Showing Why Math Matters

Rockswold teaches algebra in context, answering the question,
"Why am I learning this?"

Creating a Social Network

Although it may not be obvious, math is essential for social networks to operate properly. Matrixes are used to keep track of relationships between people on Facebook, Twitter, or Spotify. Also, a matrix can be used to describe links to and from websites on the Internet. (See the Chapter 6 opener on page 452, Example 2 on page 507, and Example 9 page 514.)

Starting Up Your Own Company

If you are starting up a small business, you might be interested in a payment startup called Square that allows businesses to swipe credit cards on iPhones and Android devices. Square's dramatic growth during recent years can be analyzed with the aid of a linear inequality. (See the Introduction to Section 2.3 on page 104 and Example 3 on pages 107-108.)

Getting the Jitters?

The side effects of caffeine include either headaches or the jitters. However, with the aid of a system of linear inequalities, we can identify the levels of caffeine intake where neither side effect occurs. (See the Introduction to Section 6.2 on page 471.)

Determining a Margin of Error

Whether a person is being shot out of cannon or manufacturing an iPhone, the concept of a margin of error is essential. To determine accurate margins of error, we need the concept of an absolute value inequality. (See the Introduction to Section 2.5 on page 134 and Example 7 on page 140.)

Modeling Half-Life of a Facebook Link

A typical Facebook link experiences half of its engagements, or hits, during the first 3 hours. By using exponential functions we can estimate how many hits a Facebook link might experience in a given period of time. (See the Chapter 5 opener on page 348 and Example 12 in Section 5.3 on pages 390-391.)

Classifying Tornados

The intensity of a tornado is often classified using the Fujita scale: the greater the wind speed of a tornado, the greater its Fujita number. To use this scale we need the concept of a piecewise-constant function. (See the discussion in Section 2.4 on page 121.)

Diminishing Returns and Overfishing

If there are only a few fishing boats in a large body of water, each boat might catch its limit. However, as the number of boats increases, there is a point of diminishing returns, where each boat starts to catch fewer and fewer fish. We can analyze this situation with a piecewise-polynomial function. (See Example 7 in Section 4.2 on pages 250-251.)

Understanding Size in Biology

Larger animals tend to have slower heart rates and larger birds tend to have bigger wings. To understand size and physical characteristics in nature, we need to study power functions. (See the discussion and Example 10 in Section 4.8 on pages 326-328.)

5th edition

Algebra and Trigonometry

with Modeling & Visualization

Gary K. Rockswold

Minnesota State University, Mankato

with

Terry A. Krieger

Rochester Community and Technical College

and

Jessica C. Rockswold

PEARSON

Boston Columbus Indianapolis New York San Francisco Upper Saddle River
Amsterdam Cape Town Dubai London Madrid Milan Munich Paris Montréal Toronto
Delhi Mexico City São Paulo Sydney Hong Kong Seoul Singapore Taipei Tokyo

Editorial Director Chris Hoag
Editor in Chief Anne Kelly
Executive Content Editor Christine O'Brien
Editorial Assistant Judith Garber
Executive Director of Development Carol Trueheart
Senior Managing Editor Karen Wernholm
Associate Managing Editor Tamela Ambush
Senior Production Project Manager Sheila Spinney
Digital Assets Manager Marianne Groth
Supplements Production Coordinator Kerri Consalvo
Media Producer Tracy Menoza
MathXL Content Supervisor Kristina Evans
Senior Content Developer Mary Durnwald
Marketing Manager Peggy Sue Lucas
Marketing Assistant Justine Goulart
Senior Author Support/Technology Specialist Joe Vetere
Image Manager Rachel Youdelman
Procurement Manager Evelyn Beaton
Procurement Specialist Debbie Rossi
Media Procurement Specialist Ginny Michaud
Associate Director of Design, USHE/HSC/EDU Andrea Nix
Senior Design Specialist and Cover Designer Heather Scott
Text Design, Production Coordination, Composition, and Illustrations Cenveo Publisher
     Services/Nesbitt Graphics, Inc.
Cover Image Sommai/Shutterstock

For permission to use copyrighted material, grateful acknowledgment is made to the copyright holders on page P-1, which is hereby made part of this copyright page.

Many of the designations used by manufacturers and sellers to distinguish their products are claimed as trademarks. Where those designations appear in this book, and Pearson was aware of a trademark claim, the designations have been printed in initial caps or all caps.

Library of Congress Cataloging-in-Publication Data

Rockswold, Gary K.
 Algebra and trigonometry with modeling and visualization / Gary K. Rockswold,
Minnesota State University, Mankato. -- 5th edition.
    pages cm
  Includes index.
  ISBN 0-321-82612-4 (student edition) --
 1. Algebra--Textbooks. 2. Trigonometry--Textbooks. 3. Functions--Textbooks.
I. Title.
  QA154.3.R63 2014
  512'.13--dc23

                                                      2012024554

2 3 4 5 6—V057— 16 15

www.pearsonhighered.com

ISBN 13: 978-0-321-82612-1
ISBN 10: 0-321-82612-4

In memory of a kind man who said to me,

"Have joy wherever you go."

Marvin, 1914–2010

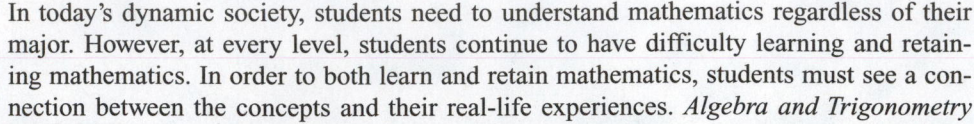

Foreword

In today's dynamic society, students need to understand mathematics regardless of their major. However, at every level, students continue to have difficulty learning and retaining mathematics. In order to both learn and retain mathematics, students must see a connection between the concepts and their real-life experiences. *Algebra and Trigonometry with Modeling and Visualization,* Fifth Edition, addresses these issues by appropriately connecting applications, modeling, and visualization to mathematical concepts and skills. This text consistently gives meaning to the equations and demonstrates that mathematics *is* relevant. It allows students to learn mathematics in the context of their experiences. Students learn mathematics more fully when concepts are presented not only symbolically but also visually. By complementing a symbolic approach with an emphasis on visual presentations, this text allows students to absorb information faster and more intuitively.

The concept of a function is the unifying theme in this text with an emphasis on the rule of four (verbal, graphical, numerical, and symbolic representations). A flexible approach allows instructors to strike their own balance of skills, rule of four, applications, modeling, and technology. Rather than reviewing all of the necessary intermediate algebra skills in the first chapter, this text integrates required math skills seamlessly by referring students "just in time" to Chapter R, "Basic Concepts from Algebra and Geometry." Instructors are free to assign supplemental homework from this chapter. Students also have additional opportunities to review their skills in the MyMathLab® course when needed. Here personalized homework and quizzes are readily available on a wide variety of review topics.

Students frequently do not realize that mathematics is transforming our society. To communicate this fact, the author has established a website at www.garyrockswold.net. Here, several resources are available, including a number of invited addresses given by the author. These presentations are accessible to students and allow them to understand the big picture of how mathematics influences everyone's life.

Contents

3 Quadratic Functions and Equations *155*

6 Trigonometric Functions *452*

7 Trigonometric Identities and Equations *553*

8 Further Topics in Trigonometry *619*

R Reference: Basic Concepts from Algebra and Geometry *R-1*

Changes to the Fifth Edition

The Fifth Edition has an exciting new look that makes mathematics more visual and easier for students to understand. The following changes are the result of suggestions made by students, instructors, and reviewers.

- Several features have been added that allow graphs and tables to be labeled in a way that explains topics visually with fewer words.
- Hundreds of new real-world applications that relate to students' lives have been added.
- Approximately 1000 examples and exercises have been replaced or modified to better meet student needs.
- Real-world data has been added and updated to be more current and meaningful to students.
- Chapter 1 has been streamlined from five to four sections at the request of reviewers. As a result, the first two chapters can be covered more efficiently. Also at the request of reviewers, the definition of increasing and decreasing has been modified.
- Chapter 2 has been reorganized so that it begins with equations of lines in Section 2.1. Now additional modeling with linear functions occurs in Section 2.4.
- Chapter 3 includes new visual presentations and explanations for quadratic equations, quadratic inequalities, and transformations of graphs. Rules for the order of transformations have also been included.
- Chapter 4 includes a new subsection covering radical functions and their transformations. Several visuals have been added to help students understand polynomial behavior and graphs of rational functions.
- Chapter 5 has increased emphasis on transformations of exponential and logarithmic functions. It includes more visual explanations of logarithms.
- Chapter 6 has new visuals to explain angular velocity and transformations of sinusoidal graphs. Additional material covering other inverse trigonometric functions has also been added.
- Chapter 7 has improved clarity for students when they verify identities, simplify expressions, find reference angles, and solve trigonometric equations.
- Chapter 8 has additional examples for solving triangles using the law of sines and the law of cosines. New visuals have been added to the topic of vectors to help students understand this important concept better.
- Chapter 9 has a new subsection on social networks and matrices. Several application topics that relate to students' lives, such as the Internet, have been included.
- Chapter 10 has several new visuals that help students understand conic sections.
- Chapter 11 has an increased emphasis on explaining the distinction between sequences and series. More explanation of conditional probability is also given.
- Appendix D is new and explains how percentages, constant percent change, and exponential functions are related. These topics are important to students in their everyday lives.

Features

 NEW!

■ **See the Concept**

This new and exciting feature allows students to make important connections by walking them through detailed visualizations. Students use graphs, tables, and diagrams to learn new concepts in a concise and efficient way. This feature also promotes multiple learning styles and deepens every student's understanding of mathematics. (See pages 32, 109, 233, 290, and 373.)

NEW!

■ **Comment Boxes**

This new feature allows graphs, tables, and symbolic explanations to be labeled in such a way that a *concept is easier to understand*. The explanation is now tied closely to a graph, table, or equation. (See pages 26, 73, 214, 271, and 294.)

■ **Chapter and Section Introductions**

Many algebra and trigonometry students have little or no understanding of mathematics beyond basic computation. To motivate students, chapter and section introductions explain some of the reasons for studying mathematics. (See pages 1, 67, 134, and 232.)

■ **Now Try**

This feature occurs after each example. It suggests a similar exercise students can work to see if they understand the concept presented in the example. (See pages 17, 74, and 106.)

■ **Getting Started**

This feature occurs in selected examples that require multistep solutions. Getting Started helps students develop an overall problem-solving strategy before they begin writing a detailed solution. (See pages 6, 71, and 390.)

■ **Algebra and Geometry Review Notes**

Throughout the text, Algebra and Geometry Review Notes, located in the margins, direct students "just in time" to Chapter R, where important topics in algebra and geometry are reviewed. Instructors can use this chapter for extra review or refer students to it as needed. This feature *frees* instructors from having to frequently review material from intermediate algebra and geometry. (See pages 97 and 160.) In addition, quizzes and personalized homework on review skills are now embedded in MyMathLab.

■ **Calculator Help Notes**

The Calculator Help Notes in the margins direct students "just in time" to Appendix A, "Using the Graphing Calculator." This appendix shows students the keystrokes necessary to complete specific examples from the text. This feature *frees* instructors from having to teach the specifics of the graphing calculator and gives students a convenient reference written specifically for this text. (See pages 6, 20, and 93.)

■ **Class Discussion**

This feature, included in most sections, poses a question that can be used for either classroom discussion or homework. (See pages 47, 157, and 236.)

■ **Making Connections**

This feature, which occurs throughout the text, shows students how concepts covered previously are related to new concepts being presented. (See pages 29, 94, 110, 136, and 175.)

■ **Putting It All Together**

This helpful feature at the end of each section summarizes techniques and reinforces the mathematical concepts presented in the section. It is given in an easy-to-follow grid. (See pages 98–99, 331–332, and 392–393.)

■ **Checking Basic Concepts**

This feature, included after every two sections, provides a small set of exercises that can be used as mixed review. These exercises require about 15 or 20 minutes to complete and can be used for collaborative learning exercises if time permits. (See pages 104, 133, and 188.)

■ **Exercise Sets**

The exercise sets are the heart of any mathematics text, and this text includes a large variety of instructive exercises. Each set of exercises covers skill building, mathematical concepts, and applications. Graphical interpretation and tables of data are often used to extend students' understanding of mathematical concepts. The exercise sets are graded carefully and categorized according to topic, making it easy for an instructor to select appropriate assignments. (See pages 80–85 and 183–188.)

■ **Chapter Summaries**

Chapter summaries are presented in an easy-to-read grid. They allow students to quickly review key concepts from the chapter. (See pages 224–227 and 337–341.)

■ **Chapter Review Exercises**

This exercise set contains both skill-building and applied exercises. These exercises stress different techniques for solving problems and provide students with the review necessary to pass a chapter test. (See pages 63–66 and 341–345.)

■ **Extended and Discovery Exercises**

Extended and Discovery Exercises occur at the end of selected sections and at the end of every chapter. These exercises are usually more complex and challenging than the rest of the exercises and often require extension of a topic presented or exploration of a new topic. They can be used for either collaborative learning or extra homework assignments. (See pages 65–66, 230, and 344–345.)

■ **Cumulative Review Exercises**

These comprehensive exercise sets, which occur after every two chapters, give students an opportunity to review previous material. (See pages 152–154 and 345–347.)

Instructor Supplements

ANNOTATED INSTRUCTOR'S EDITION

- Includes sample homework assignments indicated by problem numbers underlined in blue within each end-of-section exercise set.
- Sample homework assignments assignable in MyMathLab.
- Includes Teaching Examples, an extra set of examples for instructors to present in class, doubling the number of examples available for instructors. Solutions and Power Point Slides are available for these.
- Includes Teaching Tips, helpful ideas about presenting topics or teaching from the text
- Includes all the answers to the exercise sets, usually right on the page where the exercise appears

ISBN: 0-321-83679-0 / 978-0-321-83679-3

INSTRUCTOR'S SOLUTIONS MANUAL

- By David Atwood, *Rochester Community and Technical College*
- Provides complete solutions to all text exercises, excluding Writing about Mathematics

ISBN: 0-321-82619-1 / 978-0-321-82619-0

INSTRUCTOR'S TESTING MANUAL (DOWNLOAD ONLY)

- By David Atwood, *Rochester Community and Technical College*
- Provides prepared tests for each chapter of the text, as well as answers
- Available in MyMathLab or downloadable from Pearson Education's online catalog.

TESTGEN® (DOWNLOAD ONLY)

- Enables instructors to build, edit, print, and administer tests using a computerized bank of questions that cover all the objectives of the text
- Using algorithmically based questions, allows instructors to create multiple but equivalent versions of the same question or test with the click of a button
- Lets instructors modify test bank questions or add new questions
- Provides printable or online tests
- Available in MyMathLab or downloadable from Pearson Education's online catalog

INSIDER'S GUIDE

- Includes resources to help faculty with course preparation and classroom management
- Provides helpful teaching tips correlated to the sections of text, as well as general teaching advice

ISBN: 0-321-57717-5 / 978-0-321-57717-7

POWERPOINT PRESENTATION (DOWNLOAD ONLY)

- Classroom presentation software correlated specifically to this textbook sequence
- Available for download within MyMathLab or from Pearson Education's online catalog

Student Supplements

STUDENT'S SOLUTIONS MANUAL

- By David Atwood, *Rochester Community and Technical College*
- Provides complete solutions to all odd-numbered text exercises, excluding Writing about Mathematics and Extended and Discovery Exercises

ISBN: 0-321-83307-4 / 978-0-321-83307-5

Technology Resources

MyMathLab®

MyMathLab Online Course (access code required)

MyMathLab delivers **proven results** in helping individual students succeed.

- MyMathLab has a consistently positive impact on the quality of learning in higher education math instruction. MyMathLab can be successfully implemented in any environment—lab-based, hybrid, fully online, traditional—and demonstrates the quantifiable difference that integrated usage has on student retention, subsequent success, and overall achievement.

- MyMathLab's comprehensive online gradebook automatically tracks your students' results on tests, quizzes, homework, and in the study plan. You can use the gradebook to quickly intervene if your students have trouble or to provide positive feedback on a job well done. The data within MyMathLab is easily exported to a variety of spreadsheet programs, such as Microsoft Excel. You can determine which points of data you want to export and then analyze the results to determine success.

MyMathLab provides **engaging experiences** that personalize, stimulate, and measure learning for each student.

- **Exercises:** The homework and practice exercises in MyMathLab are correlated to the exercises in the textbook, and they regenerate algorithmically to give students unlimited opportunity for practice and mastery. The software offers immediate, helpful feedback when students enter incorrect answers.

- **Multimedia Learning Aids:** Exercises include guided solutions, sample problems, videos, and eText clips for extra help at point-of-use.

- **Expert Tutoring:** Although many students describe the whole of MyMathLab as "like having your own personal tutor," students using MyMathLab do have access to live tutoring from Pearson, from qualified math and statistics instructors.

And, MyMathLab comes from a **trusted partner** with educational expertise and an eye on the future.

- Knowing that you are using a Pearson product means knowing that you are using quality content. That means that our eTexts are accurate and our assessment tools work. Whether you are just getting started with MyMathLab, or have a question along the way, we're here to help you learn about our technologies and how to incorporate them into your course.

Rockswold's MyMathLab course engages students and keeps them thinking.

- Author designated preassigned homework assignments are provided.

- Integrated Review provides optional quizzes throughout the course that test prerequisite knowledge. After taking each quiz, students receive a personalized, just-in-time review assignment to help them refresh forgotten skills.

- Interactive figures are available, enabling users to manipulate figures to bring hard-to-convey math concepts to life.

- Section-Lecture Videos provide lectures for each section of the text to help students review important concepts and procedures 24/7. Assignable questions are available to check students' video comprehension.

To learn more about how MyMathLab combines proven learning applications with powerful assessment, visit **www.mymathlab.com** or contact your Pearson representative.

MyMathLab Ready to Go Course (access code required)

These new Ready to Go courses provide students with all the same great MyMathLab features, but make it easier for instructors to get started. Each course includes pre-assigned homework and quizzes to make creating a course even simpler.

Ask your Pearson representative about the details for this particular course or to see a copy of this course.

MyMathLab Plus/MyLabsPlus

MyLabsPlus combines proven results and engaging experiences from MyMathLab with convenient management tools and a dedicated services team. Designed to support growing math and statistics programs, it includes additional features such as

- **Batch Enrollment:** Your school can create the login name and password for every student and instructor, so everyone can be ready to start class on the first day. Automation of this process is also possible through integration with your school's Student Information System.
- **Login from your campus portal:** You and your students can link directly from your campus portal into your MyLabsPlus courses. A Pearson service team works with your institution to create a single sign-on experience for instructors and students.
- **Advanced Reporting:** MyLabsPlus's advanced reporting allows instructors to review and analyze students' strengths and weaknesses by tracking their performance on tests, assignments, and tutorials. Administrators can review grades and assignments across all courses on your MyLabsPlus campus for a broad overview of program performance.
- **24/7 Support:** Students and instructors receive 24/7 support, 365 days a year, by email or online chat.

MyLabsPlus is available to qualified adopters. For more information, visit our website at www.mylabsplus.com or contact your Pearson representative.

MathXL® MathXL Online Course (access code required)

MathXL is the homework and assessment engine that runs MyMathLab. (MyMathLab is MathXL plus a learning management system.)

With MathXL, instructors can
- Create, edit, and assign online homework and tests using algorithmically generated exercises correlated at the objective level to the textbook.
- Create and assign their own online exercises and import TestGen tests for added flexibility.
- Maintain records of all student work tracked in MathXL's online gradebook.

With MathXL, students can:
- Take chapter tests in MathXL and receive personalized study plans and/or personalized homework assignments based on their test results.
- Use the study plan and/or the homework to link directly to tutorial exercises for the objectives they need to study.
- Access supplemental animations and video clips directly from selected exercises.

MathXL is available to qualified adopters. For more information, visit our website at www.mathxl.com or contact your Pearson representative.

Acknowledgments

Many individuals contributed to the development of this textbook. I would like to thank the following reviewers, whose comments and suggestions were invaluable in preparing this edition of the text.

Dawit Aberra	*Fort Valley State University*
Dr. Josephine D. Davis	*Fort Valley State University*
Christy Dittmar	*Austin Community College*
Chi Giang	*Westchester Community College*
Christian Mason	*Virginia Commonwealth University*
Val Mohanakumar	*Hillsborough Community College*
Nancy Pevey	*Pellissippi State Community College*
Carolynn Reed	*Austin Community College*
Tracy Romesser	*Erie Community College North*
Jeffrey Saikali	*San Diego Miramar College*
Meredith Watts	*Massachusetts Bay Community College*
Cathleen Zucco-Teveloff	*Rowan University*

I would like to welcome Terry Krieger and Jessica Rockswold to the team for the fifth edition. They have provided invaluable help with developing new applications, visualizations, examples, and exercises. Terry and Jessica have contributed at all levels in the development of this new and exciting edition.

I would like to thank Paul Lorczak, Lynn Baker, Namyong Lee at Minnesota State University, Mankato, Mark Rockswold at Denver Community College, and David Atwood at Rochester Community and Technical College for their superb work with proofreading and accuracy checking.

Without the excellent cooperation from the professional staff at Pearson, this project would have been impossible. They are, without a doubt, the best. Thanks go to Greg Tobin for his support of this project. Particular recognition is due Anne Kelly and Christine O'Brien, who gave advice, support, assistance, and encouragement. The outstanding contributions of Sheila Spinney, Judith Garber, Heather Scott, Peggy Sue Lucas, Justine Goulart, and Joe Vetere are much appreciated. The outstanding work of Kathy Diamond was instrumental to the success of this project.

Thanks go to Wendy Rockswold, who gave invaluable assistance and encouragement throughout the project. She also supplied several of the photographs found throughout the text.

A special thank you goes to the many students and instructors who used the first four editions of this text. Their suggestions were insightful. Please feel free to contact me at either *gary.rockswold@mnsu.edu* or *www.garyrockswold.net* with your comments. Your opinion is important.

Gary Rockswold

1 Introduction to Functions and Graphs

> The essence of mathematics is not to make simple things complicated, but to make complicated things simple.
>
> —Stanley Gudder

Have you ever thought about how we "live by the numbers"? Money, sports, digital televisions, speed limits, grade point averages, gas mileages, and temperatures are all based on numbers. When we are told what our weight, blood pressure, body mass index, and cholesterol levels are, it can even affect how we feel about ourselves. Numbers permeate our society.

Numbers are part of mathematics, but mathematics is *much more* than numbers. Mathematics also includes techniques to analyze these numbers and to guide our decisions about the future. Mathematics is used not only in science and technology; it is also used to describe almost every facet of life, including consumer behavior, social networks, and the Internet. Mathematics gives people the reasoning skills to solve problems from work and life.

In this chapter we discuss numbers and how functions are used to do computations with these numbers. Understanding numbers and mathematical concepts is essential to understanding and dealing with the many changes that will occur in our lifetimes. Mathematics makes life easier!

1.1 Numbers, Data, and Problem Solving

- Recognize common sets of numbers
- Evaluate expressions by applying the order of operations
- Learn scientific notation and use it in applications
- Apply problem-solving strategies

Introduction

Because society is becoming more complex and diverse, our need for mathematics is increasing dramatically each year. Numbers are essential to our everyday lives. For example, the iPhone 4Gs is 4.5 inches in height, 2.31 inches in width, and 0.37 inch in thickness. It has an 8-gigabyte flash drive and a 5-megapixel camera, and it can operate at temperatures between 32° and 95°F. (**Source:** Apple Corporation.)

Mathematics not only provides numbers to describe new products but also gives us problem-solving strategies. This section discusses basic sets of numbers and introduces some essential problem-solving strategies.

Sets of Numbers

One important set of numbers is the set of **natural numbers**. This set comprises the *counting numbers* $N = \{1, 2, 3, 4, \dots\}$.

The **integers** $I = \{\dots, -3, -2, -1, 0, 1, 2, 3, \dots\}$ are a set of numbers that contains the natural numbers, their additive inverses (negatives), and 0.

A **rational number** can be expressed as the *ratio* of two integers $\frac{p}{q}$, where $q \neq 0$. A rational number results when an integer is divided by a nonzero integer. Thus rational numbers include fractions and the integers. Examples of rational numbers are

$$\frac{2}{1}, \frac{1}{3}, -\frac{1}{4}, \frac{-50}{2}, \frac{22}{7}, 0, \sqrt{25}, \text{ and } 1.2.$$

Rational Numbers

Note that 0 and 1.2 are both rational numbers. They can be represented by the fractions $\frac{0}{1}$ and $\frac{12}{10}$. Because two fractions that look different can be equivalent, rational numbers have more than one form. A rational number can always be expressed in a decimal form that either *repeats* or *terminates*. For example, $\frac{2}{3} = 0.\overline{6}$, a repeating decimal, and $\frac{1}{4} = 0.25$, a terminating decimal. The overbar indicates that $0.\overline{6} = 0.6666666\dots.$

CLASS DISCUSSION

The number 0 was invented well after the natural numbers. Many societies did not have a zero—for example, there is no Roman numeral for 0. Discuss some possible reasons for this.

Real numbers can be represented by decimal numbers. Since every rational number has a decimal form, real numbers include rational numbers. However, some real numbers cannot be expressed as a ratio of two integers. These numbers are called **irrational numbers**. The numbers $\sqrt{2}$, $\sqrt{15}$, and π are examples of irrational numbers. They can be represented by nonrepeating, nonterminating decimals. Note that for any positive integer a, if $\sqrt{a}$ is not an integer, then $\sqrt{a}$ is an irrational number.

Real numbers are either rational or irrational numbers and can always be *approximated* by a terminating decimal. Examples of real numbers include

$$2, -10, -131.3337, \frac{1}{3} = 0.\overline{3}, -\sqrt{5} \approx -2.2361, \text{ and } \sqrt{11} \approx 3.3166.$$

Real Numbers

NOTE The symbol $\approx$ means **approximately equal**. This symbol is used in place of an equals sign whenever two unequal quantities are close in value. For example, $\frac{1}{2} = 0.5$, whereas $\frac{1}{3} \approx 0.3333$.

Figure 1.1 illustrates how the different sets of numbers are related.

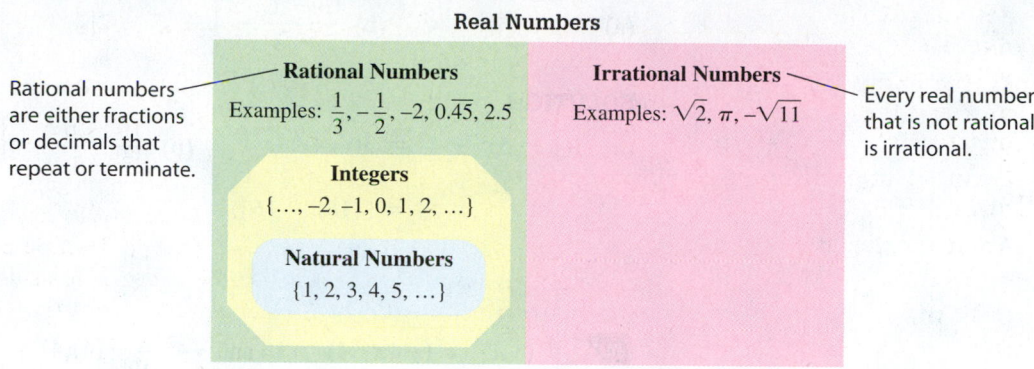

Rational numbers are either fractions or decimals that repeat or terminate.

Every real number that is not rational is irrational.

Figure 1.1

EXAMPLE 1   **Classifying numbers**

Classify each real number as one or more of the following: natural number, integer, rational number, or irrational number.

$$5, -1.2, \frac{13}{7}, -\sqrt{7}, -12, \sqrt{16}$$

SOLUTION

5: natural number, integer, and rational number

-1.2: rational number

$\frac{13}{7}$: rational number

$-\sqrt{7}$: irrational number

-12: integer and rational number

$\sqrt{16} = 4$: natural number, integer, and rational number

Now Try Exercise 7

Order of Operations

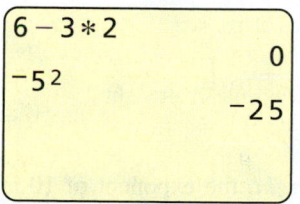

Figure 1.2

Does $6 - 3 \cdot 2$ equal 0 or 6? Does -5^2 equal 25 or -25? Figure 1.2 correctly shows that $6 - 3 \cdot 2 = 0$ and that $-5^2 = -25$. Because multiplication is performed before subtraction, $6 - 3 \cdot 2 = 0$. Similarly, because exponents are evaluated before performing negation, $-5^2 = -25$. It is essential that algebraic expressions be evaluated consistently, so the following rules have been established.

ORDER OF OPERATIONS

Using the following order of operations, perform all calculations within parentheses, square roots, and absolute value bars and above and below fraction bars. Then use the same order of operations to perform any remaining calculations.

1. Evaluate all exponents. Then do any negation *after* evaluating exponents.
2. Do all multiplication and division from *left to right*.
3. Do all addition and subtraction from *left to right*.

EXAMPLE 2   **Evaluating arithmetic expressions by hand**

Evaluate each expression by hand.

(a) $3(1 - 5)^2 - 4^2$ **(b)** $\dfrac{10 - 6}{5 - 3} - 4 - |7 - 2|$

SOLUTION

(a) $3(1 - 5)^2 - 4^2 = 3(-4)^2 - 4^2$

$= 3(16) - 16$

$= 48 - 16$

$= 32$

(b) $\dfrac{10 - 6}{5 - 3} - 4 - |7 - 2| = \dfrac{4}{2} - 4 - |5|$

$= 2 - 4 - 5$

$= -2 - 5$

$= -7$

NOTE $(-4)^2 = (-4)(-4) = 16$ and $-4^2 = -(4)(4) = -16.$

Now Try Exercises 19 and 21

Scientific Notation

Numbers that are large or small in absolute value are often expressed in scientific notation. Table 1.1 lists examples of numbers in **standard (decimal) form** and in **scientific notation**.

Applications of Scientific Notation

Standard Form	Scientific Notation	Application
93,000,000 mi	9.3×10^7 mi	Distance to the sun
256,000	2.56×10^5	Number of cell towers in 2010
9,000,000,000	9×10^9	Estimated world population in 2050
0.00000538 sec	5.38×10^{-6} sec	Time for light to travel 1 mile
0.000005 cm	5×10^{-6} cm	Size of a typical virus

Table 1.1

EXAMPLE 3   **Writing a number in scientific notation**

Write 0.000578 in scientific notation.

SOLUTION

To write 0.000578 in scientific notation, start by moving the decimal point to the right of the first nonzero digit, 5, to obtain 5.78.

Decimal Form   *Scientific Notation*

$0 . \underset{1\ 2\ 3\ 4}{0\ 0\ 0\ 5} . 7\ 8 \quad \rightarrow \quad 5.78 \times 10^{-4}$ *Move the decimal point right.*

Since the decimal point was moved four places to the *right*, the exponent of 10 is *negative* 4, or -4. If the decimal point had been moved to the *left*, the exponent of 10 would be *positive* 4.

Now Try Exercise 35

Here is a formal definition of scientific notation.

SCIENTIFIC NOTATION

A real number r is in **scientific notation** when r is written as $c \times 10^n$, where $1 \leq |c| < 10$ and n is an integer.

Calculator Help

To display numbers in scientific notation, see Appendix A (page AP-2).

An Application The next example demonstrates how scientific notation appears in the description of a new technology.

EXAMPLE 4   Analyzing the energy produced by your body

Nanotechnology is a technology of the very small: on the order of one billionth of a meter. Researchers are using nanotechnology to power tiny devices with energy from the human body. (**Source:** Z. Wang, "Self-Powered Nanotech," *Scientific American*, January 2008.)

(a) Write one billionth in scientific notation.

(b) While typing, a person's fingers generate about 2.2×10^{-3} watt of electrical energy. Write this number in standard (decimal) form.

SOLUTION

(a) One billionth can be written as $\frac{1}{1,000,000,000} = \frac{1}{10^9} = 1 \times 10^{-9}$.

(b) Move the decimal point in 2.2 three places to the left: $2.2 \times 10^{-3} = 0.0022$.

> **Now Try Exercise 77**

The next example illustrates how to evaluate expressions in scientific notation.

EXAMPLE 5   Evaluating expressions by hand

Evaluate each expression. Write your result in scientific notation and standard form.

(a) $(3 \times 10^3)(2 \times 10^4)$ **(b)** $(5 \times 10^{-3})(6 \times 10^5)$ **(c)** $\dfrac{4.6 \times 10^{-1}}{2 \times 10^2}$

SOLUTION

(a) $(3 \times 10^3)(2 \times 10^4) = 3 \times 2 \times 10^3 \times 10^4$ Commutative property

$\qquad\qquad\qquad\qquad\quad = 6 \times 10^{3+4}$ Add exponents.

$\qquad\qquad\qquad\qquad\quad = 6 \times 10^7$ Scientific notation

$\qquad\qquad\qquad\qquad\quad = 60{,}000{,}000$ Standard form

(b) $(5 \times 10^{-3})(6 \times 10^5) = 5 \times 6 \times 10^{-3} \times 10^5$ Commutative property

$\qquad\qquad\qquad\qquad\qquad = 30 \times 10^2$ Add exponents.

$\qquad\qquad\qquad\qquad\qquad = 3 \times 10^3$ Scientific notation

$\qquad\qquad\qquad\qquad\qquad = 3000$ Standard form

Algebra Review

To review exponents, see Chapter R (page R-8).

(c) $\dfrac{4.6 \times 10^{-1}}{2 \times 10^2} = \dfrac{4.6}{2} \times \dfrac{10^{-1}}{10^2}$ Multiplication of fractions

$\qquad\qquad\qquad\quad = 2.3 \times 10^{-1-2}$ Subtract exponents.

$\qquad\qquad\qquad\quad = 2.3 \times 10^{-3}$ Scientific notation

$\qquad\qquad\qquad\quad = 0.0023$ Standard form

> **Now Try Exercises 53, 55, and 57**

Calculators Calculators often use **E** to express powers of 10. For example, 4.2×10^{-3} might be displayed as 4.2E–3. On some calculators, numbers can be entered in scientific notation with the (EE) key, which you can find by pressing (2nd) (,).

EXAMPLE 6   Computing in scientific notation with a calculator

Approximate each expression. Write your answer in scientific notation.

(a) $\left(\dfrac{6 \times 10^3}{4 \times 10^6}\right)(1.2 \times 10^2)$ **(b)** $\sqrt{4500\pi}\left(\dfrac{103 + 450}{0.233}\right)^3$

```
(6*10^3)/(4*10^6
)*(1.2*10^2)
                .18
(6E3)/(4E6)*(1.2
E2)
                .18
```

Figure 1.3

SOLUTION

(a) The given expression is entered in two ways in Figure 1.3. Note that in both cases

$$\left(\dfrac{6 \times 10^3}{4 \times 10^6}\right)(1.2 \times 10^2) = 0.18 = 1.8 \times 10^{-1}.$$

Calculator Help

To enter numbers in scientific notation, see Appendix A (page AP-2).

(b) Be sure to insert parentheses around 4500π and around the numerator, $103 + 450$, in the expression $\sqrt{4500\pi}\left(\frac{103 + 450}{0.233}\right)^3$. From Figure 1.4 we can see that the result is approximately 1.59×10^{12}.

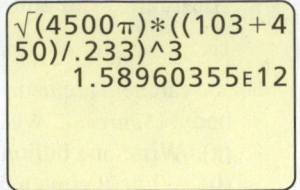

```
√(4500π)*((103+4
50)/.233)^3
        1.58960355E12
```

Figure 1.4

Now Try Exercises 61 and 63

EXAMPLE 7   **Computing with a calculator**

Use a calculator to evaluate each expression. Round answers to the nearest thousandth.

(a) $\sqrt[3]{131}$ **(b)** $\pi^3 + 1.2^2$ **(c)** $\dfrac{1 + \sqrt{2}}{3.7 + 9.8}$ **(d)** $|\sqrt{3} - 6|$

SOLUTION

(a) On some calculators the cube root can be found by using the MATH menu. If your calculator does not have a cube root key, enter $131^{\wedge}(1/3)$. From the first two lines in Figure 1.5, we see that $\sqrt[3]{131} \approx 5.079$.

(b) Do *not* use 3.14 for the value of π. Instead, use the built-in key to obtain a more accurate value of π. From the bottom two lines in Figure 1.5, $\pi^3 + 1.2^2 \approx 32.446$.

(c) When evaluating this expression be sure to include parentheses around the numerator and around the denominator. Most calculators have a special square root key that can be used to evaluate $\sqrt{2}$. From the first three lines in Figure 1.6, $\frac{1 + \sqrt{2}}{3.7 + 9.8} \approx 0.179$.

(d) The absolute value can be found on some calculators by using the MATH NUM menus. From the bottom two lines in Figure 1.6, $|\sqrt{3} - 6| \approx 4.268$.

Algebra Review

To review cube roots, see Chapter R (page R-37).

Calculator Help

To enter expressions such as $\sqrt[3]{131}$, $\sqrt{2}$, π, and $|\sqrt{3} - 6|$, see Appendix A (page AP-2).

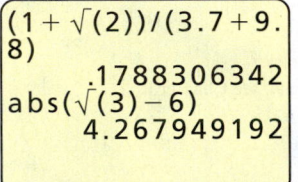

```
³√(131)
        5.078753078
π^3+1.2²
        32.44627668
```

Figure 1.5

```
(1+√(2))/(3.7+9.
8)
        .1788306342
abs(√(3)−6)
        4.267949192
```

Figure 1.6

Now Try Exercises 67, 69, 71, and 73

Problem Solving

Many problem-solving strategies are used in algebra. However, in this subsection we focus on two important strategies that are used frequently: making a sketch and applying one or more formulas. These strategies are illustrated in the next two examples.

EXAMPLE 8   **Finding the speed of Earth**

Earth travels around the sun in an approximately circular orbit with an average radius of 93 million miles. If Earth takes 1 year, or about 365 days, to complete one orbit, estimate the orbital speed of Earth in miles per hour.

SOLUTION

Getting Started Speed S equals distance D divided by time T, $S = \frac{D}{T}$. We need to find the number of miles Earth travels in 1 year and then divide it by the number of hours in 1 year. ▶

Geometry Review
To find the circumference of a circle, see Chapter R (page R-2).

Earth's Orbit

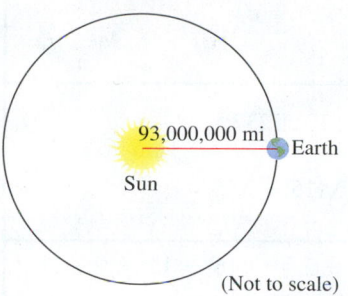

93,000,000 mi

Earth

Sun

(Not to scale)

Figure 1.7

Distance Traveled Make a sketch of Earth orbiting the sun, as shown in Figure 1.7. In 1 year Earth travels the circumference of a circle with a radius of 93 million miles. The circumference of a circle is $2\pi r$, where r is the radius, so the distance D is

$$D = 2\pi r = 2\pi(93{,}000{,}000) \approx 584{,}300{,}000 \text{ miles.}$$

Hours in 1 Year The number of hours T in 1 year, or 365 days, equals

$$T = 365 \times 24 = 8760 \text{ hours.}$$

Speed of Earth $S = \dfrac{D}{T} = \dfrac{584{,}300{,}000}{8760} \approx 66{,}700$ miles per hour.

> **Now Try Exercise 79**

Many times in geometry we evaluate formulas to determine quantities, such as perimeter, area, and volume. In the next example we use a formula to determine the number of fluid ounces in a soda can.

EXAMPLE 9 **Finding the volume of a soda can**

A Soda Can

Cola

16oz serving

Figure 1.8

The volume V of the cylindrical soda can in Figure 1.8 is given by $V = \pi r^2 h$, where r is its radius and h is its height.
(a) If $r = 1.4$ inches and $h = 5$ inches, find the volume of the can in cubic inches.
(b) Could this can hold 16 fluid ounces? (*Hint:* 1 cubic inch equals 0.55 fluid ounce.)

SOLUTION
(a) $V = \pi r^2 h = \pi(1.4)^2(5) = 9.8\pi \approx 30.8$ cubic inches.
(b) To find the number of fluid ounces, multiply the number of cubic inches by 0.55.

$$30.8 \times 0.55 = 16.94$$

Yes, the can could hold 16 fluid ounces.

> **Now Try Exercise 85**

1.1 Putting It All Together

Numbers play a central role in our society. Without numbers, data could be described qualitatively but not quantitatively. For example, we could say that the day seems hot but would not be able to give an actual number for the temperature. Mathematics provides problem-solving strategies that are used in almost every facet of our lives.

CONCEPT	COMMENTS	EXAMPLES
Natural numbers	Sometimes referred to as the *counting numbers*	$1, 2, 3, 4, 5, \ldots$
Integers	Include the natural numbers, their opposites, and 0	$\ldots, -2, -1, 0, 1, 2, \ldots$
Rational numbers	Include integers; all fractions $\frac{p}{q}$, where p and q are integers with $q \neq 0$; all repeating and all terminating decimals	$\dfrac{1}{2}, -3, \dfrac{128}{6}, -0.335, 0,$ $0.25 = \dfrac{1}{4}, 0.\overline{33} = \dfrac{1}{3}$

continued on next page

CONCEPT	COMMENTS	EXAMPLES
Irrational numbers	Can be written as nonrepeating, nonterminating decimals; cannot be a rational number; if a square root of a positive integer is not an integer, it is an irrational number.	$\pi, \sqrt{2}, -\sqrt{5}, \sqrt[3]{7}, \pi^4$
Real numbers	Any number that can be expressed in standard (decimal) form Include the rational numbers and irrational numbers	$\pi, \sqrt{7}, -\dfrac{4}{7}, 0, -10, 1.237$ $0.\overline{6} = \dfrac{2}{3}, 1000, \sqrt{15}, -\sqrt{5}$
Order of operations	Using the following order of operations, perform all calculations within parentheses, square roots, and absolute value bars and above and below fraction bars. Then perform any remaining calculations. 1. Evaluate all exponents. Then do any negation *after* evaluating exponents. 2. Do all multiplication and division from *left to right*. 3. Do all addition and subtraction from *left to right*.	$\begin{aligned} -4^2 - 12 \div 2 - 2 &= -16 - 12 \div 2 - 2 \\ &= -16 - 6 - 2 \\ &= -22 - 2 \\ &= -24 \end{aligned}$ $\begin{aligned} \dfrac{2 + 4^2}{3 - 3 \cdot 5} &= \dfrac{2 + 16}{3 - 15} \\ &= \dfrac{18}{-12} \\ &= -\dfrac{3}{2} \end{aligned}$
Scientific notation	A number in the form $c \times 10^n$, where $1 \le |c| < 10$ and n is an integer Used to represent numbers that are large or small in absolute value	$3.12 \times 10^4 = 31{,}200$ $-1.4521 \times 10^{-2} = -0.014521$ $5 \times 10^9 = 5{,}000{,}000{,}000$ $1.5987 \times 10^{-6} = 0.0000015987$

1.1 Exercises

Classifying Numbers

Exercises 1–6: Classify the number as one or more of the following: natural number, integer, rational number, or real number.

1. $\frac{21}{24}$ (Fraction of people in the United States completing at least 4 years of high school)

2. 695,000 (Number of Facebook status updates every 60 seconds)

3. 7.5 (Average number of gallons of water used each minute while taking a shower)

4. 8.4 (Neilsen rating of the TV show *Modern Family* the week of January 2, 2012)

5. $90\sqrt{2}$ (Distance in feet from home plate to second base on a baseball field)

6. -71 (Wind chill when the temperature is $-30°F$ and the wind speed is 40 mi/hr)

Exercises 7–10: Classify each number as one or more of the following: natural number, integer, rational number, or irrational number.

7. $\pi, -3, \frac{2}{9}, \sqrt{9}, 1.\overline{3}, -\sqrt{2}$

8. $\frac{3}{1}, -\frac{5}{8}, \sqrt{7}, 0.\overline{45}, 0, 5.6 \times 10^3$

9. $\sqrt{13}, \frac{1}{3}, 5.1 \times 10^{-6}, -2.33, 0.\overline{7}, -\sqrt{4}$

10. $-103, \frac{21}{25}, \sqrt{100}, -\frac{5.7}{10}, \frac{2}{9}, -1.457, \sqrt{3}$

Exercises 11–16: For the measured quantity, state the set of numbers that most appropriately describes it. Choose from the natural numbers, integers, and rational numbers. Explain your answer.

11. Shoe sizes

12. Populations of states

13. Speed limits

14. Gallons of gasoline

15. Temperatures in a winter weather forecast in Montana

16. Numbers of compact disc sales

Order of Operations
Exercises 17–28: Evaluate by hand.

17. $|5 - 8 \cdot 7|$

18. $-2(16 - 3 \cdot 5) \div 2$

19. $-6^2 - 3(2 - 4)^4$

20. $(4 - 5)^2 - 3^2 - 3\sqrt{9}$

21. $\sqrt{9 - 5} - \dfrac{8 - 4}{4 - 2}$

22. $\dfrac{6 - 4^2 \div 2^3}{3 - 4}$

23. $\sqrt{13^2 - 12^2}$

24. $\dfrac{13 - \sqrt{9 + 16}}{|5 - 7|^2}$

25. $\dfrac{4 + 9}{2 + 3} - \dfrac{-3^2 \cdot 3}{5}$

26. $10 \div 2 \div \dfrac{5 + 10}{5}$

27. $-5^2 - 20 \div 4 - 2$

28. $5 - (-4)^3 - (4)^3$

Scientific Notation
Exercises 29–40: Write the number in scientific notation.

29. 40 (Percent of smartphones that run the Android operating system in 2012)

30. 11,700,000 (Number of U. S. cancer survivors in 2007)

31. 0.00365 (Proportion of cosmetic surgeries performed on 13–19 year olds)

32. 0.62 (Number of miles in 1 kilometer)

33. 2450

34. 105.6

35. 0.56

36. −0.00456

37. −0.0087

38. 1,250,000

39. 206.8

40. 0.00007

Exercises 41–52: Write the number in standard form.

41. 1×10^{-6} (Wavelength of visible light in meters)

42. 9.11×10^{-31} (Weight of an electron in kilograms)

43. 2×10^8 (Years required for the sun to orbit our galaxy)

44. 8×10^9 (Annual dollars spent in the United States on cosmetics)

45. 1.567×10^2

46. -5.68×10^{-1}

47. 5×10^5

48. 3.5×10^3

49. 0.045×10^5

50. -5.4×10^{-5}

51. 67×10^3

52. 0.0032×10^{-1}

Exercises 53–60: Evaluate the expression by hand. Write your result in scientific notation and standard form.

53. $(4 \times 10^3)(2 \times 10^5)$

54. $(3 \times 10^1)(3 \times 10^4)$

55. $(5 \times 10^2)(7 \times 10^{-4})$

56. $(8 \times 10^{-3})(7 \times 10^1)$

57. $\dfrac{6.3 \times 10^{-2}}{3 \times 10^1}$

58. $\dfrac{8.2 \times 10^2}{2 \times 10^{-2}}$

59. $\dfrac{4 \times 10^{-3}}{8 \times 10^{-1}}$

60. $\dfrac{2.4 \times 10^{-5}}{4.8 \times 10^{-7}}$

Exercises 61–66: Use a calculator to approximate the expression. Write your result in scientific notation.

61. $\dfrac{8.947 \times 10^7}{0.00095}(4.5 \times 10^8)$

62. $(9.87 \times 10^6)(3.4 \times 10^{12})$

63. $\left(\dfrac{101 + 23}{0.42}\right)^2 + \sqrt{3.4 \times 10^{-2}}$

64. $\sqrt[3]{(2.5 \times 10^{-8}) + 10^{-7}}$

65. $(8.5 \times 10^{-5})(-9.5 \times 10^7)^2$

66. $\sqrt{\pi(4.56 \times 10^4) + (3.1 \times 10^{-2})}$

Exercises 67–76: Use a calculator to evaluate the expression. Round your result to the nearest thousandth.

67. $\sqrt[3]{192}$

68. $\sqrt{(32 + \pi^3)}$

69. $|\pi - 3.2|$

70. $\dfrac{1.72 - 5.98}{35.6 + 1.02}$

71. $\dfrac{0.3 + 1.5}{5.5 - 1.2}$

72. $3.2(1.1)^2 - 4(1.1) + 2$

73. $\dfrac{1.5^3}{\sqrt{2} + \pi - 5}$

74. $4.3^2 - \dfrac{5}{17}$

75. $15 + \dfrac{4 + \sqrt{3}}{7}$

76. $\dfrac{5 + \sqrt{5}}{2}$

Applications

77. **Nanotechnology** (Refer to Example 4.) During inhalation, the typical body generates 0.14 watt of electrical power, which could be used to power tiny electrical circuits. Write this number in scientific notation. (*Source: Scientific American,* January 2008.)

78. Movement of the Pacific Plate The Pacific plate (the floor of the Pacific Ocean) near Hawaii is moving at about 0.000071 kilometer per year. This is about the speed at which a fingernail grows. Use scientific notation to determine how many kilometers the Pacific plate travels in 1 million years.

79. Orbital Speed (Refer to Example 8.) The planet Mars travels around the sun in a nearly circular orbit with a radius of 141 million miles. If it takes 1.88 years for Mars to complete one orbit, estimate the orbital speed of Mars in miles per hour.

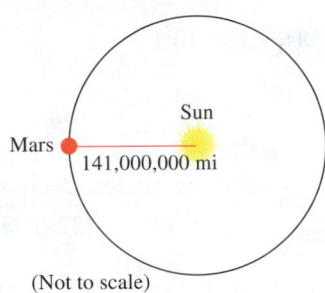

(Not to scale)

80. Size of the Milky Way The speed of light is about 186,000 miles per second. The Milky Way galaxy has an approximate diameter of 6×10^{17} miles. Estimate, to the nearest thousand, the number of years it takes for light to travel across the Milky Way. (*Source:* C. Ronan, *The Natural History of the Universe.*)

81. Living with Cancer The number of people living with cancer (cancer survivors) has increased dramatically from 1971 to 2007. (*Source:* CDC.)
   (a) In 1971 the population of the United States was 208 million and the number living with cancer was 3,000,000. To the nearest tenth, approximate the percentage of the population living with cancer in 1971.

   (b) In 2007 the population of the United States was 300 million and the number living with cancer was 11,700,000. To the nearest tenth, approximate the percentage of the population living with cancer in 2007.

82. Discharge of Water The Amazon River discharges water into the Atlantic Ocean at an average rate of 4,200,000 cubic feet per second, the highest rate of any river in the world. Is this more or less than 1 cubic mile of water per day? Explain your calculations. (*Source: The Guinness Book of Records 1993.*)

83. Analyzing Debt A 1-inch-high stack of $100 bills contains about 250 bills. In 2010 the gross federal debt was approximately 13.5 trillion dollars.
   (a) If the entire federal debt were converted into a stack of $100 bills, how many feet high would it be?

   (b) The distance between Los Angeles and New York is approximately 2500 miles. Could this stack of $100 bills reach between these two cities?

84. Volume of a Cone The volume V of a cone is given by $V = \frac{1}{3}\pi r^2 h$, where r is its radius and h is its height. Find V when $r = 4$ inches and $h = 1$ foot. Round your answer to the nearest hundredth.

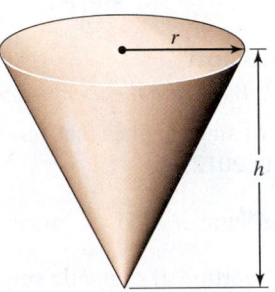

85. Size of a Soda Can (Refer to Example 9.) The volume V of a cylindrical soda can is given by $V = \pi r^2 h$, where r is its radius and h is its height.
   (a) If $r = 1.3$ inches and $h = 4.4$ inches, find the volume of the can in cubic inches.

   (b) Could this can hold 12 fluid ounces? (*Hint:* 1 cubic inch equals about 0.55 fluid ounce.)

86. Volume of a Sphere The volume of a sphere is given by $V = \frac{4}{3}\pi r^3$, where r is the radius of the sphere. Calculate the volume if the radius is 3 feet. Approximate your answer to the nearest tenth.

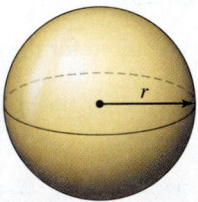

Writing about Mathematics

87. Describe some basic sets of numbers that are used in mathematics and where you find them in everyday life.

88. Suppose that a positive number a is written in scientific notation as $a = c \times 10^n$, where n is an integer and $1 \le c < 10$. Explain what n indicates about the size of a.

Extended and Discovery Exercise

Exercises 1–4: Measuring the thickness of a very thin layer of material can be difficult to do directly. For example, it would be difficult to measure the thickness of an oil film on water or a coat of paint with a ruler. However, it can be done indirectly using the following formula.

$$\text{Thickness} = \frac{\text{Volume}}{\text{Area}}$$

That is, the thickness of a thin layer equals the volume of the substance divided by the area that it covers. For example, if a volume of 1 cubic inch of paint is spread over an area of 100 square inches, then the thickness of the paint equals $\frac{1}{100}$ inch. Use this formula in the following exercises.

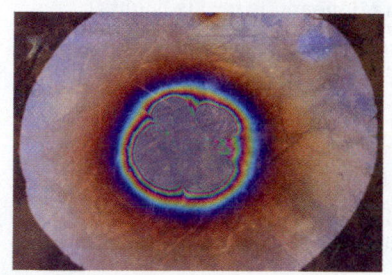

1. **Thickness of an Oil Film** A drop of oil measuring 0.12 cubic centimeter in volume is spilled onto a lake. The oil spreads out in a circular shape having a *diameter* of 23 centimeters. Approximate the thickness of the oil film.

2. **Thickness of Gold Foil** A flat, rectangular sheet of gold foil measures 20 centimeters by 30 centimeters and has a mass of 23.16 grams. If 1 cubic centimeter of gold has a mass of 19.3 grams, find the thickness of the gold foil. (*Source:* U. Haber-Schaim, *Introductory Physical Science.*)

3. **Thickness of Cement** A 100-foot-long sidewalk is 5 feet wide. If 125 cubic feet of cement are evenly poured to form the sidewalk, find the thickness of the sidewalk.

4. **Depth of a Lake** A lake covers 2.5×10^7 square feet and contains 7.5×10^8 cubic feet of water. Find the average depth of the lake.

1.2 Visualizing and Graphing Data

- Analyze one-variable data
- Find the domain and range of a relation
- Graph in the *xy*-plane
- Calculate distance
- Find the midpoint
- Learn the standard equation of a circle
- Learn to graph equations with a calculator (optional)

Introduction

There is a wealth of information in data, but the challenge is to convert data into meaningful information that can be used to solve problems. Visualization can be a powerful tool to analyze data, as pictures and graphs are often easier to understand than a list of numbers. For example, looking at Table 1.2, it is difficult to identify trends in the data for *Popular Science* iPad subscriptions. On the other hand, the *line graph* in Figure 1.9 makes it easy to see that there is a sudden increase in subscriptions in October 2011. This increase could be an indicator for Apple that its new product Newsstand, which makes it easier to subscribe to magazines, is a success. Insights like this are important to businesses and therefore visualization is an effective business strategy.

Popular Science iPad Subscription Sales 2011

Date	Subscriptions
27 Feb	7,100
24 Apr	15,400
19 Jun	20,100
14 Aug	25,000
9 Oct	28,700
30 Oct	36,000
27 Nov	40,700

Table 1.2

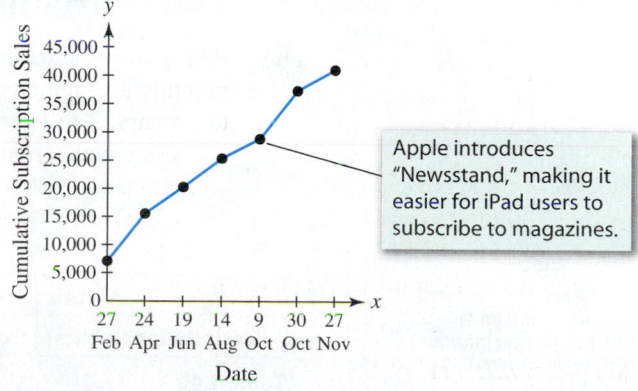

Figure 1.9

One-Variable Data

Data often occur in the form of a list. A list of test scores without names is an example; the only variable is the score. Data of this type are referred to as **one-variable data.** If the values in a list are unique, they can be represented visually on a number line.

Means and medians can be found for one-variable data sets. To calculate the **mean** (or **average**) of a set of n numbers, we add the n numbers and then divide the sum by n. The **median** is equal to the value that is located in the middle of a *sorted* list. If there is an odd number of data items, the median is the middle data item. If there is an even number of data items, the median is the average of the two middle items.

EXAMPLE 1 **Analyzing a list of temperatures**

Table 1.3 lists the low temperatures T in degrees Fahrenheit that occurred in Minneapolis, Minnesota, for six consecutive nights during January 2012.

Low Temperatures in Minneapolis for Six Nights

T	−12	−4	−8	21	18	9

Table 1.3

(a) Plot these temperatures on a number line.
(b) Find the maximum and minimum of these temperatures.
(c) Determine the mean of these six temperatures.
(d) Find the median and interpret the result.

SOLUTION
(a) In Figure 1.10 the numbers in Table 1.3 are plotted on a number line.

Number Line Graph of Low Temperatures

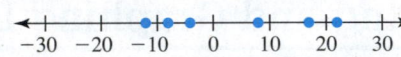

Figure 1.10

(b) The maximum is 21°F and the minimum is −12°F.
(c) The mean temperature is calculated as follows:

$$\frac{-12 + (-4) + (-8) + 21 + 18 + 9}{6} = \frac{24}{6} = 4.$$

Thus the average low temperature was 4°F.
(d) In the ordered list −12, −8, **−4**, **9**, 18, 21, the median is the average of the two middle temperatures, **−4**°F and **9**°F. The median is $\frac{-4 + 9}{2} = 2.5$°F. This means that half of the temperatures are above 2.5°F and half are below.

Now Try Exercises 1 and 5

Two-Variable Data

Relations Sometimes a relationship exists between two lists of data. Table 1.4 lists the monthly average precipitation in inches for Portland, Oregon. In this table, 1 corresponds to January, 2 to February, and so on. We show the relationship between a month and its average precipitation by combining the two lists so that corresponding months and precipitations are visually paired.

> This table forms a relation with ordered pairs in the form (month, precipitation).

Average Precipitation for Portland, Oregon

Month	1	2	3	4	5	6	7	8	9	10	11	12
Precipitation (inches)	6.2	3.9	3.6	2.3	2.0	1.5	0.5	1.1	1.6	3.1	5.2	6.4

Table 1.4

If x is the month and y is the precipitation in inches, then the **ordered pair** (x, y) represents the average amount of precipitation y during month x. For example, the ordered pair $(5, 2.0)$ indicates that the average precipitation in May is 2.0 inches, whereas the ordered pair $(2, 3.9)$ indicates that the average precipitation in February is 3.9 inches. *Order is important* in an ordered pair.

Since the data in Table 1.4 involve two variables, the month and precipitation, we refer to them as **two-variable data**. It is important to realize that a relation established by two-variable data is between two lists rather than within a single list. January is not related to August, and 6.2 inches of precipitation is not associated with 1.1 inches of precipitation. Instead, January is paired with 6.2 inches, and August is paired with 1.1 inches. We now define the mathematical concept of a relation.

RELATION

A **relation** is a set of ordered pairs.

If we denote the ordered pairs in a relation by (x, y), then the set of all x-values is called the **domain** of the relation and the set of all y-values is called the **range**. The relation shown in Table 1.4 has domain

$$D = \{1, 2, 3, 4, 5, 6, 7, 8, 9, 10, 11, 12\} \qquad \textit{x-values}$$

The domain is the set of months.

and range

$$R = \{0.5, 1.1, 1.5, 1.6, 2.0, 2.3, 3.1, 3.6, 3.9, 5.2, 6.2, 6.4\}. \qquad \textit{y-values}$$

The range is the set of average precipitations.

EXAMPLE 2 **Finding the domain and range of a relation**

A physics class measured the time y that it takes for an object to fall x feet, as shown in Table 1.5. The object was dropped twice from each height.

Falling Object

x (feet)	20	20	40	40
y (seconds)	1.2	1.1	1.5	1.6

Table 1.5

(a) Express the data as a relation S.
(b) Find the domain and range of S.

SOLUTION
(a) A relation is a set of ordered pairs, so we can write

$$S = \{(20, 1.2), (20, 1.1), (40, 1.5), (40, 1.6)\}.$$

(b) The domain is the set of x-values of the ordered pairs, or $D = \{20, 40\}$. The range is the set of y-values of the ordered pairs, or $R = \{1.1, 1.2, 1.5, 1.6\}$.

> **NOTE** If an element in the domain or range occurs more than once in a data table, it is listed only once in the set for the domain or the range.

Now Try Exercise 13

Graphing Relations To visualize a relation, we often use the **Cartesian (rectangular) coordinate plane**, or *xy-plane*. The horizontal axis is the *x*-axis and the vertical axis is the *y*-axis. The axes intersect at the **origin** and determine four regions called **quadrants**, numbered I, II, III, and IV, counterclockwise, as shown in Figure 1.11. We can plot the ordered pair (x, y) using the x- and y-axes. A grid is sometimes helpful when plotting points, as shown in Figure 1.12. A point lying on the x-axis or y-axis does not belong to any quadrant. The point $(-3, 0)$ is located on the x-axis, whereas the point $(0, -3)$ lies on the y-axis. Neither point belongs to a quadrant.

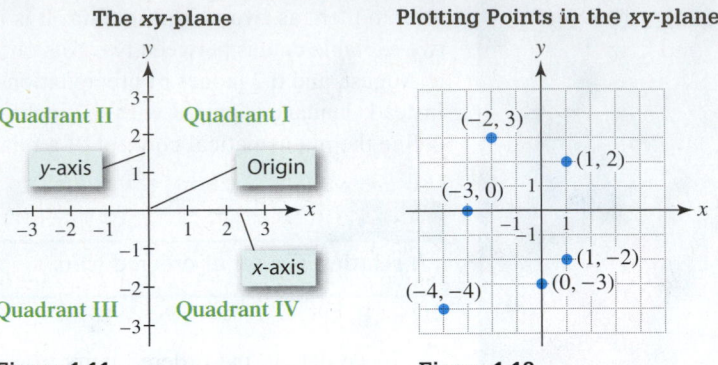

Figure 1.11   **Figure 1.12**

The term **scatterplot** is given to a graph in the xy-plane where distinct points are plotted. Figure 1.12 is an example of a scatterplot.

EXAMPLE 3 Graphing a relation

Complete the following for the relation

$$S = \{(5, 10), (5, -5), (-10, 10), (0, 15), (-15, -10)\}.$$

(a) Find the domain and range of the relation.
(b) Determine the maximum and minimum of the x-values and then of the y-values.
(c) Label appropriate scales on the xy-axes.
(d) Plot the relation as a scatterplot.

SOLUTION

(a) The elements of the domain D correspond to the first number in each ordered pair.

$$D = \{-15, -10, 0, 5\}$$

The elements of the range R correspond to the second number in each ordered pair.

$$R = \{-10, -5, 10, 15\}$$

(b) x-minimum: -15; x-maximum: 5; y-minimum: -10; y-maximum: 15
(c) An appropriate scale for both the x-axis and the y-axis might be -20 to 20, with each tick mark representing a distance of 5. This scale is shown in Figure 1.13.

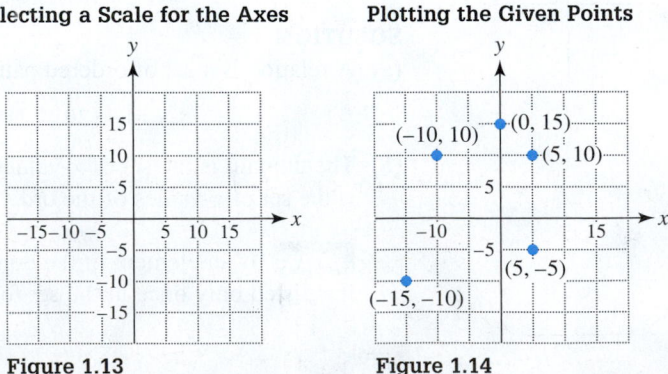

Figure 1.13   **Figure 1.14**

(d) The points in S are plotted in Figure 1.14.

Now Try Exercise 17

Sometimes it is helpful to connect consecutive data points in a scatterplot with straight-line segments. This type of graph, which visually emphasizes changes in the data, is called a **line graph**.

EXAMPLE 4   Making a scatterplot and a line graph

Use Table 1.4 on page 12 to make a scatterplot of average monthly precipitation in Portland, Oregon. Then make a line graph.

SOLUTION

Use the x-axis for the months and the y-axis for the precipitation amounts. To make a scatterplot, simply graph the ordered pairs $(1, 6.2)$, $(2, 3.9)$, $(3, 3.6)$, $(4, 2.3)$, $(5, 2.0)$, $(6, 1.5)$, $(7, 0.5)$, $(8, 1.1)$, $(9, 1.6)$, $(10, 3.1)$, $(11, 5.2)$, and $(12, 6.4)$ in the xy-plane, as shown in Figure 1.15. Then connect consecutive data points to make a line graph, as shown in Figure 1.16.

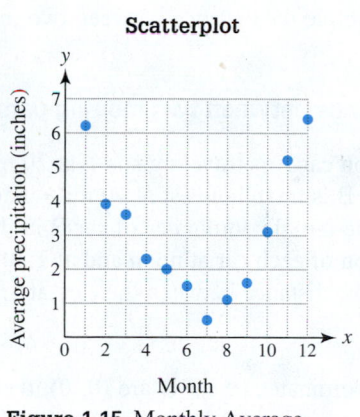

Figure 1.15 Monthly Average Precipitation

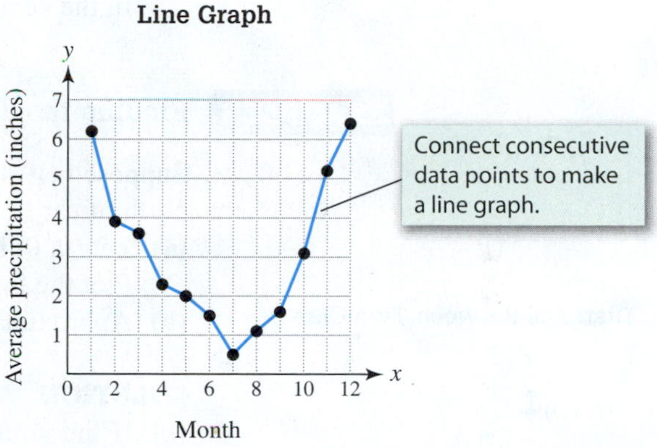

Figure 1.16

Connect consecutive data points to make a line graph.

Now Try Exercise 23

Geometry Review

To review the Pythagorean theorem, see Chapter R (page R-2).

Distance Between Two Points

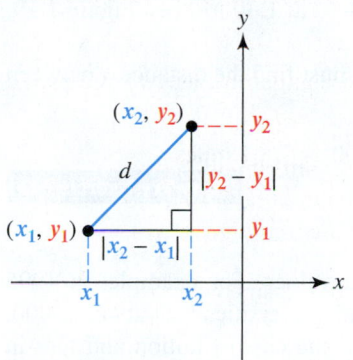

Figure 1.17

The Distance Formula

In the xy-plane, the length of a line segment with endpoints (x_1, y_1) and (x_2, y_2) can be calculated by using the **Pythagorean theorem**. See Figure 1.17.

The lengths of the legs of the right triangle are $|x_2 - x_1|$ and $|y_2 - y_1|$. The distance d is the hypotenuse of the right triangle. Applying the Pythagorean theorem to this triangle gives $d^2 = (x_2 - x_1)^2 + (y_2 - y_1)^2$. Because distance is nonnegative, we can solve this equation for d to get $d = \sqrt{(x_2 - x_1)^2 + (y_2 - y_1)^2}$.

DISTANCE FORMULA

The **distance** d between the points (x_1, y_1) and (x_2, y_2) in the xy-plane is

$$d = \sqrt{(x_2 - x_1)^2 + (y_2 - y_1)^2}.$$

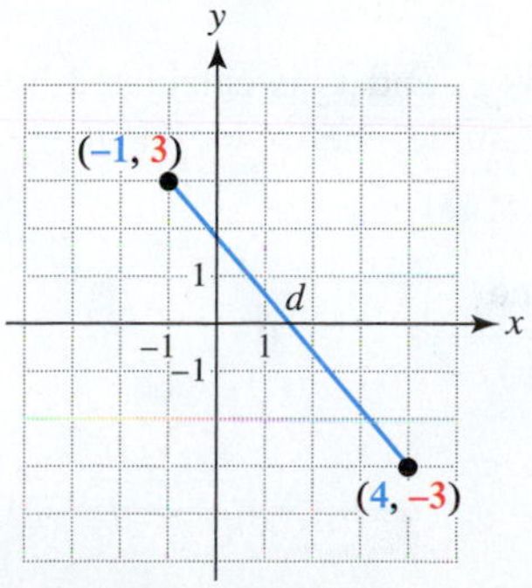

Figure 1.18 shows a line segment connecting the points $(-1, 3)$ and $(4, -3)$. Its length is

$$d = \sqrt{(4 - (-1))^2 + (-3 - 3)^2} \quad \text{Distance formula}$$
$$= \sqrt{61} \quad \text{Exact length}$$
$$\approx 7.81. \quad \text{Approximate.}$$

Figure 1.18

EXAMPLE 5 **Finding the distance between two points**

Find the exact distance between $(3, -4)$ and $(-2, 7)$. Then approximate this distance to the nearest hundredth.

SOLUTION

In the distance formula, let (x_1, y_1) be $(3, -4)$ and (x_2, y_2) be $(-2, 7)$.

$$\sqrt{(x_2 - x_1)^2 + (y_2 - y_1)^2} = \sqrt{(-2 - 3)^2 + (7 - (-4))^2} \qquad \text{Distance formula}$$
$$= \sqrt{(-5)^2 + 11^2} \qquad \text{Subtract.}$$
$$= \sqrt{146} \qquad \text{Simplify.}$$
$$\approx 12.08 \qquad \text{Approximate.}$$

The *exact* distance is $\sqrt{146}$, and the *approximate* distance, rounded to the nearest hundredth, is 12.08. Note that we would obtain the same answer if we let (x_1, y_1) be $(-2, 7)$ and (x_2, y_2) be $(3, -4)$.

Now Try Exercise 27

In the next example the distance between two moving cars is found.

EXAMPLE 6 **Finding the distance between two moving cars**

Suppose that at noon car A is traveling south at 20 miles per hour and is located 80 miles north of car B. Car B is traveling east at 40 miles per hour.
(a) Let $(0, 0)$ be the initial coordinates of car B in the xy-plane, where units are in miles. Plot the location of each car at noon and at 1:30 P.M.
(b) Approximate the distance between the cars at 1:30 P.M.

Distance Between Two Cars

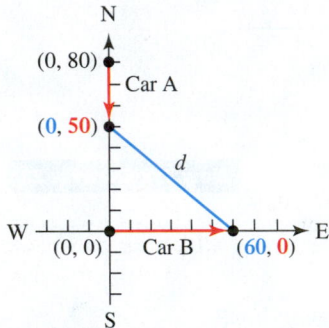

Figure 1.19

SOLUTION

(a) If the initial coordinates of car B are $(0, 0)$, then the initial coordinates of car A are $(0, 80)$, because car A is 80 miles north of car B. After 1 hour and 30 minutes, or 1.5 hours, car A has traveled $1.5 \times 20 = 30$ miles south, and so it is located 50 miles north of the initial location of car B. Thus its coordinates are $(0, 50)$ at 1:30 PM. Car B traveled $1.5 \times 40 = 60$ miles east, so its coordinates are $(60, 0)$ at 1:30 PM. See Figure 1.19, where these points are plotted.
(b) To find the distance between the cars at 1:30 P.M. we must find the distance d between the points $(0, 50)$ and $(60, 0)$.

$$d = \sqrt{(60 - 0)^2 + (0 - 50)^2} = \sqrt{6100} \approx 78.1 \text{ miles}$$

Now Try Exercise 43

The Midpoint Formula

A common way to make estimations is to average data values. For example, in 1995 the average cost of tuition and fees at public colleges and universities was about \$4000, whereas in 2005 it was about \$6000. One might estimate the cost of tuition and fees in 2000 to be \$5000. This type of averaging is referred to as finding the midpoint. If a line segment is drawn between two data points, then its *midpoint* is the unique point on the line segment that is equidistant from the endpoints.

On a real number line, the midpoint M of two data points x_1 and x_2 is calculated by averaging their coordinates, as shown in Figure 1.20. For example, the midpoint of -3 and 5 is $M = \frac{-3 + 5}{2} = 1$.

Finding the Midpoint on a Number Line

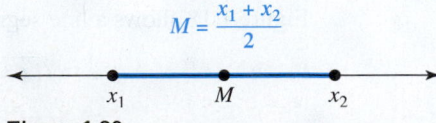

Figure 1.20

Finding the Midpoint

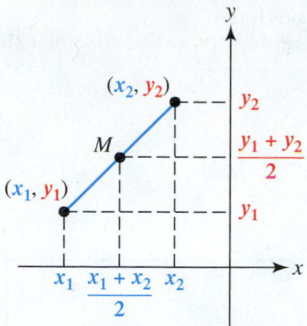

Figure 1.21

The midpoint formula in the xy-plane is similar to the formula for the real number line, except that both coordinates are averaged. Figure 1.21 shows midpoint M located on the line segment connecting the two data points (x_1, y_1) and (x_2, y_2). The x-coordinate of the midpoint M is located halfway between x_1 and x_2 and is $\frac{x_1 + x_2}{2}$. Similarly, the y-coordinate of M is the average of y_1 and y_2. For example, if we let (x_1, y_1) be $(-3, 1)$ and (x_2, y_2) be $(-1, 3)$ in Figure 1.21, then the midpoint is computed by

$$\left(\frac{-3 + -1}{2}, \frac{1 + 3}{2} \right) = (-2, 2).$$

MIDPOINT FORMULA IN THE XY-PLANE

The **midpoint** of the line segment with endpoints (x_1, y_1) and (x_2, y_2) in the xy-plane is

$$\left(\frac{x_1 + x_2}{2}, \frac{y_1 + y_2}{2} \right).$$

EXAMPLE 7   Finding the midpoint

Find the midpoint of the line segment connecting the points $(6, -7)$ and $(-4, 6)$.

SOLUTION

In the midpoint formula, let (x_1, y_1) be $(6, -7)$ and (x_2, y_2) be $(-4, 6)$. Then the midpoint M can be found as follows.

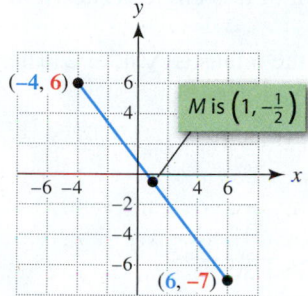

Figure 1.22

$$M = \left(\frac{x_1 + x_2}{2}, \frac{y_1 + y_2}{2} \right) \qquad \text{Midpoint formula}$$

$$= \left(\frac{6 + (-4)}{2}, \frac{-7 + 6}{2} \right) \qquad \text{Substitute.}$$

$$= \left(1, -\frac{1}{2} \right) \qquad \text{Simplify}$$

The midpoint M of this line segment is shown in Figure 1.22

Now Try Exercise 49

EXAMPLE 8   Estimating U.S. population

In 1990 the population of the United States was 249 million, and by 2010 it had increased to 308 million. Use the midpoint formula to estimate the population in 2000. (**Source:** Bureau of the Census.)

SOLUTION

The U.S. populations in 1990 and 2010 are given by the points $(1990, 249)$ and $(2010, 308)$. The midpoint M of the line segment connecting these points is

$$M = \left(\frac{1990 + 2010}{2}, \frac{249 + 308}{2} \right) = (2000, 278.5).$$

The midpoint formula estimates a population of 278.5 million in 2000 (The actual population was 281 million.)

Now Try Exercise 45

Circles

Applying the Distance Formula to Circles A **circle** consists of the set of points in a plane that are equidistant from a fixed point. The distance is called the **radius** of the circle, and the fixed point is called the **center**. If we let the center of the circle be (h, k),

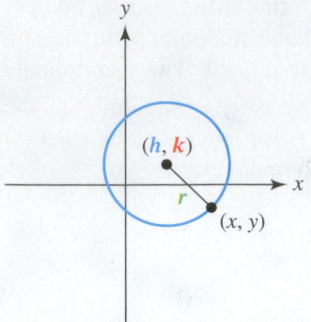

Figure 1.23

the radius be r, and (x, y) be any point on the circle, then the distance between (x, y) and (h, k) must equal r. See Figure 1.23. By the distance formula we have

$$\sqrt{(x - h)^2 + (y - k)^2} = r.$$

Squaring each side gives

$$(x - h)^2 + (y - k)^2 = r^2.$$

STANDARD EQUATION OF A CIRCLE

The circle with center (h, k) and radius r has equation

$$(x - h)^2 + (y - k)^2 = r^2.$$

NOTE If the center of a circle is $(0, 0)$, then the equation simplifies to $x^2 + y^2 = r^2$. For example, the equation of the circle with center $(0, 0)$ and radius 7 is $x^2 + y^2 = 49$.

EXAMPLE 9 Finding the center and radius of a circle

Find the center and radius of the circle with the given equation. Graph each circle.
(a) $x^2 + y^2 = 9$ **(b)** $(x - 1)^2 + (y + 2)^2 = 4$

SOLUTION
(a) Because the equation $x^2 + y^2 = 9$ can be written as $(x - \mathbf{0})^2 + (y - \mathbf{0})^2 = 3^2$, the center is $(\mathbf{0}, \mathbf{0})$ and the radius is $\sqrt{9}$, or $\mathbf{3}$. The graph of this circle is shown in Figure 1.24.
(b) For $(x - 1)^2 + (y + 2)^2 = 4$, the center is $(1, -2)$ and the radius is $\sqrt{4}$, or 2. The graph is shown in Figure 1.25.

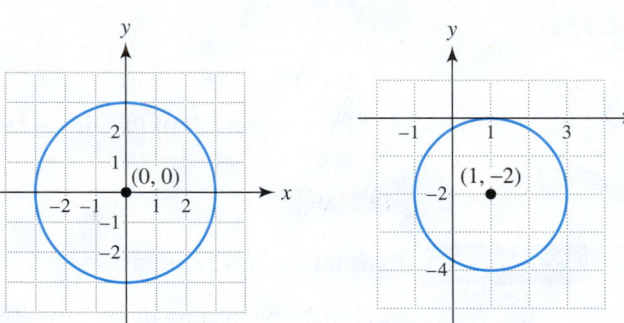

Figure 1.24   **Figure 1.25**

Now Try Exercises 59, 63, 81, and 83

EXAMPLE 10 Finding the equation of a circle

Find the equation of the circle that satisfies the conditions. Graph each circle.
(a) Radius 4, center $(-3, 5)$
(b) Center $(6, -3)$ with the point $(1, 2)$ on the circle

SOLUTION
(a) Let $r = \mathbf{4}$ and $(h, k) = (\mathbf{-3, 5})$. The equation of this circle is

$$(x - (\mathbf{-3}))^2 + (y - \mathbf{5})^2 = \mathbf{4}^2 \quad \text{or} \quad (x + 3)^2 + (y - 5)^2 = 16.$$

A graph of the circle is shown in Figure 1.26.
(b) First we must find the distance between the points $(\mathbf{6, -3})$ and $(\mathbf{1, 2})$ to determine r.

$$r = \sqrt{(\mathbf{6} - 1)^2 + (\mathbf{-3} - 2)^2} = \sqrt{50} \approx 7.1$$

Since $r^2 = 50$, the equation of the circle is $(x - 6)^2 + (y + 3)^2 = 50$. Its graph is shown in Figure 1.27.

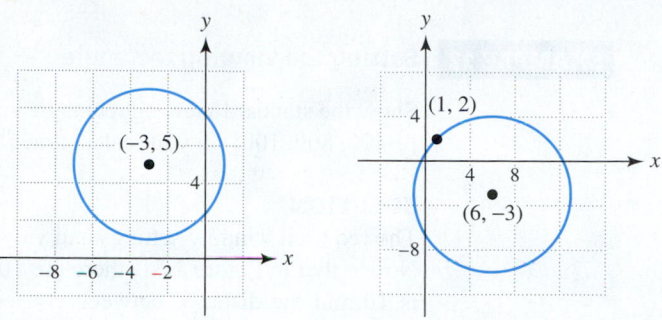

Figure 1.26   Figure 1.27

Now Try Exercises 71, 75, 85, and 87

EXAMPLE 11 **Finding the equation of a circle**

The diameter of a circle is shown in Figure 1.28. Find the standard equation of the circle.

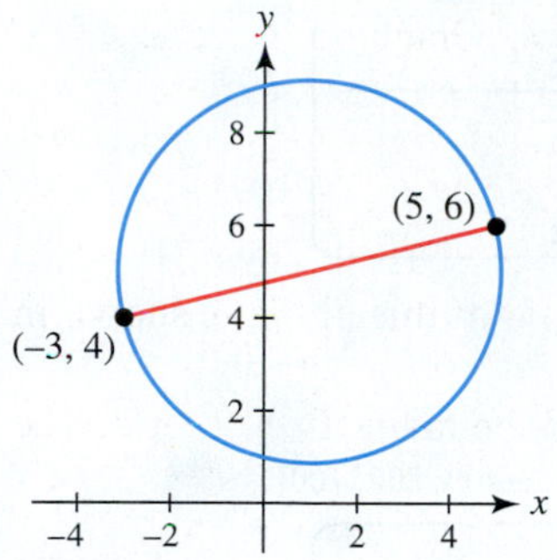

SOLUTION

Getting Started First find the center of the circle, which is the midpoint of a diameter. Then find the radius by calculating the distance between the center and one of the given endpoints of the diameter. ▶

Find the center C of the circle by applying the midpoint formula to the endpoints of the diameter $(-3, 4)$ and $(5, 6)$.

$$C = \left(\frac{-3 + 5}{2}, \frac{4 + 6}{2}\right) = (1, 5)$$

Use the distance formula to find the radius, which equals the distance from the center $(1, 5)$ to the endpoint $(5, 6)$.

$$r = \sqrt{(5 - 1)^2 + (6 - 5)^2} = \sqrt{17}$$

Thus r^2 equals 17, and the standard equation is $(x - 1)^2 + (y - 5)^2 = 17$.

Now Try Exercise 79

Figure 1.28

Graphing with a Calculator (Optional)

Graphing calculators can be used to create tables, scatterplots, line graphs, and other types of graphs. The **viewing rectangle**, or **window**, on a graphing calculator is similar to the view finder in a camera. A camera cannot take a picture of an entire scene; it must be centered on a portion of the available scenery and then it can capture different views of the same scene by zooming in and out. Graphing calculators have similar capabilities. The calculator screen can show only a finite, rectangular region of the xy-plane, which is infinite. The viewing rectangle must be specified by setting minimum and maximum values for both the x- and y-axes before a graph can be drawn.

We will use the following terminology to describe a viewing rectangle. **Xmin** is the minimum x-value and **Xmax** is the maximum x-value along the x-axis. Similarly, **Ymin** is the minimum y-value and **Ymax** is the maximum y-value along the y-axis. Most graphs show an x-scale and a y-scale using tick marks on the respective axes. The distance represented by consecutive tick marks on the x-axis is called **Xscl**, and the distance represented by consecutive tick marks on the y-axis is called **Yscl**. See Figure 1.29. This information can be written concisely as

Identifying the Window Size

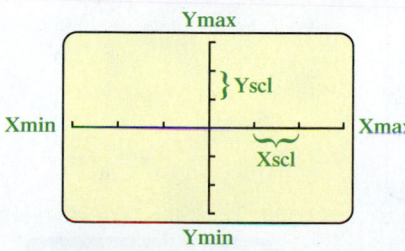

Figure 1.29

$$[\textbf{Xmin, Xmax, Xscl}] \text{ by } [\textbf{Ymin, Ymax, Yscl}].$$

For example, $[-10, 10, 1]$ by $[-10, 10, 1]$ means that Xmin $= -10$, Xmax $= 10$, Xscl $= 1$, Ymin $= -10$, Ymax $= 10$, and Yscl $= 1$. This setting is referred to as the **standard viewing rectangle**.

EXAMPLE 12   Setting the viewing rectangle

Show the standard viewing rectangle and the viewing rectangle given by $[-30, 40, 10]$ by $[-400, 800, 100]$ on your calculator.

SOLUTION

The required window settings and viewing rectangles are displayed in Figures 1.30–1.33. Notice that in Figure 1.31, there are 10 tick marks on the positive x-axis, since its length is 10 and the distance between consecutive tick marks is 1. Note that **Xres** is usually equal to 1.

The Standard Viewing Rectangle

$[-10, 10, 1]$ by $[-10, 10, 1]$

Calculator Help

To set a viewing rectangle or window, see Appendix A (page AP-3).

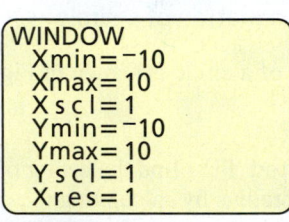

Figure 1.30

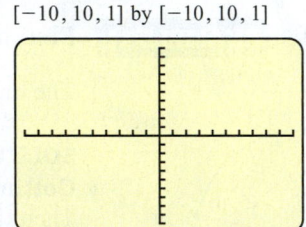

Figure 1.31

Setting a Different Window

$[-30, 40, 10]$ by $[-400, 800, 100]$

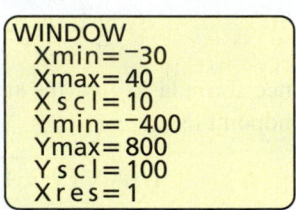

Figure 1.32

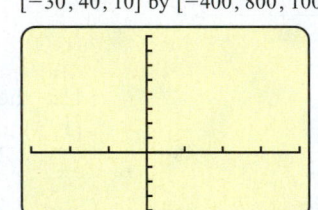

Figure 1.33

Now Try Exercise 93

EXAMPLE 13   Making a scatterplot with a graphing calculator

Plot the points $(-5, -5)$, $(-2, 3)$, $(1, -7)$, and $(4, 8)$ in the standard viewing rectangle.

SOLUTION

The standard viewing rectangle is given by $[-10, 10, 1]$ by $[-10, 10, 1]$. The points $(-5, -5)$, $(-2, 3)$, $(1, -7)$, and $(4, 8)$ are plotted in Figure 1.34.

$[-10, 10, 1]$ by $[-10, 10, 1]$

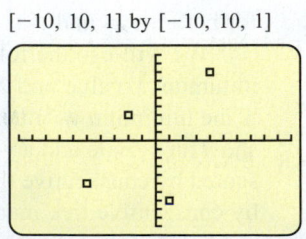

Figure 1.34

Calculator Help

To make the scatterplot in Figure 1.34, see Appendix A (page AP-3).

Now Try Exercise 101

EXAMPLE 14 Creating a line graph with a graphing calculator

Table 1.6 lists the percentage of music album sales accounted for by compact discs (CDs) from 1990 to 2010. Make a line graph of these sales in [1988, 2012, 2] by [0, 100, 20].

CD Album Sales

Year	1990	1995	2000	2005	2010
CDs (% share)	31	65	89	90	74

Source: Recording Industry Association of America.

Table 1.6

SOLUTION

Enter the data points (1990, 31), (1995, 65), (2000, 89), (2005, 90), and (2010, 74). A line graph can be created by selecting this option on your graphing calculator. The graph is shown in Figure 1.35.

Calculator Help

To make a line graph, see Appendix A (page AP-3).

[1988, 2012, 2] by [0, 100, 20]

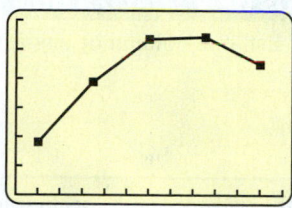

Figure 1.35

Now Try Exercise 105

1.2 Putting It All Together

CONCEPT	EXPLANATION	EXAMPLES
Mean, or average	To find the mean, or average, of n numbers, divide their sum by n.	The mean of the four numbers $-3, 5, 6, 9$ is $$\frac{-3 + 5 + 6 + 9}{4} = 4.25.$$
Median	The median of a sorted list of numbers equals the value that is located in the middle of the list. Half the data are greater than or equal to the median, and half the data are less than or equal to the median.	The median of 2, 3, 6, 9, 11 is 6, the middle data item. The median of 2, 3, 6, 9 is the average of the two middle values: 3 and 6. Therefore the median is $\frac{3+6}{2} = 4.5$.
Relation, domain, and range	A relation is a set of ordered pairs (x, y). The set of x-values is called the domain, and the set of y-values is called the range.	The relation $S = \{(1, 3), (2, 5), (1, 6)\}$ has domain $D = \{1, 2\}$ and range $R = \{3, 5, 6\}$.
Distance formula	The distance between (x_1, y_1) and (x_2, y_2) is $$d = \sqrt{(x_2 - x_1)^2 + (y_2 - y_1)^2}.$$	The distance between $(2, -1)$ and $(-1, 3)$ is $$d = \sqrt{(-1 - 2)^2 + (3 - (-1))^2} = 5.$$

continued on next page

CONCEPT	EXPLANATION	EXAMPLES
Midpoint formula	The midpoint of the line segment connecting (x_1, y_1) and (x_2, y_2) is $$M = \left(\frac{x_1 + x_2}{2}, \frac{y_1 + y_2}{2}\right).$$	The midpoint of the line segment connecting $(4, 3)$ and $(-2, 5)$ is $$M = \left(\frac{4 + (-2)}{2}, \frac{3 + 5}{2}\right) = (1, 4).$$
Standard equation of a circle	The circle with center (h, k) and radius r has the equation $$(x - h)^2 + (y - k)^2 = r^2.$$	The circle with center $(-3, 4)$ and radius 5 has the equation $$(x + 3)^2 + (y - 4)^2 = 25.$$

The next table summarizes some basic concepts related to one-variable and two-variable data.

TYPE OF DATA	METHODS OF VISUALIZATION	COMMENTS
One-variable data	Number line, list, one-column or one-row table	The data items are the same type and can be described using x-values. Computations of the mean and median are performed on one-variable data.
Two-variable data	Two-column or two-row table, scatterplot, line graph or other type of graph in the xy-plane	Two types of data are related, can be described by using ordered pairs (x, y), and are often called a relation.

1.2 Exercises

Data Involving One Variable

Exercises 1–4: For the table of data, complete the following.

(a) *Plot the numbers on a number line.*
(b) *Find the maximum and minimum of the data.*
(c) *Determine the mean of the data.*

1. | 3 | −2 | 5 | 0 | 6 | −1 |
|---|---|---|---|---|---|

2. | 5 | −3 | 4 | −2 | 1 | 6 |
|---|---|---|---|---|---|

3. | −10 | 20 | 30 | −20 | 0 | 10 |
|---|---|---|---|---|---|

4. | 0.5 | −1.5 | 2.0 | 4.5 | −3.5 | −1.0 |
|---|---|---|---|---|---|

Exercises 5–8: Sort the list of numbers from smallest to largest and display the result in a table.

(a) *Determine the maximum and minimum values.*
(b) *Calculate the mean and median. Round each result to the nearest hundredth when appropriate.*

5. $-10, 25, 15, -30, 55, 61, -30, 45, 5$

6. $-1.25, 4.75, -3.5, 1.5, 2.5, 4.75, 1.5$

7. $\sqrt{15}, 2^{2.3}, \sqrt[3]{69}, \pi^2, 2^\pi, 4.1$

8. $\frac{22}{7}, 3.14, \sqrt[3]{28}, \sqrt{9.4}, 4^{0.9}, 3^{1.2}$

Exercises 9 and 10: **Geography**   *The set of numbers contains data about geographic features of the world.*

(a) *Plot the numbers on a number line.*
(b) *Calculate the mean and median for the set of numbers. Interpret your results.*
(c) *Try to identify the geographic feature associated with the largest number in the set.*

9. $\{31.7, 22.3, 12.3, 26.8, 24.9, 23.0\}$ (Areas of largest freshwater lakes in thousands of square miles) (**Source:** U.S. National Oceanic and Atmospheric Administration.)

10. $\{19.3, 18.5, 29.0, 7.31, 16.1, 22.8, 20.3\}$ (Highest elevations of the continents in thousands of feet) (**Source:** National Geographic Atlas of the World.)

11. **Designing a Data Set**   Find a set of three numbers with a mean of 20 and a median of 18. Is your answer unique?

12. **Designing a Data Set**   Find a set of five numbers with a mean of 10 and a median of 9. Is your answer unique?

Data Involving Two Variables

Exercises 13–16: For the table of data, complete the following.

(a) *Express the data as a relation S.*
(b) *Find the domain and range of S.*

13.
x	−1	2	3	5	9
y	5	2	−1	−4	−5

14.
x	−2	0	2	4	6
y	−4	−2	−1	0	4

15.
x	1	4	5	4	1
y	5	5	6	6	5

16.
x	−1	0	3	−1	−2
y	$\frac{1}{2}$	1	$\frac{3}{4}$	3	$-\frac{5}{6}$

Exercises 17–22: Complete the following.

(a) *Find the domain and range of the relation.*
(b) *Determine the maximum and minimum of the x-values and then of the y-values.*
(c) *Label appropriate scales on the xy-axes.*
(d) *Plot the relation.*

17. $\{(0, 5), (-3, 4), (-2, -5), (7, -3), (0, 0)\}$

18. $\{(1, 1), (3, 0), (-5, -5), (8, -2), (0, 3)\}$

19. $\{(2, 2), (-3, 1), (-4, -1), (-1, 3), (0, -2)\}$

20. $\{(1, 1), (2, -3), (-1, -1), (-1, 2), (-1, 0)\}$

21. $\{(10, 50), (-35, 45), (0, -55), (75, 25), (-25, -25)\}$

22. $\{(-1.2, 1.5), (1.0, 0.5), (-0.3, 1.1), (-0.8, -1.3)\}$

Exercises 23 and 24: **Plotting Real Data**   *Use the table to make a scatterplot and line graph of the data.*

23. Global Cell Phone Subscribers (millions)

Year	1990	1995	2000	2005	2011
Subscribers	5	34	109	208	324

Source: CTIA–The Wireless Association.

24. Atmospheric CO_2 Levels (parts per million)

Year	1958	1975	1990	2005	2011
CO_2 Amounts	315	335	355	380	392

Source: Mauna Loa Observatory.

Distance Formula

Exercises 25–40: Find the exact distance between the two points. Where appropriate, also give approximate results to the nearest hundredth.

25. $(2, -2), (5, 2)$

26. $(0, -3), (12, -8)$

27. $(7, -4), (9, 1)$

28. $(-1, -6), (-8, -5)$

29. $(3.6, 5.7), (-2.1, 8.7)$

30. $(-6.5, 2.7), (3.6, -2.9)$

31. $(-3, 2), (-3, 10)$

32. $(7, 9), (-1, 9)$

33. $\left(\frac{1}{2}, -\frac{1}{2}\right), \left(\frac{3}{4}, \frac{1}{2}\right)$

34. $\left(-\frac{1}{3}, \frac{2}{3}\right), \left(\frac{1}{3}, -\frac{4}{3}\right)$

35. $\left(\frac{2}{5}, \frac{3}{10}\right), \left(-\frac{1}{10}, \frac{4}{5}\right)$

36. $\left(-\frac{1}{2}, \frac{2}{3}\right), \left(\frac{1}{3}, -\frac{5}{2}\right)$

37. $(20, 30), (-30, -90)$

38. $(40, 6), (-20, 17)$

39. $(a, 0), (0, -b)$

40. $(x, y), (1, 2)$

41. **Geometry**   An **isosceles triangle** has at least two sides of equal length. Determine whether the triangle with vertices $(0, 0)$, $(3, 4)$, $(7, 1)$ is isosceles.

42. **Geometry**   An **equilateral triangle** has sides of equal length. Determine whether the triangle with vertices $(-1, -1)$, $(2, 3)$, $(-4, 3)$ is equilateral.

43. **Distance between Cars**   (Refer to Example 6.) At 9:00 A.M. car A is traveling north at 50 miles per hour and is located 50 miles south of car B. Car B is traveling west at 20 miles per hour.
    (a) Let $(0, 0)$ be the initial coordinates of car B in the xy-plane, where units are in miles. Plot the locations of each car at 9:00 A.M. and at 11:00 A.M.

    (b) Find the distance d between the cars at 11:00 A.M.

44. **Distance between Ships**   Two ships leave a harbor at the same time. The first ship heads north at 20 miles per hour, and the second ship heads west at 15 miles per hour. Write an expression that gives the distance d between the ships after t hours.

Midpoint Formula

Exercises 45–48: Use the midpoint formula for the following.

45. **Nintendo Wii**   Six months after Nintendo Wii was introduced it had sold 10 million units, and after 24 months it had sold 44 million units. Estimate the number of units sold 15 months after the Nintendo Wii was introduced. (The actual value was 25 million units.) (*Source:* Company Reports.)

46. World Population In 1874 the world population was 1.24 billion. By 2050 this number is expected to be 9 billion. Estimate the world population in 1962. (The actual value was 3.14 billion.) (*Source:* United Nations Data.)

47. Olympic Times In the Olympic Games, the 200-meter dash is run in approximately 19 seconds. Estimate the time to run the 100-meter dash.

48. Real Numbers Between any two real numbers a and b there is always another real number. How could such a number be found?

Exercises 49–58: Find the midpoint of the line segment connecting the points.

49. $(1, 2)$, $(5, -3)$ **50.** $(-6, 7)$, $(9, -4)$

51. $(-30, 50)$, $(50, -30)$ **52.** $(28, -33)$, $(52, 38)$

53. $(1.5, 2.9)$, $(-5.7, -3.6)$ **54.** $(9.4, -4.5)$, $(-7.7, 9.5)$

55. $(\sqrt{2}, \sqrt{5})$, $(\sqrt{2}, -\sqrt{5})$

56. $(\sqrt{7}, 3\sqrt{3})$, $(-\sqrt{7}, -\sqrt{3})$

57. (a, b), $(-a, 3b)$ **58.** $(-a, b)$, $(3a, b)$

Circles

Exercises 59–66: Find the center and radius of the circle.

59. $x^2 + y^2 = 25$ **60.** $x^2 + y^2 = 100$

61. $x^2 + y^2 = 7$ **62.** $x^2 + y^2 = 20$

63. $(x - 2)^2 + (y + 3)^2 = 9$

64. $(x + 1)^2 + (y - 1)^2 = 16$

65. $x^2 + (y + 1)^2 = 100$ **66.** $(x - 5)^2 + y^2 = 19$

Exercises 67–70: Find the standard equation of the circle.

67.
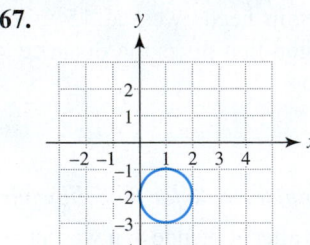

68.
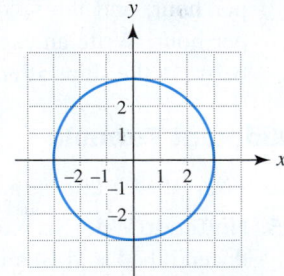

69.
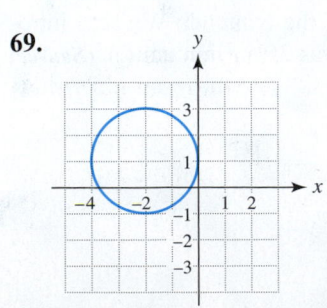

70.
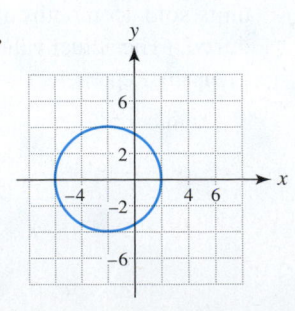

Exercises 71–78: Find the standard equation of a circle that satisfies the conditions.

71. Radius 8, center $(3, -5)$

72. Radius 5, center $(-1, 4)$

73. Radius 7, center $(3, 0)$

74. Radius 1, center $(0, 0)$

75. Center $(3, -5)$ with the point $(4, 2)$ on the circle

76. Center $(0, 0)$ with the point $(-3, -1)$ on the circle

77. Endpoints of a diameter $(-5, -7)$ and $(1, 1)$

78. Endpoints of a diameter $(-3, -2)$ and $(1, -4)$

Exercises 79 and 80: (Refer to Example 11.) Use the diameter to find the standard equation of the circle shown.

79.
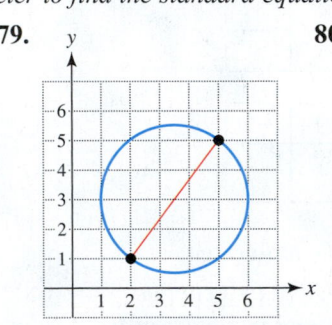

80.
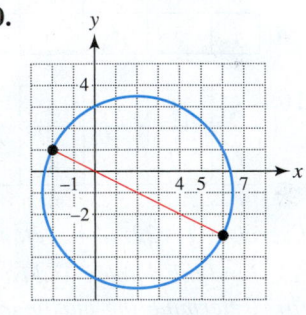

Exercises 81–90: Graph the equation.

81. $x^2 + y^2 = 25$ **82.** $x^2 + y^2 = 1$

83. $(x - 2)^2 + (y + 1)^2 = 25$

84. $(x + 2)^2 + (y + 3)^2 = 1$

85. $(x + 2)^2 + y^2 = 16$ **86.** $x^2 + (y - 2)^2 = 9$

87. $(x + 3)^2 + (y + 1)^2 = 9$

88. $(x - 1)^2 + (y - 2)^2 = 4$

89. $\left(x - \frac{1}{2}\right)^2 + (y - 1)^2 = 10$

90. $(x + 1)^2 + \left(y + \frac{3}{2}\right)^2 = 6$

Graphing Calculators

Exercises 91–96: Predict the number of tick marks on the positive x-axis and the positive y-axis. Then show the viewing rectangle on your graphing calculator.

91. Standard viewing rectangle

92. $[-4.7, 4.7, 1]$ by $[-3.1, 3.1, 1]$

93. $[0, 100, 10]$ by $[-50, 50, 10]$

94. [−30, 30, 5] by [−20, 20, 5]

95. [1980, 1995, 1] by [12000, 16000, 1000]

96. [1800, 2000, 20] by [5, 20, 5]

Exercises 97–100: Match the settings for a viewing rectangle with the correct figure (a–d).

97. [−9, 9, 1] by [−6, 6, 1]

98. [−6, 6, 1] by [−9, 9, 1]

99. [−2, 2, 0.5] by [−4.5, 4.5, 0.5]

100. [−4, 8, 1] by [−600, 600, 100]

a. 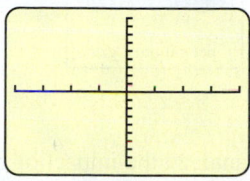   **b.**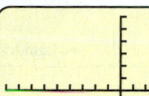

c. 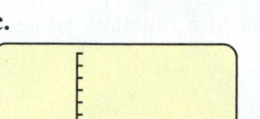   **d.**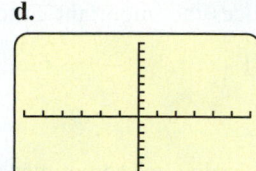

Exercises 101–104: Make a scatterplot of the relation.

101. $\{(1, 3), (−2, 2), (−4, 1), (−2, −4), (0, 2)\}$

102. $\{(6, 8), (−4, −10), (−2, −6), (2, −5)\}$

103. $\{(10, −20), (−40, 50), (30, 60), (−50, −80), (70, 0)\}$

104. $\{(−1.2, 0.6), (1.0, −0.5), (0.4, 0.2), (−2.8, 1.4)\}$

Exercises 105–108: The table contains real data.

(a) Determine the maximum and minimum values for each variable in the table.

(b) Use your results from part (a) to find an appropriate viewing rectangle.

(c) Make a scatterplot of the data.

(d) Make a line graph of the data.

105. Netflix Subscriptions (millions)

x	2006	2007	2008	2009	2010
y	6.1	7.3	9.1	11.9	19.4

Source: Company Reports.

106. Global Vehicle Sales That Are Electric/Hybrid

x	2011	2015	2019	2022	2025
y	0%	1%	2%	4.5%	8%

Source: Business Insider (projected).

107. MySpace U.S. Advertising Revenue ($ millions)

x	2006	2007	2008	2009	2010	2011
y	225	475	590	430	270	180

Source: eMarketer.

108. U.S. College Students Who Study Chinese (thousands)

x	1995	1998	2002	2006	2009
y	26	29	34	51	60

Source: Institute of International Education.

Writing about Mathematics

109. Give an example of a relation that has meaning in the real world. Give an example of an ordered pair (x, y) that is in your relation. Does the ordered pair (y, x) also have meaning? Explain your answers.

110. Do the mean and median represent the same thing? Explain your answer and give an example.

CHECKING BASIC CONCEPTS FOR SECTIONS 1.1 AND 1.2

1. Approximate each expression to the nearest hundredth.

(a) $\sqrt{4.2(23.1 + 0.5^3)}$ (b) $\dfrac{23 + 44}{85.1 − 32.9}$

2. Evaluate the expression by hand.
(a) $5 − (−4)^2 \cdot 3$ (b) $5 ÷ 5\sqrt{2} + 2$

3. Write each number using scientific notation.
(a) 348,500,000   (b) −1237.4

(c) 0.00198

4. Find the exact distance between the points $(−3, 1)$ and $(3, −5)$. Then round this distance to the nearest hundredth.

5. Find the midpoint of the line segment connecting the points $(−2, 3)$ and $(4, 2)$.

6. Find the standard equation of a circle with center $(−4, 5)$ and radius 8.

7. The average depths in feet of four oceans are 13,215, 12,881, 13,002, and 3953. Calculate the mean and median of these depths.

8. Make a scatterplot and a line graph with the four points $(−5, −4)$, $(−1, 2)$, $(2, −2)$, and $(3, 6)$. State the quadrant in which each point lies.

1.3 Functions and Their Representations

- Learn function notation
- Represent a function four different ways
- Define a function formally
- Identify the domain and range of a function
- Use calculators to represent functions (optional)
- Identify functions
- Represent functions with diagrams and equations

Introduction

Because there are more than 300 million people in the United States who consume many natural resources, *going green* has become an important social and environmental issue. Figure 1.36 gives some information about U.S. consumption.

Population and Consumption in the United States

313,000,000 People	2,200 Napkins	40 Gallons H_2O	19.4 Pounds CO_2
in 2012	per person per year	per laundry load	per gallon of gas

Figure 1.36

The mathematical concept of a function can be used to analyze the impact of human consumption (see Exercises 101 and 102) and also to describe natural phenomena, such as lightning. This section introduces the important concept of a function, which is used throughout the course.

Basic Concepts

Although thunder is caused by lightning, we sometimes see a flash of lightning before we hear the thunder. The farther away lightning is, the greater the time lapse between seeing the flash of lightning and hearing the thunder. Table 1.7 lists the *approximate* distance y in miles between a person and a bolt of lightning when there is a time lapse of x seconds between seeing the lightning and hearing the thunder.

Time lapse between seeing lightning and hearing thunder

Distance from lightning

Distance from Lightning

x (seconds)	5	10	15	20	25
y (miles)	1	2	3	4	5

Divide x by 5 to get y.

Table 1.7

Table 1.7 established a special type of relation between x and y, called a *function*.

- Each x determines *exactly* one y, so we say that Table 1.7 *represents* or *defines* a function f.
- Function f *computes* the distance y between an observer and a lightning bolt, given the time lapse x. We say that y is a function of x.
- This computation is denoted $y = f(x)$, which is called **function notation** and is read "y equals f of x." It means that function f with **input** x produces **output** y. That is,

$$f(\textbf{Input}) = \textbf{Output}$$

or

$$f(\textbf{Time lapse in seconds}) = \textbf{Miles from lightning}.$$

We can represent the five values in Table 1.7 in function notation as

$$f(5) = 1, \ f(10) = 2, \ f(15) = 3, \ f(20) = 4, \ f(25) = 5,$$

A **5** second delay means the lightning bolt was about **1** mile away.

or more generally as,

$$f(x) = y.$$

> The expression $f(x)$ represents the **output y** from f when the **input** is x.

Function f calculates a set of ordered pairs (x, y), where $y = f(x)$. From Table 1.7, the five ordered pairs $(5, 1)$, $(10, 2)$, $(15, 3)$, $(20, 4)$, and $(25, 5)$ belong to the relation computed by f. We can think of these ordered pairs as input-output pairs in the form (**input**, **output**).

A given x-value determines exactly one y-value. For example, if x equals 30 seconds, then y equals $\frac{30}{5} = 6$ miles. Thus the y-values (outputs) *depend* on the x-values (inputs). We call x the **independent variable** and y the **dependent variable**.

Computation Performed by Function f

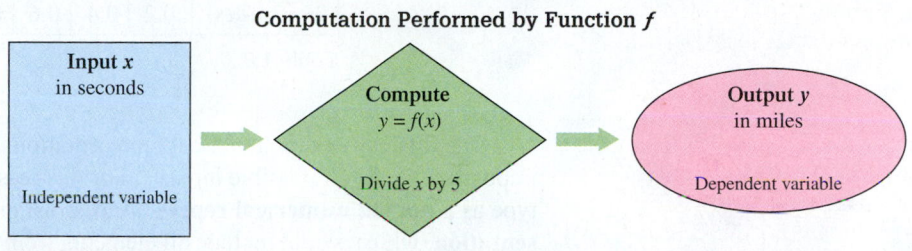

Figure 1.37

In Section 1.2 we discussed the domain and range of a relation. For a function, the set of valid or meaningful inputs x is called the **domain**, and the set of corresponding outputs y is the **range**. For example, suppose that a function f computes the height after x seconds of a ball that was thrown into the air. Then the domain of f consists of all times that the ball was in flight, and the range includes all heights attained by the ball.

This discussion about function notation is summarized as follows.

See the Concept: Function Notation

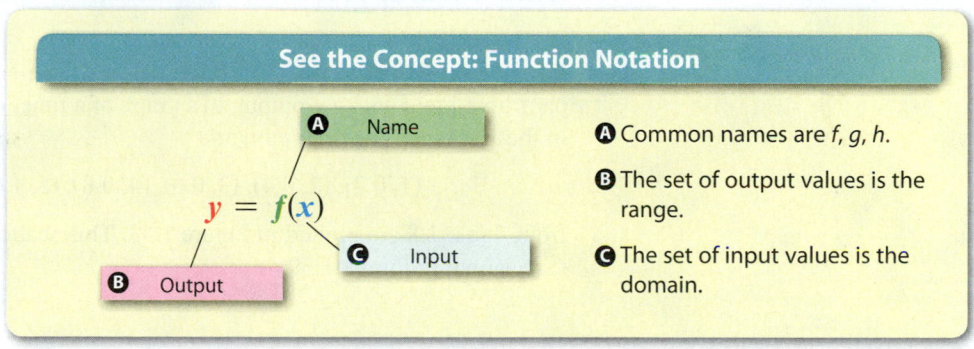

A Name
B Output
C Input

A Common names are f, g, h.
B The set of output values is the range.
C The set of input values is the domain.

$$y = f(x)$$

NOTE Every function is a *relation* that calculates *exactly* one output for each valid input.

Representations of Functions

Functions can be represented by verbal descriptions, tables, symbols, and graphs.

Verbal Representation (Words) If function f gives the distance between an observer and a bolt of lightning, then we can verbally describe f with the following sentence: "Divide x seconds by 5 to obtain y miles." We call this a **verbal representation** of f.

Sometimes when the computation performed by a function has meaning, we can interpret this computation verbally. For example, a verbal description of function f is "f calculates the number of miles from a lightning bolt when the delay between thunder and lightning is x seconds."

Numerical Representation (Table of Values)   Table 1.7 gave a numerical representation for the function f that calculates the distance between a lightning bolt and an observer. A **numerical representation** is a *table of values* that lists input-output pairs for a function. A different numerical representation for f is shown in Table 1.8.

Distance from a Bolt of Lightning

x (seconds)	1	2	3	4	5	6	7
y (miles)	0.2	0.4	0.6	0.8	1.0	1.2	1.4

Table 1.8

One difficulty with numerical representations is that it is often either inconvenient or impossible to list all possible inputs x. For this reason we sometimes refer to a table of this type as a **partial numerical representation** as opposed to a **complete numerical representation**, which would include all elements from the domain of a function. For example, many valid inputs do not appear in Table 1.8, such as $x = 11$ or $x = 0.75$.

Symbolic Representation (Formula)   A formula gives a **symbolic representation** of a function. The computation performed by f is expressed by

$$f(x) = \frac{x}{5},$$ Formula for f

where $y = f(x)$. We say that function f is *represented by, defined by,* or *given by* $f(x) = \frac{x}{5}$. For example, if the elapsed time is **6** seconds, we write $f(\mathbf{6}) = \frac{6}{5} = \mathbf{1.2}$ miles. A formula is an efficient and complete, but less visual, way to define a function.

Graphical Representation (Graph)   A **graphical representation**, or **graph**, visually pairs an x-input with a y-output. In a graph of a function, the ordered pairs (x, y) are plotted in the xy-plane. The ordered pairs

$$(1, 0.2), (2, 0.4), (3, 0.6), (4, 0.8), (5, 1.0), (6, 1.2), \text{ and } (7, 1.4)$$

from Table 1.8 are plotted in Figure 1.38. This scatterplot suggests a line for the graph of f, as shown in Figure 1.39.

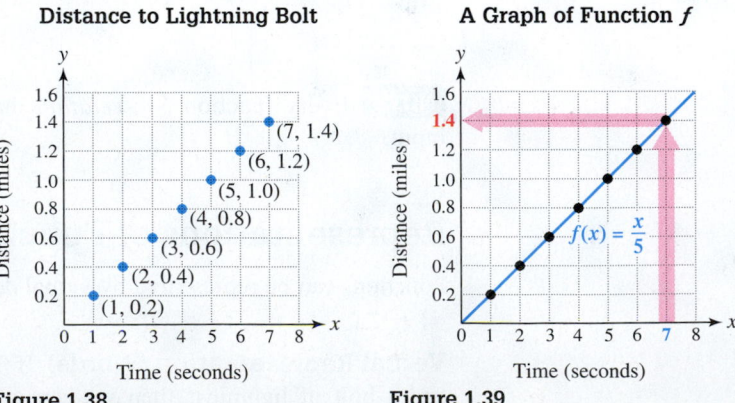

Figure 1.38   **Figure 1.39**

Evaluating _f(a)_ Graphically

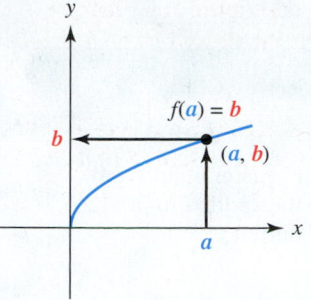

Figure 1.40

From Table 1.8 $f(5) = 1$, so the point $(5, 1)$ lies on the graph of f. Similarly, since $(7, 1.4)$ lies on the graph of f, it follows that $f(7) = 1.4$. See Figure 1.39 where the red arrows illustrate that $f(7) = 1.4$ because the point $(7, 1.4)$ lies on the graph of f.

> **MAKING CONNECTIONS**
>
> **Functions, Points, and Graphs** If $f(a) = b$, then the point (a, b) lies on the graph of f. Conversely, if the point (a, b) lies on the graph of f, then $f(a) = b$. Thus each point on the graph of f can be written in the form $(a, f(a))$. See Figure 1.40.

The next example shows how we often use symbolic and numerical representations (formulas and tables) to graph a function by hand.

EXAMPLE 1   **Graphing the absolute value function by hand**

Graph $f(x) = |x|$ by hand.

SOLUTION

Getting Started Unless you already know what the graph of a given function looks like, a good technique to use when graphing by hand is to first make a table of values. Then plot the points in the table and sketch a smooth curve (or line) between these points. ▶

Start by selecting convenient x-values and then substitute them into $f(x) = |x|$, as shown in Table 1.9. For example, when $x = -2$, then $f(-2) = |-2| = 2$, so the point $(-2, 2)$ is located on the graph of $y = f(x)$.

Next, plot the points as shown in Figure 1.41. The points appear to be V-shaped, and the graph of f shown in Figure 1.42 results if all possible real number ordered pairs $(x, |x|)$ are plotted.

Make a Table

x	$|x|$
-2	2
-1	1
0	0
1	1
2	2

Table 1.9

Plot Points

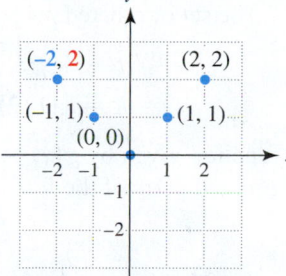

Figure 1.41

Sketch a Graph

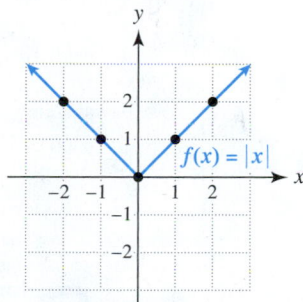

Figure 1.42

> **Now Try Exercise 13**

> **MAKING CONNECTIONS**
>
> **The Expressions f and $f(x)$** The italic letter f represents the *name* of a function, whereas the expression $f(x)$ represents the function f evaluated for input x. That is, $f(x)$ typically represents a formula for function f that can be used to evaluate f for various values of x.

The following See the Concept illustrates four representations for a function.

See the Concept: Four Representations of a Function

| **Symbolic** | **Numerical** | **Graphical** | **Verbal** |

Symbolic

$f(x) = x^2 + 1$

To get the table, evaluate $f(x) = x^2 + 1$ for x equal to $-2, -1, 0, 1,$ and 2.

Numerical

x	y
-2	5
-1	2
0	1
1	2
2	5

Graphical

$f(x) = x^2 + 1$

Plot the points in the table, and sketch a smooth graph.

Verbal

f squares input x and then adds 1 to produce output y.

Formal Definition of a Function

Because the idea of a function is a fundamental concept in mathematics, it is important that we define a function precisely. This definition should allow for *all* representations of a function. The commonality among representations is the concept of an ordered pair. A relation is a set of ordered pairs, and a function is a special type of relation.

> **FUNCTION**
>
> A **function** is a relation in which *each* element in the domain corresponds to *exactly one* element in the range.

The set of ordered pairs for a function can be either finite or infinite.

$$f = \{(-3, 5), (0, 6), (3, 2)\} \qquad \textit{Finite Set}$$
$$h = \{(1, 2), (2, 4), (3, 6), (4, 8), \dots\} \qquad \textit{Infinite Set}$$

A function given by $g(x) = x^2$, where x can be any real number, is another infinite set of ordered pairs of the form (x, x^2). Examples include $(-3, 9)$, $(4, 16)$, and $\left(\frac{1}{2}, \frac{1}{4}\right)$.

EXAMPLE 2   Finding the domain and range

Let a function f be defined by $f(-1) = 4$, $f(0) = 3$, $f(1) = 4$, and $f(2) = -2$. Write f as a set of ordered pairs. Give the domain and range.

SOLUTION

Because $f(-1) = 4$, the ordered pair $(-1, 4)$ is in the set. It follows that

$$f = \{(-1, 4), (0, 3), (1, 4), (2, -2)\}.$$

The domain D of f is the set of x-values, and the range R of f is the set of y-values. Thus

$$D = \{-1, 0, 1, 2\} \quad \text{and} \quad R = \{-2, 3, 4\}.$$

Now Try Exercise 21

EXAMPLE 3   Computing revenue of technology companies

The function f computes the revenue in dollars per unique user for different technology companies. This function is defined by $f(A) = 189$, $f(G) = 24$, $f(Y) = 8$, $f(F) = 4$, where A is Amazon, G is Google, Y is Yahoo, and F is Facebook. (**Source:** *Business Insider.*)
(a) Write f as a set of ordered pairs.
(b) Give the domain and range of f.

SOLUTION
(a) $f = \{(A, 189), (G, 24), (Y, 8), (F, 4)\}$
(b) The domain D and range R of f are

$$D = \{A, F, G, Y\} \quad \text{and} \quad R = \{4, 8, 24, 189\}.$$

Now Try Exercise 97

MAKING CONNECTIONS

Relations and Functions   Every function is a relation, whereas not every relation is a function. A function has exactly one output for each valid input.

Finding Domain, Range, and Function Values

Implied Domain   Unless stated otherwise, the domain of a function f is the set of all *real* numbers for which its symbolic representation (formula) is defined. The domain can be thought of as the set of all valid inputs that make sense in the expression for $f(x)$. In this case the domain is sometimes referred to as the **implied domain**.

When determining the domain of a function we must exclude values of x that result in division by 0. Division by 0 is *always* undefined. Also, unless stated otherwise, we will exclude values of x that result in taking the square root of a negative number. Examples of three functions and their domains D are the following.

The Domains D of Three Functions

1. $f(x) = 2x$

D is all real numbers because $2x$ is defined for all real numbers x.

2. $g(x) = \frac{1}{x}$

D is all real numbers except 0 ($x \neq 0$).

3. $h(x) = \sqrt{x}$

D is all nonnegative real numbers ($x \geq 0$).

EXAMPLE 4   Evaluating a function and finding its domain

Let $f(x) = \frac{x}{x - 1}$.
(a) If possible, evaluate $f(2)$, $f(1)$, and $f(a + 1)$.   (b) Find the domain of f.

SOLUTION
(a) To evaluate $f(2)$, substitute 2 for x in the formula $f(x) = \frac{x}{x - 1}$.

$$f(2) = \frac{2}{2 - 1} = 2 \qquad \textit{Let } x = 2.$$

To evaluate $f(1)$, let x be 1 in the formula.

$$f(1) = \frac{1}{1 - 1} = \frac{1}{0}, \qquad \textit{Division by 0 is undefined.}$$

which is undefined. Thus **1** is not in the domain of f.
To evaluate $f(a + 1)$, let x be $a + 1$ in the formula.

$$f(a + 1) = \frac{a + 1}{a + 1 - 1} = \frac{a + 1}{a} \qquad \textit{Let } x = a + 1.$$

(b) The formula $f(x) = \frac{x}{x-1}$ is undefined whenever $x - 1$ equals 0. Thus we exclude 1 from the domain ($x \neq 1$). In *set-builder notation* the domain is written $\{x \mid x \neq 1\}$.

Now Try Exercise 31

> **SET-BUILDER NOTATION**
>
> The expression $\{x \mid x \neq 1\}$ is written in **set-builder notation** and represents the set of all real numbers x such that x does not equal 1. Another example is $\{y \mid 1 < y < 5\}$, which represents the set of all real numbers y such that y is greater than 1 *and* less than 5.

EXAMPLE 5 **Evaluating a function symbolically and graphically**

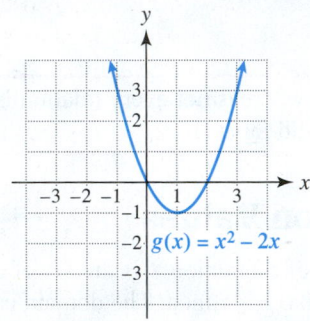

Figure 1.43

A function g is given by $g(x) = x^2 - 2x$, and its graph is shown in Figure 1.43.
(a) Find the domain and range of g.
(b) Use $g(x)$ to evaluate $g(-1)$.
(c) Use the graph of g to evaluate $g(-1)$.

SOLUTION
(a) The domain for $g(x) = x^2 - 2x$ includes all real numbers because the formula is defined for all real number inputs x.

   The minimum y-value on the graph is -1. The arrows on the graph point upward, so there is no maximum y-value. The range is all real numbers greater than or equal to -1, or $\{y \mid y \geq -1\}$.
(b) To evaluate $g(-1)$, substitute -1 for x in $g(x) = x^2 - 2x$.

$$g(-1) = (-1)^2 - 2(-1) = 1 + 2 = 3$$

(c) Refer to Figure 1.44 in the following See the Concept where $g(-1) = 3$.

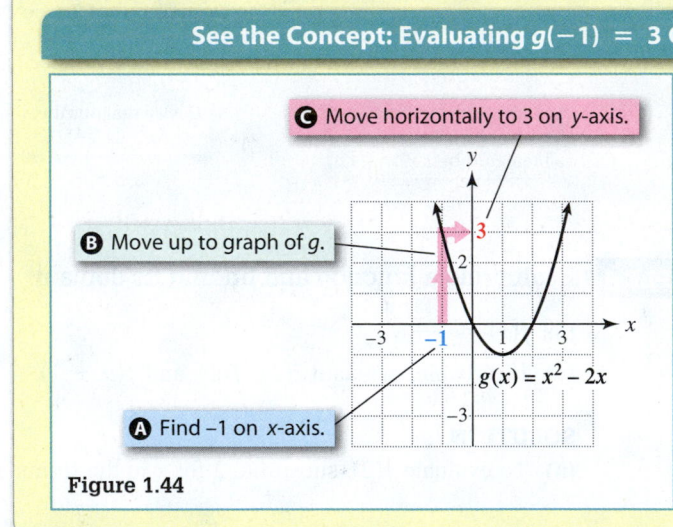

See the Concept: Evaluating $g(-1) = 3$ Graphically

C Move horizontally to 3 on y-axis.
B Move up to graph of g.
A Find -1 on x-axis.

Figure 1.44

A To evaluate $g(-1)$, begin by finding $x = -1$ on the x-axis.

B Next, move upward until the graph of g is reached.

C Finally, move to the right until the y-axis is reached. The y-value corresponding to an x-value of -1 is 3. Thus $g(-1) = 3$.

Now Try Exercise 37

EXAMPLE 6 **Finding the domain and range graphically**

A graph of $f(x) = \sqrt{x - 2}$ is shown in Figure 1.45.
(a) Evaluate $f(1)$.
(b) Find the domain and range of f.

SOLUTION
(a) Start by finding 1 on the x-axis. Move up and down on the grid. Note that we do not intersect the graph of f. See Figure 1.46. Thus $f(1)$ is *undefined*.

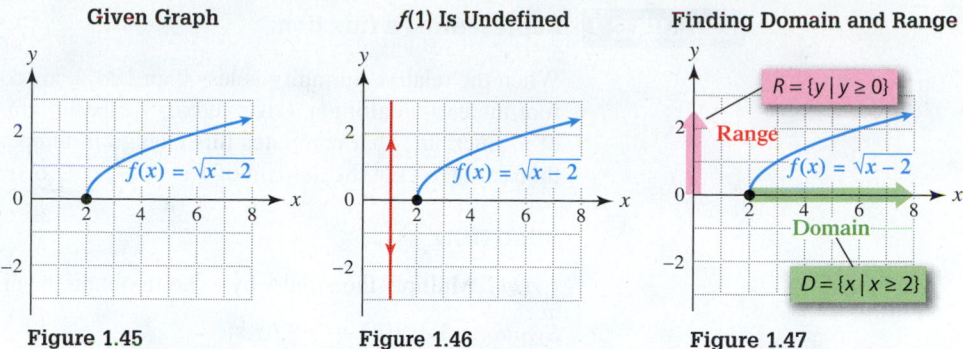

Figure 1.45   **Figure 1.46**   **Figure 1.47**

(b) In Figure 1.45 the arrow on the graph of f indicates that both the x-values and the y-values increase without reaching a maximum value. In Figure 1.47 the domain and range of f have been labeled by an arrow on each axis. Note that points appear on the graph for all x greater than or equal to 2. Thus the domain is $D = \{x \mid x \geq 2\}$. The minimum y-value on the graph of f is 0 and it occurs at the point $(2, 0)$. There is no maximum y-value on the graph, so the range is $R = \{y \mid y \geq 0\}$.

Now Try Exercise 45

Graphing Calculators and Functions (Optional)

Graphing calculators can be used to create graphs and tables of a function—usually more efficiently and reliably than pencil-and-paper techniques. However, a graphing calculator uses the same basic method that we might use to draw a graph. For example, one way to sketch a graph of $y = x^2$ is to first make a table of values, such as Table 1.10.

Making a Table for $y = x^2$

x	-3	-2	-1	0	1	2	3
y	9	4	1	0	1	4	9

Table 1.10

We can plot these points in the xy-plane, as shown in Figure 1.48. Next we might connect the points with a smooth curve, as shown in Figure 1.49. A graphing calculator typically plots numerous points and connects them to make a graph. In Figure 1.50, a graphing calculator has been used to graph $y = x^2$.

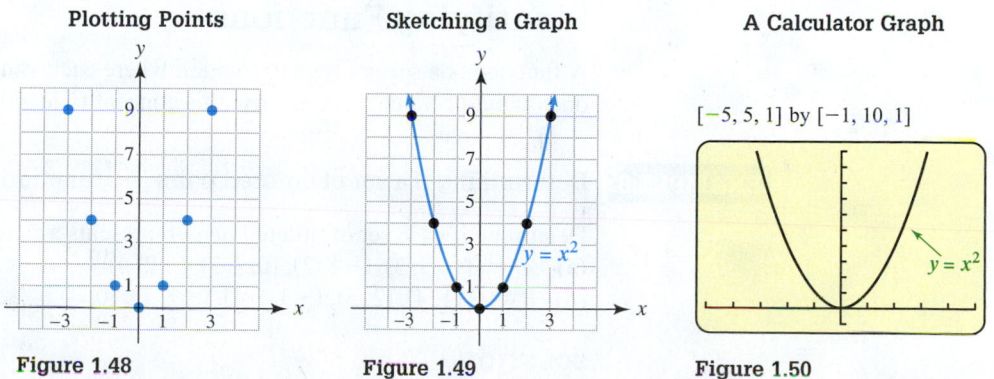

Figure 1.48   **Figure 1.49**   **Figure 1.50**

An Application In the next example we use a graphing calculator to represent a function that computes the decrease in air temperature as altitude increases.

EXAMPLE 7 Representing a function

When the relative humidity is less than 100%, air cools at a rate of 3.6°F for every 1000-foot increase in altitude. Give verbal, symbolic, graphical, and numerical representations of a function f that computes this change in temperature for an increase in altitude of x thousand feet. Let the domain of f be $0 \leq x \leq 6$. (***Source:*** L. Battan, *Weather in Your Life*.)

SOLUTION

Verbal Multiply the input x by -3.6 to obtain the change y in temperature.

Symbolic Let $f(x) = -3.6x$.

Graphical Since $f(x) = -3.6x$, enter $y_1 = -3.6x$, as shown in Figure 1.51. Graph y_1 in a viewing rectangle such as $[0, 6, 1]$ by $[-25, 10, 5]$, used in Figure 1.52.

Graphing with a Calculator

$[0, 6, 1]$ by $[-25, 10, 5]$

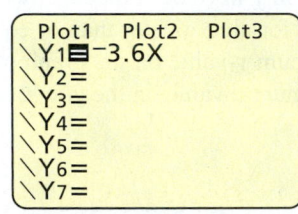

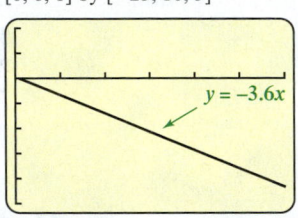

Calculator Help

To enter a formula and create a graph, see Appendix A (pages AP-4 and AP-5).

Figure 1.51   **Figure 1.52**

Calculator Help

To create a table similar to Figure 1.54, see Appendix A (page AP-5).

Numerical It is impossible to list all inputs x, since $0 \leq x \leq 6$. However, Figures 1.53 and 1.54 show how to create a table for $y_1 = -3.6x$ with $x = 0, 1, 2, 3, 4, 5, 6$. Other values for x in the domain of f are possible.

Making a Table of Values

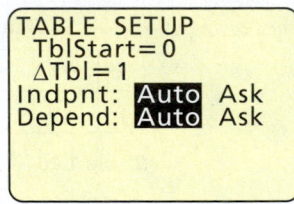

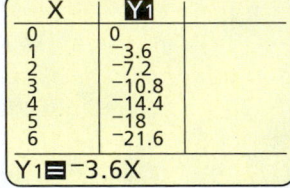

Figure 1.53   **Figure 1.54**

Now Try Exercise 103

NOTE Each of the four representations presented in Example 7 represent the *same* function f.

Identifying Functions

A function is a special type of relation where each valid input (x-value) produces exactly one output (y-value). We can use this concept to identify functions.

EXAMPLE 8 Determining if a set of ordered pairs is a function

Determine if each set of ordered pairs represents a function.
(a) $A = \{(-2, 3), (-1, 2), (0, -3), (-2, 4)\}$
(b) $B = \{(1, 4), (2, 5), (-3, -4), (-1, 7), (0, 4)\}$

SOLUTION

(a) Set A does not represent a function because input -2 results in two outputs: 3 and 4.
(b) Inputs 1 and 0 have the same output 4. However, set B represents a function because each input (x-value) results in one output (y-value).

Now Try Exercises 77 and 79

The following operations can be carried out by functions because they result in *one* output for each valid input.

- Calculating the square of a number x
- Finding the sale price when an item with regular price x is discounted 25%
- Naming the biological mother of person x

NOTE Not all computations can be done by functions. Suppose that we were given an eye color as an input and asked to output the name of each person in the class having this eye color. Typically we would *not* be computing a function. If, for example, several people in the class had brown eyes, there would not be a unique output (name) for each input (eye color).

Some tables do not represent a function. For example, Table 1.11 represents a relation but *not a function* because input **1** produces two outputs, **3** and **12**.

Points on a Vertical Line Have the Same *x*-Values

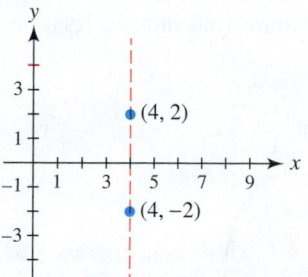

Figure 1.55

A Relation That Is Not a Function

x	1	2	3	1	4
y	3	6	9	12	2

Table 1.11

Vertical Line Test To conclude that a graph represents a function, we must be convinced that it is impossible for two distinct points with the same x-coordinate to lie on the graph. For example, the ordered pairs (**4**, **2**) and (**4**, **−2**) are distinct points with the same x-coordinate. These two points could not lie on the graph of the same function because input **4** would result in two outputs: **2** and **−2**. When the points (4, 2) and (4, −2) are plotted, they lie on the same vertical line, as shown in Figure 1.55. A graph passing through these points intersects the line twice, as illustrated in Figure 1.56. Therefore the graph in Figure 1.56 does *not* represent a function.

To determine if a graph represents a function, simply visualize vertical lines in the xy-plane. If every vertical line intersects a graph at no more than one point, then it is a graph of a function. This procedure is called the **vertical line test** for a function.

Not a Function

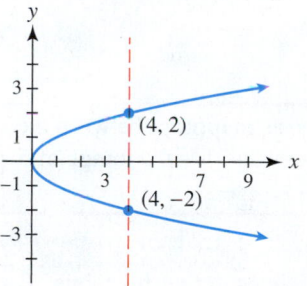

Figure 1.56

> ### VERTICAL LINE TEST
> If every vertical line intersects a graph at no more than one point, then the graph represents a function.

NOTE If a vertical line intersects a graph more than once, then the graph does *not* represent a function.

EXAMPLE 9 **Identifying a function graphically**

Use the vertical line test to determine if the graph represents a function.

(a)

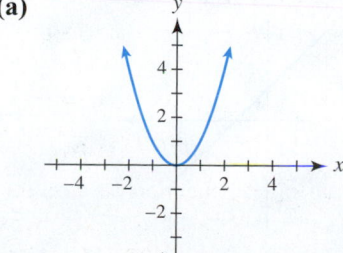

(b)

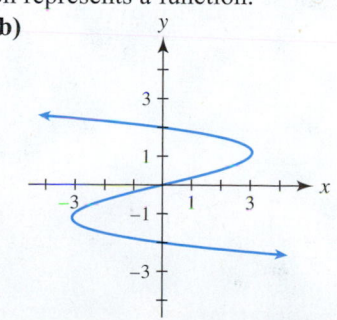

Function

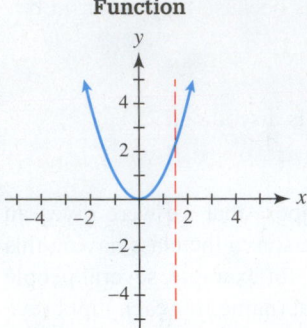

Figure 1.57

Not a Function

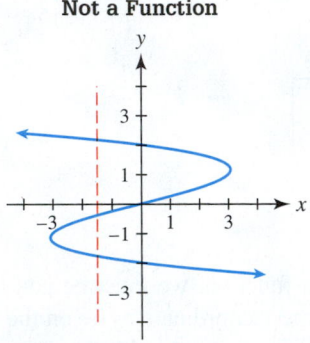

Figure 1.58

SOLUTION

(a) Note in Figure 1.57 that every vertical line that could be visualized would intersect the graph at most once. Therefore the graph represents a function.

(b) The graph in Figure 1.58 does not represent a function because it is possible for a vertical line to intersect the graph more than once.

> **Now Try Exercises 65 and 67**

Functions Represented by Diagrams and Equations

Thus far we have discussed four representations of a function: verbal, numerical, symbolic, and graphical. Two other ways that we can represent, or define, a function are with diagrams and equations.

Diagrammatic Representation (Diagram) Functions are sometimes represented using **diagrammatic representations,** or **diagrams.** Figure 1.59 is a diagram of a function with domain $D = \{5, 10, 15, 20\}$ and range $R = \{1, 2, 3,\}$. An arrow is used to show that input x produces output y. For example, in Figure 1.59 an arrow points from **5** to **1**, so $f(\mathbf{5}) = \mathbf{1}$ and the point $(\mathbf{5}, \mathbf{1})$ lies on the graph of f. Figure 1.60 shows a relation, but not a function.

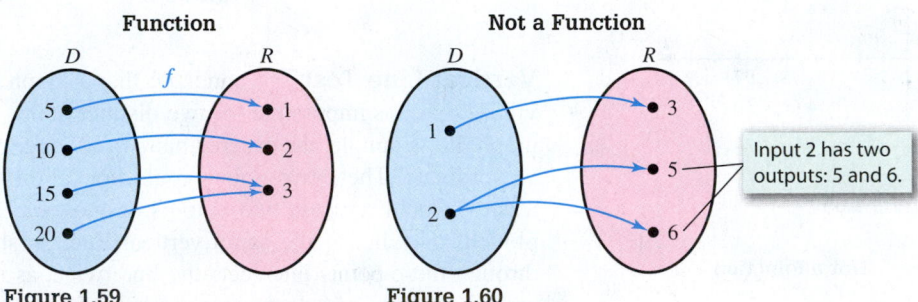

Figure 1.59   **Figure 1.60**

MAKING CONNECTIONS

Functions as Mappings Functions are sometimes referred to as **mappings** between the domain and the range. If $f(\mathbf{5}) = \mathbf{1}$, then we say that the range value **1** is the **image** of **5** and that the domain value **5** is the **preimage** of **1**.

Functions Defined by Equations Equations can sometimes define functions. For example, the equation $x + y = 1$ defines a function f given by $f(x) = 1 - x$, where $y = f(x)$. Notice that for each input x, there is exactly one y output determined by $y = 1 - x$. In Figure 1.61 the graph passes the vertical line test. However, the graph of the equation $x^2 + y^2 = 4$ is a circle with center $(0, 0)$ and radius 2. Because a circle does not pass the vertical line test, this equation does not represent a function. See Figure 1.62.

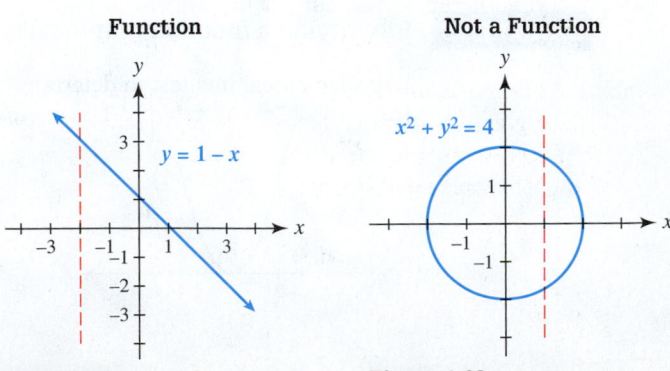

Figure 1.61   **Figure 1.62**

EXAMPLE 10   Identifying a function

Determine if y is a function of x.
(a) $x = y^2$
(b) $y = x^2 - 2$

SOLUTION

(a) For y to be a function of x in the equation $x = y^2$, each valid x-value must result in one y-value. If we let $x = 4$, then y could be either -2 or 2 since

$$4 = (-2)^2 \quad \text{and} \quad 4 = (2)^2. \qquad x = y^2$$

Therefore y is not a function of x. A graph of $x = y^2$ is shown in Figure 1.63. Note that this graph fails the vertical line test.

(b) In the equation $y = x^2 - 2$ each x-value determines exactly one y-value, and so y is a function of x. A graph of this equation is shown in Figure 1.64. This graph passes the vertical line test.

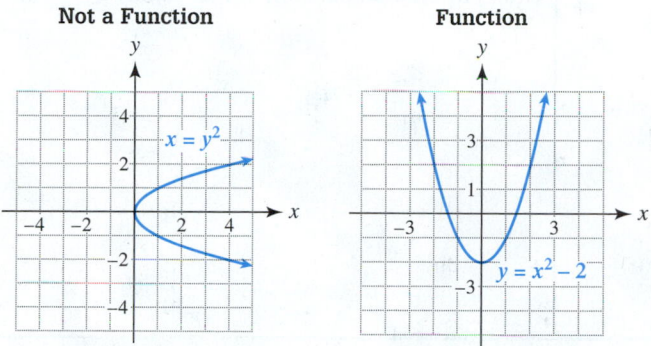

Not a Function   Function

Figure 1.63   Figure 1.64

Now Try Exercises 83 and 89

1.3 Putting It All Together

CONCEPT	EXPLANATION	EXAMPLES
Function	A function is a *relation* in which each valid input results in one output. The *domain* of a function is the set of valid inputs (x-values), and the *range* is the set of resulting outputs (y-values).	$f = \{(1, 3), (2, 6), (3, 9), (4, 9)\}$ The domain is $D = \{1, 2, 3, 4\}$, and the range is $R = \{3, 6, 9\}$.
Implied domain	When a function is represented by a formula, its domain, unless otherwise stated, is the set of all valid inputs (x-values) that are defined or make sense in the formula.	$f(x) = \dfrac{1}{x + 4}$ Domain of f: $\{x \mid x \neq -4\}$
Identifying graphs of functions	*Vertical Line Test*: If every vertical line intersects a graph at no more than one point, then the graph represents a function. (Otherwise the graph does not represent a function.)	Not a function

continued on next page

CONCEPT	EXPLANATION	EXAMPLES
Verbal representation of a function	Words describe precisely what is computed.	A verbal representation of $f(x) = x^2$ is "Square the input x to obtain the output."
Symbolic representation of a function	Mathematical formula	The squaring function is given by $f(x) = x^2$, and the square root function is given by $g(x) = \sqrt{x}$.
Numerical representation of a function	Table of values	A *partial* numerical representation of $f(x) = 3x$ is shown. | x	0	1	2	3
---	---	---	---	---
$f(x)$	0	3	6	9	
Graphical representation of a function	Graph of ordered pairs (x, y) that satisfy $y = f(x)$	Each point on the graph satisfies $y = 2x$.

1.3 Exercises

Evaluating and Representing Functions

1. If $f(-2) = 3$, identify a point on the graph of f.

2. If $f(3) = -9.7$, identify a point on the graph of f.

3. If $(7, 8)$ lies on the graph of f, then $f(\underline{}) = \underline{}$.

4. If $(-3, 2)$ lies on the graph of f, then $f(\underline{}) = \underline{}$.

Exercises 5–20: Graph $y = f(x)$ by hand by first plotting points to determine the shape of the graph.

5. $f(x) = 3$　　　**6.** $f(x) = -2$

7. $f(x) = 2x$　　　**8.** $f(x) = x + 1$

9. $f(x) = 4 - x$　　　**10.** $f(x) = 3 + 2x$

11. $f(x) = \frac{1}{2}x - 2$　　　**12.** $f(x) = 2 - 2x$

13. $f(x) = |x - 1|$　　　**14.** $f(x) = |0.5x|$

15. $f(x) = |3x|$　　　**16.** $f(x) = |2x - 1|$

17. $f(x) = \frac{1}{2}x^2$　　　**18.** $f(x) = 2x^2$

19. $f(x) = x^2 - 2$　　　**20.** $f(x) = x^2 + 1$

Exercises 21–26: A function g is defined.
(a) Write g as a set of ordered pairs.
(b) Give the domain and range of g.

21. $g(-1) = 0, g(2) = -2, g(5) = 7$

22. $g(-2) = 5, g(3) = 9, g(4) = -4$

23. $g(1) = 8, g(2) = 8, g(3) = 8$

24. $g(-5) = 0, g(0) = -5, g(5) = 0$

25. $g(-1) = 2, g(0) = 4, g(1) = -3, g(2) = 2$

26. $g(-4) = 5, g(0) = -5, g(4) = 5, g(8) = 0$

Exercises 27–36: Complete the following.

(a) *Find $f(x)$ for the indicated values of x, if possible.*

(b) *Find the domain of f.*

27. $f(x) = x^3$ for $x = -2, 5$

28. $f(x) = 2x - 1$ for $x = 8, -1$

29. $f(x) = \sqrt{x}$ for $x = -1, a + 1$

30. $f(x) = \sqrt{1 - x}$ for $x = -2, a + 2$

31. $f(x) = 6 - 3x$ for $x = -1, a + 1$

32. $f(x) = -7$ for $x = 6, a - 1$

33. $f(x) = \dfrac{3x - 5}{x + 5}$ for $x = -1, a$

34. $f(x) = x^2 - x + 1$ for $x = 1, -2$

35. $f(x) = \dfrac{1}{x^2}$ for $x = 4, -7$

36. $f(x) = \dfrac{1}{x - 9}$ for $x = 4, a + 9$

Exercises 37–42: (Refer to Example 5.) Use the graph to complete the following.

(a) *Estimate the domain and range of g.*

(b) *Use the formula to evaluate $g(-1)$ and $g(2)$.*

(c) *Use the graph of g to evaluate $g(-1)$ and $g(2)$.*

37.
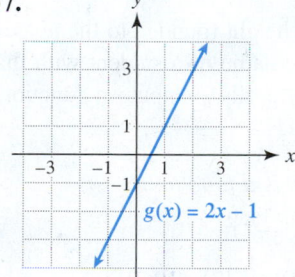
$g(x) = 2x - 1$

38.
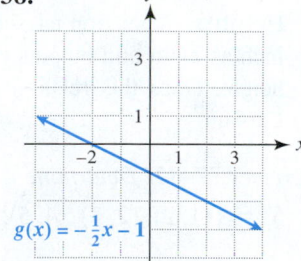
$g(x) = -\frac{1}{2}x - 1$

39.
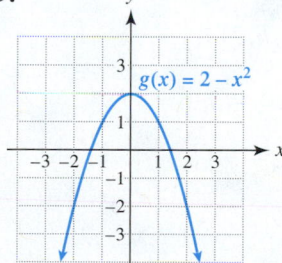
$g(x) = 2 - x^2$

40.
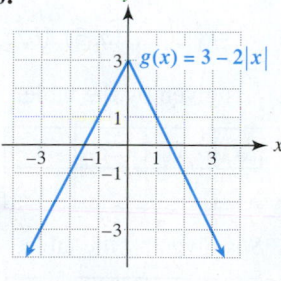
$g(x) = 3 - 2|x|$

41.
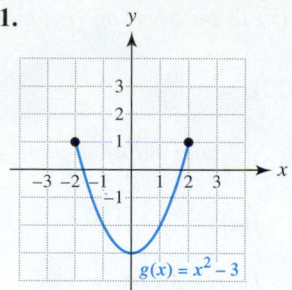
$g(x) = x^2 - 3$

42.
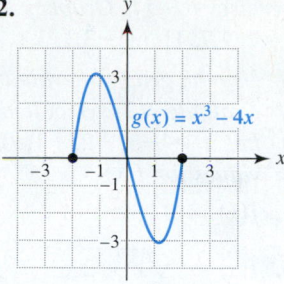
$g(x) = x^3 - 4x$

Exercises 43–48: Use the graph of the function f to estimate its domain and range. Evaluate $f(0)$.

43.
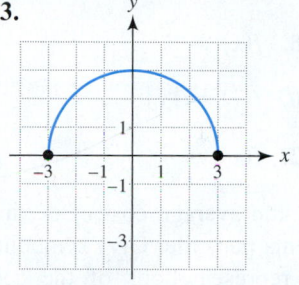

44.
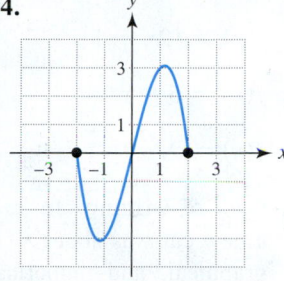

45.
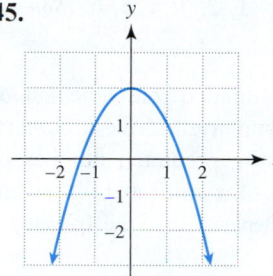

46.
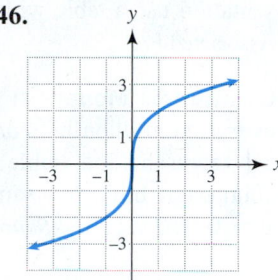

47.
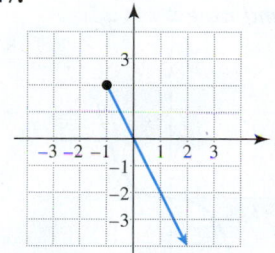

48.
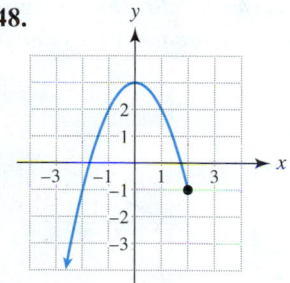

Exercises 49 and 50: **Diagrams** *Complete the following.*

(a) *Evaluate $f(2)$.*

(b) *Write f as a set of ordered pairs.*

(c) *Find the domain and range of f.*

49.
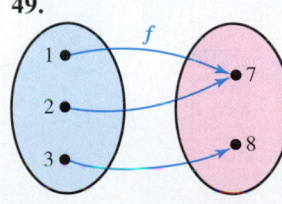

50.
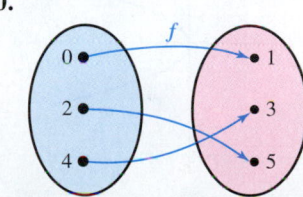

Exercises 51–54: Graph $y = f(x)$ in the viewing rectangle $[-4.7, 4.7, 1]$ by $[-3.1, 3.1, 1]$.

(a) *Use the graph to evaluate $f(2)$.*

(b) *Evaluate $f(2)$ symbolically.*

(c) *Let $x = -3, -2, -1, 0, 1, 2, 3$ and make a table of values for $f(x)$.*

51. $f(x) = 0.25x^2$ **52.** $f(x) = 3 - 1.5x^2$

53. $f(x) = \sqrt{x + 2}$ **54.** $f(x) = |1.6x - 2|$

Exercises 55–62: Use $f(x)$ to determine verbal, graphical, and numerical representations. For the numerical representation use a table with $x = -2, -1, 0, 1, 2$. Evaluate $f(2)$.

55. $f(x) = x^2$ **56.** $f(x) = 2x - 5$

57. $f(x) = |2x + 1|$ **58.** $f(x) = 8$

59. $f(x) = 5 - x$ **60.** $f(x) = |x|$

61. $f(x) = \sqrt{x + 1}$ **62.** $f(x) = x^2 - 1$

63. Cost of Driving In 2012 the average cost of driving a new car was about 50 cents per mile. Give symbolic, graphical, and numerical representations of the cost in dollars of driving x miles. For the numerical representation use a table with $x = 1, 2, 3, 4, 5, 6$. (*Source:* Associated Press.)

64. Counterfeit Money It is estimated that nine out of every one million bills are counterfeit. Give a numerical representation (table) of the predicted number of counterfeit bills in a sample of x million bills where $x = 0, 1, 2, \ldots, 6$. (*Source:* Department of the Treasury.)

Identifying Functions

Exercises 65–70: Does the graph represent a function? If so, determine the function's domain and range.

65.
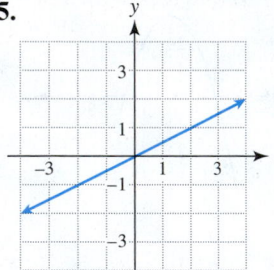

66.
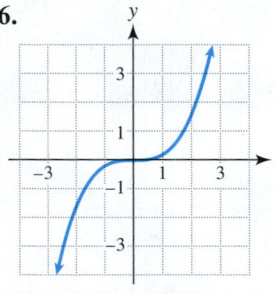

67.
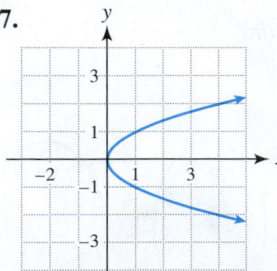

68.
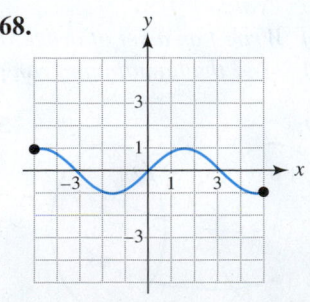

69.
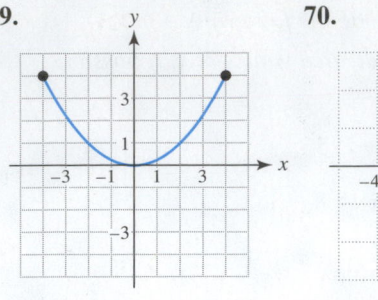

70.
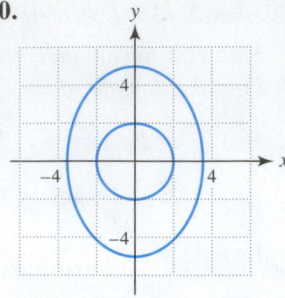

Exercises 71–74: Complete the following.

(a) *Determine if the following can be calculated with a function.*

(b) *Explain your answer.*

71. Input is a real number x; output is its real cube root.

72. Input is a Social Security number x; output is the age of the person with Social Security number x.

73. Input x is a score on a math exam; output is the ID numbers for all students having score x.

74. Input is a Social Security number x; output is the full names of the children of the person with Social Security number x.

75. Identification Numbers A relation takes a student's identification number at your college as input and outputs the student's name. Does this relation compute a function? Explain.

76. Heights A relation takes a height rounded to the nearest inch as input and outputs the name of a student with that height. Does this relation typically compute a function? Explain.

Exercises 77–82: Determine if S is a function.

77. $S = \{(1, 2), (2, 3), (4, 5), (1, 3)\}$

78. $S = \{(-3, 7), (-1, 7), (3, 9), (6, 7), (10, 0)\}$

79. $S = \{(a, 2), (b, 3), (c, 3), (d, 3), (e, 2)\}$

80. $S = \{(a, 2), (a, 3), (b, 5), (-b, 7)\}$

81. S is given by the table.

x	1	3	1
y	10.5	2	-0.5

82. S is given by the table.

x	1	2	3
y	1	1	1

Exercises 83–90: Determine if y is a function of x.

83. $x = y^4$

84. $y^2 = x + 1$

85. $\sqrt{x + 1} = y$

86. $x^2 = y - 7$

87. $x^2 + y^2 = 70$

88. $(x - 1)^2 + y^2 = 1$

89. $x + y = 2$

90. $y = |x|$

Exercises 91–96: **Formulas** *Write a symbolic representation (formula) for a function g that calculates the given quantity. Then evaluate g(10) and interpret the result.*

91. The number of inches in x feet

92. The number of quarts in x gallons

93. The number of dollars in x quarters

94. The number of quarters in x dollars

95. The number of seconds in x days

96. The number of feet in x miles

Applications

97. **DVD Video Rentals** (Refer to Example 3.) The function V computes the percent share of disc DVD rentals accounted for by various companies. This function is defined by $V(R) = 37$, $V(N) = 30$, and $V(S) = 17$, where R is Redbox, N is Netflix, and S is rental stores. (*Source: Business Insider.*)
(a) Write V as a set of ordered pairs.

(b) Give the domain and range of V.

98. **Food Insecurity** Function P computes the percentage of U.S. households that were food insecure during a selected year. This function is defined by $P(2006) = 10.9$, $P(2007) = 11.1$, $P(2008) = 14.6$, $P(2009) = 14.7$, and $P(2010) = 14.5$. (*Source:* U.S. Census Bureau.)
(a) Write P as a set of ordered pairs.

(b) Give the domain and range of P.

99. **Electronic Waste** As technology advances there are more and more obsolete electronic devices that need to be disposed of. The function $f(x) = 40x$ estimates the millions of tons of electronic waste that will accumulate worldwide after x years. Evaluate $f(5)$ and interpret the result. (*Source: Environmental Protection Agency.*)

100. **Portion Size** If the average American always ordered the larger portion while eating out, then $W(x) = 8x$ would estimate the resulting weight gain in pounds after x years. Evaluate $W(3)$ and interpret the result. (*Source:* FDA.)

101. **Going Green** The average person uses 2200 paper napkins in one year. Write the formula for a function N that calculates the number of paper napkins that the average person uses in x years. Evaluate $N(3)$ and interpret your result.

102. **Going Green** The average top-loading washing machine uses about 40 gallons of water per load of clothes. Write the formula for a function W that calculates the number of gallons of water used while washing x loads of clothes. Evaluate $W(30)$ and interpret your result.

103. **Air Temperature** When the relative humidity is 100%, air cools 5.8°F for every 1-mile increase in altitude. Give verbal, symbolic, graphical, and numerical representations of a function f that computes this change in temperature for an increase in altitude of x miles for $0 \le x \le 3$. (*Source:* L. Battan.)

104. **Crutch Length** Each year 15 million people have foot and ankle problems. Many times they need crutches. The formula $f(x) = 0.72x + 2$ calculates the appropriate crutch length in inches for a person with a height of x inches. (*Source: Journal of the American Physical Therapy Association.*)
(a) Find the crutch length for a person 6 feet 3 inches tall.

(b) For each 1-inch increase in height, by how much does the recommended crutch length increase?

Writing about Mathematics

105. Explain how you could use a complete numerical representation (table) for a function to determine its domain and range.

106. Explain in your own words what a function is. How is a function different from a relation?

1.4 Types of Functions and Their Rates of Change

- Identify linear functions
- Interpret slope as a rate of change
- Identify nonlinear functions
- Identify where a function is increasing or decreasing
- Use interval notation
- Use and interpret average rate of change
- Calculate the difference quotient

Introduction

Functions are used to describe, or *model*, everything from weather to new product "specs," global warming, and U.S. population. New functions are created each day in the *dynamic field* of mathematics. Finding new functions, such as one that calculates memory requirements for an iPod (Example 3), requires creativity. This section discusses two basic types of functions: *linear* and *nonlinear*. We also discuss *constant* functions, which are a special type of linear function.

Linear Functions

A car is initially located 30 miles north of the Texas border and is traveling north on Interstate 35 at 60 miles per hour. The distances between the automobile and the border are listed in Table 1.12 for various times. A scatterplot of the data is shown in Figure 1.65. It suggests that a line that rises from left to right might model these data.

Distance from Texas Border

Elapsed time (hours)	0	1	2	3	4	5	
Distance (miles)	30	90	150	210	270	330	Increases by 60 miles each hour

Table 1.12

A Scatterplot of Table 1.12

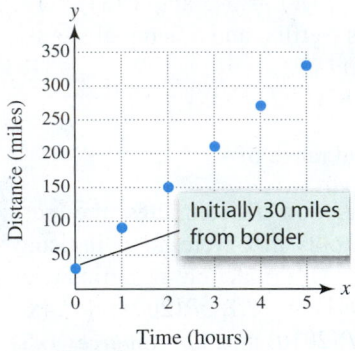

Initially 30 miles from border

Figure 1.65

A Linear Model

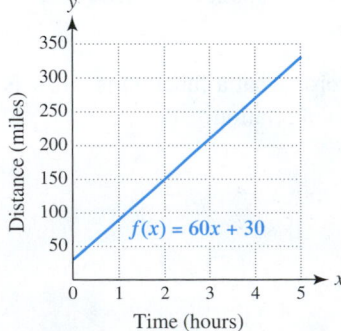

$f(x) = 60x + 30$

Figure 1.66

If the car travels for x hours, the distance traveled can be found by multiplying **60** times x and adding the initial distance of **30** miles. This computation can be expressed as $f(x) = 60x + 30$. For example, $f(1.5) = 60(1.5) + 30 = 120$ means that the car is **120** miles from the border after **1.5** hours. The formula is valid for nonnegative x. The graph of $f(x) = 60x + 30$ is a *line* (ray), shown in Figure 1.66. We call f a *linear function*.

> **LINEAR FUNCTION**
>
> A function f represented by $f(x) = mx + b$, where m and b are constants, is a **linear function**.

Recognizing Linear Functions In the example of the moving car, $f(x) = 60x + 30$, so $m = 60$ and $b = 30$. The value of m represents the **speed** of the car, and b is the **initial distance** of the car from the border.

A distinguishing feature of a linear function f is that each time x increases by one unit, the value of $f(x)$ always changes by an amount equal to m. That is, a linear function has a **constant rate of change**. (The constant rate of change m is equal to the slope of the graph of f.) The following applications are modeled by linear functions. Try to determine the value of the constant m in each case.

- The wages earned by an individual working x hours at \$9.25 per hour
- The amount of tuition and fees due when registering for x credits if each credit costs \$350 and the fees are fixed at \$560

Constant Functions If $m = 0$, then a linear function can be written as $f(x) = b$. In this case, f is a *constant function*.

> **CONSTANT FUNCTION**
>
> A function f represented by $f(x) = b$, where b is a constant (fixed number), is a **constant function**.

For example, if $f(x) = \mathbf{2}$, then every input x always results in an output of 2. Thus every point on the graph of f has a y-coordinate of **2** and its graph is a horizontal line. See Table 1.13 and Figure 1.67.

A Constant Function

For every input x, the output y is always **2**.

x	y
-2	2
-1	2
1	2
2	2

Table 1.13

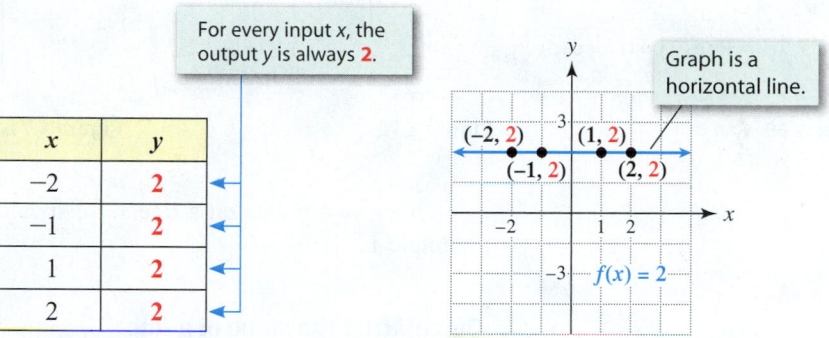

Graph is a horizontal line.

$(-2, 2)$ $(1, 2)$
$(-1, 2)$ $(2, 2)$

$f(x) = 2$

Figure 1.67

Slope as a Rate of Change

The graph of a linear function is a line. The slope m is a real number that measures the "tilt" of a line in the xy-plane. If the input x to a linear function increases by 1 unit, then the output y changes by a constant amount that is equal to the slope of its graph. In Figure 1.68 a line passes through the points (x_1, y_1) and (x_2, y_2). The *change in y* is $y_2 - y_1$, and the *change in x* is $x_2 - x_1$. The ratio of the change in y to the change in x is called the *slope*. We sometimes denote the change in y by Δy (delta y) and the change in x by Δx (delta x). That is, $\Delta y = y_2 - y_1$ and $\Delta x = x_2 - x_1$.

Finding Slope Given Two Points

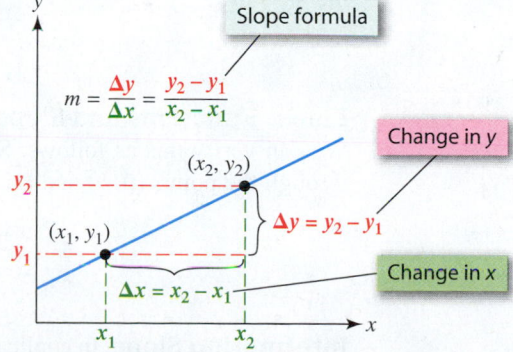

Slope formula

$$m = \frac{\Delta y}{\Delta x} = \frac{y_2 - y_1}{x_2 - x_1}$$

Change in y

(x_2, y_2)

$\Delta y = y_2 - y_1$

(x_1, y_1)

Change in x

$\Delta x = x_2 - x_1$

Figure 1.68

The following is a definition of slope.

SLOPE

The **slope** m of the line passing through the points (x_1, y_1) and (x_2, y_2) is

$$m = \frac{\Delta y}{\Delta x} = \frac{y_2 - y_1}{x_2 - x_1}, \qquad \text{where } x_1 \neq x_2.$$

Figures 1.69–1.72 summarize some basic concepts about slope.

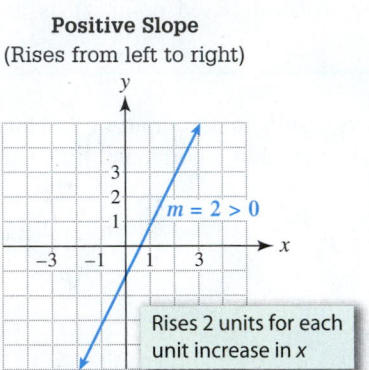

Positive Slope
(Rises from left to right)

$m = 2 > 0$

Rises 2 units for each unit increase in x

Figure 1.69

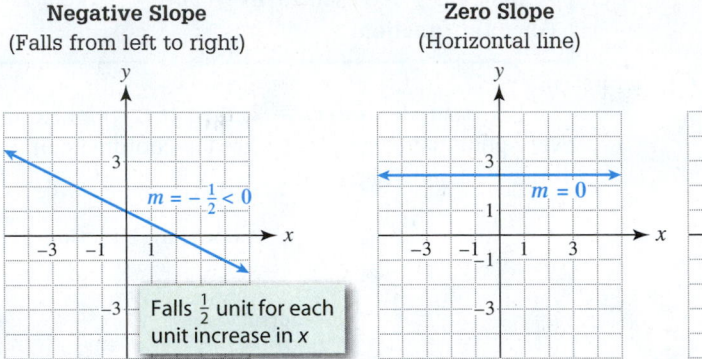

Negative Slope
(Falls from left to right)

$m = -\frac{1}{2} < 0$

Falls $\frac{1}{2}$ unit for each unit increase in x

Figure 1.70

Zero Slope
(Horizontal line)

$m = 0$

Figure 1.71

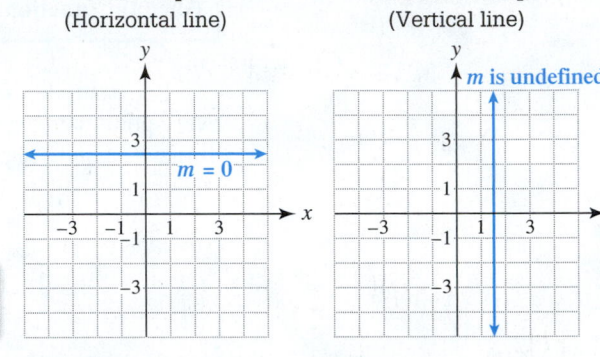

Undefined Slope
(Vertical line)

m is undefined

Figure 1.72

Given two points on a line we can calculate the slope of the line, as illustrated in Example 1.

EXAMPLE 1 Calculating the slope of a line

Find the slope of the line passing through the points $(-2, 3)$ and $(1, -2)$. Plot these points together with the line. Explain what the slope indicates about the line.

SOLUTION

Let $(x_1, y_1) = (-2, 3)$ and $(x_2, y_2) = (1, -2)$. The slope is

$$m = \frac{y_2 - y_1}{x_2 - x_1} = \frac{-2 - 3}{1 - (-2)} = -\frac{5}{3}.$$

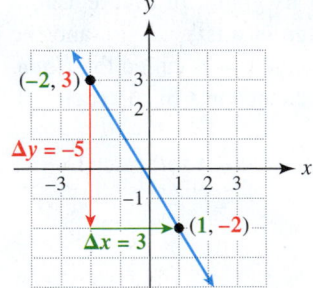

Figure 1.73

A line passing through these two points is shown in Figure 1.73. The change in y is $\Delta y = -5$ and the change in x is $\Delta x = 3$, so $m = \frac{\Delta y}{\Delta x} = -\frac{5}{3}$ indicates that the line falls $\frac{5}{3}$ units for each unit increase in x, or equivalently, the line falls 5 units for each 3-unit increase in x.

Now Try Exercise 5

Linear Functions and Slope The graph of $f(x) = mx + b$ is a line that has slope m. We can verify this as follows. Since $f(0) = b$ and $f(1) = m + b$, the graph of f passes through the points $(0, b)$ and $(1, m + b)$. The slope of this line is

$$\frac{y_2 - y_1}{x_2 - x_1} = \frac{m + b - b}{1 - 0} = \frac{m}{1} = m.$$

Interpreting Slope In applications involving linear functions, slope sometimes is interpreted as a (*constant*) *rate of change* as in the next two examples.

| EXAMPLE 2 | Interpreting slope and going green |

The function given by $P(x) = 19.4x$ calculates the pounds of CO_2 (carbon dioxide) released into the atmosphere by a car burning x gallons of gasoline.
(a) Calculate $P(5)$ and interpret the result.
(b) Find the slope of the graph of P. Interpret this slope as a rate of change.

SOLUTION
(a) $P(5) = 19.4(5) = 97$, so burning 5 gallons of gasoline releases 97 pounds of CO_2.
(b) The slope of the graph of P is 19.4. This means that 19.4 pounds of CO_2 are released for every gallon of gasoline burned by a car.

Now Try Exercise 27

| EXAMPLE 3 | Interpreting slope and iPod memory |

Figure 1.74 shows the (approximate) number of songs that can be stored on x gigabytes of Classic iPod memory.   (*Source:* Apple Corporation.)
(a) Why is it reasonable for the graph to pass through the origin?
(b) Find the slope of the line segment.
(c) Interpret the slope as a rate of change.

iPod Memory

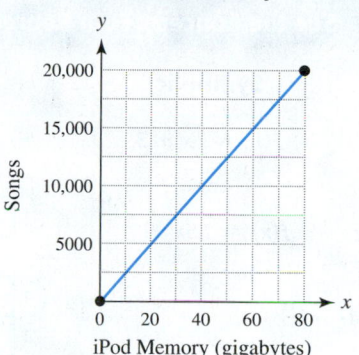

Figure 1.74

SOLUTION
(a) Because 0 songs require no memory, the graph passes through the point $(0, 0)$.
(b) The graph passes through the points $(0, 0)$ and $(80, 20,000)$. The slope of the line is

$$m = \frac{20,000 - 0}{80 - 0} = 250.$$

(c) An iPod holds 250 songs per gigabyte.

Now Try Exercise 25

MAKING CONNECTIONS

Units for Rates of Change   The units for a rate of change can be found from a graph by placing the units from the vertical axis over the units from the horizontal axis. For example, in Figure 1.74 the units on the y-axis are *songs* and the units on the x-axis are *gigabytes*. Thus the units for the slope, or rate of change, are *songs per gigabyte*.

Representations of Linear Functions   Any linear function can be written as $f(x) = mx + b$, where m equals the slope of the graph of f. Also, because $f(0) = m(0) + b = b$, the point $(0, b)$ lies on the graph of f and the value of b is the **y-intercept** of the graph of f. A function can have at most one y-intercept because $f(0)$ can have at most one value.

**Linear Function
with $m = \frac{3}{2}$ and $b = 3$**

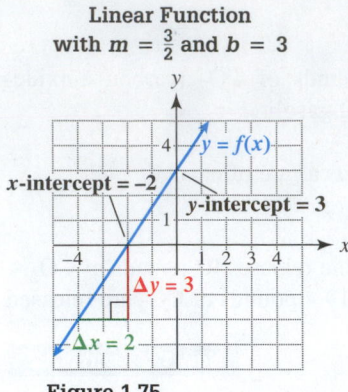

Figure 1.75

Consider the graph of the linear function f shown in Figure 1.75. The graph is a line that intersects each axis once. From the graph we can see that when y increases by 3 units, x increases by 2 units. Thus the change in y is $\Delta y = 3$, the change in x is $\Delta x = 2$, and the slope is $\frac{\Delta y}{\Delta x} = \frac{3}{2}$. The graph f intersects the y-axis at the point $(0, 3)$, and so the y-intercept is 3. We can write the formula for f as:

$$f(x) = \underset{\text{slope}}{\frac{3}{2}}x + \underset{\text{y-intercept}}{3}.$$

The graph of f in Figure 1.75 intersects the x-axis at the point $(-2, 0)$. We say that the **x-intercept** on the graph of f is -2. When we evaluate $f(-2)$, we obtain

$$f(-2) = \frac{3}{2}(-2) + 3 = 0.$$

An x-intercept corresponds to an input that results in an output of 0. We also say that -2 is a *zero* of the function f, since $f(-2) = 0$. A **zero** of a function f corresponds to an x-intercept on the graph of f. If the slope of the graph of a linear function f is not 0, then the graph of f has exactly one x-intercept.

The following See the Concept shows four representations of a linear function.

See the Concept: Four Representations of a Linear Function f

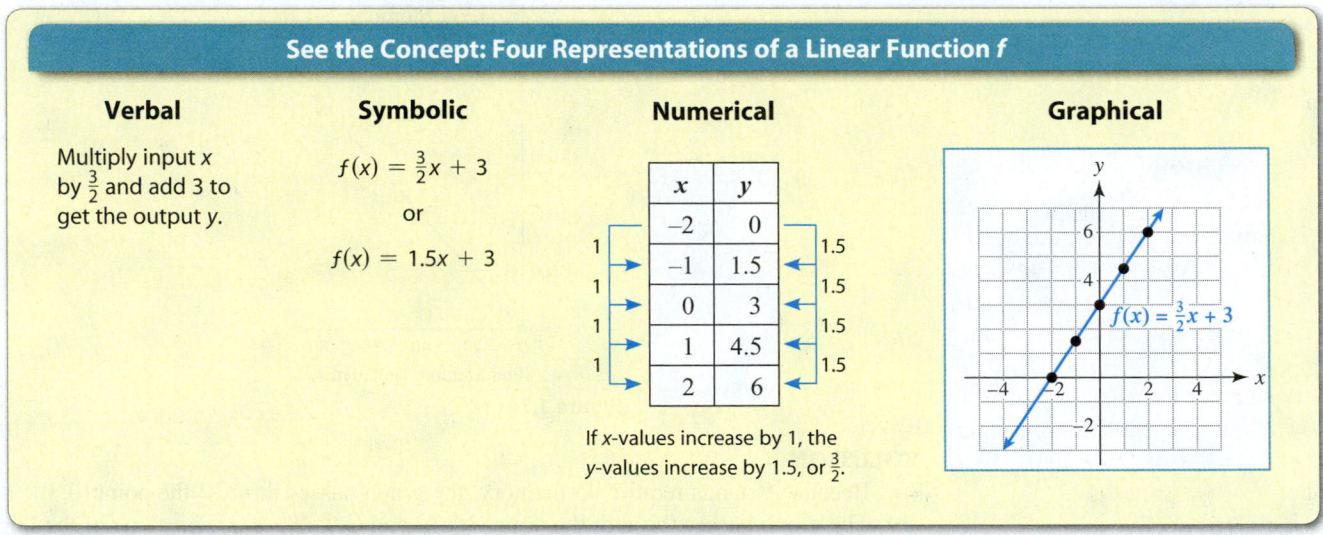

Verbal	Symbolic	Numerical	Graphical
Multiply input x by $\frac{3}{2}$ and add 3 to get the output y.	$f(x) = \frac{3}{2}x + 3$ or $f(x) = 1.5x + 3$		

x	y
-2	0
-1	1.5
0	3
1	4.5
2	6

If x-values increase by 1, the y-values increase by 1.5, or $\frac{3}{2}$.

EXAMPLE 4 Finding a formula from a graph

Use the graph of a linear function f in Figure 1.76 to complete the following.
(a) Find the slope, y-intercept, and x-intercept.
(b) Write a formula for f.
(c) Find any zeros of f.

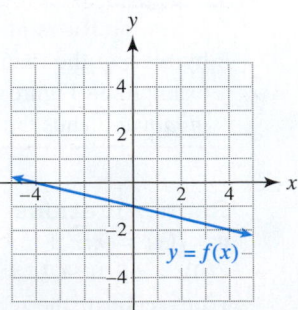

Figure 1.76

SOLUTION

(a) The line **falls 1** unit each time the x-values increase by **4** units. Therefore the slope is $-\frac{1}{4}$. The graph intersects the y-axis at the point $(0, -1)$ and intersects the x-axis at the point $(-4, 0)$. Therefore the y-intercept is -1, and the x-intercept is -4.

(b) Because the slope is $-\frac{1}{4}$ and the y-intercept is -1, it follows that

$$f(x) = -\frac{1}{4}x - 1.$$

(c) Zeros of f correspond to x-intercepts, so the only zero is -4.

> **Now Try Exercise 43**

Nonlinear Functions

We have discussed linear and constant functions. *Nonlinear functions* are another type of function.

Recognizing Nonlinear Functions If a function is not linear, then it is called a **nonlinear function.** *The graph of a nonlinear function is not a (straight) line.* With a nonlinear function, it is possible for the input x to increase by 1 unit and the output y to change by different amounts. Nonlinear functions *cannot* be written in the form $f(x) = mx + b$.

One example of a nonlinear function is $f(x) = x^2$. In Figure 1.77 (below), its graph is *not* a line. In Table 1.14 we see that $f(x)$ does not increase by a constant amount for each unit increase in x.

The Function $f(x) = x^2$

x	0	1	2	3	4
$f(x)$	0	1	4	9	16

Table 1.14

1   3   5   7

Increase is not constant.

> The increase in $f(x)$ is not the same for each unit increase in x, so $f(x) = x^2$ is nonlinear.

CLASS DISCUSSION

The time required to drive 100 miles depends on the average speed x. Let $f(x)$ compute this time, given x as input. For example, $f(50) = 2$, because it would take 2 hours to travel 100 miles at an average speed of 50 miles per hour. Find a formula for f. Is f linear or nonlinear?

Real-world phenomena often are modeled by using nonlinear functions. The following are two examples of quantities that can be described by nonlinear functions.

- The monthly average temperature in Chicago (Monthly average temperatures increase and decrease throughout the year.)
- The height of a child between the ages of 2 and 18 (A child grows faster at certain ages.)

Graphs of Nonlinear Functions There are many nonlinear functions. In Figures 1.77–1.80 graphs and formulas are given for four common nonlinear functions. Note that each graph is not a line. See Appendix B for more examples of functions.

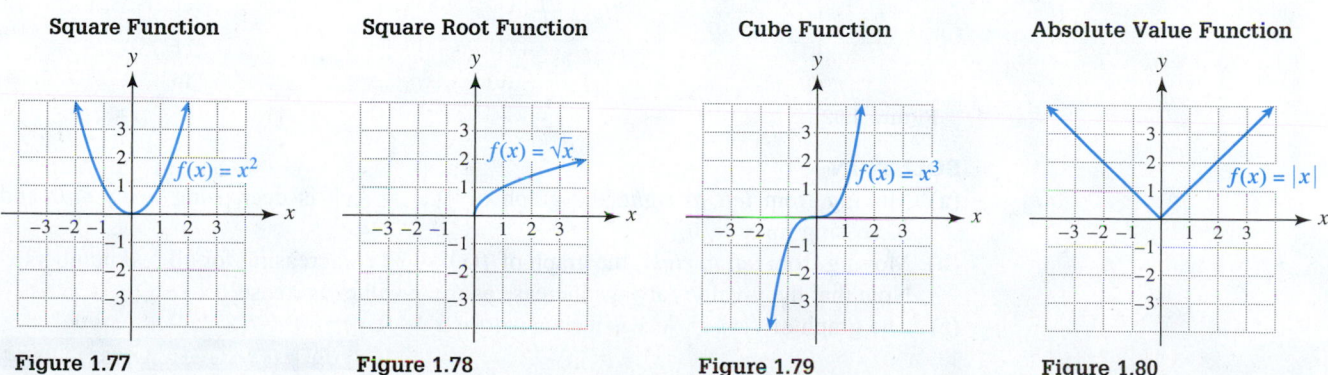

Square Function	Square Root Function	Cube Function	Absolute Value Function
$f(x) = x^2$	$f(x) = \sqrt{x}$	$f(x) = x^3$	$f(x) = |x|$
Figure 1.77	Figure 1.78	Figure 1.79	Figure 1.80

Increasing and Decreasing Functions

Sales of rock music have not remained constant during the past two decades. In 1990, rock music accounted for 36% of all U.S. music sales. This percentage decreased to a low of 24% in 2001 and then increased to 34% in 2006.   (*Source:* Recording Industry Association of America.)

A linear function cannot be used to describe these data because the graph of a (nonconstant) linear function either always rises or always falls. The concepts of increasing and decreasing are important to nonlinear functions. Figure 1.81 illustrates the concepts of an increasing and decreasing function f that models rock music sales.

See the Concept: Rock Music's Share of All U.S. Sales (Percentage)

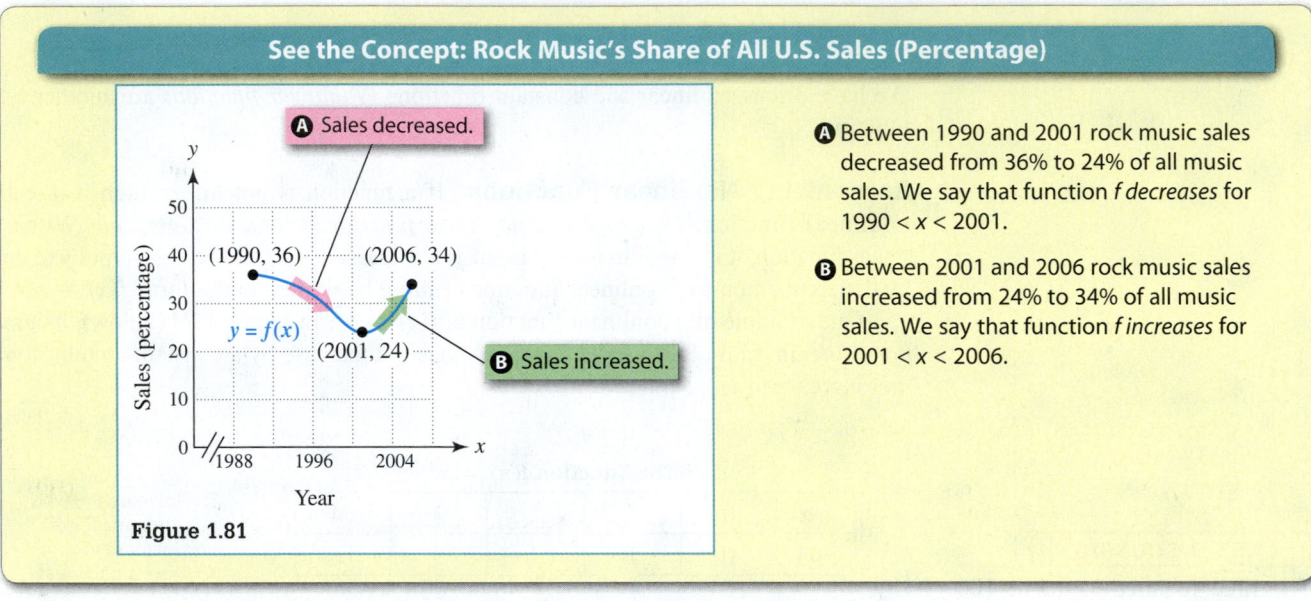

Ⓐ Between 1990 and 2001 rock music sales decreased from 36% to 24% of all music sales. We say that function f *decreases* for $1990 < x < 2001$.

Ⓑ Between 2001 and 2006 rock music sales increased from 24% to 34% of all music sales. We say that function f *increases* for $2001 < x < 2006$.

Figure 1.81

NOTE   The inequality $a < x < b$ means that $x > a$ *and* $x < b$.

EXAMPLE 5   Recognizing increasing and decreasing graphs

The graphs of three functions are shown in Figure 1.82. Determine intervals where each function is increasing or decreasing.

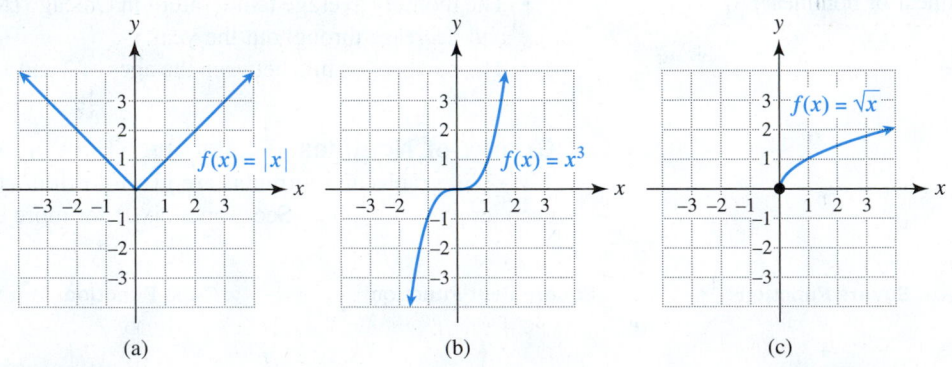

(a)   (b)   (c)

Figure 1.82

SOLUTION

(a) Moving from *left to right*, the graph of $f(x) = |x|$ is decreasing for $x < 0$ and increasing for $x > 0$.

(b) Moving from *left to right*, the graph of $f(x) = x^3$ is increasing for all real numbers x. Note that the y-values always increase as the x-values increase.

(c) The graph of $f(x) = \sqrt{x}$ is increasing for $x > 0$.

Now Try Exercises 65, 66, and 67

The concepts of increasing and decreasing are defined as follows.

> **INCREASING AND DECREASING FUNCTIONS**
>
> Suppose that a function f is defined over an *open* interval I on the number line. If x_1 and x_2 are in I,
> **(a)** f **increases** on I if, whenever $x_1 < x_2$, $f(x_1) < f(x_2)$;
> **(b)** f **decreases** on I if, whenever $x_1 < x_2$, $f(x_1) > f(x_2)$.

Figures 1.83–1.85 in See the Concept illustrate increasing and decreasing functions.

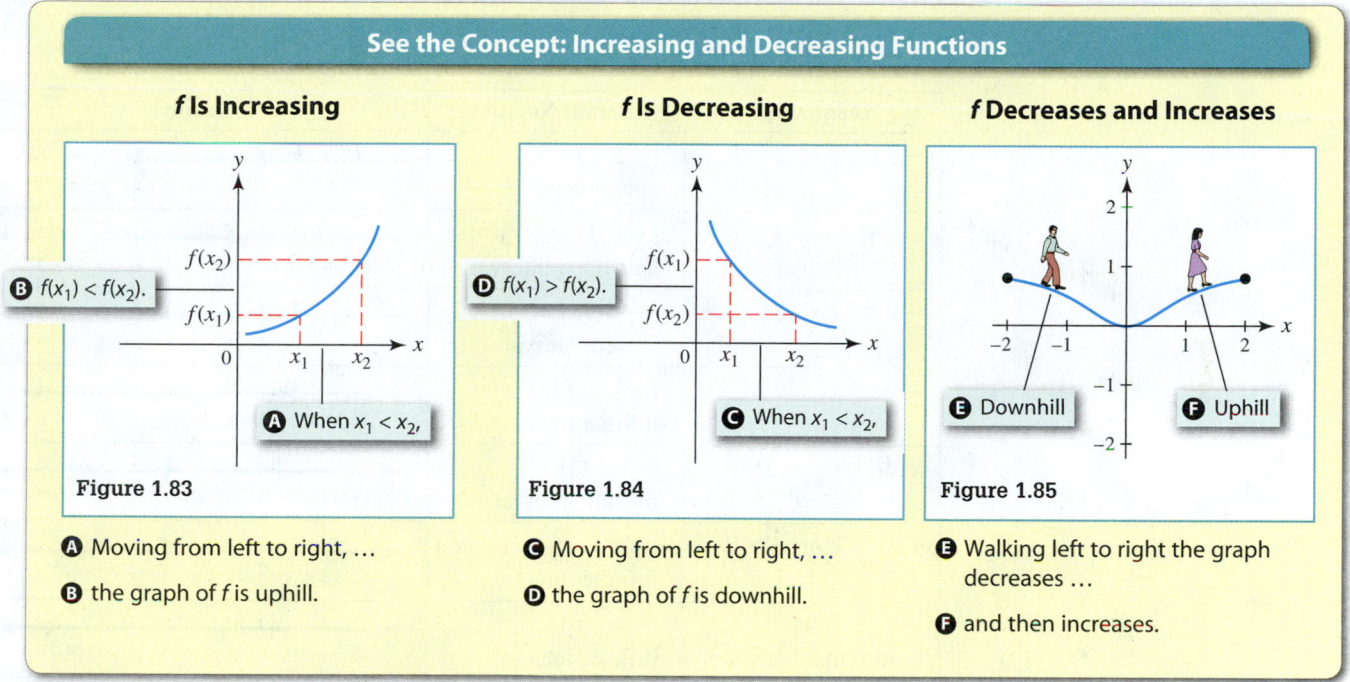

See the Concept: Increasing and Decreasing Functions

f Is Increasing

B $f(x_1) < f(x_2)$.

A When $x_1 < x_2$,

Figure 1.83

A Moving from left to right, …
B the graph of f is uphill.

f Is Decreasing

D $f(x_1) > f(x_2)$.

C When $x_1 < x_2$,

Figure 1.84

C Moving from left to right, …
D the graph of f is downhill.

f Decreases and Increases

E Downhill   **F** Uphill

Figure 1.85

E Walking left to right the graph decreases …
F and then increases.

> **NOTE**   When stating where a function is increasing and where it is decreasing, it is important to give x-intervals and not y-intervals. These x-intervals do *not* include the endpoints.

Interval Notation

To describe intervals where functions are increasing or decreasing, a number line graph is sometimes used. The set $\{x \mid x > 2\}$, which includes all real numbers greater than 2, is graphed in Figure 1.86. Note that a parenthesis at $x = 2$ indicates that the endpoint *is not included*. The set $\{x \mid -1 \leq x \leq 4\}$ is shown in Figure 1.87 and the set $\{x \mid -\frac{7}{2} < x < -\frac{1}{2}\}$ is shown in Figure 1.88. Note that brackets, either $[$ or $]$, are used when endpoints *are included*.

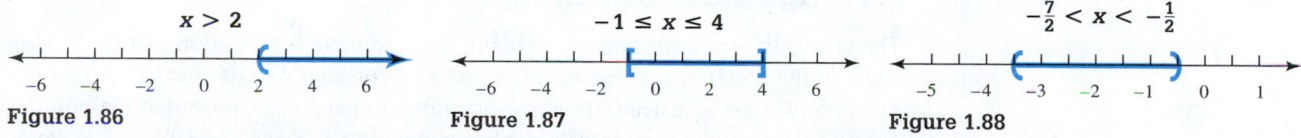

$x > 2$

Figure 1.86

$-1 \leq x \leq 4$

Figure 1.87

$-\frac{7}{2} < x < -\frac{1}{2}$

Figure 1.88

A convenient notation is called **interval notation**. Instead of drawing the entire number line, as in Figure 1.87, we can express this set of real numbers as $[-1, 4]$. Because the set includes the endpoints −1 and 4, the interval is a **closed interval** and brackets are used. A set that included all real numbers satisfying $-\frac{7}{2} < x < -\frac{1}{2}$ would be expressed as the **open interval** $\left(-\frac{7}{2}, -\frac{1}{2}\right)$. Parentheses indicate that the endpoints are not included in the set. An example of a **half-open interval** is [0, 4), which represents the set of real numbers satisfying $0 \leq x < 4$.

Table 1.15 provides some examples of interval notation. The symbol ∞ refers to **infinity**; it does not represent a real number. The notation $(1, \infty)$ means $\{x \mid x > 1\}$, or simply $x > 1$. Since this interval has no maximum x-value, ∞ is used in the position of the right endpoint. A similar interpretation holds for the symbol $-\infty$, which represents **negative infinity**.

NOTE An inequality in the form $x < 1$ or $x > 3$ indicates the set of real numbers that are either less than 1 or greater than 3. The **union symbol** $\cup$ can be used to write this inequality in interval notation as $(-\infty, 1) \cup (3, \infty)$.

Interval Notation

Inequality	Interval Notation	Graph
$-2 < x < 2$	$(-2, 2)$ open interval	
$-1 < x \leq 3$	$(-1, 3]$ half-open interval	
$-3 \leq x \leq 2$	$[-3, 2]$ closed interval	
$x > -3$	$(-3, \infty)$ infinite interval	
$x \leq 1$	$(-\infty, 1]$ infinite interval	
$x \leq -2$ or $x > 1$	$(-\infty, -2] \cup (1, \infty)$ infinite intervals	
$-\infty < x < \infty$ (entire number line)	$(-\infty, \infty)$ infinite interval	

Table 1.15

EXAMPLE 6 **Determining where a function is increasing or decreasing**

Use the graph of $f(x) = 4x - \frac{1}{3}x^3$ (shown in Figure 1.89) and interval notation to identify where f is increasing or decreasing.

SOLUTION Moving from left to right on the graph of f, the y-values decrease until $x = -2$, increase until $x = 2$, and decrease thereafter. Thus $f(x) = 4x - \frac{1}{3}x^3$ is decreasing on $(-\infty, -2)$, increasing on $(-2, 2)$, and decreasing again on $(2, \infty)$.

Now Try Exercise 69

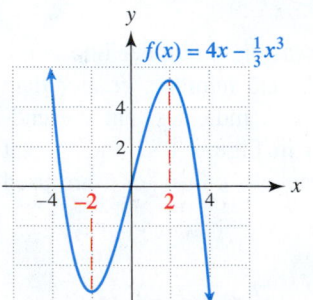

Figure 1.89

Average Rate of Change

The graphs of nonlinear functions are not lines, so there is no notion of a single slope. The slope of the graph of a linear function gives its constant rate of change. With a nonlinear function, we speak instead of an *average* rate of change. Suppose that the points (x_1, y_1) and (x_2, y_2) lie on the graph of a nonlinear function f. See Figure 1.90. The slope of the line L passing through these two points represents the *average rate of change of f from x_1 to x_2*. The line L is referred to as a **secant line**. If different values for x_1 and x_2 are selected, then a different secant line and a different average rate of change (slope) usually result.

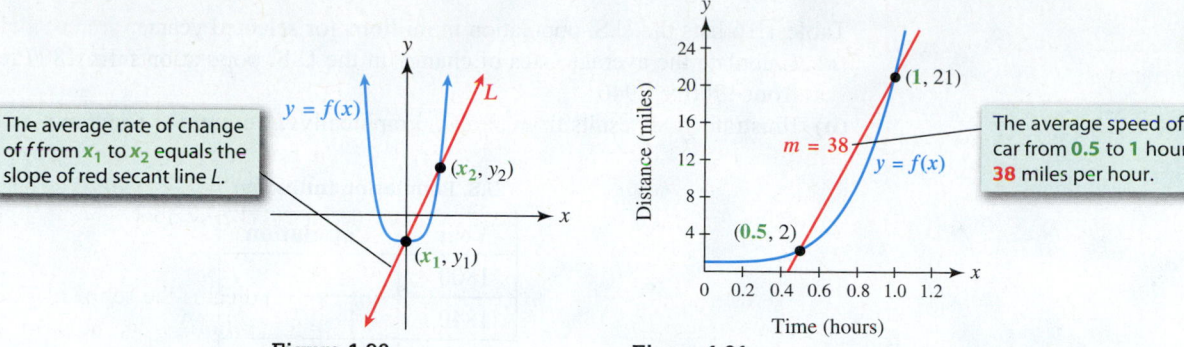

Slope of _L_ Is Average Rate of Change

The average rate of change of _f_ from x_1 to x_2 equals the slope of red secant line _L_.

Figure 1.90

Distance Traveled by a Car

The average speed of the car from **0.5** to **1** hour is **38** miles per hour.

Figure 1.91

In applications the average rate of change measures how fast a quantity is changing over an interval of its domain, *on average*. For example, suppose the graph of the function _f_ in Figure 1.91 represents the distance _y_ in miles that a car has traveled on a straight highway (under construction) after _x_ hours. The points $(\mathbf{0.5}, \mathbf{2})$ and $(\mathbf{1}, \mathbf{21})$ lie on this graph. Thus after 0.5 hour the car has traveled 2 miles and after 1 hour the car has traveled 21 miles. The slope of the red line passing through these two points is

$$m = \frac{y_2 - y_1}{x_2 - x_1} = \frac{21 - 2}{1 - 0.5} = \mathbf{38}.$$

This means that during the half hour from **0.5** to **1** hour the average rate of change, or average *speed*, is **38** miles per hour.

AVERAGE RATE OF CHANGE

Let (x_1, y_1) and (x_2, y_2) be distinct points on the graph of a function _f_. The **average rate of change of _f_ from x_1 to x_2** is

$$\frac{y_2 - y_1}{x_2 - x_1}.$$

That is, the average rate of change from x_1 to x_2 equals the slope of the line passing through (x_1, y_1) and (x_2, y_2).

NOTE If $y = f(x)$, then average rate of change equals $\frac{f(x_2) - f(x_1)}{x_2 - x_1}$.

EXAMPLE 7   **Finding an average rate of change**

Let $f(x) = 2x^2$. Find the average rate of change from $x = 1$ to $x = 3$.

SOLUTION First calculate $f(1) = 2(1)^2 = \mathbf{2}$ and $f(3) = 2(3)^2 = \mathbf{18}$. The average rate of change equals the slope of the line passing through the points $(\mathbf{1}, \mathbf{2})$ and $(\mathbf{3}, \mathbf{18})$.

$$\text{Average rate of change} = \frac{\mathbf{18} - \mathbf{2}}{\mathbf{3} - \mathbf{1}} = \frac{16}{2} = \mathbf{8}.$$

The average rate of change from $x = 1$ to $x = 3$ is **8**.

Now Try Exercise 97

NOTE If _f_ is a constant function, its average rate of change is always 0. For a linear function defined by $f(x) = mx + b$, the average rate of change is always _m_, the slope of its graph. The average rate of change for a nonlinear function varies.

EXAMPLE 8 **Calculating and interpreting average rates of change**

Table 1.16 lists the U.S. population in millions for selected years.
(a) Calculate the average rates of change in the U.S. population from 1800 to 1840 and from 1900 to 1940.
(b) Illustrate your results from part (a) graphically. Interpret the results.

U.S. Population (millions)

Year	Population
1800	5
1840	17
1900	76
1940	132

Table 1.16

SOLUTION

(a) In **1800** the population was **5** million, and in **1840** it was **17** million. Therefore the average rate of change in the population from 1800 to 1840 was

$$\frac{17 - 5}{1840 - 1800} = 0.3.$$

In **1900** the population was **76** million, and in **1940** it was **132** million. Therefore the average rate of change in the population from 1900 to 1940 was

$$\frac{132 - 76}{1940 - 1900} = 1.4.$$

(b) These average rates of change are illustrated graphically and interpreted in Figure 1.92.

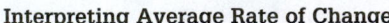

Interpreting Average Rate of Change

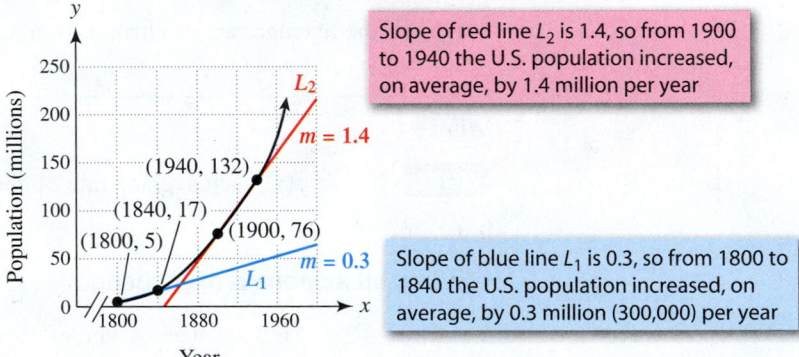

Slope of red line L_2 is 1.4, so from 1900 to 1940 the U.S. population increased, on average, by 1.4 million per year

Slope of blue line L_1 is 0.3, so from 1800 to 1840 the U.S. population increased, on average, by 0.3 million (300,000) per year

Figure 1.92

Now Try Exercise 99

The Difference Quotient

The difference quotient is often used in calculus to calculate rates of change and is explained in the following See the Concept. In Figure 1.93, the red secant line L passes through two points, $(x, f(x))$ and $(x + h, f(x + h))$, on the graph of $y = f(x)$. (In Figure 1.93, we assume that $h > 0$.)

See the Concept: Understanding the Difference Quotient

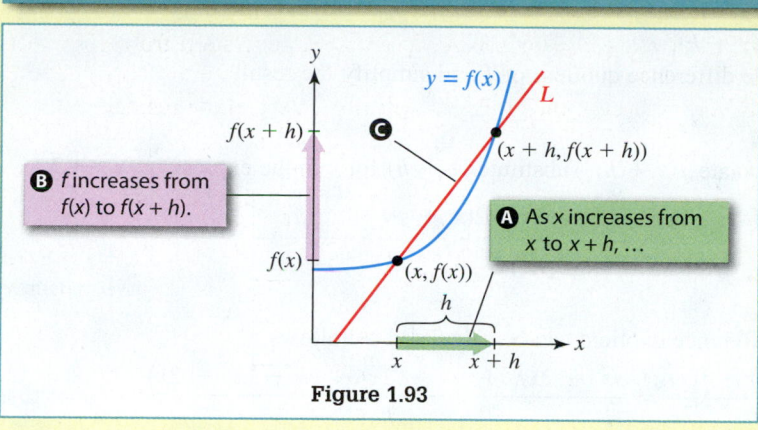

B *f* increases from $f(x)$ to $f(x + h)$.

A As *x* increases from *x* to $x + h$, ...

Figure 1.93

C The slope of the red line *L* represents the average rate of change of *f* from *x* to $x + h$.

The slope of *L* is $\dfrac{f(x + h) - f(x)}{(x + h) - x}$ and this

formula simplifies to the *difference quotient* of *f*.

The following box gives a definition of the difference quotient.

DIFFERENCE QUOTIENT

The **difference quotient of a function *f*** is an expression of the form

$$\frac{f(x + h) - f(x)}{h}, \qquad \text{where } h \neq 0.$$

EXAMPLE 9   Finding a difference quotient

Let $f(x) = 3x - 2$.
(a) Find $f(x + h)$.
(b) Find the difference quotient of *f* and simplify the result.

SOLUTION
(a) To find $f(x + h)$, substitute $(x + h)$ for *x* in the expression $3x - 2$.

$$f(x + h) = 3(x + h) - 2 \qquad\qquad f(x) = 3x - 2$$
$$= 3x + 3h - 2 \qquad\qquad \text{Distributive property}$$

(b) The difference quotient can be calculated as follows.

> Include parentheses when $f(x)$ is more than one term.

$$\underbrace{\frac{f(x + h) - f(x)}{h}}_{\text{Difference quotient}} = \frac{(3x + 3h - 2) - (3x - 2)}{h} \qquad \text{Substitute.}$$

$$= \frac{3x + 3h - 2 - 3x + 2}{h} \qquad \text{Distributive property}$$

$$= \frac{3h}{h} \qquad \text{Combine like terms.}$$

$$= 3 \qquad \text{Simplify.}$$

NOTE   The difference quotient for a linear function $f(x) = mx + b$ always equals the slope *m* of the graph of *f*.

Now Try Exercise 107

EXAMPLE 10   Calculating a difference quotient

Let $f(x) = x^2 - 2x$.
(a) Find $f(x + h)$.
(b) Find the difference quotient of f and simplify the result.

SOLUTION
(a) To calculate $f(x + h)$, substitute $(x + h)$ for x in the expression $x^2 - 2x$.

$$f(x + h) = (x + h)^2 - 2(x + h) \qquad f(x) = x^2 - 2x$$
$$= x^2 + 2xh + h^2 - 2x - 2h \qquad \text{Square the binomial;}$$
$$\text{distributive property.}$$

Algebra Review
To square a binomial, see Chapter R
(page R-17)

(b) The difference quotient can be calculated as follows.

$$\frac{f(x + h) - f(x)}{h} = \frac{x^2 + 2xh + h^2 - 2x - 2h - (x^2 - 2x)}{h} \qquad \text{Substitute.}$$

$$= \frac{2xh + h^2 - 2h}{h} \qquad \text{Combine like terms.}$$

$$= \frac{h(2x + h - 2)}{h} \qquad \text{Factor out } h.$$

$$= 2x + h - 2 \qquad \text{Simplify.}$$

Now Try Exercise 109

1.4 Putting It All Together

CONCEPT	FORMULA	EXAMPLES
Slope of a line passing through (x_1, y_1) and (x_2, y_2)	$m = \dfrac{\Delta y}{\Delta x} = \dfrac{y_2 - y_1}{x_2 - x_1}$ $\Delta y = y_2 - y_1$ denotes the change in y. $\Delta x = x_2 - x_1$ denotes the change in x.	A line passing through $(-1, 3)$ and $(1, 7)$ has slope $m = \dfrac{7 - 3}{1 - (-1)} = \dfrac{4}{2} = 2$. This slope indicates that the line rises 2 units for each unit increase in x.
Constant function	$f(x) = b$, where b is a fixed number, or constant.	$f(x) = 12, g(x) = -2.5$, and $h(x) = 0$. Every constant function is also linear.
Linear function	$f(x) = mx + b$, where m and b are constants. The graph of f has slope m.	$f(x) = 3x - 1, g(x) = -5$, and $h(x) = \frac{1}{2} - \frac{3}{4}x$. Their graphs have slopes 3, 0, and $-\frac{3}{4}$.
Nonlinear function	A nonlinear function cannot be expressed in the form $f(x) = mx + b$.	$f(x) = \sqrt{x + 1}, g(x) = 4x^3$, and $h(x) = x^{1.01} + 2$.

CONCEPT	CONSTANT FUNCTION	LINEAR FUNCTION	NONLINEAR FUNCTION
Slope of graph	Always zero	Always constant	No notion of one slope
Graph	Horizontal line	Nonvertical line	Not a line
Examples			

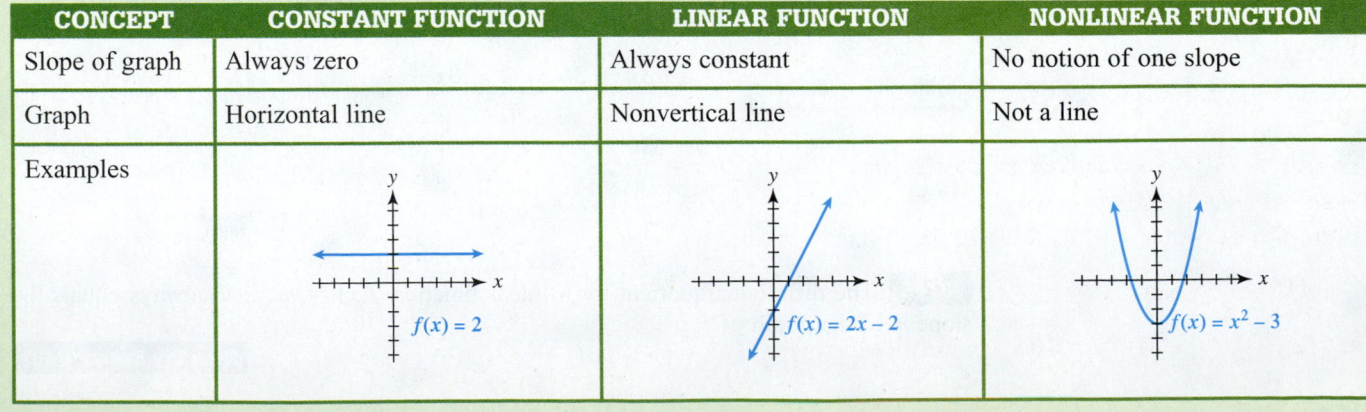

CONCEPT	EXPLANATION	EXAMPLES
Interval notation	An efficient notation for writing inequalities	$x \leq 6$ is equivalent to $(-\infty, 6]$. $x > 3$ is equivalent to $(3, \infty)$. $2 < x \leq 5$ is equivalent to $(2, 5]$.
Increasing and decreasing	f increases on an *open* interval if, whenever $x_1 < x_2$, then $f(x_1) < f(x_2)$. f decreases on an *open* interval if, whenever $x_1 < x_2$, then $f(x_1) > f(x_2)$.	 f is increasing on $(-\infty, -2)$ and on $(2, \infty)$. f is decreasing on $(-2, 2)$.
Average rate of change of f from x_1 to x_2	If (x_1, y_1) and (x_2, y_2) are distinct points on the graph of f, then the average rate of change from x_1 to x_2 equals $$\frac{y_2 - y_1}{x_2 - x_1}.$$	If $f(x) = 3x^2$, then the average rate of change from $x = 1$ to $x = 3$ is $$\frac{27 - 3}{3 - 1} = 12$$ because $f(3) = 27$ and $f(1) = 3$. This means that, on average, $f(x)$ increases by 12 units for each unit increase in x from 1 to 3.
Difference quotient	Calculates average rate of change of f from x to $x + h$. $$\frac{f(x + h) - f(x)}{h}, h \neq 0$$	If $f(x) = 2x$, then the difference quotient equals $$\frac{2(x + h) - 2x}{h} = \frac{2h}{h} = 2.$$

1.4 Exercises

Formulas for Linear Functions

Exercises 1–4: A linear function f can be written in the form
$f(x) = mx + b$. Identify m and b for the given $f(x)$.

1. $f(x) = 5 - 2x$ **2.** $f(x) = 3 - 4x$

3. $f(x) = -8x$ **4.** $f(x) = -6$

Slope

Exercises 5–18: If possible, find the slope of the line passing through each pair of points.

5. $(4, 6), (2, 5)$ **6.** $(-8, 5), (-3, -7)$

7. $(-1, 4), (5, -2)$ **8.** $(10, -4), (-15, 7)$

9. $(12, -8), (7, -8)$ **10.** $(8, -5), (8, 2)$

11. $(0.2, -0.1), (-0.3, 0.4)$ **12.** $(-0.3, 0.6), (-0.2, 1.1)$

13. $(-0.5, 9.2), (-0.3, 7.6)$ **14.** $(1.6, 12), (1.6, 5)$

15. $(-5, 6), (-5, 8)$ **16.** $(17, 7), (19, 7)$

17. $\left(\frac{1}{3}, -\frac{3}{5}\right), \left(-\frac{5}{6}, \frac{7}{10}\right)$ **18.** $\left(-\frac{13}{15}, -\frac{7}{8}\right), \left(\frac{1}{10}, \frac{3}{16}\right)$

Exercises 19–24: State the slope of the graph of f. Explain what the slope indicates about the graph.

19. $f(x) = 2x + 7$ **20.** $f(x) = 6 - x$

21. $f(x) = -\frac{3}{4}x$ **22.** $f(x) = \frac{2}{3}x$

23. $f(x) = 9 - x$ **24.** $f(x) = 23$

Slope as a Rate of Change

25. Price of Carpet The graph shows the price of x square feet of carpeting.

(a) Why is it reasonable for the graph to pass through the origin?

(b) Find the slope of the graph.

(c) Interpret the slope as a rate of change.

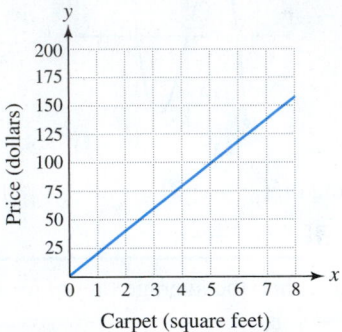

Carpet (square feet)

26. Landscape Rock The figure shows the price of x tons of landscape rock.

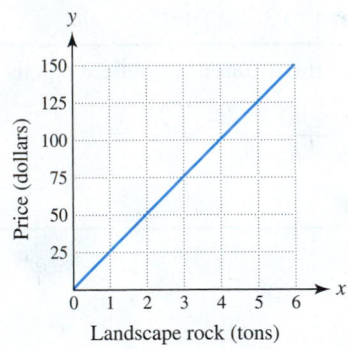

Landscape rock (tons)

(a) Why is it reasonable for the graph to pass through the origin?

(b) Find the slope of the graph.

(c) Interpret the slope as a rate of change.

27. Velocity of a Train The distance D in miles that a train is from a station after x hours is given by the formula $D(x) = 150 - 20x$.

(a) Calculate $D(5)$ and interpret the result.

(b) Find the slope of the graph of D. Interpret this slope as a rate of change.

28. Cost of Paint The cost C in dollars of purchasing x gallons of paint is given by $C(x) = 29x$.

(a) Evaluate $C(5)$ and interpret your result.

(b) Find the slope of the graph of C. Interpret this slope as a rate of change.

29. Velocity of a Car A driver's distance D in miles from a rest stop after x hours is given by $D(x) = 75x$.

(a) How far is the driver from the rest stop after 2 hours?

(b) Find the slope of the graph of D. Interpret this slope as a rate of change.

30. Age in the United States The median age of the U.S. population for each year t between 1970 and 2010 can be approximated by the formula $A(t) = 0.243t - 450.8$. (*Source:* Bureau of the Census.)

(a) Compute the median ages in 1980 and 2000.

(b) What is the slope of the graph of A? Interpret the slope.

Linear and Nonlinear Functions

Exercises 31–38: Determine if f is a linear or nonlinear function. If f is a linear function, determine if f is a constant function. Support your answer by graphing f.

31. $f(x) = -2x + 5$ **32.** $f(x) = 3x - 2$

33. $f(x) = 1$ **34.** $f(x) = -2$

35. $f(x) = |x + 1|$ **36.** $f(x) = |2x - 1|$

37. $f(x) = x^2 - 1$ **38.** $f(x) = \sqrt{x - 1}$

Recognizing Linear Data

Exercises 39–42: Decide whether a line can pass through the data points. If it can, determine the slope of the line.

39.

x	0	1	2	3	4
y	-1	3	7	11	15

40.

x	-4	-2	0	2	4
y	1	$-\frac{1}{2}$	-2	$-\frac{7}{2}$	-5

41.

x	-5	-3	1	3	5
y	-5	-2	1	4	7

42.

x	10	20	25	35	40
y	40	190	300	600	790

Linear Functions

Exercises 43–46: The graph of a linear function f is shown.

(a) *Identify the slope, y-intercept, and x-intercept.*
(b) *Write a formula for f.*
(c) *Estimate the zero of f.*

43.

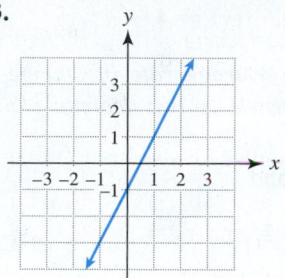

44.

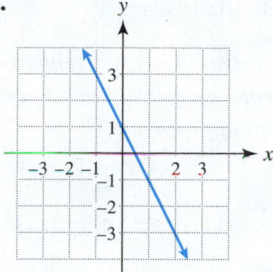

45.

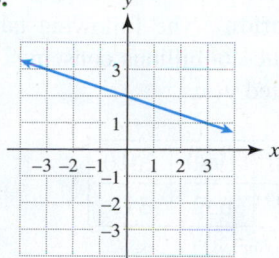

46.

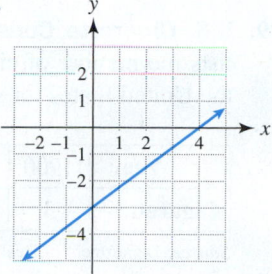

Exercises 47–50: Write a formula for a linear function f whose graph satisfies the conditions.

47. Slope $-\frac{3}{4}$, y-intercept $\frac{1}{3}$

48. Slope -122, y-intercept 805

49. Slope 15, passing through the origin

50. Slope 1.68, passing through (0, 1.23)

Interval Notation

Exercises 51–64: Express each of the following in interval notation.

51. $x \geq 5$

52. $x < 100$

53. $4 \leq x < 19$

54. $-4 < x < -1$

55. $\{x \mid -1 \leq x\}$

56. $\{x \mid x \leq -3\}$

57. $\{x \mid x < 1 \text{ or } x \geq 3\}$

58. $\{x \mid x \leq -2 \text{ or } x \geq 0\}$

59.

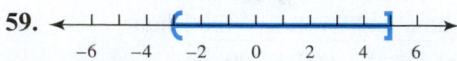

60.

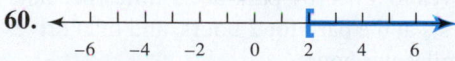

61.

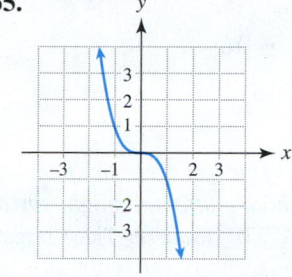

62.

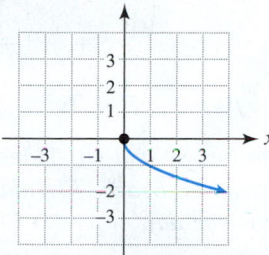

63.

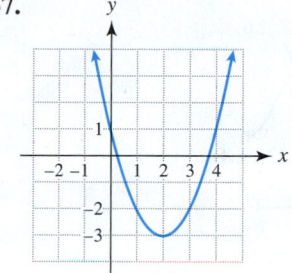

64.

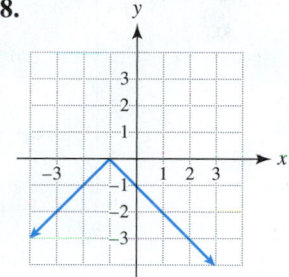

Increasing and Decreasing Functions

Exercises 65–72: Use the graph of f to determine intervals where f is increasing and where f is decreasing.

65.

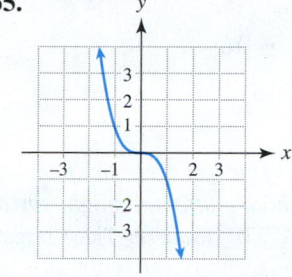

66.

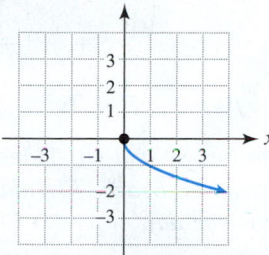

67.

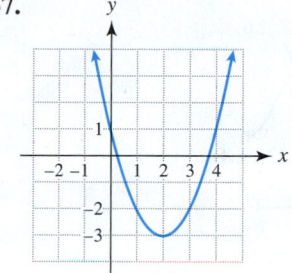

68.

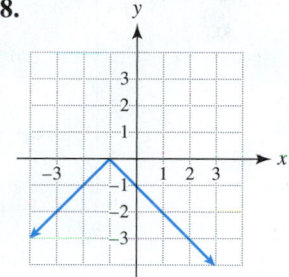

69.

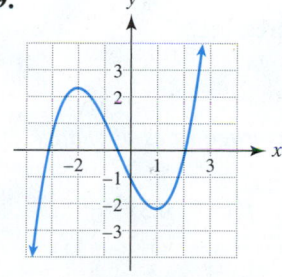

70.

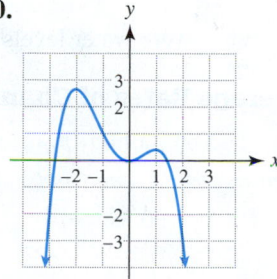

71.

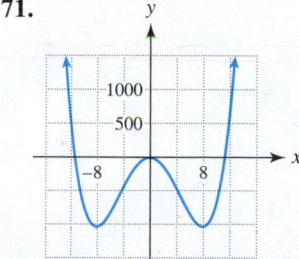

72.

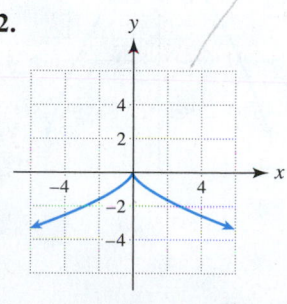

Exercises 73–88: Identify where f is increasing and where f is decreasing. (Hint: Consider the graph $y = f(x)$.)

73. $f(x) = 2x - 1$ **74.** $f(x) = 4 - x$

75. $f(x) = x^2 - 2$ **76.** $f(x) = -\frac{1}{2}x^2$

77. $f(x) = 2x - x^2$ **78.** $f(x) = x^2 - 4x$

79. $f(x) = \sqrt{x - 1}$ **80.** $f(x) = -\sqrt{x + 1}$

81. $f(x) = |x + 3|$ **82.** $f(x) = |x - 1|$

83. $f(x) = x^3$ **84.** $f(x) = \sqrt[3]{x}$

85. $f(x) = \frac{1}{3}x^3 - 4x$ **86.** $f(x) = x^3 - 3x$

87. $f(x) = -\frac{1}{4}x^4 + \frac{1}{3}x^3 + x^2$

88. $f(x) = \frac{1}{4}x^4 - 2x^2$

Exercises 89 and 90: **Tides** *The graph gives the tides at Clearwater Beach, Florida, x hours after midnight on a particular day, where $0 \le x \le 27$.* (**Source:** Tide/Current Predictor.)

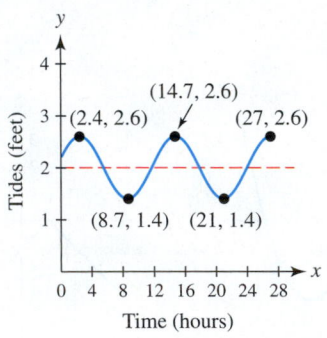

89. When were water levels increasing?

90. When were water levels decreasing?

Average Rates of Change

Exercises 91 and 92: Find the average rates of change of f from −3 to −1 and from 1 to 3.

91. $f(x) = -0.3x^2 + 4$ **92.** $f(x) = 0.3x^2 - 4$

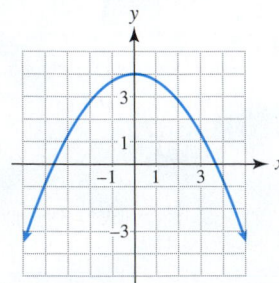 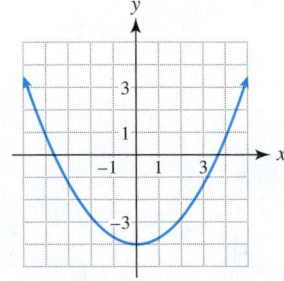

Exercises 93 and 94: (Refer to Examples 7 and 8.) Use the given f(x) to complete the following.

(a) Calculate the average rate of change of f from $x = 1$ to $x = 2$.

(b) Illustrate your result from part (a) graphically.

93. $f(x) = x^2$ **94.** $f(x) = 4 - x^2$

Exercises 95–98: Compute the average rate of change of f from x_1 to x_2. Round your answer to two decimal places when appropriate.

95. $f(x) = 7x - 2, x_1 = 1$, and $x_2 = 4$

96. $f(x) = -8x + 5, x_1 = -2$, and $x_2 = 0$

97. $f(x) = \sqrt{2x - 1}, x_1 = 1$, and $x_2 = 3$

98. $f(x) = 0.5x^2 - 5, x_1 = -1$, and $x_2 = 4$

99. **U.S. Cigarette Consumption** The following table lists the number of cigarettes in billions consumed in the United States for selected years.

Year	1900	1940	1980	2010
Cigarettes	3	182	632	315

Source: Department of Health and Human Services.

(a) Find the average rate of change during each time period.

(b) Interpret the results.

100. **Torricelli's Law** A cylindrical tank contains 100 gallons of water. A plug is pulled from the bottom of the tank and the amount of water in gallons remaining in the tank after x minutes is given by

$$A(x) = 100\left(1 - \frac{x}{5}\right)^2.$$

(a) Calculate the average rate of change of A from 1 to 1.5 and from 2 to 2.5. Interpret your results.

(b) Are the two average rates of change the same or different? Explain why.

Curve Sketching

Exercises 101 and 102: Sketch a graph that illustrates the motion of the person described. Let the x-axis represent time and the y-axis represent distance from home. Be sure to label each axis.

101. A person drives a car away from home for 2 hours at 50 miles per hour and then stops for 1 hour.

102. A person drives to a nearby park at 25 miles per hour for 1 hour, rests at the park for 2 hours, and then drives home at 50 miles per hour.

Exercises 103 and 104: **Critical Thinking** *Do not use a graphing calculator.*

103. On the same coordinate axes, sketch the graphs of a constant function f and a nonlinear function g that intersect exactly twice.

104. Sketch a graph of a function that has only positive average rates of change for $x \geq 1$ and only negative average rates of change for $x \leq 1$.

The Difference Quotient

Exercises 105–116: (Refer to Examples 9 and 10.) Complete the following for the given $f(x)$.

(a) Find $f(x + h)$.
(b) Find the difference quotient of f and simplify.

105. $f(x) = 3$

106. $f(x) = -5$

107. $f(x) = 2x + 1$

108. $f(x) = -3x + 4$

109. $f(x) = 3x^2 + 1$

110. $f(x) = x^2 - 2$

111. $f(x) = -x^2 + 2x$

112. $f(x) = -4x^2 + 1$

113. $f(x) = 2x^2 - x + 1$

114. $f(x) = x^2 + 3x - 2$

115. $f(x) = x^3$

116. $f(x) = 1 - x^3$

117. **Speed of a Car** Let the distance in feet that a car travels in t seconds be given by $d(t) = 8t^2$ for $0 \leq t \leq 6$.

  (a) Find $d(t + h)$.

  (b) Find the difference quotient for d and simplify.

  (c) Evaluate the difference quotient when $t = 4$ and $h = 0.05$. Interpret your result.

118. **Draining a Pool** Let the number of gallons G of water in a pool after t hours be given by $G(t) = 4000 - 100t$ for $0 \leq t \leq 40$.

  (a) Find $G(t + h)$.

  (b) Find the difference quotient. Interpret your result.

Writing about Mathematics

119. What does the average rate of change represent for a linear function? What does it represent for a nonlinear function? Explain your answers.

120. Suppose you are given a graphical representation of a function f. Explain how you would determine whether f is constant, linear, or nonlinear. How would you determine the type if you were given a numerical or symbolic representation? Give examples.

121. Suppose that a function f has a positive average rate of change from 1 to 4. Is it correct to assume that function f only increases on the interval $(1, 4)$? Make a sketch to support your answer.

122. If $f(x) = mx + b$, what does the difference quotient for function f equal? Explain your reasoning.

Extended and Discovery Exercise

1. **Geometry** Suppose that the radius of a circle on a computer monitor is increasing at a constant rate of 1 inch per second.

  (a) Does the circumference of the circle increase at a constant rate? If it does, find this rate.

  (b) Does the area of the circle increase at a constant rate? Explain.

2. **Velocity** If the distance in feet run by a racehorse in t seconds is given by $d(t) = 2t^2$, then the difference quotient for d is $4t + 2h$. How could you estimate the velocity of the racehorse at exactly 7 seconds?

CHECKING BASIC CONCEPTS FOR SECTIONS 1.3 AND 1.4

1. Give symbolic, numerical, and graphical representations of a function f that computes the number of feet in x miles. For the numerical representation use a table and let $x = 1, 2, 3, 4, 5$.

2. Let $f(x) = \frac{2x}{x - 4}$.
  (a) Find $f(2)$ and $f(a + 4)$.

  (b) Find the domain of f.

3. Find the slope of the line passing through the points $(-2, 4)$ and $(4, -5)$.

4. Identify each function f as constant, linear, or nonlinear.
  (a) $f(x) = -1.4x + 5.1$

  (b) $f(x) = 25$

  (c) $f(x) = 2x^2 - 5$

5. Write each expression in interval notation.
  (a) $x \leq 5$ **(b)** $1 \leq x < 6$

6. Determine where $f(x) = x^2 - 2$ is increasing and where it is decreasing.

7. Find the average rate of change of $f(x) = x^2 - 3x$ from $x = -3$ to $x = -1$.

8. Find the difference quotient for $f(x) = 4x^2$.

1 Summary

CONCEPT	EXPLANATION AND EXAMPLES

Section 1.1 Numbers, Data, and Problem Solving

Sets of Numbers

Natural numbers: $N = \{1, 2, 3, 4, \dots\}$

Integers: $I = \{\dots, -3, -2, -1, 0, 1, 2, 3, \dots\}$

Rational numbers: $\frac{p}{q}$, where p and q are integers with $q \neq 0$; includes fractions, repeating and terminating decimals

Irrational numbers: Includes nonrepeating, nonterminating decimals

Real numbers: Any number that can be expressed in decimal form; includes rational and irrational numbers

Order of Operations

Using the following order of operations, perform all calculations within parentheses, square roots, and absolute value bars and above and below fraction bars. Then use the same order of operations to perform any remaining calculations.

1. Evaluate all exponents. Then do any negation *after* evaluating exponents.
2. Do all multiplication and division from *left to right*.
3. Do all addition and subtraction from *left to right*.

Example: $5 + 3 \cdot 2^3 = 5 + 3 \cdot 8 = 5 + 24 = 29$

Scientific Notation

A real number r is written as $c \times 10^n$, where $1 \leq |c| < 10$.

Examples: $1234 = 1.234 \times 10^3$ $0.054 = 5.4 \times 10^{-2}$

Section 1.2 Visualizing and Graphing Data

Mean (Average) and Median

The mean represents the average of a set of numbers, and the median represents the middle of a sorted list.

Example: $4, 6, 9, 13, 15$; $\text{Mean} = \dfrac{4 + 6 + 9 + 13 + 15}{5} = 9.4$; $\text{Median} = 9$

Relation, Domain, and Range

A relation S is a set of ordered pairs. The domain D is the set of x-values, and the range R is the set of y-values.

Example: $S = \{(-1, 2), (4, -5), (5, 9)\}$; $D = \{-1, 4, 5,\}, R = \{-5, 2, 9\}$

Cartesian (Rectangular) Coordinate System, or xy-Plane

The xy-plane has four quadrants and is used to graph ordered pairs.

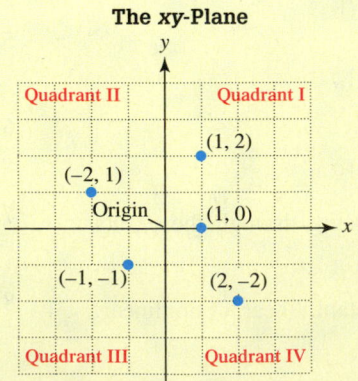

The xy-Plane

CONCEPT	EXPLANATION AND EXAMPLES

Section 1.2 Visualizing and Graphing Data (CONTINUED)

Distance Formula

The distance d between the points (x_1, y_1) and (x_2, y_2) is

$$d = \sqrt{(x_2 - x_1)^2 + (y_2 - y_1)^2}.$$

Example: The distance between $(-3, 5)$ and $(2, -7)$ is

$$d = \sqrt{(2 - (-3))^2 + (-7 - 5)^2} = \sqrt{5^2 + (-12)^2} = 13.$$

Midpoint Formula

The midpoint M of the line segment with endpoints (x_1, y_1) and (x_2, y_2) is

$$M = \left(\frac{x_1 + x_2}{2}, \frac{y_1 + y_2}{2} \right).$$

Example: The midpoint of the line segment connecting $(1, 2)$ and $(-3, 5)$ is

$$M = \left(\frac{1 + (-3)}{2}, \frac{2 + 5}{2} \right) = \left(-1, \frac{7}{2} \right).$$

Standard Equation of a Circle

The circle with center (h, k) and radius r has the equation

$$(x - h)^2 + (y - k)^2 = r^2.$$

Example: A circle with center $(-2, 5)$ and radius 6 has the equation

$$(x + 2)^2 + (y - 5)^2 = 36.$$

Scatterplot and Line Graph

A scatterplot consists of a set of ordered pairs plotted in the xy-plane. When consecutive points are connected with line segments, a line graph results.

Section 1.3 Functions and Their Representations

Function

A function computes exactly one output for each valid input. The set of valid inputs is called the domain D, and the set of outputs is called the range R.

Examples: $f(x) = \sqrt{1 - x}$

$D = \{x \mid x \leq 1\}, R = \{y \mid y \geq 0\}$

$g = \{(-1, 0.5), (0, 4), (2, 4), (6, \pi)\}$

$D = \{-1, 0, 2, 6\}, R = \{0.5, \pi, 4\}$

Function Notation

Examples: $f(x) = x^2 - 4$; $f(3) = 3^2 - 4 = 5$ and

$f(a + 1) = (a + 1)^2 - 4 = a^2 + 2a - 3$

Representations of Functions

A function can be represented symbolically (formula), graphically (graph), numerically (table of values), and verbally (words). Other representations are possible.

Symbolic Representation $f(x) = x^2 - 1$

CONCEPT	EXPLANATION AND EXAMPLES

Section 1.3 Functions and Their Representations (CONTINUED)

Representations of Functions

Numerical Representation

x	y
-2	3
-1	0
0	-1
1	0
2	3

Graphical Representation

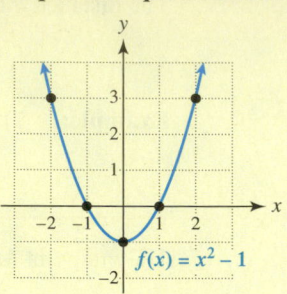

Verbal Representation f computes the square of the input x and then subtracts 1.

Vertical Line Test

If every vertical line intersects a graph at most once, then the graph represents a function.

Section 1.4 Types of Functions and Their Rates of Change

Slope

The slope m of the line passing through (x_1, y_1) and (x_2, y_2) is

$$m = \frac{\Delta y}{\Delta x} = \frac{y_2 - y_1}{x_2 - x_1}.$$

Example: The slope of the line passing through $(1, -1)$ and $(-2, 3)$ is

$$m = \frac{3 - (-1)}{-2 - 1} = -\frac{4}{3}.$$

Constant Function

Given by $f(x) = b$, where b is a constant; its graph is a horizontal line.

Linear Function

Given by $f(x) = mx + b$; its graph is a nonvertical line; the slope of its graph is equal to m, which is also equal to its constant rate of change.

Examples: The graph of $f(x) = -8x + 100$ has slope -8.
If $G(t) = -8t + 100$ calculates the number of gallons of water in a tank after t minutes, then water is *leaving* the tank at 8 gallons per minute.

Nonlinear Function

A nonlinear function *cannot* be written as $f(x) = mx + b$ and its graph is not a line.

Examples: $f(x) = x^2 - 4$; $g(x) = \sqrt[3]{x} - 2$; $h(t) = \dfrac{1}{t + 1}$

Interval Notation

A concise way to express intervals on the number line

Example: $x < 4$ is expressed as $(-\infty, 4)$.
$-3 \le x < 1$ is expressed as $[-3, 1)$.
$x \le 2$ or $x \ge 5$ is expressed as $(-\infty, 2] \cup [5, \infty)$.

Increasing/Decreasing

f increases on an *open* interval I if, whenever $x_1 < x_2$, $f(x_1) < f(x_2)$.
f decreases on an *open* interval I if, whenever $x_1 < x_2$, $f(x_1) > f(x_2)$.

Example: $f(x) = |x|$ increases on $(0, \infty)$ and decreases on $(-\infty, 0)$.

CONCEPT	EXPLANATION AND EXAMPLES

Section 1.4 Types of Functions and Their Rates of Change (CONTINUED)

Average Rate of Change

If (x_1, y_1) and (x_2, y_2) are distinct points on the graph of f, then the average rate of change from x_1 to x_2 equals the slope of the line passing through these two points, given by

$$\frac{y_2 - y_1}{x_2 - x_1}.$$

Example: $f(x) = x^2$; because $f(2) = 4$ and $f(3) = 9$, the graph of f passes through the points $(2, 4)$ and $(3, 9)$, and the average rate of change from 2 to 3 is given by $\frac{9 - 4}{3 - 2} = 5$.

Difference Quotient

The difference quotient of a function f is an expression of the form

$$\frac{f(x + h) - f(x)}{h}, \quad \text{where } h \neq 0,$$

and is the average rate of change from x to $x + h$.

Example: Let $f(x) = x^2$. The difference quotient of f is

$$\frac{(x + h)^2 - x^2}{h} = \frac{x^2 + 2xh + h^2 - x^2}{h} = 2x + h.$$

1 Review Exercises

Exercises 1 and 2: Classify each number listed as one or more of the following: natural number, integer, rational number, or real number.

1. $-2, \frac{1}{2}, 0, 1.23, \sqrt{7}, \sqrt{16}$

2. $55, 1.5, \frac{104}{17}, 2^3, \sqrt{3}, -1000$

Exercises 3 and 4: Write each number in scientific notation.

3. $1,891,000$ **4.** 0.0001001

Exercises 5 and 6: Write each number in standard form.

5. 1.52×10^4 **6.** -7.2×10^{-3}

7. Evaluate each expression with a calculator. Round answers to the nearest hundredth.

(a) $\sqrt[3]{1.2} + \pi^3$ (b) $\dfrac{3.2 + 5.7}{7.9 - 4.5}$

(c) $\sqrt{5^2 + 2.1}$ (d) $1.2(6.3)^2 + \dfrac{3.2}{\pi - 1}$

8. Evaluate each expression. Write your answer in scientific notation and in standard form.

(a) $(4 \times 10^3)(5 \times 10^{-5})$ (b) $\dfrac{3 \times 10^{-5}}{6 \times 10^{-2}}$

Exercises 9 and 10: Evaluate by hand.

9. $4 - 3^2 \cdot 5$ **10.** $3 \cdot 3^2 \div \dfrac{3 - 5}{6 + 2}$

Exercises 11 and 12: Sort the list of numbers from smallest to largest and display the result in a table.

(a) Determine the maximum and minimum values.

(b) Calculate the mean and median.

11. $-5, 8, 19, 24, -23$

12. $8.9, -1.2, -3.8, 0.8, 1.7, 1.7$

Exercises 13 and 14: Complete the following.

(a) Express the data as a relation S.

(b) Find the domain and range of S.

13.

x	-15	-10	0	5	20
y	-3	-1	1	3	5

14.

x	-0.6	-0.2	0.1	0.5	1.2
y	10	20	25	30	80

Exercises 15 and 16: Make a scatterplot of the relation. Determine if the relation is a function.

15. $\{(10, 13), (-12, 40), (-30, -23), (25, -22), (10, 20)\}$

16. $\{(1.5, 2.5), (0, 2.1), (-2.3, 3.1), (0.5, -0.8), (-1.1, 0)\}$

Exercises 17 and 18: Find the distance between the points.

17. $(-4, 5), (2, -3)$ **18.** $(1.2, -4), (0.2, 6)$

Exercises 19 and 20: Find the midpoint of the line segment with the given endpoints.

19. $(24, -16), (-20, 13)$ **20.** $\left(\frac{1}{2}, \frac{5}{4}\right)\left(\frac{1}{2}, -\frac{5}{2}\right)$

21. Determine if the triangle with vertices $(1, 2), (-3, 5)$, and $(0, 9)$ is isosceles.

22. Find the standard equation of a circle with center $(-5, 3)$ and radius 9.

23. A diameter of a circle has endpoints $(-2, 4)$ and $(6, 6)$. Find the standard equation of the circle.

24. Use the graph to determine the domain and range of each function. Evaluate $f(-2)$.

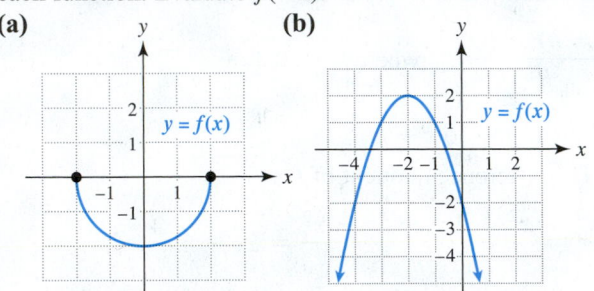

(a)   **(b)**

Exercises 25–32: Graph $y = f(x)$ by first plotting points to determine the shape of the graph.

25. $f(x) = -2$ **26.** $f(x) = 3x$

27. $f(x) = -x + 1$ **28.** $f(x) = 2x - 3$

29. $f(x) = 4 - 2x^2$ **30.** $f(x) = \frac{1}{2}x^2 - 1$

31. $f(x) = |x + 3|$ **32.** $f(x) = \sqrt{3 - x}$

Exercises 33 and 34: Use the verbal representation to express the function f symbolically, graphically, and numerically. Let $y = f(x)$ with $0 \le x \le 100$. For the numerical representation, use a table with $x = 0, 25, 50, 75, 100$.

33. To convert x pounds to y ounces, multiply x by 16.

34. To find the area y of a square, multiply the length x of a side by itself.

Exercises 35–40: Complete the following for the function f.

(a) Evaluate $f(x)$ at the indicated values of x.
(b) Find the domain of f.

35. $f(x) = 5$ for $x = -3, 1.5$

36. $f(x) = 4 - 5x$ for $x = -5, 6$

37. $f(x) = x^2 - 3$ for $x = -10, a + 2$

38. $f(x) = x^3 - 3x$ for $x = -10, a + 1$

39. $f(x) = \frac{1}{x - 4}$ for $x = -3, a + 1$

40. $f(x) = \sqrt{x + 3}$ for $x = 1, a - 3$

41. Determine if y is a function of x in $x = y^2 + 5$.

42. Write $5 \le x < 10$ in interval notation.

Exercises 43 and 44: The graph of a linear function f is shown.

(a) Identify the slope, y-intercept, and x-intercept.
(b) Write a formula for f.
(c) Find any zeros of f.

43.   **44.**

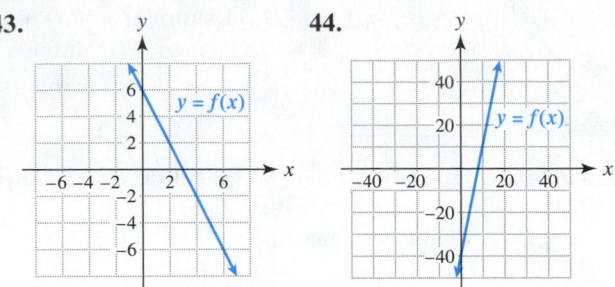

Exercises 45 and 46: Determine if the graph represents a function.

45.   **46.**

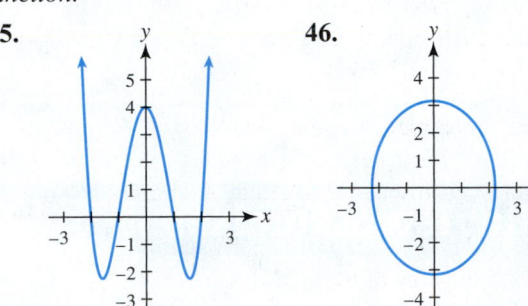

Exercises 47 and 48: Determine if S represents a function.

47. $S = \{(-3, 4), (-1, 2), (3, -5), (4, 2)\}$

48. $S = \{(-1, 3), (0, 2), (-1, 7), (3, -3)\}$

Exercises 49 and 50: State the slope of the graph of f.

49. $f(x) = 7$ **50.** $f(x) = \frac{1}{3}x - \frac{2}{3}$

Exercises 51–54: If possible, find the slope of the line passing through each pair of points.

51. $(-1, 7), (3, 4)$ **52.** $(1, -4), (2, 10)$

53. $(8, 4), (-2, 4)$ **54.** $\left(-\frac{1}{3}, \frac{2}{3}\right), \left(-\frac{1}{3}, -\frac{5}{6}\right)$

Exercises 55–58: Decide whether the function f is constant, linear, or nonlinear.

55. $f(x) = 8 - 3x$ **56.** $f(x) = 2x^2 - 3x - 8$

57. $f(x) = |x + 2|$ **58.** $f(x) = 6$

59. Sketch a graph for a 2-hour period showing the distance between two cars meeting on a straight highway, each traveling 60 miles per hour. Assume that the cars are initially 120 miles apart.

60. Determine where the graph of $f(x) = |x - 3|$ is increasing and where it is decreasing.

61. Determine if a line passes through every point in the table. If it does, give its slope.

x	−2	0	2	4
y	50	42	34	26

62. Find the average rate of change of $f(x) = x^2 - x + 1$ from $x_1 = 1$ to $x_2 = 3$.

Exercises 63 and 64: Find the difference quotient for $f(x)$.

63. $f(x) = 5x + 1$

64. $f(x) = 3x^2 - 2$

Applications

65. Speed of Light The average distance between the planet Mars and the sun is approximately 228 million kilometers. Estimate the time required for sunlight, traveling at 300,000 kilometers per second, to reach Mars. (*Source:* C. Ronan, *The Natural History of the Universe.*)

66. Geometry Suppose that 0.25 cubic inch of paint is applied to a circular piece of plastic with a diameter of 20 inches. Estimate the thickness of the paint. (*Hint:* Thickness equals volume divided by area.)

67. Enclosing a Pool A rectangular swimming pool that is 25 feet by 50 feet has a 6-foot-wide sidewalk around it.
(a) How much fencing would be needed to enclose the sidewalk?

(b) Find the area of the sidewalk.

68. Distance A driver's distance D in miles from a rest stop after t hours is given by $D(t) = 280 - 70t$.
(a) How far is the driver from the rest stop after 2 hours?

(b) Find the slope of the graph of D. Interpret this slope as a rate of change.

69. Survival Rates The survival rates for song sparrows are shown in the table. The values listed are the numbers of song sparrows that attain a given age from 100 eggs. For example, 6 sparrows reach an age of 2 years from 100 eggs laid in the wild. (*Source:* S. Kress, *Bird Life.*)

Age	0	1	2	3	4
Number	100	10	6	3	2

(a) Make a line graph of the data. Interpret the data.

(b) Does this line graph represent a function?

(c) Calculate and interpret the average rate of change for each 1-year period.

70. Cost of Tuition The graph shows the cost of taking x credits at a university.
(a) Why is it reasonable for the graph to pass through the origin?

(b) Find the slope of the graph.

(c) Interpret the slope as a rate of change.

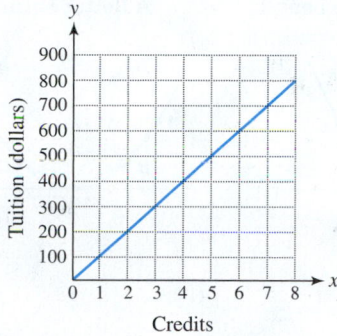

Credits

71. Average Rate of Change Let $f(x) = 0.5x^2 + 50$ represent the outside temperature in degrees Fahrenheit at x P.M., where $1 \le x \le 5$.
(a) Graph f. Is f linear or nonlinear?

(b) Calculate the average rate of change of f from 1 P.M. to 4 P.M.

(c) Interpret this average rate of change.

72. Distance At noon car A is traveling north at 30 miles per hour and is located 20 miles north of car B. Car B is traveling west at 50 miles per hour. Approximate the distance between the cars at 12:45 P.M. to the nearest mile.

Extended and Discovery Exercises

Because a parabolic curve becomes sharp gradually, as shown in the first figure on the next page, curves designed by engineers for highways and railroads frequently have parabolic, rather than circular, shapes. If railroad tracks changed abruptly from straight to circular, the momentum of the locomotive could cause a derailment. The second figure on the next page illustrates straight tracks connecting to a circular curve. (*Source:* F. Mannering and W. Kilareski, *Principles of Highway Engineering and Traffic Analysis.*)

In order to design a curve and estimate its cost, engineers determine the distance around the curve before it is built. In the third figure on the next page the distance along a parabolic curve from A to C is approximated by two line segments AB and BC. The distance formula can be used to calculate the length of each segment. The sum of these two lengths gives a crude estimate of the length of the curve.

A better estimate can be made using four line segments, as shown in the fourth figure. As the number of segments increases, so does the accuracy of the approximation.

A Parabolic Curve   **A Circular Curve**

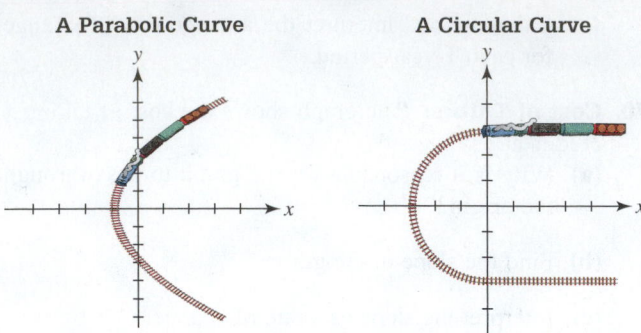

An Estimate of Curve Length   **A Better Estimate**

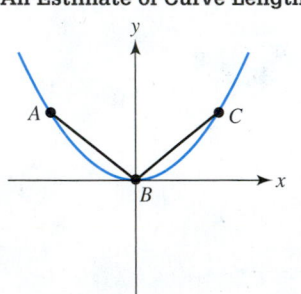   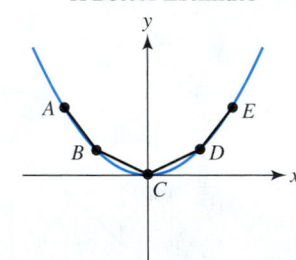

1. **Curve Length** Suppose that a curve designed for railroad tracks is represented by the equation $y = 0.2x^2$, where the units are in kilometers. The points $(-3, 1.8)$, $(-1.5, 0.45)$, $(0, 0)$, $(1.5, 0.45)$, and $(3, 1.8)$ lie on the graph of $y = 0.2x^2$. Approximate the length of the curve from $x = -3$ to $x = 3$ by using line segments connecting these points.

Exercises 2–5: **Curve Length** *Use three line segments connecting the four points to estimate the length of the curve on the graph of f from x = -1 to x = 2. Graph f and a line graph of the four points in the indicated viewing rectangle.*

2. $f(x) = x^2$; $(-1, 1)$, $(0, 0)$, $(1, 1)$, $(2, 4)$;
   $[-4.5, 4.5, 1]$ by $[-1, 5, 1]$

3. $f(x) = \sqrt[3]{x}$; $(-1, -1)$, $(0, 0)$, $(1, 1)$, $(2, \sqrt[3]{2})$;
   $[-3, 3, 1]$ by $[-2, 2, 1]$

4. $f(x) = 0.5x^3 + 2$; $(-1, 1.5)$, $(0, 2)$, $(1, 2.5)$, $(2, 6)$;
   $[-4.5, 4.5, 1]$ by $[0, 6, 1]$

5. $f(x) = 2 - 0.5x^2$; $(-1, 1.5)$, $(0, 2)$, $(1, 1.5)$, $(2, 0)$;
   $[-3, 3, 1]$ by $[-1, 3, 1]$

6. The distance along the curve of $y = x^2$ from $(0, 0)$ to $(3, 9)$ is about 9.747. Use this fact to estimate the distance along the curve of $y = 9 - x^2$ from $(0, 9)$ to $(3, 0)$.

7. Estimate the distance along the curve of $y = \sqrt{x}$ from $(1, 1)$ to $(4, 2)$. (The actual value is approximately 3.168.)

8. **Endangered Species** The Florida scrub-jay is an endangered species that prefers to live in open landscape

with short vegetation. NASA has attempted to create a habitat for these birds near Kennedy Space Center. The following table lists their population for selected years, where $x = 0$ corresponds to 1980, $x = 1$ to 1981, $x = 2$ to 1982, and so on.

x (1980 $\leftrightarrow$ 0)	0	5	9
y (population)	3697	2512	2176

x (1980 $\leftrightarrow$ 0)	11	15	19
y (population)	2100	1689	1127

Source: Mathematics Explorations II, NASA–AMATYC–NSF.

(a) Make a scatterplot of the data.

(b) Find a linear function f that models the data.

(c) Graph the data and f in the same viewing rectangle.

(d) Estimate the scrub-jay population in 1987 and in 2003.

9. **Rise in Sea Level** If the global climate were to warm significantly, the Arctic ice cap would melt. It is estimated that this ice cap contains the equivalent of 680,000 cubic miles of water. Over 200 million people currently live on soil that is less than 3 feet above sea level. In the United States, several large cities have low average elevations, such as Boston (14 feet), New Orleans (4 feet), and San Diego (13 feet). (*Sources:* Department of the Interior, Geological Survey.)

(a) Devise a plan to determine how much sea level would rise if the Arctic cap melted. (*Hint:* The radius of Earth is 3960 miles and 71% of its surface is covered by oceans.)

(b) Use your plan to estimate this rise in sea level.

(c) Discuss the implications of your calculation.

(d) Estimate how much sea level would rise if the 6,300,000 cubic miles of water in the Antarctic ice cap melted.

10. Prove that $\sqrt{2}$ is irrational by assuming that $\sqrt{2}$ is rational and arriving at a contradiction.

2 Linear Functions and Equations

For over two centuries people have been transferring carbon from below the surface of the earth into the atmosphere. In 2006, the burning of coal, oil, and natural gas released 7 billion tons of carbon into the atmosphere. If the rate of growth continues, the amount could double to 14 billion tons by 2056. This increase is modeled by a linear function C in the figure on the left below. The green horizontal line $y = 7$ represents the 2006 level of emissions, and the red horizontal line $y = 14$ represents a doubling of carbon emissions. Their points of intersection with the graph of C represent when these levels of emission could

No Restrictions on Carbon Emissions

Restrictions on Carbon Emissions

occur. The figure on the right illustrates what might happen if levels of carbon emission could be held at 7 billion tons per year for the next 50 years. In this case, emissions are expected to decline after 50 years. This graph, made up of line segments, is called a *piecewise-linear function*.

Whatever your point of view, mathematics plays an essential role in understanding the future of carbon emissions; without mathematical support, predictions lack credibility. To model carbon emissions, we need constant, linear, and piecewise-defined functions. All of these important concepts are discussed in this chapter.

Source: R. Socolow and S. Pacala, "A Plan to Keep Carbon in Check," *Scientific American,* September, 2006.

2.1 Equations of Lines

- Write the point-slope and slope-intercept forms
- Find the intercepts of a line
- Write equations for horizontal, vertical, parallel, and perpendicular lines
- Model data with lines and linear functions (optional)
- Use linear regression to model data (optional)

Introduction

The graph of a linear function is a line. One way to determine a linear function is to find the equation of this line. Once this equation is known, we can easily write the symbolic representation for the linear function.

For example, Apple Corporation sold approximately 55 million iPods in fiscal 2008, making the iPod the fastest-selling music player in history. However, sales then decreased to about 43 million in 2011. (*Source:* Apple Corporation.) This decline can be modeled by a line. The equation of this line determines a linear function that we can use to estimate iPod sales for other years. See Example 4. This section discusses how to use data points to find equations of lines and linear functions.

Forms for Equations of Lines

Point-Slope Form   Suppose that a nonvertical line with slope m passes through the point (x_1, y_1). If (x, y) is any point on this *nonvertical* line with $x \neq x_1$, then the change in y is $\Delta y = y - y_1$, the change in x is $\Delta x = x - x_1$, and the slope is $m = \frac{\Delta y}{\Delta x} = \frac{y - y_1}{x - x_1}$, as illustrated in Figure 2.1.

Two Points Determine a Line

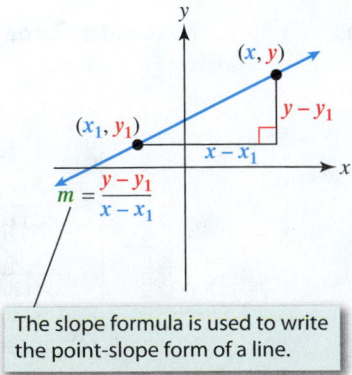

The slope formula is used to write the point-slope form of a line.

Figure 2.1

With this slope formula, the equation of the line can be found.

$$m = \frac{y - y_1}{x - x_1}$$ Slope formula

$$y - y_1 = m(x - x_1)$$ Cross multiply.

$$y = m(x - x_1) + y_1$$ Add y_1 to each side.

The equation $y - y_1 = m(x - x_1)$ is traditionally called the *point-slope form* of the equation of a line. Since we think of y as being a function of x, written $y = f(x)$, the equivalent form $y = m(x - x_1) + y_1$ will also be referred to as the point-slope form. The point-slope form is not unique, as any point on the line can be used for (x_1, y_1). However, these point-slope forms are *equivalent*, meaning their graphs are identical.

POINT-SLOPE FORM

The line with slope m passing through the point (x_1, y_1) has an equation

$$y = m(x - x_1) + y_1, \quad \text{or} \quad y - y_1 = m(x - x_1),$$

the **point-slope form** of the equation of a line.

In the next example we find the equation of a line given two points.

EXAMPLE 1   Determining a point-slope form

Find an equation of the line passing through the points $(-2, -3)$ and $(1, 3)$. Plot the points and graph the line by hand.

SOLUTION   Begin by finding the slope of the line.

$$m = \frac{3 - (-3)}{1 - (-2)} = \frac{6}{3} = 2$$

Substituting $(x_1, y_1) = (1, 3)$ and $m = 2$ into the point-slope form results in

First point-slope form $y = 2(x - 1) + 3.$ $y = m(x - x_1) + y_1$

If we use the point $(-2, -3)$, the point-slope form is

Second point-slope form $y = 2(x + 2) - 3.$ Note that $(x - (-2)) = (x + 2)$.

This line and the two points are shown in Figure 2.2.

Now Try Exercise 1

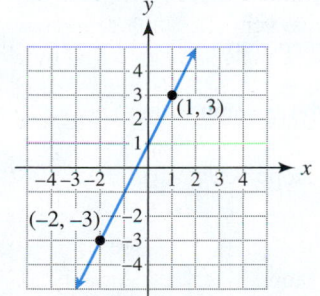

Figure 2.2

Slope-Intercept Form   The two point-slope forms found in Example 1 are equivalent.

First point-slope form	Second point-slope form	
$y = 2(x - 1) + 3$	$y = 2(x + 2) - 3$	Point-slope form
$y = 2x - 2 + 3$	$y = 2x + 4 - 3$	Distributive property
$y = 2x + 1$	$y = 2x + 1$	Simplify.

Both point-slope forms simplify to the same equation.

The form $y = mx + b$ is called the *slope-intercept form*. Unlike the point-slope form, the slope-intercept form is *unique*. The real number m represents the slope and the real number b represents the y-intercept, as illustrated in Figure 2.3.

Slope-Intercept Form
$$y = mx + b$$

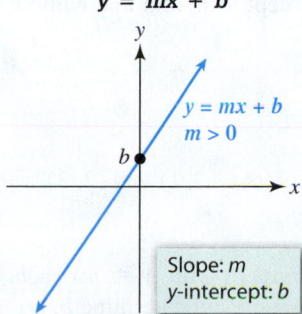

Figure 2.3

SLOPE-INTERCEPT FORM

The line with slope m and y-intercept b is given by

$$y = mx + b,$$

the **slope-intercept form** of the equation of a line.

EXAMPLE 2 Finding equations of lines

Find the point-slope form for the line that satisfies the conditions. Then convert this equation into slope-intercept form and write the formula for a function f whose graph is the line.
(a) Slope $-\frac{1}{2}$, passing through the point $(-3, -7)$
(b) x-intercept -4, y-intercept 2

SOLUTION
(a) Let $m = -\frac{1}{2}$ and $(x_1, y_1) = (-3, -7)$ in the point-slope form.

$$y = m(x - x_1) + y_1 \qquad \text{Point-slope form}$$

$$y = -\frac{1}{2}(x + 3) - 7 \qquad \text{Substitute.}$$

The slope-intercept form can be found by simplifying.

$$y = -\frac{1}{2}(x + 3) - 7 \qquad \text{Point-slope form}$$

$$y = -\frac{1}{2}x - \frac{3}{2} - 7 \qquad \text{Distributive property}$$

$$y = -\frac{1}{2}x - \frac{17}{2} \qquad \text{Slope-intercept form}$$

Thus $f(x) = -\frac{1}{2}x - \frac{17}{2}$ is the formula for the function whose graph is this line.

(b) The line passes through the points $(-4, 0)$ and $(0, 2)$. Its slope is

$$m = \frac{2 - 0}{0 - (-4)} = \frac{1}{2}.$$

Thus a point-slope form for the line is $y = \frac{1}{2}(x + 4) + 0$, where the point $(-4, 0)$ is used for (x_1, y_1). The slope-intercept form is $y = \frac{1}{2}x + 2$ and $f(x) = \frac{1}{2}x + 2$ is the formula for the function whose graph is this line.

> **Now Try Exercises 5 and 9**

The next example demonstrates how to find the slope-intercept form of a line without first finding the point-slope form.

EXAMPLE 3 Finding slope-intercept form

Find the slope-intercept form of the line passing through the points $(-2, 1)$ and $(2, 3)$.

SOLUTION
Getting Started We need to determine m and b in the slope-intercept form, $y = mx + b$. First find the slope m. Then substitute *either* point into the equation and determine b. ▶

$$m = \frac{3 - 1}{2 - (-2)} = \frac{2}{4} = \frac{1}{2}$$

Thus $y = \frac{1}{2}x + b$. To find b, we substitute $(2, 3)$ in this equation.

$$3 = \frac{1}{2}(2) + b \qquad \text{Let } x = 2 \text{ and } y = 3.$$
$$3 = 1 + b \qquad \text{Multiply.}$$
$$2 = b \qquad \text{Determine } b.$$

Thus $y = \frac{1}{2}x + 2$.

> **Now Try Exercise 21**

An Application   In the next example we model the data about iPods discussed in the introduction to this section.

EXAMPLE 4   **Estimating iPod sales**

Apple Corporation sold approximately 55 million iPods in fiscal 2008 and 43 million iPods in fiscal 2011.

(a) Find the point-slope form of the line passing through $(2008, 55)$ and $(2011, 43)$. Interpret the slope of the line as a rate of change.

(b) Sketch a graph of the data and the line connecting these points.

(c) Estimate sales in 2010 and compare the estimate to the true value of 50 million.

(d) Estimate sales in 2023. Discuss the accuracy of your answers.

SOLUTION

Getting Started   First find the slope m of the line connecting the data points, and then substitute this value for m and either of the two data points in the point-slope form. We can use this equation to estimate sales by substituting the required year for x in the equation. ▶

(a)   The slope of the line passing through $(2008, 55)$ and $(2011, 43)$ is

$$m = \frac{43 - 55}{2011 - 2008} = -4.$$

Thus sales of iPods decreased, on average, by **4** million iPods per year from **2008** to **2011**. If we substitute -4 for m and $(2008, 55)$ for (x_1, y_1), the point-slope form is

$$y = -4(x - 2008) + 55.$$

(b)   The requested line passing through the data points is shown in Figure 2.4.

(c)   If $x = 2010$, then $y = -4(2010 - 2008) + 55 = 47$ million. This estimated value is 3 million lower than the true value of 50 million.

(d)   We can use the equation to estimate **2023** sales as follows.

$$y = -4(2023 - 2008) + 55 = -5 \text{ million} \qquad \text{Let } x = 2023.$$

The 2023 value is clearly incorrect because sales cannot be negative.

> **Now Try Exercise 81**

iPod Sales

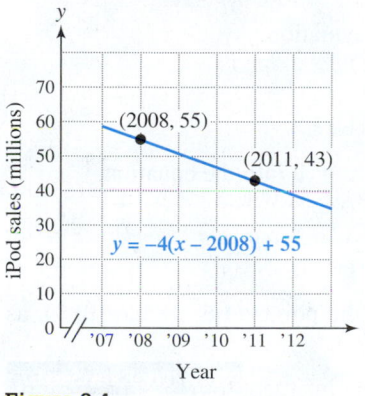

Figure 2.4

Finding Intercepts

The point-slope form and the slope-intercept form are not the only forms for the equation of a line. An equation of a line is in **standard form** when it is written as

$$ax + by = c,$$

where a, b, and c are constants and a and b are not *both* zero. By using standard form, we can write the equation of any line, including vertical lines (which are discussed later in this section). Examples of equations of lines in standard form include the following.

Equations in Standard Form: $ax + by = c$

$$2x - 3y = -6, \qquad y = \frac{1}{4}, \qquad x = -3, \qquad \text{and} \qquad -3x + y = \frac{1}{2}$$

$$\boxed{a = 0} \qquad \boxed{b = 0}$$

The following box gives an example of how to find the intercepts for a line given in standard form.

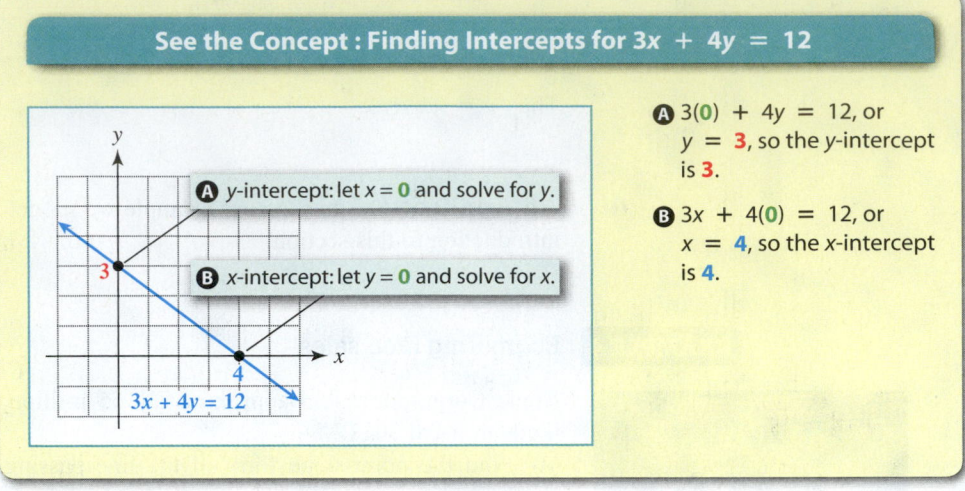

See the Concept : Finding Intercepts for $3x + 4y = 12$

Ⓐ y-intercept: let $x = 0$ and solve for y.

Ⓑ x-intercept: let $y = 0$ and solve for x.

$3x + 4y = 12$

Ⓐ $3(0) + 4y = 12$, or
$y = 3$, so the y-intercept is **3**.

Ⓑ $3x + 4(0) = 12$, or
$x = 4$, so the x-intercept is **4**.

NOTE To solve $ax = b$, divide each side by a to obtain $x = \frac{b}{a}$. Thus $3x = 12$ implies that $x = \frac{12}{3} = 4$. Linear equations are solved in general in the next section.

FINDING INTERCEPTS

To find any x-intercepts, let $y = 0$ in the equation and solve for x.
To find any y-intercepts, let $x = 0$ in the equation and solve for y.

EXAMPLE 5 Finding intercepts

Locate the x- and y-intercepts for the line whose equation is $4x + 3y = 6$. Use the intercepts to graph the equation.

SOLUTION To locate the x-intercept, let $y = 0$ in the equation.

$$4x + 3(0) = 6 \qquad \text{Let } y = 0.$$
$$x = 1.5 \qquad \text{Divide each side by 4.}$$

The x-intercept is **1.5**. To find the y-intercept, substitute $x = 0$ into the equation.

$$4(0) + 3y = 6 \qquad \text{Let } x = 0.$$
$$y = 2 \qquad \text{Divide each side by 3.}$$

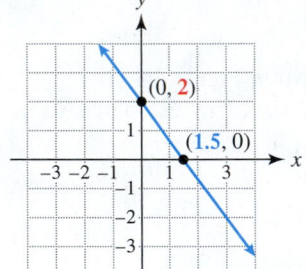

Figure 2.5

The y-intercept is **2**. Therefore the line passes through the points (**1.5**, 0) and (0, **2**), as shown in Figure 2.5.

Now Try Exercise 57

Horizontal, Vertical, Parallel, and Perpendicular Lines

Horizontal and Vertical Lines The graph of a constant function f, defined by the formula $f(x) = b$, is a *horizontal line* having slope 0 and y-intercept b.

A *vertical line* cannot be represented by a function because distinct points on a vertical line have the same x-coordinate. In fact, this is the distinguishing feature of points on a vertical line—they all have the same x-coordinate. See Figures 2.6 and 2.7.

CLASS DISCUSSION

Why do you think that a vertical line sometimes is said to have "infinite slope?" What are some problems with taking this phrase too literally?

Equations of Vertical Lines

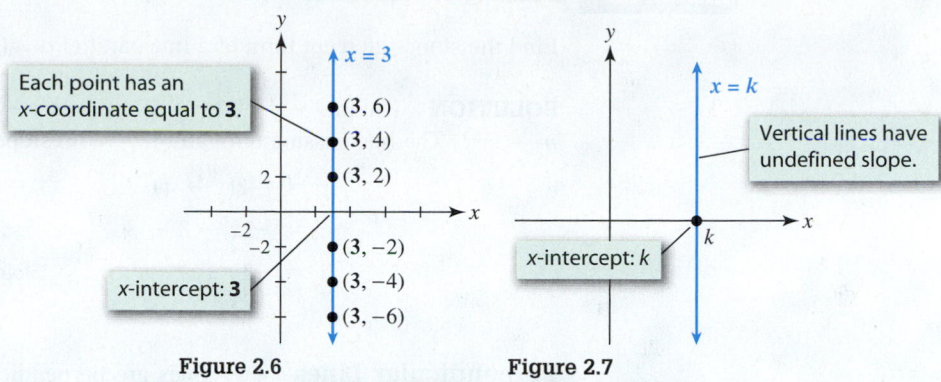

Figure 2.6   Figure 2.7

EQUATIONS OF HORIZONTAL AND VERTICAL LINES

An equation of the horizontal line with y-intercept b is $y = b$. An equation of the vertical line with x-intercept k is $x = k$.

EXAMPLE 6   Finding equations of horizontal and vertical lines

Find equations of vertical and horizontal lines passing through the point $(8, 5)$. If possible, for each line write a formula for a linear function whose graph is the line.

SOLUTION   The x-coordinate of the point $(8, 5)$ is 8. The vertical line $x = 8$ passes through every point in the xy-plane with an x-coordinate of 8, including the point $(8, 5)$. Similarly, the horizontal line $y = 5$ passes through every point with a y-coordinate of 5, including $(8, 5)$. See Figure 2.8, where each line is shown.

Vertical and Horizontal Lines

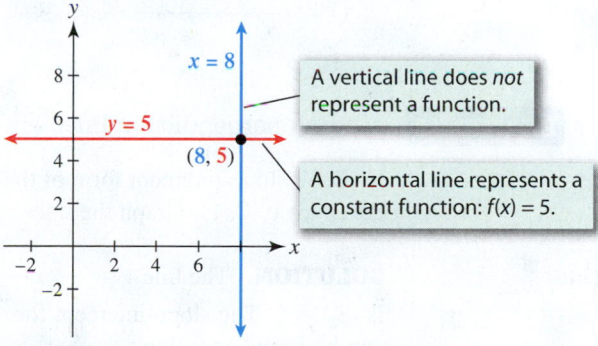

Figure 2.8

The horizontal line $y = 5$ represents the constant function $f(x) = 5$. The vertical line $x = 8$ does *not* represent a function because it does *not* pass the vertical line test.

Now Try Exercises 49 and 51

Parallel Lines   Two nonvertical parallel lines have equal slopes.

PARALLEL LINES

Two lines with slopes m_1 and m_2, neither of which is vertical, are parallel if and only if their slopes are equal; that is, $m_1 = m_2$.

NOTE   The phrase "if and only if" is used when two statements are mathematically equivalent. If two nonvertical lines are parallel, then $m_1 = m_2$. Conversely, if two nonvertical lines have equal slopes, then they are parallel. Either condition implies the other.

EXAMPLE 7 **Finding parallel lines**

Find the slope-intercept form of a line parallel to $y = -2x + 5$, passing through $(4, 3)$.

SOLUTION The line $y = -2x + 5$ has slope -2, so any parallel line also has slope $m = -2$. The line passing through $(4, 3)$ with slope -2 is determined as follows.

$$y = -2(x - 4) + 3 \qquad \text{Point-slope form}$$
$$y = -2x + 8 + 3 \qquad \text{Distributive property}$$
$$y = -2x + 11 \qquad \text{Slope-intercept form}$$

> **Now Try Exercise 35**

Perpendicular Lines Two lines are perpendicular if the product of their slopes is equal to -1.

> ### PERPENDICULAR LINES
>
> Two lines with nonzero slopes m_1 and m_2 are perpendicular if and only if their slopes have product -1; that is, $m_1 m_2 = -1$.

For perpendicular lines, m_1 and m_2 are *negative reciprocals*. That is, $m_1 = -\frac{1}{m_2}$ and $m_2 = -\frac{1}{m_1}$. Table 2.1 shows some slopes of perpendicular lines.

Slopes of Perpendicular Lines

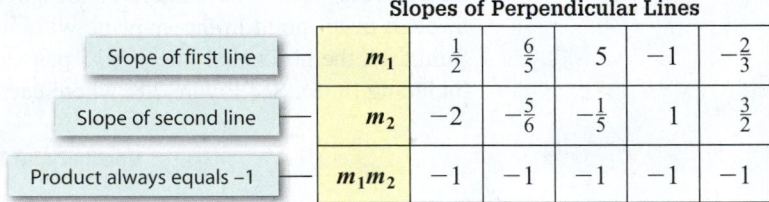

Slope of first line — m_1	$\frac{1}{2}$	$\frac{6}{5}$	5	-1	$-\frac{2}{3}$
Slope of second line — m_2	-2	$-\frac{5}{6}$	$-\frac{1}{5}$	1	$\frac{3}{2}$
Product always equals –1 — $m_1 m_2$	-1	-1	-1	-1	-1

Table 2.1

EXAMPLE 8 **Finding perpendicular lines**

Find the slope-intercept form of the line perpendicular to $y = -\frac{2}{3}x + 2$, passing through the point $(-2, 1)$. Graph the lines.

Perpendicular Lines

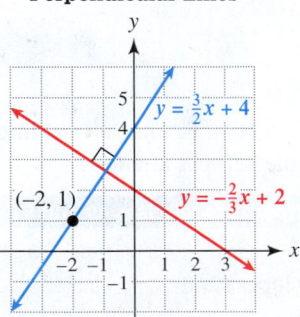

Figure 2.9

SOLUTION The line $y = -\frac{2}{3}x + 2$ has slope $-\frac{2}{3}$. The negative reciprocal of $m_1 = -\frac{2}{3}$ is $m_2 = \frac{3}{2}$. The slope-intercept form of a line having slope $\frac{3}{2}$ and passing through $(-2, 1)$ can be found as follows.

$$y = m(x - x_1) + y_1 \qquad \text{Point-slope form}$$
$$y = \frac{3}{2}(x + 2) + 1 \qquad \text{Let } m = \tfrac{3}{2}, x_1 = -2, \text{ and } y_1 = 1.$$
$$y = \frac{3}{2}x + 3 + 1 \qquad \text{Distributive property}$$
$$y = \frac{3}{2}x + 4 \qquad \text{Slope-intercept form}$$

Figure 2.9 shows graphs of these perpendicular lines.

> **Now Try Exercise 41**

Calculator Help

To set a square viewing rectangle, see Appendix A (page AP-5).

NOTE If a graphing calculator is used to graph these lines, a square viewing rectangle must be used for the lines to appear perpendicular.

EXAMPLE 9   Determining a rectangle

In Figure 2.10 a rectangle is outlined by four lines denoted y_1, y_2, y_3, and y_4. Find the equation of each line.

CLASS DISCUSSION

Check the results from Example 9 by graphing the four equations in the same viewing rectangle. How does your graph compare with Figure 2.10? Why is it important to use a square viewing rectangle?

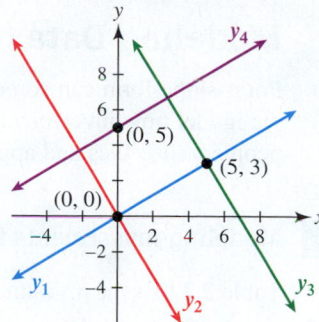

Figure 2.10

SOLUTION

Line y_1: This line passes through the points $(0, 0)$ and $(5, 3)$, so $m = \frac{3}{5}$ and the y-intercept is 0. Its equation is $y_1 = \frac{3}{5}x$.

Line y_2: This line passes through the point $(0, 0)$ and is perpendicular to y_1, so its slope is given by $m = -\frac{5}{3}$ and the y-intercept is 0. Its equation is $y_2 = -\frac{5}{3}x$.

Line y_3: This line passes through the point $(5, 3)$ and is parallel to y_2, so its slope is given by $m = -\frac{5}{3}$. In a point-slope form, its equation is $y_3 = -\frac{5}{3}(x - 5) + 3$, which is equivalent to $y_3 = -\frac{5}{3}x + \frac{34}{3}$.

Line y_4: This line passes through the point $(0, \mathbf{5})$, so its y-intercept is **5**. It is parallel to y_1, so its slope is given by $m = \frac{3}{5}$. Its equation is $y_4 = \frac{3}{5}x + \mathbf{5}$.

Now Try Exercise 95

Interpolation and Extrapolation

In **2005** about **\$350** million was spent on U.S. digital music single downloads. This amount reached **\$1350** million in **2010**. (*Source:* RIAA.) These data can be modeled with a line, and the slope of this line is

$$m = \frac{\mathbf{1350} - \mathbf{350}}{\mathbf{2010} - \mathbf{2005}} = \mathbf{200}.$$

This value means that the increase in the amount spent on U.S. music downloads was, on average, \$200 million per year from 2005 to 2010.

A point-slope form of the line passing through (**2005**, **350**) with slope **200** is

$$y = \mathbf{200}(x - \mathbf{2005}) + \mathbf{350}.$$

We can easily write a formula for a linear function

$$D(x) = 200(x - 2005) + 350,$$

whose graph is this line. See Figure 2.11.

This formula can be used to estimate the value of downloads in 2008 as follows.

$$D(\mathbf{2008}) = 200(\mathbf{2008} - 2005) + 350 = \mathbf{950}$$

Thus $D(x)$ estimates that the value of digital music single downloads was \$950 million in 2008. (The actual value was \$1000 million.) Because 2008 is *between* 2005 and 2010, we say that this estimation involves **interpolation**. Interpolation occurs when we estimate between given data points. However, if we use $D(x)$ to estimate this value in 2003 we obtain the following.

$$D(\mathbf{2003}) = 200(\mathbf{2003} - 2005) + 350 = \mathbf{-50}$$

U.S. Digital Single Download Sales

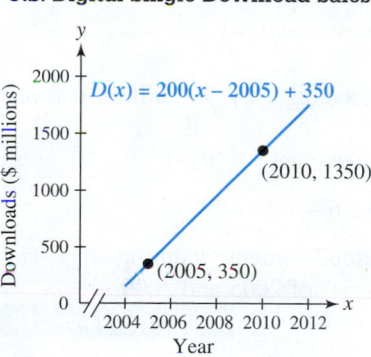

Figure 2.11

This estimate is incorrect because sales cannot be negative. Because 2003 is *not between* 2005 and 2010, we say that this estimation involves **extrapolation.** Extrapolation occurs when we estimate "outside" of the given data. Interpolation is usually more accurate than extrapolation.

Modeling Data (Optional)

Point-slope form can sometimes be useful when modeling real data. In the next example we model how investments in cloud computing have increased. Cloud computing allows people to use files and applications on the Internet.

Calculator Help

To make a scatterplot, see Appendix A (page AP-3). To plot data and graph an equation, see Appendix A (page AP-6).

[2004, 2010, 1] by [0, 500, 100]

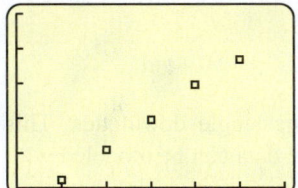

Figure 2.12

Modeling Cloud Investments
[2004, 2010, 1] by [0, 500, 100]

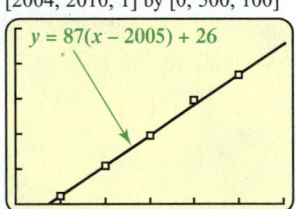

Figure 2.13

| EXAMPLE 10 | Modeling investments for cloud computing |

Table 2.2 lists the investments in billions of dollars for cloud computing for selected years.

Investments in Cloud Computing ($ billions)

Year	2005	2006	2007	2008	2009
Investments	26	113	195	299	374

Source: Thomson Reuters.

Table 2.2

(a) Make a scatterplot of the data.
(b) Find a formula in point-slope form for a linear function f that models the data.
(c) Graph the data and $y = f(x)$ in the same xy-plane.
(d) Interpret the slope of the graph of $y = f(x)$.
(e) Estimate the investment in cloud computing in 2014. Does your answer involve interpolation or extrapolation?

SOLUTION

(a) See Figure 2.12
(b) Because the data are nearly linear, we could require that the line pass through the first data point (**2005**, **26**) and the last data point (**2009**, **374**). The slope of this line is

$$m = \frac{374 - 26}{2009 - 2005} = 87. \quad \text{Slope} = \frac{y_2 - y_1}{x_2 - x_1}$$

Thus we can write a formula for f as

$$f(x) = 87(x - 2005) + 26.$$

NOTE When modeling real data, answers for $f(x)$ may vary, depending on the points that are used to determine the line.

(c) See Figure 2.13
(d) Slope 87 indicates that investments increased, on average, by $87 billion per year between 2005 and 2009.
(e) To estimate the investment amount in 2014, we can evaluate $f(2014)$.

$$f(\mathbf{2014}) = 87(\mathbf{2014} - 2005) + 26 = \mathbf{809}$$

This model predicts an **$809** billion investment in cloud computing during **2014**. This result involves extrapolation because 2014 is "outside" of 2005 and 2009.

Now Try Exercise 87

Linear Regression (Optional)

We have used linear functions to model data involving the variables x and y. Problems where one variable is used to predict the behavior of a second variable are called **regression** problems. If a linear function or line is used to approximate the data, then the technique is referred to as **linear regression.**

We have already solved problems by selecting a line that *visually* fits the data in a scatterplot. See Example 10. However, this technique has some disadvantages. First, it does not produce a unique line. Different people may arrive at different lines to fit the same data. Second, the line is not determined automatically by a calculator or computer. A statistical method used to determine a unique linear function or line is based on the method of **least squares**.

Correlation Coefficient   Most graphing calculators have the capability to calculate the least-squares regression line automatically. When determining the least-squares line, calculators often compute a real number r, called the **correlation coefficient**, where $-1 \le r \le 1$. When r is positive and near 1, low x-values correspond to low y-values and high x-values correspond to high y-values. For example, there is a positive correlation between years of education x and income y. More years of education correlate with higher income. When r is near -1, the reverse is true. Low x-values correspond to high y-values and high x-values correspond to low y-values. If $r \approx 0$, then there is little or no correlation between the data points. In this case, a linear function does not provide a suitable model. A summary of these concepts is shown in the following.

Correlation Coefficient r ($-1 \le r \le 1$)

Value of r	Comments	Sample Scatterplot
$r = 1$	There is an exact linear fit. The line passes through all data points and has a positive slope.	
$r = -1$	There is an exact linear fit. The line passes through all data points and has a negative slope.	
$0 < r < 1$	There is a positive correlation. As the x-values increase, so do the y-values. The fit is not exact.	
$-1 < r < 0$	There is a negative correlation. As the x-values increase, the y-values decrease. The fit is not exact.	
$r = 0$	There is no correlation. The data has no tendency toward being linear. A regression line predicts poorly.	

MAKING CONNECTIONS

Correlation and Causation
When geese begin to fly north, summer is coming and the weather becomes warmer. Geese flying north correlate with warmer weather. However, geese flying north clearly do not *cause* warmer weather. It is important to remember that correlation does not always indicate causation.

Calculator Help

To find a line of least-squares fit, see
Appendix A (page AP-6).

In the next example we use a graphing calculator to find the line of least-squares fit
that models three data points.

EXAMPLE 11   Determining a line of least-squares fit

Find the line of least-squares fit for the data points $(1, 1)$, $(2, 3)$, and $(3, 4)$. What is the
correlation coefficient? Plot the data and graph the line.

SOLUTION   Begin by entering the three data points into the STAT EDIT menu. Refer
to Figures 2.14–2.17. Select the LinReg (ax + b) option from the STAT CALC menu.
From the home screen we can see that the line (linear function) of least squares is given
by the formula $y = \frac{3}{2}x - \frac{1}{3}$. The correlation coefficient is $r \approx 0.98$. Since $r \neq 1$, the
line does not provide an *exact* model of the data.

Enter Data into Two Lists	Select Linear Regression	Equation for Regression Line	Data and Regression Line [0, 5, 1] by [0, 5, 1]
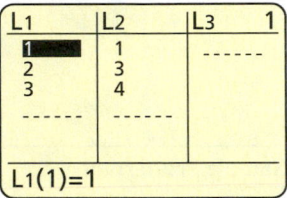	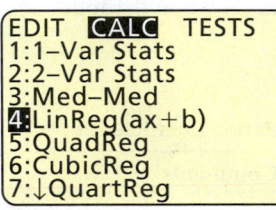	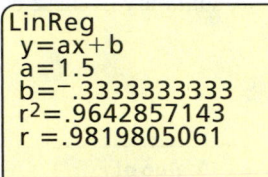	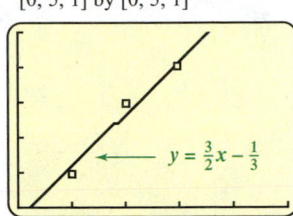
Figure 2.14	**Figure 2.15**	**Figure 2.16**	**Figure 2.17**

Now Try Exercise 99

An Application   In the next example, we find the regression line that models the cloud
computing data found in Example 10.

EXAMPLE 12   Modeling data with a regression line

Refer to the data from Table 2.2 in Example 10.
(a) Find the least-squares regression line that models this data.
(b) Compare the regression line with the one that was found in Example 10 by writing
both lines in slope-intercept form.

SOLUTION
(a) Figures 2.18–2.21 show how to find this regression line. Its equation is

$$y = 88.2x - 176{,}816.$$

Enter Data into Two Lists	Select Linear Regression	Equation for Regression Line	Data and Regression Line [2004, 2010, 1] by [0, 500, 100]
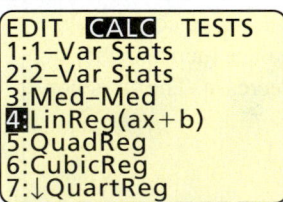	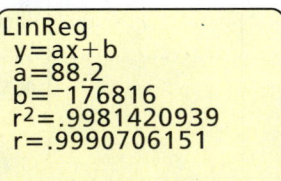	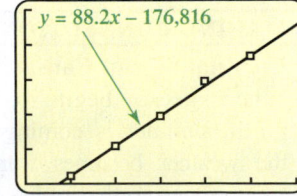	
Figure 2.18	**Figure 2.19**	**Figure 2.20**	**Figure 2.21**

(b) The equation of the line found in Example 10 can be written in slope-intercept form
as follows.

$$y = 87(x - 2005) + 26 \qquad \text{Example 10 equation}$$
$$y = 87x - 174{,}435 + 26 \qquad \text{Distributive property}$$
$$y = 87x - 174{,}409$$

Calculator Help

To copy the regression equation directly
into Y_1, see Appendix A (page AP-11).

Notice that the slope-intercept forms for these two lines are not exactly alike. However, both lines model the data reasonably well, as shown in Figures 2.13 and 2.21.

Now Try Exercise 107

2.1 Putting It All Together

The following table summarizes three forms for equations of a line and how to find the intercepts.

CONCEPT	COMMENTS	EXAMPLES
Point-slope form $$y = m(x - x_1) + y_1$$ or $$y - y_1 = m(x - x_1)$$	Used to find the equation of a line, given two points or one point and the slope	Given two points $(5, 1)$ and $(4, 3)$, first compute $m = \frac{3 - 1}{4 - 5} = -2$. An equation of this line is $$y = -2(x - 5) + 1.$$
Slope-intercept form $$y = mx + b$$	A unique equation for a line, determined by the slope m and the y-intercept b	An equation of the line with slope 5 and y-intercept -4 is $y = 5x - 4$.
Standard form $$ax + by = c$$	Any line can be written in this form.	$3x + 5y = 15$ $x = -4$ Vertical line $y = 7$ Horizontal line
Finding intercepts	1. To find x-intercepts, let $y = 0$ and solve for x. 2. To find y-intercepts, let $x = 0$ and solve for y.	1. In $3x + 5y = 15$ let $y = 0$ to obtain $3x = 15$, or $x = 5$. The x-intercept is 5. 2. In $3x + 5y = 15$ let $x = 0$ to obtain $5y = 15$, or $y = 3$. The y-intercept is 3.

The following table summarizes special types of lines.

CONCEPT	EQUATION(S)	EXAMPLES
Horizontal line	$y = b$, where b is a constant	A horizontal line with y-intercept 7 has the equation $y = 7$.
Vertical line	$x = k$, where k is a constant	A vertical line with x-intercept -8 has the equation $x = -8$.
Parallel lines	$y = m_1 x + b_1$ and $y = m_2 x + b_2$, where $m_1 = m_2$	The lines given by $y = -3x - 1$ and $y = -3x + 5$ are parallel because they both have slope -3.
Perpendicular lines	$y = m_1 x + b_1$ and $y = m_2 x + b_2$, where $m_1 m_2 = -1$	The lines $y = 2x - 5$ and $y = -\frac{1}{2}x + 2$ are perpendicular because $m_1 m_2 = 2\left(-\frac{1}{2}\right) = -1$.

continued on next page

The following table summarizes linear regression.

CONCEPT	DESCRIPTION
Correlation coefficient r	The values of r satisfy $-1 \le r \le 1$, where a line fits the data better if r is near -1 or 1. A value near 0 indicates a poor fit.
Least-squares regression line	The line of least-squares fit for the points $(1, 3)$, $(2, 5)$, and $(3, 6)$ is $y = \frac{3}{2}x + \frac{5}{3}$ and $r \approx 0.98$. Try verifying this with a calculator.

2.1 Exercises

Equations of Lines

Exercises 1–4: Find the point-slope form of the line passing through the given points. Use the first point as (x_1, y_1). Plot the points and graph the line by hand.

1. $(1, 2), (3, -2)$　　　　**2.** $(-2, 3), (1, 0)$

3. $(-3, -1), (1, 2)$　　　　**4.** $(-1, 2), (-2, -3)$

Exercises 5–10: Find a point-slope form of the line satisfying the conditions. Use the first point given for (x_1, y_1). Then convert the equation to slope-intercept form, and write the formula for a function f whose graph is the line.

5. Slope -2.4, passing through $(4, 5)$

6. Slope 1.7, passing through $(-8, 10)$

7. Passing through $(1, -2)$ and $(-9, 3)$

8. Passing through $(-6, 10)$ and $(5, -12)$

9. x-intercept 4, y-intercept -3

10. x-intercept -2, y-intercept 5

Exercises 11–14: Find the slope-intercept form for the line in the figure.

11.
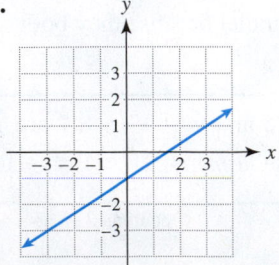

12.
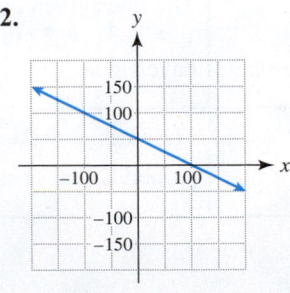

13.
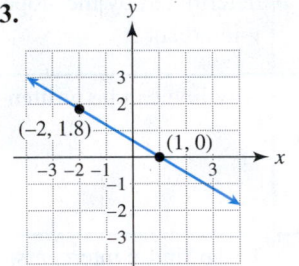

14.
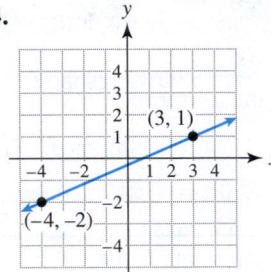

Exercises 15–20: **Concepts**　*Match the given equation to its graph (a–f) shown.*

15. $y = m(x - x_1) + y_1, m > 0$

16. $y = m(x - x_1) + y_1, m < 0$

17. $y = mx, m > 0$

18. $y = mx + b, m < 0$ and $b > 0$

19. $x = k, k > 0$

20. $y = b, b < 0$

(a)

(b)
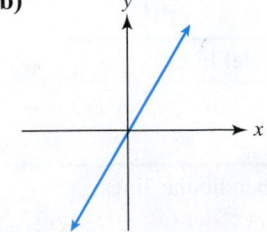

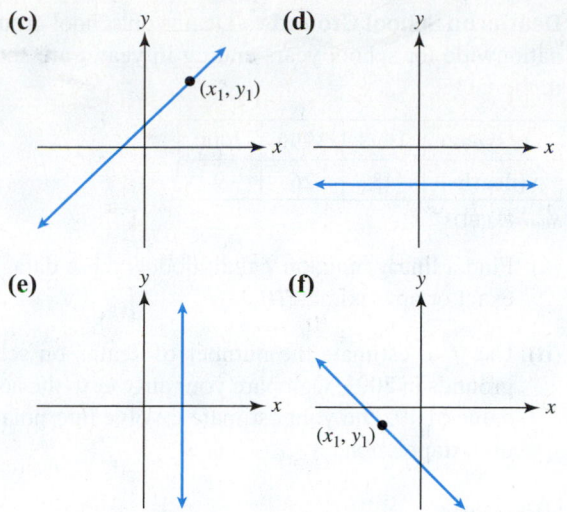

(c)

(d)

(e)

(f)

Exercises 21–48: Find the slope-intercept form for the line satisfying the conditions.

21. Passing through $(-1, -4)$ and $(1, 2)$

22. Passing through $(-1, 6)$ and $(2, -3)$

23. Passing through $(4, 5)$ and $(1, -3)$

24. Passing through $(8, -2)$ and $(-2, 3)$

25. y-intercept 5, slope -7.8

26. y-intercept -155, slope 5.6

27. y-intercept 45, x-intercept 90

28. x-intercept -6, y-intercept -8

29. Slope -3, passing through $(0, 5)$

30. Slope $\frac{1}{3}$, passing through $\left(\frac{1}{2}, -2\right)$

31. Passing through $(0, -6)$ and $(4, 0)$

32. Passing through $\left(\frac{3}{4}, -\frac{1}{4}\right)$ and $\left(\frac{5}{4}, \frac{7}{4}\right)$

33. Passing through $\left(\frac{1}{2}, \frac{3}{4}\right)$ and $\left(\frac{1}{5}, \frac{2}{3}\right)$

34. Passing through $\left(-\frac{7}{3}, \frac{5}{3}\right)$ and $\left(\frac{5}{6}, -\frac{7}{6}\right)$

35. Parallel to $y = 4x + 16$, passing through $(-4, -7)$

36. Parallel to the line $y = -\frac{3}{4}(x - 100) - 99$, passing through $(1, 3)$

37. Perpendicular to the line $y = -\frac{2}{3}(x - 1980) + 5$, passing through $(1980, 10)$

38. Perpendicular to $y = 6x - 10$, passing through $(15, -7)$

39. Parallel to $y = \frac{2}{3}x + 3$, passing through $(0, -2.1)$

40. Parallel to $y = -4x - \frac{1}{4}$, passing through $(2, -5)$

41. Perpendicular to $y = -2x$, passing through $(-2, 5)$

42. Perpendicular to $y = -\frac{6}{7}x + \frac{3}{7}$, passing through $(3, 8)$

43. Perpendicular to $x + y = 4$, passing through $(15, -5)$

44. Parallel to $2x - 3y = -6$, passing through $(4, -9)$

45. Passing through $(5, 7)$ and parallel to the line passing through $(1, 3)$ and $(-3, 1)$

46. Passing through $(1990, 4)$ and parallel to the line passing through $(1980, 3)$ and $(2000, 8)$

47. Passing through $(-2, 4)$ and perpendicular to the line passing through $\left(-5, \frac{1}{2}\right)$ and $\left(-3, \frac{2}{3}\right)$

48. Passing through $\left(\frac{3}{4}, \frac{1}{4}\right)$ and perpendicular to the line passing through $(-3, -5)$ and $(-4, 0)$

Horizontal and Vertical Lines

Exercises 49–56: Find an equation of the line satisfying the conditions. If possible, for each line write a formula for a linear function whose graph is the line.

49. Vertical, passing through $(-5, 6)$

50. Vertical, passing through $(1.95, 10.7)$

51. Horizontal, passing through $(-5, 6)$

52. Horizontal, passing through $(1.95, 10.7)$

53. Perpendicular to $y = 15$, passing through $(4, -9)$

54. Perpendicular to $x = 15$, passing through $(1.6, -9.5)$

55. Parallel to $x = 4.5$, passing through $(19, 5.5)$

56. Parallel to $y = -2.5$, passing through $(1985, 67)$

Finding Intercepts

Exercises 57–68: Determine the x- and y-intercepts on the graph of the equation. Graph the equation.

57. $4x - 5y = 20$ **58.** $-3x - 5y = 15$

59. $x - y = 7$ **60.** $15x - y = 30$

61. $6x - 7y = -42$ **62.** $5x + 2y = -20$

63. $y - 3x = 7$ **64.** $4x - 3y = 6$

65. $0.2x + 0.4y = 0.8$ **66.** $\frac{2}{3}y - x = 1$

67. $y = 8x - 5$ **68.** $y = -1.5x + 15$

*Exercises 69–72: The **intercept form of a line** is $\frac{x}{a} + \frac{y}{b} = 1$. Determine the x- and y-intercepts on the graph of the equation. Draw a conclusion about what the constants a and b represent in this form.*

69. $\frac{x}{5} + \frac{y}{7} = 1$ **70.** $\frac{x}{2} + \frac{y}{3} = 1$

71. $\frac{2x}{3} + \frac{4y}{5} = 1$ **72.** $\frac{5x}{6} - \frac{y}{2} = 1$

Exercises 73 and 74: (Refer to Exercises 69–72.) Write the intercept form for the line with the given intercepts.

73. x-intercept 5, y-intercept 9

74. x-intercept $\frac{2}{3}$, y-intercept $-\frac{5}{4}$

Interpolation and Extrapolation

Exercises 75–78: The table lists data that are exactly linear.

(a) Find the slope-intercept form of the line that passes through these data points.

(b) Predict y when x = −2.7 and 6.3. Decide if these calculations involve interpolation or extrapolation.

75.

x	−3	−2	−1	0	1
y	−7.7	−6.2	−4.7	−3.2	−1.7

76.

x	−2	−1	0	1	2
y	10.2	8.5	6.8	5.1	3.4

77.

x	5	23	32	55	61
y	94.7	56.9	38	−10.3	−22.9

78.

x	−11	−8	−7	−3	2
y	−16.1	−10.4	−8.5	−0.9	8.6

79. iPhone Revenues The percentage of smartphone revenues that were due to iPhones is shown in the table.

Year	2008	2009	2010	2011
Percentage	3	10	16	24

Source: Business Insider.

(a) Find a linear function *f* that models these data. Does *f* model these data exactly or approximately?

(b) Use *f* to estimate this percentage in 2007.

(c) Did your answer use interpolation or extrapolation? Comment on your result.

80. Deaths on School Grounds Deaths on school grounds nationwide for school years ending in year *x* are shown in the table.

x (year)	1998	1999	2000
y (deaths)	43	26	9

Source: FBI.

(a) Find a linear function *f* that models these data. Is *f* exact or approximate?

(b) Use *f* to estimate the number of deaths on school grounds in 2003. Compare your answer to the actual value of 49. Did your estimate involve interpolation or extrapolation?

Applications

81. Projected Cost of College In 2003 the average annual cost of attending a private college or university, including tuition, fees, room, and board, was $25,000. This cost rose to about $37,000 in 2010, as illustrated in the figure. (*Source: Cerulli Associates.*)

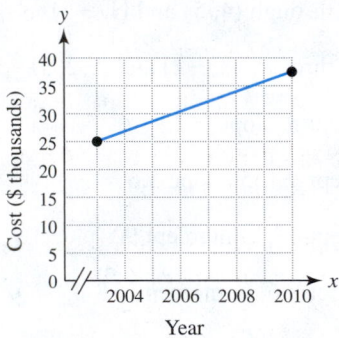

(a) Find a point-slope form of the line passing through the points (2003, 25,000) and (2010, 37,000). Interpret the slope.

(b) Use the equation to estimate the cost of attending a private college in 2007.

82. Distance A person is riding a bicycle. The graph shows the rider's distance *y* in miles from an interstate highway after *x* hours.

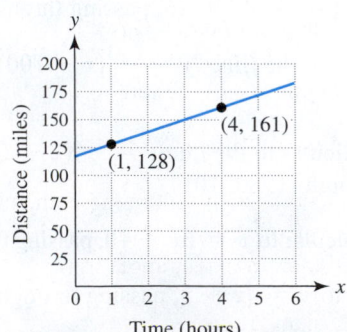

(a) How fast is the bicyclist traveling?

(b) Find the slope-intercept form of the line.

(c) How far was the bicyclist from the interstate highway initially?

(d) How far was the bicyclist from the interstate highway after 1 hour and 15 minutes?

83. Water in a Tank   The graph shows the amount of water in a 100-gallon tank after x minutes have elapsed.

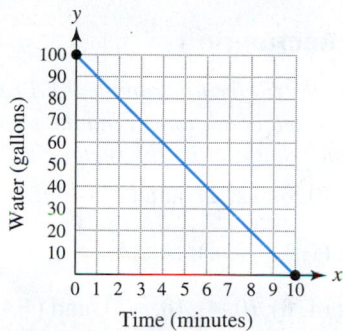

(a) Is water entering or leaving the tank? How much water is in the tank after 3 minutes?

(b) Find the x- and y-intercepts. Interpret each intercept.

(c) Find the slope-intercept form of the equation of the line. Interpret the slope.

(d) Use the graph to estimate the x-coordinate of the point $(x, 50)$ that lies on the line.

84. Cost of Driving   The cost of driving a car includes both fixed costs and mileage costs. Assume that insurance and car payments cost $350 per month and gasoline, oil, and routine maintenance cost $0.29 per mile.
(a) Find a linear function f that gives the *annual* cost of driving this car x miles.

(b) What does the y-intercept on the graph of f represent?

85. U.S. Music Sales   In 1999 music sales from vinyl, CDs, and downloads were $15 billion and in 2013 they were $9 billion. (*Source:* RIAA.)
(a) Find the point-slope form for a line passing through the points (1999, 15) and (2013, 9).

(b) Interpret the slope of this line.

(c) Estimate sales in 2008 and compare them with the true value of $10.4 billion. Did your answer use interpolation or extrapolation?

86. Toyota Vehicle Sales   In 1998 Toyota sold 1.4 million vehicles and in 2007 it sold 2.3 million vehicles. (*Source:* Toyota Motor Division.)
(a) Find the point-slope form for a line passing through the points (1998, 1.4) and (2007, 2.3).

(b) Interpret the slope of this line.

(c) Estimate sales in 2004 and compare them with the true value of 2.0 million vehicles. Did your answer use interpolation or extrapolation?

87. Bankruptcies   The table lists the number of bankruptcies filed in thousands for selected years.

Year	2006	2007	2008	2009	2010
Bankruptcies	160	220	290	380	425

Source: Administrative Office of the United States.

(a) Make a scatterplot of the data.

(b) Find a formula in point-slope form for a linear function f that models the data.

(c) Graph the data and $y = f(x)$ in the same xy-plane.

(d) Interpret the slope of the graph of $y = f(x)$.

(e) Estimate the number of bankruptcies in 2014. Did your answer involve interpolation or extrapolation?

88. College Tuition   The table lists average tuition and fees in dollars at private colleges for selected years.

Year	1995	2000	2005	2010
Cost	12,432	16,233	21,235	26,273

Source: The College Board.

(a) Make a scatterplot of the data.

(b) Find a formula in point-slope form for a linear function f that models the data.

(c) Graph the data and $y = f(x)$ in the same xy-plane.

(d) Interpret the slope of the graph of $y = f(x)$.

(e) Estimate the cost in 2014. Did your answers involve interpolation or extrapolation?

89. Hours Worked in Europe   The table lists the annual hours worked by the average worker in Europe for selected years.

Year	1970	1980	1990	2000	2010
Hours	2000	1860	1750	1690	1590

Source: Gallup Research.

continued on next page

(a) Let x represent the number of years *after* 1970. Find a formula in slope-intercept form for a linear function f that models the data.

(b) Interpret the slope of the graph of $y = f(x)$.

(c) Estimate the annual hours worked in 2014.

90. Green Building Material The table lists U.S. demand for green building materials in billions of dollars for selected years.

Year	2010	2011	2012	2013
Demand	65	70	75	80

Source: Freedonia Group.

(a) Let x represent the number of years *after* 2010. Find a formula in slope-intercept form for a linear function f that models the data.

(b) Interpret the slope of the graph of $y = f(x)$.

(c) Estimate the demand in 2020.

Perspectives and Viewing Rectangles

91. Graph $y = \frac{1}{1024}x + 1$ in [0, 3, 1] by [−2, 2, 1].
(a) Is the graph a horizontal line?

(b) Why does the calculator screen appear as it does?

92. Graph $y = 1000x + 1000$ in the standard window.
(a) Is the graph a vertical line?

(b) Explain why the calculator screen appears as it does.

93. Square Viewing Rectangle Graph the lines $y = 2x$ and $y = -\frac{1}{2}x$ in the standard viewing rectangle.
(a) Do the lines appear to be perpendicular?

(b) Graph the lines in the following viewing rectangles.
  i. [−15, 15, 1] by [−10, 10, 1]
  ii. [−10, 10, 1] by [−3, 3, 1]
  iii. [−3, 3, 1] by [−2, 2, 1]
  Do the lines appear to be perpendicular in any of these viewing rectangles?

(c) Determine the viewing rectangles where perpendicular lines will appear perpendicular. (Answers may vary.)

94. Square Viewing Rectangle Continuing with Exercise 93, make a conjecture about which viewing rectangles result in the graph of a circle with radius 5 and center at the origin appearing circular.
  i. [−9, 9, 1] by [−6, 6, 1]
  ii. [−5, 5, 1] by [−10, 10, 1]
  iii. [−5, 5, 1] by [−5, 5, 1]
  iv. [−18, 18, 1] by [−12, 12, 1]
Test your conjecture by graphing this circle in each viewing rectangle. (*Hint:* Graph $y_1 = \sqrt{25 - x^2}$ and $y_2 = -\sqrt{25 - x^2}$ to create the circle.)

Finding a Rectangle

Exercises 95–98: (Refer to Example 9.) A rectangle is determined by the stated conditions. Find the slope-intercept form of the four lines that outline the rectangle.

95. Vertices (0, 0), (2, 2), and (1, 3)

96. Vertices (1, 1), (5, 1), and (5, 5)

97. Vertices (4, 0), (0, 4), (0, −4), and (−4, 0)

98. Vertices (1, 1) and (2, 3); the point (3.5, 1) lies on a side of the rectangle.

Linear Regression

Exercises 99 and 100: Find the line of least-squares fit for the given data points. What is the correlation coefficient? Plot the data and graph the line.

99. (−2, 2), (1, 0), (3, −2)

100. (−1, −1), (1, 4), (2, 6)

Exercises 101–104: Complete the following.
(a) *Conjecture whether the correlation coefficient r for the data will be positive, negative, or zero.*
(b) *Use a calculator to find the equation of the least-squares regression line and the value of r.*
(c) *Use the regression line to predict y when x = 2.4.*

101.

x	−1	0	1	2	3
y	−5.7	−2.6	1.1	3.9	7.3

102.

x	−4	−2	0	2	4
y	1.2	2.8	5.3	6.7	9.1

103.

x	1	3	5	7	10
y	5.8	−2.4	−10.7	−17.8	−29.3

104.

x	−4	−3	−1	3	5
y	37.2	33.7	27.5	16.4	9.8

105. Distant Galaxies   In the late 1920s Edwin P. Hubble (1889–1953) determined both the distance to several galaxies and the velocity at which they were receding from Earth. Four galaxies with their distances in light-years and velocities in miles per second are listed in the table.

Galaxy	Distance	Velocity
Virgo	50	990
Ursa Minor	650	9,300
Corona Borealis	950	15,000
Bootes	1700	25,000

Source: A. Sharov and 1. Novikov, *Edwin Hubble: The Discoverer of the Big Bang Universe.*

(a) Let x be distance and y be velocity. Plot the points in [−100, 1800, 100] by [−1000, 28000, 1000].

(b) Find the least-squares regression line.

(c) If the galaxy Hydra is receding at a speed of 37,000 miles per second, estimate its distance.

106. Airline Travel   The table at the top of the next column lists the numbers of airline passengers in millions at some of the largest airports in the United States during 2002 and 2006.

(a) Graph the data by using the 2002 data for x-values and the corresponding 2006 data for y-values. Predict whether the correlation coefficient will be positive or negative.

(b) Use a calculator to find the linear function f based on least-squares regression that models the data. Graph y = f(x) and the data in the same viewing rectangle.

(c) In 2002 Newark International Airport had 29.0 million passengers. Assuming that this airport followed a trend similar to that of the five airports listed in the table, use f(x) to estimate the number of passengers at Newark International in 2006. Compare this result to the actual value of 36.7 million passengers.

Airport	2002	2006
Atlanta (Hartsfield)	76.9	84.4
Chicago (O'Hare)	66.5	77.0
Los Angeles (LAX)	56.2	61.0
Dallas/Fort Worth	52.8	60.2
Denver	35.7	47.3

Source: Airports Association Council International.

107. Passenger Travel   The table shows the number of miles (in trillions) traveled by passengers of all types x years after 1970.

Year (1970 ↔ 0)	0	10	20	30	40
Miles (trillions)	2.2	2.8	3.7	4.7	5.5

Source: Department of Transportation.

(a) Make a scatterplot of the data. Predict whether the correlation coefficient will be positive or negative.

(b) Use least-squares regression to find f(x) = ax + b so that f models the data.

(c) Graph f and the data. Interpret the slope.

(d) Predict the number of passenger miles in 2015.

108. High School Enrollment   The table lists the number of students (in millions) attending U.S. public school (grades 9–12) x years after 2000.

x (year)	0	3	5	7
y (students)	13.5	14.3	14.8	15.1

Source: National Center for Education Statistics.

(a) Use least-squares regression to find f(x) = ax + b so that f models the data.

(b) Graph f and the data. Interpret the slope.

(c) Estimate enrollment in 2002 and compare the estimate to the actual value of 14.1 million.

Writing about Mathematics

109. Compare the slope-intercept form with the point-slope form. Give examples of each.

110. Explain how you would find the equation of a line passing through two points. Give an example.

2.2 Linear Equations

- Learn about equations and recognize a linear equation
- Solve linear equations symbolically
- Solve linear equations graphically and numerically
- Solve problems involving percentages
- Apply problem-solving strategies

Introduction

In Example 4 of Section 2.1, we modeled iPod sales y in millions during year x with the equation of a line, or *linear function*, given by

$$f(x) = -4(x - 2008) + 55.$$

To predict the year when iPod sales might decrease to **27** million, we could set the formula for $f(x)$ equal to **27** and solve the following *linear equation* for x.

$$27 = -4(x - 2008) + 55 \qquad \text{Linear equation}$$

This section discusses linear equations and their solutions. See Example 5 in this section.

Equations

An **equation** is a statement that two mathematical expressions are equal. Equations always contain an equals sign.

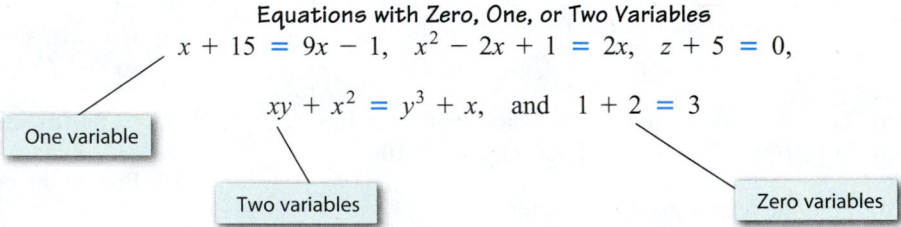

Equations with Zero, One, or Two Variables

$$x + 15 = 9x - 1, \quad x^2 - 2x + 1 = 2x, \quad z + 5 = 0,$$

$$xy + x^2 = y^3 + x, \quad \text{and} \quad 1 + 2 = 3$$

One variable

Two variables

Zero variables

We will concentrate on equations with one variable.

To **solve** an equation means to find all values for the variable that make the equation a true statement. Such values are called **solutions.** The set of all solutions is the **solution set.** For example, the solutions to the equation $x^2 - 1 = 0$ are 1 or -1, written as ± 1. Either value for x **satisfies** the equation. The solution set is $\{-1, 1\}$. Two equations are **equivalent** if they have the same solution set. For example, the equations $x + 2 = 5$ and $x = 3$ are equivalent.

If an equation has no solutions, then its solution set is empty and the equation is called a **contradiction.** The equation $x + 2 = x$ has no solutions and is a contradiction. However, if every (meaningful) value for the variable is a solution, then the equation is an **identity.** The equation $x + x = 2x$ is an identity because every value for x makes the equation true. Any equation that is satisfied by some but not all values of the variable is a **conditional equation.** The equation $x^2 - 1 = 0$ is a conditional equation. Only the values -1 and 1 for x make this equation a true statement.

Like functions, equations can be either *linear* or *nonlinear.* A linear equation is one of the simplest types of equations.

LINEAR EQUATION IN ONE VARIABLE

A **linear equation** in one variable is an equation that can be written in the form

$$ax + b = 0,$$

where a and b are constants with $a \neq 0$.

If an equation is not linear, then we say that it is a **nonlinear equation.** The following gives examples of linear and nonlinear equations.

Example	Explanation
$2x + 5 = 0$	*Linear:* Is written as $ax + b = 0$ with $a = 2$ and $b = 5$
$3x = -7$	*Linear:* Can be written as $3x + 7 = 0$
$5(x - 1) = 4$	*Linear:* Can be written as $5x - 9 = 0$
$x^2 + 2x = 1$	*Nonlinear:* Contains x^2
$\sqrt{x} = 7$	*Nonlinear:* Contains $\sqrt{x}$

Symbolic Solutions

A linear equation can be solved symbolically, and the solution is *always exact.* To solve a linear equation symbolically, we usually apply the *properties of equality* to the given equation and transform it into an equivalent equation that is simpler.

PROPERTIES OF EQUALITY

Addition Property of Equality

If a, b, and c are real numbers, then

$$a = b \quad \text{is equivalent to} \quad a + c = b + c.$$

Multiplication Property of Equality

If a, b, and c are real numbers with $c \neq 0$, then

$$a = b \quad \text{is equivalent to} \quad ac = bc.$$

Loosely speaking, the addition property states that "if equals are added to equals, the results are equal." For example, if $x + 5 = 15$, then we can add -5 to each side of the equation, or equivalently subtract 5 from each side, to obtain the following.

	$x + 5 = 15$	Given equation
Equivalent equations	$x + 5 - 5 = 15 - 5$	Subtract 5 from each side.
	$x = 10$	Simplify.

Similarly, the multiplication property states that "if equals are multiplied by nonzero equals, the results are equal." For example, if $5x = 20$, then we can multiply each side by $\frac{1}{5}$, or equivalently divide each side by 5, to obtain the following.

	$5x = 20$	Given equation
Equivalent equations	$\dfrac{5x}{5} = \dfrac{20}{5}$	Divide each side by 5.
	$x = 4$	Simplify.

These two properties along with the distributive property are applied in the next two examples. We also check our answers.

EXAMPLE 1 Solving a linear equation symbolically

Solve the equation $3(x - 4) = 2x - 1$. Check your answer.

SOLUTION

Getting Started First we apply the distributive property: $a(b - c) = ab - ac$. Thus

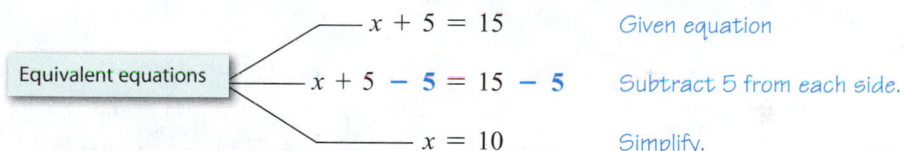

$$3(x - 4) = 3 \cdot x - 3 \cdot 4 = 3x - 12. \blacktriangleright$$

Algebra Review
To review the distributive properties, see Chapter R (page R-15).

Solving the given equation results in the following.

$$3(x - 4) = 2x - 1 \qquad \text{Given equation}$$
$$3x - 12 = 2x - 1 \qquad \text{Distributive property}$$
$$3x - 2x - 12 + 12 = 2x - 2x - 1 + 12 \qquad \text{Subtract 2x and add 12.}$$
$$3x - 2x = 12 - 1 \qquad \text{Simplify.}$$
$$x = 11 \qquad \text{Simplify.}$$

The solution is 11. We can check our answer as follows.

$$3(x - 4) = 2x - 1 \qquad \text{Given equation}$$
$$3(11 - 4) \stackrel{?}{=} 2 \cdot 11 - 1 \qquad \text{Let x = 11.}$$
$$21 = 21 \ \checkmark \qquad \text{The answer checks.}$$

Now Try Exercise 23

EXAMPLE 2 **Solving a linear equation symbolically**

Solve $3(2x - 5) = 10 - (x + 5)$. Check your answer.

SOLUTION

Getting Started In this problem subtraction must be distributed over the quantity $(x + 5)$. Thus

$$10 - (x + 5) = 10 - 1(x + 5) = 10 - x - 5. \ \blacktriangleright$$

Solving the given equation results in the following.

$$3(2x - 5) = 10 - (x + 5) \qquad \text{Given equation}$$
$$6x - 15 = 10 - x - 5 \qquad \text{Distributive property}$$
$$6x - 15 = 5 - x \qquad \text{Simplify.}$$
$$7x - 15 = 5 \qquad \text{Add x to each side.}$$
$$7x = 20 \qquad \text{Add 15 to each side.}$$
$$x = \frac{20}{7} \qquad \text{Divide each side by 7.}$$

The solution is $\frac{20}{7}$. To check this answer, let $x = \frac{20}{7}$ and simplify.

$$3(2x - 5) = 10 - (x + 5) \qquad \text{Given equation}$$
$$3\left(2 \cdot \frac{20}{7} - 5\right) \stackrel{?}{=} 10 - \left(\frac{20}{7} + 5\right) \qquad \text{Let x = } \frac{20}{7}.$$
$$3\left(\frac{5}{7}\right) \stackrel{?}{=} 10 - \frac{55}{7} \qquad \text{Simplify.}$$
$$\frac{15}{7} = \frac{15}{7} \ \checkmark \qquad \text{The answer checks.}$$

Now Try Exercise 27

Fractions and Decimals When fractions or decimals appear in an equation, we sometimes can make our work simpler by multiplying each side of the equation by the least common denominator (LCD) or a common denominator of all fractions in the equation. This method is illustrated in the next two examples.

EXAMPLE 3   **Eliminating fractions**

Solve each linear equation.

(a) $\dfrac{x}{3} + 1 = \dfrac{2}{3}$ **(b)** $\dfrac{t - 2}{4} - \dfrac{1}{3}t = 5 - \dfrac{1}{12}(3 - t)$

SOLUTION

(a) To eliminate fractions, multiply each side (or term in the equation) by the LCD, 3.

$$\dfrac{x}{3} + 1 = \dfrac{2}{3} \qquad\qquad \text{Given equation}$$

$$3 \cdot \dfrac{x}{3} + 3 \cdot 1 = 3 \cdot \dfrac{2}{3} \qquad \text{Multiply each side (term) by 3.}$$

$$x + 3 = 2 \qquad\qquad \text{Simplify.}$$

$$x + 3 - 3 = 2 - 3 \qquad \text{Subtract 3 from each side.}$$

$$x = -1 \qquad\qquad \text{Simplify.}$$

The solution is -1.

(b) To eliminate fractions, multiply each side (or term in the equation) by the LCD, 12.

Algebra Review

To review least common multiples and least common denominators, see Chapter R (page R-30).

$$\dfrac{t - 2}{4} - \dfrac{1}{3}t = 5 - \dfrac{1}{12}(3 - t) \qquad \text{Given equation}$$

$$\dfrac{12(t - 2)}{4} - \dfrac{12}{3}t = 12(5) - \dfrac{12}{12}(3 - t) \qquad \text{Multiply each side (term) by 12.}$$

$$3(t - 2) - 4t = 60 - (3 - t) \qquad \text{Simplify.}$$

$$3t - 6 - 4t = 60 - 3 + t \qquad \text{Distributive property}$$

$$-t - 6 = 57 + t \qquad \text{Combine like terms on each side.}$$

$$-2t = 63 \qquad \text{Add } -t \text{ and 6 to each side.}$$

$$t = -\dfrac{63}{2} \qquad \text{Divide each side by } -2.$$

The solution is $-\dfrac{63}{2}$.

Now Try Exercises 29 and 33

EXAMPLE 4   **Eliminating decimals**

Solve each linear equation.

(a) $5.1x - 2 = 3.7$ **(b)** $0.03(z - 3) - 0.5(2z + 1) = 0.23$

SOLUTION

(a) To eliminate decimals, multiply each side (or term in the equation) by 10.

$$5.1x - 2 = 3.7 \qquad \text{Given equation}$$

$$10(5.1x) - 10(2) = 10(3.7) \qquad \text{Multiply each side (term) by 10.}$$

$$51x - 20 = 37 \qquad \text{Simplify.}$$

$$51x - 20 + 20 = 37 + 20 \qquad \text{Add 20 to each side.}$$

$$51x = 57 \qquad \text{Simplify.}$$

$$x = \dfrac{57}{51} \text{ or } \dfrac{19}{17} \qquad \text{Divide each side by 51; simplify.}$$

The solution is $\dfrac{19}{17}$.

(b) To eliminate decimals, multiply each side (or term in the equation) by 100.

$$0.03(z - 3) - 0.5(2z + 1) = 0.23 \qquad \textit{Given equation}$$

$$3(z - 3) - 50(2z + 1) = 23 \qquad \textit{Multiply each side (term) by 100.}$$

$$3z - 9 - 100z - 50 = 23 \qquad \textit{Distributive property}$$

$$-97z - 59 = 23 \qquad \textit{Combine like terms.}$$

$$-97z = 82 \qquad \textit{Add 59 to each side.}$$

$$z = -\frac{82}{97} \qquad \textit{Divide each side by −97.}$$

The solution is $-\frac{82}{97}$.

Now Try Exercises 35 and 37

An Application   In the next example we solve the equation presented in the introduction to this section.

EXAMPLE 5   Analyzing iPod sales

The linear function defined by $f(x) = -4(x - 2008) + 55$ estimates iPod sales (in millions of units) during fiscal year x. Use $f(x)$ to estimate when iPod sales could reach 27 million.

SOLUTION   We need to find the year x when iPod sales, given by $f(x)$, equal 27 million.

$$-4(x - 2008) + 55 = 27 \qquad \textit{Given equation}$$

$$-4(x - 2008) + 55 - 55 = 27 - 55 \qquad \textit{Subtract 55 from each side.}$$

$$-4(x - 2008) = -28 \qquad \textit{Simplify.}$$

$$\frac{-4(x - 2008)}{-4} = \frac{-28}{-4} \qquad \textit{Divide each side by −4.}$$

> Rather than dividing by −4, the distributive property could be used.

$$x - 2008 = 7 \qquad \textit{Simplify.}$$

$$x = 2015 \qquad \textit{Add 2008 to each side.}$$

This model predicts that iPod sales could decrease to **27** million in **2015**.

Now Try Exercise 97

Contradictions, Identities, and Conditional Equations   The next example illustrates how an equation can have no solutions (contradiction), one solution (conditional equation), or infinitely many solutions (identity).

EXAMPLE 6   Identifying contradictions, identities, and conditional equations

Identify each equation as a contradiction, identity, or conditional equation.
(a) $7 + 6x = 2(3x + 1)$ \qquad\qquad **(b)** $2x - 5 = 3 - (1 + 2x)$
(c) $2(5 - x) - 25 = 3(x - 5) - 5x$

SOLUTION
(a)

$$7 + 6x = 2(3x + 1) \qquad \textit{Given equation}$$

$$7 + 6x = 6x + 2 \qquad \textit{Distributive property}$$

$$7 = 2 \qquad \textit{Subtract 6x from each side.}$$

The statement $7 = 2$ is false and there are *no* solutions. The equation is a contradiction.

(b)

$$2x - 5 = 3 - (1 + 2x) \quad \text{Given equation}$$
$$2x - 5 = 3 - 1 - 2x \quad \text{Distributive property}$$
$$4x = 7 \quad \text{Add 2x and 5 to each side.}$$
$$x = \frac{7}{4} \quad \text{Divide each side by 4.}$$

There is one solution: $\frac{7}{4}$. This is a conditional equation.

(c)

$$2(5 - x) - 25 = 3(x - 5) - 5x \quad \text{Given equation}$$
$$10 - 2x - 25 = 3x - 15 - 5x \quad \text{Distributive property}$$
$$-2x - 15 = -2x - 15 \quad \text{Simplify each side.}$$
$$\mathbf{0 = 0} \quad \text{Add 2x and 15 to each side.}$$

The statement $\mathbf{0 = 0}$ is true and the solution set includes *all real numbers*. The equation is an identity.

Now Try Exercises 39, 41, and 43

Graphical and Numerical Solutions

Graphical Solutions   The equation $f(x) = g(x)$ results whenever the formulas for two functions f and g are set equal to each other. A solution to this equation corresponds to the x-coordinate of a point where the graphs of f and g intersect. We call this graphical technique the **intersection-of-graphs** method and it is used to solve the linear equation $2x + 1 = -x + 4$ in the following box.

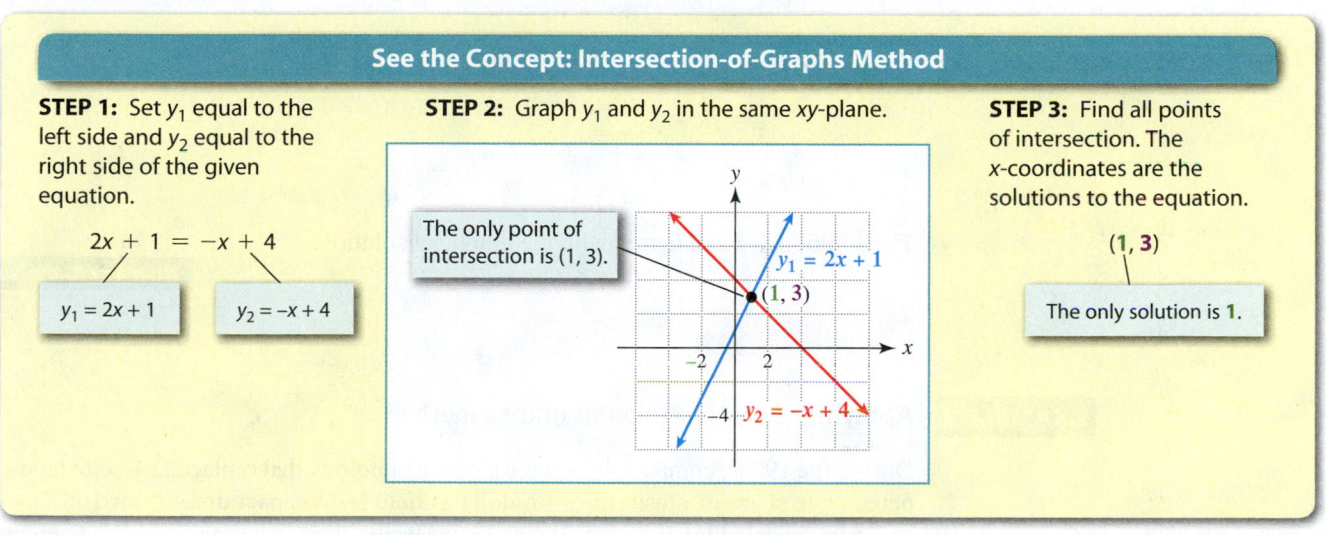

See the Concept: Intersection-of-Graphs Method

STEP 1: Set y_1 equal to the left side and y_2 equal to the right side of the given equation.

$$2x + 1 = -x + 4$$

$y_1 = 2x + 1$ $y_2 = -x + 4$

STEP 2: Graph y_1 and y_2 in the same xy-plane.

The only point of intersection is (1, 3).

$y_1 = 2x + 1$
$(1, 3)$
$y_2 = -x + 4$

STEP 3: Find all points of intersection. The x-coordinates are the solutions to the equation.

$(1, 3)$

The only solution is **1**.

We can check our work from the example above by substituting **1** for x in the given equation.

$$2x + 1 = -x + 4 \quad \text{Given equation}$$
$$2(1) + 1 \stackrel{?}{=} -(1) + 4 \quad \text{Let } x = 1.$$
$$3 = 3 \ \checkmark \quad \text{It checks.}$$

Notice that the point of intersection is $(\mathbf{1, 3})$, and that when we substitute **1** for x, each side of the equation evaluates to **3**.

EXAMPLE 7   Solving an equation graphically and symbolically

Solve $2x - 1 = \frac{1}{2}x + 2$ graphically and symbolically.

SOLUTION

Graphical Solution Graph $y_1 = 2x - 1$ and $y_2 = \frac{1}{2}x + 2$. Their graphs intersect at the point $(2, 3)$, as shown in Figure 2.22, so the solution is **2**. Figure 2.23 shows these graphs as created by a graphing calculator.

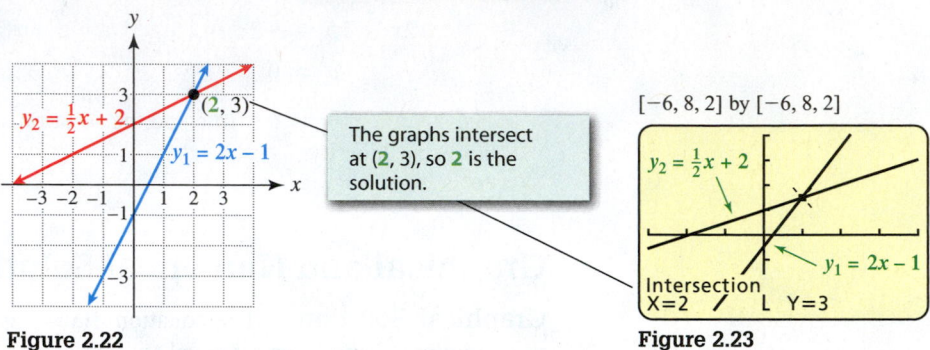

Intersection-of-Graphs Method

The graphs intersect at (**2**, **3**), so **2** is the solution.

$[-6, 8, 2]$ by $[-6, 8, 2]$

Figure 2.22   **Figure 2.23**

Symbolic Solution $2x - 1 = \dfrac{1}{2}x + 2$ Given equation

$2x = \dfrac{1}{2}x + 3$ Add 1 to each side.

$\dfrac{3}{2}x = 3$ Subtract $\frac{1}{2}x$ from each side.

$\dfrac{2}{3} \cdot \dfrac{3}{2}x = \dfrac{2}{3} \cdot 3$ Multiply each side by $\frac{2}{3}$.

$x = 2$ Multiply fractions.

The solution is **2** and agrees with the graphical solution.

Now Try Exercise 53

EXAMPLE 8   Applying the intersection-of-graphs method

During the 1990s, compact discs were a new technology that replaced cassette tapes. The percentage share of music sales (in dollars) held by compact discs from 1987 to 1998 could be modeled by $f(x) = 5.91x + 13.7$. During the same time period the percentage share of music sales held by cassette tapes could be modeled by $g(x) = -4.71x + 64.7$. In these formulas $x = 0$ corresponds to 1987, $x = 1$ to 1988, and so on. Use the intersection-of-graphs method to estimate the year when the percentage share of CDs equaled the percentage share of cassettes.   (**Source:** Recording Industry Association of America.)

SOLUTION   We must solve the linear equation $f(x) = g(x)$, or equivalently,

$$5.91x + 13.7 = -4.71x + 64.7.$$

Graph $y_1 = 5.91x + 13.7$ and $y_2 = -4.71x + 64.7$, as in Figure 2.24. In Figure 2.25 their graphs intersect near the point $(\mathbf{4.8}, \mathbf{42.1})$. Since $x = 0$ corresponds to 1987 and $1987 + \mathbf{4.8} \approx 1992$, it follows that in 1992 sales of CDs and cassette tapes were approximately equal. Each had about **42.1%** of the sales in 1992.

Calculator Help

To find the point of intersection in Figure 2.25, see Appendix A (page AP-7).

Determining When Sales Were Equal

[0, 12, 2] by [0, 100, 10]   [0, 12, 2] by [0, 100, 10]

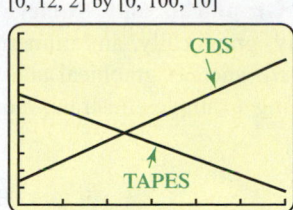   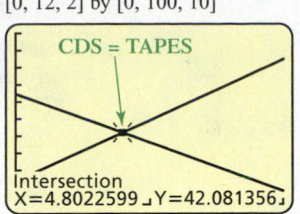

Figure 2.24   Figure 2.25

Now Try Exercise 99

Numerical Solutions Sometimes it is also possible to find a numerical solution to an equation by using a table of values. When the solution is an integer or a convenient fraction, we can usually find it by using the table feature on a graphing calculator. However, when the solution is a fraction with a repeating decimal representation or an irrational number, a numerical method gives only an approximate solution. The formula

$$f(x) = 755.7(x - 1985) + 6121 \qquad \textit{Tuitions and fees after 1985}$$

can be used to model tuition and fees at private colleges and universities during year x. See Exercise 98. To determine when tuition and fees were about \$23,500, we could solve the equation

$$755.7(x - 1985) + 6121 = 23{,}500$$

for x by making a table of values for $f(x)$, as shown in Figure 2.26. By scrolling through the x-values, we can see that tuition and fees at private schools were about \$23,500 in 2008.

$f(2008) \approx 23,500$

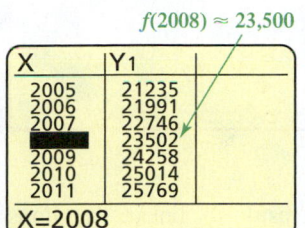

Figure 2.26

Calculator Help

To make a table of values, see Appendix A (page AP-5).

NOTE Regardless of whether we use a symbolic, graphical, or numerical method to solve an equation, we should find the same solution set. However, our answers may differ slightly because of rounding.

A numerical technique is used in the next example, where a sequence of tables are generated to obtain an increasingly accurate answer.

EXAMPLE 9 Solving an equation numerically

Solve $\sqrt{3}(2x - \pi) + \frac{1}{3}x = 0$ numerically to the nearest tenth.

SOLUTION Make a table of $y_1 = \sqrt{3}(2x - \pi) + \frac{1}{3}x$, incrementing x, as shown in Figures 2.27–2.29. The solution to $y_1 = 0$, lies between x-values where y_1 changes from negative to positive. This process shows that the solution is 1.4 (to the nearest tenth).

Increment x by 1   Increment x by 0.1   Increment x by 0.01

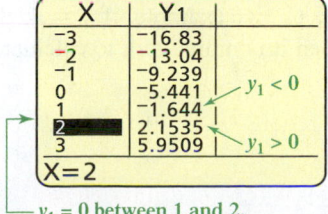   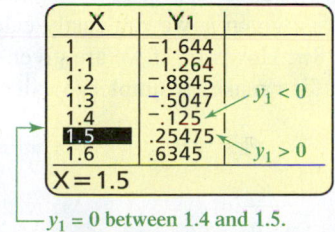   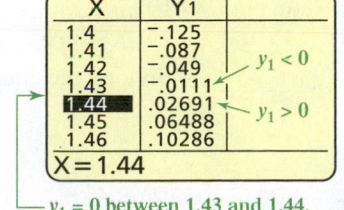

To the nearest tenth, the solution is 1.4.

$y_1 = 0$ between 1 and 2. $y_1 = 0$ between 1.4 and 1.5. $y_1 = 0$ between 1.43 and 1.44.

Figure 2.27   Figure 2.28   Figure 2.29

Now Try Exercise 71

Symbolic, Graphical, and Numerical Solutions Linear equations can be solved symbolically, graphically, and numerically. Symbolic solutions to linear equations are *always exact*, whereas graphical and numerical solutions are *sometimes approximate*. The following example illustrates how to solve the equation $2x - 1 = 3$ with each method.

Symbolic Solution

$$2x - 1 = 3$$
$$2x = 4$$
$$x = 2$$

Check:

$$2(2) - 1 \overset{?}{=} 3$$
$$3 = 3 \checkmark$$

It checks.

Graphical Solution

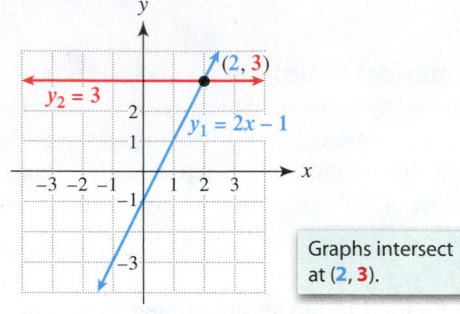

Graphs intersect at (**2, 3**).

The solution is **2**.

Numerical Solution

x	0	1	**2**	3
$2x - 1$	−1	1	**3**	5

Because $2x - 1$ equals **3** when $x = $ **2**, the solution to $2x - 1 = 3$ is **2**.

Percentages Applications involving percentages often result in linear equations because percentages can be computed by linear functions. A function for taking P percent of x is given by $f(x) = \frac{P}{100}x$, where $\frac{P}{100}$ is the decimal form for P percent. For example, to calculate 35% of x, let $f(x) = 0.35x$. Then 35% of $150 is $f(150) = 0.35(150) = 52.5$, or $52.50.

EXAMPLE 10 Solving an application involving percentages

A survey found that 76% of bicycle riders do not wear helmets. (*Source:* Opinion Research Corporation for Glaxo Wellcome, Inc.)
(a) Find a formula $f(x)$ for a function that computes the number of people who do not wear helmets among x bicycle riders.
(b) There are approximately 38.7 million riders of all ages who do not wear helmets. Find the total number of bicycle riders.

SOLUTION
(a) A function f that computes 76% of x is given by $f(x) = 0.76x$.
(b) We must find the x-value for which $f(x) = 38.7$ million, or solve the equation $0.76x = 38.7$. Solving gives $x = \frac{38.7}{0.76} \approx 50.9$ million bike riders.

Now Try Exercise 105

Solving for a Variable

The circumference C of a circle is given by $C = 2\pi r$, where r is the radius. This equation is *solved* for C. That is, given r, we can easily calculate C. For example, if $r = 4$, then $C = 2\pi(4)$, or $C = 8\pi$. However, if we are given C, then it is more work to calculate r. Solving the equation for r makes it simpler to calculate r.

$$C = 2\pi r \qquad \textit{Given equation}$$
$$\frac{C}{2\pi} = r \qquad \textit{Divide each side by } 2\pi.$$

The equation $r = \frac{C}{2\pi}$ is solved for r.

EXAMPLE 11   Solving for a variable

The area of a trapezoid with bases a and b and height h is given by $A = \frac{1}{2}h(a + b)$. Solve this equation for b.

SOLUTION

Getting Started If we multiply each side by 2 and divide each side by h, the right side of the equation becomes $a + b$. Subtracting a from each side isolates b. ▶

$$A = \frac{1}{2}h(a + b) \qquad \textit{Given equation}$$

$$2A = h(a + b) \qquad \textit{Multiply each side by 2.}$$

$$\frac{2A}{h} = a + b \qquad \textit{Divide each side by h.}$$

$$\frac{2A}{h} - a = b \qquad \textit{Subtract a from each side.}$$

The equation $b = \frac{2A}{h} - a$ is solved for b.

Now Try Exercise 85

Problem-Solving Strategies

To become more proficient at solving problems, we need to establish a procedure to guide our thinking. The following steps may be helpful in solving application problems.

SOLVING APPLICATION PROBLEMS

STEP 1: Read the problem and make sure you understand it. Assign a variable to what you are being asked to find. If necessary, write other quantities in terms of this variable.

STEP 2: Write an equation that relates the quantities described in the problem. You may need to sketch a diagram and refer to known formulas.

STEP 3: Solve the equation and determine the solution.

STEP 4: Look back and check your solution. Does it seem reasonable?

These steps are applied in the next four examples.

EXAMPLE 12   Working together

A large pump can empty a tank of gasoline in 5 hours, and a smaller pump can empty the same tank in 9 hours. If both pumps are used to empty the tank, how long will it take?

SOLUTION

STEP 1: We are asked to find the time it takes for *both* pumps to empty the tank. Let this time be t.

t: Time to empty the tank

STEP 2: In 1 hour the large pump will empty $\frac{1}{5}$ of the tank and the smaller pump will empty $\frac{1}{9}$ of the tank. The fraction of the tank that they will empty together in 1 hour is given by $\frac{1}{5} + \frac{1}{9}$. In 2 hours the large pump will empty $\frac{2}{5}$ of the tank and the smaller pump will empty $\frac{2}{9}$ of the tank. The fraction of the tank that they will empty together in 2 hours is $\frac{2}{5} + \frac{2}{9}$. Similarly, in t hours the fraction of the tank that the two pumps can empty is $\frac{t}{5} + \frac{t}{9}$. Since the tank is empty when this fraction reaches 1, we must solve the following equation.

$$\frac{t}{5} + \frac{t}{9} = 1$$

STEP 3: Multiply by the LCD, 45, to eliminate fractions.

$$\frac{45t}{5} + \frac{45t}{9} = 45 \qquad \text{Multiply by LCD.}$$

$$9t + 5t = 45 \qquad \text{Simplify.}$$

$$14t = 45 \qquad \text{Add like terms.}$$

$$t = \frac{45}{14} \approx 3.21 \qquad \text{Divide by 14 and approximate.}$$

Working together, the two pumps can empty the tank in about 3.21 hours.

STEP 4: This sounds reasonable. Working together the two pumps should be able to empty the tank faster than the large pump working alone, but not twice as fast. Note that $\frac{3.21}{5} + \frac{3.21}{9} \approx 1$.

> **Now Try Exercise 107**

EXAMPLE 13	**Solving an application involving motion**

In 1 hour an athlete traveled 10.1 miles by running first at 8 miles per hour and then at 11 miles per hour. How long did the athlete run at each speed?

SOLUTION

STEP 1: We are asked to find the time spent running at each speed. If we let x represent the time in hours spent running at 8 miles per hour, then $1 - x$ represents the time spent running at 11 miles per hour because the total running time was 1 hour.

$$x: \text{Time spent running at 8 miles per hour}$$

$$1 - x: \text{Time spent running at 11 miles per hour}$$

STEP 2: Distance d equals rate r times time t, that is, $d = rt$. In this example we have two rates (or speeds) and two times. The total distance must sum to 10.1 miles.

$$d = r_1 t_1 + r_2 t_2 \qquad \text{General equation}$$

$$10.1 = 8x + 11(1 - x) \qquad \text{Substitute.}$$

STEP 3: We can solve this equation symbolically.

$$10.1 = 8x + 11 - 11x \qquad \text{Distributive property}$$

$$10.1 = 11 - 3x \qquad \text{Combine like terms.}$$

$$3x = 0.9 \qquad \text{Add } 3x; \text{subtract } 10.1.$$

$$x = 0.3 \qquad \text{Divide by 3.}$$

The athlete runs 0.3 hour (18 minutes) at 8 miles per hour and 0.7 hour (42 minutes) at 11 miles per hour.

STEP 4: We can check this solution as follows.

$$8(0.3) + 11(0.7) = 10.1 \checkmark \quad \text{It checks.}$$

This sounds reasonable. The runner's average speed was 10.1 miles per hour, so the runner must have run longer at 11 miles per hour than at 8 miles per hour.

> **Now Try Exercise 109**

Geometry Review
To review similar triangles, see Chapter R (page R-5).

Similar triangles are often used in applications involving geometry and are used to solve the next application.

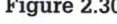

EXAMPLE 14   Solving an application involving similar triangles

A person 6 feet tall stands 17 feet from the base of a streetlight, as illustrated in Figure 2.30. If the person's shadow is 8 feet, estimate the height of the streetlight.

SOLUTION

STEP 1: We are asked to find the height of the streetlight in Figure 2.30. Let x represent this height.

$$x: \text{Height of the streetlight}$$

STEP 2: In Figure 2.31, triangle ACD is similar to triangle BCE. Thus ratios of corresponding sides are equal.

$$\frac{AD}{BE} = \frac{DC}{EC} \qquad \textcolor{blue}{\text{Similar triangles}}$$

$$\frac{x}{6} = \frac{17 + 8}{8}$$

Figure 2.30

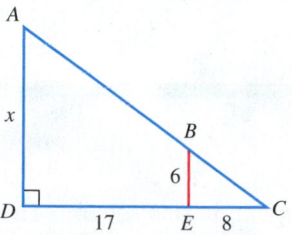

Figure 2.31

STEP 3: We can solve this equation symbolically.

$$\frac{x}{6} = \frac{25}{8} \qquad \textcolor{blue}{\text{Simplify.}}$$

$$x = \frac{6 \cdot 25}{8} \qquad \textcolor{blue}{\text{Multiply by 6.}}$$

$$x = 18.75 \qquad \textcolor{blue}{\text{Simplify.}}$$

The height of the streetlight is 18.75 feet.

STEP 4: One way to check this answer is to form a different proportion. Note that x is to $17 + 8$ in triangle ADC as 6 is to 8 in triangle BEC. If $x = 18.75$, then $\frac{18.75}{25} = \frac{6}{8}$, which is true, and our answer checks.

Now Try Exercise 113

EXAMPLE 15   Mixing acid in chemistry

Pure water is being added to 153 milliliters of a 30% solution of hydrochloric acid. How much water should be added to dilute the solution to a 13% mixture?

SOLUTION

STEP 1: We are asked to find the amount of water that should be added to 153 milliliters of 30% acid to make a 13% solution. Let this amount of water be equal to x. See Figure 2.32.

$$x: \text{Amount of pure water to be added}$$

$$x + 153: \text{Final volume of 13\% solution}$$

Figure 2.32

STEP 2: Since only water is added, the total amount of acid in the solution after the water is added must equal the amount of acid before the water is added. The volume of pure acid after the water is added equals **13%** of $x + 153$ milliliters, and the volume of pure acid before the water is added equals **30%** of **153** milliliters. We must solve the following equation.

$$\textbf{0.13}(\boldsymbol{x} + \textbf{153}) = \textbf{0.30}(\textbf{153}) \qquad \textcolor{blue}{\text{Pure acid after equals pure acid before.}}$$

STEP 3: Begin by dividing each side by 0.13.

$$0.13(x + 153) = 0.30(153) \qquad \text{Equation to solve}$$

$$x + 153 = \frac{0.30(153)}{0.13} \qquad \text{Divide by 0.13.}$$

$$x = \frac{0.30(153)}{0.13} - 153 \qquad \text{Subtract 153.}$$

$$x \approx 200.08 \qquad \text{Approximate.}$$

We should add about 200 milliliters of pure water.

STEP 4: Initially the solution contains $0.30(153) = 45.9$ milliliters of pure acid. If we add 200 milliliters of water to the 153 milliliters, the final solution is 353 milliliters, which includes 45.9 milliliters of pure acid. Its concentration is $\frac{45.9}{353} \approx 0.13$, or about 13%.

Now Try Exercise 117

2.2 Putting It All Together

A general four-step procedure for solving applications is given in this section. The following table summarizes some of the important concepts.

CONCEPT	EXPLANATION	EXAMPLES
Linear equation	A linear equation can be written as $ax + b = 0,\ a \neq 0.$	$4x + 5 = 0$ $3x - 1 = x + 2$
Addition property	$a = b$ is equivalent to $$a + c = b + c.$$	$x - 7 = 25$ $x - 7 + 7 = 25 + 7$ $x = 32$
Multiplication property	$a = b$ is equivalent to $$ac = bc,\ c \neq 0.$$	$\frac{1}{2}x = 4$ $2 \cdot \frac{1}{2}x = 4 \cdot 2$ $x = 8$
Distributive property	$a(b + c) = ab + ac$ $a(b - c) = ab - ac$	$2(5 + x) = 10 + 2x$ $-(2 - x) = -1(2 - x) = -2 + x$
Identity	An equation that is true for all (meaningful) values of the variable	$3(x - 2) = 3x - 6$ $2x + 3x = (2 + 3)x = 5x$
Contradiction	An equation that has no solutions	$x + 5 = x$ $2x - 2x = 5$
Conditional equation	An equation that is satisfied by some, but not all, of the values of the variable	$2x - 1 = 5$ Given equation $2x = 6$ Add 1. $x = 3$ Divide by 2.
Percentages	P percent of x equals $\frac{P}{100}x$, where $\frac{P}{100}$ is the decimal form for P percent.	35% of x is calculated by $f(x) = \frac{35}{100}x$, or $f(x) = 0.35x$.

The following example illustrates three ways to solve $5x - 1 = 3$.

Symbolic Solution

$$5x - 1 = 3$$
$$5x = 4$$
$$x = \frac{4}{5}$$

Check:

$$5\left(\frac{4}{5}\right) - 1 \overset{?}{=} 3$$
$$4 - 1 \overset{?}{=} 3$$
$$3 \overset{?}{=} 3 \checkmark$$

It checks.

Graphical Solution

$[-9, 9, 1]$ by $[-6, 6, 1]$

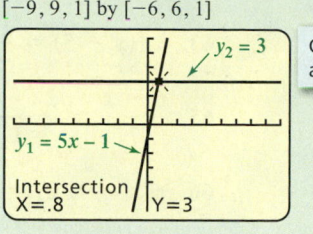

$y_2 = 3$

Graphs intersect at $(0.8, 3)$.

$y_1 = 5x - 1$

Intersection
X=.8 Y=3

The solution is **0.8**.

Numerical Solution

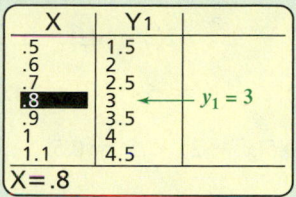

In the table, $y_1 = 3$ when $x = 0.8$.

2.2 Exercises

Concepts about Linear Equations

1. How many solutions are there to $ax + b = 0$ with $a \neq 0$?

2. How many times does the graph of $y = ax + b$ with $a \neq 0$ intersect the x-axis?

3. Apply the distributive property to $4 - (5 - 4x)$.

4. What property is used to solve $15x = 5$?

5. If $f(x) = ax + b$ with $a \neq 0$, how are the zero of f and the x-intercept of the graph of f related?

6. Distinguish between a contradiction and an identity.

Identifying Linear and Nonlinear Equations

Exercises 7–12: Determine whether the equation is linear or nonlinear by trying to write it in the form $ax + b = 0$.

7. $3x - 1.5 = 7$

8. $100 - 23x = 20x$

9. $2\sqrt{x} + 2 = 1$

10. $4x^3 - 7 = 0$

11. $7x - 5 = 3(x - 8)$

12. $2(x - 3) = 4 - 5x$

Solving Linear Equations Symbolically

Exercises 13–38: Solve the equation and check your answer.

13. $2x - 8 = 0$

14. $4x - 8 = 0$

15. $-5x + 3 = 23$

16. $-9x - 3 = 24$

17. $4(z - 8) = z$

18. $-3(2z - 1) = 2z$

19. $-5(3 - 4t) = 65$

20. $6(5 - 3t) = 66$

21. $k + 8 = 5k - 4$

22. $2k - 3 = k + 3$

23. $2(1 - 3x) + 1 = 3x$

24. $5(x - 2) = -2(1 - x)$

25. $-5(3 - 2x) - (1 - x) = 4(x - 3)$

26. $-3(5 - x) - (x - 2) = 7x - 2$

27. $-4(5x - 1) = 8 - (x + 2)$

28. $6(3 - 2x) = 1 - (2x - 1)$

29. $\frac{2}{7}n + 2 = \frac{4}{7}$

30. $\frac{6}{11} - \frac{2}{33}n = \frac{5}{11}n$

31. $\frac{1}{2}(d - 3) - \frac{2}{3}(2d - 5) = \frac{5}{12}$

32. $\frac{7}{3}(2d - 1) - \frac{2}{5}(4 - 3d) = \frac{1}{5}d$

33. $\dfrac{x - 5}{3} + \dfrac{3 - 2x}{2} = \dfrac{5}{4} - 2(1 - x)$

34. $\dfrac{3x - 1}{5} - 2 = \dfrac{2 - x}{3}$

35. $0.1z - 0.05 = -0.07z$

36. $1.1z - 2.5 = 0.3(z - 2)$

37. $0.15t + 0.85(100 - t) = 0.45(100)$

38. $0.35t + 0.65(10 - t) = 0.55(10)$

Exercises 39–48: Complete the following.

(a) *Solve the equation symbolically.*

(b) *Classify the equation as a contradiction, an identity, or a conditional equation.*

39. $5x - 1 = 5x + 4$

40. $7 - 9z = 2(3 - 4z) - z$

41. $3(x - 1) = 5$ **42.** $22 = -2(2x + 1.4)$

43. $0.5(x - 2) + 5 = 0.5x + 4$

44. $\frac{1}{2}x - 2(x - 1) = -\frac{3}{2}x + 2$

45. $\frac{t + 1}{2} = \frac{3t - 2}{6}$ **46.** $\frac{2x + 1}{3} = \frac{2x - 1}{3}$

47. $\frac{1 - 2x}{4} = \frac{3x - 1.5}{-6}$

48. $0.5(3x - 1) + 0.5x = 2x - 0.5$

Solving Linear Equations Graphically

Exercises 49 and 50: A linear equation is solved by using the intersection-of-graphs method. Find the solution by interpreting the graph. Assume that the solution is an integer.

49.
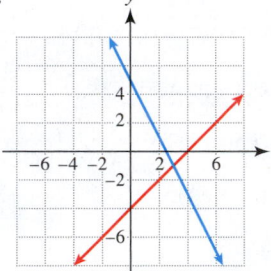

50.
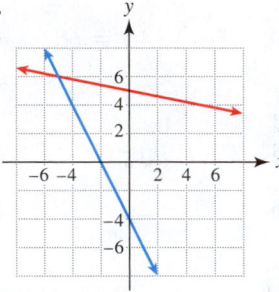

Exercises 51 and 52: Use the graph of $y = f(x)$ to solve each equation.

(a) $f(x) = -1$ (b) $f(x) = 0$ (c) $f(x) = 2$

51.
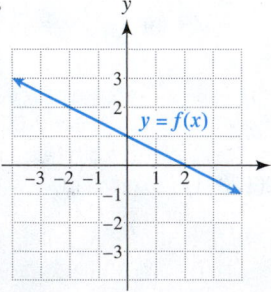

52.
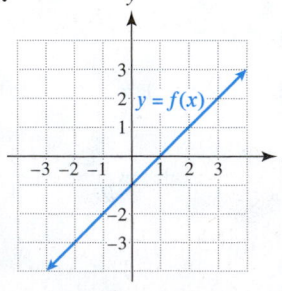

Exercises 53–58: Use the intersection-of-graphs method to solve the equation. Then solve symbolically.

53. $x + 4 = 1 - 2x$ **54.** $2x = 3x - 1$

55. $-x + 4 = 3x$ **56.** $1 - 2x = x + 4$

57. $2(x - 1) - 2 = x$ **58.** $-(x + 1) - 2 = 2x$

Exercises 59–66: Solve the linear equation with the intersection-of-graphs method. Approximate the solution to the nearest thousandth whenever appropriate.

59. $5x - 1.5 = 5$ **60.** $8 - 2x = 1.6$

61. $3x - 1.7 = 1 - x$ **62.** $\sqrt{2}x = 4x - 6$

63. $3.1(x - 5) = \frac{1}{5}x - 5$ **64.** $65 = 8(x - 6) - 5.5$

65. $\frac{6 - x}{7} = \frac{2x - 3}{3}$

66. $\pi(x - \sqrt{2}) = 1.07x - 6.1$

Solving Linear Equations Numerically

Exercises 67–74: Use tables to solve the equation numerically to the nearest tenth.

67. $2x - 7 = -1$ **68.** $1 - 6x = 7$

69. $2x - 7.2 = 10$ **70.** $5.8x - 8.7 = 0$

71. $\sqrt{2}(4x - 1) + \pi x = 0$

72. $\pi(0.3x - 2) + \sqrt{2}x = 0$

73. $0.5 - 0.1(\sqrt{2} - 3x) = 0$

74. $\sqrt{5} - \pi(\pi + 0.3x) = 0$

Solving Linear Equations by More Than One Method

Exercises 75–82: Solve the equation (to the nearest tenth)

(a) *symbolically,*

(b) *graphically, and*

(c) *numerically.*

75. $5 - (x + 1) = 3$ **76.** $7 - (3 - 2x) = 1$

77. $x - 3 = 2x + 1$ **78.** $3(x - 1) = 2x - 1$

79. $\sqrt{3}(2 - \pi x) + x = 0$

80. $3(\pi - x) + \sqrt{2} = 0$

81. $6x - 8 = -7x + 18$

82. $5 - 8x = 3(x - 7) + 37$

Solving for a Variable

Exercises 83–90: Solve the equation for the specified variable.

83. $A = LW$ for W

84. $E = IR + 2$ for R

85. $P = 2L + 2W$ for L

86. $V = 2\pi rh + \pi r^2$ for h

87. $3x + 2y = 8$ for y

88. $5x - 4y = 20$ for y

89. $y = 3(x - 2) + x$ for x

90. $y = 4 - (8 - 2x)$ for x

Exercises 91–96: The equation of a line is written in standard form.

(a) *Solve the equation for y.*

(b) *Write a formula f(x) for a function whose graph is the given line.*

91. $2x + y = 8$ **92.** $3x - y = 5$

93. $2x - 4y = -1$ **94.** $7x + 3y = -4$

95. $-9x + 8y = 9$ **96.** $-2x - 6y = 3$

Applications

97. Income The per capita (per person) income from 1980 to 2010 can be modeled by

$$f(x) = 1000(x - 1980) + 10,000,$$

where x is the year. Determine the year when the per capita income was \$19,000. (*Source:* Bureau of the Census.)

98. Tuition and Fees Tuition and fees during year x at private colleges can be modeled by

$$f(x) = 755.7(x - 1985) + 6121.$$

Use $f(x)$ to determine when tuition and fees might reach \$28,000.

99. Vinyl and CD Sales During the 1980s, sales of compact discs surpassed vinyl record sales. From 1985 to 1990, sales of compact discs in millions can be modeled by the formula $f(x) = 51.6(x - 1985) + 9.1$, whereas sales of vinyl LP records in millions can be modeled by $g(x) = -31.9(x - 1985) + 167.7$. Approximate the year x when sales of LP records and compact discs were equal by using the intersection-of-graphs method. (*Source:* Recording Industry Association of America.)

100. Median Age The median age A in the United States during year x, where $2000 \leq x \leq 2050$, is projected to be

$$A(x) = 0.07(x - 2000) + 35.3.$$

Use $A(x)$ to estimate when the median age may reach 37 years. (*Source:* Bureau of the Census.)

101. Legalize of Not? The attitudes toward legalization of marijuana changed between 1995 and 2012. The formula $A(x) = -1.6x + 3265$ gives the percentage of people in year x who were against legalization, and the formula $F(x) = 1.5x - 2968$ gives the percentage of people in year x who were in favor of legalization. (*Source:* Gallup Poll.)

(a) Interpret the slope of the graphs of A and F.

(b) Estimate the year when the percentages for and against were equal.

102. Population Density In 2000 the population density of the United States was 80 people per square mile, and in 2011 it was 88 people per square mile. Use a linear function to estimate when the U.S. population density will reach 91 people per square mile.

103. Sale Price A store is discounting all regularly priced merchandise by 25%. Find a function f that computes the sale price of an item having a regular price of x. If an item normally costs \$56.24, what is its sale price?

104. Sale Price Continuing Exercise 103, use f to find the regular price of an item that costs \$19.62 on sale.

105. Skin Cancer Approximately 4.8% of all cancer cases diagnosed in 2011 were skin cancer. (*Source:* American Cancer Society.)

(a) If x cases of cancer were diagnosed, how many of these were skin cancer?

(b) There were 76,000 cases of skin cancer diagnosed in 2011. Find the total number of cancer cases in 2011.

106. Grades In order to receive an A in a college course it is necessary to obtain an average of 90% correct on three 1-hour exams of 100 points each and on one final exam of 200 points. If a student scores 82, 88, and 91 on the 1-hour exams, what is the minimum score that the person can receive on the final exam and still earn an A?

107. Working Together Suppose that a lawn can be raked by one gardener in 3 hours and by a second gardener in 5 hours.

(a) Mentally estimate how long it will take the two gardeners to rake the lawn working together.

(b) Solve part (a) symbolically.

108. Pumping Water Suppose that a large pump can empty a swimming pool in 50 hours and a small pump can empty the pool in 80 hours. How long will it take to empty the pool if both pumps are used?

109. Motion A car went 372 miles in 6 hours, traveling part of the time at 55 miles per hour and part of the time at 70 miles per hour. How long did the car travel at each speed?

110. Mixing Candy Two types of candy sell for \$2.50 per pound and \$4.00 per pound. A store clerk is trying to make a 5-pound mixture worth \$17.60. How much of each type of candy should be included in the mixture?

111. **Running** At 2:00 P.M. a runner heads north on a highway, jogging at 10 miles per hour. At 2:30 P.M. a driver heads north on the same highway to pick up the runner. If the car travels at 55 miles per hour, how long will it take the driver to catch the runner?

112. **Investments** A total of $5000 was invested in two accounts. One pays 5% annual interest, and the second pays 7% annual interest. If the first-year interest is $325, how much was invested in each account?

113. **Height of a Tree** In the accompanying figure, a person 5 feet tall casts a shadow 4 feet long. A nearby tree casts a shadow 33 feet long. Find the height of the tree by solving a linear equation.

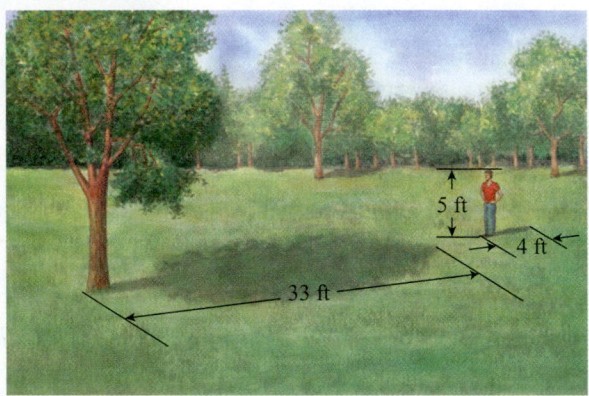

114. **Shadow Length** A person 66 inches tall is standing 15 feet from a streetlight. If the person casts a shadow 84 inches long, how tall is the streetlight?

115. **Conical Water Tank** A water tank in the shape of an inverted cone has a height of 11 feet and a radius of 3.5 feet, as illustrated in the figure. If the volume of the cone is $V = \frac{1}{3}\pi r^2 h$, find the volume of the water in the tank when the water is 7 feet deep. (*Hint:* Consider using similar triangles.)

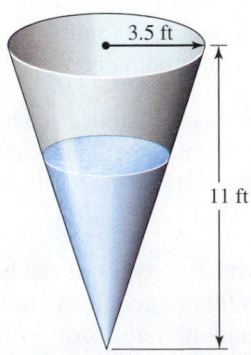

116. **Dimension of a Cone** (Refer to Exercise 115.) A conical water tank holds 100 cubic feet of water and has a diameter of 6 feet. Estimate its height to the nearest tenth of a foot.

117. **Chemistry** Determine how much pure water should be mixed with 5 liters of a 40% solution of sulfuric acid to make a 15% solution of sulfuric acid.

118. **Mixing Antifreeze** A radiator holds 5 gallons of fluid. If it is full with a 15% solution, how much fluid should be drained and replaced with a 65% antifreeze mixture to result in a 40% antifreeze mixture?

119. **Window Dimensions** A rectangular window has a length that is 18 inches more than its width. If its perimeter is 180 inches, find its dimensions.

120. **Sales of CRT and LCD Screens** In 2002, 75 million CRT (cathode ray tube) monitors were sold and 29 million flat LCD (liquid crystal display) monitors were sold. In 2006 the numbers were 45 million for CRT monitors and 88 million for LCD monitors. (*Source:* International Data Corporation.)
    (a) Find a linear function C that models these data for CRT monitors and another linear function L that models these data for LCD monitors. Let x be the year.

    (b) Interpret the slopes of the graphs of C and of L.

    (c) Determine graphically the year when sales of these two types of monitors were equal.

    (d) Solve part (c) symbolically.

    (e) Solve part (c) numerically.

121. **Online Shopping** In 2011 online sales were $192 billion, and in 2014 they are predicted to be $249 billion. (*Source:* Forrestor Forecast.)
    (a) Find a linear function S that models these data. Write $S(x)$ in slope-intercept form.

    (b) Interpret the slope of the graph of S.

    (c) Determine when online sales were $230 billion.

122. **Geometry** A 174-foot-long fence is being placed around the perimeter of a rectangular swimming pool that has a 3-foot-wide sidewalk around it. The actual swimming pool without the sidewalk is twice as long as it is wide. Find the dimensions of the pool without the sidewalk.

123. **Temperature Scales** The Celsius and Fahrenheit scales are related by the equation $C = \frac{5}{9}(F - 32)$. These scales have the same temperature reading at a unique value where $F = C$. Find this temperature.

124. Business A company manufactures compact discs with recorded music. The master disc costs $2000 to produce and copies cost $0.45 each. If a company spent $2990 producing compact discs, how many copies did the company manufacture?

125. Two-Cycle Engines Two-cycle engines, used in snowmobiles, chain saws, and outboard motors, require a mixture of gasoline and oil. For certain engines the amount of oil in pints that should be added to x gallons of gasoline is computed by $f(x) = 0.16x$. (**Source:** Johnson Outboard Motor Company.)

(a) Why is it reasonable to expect f to be linear?

(b) Evaluate $f(3)$ and interpret the answer.

(c) How much gasoline should be mixed with 2 pints of oil?

126. Perimeter Find the length of the longest side of the rectangle if its perimeter is 25 feet.

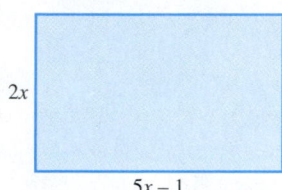

2x

5x − 1

Modeling Data with Linear Functions

Exercises 127 and 128: The following data can be modeled by a linear function. Estimate the value of x when y = 2.99.

127.

x	2	4	6	8
y	0.51	1.23	1.95	2.67

128.

x	1	2	3	4
y	−1.66	2.06	5.78	9.50

Writing about Mathematics

129. Describe a basic graphical method used to solve a linear equation. Give an example.

130. Describe verbally how to solve $ax + b = 0$. What assumptions have you made about the value of a?

Extended and Discovery Exercises

1. Geometry Suppose that two rectangles are similar and the sides of the first rectangle are twice as long as the corresponding sides of the second rectangle.

(a) Is the perimeter of the first rectangle twice the perimeter of the second rectangle? Explain.

(b) Is the area of the first rectangle twice the area of the second rectangle? Explain.

2. Geometry Repeat the previous exercise for an equilateral triangle. Try to make a generalization. (*Hint:* The area of an equilateral triangle is $A = \frac{\sqrt{3}}{4}x^2$, where x is the length of a side.) What will happen to the circumference of a circle if the radius is doubled? What will happen to its area?

3. Indoor Air Pollution Formaldehyde is an indoor air pollutant formerly found in plywood, foam insulation, and carpeting. When concentrations in the air reach 33 micrograms per cubic foot ($\mu g/ft^3$), eye irritation can occur. One square foot of new plywood could emit 140 μg per hour. (*Source:* A. Hines, *Indoor Air Quality & Control.*)

(a) A room has 100 square feet of new plywood flooring. Find a linear function f that computes the amount of formaldehyde in micrograms that could be emitted in x hours.

(b) The room contains 800 cubic feet of air and has no ventilation. Determine how long it would take for concentrations to reach 33 $\mu g/ft^3$.

4. Temperature and Volume The table shows the relationship between the temperature of a sample of helium and its volume.

Temperature (°C)	0	25	50	75	100
Volume (in³)	30	32.7	35.4	38.1	40.8

(a) Make a scatterplot of the data.

(b) Write a formula for a function f that receives the temperature x as input and outputs the volume y of the helium.

(c) Find the volume when the temperature is 65°C.

(d) Find the temperature if the volume is 25 cubic inches. Did your answer involve interpolation or extrapolation? Do you believe that your answer is accurate?

CHECKING BASIC CONCEPTS FOR SECTIONS 2.1 AND 2.2

1. Find an equation of the line passing through the points $(-3, 4)$ and $(5, -2)$. Give equations of lines that are parallel and perpendicular to this line.

2. Find equations of horizontal and vertical lines that pass through the point $(-4, 7)$.

3. Write the slope-intercept form of the line.

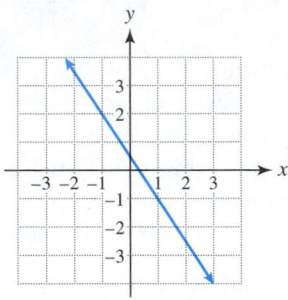

4. Find the x- and y-intercepts of the graph of the equation $-3x + 2y = -18$.

5. Solve $5x + 2 = 2x - 3$ and check your answer.

6. Solve each equation.
   (a) $\frac{3x}{4} + 2 = \frac{2x}{3}$

   (b) $-4.1x + 5.45 = 1.05(5 - 2x)$

7. Solve the linear equation $4(x - 2) = 2(5 - x) - 3$ by using each method. Compare your results.
   (a) Graphical

   (b) Numerical

   (c) Symbolic

2.3   Linear Inequalities

- **Understand basic terminology related to inequalities**
- **Solve linear inequalities symbolically**
- **Solve linear inequalities graphically and numerically**
- **Solve compound inequalities**

Introduction

Twitter cofounder Jack Dorsey has started another company called Square. Square is a payment startup that allows businesses to swipe credit cards on their iPhones and Android devices. During a 12-month period from March 2011 to March 2012, Square's *daily* payment volume grew from \$1 million to \$11 million. In Example 3 we make estimations about this growth by using linear inqualities. (***Source:*** Business Insider.)

Inequalities

Inequalities result whenever the equals sign in an equation is replaced with any one of the symbols $<$, $\leq$, $>$, or $\geq$.

Inequalities with Zero, One, and Two Variables

$$x + 15 < 9x - 1, \qquad x^2 - 2x + 1 \geq 2x, \qquad z + 5 > 0,$$
$$xy + x^2 \leq y^3 + x, \qquad \text{and} \qquad 2 + 3 > 1$$

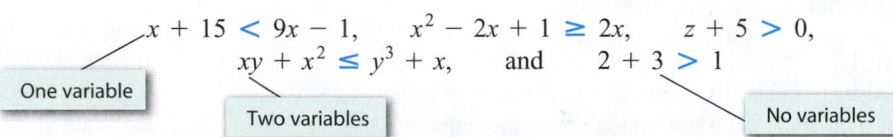

One variable

Two variables

No variables

We will concentrate on inequalities with one variable.

To **solve** an inequality means to find all values for the variable that make the inequality a true statement. Such values are **solutions**, and the set of all solutions is the **solution set** to the inequality. Two inequalities are **equivalent** if they have the same solution set. It is common for an inequality to have infinitely many solutions. For instance, the inequality $x - 1 > 0$ has infinitely many solutions because any real number x satisfying $x > 1$ is a solution. The solution set is $\{x \mid x > 1\}$.

Like functions and equations, inequalities in one variable can be classified as *linear* or *nonlinear*.

> **LINEAR INEQUALITY IN ONE VARIABLE**
>
> A **linear inequality** in one variable is an inequality that can be written in the form
>
> $$ax + b > 0,$$
>
> where $a \neq 0$. (The symbol $>$ may be replaced by $\geq$, $<$, or $\leq$.)

Linear Inequalities

$$3x - 4 < 0, \qquad 7x + 5 \geq x, \qquad x + 6 > 23, \qquad \text{and} \qquad 7x + 2 \leq -3x + 6.$$

Using techniques from algebra, we can transform these inequalities into one of the forms $ax + b > 0$, $ax + b \geq 0$, $ax + b < 0$, or $ax + b \leq 0$. For example, by subtracting x from each side of $7x + 5 \geq x$, we obtain the equivalent inequality $6x + 5 \geq 0$. If an inequality is not a linear inequality, it is called a **nonlinear inequality**.

> **PROPERTIES OF INEQUALITIES**
>
> Let a, b, and c be real numbers.
>
> 1. $a < b$ and $a + c < b + c$ are equivalent.
>   (The same number may be added to or subtracted from each side of an inequality.)
> 2. If $c > 0$, then $a < b$ and $ac < bc$ are equivalent.
>   (Each side of an inequality may be multiplied or divided by the same *positive* number.)
> 3. If $c < 0$, then $a < b$ and $ac > bc$ are equivalent.
>   (Each side of an inequality may be multiplied or divided by the same *negative* number provided the inequality symbol is *reversed*.)
>
> Replacing $<$ with $\leq$ and $>$ with $\geq$ results in similar properties.

The following examples illustrate each property.

Property 1: To solve $x - 5 < 6$, add 5 to each side to obtain $x < 11$.

Property 2: To solve $5x < 10$, divide each side by 5 to obtain $x < 2$.

Property 3: To solve $-5x < 10$, divide each side by -5 to obtain $x > -2$. (Whenever you multiply or divide an inequality by a *negative* number, *reverse* the inequality symbol.)

> **MAKING CONNECTIONS**
>
> **Linear Functions, Equations, and Inequalities**   These concepts are closely related.
>
> $$f(x) = \boldsymbol{ax + b} \qquad \text{Linear function } (a = m)$$
> $$\boldsymbol{ax + b} = 0, a \neq 0 \qquad \text{Linear equation}$$
> $$\boldsymbol{ax + b} > 0, a \neq 0 \qquad \text{Linear inequality}$$

Review of Interval Notation   In Section 1.4 interval notation was introduced as an efficient way to express intervals on the real number line. For example, the interval $3 \leq x \leq 5$ is written as $[3, 5]$, whereas the interval $3 < x < 5$ is written as $(3, 5)$. A bracket, $[$ or $]$, is used when an endpoint is included, and a parenthesis, $($ or $)$, is used when an endpoint is not included. The interval $x \geq 2$ is written as $[2, \infty)$, where ∞ denotes infinity, and the interval $x < 2$ is written as $(-\infty, 2)$

Symbolic Solutions   In the next example we solve linear inequalities symbolically and express the solution set in both set-builder and interval notation.

EXAMPLE 1   **Solving linear inequalities symbolically**

Solve each inequality. Write the solution set in set-builder and interval notation.

(a) $2x - 3 < \dfrac{x + 2}{-3}$

(b) $-3(4z - 4) \geq 4 - (z - 1)$

SOLUTION

(a) Use Property 3 by multiplying each side by -3 to clear fractions.

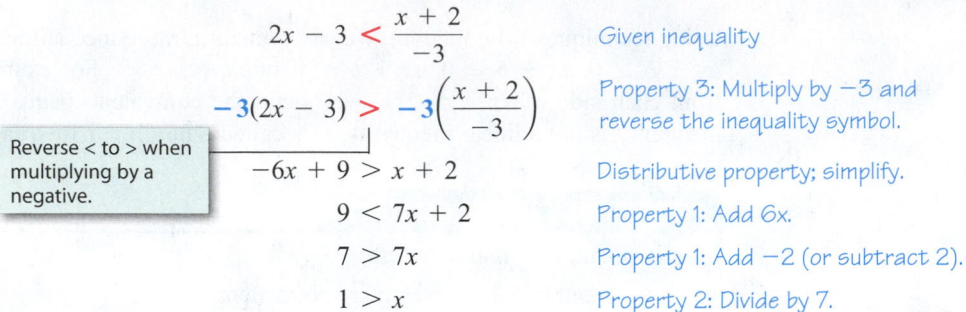

$2x - 3 < \dfrac{x + 2}{-3}$	*Given inequality*
$-3(2x - 3) > -3\left(\dfrac{x + 2}{-3}\right)$	*Property 3: Multiply by -3 and reverse the inequality symbol.*
$-6x + 9 > x + 2$	*Distributive property; simplify.*
$9 < 7x + 2$	*Property 1: Add 6x.*
$7 > 7x$	*Property 1: Add -2 (or subtract 2).*
$1 > x$	*Property 2: Divide by 7.*

Reverse < to > when multiplying by a negative.

The solution set is $\{x \mid x < 1\}$, and in interval notation it is written as $(-\infty, 1)$.

(b) Begin by applying the distributive property.

$-3(4z - 4) \geq 4 - (z - 1)$	*Given inequality*
$-12z + 12 \geq 4 - z + 1$	*Distributive property*
$-12z + 12 \geq -z + 5$	*Simplify.*
$-12z + z \geq 5 - 12$	*Property 1: Add z and -12.*
$-11z \geq -7$	*Simplify.*
$z \leq \dfrac{7}{11}$	*Property 3: Divide by -11 and reverse inequality symbol.*

The solution set is $\left\{z \mid z \leq \frac{7}{11}\right\}$, and in interval notation it is written as $\left(-\infty, \frac{7}{11}\right]$.

Now Try Exercises 15 and 17

Graphical Solutions   The intersections-of-graphs method can be extended to solve inequalities. The following box shows the actual and projected percentages of college degrees conferred to females F and to males M.   (***Source:*** Bureau of Labor Statistics.)

See the Concept: Solving an Inequality Graphically

College Degrees Conferred, 1975–2017 (Female versus Male)

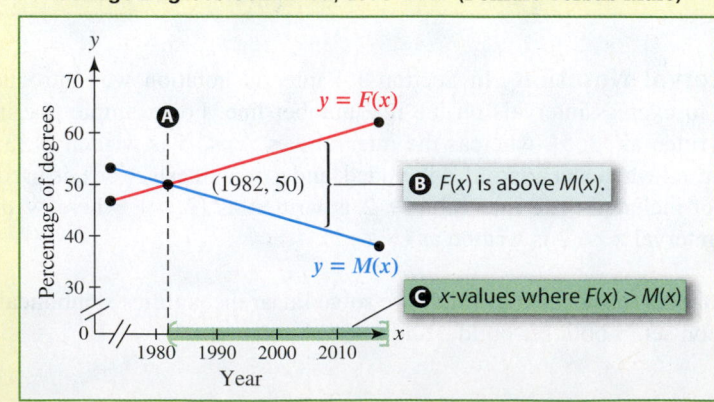

Ⓐ In 1982 half, or 50%, of all college degrees conferred were given to females.

Ⓑ There are more female graduates after 1982.

Ⓒ $F(x) > M(x)$ when $1982 < x \leq 2017$.

The preceding box illustrates how we can solve the inequality $F(x) > M(x)$ graphically. This graphical technique is used in the next example.

EXAMPLE 2   Solving a linear inequality graphically

Graph $y_1 = \frac{1}{2}x + 2$ and $y_2 = 2x - 1$ by hand. Use the graph to solve the linear inequality $\frac{1}{2}x + 2 > 2x - 1$.

SOLUTION   The graphs of $y_1 = \frac{1}{2}x + 2$ and $y_2 = 2x - 1$ are shown in Figure 2.33. The graphs intersect at the point $(2, 3)$. The graph of $y_1 = \frac{1}{2}x + 2$ is above the graph of $y_2 = 2x - 1$ to the left of the point of intersection, or when $x < 2$. Thus the solution set to the inequality $\frac{1}{2}x + 2 > 2x - 1$ is $\{x \mid x < 2\}$, or $(-\infty, 2)$.

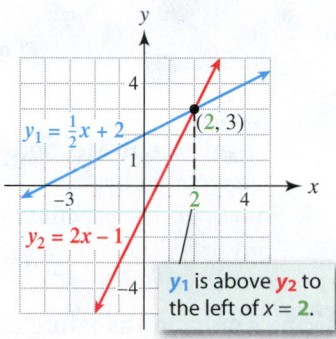

Figure 2.33

Now Try Exercise 39

An Application   In the next example we use graphical and symbolic techniques to analyze the growth of Square, which is discussed in the introduction to this section.

EXAMPLE 3   Analyzing the growth of Square

The *daily* payment processing volume for the company Square grew from \$1 million in March 2011 to \$11 million in March 2012. Its growth was approximately linear.
(a) Write a formula $S(x)$ for a linear function that models this growth in millions of dollars, where x represents months after March 2011. Assume that $0 \leq x \leq 12$.
(b) Determine symbolically and graphically when Square's daily volume was \$8.5 million or less.

SOLUTION
(a) The graph of S must pass through the point $(0, 1)$ and $(12, 11)$. The slope of this graph is

$$m = \frac{11 - 1}{12 - 0} = \frac{10}{12} = \frac{5}{6}.$$

The graph passes through $(0, 1)$, so the y-intercept is **1**. Thus $S(x) = \frac{5}{6}x + 1$.

(b) *Symbolic Solution*   We must determine when $S(x) \leq 8.5$.

$$\frac{5}{6}x + 1 \leq 8.5 \qquad \text{Substitute for } S(x).$$

$$\frac{5}{6}x \leq 7.5 \qquad \text{Subtract 1 from each side.}$$

$$x \leq 9 \qquad \text{Multiply each side by } \frac{6}{5}, \text{ or 1.2.}$$

The daily payment processing volume for Square was \$8.5 million or less from March 2011 to December 2011. (Note that $x = 0$ corresponds to March 2011.)

Graphical Solution   A graphical solution to $\frac{5}{6}x + 1 \le 8.5$ is shown in Figure 2.34.

Daily Volume for Square

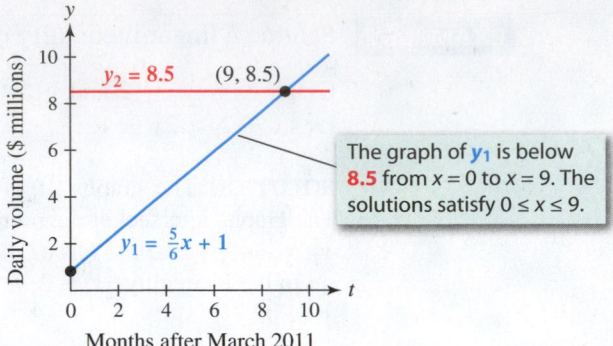

Months after March 2011

Figure 2.34

Now Try Exercise 89

x-Intercept Method   If a linear inequality can be written as $y_1 > 0$, where $>$ may be replaced by $\ge$, $\le$, or $<$, then we can solve this inequality by using the ***x*-intercept method**. To apply this method for $y_1 > 0$, graph y_1 and find the *x*-intercept. The solution set includes *x*-values where the graph of y_1 is **above** the *x*-axis.

EXAMPLE 4   **Applying the *x*-intercept method**

Solve the inequality $1 - x > \frac{1}{2}x - 2$ by using the *x*-intercept method. Write the solution set in set-builder and interval notation.

SOLUTION

Graphing by Hand   First, use properties of inequalities to rewrite the given inequality so that 0 is on the right side.

$$1 - x > \frac{1}{2}x - 2 \qquad \text{Given inequality}$$

$$1 - x - \frac{1}{2}x + 2 > 0 \qquad \text{Subtract } \tfrac{1}{2}x \text{ and add 2.}$$

$$3 - \frac{3}{2}x > 0 \qquad \text{Simplify.}$$

Next, graph $y_1 = 3 - \frac{3}{2}x$ and locate the *x*-intercept **2**. From Figure 2.35(a), $y_1 > 0$ when $x < $ **2** and the solution set is $\{x \mid x < $ **2**$\}$, or $(-\infty, $ **2**$)$.

Graphing with a Calculator   With a calculator graph $y_1 = 1 - x - \frac{1}{2}x + 2$ without simplifying further, as shown in Figure 2.35(b). The calculator can locate the zero, or *x*-intercept, of 2.

Calculator Help

To locate a zero or *x*-intercept on the graph of a function, see Appendix A (page AP-7).

x-Intercept Method

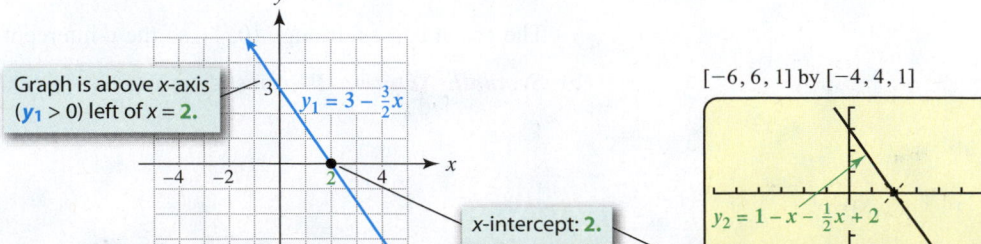

Figure 2.35(a)

Figure 2.35(b)

Now Try Exercise 51

MAKING CONNECTIONS

Equations and Inequalities   In mathematics a lot of time is spent solving equations and determining equality. One reason is that equality is frequently a boundary between *greater than* and *less than*. The solution set to an inequality often can be found by first locating where two expressions are equal. Since equality and inequality are closely connected, many of the techniques used to solve equations can also be applied to inequalities.

Visualizing Solutions   We can visualize the solution set to the linear inequality $ax + b > 0$ with $a \neq 0$, as shown in the following See the Concept. (The inequality $ax + b < 0$ can be solved similarly.)

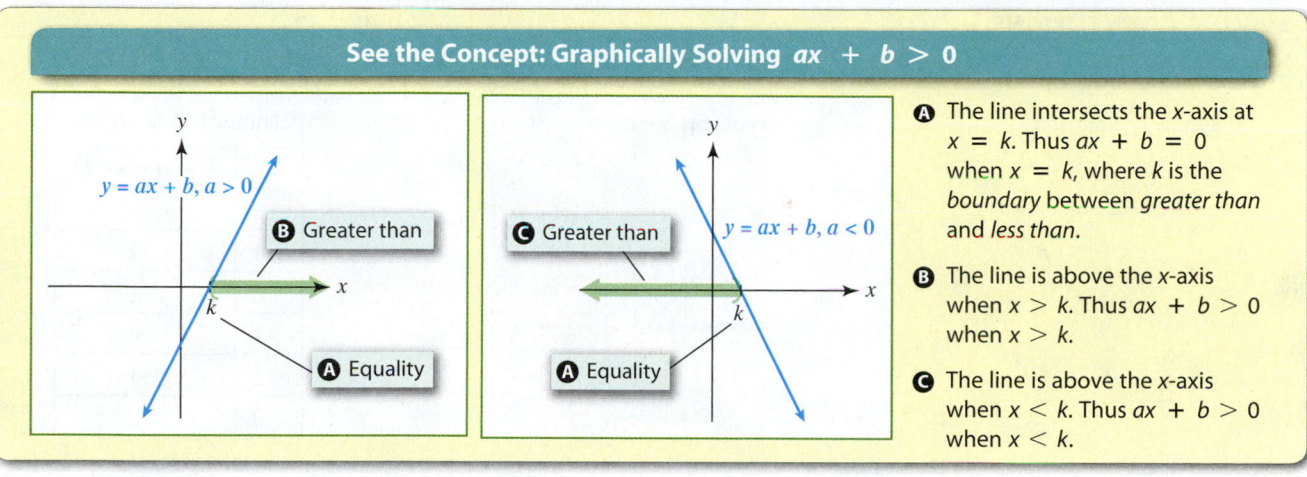

See the Concept: Graphically Solving $ax + b > 0$

Ⓐ The line intersects the x-axis at $x = k$. Thus $ax + b = 0$ when $x = k$, where k is the *boundary* between *greater than* and *less than*.

Ⓑ The line is above the x-axis when $x > k$. Thus $ax + b > 0$ when $x > k$.

Ⓒ The line is above the x-axis when $x < k$. Thus $ax + b > 0$ when $x < k$.

Numerical Solutions   Suppose that it costs a company $5x + 200$ dollars to produce x earbuds and the company receives $15x$ dollars for selling x earbuds. Then the profit P from selling x earbuds is $P = 15x - (5x + 200) = 10x - 200$. A value of $x = 20$ results in $P = 0$, and so $x = 20$ is called the **boundary number** because selling 20 earbuds represents the boundary between making money and losing money (the break-even point). To make money, the profit P must be positive, and the inequality

$$10x - 200 > 0$$

must be satisfied. The table of values for $y_1 = 10x - 200$ in Table 2.3 shows the boundary number $x = 20$ along with several **test values**. The test values of $x = 17$, 18, and 19 result in a loss. The test values of $x = 21$, 22, and 23 result in a profit. Therefore the solution set to $10x - 200 > 0$ is $\{x \mid x > 20\}$.

Profit from Sales

x	17	18	19	**20**	21	22	23
$10x - 200$	−30	−20	−10	**0**	10	20	30

Table 2.3   Less than 0   Greater than 0
                                      Boundary number

EXAMPLE 5   **Solving a linear inequality with test values**

Solve $3(6 - x) + 5 - 2x < 0$ numerically.

SOLUTION Make a table of $y_1 = 3(6 - x) + 5 - 2x$ as shown in Figure 2.36 on the next page. The boundary number for this inequality lies between $x = 4$ and $x = 5$. Changing the increment from 1 to 0.1 in Figure 2.37 shows that the boundary number for the inequality is $x = 4.6$. The test values of $x = 4.7, 4.8$, and 4.9 indicate that when $x > 4.6$, the inequality $y_1 < 0$ is true. The solution set to $3(6 - x) + 5 - 2x < 0$ is $\{x \mid x > 4.6\}$.

Incrementing x by 1

X	Y1	
1	18	Greater
2	13	than 0
3	8	
4	3	
5	−2	Less
6	−7	than 0
7	−12	

X=4

Figure 2.36

Incrementing x by 0.1

X	Y1	
4.3	1.5	Greater
4.4	1	than 0
4.5	.5	
4.6	0	
4.7	−.5	Less
4.8	−1	than 0
4.9	−1.5	

X=4.6

Figure 2.37

Now Try Exercise 71

MAKING CONNECTIONS

Symbolic, Graphical, and Numerical Solutions Each method is used to solve the inequality $3 - (x + 2) > 0$.

Symbolic Solution

$$3 - (x + 2) > 0$$
$$-(x + 2) > -3$$
$$x + 2 < 3$$
$$x < 1$$

Graphical Solution

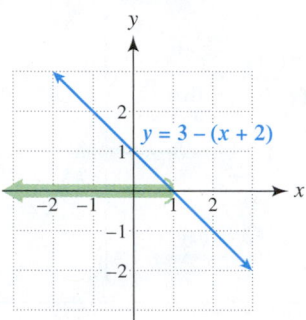

The graph of $y = 3 - (x + 2)$ is above the x-axis when **$x < 1$**.

Numerical Solution

x	$3 - (x + 2)$	
−2	3	
−1	2	Greater than 0
0	1	
1	0	Equals 0
2	−1	Less than 0
3	−2	

The values of $3 - (x + 2)$ are greater than 0 when **$x < 1$**.

Compound Inequalities

Sometimes a variable must satisfy two inequalities. For example, on a freeway there may be a minimum speed limit of 40 miles per hour and a maximum speed limit of 70 miles per hour. If x represents the speed of a vehicle, then x must satisfy the compound inequality

$$x \geq 40 \quad \text{and} \quad x \leq 70.$$

A **compound inequality** occurs when two inequalities are connected by the word *and* or *or*. When the word *and* connects two inequalities, the two inequalities can sometimes be written as a **three-part inequality**. For example, the previous compound inequality may be written as the three-part inequality

$$40 \leq x \leq 70.$$

Compound inequalities involving the word *or* are discussed in the next section.

EXAMPLE 6 Solving a three-part inequality symbolically

Solve the inequality. Write the solution set in set-builder and interval notation.

(a) $-4 \leq 5x + 1 < 21$ **(b)** $\dfrac{1}{2} < \dfrac{1 - 2t}{4} < 2$

SOLUTION

(a) Use properties of inequalities to simplify the three-part inequality.

$-4 \leq 5x + 1 < 21$	Given inequality
$-5 \leq 5x < 20$	Add −1 to each part.
$-1 \leq x < 4$	Divide each part by 5.

The solution set is $\{x \mid -1 \leq x < 4\}$, or $[-1, 4)$.

(b) Begin by multiplying each part by 4 to clear fractions.

$$\frac{1}{2} < \frac{1 - 2t}{4} < 2 \qquad \textcolor{blue}{\text{Given inequality}}$$

$$2 < 1 - 2t < 8 \qquad \textcolor{blue}{\text{Multiply each part by 4.}}$$

$$1 < -2t < 7 \qquad \textcolor{blue}{\text{Add } -1 \text{ to each part.}}$$

$$-\frac{1}{2} > t > -\frac{7}{2} \qquad \textcolor{blue}{\text{Divide by } -2; \text{ reverse inequalities.}}$$

$$-\frac{7}{2} < t < -\frac{1}{2} \qquad \textcolor{blue}{\text{Rewrite the inequality.}}$$

The solution set is $\left\{ t \mid -\frac{7}{2} < t < -\frac{1}{2} \right\}$, or $\left(-\frac{7}{2}, -\frac{1}{2} \right)$.

Now Try Exercises 23 and 35

NOTE In Example 6, it is correct to write a three-part inequality as either $-\frac{1}{2} > t > -\frac{7}{2}$ or $-\frac{7}{2} < t < -\frac{1}{2}$. However, we usually write the smaller number on the left side and the larger number on the right side.

Three-part inequalities occur in many applications and can often be solved symbolically and graphically. This is demonstrated in the next example.

EXAMPLE 7   Modeling sunset times

In Boston, on the 82nd day (March 22) the sun set at 7:00 P.M., and on the 136th day (May 15) the sun set at 8:00 P.M. Use a linear function S to estimate the days when the sun set between 7:15 P.M. and 7:45 P.M., inclusive. Do not consider any days of the year after May 15. (**Source:** R. Thomas, *The Old Farmer's 2012 Almanac.*)

SOLUTION

Getting Started First find a linear function S whose graph passes through the points $(82, 7)$ and $(136, 8)$. Then solve the compound inequality $7.25 \leq S(x) \leq 7.75$. Note that 7.25 hours past noon corresponds to 7:15 P.M. and 7.75 hours past noon corresponds to 7:45 P.M. ▶

Symbolic Solution   The slope of the line passing through $(82, 7)$ and $(136, 8)$ is given by $\frac{8 - 7}{136 - 82} = \frac{1}{54}$. The point-slope form of the line passing through $(\textcolor{green}{82}, \textcolor{red}{7})$ with slope $\frac{1}{54}$ is

$$S(x) = \frac{1}{54}(x - \textcolor{green}{82}) + \textcolor{red}{7}. \qquad \textcolor{blue}{\text{Point-slope form}}$$

Now solve the following compound inequality.

$$7.25 \leq \frac{1}{54}(x - 82) + 7 \leq 7.75 \qquad \textcolor{blue}{\text{Given inequality}}$$

$$0.25 \leq \frac{1}{54}(x - 82) \leq 0.75 \qquad \textcolor{blue}{\text{Subtract 7 from each part.}}$$

$$13.5 \leq x - 82 \leq 40.5 \qquad \textcolor{blue}{\text{Multiply each part by 54.}}$$

$$95.5 \leq x \leq 122.5 \qquad \textcolor{blue}{\text{Add 82 to each part.}}$$

If we round 95.5 and 122.5 up to 96 and 123, then this model predicts that the sun set between 7:15 P.M. and 7:45 P.M. from the 96th day (April 5) to the 123rd day (May 2). (Note that the actual days were April 5 and May 1.)

[80, 150, 10] by [6.5, 8.5, 1]

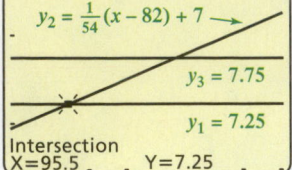

Figure 2.38

[80, 150, 10] by [6.5, 8.5, 1]

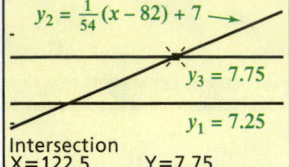

Figure 2.39

Graphical Solution   Graph $y_1 = 7.25$, $y_2 = \frac{1}{54}(x - 82) + 7$, and $y_3 = 7.75$ and determine their points of intersection, (**95.5**, 7.25) and (**122.5**, 7.75), as shown in Figures 2.38 and 2.39. The graph of y_2 is between the graphs of y_1 and y_3 for **95.5** $\leq x \leq$ **122.5**. This agrees with the symbolic solution. A different graph showing this solution appears in Figure 2.40.

Modeling Sunset Times in Boston

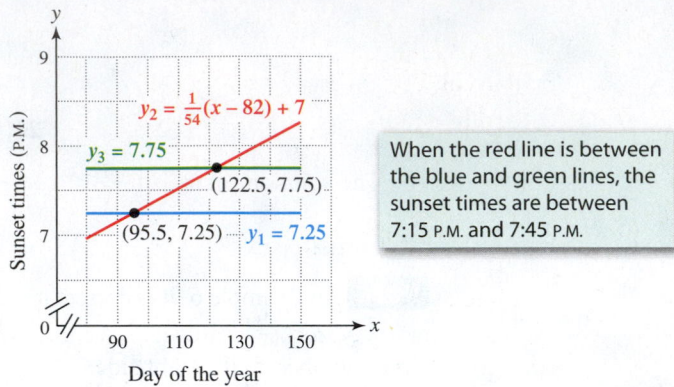

When the red line is between the blue and green lines, the sunset times are between 7:15 P.M. and 7:45 P.M.

Figure 2.40

Now Try Exercise 93

EXAMPLE 8   **Solving inequalities symbolically**

Solve the linear inequalities symbolically. Express the solution set using interval notation.

(a) $-\dfrac{x}{2} + 1 \leq 3$ **(b)** $-8 < \dfrac{3x - 1}{2} \leq 5$ **(c)** $5(x - 6) < 2x - 2(1 - x)$

SOLUTION
(a) Simplify the inequality as follows.

Reverse the inequality when multiplying by a negative.

$$-\frac{x}{2} + 1 \leq 3 \qquad \text{Given inequality}$$

$$-\frac{x}{2} \leq 2 \qquad \text{Add } -1, \text{ or subtract 1.}$$

$$x \geq -4 \qquad \text{Multiply by } -2.$$

In interval notation the solution set is $[-4, \infty)$.
(b) The parts of this compound inequality can be solved simultaneously.

$$-8 < \frac{3x - 1}{2} \leq 5 \qquad \text{Given inequality}$$

$$-16 < 3x - 1 \leq 10 \qquad \text{Multiply by 2.}$$

$$-15 < 3x \leq 11 \qquad \text{Add 1.}$$

$$-5 < x \leq \frac{11}{3} \qquad \text{Divide by 3.}$$

The solution set is $\left(-5, \frac{11}{3}\right]$.
(c) Start by applying the distributive property to each side of the inequality.

$$5(x - 6) < 2x - 2(1 - x) \qquad \text{Given inequality}$$

$$5x - 30 < 2x - 2 + 2x \qquad \text{Distributive property}$$

$$5x - 30 < 4x - 2 \qquad \text{Simplify.}$$

$$x - 30 < -2 \qquad \text{Subtract } 4x.$$

$$x < 28 \qquad \text{Add 30.}$$

The solution set is $(-\infty, 28)$.

Now Try Exercises 13, 27, and 33

2.3 Putting It All Together

$\mathbf{A}$ny linear inequality can be written as $ax + b > 0$ with $a \neq 0$, where $>$ can be replaced by $\geq$, $<$, or $\leq$. The following table includes methods for solving linear inequalities.

CONCEPT	EXPLANATION	EXAMPLES
Compound inequality	Two inequalities connected by the word *and* or *or*	$x \leq 4$ or $x \geq 10$ $x \geq -3$ and $x < 4$ $x > 5$ and $x \leq 20$ can be written as the three-part inequality $5 < x \leq 20$.
Symbolic method	Use properties of inequalities to simplify $f(x) > g(x)$ to either $x > k$ or $x < k$ for some real number k.	$\begin{aligned} \frac{1}{2}x + 1 &> 3 - \frac{3}{2}x & &\text{Given inequality}\\ 2x + 1 &> 3 & &\text{Add } \tfrac{3}{2}x.\\ 2x &> 2 & &\text{Subtract 1.}\\ x &> 1 & &\text{Divide by 2.} \end{aligned}$
Intersection-of-graphs method	To solve $f(x) > g(x)$, graph $y_1 = f(x)$ and $y_2 = g(x)$. Find the point of intersection. The solution set includes x-values where the graph of y_1 is above the graph of y_2.	$\frac{1}{2}x + 1 > 3 - \frac{3}{2}x$ Graph $y_1 = \frac{1}{2}x + 1$ and $y_2 = 3 - \frac{3}{2}x$. The solution set for $y_1 > y_2$ is $\{x \mid x > 1\}$.
The x-intercept method	Write the inequality as $h(x) > 0$. Graph $y_1 = h(x)$. Solutions occur where the graph is above the x-axis.	$\frac{1}{2}x + 1 > 3 - \frac{3}{2}x$ Graph $y_1 = \frac{1}{2}x + 1 - \left(3 - \frac{3}{2}x\right)$. The solution set for $y_1 > 0$ is $\{x \mid x > 1\}$.
Numerical method	Write the inequality as $h(x) > 0$. Create a table for $y_1 = h(x)$ and find the boundary number $x = k$ such that $h(k) = 0$. Use the test values in the table to determine if the solution set is $x > k$ or $x < k$.	$\frac{1}{2}x + 1 > 3 - \frac{3}{2}x$ Table $y_1 = \frac{1}{2}x + 1 - \left(3 - \frac{3}{2}x\right)$. The solution set for $y_1 > 0$ is $\{x \mid x > 1\}$. <table below>

x	-1	0	**1**	2	3
y_1	-4	-2	**0**	2	4

Less than 0   Greater than 0

2.3 Exercises

Review of Interval Notation

Exercises 1–8: Express the following in interval notation.

1. $x < 2$

2. $x > -3$

3. $x \geq -1$

4. $x \leq 7$

5. $\{x \mid 1 \leq x < 8\}$

6. $\{x \mid -2 < x \leq 4\}$

7. $\{x \mid x \leq 1\}$

8. $\{x \mid x > 5\}$

Solving Linear Inequalities Symbolically

Exercises 9–38: Solve the inequality symbolically. Express the solution set in set-builder or interval notation.

9. $2x + 6 \geq 10$

10. $-4x - 3 < 5$

11. $-2(x - 10) + 1 > 0$

12. $3(x + 5) \leq 0$

13. $\dfrac{t + 2}{3} \geq 5$

14. $\dfrac{2 - t}{6} < 0$

15. $4x - 1 < \dfrac{3 - x}{-3}$

16. $\dfrac{x + 5}{-10} > 2x + 3$

17. $-3(z - 4) \geq 2(1 - 2z)$

18. $-\frac{1}{4}(2z - 6) + z \geq 5$

19. $\dfrac{1 - x}{4} < \dfrac{2x - 2}{3}$

20. $\dfrac{3x}{4} < x - \dfrac{x + 2}{2}$

21. $2x - 3 > \frac{1}{2}(x + 1)$

22. $5 - (2 - 3x) \leq -5x$

23. $5 < 4t - 1 \leq 11$

24. $-1 \leq 2t \leq 4$

25. $3 \leq 4 - x \leq 20$

26. $-5 < 1 - 2x < 40$

27. $-7 \leq \dfrac{1 - 4x}{7} < 12$

28. $0 < \dfrac{7x - 5}{3} \leq 4$

29. $5 > 2(x + 4) - 5 > -5$

30. $\frac{8}{3} \geq \frac{4}{3} - (x + 3) \geq \frac{2}{3}$

31. $3 \leq \frac{1}{2}x + \frac{3}{4} \leq 6$

32. $-4 \leq 5 - \frac{4}{5}x < 6$

33. $5x - 2(x + 3) \geq 4 - 3x$

34. $3x - 1 < 2(x - 3) + 1$

35. $\dfrac{1}{2} \leq \dfrac{1 - 2t}{3} < \dfrac{2}{3}$

36. $-\dfrac{3}{4} < \dfrac{2 - t}{5} < \dfrac{3}{4}$

37. $\frac{1}{2}z + \frac{2}{3}(3 - z) - \frac{5}{4}z \geq \frac{3}{4}(z - 2) + z$

38. $\frac{2}{3}(1 - 2z) - \frac{3}{2}z + \frac{5}{6}z \geq \dfrac{2z - 1}{3} + 1$

Solving Linear Inequalities Graphically

Exercises 39–46: (Refer to Example 2.) Solve the inequality graphically. Use set-builder notation.

39. $x + 2 \geq 2x$

40. $2x - 1 \leq x$

41. $\frac{2}{3}x - 2 > -\frac{4}{3}x + 4$

42. $-2x \geq -\frac{5}{3}x + 1$

43. $-1 \leq 2x - 1 \leq 3$

44. $-2 < 1 - x < 2$

45. $-3 < x - 2 \leq 2$

46. $-1 \leq 1 - 2x < 5$

Exercises 47–50: Use the given graph of $y = ax + b$ to solve each equation and inequality. Write the solution set to each inequality in set-builder or interval notation.

(a) $ax + b = 0$ (b) $ax + b < 0$ (c) $ax + b \geq 0$

47.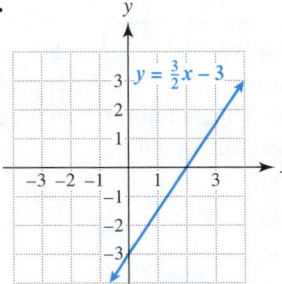

48.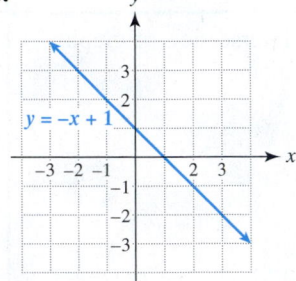

49.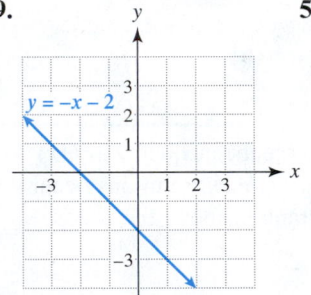

50.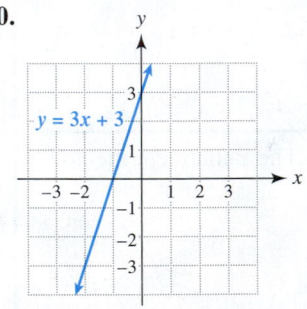

Exercises 51–54: **x-Intercept Method** *(Refer to Example 4.) Use the x-intercept method to solve the inequality. Write the solution set in set-builder or interval notation. Then check your answer by solving the inequality symbolically.*

51. $x - 3 \leq \frac{1}{2}x - 2$

52. $x - 2 \leq \frac{1}{3}x$

53. $2 - x < 3x - 2$

54. $\frac{1}{2}x + 1 > \frac{3}{2}x - 1$

Exercises 55–60: Solve the linear inequality graphically. Write the solution set in set-builder notation. Approximate endpoints to the nearest hundredth whenever appropriate.

55. $5x - 4 > 10$

56. $-3x + 6 \leq 9$

57. $-2(x - 1990) + 55 \geq 60$

58. $\sqrt{2}x > 10.5 - 13.7x$

59. $\sqrt{5}(x - 1.2) - \sqrt{3}x < 5(x + 1.1)$

60. $1.238x + 0.998 \leq 1.23(3.987 - 2.1x)$

▣ *Exercises 61–66; Solve the compound linear inequality graphically. Write the solution set in set-builder or interval notation, and approximate endpoints to the nearest tenth whenever appropriate.*

61. $3 \leq 5x - 17 < 15$ **62.** $-4 < \dfrac{55 - 3.1x}{4} < 17$

63. $1.5 \leq 9.1 - 0.5x \leq 6.8$ **64.** $0.2x < \dfrac{2x - 5}{3} < 8$

65. $x - 4 < 2x - 5 < 6$ **66.** $-3 \leq 1 - x \leq 2x$

67. The graphs of two linear functions f and g are shown.
(a) Solve the equation $g(x) = f(x)$.

(b) Solve the inequality $g(x) > f(x)$.

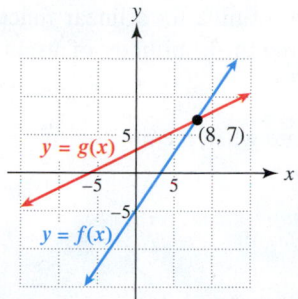

68. Use the figure to solve each equation or inequality.
(a) $f(x) = g(x)$ (b) $g(x) = h(x)$

(c) $f(x) < g(x) < h(x)$ (d) $g(x) > h(x)$

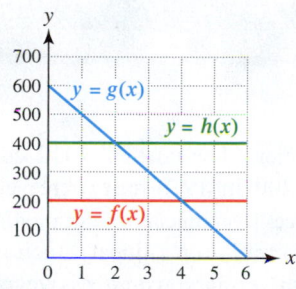

Solving Linear Inequalities Numerically

Exercises 69 and 70: Assume y_1 represents a linear function with the set of real numbers for its domain. Use the table to solve the inequalities. Use set-builder notation.

69. $y_1 > 0, y_1 \leq 0$ **70.** $y_1 < 0, y_1 \geq 0$

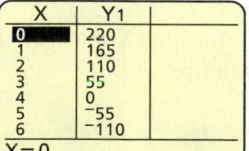   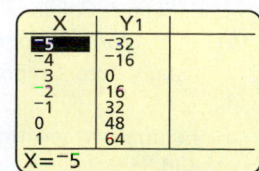

Exercises 71–78: Solve each inequality numerically. Write the solution set in set-builder or interval notation, and approximate endpoints to the nearest tenth when appropriate.

71. $-4x - 6 > 0$ **72.** $1 - 2x \geq 9$

73. $1 \leq 3x - 2 \leq 10$ **74.** $-5 < 2x - 1 < 15$

▣ **75.** $-\dfrac{3}{4} < \dfrac{2 - 5x}{3} \leq \dfrac{3}{4}$ ▣ **76.** $\dfrac{3x - 1}{5} < 15$

▣ **77.** $(\sqrt{11} - \pi)x - 5.5 \leq 0$

▣ **78.** $1.5(x - 0.7) + 1.5x < 1$

You Decide the Method

Exercises 79–82: Solve the inequality. Approximate the endpoints to the nearest thousandth when appropriate.

79. $2x - 8 > 5$ **80.** $5 < 4x - 2.5$

81. $\pi x - 5.12 \leq \sqrt{2}x - 5.7(x - 1.1)$

82. $5.1x - \pi \geq \sqrt{3} - 1.7x$

Applications

83. Distance Between Cars Cars A and B are both traveling in the same direction. Their distances in miles north of St. Louis after x hours are computed by the functions f_A and f_B, respectively. The graphs of f_A and f_B are shown in the figure for $0 \leq x \leq 10$.
(a) Which car is traveling faster? Explain.

(b) How many hours elapse before the two cars are the same distance from St. Louis? How far are they from St. Louis when this occurs?

(c) During what time interval is car B farther from St. Louis than car A?

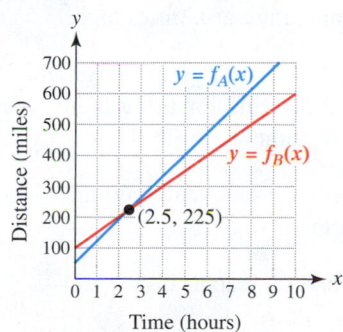

84. Distance Function f computes the distance y in miles between a car and the city of Omaha after x hours, where $0 \leq x \leq 6$. The graphs of f and the horizontal lines $y = 100$ and $y = 200$ are shown in the figure on the next page.
(a) Is the car moving toward or away from Omaha? Explain.

(b) Determine the times when the car is 100 miles and 200 miles from Omaha.

(c) Determine when the car is from 100 to 200 miles from Omaha.

(d) When is the car's distance from Omaha greater than 100 miles?

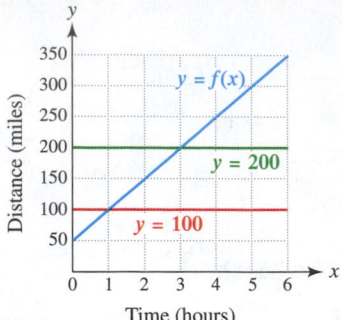

85. **Clouds and Temperature** As the altitude increases, both the air temperature and the dew point decrease. As long as the air temperature is greater than the dew point, clouds will not form. Typically, the air temperature T cools at 19°F for each 1-mile increase in altitude and the dew point D decreases by 5.8°F for each 1-mile increase in altitude. (Source: A. Miller and R. Anthes,

Meteorology.)

(a) Suppose the air temperature at ground level is 65°F. Write a formula for a linear function T that gives the air temperature at x miles high.

(b) Suppose the dew point at ground level is 50°F. Write a formula for a linear function D that gives the dew point at x miles high.

(c) Determine symbolically altitudes where clouds will not form.

(d) Solve part (c) graphically.

86. **Temperature and Altitude** Suppose the Fahrenheit temperature x miles above ground level is given by the formula $T(x) = 85 - 19x$.

(a) Use the intersection-of-graphs method to estimate the altitudes where the temperature is below freezing. Assume that the domain of T is $0 \le x \le 6$.

(b) What does the x-intercept on the graph of $y = T(x)$ represent?

(c) Solve part (a) symbolically.

87. **U.S. Social Gaming** Social gaming revenues (via the Internet) in billions of dollars from 2010 to 2015 are projected to be modeled by $R(x) = 0.86x + 1.2$, where x represents years after 2010. (Source: BI Intelligence.)

(a) Interpret the slope of the graph of R.

(b) Estimate when this revenue is expected to be more than $3 billion.

88. **Overweight Americans** If trends continue, the past and future percentages of the population who are overweight can be estimated by $W(x) = x + 45$, where x represents years after 1980. (Source: New York Times.)

(a) Interpret the slope of the graph of W.

(b) Estimate when this percentage was between 70% and 77%.

89. **Facebook Users** The number of active Facebook users increased from 100 million in 2008 to 1 billion in 2012. (Source: Business Insider.)

(a) Find a formula for a linear function U that models this growth in millions of users, where x is years after 2008.

(b) Estimate when the number of users was 550 million or more.

90. **Cost of Education** The cost of K–12 education per student was $2200 in 1978 and increased to $10,300 in 2008. (Source: Department of Education.)

(a) Find a formula for a linear function C that models this cost in dollars, where x is years after 1978.

(b) Estimate when this cost was from $4900 to $7600.

91. **Video Sharing** In 2006 about 33% of Americans reported watching online videos via sharing sites such as YouTube and Vimeo. By 2011 this number increased to 71% (Source: Column Five.)

(a) Find a formula for a linear function V that models the data, where x represents the year.

(b) Estimate when this percentage was between 40% and 55%.

92. Consumer Spending In 2005 consumers used credit and debit cards to pay for 40% of all purchases. This percentage was 55% in 2011. (*Source:* Bloomburg.)

(a) Find a linear function P that models the data.

(b) Estimate when this percentage was between 45% and 50%.

93. Modeling Sunrise Times In Boston, on the 90th day (March 30) the sun rose at 6:30 A.M., and on the 129th day (May 8) the sun rose at 5:30 A.M. Use a linear function to estimate the days when the sun rose between 5:45 A.M. and 6:00 A.M. Do not consider days after May 8. (*Source:* R Thomas.)

94. Modeling Sunrise Times In Denver, on the 77th day (March 17) the sun rose at 7:00 A.M., and on the 112th day (April 21) the sun rose at 6:00 A.M. Use a linear function to estimate the days when the sun rose between 6:10 A.M. and 6:40 A.M. Do not consider days after April 21. (*Source:* R. Thomas.)

95. Error Tolerances Suppose that an aluminum can is manufactured so that its radius r can vary from 1.99 inches to 2.01 inches. What range of values is possible for the circumference C of the can? Express your answer by using a three-part inequality.

96. Error Tolerances Suppose that a square picture frame has sides that vary between 9.9 inches and 10.1 inches. What range of values is possible for the perimeter P of the picture frame? Express your answer by using a three-part inequality.

97. Modeling Data The following data are exactly linear.

x	0	2	4	6
y	−1.5	4.5	10.5	16.5

(a) Find a linear function f that models the data.

(b) Solve the inequality $f(x) > 2.25$.

98. Modeling Data The following data are exactly linear.

x	1	2	3	4	5
y	0.4	3.5	6.6	9.7	12.8

(a) Find a linear function f that models the data.

(b) Solve the inequality $2 \leq f(x) \leq 8$.

Linear Regression

99. Female College Graduates The table in the next column lists the percentage P of the female population who had college degrees in selected years.

x	1980	1990	2000	2010
$P(\%)$	12.8	18.4	23.6	33.0

Source: Bureau of the Census.

(a) Find a linear function P that models the data.

(b) Estimate when this percentage was between 18% and 28.5%.

(c) Did your estimate involve interpolation or extrapolation?

100. Tablet Sales The table lists the actual and projected number N of tablets, such as the iPad, sold worldwide in millions for selected years x.

x	2011	2012	2013	2014	2015
N	90	165	250	340	470

Source: Business Insider.

(a) Find a linear function N that models the data.

(b) Estimate when this number is expected to be more than 637 million.

(c) Did your estimate involve interpolation or extrapolation?

Writing about Mathematics

101. Suppose the solution to the equation $ax + b = 0$ with $a > 0$ is $x = k$. Discuss how the value of k can be used to help solve the linear inequalities $ax + b > 0$ and $ax + b < 0$. Illustrate this process graphically. How would the solution sets change if $a < 0$?

102. Describe how to numerically solve the linear inequality $ax + b \leq 0$. Give an example.

103. If you multiply each part of a three-part inequality by the same negative number, what must you make sure to do? Explain by using an example.

104. Explain how a linear function, a linear equation, and a linear inequality are related. Give an example.

Extended and Discovery Exercises

1. Arithmetic Mean The **arithmetic mean** of two numbers a and b is given by $\frac{a + b}{2}$. Use properties of inequalities to show that if $a < b$, then $a < \frac{a + b}{2} < b$.

2. Geometric Mean The **geometric mean** of two numbers a and b is given by $\sqrt{ab}$. Use properties of inequalities to show that if $0 < a < b$, then $a < \sqrt{ab} < b$.

2.4 More Modeling with Functions

- Model data with a linear function
- Evaluate and graph piecewise-defined functions
- Evaluate and graph the greatest integer function
- Use direct variation to solve problems

Introduction

Throughout history, people have attempted to explain the world around them by creating models. A model is based on observations. It can be a diagram, a graph, an equation, a verbal expression, or some other form of communication. Models are used in diverse areas such as economics, chemistry, astronomy, religion, and mathematics. Regardless of where it is used, a **model** is an *abstraction* with the following two characteristics:

1. A model is able to explain present phenomena. It should not contradict data and information already known to be correct.
2. A model is able to make predictions about data or results. It should use current information to forecast phenomena or create new information.

Mathematical models are used to forecast business trends, design social networks, estimate ecological trends, control highway traffic, describe the Internet, predict weather, and discover new information when human knowledge is inadequate.

Modeling with Linear Functions

Linear functions can be used to model things that change at a constant rate. For example, the distance traveled by a car can be modeled by a linear function *if* the car is traveling at a constant speed.

MODELING WITH A LINEAR FUNCTION

To model a quantity that is changing at a constant rate with $f(x) = mx + b$, the following formula may be used.

$$f(x) = (\text{constant rate of change}) \, x + (\text{initial amount})$$

The constant rate of change corresponds to the slope of the graph of f, and the initial amount corresponds to the y-intercept.

This method is illustrated in the next two examples.

EXAMPLE 1   Writing formulas for functions

Write the formula for a linear function that models each situation. Choose both an appropriate name and an appropriate variable for the function. State what the input variable represents and the domain of the function.

(a) In 2011 the average cost of attending a public college was $8200, and it is projected to increase, on average, by $600 per year until 2014.   (*Source:* The College Board.)

(b) A car's speed is 50 miles per hour, and it begins to slow down at a constant rate of 10 miles per hour each second.

SOLUTION

(a) **Getting Started** To model cost with a linear function, we need to find two quantities: the initial amount and the rate of change. In this example the initial amount is $8200 and the rate of change is $600 per year. ▶

Let C be the name of the function and x be the number of years after 2011. Then

$$C(x) = (\textcolor{blue}{\textbf{constant rate of change}})x + (\textcolor{red}{\textbf{initial amount}})$$

$$= \textcolor{blue}{600x} + \textcolor{red}{8200}$$

models the cost in dollars of attending a public college x years after 2011. Because this projection is valid only until 2014, or for 3 years past 2011, the domain D of function C is

$$D = \{x \mid x = 0, 1, 2, \text{ or } 3\}.$$

Note that x represents a year, so it may be most appropriate to restrict the domain to integer values for x.

(b) Let S be the name of the function and t be the elapsed time in seconds that the car has been slowing down. Then

$$S(t) = (\textbf{constant rate of change})t + (\textbf{initial speed})$$
$$= -\mathbf{10}t + \mathbf{50}$$

models the speed of the car after an elapsed time of t seconds. Because the car's initial speed is **50** miles per hour and it **slows at 10** miles per hour per second, the car can slow down for at most 5 seconds before it comes to a stop. Thus the domain D of S is

$$D = \{t \mid 0 \leq t \leq 5\}.$$

Note that t represents time in seconds and does not need to be restricted to an integer.

Now Try Exercises 11 and 13

EXAMPLE 2 **Finding a symbolic representation**

A 100-gallon tank, initially full of water, is being drained at a rate of 5 gallons per minute.
(a) Write a formula for a linear function f that models the number of gallons of water in the tank after x minutes.
(b) How much water is in the tank after 4 minutes?
(c) Graph f. Identify the x- and y-intercepts and interpret each.
(d) Discuss the domain of f.

SOLUTION
(a) The amount of water in the tank is *decreasing* at 5 gallons per minute, so the constant rate of change is −5. The initial amount of water is 100 gallons.

$$f(x) = (\textbf{constant rate of change})x + (\textbf{initial amount})$$
$$= -\mathbf{5}x + \mathbf{100}$$

(b) After **4** minutes the tank contains $f(\mathbf{4}) = -5(\mathbf{4}) + 100 = 80$ gallons.
(c) $f(x) = -5x + 100$, so the graph has y-intercept 100 and slope −5. See Figure 2.41.

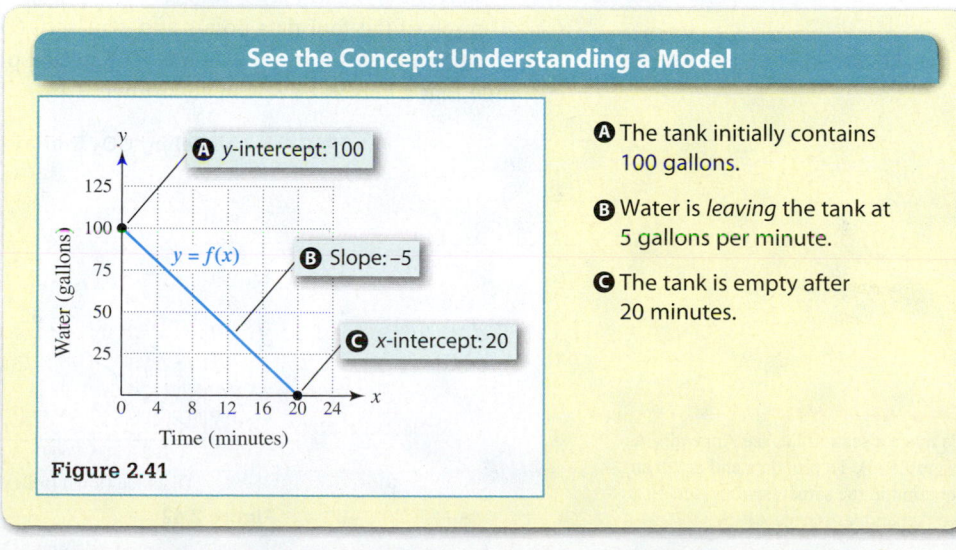

See the Concept: Understanding a Model

Ⓐ y-intercept: 100
$y = f(x)$
Ⓑ Slope: −5
Ⓒ x-intercept: 20

Ⓐ The tank initially contains 100 gallons.

Ⓑ Water is *leaving* the tank at 5 gallons per minute.

Ⓒ The tank is empty after 20 minutes.

Water (gallons)
Time (minutes)

Figure 2.41

(d) From the graph we see that the domain of f must be restricted to $0 \le x \le 20$. For example, 21 is not in the domain of f because $f(21) = -5(21) + 100 = -5$; the tank cannot hold -5 gallons. Similarly, -1 is not in the domain of f because $f(-1) = -5(-1) + 100 = 105$; the tank holds *at most* 100 gallons.

Now Try Exercise 17

If the slopes between consecutive pairs of data points are always the same, the data can be modeled exactly by a linear function. If the slopes between consecutive pairs of data points are nearly the same, then the data can be modeled approximately by a linear function. In the next example we model data approximately.

EXAMPLE 3 Modeling airliner CO_2 emissions

Airliners emit carbon dioxide into the atmosphere when they burn jet fuel. Table 2.4 shows the *average* number y of pounds of carbon dioxide (CO_2) emitted by an airliner for each passenger who flies a distance of x miles.

Carbon Dioxide Emissions

x (miles)	240	360	680	800
y (pounds)	150	230	435	510

Source: E. Rogers and T. Kostigen, *The Green Book.*
Table 2.4

(a) Calculate the slopes of the line segments that connect consecutive data points.
(b) Find a linear function f that models the data.
(c) Graph f and the data. What does the slope of the graph of f indicate?
(d) Calculate $f(1000)$ and interpret the result.

SOLUTION
(a) The slopes of the lines passing through the points $(240, 150)$, $(360, 230)$, $(680, 435)$, and $(800, 510)$ are as follows.

$$m_1 = \frac{230 - 150}{360 - 240} \approx 0.67, \quad m_2 = \frac{435 - 230}{680 - 360} \approx 0.64, \quad \text{and}$$

$$m_3 = \frac{510 - 435}{800 - 680} \approx 0.63$$

(b) One possibility for m is to find the average of m_1, m_2, and m_3. The average of 0.67, 0.64, and 0.63 is 0.65, rounded to the nearest hundredth. Because traveling 0 miles produces 0 pounds of carbon dioxide, let the graph of f pass through $(0, 0)$. Thus the y-intercept is 0 and $f(x) = 0.65x + 0$.

(c) A graph of the four data points and $f(x) = 0.65x$ is shown in Figure 2.42. The slope of 0.65 indicates that, on average, 0.65 pound of carbon dioxide is produced for each mile that a person travels in an airliner.

Modeling CO_2 Emissions

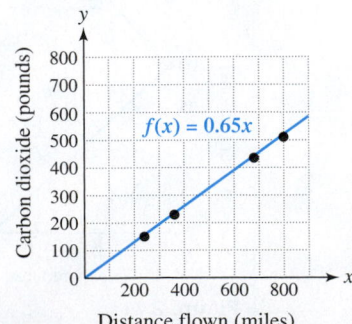

Figure 2.42

Calculator Help

To make a scatterplot, see Appendix A (page AP-3). To plot data and graph an equation in the same viewing rectangle, see Appendix A (page AP-6).

(d) $f(\textbf{1000}) = 0.65(\textbf{1000}) = \textbf{650}$; thus **650** pounds of carbon dioxide are emitted into the atmosphere, on average, when a person flies **1000** miles.

Now Try Exercise 23

MAKING CONNECTIONS

Slope and Approximately Linear Data   Another way to obtain an initial value for m is to calculate the slope between the first and last data point in the table. The value for m can then be adjusted visually by graphing f and the data. In Example 3 this would have resulted in

$$m = \frac{510 - 150}{800 - 240} \approx 0.64,$$

which compares favorably with our decision to let $m = 0.65$.

Piecewise-Defined Functions

When a function f models data, there may not be one formula for $f(x)$ that works. In this case, the function is sometimes defined on pieces of its domain and is therefore called a **piecewise-defined function**. If each piece is linear, the function is a **piecewise-linear function**. An example of a piecewise-defined function is the *Fujita scale,* which classifies tornadoes by intensity. If a tornado has wind speeds between 40 and 72 miles per hour, it is an F1 tornado. Tornadoes with wind speeds *greater than* 72 miles per hour but not more than 112 miles per hour are F2 tornadoes. The Fujita scale is represented by the following function F, where the input x represents the maximum wind speed of a tornado and the output is the F-scale number from 1 to 5.

Fujita Scale $(F1 \rightarrow F5)$

$$F(x) = \begin{cases} 1 & \text{if } 40 \le x \le 72 \\ 2 & \text{if } 72 < x \le 112 \\ 3 & \text{if } 112 < x \le 157 \\ 4 & \text{if } 157 < x \le 206 \\ 5 & \text{if } 206 < x \le 260 \end{cases}$$

> F is defined in five pieces over intervals of its domain.

> $F(180) = 4$

For example, if the maximum wind speed is 180 miles per hour, then $F(180) = 4$ because **180** is between **157** and **206**; that is, $\textbf{157} < \textbf{180} \le \textbf{206}$. Thus a tornado with a maximum wind speed of 180 miles per hour is an F4 tornado.

A graph of $y = f(x)$ is shown in Figure 2.43. It is composed of horizontal line segments. Because each piece is constant, F is sometimes called a **piecewise-constant function** or a **step function**. A solid dot occurs at the point $(72, 1)$ and an open circle occurs at the point $(72, 2)$, because technically a tornado with 72-mile-per-hour winds is an F1 tornado, not an F2 tornado.

Tornado Intensities

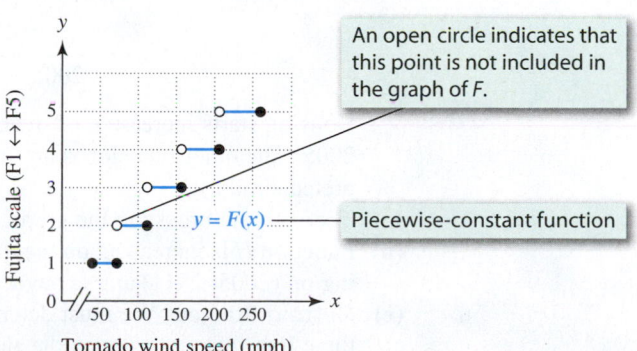

> An open circle indicates that this point is not included in the graph of F.

> $y = F(x)$

> Piecewise-constant function

Figure 2.43

Continuous Functions The graph of a **continuous function** can be sketched without picking up the pencil. There are no breaks in the graph of a continuous function. Because there are breaks in the graph shown in Figure 2.43, function F is not continuous; rather it is **discontinuous** at $x = 72$, 112, 157, and 206.

An Application The housing market peaked in 2005. Shortly after, the housing bubble burst and continued to plummet, hitting a low in 2011. In the next example we use a piecewise-defined function to model housing starts from 2005 to 2011.

EXAMPLE 4 **Analyzing housing starts**

Table 2.5 lists numbers of single residential homes built during selected years.

Housing Starts in Millions

Year	2000	2005	2011	2012
Homes	1.3	1.7	0.4	0.5

Source: Bureau of the Census.

Table 2.5

(a) Plot a line graph of these data. Let this graph define a function H.
(b) Find and interpret the slope of each line segment.
(c) Is H continuous on its domain?
(d) Identify where H is increasing, decreasing, or constant.
(e) Write a piecewise-defined formula for H.

SOLUTION
(a) Plot the points (2000, 1.3), (2005, 1.7), (2011, 0.4), and (2012, 0.5). Connect these points with three line segments as shown in Figure 2.44. (*Source:* Bureau of the Census.)

Housing Starts

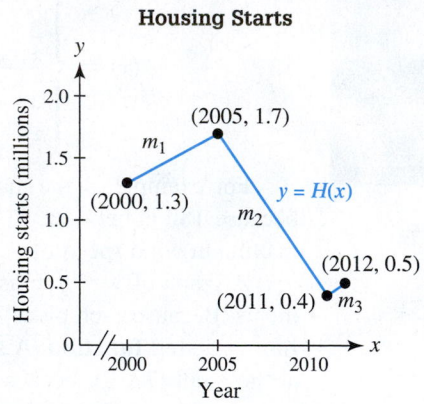

Figure 2.44

(b) The first line segment passes through (2000, 1.3) and (2005, 1.7) with slope

$$m_1 = \frac{1.7 - 1.3}{2005 - 2000} = \frac{0.4}{5} = \frac{2}{25} = 0.08.$$

Housing starts increased, on average, by 0.08 million (80,000) per year from 2000 to 2005. The other two slopes are $m_2 = -\frac{13}{60} \approx -0.22$ and $m_3 = \frac{1}{10}$ and can be interpreted similarly.
(c) There are no breaks in the graph so H is continuous on its domain, [2000, 2012].
(d) Function H is increasing on the intervals (2000, 2005) and (2011, 2012). It is decreasing on (2005, 2011) and is never constant.
(e) **Getting Started** We must determine three equations for the lines that represent the three line segments. Given the slope and one point we can write the point-slope form of each line. ▶

The first line segment passes through (2000, 1.3) and has slope $\frac{2}{25}$. Therefore $H(x) = \frac{2}{25}(x - 2000) + 1.3$, if $2000 \le x \le 2005$. Continuing in this manner, we can write a piecewise-defined formula for $H(x)$.

$$H(x) = \begin{cases} \frac{2}{25}(x - 2000) + 1.3, & \text{if } 2000 \le x \le 2005 \\ -\frac{13}{60}(x - 2005) + 1.7, & \text{if } 2005 < x \le 2011 \\ \frac{1}{10}(x - 2011) + 0.4, & \text{if } 2011 < x \le 2012 \end{cases}$$

> **Now Try Exercise 43**

EXAMPLE 5   **Evaluating and graphing a piecewise-defined function**

Use $f(x)$ to complete the following.

$$f(x) = \begin{cases} x - 1 & \text{if } -4 \le x < 2 \\ -2x & \text{if } 2 \le x \le 4 \end{cases}$$

(a) What is the domain of f?
(b) Evaluate $f(-3)$, $f(2)$, $f(4)$, and $f(5)$.
(c) Sketch a graph of f.
(d) Is f a continuous function on its domain?

SOLUTION
(a) Function f is defined for x-values satisfying either $-4 \le x < 2$ or $2 \le x \le 4$. Thus the domain of f is $D = \{x \mid -4 \le x \le 4\}$, or $[-4, 4]$.
(b) For x-values satisfying $-4 \le x < 2$, $f(x) = x - 1$ and so $f(-3) = -4$. Similarly, if $2 \le x \le 4$, then $f(x) = -2x$. Thus $f(2) = -4$ and $f(4) = -8$. The expression $f(5)$ is undefined because 5 is not in the domain of f.
(c) **Getting Started** Because each piece of $f(x)$ is linear, the graph of $y = f(x)$ consists of two line segments. Therefore we can find the endpoints of each line segment and then sketch the graph. ▶

Graph the First Piece

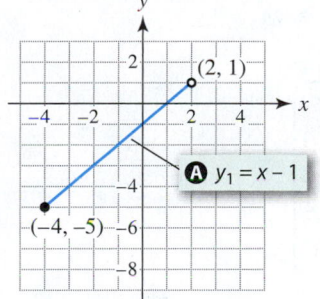

Figure 2.45

Graph the Second Piece

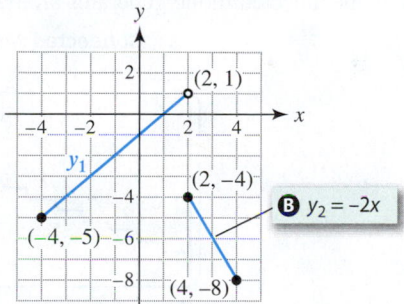

Figure 2.46

Ⓐ Evaluate $y_1 = x - 1$ at $x = -4$ and $x = 2$. Place a dot at $(-4, -5)$ and an open circle at $(2, 1)$, because $y_1 = x - 1$ applies for $x < 2$. Sketch a line segment between these points.

Ⓑ Evaluate $y_2 = -2x$ at $x = 2$ and $x = 4$. Place dots at $(2, -4)$ and $(4, -8)$. Sketch a line segment between these points.

(d) The function f is *not* continuous because in Figure 2.46 there is a break in its graph at $x = 2$.

> **Now Try Exercise 33**

The Greatest Integer Function

A common piecewise-defined function used in mathematics is the greatest integer function, denoted $f(x) = [\![x]\!]$. The **greatest integer function** is defined as follows.

$[\![x]\!]$ **is the greatest integer less than or equal to x.**

Some examples of the evaluation of $[\![x]\!]$ include

$$[\![6.7]\!] = 6, \quad [\![3]\!] = 3, \quad [\![-2.3]\!] = -3, \quad [\![-10]\!] = -10, \quad \text{and} \quad [\![-\pi]\!] = -4.$$

The graph of $y = [\![x]\!]$ is shown in Figure 2.47. The greatest integer function is both a piecewise-constant function and a step function.

The Greatest Integer Function

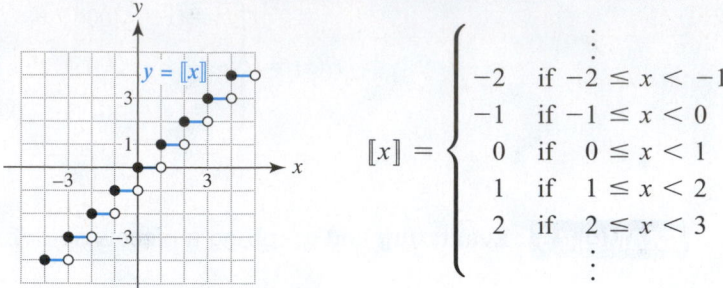

$$[\![x]\!] = \begin{cases} \vdots \\ -2 & \text{if } -2 \le x < -1 \\ -1 & \text{if } -1 \le x < 0 \\ 0 & \text{if } 0 \le x < 1 \\ 1 & \text{if } 1 \le x < 2 \\ 2 & \text{if } 2 \le x < 3 \\ \vdots \end{cases}$$

Figure 2.47

Calculator Help

To access the greatest integer function or to set a calculator in dot mode, see Appendix A (pages AP-6 and AP-8).

In some applications, fractional parts are either not allowed or ignored. Framing lumber for houses is measured in 2-foot multiples, and mileage charges for rental cars may be calculated to the mile.

Suppose a car rental company charges \$31.50 per day plus \$0.25 for each mile driven, where fractions of a mile are ignored. The function given by $f(x) = 0.25[\![x]\!] + 31.50$ calculates the cost of driving x miles in one day. For example, the cost of driving 100.4 miles is

$$f(100.4) = 0.25[\![\textbf{100.4}]\!] + 31.50 = 0.25(\textbf{100}) + 31.50 = \$56.50.$$

On some calculators and computers, the greatest integer function is denoted int(X). A graph of $Y_1 = 0.25*\text{int}(X) + 31.5$ is shown in Figure 2.48.

Dot Mode

[0, 10, 1] by [31, 35, 1]

$y = 0.25[\![x]\!] + 31.5$

Figure 2.48

MAKING CONNECTIONS

Connected and Dot Modes Graphing calculators often connect points to make a graph look continuous. However, if a graph has breaks in it, a graphing calculator may connect points where there should be breaks. In *dot mode*, points are plotted but not connected. Figure 2.49 is the same graph shown in Figure 2.48, except that it is plotted in *connected mode*. Note that connected mode generates an inaccurate graph of this step function.

Connected Mode

[0, 10, 1] by [31, 35, 1]

$y = 0.25[\![x]\!] + 31.5$

Inaccurate graph

Figure 2.49

Direct Variation

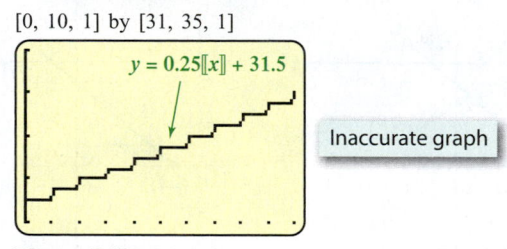

When a change in one quantity causes a proportional change in another quantity, the two quantities are said to *vary directly* or to *be directly proportional*. For example, if we work for \$8 per hour, our pay is proportional to the number of hours that we work. Doubling the hours doubles the pay, tripling the hours triples the pay, and so on.

DIRECT VARIATION

Let x and y denote two quantities. Then y is **directly proportional** to x, or y **varies directly** with x, if there exists a nonzero number k such that

$$y = kx.$$

The number k is called the **constant of proportionality** or the **constant of variation**.

If a person earns $**57.75** working for **7** hours, the constant of proportionality k is the hourly pay rate. If y represents the pay in dollars and x the hours worked, then k is found by substituting values for x and y into the equation $y = kx$ and solving for k. That is,

$$57.75 = k(7), \quad \text{or} \quad k = \frac{57.75}{7} = 8.25,$$

so the hourly pay rate is $8.25 and, in general, $y = 8.25x$.

A Spring Being Stretched

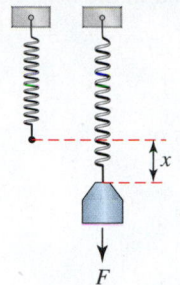

Figure 2.50

An Application   Hooke's law states that the distance that an elastic spring stretches beyond its natural length is *directly proportional* to the amount of weight hung on the spring, as illustrated in Figure 2.50. This law is valid whether the spring is stretched or compressed. The constant of proportionality is called the **spring constant**. Thus if a weight or force F is applied and the spring stretches a distance x beyond its natural length, then the equation $F = kx$ models this situation, where k is the spring constant.

EXAMPLE 6   **Working with Hooke's law**

A 12-pound weight is hung on a spring, and it stretches 2 inches.
(a) Find the spring constant.
(b) Determine how far the spring will stretch when a 19-pound weight is hung on it.

SOLUTION
(a) Let $F = kx$, given that $F = 12$ pounds and $x = 2$ inches. Thus

$$12 = k(2), \quad \text{or} \quad k = 6,$$

and the spring constant equals **6**.
(b) If $F = 19$ and $F = 6x$, then $19 = 6x$, or $x = \frac{19}{6} \approx 3.17$ inches.

Now Try Exercise 63

The following four-step method can often be used to solve variation problems.

SOLVING A VARIATION PROBLEM

When solving a variation problem, the following steps can be used.

STEP 1: Write the general equation for the type of variation problem that you are solving.

STEP 2: Substitute given values in this equation so the constant of variation k is the only unknown value in the equation. Solve for k.

STEP 3: Substitute the value of k in the general equation in Step 1.

STEP 4: Use this equation to find the requested quantity.

EXAMPLE 7   **Solving a direct variation problem**

Let T vary directly with x, and suppose that $T = 33$ when $x = 5$. Find T when $x = 31$.

SOLUTION
STEP 1: The equation for direct variation is $T = kx$.
STEP 2: Substitute 33 for T and 5 for x. Then solve for k.

$$T = kx \qquad \textit{Direct variation equation}$$
$$33 = k(5) \qquad \textit{Let T = 33 and x = 5.}$$
$$\frac{33}{5} = k \qquad \textit{Divide each side by 5.}$$

STEP 3: Thus $T = \frac{33}{5}x$, or $T = 6.6x$.
STEP 4: When $x = 31$, we have $T = 6.6(31) = 204.6$.

Now Try Exercise 51

Suppose that, for each point (x, y) in a data set, the ratios $\frac{y}{x}$ are equal to some constant k. That is, $\frac{y}{x} = k$ for each data point. Then $y = kx$, so y varies directly with x and the constant of variation is k. In addition, the data points (x, y) all lie on the line $y = kx$, which has slope k and passes through the origin. These concepts are used in the next example.

EXAMPLE 8 Modeling memory requirements

Table 2.6 lists the megabytes (MB) x needed to record y seconds of music.

Recording Digital Music

x (MB)	0.23	0.49	1.16	1.27
y (sec)	10.7	22.8	55.2	60.2

Table 2.6

(a) Compute the ratios $\frac{y}{x}$ for the four data points. Does y vary directly with x? If it does, what is the constant of variation k?

(b) Estimate the seconds of music that can be stored on 5 megabytes.

(c) Graph the data in Table 2.6 and the line $y = kx$.

SOLUTION

(a) The four ratios $\frac{y}{x}$ from Table 2.6 are

$$\frac{10.7}{0.23} \approx 46.5, \quad \frac{22.8}{0.49} \approx 46.5, \quad \frac{55.2}{1.16} \approx 47.6, \quad \text{and} \quad \frac{60.2}{1.27} \approx 47.4.$$

Because the ratios are nearly equal, it is reasonable to say that y is directly proportional to x. The constant of proportionality is about **47**, the average of the four ratios. This means that $y = 47x$ and we can store about **47** seconds of music per megabyte.

(b) Let $x = 5$ in the equation $y = 47x$, to obtain $y = 47(5) = 235$ seconds.

(c) Graphs of the data and the line $y = 47x$ are shown in Figure 2.51.

Now Try Exercise 65

Music and Memory Requirements
[0, 1.5, 0.5] by [0, 70, 10]

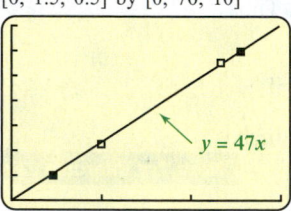

$y = 47x$

Figure 2.51

2.4 Putting It All Together

The following table summarizes important concepts.

CONCEPT	DESCRIPTION
Models	A good model describes and explains current data. It should also make predictions and forecast phenomena.
Linear model	If a quantity experiences a constant rate of change, then it can be modeled by a linear function in the form $f(x) = mx + b$. $f(x) = (\text{constant rate of change})x + (\text{initial value})$
Piecewise-defined function	A function is piecewise-defined if it has different formulas on different intervals of its domain. Many times the domain is restricted. $$f(x) = \begin{cases} 2x - 3 & \text{if } -3 \leq x < 1 \\ x + 5 & \text{if } 1 \leq x \leq 5 \end{cases}$$ When $x = 2$ then $f(x) = x + 5$, so $f(2) = 2 + 5 = 7$. The domain of f is $[-3, 5]$.

CONCEPT	DESCRIPTION
The greatest integer function	$[\![x]\!]$ is the greatest integer less than or equal to x. $[\![-5.9]\!] = -6,\quad [\![8.7]\!] = 8,\quad \text{and}\quad [\![-5]\!] = -5$
Direct variation	The variable y is directly proportional to x or varies directly with x if $y = kx$ for some nonzero constant k. Constant k is the constant of proportionality or the constant of variation.

2.4 Exercises

Modeling with Linear Functions

Exercises 1 and 2: Write a symbolic representation (formula) for a function f that computes the following.

1. (a) The number of pounds in x ounces

   (b) The number of dimes in x dollars

   (c) The monthly electric bill in dollars if x kilowatt-hours are used at 6 cents per kilowatt-hour and there is a fee of $6.50

   (d) The cost of skiing x times with a $500 season pass

2. (a) The distance traveled by a car moving at 50 miles per hour for x hours

   (b) The total number of hours in day x

   (c) The distance in miles between a runner and home after x hours if the runner starts 1 mile from home and jogs *away* from home at 6 miles per hour

   (d) A car's speed in feet per second after x seconds if its tires are 2 feet in diameter and rotating 14 times per second

Exercises 3–6: Find the formula for a linear function f that models the data in the table exactly.

3.
x	-2	0	4
$f(x)$	4	3	1

4.
x	-6	0	3
$f(x)$	-5	-1	1

5.
x	1	2	3
$f(x)$	7	9	11

6.
x	15	30	45
$f(x)$	40	30	20

Exercises 7–10: Match the situation with the graph (a–d) that models it best, where x-values represent time.

7. Height of the Empire State Building from 1990 to 2010

8. Average cost of a new car from 1980 to 2010

9. Distance between a runner in a race and the finish line

10. Amount of money earned after x hours when working at an hourly rate of pay

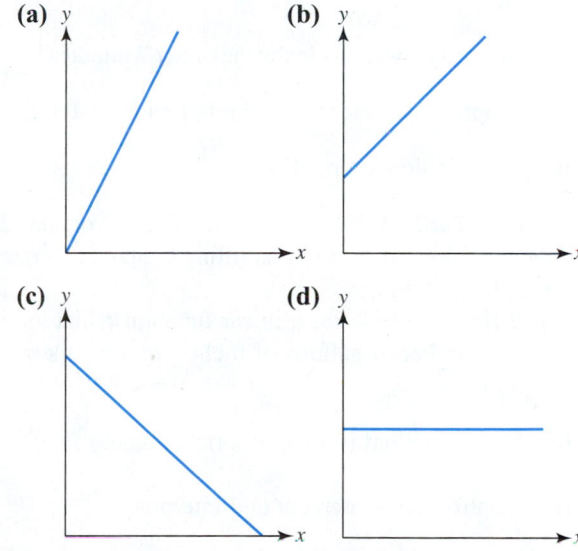

Exercises 11–16: Write a formula for a linear function that models the situation. Choose both an appropriate name and an appropriate variable for the function. State what the input variable represents and the domain of the function. Assume that the domain is an interval of the real numbers.

11. **U.S. Homes with Broadband** In 2010 about 27% of U.S. homes had broadband Internet access. This percentage is expected to increase, on average, by 1.2 percentage points per year for the next 4 years. (*Source:* Bureau of the Census.)

12. **Velocity of a Falling Object** A stone is dropped from a water tower and its velocity increases at a rate of 32 feet per second. The stone hits the ground with a velocity of 96 feet per second.

13. **Text Messages** In the first month of 2010 about 0.4 trillion text messages were sent. The number of text messages being sent during 2010 grew by 0.5 trillion per month.   (*Source:* CTIA.)

14. **Speed of a Car** A car is traveling at 30 miles per hour, and then it begins to slow down at a constant rate of 6 miles per hour every 4 seconds.

15. **Population Density** In 1900 the average number of people per square mile in the United States was 21.5, and it increased, on average, by 6 people every 10 years until 2011.   (*Source:* Bureau of the Census.)

16. **Injury Rate** In 1992 the number of injury cases recorded in private industry per 100 full-time workers was 8.3, and it decreased, on average, by 0.27 injury every year until 2010.   (*Source:* Bureau of Labor Statistics.)

17. **Draining a Water Tank** A 300-gallon tank is initially full of water and is being drained at a rate of 10 gallons per minute.
    (a) Write a formula for a function W that gives the number of gallons of water in the tank after t minutes.

    (b) How much water is in the tank after 7 minutes?

    (c) Graph W and identify and interpret the intercepts.

    (d) Find the domain of W.

18. **Filling a Tank** A 500-gallon tank initially contains 200 gallons of fuel oil. A pump is filling the tank at a rate of 6 gallons per minute.
    (a) Write a formula for a linear function f that models the number of gallons of fuel oil in the tank after x minutes.

    (b) Graph f. What is an appropriate domain for f?

    (c) Identify the y-intercept and interpret it.

    (d) Does the x-intercept of the graph of f have any physical meaning in this problem? Explain.

19. **Living with HIV/AIDS** In 2007 there were 1.2 million people in the United States who had been infected with HIV. At that time the infection rate was 40,000 people per year.   (*Source:* CDC.)
    (a) Write a formula for a linear function f that models the total number of people in millions who were living with HIV/AIDS x years after 2007.

    (b) Estimate the number of people who may have been infected by the year 2014.

20. **Birth Rate** In 1990 the number of births per 1000 people in the United States was 16.7 and decreasing, on average, at 0.136 birth per 1000 people each year. (*Source:* National Center for Health Statistics.)
    (a) Write a formula for a linear function f that models the birth rate x years after 1990.

    (b) Estimate the birth rate in 2012 and compare this estimate to the actual value of 13.7.

21. **Ice Deposits** A roof has a 0.5-inch layer of ice on it from a previous storm. Another ice storm begins to deposit ice at a rate of 0.25 inch per hour.
    (a) Find a formula for a linear function f that models the thickness of the ice on the roof x hours after the second ice storm started.

    (b) How thick is the ice after 2.5 hours?

22. **Rainfall** Suppose that during a storm rain is falling at a rate of 1 inch per hour. The water coming from a circular roof with a radius of 20 feet is running down a downspout that can accommodate 400 gallons of water per hour. See the figure.
    (a) Determine the number of cubic inches of water falling on the roof in 1 hour.

    (b) One gallon equals about 231 cubic inches. Write a formula for a function g that computes the gallons of water landing on the roof in x hours.

    (c) How many gallons of water land on the roof during a 2.5-hour rain storm?

    (d) Will one downspout be sufficient to handle this type of rainfall? How many downspouts should there be?

Exercises 23 and 24: **Modeling Fuel Consumption** *The table shows the distance y in miles traveled by a vehicle using x gallons of gasoline.*

(a) *Calculate the slopes of the line segments that connect consecutive points.*

(b) *Find a linear function that models the data.*

(c) *Graph f and the data together. What does the slope indicate?*

(d) *Evaluate f(30) and interpret the result.*

23.

x (gallons)	5	10	15	20
y (miles)	84	169	255	338

24.

x (gallons)	5	10	15	20
y (miles)	194	392	580	781

Piecewise-Defined Functions

25. Speed Limits The graph of $y = f(x)$ gives the speed limit y along a rural highway x miles from its starting point.

(a) What are the maximum and minimum speed limits along this stretch of highway?

(b) Estimate the miles of highway with a speed limit of 55 miles per hour.

(c) Evaluate $f(4)$, $f(12)$, and $f(18)$.

(d) At what x-values is the graph discontinuous? Interpret each discontinuity.

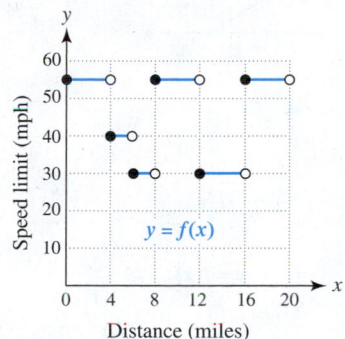

26. ATM The graph of $y = f(x)$ depicts the amount of money y in dollars in an automatic teller machine (ATM) after x minutes.

(a) Determine the initial and final amounts of money in the ATM.

(b) Evaluate $f(10)$ and $f(50)$. Is f continuous?

(c) How many *withdrawals* occurred?

(d) When did the largest withdrawal occur? How much was it?

(e) How much was deposited into the machine?

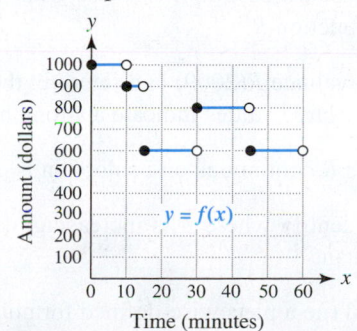

27. First-Class Mail In January 2012, the retail flat rate in dollars for first-class mail weighing up to 5 ounces could be computed by the piecewise-constant function P, where x is the number of ounces.

$$P(x) = \begin{cases} 0.90 & \text{if } 0 < x \le 1 \\ 1.10 & \text{if } 1 < x \le 2 \\ 1.30 & \text{if } 2 < x \le 3 \\ 1.50 & \text{if } 3 < x \le 4 \\ 1.70 & \text{if } 4 < x \le 5 \end{cases}$$

(a) Evaluate $P(1.5)$ and $P(3)$. Interpret your results.

(b) Sketch a graph of P. What is the domain of P?

(c) Where is P discontinuous on its domain?

28. Swimming Pool Levels The graph of $y = f(x)$ shows the amount of water y in thousands of gallons remaining in a swimming pool after x days.

(a) Estimate the initial and final amounts of water in the pool.

(b) When did the amount of water in the pool remain constant?

(c) Approximate $f(2)$ and $f(4)$.

(d) At what rate was water being drained from the pool when $1 \le x \le 3$?

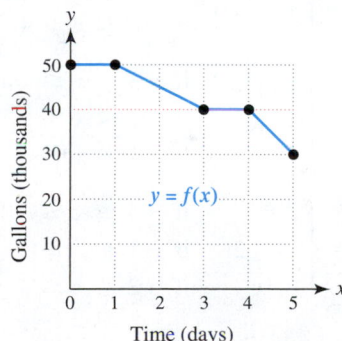

Exercises 29 and 30: An individual is driving a car along a straight road. The graph shows the driver's distance from home after x hours.

(a) *Use the graph to evaluate $f(1.5)$ and $f(4)$.*

(b) *Interpret the slope of each line segment.*

(c) *Describe the motion of the car.*

(d) *Identify where f is increasing, decreasing, or constant.*

29.

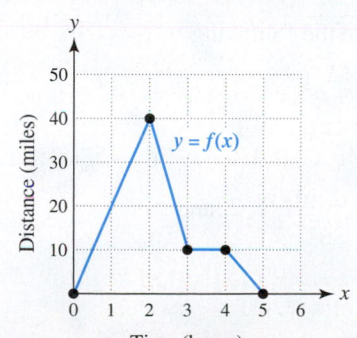

30.

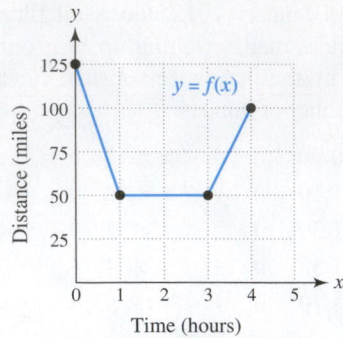

Exercises 31–36: Complete the following for f(x).

(a) *Determine the domain of f.*

(b) *Evaluate f(−2), f(0), and f(3).*

(c) *Graph f.*

(d) *Is f continuous on its domain?*

31. $f(x) = \begin{cases} 2 & \text{if } -5 \le x \le -1 \\ x + 3 & \text{if } -1 < x \le 5 \end{cases}$

32. $f(x) = \begin{cases} 2x + 1 & \text{if } -3 \le x < 0 \\ x - 1 & \text{if } \;\;\; 0 \le x \le 3 \end{cases}$

33. $f(x) = \begin{cases} 3x & \text{if } -1 \le x < 1 \\ x + 1 & \text{if } \;\;\; 1 \le x \le 2 \end{cases}$

34. $f(x) = \begin{cases} -2 & \text{if } -6 \le x < -2 \\ 0 & \text{if } -2 \le x < 0 \\ 3x & \text{if } \;\;\; 0 \le x \le 4 \end{cases}$

35. $f(x) = \begin{cases} x & \text{if } -3 \le x \le -1 \\ 1 & \text{if } -1 < x < 1 \\ 2 - x & \text{if } \;\;\; 1 \le x \le 3 \end{cases}$

36. $f(x) = \begin{cases} 3 & \text{if } -4 \le x \le -1 \\ x - 2 & \text{if } -1 < x \le 2 \\ 0.5x & \text{if } \;\;\; 2 < x \le 4 \end{cases}$

Exercises 37 and 38: Graph f.

37. $f(x) = \begin{cases} -\frac{1}{2}x + 1 & \text{if } -4 \le x \le -2 \\ 1 - 2x & \text{if } -2 < x \le 1 \\ \frac{2}{3}x + \frac{4}{3} & \text{if } \;\;\; 1 < x \le 4 \end{cases}$

38. $f(x) = \begin{cases} \frac{3}{2} - \frac{1}{2}x & \text{if } -3 \le x < -1 \\ -2x & \text{if } -1 \le x \le 2 \\ \frac{1}{2}x - 5 & \text{if } \;\;\; 2 < x \le 3 \end{cases}$

39. Use $f(x)$ to complete the following.

$$f(x) = \begin{cases} 3x - 1 & \text{if } -5 \le x < 1 \\ 4 & \text{if } \;\;\; 1 \le x \le 3 \\ 6 - x & \text{if } \;\;\; 3 < x \le 5 \end{cases}$$

(a) Evaluate f at $x = -3, 1, 2,$ and 5.

(b) On what interval is f constant?

(c) Sketch a graph of f. Is f continuous on its domain?

40. Use $g(x)$ to complete the following.

$$g(x) = \begin{cases} -2x - 6 & \text{if } -8 \le x \le -2 \\ x & \text{if } -2 < x < 2 \\ 0.5x + 1 & \text{if } \;\;\; 2 \le x \le 8 \end{cases}$$

(a) Evaluate g at $x = -8, -2, 2,$ and 8.

(b) For what x-values is g increasing?

(c) Sketch a graph of g. Is g continuous on its domain?

Exercises 41 and 42: Use the graph of $y = f(x)$ to write a piecewise-defined formula for f. Write each piece in slope-intercept form.

41.

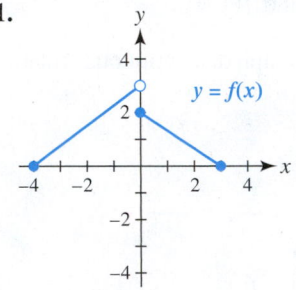

42.

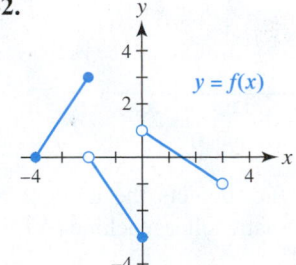

43. Housing Market One way to describe the housing market is to divide the population of the United States by the number of new housing starts. This ratio is listed in the following table for selected years.

Year	2000	2005	2009	2011
Ratio	225	180	700	727

Source: Bureau of the Census.

(a) Plot a line graph of these data. Let this graph be function R.

(b) Evaluate $R(2009)$ and interpret the result. Do small or large values indicate a strong housing market?

(c) Is R continuous on its domain?

(d) Identify where R is increasing, decreasing, or constant.

(e) Write a piecewise-defined formula for R.

44. High School Dropouts The table lists the percentage of the population that did not have a high school diploma for selected years.

Year	1960	1980	2000	2010
Percentage	28	14	11	8

Source: Bureau of the Census.

(a) Plot a line graph of these data. Let this graph be function D.

(b) Interpret the slope of each line segment.

(c) Is D continuous on its domain?

(d) Identify where D is increasing, decreasing, or constant.

(e) Write a piecewise-defined formula for D.

Greatest Integer Function

Exercises 45–48: Complete the following.

(a) *Use dot mode to graph the function f in the standard viewing rectangle.*

(b) *Evaluate $f(-3.1)$ and $f(1.7)$.*

45. $f(x) = [\![2x - 1]\!]$ **46.** $f(x) = [\![x + 1]\!]$

47. $f(x) = 2[\![x]\!] + 1$ **48.** $f(x) = [\![-x]\!]$

49. Lumber Costs The lumber used to frame walls of houses is frequently sold in multiples of 2 feet. If the length of a board is not exactly a multiple of 2 feet, there is often no charge for the additional length. For example, if a board measures at least 8 feet but less than 10 feet, then the consumer is charged for only 8 feet.

(a) Suppose that the cost of lumber is $0.80 for every 2 feet. Find a formula for a function f that computes the cost of a board x feet long for $6 \le x \le 18$.

(b) Graph f.

(c) Determine the costs of boards with lengths of 8.5 feet and 15.2 feet.

50. Cost of Carpet Each foot of carpet purchased from a 12-foot-wide roll costs $36. If a fraction of a foot is purchased, a customer does not pay for the extra amount. For example, if a customer wants 14 feet of carpet and the salesperson cuts off 14 feet 4 inches, the customer does not pay for the extra 4 inches.

(a) How much does 9 feet 8 inches of carpet from this roll cost?

(b) Using the greatest integer function, write a formula for the price P of x feet of carpet.

Direct Variation

Exercises 51–54: Let y be directly proportional to x. Complete the following.

51. Find y when $x = 5$, if $y = 7$ when $x = 14$.

52. Find y when $x = 2.5$, if $y = 13$ when $x = 10$.

53. Find y when $x = \frac{1}{2}$, if $y = \frac{3}{2}$ when $x = \frac{2}{3}$.

54. Find y when $x = 1.3$, if $y = 7.2$ when $x = 5.2$.

Exercises 55–58: Find the constant of proportionality k and the undetermined value in the table if y is directly proportional to x.

55.

x	3	5	6	8
y	7.5	12.5	15	?

56.

x	1.2	4.3	5.7	?
y	3.96	14.19	18.81	23.43

57. Sales tax y on a purchase of x dollars

x	$25	$55	?
y	$1.50	$3.30	$5.10

58. Cost y of buying x compact discs at the same price

x	3	4	5
y	$41.97	$55.96	?

59. Cost of Tuition The cost of tuition is directly proportional to the number of credits taken. If 11 credits cost $720.50, find the cost of taking 16 credits. What is the constant of proportionality?

60. Strength of a Beam The maximum load that a horizontal beam can carry is directly proportional to its width. If a beam 1.5 inches wide can support a load of 250 pounds, find the load that a beam of the same type can support if its width is 3.5 inches.

61. Antarctic Ozone Layer Stratospheric ozone occurs in the atmosphere between altitudes of 12 and 18 miles. Ozone in the stratosphere is frequently measured in Dobson units, where 300 Dobson units corresponds to an ozone layer 3 millimeters thick. In 1991 the reported minimum in the Antarctic *ozone hole* was about 110 Dobson units. (*Source:* R. Huffman, *Atmospheric Ultraviolet Remote Sensing.*)

(a) The thickness y of the ozone layer is directly proportional to the number of Dobson units x. Find the constant of proportionality k.

(b) How thick was the ozone layer in 1991?

62. **Weight on Mars** The weight of an object on Earth is directly proportional to the weight of an object on Mars. If a 25-pound object on Earth weighs 10 pounds on Mars, how much would a 195-pound astronaut weigh on Mars?

63. **Hooke's Law** Suppose a 15-pound weight stretches a spring 8 inches, as shown in the figure.

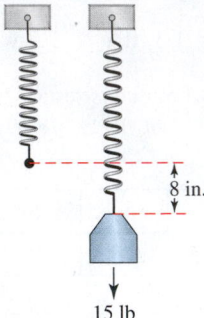

8 in.

15 lb

   (a) Find the spring constant.

   (b) How far will a 25-pound weight stretch this spring?

64. **Hooke's Law** If an 80-pound force compresses a spring 3 inches, how much force must be applied to compress the spring 7 inches?

65. **Force of Friction** The table lists the force F needed to push a cargo box weighing x pounds on a wood floor.

x (lb)	150	180	210	320
F (lb)	26	31	36	54

   (a) Compute the ratio $\frac{F}{x}$ for each data pair in the table. Interpret these ratios.

   (b) Approximate a constant of proportionality k satisfying $F = kx$. (k is the *coefficient of friction*.)

   (c) Graph the data and the equation together.

   (d) Estimate the force needed to push a 275-pound cargo box on the floor.

66. **Electrical Resistance** The electrical resistance of a wire varies directly with its length. If a 255-foot wire has a resistance of 1.2 ohms, find the resistance of 135 feet of the same type of wire. Interpret the constant of proportionality in this situation.

Linear Regression

67. **Ring Size** The table lists ring size S for a finger with circumference x in centimeters.

x (cm)	4.65	5.40	5.66	6.41
S (size)	4	7	8	11

Source: Overstock.

   (a) Find a linear function S that models the data.

   (b) Find the circumference of a finger with a ring size of 6.

68. **Hat Size** The table lists hat size H for a head with circumference x in inches.

x (in.)	$21\frac{1}{8}$	$21\frac{7}{8}$	$22\frac{5}{8}$	25
S (size)	$6\frac{3}{4}$	7	$7\frac{1}{4}$	8

   (a) Find a linear function S that models the data.

   (b) Find the circumference of a head with a hat size of $7\frac{1}{2}$.

69. **Super Bowl Ads** The table lists the cost in millions of dollars for a 30-second Super Bowl commercial for selected years.

Year	1994	1998	2004	2008	2012
Cost	1.2	1.6	2.3	2.7	3.5

Source: MSNBC.

   (a) Find a linear function f that models the data.

   (b) Estimate the cost in 2009 and compare the estimate to the actual value of $3.0 million. Did your estimate involve interpolation or extrapolation?

   (c) Use f to predict the year when the cost could reach $4.0 million.

70. **Women in Politics** The table lists the percentage P of women in state legislatures during year x.

x	1993	1997	2001	2005	2007
P	20.5	21.6	22.4	22.7	23.5

Source: National Women's Political Caucus.

   (a) Find a linear function P that models the data.

   (b) Estimate this percentage in 2003 and compare the estimate to the actual value of 22.4%. Did your estimate involve interpolation or extrapolation?

   (c) Use P to predict the year when this percentage could reach 25%.

Writing about Mathematics

71. Explain what a piecewise-defined function is and why it is used. Sketch a graph of a continuous piecewise-linear function f that increases, decreases, and is constant. Let the domain of f be $-4 \le x \le 4$.

72. Find a real data set on the Internet that can be modeled by a linear function. Find the linear modeling function. Is your model exact or approximate? Explain.

73. How can you recognize a symbolic representation (formula) of a linear function? How can you recognize a graph or table of values of a linear function?

74. Explain how you determine whether a linear function is increasing, decreasing, or constant. Give an example of each.

Extended and Discovery Exercises

Exercises 1 and 2: **Estimating Populations** *Biologists sometimes use direct variation to estimate the number of fish in small lakes. They start by tagging a small number of fish and then releasing them. They assume that over a period of time, the tagged fish distribute themselves evenly throughout the lake. Later, they collect a second sample. The total number of fish and the number of tagged fish in the second sample are counted. To determine the total population of fish in the lake, biologists assume that the proportion of tagged fish in the second sample is equal to the proportion of tagged fish in the entire lake. This technique can also be used to count other types of animals, such as birds, when they are not migrating.*

1. Eighty-five fish are tagged and released into a pond. A later sample of 94 fish from the pond contains 13 tagged fish. Estimate the number of fish in the pond.

2. Sixty-three blackbirds are tagged and released. Later it is estimated that out of a sample of 32 blackbirds, only 8 are tagged. Estimate the population of blackbirds in the area.

3. **Height and Shoe Size** In this exercise you will determine if there is a relationship between height and shoe size.

(a) Have classmates write their sex, shoe size, and height in inches on a slip of paper. When you have enough information, complete the following table—one for adult males and one for adult females.

Height (inches)				
Shoe size				

(b) Make a scatterplot of each table, with height on the x-axis and shoe size on the y-axis. Is there any relationship between height and shoe size? Explain.

(c) Try to find a linear function that models each data set.

Exercises 4 and 5: **Linear Approximation** *Graph the function f in the standard viewing rectangle.*

(a) *Choose any curved portion of the graph of f and repeatedly zoom in. Describe how the graph appears. Repeat this process on different portions of the graph.*

(b) *Under what circumstances could a linear function be used to accurately model a nonlinear graph?*

4. $f(x) = 4x - x^3$

5. $f(x) = x^4 - 5x^2$

CHECKING BASIC CONCEPTS FOR SECTIONS 2.3 AND 2.4

1. Solve the inequality $2(x - 4) > 1 - x$. Express the solution set in set-builder notation.

2. Solve the compound inequality $-2 \leq 1 - 2x \leq 3$. Use set-builder or interval notation.

3. Use the graph to solve each equation and inequality. Then solve each part symbolically. Use set-builder or interval notation when possible.

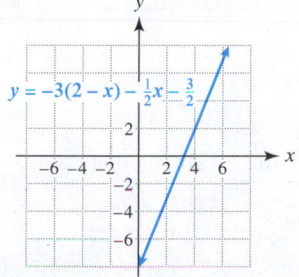

(a) $-3(2 - x) - \frac{1}{2}x - \frac{3}{2} = 0$

(b) $-3(2 - x) - \frac{1}{2}x - \frac{3}{2} > 0$

(c) $-3(2 - x) - \frac{1}{2}x - \frac{3}{2} \leq 0$

4. The death rate from heart disease for people ages 15 through 24 is 2.7 per 100,000 people.

(a) Write a function f that models the number of deaths in a population of x million people 15 to 24 years old.

(b) There are about 39 million people in the United States who are 15 to 24 years old. Estimate the number of deaths from heart disease in this age group.

5. A driver of a car is initially 50 miles south of home, driving 60 miles per hour south. Write a function f that models the distance between the driver and home.

2.5 Absolute Value Equations and Inequalities

- Evaluate and graph the absolute value function
- Solve absolute value equations
- Solve absolute value inequalities

Introduction

A margin of error can be very important in many aspects of life, including being fired out of a cannon. The most dangerous part of the feat, first done by a human in 1875, is to land squarely on a net. For a human cannonball who wants to fly 180 feet in the air and then land in the center of a net with a 60-foot-long safe zone, there is a margin of error of ± 30 feet. That is, the horizontal distance D traveled by the human cannonball can vary between **180** $-$ **30** $= 150$ feet and **180** $+$ **30** $= 210$ feet. (*Source:* Ontario Science Center.)

This margin of error can be expressed mathematically by using the *absolute value inequality*

$$|D - \mathbf{180}| \leq \mathbf{30}.$$

The absolute value is necessary because D can be either less than or greater than 180, but by not more than 30 feet.

The Absolute Value Function

The absolute value function is defined by $f(x) = |x|$. The following See the Concept describes many of the properties of this function.

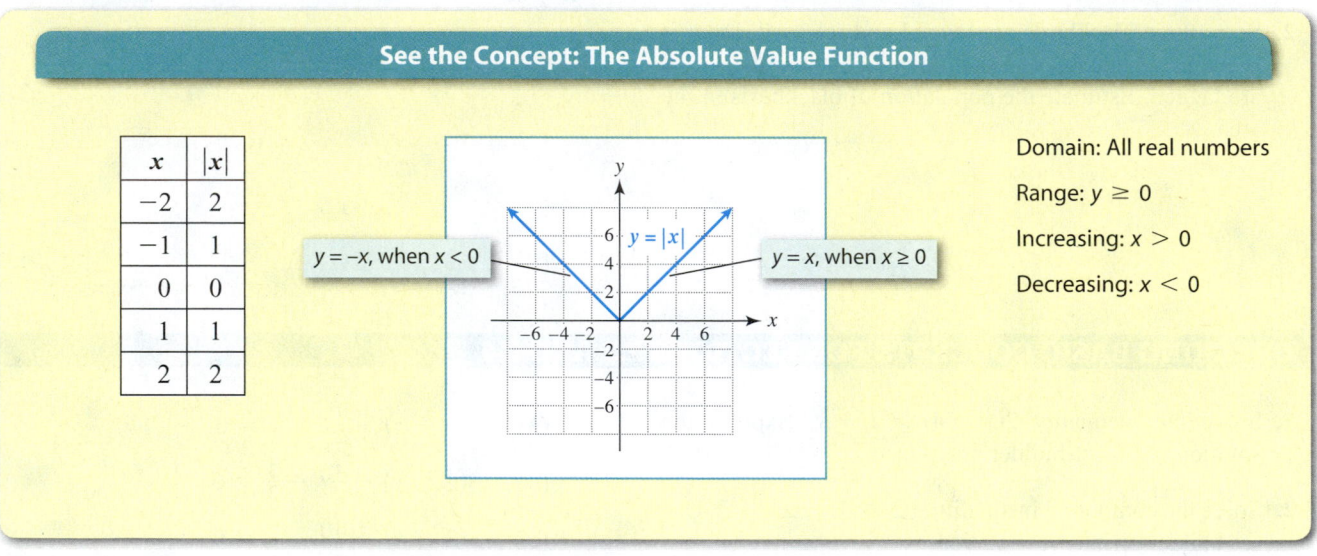

See the Concept: The Absolute Value Function

x	$|x|$
-2	2
-1	1
0	0
1	1
2	2

$y = -x$, when $x < 0$ $y = |x|$ $y = x$, when $x \geq 0$

Domain: All real numbers

Range: $y \geq 0$

Increasing: $x > 0$

Decreasing: $x < 0$

Symbolic Representations (Formulas) The graph of the absolute value function suggests that the absolute value function can be defined symbolically using the following piecewise-linear function.

$$|x| = \begin{cases} -x & \text{if } x < 0 \\ x & \text{if } x \geq 0 \end{cases} \qquad \textcolor{teal}{\textit{Piecewise-linear function}}$$

There is another formula for $|x|$. Consider the following examples.

$$\sqrt{3^2} = \sqrt{9} = 3 \quad \text{and} \quad \sqrt{(-3)^2} = \sqrt{9} = 3$$
$$\sqrt{7^2} = \sqrt{49} = 7 \quad \text{and} \quad \sqrt{(-7)^2} = \sqrt{49} = 7$$

That is, regardless of whether a real number x is positive or negative, the expression $\sqrt{x^2}$ equals the *absolute value* of x. This statement is summarized by

$$\sqrt{x^2} = |x| \qquad \text{for all real numbers } x.$$

Calculator Help

To access the absolute value function, see Appendix A (page AP-8).

For example, $\sqrt{y^2} = |y|$, $\sqrt{(x-1)^2} = |x-1|$, and $\sqrt{(2x)^2} = |2x|$.

EXAMPLE 1   **Analyzing the graph of $y = |ax + b|$**

Graph $y = f(x)$ and $y = |f(x)|$ separately. Discuss how the absolute value affects the graph of f.

(a) $f(x) = x + 2$ (b) $f(x) = -2x + 4$

SOLUTION

(a) The line $y = x + 2$ is shown in Figure 2.52. The graph of $y = |x + 2|$ in Figure 2.53 is V-shaped and *never* dips below the x-axis because an absolute value is *never* negative.

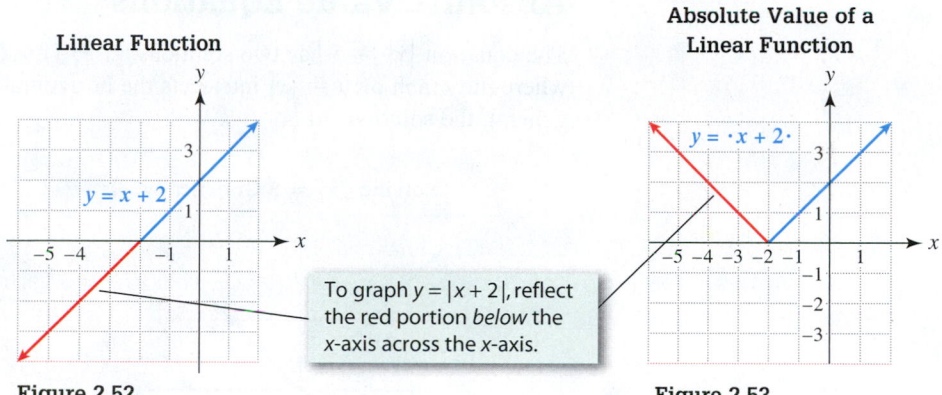

Figure 2.52   Figure 2.53

(b) The graphs of $y_1 = -2x + 4$ and $y_2 = |-2x + 4|$ are shown in Figures 2.54 and 2.55, respectively.

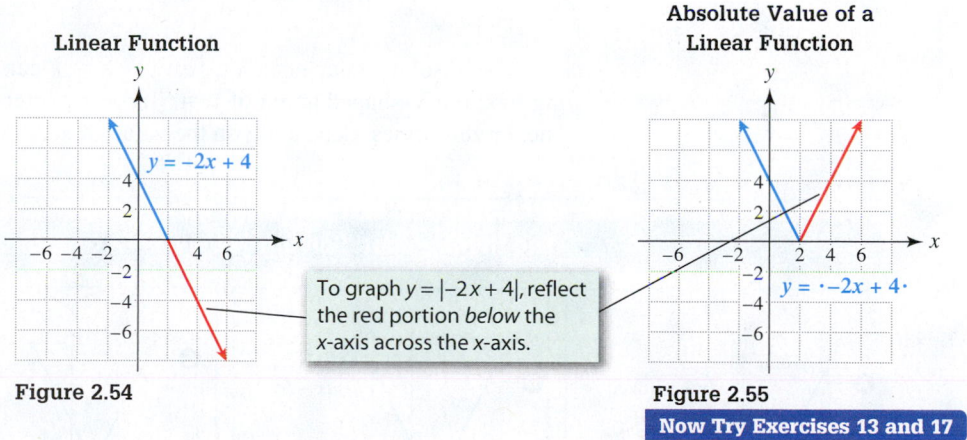

Figure 2.54   Figure 2.55

Now Try Exercises 13 and 17

NOTE   Example 1 illustrates the fact that the graph of $y = |ax + b|$ with $a \neq 0$ is V-shaped and is never located below the x-axis. The vertex (or point) of the V-shaped graph corresponds to the x-intercept, which can be found by solving the linear equation $ax + b = 0$.

Graphs and the Absolute Value In general, the graph of $y = |f(x)|$ is a reflection of the graph of $y = f(x)$ across the x-axis whenever $f(x) < 0$. Otherwise (whenever $f(x) > 0$), their graphs are identical. The following graphs illustrate this connection.

The Effect of the Absolute Value on a Graph

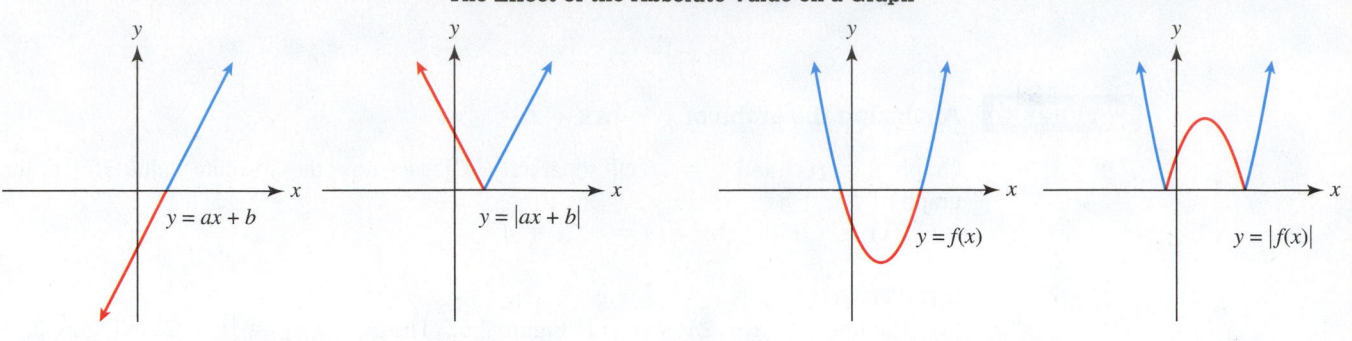

Absolute Value Equations

The equation $|x| = 5$ has two solutions: ± 5. This fact is shown visually in Figure 2.56, where the graph of $y = |x|$ intersects the horizontal line $y = 5$ at the points $(\pm 5, 5)$. In general, the solutions to $|x| = k$ with $k > 0$ are given by $x = \pm k$. See Figure 2.57.

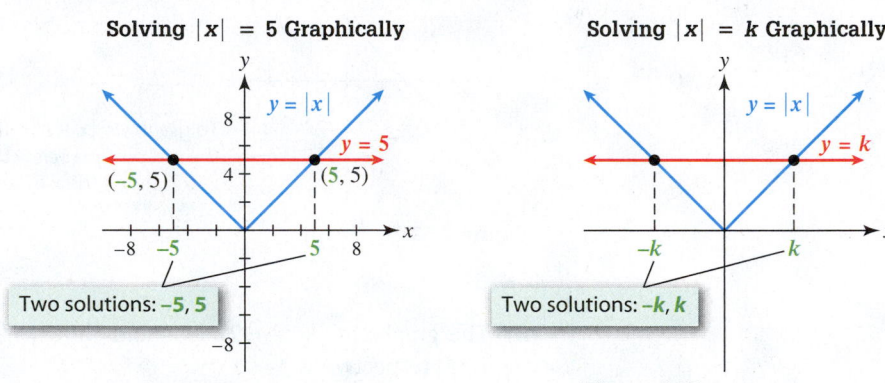

Solving $|x| = 5$ Graphically

Two solutions: **−5, 5**

Figure 2.56

Solving $|x| = k$ Graphically

Two solutions: **−k, k**

Figure 2.57

The absolute value equation $|ax + b| = k$ can be solved graphically. In the following box, the V-shaped graph of $y = |ax + b|$ intersects the horizontal line ($y = k$) two, one, or zero times, depending on the value of k.

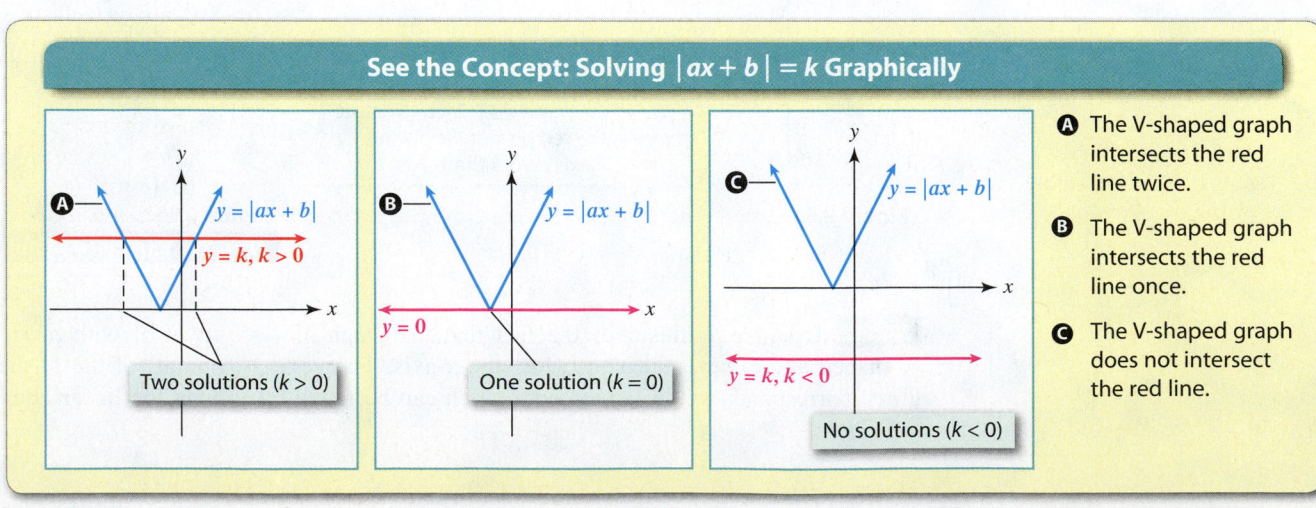

See the Concept: Solving $|ax + b| = k$ Graphically

A $y = |ax + b|$

$y = k, k > 0$

Two solutions ($k > 0$)

B $y = |ax + b|$

$y = 0$

One solution ($k = 0$)

C $y = |ax + b|$

$y = k, k < 0$

No solutions ($k < 0$)

A The V-shaped graph intersects the red line twice.

B The V-shaped graph intersects the red line once.

C The V-shaped graph does not intersect the red line.

Because the solution to $|x| = k$ are given by $x = \pm k$, it follows that the solutions to $|ax + b| = k$ are given by $ax + b = \pm k$. This concept is used to solve absolute value equations symbolically.

> ### ABSOLUTE VALUE EQUATIONS
>
> Let k be a positive number. Then
>
> $$|ax + b| = k \quad \text{is equivalent to} \quad ax + b = \pm k.$$

EXAMPLE 2 Solving absolute value equations

Solve each equation.
(a) $\left|\frac{3}{4}x - 6\right| = 15$ **(b)** $|1 - 2x| = -3$ **(c)** $|3x - 2| - 5 = -2$

SOLUTION
(a) The equation $\left|\frac{3}{4}x - 6\right| = 15$ is satisfied when $\frac{3}{4}x - 6 = \pm 15$.

$\frac{3}{4}x - 6 = 15$	or	$\frac{3}{4}x - 6 = -15$	*Equations to solve*
$\frac{3}{4}x = 21$	or	$\frac{3}{4}x = -9$	*Add 6 to each side.*
$x = 28$	or	$x = -12$	*Multiply by $\frac{4}{3}$.*

The solutions are -12 and 28.

(b) Because an absolute value is never negative, $|1 - 2x| \geq 0$ for all x and can never equal -3. There are no solutions. This is illustrated graphically in Figure 2.58.

(c) Because the right side of the equation is a negative number, it might appear at first glance that there were no solutions. However, if we add 5 to each side of the equation,

$$|3x - 2| - 5 = -2 \quad \text{becomes} \quad |3x - 2| = 3.$$

This equation is equivalent to $3x - 2 = \pm 3$ and has two solutions.

$3x - 2 = 3$	or	$3x - 2 = -3$	*Equations to solve*
$3x = 5$	or	$3x = -1$	*Add 2 to each side.*
$x = \dfrac{5}{3}$	or	$x = -\dfrac{1}{3}$	*Divide by 3.*

The solutions are $-\frac{1}{3}$ and $\frac{5}{3}$.

Now Try Exercises 21, 29, and 31

No Solutions

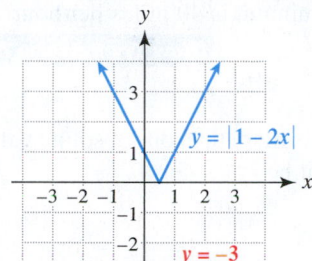

Figure 2.58

EXAMPLE 3 Solving an equation with technology

Solve the equation $|2x + 5| = 2$ graphically, numerically, and symbolically.

SOLUTION
Graphical Solution Graph $y_1 = |2x + 5|$ and $y_2 = 2$. The V-shaped graph of y_1 intersects the horizontal line at the points $(-3.5, 2)$ and $(-1.5, 2)$, as shown in Figures 2.59 and 2.60. The solutions are -3.5 and -1.5.

Numerical Solution Table y_1 and y_2, as shown in Figure 2.61. The solutions to $y_1 = y_2$ are -3.5 and -1.5.

Calculator Help
To find a point of intersection, see Appendix A (page AP-7).

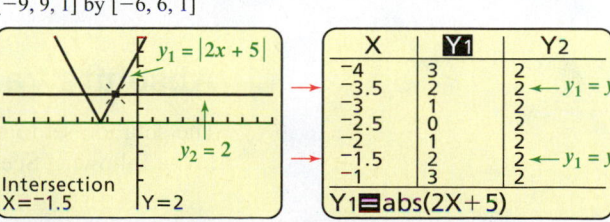

First Graphical Solution   **Second Graphical Solution**   **Numerical Solutions**

Figure 2.59   **Figure 2.60**   **Figure 2.61**

Symbolic Solution The equation $|2x + 5| = 2$ is satisfied when $2x + 5 = \pm 2$.

$$2x + 5 = 2 \qquad \text{or} \qquad 2x + 5 = -2 \qquad \textcolor{teal}{\text{Equations to solve}}$$

$$2x = -3 \qquad \text{or} \qquad 2x = -7 \qquad \textcolor{teal}{\text{Subtract 5 from each side.}}$$

$$x = -\frac{3}{2} \qquad \text{or} \qquad x = -\frac{7}{2} \qquad \textcolor{teal}{\text{Divide by 2.}}$$

> **Now Try Exercises 45(a), (b), (c)**

EXAMPLE 4 **Describing speed limits with absolute values**

The maximum and minimum lawful speeds S on an interstate highway satisfy the equation $|S - 55| = 15$. Find the maximum and minimum speed limits.

SOLUTION The equation $|S - 55| = 15$ is equivalent to $S - 55 = \pm 15$.

$$S - 55 = 15 \qquad \text{or} \qquad S - 55 = -15 \qquad \textcolor{teal}{\text{Equations to solve}}$$

$$S = 70 \qquad \text{or} \qquad S = 40 \qquad \textcolor{teal}{\text{Add 55 to each side.}}$$

The maximum speed limit is 70 miles per hour and the minimum is 40 miles per hour.

> **Now Try Exercise 73**

An Equation with Two Absolute Values Sometimes more than one absolute value sign occurs in an equation. For example, an equation might be in the form

$$|ax + b| = |cx + d|.$$

In this case there are two possibilities:

$$\text{either} \qquad ax + b = cx + d \qquad \text{or} \qquad ax + b = -(cx + d).$$

This symbolic technique is demonstrated in the next example.

EXAMPLE 5 **Solving an equation involving two absolute values**

Solve the equation $|x - 2| = |1 - 2x|$.

SOLUTION We must solve both of the following equations.

$$x - 2 = 1 - 2x \qquad \text{or} \qquad x - 2 = -(1 - 2x)$$

$$3x = 3 \qquad \text{or} \qquad x - 2 = -1 + 2x$$

$$x = 1 \qquad \text{or} \qquad -1 = x$$

There are two solutions: -1 and 1.

> **Now Try Exercise 35**

Absolute Value Inequalities

The solution set to an absolute value inequality can be understood graphically, as shown in the following See the Concept.

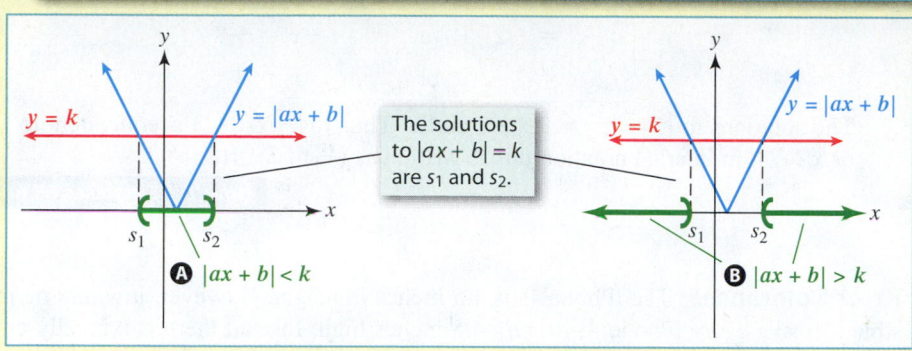

See the Concept: Solving $|ax + b| < k$ and $|ax + b| > k$ Graphically

Ⓐ The V-shaped graph is *below* the red line between s_1 and s_2. The solutions to $|ax + b| < k$ satisfy $s_1 < x < s_2$.

Ⓑ The V-shaped graph is *above* the red line to the left of s_1 and to the right of s_2. The solutions to $|ax + b| > k$ satisfy $x < s_1$ or $x > s_2$.

> **NOTE** In both of the above figures, equality (determined by s_1 and s_2) is the boundary between *greater than* and *less than*. For this reason, s_1 and s_2 are called *boundary numbers.*

ABSOLUTE VALUE INEQUALITIES

Let the solutions to $|ax + b| = k$ be s_1 and s_2, where $s_1 < s_2$ and $k > 0$.

1. $|ax + b| < k$ is equivalent to $s_1 < x < s_2$.
2. $|ax + b| > k$ is equivalent to $x < s_1$ or $x > s_2$.

Similar statements can be made for inequalities involving $\leq$ or $\geq$.

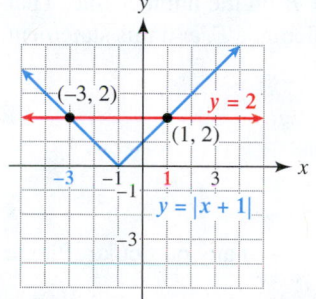

Figure 2.62

For example, the graphs of $y = |x + 1|$ and $y = 2$ are shown in Figure 2.62. These graphs intersect at the points $(-3, 2)$ and $(1, 2)$. It follows that the two solutions to

$$|x + 1| = 2$$

are $s_1 = -3$ and $s_2 = 1$. The solutions to $|x + 1| < 2$ lie between $s_1 = -3$ and $s_2 = 1$, which can be written as $-3 < x < 1$. Furthermore, the solutions to $|x + 1| > 2$ lie "outside" $s_1 = -3$ and $s_2 = 1$. This can be written as $x < -3$ or $x > 1$.

> **NOTE** The **union symbol** $\cup$ may be used to write $x < s_1$ or $x > s_2$ in interval notation. For example, $x < -3$ or $x > 1$ is written as $(-\infty, -3) \cup (1, \infty)$ in interval notation. This indicates that the solution set includes all real numbers in either $(-\infty, -3)$ or $(1, \infty)$.

EXAMPLE 6 Solving inequalities involving absolute values symbolically

Solve each inequality symbolically. Write the solution set in interval notation.
(a) $|2x - 5| \leq 6$ (b) $|5 - x| > 3$

SOLUTION
(a) Begin by solving $|2x - 5| = 6$, or equivalently, $2x - 5 = \pm 6$.

$$2x - 5 = 6 \qquad \text{or} \qquad 2x - 5 = -6$$
$$2x = 11 \qquad \text{or} \qquad 2x = -1$$
$$x = \frac{11}{2} \qquad \text{or} \qquad x = -\frac{1}{2}$$

The solutions to $|2x - 5| = 6$ are $-\frac{1}{2}$ and $\frac{11}{2}$. The solution set for the inequality $|2x - 5| \leq 6$ includes all real numbers x satisfying $-\frac{1}{2} \leq x \leq \frac{11}{2}$. In interval notation this is written as $\left[-\frac{1}{2}, \frac{11}{2}\right]$.

(b) To solve $|5 - x| > 3$, begin by solving $|5 - x| = 3$, or equivalently, $5 - x = \pm 3$.

$$5 - x = 3 \qquad \text{or} \qquad 5 - x = -3$$
$$-x = -2 \qquad \text{or} \qquad -x = -8$$
$$x = 2 \qquad \text{or} \qquad x = 8$$

The solutions to $|5 - x| = 3$ are 2 and 8. Thus $|5 - x| > 3$ is equivalent to $x < 2$ or $x > 8$. In interval notation this is written as $(-\infty, 2) \cup (8, \infty)$.

Now Try Exercises 55 and 63

Error Tolerances The iPhone 4s is 4.5 inches in height. However, it would be impossible to make every iPhone 4s *exactly* 4.5 inches high. Instead there is typically an error tolerance, where the height of an actual iPhone must be within a certain range.

Suppose that the actual height of a particular phone is A inches, but it is specified that this phone should be S inches high. If the maximum error tolerance is E, then Figure 2.63 illustrates this situation.

Distance Between A and S Must Be Less Than E

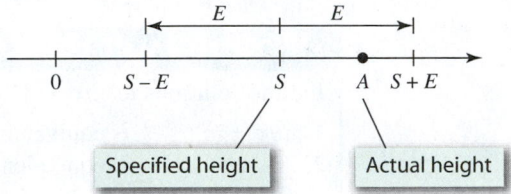

Figure 2.63

The value of A must be located between $S - E$ and $S + E$ on the number line. That is, the distance between A and S must be less than the error tolerance E, and this statement can be written as $|A - S| < E$.

EXAMPLE 7   **Finding error tolerances on the iPhone 4s**

The iPhone 4s is 4.5 inches high. Suppose that the actual height A of any particular iPhone has a maximum error tolerance that is less than 0.005 inch.
(a) Write an absolute value statement that describes this situation.
(b) Solve this inequality for A and interpret your result.

SOLUTION
(a) The distance between A and 4.5 on the number line must be less than 0.005. This statement can be written as $|A - 4.5| < 0.005$.
(b) To solve this inequality, we first solve $|A - 4.5| = 0.005$.

$$A - 4.5 = -0.005 \quad \text{or} \quad A - 4.5 = 0.005 \qquad \textit{First solve equality.}$$
$$A = 4.495 \qquad \text{or} \qquad A = 4.505 \qquad \textit{Add 4.5 to each side.}$$

Thus $4.495 < A < 4.505$. The actual height must be greater than 4.495 inches *and* less than 4.505 inches.

Now Try Exercise 83

An Alternative Method There is a second symbolic method that can be used to solve absolute value inequalities. This method is often used in advanced mathematics courses, such as calculus. It is based on the following two properties.

> ## ABSOLUTE VALUE INEQUALITIES (ALTERNATIVE METHOD)
>
> Let k be a positive number.
> 1. $|ax + b| < k$ is equivalent to $-k < ax + b < k$.
> 2. $|ax + b| > k$ is equivalent to $ax + b < -k$ or $ax + b > k$.
>
> Similar statements can be made for inequalities involving $\leq$ or $\geq$.

EXAMPLE 8   Using an alternative method

Solve each absolute value inequality. Write your answer in interval notation.
(a) $|4 - 5x| \leq 3$
(b) $|-4x - 6| > 2$

SOLUTION
(a) $|4 - 5x| \leq 3$ is equivalent to the following three-part inequality.

$$-3 \leq 4 - 5x \leq 3 \qquad \text{Equivalent inequality}$$
$$-7 \leq -5x \leq -1 \qquad \text{Subtract 4 from each part.}$$
$$\frac{7}{5} \geq x \geq \frac{1}{5} \qquad \text{Divide each part by } -5; \text{ reverse the inequality.}$$

In interval notation the solution is $\left[\frac{1}{5}, \frac{7}{5}\right]$.

(b) $|-4x - 6| > 2$ is equivalent to the following compound inequality.

$$-4x - 6 < -2 \quad \text{or} \quad -4x - 6 > 2 \qquad \text{Equivalent compound inequality}$$
$$-4x < 4 \quad \text{or} \quad -4x > 8 \qquad \text{Add 6 to each side.}$$
$$x > -1 \quad \text{or} \quad x < -2 \qquad \text{Divide each by } -4; \text{ reverse the inequality.}$$

In interval notation the solution set is $(-\infty, -2) \cup (-1, \infty)$.

Now Try Exercises 57 and 61

> **CLASS DISCUSSION**
>
> Sketch the graphs of $y = ax + b$, $y = |ax + b|$, $y = -k$, and $y = k$ on one xy-plane. Now use these graphs to explain why the alternative method for solving absolute value inequalities is correct.

2.5 Putting It All Together

The following table summarizes some important concepts from this section.

CONCEPT	EXPLANATION	EXAMPLES
Absolute value function	$f(x) = |x|$: The output from the absolute value function is never negative. $|x|$ is equivalent to $\sqrt{x^2}$. $|x| = \begin{cases} -x & \text{if } x < 0 \\ x & \text{if } x \geq 0 \end{cases}$	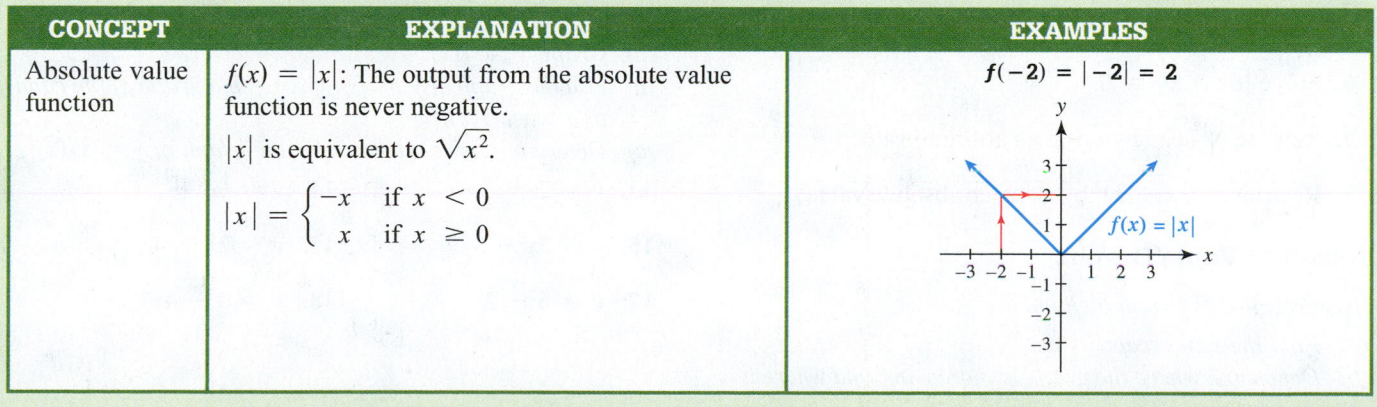

$f(-2) = |-2| = 2$

continued on next page

CONCEPT	EXPLANATION	EXAMPLES
Absolute value equations	**1.** If $k > 0$, then $\lvert ax + b \rvert = k$ has two solutions, given by $ax + b = \pm k$.	**1.** Solve $\lvert 3x - 5 \rvert = 4$. $$3x - 5 = -4 \quad \text{or} \quad 3x - 5 = 4$$ $$3x = 1 \quad \text{or} \quad 3x = 9$$ $$x = \tfrac{1}{3} \quad \text{or} \quad x = 3$$
	2. If $k = 0$, then $\lvert ax + b \rvert = k$ has one solution, given by $ax + b = 0$.	**2.** Solve $\lvert x - 1 \rvert = 0$. $$x - 1 = 0 \quad \text{implies} \quad x = 1.$$
	3. If $k < 0$, then $\lvert ax + b \rvert = k$ has no solutions.	**3.** $\lvert 4x - 9 \rvert = -2$ has no solutions.
Absolute value inequalities	To solve $\lvert ax + b \rvert < k$ or $\lvert ax + b \rvert > k$ with $k > 0$, first solve $\lvert ax + b \rvert = k$. Let these solutions be s_1 and s_2, where $s_1 < s_2$. **1.** $\lvert ax + b \rvert < k$ is equivalent to $s_1 < x < s_2$. **2.** $\lvert ax + b \rvert > k$ is equivalent to $x < s_1$ or $x > s_2$.	To solve $\lvert x - 5 \rvert < 4$ or $\lvert x - 5 \rvert > 4$, first solve $\lvert x - 5 \rvert = 4$ to obtain the solutions $s_1 = 1$ and $s_2 = 9$. **1.** $\lvert x - 5 \rvert < 4$ is equivalent to $1 < x < 9$. **2.** $\lvert x - 5 \rvert > 4$ is equivalent to $x < 1$ or $x > 9$.
Alternative method	**1.** $\lvert ax + b \rvert < k$ with $k > 0$ is equivalent to $$-k < ax + b < k.$$ **2.** $\lvert ax + b \rvert > k$ with $k > 0$ is equivalent to $$ax + b < -k \text{ or } ax + b > k.$$	**1.** $\lvert x - 1 \rvert < 5$ is solved as follows. $$-5 < x - 1 < 5$$ $$-4 < x < 6$$ **2.** $\lvert x - 1 \rvert > 5$ is solved as follows. $$x - 1 < -5 \quad \text{or} \quad x - 1 > 5$$ $$x < -4 \quad \text{or} \quad x > 6$$

2.5 Exercises

Basic Concepts

Exercises 1–8: Let $a \neq 0$.

1. Solve $\lvert x \rvert = 3$.

2. Solve $\lvert x \rvert \leq 3$.

3. Solve $\lvert x \rvert > 3$.

4. Solve $\lvert ax + b \rvert \leq -2$.

5. Describe the graph of $y = \lvert ax + b \rvert$.

6. Solve $\lvert ax + b \rvert = 0$.

7. Rewrite $\sqrt{36a^2}$ by using an absolute value.

8. Rewrite $\sqrt{(ax + b)^2}$ by using an absolute value.

Absolute Value Graphs

Exercises 9–12: Graph by hand.

(a) Find the x-intercept.

(b) Determine where the graph is increasing and where it is decreasing.

9. $y = \lvert x + 1 \rvert$

10. $y = \lvert 1 - x \rvert$

11. $y = \lvert 2x - 3 \rvert$

12. $y = \lvert \tfrac{1}{2}x + 1 \rvert$

Exercises 13–18: (Refer to Example 1.) Do the following.

(a) Graph $y = f(x)$.

(b) Use the graph of $y = f(x)$ to sketch a graph of the equation $y = \lvert f(x) \rvert$.

(c) Determine the x-intercept for the graph of $y = \lvert f(x) \rvert$.

13. $y = 2x$

14. $y = \tfrac{1}{2}x$

15. $y = 3x - 3$

16. $y = 2x - 4$

17. $y = 6 - 2x$

18. $y = 2 - 4x$

Absolute Value Equations and Inequalities

Exercises 19–40: Solve the absolute value equation.

19. $|-2x| = 4$ **20.** $|3x| = -6$

21. $|5x - 7| = 2$ **22.** $|-3x - 2| = 5$

23. $|3 - 4x| = 5$ **24.** $|2 - 3x| = 1$

25. $|-6x - 2| = 0$ **26.** $|6x - 9| = 0$

27. $|7 - 16x| = 0$ **28.** $|-x - 4| = 0$

29. $|17x - 6| = -3$ **30.** $|-8x - 11| = -7$

31. $|1.2x - 1.7| - 5 = -1$

32. $|3 - 3x| - 2 = 2$

33. $|4x - 5| + 3 = 2$ **34.** $|4.5 - 2x| + 1.1 = 9.7$

35. $|2x - 9| = |8 - 3x|$ **36.** $|x - 3| = |8 - x|$

37. $\left|\frac{3}{4}x - \frac{1}{4}\right| = \left|\frac{3}{4} - \frac{1}{4}x\right|$ **38.** $\left|\frac{1}{2}x + \frac{3}{2}\right| = \left|\frac{3}{2}x - \frac{7}{2}\right|$

39. $|15x - 5| = |35 - 5x|$

40. $|20x - 40| = |80x - 20|$

Exercises 41 and 42. The graphs of f and g are shown. Solve each equation and inequality.

41. (a) $f(x) = g(x)$

(b) $f(x) < g(x)$

(c) $f(x) > g(x)$

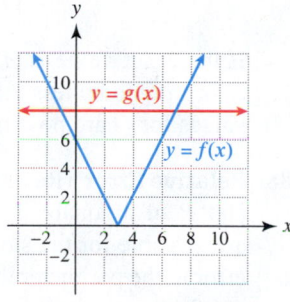

42. (a) $f(x) = g(x)$

(b) $f(x) \leq g(x)$

(c) $f(x) \geq g(x)$

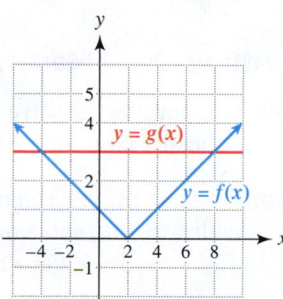

Exercises 43 and 44: Solve each equation or inequality.

43. (a) $|2x - 3| = 1$ **44. (a)** $|5 - x| = 2$

(b) $|2x - 3| < 1$ **(b)** $|5 - x| \leq 2$

(c) $|2x - 3| > 1$ **(c)** $|5 - x| \geq 2$

Exercises 45–48: Solve the equation
(a) graphically,
(b) numerically, and
(c) symbolically.
Then solve the related inequality.

45. $|2x - 5| = 10$, $|2x - 5| < 10$

46. $|3x - 4| = 8$, $|3x - 4| \leq 8$

47. $|5 - 3x| = 2$, $|5 - 3x| > 2$

48. $|4x - 7| = 5$, $|4x - 7| \geq 5$

Exercises 49–54: Solve the equation symbolically. Then solve the related inequality.

49. $|2.1x - 0.7| = 2.4$, $|2.1x - 0.7| \geq 2.4$

50. $\left|\frac{1}{2}x - \frac{3}{4}\right| = \frac{7}{4}$, $\left|\frac{1}{2}x - \frac{3}{4}\right| \leq \frac{7}{4}$

51. $|3x| + 5 = 6$, $|3x| + 5 > 6$

52. $|x| - 10 = 25$, $|x| - 10 < 25$

53. $\left|\frac{2}{3}x - \frac{1}{2}\right| = -\frac{1}{4}$, $\left|\frac{2}{3}x - \frac{1}{2}\right| \leq -\frac{1}{4}$

54. $|5x - 0.3| = -4$, $|5x - 0.3| > -4$

You Decide the Method

Exercises 55–66: Solve the inequality. Write the solution in interval notation.

55. $|3x - 1| < 8$ **56.** $|15 - x| < 7$

57. $|7 - 4x| \leq 11$ **58.** $|-3x + 1| \leq 5$

59. $|0.5x - 0.75| < 2$ **60.** $|2.1x - 5| \leq 8$

61. $|2x - 3| > 1$ **62.** $|5x - 7| > 2$

63. $|-3x + 8| \geq 3$ **64.** $|-7x - 3| \geq 5$

65. $|0.25x - 1| > 3$ **66.** $|-0.5x + 5| \geq 4$

Domain and Range

67. If $f(k) = -6$, what is the value of $|f(k)|$?

68. If $f(k) = 17$, what is the value of $|f(k)|$?

69. If the domain of $f(x)$ is given by $[-2, 4]$, what is the domain of $|f(x)|$?

70. If the domain of $f(x)$ is given by $(-\infty, 0]$, what is the domain of $|f(x)|$?

71. If the range of $f(x)$ is given by $(-\infty, 0]$, what is the range of $|f(x)|$?

72. If the range of $f(x)$ is given by $(-4, 5)$, what is the range of $|f(x)|$?

Applications

73. **Speed Limits** The maximum and minimum lawful speeds S on an interstate highway satisfy the equation $|S - 57.5| = 17.5$. Find the maximum and minimum speed limits.

74. **Human Cannonball** A human cannonball plans to travel 180 feet and land squarely on a net with a 70-foot-long safe zone.
    (a) What distances D can this performer travel and still land safely on the net?

    (b) Use an absolute value inequality to describe the restrictions on D.

75. **Temperature and Altitude** Air temperature decreases as altitude increases. If the ground temperature is 80°F, then the air temperature x miles high is $T = 80 - 19x$.
    (a) Determine the altitudes x where the air temperature T is between 0°F and 32°F, inclusive.

    (b) Use an absolute value inequality to describe these altitudes.

76. **Dew Point and Altitude** The dew point decreases as altitude increases. If the dew point on the ground is 80°F, then the dew point x miles high is $D = 80 - \frac{29}{5}x$.
    (a) Determine the altitudes x where the dew point D is between 50°F and 60°F, inclusive.

    (b) Use an absolute value inequality to describe these altitudes.

Exercises 77–82: **Average Temperatures** *The inequality describes the range of monthly average temperatures T in degrees Fahrenheit at a certain location.*

(a) Solve the inequality.
(b) If the high and low monthly average temperatures satisfy equality, interpret the inequality.

77. $|T - 43| \leq 24$, Marquette, Michigan

78. $|T - 62| \leq 19$, Memphis, Tennessee

79. $|T - 50| \leq 22$, Boston, Massachusetts

80. $|T - 10| \leq 36$, Chesterfield, Canada

81. $|T - 61.5| \leq 12.5$, Buenos Aires, Argentina

82. $|T - 43.5| \leq 8.5$, Punta Arenas, Chile

83. **Classic iPod Dimensions** The classic iPod is 10.5 millimeters thick. Suppose that the actual thickness T of any particular iPod has a maximum error tolerance that is less than 0.05 millimeter.
    (a) Write an absolute value statement that describes this situation.

    (b) Solve this inequality for T and interpret your result.

84. **Error In Measurements** An aluminum can should have a diameter D of 3 inches with a maximum error tolerance that is less than 0.004 inch.
    (a) Write an absolute value statement that describes this situation.

    (b) Solve this inequality for D and interpret your result.

85. **Machine Parts** A part for a machine must fit into a hole and must have a diameter D that is less than or equal to 2.125 inches. The diameter D cannot be less than this maximum diameter of 2.125 inches by more than 0.014 inch.
    (a) Write an absolute value statement that describes this situation.

    (b) Solve this inequality for D and interpret your result.

86. **Error in Measurements** Suppose that a 12-inch ruler must have the correct length L to within 0.0002 inch.
    (a) Write an absolute value inequality for L that describes this requirement.

    (b) Solve this inequality and interpret the results.

87. **Relative Error** If a quantity is measured to be Q and its exact value is A, then the relative error in Q is

    $$\left| \frac{Q - A}{A} \right|.$$

    If the exact value is $A = 35$ and you want the relative error in Q to be less than or equal to 0.02 (or 2%), what values for Q are possible?

88. **Relative Error** (Refer to Exercise 87.) The exact perimeter P of a square is 50 feet. What measured lengths are possible for the side S of the square to have relative error in the perimeter that is less than or equal to 0.04 (or 4%)?

Writing about Mathematics

89. Explain how to solve $|ax + b| = k$ with $k > 0$ symbolically. Give an example.

90. Explain how you can use the solutions to $|ax + b| = k$ with $k > 0$ to solve the inequalities $|ax + b| < k$ and $|ax + b| > k$. Give an example.

Extended and Discovery Exercises

1. Let δ be a positive number and let x and c be real numbers. Write an absolute value inequality that expresses that the distance between x and c on the number line is less than δ.

2. Let ε be a positive number, L be a real number, and f be a function. Write an absolute value inequality that expresses that the distance between $f(x)$ and L on the number line is less than ε.

CHECKING BASIC CONCEPTS FOR SECTION 2.5

1. Rewrite $\sqrt{4x^2}$ by using an absolute value.

2. Graph $y = |3x - 2|$ by hand.

3. (a) Solve the equation $|2x - 1| = 5$.

   (b) Use part (a) to solve the absolute value inequalities $|2x - 1| \le 5$ and $|2x - 1| > 5$.

4. Solve each equation or inequality. For each inequality, write the solution set in interval notation.
   (a) $|2 - 5x| - 4 = -1$

   (b) $|3x - 5| \le 4$

   (c) $\left|\frac{1}{2}x - 3\right| > 5$

5. Solve $|x + 1| = |2x|$.

2 | Summary

CONCEPT	EXPLANATION AND EXAMPLES
Section 2.1 Equations of Lines	
Point-Slope Form	If a line with slope m passes through (x_1, y_1), then $$y = m(x - x_1) + y_1 \quad \text{or} \quad y - y_1 = m(x - x_1).$$ **Example:** $y = -\frac{3}{4}(x + 4) + 5$ has slope $-\frac{3}{4}$ and passes through $(-4, 5)$.
Slope-Intercept Form	If a line has slope m and y-intercept b, then $y = mx + b$. **Example:** $y = 3x - 4$ has slope 3 and y-intercept -4.
Determining Intercepts	To find the x-intercept(s), let $y = 0$ in the equation and solve for x. To find the y-intercept(s), let $x = 0$ in the equation and solve for y. **Examples:** The x-intercept on the graph of $3x - 4y = 12$ is 4 because $3x - 4(0) = 12$ implies that $x = 4$. The y-intercept on the graph of $3x - 4y = 12$ is -3 because $3(0) - 4y = 12$ implies that $y = -3$.
Horizontal and Vertical Lines	A horizontal line passing through the point (a, b) is given by $y = b$, and a vertical line passing through (a, b) is given by $x = a$. **Examples:** The horizontal line $y = -3$ passes through $(6, -3)$. The vertical line $x = 4$ passes through $(4, -2)$.
Parallel and Perpendicular Lines	Parallel lines have equal slopes satisfying $m_1 = m_2$, and perpendicular lines have slopes satisfying $m_1 m_2 = -1$, provided neither line is vertical. **Examples:** The lines $y_1 = 3x - 1$ and $y_2 = 3x + 4$ are parallel. The lines $y_1 = 3x - 1$ and $y_2 = -\frac{1}{3}x + 4$ are perpendicular.

CONCEPT	EXPLANATION AND EXAMPLES

Section 2.1 Equations of Lines (CONTINUED)

Linear Regression

One way to determine a linear function or a line that models data is to use the method of least squares. This method determines a unique line that can be found with a calculator. The correlation coefficient r $(-1 \leq r \leq 1)$ measures how well a line fits the data.

Example: The line of least squares modeling the data $(1, 1)$, $(3, 4)$, and $(4, 6)$ is given by $y \approx 1.643x - 0.714$, with $r \approx 0.997$.

Section 2.2 Linear Equations

Linear Equation

Can be written as $ax + b = 0$ with $a \neq 0$ and has one solution

Example: The solution to $2x - 4 = 0$ is 2 because $2(2) - 4 = 0$.

Properties of Equality

Addition: $a = b$ is equivalent to $a + c = b + c$.
Multiplication: $a = b$ is equivalent to $ac = bc$, provided $c \neq 0$.

Example:

$\frac{1}{2}x - 4 = 3$	Given equation
$\frac{1}{2}x = 7$	Addition property; add 4.
$x = 14$	Multiplication property; multiply by 2.

Contradiction, Identity, and Conditional Equation

A contradiction has no solutions, an identity is true for all (meaningful) values of the variable, and a conditional equation is true for some, but not all, values of the variable.

Examples:

$3(1 - 2x) = 3 - 6x$	Identity
$x + 5 = x$	Contradiction
$x - 1 = 4$	Conditional equation

Intersection-of-Graphs and x-Intercept Methods

Intersection-of-graphs method: Set y_1 equal to the left side of the equation and set y_2 equal to the right side. The x-coordinate of a point of intersection is a solution.

Example: The graphs of $y_1 = 2x$ and $y_2 = x + 1$ intersect at $(1, 2)$, so the solution to the linear equation $2x = x + 1$ is 1. See the figure below on the left.

x-intercept method: Move all terms to the left side of the equation. Set y_1 equal to the left side of the equation. The solutions are the x-intercepts. (See Section 2.3.)

Example: Write $2x = x + 1$ as $2x - (x + 1) = 0$. Graph $y_1 = 2x - (x + 1)$. The only x-intercept is 1, as shown in the figure on the right.

Point of Intersection (**1**, **2**)   x-intercept: **1**

CONCEPT	EXPLANATION AND EXAMPLES

Section 2.2 Linear Equations (CONTINUED)

Problem-Solving Strategies

STEP 1: Read the problem and make sure you understand it. Assign a variable to what you are being asked to find. If necessary, write other quantities in terms of this variable.

STEP 2: Write an equation that relates the quantities described in the problem. You may need to sketch a diagram and refer to known formulas.

STEP 3: Solve the equation and determine the solution.

STEP 4: Look back and check your solution. Does it seem reasonable?

Section 2.3 Linear Inequalities

Linear Inequality

Can be written as $ax + b > 0$ with $a \neq 0$, where $>$ can be replaced by $<$, $\leq$, or $\geq$. If the solution to $ax + b = 0$ is k, then the solution to the linear inequality $ax + b > 0$ is either the interval $(-\infty, k)$ or the interval (k, ∞).

Example: $3x - 1 < 2$ is linear since it can be written as $3x - 3 < 0$. The solution set is $\{x \mid x < 1\}$, or $(-\infty, 1)$.

Properties of Inequalities

Addition: $a < b$ is equivalent to $a + c < b + c$.
Multiplication: $a < b$ is equivalent to $ac < bc$ when $c > 0$.
$a < b$ is equivalent to $ac > bc$ when $c < 0$.

Example: $-3x - 4 < 14$ *Given equation*

$-3x < 18$ *Addition property; add 4.*

$x > -6$ *Multiplication property; divide by -3. Reverse the inequality symbol.*

Compound Inequality

Example: $x \geq -2$ and $x \leq 4$ is equivalent to $-2 \leq x \leq 4$. This is called a three-part inequality.

Section 2.4 More Modeling with Functions

Linear Model

If a quantity increases or decreases by a constant amount for each unit increase in x, then it can be modeled by a linear function given by

$$f(x) = (\text{constant rate of change})x + (\text{initial amount}).$$

Example: If water is pumped from a full 100-gallon tank at 7 gallons per minute, then $A(t) = 100 - 7t$ gives the gallons of water in the tank after t minutes.

Piecewise-Defined Function

A function defined by more than one formula on its domain

Examples: Step function, greatest integer function, absolute value function, and

$$f(x) = \begin{cases} 4 - x & \text{if } -4 \leq x < 1 \\ 3x & \text{if } 1 \leq x \leq 5 \end{cases}$$

It follows that $f(2) = 6$ because if $1 \leq x \leq 5$ then $f(x) = 3x$. Note that f is continuous on its domain of $[-4, 5]$.

CONCEPT	EXPLANATION AND EXAMPLES

Section 2.4 More Modeling with Functions (CONTINUED)

Direct Variation

A quantity y is directly proportional to a quantity x, or y varies directly with x, if $y = kx$, where $k \neq 0$. If data vary directly, the ratios $\frac{y}{x}$ are equal to the constant of variation k.

Example: If a person works for $8 per hour, then that person's pay P is directly proportional to, or varies directly with, the number of hours H that the person works by the equation $P = 8H$, where the constant of variation is $k = 8$.

Section 2.5 Absolute Value Equations and Inequalities

Absolute Value Function

An absolute value function is defined by $f(x) = |x|$. Its graph is V-shaped. An equivalent formula is $f(x) = \sqrt{x^2}$.

Examples: $f(-9) = |-9| = 9$; $\sqrt{(2x+1)^2} = |2x+1|$

Absolute Value Equations

$|ax + b| = k$ with $k > 0$ is equivalent to $ax + b = \pm k$.

Example: $|2x - 3| = 4$ is equivalent to $2x - 3 = 4$ or $2x - 3 = -4$.

The solutions are $\frac{7}{2}$ and $-\frac{1}{2}$.

Absolute Value Inequalities

Let the solutions to $|ax + b| = k$ be s_1 and s_2, where $s_1 < s_2$ and $k > 0$.

1. $|ax + b| < k$ is equivalent to $s_1 < x < s_2$.

2. $|ax + b| > k$ is equivalent to $x < s_1$ or $x > s_2$.

Similar statements can be made for inequalities involving $\leq$ or $\geq$.

Example: The solutions to $|2x + 1| = 5$ are given by $x = -3$ and $x = 2$.
The solutions to $|2x + 1| < 5$ are given by $-3 < x < 2$.
The solutions to $|2x + 1| > 5$ are given by $x < -3$ or $x > 2$.

2 Review Exercises

Exercises 1 and 2: Find the point-slope form of the line passing through the given points. Use the first point as (x_1, y_1).

1. $(-3, 4), (2, 5)$ **2.** $(1, -6), (-7, 5)$

Exercises 3–6: Find the point-slope form of the line passing through the given points. Use the first point as (x_1, y_1). Then convert the equation to slope-intercept form and write a formula for a function f whose graph is the line.

3. Slope $-\frac{7}{5}$, passing through $(-5, 6)$

4. Passing through $(-4, -7)$ and $(-2, 3)$

5. x-intercept 5, y-intercept -2

6. y-intercept 6, passing through $(8, -2)$

7. Write a formula for a linear function f whose graph has slope -2 and passes through $(-2, 3)$.

8. Find the average rate of change of $f(x) = -3x + 8$ from -2 to 3.

Exercises 9–14: Find the slope-intercept form of the equation of a line satisfying the conditions.

9. Slope 7, passing through $(-3, 9)$

10. Passing through $(2, -4)$ and $(7, -3)$

11. Passing through $(1, -1)$, parallel to $y = -3x + 1$

12. Passing through the point $(-2, 1)$, perpendicular to the line $y = 2(x + 5) - 22$

13. Parallel to the line segment connecting $(0, 3.1)$ and $(5.7, 0)$, passing through $(1, -7)$

14. Perpendicular to $y = -\frac{5}{7}x$, passing through $\left(\frac{6}{7}, 0\right)$

Exercises 15–20: Find an equation of the specified line.

15. Parallel to the y-axis, passing through $(6, -7)$

16. Parallel to the x-axis, passing through $(-3, 4)$

17. Horizontal, passing through $(1, 3)$

18. Vertical, passing through $(1.5, 1.9)$

19. Vertical with x-intercept 2.7

20. Horizontal with y-intercept -8

Exercises 21 and 22: Determine the x- and y-intercepts for the graph of the equation. Graph the equation.

21. $5x - 4y = 20$　　**22.** $\dfrac{x}{3} - \dfrac{y}{2} = 1$

Exercises 23–28: Solve the linear equation either symbolically or graphically.

23. $5x - 22 = 10$　　**24.** $5(4 - 2x) = 16$

25. $-2(3x - 7) + x = 2x - 1$

26. $5x - \frac{1}{2}(4 - 3x) = \frac{3}{2} - (2x + 3)$

27. $\pi x + 1 = 6$

28. $\dfrac{x - 4}{2} = x + \dfrac{1 - 2x}{3}$

Exercises 29 and 30: Use a table to solve each linear equation numerically to the nearest tenth.

29. $3.1x - 0.2 - 2(x - 1.7) = 0$

30. $\sqrt{7} - 3x - 2.1(1 + x) = 0$

Exercises 31–34: Complete the following.

(a) Solve the equation symbolically.
(b) Classify the equation as a contradiction, an identity, or a conditional equation.

31. $4(6 - x) = -4x + 24$

32. $\frac{1}{2}(4x - 3) + 2 = 3x - (1 + x)$

33. $5 - 2(4 - 3x) + x = 4(x - 3)$

34. $\dfrac{x - 3}{4} + \dfrac{3}{4}x - 5(2 - 7x) = 36x - \dfrac{43}{4}$

Exercises 35–38: Express the inequality in interval notation.

35. $x > -3$　　**36.** $x \le 4$

37. $-2 \le x < \frac{3}{4}$　　**38.** $x \le -2$ or $x > 3$

Exercises 39–44: Solve the linear inequality. Write the solution set in set-builder or interval notation.

39. $3x - 4 \le 2 + x$　　**40.** $-2x + 6 \le -3x$

41. $\dfrac{2x - 5}{2} < \dfrac{5x + 1}{5}$

42. $-5(1 - x) > 3(x - 3) + \frac{1}{2}x$

43. $-2 \le 5 - 2x < 7$　　**44.** $-1 < \dfrac{3x - 5}{-3} < 3$

Exercises 45 and 46: Solve the inequality graphically.

45. $2x > x - 1$　　**46.** $-1 \le 1 + x \le 2$

47. The graphs of two linear functions f and g are shown in the figure. Solve each equation or inequality.

　　(a) $f(x) = g(x)$

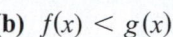

　　(b) $f(x) < g(x)$

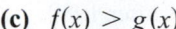

　　(c) $f(x) > g(x)$

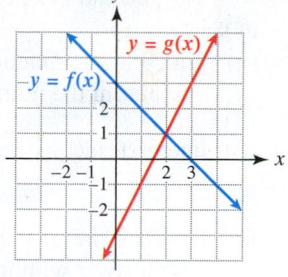

48. The graphs of three linear functions f, g, and h with domains $D = \{x \,|\, 0 \le x \le 7\}$ are shown in the figure. Solve each equation or inequality.

　　(a) $f(x) = g(x)$

　　(b) $g(x) = h(x)$

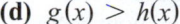

　　(c) $f(x) < g(x) < h(x)$

　　(d) $g(x) > h(x)$

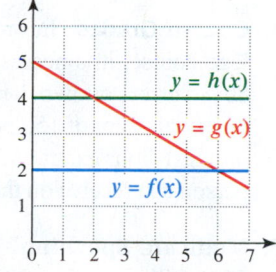

49. Use $f(x)$ to complete the following.

$$f(x) = \begin{cases} 8 + 2x & \text{if } -3 \le x \le -1 \\ 5 - x & \text{if } -1 < x \le 2 \\ x + 1 & \text{if } 2 < x \le 5 \end{cases}$$

　　(a) Evaluate f at $x = -2, -1, 2,$ and 3.

　　(b) Sketch a graph of f. Is f continuous on its domain?

　　(c) Determine the x-value(s) where $f(x) = 3$.

50. If $f(x) = [\![2x - 1]\!]$, evaluate $f(-3.1)$ and $f(2.5)$.

Exercises 51–54: Solve the equation.

51. $|2x - 5| - 1 = 8$ **52.** $|3 - 7x| = 10$

53. $|6 - 4x| = -2$ **54.** $|9 + x| = |3 - 2x|$

Exercises 55–58: Solve the equation. Use the solutions to help solve the related inequality.

55. $|x| = 3$, $|x| > 3$

56. $|-3x + 1| = 2, |-3x + 1| < 2$

57. $|3x - 7| = 10, |3x - 7| > 10$

58. $|4 - x| = 6$, $|4 - x| \leq 6$

Exercises 59–62: Solve the inequality.

59. $|3 - 2x| < 9$ **60.** $|-2x - 3| > 3$

61. $\left|\frac{1}{3}x - \frac{1}{6}\right| \geq 1$ **62.** $\left|\frac{1}{2}x\right| - 3 \leq 5$

Applications

63. U.S. Median Income In 1980 the median family income was \$17,700 and in 2010 it was \$49,500. (*Source:* Bureau of the Census.)
  (a) Find a linear function f that models these data. Let x represent the number of years after 1980.

  (b) Interpret the slope and y-intercept for the graph of f.

  (c) Estimate the median income in 1992 and compare your answer with the true value of \$30,600.

  (d) Predict the year when median income might reach \$60,000. Did your answer involve interpolation or extrapolation?

64. Course Grades In order to receive a B grade in a college course, it is necessary to have an overall average of 80% correct on two 1-hour exams of 75 points each and one final exam of 150 points. If a person scores 55 and 72 on the 1-hour exams, what is the minimum score that the person can receive on the final exam and still earn a B?

65. Medicare Spending In 2010 Medicare spending was \$524 billion and in 2020 it is projected to be \$949 billion. (*Source:* Congressional Budget Office.)
  (a) Find a linear function f that models these data. Let x represent the number of years after 2010.

  (b) Interpret the slope and y-intercept for the graph of f.

  (c) Estimate Medicare spending in 2016. Did your answer involve interpolation or extrapolation?

  (d) Predict the years when Medicare spending could be between \$694 billion and \$864 billion.

66. Temperature Scales The table shows equivalent temperatures in degrees Celsius and degrees Fahrenheit.

°F	−40	32	59	95	212
°C	−40	0	15	35	100

(a) Plot the data with Fahrenheit temperature on the x-axis and Celsius temperature on the y-axis. What type of relation exists between the data?

  (b) Find a function C that receives the Fahrenheit temperature x as input and outputs the corresponding Celsius temperature. Interpret the slope.

  (c) If the temperature is 83°F, what is it in degrees Celsius?

67. Distance from Home The graph depicts the distance y that a person driving a car on a straight road is from home after x hours. Interpret the graph. What speeds did the car travel?

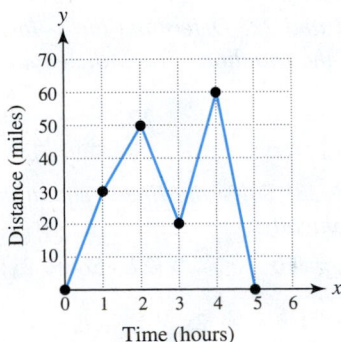

68. Piecewise-Linear Function Given the data points $(1, 2)$, $(4, 9)$, and $(6, 3)$, complete the following.
  (a) Write the formula for a piecewise-linear function f that passes through these data points whose domain is $1 \leq x \leq 6$.

  (b) Evaluate $f(5)$.

  (c) Is f continuous on its domain?

69. Population Estimates In 2008 the population of a city was 143,247, and in 2012 it was 167,933. Estimate the population in 2010.

70. Distance A driver of a car is initially 455 miles from home, traveling toward home on a straight freeway at 70 miles per hour.
  (a) Write a formula for a linear function f that models the distance between the driver and home after x hours.

  (b) Graph f. What is an appropriate domain?

  (c) Identify the x- and y-intercepts. Interpret each.

71. Working Together Suppose that one worker can shovel snow from a storefront sidewalk in 50 minutes and another worker can shovel it in 30 minutes. How long will it take if they work together?

72. Antifreeze Initially, a tank contains 20 gallons of a 30% antifreeze solution. How many gallons of an 80% antifreeze solution should be added to the tank in order to increase the concentration of the antifreeze in the tank to 50%?

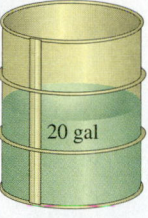

73. Running An athlete traveled 13.5 miles in 1 hour and 48 minutes, jogging at 7 miles per hour and then at 8 miles per hour. How long did the runner jog at each speed?

74. Least-Squares Fit The table lists the actual annual cost y to drive a midsize car 15,000 miles per year for selected years x.

x	1970	1980	1990	2000	2010
y	$1763	$3176	$5136	$6880	$8595

Source: Runzheimer International.

(a) Predict whether the correlation coefficient is positive, negative, or zero.

(b) Find a least-squares regression line that models these data. What is the correlation coefficient?

(c) Estimate the cost of driving a midsize car in 2005.

(d) Estimate the year when the cost to drive a car could reach $10,000.

75. Modeling The table lists data that are exactly linear.

x	−3	−2	−1	1	2
y	6.6	5.4	4.2	1.8	0.6

(a) Determine the slope-intercept form of the line that passes through these data points.

(b) Predict y when $x = -1.5$ and 3.5. State whether these calculations involve interpolation or extrapolation.

(c) Predict x when $y = 1.3$.

76. Geometry A rectangle is twice as long as it is wide and has a perimeter of 78 inches. Find the width and length of this rectangle.

77. Flow Rates A water tank has an inlet pipe with a flow rate of 5 gallons per minute and an outlet pipe with a flow rate of 3 gallons per minute. A pipe can be either closed or completely open. The graph shows the number of gallons of water in the tank after x minutes have

elapsed. Use the concept of slope to interpret each piece of this graph.

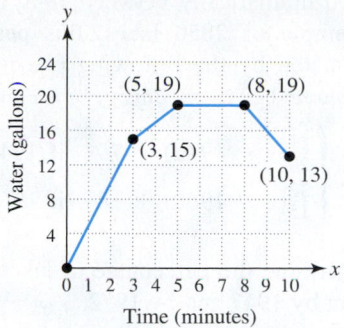

78. Flow Rates (Refer to Exercise 77.) Suppose the tank is modified so that it has a second inlet pipe, which flows at a rate of 2 gallons per minute. Interpret the graph by determining when each inlet and outlet pipe is open or closed.

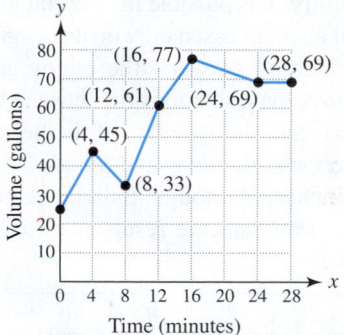

79. Air Temperature For altitudes up to 4 kilometers, moist air will cool at a rate of about 6°C per kilometer. If the ground temperature is 25°C, at what altitudes would the air temperature be from 5°C to 15°C? (*Source:* A. Miller and R. Anthes, *Meteorology.*)

80. Water Pollution At one time the Thames River in England supported an abundant community of fish. Pollution then destroyed all the fish in a 40-mile stretch near its mouth for a 45-year period beginning in 1915. Since then, improvement of sewage treatment facilities and other ecological steps have resulted in a dramatic increase in the number of different fish present. The number of species present from 1967 to 1978 can be modeled by $f(x) = 6.15x - 12{,}059$, where x is the year.
(a) Estimate the year when the number of species first exceeded 70.

(b) Estimate the years when the number of species was between 50 and 100.

81. Relative Error The actual length of a side of a building is 52.3 feet. How accurately must an apprentice carpenter measure this side to have the relative error in the measurement be less than 0.003 (0.3%)? (*Hint:* Use $\left|\frac{C - A}{A}\right|$, where C is the carpenter's measurement and A is the actual length.)

82. Brown Trout Due to acid rain, the percentage of lakes in Scandinavia that lost their population of brown trout increased dramatically between 1940 and 1975. Based on a sample of 2850 lakes, this percentage can be approximated by the following piecewise-linear function. (*Source:* C. Mason, *Biology of Freshwater Pollution.*)

$$f(x) = \begin{cases} \frac{11}{20}(x - 1940) + 7 & \text{if } 1940 \leq x < 1960 \\ \frac{32}{15}(x - 1960) + 18 & \text{if } 1960 \leq x \leq 1975 \end{cases}$$

(a) Determine the percentage of lakes that lost brown trout by 1947 and by 1972.

(b) Sketch a graph of f.

(c) Is f a continuous function on its domain?

Extended and Discovery Exercises

1. Archeology It is possible for archeologists to estimate the height of an adult based only on the length of the humerus, a bone located between the elbow and the shoulder. The approximate relationship between the height y of an individual and the length x of the humerus is shown in the table for both males and females. All measurements are in inches. Although individual values may vary, tables like this are the result of measuring bones from many skeletons.

x	8	9	10	11
y (females)	50.4	53.5	56.6	59.7
y (males)	53.0	56.0	59.0	62.0

x	12	13	14
y (females)	62.8	65.9	69.0
y (males)	65.0	68.0	71.0

(a) Find the estimated height of a female with a 12-inch humerus.

(b) Plot the ordered pairs (x, y) for both sexes. What type of relation exists between the data?

(c) For each 1-inch increase in the length of the humerus, what are the corresponding increases in the heights of females and of males?

(d) Determine linear functions f and g that model these data for females and males, respectively.

(e) Suppose a humerus from a person of unknown sex is estimated to be between 9.7 and 10.1 inches long. Use f and g to approximate the range for the height of a female and a male.

2. Archeology Continuing with Exercise 1, have members of the class measure their heights and the lengths of their humeri (plural of *humerus*) in inches.
(a) Make a table of the results.

(b) Find regression lines that fit the data points for males and females.

(c) Compare your results with the table in Exercise 1.

3. A Puzzle Three people leave for a city 15 miles away. The first person walks 4 miles per hour, and the other two people ride in a car that travels 28 miles per hour. After some time, the second person gets out of the car and walks 4 miles per hour to the city while the driver goes back and picks up the first person. The driver takes the first person to the city. If all three people arrive in the city at the same time, how far did each person walk?

4. Limit Notation Let ε and δ be positive numbers; let x, c, and L be real numbers; and let f be a function. Consider the following: "If the distance between x and c is less than δ, then the distance between $f(x)$ and L is less than ε." Rewrite this sentence by using two absolute value inequalities.

1–2 Cumulative Review Exercises

1. Write 123,000 and 0.0051 in scientific notation.

2. Write 6.7×10^6 and 1.45×10^{-4} in standard form.

3. Evaluate $\dfrac{4 + \sqrt{2}}{4 - \sqrt{2}}$. Round your answer to the nearest hundredth.

4. The table represents a relation S.

x	−1	0	1	2	3
y	6	4	3	0	0

(a) Does S represent a function?

(b) Determine the domain and range of S.

5. Find the standard equation of a circle with center $(-2, 3)$ and radius 7.

6. Evaluate $-5^2 - 2 - \frac{10 - 2}{5 - 1}$ by hand.

7. Find the exact distance between $(-3, 5)$ and $(2, -3)$.

8. Find the midpoint of the line segment with endpoints $(5, -2)$ and $(-3, 1)$.

9. Find the domain and range of the function shown in the graph. Evaluate $f(-1)$.

(a) 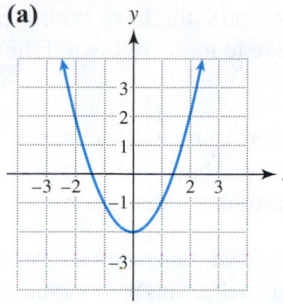   (b)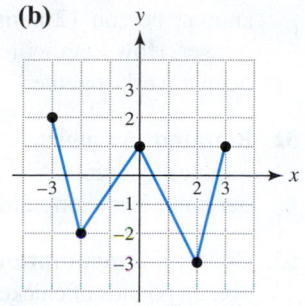

10. Graph f by hand.
(a) $f(x) = 3 - 2x$ (b) $f(x) = |x + 1|$

Exercises 11 and 12: Complete the following.

(a) *Evaluate $f(2)$ and $f(a - 1)$.*
(b) *Determine the domain of f.*

11. $f(x) = 5x - 3$

12. $f(x) = \sqrt{2x - 1}$

13. Determine if the graph represents a function. Explain your answer.

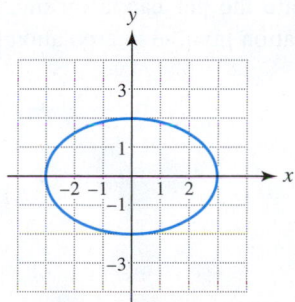

14. Write a formula for a function f that computes the cost of taking x credits if credits cost \$80 each and fees are fixed at \$89.

15. Find the average rate of change of $f(x) = x^2 - 2x + 1$ from $x = 1$ to $x = 2$.

16. Find the difference quotient for $f(x) = 2x^2 - x$.

Exercises 17 and 18: The graph of a linear function f is shown.

(a) *Identify the slope, y-intercept, and x-intercept.*
(b) *Write a formula for f.*
(c) *Find any zeros of f.*

17. 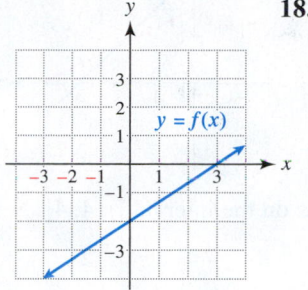   **18.**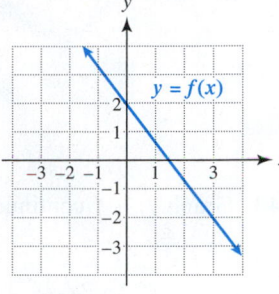

Exercises 19–24: Write an equation of a line satisfying the given conditions. Use slope-intercept form whenever possible.

19. Passing through $(1, -5)$ and $\left(-3, \frac{1}{2}\right)$

20. Passing through the point $(-3, 2)$ and perpendicular to the line $y = \frac{2}{3}x - 7$

21. Parallel to the y-axis and passing through $(-1, 3)$

22. Slope 30, passing through $(2002, 50)$

23. Passing through $(-3, 5)$ and parallel to the line segment connecting $(2.4, 5.6)$ and $(3.9, 8.6)$

24. Perpendicular to the y-axis and passing through the origin

Exercises 25 and 26: Determine the x- and y-intercepts on the graph of the equation. Graph the equation.

25. $-2x + 3y = 6$ **26.** $x = 2y - 3$

Exercises 27–30: Solve the equation.

27. $4x - 5 = 1 - 2x$ **28.** $\frac{2x - 4}{2} = \frac{3x}{7} - 1$

29. $\frac{2}{3}(x - 2) - \frac{4}{5}x = \frac{4}{15} + x$

30. $-0.3(1 - x) - 0.1(2x - 3) = 0.4$

31. Solve $x + 1 = 2x - 2$ graphically and numerically.

32. Solve $2x - (5 - x) = \frac{1 - 4x}{2} + 5(x - 2)$. Is this equation either an identity or a contradiction?

Exercises 33–36: Express each inequality in interval notation.

33. $x < 5$ **34.** $-2 \le x \le 5$

35. $x < -2$ or $x > 2$ **36.** $x \ge -3$

Exercises 37 and 38: Solve the inequality. Write the solution set in set-builder or interval notation.

37. $-3(1 - 2x) + x \le 4 - (x + 2)$

38. $\frac{1}{3} \le \frac{2 - 3x}{2} < \frac{4}{3}$

39. The graphs of two linear functions f and g are shown. Solve each equation or inequality.

(a) $f(x) = g(x)$

(b) $f(x) > g(x)$

(c) $f(x) \leq g(x)$

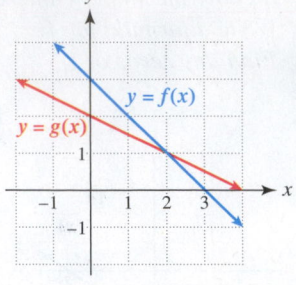

40. Graph f. Is f continuous on the interval $[-4, 4]$?

$$f(x) = \begin{cases} 2 - x & \text{if } -4 \leq x < -2 \\ \frac{1}{2}x + 5 & \text{if } -2 \leq x < 2 \\ 2x + 1 & \text{if } 2 \leq x \leq 4 \end{cases}$$

Exercises 41–44: Solve the equation.

41. $|d + 1| = 5$

42. $|3 - 2x| = 7$

43. $|2t| - 4 = 10$

44. $|11 - 2x| = |3x + 1|$

Exercises 45 and 46: Solve the inequality.

45. $|2t - 5| \leq 5$

46. $|5 - 5t| > 7$

Applications

47. Cost A company's cost C in dollars for making x computers is $C(x) = 500x + 20,000$.

(a) Evaluate $C(1500)$. Interpret the result.

(b) Find the slope of the graph of C. Interpret the slope and y-intercept.

48. Distance At midnight car A is traveling north at 60 miles per hour and is located 40 miles south of car B. Car B is traveling west at 70 miles per hour. Approximate the distance between the cars at 1:15 A.M. to the nearest tenth of a mile.

49. Average Rate of Change On a warm summer day the Fahrenheit temperature x hours past noon is given by the formula $T(x) = 70 + \frac{3}{2}x^2$.

(a) Find the average rate of change of T from 2:00 P.M. to 4:00 P.M.

(b) Interpret this average rate of change.

50. Distance from Home A driver is initially 270 miles from home, traveling toward home on a straight interstate at 72 miles per hour.

(a) Write a formula for a function D that models the distance between the driver and home after x hours.

(b) What is an appropriate domain for D? Graph D.

(c) Identify the x- and y-intercepts. Interpret each.

51. Working Together Suppose one person can mow a large lawn in 5 hours with a riding mower and it takes another person 12 hours to mow the lawn with a push mower. How long will it take to mow the lawn if the two people work together?

52. Running An athlete traveled 15 miles in 1 hour and 45 minutes, running at 8 miles per hour and then 10 miles per hour. How long did the athlete run at each speed?

53. Chicken Consumption In 2001 Americans ate, on average, 56 pounds of chicken annually. This amount increased to 84 pounds in 2010. (*Source:* Department of Agriculture.)

(a) Determine a formula

$$f(x) = m(x - x_1) + y_1$$

that models these data. Let x be the year.

(b) Estimate the annual chicken consumption in 2015.

54. Income The table lists per capita income.

Year	1980	1990	2000	2010
Income	$10,183	$19,572	$29,760	$40,584

Source: Bureau of Economic Analysis.

(a) Find the least-squares regression line for the data.

(b) Estimate the per capita income in 1995. Did this calculation involve interpolation or extrapolation?

3

Quadratic Functions and Equations

The last basketball player in the NBA to shoot foul shots underhand was Rick Barry, who retired in 1980. On average, he was able to make about 9 out of 10 shots. Since then, every NBA player has used the overhand style of shooting foul shots—even though this style has often resulted in lower free-throw percentages.

According to Dr. Peter Brancazio, a physics professor emeritus from Brooklyn College and author of *Sports Science*, there are good reasons for shooting underhand. An underhand shot obtains a higher arc, and as the ball approaches the hoop, it has a better chance of going through the hoop than does a ball with a flatter arc. Lower release points require steeper arcs and increase the chances of the ball passing through the hoop. (See the Extended and Discovery Exercise at the end of this chapter to model the arc of a basketball.)

Rick Barry

Mathematics plays an important role in analyzing applied problems such as shooting foul shots. Whether NBA players choose to listen to Professor Brancazio is another question, but mathematics tells us that steeper arcs are necessary for accurate foul shooting.

Source: Curtis Rist, "The Physics of Foul Shots," *Discover*, October 2000. (Photograph reprinted with permission.)

3.1 Quadratic Functions and Models

- **Learn basic concepts about quadratic functions and their graphs**
- **Complete the square and apply the vertex formula**
- **Graph a quadratic function by hand**
- **Solve applications and model data**
- **Use quadratic regression to model data (optional)**

Introduction

Sometimes when data lie on or nearly on a line, they can be modeled with a linear function ($f(x) = mx + b$) and are called *linear data*. Data that are not linear are called **nonlinear data** and must be modeled with a nonlinear function. One of the simplest types of nonlinear functions is a *quadratic function,* which can be used to model the flight of a baseball, an athlete's heart rate, or MySpace advertising revenue. Figures 3.1–3.3 illustrate three sets of nonlinear data that can be modeled by a quadratic function. Although the complete graph of a quadratic function is either ∪-shaped or ∩-shaped, we often use only one side or a portion of the graph to model real-world data.

Data Modeled with a Parabola

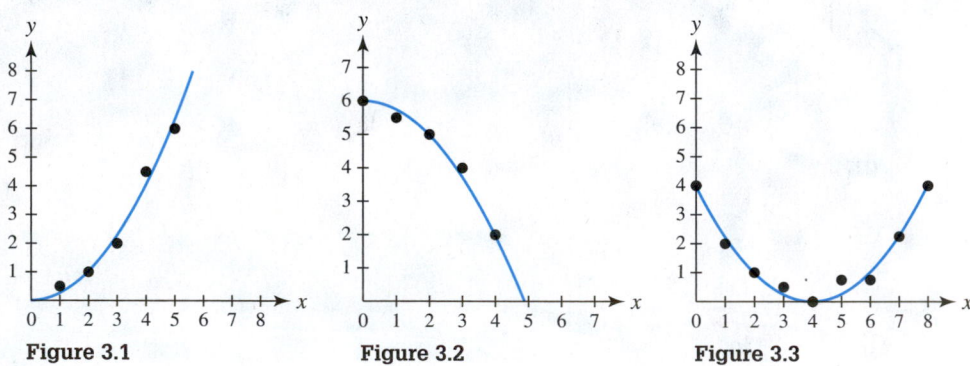

Figure 3.1　　　　　Figure 3.2　　　　　Figure 3.3

Basic Concepts

The formula for a quadratic function is different from that of a linear function because it contains an x^2**-term**. Examples of quadratic functions include the following.

Quadratic Functions

$$f(x) = 2x^2 - 4x - 1, \quad g(x) = 4 - x^2, \quad and \quad h(x) = \frac{1}{3}x^2 + \frac{2}{3}x + 1$$

The following box defines a *general form* for a quadratic function.

> **QUADRATIC FUNCTION**
>
> Let a, b, and c be constants with $a \neq 0$. A function represented by
> $$f(x) = ax^2 + bx + c$$
> is a **quadratic function**.

The following are some basics of quadratic functions.

- The domain is *all* real numbers.
- The leading coefficient is a: $f(x) = ax^2 + bx + c$.

    Examples: $f(x) = 2x^2 - 4x - 1, \quad g(x) = 4 - 1x^2, \quad h(x) = \frac{1}{3}x^2 + \frac{2}{3}x + 1$

    $a = 2$　　　　　$a = -1$　　　　　$a = \frac{1}{3}$

- The graph is a **parabola**, or ∪-shaped, that opens upward if a is positive, and opens downward if a is negative.

See the Concept: Graphs of Quadratic Functions

$a > 0$: Opens Upward

$a < 0$: Opens Downward

Ⓐ The highest point on a parabola that opens downward or the lowest point on a parabola that opens upward is called the **vertex**.

Ⓑ Axis of symmetry: $x = 1$

Ⓐ Vertex

$f(x) = 2x^2 - 4x - 1$

$(1, -3)$

f decreases for $x < 1$ and increases for $x > 1$

Ⓐ Vertex $(0, 4)$

$g(x) = 4 - x^2$

Ⓑ Axis of symmetry: $x = 0$

g increases for $x < 0$ and decreases for $x > 0$

Ⓑ The vertical line passing through the vertex is called the **axis of symmetry**.

The leading coefficient a of a quadratic function not only determines whether its graph opens upward or downward but also controls the width of the parabola. This concept is illustrated in Figures 3.4 and 3.5.

The Effect of a on the Graph of $y = ax^2$

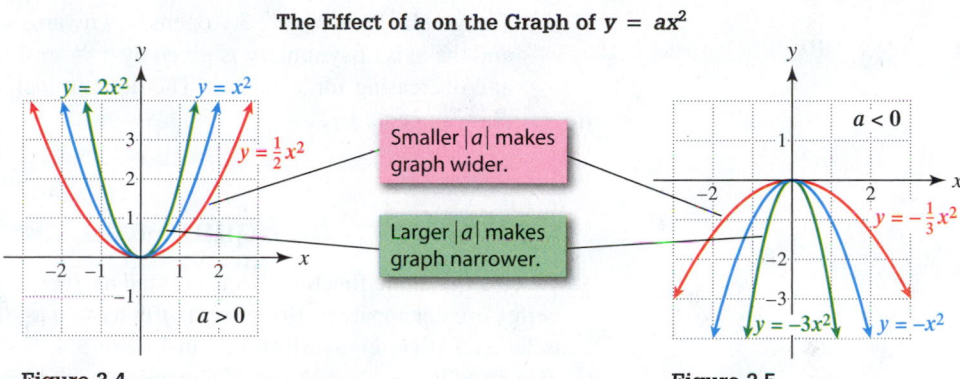

$y = 2x^2$ $y = x^2$ $y = \frac{1}{2}x^2$

Smaller $|a|$ makes graph wider.

Larger $|a|$ makes graph narrower.

$a > 0$

$a < 0$

$y = -\frac{1}{3}x^2$

$y = -3x^2$ $y = -x^2$

Figure 3.4

Figure 3.5

CLASS DISCUSSION
What does the graph of a quadratic function resemble if its leading coefficient is nearly 0?

EXAMPLE 1 **Identifying quadratic functions**

Identify the function as linear, quadratic, or neither. If it is quadratic, identify the leading coefficient and evaluate the function at $x = 2$.

(a) $f(x) = 3 - 2^2x$ **(b)** $g(x) = 5 + x - 3x^2$ **(c)** $h(x) = \dfrac{3}{x^2 + 1}$

SOLUTION

Getting Started The formula for a quadratic function always has an x^2-term and may have an x-term and a constant. It does not have a variable raised to any other power or a variable in a denominator. ▶

(a) Because $f(x) = 3 - 2^2x$ can be written as $f(x) = -4x + 3$, f is linear.
(b) Because $g(x) = 5 + x - 3x^2$ can be written as $g(x) = -3x^2 + x + 5$, g is a quadratic function with $a = -3$, $b = 1$, and $c = 5$. The leading coefficient is $a = -3$ and

$$g(2) = 5 + 2 - 3(2)^2 = -5.$$

(c) $h(x) = \dfrac{3}{x^2 + 1}$ cannot be written as $h(x) = ax + b$ or as $h(x) = ax^2 + bx + c$, so h is neither a linear nor a quadratic function.

Now Try Exercises 1, 3, and 5

EXAMPLE 2 Analyzing graphs of quadratic functions

Use the graph of each quadratic function to determine the sign of the leading coefficient a, the vertex, and the equation of the axis of symmetry. Give the intervals where the function is increasing and where it is decreasing. Give the domain and range.

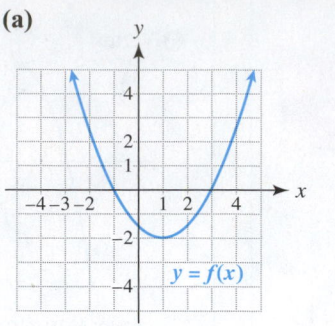

(a)

Figure 3.6

(b)

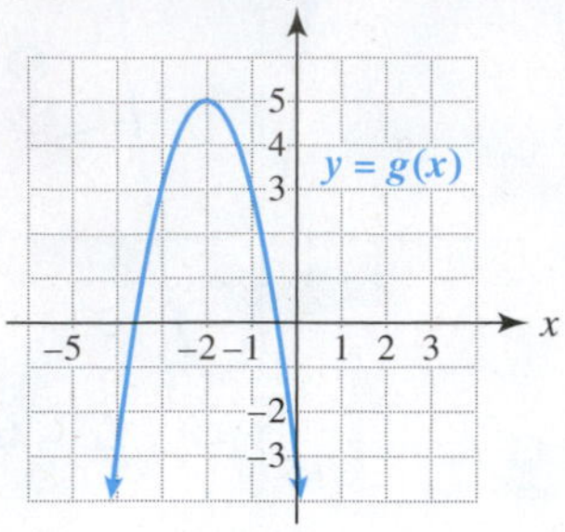

Figure 3.7

SOLUTION

(a) The graph of f in Figure 3.6 opens upward, so a is positive. The vertex is the lowest point on the graph and is $(1, -2)$. The axis of symmetry is a vertical line passing through the vertex with equation $x = 1$. Function f increases to the right of the vertex and decreases to the left of the vertex. Thus f is increasing for $x > 1$ and decreasing for $x < 1$. The domain includes all real numbers and the range is $R = \{y \mid y \geq -2\}$.

(b) The graph of g in Figure 3.7 opens downward, so a is negative. The vertex is $(-2, 5)$, and the axis of symmetry is given by $x = -2$. Function g is increasing for $x < -2$ and decreasing for $x > -2$. The domain includes all real numbers and the range is $R = \{y \mid y \leq 5\}$.

Now Try Exercises 7 and 9

Completing the Square and the Vertex Formula

When a quadratic function f is expressed as $f(x) = ax^2 + bx + c$, the coordinates of the vertex are not apparent. However, if f is written as $f(x) = a(x - h)^2 + k$, then the vertex is located at (h, k), as illustrated in Figure 3.8. For example, in Figure 3.9 the graph of $f(x) = -2(x - 1)^2 + 2$ opens downward with vertex $(1, 2)$.

A Parabola with Vertex (h, k)

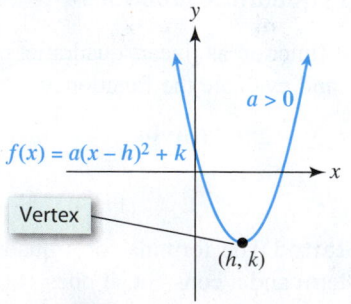

Figure 3.8

A Parabola with Vertex $(1, 2)$

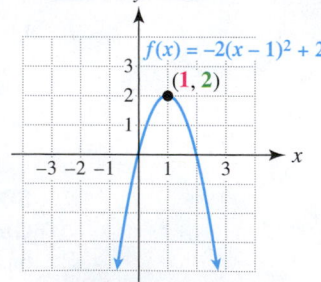

Figure 3.9

To justify that the vertex is indeed (h, k), consider the following. If $a > 0$ in the form $f(x) = a(x - h)^2 + k$, then the term $a(x - h)^2$ is never negative and the minimum value of $f(x)$ is k. This value occurs when $x = h$ because

$$f(h) = a(h - h)^2 + k = 0 + k = k.$$

Thus the lowest point on the graph of $f(x) = a(x - h)^2 + k$ with $a > 0$ is (h, k), and because this graph is a parabola that opens upward, the vertex must be (h, k). A similar discussion can be used to justify that (h, k) is the vertex when $a < 0$.

The formula $f(x) = a(x - h)^2 + k$ is sometimes called the **standard form for a parabola with a vertical axis**. Because the vertex is apparent in this formula, we will call it simply the **vertex form**.

> **VERTEX FORM**
>
> The parabolic graph of $f(x) = a(x - h)^2 + k$ with $a \neq 0$ has vertex (h, k). Its graph opens upward when $a > 0$ and opens downward when $a < 0$.

The next example illustrates how to determine the vertex form given the graph of a parabola.

EXAMPLE 3 Writing the equation of a parabola

Find the vertex form for the graph shown in Figure 3.10.

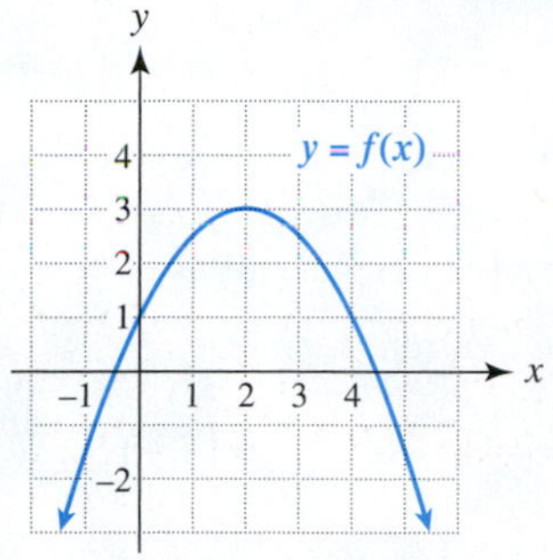

SOLUTION

Getting Started We must determine a, h, and k in $f(x) = a(x - h)^2 + k$. The coordinates of the vertex correspond to the values of h and k. To find a, substitute the coordinates of a point on the graph in the equation and solve for a. ▶

From Figure 3.10, the vertex is $(2, 3)$. Thus $h = 2$ and $k = 3$ and $f(x) = a(x - 2)^2 + 3$. The point $(0, 1)$ lies on the graph, so $f(0) = 1$. (Any point on the graph of f other than the vertex could be used.)

$$f(x) = a(x - 2)^2 + 3 \qquad \text{Vertex form}$$
$$1 = a(0 - 2)^2 + 3 \qquad \text{Let } x = 0 \text{ and } f(0) = 1. \text{ Solve for } a.$$
$$1 = 4a + 3 \qquad \text{Simplify.}$$
$$-2 = 4a \qquad \text{Subtract 3 from each side.}$$
$$a = -\frac{1}{2} \qquad \text{Divide by 4 and rewrite equation.}$$

Thus $f(x) = -\frac{1}{2}(x - 2)^2 + 3$.

Now Try Exercise 25

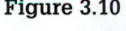

Figure 3.10

EXAMPLE 4 Converting to $f(x) = ax^2 + bx + c$

Write $f(x) = 2(x - 1)^2 + 4$ in the form $f(x) = ax^2 + bx + c$.

SOLUTION Begin by expanding the expression $(x - 1)^2$.

$$2(x - 1)^2 + 4 = 2(x^2 - 2x + 1) + 4 \qquad \text{Square binomial.}$$
$$= 2x^2 - 4x + 2 + 4 \qquad \text{Distributive property}$$
$$= 2x^2 - 4x + 6 \qquad \text{Add terms.}$$

Algebra Review
To review squaring a binomial, see Chapter R (page R-17).

Thus $f(x) = 2x^2 - 4x + 6$.

Now Try Exercise 15

Completing the Square We can convert the general form $f(x) = ax^2 + bx + c$ to vertex form by **completing the square**. If a quadratic expression can be written as $x^2 + kx + \left(\frac{k}{2}\right)^2$, then it is a perfect square trinomial and can be factored as

$$x^2 + kx + \left(\frac{k}{2}\right)^2 = \left(x + \frac{k}{2}\right)^2.$$

Note that the k used to complete the square is different from the k found in the vertex form. (A variable can have different meanings in different situations.)

This technique of converting to vertex form by completing the square is illustrated in the next example.

| EXAMPLE 5 | Converting to vertex form |

Write each formula in vertex form by completing the square. Identify the vertex.

(a) $f(x) = x^2 + 6x - 3$　　**(b)** $f(x) = \frac{1}{3}x^2 - x + 2$

SOLUTION

(a) Start by letting $y = f(x)$.

Algebra Review
To review perfect square trinomials, see Chapter R (page R-25).

$$y = x^2 + 6x - 3 \qquad \text{Given formula}$$

$$y + 3 = x^2 + 6x \qquad \text{Add 3 to each side.}$$

$$y + 3 + 9 = x^2 + 6x + 9 \qquad \text{Let } k = 6; \text{ add } \left(\frac{k}{2}\right)^2 = \left(\frac{6}{2}\right)^2 = 9.$$

$$y + 12 = (x + 3)^2 \qquad \text{Factor perfect square trinomial.}$$

$$y = (x + 3)^2 - 12 \qquad \text{Subtract 12.}$$

The required form is $f(x) = (x + 3)^2 - 12$. The vertex is $(-3, -12)$.

(b) Start by letting $y = f(x)$.

$$y = \frac{1}{3}x^2 - x + 2 \qquad \text{Given formula}$$

Multiply each side by 3.

$$3y = x^2 - 3x + 6 \qquad \text{Make leading coefficient 1.}$$

$$3y - 6 = x^2 - 3x \qquad \text{Subtract 6 from each side.}$$

$$3y - 6 + \frac{9}{4} = x^2 - 3x + \frac{9}{4} \qquad \text{Let } k = -3; \text{ add} \left(\frac{k}{2}\right)^2 = \left(\frac{-3}{2}\right)^2 = \frac{9}{4}.$$

$$3y - \frac{15}{4} = \left(x - \frac{3}{2}\right)^2 \qquad \text{Factor perfect square trinomial.}$$

$$3y = \left(x - \frac{3}{2}\right)^2 + \frac{15}{4} \qquad \text{Add } \frac{15}{4} \text{ to each side.}$$

$$y = \frac{1}{3}\left(x - \frac{3}{2}\right)^2 + \frac{5}{4} \qquad \text{Multiply each side by } \frac{1}{3}.$$

The required form is $f(x) = \frac{1}{3}\left(x - \frac{3}{2}\right)^2 + \frac{5}{4}$. The vertex is $\left(\frac{3}{2}, \frac{5}{4}\right)$.

Now Try Exercises 29 and 35

Derivation of the Vertex Formula　The procedure above of completing the square can be done in general to derive a formula for determining the vertex of any parabola.

$$y = ax^2 + bx + c \qquad \text{General equation for a parabola}$$

$$\frac{y}{a} = x^2 + \frac{b}{a}x + \frac{c}{a} \qquad \text{Divide each side by } a \text{ to make leading coefficient 1.}$$

$$\frac{y}{a} - \frac{c}{a} = x^2 + \frac{b}{a}x \qquad \text{Subtract } \frac{c}{a} \text{ from each side.}$$

$$\frac{y}{a} - \frac{c}{a} + \frac{b^2}{4a^2} = x^2 + \frac{b}{a}x + \frac{b^2}{4a^2} \qquad \text{Add } \left(\frac{b/a}{2}\right)^2 = \frac{b^2}{4a^2}.$$

$$\frac{y}{a} + \frac{b^2 - 4ac}{4a^2} = \left(x + \frac{b}{2a}\right)^2 \qquad \text{Combine left terms; factor perfect square trinomial.}$$

$$\frac{y}{a} = \left(x + \frac{b}{2a}\right)^2 - \frac{b^2 - 4ac}{4a^2} \qquad \text{Isolate } y\text{-term on the left side.}$$

$$y = a\left(x + \frac{b}{2a}\right)^2 - \frac{b^2 - 4ac}{4a} \qquad \text{Multiply by } a.$$

$$y = a\left(x - \left(-\frac{b}{2a}\right)\right)^2 + \frac{4ac - b^2}{4a} \qquad \text{Write } y = a(x - h)^2 + k.$$
$$\underbrace{\phantom{x - \left(-\frac{b}{2a}\right)}}_{h} \qquad \underbrace{\phantom{\frac{4ac - b^2}{4a}}}_{k}$$

Because the coordinates of the vertex are (h, k), the x-coordinate is $-\frac{b}{2a}$. Note that it is *not* necessary to memorize the expression for k, because the y-coordinate can be found by evaluating $y = f(x)$ for $x = -\frac{b}{2a}$. This derivation of the *vertex formula* is now summarized.

VERTEX FORMULA

The *vertex* of the graph of $f(x) = ax^2 + bx + c$ with $a \neq 0$ is the point $\left(-\frac{b}{2a}, f\left(-\frac{b}{2a}\right)\right)$.

NOTE If a parabola has two x-intercepts, then the x-coordinate of the vertex is equal to the midpoint of these two x-intercepts. For example, the x-intercepts in Figure 3.9 on page 158 are 0 and 2. Their midpoint is $\frac{0 + 2}{2} = 1$, which is the x-coordinate of the vertex.

EXAMPLE 6 Converting to $f(x) = a(x - h)^2 + k$

Use the vertex formula to write $f(x) = 3x^2 + 12x + 7$ in vertex form.

SOLUTION Begin by finding the vertex. Let $a = 3$ and $b = 12$.

$$x = -\frac{b}{2a} = -\frac{12}{2(3)} = -2$$

Since $f(-2) = 3(-2)^2 + 12(-2) + 7 = -5$, the vertex is $(-2, -5)$. Because $a = 3$, $f(x)$ can be written as $f(x) = 3(x + 2)^2 - 5$.

> **Now Try Exercise 37**

Graphing Quadratic Functions When sketching a parabola, it is important to determine the vertex, the axis of symmetry, and whether the parabola opens upward or downward. In the next example we sketch the graphs of two quadratic functions by hand.

EXAMPLE 7 Graphing quadratic functions by hand

Graph each quadratic function. Find the intervals where the function is increasing and where it is decreasing.

(a) $g(x) = 2(x - 1)^2 - 3$ **(b)** $h(x) = -\frac{1}{2}x^2 - x + 2$

SOLUTION
(a) The vertex is $(1, -3)$ and the axis of symmetry is $x = 1$. The parabola opens upward because $a = 2$ is positive. In Table 3.1 we list the vertex and a few other points located on either side of the vertex. Note the symmetry of the y-values on each side of the vertex. These points and a smooth $\cup$-shaped curve are plotted in Figure 3.11. When $x = 0, y = -1$, and so the y-intercept is -1. Function g is decreasing when $x < 1$ and increasing when $x > 1$.

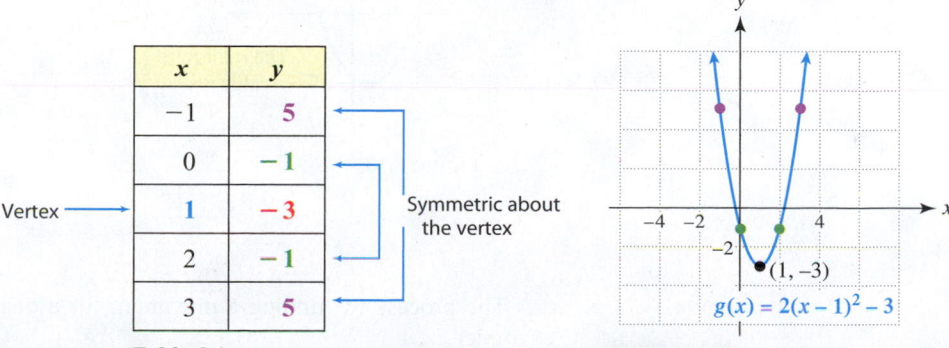

x	y
-1	5
0	-1
1	-3
2	-1
3	5

Vertex → 1, -3

Symmetric about the vertex

Table 3.1

$g(x) = 2(x - 1)^2 - 3$

$(1, -3)$

Figure 3.11

(b) The formula $h(x) = -\frac{1}{2}x^2 - x + 2$ is not in vertex form, but we can find the vertex.

$$x = -\frac{b}{2a} = -\frac{-1}{2\left(-\frac{1}{2}\right)} = -1$$

The y-coordinate of the vertex is $h(-1) = -\frac{1}{2}(-1)^2 - (-1) + 2 = \frac{5}{2}$. Thus the vertex is $\left(-1, \frac{5}{2}\right)$, the axis of symmetry is $x = -1$, and the parabola opens downward because $a = -\frac{1}{2}$ is negative. In Table 3.2 we list the vertex and a few other points located on either side of the vertex. These points and a smooth ∩-shaped curve are plotted in Figure 3.12. When $x = 0$, $y = 2$, and so the y-intercept is 2. Function h is increasing when $x < -1$ and decreasing when $x > -1$.

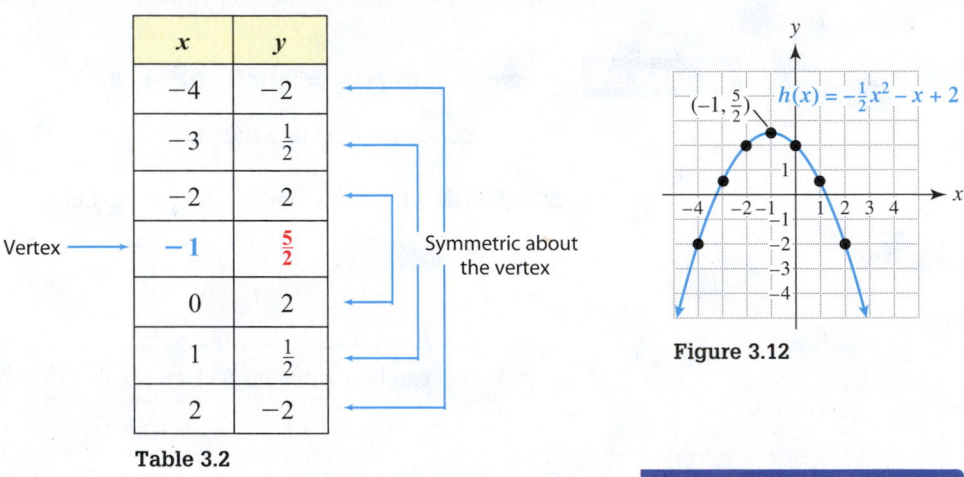

x	y
-4	-2
-3	$\frac{1}{2}$
-2	2
-1	$\frac{5}{2}$
0	2
1	$\frac{1}{2}$
2	-2

Vertex → -1 Symmetric about the vertex

Table 3.2

Figure 3.12

Now Try Exercises 71 and 83

Applications and Models

Min-Max Values Sometimes when a quadratic function f is used in applications, the vertex provides important information. The reason is that the y-coordinate of the vertex is the minimum value of $f(x)$ when its graph opens upward and is the maximum value of $f(x)$ when its graph opens downward. See Figure 3.13 and 3.14.

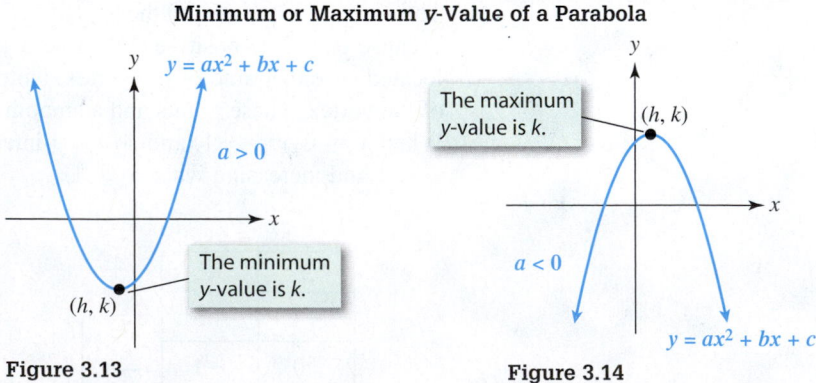

Minimum or Maximum y-Value of a Parabola

$y = ax^2 + bx + c$

$a > 0$

The minimum y-value is k.

(h, k)

Figure 3.13

The maximum y-value is k.

(h, k)

$a < 0$

$y = ax^2 + bx + c$

Figure 3.14

The process for finding a maximum for a quadratic function is applied in the next example.

EXAMPLE 8   **Maximizing area**

A rancher is fencing a rectangular area for cattle using the straight portion of a river as one side of the rectangle, as illustrated in Figure 3.15. If the rancher has 2400 feet of fence, find the dimensions of the rectangle that give the maximum area for the cattle.

Figure 3.15

SOLUTION

Getting Started Because the goal is to maximize the area, first write a formula for the area. If the formula is quadratic with $a < 0$ (parabola opening downward), then the maximum area will be the y-coordinate of the vertex. ▶

Let W be the width and L be the length of the rectangle. Because the 2400-foot fence does not go along the river, it follows that

$$W + L + W = 2400, \quad \text{or} \quad L = 2400 - 2W.$$

Area A of a rectangle equals length times width, so

$$\begin{aligned} A &= LW & \text{Area of rectangle} \\ &= (2400 - 2W)W & \text{Substitute for L.} \\ &= 2400W - 2W^2. & \text{Distributive property} \end{aligned}$$

Thus the graph of $A = -2W^2 + 2400W$ is a parabola opening downward, and by the vertex formula, maximum area occurs when

$$W = -\frac{b}{2a} = -\frac{2400}{2(-2)} = 600 \text{ feet.}$$

The corresponding length is $L = 2400 - 2W = 2400 - 2(600) = 1200$ feet. The dimensions that maximize area are 600 feet by 1200 feet. The maximum area is 720,000 square feet.

Now Try Exercise 105

Modeling   Another application of quadratic functions occurs in projectile motion, such as when a baseball is hit up in the air. If air resistance is ignored, then the formula

$$s(t) = -16t^2 + v_0 t + h_0$$

calculates the height s of the object above the ground in feet after t seconds. In this formula h_0 represents the initial height of the object in feet and v_0 represents its initial *vertical* velocity in feet per second. If the initial velocity is upward, then $v_0 > 0$, and if the initial velocity is downward, then $v_0 < 0$.

EXAMPLE 9   **Modeling the flight of a baseball**

A baseball is hit straight up with an initial velocity of $v_0 = 80$ feet per second (or about 55 miles per hour) and leaves the bat with an initial height of $h_0 = 3$ feet, as shown in Figure 3.16.
(a) Write a formula $s(t)$ that models the height of the baseball after t seconds.
(b) How high is the baseball after 2 seconds?
(c) Find the maximum height of the baseball. Support your answer graphically.

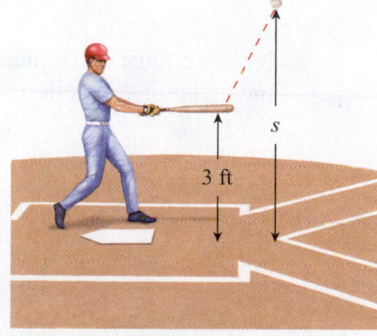

Figure 3.16

Calculator Help

To find a maximum or minimum point on a graph, see Appendix A (page AP-8).

SOLUTION

(a) Because $v_0 = 80$ and the initial height is $h_0 = 3$,

$$s(t) = -16t^2 + v_0 t + h_0 = -16t^2 + 80t + 3.$$

(b) $s(2) = -16(2)^2 + 80(2) + 3 = 99$, so the baseball is 99 feet high after 2 seconds.

(c) Because $a = -16$, the graph of s is a parabola opening downward. The vertex is the highest point on the graph, with a t-coordinate of

Height of a Baseball

[0, 5, 1] by [−20, 120, 20]

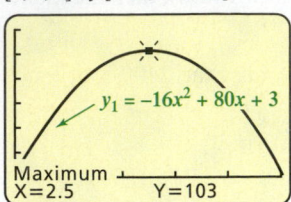

$y_1 = -16x^2 + 80x + 3$

Maximum
X=2.5   Y=103

Figure 3.17

$$t = -\frac{b}{2a} = -\frac{80}{2(-16)} = 2.5.$$

The corresponding y-coordinate of the vertex is

$$s(2.5) = -16(2.5)^2 + 80(2.5) + 3 = 103 \text{ feet.}$$

Thus the vertex is $(2.5, 103)$ and the maximum height of the baseball is 103 feet after 2.5 seconds. Graphical support is shown in Figure 3.17, where the vertex is (2.5, 103).

Now Try Exercise 101

A well-conditioned athlete's heart rate can reach 200 beats per minute (bpm) during strenuous physical activity. Upon stopping an activity, a typical heart rate decreases rapidly at first and then gradually levels off, as shown in Table 3.3.

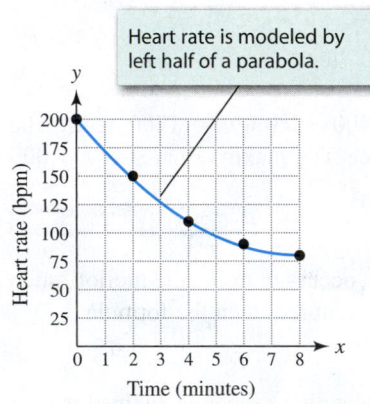

Heart rate is modeled by left half of a parabola.

Figure 3.18

Athlete's Heart Rate After Exercise

Time (min)	0	2	4	6	8
Heart rate (bpm)	200	150	110	90	80

Source: Adapted from: V. Thomas, *Science and Sport.*

Table 3.3

The data are not linear because for each 2-minute interval the heart rate does not decrease by a fixed amount. In Figure 3.18 the data are modeled with a nonlinear function. Note that the graph resembles the left half of a parabola that opens upward.

EXAMPLE 10 **Modeling an athlete's heart rate**

Find a quadratic function f expressed in vertex form that models the data in Table 3.3. Support your result by graphing f and the data in the same xy-plane. What is the domain of your function?

SOLUTION To model the data we use the left half of a parabola. Since the minimum heart rate of 80 beats per minute occurs when $x = 8$, let $(8, 80)$ be the vertex and write

$$f(x) = a(x - 8)^2 + 80.$$

Next we must determine a value for the leading coefficient a. One possibility is to have the graph of f pass through the first data point $(0, 200)$, or equivalently, let $f(0) = 200$.

$$f(0) = 200 \qquad \text{Have the graph pass through } (0, 200).$$

$$a(0 - 8)^2 + 80 = 200 \qquad \text{Let } x = 0 \text{ in } f(x). \text{ Solve for } a.$$

$$a(0 - 8)^2 = 120 \qquad \text{Subtract 80.}$$

$$a = \frac{120}{64} \qquad \text{Divide by } (0 - 8)^2 = 64.$$

$$a = 1.875 \qquad \text{Write as a decimal.}$$

Thus $f(x) = \mathbf{1.875}(x - 8)^2 + 80$ can be used to model the athlete's heart rate. A graph of f and the data are shown in Figure 3.19, which is similar to Figure 3.18. Figure 3.20 shows a table of $f(x)$. Although the table in Figure 3.20 does not match Table 3.3 exactly, it gives reasonable approximations. (Note that formulas for $f(x)$ may vary. For example, if we had selected the point (2, 150) rather than (0, 200), then $a \approx 1.94$. You may wish to verify this result.)

The domain D of $f(x) = 1.875(x - 8)^2 + 80$ needs to be restricted to $0 \le x \le 8$ because this interval corresponds to the domain of the data in Table 3.3.

Calculator Help

To create a table similar to Figure 3.20, see Appendix A (page AP-9).

Modeling an Athlete's Heart Rate

$[-0.5, 8.5, 2]$ by $[0, 220, 20]$

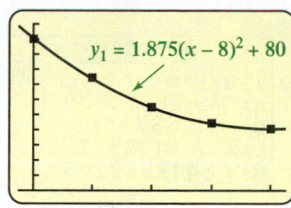

Figure 3.19

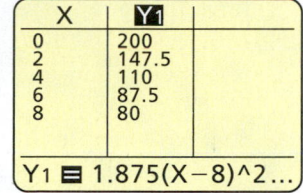

Figure 3.20

> **Now Try Exercise 115**

MAKING CONNECTIONS

General Form, Vertex Form, and Modeling   When modeling quadratic data by hand, it is often easier to use the vertex form, $f(x) = a(x - h)^2 + k$, rather than the general form, $f(x) = ax^2 + bx + c$. Because (h, k) corresponds to the vertex of a parabola, it may be appropriate to let (h, k) correspond to either the highest data point or the lowest data point in the scatterplot. Then a value for a can be found by substituting a data point into the formula for $f(x)$. In the next subsection least-squares regression provides a quadratic modeling function in general form.

Quadratic Regression (Optional)

In Chapter 2 we discussed how a regression line could be found by the method of least squares. This method can also be applied to quadratic data; the process is illustrated next.

EXAMPLE 11   Finding a quadratic regression model

Table 3.4 lists MySpace U.S. advertising revenue in millions of dollars for 2006 through 2011, where x corresponds to years after 2006.
(a) Plot the data. Discuss reasons why a quadratic function might model the data.
(b) Find a least-squares function f given by $f(x) = ax^2 + bx + c$ that models the data.
(c) Graph f and the data in $[-2, 7, 1]$ by $[0, 1000, 100]$. Discuss the fit.

MySpace Ad Revenue ($ millions)

Year	0	1	2	3	4	5
Revenue	225	450	590	435	285	190

Table 3.4

$[-2, 7, 1]$ by $[0, 1000, 100]$

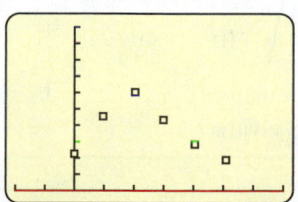

Figure 3.21

SOLUTION
(a) A plot of the data is shown in Figure 3.21. The y-values (ad revenue) first increase and then decrease as the year x increases. The data suggest a parabolic shape opening downward.

(b) Enter the data into your calculator, and then select quadratic regression from the menu, as shown in Figure 3.22 and Figure 3.23. In Figure 3.24 the modeling function f is given (approximately) by $f(x) = -49.29x^2 + 222.9x + 257$.

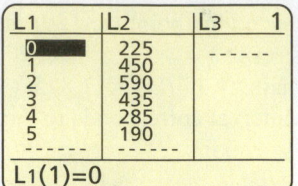

Figure 3.22

```
EDIT  CALC  TESTS
1:1-Var Stats
2:2-Var Stats
3:Med-Med
4:LinReg(ax+b)
5:QuadReg
6:CubicReg
7↓QuartReg
```

Figure 3.23

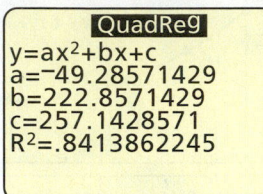

Figure 3.24

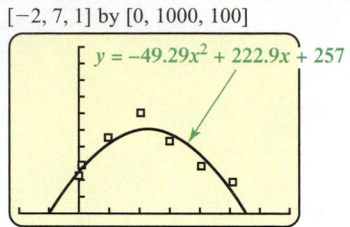

$[-2, 7, 1]$ by $[0, 1000, 100]$

Figure 3.25

Calculator Help

To find an equation of least-squares fit, see Appendix A (page AP-9).

(c) The graph of $y_1 = -49.29x^2 + 222.9x + 257$ with the data is shown in Figure 3.25. Although the model is not exact, the parabola describes the general trend in the data. The parabola opens downward because $a = -49.29$ is negative.

Now Try Exercise 125

3.1 Putting It All Together

The following table summarizes some important topics from this section.

CONCEPT	SYMBOLIC REPRESENTATION	COMMENTS AND EXAMPLES
Quadratic function	$f(x) = ax^2 + bx + c$, where a, b, and c are constants with $a \neq 0$ (general form)	It models data that are not linear. Its graph is a parabola ($\cup$-shaped) that opens either upward ($a > 0$) or downward ($a < 0$).
Parabola	The graph of $y = ax^2 + bx + c$, $a \neq 0$, is a parabola.	Vertex: (h, k); axis of symmetry: $x = h$ Maximum (or minimum) y-value: k

CONCEPT	SYMBOLIC REPRESENTATION	COMMENTS AND EXAMPLES
Completing the square to find vertex form	To complete the square for $x^2 + kx$, add $\left(\frac{k}{2}\right)^2$ to make a perfect square trinomial.	$y = x^2 - 2x + 3$ $y - 3 = x^2 - 2x$ $k = -2$ $y - 3 + 1 = x^2 - 2x + 1$ Add $\left(\frac{-2}{2}\right)^2 = 1.$ $y - 2 = (x - 1)^2$ $y = (x - 1)^2 + 2$
Vertex formula	The vertex for $f(x) = ax^2 + bx + c$ is the point $$\left(-\frac{b}{2a}, f\left(-\frac{b}{2a}\right)\right).$$	If $f(x) = x^2 - 2x + 3$, then $$x = -\frac{-2}{2(1)} = 1.$$ y-value of vertex: $f(1) = 2$ Vertex: $(1, 2)$; axis of symmetry: $x = 1$
Vertex form (standard form for a parabola with vertical axis)	The vertex form for a quadratic function is $f(x) = a(x - h)^2 + k$, with vertex (h, k).	Let $f(x) = 2(x + 3)^2 - 5$. Parabola opens upward: $a > 0$ $a = 2$ Vertex: $(-3, -5)$ Axis of symmetry: $x = -3$ Minimum y-value on graph: -5

3.1 Exercises

Basics of Quadratic Functions

Exercises 1–6: Identify f as being linear, quadratic, or neither. If f is quadratic, identify the leading coefficient a and evaluate f(−2).

1. $f(x) = 1 - 2x + 3x^2$

2. $f(x) = -5x + 11$

3. $f(x) = \dfrac{1}{x^2 - 1}$

4. $f(x) = (x^2 + 1)^2$

5. $f(x) = \frac{1}{2} - \frac{3}{10}x$

6. $f(x) = \frac{1}{5}x^2$

Exercises 7–10: Use the graph to find the following.

(a) *Sign of the leading coefficient*
(b) *Vertex*
(c) *Axis of symmetry*
(d) *Intervals where f is increasing and where f is decreasing*
(e) *Domain and range*

7.

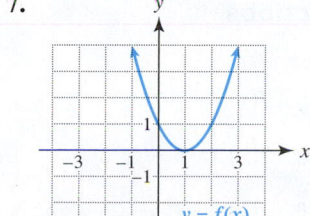

8.

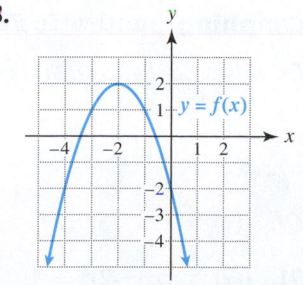

9.

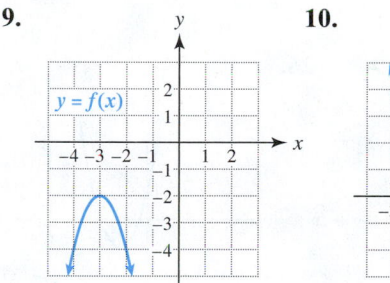

10.

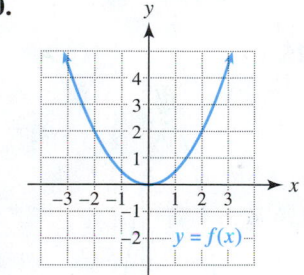

Exercises 11–14: The formulas for f(x) and g(x) are identical except for their leading coefficients a. Compare the graphs of f and g. You may want to support your answers by graphing f and g together.

11. $f(x) = x^2$, $g(x) = 2x^2$

12. $f(x) = \frac{1}{2}x^2$, $g(x) = -\frac{1}{2}x^2$

13. $f(x) = 2x^2 + 1$, $g(x) = -\frac{1}{3}x^2 + 1$

14. $f(x) = x^2 + x$, $g(x) = \frac{1}{4}x^2 + x$

Vertex Formula

Exercises 15–20: Identify the vertex and leading coefficient. Then write the expression as $f(x) = ax^2 + bx + c$.

15. $f(x) = -3(x - 1)^2 + 2$

16. $f(x) = 5(x + 2)^2 - 5$

17. $f(x) = 5 - 2(x - 4)^2$ **18.** $f(x) = \frac{1}{2}(x + 3)^2 - 5$

19. $f(x) = \frac{3}{4}(x + 5)^2 - \frac{7}{4}$ **20.** $f(x) = -5(x - 4)^2$

Exercises 21–28: Use the graph of the quadratic function f to write its formula as $f(x) = a(x - h)^2 + k$.

21.
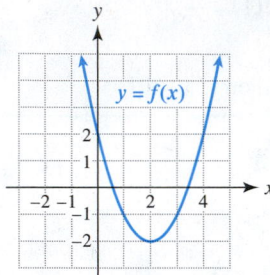

22.
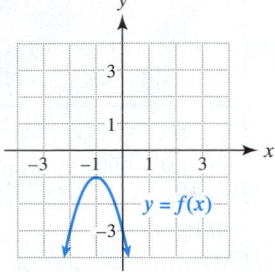

23.
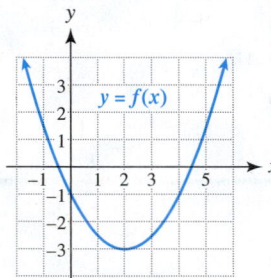

24.
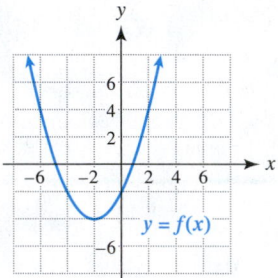

25.
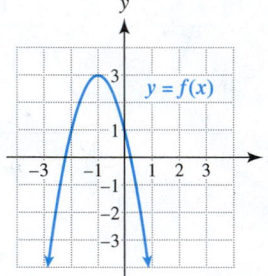

26.
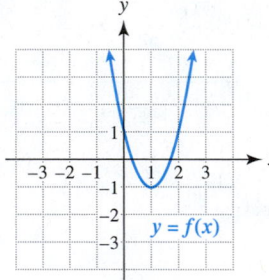

27.
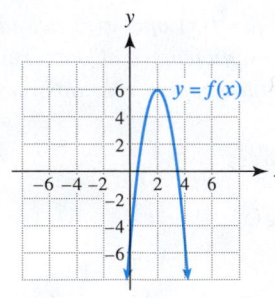

28.
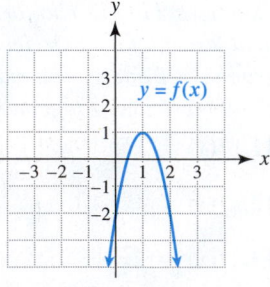

Exercises 29–40: Write the given expression in the form $f(x) = a(x - h)^2 + k$. Identify the vertex.

29. $f(x) = x^2 + 4x - 5$ **30.** $f(x) = x^2 + 10x + 7$

31. $f(x) = x^2 - 3x$ **32.** $f(x) = x^2 - 7x + 5$

33. $f(x) = 2x^2 - 5x + 3$ **34.** $f(x) = 3x^2 + 6x + 2$

35. $f(x) = \frac{1}{3}x^2 + x + 1$ **36.** $f(x) = -\frac{1}{2}x^2 - \frac{3}{2}x + 1$

37. $f(x) = 2x^2 - 8x - 1$ **38.** $f(x) = -\frac{1}{2}x^2 - x$

39. $f(x) = 2 - 9x - 3x^2$

40. $f(x) = 6 + 5x - 10x^2$

Exercises 41–52: Complete the following.

(a) Use the vertex formula to find the vertex.

(b) Find the intervals where f is increasing and where f is decreasing.

41. $f(x) = 6 - x^2$ **42.** $f(x) = 2x^2 - 2x + 1$

43. $f(x) = x^2 - 6x$ **44.** $f(x) = -2x^2 + 4x + 5$

45. $f(x) = 2x^2 - 4x + 1$ **46.** $f(x) = -3x^2 + x - 2$

47. $f(x) = \frac{1}{2}x^2 + 10$ **48.** $f(x) = \frac{9}{10}x^2 - 12$

49. $f(x) = -\frac{3}{4}x^2 + \frac{1}{2}x - 3$ **50.** $f(x) = -\frac{4}{5}x^2 - \frac{1}{5}x + 1$

51. $f(x) = 1.5 - 3x - 6x^2$ **52.** $f(x) = -4x^2 + 16x$

Min-Max

Exercises 53–58: Find the minimum y-value on the graph of $y = f(x)$.

53. $f(x) = x^2 + 4x - 2$ **54.** $f(x) = x^2 - 6x$

55. $f(x) = 3x^2 - 4x + 2$ **56.** $f(x) = 2x^2 + 6x$

57. $f(x) = x^2 + 3x + 5$ **58.** $f(x) = 2x^2 - x + 1$

Exercises 59–64: Find the maximum y-value on the graph of $y = f(x)$.

59. $f(x) = -x^2 + 3x - 2$ **60.** $f(x) = -x^2 + 4x + 5$

61. $f(x) = 5x - x^2$ **62.** $f(x) = -2x^2 - 2x - 5$

63. $f(x) = 2x - 3x^2$ **64.** $f(x) = -4x^2 + 6x - 9$

Graphing Quadratic Functions

Exercises 65–84: Sketch a graph of f.

65. $f(x) = x^2$ **66.** $f(x) = -2x^2$

67. $f(x) = -\frac{1}{2}x^2$ **68.** $f(x) = 4 - x^2$

69. $f(x) = x^2 - 3$ **70.** $f(x) = x^2 + 2$

71. $f(x) = (x - 2)^2 + 1$ **72.** $f(x) = (x + 1)^2 - 2$

73. $f(x) = -3(x + 1)^2 + 3$

74. $f(x) = -2(x - 1)^2 + 1$

75. $f(x) = x^2 - 2x - 2$ **76.** $f(x) = x^2 - 4x$

77. $f(x) = -x^2 + 4x - 2$ **78.** $f(x) = -x^2 + 2x + 1$

79. $f(x) = 2x^2 - 4x - 1$ **80.** $f(x) = 3x^2 + 6x$

81. $f(x) = -3x^2 - 6x + 1$ **82.** $f(x) = -2x^2 + 4x - 1$

83. $f(x) = -\frac{1}{2}x^2 + x + 1$ **84.** $f(x) = \frac{1}{2}x^2 - 2x + 2$

Exercises 85 and 86: **Average Rate of Change**   *Find the average rate of change of f from 1 to 3.*

85. $f(x) = -3x^2 + 5x$ **86.** $f(x) = 4x^2 - 3x + 1$

Exercises 87 and 88: **Difference Quotient**   *Find the difference quotient of f.*

87. $f(x) = 3x^2 - 2x$ **88.** $f(x) = 5 - 4x^2$

Exercises 89 and 90: Find the formula for a quadratic function that satisfies the given conditions.

89. Axis of symmetry $x = 3$, passing through the points $(3, 1)$ and $(1, 9)$

90. Vertex $(-3, 4)$, passing through $(-2, 1)$

Graphs and Models

Exercises 91–94: Match the situation with the graph of the quadratic function (a–d) that models it best.

91. The height y of a stone thrown from ground level after x seconds

92. The number of people attending a popular movie x weeks after its opening

93. The temperature after x hours in a house where the furnace quits and a repair person fixes it

94. The cumulative number of reported AIDS cases in year x, where $1982 \leq x \leq 1994$

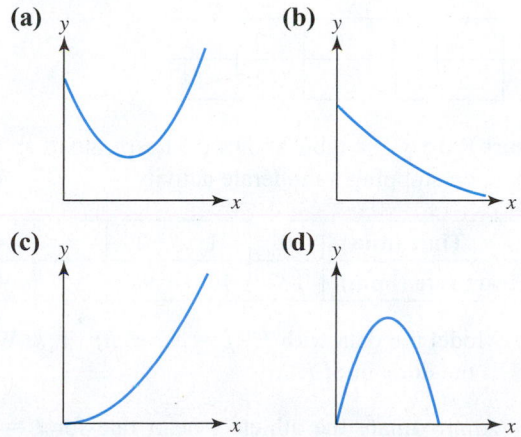

(a)   **(b)**

(c)   **(d)**

Applications and Models

95. **Maximizing Area**   A farmer has 1000 feet of fence to enclose a rectangular area. What dimensions for the rectangle result in the maximum area enclosed by the fence?

96. **Maximizing Area**   A homeowner has 80 feet of fence to enclose a rectangular garden. What dimensions for the garden give the maximum area?

97. **Maximizing Revenue**   Suppose the revenue R in thousands of dollars that a company receives from producing x thousand DVD players is given by the formula $R(x) = x(40 - 2x)$.
 (a) Evaluate $R(2)$ and interpret the result.

 (b) How many DVD players should the company produce to maximize its revenue?

 (c) What is the maximum revenue?

98. **Maximizing Revenue**   A large hotel is considering giving the following group discount on room rates: the regular price of $120 decreases by $2 for each room rented. For example, one room costs $118, two rooms cost $116 \times 2 = $232, three rooms cost $114 \times 3 = $342, and so on.
 (a) Write a formula for a function R that gives the revenue for renting x rooms.

 (b) Sketch a graph of R. What is a reasonable domain?

 (c) Determine the maximum revenue and the corresponding number of rooms rented.

99. **Minimizing Cost**   A business that produces color copies is trying to minimize its average cost per copy (total cost divided by the number of copies). This average cost in cents is given by

$$f(x) = 0.00000093x^2 - 0.0145x + 60,$$

where x represents the total number of copies produced.
 (a) Describe the graph of f.

 (b) Find the minimum average cost per copy and the corresponding number of copies made.

100. **Minimizing Cost**   A publisher is trying to minimize its average cost per book printed (total cost divided by the number of books printed). This average cost in dollars is given by

$$f(x) = 0.000000015x^2 - 0.0007x + 26,$$

where x represents the total number of books printed.
 (a) Describe the graph of f.

 (b) Find the minimum average cost per book and the corresponding number of books printed.

101. Hitting a Baseball A baseball is hit so that its height in feet after t seconds is $s(t) = -16t^2 + 44t + 4$.
(a) How high is the baseball after 1 second?

(b) Find the maximum height of the baseball. Support your answer graphically.

102. Flight of a Baseball (Refer to Example 9.) A baseball is hit straight up with an initial velocity of $v_0 = 96$ feet per second (about 65 miles per hour) and leaves the bat with an initial height of $h_0 = 2.5$ feet.
(a) Write a formula $s(t)$ that models the height after t seconds.

(b) How high is the baseball after 4 seconds?

(c) Find the maximum height of the baseball. Support your answer graphically.

103. Throwing a Stone (Refer to Example 9.) A stone is thrown *downward* with a velocity of 66 feet per second (45 miles per hour) from a bridge that is 120 feet above a river, as illustrated in the figure.
(a) Write a formula $s(t)$ that models the height of the stone after t seconds.

(b) Does the stone hit the water within the first 2 seconds? Explain.

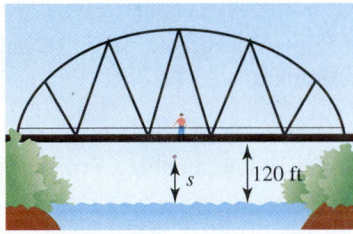

104. Hitting a Golf Ball A golf ball is hit so that its height h in feet after t seconds is $h(t) = -16t^2 + 60t$.
(a) What is the initial height of the golf ball?

(b) How high is the golf ball after 1.5 seconds?

(c) Find the maximum height of the golf ball.

105. Maximizing Area (Refer to Example 8.) A farmer wants to fence a rectangular area by using the wall of a barn as one side of the rectangle and then enclosing the other three sides with 160 feet of fence. Find the dimensions of the rectangle that give the maximum area inside.

106. Maximizing Area A rancher plans to fence a rectangular area for cattle using the straight portion of a river as one side of the rectangle. If the farmer has P feet of fence, find the dimensions of the rectangle that give the maximum area for the cattle.

Exercises 107–110: **Maximizing Altitude** *If air resistance is ignored, the height h of a projectile above the ground after x seconds is given by $h(x) = -\frac{1}{2}gx^2 + v_0x + h_0$, where g is the acceleration due to gravity. This formula is also valid for other celestial bodies. Suppose a ball is thrown straight up at 88 feet per second from a height of 25 feet.*

(a) *For the given g, graphically estimate both the maximum height and the time when it occurs.*

(b) *Solve part (a) symbolically.*

107. $g = 32$ (Earth)   **108.** $g = 5.1$ (Moon)

109. $g = 13$ (Mars)   **110.** $g = 88$ (Jupiter)

111. Suspension Bridge The cables that support a suspension bridge, such as the Golden Gate Bridge, can be modeled by parabolas. Suppose that a 300-foot-long suspension bridge has at each end a tower that is 120 feet tall, as shown in the figure. If the cable comes within 20 feet of the road at the center of the bridge, find a function that models the height of the cable above the road a distance of x feet from the center of the bridge.

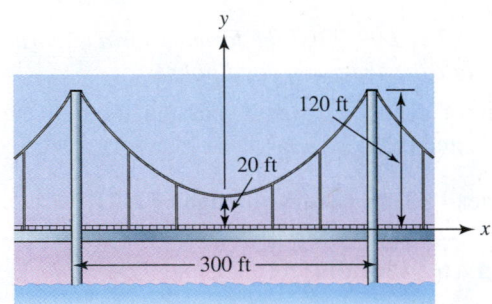

112. Suspension Bridge Repeat Exercise 111 for a suspension bridge that has 100-foot towers, a length of 200 feet, and a cable that comes within 15 feet of the road at the center of the bridge.

Exercises 113 and 114: Find $f(x) = a(x - h)^2 + k$ so that f models the data exactly.

113.

x	-1	0	1	2	3
y	5	-1	-3	-1	5

114.

x	-2	-1	0	1	2
y	2	4	2	-4	-14

115. Heart Rate The table shows the heart rate of an athlete upon stopping a moderate activity.

Time (min)	0	1	2	3	4
Heart rate (bpm)	122	108	98	92	90

(a) Model the data with $H(t) = a(t - h)^2 + k$. What is the domain of H?

(b) Approximate the athlete's heart rate for $t = 1.5$ minutes.

116. Heart Rate The heart rate of an athlete while weight training is recorded for 4 minutes. The table lists the heart rate after x minutes.

Time (min)	0	1	2	3	4
Heart rate (bpm)	84	111	120	110	85

(a) Explain why the data are not linear.

(b) Find a quadratic function f that models the data.

(c) What is the domain of your function?

117. Slideshare Slideshare is a social sharing site where users share professional materials such as PowerPoint presentations. The table shows the number of Slideshare visitors in millions for selected years, where x represents years after 2007.

Year	0	1	2	3	4
Visitors	1.5	6.1	17	33.2	60.1

Source: Business Insider.

(a) Model the data with $V(x) = a(x - h)^2 + k$.

(b) Approximate the number of visitors to Slideshare in 2012 if trends were to continue.

118. iPhone Sales The table lists the number of iPhones sold in millions in the first 5 years on the market. Find a quadratic function f that models the data.

Year	1	2	3	4	5
Units sold	1	20	50	100	180

Source: Business Insider.

Exercises 119 and 120: **AIDS in America (1982–1994)** *In the early years of AIDS, the numbers of both AIDS cases and AIDS deaths could be modeled with quadratic functions. The tables list cumulative numbers for selected years.*
(a) Find a quadratic function f that models the data.
(b) Graph the data and f together.
(c) Evaluate f(1991) and interpret the result.

119. Cumulative AIDS cases in thousands

Year	1982	1986	1990	1994
Cases	1.6	41.9	197	442

Source: Department of Health and Human Services.

120. Cumulative AIDS deaths in thousands

Year	1982	1986	1990	1994
Deaths	0.6	24.8	122	298

Source: Department of Health and Human Services.

Quadratic Regression

Exercises 121 and 122: **Quadratic Models** *Use least-squares regression to find a quadratic function f that models the data given in the table. Estimate f(3.5) to the nearest hundredth.*

121.

x	0	2	4	6
$f(x)$	−1	16	57	124

122.

x	10	20	30	40
$f(x)$	4.2	24.3	84.1	184

123. China's Rise The economic rise of China has greatly increased American interest in learning Chinese. The following table lists the U.S. college and university enrollments in thousands to study Chinese.

Year	1980	1986	1995	2002	2009
Enrollment	11	17	27	34	61

Source: The Economist

(a) Find a quadratic function that models the data. Support your result graphically.

(b) Estimate the enrollment to study Chinese in 2006 and compare to the actual value of 51 thousand.

124. MySpace Visitors The table lists the number of unique MySpace visitors per month in millions for various years.

Year	2006	2007	2008	2009	2010	2011
Visitors	55	70	77	65	55	35

Source: comScore

(a) Find a quadratic function that models the data. Support your result graphically.

(b) Estimate the number of MySpace visitors in 2012 if trends continued.

125. Head Start Enrollment Head Start provides a wide range of services to children of low-income families. The table lists Head Start participation in thousands.

Year	1970	1980	1990	2006
Enrollment	447	376	541	909

Source: Department of Health and Human Services.

(a) Find a quadratic function that models the data.

(b) Estimate enrollment in 1985.

126. Photosynthesis In one study the efficiency of photosynthesis in an Antarctic species of grass was investigated. The table lists results for various temperatures. The temperature x is in degrees Celsius and the efficiency y is given as a percent. (*Source:* D. Brown and P. Rothery, *Models in Biology: Mathematics, Statistics and Computing.*)

x (°C)	−1.5	0	2.5	5	7	10	12
y (%)	33	46	55	80	87	93	95
x (°C)	15	17	20	22	25	27	30
y (%)	91	89	77	72	54	46	34

(a) Plot the data. Discuss reasons why a quadratic function might model the data.

(b) Find a least-squares quadratic function f given by $f(x) = ax^2 + bx + c$ that models the data.

(c) Graph f and the data in the window $[-5, 35, 5]$ by $[20, 110, 10]$. Discuss the fit.

Writing about Mathematics

127. How do the values of a, h, and k affect the graph of $f(x) = a(x - h)^2 + k$?

128. Explain why the vertex is important when you are trying to find either the maximum y-value or the minimum y-value on the graph of a quadratic function.

Extended and Discovery Exercises

Exercises 1–4: **Difference Quotient**   *Complete the following.*

(a) *Evaluate $f(x)$ for each x-value in the table.*

x	1	2	3	4	5
$f(x)$					

(b) *Calculate the average rate of change of f between consecutive data points in the table.*

(c) *Find the difference quotient for $f(x)$. Then let $h = 1$ in the difference quotient.*

(d) *Evaluate this difference quotient for $x = 1, 2, 3,$ and 4. Compare these results to your results in part (b).*

1. $f(x) = x^2 - 3$

2. $f(x) = 2x - x^2$

3. $f(x) = -2x^2 + 3x - 1$

4. $f(x) = 3x^2 + x + 2$

3.2   Quadratic Equations and Problem Solving

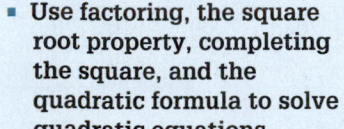

- **Understand basic concepts about quadratic equations**
- **Use factoring, the square root property, completing the square, and the quadratic formula to solve quadratic equations**
- **Understand the discriminant**
- **Solve problems involving quadratic equations**

Introduction

In Example 10 of Section 3.1 we modeled an athlete's heart rate x minutes after exercise stopped by using $f(x) = 1.875(x - 8)^2 + 80$. This vertex form can easily be changed to *general form*.

$$f(x) = 1.875(x - 8)^2 + 80 \qquad \text{Vertex form}$$
$$= 1.875(x^2 - 16x + 64) + 80 \qquad \text{Square the binomial.}$$
$$= 1.875x^2 - 30x + 200 \qquad \text{General form}$$

To determine the length of time needed for the athlete's heart rate to slow from **200** beats per minute to **110** beats per minute, we can solve the *quadratic equation*

$$1.875x^2 - 30x + 200 = 110, \quad \text{or} \qquad \text{Quadratic equation}$$
$$1.875x^2 - 30x + 90 = 0 \qquad \text{Subtract 110 from each side.}$$

(See Example 7.) A quadratic equation results when the formula for a quadratic function is set equal to a constant.

Quadratic Equations

A quadratic equation can be defined as follows.

> **QUADRATIC EQUATION**
>
> A **quadratic equation** in one variable is an equation that can be written in the form
> $$ax^2 + bx + c = 0,$$
> where a, b, and c are constants with $a \neq 0$.

Examples of quadratic equations include

$$2x^2 - 3x - 4 = 0, \quad x^2 = 3, \quad -5x^2 + x = 0, \quad \text{and} \quad 3x + 1 = x^2.$$

The following See the Concept explains that quadratic equations can have zero, one, or two *real* solutions by showing that a parabola can intersect the *x*-axis zero, one, or two times. (Complex solutions are discussed in the next section.)

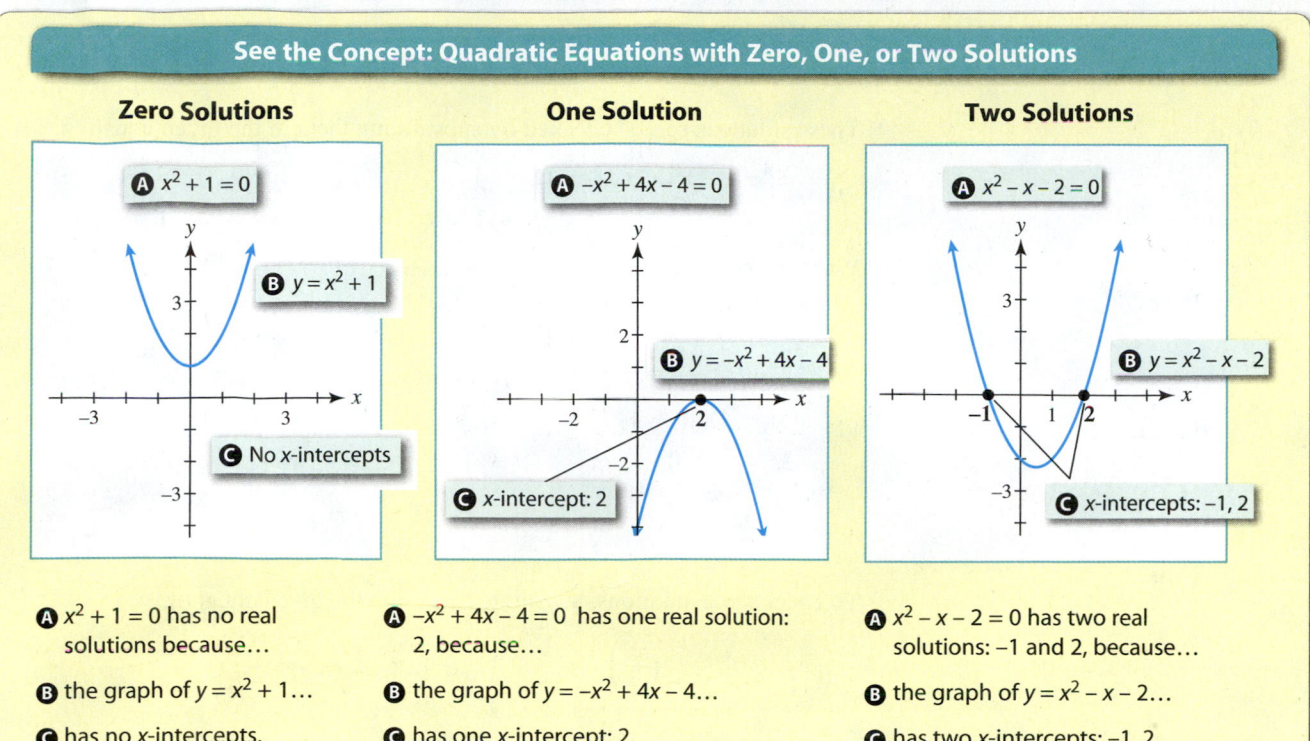

See the Concept: Quadratic Equations with Zero, One, or Two Solutions

Zero Solutions

Ⓐ $x^2 + 1 = 0$
Ⓑ $y = x^2 + 1$
Ⓒ No *x*-intercepts

One Solution

Ⓐ $-x^2 + 4x - 4 = 0$
Ⓑ $y = -x^2 + 4x - 4$
Ⓒ *x*-intercept: 2

Two Solutions

Ⓐ $x^2 - x - 2 = 0$
Ⓑ $y = x^2 - x - 2$
Ⓒ *x*-intercepts: −1, 2

Ⓐ $x^2 + 1 = 0$ has no real solutions because...

Ⓑ the graph of $y = x^2 + 1$...

Ⓒ has no *x*-intercepts.

Ⓐ $-x^2 + 4x - 4 = 0$ has one real solution: 2, because...

Ⓑ the graph of $y = -x^2 + 4x - 4$...

Ⓒ has one *x*-intercept: 2.

Ⓐ $x^2 - x - 2 = 0$ has two real solutions: −1 and 2, because...

Ⓑ the graph of $y = x^2 - x - 2$...

Ⓒ has two *x*-intercepts: −1, 2.

Quadratic equations can be solved symbolically by a variety of methods: factoring, the square root property, completing the square, and the quadratic formula. They can also be solved graphically and numerically; however, the exact solution can *always* be obtained symbolically.

Factoring

Factoring to solve an equation is based on the **zero-product property**, which states that if $ab = 0$, then $a = 0$ or $b = 0$ or both. It is important to remember that this property works only for 0. For example, if $ab = 1$, then this equation does *not* imply that either $a = 1$ or $b = 1$. For example, $a = \frac{1}{2}$ and $b = 2$ also satisfies $ab = 1$ and neither a nor b is 1.

EXAMPLE 1 **Solving quadratic equations with factoring**

Solve each quadratic equation. Check your results.
(a) $x^2 - 2x + 1 = 0$ **(b)** $2x^2 + 2x - 11 = 1$ **(c)** $12t^2 = t + 1$

SOLUTION
(a) Begin by factoring and applying the zero-product property.

$$x^2 - 2x + 1 = 0 \qquad \text{Given equation}$$
$$(x - 1)(x - 1) = 0 \qquad \text{Factor.}$$
$$x - 1 = 0 \quad \text{or} \quad x - 1 = 0 \qquad \text{Zero-product property}$$
$$x = \mathbf{1} \quad \text{or} \quad x = \mathbf{1} \qquad \text{Solve.}$$

The only solution is **1**. We can check our answer.

$$(\mathbf{1})^2 - 2(\mathbf{1}) + 1 = 0 \qquad \text{Let } x = 1.$$
$$0 = 0 \ \checkmark \qquad \text{It checks.}$$

Algebra Review
To review factoring trinomials, see Chapter R (pages R-22–R-23).

(b) Start by writing the equation in the form $ax^2 + bx + c = 0$.

$$2x^2 + 2x - 11 = 1 \qquad \text{Given equation}$$
$$2x^2 + 2x - 12 = 0 \qquad \text{Subtract 1 from each side.}$$
$$x^2 + x - 6 = 0 \qquad \text{Divide each side by 2.}$$
$$(x + 3)(x - 2) = 0 \qquad \text{Factor.}$$
$$x + 3 = 0 \quad \text{or} \quad x - 2 = 0 \qquad \text{Zero-product property}$$
$$x = -3 \quad \text{or} \quad x = 2 \qquad \text{Solve.}$$

These solutions can be checked by substituting them in the given equation.

$$2(-3)^2 + 2(-3) - 11 \overset{?}{=} 1 \qquad 2(2)^2 + 2(2) - 11 \overset{?}{=} 1$$
$$1 = 1 \checkmark \qquad \qquad 1 = 1 \checkmark$$

(c) Write the equation in the form $at^2 + bt + c = 0$.

$$12t^2 = t + 1 \qquad \text{Given equation}$$
$$12t^2 - t - 1 = 0 \qquad \text{Subtract } t \text{ and 1.}$$
$$(3t - 1)(4t + 1) = 0 \qquad \text{Factor.}$$
$$3t - 1 = 0 \quad \text{or} \quad 4t + 1 = 0 \qquad \text{Zero-product property}$$
$$t = \frac{1}{3} \quad \text{or} \quad t = -\frac{1}{4} \qquad \text{Solve.}$$

To check these solutions, substitute them into the given equation.

$$12\left(\frac{1}{3}\right)^2 = \frac{1}{3} + 1 \qquad 12\left(-\frac{1}{4}\right)^2 = -\frac{1}{4} + 1$$
$$\frac{4}{3} = \frac{4}{3} \checkmark \qquad \qquad \frac{3}{4} = \frac{3}{4} \checkmark$$

> **Now Try Exercises 1, 7, and 9**

We can also use factoring to find the x-intercepts of the graph of a quadratic function, because the x-intercepts of the graph of $y = ax^2 + bx + c$ correspond to the solutions to $ax^2 + bx + c = 0$. The y-intercept is c, the value of y when $x = 0$.

> **EXAMPLE 2** **Finding intercepts**

Find the exact values for both the x-intercepts and the y-intercept shown in Figure 3.26.

SOLUTION From the graph it is difficult to determine the *exact* x-intercepts. However, they can be determined symbolically.

$$24x^2 + 7x - 6 = 0 \qquad \text{Set expression equal to 0.}$$
$$(3x + 2)(8x - 3) = 0 \qquad \text{Factor.}$$
$$3x + 2 = 0 \quad \text{or} \quad 8x - 3 = 0 \qquad \text{Zero-product property}$$
$$x = -\frac{2}{3} \quad \text{or} \quad x = \frac{3}{8} \qquad \text{Solve.}$$

The x-intercepts are $-\frac{2}{3}$ and $\frac{3}{8}$.

To find the y-intercept, let $x = 0$ in $y = 24x^2 + 7x - 6$.

$$24(0)^2 + 7(0) - 6 = -6$$

The y-intercept is -6.

> **Now Try Exercise 33**

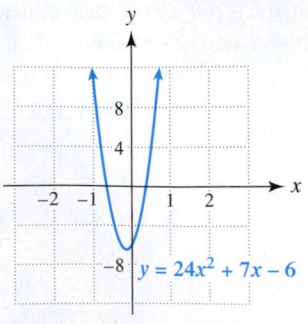

Figure 3.26

$y = 24x^2 + 7x - 6$

Symbolic, Numerical, and Graphical Solutions   Quadratic equations can be solved symbolically, numerically, and graphically. The following example illustrates each technique for the equation $x(x - 2) = 3$. (Also, see Example 7.)

Symbolic Solution	Numerical Solution	Graphical Solution

Symbolic Solution

$$x(x - 2) = 3$$
$$x^2 - 2x = 3$$
$$x^2 - 2x - 3 = 0$$
$$(x + 1)(x - 3) = 0$$

The solutions are **−1** and **3**.

Numerical Solution

x	$x(x - 2)$
−2	8
−1	**3**
0	0
1	−1
2	0
3	**3**
4	8

Let $y = x(x - 2)$. In the table $y = 3$ when $x = -1$ or $x = 3$.

Graphical Solution

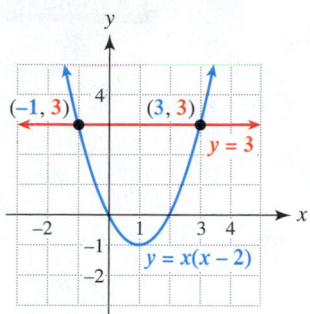

The graph of $y = x(x - 2)$ intersects the graph of $y = 3$ at ($-$**1**, **3**) and (**3**, **3**). The solutions are $-$**1** and **3**.

The Square Root Property

Some quadratic equations can be written as $x^2 = k$, where k is a nonnegative number. The solutions are $\pm \sqrt{k}$. (Recall that the symbol $\pm$ represents *plus or minus*.) For example, $x^2 = 16$ has two solutions: ± 4. We refer to this as the **square root property**. See Extended and Discovery Exercise 7 in this section for a justification of this property.

> **SQUARE ROOT PROPERTY**
>
> Let k be a nonnegative number. Then the solutions to the equation
> $$x^2 = k$$
> are given by $x = \pm \sqrt{k}$.

EXAMPLE 3   Using the square root property

If a metal ball is dropped 100 feet from a water tower, its height h in feet above the ground after t seconds is given by $h(t) = 100 - 16t^2$. Determine how long it takes the ball to hit the ground.

SOLUTION   The ball strikes the ground when the equation $100 - 16t^2 = 0$ is satisfied.

$$100 - 16t^2 = 0$$

$$100 = 16t^2 \qquad \text{Add } 16t^2 \text{ to each side.}$$

$$t^2 = \frac{100}{16} \qquad \text{Divide each side by 16. Rewrite.}$$

$$t = \pm\sqrt{\frac{100}{16}} \qquad \text{Square root property}$$

$$t = \pm\frac{10}{4} \qquad \text{Simplify.}$$

In this example only positive values for time are valid, so the ball strikes the ground after $\frac{10}{4}$, or 2.5, seconds.

Now Try Exercise 105

Functions can be defined by formulas, graphs, tables, and diagrams. Functions can also be defined by equations. In the next example we use the square root property to solve equations for y and then determine if y is a function of x, where $y = f(x)$.

EXAMPLE 4 Determining if equations represent functions

Solve each equation for y. Determine if y is a function of x.

(a) $x^2 + (y - 1)^2 = 4$ **(b)** $2y = \dfrac{x + y}{2}$

SOLUTION

(a) Start by subtracting x^2 from each side of the equation.

$$(y - 1)^2 = 4 - x^2 \qquad \text{Subtract } x^2 \text{ from each side.}$$
$$y - 1 = \pm\sqrt{4 - x^2} \qquad \text{Square root property}$$
$$y = 1 \pm \sqrt{4 - x^2} \qquad \text{Add 1 to each side.}$$

There are two formulas, $y = 1 + \sqrt{4 - x^2}$ and $y = 1 - \sqrt{4 - x^2}$, which indicates that y is *not* a function of x. That is, one x-input can produce two y-outputs.

NOTE The equation $x^2 + (y - 1)^2 = 4$ is in standard form for a circle that has center $(0, 1)$ and radius 2. A circle does not pass the vertical line test, so the equation does not represent a function.

Algebra Review
To clear fractions, see Chapter R (page R-33).

(b) Clear fractions by multiplying each side of $2y = \dfrac{x + y}{2}$ by 2.

$$4y = x + y \qquad \text{Multiply each side by 2.}$$
$$3y = x \qquad \text{Subtract } y \text{ from each side.}$$
$$y = \frac{x}{3} \qquad \text{Divide each side by 3.}$$

The equation $y = \frac{x}{3}$ defines a linear function, $f(x) = \frac{1}{3}x$, so y is a function of x.

Now Try Exercises 71 and 73

Completing the Square

Another technique that can be used to solve a quadratic equation is *completing the square*. If a quadratic equation is written in the form $x^2 + kx = d$, where k and d are constants, then the equation can be solved using

$$x^2 + kx + \left(\frac{k}{2}\right)^2 = \left(x + \frac{k}{2}\right)^2.$$

Algebra Review
To review factoring perfect square trinomials, see Chapter R (page R-25)

For example, $k = 6$ in $x^2 + 6x = 7$, so add $\left(\frac{k}{2}\right)^2 = \left(\frac{6}{2}\right)^2 = 9$ to each side.

$$x^2 + 6x = 7 \qquad \text{Given equation}$$
$$x^2 + 6x + 9 = 7 + 9 \qquad \text{Add 9 to each side.}$$
$$(x + 3)^2 = 16 \qquad \text{Factor the perfect square.}$$
$$x + 3 = \pm 4 \qquad \text{Square root property}$$
$$x = -3 \pm 4 \qquad \text{Add } -3 \text{ to each side.}$$
$$x = 1 \quad \text{or} \quad x = -7 \qquad \text{Simplify.}$$

NOTE If the coefficient a of the x^2-term is not 1, we can divide each side of the equation by a so that it becomes 1. See Example 5(b).

Completing the square is useful when solving quadratic equations that do not factor easily.

EXAMPLE 5   Completing the square

Solve each equation.
(a) $x^2 - 8x + 9 = 0$ **(b)** $2x^2 - 8x = 7$

SOLUTION
(a) Start by writing the equation in the form $x^2 + kx = d$ with $k = -8$ and $d = -9$.

$x^2 - 8x + 9 = 0$	*Given equation*
$x^2 - 8x = -9$	*Subtract 9 from each side.*
$x^2 - 8x + 16 = -9 + 16$	*Add $\left(\frac{k}{2}\right)^2 = \left(\frac{-8}{2}\right)^2 = 16$.*
$(x - 4)^2 = 7$	*Factor the perfect square.*
$x - 4 = \pm\sqrt{7}$	*Square root property*
$x = 4 \pm \sqrt{7}$	*Add 4 to each side.*

(b) Divide each side by 2 to obtain a 1 for the leading coefficient.

$2x^2 - 8x = 7$	*Given equation*
$x^2 - 4x = \frac{7}{2}$	*Divide each side by 2.*
$x^2 - 4x + 4 = \frac{7}{2} + 4$	*Add $\left(\frac{-4}{2}\right)^2 = 4$ to each side.*
$(x - 2)^2 = \frac{15}{2}$	*Factor the perfect square.*
$x - 2 = \pm\sqrt{\frac{15}{2}}$	*Square root property*
$x = 2 \pm \sqrt{\frac{15}{2}}$	*Add 2 to each side.*

Now Try Exercises 53 and 55

The Quadratic Formula

The quadratic formula can be used to find the solutions to *any* quadratic equation.

QUADRATIC FORMULA

The solutions to the quadratic equation $ax^2 + bx + c = 0$, where $a \neq 0$, are given by

$$x = \frac{-b \pm \sqrt{b^2 - 4ac}}{2a}.$$

EXAMPLE 6   Using the quadratic formula

Solve the equation $3x^2 - 6x + 2 = 0$.

SOLUTION   Let $a = 3$, $b = -6$, and $c = 2$.

$x = \dfrac{-b \pm \sqrt{b^2 - 4ac}}{2a}$	*Quadratic formula*
$x = \dfrac{-(-6) \pm \sqrt{(-6)^2 - 4(3)(2)}}{2(3)}$	*Substitute for a, b, and c.*
$x = \dfrac{6 \pm \sqrt{12}}{6}$	*Simplify.*
$x = 1 \pm \dfrac{1}{6}\sqrt{12}$	*Divide: $\frac{a \pm b}{c} = \frac{a}{c} \pm \frac{b}{c}$.*

Algebra Review
To review simplifying square roots, see Chapter R (page R-43).

NOTE   Because $\sqrt{12} = \sqrt{4} \cdot \sqrt{3} = 2\sqrt{3}$, we can write $1 \pm \frac{1}{6}\sqrt{12}$ as $1 \pm \frac{1}{3}\sqrt{3}$.

Now Try Exercise 25

> **NOTE** When solving quadratic equations, a common strategy is to first write the equation in the form $ax^2 + bx + c = 0$ and to then try to factor the left side of the equation. If the factors are not easily found, the quadratic formula is used.

EXAMPLE 7 Estimating an athlete's heart rate

An athlete's heart rate is given by $1.875x^2 - 30x + 200$. Determine when the heart rate was 110 beats per minute by solving the quadratic equation $1.875x^2 - 30x + 90 = 0$ symbolically, graphically, and numerically, where $0 \le x \le 8$. (This equation was discussed in the introduction to this section.)

SOLUTION

Symbolic Solution Let $a = 1.875$, $b = -30$, and $c = 90$ in the quadratic formula.

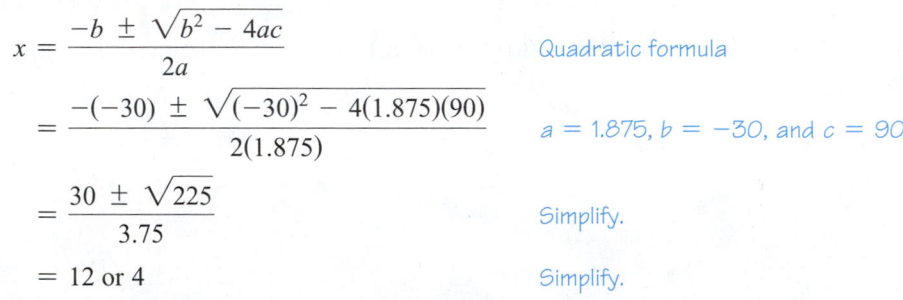

$$x = \frac{-b \pm \sqrt{b^2 - 4ac}}{2a} \qquad \text{Quadratic formula}$$

$$= \frac{-(-30) \pm \sqrt{(-30)^2 - 4(1.875)(90)}}{2(1.875)} \qquad a = 1.875,\ b = -30,\ \text{and } c = 90$$

$$= \frac{30 \pm \sqrt{225}}{3.75} \qquad \text{Simplify.}$$

$$= 12 \text{ or } 4 \qquad \text{Simplify.}$$

The x-values are restricted to $0 \le x \le 8$, so the only valid solution is 4. Thus the athlete's heart rate reached 110 beats per minute 4 minutes after the athlete stopped exercising.

Graphical Solution To use the *x-intercept method* to solve this quadratic equation, graph $y_1 = 1.875x^2 - 30x + 90$ and locate the x-intercepts, as shown in Figures 3.27 and 3.28. The x-intercepts are 4 and 12, in agreement with the symbolic solution.

Numerical Solution Make a table of $y_1 = 1.875x^2 - 30x + 90$, as shown in Figure 3.29. The numerical solution agrees with the symbolic and graphical solutions because $y_1 = 0$ when $x = 4$ or $x = 12$.

Calculator Help

To find a zero, or *x*-intercept, see Appendix A (page AP-7).

Graphical Solutions   **Numerical Solutions**

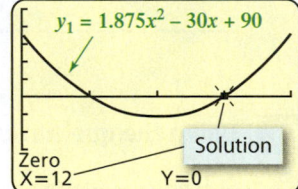   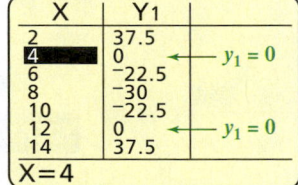

X	Y1
2	37.5
4	0
6	-22.5
8	-30
10	-22.5
12	0
14	37.5

X=4

Figure 3.27   **Figure 3.28**   **Figure 3.29**

Now Try Exercise 107

The Discriminant

The quantity $b^2 - 4ac$ in the quadratic formula is called the **discriminant.** It provides information about the number of real solutions to a quadratic equation.

QUADRATIC EQUATIONS AND THE DISCRIMINANT

To determine the number of real solutions to $ax^2 + bx + c = 0$ with $a \ne 0$, evaluate the discriminant $b^2 - 4ac$.

1. If $b^2 - 4ac > 0$, there are two real solutions.
2. If $b^2 - 4ac = 0$, there is one real solution.
3. If $b^2 - 4ac < 0$, there are no real solutions.

NOTE When $b^2 - 4ac < 0$, the solutions to a quadratic equation may be expressed as two complex numbers. Complex numbers are discussed in the next section.

In Example 6 the discriminant is $b^2 - 4ac = (-6)^2 - 4(3)(2) = 12$. Because the discriminant is positive, there are two real solutions.

EXAMPLE 8 **Using the discriminant**

Use the discriminant to find the number of solutions to $4x^2 - 12x + 9 = 0$. Then solve the equation by using the quadratic formula. Support your result graphically.

SOLUTION

Symbolic Solution Let $a = 4, b = -12$, and $c = 9$. The discriminant is given by

$$b^2 - 4ac = (-12)^2 - 4(4)(9) = 0.$$

Since the discriminant is 0, there is one solution.

$$x = \frac{-b \pm \sqrt{b^2 - 4ac}}{2a} \qquad \textcolor{blue}{\text{Quadratic formula}}$$

$$x = \frac{-(-12) \pm \sqrt{0}}{8} \qquad \textcolor{blue}{\text{Substitute.}}$$

$$x = \frac{3}{2} \qquad \textcolor{blue}{\text{Simplify.}}$$

The only solution is $\frac{3}{2}$.

Graphical Solution A graph of $y = 4x^2 - 12x + 9$ is shown in Figure 3.30. The graph suggests that there is one solution because there is one x-intercept: $\frac{3}{2}$.

Now Try Exercise 87

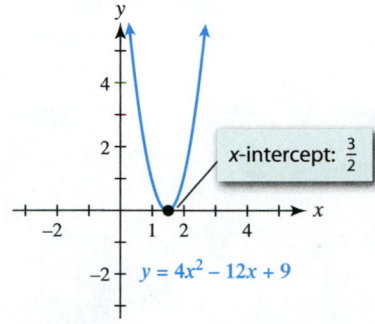

One Solution: $b^2 - 4ac = 0$

x-intercept: $\frac{3}{2}$

$y = 4x^2 - 12x + 9$

Figure 3.30

See the Concept: The Discriminant and Solutions to $ax^2 + bx + c = 0$

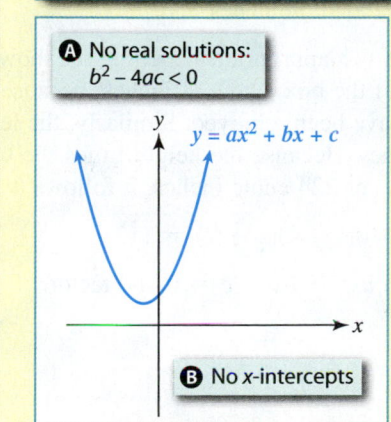

A No real solutions: $b^2 - 4ac < 0$

$y = ax^2 + bx + c$

B No x-intercepts

A Discriminant is negative because the graph…

B has no x-intercepts.

A One real solution: $b^2 - 4ac = 0$

$y = ax^2 + bx + c$

B One x-intercept

A Discriminant equals 0 because the graph…

B has one x-intercept.

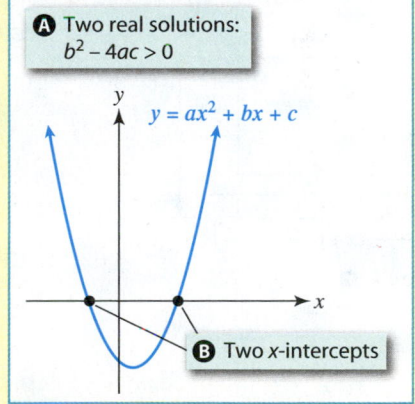

A Two real solutions: $b^2 - 4ac > 0$

$y = ax^2 + bx + c$

B Two x-intercepts

A Discriminant is positive because the graph…

B has two x-intercepts.

Factoring and the Discriminant If a, b, and c are *integers* and $b^2 - 4ac$ is a perfect square, then the trinomial $ax^2 + bx + c$ can be factored using only integer coefficients. For example, if $6x^2 + x - 2 = 0$, then

$$b^2 - 4ac = 1^2 - 4(6)(-2) = 49,$$

which *is* a perfect square ($49 = 7^2$). Thus we can factor $6x^2 + x - 2$ as $(2x - 1)(3x + 2)$ to solve $6x^2 + x - 2 = 0$. However, if $3x^2 - x - 1 = 0$, then $b^2 - 4ac = 13$, which is *not* a perfect square. This trinomial cannot be factored (by using traditional methods with integer coefficients), so either the quadratic formula or completing the square should be used to solve $3x^2 - x - 1 = 0$. See Extended and Discovery Exercises 1–4 at the end of this section.

Problem Solving and Modeling

Many types of applications involve quadratic equations. To solve the next two problems, we use the steps for "Solving Application Problems" from Section 2.2.

EXAMPLE 9 Solving a construction problem

A box is being constructed by cutting 2-inch squares from the corners of a rectangular piece of cardboard that is 6 inches longer than it is wide, as illustrated in Figure 3.31. If the box is to have a volume of 224 cubic inches, find the dimensions of the piece of cardboard.

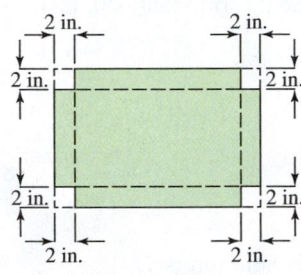
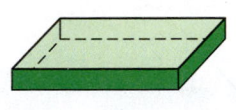

Figure 3.31

SOLUTION

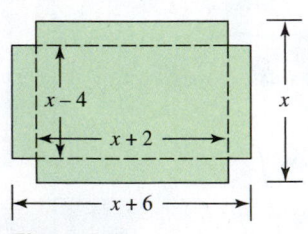

Geometry Review
To review formulas related to boxes, see Chapter R (page R-3).

Figure 3.32

STEP 1: The rectangular piece of cardboard is 6 inches longer than it is wide. Let x be its width and $x + 6$ be its length.

$$x\text{: Width of the cardboard in inches}$$

$$x + 6\text{: Length of the cardboard in inches}$$

STEP 2: First make a drawing of the box with the appropriate labeling, as shown in Figure 3.32. The width of the *bottom* of the box is $x - 4$ inches, because two square corners with sides of 2 inches have been removed. Similarly, the length of the bottom of the box is $x + 2$ inches. Because the height times the width times the length must equal the volume, or 224 cubic inches, it follows that

$$2(x - 4)(x + 2) = 224, \quad \text{or} \quad (x - 4)(x + 2) = 112.$$

STEP 3: Write the quadratic equation in the form $ax^2 + bx + c = 0$ and factor.

$$x^2 - 2x - 8 = 112 \qquad \text{Equation to be solved}$$
$$x^2 - 2x - 120 = 0 \qquad \text{Subtract 112.}$$
$$(x - 12)(x + 10) = 0 \qquad \text{Factor.}$$
$$x = 12 \quad \text{or} \quad x = -10 \qquad \text{Zero-product property}$$

Since the dimensions cannot be negative, the width of the cardboard is 12 inches and the length is 6 inches more, or 18 inches.

STEP 4: After the **2**-inch-square corners are cut out, the dimensions of the bottom of the box are $12 - 4 = $ **8** inches by $18 - 4 = $ **14** inches. The volume of the box is then $2 \cdot 8 \cdot 14 = 224$ cubic inches, which checks.

Now Try Exercise 111

When items are sold, a discount is sometimes given to a customer who makes a large order, which affects the revenue that a company receives. We discuss this situation in the next example.

EXAMPLE 10 **Determining revenue**

A company charges \$5 for earbud headphones, but it reduces this cost by \$0.05 for each additional pair ordered, up to a maximum of 50 earbuds. For example, the price for one pair is \$5, the price for two pair is $2(\$4.95) = \$9.90$, the price for 3 pair is $3(\$4.90) = \$14.70$, and so on. If the total price is \$95, how many earbuds were ordered?

SOLUTION

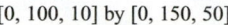
[0, 100, 10] by [0, 150, 50]

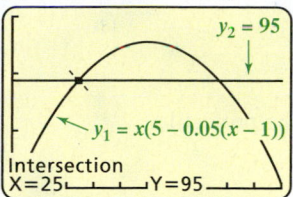
$y_2 = 95$

$y_1 = x(5 - 0.05(x - 1))$

Intersection
X=25   Y=95

Figure 3.33

[0, 100, 10] by [0, 150, 50]

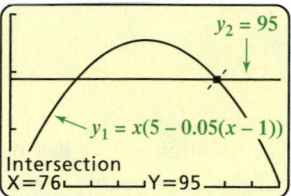
$y_2 = 95$

$y_1 = x(5 - 0.05(x - 1))$

Intersection
X=76   Y=95

Figure 3.34

STEP 1: We are asked to find the number of earbuds that results in an order costing \$95. Let this number be x.

$$x: \text{Number of earbuds ordered}$$

STEP 2: Revenue equals the number of earbuds sold times the price of each pair. If x earbuds are sold, then the price in dollars of each pair is $5 - 0.05(x - 1)$. (Note that when $x = 1$ the price is $5 - 0.05(1 - 1) = \$5$.) The revenue R is given by

$$R(x) = x(5 - 0.05(x - 1)),$$

Revenue = Number Sold × Price

and we must solve the equation

$$x(5 - 0.05(x - 1)) = 95, \quad \text{or} \quad -0.05x^2 + 5.05x - 95 = 0.$$

STEP 3: We can solve $-0.05x^2 + 5.05x - 95$ by using the quadratic formula with $a = -0.05$, $b = 5.05$, and $c = -95$.

$$x = \frac{-5.05 \pm \sqrt{(5.05)^2 - 4(-0.05)(-95)}}{2(-0.05)} \qquad \textit{Quadratic formula}$$

$$= \frac{-5.05 \pm 2.55}{-0.1} \qquad \textit{Use a calculator.}$$

$$= 25 \text{ or } 76 \qquad \textit{Simplify.}$$

This discount only applies to orders of 50 earbuds or fewer, so the answer is 25 earbuds. A graphical solution is shown in Figure 3.33 and 3.34.

STEP 4: If 25 earbuds are ordered, then the cost of each pair is $5 - 0.05(24) = \$3.80$ and the total revenue is $25(3.80) = \$95$.

Now Try Exercise 117

Figure 3.35

A Modeling Application In Section 3.1 we learned that the height s of an object propelled into the air is modeled by $s(t) = -16t^2 + v_0t + h_0$, where v_0 represents the object's initial (vertical) velocity in feet per second and h_0 represents the object's initial height in feet, as illustrated in Figure 3.35. In the next example we model the position of a projectile.

EXAMPLE 11 **Modeling projectile motion**

Table 3.5 shows the height of a projectile shot into the air.

Height of a Projectile

t (seconds)	0	2	4	6	8
$s(t)$ (feet)	96	400	576	624	544

Table 3.5

(a) Use $s(t) = -16t^2 + v_0t + h_0$ to model the data.
(b) After how many seconds did the projectile strike the ground?

t (seconds)	$s(t)$ (feet)
0	96
2	400
4	576
6	624
8	544

Table 3.5 (repeated)

SOLUTION

Getting Started The value for h_0 can be determined by noting that $s(0) = 96$. The value for v_0 can be determined by using any other value in Table 3.5. We use $s(2) = 400$ to determine v_0. ▶

(a) Because $s(0) = 96$, $h_0 = 96$ and $s(t) = -16t^2 + v_0 t + 96$. Substituting $s(2) = 400$ gives the following result.

$$-16(2)^2 + v_0(2) + 96 = 400 \qquad \text{\small\color{teal}$s(2) = 400$}$$

$$2v_0 + 32 = 400 \qquad \text{\small\color{teal}Simplify.}$$

$$2v_0 = 368 \qquad \text{\small\color{teal}Subtract 32 from each side.}$$

$$v_0 = 184 \qquad \text{\small\color{teal}Divide each side by 2.}$$

Thus $s(t) = -16t^2 + 184t + 96$ models the height of the projectile.

(b) The projectile strikes the ground when $s(t) = 0$.

$$-16t^2 + 184t + 96 = 0 \qquad \text{\small\color{teal}$s(t) = 0$; equation to be solved}$$

$$2t^2 - 23t - 12 = 0 \qquad \text{\small\color{teal}Divide each term by -8.}$$

$$(2t + 1)(t - 12) = 0 \qquad \text{\small\color{teal}Factor the trinomial.}$$

$$t = -\frac{1}{2} \quad \text{or} \quad t = 12 \qquad \text{\small\color{teal}Solve the equation.}$$

Thus the projectile strikes the ground after 12 seconds. The solution of $-\frac{1}{2}$ has no meaning in this problem because it corresponds to a time before the projectile is shot into the air.

Now Try Exercise 119

3.2 Putting It All Together

The following table summarizes important topics related to quadratic equations.

CONCEPT	EXPLANATION	EXAMPLES
Quadratic equation	$ax^2 + bx + c = 0$, where a, b, and c are constants with $a \neq 0$	A quadratic equation can have zero, one, or two real solutions. $x^2 = -5$ No real solutions $(x - 2)^2 = 0$ One real solution: 2 $x^2 - 4 = 0$ Two real solutions: ± 2
Factoring	A symbolic technique for solving equations, based on the zero-product property: if $ab = 0$, then either $a = 0$ or $b = 0$.	$x^2 - 3x = -2$ $x^2 - 3x + 2 = 0$ $(x - 1)(x - 2) = 0$ $x - 1 = 0$ or $x - 2 = 0$ $x = 1$ or $x = 2$

CONCEPT	EXPLANATION	EXAMPLES
Square root property	The solutions to $x^2 = k$ are $x = \pm\sqrt{k}$, where $k \geq 0$.	$x^2 = 9$ is equivalent to $x = \pm 3$. $x^2 = 11$ is equivalent to $x = \pm\sqrt{11}$.
Completing the square	To solve $x^2 + kx = d$ symbolically, add $\left(\frac{k}{2}\right)^2$ to each side to obtain a perfect square trinomial. Then apply the square root property.	$$x^2 - 6x = 1$$ $$x^2 - 6x + 9 = 1 + 9 \qquad \left(\frac{-6}{2}\right)^2 = 9$$ $$(x-3)^2 = 10$$ $$x - 3 = \pm\sqrt{10}$$ $$x = 3 \pm \sqrt{10}$$
Quadratic formula	The solutions to $ax^2 + bx + c = 0$ are given by $$x = \frac{-b \pm \sqrt{b^2 - 4ac}}{2a}.$$ Always gives the *exact* solutions	To solve $2x^2 - x - 4 = 0$, let $a = 2$, $b = -1$, and $c = -4$. $$x = \frac{-(-1) \pm \sqrt{(-1)^2 - 4(2)(-4)}}{2(2)}$$ $$= \frac{1 \pm \sqrt{33}}{4} \approx 1.69 \text{ or } -1.19$$
Graphical solution	To use the *x-intercept method* to solve $ax^2 + bx + c = 0$, let y_1 equal the left side of the equation and graph y_1. The real solutions correspond to the x-intercepts. The *intersection-of-graphs* method can also be used when one side of the equation is *not* equal to 0. Let y_1 equal the left side of the equation and y_2 equal the right side of the equation. The real solutions correspond to the x-coordinates of any points of intersection.	To solve $x^2 + x - 2 = 0$, graph $y = x^2 + x - 2$. The solutions are -2 and 1. Solutions: $-2, 1$
Numerical solution	To solve $ax^2 + bx + c = 0$, let y_1 equal the left side of the equation and create a table for y_1. The zeros of y_1 are the real solutions. May *not* be a good method when solutions are fractions or irrational numbers	To solve $2x^2 + x - 1 = 0$, make a table. The solutions are -1 and 0.5.

3.2 Exercises

Quadratic Equations

Exercises 1–32: Solve the quadratic equation. Check your answers for Exercises 1–16.

1. $x^2 + x - 11 = 1$

2. $x^2 - 9x + 10 = -8$

3. $t^2 = 2t$

4. $t^2 - 7t = 0$

5. $3x^2 - 7x = 0$

6. $5x = 9x^2$

7. $2z^2 = 13z + 15$

8. $4z^2 = 7 - 27z$

9. $x^2 + 6x + 9 = 0$

10. $x^2 - 8x + 16 = 0$

11. $4x^2 + 1 = 4x$

12. $9x^2 + 4 = 12x$

13. $x(3x + 14) = 5$

14. $x(5x + 19) = 4$

15. $6x^2 + \frac{5}{2} = 8x$

16. $8x^2 + 63 = -46x$

17. $(t + 3)^2 = 5$

18. $(t - 2)^2 = 11$

19. $4x^2 - 13 = 0$

20. $9x^2 - 11 = 0$

21. $2(x - 1)^2 + 4 = 0$

22. $-3(x + 5)^2 - 6 = 0$

23. $\frac{1}{2}x^2 - 3x + \frac{1}{2} = 0$

24. $\frac{3}{4}x^2 + \frac{1}{2}x - \frac{1}{2} = 0$

25. $-3z^2 - 2z + 4 = 0$

26. $-4z^2 + z + 1 = 0$

27. $25k^2 + 1 = 10k$

28. $49k^2 + 4 = -28k$

29. $-0.3x^2 + 0.1x = -0.02$

30. $-0.1x^2 + 1 = 0.5x$

31. $2x(x + 2) = (x - 1)(x + 2)$

32. $(2x - 1)(x + 2) = (x + 3)(x + 1)$

Exercises 33–38: Find the exact values of all intercepts.

33.
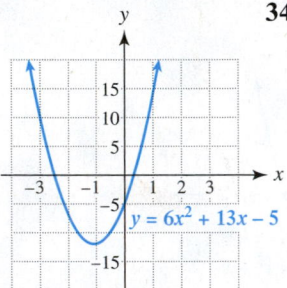
$y = 6x^2 + 13x - 5$

34.
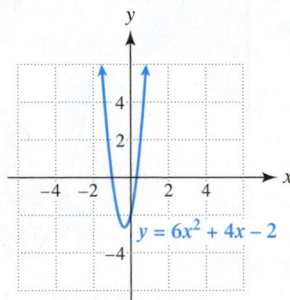
$y = 6x^2 + 4x - 2$

35.
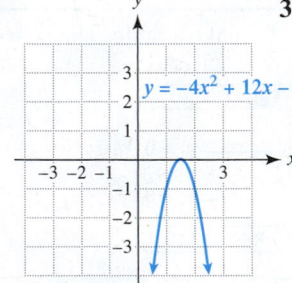
$y = -4x^2 + 12x - 9$

36.
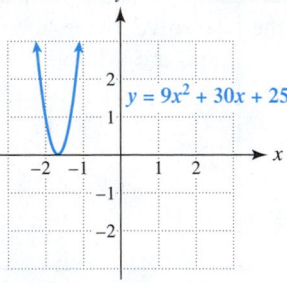
$y = 9x^2 + 30x + 25$

37.
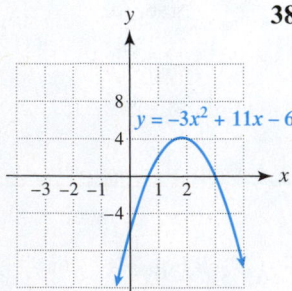
$y = -3x^2 + 11x - 6$

38.
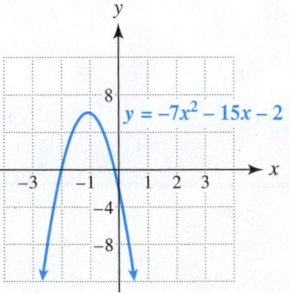
$y = -7x^2 - 15x - 2$

Graphical and Numerical Solutions

Exercises 39–46: Solve each quadratic equation (a) graphically, (b) numerically, and (c) symbolically. Express graphical and numerical solutions to the nearest tenth when appropriate.

39. $x^2 + 2x = 0$

40. $x^2 - 4 = 0$

41. $x^2 - x - 6 = 0$

42. $2x^2 + 5x - 3 = 0$

43. $2x^2 = 6$

44. $x^2 - 225 = 0$

45. $4x(x - 3) = -9$

46. $-4x(x - 1) = 1$

Exercises 47–50: Solve the quadratic equation graphically.

47. $20x^2 + 11x = 3$

48. $-2x^2 + 4x = 1.595$

49. $2.5x^2 = 4.75x - 2.1$

50. $x(x + 24) = 6912$

Completing the Square

Exercises 51–64: Solve the equation by completing the square.

51. $x^2 + 4x - 6 = 0$

52. $x^2 - 10x = 1$

53. $x^2 + 5x = 4$

54. $x^2 + 6x - 5 = 0$

55. $3x^2 - 6x = 2$

56. $2x^2 - 3x + 1 = 0$

57. $x^2 - 8x = 10$

58. $x^2 - 2x = 2$

59. $\frac{1}{2}t^2 - \frac{3}{2}t = 1$

60. $\frac{1}{3}t^2 + \frac{1}{2}t = 2$

61. $-2z^2 + 3z + 1 = 0$

62. $-3z^2 - 5z + 3 = 0$

63. $-\frac{3}{2}z^2 - \frac{1}{4}z + 1 = 0$

64. $-\frac{1}{5}z^2 - \frac{1}{2}z + 2 = 0$

Finding Domains

Exercises 65–68: Find the domain of the function. Write your answer in set-builder notation.

65. $f(x) = \dfrac{1}{x^2 - 5}$

66. $f(x) = \dfrac{4x}{7 - x^2}$

67. $g(t) = \dfrac{5 - t}{t^2 - t - 2}$

68. $g(t) = \dfrac{t + 1}{2t^2 - 11t - 21}$

Solving for a Variable

Exercises 69–76: (Refer to Example 4.) Solve the equation for y. Determine if y is a function of x.

69. $4x^2 + 3y = \dfrac{y + 1}{3}$

70. $\dfrac{x^2 + y}{2} = y - 2$

71. $3y = \dfrac{2x - y}{3}$

72. $\dfrac{5 - y}{3} = \dfrac{x + 3y}{4}$

73. $x^2 + (y - 3)^2 = 9$

74. $(x + 2)^2 + (y + 1)^2 = 1$

75. $3x^2 + 4y^2 = 12$

76. $x - 25y^2 = 50$

Exercises 77–84: Solve for the specified variable.

77. $V = \frac{1}{3}\pi r^2 h$ for r **78.** $V = \frac{1}{2}gt^2 + h$ for t

79. $K = \frac{1}{2}mv^2$ for v **80.** $W = I^2 R$ for I

81. $a^2 + b^2 = c^2$ for b **82.** $S = 4\pi r^2 + x^2$ for r

83. $s = -16t^2 + 100t$ for t

84. $T^2 - kT - k^2 = 0$ for T

The Discriminant

Exercises 85–100: Complete the following.

(a) Write the equation as $ax^2 + bx + c = 0$ with $a > 0$.
(b) Calculate the discriminant $b^2 - 4ac$ and determine the number of real solutions.
(c) Solve the equation.

85. $3x^2 = 12$ **86.** $8x^2 - 2 = 14$

87. $x^2 - 2x = -1$ **88.** $6x^2 = 4x$

89. $4x = x^2$ **90.** $16x^2 + 9 = 24x$

91. $x^2 + 1 = x$ **92.** $2x^2 + x = 2$

93. $2x^2 + 3x = 12 - 2x$ **94.** $3x^2 + 3 = 5x$

95. $9x(x - 4) = -36$ **96.** $\frac{1}{4}x^2 + 3x = x - 4$

97. $x\left(\frac{1}{2}x + 1\right) = -\frac{13}{2}$ **98.** $4x = 6 + x^2$

99. $3x^2 = 1 - x$ **100.** $x(5x - 3) = 1$

Exercises 101–104: The graph of $f(x) = ax^2 + bx + c$ is shown in the figure.

(a) State whether $a > 0$ or $a < 0$.
(b) Solve the equation $ax^2 + bx + c = 0$.
(c) Is the discriminant positive, negative, or zero?

101. 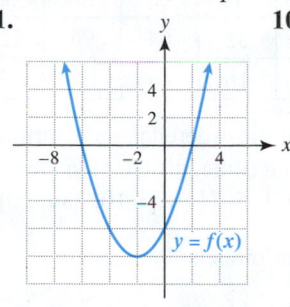   **102.**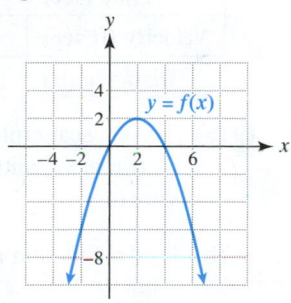

103. 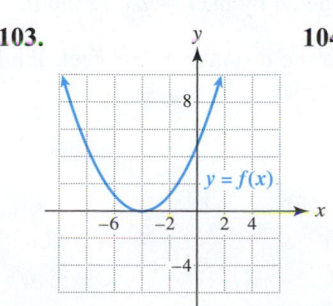   **104.**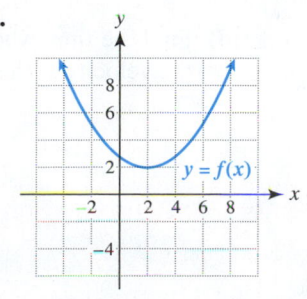

Applications and Models

105. Height of a Baseball A baseball is dropped from a stadium seat that is 75 feet above the ground. Its height s in feet after t seconds is given by $s(t) = 75 - 16t^2$. Estimate to the nearest tenth of a second how long it takes for the baseball to strike the ground.

106. Height of a Baseball A baseball is thrown *downward* with an initial velocity of 30 feet per second from a stadium seat that is 80 feet above the ground. Estimate to the nearest tenth of a second how long it takes for the baseball to strike the ground.

107. Facebook Visitors The number of unique monthly visitors in millions to Facebook can be approximated by

$$V(x) = 16x^2 + 7x + 32,$$

where x is the number of years after 2008. Estimate the year when Facebook averaged 55 million unique monthly visitors.

108. U.S. AIDS Cases From 1984 to 1994 the cumulative number of AIDS cases can be modeled by the equation

$$C(x) = 3034x^2 + 14{,}018x + 6400,$$

where x represents years after 1984. Estimate the year when 200,000 AIDS cases had been diagnosed.

109. Screen Dimensions The width of a rectangular computer screen is 2.5 inches more than its height. If the area of the screen is 93.5 square inches, determine its dimensions symbolically, graphically, and numerically. Do your answers agree?

110. Maximizing Area A rectangular pen for a pet is under construction using 100 feet of fence.
(a) Find the dimensions that give an area of 576 square feet.

(b) Find the dimensions that give maximum area.

111. Construction (Refer to Example 9.) A box is being constructed by cutting 4-inch squares from the corners of a rectangular sheet of metal that is 10 inches longer than it is wide. If the box is to have a volume of 476 cubic inches, find the dimensions of the metal sheet.

112. **Construction** A box is being constructed by cutting 2-inch squares from the corners of a *square* sheet of metal. If the box is to have a volume of 1058 cubic inches, find the dimensions of the metal sheet.

113. **Geometry** A cylindrical aluminum can is being constructed to have a height h of 4 inches. If the can is to have a volume of 28 cubic inches, approximate its radius r. (*Hint: $V = \pi r^2 h$.*)

114. **Braking Distance** Braking distance for cars on level pavement can be approximated by $D(x) = \frac{x^2}{30k}$. The input x is the car's velocity in miles per hour and the output $D(x)$ is the braking distance in feet. The positive constant k is a measure of the traction of the tires. Small values of k indicate a slippery road or worn tires. (*Source:* L. Haefner, *Introduction to Transportation Systems.*)

    (a) Let $k = 0.3$. Evaluate $D(60)$ and interpret the result.

    (b) If $k = 0.25$, find the velocity x that corresponds to a braking distance of 300 feet.

115. **Window Dimensions** A window comprises a square with sides of length x and a semicircle with diameter x, as shown in the figure. If the total area of the window is 463 square inches, estimate the value of x to the nearest hundredth of an inch.

116. **Picture Frame** A frame for a picture is 2 inches wide. The picture inside the frame is 4 inches longer than it is wide. See the figure. If the area of the picture is 320 square inches, find the outside dimensions of the picture frame.

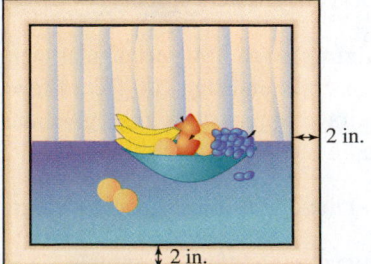

117. **Cost** (Refer to Example 10.) A company charges $20 to make one monogrammed shirt but reduces this cost by $0.10 per shirt for each *additional* shirt ordered up to 100 shirts. If the cost of an order is $989, how many shirts were ordered?

118. **Ticket Prices** The price of one airline ticket is $250. For each additional ticket sold to a group, the price of every ticket is reduced by $2. For example, 2 tickets cost $2 \cdot 248 = \$496$ and 3 tickets cost $3 \cdot 246 = \$738$.

    (a) Write a quadratic function that gives the total cost of buying x tickets.

    (b) What is the cost of 5 tickets?

    (c) How many tickets were sold if the cost is $5200?

    (d) What number of tickets sold gives the greatest cost?

Modeling Quadratic Data

119. **Projectile Motion** The table shows the height of a projectile that is shot into the air.

t (seconds)	0	1	2	3	4
s (feet)	32	176	288	368	416

    (a) Use $s(t) = -16t^2 + v_0 t + h_0$ to model the data.

    (b) After how long did the projectile strike the ground?

120. **Falling Object** The table lists the velocity and distance traveled by a falling object for various elapsed times.

Time (sec)	0	1	2	3	4
Velocity (ft/sec)	0	32	64	96	128
Distance (ft)	0	16	64	144	256

    (a) Make a scatterplot of the ordered pairs determined by (time, velocity) and (time, distance) in the same viewing rectangle $[-1, 5, 1]$ by $[-10, 280, 20]$.

    (b) Find a function v that models the velocity.

    (c) The distance is modeled by $d(x) = ax^2$. Find a.

    (d) Find the time when the distance is 200 feet. Find the velocity at this time.

121. Pedestrian and Bicycle Programs   The Department of Transportation's budget for pedestrian and bicycle programs in millions of dollars for selected years is given in the table.

Year	1995	2001	2009
Budget	180	340	1200

Source: DOT.

(a) Determine a quadratic function B whose vertex is (1995, 180) and whose graph passes through the point (2009, 1200). Write $B(x)$ in vertex form.

(b) Use $B(x)$ to determine when the budget was $700 million.

122. Safe Runway Speed   The taxiway used by an aircraft to exit a runway should not have sharp curves. The safe radius for any curve depends on the speed of the airplane. The table lists the minimum radius R of the exit curves, where the taxiing speed of the airplane is x miles per hour.

x(mi/hr)	10	20	30	40	50
R(ft)	50	200	450	800	1250

Source: Federal Aviation Administration.

(a) If the taxiing speed x of the plane doubles, what happens to the minimum radius R of the curve?

(b) The FAA used $R(x) = ax^2$ to compute the values in the table. Determine a.

(c) If $R = 500$, find x. Interpret your results.

Quadratic Regression

123. iPod Sales   The table shows the number of iPods sold in millions of units for various years.

Year	2006	2007	2008	2009	2010	2011
Sales	39.4	51.6	54.8	54.1	50.3	42.6

Source: Apple Corporation.

(a) Use regression to find a quadratic function I that models the data. Let $x = 0$ correspond to 2006.

(b) Use I to estimate when sales were 28 million units.

124. Biology   Some types of worms have a remarkable capacity to live without moisture. The table shows the number of worms y surviving after x days in one study.

x (days)	0	20	40	80	120	160
y (worms)	50	48	45	36	20	3

Source: D. Brown and P. Rothery, Models in Biology.

(a) Use regression to find a quadratic function f that models these data.

(b) Graph f and the data in the same window.

(c) Solve the quadratic equation $f(x) = 0$ graphically. Do both solutions have meaning? Explain.

125. Walmart Employees   The table lists numbers of Walmart employees E in millions, x years after 1987.

x	0	5	10	15	20
E	0.20	0.38	0.68	1.4	2.2

Source: Walmart.

(a) Evaluate $E(15)$ and interpret the result.

(b) Find a quadratic function f that models these data.

(c) Graph the data and quadratic function f in the same xy-plane.

(d) Use f to estimate the year when the number of employees reached 3 million.

126. Women in the Workforce   The number N of women in millions who were gainfully employed in the workforce in selected years is shown in the table.

Year	1900	1910	1920	1930	1940	1950
N	5.3	7.4	8.6	10.8	12.8	18.4

Year	1960	1970	1980	1990	2000	2010
N	23.2	31.5	45.5	56.6	65.6	74.8

Source: Department of Labor.

(a) Use regression to find a quadratic function f that models the data. Support your result graphically.

(b) Predict the number of women in the labor force in 2020.

Writing about Mathematics

127. Discuss three symbolic methods for solving a quadratic equation. Make up a quadratic equation and use each method to find the solution set.

128. Explain how to solve a quadratic equation graphically.

EXTENDED AND DISCOVERY EXERCISES

Exercises 1–4: **Discriminant and Factoring**   *(Refer to pages 179 and 180.) For each equation, calculate the discriminant. Use the discriminant to decide whether the equation can be solved by factoring. If it can, solve the equation by factoring. Otherwise, use the quadratic formula.*

1. $8x^2 + 14x - 15 = 0$ **2.** $15x^2 - 17x - 4 = 0$

3. $5x^2 - 3x - 3 = 0$ **4.** $3x^2 - 2x - 4 = 0$

5. Quadratic Formula Prove the quadratic formula by completing the following.

(a) Write $ax^2 + bx + c = 0$ as $x^2 + \frac{b}{a}x = -\frac{c}{a}$.

(b) Complete the square to obtain $\left(x + \frac{b}{2a}\right)^2 = \frac{b^2 - 4ac}{4a^2}$.

(c) Use the square root property and solve for x.

6. Difference Quotient If the difference quotient for the function $f(x) = ax^2 - bx + 1$ equals $2x + h - 4$, find values for a and b.

7. Square Root Property Use the fact that $\sqrt{x^2} = |x|$ to show that the solutions to $x^2 = k$ with $k > 0$ are given by $x = \pm\sqrt{k}$.

CHECKING BASIC CONCEPTS FOR SECTIONS 3.1 AND 3.2

1. Graph $f(x) = (x - 1)^2 - 4$. Identify the vertex, axis of symmetry, and x-intercepts.

2. A graph of $f(x) = ax^2 + bx + c$ is shown.

   (a) Is a positive, negative, or zero?

   (b) Find the vertex and axis of symmetry.

   (c) Solve $ax^2 + bx + c = 0$.

   (d) Is the discriminant positive, negative, or zero?

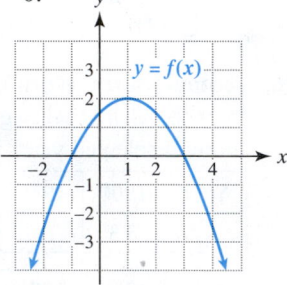

3. Use $f(x) = a(x - h)^2 + k$ to model the data exactly.

x	-3	-2	-1	0	1
$f(x)$	11	5	3	5	11

4. Find the vertex on the graph of $y = 3x^2 - 9x - 2$.

5. Write $f(x) = x^2 + 4x - 3$ as $f(x) = a(x - h)^2 + k$. What are the coordinates of the vertex? What is the minimum y-value of the graph of f?

6. Solve the quadratic equations.

   (a) $16x^2 = 81$ (b) $2x^2 + 3x = 2$

   (c) $x^2 = x - 3$ (d) $2x^2 = 3x + 4$

7. **Dimensions of a Rectangle** A rectangle is 4 inches longer than it is wide and has an area of 165 square inches. Find its dimensions.

8. **Height of a Baseball** The height s of a baseball in feet after t seconds is given by $s(t) = -16t^2 + 96t + 2$.

   (a) Find the height of the baseball after 1 second.

   (b) After how long is the baseball 142 feet high?

   (c) Find the maximum height of the baseball.

   (d) How long is the baseball in the air?

3.3 Complex Numbers

- Perform arithmetic operations on complex numbers
- Solve quadratic equations having complex solutions

Introduction

Throughout history, people have invented (or discovered) new numbers to solve equations and describe data. New numbers were often met with resistance and were seen as imaginary and useless. There was no Roman numeral for 0, as many skeptics probably wondered why a number was needed to represent nothing. Negative numbers also met strong resistance. After all, how could one possibly have -6 apples? Complex numbers were no different. In this section we introduce complex numbers and the "imaginary" unit i. Ultimately, complex numbers are no more imaginary than any other number. Society needs and uses complex numbers just as much as other numbers.

Basic Concepts

Graphing a quadratic equation is helpful only in finding the *real* solutions to the equation. Real solutions to the equation $x^2 + 1 = 0$ correspond to x-intercepts. See Figure 3.36. We can also solve this equation for real solutions symbolically, as shown next to Figure 3.36.

Solving $x^2 + 1 = 0$ Graphically and Symbolically

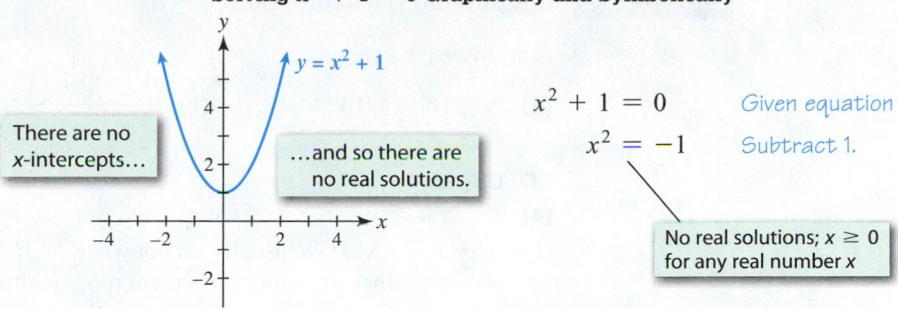

$$x^2 + 1 = 0 \qquad \textit{Given equation}$$
$$x^2 = -1 \qquad \textit{Subtract 1.}$$

No real solutions; $x \geq 0$ for any real number x

Figure 3.36

Above we showed that there are no *real* solutions to the equation $x^2 + 1 = 0$. However, we can *invent* two solutions.

$$x^2 = -1, \quad \text{or} \quad x = \pm\sqrt{-1}$$

We now define a new number called the **imaginary unit**, denoted i.

PROPERTIES OF THE IMAGINARY UNIT i

$$i = \sqrt{-1}, \qquad i^2 = -1$$

Solving $x^2 + 1 = 0$ using the imaginary unit

$$x = \pm\sqrt{-1} \qquad \text{or}$$
$$x = \pm i$$

Using the real numbers and the imaginary unit i, complex numbers can be defined. A **complex number** can be written in **standard form** as $a + bi$, where a and b are real numbers. The **real part** is a and the **imaginary part** is b. Every real number a is also a complex number because it can be written as $a + 0i$. A complex number $a + bi$ with $b \neq 0$ is an **imaginary number**. A complex number $a + bi$ with $a = 0$ and $b \neq 0$ is sometimes called a **pure imaginary number**. Examples of pure imaginary numbers include $3i$ and $-i$. Table 3.6 lists several complex numbers with their real and imaginary parts.

Complex Numbers

Example	$a + bi$	a	b
5	$5 + 0i$	5	0
$-5 - 2i$	$-5 - 2i$	-5	-2
$-3i$	$0 - 3i$	0	-3
$4 + 6i$	$4 + 6i$	4	6
	Standard form	Real part	Imaginary part

Table 3.6

Using the imaginary unit i, square roots of negative numbers can be written as complex numbers.

THE EXPRESSION $\sqrt{-a}$

If $a > 0$, then $\sqrt{-a} = i\sqrt{a}$.

EXAMPLE 1 **Simplifying the expression $\sqrt{-a}$**

Simplify each expression.

(a) $\sqrt{-16}$ **(b)** $\sqrt{-3}$ **(c)** $\dfrac{2 \pm \sqrt{-24}}{2}$

SOLUTION

(a) When $a > 0$, $\sqrt{-a} = i\sqrt{a}$. Thus $\sqrt{-16} = i\sqrt{16} = 4i$.

(b) $\sqrt{-3} = i\sqrt{3}$; we usually do not write $\sqrt{3}i$ because of the possibility of confusion about whether i is under the square root symbol.

(c) First note that $\sqrt{-24} = i\sqrt{24} = i\sqrt{4} \cdot \sqrt{6} = 2i\sqrt{6}$.

$$\frac{2 \pm \sqrt{-24}}{2} = \frac{2 \pm 2i\sqrt{6}}{2} \qquad \text{Simplify } \sqrt{-24}.$$

$$= \frac{2}{2} \pm \frac{2i\sqrt{6}}{2} \qquad \tfrac{a \pm b}{c} = \tfrac{a}{c} \pm \tfrac{b}{c}$$

$$= 1 \pm i\sqrt{6} \qquad \text{Simplify fractions.}$$

Now Try Exercises 1, 5, and 13

The property $\sqrt{a} \cdot \sqrt{b} = \sqrt{ab}$ is true only when *both* a and b are positive. When simplifying products containing square roots of negative numbers, it is important to first apply the property $\sqrt{-a} = i\sqrt{a}$, where $a > 0$. This technique is illustrated in the next example.

EXAMPLE 2 **Simplifying complex expressions**

Simplify each expression.

(a) $\sqrt{-3} \cdot \sqrt{-3}$ **(b)** $\sqrt{-2} \cdot \sqrt{-8}$

SOLUTION

(a) $\sqrt{-3} \cdot \sqrt{-3} = i\sqrt{3} \cdot i\sqrt{3} = i^2(\sqrt{3})^2 = -1(3) = -3$ $i^2 = -1$

> **NOTE** $\sqrt{-3} \cdot \sqrt{-3} \neq \sqrt{(-3) \cdot (-3)} = \sqrt{9} = 3$

(b) $\sqrt{-2} \cdot \sqrt{-8} = i\sqrt{2} \cdot i\sqrt{8} = i^2\sqrt{16} = -1(4) = -4$

Now Try Exercises 15 and 17

Calculator Help

To perform arithmetic on complex numbers, see Appendix A (page AP-10).

Arithmetic Operations on Complex Numbers

Arithmetic operations are also defined for complex numbers.

Addition and Subtraction   To add the complex numbers $(-2 + 3i)$ and $(4 - 6i)$, simply combine the real parts and the imaginary parts.

$$(-2 + 3i) + (4 - 6i) = -2 + 4 + 3i - 6i$$

$$= 2 - 3i$$

This same process works for subtraction.

$$(5 - 7i) - (8 + 3i) = 5 - 8 - 7i - 3i$$

$$= -3 - 10i$$

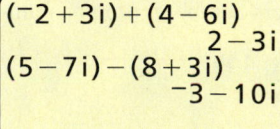

Figure 3.37

These operations are performed on a calculator in Figure 3.37.

Algebra Review
Before multiplying complex numbers, you may want to review multiplication of binomials in Chapter R (page R-15).

Multiplication   Two complex numbers can be multiplied. The property $i^2 = -1$ is applied when appropriate.

$$(-5 + i)(7 - 9i) = -5(7) + (-5)(-9i) + (i)(7) + (i)(-9i)$$
$$= -35 + 45i + 7i - 9i^2$$
$$= -35 + 52i - 9(\mathbf{-1})$$
$$= -26 + 52i$$

NOTE   Express your results in the standard form $a + bi$.

Division   The **conjugate** of $a + bi$ is $a - bi$. To find the conjugate, change the sign of the imaginary part b. Table 3.7 lists examples of complex numbers and their conjugates.

Complex Conjugates

$a + bi$	$2 + 5i$	$6 - 3i$	$-2 + 7i$	$-1 - i$	5	$-4i$
$a - bi$	$2 - 5i$	$6 + 3i$	$-2 - 7i$	$-1 + i$	5	$4i$

Change the sign on b.

Table 3.7

To simplify the quotient $\frac{3 + 2i}{5 - i}$, first multiply both the numerator and the denominator by the conjugate of the *denominator*.

$$\frac{3 + 2i}{5 - i} = \frac{(3 + 2i)(\mathbf{5 + i})}{(5 - i)(\mathbf{5 + i})} \qquad \text{Multiply by } \frac{conjugate}{conjugate}.$$

$$= \frac{3(5) + (3)(i) + (2i)(5) + (2i)(i)}{(5)(5) + (5)(i) + (-i)(5) + (-i)(i)} \qquad \text{Expand.}$$

$$= \frac{15 + 3i + 10i + 2i^2}{25 + 5i - 5i - i^2} \qquad \text{Simplify.}$$

$$= \frac{15 + 13i + 2(\mathbf{-1})}{25 - (\mathbf{-1})} \qquad i^2 = -1$$

$$= \frac{13 + 13i}{26} \qquad \text{Simplify.}$$

$$= \frac{1}{2} + \frac{1}{2}i \qquad \frac{a + bi}{c} = \frac{a}{c} + \frac{b}{c}i$$

The last step expresses the quotient as a complex number in standard form. The evaluation of the previous multiplication and division examples is shown in Figure 3.38.

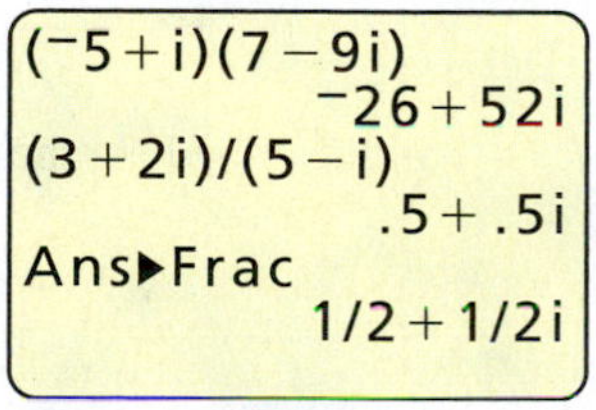

```
(-5+i)(7-9i)
            -26+52i
(3+2i)/(5-i)
             .5+.5i
Ans▶Frac
          1/2+1/2i
```

Figure 3.38

EXAMPLE 3   **Performing complex arithmetic**

Write each expression in standard form. Support your results using a calculator.
(a) $(-3 + 4i) + (5 - i)$ **(b)** $(-7i) - (6 - 5i)$

(c) $(-3 + 2i)^2$ **(d)** $\dfrac{17}{4 + i}$

SOLUTION
(a) $(\mathbf{-3} + \mathbf{4i}) + (\mathbf{5} - \mathbf{i}) = \mathbf{-3} + \mathbf{5} + \mathbf{4i} - \mathbf{i} = 2 + 3i$
(b) $(-7i) - (6 - 5i) = -6 - 7i + 5i = -6 - 2i$
(c) $(-3 + 2i)^2 = (-3 + 2i)(-3 + 2i)$
$$= 9 - 6i - 6i + 4i^2$$
$$= 9 - 12i + 4(\mathbf{-1})$$
$$= 5 - 12i$$

(d) $\dfrac{17}{4+i} = \dfrac{17}{4+i} \cdot \dfrac{4-i}{4-i}$ Multiply by $\dfrac{conjugate}{conjugate}$.

$\qquad\quad = \dfrac{68-17i}{16-i^2}$

$\qquad\quad = \dfrac{68-17i}{17}$ $i^2 = -1$

$\qquad\quad = 4 - i$

Standard forms can be found using a calculator. See Figures 3.39 and 3.40.

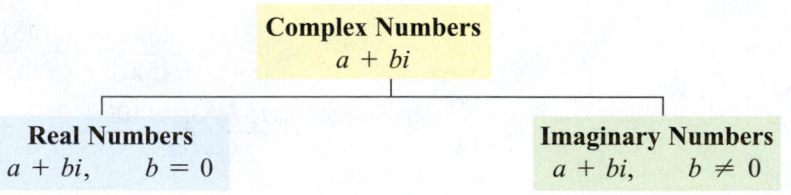

Figure 3.39   Figure 3.40

Now Try Exercises 23, 25, 35, and 39

MAKING CONNECTIONS

Complex, Real, and Imaginary Numbers   The following diagram illustrates the relationship among complex, real, and imaginary numbers, where a and b are real numbers. Note that complex numbers comprise two disjoint sets of numbers: the real numbers and the imaginary numbers.

Complex Numbers
$a + bi$

Real Numbers
$a + bi, \quad b = 0$

Imaginary Numbers
$a + bi, \quad b \neq 0$

NOTE   *Every* real number is a complex number.

Powers of i   We can simplify powers of i using the fact that $i^1 = i$ and $i^2 = -1$. In the next example we use these ideas.

EXAMPLE 4   **Simplifying Powers of i**

Simplify each power of i.
(a) i^8 **(b)** i^{19}

SOLUTION
(a) $i^8 = (i^2)^4 = (-1)^4 = 1$. Another way to simplify is the following.

$$i^8 = (i^2)(i^2)(i^2)(i^2) = (-1)(-1)(-1)(-1) = 1$$

(b) Write $i^{19} = i^{18} \cdot i$. Then $i^{19} = (i^2)^9 \cdot i = (-1)^9 i = -1 \cdot i = -i$.

Now Try Exercises 61 and 63

NOTE   In the Extended and Discovery Exercise in this section we discover a pattern for calculating the powers of i.

Quadratic Equations with Complex Solutions

We can use the quadratic formula to solve any quadratic equation $ax^2 + bx + c = 0$. If the discriminant $b^2 - 4ac$ is negative, then there are no real solutions, and the graph of

$y = ax^2 + bx + c$ does not intersect the x-axis. However, there are solutions that can be expressed as imaginary numbers. This is illustrated in the next example.

EXAMPLE 5 **Solving quadratic equations with imaginary solutions**

Solve the quadratic equation. Support your results graphically.

(a) $x^2 + 3x + 5 = 0$ **(b)** $\frac{1}{2}x^2 + 17 = 5x$ **(c)** $-2x^2 = 3$

SOLUTION

Getting Started Make sure each equation is in the form $ax^2 + bx + c = 0$, and then apply the quadratic formula, which always "works." ▶

(a) Let $a = 1$, $b = 3$, and $c = 5$ and apply the quadratic formula.

$$
\begin{aligned}
x &= \frac{-b \pm \sqrt{b^2 - 4ac}}{2a} \\
&= \frac{-3 \pm \sqrt{3^2 - 4(1)(5)}}{2(1)} \\
&= \frac{-3 \pm \sqrt{-11}}{2} \\
&= \frac{-3 \pm i\sqrt{11}}{2} \\
&= -\frac{3}{2} \pm \frac{i\sqrt{11}}{2}
\end{aligned}
$$

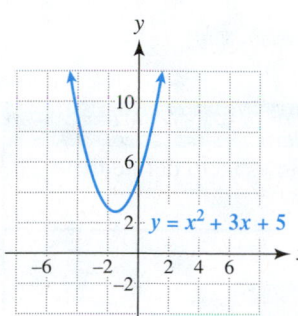

Figure 3.41

In Figure 3.41 the graph of $y_1 = x^2 + 3x + 5$ does not intersect the x-axis, which indicates that the equation $x^2 + 3x + 5 = 0$ has no real solutions. However, there are two complex solutions that are imaginary.

(b) Rewrite the equation as $\frac{1}{2}x^2 - 5x + 17 = 0$, and let $a = \frac{1}{2}$, $b = -5$, and $c = 17$.

$$
\begin{aligned}
x &= \frac{5 \pm \sqrt{(-5)^2 - 4(0.5)(17)}}{2(0.5)} \\
&= 5 \pm \sqrt{-9} \\
&= 5 \pm 3i
\end{aligned}
$$

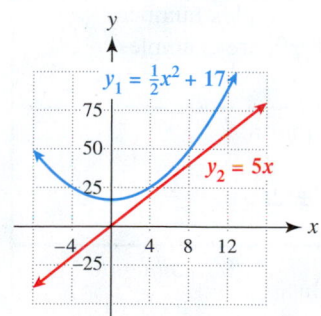

Figure 3.42

In Figure 3.42 the graphs of $y_1 = \frac{1}{2}x^2 + 17$ and $y_2 = 5x$ do not intersect, which indicates that the equation $\frac{1}{2}x^2 + 17 = 5x$ has no real solutions. However, there are two complex solutions that are imaginary.

(c) Rather than use the quadratic formula for this equation, we apply the square root property because the equation contains no x-term.

$$
\begin{aligned}
-2x^2 &= 3 && \text{\color{teal}Given equation} \\
x^2 &= -\frac{3}{2} && \text{\color{teal}Divide each side by -2.} \\
x &= \pm\sqrt{-\frac{3}{2}} && \text{\color{teal}Square root property} \\
x &= \pm i\sqrt{\frac{3}{2}} && \text{\color{teal}$\sqrt{-a} = i\sqrt{a}$}
\end{aligned}
$$

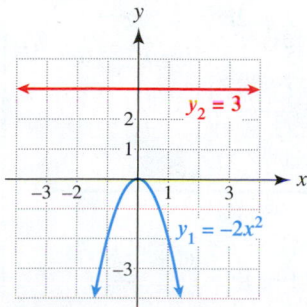

Figure 3.43

In Figure 3.43 the graphs of $y_1 = -2x^2$ and $y_2 = 3$ do not intersect, which indicates that the equation $-2x^2 = 3$ has no real solutions. However, there are two complex solutions that are imaginary.

Now Try Exercises 67, 73, and 75

What is the result if the expression is evaluated? (See Example 5(a).)

$$\left(-\frac{3}{2} + \frac{i\sqrt{11}}{2}\right)^2 + 3\left(-\frac{3}{2} + \frac{i\sqrt{11}}{2}\right) + 5$$

3.3 Putting It All Together

Some of the important topics in this section are summarized in the following table.

CONCEPT	EXPLANATION	COMMENTS AND EXAMPLES
Imaginary unit	$i = \sqrt{-1}, i^2 = -1$	The imaginary unit i allows us to define the complex numbers.
The expression $\sqrt{-a}$ with $a > 0$	$\sqrt{-a} = i\sqrt{a}$	$\sqrt{-4} = 2i$ $\sqrt{-5} = i\sqrt{5}$ $\sqrt{-32} = i\sqrt{32} = i\sqrt{16}\sqrt{2} = 4i\sqrt{2}$ $\sqrt{-5}\sqrt{-20} = i\sqrt{5}i\sqrt{20} = i^2\sqrt{100} = -10$
Complex number	$a + bi$, where a and b are real numbers	Every real number is a complex number. $5 - 4i, 5, 2 + i$, and $-9i$ are examples of complex numbers.
Standard form of a complex number	$a + bi$, where a and b are real numbers	Converting to standard form: $\dfrac{3 \pm 4i}{2} = \dfrac{3}{2} + 2i \quad \text{or} \quad \dfrac{3}{2} - 2i$
Conjugates	The conjugate of $a + bi$ is $a - bi$.	*Number* *Conjugate* $5 - 6i$ $5 + 6i$ $-12i$ $12i$ -7 -7 $2 + 3i$ $2 - 3i$
Arithmetic operations on complex numbers	Complex numbers may be added, subtracted, multiplied, or divided.	$(2 + 3i) + (-3 - i) = -1 + 2i$ $(5 + i) - (3 - i) = 2 + 2i$ $(1 + i)(5 - i) = 5 - i + 5i - i^2 = 6 + 4i$ $\dfrac{3 + i}{1 - i} = \dfrac{(3 + i)(1 + i)}{(1 - i)(1 + i)}$ $= 1 + 2i$
Powers of i	Use the fact that $i^2 = -1$ to simplify powers of i.	$i^{27} = i^{26} \cdot i$ $= (i^2)^{13} \cdot i$ $= (-1)^{13} \cdot i$ $= -i$

CONCEPT	EXPLANATION	COMMENTS AND EXAMPLES
Complex solutions to equations	Complex numbers $a + bi$ with $b \neq 0$ can be solutions to equations that cannot be solved with only real numbers.	$x^2 + 5 = 0$ implies $x^2 = -5$; there are no real solutions, but $x = \pm i\sqrt{5}$ are two complex solutions. Note that the graph of $y = x^2 + 5$ has no x-intercepts. The quadratic formula can be used to find complex solutions.

3.3 Exercises

Complex Numbers

Exercises 1–20: Simplify by using the imaginary unit i.

1. $\sqrt{-4}$
2. $\sqrt{-16}$
3. $\sqrt{-100}$
4. $\sqrt{-49}$
5. $\sqrt{-23}$
6. $\sqrt{-11}$
7. $\sqrt{-12}$
8. $\sqrt{-32}$
9. $\sqrt{-54}$
10. $\sqrt{-28}$
11. $\dfrac{4 \pm \sqrt{-16}}{2}$
12. $\dfrac{-2 \pm \sqrt{-36}}{6}$
13. $\dfrac{-6 \pm \sqrt{-72}}{3}$
14. $\dfrac{2 \pm \sqrt{-8}}{4}$
15. $\sqrt{-5} \cdot \sqrt{-5}$
16. $\sqrt{-8} \cdot \sqrt{-8}$
17. $\sqrt{-18} \cdot \sqrt{-2}$
18. $\sqrt{-20} \cdot \sqrt{-5}$
19. $\sqrt{-3} \cdot \sqrt{-6}$
20. $\sqrt{-15} \cdot \sqrt{-5}$

Exercises 21–48: Write the expression in standard form.

21. $3i + 5i$
22. $-7i + 5i$
23. $(3 + i) + (-5 - 2i)$
24. $(-4 + 2i) + (7 + 35i)$
25. $2i - (-5 + 23i)$
26. $(12 - 7i) - (-1 + 9i)$
27. $3 - (4 - 6i)$
28. $(7 + i) - (-8 + 5i)$
29. $(2)(2 + 4i)$
30. $(-5)(-7 + 3i)$
31. $(1 + i)(2 - 3i)$
32. $(-2 + i)(1 - 2i)$
33. $(-3 + 2i)(-2 + i)$
34. $(2 - 3i)(1 + 4i)$
35. $(-2 + 3i)^2$
36. $(2 - 3i)^2$
37. $2i(1 - i)^2$
38. $-i(5 - 2i)^2$

39. $\dfrac{1}{1 + i}$
40. $\dfrac{1 - i}{2 + 3i}$
41. $\dfrac{4 + i}{5 - i}$
42. $\dfrac{10}{1 - 4i}$
43. $\dfrac{2i}{10 - 5i}$
44. $\dfrac{3 - 2i}{1 + 2i}$
45. $\dfrac{3}{-i}$
46. $\dfrac{4 - 2i}{i}$
47. $\dfrac{-2 + i}{(1 + i)^2}$
48. $\dfrac{3}{(2 - i)^2}$

Exercises 49–54: Evaluate the expression with a calculator.

49. $(23 - 5.6i) + (-41.5 + 93i)$
50. $(-8.05 - 4.67i) + (3.5 + 5.37i)$
51. $(17.1 - 6i) - (8.4 + 0.7i)$
52. $\left(\frac{3}{4} - \frac{1}{10}i\right) - \left(-\frac{1}{8} + \frac{4}{25}i\right)$
53. $(-12.6 - 5.7i)(5.1 - 9.3i)$
54. $(7.8 + 23i)(-1.04 + 2.09i)$

Exercises 55 and 56: Evaluate with a calculator. Round values to the nearest thousandth.

55. $\dfrac{17 - 135i}{18 + 142i}$
56. $\dfrac{141 + 52i}{102 - 31i}$

Powers of *i*

Exercises 57–64: Simplify the powers of i.

57. i^{50}
58. i^{28}
59. i^{31}
60. i^{21}
61. i^{12}
62. i^{103}
63. i^{57}
64. i^{30}

Quadratic Equations with Complex Solutions

Exercises 65–82: Solve. Write answers in standard form.

65. $x^2 + 5 = 0$

66. $4x^2 + 3 = 0$

67. $5x^2 + 1 = 3x^2$

68. $x(3x + 1) = -1$

69. $3x = 5x^2 + 1$

70. $4x^2 = x - 1$

71. $x(x - 4) = -5$

72. $2x^2 + x + 1 = 0$

73. $x^2 = 3x - 5$

74. $3x - x^2 = 5$

75. $x^2 + 2x + 4 = 0$

76. $x(x - 4) = -8$

77. $3x^2 - 4x = x^2 - 3$

78. $2x^2 + 3 = 1 - x$

79. $2x(x - 2) = x - 4$

80. $3x^2 + x = x(5 - x) - 2$

81. $3x(3 - x) - 8 = x(x - 2)$

82. $-x(7 - 2x) = -6 - (3 - x)$

Zeros of Quadratic Functions

Exercises 83–88: The graph of a function is given.

(a) Use the graph to predict the number of real zeros and the number of imaginary zeros.

(b) Find these zeros using the quadratic formula.

83.
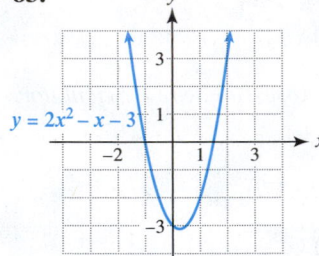

84.
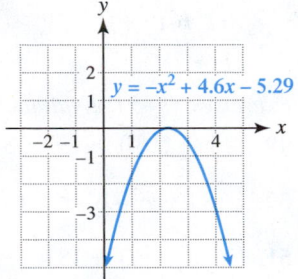

85.
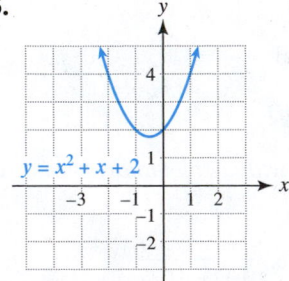

86.
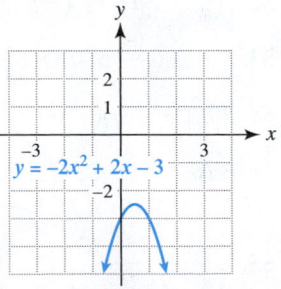

87.
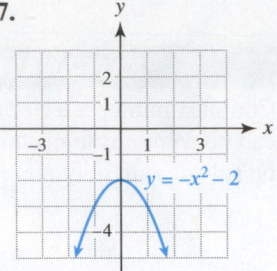

88.
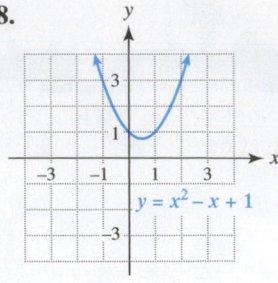

Applications

Exercises 89–94: **Electricity**   *Complex numbers are used in the study of electrical circuits. Impedance Z (or the opposition to the flow of electricity), voltage V, and current I can all be represented by complex numbers. They are related by the equation $Z = \frac{V}{I}$. Find the value of the missing variable.*

89. $V = 50 + 98i$ $I = 8 + 5i$

90. $V = 30 + 60i$ $I = 8 + 6i$

91. $I = 1 + 2i$ $Z = 3 - 4i$

92. $I = \frac{1}{2} + \frac{1}{4}i$ $Z = 8 - 9i$

93. $Z = 22 - 5i$ $V = 27 + 17i$

94. $Z = 10 + 5i$ $V = 10 + 8i$

Writing about Mathematics

95. Could a quadratic function have one real zero and one imaginary zero? Explain.

96. Give an example of a quadratic function that has only real zeros and an example of one that has only imaginary zeros. How do their graphs compare? Explain how to determine from a graph whether a quadratic function has real zeros.

Extended and Discovery Exercise

1. Powers of i   The properties of the imaginary unit are $i = \sqrt{-1}$ and $i^2 = -1$.

   (a) Begin simplifying the expressions $i, i^2, i^3, i^4, i^5, \ldots,$ until a simple pattern is discovered. For example, $i^3 = i \cdot i^2 = i \cdot (-1) = -i$.

   (b) Summarize your findings by describing how to simplify i^n for any natural number n.

3.4 Quadratic Inequalities

- **Understand basic concepts about quadratic inequalities**
- **Solve quadratic inequalities graphically**
- **Solve quadratic inequalities symbolically**

Introduction

Highway engineers often use quadratic functions to model safe stopping distances for cars. For example, $f(x) = \frac{1}{12}x^2 + \frac{11}{5}x$ is sometimes used to model the stopping distance for a car traveling at x miles per hour on dry, level pavement. If a driver can see only **200** feet ahead on a highway with a sharp curve, then safe driving speeds x satisfy the *quadratic inequality*

$$\frac{1}{12}x^2 + \frac{11}{5}x \leq 200, \qquad \text{Quadratic inequality}$$

or equivalently,

$$\frac{1}{12}x^2 + \frac{11}{5}x - 200 \leq 0. \qquad \text{Subtract 200.}$$

(See Example 4.) This section discusses methods for solving quadratic inequalities.

Basic Concepts

A quadratic equation can be written as $ax^2 + bx + c = 0$ with $a \neq 0$. If the equals sign is replaced by $>$, $\geq$, $<$, or $\leq$, a **quadratic inequality** results. Examples of *quadratic equations* include

$$x^2 + 2x - 1 = 0, \qquad 4x^2 = 1, \qquad 2x^2 = 1 - 3x,$$

and examples of *quadratic inequalities* include

$$x^2 + 2x - 1 \geq 0, \qquad 4x^2 < 1, \quad \text{and} \quad 2x^2 \leq 1 - 3x.$$

MAKING CONNECTIONS

Quadratic Function, Equation, and Inequality   The three concepts are closely related.

$$f(x) = ax^2 + bx + c, a \neq 0 \qquad \text{Quadratic function}$$
$$ax^2 + bx + c = 0, a \neq 0 \qquad \text{Quadratic equation}$$
$$ax^2 + bx + c > 0, a \neq 0 \qquad \text{Quadratic inequality}$$

Because equality is (usually) the boundary between *greater than* and *less than*, a first step in solving a quadratic inequality is to determine the x-values where equality occurs. These x-values are *boundary numbers*. We begin by discussing graphical solutions to quadratic inequalities.

Graphical and Numerical Solutions

Quadratic inequalities can be solved graphically, as illustrated in the next example.

EXAMPLE 1   **Solving quadratic inequalities graphically**

Solve each inequality graphically.
(a) $x^2 - 4 < 0$ **(b)** $x^2 - 4 > 0$

SOLUTION
(a) The graph of $y = x^2 - 4$ opens upward with x-intercepts -2 and 2, as shown in Figure 3.44 on the next page. Because the x-axis represents where $y = 0$, the solutions

to $x^2 - 4 < 0$ correspond to the red portion where the graph of $y = x^2 - 4$ is **below** the x-axis, which is between the x-intercepts ± 2. Thus the solution set is $\{x \mid -2 < x < 2\}$ or $(-2, 2)$.

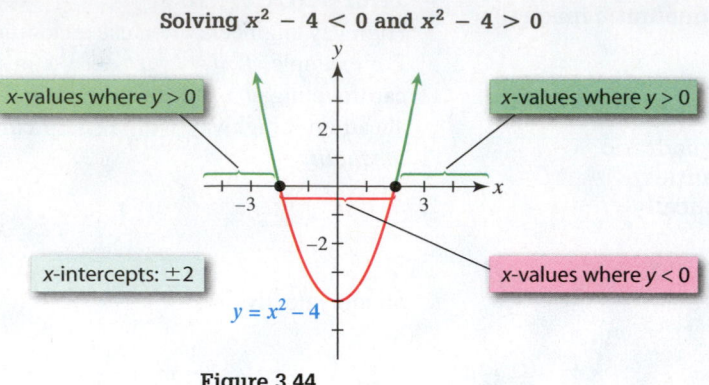

Figure 3.44

(b) From Figure 3.44 the green portions of the graph of $y = x^2 - 4$ are **above** the x-axis left of x-intercept -2 or right of x-intercept 2. Thus the solution set can be written as $\{x \mid x < -2 \text{ or } x > 2\}$, or $(-\infty, -2) \cup (2, \infty)$.

> **Now Try Exercise 1**

The following steps summarize how to solve a quadratic inequality graphically.

SOLVING QUADRATIC INEQUALITIES GRAPHICALLY

STEP 1: If necessary, rewrite the inequality as $ax^2 + bx + c > 0$, where $>$ may be replaced by $<$, $\leq$, or $\geq$.

STEP 2: Graph $y = ax^2 + bx + c$ and determine any x-intercepts.

STEP 3: Depending on the inequality symbol, locate the x-values where the graph is either above or below the x-axis.

STEP 4: Use the information from Step 3 to solve the inequality.

In the next example we apply these steps to solve a quadratic inequality.

EXAMPLE 2   Solving a quadratic inequality graphically

Solve the inequality $2 + x \geq x^2$ graphically.

SOLUTION

STEP 1: Rewrite $2 + x \geq x^2$ as $-x^2 + x + 2 \geq 0$.

STEP 2: Figure 3.45 shows a graph of $y = -x^2 + x + 2$ where the x-intercepts are -1 and 2. When graphing by hand, it may be helpful to locate the x-intercepts symbolically.

$$-x^2 + x + 2 = 0 \qquad \text{\textcolor{blue}{Let $y = 0$ to find x-intercepts.}}$$
$$x^2 - x - 2 = 0 \qquad \text{\textcolor{blue}{Multiply each side by -1.}}$$
$$(x + 1)(x - 2) = 0 \qquad \text{\textcolor{blue}{Factor.}}$$
$$x = -1 \quad \text{or} \quad x = 2 \qquad \text{\textcolor{blue}{Zero-product property.}}$$

This confirms that the x-intercepts are -1 and 2.

STEP 3: The inequality symbol in $-x^2 + x + 2 \geq 0$ is $\geq$, so locate x-values where the parabola is **above** the x-axis. This occurs between the x-intercepts: $-1, 2$.

Solving $-x^2 + x + 2 \geq 0$ Graphically

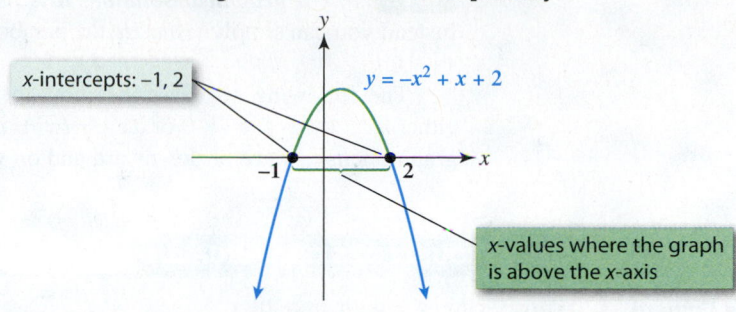

x-intercepts: $-1, 2$

$y = -x^2 + x + 2$

x-values where the graph is above the x-axis

Figure 3.45

STEP 4: Because $\geq$ includes equality, the solution set includes -1 and 2 and can be written as $[-1, 2]$, or $\{x \mid -1 \leq x \leq 2\}$.

Now Try Exercise 35

In the next example we use symbolic, graphical, and numerical methods.

EXAMPLE 3 **Solving quadratic equations and inequalities**

Solve each equation or inequality.
(a) $2x^2 - 3x - 2 = 0$ **(b)** $2x^2 - 3x - 2 < 0$ **(c)** $2x^2 - 3x - 2 > 0$

SOLUTION
(a) *Symbolic Solution* The equation $2x^2 - 3x - 2 = 0$ can be solved by factoring.

$$(2x + 1)(x - 2) = 0 \qquad \text{Factor trinomial.}$$

$$x = -\frac{1}{2} \quad \text{or} \quad x = 2 \qquad \text{Zero-product property}$$

The solutions are $-\frac{1}{2}$ and 2.

(b) *Graphical Solution* The graph of $y = 2x^2 - 3x - 2$ is a parabola opening upward. Its x-intercepts are $-\frac{1}{2}$ and 2. See Figure 3.46. This parabola is below the x-axis ($y < 0$) for x-values between $-\frac{1}{2}$ and 2, so the solution set is $\left(-\frac{1}{2}, 2\right)$, or $\left\{x \mid -\frac{1}{2} < x < 2\right\}$.

(c) *Graphical Solution* In Figure 3.47, the graph of $y = 2x^2 - 3x - 2$ is above the x-axis ($y > 0$) for x-values less than $-\frac{1}{2}$ or greater than 2, so the solution set is $\left(-\infty, -\frac{1}{2}\right) \cup (2, \infty)$, or $\left\{x \mid x < -\frac{1}{2} \text{ or } x > 2\right\}$.

Numerical Solution The table of values in Figure 3.48 supports these graphical results. Note that $y_1 = 0$ for $x = -\frac{1}{2}$ and $x = 2$. For $-\frac{1}{2} < x < 2$, we see that $y_1 < 0$, and for $x < -\frac{1}{2}$ or $x > 2$, we see that $y_1 > 0$.

Graphical Solutions

Numerical Solutions

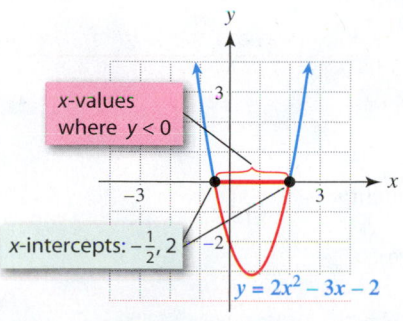

x-values where $y < 0$

x-intercepts: $-\frac{1}{2}, 2$

$y = 2x^2 - 3x - 2$

Figure 3.46

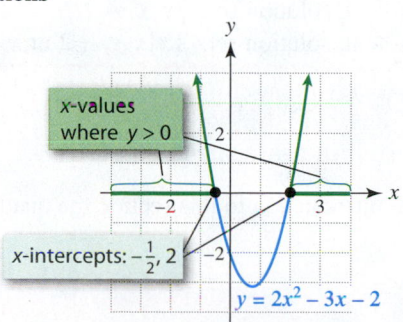

x-values where $y > 0$

x-intercepts: $-\frac{1}{2}, 2$

$y = 2x^2 - 3x - 2$

Figure 3.47

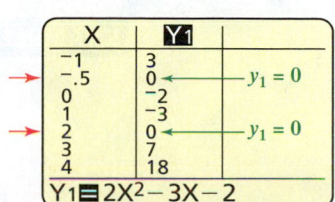

Figure 3.48

Now Try Exercise 7

NOTE *For a graphical solution, it is not necessary to graph the parabola precisely. Instead you can simply* visualize *the parabola opening upward with x-intercepts* $-\frac{1}{2}$ *and 2.*

The following box describes how to use the graph of $y = ax^2 + bx + c$ to solve either $ax^2 + bx + c > 0$ or $ax^2 + bx + c < 0$. The solution set depends on whether the graph opens upward or downward and on whether there are zero, one, or two intercepts.

SOLVING QUADRATIC INEQUALITIES GRAPHICALLY

Opens Upward $(y = ax^2 + bx + c$ with $a > 0)$

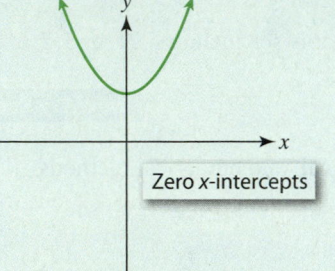

Zero x-intercepts

$y > 0$ for $(-\infty, \infty)$
$y < 0$: never

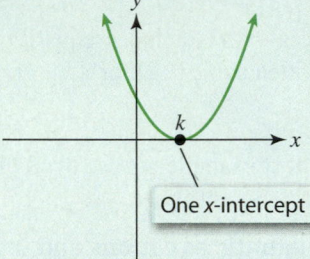

One x-intercept

$y > 0$ whenever $x \neq k$
$y < 0$: never

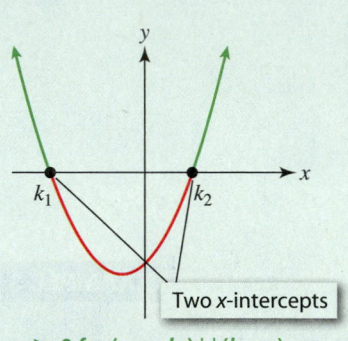

Two x-intercepts

$y > 0$ for $(-\infty, k_1) \cup (k_2, \infty)$
$y < 0$ for (k_1, k_2)

Opens Downward $(y = ax^2 + bx + c$ with $a < 0)$

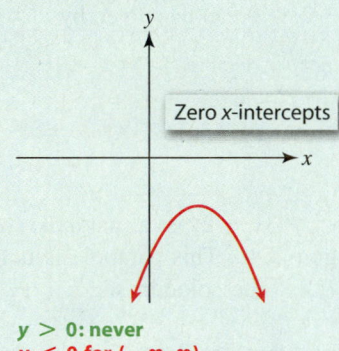

Zero x-intercepts

$y > 0$: never
$y < 0$ for $(-\infty, \infty)$

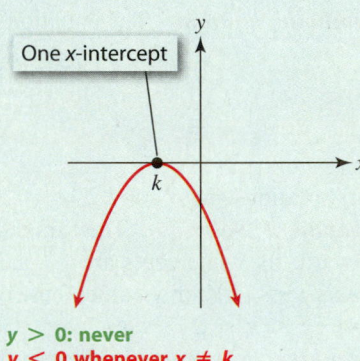

One x-intercept

$y > 0$: never
$y < 0$ whenever $x \neq k$

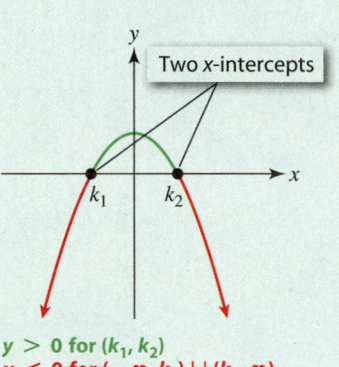

Two x-intercepts

$y > 0$ for (k_1, k_2)
$y < 0$ for $(-\infty, k_1) \cup (k_2, \infty)$

CLASS DISCUSSION

Sketch a graph of $y = ax^2 + bx + c$ if the quadratic inequality $ax^2 + bx + c < 0$ satisfies the following conditions.
(a) $a > 0$, solution set: $\{x \mid -1 < x < 3\}$
(b) $a < 0$, solution set: $\{x \mid x \neq 1\}$
(c) $a < 0$, solution set: $\{x \mid x < -2 \text{ or } x > 2\}$

EXAMPLE 4   **Determining safe speeds**

In the introduction to this section the quadratic inequality

$$\frac{1}{12}x^2 + \frac{11}{5}x \leq 200$$

was explained. Solve this inequality to determine safe speeds on a curve where a driver can see the road ahead for only 200 feet. What might be a safe speed limit for this curve?

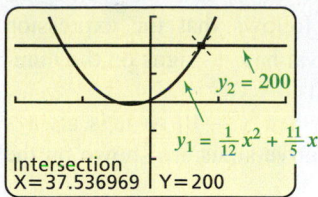

[−100, 100, 50] by [−300, 300, 100]

Intersection
X=37.536969 Y=200

Figure 3.49

SOLUTION Graph $y_1 = \frac{1}{12}x^2 + \frac{11}{5}x$ and $y_2 = 200$, as shown in Figure 3.49. Since we are interested in *positive* speeds, we need to locate only the point of intersection where x is positive. This is where $x \approx 37.5$. For positive x-values to the left of $x \approx 37.5$, $y_1 < 200$. Thus safe speeds are less than 37.5 miles per hour. A reasonable speed limit might be 35 miles per hour.

Now Try Exercise 67

Symbolic Solutions

Although it is usually easier to solve a quadratic inequality graphically by visualizing a parabola and its x-intercepts, we can also solve a quadratic inequality symbolically without the aid of a graph. The symbolic method, which involves a table or a number line, is often used to solve more complicated inequalities. (See Section 4.7.)

SOLVING QUADRATIC INEQUALITIES SYMBOLICALLY

STEP 1: If necessary, write the inequality as $ax^2 + bx + c < 0$, where $<$ may be replaced by $>$, $\leq$, or $\geq$.

STEP 2: Solve the equation $ax^2 + bx + c = 0$. The solutions are called boundary numbers.

STEP 3: Use the boundary numbers to separate the number line into disjoint open intervals. Note that on each open interval, $y = ax^2 + bx + c$ is either always positive or always negative.

STEP 4: To solve the inequality, choose a convenient test value (an x-value) from each disjoint interval in Step 3. Evaluate $y = ax^2 + bx + c$ at each test point. If the result is positive, then $y > 0$ over that interval. If the result is negative, then $y < 0$ over that interval. You may want to use either a number line or a table of values to organize your work.

Algebra Review
To review factoring, see Chapter R (pages R-22–R-23).

NOTE Do not pick a boundary number for a test value, because the result will be $y = 0$.

For example, to solve $2x^2 - 5x - 12 < 0$ symbolically, replace $<$ with $=$ and solve the equation by factoring.

$$2x^2 - 5x - 12 = 0 \qquad \text{Quadratic equation}$$
$$(2x + 3)(x - 4) = 0 \qquad \text{Factor.}$$
$$2x + 3 = 0 \quad \text{or} \quad x - 4 = 0 \qquad \text{Zero-product property}$$
$$x = -\frac{3}{2} \quad \text{or} \quad x = 4 \qquad \text{Solve.}$$

The boundary numbers $-\frac{3}{2}$ and 4 separate the number line into three disjoint intervals:

$$\left(-\infty, -\frac{3}{2}\right), \left(-\frac{3}{2}, 4\right), \quad \text{and} \quad (4, \infty),$$

as illustrated in Figure 3.50.

$2x^2 - 5x - 12 = 0$

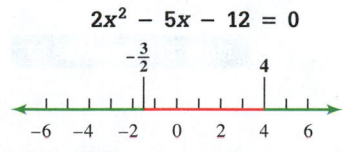

Figure 3.50

The expression $2x^2 - 5x - 12$ is either always positive or always negative on a particular interval. To determine where $2x^2 - 5x - 12 < 0$, we can use the **test values** shown in Table 3.8.

Checking Test Values

Interval	Test Value x	$2x^2 - 5x - 12$	Positive or Negative?
$\left(-\infty, -\frac{3}{2}\right)$	-2	6	Positive
$\left(-\frac{3}{2}, 4\right)$	0	-12	Negative
$(4, \infty)$	6	30	Positive

Table 3.8

$2x^2 - 5x - 12$

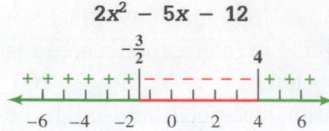

Figure 3.51

For example, since the test value -2 lies in the interval $\left(-\infty, -\frac{3}{2}\right)$ and $2x^2 - 5x - 12$ evaluated at $x = -2$ equals **6**, which is greater than 0, it follows that the expression $2x^2 - 5x - 12$ is always **positive** for $\left(-\infty, -\frac{3}{2}\right)$. This interval has $+$ signs on the number line in Figure 3.51. (See Table 3.8 on the previous page.)

See that the expression $2x^2 - 5x - 12$ equals -12 when $x = 0$, so it is always **negative** between the boundary numbers of $-\frac{3}{2}$ and 4. Negative signs are shown on the real number line in Figure 3.51 in the interval $\left(-\frac{3}{2}, 4\right)$.

Finally, when $x = 6$, the expression $2x^2 - 5x - 12$ equals **30**, which is **positive**. Thus $+$ signs are placed along the x-axis in Figure 3.51 for $x > 4$. Therefore the solution set for $2x^2 - 5x - 12 < 0$ is $\left(-\frac{3}{2}, 4\right)$.

Note that it is important to choose one test value less than $-\frac{3}{2}$, one test value between $-\frac{3}{2}$ and 4, and one test value greater than 4. You do *not* need to use both a table and a number line.

EXAMPLE 5 Solving a quadratic inequality

Solve $x^2 \geq 2 - x$ symbolically. Write the solution set in interval notation.

SOLUTION

STEP 1: Rewrite the inequality as $x^2 + x - 2 \geq 0$.

STEP 2: Solve the quadratic equation $x^2 + x - 2 = 0$.

$$(x + 2)(x - 1) = 0 \qquad \text{Factor.}$$
$$x = -2 \quad \text{or} \quad x = 1 \qquad \text{Zero-product property}$$

$x^2 - x - 2$

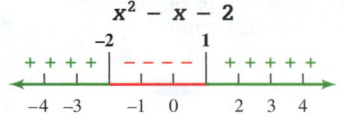

Figure 3.52

STEP 3: These two boundary numbers separate the number line into three disjoint intervals:

$$(-\infty, -2), \quad (-2, 1), \quad \text{and} \quad (1, \infty).$$

STEP 4: We choose the test values $x = -3, x = 0$, and $x = 2$. From Table 3.9 or Figure 3.52 the expression $x^2 + x - 2$ is positive when $x = -3$ and $x = 2$. Thus the solution set is $(-\infty, -2] \cup [1, \infty)$. The boundary numbers, -2 and 1, are included because the inequality involves $\geq$ rather than $>$.

Interval	Test Value x	$x^2 + x - 2$	Positive or Negative?
$(-\infty, -2)$	-3	4	Positive
$(-2, 1)$	0	-2	Negative
$(1, \infty)$	2	4	Positive

Table 3.9

Now Try Exercise 61

3.4 Putting It All Together

The following table summarizes concepts related to solving quadratic inequalities.

CONCEPT	DESCRIPTION
Quadratic inequality	Can be written as $ax^2 + bx + c < 0$, where $<$ may be replaced by $>$, $\leq$, or $\geq$. **Example:** $-x^2 + x < -2$ is a quadratic inequality; it can be written as $-x^2 + x + 2 < 0$.

CONCEPT	DESCRIPTION	
Graphical solution	Write the inequality as $ax^2 + bx + c < 0$, where $<$ may be $>$, $\leq$, or $\geq$. Graph $y = ax^2 + bx + c$, and use the x-intercepts, or boundary numbers, to determine x-values where the graph is below (above) the x-axis. In the figure, the inequality $-x^2 + x + 2 < 0$ is satisfied ($y < 0$) when either $x < -1$ or $x > 2$.	

Symbolic solution	Write the inequality as $ax^2 + bx + c < 0$, where $<$ may be $>$, $\leq$, or $\geq$. Solve the equation $ax^2 + bx + c = 0$. To determine where $y = ax^2 + bx + c$ is positive or negative, use a table of test values or a number line.

Example: Solve $-x^2 + x + 2 < 0$.
Solving $-x^2 + x + 2 = 0$ results in $x = -1$ or $x = 2$. From the table the solution set is $\{x \mid x < -1 \text{ or } x > 2\}$ or $(-\infty, -1) \cup (2, \infty)$.

Interval	Test Value x	$-x^2 + x + 2$	Positive or Negative?
$(-\infty, -1)$	-2	-4	Negative
$(-1, 2)$	0	2	Positive
$(2, \infty)$	3	-4	Negative

3.4 Exercises

Quadratic Inequalities

Exercises 1–6: Solve each inequality.

1. (a) $x^2 - 1 > 0$ (b) $x^2 - 1 < 0$

2. (a) $x^2 - 9 < 0$ (b) $x^2 - 9 > 0$

3. (a) $x^2 - 16 \leq 0$ (b) $x^2 - 16 \geq 0$

4. (a) $x^2 - 25 \geq 0$ (b) $x^2 - 25 \leq 0$

5. (a) $4 - x^2 > 0$ (b) $4 - x^2 < 0$

6. (a) $\frac{1}{4} - x^2 < 0$ (b) $\frac{1}{4} - x^2 > 0$

Exercises 7–18: Solve each equation and inequality. Use set-builder or interval notation to write solution sets to the inequalities.

7. (a) $x^2 - x - 12 = 0$ **8.** (a) $x^2 - 8x + 12 = 0$

   (b) $x^2 - x - 12 < 0$ (b) $x^2 - 8x + 12 < 0$

   (c) $x^2 - x - 12 > 0$ (c) $x^2 - 8x + 12 > 0$

9. (a) $k^2 - 5 = 0$

   (b) $k^2 - 5 \leq 0$

   (c) $k^2 - 5 \geq 0$

10. (a) $n^2 - 17 = 0$

   (b) $n^2 - 17 \leq 0$

   (c) $n^2 - 17 \geq 0$

11. (a) $3x^2 + 8x = 0$

   (b) $3x^2 + 8x \leq 0$

   (c) $3x^2 + 8x \geq 0$

12. (a) $7x^2 - 4x = 0$

   (b) $7x^2 - 4x \leq 0$

   (c) $7x^2 - 4x \geq 0$

13. (a) $-4x^2 + 12x - 9 = 0$

   (b) $-4x^2 + 12x - 9 < 0$

   (c) $-4x^2 + 12x - 9 > 0$

14. (a) $x^2 + 2x + 1 = 0$

   (b) $x^2 + 2x + 1 < 0$

   (c) $x^2 + 2x + 1 > 0$

15. (a) $12z^2 - 23z + 10 = 0$

  (b) $12z^2 - 23z + 10 \le 0$

  (c) $12z^2 - 23z + 10 \ge 0$

16. (a) $18z^2 + 9z - 20 = 0$

  (b) $18z^2 + 9z - 20 \le 0$

  (c) $18z^2 + 9z - 20 \ge 0$

17. (a) $x^2 + 2x - 1 = 0$ **18. (a)** $x^2 + 4x - 3 = 0$

  (b) $x^2 + 2x - 1 < 0$ **(b)** $x^2 + 4x - 3 < 0$

  (c) $x^2 + 2x - 1 > 0$ **(c)** $x^2 + 4x - 3 > 0$

Exercises 19–24: The graph of $f(x) = ax^2 + bx + c$ is shown in the figure. Solve each inequality.

19. (a) $f(x) < 0$ **20. (a)** $f(x) > 0$

  (b) $f(x) \ge 0$ **(b)** $f(x) < 0$

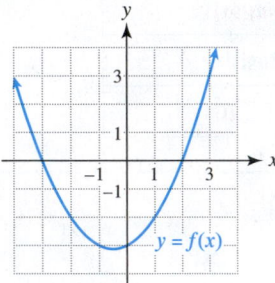
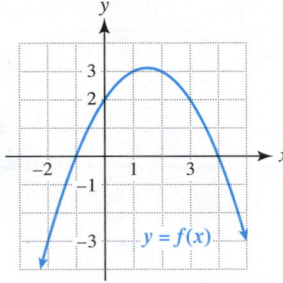

21. (a) $f(x) \le 0$ **22. (a)** $f(x) \ge 0$

  (b) $f(x) > 0$ **(b)** $f(x) \le 0$

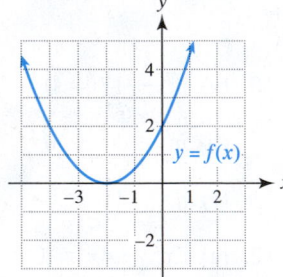
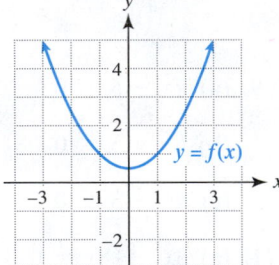

23. (a) $f(x) > 0$ **24. (a)** $f(x) \ge 0$

  (b) $f(x) < 0$ **(b)** $f(x) < 0$

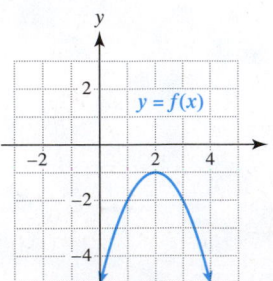
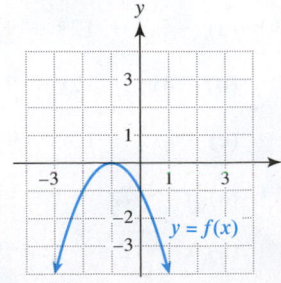

Exercises 25–28: Use the graph of $y = f(x)$ to solve each equation or inequality. Use set-builder or interval notation to write solution sets to the inequalities.

(a) $f(x) = 0$ (b) $f(x) < 0$ (c) $f(x) > 0$

25.
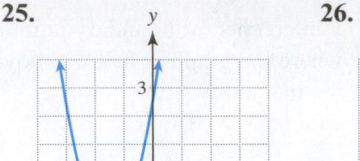

26.
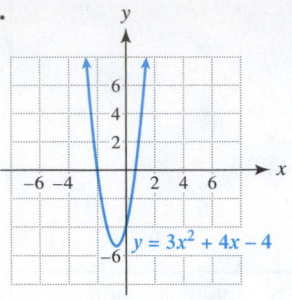

27.
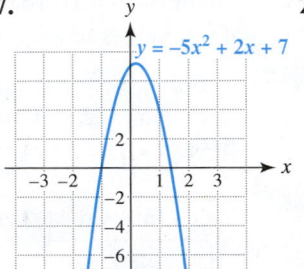

28.
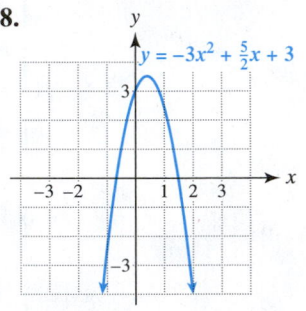

Exercises 29–32: The table contains test values for a quadratic function $f(x) = ax^2 + bx + c$. Solve each inequality.

(a) $f(x) > 0$ (b) $f(x) \le 0$

29.

x	-2	-1	0	1	2
$f(x)$	3	0	-1	0	3

30.

x	-6	-4	1	5	8
$f(x)$	-22	0	20	0	-36

31.

x	-6	-4	-2	0	2
$f(x)$	0	4	0	-12	-32

32.

x	-4	-2	0	2	4
$f(x)$	16	0	-8	-8	0

Exercises 33–58: Solve the inequality.

33. $2x^2 + 5x + 2 \le 0$ **34.** $x^2 - 3x - 4 < 0$

35. $x^2 + x > 6$ **36.** $-3x \ge 9 - 12x^2$

37. $x^2 \le 4$ **38.** $2x^2 > 16$

39. $x(x - 4) \ge -4$ **40.** $x^2 - 3x - 10 < 0$

41. $-x^2 + x + 6 \le 0$ **42.** $-x^2 - 2x + 8 > 0$

43. $6x^2 - x < 1$ **44.** $5x^2 \le 10 - 5x$

45. $(x + 4)(x - 10) \le 0$ **46.** $(x - 3.1)(x + 2.7) > 0$

47. $2x^2 + 4x + 3 < 0$ 48. $2x^2 + x + 4 < 0$

49. $9x^2 + 4 > 12x$ 50. $x^2 + 2x \geq 35$

51. $x^2 \geq x$ 52. $x^2 \geq -3$

53. $x(x - 1) \geq 6$ 54. $x^2 - 9 < 0$

55. $x^2 - 5 \leq 0$

56. $0.5x^2 - 3x > -1$

57. $7x^2 + 515.2 \geq 179.8x$

58. $-10 < 3x - x^2$

Exercises 59–66: (Refer to Example 5.) Use a table to solve.

59. $x^2 - 9x + 14 \leq 0$ 60. $x^2 + 10x + 21 > 0$

61. $x^2 \geq 3x + 10$ 62. $x^2 < 3x + 4$

63. $\frac{1}{8}x^2 + x + 2 \geq 0$ 64. $x^2 - \frac{1}{2}x - 5 < 0$

65. $x^2 > 3 - 4x$

66. $2x^2 \leq 1 - 4x$

Applications

67. **Stopping Distance** The stopping distance D in feet for a car traveling at x miles per hour on *wet* level pavement can be estimated by $D(x) = \frac{1}{9}x^2 + \frac{11}{3}x$. If a driver can see only 300 feet ahead on a curve, find a safe speed limit.

68. **Safe Driving Speeds** The stopping distance d in feet for a car traveling at x miles per hour is given by $d(x) = \frac{1}{12}x^2 + \frac{11}{9}x$. Determine the driving speeds that correspond to stopping distances between 300 and 500 feet, inclusive. Round speeds to the nearest mile per hour.

69. **Geometry** The volume of a cylinder is given by $V = \pi r^2 h$, where r is the radius and h is the height. If the height of a cylindrical can is 6 inches and the volume must be between 24π and 54π cubic inches, inclusive, find the possible values for the radius of the can.

70. **Geometry** A rectangle is 4 feet longer than it is wide. If the area of the rectangle must be less than or equal to 672 square feet, find the possible values for the width x.

71. **Heart Rate** Suppose that a person's heart rate, x minutes after vigorous exercise has stopped, can be modeled

by $f(x) = \frac{4}{5}(x - 10)^2 + 80$. The output is in beats per minute, where the domain of f is $0 \leq x \leq 10$.
(a) Evaluate $f(0)$ and $f(2)$. Interpret the result.

(b) Estimate the times when the person's heart rate was between 100 and 120 beats per minute, inclusive.

72. **Carbon Monoxide Exposure** When a person breathes carbon monoxide (CO), it enters the bloodstream to form carboxyhemoglobin (COHb), which reduces the transport of oxygen to tissues. The formula given by $T(x) = 0.0079x^2 - 1.53x + 76$ approximates the number of hours T that it takes for a person's bloodstream to reach the 5% COHb level, where x is the concentration of CO in the air in parts per million (ppm) and $50 \leq x \leq 100$. (Smokers routinely have a 5% concentration.) Estimate the CO concentration x necessary for a person to reach the 5% COHb level in 4–5 hours. (*Source:* *Indoor Air Quality Environmental Information Handbook.*)

73. **AIDS Deaths** Let $f(x) = 2375x^2 + 5134x + 5020$ estimate the number of U.S. AIDS deaths x years after 1984, where $0 \leq x \leq 10$. Estimate when the number of AIDS deaths was from 90,000 to 200,000.

74. **Air Density** As the altitude increases, air becomes thinner, or less dense. An approximation of the density of air at an altitude of x meters above sea level is given by

$$d(x) = (3.32 \times 10^{-9})x^2 - (1.14 \times 10^{-4})x + 1.22.$$

The output is the density of air in kilograms per cubic meter. The domain of d is $0 \leq x \leq 10,000$. (*Source:* A. Miller and J. Thompson, *Elements of Meteorology.*)
(a) Denver is sometimes referred to as the mile-high city. Compare the density of air at sea level and in Denver. (*Hint:* 1 ft $\approx$ 0.305 m.)

(b) Determine the altitudes where the density is greater than 1 kilogram per cubic meter.

75. **iPod Sales** Sales of iPods in millions x years after 2006 can be modeled by

$$I(x) = -2.277x^2 + 11.71x + 40.4.$$

To the nearest year, estimate when sales were between 50 and 55 million iPods. (*Source:* Apple Corporation.)

76. **Heart Rate** The table shows a person's heart rate after exercise has stopped.
(a) Find values for the constants a, h, and k so that the formula $f(x) = a(x - h)^2 + k$ models the data, where x represents time and $0 \leq x \leq 4$.

Time (min)	0	2	4
Heart rate (bpm)	154	106	90

(b) Evaluate $f(1)$ and interpret the result.

(c) Estimate the times when the heart rate was from 115 to 125 beats per minute.

Writing about Mathematics

77. Explain how a table of values can be used to help solve a quadratic inequality, provided that the boundary numbers are listed in the table.

78. Explain how to determine the solution set for the inequality $ax^2 + bx + c < 0$, where $a > 0$. How would the solution set change if $a < 0$?

CHECKING BASIC CONCEPTS FOR SECTIONS 3.3 AND 3.4

1. Simplify by using the imaginary unit i.
   (a) $\sqrt{-25}$ (b) $\sqrt{-3} \cdot \sqrt{-18}$
   (c) $\dfrac{7 \pm \sqrt{-98}}{14}$

2. Write each expression in standard form.
   (a) $-3i - (5 - 2i)$ (b) $(6 - 7i) + (-1 + i)$
   (c) $i(1 - i)(1 + i)$ (d) $\dfrac{1 + 2i}{4 - i}$

3. Use the graph of $y = f(x)$ to solve $f(x) \le 0$ and $f(x) > 0$. Write your answer in set-builder or interval notation.
   (a)   (b)

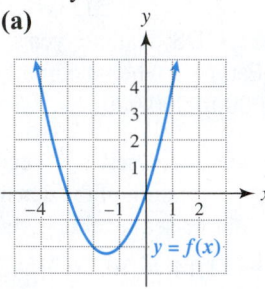
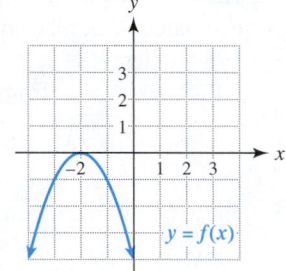

4. Solve each equation and inequality. Write the solution set for each inequality in set-builder or interval notation.
   (a) $2x^2 + 7x - 4 = 0$
   (b) $2x^2 + 7x - 4 < 0$
   (c) $2x^2 + 7x - 4 > 0$

5. Solve each inequality. Use set-builder or interval notation.
   (a) $x^2 - 36 \ge 0$
   (b) $4x^2 + 9 > 9x$
   (c) $2x(x - 1) \le 2$

6. **Safe Driving Speeds** The stopping distance d in feet for a car traveling x miles per hour on *wet* level pavement can be estimated by $d(x) = \frac{1}{9}x^2 + \frac{11}{3}x$. Determine the driving speeds that correspond to stopping distances between 80 and 180 feet, inclusive.

3.5 Transformations of Graphs

- **Graph functions using vertical and horizontal shifts**
- **Graph functions using stretching and shrinking**
- **Graph functions using reflections**
- **Combine transformations**
- **Model data with transformations (optional)**

Introduction

Graphs are often used to model different types of phenomena. For example, when a cold front moves across the United States, we might use a circular arc on a weather map to describe its shape. (See Exercise 7 in the Extended and Discovery Exercises on page 230.) If the front does not change its shape significantly, we could model the movement of the front on a television weather map by translating the circular arc. Before we can portray a cold front on a weather map, we need to discuss how to transform graphs of functions. (*Sources:* S. Hoggar, *Mathematics for Computer Graphics;* A. Watt, *3D Computer Graphics.*)

Vertical and Horizontal Shifts

Graphs of $f(x) = x^2$ and $g(x) = \sqrt{x}$ will be used to demonstrate shifts, or translations, in the xy-plane. Basic functions such as these are sometimes referred to as **parent functions**. Symbolic, numerical, and graphical representations of f and g are shown in Figures 3.53 and 3.54, respectively. Points listed in the table are plotted on the graph.

Vertical Shifts If 2 is added to the formula for each parent function, their graphs are shifted upward 2 units. The graphs of $y = x^2 + 2$ and $y = \sqrt{x} + 2$ are shown in Figures 3.55 and 3.56. Notice that the y-values in the graphs and tables increase by 2 units.

If 2 is subtracted from each of the parent formulas, the graphs are shifted downward 2 units. Verify this by graphing $y = x^2 - 2$ and $y = \sqrt{x} - 2$. Translations of this type are called **vertical shifts**, or **vertical translations**. They do not alter the shape of the graph, only its position. The parent and shifted graphs are congruent.

Parent Functions

x	−2	−1	0	1	2
y	4	1	0	1	4

x	0	1	4
y	0	1	2

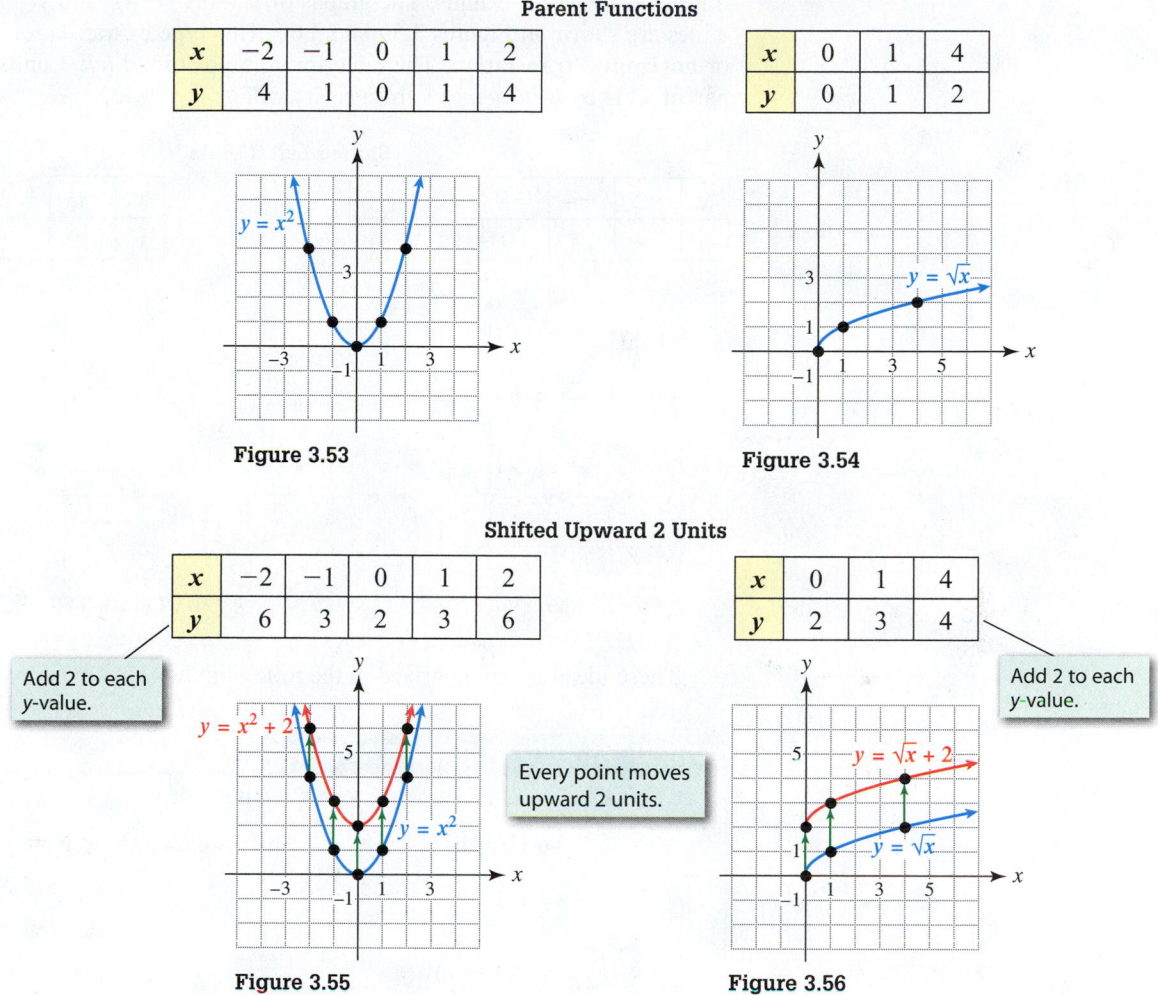

Figure 3.53   Figure 3.54

Shifted Upward 2 Units

x	−2	−1	0	1	2
y	6	3	2	3	6

x	0	1	4
y	2	3	4

Add 2 to each y-value.

Add 2 to each y-value.

Every point moves upward 2 units.

Figure 3.55   Figure 3.56

Horizontal Shifts   If the variable x is replaced by $(x - 2)$ in the formulas for f and g, a different type of shift results. Figures 3.57 and 3.58 show the graphs and tables of $y = (x - 2)^2$ and $y = \sqrt{(x - 2)}$, together with the graphs of the parent functions.

Shifted Right 2 Units

x	0	1	2	3	4
y	4	1	0	1	4

x	2	3	6
y	0	1	2

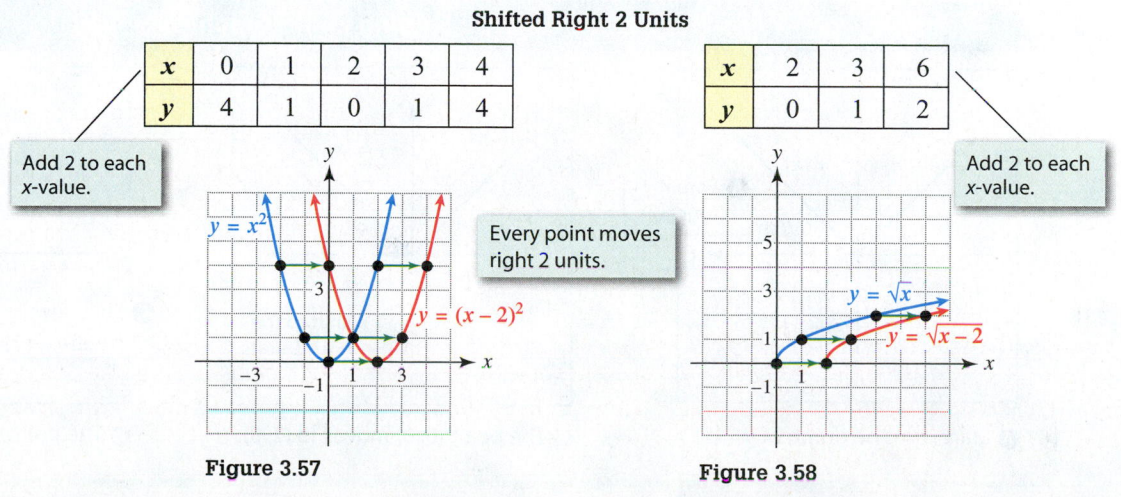

Add 2 to each x-value.

Add 2 to each x-value.

Every point moves right 2 units.

Figure 3.57   Figure 3.58

Each new graph shows a shift of the parent graph to the *right* by 2 units. Notice that a table for a graph shifted *right* 2 units can be obtained from the parent table by *adding* 2 to each *x*-value.

If the variable x is replaced by $(x + 3)$ in each equation, the parent graphs are translated to the *left* 3 units. The graphs of $y = (x + 3)^2$ and $y = \sqrt{(x + 3)}$ and their tables are shown in Figures 3.59 and 3.60. This type of translation is a **horizontal shift**, or **horizontal translation**. The table for a graph shifted *left* 3 units is obtained from the parent table by *subtracting* 3 from each *x*-value.

Shifted Left 3 Units

Subtract 3 from each *x*-value.

x	-5	-4	-3	-2	-1
y	4	1	0	1	4

x	-3	-2	1
y	0	1	2

Subtract 3 from each *x*-value.

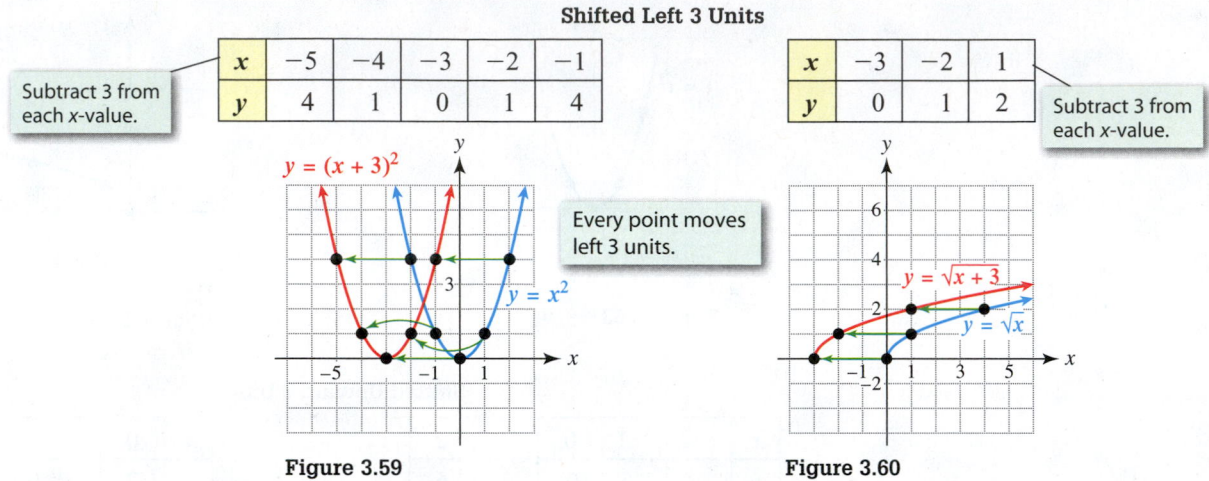

Every point moves left 3 units.

Figure 3.59   **Figure 3.60**

These ideas are summarized in the following box.

VERTICAL AND HORIZONTAL SHIFTS

Let f be a function, and let c be a positive number.

To Graph	Shift the Graph of $y = f(x)$ by c Units
$y = f(x) + c$	upward
$y = f(x) - c$	downward
$y = f(x - c)$	right
$y = f(x + c)$	left

Shifts can be combined to translate a graph of $y = f(x)$. For example, we can shift the graph of $y = |x|$ to the right 2 units and downward 4 units, as follows.

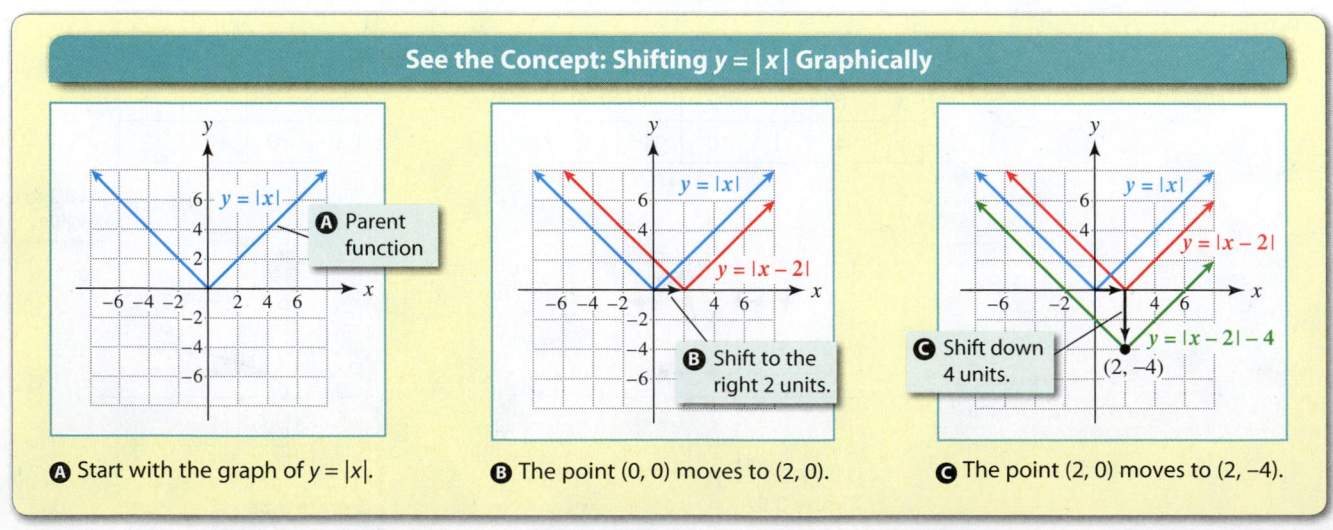

See the Concept: Shifting $y = |x|$ Graphically

Ⓐ Parent function

Ⓑ Shift to the right 2 units.

Ⓒ Shift down 4 units.

Ⓐ Start with the graph of $y = |x|$.   Ⓑ The point $(0, 0)$ moves to $(2, 0)$.   Ⓒ The point $(2, 0)$ moves to $(2, -4)$.

Shifting y = |x| Symbolically

$$y = |x| \xrightarrow[\text{right 2 units}]{\text{Shift to the}} y = |x - 2| \xrightarrow[\text{4 units}]{\text{Shift down}} y = |x - 2| - 4$$

EXAMPLE 1 **Combining vertical and horizontal shifts**

Complete the following.
(a) Write an equation that shifts the graph of $f(x) = x^2$ left 2 units. Graph your equation.
(b) Write an equation that shifts the graph of $f(x) = x^2$ left 2 units and downward 3 units. Graph your equation.

SOLUTION
(a) To shift $f(x) = x^2$ left 2 units, replace x with $x + 2$ to obtain $y = f(x + 2)$, or $y = (x + 2)^2$. Its graph is shown in Figure 3.61.
(b) To shift $f(x) = x^2$ left 2 units and downward 3 units, we subtract 3 from the equation found in part (a) to obtain $y = f(x + 2) - 3$, or $y = (x + 2)^2 - 3$. Its graph is shown in Figure 3.62. The graph does not change shape.

Shifting y = x² Left 2 Units

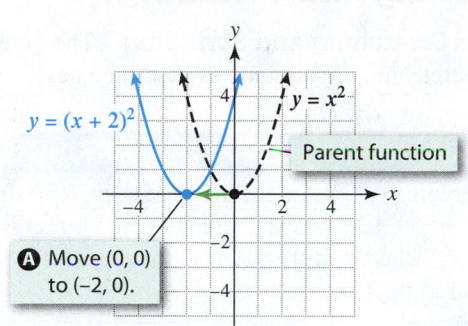

Figure 3.61

Shifting y = x² Left 2 Units and Downward 3 Units

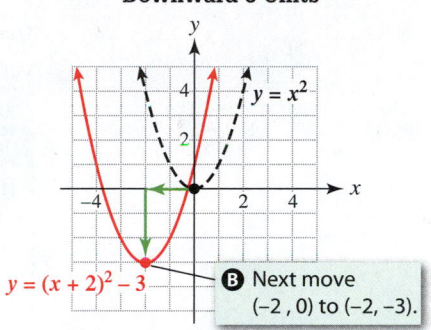

Figure 3.62

Now Try Exercise 9

EXAMPLE 2 **Writing formulas**

Write a formula for a function g whose graph is similar to that of $f(x) = 4x^2 - 2x + 1$ but is shifted right 1980 units and upward 50 units. Do not simplify the formula.

SOLUTION   Replace x with $(x - 1980)$ in the formula for $f(x)$ and then add **50**.

$$g(x) = f(x - 1980) + 50$$
$$= 4(x - 1980)^2 - 2(x - 1980) + 1 + 50$$
$$= 4(x - 1980)^2 - 2(x - 1980) + 51$$

Now Try Exercise 19

In the next example we translate a circle that is centered at the origin.

EXAMPLE 3 **Translating a circle**

The equation of a circle having radius 3 and center $(0, 0)$ is $x^2 + y^2 = 9$. Write an equation that shifts this circle to the right 4 units and upward 2 units. What are the center and radius of this circle? (Note that a circle is not a function.)

SOLUTION

Getting Started The standard equation for a circle with center (h, k) and radius r is

$$(x - h)^2 + (y - k)^2 = r^2.$$

If we determine the new center and radius, we can apply the standard equation. ▶

Translating a Circle

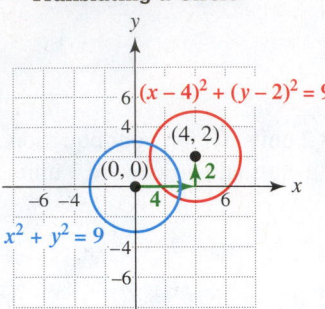

Figure 3.63

If a circle with center $(0, 0)$ and radius 3 is translated to the right 4 units and upward 2 units, then the center of the new circle is $(0 + 4, 0 + 2)$, or $(4, 2)$; the radius remains the same. The standard equation for a circle with center $(\mathbf{4}, \mathbf{2})$ and radius 3 is

$$(x - \mathbf{4})^2 + (y - \mathbf{2})^2 = 9.$$

Figure 3.63 illustrates this translation.

Now Try Exercise 23

NOTE Example 3 illustrates that to translate a circle horizontally c units, replace x with $(x - c)$, and to translate a circle vertically c units, replace y with $(y - c)$. If c is positive, the translation is either to the right or upward. If c is negative, the translation is either to the left or downward.

Stretching and Shrinking

Vertical Stretching and Shrinking The graph of a function can be transformed by vertical stretching or shrinking as described next.

> ### VERTICAL STRETCHING AND SHRINKING
>
> If the point (x, y) lies on the graph of $y = f(x)$, then the point (x, cy) lies on the graph of $y = cf(x)$. If $c > 1$, the graph of $y = cf(x)$ is a vertical stretching of the graph of $y = f(x)$, whereas if $0 < c < 1$, the graph of $y = cf(x)$ is a vertical shrinking of the graph of $y = f(x)$.

For example, if the point $(\mathbf{4}, \mathbf{2})$ is on the graph of $y = f(x)$, then the point $(\mathbf{4}, \mathbf{4})$ is on the graph of $y = 2f(x)$ and the point $(\mathbf{4}, \mathbf{1})$ is on the graph of $y = \frac{1}{2}f(x)$. The graph of $f(x) = \sqrt{x}$ in Figure 3.64 can be stretched or shrunk *vertically*. In Figure 3.65, the graph of $y = 2f(x)$, or $y = 2\sqrt{x}$, represents a vertical stretching of the graph of $y = \sqrt{x}$. In Figure 3.66, the graph of $y = \frac{1}{2}f(x)$, or $y = \frac{1}{2}\sqrt{x}$, represents a vertical shrinking of the graph of $y = \sqrt{x}$. Compared to the y-values in the table for $y = \sqrt{x}$, the y-values in the tables for $y = 2f(x)$ and $y = \frac{1}{2}f(x)$ have been multiplied by 2 and $\frac{1}{2}$, respectively. The x-values have not changed.

Parent Function

x	0	1	4
$f(x)$	0	1	2

Vertical Stretching

x	0	1	4
$2f(x)$	0	2	4

Vertical Shrinking

x	0	1	4
$\frac{1}{2}f(x)$	0	$\frac{1}{2}$	1

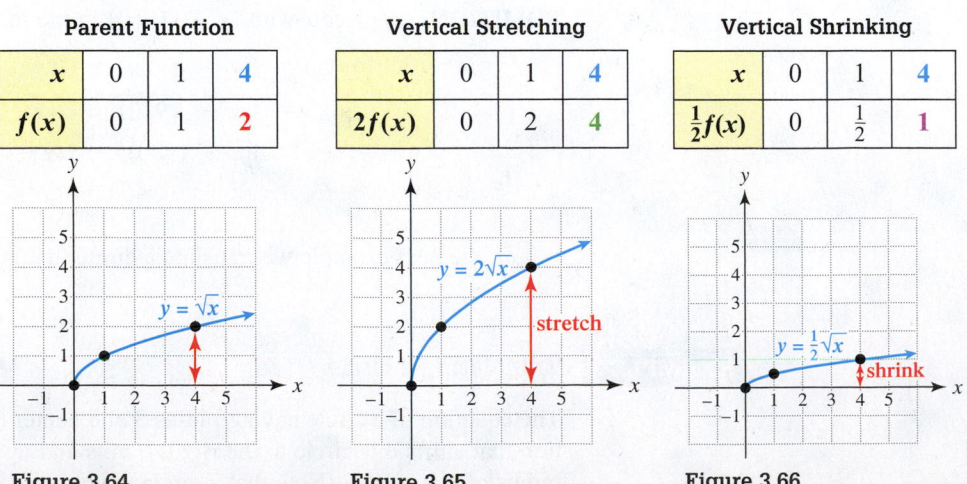

Figure 3.64   **Figure 3.65**   **Figure 3.66**

Horizontal Stretching and Shrinking The line graph in Figure 3.67 can be stretched or shrunk *horizontally*. On one hand, if the line graph represents the graph of a function f, then the graph of $y = f\left(\frac{1}{2}x\right)$ in Figure 3.68 is a horizontal stretching of the graph of $y = f(x)$. On the other hand, the graph of $y = f(2x)$ in Figure 3.69 represents a horizontal shrinking of the graph of $y = f(x)$. Compared to the x-values in the table for $y = f(x)$, the x-values in the table for $y = f\left(\frac{1}{2}x\right)$ have been multiplied by 2 and the x-values in the table for $y = f(2x)$ have been multiplied by $\frac{1}{2}$. The y-values have not changed.

Parent Function

x	-2	-1	1	2
$f(x)$	3	-3	3	-3

Horizontal Stretching

x	-4	-2	2	4
$f\left(\frac{1}{2}x\right)$	3	-3	3	-3

Horizontal Shrinking

x	-1	$-\frac{1}{2}$	$\frac{1}{2}$	1
$f(2x)$	3	-3	3	-3

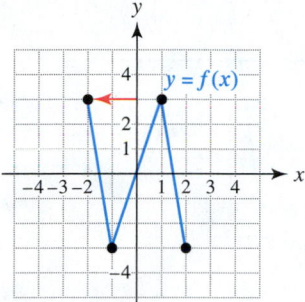

Figure 3.67

Figure 3.68

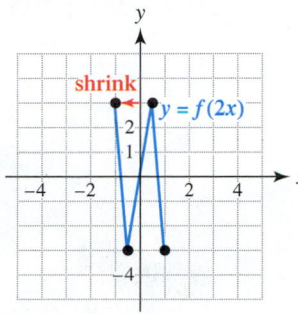

Figure 3.69

NOTE Horizontal stretching or shrinking does not change the height (maximum or minimum y-values) of the graph, nor does it change the y-intercept. Horizontal stretching and shrinking can be generalized for any function f.

HORIZONTAL STRETCHING AND SHRINKING

If the point (x, y) lies on the graph of $y = f(x)$, then the point $\left(\frac{x}{c}, y\right)$ lies on the graph of $y = f(cx)$. If $c > 1$, the graph of $y = f(cx)$ is a horizontal shrinking of the graph of $y = f(x)$, whereas if $0 < c < 1$, the graph of $y = f(cx)$ is a horizontal stretching of the graph of $y = f(x)$.

For example, if the point $(-2, 3)$ is on the graph of $y = f(x)$, then the point $(-1, 3)$ is on the graph of $y = f(2x)$ and the point $(-4, 3)$ is on the graph of $y = f\left(\frac{1}{2}x\right)$.

MAKING CONNECTIONS

Horizontal Stretching and Shrinking

Horizontal stretching "pulls" the graph away from the y-axis.

Horizontal shrinking "pushes" the graph towards the y-axis. In both cases the y-intercepts do not change.

EXAMPLE 4 Stretching and shrinking of a graph

Use the graph and table of $y = f(x)$ in Figure 3.70 on the next page to sketch a graph of each equation.

(a) $y = 3f(x)$
(b) $y = f\left(\frac{1}{2}x\right)$

SOLUTION

(a) The graph of $y = 3f(x)$, shown in Figure 3.71, is a vertical stretching of the graph of the given function $y = f(x)$, shown in Figure 3.70, and can be obtained by multiplying each y-coordinate on the graph of $y = f(x)$ by 3.

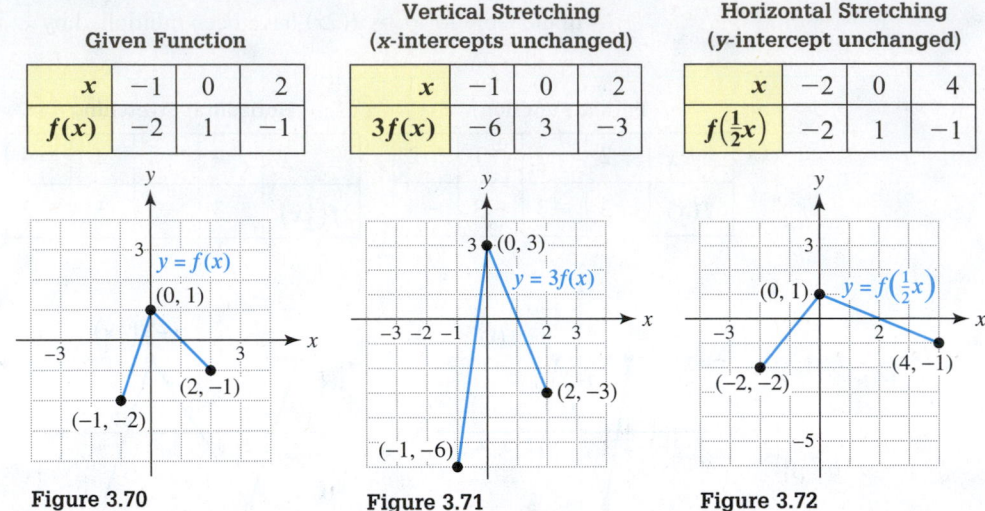

Given Function

x	-1	0	2
$f(x)$	-2	1	-1

Vertical Stretching
(x-intercepts unchanged)

x	-1	0	2
$3f(x)$	-6	3	-3

Horizontal Stretching
(y-intercept unchanged)

x	-2	0	4
$f\left(\frac{1}{2}x\right)$	-2	1	-1

Figure 3.70   **Figure 3.71**   **Figure 3.72**

(b) The graph of $y = f\left(\frac{1}{2}x\right)$, shown in Figure 3.72, is a horizontal stretching of the graph of $y = f(x)$, shown in Figure 3.70, and can be obtained by *dividing* each x-coordinate on the graph of $y = f(x)$ by $\frac{1}{2}$, which is equivalent to multiplying each x-coordinate by 2.

Now Try Exercises 33(a) and (b)

Reflection of Graphs

Another type of translation is called a **reflection**. The reflection of the blue graph of $y = f(x)$ across the x-axis is shown in Figure 3.73 as a red curve. This reflection can be thought of as flipping the graph of $y = f(x)$ across the x-axis.

If (x, y) is a point on the graph of f, then $(x, -y)$ lies on the graph of its reflection across the x-axis, as shown in Figure 3.73. Thus a reflection of $y = f(x)$ is given by the equation $-y = f(x)$, or equivalently, $y = -f(x)$.

If a point (x, y) lies on the graph of a function f, then the point $(-x, y)$ lies on the graph of its reflection across the y-axis, as shown in Figure 3.74. Thus a reflection of $y = f(x)$ across the y-axis is given by $y = f(-x)$. This reflection can be thought of as flipping the graph of $y = f(x)$ across the y-axis. Another example is shown in Figure 3.75.

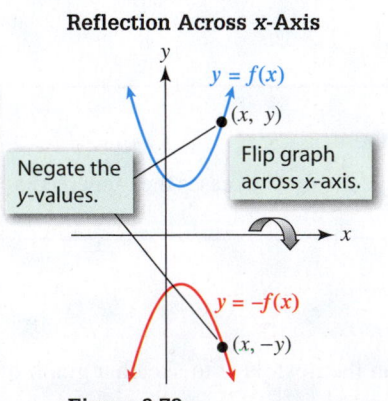

Reflection Across x-Axis

Figure 3.73

Reflections Across y-Axis

Figure 3.74   **Figure 3.75**

These results are summarized in the following box.

> **REFLECTIONS OF GRAPHS ACROSS THE x-AND y-AXES**
>
> 1. The graph of $y = -f(x)$ is a reflection of the graph of $y = f(x)$ across the x-axis.
> 2. The graph of $y = f(-x)$ is a reflection of the graph of $y = f(x)$ across the y-axis.

EXAMPLE 5   Reflecting graphs of functions

Complete the following.
(a) Write an equation that reflects the graph of $f(x) = x^2 + 1$ across the x-axis. Graph your equation.
(b) Write an equation that reflects the graph of $f(x) = \sqrt{x}$ across the y-axis. Graph your equation.

SOLUTION
(a) To reflect $f(x) = x^2 + 1$ across the x-axis, replace $f(x)$ with $-f(x)$ to obtain $y = -f(x)$, or $y = -x^2 - 1$. See Figure 3.76.
(b) To reflect $f(x) = \sqrt{x}$ across the y-axis, replace x with $-x$ to obtain $y = f(-x)$, or $y = \sqrt{-x}$. See Figure 3.77.

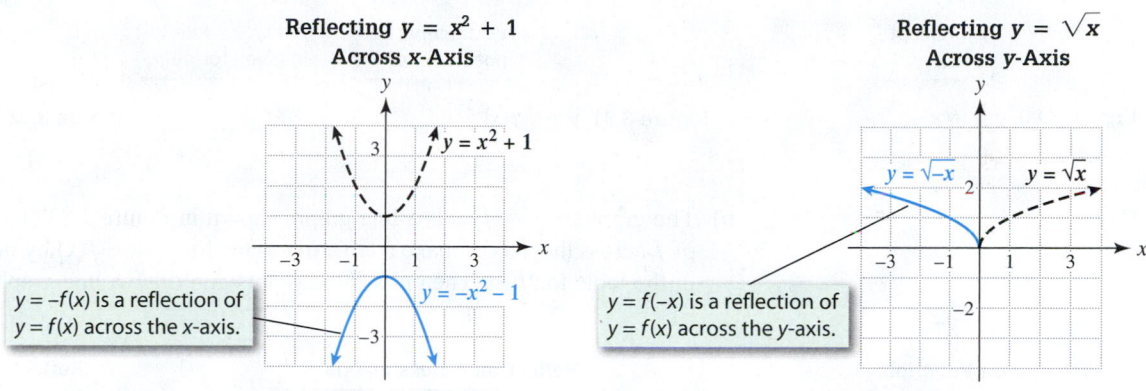

Figure 3.76

Figure 3.77

> **Now Try Exercises 35 and 37**

Calculator Help

To access the variable Y_1 as shown in Figure 3.78, see Appendix A (page AP-10).

Graphing Calculators (Optional)   On a graphing calculator capable of using function notation, entering equations for reflections of a function f is easy. For example, if $f(x) = (x - 4)^2$, let $Y_1 = (X - 4)^2$, $Y_2 = -Y_1$, and $Y_3 = Y_1(-X)$. See Figures 3.78 and 3.79.

$[-10, 10, 1]$ by $[-10, 10, 1]$

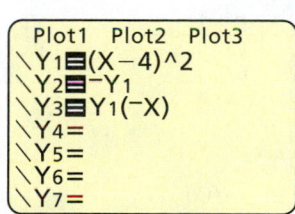

Figure 3.78

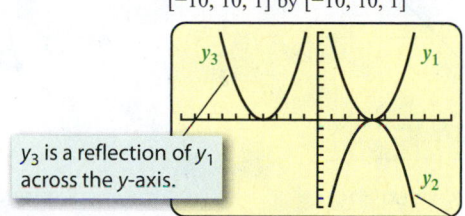

Figure 3.79

EXAMPLE 6   Reflecting graphs of functions

For each representation of f, graph the reflection across the x-axis and across the y-axis.
(a) $f(x) = x^2 + 2x - 3$
(b) The graph of f is a line graph determined by Table 3.10.

x	-2	-1	0	3
$f(x)$	1	-3	-1	2

Table 3.10

SOLUTION

(a) The graph of $f(x) = x^2 + 2x - 3$ is shown in Figure 3.80. To obtain its reflection across the x-axis, graph $y = -f(x)$, or $y = -x^2 - 2x + 3$, as shown in Figure 3.81. The vertex is now $(-1, 4)$.

For the reflection across the y-axis, let $y = f(-x)$, or $y = (-x)^2 + 2(-x) - 3$, and graph, as shown in Figure 3.82. The vertex is now $(1, -4)$.

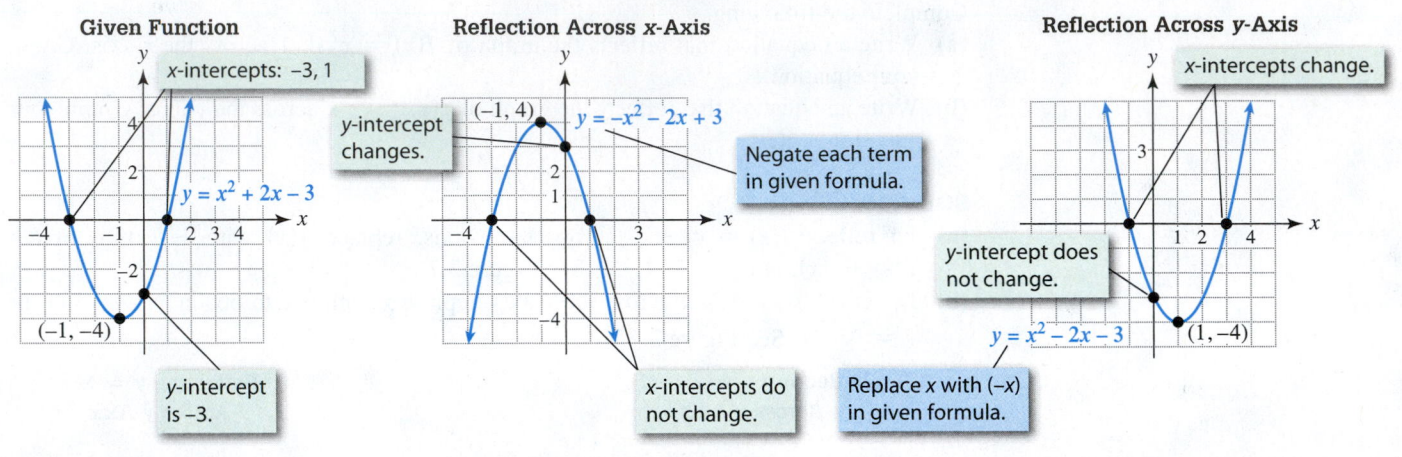

Figure 3.80 $y = f(x)$ **Figure 3.81** $y = -f(x)$ **Figure 3.82** $y = f(-x)$

(b) The graph of $y = f(x)$ is a line graph, shown in Figure 3.83. To graph the reflection of f across the x-axis, make a table of values for $y = -f(x)$ by negating each y-value in the table for $f(x)$. Then plot these points and draw a line graph, as in Figure 3.84.

Given Function

x	-2	-1	0	3
$f(x)$	1	-3	-1	2

Reflection Across x-Axis

x	-2	-1	0	3
$-f(x)$	-1	3	1	-2

Reflection Across y-Axis

x	2	1	0	-3
$f(-x)$	1	-3	-1	2

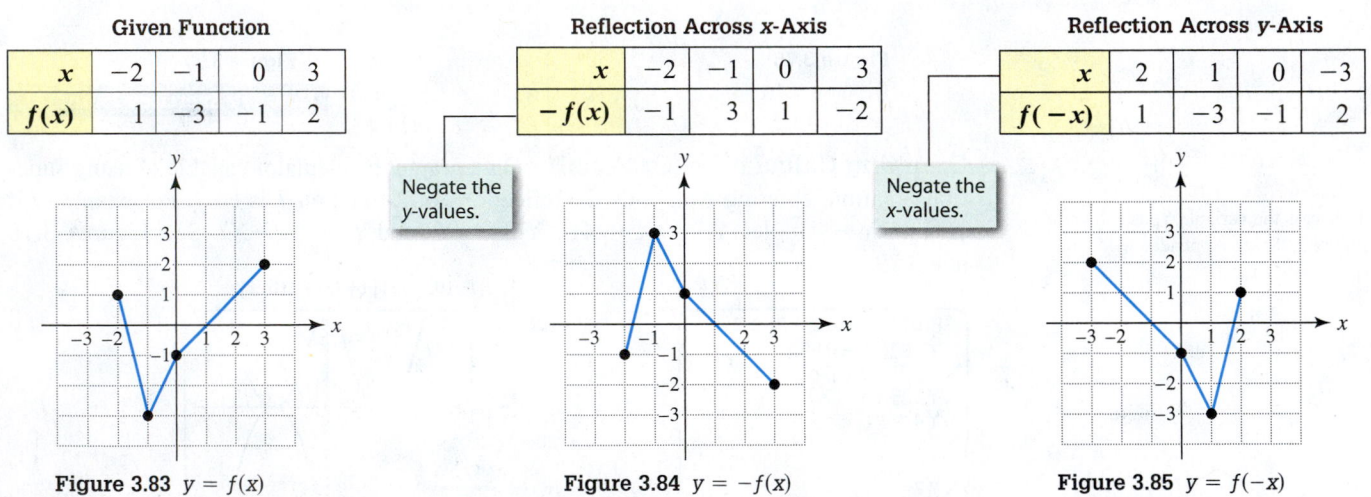

Figure 3.83 $y = f(x)$ **Figure 3.84** $y = -f(x)$ **Figure 3.85** $y = f(-x)$

To graph the reflection of f across the y-axis, make a table of values for $y = f(-x)$ by negating each x-value in the table for $f(x)$. Then plot these points and draw a line graph, as in Figure 3.85.

Now Try Exercises 39 and 43

NOTE Compared to the y-values in a table of values for $y = f(x)$, the y-values in a table of values for $y = -f(x)$ are negated; the x-values do *not* change. Compared to the x-values in a table of values for $y = f(x)$, the x-values in a table of values for $y = f(-x)$ are negated; the y-values do *not* change.

Combining Transformations

CLASS DISCUSSION

Given a table of values for $y = f(x)$, how would you make a table of values for $y = -f(-x)$?

Transformation of graphs can be combined to create new graphs. For example, the graph of $y = -2(x - 1)^2 + 3$ can be obtained by performing four transformations on the graph of $y = x^2$.

1. Shift the graph of $y = x^2$ to the right 1 unit: $y = (x - 1)^2$.
2. Vertically stretch the graph of $y = (x - 1)^2$ by a factor of 2: $y = 2(x - 1)^2$.
3. Reflect the graph of $y = 2(x - 1)^2$ across the x-axis: $y = -2(x - 1)^2$.
4. Shift the graph of $y = -2(x - 1)^2$ upward 3 units: $y = -2(x - 1)^2 + 3$.

These steps are summarized as follows.

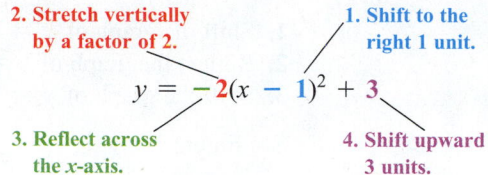

The resulting sequence of graphs is shown in Figures 3.86–3.89.

1. Shift Right

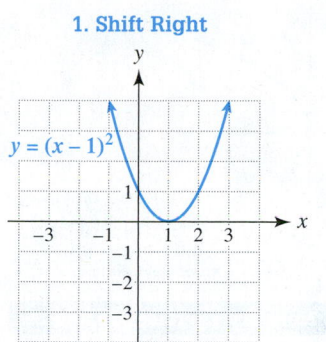

Figure 3.86

2. Vertical Stretch

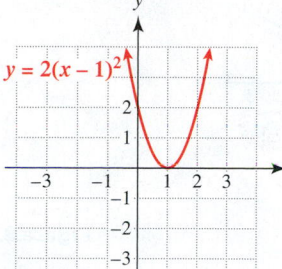

Figure 3.87

3. Reflect Across x-Axis

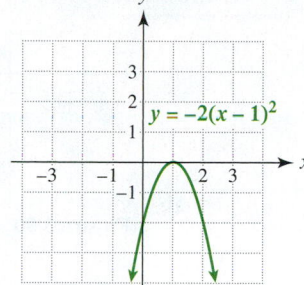

Figure 3.88

4. Shift Upward

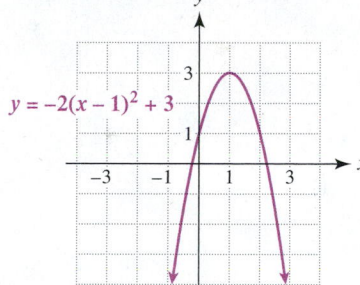

Figure 3.89

NOTE *The order in which transformations are made can be important.* For example, changing the order of a stretch and shift can result in a different equation and graph.

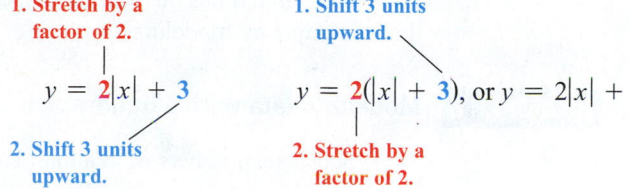

Also be careful when performing reflections and shifts. On the one hand, if we reflect the graph of $y = \sqrt{x}$ across the y-axis to obtain $y = \sqrt{-x}$ and then shift it right 2 units, we obtain $y = \sqrt{-(x - 2)}$. On the other hand, if we shift the graph of $y = \sqrt{x}$ right 2 units to obtain $y = \sqrt{x - 2}$ and then reflect it across the y-axis, we obtain $y = \sqrt{-x - 2}$. The final equations are different and so are their graphs. (Try sketching each graph.)

The following order can be used to graph the functions that we will encounter in this section.

COMBINING TRANSFORMATIONS

To graph a function by applying more than one transformation, use the following order.

1. Horizontal transformations
2. Stretching, shrinking, and reflections
3. Vertical transformations

EXAMPLE 7 Combining transformations of graphs

Describe how the graph of each equation can be obtained by transforming the parent graph of $y = \sqrt{x}$. Then graph the equation.

(a) $y = -\frac{1}{2}\sqrt{x}$ **(b)** $y = \sqrt{-x - 2} - 1$

SOLUTION

(a) Vertically shrink the graph of $y = \sqrt{x}$ by a factor of $\frac{1}{2}$ and then reflect it across the x-axis. See Figure 3.90.

(b) The following transformations can be used to obtain the graph of the equation $y = \sqrt{-x - 2} - 1$ from $y = \sqrt{x}$.

  1. Shift the graph of $y = \sqrt{x}$ right 2 units: $y = \sqrt{x - 2}$.
  2. Reflect the graph of $y = \sqrt{x - 2}$ across the y-axis: $y = \sqrt{-x - 2}$.
  3. Shift the graph of $y = \sqrt{-x - 2}$ down 1 unit: $y = \sqrt{-x - 2} - 1$.

See Figure 3.91.

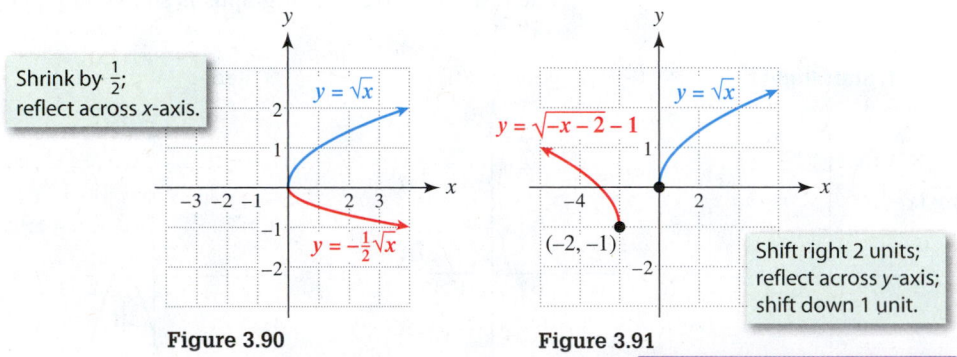

Shrink by $\frac{1}{2}$; reflect across x-axis.

Figure 3.90

Shift right 2 units; reflect across y-axis; shift down 1 unit.

Figure 3.91

Now Try Exercises 51 and 65

Modeling with Transformations (Optional)

Transformations of the graph of $y = x^2$ can be used to model some types of nonlinear data. By shifting, stretching, and shrinking this graph, we can transform it into a *portion of a parabola* that has the desired shape and location. In the next example we demonstrate this technique by modeling numbers of Walmart employees.

EXAMPLE 8 Modeling data with a quadratic function

Table 3.11 lists numbers of Walmart employees in millions for selected years.

(a) Make a scatterplot of the data.

(b) Use transformations to determine $f(x) = a(x - h)^2 + k$ so that $f(x)$ models the data. Graph $y = f(x)$ together with the data.

(c) Use $f(x)$ to estimate the number of Walmart employees in 2010. Compare it with the actual value of 2.1 million employees.

Walmart Employees (in millions)

Year	Employees
1987	0.20
1992	0.37
1997	0.68
2002	1.4
2007	2.2

Source: Walmart.

Table 3.11

SOLUTION

(a) A scatterplot of the data is shown in Figure 3.92. This plot suggests that the data could be modeled by the *right half* of a parabola that opens upward.

(b) Because the parabola opens upward, it follows that $a > 0$ and the vertex is the lowest point on the parabola. The minimum number of employees is 0.20 million in 1987. One possibility for the vertex (h, k) is $(1987, 0.20)$. Translate the graph of $y = x^2$ right 1987 units and upward 0.20 unit. Thus $f(x) = a(x - 1987)^2 + 0.20$.

To determine a, graph the data and $y = f(x)$ for different values of a. See Figures 3.93 and 3.94. With a little experimentation, a reasonable value for a near 0.005 can be found.

A scatterplot of the data and graph of $f(x) = 0.005(x - 1987)^2 + 0.2$ are shown in Figure 3.94. (Answers may vary.) Note that this equation is in *vertex form*.

Testing Values Graphically

[1985, 2010, 5] by [0, 3, 0.5]

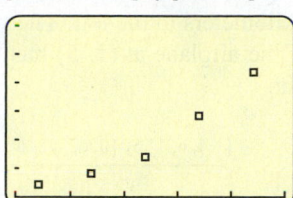

Figure 3.92

[1985, 2010, 5] by [0, 3, 0.5]

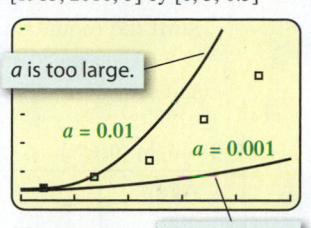

a is too large.

$a = 0.01$

$a = 0.001$

Figure 3.93   *a* is too small.

[1985, 2010, 5] by [0, 3, 0.5]

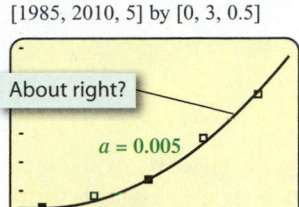

About right?

$a = 0.005$

Figure 3.94

(c) To estimate the number of employees in 2010, evaluate $f(2010)$.

$$f(2010) = 0.005(2010 - 1987)^2 + 0.2 = 2.845$$

This model provides an estimate of about 2.8 million Walmart employees in 2010. The calculation involves extrapolation and is not accurate.

Now Try Exercise 95

Translations and Computer Graphics   In video games, the background is often translated to give the illusion that the player in the game is moving. A simple scene of a mountain and an airplane is shown in Figure 3.95. To make it appear to the player as though the airplane were flying to the right, the image of the mountain could be translated horizontally to the left, as shown in Figure 3.96. Note that the position of the plane does not change.   (*Source:* C. Pokorny and C. Gerald, *Computer Graphics.*)

Shifting the Background to Show Motion

Figure 3.95

Figure 3.96

EXAMPLE 9   **Using translations to model movement**

Suppose that the mountain in Figure 3.95 can be described by $f(x) = -0.4x^2 + 4$ and that the airplane is located at the point $(1, 5)$.

(a) Graph f in $[-4, 4, 1]$ by $[0, 6, 1]$, where the units are in kilometers. Plot a point (a scatterplot with one point) to mark the position of the airplane.

(b) Assume that the airplane is moving horizontally to the right at 0.4 kilometer per second. To give a video player the illusion that the airplane is moving, graph the image of the mountain and the position of the airplane after 5 seconds and then after 10 seconds.

SOLUTION

(a) The graph of $y = f(x) = -0.4x^2 + 4$ and the position of the airplane at $(1, 5)$ are shown in Figure 3.97 on the next page. The "mountain" has been shaded to emphasize its position.

(b) Five seconds later, the airplane has moved $5(0.4) = 2$ kilometers right. In 10 seconds it has moved $10(0.4) = 4$ kilometers right. To graph these new positions, translate the

graph of the mountain 2 and 4 kilometers (units) to the left. First, shift the mountain 2 kilometers to the left by replacing x with $(x + 2)$ and graphing

$$y = f(x + 2) = -0.4(x + 2)^2 + 4$$

together with the point $(1, 5)$. Next, graph $y = f(x + 4) = -0.4(x + 4)^2 + 4$ to shift the mountain 4 kilometers to the left. The results are shown in Figures 3.98 and 3.99. The position of the airplane at $(1, 5)$ has not changed. However, it appears to have flown to the right.

Calculator Help

To shade below a parabola, see Appendix A (page AP-10).

[−4, 4, 1] by [0, 6, 1]   [−4, 4, 1] by [0, 6, 1]   [−4, 4, 1] by [0, 6, 1]

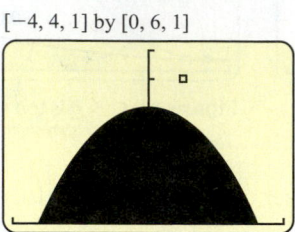   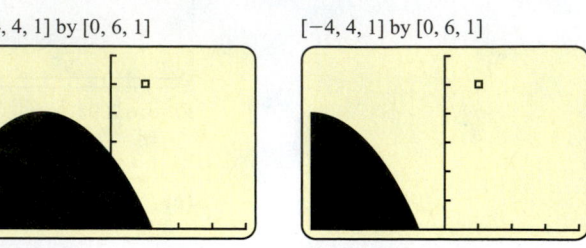

Figure 3.97   **Figure 3.98**   **Figure 3.99**

Now Try Exercise 101

CLASS DISCUSSION

Discuss how one might create the illusion of the airplane moving to the left and *gaining altitude* as it passes over the mountain.

3.5 Putting It All Together

In this section we discussed several transformations of graphs. The following table summarizes how these transformations affect the graph of $y = f(x)$.

EQUATION	EFFECT ON GRAPH OF $y = f(x)$
Let $c > 0$.	
$y = f(x) + c$	The graph of $y = f(x)$ is shifted upward c units.
$y = f(x) - c$	The graph of $y = f(x)$ is shifted downward c units.
$y = f(x + c)$	The graph of $y = f(x)$ is shifted to the left c units.
$y = f(x - c)$	The graph of $y = f(x)$ is shifted to the right c units.

Examples:

Shifted Down

$y = f(x)$

$y = f(x) - 1$

Shifted Left

$y = f(x + 1)$

$y = f(x)$

EQUATION	EFFECT ON GRAPH OF $y = f(x)$
Let $c > 0$. $y = cf(x)$	If (x, y) lies on the graph of $y = f(x)$, then (x, cy) lies on the graph of $y = cf(x)$. The graph is vertically stretched if $c > 1$ and vertically shrunk if $0 < c < 1$. **Examples:** Vertically Stretched 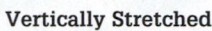 Vertically Shrunk 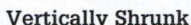 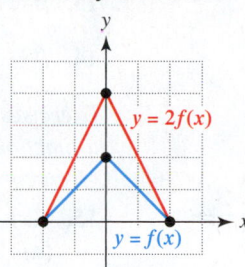 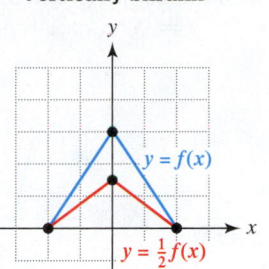
Let $c > 0$. $y = f(cx)$	If (x, y) lies on the graph of $y = f(x)$, then $\left(\frac{x}{c}, y\right)$ lies on the graph of $y = f(cx)$. The graph is horizontally shrunk if $c > 1$ and horizontally stretched if $0 < c < 1$. **Examples:** Horizontally Shrunk Horizontally Stretched  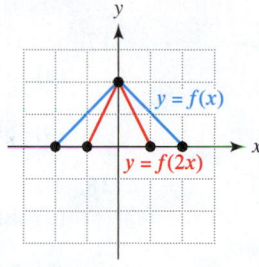 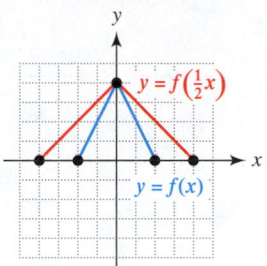
$y = -f(x)$ $y = f(-x)$	The graph of $y = f(x)$ is reflected across the x-axis. The graph of $y = f(x)$ is reflected across the y-axis. **Examples:** Reflected Across x-Axis Reflected Across y-Axis 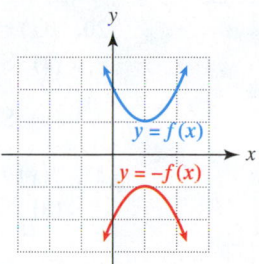 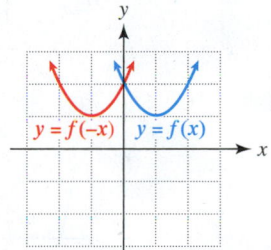

3.5 Exercises

Vertical and Horizontal Translations

Exercises 1–8: Write the equation of the graph. (Note: The given graph is a translation of the graph of one of the following equations: $y = x^2$, $y = \sqrt{x}$, or $y = |x|$.)

1.

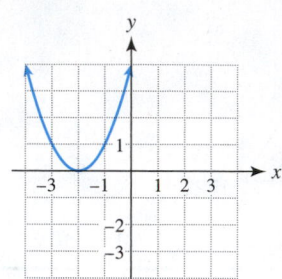

2.

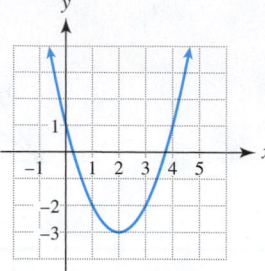

3.

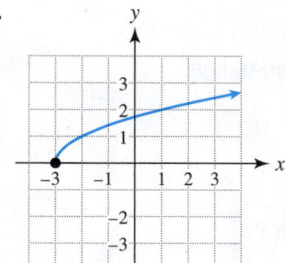

4.

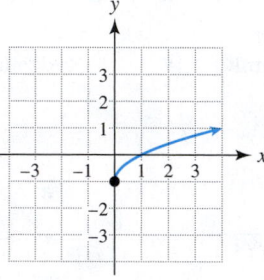

5.

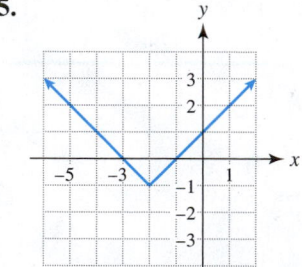

6.

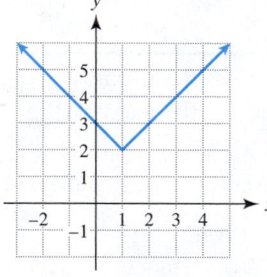

7.

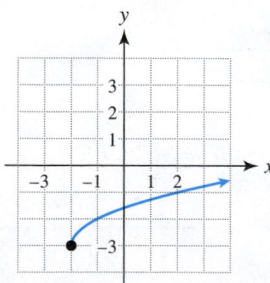

8.

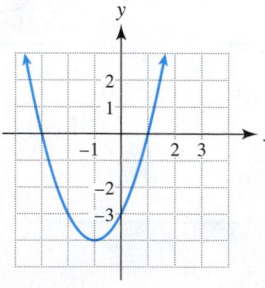

Exercises 9–14: Find an equation that shifts the graph of f by the desired amounts. Do not simplify. Graph f and the shifted graph in the same xy-plane.

9. $f(x) = x^2$; right 2 units, downward 3 units

10. $f(x) = 3x - 4$; left 3 units, upward 1 unit

11. $f(x) = x^2 - 4x + 1$; left 6 units, upward 4 units

12. $f(x) = x^2 - x - 2$; right 2 units, upward 3 units

13. $f(x) = \frac{1}{2}x^2 + 2x - 1$; left 3 units, downward 2 units

14. $f(x) = 5 - 3x - \frac{1}{2}x^2$; right 5 units, downward 8 units

Exercises 15–22: (Refer to Example 2.) Write a formula for a function g whose graph is similar to f(x) but satisfies the given conditions. Do not simplify the formula.

15. $f(x) = 3x^2 + 2x - 5$
   (a) Shifted left 3 units

   (b) Shifted downward 4 units

16. $f(x) = 2x^2 - 3x + 2$
   (a) Shifted right 8 units

   (b) Shifted upward 2 units

17. $f(x) = 2x^2$
   (a) Shifted right 2 units and upward 4 units

   (b) Shifted left 8 units and downward 5 units

18. $f(x) = 5x^2$
   (a) Shifted left 10 units and downward 6 units

   (b) Shifted right 1 unit and upward 10 units

19. $f(x) = 3x^2 - 3x + 2$
   (a) Shifted right 2000 units and upward 70 units

   (b) Shifted left 300 units and downward 30 units

20. $f(x) = |x|$
   (a) Shifted right 4 units and downward 3 units

   (b) Shifted left 5 units and upward 2 units

21. $f(x) = \sqrt{x}$
   (a) Shifted right 4 units, reflected about the x-axis

   (b) Shifted left 2 units, reflected about the y-axis

22. $f(x) = \sqrt{x}$
   (a) Reflected about the x-axis, shifted left 2 units

   (b) Reflected about the y-axis, shifted right 3 units

Exercises 23–26: **Translating Circles** *Write an equation that shifts the given circle in the specified manner. State the center and radius of the translated circle.*

23. $x^2 + y^2 = 4$; right 3 units, downward 4 units

24. $x^2 + y^2 = 9$; right 2 units, downward 6 units

25. $x^2 + y^2 = 5$; left 5 units, upward 3 units

26. $x^2 + y^2 = 7$; left 3 units, downward 7 units

Transforming Graphical Representations

Exercises 27–34: Use the accompanying graph of $y = f(x)$ to sketch a graph of each equation.

27. (a) $y = f(x) + 2$ **28.** (a) $y = f(x + 1)$

  (b) $y = f(x - 2) - 1$ (b) $y = -f(x)$

  (c) $y = -f(x)$ (c) $y = 2f(x)$

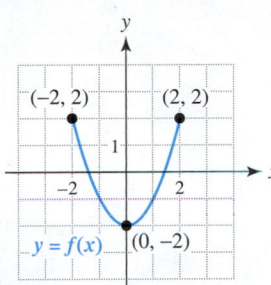
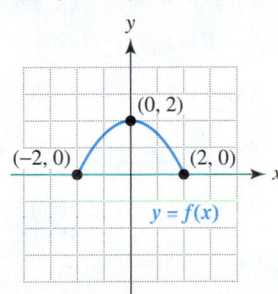

29. (a) $y = f(x + 3) - 2$ **30.** (a) $y = f(x - 1) - 2$

  (b) $y = f(-x)$ (b) $y = -f(x) + 1$

  (c) $y = \frac{1}{2}f(x)$ (c) $y = f\left(\frac{1}{2}x\right)$

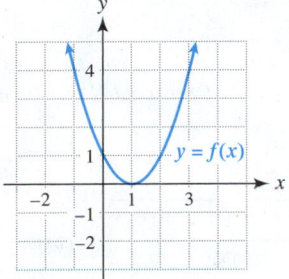
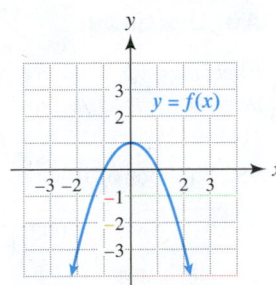

31. (a) $y = f(x + 1) + 1$ **32.** (a) $y = f(x) - 2$

  (b) $y = -f(x) - 1$ (b) $y = f(x - 1) + 2$

  (c) $y = \frac{1}{2}f(2x)$ (c) $y = 2f(-x)$

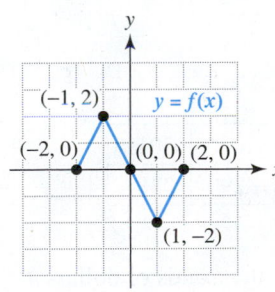
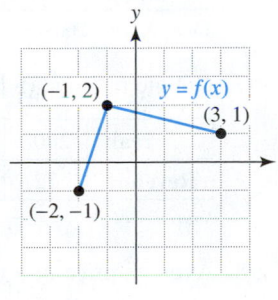

33. (a) $y = f(2x) + 1$ **34.** (a) $y = f(2x)$

  (b) $y = 2f\left(\frac{1}{2}x\right) + 1$ (b) $y = f\left(\frac{1}{2}x\right) - 1$

  (c) $y = \frac{1}{2}f(2 - x)$ (c) $y = 2f(1 - x)$

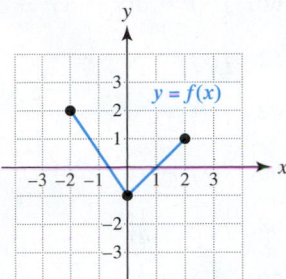
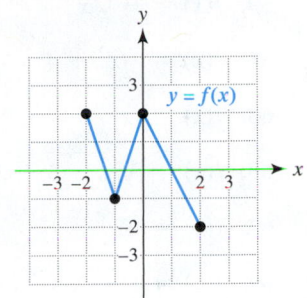

Reflections of Graphs

Exercises 35–38: Write the requested equation. Then graph the given equation and your equation.

35. Reflect $f(x) = \sqrt{x} + 1$ across the x-axis.

36. Reflect $f(x) = |x| - 1$ across the x-axis.

37. Reflect $f(x) = x^2 - x$ across the y-axis.

38. Reflect $f(x) = \sqrt{-x} + 1$ across the y-axis.

Exercises 39–44: (Refer to Example 6.) For the given representation of a function f, graph the reflection across the x-axis and graph the reflection across the y-axis.

39. $f(x) = x^2 - 2x - 3$ **40.** $f(x) = 4 - 7x - 2x^2$

41. $f(x) = |x + 1| - 1$ **42.** $f(x) = \frac{1}{2}|x - 2| + 2$

43. Line graph determined by the table

x	-3	-1	1	2
$f(x)$	2	3	-1	-2

44. Line graph determined by the table

x	-4	-2	0	1
$f(x)$	-1	-4	2	2

Graphing Transformations of Functions

Exercises 45–54. Use transformations to explain how the graph of f can be found by using the graph of $y = x^2$, $y = \sqrt{x}$, or $y = |x|$. You do not need to graph $y = f(x)$.

45. $f(x) = (x - 3)^2 + 1$ **46.** $f(x) = (x + 2)^2 - 3$

47. $f(x) = \frac{1}{4}(x + 1)^2$ **48.** $f(x) = 2(x - 4)^2$

49. $f(x) = -\sqrt{x + 5}$ **50.** $f(x) = -\sqrt{x} - 3$

51. $f(x) = 2\sqrt{-x}$ **52.** $f(x) = \sqrt{-\frac{1}{2}x}$

53. $f(x) = |-(x + 1)|$ **54.** $f(x) = |4 - x|$

Exercises 55–78: Use transformations to sketch a graph of f.

55. $f(x) = x^2 - 3$

56. $f(x) = -x^2$

57. $f(x) = (x - 5)^2 + 3$

58. $f(x) = (x + 4)^2$

59. $f(x) = -\sqrt{x}$

60. $f(x) = 2(x - 1)^2 + 1$

61. $f(x) = -x^2 + 4$

62. $f(x) = \sqrt{-x}$

63. $f(x) = |x| - 4$

64. $f(x) = \sqrt{x} + 1$

65. $f(x) = \sqrt{x - 3} + 2$

66. $f(x) = |x + 2| - 3$

67. $f(x) = |2x|$

68. $f(x) = \frac{1}{2}|x|$

69. $f(x) = 1 - \sqrt{x}$

70. $f(x) = 2\sqrt{x - 2} - 1$

71. $f(x) = -\sqrt{1 - x}$

72. $f(x) = \sqrt{-x} - 1$

73. $f(x) = \sqrt{-(x + 1)}$

74. $f(x) = 2 + \sqrt{-(x - 3)}$

75. $f(x) = (x - 1)^3$

76. $f(x) = (x + 2)^3$

77. $f(x) = -x^3$

78. $f(x) = (-x)^3 + 1$

Transforming Numerical Representations

Exercises 79–86: Two functions, f and g, are related by the given equation. Use the numerical representation of f to make a numerical representation of g.

79. $g(x) = f(x) + 7$

x	1	2	3	4	5	6
f(x)	5	1	6	2	7	9

80. $g(x) = f(x) - 10$

x	0	5	10	15	20
f(x)	−5	11	21	32	47

81. $g(x) = f(x - 2)$

x	−4	−2	0	2	4
f(x)	5	2	−3	−5	−9

82. $g(x) = f(x + 50)$

x	−100	−50	0	50	100
f(x)	25	80	120	150	100

83. $g(x) = f(x + 1) - 2$

x	1	2	3	4	5	6
f(x)	2	4	3	7	8	10

84. $g(x) = f(x - 3) + 5$

x	−3	0	3	6	9
f(x)	3	8	15	27	31

85. $g(x) = f(-x) + 1$

x	−2	−1	0	1	2
f(x)	11	8	5	2	−1

86. $g(x) = -f(x + 2)$

x	−4	−2	0	2	4
f(x)	5	8	10	8	5

Exercises 87–94: The points $(-12, 6)$, $(0, 8)$, and $(8, -4)$ lie on the graph of $y = f(x)$. Determine three points that lie on the graph of $y = g(x)$.

87. $g(x) = f(x) + 2$

88. $g(x) = f(x) - 3$

89. $g(x) = f(x - 2) + 1$

90. $g(x) = f(x + 1) - 1$

91. $g(x) = -\frac{1}{2}f(x)$

92. $g(x) = -2f(x)$

93. $g(x) = f(-2x)$

94. $g(x) = f\left(-\frac{1}{2}x\right)$

Applications

Exercises 95–98: (Refer to Example 8.) Use transformations of graphs to model the table of data with the formula $f(x) = a(x - h)^2 + k$. (Answers may vary.)

95. Number of iPhones sold (millions)

Year	2008	2009	2010	2011
iPhones	11.6	20.7	40.0	72.3

Source: Apple Corporation.

96. Apple apps downloaded (billions)

Year	2009	2010	2011
Apps	1.1	2.9	6.5

Source: Apple Corporation.

97. Google revenue ($ billions)

Year	2008	2009	2010	2011
Revenue	22	24	29	38

Source: Google.

98. Average price of a home in thousands of dollars

Year	1970	1980	1990	2000	2005
Price	30	80	150	210	300

Source: Bureau of the Census.

99. U.S. Home Ownership The general trend in the percentage P of homes lived in by owners rather than renters between 1990 and 2006 is modeled by

$$P(x) = 0.00075x^2 + 0.17x + 44,$$

where $x = 0$ corresponds to 1990, $x = 1$ to 1991, and so on. Determine a function g that computes P, where x is the actual year. For example, $P(0) = 44$, so $g(1990) = 44$.

100. U.S. AIDS Deaths The function D defined by

$$D(x) = 2375x^2 + 5134x + 5020$$

models AIDS deaths x years after 1984. Write a formula $g(x)$ that computes AIDS deaths during year x, where x is the actual year.

Using Transformations to Model Motion

101. Computer Graphics (Refer to Example 9.) Suppose that the airplane in Figure 3.95 is flying at 0.2 kilometer per second to the left, rather than to the right. If the position of the airplane is fixed at $(-1, 5)$, graph the image of the mountain and the position of the airplane after 15 seconds.

102. Computer Graphics (Refer to Example 9.) Suppose that the airplane in Figure 3.95 is traveling to the right at 0.1 kilometer per second and gaining altitude at 0.05 kilometer per second. If the airplane's position is fixed at $(-1, 5)$, graph the image of the mountain and the position of the airplane after 20 seconds.

103. Modeling a Weather Front Suppose a cold front passing through the United States at noon, has a shape described by the function $y = \frac{1}{20}x^2$. Each unit represents 100 miles. Des Moines, Iowa, is located at $(0, 0)$, and the positive y-axis points north. See the figure at the top of the next column.

(a) If the cold front moves south at 40 miles per hour and retains its present shape, graph its new location at 4 P.M.

(b) Suppose that by midnight the vertex of the front, which is maintaining the same shape, has moved 250 miles south and 210 miles east of Des Moines. Columbus, Ohio, is located approximately 550 miles east and 80 miles south of Des Moines. Plot the locations of Des Moines and Columbus together with the new position of the cold front. Determine whether the cold front has reached Columbus by midnight.

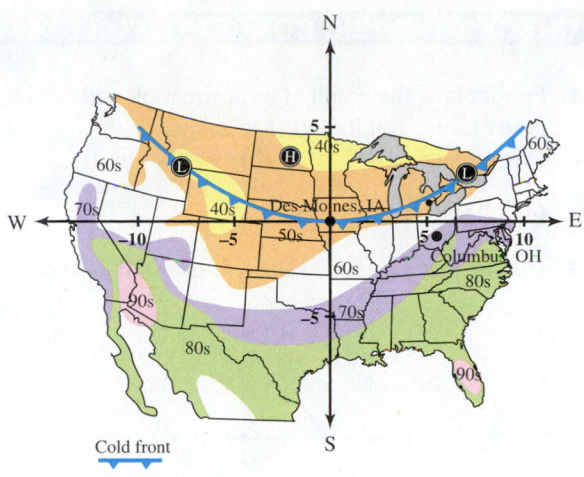

Cold front

104. Modeling Motion The first figure below is a picture composed of lines and curves. In this exercise we will model only the semicircle that outlines the top of the silo. In order to make it appear that the person is walking to the right, the background must be translated horizontally to the left, as shown in the second figure.

The semicircle at the top of the silo in the first figure is described by $f(x) = \sqrt{9 - x^2} + 12$.

(a) Graph f in the window $[-12, 12, 1]$ by $[0, 16, 1]$.

(b) To give the illusion that the person is walking to the right at 2 units per second, graph the top of the silo after 1 second and after 4 seconds.

Writing about Mathematics

105. Explain how to graph the reflection of $y = f(x)$ across the x-axis. Give an example.

106. Let c be a positive number. Explain how to shift the graph of $y = f(x)$ upward, downward, left, or right c units. Give examples.

107. If the graph of $y = f(x)$ undergoes a vertical stretch or shrink to become the graph of $y = g(x)$, do these two graphs have the same x-intercepts? y-intercepts? Explain your answers.

108. If the graph of $y = f(x)$ undergoes a horizontal stretch or shrink to become the graph of $y = g(x)$, do these two graphs have the same x-intercepts? y-intercepts? Explain your answers.

CHECKING BASIC CONCEPTS FOR SECTION 3.5

1. Predict how the graph of each equation will appear compared to the graph of $f(x) = x^2$.
   (a) $y = (x + 4)^2$ (b) $y = x^2 - 3$

   (c) $y = (x - 5)^2 + 3$

2. Use the graph shown to sketch a graph of each equation.

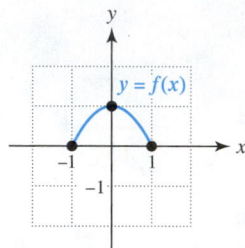

   (a) $y = -2f(x)$ (b) $y = f\left(-\frac{1}{2}x\right)$

   (c) $y = f(x - 1) + 1$

3. Write an equation that transforms the graph of $f(x) = x^2$ in the desired ways. Do not simplify.
   (a) Right 3 units, downward 4 units

   (b) Reflected about the x-axis

   (c) Shifted left 6 units, reflected about the y-axis

   (d) Reflected about the y-axis, shifted left 6 units

4. Use transformations to sketch a graph of the equation $y = \sqrt{x + 1} - 2$.

5. Use the table for $f(x)$ to make tables for $g(x)$ and $h(x)$.
   (a) $g(x) = f(x - 2) + 3$ (b) $h(x) = -2f(x + 1)$

x	-4	-2	0	2	4
$f(x)$	1	3	6	8	9

3 Summary

CONCEPT	EXPLANATION AND EXAMPLES
Section 3.1 Quadratic Functions and Models	

Quadratic Function

General form: $f(x) = ax^2 + bx + c, \quad a \neq 0$

Examples: $f(x) = x^2$ and $f(x) = -3x^2 + x + 5$

Parabola

The graph of a quadratic function is a parabola.

Vertex form: $f(x) = a(x - h)^2 + k$ (standard form for a parabola with a vertical axis)
Leading coefficient: a;   *vertex*: (h, k);   *axis of symmetry*: $x = h$

Example:

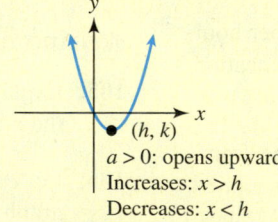

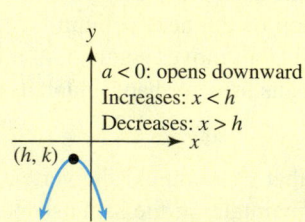

Completing the Square to Find the Vertex

The vertex of a parabola can be found by completing the square.

Example:

$$y = x^2 - 4x + 1$$
$$y - 1 = x^2 - 4x \qquad \text{Subtract 1.}$$
$$y - 1 + 4 = x^2 - 4x + 4 \qquad \text{Add } \left(\frac{-4}{2}\right)^2 = 4.$$
$$y + 3 = (x - 2)^2 \qquad \text{Perfect square trinomial}$$
$$y = (x - 2)^2 - 3 \qquad \text{Subtract 3.}$$

The vertex is $(2, -3)$.

CONCEPT	EXPLANATION AND EXAMPLES

Section 3.1 Quadratic Functions and Models (CONTINUED)

Vertex Formula

x-coordinate: $x = -\dfrac{b}{2a}$; y-coordinate: $y = f\left(-\dfrac{b}{2a}\right)$

Example: $f(x) = 2x^2 + 4x - 4$

$$x = -\frac{4}{2(2)} = -1, \quad y = 2(-1)^2 + 4(-1) - 4 = -6$$

Vertex: $(-1, -6)$

Section 3.2 Quadratic Equations and Problem Solving

Quadratic Equation

Can be written as $ax^2 + bx + c = 0, \quad a \neq 0$
A quadratic equation can have zero, one, or two real solutions.

Examples: $x^2 + 1 = 0,$ $x^2 + 2x + 1 = 0,$ and $x(x - 1) = 20$
                Zero solutions   *One solution*   *Two solutions*

Factoring

Write an equation in the form $ab = 0$ and apply the zero-product property.

Example:

$x^2 - 3x = -2$	*Given equation*
$x^2 - 3x + 2 = 0$	*Set equal to 0.*
$(x - 1)(x - 2) = 0$	*Factor.*
$x = 1$ or $x = 2$	*Zero-product property*

Square Root Property

If $x^2 = k$ and $k \geq 0$, then $x = \pm\sqrt{k}$.

Example: $x^2 = 16$ implies $x = \pm 4$.

Completing the Square

If $x^2 + kx = d$, then add $\left(\dfrac{k}{2}\right)^2$ to each side.

Example:

$x^2 - 4x = 2$	$k = -4$
$x^2 - 4x + 4 = 2 + 4$	Add $\left(\dfrac{-4}{2}\right)^2 = 4.$
$(x - 2)^2 = 6$	*Perfect square trinomial*
$x - 2 = \pm\sqrt{6}$	*Square root property*
$x = 2 \pm \sqrt{6}$	*Add 2.*

Quadratic Formula

$$x = \frac{-b \pm \sqrt{b^2 - 4ac}}{2a}$$

Always works to solve
$ax^2 + bx + c = 0$

Example: $2x^2 - 5x - 3 = 0$

$$x = \frac{-(-5) \pm \sqrt{(-5)^2 - 4(2)(-3)}}{2(2)} = \frac{5 \pm 7}{4} = 3, -\frac{1}{2}$$

CONCEPT	EXPLANATION AND EXAMPLES

Section 3.2 Quadratic Equations and Problem Solving (CONTINUED)

Discriminant

The number of real solutions to $ax^2 + bx + c = 0$ with $a \neq 0$ can be found by evaluating the discriminant, $b^2 - 4ac$.

1. If $b^2 - 4ac > 0$, there are two real solutions.

2. If $b^2 - 4ac = 0$, there is one real solution.

3. If $b^2 - 4ac < 0$, there are no real solutions (two complex solutions).

Section 3.3 Complex Numbers

Imaginary Unit

$$i = \sqrt{-1}, \qquad i^2 = -1$$

Examples: $\sqrt{-4} = 2i, \ \sqrt{-7} = i\sqrt{7}$

$$\sqrt{-3} \cdot \sqrt{-27} = i\sqrt{3} \cdot i\sqrt{27} = i^2\sqrt{81} = -9$$

Complex Number

$a + bi$, where a and b are real numbers (standard form)

Complex numbers include all real numbers. We can add, subtract, multiply, and divide complex numbers.

Examples:

$(2 - 3i) + (1 + 5i) = (2 + 1) + (-3 + 5)i = 3 + 2i$ (Add)

$3i - (2 + i) = -2 + (3 - 1)i = -2 + 2i$ (Subtract)

$(3 - i)(1 + 2i) = 3(1) + 3(2i) - i(1) - i(2i) = 5 + 5i$ (Multiply)

$\dfrac{1 - i}{2 + i} = \dfrac{(1 - i)(2 - i)}{(2 + i)(2 - i)} = \dfrac{1 - 3i}{5} = \dfrac{1}{5} - \dfrac{3}{5}i$ (Divide)

Complex Conjugate

The conjugate of $a + bi$ is $a - bi$.

Examples:

Number	$5 - 2i$	$5i$	-7	$-1 + 4i$
Conjugate	$5 + 2i$	$-5i$	-7	$-1 - 4i$

Complex Solutions

The quadratic formula can be used to solve quadratic equations with complex solutions.

Example: The solutions to $x^2 - x + 2 = 0$ are

$$x = \frac{1 \pm \sqrt{(-1)^2 - 4(1)(2)}}{2(1)} = \frac{1}{2} \pm i\frac{\sqrt{7}}{2}.$$

Section 3.4 Quadratic Inequalities

Quadratic Inequality

$ax^2 + bx + c < 0$ with $a \neq 0$, where $<$ may be replaced by $\leq$, $>$, or $\geq$.

Example: $3x^2 - x + 1 \leq 0$

CONCEPT	EXPLANATION AND EXAMPLES

Section 3.4 Quadratic Inequalities (CONTINUED)

Graphical Solution

Graph $y = ax^2 + bx + c$ and find the x-intercepts; then determine x-values where the inequality is satisfied.

Example: Solve $-x^2 - x + 2 > 0$.

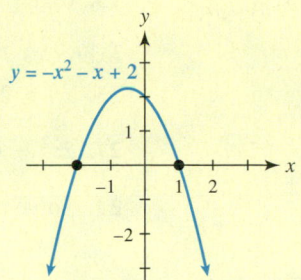

The x-intercepts are -2 and 1.

Solution set is $\{x \mid -2 < x < 1\}$, or $(-2, 1)$ in interval notation.

Symbolic Solution

First solve $ax^2 + bx + c = 0$ and use a table of values or a number line to determine the x-intervals where the inequality is satisfied.

Example: Solve $x^2 - 4 \geq 0$.

$x^2 - 4 = 0$ implies $x = \pm 2$.

Solution set is $\{x \mid x \leq -2 \text{ or } x \geq 2\}$, or $(-\infty, -2] \cup [2, \infty)$.

Interval	Test Value x	$x^2 - 4$	Positive or Negative?
$(-\infty, -2)$	-3	5	Positive
$(-2, 2)$	0	-4	Negative
$(2, \infty)$	3	5	Positive

Section 3.5 Transformations of Graphs

Vertical Shifts with $c > 0$

$y = f(x) + c$ shifts the graph of $y = f(x)$ upward c units.
$y = f(x) - c$ shifts the graph of $y = f(x)$ downward c units.

Horizontal Shifts with $c > 0$

$y = f(x - c)$ shifts the graph of $y = f(x)$ to the right c units.
$y = f(x + c)$ shifts the graph of $y = f(x)$ to the left c units.

Vertical Stretching and Shrinking

$y = cf(x)$ vertically stretches the graph of $y = f(x)$ when $c > 1$ and shrinks the graph when $0 < c < 1$.

Horizontal Stretching and Shrinking

$y = f(cx)$ horizontally shrinks the graph of $y = f(x)$ when $c > 1$ and stretches the graph when $0 < c < 1$.

Reflections

$y = -f(x)$ is a reflection of $y = f(x)$ across the x-axis.
$y = f(-x)$ is a reflection of $y = f(x)$ across the y-axis.

Review Exercises

Exercises 1 and 2: Use the graph to find the following.

(a) *Sign of the leading coefficient*
(b) *Vertex*
(c) *Axis of symmetry*
(d) *Intervals where f is increasing and where f is decreasing*

1.

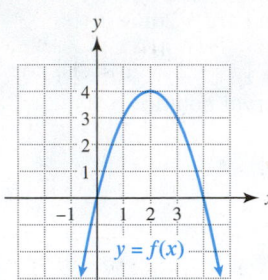

2.

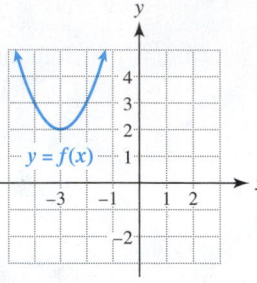

Exercises 3 and 4: Write f(x) in the form f(x) = ax² + bx + c, and identify the leading coefficient.

3. $f(x) = -2(x - 5)^2 + 1$ **4.** $f(x) = \frac{1}{3}(x + 1)^2 - 2$

Exercises 5 and 6: Use the graph of the quadratic function f to write it as f(x) = a(x − h)² + k.

5.

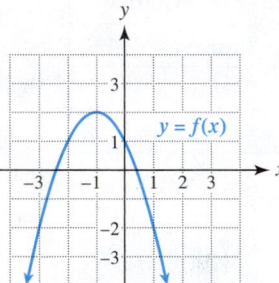

6.

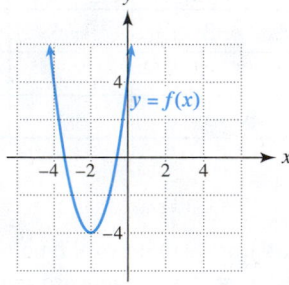

Exercises 7 and 8: Write f(x) in the form f(x) = a(x − h)² + k, and identify the vertex.

7. $f(x) = x^2 + 6x - 1$ **8.** $f(x) = 2x^2 + 4x - 5$

Exercises 9 and 10: Use the vertex formula to determine the vertex on the graph of f.

9. $f(x) = -3x^2 + 2x - 4$ **10.** $f(x) = x^2 + 8x - 5$

Exercises 11–14: Sketch a graph of the function.

11. $f(x) = -3x^2 + 3$ **12.** $g(x) = 2(x - 1)^2 - 3$

13. $f(x) = -|x + 3|$ **14.** $f(x) = \sqrt{2 - x}$

15. Average Rate of Change Find the average rate of change of $f(x) = -6x^2 + 7x + 5$ from 2 to 4.

16. Difference Quotient Find the difference quotient for $f(x) = x^2 - 2x$.

Exercises 17–24: Solve the quadratic equation.

17. $x^2 - x - 20 = 0$ **18.** $-5x^2 - 3x = 0$

19. $4z^2 - 7 = 0$ **20.** $25z^2 = 9$

21. $-2t^2 - 3t + 14 = 0$ **22.** $x(6 - x) = -16$

23. $0.1x^2 - 0.3x = 1$ **24.** $(k + 2)^2 = 7$

Exercises 25–28: Solve by completing the square.

25. $x^2 + 2x = 5$ **26.** $x^2 - 3x = 3$

27. $2z^2 - 6z - 1 = 0$ **28.** $-\frac{1}{4}x^2 - \frac{1}{2}x + 1 = 0$

29. Solve the equation $2x^2 - 3y^2 = 6$ for y. Is y a function of x?

30. Solve $h = -\frac{1}{2}gt^2 + 100$ for t.

31. Use the imaginary unit i to simplify each expression.

(a) $\sqrt{-16}$ (b) $\sqrt{-48}$ (c) $\sqrt{-5} \cdot \sqrt{-15}$

32. Write each expression in standard form.

(a) $(2 - 3i) + (-3 + 3i)$ (c) $(3 + 2i)(-4 - i)$

(b) $(-5 + 3i) - (-3 - 5i)$ (d) $\dfrac{3 + 2i}{2 - i}$

Exercises 33 and 34 Use the graph and the given f(x) to complete the following.

(a) *Find any x-intercepts.*
(b) *Find the complex zeros of f.*

33.

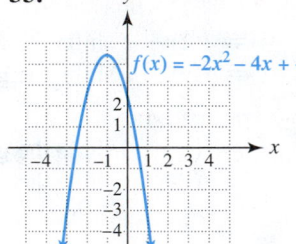

34.

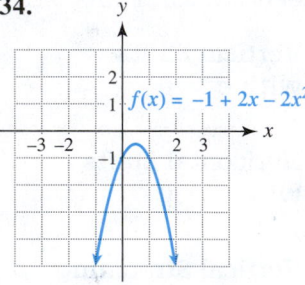

Exercises 35 and 36: Find all complex solutions.

35. $4x^2 + 9 = 0$ **36.** $2x^2 + 3 = 2x$

37. Use the graph of $y = f(x)$ to solve the inequality.
(a) $f(x) > 0$ (b) $f(x) \le 0$

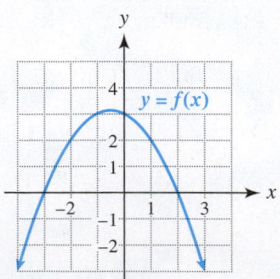

38. Solve the equation or inequality.
  (a) $x^2 - 3x + 2 = 0$ (b) $x^2 - 3x + 2 < 0$
  (c) $x^2 - 3x + 2 > 0$

Exercises 39–42: Solve the inequality. Use set-builder or interval notation to write a solution set to the inequality.

39. $x^2 - 3x + 2 \leq 0$ **40.** $9x^2 - 4 > 0$

41. $n(n - 2) \geq 15$ **42.** $n^2 + 4 \leq 6n$

43. If $f(x) = 2x^2 - 3x + 1$, use transformations to graph $y = -f(x)$ and $y = f(-x)$.

44. Use the given graph of $y = f(x)$ to sketch a graph of each expression.
  (a) $y = f(x + 1) - 2$
  (b) $y = -2f(x)$
  (c) $y = f(2x)$

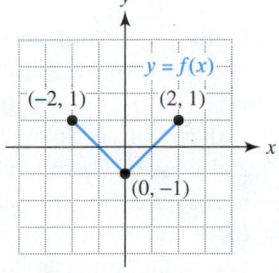

Exercises 45–48: Use transformations to sketch a graph of f.

45. $f(x) = x^2 - 4$ **46.** $f(x) = -4\sqrt{-x}$

47. $f(x) = -2(x - 2)^2 + 3$ **48.** $f(x) = -|x - 3|$

Applications

49. Maximizing Area A homeowner has 44 feet of fence to enclose a rectangular garden. One side of the garden needs no fencing because it is along the wall of the house. What dimensions will maximize area?

50. Maximizing Revenue The revenue R in dollars received from selling x radios is $R(x) = x(90 - x)$.
  (a) Evaluate $R(20)$ and interpret the result.
  (b) What number of radios sold will maximize revenue?
  (c) What is the maximum revenue?
  (d) What number of radios should be sold for revenue to be $2000 or more?

51. Projectile A slingshot is used to propel a stone upward so that its height h in feet after t seconds is given by $h(t) = -16t^2 + 88t + 5$.
  (a) Evaluate $h(0)$ and interpret the result.
  (b) How high was the stone after 2 seconds?
  (c) Find the maximum height of the stone.
  (d) At what time(s) was the stone 117 feet high?

52. World Population The function given by the formula $f(x) = 0.000478x^2 - 1.813x + 1720.1$ models world population in billions from 1950 to 2010 during year x.
  (a) Evaluate $f(1985)$ and interpret the result.
  (b) Estimate world population during the year 2000.
  (c) According to this model, when did world population reach 7 billion?

53. Construction A box is being constructed by cutting 3-inch squares from the corners of a rectangular sheet of metal that is 4 inches longer than it is wide. If the box is to have a volume of 135 cubic inches, find the dimensions of the metal sheet.

54. Room Prices Room prices are regularly $100, but for each additional room rented by a group, the price is reduced by $3 for each room. For example, 1 room costs $100, 2 rooms cost $2 \times \$97 = \$194$, and so on.
  (a) Write a quadratic function C that gives the total cost of renting x rooms.
  (b) What is the total cost of renting 6 rooms?
  (c) How many rooms are rented if the cost is $730?
  (d) What number of rooms rented gives the greatest cost?

55. Minutes Spent on Facebook The following table gives estimates for the total number of minutes in billions spent on Facebook per year. Find a function in the form $M(x) = a(x - h)^2 + k$ that models this data.

Year	2007	2008	2009
Minutes	60	84	184

Source: Business Insider.

56. Irrigation and Yield The table shows how irrigation of rice crops affects yield, where x represents the percent of total area that is irrigated and y is the rice yield in tons per hectare. (1 hectare ≈ 2.47 acres.)

x	0	20	40	60	80	100
y	1.6	1.8	2.2	3.0	4.5	6.1

Source: D. Grigg, The World Food Problem.

  (a) Use least-squares regression to find a quadratic function that models the data.
  (b) Solve the equation $f(x) = 3.7$. Interpret the results.

Extended and Discovery Exercises

1. **Shooting a Foul Shot** (Refer to the introduction to this chapter.) When a basketball player shoots a foul shot, the ball follows a parabolic arc. This arc depends on both the angle and velocity with which the basketball is released. If a person shoots the basketball overhand from a position 8 feet above the floor, then the path can sometimes be modeled by the parabola

$$y = \frac{-16x^2}{0.434v^2} + 1.15x + 8,$$

where v is the velocity of the ball in feet per second, as illustrated in the figure. (**Source:** C. Rist, "The Physics of Foul Shots.")

(a) If the basketball hoop is 10 feet high and located 15 feet away, what initial velocity v should the basketball have?

(b) Check your answer from part (a) graphically. Plot the point $(0, 8)$ where the ball is released and the point $(15, 10)$ where the basketball hoop is. Does your graph pass through both points?

(c) What is the maximum height of the basketball?

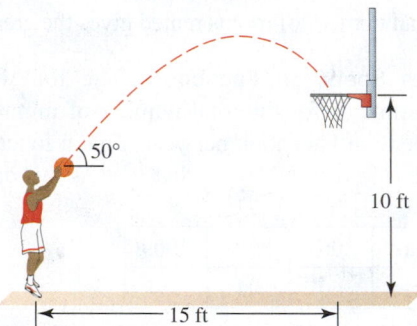

2. **Shooting a Foul Shot** (Continuation of Exercise 1) If a person releases a basketball underhand from a position 3 feet above the floor, it often has a steeper arc than if it is released overhand and the path sometimes may be modeled by

$$y = \frac{-16x^2}{0.117v^2} + 2.75x + 3.$$

See the figure below. Complete parts (a), (b), and (c) from Exercise 1. Then compare the paths for an overhand shot and an underhand shot.

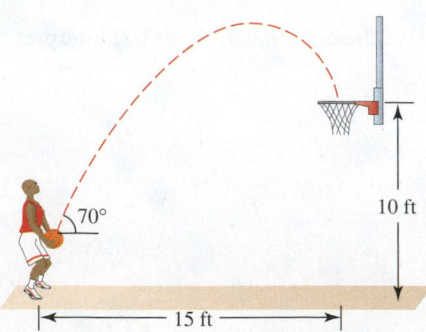

Exercises 3–6: **Reflecting Functions** *Computer graphics frequently use reflections. Reflections can speed up the generation of a picture or create a figure that appears perfectly symmetrical.* (**Source:** S. Hoggar, *Mathematics for Computer Graphics.*)

(a) *For the given $f(x)$, constant k, and viewing rectangle, graph $x = k$, $y = f(x)$, and $y = f(2k - x)$.*

(b) *Generalize how the graph of $y = f(2k - x)$ compares to the graph of $y = f(x)$.*

3. $f(x) = \sqrt{x}$, $k = 2$, $[-1, 8, 1]$ by $[-4, 4, 1]$

4. $f(x) = x^2$, $k = -3$, $[-12, 6, 1]$ by $[-6, 6, 1]$

5. $f(x) = x^4 - 2x^2 + 1$, $k = -6$, $[-15, 3, 1]$ by $[-3, 9, 1]$

6. $f(x) = 4x - x^3$, $k = 5$, $[-6, 18, 1]$ by $[-8, 8, 1]$

7. **Modeling a Cold Front** A weather map of the United States on April 22, 1996, is shown in the figure. There was a cold front roughly in the shape of a circular arc, with a radius of about 750 miles, passing north of Dallas and west of Detroit. The center of the arc was located near Pierre, South Dakota. If Pierre has the coordinates $(0, 0)$ and the positive y-axis points north, then the equation of the front can be modeled by

$$f(x) = -\sqrt{750^2 - x^2},$$

where $0 \leq x \leq 750$. (**Source:** AccuWeather, Inc.)

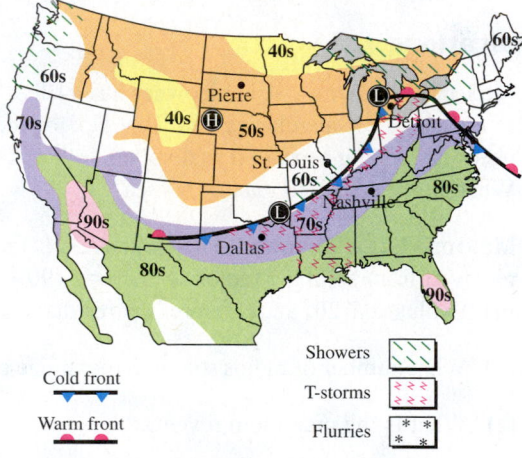

(a) St. Louis is located at $(535, -400)$ and Nashville is at $(730, -570)$, where units are in miles. Plot these points and graph f in the window $[0, 1200, 100]$ by $[-800, 0, 100]$. Did the cold front reach these cities?

(b) During the next 12 hours, the center of the front moved approximately 110 miles south and 160 miles east. Assuming the cold front did not change shape, use transformations of graphs to determine an equation that models its new location.

(c) Use graphing to determine visually if the cold front reached both cities.

4

More Nonlinear Functions and Equations

Mathematics can be both abstract and applied. Abstract mathematics is focused on axioms, theorems, and proofs that can be derived independently of empirical evidence. Theorems that were proved centuries ago are still valid today. In this sense, abstract mathematics transcends time. Yet, even though mathematics can be developed in an abstract setting—separate from science and all measured data—it also has countless applications.

There is a common misconception that theoretical mathematics is unimportant, yet many of the ideas that eventually had great practical importance were first born in the abstract. For example, in 1854 George Boole published *Laws of Thought,* which outlined the basis for Boolean algebra. This was 85 years before the invention of the first digital computer. However, Boolean algebra became the basis on which modern computer hardware operates.

Much like Boolean algebra, the topic of complex numbers was at first theoretical. However, today complex numbers are used in the design of electrical circuits, ships, and airplanes.

In this chapter we discuss some important topics in algebra that have had an impact on society. We are privileged to read in a few hours what took people centuries to discover. To ignore either the abstract beauty or the profound applicability of mathematics is like seeing a rose but never smelling one.

4.1 More Nonlinear Functions and Their Graphs

- Learn terminology about polynomial functions
- Find extrema of a function
- Identify symmetry in a graph of a function
- Determine if a function is odd, even, or neither

Introduction

Monthly average high temperatures at Daytona Beach are shown in Table 4.1.

Monthly Average High Temperatures at Daytona Beach

Month	1	2	3	4	5	6	7	8	9	10	11	12
Temperature (°F)	69	70	75	80	85	88	90	89	87	81	76	70

Table 4.1 *Source:* J. Williams, *The USA Weather Almanac.*

Figure 4.1 shows a scatterplot of the data. A linear function would not model these data because these data do not lie on a line. One possibility is to model the data with a quadratic function, as shown in Figure 4.2. However, a better fit can be obtained with the nonlinear function f whose graph is shown in Figure 4.3 and is given by

$$f(x) = 0.0145x^4 - 0.426x^3 + 3.53x^2 - 6.23x + 72,$$

A Polynomial Function

where $x = 1$ corresponds to January, $x = 2$ to February, and so on. (Least-squares regression was used to determine $f(x)$.) Function f is a *polynomial function* with degree 4.

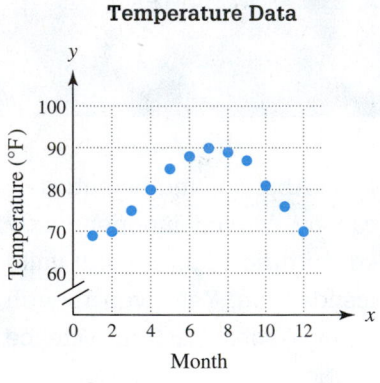

Figure 4.1

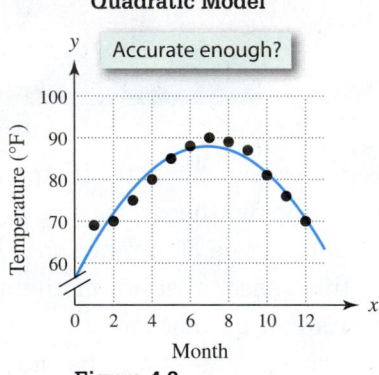

Figure 4.2

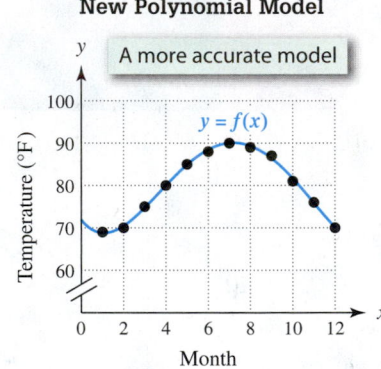

Figure 4.3

Polynomial Functions

The domain of a polynomial function is all real numbers, and its graph is continuous and smooth without breaks or sharp edges.

POLYNOMIAL FUNCTION

A **polynomial function f of degree n in the variable x** can be represented by

$$f(x) = a_n x^n + a_{n-1}x^{n-1} + \cdots + a_2 x^2 + a_1 x + a_0,$$

where each coefficient a_k is a real number, $a_n \neq 0$, and n is a nonnegative integer. The **leading coefficient** is a_n and the **degree** is n.

Algebra Review
To review polynomials, see Chapter R
(page R-12).

Examples of polynomial functions include the following.

Formula	Degree	Leading Coefficient
$f(x) = 10$	0	$a_0 = 10$
$g(x) = 2x - 3.7$	1	$a_1 = 2$
$h(x) = 1 - 1.4x + 3x^2$	2	$a_2 = 3$
$k(x) = -\dfrac{1}{2}x^6 + 4x^4 + x$	6	$a_6 = -\dfrac{1}{2}$

A polynomial function of degree 2 or higher is a *nonlinear* function. Functions f and g are linear, whereas functions h and k are nonlinear.

NOTE Quadratic functions, which were discussed in Chapter 3, are examples of nonlinear functions. This chapter introduces *more* nonlinear functions.

Functions that contain radicals, ratios, or absolute values of variables are not polynomials. For example, $f(x) = 2\sqrt{x}$, $g(x) = \frac{1}{x-1}$, and $h(x) = |2x + 5|$ are *not* polynomials. However, they *are* nonlinear functions.

Identifying Extrema

In Figure 4.3 the minimum monthly average temperature of 69°F occurs in January ($x = 1$) and the maximum monthly average temperature of 90°F occurs in July ($x = 7$). Minimum and maximum y-values on the graph of a function often represent important data.

Graphs of two polynomial functions with "hills" and "valleys" are shown in Figures 4.4 and 4.5. These hills and valleys are associated with maximum and minimum y-values on the graphs. The following See the Concept shows how these minimum and maximum y-values are classified.

See the Concept: Absolute and Local Maximums and Minimums

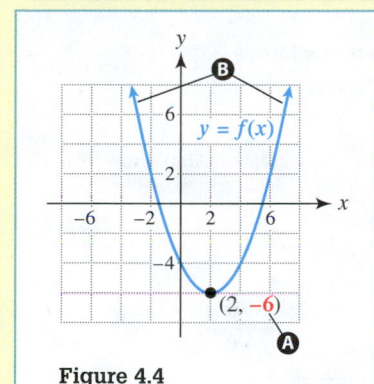

Figure 4.4

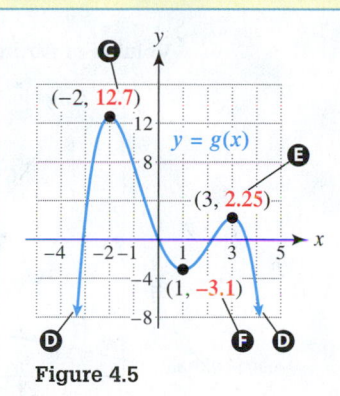

Figure 4.5

Ⓐ Absolute minimum: **−6**

Ⓑ Absolute maximum: None

Ⓒ Absolute maximum: **12.7**

Ⓓ Absolute minimum: None

Ⓔ Local maximum: **2.25**

Ⓕ Local minimum: **−3.1**

Note that the absolute minimum of −6 in Figure 4.4 is also a local minimum, and that the absolute maximum of 12.7 in Figure 4.5 is also a local maximum.

In Figure 4.4, the minimum y-value on the graph of f is **−6**. It is called the *absolute minimum* of f. Function f has no *absolute maximum* because there is no largest y-value on a parabola opening upward.

In Figure 4.5 the peak of the highest "hill" on the graph of g is (−2, **12.7**). Therefore the absolute maximum of g is **12.7**. There is a smaller peak located at the point (3, **2.25**). In a small open interval near $x = 3$, the y-value of 2.25 is locally the largest. We say that g has a *local maximum* of **2.25**. Similarly, a "valley" occurs on the graph of g, where the lowest point is (1, **−3.1**). The value **−3.1** is not the smallest y-value on the entire graph of g. Therefore it is not an absolute minimum. Rather, −3.1 is a *local minimum*.

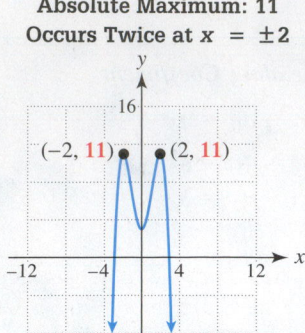

Absolute Maximum: 11
Occurs Twice at x = ±2

Figure 4.6

Maximum and minimum values that are either absolute or local are called **extrema** (plural of extremum). A function may have several local extrema, but at most one absolute maximum and one absolute minimum. However, it is possible for a function to assume an absolute extremum at two values of x. In Figure 4.6 the absolute maximum is **11**. It occurs at $x = \pm 2$.

NOTE In Figure 4.6, the absolute maximum of 11 is *also* a local maximum because it is the largest y-value in a small interval near either $x = -2$ or $x = 2$.

Sometimes an absolute maximum (minimum) is called a *global maximum* (*minimum*). Similarly, sometimes a local maximum (minimum) is called a *relative maximum* (*minimum*).

ABSOLUTE AND LOCAL EXTREMA

Let c be in the domain of f.

$f(c)$ is an **absolute (global) maximum** if $f(c) \geq f(x)$ *for all x in the domain of f.*
$f(c)$ is an **absolute (global) minimum** if $f(c) \leq f(x)$ *for all x in the domain of f.*
$f(c)$ is a **local (relative) maximum** if $f(c) \geq f(x)$ *when x is near c.*
$f(c)$ is a **local (relative) minimum** if $f(c) \leq f(x)$ *when x is near c.*

NOTE The expression "near c" means that there is an open interval in the domain of f containing c where $f(c)$ satisfies the inequality.

EXAMPLE 1   Identifying and interpreting extrema

Figure 4.7 shows the graph of a function f that models the volume of air in a person's lungs, measured in liters, after x seconds. (*Source:* V. Thomas, *Science and Sport.*)
(a) Find the absolute maximum and the absolute minimum of f. Interpret the results.
(b) Identify two local maximums and two local minimums of f. Interpret the results.

Volume of Air in a Person's Lungs

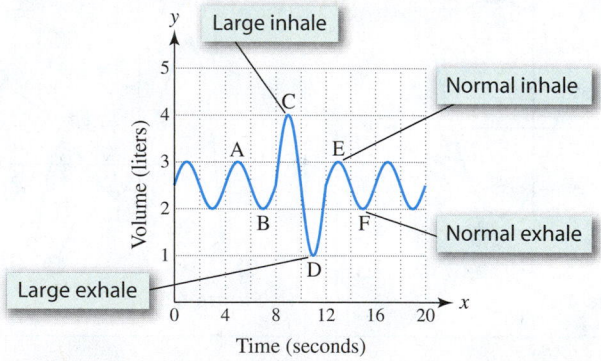

Figure 4.7

SOLUTION
(a) The absolute maximum is 4 liters and occurs at C. The absolute minimum is 1 liter and occurs at D. At C a deep breath has been taken and the lungs are more inflated. After C, the person exhales until the lungs contain only 1 liter of air at D.
(b) One local maximum is 3 liters. It occurs at A and E and represents the amount of air in a person's lungs after inhaling normally. One local minimum is 2 liters. It occurs at B and F and represents the amount of air after exhaling normally. Another local maximum is 4 liters, which is also the absolute maximum. Similarly, 1 liter is a local minimum and also the absolute minimum.

Now Try Exercise 91

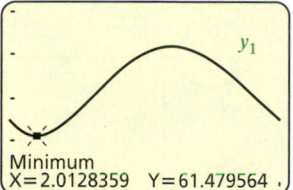

Figure 4.8

EXAMPLE 2   Identifying extrema

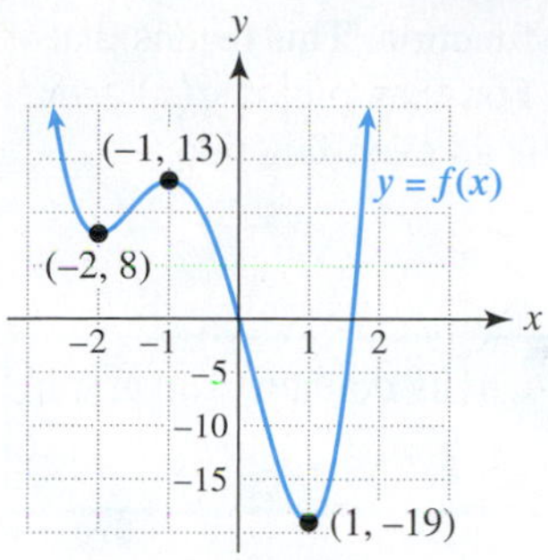

Use the graph of f in Figure 4.8 to estimate any local and absolute extrema.

SOLUTION

Local Extrema   The points $(-2, \mathbf{8})$ and $(1, -\mathbf{19})$ on the graph of f correspond to the lowest point in a "valley." Thus there are local minimums of $\mathbf{8}$ and $-\mathbf{19}$. The point $(-1, \mathbf{13})$ corresponds to the highest point on a "hill." Thus there is a local maximum of $\mathbf{13}$.

Absolute Extrema   Because the arrows point upward, there is no maximum y-value on the graph. Thus there is no absolute maximum. However, the minimum y-value on the graph of f occurs at the point $(1, -\mathbf{19})$. The absolute minimum is $-\mathbf{19}$.

> **Now Try Exercise 15**

> **NOTE**   Extrema are y-values on the graph of a function, not x-values.

EXAMPLE 3   Modeling ocean temperatures

The monthly average ocean temperature in degrees Fahrenheit at Bermuda can be modeled by $f(x) = 0.0215x^4 - 0.648x^3 + 6.03x^2 - 17.1x + 76.4$, where $x = 1$ corresponds to January and $x = 12$ to December. The domain of f is $D = \{x \mid 1 \le x \le 12\}$.
(a) Graph f in $[1, 12, 1]$ by $[50, 90, 10]$.
(b) Estimate the absolute extrema. Interpret the results.

[1, 12, 1] by [50, 90, 10]

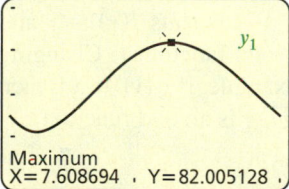

Figure 4.9

[1, 12, 1] by [50, 90, 10]

Figure 4.10

SOLUTION
(a) The graph of $y_1 = f(x)$ is shown in Figure 4.9.
(b) Many graphing calculators have the capability to find maximum and minimum y-values. The points associated with absolute extrema are shown in Figures 4.9 and 4.10. An absolute minimum of about 61.5 corresponds to the point $(2.01, 61.5)$. This means that the monthly average ocean temperature is coldest during the month of February ($x \approx 2$) when it reaches a minimum of about 61.5°F.

  An absolute maximum of approximately 82 corresponds to $(7.61, 82.0)$. Rounding, we might say that the warmest average temperature occurs during August ($x \approx 8$) when it reaches a maximum of 82°F. (Or we might say that this maximum occurs in late July, since $x \approx 7.61$.)

> **Now Try Exercise 95**

Symmetry

Even Functions   Symmetry is used frequently in art, mathematics, science, and computer graphics. Many objects and animals are symmetric along a line so that the left and right sides are mirror images. For example, if a butterfly is viewed from the top, the left wing is typically a mirror image of the right wing. Graphs of functions may also exhibit this type of symmetry, as shown in Figures 4.11–4.13. These graphs are **symmetric with respect to the y-axis**. A function whose graph satisfies this characteristic is called an *even function*.

Even Functions: Symmetry with Respect to the y-Axis

If these graphs are folded on the y-axis, the two halves match.

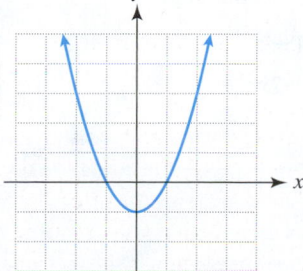

Figure 4.11

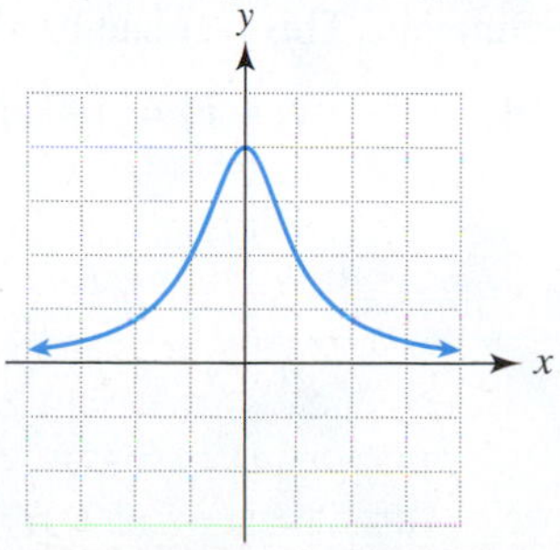

Figure 4.12

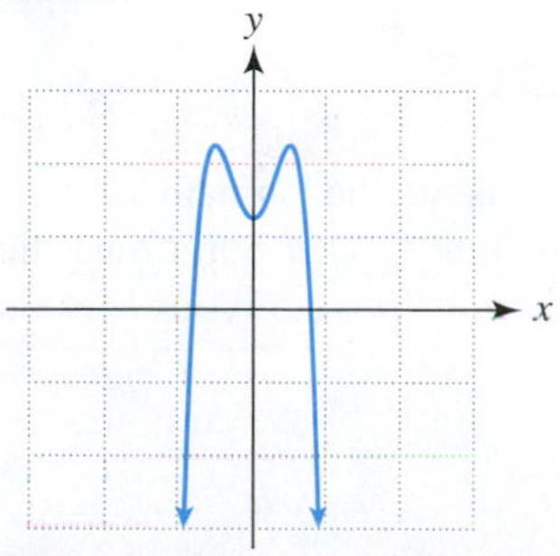

Figure 4.13

An Even Function

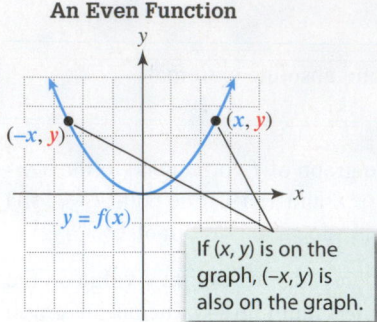

If (x, y) is on the graph, $(-x, y)$ is also on the graph.

Figure 4.14

Figure 4.14 shows a graph of an even function f. Since the graph is symmetric with respect to the y-axis, the points (x, y) and $(-x, y)$ both lie on the graph of f. Thus $f(x) = y$ and $f(-x) = y$, and so $f(x) = f(-x)$ for an even function. This means that if we change the sign of the input, the output does not change. For example, if $g(x) = x^2$, then $g(2) = g(-2) = 4$. Since this is true for *every input*, g is an even function.

> **EVEN FUNCTION**
>
> A function f is an **even function** if $f(-x) = f(x)$ for every x in its domain. The graph of an even function is symmetric with respect to the y-axis.

Odd Functions A second type of symmetry is shown in Figures 4.15–4.17. If we could spin or rotate the graph about the origin, the original graph would reappear after half a turn. These graphs are **symmetric with respect to the origin** and represent *odd functions*.

Odd Functions: Symmetry with Respect to the Origin

If these graphs are spun 180° around the origin, they match.

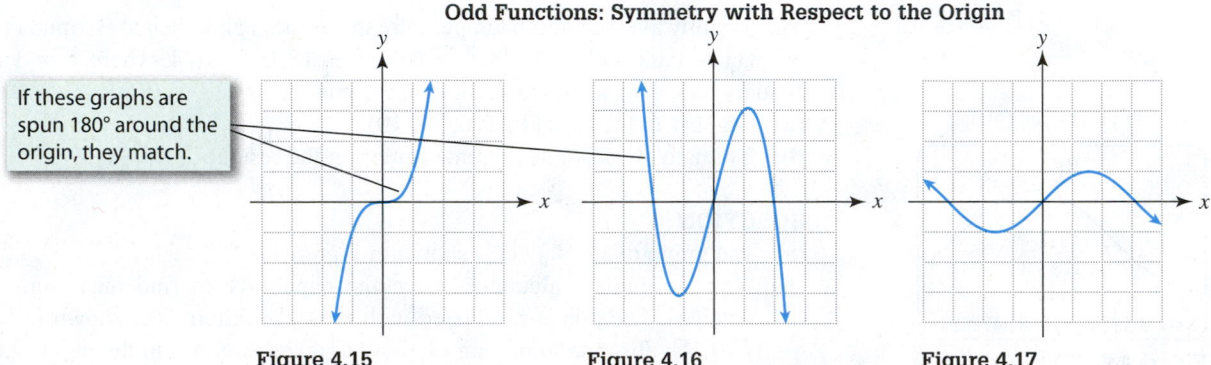

Figure 4.15   **Figure 4.16**   **Figure 4.17**

An Odd Function

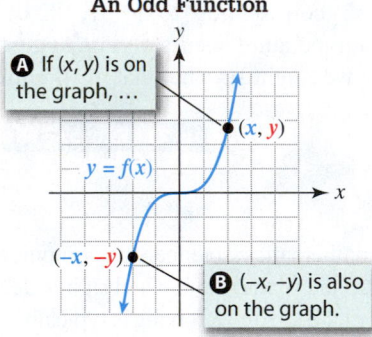

Ⓐ If (x, y) is on the graph, …

Ⓑ $(-x, -y)$ is also on the graph.

Figure 4.18

In Figure 4.18 the point (x, y) lies on the graph of an odd function f. If this point spins half a turn, or 180°, around the origin, its new location is $(-x, -y)$. Thus $f(x) = y$ and $f(-x) = -y$. It follows that $f(-x) = -y = -f(x)$ for any odd function f. Changing the sign of the input only changes the sign of the output. For example, if $g(x) = x^3$, then $g(3) = 27$ and $g(-3) = -27$. Since this is true for *every input*, g is an odd function.

> **ODD FUNCTION**
>
> A function f is an **odd function** if $f(-x) = -f(x)$ for every x in its domain. The graph of an odd function is symmetric with respect to the origin.

Identifying Odd and Even Functions The terms *odd* and *even* have special meaning when they are applied to a polynomial function f. If $f(x)$ contains terms that have only odd powers of x, then f is an odd function. Similarly, if $f(x)$ contains terms that have only even powers of x (and possibly a constant term), then f is an even function. For example, $f(x) = x^6 - 4x^4 - 2x^2 + 5$ is an **even** function, whereas $g(x) = x^5 + 4x^3$ is an **odd** function. This fact can be shown *symbolically*.

$$f(-x) = (-x)^6 - 4(-x)^4 - 2(-x)^2 + 5 \qquad \text{Substitute } -x \text{ for } x.$$
$$= x^6 - 4x^4 - 2x^2 + 5 \qquad \text{Simplify.}$$
$$= f(x) \qquad f \text{ is an even function.}$$

$$g(-x) = (-x)^5 + 4(-x)^3 \qquad \text{Substitute } -x \text{ for } x.$$
$$= -x^5 - 4x^3 \qquad \text{Simplify.}$$
$$= -g(x) \qquad g \text{ is an odd function.}$$

CLASS DISCUSSION

If 0 is in the domain of an odd function f, what point must lie on its graph? Explain your reasoning.

NOTE It is important to remember that the graphs of many functions exhibit *no symmetry* with respect to either the *y*-axis or the origin. These functions are *neither* odd *nor* even.

EXAMPLE 4 Identifying odd and even functions

For each representation of a function f, identify whether f is odd, even, or neither.

(a)

x	-3	-2	-1	0	1	2	3
$f(x)$	10.5	2	-0.5	-2	-0.5	2	10.5

(b)

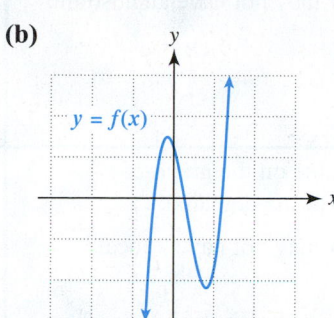

(c) $f(x) = x^3 - 5x$

(d) f is the cube root function.

SOLUTION

Getting Started If either $f(-x) = f(x)$ or its graph is symmetric with respect to the *y*-axis, then f is even. If either $f(-x) = -f(x)$ or its graph is symmetric with respect to the origin, then f is odd. Otherwise, f is neither even nor odd. ▶

(a) The function defined by the table has domain $D = \{-3, -2, -1, 0, 1, 2, 3\}$. Notice that $f(-3) = 10.5 = f(3)$, $f(-2) = 2 = f(2)$, and $f(-1) = -0.5 = f(1)$. The function f satisfies the statement $f(-x) = f(x)$ for every x in D. Thus f is an even function.

(b) If we fold the graph on the *y*-axis, the two halves do not match, so f is *not* an even function. Similarly, f is *not* an odd function since spinning its graph half a turn about the origin does not result in the same graph. The function f is neither odd nor even.

(c) Since f is a polynomial containing only odd powers of x, it is an odd function. This fact can also be shown symbolically.

$$f(-x) = (-x)^3 - 5(-x) \qquad \text{Substitute } -x \text{ for } x.$$
$$= -x^3 + 5x \qquad \text{Simplify.}$$
$$= -(x^3 - 5x) \qquad \text{Distributive property}$$
$$= -f(x) \qquad f \text{ is an odd function.}$$

(d) Note that $\sqrt[3]{-8} = -2$ and that $\sqrt[3]{8} = 2$. In general, $\sqrt[3]{-x} = -\sqrt[3]{x}$, which indicates that $f(-x) = -f(x)$, where $f(x) = \sqrt[3]{x}$. Thus f is an odd function. This fact can also be seen by graphing $f(x) = \sqrt[3]{x}$, as shown in Figure 4.19. Spinning the graph of $f(x) = \sqrt[3]{x}$ a half a turn about the origin results in the same graph.

Now Try Exercises 47, 59, 63, and 71

Algebra Review

To review cube roots, see Chapter R (page R-37).

Cube Root Function

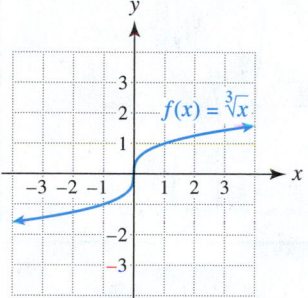

Figure 4.19

CLASS DISCUSSION

Discuss the possibility of the graph of a *function* being symmetric with respect to the *x*-axis.

4.1 Putting It All Together

The following table summarizes some important concepts related to the graphs of nonlinear functions.

CONCEPT	EXPLANATION	GRAPHICAL EXAMPLE
Absolute, or global, maximum (minimum)	The maximum (minimum) y-value on the graph of $y = f(x)$ A graph of a function may or may not have an absolute maximum (minimum).	
Local, or relative, maximum (minimum)	A maximum (minimum) y-value on the graph of $y = f(x)$ in an open interval of the domain of f A graph of a function may or may not have a local maximum (minimum). Note that it is possible for a y-value on the graph of f to be both an absolute maximum (minimum) *and* a local maximum (minimum).	
Even function	$f(-x) = f(x)$ The graph is symmetric with respect to the y-axis. If the graph is folded on the y-axis, the left and right halves match. Changing the sign of the input does not change the output.	
Odd function	$f(-x) = -f(x)$ The graph is symmetric with respect to the origin. If the graph is spun about the origin, the graph reappears after half a turn, or 180°. Changing the sign of the input only changes the sign of the output.	

4.1 Exercises

Note: Many of the answers in this section involve estimations. Your answers may vary slightly, particularly when you are reading a graph.

Polynomials

Exercises 1–10: Determine if the function is a polynomial function. If it is, state its degree and leading coefficient a.

1. $f(x) = 2x^3 - x + 5$ **2.** $f(x) = -x^4 + 1$

3. $f(x) = \sqrt{x}$ **4.** $f(x) = 2x^3 - \sqrt[3]{x}$

5. $f(x) = 1 - 4x - 5x^4$ **6.** $f(x) = 5 - 4x$

7. $g(t) = \dfrac{1}{t^2 + 3t - 1}$ **8.** $g(t) = \dfrac{1}{1 - t}$

9. $g(t) = 22$ **10.** $g(t) = |2t|$

Finding Extrema of Polynomials

Exercises 11–26: Use the graph of f to estimate the

(a) *local extrema and*

(b) *absolute extrema.*

11.
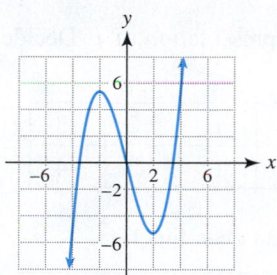

12.
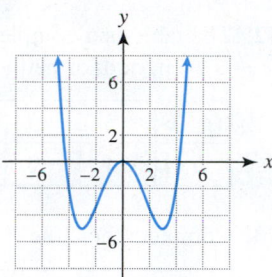

13.
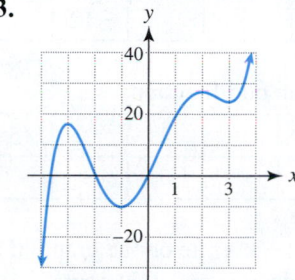

14.
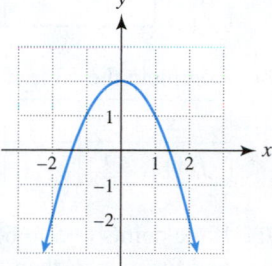

15.
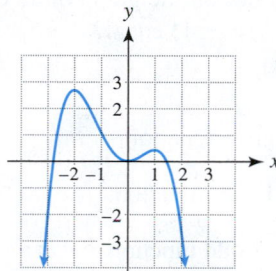

16.
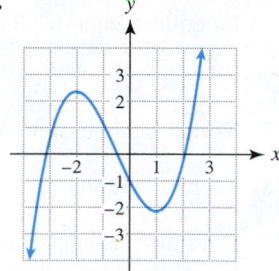

17.
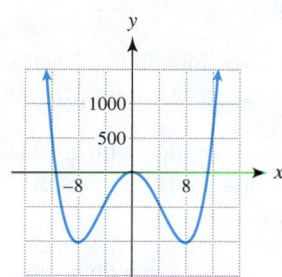

18.
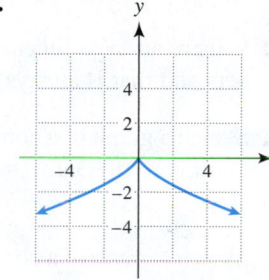

19.
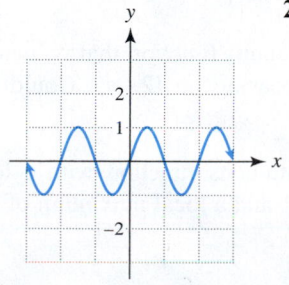

20.
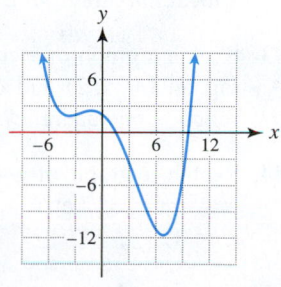

(Hint: *Exercises 21–26: Local extrema cannot occur at end-points because they only occur on open intervals.*)

21.
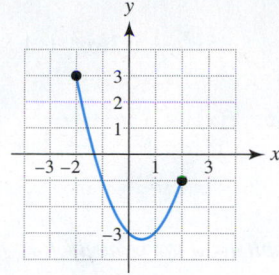

22.
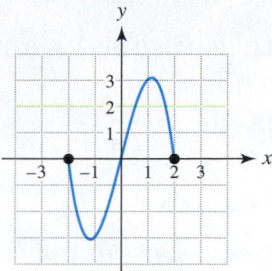

23.
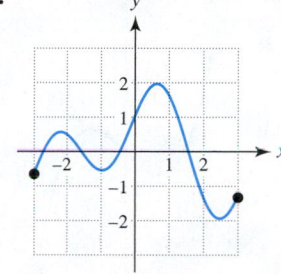

24.
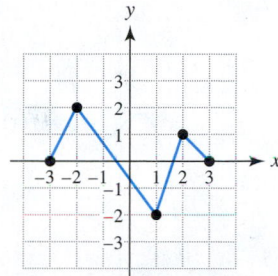

25.
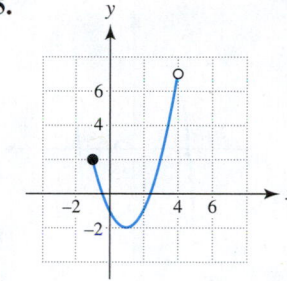

26.
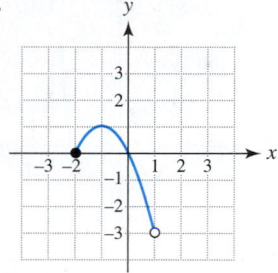

Exercises 27–38: Determine any

(a) *local extrema and*

(b) *absolute extrema.*

(Hint: *Consider the graph* $y = g(x)$.)

27. $g(x) = 1 - 3x$ **28.** $g(x) = \frac{1}{4}x$

29. $g(x) = x^2 + 1$ **30.** $g(x) = 1 - x^2$

31. $g(x) = -2(x + 3)^2 + 4$

32. $g(x) = \frac{1}{3}(x - 1)^2 - 2$

33. $g(x) = 2x^2 - 3x + 1$ **34.** $g(x) = -3x^2 + 4x - 1$

35. $g(x) = |x + 3|$ **36.** $g(x) = -|x| + 2$

37. $g(x) = \sqrt[3]{x}$ **38.** $g(x) = -x^3$

Exercises 39–46: Determine graphically any

(a) *local extrema and*

(b) *absolute extrema.*

39. $g(x) = 3x - x^3$ **40.** $g(x) = \dfrac{1}{1 + |x|}$

41. $f(x) = -3x^4 + 8x^3 + 6x^2 - 24x$

42. $f(x) = -x^4 + 4x^3 - 4x^2$

43. $f(x) = 0.5x^4 - 5x^2 + 4.5$

44. $f(x) = 0.01x^5 + 0.02x^4 - 0.35x^3 - 0.36x^2 + 1.8x$

45. $f(x) = \dfrac{8}{1 + x^2}$ 46. $f(x) = \dfrac{6}{x^2 + 2x + 2}$

Symmetry

Exercises 47–50: Use the graph to determine if f is odd, even, or neither.

47.
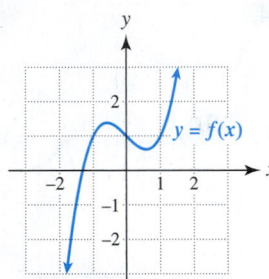

48.
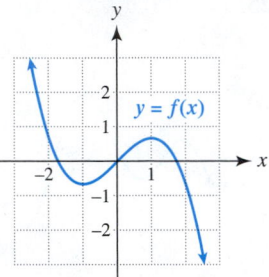

49.
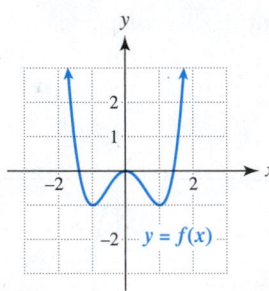

50.
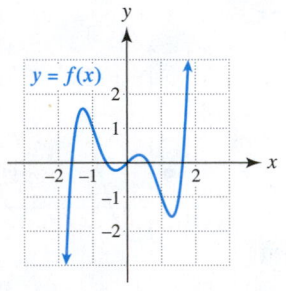

Exercises 51–70: Determine if f is odd, even, or neither.

51. $f(x) = 5x$ 52. $f(x) = -3x$

53. $f(x) = x + 3$ 54. $f(x) = 2x - 1$

55. $f(x) = x^2 - 10$ 56. $f(x) = 8 - 2x^2$

57. $f(x) = x^4 - 6x^2 + 2$ 58. $f(x) = -x^6 + 5x^2$

59. $f(x) = x^3 - 2x$ 60. $f(x) = -x^5$

61. $f(x) = x^2 - x^3$ 62. $f(x) = 3x^3 - 1$

63. $f(x) = \sqrt[3]{x^2}$ 64. $f(x) = \sqrt{-x}$

65. $f(x) = \sqrt{1 - x^2}$ 66. $f(x) = \sqrt{x^2}$

67. $f(x) = \dfrac{1}{1 + x^2}$ 68. $f(x) = \dfrac{1}{x}$

69. $f(x) = |x + 2|$ 70. $f(x) = \dfrac{1}{x + 1}$

71. The table is a complete representation of f. Decide if f is even, odd, or neither.

x	-100	-10	-1	0	1	10	100
$f(x)$	56	-23	5	0	-5	23	-56

72. The table is a complete representation of f. Decide if f is even, odd, or neither.

x	-5	-3	-1	1	2	3
$f(x)$	-4	-2	1	1	-2	-4

73. Complete the table if f is an even function.

x	-3	-2	-1	0	1	2	3
$f(x)$	21		-25			-12	

74. Complete the table if f is an odd function.

x	-5	-3	-2	0	2	3	5
$f(x)$	13		-5			-1	

75. If the points $(-5, -6)$ and $(-3, 4)$ lie on the graph of an odd function f, then what do $f(5)$ and $f(3)$ equal?

76. If the point $(1 - a, b + 1)$ lies on the graph of an even function f, then what does $f(a - 1)$ equal?

Concepts

77. Sketch a graph of an odd linear function.

78. Sketch a graph of an even linear function.

79. Does there exist a continuous odd function that is always increasing and whose graph passes through the points $(-3, -4)$ and $(2, 5)$? Explain.

80. Is there an even function whose domain is all real numbers and that is always decreasing? Explain.

81. Sketch a graph of a continuous function with an absolute minimum of -3 at $x = -2$ and a local minimum of -1 at $x = 2$.

82. Sketch a graph of a continuous function with no absolute extrema but with a local minimum of -2 at $x = -1$ and a local maximum of 2 at $x = 1$.

83. Sketch a graph of a continuous function that is increasing on $(-\infty, 2)$ and decreasing on $(2, \infty)$. Could this function be quadratic?

84. Sketch a graph of a continuous function with a local maximum of 2 at $x = -1$ and a local maximum of 0 at $x = 1$.

Translations of Graphs

Exercises 85–88: Use the graph of $f(x) = 4x - \frac{1}{3}x^3$ and translations of graphs to sketch the graph of the equation.

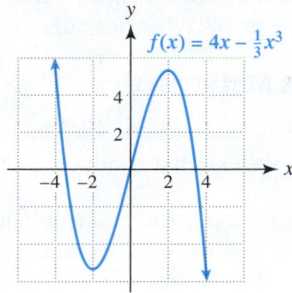

85. $y = f(x + 1)$

86. $y = f(x) - 2$

87. $y = 2f(x)$

88. $y = f\left(\frac{1}{2}x\right)$

89. If the graph of $y = f(x)$ is increasing on $(1, 4)$, then where is the graph of $y = f(x + 1) - 2$ increasing? Where is the graph of $y = -f(x - 2)$ decreasing?

90. If the graph of f is decreasing on $(0, \infty)$, then what can be said about the graph of $y = f(-x) + 1$? The graph of $y = -f(x) - 1$?

Applications

91. Temperature in Sunlight   The graph shows the temperature readings of a thermometer (on a partly cloudy day) x hours past noon.

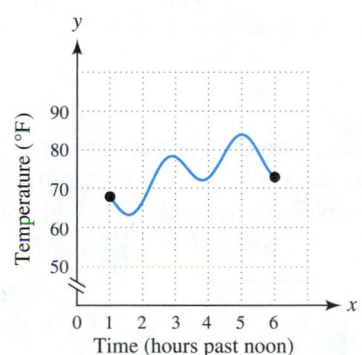

(a) Identify the absolute maximum and minimum. Interpret each.

(b) Identify any local maximums and minimums. (Do not consider the endpoints.)

(c) For what x-values was the temperature increasing?

92. Daytona Beach (Refer to the introduction to this section.) The graph at the top of the next column shows the monthly average high temperatures at Daytona Beach.

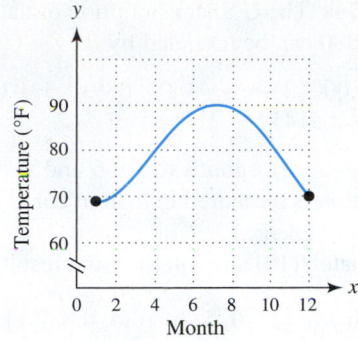

(a) Identify the absolute maximum and minimum.

(b) Identify a local maximum.

(c) For what x-values was the temperature increasing?

93. Facebook Versus Yahoo During 2009, Facebook surpassed Yahoo for the greatest number of unique monthly users. The formula

$$F(x) = 0.0484x^3 - 1.504x^2 + 17.7x + 53$$

models these numbers in millions of unique monthly users x months after December 2008.

(a) Evaluate $F(1)$ and $F(12)$. Interpret the result.

(b) Graph $y = F(x)$ in $[0, 13, 2]$ by $[0, 150, 25]$. Does F have any local extrema for $1 \leq x \leq 12$?

(c) Where is F increasing if its domain is $1 \leq x \leq 12$?

94. Google + Users Google+ experienced an increase in the number of users between July 1, 2011 and February 1, 2012. The total number of users in millions can be modeled by the polynomial function

$$G(x) = 0.000014437x^3 - 0.00406x^2 + 0.603x + 3.7,$$

where x represents days after July 1, 2011.

(a) Evaluate $G(31)$ and interpret the result.

(b) Graph $y = G(x)$ in $[0, 215, 20]$ by $[0, 90, 10]$. Does G have any local extrema for $0 \leq x \leq 215$?

(c) Where is G increasing if its domain is $0 \leq x \leq 215$?

95. Heating Costs   In colder climates the cost for natural gas to heat homes can vary from one month to the next. The polynomial function given by

$$f(x) = -0.1213x^4 + 3.462x^3 - 29.22x^2 + 64.68x + 97.69$$

models the monthly cost in dollars of heating a typical home. The input x represents the month, where $x = 1$ corresponds to January and $x = 12$ to December.

(a) Where might the absolute extrema occur for $1 \leq x \leq 12$?

(b) Graph f in $[1, 12, 1]$ by $[0, 150, 10]$. Find the absolute extrema and interpret the results.

96. Natural Gas The U.S. consumption of natural gas from 1965 to 1980 can be modeled by

$$f(x) = 0.0001234x^4 - 0.005689x^3 + 0.08792x^2 - 0.5145x + 1.514,$$

where $x = 6$ corresponds to 1966 and $x = 20$ to 1980. Consumption is measured in trillion cubic feet. (*Source:* Department of Energy.)

(a) Evaluate $f(10)$ and interpret the result.

(b) Graph f in [6, 20, 5] by [0.4, 0.8, 0.1]. Describe the energy usage during this time period.

(c) Determine the local extrema and interpret the results.

97. Average Temperature The graph approximates the monthly *average* temperatures in degrees Fahrenheit in Austin, Texas. In this graph x represents the month, where $x = 0$ corresponds to July.

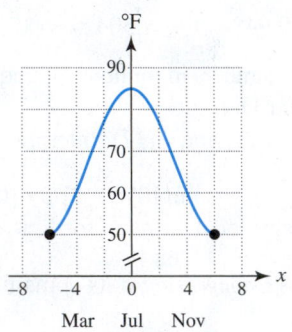

(a) Is this a graph of an odd or even function?

(b) June corresponds to $x = -1$ and August to $x = 1$. The average temperature in June is 83°F. What is the average temperature in August?

(c) March corresponds to $x = -4$ and November to $x = 4$. According to the graph, how do their average temperatures compare?

(d) Interpret what this type of symmetry implies about average temperatures in Austin.

98. Height of a Projectile When a projectile is shot into the air, it attains a maximum height and then falls back to the ground. Suppose that $x = 0$ corresponds to the time when the projectile's height is maximum. If air resistance is ignored, its height h above the ground at any time x may be modeled by $h(x) = -16x^2 + h_{max}$, where h_{max} is the projectile's maximum height above the ground. Height is measured in feet and time in seconds. Let $h_{max} = 400$ feet.

(a) Evaluate $h(-2)$ and $h(2)$. Interpret these results.

(b) Evaluate $h(-5)$ and $h(5)$. Interpret these results.

(c) Graph h for $-5 \le x \le 5$. Is h even or odd?

(d) How do the values of $h(x)$ and $h(-x)$ compare for $-5 \le x \le 5$? What does this result indicate?

Writing About Mathematics

99. Explain the difference between a local and an absolute maximum. Are extrema x-values or y-values?

100. Describe ways to determine if a polynomial function is odd, even, or neither. Give examples.

101. If an odd function f has one local maximum of 5 at $x = 3$, then what else can be said about f? Explain.

102. If an even function f has an absolute minimum of -6 at $x = -2$, then what else can be said about f? Explain.

Extended and Discovery Exercises

1. Maximizing Area Find the dimensions of the rectangle of maximum area that can be inscribed in a semicircle with radius 3. Assume that the rectangle is positioned as shown in the accompanying figure.

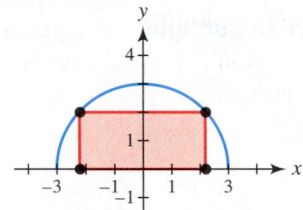

2. Minimizing Area A piece of wire 20 inches long is cut into two pieces. One piece is bent into a square and the other is bent into an equilateral triangle, as illustrated.

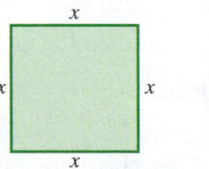 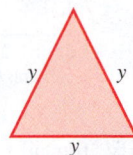

(a) Write a formula that gives the area A of the two shapes in terms of x.

(b) Find the length of wire (to the nearest tenth of an inch) that should be used for the square if the combined area of the two shapes is to be minimized?

3. Minimizing Time A person is in a rowboat 3 miles from the closest point on a straight shoreline, as illustrated in the figure. The person would like to reach a cabin that is 8 miles down the shoreline. The person can row at 4 miles per hour and jog at 7 miles per hour.

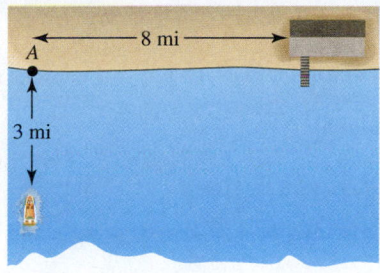

(a) How long will it take to reach the cabin if the person rows straight toward shore at point A and then jogs to the cabin?

(b) How long will it take to reach the cabin if the person rows straight to the cabin and does no jogging?

(c) Find the minimum time to reach the cabin.

4.2 Polynomial Functions and Models

- Understand the graphs of polynomial functions
- Evaluate and graph piecewise-defined functions
- Use polynomial regression to model data (optional)

Introduction

The consumption of natural gas by the United States has varied in the past. As shown in Table 4.2, energy consumption (in quadrillion Btu) increased, decreased, and then increased again. A scatterplot of the data is shown in Figure 4.20, and one possibility for a polynomial modeling function f is shown in Figure 4.21. Notice that f is neither linear nor quadratic. What degree of polynomial might we use to model these data? This question is answered in Example 4.

Natural Gas Consumption

Year	1960	1970	1980	1990	2000
Consumption	12.4	21.8	20.4	19.3	24.0

Table 4.2   *Source:* Department of Energy.

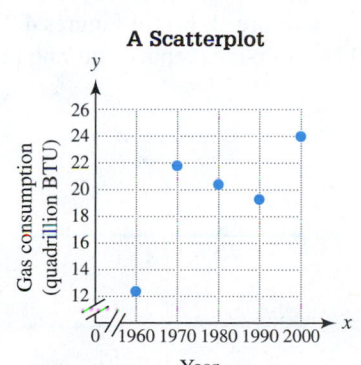

Figure 4.20

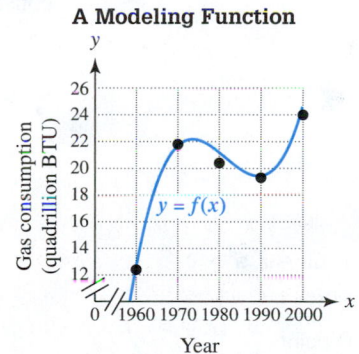

Figure 4.21

Graphs of Polynomial Functions

In Section 4.1 polynomial functions were defined. The following Making Connections helps show the relationship among polynomials, polynomial functions, and polynomial equations with one variable.

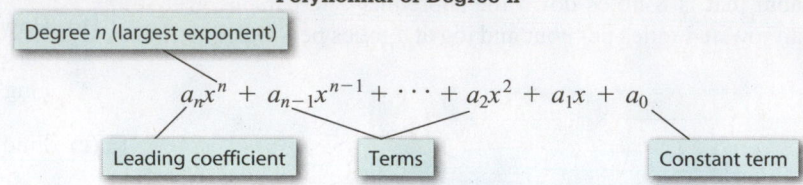

MAKING CONNECTIONS

Polynomials, Functions, and Equations

Polynomial of Degree n

Degree n (largest exponent)

$$a_n x^n + a_{n-1} x^{n-1} + \cdots + a_2 x^2 + a_1 x + a_0$$

Leading coefficient   Terms   Constant term

Example: $x^3 - 3x^2 + x - 5$ is a polynomial.

Polynomial Function of Degree n

$$f(x) = a_n x^n + a_{n-1} x^{n-1} + \cdots + a_2 x^2 + a_1 x + a_0$$

Domain: All real numbers

Graph is continuous and smooth; no breaks or sharp edges.

Example: $f(x) = x^3 - 3x^2 + x - 5$ defines a polynomial function.

Polynomial Equation of Degree n

$$a_n x^n + a_{n-1} x^{n-1} + \cdots + a_2 x^2 + a_1 x + a_0 = 0$$

At most n real solutions (See Section 4.4)

Example: $x^3 - 3x^2 + x - 5 = 0$ is a polynomial equation.

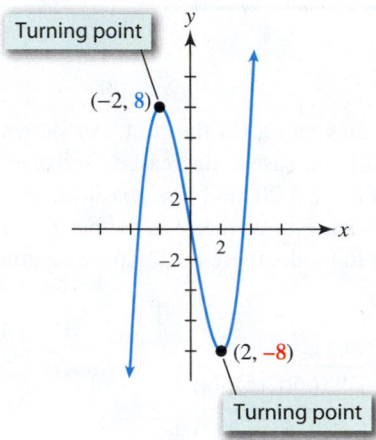

Turning point

(−2, 8)

(2, −8)

Turning point

Figure 4.22

A **turning point** occurs whenever the graph of a polynomial function changes from increasing to decreasing or from decreasing to increasing. Turning points are associated with "hills" or "valleys" on a graph. The y-value at a turning point is either a local maximum or a local minimum of the function. In Figure 4.22 the graph has two turning points, $(-2, 8)$ and $(2, -8)$. A local maximum is 8 and a local minimum is -8.

We discuss the graphs of polynomial functions, starting with degree 0 and continuing to degree 5. Look for patterns in the graphs of these polynomial functions.

Constant Polynomial Functions   If $f(x) = a$ and $a \neq 0$, then f is both a constant function and a polynomial function of **degree 0**. (If $a = 0$, then f has an **undefined degree**.) Its graph is a horizontal line that does not coincide with the x-axis. Graphs of two constant functions are shown in Figures 4.23 and 4.24. A graph of a polynomial function of degree 0 has no x-intercepts or turning points.

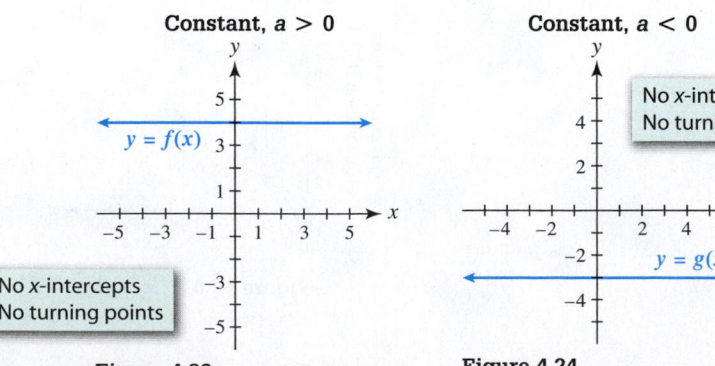

Constant, $a > 0$

$y = f(x)$

No x-intercepts
No turning points

Figure 4.23

Constant, $a < 0$

No x-intercepts
No turning points

$y = g(x)$

Figure 4.24

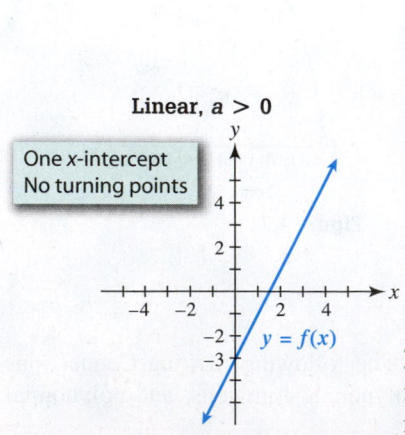

Linear, $a > 0$

One x-intercept
No turning points

$y = f(x)$

Figure 4.25

Linear Polynomial Functions   If $f(x) = ax + b$ and $a \neq 0$, then f is both a linear function and a polynomial function of **degree 1**. Its graph is a line that is neither horizontal nor vertical. The graphs of two linear functions are shown in Figures 4.25 and 4.26. A polynomial function of degree 1 has one x-intercept and no turning points.

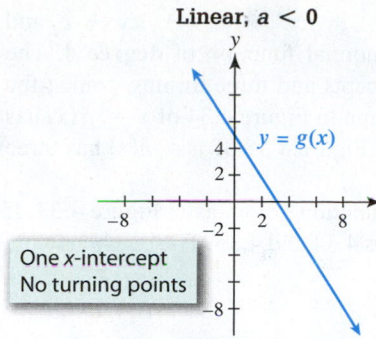

Linear, a < 0

One x-intercept
No turning points

Figure 4.26

The graph of $f(x) = ax + b$ with $a > 0$ is a line sloping upward from left to right. As one traces from left to right, the y-values become larger without a maximum. We say that the **end behavior** of the graph tends to $-\infty$ on the left and ∞ on the right. (Strictly speaking, the graph of a polynomial has infinite length and does not have an end.) More formally, we say that $f(x) \rightarrow -\infty$ as $x \rightarrow -\infty$ and $f(x) \rightarrow \infty$ as $x \rightarrow \infty$.

If $a < 0$, then the end behavior is reversed. The line slopes downward from left to right. The y-values on the left side of the graph become large positive values without a maximum and the y-values on the right side become negative without a minimum. The end behavior tends to ∞ on the left and $-\infty$ on the right, or $f(x) \rightarrow \infty$ as $x \rightarrow -\infty$ and $f(x) \rightarrow -\infty$ as $x \rightarrow \infty$.

Quadratic Polynomial Functions If $f(x) = ax^2 + bx + c$ and $a \neq 0$, then f is both a quadratic function and a polynomial function of **degree 2.** Its graph is a parabola that opens either upward ($a > 0$) or downward ($a < 0$). The graphs of three quadratic functions are shown in Figures 4.27–4.29, respectively. Quadratic functions can have zero, one, or two x-intercepts. A parabola has *exactly one* turning point, which is also the vertex.

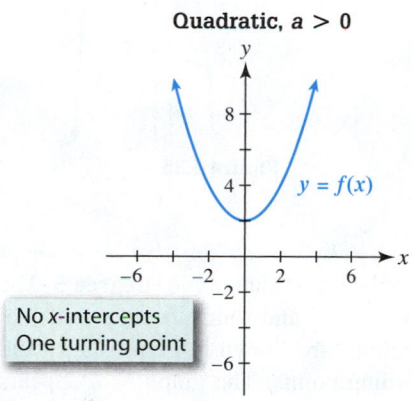

Quadratic, a > 0

No x-intercepts
One turning point

Figure 4.27

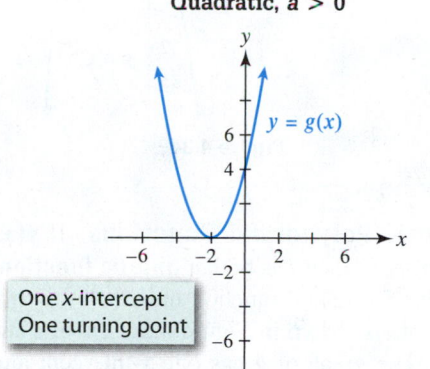

Quadratic, a > 0

One x-intercept
One turning point

Figure 4.28

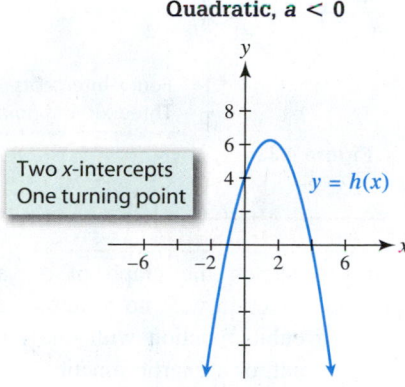

Quadratic, a < 0

Two x-intercepts
One turning point

Figure 4.29

If $a > 0$, as in Figure 4.27, then both sides of the graph go up. The end behavior tends to ∞ on both sides, or $f(x) \rightarrow \infty$ as $x \rightarrow \pm\infty$. If $a < 0$, as in Figure 4.29, then the end behavior is reversed and tends to $-\infty$ on both sides, or $f(x) \rightarrow -\infty$ as $x \rightarrow \pm\infty$.

Cubic Polynomial Functions If $f(x) = ax^3 + bx^2 + cx + d$ and $a \neq 0$, then f is both a **cubic function** and a polynomial function of **degree 3.** The graph of a cubic function can have zero or two turning points. The graph of $y = f(x)$ in Figure 4.30 has two turning points, whereas the graph of $y = g(x)$ in Figure 4.31 has no turning points.

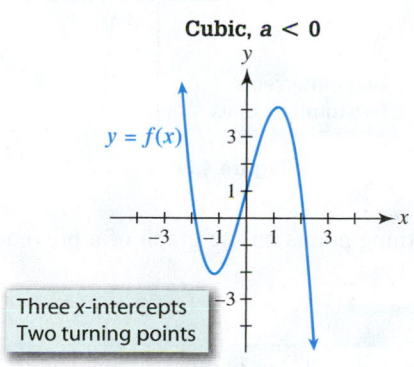

Cubic, a < 0

Three x-intercepts
Two turning points

Figure 4.30

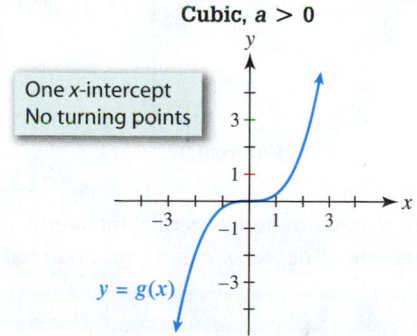

Cubic, a > 0

One x-intercept
No turning points

Figure 4.31

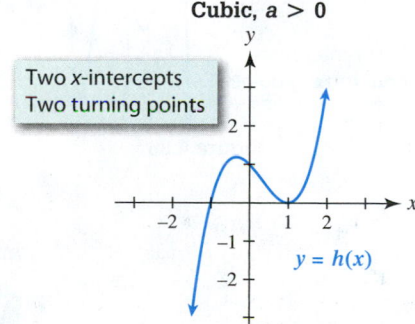

Cubic, a > 0

Two x-intercepts
Two turning points

Figure 4.32

If $a > 0$, the graph of a cubic function falls to the left and rises to the right, as in Figure 4.31. If $a < 0$, its graph rises to the left and falls to the right, as in Figure 4.30. The end behavior of a cubic function is similar to that of a linear function, tending to ∞ on one side and $-\infty$ on the other. Therefore its graph must cross the x-axis at least once. A cubic function can have up to three x-intercepts. The graph of $y = h(x)$ in Figure 4.32 has two x-intercepts.

Quartic Polynomial Functions   If $f(x) = ax^4 + bx^3 + cx^2 + dx + e$ and $a \neq 0$, then f is both a **quartic function** and a polynomial function of **degree 4**. The graph of a quartic function can have up to four x-intercepts and three turning points; the graph of $y = f(x)$ in Figure 4.33 is an example. The graph in Figure 4.34 of $y = g(x)$ has one turning point and two x-intercepts, and the graph in Figure 4.35 of $y = h(x)$ has three turning points and three x-intercepts.

If $a > 0$, then both ends of the graph of a quartic function go up, as in Figure 4.33. If $a < 0$, then both ends of its graph go down, as in Figures 4.34 and 4.35. The end behaviors of quartic and quadratic functions are similar.

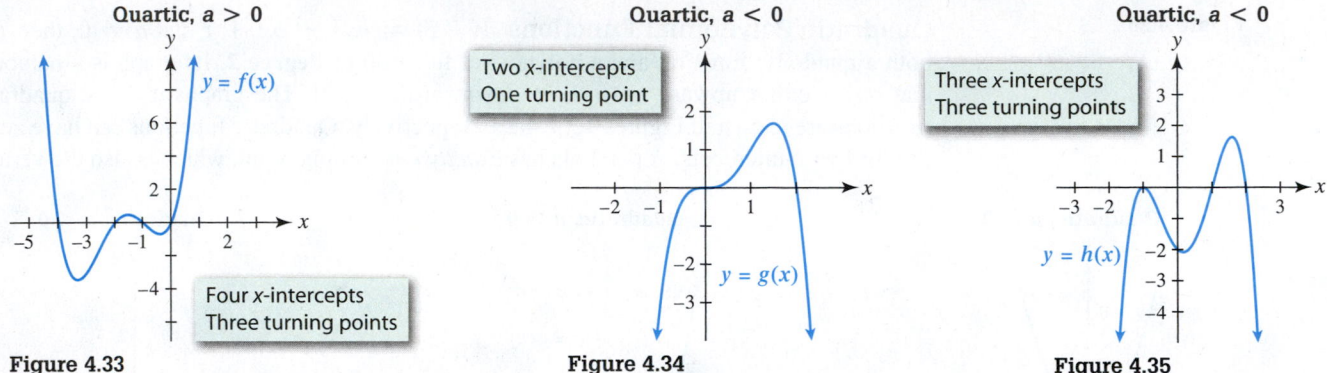

Figure 4.33   Figure 4.34   Figure 4.35

Quintic Polynomial Functions   If $f(x) = ax^5 + bx^4 + cx^3 + dx^2 + ex + k$ and $a \neq 0$, then f is both a **quintic function** and a polynomial function of **degree 5**. The graph of a quintic function may have up to five x-intercepts and four turning points. An example is shown in Figure 4.36. Other quintic functions are shown in Figures 4.37 and 4.38. The graph of g has one x-intercept and no turning points. The graph of h appears to have two x-intercepts and two turning points. Notice that the end behavior of a quintic function is similar to that of linear and cubic functions.

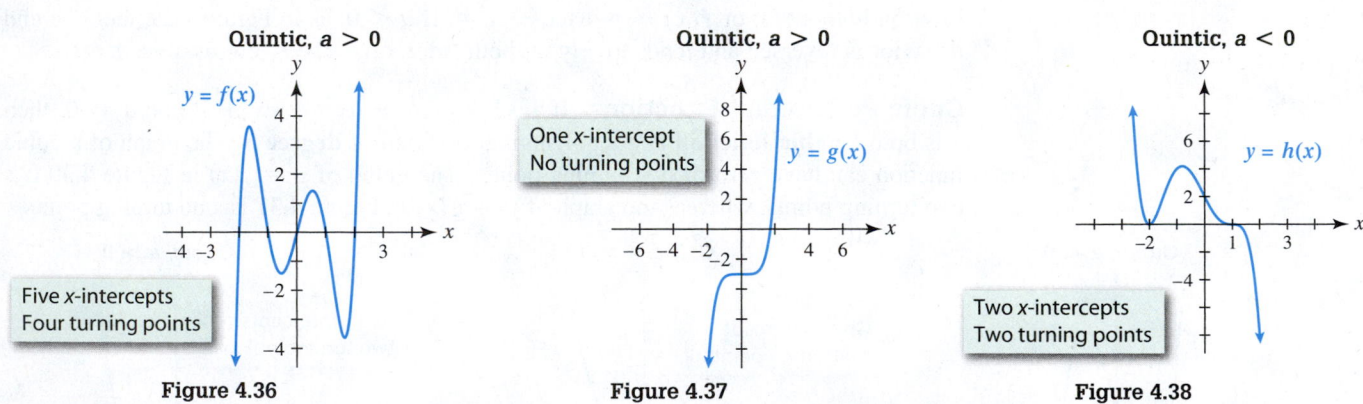

Figure 4.36   Figure 4.37   Figure 4.38

The maximum numbers of x-intercepts and turning points on the graph of a polynomial function of degree n can be summarized as follows.

DEGREE, x-INTERCEPTS, AND TURNING POINTS

The graph of a polynomial function of degree n, with $n \geq 1$, has at most n x-intercepts and at most $n - 1$ turning points.

The end behavior of a polynomial function depends on whether its degree is even or odd and whether its leading coefficient is positive or negative. The following See the Concept summarizes this discussion. The region inside the oval indicates where the graph might have x-intercepts and turning points.

See the Concept: End Behavior of Polynomial Functions

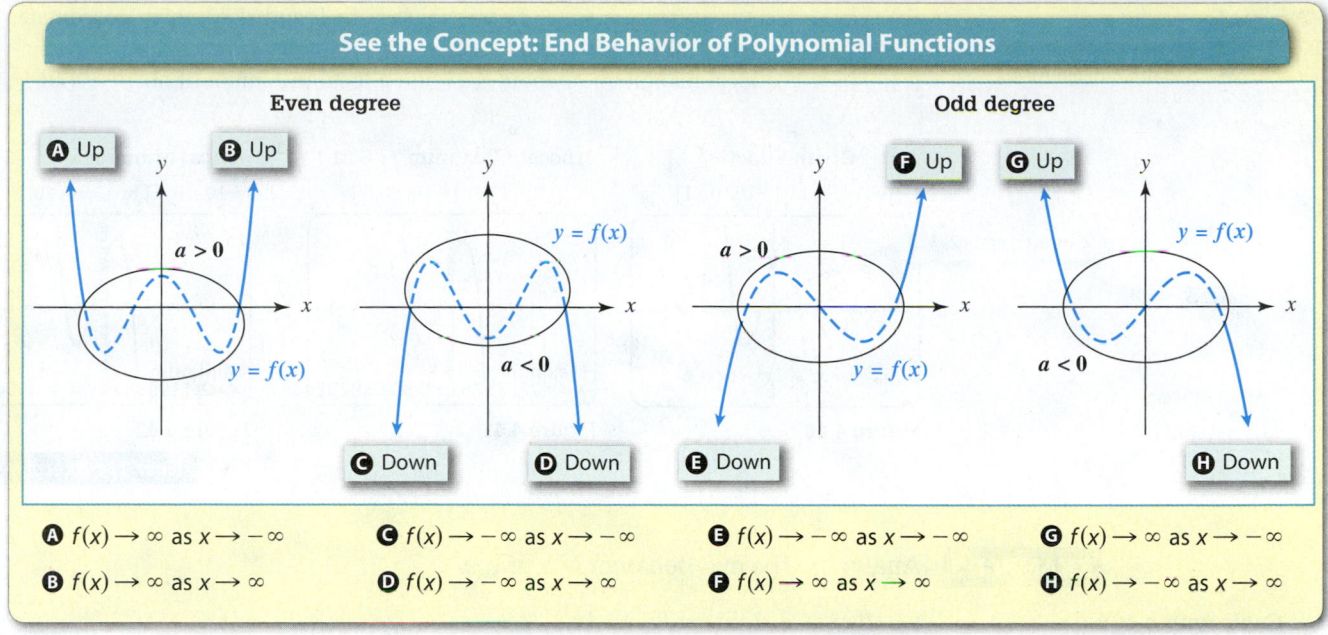

Ⓐ $f(x) \rightarrow \infty$ as $x \rightarrow -\infty$

Ⓑ $f(x) \rightarrow \infty$ as $x \rightarrow \infty$

Ⓒ $f(x) \rightarrow -\infty$ as $x \rightarrow -\infty$

Ⓓ $f(x) \rightarrow -\infty$ as $x \rightarrow \infty$

Ⓔ $f(x) \rightarrow -\infty$ as $x \rightarrow -\infty$

Ⓕ $f(x) \rightarrow \infty$ as $x \rightarrow \infty$

Ⓖ $f(x) \rightarrow \infty$ as $x \rightarrow -\infty$

Ⓗ $f(x) \rightarrow -\infty$ as $x \rightarrow \infty$

EXAMPLE 1   **Analyzing the graph of a polynomial function**

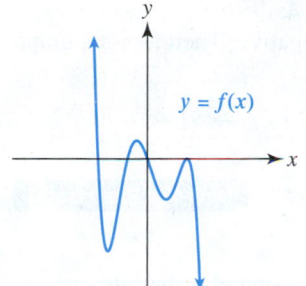

Figure 4.39

Figure 4.39 shows the graph of a polynomial function f.
(a) How many turning points and how many x-intercepts are there?
(b) Is the leading coefficient a positive or negative? Is the degree odd or even?
(c) Determine the minimum possible degree of f.

SOLUTION
(a) There are four turning points corresponding to the two "hills" and two "valleys." There appear to be four x-intercepts.
(b) The left side of the graph rises and the right side falls. Therefore $a < 0$ and the polynomial function has odd degree.
(c) The graph has four turning points. A polynomial of degree n can have at most $n - 1$ turning points. Therefore f must be *at least* degree 5.

> **Now Try Exercise 7**

NOTE   More examples of graphs of polynomials and their characteristics are found in the "Putting It All Together" for this section.

EXAMPLE 2   **Analyzing the graph of a polynomial function**

Graph $f(x) = x^3 - 2x^2 - 5x + 6$, and then complete the following.
(a) Identify the x-intercepts.
(b) Approximate the coordinates of any turning points to the nearest hundredth.
(c) Use the turning points to approximate any local extrema.

SOLUTION
(a) A calculator graph of f, shown in Figure 4.40 on the next page, *appears* to intersect the x-axis at the points $(-2, 0)$, $(1, 0)$, and $(3, 0)$. Because $f(-2) = 0$, $f(1) = 0$, and $f(3) = 0$, the x-intercepts are -2, 1, and 3.

Calculator Help

To find a minimum or maximum point on a graph, see Appendix A (page AP-8).

(b) There are two turning points. From Figures 4.41 and 4.42 their coordinates are approximately $(-0.79, 8.21)$ and $(2.12, -4.06)$.

(c) There is a local maximum of about 8.21 and a local minimum of about -4.06.

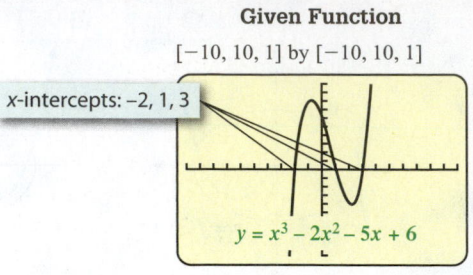

Given Function

$[-10, 10, 1]$ by $[-10, 10, 1]$

x-intercepts: –2, 1, 3

$y = x^3 - 2x^2 - 5x + 6$

Figure 4.40

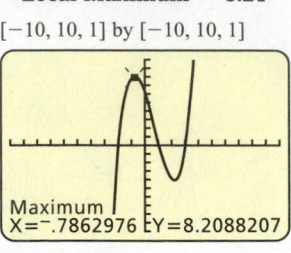

Local Maximum ≈ 8.21

$[-10, 10, 1]$ by $[-10, 10, 1]$

Maximum
X=⁻.7862976 Y=8.2088207

Figure 4.41

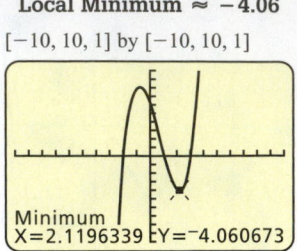

Local Minimum ≈ −4.06

$[-10, 10, 1]$ by $[-10, 10, 1]$

Minimum
X=2.1196339 Y=⁻4.060673

Figure 4.42

Now Try Exercise 25

EXAMPLE 3 Analyzing the end behavior of a graph

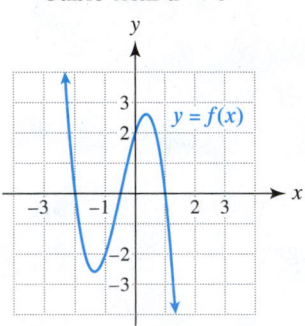

Cubic with *a* < 0

$y = f(x)$

Figure 4.43

Let $f(x) = 2 + 3x - 3x^2 - 2x^3$.
(a) Give the degree and leading coefficient.
(b) State the end behavior of the graph of f.

SOLUTION

(a) Rewriting gives $f(x) = -2x^3 - 3x^2 + 3x + 2$. The term with highest degree is $-2x^3$, so the degree is **3** and the leading coefficient is -2.

(b) The degree of $f(x)$ is odd, and the leading coefficient is negative. Therefore the graph of f rises to the left and falls to the right. More formally,

$$f(x) \to \infty \text{ as } x \to -\infty \quad \text{and} \quad f(x) \to -\infty \text{ as } x \to \infty.$$

This conclusion is supported by Figure 4.43.

Now Try Exercise 33

An Application In the next example, we analyze the data presented in the introduction to this section.

EXAMPLE 4 Modeling natural gas consumption

Figure 4.20, which shows natural gas consumption from 1960 to 2000, is repeated in the margin.
(a) Could a linear or quadratic function model the data?
(b) What minimum degree polynomial might be appropriate to model the data?
(c) Should the leading coefficient a be positive or negative?

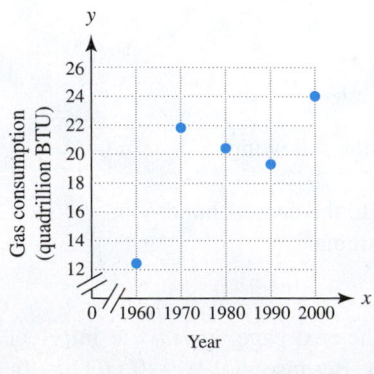

Gas consumption (quadrillion BTU)

Year

SOLUTION

(a) The data clearly do not lie on a line, so a linear function is not appropriate. Because natural gas consumption increases, decreases, and then increases, a quadratic function would not be a good choice either. The data are not ∪-shaped or ∩-shaped.

(b) Because the data increase, decrease, and then increase, a polynomial with at least two turning points would be appropriate. A cubic, or degree 3, polynomial is a possibility for a modeling function.

(c) The leading coefficient a should be positive because the data fall to the left and rise to the right.

Now Try Exercises 39(a), (b), and (c)

Concavity (Optional)   Graphs of polynomial functions with degree 2 or greater are curves. **Concavity** is a mathematical description of how a curve bends. A line exhibits no, or zero, concavity because it is straight. A parabola that opens upward is said to be **concave upward** everywhere on its domain. See Figure 4.44. A parabola that opens downward is said to be **concave downward** everywhere on its domain. See Figure 4.45. A graph of a higher degree polynomial can be concave upward on one interval of its domain and concave downward on a different interval of its domain. See Figure 4.46, where the graph of $f(x) = 4x - x^3$ is concave upward on the interval $(-\infty, 0)$, shown in blue, and concave downward on $(0, \infty)$, shown in red. Concavity is usually defined for *open* intervals. Determining the exact x-value where a graph switches from concave upward to downward or vice versa can be difficult to do visually and often requires techniques learned in calculus. This point is called the **point of inflection**. (See Extended and Discovery Exercises 2–7 for this section.)

Graphs with Concavity

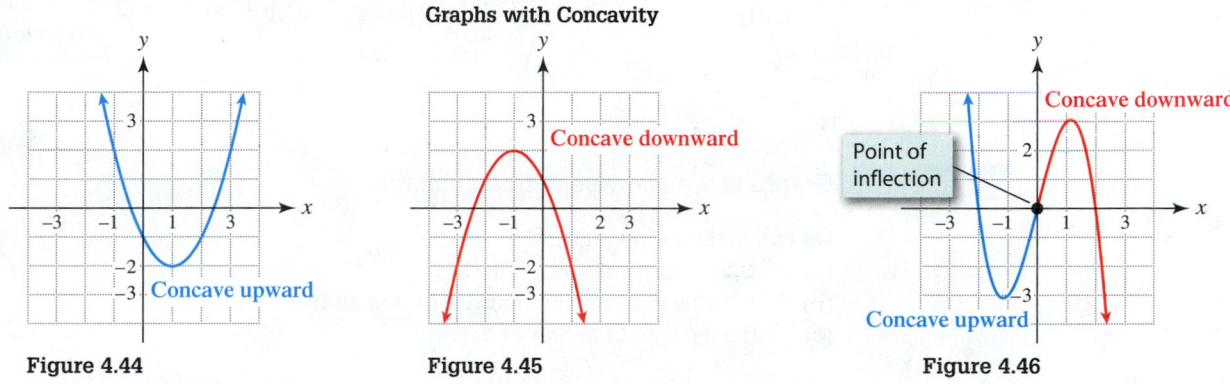

Figure 4.44   Figure 4.45   Figure 4.46

Piecewise-Defined Polynomial Functions

In Section 2.4 piecewise-defined functions were discussed. If each piece is a polynomial, then the function is a **piecewise-defined polynomial function** or **piecewise-polynomial function**. An example is given by $f(x)$.

$$f(x) = \begin{cases} x^3 & \text{if } x < 1 \quad \text{— First piece} \\ x^2 - 1 & \text{if } x \geq 1 \quad \text{— Second piece} \end{cases}$$

One way to graph f is to first graph $y = x^3$ and $y = x^2 - 1$, as shown in Figures 4.47 and 4.48. Then the graph of f is found by using the blue portion of $y = x^3$ for $x < 1$ and the blue portion of $y = x^2 - 1$ for $x \geq 1$, as illustrated in Figure 4.49. At $x = 1$ there is a break in the graph, where the graph of f is discontinuous.

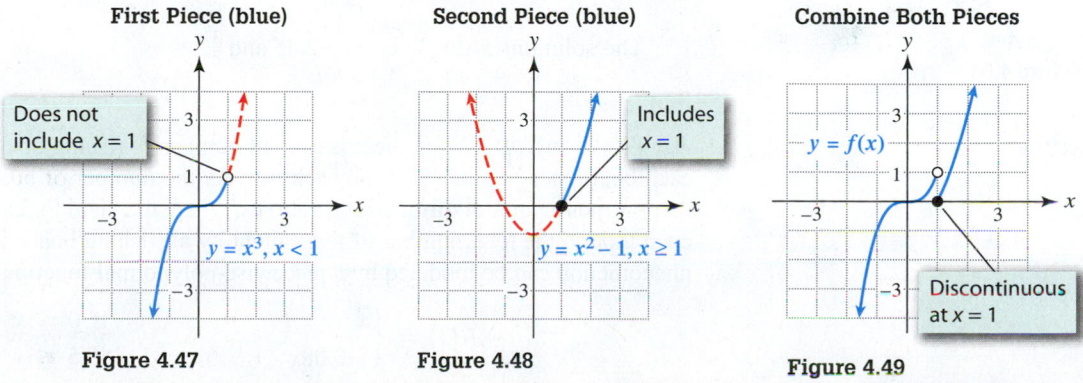

Figure 4.47   Figure 4.48   Figure 4.49

EXAMPLE 5 Evaluating a piecewise-defined polynomial function

Evaluate $f(x)$ at $x = -3, -2, 1,$ and 2.

$$f(x) = \begin{cases} x^2 - x & \text{if } -5 \le x < -2 \\ -x^3 & \text{if } -2 \le x < 2 \\ 4 - 4x & \text{if } 2 \le x \le 5 \end{cases}$$

SOLUTION To evaluate $f(-3)$ we use the formula $f(x) = x^2 - x$, because -3 is in the interval $-5 \le x < -2$.

$$f(-3) = (-3)^2 - (-3) = 12$$

To evaluate $f(-2)$ we use $f(x) = -x^3$, because -2 is in the interval $-2 \le x < 2$.

$$f(-2) = -(-2)^3 = -(-8) = 8$$

Similarly, $f(1) = -1^3 = -1$ and $f(2) = 4 - 4(2) = -4$.

> Now Try Exercise 73

EXAMPLE 6 Graphing a piecewise-defined function

Complete the following.
(a) Sketch a graph of f.
(b) Determine if f is continuous on its domain.
(c) Solve the equation $f(x) = 1$.

$$f(x) = \begin{cases} \frac{1}{2}x^2 - 2 & \text{if } -4 \le x \le 0 \\ 2x - 2 & \text{if } 0 < x < 2 \\ 2 & \text{if } 2 \le x \le 4 \end{cases}$$

SOLUTION

(a) For the first piece, graph the parabola determined by $y = \frac{1}{2}x^2 - 2$ on the interval $-4 \le x \le 0$. Place dots at the endpoints, which are $(-4, 6)$ and $(0, -2)$. See Figure 4.50. For the second piece, graph the line determined by $y = 2x - 2$ *between* the points $(0, -2)$ and $(2, 2)$. Note that the left endpoint of the middle piece coincides with the right endpoint of the first piece. Finally, graph the horizontal line $y = 2$ from the points $(2, 2)$ to $(4, 2)$. Note that the left endpoint of the third piece coincides with the right endpoint of the middle piece.

(b) The domain of f is $-4 \le x \le 4$. Because there are no breaks in the graph of f on its domain, the graph of f is continuous.

(c) The red horizontal line $y = 1$ intersects the blue graph of $y = f(x)$ at two points, as shown in Figure 4.51. The x-coordinates of these two points of intersection coincide with the green segments and can be found by solving the equations

$$\frac{1}{2}x^2 - 2 = 1 \quad \text{and} \quad 2x - 2 = 1.$$

The solutions are $-\sqrt{6} \approx -2.45$ and $\frac{3}{2}$.

> Now Try Exercise 79

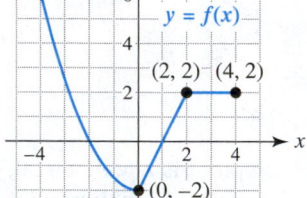

Figure 4.50

Solving $f(x) = 1$

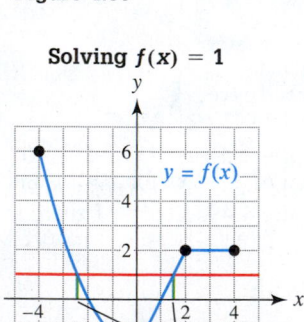

Figure 4.51

An Application When there is a small number of fishing boats in a large area of water, each boat tends to catch its limit each trip. As the number of boats increases dramatically, there comes a point of **diminishing returns**, where the yield for each boat begins to decrease even though the total number of fish caught by all fishing boats continues to increase. This phenomenon can be modeled by a piecewise-polynomial function F defined by

$$F(x) = \begin{cases} x & \text{if } 0 \le x \le 5 \\ -0.08x^2 + 1.6x - 1 & \text{if } 5 \le x \le 15 \end{cases}$$

where x is the number of fishing boats in hundreds and $F(x)$ outputs thousands of tons of fish harvested. A graph of F is shown in Figure 4.52.

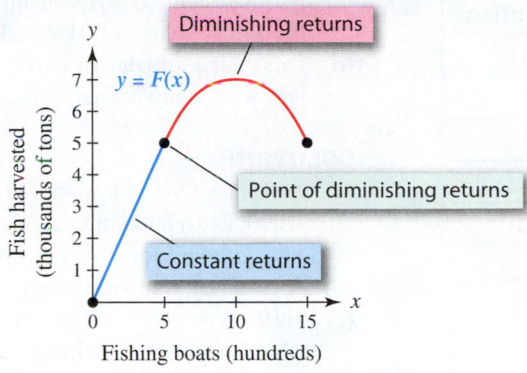

Figure 4.52

EXAMPLE 7 **Analyzing diminishing returns**

Use the preceding discussion to complete the following.
(a) Evaluate $F(2)$ and interpret the result.
(b) Find the absolute maximum on the graph of F and interpret the result.

SOLUTION
(a) For $0 \leq x \leq 5$, $F(x) = x$, so $F(2) = 2$. When 200 fishing boats are used, 2 thousand tons of fish are harvested.
(b) The second piece is $F(x) = -0.08x^2 + 1.6x - 1$, so its graph is a parabola opening downward. To find the absolute maximum, we first need to find its vertex.

$$x = -\frac{b}{2a} \qquad \text{Vertex formula}$$

$$= -\frac{1.6}{2(-0.08)} \qquad \text{Substitute.}$$

$$= 10 \qquad \text{x-coordinate of vertex}$$

Because $F(10) = -0.08(10)^2 + 1.6(10) - 1 = 7$, the vertex is $(10, 7)$. This result means that 7 thousand tons of fish is the maximum amount of fish that can be caught in these waters and it occurs when 1000 fishing vessels are used. If more boats are used, there is a decrease in the overall catch.

Now Try Exercise 83

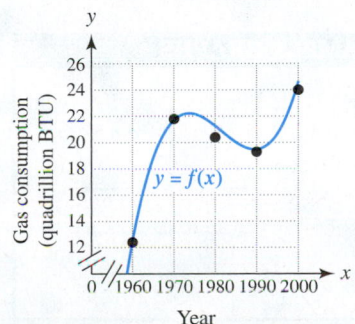

Natural Gas Usage

Polynomial Regression (Optional)

We now have the mathematical understanding to model the data presented in the introduction to this section. The polynomial modeling function f (shown in Figure 4.21 and repeated in the margin) falls to the left and rises to the right, so it has odd degree and the leading coefficient is positive. Since the graph of f has two turning points, it must be at least degree 3. A cubic polynomial $f(x)$ is a possible choice, where

$$f(x) = ax^3 + bx^2 + cx + d.$$

Trial and error would be a difficult way to find values for a, b, c, and d. Instead, we can use least-squares regression, which was also discussed in Sections 2.1 and 3.1, for linear and quadratic functions. The next example illustrates *cubic regression*.

EXAMPLE 8 Determining a cubic modeling function

The data in Table 4.2 (repeated in the margin) lists natural gas consumption.
(a) Find a polynomial function of degree 3 that models the data.
(b) Graph f and the data together.
(c) Estimate natural gas consumption in 1974 and in 2010. Compare these estimates to the actual values of 21.2 and 24.9 quadrillion Btu, respectively.
(d) Did your estimates in part (c) involve interpolation or extrapolation? Is there a problem with using higher degree polynomials ($n \geq 3$) for extrapolation? Explain.

Year	Consumption
1960	12.4
1970	21.8
1980	20.4
1990	19.3
2000	24.0

SOLUTION
(a) Enter the five data points (1960, 12.4), (1970, 21.8), (1980, 20.4), (1990, 19.3), and (2000, 24.0) into your calculator. Then select cubic regression, as shown in Figure 4.53. The equation for $f(x)$ is shown in Figure 4.54.
(b) A graph of f and a scatterplot of the data are shown in Figure 4.55.
(c) $f(1974) \approx 21.9$ and $f(2010) \approx 44.5$; the 1974 estimate is reasonably close to 21.2, whereas the 2010 estimate is not close to 24.9.
(d) The 1974 estimate uses interpolation, and the 2010 estimate uses extrapolation. Because the end behavior of a higher degree polynomial rapidly tends to either ∞ or $-\infty$, extrapolation-based estimates are usually inaccurate with higher degree polynomials.

Calculator Help

To find an equation of least-squares fit, see Appendix A (page AP-9). To copy a regression equation into Y_1, see Appendix A (page AP-11.)

Select Cubic Regression

```
EDIT  CALC  TESTS
1:1-Var Stats
2:2-Var Stats
3:Med-Med
4:LinReg(ax+b)
5:QuadReg
6:CubicReg
7↓QuartReg
```

Figure 4.53

Regression Equation

```
CubicReg
y=ax³+bx²+cx+d
a=.0013833333
b=-8.2235
c=16295.13667
d=-10762915.92
R²=.9941642351
```

Figure 4.54

Graph of f and Data
[1955, 2005, 5] by [10, 25, 5]

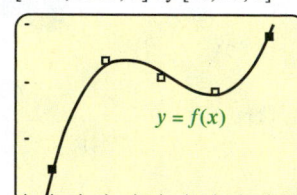

$y = f(x)$

Figure 4.55

Now Try Exercise 91

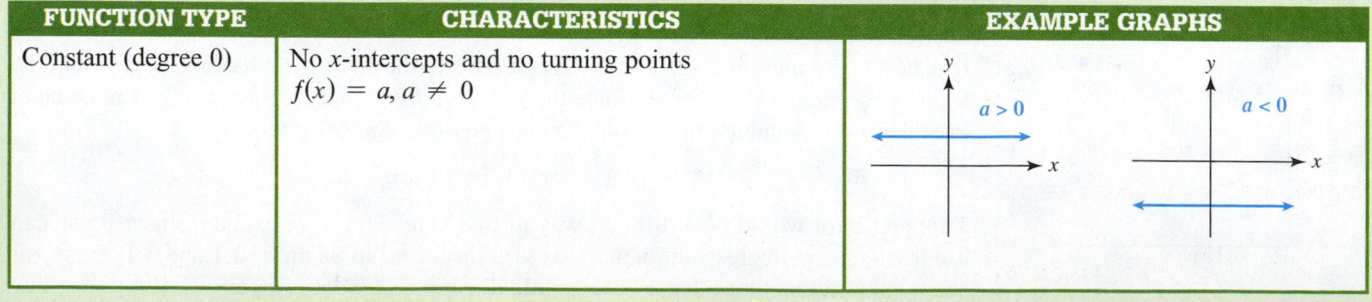

4.2 Putting It All Together

Higher degree polynomials generally have more complicated graphs. Each additional degree allows the graph to have one more possible turning point and x-intercept. The graph of a polynomial function is continuous and smooth; it has no breaks or sharp edges. Its domain includes all real numbers. The end behavior of a polynomial always tends to either ∞ or $-\infty$. End behavior describes what happens to the y-values as $|x|$ becomes large.

FUNCTION TYPE	CHARACTERISTICS	EXAMPLE GRAPHS
Constant (degree 0)	No x-intercepts and no turning points $f(x) = a, a \neq 0$	

FUNCTION TYPE	CHARACTERISTICS	EXAMPLE GRAPHS
Linear (degree 1)	One x-intercept and no turning points $f(x) = ax + b, a \neq 0$	
Quadratic (degree 2)	At most two x-intercepts and exactly one turning point $f(x) = ax^2 + bx + c, a \neq 0$	
Cubic (degree 3)	At most three x-intercepts and up to two turning points $f(x) = ax^3 + bx^2 + cx + d, a \neq 0$	
Quartic (degree 4)	At most four x-intercepts and up to three turning points $f(x) = ax^4 + bx^3 + cx^2 + dx + e, a \neq 0$	
Quintic (degree 5)	At most five x-intercepts and up to four turning points $f(x) = ax^5 + bx^4 + cx^3 + dx^2 + ex + k,$ $a \neq 0$	

4.2 Exercises

Note: Many of the answers in this section involve estimations. Your answers may vary slightly, particularly when you are reading a graph.

Graphs of Polynomial Functions

1. A runner is working out on a straight track. The graph on the next page shows the runner's distance y in hundreds of feet from the starting line after t minutes.

(a) Estimate the turning points.

(b) Interpret each turning point.

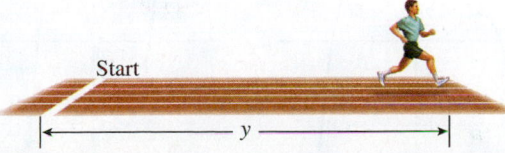

Start

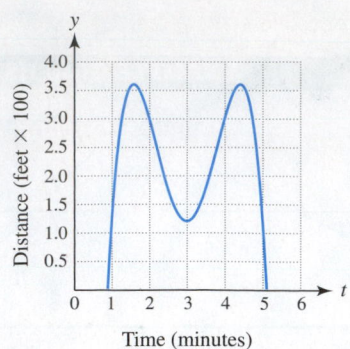

2. A stone is thrown into the air. Its height y in feet after t seconds is shown in the graph. Use the graph to complete the following.

   (a) Estimate the turning point.

   (b) Interpret this point.

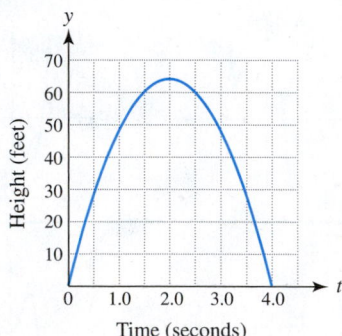

Exercises 3–12: Use the graph of the polynomial function f to complete the following. Let a be the leading coefficient of the polynomial $f(x)$.

(a) Determine the number of turning points and estimate any x-intercepts.

(b) State whether $a > 0$ or $a < 0$.

(c) Determine the minimum degree of f.

3.
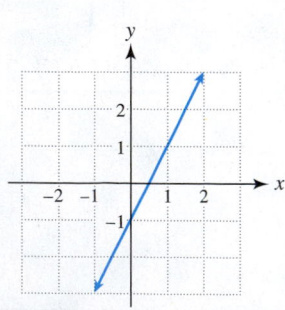

4.
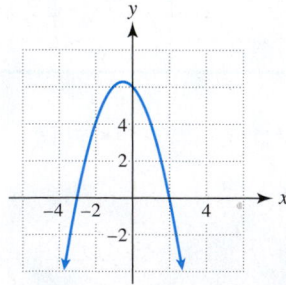

5.
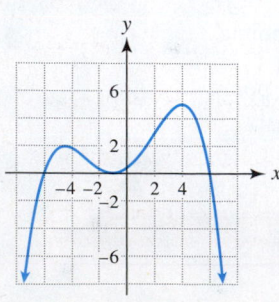

6.
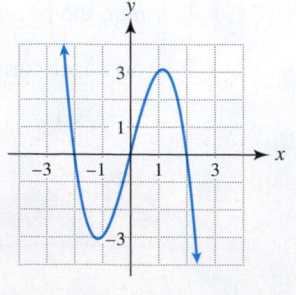

7.
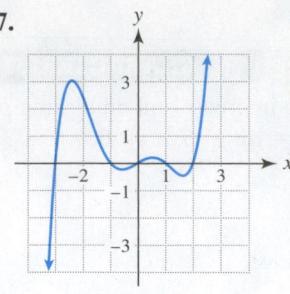

8.
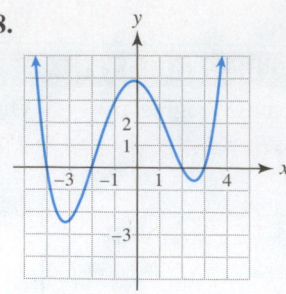

9.
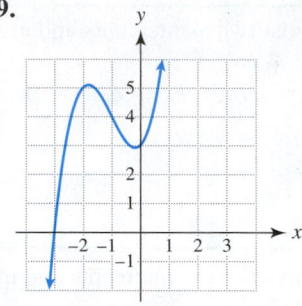

10.
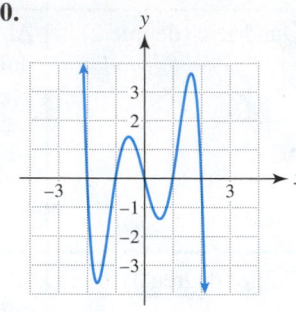

11.
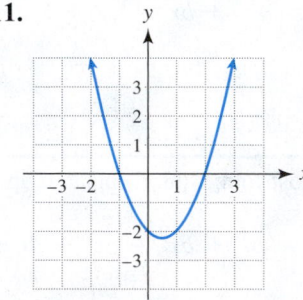

12.
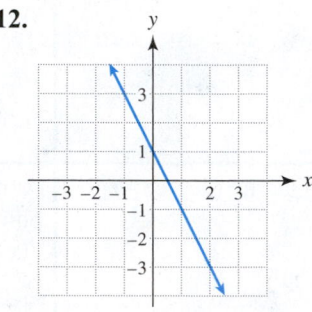

Exercises 13–18: Complete the following without a calculator.

(a) Match the equation with its graph (a–f).

(b) Identify the turning points.

(c) Estimate the x-intercepts.

(d) Estimate any local extrema.

(e) Estimate any absolute extrema.

13. $f(x) = 1 - 2x + x^2$

14. $f(x) = 3x - x^3$

15. $f(x) = x^3 + 3x^2 - 9x$

16. $f(x) = x^4 - 8x^2$

17. $f(x) = 8x^2 - x^4$

18. $f(x) = x^5 + \frac{5}{2}x^4 - \frac{5}{3}x^3 - 5x^2$

a.
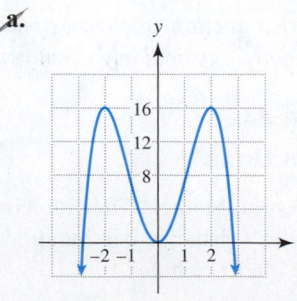

b.
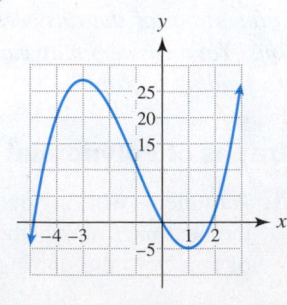

c.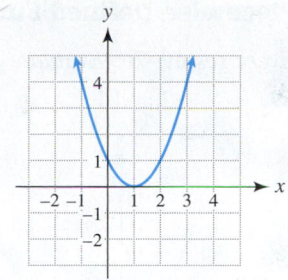

d.

e.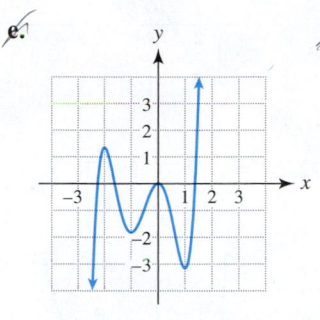

f.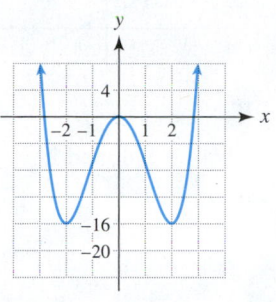

Exercises 19–26: *Complete the following.*

(a) *Graph* $y = f(x)$ *in the standard viewing rectangle.*

(b) *Approximate the coordinates of each turning point.*

(c) *Estimate any local extrema.*

19. $f(x) = \frac{1}{9}x^3 - 3x$

20. $f(x) = x^2 - 4x - 3$

21. $f(x) = 0.025x^4 - 0.45x^2 - 5$

22. $f(x) = -\frac{1}{8}x^4 + \frac{1}{3}x^3 + \frac{5}{4}x^2 - 3x + 3$

23. $f(x) = 1 - 2x + 3x^2$

24. $f(x) = 4x - \frac{1}{3}x^3$

25. $f(x) = \frac{1}{3}x^3 + \frac{1}{2}x^2 - 2x$

26. $f(x) = \frac{1}{4}x^4 + \frac{2}{3}x^3 - \frac{1}{2}x^2 - 2x + 1$

Exercises 27–38: *Complete the following.*

(a) *State the degree and leading coefficient of f.*

(b) *State the end behavior of the graph of f.*

27. $f(x) = -2x + 3$ **28.** $f(x) = \frac{2}{3}x - 2$

29. $f(x) = x^2 + 4x$ **30.** $f(x) = 5 - \frac{1}{2}x^2$

31. $f(x) = -2x^3$ **32.** $f(x) = 4x - \frac{1}{3}x^3$

33. $f(x) = x^2 - x^3 - 4$

34. $f(x) = x^4 - 4x^3 + 3x^2 - 3$

35. $f(x) = 0.1x^5 - 2x^2 - 3x + 4$

36. $f(x) = 3x^3 - 2 - x^4$

37. $f(x) = 4 + 2x - \frac{1}{2}x^2$

38. $f(x) = -0.2x^5 + 4x^2 - 3$

Modeling Data with Polynomials

Exercises 39–44: *The data are modeled exactly by a linear, quadratic, cubic, or quartic function f with leading coefficient a. All zeros of f are real numbers located in the interval* $[-3, 3]$.

(a) *Make a line graph of the data.*

(b) *State the minimum degree of f.*

(c) *Is* $a > 0$ *or is* $a < 0$?

(d) *Find a formula for* $f(x)$.

39.

x	-3	-2	-1	0	1	2	3
$f(x)$	3	-8	-7	0	7	8	-3

40.

x	-3	-2	-1	0	1	2	3
$f(x)$	11	9	7	5	3	1	-1

41.

x	-3	-2	-1	0	1	2	3
$f(x)$	14	7	2	-1	-2	-1	2

42.

x	-3	-2	-1	0	1	2	3
$f(x)$	-13	-6	-1	2	3	2	-1

43.

x	-3	-2	-1	0	1	2	3
$f(x)$	-55	-5	1	-1	1	-5	-55

44.

x	-3	-2	-1	0	1	2	3
$f(x)$	-15	0	3	0	-3	0	15

Sketching Graphs of Polynomials

Exercises 45–56: *If possible, sketch a graph of a polynomial that satisfies the conditions. Let a be the leading coefficient.*

45. Degree 3 with three real zeros and $a > 0$

46. Degree 4 with four real zeros and $a < 0$

47. Linear with $a < 0$

48. Cubic with one real zero and $a > 0$

49. Degree 4 and an even function with four turning points

50. Degree 5 and symmetric with respect to the y-axis

51. Degree 3 and an odd function with no x-intercepts

52. Degree 6 and an odd function with five turning points

53. Degree 3 with turning points $(-1, 2)$ and $\left(1, \frac{2}{3}\right)$

54. Degree 4 with turning points $(-1, -1)$, $(0, 0)$, and $(1, -1)$

55. Degree 2 with turning point $(-1, 2)$, passing through $(-3, 4)$ and $(1, 4)$

56. Degree 5 and an odd function with five x-intercepts and a negative leading coefficient.

Dominant Term of a Polynomial

Exercises 57 and 58: Graph the functions f, g, and h in the same viewing rectangle. What happens to their graphs as the size of the viewing rectangle increases? Explain why the term of highest degree in a polynomial is sometimes called the **dominant term.**

57. $f(x) = 2x^4$, $g(x) = 2x^4 - 5x^2 + 1$, and $h(x) = 2x^4 + 3x^2 - x - 2$
 (a) $[-4, 4, 1]$ by $[-4, 4, 1]$

 (b) $[-10, 10, 1]$ by $[-100, 100, 10]$

 (c) $[-100, 100, 10]$ by $[-10^6, 10^6, 10^5]$

58. $f(x) = -x^3$, $g(x) = -x^3 + x^2 + 2$, and $h(x) = -x^3 - 2x^2 + x - 1$
 (a) $[-4, 4, 1]$ by $[-4, 4, 1]$

 (b) $[-10, 10, 1]$ by $[-100, 100, 10]$

 (c) $[-100, 100, 10]$ by $[-10^5, 10^5, 10^4]$

Average Rates of Change

59. Compare the average rates of change from 0 to $\frac{1}{2}$ for $f(x) = x$, $g(x) = x^2$, and $h(x) = x^3$.

60. Compare the average rates of change from 1 to $\frac{3}{2}$ for $f(x) = x$, $g(x) = x^2$, and $h(x) = x^3$.

Exercises 61–64: Calculate the average rate of change of f on each interval. What happens to this average rate of change as the interval decreases in length?

(a) $[1.9, 2.1]$
(b) $[1.99, 2.01]$
(c) $[1.999, 2.001]$

61. $f(x) = x^3$ **62.** $f(x) = 4x - \frac{1}{3}x^3$

63. $f(x) = \frac{1}{4}x^4 - \frac{1}{3}x^3$ **64.** $f(x) = 4x^2 - \frac{1}{2}x^4$

Exercises 65–68: Find the difference quotient of g.

65. $g(x) = 3x^3$ **66.** $g(x) = -2x^3$

67. $g(x) = 1 + x - x^3$ **68.** $g(x) = \frac{1}{2}x^3 - 2x$

Piecewise-Defined Functions

Exercises 69–76: Evaluate f(x) at the given values of x.

69. $x = -2$ and 1   **70.** $x = -1, 0,$ and 3

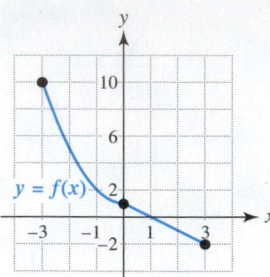 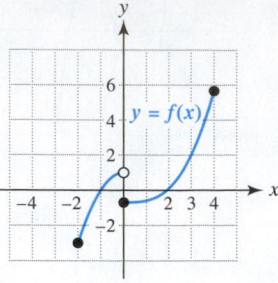

71. $x = -1, 1,$ and 2   **72.** $x = -2, 0,$ and 2

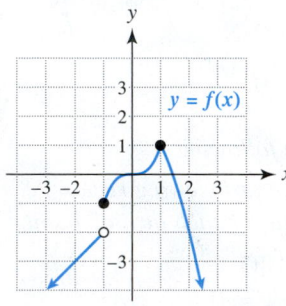 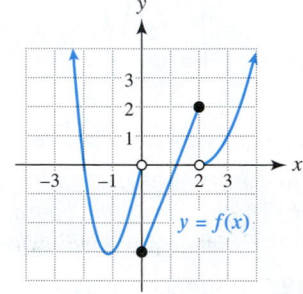

73. $x = -3, 1,$ and 4
$$f(x) = \begin{cases} x^3 - 4x^2 & \text{if} & x \le -3 \\ 3x^2 & \text{if} -3 < x < 4 \\ x^3 - 54 & \text{if} & x \ge 4 \end{cases}$$

74. $x = -4, 0,$ and 4
$$f(x) = \begin{cases} -4x & \text{if} & x \le -4 \\ x^3 + 2 & \text{if} -4 < x \le 2 \\ 4 - x^2 & \text{if} & x > 2 \end{cases}$$

75. $x = -2, 1,$ and 2
$$f(x) = \begin{cases} x^2 + 2x + 6 & \text{if} -5 \le x < 0 \\ x + 6 & \text{if} & 0 \le x < 2 \\ x^3 + 1 & \text{if} & 2 \le x \le 5 \end{cases}$$

76. $x = 1975, 1980,$ and 1998
$$f(x) = \begin{cases} 0.2(x - 1970)^3 + 60 & \text{if } 1970 \le x < 1980 \\ 190 - (x - 1980)^2 & \text{if } 1980 \le x < 1990 \\ 2(x - 1990) + 100 & \text{if } 1990 \le x \le 2000 \end{cases}$$

Exercises 77–82: Complete the following.

(a) Sketch a graph of f.
(b) Determine if f is continuous on its domain.
(c) Solve f(x) = 0.

77. $f(x) = \begin{cases} 4 - x^2 & \text{if } -3 \le x \le 0 \\ x^2 - 4 & \text{if } 0 < x \le 3 \end{cases}$

78. $f(x) = \begin{cases} x^2 & \text{if } -2 \leq x < 0 \\ x+1 & \text{if } 0 \leq x \leq 2 \end{cases}$

79. $f(x) = \begin{cases} 2x & \text{if } -5 \leq x < -1 \\ -2 & \text{if } -1 \leq x < 0 \\ x^2 - 2 & \text{if } 0 \leq x \leq 2 \end{cases}$

80. $f(x) = \begin{cases} 0.5x^2 & \text{if } -4 \leq x \leq -2 \\ x & \text{if } -2 < x < 2 \\ x^2 - 4 & \text{if } 2 \leq x \leq 4 \end{cases}$

81. $f(x) = \begin{cases} x^3 + 3 & \text{if } -2 \leq x \leq 0 \\ x + 3 & \text{if } 0 < x < 1 \\ 4 + x - x^2 & \text{if } 1 \leq x \leq 3 \end{cases}$

82. $f(x) = \begin{cases} -2x & \text{if } -3 \leq x < -1 \\ x^2 + 1 & \text{if } -1 \leq x \leq 2 \\ \frac{1}{2}x^3 + 1 & \text{if } 2 < x \leq 3 \end{cases}$

Applications

83. Diminishing Returns   (Refer to Example 7.) Let the function F, defined by

$$F(x) = \begin{cases} 2x & \text{if } 0 \leq x \leq 4 \\ -\frac{1}{4}x^2 + 4x - 4 & \text{if } 4 \leq x \leq 12, \end{cases}$$

calculate a fish harvest in thousands of tons when x hundred fishing boats are used in a region.

(a) Evaluate $F(3)$ and interpret the result.

(b) Find the absolute maximum on the graph of F and interpret the result.

84. Diminishing Returns   The function F, defined by

$$F(x) = \begin{cases} x & \text{if } 0 \leq x \leq 5 \\ -0.08x^2 + 1.6x - 1 & \text{if } 5 \leq x \leq 15, \end{cases}$$

was used in Example 7 to describe a fish harvest in a region.

(a) Evaluate $F(6)$ and interpret your answer.

(b) Solve $F(x) = 6$ and interpret the result.

(c) Use Figure 4.52 to find the point of diminishing returns and interpret its meaning.

85. Electronics   The **Heaviside function** H, used in the study of electrical circuits, is defined by

$$H(t) = \begin{cases} 0 & \text{if } t < 0 \\ 1 & \text{if } t \geq 0. \end{cases}$$

(a) Evaluate $H(-2)$, $H(0)$, and $H(3.5)$.

(b) Graph $y = H(t)$.

86. A Strange Graph   The following definition is discussed in advanced mathematics courses.

$$f(x) = \begin{cases} 0 & \text{if } x \text{ is a rational number} \\ 1 & \text{if } x \text{ is an irrational number} \end{cases}$$

(a) Evaluate $f\left(-\frac{3}{4}\right)$, $f(-\sqrt{2})$, and $f(\pi)$.

(b) Is f a function? Explain.

(c) Discuss the difficulty with graphing $y = f(x)$.

87. Modeling Temperature   In the figure the monthly average temperature in degrees Fahrenheit from January to December in Minneapolis is modeled by a polynomial function f, where $x = 1$ corresponds to January and $x = 12$ to December.   (*Source:* A. Miller and J. Thompson, *Elements of Meteorology.*)

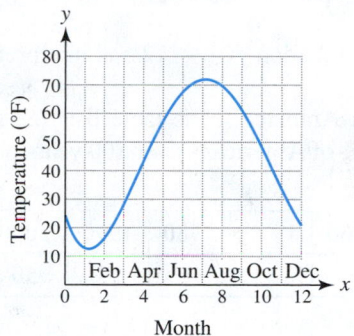

(a) Estimate the turning points.

(b) Interpret each turning point.

88. Natural Gas Consumption   Refer to Example 8 and the introduction to this section.
(a) Solve the equation $f(x) = 20$ graphically. Interpret the solution set.

(b) Calculate the average rate of change in natural gas consumption from 1970 to 1980. Interpret the result.

89. Modeling   An object is lifted rapidly into the air at a constant speed and then dropped. Its height h in feet after x seconds is listed in the table.

x (sec)	0	1	2	3	4	5	6	7
h (ft)	0	36	72	108	144	128	80	0

(a) At what time does it appear that the object was dropped?

(b) Identify the time interval when the height could be modeled by a linear function. When could it be modeled by a nonlinear function?

(*continued*)

(c) Determine values for the constants m, a, and b so that f models the data.

$$f(x) = \begin{cases} mx & \text{if } 0 \leq x \leq 4 \\ a(x-4)^2 + b & \text{if } 4 < x \leq 7 \end{cases}$$

(d) Solve $f(x) = 100$ and interpret your answer.

90. **Modeling**   A water tank is filled with a hose and then drained. The table shows the number of gallons y in the tank after t minutes.

t (min)	0	1	2	3	4	5	6	7
y (gal)	0	9	18	27	36	16	4	0

The following function f models the data in the table.

$$f(t) = \begin{cases} 9t & \text{if } 0 \leq t \leq 4 \\ 4t^2 - 56t + 196 & \text{if } 4 < t \leq 7 \end{cases}$$

Solve the equation $f(t) = 12$ and interpret the results.

91. **Aging in America**   The table lists the number N (in thousands) of Americans over 100 years old for selected years x.

x	1960	1970	1980	1990	2000	2010
N	3	5	15	37	50	70

(a) Use regression to find a polynomial of degree 3 that models the data. Let $x = 0$ correspond to 1960.

(b) Graph f and the data.

(c) Estimate N in 1994 and in 2020.

(d) Did your estimates in part (c) involve interpolation or extrapolation?

92. **Modeling Water Flow**   A cylindrical container has a height of 16 centimeters. Water entered the container at a constant rate until it was completely filled. Then water was allowed to leak out through a small hole in the bottom. The height of the water in the container was recorded every half minute over a 5-minute period.

Time (min)	0	0.5	1.0	1.5	2.0	2.5
Height (cm)	0	4	8	12	16	11.6

Time (min)	3.0	3.5	4.0	4.5	5.0
Height (cm)	8.1	5.3	3.1	1.4	0.5

(a) Plot the data.

(b) Find a piecewise-defined function that models the data. (*Hint:* Use regression.)

(c) Approximate the water level after 1.25 minutes and after 3.2 minutes.

(d) Estimate the time when water was flowing out of the tank and the water level was 5 centimeters.

Writing About Mathematics

Exercises 93–96: Discuss possible local extrema and absolute extrema on the graph of f. Assume that a > 0.

93. $f(x) = ax + b$

94. $f(x) = ax^2 + bx + c$

95. $f(x) = ax^3 + bx^2 + cx + d$

96. $f(x) = a|x|$

Extended and Discovery Exercises

1. **Torricelli's Law**   A cylindrical tank contains 500 gallons of water. A plug is pulled from the bottom of the tank, and it takes 10 minutes to drain the tank. The amount A of water in gallons remaining in the tank after t minutes is approximated by

$$A(t) = 500\left(1 - \frac{t}{10}\right)^2.$$

(a) What is a reasonable domain for A?

(b) Evaluate $A(1)$ and interpret the result.

(c) What are the degree and leading coefficient of $A(t)$?

(d) Has half the water drained from the tank after 5 minutes? Does this agree with your intuition? Explain.

Exercises 2–7: **Concavity**   *Estimate the intervals where the graph of f is concave upward and where the graph is concave downward. Use interval notation.*

2.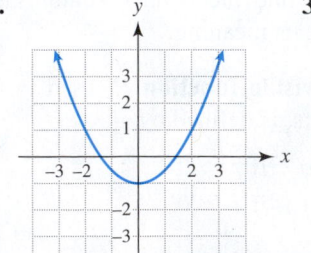

3.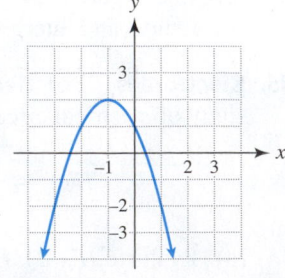

4.

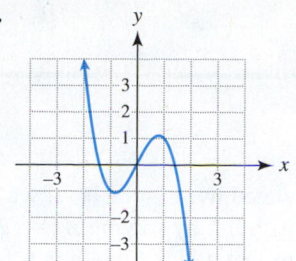

5.

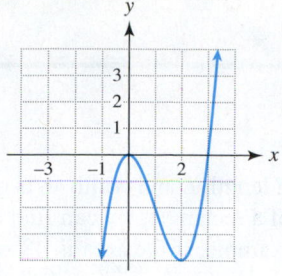

6.

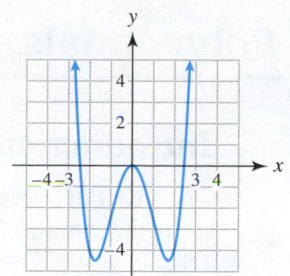

7.

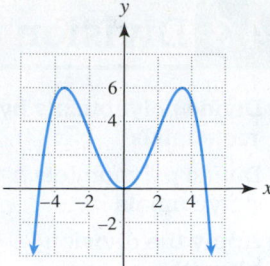

CHECKING BASIC CONCEPTS FOR SECTIONS 4.1 AND 4.2

1. Use the graph of f to complete the following.

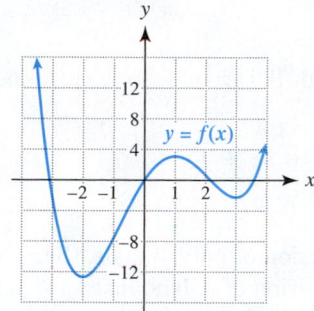

   (a) Determine where f is increasing or decreasing.

   (b) Identify any local extrema.

   (c) Identify any absolute extrema.

   (d) Approximate the x-intercepts and zeros of f. Then solve $f(x) = 0$. How are the x-intercepts, zeros, and solutions to $f(x) = 0$ related?

2. Use the graph to complete the following.

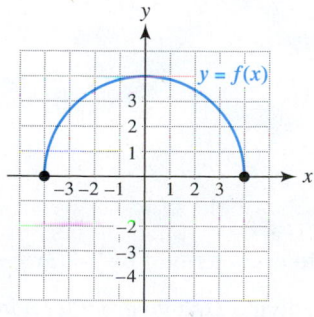

(a) Evaluate $f(-4)$, $f(0)$, and $f(4)$.

(b) What type of symmetry does the graph of f exhibit?

(c) Is f an odd function or an even function? Why?

(d) Find the domain and range of f.

3. If possible, sketch a graph of a cubic polynomial with a negative leading coefficient that satisfies each of the following conditions.

   (a) Zero x-intercepts

   (b) One x-intercept

   (c) Two x-intercepts

   (d) Four x-intercepts

4. Plot the data in the table.

x	-3.2	-2	0	2	3.2
y	-11	15	-10	15	-11

   (a) What is the minimum degree of the polynomial function f that would be needed to model these data? Explain.

   (b) Should function f be odd, even, or neither? Explain.

   (c) Should the leading coefficient of f be positive or negative? Explain.

5. Use least-squares regression to find a polynomial that models the data in Exercise 4.

4.3 Division of Polynomials

- Divide polynomials by monomials
- Divide polynomials by polynomials
- Apply the division algorithm
- Learn synthetic division
- Understand the remainder theorem

Introduction

The area A of a rectangle with length L and width W is calculated by $A = LW$. If we are given the area and the width of a rectangle, we can find the length L by solving $A = LW$ for L to obtain $L = \frac{A}{W}$. For example, if the area is 48 square feet and the width is 6 feet, then the length of the rectangle equals $L = \frac{48}{6} = 8$ feet. Now consider the more general situation shown in Figure 4.56.

$$x + 3 \qquad A = x^2 + 8x + 15$$

$$L$$

Figure 4.56

The area A is $x^2 + 8x + 15$ and the width W is $x + 3$, so we can find an expression for L in the same way by calculating

$$L = \frac{A}{W} = \frac{x^2 + 8x + 15}{x + 3}.$$

In this case, the calculation of L involves division of polynomials. (See Example 5.) This section discusses basic concepts related to division of polynomials.

Division by Monomials

Adding (or subtracting) fractions having like denominators is straightforward. For example,

$$\frac{3}{17} + \frac{7}{17} = \frac{3 + 7}{17}, \qquad \text{and so} \qquad \frac{3 + 7}{17} = \frac{3}{17} + \frac{7}{17}.$$

Note that when we reverse the process, the denominator of 17 is divided into each term in the numerator. By reversing the process, we can sometimes simplify expressions.

$$\frac{3x^4 + 5x^3}{5x^3} = \frac{3x^4}{5x^3} + \frac{5x^3}{5x^3} = \frac{3}{5}x + 1 \qquad \textcolor{blue}{\text{Subtract exponents: } \frac{3x^4}{5x^3} = \frac{3}{5}x.}$$

This process is used in the next example.

EXAMPLE 1 **Dividing by a monomial**

Divide $6x^3 - 3x^2 + 2$ by $2x^2$.

SOLUTION

Getting Started Remember to divide $2x^2$ into *every* term of $6x^3 - 3x^2 + 2$. ▶

Write the problem as $\frac{6x^3 - 3x^2 + 2}{2x^2}$. Then divide $2x^2$ into *every* term in the numerator.

$$\frac{6x^3 - 3x^2 + 2}{2x^2} = \frac{6x^3}{2x^2} - \frac{3x^2}{2x^2} + \frac{2}{2x^2} \qquad \textcolor{blue}{\text{Write as three terms.}}$$

$$= 3x - \frac{3}{2} + \frac{1}{x^2}$$

Algebra Review

To review simplification of rational expressions, see Chapter R (page R-28).

Now Try Exercise 3

Division by Polynomials

Before dividing a polynomial by a binomial, we review division of natural numbers.

$$\begin{array}{r} \text{quotient} \rightarrow 58 \\ \text{divisor} \rightarrow 3\overline{)175} \leftarrow \text{dividend} \\ \underline{15} \\ 25 \\ \underline{24} \\ 1 \leftarrow \text{remainder} \end{array}$$

This result is checked as follows: $3 \cdot 58 + 1 = 175$. That is,

(Divisor) (Quotient) + (Remainder) = (Dividend).

The quotient and remainder can also be expressed as $58\frac{1}{3}$. Since 3 does not divide into 175 evenly, 3 is *not* a factor of 175. When the remainder is 0, the divisor is a *factor* of the dividend. Division of polynomials is similar to division of natural numbers.

EXAMPLE 2 Dividing polynomials

Divide $2x^3 - 3x^2 - 11x + 7$ by $x - 3$. Check the result.

SOLUTION Begin by dividing x into $2x^3$.

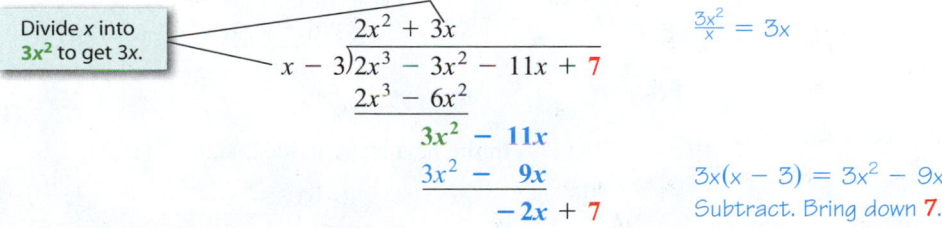

Start by dividing x into $2x^3$ to get $2x^2$.

$$\frac{2x^3}{x} = 2x^2$$

$2x^2(x - 3) = 2x^3 - 6x^2$
Subtract. Bring down $-11x$.

In the next step, divide x into $3x^2$.

Divide x into $3x^2$ to get $3x$.

$$\frac{3x^2}{x} = 3x$$

$3x(x - 3) = 3x^2 - 9x$
Subtract. Bring down 7.

Now divide x into $-2x$.

Divide x into $-2x$ to get -2.

$$\frac{-2x}{x} = -2$$

$-2(x - 3) = -2x + 6$
Subtract. Remainder is 1.

The quotient is $2x^2 + 3x - 2$ and the remainder is 1. Polynomial division is also checked by multiplying the divisor and quotient and then adding the remainder.

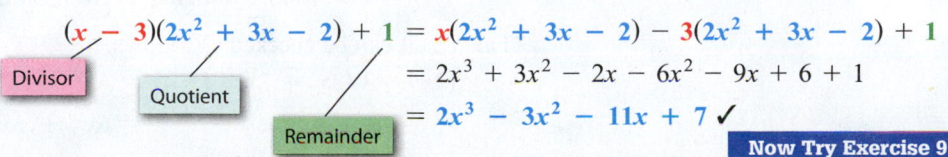

$$(x - 3)(2x^2 + 3x - 2) + 1 = x(2x^2 + 3x - 2) - 3(2x^2 + 3x - 2) + 1$$
$$= 2x^3 + 3x^2 - 2x - 6x^2 - 9x + 6 + 1$$
$$= 2x^3 - 3x^2 - 11x + 7 \checkmark$$

Divisor   Quotient   Remainder

Now Try Exercise 9

Algebra Review
To review multiplication of polynomials, see Chapter R (page R-16).

A division problem, such as $\frac{175}{3}$, is typically given in the form $\frac{\text{(Dividend)}}{\text{(Divisor)}}$. If we divide each term in the equation

$$\text{(Dividend)} = \text{(Divisor)}\,\text{(Quotient)} + \text{(Remainder)}$$

by **(Divisor)**, we obtain the equation

$$\frac{\text{(Dividend)}}{\text{(Divisor)}} = \text{(Quotient)} + \frac{\text{(Remainder)}}{\text{(Divisor)}}.$$

For example, because **175** divided by **3** equals **58** remainder **1**, we can use this equation to justify writing $\frac{175}{3} = 58 + \frac{1}{3}$, or $58\frac{1}{3}$. We can use the results from Example 2 to write

$$\frac{2x^3 - 3x^2 - 11x + 7}{x - 3} = 2x^2 + 3x - 2 + \frac{1}{x - 3}.$$

This process is applied in the next example.

EXAMPLE 3 **Dividing polynomials**

Divide each expression. Check your answer.

(a) $\dfrac{6x^2 + 5x - 10}{2x + 3}$ **(b)** $(5x^3 - 4x^2 + 7x - 2) \div (x^2 + 1)$

SOLUTION

(a) Begin by dividing $2x$ into $6x^2$.

$$
\begin{array}{r}
3x \\
2x + 3\overline{)6x^2 + 5x - 10} \\
\underline{6x^2 + 9x} \\
-4x - 10
\end{array}
$$

$\dfrac{6x^2}{2x} = 3x$

$3x(2x + 3) = 6x^2 + 9x$

Subtract: $5x - 9x = -4x$.
Bring down the -10.

In the next step, divide $2x$ into $-4x$.

$$
\begin{array}{r}
3x - 2 \\
2x + 3\overline{)6x^2 + 5x - 10} \\
\underline{6x^2 + 9x} \\
-4x - 10 \\
\underline{-4x - 6} \\
-4
\end{array}
$$

$\dfrac{-4x}{2x} = -2$

$-2(2x + 3) = -4x - 6$

Subtract: $-10 - (-6) = -4$.

The quotient is $3x - 2$ with remainder -4. This result can also be written as follows.

$$3x - 2 + \frac{-4}{2x + 3} \qquad \text{(Quotient)} + \frac{\text{(Remainder)}}{\text{(Divisor)}}$$

To check this result use the equation

$$\text{(Divisor)}\text{(Quotient)} + \text{(Remainder)} = \text{(Dividend)}.$$

This result can be checked as follows.

$$(2x + 3)(3x - 2) + (-4) = 6x^2 + 5x - 6 - 4$$

$$= 6x^2 + 5x - 10 \checkmark \qquad \text{The result checks.}$$

(b) Begin by writing $x^2 + 1$ as $x^2 + 0x + 1$.

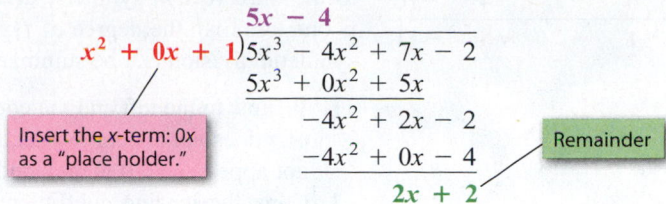

$$
\begin{array}{r}
5x - 4 \\
x^2 + 0x + 1 \overline{)5x^3 - 4x^2 + 7x - 2} \\
\underline{5x^3 + 0x^2 + 5x} \\
-4x^2 + 2x - 2 \\
\underline{-4x^2 + 0x - 4} \\
2x + 2
\end{array}
$$

Insert the x-term: $0x$ as a "place holder."

Remainder

The quotient is $5x - 4$ with remainder of $2x + 2$. This result can also be written as

$$5x - 4 + \frac{2x + 2}{x^2 + 1}. \qquad \text{(Quotient)} + \frac{\text{(Remainder)}}{\text{(Divisor)}}$$

This result can be checked as follows.

$$(x^2 + 1)(5x - 4) + 2x + 2 = 5x^3 - 4x^2 + 5x - 4 + 2x + 2$$
$$= 5x^3 - 4x^2 + 7x - 2 \checkmark \qquad \text{The result checks.}$$

Now Try Exercises 21 and 25

This process is summarized by the following *division algorithm for polynomials.*

DIVISION ALGORITHM FOR POLYNOMIALS

Let $f(x)$ and $d(x)$ be two polynomials, with the degree of $d(x)$ greater than zero and less than the degree of $f(x)$. Then there exist unique polynomials $q(x)$ and $r(x)$ such that

$$f(x) \quad = \quad d(x) \quad \cdot \quad q(x) \quad + \quad r(x),$$

(Dividend) = (Divisor) · (Quotient) + (Remainder)

where either $r(x) = 0$ or the degree of $r(x)$ is less than the degree of $d(x)$. The polynomial $r(x)$ is called the remainder.

The Division Algorithm in Real Life   Although the division algorithm may seem unrelated to everyday life, it is actually expressing something that is very common. For example, if there are 5 calculators for 16 students to use, then there should be 3 students to a group leaving 1 extra student, who could join a group of 3. That is, $\frac{16}{5} = 3$ with remainder 1.

Synthetic Division

A shortcut called **synthetic division** can be used to divide $x - k$ into a polynomial. For example, to divide $x - 2$ into $3x^4 - 7x^3 - 4x + 5$, we perform the following steps with $k = 2$. The equivalent steps involving long division are shown to the right.

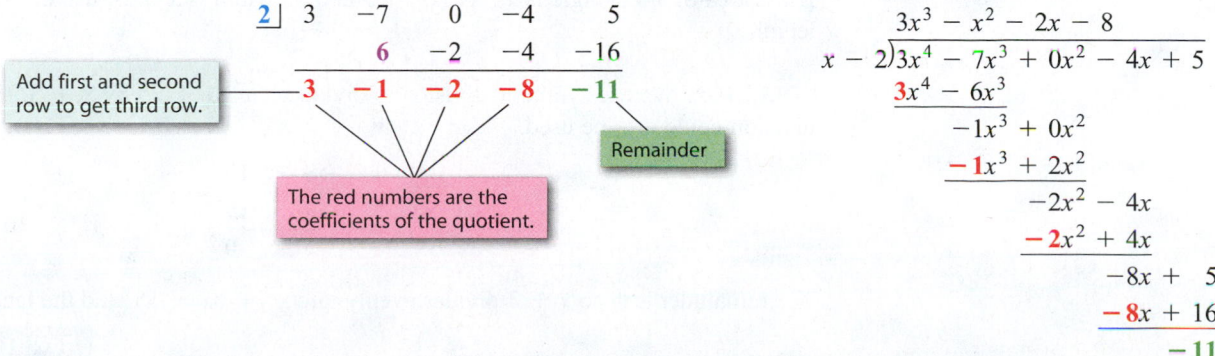

$$
\begin{array}{r|rrrrr}
2 & 3 & -7 & 0 & -4 & 5 \\
  &   & 6 & -2 & -4 & -16 \\
\hline
  & 3 & -1 & -2 & -8 & -11
\end{array}
$$

Add first and second row to get third row.

The red numbers are the coefficients of the quotient.

Remainder

$$
\begin{array}{r}
3x^3 - x^2 - 2x - 8 \\
x - 2 \overline{)3x^4 - 7x^3 + 0x^2 - 4x + 5} \\
\underline{3x^4 - 6x^3} \\
-1x^3 + 0x^2 \\
\underline{-1x^3 + 2x^2} \\
-2x^2 - 4x \\
\underline{-2x^2 + 4x} \\
-8x + 5 \\
\underline{-8x + 16} \\
-11
\end{array}
$$

$$\begin{array}{r|rrrr} 2 & 3 & -7 & 0 & -4 & 5 \\ & & 6 & -2 & -4 & -16 \\ \hline & 3 & -1 & -2 & -8 & -11 \end{array}$$

Notice how the red and green numbers in the expression for long division correspond to the third row in synthetic division. The degree of the quotient, $3x^3 - x^2 - 2x - 8$, is one less than the degree of $f(x)$. The steps to divide a polynomial $f(x)$ by $x - k$ using synthetic division can be summarized as follows.

1. Write k to the left and the coefficients of $f(x)$ to the right in the top row. If any power of x does *not* appear in $f(x)$, include a 0 for that term. In this example, an x^2-term did not appear, so a 0 is included in the first row.
2. Copy the leading coefficient of $f(x)$ into the third row and multiply it by k. Write the result below the next coefficient of $f(x)$ in the second row. Add the numbers in the second column and place the result in the third row. Repeat the process. In this example, the leading coefficient is **3** and $k = $ **2**. Since **3 · 2 = 6**, **6** is placed below the -7. Then add to obtain $-7 + 6 = $ **−1**. Multiply **−1** by **2** and repeat.
3. The last number in the third row is the remainder. If the remainder is 0, then the binomial $x - k$ is a factor of $f(x)$. The other red numbers in the third row are the coefficients of the quotient, with terms written in descending powers.

EXAMPLE 4 Performing synthetic division

Use synthetic division to divide $2x^3 + 4x^2 - x + 5$ by $x + 2$.

SOLUTION
Getting Started To find the value of k, write the divisor as $x - k$. Because $x + 2$ equals $x - (-2)$, the value of k is -2 ▶

Let $k = $ **−2** and perform synthetic division on the problem $\frac{2x^3 + 4x^2 - x + 5}{x + 2}$.

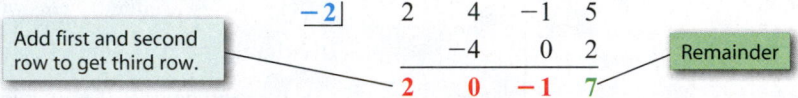

Add first and second row to get third row.

Remainder

$$\begin{array}{r|rrrr} -2 & 2 & 4 & -1 & 5 \\ & & -4 & 0 & 2 \\ \hline & 2 & 0 & -1 & 7 \end{array}$$

The remainder is **7** and the quotient is $2x^2 + 0x - 1 = 2x^2 - 1$. This result is expressed by the equation

$$\frac{2x^3 + 4x^2 - x + 5}{x + 2} = 2x^2 - 1 + \frac{7}{x + 2}.$$

Now Try Exercise 41

An Application from Geometry In the final example, we use division to solve the problem presented in the introduction to this section.

EXAMPLE 5 Finding the length of a rectangle

If the area of a rectangle is $x^2 + 8x + 15$ and its width is $x + 3$, use division to find its length.

SOLUTION We use synthetic division to divide $x^2 + 8x + 15$ by $x + 3$. However, long division could also be used.

$$\begin{array}{r|rrr} -3 & 1 & 8 & 15 \\ & & -3 & -15 \\ \hline & 1 & 5 & 0 \end{array}$$

The remainder is **0**, so $x + 3$ divides evenly into $x^2 + 8x + 15$, and the length is $x + 5$.

Now Try Exercise 51

Remainder Theorem   If the divisor $d(x)$ is $x - k$, then the *division algorithm for polynomials* simplifies to

$$f(x) = (x - k)q(x) + r,$$

where r is a constant. If we let $x = k$ in this equation, then

$$f(k) = (k - k)q(k) + r = r.$$

Thus $f(k)$ is equal to the remainder obtained in synthetic division. In Example 4, when $f(x) = 2x^3 + 4x^2 - x + 5$ is divided by $x + 2$, the remainder is 7. It follows that $k = -2$ and $f(-2) = 2(-2)^3 + 4(-2)^2 - (-2) + 5 = 7$. This result is summarized by the *remainder theorem*.

> **REMAINDER THEOREM**
>
> If a polynomial $f(x)$ is divided by $x - k$, the remainder is $f(k)$.

4.3 Putting It All Together

The following table lists some important concepts related to division of polynomials.

CONCEPT	EXPLANATION	EXAMPLE
Division by a monomial	Be sure to divide the denominator into *every term* in the numerator.	$\dfrac{5a^3 - 10a^2}{5a^2} = \dfrac{5a^3}{5a^2} - \dfrac{10a^2}{5a^2} = a - 2$
Division by a polynomial	Division by a polynomial can be done in a manner similar to long division of natural numbers. See Examples 2 and 3.	When $6x^3 + 5x^2 - 8x + 4$ is divided by $2x - 1$, the quotient is $3x^2 + 4x - 2$ with remainder 2 and can be written as $$\dfrac{6x^3 + 5x^2 - 8x + 4}{2x - 1} = 3x^2 + 4x - 2 + \dfrac{2}{2x - 1}.$$
Division algorithm	(Dividend) = (Divisor)(Quotient) + (Remainder) This equation can be written as $$\dfrac{\text{(Dividend)}}{\text{(Divisor)}} = \text{(Quotient)} + \dfrac{\text{(Remainder)}}{\text{(Divisor)}}.$$	$\dfrac{x^3 - 1}{x + 1} = x^2 - x + 1 + \dfrac{-2}{x + 1}$ Dividend: $x^3 - 1$ Divisor: $x + 1$ Quotient: $x^2 - x + 1$ Remainder: -2
Synthetic division	An efficient method for dividing $x - k$ into a polynomial	Divide $2x^3 - 3x^2 + x + 2$ by $x + 1$. $$\begin{array}{r} -1 \,\rvert\ \begin{array}{rrrr} 2 & -3 & 1 & 2 \\ & -2 & 5 & -6 \\ \hline 2 & -5 & 6 & -4 \end{array} \end{array}$$ $k = -1$ The quotient is $2x^2 - 5x + 6$, and the remainder is -4.
Remainder theorem	If a polynomial $f(x)$ is divided by $x - k$, the remainder is $f(k)$.	If $f(x) = 3x^2 - 2x + 6$ is divided by $x - 2$, the remainder is $f(2) = 3(2)^2 - 2(2) + 6 = 14$.

4.3 Exercises

Division by Monomials

Exercises 1–8: Divide the expression.

1. $\dfrac{5x^4 - 15}{10x}$

2. $\dfrac{x^2 - 5x}{5x}$

3. $\dfrac{3x^4 - 2x^2 - 1}{3x^3}$

4. $\dfrac{5x^3 - 10x^2 + 5x}{15x^2}$

5. $\dfrac{x^3 - 4}{4x^3}$

6. $\dfrac{2x^4 - 3x^2 + 4x - 7}{-4x}$

7. $\dfrac{5x(3x^2 - 6x + 1)}{3x^2}$

8. $\dfrac{(1 - 5x^2)(x + 1) + x^2}{2x}$

Division by Polynomials

Exercises 9–14: Divide the first polynomial by the second. State the quotient and remainder.

9. $x^3 - 2x^2 - 5x + 6$ $x - 3$

10. $3x^3 - 10x^2 - 27x + 10$ $x + 2$

11. $2x^4 - 7x^3 - 5x^2 - 19x + 17$ $x + 1$

12. $x^4 - x^3 - 4x + 1$ $x - 2$

13. $3x^3 - 7x + 10$ $x - 1$

14. $x^4 - 16x^2 + 1$ $x + 4$

Exercises 15–22: Divide. Check your answer.

15. $\dfrac{x^4 - 3x^3 - x + 3}{x - 3}$

16. $\dfrac{x^3 - 2x^2 - x + 3}{x + 1}$

17. $\dfrac{4x^3 - x^2 - 5x + 6}{x - 1}$

18. $\dfrac{x^4 + 3x^3 - 4x + 1}{x + 2}$

19. $\dfrac{x^3 + 1}{x + 1}$

20. $\dfrac{x^5 + 3x^4 - x - 3}{x + 3}$

21. $\dfrac{6x^3 + 5x^2 - 8x + 4}{2x - 1}$

22. $\dfrac{12x^3 - 14x^2 + 7x - 7}{3x - 2}$

Exercises 23–30: Divide the expression.

23. $\dfrac{3x^4 - 7x^3 + 6x - 16}{3x - 7}$

24. $\dfrac{20x^4 + 6x^3 - 2x^2 + 15x - 2}{5x - 1}$

25. $\dfrac{5x^4 - 2x^2 + 6}{x^2 + 2}$

26. $\dfrac{x^3 - x^2 + 2x - 3}{x^2 + 3}$

27. $\dfrac{8x^3 + 10x^2 - 12x - 15}{2x^2 - 3}$

28. $\dfrac{3x^4 - 2x^2 - 5}{3x^2 - 5}$

29. $\dfrac{2x^4 - x^3 + 4x^2 + 8x + 7}{2x^2 + 3x + 2}$

30. $\dfrac{3x^4 + 2x^3 - x^2 + 4x - 3}{x^2 + x - 1}$

Division Algorithm

Exercises 31 and 32: Use the equation

  (Dividend) = (Divisor)(Quotient) + (Remainder)

to complete the following.

31. $\dfrac{x^3 - 8x^2 + 15x - 6}{x - 2} = x^2 - 6x + 3$ implies
$(x - 2)(x^2 - 6x + 3) = \underline{\ \ ?\ \ }$.

32. $\dfrac{x^4 - 15}{x + 2} = x^3 - 2x^2 + 4x - 8 + \dfrac{1}{x + 2}$
implies $x^4 - 15 = (x + 2) \times \underline{\ \ ?\ \ } + \underline{\ \ ?\ \ }$.

Exercises 33–38: Use division to express the (Dividend) as (Divisor)(Quotient) + (Remainder).

33. $\dfrac{x^2 - 3x + 1}{x - 2}$

34. $\dfrac{2x^2 - x + 2}{x + 4}$

35. $\dfrac{2x^3 + x^2 - 2x}{2x + 1}$

36. $\dfrac{1 - x^2 + x^3}{x - 1}$

37. $\dfrac{x^3 - x^2 + x + 1}{x^2 + 1}$

38. $\dfrac{2x^3 + x^2 - x + 4}{x^2 + x}$

Synthetic Division

Exercises 39–46: Use synthetic division to divide the first polynomial by the second.

39. $x^3 + 2x^2 - 17x - 10$ $x + 5$

40. $x^3 - 2x + 1$ $x + 4$

41. $3x^3 - 11x^2 - 20x + 3$ $x - 5$

42. $x^4 - 3x^3 - 5x^2 + 2x - 16$ $x - 3$

43. $x^4 - 3x^3 - 4x^2 + 12x$ $x - 2$

44. $x^5 + \frac{1}{4}x^4 - x^3 - \frac{1}{4}x^2 + 3x - \frac{5}{4}$ $x + \frac{1}{4}$

45. $2x^5 - x^4 - x^3 + 4x + 3 \quad x + \frac{1}{2}$

46. $x^4 - \frac{1}{2}x^3 + 3x^2 - \frac{5}{2}x + \frac{9}{2} \quad x - \frac{1}{2}$

Remainder Theorem

Exercises 47–50: Use the remainder theorem to find the remainder when $f(x)$ is divided by the given $x - k$.

47. $f(x) = 5x^2 - 3x + 1 \qquad x - 1$

48. $f(x) = -4x^2 + 6x - 7 \qquad x + 4$

49. $f(x) = 4x^3 - x^2 + 4x + 2 \quad x + 2$

50. $f(x) = -x^4 + 4x^3 - x + 3 \quad x - 3$

Applications

51. Area of a Rectangle Use the figure to find the length L of the rectangle from its width and area A. Determine the value of L when $x = 10$ feet.

$3x + 1$ | $A = 12x^2 + 13x + 3$

L

52. Area of a Rectangle Use the figure to find the width W of the rectangle from its length and area A. Determine the value of W when $x = 5$ inches.

W | $A = 3x^3 - 5x^2 + 3x - 5$

$x^2 + 1$

Writing About Mathematics

53. Compare division of integers to division of polynomials. Give examples.

54. When can you use synthetic division to divide two polynomials? Give one example where synthetic division can be used and one example where it cannot be used.

4.4 Real Zeros of Polynomial Functions

- **Understand the factor theorem**
- **Factor higher degree polynomials**
- **Analyze polynomials having multiple zeros**
- **Solve higher degree polynomial equations**
- **Understand the rational zeros test, Descartes' rule of signs, and the intermediate value property**

Introduction

Some species of birds, such as robins, have two nesting periods each summer. Because the survival rate for young birds is low, bird populations can fluctuate greatly during the summer months.   (*Source:* S. Kress, *Bird Life.*)

The graph of $f(x) = x^3 - 61x^2 + 839x + 4221$ shown in Figure 4.57 models a population of birds in a small county, x days after May 31.

A Summer Bird Population

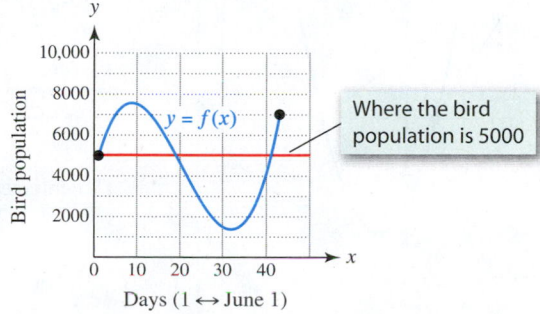

Where the bird population is 5000

Figure 4.57

If we want to determine the dates when the population was **5000**, we can solve

$$x^3 - 61x^2 + 839x + 4221 = 5000.$$

From the graph of f on the preceding page it appears that there were 5000 birds around June 1 ($x = 1$), June 20 ($x = 20$), and July 10 ($x = 40$). See Example 5.

Factoring Polynomials

The polynomial $f(x) = x^2 - 3x + 2$ can be factored as $f(x) = (x - 1)(x - 2)$. Note that $f(\mathbf{1}) = 0$ and $(x - \mathbf{1})$ is a factor of $f(x)$. Similarly, $f(\mathbf{2}) = 0$ and $(x - \mathbf{2})$ is a factor. This discussion can be generalized. By the remainder theorem we know that

$$f(x) = (x - k)q(x) + r,$$

where r is the remainder. If $r = 0$, then $f(x) = (x - k)q(x)$ and $(x - k)$ is a factor of $f(x)$. Similarly, if $(x - k)$ is a factor of $f(x)$, then $r = 0$. That is,

$$f(x) = (x - k)q(x)$$

and $f(k) = (k - k)q(k) = 0 \cdot q(k) = 0$. This discussion justifies the *factor theorem*.

> **FACTOR THEOREM**
>
> A polynomial $f(x)$ has a factor $x - k$ if and only if $f(k) = 0$.

EXAMPLE 1 Applying the factor theorem

Use the graph in Figure 4.58 and the factor theorem to list the factors of $f(x)$.

SOLUTION
Figure 4.58 shows that the **zeros** (or x-intercepts) of f are $-\mathbf{2}, \mathbf{1}$, and $\mathbf{3}$. Since $f(-\mathbf{2}) = 0$, the factor theorem states that $(x + 2)$ is a factor of $f(x)$. Similarly, $f(\mathbf{1}) = 0$ implies that $(x - 1)$ is a factor, and $f(\mathbf{3}) = 0$ implies that $(x - 3)$ is a factor. Thus the factors of $f(x)$ are $(x + 2)$, $(x - 1)$, and $(x - 3)$.

NOTE If $f(-2) = 0$, then $(x - (-2))$, or $(x + 2)$, is a factor.

Now Try Exercise 1

Figure 4.58

See the Concept: x-Intercepts, Zeros, and Factors

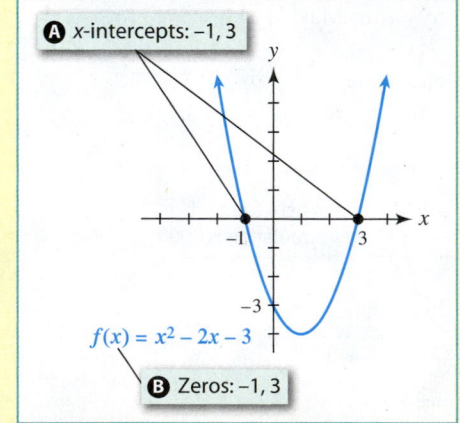

Ⓐ x-intercepts: $-1, 3$

$f(x) = x^2 - 2x - 3$

Ⓑ Zeros: $-1, 3$

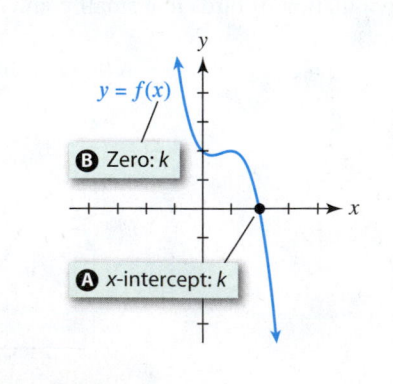

Ⓑ Zero: k

Ⓐ x-intercept: k

Equivalent Concepts:

- The graph of $y = f(x)$ has x-intercept k.

- A real zero of $f(x)$ is k.

- A factor of $f(x)$ is $(x - k)$.

Ⓑ $f(-1) = 0$ and $f(3) = 0$

Ⓒ Factors of $x^2 - 2x - 3$ are $(x + 1)$ and $(x - 3)$.

Ⓑ $f(k) = 0$

Ⓒ Because k is a zero, $(x - k)$ is a factor of $f(x)$.

Zeros with Multiplicity If $f(x) = (x + 2)^2$, then the factor $(x + 2)$ occurs twice and the zero -2 is called a **zero of multiplicity** **2**. See Figure 4.59. The polynomial $g(x) = (x + 1)^3(x - 2)$ has zeros -1 and 2 with *multiplicities* **3** and **1**, respectively. See Figure 4.60. *Counting multiplicities*, a polynomial of degree n has at most n real zeros. For $g(x)$, the sum of the multiplicities is $3 + 1 = 4$, which equals its degree.

See the Concept: Zeros, x-Intercepts, and Multiplicities

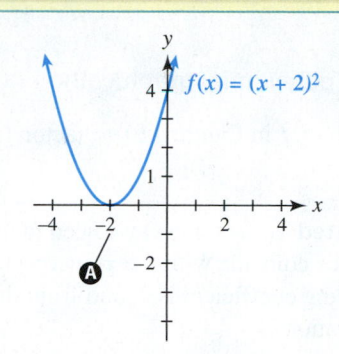

Figure 4.59

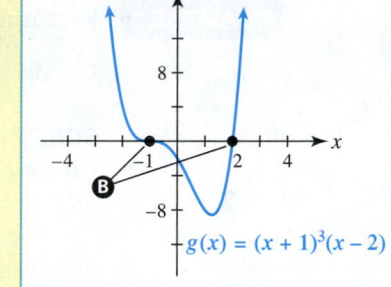

Figure 4.60

🅐 The *x*-intercept corresponds to the zero -2, which has multiplicity **2**.

🅑 The two *x*-intercepts correspond to zeros -1 and 2, which have multiplicities **3** and **1**, respectively.

Complete Factored Form The concepts discussed above, together with the factor theorem, can be used to find the *complete factored form* of a polynomial. The complete factored form of a polynomial is *unique*.

COMPLETE FACTORED FORM

Suppose a polynomial

$$f(x) = a_n x^n + \cdots + a_2 x^2 + a_1 x + a_0$$

has n real zeros $c_1, c_2, c_3, \ldots, c_n$, where a distinct zero is listed as many times as its multiplicity. Then $f(x)$ can be written in **complete factored form** as

$$f(x) = a_n(x - c_1)(x - c_2)(x - c_3) \cdots (x - c_n).$$

EXAMPLE 2 **Finding a complete factorization**

Write the complete factorization for each polynomial with the given zeros.
(a) $f(x) = 13x^2 - 13x - 26$;   zeros: -1 and 2
(b) $f(x) = 7x^3 - 21x^2 - 7x + 21$;   zeros: -1, 1, and 3

SOLUTION
(a) The leading coefficient is **13** and the zeros are **-1** and **2**. By the factor theorem, $(x + 1)$ and $(x - 2)$ are factors. The complete factorization is

$$f(x) = \mathbf{13}(x - (\mathbf{-1}))(x - \mathbf{2}) = 13(x + 1)(x - 2).$$

> *f* is degree 2 and has two factors.

(b) The leading coefficient is 7 and the zeros are -1, 1, and 3. The complete factorization is

$$f(x) = \mathbf{7}(x + 1)(x - 1)(x - 3).$$

> *f* is degree 3 and has three factors.

Now Try Exercises 5 and 7

MAKING CONNECTIONS

Types of Factored Forms   If the leading coefficient of a polynomial $f(x)$ is 6 and the only zeros are $\frac{1}{3}$ and $\frac{1}{2}$, then the complete factored form is $f(x) = 6\left(x - \frac{1}{3}\right)\left(x - \frac{1}{2}\right)$ and it is *unique*. Sometimes we factor 6 as $3 \cdot 2$ and distribute the 3 over the first factor and the 2 over the second factor to obtain the slightly different factored form of $f(x) = (3x - 1)(2x - 1)$.

EXAMPLE 3 | **Factoring a polynomial graphically**

Use the graph of f in Figure 4.61 to factor $f(x) = 2x^3 - 4x^2 - 10x + 12$.

SOLUTION

Getting Started To factor $f(x)$ we need to determine the leading coefficient and the zeros of f. The zeros coincide with the x-intercepts of the graph of f. ▶

The leading coefficient is **2**, and from the graph the zeros are -2, 1, and 3. The complete factorization is

$$f(x) = \mathbf{2}(x + 2)(x - 1)(x - 3).$$

> f is degree 3 and has three factors.

> **Now Try Exercise 13**

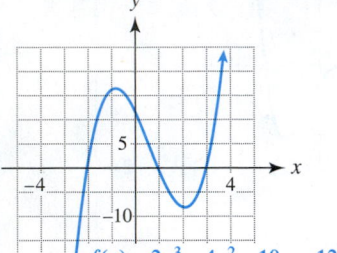

Figure 4.61

When factoring polynomials by hand, it is sometimes helpful to use the techniques of division discussed in Section 4.3.

EXAMPLE 4 | **Factoring a polynomial symbolically**

One of the zeros of the polynomial $f(x) = 2x^3 - 2x^2 - 34x - 30$ is -1. Express $f(x)$ in complete factored form.

SOLUTION

If -1 is a zero, then by the factor theorem $(x + 1)$ is a factor. To factor $f(x)$, divide $x + 1$ into $2x^3 - 2x^2 - 34x - 30$ by using synthetic division.

$$
\begin{array}{r|rrrr}
-1 & 2 & -2 & -34 & -30 \\
   &   & -2 & 4 & 30 \\
\hline
   & 2 & -4 & -30 & 0
\end{array}
$$

> Long division could also be used.

Algebra Review

To review factoring trinomials, see Chapter R (pages R22–R23).

The remainder is **0**, so $x + 1$ divides evenly into the dividend. By the division algorithm,

$$2x^3 - 2x^2 - 34x - 30 = (x + 1)(2x^2 - 4x - 30).$$

The quotient $2x^2 - 4x - 30$ can be factored further.

$$2x^2 - 4x - 30 = 2(x^2 - 2x - 15) \qquad \text{Factor out 2.}$$
$$= \mathbf{2}(x + 3)(x - 5) \qquad \text{Factor trinomial.}$$

The complete factored form is $f(x) = \mathbf{2}(x + 1)(x + 3)(x - 5)$.

> **Now Try Exercise 31**

An Application   In the introduction to this section we presented the equation

$$x^3 - 61x^2 + 839x + 4221 = 5000, \qquad \textit{Equation to be solved}$$

which can be rewritten as

$$x^3 - 61x^2 + 839x - 779 = 0. \qquad \textit{Subtract 5000.}$$

This second equation gives the days when a summer bird population was 5000 and is solved in the next example.

EXAMPLE 5 Factoring a polynomial

Factor $g(x) = x^3 - 61x^2 + 839x - 779$. Use the zeros of $g(x)$ to determine when the bird population was 5000.

A Summer Bird Population

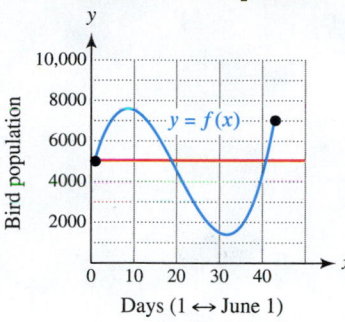

Days (1 ↔ June 1)

SOLUTION

From Figure 4.57, repeated in the margin, it appears that the bird population was 5000 when $x = 1$. If we substitute $x = 1$ in this polynomial, the result is **0**.

$$g(1) = 1^3 - 61(1)^2 + 839(1) - 779 = 0$$

By the factor theorem, $(x - 1)$ is a factor of $g(x)$. We can use synthetic division to divide $x^3 - 61x^2 + 839x - 779$ by $x - 1$.

$$\begin{array}{r|rrrr} 1\rfloor & 1 & -61 & 839 & -779 \\ & & 1 & -60 & 779 \\ \hline & 1 & -60 & 779 & 0 \end{array}$$ *The remainder is 0.*

By the division algorithm,

$$x^3 - 61x^2 + 839x - 779 = (x - 1)(x^2 - 60x + 779).$$

Since it is not obvious how to factor $x^2 - 60x + 779$, we can use the quadratic formula with $a = 1$, $b = -60$, and $c = 779$ to find its zeros.

$$\begin{aligned} x &= \frac{-b \pm \sqrt{b^2 - 4ac}}{2a} & \text{\textit{Quadratic formula}} \\ &= \frac{-(-60) \pm \sqrt{(-60)^2 - 4(1)(779)}}{2(1)} & \text{\textit{a = 1, b = -60, c = 779}} \\ &= \frac{60 \pm 22}{2} & \text{\textit{Simplify.}} \\ &= 41 \text{ or } 19 & \text{\textit{Two zeros}} \end{aligned}$$

The zeros of $g(x) = x^3 - 61x^2 + 839x - 779$ are 1, 19, and 41, and its leading coefficient is 1. The complete factorization is $g(x) = (x - 1)(x - 19)(x - 41)$. The bird population equals 5000 on June 1 ($x = 1$), June 19 ($x = 19$), and July 11 ($x = 41$).

Now Try Exercise 113

Graphs and Multiple Zeros

The following See the Concept explains graphs of polynomials with multiple zeros. See Figures 4.62 and 4.63.

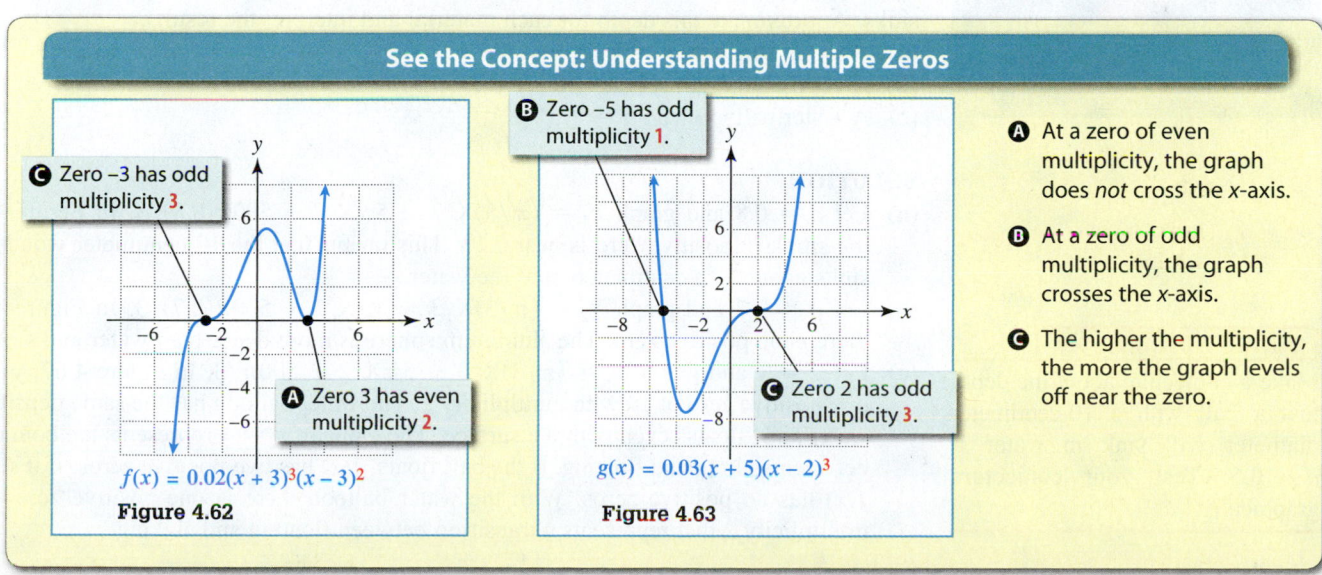

See the Concept: Understanding Multiple Zeros

Ⓒ Zero −3 has odd multiplicity **3**.

Ⓐ Zero 3 has even multiplicity **2**.

$f(x) = 0.02(x + 3)^3(x - 3)^2$

Figure 4.62

Ⓑ Zero −5 has odd multiplicity **1**.

Ⓒ Zero 2 has odd multiplicity **3**.

$g(x) = 0.03(x + 5)(x - 2)^3$

Figure 4.63

Ⓐ At a zero of even multiplicity, the graph does *not* cross the x-axis.

Ⓑ At a zero of odd multiplicity, the graph crosses the x-axis.

Ⓒ The higher the multiplicity, the more the graph levels off near the zero.

EXAMPLE 6 Finding multiplicities graphically

Figure 4.64 shows the graph of a sixth-degree polynomial $f(x)$ with leading coefficient 1. All zeros are integers. Write $f(x)$ in complete factored form.

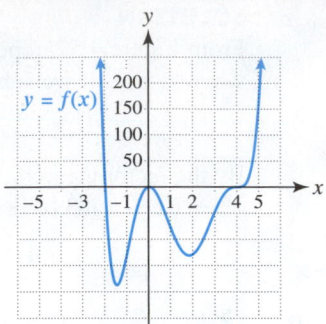

Figure 4.64

SOLUTION

The x-intercepts or zeros of f are -2, 0, and 4. Since the graph crosses the x-axis at -2 and 4, these zeros have odd multiplicity. The graph of f levels off more at $x = 4$ than at $x = -2$, so 4 has a higher multiplicity than -2. At $x = 0$ the graph of f does not cross the x-axis. Thus 0 has even multiplicity. To make $f(x)$ a sixth degree polynomial, -2 has multiplicity 1, 0 has multiplicity 2, and 4 has multiplicity 3. Then the sum of the multiplicities is given by $1 + 2 + 3 = 6$, which equals the degree of $f(x)$. List the zeros as $-2, 0, 0, 4, 4,$ and 4. The leading coefficient is 1, so the complete factorization of $f(x)$ is

$$f(x) = 1(x + 2)(x - 0)(x - 0)(x - 4)(x - 4)(x - 4), \quad \text{or}$$
$$f(x) = x^2(x + 2)(x - 4)^3.$$

Now Try Exercise 47

An Application: "Will It Float?" Multiple zeros can have physical significance. The next example shows how a multiple zero represents the boundary between an object floating and sinking.

EXAMPLE 7 Interpreting a multiple zero

The polynomial $f(x) = \frac{\pi}{3}x^3 - 5\pi x^2 + \frac{500\pi d}{3}$ can be used to find the depth that a ball, 10 centimeters in diameter, sinks in water. The constant d is the density of the ball, where the density of water is 1. The smallest *positive* zero of $f(x)$ equals the depth that the sphere sinks. Approximate this depth for each material and interpret the results.
 (a) A wood ball with $d = 0.8$
 (b) A solid aluminum sphere with $d = 2.7$
 (c) A water balloon with $d = 1$

SOLUTION

(a) Let $d = 0.8$ and graph $Y_1 = (\pi/3)X^3 - 5\pi X^2 + 500\pi(0.8)/3$. In Figure 4.65 the smallest positive zero is near 7.13. This means that the 10-centimeter wood ball sinks about 7.13 centimeters into the water.

(b) Let $d = 2.7$ and graph $Y_2 = (\pi/3)X^3 - 5\pi X^2 + 500\pi(2.7)/3$. In Figure 4.66 there is no positive zero. The aluminum sphere is more dense than water and sinks.

(c) Let $d = 1$ and graph $Y_3 = (\pi/3)X^3 - 5\pi X^2 + 500\pi/3$. In Figure 4.67, y_3 has one positive zero of 10 with multiplicity 2. The water balloon has the same density as water and "floats" even with the surface. The value of $d = 1$ represents the boundary between sinking and floating. If the ball floats, $f(x)$ has two positive zeros; if it sinks, $f(x)$ has no positive zeros. With the water balloon there is one positive zero with multiplicity 2 that represents a transition between floating and sinking.

> **CLASS DISCUSSION**
>
> Make a conjecture about the depth that a ball with a 10-centimeter diameter will sink in water if $d = 0.5$. Test your conjecture graphically.

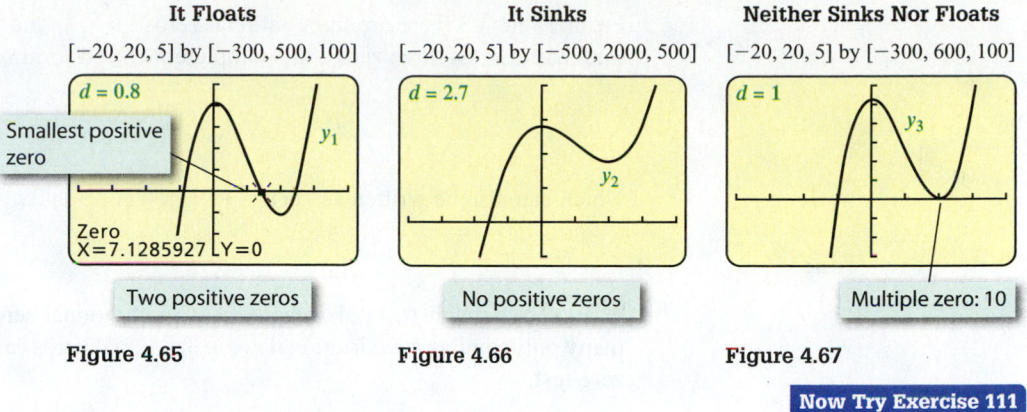

It Floats	It Sinks	Neither Sinks Nor Floats
$[-20, 20, 5]$ by $[-300, 500, 100]$	$[-20, 20, 5]$ by $[-500, 2000, 500]$	$[-20, 20, 5]$ by $[-300, 600, 100]$

Figure 4.65   Figure 4.66   Figure 4.67

Now Try Exercise 111

Rational Zeros

If a polynomial has a rational zero, it can be found by using the **rational zero test**.

RATIONAL ZERO TEST

Let $f(x) = a_n x^n + \cdots + a_2 x^2 + a_1 x + a_0$, where $a_n \neq 0$, represent a polynomial function f with *integer* coefficients. If $\frac{p}{q}$ is a rational number written in lowest terms and if $\frac{p}{q}$ is a zero of f, then p is a factor of the constant term a_0 and q is a factor of the leading coefficient a_n.

The following example illustrates how to find rational zeros by using this test.

EXAMPLE 8   **Finding rational zeros of a polynomial**

Find all rational zeros of $f(x) = 6x^3 - 5x^2 - 7x + 4$ and factor $f(x)$.

SOLUTION

If $\frac{p}{q}$ is a rational zero in lowest terms, then p is a factor of the constant term 4 and q is a factor of the leading coefficient 6. The possible values for p and q are as follows.

$$p: \quad \pm 1, \quad \pm 2, \quad \pm 4$$
$$q: \quad \pm 1, \quad \pm 2, \quad \pm 3, \quad \pm 6$$

As a result, any rational zero of $f(x)$ in the form $\frac{p}{q}$ must occur in the list

$$\pm \frac{1}{6}, \quad \pm \frac{1}{3}, \quad \pm \frac{1}{2}, \quad \pm \frac{2}{3}, \quad \pm \frac{1}{1}, \quad \pm \frac{4}{3}, \quad \pm \frac{2}{1}, \quad \text{or} \quad \pm \frac{4}{1}.$$

Evaluate $f(x)$ at each value in the list. See Table 4.3.

x	$f(x)$	x	$f(x)$	x	$f(x)$	x	$f(x)$
$\frac{1}{6}$	$\frac{49}{18}$	$\frac{1}{2}$	0	1	-2	2	18
$-\frac{1}{6}$	5	$-\frac{1}{2}$	$\frac{11}{2}$	-1	0	-2	-50
$\frac{1}{3}$	$\frac{4}{3}$	$\frac{2}{3}$	$-\frac{10}{9}$	$\frac{4}{3}$	0	4	280
$-\frac{1}{3}$	$\frac{50}{9}$	$-\frac{2}{3}$	$\frac{14}{3}$	$-\frac{4}{3}$	$-\frac{88}{9}$	-4	-432

Table 4.3

From Table 4.3 there are three rational zeros: $-1, \frac{1}{2}$, and $\frac{4}{3}$. Since a third-degree polynomial has at most three zeros, the complete factored form of $f(x)$ is

$$f(x) = 6(x + 1)\left(x - \frac{1}{2}\right)\left(x - \frac{4}{3}\right),$$

which can also be written as $f(x) = (x + 1)(2x - 1)(3x - 4)$.

Now Try Exercise 57

> **NOTE** Although $f(x)$ in Example 8 had only rational zeros, it is important to realize that many polynomials have irrational zeros. Irrational zeros cannot be found using the rational zero test.

Descartes' Rule of Signs

Descartes' rule of signs helps to determine the numbers of positive and negative real zeros of a polynomial function.

DESCARTES' RULE OF SIGNS

Let $P(x)$ define a polynomial function with real coefficients and a nonzero constant term, with terms in descending powers of x.
(a) The number of positive real zeros either equals the number of variations in sign occurring in the coefficients of $P(x)$ or is less than the number of variations by a positive even integer.
(b) The number of negative real zeros either equals the number of variations in sign occurring in the coefficients of $P(-x)$ or is less than the number of variations by a positive even integer.

A **variation in sign** is a change from positive to negative or negative to positive in successive terms of the polynomial when written in descending powers of the variable. Missing terms (those with 0 coefficients) can be ignored.

EXAMPLE 9 Applying Descartes' rule of signs

Determine the possible numbers of positive real zeros and negative real zeros of $P(x) = x^4 - 6x^3 + 8x^2 + 2x - 1$.

SOLUTION

We first consider the possible number of positive zeros by observing that $P(x)$ has three variations in sign.

$$+x^4 - 6x^3 + 8x^2 + 2x - 1 \qquad \text{\color{blue}Three variations in sign.}$$
$$\color{blue}1 \qquad 2 \qquad 3$$

Thus by Descartes' rule of signs, $P(x)$ has either **3** or $3 - 2 = \mathbf{1}$ positive real zeros.
For negative zeros, consider the variations in sign for $P(-x)$.

$$P(-x) = (-x)^4 - 6(-x)^3 + 8(-x)^2 + 2(-x) - 1$$
$$= x^4 + 6x^3 + 8x^2 - 2x - 1 \qquad \text{\color{blue}One variation in sign.}$$
$$\color{blue}1$$

Since there is only one variation in sign, $P(x)$ has only **1** negative real zero.

Now Try Exercise 65

Polynomial Equations

In Section 3.2, factoring was used to solve quadratic equations. Factoring also can be used to solve polynomial equations with degree greater than 2.

EXAMPLE 10 Solving a cubic equation

Solve $x^3 + 3x^2 - 4x = 0$ symbolically. Support your answer graphically and numerically.

SOLUTION
Symbolic Solution

$$
\begin{array}{ll}
x^3 + 3x^2 - 4x = 0 & \text{Given equation} \\
x(x^2 + 3x - 4) = 0 & \text{Factor out x.} \\
x(x + 4)(x - 1) = 0 & \text{Factor the quadratic expression} \\
x = 0, \quad x + 4 = 0, \quad \text{or} \quad x - 1 = 0 & \text{Zero-product property} \\
x = 0, -4, \text{ or } 1 & \text{Solve.}
\end{array}
$$

Graphical Solution   Graph $y = x^3 + 3x^2 - 4x$ as in Figure 4.68. The x-intercepts are -4, 0, and 1, which correspond to the solutions.

Numerical Solution   Table $y = x^3 + 3x^2 - 4x$ as in Figure 4.69. The zeros of y occur at $x = -4$, 0, and 1.

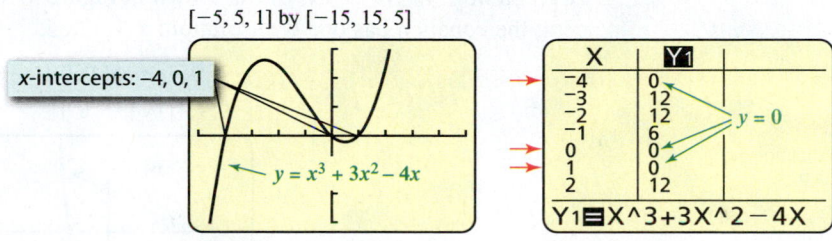

Solving $x^3 + 3x^2 - 4x = 0$

Figure 4.68   **Figure 4.69**

Now Try Exercise 71

EXAMPLE 11 Solving a polynomial equation

Find all real solutions to each equation symbolically.
(a) $4x^4 - 5x^2 - 9 = 0$ **(b)** $2x^3 + 12 = 3x^2 + 8x$

SOLUTION
(a) The expression $4x^4 - 5x^2 - 9$ can be factored in a manner similar to the way quadratic expressions are factored.

$$
\begin{array}{ll}
4x^4 - 5x^2 - 9 = 0 & \text{Given equation} \\
(4x^2 - 9)(x^2 + 1) = 0 & \text{Factor.} \\
4x^2 - 9 = 0 \quad \text{or} \quad x^2 + 1 = 0 & \text{Zero-product property} \\
4x^2 = 9 \quad \text{or} \quad x^2 = -1 & \text{Add 9 or subtract 1.} \\
x^2 = \dfrac{9}{4} \quad \text{or} \quad x^2 = -1 & \text{Divide left equation by 4.} \\
x = \pm\dfrac{3}{2} \quad \text{or} \quad x^2 = -1 & \text{Square root property}
\end{array}
$$

The equation $x^2 = -1$ has no *real* solutions. The solutions are $-\frac{3}{2}$ and $\frac{3}{2}$.

Algebra Review

To review factoring a cubic polynomial by grouping, see Chapter R (page R-20).

(b) First transpose each term on the right side of the equation to the left side of the equation. Then use *grouping* to factor the polynomial.

$$2x^3 + 12 = 3x^2 + 8x \qquad \text{Given equation}$$
$$2x^3 - 3x^2 - 8x + 12 = 0 \qquad \text{Rewrite the equation.}$$
$$(2x^3 - 3x^2) + (-8x + 12) = 0 \qquad \text{Associative property}$$
$$x^2(2x - 3) - 4(2x - 3) = 0 \qquad \text{Factor.}$$
$$(x^2 - 4)(2x - 3) = 0 \qquad \text{Factor out } 2x - 3.$$
$$x^2 - 4 = 0 \quad \text{or} \quad 2x - 3 = 0 \qquad \text{Zero-product property}$$
$$x = \pm 2 \quad \text{or} \qquad x = \frac{3}{2} \qquad \text{Solve each equation.}$$

The solutions are -2, $\frac{3}{2}$, and 2.

Now Try Exercises 79 and 89

Some types of polynomial equations cannot be solved easily by factoring. The next example illustrates how we can obtain an approximate solution graphically.

EXAMPLE 12   **Finding a solution graphically**

Solve the equation $\frac{1}{2}x^3 - 2x - 4 = 0$ graphically. Round to the nearest hundredth.

SOLUTION

A graph of $y = \frac{1}{2}x^3 - 2x - 4$ is shown in Figure 4.70. Since there is only one x-intercept, the equation has one *real* solution: $x \approx 2.65$.

Graphical Solution

$[-9, 9, 1]$ by $[-6, 6, 1]$

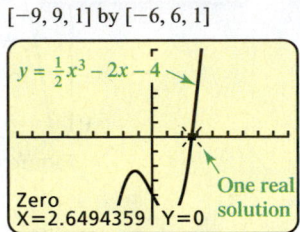

Figure 4.70

Calculator Help

To find a zero of a function, see Appendix A (page AP-7).

Now Try Exercise 97

Intermediate Value Property

In Example 12, we approximated a solution to $\frac{1}{2}x^3 - 2x - 4 = 0$ to be 2.65. How do we know for sure that there is indeed such a solution? The *intermediate value property* helps answer this question.

> **INTERMEDIATE VALUE PROPERTY**
>
> Let (x_1, y_1) and (x_2, y_2), with $y_1 \neq y_2$ and $x_1 < x_2$, be two points on the graph of a continuous function f. Then, on the interval $x_1 \leq x \leq x_2$, f assumes every value between y_1 and y_2 at least once.

From Example 12, let $f(x) = \frac{1}{2}x^3 - 2x - 4$. Because $f(0) = -4$ and $f(3) = 3.5$, the points $(0, -4)$ and $(3, 3.5)$ lie on the graph of f in Figure 4.70. By the intermediate value property, $f(x) = 0$ for at least one x-value, because 0 is between -4 and 3.5. Although we have not found the exact x-value, we know that a real zero of f does exist. Loosely speaking, the intermediate value property says that if $y_1 < 0$ and $y_2 > 0$, then we cannot draw a continuous graph of f without crossing the x-axis. The only way not to cross the x-axis would be to pick up the pencil, but this would create a discontinuous graph.

Applications There are many examples of the intermediate value property. Physical motion is usually considered to be continuous. Suppose at one time a car is traveling at 20 miles per hour and at another time it is traveling at 40 miles per hour. It is logical to assume that the car traveled 30 miles per hour at least once between these times. In fact, by the intermediate value property, the car must have assumed all speeds between 20 and 40 miles per hour at least once. Similarly, if a jet airliner takes off and flies at an altitude of 30,000 feet, then by the intermediate value property we may conclude that the airliner assumed all altitudes between ground level and 30,000 feet at least once.

4.4 Putting It All Together

$\mathbf{T}$he following table summarizes important topics about real zeros of polynomial functions.

CONCEPT	EXPLANATION	EXAMPLES
Factor theorem	$(x - k)$ is a factor of $f(x)$ if and only if $f(k) = 0$.	$f(x) = x^2 + 3x - 4$, $f(-4) = 0$, and $f(1) = 0$ imply that $(x + 4)$ and $(x - 1)$ are factors of $f(x)$. That is, $f(x) = (x + 4)(x - 1)$.
x-intercepts, zeros, and factors	The following are *equivalent:* **1.** The graph of f has x-intercept k. **2.** A real zero of f is k. That is, $f(k) = 0$. **3.** A factor of f is $(x - k)$.	Let $f(x) = x^2 - 2x - 3$. See the graph below. **1.** The graph of f has x-intercepts -1 and 3. **2.** $f(-1) = 0$ and $f(3) = 0$ **3.** $f(x) = (x + 1)(x - 3)$
Complete factored form	$f(x) = a_n(x - c_1) \cdots (x - c_n)$, where the c_k are zeros of f, with a distinct zero listed as many times as its multiplicity. This form is unique.	$\begin{aligned} f(x) &= 3(x - 5)(x + 3)(x + 3) \\ &= 3(x - 5)(x + 3)^2 \end{aligned}$ $a_n = 3$, $c_1 = 5$, $c_2 = -3$, $c_3 = -3$
Factoring a polynomial graphically (only real zeros)	Graph $y = f(x)$ and locate all the zeros or x-intercepts. If the leading coefficient is a and the zeros are $c_1, c_2,$ and c_3, then $f(x) = a(x - c_1)(x - c_2)(x - c_3)$.	$f(x) = 2x^3 + 4x^2 - 2x - 4$ has zeros -2, -1, and 1 and leading coefficient 2. Thus $$f(x) = 2(x + 2)(x + 1)(x - 1).$$
Solving polynomial equations	Polynomial equations can be solved symbolically, graphically, and numerically. Factoring is a useful symbolic technique.	Solve $x^3 - 4x^2 - 5x = 0$. $\begin{aligned} x(x^2 - 4x - 5) &= 0 \\ x(x - 5)(x + 1) &= 0 \\ x = 0, 5, \text{ or } -1 \end{aligned}$ See also Examples 10 and 11.

continued on next page

The following explains the concepts of multiple zeros and graphs of polynomials.

CONCEPT	DESCRIPTION
Zeros and multiplicity	If a zero of a polynomial has odd multiplicity, then the graph crosses the x-axis at the zero. If a polynomial has a zero of even multiplicity, then the graph intersects, but does not cross, the x-axis at the zero. The higher the multiplicity of a zero, the more the graph levels off near the zero.

Multiplicity 1 (odd) **Multiplicity 2 (even)** **Multiplicity 3 (odd)** **Multiplicity 4 (even)**

4.4 Exercises

Factoring Polynomials

Exercises 1–4: Use the graph and the factor theorem to list the factors of $f(x)$.

1.
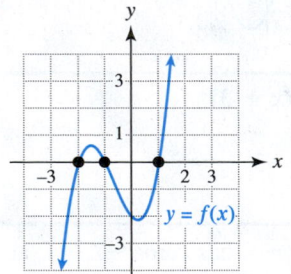

2.
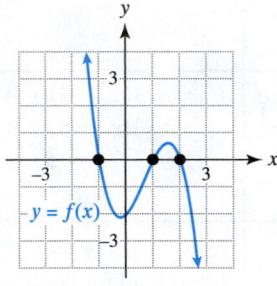

3.
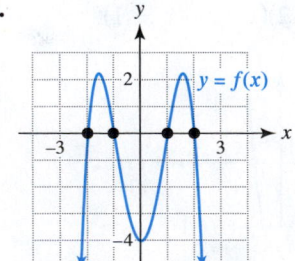

4.
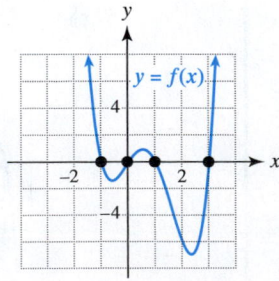

Exercises 5–12: Use the given zeros to write the complete factored form of $f(x)$.

5. $f(x) = 2x^2 - 25x + 77$; zeros: $\frac{11}{2}$ and 7

6. $f(x) = 6x^2 + 21x - 90$; zeros: -6 and $\frac{5}{2}$

7. $f(x) = x^3 - 2x^2 - 5x + 6$; zeros: -2, 1, and 3

8. $f(x) = x^3 + 6x^2 + 11x + 6$; zeros: -3, -2, and -1

9. $f(x) = -2x^3 + 3x^2 + 59x - 30$; zeros: $-5, \frac{1}{2},$ and 6

10. $f(x) = 3x^4 - 8x^3 - 67x^2 + 112x + 240$; zeros: $-4, -\frac{4}{3}, 3,$ and 5

11. Let $f(x)$ be a quadratic polynomial with leading coefficient 7. Suppose that $f(-3) = 0$ and $f(2) = 0$. Write the complete factored form of $f(x)$.

12. Let $g(x)$ be a cubic polynomial with leading coefficient -4. Suppose that $g(-2) = 0, g(1) = 0,$ and $g(4) = 0$. Write the complete factored form of $g(x)$.

Exercises 13 and 14: Use the graph to factor $f(x)$.

13. $f(x) = -2x^3 + 2x$

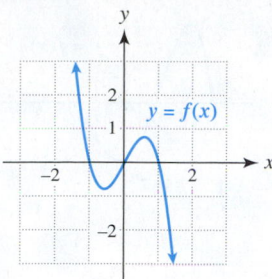

14. $f(x) = \frac{1}{4}x^4 - \frac{3}{2}x^3 + \frac{3}{4}x^2 + \frac{13}{2}x - 6$

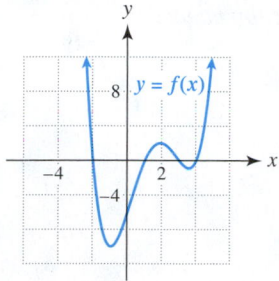

Exercises 15–18: The graph of a polynomial $f(x)$ with leading coefficient ± 1 and integer zeros is shown in the figure. Write its complete factored form.

15.

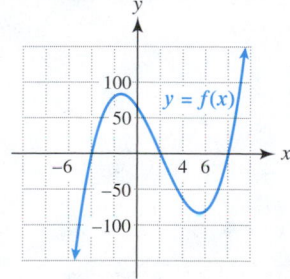

16.

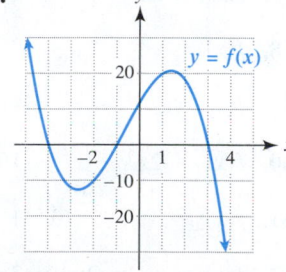

17.

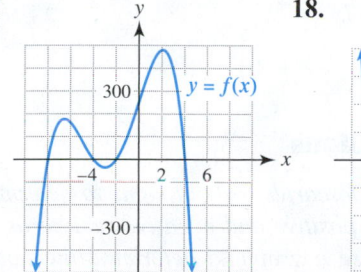

18.

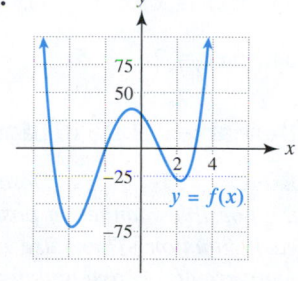

19. Let $f(x)$ be a cubic polynomial with zeros -1, 2, and 3. If the graph of f passes through the point $(0, 3)$, write the complete factored form of $f(x)$.

20. Let $g(x)$ be a quartic polynomial with zeros -2, -1, 1, and 2. If the graph of g passes through the point $(0, 8)$, write the complete factored form of $g(x)$.

Exercises 21–24: The graph of a polynomial $f(x)$ with integer zeros is shown in the figure. Write its complete factored form. Note that the leading coefficient of $f(x)$ is not ± 1.

21.

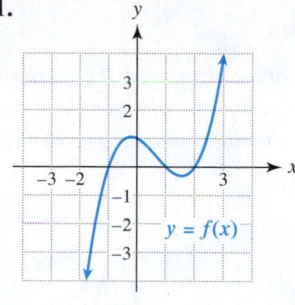

22.

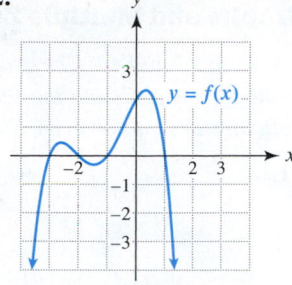

23.

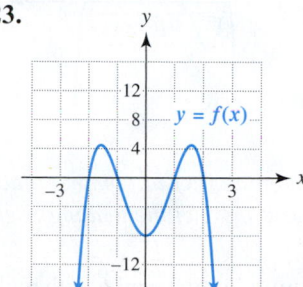

24.

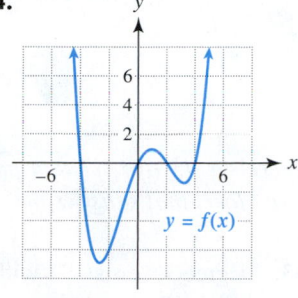

Exercises 25–30: Use graphing to factor $f(x)$.

25. $f(x) = 10x^2 + 17x - 6$

26. $f(x) = 2x^3 + 7x^2 + 2x - 3$

27. $f(x) = -3x^3 - 3x^2 + 18x$

28. $f(x) = \frac{1}{2}x^3 + \frac{5}{2}x^2 + x - 4$

29. $f(x) = x^4 + \frac{5}{2}x^3 - 3x^2 - \frac{9}{2}x$

30. $f(x) = 10x^4 + 7x^3 - 27x^2 + 2x + 8$

Exercises 31–36: (Refer to Example 4.) Write the complete factored form of the polynomial $f(x)$, given that k is a zero.

31. $f(x) = x^3 - 9x^2 + 23x - 15$ $\qquad k = 1$

32. $f(x) = 2x^3 + x^2 - 11x - 10$ $\qquad k = -2$

33. $f(x) = -4x^3 - x^2 + 51x - 36$ $\qquad k = -4$

34. $f(x) = 3x^3 - 11x^2 - 35x + 75$ $\qquad k = 5$

35. $f(x) = 2x^4 - x^3 - 13x^2 - 6x$ $\qquad k = -2$

36. $f(x) = 35x^4 + 48x^3 - 41x^2 + 6x$ $\qquad k = \frac{3}{7}$

Factor Theorem

Exercises 37–40: Use the factor theorem to decide if $x - k$ is a factor of $f(x)$ for the given k.

37. $f(x) = x^3 - 6x^2 + 11x - 6$ $\qquad k = 2$

38. $f(x) = x^3 + x^2 - 14x - 24$ $\qquad k = -3$

39. $f(x) = x^4 - 2x^3 - 13x^2 - 10x$ $k = 3$

40. $f(x) = 2x^4 - 11x^3 + 9x^2 + 14x$ $k = \frac{1}{2}$

Graphs and Multiple Zeros

Exercises 41 and 42: The graph of a polynomial f(x) is shown in the figure. Estimate the zeros and state whether their multiplicities are odd or even. State the minimum degree of f(x).

41.
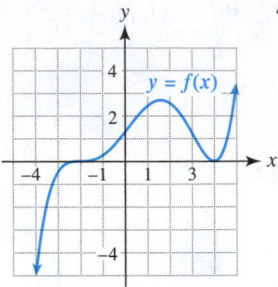

42.
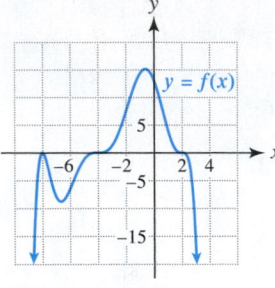

Exercises 43–46: Write a polynomial f(x) in complete factored form that satisfies the conditions. Let the leading coefficient be 1.

43. Degree 3; zeros: −1 with multiplicity 2, and 6 with multiplicity 1

44. Degree 4; zeros: 5 and 7, both with multiplicity 2

45. Degree 4; zeros: 2 with multiplicity 3, and 6 with multiplicity 1

46. Degree 5; zeros: −2 with multiplicity 2, and 4 with multiplicity 3

Exercises 47–52: The graph of either a cubic, quartic, or quintic polynomial f(x) with integer zeros is shown. Write the complete factored form of f(x). (Hint: In Exercises 51 and 52 the leading coefficient is not ±1.)

47.
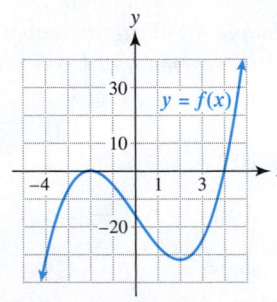

48.
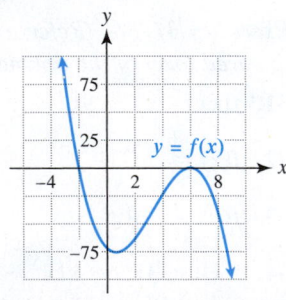

49.
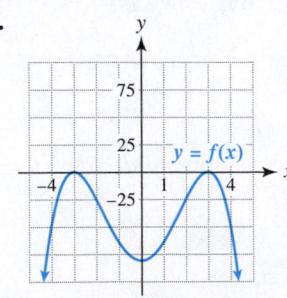

50.
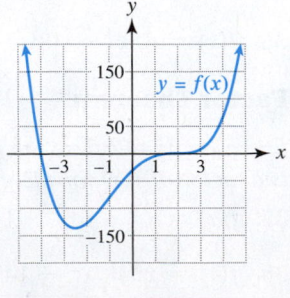

51.   **52.**
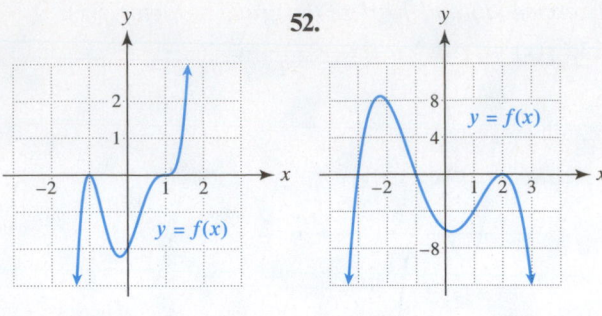

Exercises 53–56: Complete the following.

(a) Find the x- and y-intercepts.

(b) Determine the multiplicity of each zero of f.

(c) Sketch a graph of y = f(x) by hand.

53. $f(x) = 2(x + 2)(x + 1)^2$

54. $f(x) = -(x + 1)(x - 1)(x - 2)$

55. $f(x) = x^2(x + 2)(x - 2)$

56. $f(x) = -\frac{1}{2}(x + 2)^2(x - 1)^3$

Rational Zeros

Exercises 57–64: (Refer to Example 8.)

(a) Use the rational zero test to find any rational zeros of the polynomial f(x).

(b) Write the complete factored form of f(x).

57. $f(x) = 2x^3 + 3x^2 - 8x + 3$

58. $f(x) = x^3 - 7x + 6$

59. $f(x) = 2x^4 + x^3 - 8x^2 - x + 6$

60. $f(x) = 2x^4 + x^3 - 19x^2 - 9x + 9$

61. $f(x) = 3x^3 - 16x^2 + 17x - 4$

62. $f(x) = x^3 + 2x^2 - 3x - 6$

63. $f(x) = x^3 - x^2 - 7x + 7$

64. $f(x) = 2x^3 - 5x^2 - 4x + 10$

Descartes' Rule of Signs

Exercises 65–70: Use Descartes' rule of signs to determine the possible number of positive and negative real zeros for each function. Then, use a graph to determine the actual numbers of positive and negative real zeros.

65. $P(x) = 2x^3 - 4x^2 + 2x + 7$

66. $P(x) = x^3 + 2x^2 + x - 10$

67. $P(x) = 5x^4 + 3x^2 + 2x - 9$

68. $P(x) = 3x^4 + 2x^3 - 8x^2 - 10x - 1$

69. $P(x) = x^5 + 3x^4 - x^3 + 2x + 3$

70. $P(x) = 2x^5 - x^4 + x^3 - x^2 + x + 5$

Polynomial Equations

Exercises 71–76: Solve the equation

(a) *symbolically,*
(b) *graphically, and*
(c) *numerically.*

71. $x^3 + x^2 - 6x = 0$ **72.** $2x^2 - 8x + 6 = 0$

73. $x^4 - 1 = 0$ **74.** $x^4 - 5x^2 + 4 = 0$

75. $-x^3 + 4x = 0$ **76.** $6 - 4x - 2x^2 = 0$

Exercises 77–96: Solve the equation.

77. $x^3 - 25x = 0$ **78.** $x^4 - x^3 - 6x^2 = 0$

79. $x^4 - x^2 = 2x^2 + 4$ **80.** $x^4 + 5 = 6x^2$

81. $x^3 - 3x^2 - 18x = 0$ **82.** $x^4 - x^2 = 0$

83. $2x^3 = 4x^2 - 2x$ **84.** $x^3 = x$

85. $12x^3 = 17x^2 + 5x$ **86.** $3x^3 + 3x = 10x^2$

87. $9x^4 + 4 = 13x^2$ **88.** $4x^4 + 7x^2 - 2 = 0$

89. $4x^3 + 4x^2 - 3x - 3 = 0$

90. $9x^3 + 27x^2 - 2x - 6 = 0$

91. $2x^3 + 4 = x(x + 8)$ **92.** $3x^3 + 18 = x(2x + 27)$

93. $8x^4 = 30x^2 - 27$ **94.** $4x^4 - 21x^2 + 20 = 0$

95. $x^6 - 19x^3 = 216$ **96.** $x^6 = 7x^3 + 8$

Exercises 97–102: (Refer to Example 12.) Solve the equation graphically. Round your answers to the nearest hundredth.

97. $x^3 - 1.1x^2 - 5.9x + 0.7 = 0$

98. $x^3 + x^2 - 18x + 13 = 0$

99. $-0.7x^3 - 2x^2 + 4x + 2.5 = 0$

100. $3x^3 - 46x^2 + 180x - 99 = 0$

101. $2x^4 - 1.5x^3 + 13 = 24x^2 + 10x$

102. $-x^4 + 2x^3 + 20x^2 = 22x + 41$

Intermediate Value Property

Exercises 103–106: Use the intermediate value property to show that $f(x) = 0$ for some x on the given interval.

103. $f(x) = x^2 - 5, 2 \le x \le 3$ (*Hint:* Evaluate $f(2)$ and $f(3)$ and then apply the intermediate value property.)

104. $f(x) = x^3 - x - 1, 1 \le x \le 2$

105. $f(x) = 2x^3 - 1, 0 \le x \le 1$

106. $f(x) = 4x^2 - x - 1, -1 \le x \le 0$

107. Let $f(x) = x^5 - x^2 + 4$. Evaluate $f(1)$ and $f(2)$. Is there a real number k such that $f(k) = 20$? Explain your answer.

108. Sketch a graph of a function f that passes through the points $(-2, 3)$ and $(1, -2)$ but never assumes a value of 0. What must be true about the graph of f?

Applications

109. **Winter Temperature**   The temperature T in degrees Fahrenheit on a cold night x hours past midnight can be approximated by $T(x) = x^3 - 6x^2 + 8x$, where $0 \le x \le 4$. Determine when the temperature was 0°F.

110. **Geometry**   A rectangular box has sides with lengths $x, x + 1$, and $x + 2$. If the volume of the box is 504 cubic inches, find the dimensions of the box.

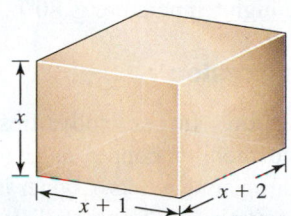

111. **Floating Ball**   (Refer to Example 7.) If a ball has a 20-centimeter diameter, then

$$f(x) = \frac{\pi}{3}x^3 - 10\pi x^2 + \frac{4000\pi d}{3}$$

determines the depth that it sinks in water. Find the depth that this size ball sinks when $d = 0.6$.

112. **Floating Ball**   (Refer to Example 7.) Determine the depth that a pine ball with a 10-centimeter diameter sinks in water if $d = 0.55$.

113. **Bird Populations**   (Refer to Example 5.) A bird population can be modeled by

$$f(x) = x^3 - 66x^2 + 1052x + 1652,$$

where $x = 1$ corresponds to June 1, $x = 2$ to June 2, and so on. Find the days when f estimates that there were 3500 birds.

114. **Insect Population**   An insect population P in thousands per acre x days past May 31 is approximated by $P(x) = 2x^3 - 18x^2 + 46x$, where $0 \le x \le 6$. Determine the dates when the insect population equaled 30 thousand per acre.

115. **Modeling Temperature**   Complete the following.
(a) Approximate the complete factored form of $f(x) = -0.184x^3 + 1.45x^2 + 10.7x - 27.9$.

(b) The cubic polynomial $f(x)$ models monthly average temperature at Trout Lake, Canada, in degrees Fahrenheit, where $x = 1$ corresponds to January and $x = 12$ represents December. Interpret the zeros of f.

116. Average High Temperatures   The monthly average high temperatures in degrees Fahrenheit at Daytona Beach can be modeled by

$$f(x) = 0.0151x^4 - 0.438x^3 + 3.60x^2 - 6.49x + 72.5,$$

where $x = 1$ corresponds to January and $x = 12$ represents December.

(a) Find the average high temperature during March and July.

(b) Graph f in [0.5, 12.5, 1] by [60, 100, 10]. Interpret the graph.

(c) Estimate graphically and numerically when the average high temperature is 80°F.

Polynomial Regression

117. Water Pollution   In one study, freshwater mussels were used to monitor copper discharge into a river from an electroplating works. Copper in high doses can be lethal to aquatic life. The table lists copper concentrations in mussels after 45 days at various distances downstream from the plant. The concentration C is measured in micrograms of copper per gram of mussel x kilometers downstream.

x	5	21	37	53	59
C	20	13	9	6	5

Source: R. Foster and J. Bates, "Use of mussels to monitor point source industrial discharges."

(a) Describe the relationship between x and C.

(b) Use regression to find a cubic polynomial function $f(x)$ that models the data.

(c) Graph C and the data.

(d) Concentrations above 10 are lethal to mussels. Locate this region in the river.

118. Dog Years   There is a saying that every year of a dog's life is equal to 7 years for a human. A more accurate approximation is given by the graph of f. Given a dog's age x, where $x \geq 1$, $f(x)$ models the equivalent age in human years. According to the Bureau of the Census, middle age for people begins at age 45.
(*Source:* J. Brearley and A. Nicholas, *This Is the Bichon Frise.*)

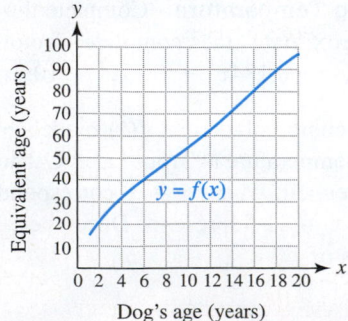

(a) Use the graph of f to estimate the equivalent age for dogs.

(b) Estimate $f(x)$ at $x = 2, 6, 10, 14,$ and 18.

(c) Use regression and the points you estimated to find a quartic polynomial function f that models the data points.

(d) Use $f(x)$ to solve part (a) either graphically or numerically.

Writing About Mathematics

119. Suppose that $f(x)$ is a quintic polynomial with distinct real zeros. Assuming you have access to technology, explain how to factor $f(x)$ approximately. Have you used the factor theorem? Explain.

120. Explain how to determine graphically whether a zero of a polynomial is a multiple zero. Sketch examples.

Extended and Discovery Exercises

Exercises 1–6: **Boundedness Theorem**   *The boundedness theorem shows how the bottom row of a synthetic division is used to place upper and lower bounds on possible real zeros of a polynomial function.*

Let P(x) define a polynomial function of degree n ≥ 1 with real coefficients and with a positive leading coefficient. If P(x) is divided synthetically by x − c and

(a) if c > 0 and all numbers in the bottom row of the synthetic division are nonnegative, then P(x) has no zero greater than c;

(b) if c < 0 and the numbers in the bottom row of the synthetic division alternate in sign (with 0 considered positive or negative, as needed), then P(x) has no zero less than c.

Use the boundedness theorem to show that the real zeros of each polynomial function satisfy the given conditions.

1. $P(x) = x^4 - x^3 + 3x^2 - 8x + 8;$
no real zero greater than 2

2. $P(x) = 2x^5 - x^4 + 2x^3 - 2x^2 + 4x - 4;$
no real zero greater than 1

3. $P(x) = x^4 + x^3 - x^2 + 3;$
no real zero less than −2

4. $P(x) = x^5 + 2x^3 - 2x^2 + 5x + 5;$
no real zero less than −1

5. $P(x) = 3x^4 + 2x^3 - 4x^2 + x - 1;$
no real zero greater than 1

6. $P(x) = 3x^4 + 2x^3 - 4x^2 + x - 1;$
no real zero less than −2

CHECKING BASIC CONCEPTS FOR SECTIONS 4.3 AND 4.4

1. Simplify the expression $\dfrac{5x^4 - 10x^3 + 5x^2}{5x^2}$.

2. Divide the expression.

   (a) $\dfrac{x^3 - x^2 + 4x - 4}{x - 1}$

   (b) $\dfrac{2x^3 - 3x^2 + 4x + 4}{2x + 1}$

   (c) $\dfrac{x^4 - 3x^3 + 6x^2 - 13x + 9}{x^2 + 4}$

3. Use the graph of the cubic polynomial $f(x)$ in the next column to determine its complete factored form. State the multiplicity of each zero. Assume that all zeros are integers and that the leading coefficient is *not* ± 1.

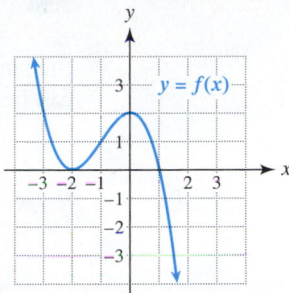

4. Solve $x^3 - 2x^2 - 15x = 0$.

5. Determine graphically the zeros of

   $$f(x) = x^4 - x^3 - 18x^2 + 16x + 32.$$

   Write $f(x)$ in complete factored form.

4.5 The Fundamental Theorem of Algebra

- Apply the fundamental theorem of algebra
- Factor polynomials having complex zeros
- Solve polynomial equations having complex solutions

Introduction

In Section 3.2 the quadratic formula was used to solve $ax^2 + bx + c = 0$. Are there similar formulas for higher degree polynomial equations? One of the most spectacular mathematical achievements during the sixteenth century was the discovery of formulas for solving cubic and quartic equations. This was accomplished by the Italian mathematicians Tartaglia, Cardano, Fior, del Ferro, and Ferrari between 1515 and 1545. These formulas are quite complicated and typically used only in computer software. In about 1805, the Italian physician Ruffini proved that finding formulas for quintic or higher degree equations was impossible. Another spectacular result came from Carl Friedrich Gauss. He proved that all polynomials can be completely factored by using complex numbers. This result is called the **fundamental theorem of algebra**. (*Source:* H. Eves, *An Introduction to the History of Mathematics.*)

Fundamental Theorem of Algebra

One of the most brilliant mathematicians of all time, Carl Friedrich Gauss proved the fundamental theorem of algebra as part of his doctoral thesis at age 20. Although his theorem and proof were completed in 1797, they are still valid today.

> **FUNDAMENTAL THEOREM OF ALGEBRA**
>
> A polynomial $f(x)$ of degree n, with $n \geq 1$, has at least one complex zero.

NOTE The fundamental theorem of algebra guarantees that *every polynomial has a complete factorization*, provided we are allowed to use complex numbers.

Justification for Complete Factorization of Polynomials If $f(x)$ is a polynomial of degree 1 or higher, then by the fundamental theorem of algebra there is a zero c_1 such that $f(c_1) = 0$. By the factor theorem, $(x - c_1)$ is a factor of $f(x)$ and $f(x) = (x - c_1) q_1(x)$ for some polynomial $q_1(x)$. If $q_1(x)$ has positive degree, then by the fundamental theorem of algebra there exists a zero c_2 of $q_1(x)$. By the factor theorem, $q_1(x)$ can be written as $q_1(x) = (x - c_2) q_2(x)$. Then

$$f(x) = (x - c_1)q_1(x) = (x - c_1)(x - c_2)q_2(x).$$

If $f(x)$ has degree n, this process can be continued until $f(x)$ is written in the complete factored form

$$f(x) = a_n(x - c_1)(x - c_2) \cdots (x - c_n),$$

where a_n is the leading coefficient and the c_k are complex zeros of $f(x)$. If each c_k is distinct, then $f(x)$ has n zeros. However, in general the c_k may not be distinct since multiple zeros are possible.

NUMBER OF ZEROS THEOREM

A polynomial of degree n has at most n distinct zeros.

EXAMPLE 1   **Classifying zeros**

All zeros for the given polynomials are distinct. Use Figures 4.71–4.73 to determine graphically the number of real zeros and the number of imaginary zeros.

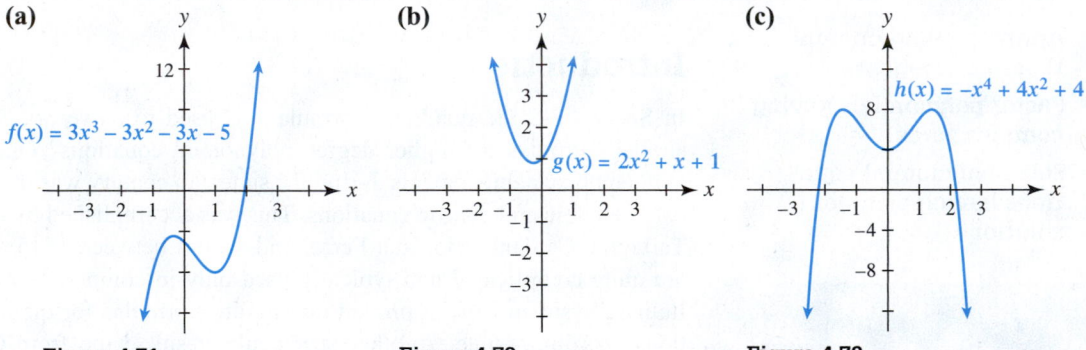

Figure 4.71   **Figure 4.72**   **Figure 4.73**

SOLUTION

Getting Started Each (distinct) real zero corresponds to an x-intercept. Imaginary zeros do *not* correspond to x-intercepts, but their number can be determined after the number of real zeros is known. ▶

(a) The graph of $f(x)$ in Figure 4.71 crosses the x-axis once, so there is one real zero. Since f is degree 3 and all zeros are distinct, there are two imaginary zeros.

(b) The graph of $g(x)$ in Figure 4.72 never crosses the x-axis. Since g is degree 2, there are no real zeros and two imaginary zeros.

(c) The graph of $h(x)$ is shown in Figure 4.73. Since h is degree 4, there are two real zeros and the remaining two zeros are imaginary.

Now Try Exercises 1, 3, and 5

NOTE   The sum of the imaginary zeros and the real zeros (counting multiplicities) equals the degree n of the polynomial. A polynomial with real coefficients always has an even number of imaginary zeros.

EXAMPLE 2 Constructing a polynomial with prescribed zeros

Determine a polynomial $f(x)$ of degree 4 with leading coefficient 2 and zeros -3, 5, i, and $-i$ in **(a)** complete factored form and **(b)** expanded form.

SOLUTION

(a) Let $a_n = 2$, $c_1 = -3$, $c_2 = 5$, $c_3 = i$, and $c_4 = -i$. Then

$$f(x) = 2(x + 3)(x - 5)(x - i)(x + i).$$

(b) To expand this expression for $f(x)$, perform the following steps.

$$2(x + 3)(x - 5)(x - i)(x + i) = 2(x + 3)(x - 5)(x^2 + 1)$$
$$= 2(x + 3)(x^3 - 5x^2 + x - 5)$$
$$= 2(x^4 - 2x^3 - 14x^2 - 2x - 15)$$
$$= 2x^4 - 4x^3 - 28x^2 - 4x - 30$$

Thus $f(x) = 2x^4 - 4x^3 - 28x^2 - 4x - 30$.

> **Now Try Exercise 13**

EXAMPLE 3 Factoring a cubic polynomial with imaginary zeros

Determine the complete factored form for $f(x) = x^3 + 2x^2 + 4x + 8$.

SOLUTION

We can use factoring by grouping to determine the complete factored form.

$$x^3 + 2x^2 + 4x + 8 = (x^3 + 2x^2) + (4x + 8) \qquad \text{Associative property}$$
$$= x^2(x + 2) + 4(x + 2) \qquad \text{Distributive property}$$
$$= (x^2 + 4)(x + 2) \qquad \text{Factor out } x + 2.$$

Algebra Review

To review factoring a cubic polynomial by grouping, see Chapter R (page R-20).

To factor $x^2 + 4$, first find its zeros.

$$x^2 + 4 = 0$$
$$x^2 = -4$$
$$x = \pm\sqrt{-4}$$
$$x = \pm 2i$$

The zeros of $f(x)$ are -2, $2i$, and $-2i$. Its complete factored form is

$$f(x) = (x + 2)(x - 2i)(x + 2i).$$

> **Now Try Exercise 29**

Conjugate Zeros Notice that in Example 3 both $2i$ and $-2i$ were zeros of $f(x)$. The numbers $2i$ and $-2i$ are *conjugates*. See Section 3.3. This result can be generalized.

> **CONJUGATE ZEROS THEOREM**
>
> If a polynomial $f(x)$ has only real coefficients and if $a + bi$ is a zero of $f(x)$, then the conjugate $a - bi$ is also a zero of $f(x)$.

EXAMPLE 4 Constructing a polynomial with prescribed zeros

Determine a cubic polynomial $f(x)$ with real coefficients, leading coefficient 2, and zeros 3 and $5i$. Express $f(x)$ in **(a)** complete factored form and **(b)** expanded form.

SOLUTION

(a) Since $f(x)$ has real coefficients, it must also have a third zero of $-5i$, the conjugate of $5i$. Let $c_1 = 3$, $c_2 = 5i$, $c_3 = -5i$, and $a_n = 2$. The complete factored form is

$$f(x) = 2(x - 3)(x - 5i)(x + 5i).$$

(b) To expand $f(x)$, perform the following steps.

$$2(x - 3)(x - 5i)(x + 5i) = 2(x - 3)(x^2 + 25)$$
$$= 2(x^3 - 3x^2 + 25x - 75)$$
$$= 2x^3 - 6x^2 + 50x - 150$$

> **Now Try Exercise 15**

EXAMPLE 5 Finding imaginary zeros of a polynomial

Find the zeros of $f(x) = x^4 + x^3 + 2x^2 + x + 1$, given that one zero is $-i$.

SOLUTION

By the conjugate zeros theorem, it follows that i must also be a zero of $f(x)$. Therefore $(x - i)$ and $(x + i)$ are factors of $f(x)$. Because $(x - i)(x + i) = x^2 + 1$, we can use long division to find another quadratic factor of $f(x)$.

$$
\begin{array}{r}
x^2 + x + 1 \\
x^2 + 0x + 1 \overline{)x^4 + x^3 + 2x^2 + x + 1} \\
\underline{x^4 + 0x^3 + x^2} \\
x^3 + x^2 + x \\
\underline{x^3 + 0x^2 + x} \\
x^2 + 0x + 1 \\
\underline{x^2 + 0x + 1} \\
0
\end{array}
$$

The quotient is $x^2 + x + 1$ with remainder 0. By the division algorithm,

$$x^4 + x^3 + 2x^2 + x + 1 = (x^2 + 1)(x^2 + x + 1).$$

We can use the quadratic formula to find the zeros of $x^2 + x + 1$.

$$x = \frac{-b \pm \sqrt{b^2 - 4ac}}{2a}$$
$$= \frac{-1 \pm \sqrt{1^2 - 4(1)(1)}}{2(1)}$$
$$= -\frac{1}{2} \pm i\frac{\sqrt{3}}{2}$$

The four zeros of $f(x)$ are $\pm i$ and $-\frac{1}{2} \pm i\frac{\sqrt{3}}{2}$.

> **Now Try Exercise 21**

Polynomial Equations with Complex Solutions

Every polynomial equation of degree n can be written in the form

$$a_n x^n + \cdots + a_2 x^2 + a_1 x + a_0 = 0.$$

If we let $f(x) = a_n x^n + \cdots + a_2 x^2 + a_1 x + a_0$ and write $f(x)$ in complete factored form as

$$f(x) = a_n(x - c_1)(x - c_2) \cdots (x - c_n),$$

then the solutions to the polynomial equation are the zeros $c_1, c_2, \ldots, c_n$ of $f(x)$. Solving cubic and quartic polynomial equations with this technique is illustrated in the next two examples.

EXAMPLE 6 Solving a polynomial equation

Solve $x^3 = 3x^2 - 7x + 21$.

SOLUTION

Write the equation as $f(x) = 0$, where $f(x) = x^3 - 3x^2 + 7x - 21$. Although we could use factoring by grouping, as is done in Example 3, we use graphing instead to find one real zero of $f(x)$. Figure 4.74 shows that 3 is a zero of $f(x)$. By the factor theorem, $x - 3$ is a factor of $f(x)$. Using synthetic division, we divide $x - 3$ into $f(x)$.

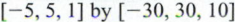

$[-5, 5, 1]$ by $[-30, 30, 10]$

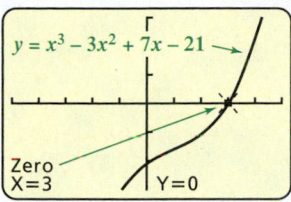

Figure 4.74

$$\begin{array}{r|rrrr} 3 & 1 & -3 & 7 & -21 \\ & & 3 & 0 & 21 \\ \hline & 1 & 0 & 7 & 0 \end{array}$$

Thus $x^3 - 3x^2 + 7x - 21 = (x - 3)(x^2 + 7)$, and we can solve as follows.

$$x^3 - 3x^2 + 7x - 21 = 0 \qquad \text{Solve } f(x) = 0.$$
$$(x - 3)(x^2 + 7) = 0 \qquad \text{Factor.}$$
$$x - 3 = 0 \quad \text{or} \quad x^2 + 7 = 0 \qquad \text{Zero-product property}$$
$$x = 3 \quad \text{or} \quad x^2 = -7 \qquad \text{Solve.}$$
$$x = 3 \quad \text{or} \quad x = \pm i\sqrt{7} \qquad \text{Property of } i$$

The solutions are 3 and $\pm i\sqrt{7}$.

Now Try Exercise 33

EXAMPLE 7 Solving a polynomial equation

Solve $x^4 + x^2 = x^3$.

SOLUTION

Write the equation as $f(x) = 0$, where $f(x) = x^4 - x^3 + x^2$.

$$x^4 - x^3 + x^2 = 0 \qquad f(x) = 0$$
$$x^2(x^2 - x + 1) = 0 \qquad \text{Factor out } x^2.$$
$$x^2 = 0 \quad \text{or} \quad x^2 - x + 1 = 0 \qquad \text{Zero-product property}$$

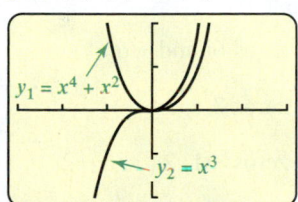

$[-3, 3, 1]$ by $[-2, 2, 1]$

Figure 4.75

The only solution to $x^2 = 0$ is 0. To solve $x^2 - x + 1 = 0$, use the quadratic formula, as in Example 5. The solutions are 0 and $\frac{1}{2} \pm i\frac{\sqrt{3}}{2}$.

The graphs of $y_1 = x^4 + x^2$ and $y_2 = x^3$ are shown in Figure 4.75. Notice that they appear to intersect only at the origin. This indicates that the only *real* solution is 0.

Now Try Exercise 37

4.5 Putting It All Together

Some of the important topics in this section are summarized in the following table.

CONCEPT	EXPLANATION	COMMENTS AND EXAMPLES
Number of zeros theorem	A polynomial of degree n has at most n distinct zeros. These zeros can be real or imaginary numbers.	The cubic polynomial, $$ax^3 + bx^2 + cx + d,$$ has *at most* three distinct zeros.
Fundamental theorem of algebra	A polynomial of degree n, with $n \geq 1$, has at least one complex zero.	This theorem guarantees that we can always factor a polynomial $f(x)$ into complete factored form: $$f(x) = a_n(x - c_1) \cdots (x - c_n),$$ where the c_k are complex numbers.

continued on next page

CONCEPT	EXPLANATION	COMMENTS AND EXAMPLES
Conjugate zeros theorem	If a polynomial has *real* coefficients and $a + bi$ is a zero, then $a - bi$ is also a zero.	Since $\frac{1}{2} + \frac{1}{2}i$ is a zero of $2x^2 - 2x + 1$, it follows that $\frac{1}{2} - \frac{1}{2}i$ is also a zero.

4.5 Exercises

Zeros of Polynomials

Exercises 1–8: The graph and degree of a polynomial with real coefficients $f(x)$ are given. Determine the number of real zeros and the number of imaginary zeros. Assume that all zeros of $f(x)$ are distinct.

1. Degree 2

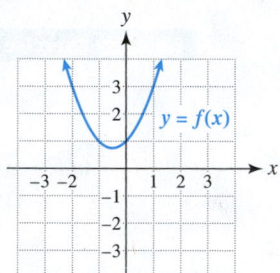

2. Degree 2

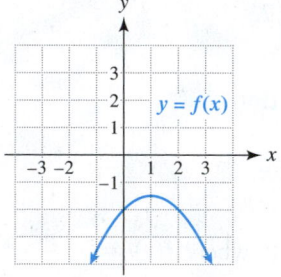

3. Degree 3

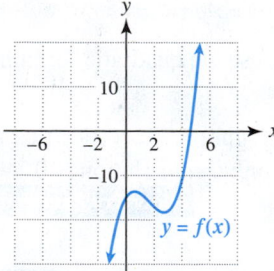

4. Degree 3

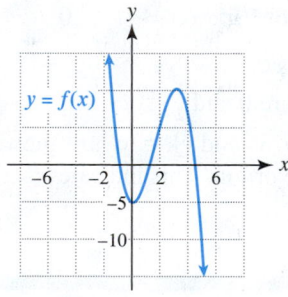

5. Degree 4

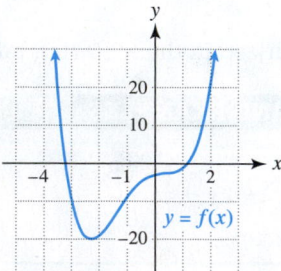

6. Degree 4

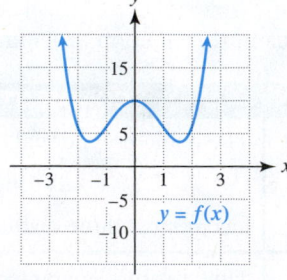

7. Degree 5

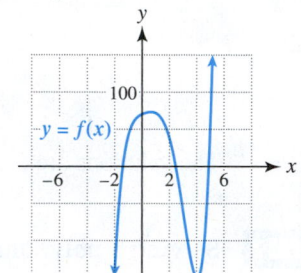

8. Degree 5

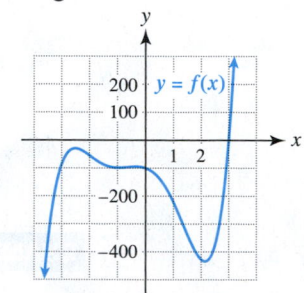

Exercises 9–18: Let a_n be the leading coefficient.

(a) *Find the complete factored form of a polynomial with real coefficients $f(x)$ that satisfy the conditions.*

(b) *Express $f(x)$ in expanded form.*

9. Degree 2; $a_n = 1$;   zeros 6i and $-6i$

10. Degree 3; $a_n = 5$;   zeros 2, i, and $-i$

11. Degree 3; $a_n = -1$;   zeros -1, 2i, and $-2i$

12. Degree 4; $a_n = 3$;   zeros -2, 4, i, and $-i$

13. Degree 4; $a_n = 10$;   zeros 1, -1, 3i, and $-3i$

14. Degree 2; $a_n = -5$;   zeros $1 + i$ and $1 - i$

15. Degree 4; $a_n = \frac{1}{2}$;   zeros $-i$ and 2i

16. Degree 3; $a_n = -\frac{3}{4}$;   zeros $-3i$ and $\frac{2}{5}$

17. Degree 3; $a_n = -2$;   zeros $1 - i$ and 3

18. Degree 4; $a_n = 7$;   zeros 2i and 3i

Exercises 19–22: (Refer to Example 5.) Find the zeros of $f(x)$, given that one zero is k.

19. $f(x) = 3x^3 - 5x^2 + 75x - 125$ $k = \frac{5}{3}$

20. $f(x) = x^4 + 2x^3 + 8x^2 + 8x + 16$ $k = 2i$

21. $f(x) = 2x^4 - x^3 + 19x^2 - 9x + 9$ $k = -3i$

22. $f(x) = 7x^3 + 5x^2 + 12x - 4$ $k = \frac{2}{7}$

Exercises 23–30: Complete the following.

(a) Find all zeros of $f(x)$.
(b) Write the complete factored form of $f(x)$.

23. $f(x) = x^2 + 25$ **24.** $f(x) = x^2 + 11$

25. $f(x) = 3x^3 + 3x$ **26.** $f(x) = 2x^3 + 10x$

27. $f(x) = x^4 + 5x^2 + 4$ **28.** $f(x) = x^4 + 4x^2$

29. $f(x) = x^3 + 2x^2 + 16x + 32$

30. $f(x) = x^4 + 2x^3 + x^2 + 8x - 12$

Exercises 31–42: Solve the polynomial equation.

31. $x^3 + x = 0$ **32.** $2x^3 - x + 1 = 0$

33. $x^3 = 2x^2 - 7x + 14$ **34.** $x^2 + x + 2 = x^3$

35. $x^4 + 5x^2 = 0$ **36.** $x^4 - 2x^3 + x^2 - 2x = 0$

37. $x^4 = x^3 - 4x^2$ **38.** $x^5 + 9x^3 = x^4 + 9x^2$

39. $x^4 + x^3 = 16 - 8x - 6x^2$

40. $x^4 + 2x^2 = x^3$ **41.** $3x^3 + 4x^2 + 6 = x$

42. $2x^3 + 5x^2 + x + 12 = 0$

Writing About Mathematics

43. Could a cubic function with real coefficients have only imaginary zeros? Explain.

44. Give an example of a polynomial function that has only imaginary zeros and a polynomial function that has only real zeros. Explain how to determine graphically if a function has only imaginary zeros.

4.6 Rational Functions and Models

- **Identify a rational function and state its domain**
- **Identify asymptotes**
- **Interpret asymptotes**
- **Graph a rational function by using transformations**
- **Graph a rational function by hand (optional)**

Introduction

Rational functions are (typically) nonlinear functions that frequently occur in applications. For example, rational functions are used to model postings on social networks, design curves for railroad tracks, determine stopping distances on hills, and calculate the average number of people waiting in a line.

Rational Functions

A *rational* number can be expressed as a *ratio* $\frac{p}{q}$, where p and q are integers with $q \neq 0$. A rational function is defined similarly by using the concept of a polynomial.

> **RATIONAL FUNCTION**
>
> A function f represented by $f(x) = \frac{p(x)}{q(x)}$, where $p(x)$ and $q(x)$ are polynomials and $q(x) \neq 0$, is a **rational function**.

The domain of a rational function includes all real numbers *except* the zeros of the denominator $q(x)$. The graph of a rational function is continuous except at x-values where $q(x) = 0$.

EXAMPLE 1   **Identifying rational functions**

Algebra Review

To review rational expressions, see Chapter R (page R-28).

Determine if the function is rational and state its domain.

(a) $f(x) = \dfrac{2x - 1}{x^2 + 1}$ **(b)** $g(x) = \dfrac{1}{\sqrt{x}}$ **(c)** $h(x) = \dfrac{x^3 - 2x^2 + 1}{x^2 - 3x + 2}$

SOLUTION

(a) Both the numerator, $2x - 1$, and the denominator, $x^2 + 1$, are polynomials, so f is a rational function. The domain of f includes all real numbers because its denominator $x^2 + 1 \neq 0$ for any real number x.

(b) The expression $\sqrt{x}$ is not a polynomial, so $g(x) = \frac{1}{\sqrt{x}}$ is not a rational function. The domain is $\{x \mid x > 0\}$.

(c) Both the numerator and the denominator are polynomials, so $h(x) = \frac{x^3 - 2x^2 + 1}{x^2 - 3x + 2}$ is a rational function. Because

$$x^2 - 3x + 2 = (x - 1)(x - 2) = 0$$

when $x = 1$ or $x = 2$, the domain of h is $\{x \mid x \neq 1, x \neq 2\}$.

Now Try Exercises 1, 7, and 9

CLASS DISCUSSION

Is an integer a rational number? Is a polynomial function a rational function?

Vertical Asymptotes

A rational function given by $f(x) = \frac{p(x)}{q(x)}$ is undefined whenever $q(x) = 0$. If $q(k) = 0$ for some k, then a *vertical asymptote* of the graph of f *may* occur at $x = k$. Near a vertical asymptote, the y-values on the graph of f become very large (unbounded) in absolute value.

For example, Figure 4.76 shows the graph of a rational function defined by $f(x) = \frac{1}{x - 2}$. It has a vertical asymptote, which is shown as a dashed red line at $x = 2$. Note that $x = 2$ is *not* in the domain of f.

See the Concept: Rational Functions and Vertical Asymptotes

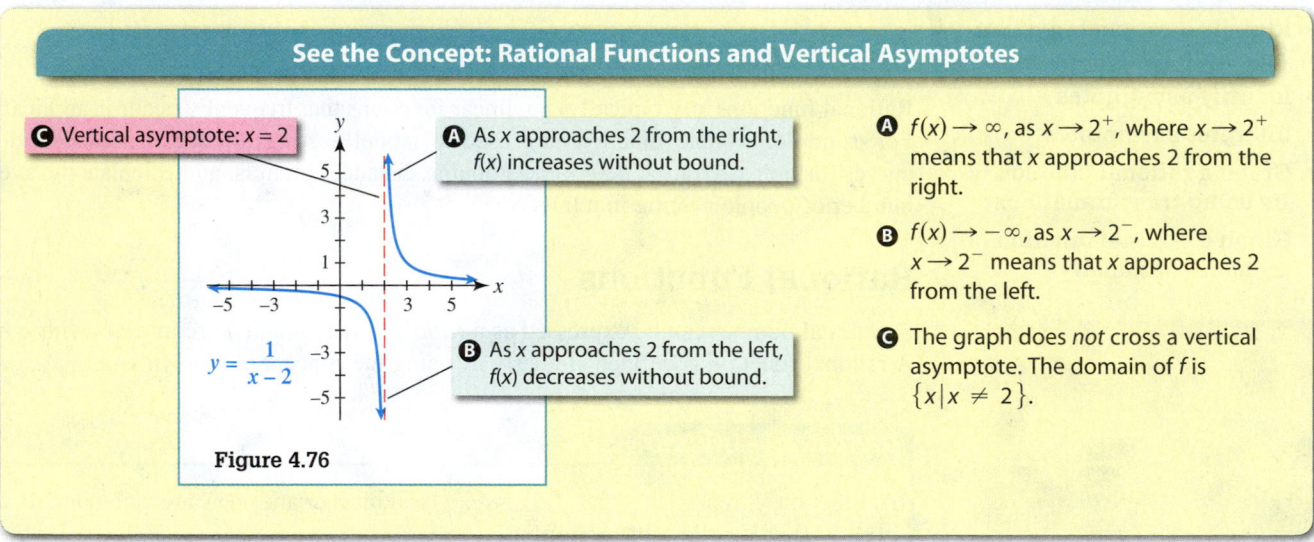

C Vertical asymptote: $x = 2$

A As x approaches 2 from the right, $f(x)$ increases without bound.

B As x approaches 2 from the left, $f(x)$ decreases without bound.

$y = \dfrac{1}{x - 2}$

A $f(x) \to \infty$, as $x \to 2^+$, where $x \to 2^+$ means that x approaches 2 from the right.

B $f(x) \to -\infty$, as $x \to 2^-$, where $x \to 2^-$ means that x approaches 2 from the left.

C The graph does *not* cross a vertical asymptote. The domain of f is $\{x \mid x \neq 2\}$.

Figure 4.76

NOTE   A vertical asymptote is not part of the graph of a rational function; rather, it is an aid that is used to sketch and to better understand the graph of a rational function.

VERTICAL ASYMPTOTE

The line $x = k$ is a **vertical asymptote** of the graph of f if $f(x) \to \infty$ or $f(x) \to -\infty$ as x approaches k from either the left or the right.

An Application   If cars leave a parking garage randomly and stop to pay the parking attendant on the way out, then the average length of the line depends on two factors: the average traffic rate at which cars are exiting the ramp and the average rate at which the parking attendant can wait on cars. For instance, if the average traffic rate is **three** cars per minute and the parking attendant can serve **four** cars per minute, then at times a line may form *if* cars arrive in a *random* manner. The **traffic intensity** x is the ratio of the average traffic rate to the average working rate of the attendant. In this example, $x = \frac{3}{4}$.   (*Source:* F. Mannering and W. Kilareski, *Principles of Highway Engineering and Traffic Control*.)

EXAMPLE 2   **Estimating the length of parking garage lines**

If the traffic intensity is x, then the average number of cars waiting in line to exit a parking garage can be estimated by $N(x) = \frac{x^2}{2 - 2x}$, where $0 \le x < 1$.

(a) Evaluate $N(0.5)$ and $N(0.9)$. Interpret the results.

(b) Use the graph of $y = N(x)$ in Figure 4.77 to explain what happens to the length of the line as the traffic intensity x increases to 1 from the left (denoted $x \to 1^-$.)

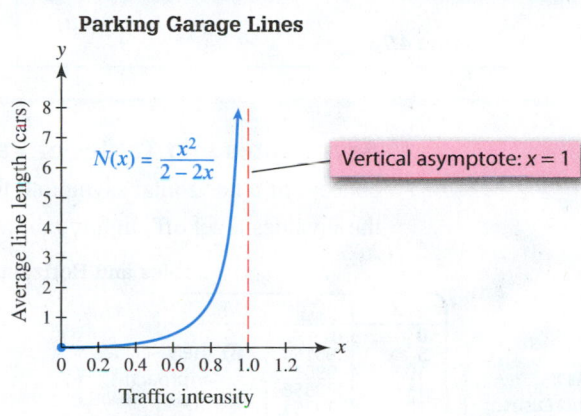

Figure 4.77

SOLUTION

(a) $N(0.5) = \frac{0.5^2}{2 - 2(0.5)} = 0.25$ and $N(0.9) = \frac{0.9^2}{2 - 2(0.9)} = 4.05$. This means that if the traffic intensity is 0.5, there is little waiting in line. As the traffic intensity increases to 0.9, the average line has more than four cars.

(b) As the traffic intensity x approaches 1 from the left in Figure 4.77 the graph of f increases rapidly without bound. Numerical support is given in Table 4.4. With a traffic intensity slightly less than 1, the attendant has difficulty keeping up. If cars occasionally arrive in groups, long lines will form. At $x = 1$ the denominator, $2 - 2x$, equals 0 and $N(x)$ is undefined.

Traffic Intensity Approaching 1

x	0.94	0.95	0.96	0.97	0.98	0.99	1
$\frac{x^2}{2-2x}$	7.36	9.03	11.52	15.68	24.01	49.01	—

Table 4.4

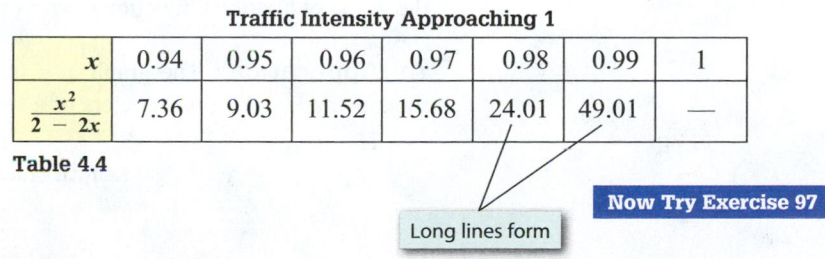

Long lines form

Now Try Exercise 97

Horizontal Asymptotes

If the absolute value of x becomes large in the formula $f(x)$ for a rational function, then the graph of f *may* level off and begin to approximate a horizontal line. This horizontal line is called a *horizontal asymptote*.

In See the Concept on the next page, Figure 4.78 shows the graph of the rational function $f(x) = \frac{x^2}{x^2 + 1}$. This rational function does not have a vertical asymptote because the denominator, $x^2 + 1$, has no real zeros. However, the graph does have a horizontal asymptote, which is shown as a dashed red line at $y = 1$.

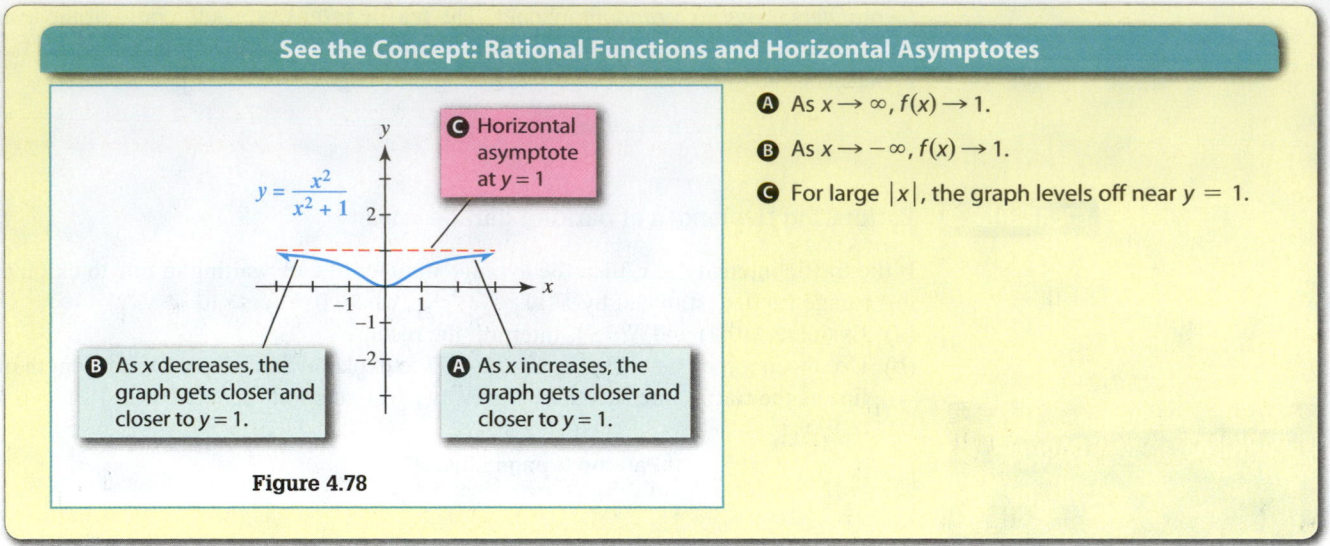

See the Concept: Rational Functions and Horizontal Asymptotes

Ⓐ As $x \to \infty$, $f(x) \to 1$.

Ⓑ As $x \to -\infty$, $f(x) \to 1$.

Ⓒ For large $|x|$, the graph levels off near $y = 1$.

Figure 4.78

Asymptotes and Tables of Values   A table of values can be used to illustrate the concept of a horizontal asymptote for $y = \dfrac{x^2}{x^2 + 1}$. See Figures 4.79 and 4.80. Notice that the y-values level off slightly below 1, which agrees with the graph in Figure 4.78.

Tables and Horizontal Asymptotes

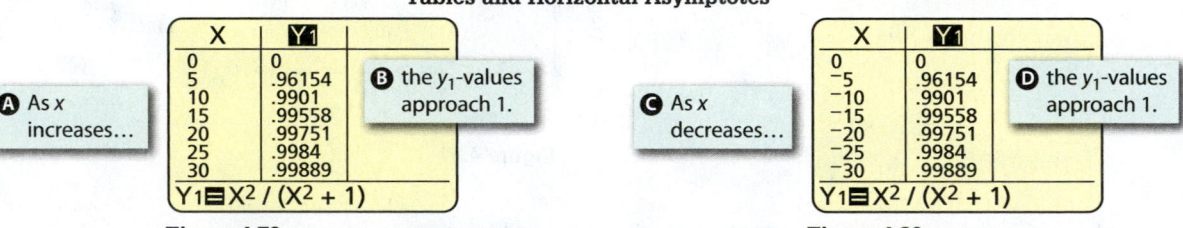

Ⓐ As x increases...

Ⓑ the y_1-values approach 1.

X	Y1
0	0
5	.96154
10	.9901
15	.99558
20	.99751
25	.9984
30	.99889

Y1◼X² / (X² + 1)

Figure 4.79

Ⓒ As x decreases...

Ⓓ the y_1-values approach 1.

X	Y1
0	0
−5	.96154
−10	.9901
−15	.99558
−20	.99751
−25	.9984
−30	.99889

Y1◼X² / (X² + 1)

Figure 4.80

HORIZONTAL ASYMPTOTE

The line $y = b$ is a **horizontal asymptote** of the graph of f if $f(x) \to b$ as x approaches either ∞ or $-\infty$.

NOTE   Like a vertical asymptote, a horizontal asymptote is *not* part of the graph of a rational function; rather, it is an aid that is used to sketch and to better understand the graph of a rational function. However, unlike in the case of vertical asymptotes, it *is possible* for the graph of a rational function to *cross* a horizontal asymptote. See Examples 4(b) and 11.

An Application   The graph of f in Figure 4.81 is an example of a growth curve. It models the length in millimeters of a small fish after x weeks.

Length of a Small Fish

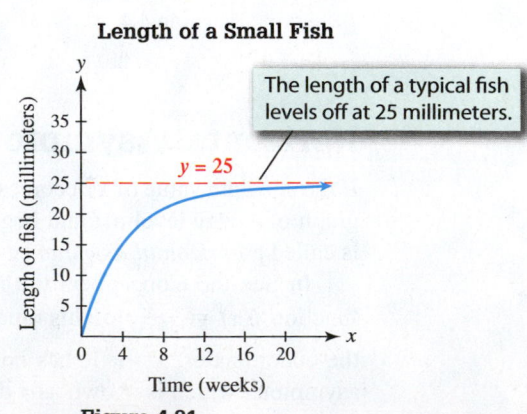

The length of a typical fish levels off at 25 millimeters.

$y = 25$

Figure 4.81

After several weeks the length of the fish begins to level off near 25 millimeters. Thus $y = 25$ is a horizontal asymptote of the graph of f. This is denoted by $f(x) \to 25$ as $x \to \infty$. (*Source:* D. Brown and P. Rothery, *Models in Biology.*)

In real-life applications, time does not actually approach infinity. For example, a fish does not live forever. However, the asymptote $y = 25$ does tend to model the length of the fish as it becomes older.

Identifying Asymptotes

Asymptotes can be found visually from a graph and symbolically from a formula. The next two examples discuss these techniques.

EXAMPLE 3   Determining horizontal and vertical asymptotes visually

Use the graph of each rational function to determine any vertical or horizontal asymptotes.

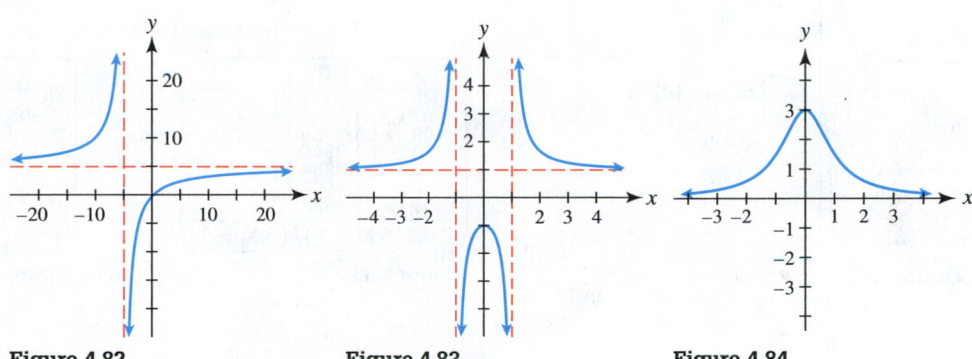

Figure 4.82   Figure 4.83   Figure 4.84

SOLUTION
In Figure 4.82, $x = -5$ is a vertical asymptote and $y = 5$ is a horizontal asymptote. In Figure 4.83, $x = \pm 1$ are vertical asymptotes and $y = 1$ is a horizontal asymptote. In Figure 4.84, there are no vertical asymptotes. The x-axis ($y = 0$) is a horizontal asymptote.

Now Try Exercises 13, 15, and 17

The following technique can be used for rational functions to find vertical and horizontal asymptotes symbolically.

FINDING VERTICAL AND HORIZONTAL ASYMPTOTES

Let f be a rational function given by $f(x) = \frac{p(x)}{q(x)}$ written in *lowest* terms.

Vertical Asymptote

To find a vertical asymptote, set the denominator, $q(x)$, equal to 0 and solve. If k is a zero of $q(x)$, then $x = k$ is a vertical asymptote. *Caution:* If k is a zero of both $q(x)$ *and* $p(x)$, then $f(x)$ is not written in lowest terms, and $x - k$ is a common factor.

Horizontal Asymptote

(a) If the degree of the numerator is less than the degree of the denominator, then $y = 0$ (the x-axis) is a horizontal asymptote.

(b) If the degree of the numerator equals the degree of the denominator, then $y = \frac{a}{b}$ is a horizontal asymptote, where a is the leading coefficient of the numerator and b is the leading coefficient of the denominator.

(c) If the degree of the numerator is greater than the degree of the denominator, then there are no horizontal asymptotes.

EXAMPLE 4   Finding asymptotes

For each rational function, determine any horizontal or vertical asymptotes.

(a) $f(x) = \dfrac{6x - 1}{3x + 3}$ **(b)** $g(x) = \dfrac{x + 1}{x^2 - 4}$ **(c)** $h(x) = \dfrac{x^2 - 1}{x + 1}$

SOLUTION

(a) The numerator and denominator have no common factors, so the expression is in lowest terms.

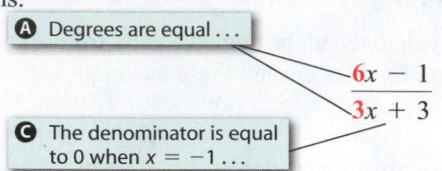

A Degrees are equal . . .   B so $y = \frac{6}{3}$, or $y = 2$, is a horizontal asymptote. See Figure 4.85.

$$\dfrac{6x - 1}{3x + 3}$$

C The denominator is equal to 0 when $x = -1$ . . .   D so $x = -1$ is a vertical asymptote.

A graph of $f(x) = \dfrac{6x - 1}{3x + 3}$ is shown in Figure 4.85. In Figures 4.86 and 4.87, the tables support that $y = 2$ is a horizontal asymptote.

Horizontal Asymptote: $y = 2$

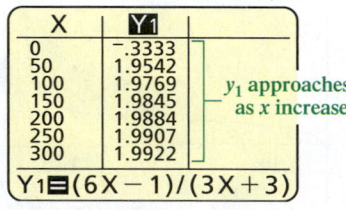

X	Y1
0	-.3333
50	1.9542
100	1.9769
150	1.9845
200	1.9884
250	1.9907
300	1.9922

y_1 approaches 2 as x increases.

Y1◨(6X−1)/(3X+3)

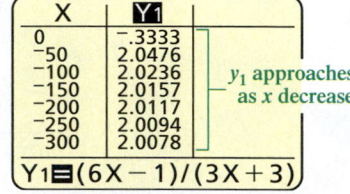

X	Y1
0	-.3333
-50	2.0476
-100	2.0236
-150	2.0157
-200	2.0117
-250	2.0094
-300	2.0078

y_1 approaches 2 as x decreases.

Y1◨(6X−1)/(3X+3)

Figure 4.86   **Figure 4.87**

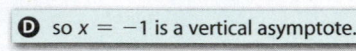

y

8

$f(x) = \dfrac{6x - 1}{3x + 3}$

4

−4   4   8   *x*

$y = 2$

−4

$x = -1$

−8

Figure 4.85

Algebra Review
To review simplifying rational expressions, see Chapter R (page R-28).

(b) In the expression $\dfrac{x + 1}{x^2 - 4}$, the degree of the numerator is one less than the degree of the denominator, so the x-axis, or $y = 0$, is a horizontal asymptote. When $x = \pm 2$, the denominator, $x^2 - 4$, equals 0 and the numerator, $x + 1$, does not equal 0. Thus $x = \pm 2$ are vertical asymptotes. See Figure 4.88. Note that the graph crosses the horizontal asymptote $y = 0$ but does not cross either vertical asymptote.

(c) The degree of the numerator, $x^2 - 1$, is greater than the degree of the denominator, $x + 1$, so there are no horizontal asymptotes. When $x = -1$, both numerator and denominator equal 0, so the expression is *not* in lowest terms. We can simplify $h(x)$ as follows.

$$h(x) = \dfrac{x^2 - 1}{x + 1} = \dfrac{(x + 1)(x - 1)}{x + 1} = x - 1, \qquad x \neq -1$$

The graph of $h(x)$ is the line $y = x - 1$ with the point $(-1, -2)$ missing. There are no vertical asymptotes. See Figure 4.89.

Graph with Asymptotes

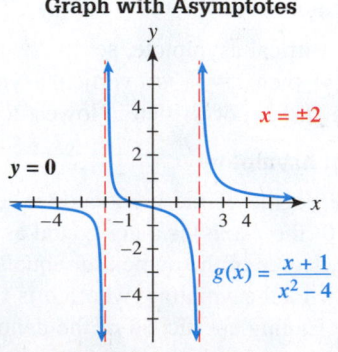

y

4

$x = \pm 2$

2

$y = 0$

−4   −1   3   4   *x*

−2

−4

$g(x) = \dfrac{x + 1}{x^2 - 4}$

Figure 4.88

Graph with a "Hole"

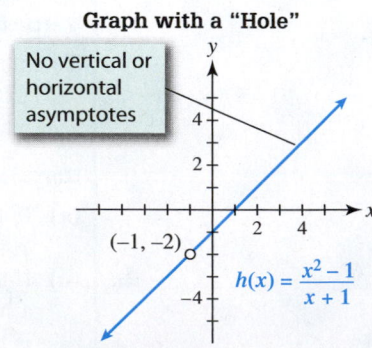

No vertical or horizontal asymptotes

y

4

2

2   4   *x*

$(-1, -2)$

−4

$h(x) = \dfrac{x^2 - 1}{x + 1}$

Figure 4.89

Now Try Exercises 21, 23, and 31

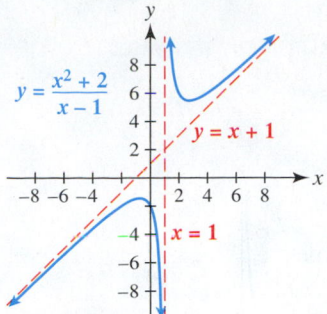

Slant Asymptote

$y = \dfrac{x^2 + 2}{x - 1}$

$y = x + 1$

$x = 1$

Figure 4.90

Slant, or Oblique, Asymptotes A third type of asymptote, which is neither vertical nor horizontal, occurs when the numerator of a rational function has degree *one more* than the degree of the denominator. For example, let $f(x) = \dfrac{x^2 + 2}{x - 1}$. If $x - 1$ is divided into $x^2 + 2$, the quotient is $x + 1$ with remainder 3. Thus

$$f(x) = x + 1 + \frac{3}{x - 1}$$

is an equivalent representation of $f(x)$. For large values of $|x|$, the ratio $\dfrac{3}{x - 1}$ approaches 0 and the graph of f approaches $y = x + 1$. The line $y = x + 1$ is called a **slant asymptote,** or **oblique asymptote**, of the graph of f. A graph of $f(x) = \dfrac{x^2 + 2}{x - 1}$ with vertical asymptote $x = 1$ and slant asymptote $y = x + 1$ is shown in Figure 4.90.

MAKING CONNECTIONS

Division Algorithm and Asymptotes Suppose that the division algorithm is used to write a rational function f in the form

$$f(x) = \textbf{(Quotient)} + \frac{(\textbf{Remainder})}{(\textbf{Divisor})}.$$

1. If the quotient equals a constant k, then $y = k$ is a horizontal asymptote.

   Example: $f(x) = \dfrac{2x - 1}{x - 1} = 2 + \dfrac{1}{x - 1}$, so $y = 2$ is a horizontal asymptote.

2. If the quotient equals $ax + b$ with $a \neq 0$ (linear), then $y = ax + b$ is a slant asymptote.

   Example: $f(x) = \dfrac{x^2 + 2}{x - 1} = x + 1 + \dfrac{3}{x - 1}$, so $y = x + 1$ is a slant asymptote.

Graphs and Transformations of Rational Functions

Graphs of rational functions can vary greatly in complexity. We begin by graphing $y = \frac{1}{x}$ and then use transformations to graph other rational functions.

EXAMPLE 5 **Analyzing the graph of $y = \frac{1}{x}$**

Sketch a graph of $y = \frac{1}{x}$ and identify any asymptotes.

SOLUTION

Note that when $x = 0$, the denominator is 0 but the numerator is not. Thus $x = 0$ (the y-axis) is a vertical asymptote. Also, the degree of the numerator is less than the degree of the denominator, so $y = 0$ (the x-axis) is a horizontal asymptote. Table 4.5 lists points that lie on the graph of $y = \frac{1}{x}$. These points and the graph are shown in Figure 4.91.

x	$y = \frac{1}{x}$
-2	$-\frac{1}{2}$
-1	-1
$-\frac{1}{2}$	-2
0	—
$\frac{1}{2}$	2
1	1
2	$\frac{1}{2}$

Table 4.5

Graph of $y = \frac{1}{x}$

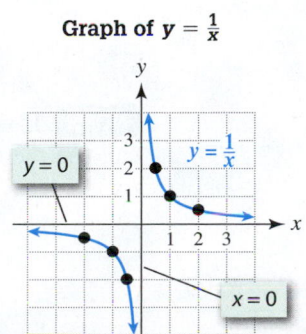

Figure 4.91

Now Try Exercise 41

Transformations   Transformations of graphs can be used to graph some types of rational functions by hand, as illustrated in Figures 4.92 and 4.93.

Transformations of $y = \frac{1}{x}$

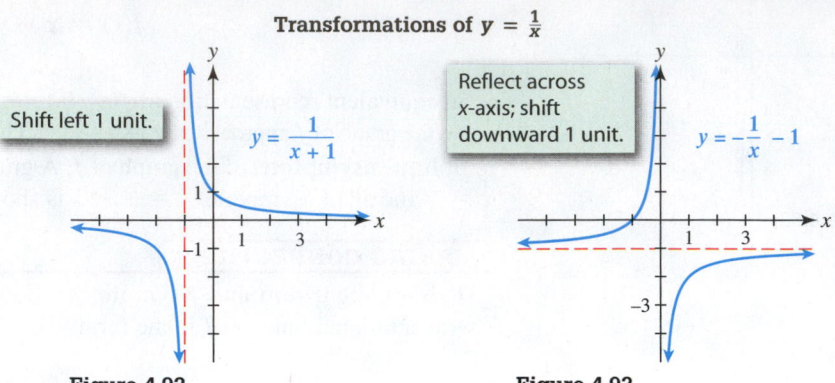

Figure 4.92   **Figure 4.93**

EXAMPLE 6   **Graphing with transformations**

Graph $y = \dfrac{1}{x - 1} + 2$.

SOLUTION

We can graph $g(x) = \frac{1}{x-1} + \mathbf{2}$ by translating the graph of $f(x) = \frac{1}{x}$ right **1** unit and upward **2** units. That is, $g(x)$ can be written in terms of $f(x)$ using $g(x) = f(x - \mathbf{1}) + \mathbf{2}$. Because the graph of $y = \frac{1}{x}$ in Figure 4.91 has vertical asymptote $x = 0$ and horizontal asymptote $y = 0$, the graph of g in Figure 4.94 has vertical asymptote $x = \mathbf{1}$ and horizontal asymptote $y = \mathbf{2}$.

Translating $y = \frac{1}{x}$

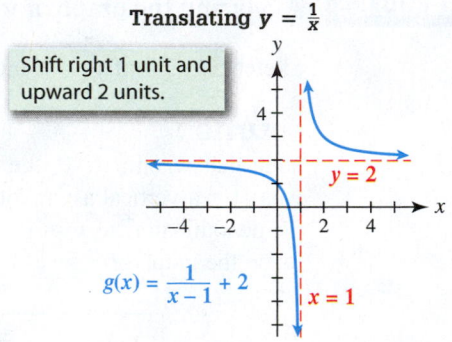

Figure 4.94

Now Try Exercise 49

NOTE   If we are given function g from Example 6 in the form $g(x) = \frac{2x - 1}{x - 1}$, then we can use the division algorithm to divide $x - 1$ into $2x - 1$ and obtain quotient 2 with remainder 1.

$$
\begin{array}{r}
2 \\
x - 1 \overline{)\,2x - 1} \\
\underline{2x - 2} \\
1
\end{array}
$$

$\frac{2x}{x} = 2$

$2(x - 1) = 2x - 2$

Subtract: $-1 - (-2) = 1$.

Thus $g(x) = 2 + \frac{1}{x - 1}$, and we can graph g as in Figure 4.94.

EXAMPLE 7   Using transformations to graph a rational function

Use the graph of $f(x) = \frac{1}{x^2}$ in Figure 4.95 to sketch a graph of $g(x) = -\frac{1}{(x+2)^2}$. Include all asymptotes in your graph. Write $g(x)$ in terms of $f(x)$.

Graph of $y = \dfrac{1}{x^2}$

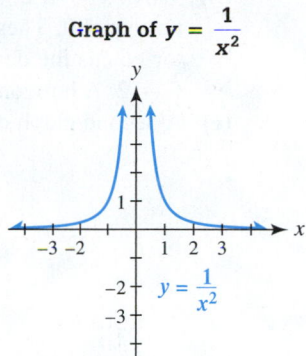

Figure 4.95

SOLUTION

The graph of $y = \frac{1}{x^2}$ has vertical asymptote $x = 0$ and horizontal asymptote $y = 0$. The graph of $g(x) = -\frac{1}{(x+2)^2}$ is a translation of the graph of $f(x) = \frac{1}{x^2}$ left 2 units and then a reflection across the x-axis. The vertical asymptote for $y = g(x)$ is $x = -2$ and the horizontal asymptote is $y = 0$, as shown in Figure 4.96. We can write $g(x)$ in terms of $f(x)$ as $g(x) = -f(x + 2)$.

Translating $y = \dfrac{1}{x^2}$

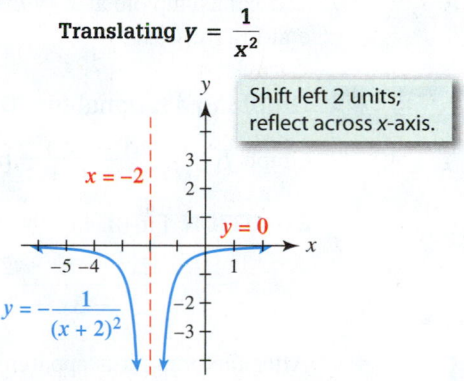

Shift left 2 units; reflect across x-axis.

Figure 4.96

Now Try Exercise 51

Graphing with Technology   Although many graphing calculators have difficulty accurately showing some features of the graph of a rational function, they can be helpful when they are used in conjunction with symbolic techniques.

NOTE   Calculators often graph in connected or dot mode. (See Figure 4.97 on the next page.) If connected mode is used to graph a rational function, it may appear as though the calculator is graphing vertical asymptotes automatically. However, in most instances the calculator is connecting points inappropriately. Sometimes rational functions can be graphed in connected mode using a *decimal* or *friendly* viewing rectangle.

Calculator Help

To set dot mode, see Appendix A (page AP-8). To set a decimal window, see Appendix A (page AP-11).

EXAMPLE 8   Analyzing a rational function with technology

Let $f(x) = \frac{2x^2 + 1}{x^2 - 4}$.

(a) Use a calculator to graph f. Find the domain of f.
(b) Identify any vertical or horizontal asymptotes.
(c) Sketch a graph of f that includes the asymptotes.

Graphing in Dot Mode

[−6, 6, 1] by [−6, 6, 1]

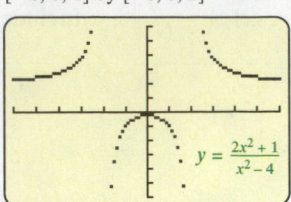

Figure 4.97

SOLUTION

(a) A calculator graph of $f(x) = \frac{2x^2 + 1}{x^2 - 4}$ using dot mode is shown in Figure 4.97. The function is undefined when $x^2 - 4 = 0$, or when $x = \pm 2$. The domain of function f is given by $D = \{x \mid x \neq 2, x \neq -2\}$.

(b) When $x = \pm 2$, the denominator, $x^2 - 4$, equals 0 and the numerator, $2x^2 + 1$, does not equal 0. Therefore $x = \pm 2$ are vertical asymptotes. The degree of the numerator equals the degree of the denominator, and the ratio of the leading coefficients is $\frac{2}{1} = 2$. A horizontal asymptote of the graph of f is $y = 2$.

(c) A second graph of f and its asymptotes is shown in Figure 4.98.

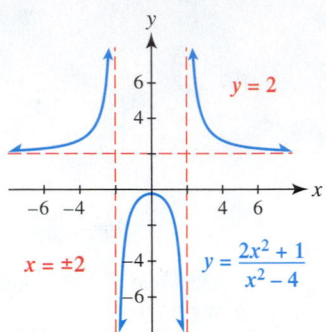

Figure 4.98

Now Try Exercise 57

Graphs with "Holes" If $f(x) = \frac{p(x)}{q(x)}$ is *not* in lowest terms, then it is possible that for some number k both $p(k) = 0$ and $q(k) = 0$. In this case, the graph of f may *not* have a vertical asymptote at $x = k$; rather, it may have a "hole" at $x = k$. See the next example and Figure 4.89.

EXAMPLE 9 **Graphing a rational function having a "hole"**

Graph $f(x) = \frac{2x^2 - 5x + 2}{x^2 - 3x + 2}$ by hand.

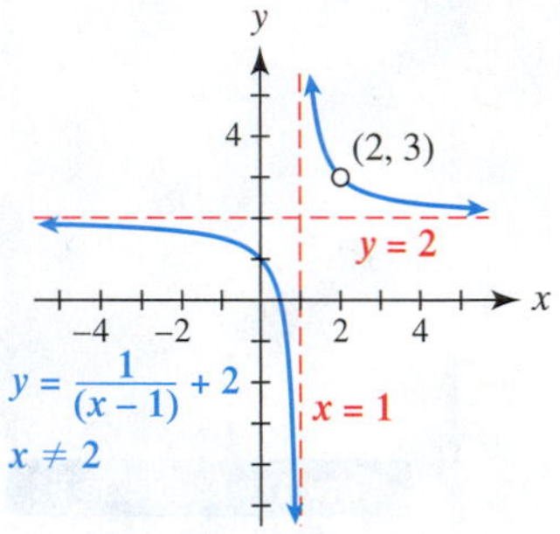

SOLUTION First factor the numerator and the denominator.

$$\frac{2x^2 - 5x + 2}{x^2 - 3x + 2} = \frac{(2x - 1)(x - 2)}{(x - 1)(x - 2)} = \frac{2x - 1}{x - 1}, \qquad x \neq 2$$

After factoring, it is apparent that both the numerator and the denominator equal 0 when $x = 2$. Therefore we simplify the rational expression to lowest terms and restrict the domain to $x \neq 2$.

From our previous work, long division can be used to show that $f(x) = \frac{2x - 1}{x - 1}$ is equivalent to $f(x) = 2 + \frac{1}{x - 1}$. Thus the graph of f is similar to Figure 4.94 except that the point $(2, 3)$ is missing and an open circle appears in its place. See Figure 4.99.

Now Try Exercise 69

Figure 4.99

Graphing Rational Functions by Hand (Optional)

To graph a rational function by hand, we sometimes need to solve a rational equation of the form $\frac{a}{b} = \frac{c}{d}$. One way to solve this equation is to **cross multiply** and obtain $ad = bc$, provided b and d are nonzero. Consider the following example.

$$\frac{2x - 1}{3x + 2} = \frac{5}{4} \qquad \text{Given equation}$$

$$4(2x - 1) = 5(3x + 2) \qquad \text{Cross multiply.}$$

$$8x - 4 = 15x + 10 \qquad \text{Simplify.}$$

$$-7x = 14 \qquad \text{Subtract 15x; add 4.}$$

$$x = -2 \qquad \text{Divide by −7. The answer checks.}$$

The following guidelines can be used to graph a rational function by hand.

GRAPHING A RATIONAL FUNCTION

Let $f(x) = \dfrac{p(x)}{q(x)}$ define a rational function in *lowest* terms. To sketch its graph, follow these steps:

STEP 1: Find all vertical asymptotes.

STEP 2: Find all horizontal or oblique asymptotes.

STEP 3: Find the y-intercept, if possible, by evaluating $f(0)$.

STEP 4: Find the x-intercepts, if any, by solving $f(x) = 0$. (These will be the zeros of the numerator $p(x)$.)

STEP 5: Determine whether the graph will intersect its nonvertical asymptote $y = b$ by solving $f(x) = b$, where b is the y-value of the horizontal asymptote, or by solving $f(x) = mx + b$, where $y = mx + b$ is the equation of the oblique asymptote.

STEP 6: Plot selected points as necessary. Choose an x-value in each interval of the domain determined by the vertical asymptotes and x-intercepts.

STEP 7: Complete the sketch.

EXAMPLE 10 Graphing a rational function by hand

Graph $f(x) = \dfrac{2x + 1}{x - 3}$.

SOLUTION

Getting Started As you go through Steps 1 through 7, be sure to sketch all asymptotes first and then plot some key points. Finally, sketch the entire graph. ▶

STEP 1: The vertical asymptote has equation $x = 3$.

STEP 2: The horizontal asymptote has equation $y = 2$.

STEP 3: $f(0) = -\frac{1}{3}$, so the y-intercept is $-\frac{1}{3}$.

STEP 4: Solve $f(x) = 0$ to find any x-intercepts.

$$\frac{2x + 1}{x - 3} = 0$$

$$2x + 1 = 0 \qquad \text{\textit{If a fraction equals 0, its}}$$
$$\text{\textit{numerator must be 0.}}$$

$$x = -\frac{1}{2}$$

The x-intercept is $-\frac{1}{2}$.

STEP 5: The graph does not intersect its horizontal asymptote, since $f(x) = 2$ has no solutions. (Verify this.)

STEP 6 AND 7: The points $(-4, 1)$, $\left(1, -\frac{3}{2}\right)$, and $\left(6, \frac{13}{3}\right)$ are on the graph and can be used to complete the sketch, shown in Figure 4.100.

Now Try Exercise 87

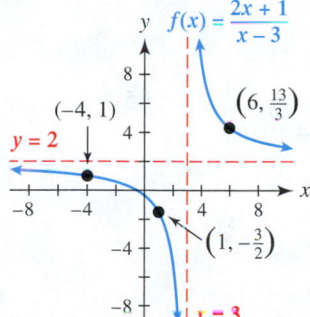

Figure 4.100

EXAMPLE 11 Graphing a function that intersects its horizontal asymptote

Graph $f(x) = \dfrac{3x^2 - 3x - 6}{x^2 + 8x + 16}$.

SOLUTION

Let $f(x) = \dfrac{3x^2 - 3x - 6}{x^2 + 8x + 16}$.

STEP 1: To find the vertical asymptote(s), solve $x^2 + 8x + 16 = 0$.

$$x^2 + 8x + 16 = 0$$
$$(x + 4)^2 = 0$$
$$x = -4$$

Since the numerator is not 0 when $x = -4$, the only vertical asymptote has equation $x = -4$.

STEP 2: Because the degrees of the numerator and denominator both equal 2, the ratio of the leading coefficients gives the horizontal asymptote, which is,

$$y = \frac{3}{1},$$ ← Leading coefficient of numerator
← Leading coefficient of denominator

or $y = 3$.

STEP 3: The y-intercept is $f(0) = -\frac{3}{8}$.

STEP 4: To find the x-intercept(s), if any, solve $f(x) = 0$.

$$\frac{3x^2 - 3x - 6}{x^2 + 8x + 16} = 0 \qquad \text{Set } f(x) \text{ equal to 0.}$$
$$3x^2 - 3x - 6 = 0 \qquad \text{Set the numerator equal to 0.}$$
$$x^2 - x - 2 = 0 \qquad \text{Divide by 3.}$$
$$(x - 2)(x + 1) = 0 \qquad \text{Factor.}$$
$$x = 2 \quad \text{or} \quad x = -1 \qquad \text{Zero-product property}$$

The x-intercepts are -1 and 2.

STEP 5: Because the horizontal asymptote is $y = 3$, set $f(x) = 3$ and solve to locate the point where the graph intersects the horizontal asymptote.

$$\frac{3x^2 - 3x - 6}{x^2 + 8x + 16} = 3$$
$$3x^2 - 3x - 6 = 3x^2 + 24x + 48 \qquad \text{Let } 3 = \tfrac{3}{1}; \text{ cross multiply.}$$
$$-3x - 6 = 24x + 48 \qquad \text{Subtract } 3x^2.$$
$$-27x = 54 \qquad \text{Subtract } 24x, \text{ add 6.}$$
$$x = -2 \qquad \text{Divide by } -27.$$

The graph intersects its horizontal asymptote at $(-2, 3)$.

STEP 6 AND 7: Some other points that lie on the graph are $(-10, 9)$, $\left(-8, 13\frac{1}{8}\right)$, and $\left(5, \frac{2}{3}\right)$. These can be used to complete the graph, shown in Figure 4.101.

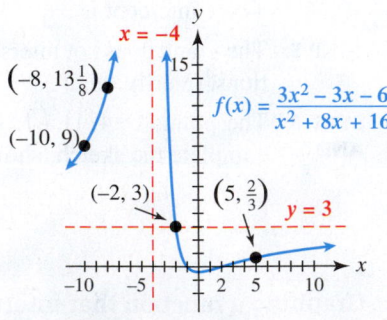

Figure 4.101

Now Try Exercise 91

4.6 Putting It All Together

$\mathbf{T}$he following table summarizes some concepts about rational functions (in lowest terms) and rational equations. To determine vertical and horizontal asymptotes see the box on page 293.

CONCEPT	EXPLANATION	EXAMPLES
Rational function	$f(x) = \frac{p(x)}{q(x)}$, where $p(x)$ and $q(x)$ are polynomials with $q(x) \neq 0$.	$f(x) = \dfrac{x - 1}{x^2 + 2x + 1}$ $g(x) = 1 + \dfrac{1}{x}$ $\left(\textit{Note: } 1 + \dfrac{1}{x} = \dfrac{x + 1}{x}.\right)$
Vertical asymptote	If k is a zero of the denominator, but not of the numerator, then $x = k$ is a vertical asymptote.	The graph of $f(x) = \frac{2x + 1}{x - 2}$ has a vertical asymptote at $x = 2$ because 2 is a zero of $x - 2$, but not a zero of $2x + 1$.
Horizontal asymptote	A horizontal asymptote occurs when the degree of the numerator is less than or equal to the degree of the denominator.	$$f(x) = \frac{1 - 4x^2}{3x^2 - x}$$ Horizontal asymptote: $y = -\frac{4}{3}$ $$g(x) = \frac{x}{4x^2 + 2x}$$ Horizontal asymptote: $y = 0$
Graph of a rational function	The graph of a rational function is continuous, except at x-values where the denominator equals zero.	The graph of $f(x) = \frac{3x^2 + 1}{x^2 - 4}$ is discontinuous at $x = \pm 2$. It has vertical asymptotes of $x = \pm 2$ and a horizontal asymptote of $y = 3$.
Basic rational equation	$\frac{a}{b} = \frac{c}{d}$ is equivalent to $ad = bc$, provided b and d are nonzero. Check your answer. Can be used to graph a rational function by hand	To solve the rational equation $$\frac{4}{2x - 1} = 8,$$ write 8 as $\frac{8}{1}$ and cross multiply. $$4 = 8(2x - 1)$$ $$12 = 16x$$ $$x = \frac{3}{4}$$

4.6 Exercises

Rational Functions

Exercises 1–12: Determine whether f is a rational function and state its domain.

1. $f(x) = \dfrac{x^3 - 5x + 1}{4x - 5}$

2. $f(x) = \dfrac{6}{x^2}$

3. $f(x) = x^2 - x - 2$

4. $f(x) = \dfrac{x^2 + 1}{\sqrt{x - 8}}$

5. $f(x) = \dfrac{|x - 1|}{x + 1}$

6. $f(x) = \dfrac{4}{x} + 1$

7. $f(x) = \dfrac{3x}{x^2 + 1}$

8. $f(x) = \dfrac{|x + 1|}{x + 1}$

9. $f(x) = \dfrac{3 - \sqrt{x}}{x^2 + x}$

10. $f(x) = \dfrac{x^3 - 3x + 1}{x^2 - 5}$

11. $f(x) = 4 - \dfrac{3}{x + 1}$

12. $f(x) = 5x^3 - 4x$

Asymptotes and Graphs

Exercises 13–18: Identify any horizontal or vertical asymptotes in the graph. State the domain of f.

13.
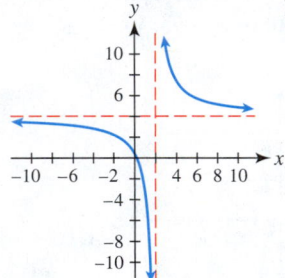

14.
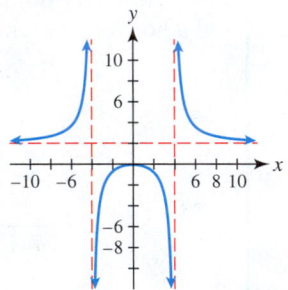

15.
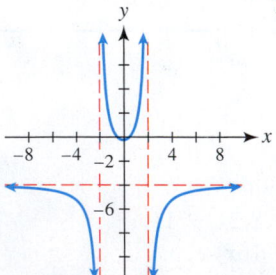

16.
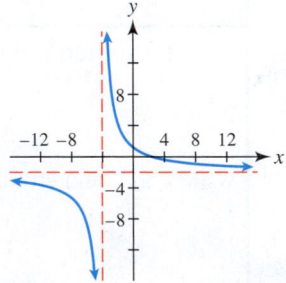

17.
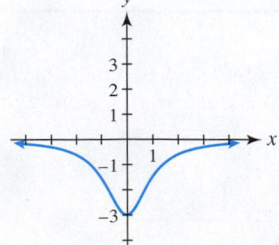

18.
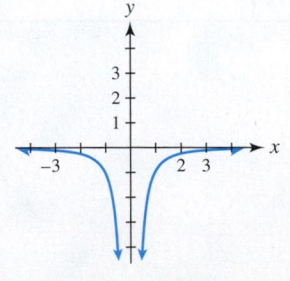

Exercises 19 and 20: In the table, Y_1 is a rational function. Give a possible equation for a horizontal asymptote.

19.

X	Y1
50	2.8654
100	2.9314
150	2.9539
200	2.9653
250	2.9722
300	2.9768
350	2.9801

X=50

20.

X	Y1
−10	4.8922
−20	4.9726
−30	4.9878
−40	4.9931
−50	4.9956
−60	4.9969
−70	4.9978

X=−10

Exercises 21–32: Find any horizontal or vertical asymptotes.

21. $f(x) = \dfrac{4x + 1}{2x - 6}$

22. $f(x) = \dfrac{x + 6}{5 - 2x}$

23. $f(x) = \dfrac{3}{x^2 - 5}$

24. $f(x) = \dfrac{3x^2}{x^2 - 9}$

25. $f(x) = \dfrac{x^4 + 1}{x^2 + 3x - 10}$

26. $f(x) = \dfrac{4x^3 - 2}{x + 2}$

27. $f(x) = \dfrac{x^2 + 2x + 1}{2x^2 - 3x - 5}$

28. $f(x) = \dfrac{6x^2 - x - 2}{2x^2 + x - 6}$

29. $f(x) = \dfrac{3x(x + 2)}{(x + 2)(x - 1)}$

30. $f(x) = \dfrac{x}{x^3 - x}$

31. $f(x) = \dfrac{x^2 - 9}{x + 3}$

32. $f(x) = \dfrac{2x^2 - 3x + 1}{2x - 1}$

Exercises 33–36: Let a be a positive constant. Match $f(x)$ with its graph (a–d) without using a calculator.

33. $f(x) = \dfrac{a}{x - 1}$

34. $f(x) = \dfrac{2x + a}{x - 1}$

35. $f(x) = \dfrac{x - a}{x + 2}$

36. $f(x) = \dfrac{-2x}{x^2 - a}$

a.
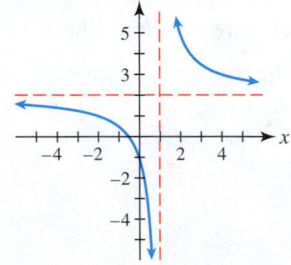

b.
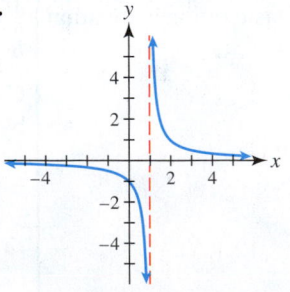

c.
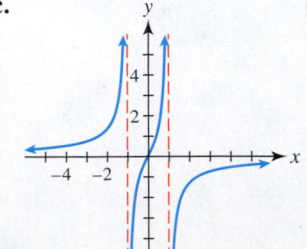

d.
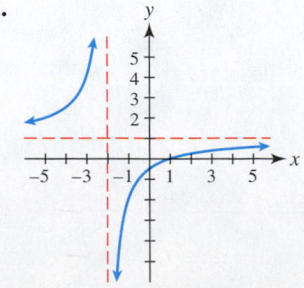

Exercises 37–40: Write a formula $f(x)$ for a rational function so that its graph has the specified asymptotes.

37. Vertical: $x = -3$; horizontal: $y = 1$

38. Vertical: $x = 4$; horizontal: $y = -3$

39. Vertical: $x = \pm 3$; horizontal: $y = 0$

40. Vertical: $x = -2$ and $x = 4$; horizontal: $y = 5$

Graphing Rational Functions

Exercises 41–44: Graph f and identify any asymptotes.

41. $f(x) = \dfrac{1}{x^2}$

42. $f(x) = -\dfrac{1}{x}$

43. $f(x) = -\dfrac{1}{2x}$

44. $f(x) = \dfrac{2}{x^2}$

Exercises 45–54: **Transformations** *Use transformations of the graph of either $f(x) = \frac{1}{x}$ or $h(x) = \frac{1}{x^2}$ to sketch a graph of $y = g(x)$ by hand. Show all asymptotes. Write $g(x)$ in terms of either $f(x)$ or $h(x)$.*

45. $g(x) = \dfrac{1}{x - 3}$

46. $g(x) = \dfrac{1}{x + 2}$

47. $g(x) = \dfrac{1}{x} + 2$

48. $g(x) = 1 - \dfrac{2}{x}$

49. $g(x) = \dfrac{1}{x + 1} - 2$

50. $g(x) = \dfrac{1}{x - 2} + 1$

51. $g(x) = -\dfrac{2}{(x - 1)^2}$

52. $g(x) = \dfrac{1}{x^2} - 1$

53. $g(x) = \dfrac{1}{(x + 1)^2} - 2$

54. $g(x) = 1 - \dfrac{1}{(x - 2)^2}$

Exercises 55–62: Complete the following.

(a) Find the domain of f.

(b) Graph f in an appropriate viewing rectangle.

(c) Find any horizontal or vertical asymptotes.

(d) Sketch a graph of f that includes any asymptotes.

55. $f(x) = \dfrac{x + 3}{x - 2}$

56. $f(x) = \dfrac{6 - 2x}{x + 3}$

57. $f(x) = \dfrac{4x + 1}{x^2 - 4}$

58. $f(x) = \dfrac{0.5x^2 + 1}{x^2 - 9}$

59. $f(x) = \dfrac{4}{1 - 0.25x^2}$

60. $f(x) = \dfrac{x^2}{1 + 0.25x^2}$

61. $f(x) = \dfrac{x^2 - 4}{x - 2}$

62. $f(x) = \dfrac{4(x - 1)}{x^2 - x - 6}$

Exercises 63–72: Graph $y = f(x)$. You may want to use division, factoring, or transformations as an aid. Show all asymptotes and "holes."

63. $f(x) = \dfrac{x^2 - 2x + 1}{x - 1}$

64. $f(x) = \dfrac{4x^2 + 4x + 1}{2x + 1}$

65. $f(x) = \dfrac{x + 2}{x + 1}$

66. $f(x) = \dfrac{2x + 3}{x + 1}$

67. $f(x) = \dfrac{2x^2 - 3x - 2}{x^2 - 4x + 4}$

68. $f(x) = \dfrac{x^2 - x - 2}{x^2 - 2x - 3}$

69. $f(x) = \dfrac{2x^2 + 9x + 9}{2x^2 + 7x + 6}$

70. $f(x) = \dfrac{x^2 - 4}{x^2 - x - 6}$

71. $f(x) = \dfrac{-2x^2 + 11x - 14}{x^2 - 5x + 6}$

72. $f(x) = \dfrac{2x^2 - 3x - 14}{x^2 - 2x - 8}$

Slant Asymptotes

Exercises 73–80: Complete the following.

(a) Find any slant or vertical asymptotes.

(b) Graph $y = f(x)$. Show all asymptotes.

73. $f(x) = \dfrac{x^2 + 1}{x + 1}$

74. $f(x) = \dfrac{2x^2 - 5x - 2}{x - 2}$

75. $f(x) = \dfrac{0.5x^2 - 2x + 2}{x + 2}$

76. $f(x) = \dfrac{0.5x^2 - 5}{x - 3}$

77. $f(x) = \dfrac{x^2 + 2x + 1}{x - 1}$

78. $f(x) = \dfrac{2x^2 + 3x + 1}{x - 2}$

79. $f(x) = \dfrac{4x^2}{2x - 1}$

80. $f(x) = \dfrac{4x^2 + x - 2}{4x - 3}$

Basic Rational Equations

Exercises 81–86: Solve the equation.

81. $\dfrac{4}{x + 2} = -4$

82. $\dfrac{3}{2x + 1} = -1$

83. $\dfrac{x + 1}{x} = 2$

84. $\dfrac{2x}{x - 3} = -4$

85. $\dfrac{1 - x}{3x - 1} = -\dfrac{3}{5}$

86. $\dfrac{3 - 2x}{x + 2} = 12$

Graphing Rational Functions by Hand

Exercises 87–92: (Refer to Examples 10 and 11.) Graph f. Use the steps for graphing a rational function described in this section.

87. $f(x) = \dfrac{2x - 4}{x - 1}$

88. $f(x) = \dfrac{x + 3}{2x - 4}$

89. $f(x) = \dfrac{x^2 - 2x}{x^2 + 6x + 9}$ **90.** $f(x) = \dfrac{2x + 1}{x^2 + 6x + 8}$

91. $f(x) = \dfrac{x^2 + 2x + 1}{x^2 - x - 6}$ **92.** $f(x) = \dfrac{3x^2 + 3x - 6}{x^2 - x - 12}$

Applications

93. Social Network Participation You may have noticed that a relatively small percentage of people do the vast majority of postings on social networks. The rational function defined by

$$f(x) = \frac{100}{101 - x}, \qquad 5 \le x \le 100,$$

models this participation inequality. In this formula, $f(x)$ outputs the percentage of the postings done by the least active (bottom) x percent of the population.
(a) Evaluate $f(98)$. Interpret your answer.

(b) Evaluate $f(100) - f(98)$. Interpret your answer.

94. Probability A container holds x balls numbered 1 through x. Only one ball has the winning number.
(a) Find a function f that computes the probability, or likelihood, of *not* drawing the winning ball.

(b) What is the domain of f?

(c) What happens to the probability of *not* drawing the winning ball as the number of balls increases?

(d) Interpret the horizontal asymptote of the graph of f.

95. Abandoning Websites The first minute is critical to a visitor's decision whether to stay or leave a website. The longer a person visits a website, the less likely it is that he or she will leave the page. If x represents the number of seconds that a visitor has been visiting a website, then the likelihood as a percentage that this visitor is abandoning the website at x seconds is modeled by

$$P(x) = \frac{5}{0.03x + 0.97}, \qquad 1 \le x \le 60.$$

(a) Approximate $P(1)$ and interpret your answer.

(b) Approximate $P(60)$ and interpret your answer.

96. Concentration of a Drug The concentration of a drug in a medical patient's bloodstream is given by the formula $f(t) = \dfrac{5}{t^2 + 1}$, where the input t is in hours, $t \ge 0$, and the output is in milligrams per liter.
(a) Does the concentration of the drug increase or decrease? Explain.

(b) The patient should not take a second dose until the concentration is below 1.5 milligrams per liter. How long should the patient wait before taking a second dose?

97. Time Spent in Line If two parking attendants can wait on 8 vehicles per minute and vehicles are leaving the parking garage randomly at an average rate of x vehicles per minute, then the average time T in minutes spent waiting in line *and* paying the attendant is given by the formula $T(x) = -\dfrac{1}{x - 8}$, where $0 \le x < 8$. A graph of T is shown in the figure.

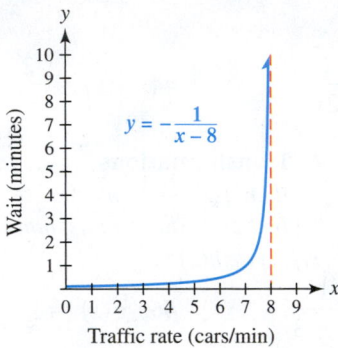

(a) Evaluate $T(4)$ and $T(7.5)$. Interpret the results.

(b) What happens to the wait as vehicles arrive at an average rate that approaches 8 cars per minute?

98. Time Spent in Line (Refer to Exercise 97.) If the parking attendants can wait on 5 vehicles per minute, the average time T in minutes spent waiting in line *and* paying the attendant becomes $T(x) = -\dfrac{1}{x - 5}$.
(a) What is a reasonable domain for T?

(b) Graph $y = T(x)$. Be sure to include any vertical asymptotes.

(c) Explain what happens to $T(x)$ as $x \to 5^-$.

99. Length of Lines (Refer to Example 2.) Suppose that a parking attendant can wait on 40 cars per hour and that cars arrive randomly at a rate of x cars per hour. Then the average number of cars waiting in line can be estimated by

$$N(x) = \frac{x^2}{1600 - 40x}.$$

(a) Evaluate $N(20)$ and $N(39)$.

(b) Explain what happens to the length of the line as x approaches 40.

(c) Find any vertical asymptotes of the graph of N.

100. Construction Zone Suppose that a construction zone can allow 50 cars per hour to pass through and that cars arrive randomly at a rate of x cars per hour. Then the average number of cars waiting in line to get through the construction zone can be estimated by

$$N(x) = \frac{x^2}{2500 - 50x}.$$

(a) Evaluate $N(20)$, $N(40)$, and $N(49)$.

(b) Explain what happens to the length of the line as x approaches 50.

(c) Find any vertical asymptotes of the graph of N.

101. Interpreting an Asymptote Suppose that an insect population in millions is modeled by $f(x) = \frac{10x + 1}{x + 1}$, where $x \geq 0$ is in months.

(a) Graph f in $[0, 14, 1]$ by $[0, 14, 1]$. Find the equation of the horizontal asymptote.

(b) Determine the initial insect population.

(c) What happens to the population over time?

(d) Interpret the horizontal asymptote.

102. Interpreting an Asymptote Suppose that the population of a species of fish (in thousands) is modeled by $f(x) = \frac{x + 10}{0.5x^2 + 1}$, where $x \geq 0$ is in years.

(a) Graph f in $[0, 12, 1]$ by $[0, 12, 1]$. What is the horizontal asymptote?

(b) Determine the initial population.

(c) What happens to the population of this fish?

(d) Interpret the horizontal asymptote.

103. Train Curves When curves are designed for trains, sometimes the outer rail is elevated or banked, so that a locomotive can safely negotiate the curve at a higher speed. See the figure at the top of the next column. Suppose a circular curve is designed for 60 miles per hour. The formula $f(x) = \frac{2540}{x}$ computes the elevation y in inches of the outer track for a curve having a radius of x feet, where $y = f(x)$. (***Source:*** L. Haefner, *Introduction to Transportation Systems.*)

(a) Evaluate $f(400)$ and interpret the result.

(b) Graph f in $[0, 600, 100]$ by $[0, 50, 5]$. How does the elevation change as the radius increases?

(c) Interpret the horizontal asymptote.

(d) Find the radius if the elevation is 12.7 inches.

104. Slippery Roads If a car is moving at 50 miles per hour on a level highway, then its braking distance depends on the road conditions. This distance in feet can be computed by $D(x) = \frac{250}{30x}$, where x is the coefficient of friction between the tires and the road and $0 < x \leq 1$. A smaller value of x indicates that the road is more slippery.

(a) Identify and interpret the vertical asymptote.

(b) Estimate the coefficient of friction associated with a braking distance of 340 feet.

Writing About Mathematics

105. Let $f(x)$ be the formula for a rational function.

(a) Explain how to find any vertical or horizontal asymptotes of the graph of f.

(b) Discuss what a horizontal asymptote represents.

106. Discuss how to find the domain of a rational function symbolically and graphically.

Extended and Discovery Exercises

Exercises 1–4: **Rate of Change/Difference Quotient** *Find the average rate of change of f from $x_1 = 1$ to $x_2 = 3$. Then find the difference quotient of f.*

1. $f(x) = \frac{1}{x}$

2. $f(x) = \frac{1}{x^2}$

3. $f(x) = \frac{3}{2x}$

4. $f(x) = \frac{1}{5 - x}$

CHECKING BASIC CONCEPTS FOR SECTIONS 4.5 AND 4.6

1. Find a quadratic polynomial $f(x)$ with zeros $\pm 4i$ and leading coefficient 3. Write $f(x)$ in complete factored form and expanded form.

2. Sketch a graph of a quartic function (degree 4) with a negative leading coefficient, two real zeros, and two imaginary zeros.

3. Write $x^3 - x^2 + 4x - 4$ in complete factored form.

4. Solve each equation.
   (a) $2x^3 + 45 = 5x^2 - 18x$

   (b) $x^4 + 5x^2 = 36$

5. Let $f(x) = \frac{1}{x-1} + 2$.
   (a) Find the domain of f.

   (b) Identify any vertical or horizontal asymptotes.

   (c) Sketch a graph of f that includes all asymptotes.

6. Find any vertical or horizontal asymptotes for the graph of $f(x) = \frac{4x^2}{x^2 - 4}$. State the domain of f.

7. Sketch a graph of each rational function f. Include all asymptotes and any "holes" in your graph.
   (a) $f(x) = \frac{3x-1}{2x-2}$ (b) $f(x) = \frac{1}{(x+1)^2}$

   (c) $f(x) = \frac{x+2}{x^2-4}$ (d) $f(x) = \frac{x^2+1}{x^2-1}$

4.7 More Equations and Inequalities

- Solve rational equations
- Solve variation problems
- Solve polynomial inequalities
- Solve rational inequalities

Introduction

Waiting in line is a part of almost everyone's life. When people arrive randomly at a line, rational functions can be used to estimate the average number of people standing in line. For example, if an attendant at a ticket booth can wait on 30 customers per hour and if customers arrive at an average rate of x per hour, then the average number of customers waiting in line is computed by

$$f(x) = \frac{x^2}{900 - 30x},\qquad \text{Rational function}$$

where $0 \le x < 30$. Thus $f(28) \approx 13$ indicates that if customers arrive, on average, at 28 per hour, then the average number of people in line is 13. If a line length of **8** customers or fewer is acceptable, then we can use $f(x)$ to estimate customer arrival rates x that one attendant can accommodate by solving the *rational inequality*

$$\frac{x^2}{900 - 30x} \le \mathbf{8}.\qquad \text{Rational inequality}$$

(See Example 8.) Rational inequalities are discussed in this section along with other types of inequalities and equations.   (**Source:** N. Garber and L. Hoel, *Traffic and Highway Engineering.*)

Rational Equations

If $f(x)$ represents a rational function, then an equation that can be written in the form $f(x) = k$ for some constant k is a **rational equation**.

$$\frac{x^2 - 1}{x^2 + x + 3} = 0, \quad \frac{3x}{x^3 + x} = \frac{3}{2}, \quad \text{and} \quad \frac{2}{x-1} + \frac{1}{x} = -2$$

Rational Equations

Rational equations can be solved symbolically, graphically, and numerically.

| EXAMPLE 1 | Solving a rational equation |

Solve $\frac{4x}{x-1} = 6$ symbolically, graphically, and numerically.

SOLUTION

Getting Started The equation $\frac{a}{b} = \frac{c}{d}$ with $b \ne 0$ and $d \ne 0$ is equivalent to $ad = bc$. In this example, you can think of 6 as the ratio $\frac{6}{1}$. This technique is sometimes called *cross multiplying.* ▶

Symbolic Solution

$$\frac{4x}{x-1} = 6 \qquad \text{Given equation}$$

$$4x = 6(x-1) \qquad \text{Cross multiply: } \tfrac{a}{b} = \tfrac{c}{d} \text{ implies } ad = bc.$$

$$4x = 6x - 6 \qquad \text{Distributive property}$$

$$-2x = -6 \qquad \text{Subtract 6x.}$$

$$x = 3 \qquad \text{Divide by } -2. \text{ (Check this answer.)}$$

The only solution is 3.

Graphical Solution Graph $Y_1 = 4X/(X-1)$ and $Y_2 = 6$. Their graphs intersect at $(\mathbf{3}, 6)$, so the solution is **3**. See Figure 4.102.

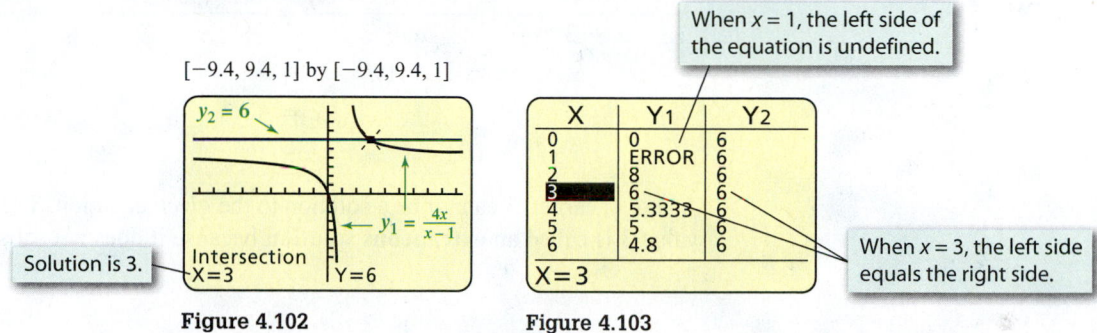

$[-9.4, 9.4, 1]$ by $[-9.4, 9.4, 1]$

X	Y₁	Y₂
0	0	6
1	ERROR	6
2	8	6
3	6	6
4	5.3333	6
5	5	6
6	4.8	6

When $x = 1$, the left side of the equation is undefined.

When $x = 3$, the left side equals the right side.

Solution is 3.

Figure 4.102   **Figure 4.103**

Numerical Solution In Figure 4.103, $y_1 = y_2$ when $x = 3$.

Now Try Exercise 1

Algebra Review

To review clearing fractions, see Chapter R (page R-33).

A common approach to solving rational equations symbolically is to multiply each side of the equation by a common denominator. This technique, which clears fractions from an equation, is used in Examples 2 and 3.

EXAMPLE 2 **Solving a rational equation**

Solve $\frac{6}{x^2} - \frac{5}{x} = 1$ symbolically.

SOLUTION

The least common denominator for x^2 and x is x^2. Multiply each side of the equation by x^2.

Algebra Review

To review finding a least common denominator, see Chapter R (page R-30).

$$\frac{6}{x^2} - \frac{5}{x} = 1 \qquad \text{Given equation}$$

$$\frac{6}{x^2} \cdot x^2 - \frac{5}{x} \cdot x^2 = 1 \cdot x^2 \qquad \text{Multiply each term by } x^2.$$

$$6 - 5x = x^2 \qquad \text{Simplify.}$$

$$0 = x^2 + 5x - 6 \qquad \text{Add 5x and subtract 6.}$$

$$0 = (x+6)(x-1) \qquad \text{Factor.}$$

$$x + 6 = 0 \quad \text{or} \quad x - 1 = 0 \qquad \text{Zero-product property}$$

$$x = -6 \quad \text{or} \quad x = 1 \qquad \text{Solve.}$$

Check to verify that each answer is correct.

Now Try Exercise 17

The next example illustrates the importance of checking possible solutions.

EXAMPLE 3 Solving a rational equation

Solve $\frac{1}{x+3} + \frac{1}{x-3} = \frac{6}{x^2-9}$ symbolically. Check the result.

SOLUTION

The least common denominator is $(x+3)(x-3)$, or x^2-9.

NOTE If either -3 or 3 is substituted into the given equation, two of the expressions are undefined. (Their denominators equal 0.) Thus neither of these values can be a solution to the *given* equation.

$$\frac{1}{x+3} + \frac{1}{x-3} = \frac{6}{x^2-9} \qquad \text{Given equation}$$

$$\frac{(x+3)(x-3)}{x+3} + \frac{(x+3)(x-3)}{x-3} = \frac{6(x+3)(x-3)}{x^2-9} \qquad \text{Multiply by } (x+3)(x-3).$$

$$(x-3) + (x+3) = 6 \qquad \text{Simplify.}$$

$$2x = 6 \qquad \text{Combine terms.}$$

$$x = 3 \qquad \text{Divide by 2.}$$

As noted earlier, 3 cannot be a solution to the *given* equation. There are no solutions. (The value 3 is called an **extraneous solution** because it does not satisfy the given equation.)

Now Try Exercise 21

An Application Rational equations are used in real-world applications such as the construction problem in the next example. Steps for solving application problems (see page 95) have been used to structure the solution.

EXAMPLE 4 Designing a box

A box with rectangular sides and a top is being designed to hold 324 cubic inches and to have a surface area of 342 square inches. If the length of the box is four times the height, find possible dimensions of the box.

SOLUTION

STEP 1: We are asked to find the dimensions of a box. If x is the height of the box and y is the width, then the length of the box is $4x$, or four times the height.

x: Height of the box y: Width of the box $4x$: Length of the box

STEP 2: To relate these variables to an equation, sketch a box as shown in Figure 4.104. The volume V of the box is **height** times **width** times **length**.

$$V = xy(4x) = 4x^2y$$

The surface area A of this box is determined by finding the area of the 6 rectangular sides: left and right sides, front and back, and top and bottom.

$$A = 2(4x \cdot x) + 2(x \cdot y) + 2(4x \cdot y)$$
$$= 8x^2 + 10xy$$

If we solve $V = 4x^2y$ for y and let $V = 324$, we obtain

$$y = \frac{V}{4x^2} = \frac{324}{4x^2} = \frac{81}{x^2}.$$

Substituting $y = \frac{81}{x^2}$ in the formula for A eliminates the y variable.

$$A = 8x^2 + 10xy \qquad \text{Area formula}$$

$$= 8x^2 + 10x \cdot \frac{81}{x^2} \qquad \text{Let } y = \frac{81}{x^2}.$$

$$= 8x^2 + \frac{810}{x} \qquad \text{Simplify.}$$

Designing a Box

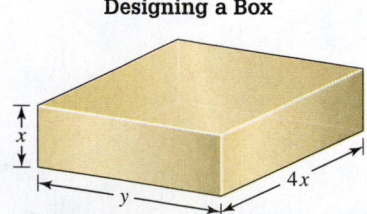

Figure 4.104

Geometry Review
To review formulas related to box shapes, see Chapter R (page R-3).

Since the surface area is $A = 342$ square inches, the height x can be determined by solving the rational equation

$$8x^2 + \frac{810}{x} = 342. \qquad \text{Area} = 342$$

STEP 3: Figures 4.105 and 4.106 show the graphs of $Y_1 = 8X^2 + 810/X$ and $Y_2 = 342$. There are two *positive* solutions: $x = 3$ and $x = 4.5$.

Locating Solutions Graphically

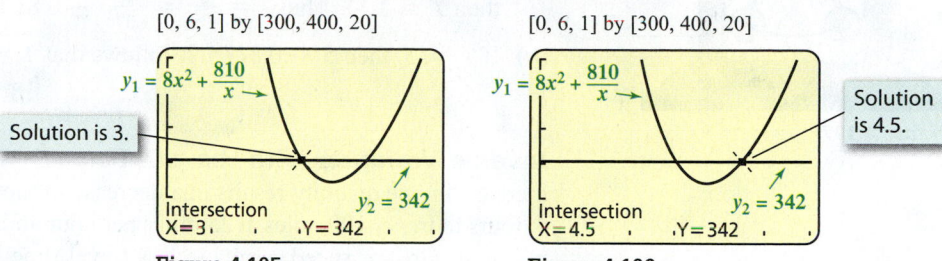

Figure 4.105   **Figure 4.106**

Calculator Help
To find a point of intersection, see Appendix A (page AP-7).

NOTE This equation can be written as $8x^3 - 342x + 810 = 0$, and the rational zeros test or factoring could be used to find the solutions, 3 and $\frac{9}{2}$.

For the first solution, the height is $x = 3$ inches, so the length is $4 \cdot 3 = 12$ inches and the width is $y = \frac{81}{3^2} = 9$ inches. (Note that $y = \frac{81}{x^2}$.) For the second solution, the height is 4.5 inches, so the length is $4 \cdot 4.5 = 18$ inches and the width is $y = \frac{81}{4.5^2} = 4$ inches. Thus the dimensions of the box in inches can be either $3 \times 9 \times 12$ or $4.5 \times 4 \times 18$.

STEP 4: We can check our results directly. If the dimensions are $3 \times 9 \times 12$, then

$$V = 3 \cdot 9 \cdot 12 = 324 \; \checkmark \text{ and}$$
$$S = 2(3 \cdot 9) + 2(3 \cdot 12) + 2(9 \cdot 12) = 342. \; \checkmark$$

If the dimensions are $4.5 \times 4 \times 18$, then

$$V = 4.5 \cdot 4 \cdot 18 = 324 \; \checkmark \text{ and}$$
$$S = 2(4.5 \cdot 4) + 2(4.5 \cdot 18) + 2(4 \cdot 18) = 342. \; \checkmark$$

In both cases our results check.

Now Try Exercise 79

Variation

Direct Variation with the *n*th Power   In Section 2.4 direct variation was discussed. Sometimes a quantity y varies directly with a *power* of a variable. For example, the area A of a circle varies directly with the second power (square) of the radius r. That is, $A = \pi r^2$.

> **DIRECT VARIATION WITH THE *n*TH POWER**
>
> Let x and y denote two quantities and n be a positive real number. Then y is **directly proportional to the *n*th power** of x, or y **varies directly with the *n*th power** of x, if there exists a nonzero number k such that
>
> $$y = kx^n.$$

The number k is called the *constant of variation* or the *constant of proportionality*. In the formula $A = \pi r^2$, the constant of variation is π.

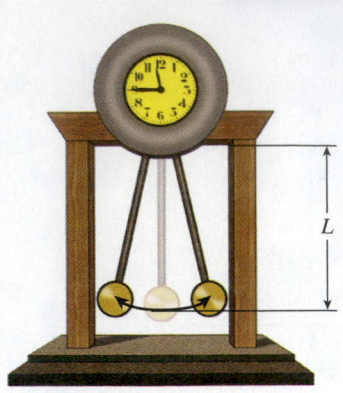

Figure 4.107

EXAMPLE 5   **Modeling a pendulum**

The time T required for a pendulum to swing back and forth once is called its *period*. The length L of a pendulum is directly proportional to the square of T. See Figure 4.107. A 2-foot pendulum has a 1.57-second period.

(a) Find the constant of proportionality k.

(b) Predict T for a pendulum having a length of 5 feet.

SOLUTION

(a) Because L is directly proportional to the **square** of T, we can write $L = kT^2$. If $L = 2$, then $T = \textbf{1.57}$. Thus $k = \frac{L}{T^2} = \frac{2}{1.57^2} \approx 0.81$ and $L = 0.81T^2$.

(b) If $L = 5$, then $5 = 0.81T^2$. It follows that $T = \sqrt{5/0.81} \approx 2.48$ seconds.

> **Now Try Exercise 107**

Inverse Variation with the nth Power   When two quantities vary inversely, an increase in one quantity results in a decrease in the second quantity. For example, it takes 4 hours to travel 100 miles at 25 miles per hour and 2 hours to travel 100 miles at 50 miles per hour. Greater speed results in less travel time. If s represents the average speed of a car and t is the time to travel 100 miles, then $s \cdot t = 100$, or $t = \frac{100}{s}$. Doubling the speed cuts the time in half; tripling the speed reduces the time by one-third. The quantities t and s are said to *vary inversely*. The constant of variation is 100.

INVERSE VARIATION WITH THE nTH POWER

Let x and y denote two quantities and n be a positive real number. Then y is **inversely proportional to the nth power** of x, or y **varies inversely with the nth power** of x, if there exists a nonzero number k such that

$$y = \frac{k}{x^n}.$$

If $y = \frac{k}{x}$, then y is **inversely proportional** to x or y **varies inversely** with x.

NOTE   To review steps for solving variation problems see the box on page 125.

Inverse variation occurs in measuring the intensity of light. If we increase our distance from a lightbulb, the intensity of the light decreases. Intensity I is inversely proportional to the second power of the distance d. The equation $I = \frac{k}{d^2}$ models this phenomenon.

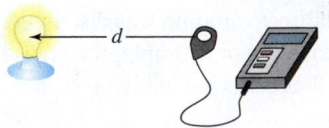

EXAMPLE 6   **Modeling the intensity of light**

At a distance of 3 meters, a 100-watt bulb produces an intensity of 0.88 watt per square meter.   (*Source*: R. Weidner and R. Sells, *Elementary Classical Physics*, Volume 2.)

(a) Find the constant of proportionality k.

(b) Determine the intensity at a distance of 2 meters.

SOLUTION

(a) Substitute $d = 3$ and $I = 0.88$ in the equation $I = \frac{k}{d^2}$. Solve for k.

$$0.88 = \frac{k}{3^2}, \quad \text{or} \quad k = \textbf{7.92}$$

(b) Let $I = \frac{7.92}{d^2}$ and $d = 2$. Then $I = \frac{7.92}{2^2} = 1.98$. The intensity at 2 meters is 1.98 watts per square meter.

> **Now Try Exercise 111**

Polynomial Inequalities

Algebra Review
To review interval notation, see Section 1.4.

Graphical Solutions In Section 3.4 a strategy for solving quadratic inequalities was presented. This strategy involves first finding boundary numbers (x-values) where equality holds. Once the boundary numbers are known, a graph or a table of test values can be used to determine the intervals where inequality holds. This strategy can be applied to other types of inequalities.

The following See the Concept illustrates how to solve polynomial inequalities, where $p(x) = -x^4 + 5x^2 - 4$. Note that the solution sets are also shown on the number line above each graph.

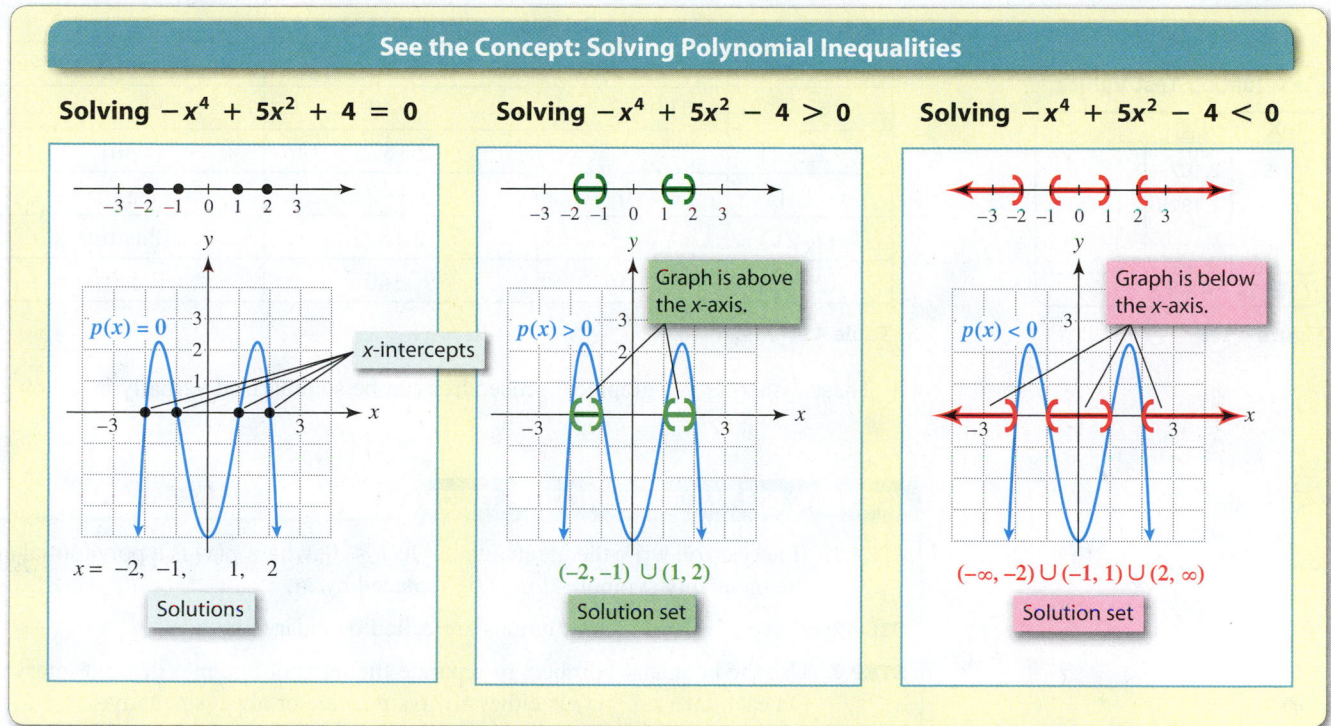

See the Concept: Solving Polynomial Inequalities

Solving $-x^4 + 5x^2 + 4 = 0$

$x = -2, -1, \quad 1, \quad 2$

Solutions

Solving $-x^4 + 5x^2 - 4 > 0$

Graph is above the x-axis.

$p(x) > 0$

$(-2, -1) \cup (1, 2)$

Solution set

Solving $-x^4 + 5x^2 - 4 < 0$

Graph is below the x-axis.

$p(x) < 0$

$(-\infty, -2) \cup (-1, 1) \cup (2, \infty)$

Solution set

MAKING CONNECTIONS

Visualization and Inequalities A *precise* graph of p is *not* necessary to solve the polynomial inequality $p(x) > 0$ or $p(x) < 0$. Once the x-intercepts have been determined, we can use our knowledge about graphs of quartic polynomials (see Section 4.2) to visualize the graph of p. Then we can determine where the graph of p is above the x-axis and where it is below the x-axis.

Symbolic Solutions Polynomial inequalities can also be solved symbolically. For example, to solve the inequality $-x^4 + 5x^2 - 4 > 0$, begin by finding the boundary numbers.

$$-x^4 + 5x^2 - 4 = 0 \qquad \text{Replace } > \text{ with } =.$$
$$x^4 - 5x^2 + 4 = 0 \qquad \text{Multiply by } -1.$$
$$(x^2 - 4)(x^2 - 1) = 0 \qquad \text{Factor.}$$
$$x^2 - 4 = 0 \quad \text{ or } \quad x^2 - 1 = 0 \qquad \text{Zero-product property}$$
$$x = \pm 2 \quad \text{ or } \quad x = \pm 1 \qquad \text{Square root property}$$

The boundary numbers $-2, -1, 1,$ and 2 separate the number line into five intervals:

$$(-\infty, -2), (-2, -1), (-1, 1), (1, 2), \text{ and } (2, \infty),$$

as illustrated in Figure 4.108.

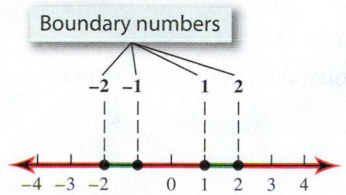

Boundary numbers

Figure 4.108

The polynomial $p(x) = -x^4 + 5x^2 - 4$ is either always positive or always negative on each of these intervals. To determine which is the case, we can choose one test value (x-value) from each interval and substitute this test value in $p(x)$. From Table 4.6 we see that the test value $x = -1.5$ results in

$$p(-1.5) = -(-1.5)^4 + 5(-1.5)^2 - 4 = 2.1875 > 0.$$

Figure 4.109, which was created by a graphing calculator, is similar to Table 4.6. Since $x = -1.5$ is in the interval $(-2, -1)$, it follows that $p(x)$ is **positive** on this interval. The sign of $p(x)$ on the other intervals is determined similarly. Thus the solution set to $p(x) = -x^4 + 5x^2 - 4 > 0$ is $(-2, -1) \cup (1, 2)$.

Evaluating Test Values

X	Y1
-3	-40
-1.5	2.1875
0	-4
1.5	2.1875
3	-40

Y1▪-X^4+5X²-4

Figure 4.109

Solving $-x^4 + 5x^2 - 4 > 0$

Interval	Test Value x	$-x^4 + 5x^2 - 4$	Positive or Negative?
$(-\infty, -2)$	-3	-40	**Negative**
$(-2, -1)$	-1.5	2.1875	**Positive**
$(-1, 1)$	0	-4	**Negative**
$(1, 2)$	1.5	2.1875	**Positive**
$(2, \infty)$	3	-40	**Negative**

Table 4.6

These symbolic and graphical procedures can be summarized verbally as follows.

SOLVING POLYNOMIAL INEQUALITIES

STEP 1: If necessary, write the inequality as $p(x) < 0$, where $p(x)$ is a polynomial and the inequality symbol $<$ may be replaced by $>$, $\leq$, or $\geq$.

STEP 2: Solve $p(x) = 0$. The solutions are called boundary numbers.

STEP 3: Use the boundary numbers to separate the number line into disjoint intervals. On each interval, $p(x)$ is either always positive or always negative.

STEP 4: To solve the inequality, either make a table of test values for $p(x)$ or use a graph of $y = p(x)$. For example, the solution set for $p(x) < 0$ corresponds to intervals where test values result in negative outputs or to intervals where the graph of $y = p(x)$ is below the x-axis.

EXAMPLE 7 **Solving a polynomial inequality**

Solve $x^3 \geq 2x^2 + 3x$ symbolically and graphically.

SOLUTION
Symbolic Solution

STEP 1: Begin by writing the inequality as $x^3 - 2x^2 - 3x \geq 0$.

STEP 2: Replace the $\geq$ symbol with an equals sign and solve the resulting equation.

$$x^3 - 2x^2 - 3x = 0 \qquad \textcolor{blue}{\text{Replace } \geq \text{ with } =.}$$
$$x(x^2 - 2x - 3) = 0 \qquad \textcolor{blue}{\text{Factor out } x.}$$
$$x(x + 1)(x - 3) = 0 \qquad \textcolor{blue}{\text{Factor the trinomial.}}$$
$$x = 0 \quad \text{or} \quad x = -1 \quad \text{or} \quad x = 3 \qquad \textcolor{blue}{\text{Zero-product property}}$$

Algebra Review
To review factoring, see Chapter R, (page R-23).

The boundary numbers are -1, 0, and 3.

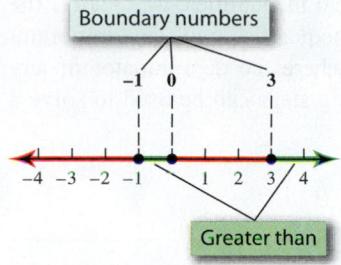

Figure 4.110

Evaluating Test Values

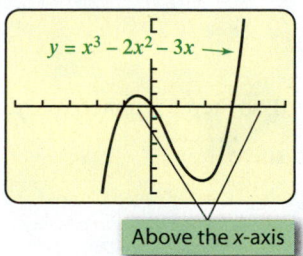

Figure 4.111

$[-5, 5, 1$ by $[-7, 7, 1]$

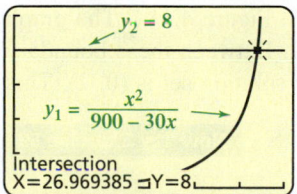

Figure 4.112

STEP 3: The boundary numbers separate the number line into four disjoint intervals:

$$(-\infty, -1), (-1, 0), (0, 3), \text{ and } (3, \infty),$$

as illustrated in Figure 4.110.

STEP 4: In Table 4.7 the expression $x^3 - 2x^2 - 3x$ is evaluated at a test value from each interval. The solution set is $[-1, 0] \cup [3, \infty)$. In Figure 4.111 a graphing calculator has been used to evaluate the same test values. (*Note:* The boundary numbers are included in the solution set because the inequality involves $\geq$ rather than $>$.)

Interval	Test Value x	$x^3 - 2x^2 - 3x$	Positive or Negative?
$(-\infty, -1)$	-2	-10	**Negative**
$(-1, 0)$	-0.5	0.875	**Positive**
$(0, 3)$	1	-4	**Negative**
$(3, \infty)$	4	20	**Positive**

Table 4.7

Graphical Solution Graph $y = x^3 - 2x^2 - 3x$, as shown in Figure 4.112. The zeros or x-intercepts are located at $-1, 0$, and 3. The graph of y is positive (or above the x-axis) for $-1 < x < 0$ or $3 < x < \infty$. If we include the boundary numbers, this result agrees with the symbolic solution.

Now Try Exercise 43

MAKING CONNECTIONS

Functions, Equations, and Inequalities   The three concepts are related. For example,

$$f(x) = ax^3 + bx^2 + cx + d \qquad \text{Cubic function}$$
$$ax^3 + bx^2 + cx + d = 0 \qquad \text{Cubic equation}$$
$$ax^3 + bx^2 + cx + d < 0 \qquad \text{Cubic inequality}$$

where $a \neq 0$. These concepts also apply to higher degree polynomials and to rational expressions.

Rational Inequalities

An Application   In the introduction we looked at how a rational inequality can be used to estimate the number of people standing in line at a ticket booth. In the next example, this inequality is solved graphically.

EXAMPLE 8   Modeling customers in a line

$[0, 30, 5]$ by $[0, 10, 2]$

A ticket booth attendant can wait on 30 customers per hour. To keep the time waiting in line reasonable, the line length should not exceed 8 customers on average. Solve the inequality $\frac{x^2}{900 - 30x} \leq 8$ to determine the rates x at which customers can arrive before a second attendant is needed. Note that the x-values are limited to $0 \leq x < 30$.

SOLUTION Graph $Y_1 = X^2/(900 - 30X)$ and $Y_2 = 8$ for $0 \leq x \leq 30$, as shown in Figure 4.113. The only point of intersection on this interval is near $(26.97, 8)$. The graph of y_1 is below the graph of y_2 for x-values to the left of this point. We conclude that if the arrival rate is about 27 customers per hour *or less*, then the line length does not exceed 8 customers on average. If the arrival rate is more than 27 customers per hour, a second ticket booth attendant is needed.

Figure 4.113

Now Try Exercise 85

Graphical and Symbolic Solutions To solve rational inequalities, we can use the same basic techniques that we used to solve polynomial inequalities, with one important modification: boundary numbers also occur at x-values where the denominator of any rational expression in the inequality equals 0. The following steps can be used to solve a rational inequality.

SOLVING RATIONAL INEQUALITIES

STEP 1: If necessary, write the inequality in the form $\frac{p(x)}{q(x)} > 0$, where $p(x)$ and $q(x)$ are polynomials. Note that $>$ may be replaced by $<$, $\leq$, or $\geq$.

STEP 2: Solve $p(x) = 0$ and $q(x) = 0$. The solutions are boundary numbers.

STEP 3: Use the boundary numbers to separate the number line into disjoint intervals. On each interval, $\frac{p(x)}{q(x)}$ is either always positive or always negative.

STEP 4: Use a table of test values or a graph to solve the inequality in Step 1.

EXAMPLE 9 Solving a rational inequality

Solve $\frac{2-x}{2x} > 0$ symbolically. Support your answer graphically.

SOLUTION

Symbolic Solution The inequality is written in the form $\frac{p(x)}{q(x)} > 0$, so Step 1 is unnecessary.

STEP 2: Set the numerator and the denominator equal to 0 and solve.

Numerator	Denominator
$2 - x = 0$	$2x = 0$
$x = 2$	$x = 0$

STEP 3: The boundary numbers are 0 and 2, which separate the number line into three disjoint intervals: $(-\infty, 0)$, $(0, 2)$, and $(2, \infty)$.

STEP 4: Table 4.8 shows that the expression is positive between the two boundary numbers or when $0 < x < 2$. In interval notation the solution set is $(0, 2)$.

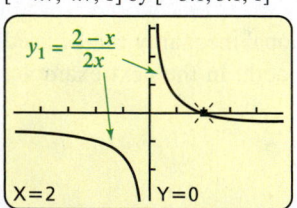

$[-4.7, 4.7, 1]$ by $[-3.1, 3.1, 1]$

$y_1 = \dfrac{2-x}{2x}$

X=2   Y=0

Figure 4.114

Solving $\dfrac{2-x}{2x} > 0$

Interval	Test Value x	$(2-x)/(2x)$	Positive or Negative?
$(-\infty, 0)$	-1	-1.5	**Negative**
$(0, 2)$	1	0.5	**Positive**
$(2, \infty)$	4	-0.25	**Negative**

Table 4.8

Graphical Solution Graph $Y_1 = (2 - X)/(2X)$, as shown in Figure 4.114. The graph has a vertical asymptote at $x = 0$ and an x-intercept at $x = 2$. Between these boundary numbers the graph of y_1 is positive (or above the x-axis). The solution set is $(0, 2)$. This agrees with our symbolic solution.

Now Try Exercise 59

EXAMPLE 10 Solving a rational inequality symbolically

Solve $\frac{1}{x} \leq \frac{2}{x+1}$.

SOLUTION

STEP 1: Begin by writing the inequality in the form $\frac{p(x)}{q(x)} \leq 0$.

$$\frac{1}{x} - \frac{2}{x+1} \leq 0 \qquad \textsf{Subtract } \frac{2}{x+1}.$$

$$\frac{1}{x} \cdot \frac{(x+1)}{(x+1)} - \frac{2}{x+1} \cdot \frac{x}{x} \leq 0 \qquad \textsf{Common denominator is } x(x+1).$$

$$\frac{x+1}{x(x+1)} - \frac{2x}{x(x+1)} \leq 0 \qquad \textsf{Multiply.}$$

$$\frac{1-x}{x(x+1)} \leq 0 \qquad \textsf{Subtract numerators:}$$
$$\textsf{x + 1 − 2x = 1 − x.}$$

Algebra Review

To review subtraction of rational expressions, see Chapter R (page R-32).

STEP 2: Find the zeros of the numerator and the denominator.

Numerator	*Denominator*
$1 - x = 0$	$x(x+1) = 0$
$x = 1$	$x = 0 \quad \text{or} \quad x = -1$

STEP 3: The boundary numbers are -1, 0, and 1, which separate the number line into four disjoint intervals: $(-\infty, -1)$, $(-1, 0)$, $(0, 1)$, and $(1, \infty)$.

STEP 4: Table 4.9 can be used to solve the inequality $\frac{1-x}{x(x+1)} \leq 0$. The solution set is $(-1, 0) \cup [1, \infty)$.

> **NOTE** The boundary numbers -1 and 0 are not included in the solution set because the given inequality is undefined when $x = -1$ or $x = 0$.

$$\text{Solving } \frac{1-x}{x(x+1)} \leq 0$$

Interval	Test Value x	$(1-x)/(x(x+1))$	Positive or Negative?
$(-\infty, -1)$	-2	1.5	**Positive**
$(-1, 0)$	-0.5	-6	**Negative**
$(0, 1)$	0.5	$0.\overline{6}$	**Positive**
$(1, \infty)$	2	$-0.1\overline{6}$	**Negative**

Table 4.9

Now Try Exercise 75

Multiplying an Inequality by a Variable

When solving a rational inequality, it is essential *not* to multiply or divide each side of the inequality by the LCD (least common denominator) if the LCD contains a *variable*. This technique often leads to an incorrect solution set.

For example, if each side of the rational inequality

$$\frac{1}{x} < 2$$

is multiplied by x to clear fractions, the inequality becomes

$$1 < 2x \quad \text{or} \quad x > \frac{1}{2}.$$

However, this solution set is clearly incomplete because $x = -1$ is also a solution to the given inequality. In general, the variable x can be either negative or positive. If $x < 0$, the inequality symbol should be reversed, whereas if $x > 0$, the inequality symbol should not be reversed. Because we have no way of knowing ahead of time which is the case, this technique of multiplying by a variable should be *avoided*.

4.7 Putting It All Together

$\mathbf{T}$he following table outlines basic concepts for this section.

CONCEPT	DESCRIPTION
Rational equation	Can be written as $f(x) = k$, where f is a rational function. To solve a rational equation, first clear fractions by multiplying each side by the LCD. *Check your answers.* **Example:** Solve $\frac{1}{x} + \frac{1}{x+2} = \frac{1}{x(x+2)}$. Multiply each side of the equation by the LCD: $x(x+2)$. $$\frac{x(x+2)}{x} + \frac{x(x+2)}{x+2} = \frac{x(x+2)}{x(x+2)}$$ $$(x+2) + x = 1$$ $$x = -\frac{1}{2} \qquad \text{It checks.}$$
Variation	*y varies directly with the nth power of x:* $y = kx^n$. **Example:** Let y vary directly with the cube (third power) of x. If the constant of variation is 5, then the variation equation is $y = 5x^3$. *y varies inversely with the nth power of x:* $y = \dfrac{k}{x^n}$. **Example:** Let y vary inversely with the square (second power) of x. If the constant of variation is 3, then the variation equation is $y = \dfrac{3}{x^2}$.
Polynomial inequality	Can be written as $p(x) < 0$, where $p(x)$ is a polynomial and $<$ may be replaced by $>$, $\leq$, or $\geq$ **Examples:** $x^3 - x \leq 0, \quad 2x^4 - 3x^2 \geq 5x + 1$
Solving a polynomial inequality	Follow the steps for solving a polynomial inequality presented on page 312. Either graphical or symbolic methods can be used. **Example:** A graph of $y = x^3 - 2x^2 - 5x + 6$ is shown. The boundary numbers are -2, 1, and 3. The solution set to $x^3 - 2x^2 - 5x + 6 > 0$ is $(-2, 1) \cup (3, \infty)$, because the graph is above the x-axis on these intervals.
Rational inequality	Can be written as $\dfrac{p(x)}{q(x)} < 0$, where $p(x)$ and $q(x) \neq 0$ are polynomials and $<$ may be replaced by $>$, $\leq$, or $\geq$ **Examples:** $\dfrac{x-3}{x+2} \geq 0, \quad 2x - \dfrac{2}{x^2-1} > 5$

CONCEPT	DESCRIPTION
Solving a rational inequality	Follow the steps for solving a rational inequality presented on page 314. Either graphical or symbolic methods can be used.

Example: A graph of $y = \frac{x-1}{2x+1}$ is shown. Boundary numbers occur where either the numerator or denominator equals zero: $x = 1$ or $x = -\frac{1}{2}$. The solution set to $\frac{x-1}{2x+1} < 0$ is $\left(-\frac{1}{2}, 1\right)$, because the graph is below the x-axis on this interval.

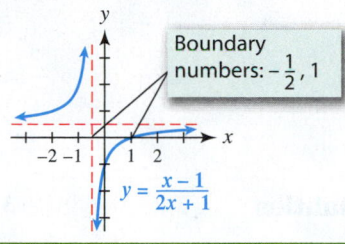

4.7 Exercises

Rational Equations

Exercises 1–6: Solve the rational equation

(a) *symbolically,*

(b) *graphically, and*

(c) *numerically.*

1. $\dfrac{2x}{x+2} = 6$

2. $\dfrac{3x}{2x-1} = 3$

3. $2 - \dfrac{5}{x} + \dfrac{2}{x^2} = 0$

4. $\dfrac{1}{x^2} + \dfrac{1}{x} = 2$

5. $\dfrac{1}{x+1} + \dfrac{1}{x-1} = \dfrac{1}{x^2-1}$

6. $\dfrac{4}{x-2} = \dfrac{3}{x-1}$

Exercises 7–28: Find all real solutions. Check your results.

7. $\dfrac{x+1}{x-5} = 0$

8. $\dfrac{x-2}{x+3} = 1$

9. $\dfrac{6(1-2x)}{x-5} = 4$

10. $\dfrac{2}{5(2x+5)} + 3 = -1$

11. $\dfrac{1}{x+2} + \dfrac{1}{x} = 1$

12. $\dfrac{2x}{x-1} = 5 + \dfrac{2}{x-1}$

13. $\dfrac{1}{x} - \dfrac{2}{x^2} = 5$

14. $\dfrac{1}{x^2-2} = \dfrac{1}{x}$

15. $\dfrac{x^3-4x}{x^2+1} = 0$

16. $\dfrac{1}{x+2} + \dfrac{1}{x+3} = \dfrac{2}{x^2+5x+6}$

17. $\dfrac{35}{x^2} = \dfrac{4}{x} + 15$

18. $6 - \dfrac{35}{x} + \dfrac{36}{x^2} = 0$

19. $\dfrac{x+5}{x+2} = \dfrac{x-4}{x-10}$

20. $\dfrac{x-1}{x+1} = \dfrac{x+3}{x-4}$

21. $\dfrac{1}{x-2} - \dfrac{2}{x-3} = \dfrac{-1}{x^2-5x+6}$

22. $\dfrac{1}{x-1} + \dfrac{3}{x+1} = \dfrac{4}{x^2-1}$

23. $\dfrac{2}{x-1} + 1 = \dfrac{4}{x^2-1}$

24. $\dfrac{1}{x} + 2 = \dfrac{1}{x^2 + x}$

25. $\dfrac{1}{x + 2} = \dfrac{4}{4 - x^2} - 1$

26. $\dfrac{1}{x - 3} + 1 = \dfrac{6}{x^2 - 9}$

27. $\dfrac{1}{x - 1} + \dfrac{1}{x + 1} = \dfrac{2}{x^2 - 1}$

28. $\dfrac{1}{2x + 1} + \dfrac{1}{2x - 1} = \dfrac{2}{4x^2 - 1}$

Graphical Solutions to Inequalities

Exercises 29–34: Solve the equation and inequalities.

(a) $f(x) = 0$ *(b)* $f(x) > 0$ *(c)* $f(x) < 0$

29.
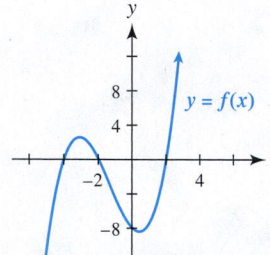

30.
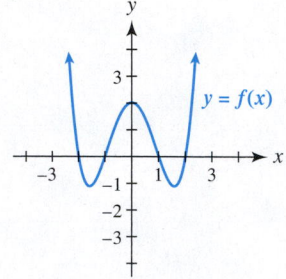

31.
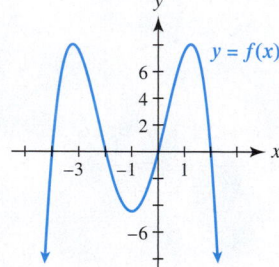

32.
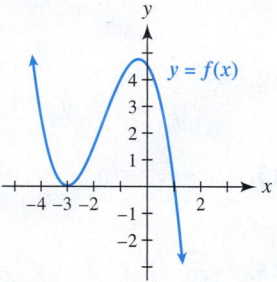

33.
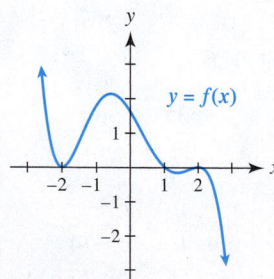

34.
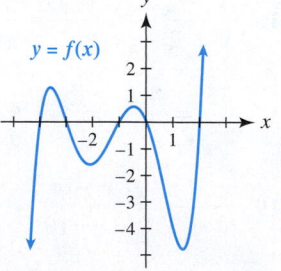

Exercises 35–40: Complete the following.

(a) Identify where $f(x)$ is undefined or $f(x) = 0$.

(b) Solve $f(x) > 0$.

(c) Solve $f(x) < 0$.

35.
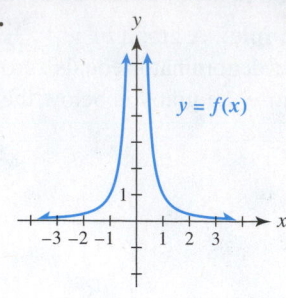

36.
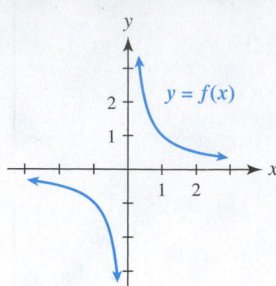

37.
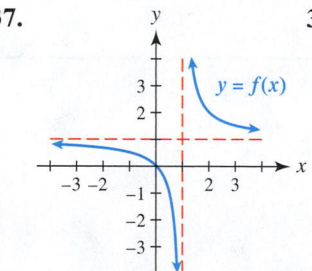

38.
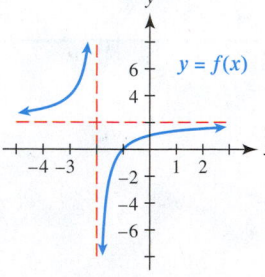

39.
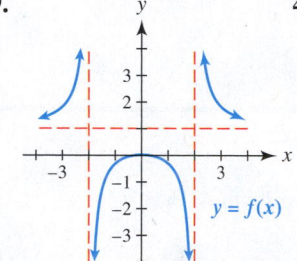

40.
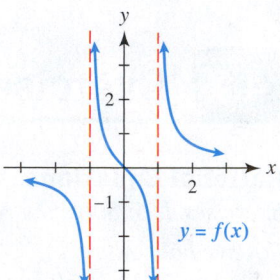

Polynomial Inequalities

Exercises 41–46: Solve the polynomial inequality

(a) symbolically and   *(b)* graphically.

41. $x^3 - x > 0$ **42.** $8x^3 < 27$

43. $x^3 + x^2 \geq 2x$ **44.** $2x^3 \leq 3x^2 + 5x$

45. $x^4 - 13x^2 + 36 < 0$

46. $4x^4 - 5x^2 - 9 \geq 0$

Exercises 47–52: Solve the polynomial inequality.

47. $7x^4 > 14x^2$

48. $3x^4 - 4x^2 < 7$

49. $(x - 1)(x - 2)(x + 2) \geq 0$

50. $-(x + 1)^2(x - 2) \geq 0$

51. $2x^4 + 2x^3 \leq 12x^2$

52. $x^3 + 6x^2 + 9x > 0$

Exercises 53–56: Solve the polynomial inequality graphically.

53. $x^3 - 7x^2 + 14x \leq 8$

54. $2x^3 + 3x^2 - 3x < 2$

55. $3x^4 - 7x^3 - 2x^2 + 8x > 0$

56. $x^4 - 5x^3 \leq 5x^2 + 45x + 36$

Rational Inequalities

Exercises 57–62: Solve the rational inequality

(a) symbolically and   (b) graphically.

57. $\dfrac{1}{x} < 0$

58. $\dfrac{1}{x^2} > 0$

59. $\dfrac{4}{x + 3} \geq 0$

60. $\dfrac{x - 1}{x + 1} < 0$

61. $\dfrac{5}{x^2 - 4} < 0$

62. $\dfrac{x}{x^2 - 1} \geq 0$

Exercises 63–76: Solve the rational inequality.

63. $\dfrac{(x + 1)^2}{x - 2} \leq 0$

64. $\dfrac{2x}{(x - 2)^2} > 0$

65. $\dfrac{3 - 2x}{1 + x} < 0$

66. $\dfrac{x + 1}{4 - 2x} \geq 1$

67. $\dfrac{(x + 1)(x - 2)}{(x + 3)} < 0$

68. $\dfrac{x(x - 3)}{x + 2} \geq 0$

69. $\dfrac{2x - 5}{x^2 - 1} \geq 0$

70. $\dfrac{5 - x}{x^2 - x - 2} < 0$

71. $\dfrac{1}{x - 3} \leq \dfrac{5}{x - 3}$

72. $\dfrac{3}{2 - x} > \dfrac{x}{2 + x}$

73. $2 - \dfrac{5}{x} + \dfrac{2}{x^2} \geq 0$

74. $\dfrac{1}{x - 1} + \dfrac{1}{x + 1} > \dfrac{3}{4}$

75. $\dfrac{1}{x} \leq \dfrac{2}{x + 2}$

76. $\dfrac{1}{x + 1} < \dfrac{1}{x} + 1$

Applications

77. Time Spent in Line   Suppose the average number of vehicles arriving at the main gate of an amusement park is equal to 10 per minute, while the average number of vehicles being admitted through the gate per minute is equal to x. Then the average waiting time in minutes for each

vehicle at the gate can be computed by $f(x) = \dfrac{x - 5}{x^2 - 10x}$, where $x > 10$.   (*Source:* F. Mannering.)

(a) Estimate the admittance rate x that results in an average wait of 15 seconds.

(b) If one attendant can serve 5 vehicles per minute, how many attendants are needed to keep the average wait to 15 seconds or less?

78. Length of Lines   (Refer to Example 2 in Section 4.6.) Solve $\dfrac{x^2}{2 - 2x} = 3$ to determine the traffic intensity x when the average number of vehicles in line equals 3.

79. Construction   (Refer to Example 4.) Find possible dimensions for a box with a volume of 196 cubic inches, a surface area of 280 square inches, and a length that is twice the width.

80. Minimizing Surface Area   An aluminum can is being designed to hold a volume of 100π cubic centimeters.

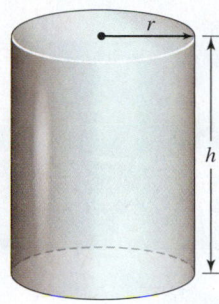

Geometry Review
To review formulas for cylinders, see Chapter R (page R-4).

(a) Find a formula for the volume V in terms of r and h.

(b) Write a formula for a function S that calculates the outside surface area of the can in terms of only r. Evaluate $S(2)$ and interpret the result.

(c) Find the dimensions that result in the least amount of aluminum being used in its construction.

81. Minimizing Cost   A cardboard box with no top and a square base is being constructed and must have a volume of 108 cubic inches. Let x be the length of a side of its base in inches.

(a) Write a formula $A(x)$ that calculates the outside surface area in square feet of the box.

(b) If cardboard costs $0.10 per square foot, write a formula $C(x)$ that gives the cost in dollars of the cardboard in the box.

(c) Find the dimensions of the box that would minimize the cost of the cardboard.

82. Cost-Benefit A cost-benefit function C computes the cost in millions of dollars of implementing a city recycling project when x percent of the citizens participate, where $C(x) = \frac{1.2x}{100 - x}$.

(a) Graph C in [0, 100, 10] by [0, 10, 1]. Interpret the graph as x approaches 100.

(b) If 75% participation is expected, determine the cost for the city.

(c) The city plans to spend $5 million on this recycling project. Estimate the percentage of participation that can be expected.

83. Braking Distance The *grade* x of a hill is a measure of its steepness. For example, if a road rises 10 feet for every 100 feet of horizontal distance, then it has an uphill grade of $x = \frac{10}{100}$, or 10%. See the figure. The braking distance D for a car traveling at 50 miles per hour on a wet uphill grade is given by the formula $D(x) = \frac{2500}{30(0.3 + x)}$. (*Source:* L. Haefner.)

(a) Evaluate $D(0.05)$ and interpret the result.

(b) Describe what happens to the braking distance as the hill becomes steeper. Does this agree with your driving experience?

(c) Estimate the grade associated with a braking distance of 220 feet.

84. Braking Distance (Refer to Exercise 83.) If a car is traveling 50 miles per hour downhill, then the car's braking distance on a wet pavement is given by

$$D(x) = \frac{2500}{30(0.3 + x)},$$

where $x < 0$ for a downhill grade.

(a) Evaluate $D(-0.1)$ and interpret the result.

(b) What happens to the braking distance as the downhill grade becomes steeper? Does this agree with your driving experience?

(c) The graph of D has a vertical asymptote at $x = -0.3$. Give the physical significance of this asymptote.

(d) Estimate the grade associated with a braking distance of 350 feet.

85. Waiting in Line (Refer to Example 8.) A parking garage attendant can wait on 40 cars per hour. If cars arrive randomly at a rate of x cars per hour, then the average line length is given by

$$f(x) = \frac{x^2}{1600 - 40x},$$

where the x-values are limited to $0 \le x < 40$.

(a) Solve the inequality $f(x) \le 8$.

(b) Interpret your answer from part (a).

86. Time Spent in Line If a parking garage attendant can wait on 3 vehicles per minute and vehicles are leaving the ramp at x vehicles per minute, then the average wait in minutes for a car trying to exit is given by the formula $f(x) = \frac{1}{3 - x}$.

(a) Solve the three-part inequality $5 \le \frac{1}{3 - x} \le 10$.

(b) Interpret your result from part (a).

87. Slippery Roads The coefficient of friction x measures the friction between the tires of a car and the road, where $0 < x \le 1$. A smaller value of x indicates that the road is more slippery. If a car is traveling at 60 miles per hour, then the braking distance D in feet is given by the formula $D(x) = \frac{120}{x}$.

(a) What happens to the braking distance as the coefficient of friction becomes smaller?

(b) Find values for the coefficient of friction x that correspond to a braking distance of 400 feet or more.

88. Average Temperature The monthly average high temperature in degrees Fahrenheit at Daytona Beach, Florida, can be approximated by

$$f(x) = 0.0145x^4 - 0.426x^3 + 3.53x^2 - 6.22x + 72,$$

where $x = 1$ corresponds to January, $x = 2$ to February, and so on. Estimate graphically when the monthly average high temperature is 75°F or more.

89. Geometry A cubical box is being manufactured to hold 213 cubic inches. If this measurement can vary between 212.8 cubic inches and 213.2 cubic inches inclusive, by how much can the length x of a side of the cube vary?

90. Construction A cylindrical aluminum can is being manufactured so that its height h is 8 centimeters more than its radius r. Estimate values for the radius (to the nearest hundredth) that result in the can having a volume between 1000 and 1500 cubic centimeters inclusive.

Variation

Exercises 91–94: Find the constant of proportionality k.

91. $y = \frac{k}{x}$, and $y = 2$ when $x = 3$

92. $y = \frac{k}{x^2}$, and $y = \frac{1}{4}$ when $x = 8$

93. $y = kx^3$, and $y = 64$ when $x = 2$

94. $y = kx^{3/2}$, and $y = 96$ when $x = 16$

Exercises 95–98: Solve the variation problem.

95. Suppose T varies directly with the $\frac{3}{2}$ power of x. When $x = 4$, $T = 20$. Find T when $x = 16$.

96. Suppose y varies directly with the second power of x. When $x = 3$, $y = 10.8$. Find y when $x = 1.5$.

97. Let y be inversely proportional to x. When $x = 6$, $y = 5$. Find y when $x = 15$.

98. Let z be inversely proportional to the third power of t. When $t = 5$, $z = 0.08$. Find z when $t = 2$.

Exercises 99–102: Assume that the constant of proportionality is positive.

99. Let y be inversely proportional to x. If x doubles, what happens to y?

100. Let y vary inversely with the second power of x. If x doubles, what happens to y?

101. Suppose y varies directly with the third power of x. If x triples, what happens to y?

102. Suppose y is directly proportional to the second power of x. If x is halved, what happens to y?

Exercises 103 and 104: The data satisfy the equation $y = kx^n$, where n is a positive integer. Determine k and n.

103.

x	2	3	4	5
y	2	4.5	8	12.5

104.

x	3	5	7	9
y	32.4	150	411.6	874.8

Exercises 105 and 106: The data in the table satisfy the equation $y = \frac{k}{x^n}$, where n is a positive integer. Determine k and n.

105.

x	2	3	4	5
y	1.5	1	0.75	0.6

106.

x	2	4	6	8
y	9	2.25	1	0.5625

107. Fiddler Crab Growth The weight y of a fiddler crab is directly proportional to the 1.25 power of the weight x of its claws. A crab with a body weight of 1.9 grams has claws weighing 1.1 grams. Estimate the weight of a fiddler crab with claws weighing 0.75 gram. (**Source:** D. Brown.)

108. Gravity The weight of an object varies inversely with the second power of the distance from the *center* of Earth. The radius of Earth is approximately 4000 miles. If a person weighs 160 pounds on Earth's surface, what would this individual weigh 8000 miles above the surface of Earth?

109. Hubble Telescope The brightness, or intensity, of starlight varies inversely with the square of its distance from Earth. The Hubble Telescope can see stars whose intensities are $\frac{1}{50}$ that of the faintest star now seen by ground-based telescopes. Determine how much farther the Hubble Telescope can see into space than ground-based telescopes. (**Source:** National Aeronautics and Space Administration.)

110. Volume The volume V of a cylinder with a fixed height is directly proportional to the square of its radius r. If a cylinder with a radius of 10 inches has a volume of 200 cubic inches, what is the volume of a cylinder with the same height and a radius of 5 inches?

111. Electrical Resistance The electrical resistance R of a wire varies inversely with the square of its diameter d. If a 25-foot wire with a diameter of 2 millimeters has a resistance of 0.5 ohm, find the resistance of a wire having the same length and a diameter of 3 millimeters.

112. Strength of a Beam The strength of a rectangular wood beam varies directly with the square of the depth of its cross section. If a beam with a depth of 3.5 inches can support 1000 pounds, how much weight can the same type of beam hold if its depth is 12 inches?

Exercises 113 and 114: **Violin String** *The frequency F of a vibrating string is directly proportional to the square root of the tension T on the string and inversely proportional to the length L of the string.*

113. If both the tension and the length are doubled, what happens to *F*?

114. Give two ways to double the frequency *F*.

Writing About Mathematics

115. Describe the steps to graphically solve a polynomial inequality in the form $p(x) > 0$.

116. Describe the steps to symbolically solve a rational inequality in the form $f(x) > 0$.

4.8 Radical Equations and Power Functions

- Learn properties of rational exponents
- Learn radical notation
- Use radical functions and solve radical equations
- Understand properties and graphs of power functions
- Use power functions to model data
- Solve equations involving rational exponents
- Use power regression to model data (optional)

Introduction

Johannes Kepler (1571–1630) was the first to recognize that a power function models the relationship between a planet's distance from the sun and its period of revolution. Table 4.10 lists the average distance x from the sun and the time y in years for several planets to orbit the sun. The distance x has been normalized so that Earth is one unit away from the sun. For example, Jupiter is 5.2 times farther from the sun than Earth and requires 11.9 years to orbit the sun.

Planets and Their Orbits

Planet	Mercury	Venus	Earth	Mars	Jupiter	Saturn
x (distance)	0.387	0.723	1.00	1.52	5.20	9.54
y (period)	0.241	0.615	1.00	1.88	11.9	29.5

Table 4.10 *Source:* C. Ronan, *The Natural History of the Universe.*

A scatterplot of the data in Table 4.10 is shown in Figure 4.115. To model these data, we might try a polynomial, such as $f(x) = x$ or $g(x) = x^2$. Figure 4.116 shows that $f(x) = x$ increases too slowly and $g(x) = x^2$ increases too fast. To model these data, a new type of function is required. That is, we need a function in the form $h(x) = x^b$, where $1 < b < 2$. Polynomials allow the exponent b to be only a nonnegative integer, whereas *power functions* allow b to be any real number. See Figure 4.117 and Example 11.

Orbital Data

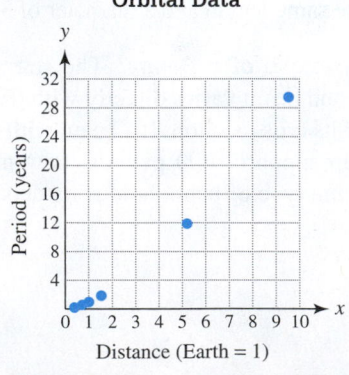

Figure 4.115

Polynomial Models

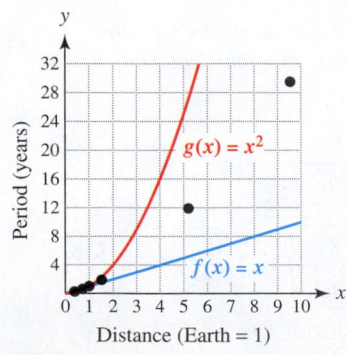

Figure 4.116

Power Model

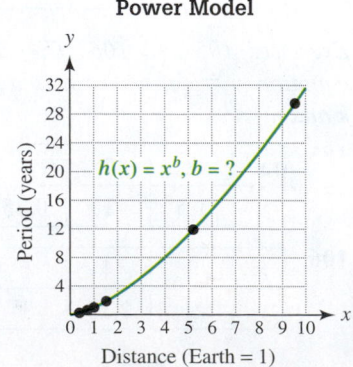

Figure 4.117

Rational Exponents and Radical Notation

The following properties can be used to simplify expressions with rational exponents.

Algebra Review

To review integer exponents, see Chapter R (page R-7). To review radical notation and rational exponents, see Chapter R (page R-38).

> **PROPERTIES OF RATIONAL EXPONENTS**
>
> Let m and n be positive integers with $\frac{m}{n}$ in *lowest* terms and $n \geq 2$. Let r and p be rational numbers. Assume that b is a nonzero real number and that each expression is a real number.
>
> | *Property* | *Example* |
> |---|---|
> | **1.** $b^{m/n} = (b^m)^{1/n} = (b^{1/n})^m$ | $4^{3/2} = (4^3)^{1/2} = (4^{1/2})^3 = 2^3 = 8$ |
> | **2.** $b^{m/n} = \sqrt[n]{b^m} = (\sqrt[n]{b})^m$ | $8^{2/3} = \sqrt[3]{8^2} = (\sqrt[3]{8})^2 = 2^2 = 4$ |
> | **3.** $(b^r)^p = b^{rp}$ | $(2^{3/2})^4 = 2^{(3/2)4} = 2^6 = 64$ |
> | **4.** $b^{-r} = \dfrac{1}{b^r}$ | $4^{-1/2} = \dfrac{1}{4^{1/2}} = \dfrac{1}{2}$ |
> | **5.** $b^r b^p = b^{r+p}$ | $3^{5/2} \cdot 3^{3/2} = 3^{(5/2)+(3/2)} = 3^4 = 81$ |
> | **6.** $\dfrac{b^r}{b^p} = b^{r-p}$ | $\dfrac{5^{5/4}}{5^{3/4}} = 5^{(5/4)-(3/4)} = 5^{1/2}$ |

EXAMPLE 1   Applying properties of exponents

Simplify each expression by hand.

(a) $16^{3/4}$ (b) $\dfrac{4^{1/3}}{4^{5/6}}$ (c) $27^{-2/3} \cdot 27^{1/3}$ (d) $(5^{3/4})^{2/3}$ (e) $(-125)^{-4/3}$

SOLUTION

(a) $16^{3/4} = (\sqrt[4]{16})^3 = (2)^3 = 8$ $\qquad b^{m/n} = \sqrt[n]{b^m} = (\sqrt[n]{b})^m$

(b) $\dfrac{4^{1/3}}{4^{5/6}} = 4^{(1/3)-(5/6)} = 4^{-1/2} = \dfrac{1}{\sqrt{4}} = \dfrac{1}{2}$ $\qquad \frac{b^r}{b^p} = b^{r-p}$

(c) $27^{-2/3} \cdot 27^{1/3} = 27^{(-2/3)+(1/3)} = 27^{-1/3} = \dfrac{1}{\sqrt[3]{27}} = \dfrac{1}{3}$ $\qquad b^r b^p = b^{r+p}$

(d) $(5^{3/4})^{2/3} = 5^{(3/4)(2/3)} = 5^{1/2}$ or $\sqrt{5}$ $\qquad (b^r)^p = b^{rp}$

(e) $(-125)^{-4/3} = \dfrac{1}{(\sqrt[3]{-125})^4} = \dfrac{1}{(-5)^4} = \dfrac{1}{625}$ $\qquad b^{-r} = \frac{1}{b^r}$

> **Now Try Exercises 1, 7, 9, 11, and 13**

EXAMPLE 2   Writing radicals with rational exponents

Use positive rational exponents to write each expression.

(a) $\sqrt{x}$ (b) $\sqrt[3]{x^2}$ (c) $(\sqrt[4]{z})^{-5}$ (d) $\sqrt{\sqrt[3]{y} \cdot \sqrt[4]{y}}$

SOLUTION

(a) $\sqrt{x} = x^{1/2}$ (b) $\sqrt[3]{x^2} = (x^2)^{1/3} = x^{2/3}$

(c) $(\sqrt[4]{z})^{-5} = (z^{1/4})^{-5} = z^{-5/4} = \dfrac{1}{z^{5/4}}$

(d) $\sqrt{\sqrt[3]{y} \cdot \sqrt[4]{y}} = (y^{1/3} \cdot y^{1/4})^{1/2} = (y^{(1/3)+(1/4)})^{1/2} = (y^{7/12})^{1/2} = y^{7/24}$

> **Now Try Exercises 19, 21, 23, and 27**

Functions Involving Radicals

Expressions involving radicals are sometimes used to define **radical functions**. Two common radical functions are the **square root function** and the **cube root function**. Their graphs and domains are shown in Figure 4.118 and 4.119. The fourth root function is also shown in Figure 4.120.

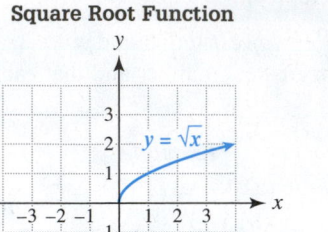

Square Root Function

Domain: $[0, \infty)$

Figure 4.118

Cube Root Function

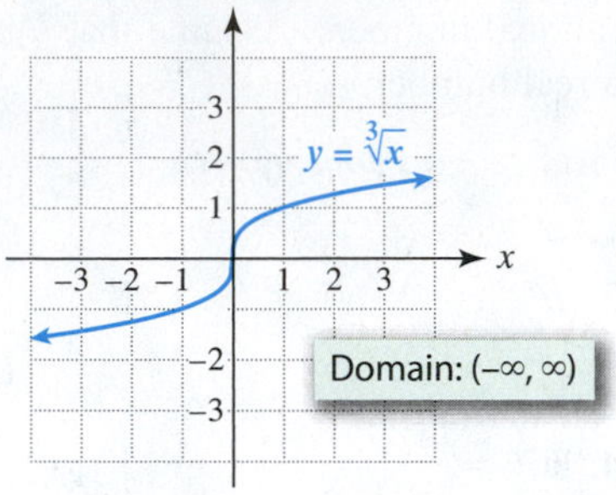

Domain: $(-\infty, \infty)$

Figure 4.119

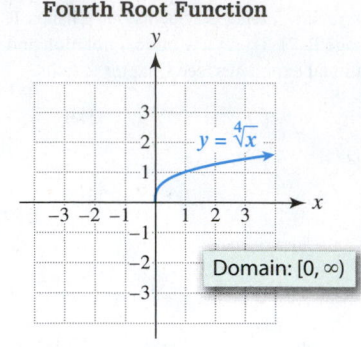

Fourth Root Function

Domain: $[0, \infty)$

Figure 4.120

EXAMPLE 3 **Evaluating radical functions**

Evaluate each function for the given value of x.

(a) $f(x) = 2\sqrt[3]{x}$ for $x = 27$

(b) $g(x) = \sqrt{x^3}$ for $x = 4$

(c) $h(x) = \sqrt[4]{2x}$ for $x = 8$

SOLUTION

(a) $f(\mathbf{27}) = 2\sqrt[3]{\mathbf{27}} = 2(3) = 6$

(b) $g(\mathbf{4}) = \sqrt{\mathbf{4}^3} = \sqrt{64} = 8$

(c) $h(\mathbf{8}) = \sqrt[4]{2(\mathbf{8})} = \sqrt[4]{16} = 2$

> **Now Try Exercises 33, 35, and 37**

In the next example, we use transformations to graph a function.

EXAMPLE 4 **Using transformations to graph a function**

Graph $g(x) = -\sqrt[3]{x + 1}$.

SOLUTION

To obtain the graph of g we can reflect the graph of $f(x) = \sqrt[3]{x}$ (see Figure 4.119) across the x-axis to obtain $y = -\sqrt[3]{x}$, and then shift this new graph left 1 unit to obtain $y = -\sqrt[3]{(x + 1)}$, as shown in Figures 4.121 and 4.122. (In this example, we could also shift left 1 unit and then reflect across the x-axis.)

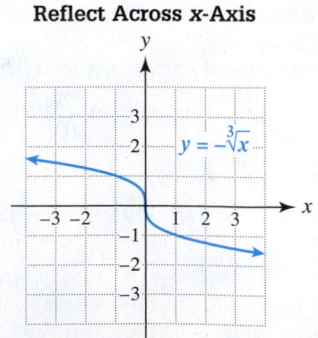

Reflect Across x-Axis

$y = -\sqrt[3]{x}$

Figure 4.121

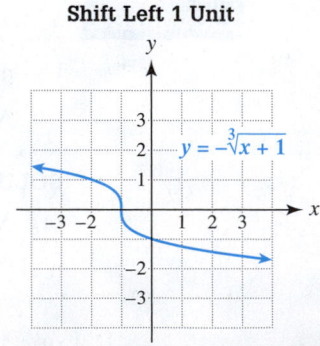

Shift Left 1 Unit

$y = -\sqrt[3]{x + 1}$

Figure 4.122

> **Now Try Exercise 43**

Functions involving radicals have many applications. In the next example we use the cube root function to model numbers of plant species on different Galápagos islands.

EXAMPLE 5 Estimating numbers of plant species

The number N of different plant species that live on a Galápagos island is sometimes related to the island's area A in square miles by the function

$$N(A) = 28.6\sqrt[3]{A}.$$

(a) Approximate N to the nearest whole number for islands of 100 square miles and 200 square miles. Interpret your answers.
(b) Does N double if A doubles?

SOLUTION
(a) $N(100) = 28.6\sqrt[3]{100} \approx 133$ and $N(200) = 28.6\sqrt[3]{200} \approx 167$. If the island is 100 square miles, there are about 133 different species of plants. This number increases to 167 on an island of 200 square miles.
(b) From part (a), we see that N does not double if A doubles. It increases by a factor of $\sqrt[3]{2} \approx 1.26$, not 2.

> **Now Try Exercise 111**

Equations Involving Radicals

In solving equations that contain square roots, it is common to square each side of an equation and then check the results. This is done in the next example.

EXAMPLE 6 Solving an equation containing a square root

Solve $x = \sqrt{15 - 2x}$. Check your answers.

SOLUTION
Begin by squaring each side of the equation.

$$x = \sqrt{15 - 2x} \qquad \textit{Given equation}$$
$$x^2 = \left(\sqrt{15 - 2x}\right)^2 \qquad \textit{Square each side.}$$
$$x^2 = 15 - 2x \qquad \textit{Simplify.}$$
$$x^2 + 2x - 15 = 0 \qquad \textit{Add 2x and subtract 15.}$$
$$(x + 5)(x - 3) = 0 \qquad \textit{Factor.}$$
$$x = -5 \quad \text{or} \quad x = 3 \qquad \textit{Solve.}$$

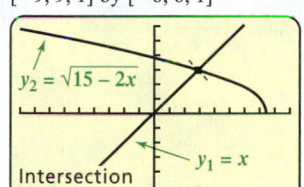

$[-9, 9, 1]$ by $[-6, 6, 1]$

Figure 4.123

Check: Now substitute these values in the *given* equation $x = \sqrt{15 - 2x}$.

$$-5 \neq \sqrt{15 - 2(-5)} = 5, \qquad 3 = \sqrt{15 - 2(3)} \; \checkmark$$

Thus 3 is the only solution. This result is supported graphically in Figure 4.123. Notice that *no* point of intersection occurs when $x = -5$.

> **Now Try Exercise 57**

The value -5 in Example 6 is called an **extraneous solution** because it does not satisfy the given equation. It is important to check results whenever *squaring* has been used to solve an equation. (Graphical solutions do not have extraneous solutions.)

NOTE If each side of an equation is raised to the same positive integer power, then any solutions to the given equation are *among* the solutions to the new equation. That is, the solutions to the equation $a = b$ are *among* the solutions to $a^n = b^n$. For this reason, we *must* check our answers.

In the next example, we cube each side of an equation that contains a cube root.

EXAMPLE 7 Solving an equation containing a cube root

Solve $\sqrt[3]{2x + 5} - 2 = 1$.

SOLUTION Start by adding 2 to each side. Then cube each side.

$$\sqrt[3]{2x + 5} = 3 \qquad \text{Add 2 to each side.}$$
$$\left(\sqrt[3]{2x + 5}\right)^3 = 3^3 \qquad \text{Cube each side.}$$
$$2x + 5 = 27 \qquad \text{Simplify.}$$
$$2x = 22 \qquad \text{Subtract 5 from each side.}$$
$$x = 11 \qquad \text{Divide each side by 2.}$$

The only solution is 11. The answer checks.

Now Try Exercise 65

Squaring Twice In the next example, we need to square twice to solve a radical equation.

EXAMPLE 8 Squaring twice

Solve $\sqrt{2x + 3} - \sqrt{x + 1} = 1$.

SOLUTION

Getting Started When an equation contains two square root expressions, we frequently need to square twice. Start by isolating the more complicated radical expression and then square each side. ▶

$$\sqrt{2x + 3} - \sqrt{x + 1} = 1 \qquad \text{Given equation}$$
$$\sqrt{2x + 3} = 1 + \sqrt{x + 1} \qquad \text{Isolate } \sqrt{2x + 3}.$$
$$\left(\sqrt{2x + 3}\right)^2 = \left(1 + \sqrt{x + 1}\right)^2 \qquad \text{Square each side.}$$
$$2x + 3 = 1 + 2\sqrt{x + 1} + x + 1 \qquad \text{Simplify.}$$
$$x + 1 = 2\sqrt{x + 1} \qquad \text{Isolate the remaining radical.}$$
$$(x + 1)^2 = \left(2\sqrt{x + 1}\right)^2 \qquad \text{Square each side again.}$$
$$x^2 + 2x + 1 = 4(x + 1) \qquad \text{Simplify.}$$
$$x^2 + 2x + 1 = 4x + 4 \qquad \text{Distributive property}$$
$$x^2 - 2x - 3 = 0 \qquad \text{Subtract } 4x + 4.$$
$$(x - 3)(x + 1) = 0 \qquad \text{Factor.}$$
$$x - 3 = 0 \quad \text{or} \quad x + 1 = 0 \qquad \text{Zero-product property}$$
$$x = 3 \quad \text{or} \qquad x = -1 \qquad \text{Solve each equation.}$$

Checking reveals that both -1 and 3 are solutions to the *given* equation.

Now Try Exercise 63

Power Functions and Models

Functions with rational exponents are often used to model physical characteristics of living organisms. For example, larger animals tend to have slower heart rates and larger birds tend to have bigger wings. There is a relationship between a bird's weight and its wing size. (See Example 10.)

Power functions often have rational exponents, and can be written in radical notation. A special type of power function is a root function.

POWER FUNCTION

A function f given by $f(x) = x^b$, where b is a constant, is a **power function**. If $b = \frac{1}{n}$ for some integer $n \geq 2$, then f is a **root function** given by $f(x) = x^{1/n}$, or equivalently, $f(x) = \sqrt[n]{x}$.

Cube Root Square

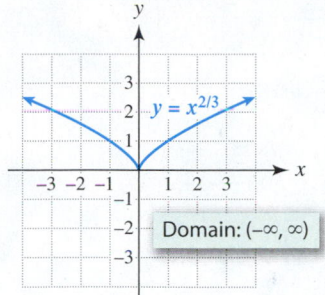

Figure 4.124

Examples of power functions include

$$f_1(x) = x^2, \qquad f_2(x) = x^{3/4}, \qquad f_3(x) = x^{0.4}, \qquad \text{and} \qquad f_4(x) = \sqrt[3]{x^2}.$$
Power Functions

Domains of Power Functions   Suppose a positive rational number $\frac{p}{q}$ is written in lowest terms. Then the domain of $f(x) = x^{p/q}$ is all real numbers whenever q is odd and all *nonnegative* real numbers whenever q is even. A graph of a common power function is shown in Figure 4.124.

EXAMPLE 9   Graphing power functions

Graph $f(x) = x^b$, where $b = 0.3, 1,$ and 1.7, for $x \geq 0$. Discuss the effect that b has on the graph of f for $x \geq 1$.

SOLUTION
The graphs of $y = x^{0.3}$, $y = x^1$, and $y = x^{1.7}$ are shown in Figure 4.125. For $x \geq 1$, larger values of b cause the graph of f to increase faster. Note that each graph passes through the point $(1, 1)$. Why?

$$f(x) = x^b, \, x \geq 0$$

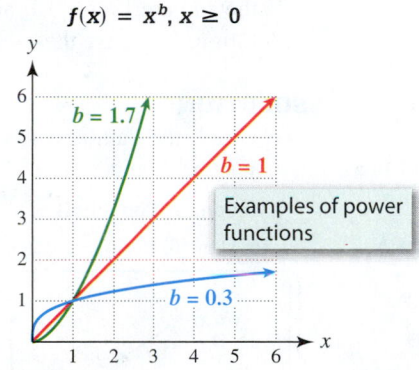

Figure 4.125

Now Try Exercises 73 and 74

Modeling   In the next two examples, we use power functions to determine weights of birds based on wing size and to describe planetary motion.

EXAMPLE 10   Modeling wing size of a bird

Heavier birds have larger wings with more surface area than do lighter birds. For some species of birds, this relationship can be modeled by $S(w) = 0.2w^{2/3}$, where w is the weight of the bird in kilograms, with $0.1 \leq w \leq 5$, and S is the surface area of the wings in square meters.   (*Source:* C. Pennycuick, *Newton Rules Biology.*)
(a) Approximate $S(0.5)$ and interpret the result.
(b) What weight corresponds to a surface area of 0.25 square meter?

SOLUTION
(a) $S(0.5) = 0.2(0.5)^{2/3} \approx 0.126$. The wings of a bird that weighs 0.5 kilogram have a surface area of about 0.126 square meter.

(b) To determine the weight that corresponds to a surface area of 0.25 square meter, we must solve the equation $0.2w^{2/3} = 0.25$.

$$0.2w^{2/3} = 0.25 \qquad \text{Equation to solve}$$

$$w^{2/3} = \frac{0.25}{0.2} \qquad \text{Divide by 0.2.}$$

$$(w^{2/3})^3 = \left(\frac{0.25}{0.2}\right)^3 \qquad \text{Cube each side.}$$

$$w^2 = \left(\frac{0.25}{0.2}\right)^3 \qquad \text{Simplify.}$$

$$w = \pm\sqrt{\left(\frac{0.25}{0.2}\right)^3} \qquad \text{Square root property}$$

$$w \approx \pm 1.4 \qquad \text{Approximate.}$$

Since w must be positive, the wings of a 1.4-kilogram bird have a surface area of about 0.25 square meter.

> **Now Try Exercise 105**

EXAMPLE 11 **Modeling the period of planetary orbits**

Use the data in Table 4.10 on page 322 to complete the following.
(a) Make a scatterplot of the data. Graphically estimate a value for b so that $f(x) = x^b$ models the data.
(b) Check the accuracy of $f(x)$.
(c) The average distances of Uranus, Neptune, and Pluto (no longer a major planet) from the sun are 19.2, 30.1, and 39.5, respectively. Use f to estimate their periods of revolution. Compare these estimates to the actual values of 84.0, 164.8, and 248.5 years.

SOLUTION
(a) Graph the data and $y = x^b$ for different values of b. From the graphs of $y = x^{1.4}$, $y = x^{1.5}$, and $y = x^{1.6}$ in Figures 4.126–4.128, it can be seen that $b \approx 1.5$.

Calculator Help

To make a scatterplot, see Appendix A (page AP-3). To make a table like Figure 4.129, see Appendix A (page AP-9).

Exponent Too Small	Exponent About Right?	Exponent Too Large
[0, 10, 1] by [0, 30, 10]	[0, 10, 1] by [0, 30, 10]	[0, 10, 1] by [0, 30, 10]

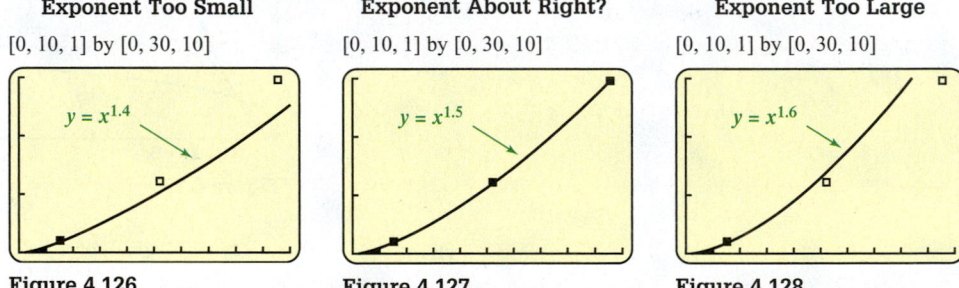

$y = x^{1.4}$	$y = x^{1.5}$	$y = x^{1.6}$

Figure 4.126 **Figure 4.127** **Figure 4.128**

(b) Let $y_1 = x^{1.5}$. The values shown in Figure 4.129 model the data in Table 4.10 remarkably well.

Comparing Model with True Values

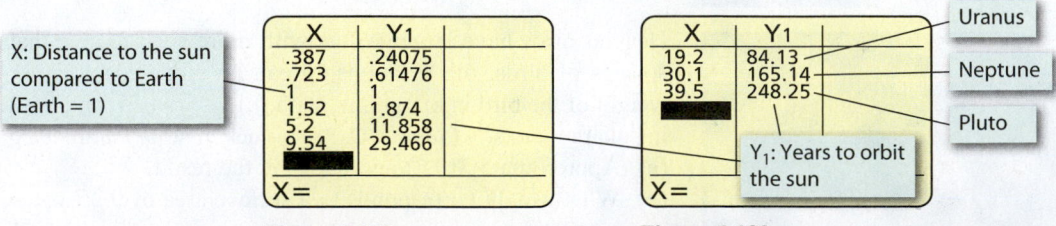

X: Distance to the sun compared to Earth (Earth = 1)

Y_1: Years to orbit the sun

Uranus

Neptune

Pluto

Figure 4.129 **Figure 4.130**

(c) To approximate the number of years required for Uranus, Neptune, and Pluto to orbit the sun, evaluate $y_1 = x^{1.5}$ at $x = 19.2$, 30.1, and 39.5, as shown in Figure 4.130. These values are close to the actual values.

> **Now Try Exercise 109**

Equations Involving Rational Exponents

The next example demonstrates a basic technique that can be used to solve equations with rational exponents.

EXAMPLE 12 **Solving an equation with rational exponents**

Solve $2x^{5/2} - 7 = 23$. Round to the nearest hundredth and give graphical support.

SOLUTION

Symbolic Solution   Start by adding 7 to each side of the given equation.

$2x^{5/2} = 30$	Add 7 to each side.
$x^{5/2} = 15$	Divide each side by 2.
$(x^{5/2})^2 = 15^2$	Square each side.
$x^5 = 225$	Properties of exponents
$x = 225^{1/5}$	Take the fifth root of each side.
$x \approx 2.95$	Approximate.

Graphic Solution   Graphical support is shown in Figure 4.131, where the graphs of $Y_1 = 2X^{\wedge}(5/2) - 7$ and $Y_2 = 23$ intersect near (2.95, 23).

Now Try Exercise 89

$[-5, 5, 1]$ by $[-40, 40, 10]$

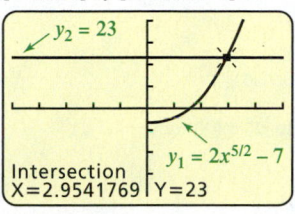

Figure 4.131

MAKING CONNECTIONS

Solutions to $x^n = k$ (k a constant and $n \geq 2$ an integer)

n odd: The real solution to $x^n = k$ is $x = \sqrt[n]{k}$, or $x = k^{1/n}$, for all k. See Example 12.

n even: The real solutions to $x^n = k$ are $x = \pm\sqrt[n]{k}$, or $x = \pm k^{1/n}$, for $k \geq 0$. See Example 10.

Equations that have rational exponents are sometimes reducible to quadratic form.

EXAMPLE 13 **Solving an equation having negative exponents**

Solve $15n^{-2} - 19n^{-1} + 6 = 0$.

SOLUTION   Two methods for solving this equation are presented.

Method 1:   Use the substitution $u = n^{-1} = \frac{1}{n}$ and $u^2 = n^{-2} = \frac{1}{n^2}$.

$15n^{-2} - 19n^{-1} + 6 = 0$	Given equation
$15u^2 - 19u + 6 = 0$	Let $u = n^{-1}$ and $u^2 = n^{-2}$.
$(3u - 2)(5u - 3) = 0$	Factor.
$u = \dfrac{2}{3}$ or $u = \dfrac{3}{5}$	Zero-product property

Because $u = \frac{1}{n}$, it follows that $n = \frac{1}{u}$. Thus $n = \frac{3}{2}$ or $n = \frac{5}{3}$.

Method 2:   Another way to solve this equation is to multiply each side by n^2 to eliminate negative exponents.

$15n^{-2} - 19n^{-1} + 6 = 0$	Given equation
$n^2(15n^{-2} - 19n^{-1} + 6) = n^2(0)$	Multiply each side by n^2.
$15n^2n^{-2} - 19n^2n^{-1} + 6n^2 = 0$	Distributive property
$15 - 19n + 6n^2 = 0$	Properties of exponents

$$6n^2 - 19n + 15 = 0 \qquad \text{Rewrite the equation.}$$

$$(2n - 3)(3n - 5) = 0 \qquad \text{Factor.}$$

$$n = \frac{3}{2} \quad \text{or} \quad n = \frac{5}{3} \qquad \text{Zero-product property}$$

> **Now Try Exercise 93**

In the next example, we solve an equation with fractional exponents that can be written in quadratic form by using substitution.

EXAMPLE 14 Solving an equation having fractions for exponents

Solve $2x^{2/3} + 5x^{1/3} - 3 = 0$.

SOLUTION
To solve this equation, use the substitution $u = x^{1/3}$.

$$2x^{2/3} + 5x^{1/3} - 3 = 0 \qquad \text{Given equation}$$

$$2(x^{1/3})^2 + 5(x^{1/3}) - 3 = 0 \qquad \text{Properties of exponents}$$

$$2u^2 + 5u - 3 = 0 \qquad \text{Let } u = x^{1/3}.$$

$$(2u - 1)(u + 3) = 0 \qquad \text{Factor.}$$

$$u = \frac{1}{2} \quad \text{or} \quad u = -3 \qquad \text{Zero-product property}$$

Because $u = x^{1/3}$, it follows that $x = u^3$. Thus $x = \left(\frac{1}{2}\right)^3 = \frac{1}{8}$ or $x = (-3)^3 = -27$.

> **Now Try Exercise 97**

Power Regression (Optional)

Rather than visually fit a curve to data, as was done in Example 11, we can use least-squares regression to fit the data. Least-squares regression was introduced in Section 2.1. In the next example, we apply this technique to data from biology.

EXAMPLE 15 Modeling the length of a bird's wing

Table 4.11 lists the weight W and the wingspan L for birds of a particular species.

Weights and Wingspans

W (kilograms)	0.5	1.5	2.0	2.5	3.0
L (meters)	0.77	1.10	1.22	1.31	1.40

Table 4.11   *Source:* C. Pennycuick.

(a) Use power regression to model the data with $L = aW^b$. Graph the data and the equation.
(b) Approximate the wingspan for a bird weighing 3.2 kilograms.

SOLUTION
(a) Let x be the weight W and y be the length L. Enter the data, and then select power regression (PwrReg), as shown in Figures 4.132 and 4.133. The results are shown in Figure 4.134. Let

$$y = 0.9674x^{0.3326} \quad \text{or} \quad L = 0.9674W^{0.3326}.$$

The data and equation are graphed in Figure 4.135.
(b) If a bird weighs **3.2** kilograms, this model predicts the wingspan to be

$$L = 0.9674(\mathbf{3.2})^{0.3326} \approx 1.42 \text{ meters.}$$

Calculator Help

To find an equation of least-squares fit,
see Appendix A (page AP-9).

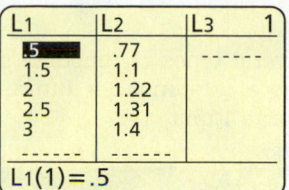

Figure 4.132

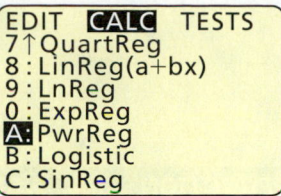

Figure 4.133

[0, 4, 1] by [0.5, 1.5, 0.5]

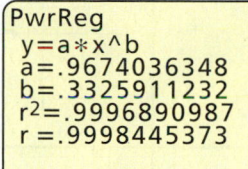

Figure 4.134

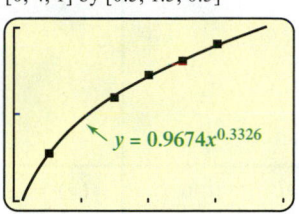

Figure 4.135

Now Try Exercise 115

4.8 Putting It All Together

The following table outlines important concepts in this section.

CONCEPT	EXPLANATION	EXAMPLES
Rational exponents	$x^{m/n} = (x^m)^{1/n}$ $= (x^{1/n})^m$	$9^{3/2} = (9^3)^{1/2} = (729)^{1/2} = 27$ $9^{3/2} = (9^{1/2})^3 = (3)^3 = 27$
Radical notation	$x^{1/2} = \sqrt{x}$ $x^{1/3} = \sqrt[3]{x}$ $x^{m/n} = \sqrt[n]{x^m}$ $= (\sqrt[n]{x})^m$	$25^{1/2} = \sqrt{25} = 5$ $27^{1/3} = \sqrt[3]{27} = 3$ $8^{2/3} = \sqrt[3]{8^2} = \sqrt[3]{64} = 4$ $4^{3/2} = \sqrt{4^3} = (\sqrt{4})^3 = (2)^3 = 8$
Solving radical equations	The solutions to $a = b$ are among the solutions to $a^n = b^n$ when n is a positive integer. *Check your results.*	Solve $\sqrt{2x + 3} = x$. $\qquad 2x + 3 = x^2$ Square each side. $\qquad x^2 - 2x - 3 = 0$ Rewrite equation. $\qquad (x + 1)(x - 3) = 0$ Factor. $\qquad x = -1$ or $x = 3$ Solve. Checking reveals that 3 is the only solution.
Power function	$f(x) = x^b$, where b is a constant	$f(x) = x^{5/4}$ $g(x) = x^{-3.14}$ $h(x) = x^{1/3}$
Root function	$f(x) = x^{1/n}$, where $n \geq 2$ is an integer	$f(x) = x^{1/2}$ or $f(x) = \sqrt{x}$ $g(x) = x^{1/5}$ or $g(x) = \sqrt[5]{x}$

continued on next page

Several types of functions are listed in the following summary, which may be used as a reference for future work. Unless specified otherwise, each tick mark represents 1 unit.

TYPE OF FUNCTION	EXAMPLES	GRAPHS
Linear function $f(x) = ax + b$	$f(x) = 0.5x - 1$ $g(x) = -3x + 2$ $h(x) = 2$	
Polynomial function $f(x) = a_n x^n + \cdots + a_2 x^2 + a_1 x + a_0$	$f(x) = x^2 - 1$ $g(x) = x^3 - 4x - 1$ $h(x) = -x^4 + 4x^2 - 2$	
Rational function $f(x) = \frac{p(x)}{q(x)}$, where $p(x)$ and $q(x)$ are polynomials with $q(x) \neq 0$	$f(x) = \dfrac{1}{x}$ $g(x) = \dfrac{2x - 1}{x + 2}$ $h(x) = \dfrac{1}{x^2 - 1}$	
Root function $f(x) = x^{1/n}$, where $n \geq 2$ is an integer	$f(x) = x^{1/2} = \sqrt{x}$ $g(x) = x^{1/3} = \sqrt[3]{x}$ $h(x) = x^{1/4} = \sqrt[4]{x}$	
Power function $f(x) = x^b$, where b is a constant	$f(x) = x^{2/3}$ $g(x) = x^{1.41}$ $h(x) = x^3$	

4.8 Exercises

Properties of Exponents

Exercises 1–18: Evaluate the expression by hand.

1. $8^{2/3}$

2. $-16^{3/2}$

3. $16^{-3/4}$

4. $25^{-3/2}$

5. $-81^{0.5}$

6. $32^{1/5}$

7. $(9^{3/4})^2$

8. $(4^{-1/2})^{-4}$

9. $\dfrac{8^{5/6}}{8^{1/2}}$

10. $\dfrac{4^{-1/2}}{4^{3/2}}$

11. $27^{5/6} \cdot 27^{-1/6}$

12. $16^{2/3} \cdot 16^{-1/6}$

13. $(-27)^{-5/3}$

14. $(-32)^{-3/5}$

15. $(0.5^{-2})^2$

16. $(2^{-2})^{-3/2}$

17. $\left(\frac{2}{3}\right)^{-2}$

18. $(8^{-1/3} + 27^{-1/3})^2$

Exercises 19–28: Use positive exponents to rewrite.

19. $\sqrt{2x}$

20. $\sqrt{x+1}$

21. $\sqrt[3]{z^5}$

22. $\sqrt[5]{x^2}$

23. $(\sqrt[4]{y})^{-3}$

24. $(\sqrt[3]{y^2})^{-5}$

25. $\sqrt{x} \cdot \sqrt[3]{x}$

26. $(\sqrt[5]{z})^{-3}$

27. $\sqrt{y} \cdot \sqrt{y}$

28. $\dfrac{\sqrt[3]{x}}{\sqrt{x}}$

Exercises 29–32: Use radical notation to rewrite.

29. $a^{-3/4}b^{1/2}$

30. $a^{-2/3}b^{3/5}$

31. $(a^{1/2} + b^{1/2})^{1/2}$

32. $(a^{3/4} - b^{3/2})^{1/3}$

Functions Involving Radicals

Exercises 33–40: Evaluate the function for the given value of x.

33. $f(x) = \sqrt[3]{2x}$ for $x = 32$

34. $f(x) = 5\sqrt[3]{-x}$ for $x = 8$

35. $f(x) = 2\sqrt{x^5}$ for $x = 4$

36. $f(x) = \sqrt{2x^3}$ for $x = 2$

37. $f(x) = \sqrt[4]{5x}$ for $x = 125$

38. $f(x) = 2\sqrt[4]{-x}$ for $x = -81$

39. $f(x) = \sqrt[5]{32x}$ for $x = -1$

40. $f(x) = 3\sqrt[5]{-x}$ for $x = 32$

Exercises 41–52: Use transformations of $y = \sqrt{x}$, $y = \sqrt[3]{x}$, or $y = \sqrt[4]{x}$ to graph $y = f(x)$.

41. $f(x) = \sqrt[3]{-x}$

42. $f(x) = \sqrt[3]{x+1} - 2$

43. $f(x) = \sqrt[3]{x-1} + 1$

44. $f(x) = -2\sqrt[3]{x}$

45. $f(x) = \sqrt[4]{x+2} - 1$

46. $f(x) = \sqrt[4]{-x-1}$

47. $f(x) = 2\sqrt[4]{x-1}$

48. $f(x) = \sqrt[4]{2x}$

49. $f(x) = \sqrt{x+3} + 2$

50. $f(x) = -\sqrt{-x}$

51. $f(x) = 2\sqrt{x}$

52. $f(x) = \sqrt{-x} + 1$

Equations Involving Radicals

Exercises 53–70: Solve the equation. Check your answers.

53. $\sqrt{x+2} = x - 4$

54. $\sqrt{2x+1} = 13$

55. $\sqrt{3x+7} = 3x + 5$

56. $\sqrt{1-x} = x + 5$

57. $\sqrt{5x-6} = x$

58. $x - 5 = \sqrt{5x-1}$

59. $\sqrt{x+5} + 1 = x$

60. $\sqrt{4-3x} = x + 8$

61. $\sqrt{x+1} + 3 = \sqrt{3x+4}$

62. $\sqrt{x} = \sqrt{x-5} + 1$

63. $\sqrt{2x} - \sqrt{x+1} = 1$

64. $\sqrt{2x-4} + 2 = \sqrt{3x+4}$

65. $\sqrt[3]{z+1} = -3$

66. $\sqrt[3]{z} + 5 = 4$

67. $\sqrt[3]{x+1} = \sqrt[3]{2x-1}$

68. $\sqrt[3]{2x^2 + 1} = \sqrt[3]{1-x}$

69. $\sqrt[4]{x-2} + 4 = 20$

70. $\sqrt[4]{2x+3} = \sqrt{x+1}$

Power Functions

Exercises 71 and 72: Evaluate each $f(x)$ at the given x. Approximate each result to the nearest hundredth.

71. $f(x) = x^{3/2} - x^{1/2}$, $x = 50$

72. $f(x) = x^{5/4} - x^{-3/4}$, $x = 7$

Exercises 73 and 74: *Match f(x) with its graph. Assume that a and b are constants with $0 < a < 1 < b$.*

73. $f(x) = x^a$

(a)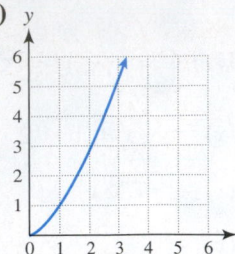

74. $f(x) = x^b$

(b)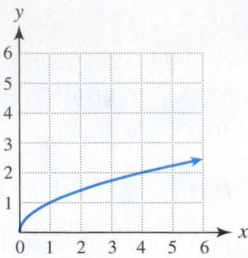

Exercises 75–80: *Use translations to graph f.*

75. $f(x) = x^{1/2} + 1$

76. $f(x) = (x - 1)^{1/3}$

77. $f(x) = x^{2/3} - 1$

78. $f(x) = (x - 1)^{1/4}$

79. $f(x) = (x + 1)^{2/3} - 2$

80. $f(x) = (x - 1)^{2/3}$

Equations Involving Rational Exponents

Exercises 81–102: *Solve the equation. Check your answers.*

81. $x^3 = 8$

82. $x^4 = \frac{1}{81}$

83. $x^{1/4} = 3$

84. $x^{1/3} = \frac{1}{5}$

85. $x^{2/5} = 4$

86. $x^{2/3} = 16$

87. $x^{4/3} = 16$

88. $x^{4/5} = 16$

89. $4x^{3/2} + 5 = 21$

90. $2x^{1/3} - 5 = 1$

91. $n^{-2} + 3n^{-1} + 2 = 0$

92. $2n^{-2} - n^{-1} = 3$

93. $5n^{-2} + 13n^{-1} = 28$

94. $3n^{-2} - 19n^{-1} + 20 = 0$

95. $x^{2/3} - x^{1/3} - 6 = 0$

96. $x^{2/3} + 9x^{1/3} + 14 = 0$

97. $6x^{2/3} - 11x^{1/3} + 4 = 0$

98. $10x^{2/3} + 29x^{1/3} + 10 = 0$

99. $x^{3/4} - x^{1/2} - x^{1/4} + 1 = 0$

100. $x^{3/4} - 2x^{1/2} - 4x^{1/4} + 8 = 0$

101. $x^{-2/3} - 2x^{-1/3} - 3 = 0$

102. $6x^{-2/3} - 13x^{-1/3} - 5 = 0$

Exercises 103 and 104: **Average Rate of Change** *Let the distance from home in miles of a person after t hours on a straight path be given by s(t). Approximate the average rate of change of s from $t_1 = \frac{1}{2}$ to $t_2 = \frac{9}{2}$ to the nearest tenth and interpret the result.*

103. $s(t) = \sqrt{96t}$

104. $s(t) = 3t^{3/4}$

Applications and Models

105. **Modeling Wing Size** Suppose that the surface area S of a bird's wings in square feet can be modeled by $S(w) = 1.27w^{2/3}$, where w is the weight of the bird in pounds, with $1 \le w \le 10$. Estimate the weight of a bird with wings having a surface area of 3 square feet.

106. **Modeling Wingspan** The wingspan L in feet of a bird weighing W pounds is given by $L = 2.43W^{0.3326}$. Estimate the wingspan of a bird that weighs 5.2 pounds.

107. **Modeling Planetary Orbits** The formula $f(x) = x^{1.5}$ calculates the number of years it takes for a planet to orbit the sun if its average distance from the sun is x times that of Earth. If there were a planet located 15 times as far from the sun as Earth, how many years would it take for the planet to orbit the sun?

108. **Modeling Planetary Orbits** (Refer to Exercise 107.) If there were a planet that took 200 years to orbit the sun, what would be its average distance x from the sun compared to that of Earth?

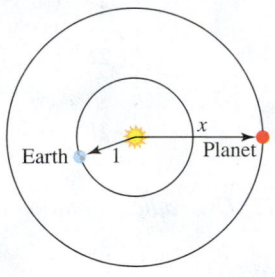

(Not to scale)

109. **Trout and Pollution** Rainbow trout are sensitive to zinc ions in the water. High concentrations are lethal. The average survival times x in minutes for trout in various concentrations of zinc ions y in milligrams per liter (mg/L) are listed in the table.

x (min)	0.5	1	2	3
y (mg/L)	4500	1960	850	525

Source: C. Mason, *Biology of Freshwater Pollution.*

(a) These data can be modeled by $f(x) = ax^b$, where a and b are constants. Determine an appropriate value for a. (*Hint:* Let $f(1) = 1960$.)

(b) Estimate b.

(c) Evaluate $f(4)$ and interpret the result.

110. Lunar Orbits for Jupiter Use the data in the table to complete the following.

Moons of Jupiter	Distance (10^3 km)	Period (days)
Metis	128	0.29
Almathea	181	0.50
Thebe	222	0.67
Europa	671	3.55
Ganymede	1070	7.16
Callisto	1883	16.69

(a) Make a scatterplot of the data. Estimate a value for b so that $f(x) = 0.0002x^b$ models the data.

(b) Check the accuracy of $f(x)$.

(c) The moon Io is 422 thousand kilometers from Jupiter. Estimate its period and compare the estimate to the actual value of 1.77 days. Did your estimate involve interpolation or extrapolation?

111. Fiddler Crab Size One study showed that for a male fiddler crab weighing over 0.75 gram, the weight of its claws can be estimated by $f(x) = 0.445x^{5/4}$. The input x is the weight of the crab in grams, and the output $f(x)$ is the weight of the claws in grams. (*Source:* J. Huxley, *Problems of Relative Growth.*)
(a) Predict the weight of the claws of a 2-gram crab.

(b) Approximate graphically the weight of a crab that has 0.5-gram claws.

(c) Solve part (b) symbolically.

112. Weight and Height The average weight in pounds for men and women can sometimes be estimated by $f(x) = ax^{1.7}$, where x is a person's height in inches and a is a constant determined by the sex of the individual.
(a) If the average weight of a 68-inch-tall man is 152 pounds, approximate a. Use f to estimate the average weight of a 66-inch-tall man.

(b) If the average weight of a 68-inch-tall woman is 137 pounds, approximate a. Use f to estimate the average weight of a 70-inch-tall woman.

Power Regression

Exercises 113 and 114: The table contains data that can be modeled by a function of the form $f(x) = ax^b$. Use regression to find the constants a and b to the nearest hundredth. Graph f and the data.

113.

x	2	4	6	8
f(x)	3.7	4.2	4.6	4.9

114.

x	3	6	9	12
f(x)	23.8	58.5	99.2	144

115. Walmart Employees The table lists numbers N of Walmart employees (in millions) x years after 1980.

x	7	12	17	22	27
N	0.20	0.37	0.68	1.4	2.2

Source: Walmart.

(a) Find a power function f that models the data in the table.

(b) Use f to predict the number of employees in the year 2012. Compare to the true value of 2.1 million employees. Did your answer involve interpolation or extrapolation?

(c) When did the number of employees first reach 1 million?

116. DVD Rentals The table lists numbers of titles T released for DVD rentals x years after 1995.

x	3	4	5	6	7
T	2049	4787	8723	14,321	21,260

Source: DVD Release Report.

(a) Find a power function f that models the data in the table.

(b) Use f to estimate the number of titles released in 2006. Did your answer involve interpolation or extrapolation?

(c) When did the number of releases first surpass 45,000?

117. Pulse Rate and Weight According to one model, the rate at which an animal's heart beats varies with its weight. Smaller animals tend to have faster pulses, whereas larger animals tend to have slower pulses. The table lists average pulse rates in beats per minute (bpm) for animals with various weights in pounds (lb). Use regression (or some other method) to find values for a and b so that $f(x) = ax^b$ models these data.

Weight (lb)	40	150	400	1000	2000
Pulse (bpm)	140	72	44	28	20

Source: C. Pennycuick.

118. Pulse Rate and Weight (Continuation of Exercise 117) Use the results in the previous exercise to calculate the pulse rates for a 60-pound dog and a 2-ton whale.

Writing About Mathematics

119. Can a function be both a polynomial function and a power function? Explain.

120. Explain the basic steps needed to solve equations that contain square roots of variables.

Extended and Discovery Exercises

1. Odd Root Functions Graph $y = \sqrt[n]{x}$ for $n = 3, 5,$ and 7. State some generalizations about a graph of an odd root function.

2. Even Root Functions Graph $y = \sqrt[n]{x}$ for $n = 2, 4,$ and 6. State some generalizations about a graph of an even root function.

Exercises 3 and 4: **Difference Quotient** *Find the difference quotient of f.*

3. $f(x) = \sqrt{x}$ **4.** $f(x) = \frac{1}{x}$

Exercises 5–8: **Negative Rational Exponents** *Write the expression as one ratio without any negative exponents.*

5. $\dfrac{x^{-2/3} + x^{1/3}}{x}$

6. $\dfrac{x^{1/4} - x^{-3/4}}{x}$

7. $\dfrac{\frac{2}{3}(x + 1)x^{-1/3} - x^{2/3}}{(x + 1)^2}$

8. $\dfrac{(x^2 + 1)^{1/2} - \frac{1}{2}x(x^2 + 1)^{-1/2}(2x)}{x^2 + 1}$

9. Wire Between Two Poles Two vertical poles of lengths 12 feet and 16 feet are situated on level ground, 20 feet apart, as shown in the figure. A piece of wire is to be strung from the top of the 12-foot pole to the top of the 16-foot pole, attached to a stake in the ground at a point P on a line connecting the bottoms of the vertical poles. Let x represent the distance from P to D.

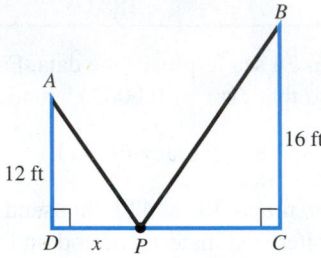

(a) Express the distance from P to C in terms of x.

(b) Express the lengths AP and BP in terms of x.

(c) Give a function f that expresses the total length of the wire used.

(d) Graph f in the window $[0, 20, 5]$ by $[0, 50, 10]$.

(e) Approximate the value of x that will minimize the amount of wire used. What is this minimum?

CHECKING BASIC CONCEPTS FOR SECTIONS 4.7 AND 4.8

1. Solve.

(a) $\dfrac{3x - 1}{1 - x} = 1$

(b) $3 + \dfrac{8}{x} = \dfrac{35}{x^2}$

(c) $\dfrac{1}{x - 1} - \dfrac{1}{3(x + 2)} = \dfrac{1}{x^2 + x - 2}$

2. Solve $2x^3 + x^2 - 6x < 0$.

3. Solve $\dfrac{x^2 - 1}{x + 2} \geq 0$.

4. Let y vary inversely with the cube of x. If $x = \frac{1}{5}$, then $y = 150$. Find y if $x = \frac{1}{2}$.

5. Simplify each expression without a calculator.

(a) $-4^{3/2}$ **(b)** $(8^{-2})^{1/3}$ **(c)** $\sqrt[3]{27^2}$

6. Solve the equation $4x^{3/2} - 3 = 29$.

7. Solve the equation $\sqrt{5x - 4} = x - 2$.

8. Solve each equation.
(a) $n^{-2} + 6n^{-1} = 16$

(b) $2x^{2/3} + 5x^{1/3} - 12 = 0$

9. Find a and b so that $f(x) = ax^b$ models the data.

x	1	2	3	4
$f(x)$	2	2.83	3.46	4

4 Summary

CONCEPT	EXPLANATION AND EXAMPLES

Section 4.1 More Nonlinear Functions and Their Graphs

Polynomial Function

Can be represented by $f(x) = a_n x^n + a_{n-1} x^{n-1} + \cdots + a_2 x^2 + a_1 x + a_0$
The leading coefficient is $a_n \neq 0$ and the degree is n.

Example: $f(x) = -4x^3 - 2x^2 + 6x + \frac{1}{2}$; $a_n = -4$; $n = 3$

Absolute and Local Extrema

The accompanying graph has the following extrema.

Absolute maximum: none

Absolute minimum: -14.8

Local maximum: 1

Local minimums: -4.3, -14.8

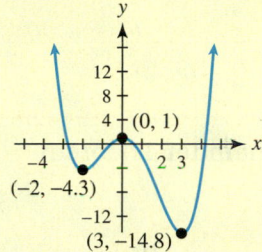

Symmetry

Even function: $f(-x) = f(x)$; the graph is symmetric with respect to the y-axis.

Odd function: $f(-x) = -f(x)$; the graph is symmetric with respect to the origin.

Examples: $f(x) = x^4 - 3x^2$ $f(x) = x - x^3$

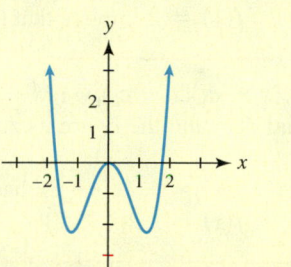   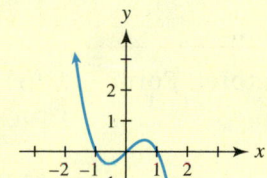

Even Function   Odd Function

Section 4.2 Polynomial Functions and Models

Graphs of Polynomial Functions

Their graphs are continuous with no breaks, and their domains include all real numbers. The graph of a polynomial function of degree $n \geq 1$ has at most n x-intercepts and at most $n - 1$ turning points. For a discussion of the end behavior for graphs of polynomial functions, see page 247.

Examples: See "Putting It All Together" for Section 4.2.

Piecewise-Polynomial Functions

Example:

$$f(x) = \begin{cases} x^2 - 2 & \text{if } x < 0 \\ 1 - 2x & \text{if } x \geq 0 \end{cases}$$

$f(-2) = (-2)^2 - 2 = 2$

$f(0) = 1 - 2(0) = 1$

$f(2) = 1 - 2(2) = -3$

f is discontinuous at $x = 0$.

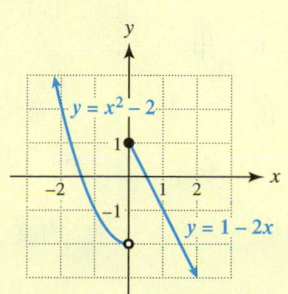

CONCEPT	EXPLANATION AND EXAMPLES

Section 4.3 Division of Polynomials

Division Algorithm

Let $f(x)$ and $d(x)$ be two polynomials with the degree of $d(x)$ greater than zero and less than the degree of $f(x)$. Then there exist unique polynomials $q(x)$ and $r(x)$ such that

$$f(x) = d(x) \cdot q(x) + r(x),$$

where either $r(x) = 0$ or the degree of $r(x)$ is less than the degree of $d(x)$. That is,

(Dividend) = (Divisor) · (Quotient) + (Remainder).

Example: $\dfrac{x^2 - 4x + 5}{x - 1} = x - 3 + \dfrac{2}{x - 1}$. That is,

$$x^2 - 4x + 5 = (x - 1)(x - 3) + 2.$$

Remainder Theorem

If a polynomial $f(x)$ is divided by $x - k$, the remainder is $f(k)$.

Example: If $x^2 - 4x + 5$ is divided by $x - 1$, the remainder is

$$f(1) = 1^2 - 4(1) + 5 = 2.$$

Section 4.4 Real Zeros of Polynomial Functions

Factor Theorem

A polynomial $f(x)$ has a factor $x - k$ if and only if $f(k) = 0$.

Example: $f(x) = x^2 - 3x + 2$;
$f(1) = 0$ implies that $(x - 1)$ is a factor of $x^2 - 3x + 2$.

Complete Factored Form

$f(x) = a_n(x - c_1)(x - c_2) \cdots (x - c_n)$, where a_n is the leading coefficient of the polynomial $f(x)$ and the c_k are its zeros. This form is unique.

Example: $f(x) = -2x^3 + 8x$ has zeros of $-2, 0,$ and 2; therefore
$f(x) = -2(x + 2)(x - 0)(x - 2)$.

Multiple Real Zeros

Odd multiplicity: Graph crosses the x-axis.

Even multiplicity: Graph touches but does not cross the x-axis.

Higher multiplicities: Graph levels off more at a zero of higher multiplicity.

Example: Let $f(x) = -2(x + 1)^3(x - 4)^2$; $f(x)$ has a zero of -1 with odd multiplicity 3 and a zero of 4 with even multiplicity 2.

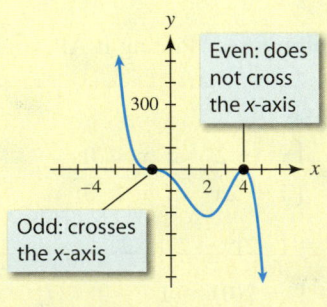

Even: does not cross the x-axis

Odd: crosses the x-axis

CONCEPT	EXPLANATION AND EXAMPLES

Section 4.4 Real Zeros of Polynomial Functions (CONTINUED)

Polynomial Equations

Polynomial equations can be solved symbolically, graphically, and numerically. A common symbolic technique is factoring.

Example: Solve $x^3 - 4x = 0$ symbolically and graphically.

Symbolic Solution

$$x(x^2 - 4) = 0$$
$$x(x - 2)(x + 2) = 0$$
$$x = 0,\ x = 2,\ \text{ or }\ x = -2$$

Graphical Solution

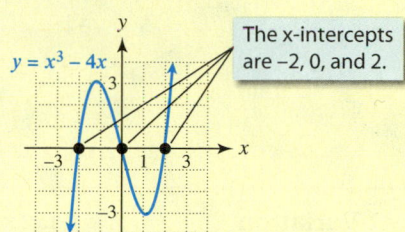

The x-intercepts are –2, 0, and 2.

Section 4.5 The Fundamental Theorem of Algebra

Fundamental Theorem of Algebra

A polynomial $f(x)$ of degree $n \geq 1$ has at least one complex zero.
Explanation: With complex numbers, any polynomial can be written in complete factored form.

Examples: $x^2 + 1 = (x + i)(x - i)$
$3x^2 - 3x - 6 = 3(x + 1)(x - 2)$

Number of Zeros Theorem

A polynomial of degree n has at most n distinct zeros.

Example: A cubic polynomial has at most three distinct zeros.

Polynomial Equations with Complex Solutions

Polynomial equations can have both real and imaginary solutions.

Example: Solve $x^4 - 1 = 0$.
$$(x^2 - 1)(x^2 + 1) = 0$$
$$(x - 1)(x + 1)(x^2 + 1) = 0$$
$$x = 1,\ x = -1,\ \text{ or }\ x^2 = -1$$
$$x = \pm 1,\ \pm i$$

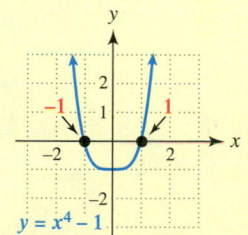

The x-intercepts are -1 and 1. The imaginary solutions $\pm i$ cannot be found from a graph.

Section 4.6 Rational Functions and Models

Rational Functions

$f(x) = \dfrac{p(x)}{q(x)}$, where $p(x)$ and $q(x) \neq 0$ are polynomials

Example: $f(x) = \dfrac{2x - 3}{x - 1}$

Horizontal asymptote: $y = 2$
Vertical asymptote: $x = 1$

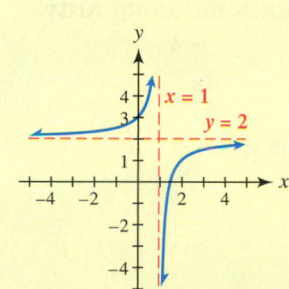

To find vertical and horizontal asymptotes, see page 293.
To graph rational functions by hand, see page 299.

CONCEPT	EXPLANATION AND EXAMPLES

Section 4.7 More Equations and Inequalities

Solving Rational Equations

Multiply each side by the LCD. Check your results.

Example:

$$\frac{-24}{x-3} - 4 = x + 3 \qquad \text{LCD is } x - 3.$$

$$-24 - 4(x - 3) = (x + 3)(x - 3) \qquad \text{Multiply each term by } x - 3.$$

$$-24 - 4x + 12 = x^2 - 9 \qquad \text{Multiply.}$$

$$0 = x^2 + 4x + 3 \qquad \text{Combine terms.}$$

$$0 = (x + 3)(x + 1) \qquad \text{Factor.}$$

$$x = -3 \quad \text{or} \quad x = -1 \qquad \text{Both solutions check.}$$

Direct Variation

Let x and y denote two quantities and n be a positive number. Then y is *directly proportional to the nth power of x*, or *y varies directly with the nth power of x*, if there exists a nonzero number k such that $y = kx^n$.

Example: Because $V = \frac{4}{3}\pi r^3$, the volume of a sphere varies directly with the third power of the radius. The constant of variation is $\frac{4}{3}\pi$.

Inverse Variation

Let x and y denote two quantities and n be a positive number. Then y is *inversely proportional to the nth power of x*, or *y varies inversely with the nth power of x*, if there exists a nonzero number k such that $y = \frac{k}{x^n}$.

Example: Because $I = \frac{k}{d^2}$, the intensity of a light source varies inversely with the square of the distance from the light source.

Polynomial Inequality

Write the inequality as $p(x) > 0$, where $>$ may be replaced by $\geq$, $<$, or $\leq$. Replace the inequality sign with an equals sign, and solve this equation. The solutions are called boundary numbers. Then use a graph or table to find the solution set to the given inequality.

Example: $4x - x^3 > 0$; Boundary numbers: $-2, 0, 2$
The solution set is $(-\infty, -2) \cup (0, 2)$ because the graph is above the x-axis for these intervals of x-values.

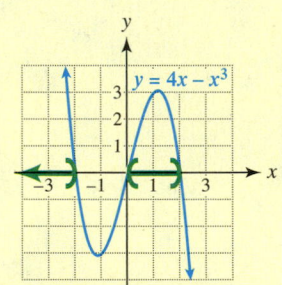

Rational Inequality

As with polynomial inequalities, find the boundary numbers, including x-values where any expressions are undefined.

Example: $\dfrac{(x + 2)(x - 3)}{x} \geq 0$

Boundary numbers: $-2, 0, 3$
Solution set: $[-2, 0) \cup [3, \infty)$

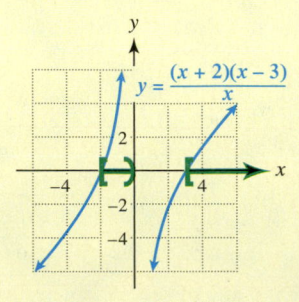

CONCEPT	EXPLANATION AND EXAMPLES

Section 4.8 Radical Equations and Power Functions

Rational Exponents

$$x^{m/n} = \sqrt[n]{x^m} = (\sqrt[n]{x})^m$$

Example: $25^{3/2} = \sqrt{25^3} = (\sqrt{25})^3 = 5^3 = 125$

Functions Involving Radicals

Functions defined by radical expressions

Examples: $f(x) = \sqrt[3]{x}; \; f(-8) = \sqrt[3]{-8} = -2$

$f(x) = \sqrt[4]{x^3} + 1; \; f(81) = \sqrt[4]{81^3} + 1 = 28$

Solving Radical Equations

When an equation contains a square root, isolate the square root and then square each side. *Be sure to check your results.*

Example:

$x + \sqrt{3x - 3} = 1$	Given equation
$\sqrt{3x - 3} = 1 - x$	Subtract x.
$3x - 3 = (1 - x)^2$	Square each side.
$3x - 3 = 1 - 2x + x^2$	Square binomial.
$0 = x^2 - 5x + 4$	Combine terms.
$0 = (x - 1)(x - 4)$	Factor.
$x = 1 \quad \text{or} \quad x = 4$	Solve.

Check: 1 is a solution, but 4 is not.

$$1 + \sqrt{3(1) - 3} = 1 \;\checkmark \quad 4 + \sqrt{3(4) - 3} \neq 1$$

Power Function

$f(x) = x^b$, where b is a constant

Example: $f(x) = x^{4/3}, \quad g(x) = x^{1.72}$

4   Review Exercises

1. State the degree and leading coefficient of the polynomial $f(x) = 4 + x - 2x^2 - 7x^3$.

Exercises 2 and 3: Use the graph of f to estimate any
(a) local extrema and   (b) absolute extrema.

2.

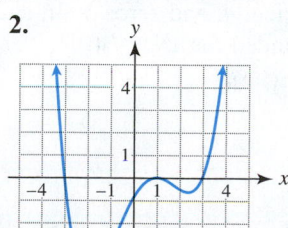

3.

4. Graph $f(x) = -0.25x^4 + 0.67x^3 + 9.5x^2 - 20x - 50$.
   (a) Approximate any local extrema.

   (b) Approximate any absolute extrema.

   (c) Determine where f is increasing or decreasing.

Exercises 5–8: Determine if f is even, odd, or neither.

5. $f(x) = 2x^6 - 5x^4 - x^2$ 6. $f(x) = -5x^3 - 18$

7. $f(x) = 7x^5 + 3x^3 - x$ 8. $f(x) = \dfrac{1}{1 + x^2}$

Exercises 9 and 10: The table is a complete representation of f. Decide if f is even, odd, or neither.

9.

x	−4	−2	0	2	−4
f(x)	13	7	0	−7	−13

10.

x	−5	−3	−1	1	3	5
f(x)	−6	2	7	7	2	−6

Exercises 11 and 12: Sketch a graph of a polynomial function that satisfies the given conditions.

11. Cubic polynomial, two x-intercepts, and a positive leading coefficient

12. Degree 4 with a positive leading coefficient, three turning points, and one x-intercept

Exercises 13 and 14: Use the graph of the polynomial function f to complete the following.

(a) Determine the number of turning points and estimate any x-intercepts.
(b) State whether $a > 0$ or $a < 0$.
(c) Determine the minimum degree of f.

13.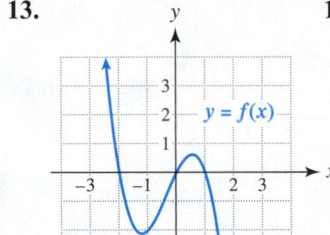

14.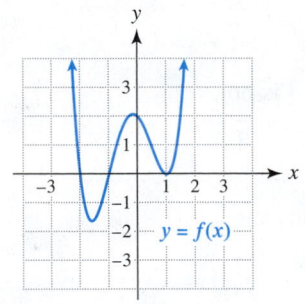

Exercises 15 and 16: State the end behavior of f.

15. $f(x) = -2x^3 + 4x - 2$

16. $f(x) = 1 - 2x - x^4$

17. Find the average rate of change of $f(x) = x^3 + 1$ from $x = -2$ to $x = -1$.

18. Find the difference quotient for $g(x) = 4x^3$.

19. Let $f(x)$ be given by

$$f(x) = \begin{cases} 2x & \text{if } 0 \le x < 2 \\ 8 - x^2 & \text{if } 2 \le x \le 4. \end{cases}$$

(a) Sketch a graph of f. Is f continuous on its domain?

(b) Evaluate $f(1)$ and $f(3)$.

(c) Solve the equation $f(x) = 2$.

20. Determine the type of symmetry that the graph of $g(x) = x^5 - 4x^3$ exhibits.

Exercises 21–24: Divide the expression.

21. $\dfrac{14x^3 - 21x^2 - 7x}{7x}$

22. $\dfrac{2x^3 - x^2 - 4x + 1}{x + 2}$

23. $\dfrac{4x^3 - 7x + 4}{2x + 3}$

24. $\dfrac{3x^3 - 5x^2 + 13x - 18}{x^2 + 4}$

25. The polynomial given by $f(x) = \frac{1}{2}x^3 - 3x^2 + \frac{11}{2}x - 3$ has zeros 1, 2, and 3. Write its complete factored form.

26. Write a complete factored form of a quintic (degree 5) polynomial $f(x)$ that has zeros −2 and 2 with multiplicities 2 and 3, respectively.

27. Use the graph of $y = f(x)$ to write its complete factored form. (Do not assume that the leading coefficient is ± 1.)

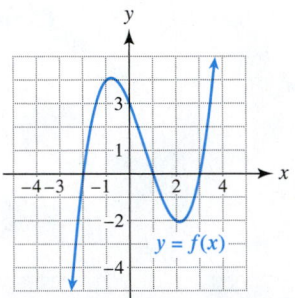

28. What is the maximum number of times that a horizontal line can intersect the graph of each type of polynomial?
   (a) linear (degree 1)   **(b)** quadratic   **(c)** cubic

Exercises 29 and 30: Use the rational zero test to determine any rational zeros of f(x).

29. $f(x) = 2x^3 + x^2 - 13x + 6$

30. $f(x) = x^3 + x^2 - 11x - 11$

Exercises 31–32: Solve the equation.

31. $9x = 3x^3$

32. $x^4 - 3x^2 + 2 = 0$

Exercises 33 and 34: Solve the equation graphically. Round your answers to the nearest hundredth.

33. $x^3 - 3x + 1 = 0$ 34. $x^4 - 2x = 2$

Exercises 35 and 36: Find all real and imaginary solutions.

35. $x^3 + x = 0$ 36. $x^4 + 3x^2 + 2 = 0$

37. Write a polynomial $f(x)$ in complete factored form that has degree 3, leading coefficient 4, and zeros 1, 3i, and −3i. Then write $f(x)$ in expanded form.

38. Use the graph of $f(x) = 2x^4 - x^2 - 1$ to predict the number of real zeros and the number of imaginary zeros of f. Find these zeros symbolically.

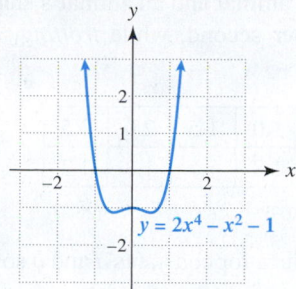

Exercises 53 and 54: Use the graph of f to solve each inequality.

(a) $f(x) > 0$ *(b)* $f(x) < 0$

53.   **54.**

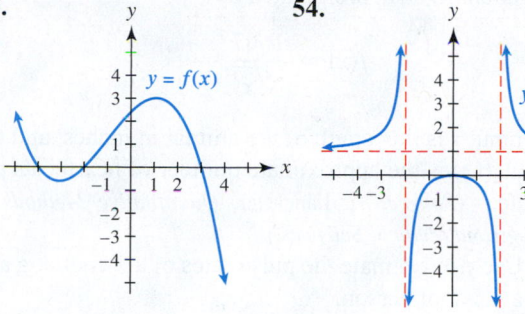

39. If a zero of f is i, find the complete factored form of $f(x) = x^4 + x^3 + 2x^2 + x + 1$.

40. State the domain of $f(x) = \frac{3x - 2}{5x + 4}$. Identify any horizontal or vertical asymptotes in the graph of f.

41. Find any horizontal or vertical asymptotes in the graph of
$$f(x) = \frac{2x^2 + x - 15}{3x^2 + 8x - 3}.$$

42. Let $f(x) = \frac{2x^2}{x^2 - 4}$.

   (a) Find the domain of f.

   (b) Identify any horizontal or vertical asymptotes.

   (c) Graph f with a graphing calculator.

   (d) Sketch a graph of f that includes all asymptotes.

Exercises 43–46: Graph $y = g(x)$ by hand.

43. $g(x) = \frac{1}{x + 1} - 2$ **44.** $g(x) = \frac{x}{x - 1}$

45. $g(x) = \frac{x^2 - 1}{x^2 + 2x + 1}$ **46.** $g(x) = \frac{2x - 3}{2x^2 + x - 6}$

47. Sketch a graph of a function f with vertical asymptote $x = -2$ and horizontal asymptote $y = 2$.

48. Solve the equation $\frac{3x}{x - 2} = 2$ symbolically, graphically, and numerically.

Exercises 49–52: Solve the equation. Check your results.

49. $\frac{5x + 1}{x + 3} = 3$ **50.** $\frac{1}{x} - \frac{1}{x^2} + 2 = 0$

51. $\frac{1}{x + 2} + \frac{1}{x - 2} = \frac{4}{x^2 - 4}$

52. $\frac{x + 5}{x - 2} = \frac{x - 1}{x + 1}$

Exercises 55–58: Solve the inequality.

55. $x^3 + x^2 - 6x > 0$ **56.** $x^4 + 4 < 5x^2$

57. $\frac{2x - 1}{x + 2} > 0$ **58.** $\frac{1}{x} + \frac{1}{x + 2} \le \frac{4}{3}$

Exercises 59–62: Evaluate the radical expression by hand.

59. $(36^{3/4})^2$ **60.** $(9^{-3/2})^{-2}$

61. $(2^{-3/2} \cdot 2^{1/2})^{-3}$ **62.** $\left(\frac{4}{9}\right)^{-3/2}$

Exercises 63–66: Write the radical expression using positive exponents.

63. $\sqrt[3]{x^4}$ **64.** $\left(\sqrt[4]{z}\right)^{-1/2}$

65. $\sqrt[3]{y} \cdot \sqrt{y}$ **66.** $\sqrt{x} \cdot \sqrt[3]{x^2} \cdot \sqrt[4]{x^3}$

Exercises 67 and 68: Give the domain of the power function. Approximate $f(3)$ to the nearest hundredth.

67. $f(x) = x^{5/2}$ **68.** $f(x) = x^{-2/3}$

Exercises 69–80: Solve the equation. Check your results.

69. $x^5 = 1024$ **70.** $x^{1/3} = 4$

71. $\sqrt{x - 2} = x - 4$ **72.** $x^{3/2} = 27$

73. $2x^{1/4} + 3 = 6$ **74.** $x^{1/3} + 3x^{1/3} = -2$

75. $\sqrt[3]{2x - 3} + 1 = 4$

76. $m^{-3} + 2m^{-2} + m^{-1} = 0$

77. $2n^{-2} - 5n^{-1} = 3$

78. $x^{3/4} - 16x^{1/4} = 0$

79. $k^{2/3} - 4k^{1/3} - 5 = 0$

80. $\sqrt{x - 2} = 5 - \sqrt{x + 3}$

Applications

81. Pulse Rate and Length During the eighteenth century Bryan Robinson found that the pulse rate of an animal could be approximated by

$$f(x) = \frac{1607}{\sqrt[4]{x^3}}.$$

The input x is the length of the animal in inches, and the output $f(x)$ is the approximate number of heartbeats per minute. (*Source*: H. Lancaster, *Quantitative Methods in Biology and Medical Sciences*.)

(a) Use f to estimate the pulse rates of a 2-foot dog and a 5.5-foot person.

(b) What length corresponds to a pulse rate of 400 beats per minute?

82. Time Spent in Line Suppose a parking garage attendant can wait on 4 vehicles per minute and vehicles are leaving the ramp randomly at an average rate of x vehicles per minute. Then the average time T in minutes spent waiting in line and paying the attendant is given by

$$T(x) = \frac{1}{4 - x},$$

where $0 \le x < 4$. (*Source*: N. Garber and L. Hoel, *Traffic and Highway Engineering*.)

(a) Evaluate $T(2)$ and interpret the result.

(b) Graph T for $0 \le x < 4$.

(c) What happens to the waiting time as x increases from 0 to (nearly) 4?

(d) Find x if the waiting time is 5 minutes.

83. Modeling Ocean Temperatures The formula

$$T(m) = -0.064m^3 + 0.56m^2 + 2.9m + 61$$

approximates the ocean temperature in degrees Fahrenheit at Naples, Florida. In this formula m is the month, with $m = 1$ corresponding to January.

(a) What is the average ocean temperature in May?

(b) Estimate the absolute maximum of T on the closed interval $[1, 12]$ and interpret the result.

84. Minimizing Surface Area Find possible dimensions that minimize the surface area of a box with no top that has a volume of 96 cubic inches and a length that is three times the width.

85. Falling Object If an object is dropped from a height h, then the time t required for the object to strike the ground is directly proportional to the square root of h. If it requires 1 second for an object to fall 16 feet, how long does it take for an object to fall 256 feet?

86. Animals and Trotting Speeds Taller animals tend to take longer, but fewer, steps per second than shorter animals. The relationship between the shoulder height h in meters of an animal and an animal's stepping frequency F in steps per second, while *trotting*, is shown in the table.

h	0.5	1.0	1.5	2.0	2.5
F	2.6	1.8	1.5	1.3	1.2

Source: C. Pennycuick, *Newton Rules Biology*.

(a) Find values for constants a and b so that the formula $f(x) = ax^b$ models the data.

(b) Estimate the stepping frequency for an elephant with a 3-meter shoulder height.

Extended and Discovery Exercises

Exercises 1 and 2: **Velocity** *Suppose that a person is riding a bicycle on a straight road and that $f(t)$ computes the total distance in feet that the rider has traveled after t seconds. To calculate the person's average velocity between time t_1 and time t_2, we can evaluate the difference quotient*

$$\frac{f(t_2) - f(t_1)}{t_2 - t_1}.$$

(a) *For the given $f(t)$ and the indicated values of t_1 and t_2, calculate the average velocity of the bike rider. Make a table to organize your work.*

(b) *Make a conjecture about the velocity of the bike rider precisely at time t_1.*

1. $f(t) = t^2, t_1 = 10$
(i) $t_2 = 11$
(ii) $t_2 = 10.1$
(iii) $t_2 = 10.01$
(iv) $t_2 = 10.001$

2. $f(t) = \sqrt{t}, t_1 = 4$
(i) $t_2 = 5$
(ii) $t_2 = 4.1$
(iii) $t_2 = 4.01$
(iv) $t_2 = 4.001$

Exercises 3–6: **Average Rates of Change** *These exercises investigate the relationship between polynomial functions and their average rates of change. For example, the average rate of change of $f(x) = x^2$ from x to $x + 0.001$ for any x can be calculated and graphed as shown in the figures. The graph of f is a parabola, and the graph of its average rate of change is a line. Try to discover what this relationship is by completing the following.*

$[-10, 10, 1]$ by $[-10, 10, 1]$

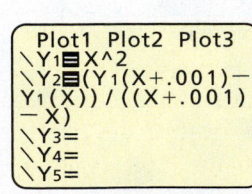

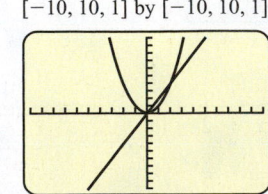

(a) *Graph each function and its average rate of change from x to x + 0.001.*

(b) *Compare the graphs. How are turning points on the graph of a function related to its average rate of change?*

(c) *Generalize your results. Test your generalization.*

3. Linear Functions
$f_1(x) = 3x + 1$ $f_2(x) = -2x + 6$
$f_3(x) = 1.5x - 5$ $f_4(x) = -4x - 2.5$

4. Quadratic Functions
$f_1(x) = 2x^2 - 3x + 1$ $f_2(x) = -0.5x^2 + 2x + 2$
$f_3(x) = x^2 + x - 2$ $f_4(x) = -1.5x^2 - 4x + 6$

5. Cubic Functions
$f_1(x) = 0.5x^3 - x^2 - 2x + 1$
$f_2(x) = -x^3 + x^2 + 3x - 5$
$f_3(x) = 2x^3 - 5x^2 + x - 3$
$f_4(x) = -x^3 + 3x - 4$

6. Quartic Functions
$f_1(x) = 0.05x^4 + 0.2x^3 - x^2 - 2.4x$
$f_2(x) = -0.1x^4 + 0.1x^3 + 1.3x^2 - 0.1x - 1.2$
$f_3(x) = 0.1x^4 + 0.4x^3 - 0.2x^2 - 2.4x - 2.4$

1–4 Cumulative Review Exercises

1. Let $S = \{(-3, 4), (-1, -2), (0, 4), (1, -2), (-1, 5)\}$.
   (a) Find the domain and range of S.

   (b) Is S a function?

2. Find the exact distance between $(-1, 4)$ and $(3, -9)$.

3. Use the graph to express the domain and range of f. Then evaluate $f(0)$.

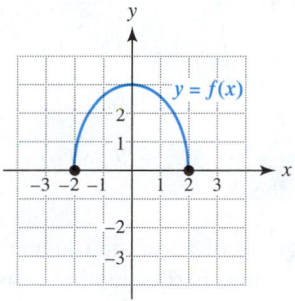

4. Graph $y = g(x)$ by hand.
   (a) $g(x) = 2 - 3x$ (b) $g(x) = |2x - 1|$

   (c) $g(x) = \frac{1}{2}(x - 2)^2 + 2$ (d) $g(x) = x^3 - 1$

   (e) $g(x) = \sqrt{-x}$ (f) $g(x) = \sqrt[3]{x}$

   (g) $g(x) = \dfrac{1}{x - 4} + 2$ (h) $g(x) = x^2 - x$

5. The monthly cost of driving a car is $200 for maintenance plus $0.25 a mile. Write a formula for a function C that calculates the monthly cost of driving a car x miles. Evaluate $C(2000)$ and interpret the result.

6. The graph of a linear function f is shown.

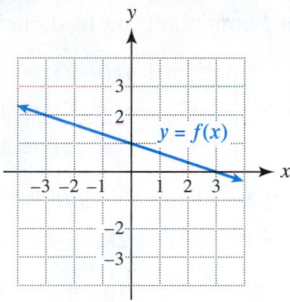

   (a) Identify the slope, y-intercept, and x-intercept.

   (b) Write a formula for $f(x)$.

   (c) Evaluate $f(-3)$ symbolically and graphically.

   (d) Find any zeros of f.

7. Find the average rate of change of $f(x) = x^3 - x$ from $x = -3$ to $x = -2$.

8. Find the difference quotient for $f(x) = x^2 + 6x$.

9. Write the slope-intercept form for a line that passes through $(-2, 5)$ and $(3, -4)$.

10. Write the slope-intercept form for a line that passes through $(-1, 4)$ and is perpendicular to the line $3x - 4y = 12$.

11. Write an equation of a line that is parallel to the x-axis and passes through $(4, -5)$.

12. Determine the x- and y-intercepts on the graph of $5x - 4y = 10$. Graph the equation.

$C(x) = 15x + 2000$ calculates the cost in dollars of producing x radios, interpret the numbers 15 and 2000 in the formula for $C(x)$.

14. Solve $-2.4x - 2.1 = \sqrt{3}x + 1.7$ both graphically and numerically. Round your answer to the nearest tenth.

Exercises 15–24: Solve the equation.

15. $-3(2 - 3x) - (-x - 1) = 1$

16. $x^3 + 5 = 5x^2 + x$ 17. $|3x - 4| + 1 = 5$

18. $2x^2 + x + 2 = 0$ 19. $7x^2 + 9x = 10$

20. $x^4 + 9 = 10x^2$ 21. $3x^{2/3} + 5x^{1/3} - 2 = 0$

22. $\sqrt{5 + 2x} + 4 = x + 5$

23. $\dfrac{2x - 3}{5 - x} = \dfrac{4x - 3}{1 - 2x}$ 24. $\sqrt[3]{x - 4} - 1 = 3$

25. Solve $\frac{1}{2}x - (4 - x) + 1 = \frac{3}{2}x - 5$. Is this equation either an identity or a contradiction?

26. Graph f. Is f continuous on its domain? Evaluate $f(1)$.

$$f(x) = \begin{cases} x^2 - 1 & \text{if } -3 \le x \le -1 \\ x + 1 & \text{if } -1 < x < 1 \\ 1 - x^2 & \text{if } 1 \le x \le 3 \end{cases}$$

Exercises 27–32: Solve the inequality.

27. $-\frac{1}{3}x - (1 + x) > \frac{2}{3}x$ 28. $-4 \le 4x - 6 < \frac{5}{2}$

29. $|5x - 7| \ge 3$ 30. $5x^2 + 13x - 6 < 0$

31. $x^3 - 9x \le 0$ 32. $\dfrac{4x - 3}{x + 2} > 0$

range?

33. The graph of a nonlinear function f is shown. Solve each equation or inequality.
    (a) $f(x) = 0$ (b) $f(x) > 0$ (c) $f(x) \le 0$

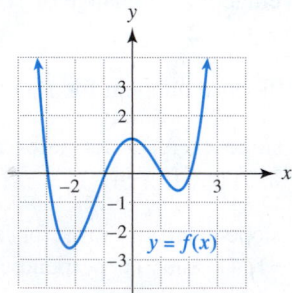

34. Write the quadratic polynomial $f(x) = 2x^2 - 4x + 1$ in the form $f(x) = a(x - h)^2 + k$.

35. Use the given graph of $y = f(x)$ at the top of the next column to sketch a graph of each equation.

(a) $y = f(x + 2) - 1$ (b) $y = -2f(x)$

(c) $y = f(-x) + 1$ (d) $y = f\left(\frac{1}{2}x\right)$

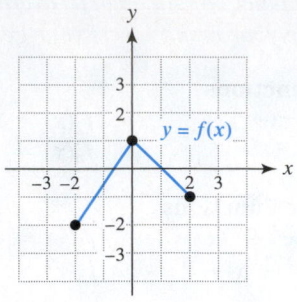

36. Use transformations of graphs to sketch a graph of $y = 2\sqrt{x} + 1$.

37. Use the graph of f to estimate each of the following.
    (a) Where f is increasing or decreasing

    (b) The zeros of f

    (c) The coordinates of any turning points

    (d) Any local extrema

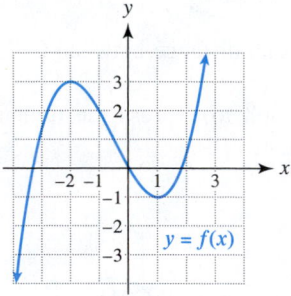

38. Are $f(x) = x^4 - 5x^3 - 7$ and $g(x) = \sqrt{9 - x^2}$ even, odd, or neither?

39. Sketch a graph of a quartic (degree 4) function with a negative leading coefficient, three x-intercepts, and three turning points.

40. State the end behavior of $f(x) = 4 + 3x - x^3$.

41. Divide each expression.

    (a) $\dfrac{4a^3 - 8a^2 + 12}{4a^2}$ (b) $\dfrac{2x^3 - 4x + 1}{x - 1}$

42. A quintic (degree 5) function f with real coefficients has leading coefficient $\frac{1}{2}$ and zeros -2, i, and $-2i$. Write $f(x)$ in complete factored form and expanded form.

43. A degree 6 function f has zeros -3, 1, and 4 with multiplicities 1, 2, and 3, respectively. If the leading coefficient is 4, write the complete factored form of $f(x)$.

44. Use the graph to write the complete factored form of the cubic polynomial $f(x)$.

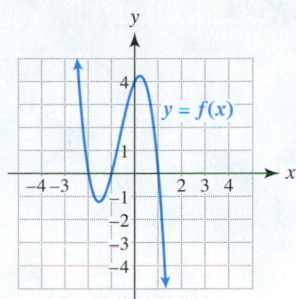

45. Write $\frac{3 + 4i}{1 - i}$ in standard form.

46. Find all solutions, real or imaginary, to $x^4 - 25 = 0$.

47. State the domain of $f(x) = \frac{2x - 5}{x^2 - 3x - 4}$. Find any vertical or horizontal asymptotes.

48. Write $\sqrt[3]{x^5}$ using rational exponents. Evaluate the expression for $x = 8$.

Applications

49. Water in a Pool The graph shows the amount of water in a swimming pool x hours past noon. Find the slope of each line segment and interpret each slope.

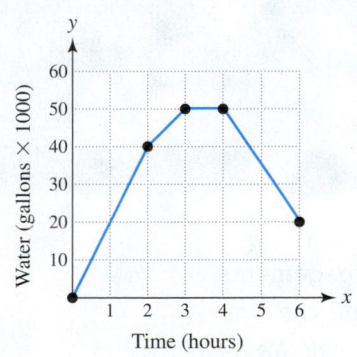

Time (hours)

50. Distance At noon, one runner is heading south at 8 miles per hour and is located 2 miles north of a second runner, who is heading west at 7 miles per hour. Approximate the distance between the runners to the nearest tenth of a mile at 12:30 P.M.

51. Mixing Acid Two liters of a 35% sulfuric acid solution need to be diluted to a 20% solution. How many liters of a 12% sulfuric acid solution should be mixed with the 2-liter solution?

52. Working Together Suppose one person can paint a room in 10 hours and another person can paint the same room in 8 hours. How long will it take to paint the room if they work together?

53. Maximizing Revenue The revenue R in dollars from selling x thousand toy figures is given by the formula $R(x) = x(800 - x)$. How many toy figures should be sold to maximize revenue?

54. Construction A box is being constructed by cutting 2-inch squares from the corners of a rectangular sheet of metal that is 6 inches longer than it is wide. If the box is to have a volume of 270 cubic inches, find the dimensions of the metal sheet.

55. Group Rates Round-trip airline tickets to Hawaii are regularly $800, but for each additional ticket purchased the price is reduced by $5. For example, 1 ticket costs $800, 2 tickets cost $2(795) = 1590, and 3 tickets cost $3(790) = 2370.

  (a) Write a quadratic function C that gives the total cost of purchasing t tickets.

  (b) Solve $C(t) = 17,000$ and interpret the result.

  (c) Find the absolute maximum for C and interpret your result. Assume that t must be an integer.

56. Modeling Data Find a quadratic function in the form $f(x) = a(x - h)^2 + k$ that models the data in the table. Graph $y = f(x)$ and the data if a graphing calculator is available.

x	4	6	8	10
y	6	15	37	80

57. Minimizing Surface Area A cylindrical can is being constructed to have a volume of 10π cubic inches. Find the dimensions of the can that result in the least amount of aluminum being used in its construction.

5 Exponential and Logarithmic Functions

When a link is posted on a social network, the majority of the engagements with this link occur within the first few hours. For example, within the first 3 hours a link on Facebook will have received half of its "hits." This *half life for* Twitter is even shorter, 2.8 hours, but much longer, 400 hours, for StumbleUpon. (See Section 5.3, Example 12.)

PGA golfers make about 99.2% of their putts that are 3 feet or less. However, every time a putt's distance increases by 6.6 feet, the chances of a pro making it decrease by about half. (See Section 5.3, Exercise 99.)

One reason for the explosive growth in mobile communication is the ability of researchers to increase the electrical efficiency of electronic devices. Since the era of the vacuum tube, the number of calculations that can be made using 1 kilowatt-hour of electricity has doubled every 1.6 years. (See Section 5.6, Example 4.)

Each of the examples above makes use of the mathematical concept of an exponential function. This chapter discusses this important topic.

The important thing is not to stop questioning.
—Albert Einstein

Sources: "The Half-Life of a Link", Column Five; "A Deeper Law than Moore's," *The Economist.*

5.1 Combining Functions

- Perform arithmetic operations on functions
- Review function notation
- Perform composition of functions

Introduction

Addition, subtraction, multiplication, and division can be performed on numbers and variables. These arithmetic operations can also be used to combine functions. For example, to model the stopping distance of a car traveling at x miles per hour, we compute two functions. The first function is the *reaction distance*, $r(x)$, which is the distance that a car travels between the time when a driver first recognizes a hazard and the time when the brakes are applied. The second function is *braking distance, $b(x)$*, which is the distance that a car travels after the brakes have been applied. *Stopping distance, $s(x)$*, is equal to the sum of $r(x)$ and $b(x)$. Figure 5.1 illustrates this example of addition of functions.

Stopping Distance Is a Sum of Functions:

$$s(x) = r(x) + b(x)$$

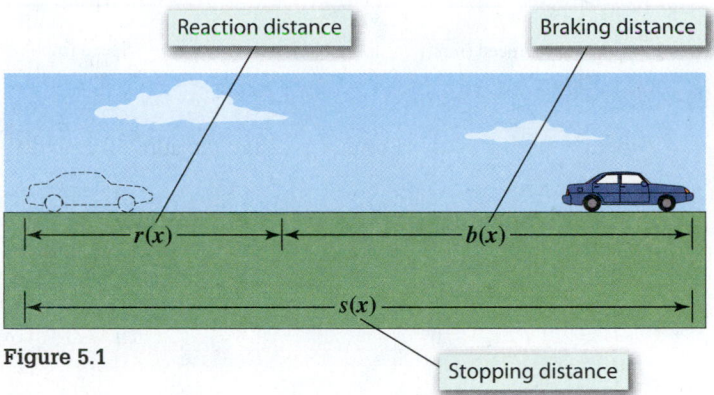

Reaction distance   Braking distance

$r(x)$ $b(x)$ $s(x)$

Figure 5.1   Stopping distance

Arithmetic Operations on Functions

The concept of finding the sum of two functions can be represented symbolically, graphically, and numerically, as illustrated in the next example.

EXAMPLE 1   **Representing stopping distance**

For wet, level pavement, highway engineers sometimes let $r(x) = \frac{11}{3}x$ and $b(x) = \frac{1}{9}x^2$. (*Source:* L. Haefner, *Introduction to Transportation Systems.*)

(a) Evaluate $r(60)$ and $b(60)$ and interpret each result.
(b) Write a formula for $s(x)$ and evaluate $s(60)$. Interpret the result.
(c) Graph r, b, and s. Interpret the graphs.
(d) Illustrate the relationship among r, b, and s numerically.

SOLUTION
(a) Substitute 60 into each formula.

$$r(60) = \frac{11}{3}(60) = 220 \quad \text{and} \quad b(60) = \frac{1}{9}(60)^2 = 400.$$

At 60 miles per hour the reaction distance is 220 feet and the braking distance is 400 feet.

(b) *Symbolic Representation* Let $s(x)$ be the sum of $r(x)$ and $b(x)$.

$$s(x) = r(x) + b(x) = \frac{11}{3}x + \frac{1}{9}x^2$$

Stopping Distance = Reaction Distance + Braking Distance

The stopping distance for a car traveling at 60 miles per hour is

$$s(60) = \frac{11}{3}(60) + \frac{1}{9}(60)^2 = 620 \text{ feet.}$$

(c) *Graphical Representation* Graph $y_1 = \frac{11}{3}x$, $y_2 = \frac{1}{9}x^2$, and $y_3 = \frac{11}{3}x + \frac{1}{9}x^2$, as shown in Figures 5.2–5.4.

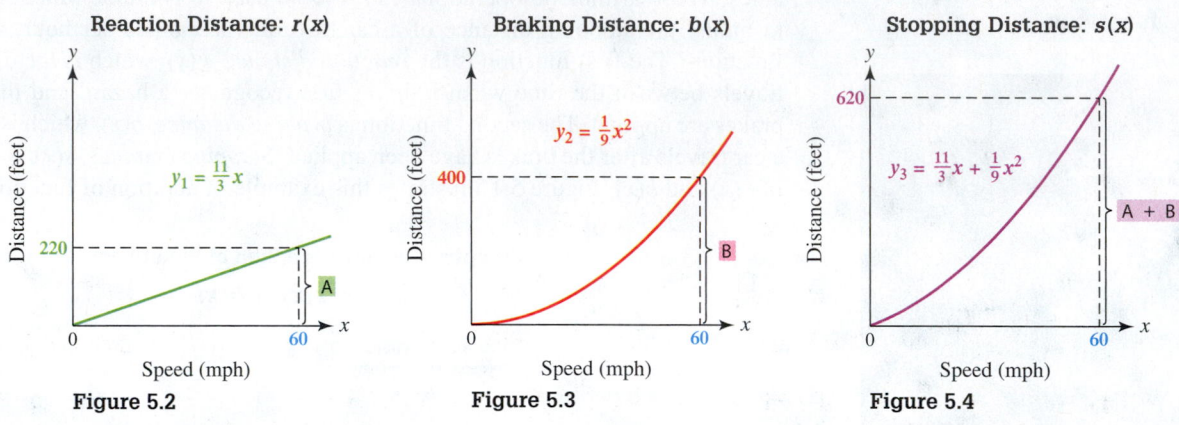

Figure 5.2   Figure 5.3   Figure 5.4

For any x-value, the sum of y_1 and y_2 equals y_3. For example, if $x = 60$, then

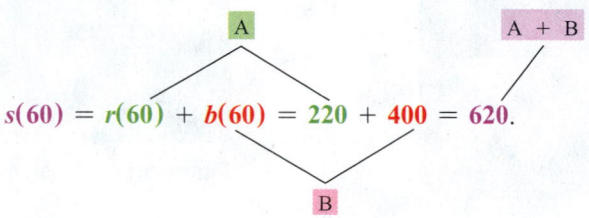

$$s(60) = r(60) + b(60) = 220 + 400 = 620.$$

(d) *Numerical Representation* Table 5.1 shows $r(x)$, $b(x)$, and $s(x) = r(x) + b(x)$. Values for $s(x)$ can be found by adding $r(x)$ and $b(x)$. For example, $s(60) = 220 + 400 = 620$.

Adding Functions

x	0	12	24	36	48	60	— Speed
$r(x)$	0	44	88	132	176	220	— Reaction distance
$b(x)$	0	16	64	144	256	400	— Braking distance
$s(x)$	0	60	152	276	432	620	— Stopping distance: $s(x) = r(x) + b(x)$

Table 5.1

Now Try Exercise 99

We now formally define arithmetic operations on functions.

OPERATIONS ON FUNCTIONS

If $f(x)$ and $g(x)$ both exist, the sum, difference, product, and quotient of two functions f and g are defined by

$$(f + g)(x) = f(x) + g(x),$$
$$(f - g)(x) = f(x) - g(x),$$
$$(fg)(x) = f(x) \cdot g(x), \quad \text{and}$$
$$\left(\frac{f}{g}\right)(x) = \frac{f(x)}{g(x)}, \quad \text{where } g(x) \neq 0.$$

Operations on Functions and Domains   The domains of the sum, difference, and product of f and g include x-values that are in *both* the domain of f and the domain of g. The domain of the quotient f/g includes all x-values in both the domain of f and the domain of g, where $g(x) \neq 0$.

Graphical, Numerical, and Symbolic Evaluation   In the next example, we evaluate the sum, difference, product, and quotient of two functions in three ways: graphically, numerically, and symbolically.

EXAMPLE 2   **Evaluating combinations of functions**

If possible, use each representation of f and g to evaluate $(f + g)(4)$, $(f - g)(-2)$, $(fg)(1)$, and $(f/g)(0)$.

(a)

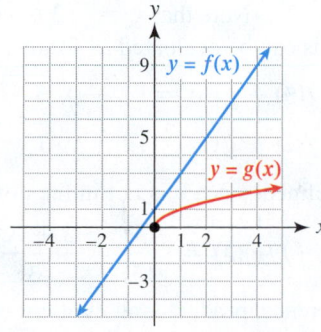

(b)

x	-2	0	1	4
$f(x)$	-3	1	3	9

x	-2	0	1	4
$g(x)$	—	0	1	2

(c) $f(x) = 2x + 1,\ g(x) = \sqrt{x}$

SOLUTION

(a) *Graphical Evaluation* From Figure 5.5, $f(4) = 9$ and $g(4) = 2$. Thus

$$(f + g)(4) = f(4) + g(4) = 9 + 2 = 11.$$

Although $f(-2) = -3$, $g(-2)$ is undefined because -2 is not in the domain of g. Thus $(f - g)(-2)$ is undefined. The domains of f and g include 1, and it follows that

$$(fg)(1) = f(1)g(1) = 3(1) = 3.$$

The graph of g intersects the origin, so $g(0) = 0$. Thus $(f/g)(0) = \dfrac{f(0)}{g(0)}$ is undefined.

(b) *Numerical Evaluation* From the tables, $f(4) = 9$ and $g(4) = 2$. As in part (a),

$$(f + g)(4) = f(4) + g(4) = 9 + 2 = 11.$$

A dash in the table indicates that $g(-2)$ is undefined, so $(f - g)(-2)$ is also undefined. The calculations of $(fg)(1)$ and $(f/g)(0)$ are done in a similar manner.

(c) *Symbolic Evaluation* Use the formulas $f(x) = 2x + 1$ and $g(x) = \sqrt{x}$.

$$(f + g)(4) = f(4) + g(4) = (2 \cdot 4 + 1) + \sqrt{4} = 9 + 2 = 11$$

$$(f - g)(-2) = f(-2) - g(-2) = (2 \cdot (-2) + 1) - \sqrt{-2} \text{ is undefined.}$$

$$(fg)(1) = f(1)g(1) = (2 \cdot 1 + 1)\sqrt{1} = 3(1) = 3$$

$$\left(\frac{f}{g}\right)(0) = \frac{f(0)}{g(0)} \text{ is undefined, since } g(0) = 0.$$

Now Try Exercises 7, 31, and 35

$f(4) = 9$ and $g(4) = 2$

Figure 5.5

EXAMPLE 3 Performing arithmetic operations on functions symbolically

Let $f(x) = 2 + \sqrt{x - 1}$ and $g(x) = x^2 - 4$.
(a) Find the domains of $f(x)$ and $g(x)$. Then find the domains of $(f + g)(x)$, $(f - g)(x)$, $(fg)(x)$, and $(f/g)(x)$.
(b) If possible, evaluate $(f + g)(5)$, $(f - g)(1)$, $(fg)(0)$, and $(f/g)(3)$.
(c) Write expressions for $(f + g)(x)$, $(f - g)(x)$, $(fg)(x)$, and $(f/g)(x)$.

SOLUTION
(a) Whenever $x \geq 1$, $f(x) = 2 + \sqrt{x - 1}$ is defined. Therefore the domain of $f(x)$ is $\{x \mid x \geq 1\}$. The domain of $g(x) = x^2 - 4$ is all real numbers. The domains of $(f + g)(x)$, $(f - g)(x)$, and $(fg)(x)$ include all x-values in *both* the domain of $f(x)$ and the domain of $g(x)$. Thus their domains are $\{x \mid x \geq 1\}$.

To determine the domain of $(f/g)(x)$, we must also exclude x-values for which $g(x) = x^2 - 4 = 0$. This occurs when $x = \pm 2$. Thus the domain of $(f/g)(x)$ is $\{x \mid x \geq 1, x \neq 2\}$. (Note that $x \neq -2$ is satisfied if $x \geq 1$.)

(b) The expressions can be evaluated as follows.

$$(f + g)(5) = f(5) + g(5) = \left(2 + \sqrt{5 - 1}\right) + (5^2 - 4) = 4 + 21 = 25$$

$$(f - g)(1) = f(1) - g(1) = \left(2 + \sqrt{1 - 1}\right) - (1^2 - 4) = 2 - (-3) = 5$$

$(fg)(0)$ is undefined, since 0 is not in the domain of $f(x)$.

$$(f/g)(3) = \frac{f(3)}{g(3)} = \frac{2 + \sqrt{3 - 1}}{3^2 - 4} = \frac{2 + \sqrt{2}}{5}$$

(c) The sum, difference, product, and quotient of f and g are calculated as follows.

$$(f + g)(x) = f(x) + g(x) = \left(2 + \sqrt{x - 1}\right) + (x^2 - 4) = \sqrt{x - 1} + x^2 - 2$$

$$(f - g)(x) = f(x) - g(x) = \left(2 + \sqrt{x - 1}\right) - (x^2 - 4) = \sqrt{x - 1} - x^2 + 6$$

$$(fg)(x) = f(x) \cdot g(x) = \left(2 + \sqrt{x - 1}\right)(x^2 - 4)$$

$$\left(\frac{f}{g}\right)(x) = \frac{f(x)}{g(x)} = \frac{2 + \sqrt{x - 1}}{x^2 - 4}$$

Now Try Exercises 9 and 15

An Application The next example is an application from business involving the difference between two functions.

EXAMPLE 4 Finding the difference of two functions

The expenses for a band to produce the master sound track for an album include renting a music studio and hiring a music engineer. These *fixed costs* are $12,000, and the cost to produce each album with packaging is $5.
(a) Assuming no other expenses, find a function C that outputs the cost of producing the master sound track plus x albums. Find the cost of making the master track and 3000 albums.
(b) Suppose that each album is sold for $12. Find a function R that computes the revenue from selling x albums. Find the revenue from selling 3000 albums.
(c) Determine a function P that outputs the profit from selling x albums. How much profit is there from selling 3000 albums?

SOLUTION
(a) The cost of producing the master sound track for $12,000 plus x albums at $5 each is given by $C(x) = 5x + 12,000$. The cost of manufacturing the master sound track and 3000 albums is

$$C(3000) = 5(3000) + 12,000 = \$27,000.$$

(b) The revenue from x albums at **$12** each is computed by $R(x) = 12x$. The revenue from selling 3000 albums is $R(3000) = 12(3000) = \$36,000$.

(c) Profit P is equal to revenue minus cost. This can be written using function notation.

$$P(x) = R(x) - C(x)$$
$$= 12x - (5x + 12,000)$$
$$= 12x - 5x - 12,000$$
$$= 7x - 12,000$$

If $x = 3000$, $P(3000) = 7(3000) - 12,000 = \$9000$.

> **Now Try Exercise 95**

Review of Function Notation

In the next example, we review how to evaluate function notation before we discuss composition of functions.

EXAMPLE 5 **Evaluating function notation**

Let $g(x) = 3x^2 - 6x + 2$. Evaluate each expression.
(a) $g(2)$ **(b)** $g(k)$ **(c)** $g(x^2)$ **(d)** $g(x + 2)$

SOLUTION
(a) $g(2) = 3(2)^2 - 6(2) + 2 = 12 - 12 + 2 = 2$
(b) $g(k) = 3k^2 - 6k + 2$
(c) $g(x^2) = 3(x^2)^2 - 6(x^2) + 2 = 3x^4 - 6x^2 + 2$
(d) $g(x + 2) = 3(x + 2)^2 - 6(x + 2) + 2$
$$= 3(x^2 + 4x + 4) - 6(x + 2) + 2$$
$$= 3x^2 + 12x + 12 - 6x - 12 + 2$$
$$= 3x^2 + 6x + 2$$

Algebra Review
To review squaring a binomial, see Chapter R (page R-17).

> **Now Try Exercise 45**

Composition of Functions

Many tasks in life are performed in sequence. For example, to go to a movie we might get into a car, drive to the movie theater, and get out of the car. A similar situation occurs with functions. For example, to convert miles to inches we might first convert miles to feet and then convert feet to inches in sequence. Since there are 5280 feet in a mile, $f(x) = 5280x$ converts x miles to an equivalent number of feet. Then $g(x) = 12x$ changes feet to inches. To convert x miles to inches, we combine the functions f and g in sequence. Figure 5.6 illustrates how to convert 5 miles to inches. First, $f(5) = 5280(5) = \mathbf{26,400}$. Then the output of 26,400 feet from f is used as input for g. The number of inches in 26,400 feet is $g(\mathbf{26,400}) = 12(\mathbf{26,400}) = \mathbf{316,800}$. This computation is called the *composition* of g and f.

The composition of g and f shown in Figure 5.6 can be expressed symbolically. The symbol $\circ$ is used to denote composition of two functions.

$$(g \circ f)(5) = g(f(5)) \qquad \text{First compute } f(5).$$
$$= g(5280 \cdot 5) \qquad f(x) = 5280x$$
$$= g(\mathbf{26,400}) \qquad \text{Simplify.}$$
$$= 12(\mathbf{26,400}) \qquad g(x) = 12x$$
$$= \mathbf{316,800} \qquad \text{Simplify.}$$

Composition of g and f

Composition of Functions
(converting 5 miles to inches)

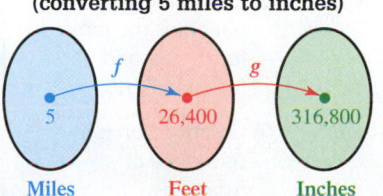

Miles   Feet   Inches

Figure 5.6

A distance of **5** miles is equivalent to **316,800** inches. The concept of composition of two functions relates to applying functions in sequence and is now defined formally.

COMPOSITION OF FUNCTIONS

If f and g are functions, then the **composite function** $g \circ f$, or **composition** of g and f is defined by

$$(g \circ f)(x) = g(f(x)).$$

We read $g(f(x))$ as "g of f of x."

The domain of $g \circ f$ is all x in the domain of f such that $f(x)$ is in the domain of g.

Symbolic Evaluation of Composite Functions   The next three examples discuss how to evaluate composite functions and find their domains symbolically.

EXAMPLE 6   **Finding a symbolic representation of a composite function**

Find a formula for the composite function $g \circ f$ that converts x miles into inches.

SOLUTION   Let $f(x) = 5280x$ and $g(x) = 12x$.

$$\begin{aligned}
(g \circ f)(x) &= g(\mathbf{f(x)}) &&\text{Definition of composition} \\
&= g(\mathbf{5280x}) &&f(x) = 5280x \text{ is the input for } g. \\
&= 12(5280x) &&g \text{ multiplies the input by 12.} \\
&= 63{,}360x &&\text{Simplify.}
\end{aligned}$$

Thus $(g \circ f)(x) = 63{,}360x$ converts x miles into inches.

Now try Exercies 97

NOTE   Converting units is only one application of composition of functions. Another application involves examining how a decrease in the ozone layer causes an increase in ultraviolet sunlight, which in turn causes increases in the number of skin cancer cases. See Example 10.

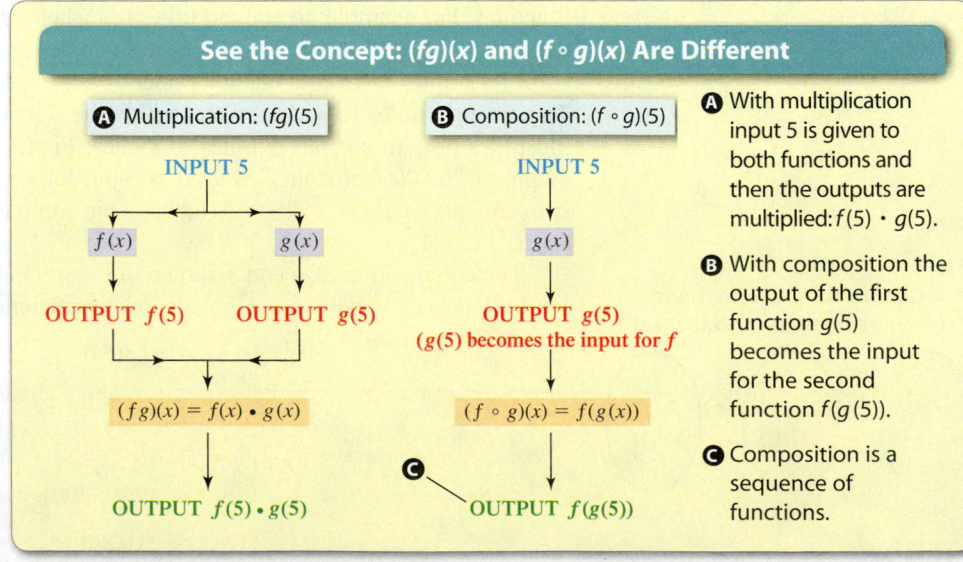

See the Concept: $(fg)(x)$ and $(f \circ g)(x)$ Are Different

A Multiplication: $(fg)(5)$

INPUT 5

$f(x)$ $g(x)$

OUTPUT $f(5)$ OUTPUT $g(5)$

$(fg)(x) = f(x) \cdot g(x)$

OUTPUT $f(5) \cdot g(5)$

B Composition: $(f \circ g)(5)$

INPUT 5

$g(x)$

OUTPUT $g(5)$
($g(5)$ becomes the input for f)

$(f \circ g)(x) = f(g(x))$

C

OUTPUT $f(g(5))$

A With multiplication input 5 is given to both functions and then the outputs are multiplied: $f(5) \cdot g(5)$.

B With composition the output of the first function $g(5)$ becomes the input for the second function $f(g(5))$.

C Composition is a sequence of functions.

EXAMPLE 7   **Evaluating a composite function symbolically**

Let $f(x) = x^2 + 3x + 2$ and $g(x) = \frac{1}{x}$.
(a) Evaluate $(f \circ g)(2)$ and $(g \circ f)(2)$. How do they compare?
(b) Find the composite functions defined by $(f \circ g)(x)$ and $(g \circ f)(x)$. Are they equivalent expressions?
(c) Find the domains of $(f \circ g)(x)$ and $(g \circ f)(x)$.

SOLUTION
(a) $(f \circ g)(2) = f(g(2)) = f\left(\frac{1}{2}\right) = \left(\frac{1}{2}\right)^2 + 3\left(\frac{1}{2}\right) + 2 = \frac{15}{4} = 3.75$

$(g \circ f)(2) = g(f(2)) = g(2^2 + 3 \cdot 2 + 2) = g(12) = \frac{1}{12} \approx 0.0833$
The results are *not* equal.

(b) $(f \circ g)(x) = f(g(x)) = f\left(\frac{1}{x}\right) = \left(\frac{1}{x}\right)^2 + 3\left(\frac{1}{x}\right) + 2 = \frac{1}{x^2} + \frac{3}{x} + 2$

$(g \circ f)(x) = g(f(x)) = g(x^2 + 3x + 2) = \dfrac{1}{x^2 + 3x + 2}$

The expressions for $(f \circ g)(x)$ and $(g \circ f)(x)$ are *not* equivalent.

(c) The domain of f is all real numbers, and the domain of g is $\{x \mid x \neq 0\}$. The domain of $(f \circ g)(x) = f(g(x))$ consists of all x in the domain of g such that $g(x)$ is in the domain of f. Thus the domain of $(f \circ g)(x) = \frac{1}{x^2} + \frac{3}{x} + 2$ is $\{x \mid x \neq 0\}$.

The domain of $(g \circ f)(x) = g(f(x))$ consists of all x in the domain of f such that $f(x)$ is in the domain of g. Since $x^2 + 3x + 2 = 0$ when $x = -1$ or $x = -2$, the domain of $(g \circ f)(x) = \frac{1}{x^2 + 3x + 2}$ is $\{x \mid x \neq -1, x \neq -2\}$.

> **Now try Exercises 53 and 57**

MAKING CONNECTIONS

Composition and Domains   To find the domain of a composition of two functions, it is sometimes helpful not to immediately simplify the resulting expression. For example, if $f(x) = x^2$ and $g(x) = \sqrt{x - 1}$, then

$$(f \circ g)(x) = \left(\sqrt{x - 1}\right)^2.$$

From this unsimplified expression, we can see that the domain (input) of $f \circ g$ must be restricted to $x \geq 1$ for the output to be a real number. As a result,

$$(f \circ g)(x) = x - 1, \quad x \geq 1.$$

EXAMPLE 8   **Finding symbolic representations for composite functions**

Find $(f \circ g)(x)$ and $(g \circ f)(x)$.
(a) $f(x) = x + 2$, $g(x) = x^3 - 2x^2 - 1$

(b) $f(x) = \sqrt{2x}$, $g(x) = \dfrac{1}{x + 1}$

(c) $f(x) = 2x - 3$, $g(x) = x^2 + 5$

SOLUTION

Getting Started When finding a composition of functions, the first step is often to write $(f \circ g)(x) = f(g(x))$ or $(g \circ f)(x) = g(f(x))$. ▶
(a) Begin by writing $(f \circ g)(x) = f(g(x)) = f(x^3 - 2x^2 - 1)$. Function f adds 2 to the input. That is, $f(\textbf{input}) = (\textbf{input}) + \textbf{2}$ because $f(x) = x + \textbf{2}$. Thus

$$f(x^3 - 2x^2 - 1) = (x^3 - 2x^2 - 1) + 2 = x^3 - 2x^2 + 1.$$

To find the composition $(g \circ f)(x)$, write $(g \circ f)(x) = g(f(x)) = g(x + 2)$. Because $g(x) = x^3 - 2x^2 - 1$, it follows that $g(\text{input}) = (\text{input})^3 - 2(\text{input})^2 - 1$. Thus

$$g(x + 2) = (x + 2)^3 - 2(x + 2)^2 - 1.$$

(b) $(f \circ g)(x) = f(g(x)) = f\left(\dfrac{1}{x + 1}\right) = \sqrt{2 \cdot \dfrac{1}{x + 1}} = \sqrt{\dfrac{2}{x + 1}}$ $f(x) = \sqrt{2x}$

$(g \circ f)(x) = g(f(x)) = g(\sqrt{2x}) = \dfrac{1}{\sqrt{2x} + 1}$ $g(x) = \dfrac{1}{x + 1}$

(c) $(f \circ g)(x) = f(g(x)) = f(x^2 + 5) = 2(x^2 + 5) - 3 = 2x^2 + 7$
$(g \circ f)(x) = g(f(x)) = g(2x - 3) = (2x - 3)^2 + 5 = 4x^2 - 12x + 14$

> **Now Try Exercises 59, 63, and 65**

> **NOTE** A composition of f with itself is denoted $(f \circ f)(x)$. For example, if $f(x) = 3x$, then
>
> $$(f \circ f)(x) = f(3x) = 3(3x) = 9x.$$

That is, f multiplies the input by 3, so $(f \circ f)(x)$ multiplies the input by 3 twice, which is equivalent to multiplying the input by 9.

Graphical Evaluation of Composite Functions   The next example shows how to evaluate a composition of functions graphically.

EXAMPLE 9   Evaluating a composite function graphically

Use the graphs of f and g shown in Figure 5.7 to evaluate each expression.
(a) $(f \circ g)(2)$ **(b)** $(g \circ f)(-3)$ **(c)** $(f \circ f)(-3)$

SOLUTION
(a) Because $(f \circ g)(2) = f(g(2))$, first evaluate $g(2)$. From Figure 5.8, $g(2) = 1$ and

$$(f \circ g)(2) = f(g(2)) = f(1).$$

To complete the evaluation of $(f \circ g)(2)$, use Figure 5.9 to determine that $f(1) = 3$. Thus $(f \circ g)(2) = 3$.

(b) Because $(g \circ f)(-3) = g(f(-3))$, first evaluate the expression $f(-3)$. Figure 5.10 shows that $f(-3) = -1$, so

$$(g \circ f)(-3) = g(f(-3)) = g(-1).$$

From Figure 5.11, $g(-1) = -2$. Thus $(g \circ f)(-3) = -2$.

(c) Similarly, $(f \circ f)(-3) = f(f(-3)) = f(-1) = 1$.

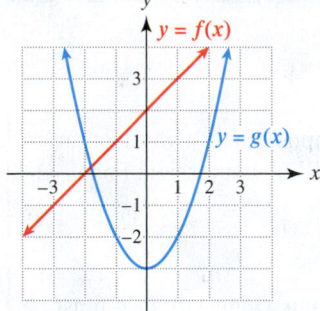

Figure 5.7

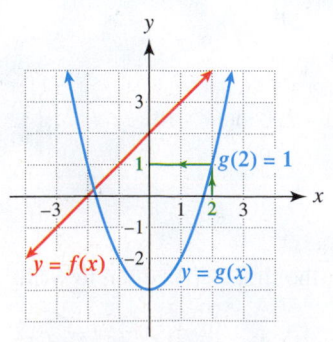

Figure 5.8

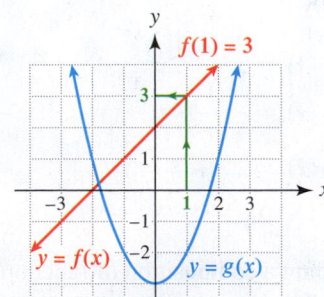

Figure 5.9

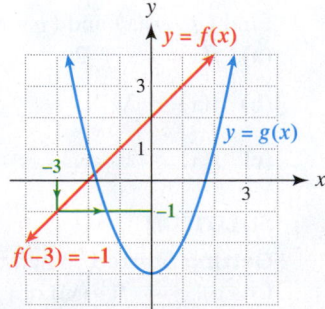

Figure 5.10

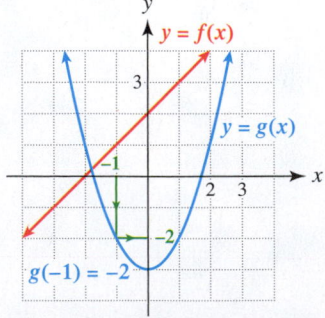

Figure 5.11

> **Now Try Exercise 75**

Numerical Evaluation of Composite Functions   Tables 5.2 and 5.3 represent two functions f and g.

x	1	2	3	4
$f(x)$	2	3	4	1

Table 5.2

x	1	2	3	4
$g(x)$	4	3	2	1

Table 5.3

We can use these tables to evaluate expressions, such as $(g \circ f)(3)$ and $(f \circ g)(3)$.

$$(g \circ f)(3) = g(f(3)) \qquad \textit{Definition of composition}$$
$$= g(4) \qquad f(3) = 4 \text{ in Table 5.2.}$$
$$= 1 \qquad g(4) = 1 \text{ in Table 5.3.}$$

We see that $(f \circ g)(3) = f(g(3)) = f(2) = 3$ by using Table 5.3 and then Table 5.2.

NOTE   Composition of functions is *not* commutative. That is, $(g \circ f)(x) \neq (f \circ g)(x)$ in general. For example, from above, $(g \circ f)(3) \neq (f \circ g)(3)$.

An Application   The next example illustrates how composition of functions occurs in the analysis of the ozone layer, ultraviolet (UV) radiation, and cases of skin cancer.

EXAMPLE 10   **Evaluating a composite function numerically**

Depletion of the ozone layer can cause an increase in the amount of UV radiation reaching the surface of the earth. An increase in UV radiation is associated with skin cancer. In Table 5.4 the function f computes the approximate percent *increase* in UV radiation resulting from an x percent *decrease* in the thickness of the ozone layer. The function g shown in Table 5.5 computes the expected percent increase in cases of skin cancer resulting from an x percent increase in UV radiation. (*Source:* R. Turner, D. Pearce, and I. Bateman, *Environmental Economics.*)

Percent Increase in UV Radiation

x	0	1	2	3	4	5	6
$f(x)$	0	1.5	3.0	4.5	6.0	7.5	9.0

Table 5.4

Percent decrease in the ozone layer

Percent increase in UV radiation

Percent Increase in Skin Cancer

x	0	1.5	3.0	4.5	6.0	7.5	9.0
$g(x)$	0	5.25	10.5	15.75	21.0	26.25	31.5

Table 5.5

Percent increase in UV radiation

Percent increase in cases of skin cancer

(a) Find $(g \circ f)(2)$ and interpret this calculation.
(b) Create a table for $g \circ f$. Describe what $(g \circ f)(x)$ computes.

SOLUTION
(a) $(g \circ f)(2) = g(f(2)) = g(3.0) = 10.5$. This means that a 2% decrease in the thickness of the ozone layer results in a 3% increase in UV radiation, which could cause a 10.5% increase in skin cancer.
(b) The values for $(g \circ f)(x)$ can be found in a similar manner. See Table 5.6.

x	0	1	2	3	4	5	6
$(g \circ f)(x)$	0	5.25	10.5	15.75	21.0	26.25	31.5

Table 5.6

Percent decrease in the ozone layer

Percent increase in cases of skin cancer

The composition $(g \circ f)(x)$ computes the percent increase in cases of skin cancer resulting from an x percent decrease in the ozone layer.

Now Try Exercise 101

Writing Compositions When you are solving problems, it is sometimes helpful to recognize a function as the composition of two simpler functions. For example, $h(x) = \sqrt[3]{x^2}$ can be thought of as the composition of the cube root function, $g(x) = \sqrt[3]{x}$, and the squaring function, $f(x) = x^2$. Then h can be written as $h(x) = g(f(x)) = \sqrt[3]{x^2}$. This concept is demonstrated in the next example. Note that answers may vary.

EXAMPLE 11 Writing a function as a composition of two functions

Find $f(x)$ and $g(x)$ so that $h(x) = (g \circ f)(x)$.

(a) $h(x) = (x + 3)^2$ **(b)** $h(x) = \sqrt{2x - 7}$ **(c)** $h(x) = \dfrac{1}{x^2 + 2x}$

SOLUTION

(a) Let $f(x) = x + 3$ and $g(x) = x^2$. Then
$$(g \circ f)(x) = g(f(x)) = g(x + 3) = (x + 3)^2.$$

(b) Let $f(x) = 2x - 7$ and $g(x) = \sqrt{x}$. Then
$$(g \circ f)(x) = g(f(x)) = g(2x - 7) = \sqrt{2x - 7}.$$

(c) Let $f(x) = x^2 + 2x$ and $g(x) = \frac{1}{x}$. Then
$$(g \circ f)(x) = g(f(x)) = g(x^2 + 2x) = \dfrac{1}{x^2 + 2x}.$$

> Now try Exercises 81, 85, and 91

5.1 Putting It All Together

The following table summarizes some concepts involved with combining functions.

CONCEPT	NOTATION	EXAMPLES
Sum of two functions	$(f + g)(x) = f(x) + g(x)$	$f(x) = x^2, g(x) = 2x + 1$ $(f + g)(3) = f(3) + g(3) = 9 + 7 = 16$ $(f + g)(x) = f(x) + g(x) = x^2 + 2x + 1$
Difference of two functions	$(f - g)(x) = f(x) - g(x)$	$f(x) = 3x, g(x) = 2x + 1$ $(f - g)(1) = f(1) - g(1) = 3 - 3 = 0$ $(f - g)(x) = f(x) - g(x) = 3x - (2x + 1)$ $\quad = x - 1$
Product of two functions	$(fg)(x) = f(x) \cdot g(x)$	$f(x) = x^3, g(x) = 1 - 3x$ $(fg)(-2) = f(-2) \cdot g(-2) = (-8)(7) = -56$ $(fg)(x) = f(x) \cdot g(x) = x^3(1 - 3x) = x^3 - 3x^4$
Quotient of two functions	$\left(\dfrac{f}{g}\right)(x) = \dfrac{f(x)}{g(x)}, g(x) \neq 0$	$f(x) = x^2 - 1, g(x) = x + 2$ $\left(\dfrac{f}{g}\right)(2) = \dfrac{f(2)}{g(2)} = \dfrac{3}{4}$ $\left(\dfrac{f}{g}\right)(x) = \dfrac{f(x)}{g(x)} = \dfrac{x^2 - 1}{x + 2}, x \neq -2$
Composition of two functions	$(g \circ f)(x) = g(f(x))$	$f(x) = x^3, g(x) = x^2 - 2x + 1$ $(g \circ f)(2) = g(f(2)) = g(8)$ $\quad = 64 - 16 + 1 = 49$ $(g \circ f)(x) = g(f(x)) = g(x^3)$ $\quad = (x^3)^2 - 2(x^3) + 1$ $\quad = x^6 - 2x^3 + 1$

Concepts

1. If $f(3) = 2$ and $g(3) = 5$, $(f + g)(3) =$ ___.

2. If $f(3) = 2$ and $g(2) = 5$, $(g \circ f)(3) =$ ___.

3. If $f(x) = x^2$ and $g(x) = 4x$, $(fg)(x) =$ ___.

4. If $f(x) = x^2$ and $g(x) = 4x$, $(f \circ g)(x) =$ ___.

5. Cost of Carpet If $f(x)$ calculates the number of square feet in x square yards and $g(x)$ calculates the cost in dollars of x square feet of carpet, what does $(g \circ f)(x)$ calculate?

6. Time Conversion If $f(x)$ calculates the number of days in x hours and $g(x)$ calculates the number of years in x days, what does $(g \circ f)(x)$ calculate?

Arithmetic Operations on Functions

Exercises 7–10: Use $f(x)$ and $g(x)$ to evaluate each expression symbolically.

7. $f(x) = 2x - 3, g(x) = 1 - x^2$

  (a) $(f + g)(3)$ **(b)** $(f - g)(-1)$

  (c) $(fg)(0)$ **(d)** $(f/g)(2)$

8. $f(x) = 4x - x^3, g(x) = x + 3$

  (a) $(g + g)(-2)$ **(b)** $(f - g)(0)$

  (c) $(gf)(1)$ **(d)** $(g/f)(-3)$

9. $f(x) = 2x + 1, g(x) = \dfrac{1}{x}$

  (a) $(f + g)(2)$ **(b)** $(f - g)\left(\tfrac{1}{2}\right)$

  (c) $(fg)(4)$ **(d)** $(f/g)(0)$

10. $f(x) = \sqrt[3]{x^2}, g(x) = |x - 3|$

  (a) $(f + g)(-8)$ **(b)** $(f - g)(-1)$

  (c) $(fg)(0)$ **(d)** $(f/g)(27)$

Exercises 11–30: Use $f(x)$ and $g(x)$ to find a formula for each expression. Identify its domain.

  (a) $(f + g)(x)$ *(b)* $(f - g)(x)$

  (c) $(fg)(x)$ *(d)* $(f/g)(x)$

11. $f(x) = 2x,$ $g(x) = x^2$

12. $f(x) = 1 - 4x,$ $g(x) = 3x + 1$

13. $f(x) = x^2 - 1,$ $g(x) = x^2 + 1$

14. $f(x) = 4x^3 - 8x^2,$ $g(x) = 4x^2$

15. $f(x) = x - \sqrt{x - 1},$ $g(x) = x + \sqrt{x - 1}$

16. $f(x) = 3 + \sqrt{2x + 9},$ $g(x) = 3 - \sqrt{2x + 9}$

17. $f(x) = \sqrt{x} - 1,$ $g(x) = \sqrt{x} + 1$

18. $f(x) = \sqrt{1 - x},$ $g(x) = x^3$

19. $f(x) = \dfrac{1}{x + 1},$ $g(x) = \dfrac{3}{x + 1}$

20. $f(x) = x^{1/2},$ $g(x) = 3$

21. $f(x) = \dfrac{1}{2x - 4},$ $g(x) = \dfrac{x}{2x - 4}$

22. $f(x) = \dfrac{1}{x},$ $g(x) = x^3$

23. $f(x) = x^2 - 1,$ $g(x) = |x + 1|$

24. $f(x) = |2x - 1|,$ $g(x) = |2x + 1|$

25. $f(x) = \dfrac{x^2 - 3x + 2}{x + 1},$ $g(x) = \dfrac{x^2 - 1}{x - 2}$

26. $f(x) = \dfrac{4x - 2}{x + 2},$ $g(x) = \dfrac{2x - 1}{3x + 6}$

27. $f(x) = \dfrac{2}{x^2 - 1},$ $g(x) = \dfrac{x + 1}{x^2 - 2x + 1}$

28. $f(x) = \dfrac{1}{x + 2},$ $g(x) = x^2 + x - 2$

29. $f(x) = x^{5/2} - x^{3/2},$ $g(x) = x^{1/2}$

30. $f(x) = x^{2/3} - 2x^{1/3} + 1,$ $g(x) = x^{1/3} - 1$

Exercises 31–34: Use the graph to evaluate each expression.

31.

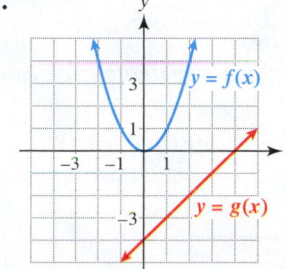

  (a) $(f + g)(2)$

  (b) $(f - g)(1)$

  (c) $(fg)(0)$

  (d) $(f/g)(1)$

32.

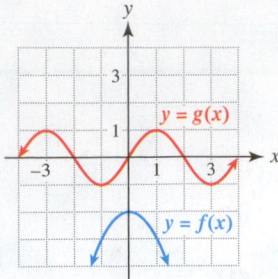

(a) $(f + g)(1)$

(b) $(f - g)(0)$

(c) $(fg)(-1)$

(d) $(f/g)(1)$

33.

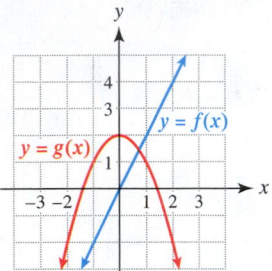

(a) $(f + g)(0)$

(b) $(f - g)(-1)$

(c) $(fg)(1)$

(d) $(f/g)(2)$

34.

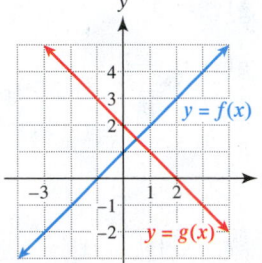

(a) $(f + g)(-1)$

(b) $(f - g)(-2)$

(c) $(fg)(0)$

(d) $(f/g)(2)$

Exercises 35 and 36: Use the tables to evaluate each expression, if possible.

(a) $(f + g)(-1)$ (b) $(g - f)(0)$

(c) $(gf)(2)$ (d) $(f/g)(2)$

35.

x	-1	0	2
$f(x)$	-3	5	1

x	-1	0	2
$g(x)$	-2	3	0

36.

x	-1	0	2
$f(x)$	4	1	3

x	-1	0	2
$g(x)$	2	0	1

Exercises 37 and 38: Use the table to evaluate each expression, if possible.

(a) $(f + g)(2)$ (b) $(f - g)(4)$

(c) $(fg)(-2)$ (d) $(f/g)(0)$

37.

x	-2	0	2	4
$f(x)$	0	5	7	10
$g(x)$	6	0	-2	5

38.

x	-2	0	2	4
$f(x)$	-4	8	5	0
$g(x)$	2	-1	4	0

39. Use the table in Exercise 37 to complete the following table.

x	-2	0	2	4
$(f + g)(x)$				
$(f - g)(x)$				
$(fg)(x)$				
$(f/g)(x)$				

40. Use the table in Exercise 38 to complete the table in Exercise 39.

Review of Function Notation

Exercises 41–52: For the given $g(x)$, evaluate each of the following.

(a) $g(-3)$ (b) $g(b)$ (c) $g(x^3)$ (d) $g(2x - 3)$

41. $g(x) = 2x + 1$ **42.** $g(x) = 5 - \frac{1}{2}x$

43. $g(x) = 2(x + 3)^2 - 4$ **44.** $g(x) = -(x - 1)^2$

45. $g(x) = \frac{1}{2}x^2 + 3x - 1$ **46.** $g(x) = 2x^2 - x - 9$

47. $g(x) = \sqrt{x + 4}$ **48.** $g(x) = \sqrt{2 - x}$

49. $g(x) = |3x - 1| + 4$ **50.** $g(x) = 2|1 - x| - 7$

51. $g(x) = \dfrac{4x}{x + 3}$ **52.** $g(x) = \dfrac{x + 3}{2}$

Composition of Functions

Exercises 53–56: Use the given $f(x)$ and $g(x)$ to evaluate each expression.

53. $f(x) = \sqrt{x + 5}, \quad g(x) = x^2$

  (a) $(f \circ g)(2)$ (b) $(g \circ f)(-1)$

54. $f(x) = |x^2 - 4|, \quad g(x) = 2x^2 + x + 1$

  (a) $(f \circ g)(1)$ (b) $(g \circ f)(-3)$

55. $f(x) = 5x - 2, \quad g(x) = |x|$

  (a) $(f \circ g)(-4)$ (b) $(g \circ f)(5)$

56. $f(x) = \dfrac{1}{x - 4}, \quad g(x) = 5$

  (a) $(f \circ g)(3)$ (b) $(g \circ f)(8)$

Exercises 57–72: Use the given $f(x)$ and $g(x)$ to find each of the following. Identify its domain.

    (a) $(f \circ g)(x)$ *(b)* $(g \circ f)(x)$ *(c)* $(f \circ f)(x)$

57. $f(x) = x^3,$ $g(x) = x^2 + 3x - 1$

58. $f(x) = 2 - x,$ $g(x) = \dfrac{1}{x^2}$

59. $f(x) = x + 2,$ $g(x) = x^4 + x^2 - 3x - 4$

60. $f(x) = x^2,$ $g(x) = \sqrt{1 - x}$

61. $f(x) = 2 - 3x,$ $g(x) = x^3$

62. $f(x) = \sqrt{x},$ $g(x) = 1 - x^2$

63. $f(x) = \dfrac{1}{x + 1},$ $g(x) = 5x$

64. $f(x) = \dfrac{1}{3x},$ $g(x) = \dfrac{2}{x - 1}$

65. $f(x) = x + 4,$ $g(x) = \sqrt{4 - x^2}$

66. $f(x) = 2x + 1,$ $g(x) = 4x^3 - 5x^2$

67. $f(x) = \sqrt{x - 1},$ $g(x) = 3x$

68. $f(x) = \dfrac{x - 3}{2},$ $g(x) = 2x + 3$

69. $f(x) = 1 - 5x,$ $g(x) = \dfrac{1 - x}{5}$

70. $f(x) = \sqrt[3]{x - 1},$ $g(x) = x^3 + 1$

71. $f(x) = \dfrac{1}{kx}, k > 0,$ $g(x) = \dfrac{1}{kx}, k > 0$

72. $f(x) = ax^2, a > 0,$ $g(x) = \sqrt{ax}, a > 0$

Exercises 73–76: Use the graph to evaluate each expression.

73.

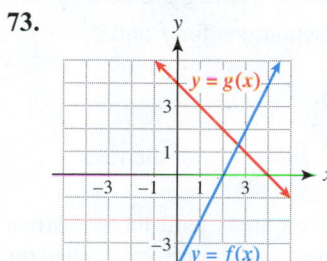

  (a) $(f \circ g)(4)$

  (b) $(g \circ f)(3)$

  (c) $(f \circ f)(2)$

74.

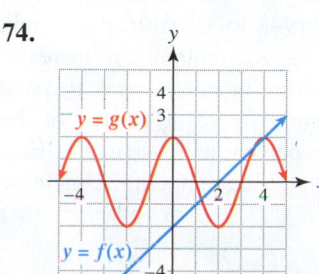

  (a) $(f \circ g)(2)$

  (b) $(g \circ g)(0)$

  (c) $(g \circ f)(4)$

75.

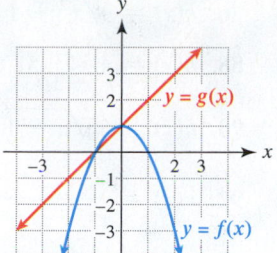

  (a) $(f \circ g)(1)$

  (b) $(g \circ f)(-2)$

  (c) $(g \circ g)(-2)$

76.

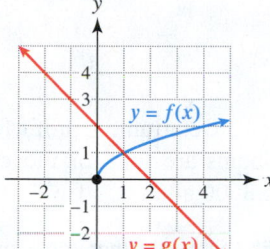

  (a) $(f \circ g)(-2)$

  (b) $(g \circ f)(1)$

  (c) $(f \circ f)(0)$

Exercises 77 and 78: Tables for the functions f and g are given. Evaluate the expression, if possible.

   (a) $(g \circ f)(1)$ *(b)* $(f \circ g)(4)$ *(c)* $(f \circ f)(3)$

77.

x	1	2	3	4
$f(x)$	4	3	1	2

x	1	2	3	4
$g(x)$	2	3	4	5

78.

x	1	3	4	6
$f(x)$	2	6	5	7

x	2	3	5	7
$g(x)$	4	2	6	0

79. Use the tables for $f(x)$ and $g(x)$ in Exercise 77 to complete the composition shown in the diagram.

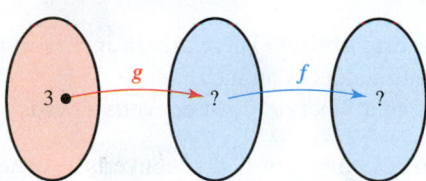

80. Use the tables for $f(x)$ and $g(x)$ in Exercise 78 to complete the composition shown in the diagram.

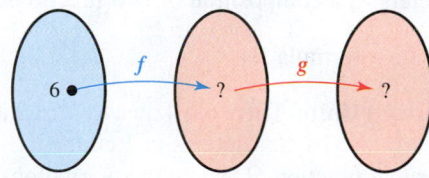

Exercises 81–94: (Refer to Example 11.) Find $f(x)$ and $g(x)$ so that $h(x) = (g \circ f)(x)$. Answers may vary.

81. $h(x) = \sqrt{x - 2}$

82. $h(x) = (x + 2)^4$

83. $h(x) = \dfrac{1}{x + 2}$

84. $h(x) = 5(x + 2)^2 - 4$

85. $h(x) = 4(2x + 1)^3$ **86.** $h(x) = \sqrt[3]{x^2 + 1}$

87. $h(x) = (x^3 - 1)^2$ **88.** $h(x) = 4(x - 5)^{-2}$

89. $h(x) = -4|x + 2| - 3$ **90.** $h(x) = 5\sqrt{x - 1}$

91. $h(x) = \dfrac{1}{(x - 1)^2}$ **92.** $h(x) = \dfrac{2}{x^2 - x + 1}$

93. $h(x) = x^{3/4} - x^{1/4}$ **94.** $h(x) = x^{2/3} - 5x^{1/3} + 4$

Applications

95. Profit (Refer to Example 4.) Determine a profit function P that results if the albums are sold for $15 each. Find the profit from selling 3000 albums.

96. Revenue, Cost, and Profit Suppose that for a major production company it costs $150,000 to produce a master track for a music video and $1.50 to produce each copy.

(a) Write a cost function C that outputs the cost of producing the master track and x copies.

(b) If the music videos are sold for $6.50 each, find a function R that outputs the revenue received from selling x music videos. What is the revenue from selling 8000 videos?

(c) Find a function P that outputs the profit from selling x music videos. What is the profit from selling 40,000 videos?

(d) How many videos must be sold to break even? That is, how many videos must be sold for the revenue to equal the cost?

97. Converting Units There are 36 inches in a yard and 2.54 centimeters in an inch.

(a) Write a function I that converts x yards to inches.

(b) Write a function C that converts x inches to centimeters.

(c) Express a function F that converts x yards to centimeters as a composition of two functions.

(d) Write a formula for F.

98. Converting Units There are 4 quarts in 1 gallon, 4 cups in 1 quart, and 16 tablespoons in 1 cup.

(a) Write a function Q that converts x gallons to quarts.

(b) Write a function C that converts x quarts to cups.

(c) Write a function T that converts x cups to tablespoons.

(d) Express a function F that converts x gallons to tablespoons as a composition of *three* functions.

(e) Write a formula for F.

99. Stopping Distance (Refer to Example 1.) A driver's reaction distance is $r(x) = \frac{11}{6}x$ and braking distance is $b(x) = \frac{1}{9}x^2$.

(a) Find a formula $s(x)$ that computes the stopping distance for this driver traveling at x miles per hour.

(b) Evaluate $s(60)$ and interpret the result.

100. Stopping Distance (Refer to Example 1.) If a driver attempts to stop while traveling at x miles per hour on dry, level pavement, the reaction distance is $r(x) = \frac{11}{5}x$ and the braking distance is $b(x) = \frac{1}{11}x^2$, where both distances are in feet.

(a) Write a formula for a function s in terms of $r(x)$ and $b(x)$ that gives the stopping distance when driving at x miles per hour. Evaluate $s(55)$.

(b) Graph r, b, and s on the same axes. Explain how the graph of s can be found using the graphs of r and b.

(c) Make tables for $r(x)$ and $b(x)$ at $x = 11, 22, 33, 44,$ and 55. Then use these tables to construct a table for $s(x)$.

101. Skin Cancer (Refer to Example 10 and Tables 5.4 and 5.5.) If possible, calculate the composition and interpret the result.

(a) $(g \circ f)(1)$ (b) $(f \circ g)(21)$

102. Skin Cancer In Example 10, f and g are both linear.

(a) Find symbolic representations for f and g.

(b) Determine $(g \circ f)(x)$.

(c) Evaluate $(g \circ f)(3.5)$ and interpret the result.

103. Urban Heat Island Urban areas tend to be warmer than the surrounding rural areas. This effect is called the *urban heat island*. In the first figure at the top of the next page, f computes the average increase in nighttime summer temperatures in degrees Celsius at Sky Harbor Airport in Phoenix from 1948 to 1990. In this graph, 1948 is the base year with a zero temperature increase. The rise in urban temperature increased peak demand for electricity. In the second figure at the top of the next page, g computes the percent increase in electrical demand for an average nighttime temperature increase of x degrees Celsius. (*Source:* W. Cotton and R. Pielke, *Human Impacts on Weather and Climate.*)

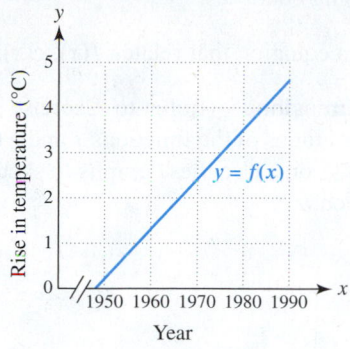

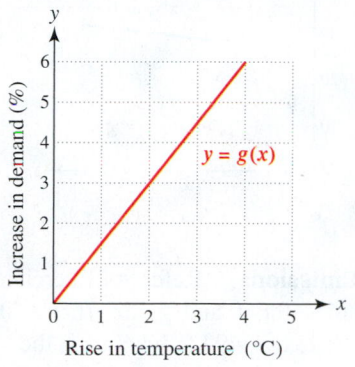

(a) Evaluate $(g \circ f)(1975)$ graphically.

(b) Interpret $(g \circ f)(x)$.

104. Urban Heat Island (Refer to Exercise 103.) If possible, calculate the composition and interpret the result.
(a) $(g \circ f)(1980)$ **(b)** $(f \circ g)(3)$

105. Urban Heat Island (Refer to Exercise 103.) The functions f and g are given by $f(x) = 0.11(x - 1948)$ and $g(x) = 1.5x$.
(a) Evaluate $(g \circ f)(1960)$.

(b) Find $(g \circ f)(x)$.

(c) What type of functions are, f, g, and $g \circ f$?

106. Swimming Pools In the figures, f computes the cubic feet of water in a pool after x days, and g converts cubic feet to gallons.

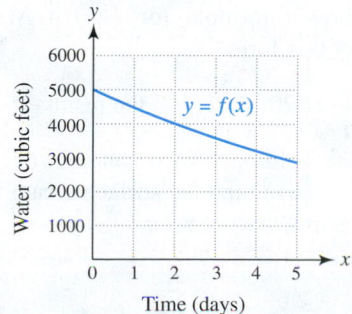

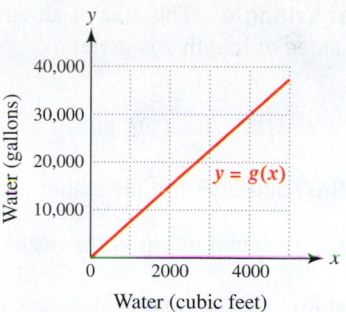

(a) Find the gallons of water in the pool after 2 days.

(b) Interpret $(g \circ f)(x)$.

107. Temperature The function f computes the temperature on a summer day after x hours, and g converts Fahrenheit temperature to Celsius temperature. See the figures.

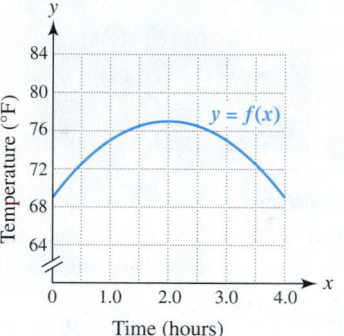

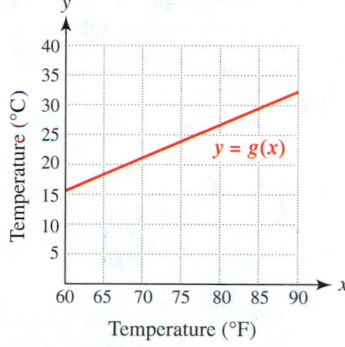

(a) Evaluate $(g \circ f)(2)$.
(b) Interpret $(g \circ f)(x)$.

108. Surface Area of a Balloon The surface area A of a balloon with radius r is given by $A(r) = 4\pi r^2$. Suppose that the radius of the balloon increases from r to $r + h$, where h is a small positive number.
(a) Find $A(r + h) - A(r)$. Interpret your answer.

(b) Evaluate your expression in part (a) for $r = 3$ and $h = 0.1$, and then for $r = 6$ and $h = 0.1$.

(c) If the radius of the balloon increases by 0.1, does the surface area always increase by a fixed amount or does the amount depend on the value of r?

109. Equilateral Triangle The area of an equilateral triangle with sides of length s is given by

$$A(s) = \frac{\sqrt{3}}{4}s^2.$$

(a) Find $A(4s)$ and interpret the result.

(b) Find $A(s + 2)$ and interpret the result.

110. Circular Wave A marble is dropped into a lake, resulting in a circular wave whose radius increases at a rate of 6 inches per second. Write a formula for C that gives the circumference of the circular wave in inches after t seconds.

111. Circular Wave (Refer to Exercise 110.) Write a function A that gives the area contained inside the circular wave in square inches after t seconds.

112. Geometry The surface area of a cone (excluding the bottom) is given by $S = \pi r \sqrt{r^2 + h^2}$, where r is its radius and h is its height, as shown in the figure. If the height is twice the radius, write a formula for S in terms of r.

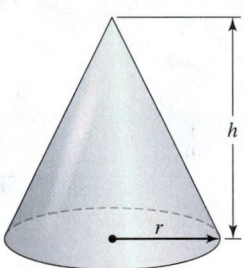

113. Methane Emissions Methane is a greenhouse gas that lets sunlight into the atmosphere but blocks heat from escaping Earth's atmosphere. In the table, $f(x)$ models the predicted methane emissions in millions of tons produced by *developed* countries. The function $g(x)$ models the same emissions for *developing* countries.

x	1990	2000	2010	2020	2030
$f(x)$	27	28	29	30	31
$g(x)$	5	7.5	10	12.5	15

Source: A. Nilsson, *Greenhouse Earth.*

(a) Make a table for a function h that models the total predicted methane emissions for developed *and* developing countries.

(b) Write an equation that relates $f(x)$, $g(x)$, and $h(x)$.

114. Methane Emissions (Refer to Exercise 113.) The figure shows graphs of the functions f and g that model methane emissions. Use these graphs to sketch a graph of the function h.

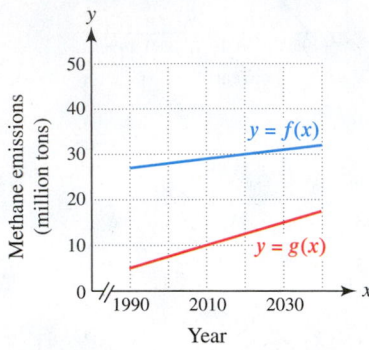

115. Methane Emissions (Refer to Exercises 113 and 114.) Formulas for f and g are $f(x) = 0.1x - 172$ and $g(x) = 0.25x - 492.5$, where x is the year. Find a symbolic representation for h.

116. Energy of a Falling Object A ball with mass m is dropped from an initial height of h_0 and lands with a final velocity of v_f. The kinetic energy of the ball is $K(v) = \frac{1}{2}mv^2$, where v is its velocity, and the potential energy of the ball is $P(h) = mgh$, where h is its height and g is a constant.

(a) Show that $P(h_0) = K(v_f)$. (*Hint:* $v_f = \sqrt{2gh_0}$.)

(b) Interpret your result from part (a).

117. Income Differences From 1990 to 2012, the 50th percentile (median) of U.S. household income in thousands of dollars could be approximated by $M(x) = 0.1x + 48$, where $M(x)$ is in 2012 dollars and x is years after 1990. The 90th percentile could be modeled by $N(x) = -0.15x^2 + 4x + 120$.

(a) Evaluate $M(20)$ and $N(20)$. Interpret the results.

(b) Determine a formula for $D(x) = N(x) - M(x)$. Interpret this formula.

(c) Evaluate $D(20)$. Interpret the results.

118. Sphere The volume V of a sphere with radius r is given by $V = \frac{4}{3}\pi r^3$, and the surface area S is given by $S = 4\pi r^2$. Show that $V = \frac{4}{3}\pi \left(\frac{S}{4\pi}\right)^{3/2}$.

Applying Concepts

119. Show that the sum of two linear functions is a linear function.

120. Show that if f and g are odd functions, then the composition $g \circ f$ is also an odd function.

121. Let $f(x) = k$ and $g(x) = ax + b$, where k, a, and b are constants.
   (a) Find $(f \circ g)(x)$. What type of function is $f \circ g$?
   (b) Find $(g \circ f)(x)$. What type of function is $g \circ f$?

122. Show that if $f(x) = ax + b$ and $g(x) = cx + d$, then $(g \circ f)(x)$ also represents a linear function. Find the slope of the graph of $(g \circ f)(x)$.

Writing about Mathematics

123. Describe differences between $(fg)(x)$ and $(f \circ g)(x)$. Give examples.
124. Describe differences between $(f \circ g)(x)$ and $(g \circ f)(x)$. Give examples.

5.2 Inverse Functions and Their Representations

- **Calculate inverse operations**
- **Identify one-to-one functions**
- **Find inverse functions symbolically**
- **Use other representations to find inverse functions**

Introduction

Many actions are reversible. A closed door can be opened—an open door can be closed. One hundred dollars can be withdrawn from and deposited into a savings account. These actions undo or cancel each other. But not all actions are reversible. Explosions and weather are two examples. In mathematics the concept of reversing a calculation and arriving at the original value is associated with an *inverse*.

Actions and their inverses occur in everyday life. Suppose a person opens a car door, gets in, and starts the engine. What are the inverse actions? The person turns off the engine, gets out, and closes the car door. Notice that we must reverse the order as well as apply the inverse operation at each step.

Inverse Operations and Inverse Functions

Inverse Operations In mathematics there are basic operations that can be considered inverse operations.

Inverse Operations: Addition and Subtraction

| Start with 10. | $10 + 5 = 15$ | $15 - 5 = 10$ | End with 10. |

Add 5.   Subtract 5.

Addition and subtraction are inverse operations. The same is true for multiplication and division, as illustrated by the following.

Inverse Operations: Multiplication and Division

| Start with 10. | $10 \times 2 = 20$ | $20 \div 2 = 10$ | End with 10. |

Multiply by 2.   Divide by 2.

In the next example, we discuss inverse operations further.

EXAMPLE 1 Finding inverse actions and operations

For each of the following, state the inverse actions or operations.
(a) Put on a coat and go outside.
(b) Subtract 5 from x and divide the result by 2.

SOLUTION
(a) To find the inverse actions, reverse the order and apply the inverse action at each step. The inverse actions would be to come inside and take off the coat.
(b) We must reverse the order and apply the inverse operation at each step. The inverse operations would be to multiply x by 2 and add 5. The original operations could be expressed as $\frac{x-5}{2}$, and the inverse operations could be written as $2x + 5$.

Now try Exercises 3 and 7

Inverse Functions Table 5.7 can be used to convert *gallons* to *pints*. There are 8 pints in a gallon, so $f(x) = 8x$ converts x gallons to an equivalent number of pints. If we want to convert *pints* to *gallons*, we need to divide by 8. This conversion is calculated by $g(x) = \frac{x}{8}$, where x is the number of pints. We say that f and g are *inverse functions* and write this as $g(x) = f^{-1}(x)$. See Table 5.8. We read f^{-1} as "f inverse."

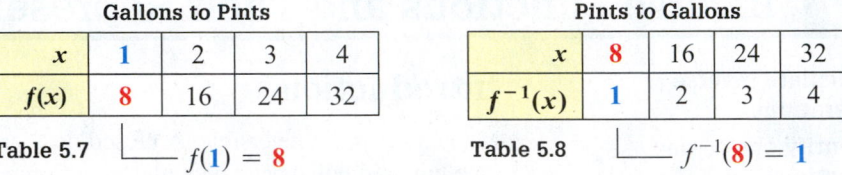

Inverse Functions

Gallons to Pints

x	1	2	3	4
$f(x)$	8	16	24	32

Table 5.7 $f(\mathbf{1}) = \mathbf{8}$

Pints to Gallons

x	8	16	24	32
$f^{-1}(x)$	1	2	3	4

Table 5.8 $f^{-1}(\mathbf{8}) = \mathbf{1}$

Multiplying by 8 and dividing by 8 are inverse operations. As discussed earlier, adding 5 and subtracting 5 are also inverse operations. If $f(x) = x + 5$, then the *inverse function* of f is given by $f^{-1}(x) = x - 5$. For example, $f(\mathbf{5}) = \mathbf{10}$, and $f^{-1}(\mathbf{10}) = \mathbf{5}$. If input x produces output y with function f, input y produces output x with function f^{-1}. This can be seen in Tables 5.9 and 5.10.

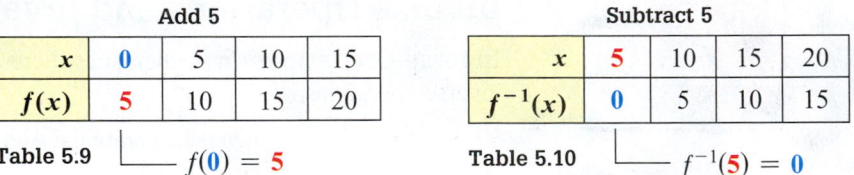

Inverse Functions

Add 5

x	0	5	10	15
$f(x)$	5	10	15	20

Table 5.9 $f(\mathbf{0}) = \mathbf{5}$

Subtract 5

x	5	10	15	20
$f^{-1}(x)$	0	5	10	15

Table 5.10 $f^{-1}(\mathbf{5}) = \mathbf{0}$

From Tables 5.9 and 5.10, if $f(a) = b$, then $f^{-1}(b) = a$. That is, if f outputs b with input a, then f^{-1} must output a with input b. *Inputs and outputs (domains and ranges) are interchanged for inverse functions.* This statement is illustrated in Figure 5.12.

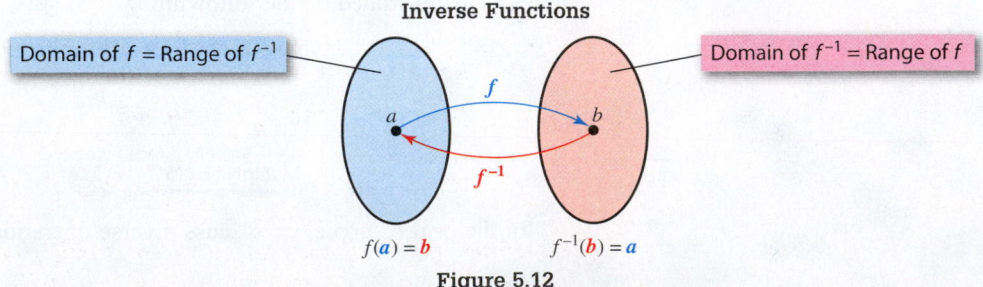

Inverse Functions

Domain of f = Range of f^{-1}

Domain of f^{-1} = Range of f

$f(a) = b$ $f^{-1}(b) = a$

Figure 5.12

When $f(x) = x + 5$ and $f^{-1}(x) = x - 5$ are applied in sequence, the output of f is used as input for f^{-1}. This is *composition* of functions.

$$(f^{-1} \circ f)(x) = f^{-1}(f(x)) \qquad \text{Definition of composition}$$
$$= f^{-1}(x + 5) \qquad f(x) = x + 5$$
$$= (x + 5) - 5 \qquad f^{-1} \text{ subtracts 5 from its input.}$$
$$= x \qquad \text{Simplify.}$$

The composition $f^{-1} \circ f$ with input x produces output x. The same action occurs when computing the composition $f \circ f^{-1}$.

$$(f \circ f^{-1})(x) = f(f^{-1}(x)) \qquad \text{Definition of composition}$$
$$= f(x - 5) \qquad f^{-1}(x) = x - 5$$
$$= (x - 5) + 5 \qquad f \text{ adds 5 to its input.}$$
$$= x \qquad \text{Simplify.}$$

Another pair of inverse functions is given by $f(x) = x^3$ and $f^{-1}(x) = \sqrt[3]{x}$. Before a formal definition of inverse functions is given, we must discuss one-to-one functions.

> **MAKING CONNECTIONS**
>
> **The Notation f^{-1} and Negative Exponents** If a represents a real number, then $a^{-1} = \frac{1}{a}$. For example, $4^{-1} = \frac{1}{4}$. On the other hand, if f represents a function, note that $f^{-1}(x) \neq \frac{1}{f(x)}$. Instead, $f^{-1}(x)$ represents the inverse function of f. For instance, if $f(x) = 5x$, then $f^{-1}(x) = \frac{x}{5} \neq \frac{1}{5x}$.

One-to-One Functions

Does every function have an inverse function? The next example answers this question.

EXAMPLE 2 **Determining if a function has an inverse function**

Table 5.11 represents a function C that computes the percentage of the time that the sky is cloudy in Augusta, Georgia, where x corresponds to the standard numbers for the months. Determine if C has an inverse function.

Cloudy Skies in Augusta

x (month)	1	2	3	4	5	6	7	8	9	10	11	12
C(x)(%)	43	40	39	29	28	26	27	25	30	26	31	39

Source: J. Williams, *The Weather Almanac.*
Table 5.11

SOLUTION

For each input (month), C computes exactly one output. For example, $C(3) = 39$ means that during March the sky is cloudy 39% of the time. If C has an inverse function, the inverse must receive 39 as input and produce exactly one output. Both March and December have cloudy skies 39% of the time. Given an input of 39, it is impossible for an inverse *function* to output both 3 and 12. Therefore C does *not* have an inverse function.

Now Try Exercise 19

If *different inputs* of a function f produce the *same output*, then an inverse function of f does *not* exist. However, if different inputs always produce different outputs, f is a *one-to-one function*. *Every one-to-one function has an inverse function.* For example, $f(x) = x^2$ is *not* one-to-one because

Different Inputs

$$f(-2) = 4 \text{ and } f(2) = 4. \qquad \text{Not one-to-one}$$

Same Output

Therefore $f(x) = x^2$ does not have an inverse function because an inverse *function* cannot receive input 4 and produce both -2 and 2 as outputs. However, $g(x) = 5x$ is one-to-one because different inputs always result in different outputs. Therefore g has an inverse function: $g^{-1}(x) = \frac{x}{5}$.

> **ONE-TO-ONE FUNCTION**
>
> A function f is a **one-to-one function** if, for elements c and d in the domain of f,
>
> $$c \neq d \quad \text{implies} \quad f(c) \neq f(d).$$
>
> That is, different inputs always result in different outputs.

NOTE A function f is one-to-one if equal outputs always have the same input. This statement can be written as $f(c) = f(d)$ **implies** $c = d$, which is an equivalent definition.

EXAMPLE 3   **Determining if a function is one-to-one graphically**

Use each graph to determine if f is one-to-one and if f has an inverse function.

(a)
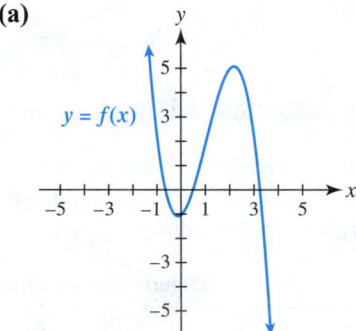

(b)
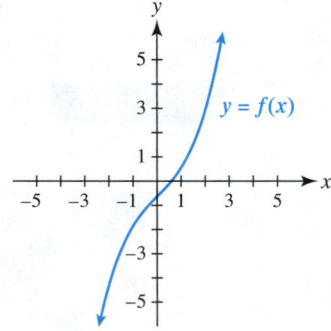

SOLUTION

(a) In Figure 5.13 the horizontal line $y = 2$ intersects the graph of f at $(-1, 2)$, $(1, 2)$, and $(3, 2)$. This means that $f(-1) = f(1) = f(3) = 2$. Three distinct inputs, $-1, 1$, and 3, produce the same output, 2. Therefore f is *not* one-to-one and does *not* have an inverse function.

(b) See Figure 5.14. Because every horizontal line intersects the graph at most once, different inputs (x-values) always result in different outputs (y-values). Therefore f *is* one-to-one and *has* an inverse function.

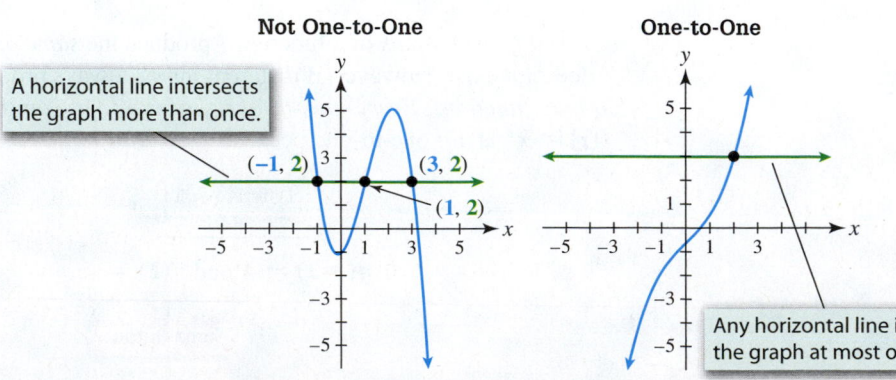

Figure 5.13   **Figure 5.14**

Now Try Exercises 13 and 15

NOTE To show that f is not one-to-one, it is not necessary to find the actual points of intersection—we have to show only that a horizontal line can intersect the graph of f more than once.

The technique of visualizing horizontal lines to determine if a graph represents a one-to-one function is called the *horizontal line test*.

CLASS DISCUSSION

Use the horizontal line test to explain why a nonconstant linear function has an inverse function, whereas a quadratic function does not.

HORIZONTAL LINE TEST

If every horizontal line intersects the graph of a function f at most once, then f is a one-to-one function.

Increasing, Decreasing, and One-to-One Functions If a continuous function f is always increasing on its domain, then every horizontal line will intersect the graph of f at most once. By the horizontal line test, f is a one-to-one function. For example, the function f shown in Example 3(b) is always increasing on its domain and so it is one-to-one. Similarly, if a continuous function g is only decreasing on its domain, then g is a one-to-one function.

Symbolic Representations of Inverse Functions

If a function f is one-to-one, then an inverse function f^{-1} exists. Therefore $f(a) = b$ implies $f^{-1}(b) = a$ for every a in the domain of f. That is,

$$(f^{-1} \circ f)(a) = f^{-1}(f(a)) = f^{-1}(b) = a.$$

Similarly, $f^{-1}(b) = a$ implies $f(a) = b$ for every b in the domain of f^{-1} and so

$$(f \circ f^{-1})(b) = f(f^{-1}(b)) = f(a) = b.$$

These two properties can be used to define an inverse function.

INVERSE FUNCTION

Let f be a one-to-one function. Then f^{-1} is the **inverse function** of f if

$(f^{-1} \circ f)(x) = f^{-1}(f(x)) = x$ for every x in the domain of f and

$(f \circ f^{-1})(x) = f(f^{-1}(x)) = x$ for every x in the domain of f^{-1}.

In the next two examples, we find an inverse function and verify that it is correct.

EXAMPLE 4   **Finding and verifying an inverse function**

Let f be the one-to-one function given by $f(x) = x^3 - 2$.
(a) Find a formula for $f^{-1}(x)$.   **(b)** Identify the domain and range of f^{-1}.
(c) Verify that your result from part (a) is correct.

SOLUTION
(a) Since $f(x) = x^3 - 2$, function f **cubes** the input x and then **subtracts 2**. To reverse this calculation, the inverse function must **add 2** to the input x and then take the **cube root**. That is, $f^{-1}(x) = \sqrt[3]{x + 2}$. An important symbolic technique for finding $f^{-1}(x)$ is to solve the equation $y = f(x)$ for x.

$$y = x^3 - 2 \qquad \textit{y = f(x); now solve for x.}$$

$$y + 2 = x^3 \qquad \textit{Add 2.}$$

$$\sqrt[3]{y + 2} = x \qquad \textit{Take the cube root.}$$

Interchange x and y to obtain $y = \sqrt[3]{x + 2}$. This gives us the formula for $f^{-1}(x)$.

(b) Both the domain and the range of the cube root function include all real numbers. The graph of $f^{-1}(x) = \sqrt[3]{x} + 2$ is the graph of the cube root function shifted left 2 units. Therefore the domain and range of f^{-1} also include all real numbers.

(c) To verify that $f^{-1}(x) = \sqrt[3]{x} + 2$ is indeed the inverse of $f(x) = x^3 - 2$, we must show that $\mathbf{f^{-1}(f(x)) = x}$ and that $\mathbf{f(f^{-1}(x)) = x}$.

$$
\begin{aligned}
f^{-1}(f(x)) &= f^{-1}(x^3 - 2) && f(x) = x^3 - 2 \\
&= \sqrt[3]{(x^3 - 2) + 2} && f^{-1}(x) = \sqrt[3]{x} + 2 \\
&= \sqrt[3]{x^3} && \text{Combine terms.} \\
&= x \;\checkmark && \text{Simplify.} \\
f(f^{-1}(x)) &= f\left(\sqrt[3]{x} + 2\right) && f^{-1}(x) = \sqrt[3]{x} + 2 \\
&= \left(\sqrt[3]{x} + 2\right)^3 - 2 && f(x) = x^3 - 2 \\
&= (x + 2) - 2 && \text{Cube the expression.} \\
&= x \;\checkmark && \text{Combine terms.}
\end{aligned}
$$

These calculations verify that our result is correct.

> **Now Try Exercise 73**

The symbolic technique used in Example 4(a) is now summarized.

FINDING A SYMBOLIC REPRESENTATION FOR f^{-1}

To find a formula for f^{-1}, perform the following steps.

STEP 1: Verify that f is a one-to-one function.

STEP 2: Solve the equation $y = f(x)$ for x, obtaining the equation $x = f^{-1}(y)$.

STEP 3: Interchange x and y to obtain $y = f^{-1}(x)$.

To verify $f^{-1}(x)$, show that $(f^{-1} \circ f)(x) = x$ and $(f \circ f^{-1})(x) = x$.

> **NOTE** One reason for interchanging the variables in Step 3 is to make it easier to graph $y = f^{-1}(x)$ in the xy-plane.

EXAMPLE 5 **Finding an inverse function**

The function $f(x) = \frac{3}{4}x + 39$ gives the percentage of China's population that may live in urban areas x years after 2000, where $0 \le x \le 40$.
(a) Explain why f is a one-to-one function.
(b) Find a formula for $f^{-1}(x)$.
(c) Evaluate and interpret the meaning of $f^{-1}(60)$.

SOLUTION
(a) Since f is a linear function, its graph is a line with a nonzero slope of $\frac{3}{4}$. Every horizontal line intersects it at most once. By the horizontal line test, f is one-to-one.
(b) To find $f^{-1}(x)$, solve the equation $y = f(x)$ for x.

$$
\begin{aligned}
y &= \frac{3}{4}x + 39 && y = f(x) \\
y - 39 &= \frac{3}{4}x && \text{Subtract 39.} \\
\frac{4}{3}(y - 39) &= x && \text{Multiply by } \tfrac{4}{3}, \text{ the reciprocal of } \tfrac{3}{4}. \\
\frac{4}{3}y - 52 &= x && \text{Distributive property}
\end{aligned}
$$

Now interchange x and y to obtain $y = \frac{4}{3}x - 52$. The formula for the inverse is, therefore,

$$f^{-1}(x) = \frac{4}{3}x - 52.$$

(c) Substitute 60 for x in $f^{-1}(x) = \frac{4}{3}x - 52$.

$$f^{-1}(\textcolor{red}{60}) = \tfrac{4}{3}(\textcolor{red}{60}) - 52 = \textcolor{blue}{28}$$

The expression $f^{-1}(x)$ predicts the number of years after 2000 when x percent of China's population will live in urban areas. In 20**28**, it is estimated that **60** percent of China's population will live in urban areas.

> **Now Try Exercise 123**

EXAMPLE 6   **Restricting the domain of a function**

Let $f(x) = (x - 1)^2$.
(a) Does f have an inverse function? Explain.
(b) Restrict the domain of f so that f^{-1} exists.
(c) Find $f^{-1}(x)$ for the restricted domain.

SOLUTION
(a) The graph of $f(x) = (x - 1)^2$, shown in Figure 5.15, does not pass the horizontal line test. Therefore f is not one-to-one and does not have an inverse function.

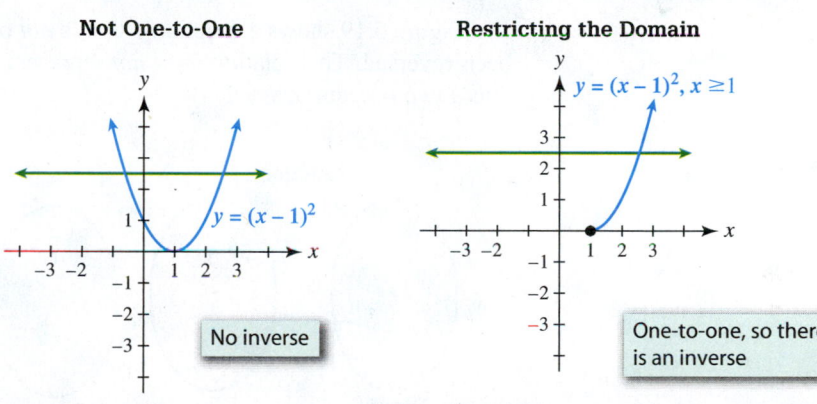

Figure 5.15   Figure 5.16

(b) If we restrict the domain of f to $D = \{x \mid x \geq 1\}$, then f becomes a one-to-one function. To illustrate this, the graph of $y = (x - 1)^2$ for $x \geq 1$ is shown in Figure 5.16. This graph passes the horizontal line test and f^{-1} exists on the restricted domain.
(c) Assume that $x \geq 1$ and solve the equation $y = f(x)$ for x.

$$y = (x - 1)^2 \qquad \textcolor{blue}{y = f(x)}$$
$$\sqrt{y} = x - 1 \qquad \textcolor{blue}{\text{Take the positive square root.}}$$
$$\textcolor{blue}{\text{Note: } x \geq 1 \text{ implies that } x - 1 \geq 0.}$$
$$\sqrt{y} + 1 = x \qquad \textcolor{blue}{\text{Add 1.}}$$

Thus $f^{-1}(x) = \sqrt{x} + 1$. $\textcolor{blue}{\text{Write the formula for } f^{-1}(x).}$

> **Now Try Exercise 63**

> **NOTE**   In Example 6 we could have restricted the domain to $x \leq 1$, rather than $x \geq 1$. In this case, we would obtain the left half of the parabola, which would also represent a one-to-one function that has an inverse.

College Graduates (%)

x	1940	1970	2010
$f(x)$	5	11	30

Table 5.12

x	5	11	30
$f^{-1}(x)$	1940	1970	2010

Table 5.13

Other Representations of Inverse Functions

Tables and graphs of a one-to-one function can also be used to find its inverse.

Numerical Representations In Table 5.12, f has domain $D = \{1940, 1970, 2010\}$ and it computes the percentage of the U.S. population with 4 or more years of college in year x.

Function f is one-to-one because different inputs always produce different outputs. Therefore f^{-1} exists. Since $f(1940) = 5$, it follows that $f^{-1}(5) = 1940$. Similarly, $f^{-1}(11) = 1970$ and $f^{-1}(30) = 2010$. Table 5.13 shows a table for f^{-1}.

The domain of f is $\{1940, 1970, 2010\}$ and the range of f is $\{5, 11, 30\}$. The domain of f^{-1} is $\{5, 11, 30\}$ and the range of f^{-1} is $\{1940, 1970, 2010\}$. The functions f and f^{-1} interchange domains and ranges. Figures 5.17 and 5.18 demonstrate this property.

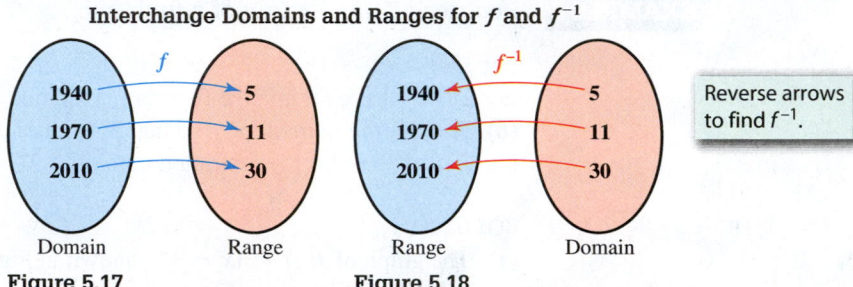

Interchange Domains and Ranges for f and f^{-1}

Reverse arrows to find f^{-1}.

| **Figure 5.17** | **Figure 5.18** |

Figure 5.19 shows a function f that is *not* one-to-one. In Figure 5.20 the arrows have been reversed. This relation does *not* represent the inverse *function* because input 4 produces two outputs, 1 and 2.

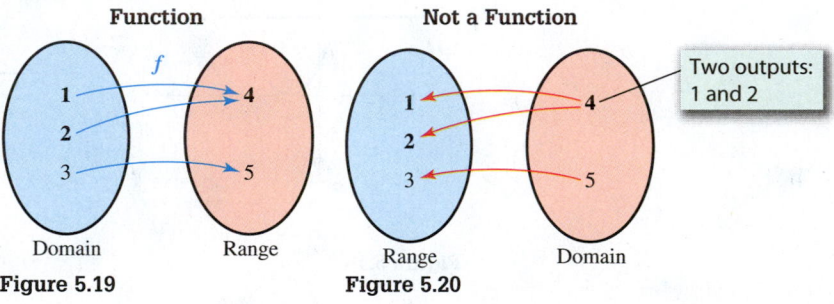

Function   **Not a Function**

Two outputs: 1 and 2

| **Figure 5.19** | **Figure 5.20** |

The relationship between domains and ranges is summarized in the following box.

DOMAINS AND RANGES OF INVERSE FUNCTIONS

The domain of f equals the range of f^{-1}.
The range of f equals the domain of f^{-1}.

Graphical Representations If the point $(2, 5)$ lies on the graph of f, then $f(2) = 5$ and $f^{-1}(5) = 2$. Therefore the point $(5, 2)$ must lie on the graph of f^{-1}. In general, if the point (a, b) lies on the graph of f, then the point (b, a) lies on the graph of f^{-1}. Refer to Figure 5.21. If a line segment is drawn between the points (a, b) and (b, a), the line $y = x$ is a perpendicular bisector of this line segment. Figure 5.22 shows pairs of points in the form (a, b) and (b, a). Figure 5.23 shows continuous graphs of f and f^{-1} passing through these points. The graph of f^{-1} is a *reflection* of the graph of f across the line $y = x$.

Reflecting to Find f^{-1}

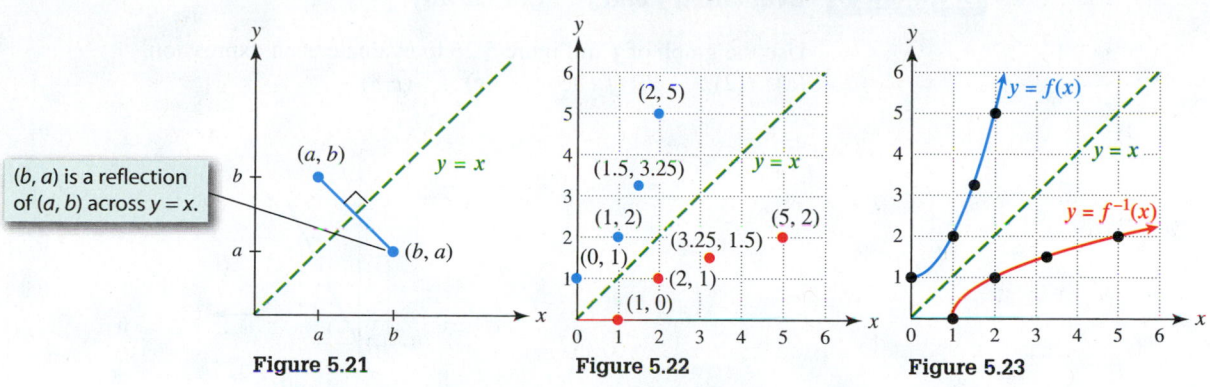

Figure 5.21   Figure 5.22   Figure 5.23

GRAPHS OF FUNCTIONS AND THEIR INVERSES

The graph of f^{-1} is a reflection of the graph of f across the line $y = x$.

EXAMPLE 7   **Representing an inverse function graphically**

Let $f(x) = x^3 + 2$. Graph f. Then sketch a graph of f^{-1}.

SOLUTION   Figure 5.24 shows a graph of f. To sketch a graph of f^{-1}, reflect the graph of f across the line $y = x$. The graph of f^{-1} appears as though it were the "reflection" of the graph of f in a mirror located along $y = x$. See Figure 5.25.

$[-5, 5, 1]$ by $[-5, 5, 1]$

Shift $y = x^3$ upward 2 units.

$y = x^3 + 2$

Figure 5.24

Calculator Help

To graph an inverse function, see Appendix A (page AP-11).

f^{-1} is a reflection of f across $y = x$.

Figure 5.25

Now Try Exercise 113

The following See the Concept shows how to represent and find inverse functions.

See the Concept: Representing Inverse Functions

Verbal

f: Multiply x by 2 and add 1.

f^{-1}: Subtract 1 from x and divide by 2.

Use the inverse operations in the reverse order.

Symbolic

$f(x) = 2x + 1$

To find $f^{-1}(x)$:

$y = 2x + 1$

$y - 1 = 2x$

$\dfrac{y - 1}{2} = x$

$f^{-1}(x) = \dfrac{x - 1}{2}$

Solve for x.

Numerical

x	$f(x)$
-2	-3
-1	-1
0	1
1	3
2	5

x	$f^{-1}(x)$
-3	-2
-1	-1
1	0
3	1
5	2

Interchange the x-and y-values.

Graphical

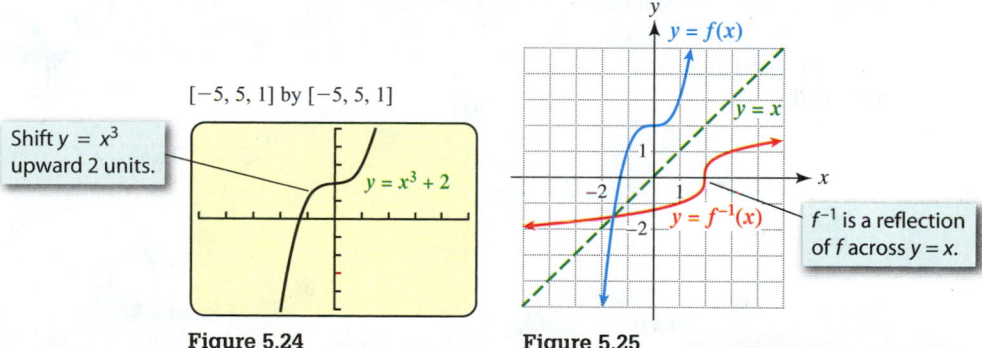

$f(x) = 2x + 1$

$y = x$

$f^{-1}(x) = \dfrac{x - 1}{2}$

Reflect the graph of f across $y = x$.

EXAMPLE 8   Evaluating f and f^{-1} graphically

Use the graph of f in Figure 5.26 to evaluate each expression.
(a) $f(2)$ **(b)** $f^{-1}(3)$ **(c)** $f^{-1}(-3)$

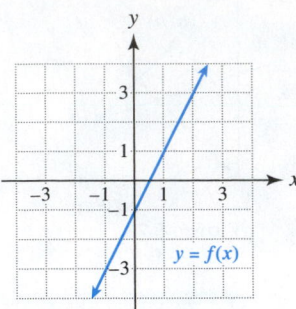

Figure 5.26

SOLUTION

Getting Started   To evaluate $f(a)$ graphically, find a on the **x-axis**. Move upward or downward to the graph of f and determine the corresponding y-value. To evaluate $f^{-1}(b)$ graphically, find b on the **y-axis**. Move left or right to the graph of f and determine the corresponding x-value. ▶

(a)   To evaluate $f(2)$, find **2** on the x-axis, move upward to the graph of f and then move left to the y-axis to obtain $f(2) = $ **3**, as shown in Figure 5.27.

$f(2) = $ **3**

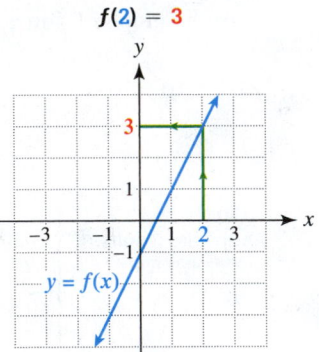

Figure 5.27

(b)   Start by finding **3** on the y-axis, move right to the graph of f, and then move downward to the x-axis to obtain $f^{-1}(3) = $ **2**, as shown in Figure 5.28. Notice that $f(2) = $ **3** from part (a) and $f^{-1}(3) = $ **2** here.

(c)   Find -3 on the y-axis, move left to the graph of f, and then move upward to the x-axis. We can see from Figure 5.29 that $f^{-1}(-3) = $ **−1**.

$f^{-1}(3) = $ **2**

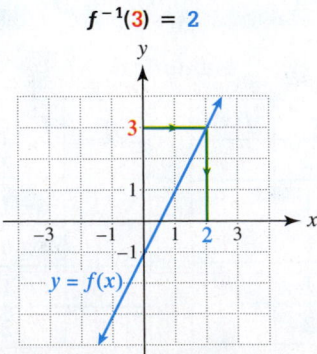

Figure 5.28

$f^{-1}(-3) = $ **−1**

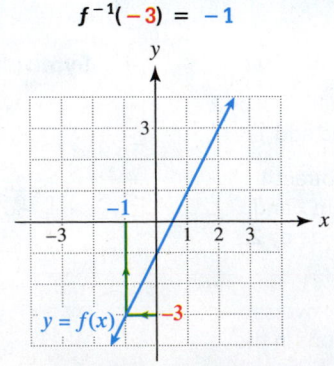

Figure 5.29

Now Try Exercise 101

5.2 Putting It All Together

The following table summarizes some important concepts about inverse functions.

CONCEPT	COMMENTS	EXAMPLES
One-to-one function	f is one-to-one if different inputs always result in different outputs. That is, $a \neq b$ implies $f(a) \neq f(b)$.	$f(x) = x^2 - 4x$ is not one-to-one because $f(0) = 0$ and $f(4) = 0$. With this function, *different* inputs can result in the *same* output.
Horizontal line test	If every horizontal line intersects the graph of f at most once, then f is one-to-one.	Not one-to-one
Inverse function	If a function f is one-to-one, it has an inverse function f^{-1} that satisfies both $$(f^{-1} \circ f)(x) = f^{-1}(f(x)) = x$$ and $$(f \circ f^{-1})(x) = f(f^{-1}(x)) = x.$$	$f(x) = 3x - 1$ is one-to-one and has inverse function $f^{-1}(x) = \frac{x+1}{3}$. $$f^{-1}(f(x)) = f^{-1}(3x - 1)$$ $$= \frac{(3x - 1) + 1}{3}$$ $$= x$$ Similarly, $$f(f^{-1}(x)) = x.$$
Domains and ranges of inverse functions	The domain of f equals the range of f^{-1}. The range of f equals the domain of f^{-1}.	Let $f(x) = (x + 2)^2$ with restricted domain $x \geq -2$ and range $y \geq 0$. It follows that $f^{-1}(x) = \sqrt{x} - 2$ with domain $x \geq 0$ and range $y \geq -2$.

5.2 Exercises

Inverse Operations

Exercises 1–4: State the inverse action or actions.

1. Opening a window

2. Climbing up a ladder

3. Walking into a classroom, sitting down, and opening a book

4. Opening the door and turning on the lights

Exercises 5–12: Describe verbally the inverse of the statement. Then express both the given statement and its inverse symbolically.

5. Add 2 to x.

6. Multiply x by 5.

7. Subtract 2 from x and multiply the result by 3.

8. Divide x by 20 and then add 10.

9. Take the cube root of x and add 1.

10. Multiply x by -2 and add 3.

11. Take the reciprocal of a nonzero number x.

12. Take the square root of a positive number x.

One-to-One Functions

Exercises 13–18: Use the graph of $y = f(x)$ to determine if f is one-to-one.

13.
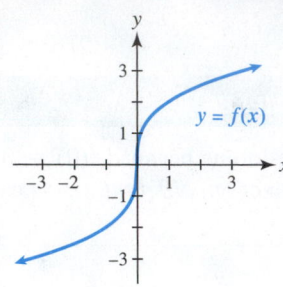

14.
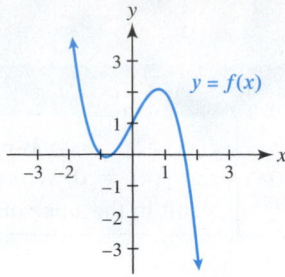

15.
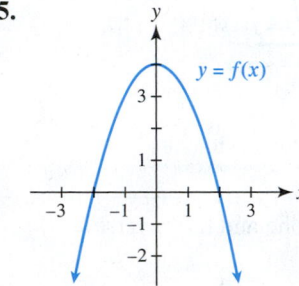

16.
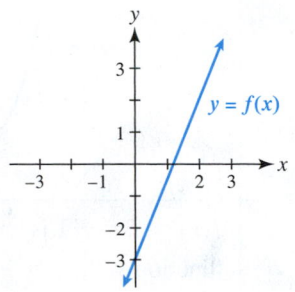

17.
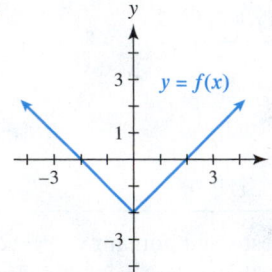

18.
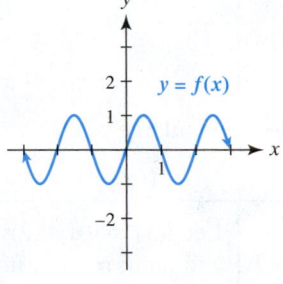

Exercises 19–22: The table is a complete representation of f. Use the table to determine if f is one-to-one and has an inverse.

19.

x	1	2	3	4
$f(x)$	4	3	3	5

20.

x	-2	0	2	4
$f(x)$	4	2	0	-2

21.

x	0	2	4	6	8
$f(x)$	-1	0	4	1	-3

22.

x	-2	-1	0	1	2
$f(x)$	4	1	0	1	4

Exercises 23–36: Determine if f is one-to-one. You may want to graph $y = f(x)$ and apply the horizontal line test.

23. $f(x) = 2x - 7$

24. $f(x) = x^2 - 1$

25. $f(x) = -2x^2 + x$

26. $f(x) = 4 - \frac{3}{4}x$

27. $f(x) = x^4$

28. $f(x) = |2x - 5|$

29. $f(x) = |x - 1|$

30. $f(x) = x^3$

31. $f(x) = \dfrac{1}{1 + x^2}$

32. $f(x) = \dfrac{1}{x}$

33. $f(x) = 3x - x^3$

34. $f(x) = x^{2/3}$

35. $f(x) = x^{1/2}$

36. $f(x) = x^3 - 4x$

Exercises 37–40: **Modeling** *Decide if the situation could be modeled by a one-to-one function.*

37. The distance between the ground and a person who is riding a Ferris wheel after x seconds

38. The cumulative numbers of AIDS cases from 1980 to 2010

39. The population of the United States from 1980 to 2010

40. The height y of a stone thrown upward after x seconds

Symbolic Representations of Inverse Functions

Exercises 41–62: Find a symbolic representation for $f^{-1}(x)$.

41. $f(x) = \sqrt[3]{x}$

42. $f(x) = 2x$

43. $f(x) = -2x + 10$

44. $f(x) = x^3 + 2$

45. $f(x) = 3x - 1$

46. $f(x) = \dfrac{x - 1}{2}$

47. $f(x) = 2x^3 - 5$

48. $f(x) = 1 - \frac{1}{2}x^3$

49. $f(x) = x^2 - 1, x \geq 0$

50. $f(x) = (x + 2)^2, x \leq -2$

51. $f(x) = \dfrac{1}{2x}$

52. $f(x) = \dfrac{2}{\sqrt{x}}$

53. $f(x) = \frac{1}{2}(4 - 5x) + 1$

54. $f(x) = 6 - \frac{3}{4}(2x - 4)$

55. $f(x) = \dfrac{x}{x + 2}$

56. $f(x) = \dfrac{3x}{x - 1}$

57. $f(x) = \dfrac{2x + 1}{x - 1}$

58. $f(x) = \dfrac{1 - x}{3x + 1}$

59. $f(x) = \dfrac{1}{x} - 3$

60. $f(x) = \dfrac{1}{x + 5} + 2$

61. $f(x) = \dfrac{1}{x^3 - 1}$

62. $f(x) = \dfrac{2}{2 - x^3}$

Exercises 63–70: Restrict the domain of f(x) so that f is one-to-one. Then find $f^{-1}(x)$. Answers may vary.

63. $f(x) = 4 - x^2$

64. $f(x) = 2(x + 3)^2$

65. $f(x) = (x - 2)^2 + 4$

66. $f(x) = x^4 - 1$

67. $f(x) = x^{2/3} + 1$

68. $f(x) = 2(x + 3)^{2/3}$

69. $f(x) = \sqrt{9 - 2x^2}$

70. $f(x) = \sqrt{25 - x^2}$

Exercises 71–84: Find a formula for $f^{-1}(x)$. Identify the domain and range of f^{-1}. Verify that f and f^{-1} are inverses.

71. $f(x) = 5x - 15$

72. $f(x) = (x + 3)^2, x \geq -3$

73. $f(x) = \sqrt[3]{x - 5}$

74. $f(x) = 6 - 7x$

75. $f(x) = \dfrac{x - 5}{4}$

76. $f(x) = \dfrac{x + 2}{9}$

77. $f(x) = \sqrt{x - 5}, x \geq 5$

78. $f(x) = \sqrt{5 - 2x}, x \leq \frac{5}{2}$

79. $f(x) = \dfrac{1}{x + 3}$

80. $f(x) = \dfrac{2}{x - 1}$

81. $f(x) = 2x^3$

82. $f(x) = 1 - 4x^3$

83. $f(x) = x^2, x \geq 0$

84. $f(x) = \sqrt[3]{1 - x}$

Numerical Representations of Inverse Functions

Exercises 85–88: Use the table for f(x) to find a table for $f^{-1}(x)$. Identify the domains and ranges of f and f^{-1}.

85.

x	1	2	3
f(x)	5	7	9

86.

x	1	10	100
f(x)	0	1	2

87.

x	0	2	4
f(x)	0	4	16

88.

x	0	1	2
f(x)	1	2	4

Exercises 89 and 90: Use f(x) to complete the table.

89. $f(x) = 4x$

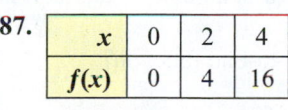

x	0	2	4	6
$f^{-1}(x)$				

90. $f(x) = x^3$

x	−8	−1	8	27
$f^{-1}(x)$				

Exercises 91–98: Use the tables to evaluate the following.

x	0	1	2	3	4
f(x)	1	3	5	4	2

x	−1	1	2	3	4
g(x)	0	2	1	4	5

91. $f^{-1}(3)$

92. $f^{-1}(5)$

93. $g^{-1}(4)$

94. $g^{-1}(0)$

95. $(f \circ g^{-1})(1)$

96. $(g^{-1} \circ g^{-1})(2)$

97. $(g \circ f^{-1})(5)$

98. $(f^{-1} \circ g)(4)$

Graphs and Inverse Functions

99. Interpreting an Inverse The graph of f computes the balance in a savings account after x years. Estimate each expression. Interpret what $f^{-1}(x)$ computes.

(a) $f(1)$ (b) $f^{-1}(110)$ (c) $f^{-1}(160)$

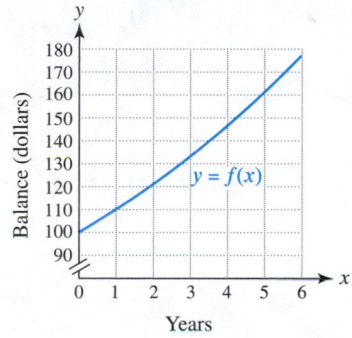

100. Interpreting an Inverse The graph of f computes the Celsius temperature of a pan of water after x minutes. Estimate each expression. Interpret what the expression $f^{-1}(x)$ computes.

(a) $f(4)$ (b) $f^{-1}(90)$ (c) $f^{-1}(80)$

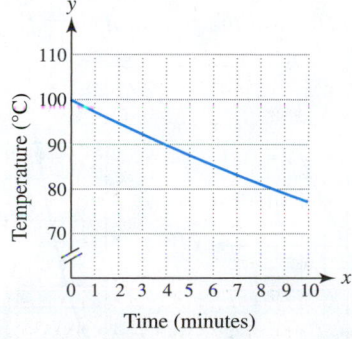

Exercises 101–104: Use the graph to evaluate the expression.

101.

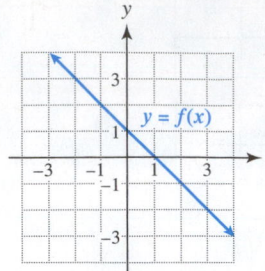

(a) $f(-1)$

(b) $f^{-1}(-2)$

(c) $f^{-1}(0)$

(d) $(f^{-1} \circ f)(3)$

102.

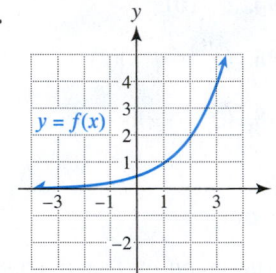

(a) $f(1)$

(b) $f^{-1}(1)$

(c) $f^{-1}(4)$

(d) $(f \circ f^{-1})(2.5)$

103.

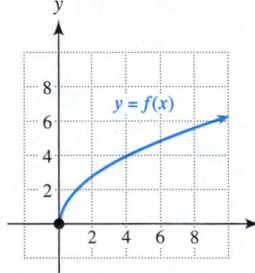

(a) $f(4)$

(b) $f^{-1}(0)$

(c) $f^{-1}(6)$

(d) $(f \circ f^{-1})(4)$

104.

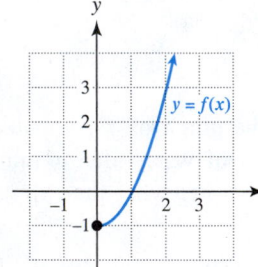

(a) $f(1)$

(b) $f^{-1}(-1)$

(c) $f^{-1}(3)$

(d) $(f \circ f^{-1})(1)$

Exercises 105–110: Use the graph of $y = f(x)$ to sketch a graph of $y = f^{-1}(x)$.

105.

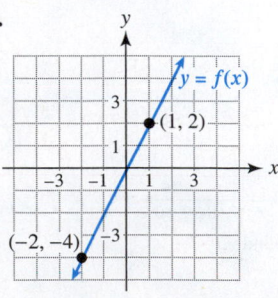

106.

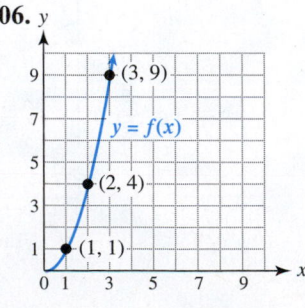

107.

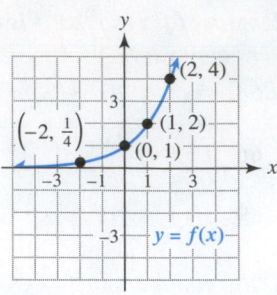

108.

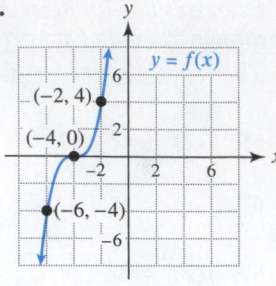

109.

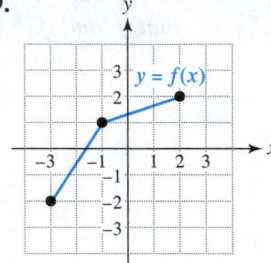

110.

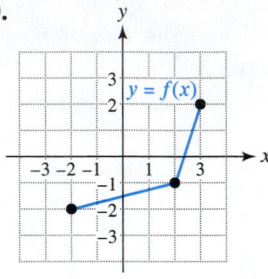

Exercises 111–116: Graph $y = f(x)$ and $y = x$. Then graph $y = f^{-1}(x)$

111. $f(x) = 2x - 1$

112. $f(x) = -\frac{1}{2}x + 1$

113. $f(x) = x^3 - 1$

114. $f(x) = \sqrt[3]{x - 1}$

115. $f(x) = (x + 1)^2, x \geq -1$

116. $f(x) = \sqrt{x + 1}$

Exercises 117–120: Graph $y = f(x)$, $y = f^{-1}(x)$, and $y = x$ in a square viewing rectangle such as $[-4.7, 4.7, 1]$ by $[-3.1, 3.1, 1]$.

117. $f(x) = 3x - 1$

118. $f(x) = \dfrac{3 - x}{2}$

119. $f(x) = \frac{1}{3}x^3 - 1$

120. $f(x) = \sqrt[3]{x - 1}$

Applications

121. **Volume** The volume V of a sphere with radius r is given by $V = \frac{4}{3}\pi r^3$.

(a) Does V represent a one-to-one function?

(b) What does the inverse of V compute?

(c) Find a formula for the inverse.

(d) Normally we interchange x and y to find the inverse function. Does it make sense to interchange V and r in part (c) of this exercise? Explain.

122. **Temperature** The formula $F = \frac{9}{5}C + 32$ converts a Celsius temperature to Fahrenheit temperature.

(a) Find a formula for the inverse.

(b) Normally we interchange x and y to find the inverse function. Does it makes sense to interchange F and C in part (a) of this exercise? Explain.

(c) What Celsius temperature is equivalent to $68°F$?

123. Height and Weight   The formula $W = \frac{25}{7}h - \frac{800}{7}$ approximates the recommended minimum weight for a person h inches tall, where $62 \le h \le 76$.
(a) What is the recommended minimum weight for someone 70 inches tall?

(b) Does W represent a one-to-one function?

(c) Find a formula for the inverse.

(d) Evaluate the inverse for 150 pounds and interpret the result.

(e) What does the inverse compute?

124. Planetary Orbits   The formula $T(x) = x^{3/2}$ calculates the time in years that it takes a planet to orbit the sun if the planet is x times farther from the sun than Earth is.
(a) Find the inverse of T.

(b) What does the inverse of T calculate?

125. Converting Units   The tables represent a function F that converts yards to feet and a function Y that converts miles to yards. Evaluate each expression and interpret the results.

x (yd)	1760	3520	5280	7040	8800
$F(x)$ (ft)	5280	10,560	15,840	21,120	26,400

x (mi)	1	2	3	4	5
$Y(x)$ (yd)	1760	3520	5280	7040	8800

(a) $(F \circ Y)(2)$

(b) $F^{-1}(26,400)$

(c) $(Y^{-1} \circ F^{-1})(21,120)$

126. Converting Units   (Refer to Exercise 125.)
(a) Find formulas for $F(x)$, $Y(x)$, and $(F \circ Y)(x)$.

(b) Find a formula for $(Y^{-1} \circ F^{-1})(x)$. What does this function compute?

127. Converting Units   The tables at the top of the next column represent a function C that converts tablespoons to cups and a function Q that converts cups to quarts. Evaluate each expression and interpret the results.

x (tbsp)	32	64	96	128
$C(x)$ (c)	2	4	6	8

x (c)	2	4	6	8
$Q(x)$ (qt)	0.5	1	1.5	2

(a) $(Q \circ C)(96)$

(b) $Q^{-1}(2)$

(c) $(C^{-1} \circ Q^{-1})(1.5)$

128. Rise in Sea Level   The global sea level could rise due to partial melting of the polar ice caps. The table represents a function R that models this expected rise in sea level in centimeters for the year t. (This model assumes no changes in current trends.)

t (yr)	1990	2000	2030	2070	2100
$R(t)$ (cm)	0	1	18	44	66

Source: A. Nilsson, Greenhouse Earth.

(a) Is R a one-to-one function? Explain.

(b) Use $R(t)$ to find a table for $R^{-1}(t)$. Interpret R^{-1}.

Writing about Mathematics

129. Explain how to find verbal, numerical, graphical, and symbolic representations of an inverse function. Give examples.

130. Can a one-to-one function have more than one x-intercept or more than one y-intercept? Explain.

131. If the graphs of $y = f(x)$ and $y = f^{-1}(x)$ intersect at a point (a, b), what can be said about these graphs? Explain.

132. If $f(x) = ax^2 + bx + c$ with $a \ne 0$, does $f^{-1}(x)$ exist? Explain.

Extended and Discovery Exercises

1. Interpreting an Inverse   Let $f(x)$ compute the height in feet of a rocket after x seconds of upward flight.
(a) Explain what $f^{-1}(x)$ computes.

(b) Interpret the solution to the equation $f(x) = 5000$.

(c) Explain how to solve the equation in part (b) using $f^{-1}(x)$.

2. If the graph of f lies entirely in quadrants I and II, in which quadrant(s) does the graph of f^{-1} lie?

CHECKING BASIC CONCEPTS FOR SECTIONS 5.1 AND 5.2

1. Use the table to evaluate each expression, if possible.

x	-2	-1	0	1	2
$f(x)$	0	1	-2	-1	2
$g(x)$	1	-2	-1	2	0

(a) $(f + g)(1)$

(b) $(f - g)(-1)$

(c) $(fg)(0)$

(d) $(f/g)(2)$

(e) $(f \circ g)(2)$

(f) $(g \circ f)(-2)$

2. Use the graph to evaluate each expression, if possible.

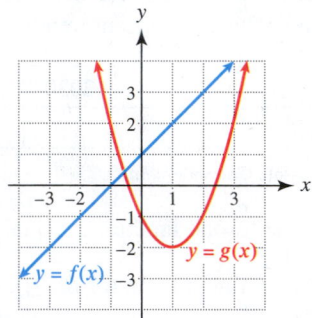

(a) $(f + g)(1)$

(b) $(g - f)(0)$

(c) $(fg)(2)$

(d) $(g/f)(-1)$

(e) $(f \circ g)(2)$

(f) $(g \circ f)(1)$

3. Let $f(x) = x^2 + 3x - 2$ and $g(x) = 3x - 1$. Find each expression.

(a) $(f + g)(x)$ **(b)** $(f/g)(x)$ **(c)** $(f \circ g)(x)$

4. If $f(x) = 5 - 2x$, find $f^{-1}(x)$.

5. Use the graph in Exercise 2 to answer the following.
(a) Is f one-to-one? Does f^{-1} exist? If so, find it.

(b) Is g one-to-one? Does g^{-1} exist? If so, find it.

6. Graph $f(x) = \sqrt[3]{x}$ and $y = x$. Then graph $y = f^{-1}(x)$.

7. Use the table in Exercise 1 to evaluate the following. (Assume that f^{-1} exists.)
(a) $f^{-1}(-2)$ **(b)** $(f^{-1} \circ g)(1)$

8. Use the graph in Exercise 2 to evaluate the following.
(a) $f^{-1}(2)$ **(b)** $(f^{-1} \circ g)(0)$

5.3 Exponential Functions and Models

- **Distinguish between linear and exponential growth**
- **Recognize exponential growth and decay**
- **Calculate compound interest**
- **Use the natural exponential function in applications**
- **Model data with exponential functions**

Introduction

Suppose that we deposit $500 into a savings account that pays 2% annual interest. If none of the money or interest is withdrawn, then the account increases in value by 2% each year. When an amount A changes by a constant percentage over each fixed time period, such as a year or a month, then the growth (or possibly decay) in A can be described by an exponential model. In this section we use exponential functions to calculate interest and model real data.

Linear and Exponential Functions

A linear function g can be written as $g(x) = mx + b$, where m represents the rate of change. For example, Table 5.14 shows that each time x increases by 1 unit, $g(x)$ increases by 2 units. We can write a formula $g(x) = 2x + 3$ because the rate of change is **2** and $g(0) = 3$.

A Linear Function

x	0	1	2	3	4	5
$y = g(x)$	3	5	7	9	11	13

> Each time x increases by 1 unit, y increases by 2 units.

Table 5.14

An exponential function is different. Rather than *adding* a fixed amount to the previous y-value for each unit increase in x, an exponential function *multiplies* the previous y-value by a fixed amount for each unit increase in x. Table 5.15 shows an exponential function f, where consecutive y-values are found by multiplying the previous y-value by 2. (The y-values increase by 100% for each unit increase in x.)

An Exponential Function

x	0	1	2	3	4	5
$y = f(x)$	3	6	12	24	48	96

Each time x increases by 1 unit, y doubles.

Table 5.15

Compare the following patterns for calculating $g(x)$ and $f(x)$.

Linear Growth

Add 2 each step.

$g(0) = 3$

$g(1) = \underbrace{3}_{g(0)} + \mathbf{2} = 3 + 2 \cdot \mathbf{1} = 5$

$g(2) = \underbrace{3 + 2}_{g(1)} + \mathbf{2} = 3 + 2 \cdot \mathbf{2} = 7$

$g(3) = \underbrace{3 + 2 + 2}_{g(2)} + \mathbf{2} = 3 + 2 \cdot \mathbf{3} = 9$

$g(4) = \underbrace{3 + 2 + 2 + 2}_{g(3)} + \mathbf{2} = 3 + 2 \cdot \mathbf{4} = 11$

$g(5) = \underbrace{3 + 2 + 2 + 2 + 2}_{g(4)} + \mathbf{2} = 3 + 2 \cdot \mathbf{5} = 13$

Exponential Growth

Multiply by 2 each step.

$f(0) = 3$

$f(1) = \underbrace{3}_{f(0)} \cdot \mathbf{2} = 3 \cdot 2^1 = 6$

$f(2) = \underbrace{3 \cdot 2}_{f(1)} \cdot \mathbf{2} = 3 \cdot 2^2 = 12$

$f(3) = \underbrace{3 \cdot 2 \cdot 2}_{f(2)} \cdot \mathbf{2} = 3 \cdot 2^3 = 24$

$f(4) = \underbrace{3 \cdot 2 \cdot 2 \cdot 2}_{f(3)} \cdot \mathbf{2} = 3 \cdot 2^4 = 48$

$f(5) = \underbrace{3 \cdot 2 \cdot 2 \cdot 2 \cdot 2}_{f(4)} \cdot \mathbf{2} = 3 \cdot 2^5 = 96$

Notice that if x is a positive integer then

$$g(x) = 3 + \underbrace{(\mathbf{2} + \mathbf{2} + \cdots + \mathbf{2})}_{x \text{ terms}} \quad \text{and} \quad f(x) = 3 \cdot \underbrace{(\mathbf{2} \cdot \mathbf{2} \cdots \cdots \mathbf{2})}_{x \text{ factors}}.$$

Using these patterns, we can write formulas for $g(x)$ and $f(x)$ as follows.

$$g(x) = 3 + \mathbf{2}x \quad \text{and} \quad f(x) = 3 \cdot \mathbf{2}^x$$

Linear function

Exponential function

This discussion gives motivation for the following definition.

> **EXPONENTIAL FUNCTION**
>
> A function f represented by
> $$f(x) = Ca^x, \quad \text{with} \quad a > 0, \quad a \neq 1, \quad \text{and} \quad C > 0,$$
> is an **exponential function with base a and coefficient C**.

Examples of exponential functions include

$$f(x) = \mathbf{3}^x, \quad g(x) = \mathbf{5}(\mathbf{1.7})^x, \quad \text{and} \quad h(x) = \mathbf{4}\left(\frac{\mathbf{1}}{\mathbf{2}}\right)^x.$$

$C = 1; a = 3$ $\qquad$ $C = 5; a = 1.7$ $\qquad$ $C = 4; a = \frac{1}{2}$

NOTE Some definitions for an exponential function require that $C = 1$. In this case, an exponential function is defined as $f(x) = a^x$.

The following Seeing the Concept compares the graph of the exponential function $f(x) = 2^x$ to the graph of the linear function $g(x) = 2x$. See Figures 5.30 and 5.31.

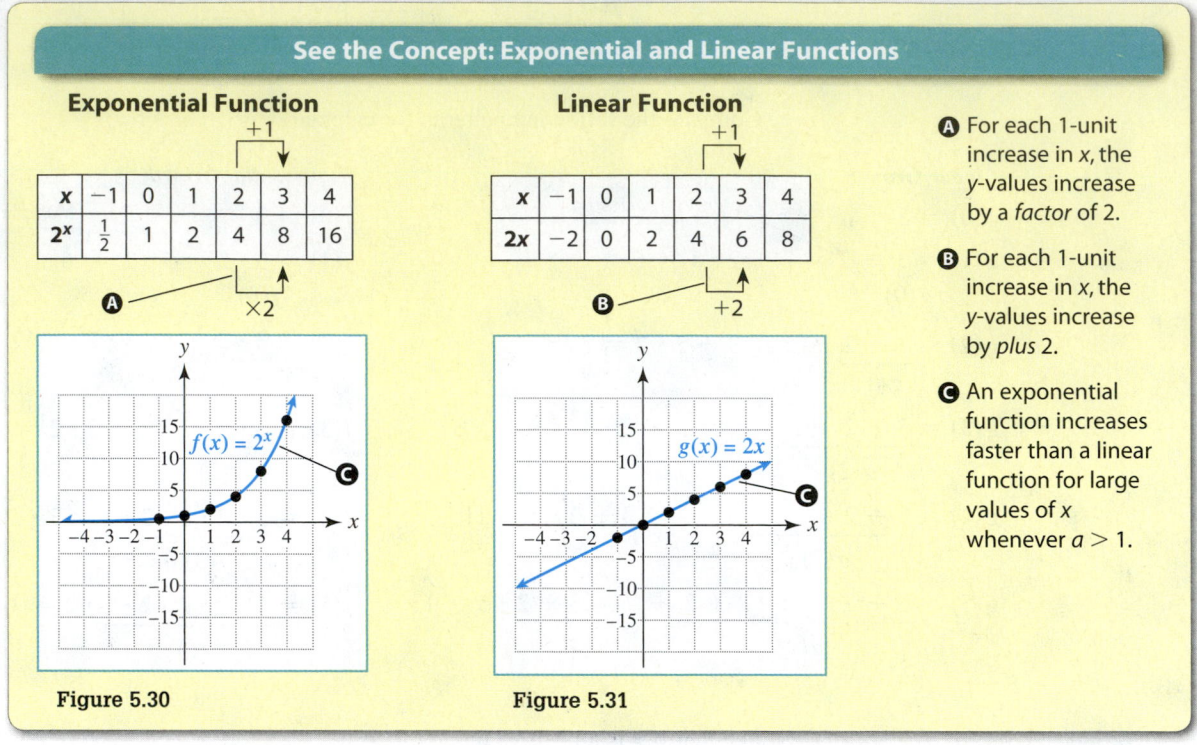

See the Concept: Exponential and Linear Functions

Exponential Function

x	-1	0	1	2	3	4
2^x	$\frac{1}{2}$	1	2	4	8	16

Linear Function

x	-1	0	1	2	3	4
$2x$	-2	0	2	4	6	8

A For each 1-unit increase in x, the y-values increase by a *factor* of 2.

B For each 1-unit increase in x, the y-values increase by *plus* 2.

C An exponential function increases faster than a linear function for large values of x whenever $a > 1$.

Figure 5.30

Figure 5.31

NOTE Any *nonzero* number raised to the 0 power equals 1. If $f(x) = Ca^x$, then $f(0) = Ca^0 = C(1) = C$. This means that

- C equals the value of $f(x)$ at $x = 0$,
- C equals the y-intercept on the graph of f, and
- C equals the *initial* value of $f(x)$ when x is time.

For example, if $g(x) = 5(1.7)^x$, then $g(0) = 5$, the y-intercept is 5, and the *initial* value of $g(x)$ is 5 when x is time.

> **MAKING CONNECTIONS**
>
> **Exponential Functions and Polynomial Functions**   An exponential function has a **variable** for an exponent, whereas a polynomial function has **constants** for exponents. For example, $f(x) = 3^x$ represents an exponential function, and $g(x) = x^3$ represents a polynomial function.

EXAMPLE 1   Recognizing linear and exponential data

For each table, find either a linear or an exponential function that models the data.

(a)

x	0	1	2	3
y	-3	-1.5	0	1.5

(b)

x	0	1	2	3	4
y	16	4	1	$\frac{1}{4}$	$\frac{1}{16}$

(c)

x	0	1	2	3
y	3	4.5	6.75	10.125

(d)

x	0	1	2	3	4
y	16	12	8	4	0

SOLUTION

(a) For each unit increase in x, the y-values increase by **1.5**, so the data are linear. Because $y = -3$ when $x = 0$, it follows that the data can be modeled by $f(x) = 1.5x - 3$.

(b) For each unit increase in x, the y-values are multiplied by $\frac{1}{4}$. This is an exponential function given by $f(x) = Ca^x$ with $C = f(0) = \mathbf{16}$ and $a = \frac{1}{4}$, so $f(x) = \mathbf{16}\left(\frac{1}{4}\right)^x$.

(c) Since the data do not change by a fixed amount *for each unit increase in x*, the data are not linear. To determine if the data are exponential, calculate ratios of consecutive y-values.

$$\frac{4.5}{3} = \mathbf{1.5}, \quad \frac{6.75}{4.5} = \mathbf{1.5}, \quad \frac{10.125}{6.75} = \mathbf{1.5}$$

For each unit increase in x, the next y-value in the table can be found by multiplying the previous y-value by $\mathbf{1.5}$, so let $a = \mathbf{1.5}$. Since $y = 3$ when $x = 0$, let $C = \mathbf{3}$. Thus $f(x) = \mathbf{3}(\mathbf{1.5})^x$.

(d) Note that $y = \mathbf{16}$ when $x = 0$. For each unit increase in x, the next y-value is found by adding -4 to the previous y-value, so $f(x) = -\mathbf{4}x + \mathbf{16}$.

> **Now Try Exercises 13 and 15**

Determining an Exponential Function Two points can be used to determine a line. In a similar way, the values of $f(x)$ at two points can be used to determine C and a for an exponential function. This technique is demonstrated in the next example.

EXAMPLE 2 **Finding exponential functions**

Find values for C and a so that $f(x) = Ca^x$ satisfies the conditions.
(a) $f(0) = 4$ and $f(1) = 8$ **(b)** $f(-1) = 8$ and $f(2) = 1$

SOLUTION
(a) $f(0) = 4$, so $C = \mathbf{4}$. Because $\frac{f(1)}{f(0)} = \frac{8}{4} = \mathbf{2}$, it follows that for each unit increase in x, the output is multiplied by $\mathbf{2}$. Thus $a = \mathbf{2}$ and $f(x) = \mathbf{4}(\mathbf{2})^x$.

(b) **Getting Started** Because we are not given $f(0)$, we cannot immediately determine C. Instead, first find a by evaluating the following ratios. ▶

$$\frac{f(2)}{f(-1)} = \frac{1}{8} \qquad \text{and} \qquad \frac{f(2)}{f(-1)} = \frac{Ca^2}{Ca^{-1}} = a^3 \qquad \color{blue}{f(x) = Ca^x;\ \text{subtract exponents.}}$$

It follows that $a^3 = \frac{1}{8}$, so $a = \frac{1}{2}$. Thus $f(x) = C\left(\frac{1}{2}\right)^x$. Next determine C by using the fact that $f(\mathbf{2}) = \mathbf{1}$.

$$f(\mathbf{2}) = C\left(\tfrac{1}{2}\right)^2 = \tfrac{1}{4}C = \mathbf{1}, \qquad \text{or} \qquad C = \mathbf{4}$$

Thus $f(x) = \mathbf{4}\left(\frac{1}{2}\right)^x$.

> **Now Try Exercises 29 and 33**

EXAMPLE 3 **Finding an exponential model**

The number of centenarians (people who are 100 years old or more) in India is projected to increase by a factor of 1.04 each year after 2010 until 2100.
(*Source:* UN Population Prospects, 2010 Revision.)
(a) If there were 50,000 centenarians in India in 2010, write a function f that gives this number x years after 2010.
(b) Estimate the number of centenarians in 2050.

SOLUTION
(a) Because the number of centenarians is expected to increase by a *factor* of 1.04 each year, function f is *exponential* with $a = \mathbf{1.04}$. The variable x represents time and the initial number of centenarians was 50,000, so $C = \mathbf{50,000}$. Thus $f(x) = \mathbf{50,000}(\mathbf{1.04})^x$.

(b) The year 2050 is 40 years after 2010, so let $x = 40$.

$$f(\mathbf{40}) = 50,000(1.04)^{40} \approx 240,000$$

This model projects that in 2050 India will have about 240,000 centenarians.

> **Now Try Exercise 85**

Exponential Growth and Decay

Graphs of Exponential Functions If $a > 1$ for the exponential function $f(x) = Ca^x$, then $f(x)$ experiences **exponential growth** with **growth factor** a. If $0 < a < 1$, then $f(x)$ experiences **exponential decay** with **decay factor** a. The following See the Concept illustrates these ideas.

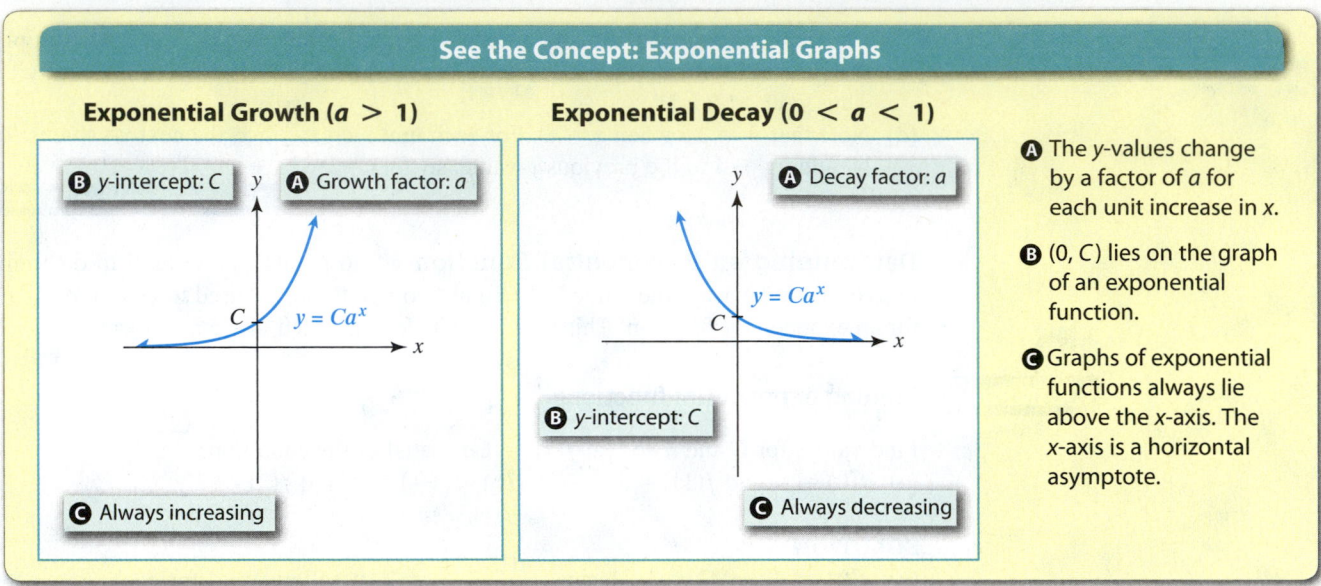

See the Concept: Exponential Graphs

Exponential Growth ($a > 1$)

B y-intercept: C **A** Growth factor: a

$y = Ca^x$

C Always increasing

Exponential Decay ($0 < a < 1$)

A Decay factor: a

$y = Ca^x$

B y-intercept: C

C Always decreasing

A The y-values change by a factor of a for each unit increase in x.

B $(0, C)$ lies on the graph of an exponential function.

C Graphs of exponential functions always lie above the x-axis. The x-axis is a horizontal asymptote.

Figures 5.32 and 5.33 show graphs having exponential growth or decay. Note that when $a > 1$, the graphs exhibit growth and when $0 < a < 1$, the graphs exhibit decay.

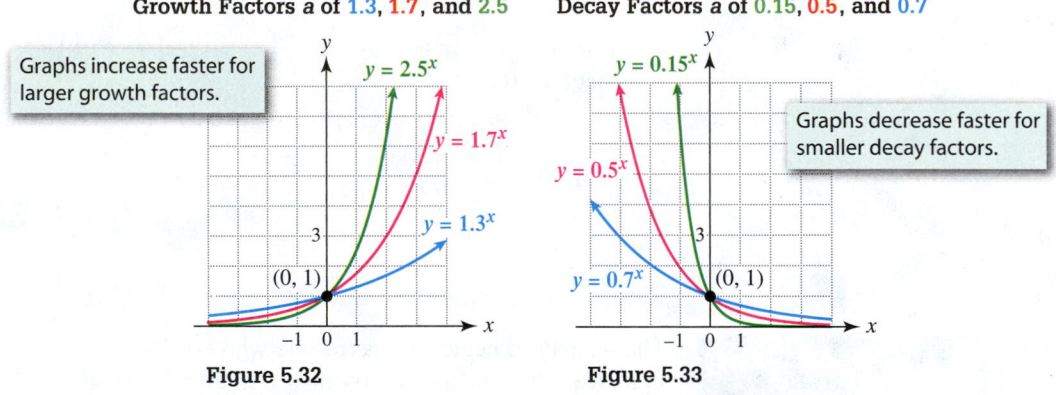

Growth Factors a of 1.3, 1.7, and 2.5

Graphs increase faster for larger growth factors.

$y = 2.5^x$

$y = 1.7^x$

$y = 1.3^x$

$(0, 1)$

Figure 5.32

Decay Factors a of 0.15, 0.5, and 0.7

$y = 0.15^x$

Graphs decrease faster for smaller decay factors.

$y = 0.5^x$

$y = 0.7^x$

$(0, 1)$

Figure 5.33

The following box summarizes some properties of exponential functions and their graphs.

PROPERTIES OF EXPONENTIAL FUNCTIONS

An *exponential function* f, defined by $f(x) = Ca^x$ with $a > 0$, $a \neq 1$, and $C > 0$, has the following properties.

1. The domain of f is $(-\infty, \infty)$ and the range of f is $(0, \infty)$.
2. The graph of f is continuous with no breaks. The x-axis is a horizontal asymptote. There are no x-intercepts and the y-intercept is C.
3. If $a > 1$, f is increasing on its domain; if $0 < a < 1$, f is decreasing on its domain.
4. f is one-to-one and therefore has an inverse. (See Section 5.4.)

Transformations of Exponential Graphs   In Section 3.5 we discussed ways to shift, reflect, stretch, and shrink graphs of functions. These transformations can also be applied to exponential functions as demonstrated in the next example.

EXAMPLE 4   **Transformations of exponential graphs**

Explain how to obtain the graph of g from the graph of $f(x) = 2^x$. Then graph both f and g in the same xy-plane.

(a) $g(x) = 2^{x-1} - 2$ **(b)** $g(x) = 2^{-x}$

SOLUTION

(a) If we replace x with $x - 1$ and subtract 2 in the formula $f(x) = 2^x$, we obtain
$f(x - 1) - 2 = 2^{x-1} - 2$, which is $g(x)$. Thus if we shift the graph of $y = 2^x$ right 1 unit and downward 2 units, we obtain the graph of g. See Figure 5.34.

(b) If we replace x with $-x$ in the formula $f(x) = 2^x$, we obtain $f(-x) = 2^{-x}$, which is $g(x)$. Thus if we reflect the graph of $y = 2^x$ across the y-axis, we obtain the graph of g. See Figure 5.35. Note that by properties of exponents, $2^{-x} = \left(\frac{1}{2}\right)^x$.

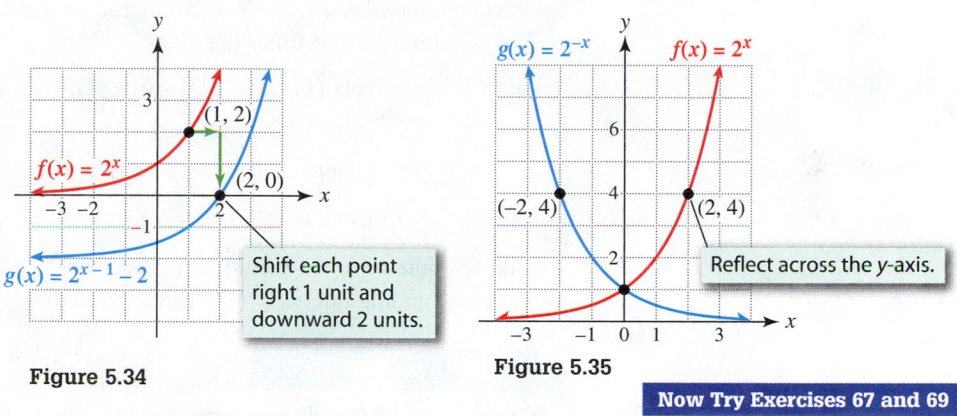

Figure 5.34   **Figure 5.35**

Now Try Exercises 67 and 69

Compound Interest

We can use exponential functions to calculate interest on money. For example, suppose a **principal** P of $1000 is deposited in an account paying 3% annual interest. At the end of 1 year, the account will contain $1000 plus 3% of $1000, or $30, which totals to $1030. In general, the amount A in an account at the end of 1 year with annual interest rate r (in decimal form) and principal P is

$$A = P + rP \qquad \textit{Amount after 1 year}$$
$$= P(1 + r). \qquad \textit{Factor out P.}$$

That is, the amount A after 1 year equals the principal P times $1 + r$. If no money is withdrawn, the amount in the account after 2 years is found by multiplying P by $1 + r$ a second time.

$$A = P(1 + r)(1 + r) \qquad \textit{Amount after 2 years}$$
$$= P(1 + r)^2 \qquad \textit{Properties of exponents}$$

This pattern continues. After t years, the amount A in the account is found by multiplying the principal P by t factors of $(1 + r)$.

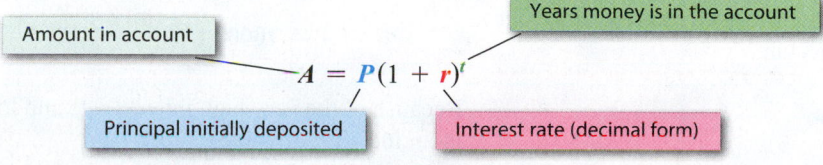

This type of interest is said to be **compounded annually.**

EXAMPLE 5 Calculating an account balance

If the principal is $2000 and the interest rate is 8% compounded annually, calculate the account balance after 4 years.

SOLUTION
The principal is $P = 2000$, the interest rate is $r = 0.08$, and the number of years is $t = 4$.

$$A = P(1 + r)^t = 2000(1 + 0.08)^4 \approx 2720.98$$

After 4 years, the account contains $2720.98.

Now Try Exercise 73

NOTE The value of r can be negative in some applications. For example, if a stock is initially worth $150 and decreases 3% each year for 4 years, then its new value is given by $150(1 - 0.03)^4 \approx \$132.79$.

In most savings accounts, interest is paid more than once a year, where a smaller amount of interest is paid more frequently. For example, if $1000 is deposited at 8% interest *compounded quarterly*, then instead of paying 8% interest once, the account pays $\frac{8\%}{4} = 2\%$ interest four times per year.

$A = 1000(1 + 0.02)^1 = \$1020.00$	Amount after 3 months
$A = 1000(1 + 0.02)^2 = \$1040.40$	Amount after 6 months
$A = 1000(1 + 0.02)^3 \approx \$1061.21$	Amount after 9 months
$A = 1000(1 + 0.02)^4 \approx \$1082.43$	Amount after 1 year

Paying 8% interest once would give $1000(1 + 0.08)^1 = \$1080$. The extra $2.43 results from compounding quarterly rather than annually.

COMPOUND INTEREST

If a principal of P dollars is deposited in an account paying an annual rate of interest r (expressed in decimal form) compounded (paid) n times per year, then after t years the account will contain A dollars, where

$$A = P\left(1 + \frac{r}{n}\right)^{nt}.$$

EXAMPLE 6 Comparing compound interest

Suppose $1000 is deposited by a 20-year-old worker in an Individual Retirement Account (IRA) that pays an annual interest rate of 12%. Describe the effect on the balance after 45 years at age 65 if interest were compounded quarterly rather than annually.

SOLUTION

Compounded Annually Let $P = 1000, r = 0.12, n = 1$, and $t = 65 - 20 = 45$.

$$A = P\left(1 + \frac{r}{n}\right)^{nt} = 1000(1 + 0.12)^{45} \approx \$163,987.60$$

Compounded Quarterly Let $P = 1000, r = 0.12, n = 4$, and $t = 45$.

$$A = 1000\left(1 + \frac{0.12}{4}\right)^{4(45)} = 1000(1 + 0.03)^{180} \approx \$204,503.36$$

CLASS DISCUSSION

Make a conjecture about the effect on the IRA balance after 45 years if the interest rate in Example 6 were 6% instead of 12%. Test your conjecture.

Because of the very high interest rate and long time period, quarterly compounding results in an increase of $40,515.76!

Now Try Exercise 75

The Natural Exponential Function

In Example 6, compounding interest quarterly rather than annually made a significant difference in the balance. What would happen if interest were compounded daily or even hourly? Would there be a limit to the amount of interest that could be earned? To answer these questions, suppose \$1 was deposited in an account at the very high interest rate of 100%. Table 5.16 shows that there *is a limit*. In this example, the interest formula $A = P\left(1 + \frac{r}{n}\right)^{nt}$ simplifies to $A = \left(1 + \frac{1}{n}\right)^n$, because $P = t = r = 1$. The graph of $y = \left(1 + \frac{1}{x}\right)^x$ is shown in Figure 5.36. The graph approaches the horizontal asymptote $y \approx 2.7183$. This means that the y-values never exceed about 2.7183.

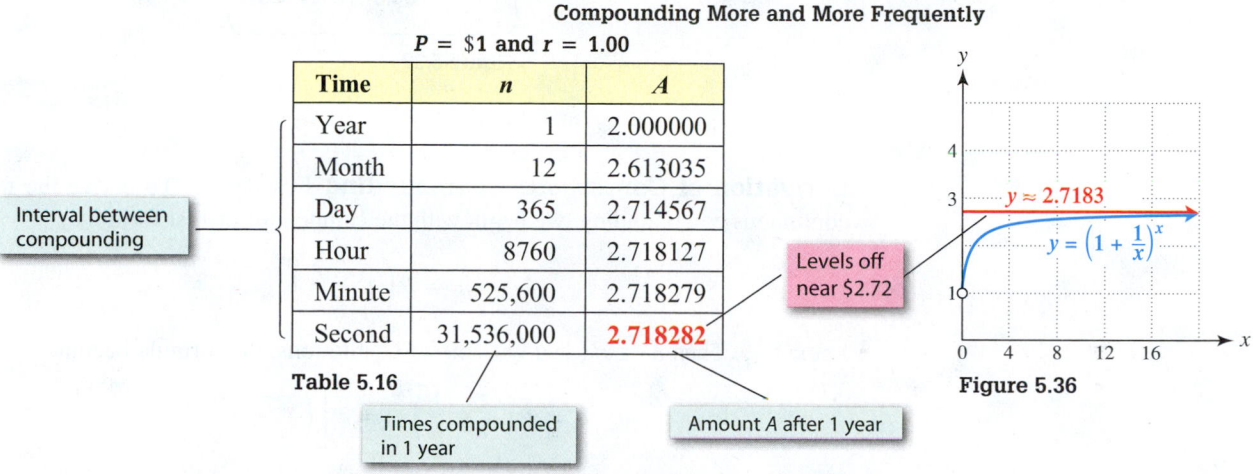

Compounding More and More Frequently

$P = \$1$ and $r = 1.00$

Interval between compounding

Time	n	A
Year	1	2.000000
Month	12	2.613035
Day	365	2.714567
Hour	8760	2.718127
Minute	525,600	2.718279
Second	31,536,000	**2.718282**

Levels off near \$2.72

Table 5.16

Times compounded in 1 year

Amount A after 1 year

Figure 5.36

Compounding that is done more frequently, by letting n become large without bound, is called **continuous compounding**. The exponential expression $\left(1 + \frac{1}{n}\right)^n$ reaches a limit of approximately 2.718281828 as $n \to \infty$. This value is so important in mathematics that it has been given its own symbol, e, sometimes called **Euler's number**.

As x becomes very large this expression approaches e.

$$\left(1 + \frac{1}{x}\right)^x \to e \quad \text{as} \quad x \to \infty.$$

The number e has many of the same characteristics as π. Its decimal expansion never terminates or repeats in a pattern. It is an irrational number.

CLASS DISCUSSION

Graph $y = 2^x$ and $y = 3^x$ on the same coordinate axes, using the viewing rectangle $[-3, 3, 1]$ by $[0, 4, 1]$. Make a conjecture about how the graph of $y = e^x$ will appear. Test your conjecture.

VALUE OF e

To eleven decimal places, $e \approx 2.71828182846$.

Continuous compounding can be applied to population growth. Compounding annually would mean that all births and deaths occurred on December 31. Similarly, compounding quarterly would mean that births and deaths occurred at the end of March, June, September, and December. In large populations, births and deaths occur *continuously* throughout the year. Compounding continuously is a *natural* way to model large populations.

Calculator Help

When evaluating e^x, be sure to use the built-in key for e^x, rather than using an approximation for e, such as 2.72.

THE NATURAL EXPONENTIAL FUNCTION

The function f, represented by $f(x) = e^x$, is the **natural exponential function**.

EXAMPLE 7 Evaluating the natural exponential function

Approximate to four decimal places $f(x) = e^x$ when $x = 1, 0.5,$ and -2.56.

SOLUTION

Figure 5.37 shows that these values are approximated as follows: $f(1) = e^1 \approx 2.7183$, $f(0.5) = e^{0.5} \approx 1.6487$, and $f(-2.56) = e^{-2.56} \approx 0.0773$.

```
e^(1)
         2.718281828
e^(.5)
         1.648721271
e^(-2.56)
          .0773047404
```

Figure 5.37

Now Try Exercise 45

Derivation of Continuous Compounding Formula To derive the formula for continuous compounding, we begin with the compound interest formula,

$$A = P\left(1 + \frac{r}{n}\right)^{nt}.$$

Let $k = \frac{n}{r}$. Then $n = rk$, and with these substitutions, the formula becomes

$$A = P\left(1 + \frac{1}{k}\right)^{rkt} = P\left[\left(1 + \frac{1}{k}\right)^k\right]^{rt}.$$

If $n \to \infty$, then $k \to \infty$ as well, and the expression $\left(1 + \frac{1}{k}\right)^k \to e$, as discussed earlier. This leads to the formula $A = Pe^{rt}$.

CONTINUOUSLY COMPOUNDED INTEREST

If a principal of P dollars is deposited in an account paying an annual rate of interest r (expressed in decimal form), compounded continuously, then after t years the account will contain A dollars, where

$$A = Pe^{rt}.$$

EXAMPLE 8 Calculating continuously compounded interest

The principal in an IRA is \$1000 and the interest rate is 12%, compounded continuously. How much money will there be after 45 years?

SOLUTION

Let $P = 1000, r = 0.12$, and $t = 45$. Then $A = 1000e^{(0.12)45} \approx \$221{,}406.42$. This is more than the \$204,503.36 that resulted from compounding quarterly in Example 6.

Now Try Exercise 77

Natural Exponential Growth and Decay The natural exponential function is often used to model growth of a quantity. If A_0 is the initial amount of a quantity A at time $t = 0$ and if k is a *positive* constant, then exponential growth of A can be modeled by

$$A(t) = A_0e^{kt}. \qquad \text{Growth } (k > 0)$$

Figure 5.38 illustrates this type of growth graphically. Similarly,

$$A(t) = A_0e^{-kt} \qquad \text{Decay } (k > 0)$$

can be used to model exponential decay provided $k > 0$. Figure 5.39 illustrates this type of decay graphically. A larger value of k causes the graph of A to increase or decrease more rapidly. That is, a larger value of k causes the average rate of change in A to be greater in absolute value.

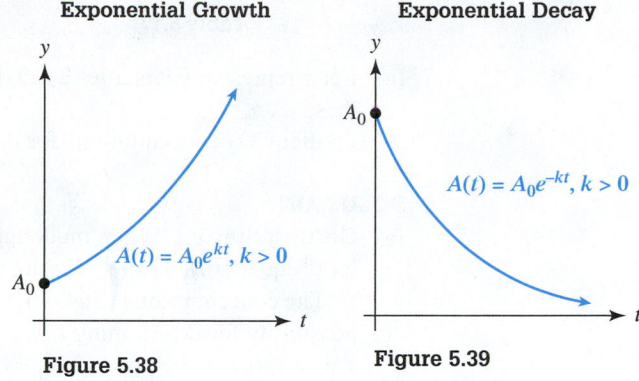

Exponential Growth

$A(t) = A_0 e^{kt}, k > 0$

Figure 5.38

Exponential Decay

$A(t) = A_0 e^{-kt}, k > 0$

Figure 5.39

NOTE In the compound interest formula $A = Pe^{rt}$, the value of r corresponds to the value of k in the formula $A = A_0 e^{kt}$. If $r > 0$, the principal P grows and if $r < 0$, the principal P decays.

EXAMPLE 9   **Modeling the growth of _E. coli_ bacteria**

E. coli (_Escherichia coli_) is a type of bacteria that inhabits the intestines of animals. These bacteria are capable of rapid growth and can be dangerous to humans—especially children. In one study, _E. coli_ bacteria were found to be capable of doubling in number about every 49.5 minutes. Their number N after t minutes could be modeled by $N(t) = N_0 e^{0.014t}$. Suppose that $N_0 = 500,000$ is the initial number of bacteria per milliliter. (**Source:** G. S. Stent, _Molecular Biology of Bacterial Viruses._)

(a) Make a conjecture about the number of bacteria per milliliter after 99 minutes. Verify your conjecture.

(b) Determine graphically the elapsed time when there were 25 million bacteria per milliliter.

SOLUTION

(a) Since the bacteria double every 49.5 minutes, there would be 1,000,000 per milliliter after 49.5 minutes and 2,000,000 after **99** minutes. This is verified by evaluating

$$N(\mathbf{99}) = 500,000 e^{0.014(\mathbf{99})} \approx 2,000,000.$$

(b) _Graphical Solution_   Solve $N(t) = 25,000,000$ by graphing $Y_1 = 500000e^{\wedge}(0.014X)$ and $Y_2 = 25000000$. Their graphs intersect near $(279.4, 25000000)$, as shown in Figure 5.40. Thus in a 1-milliliter sample, half a million _E. coli_ bacteria could increase to 25 million in approximately 279 minutes, or 4 hours and 39 minutes.

Now Try Exercise 93

$[0, 400, 100]$ by $[0, 3 \times 10^7, 1 \times 10^7]$

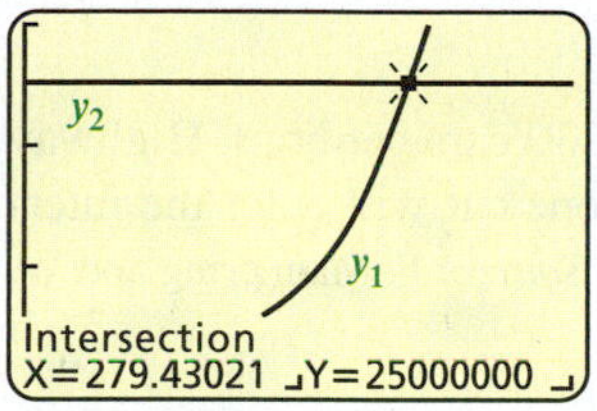

y_2

y_1

Intersection
X=279.43021 Y=25000000

Figure 5.40

NOTE You will learn how to solve the equation in Example 9(b) symbolically. See Exercise 88 in Section 5.6.

Exponential Models

The next example analyzes the increase in atmospheric carbon dioxide (CO_2).

EXAMPLE 10   **Modeling atmospheric CO_2 concentrations**

Predicted concentrations of atmospheric carbon dioxide (CO_2) in parts per million (ppm) are shown in Table 5.17 on the next page. (These concentrations assume that current trends continue.)

Concentrations of Atmospheric CO_2

Year	2000	2050	2100	2150	2200
CO_2 (ppm)	364	467	600	769	987

Source: R. Turner, *Environmental Economics.*

Table 5.17

(a) Let x represent years after 2000. Find values for C and a so that $f(x) = Ca^x$ models these data.

(b) Predict CO_2 concentrations for the year 2025.

SOLUTION

(a) Getting Started When modeling data by hand with $f(x) = Ca^x$, one strategy is to let C equal $f(0)$. Then substitute a different data point into this formula to find a. ▶

The concentration is 364 when $x = 0$, so $C = 364$. This gives $f(x) = 364a^x$. One possibility for determining a is to use the last data point and require that the graph of f pass through the point $(200, 987)$. It then follows that $f(\mathbf{200}) = \mathbf{987}$.

$$364a^{\mathbf{200}} = \mathbf{987} \qquad \color{blue}{f(200) = 987}$$

$$a^{200} = \frac{987}{364} \qquad \color{blue}{\text{Divide by 364.}}$$

$$\left(a^{200}\right)^{\mathbf{1/200}} = \left(\frac{987}{364}\right)^{\mathbf{1/200}} \qquad \color{blue}{\text{Take the } \tfrac{1}{200}\text{th power.}}$$

$$a = \left(\frac{987}{364}\right)^{1/200} \qquad \color{blue}{\text{Properties of exponents}}$$

$$a \approx 1.005 \qquad \color{blue}{\text{Approximate.}}$$

Thus $f(x) = 364(1.005)^x$. Answers for $f(x)$ may vary slightly.

(b) Since 2025 corresponds to $x = 25$, evaluate $f(25)$.

$$f(\mathbf{25}) = 364(1.005)^{\mathbf{25}} \approx 412$$

Concentration of CO_2 could reach 412 ppm by 2025.

> **Now Try Exercise 91**

EXAMPLE 11 **Modeling traffic flow**

Cars arrive randomly at an intersection with an average rate of 30 cars per hour. Highway engineers estimate the likelihood, or probability, that at least one car will enter the intersection within a period of x minutes with $f(x) = 1 - e^{-0.5x}$. (*Source:* F. Mannering and W. Kilareski, *Principles of Highway Engineering and Traffic Analysis.*)

(a) Evaluate $f(2)$ and interpret the answer.

(b) Graph f for $0 \le x \le 60$. What is the likelihood that at least one car will enter the intersection during a 60-minute period?

[0, 60, 10] by [0, 1.2, 0.2]

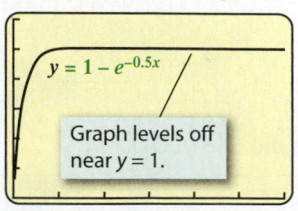

Figure 5.41

SOLUTION

(a) $f(2) = 1 - e^{-0.5(2)} = 1 - e^{-1} \approx 0.63$. There is a 63% chance that at least one car will enter the intersection during a 2-minute period.

(b) Graph $Y_1 = 1 - e^\wedge(-0.5X)$, as shown in Figure 5.41. As time progresses, the probability increases and begins to approach 1. That is, it is almost certain that at least one car will enter the intersection during a 60-minute period. (Note that a horizontal asymptote is $y = 1$.)

> **Now Try Exercise 95**

Modeling Half-Life Links on Facebook typically experience half of their engagements ("hits") during the first 3 hours. This pattern can continue over time and is sometimes referred to as a **half-life** of a link on Facebook. Because the number of engagements with a link is decreasing by a factor of $\frac{1}{2}$ over each 3-hour period, we can use an exponential function to model this situation.

EXAMPLE 12   Modeling the half-life of a Facebook link

Initially a link on Facebook has had no hits, so 100% of its hits are yet to occur. Write an exponential function F that gives the percentage of engagements yet to occur on a typical Facebook link if its half-life is 3 hours. Estimate this percentage after 4 hours.

SOLUTION
Let $F(t) = Ca^t$, where t is time in hours. Initially $F(0) = 100$, so $C = \mathbf{100}$ and $F(t) = \mathbf{100}a^t$. Next we must find the value of a. Because the half-life is 3 hours, 50% of the hits remain to occur after 3 hours, so $F(3) = 50$.

$$100a^3 = 50 \qquad \text{\textcolor{blue}{F(3) = 50}}$$
$$a^3 = \tfrac{1}{2} \qquad \text{\textcolor{blue}{Divide each side by 100.}}$$
$$(a^3)^{1/3} = \left(\tfrac{1}{2}\right)^{1/3} \qquad \text{\textcolor{blue}{Take the cube root of each side.}}$$
$$a = \left(\tfrac{1}{2}\right)^{1/3} \qquad \text{\textcolor{blue}{Simplify.}}$$

Thus we can write $F(t)$ as

$$F(t) = \mathbf{100}\left(\left(\tfrac{1}{2}\right)^{1/3}\right)^t = 100\left(\tfrac{1}{2}\right)^{t/3}.$$

After 4 hours,

$$F(\mathbf{4}) = 100\left(\tfrac{1}{2}\right)^{4/3} \approx 40\%.$$

The link has about 40% of its total hits remaining while 60% have already occurred. Note that 4 hours is more than the half-life of 3 hours, so $F(4)$ is less than 50%.

Now Try Exercise 97

These half-life results can be generalized by the following.

> **MODELING HALF-LIFE**
>
> If a quantity initially equals C and has a half-life of k, then the amount A remaining after time t is given by
> $$A(t) = C\left(\tfrac{1}{2}\right)^{t/k}.$$

Radioactive Decay   Radioactivity is an application of half-life. For example, the half-life of radium-226 is about **1600** years. After **9600** years, a **2**-gram sample decays to

$$A(\mathbf{9600}) = 2\left(\tfrac{1}{2}\right)^{9600/1600} = 0.03125 \text{ gram}.$$

EXAMPLE 13   Finding the age of a fossil

Radioactive carbon-14, which is found in all living things, has a half-life of about 5700 years and can be used to date fossils. Suppose a fossil contains 5% of the amount of carbon-14 that the organism contained when it was alive. Graphically estimate its age. The situation is illustrated in Figure 5.42.

SOLUTION
The initial amount of carbon-14 is 100% (or 1), the final amount is 5% (or 0.05), and the half-life is 5700 years, so $A = 0.05$, $C = 1$, and $k = 5700$ in $A(x) = C\left(\tfrac{1}{2}\right)^{x/k}$. To determine the age of the fossil, solve the following equation for x.

Carbon-14 (percent)

$y = C\left(\tfrac{1}{2}\right)^{x/5700}$

100%

5%

Age

Years

Figure 5.42

[0, 50000, 10000] by [0, 0.1, 0.01]

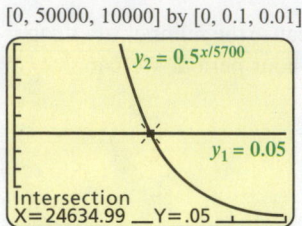

Figure 5.43

$$0.05 = 1\left(\frac{1}{2}\right)^{x/5700}$$

Graph $Y_1 = 0.05$ and $Y_2 = 0.5^{\wedge}(X/5700)$, as shown in Figure 5.43. Their graphs intersect near $(24635, 0.05)$, so the fossil is about 24,635 years old.

Now Try Exercise 101

5.3 Putting It All Together

The following table summarizes some important concepts about exponential functions and types of growth and decay.

CONCEPT	COMMENTS	EXAMPLES
Exponential function $$f(x) = Ca^x,$$ where $a > 0$, $a \neq 1$, and $C > 0$	Exponential growth occurs when $a > 1$, and exponential decay occurs when $0 < a < 1$. C often represents the initial amount present because $f(0) = C$.	$f(x) = 5(0.8)^x$ *Decay* $g(x) = 3^x$ *Growth* Decay: $0 < a < 1$ Growth: $a > 1$
Linear growth $$y = mx + b$$	If data increase by a fixed amount m for each unit increase in x, they can be modeled by a linear function.	The following data can be modeled by $f(x) = 3x + 2$ because the data increase 3 units for each unit increase in x and because $y = 2$ when $x = 0$. <table><tr><td>x</td><td>0</td><td>1</td><td>2</td><td>3</td></tr><tr><td>y</td><td>2</td><td>5</td><td>8</td><td>11</td></tr></table> $+3$ $+3$ $+3$
Exponential growth $$y = Ca^x$$	If data increase by a constant factor a for each unit increase in x, they can be modeled by an exponential function.	The following data can be modeled by $f(x) = 2(3)^x$ because the data increase by a factor of 3 for each unit increase in x and because $y = 2$ when $x = 0$. <table><tr><td>x</td><td>0</td><td>1</td><td>2</td><td>3</td></tr><tr><td>y</td><td>2</td><td>6</td><td>18</td><td>54</td></tr></table> $\times 3$ $\times 3$ $\times 3$
Interest compounded n times per year $$A = P\left(1 + \frac{r}{n}\right)^{nt}$$	P is the principal, r is the interest rate (expressed in decimal form), n is the number of times interest is paid each year, t is the number of years, and A is the amount after t years.	$500 at 8%, compounded monthly, for 3 years yields $$500\left(1 + \frac{0.08}{12}\right)^{12(3)} \approx \$635.12.$$
The number e	The number e is an irrational number that is important in mathematics, much like π.	$e \approx 2.718282$

CONCEPT	COMMENTS	EXAMPLES
Natural exponential function	This function is an exponential function with base e and $C = 1$.	$f(x) = e^x$
Interest compounded continuously $$A = Pe^{rt}$$	P is the principal, r is the interest rate (expressed in decimal form), t is the number of years, and A is the amount after t years.	\$500 at 8% compounded continuously for 3 years yields $$500e^{0.08(3)} \approx \$635.62.$$
Modeling half-life	If a quantity initially equals C and has a half-life of k, then the amount A remaining after time t is given by $$A(t) = C\left(\tfrac{1}{2}\right)^{t/k}.$$	A 5-gram sample of radioactive material with a half-life of 300 years is modeled by $$A(t) = 5\left(\frac{1}{2}\right)^{t/300}.$$

5.3 Exercises

Review of Exponents

Exercises 1–12: Simplify the expression without a calculator.

1. 2^{-3}

2. $(-3)^{-2}$

3. $3(4)^{1/2}$

4. $5\left(\frac{1}{2}\right)^{-3}$

5. $-2(27)^{2/3}$

6. $-4(8)^{-2/3}$

7. $4^{1/6}4^{1/3}$

8. $\frac{9^{5/6}}{9^{1/3}}$

9. 3^0

10. $5\left(\frac{3}{4}\right)^0$

11. $(5^{101})^{1/101}$

12. $(8^{27})^{1/27}$

Linear and Exponential Growth

Exercises 13–22: (Refer to Example 1.) Find either a linear or an exponential function that models the data in the table.

13.

x	0	1	2	3	4
y	2	0.8	-0.4	-1.6	-2.8

14.

x	0	1	2	3	4
y	2	8	32	128	512

15.

x	-3	-2	-1	0	1
y	64	32	16	8	4

16.

x	-2	-1	0	1	2
y	3	5.5	8	10.5	13

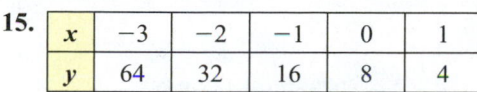

17.

x	-4	-2	0	2	4
y	0.3125	1.25	5	20	80

18.

x	-15	-5	5	15	25
y	22	24	26	28	30

19.

x	-6	-2	2	6
y	-23	-9	5	19

20.

x	-2	3	5	8
y	$\frac{3}{4}$	24	96	768

21.

x	-4	-1	2	5
y	6561	243	9	$\frac{1}{3}$

22.

x	-20	-4	16	36
y	246	234	219	204

23. Comparing Growth Which function becomes larger for $0 \le x \le 10$: $f(x) = 2^x$ or $g(x) = x^2$?

24. Comparing Growth Which function becomes larger for $0 \le x \le 10$: $f(x) = 4 + 3x$ or $g(x) = 4(3)^x$?

25. Comparing Growth Which function becomes larger for $0 \le x \le 10$: $f(x) = 2x + 1$ or $g(x) = 2^{-x}$?

26. Salaries If you were offered 1¢ for the first week of work, 3¢ for the second week, 5¢ for the third week, 7¢ for the fourth week, and so on for a year, would you accept the offer? Would you accept an offer that pays 1¢ for the first week of work, 2¢ for the second week, 4¢ for the third week, 8¢ for the fourth week, and so on for a year? Explain your answers.

Exponential Functions

Exercises 27–34: Find C and a so that $f(x) = Ca^x$ satisfies the given conditions.

27. $f(0) = 5$ and for each unit increase in x, the output is multiplied by 1.5.

28. $f(1) = 3$ and for each unit increase in x, the output is multiplied by $\frac{3}{4}$.

29. $f(0) = 10$ and $f(1) = 20$

30. $f(0) = 7$ and $f(-1) = 1$

31. $f(1) = 9$ and $f(2) = 27$

32. $f(-1) = \frac{1}{4}$ and $f(1) = 4$

33. $f(-2) = \frac{9}{2}$ and $f(2) = \frac{1}{18}$

34. $f(-2) = \frac{3}{4}$ and $f(2) = 12$

Exercises 35–38: Find a linear function f and an exponential function g whose graphs pass through the two given points.

35. $(0, 4), (1, 8)$ **36.** $\left(0, \frac{3}{2}\right), (1, 2)$

37. $(-2, 12), (1, 1.5)$ **38.** $(1, 3), (5, 48)$

Exercises 39–42: Find C and a so that $f(x) = Ca^x$ models the situation described. State what the variable x represents in your formula. (Answers may vary.)

39. Bacteria There are initially 5000 bacteria, and this sample doubles in size every hour.

40. Savings Fifteen hundred dollars is deposited in an account that triples in value every decade.

41. Home Value In 2008 a house was worth $200,000, and its value *decreases* by 5% each year thereafter.

42. Population A fish population is initially 6000 and decreases by half each year.

43. Tire Pressure The pressure in a tire with a leak is 30 pounds per square inch initially and can be modeled by $f(x) = 30(0.9)^x$ after x minutes. What is the tire's pressure after 9.5 minutes?

44. Population The population of California was about 37 million in 2010 and increasing by 1.0% each year. Estimate the population of California in 2015.

Exercises 45–48: Approximate f(x) to four decimal places.

45. $f(x) = 4e^{-1.2x}$, $x = -2.4$

46. $f(x) = -2.1e^{-0.71x}$, $x = 1.9$

47. $f(x) = \frac{1}{2}(e^x - e^{-x})$, $x = -0.7$

48. $f(x) = 4(e^{-0.3x} - e^{-0.6x})$, $x = 1.6$

Graphs of Exponential Functions

Exercises 49–56: Sketch a graph of $y = f(x)$.

49. $f(x) = 2^x$ **50.** $f(x) = 4^x$

51. $f(x) = 3^{-x}$ **52.** $f(x) = 3(2^{-x})$

53. $f(x) = 2\left(\frac{1}{3}\right)^x$ **54.** $f(x) = 2(3^x)$

55. $f(x) = \left(\frac{1}{2}\right)^x$ **56.** $f(x) = \left(\frac{1}{4}\right)^x$

Exercises 57–60: Use the graph of $y = Ca^x$ to determine values for C and a.

57.

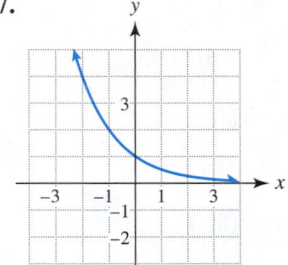

58.

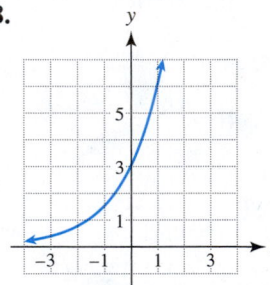

59.

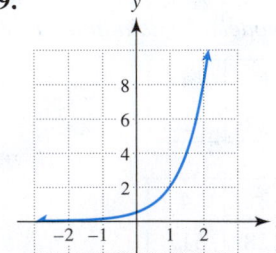

60.

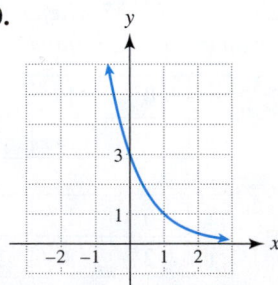

61. Let $f(x) = 7\left(\frac{1}{8}\right)^x$.

(a) What are the domain and range of f?

(b) Is f either increasing or decreasing on its domain?

(c) Find any asymptotes on the graph of f.

(d) Find any x- or y-intercepts on the graph of f.

(e) Is f a one-to-one function? Does f have an inverse?

62. Repeat Exercise 61 with $f(x) = e^x$.

63. Match the formula for f with its graph (a–d). Do *not* use a calculator.
    (i) $f(x) = e^x$ **(ii)** $f(x) = 3^{-x}$
    (iii) $f(x) = 1.5^x$ **(iv)** $f(x) = 0.99^x$

a. 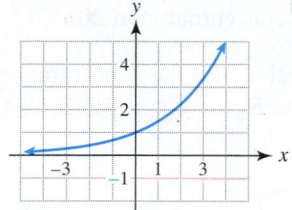   **b.**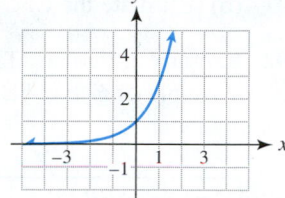

c. 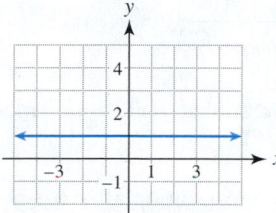   **d.**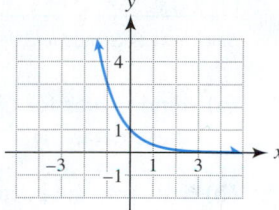

64. Modeling Phenomena Match the situation with the graph (a–d) that models it best.
  (i) Balance in an account after x years earning 10% interest compounded continuously

  (ii) Balance in an account after x years earning 5% interest compounded annually

  (iii) Air pressure in a car tire with a large hole in it after x minutes

  (iv) Air pressure in a car tire with a pinhole in it after x minutes

a. 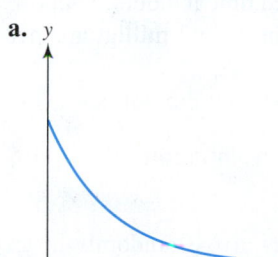   **b.**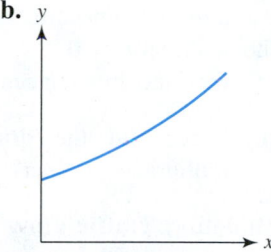

c. 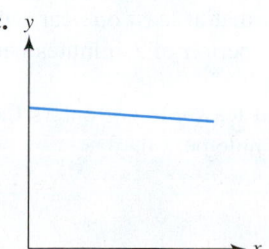   **d.**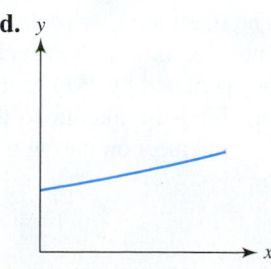

Transformations of Graphs

Exercises 65 and 66: The graph of $y = f(x)$ is shown in the figure. Sketch a graph of each equation using translations of graphs and reflections. Do not use a graphing calculator

65. $f(x) = 2^x$
  (a) $y = 2^x - 2$

  (b) $y = 2^{x-1}$

  (c) $y = 2^{-x}$

  (d) $y = -2^x$

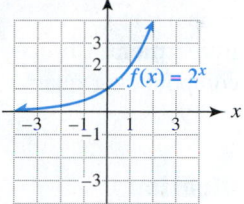

66. $f(x) = e^{-0.5x}$
  (a) $y = -e^{-0.5x}$

  (b) $y = e^{-0.5x} - 3$

  (c) $y = e^{-0.5(x-2)}$

  (d) $y = e^{0.5x}$

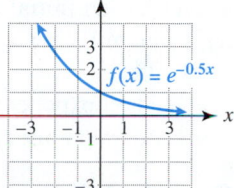

Exercises 67–72: Graph $y = f(x)$. Then use transformations to graph $y = g(x)$ on the same xy-plane.

67. $f(x) = 2^x$, $g(x) = 2^{x-1} - 2$

68. $f(x) = 2^x$, $g(x) = 2^{x+2} + 1$

69. $f(x) = 3^x$, $g(x) = 3^{-x}$

70. $f(x) = 3^x$, $g(x) = -3^x$

71. $f(x) = \left(\frac{1}{2}\right)^x$, $g(x) = 2\left(\frac{1}{2}\right)^x$

72. $f(x) = \left(\frac{1}{2}\right)^x$, $g(x) = \left(\frac{1}{2}\right)^{x-2} - 1$

Compound Interest

Exercises 73–80: Use the compound interest formula to approximate the final value of each amount.

73. $600 at 7% compounded annually for 5 years

74. $2300 at 2% compounded semiannually for 10 years

75. $950 at 3% compounded daily for 20 years

76. $3300 at 3% compounded quarterly for 2 years

77. $2000 at 10% compounded continuously for 8 years

78. $100 at 1.5% compounded continuously for 50 years

79. $1600 at 1.3% compounded monthly for 2.5 years

80. $2000 at 8.7% compounded annually for 5 years

81. College Tuition If college tuition is currently $8000 per year, inflating at 6% per year, what will be the cost of tuition in 10 years?

82. Doubling Time   How long does it take for an investment to double its value if the interest is 12% compounded annually? 6% compounded annually?

83. Investment Choice   Determine the best investment: compounding continuously at 6.0% or compounding annually at 6.3%.

84. Investment Choice   Determine the best investment: compounding quarterly at 3.1% or compounding daily at 2.9%.

Applications

85. Bacteria Growth   A sample of bacteria taken from a river has an initial concentration of 2.5 million bacteria per milliliter and its concentration triples each week.
(a) Find an exponential model that calculates the concentration after x weeks.

(b) Estimate the concentration after 1.5 weeks.

86. Fish Population   A fish population in a small lake is estimated to be 6000. Due to a change in water quality, this population is decreasing by half each year.
(a) Find an exponential model that approximates the number of fish in the lake after x years.

(b) Estimate the fish population to the nearest hundred after 3.5 years.

87. Intensity of Sunlight   The intensity I of sunlight at the surface of a lake is 300 watts per square meter. For each 1-foot increase in depth of a lake, the intensity of sunlight decreases by a factor of $\frac{9}{10}$.
(a) Estimate the intensity I of sunlight at a depth of 50 feet.

(b) Graph the intensity of sunlight to a depth of 50 feet. Interpret the y-intercept.

88. Pollution   A pollutant in a river has an initial concentration of 3 parts per million and degrades at a rate of 1.5% per year. Approximate its concentration after 20 years.

89. Population Growth   The population of Arizona was 6.6 million in 2010 and growing continuously at a 1.44% rate. Assuming this trend continues, estimate the population of Arizona in 2016.

90. Saving for Retirement   Suppose $1500 is deposited into an IRA with an interest rate of 6%, compounded continuously. How much money will there be after 30 years?

91. Greenhouse Gases   (Refer to Example 10.) Chlorofluorocarbons (CFCs) are gases that might increase the greenhouse effect. The following table lists future concentrations of CFC-12 in parts per billion (ppb) if current trends continue.

Year	2000	2005	2010	2015	2020
CFC-12 (ppb)	0.72	0.88	1.07	1.31	1.60

Source: R. Turner, *Environmental Economics.*

(a) Find values for C and a so that $f(x) = Ca^x$ models these data, where x is years after 2000.

(b) Estimate the CFC-12 concentration in 2013.

92. Bacteria Growth   The table lists the concentration of a sample of *E. coli* bacteria B (in billions per liter) after x hours.

x	0	3	5	8
B	0.5	6.2	33.3	414

Source: G. S. Stent, *Molecular Biology of Bacterial Viruses.*

(a) Find values for C and a so that $f(x) = Ca^x$ models these data.

(b) Estimate the bacteria concentration after 6.2 hours.

93. E. Coli Growth   (Refer to Example 9.)
(a) Approximate the number of *E. coli* after 3 hours.

(b) Estimate graphically the elapsed time when there are 10 million bacteria per milliliter.

94. Drug Concentrations   Sometimes after a patient takes a drug, the amount of medication A in the bloodstream can be modeled by $A = A_0 e^{-rt}$, where A_0 is the initial concentration in milligrams per liter, r is the hourly percentage decrease (in decimal form) of the drug in the bloodstream, and t is the elapsed time in hours. Suppose that a drug's concentration is initially 2 milligrams per liter and that $r = 0.2$.
(a) Find the drug concentration after 3.5 hours.

(b) When did the drug concentration reach 1.5 milligrams per liter?

95. Modeling Traffic Flow   Cars arrive randomly at an intersection with an average rate of 50 cars per hour. The likelihood, or probability, that at least one car will enter the intersection within a period of x minutes can be estimated by $P(x) = 1 - e^{-5x/6}$.
(a) Find the likelihood that at least one car enters the intersection during a 3-minute period.

(b) Graphically determine the value of x that gives a 50–50 chance of at least one car entering the intersection during an interval of x minutes.

96. Tree Density Ecologists studied the spacing between individual trees in a forest in British Columbia. The probability, or likelihood, that there is at least one tree located in a circle with a radius of x feet can be estimated by $P(x) = 1 - e^{-0.1144x}$. For example, $P(7) \approx 0.55$ means that if a person picks a point at random in the forest, there is a 55% chance that at least one tree will be located within 7 feet. See the figure. (*Source:* E. Pielou, *Populations and Community Ecology.*)

(a) Evaluate $P(2)$ and $P(20)$, and interpret the results.

(b) Graph P. Explain verbally why it is logical for P to be an increasing function. Does the graph have a horizontal asymptote?

(c) Solve $P(x) = 0.5$ and interpret the result.

97. Half-Life for Twitter (Refer to Example 12.) The half-life for a link on Twitter is 2.8 hours. Write an exponential function T that gives the percentage of engagements remaining on a typical Twitter link after t hours. Estimate this percentage after 5.5 hours.

98. Half-Life for StumbleUpon (Refer to Example 12.) The half-life for a link on StumbleUpon is 400 hours. Write an exponential function S that gives the percentage of engagements remaining on a typical StumbleUpon link after t hours. Estimate this percentage after 250 hours.

99. Putts and Pros At a distance of 3 feet, professional golfers make about 95% of their putts. For each additional foot of distance, this percentage decreases by a factor of 0.9.

(a) Write an exponential function P that gives the percentage of putts pros make at a distance of x feet beyond a 3-foot distance.

(b) Estimate this percentage for 23 feet ($x = 20$).

(c) What additional distance decreases this percentage by half?

100. Filters Impurities in water are frequently removed using filters. Suppose that a 1-inch filter allows 10% of the impurities to pass through it. The other 90% is trapped in the filter.

(a) Find a formula in the form $f(x) = 100a^x$ that calculates the percentage of impurities passing through x inches of this type of filter.

(b) Use $f(x)$ to estimate the percentage of impurities passing through 2.3 inches of the filter.

101. Radioactive Carbon-14 (Refer to Example 13.) A fossil contains 10% of the carbon-14 that the organism contained when it was alive. Graphically estimate its age.

102. Radioactive Carbon-14 A fossil contains 20% of the carbon-14 that the organism contained when it was alive. Estimate its age.

103. Radioactive Radium-226 The half-life of radium-226 is about 1600 years. After 3000 years, what percentage P of a sample of radium remains?

104. Radioactive Strontium-90 Radioactive strontium-90 has a half-life of about 28 years and sometimes contaminates the soil. After 50 years, what percentage of a sample of radioactive strontium would remain?

105. Swimming Pool Maintenance Chlorine is frequently used to disinfect swimming pools. The chlorine concentration should remain between 1.5 and 2.5 parts per million. On warm sunny days with many swimmers agitating the water, 30% of the chlorine can dissipate into the air or combine with other chemicals. (*Source:* D. Thomas, *Swimming Pool Operators Handbook.*)

(a) Find C and a so that $f(x) = Ca^x$ models the amount of chlorine in the pool after x days. Assume that the initial amount is 2.5 parts per million.

(b) What is the chlorine concentration after 2 days if no chlorine is added?

106. Thickness of Runways Heavier aircraft require runways with thicker pavement for landings and takeoffs. A pavement 6 inches thick can accommodate an aircraft weighing 80,000 pounds, whereas a 12-inch-thick pavement is necessary for a 350,000-pound plane. The relation between pavement thickness t in inches and gross weight W in thousands of pounds can be modeled by $W = Ca^t$. (*Source:* FAA.)

(a) Find values for C and a.

(b) What is the weight of the heaviest airplane a 9-inch-thick runway can accommodate?

(c) What is the minimum thickness for a 242,000-pound plane?

107. Trains The faster a locomotive travels, the more horsepower is needed. The formula

$$H(x) = 0.157(1.033)^x$$

calculates this horsepower for a level track. The input x is in miles per hour and the output $H(x)$ is the horsepower required per ton of cargo. (*Source:* L. Haefner, *Introduction to Transportation Systems.*)

(a) Evaluate $H(30)$ and interpret the result.

(b) Determine the horsepower needed to move a 5000-ton train 30 miles per hour.

(c) Some types of locomotives are rated for 1350 horsepower. How many locomotives of this type would be needed in part (b)?

108. Survival of Reindeer For all types of animals, the percentage that survive into the next year decreases. In one study, the survival rate of a sample of reindeer was modeled by $S(t) = 100(0.999993)^{t^5}$. The function S outputs the percentage of reindeer that survive t years. (*Source:* D. Brown.)

(a) Evaluate $S(4)$ and $S(15)$. Interpret the results.

(b) Graph S in [0, 15, 5] by [0, 110, 10]. Interpret the graph. Does the graph have a horizontal asymptote?

Writing about Mathematics

109. Explain how linear and exponential functions differ. Give examples.

110. Discuss the domain and range of an exponential function f. Is f one-to-one? Explain.

Extended and Discovery Exercises

Exercises 1–4: **Present Value** *In the compound interest formula $A = P(1 + r/n)^{nt}$, we can think of P as the present value of an investment and A as the future value of an investment after t years. For example, if you were saving for college and needed a future value of A dollars, then P would represent the amount needed in an account today to reach your goal in t years at an interest rate of r, compounded n times per year. If we solve the equation for P, it results in*

$$P = A(1 + r/n)^{-nt}.$$

1. Verify that the two formulas are equivalent by transforming the first equation into the second.

2. What should the present value of a savings account be to cover $30,000 of college expenses in 12.5 years, if the account pays 7.5% interest compounded quarterly?

3. Suppose you want to have $15,000 to buy a car in 3 years. What should the present value of a savings account be to reach this goal, if the account pays 5% compounded monthly?

4. A parent expects college costs to reach $40,000 in 6 years. To cover the $40,000 in future expenses, how much should the parent deposit in an account that pays 6% interest compounded *continuously*?

Exercises 5–8: **Average Rate of Change of e^x** *Complete the following. Round your answers to two decimal places.*
(a) *Find the average rate of change of $f(x) = e^x$ from x to $x + 0.001$ for the given x.*
(b) *Approximate $f(x) = e^x$ for the given x.*
(c) *Compare your answers in parts (a) and (b).*

5. $x = 0$ **6.** $x = -2$

7. $x = -0.5$ **8.** $x = 1.5$

9. Average Rate of Change What is the pattern in the results from Exercises 5–8? You may want to test your conjecture by trying different values of x.

10. Average Rate of Change (Refer to Exercises 5–8.) For any real number k, what is a good approximation for the average rate of change of $f(x) = e^x$ on a small interval near $x = k$? Explain how your answer relates to the graph of $f(x) = e^x$.

5.4 Logarithmic Functions and Models

- **Evaluate the common logarithmic function**
- **Evaluate logarithms with other bases**
- **Solve basic exponential and logarithmic equations**
- **Solve general exponential and logarithmic equations**
- **Convert between exponential and logarithmic forms**

Introduction

Bacteria growth is often exponential. For example, Table 5.18 and Figure 5.44 show the concentration B (in thousands per milliliter) in a bacteria sample after x days, where $B(x) = 10^x$. To determine how long it takes for the concentration to reach **500** thousand per milliliter, we can solve the equation $10^x = 500$. The graphical solution in Figure 5.45 is about 2.7, but finding this value symbolically requires *logarithms*.

Growth of Bacteria (1000/mL)

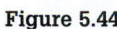

x	0	1	2	3	— Elapsed days
$B(x)$	1	10	100	1000	

Table 5.18

Exponential Growth

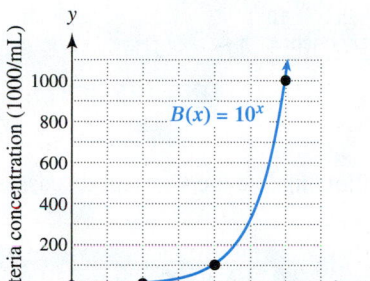

Figure 5.44

Solving $10^x = 500$

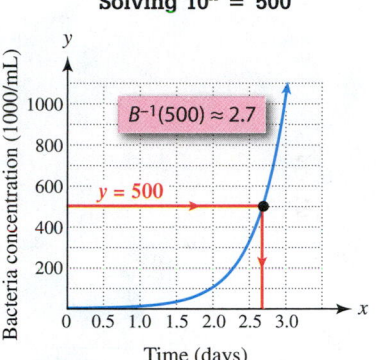

Figure 5.45

The Common Logarithmic Function

Notice that the graph of $B(x) = 10^x$ in Figure 5.44 is increasing and one-to-one. Its graph passes the horizontal line test. Therefore B has an inverse function. Because $B(1) = 10$, it follows that $B^{-1}(10) = 1$. Similarly, $B(2) = 100$ implies that $B^{-1}(100) = 2$, and in general, $B(k) = 10^k$ implies that $B^{-1}(10^k) = k$. Table 5.19 lists values for $B^{-1}(x)$. Notice that, unlike B, B^{-1} increases *very slowly* for large values of x.

Inverse Function of B

x	1	**10**	100	1000
$B^{-1}(x)$	0	**1**	2	3

Table 5.19

The solution to $10^x = 500$ represents the number of days needed for the bacteria concentration to reach 500 thousand per milliliter. To solve this equation, we must find an exponent k such that $10^k = 500$. Because $B^{-1}(10^k) = k$, this is equivalent to evaluating $B^{-1}(500)$, as shown graphically in Figure 5.45. From Table 5.19, k is between 2 and 3 because $100 \le 500 \le 1000$.

The inverse of $y = 10^x$ is defined to be the *common (base-10) logarithm,* denoted $\log x$ or $\log_{10} x$. Thus $10^2 = 100$ implies $\log 100 = 2$, $10^3 = 1000$ implies $\log 1000 = 3$, and in general, $10^a = b$ implies $\log b = a$. Thus $B^{-1}(500) = \log 500$ and $B^{-1}(x) = \log x$.

The common exponential function and the common logarithmic function, represented in Tables 5.20 and 5.21, are inverse functions.

Common Exponential Function

x	-4	-3	-2	-1	0	0.5	1	2	π
10^x	10^{-4}	10^{-3}	10^{-2}	10^{-1}	10^0	$10^{0.5}$	10^1	10^2	10^π

Table 5.20

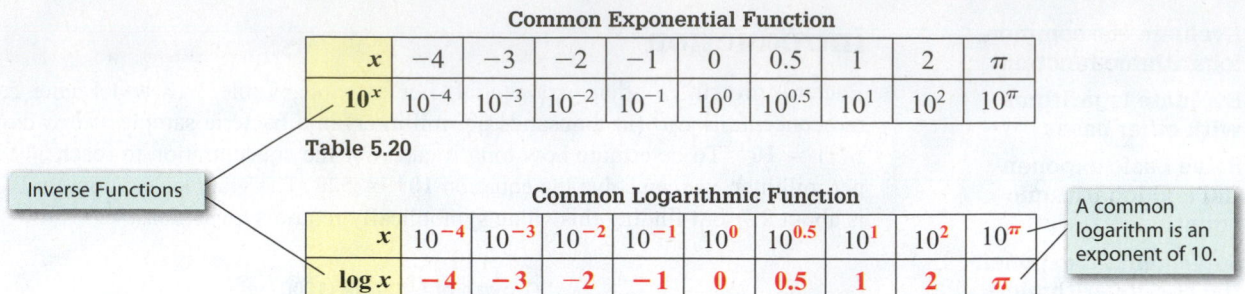

Inverse Functions

Common Logarithmic Function

x	10^{-4}	10^{-3}	10^{-2}	10^{-1}	10^0	$10^{0.5}$	10^1	10^2	10^π
$\log x$	-4	-3	-2	-1	0	0.5	1	2	π

A common logarithm is an exponent of 10.

Table 5.21

The common logarithm of a positive number x is the exponent k, such that $10^k = x$. This concept can be illustrated visually.

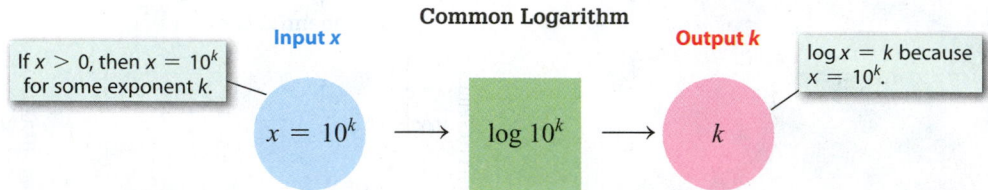

Common Logarithm

If $x > 0$, then $x = 10^k$ for some exponent k.

Input x   **Output k**

$$x = 10^k \longrightarrow \log 10^k \longrightarrow k$$

$\log x = k$ because $x = 10^k$.

The following See the Concept discusses the graph of $y = \log x$.

See the Concept: The Graph of $y = \log x$

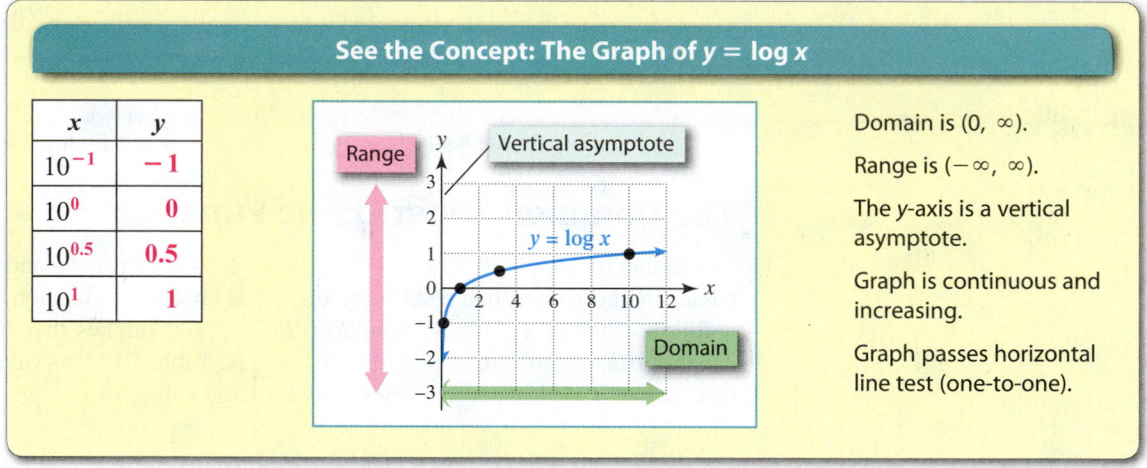

x	y
10^{-1}	-1
10^0	0
$10^{0.5}$	0.5
10^1	1

Domain is $(0, \infty)$.

Range is $(-\infty, \infty)$.

The y-axis is a vertical asymptote.

Graph is continuous and increasing.

Graph passes horizontal line test (one-to-one).

We have shown *verbal*, *numerical*, *graphical*, and *symbolic* descriptions of the common logarithmic function. A formal definition of the common logarithm is now given.

COMMON LOGARITHM

The **common logarithm of a positive number x,** denoted $\log x$, is defined by

$$\log x = k \quad \text{if and only if} \quad x = 10^k,$$

where k is a real number. The function given by

$$f(x) = \log x$$

is called the **common logarithmic function.**

NOTE The common logarithmic function outputs an *exponent k,* which may be positive, negative, or zero. However, a valid input must be positive. Thus its range is $(-\infty, \infty)$ and its domain is $(0, \infty)$.

The equation $\log x = k$ is called **logarithmic form** and the equation $x = 10^k$ is called **exponential form.** These forms are *equivalent.* That is, if one equation is true, the other equation must also be true.

Logarithmic Form   **Exponential Form**

$$\log 100 = 2 \qquad\qquad 10^2 = 100$$
$$\log 1000 = 3 \qquad\qquad 10^3 = 1000$$
$$\log \tfrac{1}{10} = -1 \qquad\qquad 10^{-1} = \tfrac{1}{10}$$
$$\log x = k \qquad\qquad 10^k = x$$

Pairs of equivalent equations

General forms

MAKING CONNECTIONS

Logarithms and Exponents *A logarithm is an exponent.* For example, to evaluate the logorithm $\log 1000$ ask, "10 to what power equals 1000?" The necessary *exponent* equals $\log 1000$. That is, $\log 1000 = \mathbf{3}$ because $10^{\mathbf{3}} = 1000$.

EXAMPLE 1   **Evaluating common logarithms**

Simplify each logarithm by hand.

(a) $\log 100{,}000$ **(b)** $\log 1$ **(c)** $\log \dfrac{1}{1000}$ **(d)** $\log \sqrt{10}$ **(e)** $\log(-2)$

SOLUTION

(a) Ask, "10 to what power equals 100,000?" Because $10^5 = 100{,}000$, it follows that $\log 100{,}000 = 5$. That is,

$$\log 100{,}000 = \log 10^5 = \mathbf{5}.$$

(b) Ask, "10 to what power equals 1?" Because $10^0 = 1$, it follows that $\log 1 = 0$. That is,

$$\log 1 = \log 10^0 = \mathbf{0}.$$

(c) $\log \dfrac{1}{1000} = \log 10^{-3} = \mathbf{-3}$

(d) $\log \sqrt{10} = \log(10^{1/2}) = \dfrac{\mathbf{1}}{\mathbf{2}}$

(e) The domain of $f(x) = \log x$ is $(0, \infty)$, so $\log(-2)$ is undefined because the input is negative.

Now Try Exercise 3

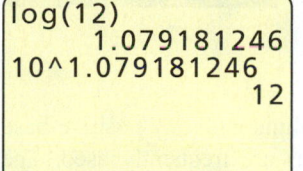

Figure 5.46

MAKING CONNECTIONS

Square Roots and Logarithms Much like the square root function, the common logarithmic function does not have an easy-to-evaluate formula. For example, $\sqrt{4} = 2$ and $\sqrt{100} = 10$ can be calculated mentally, but for $\sqrt{2}$ we usually rely on a calculator. Similarly, $\log 100 = 2$ can be found mentally since $100 = 10^2$, whereas $\log 12$ can be approximated using a calculator. See Figure 5.46. To check that $\log 12 \approx 1.079181246$, evaluate $10^{1.079181246} \approx 12$. Another similarity between the square root function and the common logarithmic function is that their domains do not include negative numbers. If outputs are restricted to real numbers, both $\sqrt{-3}$ and $\log(-3)$ are undefined expressions.

Reflection Across $y = x$

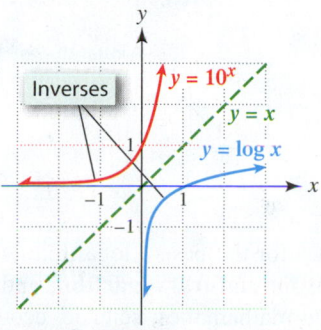

Figure 5.47

Graphs and Inverse Properties Because $\log x$ and 10^x are inverses, we can write $f(x) = \log x$ and $f^{-1}(x) = 10^x$. Their graphs are reflections across the line $y = x$, as shown in Figure 5.47. For inverse functions, $f(f^{-1}(x)) = x$ and $f^{-1}(f(x)) = x$, so the following inverse properties hold for all valid inputs x.

INVERSE PROPERTIES OF THE COMMON LOGARITHM

The following inverse properties hold for the common logarithm.

$$\log 10^x = x \quad \text{for any real number } x \text{ and}$$

$$10^{\log x} = x \quad \text{for any positive number } x$$

In Figures 5.48 and 5.49 a graphing calculator has been used to illustrate these properties. (If you have a calculator available, try some other examples.)

```
log(10^5)
                 5
log(10^1.6)
               1.6
log(10^(-2.5))
              -2.5
```

Figure 5.48

```
10^log(2)
                 2
10^log(3.7)
               3.7
10^log(0.12)
               .12
```

Figure 5.49

An Application Malaria deaths on the continent of Africa have gradually decreased during the past decade largely due to vaccines. These deaths can be modeled by a function that involves the common logarithm. (*Sources:* WHO; Malaria Vaccine Initiative.)

EXAMPLE 2 Modeling malaria deaths in Africa

The number of malaria deaths in millions x years after the year 2000 can be modeled by $D(x) = 0.9 \log\left(10 - \frac{2}{5}x\right)$.
(a) Evaluate $D(0)$ by hand and interpret the result.
(b) Approximate $D(10)$ and interpret the result.

SOLUTION
(a) $D(0) = 0.9 \log\left(10 - \frac{2}{5}(0)\right) = 0.9 \, \log 10 = 0.9(1) = 0.9$. In 2000 there were about 0.9 million deaths from malaria in Africa.
(b) $D(10) = 0.9 \log\left(10 - \frac{2}{5}(10)\right) = 0.9 \log 6 \approx 0.7$. (Use a calculator.) In 2010 there were about 0.7 million deaths from malaria in Africa.

Now Try Exercise 133(a)

Logarithms with Other Bases

Base-2 Logarithm It is possible to develop base-a logarithms with any positive base $a \neq 1$. For example, in computer science base-2 logarithms are frequently used. The base-2 logarithmic function, denoted $f(x) = \log_2 x$, is shown in Table 5.22. If x can be expressed in the form $x = 2^k$ for some k, then $\log_2 x = k$. Thus $\log_2 x$ is an *exponent*. A graph of $y = \log_2 x$ is shown in Figure 5.50.

Base-2 Logarithm

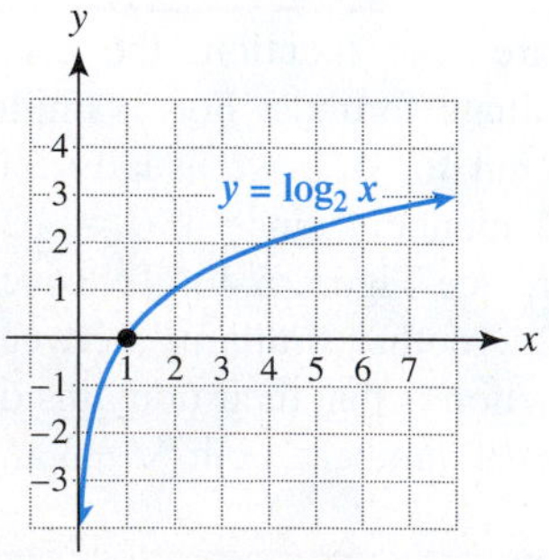

Figure 5.50

Base-2 Logarithm

x	$2^{-3.1}$	2^{-2}	$2^{-0.5}$	2^0	$2^{0.5}$	2^2	$2^{3.1}$
$\log_2 x$	-3.1	-2	-0.5	0	0.5	2	3.1

Table 5.22

A logarithm is an exponent.

Natural Logarithm In a similar manner, a table of values for the base-e logarithm is shown in Table 5.23. The base-e logarithm is referred to as the **natural logarithm** and denoted either $\log_e x$ or $\ln x$. Natural logarithms are used in mathematics, science, economics, and technology. A graph of $y = \ln x$ is shown in Figure 5.51.

Natural Logarithm

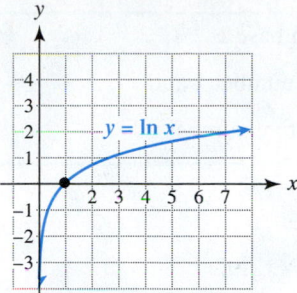

Figure 5.51

Natural Logarithm

x	$e^{-3.1}$	e^{-2}	$e^{-0.5}$	e^0	$e^{0.5}$	e^2	$e^{3.1}$
ln x	-3.1	-2	-0.5	0	0.5	2	3.1

A logarithm is an exponent.

Table 5.23

Base-a Logarithm We can define a logarithm for any positive base a, where $a \neq 1$. As with other logarithms we have discussed, any base-a logarithm has domain $(0, \infty)$ and range $(-\infty, \infty)$.

> **LOGARITHM**
>
> The **logarithm with base a of a positive number x,** denoted by $\log_a x$, is defined by
>
> $$\log_a x = k \quad \text{if and only if} \quad x = a^k,$$
>
> where $a > 0, a \neq 1$, and k is a real number. The function, given by
>
> $$f(x) = \log_a x,$$
>
> is called the **logarithmic function with base a.**

Evaluating Base-a Logarithms *A base-a logarithm is an exponent.* For example, to evaluate $\log_2 16$ ask, "2 to what power equals 16?" The required *exponent* equals $\log_2 16$. That is, $\log_2 16 = 4$ because $2^4 = 16$.

EXAMPLE 3 **Evaluating logarithms**

Evaluate each logarithm.

(a) $\log_2 8$ **(b)** $\log_5 \dfrac{1}{25}$ **(c)** $\log_7 49$ **(d)** $\ln e^{-7}$

SOLUTION

(a) Ask, "2 to what power equals 8?" Because $2^3 = 8$, it follows that $\log_2 8 = 3$. That is, $\log_2 8 = \log_2 2^3 = 3$.

(b) Ask, "5 to what power equals $\frac{1}{25}$?" Because $5^{-2} = \frac{1}{25}$, it follows that $\log_5 \frac{1}{25} = -2$. That is, $\log_5 \frac{1}{25} = \log_5 5^{-2} = -2$.

(c) $\log_7 49 = \log_7 7^2 = 2$

(d) $\ln e^{-7} = \log_e e^{-7} = -7$

> Now Try Exercises 33, 37, 39, and 49

Although the output from a base-a logarithm can be positive *or* negative, the input for a base-a logarithm must be positive. This fact can be used to determine the domain of logarithmic functions, as demonstrated in the next example.

EXAMPLE 4 **Finding the domain of a logarithmic function**

State the domain of f.

(a) $f(x) = \log_2(x - 4)$ **(b)** $f(x) = \ln(10^x)$

SOLUTION

(a) *The input for a logarithmic function must be positive.* Any x in the domain of f must satisfy $x - 4 > 0$, or equivalently, $x > 4$. Thus $D = (4, \infty)$.

(b) The expression 10^x is positive for all real numbers x. (See Figure 5.47, where the red graph of $y = 10^x$ is above the x-axis for all values of x.) Thus $D = (-\infty, \infty)$, or all real numbers.

> Now Try Exercises 13 and 17

CLASS DISCUSSION

Make a table of values for a base-4 logarithm. Evaluate $\log_4 16$.

Remember that *a logarithm is an exponent.* The expression $\log_a x$ is the exponent k such that $a^k = x$. Logarithms with base a also satisfy inverse properties.

> ### INVERSE PROPERTIES
>
> The following inverse properties hold for logarithms with base a.
>
> $$\log_a a^x = x \quad \text{for any real number } x \text{ and}$$
>
> $$a^{\log_a x} = x \quad \text{for any positive number } x$$

EXAMPLE 5   Applying inverse properties

Use inverse properties to evaluate each expression.

(a) $\log_6 6^{-1.3}$ **(b)** $5^{\log_5(x+8)}$ **(c)** $\log_{1/2}\left(\frac{1}{2}\right)^{45}$

SOLUTION

(a) $\log_a a^x = x$, so $\log_6 6^{-1.3} = \mathbf{-1.3}$.

(b) $a^{\log_a x} = x$, so $5^{\log_5(x+8)} = \mathbf{x + 8}$, provided $x > -8$.

(c) $\log_a a^x = x$, so $\log_{1/2}\left(\frac{1}{2}\right)^{45} = \mathbf{45}$. Note that the base of a logarithmic function can be a positive fraction less than 1.

> **Now Try Exercises 21, 23, and 25**

Graphs and Inverse Functions If $f(x) = a^x$, then its inverse function is given by $f^{-1}(x) = \log_a x$ and their graphs are reflections across the line $y = x$. These and other properties are illustrated for $f(x) = 2^x$ and $g(x) = e^x$ in the following See the Concept.

See the Concept: Inverses and Logarithms

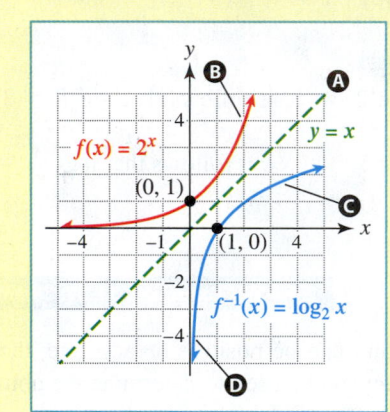

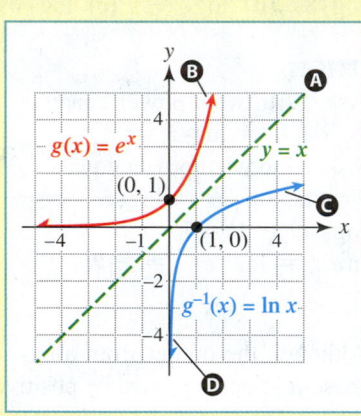

Ⓐ The graph of the inverse is a reflection across the line $y = x$.

Ⓑ Rapid exponential growth

Ⓒ Slow logarithmic growth

Ⓓ $\log_a x$ is negative (below the x-axis) for x between 0 and 1.

> **MAKING CONNECTIONS**
>
> **Exponential and Logarithmic Functions**   The inverse of an exponential function is a logarithmic function, and the inverse of a logarithmic function is an exponential function. For example,
>
> $$\text{if } f(x) = 10^x \qquad \text{then} \qquad f^{-1}(x) = \log x,$$
>
> $$\text{if } g(x) = \ln x \qquad \text{then} \qquad g^{-1}(x) = e^x, \quad \text{and}$$
>
> $$\text{if } h(x) = 2^x \qquad \text{then} \qquad h^{-1}(x) = \log_2 x.$$

Domains and Ranges of Inverses The domain of $f(x) = \log_a x$ is $(0, \infty)$ and its range is $(-\infty, \infty)$. The domain of its inverse $f^{-1}(x) = a^x$ is $(-\infty, \infty)$ and its range is $(0, \infty)$. Notice how domains and ranges are *interchanged* because they are inverse functions.

Transformations of Logarithmic Graphs

In Section 3.5 we discussed how to shift, reflect, stretch, and shrink graphs of functions. For example, to graph $y = (x - 2)^2$ we can shift the graph of $y = x^2$ right 2 units. In a similar manner, we can graph $y = \log_2(x - 2)$ by shifting the graph of $y = \log_2 x$ right 2 units. The graph of $y = \log_2 x$ has a vertical asymptote at $x = 0$, so the graph of $y = \log_2(x - 2)$ has a vertical asymptote at $x = 2$. See Figure 5.52.

Shifting a Logarithmic Graph

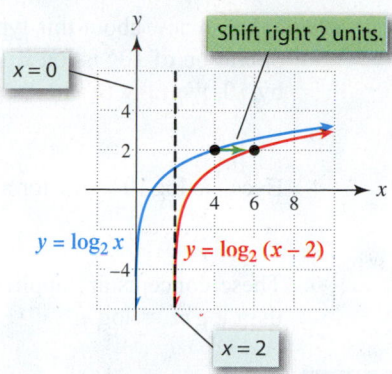

Figure 5.52

EXAMPLE 6   Transformations of a logarithmic graph

Explain how to obtain the graph of $g(x) = \log(-x) + 1$ from the graph of $f(x) = \log x$. Graph both functions in the same xy-plane.

SOLUTION

The graph of $y = \log(-x) + 1$ is similar to the graph of $f(x) = \log x$, except it is reflected about the y-axis and shifted upward 1 unit. Both graphs have a vertical asymptote at $x = 0$. See Figure 5.53.

Transforming $y = \log x$

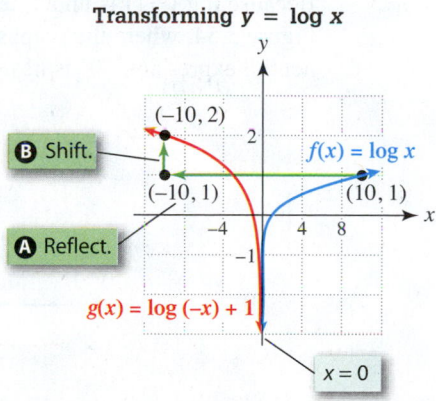

Figure 5.53

Now Try Exercise 113

Basic Equations

Base-10 Exponential Equations   To solve the equation $10 + x = 100$, we **subtract 10 from each side** because addition and subtraction are inverse operations.

$$10 + x - 10 = 100 - 10 \qquad \text{Subtract 10.}$$

$$x = 90$$

To solve the equation $10x = 100$, we **divide each side by 10** because multiplication and division are inverse operations.

$$\frac{10x}{10} = \frac{100}{10} \qquad \text{Divide by 10.}$$

$$x = 10$$

Now suppose that we want to solve the exponential equation

$$10^x = 100.$$

What is new about this type of equation is that the variable x is an *exponent*. The inverse operation of 10^x is $\log x$. Rather than subtracting 10 from each side or dividing each side by 10, we **take the base-10 logarithm of each side**. Doing so results in

$$\log 10^x = \log 100. \qquad \text{Take base-10 logarithm.}$$

Because $\log 10^x = x$ for all real numbers x, the equation becomes

$$x = \log 100, \qquad \text{or equivalently,} \qquad x = 2.$$

These concepts are applied in the next example. Note that if $m = n$ (m and n positive), then $\log m = \log n$.

EXAMPLE 7 **Solving equations of the form $10^x = k$**

Solve each equation, if possible.
(a) $10^x = 0.001$ **(b)** $10^x = 55$ **(c)** $10^x = -1$

SOLUTION
(a) Take the common logarithm of each side of the equation $10^x = 0.001$. Then

$$\log 10^x = \log 0.001, \qquad \text{or} \qquad x = \log 10^{-3} = -3.$$

(b) In a similar manner, $10^x = 55$ is equivalent to $x = \log 55 \approx 1.7404$.
(c) Taking the common logarithm of each side gives

$$\log 10^x = \log(-1).$$

Because $\log(-1)$ is undefined, there are no solutions. Graphical support is given in Figure 5.54, where the graphs of $y_1 = 10^x$ and $y_2 = -1$ do not intersect. The exponential expression 10^x is never negative.

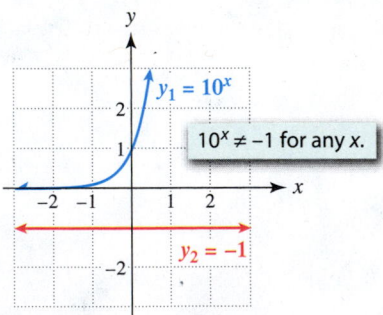

Figure 5.54

Now Try Exercise 57

EXAMPLE 8 **Solving a base-10 exponential equation**

Solve $4(10^{3x}) = 244$.

SOLUTION Begin by dividing each side by 4.

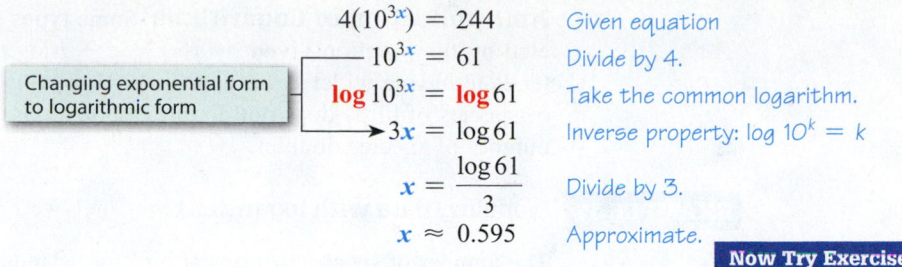

$$4(10^{3x}) = 244 \qquad \text{Given equation}$$

Changing exponential form to logarithmic form

$$10^{3x} = 61 \qquad \text{Divide by 4.}$$
$$\log 10^{3x} = \log 61 \qquad \text{Take the common logarithm.}$$
$$3x = \log 61 \qquad \text{Inverse property: } \log 10^k = k$$
$$x = \frac{\log 61}{3} \qquad \text{Divide by 3.}$$
$$x \approx 0.595 \qquad \text{Approximate.}$$

Now Try Exercise 69

Converting $10^{3x} = 61$ to $3x = \log 61$ is referred to as changing *exponential form* to *logarithmic form*. We usually include the step of taking the common logarithm of each side to emphasize the fact that *inverse properties* are being used to solve these equations.

Common Logarithmic Equations A **logarithmic equation** contains logarithms. To solve logarithmic equations we *exponentiate* each side of the equation and then apply the inverse property $10^{\log x} = x$. Note that if $m = n$, then $10^m = 10^n$.

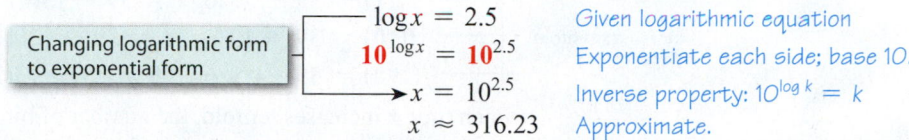

Changing logarithmic form to exponential form

$$\log x = 2.5 \qquad \text{Given logarithmic equation}$$
$$10^{\log x} = 10^{2.5} \qquad \text{Exponentiate each side; base 10.}$$
$$x = 10^{2.5} \qquad \text{Inverse property: } 10^{\log k} = k$$
$$x \approx 316.23 \qquad \text{Approximate.}$$

Converting the equation $\log x = 2.5$ to the equivalent equation $x = 10^{2.5}$ is called changing *logarithmic form* to *exponential form*.

EXAMPLE 9 **Solving equations of the form log x = k**

Solve each equation.
(a) $\log x = 3$ **(b)** $\log x = -2$ **(c)** $\log x = 2.7$

SOLUTION
(a)
$$\log x = 3 \qquad \text{Given equation}$$
$$10^{\log x} = 10^3 \qquad \text{Exponentiate each side; base 10.}$$
$$x = 10^3 \qquad \text{Inverse property: } 10^{\log k} = k$$
$$x = 1000 \qquad \text{Simplify.}$$

(b) Similarly, $\log x = -2$ is equivalent to $x = 10^{-2} = 0.01$.
(c) $\log x = 2.7$ is equivalent to $x = 10^{2.7} \approx 501.19$.

Now Try Exercise 81

EXAMPLE 10 **Solving a common logarithmic equation**

Solve $5 \log 2x = 16$.

SOLUTION Begin by dividing each side by 5.

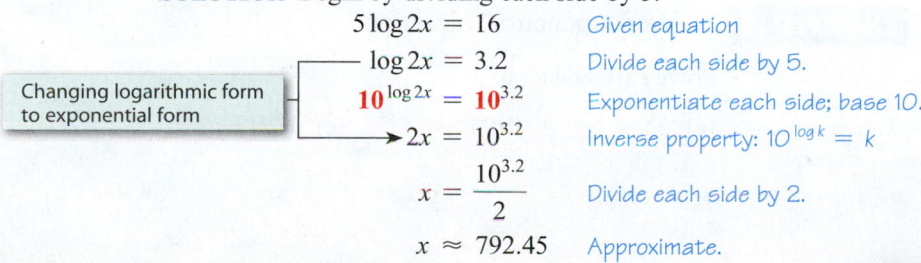

$$5 \log 2x = 16 \qquad \text{Given equation}$$

Changing logarithmic form to exponential form

$$\log 2x = 3.2 \qquad \text{Divide each side by 5.}$$
$$10^{\log 2x} = 10^{3.2} \qquad \text{Exponentiate each side; base 10.}$$
$$2x = 10^{3.2} \qquad \text{Inverse property: } 10^{\log k} = k$$
$$x = \frac{10^{3.2}}{2} \qquad \text{Divide each side by 2.}$$
$$x \approx 792.45 \qquad \text{Approximate.}$$

Now Try Exercise 91

An Application of Logarithms Some types of data grow slowly and can be modeled by the function given by $f(x) = a + b \log x$. For example, a larger area of land tends to have a wider variety of birds. However, if the land area doubles, the number of species of birds does not double; the land area has to more than double before the number of species doubles.

EXAMPLE 11 Modeling data with logarithms

The number of species of tropical birds on islands of different sizes x near New Guinea can be modeled by $f(x) = 39 + 13 \log x$, where x is in square miles. (*Source:* B. Freedman, *Environmental Ecology.*)

(a) If the number of square miles x of an island increases tenfold, how many additional species of birds are there?

(b) Determine both symbolically and graphically the size of an island that might have 50 different species of birds.

SOLUTION

(a) Start by evaluating $f(1)$, $f(10)$, and $f(100)$.

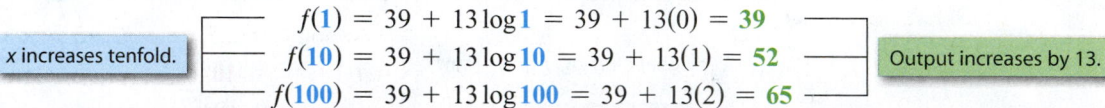

x increases tenfold.

$$f(\mathbf{1}) = 39 + 13 \log \mathbf{1} = 39 + 13(0) = \mathbf{39}$$
$$f(\mathbf{10}) = 39 + 13 \log \mathbf{10} = 39 + 13(1) = \mathbf{52}$$
$$f(\mathbf{100}) = 39 + 13 \log \mathbf{100} = 39 + 13(2) = \mathbf{65}$$

Output increases by 13.

Each time x increases tenfold, the number of bird species does not increase by a factor of 10; rather, it only increases by 13 species of birds. This slow growth is an example of *logarithmic growth*.

(b) *Symbolic Solution* We must solve the equation $39 + 13 \log x = 50$.

$39 + 13 \log x = 50$	Equation to solve
$13 \log x = 11$	Subtract 39.
$\log x = \dfrac{11}{13}$	Divide by 13.
$\mathbf{10}^{\log x} = \mathbf{10}^{11/13}$	Exponentiate each side; base 10.
$x = 10^{11/13}$	Inverse property: $10^{\log k} = k$
$x \approx 7$	Approximate.

To have 50 species of birds, an island should be about 7 square miles.

Graphical Solution The graphs of $y_1 = 39 + 13 \log x$ and $y_2 = 50$ in Figure 5.55 intersect near the point $(7.02, 50)$. This result agrees with our symbolic solution.

Now Try Exercise 139

$[0, 15, 5]$ by $[0, 70, 10]$

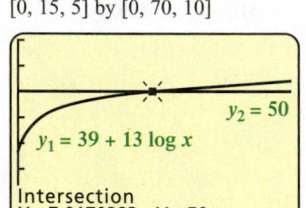

$y_2 = 50$

$y_1 = 39 + 13 \log x$

Intersection
X=7.0170383 Y=50

Figure 5.55

General Equations

Base-*a* Exponential Equations The exponential function $f(x) = a^x$ is one-to-one and therefore has an inverse function $f^{-1}(x) = \log_a x$. Because $f^{-1}(f(x)) = x$ for all real numbers x, or equivalently $\log_a a^x = x$, it follows that to solve an exponential equation we can take the base-*a* logarithm of each side. Note that if $m = n$ (m and n positive), then $\log_a m = \log_a n$.

EXAMPLE 12 Solving exponential equations

Solve each equation.

(a) $3^x = \dfrac{1}{27}$ **(b)** $e^x = 5$ **(c)** $3(2^x) - 7 = 20$

SOLUTION

(a) $\log_3 3^x = \log_3 \frac{1}{27}$ Take the base-3 logarithm of each side.

        $\log_3 3^x = \log_3 3^{-3}$ Properties of exponents

              $x = -3$ Inverse property: $\log_a a^k = k$

(b) Take the natural logarithm of each side. Then

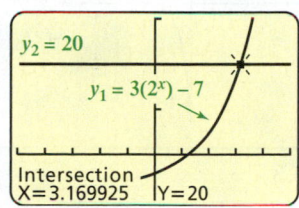

Figure 5.56

           $\ln e^x = \ln 5$ is equivalent to $x = \ln 5 \approx 1.609$.

Many calculators are able to compute natural logarithms. The evaluation of $\ln 5$ is shown in Figure 5.56. Notice that $e^{1.609437912} \approx 5$.

(c) $3(2^x) - 7 = 20$ Given equation

         $3(2^x) = 27$ Add 7 to each side.

             $2^x = 9$ Divide each side by 3.

> Changing exponential form to logarithmic form

      $\log_2 2^x = \log_2 9$ Take the base-2 logarithm of each side.

           $x = \log_2 9$ Inverse property: $\log_a a^k = k$

> **Now Try Exercises 59 and 71**

$[-5, 5, 1]$ by $[-10, 30, 10]$

$y_2 = 20$

$y_1 = 3(2^x) - 7$

Intersection
X=3.169925   Y=20

Figure 5.57

Introduction to the Change of Base Formula   Figure 5.57 gives a graphical solution of about 3.17 for the equation in Example 12(c). The exact solution of $\log_2 9$ can be approximated as 3.17 by using the following **change of base formula**. (Its derivation is given in Section 5.5.)

$$\log_a x = \frac{\log_b x}{\log_b a} \qquad \text{Change of base formula}$$

Because calculators typically have only $\log x$ and $\ln x$ keys, this formula is often written as

> Example of change of base formulas

$$\log_a x = \frac{\log x}{\log a} \quad \text{or} \quad \log_a x = \frac{\ln x}{\ln a}$$

For example, $\log_2 9 = \frac{\log 9}{\log 2} \approx 3.17$, which agrees with the graphical solution.

Base-a Logarithmic Equations   To solve the equation $\log_a x = k$, we exponentiate each side of the equation by using base a. That is, we use the inverse property: $a^{\log_a x} = x$ for all positive x. This is illustrated in the next example. Note that if $m = n$, then $a^m = a^n$.

EXAMPLE 13   Solving logarithmic equations

Solve each equation.
(a) $\log_2 x = 5$ (b) $\log_5 x = -2$ (c) $\ln x = 4.3$ (d) $3 \log_2 5x = 9$

SOLUTION

(a) $\log_2 x = 5$ Given equation

      $2^{\log_2 x} = 2^5$ Exponentiate each side; base 2.

          $x = 2^5$ Inverse property: $a^{\log_a k} = k$

          $x = 32$ Simplify.

(b) Similarly, $\log_5 x = -2$ is equivalent to $x = 5^{-2} = \frac{1}{25}$.

(c) $\ln x = 4.3$ is equivalent to $x = e^{4.3} \approx 73.7$.

(d) $3 \log_2 5x = 9$ Given equation

      $\log_2 5x = 3$ Divide each side by 3.

> Changing logarithmic form to exponential form

      $2^{\log_2 5x} = 2^3$ Exponentiate each side; base 2.

         $5x = 8$ Inverse property: $a^{\log_a k} = k$

          $x = \dfrac{8}{5}$ Divide each side by 5.

> **Now Try Exercises 83 and 87**

5.4 Putting It All Together

The following table summarizes some important concepts about base-a logarithms. Common and natural logarithms satisfy the same properties.

CONCEPT	EXPLANATION	EXAMPLES
Base-a logarithm $a > 0, a \neq 1$	The base-a logarithm of a positive number x is $$\log_a x = k \quad \text{if and only if} \quad x = a^k.$$ That is, a logarithm is an exponent k.	$\log 100 = \log 10^2 = 2$ $\log_2 8 = \log_2 2^3 = 3$ $\log_3 \sqrt[3]{3} = \log_3 3^{1/3} = \frac{1}{3}$ $\ln 5 \approx 1.609$ (using a calculator)
Graph of $y = \log_a x$	The graph of the base-a logarithmic function *always* passes through the point $(1, 0)$ because $\log_a 1 = \log_a a^0 = 0$. The y-axis is a vertical asymptote. The graph passes the horizontal line test, so $f(x) = \log_a x$ is one-to-one and has an inverse given by $f^{-1}(x) = a^x$.	$y = \log_a x, \ a > 1$ $(1, 0)$
Inverse properties	a^x and $\log_a x$ represent inverse operations. That is, $$\log_a a^x = x$$ and $$a^{\log_a x} = x \quad \text{for } x > 0.$$	$\log 10^{-3} = -3$ $\log_4 4^6 = 6$ $10^{\log x} = x, \ x > 0$ $4^{\log_4 2x} = 2x, \ x > 0$ $e^{\ln 5} = 5$
Inverse functions	The inverse function of $f(x) = a^x$ is $$f^{-1}(x) = \log_a x.$$	$f(x) = 10^x \quad \leftrightarrow \quad f^{-1}(x) = \log x$ $g(x) = e^x \quad \leftrightarrow \quad g^{-1}(x) = \ln x$ $h(x) = \log_2 x \quad \leftrightarrow \quad h^{-1}(x) = 2^x$
Exponential and logarithmic forms	$\log_a x = k$ is equivalent to $x = a^k$.	$\log_5 25 = 2$ is equivalent to $25 = 5^2$. $4^3 = 64$ is equivalent to $\log_4 64 = 3$.
Exponential equations	To solve $a^x = k$, take the base-a logarithm of each side.	$10^x = 15 \qquad\qquad e^x = 20$ $\log 10^x = \log 15 \quad \ln e^x = \ln 20$ $x = \log 15 \qquad\qquad x = \ln 20$
Logarithmic equations	To solve $\log_a x = k$, exponentiate each side; use base a.	$\log x = 3 \qquad\qquad \ln x = 5$ $10^{\log x} = 10^3 \qquad e^{\ln x} = e^5$ $x = 1000 \qquad\qquad x = e^5$

5.4 Exercises

Common Logarithms

Exercises 1 and 2: Complete the table.

1.

x	10^0	10^4	10^{-8}	$10^{1.26}$
$\log x$		4		

2.

x	10^{-2}	$10^{-\pi}$	10^5	$10^{7.89}$
$\log x$			5	

Exercises 3–8: Evaluate each expression by hand, if possible.

3. (a) $\log(-3)$ **(b)** $\log \frac{1}{100}$

(c) $\log \sqrt{0.1}$ **(d)** $\log 5^0$

4. (a) $\log 10{,}000$ **(b)** $\log(-\pi)$

(c) $\log \sqrt{0.001}$ **(d)** $\log 8^0$

5. (a) $\log 10$ **(b)** $\log 10{,}000$

(c) $20 \log 0.1$ **(d)** $\log 10 + \log 0.001$

6. (a) $\log 100$ **(b)** $\log 1{,}000{,}000$

(c) $5 \log 0.01$ **(d)** $\log 0.1 - \log 1000$

7. (a) $2 \log 0.1 + 4$ **(b)** $\log 10^{1/2}$

(c) $3 \log 100 - \log 1000$ **(d)** $\log(-10)$

8. (a) $\log(-4)$ **(b)** $\log 1$

(c) $\log 0$ **(d)** $-6 \log 100$

Exercises 9 and 10: Determine mentally an integer n so that the logarithm is between n and n + 1. Check your result with a calculator.

9. (a) $\log 79$ **(b)** $\log 500$

(c) $\log 5$ **(d)** $\log 0.5$

10. (a) $\log 63$ **(b)** $\log 5000$

(c) $\log 9$ **(d)** $\log 0.04$

Exercises 11 and 12: Find the exact value of each expression.

11. (a) $\log \sqrt{1000}$ **(b)** $\log \sqrt[3]{10}$

(c) $\log \sqrt[5]{0.1}$ **(d)** $\log \sqrt{0.01}$

12. (a) $\log \sqrt{100{,}000}$ **(b)** $\log \sqrt[3]{100}$

(c) $2 \log \sqrt{0.1}$ **(d)** $10 \log \sqrt[3]{10}$

Domains of Logarithmic Functions

Exercises 13–20: Find the domain of f and write it in set-builder or interval notation.

13. $f(x) = \log(x + 3)$ **14.** $f(x) = \ln(2x - 4)$

15. $f(x) = \log_2(x^2 - 1)$ **16.** $f(x) = \log_4(4 - x^2)$

17. $f(x) = \log_3(4^x)$ **18.** $f(x) = \log_5(5^x - 25)$

19. $f(x) = \ln(\sqrt{3 - x} - 1)$ **20.** $f(x) = \log(4 - \sqrt{2 - x})$

General Logarithms

Exercises 21–50: Simplify the expression.

21. $\log_8 8^{-5.7}$ **22.** $\log_4 4^{-1.23}$

23. $7^{\log_7 2x}$ **24.** $6^{\log_6(x+1)}$

25. $\log_{1/3}\left(\frac{1}{3}\right)^{64}$ **26.** $\log_{0.4}\left(\frac{2}{5}\right)^{-3}$

27. $\ln e^{-4}$ **28.** $2^{\log_2 k}$

29. $\log_5 5^{\pi}$ **30.** $\log_6 6^9$

31. $3^{\log_3(x-1)}$ **32.** $8^{\log_8(\pi+1)}$

33. $\log_2 64$ **34.** $\log_2 \frac{1}{4}$

35. $\log_4 2$ **36.** $\log_3 9$

37. $\ln e^{-3}$ **38.** $\ln e$

39. $\log_8 64$ **40.** $\ln \sqrt[3]{e}$

41. $\log_{1/2}\left(\frac{1}{4}\right)$ **42.** $\log_{1/3}\left(\frac{1}{27}\right)$

43. $\log_{1/6} 36$ **44.** $\log_{1/4} 64$

45. $\log_a \frac{1}{a}$ **46.** $\log_a(a^2 \cdot a^3)$

47. $\log_5 5^0$ **48.** $\ln \sqrt{e}$

49. $\log_2 \frac{1}{16}$ **50.** $\log_8 8^k$

Exercises 51 and 52: Complete the table by hand.

51. $f(x) = 2 \log_2(x - 5)$

x	6	7	21
$f(x)$			

52. $f(x) = 2 \log_3(2x)$

x	$\frac{1}{18}$	$\frac{3}{2}$	$\frac{9}{2}$
$f(x)$			

Exponential and Logarithmic Forms

Exercises 53 and 54: Change each equation to its equivalent logarithmic form.

53. (a) $7^{4x} = 4$ **(b)** $e^x = 7$ **(c)** $c^x = b$

54. (a) $5^{2x} = 9$ **(b)** $b^x = a$ **(c)** $d^{2x} = b$

Exercises 55 and 56: Change each equation to its equivalent exponential form.

55. (a) $\log_8 x = 3$ **(b)** $\log_9(2 + x) = 5$ **(c)** $\log_k b = c$

56. (a) $\log x = 4$ **(b)** $\ln 8x = 7$ **(c)** $\log_a x = b$

Solving Exponential Equations

Exercises 57–80: Solve each equation. Use the change of base formula to approximate exact answers to the nearest hundredth when appropriate.

57. (a) $10^x = 0.01$ **(b)** $10^x = 7$ **(c)** $10^x = -4$

58. (a) $10^x = 1000$ **(b)** $10^x = 5$ **(c)** $10^x = -2$

59. (a) $4^x = \frac{1}{16}$ **(b)** $e^x = 2$ **(c)** $5^x = 125$

60. (a) $2^x = 9$ **(b)** $10^x = \frac{1}{1000}$ **(c)** $e^x = 8$

61. (a) $9^x = 1$ **(b)** $10^x = \sqrt{10}$ **(c)** $4^x = \sqrt[3]{4}$

62. (a) $2^x = \sqrt{8}$ **(b)** $7^x = 1$ **(c)** $e^x = \sqrt[3]{e}$

63. $e^{-x} = 3$ **64.** $e^{-x} = \frac{1}{2}$

65. $10^x - 5 = 95$ **66.** $2 \cdot 10^x = 66$

67. $10^{3x} = 100$ **68.** $4 \cdot 10^{2x} + 1 = 21$

69. $5(10^{4x}) = 65$ **70.** $3(10^{x-2}) = 72$

71. $4(3^x) - 3 = 13$

72. $5(7^x) + 3 = 83$

73. $e^x + 1 = 24$

74. $1 - 2e^x = -5$

75. $2^x + 1 = 15$

76. $3 \cdot 5^x = 125$

77. $5e^x + 2 = 20$

78. $6 - 2e^{3x} = -10$

79. $8 - 3(2)^{0.5x} = -40$

80. $2(3)^{-2x} + 5 = 167$

Solving Logarithmic Equations

Exercises 81–102: Solve each equation. Approximate answers to four decimal places when appropriate.

81. **(a)** $\log x = 2$ **(b)** $\log x = -3$ **(c)** $\log x = 1.2$

82. **(a)** $\log x = 1$ **(b)** $\log x = -4$ **(c)** $\log x = 0.3$

83. **(a)** $\log_2 x = 6$ **(b)** $\log_3 x = -2$ **(c)** $\ln x = 2$

84. **(a)** $\log_4 x = 2$ **(b)** $\log_8 x = -1$ **(c)** $\ln x = -2$

85. $\log_2 x = 1.2$

86. $\log_4 x = 3.7$

87. $5\log_7 2x = 10$

88. $2\log_4 x = 3.4$

89. $2\log x = 6$

90. $\log 4x = 2$

91. $2\log 5x = 4$

92. $6 - \log x = 3$

93. $4\ln x = 3$

94. $\ln 5x = 8$

95. $5\ln x - 1 = 6$

96. $2\ln 3x = 8$

97. $4\log_2 x = 16$

98. $\log_3 5x = 10$

99. $5\ln(2x) + 6 = 12$

100. $16 - 4\ln 3x = 2$

101. $9 - 3\log_4 2x = 3$

102. $7\log_6(4x) + 5 = -2$

Exercises 103 and 104: Find values for a and b so that f(x) models the data exactly.

103. $f(x) = a + b\log x$

x	1	10	100
f(x)	5	7	9

104. $f(x) = a + b\log_2 x$

x	1	2	4
f(x)	3.1	6	8.9

Graphs of Logarithmic Functions

Exercises 105–108: Use the graph of f to sketch a graph of f^{-1}. Give a symbolic representation of f^{-1}.

105.

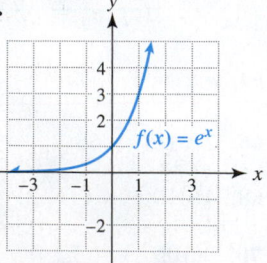

106.

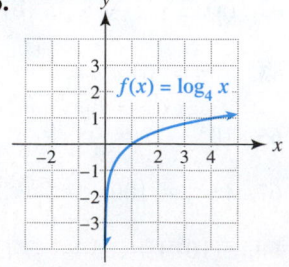

107.

108.

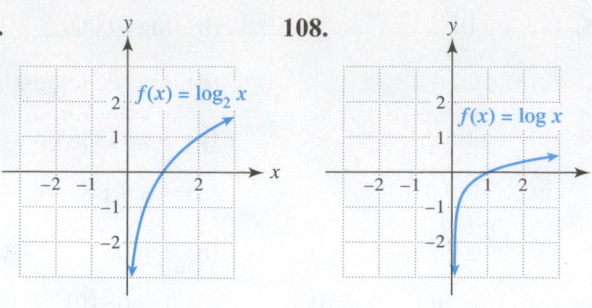

Exercises 109 and 110: Graph y = f(x). Is f increasing or decreasing on its domain?

109. $f(x) = \log_{1/2} x$

110. $f(x) = \log_{1/3} x$

Exercises 111 and 112: Complete the following.

(a) Graph $y = f(x)$, $y = f^{-1}(x)$, and $y = x$.

(b) Determine the intervals where f and f^{-1} are increasing or decreasing.

111. $f(x) = \log_3 x$

112. $f(x) = \log_{1/2} x$

Transformations of Graphs

Exercises 113–120: Use transformations to graph $y = g(x)$. Give the equation of any asymptotes.

113. $g(x) = \log(x - 2)$

114. $g(x) = \log(x + 2)$

115. $g(x) = 3\log x$

116. $g(x) = 2\log_2 x$

117. $g(x) = \log_2(-x)$

118. $g(x) = -\log_2 x$

119. $g(x) = 2 + \ln(x - 1)$

120. $g(x) = \ln(x + 1) - 1$

Exercises 121–124: Graph f and state its domain.

121. $f(x) = \log(x + 1)$

122. $f(x) = \log(x - 3)$

123. $f(x) = \ln(-x)$

124. $f(x) = \ln(x^2 + 1)$

Applications

125. Decibels Sound levels in decibels (dB) can be calculated by

$$D(x) = 10\log(10^{16}x),$$

where x is the intensity of the sound in watts per square meter. The human ear begins to hurt when the intensity reaches $x = 10^{-4}$. Find how many decibels this represents.

126. Decibels (Refer to Exercise 125.) If the intensity increases by a *factor* of 10, find the increase in decibels.

127. Runway Length There is a relation between an airplane's weight x and the runway length L required for takeoff. For some airplanes the minimum runway length L in thousands of feet is given by $L(x) = 3\log x$, where x is measured in thousands of pounds. (*Source:* L. Haefner, *Introduction to Transportation Systems.*)

(a) Evaluate $L(100)$ and interpret the result.

(b) If the weight of an airplane increases tenfold from 10,000 to 100,000 pounds, does the length of the required runway also increase by a factor of 10? Explain.

(c) Generalize your answer from part (b).

128. Runway Length   (Refer to Exercise 127.) Estimate the maximum weight of a plane that can take off from a runway that is 5 thousand feet long.

129. Acid Rain   Air pollutants frequently cause acid rain. A measure of the acidity is pH, which ranges between 1 and 14. Pure water is neutral and has a pH of 7. Acidic solutions have a pH less than 7, whereas alkaline solutions have a pH greater than 7. A pH value can be computed by $\text{pH} = -\log x$, where x represents the hydrogen ion concentration in moles per liter. In rural areas of Europe, rainwater typically has $x = 10^{-4.7}$. (**Source:** G. Howells, *Acid Rain and Acid Water*.)

(a) Find its pH.

(b) Seawater has a pH of 8.2. How many times greater is the hydrogen ion concentration in rainwater from rural Europe than in seawater?

130. Acid Rain   (Refer to Exercise 129.) Find the hydrogen ion concentration for the following pH levels of acid rain. (**Source:** G. Howells.)

(a) 4.92 (pH of rain at Amsterdam Islands)

(b) 3.9 (pH of some rain in the eastern United States)

131. Earthquakes   The Richter scale is used to measure the intensity of earthquakes, where intensity corresponds to the amount of energy released by an earthquake. If an earthquake has an intensity of x, then its *magnitude*, as computed by the Richter scale, is given by the formula $R(x) = \log \frac{x}{I_0}$, where I_0 is the intensity of a small measurable earthquake.

(a) On July 26, 1963, an earthquake in Yugoslavia had a magnitude of 6.0 on the Richter scale, and on August 19, 1977, an earthquake in Indonesia measured 8.0. Find the intensity x for each of these earthquakes if $I_0 = 1$.

(b) How many times more intense was the Indonesian earthquake than the Yugoslavian earthquake?

132. Earthquakes   (Refer to Exercise 131.) If the intensity x of an earthquake increases by a *factor* of 10^3, by how much does the Richter number R increase? Generalize your results.

133. Hurricanes   Hurricanes are some of the largest storms on Earth. They are very low pressure areas with diameters of over 500 miles. The barometric air pres-

sure in inches of mercury at a distance of x miles from the eye of a severe hurricane is modeled by the formula $f(x) = 0.48 \ln(x + 1) + 27$ (**Source:** A. Miller and R. Anthes, *Meteorology*.)

(a) Evaluate $f(0)$ and $f(100)$. Interpret the results.

(b) Graph f in $[0, 250, 50]$ by $[25, 30, 1]$. Describe how air pressure changes as one moves away from the eye of the hurricane.

(c) At what distance from the eye of the hurricane is the air pressure 28 inches of mercury?

134. Predicting Wind Speed   Wind speed typically varies in the first 20 meters above the ground. For a particular day, let the formula $f(x) = 1.2 \ln x + 2.3$ compute the wind speed in meters per second at a height x meters above the ground for $x \geq 1$. (**Source:** A. Miller.)

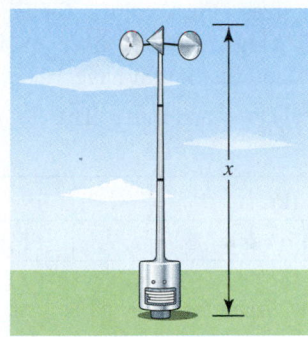

(a) Find the wind speed at a height of 5 meters.

(b) Graph f in the window $[0, 20, 5]$ by $[0, 7, 1]$. Interpret the graph.

(c) Estimate the height where the wind speed is 5 meters per second.

135. Cooling an Object   A pot of boiling water with a temperature of 100°C is set in a room with a temperature of 20°C. The temperature T of the water after x hours is given by $T(x) = 20 + 80e^{-x}$.

(a) Estimate the temperature of the water after 1 hour.

(b) How long did it take the water to cool to 60°C?

136. Warming an Object   A can of soda with a temperature of 5°C is set in a room with a temperature of 20°C. The temperature T of the soda after x minutes is given by $T(x) = 20 - 15(10)^{-0.05x}$.

(a) Estimate the temperature of the soda after 5 minutes.

(b) After how many minutes was the temperature of the soda 15°C?

137. Traffic Flow   (Refer to Example 11, Section 5.3.) Cars arrive randomly at an intersection with an average rate of 20 cars per hour. The likelihood, or probability,

(continued)

that no car enters the intersection within a period of x minutes can be estimated by $f(x) = e^{-x/3}$.
(a) What is the probability that no car enters the intersection during a 5-minute period?

(b) Determine the value of x that gives a 30% chance that no car enters the intersection during an interval of x minutes.

138. **Population Growth** The population of Nevada in millions is given by $P(x) = 2.7e^{0.014x}$, where $x = 0$ corresponds to 2010. (**Source:** Bureau of the Census.)
(a) Determine symbolically the year when the population of Nevada might be 3 million.

🖩 (b) Solve part (a) graphically.

139. **Diversity of Birds** (Refer to Example 11.) The table lists the number of species of birds on islands of various sizes. Find values for a and b so that $f(x) = a + b\log x$ models these data. Estimate the size of an island that might have 16 species of birds.

(*Hint:* Let $f(1) = 7$ and find a. Then let $f(10) = 11$ and find b.)

Area (km²)	0.1	1	10	100	1000
Species of birds	3	7	11	15	19

140. **Diversity of Insects** The table lists the number of types of insects found in wooded regions with various acreages. Find values for a and b so that $f(x) = a + b\log x$ models these data. Then use f to estimate an acreage that might have 1200 types of insects.

Area (acres)	10	100	1000	10,000
Insect Types	500	800	1100	1400

141. **Growth of Bacteria** The table lists the number of bacteria y in millions after an elapsed time of x days.

x	0	1	2	3	4
y	3	6	12	24	48

(a) Find values for C and a so that $f(x) = Ca^x$ models the data.

(b) Estimate when there were 16 million bacteria.

142. **Growth of an Investment** The growth of an investment is shown in the table in the next column.

x (years)	0	5	10	15	20
y (dollars)	100	300	900	2700	8100

(a) Find values for C and a so that $f(x) = Ca^x$ models the data.

(b) Estimate when the account contained $2000.

Writing about Mathematics

143. Describe the relationship among exponential functions and logarithmic functions. Explain why logarithms are needed to solve exponential equations.

144. Give verbal, numerical, graphical, and symbolic representations of a base-5 logarithmic function.

Extended And Discovery Exercises

1. **Average Rate of Change of ln x** Find the average rate of change of $f(x) = \ln x$ from x to $x + 0.001$ for each value of x. Round your answers to two decimal places.
(a) $x = 1$ (b) $x = 2$

(c) $x = 3$ (d) $x = 4$

2. **Average Rate of Change of ln x** (See Exercise 1.) Compare each average rate of change of $\ln x$ to x. What is the pattern? Make a generalization.

3. **Climate Change** According to one model, the future increases in average global temperatures (due to carbon dioxide levels exceeding 280 parts per million) can be estimated using $T = 6.5 \ln(C/280)$, where C is the concentration of atmospheric carbon dioxide in parts per million (ppm) and T is in degrees Fahrenheit. Let future amounts of carbon dioxide x years after 2000 be modeled by the formula $C(x) = 364(1.005)^x$. (**Source:** W. Clime, *The Economics of Global Warming.*)
(a) Use composition of functions to write T as a function of x. Evaluate T when $x = 100$ and interpret the result.

🖩 (b) Graph $C(x)$ in [0, 200, 50] by [0, 1000, 100] and $T(x)$ in [0, 200, 50] by [0, 10, 1]. Describe each graph.

(c) How does an exponential growth in carbon dioxide concentrations affect the increase in temperature?

CHECKING BASIC CONCEPTS FOR SECTIONS 5.3 AND 5.4

1. If the principal is $1200 and the interest rate is 9.5% compounded monthly, calculate the account balance after 4 years. Determine the balance if the interest is compounded continuously.

2. Find values for C and a so that $f(x) = Ca^x$ models the data in the table.

x	0	1	2	3
y	4	2	1	0.5

3. Explain verbally what $\log_2 15$ represents. Is it equal to an integer? (Do not use a calculator.)

4. Evaluate each of the following logarithms by hand.

   (a) $\log_6 36$ (b) $\log \sqrt{10} + \log 0.01$ (c) $\ln \dfrac{1}{e^2}$

5. Solve each equation.

   (a) $e^x = 5$ (b) $10^x = 25$ (c) $\log x = 1.5$

6. Solve each equation.

   (a) $2e^x + 1 = 25$ (b) $\log 2x = 2.3$

   (c) $\log x^2 = 1$

7. **Population**   The population of California in millions x years after 2010 can be modeled by $P(x) = 37.3e^{0.01x}$. (*Source:* Bureau of the Census.)

   (a) Evaluate $P(2)$. Interpret this result.

   (b) Find the y-intercept on the graph of $y = f(x)$. Interpret this result.

   (c) Estimate the annual percent increase in the population of California.

   (d) Predict when the population might reach 40 million.

8. **Growth in Salary**   Suppose that a person's salary is initially \$30,000 and is modeled by $f(x)$, where x represents the number of years of experience. Use $f(x)$ to approximate the years of experience when the salary first *exceeds* \$60,000.

   (a) $f(x) = 30{,}000(1.1)^x$

   (b) $f(x) = 30{,}000 \log(10 + x)$

   Would most people prefer that their salaries increase exponentially or logarithmically?

5.5 Properties of Logarithms

- **Apply basic properties of logarithms**
- **Expand and combine logarithmic expressions**
- **Use the change of base formula**

Introduction

The discovery of logarithms by John Napier (1550–1617) played an important role in the history of science. Logarithms were instrumental in allowing Johannes Kepler (1571–1630) to calculate the positions of the planet Mars, which led to his discovery of the laws of planetary motion. Kepler's laws were used by Isaac Newton (1643–1727) to discover the universal laws of gravitation. Although calculators and computers have made tables of logarithms obsolete, applications involving logarithms still play an important role in modern-day computation. One reason is that logarithms possess important properties. For example, the loudness of a sound can be measured in decibels by the formula $f(x) = 10 \log(10^{16}x)$, where x is the intensity of the sound. In Example 4, we use properties of logarithms to simplify this formula.

Basic Properties of Logarithms

Logarithms possess several important properties. One property of logarithms states that

   the sum of the logarithms of two numbers equals the logarithm of their product.

For example, we see in Figure 5.58 that

$$\log 5 + \log 2 = \log 10 \qquad 5 \cdot 2 = 10$$

and in Figure 5.59 that

$$\log 4 + \log 25 = \log 100. \qquad 4 \cdot 25 = 100$$

These calculations show the product rule of logarithms: $\log_a m + \log_a n = \log_a(mn)$.

Product Rule: $\log m + \log n = \log(m \cdot n)$

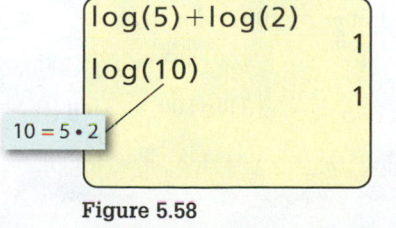

Figure 5.58

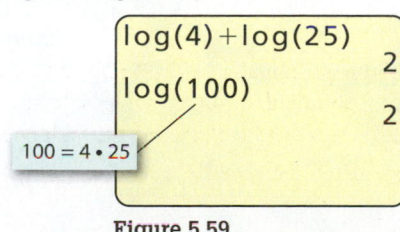

Figure 5.59

Four properties, or rules, of logarithms are as follows.

PROPERTIES OF LOGARITHMS

For positive numbers m, n, and $a \neq 1$ and any real number r:

1. $\log_a 1 = 0$ and $\log_a a = 1$ Logarithms of 1 and a

2. $\log_a m + \log_a n = \log_a (mn)$ Product rule

3. $\log_a m - \log_a n = \log_a \left(\dfrac{m}{n} \right)$ Quotient rule

4. $\log_a (m^r) = r \log_a m$ Power rule

Algebra Review

To review properties of exponents, see Section 4.8 and Chapter R (page R-7).

The properties of logarithms are a direct result of the properties of exponents and the inverse property $\log_a a^k = k$, as shown below.

Logarithms of 1 and a: This property is a direct result of the inverse property: $\log_a a^x = x$.

$$\log_a 1 = \log_a a^0 = 0 \quad \text{and} \quad \log_a a = \log_a a^1 = 1$$

Examples: $\log 1 = 0$ and $\ln e = 1$

Product Rule: If m and n are positive numbers, then we can write $m = a^c$ and $n = a^d$ for some real numbers c and d.

$$\log_a m + \log_a n = \log_a a^c + \log_a a^d = c + d$$
$$\log_a (mn) = \log_a (a^c a^d) = \log_a (a^{c+d}) = c + d$$

> Both expressions equal $c + d$.

Thus $\log_a m + \log_a n = \log_a (mn)$.

Example: Let $m = 100$ and $n = 1000$.

$$\log 100 + \log 1000 = \log 10^2 + \log 10^3 = 2 + 3 = 5$$
$$\log (100 \cdot 1000) = \log 100{,}000 = \log 10^5 = 5$$

Quotient Rule: Let $m = a^c$ and $n = a^d$ for some real numbers c and d.

$$\log_a m - \log_a n = \log_a a^c - \log_a a^d = c - d$$
$$\log_a \left(\frac{m}{n} \right) = \log_a \left(\frac{a^c}{a^d} \right) = \log_a (a^{c-d}) = c - d$$

> Both expressions equal $c - d$.

Thus $\log_a m - \log_a n = \log_a \left(\frac{m}{n} \right)$.

Example: Let $m = 100$ and $n = 1000$.

$$\log 100 - \log 1000 = \log 10^2 - \log 10^3 = 2 - 3 = -1$$
$$\log \left(\frac{100}{1000} \right) = \log \left(\frac{1}{10} \right) = \log (10^{-1}) = -1$$

Power Rule: Let $m = a^c$ and r be any real number.

$$\log_a m^r = \log_a (a^c)^r = \log_a (a^{cr}) = cr$$
$$r \log_a m = r \log_a a^c = rc = cr$$

> Both expressions equal cr.

Thus $\log_a (m^r) = r \log_a m$.

Example: Let $m = 100$ and $r = 3$.

$$\log 100^3 = \log 1{,}000{,}000 = \log 10^6 = 6$$
$$3 \log 100 = 3 \log 10^2 = 3 \cdot 2 = 6$$

Caution: $\log_a (m + n) \neq \log_a m + \log_a n; \quad \log_a (m - n) \neq \log_a m - \log_a n$

EXAMPLE 1   **Recognizing properties of logarithms**

Use a calculator to evaluate each pair of expressions. Then state which rule of logarithms this calculation illustrates.

(a) $\ln 5 + \ln 4,\ \ln 20$ **(b)** $\log 10 - \log 5,\ \log 2$ **(c)** $\log 5^2,\ 2\log 5$

SOLUTION

(a) From Figure 5.60, we see that the two expressions are equal. These calculations illustrate the product rule because $\ln 5 + \ln 4 = \ln(5 \cdot 4) = \ln 20$.

(b) The two expressions are equal in Figure 5.61, and these calculations illustrate the quotient rule because $\log 10 - \log 5 = \log \frac{10}{5} = \log 2$.

(c) The two expressions are equal in Figure 5.62, and these calculations illustrate the power rule because $\log 5^2 = 2\log 5$.

Product Rule	Quotient Rule	Power Rule

ln(5)+ln(4)
      2.995732274
ln(20)
      2.995732274
 	
$20 = 5 \cdot 4$
	log(10)−log(5)
	      .3010299957
	log(2)
	      .3010299957
	
	$2 = \frac{10}{5}$
		log(5²)
		      1.397940009
		2log(5)
		      1.397940009

Figure 5.60   Figure 5.61   Figure 5.62

Now Try Exercises 1, 3, and 5

Expanding and Combining Logarithmic Expressions

We can use the properties of logarithms to expand and combine logarithmic expressions, as illustrated by the following.

Expanding: Moving Left to Right

$\log 5b = \log 5 + \log b$ Product rule

$\log \dfrac{x}{2} = \log x - \log 2$ Quotient rule

$\log x^2 = 2\log x$ Power rule

Combining: Moving Right to Left

The next two examples demonstrate how to expand logarithmic expressions.

EXAMPLE 2   **Expanding logarithmic expressions**

Use properties of logarithms to expand each expression. Write your answers without exponents.

(a) $\log xy$ **(b)** $\ln \dfrac{5}{z}$ **(c)** $\log_4 \dfrac{\sqrt[3]{x}}{\sqrt{k}}$

SOLUTION

(a) By the product rule, $\log xy = \log x + \log y$.

(b) By the quotient rule, $\ln \frac{5}{z} = \ln 5 - \ln z$.

(c) Begin by using the quotient rule.

$$\log_4 \frac{\sqrt[3]{x}}{\sqrt{k}} = \log_4 \sqrt[3]{x} - \log_4 \sqrt{k} \qquad \text{Quotient rule}$$
$$= \log_4 x^{1/3} - \log_4 k^{1/2} \qquad \text{Properties of exponents}$$
$$= \tfrac{1}{3}\log_4 x - \tfrac{1}{2}\log_4 k \qquad \text{Power rule}$$

Now Try Exercises 7, 11, and 21

EXAMPLE 3 Expanding logarithmic expressions

Expand each expression. Write your answers without exponents.

(a) $\log_2 2x^4$ **(b)** $\ln \frac{7x^3}{k^2}$ **(c)** $\log \frac{\sqrt{x+1}}{(x-2)^3}$

SOLUTION

(a) $\log_2 2x^4 = \log_2 2 + \log_2 x^4$ Product rule

$= 1 + 4\log_2 x$ Power rule; logarithm of a

(b) $\ln \frac{7x^3}{k^2} = \ln 7x^3 - \ln k^2$ Quotient rule

$= \ln 7 + \ln x^3 - \ln k^2$ Product rule

$= \ln 7 + 3\ln x - 2\ln k$ Power rule

(c) $\log \frac{\sqrt{x+1}}{(x-2)^3} = \log \sqrt{x+1} - \log(x-2)^3$ Quotient rule

$= \log(x+1)^{1/2} - \log(x-2)^3$ Property of exponents

$= \frac{1}{2}\log(x+1) - 3\log(x-2)$ Power rule

Now Try Exercises 9, 15, and 27

An Application Sometimes properties of logarithms are used in applications to simplify a formula. This is illustrated in the next example.

EXAMPLE 4 Analyzing sound with decibels

Sound levels in decibels (dB) can be computed by $D(x) = 10\log(10^{16}x)$.
(a) Use properties of logarithms to simplify the formula for D.
(b) Ordinary conversation has an intensity of $x = 10^{-10}$ w/cm^2. Find the decibel level.

SOLUTION
(a) To simplify the formula, use the product rule.

$D(x) = 10\log(10^{16}x)$ Given formula

$= 10(\log 10^{16} + \log x)$ Product rule

$= 10(16 + \log x)$ Inverse property

$= 160 + 10\log x$ Distributive property

(b) $D(10^{-10}) = 160 + 10\log(10^{-10}) = 160 + 10(-10) = 160 - 100 = 60$

Ordinary conversation occurs at about 60 decibels.

Now Try Exercise 83

The next two examples demonstrate how properties of logarithms can be used to combine logarithmic expressions.

EXAMPLE 5 Combining terms in logarithmic expressions

Write each expression as the logarithm of a single expression.

(a) $\ln 2e + \ln \frac{1}{e}$ **(b)** $\log_2 27 + \log_2 x^3$ **(c)** $\log x^3 - \log x^2$

SOLUTION
(a) By the product rule, $\ln 2e + \ln \frac{1}{e} = \ln\left(2e \cdot \frac{1}{e}\right) = \ln 2$.

(b) By the product rule, $\log_2 27 + \log_2 x^3 = \log_2(27x^3)$.

(c) By the quotient rule, $\log x^3 - \log x^2 = \log \frac{x^3}{x^2} = \log x$.

Now Try Exercises 41, 43, and 49

EXAMPLE 6 **Combining terms in logarithmic expressions**

Write each expression as the logarithm of a single expression.

(a) $\log 5 + \log 15 - \log 3$ **(b)** $2\ln x - \frac{1}{2}\ln y - 3\ln z$

(c) $5\log_3 x + \log_3 2x - \log_3 y$

SOLUTION

(a)
$$\log 5 + \log 15 - \log 3 = \log(5 \cdot 15) - \log 3 \qquad \text{Product rule}$$
$$= \log\left(\frac{5 \cdot 15}{3}\right) \qquad \text{Quotient rule}$$
$$= \log 25 \qquad \text{Simplify.}$$

Algebra Review

To review rational exponents and radical notation, see Chapter R (page R-38).

(b)
$$2\ln x - \frac{1}{2}\ln y - 3\ln z = \ln x^2 - \ln y^{1/2} - \ln z^3 \qquad \text{Power rule}$$
$$= \ln\left(\frac{x^2}{y^{1/2}}\right) - \ln z^3 \qquad \text{Quotient rule}$$
$$= \ln \frac{x^2}{y^{1/2}z^3} \qquad \text{Quotient rule}$$
$$= \ln \frac{x^2}{z^3\sqrt{y}} \qquad \text{Properties of exponents}$$

(c)
$$5\log_3 x + \log_3 2x - \log_3 y = \log_3 x^5 + \log_3 2x - \log_3 y \qquad \text{Power rule}$$
$$= \log_3(x^5 \cdot 2x) - \log_3 y \qquad \text{Product rule}$$
$$= \log_3 \frac{2x^6}{y} \qquad \text{Quotient rule}$$

Now Try Exercises 39, 47, and 51

Change of Base Formula

Calculators usually have keys to approximate common and natural logarithms. Occasionally it is necessary to evaluate a logarithm with a base other than 10 or e. This computation can be accomplished by using a change of base formula.

CHANGE OF BASE FORMULA

Let $x, a \neq 1$, and $b \neq 1$ be positive real numbers. Then

$$\log_a x = \frac{\log_b x}{\log_b a}.$$

The change of base formula can be derived as follows.

$$y = \log_a x$$
$$a^y = a^{\log_a x} \qquad \text{Exponentiate each side; base } a.$$
$$a^y = x \qquad \text{Inverse property}$$
$$\log_b a^y = \log_b x \qquad \text{Take base-}b \text{ logarithm of each side.}$$
$$y\log_b a = \log_b x \qquad \text{Power rule}$$
$$y = \frac{\log_b x}{\log_b a} \qquad \text{Divide by } \log_b a.$$
$$\log_a x = \frac{\log_b x}{\log_b a} \qquad \text{Substitute } \log_a x \text{ for } y. \text{ (First equation)}$$

To calculate $\log_2 5$, evaluate $\frac{\log 5}{\log 2} \approx 2.322$. The change of base formula was used with $x = 5, a = 2$, and $b = 10$. We could also have evaluated $\frac{\ln 5}{\ln 2} \approx 2.322$.

EXAMPLE 7   Applying the change of base formula

Use a calculator to approximate each expression to the nearest thousandth.
(a) $\log_4 20$ (b) $\log_2 125 + \log_7 39$

SOLUTION
(a) Using the change of base formula, we have $\log_4 20 = \dfrac{\log 20}{\log 4} \approx 2.161$. We could also evaluate $\dfrac{\ln 20}{\ln 4}$, as shown in Figure 5.63.

(b) $\log_2 125 + \log_7 39 = \dfrac{\log 125}{\log 2} + \dfrac{\log 39}{\log 7} \approx 8.848$. See Figure 5.64.

Finding $\log_4 20$

```
log(20)/log(4)
            2.160964047
ln(20)/ln(4)
            2.160964047
```

Figure 5.63

Finding $\log_2 125 + \log_7 39$

```
log(125)/log(2)+
log(39)/log(7)
            8.848482542
```

Figure 5.64

Now Try Exercises 67 and 71

The change of base formula can be used to graph base-a logarithmic functions.

EXAMPLE 8   Using the change of base formula for graphing

$[-10, 10, 1]$ by $[-10, 10, 1]$

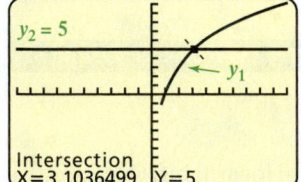

Intersection
X=3.1036499 Y=5

Figure 5.65

Solve the equation $\log_2(x^3 + x - 1) = 5$ graphically.

SOLUTION
Graph $y_1 = \log(x^3 + x - 1)/\log 2$ and $y_2 = 5$. See Figure 5.65. Their graphs intersect near the point $(3.104, 5)$. The solution is given by $x \approx 3.104$.

Now Try Exercise 77

5.5 Putting It All Together

The following table summarizes some properties of logarithms.

CONCEPT	EXPLANATION	EXAMPLES
Properties of logarithms	1. $\log_a 1 = 0$ and $\log_a a = 1$ 2. $\log_a m + \log_a n = \log_a(mn)$ 3. $\log_a m - \log_a n = \log_a(\frac{m}{n})$ 4. $\log_a(m^r) = r\log_a m$	1. $\ln 1 = 0$ and $\log_2 2 = 1$ 2. $\log 3 + \log 6 = \log(3 \cdot 6) = \log 18$ 3. $\log_3 8 - \log_3 2 = \log_3 \frac{8}{2} = \log_3 4$ 4. $\log 6^7 = 7\log 6$
Change of base formula	Let $x, a \neq 1$, and $b \neq 1$ be positive real numbers. Then $$\log_a x = \frac{\log_b x}{\log_b a}.$$	$$\log_3 6 = \frac{\log 6}{\log 3} = \frac{\ln 6}{\ln 3} \approx 1.631$$
Graphing logarithmic functions	Use the change of base formula to graph $y = \log_a x$ whenever $a \neq 10$ and $a \neq e$.	To graph $y = \log_2 x$, let $Y_1 = \log(X)/\log(2)$ or $Y_1 = \ln(X)/\ln(2)$.

5.5 Exercises

Note: When applying properties of logarithms, assume that all variables are positive.

Properties of Logarithms

Exercises 1–6: (Refer to Example 1.) Use a calculator to approximate each pair of expressions. Then state which property of logarithms this calculation illustrates.

1. $\log 4 + \log 7$, $\log 28$
2. $\ln 12 + \ln 5$, $\ln 60$
3. $\ln 72 - \ln 8$, $\ln 9$
4. $3 \log 4$, $\log 4^3$
5. $10 \log 2$, $\log 1024$
6. $\log 100 - \log 20$, $\log 5$

Exercises 7–32: (Refer to Examples 2 and 3.) Expand the expression. If possible, write your answer without exponents.

7. $\log_2 ab$
8. $\ln 3x$
9. $\ln 7a^4$
10. $\log \dfrac{a^3}{3}$
11. $\log \dfrac{6}{z}$
12. $\ln \dfrac{xy}{z}$
13. $\log \dfrac{x^2}{3}$
14. $\log 3x^6$
15. $\ln \dfrac{2x^7}{3k}$
16. $\ln \dfrac{kx^3}{5}$
17. $\log_2 4k^2x^3$
18. $\log \dfrac{5kx^2}{11}$
19. $\log_5 \dfrac{25x^3}{y^4}$
20. $\log_2 \dfrac{32}{xy^2}$
21. $\ln \dfrac{x^4}{y^2 \sqrt{z^3}}$
22. $\ln \dfrac{x \sqrt[3]{y^2}}{z^6}$
23. $\log_4 (0.25(x + 2)^3)$
24. $\log (0.001(a - b)^{-3})$
25. $\log_5 \dfrac{x^3}{(x - 4)^4}$
26. $\log_8 \dfrac{(3x - 2)^2}{x^2 + 1}$
27. $\log_2 \dfrac{\sqrt{x}}{z^2}$
28. $\log \sqrt{\dfrac{xy^2}{z}}$
29. $\ln \sqrt[3]{\dfrac{2x + 6}{(x + 1)^5}}$
30. $\log \dfrac{\sqrt{x^2 + 4}}{\sqrt[3]{x - 1}}$
31. $\log_2 \dfrac{\sqrt[3]{x^2 - 1}}{\sqrt{1 + x^2}}$
32. $\log_8 \sqrt[3]{\dfrac{x + y^2}{2z + 1}}$

Exercises 33–56: (Refer to Examples 5 and 6.) Combine the expressions by writing them as a logarithm of a single expression.

33. $\log 2 + \log 3$
34. $\log \sqrt{2} + \log \sqrt[3]{2}$
35. $\ln \sqrt{5} - \ln 25$
36. $\ln 33 - \ln 11$
37. $\log 20 + \log \dfrac{1}{10}$
38. $\log_2 24 + \log_2 \dfrac{1}{48}$
39. $\log 4 + \log 3 - \log 2$
40. $\log_3 5 - \log_3 10 - \log_3 \dfrac{1}{2}$
41. $\log_7 5 + \log_7 k^2$
42. $\log_6 45 + \log_6 b^3$
43. $\ln x^6 - \ln x^3$
44. $\log 10x^5 - \log 5x$
45. $\log \sqrt{x} + \log x^2 - \log x$
46. $\log \sqrt[4]{x} + \log x^4 - \log x^2$
47. $3 \ln x - \dfrac{3}{2} \ln y + 4 \ln z$
48. $\dfrac{2}{3} \ln y - 4 \ln x - \dfrac{1}{2} \ln z$
49. $\ln \dfrac{1}{e^2} + \ln 2e$
50. $\ln 4e^3 - \ln 2e^2$
51. $2 \ln x - 4 \ln y + \dfrac{1}{2} \ln z$
52. $\dfrac{1}{3} \log_5 (x + 1) + \dfrac{1}{3} \log_5 (x - 1)$
53. $\log 4 - \log x + 7 \log \sqrt{x}$
54. $\ln 3e - \ln \dfrac{1}{4e}$
55. $2 \log (x^2 - 1) + 4 \log (x - 2) - \dfrac{1}{2} \log y$
56. $\log_3 x + \log_3 \sqrt{x + 3} - \dfrac{1}{3} \log_3 (x - 4)$

Exercises 57–62: Complete the following.

(a) *Make a table of $f(x)$ and $g(x)$ to determine whether $f(x) = g(x)$.*

(b) *If possible, use properties of logarithms to show that $f(x) = g(x)$.*

57. $f(x) = \log 3x + \log 2x$, $\quad g(x) = \log 6x^2$
58. $f(x) = \ln 3x - \ln 2x$, $\quad g(x) = \ln x$
59. $f(x) = \ln 2x^2 - \ln x$, $\quad g(x) = \ln 2x$
60. $f(x) = \log x^2 + \log x^3$, $\quad g(x) = 5 \log x$

61. $f(x) = \ln x^4 - \ln x^2$, $g(x) = 2\ln x$

62. $f(x) = (\ln x)^2$, $g(x) = 2\ln x$

Exercises 63–66: Sketch a graph of f.

63. $f(x) = \log_2 x$ **64.** $f(x) = \log_2 x^2$

65. $f(x) = \log_3 |x|$ **66.** $f(x) = \log_4 2x$

Change of Base Formula

Exercises 67–76: Use the change of base formula to approximate the logarithm to the nearest thousandth.

67. $\log_2 25$ **68.** $\log_3 67$

69. $\log_5 130$ **70.** $\log_6 0.77$

71. $\log_2 5 + \log_2 7$ **72.** $\log_9 85 + \log_7 17$

73. $\sqrt{\log_4 46}$ **74.** $2\log_5 15 + \sqrt[3]{\log_3 67}$

75. $\dfrac{\log_2 12}{\log_2 3}$ **76.** $\dfrac{\log_7 125}{\log_7 25}$

Exercises 77–80: Solve the equation graphically. Express any solutions to the nearest thousandth.

77. $\log_2(x^3 + x^2 + 1) = 7$

78. $\log_3(1 + x^2 + 2x^4) = 4$

79. $\log_2(x^2 + 1) = 5 - \log_3(x^4 + 1)$

80. $\ln(x^2 + 2) = \log_2(10 - x^2)$

Applications

81. Runway Length (Refer to Exercise 127, Section 5.4.) Use a natural logarithm (instead of a common logarithm) to write the formula $L(x) = 3\log x$. Evaluate $L(50)$ for each formula. Do your answers agree?

82. Biology The equation $y = bx^a$ is used in applications involving biology. Another form of this equation is $\log y = \log b + a\log x$. Use properties of logarithms to obtain this second equation from the first. (*Source:* H. Lancaster, *Quantitative Methods in Biological and Medical Sciences.*)

83. Decibels (Refer to Example 4.) If the intensity x of a sound increases by a factor of 10, by how much does the decibel level increase?

84. Decibels (Refer to Example 4.) Use a natural logarithm to write the formula $f(x) = 160 + 10\log x$. Evaluate $f(5 \times 10^{-8})$ for each formula. Do your answers agree?

85. Light Absorption When sunlight passes through lake water, its initial intensity I_0 decreases to a weaker intensity I at a depth of x feet according to the formula

$$\ln I - \ln I_0 = -kx,$$

where k is a positive constant and I_0 is the sun's intensity at the surface. Solve this equation for I.

86. Dissolving Salt If C grams of salt are added to a sample of water, the amount A of undissolved salt is modeled by $A = Ca^x$, where x is time. Solve the equation for x.

87. Population Growth The population P (in millions) of California x years after 2010 can be modeled by $P = 37.3e^{0.01x}$.

    (a) Use properties of logarithms to solve this equation for x.

    (b) Use your equation to find x when $P = 40$. Interpret your answer.

88. Population Growth The population P (in millions) of Georgia x years after 2010 can be modeled by the equation $P = 9.7e^{0.017x}$.

    (a) Use properties of logarithms to solve this equation for x.

    (b) Use your equation to find x when $P = 12$. Interpret your answer.

89. Solve $A = Pe^{rt}$ for t.

90. Solve $P = P_0 e^{r(t-t_0)} + 5$ for t.

91. Write the sum

$$\log 1 + 2\log 2 + 3\log 3 + 4\log 4 + 5\log 5$$

as a logarithm of a single expression.

92. Show that

$$\log_2\left(x + \sqrt{x^2 - 4}\right) + \log_2\left(x - \sqrt{x^2 - 4}\right)$$

equals 2. What is the domain of the given expression?

Writing about Mathematics

93. A student insists that $\log(x + y)$ and $\log x + \log y$ are equal. How could you convince the student otherwise?

94. A student insists that $\log\left(\frac{x}{y}\right)$ and $\frac{\log x}{\log y}$ are equal. How could you convince the student otherwise?

5.6 Exponential and Logarithmic Equations

- Solve exponential equations
- Solve logarithmic equations

Introduction

The population of the world has grown rapidly during the past century. Near the end of 2011, world population was estimated to be **7** billion. Exponential functions and equations are often used to model this type of rapid growth, whereas logarithms are used to model slower growth.

Exponential Equations

The population P of the world was 7 billion in 2011 and can be modeled by the function $P(x) = \mathbf{7(1.01)}^{x-2011}$, where x is the year. We can use P to predict the year when world population might reach **8** billion by solving the *exponential equation*

$$\mathbf{7(1.01)}^{x-2011} = \mathbf{8}. \qquad \text{Exponential equation}$$

An equation in which the variable occurs in the *exponent* of an expression is called an **exponential equation**. In the next example, we use the power rule of logarithms, $\log_a(m^r) = r\log_a m$, to solve this equation.

EXAMPLE 1   Modeling world population

World population in billions during year x can be modeled by $P(x) = 7(1.01)^{x-2011}$, shown in Figure 5.66. Solve the equation $7(1.01)^{x-2011} = 8$ symbolically to predict the year when world population might reach 8 billion.

SOLUTION   First divide each side by 7, and then take the common logarithm of each side. (The natural logarithm could also be used.)

$$7(1.01)^{x-2011} = 8 \qquad \text{Given equation}$$

$$(1.01)^{x-2011} = \frac{8}{7} \qquad \text{Divide by 7.}$$

$$\log(1.01)^{x-2011} = \log\frac{8}{7} \qquad \text{Take the common logarithm.}$$

$$(x - 2011)\log(1.01) = \log\frac{8}{7} \qquad \log(m^r) = r\log m$$

$$x - 2011 = \frac{\log(8/7)}{\log(1.01)} \qquad \text{Divide by } \log(1.01).$$

$$x = 2011 + \frac{\log(8/7)}{\log(1.01)} \qquad \text{Add 2011.}$$

$$x \approx 2024 \qquad \text{Approximate.}$$

This model predicts that world population might reach 8 billion during 2024.

Now Try Exercise 79

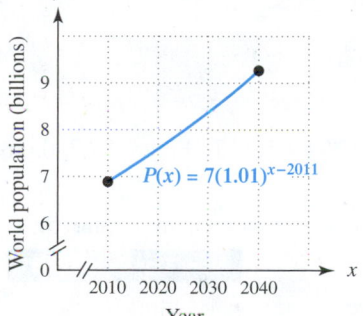

World Population

$P(x) = 7(1.01)^{x-2011}$

Year

Figure 5.66

CLASS DISCUSSION

What is the growth factor for $P(x) = 7(1.01)^{x-2011}$? By what percentage is it predicted that world population will grow, on average, each year after 2011?

The steps above are applied in Example 2 to solve the equation symbolically.

EXAMPLE 2 Calculating the thickness of a runway

Heavier aircraft require runways with thicker pavement for landings and takeoffs. The relation between the thickness of the pavement t in inches and gross weight W in thousands of pounds can be approximated by

$$W(t) = 18.29e^{0.246t}.$$

(a) Determine the required thickness of the runway for a 130,000-pound plane.
(b) Solve part (a) graphically and numerically.

SOLUTION

(a) *Symbolic Solution* Because the unit for W is thousands of pounds, we can solve the equation $W(t) = 130$ for t.

$$18.29e^{0.246t} = 130 \qquad \textcolor{blue}{W(t) = 130}$$

$$e^{0.246t} = \frac{130}{18.29} \qquad \textcolor{blue}{\text{STEP 1: Divide by 18.29.}}$$

$$\mathbf{ln}\, e^{0.246t} = \mathbf{ln}\,\frac{130}{18.29} \qquad \textcolor{blue}{\text{STEP 2: Take the natural logarithm of each side.}}$$

$$0.246t = \ln\frac{130}{18.29} \qquad \textcolor{blue}{\text{STEP 3: Inverse property: } \ln e^k = k}$$

$$t = \frac{\ln(130/18.29)}{0.246} \qquad \textcolor{blue}{\text{STEP 4: Divide by 0.246.}}$$

$$t \approx 7.97 \qquad \textcolor{blue}{\text{Approximate.}}$$

The runway should be about 8 inches thick.

Graphical Solution

[0, 10, 1] by [0, 200, 50]

$y_2 = 130$

$y_1 = 18.29e^{0.246t}$

Intersection
X=7.9722764 Y=130

Figure 5.67

Numerical Solution

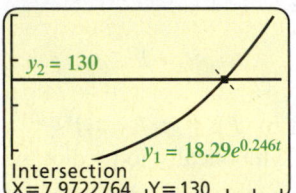

X	Y1	Y2	
5	62.574	130	
6	80.026	130	
7	102.35	130	
8	130.89	130	$\leftarrow y_1 \approx y_2$
9	167.39	130	
10	214.08	130	
11	273.79	130	

X=8

Figure 5.68

(b) *Graphical Solution* Let $Y_1 = 18.29e^{\wedge}(.246X)$ and $Y_2 = 130$. In Figure 5.67, their graphs intersect near (7.97, 130).

Numerical Solution Numerical support of this result is shown in Figure 5.68, where $y_1 \approx y_2$ when $x = 8$.

Now Try Exercise 81

To solve an exponential equation with a base other than 10 or e, the power rule of logarithms can be used. This technique is used in Examples 1 and 3.

EXAMPLE 3 Modeling the decline of bluefin tuna

Bluefin tuna are large fish that can weigh 1500 pounds and swim at speeds of 55 miles per hour. Because they are used for sushi, a prime fish can be worth over $50,000. As a result, the western Atlantic bluefin tuna have had their numbers decline exponentially. Their numbers in thousands from 1974 to 1991 can be modeled by the formula $f(x) = 230(0.881)^x$, where x is years after 1974. (In more recent years, controls have helped slow this decline.)
(*Source:* B. Freedman, *Environmental Ecology.*)

(a) Estimate the number of bluefin tuna in 1974 and 1991.
(b) Determine symbolically the year when they numbered 50 thousand.

SOLUTION

(a) To determine their numbers in 1974 and 1991, evaluate $f(0)$ and $f(17)$.

$$f(0) = 230(0.881)^0 = 230(1) = 230$$

$$f(17) = 230(0.881)^{17} \approx 26.7$$

Bluefin tuna decreased from 230 thousand in 1974 to fewer than 27 thousand in 1991.

(b) Solve the equation $f(x) = 50$ for x.

$$230(0.881)^x = 50 \qquad \textcolor{teal}{f(x) = 50}$$

$$0.881^x = \frac{5}{23} \qquad \textcolor{teal}{\text{STEP 1: Divide by 230; simplify.}}$$

> The common logarithm could also be used.

$$\ln 0.881^x = \ln \frac{5}{23} \qquad \textcolor{teal}{\text{STEP 2: Take the natural logarithm of each side.}}$$

$$x \ln 0.881 = \ln \frac{5}{23} \qquad \textcolor{teal}{\text{STEP 3: } \ln m^r = r \ln m}$$

$$x = \frac{\ln(5/23)}{\ln 0.881} \qquad \textcolor{teal}{\text{STEP 4: Divide by } \ln 0.881.}$$

$$x \approx 12.04 \qquad \textcolor{teal}{\text{Approximate.}}$$

They numbered about 50 thousand in $1974 + 12.04 \approx 1986$.

Now Try Exercise 87

Moore's law states that the processing speed and memory capacity of computers doubles every 2 years. Researchers have recently found an even more profound law relating energy efficiency and computing power. This law says that the number of computations that a computer can perform on a fixed amount of electricity (such as a kilowatt-hour) has doubled every 1.6 years since the mid-1940s, when vacuum tubes were used in computers. In the next example we use an exponential function to model this new law and make a prediction. (***Source:*** *The Economist.*)

EXAMPLE 4   **Modeling the electrical efficiency of computers**

In 1945 computers could perform about 1000 computations with 1 kilowatt-hour of electricity. This number has doubled every 1.6 years.

(a) Find an exponential function $E(x) = Ca^x$ that gives the number of computations a computer can perform on 1 kilowatt-hour, where x is years after 1945.

(b) Evaluate $E(65)$ and interpret your result.

(c) In what year did computers first perform 1,000,000 calculations per kilowatt-hour?

SOLUTION

(a) The initial value is **1000** when $x = 0$, so $C = \textbf{1000}$ and $E(x) = \textbf{1000}a^x$. When $x = 1.6$ the number of computations doubles to 2000 per kilowatt-hour, so let $E(1.6) = 2000$ and solve for a.

$$E(1.6) = 2000 \qquad \textcolor{teal}{\text{Doubles after 1.6 years}}$$

$$1000a^{1.6} = 2000 \qquad \textcolor{teal}{\text{Substitute.}}$$

> This is *not* an exponential equation. It is a power equation because the variable is the base, not the exponent.

$$a^{1.6} = 2 \qquad \textcolor{teal}{\text{Divide each side by 1000.}}$$

$$a^{8/5} = 2 \qquad \textcolor{teal}{\text{Convert 1.6 to } \tfrac{8}{5}.}$$

$$(a^{8/5})^{5/8} = 2^{5/8} \qquad \textcolor{teal}{\text{Raise each side to the } \tfrac{5}{8} \text{ power.}}$$

$$a = 2^{5/8} \qquad \textcolor{teal}{\text{Properties of exponents}}$$

Thus $E(x) = \textbf{1000}(2^{5/8})^x$.

(b) $E(65) = 1000(2^{5/8})^{65} \approx 1.7 \times 10^{15}$. In $1945 + 65 = 2010$, computers could perform about 1.7 *quadrillion* computations on 1 kilowatt-hour.

(c) We must determine a value for x, so that $E(x) = 1{,}000{,}000$.

$$1000(2^{5/8})^x = 1{,}000{,}000 \qquad \textcolor{blue}{E(x) = 1{,}000{,}000.}$$

$$(2^{5/8})^x = 1000 \qquad \textcolor{blue}{\text{STEP 1: Divide by 1000.}}$$

$$\textcolor{red}{\log}\,(2^{5/8})^x = \textcolor{red}{\log}\,1000 \qquad \textcolor{blue}{\text{STEP 2: Take common logarithm.}}$$

$$x \log 2^{5/8} = 3 \qquad \textcolor{blue}{\text{STEP 3: Power rule}}$$

$$x = \frac{3}{\log 2^{5/8}} \qquad \textcolor{blue}{\text{STEP 4: Divide by } \log 2^{5/8}.}$$

$$x \approx 16 \qquad \textcolor{blue}{\text{Approximate.}}$$

Thus in $1945 + 16 = 1961$ computers first performed 1 million computations on 1 kilowatt-hour.

> **Now Try Exercise 83**

Exponential equations can occur in many forms. Although some types of exponential equations cannot be solved symbolically, Example 5 shows four equations that can.

EXAMPLE 5 Solving exponential equations symbolically

Solve each equation.

(a) $10^{x+2} = 10^{3x}$ **(b)** $5(1.2)^x + 1 = 26$ **(c)** $\left(\dfrac{1}{4}\right)^{x-1} = \dfrac{1}{9}$ **(d)** $5^{x-3} = e^{2x}$

SOLUTION

(a) Start by taking the common logarithm of each side.

$$10^{x+2} = 10^{3x} \qquad \textcolor{blue}{\text{Given equation}}$$

$$\textcolor{red}{\log}\,10^{x+2} = \textcolor{red}{\log}\,10^{3x} \qquad \textcolor{blue}{\text{Take the common logarithm.}}$$

$$x + 2 = 3x \qquad \textcolor{blue}{\text{Inverse property: } \log 10^k = k}$$

$$2 = 2x \qquad \textcolor{blue}{\text{Subtract } x.}$$

$$x = 1 \qquad \textcolor{blue}{\text{Divide by 2; rewrite.}}$$

(b) Start by isolating the exponential term on one side of the equation.

$$5(1.2)^x = 25 \qquad \textcolor{blue}{\text{Subtract 1.}}$$

$$(1.2)^x = 5 \qquad \textcolor{blue}{\text{Divide by 5.}}$$

$$\log(1.2)^x = \log 5 \qquad \textcolor{blue}{\text{Take the common logarithm.}}$$

$$\textcolor{red}{x}\log 1.2 = \log 5 \qquad \textcolor{blue}{\log m^r = r \log m}$$

$$x = \frac{\log 5}{\log 1.2} \qquad \textcolor{blue}{\text{Divide by } \log 1.2.}$$

$$x \approx 8.827 \qquad \textcolor{blue}{\text{Approximate.}}$$

(c) Begin by taking the common logarithm of each side.

$$\left(\frac{1}{4}\right)^{x-1} = \frac{1}{9} \qquad \textcolor{blue}{\text{Given equation}}$$

$$\log\left(\frac{1}{4}\right)^{x-1} = \log\frac{1}{9} \qquad \textcolor{blue}{\text{Take the common logarithm.}}$$

$$(\textcolor{red}{x-1})\log\left(\frac{1}{4}\right) = \log\frac{1}{9} \qquad \textcolor{blue}{\log m^r = r \log m}$$

$$x - 1 = \frac{\log(1/9)}{\log(1/4)} \qquad \textcolor{blue}{\text{Divide by } \log \tfrac{1}{4}.}$$

$$x = 1 + \frac{\log(1/9)}{\log(1/4)} \approx 2.585 \quad \textcolor{blue}{\text{Add 1 and approximate.}}$$

This equation could also be solved by taking the natural logarithm of each side.

(d) Begin by taking the natural (or common) logarithm of each side.

$$5^{x-3} = e^{2x}$$ Given equation

$$\ln 5^{x-3} = \ln e^{2x}$$ Take the natural logarithm.

$$(x-3)\ln 5 = 2x$$ $\ln m^r = r \ln m$

$$x \ln 5 - 3 \ln 5 = 2x$$ Distributive property

$$x \ln 5 - 2x = 3 \ln 5$$ Subtract 2x; add 3 ln 5.

$$x(\ln 5 - 2) = 3 \ln 5$$ Factor out x.

$$x = \frac{3 \ln 5}{\ln 5 - 2}$$ Divide by ln 5 − 2.

$$x \approx -12.36$$ Approximate.

Now Try Exercises 13, 17, 23, and 25

Solving Equations Involving Like Bases If a function is one-to-one, then $f(a) = f(b)$ implies that $a = b$. In Example 5(a) we were given the equation $10^{x+2} = 10^{3x}$. Because $f(x) = 10^x$ is one-to-one, it follows that $x + 2 = 3x$, or $x = 1$. We can use this technique to solve some exponential equations if we can write the expressions on each side of the equation in terms of the same base a. Then $a^x = a^y$ implies that $x = y$.

EXAMPLE 6 **Solving an exponential equation using like bases**

Solve the equation $4^{x+1} = 8^{2-x}$.

SOLUTION

Because $4 = 2^2$ and $8 = 2^3$, we can write this equation using only base 2.

$$4^{x+1} = 8^{2-x}$$ Given equation

$$(2^2)^{x+1} = (2^3)^{2-x}$$ Substitute.

$$2^{2x+2} = 2^{6-3x}$$ Properties of exponents

$$2x + 2 = 6 - 3x$$ 2^x is one-to-one.

$$5x = 4$$ Add 3x and subtract 2.

$$x = \frac{4}{5}$$ Divide by 5.

Now Try Exercise 37

NOTE The technique used in Example 6 does not work for some exponential equations. For example, $5^{x-3} = e^{2x}$ cannot easily be written in terms of equal bases. However, this equation can be solved by taking a logarithm of each side as in Example 5(d).

An Application If a hot object is put in a room with temperature T_0, then according to **Newton's law of cooling**, the temperature T of the object after time t is modeled by

$$T(t) = T_0 + Da^t, \qquad 0 < a < 1,$$

where D is the initial temperature *difference* between the object and the room.

EXAMPLE 7 **Modeling coffee cooling**

A pot of coffee with a temperature of 100°C is set down in a room with a temperature of 20°C. The coffee cools to 60°C after 1 hour.
(a) Find values for T_0, D, and a so that $T(t) = T_0 + Da^t$ models the data.
(b) Find the temperature of the coffee after half an hour.
(c) How long did it take for the coffee to reach 50°C? Support your result graphically.

SOLUTION
(a) The room has temperature $T_0 = 20°C$, and the initial temperature difference between the coffee and the room is $D = 100 - 20 = 80°C$. Thus $T(t) = 20 + 80a^t$. To find a, use the fact that the temperature of the coffee after **1** hour was **60**°C.

Let $t = 1$ and $T = 60$ in $T(t) = 20 + 80a^t$ and solve for a.

$$T(1) = 60 \qquad \text{Temperature is 60° after 1 hour.}$$

$$20 + 80a^1 = 60 \qquad \text{Let } t = 1 \text{ in } T(t) = 20 + 80a^t.$$

$$80a = 40 \qquad \text{Subtract 20.}$$

$$a = \frac{1}{2} \qquad \text{Divide by 80.}$$

Thus $T(t) = 20 + 80\left(\frac{1}{2}\right)^t$.

(b) After half an hour, the temperature is

$$T\left(\frac{1}{2}\right) = 20 + 80\left(\frac{1}{2}\right)^{1/2} \approx 76.6°\text{C}.$$

(c) *Symbolic Solution* To determine when the coffee reached 50°C, solve $T(t) = 50$.

$$20 + 80\left(\frac{1}{2}\right)^t = 50 \qquad T(t) = 50$$

$$80\left(\frac{1}{2}\right)^t = 30 \qquad \text{Subtract 20.}$$

$$\left(\frac{1}{2}\right)^t = \frac{3}{8} \qquad \text{Divide by 80; simplify.}$$

$$\log\left(\frac{1}{2}\right)^t = \log\frac{3}{8} \qquad \text{Take the common logarithm.}$$

$$t\log\left(\frac{1}{2}\right) = \log\frac{3}{8} \qquad \text{Power rule}$$

$$t = \frac{\log(3/8)}{\log(1/2)} \qquad \text{Divide by } \log\tfrac{1}{2}.$$

$$t \approx 1.415 \qquad \text{Approximate.}$$

The temperature reaches 50°C after about 1.415 hours, or 1 hour and 25 minutes.

Graphical Solution The graphs of $Y_1 = 20 + 80(1/2)^\wedge X$ and $Y_2 = 50$ intersect near $(1.415, 50)$, as shown in Figure 5.69. This result agrees with the symbolic solution.

Now Try Exercise 91

Graphical Solution

$[0, 3, 1]$ by $[0, 100, 10]$

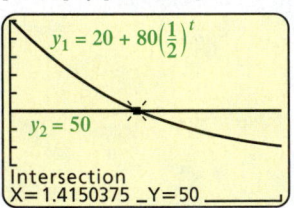

Figure 5.69

NOTE Newton's law of cooling can also model the temperature of a cold object that is brought into a warm room. In this case, the temperature difference D is *negative*.

Some exponential equations *cannot* be solved symbolically but can be solved graphically. This is demonstrated in the next example.

EXAMPLE 8 Solving an exponential equation graphically

Solve $e^{-x} + 2x = 3$ graphically. Approximate all solutions to the nearest hundredth.

SOLUTION The graphs of $Y_1 = e^\wedge(-X) + 2X$ and $Y_2 = 3$ intersect near the points $(-1.92, 3)$ and $(1.37, 3)$, as shown in Figures 5.70 and 5.71. Thus the solutions are approximately -1.92 and 1.37.

Graphical Solutions

$[-6, 6, 1]$ by $[-4, 4, 1]$ $[-6, 6, 1]$ by $[-4, 4, 1]$

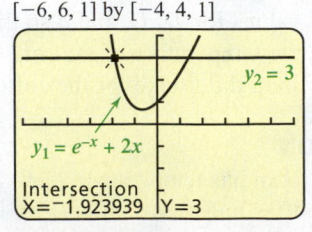 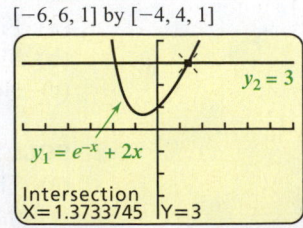

Figure 5.70   **Figure 5.71**

Now Try Exercise 73

Logarithmic Equations

Logarithmic equations contain logarithms. To solve a logarithmic equation, we use the inverse property $a^{\log_a x} = x$. This technique is illustrated in the next example.

EXAMPLE 9 Solving a logarithmic equation

Solve $3\log_3 x = 12$.

SOLUTION Begin by dividing each side by 3.

$$\log_3 x = 4 \qquad \text{Divide by 3.}$$
$$3^{\log_3 x} = 3^4 \qquad \text{Exponentiate each side; base 3.}$$
$$x = 81 \qquad \text{Inverse property: } a^{\log_a k} = k$$

> **Now Try Exercise 57**

We can solve logarithmic equations by using the following steps.

> **SOLVING LOGARITHMIC EQUATIONS SYMBOLICALLY**
>
> The following steps can be used to solve several types of logarithmic equations.
>
> **STEP 1:** Isolate the logarithmic expression on one side of the equation. (You may need to apply properties of logarithms.)
>
> **STEP 2:** Exponentiate each side of the equation with the same base as the logarithm.
>
> **STEP 3:** Apply the inverse property $a^{\log_a k} = k$.
>
> **STEP 4:** Solve for the variable. *Check your answer.*

EXAMPLE 10 Solving a logarithmic equation symbolically

In developing countries, there is a relationship between the amount of land a person owns and the average daily calories consumed. This relationship is modeled by the formula $C(x) = 280 \ln(x + 1) + 1925$, where x is the amount of land owned in acres and $0 \le x \le 4$. (***Source:*** D. Grigg, *The World Food Problem.*)
(a) Find the average caloric intake for a person who owns no land.
(b) A graph of C is shown in Figure 5.72. Interpret the graph.
(c) Determine symbolically the number of acres owned by someone whose average intake is 2000 calories per day.

Land and Calories

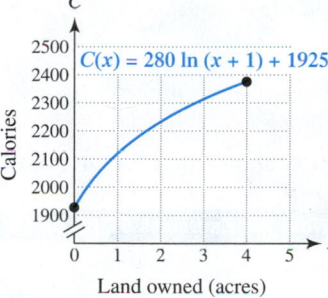

Figure 5.72

SOLUTION
(a) Since $C(0) = 280 \ln(0 + 1) + 1925 = 1925$, a person without land consumes an average of 1925 calories per day.
(b) The y-intercept of 1925 represents the caloric intake for a person who owns no land. As the amount of land x increases, the caloric intake y also increases. However, the rate of increase slows. This would be expected because there is a limit to the number of calories an average person would eat, regardless of his or her economic status.
(c) Solve the equation $C(x) = 2000$ for x.

$$280 \ln(x + 1) + 1925 = 2000 \qquad C(x) = 2000$$
$$280 \ln(x + 1) = 75 \qquad \text{STEP 1: Subtract 1925.}$$
$$\ln(x + 1) = \frac{75}{280} \qquad \text{STEP 1: Divide by 280.}$$
$$e^{\ln(x+1)} = e^{75/280} \qquad \text{STEP 2: Exponentiate each side; base } e.$$
$$x + 1 = e^{75/280} \qquad \text{STEP 3: Inverse property: } e^{\ln k} = k$$
$$x = e^{75/280} - 1 \qquad \text{STEP 4: Subtract 1.}$$
$$x \approx 0.307 \qquad \text{Approximate.}$$

A person who owns about 0.3 acre has an average intake of 2000 calories per day.

> **Now Try Exercise 95**

Like exponential equations, logarithmic equations can occur in many forms. The next example illustrates three equations that can be solved symbolically.

EXAMPLE 11 Solving logarithmic equations symbolically

Solve each equation.
(a) $\log(2x + 1) = 2$ **(b)** $\log_2 4x = 2 - \log_2 x$
(c) $2\ln(x + 1) = \ln(1 - 2x)$

SOLUTION
(a) To solve the equation, exponentiate each side of the equation using base 10.

$\log(2x + 1) = 2$	Given equation (Step 1 not needed)
$10^{\log(2x+1)} = 10^2$	STEP 2: Exponentiate each side; base 10.
$2x + 1 = 100$	STEP 3: Inverse property: $10^{\log k} = k$
$x = 49.5$	STEP 4: Solve for x.

(b) To solve this equation, apply properties of logarithms.

$\log_2 4x = 2 - \log_2 x$	Given equation
$\log_2 4x + \log_2 x = 2$	STEP 1: Add $\log_2 x$.
$\log_2 4x^2 = 2$	STEP 1: $\log_a m + \log_a n = \log_a(mn)$
$2^{\log_2 4x^2} = 2^2$	STEP 2: Exponentiate each side; base 2.
$4x^2 = 4$	STEP 3: Inverse property: $a^{\log_a k} = k$
$x = \pm 1$	STEP 4: Solve for x.

However, -1 is not a solution since $\log_2 x$ in the given equation is undefined for negative values of x. Thus the only solution is 1.

(c) For this equation we isolate a logarithmic expression on each side of the equation and then exponentiate.

$2\ln(x + 1) = \ln(1 - 2x)$	Given equation (Step 1 not needed)
$\ln(x + 1)^2 = \ln(1 - 2x)$	Power rule
$e^{\ln(x+1)^2} = e^{\ln(1-2x)}$	STEP 2: Exponentiate; base e.
$(x + 1)^2 = 1 - 2x$	STEP 3: Inverse property: $a^{\log_a k} = k$
$x^2 + 2x + 1 = 1 - 2x$	STEP 4: Expand the binomial.
$x^2 + 4x = 0$	Combine terms.
$x(x + 4) = 0$	Factor.
$x = 0$ or $x = -4$	Zero-product property

Substituting $x = 0$ and $x = -4$ in the given equation shows that 0 is a solution but -4 is not a solution.

Now Try Exercises 61, 65, and 69

MAKING CONNECTIONS

Solving Exponential and Logarithmic Equations

At some point in the process of solving an exponential equation, we often take a logarithm of each side of the equation. Similarly, when solving a logarithmic equation, we often exponentiate each side of the equation.

5.6 Putting It All Together

The following table summarizes techniques that can be used to solve some types of exponential and logarithmic equations symbolically.

CONCEPT	EXPLANATION	EXAMPLES
Exponential equations	Typical form: $Ca^x = k$ Solve for a^x. Then take a base-a logarithm of each side. Use the inverse property: $$\log_a a^x = x.$$	$4e^x = 24$ $e^x = 6$ $\ln e^x = \ln 6$ $x = \ln 6 \approx 1.79$

CONCEPT	EXPLANATION	EXAMPLES
Logarithmic equations	**Equation 1:** $C\log_a x = k$ Solve for $\log_a x$. Then exponentiate each side with base a. Use the inverse property: $$a^{\log_a x} = x.$$ **Equation 2:** $\log_a bx \pm \log_a cx = k$ When more than one logarithm with the same base occurs, use properties of logarithms to combine logarithms. *Be sure to check any solutions.*	**1.** $4\log x = 10$ $\log x = 2.5$ $10^{\log x} = 10^{2.5}$ $x = 10^{2.5} \approx 316$ **2.** $\log x + \log 4x = 2$ $\log 4x^2 = 2$ $4x^2 = 10^2$ $x^2 = 25$ $x = \pm 5$ The only solution is 5.

5.6 Exercises

Solving Exponential Equations

Exercises 1–4: The graphical and symbolic representations of f and g are shown.

(a) Use the graph to solve $f(x) = g(x)$.
(b) Solve $f(x) = g(x)$ symbolically.

1.
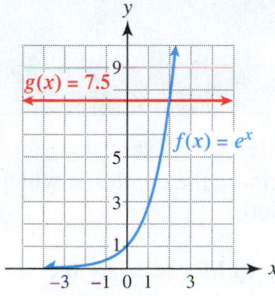

2.
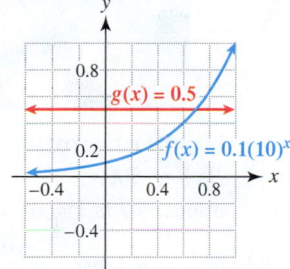

3.
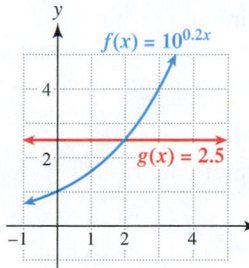

4.
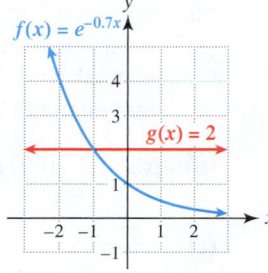

Exercises 5–32: Solve the exponential equation.

5. $4e^x = 5$

6. $2e^{-x} = 8$

7. $2(10)^x + 5 = 45$

8. $100 - 5(10)^x = 7$

9. $2.5e^{-1.2x} = 1$

10. $9.5e^{0.005x} = 19$

11. $1.2(0.9)^x = 0.6$

12. $0.05(1.15)^x = 5$

13. $4(1.1)^{x-1} = 16$

14. $3(2)^{x-2} = 99$

15. $5(1.2)^{3x-2} + 94 = 100$

16. $1.4(2)^{x+3} = 2.8$

17. $5^{3x} = 5^{1-2x}$

18. $7^{x^2} = 7^{4x-3}$

19. $10^{(x^2)} = 10^{3x-2}$

20. $e^{2x} = e^{5x-3}$

21. $\left(\frac{1}{5}\right)^x = -5$

22. $2^x = -4$

23. $\left(\frac{2}{5}\right)^{x-2} = \frac{1}{3}$

24. $\left(\frac{3}{2}\right)^{x+1} = \frac{7}{3}$

25. $4^{x-1} = 3^{2x}$

26. $3^{1-2x} = e^{0.5x}$

27. $e^{x-3} = 2^{3x}$

28. $6^{x+1} = 4^{2x-1}$

29. $3(1.4)^x - 4 = 60$

30. $2(1.05)^x + 3 = 10$

31. $5(1.015)^{x-1980} = 8$

32. $30 - 3(0.75)^{x-1} = 29$

Exercises 33–42: (Refer to Example 6.) Use the fact that $a^x = a^y$ implies $x = y$, to solve each equation.

33. $5^{2x} = 5^{x-3}$

34. $7^{-x} = 7^{2x+1}$

35. $e^{-x} = e^{x^2}$

36. $e^{x^2} = e^{2x+1}$

37. $2^{3x} = 8^{-x+2}$

38. $9^{2x} = 27^{1-x}$

39. $25^{2x} = 125^{2-x}$

40. $16^x = 4^{2-x}$

41. $32^{3x} = 16^{5x+3}$

42. $16^{x-5} = 64^{1-2x}$

Solving Logarithmic Equations

Exercises 43–72: Solve the logarithmic equation.

43. $3\log x = 2$

44. $5\ln x = 10$

45. $\ln 2x = 5$

46. $\ln 4x = 1.5$

47. $\log 2x^2 = 2$

48. $\log(2 - x) = 0.5$

49. $\log_2(3x - 2) = 4$

50. $\log_3(1 - x) = 1$

51. $\log_5(8 - 3x) = 3$

52. $\log_6(2x + 4) = 2$

53. $160 + 10\log x = 50$

54. $160 + 10\log x = 120$

55. $\ln x + \ln x^2 = 3$

56. $\log x^5 = 4 + 3\log x$

57. $2\log_2 x = 4.2$

58. $3\log_2 3x = 1$

59. $\log x + \log 2x = 2$

60. $\ln 2x + \ln 3x = \ln 6$

61. $\log(2 - 3x) = 3$

62. $\log(x^2 + 1) = 2$

63. $\ln x + \ln(3x - 1) = \ln 10$

64. $\log x + \log(2x + 5) = \log 7$

65. $2\ln x = \ln(2x + 1)$

66. $\log(x^2 + 3) = 2\log(x + 1)$

67. $\log(x + 1) + \log(x - 1) = \log 3$

68. $\ln(x^2 - 4) - \ln(x + 2) = \ln(3 - x)$

69. $\log_2 2x = 4 - \log_2(x + 2)$

70. $\log_3 x + \log_3(x + 2) = \log_3 24$

71. $\log_5(x + 1) + \log_5(x - 1) = \log_5 15$

72. $\log_7 4x - \log_7(x + 3) = \log_7 x$

▦ Solving Equations Graphically

Exercises 73–78: The following equations cannot be solved symbolically. Solve these equations graphically and round your answers to the nearest hundredth.

73. $2x + e^x = 2$

74. $xe^x - 1 = 0$

75. $x^2 - x\ln x = 2$

76. $x\ln|x| = -2$

77. $xe^{-x} + \ln x = 1$

78. $2^{x-2} = \log x^4$

Applications

79. Population Growth   World population P in billions during year x can be modeled by $P(x) = 7(1.01)^{x-2011}$. Predict the year when world population might reach 7.5 billion.

80. Population of Arizona   The population P of Arizona has been increasing at an annual rate of 2.3%. In 2010 the population of Arizona was 6.4 million.

(a) Write a formula for $P(x)$, where x is the years after 2010 and P is in millions.

(b) Estimate the population of Arizona in 2014.

81. Light Absorption   When light passes through water, its intensity I decreases according to the formula $I(x) = I_0 e^{-kx}$, where I_0 is the initial intensity of the light and x is the depth in feet. If $I_0 = 1000$ lumens per square meter and $k = 0.12$, determine the depth at which the intensity is 25% of I_0.

82. Light Absorption   (Refer to Exercise 81.) Let $I(x) = 500e^{-0.2x}$ and determine the depth x at which the intensity I is 1% of $I_0 = 500$.

83. Moore's Law   According to Moore's law the number of transistors that can be placed on an integrated circuit has doubled every 2 years. In 1971 there were only 2300 transistors on an integrated circuit.

(a) Find an exponential function $T(x) = Ca^x$ that gives the number of transistors on an integrated circuit x years after 1971.

(b) Evaluate $T(40)$ and interpret your result.

(c) Determine the year when integrated circuits first had 10 million transistors.

84. Electrical Efficiency   (Refer to Example 4.) Use the formula $E(x) = 1000(2^{5/8})^x$ to determine when 1 billion computations could first be performed with 1 kilowatt-hour.

85. Corruption and Human Development There is a relationship between perceived corruption and human development in a country that can be modeled by $H(x) = 0.3 + 0.28\ln x$. In this formula x represents the Corruption Perception Index, where 1 is very corrupt and 10 is least corrupt. The output gives the Human Development Index, which is between 0.1 and 1 with 1 being the best for human development. (*Sources:* Transparency International; UN Human Development Report.)
(a) For Britain $x = 8$. Evaluate $H(8)$ and interpret the result.

(b) The Human Development Index is 0.68 for China. Find the Corruption Perception Index for China.

86. Urbanization of Brazil The percentage of Brazil's population that lives in urban areas can be modeled by $U(x) = 72 + 4.33\ln x$, where $x \geq 1$ is the number of years after 1990.
(a) Evaluate $U(24)$ and interpret the result.

(b) Predict when urbanization will first reach 88%.

87. Bluefin Tuna The number of Atlantic bluefin tuna in thousands x years after 1974 can be modeled by $f(x) = 230(0.881)^x$. Estimate the year when the number of bluefin tuna reached 95 thousand.

88. Modeling Bacteria (Refer to Section 5.3, Example 9.) The number N of E. *coli* bacteria in millions per milliliter after t minutes can be modeled by $N(t) = 0.5e^{0.014t}$. Determine symbolically the elapsed time required for the concentration of bacteria to reach 25 million per milliliter.

89. Population Growth In 2000 the population of India reached 1 billion, and in 2025 it is projected to be 1.4 billion. (*Source:* Bureau of the Census.)
(a) Find values for C and a so that $P(x) = Ca^{x-2000}$ models the population of India in year x.

(b) Estimate India's population in 2010.

(c) Use P to determine the year when India's population might reach 1.5 billion.

90. Population of Pakistan In 2007 the population of Pakistan was 164 million, and it is expected to be 250 million in 2025. (*Source:* United Nations.)
(a) Approximate C and a so that $P(x) = Ca^{x-2007}$ models these data, where P is in millions and x is the year.

(b) Estimate the population of Pakistan in 2015, and compare your estimate to the predicted value of 204 million.

(c) Estimate when this population could reach 212 million.

91. Newton's Law of Cooling A pan of boiling water with a temperature of 212°F is set in a bin of ice with a temperature of 32°F. The pan cools to 70°F in 30 minutes.
(a) Find T_0, D, and a so that $T(t) = T_0 + Da^t$ models the data, where t is in hours.

(b) Find the temperature of the pan after 10 minutes.

(c) How long did it take the pan to reach 40°F? Support your result graphically.

92. Warming an Object A pan of cold water with a temperature of 35°F is brought into a room with a temperature of 75°F. After 1 hour, the temperature of the pan of water is 45°F.
(a) Find T_0, D, and a so that $T(t) = T_0 + Da^t$ models the data, where t is in hours.

(b) Find the temperature of the water after 3 hours.

(c) How long would it take the water to reach 60°F?

93. Warming a Soda Can Suppose that a can of soda, initially at 5°C, warms to 18°C after 2 hours in a room that has a temperature of 20°C.
(a) Find the temperature of the soda can after 1.5 hours.

(b) How long did it take for the soda to warm to 15°C?

94. Cooling a Soda Can A soda can at 80°F is put into a cooler containing ice at 32°F. The temperature of the soda after t minutes is given by $T(t) = 32 + 48(0.9)^t$.
(a) Evaluate $T(30)$ and interpret your results.

(b) How long did it take for the soda to cool to 50°F?

95. Caloric Intake (Refer to Example 10.) The formula

$$C(x) = 280\ln(x + 1) + 1925$$

models the number of calories consumed daily by a person owning x acres of land in a developing country. Estimate the number of acres owned for which average intake is 2300 calories per day.

96. Salinity The salinity of the oceans changes with latitude and with depth. In the tropics, the salinity increases on the surface of the ocean due to rapid evaporation. In the higher latitudes, there is less evaporation and rainfall causes the salinity to be less on the surface than at lower depths. The function given by

$$S(x) = 31.5 + 1.1\log(x + 1)$$

models salinity to depths of 1000 meters at a latitude of 57.5°N. The input x is the depth in meters and the output $S(x)$ is in grams of salt per kilogram of seawater. (*Source:* D. Hartman, *Global Physical Climatology.*)
(a) Evaluate $S(500)$. Interpret your result.

(b) Graph S. Discuss any trends.

(c) Find the depth where the salinity equals 33.

97. Life Span In one study, the life spans of 129 robins were monitored over a 4-year period. The equation $y = \dfrac{2 - \log(100 - x)}{0.42}$ can be used to calculate the number of years y required for x percent of the robin population to die. For example, to find the time when 40% of the robins had died, substitute $x = 40$ into the equation. The result is $y \approx 0.53$, or about half a year. Find the percentage of the robins that had died after 2 years. (*Source:* D. Lack, *The Life of a Robin.*)

98. Life Span of Sparrows (Refer to Exercise 97.) The life span of a sample of sparrows was studied. The equation $y = \dfrac{2 - \log(100 - x)}{0.37}$ calculates the number of years y required for x percent of the sparrows to die, where $0 \le x \le 95$.
(a) Find y when $x = 40$. Interpret your answer.

(b) Find x when $y = 1.5$. Interpret your answer.

99. Bacteria Growth The concentration of bacteria in a sample can be modeled by $B(t) = B_0 e^{kt}$, where t is in hours and B is the concentration in billions of bacteria per liter.
(a) If the concentration increases by 15% in 6 hours, find k.

(b) If $B_0 = 1.2$, find B after 8.2 hours.

(c) By what percentage does the concentration increase each hour?

100. Voltage The voltage in a circuit can be modeled by $V(t) = V_0 e^{kt}$, where t is in milliseconds.
(a) If the voltage decreases by 85% in 5 milliseconds, find k.

(b) If $V_0 = 4.5$ volts, find V after 2.3 milliseconds.

(c) By what percentage does the voltage decrease each millisecond?

Exercises 101 and 102: For the given annual interest rate r, estimate the time for P dollars to double.
101. $P = \$1000, r = 8.5\%$ compounded quarterly

102. $P = \$750, r = 2\%$ compounded continuously

Exercises 103 and 104: **Continuous Compounding** *Suppose that P dollars is deposited in a savings account paying 3% interest compounded continuously. After t years, the account will contain $A(t) = Pe^{0.03t}$ dollars.*
(a) Solve $A(t) = b$ for the given values of P and b.
(b) Interpret your results.

103. $P = 500$ and $b = 750$

104. $P = 1000$ and $b = 2000$

105. Radioactive Carbon-14 The percentage P of radioactive carbon-14 remaining in a fossil after t years is given by $P = 100\left(\frac{1}{2}\right)^{t/5700}$. Suppose a fossil contains 35% of the carbon-14 that the organism contained when it was alive. Estimate the age of the fossil.

106. Radioactive Radium-226 The amount A of radium in milligrams remaining in a sample after t years is given by $A(t) = 0.02\left(\frac{1}{2}\right)^{t/1600}$. How many years will it take for the radium to decay to 0.004 milligram?

107. Traffic Flow (Refer to Section 5.3, Example 11.) The probability that a car will enter an intersection within a period of x minutes is given by $P(x) = 1 - e^{-0.5x}$. Determine symbolically the elapsed time x when there is a 50–50 chance that a car has entered the intersection. (*Hint:* Solve $P(x) = 0.5$.)

108. Modeling Traffic Flow Cars arrive randomly at an intersection with an average traffic volume of one car per minute. The likelihood, or probability, that at least one car enters the intersection during a period of x minutes can be estimated by $f(x) = 1 - e^{-x}$.
(a) What is the probability that at least one car enters the intersection during a 5-minute period?

(b) Determine the value of x that gives a 40% chance that at least one car enters the intersection during an interval of x minutes.

109. Modeling Bacteria Growth Suppose that the concentration of a bacteria sample is 100,000 bacteria per milliliter. If the concentration doubles every 2 hours, how long will it take for the concentration to reach 350,000 bacteria per milliliter?

110. Modeling Bacteria Growth Suppose that the concentration of a bacteria sample is 50,000 bacteria per milliliter. If the concentration triples in 4 days, how long will it take for the concentration to reach 85,000 bacteria per milliliter?

111. Continuous Compounding Suppose that $2000 is deposited in an account and the balance increases to $2300 after 4 years. How long will it take for the account to grow to $3200? Assume continuous compounding.

112. Modeling Radioactive Decay Suppose that a 0.05-gram sample of a radioactive substance decays to 0.04 gram in 20 days. How long will it take for the sample to decay to 0.025 gram?

113. Drug Concentrations The concentration of a drug in a patient's bloodstream after t hours is modeled by the formula $C(t) = 11(0.72)^t$, where C is measured in milligrams per liter.

(a) What is the initial concentration of the drug?

(b) How long does it take for the concentration to decrease to 50% of its initial level?

114. Reducing Carbon Emissions   When fossil fuels are burned, carbon is released into the atmosphere. Governments could reduce carbon emissions by placing a tax on fossil fuels. The **cost-benefit** equation

$$\ln(1 - P) = -0.0034 - 0.0053x$$

estimates the relationship between a tax of x dollars per ton of carbon and the percent P reduction in emissions of carbon, where P is in decimal form. Determine P when $x = 60$. Interpret the result   (***Source:*** W. Clime, *The Economics of Global Warming.*)

115. Investments   The formula $A = P\left(1 + \frac{r}{n}\right)^{nt}$ can be used to calculate the future value of an investment. Solve the equation for t.

116. Decibels   The formula $D = 160 + 10\log x$ can be used to calculate loudness of a sound in decibels. Solve the equation for x.

Writing about Mathematics

117. Explain how to solve the equation $Ca^x = k$ symbolically for x. Demonstrate your method.

118. Explain how to solve the equation $b\log_a x = k$ symbolically for x. Demonstrate your method.

Extended and Discovery Exercise

1. Exponential Functions   Show that any exponential function in the form $f(x) = Ca^x$ can be written as $f(x) = Ce^{kx}$. That is, write k in terms of a. Use your method to write $g(x) = 2^x$ in the form e^{kx} for some k.

CHECKING BASIC CONCEPTS FOR SECTIONS 5.5 AND 5.6

1. Use properties of logarithms to expand $\log \frac{x^2y^3}{\sqrt[3]{z}}$. Write your answer without exponents.

2. Combine the expression $\frac{1}{2}\ln x - 3\ln y + \ln z$ as a logarithm of a single expression.

3. Solve each equation.
   (a) $5(1.4)^x - 4 = 25$ **(b)** $4^{2-x} = 4^{2x+1}$

4. Solve each equation.
   (a) $5\log_2 2x = 25$

   (b) $\ln(x + 1) + \ln(x - 1) = \ln 3$

5. The temperature T of a cooling object in degrees Fahrenheit after x minutes is given by

$$T = 80 + 120(0.9)^x.$$

   (a) What happens to T after a long time?

   (b) After how long is the object's temperature 100°F?

5.7 Constructing Nonlinear Models

- **Find an exponential model**
- **Find a logarithmic model**
- **Find a logistic model**
- **Select a model**

Introduction

If data change at a constant rate, then they can be modeled with a linear function. However, real-life data often change at a nonconstant rate. For example, a tree grows slowly when it is small and then gradually grows faster as it becomes larger. Finally, when the tree is mature, its height begins to level off. This type of growth is nonlinear.

   Three types of nonlinear data are shown in Figures 5.73–5.75 on the next page, where t represents time. In Figure 5.73 the data increase rapidly, and an *exponential function* might be an appropriate modeling function. In Figure 5.74 the data are growing, but at a slower rate than in Figure 5.73. These data could be modeled by a *logarithmic function*. Finally, in Figure 5.75 the data increase slowly, then increase faster, and finally level off. These data might represent the height of a tree over a 50-year period. To model these data, we need a new type of function called a *logistic function*.

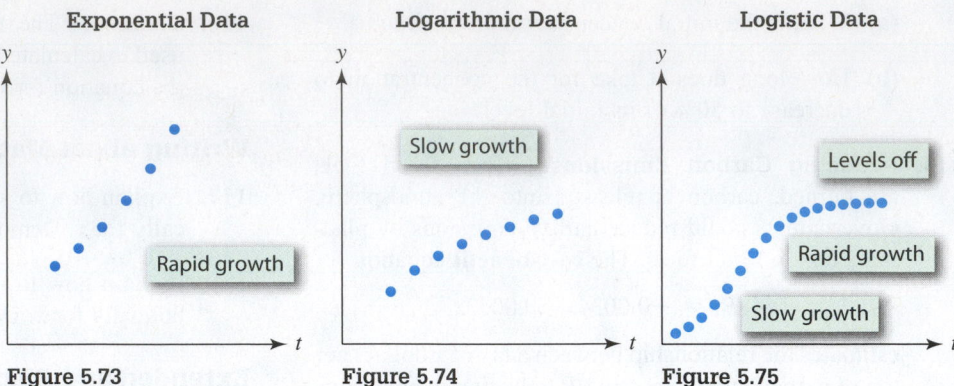

Figure 5.73   **Figure 5.74**   **Figure 5.75**

Exponential Model

Both world population and bacteria growth can sometimes be modeled by an exponential function that *increases*. Exponential functions can also be used to model data that *decrease*. In the next example, an exponential function is used to model atmospheric pressure.

EXAMPLE 1 Modeling atmospheric pressure

As altitude increases, air pressure decreases. The atmospheric pressure P in millibars (mb) at a given altitude x in meters is listed in Table 5.24.

Altitude and Air Pressure

x (m)	0	5000	10,000	15,000	20,000	25,000	30,000
P (mb)	1013	541	265	121	55	26	12

Source: A. Miller and J. Thompson, *Elements of Meteorology.*

Table 5.24

(a) Make a scatterplot of the data. What type of function might model the data?
(b) Use regression to find an exponential function given by $f(x) = ab^x$. Graph the data and f in the same viewing rectangle.
(c) Use f to estimate the air pressure at an altitude of 23,000 feet.

Calculator Help

To find an equation of least-squares fit, see Appendix A (page AP-9). To copy a regression equation directly into Y_1, see Appendix A (page AP-11).

SOLUTION
(a) The data are shown in Figure 5.76. A *decreasing* exponential function might model the data.

NOTE It is possible for a different function, such as a portion of a polynomial graph, to model the data. Answers may vary.

(b) Figures 5.77 and 5.78 show that values of $a \approx 1104.9$ and $b \approx 0.99985$ are obtained from exponential regression, where $f(x) = ab^x$. Figure 5.79 illustrates that f models the data quite accurately.
(c) $f(23,000) = 1104.9(0.99985)^{23,000} \approx 35.1$ millibars

Plot the Data

[−5000, 35000, 5000] by
[−100, 1200, 100]

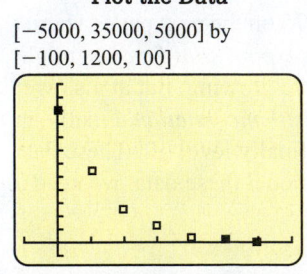

Figure 5.76

Select Exponential Regression

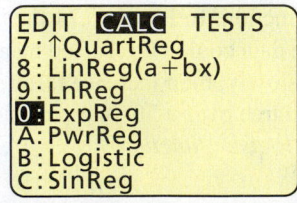

Figure 5.77

Exponential Function

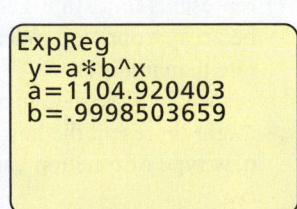

Figure 5.78

Graph Function with Data

[−5000, 35000, 5000] by
[−100, 1200, 100]

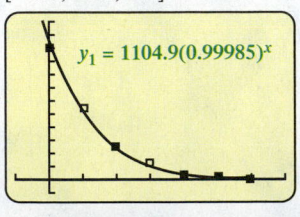

Figure 5.79

Now Try Exercise 15

Logarithmic Model

An investor buying a certificate of deposit (CD) usually gets a higher interest rate if the money is deposited for a longer period of time. However, the interest rate for a 9-month CD is not usually triple the rate for a 3-month CD. Instead, the rate of interest slowly increases with a longer-term CD. In the next example, we model interest rates with a logarithmic function.

EXAMPLE 2   **Modeling interest rates**

Table 5.25 lists the interest rates for certificates of deposit. Use the data to complete the following.

Yield on Certificates of Deposit

Time (months)	1	3	6	9	24	36	60
Yield (%)	0.25	0.39	0.74	0.80	1.25	1.40	1.50

Source: USA Today.

Table 5.25

(a) Make a scatterplot of the data. What type of function might model these data?
(b) Use least-squares regression to find a formula $f(x) = a + b \ln x$ that models the data.
(c) Graph f and the data in the same viewing rectangle.

SOLUTION

(a) Enter the data in Table 5.25 into your calculator. A scatterplot of the data is shown in Figure 5.80. The data increase but gradually level off. A logarithmic modeling function may be appropriate.
(b) In Figures 5.81 and 5.82 least-squares regression has been used to find a logarithmic function f given (approximately) by $f(x) = 0.143 + 0.334 \ln x$.
(c) A graph of f and the data are shown in Figure 5.83.

Plot the Data
[0, 70, 10] by [0, 2, 0.5]

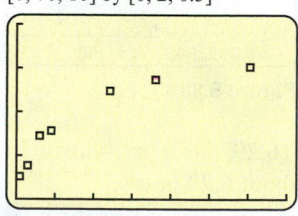

Figure 5.80

Select Logarithm Regression

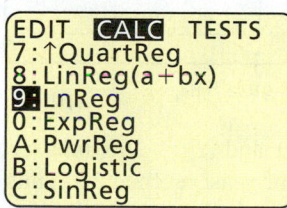

Figure 5.81

Regression Function

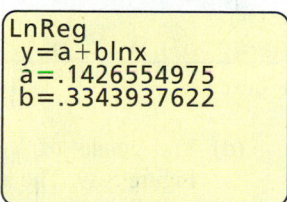

Figure 5.82

Graph Function with Data
[0, 70, 10] by [0, 2, 0.5]

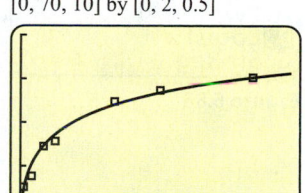

Figure 5.83

Now Try Exercise 13

Sigmoidal Curve

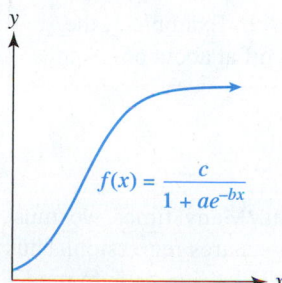

Figure 5.84

Logistic Model

In real life, populations of bacteria, insects, and animals do not continue to grow indefinitely. Initially, population growth may be slow. Then, as their numbers increase, so does the rate of growth. After a region has become heavily populated or saturated, the population usually levels off because of limited resources.

This type of growth may be modeled by a **logistic function** represented by $f(x) = \frac{c}{1 + ae^{-bx}}$, where a, b, and c are positive constants. A typical graph of a logistic function f is shown in Figure 5.84. The graph of f is referred to as a **sigmoidal curve**. The next example demonstrates how a logistic function can be used to describe the growth of a yeast culture.

EXAMPLE 3 **Modeling logistic growth**

One of the earliest studies about population growth was done in 1913 using yeast plants. A small amount of yeast was placed in a container with a fixed amount of nourishment. The units of yeast were recorded every 2 hours. The data are listed in Table 5.26.

Growth of Yeast Plants

Time	0	2	4	6	8	10	12	14	16	18
Yeast	9.6	29.0	71.1	174.6	350.7	513.3	594.8	640.8	655.9	661.8

Source: T. Carlson, *Biochem.;* D. Brown, *Models in Biology.*

Table 5.26

(a) Make a scatterplot of the data in Table 5.26. Describe the growth.
(b) Use least-squares regression to find a logistic function f that models the data.
(c) Graph f and the data in the same viewing rectangle.
(d) Approximate graphically the time when the amount of yeast was 200 units.

Plot the Data

$[-2, 20, 1]$ by $[-100, 800, 100]$

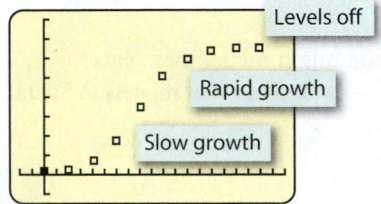

Levels off

Rapid growth

Slow growth

Figure 5.85

SOLUTION

(a) A scatterplot of the data is shown in Figure 5.85. The yeast increase slowly at first. Then they grow more rapidly until the amount of yeast gradually levels off. The limited amount of nourishment causes this leveling off.

(b) In Figure 5.86 and 5.87 we see least-squares regression being used to find a logistic function f given (approximately) by

$$f(x) = \frac{661.8}{1 + 74.46e^{-0.552x}}.$$

(c) In Figure 5.88 the data and f are graphed in the same viewing rectangle. The fit for the *real* data is remarkably good.

Select Logistic Regression

```
EDIT  CALC  TESTS
7:↑QuartReg
8:LinReg(a+bx)
9:LnReg
0:ExpReg
A:PwrReg
B:Logistic
C:SinReg
```

Figure 5.86

Regression Function

```
Logistic
y=c/(1+ae^(-bx))
a=74.46113243
b=.551931828
c=661.8044322
```

Figure 5.87

Graph Function with Data

$[-2, 20, 1]$ by $[-100, 800, 100]$

Figure 5.88

Graphical Solution

$[-2, 20, 1]$ by $[-100, 800, 100]$

y_1

$y_2 = 200$

Intersection
X=6.2932648   Y=200

Figure 5.89

(d) The graphs of $Y_1 = f(x)$ and $Y_2 = 200$ intersect near $(6.29, 200)$, as shown in Figure 5.89. The amount of yeast reached 200 units after about 6.29 hours.

Now Try Exercise 19

CLASS DISCUSSION

In Example 3, suppose that after 18 hours the experiment had been extended and more nourishment had been provided for the yeast plants. Sketch a possible graph of the amount of yeast.

MAKING CONNECTIONS

Logistic Functions and Horizontal Asymptotes If a logistic function is given by $f(x) = \frac{c}{1 + ae^{-bx}}$, where a, b, and c are positive constants, then the graph of f has a horizontal asymptote of $y = c$. (Try to explain why this is true.) In Example 3, the value of c was 661.8. This means that the amount of yeast leveled off at about 661.8 units.

Selecting a Model

In real-data applications, a modeling function is seldom given. Many times we must choose the type of modeling function and then find it using least-squares regression. Thus far in this section, we have used exponential, logarithmic, and logistic functions to model data. In the next two examples, we select a modeling function.

EXAMPLE 4   **Modeling highway design**

To allow enough distance for cars to pass on two-lane highways, engineers often calculate minimum sight distances between curves and hills. See the figure. Table 5.27 shows the minimum sight distance y in feet for a car traveling at x miles per hour.

Passing Distance

x (mph)	20	30	40	50	60	65	70
y (ft)	810	1090	1480	1840	2140	2310	2490

Source: L. Haefner, *Introduction to Transportation Systems.*
Table 5.27

(a) Find a modeling function for the data.
(b) Graph the data and your modeling function.
(c) Estimate the minimum sight distance for a car traveling at 43 miles per hour.

SOLUTION

Getting Started   One strategy is to plot the data and then decide if the data are linear or nonlinear. If the data are approximately linear, use linear regression to find the modeling function. If the data are nonlinear, think about how the data increase or decrease. You may want to try several types of modeling functions, such as quadratic, cubic, power, exponential, or logarithmic, before making a final decision. ▶

(a) A scatterplot is shown in Figure 5.90. The data appear to be (nearly) linear, so linear regression has been used to obtain $f(x) = 33.93x + 113.4$. See Figure 5.91.
(b) The data and f are graphed in Figure 5.92. Function f gives a good fit.

Plot the Data	Linear Regression Function	Graph Function with Data
[0, 80, 10] by [0, 3000, 1000]	LinReg y=ax+b a=33.92832765 b=113.4300341 r^2=.9986779453 r =.999338754	[0, 80, 10] by [0, 3000, 1000] y = 33.93x + 113.4
Figure 5.90	**Figure 5.91**	**Figure 5.92**

(c) $f(43) = 33.93(43) + 113.4 \approx 1572$ feet

Now Try Exercise 11

You may have seen or heard of asbestos being removed from buildings. Before 1960 people were generally unaware of its health hazards. As a result, insulation workers who worked with asbestos experienced higher rates of lung cancer. The following example models data from this era.

EXAMPLE 5   **Modeling asbestos and cancer**

Table 5.28 lists the number N of lung cancer cases occurring within a group of asbestos insulation workers with a cumulative total of 100,000 years of work experience, with their first date of employment x years ago.

Lung Cancer and Asbestos

x (years)	10	15	20	25	30
N (cases)	6.9	25.4	63.6	130	233

Source: A. Walker, *Observation and Inference.*
Table 5.28

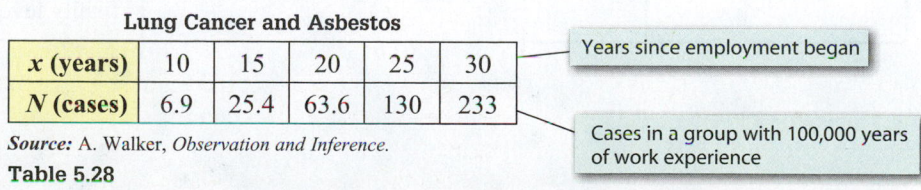

Years since employment began

Cases in a group with 100,000 years of work experience

(a) Find a modeling function for the data.

(b) Graph the data and your modeling function.

(c) Estimate the number of lung cancer cases for $x = 23$ years. Interpret your answer.

SOLUTION

Getting Started The data are nonlinear and increasing, so there are a number of functions you can try. Three possibilities are quadratic, power, and exponential. ▶

(a) A scatterplot is shown in Figure 5.93. To model the data we have used a power function given by $f(x) = 0.004334x^{3.2}$. See Figure 5.94. (Answers may vary.)

(b) The data and f are graphed in Figure 5.95. Function f gives a good fit.

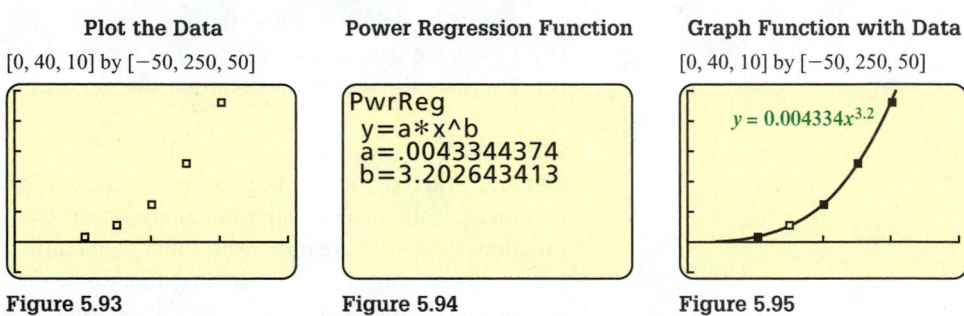

Plot the Data	Power Regression Function	Graph Function with Data
[0, 40, 10] by [−50, 250, 50]		[0, 40, 10] by [−50, 250, 50]
Figure 5.93	Figure 5.94	Figure 5.95

(c) $f(23) = 0.004334(23)^{3.2} \approx 99$ cases. If a group of asbestos workers began their employment 23 years earlier and had a cumulative work experience of 100,000 years, then the group experienced 99 cases of lung cancer.

Now Try Exercise 21

5.7 Putting It All Together

The following table summarizes the basics of exponential, logarithmic, and logistic models. Least-squares regression can be used to determine the constants a, b, c, and C.

CONCEPT	EXPLANATION	EXAMPLES
Exponential model	$f(x) = Ca^x$, $f(x) = ab^x$, or $A(t) = A_0 e^{kt}$	Exponential functions can be used to model data that increase or decrease rapidly over time.
Logarithmic model	$f(x) = a + b\log x$ or $f(x) = a + b\ln x$	Logarithmic functions can be used to model data that increase gradually over time.
Logistic model	$f(x) = \dfrac{c}{1 + ae^{-bx}}$	Logistic functions can be used to model data that at first increase slowly, then increase rapidly, and finally level off.

5.7 **Exercises**

Note: Because different functions can be used to model the same data, your answers may vary from the given answers. You can check the validity of your answer by graphing the data and your modeling function in the same viewing rectangle.

Selecting a Model

Exercises 1–4: Select an appropriate type of modeling function for the data shown in the graph. Choose from the following.

   i. Exponential   ii. Logarithmic   iii. Logistic

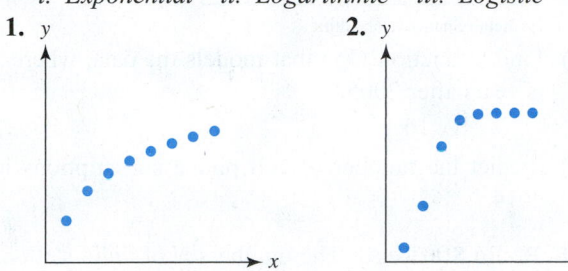

1. y

2. y

3. y

4. y

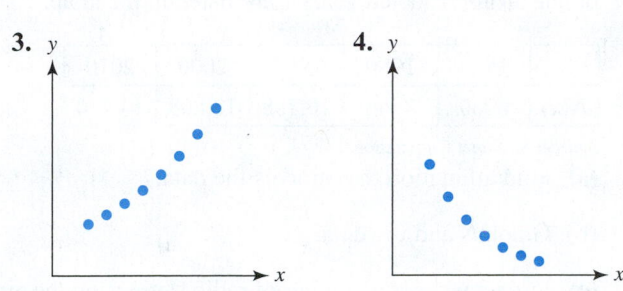

Exercises 5–10: Make a scatterplot of the data. Then find an exponential, logarithmic, or logistic function f that best models the data.

5.

x	1	2	3	4
y	2.04	3.47	5.90	10.02

6.

x	1	2	3	4	5
y	1.98	2.35	2.55	2.69	2.80

7.

x	1	2	3	4	5
y	1.1	3.1	4.3	5.2	5.8

8.

x	1	2	3	4	5	6
y	1	2	4	7	9	10

9.

x	0	1	2	3	4	5
y	0.3	1.3	4.0	7.5	9.3	9.8

10.

x	1	2	3	4	5
y	2.0	1.6	1.3	1.0	0.82

Applications

11. SlideShare Growth The following table shows the number of visitors in millions who visited the website SlideShare.

Year	2007	2008	2009	2010	2011
Visitors	1.5	6.3	17.0	33.2	60.1

Source: SlideShare.

  (a) Choose a linear, quadratic, or logarithmic function $V(x)$ that best models the data, where x is years after 2007.

  (b) Estimate the number of visitors in 2012.

12. Heart Disease Death Rates The following table contains heart disease death rates per 100,000 people for selected ages.

Age	30	40	50	60	70
Death rate	30.5	108.2	315	776	2010

Source: Department of Health and Human Services.

  (a) Make a scatterplot of the data in the viewing rectangle [25, 75, 5] by [−100, 2100, 200].

  (b) Find a function f that models the data.

  (c) Estimate the heart disease death rate for people who are 80 years old.

13. Telecommuting In the past some workers used technology such as e-mail, computers, and multiple phone lines to work at home, rather than in the office. However, because of the need for teamwork and collaboration in the workplace, fewer employees telecommuted than expected. The table lists telecommuters T in millions during year x.

x	1997	1998	1999	2000	2001
T	9.2	9.6	10.0	10.4	10.6

x	2002	2003	2004	2005	2006
T	11.0	11.1	11.2	11.3	11.4

Source: USA Today.

Find a function f that models the data, where $x = 1$ corresponds to 1997, $x = 2$ to 1998, and so on.

14. Hurricanes   The table shows the air pressure y in inches of mercury x miles from the eye of a hurricane.

x	2	4	8	15	30	100
y	27.3	27.7	28.04	28.3	28.7	29.3

Source: A. Miller and R. Anthes, *Meteorology.*

(a) Make a scatterplot of the data.

(b) Find a function f that models the data.

(c) Estimate the air pressure at 50 miles.

15. Atmospheric Density   The table lists the atmospheric density y in kilograms per cubic meter (kg/m³) at an altitude of x meters.

x (m)	0	5000	10,000	15,000
y (kg/m³)	1.2250	0.7364	0.4140	0.1948

x (m)	20,000	25,000	30,000
y (kg/m³)	0.0889	0.0401	0.0184

Source: A. Miller.

(a) Find a function f that models the data.

(b) Predict the density at 7000 meters. (The actual value is 0.59 kg/m³.)

16. Modeling Data   Use the table to complete the following.

x	1	2	5	10
y	2.5	2.1	1.6	1.2

(a) Find a function f that models the data.

(b) Solve the equation $f(x) = 1.8$.

17. Insect Population   The table shows the density y of a species of insect measured in thousands per acre after x days.

x	2	4	6	8	10	12	14
y	0.38	1.24	2.86	4.22	4.78	4.94	4.98

(a) Find a function f that models the data.

(b) Use f to estimate the insect density after a long time.

18. Heart Disease   As age increases, so does the likelihood of coronary heart disease (CHD). The percentage P of people x years old with signs of CHD is shown in the table.

x	15	25	35	45	55	65	75
P(%)	2	7	19	43	68	82	87

Source: D. Hosmer and S. Lemeshow, *Applied Logistics Regression.*

(a) Evaluate $P(25)$ and interpret the answer.

(b) Find a function that models the data.

(c) Graph P and the data.

(d) At what age does a person have a 50% chance of having signs of CHD?

19. Mobile Phones in India   The following table shows the number of cell phone subscriptions in India in billions for selected years.

Year	2005	2006	2007	2008
Cell Phones	0.05	0.12	0.21	0.40

Year	2009	2010	2011
Cell Phones	0.60	0.75	0.85

Source: Chetan Sharma Consulting.

(a) Find a function $C(x)$ that models the data, where x is years after 2005.

(b) Predict the number of cell phone subscriptions in 2014.

20. U.S. Radio Stations   The numbers N of radio stations on the air for selected years x are listed in the table.

x	1970	1980	1990	2000	2010
$N(x)$	6760	8566	10,788	12,808	14,420

Source: M. Street Corporation.

(a) Find a function that models the data.

(b) Graph N and the data.

(c) Predict when the number of radio stations on the air might reach 16,000.

21. Wing Size   Heavier birds tend to have larger wings than smaller birds. For one species of bird, the table lists the area A of the bird's wing in square inches if the bird weighs w pounds.

w (lb)	2	6	10	14	18
$A(w)$ (in²)	160	330	465	580	685

Source: C. Pennycuick, *Newton Rules Biology.*

(a) Find a function that models the data.

(b) Graph A and the data.

(c) What weight corresponds to a wing area of 500 square inches?

22. Wing Span   Heavier birds tend to have a longer wing span than smaller birds. For one species of bird, the table lists the length L of the bird's wing span in feet if the bird weighs w pounds.

w (lb)	0.22	0.88	1.76	2.42
$L(w)$ (ft)	1.38	2.19	2.76	3.07

Source: C. Pennycuick, *Newton Rules Biology.*

(a) Find a function that models the data.

(b) Graph L and the data.

(c) What weight corresponds to a wing span of 2 feet?

23. Tree Growth   (Refer to the introduction to this section.) The height H of a tree in feet after x years is listed in the table.

x (yr)	1	5	10	20	30	40
H(x) (ft)	1.3	3	8	32	47	50

(a) Evaluate $H(5)$ and interpret the answer.

(b) Find a function that models the data.

(c) Graph H and the data.

(d) What is the age of the tree when its height is 25 feet?

(e) Did your answer involve interpolation or extrapolation?

24. Bird Populations   Near New Guinea there is a relationship between the number of bird species found on an island and the size of the island. The table lists the number of species of birds y found on an island with an area of x square kilometers.

x (km²)	0.1	1	10	100	1000
y (species)	10	15	20	25	30

Source: B. Freedman, *Environmental Ecology.*

(a) Find a function f that models the data.

(b) Predict the number of bird species on an island of 5000 square kilometers.

(c) Did your answer involve interpolation or extrapolation?

25. Fertilizer Usage   Between 1950 and 1980 the use of chemical fertilizers increased. The table lists worldwide average usage y in kilograms per hectare of cropland x years after 1950. (*Note:* 1 hectare $\approx$ 2.47 acres.)

x	0	13	22	29
y	12.4	27.9	54.3	77.1

Source: D. Grigg, *The World Food Problem.*

(a) Graph the data. Are the data linear?

(b) Find a function f that models the data.

(c) Predict fertilizer usage in 1989. The actual value was 98.7 kilograms per hectare. What does this indicate about usage of fertilizer during the 1980s?

26. Social Security   If major reform occurs in the Social Security system, individuals might be able to invest some of their contributions into individual accounts. These accounts would be managed by financial firms, which often charge fees. The table lists the amount in billions of dollars that might be collected if fees are 0.93% of the assets each year.

Year	2005	2010	2015	2020
Fees ($ billions)	20	41	80	136

Source: Social Security Advisory Council.

(a) Use exponential regression to find a and b so that $f(x) = ab^x$ models the data x years after 2000.

(b) Graph f and the data.

(c) Estimate the fees in 2013.

(d) Did your answer involve interpolation or extrapolation?

Writing about Mathematics

27. How can you distinguish data that illustrate exponential growth from data that illustrate logarithmic growth?

28. Give an example of data that could be modeled by a logistic function and explain why.

Extended And Discovery Exercise

1. For medical reasons, dyes may be injected into the bloodstream to determine the health of internal organs. In one study involving animals, the dye BSP was injected to assess the blood flow in the liver. The results are listed in the table, where x represents the elapsed time in minutes and y is the concentration of the dye in the bloodstream in milligrams per milliliter.

x	1	2	3	4	5	7
y	0.102	0.077	0.057	0.045	0.036	0.023

x	9	13	16	19	22
y	0.015	0.008	0.005	0.004	0.003

Source: F. Harrison, "The measurement of liver blood flow in conscious calves."

(a) Find a function that models the data.

(b) Estimate the elapsed time when the concentration of the dye reaches 30% of its initial concentration of 0.133 mg/ml.

(c) Let $g(x) = 0.133(0.878(0.73^x) + 0.122(0.92^x))$. This formula was used by the researchers to model the data. Compare the accuracy of your formula to that of $g(x)$.

CHECKING BASIC CONCEPTS FOR SECTION 5.7

📟 *Exercises 1–4: Find a function that models the data. Choose from exponential, logarithmic, or logistic functions.*

1.
x	2	3	4	5	6	7
y	0.72	0.86	1.04	1.24	1.49	1.79

2.
x	2	3	4	5	6	7
y	0.08	1.30	2.16	2.83	3.38	3.84

3.
x	2	3	4	5	6	7
y	0.25	0.86	2.19	3.57	4.23	4.43

4. **World Population** The table lists the actual or projected world population y (in billions) for selected years x.

x	1950	1960	2000	2050	2075
y	2.5	3.0	6.1	8.9	9.2

Source: U.N. Dept. of Economic and Social Affairs.

5 Summary

CONCEPT	EXPLANATION AND EXAMPLES

Section 5.1 Combining Functions

Arithmetic Operations on Functions

Addition: $(f + g)(x) = f(x) + g(x)$

Subtraction: $(f - g)(x) = f(x) - g(x)$

Multiplication: $(fg)(x) = f(x) \cdot g(x)$

Division: $(f/g)(x) = \dfrac{f(x)}{g(x)}, g(x) \neq 0$

Examples: Let $f(x) = x^2 - 5, g(x) = x^2 - 4$.

$(f + g)(x) = (x^2 - 5) + (x^2 - 4) = 2x^2 - 9$

$(f - g)(x) = (x^2 - 5) - (x^2 - 4) = -1$

$(fg)(x) = (x^2 - 5)(x^2 - 4) = x^4 - 9x^2 + 20$

$(f/g)(x) = \dfrac{x^2 - 5}{x^2 - 4}, x \neq 2, x \neq -2$

Composition of Functions

Composition: $(g \circ f)(x) = g(f(x))$

$(f \circ g)(x) = f(g(x))$

Examples: Let $f(x) = 3x + 2, g(x) = 2x^2 - 4x + 1$.

$g(f(x)) = g(3x + 2)$

$\qquad = 2(3x + 2)^2 - 4(3x + 2) + 1$

$f(g(x)) = f(2x^2 - 4x + 1)$

$\qquad = 3(2x^2 - 4x + 1) + 2$

Section 5.2 Inverse Functions and Their Representations

Inverse Function

The inverse function of f is f^{-1} if

$f^{-1}(f(x)) = x$ for every x in the domain of f and

$f(f^{-1}(x)) = x$ for every x in the domain of f^{-1}.

Note: If $f(a) = b$, then $f^{-1}(b) = a$.

CONCEPT	EXPLANATION AND EXAMPLES

Section 5.2 Inverse Functions and Their Representations (CONTINUED)

	Example: Find the inverse function of $f(x) = 4x - 5$.
	$y = 4x - 5$ is equivalent to $\frac{y+5}{4} = x$. (Solve for x.)
	Thus $f^{-1}(x) = \frac{x+5}{4}$.

One-to-One Function	If different inputs always result in different outputs, then f is one-to-one. That is, $a \neq b$ implies $f(a) \neq f(b)$. (A function is also one-to-one if $f(a) = f(b)$ implies $a = b$.) *Note:* If f is one-to-one, then f has an inverse denoted f^{-1}.
	Example: $f(x) = x^2 + 1$ is not one-to-one because $f(2) = f(-2) = 5$.

Horizontal Line Test	If every horizontal line intersects the graph of a function f at most once, then f is a one-to-one function.

Section 5.3 Exponential Functions and Models

Exponential Function	$f(x) = Ca^x$, $a > 0$, $a \neq 1$, and $C > 0$ Exponential growth: $a > 1$; exponential decay: $0 < a < 1$
	Examples: $f(x) = 3(2)^x$ (growth); $f(x) = 1.2(0.5)^x$ (decay)

Exponential Data	For each unit increase in x, the y-values increase (or decrease) by a constant factor a.
	Example: The data in the table are modeled by $y = 5(2)^x$.

x	0	1	2	3
y	5	10	20	40

Natural Exponential Function	$f(x) = e^x$, where $e \approx 2.718282$

Compound Interest	$A = P\left(1 + \frac{r}{n}\right)^{nt}$, where P is the principal, r is the interest expressed as a decimal, n is the number of times interest is paid each year, and t is the number of years.
	Example: $A = 2000\left(1 + \frac{0.10}{12}\right)^{12(4)} \approx \$2978.71$
	calculates the future value of \$2000 invested at 10% compounded monthly for 4 years.

Continuously Compounded Interest	$A = Pe^{rt}$, where P is the principal, r is the interest expressed as a decimal, and t is the number of years
	Example: $A = 2000e^{0.10(4)} \approx \$2983.65$
	calculates the future value of \$2000 invested at 10% compounded continuously for 4 years.

CONCEPT	EXPLANATION AND EXAMPLES

Section 5.4 Logarithmic Functions and Models

Common Logarithm

$\log x = k$ if and only if $x = 10^k$

Natural Logarithm

$\ln x = k$ if and only if $x = e^k$

General Logarithm

$\log_a x = k$ if and only if $x = a^k$

Examples: $\log 100 = \mathbf{2}$ because $100 = 10^2$.

$$\ln \sqrt{e} = \frac{\mathbf{1}}{\mathbf{2}} \quad \text{because} \quad \sqrt{e} = e^{1/2}.$$

$$\log_2 \tfrac{1}{8} = \mathbf{-3} \quad \text{because} \quad \tfrac{1}{8} = 2^{-3}.$$

Inverse Properties

$\log 10^k = k, \qquad 10^{\log k} = k, \quad k > 0$
$\ln e^k = k, \qquad e^{\ln k} = k, \quad k > 0$
$\log_a a^k = k, \qquad a^{\log_a k} = k, \quad k > 0$

Examples: $10^{\log 100} = 100; \; e^{\ln 23} = 23; \; \log_4 64 = \log_4 4^3 = 3$

Inverse Functions

The inverse function of $f(x) = a^x$ is $f^{-1}(x) = \log_a x$.

Examples: If $f(x) = 10^x$, then $f^{-1}(x) = \log x$.
If $f(x) = \ln x$, then $f^{-1}(x) = e^x$.
If $f(x) = \log_5 x$, then $f^{-1}(x) = 5^x$.

Exponential and Logarithmic Forms

$\log_a x = k$ is equivalent to $x = a^k$.

Examples: $\log_2 16 = 4$ is equivalent to $16 = 2^4$.
$81 = 3^4$ is equivalent to $\log_3 81 = 4$.

Section 5.5 Properties of Logarithms

Properties of Logarithms

1. $\log_a 1 = 0$ and $\log_a a = 1$ — Logarithms of 1 and of a
2. $\log_a m + \log_a n = \log_a (mn)$ — Product rule

3. $\log_a m - \log_a n = \log_a \left(\dfrac{m}{n}\right)$ — Quotient rule

4. $\log_a (m^r) = r \log_a m$ — Power rule

Examples: 1. $\log_4 1 = 0$ and $\log_4 4 = 1$
2. $\log 2 + \log 5 = \log(2 \cdot 5) = \log 10 = 1$
3. $\log 500 - \log 5 = \log(500/5) = \log 100 = 2$
4. $\log_2 2^3 = 3 \log_2 2 = 3(1) = 3$

Change of Base Formula

$$\log_a x = \frac{\log_b x}{\log_b a}$$

Example: $\log_3 23 = \dfrac{\log 23}{\log 3} \approx 2.854$

CONCEPT	EXPLANATION AND EXAMPLES

Section 5.6 Exponential and Logarithmic Equations

Solving Exponential Equations

To solve an exponential equation we typically take the logarithm of each side.

Example:

$4e^x = 48$	*Given equation*
$e^x = 12$	*Divide by 4.*
$\ln e^x = \ln 12$	*Take the natural logarithm.*
$x = \ln 12$	*Inverse property*
$x \approx 2.485$	*Approximate.*

Solving Logarithmic Equations

To solve a logarithmic equation we typically need to exponentiate each side.

Example:

$5 \log_3 x = 10$	*Given equation*
$\log_3 x = 2$	*Divide by 5.*
$3^{\log_3 x} = 3^2$	*Exponentiate; base 3.*
$x = 9$	*Inverse property*

Section 5.7 Constructing Nonlinear Models

Exponential Model

$f(x) = Ca^x,\quad f(x) = ab^x,\quad$ or $\quad A(t) = A_0 e^{kt}$
Models data that increase or decrease rapidly

Logarithmic Model

$f(x) = a + b \ln x\quad$ or $\quad f(x) = a + b \log x.$
Models data that increase slowly

Logistic Model

$f(x) = \dfrac{c}{1 + ae^{-bx}}$, where a, b, and c are positive constants.

Models data that increase slowly at first, then increase more rapidly, and finally level off near the value of c. Its graph is a sigmoidal curve. See Figure 5.84.

5 Review Exercises

1. Use the tables to evaluate, if possible.

x	-1	0	1	3
$f(x)$	3	5	7	9

x	-1	0	1	3
$g(x)$	-2	0	1	9

(a) $(f + g)(1)$ (b) $(f - g)(3)$

(c) $(fg)(-1)$ (d) $(f/g)(0)$

2. Use the graph to evaluate each expression.
(a) $(f - g)(2)$ (b) $(fg)(0)$

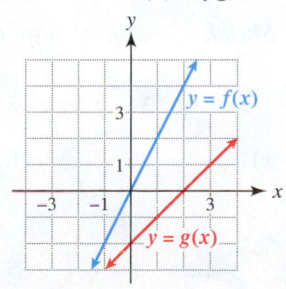

3. Let $f(x) = x^2$ and $g(x) = 1 - x$ and evaluate.
(a) $(f + g)(3)$ (b) $(f - g)(-2)$

(c) $(fg)(1)$ (d) $(f/g)(3)$

4. Use $f(x) = x^2 + 3x$ and $g(x) = x^2 - 1$ to find each expression. Identify its domain.
(a) $(f + g)(x)$ (b) $(f - g)(x)$

(c) $(fg)(x)$ (d) $(f/g)(x)$

5. Tables for f and g are given. Evaluate each expression.

x	-2	0	2	4
$f(x)$	1	4	3	2

x	1	2	3	4
$g(x)$	2	4	-2	0

(a) $(g \circ f)(-2)$ (b) $(f \circ g)(3)$ (c) $f^{-1}(3)$

6. Use the graph to evaluate each expression.
   (a) $(f \circ g)(2)$ **(b)** $(g \circ f)(0)$ **(c)** $f^{-1}(1)$

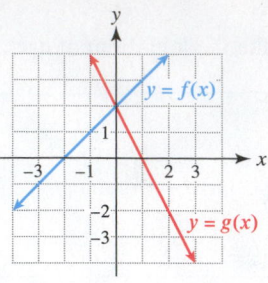

7. Let $f(x) = \sqrt{x}$ and $g(x) = x^2 + x$ and evaluate.
   (a) $(f \circ g)(2)$ **(b)** $(g \circ f)(9)$

8. Use $f(x) = x^2 + 1$ and $g(x) = x^3 - x^2 + 2x + 1$ to find each expression.
   (a) $(f \circ g)(x)$ **(b)** $(g \circ f)(x)$

Exercises 9–12: Find $(f \circ g)(x)$ and identify its domain.

9. $f(x) = x^3 - x^2 + 3x - 2$ $g(x) = x^{-1}$

10. $f(x) = \sqrt{x + 3}$ $g(x) = 1 - x^2$

11. $f(x) = \sqrt[3]{2x - 1}$ $g(x) = \frac{1}{2}x^3 + \frac{1}{2}$

12. $f(x) = \dfrac{2}{x - 5}$ $g(x) = \dfrac{1}{x + 1}$

Exercises 13 and 14: Find f and g so that $h(x) = (g \circ f)(x)$.

13. $h(x) = \sqrt{x^2 + 3}$ **14.** $h(x) = \dfrac{1}{(2x + 1)^2}$

Exercises 15 and 16: Describe the inverse operations of the given statement. Then express both the statement and its inverse symbolically.

15. Divide x by 10 and add 6.

16. Subtract 5 from x and take the cube root.

Exercises 17 and 18: Determine if f is one-to-one.

17. $f(x) = 3x - 1$ **18.** $f(x) = 3x^2 - 2x + 1$

Exercises 19 and 20: Determine if f is one-to-one.

19. 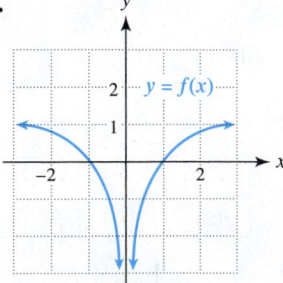   **20.**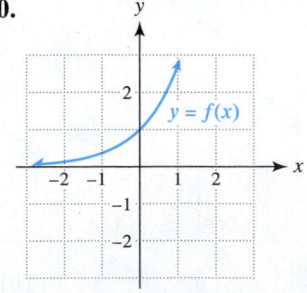

21. The table is a complete representation of f. Use the table of f to determine a table for f^{-1}. Identify the domains and ranges of f and f^{-1}.

x	-1	0	4	6
$f(x)$	6	4	3	1

22. Use the graph of f to sketch a graph of f^{-1}.

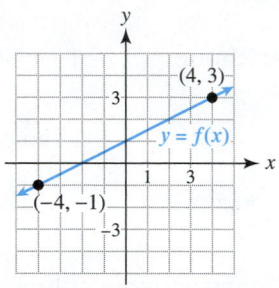

Exercises 23 and 24: Find $f^{-1}(x)$.

23. $f(x) = 3x - 5$ **24.** $f(x) = \dfrac{3x}{x + 7}$

25. Verify that $f(x) = 2x - 1$ and $f^{-1}(x) = \dfrac{x + 1}{2}$ are inverses.

26. Restrict the domain of $f(x) = 2(x - 4)^2 + 3$ so that f is one-to-one. Then find $f^{-1}(x)$.

Exercises 27 and 28: Use the tables to evaluate the given expression.

x	0	1	2	3
$f(x)$	4	3	2	1

x	0	1	2	3
$g(x)$	0	2	3	4

27. $(f \circ g^{-1})(4)$ **28.** $(g^{-1} \circ f^{-1})(1)$

29. Find $f^{-1}(x)$ if $f(x) = \sqrt{x + 1}, x \geq -1$. Identify the domain and range of f and of f^{-1}.

30. Simplify $e^x e^{-2x}$.

Exercises 31 and 32: Find C and a so that $f(x) = Ca^x$ satisfies the given conditions.

31. $f(0) = 3$ and $f(3) = 24$

32. $f(-1) = 8$ and $f(1) = 2$

Exercises 33–36: Sketch a graph of $y = f(x)$. Identify the domain of f.

33. $f(x) = 4(2)^{-x}$ **34.** $f(x) = 3^{x-1}$

35. $f(x) = \log(-x)$ **36.** $f(x) = \log(x + 1)$

Exercises 37 and 38: Use the graph of $f(x) = Ca^x$ to determine values for C and a.

37.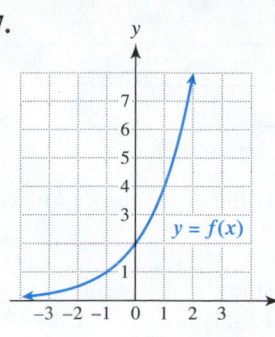

38.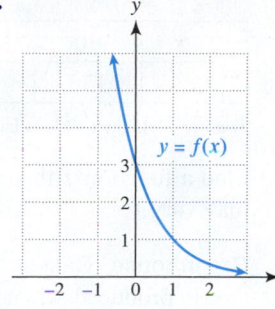

39. Determine the final value of $1200 invested at 9% compounded semiannually for 3 years.

40. Determine the final value of $500 invested at 6.5% compounded continuously for 8 years.

41. Solve $e^x = 19$ symbolically.

42. Solve $2^x - x^2 = x$ graphically. Round each solution to the nearest thousandth.

Exercises 43–46: Evaluate the expression without a calculator.

43. $\log 1000$

44. $\log 0.001$

45. $10 \log 0.01 + \log \frac{1}{10}$

46. $\log 100 + \log \sqrt[3]{10}$

Exercises 47–50: Evaluate the logarithm without a calculator.

47. $\log_3 9$

48. $\log_5 \frac{1}{25}$

49. $\ln e$

50. $\log_2 32$

Exercises 51 and 52: Approximate to the nearest thousandth.

51. $\log_3 18$

52. $\log_2 173$

Exercises 53–60: Solve the equation.

53. $10^x = 125$

54. $1.5^x = 55$

55. $e^{0.1x} = 5.2$

56. $4e^{2x} - 5 = 3$

57. $5^{-x} = 10$

58. $3(10)^{-x} = 6$

59. $50 - 3(0.78)^{x-10} = 21$

60. $5(1.3)^x + 4 = 104$

Exercises 61 and 62: Find either a linear or an exponential function that models the data in the table.

61.

x	0	1	2
y	1.5	3	6

62.

x	0	1	2
y	3	4.5	6

Exercises 63–66: Solve the equation.

63. $\log x = 1.5$

64. $\log_3 x = 4$

65. $\ln x = 3.4$

66. $4 - \ln(5 - x) = \frac{5}{2}$

Exercises 67 and 68: Use properties of logarithms to combine the expression as a logarithm of a single expression.

67. $\log 6 + \log 5x$

68. $\log \sqrt{3} - \log \sqrt[3]{3}$

69. Expand $\ln \frac{y}{x^2}$.

70. Expand $\log \frac{4x^3}{k}$.

Exercises 71–76: Solve the logarithmic equation.

71. $8 \log x = 2$

72. $\ln 2x = 2$

73. $2 \log 3x + 5 = 15$

74. $5 \log_2 x = 25$

75. $2 \log_5 (x + 2) = \log_5 (x + 8)$

76. $\ln(5 - x) - \ln(5 + x) = -\ln 9$

77. Suppose that b is the y-intercept on the graph of a one-to-one function f. What is the x-intercept on the graph of f^{-1}? Explain your reasoning.

78. Let $f(x) = ax + b$ with $a \neq 0$.
(a) Show that f^{-1} is also linear by finding $f^{-1}(x)$.

(b) How is the slope of the graph of f related to the slope of the graph of f^{-1}?

Applications

79. Bacteria Growth   There are initially 4000 bacteria per milliliter in a sample, and after 1 hour their concentration increases to 6000 bacteria per milliliter. Assume exponential growth.
(a) How many bacteria are there after 2.5 hours?

(b) After how long are there 8500 bacteria per milliliter?

80. Newton's Law of Cooling   A pan of boiling water with a temperature of 100°C is set in a room with a temperature of 20°C. The water cools to 50°C in 40 minutes.
(a) Find values for T_0, D, and a so that the formula $T(t) = T_0 + Da^t$ models the data, where t is in hours.

(b) Find the temperature of the water after 90 minutes.

(c) How long does it take the water to reach 30°C?

81. Combining Functions   The total number of gallons of water passing through a pipe after x seconds is computed by $f(x) = 10x$. Another pipe delivers $g(x) = 5x$ gallons after x seconds. Find a function h that gives the volume of water passing through both pipes in x seconds.

82. Test Scores   Let scores on a standardized test be modeled by $f(x) = 36e^{-(x-20)^2/49}$. The function f computes the number in thousands of people that received score x. Solve the equation $f(x) = 30$. Interpret your result.

83. Modeling Growth The function given by

$$W(x) = 175.6(1 - 0.66e^{-0.24x})^3$$

models the weight in milligrams of a small fish called the *Lebistes reticulatus* after x weeks, where $0 \le x \le 14$. Solve the equation $W(x) = 50$. Interpret the result. (*Source:* D. Brown and P. Rothery, *Models in Biology*.)

84. Radioactive Decay After 23 days, a 10-milligram sample of a radioactive material decays to 5 milligrams. After how many days will there be 1 milligram of the material?

85. Tire Pressure A car tire has a small leak, and the tire pressure in pounds per square inch after t minutes is given by $P(t) = 32e^{-0.2t}$. After how many minutes is the pressure 15 pounds per square inch?

86. Converting Units The figures show graphs of a function f that converts fluid ounces to pints and a function g that converts pints to quarts. Evaluate each expression. Interpret the results.

(a) $(g \circ f)(32)$

(b) $f^{-1}(1)$

(c) $(f^{-1} \circ g^{-1})(1)$

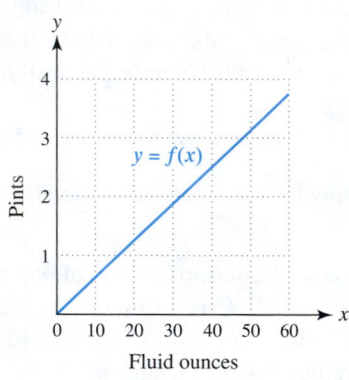

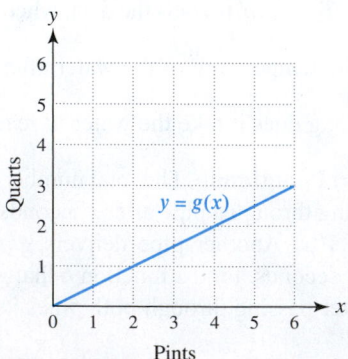

87. Modeling Epidemics In 1666 the village of Eyam, located in England, experienced an outbreak of the Great Plague. Out of 261 people in the community, only 83 people survived. The tables at the top of the next column show a function f that computes the number of people who were infected after x days.

x	0	15	30	45
$f(x)$	7	21	57	111

x	60	75	90	125
$f(x)$	136	158	164	178

Source: G. Raggett, "Modeling the Eyam plague."

Find a function f that models the given data. (Answers may vary.)

88. Greenhouse Gases Methane is a greenhouse gas that is produced when fossil fuels are burned. In 1600 methane had an atmospheric concentration of 700 parts per billion (ppb), whereas in 2000 its concentration was about 1700 ppb. (*Source:* D. Wuebbles and J. Edmonds, *Primer on Greenhouse Gases*.)

(a) Find values for C and a so that $f(x) = Ca^x$ models the data, where x is the year.

(b) Solve $f(x) = 1000$ and interpret the answer.

89. Exponential Regression The data in the table can be modeled by $f(x) = ab^x$. Use regression to estimate the constants a and b. Graph f and the data.

x	1	2	3	4
y	2.59	1.92	1.42	1.05

90. Logarithmic Regression The data in the table can be modeled by $f(x) = a + b \ln x$. Use regression to estimate the constants a and b. Graph f and the data.

x	2	3	4	5
y	2.93	3.42	3.76	4.03

Extended and Discovery Exercises

1. Modeling Data with Power Functions There is a procedure to determine whether data can be modeled by $y = ax^b$, where a and b are constants. Start by taking the natural logarithm of each side of this equation.

$\ln y = \ln(ax^b)$

$\ln y = \ln a + \ln x^b$ $\ln(mn) = \ln m + \ln n$

$\ln y = \ln a + b \ln x$ $\ln(m^r) = r \ln m$

If we let $z = \ln y$, $d = \ln a$, and $w = \ln x$, then the equation $\ln y = \ln a + b \ln x$ becomes $z = d + bw$. Thus the data points $(w, z) = (\ln x, \ln y)$ lie on the line having a slope of b and y-intercept $d = \ln a$. The following steps provide a procedure for finding the constants a and b.

Modeling Data with the Equation $y = ax^b$

If a data set (x, y) can be modeled by the (power) equation $y = ax^b$, then the following procedure can be applied to determine the constants a and b.

STEP 1: Let $w = \ln x$ and $z = \ln y$ for each data point. Graph the points (w, z). If these data are not linear, then do *not* use this procedure.

STEP 2: Find an equation of a line in the form $z = bw + d$ that models the data points (w, z). (Linear regression may be used.)

STEP 3: The slope of the line equals the constant b. The value of a is given by $a = e^d$.

Apply this procedure to the table of data for the orbital distances and periods of the moons of Jupiter. Let the distance be x and the period be y.

Moons of Jupiter	Distance (10^3 km)	Period (days)
Metis	128	0.29
Almathea	181	0.50
Thebe	222	0.67
Io	422	1.77
Europa	671	3.55
Ganymede	1070	7.16
Callisto	1883	16.69

2. **Climate Change** Greenhouse gases such as carbon dioxide trap heat from the sun. Presently, the net incoming solar radiation reaching Earth's surface is approximately 240 watts per square meter (w/m^2). Any portion of this amount that is due to greenhouse gases is called *radiative forcing*. The table lists the estimated increase in radiative forcing R above the levels in 1750.

x (year)	1800	1850	1900	1950	2000
$R(x)$ (w/m^2)	0.2	0.4	0.6	1.2	2.4

Source: A. Nilsson, *Greenhouse Earth.*

(a) Estimate constants C and k so that $R(x) = Ce^{kx}$ models the data. Let $x = 0$ correspond to 1800.

(b) Estimate the year when the additional radiative forcing could reach 3 w/m^2.

Exercises 3 and 4: Try to decide if the expression is precisely an integer. (Hint: Use computer software capable of calculating a large number of decimal places.)

3. $\left(\dfrac{1}{\pi} \ln \left(640{,}320^3 + 744\right) \right)^2$

(*Source:* I. J. Good, "What is the most amazing approximate integer in the universe?")

4. $e^{\pi \sqrt{163}}$ (*Source:* W. Cheney and D. Kincaid, *Numerical Mathematics and Computing.*)

Industrial robots comprise a multi-billion-dollar industry in the United States. Robots paint our automobiles, recognize voices, and even perform surgery that is more precise and less invasive than traditional techniques. In the future, robots will affect our quality of life by being able to assist the elderly, handle hazardous material, and help repair over 2 million miles of underground piping in our deteriorating infrastructure. Spectacular advances in microelectronic mechanical systems (MEMS) may make it possible for tiny robots to repair human tissue at the cellular level.

The trigonometry that has played a key role in the design of robots goes back to the 19th century. Some of the most important results in robotics have come from mathematicians and their interactions with engineers. Trigonometry is also used in applications involving monthly average temperatures, tidal currents, the Global Positioning System, electricity, highways, orbits of satellites, phases of the moon, and even music. In this chapter, we discuss the six basic trigonometric functions and many of their applications.

Source: National Science Foundation, *The Interplay between Mathematics and Robotics: Summary of a Workshop.*

> It is not enough to have a good mind; the main thing is to use it well.
>
> —René Descartes

6.1 Angles and Their Measure

- Learn basic concepts about angles
- Apply degree measure to problems
- Apply radian measure to problems
- Calculate arc length
- Calculate the area of a sector

Introduction

In 2007 at the French Open, Venus Williams hit the fastest serve recorded in a women's tour main-draw match, reaching 128 miles per hour. Much of the speed gained in a serve comes from flexing the wrist and rotating the shoulder joint. These rotations create *angular speed*, which transfers *linear speed* to the tennis ball. Trigonometry can be used to understand how fast it is possible to hit a tennis ball.   (*Source:* J. Cooper and R. Glassow, *Kinesiology.*)

Angles

An **angle** is formed by rotating a ray about its endpoint. The starting position of the ray is called the **initial side**, and the final position of the ray is the **terminal side**. If the rotation of the ray is counterclockwise, the angle has *positive measure*; if the rotation is clockwise, the angle has *negative measure*. For simplicity we will refer to an angle as being **positive** or **negative**. The endpoint of the ray is called the **vertex** of the angle. See Figures 6.1 and 6.2, where the Greek letter θ (theta) has been used to denote an angle.

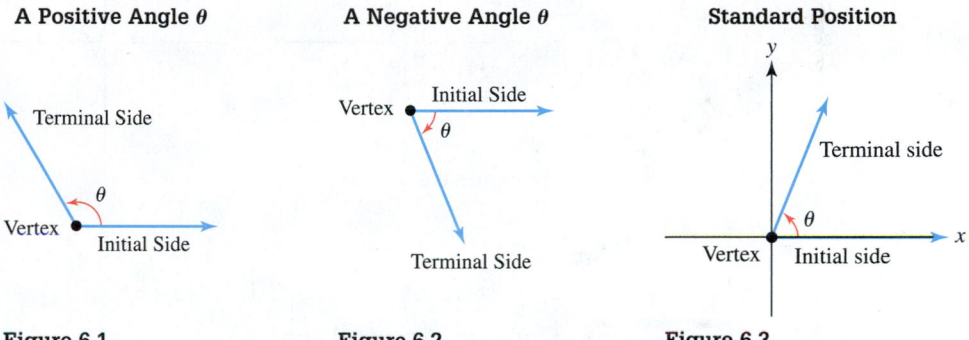

Figure 6.1   **Figure 6.2**   **Figure 6.3**

If the vertex is positioned so that it corresponds to the origin in the *xy*-plane and the initial side coincides with the positive *x*-axis, then the angle is in **standard position**, as shown in Figure 6.3.

There are two common systems for measuring the size of an angle: *degree measure* and *radian measure*.

Degree Measure

In **degree measure**, one complete rotation of a ray about its endpoint contains 360 degrees. One degree, denoted 1°, represents $\frac{1}{360}$ of a complete rotation. Figure 6.4 shows some examples of angles with their degree measure.

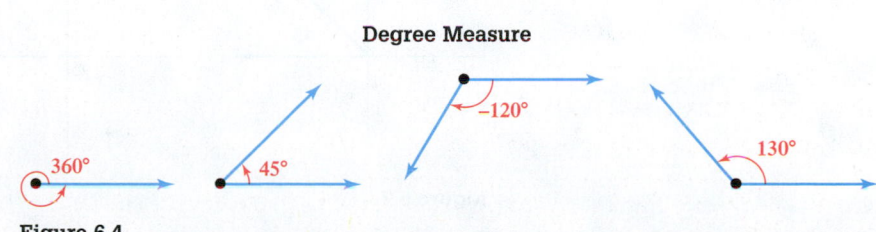

Figure 6.4

A **right angle** has measure 90°, and a **straight angle** has measure 180°. The measure of an **acute angle** is greater than 0° but less than 90°, whereas the measure of an **obtuse angle** is greater than 90° but less than 180°. Examples are shown in Figure 6.5.

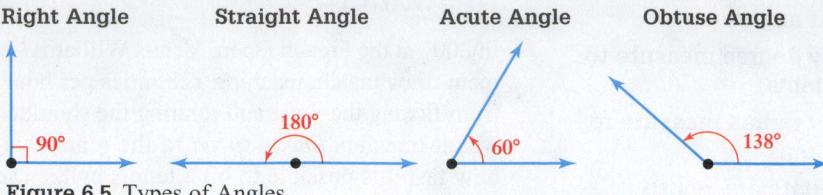

Figure 6.5 Types of Angles

We will use the Greek letters α (alpha), β (beta), γ (gamma) and θ (theta) to denote angles. For simplicity, we sometimes refer to an angle θ having measure 45° as a 45° angle, or an angle of 45°. This may be expressed as $\theta = 45°$. Two angles with the same initial and terminal sides are **coterminal angles**. Examples of coterminal angles are shown in Figure 6.6.

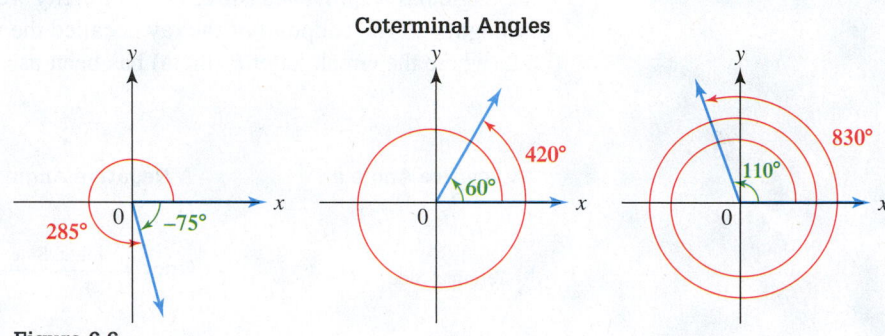

Figure 6.6

EXAMPLE 1 **Finding coterminal angles**

Find three angles coterminal with $\theta = 45°$, where θ is in standard position. Sketch these angles in standard position.

SOLUTION We can find coterminal angles by either adding multiples of 360° to θ or subtracting multiples of 360° from θ.

i. $45° + 360° = 405°$ **ii.** $45° + 2(360°) = 765°$ **iii.** $45° - 360° = -315°$

The angles **405°**, **765°**, and **−315°** are all coterminal with a **45°** angle. These three angles and θ are sketched in Figure 6.7. Note that other angles are also possible.

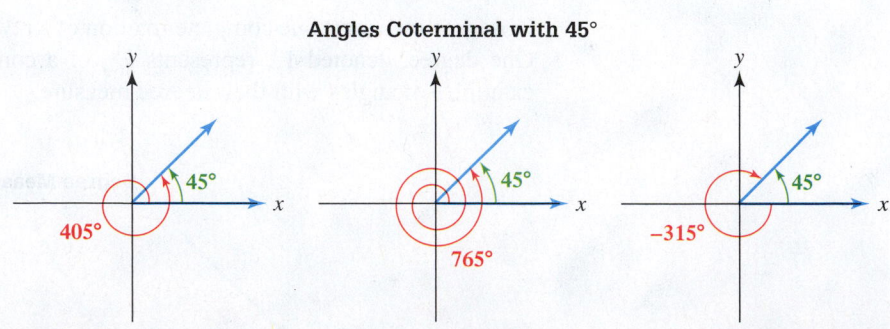

Figure 6.7

Now Try Exercise 13

Two positive angles are **complementary angles** if their sum is 90° and **supplementary angles** if their sum is 180°. For example, $\alpha = 35°$ and $\beta = 55°$ are complementary angles, whereas $\alpha = 60°$ and $\beta = 120°$ are supplementary angles. See Figures 6.8 and 6.9.

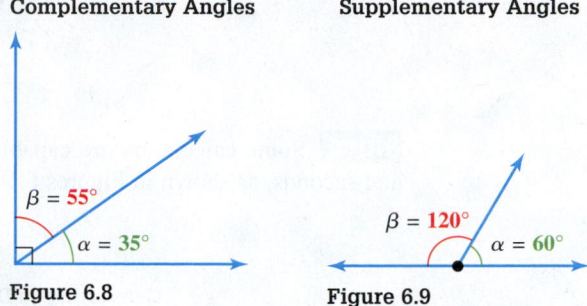

Complementary Angles   **Supplementary Angles**

$\beta = 55°$ $\beta = 120°$

$\alpha = 35°$ $\alpha = 60°$

Figure 6.8   **Figure 6.9**

Fractions of a degree may be measured using **minutes** and **seconds**. One minute, written $1'$, equals $\frac{1}{60}$ of a degree, and one second, written $1''$, equals $\frac{1}{60}$ of a minute or $\frac{1}{3600}$ of a degree. The measurement $25°45'30''$ represents 25 degrees, 45 minutes, 30 seconds. Expressed in decimal degrees, this measurement is as follows.

Converting to Decimal Degrees

$$25°45'30'' = 25° + \left(\frac{45}{60}\right)° + \left(\frac{30}{3600}\right)° = 25.7583\overline{3}°$$

$1' = \frac{1}{60}°$ $1'' = \frac{1}{3600}°$

We can convert decimal degrees to degrees, minutes, and seconds as illustrated in the next example.

EXAMPLE 2   **Converting to degrees, minutes, and seconds**

Convert 32.41° to degrees, minutes, and seconds.

SOLUTION

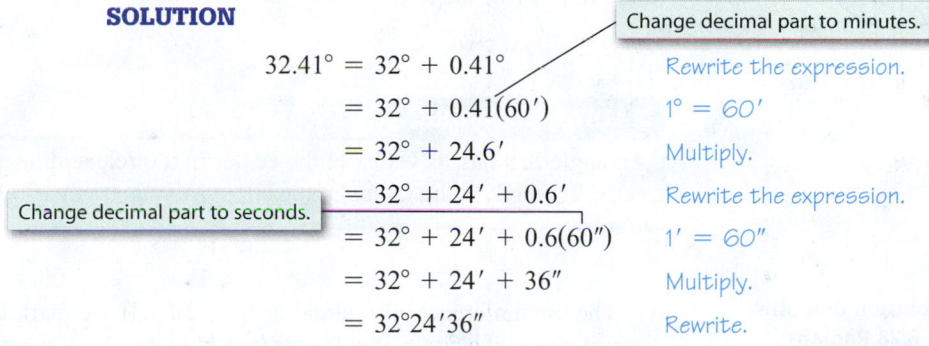

Change decimal part to minutes.

$32.41° = 32° + 0.41°$ Rewrite the expression.

$= 32° + 0.41(60')$ $1° = 60'$

$= 32° + 24.6'$ Multiply.

$= 32° + 24' + 0.6'$ Rewrite the expression.

Change decimal part to seconds.

$= 32° + 24' + 0.6(60'')$ $1' = 60''$

$= 32° + 24' + 36''$ Multiply.

$= 32°24'36''$ Rewrite.

Now Try Exercise 27

In the next example, we find complementary and supplementary angles using degree measure.

EXAMPLE 3   **Finding complementary and supplementary angles**

Find angles that are complementary and supplementary to $\alpha = 34°19'42''$.

SOLUTION If angle β is complementary to α, then $\beta = 90° - \alpha$.

$$\beta = 90° - 34°19'42''$$

$$= 89°59'60'' - 34°19'42''   \quad 90° = 89°59'60''$$

$$= 55°40'18''$$

An angle supplementary to $\alpha = 34°19'42''$ is given by $\gamma = 180° - \alpha$.

$$\gamma = 180° - 34°19'42''$$

$$= 179°59'60'' - 34°19'42'' \qquad 180° = 179°59'60''$$

$$= 145°40'18''$$

Now Try Exercise 31

NOTE Some calculators are capable of performing arithmetic using degrees, minutes, and seconds, as shown in Figures 6.10 and 6.11. (See the ANGLE menu.)

Converting to Degrees, Minutes, and Seconds

Results from Examples 2 and 3

```
32.41▶DMS
              32°24'36"
```

```
90°−34°19'42"▶DM
S
              55°40'18"
180°−34°19'42"▶D
MS
            145°40'18"
```

Figure 6.10 Degree Mode   **Figure 6.11** Degree Mode

Radian Measure

One Radian

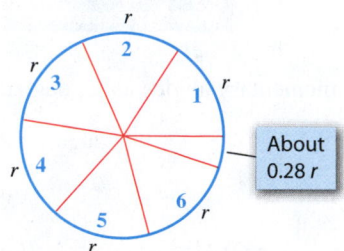

Figure 6.12

A second system of angle measure is based on *radians*. Radians are a common unit of measurement that often make formulas simpler and easier to use. Two examples are the formulas for arc length and area of a sector, which are introduced later in this section.

Angle θ in Figure 6.12 has a measure of one radian. The vertex of θ is located at the center of the circle, and its initial and terminal sides intercept an arc whose length is equal to the *radius r* of the circle.

RADIAN MEASURE

An angle that has its vertex at the center of a circle and intercepts an arc on the circle equal in length to the radius of the circle has a measure of **one radian**.

One Revolution Contains
$2\pi \approx 6.28$ **Radians**

About 0.28 r

Figure 6.13

The circumference of a circle is $C = 2\pi r$. If we mark off distances of r along the circumference of a circle, it will appear as in Figure 6.13, where $2\pi \approx 6.28$ distances of r are shown. Therefore one rotation contains $2\pi \approx 6.28$ radians, and so 360° is equivalent to 2π radians. Radian measure can be compared to degree measure using proportions. Since 180° is equivalent to π radians, it follows that

$$\frac{\text{Radian measure}}{\text{Degree measure}} = \frac{\pi}{180°}.$$

Solving for radian measure results in

$$\textbf{Radian measure} = \textbf{Degree measure} \times \frac{\pi}{180°},$$

and solving for degree measure results in

$$\textbf{Degree measure} = \textbf{Radian measure} \times \frac{180°}{\pi}.$$

The following statements summarize the preceding discussion.

CONVERTING BETWEEN DEGREES AND RADIANS

To convert *degrees to radians*, multiply a degree measure by $\frac{\pi}{180°}$.

To convert *radians to degrees*, multiply a radian measure by $\frac{180°}{\pi}$.

NOTE   One radian is equivalent to approximately 57.3°, because $1 \times \frac{180°}{\pi} \approx 57.3°$.

EXAMPLE 4   Converting degrees to radians

Convert each degree measure to radian measure.
(a) 90°   **(b)** 225°

SOLUTION
(a) To convert degrees to radians, multiply by $\frac{\pi}{180°}$.

$$90° \times \frac{\pi}{180°} = \frac{\pi}{2} \text{ radians}$$

Thus 90° is equivalent to $\frac{\pi}{2}$ radians.

(b) $225° \times \frac{\pi}{180°} = \frac{5\pi}{4}$ radians

> Now Try Exercise 39

Table 6.1 shows some equivalent measures in degrees and radians.

Degree and Radian Measure

Degrees	0°	30°	45°	60°	90°	180°	360°
Radians	0	$\frac{\pi}{6}$	$\frac{\pi}{4}$	$\frac{\pi}{3}$	$\frac{\pi}{2}$	π	2π

Table 6.1

EXAMPLE 5   Converting radians to degrees

Convert each radian measure to degree measure.
(a) $\frac{4\pi}{3}$ **(b)** $\frac{5\pi}{6}$

Converting to Radians

```
90°
        1.570796327
225°
        3.926990817
```
$\frac{\pi}{2}$
$\frac{5\pi}{4}$

Figure 6.14 Radian Mode

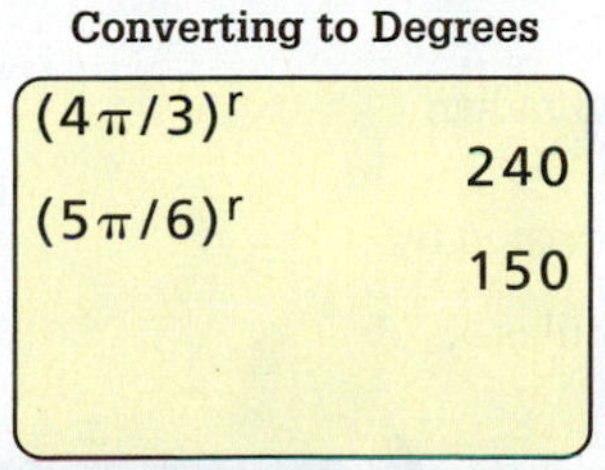

SOLUTION
(a) To convert radians to degrees, multiply by $\frac{180°}{\pi}$.

$$\frac{4\pi}{3} \times \frac{180°}{\pi} = 240°$$

Thus $\frac{4\pi}{3}$ radians is equivalent to 240°.

(b) $\frac{5\pi}{6} \times \frac{180°}{\pi} = 150°$

> Now Try Exercise 43

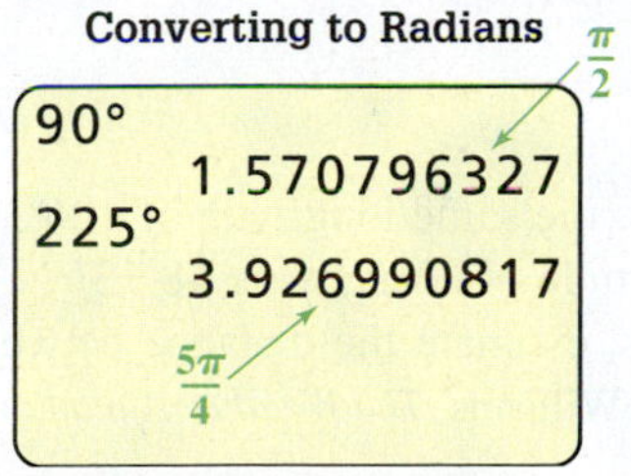

Converting to Degrees

```
(4π/3)ʳ
            240
(5π/6)ʳ
            150
```

Figure 6.15 Degree Mode

NOTE   Some calculators can convert degrees to radians and radians to degrees, as shown in Figures 6.14 and 6.15, respectively. When converting from degrees to radians, many calculators give only decimal approximations rather than exact values. For example, $\frac{\pi}{2}$ may be expressed as 1.570796327.

Arc Length $s = r\theta$

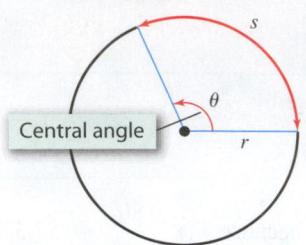

Figure 6.16

Arc Length

From geometry we know that the arc length s on a circle is proportional to the measure of the **central angle** θ. See Figure 6.16. A central angle of 2π radians corresponds to an arc length that equals the circumference $C = 2\pi r$. Using proportions yields

$$\frac{s}{\theta} = \frac{2\pi r}{2\pi},$$

which simplifies to $s = r\theta$.

> **ARC LENGTH**
>
> The **arc length** s intercepted on a circle of radius r by a central angle of θ *radians* is given by
>
> $$s = r\theta.$$

NOTE Angle θ *must* be in radian measure to use the arc length formula $s = r\theta$.

EXAMPLE 6 Finding arc length

A circle has a radius of 25 inches. Find the length of an arc intercepted by a central angle of $45°$.

SOLUTION First convert $45°$ to *radian* measure.

Calculating Arc Length

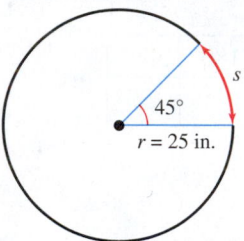

Figure 6.17

$$45° \times \frac{\pi}{180°} = \frac{\pi}{4}$$

The arc length s is given by   Must be radian measure

$$s = r\theta = 25\left(\frac{\pi}{4}\right) = 6.25\pi \text{ inches.}$$

The arc length s shown in Figure 6.17 is $6.25\pi \approx 19.6$ inches.

Now Try Exercise 47

EXAMPLE 7 Finding distance between cities

Albuquerque, New Mexico, and Glasgow, Montana, have the same longitude of $106°37'$ W. The latitude of Albuquerque is $35°03'$ N, and the latitude of Glasgow is $48°13'$ N. If the *average* radius of Earth is approximately 3955 miles, estimate the distance between Albuquerque and Glasgow. See Figure 6.18.   (*Source:* J. Williams, *The Weather Almanac.*)

Distance between Cities

Glasgow

s

Albuquerque

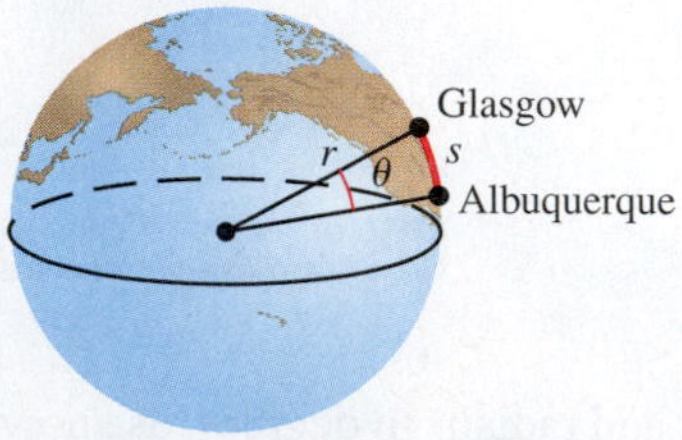

Figure 6.18 (not to scale)

SOLUTION This distance can be estimated using the arc length formula. Start by converting $\theta = 48°13' - 35°03' = 13°10'$ to radian measure.

$$\left[13° + \left(\frac{10}{60}\right)°\right] \times \frac{\pi}{180°} \approx 0.2298 \text{ radian}$$

The distance between Albuquerque and Glasgow is approximated by

$$s = r\theta \approx 3955(0.2298) \approx 909 \text{ miles.}$$

Must be radian measure

Now Try Exercise 81

Flexing the Wrist in Tennis

Figure 6.19

Rotation and Angular Speed The human joint that can be flexed the fastest is the wrist, which can rotate through 90°, or $\frac{\pi}{2}$ radians, in 0.045 second while holding a tennis racket. See Figure 6.19. **Angular speed** ω (omega) measures the speed of rotation and is defined by

$$\omega = \frac{\theta}{t}, \qquad \text{Angular speed}$$

where θ is the angle of rotation and t is time. The angular speed ω of a human wrist holding a tennis racket is

$$\frac{\theta}{t} = \frac{\pi/2}{0.045} \approx 34.9 \text{ rad/sec},$$

or about 35 radians per second. The **linear speed** v at which the tip of the racket travels as a result of flexing the wrist is given by

$$v = r\omega, \qquad \text{Linear speed}$$

where r is the radius (distance) from the tip of the racket to the wrist joint and ω is in radians per unit of time. If $r = 2$ feet, then the speed at the tip of the racket is

$$v = r\omega \approx (2)(35) = 70 \text{ ft/sec},$$

or about 48 miles per hour. In a tennis serve the arm flexes at the elbow and rotates at the shoulder, so the final speed of the racket is considerably faster. (*Source:* J. Cooper and R. Glassow, *Kinesiology.*)

EXAMPLE 8 | **Finding the speed of a GPS satellite**

GPS Satellite Coverage

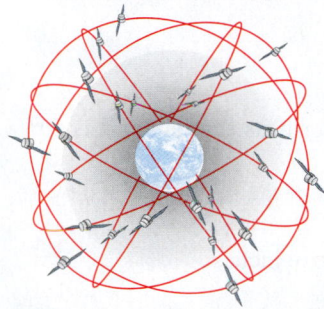

Figure 6.20

Each of the 24 satellites used in the Global Positioning System is located 16,526 miles from the *center* of Earth and has a nearly circular orbit with a period of 12 hours. See Figure 6.20. (*Source:* Y. Zhao, *Vehicle Location and Navigation Systems.*)
(a) Find the angular speed of a satellite.
(b) Estimate the linear speed of a satellite.

SOLUTION
(a) A GPS satellite circles Earth once (2π radians) every 12 hours. Its angular speed is

$$\omega = \frac{2\pi}{t} = \frac{2\pi}{12} = \frac{\pi}{6} \approx 0.5236 \text{ rad/hr}.$$

(b) Its linear speed is $v = r\omega = (16{,}526)(0.5236) \approx 8653$ miles per hour.

Now Try Exercise 79

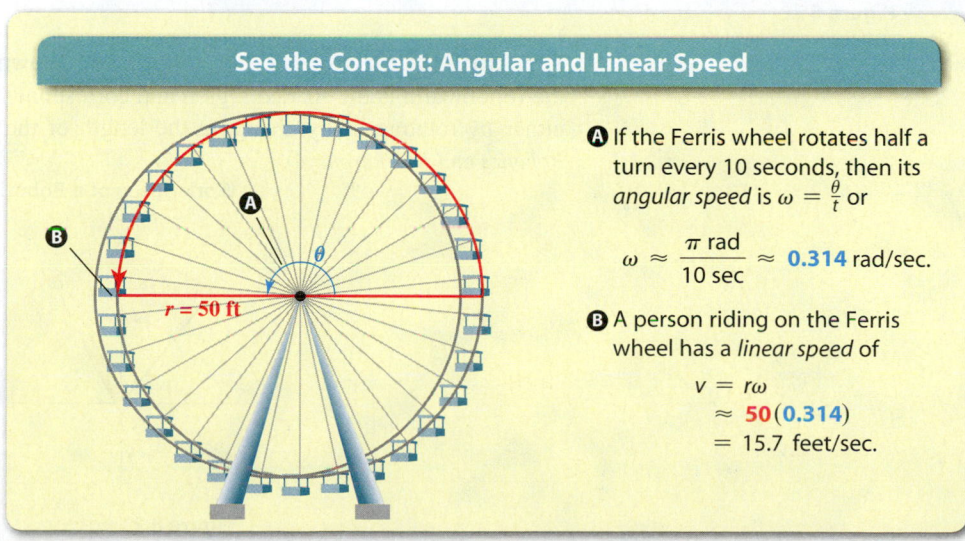

See the Concept: Angular and Linear Speed

Ⓐ If the Ferris wheel rotates half a turn every 10 seconds, then its *angular speed* is $\omega = \frac{\theta}{t}$ or

$$\omega \approx \frac{\pi \text{ rad}}{10 \text{ sec}} \approx 0.314 \text{ rad/sec}.$$

Ⓑ A person riding on the Ferris wheel has a *linear speed* of

$$v = r\omega$$
$$\approx 50(0.314)$$
$$= 15.7 \text{ feet/sec}.$$

$r = 50$ ft

A shoulder can rotate at about 25 radians per second. Estimate how much this rotation increases the speed of a racket if the total length of the (straight) arm and racket is 4 feet.

Sector of a Circle

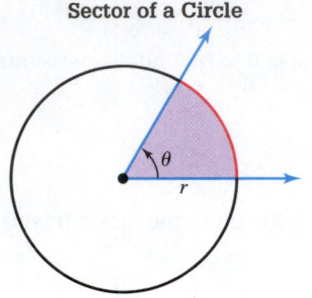

Figure 6.21

Area of a Sector

The **sector of a circle** is the portion of the interior of a circle intercepted by a central angle. The shaded region in Figure 6.21 shows a sector of a circle with radius r and central angle θ. The area of a sector is proportional to the measure of the central angle. If the central angle is 2π radians, then the area of the sector is the entire interior of the circle, which has an area of πr^2. Using proportions yields

$$\frac{\text{area of a sector}}{\theta} = \frac{\pi r^2}{2\pi}.$$

Solving the equation for the area of the sector results in

$$\text{area of a sector} = \frac{1}{2}r^2\theta.$$

AREA OF A SECTOR

The **area of a sector** A of a circle of radius r and central angle θ in *radians* is given by

$$A = \frac{1}{2}r^2\theta.$$

NOTE Angle θ *must* be in radian measure to use the area formula $A = \frac{1}{2}r^2\theta$.

EXAMPLE 9 Finding the area of a sector

A circle has a radius of 6 inches. Find the area of the sector if its central angle is 60°.

Calculating Area

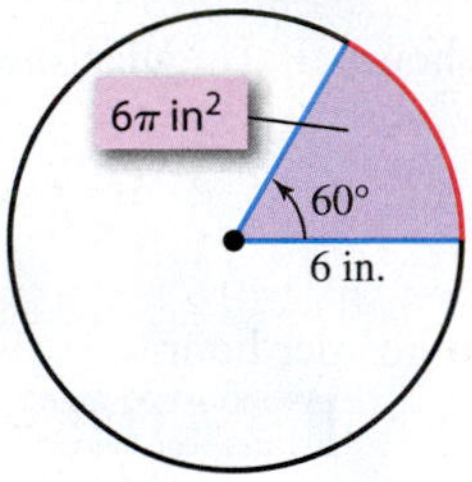

6π in^2

$60°$

6 in.

Figure 6.22

SOLUTION Since 60° is equivalent to $\frac{\pi}{3}$ radians, the area of the sector is given by

$$A = \frac{1}{2}r^2\theta = \frac{1}{2}(6)^2\left(\frac{\pi}{3}\right) = 6\pi \text{ square inches.}$$

Must be radian measure

This region of $6\pi \approx 18.8$ square inches is illustrated in Figure 6.22.

Now Try Exercise 63

An Application Consider the robotic arm shown in Figure 6.23. The *work space* of the robotic arm is the shaded region and corresponds to the places that the hand can reach either by rotating or by changing the length of the arm. (*Source:* W. Stadler, *Analytical Robotics and Mechatronics.*)

Work Space of a Robotic Arm

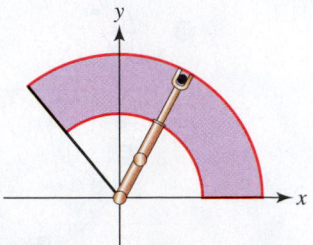

Figure 6.23

EXAMPLE 10   Finding the area of the work space for a robotic arm

Suppose that a robotic arm similar to the one in Figure 6.23 can rotate between $\theta = 10°$ and $\theta = 130°$. If the length of the robotic arm can vary between 5 inches and 20 inches, find the area of its work space.

SOLUTION The work space can be thought of as a large sector having radius $r_1 = 20$ inches with a small sector of radius $r_2 = 5$ inches removed. The arm can rotate through $130° - 10° = 120°$, or $\frac{2\pi}{3}$ radians. See Figure 6.24. The area A of the work space is computed as follows.

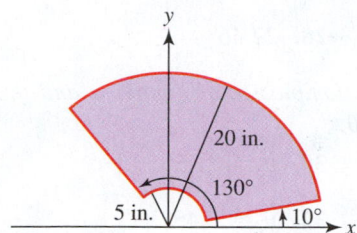

Figure 6.24

$$A = \frac{1}{2}r_1^2\theta - \frac{1}{2}r_2^2\theta \qquad \text{Area of large sector minus area of small sector}$$

$$= \frac{1}{2}\theta\left(r_1^2 - r_2^2\right) \qquad \text{Factor out } \frac{1}{2}\theta.$$

$$= \frac{1}{2}\left(\frac{2\pi}{3}\right)\left(20^2 - 5^2\right) \qquad \text{Substitute.}$$

$$= 125\pi \qquad \text{Simplify.}$$

The work space is $125\pi \approx 392.7$ square inches.

Now Try Exercise 71

6.1 Putting It All Together

Some concepts involving angles are summarized in the following table.

CONCEPT	EXPLANATION OR FORMULA	EXAMPLES
Degree measure	One complete rotation contains 360°.	A right angle contains 90°. A straight angle contains 180°. An acute angle α satisfies $0° < \alpha < 90°$. An obtuse angle β satisfies $90° < \beta < 180°$.
Radian measure	One complete rotation contains 2π radians.	2π radians is equivalent to 360°. π radians is equivalent to 180°. $\frac{\pi}{2}$ radians is equivalent to 90°. $\frac{\pi}{3}$ radians is equivalent to 60°. $\frac{\pi}{4}$ radian is equivalent to 45°. $\frac{\pi}{6}$ radian is equivalent to 30°.
Arc length	$s = r\theta$, where θ is in *radians*	If $r = 12$ feet and $\theta = 90°$, then $s = (12)\frac{\pi}{2} = 6\pi \approx 18.8$ feet.
Area of a sector	$A = \frac{1}{2}r^2\theta$, where θ is in *radians*	If $r = 6$ inches and $\theta = 45°$, then $A = \frac{1}{2}(6)^2\left(\frac{\pi}{4}\right) = 4.5\pi \approx 14.1$ square inches.
Angular speed	$\omega = \frac{\theta}{t}$, where θ is the angle of rotation and t is time	If $\theta = 5$ radians and $t = 0.1$ second, then $\omega = \frac{5}{0.1} = 50$ radians per second.
Linear speed of a rotating object	$v = r\omega$, where r is the radius and ω is the angular speed in *radians* per unit of time	If $r = 3$ feet and $\omega = 5$ radians per second, then $v = (3)(5) = 15$ feet per second.

6.1 Exercises

Angles

Exercises 1 and 2: Sketch the following angles in standard position.

1. (a) $45°$ **(b)** $-150°$
    (c) $\frac{\pi}{3}$ **(d)** $-\frac{3\pi}{4}$

2. (a) $-90°$ **(b)** $225°$
    (c) $-\frac{2\pi}{3}$ **(d)** $\frac{\pi}{6}$

Exercises 3–10: Sketch an angle θ in standard position that satisfies the conditions. Assume α is in standard position.

3. Acute   **4.** Obtuse

5. A positive straight angle   **6.** Complementary to $60°$

7. Positive and the terminal side lies in quadrant III

8. Negative and the terminal side lies in quadrant IV

9. Negative and coterminal with $\alpha = 90°$

10. Positive and coterminal with $\alpha = -135°$

11. What fraction of a complete revolution is each of the following angles?
    (a) $90°$ **(b)** $30°$ **(c)** $\frac{\pi}{3}$ **(d)** $\frac{\pi}{4}$

12. What angle is its own complement? What angle is its own supplement?

Degree Measure

Exercises 13–20: Find a positive angle and a negative angle that are coterminal with the given angle.

13. $150°$

14. $65°$

15. $-72°$

16. $-330°$

17. $\frac{\pi}{2}$

18. $\frac{5\pi}{6}$

19. $-\frac{\pi}{5}$

20. $-\frac{2\pi}{3}$

Exercises 21–24: Express the angle in decimal degrees.

21. $125°15'$ **22.** $15°30'$

23. $108°45'36''$ **24.** $256°06'12''$

Exercises 25–28: Convert the given angle to degrees, minutes, and seconds.

25. $125.3°$ **26.** $15.25°$

27. $51.36°$ **28.** $22.46°$

Exercises 29–34: Find the complementary angle α and the supplementary angle β to θ.

29. $\theta = 55.9°$ **30.** $\theta = 71.5°$

31. $\theta = 85°23'45''$ **32.** $\theta = 5°45'30''$

33. $\theta = 23°40'35''$ **34.** $\theta = 67°25'10''$

Radian Measure

Exercises 35–38: Use the figure to determine the radian measure of angle θ. Then approximate the degree measure of θ to the nearest tenth of a degree.

35. 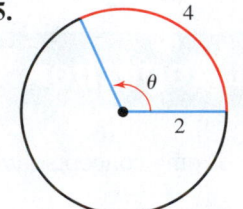   **36.**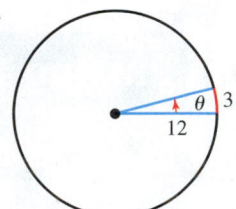

37. 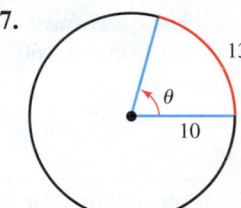   **38.**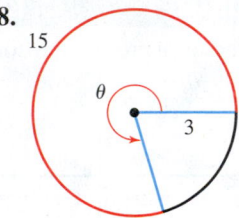

Exercises 39–42: Convert each angle from degree measure to radian measure. Round to the nearest hundredth of a radian when appropriate.

39. (a) $45°$ **(b)** $135°$ **(c)** $-120°$ **(d)** $-210°$

40. (a) $105°$ **(b)** $245°$ **(c)** $-255°$ **(d)** $-80°$

41. (a) $37°$ **(b)** $123.4°$ **(c)** $-92°25'$ **(d)** $230°17'$

42. (a) $56°$ **(b)** $88.7°$ **(c)** $122°15'$ **(d)** $-7°48'$

Exercises 43–46: Convert each angle from radian measure to degree measure. Round to the nearest hundredth of a degree when appropriate.

43. (a) $\frac{\pi}{6}$ (b) $\frac{\pi}{15}$ (c) $-\frac{5\pi}{3}$ (d) $-\frac{7\pi}{6}$

44. (a) $-\frac{\pi}{12}$ (b) $-\frac{5\pi}{2}$ (c) $\frac{17\pi}{15}$ (d) $\frac{5\pi}{6}$

45. (a) $\frac{\pi}{4}$ (b) $\frac{\pi}{7}$ (c) 3.1   (d) $-\frac{5}{2}$

46. (a) $\frac{7\pi}{2}$ (b) $\frac{2\pi}{5}$ (c) −4.1   (d) $-\frac{2}{3}$

Arc Length

Exercises 47–52: Use the formula $s = r\theta$ to determine the missing value in the figure.

47.
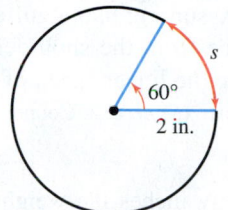

48.
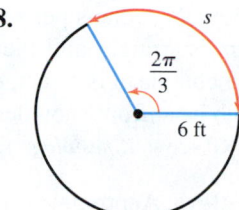

49.
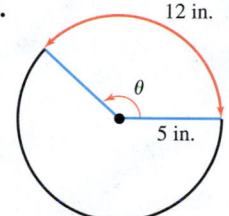

50.
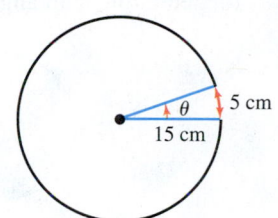

51.
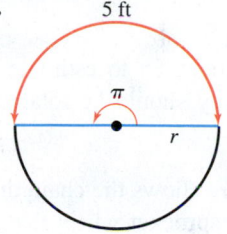

52.
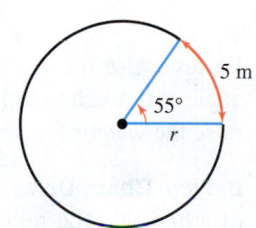

Exercises 53–58: Find the length of the arc intercepted by a central angle θ in a circle of radius r.

53. $r = 3$ m, $\theta = \frac{\pi}{12}$ **54.** $r = 7.3$ mm, $\theta = \frac{7\pi}{4}$

55. $r = 12$ ft, $\theta = 15°$ **56.** $r = 5$ cm, $\theta = 240°$

57. $r = 2$ mi, $\theta = 1°45'$ **58.** $r = 3$ mi, $\theta = 4°15'09''$

*Exercises 59–62: **Clocks** A minute hand on a clock is 4 inches long. Determine how far the tip of the minute hand travels between the given times. Find the linear speed of the tip.*

59. 10:15 A.M., 10:30 A.M.   **60.** 1:00 P.M., 1:40 P.M.

61. 3:00 P.M., 4:15 P.M.   **62.** 11:00 A.M., 1:25 P.M.

Area of a Sector

Exercises 63–66: Find the area of the shaded sector.

63.
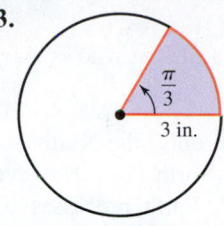

64.
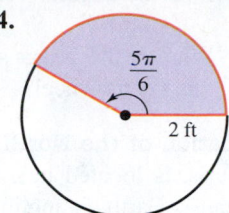

65.
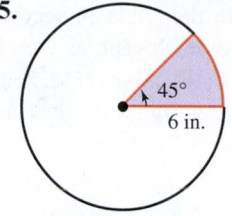

66.

Exercises 67–70: Find the area of the sector of a circle having radius r and central angle θ.

67. $r = 13.1$ cm, $\theta = \frac{\pi}{15}$ **68.** $r = 7.3$ m, $\theta = \frac{5\pi}{4}$

69. $r = 1.5$ ft, $\theta = 30°$ **70.** $r = 5.5$ in., $\theta = 225°$

*Exercises 71–74: **Robotics** (Refer to Example 10.) Find the area of the work space for a robotic arm that can rotate between angles θ_1 and θ_2 and can change its length from r_1 to r_2. See the figure.*

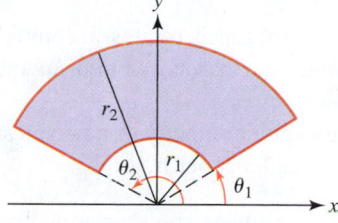

71. $\theta_1 = -45°$, $\theta_2 = 90°$, $r_1 = 6$ in., $r_2 = 26$ in.

72. $\theta_1 = -60°$, $\theta_2 = 60°$, $r_1 = 0.5$ ft, $r_2 = 2.5$ ft

73. $\theta_1 = 15°$, $\theta_2 = 195°$, $r_1 = 21$ cm, $r_2 = 95$ cm

74. $\theta_1 = 43°$, $\theta_2 = 178°$, $r_1 = 0.4$ m, $r_2 = 1.8$ m

Applications

75. Bicycle Tire A bicycle has a tire 26 inches in diameter that is rotating at 15 radians per second. Approximate the speed of the bicycle in feet per second.

76. Skateboard Wheel The wheels on a skateboard have a diameter of 2.25 inches. If a skateboarder is traveling downhill at 15 miles per hour, determine the angular velocity of the wheels in radians per second.

77. Ferris Wheel A large Ferris wheel has a diameter of 140 feet. It completes 1 revolution every 420 seconds.
(a) Find the angular velocity in radians per second.

(b) What is the linear speed of a person who is riding this Ferris wheel?

78. Location of the North Star Presently the North Star, Polaris, is located near the true North Pole. However, because Earth is inclined 23.5°, Earth precesses like a spinning top and the direction of the celestial North Pole traces out a circular path once every 26,000 years, as shown in the figure. Calculate the angle in seconds that the celestial North Pole moves each year, as viewed from the center *C* of this circular path. (*Source:* M. Zeilik et al., *Introductory Astronomy and Astrophysics.*)

79. Fan Speed The blades of a fan have a 30-inch diameter and rotate at 500 revolutions per minute.
(a) Find the angular velocity of a fan blade.

(b) Estimate the linear speed at the tip of a fan blade.

80. Earth's Rotation Earth rotates 1 complete revolution every 24 hours and has an *equatorial* radius of about 3963 miles.
(a) Find the angular velocity of a person standing at the equator in radians per hour.

(b) Estimate the linear speed in miles per hour at the equator due to Earth's rotation.

81. Distance between Cities (Refer to Example 7.) Daytona Beach, Florida, and Akron, Ohio, have nearly the same longitude of 81° W. The latitude of Daytona Beach is 29°11′, and the latitude of Akron is 40°55′. Approximate the distance between these two cities if the *average* radius of Earth is 3955 miles.

82. Nautical Miles Nautical miles are used by ships and airplanes. They are different from statute miles, which equal 5280 feet. A nautical mile is defined to be the arc length along the equator intercepted by a central angle *AOB* of 1 minute, as illustrated in the figure at the top of the next column. If the *equatorial* radius of Earth is 3963 statute miles, use the arc length formula to approximate the number of statute miles in 1 nautical mile. Round your answer to two decimal places.

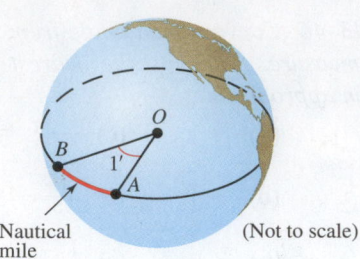

Nautical mile   (Not to scale)

83. Tennis Serve (See the discussion before Example 8.) Suppose that a tennis player rotates her wrist 85° in 0.05 second. If the tennis racket is 28 inches in length, estimate the speed of the tip of the racket in feet per second.

84. Club Speed in Golf The shoulder joint can rotate at about 25 radians per second. Assuming that a golfer's arm is straight and the distance from the shoulder to the club head is 5 feet, estimate the linear speed of the club head from shoulder rotation. (*Source:* J. Cooper and R. Glassow, *Kinesiology.*)

85. Pulleys Approximate how many inches the weight in the figure will rise if *r* = 11 inches and the pulley is rotated through an angle of 75.3°.

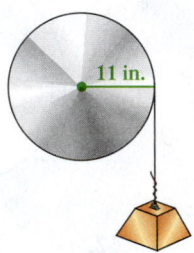

11 in.

86. Pulleys Use the figure in Exercise 85 to estimate the angle *θ* through which the pulley should be rotated to raise the weight 5 inches.

87. Bicycle Chain Drive The figure shows the chain drive of a bicycle. The radius of the sprocket wheel that the pedals are attached to is 3.75 inches, and the radius of the other sprocket wheel is 1.5 inches.
(a) Determine the number of revolutions that the bicycle tire rotates when the pedals are rotated one revolution.

(b) If the bicycle has a tire with a 26-inch diameter, determine how fast the bicycle travels in feet per second when the pedals turn through two revolutions per second.

1.5 in.   3.75 in.

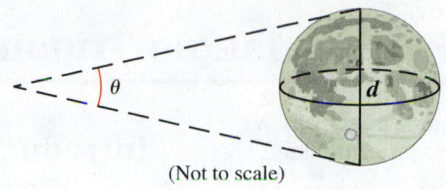

(Not to scale)

88. **Wind Speed** One of the most common ways to measure wind speed is with a *three-cup anemometer*, as shown in the figure. The cups catch the wind and cause the vertical shaft to rotate. At lower wind speeds the cups move at approximately the same speed as the wind. If the cups are rotating 5 times per second with a radius of 6 inches, estimate the wind speed in miles per hour. (**Source:** J. Navarra, *Atmosphere, Weather and Climate.*)

89. **Velocity of Planets** The average distance D in millions of miles from the sun and the orbital period P in years are given for various planets. Assuming that the orbits are circular, approximate the average orbital velocity in miles per hour for each planet. Discuss the effect that average distance from the sun has on orbital velocity. (**Source:** C. Ronan, *The Natural History of the Universe.*)

(a) Venus: $D = 67.2$, $P = 0.615$

(b) Earth: $D = 92.9$, $P = 1$

(c) Jupiter: $D = 483.6$, $P = 11.86$

(d) Neptune: $D = 2794$, $P = 164.8$

90. **Speed of a Propeller** When a 90-horsepower outboard motor is at full throttle, its propeller makes 5000 revolutions per minute. Find the angular velocity of the propeller in radians per second. What is the linear speed in inches per second of a point at the tip of the propeller if its diameter is 10 inches?

91. **Surveying** The *subtense bar method* is a technique used in surveying to measure distances. A green subtense bar is shown in the figure connecting points P and Q. If the distance d from the surveyor to the bar is large, then there is little difference between the length of the subtense bar, which is usually 2 meters, and that of the red arc connecting P and Q. Similarly, there is little difference between d and the radius r of the arc intercepted by the subtense bar. If θ is measured to be $0.835°$, approximate d using the arc length formula. (**Source:** I. Mueller and K. Ramsayer, *Introduction to Surveying.*)

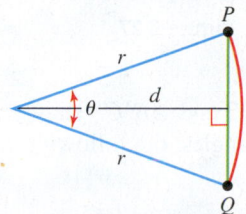

92. **Diameter of the Moon** (Refer to Exercise 91.) The distance to the moon is approximately 238,900 miles. Use the arc length formula to estimate the diameter d of the moon if angle θ in the figure at the top of the next column is measured to be $0.517°$.

93. **Solar Power Plant** A 150-megawatt solar power plant requires approximately 475,000 square meters of land to collect the required amount of energy from sunlight. (**Source:** C. Winter, *Solar Power Plants.*)

   (a) If this land area is circular, approximate its radius.

   (b) If this land area is a sector of a circle with $\theta = 70°$, approximate its radius.

94. **Measuring the Circumference of Earth** The first accurate estimate of the distance around Earth was made by the Greek astronomer Eratosthenes (276–195 B.C.), who noted that the noontime position of the sun at the summer solstice differed by $7°12'$ from the city of Syene to the city of Alexander. See the figure. The distance between these two cities is 496 miles. Use the arc length formula to estimate the radius of Earth. Then find the circumference of Earth. (**Source:** M. Zeilik et. al., *Introductory Astronomy and Astrophysics.*)

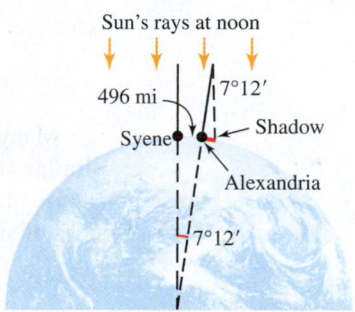

Writing about Mathematics

95. Give definitions for 1 degree and 1 radian. Compare these two units of angle measure. Which unit of measure do you prefer? Explain why.

96. Suppose a central angle θ of a circle remains fixed. Describe what happens to the arc length intercepted by θ and the area of the corresponding sector as the radius r doubles and triples.

Extended and Discovery Exercises

1. **Arc Length Formula** Modify the arc length formula $s = r\theta$ so that angle θ can be given in degrees rather than radians. Which of the two formulas is simpler?

2. **Area of a Sector Formula** Modify the area formula $A = \frac{1}{2}r^2\theta$ so that angle θ can be given in degrees rather than radians. Which of the two formulas is simpler?

6.2 Right Triangle Trigonometry

- Learn basic concepts about trigonometric functions
- Apply right triangle trigonometry
- Understand complementary angles and cofunctions

Introduction

A right triangle is a basic geometric shape that occurs in many applications such as astronomy, surveying, construction, highway design, GPS, weather, and aerial photography. Trigonometric functions are used to *solve* triangles. **Solving a triangle** involves finding the measure of each side and angle in the triangle. Like other functions that we have encountered previously, the six trigonometric functions can be used to model data and a variety of physical phenomena.

Basic Concepts of Trigonometric Functions

The **standard labeling** used to designate vertices, angles, and sides of a triangle ABC is shown in Figure 6.25.

Standard Labeling of a Triangle

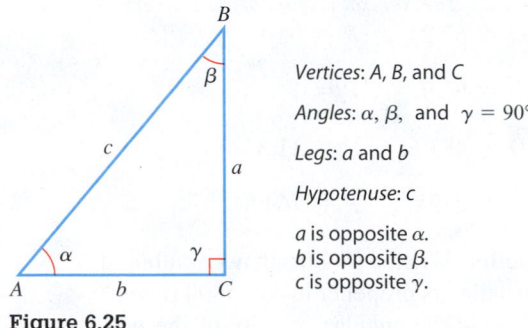

Vertices: A, B, and C

Angles: α, β, and $\gamma = 90°$

Legs: a and b

Hypotenuse: c

a is opposite α.
b is opposite β.
c is opposite γ.

Figure 6.25

Many important ideas in trigonometry depend on the properties of similar triangles. **Similar triangles** have congruent corresponding angles, but similar triangles are not necessarily the same size. Similar triangles are shown in Figures 6.26 and 6.27. Corresponding sides of similar triangles are proportional.

Similar Triangles

Geometry Review
To review similar triangles, see Chapter R (page R-5).

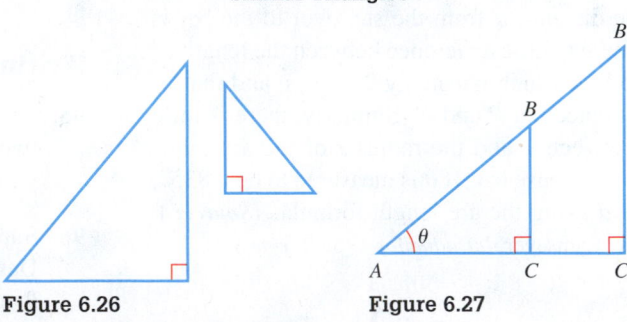

Figure 6.26   **Figure 6.27**

The Sine Function The right triangles ABC and $AB'C'$ shown in Figure 6.27 are similar triangles. By the properties of similar triangles, the following ratios are equal.

$$\frac{BC}{AB} = \frac{B'C'}{AB'}$$

That is, the ratio of the length of the side opposite angle θ to the length of the hypotenuse is constant for a given angle θ and does not depend on the size of the right triangle. If the measure of θ changes, then the ratio of the length of the side opposite θ to the length of the hypotenuse also changes. This concept can be used to define a new function called the *sine function*. That is, if θ is an acute angle in a right triangle, as shown in Figure 6.28, then we define the sine of θ as

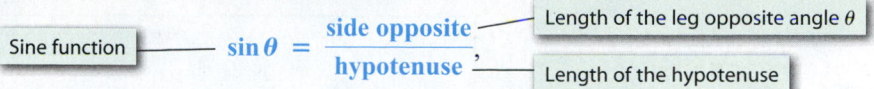

where $\sin \theta$ denotes the sine function with input θ. The three letters "sin" are used to denote the sine function, much like "log" was used to denote the common logarithmic function.

CLASS DISCUSSION

Is it possible that $\sin \theta > 1$ for some acute angle θ? Explain your reasoning.

Sides of a Right Triangle

Hypotenuse
Side opposite
θ
Side adjacent

Figure 6.28

EXAMPLE 1   **Evaluating the sine function**

Find $\sin 30°$. Support your answer by using a calculator.

SOLUTION Since the sine function depends only on the measure of θ, we can choose any size right triangle to evaluate $\sin \theta$. For convenience, let the length of the hypotenuse equal 2, as shown in Figure 6.29. From geometry we know that the length of the shortest leg in a 30°–60° right triangle is half the hypotenuse. Thus the side opposite equals 1 and

$$\sin 30° = \frac{\text{side opposite}}{\text{hypotenuse}} = \frac{1}{2}.$$

This result is supported in Figure 6.30, where the sine function has been evaluated at 30° using a calculator set in degree mode.

30°–60° Right Triangle

60°

2

1 — The shortest leg is half the hypotenuse.

30°

Figure 6.29

Evaluating sin 30°

```
sin(30)
                          .5
```

Figure 6.30 Degree Mode

Now Try Exercise 7

The Cosine Function Using Figures 6.27 and 6.28, we can define other trigonometric functions. Since

$$\frac{AC}{AB} = \frac{AC'}{AB'},$$

the ratio of the side adjacent (next) to θ to the hypotenuse is constant for a fixed angle θ and does not depend on the size of the right triangle. We define the *cosine function* to be

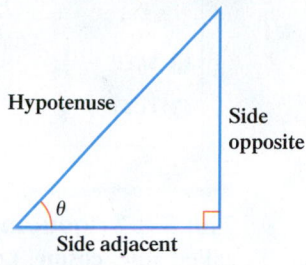

See the Concept: Evaluating Sine and Cosine

A $\sin \theta = \dfrac{\text{side opposite}}{\text{hypotenuse}}$

B $\cos \theta = \dfrac{\text{side adjacent}}{\text{hypotenuse}}$

C Side opposite θ

C Side adjacent θ

A $\sin \theta = \dfrac{4}{5}$

B $\cos \theta = \dfrac{3}{5}$

A $\sin \theta = \dfrac{3}{5}$

B $\cos \theta = \dfrac{4}{5}$

C Side opposite and side adjacent depend on how angles are labeled.

The Six Trigonometric Functions The six trigonometric functions of angle θ are called **sine, cosine, tangent, cosecant, secant,** and **cotangent**. The customary abbreviation for each trigonometric function is shown below.

RIGHT TRIANGLE-BASED DEFINITIONS OF TRIGONOMETRIC FUNCTIONS

Let θ be an acute angle in a right triangle. Then the six trigonometric functions of θ may be evaluated as follows.

$$\sin \theta = \frac{\text{side opposite}}{\text{hypotenuse}} \qquad \cos \theta = \frac{\text{side adjacent}}{\text{hypotenuse}} \qquad \tan \theta = \frac{\text{side opposite}}{\text{side adjacent}}$$

$$\csc \theta = \frac{\text{hypotenuse}}{\text{side opposite}} \qquad \sec \theta = \frac{\text{hypotenuse}}{\text{side adjacent}} \qquad \cot \theta = \frac{\text{side adjacent}}{\text{side opposite}}$$

The next example illustrates how to evaluate the trigonometric functions.

EXAMPLE 2 Evaluating trigonometric functions

Consider the right triangle shown in Figure 6.31. Find the six trigonometric functions of θ.

SOLUTION In triangle ABC, the side opposite angle θ is $a = $ **8** and the hypotenuse is $c = $ **17**. To find the adjacent side b, we apply the Pythagorean theorem.

$$c^2 = a^2 + b^2 \qquad \text{Pythagorean theorem}$$

$$b^2 = c^2 - a^2 \qquad \text{Solve for } b^2.$$

$$b^2 = 17^2 - 8^2 \qquad \text{Let } c = 17 \text{ and } a = 8.$$

$$b^2 = 225 \qquad \text{Simplify.}$$

$$b = \mathbf{15} \qquad \text{Solve for } b, \text{ where } b > 0.$$

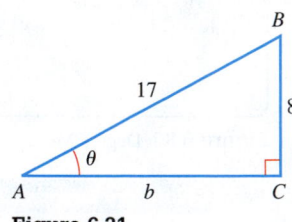

Figure 6.31

Algebra Review

To review the Pythagorean theorem, see Chapter R (page R-2).

Thus the six trigonometric functions of θ are as follows.

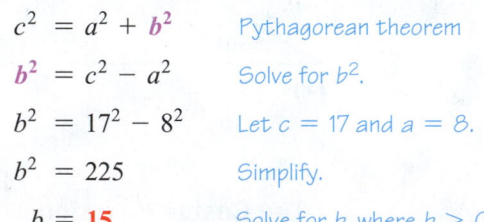

$$\sin \theta = \frac{\text{side opposite}}{\text{hypotenuse}} = \frac{8}{17} \qquad \csc \theta = \frac{\text{hypotenuse}}{\text{side opposite}} = \frac{17}{8}$$

$$\cos \theta = \frac{\text{side adjacent}}{\text{hypotenuse}} = \frac{15}{17} \qquad \sec \theta = \frac{\text{hypotenuse}}{\text{side adjacent}} = \frac{17}{15}$$

$$\tan \theta = \frac{\text{side opposite}}{\text{side adjacent}} = \frac{8}{15} \qquad \cot \theta = \frac{\text{side adjacent}}{\text{side opposite}} = \frac{15}{8}$$

Now Try Exercise 19

If an object is located above the horizontal, then the acute angle between the horizontal and the line of sight *XY* is called the **angle of elevation**. See Figure 6.32. If an object is located below the horizontal, then the acute angle between the horizontal and the line of sight *XY* is called the **angle of depression**. See Figure 6.33.

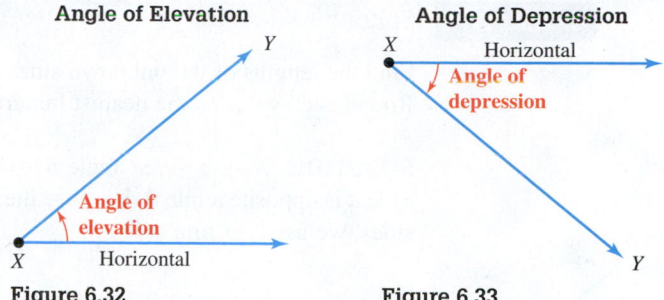

Figure 6.32   **Figure 6.33**

An Application   Trigonometry allows people to determine distances without measuring them directly. For example, the altitude of the cloud base is important at airports. Although it is not practical to measure this altitude directly, trigonometry can indirectly determine this height at nighttime. In Figure 6.34 a bright spotlight is directed vertically upward. It creates a bright spot on the cloud base. From a known horizontal distance *d* from the spotlight, the angle of elevation θ is measured. The side adjacent to θ is *d*, and the side opposite θ is *h*, where *h* represents the height of the cloud base. It follows that

θ can be measured.   *h* is the only unknown.

$$\tan \theta = \frac{\text{side opposite}}{\text{side adjacent}} = \frac{h}{d}.$$

d can be measured.

Finding the Height of the Cloud Base

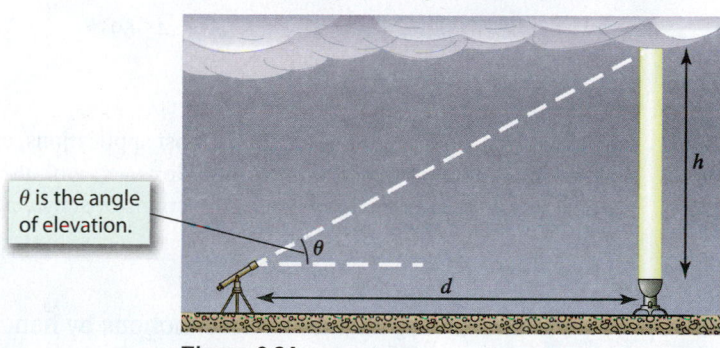

θ is the angle of elevation.

Figure 6.34

EXAMPLE 3   Determining the height of the cloud base

Suppose that $\theta = 55°$ and $d = 1150$ feet in Figure 6.34. Estimate the height of the cloud base. (Neglect the height of the telescope and spotlight in Figure 6.34.)

SOLUTION   Solve the equation $\tan \theta = \frac{h}{d}$ for *h* and then substitute values for θ and *d*.

$$\tan \theta = \frac{h}{d}$$

$h = d \tan \theta$ Multiply by *d*; rewrite.

$= 1150 \tan 55°$ Substitute for *d* and θ.

$\approx 1150 \,(1.4281)$ Approximate $\tan 55°$.

≈ 1642 feet   Multiply.

Thus the cloud base is about 1642 feet high. See Figure 6.35.

Approximating tan 55°

```
tan(55)
      1.428148007
```

Figure 6.35 Degree Mode

Now Try Exercise 59

Solving a Triangle We can use trigonometric functions to find unknown sides of a right triangle. This process, sometimes referred to as *solving a triangle*, is demonstrated in the next example.

EXAMPLE 4 Solving a triangle

Find the lengths of the unknown sides a and c for the right triangle shown in Figure 6.36. Round each value to the nearest hundredth.

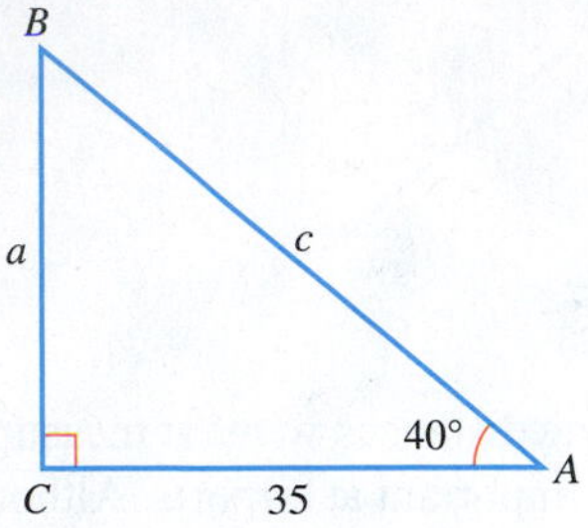

SOLUTION We are given angle $\theta = 40°$ and side $b = 35$, which is adjacent to angle θ. Side a is opposite angle θ. Because the tangent function involves the opposite and adjacent sides, we use it to find side a.

$$\tan 40° = \frac{a}{35} \qquad \tan\theta = \frac{\text{side opposite}}{\text{side adjacent}}$$

$$35 \tan 40° = a \qquad \text{Multiply by 35.}$$

$$a \approx 29.37 \qquad \text{Rewrite; approximate (if desired).}$$

To find the length of hypotenuse c, we could use the Pythagorean theorem. However, we use $\cos\theta$ instead, which involves the adjacent side and the hypotenuse.

$$\cos 40° = \frac{35}{c} \qquad \cos\theta = \frac{\text{side adjacent}}{\text{hypotenuse}}$$

$$c \cos 40° = 35 \qquad \text{Multiply by } c.$$

$$c = \frac{35}{\cos 40°} \qquad \text{Divide by } \cos 40°.$$

$$c \approx 45.69 \qquad \text{Approximate (if desired).}$$

Now Try Exercise 39

Finding Exact Values In most applications, calculators are used to approximate values of the trigonometric functions. However, with the aid of geometry we can determine exact values for the trigonometric functions of some special angles such as 30°, 45°, and 60°.

EXAMPLE 5 Evaluating trigonometric functions by hand

Evaluate the six trigonometric functions of $\theta = 45°$.

SOLUTION Begin by drawing a right triangle with a 45° angle, as shown in Figure 6.37. The lengths of the legs in this triangle are equal. Since the size of the right triangle does *not* affect the values of the trigonometric functions, let the lengths of both legs equal 1. Using the Pythagorean theorem, we can find the length of the hypotenuse as follows.

$$c^2 = a^2 + b^2 \qquad \text{Pythagorean theorem}$$
$$c^2 = 1^2 + 1^2 \qquad a = b = 1$$
$$c^2 = 2 \qquad \text{Simplify.}$$
$$c = \sqrt{2} \qquad \text{Solve for } c, \text{ where } c > 0.$$

The hypotenuse has length $\sqrt{2}$. Evaluating the six trigonometric functions gives the following.

$$\sin 45° = \frac{\text{side opposite}}{\text{hypotenuse}} = \frac{1}{\sqrt{2}} \qquad \csc 45° = \frac{\text{hypotenuse}}{\text{side opposite}} = \frac{\sqrt{2}}{1} = \sqrt{2}$$

B

a c

$40°$ A

C 35

Figure 6.36

A 45° – 45° Right Triangle

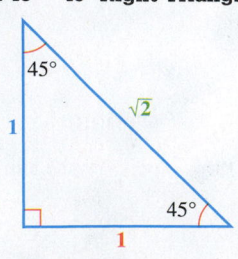

$45°$

$\sqrt{2}$

1

$45°$

1

Figure 6.37

$$\cos 45° = \frac{\text{side adjacent}}{\text{hypotenuse}} = \frac{1}{\sqrt{2}} \qquad \sec 45° = \frac{\text{hypotenuse}}{\text{side adjacent}} = \frac{\sqrt{2}}{1} = \sqrt{2}$$

$$\tan 45° = \frac{\text{side opposite}}{\text{side adjacent}} = \frac{1}{1} = 1 \qquad \cot 45° = \frac{\text{side adjacent}}{\text{side opposite}} = \frac{1}{1} = 1$$

Now Try Exercise 23

Reciprocal Trigonometric Functions In Example 5, we saw that $\sin 45° = \frac{1}{\sqrt{2}}$ and $\csc 45° = \frac{\sqrt{2}}{1}$. Because

$$\sin\theta = \frac{\text{side opposite}}{\text{hypotenuse}} \qquad \text{and} \qquad \csc\theta = \frac{\text{hypotenuse}}{\text{side opposite}},$$

> $\sin\theta$ and $\csc\theta$ are reciprocals.

it follows that $\csc\theta = \frac{1}{\sin\theta}$ in general. In a similar manner,

> $\cos\theta$ and $\sec\theta$ are reciprocals.

$$\sec\theta = \frac{1}{\cos\theta} \qquad \text{and} \qquad \cot\theta = \frac{1}{\tan\theta}.$$

> $\tan\theta$ and $\cot\theta$ are reciprocals.

For example, if $\tan\theta = \frac{2}{3}$, then $\cot\theta = \frac{3}{2}$.

Calculators Most calculators have keys to evaluate the sine, cosine, and tangent functions but do not have keys to evaluate the cosecant, secant, and cotangent functions. These three functions may be evaluated by using the following *reciprocal identities*.

Reciprocal Identities

$$\csc\theta = \frac{1}{\sin\theta}, \qquad \sec\theta = \frac{1}{\cos\theta}, \qquad \cot\theta = \frac{1}{\tan\theta}$$

Results from Example 5 are supported in Figure 6.38, where these reciprocal identities have been applied to evaluate $\csc 45°$, $\sec 45°$, and $\cot 45°$.

> **NOTE** Do not use the $\sin^{-1}$, $\cos^{-1}$, and $\tan^{-1}$ calculator keys to evaluate reciprocals. They represent inverse functions, which will be discussed in Section 6.6. You may want to use the x^{-1} key to evaluate reciprocals instead.

Using the right triangles shown in Figures 6.37 and 6.39, we can evaluate the six trigonometric functions at 30°, 45°, and 60° without the aid of a calculator. See Table 6.2.

Some Exact Values of Trigonometric Functions

θ	$\sin\theta$	$\cos\theta$	$\tan\theta$	$\csc\theta$	$\sec\theta$	$\cot\theta$
30°	$\frac{1}{2}$	$\frac{\sqrt{3}}{2}$	$\frac{1}{\sqrt{3}}$	2	$\frac{2}{\sqrt{3}}$	$\sqrt{3}$
45°	$\frac{1}{\sqrt{2}}$	$\frac{1}{\sqrt{2}}$	1	$\sqrt{2}$	$\sqrt{2}$	1
60°	$\frac{\sqrt{3}}{2}$	$\frac{1}{2}$	$\sqrt{3}$	$\frac{2}{\sqrt{3}}$	2	$\frac{1}{\sqrt{3}}$

Table 6.2

> If you can sketch Figures 6.37 and 6.39, you will be able to find the values in Table 6.2 without memorizing them.

> **NOTE** If we rationalize the denominators in Table 6.2, then

$$\frac{1}{\sqrt{2}} = \frac{1}{\sqrt{2}} \cdot \frac{\sqrt{2}}{\sqrt{2}} = \frac{\sqrt{2}}{2}.$$

Similarly, $\frac{1}{\sqrt{3}} = \frac{\sqrt{3}}{3}$ and $\frac{2}{\sqrt{3}} = \frac{2\sqrt{3}}{3}$.

Reciprocal Identities: csc 45°, sec 45°, and cot 45°

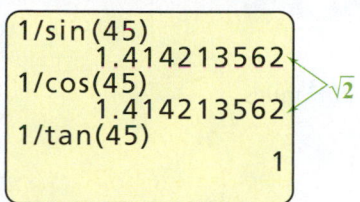

```
1/sin(45)
      1.414213562
1/cos(45)
      1.414213562
1/tan(45)
                1
```
$\sqrt{2}$

Figure 6.38 Degree Mode

A 30°–60° Right Triangle

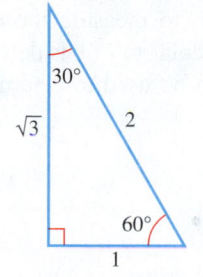

Figure 6.39

Algebra Review
To review rationalizing the denominator, see Chapter R (page R-47).

Applications of Right Triangle Trigonometry

For centuries astronomers wanted to know how far it was to the stars. Not until 1838 did the astronomer Friedrich Bessel determine the distance to a star called 61 Cygni. He used a *parallax method* that relied on the measurement of very small angles. See Figure 6.40. As Earth revolves around the sun, the observed parallax of 61 Cygni is $\theta \approx 0.0000811°$. Because stars are so distant, parallax angles are very small. (*Sources:* H. Freebury, *A History of Mathematics;* M. Zeilik et al., *Introductory Astronomy and Astrophysics.*)

Parallax of a Star

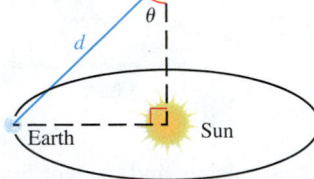

Figure 6.40 (not to scale)

EXAMPLE 6 Calculating the distance to a star

One of the nearest stars to Earth is Alpha Centauri, which has a parallax of $\theta \approx 0.000212°$. (*Source:* M. Zeilik et al.)

(a) Find the distance to Alpha Centauri if the Earth–Sun distance is 93,000,000 miles.

(b) A light-year, defined to be the distance that light travels in 1 year, equals about 5.9 trillion miles. Find the distance to Alpha Centauri in light-years.

SOLUTION

(a) Let d represent the distance between Earth and Alpha Centauri. From Figure 6.40 it can be seen that

$$\sin\theta = \frac{93,000,000}{d}, \quad \text{or} \quad d = \frac{93,000,000}{\sin\theta}.$$

Substituting for θ gives the following result.

$$d = \frac{93,000,000}{\sin 0.000212°} \approx 2.51 \times 10^{13} \text{ miles}$$

(b) This distance equals $\dfrac{2.51 \times 10^{13}}{\textcolor{red}{5.9 \times 10^{12}}} \approx 4.3$ light-years.

One light-year

Now Try Exercise 75

Figure 6.41

Applications from Surveying Water is often an obstacle to surveyors in the field when measuring distances between two points. For example, to measure the distance between points P and Q in Figure 6.41 a baseline PR, perpendicular to PQ, is determined. Angle PRQ is then measured. Right triangle trigonometry can be used to determine the length of PQ. (*Source:* P. Kissam, *Surveying Practice.*)

EXAMPLE 7 Finding distance

Suppose in Figure 6.41 the length of PR is 94.75 feet and angle PRQ has measure 41.6°. Estimate the distance between points P and Q.

SOLUTION Let angle PRQ be θ. Since $\tan\theta = \frac{PQ}{PR}$, it follows that

$$PQ = PR \tan\theta = 94.75 \tan 41.6° \approx 84.12 \text{ feet.}$$

Now Try Exercise 69

EXAMPLE 8 Finding the height of a tree

From a point A on level ground, the angle of elevation to the top of a tree is 38°. From a point B that is 46 feet farther from the tree the angle of elevation is 22°. See Figure 6.42. Find the height h of the tree to the nearest tenth of a foot.

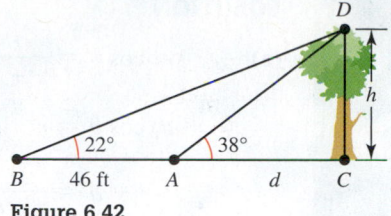

Figure 6.42

SOLUTION From triangles *ACD* and *BCD*,

$$\tan 38° = \frac{h}{d}, \qquad \text{or} \qquad h = d\tan 38°, \qquad \text{and}$$

$$\tan 22° = \frac{h}{d + 46}, \qquad \text{or} \qquad h = (d + 46)\tan 22°.$$

We can set the two expressions for *h* equal to each other and solve for *d*.

$$d\tan 38° = (d + 46)\tan 22° \qquad \text{Set expressions for } h \text{ equal.}$$

$$d\tan 38° = d\tan 22° + 46\tan 22° \qquad \text{Distributive property}$$

$$d\tan 38° - d\tan 22° = 46\tan 22° \qquad \text{Subtract } d\tan 22°.$$

$$d(\tan 38° - \tan 22°) = 46\tan 22° \qquad \text{Factor out } d.$$

$$d = \frac{46\tan 22°}{\tan 38° - \tan 22°} \qquad \text{Solve for } d.$$

From above, $h = d\tan 38°$, so the height of the tree is

$$h = \left(\frac{46\tan 22°}{\tan 38° - \tan 22°}\right)\tan 38° \approx 38.5 \text{ feet.} \qquad h = d\tan 38°$$

Expression for *d*

Now Try Exercise 71

Highway Design Next we derive a formula that is used in the design of highways.

EXAMPLE 9 **Deriving a formula for the design of highway curves**

One common type of highway curve is a *simple horizontal curve*. It consists of two straight segments of highway connected by a portion of a circular arc with radius *r*, as shown in Figure 6.43. The distance *d* is called the *external distance*. (*Source:* F. Mannering and W. Kilareski, *Principles of Highway Engineering and Traffic Analysis.*)

(a) Derive a formula for *d* that involves *r* and *θ*.
(b) Find *d* for a curve with a 750-foot radius and $θ = 36°$.

A Simple Horizontal Curve

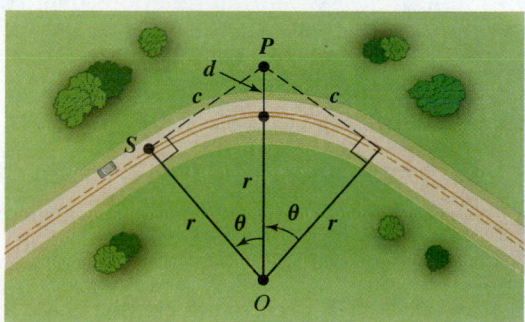

Figure 6.43

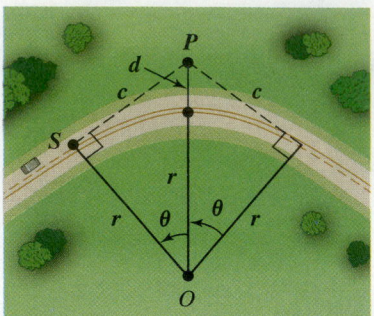

Figure 6.43 (Repeated)

SOLUTION

(a)
$$\cos \theta = \frac{r}{r + d} \qquad \text{Use triangle } OSP.$$

$$(r + d) \cos \theta = r \qquad \text{Multiply by } r + d.$$

$$r + d = \frac{r}{\cos \theta} \qquad \text{Divide by } \cos \theta.$$

$$d = \frac{r}{\cos \theta} - r \qquad \text{Subtract } r.$$

$$d = r\left(\frac{1}{\cos \theta} - 1\right) \qquad \text{Factor out } r.$$

(b) $d = 750 \left(\dfrac{1}{\cos 36°} - 1\right) \approx 177$ feet

Now Try Exercise 83

Complementary Angles and Cofunctions

Complementary Angles

$\alpha + \beta = 90°$

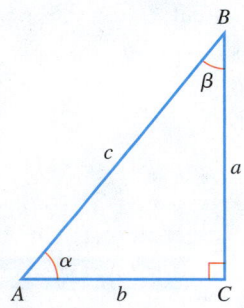

Figure 6.44

In Figure 6.44, α and β are complementary angles since their measures sum to 90°. The six trigonometric functions for α and β can be expressed as follows.

Cofunction Values ($\alpha + \beta = 90°$)

$$\sin \alpha = \frac{a}{c} = \cos \beta \qquad \cos \alpha = \frac{b}{c} = \sin \beta$$

$$\tan \alpha = \frac{a}{b} = \cot \beta \qquad \cot \alpha = \frac{b}{a} = \tan \beta$$

$$\sec \alpha = \frac{c}{b} = \csc \beta \qquad \csc \alpha = \frac{c}{a} = \sec \beta$$

Notice that the value of a trigonometric function for α equals the value of the trigonometric cofunction for β. For example, $\sin \alpha = \cos \beta$ and $\sin \beta = \cos \alpha$. This is how cofunctions were named. In 1620 Edmund Gunter combined the words "complement" and "sine" to obtain *co*sine. Similarly, the cosecant and cotangent functions are the "complementary functions" of the secant and tangent functions, respectively, and their names were shortened to *co*secant and *co*tangent.

COFUNCTION FORMULAS

$$\sin \theta = \cos (90° - \theta) \qquad \cos \theta = \sin (90° - \theta)$$

$$\tan \theta = \cot (90° - \theta) \qquad \cot \theta = \tan (90° - \theta)$$

$$\sec \theta = \csc (90° - \theta) \qquad \csc \theta = \sec (90° - \theta)$$

EXAMPLE 10 **Evaluating functions using complementary angles**

Write an equivalent expression using a cofunction. Then evaluate the expression using a calculator.
(a) $\cot 23°$ (b) $\sec 70°$ (c) $\cos 12°$

SOLUTION
(a) The complementary angle of 23° is $90° - 23° = 67°$. Thus

$$\cot 23° = \tan 67° \approx 2.3559.$$

(b) $\sec \mathbf{70°} = \csc (90° - \mathbf{70°}) = \csc 20° = \frac{1}{\sin 20°} \approx 2.9238$

(c) $\cos \mathbf{12°} = \sin (90° - \mathbf{12°}) = \sin 78° \approx 0.9781$

Now Try Exercise 55

6.2 Putting It All Together

The following table summarizes some properties of right triangle trigonometry.

CONCEPT	FORMULAS AND FIGURES	
Trigonometric functions	Let θ be an acute angle in a right triangle ABC. $$\sin\theta = \frac{\text{side opposite}}{\text{hypotenuse}} \qquad \csc\theta = \frac{\text{hypotenuse}}{\text{side opposite}}$$ $$\cos\theta = \frac{\text{side adjacent}}{\text{hypotenuse}} \qquad \sec\theta = \frac{\text{hypotenuse}}{\text{side adjacent}}$$ $$\tan\theta = \frac{\text{side opposite}}{\text{side adjacent}} \qquad \cot\theta = \frac{\text{side adjacent}}{\text{side opposite}}$$	
Cofunction formulas	Let α and β be complementary angles. $$\sin\alpha = \cos(90° - \alpha) = \cos\beta$$ $$\tan\alpha = \cot(90° - \alpha) = \cot\beta$$ $$\sec\alpha = \csc(90° - \alpha) = \csc\beta$$	

6.2 Exercises

Sketching Triangles

Exercises 1–6: Sketch a right triangle with the following properties. Label the measure of each angle and side.

1. Acute angles of 30° and 60° and a hypotenuse with length 2

2. Acute angle of 45° and a leg with length 1

3. Acute angle of 45° and a hypotenuse with length 1

4. Acute angle of 60° and a hypotenuse with length 1

5. Isosceles and a hypotenuse with length 4

6. Acute angle of 60° and the shorter leg with length 3

Evaluating Trigonometric Functions

Exercises 7–12: (Refer to Example 1.) Use a 30°–60° right triangle to find the exact value of the trigonometric expression.

7. $\sin 60°$ 8. $\tan 30°$

9. $\cos 30°$ 10. $\cot 30°$

11. $\sec 60°$ 12. $\csc 60°$

Exercises 13–18: Use a 45°–45° right triangle to find the exact value of the trigonometric expression.

13. $\tan 45°$ 14. $\sec 45°$

15. $\cot 45°$ 16. $\csc 45°$

17. $\sin 45°$ 18. $\cos 45°$

Exercises 19–22: Find the six trigonometric functions of θ.

19. 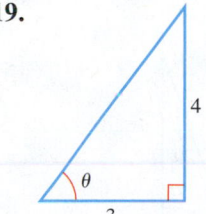   20.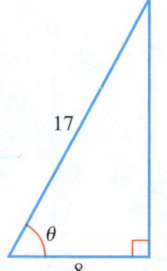

21.

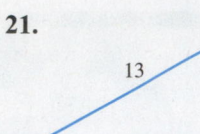

22.

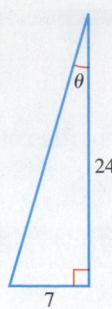

Exercises 23–32: Find the six trigonometric functions of the given angle. Approximate to three decimal places when appropriate.

23. 60°

24. 45°

25. 25°

26. 30°

27. 5°35′

28. 85°35′33″

29. 13°45′30″

30. 45°44′

31. 1.05°

32. 0.161°

Exercises 33–38: Let θ be an acute angle. Find the unknown trigonometric value, using the given information.

33. $\sec\theta$ if $\cos\theta = \frac{1}{3}$

34. $\cot\theta$ if $\tan\theta = 5$

35. $\csc\theta$ if $\sin\theta = \frac{12}{13}$

36. $\sin\theta$ if $\csc\theta = \frac{5}{4}$

37. $\tan\theta$ if $\cot\theta = \frac{7}{24}$

38. $\cos\theta$ if $\sec\theta = \frac{7}{5}$

Solving Triangles

Exercises 39–46: Find the lengths of the unknown sides in the right triangle. Round values to the nearest hundredth.

39.

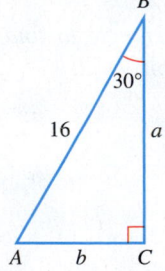

40.

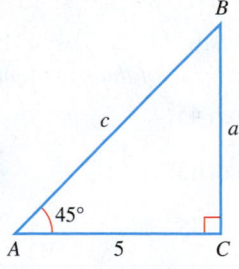

41.

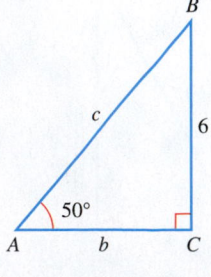

42.

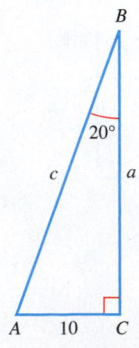

43.

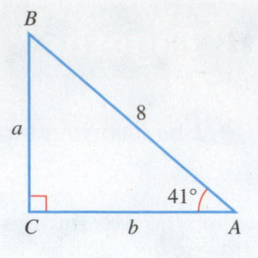

44.

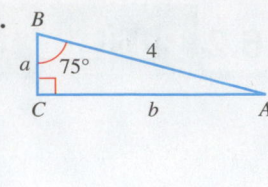

45.

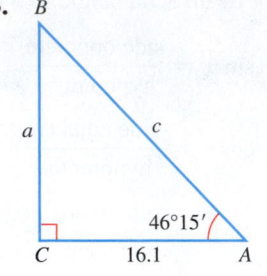

46.

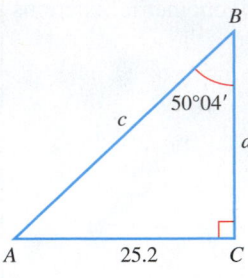

Exercises 47–50: Find the exact length of each side labeled with a variable in the figure.

47.

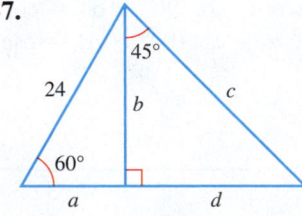

48.

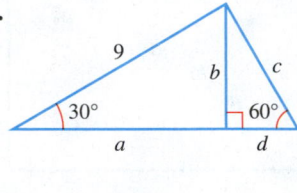

49.

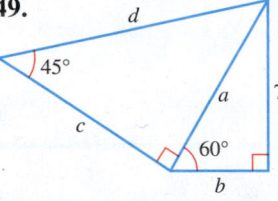

50.

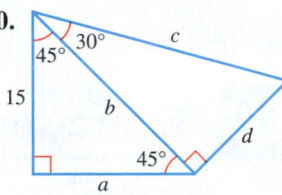

Exercises 51–54: Complete the following for right triangle ABC having the standard labeling shown in Figure 6.25. Approximate the answer to the nearest hundredth.

51. Find a if $b = 12$ and $\alpha = 60°$.

52. Find b if $c = 23$ and $\beta = 45°$.

53. Find c if $a = 100$ and $\beta = 53°43′$.

54. Find a if $b = 64$ and $\alpha = 78°15′$.

Cofunctions

Exercises 55–58: (Refer to Example 10.) Write an equivalent expression using a cofunction. Approximate the expression to four decimal places using a calculator.

55. (a) $\sin 70°$ (b) $\cos 40°$

56. (a) $\cot 23°$ (b) $\tan 48°$

57. (a) csc 49°   **(b)** sec 63°

58. (a) cot 87°   **(b)** sec 72°

Applications

59. Height of the Cloud Base (Refer to Example 3 and Figure 6.34.) From a distance of 1500 feet from the spotlight, the angle of elevation θ equals 37°30′. Find the height of the cloud base.

60. Height of a Tree One hundred feet from the trunk of a tree on level ground, the angle of elevation of the top of the tree is 35°. Find the height of the tree to the nearest foot.

61. Length of a Shadow The angle of elevation of the sun is 34°. Find the length of a shadow cast by a person who is 5 feet 3 inches tall. Round your answer to the nearest tenth of a foot.

62. Height of a Tower The shadow of a vertical tower is 40.6 meters long when the angle of elevation of the sun is 34.6°. Find the height of the tower.

63. Angle of Depression of a Floodlight A company safety committee recommended that a floodlight be mounted in a parking lot as shown in the figure, so as to illuminate the employee exit. Find h to the nearest tenth of a foot.

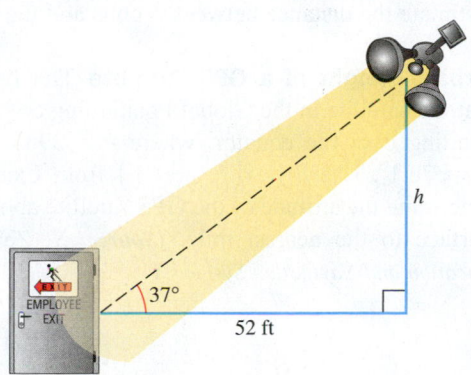

64. Aerial Photography An aerial photograph is taken directly above a building. The length of the building's shadow is 48 feet when the angle of elevation of the sun is 35.3°. Estimate the height of the building.

65. Angle of Depression An airplane is flying near a football stadium at 12,000 feet above level ground. The angle of depression from the airplane to the stadium is 13°. How far horizontally must the airplane fly to be directly over the football stadium? Round your answer to the nearest thousand feet.

66. Angle of Depression An airplane is flying 10,500 feet above the level ground. The angle of depression from the plane to the base of a tree is 13°50′. How far horizontally must the plane fly to be directly over the tree?

67. Height of a Building From a window 30 feet above the street, the angle of elevation to the top of the building across the street is 50° and the angle of depression to the base of this building is 20°. See the figure. Find the height of the building across the street.

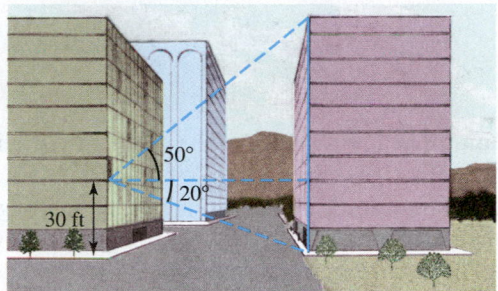

68. Height of a Building The angle of elevation from the top of a small building to the top of a nearby taller building is 46°40′, while the angle of depression to the bottom is 14°10′. If the smaller building is 28.0 meters high, find the height x of the taller building.

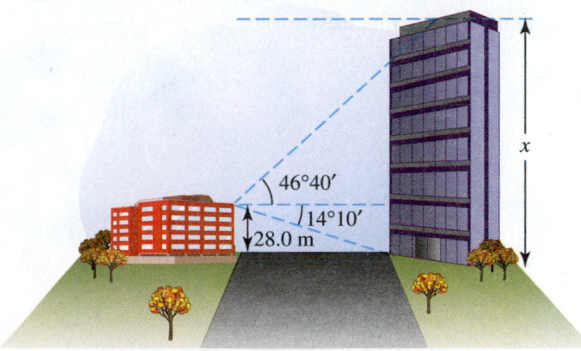

69. Surveying (Refer to Example 7 and Figure 6.41.) Find the distance from P to Q if PR is 85.62 feet and angle PQR is 23.76°.

70. Weather Tower A 410-foot weather tower used to measure wind speed has a guy wire attached to it 175 feet above the ground. The angle between the wire and the vertical tower is 57°, as shown in the figure. Approximate the length of the guy wire. (*Source:* Brookhaven National Laboratory.)

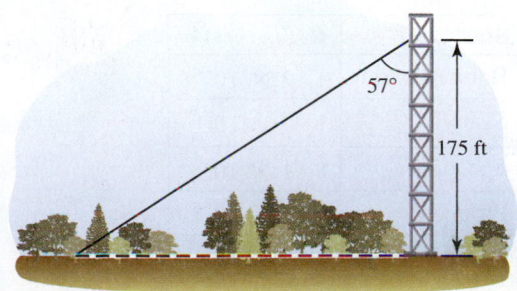

71. Height of a Mountain (Refer to Example 8.) From a point A the angle of elevation of Mount Kilimanjaro in Africa is $13.7°$, and from a point B, directly behind A, the angle of elevation is $10.4°$. See the figure. If the distance between A and B is 5 miles, approximate the height of Mount Kilimanjaro to the nearest hundred feet.

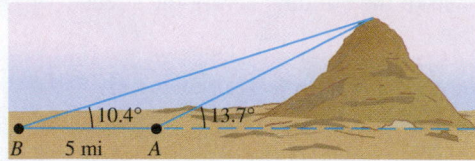

72. Height of Mt. Whitney The angle of elevation from Lone Pine to the top of Mt. Whitney is $10°50'$. Exactly 9.3 kilometers from Lone Pine along a straight, level road toward Mt. Whitney, the angle of elevation is $22°40'$. Find the height of the top of Mt. Whitney *above* the level of the road to the nearest hundredth of a kilometer.

73. Height of a Pyramid The angle of elevation from a point on the ground to the top of a pyramid is $35°30'$. The angle of elevation from a point 135 feet farther back to the top of the pyramid is $21°10'$. Find the height of the pyramid to the nearest foot.

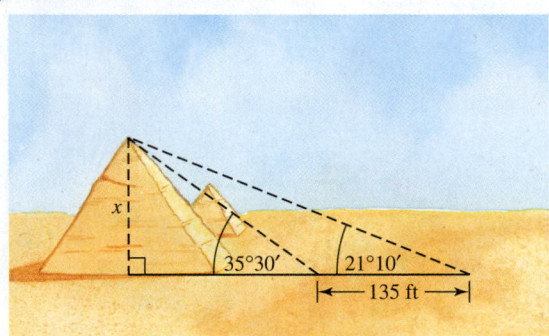

74. Height of an Antenna An antenna is on top of the center of a house. The angle of elevation from a point on the ground 28.0 meters from the center of the house to the top of the antenna is $27°10'$, and the angle of elevation to the bottom of the antenna is $18°10'$. Find the height of the antenna.

75. Distance to Nearby Stars (Refer to Example 6 and Figure 6.40.) The table lists the parallax θ in degrees for some nearby stars. Approximate the distance from Earth to each star in miles and in light-years.

Star	θ (degrees)
Barnard's Star	1.52×10^{-4}
Sirius	1.05×10^{-4}
61 Cygni	8.11×10^{-5}
Procyon	7.97×10^{-5}

Source: M. Zeilik et al., *Introductory Astronomy and Astrophysics.*

76. Parallax and Distance (Refer to Exercise 75.) When the parallax θ is equal in measure to 1 second, a star is said to have a distance from Earth of 1 parsec. If the distance between Earth and the sun is 93,000,000 miles, approximate the number of miles in 1 parsec. How many light-years is this? (*Source:* M. Zeilik et al.)

77. Observing Mercury The planet Mercury is closer to the sun than Earth. For this reason it can only be observed low in the horizon around sunset or sunrise. See the figure, where angle θ is called the *elongation*. Because Mercury's orbit is not circular, the elongation varies between $18°$ and $28°$. Approximate the minimum and maximum distances between Mercury and the sun. (*Source:* M. Zeilik et al.)

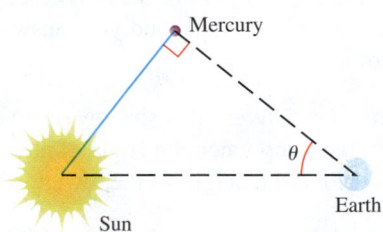

78. Observing Venus (Refer to Exercise 77.) The orbit of Venus is nearly circular with an elongation of $48°$. Estimate the distance between Venus and the sun.

79. Orbital Height of a GPS Satellite The figure illustrates a satellite in the Global Positioning System (GPS) orbiting over the equator, where $r = 3963$ miles and $\theta = 76.1°$. Use $d = r\left(\frac{1}{\cos\theta} - 1\right)$ from Example 9 to determine the altitude of the GPS satellite above Earth's surface to the nearest mile. (*Source:* Y. Zhao, *Vehicle Location and Navigation Systems.*)

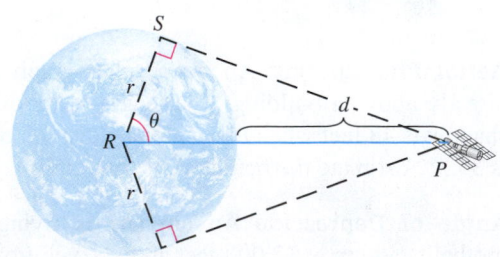

80. Heights of Lunar Mountains The lunar mountain peak Huygens has a height of 21,000 feet. The shadow of Huygens on a photograph was 2.8 mm, and the nearby mountain Bradley had a shadow of 1.8 mm on the same photograph. Use similar triangles to calculate the height of Bradley to the nearest hundred feet. (*Source:* T. Webb, *Celestial Objects for Common Telescopes.*)

81. Highway Curve Design Highway curves are sometimes banked so that the outside of the curve is slightly elevated or inclined above the inside of the curve, as shown in the figure. This inclination is called the *superelevation*. Both the curve's radius and the super-elevation must be correct for a given speed limit. The relationship among a car's velocity v in feet per second, the safe radius r of the curve in feet, and the superelevation θ in degrees is given by $r = \frac{v^2}{4.5 + 32.2 \tan\theta}$. (*Source:* F. Mannering and W. Kilareski *Principles of Highway Engineering and Traffic Analysis.*)

(a) A curve has a speed limit of 66 feet per second (45 mi/hr) and a superelevation of $\theta = 3°$. Approximate the safe radius r.

(b) Find r if $\theta = 5°$ and $v = 66$.

(c) Make a conjecture about how increasing θ affects the safe radius r. Verify your conjecture by making a table for r, starting at $\theta = 0$ and incrementing by 1. Let $v = 66$.

82. Highway Design (Refer to Exercise 81.) A highway curve has a radius of $r = 1150$ feet and a superelevation of $\theta = 2.1°$. What should be the speed limit (in miles per hour) for this curve?

83. Highway Design (Refer to Example 9 and Figure 6.43.) Find the external distance d for a highway curve with $r = 625$ feet and $\theta = 54°$.

84. Highway Design (Refer to Example 9.) A simple horizontal curve is shown in the figure at the top of the next column. The points P and S mark the beginning

and end of the curve. Let Q be the point of intersection where the two straight sections of highway leading into the curve would meet if extended. The radius of the curve is r, and the angle θ denotes how many degrees the curve turns. If $r = 765$ feet and $\theta = 83°$, find the distance between P and Q. (*Source:* F. Mannering and W. Kilareski.)

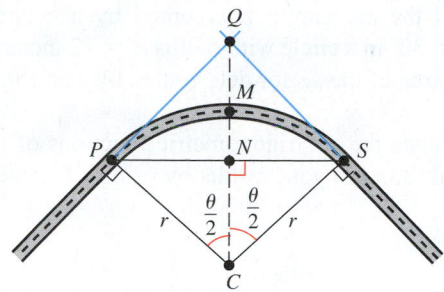

85. Area of an Equilateral Triangle Find the area of the equilateral triangle shown in the figure in terms of s.

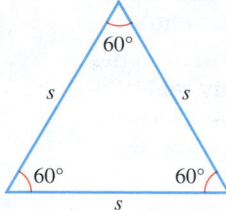

86. Area of a Hexagon Write the area of the hexagon shown in the figure in terms of x. Assume that the six triangles that comprise the hexagon are equilateral and congruent.

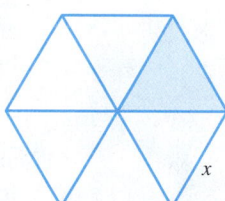

Writing about Mathematics

87. Most calculators have built-in keys to compute the sine, cosine, and tangent functions but not the secant, cosecant, and cotangent functions. Is it possible to evaluate all of the trigonometric functions with this type of calculator? Explain and include examples.

88. The sine function is defined in terms of right triangles as $\sin\theta = \frac{\text{side opposite}}{\text{hypotenuse}}$. Suppose that a fixed angle θ occurs in two right triangles. If the hypotenuse in the first triangle has twice the length of the hypotenuse in the second triangle, what can be said about the sides opposite in each triangle? How does the value $\sin\theta$ compare in each triangle? Explain.

1. Find the radian measure of each angle.
 (a) 45°   (b) 75°

2. Find the degree measure of each angle.
 (a) $\frac{\pi}{6}$ (b) $\frac{5\pi}{4}$

3. Find the arc length intercepted by a central angle of $\theta = 30°$ in a circle with radius $r = 12$ inches. Calculate the area of the sector determined by r and θ.

4. Evaluate the six trigonometric functions of $\theta = 60°$ by hand. Support your results by using a calculator.

5. Use the right triangle in the figure to find the six trigonometric functions of θ.

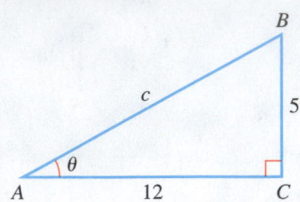

6. If $\alpha = 63°$ and $a = 9$ in right triangle ABC, approximate the length of the hypotenuse to the nearest tenth.

6.3 The Sine and Cosine Functions and Their Graphs

- **Define the sine and cosine functions for any angle**
- **Define the sine and cosine functions for any real number by using the unit circle and the wrapping function**
- **Represent the sine and cosine functions**
- **Use the sine and cosine functions in applications**
- **Model with the sine function (optional)**

Introduction

The sine and cosine functions are used not only in applications involving right triangles but also to model phenomena involving rotation and periodic motion. Extending the domains of the trigonometric functions from acute angles to angles of any measure will allow us to model a wide variety of phenomena such as biorhythms, weather, tides, electricity, robotic arms, and the design of highways.

Definitions

Robotics is a rapidly growing field that requires extensive mathematics. One basic problem in designing a robotic arm is determining the location of the robot's hand. Suppose we have a robotic arm that rotates at the shoulder and is controlled by changing the angle θ and the length of the arm r, as illustrated in Figure 6.45. We would like to find a relation between the xy-coordinates of the hand and the values for r and θ. (**Source:** W. Stadler, *Analytical Robotics and Mechatronics.*)

Notice that for a fixed angle θ, if the length of the arm is changed from r_1 to r_2, triangle ABC in Figure 6.46 and triangle DEF in Figure 6.47 are similar triangles. Thus the following ratios are equal and depend only on the measure of θ.

$$\frac{x_1}{r_1} = \frac{x_2}{r_2} \qquad \text{and} \qquad \frac{y_1}{r_1} = \frac{y_2}{r_2}$$

> Because the triangles are similar, these ratios are equal.

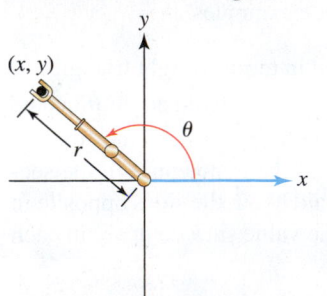

Figure 6.45

Figure 6.46

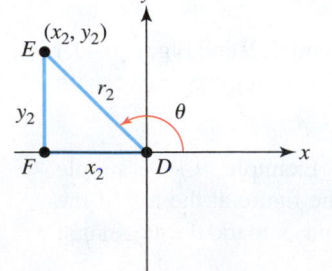

Figure 6.47

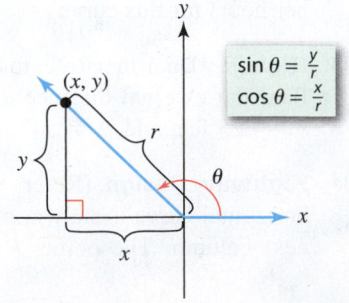

Figure 6.48

In Figure 6.48 the Pythagorean theorem gives $r^2 = x^2 + y^2$. Since $r > 0$, it follows that $r = \sqrt{x^2 + y^2}$. The ratios $\frac{y}{r}$ and $\frac{x}{r}$ depend only on θ and can be used to define the *sine* and *cosine* functions of *any* angle θ.

> ### THE SINE AND COSINE FUNCTIONS OF ANY ANGLE θ
>
> Let angle θ be in standard position with the point (x, y) lying on the angle's terminal side. If $r = \sqrt{x^2 + y^2}$, then
>
> $$\sin\theta = \frac{y}{r} \quad \text{and} \quad \cos\theta = \frac{x}{r} \ (r \neq 0).$$

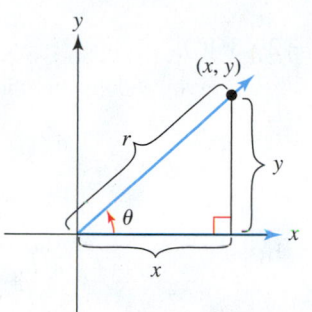

Figure 6.49

Although the terminal side of θ in Figure 6.48 is shown in the second quadrant, these definitions are valid for any angle θ having a terminal side in any of the four quadrants.

> **MAKING CONNECTIONS**
>
> **Right Triangle Trigonometry**   If $0° < \theta < 90°$, then x corresponds to the length of the adjacent side, y corresponds to the length of the opposite side, and r corresponds to the length of the hypotenuse. See Figure 6.49. These new definitions for sine and cosine are consistent with the definitions presented in Section 6.2, when θ is an acute angle.

EXAMPLE 1   **Evaluating sine and cosine for coterminal angles**

Suppose a robotic hand is located at the point $(15, -8)$, where all units are in inches.
(a) Find the length of the arm.
(b) Let α satisfy $0° \leq \alpha < 360°$ and represent the angle between the positive x-axis and the robotic arm. Find $\sin\alpha$ and $\cos\alpha$.
(c) Let β satisfy $-360° \leq \beta < 0°$ and represent the angle between the positive x-axis and the robotic arm. Find $\sin\beta$ and $\cos\beta$. How do the values for $\sin\beta$ and $\cos\beta$ compare with the values for $\sin\alpha$ and $\cos\alpha$?

SOLUTION
(a) Let $x = 15, y = -8$, and $r = \sqrt{x^2 + y^2}$. Thus $r = \sqrt{15^2 + (-8)^2} = 17$ and the length of the arm is 17 inches. See Figure 6.50.
(b) Let $x = 15, y = -8$, and $r = 17$. Then

$$\sin\alpha = \frac{y}{r} = -\frac{8}{17} \quad \text{and} \quad \cos\alpha = \frac{15}{17}.$$

(c) In Figure 6.51, β satisfies $-360° \leq \beta < 0°$. Since the values of x, y, and r do *not* change, the trigonometric values for β are the *same* as those for α in part (b).

Coterminal Angles Have Equal Sine and Cosine Values

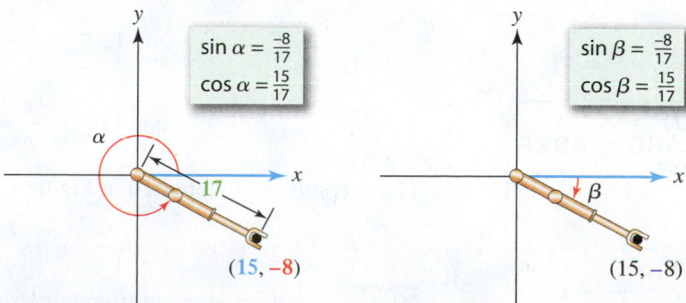

Figure 6.50   **Figure 6.51**

Now Try Exercise 1

Coterminal Angles Have Equal Trigonometric Values

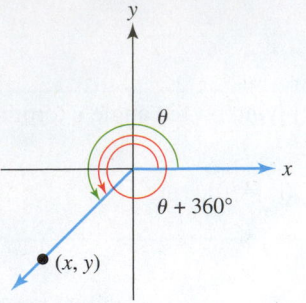

Figure 6.52

Sine and Cosine of Coterminal Angles The results of Example 1 can be generalized. If α and β are *coterminal* angles, then

$$\sin \alpha = \sin \beta \quad \text{and} \quad \cos \alpha = \cos \beta.$$

The angles θ and $\theta + 360°$ are coterminal for any θ. Their terminal sides pass through the same point (x, y), as shown in Figure 6.52. Therefore for all θ, $\sin \theta = \sin (\theta + 360°)$ and $\cos \theta = \cos (\theta + 360°)$. In general, if n is any integer, then

$$\sin \theta = \sin (\theta + n \cdot 360°) \quad \text{and} \quad \cos \theta = \cos (\theta + n \cdot 360°).$$

As a result, we say that the sine and cosine functions are **periodic** with **period 360°** (or 2π radians). For example,

$$\sin 90° = \sin (90° + 360°) = \sin (90° - 2 \cdot 360°).$$

See Figure 6.53.

Sine Has Period 360°

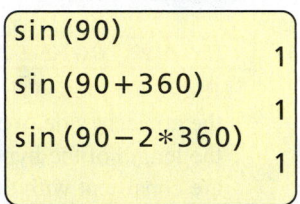

Figure 6.53 Degree Mode

Evaluating Sine and Cosine by Hand If an angle is a multiple of 30° or 45°, exact evaluation of the sine function or cosine function is possible by hand. The next example illustrates an angle for which the sine and cosine functions can be evaluated exactly by hand.

EXAMPLE 2 Finding values of $\sin \theta$ and $\cos \theta$

Find $\sin 120°$ and $\cos 120°$. Support your answer using a calculator.

Finding Exact Values

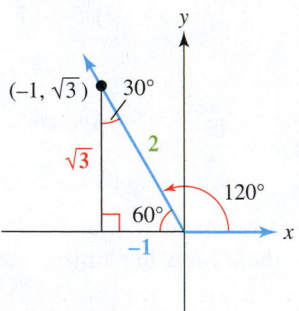

Figure 6.54

SOLUTION When evaluating the sine or cosine function by hand, we can select any positive value for r and it does not change the resulting values for $\sin \theta$ and $\cos \theta$. For convenience, we let $r = 2$, as shown in Figure 6.54. The length of the shortest leg in a 30°–60° right triangle is half the hypotenuse. Since the terminal side of θ is in the second quadrant, $x < 0$ and so $x = -1$. Next we find y.

$$x^2 + y^2 = r^2 \qquad \textcolor{blue}{\text{Pythagorean theorem}}$$
$$y^2 = r^2 - x^2 \qquad \textcolor{blue}{\text{Subtract } x^2.}$$
$$y^2 = 2^2 - (-1)^2 \qquad \textcolor{blue}{\text{Let } r = 2 \text{ and } x = -1.}$$
$$y = \pm\sqrt{3} \qquad \textcolor{blue}{\text{Square root property}}$$

Since the terminal side of θ is in the second quadrant, $y > 0$ and so $y = \sqrt{3}$. Thus

$$\sin 120° = \frac{y}{r} = \frac{\sqrt{3}}{2} \approx 0.8660 \quad \text{and}$$

$$\cos 120° = \frac{x}{r} = -\frac{1}{2} = -0.5.$$

These results are supported in Figure 6.55.

Now Try Exercise 13

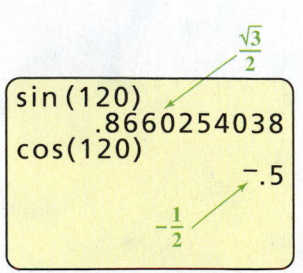

Figure 6.55 Degree Mode

NOTE When evaluating trigonometric functions by hand as in Example 2, *always* draw the perpendicular line segment from a point on the terminal side to the x-axis, *not* the y-axis.

The values of the sine and cosine functions for the **quadrantal angles** 0°, 90°, 180°, and 270° are shown in Table 6.3. Try to verify these values.

Quadrantal Angles

	0°	90°	180°	270°
$\sin \theta$	0	1	0	−1
$\cos \theta$	1	0	−1	0

Terminal sides of these angles lie on the *x*- or *y*-axis.

Table 6.3

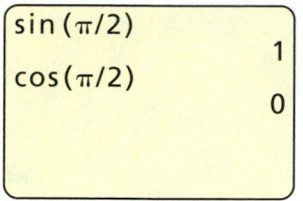

Figure 6.56 Radian Mode

NOTE If the terminal side of an angle θ in standard position lies on the *y*-axis, then $\sin \theta = \pm 1$ and $\cos \theta = 0$; if it lies on the *x*-axis, $\sin \theta = 0$ and $\cos \theta = \pm 1$.

Evaluation in Radian Mode Trigonometric functions can also be evaluated using radian mode. See Figure 6.56. Since 90° is equivalent to $\frac{\pi}{2}$ radians, it follows that

$$\sin \frac{\pi}{2} = 1 \quad \text{and} \quad \cos \frac{\pi}{2} = 0.$$

EXAMPLE 3   **Finding values of sin *t* and cos *t***

Find the exact values of $\sin\left(-\frac{3\pi}{4}\right)$ and $\cos\left(-\frac{3\pi}{4}\right)$.

SOLUTION

Getting Started Because there is no degree symbol, it is assumed that $-\frac{3\pi}{4}$ is measured in radians. Its terminal side lies in quadrant III, as shown in Figure 6.57. ▶

From a point on the terminal side, draw a perpendicular line segment to the *x*-axis, forming a 45°–45° right triangle. Label the resulting triangle conveniently so that the length of each leg equals 1. (Each side is labeled −1, because the terminal side lies in quadrant III.) By the Pythagorean theorem, the hypotenuse is $\sqrt{2}$. Thus $x = -1$, $y = -1$, and $r = \sqrt{2}$. It follows that

Finding Exact Values

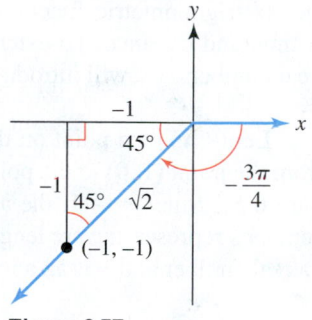

Figure 6.57

$$\sin\left(-\frac{3\pi}{4}\right) = \frac{y}{r} = -\frac{1}{\sqrt{2}} \quad \text{and} \quad \cos\left(-\frac{3\pi}{4}\right) = \frac{x}{r} = -\frac{1}{\sqrt{2}}.$$

NOTE If we rationalize the denominator, each result becomes $-\frac{\sqrt{2}}{2}$.

Now Try Exercise 27

An Application from Highway Design *Grade,* or slope, is a measure of steepness and indicates whether a highway is uphill or downhill. A 5% grade indicates that a road is increasing 5 vertical feet for each 100-foot increase in horizontal distance. *Grade resistance R* is the gravitational force acting on a vehicle and is given by

$$R = W \sin \theta,$$

where *W* is the weight of the vehicle and θ is the angle associated with the grade. See Figure 6.58. For an uphill grade $\theta > 0$, and for a downhill grade $\theta < 0$. (*Source:* F. Mannering and W. Kilareski, *Principles of Highway Engineering and Traffic Analysis.*)

Uphill and Downhill Grade

Figure 6.58

Calculating the grade resistance

A downhill highway grade is modeled by the line $y = -0.06x$ in the fourth quadrant.
(a) Find the grade of the road.
(b) Determine the grade resistance for a 3000-pound car. Interpret the result.

SOLUTION
(a) The slope of the line is -0.06, so when x *increases* by 100 feet, y *decreases* by 6 feet. See Figure 6.59. Thus this road has a grade of -6%.
(b) Because $R = W \sin\theta$, we must find $\sin\theta$. From Figure 6.59 we see that the point $(100, -6)$ lies on the terminal side of θ. Since

$$r = \sqrt{100^2 + (-6)^2} = \sqrt{10,036},$$

it follows that

$$\sin\theta = \frac{y}{r} = \frac{-6}{\sqrt{10,036}}.$$

The grade resistance is

$$R = W \sin\theta = 3000\left(\frac{-6}{\sqrt{10,036}}\right) \approx -179.7 \text{ lb}.$$

On this stretch of highway, gravity pulls a 3000-pound vehicle *downhill* with a force of about 180 pounds. Note that a downhill grade results in a negative grade resistance.

Now Try Exercise 89

A Grade of −6%

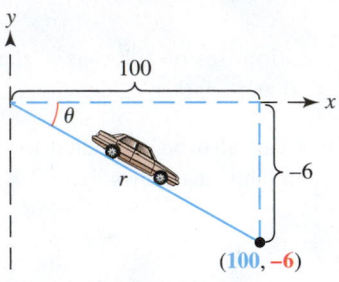

Figure 6.59 (Not to scale)

The Unit Circle

Trigonometric functions were first defined for angles. As applications became more diverse, real numbers were included in the domains of the six trigonometric functions. Real numbers were needed to represent quantities such as time and distance. To extend the domains of the sine and cosine functions to include all real numbers, we will introduce the unit circle.

The **unit circle** has radius 1 and equation $x^2 + y^2 = 1$. Let (x, y) be a point on the unit circle, and let s be the arc length along the unit circle from the point $(1, 0)$ to the point (x, y) determined by a counterclockwise rotation. See Figure 6.60. Since $r = 1$, the arc length formula $s = r\theta$ reduces to $s = \theta$. That is, the real number s representing arc length is numerically equal to the radian measure of θ. Thus if s is a real number and θ is an angle measured in radians, as shown in Figure 6.60, we define

Unit Circle ($r = 1$)

$\sin s = \sin\theta$
$\cos s = \cos\theta$

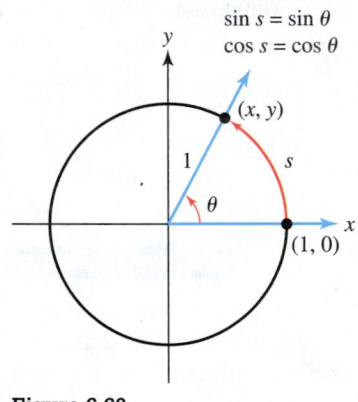

Figure 6.60

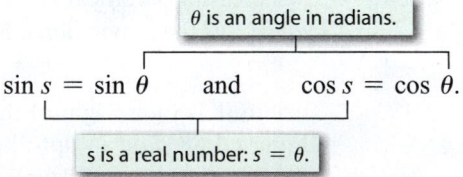

θ is an angle in radians.

$\sin s = \sin\theta$ and $\cos s = \cos\theta$.

s is a real number: $s = \theta$.

If $s < 0$, then the red arc in Figure 6.60 is wrapped around the unit circle in a *clockwise* direction and θ has negative measure. This discussion suggests how trigonometric functions of a real number are evaluated.

TRIGONOMETRIC FUNCTIONS OF REAL NUMBERS

The value of a trigonometric function for the real number s is equal to its value for s radians.

NOTE To evaluate a trigonometric function of a real number s with a calculator, use *radian* mode.

EXAMPLE 5   **Approximating the sine and cosine of a real number**

Use a calculator to approximate sin 1.78 and cos 1.78.

SOLUTION   To approximate the sine and cosine of the real number 1.78, use *radian* mode. As shown in Figure 6.61, sin 1.78 $\approx$ 0.978 and cos 1.78 $\approx$ −0.208.

Sine and Cosine of a Real Number

```
sin(1.78)
         .9781966068
cos(1.78)
        -.2076810016
```

Figure 6.61 Radian Mode

Now Try Exercise 33

Trigonometic Functions and the Unit Circle   The following See the Concept illustrates that if the terminal side of an angle θ intersects the *unit circle* at (x, y), then this point can also be written as $(\cos \theta, \sin \theta)$ since on the unit circle $x = \cos \theta$ and $y = \sin \theta$. Because trigonometric functions can be defined using the unit circle, they are also referred to as **circular functions**.

See the Concept: Circular Functions

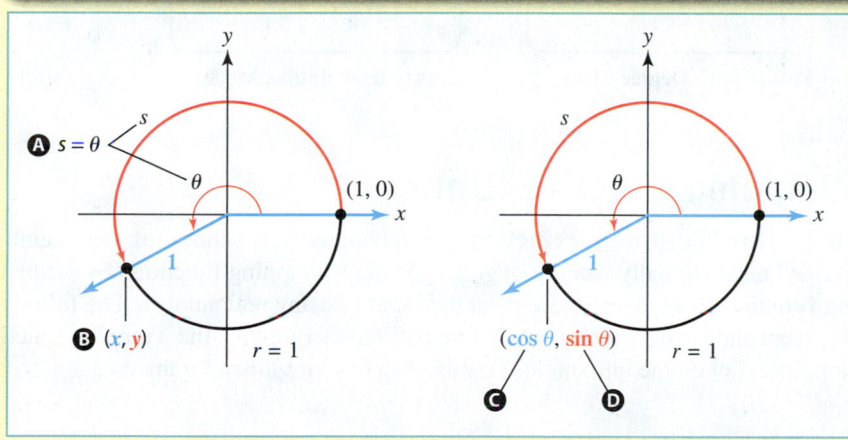

A $s = \theta$ because $s = r\theta$ and $r = 1$. (*s* is arc length; θ is in radians)

B The terminal side of θ intersects the unit circle at (x, y).

C $\cos \theta = \frac{x}{r}$ and $r = 1$, so $x = \cos \theta$.

D $\sin \theta = \frac{y}{r}$ and $r = 1$, so $y = \sin \theta$.

EXAMPLE 6   **Evaluating circular functions**

Use Figure 6.62 to find sin θ and cos θ.

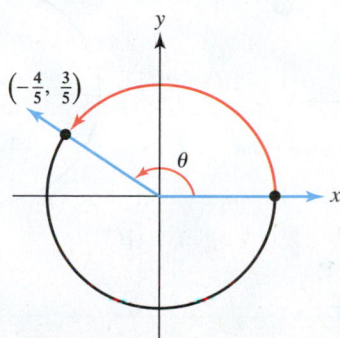

Figure 6.62

SOLUTION The terminal side of angle θ intersects the unit circle at the point $\left(-\frac{4}{5}, \frac{3}{5}\right)$. Because this point is in the form $(\cos\theta, \sin\theta)$ it follows that

$$\cos\theta = -\frac{4}{5} \quad \text{and} \quad \sin\theta = \frac{3}{5}.$$

Now Try Exercise 37

A Trigonometric Identity The unit circle is described by $x^2 + y^2 = 1$. Because $x = \cos\theta$ and $y = \sin\theta$, we have

$$(\cos\theta)^2 + (\sin\theta)^2 = 1,$$

or equivalently,

$$\sin^2\theta + \cos^2\theta = 1.$$

> Trigonometric identity
> (True for all values of θ)

This equation is an example of a **trigonometric identity**. An identity is true for all meaningful values of the variable. In Figures 6.63 and 6.64 this identity is evaluated for different values of θ. In every case the result is 1, regardless of whether θ is measured in degrees or radians.

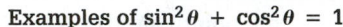

Examples of $\sin^2\theta + \cos^2\theta = 1$

```
(sin(80))²+(cos(
80))²
                1
(sin(⁻12))²+(cos
(⁻12))²
                1
```

Figure 6.63 Degree Mode

```
(sin(π/7))²+(cos
(π/7))²
                1
(sin(1.5))²+(cos
(1.5))²
                1
```

Figure 6.64 Radian Mode

The Wrapping Function (Optional)

Evaluating the Wrapping Function   Trigonometric functions of real numbers can be defined more formally *without angles* by using a wrapping function. To define the **wrapping function** W on the unit circle, let the input s be any real number. The following See the Concept shows that given any positive real number input s, the wrapping function outputs a point (x, y) on the unit circle. That is, $W(s) = (x, y)$. See Figure 6.65.

See the Concept: Evaluating the Wrapping Function

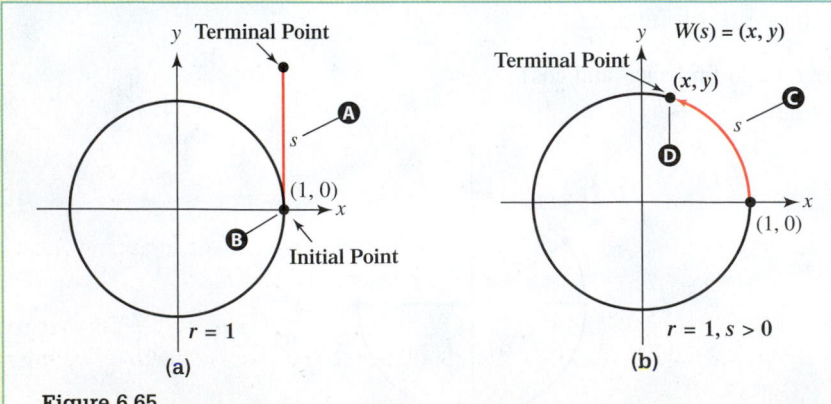

Ⓐ Cut a "string" having length s.

Ⓑ Place one end of the "string" at the initial point $(1, 0)$.

Ⓒ Wrap the "string" counterclockwise around the unit circle if $s > 0$.

Ⓓ If the "string" stops at the terminal point (x, y), then $W(s) = (x, y)$.

If s is **negative**, the string has length $|s|$ and is wrapped **clockwise**. Note that the initial point is always $(1, 0)$.

Figure 6.65

> **NOTE** The wrapping function is different from other functions that we have encountered because its output is a *point* on the unit circle.

EXAMPLE 7 **Evaluating the wrapping function**

For each real number s, evaluate $W(s)$.

(a) 2π **(b)** π **(c)** $-\dfrac{\pi}{2}$ **(d)** $-\dfrac{\pi}{4}$

SOLUTION

(a) Because the radius of the unit circle is 1, the circumference C of the unit circle is $C = 2\pi r = 2\pi$. If a string of length $s = 2\pi$ is wrapped counterclockwise around the unit circle, it will make one complete revolution and the terminal point will coincide with the initial point $(1, 0)$. Thus $W(2\pi) = (1, 0)$. See Figure 6.66.

(b) Half the circumference of the unit circle is π. The string wraps halfway around the unit circle, and the terminal point is $(-1, 0)$. Thus $W(\pi) = (-1, 0)$. See Figure 6.67.

(c) Because s is *negative*, the string wraps *clockwise* around the unit circle, as shown in Figure 6.68. A length of $\frac{\pi}{2}$ represents a fourth of the circumference of the unit circle, so the terminal point is $(0, -1)$ and $W\!\left(-\frac{\pi}{2}\right) = (0, -1)$.

(d) A distance of $\frac{\pi}{4}$ represents an eighth of the circumference of the unit circle. Thus the terminal point lies on the line $y = -x$, as shown in Figure 6.69. The 45°–45° right triangle formed has a hypotenuse with length 1 and legs both with length a.

$$a^2 + a^2 = 1 \qquad \textcolor{teal}{\text{Pythagorean theorem: } a = b}$$

$$2a^2 = 1 \qquad \textcolor{teal}{\text{Add like terms.}}$$

$$a^2 = \frac{1}{2} \qquad \textcolor{teal}{\text{Divide by 2.}}$$

$$a = \pm\frac{1}{\sqrt{2}} \qquad \textcolor{teal}{\text{Square root property}}$$

Because the terminal point is located in quadrant IV, the x-coordinate is $\frac{1}{\sqrt{2}}$ and the y-coordinate is $-\frac{1}{\sqrt{2}}$. Thus $W\!\left(-\frac{\pi}{4}\right) = \left(\frac{1}{\sqrt{2}}, -\frac{1}{\sqrt{2}}\right)$.

Evaluating $W(2\pi)$

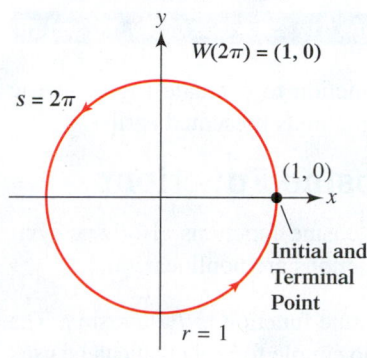

Figure 6.66

Evaluating the Wrapping Function

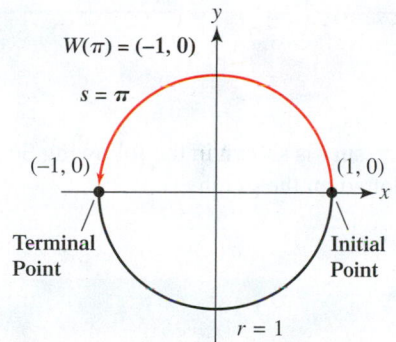

Figure 6.67

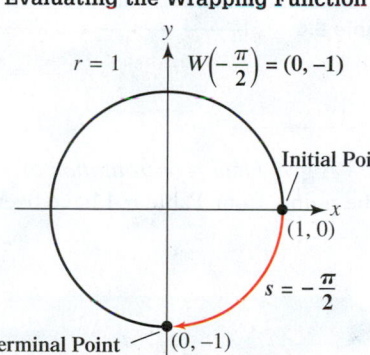

Figure 6.68

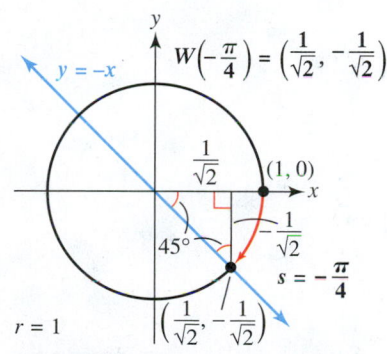

Figure 6.69

Now Try Exercises 51, 53, 55, and 59

> **NOTE** The concept of string is used only to help you visualize the wrapping function. Instead, you could visualize moving a distance s around the unit circle.

Evaluating the Sine and Cosine with the Wrapping Function The wrapping function on the unit circle can be used to define the sine and cosine functions for any real number s by letting

$$\sin s = \boldsymbol{y} \qquad \text{and} \qquad \cos s = \boldsymbol{x},$$

where $W(s) = (\boldsymbol{x}, \boldsymbol{y})$. The next example illustrates how to apply this definition.

EXAMPLE 8   **Evaluating the sine and cosine functions**

Find $\sin s$ and $\cos s$ for each real number s.

(a) 2π **(b)** π **(c)** $-\dfrac{\pi}{2}$ **(d)** $-\dfrac{\pi}{4}$

SOLUTION

Getting Started First note that the values for s are the same as those used in Example 7. Thus if $W(s) = (x, y)$, we need only apply the fact that $\sin s = y$ and $\cos s = x$. ▶

(a) $W(2\pi) = (\mathbf{1}, \mathbf{0})$ by Example 7(a), so $\sin 2\pi = \mathbf{0}$ and $\cos 2\pi = \mathbf{1}$.

(b) $W(\pi) = (\mathbf{-1}, \mathbf{0})$ by Example 7(b), so $\sin \pi = \mathbf{0}$ and $\cos \pi = \mathbf{-1}$.

(c) $W\left(-\dfrac{\pi}{2}\right) = (\mathbf{0}, \mathbf{-1})$ by Example 7(c), so $\sin\left(-\dfrac{\pi}{2}\right) = \mathbf{-1}$ and $\cos\left(-\dfrac{\pi}{2}\right) = \mathbf{0}$.

(d) $W\left(-\dfrac{\pi}{4}\right) = \left(\dfrac{1}{\sqrt{2}}, -\dfrac{1}{\sqrt{2}}\right)$ by Example 7(d), so $\sin\left(-\dfrac{\pi}{4}\right) = -\dfrac{1}{\sqrt{2}}$ and $\cos\left(-\dfrac{\pi}{4}\right) = \dfrac{1}{\sqrt{2}}$.

Now Try Exercises 67, 69, 71, and 75

NOTE The results obtained by using the wrapping function to evaluate trigonometric functions are equivalent to those arrived at by the other methods presented earlier.

Representations of the Sine and Cosine Functions

Like other functions that we have studied, the sine and cosine functions also have symbolic, numerical, and graphical representations. Both functions are nonlinear.

The Sine Function A *symbolic representation* of the sine function is $f(t) = \sin t$. The domain of the sine function is all real numbers. There is no simple formula that can be used to evaluate the sine function. Instead, we generally rely on other methods for its evaluation.

A *numerical representation* of $f(t) = \sin t$ is shown in Table 6.4. Since outputs from the sine function correspond to a y-coordinate on the unit circle, the range of the sine function is $-1 \le y \le 1$.

Evaluating the Sine Function

t	0	$\dfrac{\pi}{4}$	$\dfrac{\pi}{2}$	$\dfrac{3\pi}{4}$	π	$\dfrac{5\pi}{4}$	$\dfrac{3\pi}{2}$	$\dfrac{7\pi}{4}$	2π
$\sin t$	0	$\dfrac{1}{\sqrt{2}}$	1	$\dfrac{1}{\sqrt{2}}$	0	$-\dfrac{1}{\sqrt{2}}$	-1	$-\dfrac{1}{\sqrt{2}}$	0

Table 6.4

Increases: $\left(0, \dfrac{\pi}{2}\right)$ Decreases: $\left(\dfrac{\pi}{2}, \dfrac{3\pi}{2}\right)$ Increases: $\left(\dfrac{3\pi}{2}, 2\pi\right)$

A *graphical representation* of $f(t) = \sin t$ is shown in the following See the Concept. The points from Table 6.4 have been plotted on the graph.

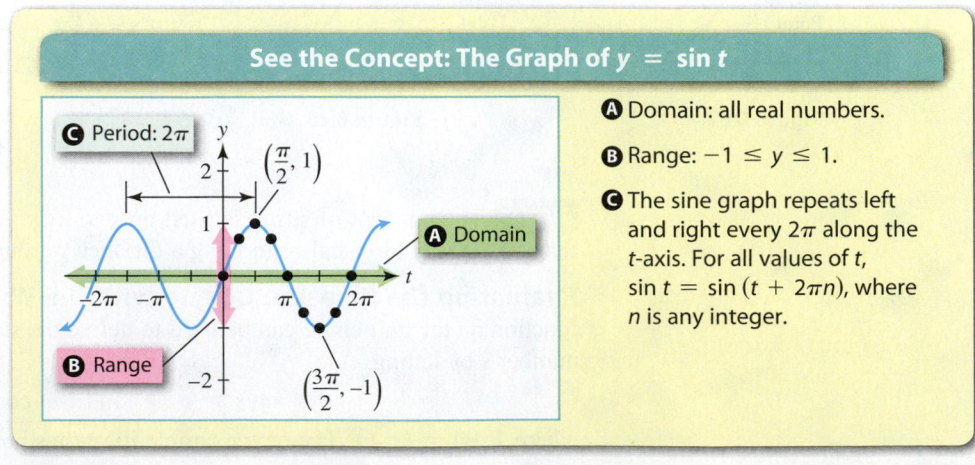

See the Concept: The Graph of $y = \sin t$

C Period: 2π $\left(\dfrac{\pi}{2}, 1\right)$ **A** Domain   **B** Range $\left(\dfrac{3\pi}{2}, -1\right)$

A Domain: all real numbers.

B Range: $-1 \le y \le 1$.

C The sine graph repeats left and right every 2π along the t-axis. For all values of t, $\sin t = \sin(t + 2\pi n)$, where n is any integer.

Unit Circle ($r = 1$)

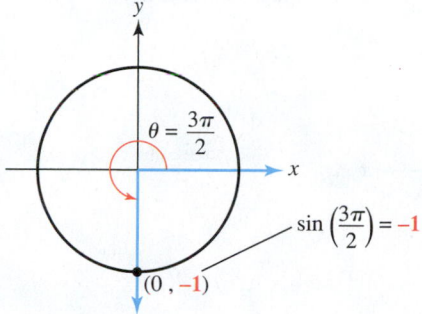

Figure 6.70

MAKING CONNECTIONS

The Unit Circle and the Sine Graph   The graph of the sine function may be easier to understand if we use a unit circle. If a point P on the unit circle starts at $(1, \mathbf{0})$ and rotates counterclockwise, then the y-coordinate of P corresponds to $\sin\theta$. Refer to Figure 6.70. As a result, $\sin\theta$ first increases from $\mathbf{0}$ to $\mathbf{1}$, then decreases from $\mathbf{1}$ to $-\mathbf{1}$, and finally increases from $-\mathbf{1}$ to $\mathbf{0}$ (see Table 6.4), where P completes one rotation and returns to the point $(1, \mathbf{0})$. If P continues to rotate, it passes through the same points a second time, and so the sine function has period 2π, or $360°$.

EXAMPLE 9   **Evaluating the sine function**

Evaluate $f(t) = \sin t$ at $t = \frac{3\pi}{2}$ by hand.

SOLUTION   An angle θ of $\frac{3\pi}{2}$ radians in standard position has a terminal side that intersects the unit circle at the point $(0, -1)$. See Figure 6.71. Therefore $\sin\left(\frac{3\pi}{2}\right) = -\mathbf{1}$.

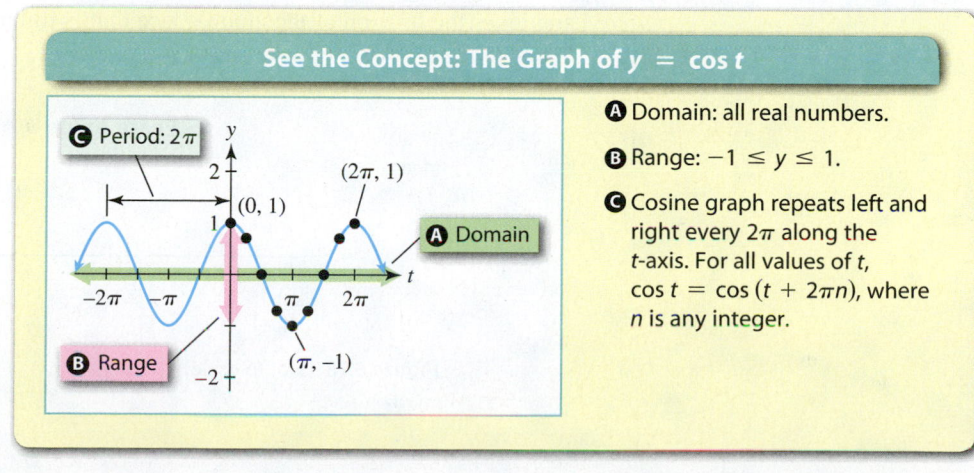

Figure 6.71

Now Try Exercise 41

The Cosine Function   The cosine function, is *represented symbolically* by $f(t) = \cos t$. Its domain is all real numbers and its range is $-1 \le y \le 1$.

A *numerical representation* of $f(t) = \cos t$ is shown in Table 6.5.

Evaluating the Cosine Function

t	0	$\frac{\pi}{4}$	$\frac{\pi}{2}$	$\frac{3\pi}{4}$	π	$\frac{5\pi}{4}$	$\frac{3\pi}{2}$	$\frac{7\pi}{4}$	2π
$\cos t$	1	$\frac{1}{\sqrt{2}}$	0	$-\frac{1}{\sqrt{2}}$	-1	$-\frac{1}{\sqrt{2}}$	0	$\frac{1}{\sqrt{2}}$	1

Table 6.5

Decreases: $(0, \pi)$ Increases: $(\pi, 2\pi)$

A *graphical representation* of $f(t) = \cos t$ is shown in the following See the Concept. The points from Table 6.5 have been plotted on the graph.

See the Concept: The Graph of $y = \cos t$

C Period: 2π

$(0, 1)$ $(2\pi, 1)$

A Domain

$(\pi, -1)$

B Range

A Domain: all real numbers.

B Range: $-1 \le y \le 1$.

C Cosine graph repeats left and right every 2π along the t-axis. For all values of t, $\cos t = \cos(t + 2\pi n)$, where n is any integer.

If a point P on the unit circle in Figure 6.70 on the preceding page starts at $(1, 0)$ and rotates counterclockwise, then the x-coordinate of P corresponds to $\cos\theta$. Explain how this relates to the graph of the cosine function.

EXAMPLE 10 **Evaluating the cosine function**

Evaluate $f(t) = \cos t$ at $t = \frac{5\pi}{6}$ by hand.

SOLUTION An angle of $\frac{5\pi}{6}$ radians in standard position has a terminal side that intersects the unit circle in the second quadrant. See Figure 6.72. To find the point of intersection (x, y), notice that the hypotenuse of the $30°$–$60°$ right triangle has length 1 and the shorter leg has length $y = \frac{1}{2}$. We can determine x symbolically.

$$x^2 + y^2 = 1 \qquad \textcolor{blue}{\text{Equation of the unit circle}}$$

$$x^2 + \left(\frac{1}{2}\right)^2 = 1 \qquad \textcolor{blue}{y = \tfrac{1}{2}}$$

$$x^2 = \frac{3}{4} \qquad \textcolor{blue}{\text{Solve for } x^2.}$$

$$x = \pm\frac{\sqrt{3}}{2} \qquad \textcolor{blue}{\text{Square root property}}$$

The point (x, y) is in the second quadrant, so choose $x = -\frac{\sqrt{3}}{2}$. Because $\cos t = x$, it follows that $\cos\frac{5\pi}{6} = -\frac{\sqrt{3}}{2} \approx -0.8660$.

Now Try Exercise 43

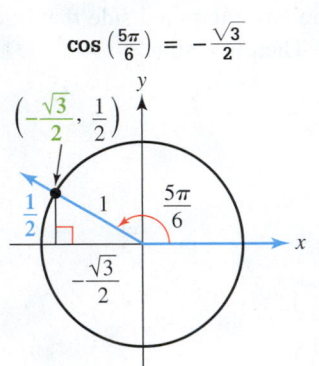

$$\cos\left(\tfrac{5\pi}{6}\right) = -\tfrac{\sqrt{3}}{2}$$

Figure 6.72

Applications of the Sine and Cosine Functions

Periodic graphs that are similar in shape to the graphs of the sine and cosine functions are **sinusoidal**. Of all the periodic graphs, sinusoidal graphs are the most important in applications because they occur in nearly every aspect of physical science.

Because the moon orbits Earth, we observe different phases of the moon during the period of a month. In Figure 6.73, angle $\boldsymbol\theta$ is called the *phase angle*. The *phase F* of the moon is computed by

$$F(\theta) = \frac{1}{2}(1 - \cos\boldsymbol\theta)$$

and gives the fraction of the moon's face that is illuminated by the sun. (***Source:*** P. Duffet-Smith, *Practical Astronomy with Your Calculator.*)

Phase Angle θ

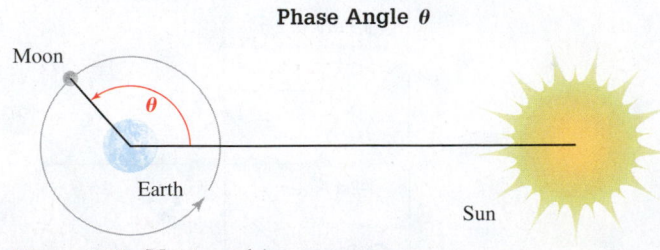

Figure 6.73 (Not to scale)

EXAMPLE 11 Modeling the phases of the moon

Let $F(\theta) = \frac{1}{2}(1 - \cos\theta)$.

(a) A graph of F is shown in Figure 6.74. Discuss how the graph relates to the phases of the moon.

(b) Evaluate $F(0)$, $F\left(\frac{\pi}{2}\right)$, $F(\pi)$, and $F\left(\frac{3\pi}{2}\right)$. Interpret each result.

Phases of the Moon

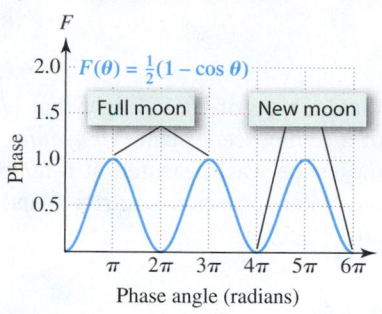

Figure 6.74

SOLUTION

(a) The phases of the moon are periodic and have a period of 2π. Each peak represents a full moon, and each valley represents a new moon.

(b) $F(0) = \frac{1}{2}(1 - \cos 0) = \frac{1}{2}(1 - 1) = 0$. When $\theta = 0$, the moon is located between Earth and the sun. Since $F = 0$, the face of the moon is not visible, which corresponds to a *new moon*.

$F\left(\frac{\pi}{2}\right) = \frac{1}{2}\left(1 - \cos\frac{\pi}{2}\right) = \frac{1}{2}(1 - 0) = \frac{1}{2}$. When $\theta = \frac{\pi}{2}$, $F = \frac{1}{2}$. Thus half the face of the moon is visible. This phase is called the *first quarter*.

$F(\pi) = \frac{1}{2}(1 - \cos\pi) = \frac{1}{2}(1 - (-1)) = 1$. When $\theta = \pi$, Earth is between the moon and the sun. Since $F = 1$, the face of the moon is completely visible, which corresponds to a *full moon*.

$F\left(\frac{3\pi}{2}\right) = \frac{1}{2}\left(1 - \cos\frac{3\pi}{2}\right) = \frac{1}{2}(1 - 0) = \frac{1}{2}$. When $\theta = \frac{3\pi}{2}$, $F = \frac{1}{2}$. Thus half the face of the moon is visible. This phase is called the *last quarter*.

Now Try Exercise 91

An Application from Electricity Common household current is called *alternating current* (AC) because the voltage and current change direction 120 times per second. Sinusoidal curves are often used to model the voltage in a common household electrical outlet, as demonstrated in the next example.

EXAMPLE 12 Analyzing household current

The voltage V in a household outlet can be modeled by $V(t) = 160\sin(120\pi t)$, where t represents time in seconds.

(a) A graph of $y = V(t)$ is shown in Figure 6.75. Describe the voltage.

(b) Evaluate $V\left(\frac{1}{240}\right)$ and interpret the result.

(c) One way to estimate the "average" voltage in a circuit is to use the **root mean square** voltage, which equals the maximum voltage divided by $\sqrt{2}$. Find this "average" or root mean square voltage.

Voltage in a Circuit

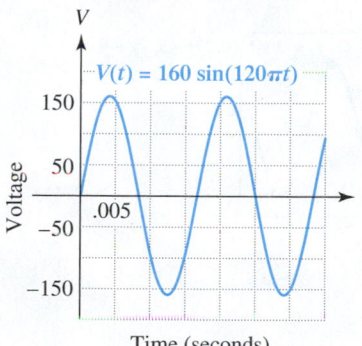

Figure 6.75

SOLUTION

(a) The graph of V in Figure 6.75 is sinusoidal and varies between -160 volts and 160 volts, corresponding to the fact that the current is changing direction in a household outlet. When the graph is above the t-axis the current flows one direction, and when it is below the t-axis the current flows in the opposite direction.

(b) $V\left(\frac{1}{240}\right) = 160\sin\frac{\pi}{2} = 160(1) = 160$. After $\frac{1}{240} \approx 0.004$ second, the voltage is 160 volts. Figure 6.75 supports this result.

(c) The maximum voltage is 160 volts. The "average" voltage is $\frac{160}{\sqrt{2}} \approx 113$ volts. Common household electricity is often rated at 110–120 volts.

Now Try Exercise 95

NOTE In Example 12, the input to the function V is time, not an angle. This example illustrates why it is necessary to extend the domain of trigonometric functions to include all real numbers.

Modeling with the Sine Function (Optional)

The study of *biological clocks* is a fascinating field. Many living organisms undergo regular biological rhythms, or circadian rhythms. (See Exercises 93 and 94.) A simple example is a flower that opens during daylight and closes at nighttime. One amazing result is that flowers often continue to open and close even when they are placed in continual darkness. (*Source:* F. Brown et al., *The Biological Clock.*)

| EXAMPLE 13 | Modeling biological rhythms |

Some types of water plants are *luminescent*—they radiate a type of "cold" light that is similar to the light emitted from a firefly. A sample of a luminescent plant (*Gonyaulax polyedra*) was put into continual dim light. The luminescence was measured at 6-hour intervals, and the data in Table 6.6 summarize the results. Noon corresponds to $t = 0$ and the y-units of luminescence are arbitrary.

Luminescence of a Plant

t (hour)	0	6	12	18	24	30	36	42	48
y (luminescence)	1	4	7	4	1	4	7	4	1

Table 6.6   *Source:* E. Bünning, *The Physiological Clock.*

(a) Make a scatterplot of the data in $[-6, 54, 6]$ by $[0, 8, 1]$. Interpret the data.
(b) Graph the data and $f(t) = 3 \sin (0.27t - 1.7) + 4$. How well does f model the data?
(c) Estimate the luminescence when $t = 33$.

SOLUTION

(a) A scatterplot of the data is shown in Figure 6.76. Luminescence appears to be periodic, increasing and decreasing at regular intervals even though the lighting was always dim. The plant was most luminescent during times that correspond to midnight: $t = 12$ and 36. It was least luminescent at times corresponding to noon: $t = 0, 24$, and 48.

(b) Let $Y_1 = 3 \sin (.27X - 1.7) + 4$; the data are graphed in Figure 6.77. The graph of f models the periodic data quite well.

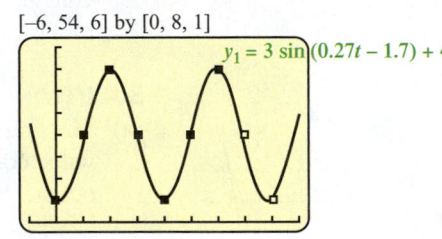

A Scatterplot

$[-6, 54, 6]$ by $[0, 8, 1]$

Figure 6.76

A Sinusoidal Model

$[-6, 54, 6]$ by $[0, 8, 1]$

$y_1 = 3 \sin (0.27t - 1.7) + 4$

Figure 6.77

(c) Evaluate $f(t)$ at $t = 33$ using radian mode.

$$f(33) = 3 \sin (0.27(33) - 1.7) + 4 \approx 6.4$$

The luminescence was about 6.4 after 33 hours, or at 9 P.M.

Calculator Help

To make a scatterplot and graph an equation, see Appendix A (pages AP-3 and AP-6).

Now Try Exercise 101

6.3 Putting It All Together

$\mathbf{I}$n this section we extended the domains of the sine and cosine functions to include any angle θ and any real number t. Both of these functions are nonlinear functions, and their ranges include values satisfying $-1 \le y \le 1$. Some concepts about the sine and cosine functions are summarized in the following table.

CONCEPT	FORMULAS AND FIGURES
Sine and cosine of any angle θ.	 $\sin \theta = \dfrac{y}{r}$ and $\cos \theta = \dfrac{x}{r}$, where $r = \sqrt{x^2 + y^2}$
The unit circle and the sine and cosine functions	$x^2 + y^2 = 1 \quad (r = 1)$ *Evaluation:* $\sin t = y,\ \cos t = x$ *Period:* 2π or $360°$ $\qquad \sin(t + 2\pi n) = \sin t$ $\qquad \cos(t + 2\pi n) = \cos t$
Graph of the sine function	**The Sine Function** *Domain:* $-\infty < t < \infty$ *Range:* $-1 \le y \le 1$
Graph of the cosine function	**The Cosine Function** *Domain:* $-\infty < t < \infty$ *Range:* $-1 \le y \le 1$

6.3 Exercises

Basic Concepts

Exercises 1–4: The xy-coordinates of the hand for a robotic arm are shown in the figure.

(a) Find the length of the arm.
(b) Find the sine and cosine functions for the angle θ.

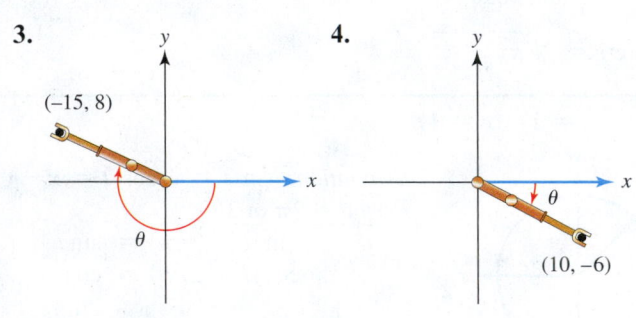

1. (12, 5)

2. (−6, −8)

3. (−15, 8)

4. (10, −6)

Exercises 5–8: Find $\sin\theta$ and $\cos\theta$.

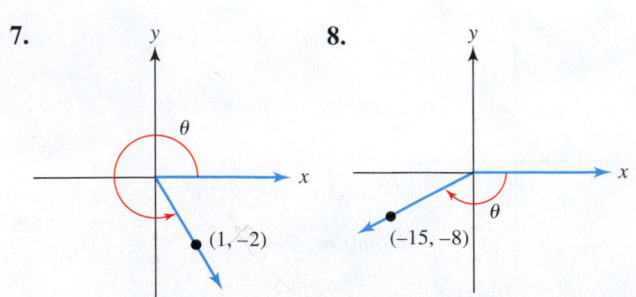

5. (4, 3)

6. (−5, 12)

7. (1, −2)

8. (−15, −8)

Exercises 9–12: The terminal side of an angle θ in standard position lies on the line in the given quadrant. Find the sine and cosine functions of θ. (Hint: Find a point (x, y) lying on the terminal side of θ.)

9. $y = 2x$,   Quadrant I

10. $y = -\frac{1}{2}x$,   Quadrant II

11. $y = -3x$,   Quadrant IV

12. $y = \frac{3}{4}x$,   Quadrant III

Evaluating Sine and Cosine

Exercises 13–28: (Refer to Examples 2 and 3.) Find the sine and cosine functions by hand for the given angle. Then support your answer using a calculator.

13. $45°$ **14.** $150°$

15. $-30°$ **16.** $-180°$

17. $225°$ **18.** $510°$

19. $-420°$ **20.** $-225°$

21. $\frac{\pi}{3}$ **22.** $\frac{5\pi}{4}$

23. $-\frac{\pi}{2}$ **24.** -2π

25. $\frac{7\pi}{6}$ **26.** $\frac{4\pi}{3}$

27. $-\frac{9\pi}{4}$ **28.** $-\frac{19\pi}{6}$

Exercises 29–36: Approximate the sine and cosine of each angle to four decimal places.

29. $93.2°$ **30.** $-43°$

31. $123°50'$ **32.** $12°40'45''$

33. -4 **34.** 1.56

35. $\frac{11\pi}{7}$ **36.** $-\frac{7\pi}{5}$

The Unit Circle

Exercises 37–40: Each figure at the top of the next page shows angle θ in standard position with its terminal side intersecting the unit circle. Evaluate sin θ and cos θ.

37.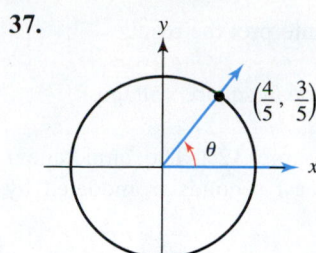

38.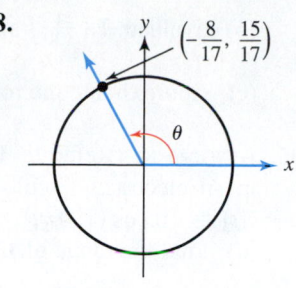

39.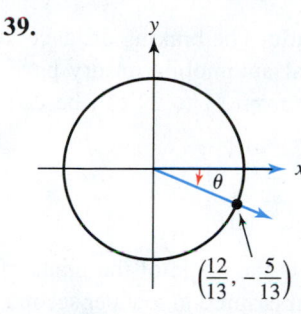

40.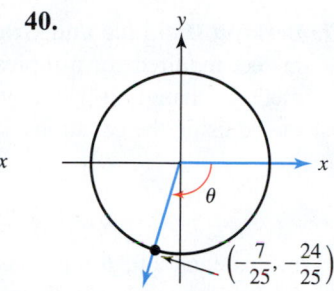

Exercises 41–50: (Refer to Examples 9 and 10.) Use a unit circle to evaluate sin t and cos t by hand.

41. $t = \frac{\pi}{2}$ **42.** $t = \pi$

43. $t = \frac{7\pi}{6}$ **44.** $t = -\frac{\pi}{4}$

45. $t = -\frac{3\pi}{4}$ **46.** $t = \frac{5\pi}{3}$

47. $t = \frac{5\pi}{2}$ **48.** $t = \frac{11\pi}{6}$

49. $t = -\frac{\pi}{3}$ **50.** $t = -\frac{3\pi}{2}$

The Wrapping Function

Exercises 51–66: (Refer to Example 7.) For each number s, evaluate W(s).

51. 3π **52.** 4π

53. -2π **54.** -5π

55. $\frac{3\pi}{2}$ **56.** $\frac{7\pi}{2}$

57. $-\frac{5\pi}{2}$ **58.** $-\frac{9\pi}{2}$

59. $\frac{5\pi}{4}$ **60.** $\frac{7\pi}{4}$

61. $-\frac{5\pi}{4}$ **62.** $-\frac{3\pi}{4}$

63. $\frac{11\pi}{6}$ **64.** $\frac{7\pi}{3}$

65. $-\frac{7\pi}{3}$ **66.** $-\frac{5\pi}{6}$

Exercises 67–82: (Refer to Example 8 and Exercises 51–66.) Use the wrapping function to evaluate sin s and cos s for each real number s.

67. 3π **68.** 4π

69. -2π **70.** -5π

71. $\frac{3\pi}{2}$ **72.** $\frac{7\pi}{2}$

73. $-\frac{5\pi}{2}$ **74.** $-\frac{9\pi}{2}$

75. $\frac{5\pi}{4}$ **76.** $\frac{7\pi}{4}$

77. $-\frac{5\pi}{4}$ **78.** $-\frac{3\pi}{4}$

79. $\frac{11\pi}{6}$ **80.** $\frac{7\pi}{3}$

81. $-\frac{7\pi}{3}$ **82.** $-\frac{5\pi}{6}$

Graphs of Trigonometric Functions

83. Sketch a graph of $y = \sin t$ for $-2\pi \le t \le 2\pi$.

84. Sketch a graph of $y = \cos t$ for $-2\pi \le t \le 2\pi$.

Exercises 85–88: Graph the function f in the viewing rectangle $[-2\pi, 2\pi, \pi/2]$ by $[-4, 4, 1]$. Identify the range of f and then evaluate $f\left(\frac{3\pi}{2}\right)$.

85. **(a)** $f(t) = 3 \sin t$ **(b)** $f(t) = \sin (3t)$

86. **(a)** $f(t) = 2 \cos t$ **(b)** $f(t) = \cos (2t)$

87. **(a)** $f(t) = 2 \cos (t) + 1$ **(b)** $f(t) = \cos (2t) - 1$

88. **(a)** $f(t) = 3 \sin (t) + 1$ **(b)** $f(t) = \sin (3t) - 1$

Applications

89. **Highway Grade** (Refer to Example 4.) Suppose an uphill grade of a highway can be modeled by the line $y = 0.03x$ in the first quadrant.
(a) Find the grade of the hill.

(b) Determine the grade resistance for a gravel truck weighing 25,000 pounds.

90. **Highway Grade** (Refer to Example 4.) A downhill graph is modeled by the line $y = -0.04x$ in the fourth quadrant.
(a) Find the grade of the road.

(b) Determine the grade resistance for a 6000-pound pickup truck. Interpret the result.

91. **Phases of the Moon** (Refer to Example 11.) Find all phase angles θ that correspond to the first quarter phase of the moon. Assume that θ can be any angle measured in radians.

92. **Phases of the Moon** (Refer to Example 11.) Find all phase angles θ that correspond to a full moon and all phase angles that correspond to a new moon. Assume that θ can be any angle measured in radians.

93. Circadian Rhythm A human body has an internal clock called *circadian rhythm* that helps determine when a person is awake and has the most energy. If a person's energy levels are determined by a scale of 0 to 100, then the function

$$C(x) = 50 - 40 \sin\left(\frac{\pi}{12}x\right),$$

where x is hours past midnight, describes a typical circadian rhythm. (*Source:* Hozumi.net)
(a) Graph $y = C(x)$ for $0 \le x \le 24$.

(b) When is a typical person's energy the greatest?

(c) When is a typical person's energy the least?

94. Circadian Rhythm Disorder (Refer to Exercise 93.) Some people experience a disorder of their circadian rhythm. An example is given by

$$C(x) = 40 + 30 \cos\left(\frac{\pi}{12}x\right),$$

where x is hours past midnight.
(a) Graph $y = C(x)$ for $0 \le x \le 24$.

(b) With this disorder, when is a person's energy the greatest?

(c) With this disorder, when is is the person's energy the least?

95. Voltage (Refer to Example 12.) Electric ranges and ovens often use a higher voltage than that found in normal household outlets. This voltage can be modeled by $V(t) = 310 \sin(120\pi t)$, where t represents time in seconds.
(a) A graph of $y = V(t)$ is shown in the given figure. Describe the voltage in the circuit.

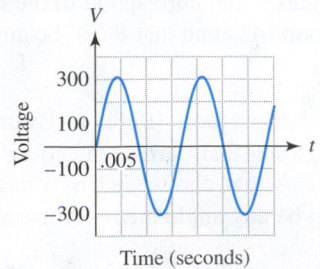

(b) Evaluate $V\left(\frac{1}{120}\right)$ and interpret the result.

(c) Approximate the root mean square voltage.

96. Amperage (Refer to Example 12.) The amperage I in an electrical circuit after t seconds is modeled by $I(t) = 10 \cos(120\pi t)$.
(a) Find the range of I.

(b) Evaluate $I\left(\frac{1}{60}\right)$ and interpret the result.

97. Braking Distance and Grade The braking distance D in feet required for a typical automobile on dry pavement to change its velocity from V_1 to V_2 can be estimated using the equation

$$D = \frac{1.05\left(V_1^2 - V_2^2\right)}{27 + 64.4 \sin\theta}.$$

In this equation, θ represents the angle of the grade of the highway and velocity is measured in feet per second. See Figure 6.58. (*Source:* F. Mannering and W. Kelareski, *Principles of Highway Engineering and Traffic Analysis*.)
(a) Compute the number of feet required to slow a car from 88 feet per second (60 mi/hr) to 44 feet per second (30 mi/hr) while it is traveling uphill with $\theta = 3°$.

(b) Repeat part (a) with $\theta = -3°$.

(c) How is braking distance affected by θ? Does this result agree with your driving experience?

98. Braking Distance (Refer to Exercise 97.) An automobile is traveling at 88 feet per second (60 mi/hr) on a highway with $\theta = -4°$. The driver sees a stalled truck in the road and applies the brakes 250 feet from the vehicle. Assuming that a collision cannot be avoided, how fast is the car traveling when it collides with the truck?

99. Highway Design When an automobile travels along a circular curve with radius r, trees and buildings situated on the inside of the curve can obstruct the driver's vision. See the figure. To ensure a safe stopping distance S, the *minimum* distance d that should be cleared on the inside of a highway curve is given by the equation $d = r\left(1 - \cos\frac{\beta}{2}\right)$, where β in radians is determined by $\beta = \frac{S}{r}$. (*Source:* F. Mannering and W. Kelareski.)

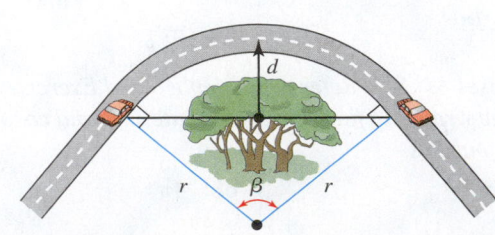

(a) At 45 miles per hour, $S = 390$ feet. If $r = 600$ feet, approximate d.

(b) At 60 miles per hour, $S = 620$ feet. Approximate d for the same curve.

(c) Discuss how the speed limit affects the amount of land that should be cleared on the inside of the curve.

100. **Dead Reckoning** Suppose an airplane flies the path shown in the figure, where distance is measured in hundreds of miles. Approximate the final coordinates of the airplane if its initial coordinates are $(0, 0)$.

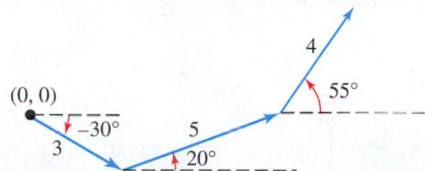

101. **Modeling Biological Rhythms** (Refer to Example 13.) The table shows the time y in the afternoon when a typical flying squirrel becomes active during month x.

x (month)	1	2	3	4
y (P.M.)	4:30	5:15	5:45	6:30

x (month)	5	6	7	8
y (P.M.)	7:00	7:30	7:45	7:30

x (month)	9	10	11	12
y (P.M.)	6:45	6:15	5:00	4:15

Source: J. Harker, The Physiology of Diurnal Rhythms.

(a) Make a scatterplot of the data. (*Hint:* Represent 4:30 P.M. by 4.5.)

(b) The formula $f(x) = 1.9 \sin(0.42x - 1.2) + 5.7$ models the time of sunset, where x represents the month. Graph f and the data together. Interpret the graph.

102. **Modeling Biological Rhythms** (Refer to Example 13.) The table lists the luminescence y of a plant after t hours, where $t = 0$ corresponds to midnight.

t	0	6	12	18	24	30	36	42	48
y	10	6	2	6	10	6	2	6	10

(a) Graph the data and $L(t) = 4 \cos(0.27t) + 6$. How well does the function model the data?

(b) Estimate the luminescence when $t = 15$.

(c) Use the midpoint formula to complete part (b). Are your answers the same? Why?

Writing about Mathematics

103. Discuss whether the sine and cosine functions are linear or nonlinear functions. Use graphical and numerical representations to justify your reasoning.

104. Describe two ways to define the sine and cosine functions. Give examples.

Extended and Discovery Exercises

1. Graph $f(t) = \sin t$ and $g(t) = \cos t$ in the same viewing rectangle. Then translate the graph of f to the left $\frac{\pi}{2}$ units by graphing $y = f\left(t + \frac{\pi}{2}\right)$. How does this translated graph compare to the graph of g?

2. (Continuation of Exercise 1) Translate the graph of $g(t) = \cos t$ to the right $\frac{3\pi}{2}$ units by graphing the function given by $y = g\left(t - \frac{3\pi}{2}\right)$. How does this translated graph compare to the graph of $f(t) = \sin t$?

6.4 Other Trigonometric Functions and Their Graphs

- Learn definitions and basic identities
- Define the other trigonometric functions for any angle and any real number
- Represent other trigonometric functions
- Solve applications

Introduction

Unlike the sine function, the tangent and cotangent functions did not originate with astronomy. Rather, they were developed as part of surveying land, finding heights of objects, and determining time. The cotangent function was computed as early as 1500 B.C. in Egypt, where sundials were used to determine time. Depending on the position of the sun in the sky, a vertical stick casts shadows of different lengths. See Figure 6.78. This is a simple device for evaluating the cotangent function. (*Source:* NCTM, *Historical Topics for the Mathematics Classroom, Thirty-first Yearbook.*)

Modeling Shadow Length

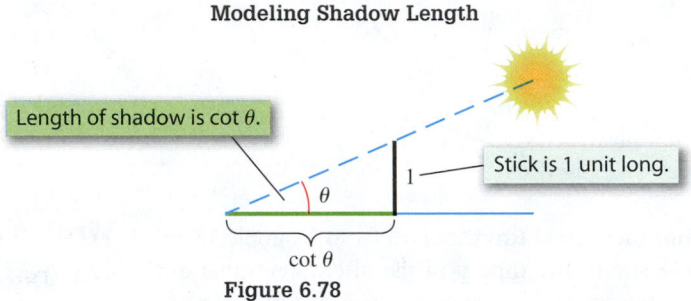

Length of shadow is $\cot\theta$.

Stick is 1 unit long.

$\cot\theta$

Figure 6.78

Definitions

Using Figure 6.79, the other trigonometric functions may be defined for any angle θ in a manner similar to the way sine and cosine functions were defined.

Finding Trig Functions of θ

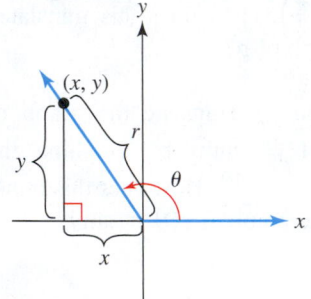

Figure 6.79

TRIGONOMETRIC FUNCTIONS OF ANY ANGLE θ

Let (x, y) be a point other than the origin on the terminal side of an angle θ in standard position. If $r = \sqrt{x^2 + y^2}$, then the six trigonometric functions are as follows.

$$\sin\theta = \frac{y}{r} \qquad\qquad \csc\theta = \frac{r}{y} \ (y \neq 0)$$

$$\cos\theta = \frac{x}{r} \qquad\qquad \sec\theta = \frac{r}{x} \ (x \neq 0)$$

$$\tan\theta = \frac{y}{x} \ (x \neq 0) \qquad \cot\theta = \frac{x}{y} \ (y \neq 0)$$

The domains of both the sine and the cosine functions include all angles. However, the cotangent and cosecant functions are undefined when $y = 0$, which corresponds to angles whose terminal sides lie on the *x-axis*. Examples include $0°$, $\pm 180°$, and $\pm 360°$. The domains of the cotangent and cosecant functions are as follows where n is an integer.

Domain of Cotangent and Cosecant

$$D = \{\theta \,|\, \theta \neq 180° \cdot n\} \qquad \text{or} \qquad D = \{t \,|\, t \neq \pi n\}$$

Degrees

Radians or real numbers

Similarly, the tangent and secant functions are undefined whenever $x = 0$. This corresponds to angles whose terminal sides lie on the *y-axis*. Examples include $\pm 90°$, $\pm 270°$, and $\pm 450°$. The domains of the tangent and secant functions are as follows, where n is an integer.

Domain of Tangent and Secant

$$D = \{\theta \mid \theta \neq 90° + 180° \cdot n\} \qquad \text{or} \qquad D = \left\{t \mid t \neq \tfrac{\pi}{2} + \pi n\right\}$$

Degrees   Radians or real numbers

In some situations we can evaluate the six trigonometric functions without a calculator, as illustrated in the next example.

EXAMPLE 1   **Finding values of trigonometric functions**

The point $(5, -12)$ is located on the terminal side of an angle θ in standard position. Find the six trigonometric functions of θ.

SOLUTION There are many coterminal angles that have the point $(5, -12)$ located on their terminal sides. However, the *trigonometric values for coterminal angles are equal.* In Figure 6.80 one possibility for θ is shown. Begin by calculating r.

$$r = \sqrt{x^2 + y^2} = \sqrt{5^2 + (-12)^2} = 13$$

Since $x = 5, y = -12$, and $r = 13$, the values of the six trigonometric functions are as follows.

$$\sin\theta = \frac{y}{r} = -\frac{12}{13} \qquad \csc\theta = \frac{r}{y} = -\frac{13}{12}$$

$$\cos\theta = \frac{x}{r} = \frac{5}{13} \qquad \sec\theta = \frac{r}{x} = \frac{13}{5}$$

$$\tan\theta = \frac{y}{x} = -\frac{12}{5} \qquad \cot\theta = \frac{x}{y} = -\frac{5}{12}$$

Now Try Exercise 1

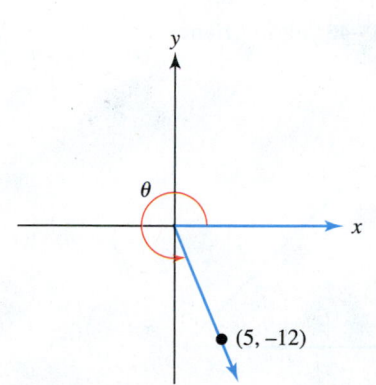

Figure 6.80

NOTE If an angle is a multiple of $30°$ or $45°$ $\left(\frac{\pi}{6} \text{ or } \frac{\pi}{4} \text{ radians}\right)$, then we can determine its six trigonometric functions by hand.

EXAMPLE 2   **Evaluating the trigonometric functions**

Find the six trigonometric functions for each angle by hand.
(a) $330°$ **(b)** $-\frac{5\pi}{4}$

SOLUTION
Getting Started First sketch the given angle in standard position and draw a perpendicular line segment from the terminal side to the *x*-axis. Choose or calculate values for x, y, or r. Use these values to determine each trigonometric function. ▶
(a) An angle of $330°$ is sketched in Figure 6.81 on the next page, with a perpendicular line segment drawn to the *x*-axis. A $30°$–$60°$ right triangle is formed. See also Figure 6.82 on the next page. From Figure 6.81 we can see that $x = \sqrt{3}, y = -1$, and $r = 2$.

Finding *x*, *y*, and *r* for 330°

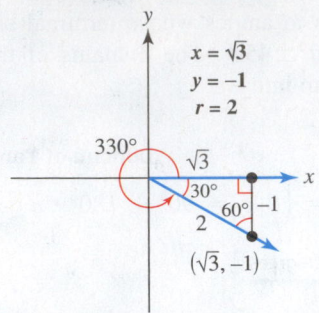

A 30°–60° Right Triangle

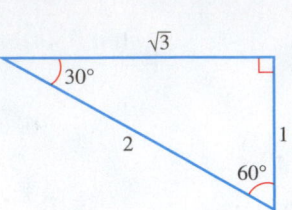

Figure 6.81   **Figure 6.82**

$$\sin 330° = \frac{y}{r} = -\frac{1}{2} \qquad \csc 330° = \frac{r}{y} = -\frac{2}{1} = -2$$

$$\cos 330° = \frac{x}{r} = \frac{\sqrt{3}}{2} \qquad \sec 330° = \frac{r}{x} = \frac{2}{\sqrt{3}}$$

$$\tan 330° = \frac{y}{x} = -\frac{1}{\sqrt{3}} \qquad \cot 330° = \frac{x}{y} = \frac{\sqrt{3}}{-1} = -\sqrt{3}$$

(b) An angle of $-\frac{5\pi}{4}$ (radians) is sketched in Figure 6.83, with a perpendicular line segment drawn to the *x*-axis. A 45°–45° right triangle is formed. See also Figure 6.84. From Figure 6.83 we can see that $x = -1$, $y = 1$, and $r = \sqrt{2}$.

Finding *x*, *y*, and *r* for $-\frac{5\pi}{4}$

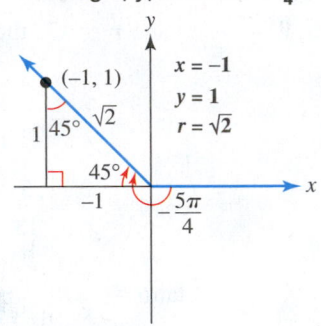

A 45°–45° Right Triangle

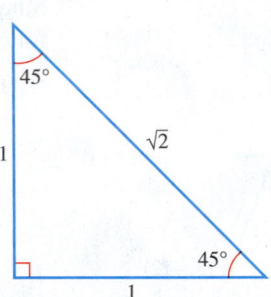

Figure 6.83   **Figure 6.84**

$$\sin\left(-\frac{5\pi}{4}\right) = \frac{y}{r} = \frac{1}{\sqrt{2}} \qquad \csc\left(-\frac{5\pi}{4}\right) = \frac{r}{y} = \frac{\sqrt{2}}{1} = \sqrt{2}$$

$$\cos\left(-\frac{5\pi}{4}\right) = \frac{x}{r} = -\frac{1}{\sqrt{2}} \qquad \sec\left(-\frac{5\pi}{4}\right) = \frac{r}{x} = -\frac{\sqrt{2}}{1} = -\sqrt{2}$$

$$\tan\left(-\frac{5\pi}{4}\right) = \frac{y}{x} = -\frac{1}{1} = -1 \qquad \cot\left(-\frac{5\pi}{4}\right) = \frac{x}{y} = -\frac{1}{1} = -1$$

> **Now Try Exercises 7 and 17**

Some Basic Trigonometric Identities

Since $\sin\theta = \frac{y}{r}$ and $\csc\theta = \frac{r}{y}$, it follows that $\sin\theta$ and $\csc\theta$ are reciprocals. This may be expressed using either of the *reciprocal identities*,

$$\sin\theta = \frac{1}{\csc\theta} \qquad \text{or} \qquad \csc\theta = \frac{1}{\sin\theta}.$$

For angle θ in Example 1, it was shown that $\sin \theta = -\frac{12}{13}$. It follows that

$$\csc \theta = \frac{1}{-12/13} = -\frac{13}{12}.$$

Identities are equations that are true for all meaningful values of a variable. Reciprocal identities also hold for $\tan \theta$ and $\cot \theta$ as well as $\cos \theta$ and $\sec \theta$.

RECIPROCAL IDENTITIES

$$\sin \theta = \frac{1}{\csc \theta} \qquad \cos \theta = \frac{1}{\sec \theta} \qquad \tan \theta = \frac{1}{\cot \theta}$$

$$\csc \theta = \frac{1}{\sin \theta} \qquad \sec \theta = \frac{1}{\cos \theta} \qquad \cot \theta = \frac{1}{\tan \theta}$$

The expressions $\cot \theta$ and $\tan \theta$ can be written in terms of $\sin \theta$ and $\cos \theta$. For example,

$$\cot \theta = \frac{x}{y} = \frac{x/r}{y/r} = \frac{\cos \theta}{\sin \theta}.$$

A *quotient identity* can be obtained for $\tan \theta$ similarly.

QUOTIENT IDENTITIES

$$\tan \theta = \frac{\sin \theta}{\cos \theta} \qquad \cot \theta = \frac{\cos \theta}{\sin \theta}$$

EXAMPLE 3   **Using identities**

If $\sin \theta = \frac{3}{5}$ and $\cos \theta = -\frac{4}{5}$, find the other four trigonometric functions of θ.

SOLUTION   To find the other four trigonometric functions, apply the quotient and reciprocal identities.

Quotient identities

$$\tan \theta = \frac{\sin \theta}{\cos \theta} = \frac{3/5}{-4/5} = -\frac{3}{4}$$

$$\cot \theta = \frac{\cos \theta}{\sin \theta} = \frac{-4/5}{3/5} = -\frac{4}{3}$$

Reciprocal identities

$$\sec \theta = \frac{1}{\cos \theta} = \frac{1}{-4/5} = -\frac{5}{4}$$

$$\csc \theta = \frac{1}{\sin \theta} = \frac{1}{3/5} = \frac{5}{3}$$

Now Try Exercise 29

In Section 6.3 we showed that $\sin^2 \theta + \cos^2 \theta = 1$ for all real numbers (or angles) θ. In the next example, we apply this identity to find the values of the trigonometric functions.

EXAMPLE 4 Applying the identity $\sin^2\theta + \cos^2\theta = 1$

If $\sin\theta = \frac{4}{5}$ and $\cos\theta < 0$, find the values of the other five trigonometric functions.

SOLUTION The identity $\sin^2\theta + \cos^2\theta = 1$ can be used to determine $\cos\theta$.

$$\sin^2\theta + \cos^2\theta = 1 \qquad \text{Identity}$$

Solve for $\cos\theta$

$$\cos^2\theta = 1 - \sin^2\theta \qquad \text{Subtract } \sin^2\theta.$$

$$\cos\theta = \pm\sqrt{1 - \sin^2\theta} \qquad \text{Square root property}$$

$$\cos\theta = \pm\sqrt{1 - \left(\frac{4}{5}\right)^2} \qquad \text{Let } \sin\theta = \frac{4}{5}.$$

$$\cos\theta = \pm\sqrt{\frac{9}{25}} \qquad \text{Simplify.}$$

$$\cos\theta = -\frac{3}{5} \qquad \cos\theta < 0$$

Now we can use $\sin\theta = \frac{4}{5}$ and $\cos\theta = -\frac{3}{5}$ to determine the other four trigonometric functions.

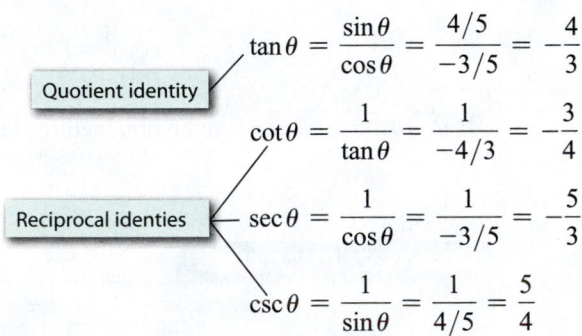

Quotient identity

$$\tan\theta = \frac{\sin\theta}{\cos\theta} = \frac{4/5}{-3/5} = -\frac{4}{3}$$

$$\cot\theta = \frac{1}{\tan\theta} = \frac{1}{-4/3} = -\frac{3}{4}$$

Reciprocal identies

$$\sec\theta = \frac{1}{\cos\theta} = \frac{1}{-3/5} = -\frac{5}{3}$$

$$\csc\theta = \frac{1}{\sin\theta} = \frac{1}{4/5} = \frac{5}{4}$$

Now Try Exercise 35

Unit Circle Evaluation Suppose the terminal side of an angle θ in standard position intersects the unit circle at the point (x, y). See Figure 6.85. In this case $r = 1$ and the definitions of the six trigonometric functions become as follows.

$$\sin\theta = y \qquad \cos\theta = x \qquad \tan\theta = \frac{y}{x}$$

$$\csc\theta = \frac{1}{y} \qquad \sec\theta = \frac{1}{x} \qquad \cot\theta = \frac{x}{y}$$

As with the sine and cosine functions, evaluating the other trigonometric functions for a real number t is equivalent to evaluating these functions for t radians.

Unit Circle ($r = 1$)

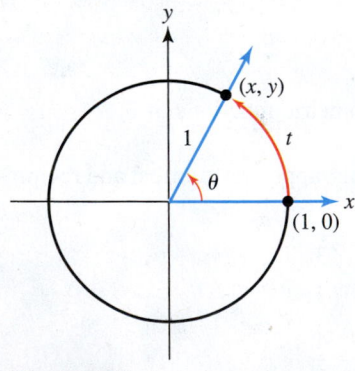

Figure 6.85

The Wrapping Function (Optional)

The six trigonometric functions can be evaluated for real numbers by using the wrapping function presented in Section 6.3. For any real number s, if $W(s) = (x, y)$, then the six trigonometric functions can be defined as follows.

$$\sin s = y \qquad \cos s = x \qquad \tan s = \frac{y}{x}$$

$$\csc s = \frac{1}{y} \qquad \sec s = \frac{1}{x} \qquad \cot s = \frac{x}{y}$$

The next example illustrates how the wrapping function can be used to evaluate the six trigonometric functions.

EXAMPLE 5   **Applying the wrapping function**

For $s = -5\pi$, evaluate $W(s)$. If possible, find the six trigonometric functions of s.

SOLUTION The circumference of the unit circle is 2π. A string of length 5π wraps (clockwise) around the unit circle two and a half times. Because the initial point is always $(1, 0)$, the terminal point is $(-1, 0)$, as shown in Figure 6.86. Thus $W(-5\pi) = (-1, 0)$.

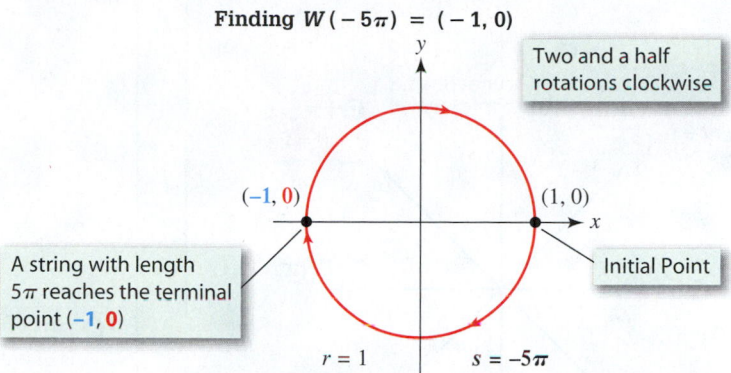

Finding $W(-5\pi) = (-1, 0)$

Figure 6.86

The six trigonometric functions can be evaluated by using $x = -1$ and $y = 0$. Note that the cosecant and cotangent are undefined because $y = 0$.

$$\sin(-5\pi) = y = 0 \qquad\qquad \csc(-5\pi) = \frac{1}{y} \text{ (undefined)}$$

$$\cos(-5\pi) = x = -1 \qquad\qquad \sec(-5\pi) = \frac{1}{x} = \frac{1}{-1} = -1$$

$$\tan(-5\pi) = \frac{y}{x} = \frac{0}{-1} = 0 \qquad\qquad \cot(-5\pi) = \frac{x}{y} \text{ (undefined)}$$

Now Try Exercise 39

Representations of Other Trigonometric Functions

Like $\sin t$ and $\cos t$, the other four trigonometric functions cannot be evaluated with simple formulas. However, these functions do have numerical and graphical representations. We begin by discussing the tangent function.

The Tangent Function Figure 6.87 shows angle θ in *standard position* with its terminal side passing through the point (x, y). The slope of this terminal side is

$$m = \frac{y - 0}{x - 0} = \frac{y}{x} = \tan\theta.$$

It can be shown that the slope of the terminal side of θ equals $\tan\theta$, regardless of the quadrant containing θ. See Figure 6.88.

$$\tan\theta = \frac{y}{x} = m$$

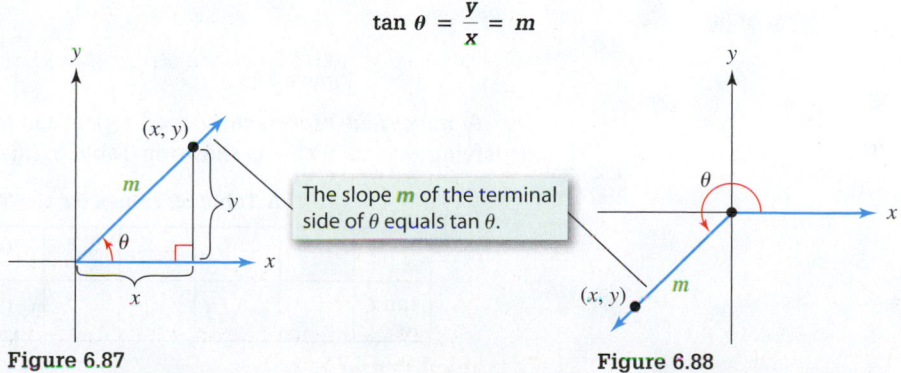

The slope m of the terminal side of θ equals $\tan\theta$.

Figure 6.87   **Figure 6.88**

The following See the Concept uses the fact that $m = \tan \theta$ to derive the graph of $y = \tan \theta$.

See the Concept: The Graph of $y = \tan \theta$.

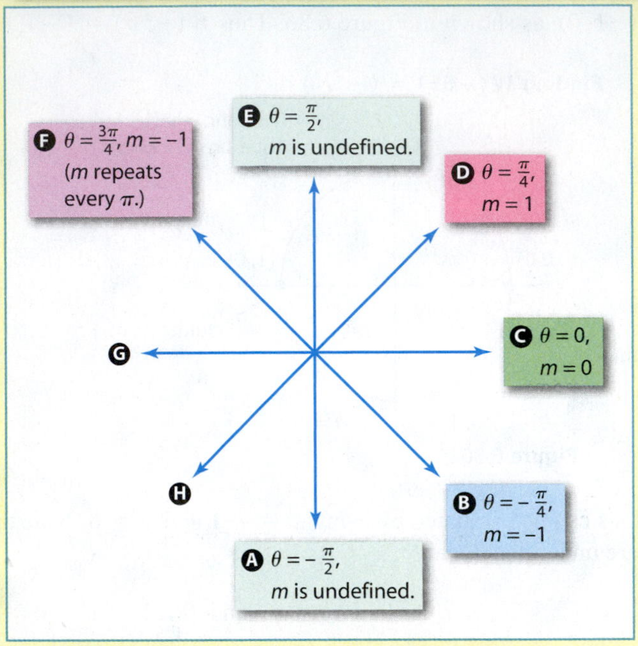

F $\theta = \frac{3\pi}{4}, m = -1$
(m repeats every π.)

E $\theta = \frac{\pi}{2}$, m is undefined.

D $\theta = \frac{\pi}{4}$, $m = 1$

C $\theta = 0$, $m = 0$

G

H

B $\theta = -\frac{\pi}{4}$, $m = -1$

A $\theta = -\frac{\pi}{2}$, m is undefined.

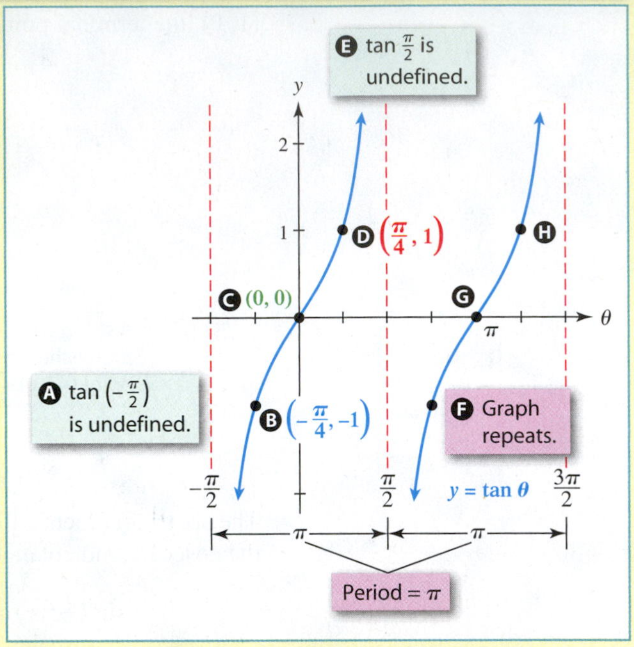

E $\tan \frac{\pi}{2}$ is undefined.

D $\left(\frac{\pi}{4}, 1\right)$

H

C $(0, 0)$

G

A $\tan\left(-\frac{\pi}{2}\right)$ is undefined.

B $\left(-\frac{\pi}{4}, -1\right)$

F Graph repeats.

$y = \tan \theta$

Period $= \pi$

A–H The slope m of the terminal side of θ (left figure) equals $\tan \theta$ (right figure). Thus all points on the graph of $y = \tan \theta$ are of the form (θ, m).

D The terminal side of $\theta = \frac{\pi}{4}$ at **D** in the left figure has slope $m = 1$, so the graph on the right passes through the point $\left(\frac{\pi}{4}, 1\right)$ at **D**.

A graph of the tangent function is shown in Figure 6.89 with period π and vertical asymptotes at $t = \pm\frac{\pi}{2}, \pm\frac{3\pi}{2}, \pm\frac{5\pi}{2}, \dots$. The domain of the tangent function is $D = \left\{t \mid t \neq \frac{\pi}{2} + \pi n\right\}$. The domain excludes x-values where vertical asymptotes occur.

The Tangent Function

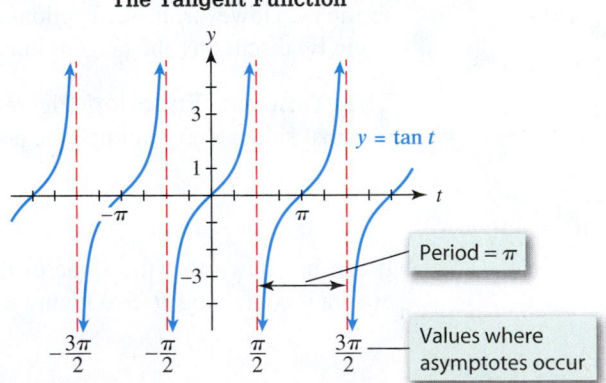

$y = \tan t$

Period $= \pi$

Values where asymptotes occur

Figure 6.89

A *numerical representation* of $f(t) = \tan t$ for selected (real number) values of t satisfying $-\frac{\pi}{2} \leq t \leq \frac{\pi}{2}$ is shown in Table 6.7. (A dash indicates that $\tan t$ is undefined.)

A Table of Values for the Tangent Function

t	$-\frac{\pi}{2}$	$-\frac{\pi}{3}$	$-\frac{\pi}{4}$	$-\frac{\pi}{6}$	0	$\frac{\pi}{6}$	$\frac{\pi}{4}$	$\frac{\pi}{3}$	$\frac{\pi}{2}$
$\tan t$	—	$-\sqrt{3}$	-1	$-\frac{1}{\sqrt{3}}$	0	$\frac{1}{\sqrt{3}}$	1	$\sqrt{3}$	—

Table 6.7

EXAMPLE 6   **Evaluating the tangent function**

Evaluate $f(t) = \tan t$ at $t = -\frac{\pi}{4}$ by hand.

SOLUTION   If angle $-\frac{\pi}{4}$ is in standard position, the terminal side intersects the unit circle at the point $\left(\frac{1}{\sqrt{2}}, -\frac{1}{\sqrt{2}} \right)$. See Figure 6.90. Therefore

$$\tan\left(-\frac{\pi}{4} \right) = \frac{y}{x} = \frac{-1/\sqrt{2}}{1/\sqrt{2}} = -1$$

Note that the 45°–45° right triangle in Figure 6.91 can be used as an aid in determining the length of the sides of the triangle in Figure 6.90. Also note that the slope of the terminal side of the angle is -1 and equals $\tan\left(-\frac{\pi}{4} \right)$.

Now Try Exercise 51

The Cosecant Function   The cosecant function can also be evaluated by hand. For example, if $t = \frac{7\pi}{6}$, then the terminal side of $\theta = \frac{7\pi}{6}$ intersects the unit circle at the point $\left(-\frac{\sqrt{3}}{2}, -\frac{1}{2} \right)$. See Figure 6.92. It follows that

$$\csc \frac{7\pi}{6} = \frac{1}{y} = \frac{1}{-1/2} = -2.$$

Note that the 30°–60° right triangle in Figure 6.93 can be used as an aid in determining the length of the sides of the triangle in Figure 6.92.

A *numerical representation* of $f(t) = \csc t$ is shown in Table 6.8 for selected values of t satisfying $-\pi \le t \le \pi$. Note that $\csc t = \frac{1}{\sin t}$. (A dash indicates that $\csc t$ is undefined.)

t	$-\pi$	$-\frac{3\pi}{4}$	$-\frac{\pi}{2}$	$-\frac{\pi}{4}$	0	$\frac{\pi}{4}$	$\frac{\pi}{2}$	$\frac{3\pi}{4}$	π
$\csc t$	—	$-\sqrt{2}$	-1	$-\sqrt{2}$	—	$\sqrt{2}$	1	$\sqrt{2}$	—

Table 6.8

A graph of $\csc t$ is given in Figure 6.94. The domain of the cosecant function is $D = \{t \mid t \ne \pi n\}$, where n is an integer and vertical asymptotes occur whenever $t = \pi n$. The period of the cosecant function equals 2π.

Since $\csc t = \frac{1}{\sin t}$, we can sketch the graph of $y = \csc t$ by using the graph of $y = \sin t$ as an aid. See Figure 6.95. When $\sin t = 0$, $\csc t$ is undefined and a vertical asymptote occurs on the graph of $\csc t$. When $\sin t = \pm 1$, $\csc t = \pm 1$. Whenever $\sin t$ increases, $\csc t$ decreases, and whenever $\sin t$ decreases, $\csc t$ increases. Because $|\sin t| \le 1$ for all t, it follows that $|\csc t| \ge 1$ for all $t \ne \pi n$.

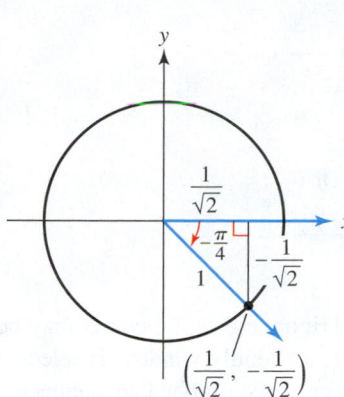

Figure 6.90

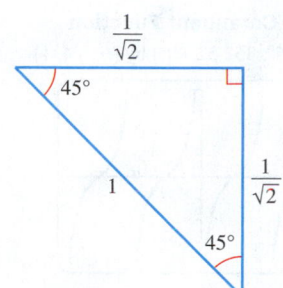

Figure 6.91

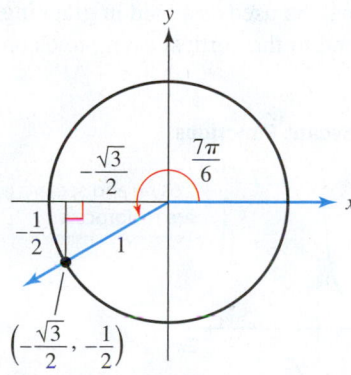

Figure 6.92

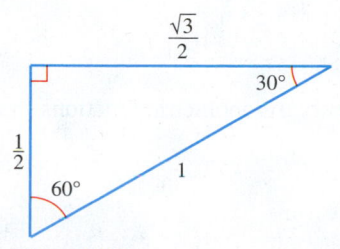

Figure 6.93

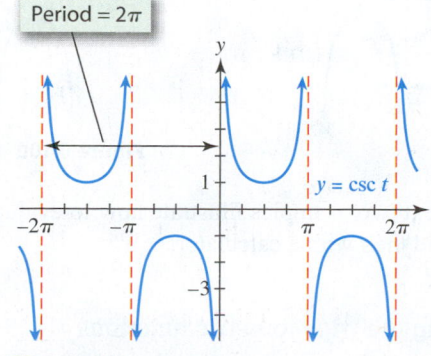

Figure 6.94

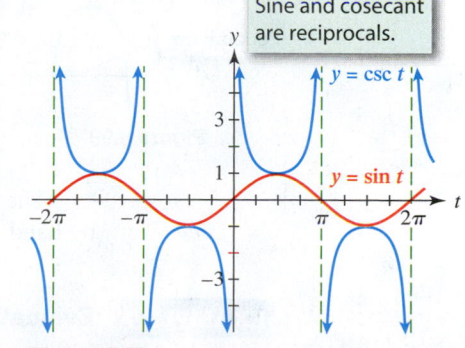

Figure 6.95

The Cotangent Function   A graph of $y = \cot t$ is shown in Figure 6.96 on the next page. Its period is π, and its domain is $D = \{t \mid t \ne \pi n\}$, where n is an integer. Vertical asymptotes occur at $t = \pi n$.

The Cotangent Function

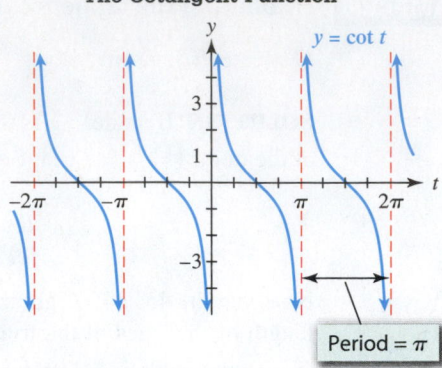

Figure 6.96

Friendly Windows in Degree Mode (Optional) Trigonometric functions may be represented graphically in either radian or degree mode. If a friendly window is selected using degree mode, a trigonometric function may be accurately graphed in connected mode. Graphs of the cosecant and cotangent functions are shown in Figures 6.97 and 6.98.

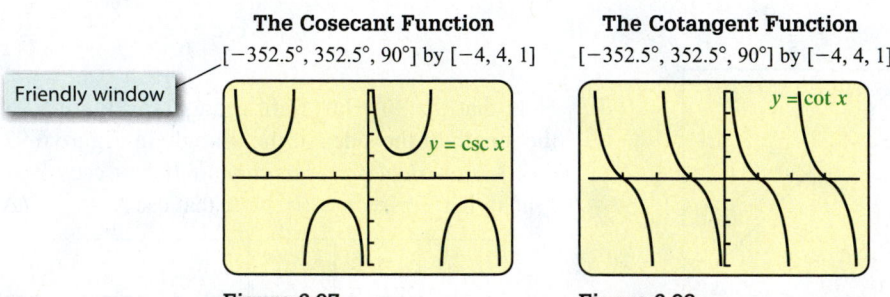

Figure 6.97   **Figure 6.98**

The Secant Function A graph of $y = \sec t$ is shown in Figure 6.99. Its period is 2π, and its domain is $D = \left\{t \mid t \neq \frac{\pi}{2} + \pi n\right\}$, where n is an integer. Vertical asymptotes occur at $t = \frac{\pi}{2} + \pi n$. The graph of the cosine function may be used as an aid in graphing the secant function, since the zeros of $y = \cos t$ correspond to the vertical asymptotes on the graph of $y = \sec t$. See Figure 6.100.

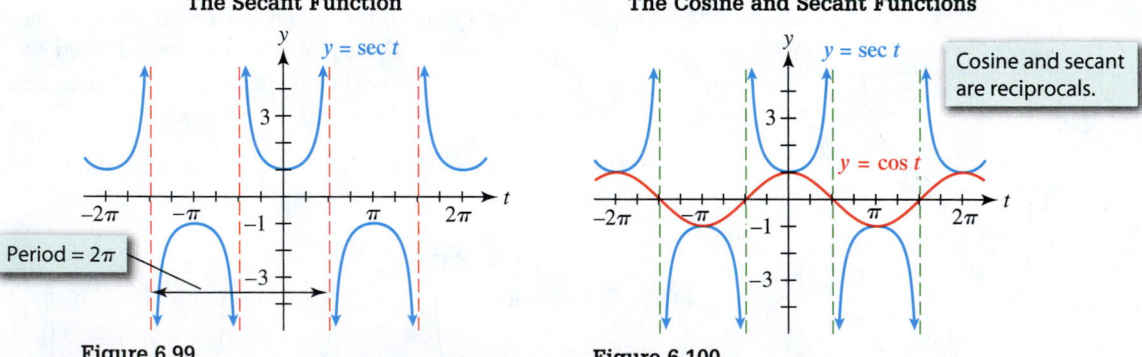

Figure 6.99   **Figure 6.100**

The next two examples illustrate how to evaluate the six trigonometric functions first by hand and then with a calculator.

EXAMPLE 7   **Evaluating the trigonometric functions**

If $\theta = \frac{3\pi}{2}$, find the six trigonometric functions of θ.

SOLUTION If $\theta = \frac{3\pi}{2}$ is in standard position, then its terminal side intersects the unit circle at the point $(0, -1)$. See Figure 6.101. Thus $x = 0$ and $y = -1$. The values of the six trigonometric functions are as follows.

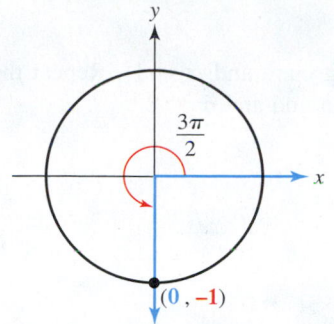

Figure 6.101

$$\sin \frac{3\pi}{2} = y = -1 \qquad\qquad \csc \frac{3\pi}{2} = \frac{1}{y} = \frac{1}{-1} = -1$$

$$\cos \frac{3\pi}{2} = x = 0 \qquad\qquad \sec \frac{3\pi}{2} = \frac{1}{x} \text{ is undefined, since } x = 0$$

$$\tan \frac{3\pi}{2} = \frac{y}{x} \text{ is undefined, since } x = 0 \qquad \cot \frac{3\pi}{2} = \frac{x}{y} = \frac{0}{-1} = 0$$

> Now Try Exercise 57

EXAMPLE 8 Using a calculator to evaluate trigonometric functions

Find values of the six trigonometric functions of θ.
(a) $\theta = 102.6°$ **(b)** $\theta = 2.56$

SOLUTION

(a) Most calculators do not have special keys for secant, cosecant, and cotangent. To evaluate these functions, we use the reciprocal identities. For example, to find sec(102.6°), evaluate $1/\cos(102.6°)$. The values of the trigonometric functions are shown in Figures 6.102 and 6.103.

(b) Since there is no degree symbol, use radian mode. With the aid of a calculator, the values of the six trigonometric functions are approximated as follows.

$$\sin 2.56 \approx 0.5494 \qquad \cos 2.56 \approx -0.8356 \qquad \tan 2.56 \approx -0.6574$$

$$\csc 2.56 \approx 1.8203 \qquad \sec 2.56 \approx -1.1968 \qquad \cot 2.56 \approx -1.5210$$

> Now Try Exercises 63 and 67

```
sin(102.6)
          .9759167619
cos(102.6)
          -.2181432414
tan(102.6)
          -4.473742829
```

Figure 6.102 Degree Mode

Applying Reciprocal Identities

```
1/sin(102.6)
          1.024677553
1/cos(102.6)
          -4.584143857
1/tan(102.6)
          -.2235264829
```

Figure 6.103 Degree Mode

Applications of Trigonometric Functions

Highway Design A *sag curve* occurs when a highway goes downhill and then uphill. Improperly designed sag curves can be dangerous at night because a vehicle's headlights point downward and may not illuminate the uphill portion of the highway, as illustrated in Figure 6.104. The minimum safe length L for a typical sag curve with a 40-mile-per-hour speed limit is given by the following.

$$L = \frac{2700}{h + 3 \tan \alpha}$$

> The tangent function is often used when designing highways.

The variable h represents the height of the headlights above the road surface, and α represents a small (upward) angle associated with the vertical alignment of the headlight, shown in Figure 6.104. (***Source:*** F. Mannering and W. Kilareski, *Principles of Highway Engineering and Traffic Analysis*.)

Designing a Safe Sag Curve

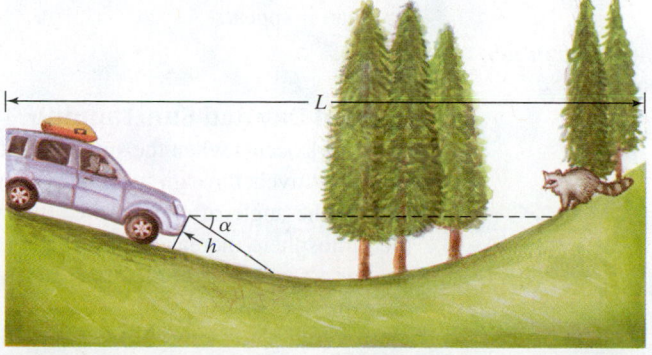

Figure 6.104 (Not to scale)

EXAMPLE 9 Designing a sag curve

Calculate L for a car with headlights 2.5 feet above the ground and $\alpha = 1°$. Repeat the calculations for a truck with headlights 4 feet above the ground and $\alpha = 2°$.

SOLUTION For the car, let $h = 2.5$ and $\alpha = 1°$.

$$L = \frac{2700}{2.5 + 3 \tan 1°} \approx 1058 \text{ feet} \qquad L = \frac{2700}{h + 3 \tan \alpha}$$

For the truck, let $h = 4$ and $\alpha = 2°$.

$$L = \frac{2700}{4 + 3 \tan 2°} \approx 658 \text{ feet} \qquad L = \frac{2700}{h + 3 \tan \alpha}$$

(Since both cars and trucks use the same highways, engineers typically use the larger of the two distances to design a safe sag curve.)

Now Try Exercise 89

Bending of Starlight When a stick is put partially into a glass of water, it appears to bend at the surface of the water. A similar phenomenon occurs when starlight enters Earth's atmosphere. To an observer on the ground, a star's apparent position α is different from its true position. This phenomenon is referred to as *atmospheric refraction*. See Figure 6.105 below. The amount that starlight is bent is given by angle θ measured in seconds, where $\theta = 57.3 \tan \alpha$ with $0° \leq \alpha \leq 45°$. (**Source:** W. Schlosser, *Challenges of Astronomy*.)

Atmospheric Refraction

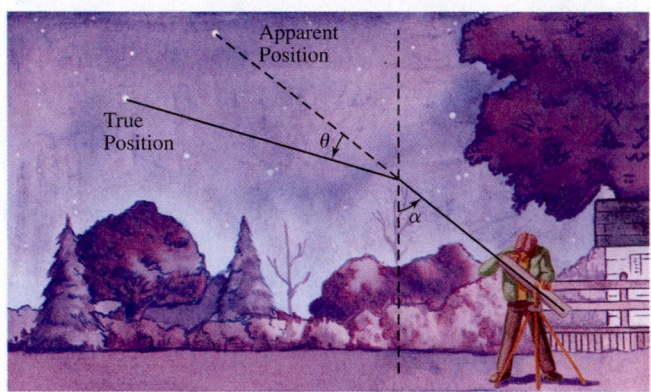

Figure 6.105 (Not to scale)

EXAMPLE 10 Calculating refraction of starlight

Use Figure 6.105 and the formula $\theta = 57.3 \tan \alpha$ to calculate θ when $\alpha = 32°35'21''$.

SOLUTION $\theta = 57.3 \tan (32°35'21'') \approx 36.6''$. The star is actually $36.6''$ lower in the sky than it appears.

Now Try Exercise 91

Sun Traveling Through the Atmosphere

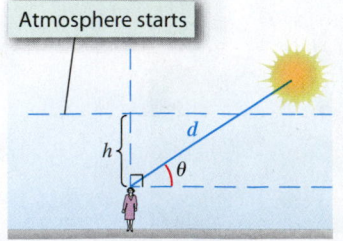

Figure 6.106 (Not to scale)

Time of Day and Sun Tanning The shortest path through Earth's atmosphere for the sun's rays occurs when the sun is directly overhead. As the sun moves lower in the horizon, sunlight travels through more atmosphere. See Figure 6.106, where $d = h \csc \theta$. *Explain why.* If the angle of elevation of the sun is θ, then the path length d of sunlight through the atmosphere increases by a factor of $\csc \theta$ from its noontime value of h. (Because of the curvature of Earth, this model is not accurate when $\theta < 20°$ and the sun is positioned near the horizon. See Exercise 95.) When sunlight passes through more atmosphere, less ultraviolet light reaches Earth's surface. This is one reason why some experts recommend sun tanning either earlier or later in the day. (**Source:** C. Winter, *Solar Power Plants.*)

EXAMPLE 11   **Measuring the intensity of the sun**

Assuming that the sun is directly overhead at noon, calculate the percent increase between the atmospheric distance that sunlight passes through at noon and at 10:00 A.M.

SOLUTION   From Figure 6.106, $\theta = 90°$ at noon and $d = h \csc \theta$. The apparent position of the sun changes 360° in the sky in 24 hours, or 15° per hour. Therefore 2 hours earlier $\theta = 60°$. At noon,

$$d = h \csc 90° = \frac{h}{\sin 90°} = \frac{h}{1} = h$$

> Apply reciprocal identity.

and at 10:00 A.M.,

$$d = h \csc 60° = \frac{h}{\sin 60°} = \frac{h}{\sqrt{3}/2} \approx 1.15h.$$

This means that sunlight travels through about 15% more atmosphere at 10:00 A.M. than at noon.

> **Now Try Exercise 93**

Putting It All Together

$\mathbf{T}$he following table summarizes the domain, range, and period of each trigonometric function. In this table n is an integer and t is a real number.

FUNCTION	DOMAIN	RANGE	PERIOD
$\sin t$	$-\infty < t < \infty$	$-1 \leq \sin t \leq 1$	2π
$\cos t$	$-\infty < t < \infty$	$-1 \leq \cos t \leq 1$	2π
$\tan t$	$t \neq \frac{\pi}{2} + \pi n$	$-\infty < \tan t < \infty$	π
$\cot t$	$t \neq \pi n$	$-\infty < \cot t < \infty$	π
$\sec t$	$t \neq \frac{\pi}{2} + \pi n$	$|\sec t| \geq 1$	2π
$\csc t$	$t \neq \pi n$	$|\csc t| \geq 1$	2π

GRAPHS OF TRIGONOMETRIC FUNCTIONS

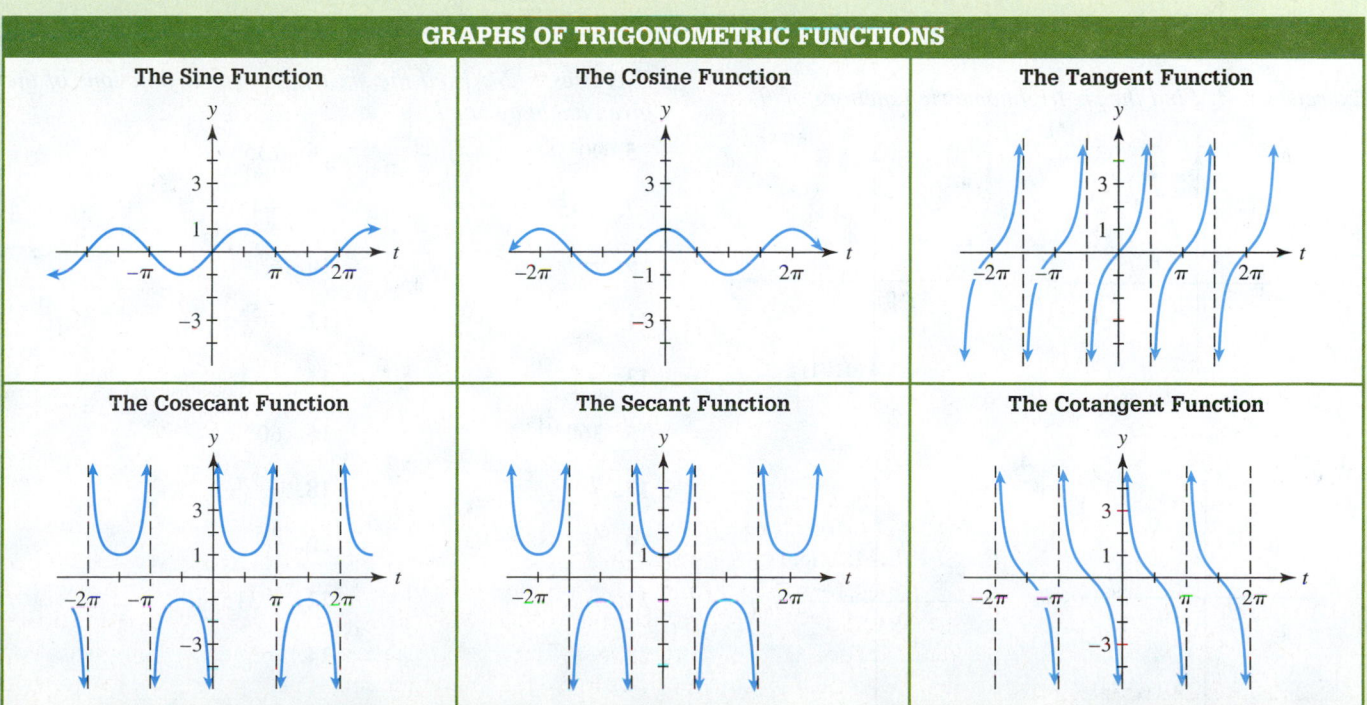

The Sine Function   The Cosine Function   The Tangent Function

The Cosecant Function   The Secant Function   The Cotangent Function

continued on next page

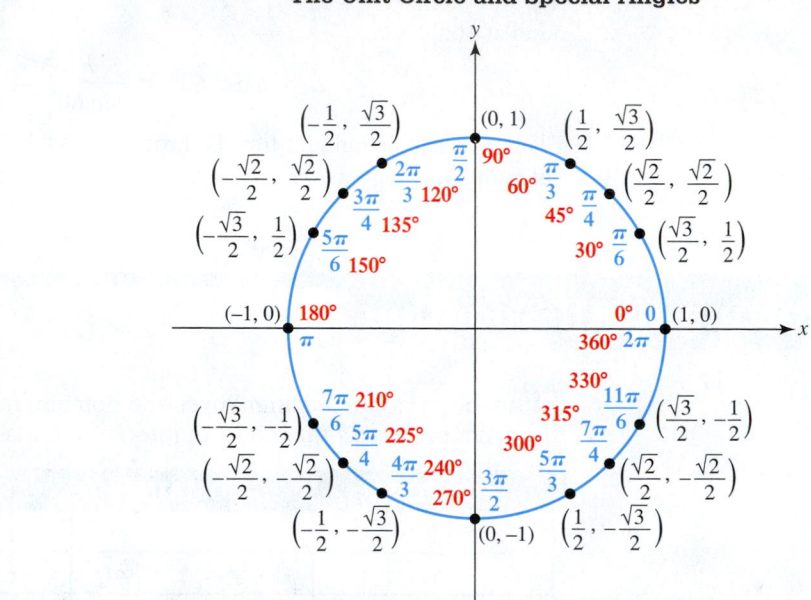

EVALUATING TRIGONOMETRIC FUNCTIONS

The following figure can be used to evaluate the trigonometric functions for certain angles. For example, if $\theta = \mathbf{120°}$ (or $\frac{2\pi}{3}$ radians) is in standard position, then its terminal side intersects the unit circle at the point $\left(-\frac{1}{2}, \frac{\sqrt{3}}{2}\right)$. It follows that $\cos 120° = -\frac{1}{2}$ and $\sin 120° = \frac{\sqrt{3}}{2}$. Other trigonometric values can also be found using these values. Note that $\frac{\sqrt{2}}{2} = \frac{1}{\sqrt{2}}$.

The Unit Circle and Special Angles

6.4 Exercises

Basic Concepts

Exercises 1–4: Find the six trigonometric functions of θ.

1.

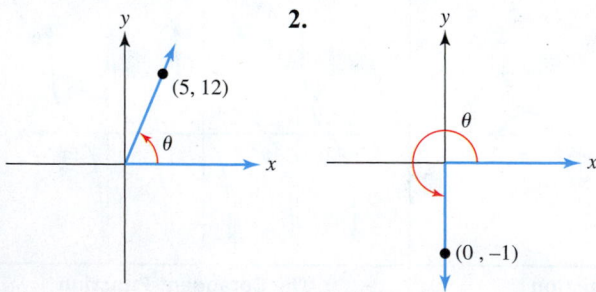

2.

Exercises 5–24: Find the six trigonometric functions of the given angle by hand.

5. 90°

6. 135°

7. −45°

8. −180°

9. π

10. $\frac{3\pi}{4}$

11. $-\frac{\pi}{3}$

12. $-\frac{5\pi}{6}$

13. $-\frac{\pi}{2}$

14. $\frac{3\pi}{2}$

15. 360°

16. 60°

3.

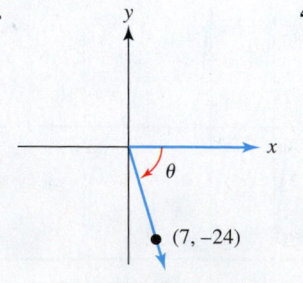

4.

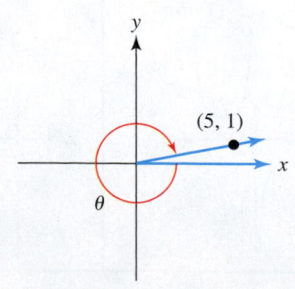

17. $\frac{\pi}{6}$

18. $-\frac{2\pi}{3}$

19. $\frac{4\pi}{3}$

20. $\frac{2\pi}{3}$

21. −225°

22. −315°

23. $-\frac{13\pi}{6}$

24. $-\frac{7\pi}{6}$

Exercises 25–28: The terminal side of an angle θ lies on the line in the given quadrant. Find the six trigonometric functions of θ. How does the slope of the line compare to tan θ?

25. $y = -4x$, Quadrant II **26.** $y = \frac{1}{2}x$, Quadrant I

27. $y = 6x$, Quadrant III **28.** $y = -\frac{2}{3}x$, Quadrant IV

Identities

Exercises 29–38: Determine the values of the trigonometric functions of θ by using the given information.

29. $\sin\theta = \frac{3}{5}$ and $\cos\theta = \frac{4}{5}$

30. $\sin\theta = -\frac{7}{25}$ and $\cos\theta = \frac{24}{25}$

31. $\csc\theta = -\frac{17}{15}$ and $\sec\theta = -\frac{17}{8}$

32. $\csc\theta = 2$ and $\sec\theta = -\frac{2}{\sqrt{3}}$

33. $\tan\theta = \frac{5}{12}$ and $\cos\theta = \frac{12}{13}$ (*Hint:* $\sin\theta = \tan\theta\cos\theta$.)

34. $\sin\theta = \frac{3}{5}$ and $\cot\theta = -\frac{4}{3}$

35. $\sin\theta = -\frac{3}{5}$ and $\cos\theta > 0$

36. $\sin\theta = -\frac{12}{13}$ and $\cos\theta < 0$

37. $\cos\theta = -\frac{4}{5}$ and $\sin\theta < 0$

38. $\cos\theta = \frac{7}{25}$ and $\sin\theta > 0$

Wrapping Function

Exercises 39–46: Use the wrapping function to determine the six trigonometric functions of the given number.

39. -7π **40.** 4π

41. $\frac{7\pi}{2}$ **42.** $-\frac{3\pi}{2}$

43. $-\frac{3\pi}{4}$ **44.** $\frac{7\pi}{4}$

45. $\frac{7\pi}{6}$ **46.** $-\frac{11\pi}{6}$

Unit Circle Evaluation

Exercises 47–50: The figure shows angle θ in standard position with its terminal side intersecting the unit circle. Evaluate the six trigonometric functions of θ.

47.

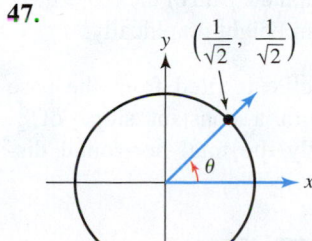

$\left(\frac{1}{\sqrt{2}}, \frac{1}{\sqrt{2}}\right)$

48.

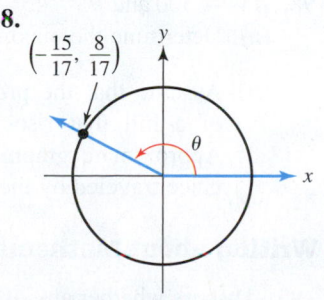

$\left(-\frac{15}{17}, \frac{8}{17}\right)$

49.

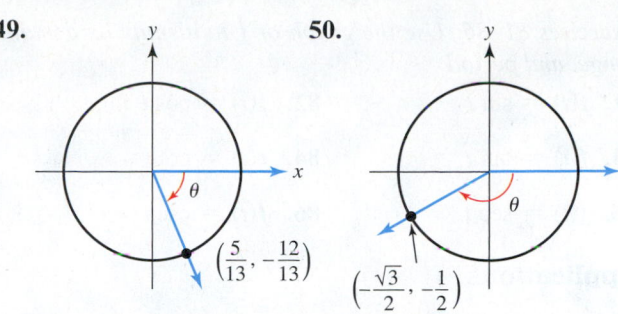

$\left(\frac{5}{13}, -\frac{12}{13}\right)$

50.

$\left(-\frac{\sqrt{3}}{2}, -\frac{1}{2}\right)$

Exercises 51–56: Evaluate the trigonometric function by using the unit circle.

51. $\tan\left(-\frac{5\pi}{4}\right)$ **52.** $\tan\left(-\frac{7\pi}{4}\right)$

53. $\csc\frac{\pi}{6}$ **54.** $\csc\frac{7\pi}{6}$

55. $\sec\frac{\pi}{6}$ **56.** $\cot\frac{7\pi}{4}$

Exercises 57–62: Evaluate the six trigonometric functions of the given number by using the unit circle.

57. $\frac{9\pi}{2}$ **58.** $-\frac{9\pi}{4}$ **59.** $-\pi$

60. 7π **61.** $\frac{4\pi}{3}$ **62.** $\frac{5\pi}{6}$

Approximations

Exercises 63–70: Approximate to four decimal places.

63. (a) $\sin 93.2°$ (b) $\csc 93.2°$

64. (a) $\cos(-43°)$ (b) $\sec(-43°)$

65. (a) $\tan 234°33'$ (b) $\cot 234°33'$

66. (a) $\sec 123°44'25''$ (b) $\cos 123°44'25''$

67. (a) $\cot(-4)$ (b) $\tan(-4)$

68. (a) $\csc 1.56$ (b) $\sin 1.56$

69. (a) $\cos\left(\frac{11\pi}{7}\right)$ (b) $\sec\left(\frac{11\pi}{7}\right)$

70. (a) $\tan\left(\frac{7\pi}{5}\right)$ (b) $\cot\left(\frac{7\pi}{5}\right)$

Exercises 71–74: **Average Rate of Change** *Approximate, to the nearest thousandth, the average rate of change of f from 0 to $\frac{\pi}{4}$.*

71. $f(x) = \sin x$ **72.** $f(x) = \cos x$

73. $f(x) = \tan\frac{1}{2}x$ **74.** $f(x) = \sin 2x$

Graphs

Exercises 75–80: Graph $y = f(t)$. Discuss the symmetry of the graph.

75. $f(t) = \sin t$ **76.** $f(t) = \cos t$

77. $f(t) = \tan t$ **78.** $f(t) = \cot t$

79. $f(t) = \sec t$ **80.** $f(t) = \csc t$

Exercises 81–86: Use the graph of f to identify its domain, range, and period.

81. $f(t) = \sin t$

82. $f(t) = \cos t$

83. $f(t) = \tan t$

84. $f(t) = \cot t$

85. $f(t) = \sec t$

86. $f(t) = \csc t$

Applications

87. Shadow Length The introduction showed how shadows can be used to compute the cotangent function.

(a) Graph $f(t) = \cot t$ for $0 \le t \le \pi$.

(b) Let $t = 0$ correspond to sunrise, $t = \frac{\pi}{2}$ to noon, and $t = \pi$ to sunset. Explain how the graph of f models the length of a shadow cast by a vertical stick with length 1.

88. Shadow Length (Refer to Figure 6.78.) Calculate the shadow length of a 2-foot stick when the angle of elevation of the sun is $\theta = 27°31'$.

89. Highway Design (Refer to Example 9.) Calculate the minimum length L of a typical sag curve on a highway with a 40-mile-per-hour speed limit for a car with headlights 2 feet above the ground and alignment set at $\alpha = 1.5°$.

90. Highway Design Repeat Exercise 89 for a truck with headlights 3.5 feet above the ground and alignment set at $\alpha = 2.5°$.

91. Refraction (Refer to Example 10.) Use the formula $\theta = 57.3 \tan \alpha$ to calculate the refraction θ in seconds when $\alpha = 17°23'43''$. Use an identity to rewrite the formula in terms of $\sin \alpha$ and $\cos \alpha$.

92. Refraction (Refer to Example 10.) Use the formula $\theta = 57.3 \tan \alpha$ to calculate the refraction θ in seconds when $\alpha = 5°15'50''$. Use an identity to rewrite the formula in terms of $\cot \alpha$.

93. Intensity of the Sun (Refer to Example 11.) If the sun is directly overhead at noon, calculate the percent increase between the atmospheric distance that sunlight must pass through at noon and at 3:00 P.M.

94. Intensity of the Sun (Refer to Example 11.) If the sun is directly overhead at 1:00 P.M., calculate the percent increase between the atmospheric distance that sunlight must pass through at 1:00 P.M. and at 2:30 P.M.

95. Intensity of the Sun (Refer to Example 11.) The formula $y_1 = \csc \theta = \frac{1}{\sin \theta}$, presented in Example 11 to calculate the path length of sunlight through the atmosphere relative to the path length at noon, is not accurate when the sun's elevation is less than 20°. A more accurate formula for small values of θ is given by

$$y_2 = \frac{1}{\sin \theta + 0.5(6° + \theta)^{-1.64}},$$

where θ is measured in degrees. Make a table of y_1 and y_2 starting at $\theta = 2°$ and incrementing by 1°. How do the values of y_1 and y_2 compare as θ increases? (**Source:** C. Winter, *Solar Power Plants.*)

96. GPS Satellite Communication Artificial satellites that orbit Earth often use VHF signals to communicate with the ground. Because VHF signals travel in straight lines, a satellite orbiting Earth can communicate with a fixed location on the ground only during certain times. The height h in miles of an orbit with communication time T is given by

$$h = 3955 (\sec (\pi T/P) - 1),$$

where P is the period for the satellite to orbit Earth. Suppose a GPS satellite orbit has a period of $P = 12$ hours and can communicate with a person at the North Pole for $T = 5.08$ hours during each orbit. Approximate the height h of its orbit. (**Sources:** W. Schlosser, *Challenges of Astronomy.*)

Exercises 97 and 98: **Projectile Flight** *If a projectile is fired with an initial velocity of v feet per second at an angle θ with the horizontal, it will follow a parabolic path described by $y = \frac{-16x^2}{v^2 (\cos \theta)^2} + x \tan \theta$. See the figure.*

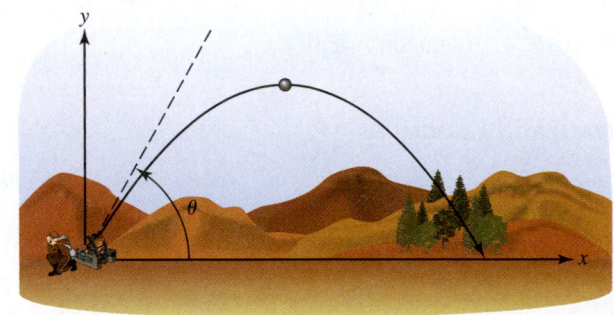

97. If $v = 750$ and $\theta = 30°$, graph the path of the projectile.

(a) Approximate the coordinates when the maximum height occurs.

(b) Assuming the ground is flat, find the total horizontal distance traveled by the projectile graphically.

98. If $v = 500$ and $\theta = 45°$, graph the path of the projectile.

(a) Determine the maximum height graphically.

(b) Assume that the projectile is fired from the base of a hill that rises with a constant slope of $\frac{1}{4}$. Approximate graphically the total horizontal distance traveled by the projectile.

Writing about Mathematics

99. Discuss whether any of the six trigonometric functions are linear functions. Justify your reasoning.

100. If α and β are coterminal angles, what can be said about the six trigonometric functions of α and β? Explain your answer using an example.

Extended and Discovery Exercise

1. The figure shows the unit circle and an acute angle θ in standard position. Explain why the trigonometric functions $\sin \theta$, $\cos \theta$, $\tan \theta$, and $\sec \theta$ are equal to the lengths of the line segments shown.

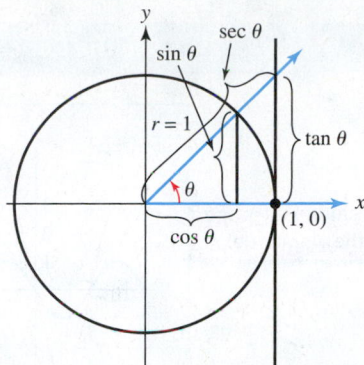

CHECKING BASIC CONCEPTS FOR SECTIONS 6.3 AND 6.4

1. Find the six trigonometric functions of θ if θ is in standard position and its terminal side passes through the point $(-7, 6)$.

2. Evaluate $\sin 45°$ by hand. Check your results with a calculator.

3. Sketch a graph of the sine, cosine, and tangent functions.

4. Evaluate the six trigonometric functions by hand at the given real number. Support your results using a calculator.
(a) $-\pi$ **(b)** $\frac{3\pi}{4}$ **(c)** $\frac{7\pi}{6}$

6.5 Graphing Trigonometric Functions

- **Learn basic transformations of trigonometric graphs**
- **Graph trigonometric functions by hand**
- **Understand simple harmonic motion**
- **Model real data with trigonometric functions (optional)**

Introduction

Trigonometric graphs can be used to model periodic data. For example, monthly average temperatures and precipitation are usually periodic. They can vary dramatically during a twelve-month period, but tend to be periodic from one year to the next. Tides are also periodic. By performing basic transformations of graphs we can model a variety of phenomena. See Example 8.

Transformations of Trigonometric Graphs

We begin by discussing how the constants a and b affect the graphs of

$$y = a \sin bx \quad \text{and} \quad y = a \cos bx.$$

The Constant a and Amplitude The constant a controls the *amplitude* of a sinusoidal wave. If $y = 3 \sin x$, then the amplitude is $|a| = 3$. The graph of $f(x) = \sin x$ oscillates between -1 and 1, whereas the graph of $g(x) = 3 \sin x$ oscillates between -3 and 3. These concepts are shown in See the Concept and Figure 6.107 on the next page. When the constant a is negative, the graphs are reflected across the x-axis. See Figure 6.108.

> **CLASS DISCUSSION**
>
> Predict how the graph of the given equation will compare with the graph of $f(x) = \cos x$. Check your prediction by graphing both f and the equation.
>
> **i.** $y = 2 \cos x$ **ii.** $y = -\frac{1}{2}\cos x$ **iii.** $y = -3 \cos x$

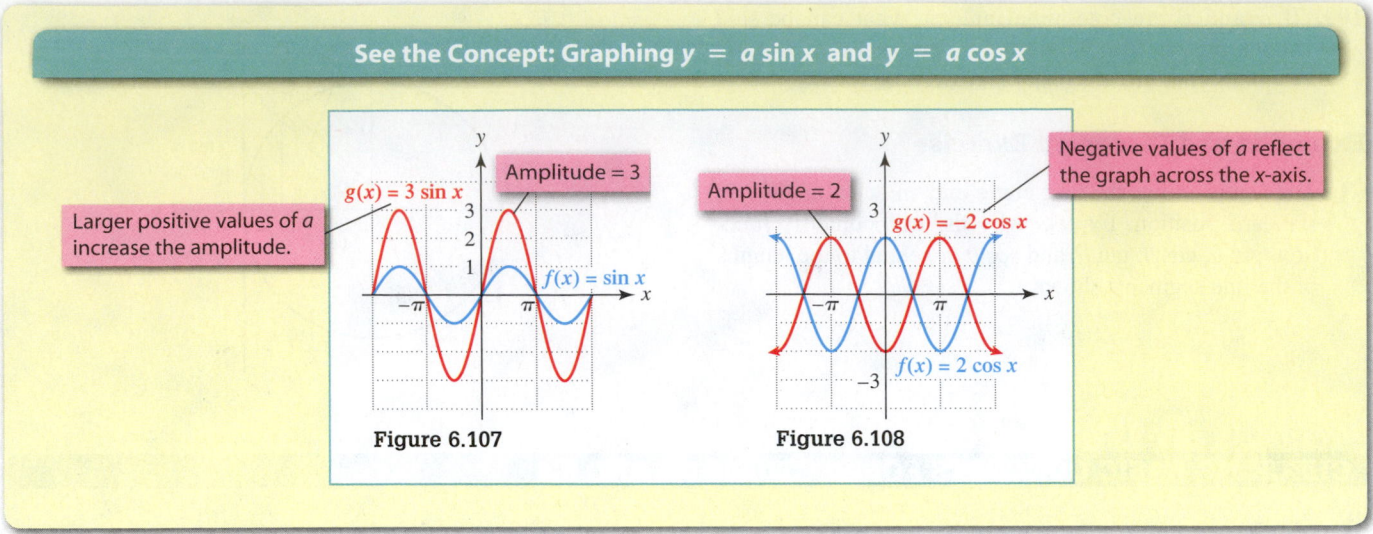

See the Concept: Graphing $y = a \sin x$ and $y = a \cos x$

Larger positive values of a increase the amplitude.

Amplitude = 3

$g(x) = 3 \sin x$

$f(x) = \sin x$

Figure 6.107

Amplitude = 2

$g(x) = -2 \cos x$

Negative values of a reflect the graph across the x-axis.

$f(x) = 2 \cos x$

Figure 6.108

The Constant b and Oscillations The constant b controls the number of oscillations in each interval of length 2π. For example, if $b = 2$, then there are two complete oscillations in every interval of length 2π. As a result, the graph repeats every π units and the period of both $y = \sin 2x$ and $y = \cos 2x$ is π. The *period P* of a sinusoidal graph can be computed by using the formula

$$P = \frac{2\pi}{b}, \qquad \text{Period formula}$$

where $b > 0$. See Figures 6.109 and 6.110.

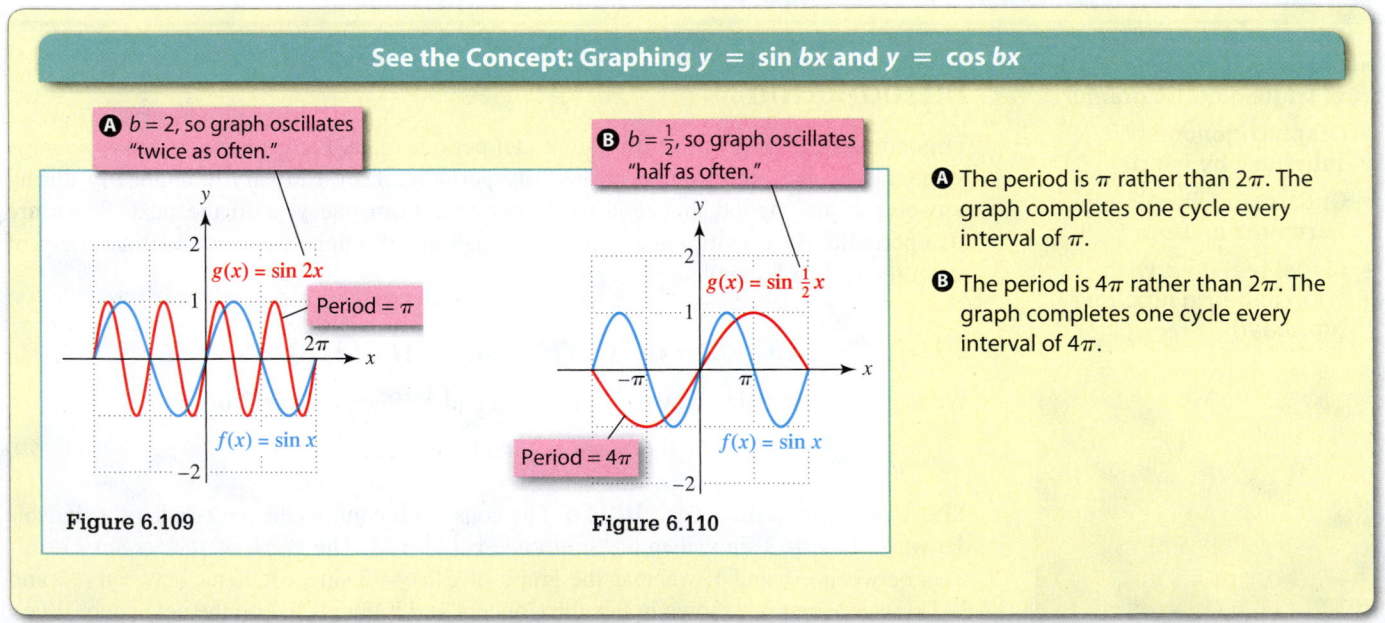

See the Concept: Graphing $y = \sin bx$ and $y = \cos bx$

A $b = 2$, so graph oscillates "twice as often."

$g(x) = \sin 2x$

Period = π

$f(x) = \sin x$

Figure 6.109

B $b = \frac{1}{2}$, so graph oscillates "half as often."

$g(x) = \sin \frac{1}{2}x$

$f(x) = \sin x$

Period = 4π

Figure 6.110

A The period is π rather than 2π. The graph completes one cycle every interval of π.

B The period is 4π rather than 2π. The graph completes one cycle every interval of 4π.

Stretching and Shrinking The preceding discussion can be understood in terms of stretching and shrinking. When $a > 1$, the graph of $y = a \sin x$ is *stretched* vertically compared to $y = \sin x$. When $0 < a < 1$, the graph of $y = a \sin x$ is vertically *shrunk* compared to $y = \sin x$. Similarly, when $b > 1$, the graph of $y = \sin bx$ is horizontally *shrunk* compared to the graph of $y = \sin x$. When $0 < b < 1$, the graph of $y = \sin bx$ is horizontally *stretched* compared to the graph of $y = \sin x$. Similar statements can be made for the graph of $y = a \cos x$ and $y = \cos bx$.

EXAMPLE 1 Sketching graphs of trigonometric functions

Sketch a graph of each equation. Identify the period and amplitude.

(a) $y = 3 \cos \frac{1}{2}x$ (b) $y = -2 \sin 3x$

SOLUTION

(a) The graph of $y = 3 \cos \frac{1}{2}x$ is similar to that of $y = \cos x$, except that its period P is

$$P = \frac{2\pi}{b} = \frac{2\pi}{1/2} = 4\pi$$

rather than 2π and its amplitude is **3** rather than 1. A graph of $y = 3 \cos \frac{1}{2}x$ is shown in Figure 6.111, and a graph $y = \cos x$ is shown in Figure 6.112 for comparison.

Transformed Graph

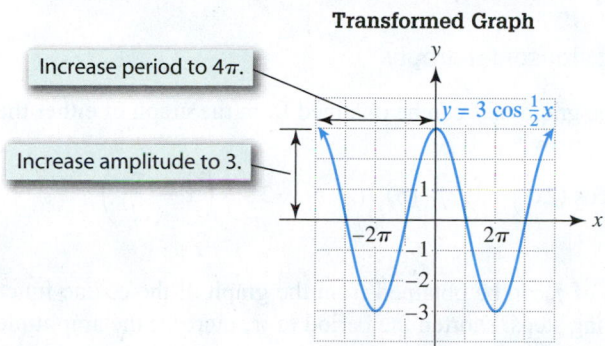

Figure 6.111

Graph of Cosine

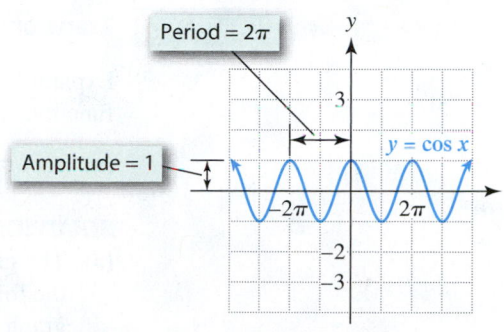

Figure 6.112

(b) The graph of $y = \sin x$ is shown in Figure 6.113. By comparison, the graph of $y = \sin 3x$, shown in Figure 6.114 has period

$$P = \frac{2\pi}{b} = \frac{2\pi}{3}.$$

It oscillates **three** times every interval of length 2π, whereas the graph of $y = \sin x$ oscillates once every interval of length 2π. The graph of $y = -2 \sin 3x$ is similar to the graph of $y = \sin 3x$, except that its amplitude is **2** rather than 1 and its graph is **reflected** across the x-axis. Figure 6.115 shows the required graph of $y = -2 \sin 3x$.

Graph of Sine

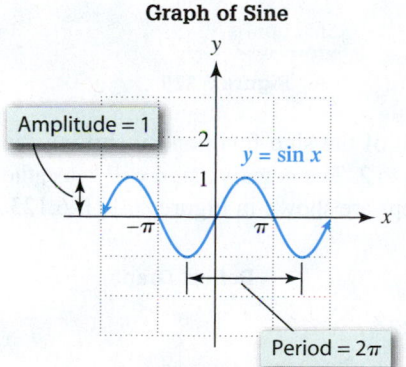

Figure 6.113

Oscillates "Three Times as Often"

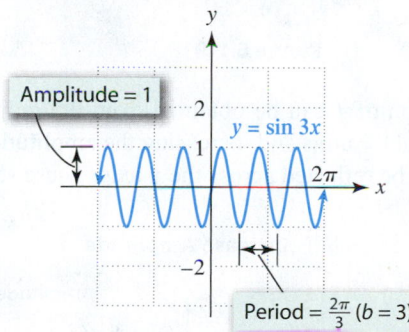

Figure 6.114

Final Transformed Graph

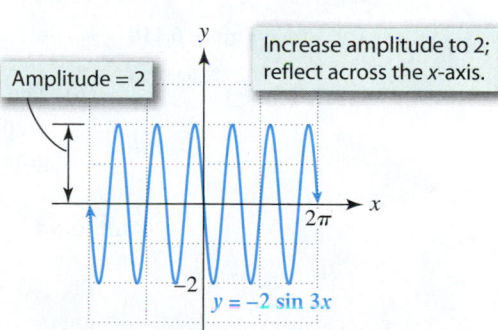

Figure 6.115

Now Try Exercises 1 and 3

Vertical and Horizontal Shifts Trigonometric graphs can be shifted vertically or horizontally. For example, to shift the graph of $f(x) = \sin x$ **upward 1** unit, graph $g(x) = \sin(x) + 1$, as shown in Figure 6.116 on the next page. Similarly, to shift the graph of $f(x) = \sin x$ right $\frac{\pi}{2}$ units, graph the equation $g(x) = \sin\left(x - \frac{\pi}{2}\right)$, as shown in Figure 6.117 on the next page. A horizontal shift of a trigonometric graph is called a *phase shift*.

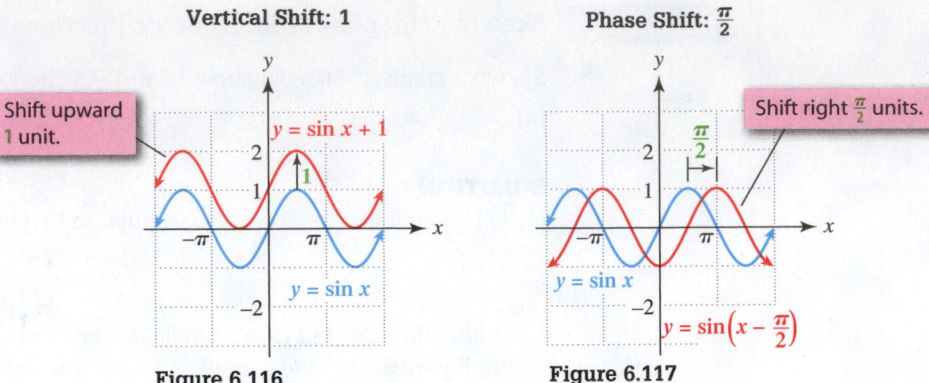

Figure 6.116

Figure 6.117

Transforming sinusoidal graphs

Explain how the graph of f can be obtained from the graph of either the sine or the cosine function. Then graph f.

(a) $f(x) = 3 \cos (2x) + 1$ **(b)** $f(x) = -2 \sin \left(x - \dfrac{\pi}{2} \right)$

SOLUTION

(a) The graph of f can be obtained from the graph of the cosine function by performing the following steps: shorten the period to π, increase the amplitude to 3, and shift the graph upward 1 unit. These three steps are shown in Figures 6.118–6.120.

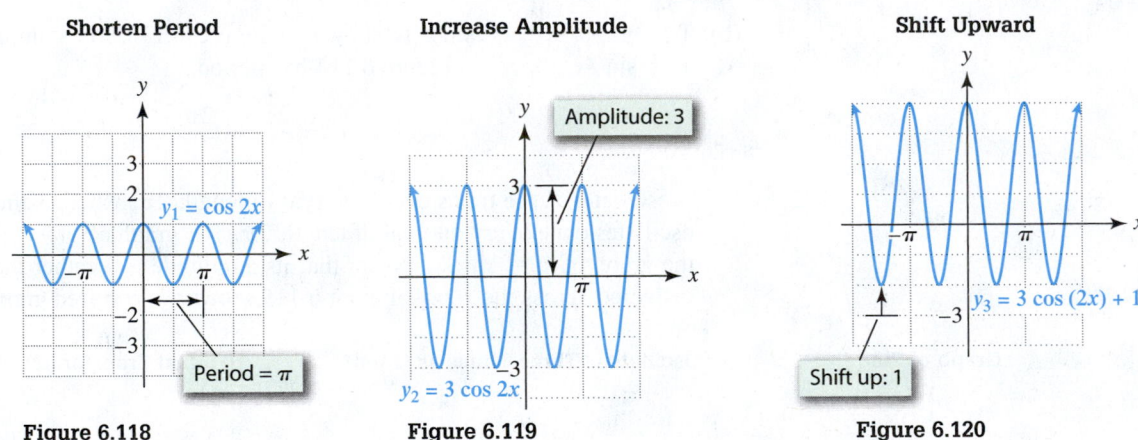

Figure 6.118   **Figure 6.119**   **Figure 6.120**

(b) The graph of f can be obtained from the graph of the sine function by shifting the graph right $\dfrac{\pi}{2}$ units and increasing the amplitude to 2. The negative sign will cause the graph to be reflected across the x-axis. These steps are shown in Figures 6.121–6.123.

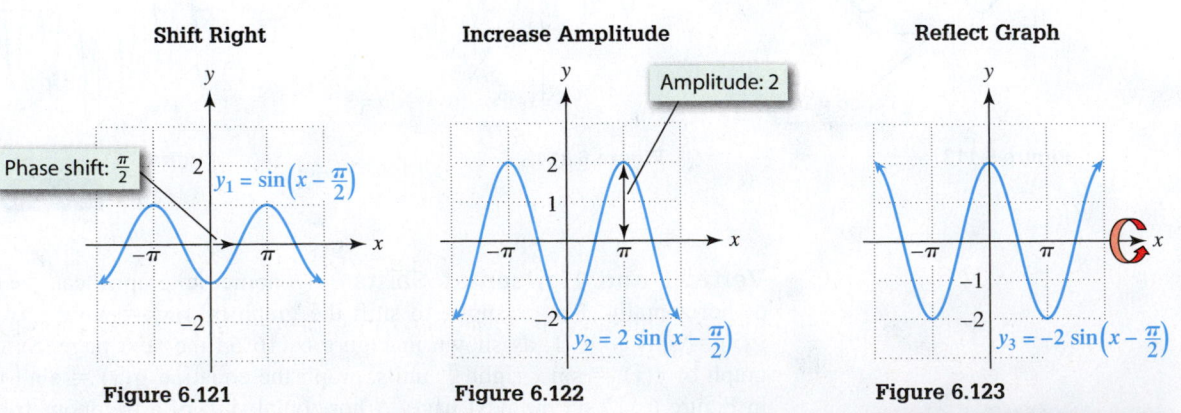

Figure 6.121   **Figure 6.122**   **Figure 6.123**

Now Try Exercises 7 and 9

The preceding discussion is summarized in the following box.

AMPLITUDE, PERIOD, PHASE SHIFT, AND VERTICAL SHIFT

The **amplitude**, **period**, **phase shift**, and **vertical shift** for the graphs of

$$y = a \sin (b(x - c)) + d \quad \text{and} \quad y = a \cos (b(x - c)) + d$$

with $b > 0$ may be determined as follows.

$$\text{Amplitude} = |a|, \quad \text{Period} = \frac{2\pi}{b}, \quad \text{Phase shift} = c, \quad \text{Vertical shift} = d$$

The vertical shift is $|d|$ units upward when $d > 0$ and $|d|$ units downward when $d < 0$.

The following See the Concept illustrates how to transform the sine graph.

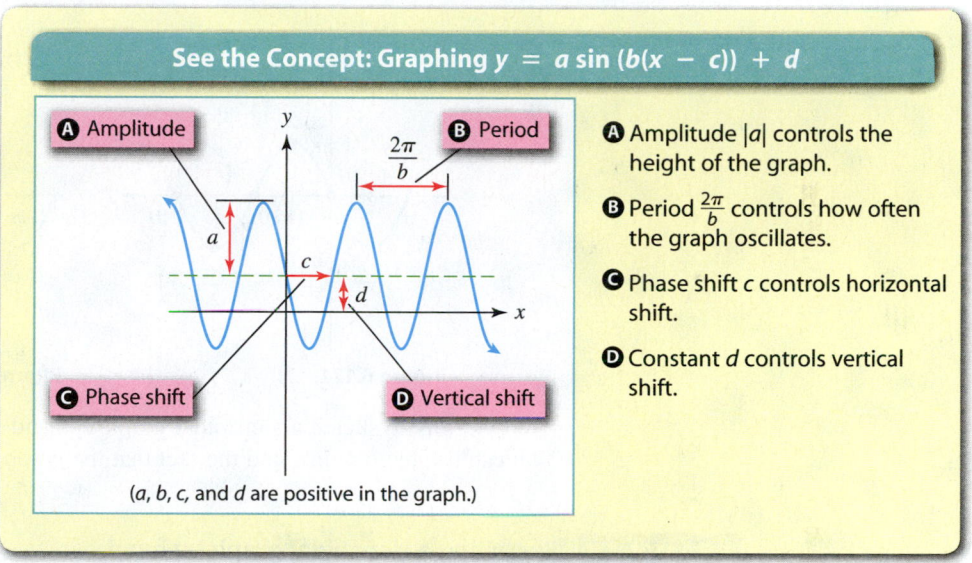

See the Concept: Graphing $y = a \sin (b(x - c)) + d$

Ⓐ Amplitude
Ⓑ Period $\frac{2\pi}{b}$
Ⓒ Phase shift
Ⓓ Vertical shift

Ⓐ Amplitude $|a|$ controls the height of the graph.

Ⓑ Period $\frac{2\pi}{b}$ controls how often the graph oscillates.

Ⓒ Phase shift c controls horizontal shift.

Ⓓ Constant d controls vertical shift.

(*a*, *b*, *c*, and *d* are positive in the graph.)

EXAMPLE 3 Identifying amplitude, period, phase shift, and vertical shift

For the graph of each equation, identify the amplitude, period, phase shift, and vertical shift.

(a) $y = 7 \sin \left(4\left(x + \frac{\pi}{3} \right) \right) + 2$ **(b)** $y = -3 \cos (5x - 3) - 6$

SOLUTION

(a) For the graph of $y = 7 \sin \left(4 \left(x - \left(-\frac{\pi}{3} \right) \right) \right) + 2$, the amplitude is 7, and the period is

$$P = \frac{2\pi}{b} = \frac{2\pi}{4} = \frac{\pi}{2}.$$

The phase shift is $-\frac{\pi}{3}$, and the vertical shift is 2.

(b) Start by applying the distributive property to rewrite the equation as

$$y = -3 \cos \left(5\left(x - \frac{3}{5} \right) \right) - 6.$$

The amplitude is 3, and the period is $P = \frac{2\pi}{b} = \frac{2\pi}{5}$. The phase shift is $\frac{3}{5}$, and the vertical shift is -6.

Now Try Exercises 13 and 17

Graphing Trigonometric Functions by Hand

Transformations can be used to graph trigonometric functions by hand, as was shown in the previous subsection. A second technique for graphing sinusoidal functions by hand is to first locate five **key points**. These key points are labeled for the graph of $y = \sin x$ in Figure 6.124. Notice that they are equally spaced along the interval $[0, 2\pi]$ on the x-axis.

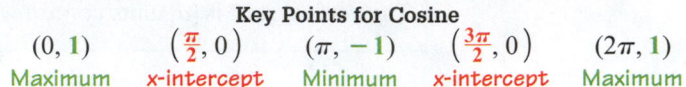

Key Points for Sine

$(\mathbf{0}, 0)$	$\left(\frac{\pi}{2}, \mathbf{1}\right)$	$(\boldsymbol{\pi}, 0)$	$\left(\frac{3\pi}{2}, -\mathbf{1}\right)$	$(\mathbf{2\pi}, 0)$
x-intercept	Maximum	x-intercept	Minimum	x-intercept

Five key points on the graph of $y = \cos x$ are shown in Figure 6.125.

Key Points for Cosine

$(0, \mathbf{1})$	$\left(\frac{\pi}{2}, 0\right)$	$(\pi, -\mathbf{1})$	$\left(\frac{3\pi}{2}, 0\right)$	$(\mathbf{2\pi}, \mathbf{1})$
Maximum	x-intercept	Minimum	x-intercept	Maximum

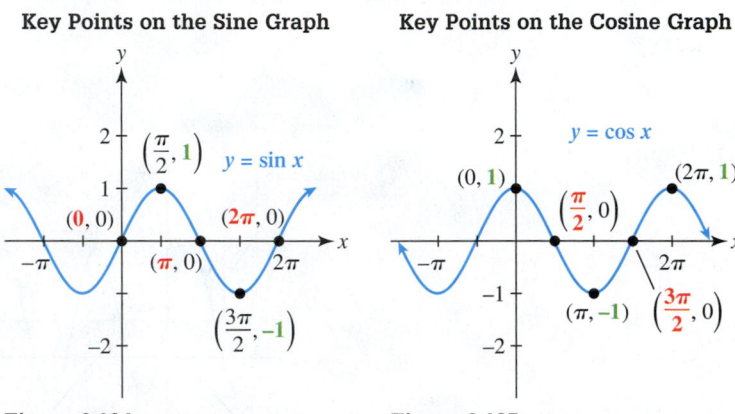

Key Points on the Sine Graph   **Key Points on the Cosine Graph**

Figure 6.124   Figure 6.125

One way to sketch a sinusoidal graph by hand is to find the *transformed* key points. You can use these points and the fact that the graph is periodic to make a sketch.

EXAMPLE 4 Sketching a sinusoidal graph by hand

Graphing with Key Points

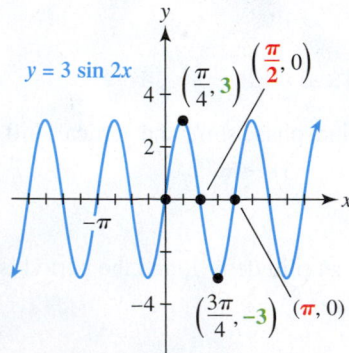

Figure 6.126

Find the amplitude, period, and phase shift of the graph of $f(x) = 3 \sin 2x$. Sketch a graph of f by hand on the interval $[-2\pi, 2\pi]$.

SOLUTION If $f(x) = 3 \sin 2x$, then the amplitude is 3 and the period is π. There is no phase or vertical shift. The graph of f passes through the point $(\mathbf{0}, 0)$ and completes one oscillation in π units. Thus the graph passes through the point $(\pi, 0)$. An amplitude of 3 will change the maximum and minimum y-values to $\mathbf{3}$ and $-\mathbf{3}$, respectively. The transformed key points for sine are equally spaced on the graph of f.

$(\mathbf{0}, 0)$	$\left(\frac{\pi}{4}, \mathbf{3}\right)$	$\left(\frac{\pi}{2}, 0\right)$	$\left(\frac{3\pi}{4}, -\mathbf{3}\right)$	$(\boldsymbol{\pi}, 0)$
x-intercept	Maximum	x-intercept	Minimum	x-intercept

These points and the graph of f are plotted in Figure 6.126.

Now Try Exercise 33

EXAMPLE 5 Sketching a graph by hand

Graph $y = 2 \cos (4x + \pi)$ by hand.

SOLUTION First write the equation in the form $y = a \cos (b(x - c))$ by factoring out 4.

$$y = 2 \cos \left(4\left(x + \frac{\pi}{4}\right)\right)$$

Graphing with Key Points

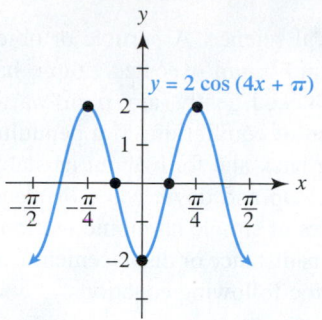

$y = 2 \cos (4x + \pi)$

Figure 6.127

The amplitude is **2**, and the period is $\frac{2\pi}{4} = \frac{\pi}{2}$. The graph is translated $\frac{\pi}{4}$ unit to the left compared to the graph of $y = 2 \cos 4x$. Because the graph is translated $\frac{\pi}{4}$ unit to the left, start the x-values in the key points at $0 - \frac{\pi}{4} = -\frac{\pi}{4}$. The first period ends at $-\frac{\pi}{4} + \frac{\pi}{2} = \frac{\pi}{4}$. The key points for this cosine graph are as follows.

$$\left(-\frac{\pi}{4}, 2\right) \quad \left(-\frac{\pi}{8}, 0\right) \quad (0, -2) \quad \left(\frac{\pi}{8}, 0\right) \quad \left(\frac{\pi}{4}, 2\right)$$

Maximum   x-intercept   Minimum   x-intercept   Maximum

Plot these key points to complete one period of the graph. The graph can be extended to show two periods, ranging from $-\frac{\pi}{2}$ to $\frac{\pi}{2}$, as shown in Figure 6.127.

Now Try Exercise 45

CLASS DISCUSSION

How would you modify the graph in Figure 6.127 to obtain the graph of

$$y = 2 \cos (4x + \pi) - 1?$$

What are the key points for this graph?

Graphs of Other Trigonometric Functions The periods of the tangent and cotangent functions are π. As a result, the graphs of $y = \tan (b(x - c))$ and $y = \cot (b(x - c))$ have a period of $P = \frac{\pi}{b}$ and a phase shift of c. Since the ranges of the tangent and cotangent functions include all real numbers, their graphs do not have an amplitude. Graphs of the secant and cosecant functions are done in Exercises 65–70.

EXAMPLE 6   Graphing other trigonometric functions

Find the period and phase shift for the graph of f. Graph f on the interval $[-2\pi, 2\pi]$. Identify where asymptotes occur in the graph.

(a) $f(x) = \tan \frac{1}{2}x$ **(b)** $f(x) = \cot \left(x + \frac{\pi}{2}\right) + 1$

SOLUTION
(a) If $f(x) = \tan \frac{1}{2}x$, then $b = \frac{1}{2}$ and $c = 0$. The period is $P = \frac{\pi}{b} = 2\pi$, and there is no phase shift. Graph $y = \tan \frac{1}{2}x$, as shown in Figure 6.128. On the interval $[-2\pi, 2\pi]$, vertical asymptotes occur at $x = \pm\pi$. The graph of f is horizontally stretched compared to the graph of the tangent function.
(b) If $f(x) = \cot \left(x + \frac{\pi}{2}\right) + 1$, then $b = 1$ and $c = -\frac{\pi}{2}$. The period is π, the phase shift is $-\frac{\pi}{2}$, and the graph is shifted upward 1 unit. Vertical asymptotes occur at $x = \pm\frac{\pi}{2}$ and $x = \pm\frac{3\pi}{2}$. The graph of $y = \cot \left(x + \frac{\pi}{2}\right) + 1$ is shown in Figure 6.129.

Increase Period to 2π

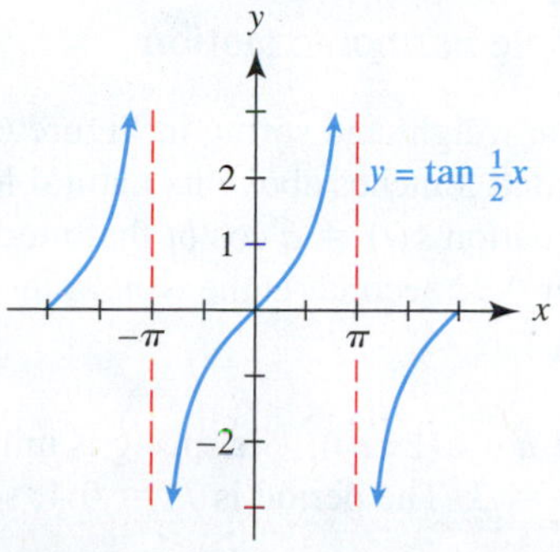

$y = \tan \frac{1}{2}x$

Figure 6.128

Shift Left $\frac{\pi}{2}$; Shift Up 1

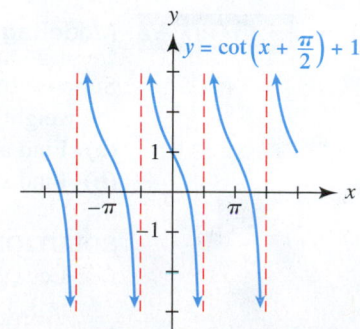

$y = \cot \left(x + \frac{\pi}{2}\right) + 1$

Figure 6.129

Now Try Exercises 57 and 63

Simple Harmonic Motion

Harmonic motion occurs frequently in nature and physical science. A particle or object undergoing small oscillations about a point of stable equilibrium executes simple harmonic motion. For example, if a string on a guitar is plucked gently, any point on the string will vibrate back and forth about its natural position of equilibrium. If a pendulum on a clock is pulled to one side, the pendulum will swing back and forth about its stable, vertical position. A small weight on a spring will bounce up and down when displaced from its natural length. All of these situations are examples of simple harmonic motion.

When an object undergoes simple harmonic motion, its distance or displacement from its stable or natural position can be modeled by either of the following equations.

> ## Modeling Simple Harmonic Motion
> $$s(t) = a \sin bt \qquad \text{or} \qquad s(t) = a \cos bt$$

In these equations, $s(t)$ represents the displacement being experienced at time t. To better understand this, consider the spring and weight shown in Figure 6.130. If $s = 0$ corresponds to the spring's natural length, then $s = -2$ indicates that the spring is stretched 2 units beyond its natural length and $s = 2$ indicates that the spring is compressed 2 units. When a spring is stretched and let go, it will oscillate up and down. If the displacement s is plotted after t seconds, a sinusoidal graph results, as shown in Figure 6.131.

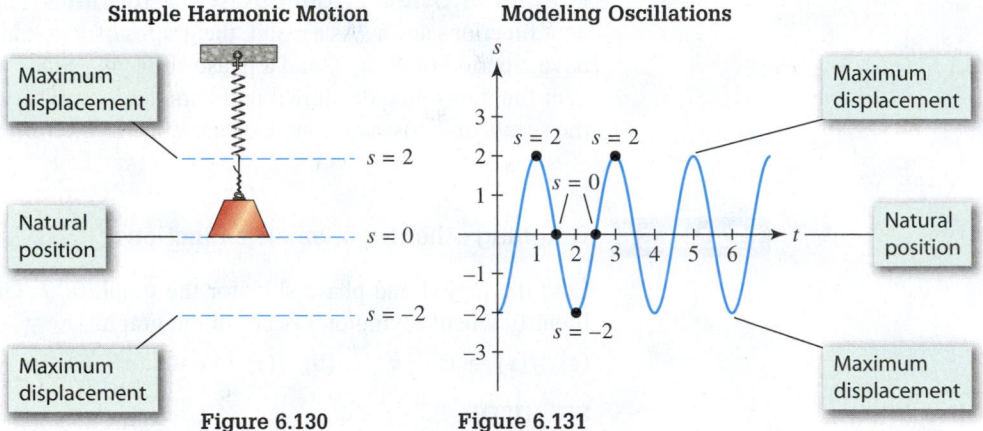

Figure 6.130 **Figure 6.131**

The amplitude of each oscillation equals $|a|$, and the period P is given by $\frac{2\pi}{b}$. The number of oscillations per unit time is the **frequency**. The frequency F equals the reciprocal of the period P. That is, $F = \frac{1}{P}$. Since $P = \frac{2\pi}{b}$, it follows that $F = \frac{b}{2\pi}$, or equivalently, $b = 2\pi F$. Substituting for b in the formulas for $s(t)$ results in the following equations.

> ## Simple Harmonic Motion with Frequency F
> $$s(t) = a \sin(2\pi Ft) \qquad \text{or} \qquad s(t) = a \cos(2\pi Ft)$$

EXAMPLE 7 **Modeling simple harmonic motion**

Suppose that the weight and spring in Figure 6.130 have a period of 0.4 second. Initially, the weight is lifted 3 inches above its natural length and then let go.
(a) Find an equation $s(t) = a \cos bt$ that models the displacement s of the weight.
(b) Find s after 0.92 second. Is the weight moving upward or downward at this time?

SOLUTION
(a) Let $s(t) = a \cos(2\pi Ft)$. The spring is initially compressed 3 inches, so the amplitude is 3. Let $a = 3$. The period is $P = 0.4$, so the frequency is

$$F = \frac{1}{P} = \frac{1}{0.4} = 2.5.$$

This indicates that the weight and spring oscillate up and down 2.5 times per second. It follows that $b = 2\pi F = 5\pi$ and $s(t) = 3\cos(5\pi t)$. Note that the initial position is $s(0) = 3$ inches.

(b) After 0.92 second, the displacement is

$$s(0.92) = 3\cos(5\pi(0.92)) \approx -0.927. \qquad \textcolor{teal}{\textit{Use radian mode.}}$$

The weight is about 0.93 inch *below* its natural position after 0.92 second.

The weight initially moves downward and oscillates with a period of 0.4 second. During the time intervals $(0, 0.2)$, $(0.4, 0.6)$, and $(0.8, 1.0)$ the weight is moving downward, and during the intervals $(0.2, 0.4)$, $(0.6, 0.8)$, and $(1.0, 1.2)$ the weight is moving upward. Thus the weight is moving *downward* when $t = 0.92$.

Now Try Exercise 71

Models Involving Trigonometric Functions (Optional)

Modeling Temperature The monthly *average* temperatures for Prince George, Canada, are shown in Table 6.9, where the months have been assigned the standard numbers.

Monthly Average Temperatures

Month	1	2	3	4	5	6	7	8	9	10	11	12
Temperature (°F)	15	19	28	41	50	55	59	57	50	41	28	19

Source: A. Miller and J. Thompson, *Elements of Meteorology.*
Table 6.9

Since the data are periodic, they are plotted over a 2-year period in Figure 6.132. For example, both $x = 1$ and $x = 13$ correspond to January. Notice that the scatterplot suggests that a sinusoidal graph might model these data.

Data of the type shown in Figure 6.132 sometimes can be modeled by

$$f(x) = a\sin(b(x - c)) + d, \qquad \textcolor{teal}{\textit{Modeling equation}}$$

where a, b, c, and d are constants. To estimate values for these constants, we will perform transformations on the graph of $y = \sin x$.

The addition of a constant d causes a vertical shift to the graph of f. The monthly average temperatures at Prince George vary from 15°F to 59°F. The average of these two temperatures is 37°F. If we let $d = 37$ and graph $y = \sin(x) + 37$, the resulting graph is as shown in Figure 6.133. Compared to the graph of the sine function, this graph is shifted upward 37 units. (Be sure to use radian mode.)

Plot Data for Two Years

[0, 25, 2] by [–5, 70, 10]

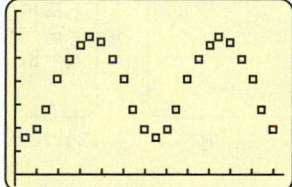

Figure 6.132

Amplitude Is Too Small

[0, 25, 2] by [–5, 70, 10]

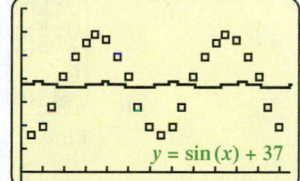

Figure 6.133

Oscillates Too Often

[0, 25, 2] by [–5, 70, 10]

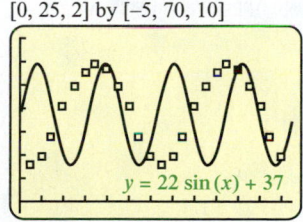

Figure 6.134

From Figure 6.133 it is apparent that the amplitude of the oscillations in the graph of $y = \sin(x) + 37$ is too small to model the data. The monthly average temperatures have a range of $59°F - 15°F = 44°F$. If we let $a = \frac{44}{2} = 22$, the peaks and valleys on the graph of $y = 22\sin(x) + 37$ will model the data better. See Figure 6.134.

In Figure 6.134 the oscillations are too frequent, which indicates that the period is too small. Since the temperature cycles every 12 months, the period is $P = \frac{2\pi}{b} = 12$. Thus

$$b = \frac{2\pi}{P} = \frac{2\pi}{12} = \frac{\pi}{6} \approx 0.524.$$

The graph of $y = 22 \sin\left(\frac{\pi}{6}x\right) + 37$ is shown in Figure 6.135.

Needs to Be Shifted Right

[0, 25, 2] by [−5, 70, 10]

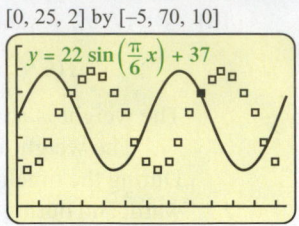

Figure 6.135

CLASS DISCUSSION

Is the value of $c = 4$ the only phase shift that can be used to model the temperature data? Explain your reasoning.

We can obtain a reasonable fit by shifting the graph horizontally to the right. The maximum of $y = 22 \sin\left(\frac{\pi}{6}x\right) + 37$ is at $x = 3$, which corresponds to March. We would like this maximum to occur in July ($x = 7$), so we translate the graph $7 - 3 = 4$ units to the right by replacing x with $x - 4$ to obtain

$$y = 22 \sin\left(\frac{\pi}{6}(x - 4)\right) + 37.$$

This equation is graphed in Figure 6.136. The phase shift is $c = 4$.

Final Model

[0, 25, 2] by [−5, 70, 10]

$y = 22 \sin\left(\frac{\pi}{6}(x-4)\right) + 37$ About right?

Figure 6.136

Sine Regression Linear and nonlinear regression were introduced in previous chapters. **Sine regression** can also be performed, using a sinusoidal function of the form

$$f(x) = a \sin(bx + c) + d.$$

Sine regression may be performed on a 2-year interval of the temperature data in Table 6.9. See Figures 6.137 and 6.138.

Calculator Help

To find an equation of least-squares fit, see Appendix A (page AP-9).

Select Sine Regression

```
EDIT  CALC  TESTS
7↑QuartReg
8:LinReg(a+bx)
9:LnReg
0:ExpReg
A:PwrReg
B:Logistic
C:SinReg
```

Figure 6.137

Regression Equation

```
SinReg
y=a*sin(bx+c)+d
a=21.7399239
b=.5207209998
c=-2.06480837
d=38.38287503
```

Figure 6.138

NOTE Sine regression gives the modeling function in the form $y = a \sin(bx + c) + d$, whereas our modeling function is in the form $f(x) = a \sin(b(x - c)) + d$. These forms have the same values for a, b, and d, but with sine regression c does not represent the phase shift. However, they are equivalent forms. If we factor out $b \approx 0.5207$ in the regression equation shown in Figure 6.138, we obtain $y \approx 21.74 \sin(0.5207(x - 3.965)) + 38.38$, which is approximately equal to our modeling equation.

Summary of Modeling Sinusoidal Data If the data appear to be sinusoidal with maximum Max, minimum Min, and period P, then the following method can be used to find values for a, b, and d for either $f(x) = a \sin(b(x - c)) + d$ or $f(x) = a \cos(b(x - c)) + d$.

$$a = \frac{\text{Max} - \text{Min}}{2}, \qquad b = \frac{2\pi}{P}, \qquad \text{and} \qquad d = \frac{\text{Max} + \text{Min}}{2}$$

If necessary, the phase shift c can be determined by shifting the graph a distance $|c|$ either left or right to fit the data. If the shift is to the right, $c > 0$, and if it is to the left, $c < 0$.

Modeling Tides Tides, which usually occur once or twice a day, represent the largest collective motion of water on Earth. We model tides in the next example.

EXAMPLE 8 Modeling tides

Figure 6.139 shows a function f that models the tides in feet at Clearwater Beach, Florida, x hours after midnight starting on August 26, 1998. (*Source: Tide and Current Predictor.*)
(a) Find the time between high tides.
(b) What is the difference in water levels between high tide and low tide?
(c) Determine a, b, c, and d so that $f(x) = a \cos(b(x - c)) + d$ models the data. Graph f and the data in the same viewing rectangle.

Tides at Clearwater Beach

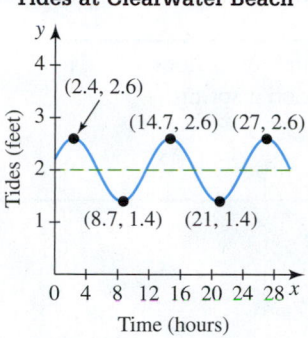

Figure 6.139

SOLUTION
(a) A high tide corresponds to a peak on the graph. The time between peaks is 12.3 hours, since $14.7 - 2.4 = 12.3$ and $27 - 14.7 = 12.3$, which is the period P.
(b) High tides were 2.6 feet, and low tides were 1.4 feet. The difference is 1.2 feet, which is twice the amplitude.
(c) *Amplitude* The amplitude of f is given by

$$a = \frac{\text{Max} - \text{Min}}{2} = \frac{2.6 - 1.4}{2} = 0.6.$$

Period Since the period is $P = 12.3$ hours, the value of b is

$$b = \frac{2\pi}{P} = \frac{2\pi}{12.3} \approx 0.511.$$

Vertical Shift The average of high tide and low tide is

$$d = \frac{\text{Max} + \text{Min}}{2} = \frac{2.6 + 1.4}{2} = 2.$$

Phase Shift Finally, a peak occurs at about 2.4 hours after midnight. Since midnight corresponds to $x = 0$ and the cosine function has a peak at $x = 0$, we translate the graph of f right **2.4** units by letting $c = 2.4$. Thus

$$f(x) = 0.6 \cos(0.511(x - 2.4)) + 2.$$

Graphs of f and the data are shown in Figure 6.140.

Final Model
[0, 28, 2] by [0, 4, 1]

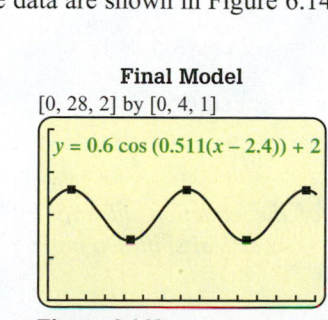

Figure 6.140

Now Try Exercise 93

6.5 Putting It All Together

Some concepts about trigonometric models are summarized in the following table.

CONCEPT	EXPLANATION	EXAMPLES
Sinusoidal Model $$f(x) = a \sin(b(x - c)) + d$$ or $$f(x) = a \cos(b(x - c)) + d$$ with $b > 0$	Amplitude $= \|a\|$ Period $= \dfrac{2\pi}{b}$ Phase shift $= c$ Vertical shift $= d$, upward if $d > 0$ and downward if $d < 0$ If $b = 2\pi F$, then F represents the frequency.	Let $f(x) = 3 \sin(2(x - \pi)) - 1$. Amplitude $= 3$ Period $= \dfrac{2\pi}{2} = \pi$ Phase shift $= \pi$ Vertical shift downward 1 unit The frequency is $F = \dfrac{b}{2\pi} = \dfrac{1}{\pi}$, which is approximately 0.32 oscillation per unit of time.
Simple Harmonic Motion $$s(t) = a \sin bt$$ or $$s(t) = a \cos bt$$	An object that oscillates about a stable equilibrium point undergoes simple harmonic motion.	A pendulum on a clock A weight on a spring

6.5 Exercises

Graphs of Trigonometric Functions

Exercises 1–6: Sketch a graph of the equation on the interval $[-4\pi, 4\pi]$. *Identify the period and amplitude.*

1. $y = 3 \sin \frac{1}{2}x$

2. $y = 2 \cos \frac{1}{3}x$

3. $y = -2 \cos 3x$

4. $y = -3 \sin 2x$

5. $y = \frac{1}{2} \sin \pi x$

6. $y = -\frac{3}{2} \cos \frac{\pi}{2}x$

Exercises 7–12: (Refer to Example 2.) Explain how the graph of f can be obtained from the graph of either the sine or the cosine function. Then sketch a graph of f on the interval $[-2\pi, 2\pi]$.

7. $y = 3 \sin(2x) - 1$

8. $y = -2 \cos(3x) + 1$

9. $y = -2 \cos\left(x + \frac{\pi}{2}\right)$

10. $y = -\frac{1}{2} \sin(x + \pi)$

11. $y = \frac{1}{2} \cos(\pi x - 1)$

12. $y = \sin\left(\frac{2}{3}x\right) + 2$

Exercises 13–18: For the graph of the equation, identify the amplitude, period, phase shift, and vertical shift. Do not graph the equation.

13. $y = 3 \sin\left(4\left(x - \frac{\pi}{4}\right)\right) - 4$

14. $y = -5 \sin\left(\frac{1}{2}(x - \pi)\right) + 7$

15. $y = -4 \cos\left(\frac{\pi}{2}(x - 1)\right) + 6$

16. $y = \frac{4}{5} \cos\left(\pi x + \frac{\pi}{3}\right) - \frac{2}{3}$

17. $y = 20 \cos\left(\frac{2}{3}x + \pi\right) + 2$

18. $y = \frac{2}{3} \sin(6x + 3\pi) - \frac{5}{2}$

Exercises 19 and 20: A graph of the equation $y = a \sin bx$ is shown, where b is a positive constant. Estimate the values for a and b.

19.

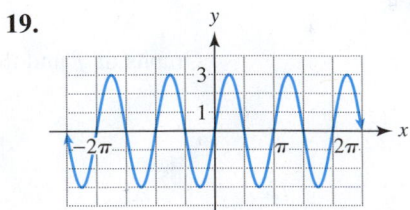

20.

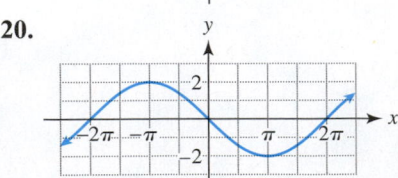

Exercises 21 and 22: The graph of a trigonometric function represented by $f(x) = a \sin(b(x - c))$ is shown, where a, b, and c are nonnegative. State the amplitude, period, and phase shift.

21.

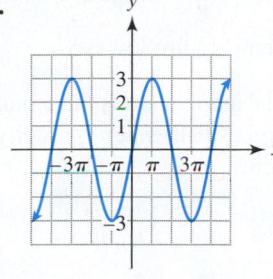

22.

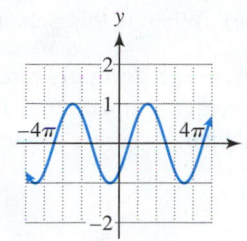

Exercises 23–28: Match the function f with its graph (a–f). Do not use a calculator.

23. $f(t) = 2 \sin\left(\frac{1}{2}t\right)$

24. $f(t) = -\sin(2t)$

25. $f(t) = 3 \cos(\pi t)$

26. $f(t) = 2 \sin\left(t - \frac{\pi}{4}\right)$

27. $f(t) = \cos\left(t + \frac{\pi}{2}\right)$

28. $f(t) = -3 \cos t$

a.

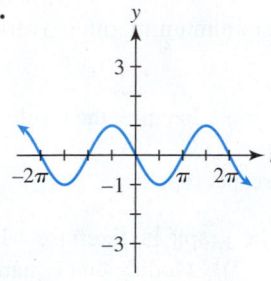

b.

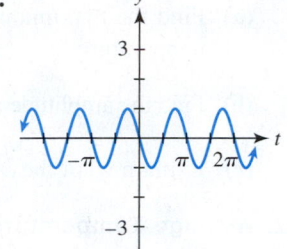

c.

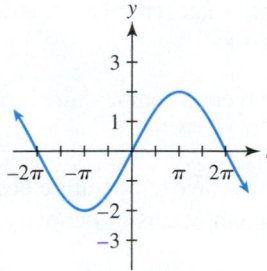

d.

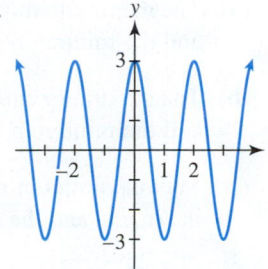

e.

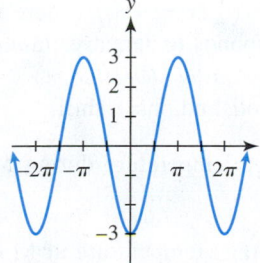

f.

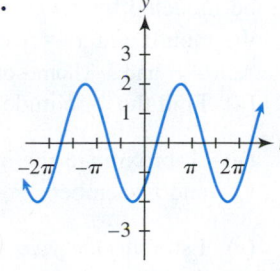

Exercises 29 and 30: Find an equation $y = a \sin(b(x - c))$ for the graph shown in the exercise. Assume that a, b, and c are nonnegative.

29. Exercise 21

30. Exercise 22

Exercises 31–48: Find the amplitude, period, and phase shift of f. Then graph f by hand on the interval $[-2\pi, 2\pi]$.

31. $f(t) = 2 \sin t$

32. $f(x) = -3 \sin x$

33. $f(x) = \sin \frac{1}{2}x$

34. $f(t) = \cos 2t$

35. $f(x) = 1 + \sin x$

36. $f(x) = -2 + 2 \cos x$

37. $f(t) = \cos(\pi t) + 2$

38. $f(t) = 2 \sin\left(t - \frac{\pi}{2}\right)$

39. $f(t) = -\sin(2(t + \pi))$

40. $f(t) = -3 \cos \frac{1}{2}t$

41. $f(x) = -\cos\left(x - \frac{\pi}{2}\right)$

42. $f(x) = -2 \cos(x + \pi)$

43. $f(x) = -\frac{1}{2} \sin 2x$

44. $f(x) = 3 \sin 4x$

45. $f(x) = 2 \cos\left(2x + \frac{\pi}{2}\right) - 1$

46. $f(x) = -\frac{1}{2} \sin(3x + \pi) + 2$

47. $f(t) = \cos\left(2\left(t - \frac{\pi}{2}\right)\right)$

48. $f(t) = 3 \sin\left(\frac{1}{2}(t - \pi)\right)$

Exercises 49–56: Graph f in $[-2\pi, 2\pi, \pi/2]$ by $[-4, 4, 1]$. State the amplitude, period, and phase shift.

49. $f(t) = 2 \sin(2t)$

50. $f(t) = -3 \sin(t - \pi)$

51. $f(t) = \frac{1}{2} \cos\left(3\left(t + \frac{\pi}{3}\right)\right)$

52. $f(t) = 1.5 \cos\left(\frac{1}{2}\left(t + \frac{\pi}{2}\right)\right)$

53. $f(t) = -2.5 \cos\left(2t + \frac{\pi}{2}\right)$

54. $f(t) = -2.5 \sin\left(\pi t + \frac{\pi}{2}\right) - 1$

55. $f(x) = -2 \cos\left(2\pi x + \frac{\pi}{4}\right) + 1$

56. $f(x) = \frac{3}{4} \sin(\pi x + \pi) - 2$

Exercises 57–64: (Refer to Example 6.) Find the period and phase shift for the graph of f. Graph f on the interval $[-2\pi, 2\pi]$. Identify where asymptotes occur in the graph.

57. $f(t) = \tan 2t$

58. $f(t) = \tan \frac{1}{2}t$

59. $f(t) = \tan\left(t - \frac{\pi}{2}\right)$

60. $f(t) = \cot\left(\frac{1}{3}\left(t - \frac{\pi}{2}\right)\right)$

61. $f(t) = -\cot 2t$

62. $f(t) = -\cot\left(t + \frac{\pi}{2}\right)$

63. $f(x) = \cot\left(2\left(x - \frac{\pi}{4}\right)\right) - 1$

64. $f(x) = \tan\left(\frac{1}{2}x - \frac{\pi}{2}\right)$

Exercises 65–70: Use the directions for Exercises 57–64 to graph f.

65. $f(t) = \sec \frac{1}{2}t$

66. $f(t) = \sec\left(2\left(t - \frac{\pi}{2}\right)\right)$

67. $f(t) = \csc(t - \pi)$

68. $f(t) = -\csc 2t$

69. $f(x) = \sec\left(\frac{1}{3}\left(x - \frac{\pi}{6}\right)\right)$

70. $f(x) = \csc(\pi(x - 1))$

Simple Harmonic Motion

Exercises 71–74: **Springs** *(Refer to Example 7.) Suppose that a weight on a spring has an initial position of s(0) and a period of P.*

(a) *Find a function s given by* $s(t) = a \cos(2\pi Ft)$ *that models the displacement of the weight.*

(b) *Evaluate s(1). Is the weight moving upward, downward, or neither when t = 1?*

71. $s(0) = 2$ inches, $P = 0.5$ second

72. $s(0) = 5$ inches, $P = 1.5$ seconds

73. $s(0) = -3$ inches, $P = 0.8$ second

74. $s(0) = -4$ inches, $P = 1.2$ seconds

Exercises 75–78: **Music** *A note on the piano has the given frequency F. Suppose the maximum displacement at the center of the piano wire is given by s(0). Find constants a and b so that* $s(t) = a \cos bt$ *models this displacement. Graph s(t) in* [0, 0.05, 0.01] *by* [-0.3, 0.3, 0.1].

75. $F = 27.5, s(0) = 0.21$ **76.** $F = 110, s(0) = 0.11$

77. $F = 55, s(0) = 0.14$ **78.** $F = 220, s(0) = 0.06$

Applications

79. Flower Opening for Sunlight Flowers tend to open in the daylight and close in the dark. The percent P of flowers open during a 24-hour period can be modeled by $P(x) = a \sin(bx) + d$, where x is hours past 6 A.M.

(a) Suppose that for P the maximum is 90%, the minimum is 10%, and the period is 24 hours. Determine $P(x)$.

(b) When are the most flowers open?

(c) When are the fewest flowers open?

80. Body Temperature A typical body temperature in degrees Fahrenheit over a 24-hour period can be modeled by the sinusoidal function

$$T(x) = a \sin(bx) + d,$$

where x is hours past 6 A.M.

(a) Suppose that for T the maximum temperature is 99.2°F, the minimum temperature is 98.0°F, and the period is 24 hours. Determine $T(x)$.

(b) When is this person's temperature highest?

(c) When is this person's temperature lowest?

81. Average Temperatures The graph models the monthly average temperature y in degrees Fahrenheit for a city in Canada, where x is the month.

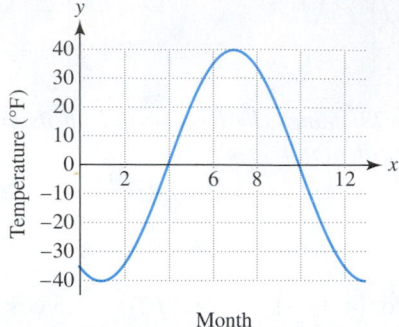

Month

(a) Find the maximum and minimum monthly average temperatures.

(b) Find the amplitude and period. Interpret the results.

(c) Explain what the x-intercepts represent.

82. Average Temperatures The graph in Exercise 81 is given by $y = 40 \cos\left(\frac{\pi}{6}(x - 7)\right)$. Modify this equation to model the following situations.

(a) The maximum monthly average temperature is 50°F and the minimum is −50°F.

(b) The maximum monthly average temperature is 60°F and the minimum is −20°F.

(c) The maximum monthly average temperature occurs in August and the minimum occurs in February.

83. Average Temperatures The monthly average temperatures in degrees Fahrenheit at Mould Bay, Canada, may be modeled by $f(x) = 34 \sin\left(\frac{\pi}{6}(x - 4.3)\right)$, where x is the month and $x = 1$ corresponds to January. (*Source:* A. Miller and J. Thompson, *Elements of Meteorology.*)

(a) Find the amplitude, period, and phase shift.

(b) Approximate the average temperature during May and December.

(c) Estimate the *yearly* average temperature at Mould Bay.

84. Average Temperatures The monthly average temperatures in degrees Fahrenheit at Austin, Texas, are given by $f(x) = 17.5 \sin\left(\frac{\pi}{6}(x - 4)\right) + 67.5$, where x is the month and $x = 1$ corresponds to January. (*Source:* A. Miller and J. Thompson, *Elements of Meteorology.*)

(a) Find the amplitude, period, phase shift, and vertical shift.

(b) Determine the maximum and minimum monthly average temperature and the months when they occur.

(c) Make a conjecture as to how the *yearly* average temperature might be related to $f(x)$.

85. Modeling Temperatures The monthly average temperatures in Vancouver, Canada, are shown in the table.

Month	1	2	3	4	5	6
Temperature (°F)	36	39	43	48	55	59

Month	7	8	9	10	11	12
Temperature (°F)	64	63	57	50	43	39

Source: A. Miller and J. Thompson, *Elements of Meteorology.*

(a) Plot these monthly average temperatures over a 24-month period by letting $x = 1$ and $x = 13$ correspond to January.

(b) Find the constants a, b, c, and d so that the function $f(x) = a \sin(b(x - c)) + d$ models the data.

(c) Graph f together with the data.

86. Modeling Temperatures The monthly average temperatures in Chicago, Illinois, are shown in the table.

Month	1	2	3	4	5	6
Temperature (°F)	25	28	36	48	61	72

Month	7	8	9	10	11	12
Temperature (°F)	74	75	66	55	39	28

Source: A. Miller and J. Thompson, *Elements of Meteorology.*

(a) Plot these monthly average temperatures over a 24-month period by letting $x = 1$ and $x = 13$ correspond to January.

(b) Find the constants a, b, c, and d so that the function $f(x) = a \sin(b(x - c)) + d$ models the data.

(c) Graph f and the data together.

87. Modeling Temperatures The monthly average high temperatures in Augusta, Georgia, are shown in the table.

Month	1	2	3	4	5	6
Temperature (°F)	58	60	68	77	82	90

Month	7	8	9	10	11	12
Temperature (°F)	92	91	83	77	68	60

Source: J. Williams, *The Weather Almanac.*

(a) Use $f(x) = a \cos(b(x - c)) + d$ to model these data.

(b) Are different values for c possible? Explain.

88. Modeling Temperatures The maximum monthly average temperature in Anchorage, Alaska, is 57°F and the minimum is 12°F.

Month	1	2	3	4	5	6
Temperature (°F)	12	18	23	36	46	55

Month	7	8	9	10	11	12
Temperature (°F)	57	55	48	36	23	16

Source: A. Miller and J. Thompson, *Elements of Meteorology.*

(a) Using only these two temperatures, determine $f(x) = a \cos(b(x - c)) + d$ so that $f(x)$ models the monthly average temperatures in Anchorage.

(b) Graph f and the actual data in the table over a 2-year period.

89. Daylight Hours The graph models the daylight hours at 60°N latitude, where $x = 1$ corresponds to January 1, $x = 2$ to February 1, and so on.

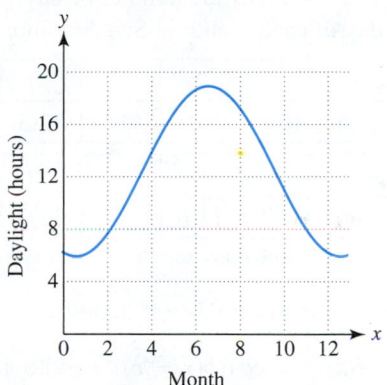

(a) Estimate the maximum number of daylight hours. When does this occur?

(b) Estimate the minimum number of daylight hours. When does this occur?

(c) Interpret the amplitude and period.

90. Daylight Hours The graph in Exercise 89 is given by $y = 6.5 \sin\left(\frac{\pi}{6}(x - 3.65)\right) + 12.4$. Modify this equation to model the following situations.

(a) At 50°N latitude, the maximum daylight is about 16.3 hours and the minimum is about 8.3 hours.

(b) The daylight hours at 60°S latitude

(c) The daylight hours at the equator

91. Average Precipitation The graph models the monthly average precipitation in inches at Mount Adams, Washington, over a 3-year period, where x is the month.

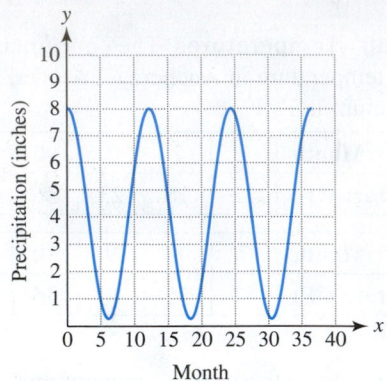

Month

(a) Find the maximum and minimum monthly average precipitation.

(b) Find the amplitude and interpret the result.

(c) Use $f(x) = a \cos(b(x - c)) + d$ to model the given graph.

92. Daylight Hours San Antonio, Texas, has a latitude of 29.5°N. The table lists the number of daylight hours on the first day of each month in San Antonio.

Month	1	2	3	4	5	6
Daylight (hr)	10.2	10.7	11.5	12.5	13.3	13.9

Month	7	8	9	10	11	12
Daylight (hr)	14.1	13.6	12.7	11.9	11.0	10.4

Source: J. Williams, *The Weather Almanac.*

(a) Plot the data over a 2-year period.

(b) Use $f(x) = a \cos(b(x - c)) + d$ to model these data.

(c) Estimate the daylight hours on February 15.

93. Average Precipitation Suppose that the monthly average precipitation at a particular location varies sinusoidally between a maximum of 6 inches in January and a minimum of 2 inches in July. Let $t = 1$ correspond to January and $t = 12$ correspond to December.
(a) Use $f(t) = a \cos(b(t - c)) + d$ to model these conditions.

(b) Evaluate $f(1)$ and $f(7)$.

94. Modeling Tidal Currents Tides cause ocean currents to flow into and out of harbors and canals. The table at the top of the next column shows the speed of the ocean current at Cape Cod Canal in bogo-knots (bk) x hours after midnight on August 26, 1998. (Note that to change bogo-knots to knots, take the square root of the absolute value of the number of bogo-knots.)

Time (hr)	3.7	6.75	9.8
Current (bk)	−18	0	18

Time (hr)	13.0	16.1	22.2
Current (bk)	0	−18	18

Source: Tide and Current Predictor.

(a) Use $f(x) = a \cos(b(x - c)) + d$ to model the data.

(b) Graph f and the data in [0, 24, 4] by [−20, 20, 5]. Interpret the graph.

95. Modeling Ocean Temperature The following table lists the monthly average ocean temperatures in degrees Fahrenheit at Veracruz, Mexico.

Month	1	2	3	4	5	6
Temperature (°F)	72	73	74	78	81	83

Month	7	8	9	10	11	12
Temperature (°F)	84	85	84	82	78	74

Source: J. Williams, *The Weather Almanac.*

(a) Make a scatterplot of the data over a 2-year period.

(b) Use $f(x) = a \sin(b(x - c)) + d$ to model the data.

96. Ocean Temperatures The water temperature in degrees Fahrenheit at St. Petersburg, Florida, can be modeled by $f(x) = 12.4 \sin\left(\frac{\pi}{6}(x - 4.2)\right) + 75$. Modify this formula to model the following situations.
(a) The monthly average water temperatures vary between 60°F and 90°F.

(b) The monthly average water temperatures vary between 50°F and 70°F.

97. Interpreting a Model Graph $y = 20 + 15 \sin\frac{\pi t}{12}$ for $0 \le t \le 12$. Let y represent the outdoor temperature in degrees Celsius at time t in hours, where $t = 0$ corresponds to 9 A.M. Interpret the graph.

98. Carbon Dioxide Levels in Hawaii At Mauna Loa, Hawaii, atmospheric carbon dioxide levels in parts per million (ppm) have been measured regularly since 1958. The equation

$$L(x) = 0.022x^2 + 0.55x + 316 + 3.5 \sin(2\pi x)$$

may be used to model these levels, where x is the year and $x = 0$ corresponds to 1960.
(a) Graph L in [20, 35, 5] by [320, 370, 10] and interpret the graph.

(b) The function L is represented by the sum of a quadratic function and a sine function. How does each function affect the shape of the graph? Discuss reasons for each function. (*Source:* A. Nilsson, *Greenhouse Earth.*)

99. Music and the Sine Function A *pure tone* can be modeled by a sine wave. Pure tones typically sound dull and uninteresting. An example of a pure tone is the sound heard when a tuning fork is lightly struck. The pure tone of the first A-note above middle C can be modeled by $f(t) = \sin(880\pi t)$. (*Source*: J. Pierce, *The Science of Musical Sound*.)

(a) In [0, 1/100, 1/880] by [−1.5, 1.5, 0.5], graph the function f.

(b) Find the period P of this tone.

(c) Frequency gives the number of vibrations or cycles per second in a sinusoidal graph. The human ear can hear frequencies from 16.4 to 16,000 cycles per second. Frequency F may be determined using the equation $f = \frac{1}{P}$, where P is the period. Find the frequency of this A-note.

100. Music (Continuation of Exercise 99.) Middle C has a frequency of 261.6 cycles per second and can be modeled by $g(t) = \sin(523.2\pi t)$. (*Source*: J. Pierce, *The Science of Musical Sound*.)

(a) Estimate the period of middle C.

(b) Graph f from Exercise 99 and g in the window [0, 1/100, 1/880] by [−1.5, 1.5, 0.5]. Compare their graphs.

Sine Regression

Exercises 101–104: **Modeling Data** *Use regression to find a formula $f(x) = a \sin(bx + c) + d$ that models the real data given in the previous exercise. Graph the data and f together.*

101. Exercise 85   **102.** Exercise 86

103. Exercise 87   **104.** Exercise 88

Writing about Mathematics

105. Discuss how the constants a, b, c, and d affect the graph of $y = a \sin(b(x - c)) + d$. Give an example.

106. Discuss some types of real data that could be modeled by $y = a \cos(b(x - c)) + d$. Give an example.

6.6 Inverse Trigonometric Functions

- **Define and use the inverse sine function**
- **Define and use the inverse cosine function**
- **Define and use the inverse tangent function**
- **Solve triangles and equations**
- **Define and use other inverse trigonometric functions**

Introduction

In construction it is sometimes necessary to determine angles. For example, the pitch, or slope, of a roof frequently is expressed as the ratio $\frac{k}{12}$, where k represents a k-foot rise for every 12 feet of run in horizontal distance. See Figure 6.141. A typical roof pitch for homes is $\frac{6}{12}$. To correctly cut the rafters, a carpenter needs to know the measure of angle θ. This problem can be solved easily using inverse trigonometric functions. See Exercise 89.

Simple Roof Truss

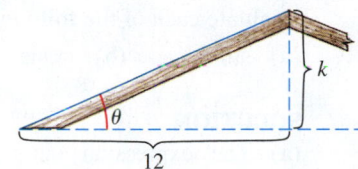

Figure 6.141

The Inverse Sine Function

A numerical representation of the sine function is shown in Table 6.10.

Sine Function

x	0	$\frac{\pi}{6}$	$\frac{\pi}{4}$	$\frac{\pi}{3}$	$\frac{\pi}{2}$	$\frac{2\pi}{3}$	$\frac{3\pi}{4}$	$\frac{5\pi}{6}$	π
$\sin x$	0	$\frac{1}{2}$	$\frac{1}{\sqrt{2}}$	$\frac{\sqrt{3}}{2}$	1	$\frac{\sqrt{3}}{2}$	$\frac{1}{\sqrt{2}}$	$\frac{1}{2}$	0

Table 6.10

$f(0) = 0$ and $f(\pi) = 0$
(not one-to-one)

Notice that different inputs do not always result in different outputs. Therefore the sine function is *not* one-to-one and so an inverse function does *not* exist. By the horizontal line test the sine function is not one-to-one. See Figure 6.142 on the next page, where a horizontal line intersects the sine graph infinitely many times.

If we restrict the domain of $f(x) = \sin x$ to $-\frac{\pi}{2} \le x \le \frac{\pi}{2}$, as shown in Figure 6.143, the graph of $y = \sin x$ is one-to-one, since a horizontal line intersects it at most once. On this restricted domain, the sine function has a unique inverse called the *inverse sine function*.

Fails Horizontal Line Test
(no inverse exists)

Restricting the Domain of Sine
(inverse exists)

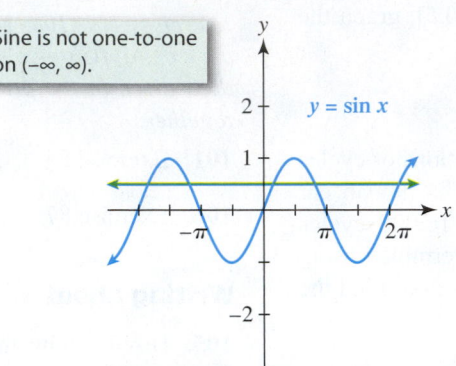

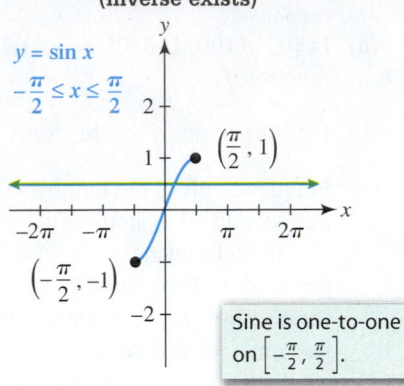

Figure 6.142

Figure 6.143

> **INVERSE SINE FUNCTION**
>
> The **inverse sine function**, denoted $\sin^{-1} x$ or arcsin x, is defined as follows. For $-1 \le x \le 1$ and y in the interval $\left[-\frac{\pi}{2}, \frac{\pi}{2}\right]$,
>
> $$y = \sin^{-1} x \quad \text{or} \quad y = \arcsin x \quad \text{means} \quad x = \sin y.$$

NOTE When evaluating the inverse sine function, it may be helpful to think of $\sin^{-1} x$ as an *angle* θ, where $\sin \theta = x$ and θ satisfies $-\frac{\pi}{2} \le \theta \le \frac{\pi}{2}$.

The next example illustrates how to evaluate the inverse sine function.

EXAMPLE 1 Evaluating the inverse sine function

Evaluate each of the following by hand and then support your results with a calculator.

(a) $\sin^{-1} 1$ **(b)** $\arcsin\left(-\frac{1}{2}\right)$

SOLUTION

(a) The expression $\sin^{-1} 1$ represents the angle (or real number) θ whose sine equals 1 and that satisfies $-\frac{\pi}{2} \le \theta \le \frac{\pi}{2}$. Thus $\theta = \sin^{-1} 1 = \frac{\pi}{2} \approx 1.57$. In degrees, $\sin^{-1} 1 = 90°$.

(b) The expression $\arcsin\left(-\frac{1}{2}\right)$ represents the angle (or real number) θ whose sine equals $-\frac{1}{2}$ and that satisfies $-\frac{\pi}{2} \le \theta \le \frac{\pi}{2}$. Thus $\theta = \arcsin\left(-\frac{1}{2}\right) = -\frac{\pi}{6} \approx -0.52$. In degrees, $\sin^{-1}\left(-\frac{1}{2}\right) = -30°$. Figures 6.144 and 6.145 support these results in both radian mode and degree mode.

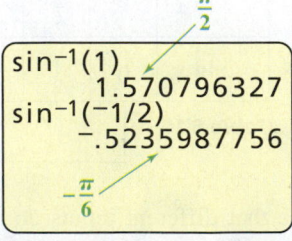

Figure 6.144 Radian Mode

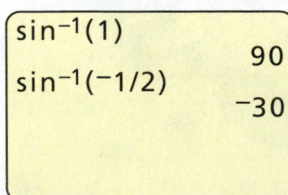

Figure 6.145 Degree Mode

Now Try Exercise 13

Domain, Range, and Inverse Properties Functions and their inverses interchange domains and ranges. The range of the sine function is $-1 \leq y \leq 1$. Therefore the domain of the inverse sine function is $-1 \leq x \leq 1$. Since the domain of the sine function has been restricted to $-\frac{\pi}{2} \leq x \leq \frac{\pi}{2}$, the range of the inverse sine function is $-\frac{\pi}{2} \leq y \leq \frac{\pi}{2}$. That is, $\sin^{-1} x$ outputs angles (numbers) only in the interval $\left[-\frac{\pi}{2}, \frac{\pi}{2}\right]$. The following are properties of the inverse sine function.

$$\sin^{-1}(\sin x) = x \quad \text{for} \quad -\frac{\pi}{2} \leq x \leq \frac{\pi}{2}$$

$$\sin(\sin^{-1} x) = x \quad \text{for} \quad -1 \leq x \leq 1$$

Inverse properties

NOTE $\sin^{-1} 2$ is undefined because $-1 \leq \sin x \leq 1$ for all angles (or numbers) x.

MAKING CONNECTIONS

Notation and Inverse Functions In Section 5.2, we learned that $f^{-1}(x) \neq \frac{1}{f(x)}$. The same is true for the inverse sine function: $\sin^{-1} x \neq \frac{1}{\sin x}$. For example, $\sin^{-1} 1 = \frac{\pi}{2} \approx 1.57$ and $\frac{1}{\sin 1} \approx 1.19$. Note that $(\sin x)^{-1} = \frac{1}{\sin x}$.

EXAMPLE 2 **Finding representations of the inverse sine function**

Represent the inverse sine function verbally, numerically, graphically, and symbolically.

SOLUTION

Verbal Representation To compute $\sin^{-1} x$ for $-1 \leq x \leq 1$, determine the angle (or real number) θ such that $\sin \theta = x$ and $-\frac{\pi}{2} \leq \theta \leq \frac{\pi}{2}$.

Numerical Representation Table 6.11 shows a numerical representation of $\sin x$ on the interval $\left[-\frac{\pi}{2}, \frac{\pi}{2}\right]$. It follows that a numerical representation of $\sin^{-1} x$ is as shown in Table 6.12. Notice that if $\sin a = b$, then $\sin^{-1} b = a$, provided $-\frac{\pi}{2} \leq a \leq \frac{\pi}{2}$.

Sine Function on $\left[-\frac{\pi}{2}, \frac{\pi}{2}\right]$

x	$-\frac{\pi}{2}$	$-\frac{\pi}{3}$	$-\frac{\pi}{4}$	$-\frac{\pi}{6}$	0	$\frac{\pi}{6}$	$\frac{\pi}{4}$	$\frac{\pi}{3}$	$\frac{\pi}{2}$
$\sin x$	-1	$-\frac{\sqrt{3}}{2}$	$-\frac{1}{\sqrt{2}}$	$-\frac{1}{2}$	0	$\frac{1}{2}$	$\frac{1}{\sqrt{2}}$	$\frac{\sqrt{3}}{2}$	1

Table 6.11

Inverse Sine Function on [−1, 1]

x	-1	$-\frac{\sqrt{3}}{2}$	$-\frac{1}{\sqrt{2}}$	$-\frac{1}{2}$	0	$\frac{1}{2}$	$\frac{1}{\sqrt{2}}$	$\frac{\sqrt{3}}{2}$	1
$\sin^{-1} x$	$-\frac{\pi}{2}$	$-\frac{\pi}{3}$	$-\frac{\pi}{4}$	$-\frac{\pi}{6}$	0	$\frac{\pi}{6}$	$\frac{\pi}{4}$	$\frac{\pi}{3}$	$\frac{\pi}{2}$

Table 6.12

Interchange domain and range.

Inverse Sine Function

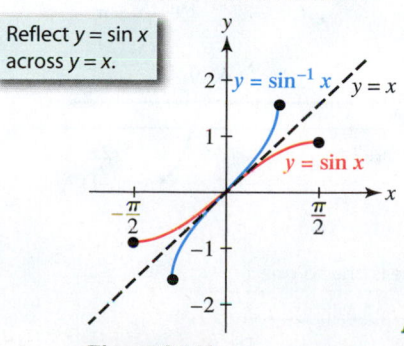

Reflect $y = \sin x$ across $y = x$.

Figure 6.146

Graphical Representation The graph of $y = \sin^{-1} x$ can be found by reflecting the graph of $y = \sin x$ for $-\frac{\pi}{2} \leq x \leq \frac{\pi}{2}$ across the line $y = x$, as shown in Figure 6.146. See also Putting It All Together at the end of this section.

Symbolic Representation A symbolic representation of the inverse sine function can be written as either $f(x) = \sin^{-1} x$ or $f(x) = \arcsin x$. There is no simple formula that can be used to evaluate $\sin^{-1} x$.

Now Try Exercise 31

An Application In track and field, when an athlete throws the shot, the distance that the shot travels depends on the angle θ that the initial direction of the shot makes with the horizontal. Angle θ in Figure 6.147 is called the *projection angle*. The optimal projection angle θ, which results in maximum distance for the shot, may be calculated by

$$\theta = \sin^{-1} \sqrt{\frac{v^2}{2v^2 + 64.4h}},$$

where v is the initial speed in feet per second of the shot and h is the height in feet of the shot when it is released.   (**Source:** J. Cooper and R. Glassow, *Kinesiology*.)

Throwing the Shot with Projection Angle θ

Figure 6.147

EXAMPLE 3   **Finding the optimal projection angle for a shot-putter**

Suppose that an athlete releases a shot 8 feet above the ground with velocity v. Give a numerical representation of the optimum projection angle θ. Interpret the results.

SOLUTION Make a table for $Y_1 = \sin^{-1}(\sqrt{(X^2/(2X^2 + 64.4*8))})$, as shown in Figure 6.148. We can see that the faster a person throws the shot, the greater the optimal projection angle θ becomes. For example, if a shot is thrown at 25 feet per second, then $\theta \approx 36.5°$, whereas if the shot is thrown at 50 feet per second, then $\theta \approx 42.3°$.

Now Try Exercise 91

Optimal Projection Angle

X	Y₁
20	33.469
25	36.515
30	38.571
35	39.997
40	41.014
45	41.76
50	42.32

X = 25

Figure 6.148 Degree Mode

The Inverse Cosine Function

By the horizontal line test, the cosine function is not one-to-one, as illustrated in Figure 6.149. If we restrict the domain of $f(x) = \cos x$ to $0 \le x \le \pi$, then the resulting function is one-to-one and has an inverse function. See Figure 6.150. This inverse function is called the *inverse cosine function*.

Fails Horizontal Line Test
(no inverse exists)

Restricting the Domain of Cosine
(inverse exists)

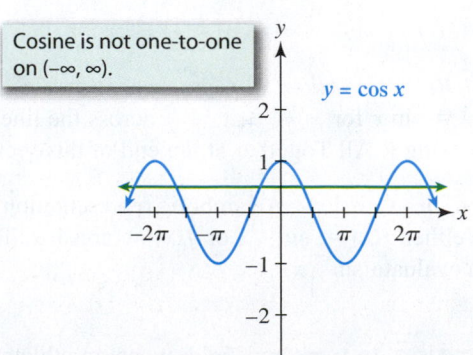

Figure 6.149

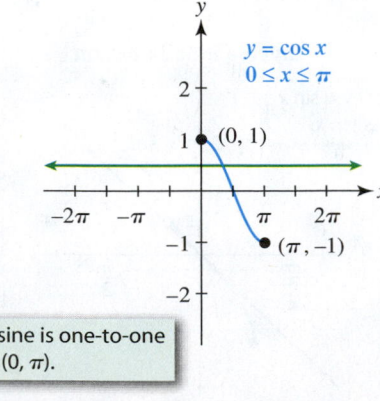

Figure 6.150

INVERSE COSINE FUNCTION

The **inverse cosine function**, denoted $\cos^{-1} x$ or arccos x, is defined as follows. For $-1 \le x \le 1$ and y in the interval $[0, \pi]$,

$$y = \cos^{-1} x \quad \text{or} \quad y = \arccos x \qquad \text{means} \qquad x = \cos y.$$

EXAMPLE 4 Evaluating the inverse cosine function

Evaluate each of the following.
(a) $\cos^{-1} 1$ **(b)** $\arccos(-0.75)$

SOLUTION

(a) The expression $\cos^{-1} 1$ represents the angle (or real number) θ whose cosine equals 1 and that satisfies $0 \le \theta \le \pi$. Thus $\theta = \cos^{-1} 1 = 0$, or $0°$.

(b) A calculator is often necessary to evaluate inverse trigonometric functions approximately. In Figure 6.151, $\cos^{-1}(-0.75) \approx 2.42$ radians, or about $138.6°$.

Now Try Exercises 15 and 19(c)

```
cos⁻¹(‾.75)
        2.418858406
Ans*180/π
        138.5903779
```
↗ Change to degrees

Figure 6.151 Radian Mode

The following are properties of the inverse cosine function.

$$\cos^{-1}(\cos x) = x \quad \text{for} \quad 0 \le x \le \pi$$

$$\cos(\cos^{-1} x) = x \quad \text{for} \quad -1 \le x \le 1$$

Inverse properties

EXAMPLE 5 Finding representations of the inverse cosine function

Represent the inverse cosine function verbally, numerically, graphically, and symbolically.

SOLUTION

Verbal Representation To compute $\cos^{-1} x$ for $-1 \le x \le 1$, determine the angle θ (or real number) such that $\cos \theta = x$ and $0 \le \theta \le \pi$.

Numerical Representation A numerical representation of $\cos x$ on the interval $[0, \pi]$ is shown in Table 6.13. A numerical representation of $\cos^{-1} x$ is shown in Table 6.14. Notice that if $\cos a = b$, then $\cos^{-1} b = a$, provided that $0 \le a \le \pi$.

Cosine Function on [0, π]

x	0	$\frac{\pi}{6}$	$\frac{\pi}{4}$	$\frac{\pi}{3}$	$\frac{\pi}{2}$	$\frac{2\pi}{3}$	$\frac{3\pi}{4}$	$\frac{5\pi}{6}$	π
$\cos x$	1	$\frac{\sqrt{3}}{2}$	$\frac{1}{\sqrt{2}}$	$\frac{1}{2}$	0	$-\frac{1}{2}$	$-\frac{1}{\sqrt{2}}$	$-\frac{\sqrt{3}}{2}$	-1

Table 6.13

Inverse Cosine Function on [–1, 1]

x	-1	$-\frac{\sqrt{3}}{2}$	$-\frac{1}{\sqrt{2}}$	$-\frac{1}{2}$	0	$\frac{1}{2}$	$\frac{1}{\sqrt{2}}$	$\frac{\sqrt{3}}{2}$	1
$\cos^{-1} x$	π	$\frac{5\pi}{6}$	$\frac{3\pi}{4}$	$\frac{2\pi}{3}$	$\frac{\pi}{2}$	$\frac{\pi}{3}$	$\frac{\pi}{4}$	$\frac{\pi}{6}$	0

Table 6.14

Interchange domain and range.

Graphical Representation The graph of $y = \cos^{-1} x$ can be found by reflecting the graph of $y = \cos x$ for $0 \le x \le \pi$ across the line $y = x$, as shown in Figure 6.152. See also Putting It All Together at the end of this section.

Symbolic Representation A symbolic representation of the inverse cosine function can be written as either $f(x) = \cos^{-1} x$ or $f(x) = \arccos x$. There is no simple formula that can be used to evaluate $\cos^{-1} x$.

Now Try Exercise 33

Inverse Cosine Function

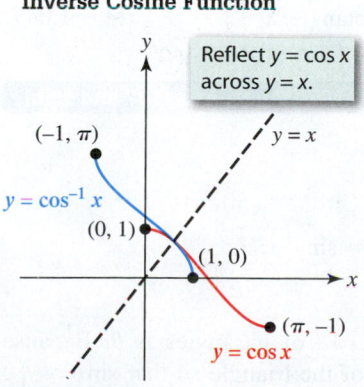

Reflect $y = \cos x$ across $y = x$.

$(-1, \pi)$
$y = x$
$y = \cos^{-1} x$
$(0, 1)$
$(1, 0)$
$(\pi, -1)$
$y = \cos x$

Figure 6.152

Explain the results in Figure 6.153.

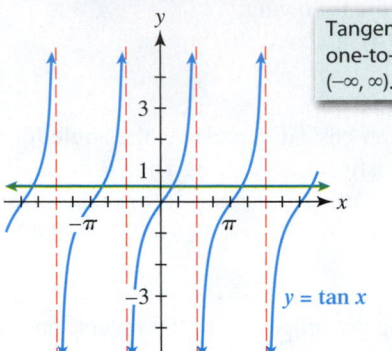

Figure 6.153 Degree Mode

The Inverse Tangent Function

By the horizontal line test, the tangent function is not one-to-one on $(-\infty, \infty)$, as shown in Figure 6.154. If we restrict the domain of $f(x) = \tan x$ to $-\frac{\pi}{2} < x < \frac{\pi}{2}$, then the resulting function is one-to-one and has an inverse function. See Figure 6.155. This inverse function is called the *inverse tangent function*.

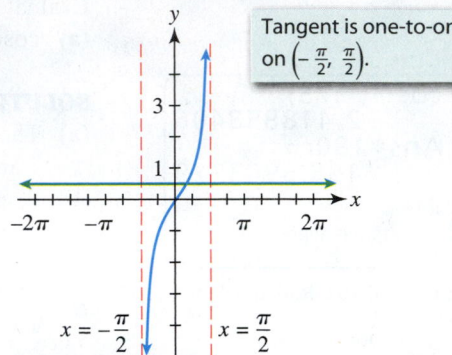

Figure 6.154   **Figure 6.155**

INVERSE TANGENT FUNCTION

The **inverse tangent function**, denoted $\tan^{-1} x$ or arctan x, is defined as follows. For y in the interval $\left(-\frac{\pi}{2}, \frac{\pi}{2}\right)$,

$$y = \tan^{-1} x \quad \text{or} \quad y = \arctan x \quad \text{means} \quad x = \tan y.$$

The following are properties of the inverse tangent function.

$$\tan^{-1}(\tan x) = x \quad \text{for} \quad -\frac{\pi}{2} < x < \frac{\pi}{2}$$

$$\tan(\tan^{-1} x) = x \quad \text{for} \quad \text{all real numbers } x$$

Inverse properties

EXAMPLE 6 **Evaluating the inverse tangent function**

Evaluate each of the following. Support your answer using a calculator.

(a) $\tan^{-1} 1$ **(b)** $\arctan\left(-\sqrt{3}\right)$

SOLUTION

(a) The expression $\tan^{-1} 1$ represents the angle (or real number) θ whose tangent equals 1 and that satisfies $-\frac{\pi}{2} < \theta < \frac{\pi}{2}$. Thus $\theta = \tan^{-1}(1) = \frac{\pi}{4} \approx 0.7854$. See Figure 6.156. In degrees, $\tan^{-1}(1) = 45°$.

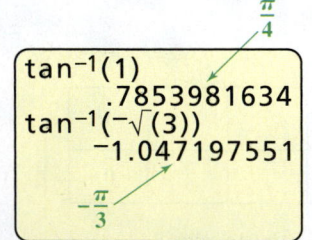

Figure 6.156 Radian Mode

(b) The expression $\arctan\left(-\sqrt{3}\right)$ represents the angle (or real number) θ whose tangent is $-\sqrt{3}$ and that satisfies $-\frac{\pi}{2} < \theta < \frac{\pi}{2}$. Thus $\theta = \arctan\left(-\sqrt{3}\right) = -\frac{\pi}{3} \approx -1.047$. Support is shown in Figure 6.156. In degrees, $\arctan\left(-\sqrt{3}\right) = -60°$.

Now Try Exercises 17 and 19(b)

EXAMPLE 7 **Evaluating inverse trigonometric functions**

Let θ be an acute angle. Write $\cos \theta$ in terms of x, if $\theta = \sin^{-1} x$.

SOLUTION

Getting Started First sketch a right triangle so that one of its angles is θ. Because $\theta = \sin^{-1} x$, it follows that $\sin \theta = x$. Label the sides of the triangle so that $\sin \theta = \frac{x}{1}$. Use the triangle to find $\cos \theta$. ▶

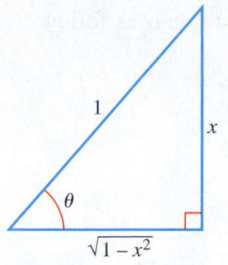

Figure 6.157

Figure 6.157 shows angle θ with $\sin\theta = \frac{x}{1}$. By the Pythagorean theorem, the other leg has length $\sqrt{1 - x^2}$. Because $\cos\theta$ equals the side adjacent divided by the hypotenuse, we have $\cos\theta = \sqrt{1 - x^2}$.

Now Try Exercise 35

NOTE Properties and graphs of the inverse sine, inverse cosine, and inverse tangent functions are given in Putting It All Together at the end of this section.

An Application The next example applies the inverse tangent function to robotics.

EXAMPLE 8 Using robots to spray paint

In industry it is common to use robots to spray paint. The robotic arm in Figure 6.158 is being used to paint a flat surface. Because the spray gun must move at a constant speed v, parallel to the surface being painted, the angle of the arm θ_1 and the angle of the spray gun θ_2 must be continually adjusted. Using Figure 6.158, it can be shown that

$$\theta_1 = \arctan\frac{h}{vt} \qquad \text{and} \qquad \theta_2 = 90° - \theta_1,$$

where $t > 0$ is time in seconds. (Try to verify this.) Let $v = 3$ inches per second and $h = 24$ inches. Determine the degree measure of θ_1 and θ_2 after 10 seconds. (*Source:* W. Stadler, *Analytical Robotics and Mechatronics.*)

CLASS DISCUSSION
Give verbal, numerical, and graphical representations of $y = \tan^{-1} x$.

Robotic Arm Painting

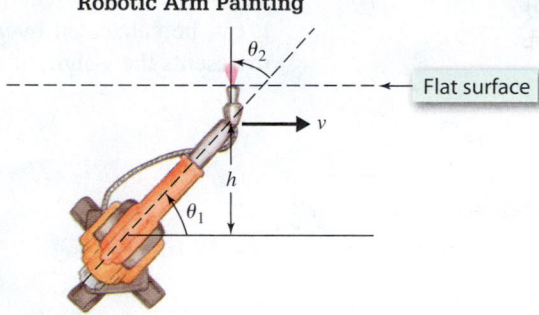

Figure 6.158

SOLUTION Substitute $h = 24$ and $v = 3$. When $t = 10$, $\theta_1 = \arctan\frac{24}{3(10)} \approx 38.7°$ and $\theta_2 = 90° - \theta_1 \approx 51.3°$.

Now Try Exercise 95

Solving Triangles and Equations

Figure 6.159 shows *standard labeling* used to denote the vertices, sides, and angles of a right triangle. The next example illustrates the process of finding the measures of the angles and sides in a triangle, called *solving a triangle*.

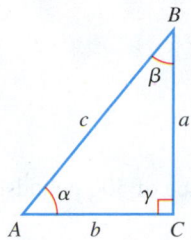

Figure 6.159

EXAMPLE 9 Solving a right triangle

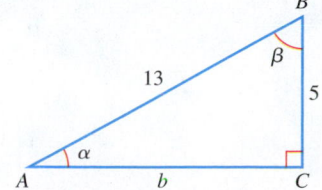

Figure 6.160

Solve triangle ABC in Figure 6.160 if $a = 5$ and $c = 13$. Round values to the nearest tenth.

SOLUTION We are given $a = 5$, $c = 13$, and $\gamma = 90°$. We must find b, α, and β. We begin by finding b using the Pythagorean theorem.

$$a^2 + b^2 = c^2 \qquad \text{Pythagorean theorem}$$
$$b^2 = c^2 - a^2 \qquad \text{Subtract } a^2.$$
$$b^2 = 13^2 - 5^2 \qquad \text{Substitute.}$$
$$b^2 = 144 \qquad \text{Simplify.}$$

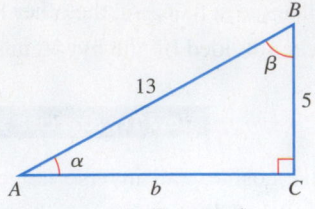

Figure 6.160 (Repeated)

Thus $b = 12$ in Figure 6.160 because $b^2 = 144$. We can find angle α as follows.

$$\sin \alpha = \frac{5}{13} \qquad \sin \alpha = \frac{\text{side opposite}}{\text{hypotenuse}}$$

$$\alpha = \sin^{-1} \frac{5}{13} \qquad \text{Solve for } \alpha.$$

$$\alpha \approx 22.6° \qquad \text{Approximate.}$$

Since β is complementary to α, $\beta \approx 90° - 22.6° = 67.4°$.

Now Try Exercise 47

NOTE There is more than one way to solve the triangle in Example 9. We could have let $\alpha = \tan^{-1} \frac{5}{12} \approx 22.6°$ and $\beta = \cos^{-1} \frac{5}{13} \approx 67.4°$. There are other possibilities.

EXAMPLE 10 Finding angles in a triangle

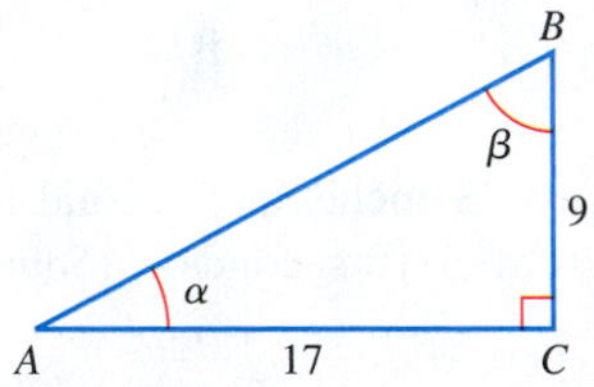

Approximate the degree measure of the angles α and β shown in Figure 6.161.

SOLUTION From Figure 6.161 we see that $\tan \alpha = \frac{9}{17}$. Thus $\alpha = \tan^{-1} \frac{9}{17} \approx 27.9°$. Since α and β are complementary angles, $\beta \approx 90° - 27.9° = 62.1°$.

Now Try Exercise 45

Figure 6.161

An Application Grade resistance is the force F that causes a car to roll down a hill. It can be calculated by $F = W \sin \theta$, where θ represents the angle of the grade and W represents the weight of the vehicle. See Figure 6.162.

Uphill and Downhill Grade

Figure 6.162

EXAMPLE 11 Calculating highway grade

Find the angle θ for which a 3000-pound car has grade resistance of 500 pounds.

SOLUTION Solve the equation $F = W \sin \theta$ for θ.

$$F = W \sin \theta \qquad \text{Given equation}$$

$$500 = 3000 \sin \theta \qquad \text{Let } W = 3000 \text{ and } F = 500.$$

$$\sin \theta = \frac{1}{6} \qquad \text{Solve for } \sin \theta.$$

$$\theta = \sin^{-1} \frac{1}{6} \qquad \text{Property of inverse sine}$$

$$\theta \approx 9.6° \qquad \text{Approximate.}$$

Thus if a road is inclined at approximately 9.6°, a 3000-pound car would experience a force of 500 pounds pulling downhill.

Now Try Exercise 85

Solving an Equation There are a wide variety of trigonometric equations. In the next example, we solve one basic type of equation. In Chapter 7, we will solve more types of trigonometric equations.

EXAMPLE 12   Solving a trigonometric equation

Solve $9\cos^2\theta = 4$, where θ is an acute angle.

SOLUTION Begin by dividing each side of the equation by 9.

$$9\cos^2\theta = 4 \qquad \text{Given equation}$$

$$\cos^2\theta = \frac{4}{9} \qquad \text{Divide by 9.}$$

$$\cos\theta = \pm\frac{2}{3} \qquad \text{Square root property}$$

Because θ is an acute angle, $\cos\theta$ must be positive. Thus $\cos\theta = \frac{2}{3}$, and the solution to the equation is $\theta = \cos^{-1}\frac{2}{3} \approx 48.2°$.

Now Try Exercise 61

Other Trigonometric Inverse Functions

Definitions and Graphs The cotangent, secant, and cosecant functions are not one-to-one functions. However, by restricting their domains, inverse trigonometric functions can be defined. The following box gives common definitions for these functions; some texts use slightly different definitions.

OTHER INVERSE TRIGONOMETRIC FUNCTIONS

Inverse Cotangent For $0 < y < \pi$,

$$y = \cot^{-1}x \quad \text{or} \quad y = \text{arccot } x \quad \text{means that} \quad x = \cot y.$$

Inverse Secant For $0 \le y \le \pi$ with $y \ne \frac{\pi}{2}$ and $|x| \ge 1$,

$$y = \sec^{-1}x \quad \text{or} \quad y = \text{arcsec } x \quad \text{means that} \quad x = \sec y$$

Inverse Cosecant For $-\frac{\pi}{2} \le y \le \frac{\pi}{2}$ with $y \ne 0$ and $|x| \ge 1$,

$$y = \csc^{-1}x \quad \text{or} \quad y = \text{arccsc } x \quad \text{means that} \quad x = \csc y.$$

The graphs of the functions $y = \cot^{-1}x$, $y = \sec^{-1}x$, and $y = \csc^{-1}x$ are shown in Figures 6.163–6.165. These inverse graphs can be found by reflecting the graphs of $y = \cot x$, $y = \sec x$, and $y = \csc x$ across the line $y = x$ on a restricted interval.

Inverse Cotangent

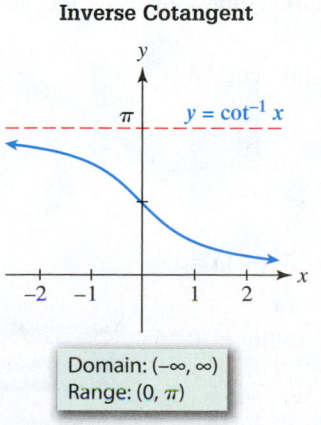

Domain: $(-\infty, \infty)$
Range: $(0, \pi)$

Figure 6.163

Inverse Secant

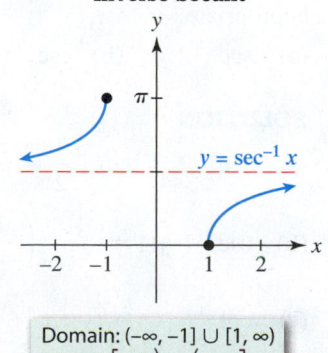

Domain: $(-\infty, -1] \cup [1, \infty)$
Range: $\left[0, \frac{\pi}{2}\right) \cup \left(\frac{\pi}{2}, \pi\right]$

Figure 6.164

Inverse Cosecant

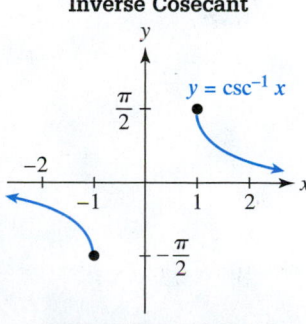

Domain: $(-\infty, -1] \cup [1, \infty)$
Range: $\left[-\frac{\pi}{2}, 0\right) \cup \left(0, \frac{\pi}{2}\right]$

Figure 6.165

Evaluating Other Inverse Functions Most calculators do not have keys for $\sec^{-1}x$, $\csc^{-1}x$, or $\cot^{-1}x$. However, we can evaluate these functions by using the functions $\cos^{-1}x$, $\sin^{-1}x$, and $\tan^{-1}x$.

For example, $y = \sec^{-1}x$ can be rewritten in terms of the inverse cosine function as follows.

$$y = \sec^{-1}x \qquad \text{Inverse secant, } |x| \geq 1$$

$$\sec y = x \qquad \text{Definition of inverse secant}$$

$$\frac{1}{\cos y} = x \qquad \text{Reciprocal identity}$$

$$\cos y = \frac{1}{x} \qquad \text{Invert each side, } x \neq 0.$$

$$\cos^{-1}(\cos y) = \cos^{-1}\frac{1}{x} \qquad \text{Take inverse cosine.}$$

$$y = \cos^{-1}\frac{1}{x} \qquad \text{Inverse properties}$$

Thus to evaluate $y = \sec^{-1}x$, we can evaluate $y = \cos^{-1}\frac{1}{x}$. The following box shows how to evaluate $\sec^{-1}x$, $\csc^{-1}x$, and $\cot^{-1}x$.

EVALUATING OTHER INVERSE TRIGONOMETRIC FUNCTIONS

Inverse Secant $\sec^{-1}x = \cos^{-1}\frac{1}{x}$ for $|x| \geq 1$.

Inverse Cosecant $\csc^{-1}x = \sin^{-1}\frac{1}{x}$ for $|x| \geq 1$.

Inverse Cotangent *Method I:* $\cot^{-1}x = \frac{\pi}{2} - \tan^{-1}x$

Method II: $\cot^{-1}x = \tan^{-1}\frac{1}{x}$ if $x > 0$

$\cot^{-1}x = \frac{\pi}{2}$ if $x = 0$

$\cot^{-1}x = \pi + \tan^{-1}\frac{1}{x}$ if $x < 0$

In the next example, we evaluate other inverse trigonometric functions.

EXAMPLE 13 **Evaluating other inverse functions**

Find each of the following in radians. Approximate to the nearest thousandth when appropriate.

(a) $\sec^{-1}2$ **(b)** $\csc^{-1}(-1)$ **(c)** $\cot^{-1}6$

SOLUTION

(a) $\sec^{-1}2 = \cos^{-1}\frac{1}{2} = \frac{\pi}{3}$

(b) $\csc^{-1}(-1) = \sin^{-1}\left(\frac{1}{-1}\right) = \sin^{-1}(-1) = -\frac{\pi}{2}$

(c) $\cot^{-1}6 = \frac{\pi}{2} - \tan^{-1}6 \approx \frac{\pi}{2} - 1.406 \approx 0.165$

Now Try Exercises 71, 73, and 81

6.6 Putting It All Together

The six trigonometric functions are not one-to-one functions. However, by *restricting* their domains to appropriate intervals, inverse trigonometric functions can be defined. The following table summarizes some properties of three important inverse trigonometric functions.

FUNCTION	EXPLANATION	EXAMPLES AND GRAPHS
Inverse sine	*Description:* $f(x) = \sin^{-1} x$ or $f(x) = \arcsin x$ computes the angle or number in $\left[-\frac{\pi}{2}, \frac{\pi}{2}\right]$ whose sine equals x, where $-1 \le x \le 1$. *Domain:* $\{x \mid -1 \le x \le 1\}$ *Range:* $\{y \mid -\frac{\pi}{2} \le y \le \frac{\pi}{2}\}$ *Inverse Properties:* $\sin^{-1}(\sin x) = x$ for $-\frac{\pi}{2} \le x \le \frac{\pi}{2}$ $\sin(\sin^{-1} x) = x$ for $-1 \le x \le 1$	$\sin^{-1} 1 = 90°$, or $\frac{\pi}{2}$ $\arcsin\left(-\frac{1}{2}\right) = -30°$, or $-\frac{\pi}{6}$ $\sin^{-1} 0 = 0°$, or 0 $\sin^{-1} 4$ is undefined. $\sin^{-1} 0.3 \approx 17.5°$, or 0.305 $y = \sin^{-1} x$
Inverse cosine	*Description:* $f(x) = \cos^{-1} x$ or $f(x) = \arccos x$ computes the angle or number in $[0, \pi]$ whose cosine equals x, where $-1 \le x \le 1$. *Domain:* $\{x \mid -1 \le x \le 1\}$ *Range:* $\{y \mid 0 \le y \le \pi\}$ *Inverse Properties:* $\cos^{-1}(\cos x) = x$ for $0 \le x \le \pi$ $\cos(\cos^{-1} x) = x$ for $-1 \le x \le 1$	$\cos^{-1} 1 = 0°$, or 0 $\arccos\left(-\frac{1}{2}\right) = 120°$, or $\frac{2\pi}{3}$ $\cos^{-1} 0 = 90°$, or $\frac{\pi}{2}$ $\cos^{-1}(-5)$ is undefined. $\cos^{-1} 0.8 \approx 36.9°$, or 0.644 $y = \cos^{-1} x$
Inverse tangent	*Description:* $f(x) = \tan^{-1} x$ or $f(x) = \arctan x$ computes the angle or number in $\left(-\frac{\pi}{2}, \frac{\pi}{2}\right)$ whose tangent equals x, where x is any real number. *Domain:* $\{x \mid -\infty < x < \infty\}$ (all real numbers) *Range:* $\{y \mid -\frac{\pi}{2} < y < \frac{\pi}{2}\}$ *Inverse Properties:* $\tan^{-1}(\tan x) = x$ for $-\frac{\pi}{2} < x < \frac{\pi}{2}$ $\tan(\tan^{-1} x) = x$ for all real numbers x *Horizontal Asymptotes:* $y = -\frac{\pi}{2}, y = \frac{\pi}{2}$	$\tan^{-1} 1 = 45°$, or $\frac{\pi}{4}$ $\tan^{-1} 0 = 0°$, or 0 $\tan^{-1} 8 \approx 82.9°$, or 1.446 $\arctan(-1) = -45°$, or $-\frac{\pi}{4}$ $y = \tan^{-1} x$

6.6 Exercises

Review of Inverses

1. For a function f to have an inverse, f must be _____.

2. A function is one-to-one if different inputs always result in _____ outputs.

3. If $f(\pi) = -1$, then $f^{-1}(-1) = $_____.

4. If $f(c) = d$, then $f^{-1}(d) = $_____.

5. If $f^{-1}(0) = 1$, then $f(1) = $_____.

6. If $f^{-1}(b) = a$, then $f(a) = $_____.

Inverse Trigonometric Functions

7. Since $\sin\frac{\pi}{2} = 1$ and $\frac{\pi}{2}$ is in the interval $\left[-\frac{\pi}{2}, \frac{\pi}{2}\right]$, $\sin^{-1} 1 = $_____.

8. Since $\cos\frac{\pi}{3} = \frac{1}{2}$ and $\frac{\pi}{3}$ is in the interval $[0, \pi]$, $\cos^{-1}\frac{1}{2} = $_____.

9. Since $\tan\left(-\frac{\pi}{4}\right) = -1$ and $-\frac{\pi}{4}$ is in the interval $\left(-\frac{\pi}{2}, \frac{\pi}{2}\right)$, $\tan^{-1}(-1) = $_____.

10. Since $\sin\left(-\frac{\pi}{6}\right) = -\frac{1}{2}$ and $-\frac{\pi}{6}$ is in the interval $\left[-\frac{\pi}{2}, \frac{\pi}{2}\right]$, $\sin^{-1}\left(-\frac{1}{2}\right) = $_____.

11. Since $\cos\left(\frac{2\pi}{3}\right) = -\frac{1}{2}$ and $\frac{2\pi}{3}$ is in the interval $[0, \pi]$, $\cos^{-1}\left(-\frac{1}{2}\right) = $_____.

12. Since $\tan\left(\frac{\pi}{3}\right) = \sqrt{3}$ and $\frac{\pi}{3}$ is in the interval $\left(-\frac{\pi}{2}, \frac{\pi}{2}\right)$, $\tan^{-1}\sqrt{3} = $_____.

Exercises 13–18: Evaluate each of the following, if possible. Give results in both radians and degrees.

13. (a) $\sin^{-1} 1$ (b) $\arcsin 4$

   (c) $\arcsin\left(-\frac{\sqrt{3}}{2}\right)$

14. (a) $\arcsin\frac{1}{2}$ (b) $\sin^{-1}(-2)$

   (c) $\sin^{-1}(-1)$

15. (a) $\cos^{-1} 0$ (b) $\arccos(-1)$

   (c) $\cos^{-1}\frac{1}{2}$

16. (a) $\arccos\frac{\sqrt{3}}{2}$ (b) $\cos^{-1}\left(-\frac{1}{2}\right)$

   (c) $\arccos 1$

17. (a) $\tan^{-1} 1$ (b) $\arctan(-1)$

   (c) $\tan^{-1}\sqrt{3}$

18. (a) $\arctan(-\sqrt{3})$ (b) $\tan^{-1} 0$

   (c) $\tan^{-1}\left(-\frac{1}{\sqrt{3}}\right)$

Exercises 19 and 20: If possible, approximate the following to a hundredth of a radian and a tenth of a degree.

19. (a) $\sin^{-1} 1.5$ (b) $\tan^{-1} 10$

   (c) $\arccos(-0.25)$

20. (a) $\cos^{-1}(-3)$ (b) $\arcsin(-0.54)$

   (c) $\arctan(-2.5)$

Exercises 21 and 22: Evaluate each expression using the figure to obtain either α or β.

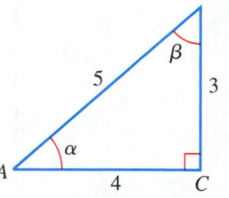

21. (a) $\tan^{-1}\frac{4}{3}$

   (b) $\sin^{-1}\frac{3}{5}$

   (c) $\arccos\frac{3}{5}$

22. (a) $\arcsin\frac{12}{13}$

   (b) $\cos^{-1}\frac{5}{13}$

   (c) $\tan^{-1}\frac{5}{12}$

Exercises 23–28: Evaluate each expression.

23. $\sin(\sin^{-1} 1)$ 24. $\sin^{-1}\left(\sin\frac{\pi}{4}\right)$

25. $\cos^{-1}\left(\cos\frac{5\pi}{4}\right)$ 26. $\cos(\cos^{-1}(-3))$

27. $\tan(\tan^{-1}(-3))$ 28. $\tan^{-1}\left(\tan\frac{\pi}{5}\right)$

Exercises 29 and 30: Evaluate the expression in degree mode. Make a generalization about the result and then test your conjecture.

29. (a) $\sin^{-1}\frac{3}{5} + \cos^{-1}\frac{3}{5}$ (b) $\sin^{-1}\frac{1}{3} + \cos^{-1}\frac{1}{3}$

   (c) $\sin^{-1}\frac{2}{7} + \cos^{-1}\frac{2}{7}$

30. (a) $\tan^{-1}\frac{3}{4} + \tan^{-1}\frac{4}{3}$ (b) $\tan^{-1}\frac{5}{12} + \tan^{-1}\frac{12}{5}$

   (c) $\tan^{-1}\frac{1}{4} + \tan^{-1} 4$

Exercises 31–34: Represent the given $f(x)$ verbally, numerically, and graphically.

31. $f(x) = \sin^{-1} 2x$ 32. $f(x) = \sin^{-1}\frac{1}{2}x$

33. $f(x) = \cos^{-1}\frac{1}{2}x$ 34. $f(x) = \cos^{-1} 2x$

Evaluating Inverses with Variables

Exercises 35–38: (Refer to Example 7.) Evaluate the indicated trigonometric function of θ, where θ is an acute angle determined by an inverse trigonometric function. (Hint: Make a sketch of a right triangle containing angle θ.)

35. $\tan\theta$, if $\theta = \sin^{-1}x$

36. $\sin\theta$, if $\theta = \tan^{-1}\dfrac{x}{\sqrt{1-x^2}}$

37. $\cos\theta$, if $\theta = \sin^{-1}\dfrac{x}{\sqrt{1+x^2}}$

38. $\tan\theta$, if $\theta = \cos^{-1}\dfrac{1}{x}$

Exercises 39–44: (Refer to Example 7.) Write the expression as an algebraic expression of u if $0 < u < 1$.

39. $\sin(\cos^{-1}u)$ **40.** $\cos(\sin^{-1}u)$

41. $\tan(\cos^{-1}u)$ **42.** $\sin(\tan^{-1}u)$

43. $\cot\left(\tan^{-1}\dfrac{1}{u}\right)$ **44.** $\sec(\sin^{-1}u)$

Solving Triangles

Exercises 45–50: (Refer to Example 9.) Solve the triangle.

45.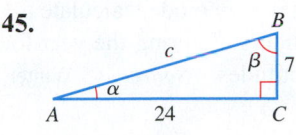

46.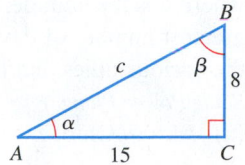

47.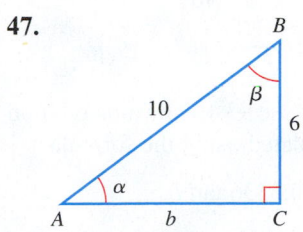

48.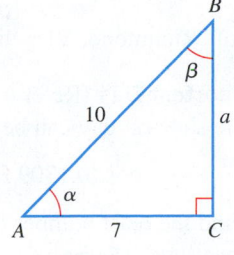

49.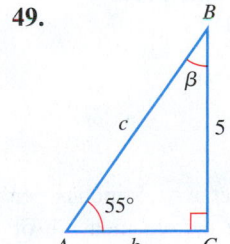

50.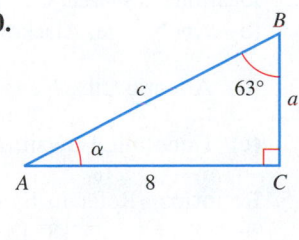

Solving Trigonometric Equations

Exercises 51–56: Solve the trigonometric equation for θ, where $0° \le \theta \le 90°$.

51. $\sin\theta = 1$ **52.** $\cos\theta = \dfrac{1}{2}$

53. $\tan\theta = 1$ **54.** $\sin\theta = \dfrac{\sqrt{3}}{2}$

55. $\cos\theta = 0$ **56.** $\tan\theta = \dfrac{1}{\sqrt{3}}$

Exercises 57–62: Solve the equation for θ, where θ is an acute angle. Approximate θ to the nearest tenth of a degree.

57. $2\cos\theta = \dfrac{1}{4}$ **58.** $3\sin\theta = \dfrac{4}{5}$

59. $\tan\theta - 1 = 5$ **60.** $4\cos\theta + 1 = 6$

61. $\sin^2\theta = 0.87$ **62.** $\tan^3\theta - 2 = 1.65$

Exercises 63–70: Solve the equation for t, where t is a real number in the given interval. Approximate t to three decimal places.

63. $\tan t = -\dfrac{1}{5}, \left(-\dfrac{\pi}{2}, \dfrac{\pi}{2}\right)$ **64.** $\sin t = -\dfrac{1}{3}, \left[-\dfrac{\pi}{2}, \dfrac{\pi}{2}\right]$

65. $\cos t = 0.452, [0, \pi]$ **66.** $\tan t = 5.67, \left(-\dfrac{\pi}{2}, \dfrac{\pi}{2}\right)$

67. $2\sin t = -0.557, \left[-\dfrac{\pi}{2}, \dfrac{\pi}{2}\right]$

68. $3\cos t + 1 = 0.333, [0, \pi]$

69. $\cos^2 t = \dfrac{1}{25}, [0, \pi]$ **70.** $\sin^2 t = \dfrac{1}{16}, \left[-\dfrac{\pi}{2}, \dfrac{\pi}{2}\right]$

Exercises 71–82: Find each of the following in radians. Approximate to the nearest thousandth when appropriate.

71. $\sec^{-1}(-1)$ **72.** $\sec^{-1}\sqrt{2}$

73. $\csc^{-1}(-\sqrt{2})$ **74.** $\csc^{-1}\dfrac{2\sqrt{3}}{3}$

75. $\cot^{-1}(-1)$ **76.** $\cot^{-1}0$

77. $\sec^{-1}3$ **78.** $\sec^{-1}(-4)$

79. $\csc^{-1}5.1$ **80.** $\csc^{-1}(-3)$

81. $\cot^{-1}1.5$ **82.** $\cot^{-1}(-7.1)$

Applications

83. Angle of Elevation Find the angle of elevation θ of the top of a 50-foot tree at a distance of 85 feet. See the figure.

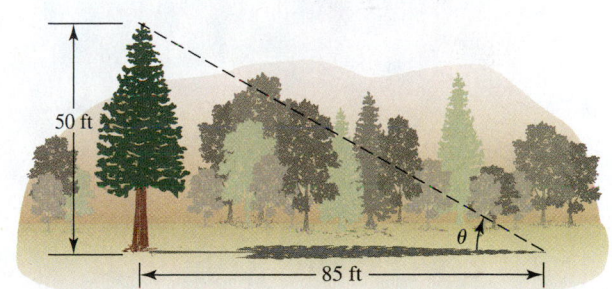

84. Angle of Elevation A 28-foot building casts a 40-foot shadow on level ground. Find the angle of elevation θ of the sun to the nearest tenth of a degree. See the figure.

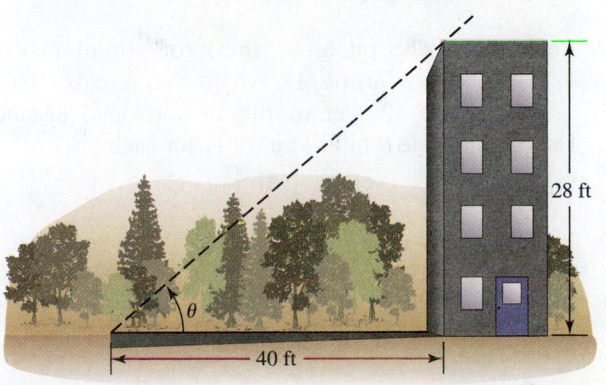

85. Grade Resistance (Refer to Example 11.) Approximate θ to the nearest tenth of a degree for the given grade resistance F and vehicle weight W. Use the equation $F = W \sin \theta$.

(a) $F = 400$ lb, $W = 5000$ lb

(b) $F = 130$ lb, $W = 3500$ lb

(c) $F = -200$ lb, $W = 4000$ lb

86. Robotics Approximate the angle θ if the robotic hand is located at the following points, where $-90° < \theta < 90°$. See the figure.

(a) $(5, 11)$ (b) $(1, -3)$

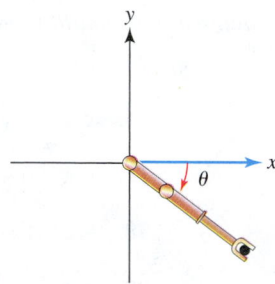

87. Designing Steps Steps are being attached to a deck as shown in the figure. The bottom of the steps should land 10 feet from the deck. If the deck is 4 feet above the ground, estimate angle θ between the ground and the side board of the step.

88. Designing Steps (Refer to Exercise 87.) If the length of the side boards for the steps is 4 feet and the deck is 2 feet above the ground, find angle θ.

89. Roof Pitch The pitch, or slope, of a roof may be expressed in the form $k/12$, where k represents a k-foot rise for every 12 feet of run in horizontal distance. Determine angle θ in Figure 6.141 for each pitch.

(a) $\frac{3}{12}$ (b) $\frac{4}{12}$ (c) $\frac{6}{12}$ (d) $\frac{12}{12}$

90. Phases of the Moon (See Example 11 in Section 6.3.) Find the phase angle θ for the given phase F. Assume that $0° \le \theta \le 180°$ and use $F(\theta) = \frac{1}{2}(1 - \cos \theta)$.

(a) $F = \frac{1}{4}$ (b) $F = \frac{3}{4}$

91. Shot Put (Refer to Example 3.) Suppose that a shot is released 7 feet above the ground with a velocity of 43 feet per second. Find the optimal projection angle.

92. Shot-Putting on the Moon Repeat Exercise 91 with $v = 50$ feet per second if the shot is thrown on the moon and the optimal projection angle is given by

$$\theta = \sin^{-1} \sqrt{\frac{v^2}{2v^2 + 10.2h}}.$$

93. Calculating Daylight Hours The ability to calculate the number of daylight hours H at any location is important for estimating potential solar energy production. The value of H on the longest day can be calculated using the formula

$$\cos (0.1309 H) = -0.4336 \tan L,$$

where L is the latitude. Using *radian* mode, calculate the greatest number of daylight hours H during the year for the various cities and their latitudes. (*Source:* C. Winter, *Solar Power Plants.*)

(a) Akron, Ohio; $L = 40°55'$

(b) Corpus Christi, Texas; $L = 27°46'$

(c) Richmond, Virginia; $L = 37°30'$

94. Shortest Day (Refer to Exercise 93.) The value of H on the shortest day can be calculated using the formula

$$\cos (0.1309 H) = 0.4336 \tan L.$$

Find the least number of daylight hours at the following locations. (*Source:* C. Winter, *Solar Power Plants.*)

(a) Anchorage, Alaska; $L = 61°10'$

(b) Atlantic City, New Jersey; $L = 39°27'$

(c) Honolulu, Hawaii; $L = 21°20'$

95. Robotics (Refer to Example 8.) Let $v = 5$ inches per second and $h = 18$. Determine the degree measure of θ_1 and θ_2 after 5 seconds.

96. Snell's Law When a ray of light enters water, it is bent. This change in direction can be calculated using Snell's law. See the figure. The angles θ_1 and θ_2 are related by the equation $n_1 \sin \theta_1 = n_2 \sin \theta_2$, where n_1 and n_2 are constants called *indexes of refraction*. For air $n_1 = 1$, and for water $n_2 = 1.33$. If a ray of light enters the water with $\theta_1 = 40°$, estimate θ_2. (*Source:* R. Weidner and R. Sells, *Elementary Classical Physics,* Vol. 2.)

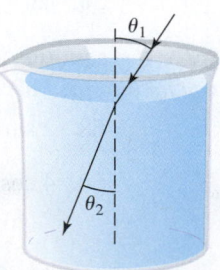

97. Landscaping Formula A shrub is planted in a 100-foot-wide space between buildings measuring 75 feet and 150 feet tall. The location of the shrub determines how much sun it receives each day. Show that if θ is the angle in the figure and x is the distance of the shrub from the taller building, then the value of θ (in radians) is given by the equation

$$\theta = \pi - \arctan\left(\frac{75}{100 - x}\right) - \arctan\left(\frac{150}{x}\right).$$

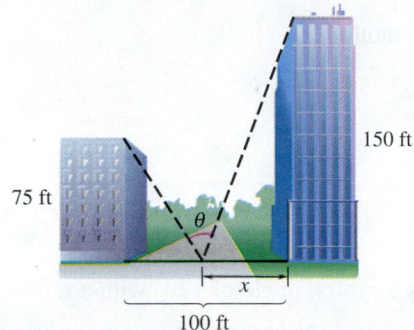

98. Communications Satellite Coverage The figure shows a stationary communications satellite positioned 20,000 miles above the equator. What percent of the equator can be seen from the satellite? The diameter of Earth is 7927 miles at the equator.

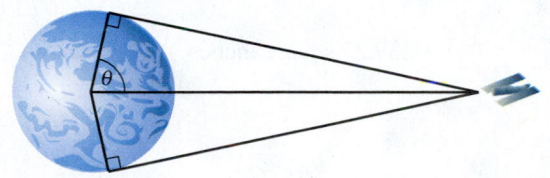

Writing about Mathematics

99. Explain verbally what each expression computes. Give examples.
(a) $\sin^{-1} x$ **(b)** $\cos^{-1} x$ **(c)** $\tan^{-1} x$

100. Explain why
$$\sin^{-1}\left(\sin\frac{\pi}{2}\right) = \frac{\pi}{2} \quad \text{but} \quad \sin^{-1}\left(\sin\frac{5\pi}{2}\right) \neq \frac{5\pi}{2}.$$
Give a similar example using $\cos x$ and $\cos^{-1} x$.

EXTENDED AND DISCOVERY EXERCISE

1. Movie Screen A 10-foot-high movie screen is mounted on a vertical wall so that the bottom of the screen is 6 feet above a horizontal floor. A person sits on a level floor x feet from the screen. If eye level is 3 feet above the floor, then angle θ shown in the figure can be expressed as

$$\theta = \tan^{-1}\left(\frac{10x}{x^2 + 39}\right).$$

Graph θ in [0, 50, 10] by [0, 50, 10] using degree mode. Determine where a person should sit to maximize θ.

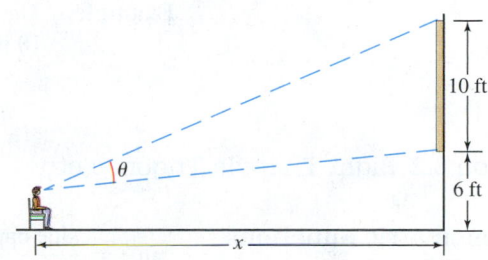

CHECKING BASIC CONCEPTS FOR SECTIONS 6.5 AND 6.6

1. Graph $f(t) = 3\sin\left(2\left(t - \frac{\pi}{4}\right)\right)$ for $-\pi \leq t \leq \pi$. State the amplitude, period, and phase shift.

2. The accompanying table contains data that can be modeled by $f(t) = a\cos(bt)$. Find values for a and b.

t	0	0.5	1.0	1.5	2.0	2.5	3.0
$f(t)$	2	0	-2	0	2	0	-2

3. Evaluate the following, expressing your answer in degrees.
(a) $\sin^{-1} 0$ **(b)** $\cos^{-1}(-1)$

(c) $\tan^{-1}(-1)$ **(d)** $\sin^{-1}\frac{1}{2}$

(e) $\tan^{-1}\sqrt{3}$ **(f)** $\cos^{-1}\frac{1}{2}$

4. Solve a right triangle if $a = 30$ and $b = 40$.

5. Use a calculator to solve each equation, where t is in the indicated interval.
(a) $\sin t = 0.55$, $\left[-\frac{\pi}{2}, \frac{\pi}{2}\right]$

(b) $\cos t = -0.35$, $[0, \pi]$

(c) $\tan t = -2.9$, $\left(-\frac{\pi}{2}, \frac{\pi}{2}\right)$

6 Summary

CONCEPT	EXPLANATION AND EXAMPLES

Section 6.1 Angles and Their Measure

Angle Measure

Degree measure: $360° =$ one revolution
$1° = 60'$ (minutes)
$1' = 60''$ (seconds)

Radian measure: 2π radians $=$ one revolution

radian measure $\times \frac{180°}{\pi} =$ degree measure

degree measure $\times \frac{\pi}{180°} =$ radian measure

Arc Length

$s = r\theta$, where θ is in *radians*

Example: The arc length intercepted by a central angle of 120° with a radius of 5 inches is

$$s = 5\left(\frac{2\pi}{3}\right) = \frac{10\pi}{3} \approx 10.47 \text{ inches.}$$

Area of Sector

$A = \frac{1}{2}r^2\theta$, where θ is in *radians*

Example: The area of the sector determined by a central angle of 45° with a radius of 10 inches is

$$A = \frac{1}{2}\left(10^2\right)\left(\frac{\pi}{4}\right) = \frac{25\pi}{2} \approx 39.27 \text{ square inches.}$$

Section 6.2 Right Triangle Trigonometry

Trigonometric Functions (Right Triangles)

$$\sin\theta = \frac{\text{side opposite}}{\text{hypotenuse}} \qquad \csc\theta = \frac{\text{hypotenuse}}{\text{side opposite}}$$

$$\cos\theta = \frac{\text{side adjacent}}{\text{hypotenuse}} \qquad \sec\theta = \frac{\text{hypotenuse}}{\text{side adjacent}}$$

$$\tan\theta = \frac{\text{side opposite}}{\text{side adjacent}} \qquad \cot\theta = \frac{\text{side adjacent}}{\text{side opposite}}$$

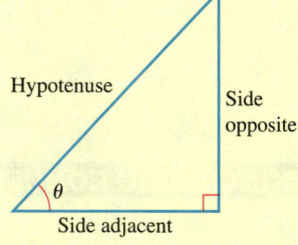

Example: The six trigonometric functions of θ in the figure are as follows.

$$\sin\theta = \frac{11}{61} \qquad \csc\theta = \frac{61}{11}$$

$$\cos\theta = \frac{60}{61} \qquad \sec\theta = \frac{61}{60}$$

$$\tan\theta = \frac{11}{60} \qquad \cot\theta = \frac{60}{11}$$

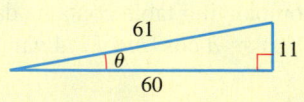

Cofunction Formulas

Let α and β be complementary angles.

$$\sin\alpha = \cos(90° - \alpha) = \cos\beta$$

$$\tan\alpha = \cot(90° - \alpha) = \cot\beta$$

$$\sec\alpha = \csc(90° - \alpha) = \csc\beta$$

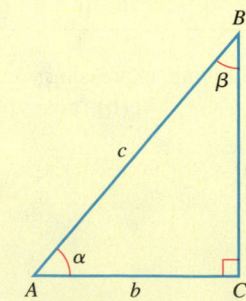

CONCEPT	EXPLANATION AND EXAMPLES

Section 6.3 The Sine and Cosine Functions and Their Graphs

Sine and Cosine

If angle θ is in standard position and its terminal side passes through the point (x, y), then

$$\sin\theta = \frac{y}{r} \quad \text{and} \quad \cos\theta = \frac{x}{r},$$

where $r = \sqrt{x^2 + y^2}$. In the figure to the right, $x = 3$, $y = -4$, and $r = \sqrt{3^2 + (-4)^2} = 5$.

$$\sin\theta = -\frac{4}{5} \quad \text{and} \quad \cos\theta = \frac{3}{5}$$

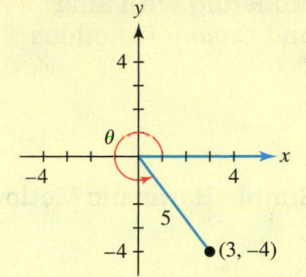

Unit Circle

If the terminal side of an angle t intersects the unit circle at the point (x, y), then $\sin t = y$ and $\cos t = x$. The domains of the sine and cosine functions include all real numbers, and their ranges include all real numbers y, such that $-1 \leq y \leq 1$.

In the figure, $x = -\frac{1}{2}$ and $y = \frac{\sqrt{3}}{2}$. Thus

$$\sin t = \frac{\sqrt{3}}{2} \quad \text{and} \quad \cos t = -\frac{1}{2}.$$

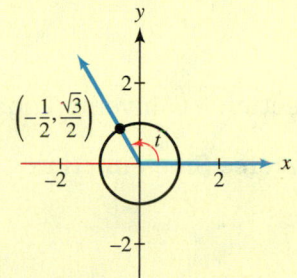

Section 6.4 Other Trigonometric Functions and Their Graphs

Trigonometric Functions

The domains, ranges, periods, and graphs of the six trigonometric functions are discussed in Putting It All Together in Section 6.4.

If angle θ is in standard position and its terminal side passes through the point (x, y), then it follows that

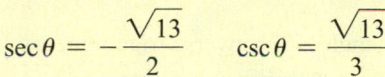

$$\tan\theta = \frac{y}{x}, \quad \cot\theta = \frac{x}{y}, \quad \sec\theta = \frac{r}{x}, \quad \text{and} \quad \csc\theta = \frac{r}{y}, \text{ where } r = \sqrt{x^2 + y^2}.$$

In the figure, $x = -2$, $y = 3$, and $r = \sqrt{(-2)^2 + 3^2} = \sqrt{13}$.

$$\tan\theta = -\frac{3}{2} \qquad \cot\theta = -\frac{2}{3}$$

$$\sec\theta = -\frac{\sqrt{13}}{2} \qquad \csc\theta = \frac{\sqrt{13}}{3}$$

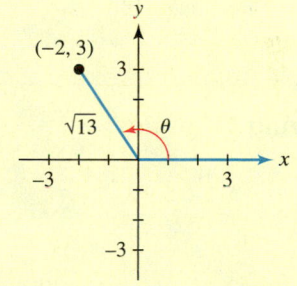

Unit Circle

If the terminal side of an angle t intersects the unit circle at the point (x, y), then

$$\tan t = \frac{y}{x}, \quad \cot t = \frac{x}{y}, \quad \sec t = \frac{1}{x}, \quad \text{and} \quad \csc t = \frac{1}{y}.$$

In the figure, $x = 0$ and $y = 1$. Thus

$\tan t$ is undefined; $\cot t = 0$;

$\sec t$ is undefined; $\csc t = 1$.

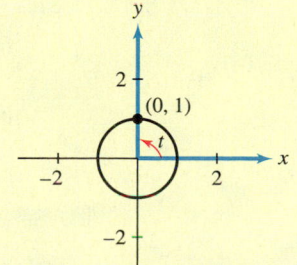

CONCEPT	EXPLANATION AND EXAMPLES

Section 6.5 Graphing Trigonometric Functions

Modeling with Sine and Cosine Functions

$f(x) = a \sin(b(x - c)) + d$ or $f(x) = a \cos(b(x - c)) + d$

Amplitude $= |a|$ Period $= \dfrac{2\pi}{b}, b > 0$

Phase shift $= c$ Vertical shift $= d$

Simple Harmonic Motion

Can be modeled by $s(t) = a \sin bt$ or $s(t) = a \cos bt$

If $b = 2\pi F$, then F represents the frequency of the sinusoidal graph.

Example: If a spring is initially compressed 3 inches and oscillates 4 times per second, then $a = 3$, $b = 2\pi F = 8\pi$, and its motion can be modeled by $s(t) = 3 \cos(8\pi t)$.

Section 6.6 Inverse Trigonometric Functions

Inverse Sine Function

$\theta = \sin^{-1} x$ implies that $\sin \theta = x$ and either $-\dfrac{\pi}{2} \leq \theta \leq \dfrac{\pi}{2}$ or $-90° \leq \theta \leq 90°$. $\sin^{-1} x$ is also denoted $\arcsin x$.

Example: $\sin^{-1} 1 = \dfrac{\pi}{2}, \quad \sin^{-1}\left(-\dfrac{1}{2}\right) = -30°, \quad \arcsin 0 = 0°$

Inverse Cosine Function

$\theta = \cos^{-1} x$ implies that $\cos \theta = x$ and either $0 \leq \theta \leq \pi$ or $0° \leq \theta \leq 180°$. $\cos^{-1} x$ is also denoted $\arccos x$.

Example: $\cos^{-1} 1 = 0, \quad \cos^{-1}(-1) = 180°, \quad \arccos \dfrac{1}{2} = 60°$

Inverse Tangent Function

$\theta = \tan^{-1} x$ implies that $\tan \theta = x$ and either $-\dfrac{\pi}{2} < \theta < \dfrac{\pi}{2}$ or $-90° < \theta < 90°$. $\tan^{-1} x$ is also denoted $\arctan x$.

Example: $\tan^{-1} 1 = \dfrac{\pi}{4}, \quad \tan^{-1}(-1) = -45°, \quad \arctan \sqrt{3} = 60°$

Solving Triangles

Inverse trigonometric functions can be used to solve equations and find angles in triangles.

Example: In triangle ABC, inverse trigonometric functions can be used to find α and β.

$\alpha = \sin^{-1} \dfrac{3}{5} \approx 36.9°$

$\alpha = \tan^{-1} \dfrac{3}{4} \approx 36.9°$

$\beta = \cos^{-1} \dfrac{3}{5} \approx 53.1°$

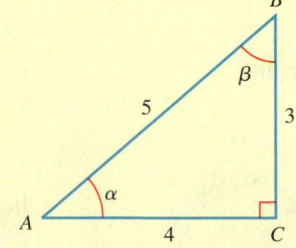

6 Review Exercises

1. Sketch the following angles in standard position.
   (a) 60° (b) $-\frac{5\pi}{6}$

2. Find the complementary angle and the supplementary angle to $\theta = 61°40'$.

3. Convert each angle from radians to degrees.
   (a) $\frac{\pi}{3}$ (b) $\frac{\pi}{36}$ (c) $-\frac{5\pi}{6}$ (d) $-\frac{7\pi}{4}$

4. Convert each angle from degrees to radians.
   (a) 30° (b) 165° (c) −90° (d) −105°

5. Find the length of the arc intercepted by a central angle $\theta = 60°$ and a radius $r = 6$ feet.

6. Find the area of the sector of a circle having a radius $r = 5$ inches and a central angle $\theta = 150°$.

Exercises 7–12: Find the exact value of each trigonometric expression by hand.

7. $\sin 30°$ 8. $\tan 45°$

9. $\cot 60°$ 10. $\cos 60°$

11. $\sec \frac{\pi}{4}$ 12. $\csc \frac{\pi}{6}$

13. Find the six trigonometric functions of θ.

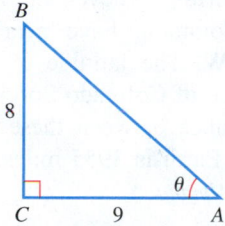

14. Solve triangle ABC.

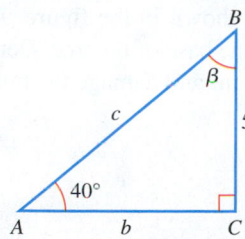

15. Find $\csc \theta$ if $\sin \theta = \frac{1}{3}$.

16. Find $\cot \theta$ if $\sin \theta = \frac{5}{13}$ and $\cos \theta = -\frac{12}{13}$.

Exercises 17 and 18: Approximate the six trigonometric functions of θ to three decimal places.

17. $\theta = 25°$ 18. $\theta = -\frac{6\pi}{7}$

Exercises 19 and 20: Find the six trigonometric functions of angle θ.

19.   20.

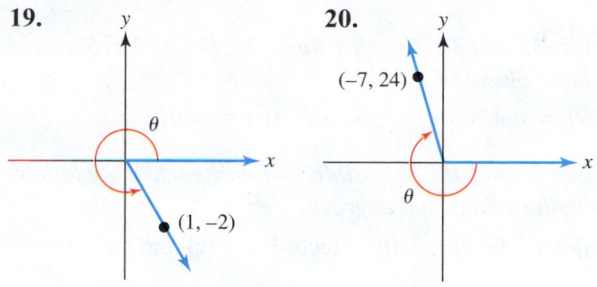

Exercises 21 and 22: The figure shows angle θ in standard position with its terminal side intersecting the unit circle. Find the six trigonometric functions of θ.

21.   22.

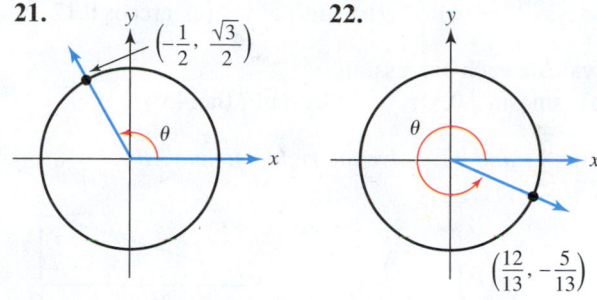

Exercises 23–26: (Wrapping function) Evaluate the function f at the given value of t.

23. $f(t) = \sin t, t = -\frac{\pi}{2}$ 24. $f(t) = \cos t, t = \pi$

25. $f(t) = \tan t, t = -3\pi$ 26. $f(t) = \csc t, t = \frac{\pi}{2}$

27. Find the other trigonometric functions of θ if $\sin \theta = -\frac{4}{5}$ and $\cos \theta = \frac{3}{5}$.

28. Convert $65°45'36''$ to decimal degrees.

Exercises 29–32: Graph f for $-2\pi \le t \le 2\pi$. State the amplitude, period, and phase shift.

29. $f(t) = 3\cos 2t$ 30. $f(t) = -2\sin(2t + \pi)$

31. $f(t) = -3\sin\left(\frac{1}{2}(t - \pi)\right) + 1$

32. $f(t) = \frac{1}{2}\cos\left(\frac{\pi}{2}(t - 1)\right) - 2$

Exercises 33 and 34: A graph of $y = a\cos bx$ is shown, where b is a positive constant. Estimate the values for a and b.

33.

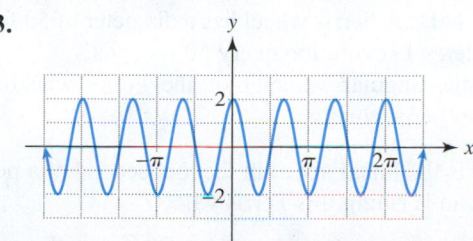

34.

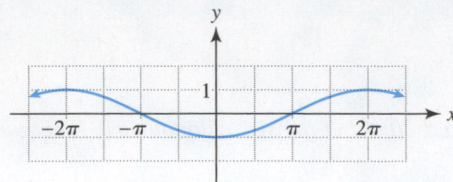

Exercises 35 and 36: Graph f for $-2\pi \le t \le 2\pi$. State the period and phase shift.

35. $f(t) = \cot 2t$ **36.** $f(t) = \sec 2t$

Exercises 37 and 38: If possible, evaluate each of the following in both radians and degrees.

37. **(a)** $\sin^{-1}(-1)$ **(b)** $\arccos \frac{1}{2}$ **(c)** $\tan^{-1} 1$

38. **(a)** $\arcsin 3$ **(b)** $\cos^{-1} 0$ **(c)** $\arctan(-\sqrt{3})$

39. Approximate the following to a hundredth of a radian and a tenth of a degree.
   (a) $\sin^{-1}(-0.6)$ **(b)** $\tan^{-1} 5$ **(c)** $\arccos 0.12$

40. Evaluate each expression.
   (a) $\sin(\sin^{-1} 0.5)$ **(b)** $\tan^{-1}(\tan 45°)$

Exercises 41 and 42: Solve the right triangle ABC.

41.

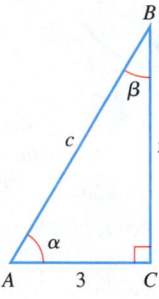

42.

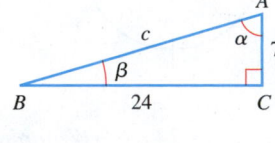

Exercises 43 and 44: Solve the equation for θ, where $0° \le \theta \le 90°$.

43. $\tan \theta = \frac{1}{\sqrt{3}}$ **44.** $\cos \theta = 1$

Exercises 45 and 46: Solve the equation for θ, where θ is an acute angle. Approximate θ to the nearest tenth of a degree.

45. $\cos \theta = \frac{1}{5}$ **46.** $3 \sin \theta = \frac{15}{13}$

Exercises 47 and 48: Solve the equation for t, where t is a real number located in the indicated interval. Approximate t to four decimal places.

47. $\tan t = -\frac{3}{4}, \left(-\frac{\pi}{2}, \frac{\pi}{2}\right)$ **48.** $\sin t = -\frac{3}{5}, \left[-\frac{\pi}{2}, \frac{\pi}{2}\right]$

Applications

49. Ferris Wheel   A Ferris wheel has a diameter of 50 feet and completes 1 revolution every 50 seconds.
   (a) Find the angular velocity of the Ferris wheel in radians per second.

   (b) What is the linear speed in feet per second of a person who is riding this Ferris wheel?

50. Fan Speed   The blades of a fan have a 25-inch diameter and rotate at 400 revolutions per minute.
   (a) Find the angular velocity of a fan blade.

   (b) Estimate the linear speed in inches per second at the tip of a fan blade.

51. Height of a Tree   Eighty feet from the trunk of a tree on level ground, the angle of elevation of the top of the tree is 48°. Estimate the height of the tree to the nearest foot.

52. Angle of Elevation   Find the angle of elevation of the top of a 35-foot building at a horizontal distance of 52 feet.

53. Grade Resistance   Approximate θ to the nearest tenth of a degree for the given grade resistance F and vehicle weight W, where $F = W \sin \theta$.
   (a) $F = 350$ lb, $W = 6000$ lb

   (b) $F = 160$ lb, $W = 4500$ lb

54. Highway Grade   (Refer to Exercise 53.) Suppose an uphill grade of a highway can be modeled by the line $y = 0.05x$.
   (a) Find the grade of the hill.

   (b) Determine the grade resistance for a gravel truck that weighs 30,000 pounds.

55. Distance Between Cities   Cheyenne, Wyoming, and Colorado Springs, Colorado, have nearly the same longitude of 104°45′W. The latitude of Cheyenne is 41°09′, and the latitude of Colorado Springs is 38°49′. Approximate the distance between these two cities if the average radius of Earth is 3955 miles. (*Source:* J. Williams, *The Weather Almanac.*)

56. Safe Distance for a Tree   From a distance of 45 feet from the base of a tree, the angle of elevation to the top of the tree is 57°, as shown in the figure. A building is located 52 feet from the base of the tree. Determine if the tree could fall in a storm and damage the building.

57. Modeling Temperatures The monthly average low temperatures in Green Bay, Wisconsin, are shown in the table.

Month	1	2	3	4	5	6
Temperature (°F)	6	10	22	35	45	52

Month	7	8	9	10	11	12
Temperature (°F)	58	56	48	38	26	11

Source: A. Miller and J. Thompson, *Elements of Meteorology.*

(a) Plot the monthly average temperature over a 24-month period by letting $x = 1$ and $x = 13$ correspond to January.

(b) Use $f(x) = a \cos(b(x - c)) + d$ to model the data.

(c) Graph f together with the data.

58. Phases of the Moon If the phase angle of the moon is θ, then the phase F of the moon is given by

$$F = \frac{1}{2}(1 - \cos\theta).$$

Solve this equation for θ.

Extended and Discovery Exercises

1. Surveying The first fundamental problem of surveying is to determine the coordinates of a point Q given the coordinates of a point P, the distance between P and Q, and the bearing θ from P to Q. See the figure. (*Source:* I. Mueller and K. Ramsayer, *Introduction to Surveying.*)

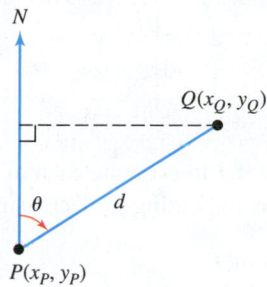

(a) Find a formula for the coordinates (x_Q, y_Q) of the point Q given θ, the coordinates (x_P, y_P) of P, and the distance d between P and Q.

(b) Use your formula to find the coordinates (x_Q, y_Q) if $(x_P, y_P) = (152, 186), \theta = 23.2°$, and $d = 208$ feet.

2. Highway Grade (Refer to Example 4, Section 6.3.) Complete the table at the top of the next column for the trigonometric values of θ to five decimal places.

(a) How do $\sin\theta$ and $\tan\theta$ compare for small values of θ?

θ	$\sin\theta$	$\tan\theta$
0°		
1°		
2°		
3°		
4°		

(b) Highway grades are usually small. Give an approximation to the grade resistance given by $F = W\sin\theta$ that uses the tangent function instead of the sine function.

(c) A stretch of highway has a 4-foot vertical rise for every 100 feet of horizontal run. Use your approximation to estimate the grade resistance for a 3000-pound car.

(d) Compare your result to the exact answer using $F = W\sin\theta$.

(*Source:* F. Mannering and W. Kilareski, *Principles of Highway Engineering and Traffic Analysis.*)

3. Average Temperature The maximum monthly average temperature in Buenos Aires, Argentina, is 74°F and the minimum monthly average temperature is 49°F.

(a) Using these two temperatures and the fact that the highest monthly average temperature occurs in January, find values for a, b, c, and d so that

$$f(x) = a \cos(b(x - c)) + d$$

models the monthly average temperature.

(b) On the same coordinate axes, graph f for a 2-year period together with the actual data values found in the table. Are your results as good as you expected? Explain.

(c) Buenos Aires is located in the Southern Hemisphere. Discuss the effect that this has on the graph of f compared to a city in the Northern Hemisphere.

Month	1	2	3	4	5	6
Temperature (°F)	74	73	69	61	55	50

Month	7	8	9	10	11	12
Temperature (°F)	49	51	55	60	66	71

Source: A. Miller and J. Thompson, *Elements of Meteorology.*

Cumulative Review Exercises

1. Write 125,000 in scientific notation and 4.67×10^{-3} in standard notation.

2. Find the midpoint of the line segment connecting the points $(-3, 2)$ and $(-1, 6)$.

3. Graph $y = g(x)$ by hand.
   (a) $g(x) = 2^x$ (b) $g(x) = |x + 2|$

   (c) $g(x) = \dfrac{1}{x + 1}$ (d) $g(x) = \sqrt{x - 2}$

Exercises 4 and 5: Complete the following.
   (a) *Determine the domain of f.*
   (b) *Evaluate $f(-1)$ and $f(2a)$.*

4. $f(x) = \sqrt{4 - x}$ 5. $f(x) = \dfrac{x - 2}{4x^2 - 16}$

6. The graph of a linear function f is shown.
   (a) Identify the slope, y-intercept, and x-intercept.

   (b) Write a formula for $f(x)$.

   (c) Evaluate $f(-2)$ symbolically and graphically.

   (d) Find any zeros of f.

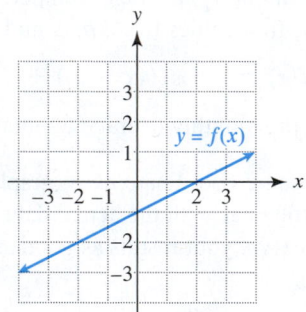

7. Find the average rate of change of $f(x) = 10^x$ from $x = 1$ to $x = 2$.

8. Find the difference quotient for $f(x) = 3x^2$.

9. Write the slope-intercept form of a line that passes through $(2, -3)$ and is parallel to the line $2x + 3y = 6$.

10. Determine the x- and y-intercepts on the graph of the equation $-2x + 5y = 20$.

11. *Solve each equation.*
    (a) $|4 - 5x| = 8$ (b) $2e^x - 1 = 27$

    (c) $x^3 - 3x^2 + 2x = 0$ (d) $\sqrt{2x - 1} = x - 2$

    (e) $x^2 - x - 2 = 0$ (f) $\log_2(x + 1) = 16$

12. Graph f. Is f continuous on its domain? Evaluate $f(1)$.
$$f(x) = \begin{cases} 1 - x & \text{if} \quad -4 \le x \le -1 \\ -2x & \text{if} \quad -1 < x < 2 \\ \frac{1}{2}x^2 & \text{if} \quad 2 \le x \le 4 \end{cases}$$

13. Solve the inequality. Write the solution set in interval notation.
    (a) $-3(2 - x) < 4 - (2x + 1)$

    (b) $-3 \le 4 - 3x < 6$ (c) $|4x - 3| \ge 9$

    (d) $x^2 - 5x + 4 \le 0$ (e) $t^3 - t > 0$

14. Write $f(x) = -2x^2 + 6x - 1$ in the vertex form given by $f(x) = a(x - h)^2 + k$.

15. Solve $2x^2 + 4x = 1$ by completing the square.

16. Use the given graph of $y = f(x)$ to sketch a graph of each expression.
    (a) $y = f(x - 1) + 2$ (b) $y = \frac{1}{2}f(x)$

    (c) $y = -f(-x)$ (d) $y = f(2x)$

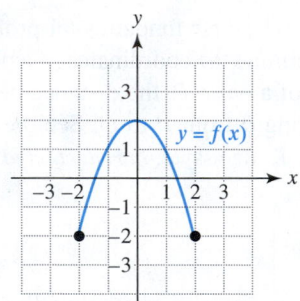

17. Use the graph of f to estimate each of the following.
    (a) Where f is increasing or decreasing

    (b) The zeros of f

    (c) The coordinates of any turning points

    (d) Any local extrema

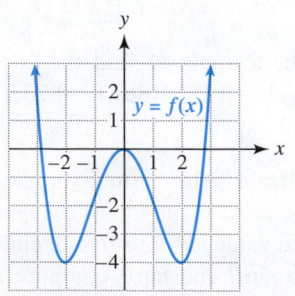

18. Write the complete factored form for the polynomial given by $f(x) = 2x^3 + x^2 - 8x - 4$.

19. Divide each expression.

(a) $\dfrac{5a^4 - 2a^2 + 4}{2a^2}$ **(b)** $\dfrac{x^3 - 3x^2 + x + 1}{x^2 + 1}$

20. Find all zeros, real or imaginary, of

$$f(x) = x^3 - x^2 + 4x - 4,$$

given that one zero is $2i$.

21. State the domain of $f(x) = \frac{2x + 5}{3x - 7}$. Find any vertical or horizontal asymptotes.

22. Write $x^{2/3}$ in radical notation. Evaluate the expression for $x = 27$.

23. Use the tables to evaluate each expression, if possible.

x	0	1	2	3	4
$f(x)$	4	3	2	1	0

x	0	1	2	3	4
$g(x)$	0	4	3	2	1

(a) $(f + g)(2)$ **(b)** $(g/f)(4)$

(c) $(f \circ g)(3)$ **(d)** $(f^{-1} \circ g)(1)$

24. Use the graphs of f and g to complete the following.

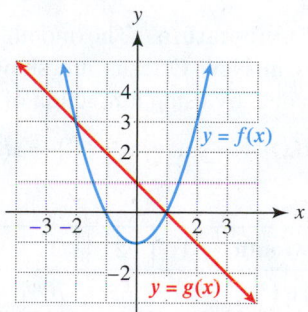

(a) $(f - g)(-1)$ **(b)** $(fg)(2)$

(c) $(g \circ f)(0)$ **(d)** $(g^{-1} \circ f)(2)$

25. Let $f(x) = x^2 + 3x - 2$ and $g(x) = x - 2$. Find the following.

(a) $(f + g)(2)$ **(b)** $(g \circ f)(1)$

(c) $(f - g)(x)$ **(d)** $(f \circ g)(x)$

26. Find $f^{-1}(x)$ if $f(x) = 2\sqrt[3]{x + 1}$.

27. Use the graph of $y = Ca^x$ to determine values for C and a.

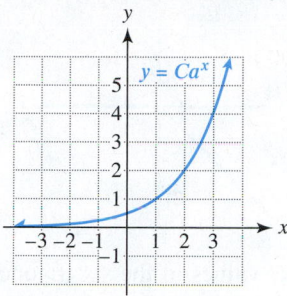

28. Five hundred dollars is deposited in an account that pays 5% annual interest compounded monthly. Find the amount in the account after 10 years.

29. Simplify each logarithm by hand.

(a) $\log 100$ **(b)** $\log_2 16$

(c) $\ln \dfrac{1}{e^2}$ **(d)** $\log 4 + \log 25$

30. Write $3 \log x - 4 \log y + \frac{1}{2} \log z$ as a logarithm of a single expression.

31. Approximate $\log_3 125$ to three decimal places.

32. Solve each equation.

(a) $3(2)^{-2x} + 4 = 100$

(b) $2 \log_3 3x = 4$

33. Convert $150°$ to radians.

34. Convert $\frac{5\pi}{4}$ radians to degrees.

35. Find the length of the arc intercepted by a central angle of $15°$ in a circle with a radius of 3 feet.

36. Find the six trigonometric functions of θ.

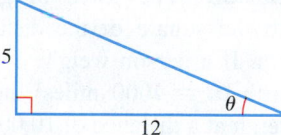

37. Find the length of a in the figure.

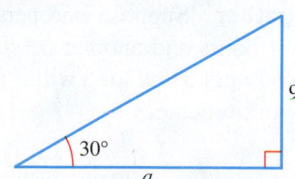

38. Approximate $\sec(1.24)$ to the nearest hundredth.

39. Find the six trigonometric functions of θ.

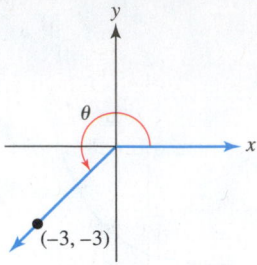

40. Find the exact values of the six trigonometric functions of $\theta = \frac{7\pi}{6}$.

41. Find the exact values of the six trigonometric functions of θ if $\cos \theta = \frac{5}{13}$ and $\sin \theta < 0$.

42. State the amplitude, period, phase shift, and vertical shift for the graph of $f(x) = 5 \sin \left(\frac{\pi}{2}(x - 1) \right) - 2$.

43. Sketch a graph of f on the interval $[-2\pi, 2\pi]$.
   (a) $f(x) = 2 \sin 3x$

   (b) $f(x) = \sec x$

   (c) $f(x) = -2 \cos \left(2x + \frac{\pi}{2} \right) + 1$

44. Evaluate $\sin^{-1} \left(-\frac{1}{2} \right)$.

45. Solve the right triangle.

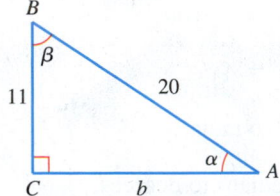

46. Solve $2 \cos \theta = -1$ if $0 \le \theta \le 180°$.

Applications

47. Inverse Variation The force of gravity F varies inversely with the square of the distance d from the *center* of Earth. If a person weighs 150 pounds on the surface of Earth ($d = 4000$ miles), how much would this person weigh at a distance of 10,000 miles from the center of Earth?

48. Working Together Suppose one person can mow a large lawn in 4 hours and another person can mow the same lawn in 6 hours. How long will it take to mow the lawn if they work together?

49. Inverse Function The function given by the formula $f(x) = \frac{5}{9}(x - 32)$ converts degrees Fahrenheit to degrees Celsius.
   (a) Find $f^{-1}(x)$.

   (b) What does f^{-1} compute?

50. Volume of a Balloon The radius r in inches of a spherical balloon after t seconds is given by $r = \sqrt{t}$.
   (a) Is the radius increasing or decreasing?

   (b) Write a formula for a function V that calculates the volume of the sphere after t seconds.

   (c) Evaluate $V(4)$ and interpret the result.

51. Bacteria Growth There are initially 200,000 bacteria per milliliter in a sample. The number of bacteria grows exponentially and reaches 300,000 per milliliter after 3 hours.
   (a) Use the formula $N(t) = N_0 e^{kt}$ to model the concentration of bacteria after t hours.

   (b) Evaluate $N(5)$ and interpret the result.

   (c) After how long did the concentration reach 500,000 per milliliter?

52. Length of a Shadow If the angle of elevation of the sun is 43°, find the length of the shadow cast by a person who is 6 feet tall. Round your answer to the nearest tenth of a foot.

53. Modeling Temperature The monthly average high temperatures in New Orleans are shown in the table. Model these data by using a function of the form

$$f(x) = a \sin (b(x - c)) + d.$$

Month	1	2	3	4	5	6
Temperature (°F)	61	65	70	78	85	89

Month	7	8	9	10	11	12
Temperature (°F)	90	89	88	80	70	63

Source: J. Williams, *The Weather Almanac.*

54. Angle of Elevation A 52-foot building cast a 63-foot shadow. Estimate the angle of elevation of the sun to the nearest tenth of a degree.

7 Trigonometric Identities and Equations

Music is both art and science. Although Pythagoras is usually associated with the Pythagorean theorem, in 500 B.C. he also discovered the mathematical ratios that governed pitch and motion. These discoveries marked the beginning of the science of musical sound. In 1862, the psychologist and scientist Hermann von Helmholtz published his classic work that opened a new direction for music using mathematics and technology. Then in 1957, Max Mathews created complex musical sounds with a computer. Mathematics has been essential in the development and reproduction of music.

The communications industry also uses the mathematics of sound. For example, each number on a touch-tone phone has a unique sound that is a combination of two different tones. These unique tones are easily distinguished when transmitted as cellular phone signals. Trigonometric functions play an important role in clear, reliable communication systems.

This chapter introduces the concept of verifying an identity. Identities are important because they allow us to write trigonometric expressions in simpler and more convenient forms.

> Music is the pleasure the human mind experiences from counting without being aware that it is counting.
> —Gottfried Leibniz

Source: J. Pierce, *The Science of Musical Sound.*

7.1 Fundamental Identities

- Learn and apply the reciprocal and quotient identities
- Learn and apply the Pythagorean identities
- Learn and apply the negative-angle identities

Introduction

Trigonometric expressions can often be written in more than one way. For example, in Section 6.4 we discussed that $\cot \theta$ is equivalent to $\frac{\cos \theta}{\sin \theta}$. The equation

$$\cot \theta = \frac{\cos \theta}{\sin \theta} \qquad \textit{Trigonometric identity}$$

is a *trigonometric identity*. This identity is true for every value of θ, provided $\sin \theta \neq 0$. Trigonometric identities are used to help solve equations and model physical phenomena such as light intensity and temperature variations. (See Example 11 and Exercises 85 and 89.) They are also used in calculus. We begin our discussion with the reciprocal and quotient identities.

Reciprocal and Quotient Identities

In Section 6.4, the following definitions were presented for the trigonometric functions of any angle θ. See Figure 7.1.

Visualizing Trigonometric Functions

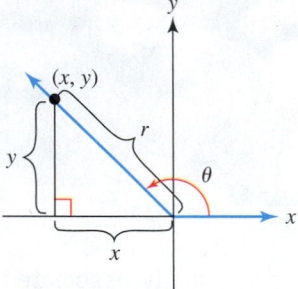

Figure 7.1

TRIGONOMETRIC FUNCTIONS OF ANY ANGLE θ

Let (x, y) be a point other than the origin on the terminal side of an angle θ in standard position. If $r = \sqrt{x^2 + y^2}$, then the six trigonometric functions are as follows.

$$\sin \theta = \frac{y}{r} \qquad\qquad \csc \theta = \frac{r}{y} \,(y \neq 0)$$

$$\cos \theta = \frac{x}{r} \qquad\qquad \sec \theta = \frac{r}{x} \,(x \neq 0)$$

$$\tan \theta = \frac{y}{x} \,(x \neq 0) \qquad \cot \theta = \frac{x}{y} \,(y \neq 0)$$

NOTE The definitions of trigonometric functions of angle θ are valid whether angle θ is measured in radians or degrees.

These definitions allow us to write several identities. For example,

$$\cos \theta = \frac{x}{r} = \frac{1}{r/x} = \frac{1}{\sec \theta} \qquad \text{and} \qquad \sec \theta = \frac{r}{x} = \frac{1}{x/r} = \frac{1}{\cos \theta}.$$

These identities are examples of *reciprocal identities*. Each of the six trigonometric functions can be written as the reciprocal of another trigonometric function.

RECIPROCAL IDENTITIES

$$\sin \theta = \frac{1}{\csc \theta} \qquad \cos \theta = \frac{1}{\sec \theta} \qquad \tan \theta = \frac{1}{\cot \theta}$$

$$\csc \theta = \frac{1}{\sin \theta} \qquad \sec \theta = \frac{1}{\cos \theta} \qquad \cot \theta = \frac{1}{\tan \theta}$$

In the next example, we use reciprocal identities to write a trigonometric expression in a more simplified form.

EXAMPLE 1 Applying reciprocal identities

Use reciprocal identities to rewrite

$$\frac{1}{\csc^2\theta} + \frac{1}{\sec^2\theta}$$

in terms of $\sin\theta$ and $\cos\theta$ and then simplify if possible.

SOLUTION

Getting Started First note that $\csc^2\theta = (\csc\theta)^2$ and $\sec^2\theta = (\sec\theta)^2$. The reciprocal identities $\sin\theta = \frac{1}{\csc\theta}$ and $\cos\theta = \frac{1}{\sec\theta}$ can be used to rewrite the given identity as follows. ▶

$$\frac{1}{\csc^2\theta} + \frac{1}{\sec^2\theta} = \left(\frac{1}{\csc\theta}\right)^2 + \left(\frac{1}{\sec\theta}\right)^2 \qquad \text{Rewrite: } \frac{1}{x^2} = \left(\frac{1}{x}\right)^2.$$

$$= (\sin\theta)^2 + (\cos\theta)^2 \qquad \text{Reciprocal identities}$$

$$= \sin^2\theta + \cos^2\theta \qquad \text{Rewrite.}$$

$$= 1 \qquad \text{Identity}$$

(The identity $\sin^2\theta + \cos^2\theta = 1$ was introduced in Section 6.3. We discuss this identity in more depth later in this section.)

Now Try Exercise 13

NOTE When writing the square of $\sin\theta$, it is common to write $\sin^2\theta$ instead of $(\sin\theta)^2$. This is also true for the other trigonometric functions.

By using reciprocal identities it is possible to write $\csc\theta$ and $\sec\theta$ in terms of $\sin\theta$ and $\cos\theta$, respectively. It is also possible to write the other two trigonometric functions, $\tan\theta$ and $\cot\theta$, in terms of $\sin\theta$ and $\cos\theta$.

$$\tan\theta = \frac{y}{x} = \frac{y/r}{x/r} = \frac{\sin\theta}{\cos\theta} \qquad \text{and} \qquad \cot\theta = \frac{x}{y} = \frac{x/r}{y/r} = \frac{\cos\theta}{\sin\theta}$$

These identities are called the *quotient identities*.

QUOTIENT IDENTITIES

$$\tan\theta = \frac{\sin\theta}{\cos\theta} \qquad \cot\theta = \frac{\cos\theta}{\sin\theta}$$

NOTE The reciprocal and quotient identities are valid not only for angles but also for any real number t. For example, $\tan t = \frac{\sin t}{\cos t}$ for all real numbers t such that $\cos t \neq 0$.

Evaluating Trigonometric Functions If $\sin\theta$ and $\cos\theta$ are known, the reciprocal and quotient identities can be used to find the other four trigonometric functions of θ, as illustrated in the next example.

EXAMPLE 2 Using reciprocal and quotient identities

If $\sin\theta = \frac{7}{25}$ and $\cos\theta = -\frac{24}{25}$, find the other four trigonometric functions of θ.

SOLUTION We can use identities to find $\tan\theta$, $\cot\theta$, $\sec\theta$, and $\csc\theta$.

$$\tan\theta = \frac{\sin\theta}{\cos\theta} = \frac{7/25}{-24/25} = -\frac{7}{24}, \qquad \cot\theta = \frac{\cos\theta}{\sin\theta} = \frac{-24/25}{7/25} = -\frac{24}{7},$$

$$\sec\theta = \frac{1}{\cos\theta} = \frac{1}{-24/25} = -\frac{25}{24}, \qquad \text{and} \qquad \csc\theta = \frac{1}{\sin\theta} = \frac{1}{7/25} = \frac{25}{7}$$

Now Try Exercise 15

> **NOTE** If the value of a trigonometric function of θ is 0, the reciprocal and quotient identities will reveal that two of the remaining five trigonometric functions are undefined at θ. For example, if $\sin\theta = 0$ then both $\csc\theta = \frac{1}{\sin\theta}$ and $\cot\theta = \frac{\cos\theta}{\sin\theta}$ are undefined.

EXAMPLE 3 Using identities to find trigonometric values

If $\tan\theta = -\frac{8}{15}$ and $\cos\theta = -\frac{15}{17}$, find the other four trigonometric functions of θ.

SOLUTION Using the reciprocal identities, we can find $\cot\theta$ and $\sec\theta$.

$$\cot\theta = \frac{1}{\tan\theta} = \frac{1}{-8/15} = -\frac{15}{8} \qquad \text{and} \qquad \sec\theta = \frac{1}{\cos\theta} = \frac{1}{-15/17} = -\frac{17}{15}$$

To find $\sin\theta$, consider the following.

$$\tan\theta\cos\theta = \frac{\sin\theta}{\cos\theta}\cdot\cos\theta = \sin\theta$$

Thus

$$\sin\theta = \tan\theta\cos\theta = \left(-\frac{8}{15}\right)\left(-\frac{15}{17}\right) = \frac{8}{17}$$

and, using a reciprocal identity, we find

$$\csc\theta = \frac{1}{\sin\theta} = \frac{1}{8/17} = \frac{17}{8}.$$

Now Try Exercise 17

Pythagorean Identities

Unit Circle

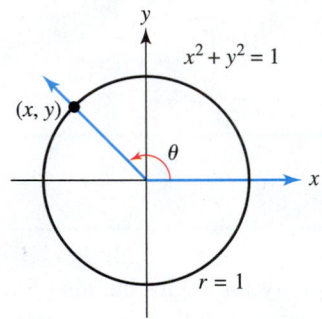

Figure 7.2

In Section 6.3, the unit circle (shown in Figure 7.2) was used to define the sine and cosine functions.

$$\sin\theta = y \qquad \text{and} \qquad \cos\theta = x$$

An equation for the unit circle is given by $x^2 + y^2 = 1$. By substitution, it follows that

$$(\cos\theta)^2 + (\sin\theta)^2 = 1,$$

or equivalently, $\cos^2\theta + \sin^2\theta = 1$. This identity can be supported graphically and numerically by letting

$$Y_1 = (\cos(X))^2 + (\sin(X))^2,$$

as shown in Figures 7.3 and 7.4. The graph of Y_1 is the horizontal line $y = 1$. Either radian or degree mode may be used. Notice that regardless of the value of x, $Y_1 = 1$.

$[-352.5, 352.5, 90]$ by $[-2, 2, 1]$

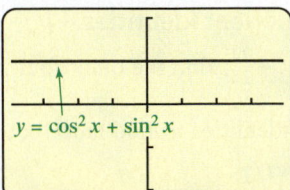

Figure 7.3 Degree Mode

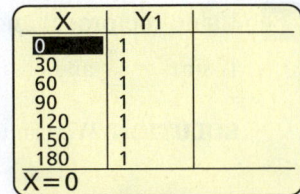

Figure 7.4 Degree Mode

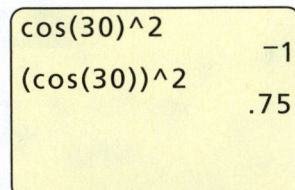

Figure 7.5 Degree Mode

> **NOTE** On some older graphing calculators, $\cos^2 x$ must be entered as $(\cos(X))^2$, rather than $\cos(X)^2$, since $\cos(X)^2$, may be interpreted as $\cos(x^2)$. See Figure 7.5.

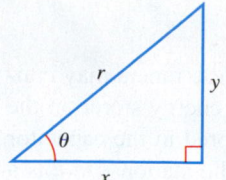

Figure 7.6

Deriving the Pythagorean Identities Using the right triangle shown in Figure 7.6, we also can derive the Pythagorean identity $\cos^2\theta + \sin^2\theta = 1$ geometrically when θ is acute.

$$x^2 + y^2 = r^2 \qquad \text{Pythagorean theorem}$$

$$\frac{x^2}{r^2} + \frac{y^2}{r^2} = \frac{r^2}{r^2} \qquad \text{Divide each side by } r^2.$$

$$\left(\frac{x}{r}\right)^2 + \left(\frac{y}{r}\right)^2 = 1 \qquad \text{Properties of exponents}$$

$$\cos^2\theta + \sin^2\theta = 1 \qquad \cos\theta = \tfrac{x}{r}, \sin\theta = \tfrac{y}{r}$$

For this reason, $\sin^2\theta + \cos^2\theta = 1$ is an example of a Pythagorean identity. Another Pythagorean identity can be derived as follows.

$$\sin^2\theta + \cos^2\theta = 1$$

$$\frac{\sin^2\theta}{\cos^2\theta} + \frac{\cos^2\theta}{\cos^2\theta} = \frac{1}{\cos^2\theta} \qquad \text{Divide each side by } \cos^2\theta.$$

$$\tan^2\theta + 1 = \sec^2\theta \qquad \tan\theta = \tfrac{\sin\theta}{\cos\theta}, \sec\theta = \tfrac{1}{\cos\theta}$$

In a similar manner, a third Pythagorean identity can be found.

$$\sin^2\theta + \cos^2\theta = 1$$

$$\frac{\sin^2\theta}{\sin^2\theta} + \frac{\cos^2\theta}{\sin^2\theta} = \frac{1}{\sin^2\theta} \qquad \text{Divide each side by } \sin^2\theta.$$

$$1 + \cot^2\theta = \csc^2\theta \qquad \cot\theta = \tfrac{\cos\theta}{\sin\theta}, \csc\theta = \tfrac{1}{\sin\theta}$$

This discussion is summarized in the following box.

CLASS DISCUSSION

Use the triangle in Figure 7.6 to justify the Pythagorean identity

$$1 + \tan^2\theta = \sec^2\theta.$$

PYTHAGOREAN IDENTITIES

$$\sin^2\theta + \cos^2\theta = 1 \qquad 1 + \tan^2\theta = \sec^2\theta \qquad 1 + \cot^2\theta = \csc^2\theta$$

In the next example, we use a reciprocal, quotient, and Pythagorean identity to simplify a trigonometric expression.

EXAMPLE 4 Applying identities to an expression

Simplify the expression $1 + \sin^2\theta\sec^2\theta$.

SOLUTION Begin by applying a reciprocal identity.

$$1 + \sin^2\theta\,\mathbf{sec^2\theta} = 1 + \sin^2\theta \cdot \frac{\mathbf{1}}{\mathbf{cos^2\,\theta}} \qquad \text{Reciprocal identity: } \sec\theta = \tfrac{1}{\cos\theta}$$

$$= 1 + \frac{\sin^2\theta}{\cos^2\theta} \qquad \text{Multiply.}$$

$$= 1 + \left(\frac{\mathbf{sin\,\theta}}{\mathbf{cos\,\theta}}\right)^2 \qquad \text{Properties of exponents}$$

$$= 1 + \mathbf{tan^2\,\theta} \qquad \text{Quotient identity}$$

$$= \sec^2\theta \qquad \text{Pythagorean identity: } 1 + \tan^2\theta = \sec^2\theta$$

Now Try Exercise 35

An Application The next example illustrates the use of identities in electronics.

EXAMPLE 5 **Applying a Pythagorean identity to radios**

Inductor L, Capacitor C

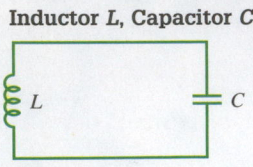

Figure 7.7

Tuners in radios select a radio station by adjusting the frequency. These tuners may contain an inductor L and a capacitor C, as illustrated in Figure 7.7. The energy stored in the inductor at time t is given by $L(t) = k \sin^2(2\pi Ft)$, and the energy stored in the capacitor is given by $C(t) = k \cos^2(2\pi Ft)$, where F is the frequency of the radio station and k is a constant. The total energy E in the circuit is given by $E(t) = L(t) + C(t)$. Show that E is a constant function. (***Source:*** R. Weidner and R. Sells, *Elementary Classical Physics,* Vol. 2.)

SOLUTION

$$
\begin{aligned}
E(t) &= L(t) + C(t) &&\text{Given equation} \\
&= k\sin^2(2\pi Ft) + k\cos^2(2\pi Ft) &&\text{Substitute.} \\
&= k\left(\mathbf{\sin^2(2\pi Ft) + \cos^2(2\pi Ft)}\right) &&\text{Factor out } k. \\
&= k(\mathbf{1}) &&\sin^2\theta + \cos^2\theta = 1 \,(\theta = 2\pi Ft) \\
&= k &&k \text{ is constant.}
\end{aligned}
$$

Now Try Exercise 87

Angle θ in Quadrant II

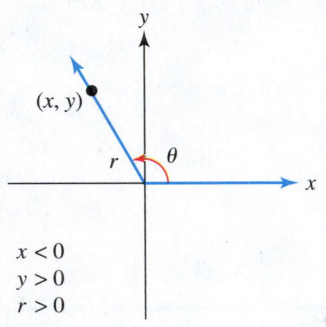

$x < 0$
$y > 0$
$r > 0$

Figure 7.8

Quadrants and Signs of Trigonometric Functions If an angle θ is in standard position and its terminal side lies in quadrant II, as shown in Figure 7.8, then we say that θ is *contained in quadrant II*, or θ is a *second quadrant angle*. Similar statements can be made for angles whose terminal sides lie in other quadrants.

If an angle θ is contained in a particular quadrant, then any point (x, y) on the terminal side of θ with $r = \sqrt{x^2 + y^2} > 0$ also lies in that quadrant. For example, the point (x, y) on the terminal side of angle θ in Figure 7.8 lies in quadrant II because θ is a second quadrant angle.

Since each of the six trigonometric functions can be defined in terms of x, y, and r, with $r > 0$, we can determine whether a trigonometric function of θ is positive or negative by simply considering the quadrant containing θ. For example, for angle θ in Figure 7.8, $\sin\theta = \frac{y}{r}$ is positive because both y and r are positive in quadrant II. Similarly, $\tan\theta = \frac{y}{x}$ is negative because x is negative and y is positive in quadrant II.

Figure 7.9 in the following See the Concept shows whether trigonometric functions of θ are positive or negative as determined by the quadrant containing θ.

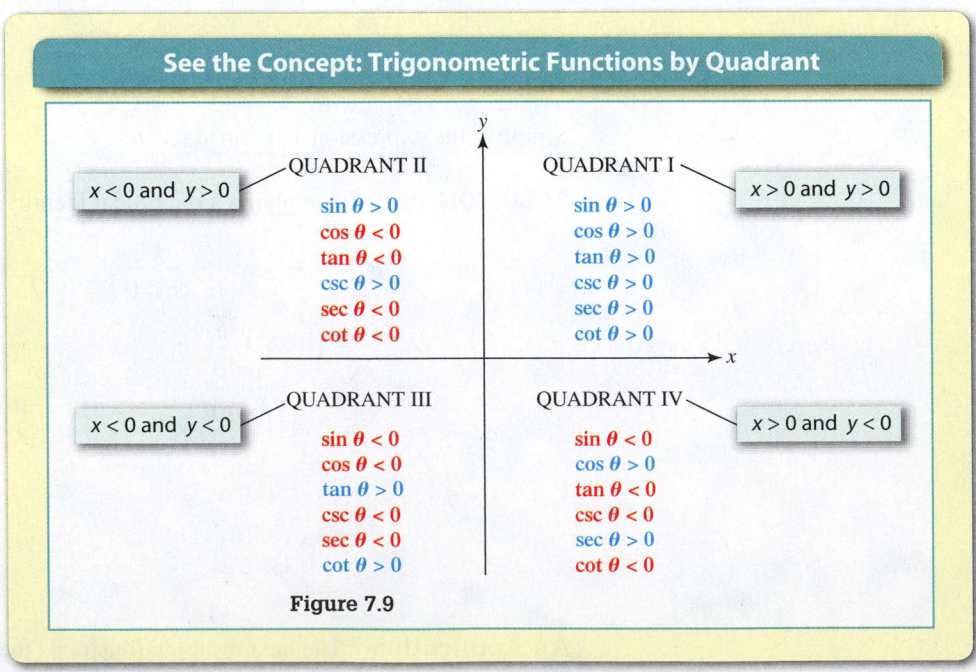

Figure 7.9

EXAMPLE 6   **Finding the quadrant containing an angle**

If $\sin \theta > 0$ and $\cos \theta < 0$, find the quadrant that contains θ. Support your results graphically and numerically.

SOLUTION   If $\sin \theta > 0$ and $\cos \theta < 0$, then any point (x, y) on the terminal side of θ must satisfy $y > 0$ and $x < 0$. Thus θ is contained in quadrant II.

Graphical Support   The graphs of $Y_1 = \sin (X)$ and $Y_2 = \cos (X)$ are shown in Figure 7.10. Notice that when $\theta = 135°$ (a second quadrant angle), the graph of $\sin \theta$ is above the x-axis and the graph of $\cos \theta$ is below the x-axis.

Numerical Support   In Figure 7.11, angles in quadrant II have positive sine values listed under Y_1 and negative cosine values listed under Y_2.

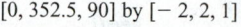

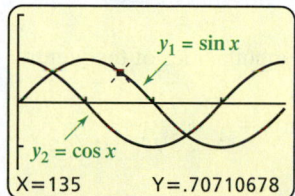

Figure 7.10 Degree Mode   **Figure 7.11** Degree Mode

Now Try Exercise 49

Evaluating Trigonometric Functions   If a trigonometric function of θ is known and the quadrant containing θ is also known, then we can find the other five trigonometric functions of θ, as illustrated in the next example.

EXAMPLE 7   **Using identities to find trigonometric values**

If $\sin \theta = -\frac{3}{5}$ and θ is a third quadrant angle, find the values of the other trigonometric functions.

SOLUTION

Getting Started   If we can find $\cos \theta$, then we can determine the other four trigonometric functions by using reciprocal and quotient identities. To determine $\cos \theta$, we can use the identity $\sin^2 \theta + \cos^2 \theta = 1$ and the fact that θ is a third quadrant angle. ▶

$$\sin^2 \theta + \cos^2 \theta = 1 \qquad \text{Pythagorean identity}$$
$$\cos^2 \theta = 1 - \sin^2 \theta \qquad \text{Subtract } \sin^2 \theta.$$
$$\cos \theta = \pm \sqrt{1 - \sin^2 \theta} \qquad \text{Square root property}$$
$$\cos \theta = \pm \sqrt{1 - (-3/5)^2} \qquad \text{Substitute for } \sin \theta.$$
$$\cos \theta = \pm \frac{4}{5} \qquad \text{Simplify.}$$

In quadrant III, $x < 0$, and so $\cos \theta = -\frac{4}{5}$. The other four trigonometric functions of θ can be found by letting $\sin \theta = -\frac{3}{5}$ and $\cos \theta = -\frac{4}{5}$ and applying identities.

$$\tan \theta = \frac{\sin \theta}{\cos \theta} = \frac{-3/5}{-4/5} = \frac{3}{4}, \qquad \cot \theta = \frac{\cos \theta}{\sin \theta} = \frac{-4/5}{-3/5} = \frac{4}{3},$$

$$\sec \theta = \frac{1}{\cos \theta} = \frac{1}{-4/5} = -\frac{5}{4}, \quad \text{and} \quad \csc \theta = \frac{1}{\sin \theta} = \frac{1}{-3/5} = -\frac{5}{3}$$

Now Try Exercise 63

EXAMPLE 8 Using identities to find trigonometric values

If $\tan\theta = -\frac{7}{3}$ and $\sin\theta > 0$, find the values of the other trigonometric functions.

SOLUTION Since $\tan\theta = -\frac{7}{3}$, it follows that $\cot\theta = -\frac{3}{7}$. Next we find $\sec\theta$.

$$\sec^2\theta = 1 + \tan^2\theta \qquad \text{Pythagorean identity}$$

$$\sec\theta = \pm\sqrt{1 + \tan^2\theta} \qquad \text{Square root property}$$

$$= \pm\sqrt{1 + (-7/3)^2} \qquad \text{Substitute for } \tan\theta.$$

$$= \pm\frac{\sqrt{58}}{3} \qquad \text{Simplify.}$$

Since $\tan\theta < 0$ and $\sin\theta > 0$, θ is a second quadrant angle. It follows that

$$\sec\theta = -\frac{\sqrt{58}}{3} \quad \text{and} \quad \cos\theta = -\frac{3}{\sqrt{58}}.$$

Multiplying both sides of the identity $\tan\theta = \frac{\sin\theta}{\cos\theta}$ by $\cos\theta$ gives $\sin\theta = \tan\theta \cos\theta$. Thus

$$\sin\theta = \left(-\frac{7}{3}\right)\left(-\frac{3}{\sqrt{58}}\right) = \frac{7}{\sqrt{58}} \quad \text{and} \quad \csc\theta = \frac{1}{\sin\theta} = \frac{\sqrt{58}}{7}.$$

Now Try Exercise 57

We can also express any trigonometric function in terms of any other trigonometric function. This fact is demonstrated in the next example.

EXAMPLE 9 Expressing one function in terms of another

Write $\cos x$ and $\cot x$ in terms of $\sin x$, if $\sec x < 0$.

SOLUTION We can write $\cos x$ in terms of $\sin x$ by applying a Pythagorean identity. Note that $\sec x < 0$ implies that $\cos x < 0$ because $\cos x = \frac{1}{\sec x}$ by a reciprocal identity.

$$\sin^2 x + \cos^2 x = 1 \qquad \text{Pythagorean identity}$$

$$\cos^2 x = 1 - \sin^2 x \qquad \text{Subtract } \sin^2 x.$$

$$\cos x = \pm\sqrt{1 - \sin^2 x} \qquad \text{Square root property}$$

Because $\cos x < 0$, we let $\cos x = -\sqrt{1 - \sin^2 x}$. Now we can write $\cot x$ in terms of $\sin x$ by using a quotient identity.

$$\cot x = \frac{\cos x}{\sin x} \qquad \text{Quotient identity}$$

$$= -\frac{\sqrt{1 - \sin^2 x}}{\sin x} \qquad \text{Substitute for } \cos x.$$

Now Try Exercise 67

EXAMPLE 10 Using identities to find trigonometric expressions

If $\sin\theta = x$ and θ is a fourth quadrant angle, find an expression for $\sec\theta$ in terms of x. Approximate $\sec\theta$ if $\sin\theta = -0.7813$.

SOLUTION We begin by writing $\sec\theta$ in terms of $\sin\theta$.

$$\sin^2\theta + \cos^2\theta = 1 \qquad \text{Pythagorean identity}$$

$$\cos^2\theta = 1 - \sin^2\theta \qquad \text{Subtract } \sin^2\theta.$$

$$\cos\theta = \pm\sqrt{1 - \sin^2\theta} \qquad \text{Square root property}$$

$$\sec\theta = \pm\frac{1}{\sqrt{1 - \sin^2\theta}} \qquad \sec\theta = \frac{1}{\cos\theta}$$

Since $\sin\theta = x$ and angle θ is in quadrant IV, where $\sec\theta > 0$,

$$\sec\theta = \frac{1}{\sqrt{1 - x^2}}. \qquad \sin\theta = x;\ \sec\theta > 0$$

Since $\sin\theta = x$, let $x = -0.7813$. Then

$$\sec\theta = \frac{1}{\sqrt{1 - (-0.7813)^2}} \approx 1.602.$$

Now Try Exercise 73

Negative-Angle Identities

In Section 4.1, we discussed odd and even functions. The graphs of odd functions are symmetric with respect to the origin, and the graphs of even functions are symmetric with respect to the y-axis. The graphs of all six trigonometric functions have symmetry, as shown in Figures 7.12–7.17.

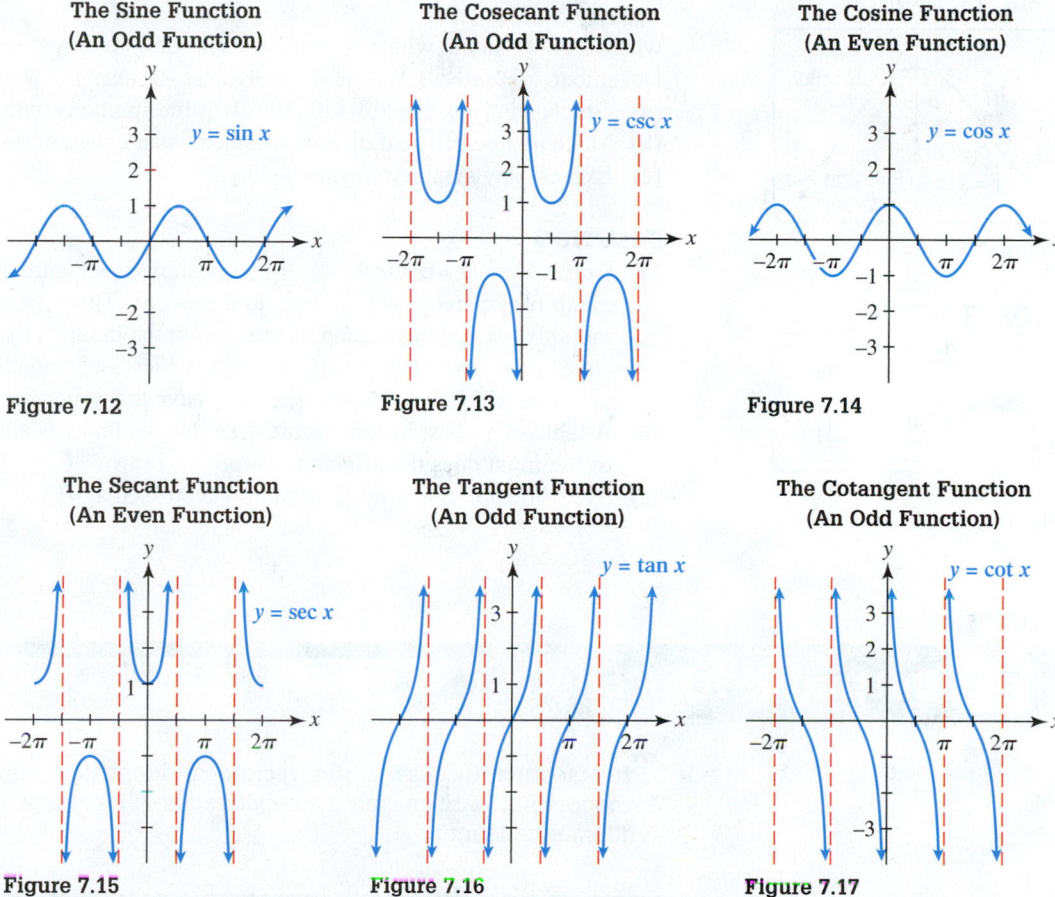

The Sine Function (An Odd Function)

$y = \sin x$

Figure 7.12

The Cosecant Function (An Odd Function)

$y = \csc x$

Figure 7.13

The Cosine Function (An Even Function)

$y = \cos x$

Figure 7.14

The Secant Function (An Even Function)

$y = \sec x$

Figure 7.15

The Tangent Function (An Odd Function)

$y = \tan x$

Figure 7.16

The Cotangent Function (An Odd Function)

$y = \cot x$

Figure 7.17

The cosine and secant functions are even functions, and the other four trigonometric functions are odd functions. For even functions the sign of the input does not affect the output, whereas for odd functions changing the sign of the input changes the sign of the output. That is, for all x in the domain of an even function f or an odd function g,

Even Function   **Odd Function**

$$f(-x) = f(x) \qquad \text{and} \qquad g(-x) = -g(x).$$

These results can be expressed using the negative-angle identities.

NEGATIVE-ANGLE IDENTITIES

$$\sin(-\theta) = -\sin\theta \qquad \cos(-\theta) = \cos\theta \qquad \tan(-\theta) = -\tan\theta$$

$$\csc(-\theta) = -\csc\theta \qquad \sec(-\theta) = \sec\theta \qquad \cot(-\theta) = -\cot\theta$$

Negative-angle identities are often used to simplify expressions. For example, the expression $\cos(-x) + \sin(-x)$ can be simplified to $\cos x - \sin x$.

An Application   In the next example, we see how the symmetry of a trigonometric function can be used to model temperature.

EXAMPLE 11   Modeling Temperature

The monthly average high temperatures in degrees Fahrenheit at Chattanooga, Tennessee, can be modeled by

$$f(x) = 21\cos\left(\frac{\pi x}{6}\right) + 70,$$

where x is the month with $x = -6$ corresponding to January, $x = 0$ to July, and $x = 5$ to December.   (*Source:* J. Williams, *The Weather Almanac.*)
(a) Graph f in $[-6, 6, 1]$ by $[40, 100, 10]$. Interpret any symmetry in the graph.
(b) Make a table of f and discuss whether f is an even or odd function.
(c) Express any symmetry symbolically.

SOLUTION
(a) Graph $Y_1 = 21\cos(\pi X/6) + 70$, as shown in Figure 7.18 using radian mode. The graph is symmetric with respect to the y-axis. This type of symmetry implies that the monthly average high temperatures x months before July and x months after July are equal. For example, April ($x = -3$), which is 3 months before July, and October ($x = 3$), which is 3 months after July, have the same average high temperature of 70°.
(b) A table of Y_1 is shown in Figure 7.19. Notice that f is an even function since the sign of the input does not affect the output.
(c) Symbolically, this symmetry can be expressed as $f(-x) = f(x)$.

Now Try Exercise 89

$[-6, 6, 1]$ by $[40, 100, 10]$

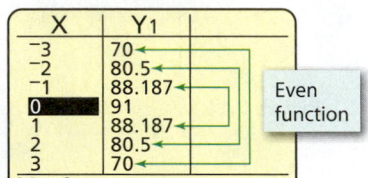

$y = 21\cos\left(\frac{\pi x}{6}\right) + 70$ Even function

Figure 7.18 Radian Mode

X	Y₁	
−3	70	
−2	80.5	
−1	88.187	Even function
0	91	
1	88.187	
2	80.5	
3	70	

X=0

Figure 7.19 Radian Mode

7.1 Putting It All Together

This section discussed the reciprocal identities, the quotient identities, the Pythagorean identities, and the negative-angle identities. Collectively, they are called fundamental identities.

IDENTITY	GENERAL FORM		
Reciprocal	$\sin\theta = \dfrac{1}{\csc\theta}$	$\cos\theta = \dfrac{1}{\sec\theta}$	$\tan\theta = \dfrac{1}{\cot\theta}$
	$\csc\theta = \dfrac{1}{\sin\theta}$	$\sec\theta = \dfrac{1}{\cos\theta}$	$\cot\theta = \dfrac{1}{\tan\theta}$
Quotient	$\tan\theta = \dfrac{\sin\theta}{\cos\theta}$	$\cot\theta = \dfrac{\cos\theta}{\sin\theta}$	

IDENTITY	GENERAL FORM		
Pythagorean	$\sin^2\theta + \cos^2\theta = 1$	$1 + \tan^2\theta = \sec^2\theta$	$1 + \cot^2\theta = \csc^2\theta$
Negative-angle	$\sin(-\theta) = -\sin\theta$ $\csc(-\theta) = -\csc\theta$	$\cos(-\theta) = \cos\theta$ $\sec(-\theta) = \sec\theta$	$\tan(-\theta) = -\tan\theta$ $\cot(-\theta) = -\cot\theta$

7.1 Exercises

Reciprocal Identities

Exercises 1–8: If possible, use a reciprocal identity to find the indicated trigonometric function of θ.

1. $\cot\theta$ if $\tan\theta = \frac{1}{2}$ **2.** $\csc\theta$ if $\sin\theta = -\frac{5}{6}$

3. $\sec\theta$ if $\cos\theta = \frac{2}{7}$ **4.** $\tan\theta$ if $\cot\theta = -\frac{3}{7}$

5. $\cos\theta$ if $\sec\theta = -4$ **6.** $\sin\theta$ if $\csc\theta = 7$

7. $\tan\theta$ if $\cos\theta = 0$ **8.** $\csc\theta$ if $\tan\theta = 0$

Exercises 9–14: Use reciprocal identities to rewrite the expression in terms of $\sin\theta$ and $\cos\theta$.

9. $\dfrac{\tan\theta}{\cot\theta}$ **10.** $\dfrac{\sec\theta}{\csc\theta}$

11. $\dfrac{\cot^2\theta}{\csc^2\theta}$ **12.** $\dfrac{\tan^2\theta}{\sec^2\theta}$

13. $\dfrac{1}{\csc\theta} + \dfrac{1}{\sec\theta}$ **14.** $\csc\theta\sec\theta\tan\theta$

Fundamental Identities

Exercises 15–22: Find the other trigonometric functions of θ.

15. $\sin\theta = \frac{3}{5}$ and $\cos\theta = -\frac{4}{5}$

16. $\tan\theta = -\frac{12}{5}$ and $\cos\theta = \frac{5}{13}$

17. $\cot\theta = \frac{7}{24}$ and $\sin\theta = -\frac{24}{25}$

18. $\sin\theta = \frac{12}{13}$ and $\cos\theta = \frac{5}{13}$

19. $\sin\theta = -\frac{60}{61}$ and $\cos\theta = -\frac{11}{61}$

20. $\sin\theta = \frac{2}{3}$ and $\cos\theta = -\frac{\sqrt{5}}{3}$

21. $\csc\theta = \sqrt{2}$ and $\sec\theta = -\sqrt{2}$

22. $\csc\theta = -\frac{13}{12}$ and $\sec\theta = -\frac{13}{5}$

Exercises 23–44: Simplify each expression.

23. $\sec\theta\cos\theta$ **24.** $\tan\theta\cot\theta$

25. $\sin\theta\csc\theta$ **26.** $\tan\theta\cos\theta$

27. $(\sin^2\theta + \cos^2\theta)^3$ **28.** $(1 + \tan^2\theta)\cos^2\theta$

29. $1 - \sin^2\theta$ **30.** $1 - \cos^2(-\theta)$

31. $\sec^2\theta - 1$ **32.** $1 + \dfrac{\cos^2\theta}{\sin^2\theta}$

33. $\dfrac{\sin(-\theta)}{\cos(-\theta)}$ **34.** $\sin\theta(\csc\theta + \sec\theta)$

35. $\dfrac{\sin^2\theta + \cos^2\theta}{\cos\theta}$ **36.** $\dfrac{1 + \tan^2\theta}{\sec^2\theta}$

37. $\dfrac{\sec^2(-\theta)}{\csc^2\theta}$ **38.** $\dfrac{1 - \cos^2\theta}{\sin^2\theta + \cos^2\theta}$

39. $\dfrac{\cot x}{\csc x}$ *cos x* **40.** $\dfrac{\tan x}{\sec x}$

41. $(\sin^2 x)(1 + \cot^2 x)$ **42.** $\dfrac{\cos x}{\sin x \cot x}$

43. $\sec(-x) + \csc(-x)$

44. $-\cos(-x)\sec(-x)$

Exercises 45–48: Determine if the equation represents an identity. If you have a graphing calculator, support your answer by making a table of the left and right sides of the equation.

45. $\sec\theta\cot\theta = \csc\theta$ **46.** $(\sin\theta + \cos\theta)^2 = 1$

47. $\cot^2\theta - \csc^2\theta = 1$ **48.** $1 + \tan^2\theta = \dfrac{1}{\cos^2\theta}$

Exercises 49–54: (Refer to Example 6.) Determine the quadrant that contains θ. If you have a graphing calculator, support your result either graphically or numerically.

49. $\sin\theta < 0$ and $\cos\theta > 0$

50. $\tan\theta > 0$ and $\cos\theta < 0$

51. $\sec\theta < 0$ and $\sin\theta < 0$

52. $\csc\theta > 0$ and $\tan\theta > 0$

53. $\cot\theta < 0$ and $\sin\theta > 0$

54. $\cos\theta > 0$ and $\cot\theta < 0$

Exercises 55–66: Find the other trigonometric functions of θ.

55. $\cos\theta = \frac{1}{2}$ and $\sin\theta < 0$

56. $\csc\theta = \sqrt{3}$ and $\cos\theta < 0$

57. $\tan\theta = -\frac{11}{60}$ and $\csc\theta < 0$

58. $\sec\theta = -\frac{5}{4}$ and $\sin\theta > 0$

59. $\sin\theta = \frac{7}{25}$ and $\cos\theta > 0$

60. $\cot\theta = \frac{12}{5}$ and $\sec\theta < 0$

61. $\sin\theta = -\frac{1}{3}$ and $\sec\theta < 0$

62. $\tan\theta = -\frac{1}{2}$ and $\sin\theta > 0$

63. $\sec\theta = \frac{37}{12}$ and θ in quadrant IV

64. $\tan\theta = \frac{3}{4}$ and θ in quadrant I

65. $\csc\theta = \frac{7}{3}$ and θ in quadrant II

66. $\cos\theta = -\frac{3}{5}$ and θ in quadrant III

Writing Trigonometric Expressions

67. Write $\sin x$ and $\tan x$ in terms of $\cos x$, if $\csc x > 0$.

68. Write $\cos x$ and $\sec x$ in terms of $\sin x$, if $\cos x < 0$.

69. Write $\sin x$ and $\sec x$ in terms of $\tan x$, if $\cos x < 0$.

70. Write $\cos x$ and $\csc x$ in terms of $\cot x$, if $\sin x > 0$.

71. Write $\cot x$ and $\cos x$ in terms of $\csc x$, if $\cot x < 0$.

72. Write $\tan x$ and $\sin x$ in terms of $\sec x$, if $\tan x > 0$.

Exercises 73–76: (Refer to Example 10.) Write the trigonometric function in terms of x. Then evaluate this trigonometric function if $x = 0.5126$.

73. $\cos\theta$ if $\sin\theta = x$ and θ is acute

74. $\sec\theta$ if $\cos\theta = x$

75. $\sin\theta$ if $\cot\theta = x$ and θ is in quadrant III

76. $\tan\theta$ if $\cos\theta = x$ and θ is in quadrant IV

77. Write $\tan\theta$ in terms of x if $\sin\theta = x$ and θ is acute.

78. Write $\sin\theta$ in terms of x if $\sec\theta = x$ and θ is acute.

Negative-Angle Identities

Exercises 79–84: Use a negative-angle identity to write an equivalent trigonometric expression involving a positive angle.

79. $\sin(-13°)$

80. $\cos\left(-\frac{\pi}{7}\right)$

81. $\tan\left(-\frac{\pi}{11}\right)$

82. $\cot(-75°)$

83. $\sec\left(-\frac{2\pi}{5}\right)$

84. $\csc(-160°)$

Applications

85. Intensity of a Lamp According to Lambert's law, the intensity of light from a single source on a flat surface at point P is given by $I = k\cos^2\theta$, where k is a constant. See the figure. (*Source:* C. Winter, *Solar Power Plants.*)

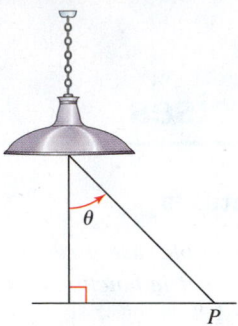

(a) Let $k = 1$ and use degree mode to graph I in the window $[-90, 90, 45]$ by $[-1, 2, 1]$. For what value of θ is I maximum?

(b) Write I in terms of the sine function.

86. Height of a Building If the angle of elevation of the sun is θ, then a building 40 feet high will cast a shadow x feet long, where $x = \frac{40}{\tan\theta}$. Use a reciprocal identity to rewrite this formula.

87. Radio Tuners (Refer to Example 5.) Let the energy stored in the inductor L be given by the formula $L(t) = 3\cos^2(6{,}000{,}000t)$ and the energy in the capacitor C be given by $C(t) = 3\sin^2(6{,}000{,}000t)$, where t is time in seconds. The total energy E in the circuit is given by $E(t) = L(t) + C(t)$.

(a) Graph L, C, and E in the window $[0, 10^{-6}, 10^{-7}]$ by $[-1, 4, 1]$. Interpret the graph.

(b) Make a table of L, C, and E, starting at $t = 0$ and incrementing by 10^{-7}. Interpret your results.

(c) Use a fundamental identity to derive a simplified expression for $E(t)$.

88. Distance to the Stars The distance d to a star can be found by using

$$d = \frac{93{,}000{,}000}{\sin\theta},$$

where θ is the parallax of the star. Use a reciprocal identity to rewrite this formula.

Exercises 89 and 90: **Temperature** *(Refer to Example 11.) Suppose that the monthly high temperature at a location is modeled by $f(x)$, where x is the month with $x = -6$ corresponding to January, $x = 0$ to July, and $x = 5$ to December.*

(a) *Graph f in $[-6, 6, 1]$ by $[0, 100, 10]$. Interpret any symmetry in the graph.*

(b) Make a table of f and discuss whether f is an even or odd function.

(c) Express any symmetry symbolically.

89. $f(x) = 40\cos\left(\frac{\pi x}{6}\right) + 50$

90. $f(x) = -15\cos\left(\frac{\pi x}{6}\right) + 60$

91. Oscillating Spring The distance or displacement y of a weight attached to an oscillating spring from its natural position is modeled by $y = 4\cos(2\pi t)$, where t is in seconds. See the figure. Potential energy is the energy of position and is given by $P = ky^2$, where k is a constant. The weight has the greatest potential energy when the spring is stretched or compressed the most. (*Source:* R. Weidner and R. Sells, *Elementary Classical Physics*, Vol. 1.)

(a) Write an expression for P that involves the cosine function.

(b) Let $k = 2$ and graph P in the viewing rectangle $[0, 2, 0.5]$ by $[-1, 40, 8]$. For $0 \le t \le 2$, at what times is P maximum and at what times is P minimum? Interpret your result.

(c) Use a fundamental identity to write P in terms of the sine function.

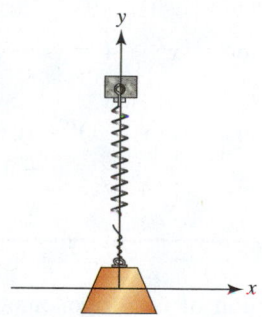

92. Energy in a Spring (Refer to Exercise 91.) Two types of mechanical energy are kinetic energy and potential energy. Kinetic energy is the energy of motion, and potential energy is the energy of position. A stretched spring has potential energy, which is converted to kinetic energy when it is released. If the potential energy of a weight on a spring is $P(t) = k\cos^2(4\pi t)$, where k is a constant and t is in seconds, then its kinetic energy is given by $K(t) = k\sin^2(4\pi t)$. The total mechanical energy E is $E(t) = P(t) + K(t)$.

(a) If $k = 2$, graph P, K, and E in $[0, 0.5, 0.25]$ by $[-1, 3, 1]$. Interpret the graph.

(b) Make a table of K, P, and E, starting at $t = 0$ and incrementing by 0.05. Interpret the results.

(c) Use a fundamental identity to derive a simplified expression for $E(t)$.

Writing about Mathematics

93. Explain in your own words what a trigonometric identity is. Give two examples.

94. Answer each of the following.
(a) Give two characteristics of an even function. Which of the trigonometric functions are even?

(b) Give two characteristics of an odd function. Which of the trigonometric functions are odd?

95. A student writes "$\cos^2 + \sin^2 = 1$." Comment on the correctness of this expression.

96. Since $\sec^2\theta = 1 + \tan^2\theta$ is an identity, does it follow that $\sec\theta = 1 + \tan\theta$? Explain your reasoning.

7.2 Verifying Identities

- Simplify trigonometric expressions
- Learn how to verify identities

Introduction

In Example 5 of the previous section, we used the identity $\sin^2\theta + \cos^2\theta = 1$ to show that the energy stored in a tuner in a radio is constant. Trigonometric identities are used in both applications and calculus. Before we can use an identity, we must verify that it is correct. Although the equality of two trigonometric expressions can be *supported* graphically and numerically, symbolic *verification* is necessary to be certain that an equation is indeed an identity.

Simplifying Trigonometric Expressions

Many of the algebraic skills that you have already learned can be used to simplify trigonometric expressions. For example, suppose we wanted to multiply the following expression.

$$(1 + \cos\theta)(1 - \cos\theta)$$

In algebra we learned that

$$(1 + x)(1 - x) = 1 - x + x - x^2$$
$$= 1 - x^2.$$

Product of a sum and a difference

If we substitute $\cos \theta$ for x, then

$$(1 + \cos \theta)(1 - \cos \theta) = 1 - \cos \theta + \cos \theta - \cos^2 \theta$$
$$= 1 - \cos^2 \theta.$$

Product of a sum and a difference

In algebra we do not simplify $1 - x^2$ further. However, since $\sin^2 \theta + \cos^2 \theta = 1$, it follows that $\sin^2 \theta = 1 - \cos^2 \theta$. As a result,

$$(1 + \cos \theta)(1 - \cos \theta) = \sin^2 \theta.$$

NOTE If you use algebraic expressions as an aid in simplifying trigonometric expressions, don't forget to check to see whether the resulting trigonometric expression can be simplified further by using fundamental identities.

MAKING CONNECTIONS

Algebraic and Trigonometric Expressions Many of the techniques used to rewrite algebraic expressions can also be used to rewrite trigonometric expressions. Here are some examples.

1. $\tan^2 \theta - 4$
   $= (\tan \theta - 2)(\tan \theta + 2)$ is similar to $x^2 - 4 = (x - 2)(x + 2)$.

2. $\cos \theta \, (\sin \theta + \cos \theta)$
   $= \cos \theta \sin \theta + \cos^2 \theta$ is similar to $x(y + x) = xy + x^2$.

3. $\dfrac{\sin \theta}{\cos \theta} + \dfrac{1}{\cos \theta} = \dfrac{\sin \theta + 1}{\cos \theta}$ is similar to $\dfrac{y}{x} + \dfrac{1}{x} = \dfrac{y + 1}{x}$.

The next example illustrates that the addition of two trigonometric expressions is often accomplished in a manner that is similar to that used to add two algebraic expressions.

EXAMPLE 1 Adding two trigonometric expressions

Write $\tan t + \cot t$ as a product of two trigonometric functions.

SOLUTION
Getting Started We begin by writing $\tan t$ and $\cot t$ as ratios involving $\sin t$ and $\cos t$.

$$\tan t + \cot t = \frac{\sin t}{\cos t} + \frac{\cos t}{\sin t}$$

In algebra we combine $\frac{y}{x} + \frac{x}{y}$ by using the common denominator xy as follows.

Algebra Review
To review addition of rational expressions, see Chapter R (page R-32).

$$\frac{y}{x} + \frac{x}{y} = \frac{y}{x} \cdot \frac{y}{y} + \frac{x}{y} \cdot \frac{x}{x} \qquad \text{Multiply each ratio by 1.}$$

$$= \frac{y^2}{xy} + \frac{x^2}{xy} \qquad \text{Simplify.}$$

Common denominator

$$= \frac{y^2 + x^2}{xy} \qquad \text{Add.}$$

Now substitute $\cos t$ for x and $\sin t$ for y. ▶

$$\tan t + \cot t = \frac{\sin t}{\cos t} + \frac{\cos t}{\sin t} \qquad \text{Quotient identities}$$

$$= \frac{\sin t}{\cos t} \cdot \frac{\sin t}{\sin t} + \frac{\cos t}{\sin t} \cdot \frac{\cos t}{\cos t} \qquad \text{Multiply each ratio by 1.}$$

$$= \frac{\sin^2 t}{\cos t \sin t} + \frac{\cos^2 t}{\cos t \sin t} \qquad \text{Simplify.}$$

Common denominator

$$= \frac{\sin^2 t + \cos^2 t}{\cos t \sin t} \qquad \text{Add.}$$

$$= \frac{1}{\cos t \sin t} \qquad \sin^2 t + \cos^2 t = 1$$

$$= \sec t \csc t \qquad \sec t = \frac{1}{\cos t}; \csc t = \frac{1}{\sin t}$$

Thus $\tan t + \cot t$ is equivalent to $\sec t \csc t$.

> **Now Try Exercise 15**

NOTE To simplify a trigonometric expression, it is *not* necessary to first write a similar algebraic expression using x and y, as was done in Example 1. However, sometimes you may find this technique helpful.

EXAMPLE 2 Factoring a trigonometric expression

Factor each expression.
(a) $\sec^2 \theta - 1$ **(b)** $2\sin^2 t + \sin t - 1$

SOLUTION
(a) In algebra we factor $x^2 - 1$ as $(x - 1)(x + 1)$. This technique can be applied to the given expression.

Difference of squares — $\sec^2 \theta - 1 = (\sec \theta - 1)(\sec \theta + 1)$

(b) Since $2y^2 + y - 1$ can be factored as $(2y - 1)(y + 1)$, it follows that

$$2\sin^2 t + \sin t - 1 = (2\sin t - 1)(\sin t + 1).$$

> **Now Try Exercises 29 and 31**

An Application from Electricity Trigonometric expressions are used in applications involving electricity, as illustrated in the next example.

EXAMPLE 3 Analyzing electromagnets

Electromagnets are used in a variety of situations, such as lifting scrap metal, ringing door bells, and opening door locks in apartments. Let the wattage W consumed by an electromagnet at t seconds be

$$W(t) = 100\sin^2(120\pi t) \qquad \text{Wattage}$$

and the voltage V in the circuit be

$$V(t) = 160\cos(120\pi t). \qquad \text{Voltage}$$

(*Source:* A. Howatson, *Electrical Circuits and Systems.*)
(a) Express $W(t)$ in terms of the cosine function. When V is maximum or minimum, what is the value of W? Explain.
(b) Support your answer in part (a) by graphing W and V in the viewing rectangle $[0, 1/30, 1/60]$ by $[-180, 180, 20]$.

[0, 1/30, 1/60] by [−180, 180, 20]

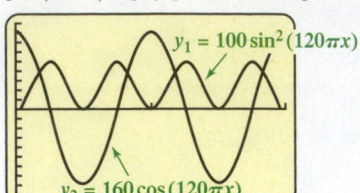

Figure 7.20 Radian Mode

SOLUTION

(a) Let $\theta = 120\pi t$ in the identity $\sin^2\theta = 1 - \cos^2\theta$ to write W as

$$W(t) = 100(1 - \cos^2(120\pi t)). \qquad \textcolor{blue}{\sin^2\theta = 1 - \cos^2\theta}$$

Since $V(t) = 160\cos(120\pi t)$, the voltage is maximum (160) or minimum (−160) whenever $\cos(120\pi t) = \pm 1$. When $\cos(120\pi t) = \pm 1$, it follows that the value of W is given by $W(t) = 100(1 - (\pm 1)^2) = 0$.

(b) Graph $Y_1 = 100(\sin(120\pi X))^\wedge 2$ and $Y_2 = 160\cos(120\pi X)$, as shown in Figure 7.20. The wattage y_1 is 0 whenever the voltage y_2 is ± 160.

> **Now Try Exercise 81**

Verification of Identities

When verifying that an equation is an identity, we usually begin with one side of the equation and write a sequence of equivalent expressions until it is transformed into the other side. This technique is illustrated in the following examples.

EXAMPLE 4 **Verifying an identity**

Verify that $(\cos\theta + \sin\theta)^2 = 1 + 2\sin\theta\cos\theta$.

SOLUTION

Getting Started In algebra, the expression $(x + y)^2$ can be expanded as follows.

$$(x + y)^2 = (x + y)(x + y)$$

Square the binomial

$$= x^2 + xy + yx + y^2$$
$$= x^2 + 2xy + y^2$$

We can perform similar steps with the left side of the trigonometric equation. ►

$$(\cos\theta + \sin\theta)^2 = (\cos\theta + \sin\theta)(\cos\theta + \sin\theta)$$

Square the binomial

$$= \cos^2\theta + \cos\theta\sin\theta + \sin\theta\cos\theta + \sin^2\theta$$
$$= \boldsymbol{\cos^2\theta} + 2\sin\theta\cos\theta + \boldsymbol{\sin^2\theta}$$
$$= \boldsymbol{1} + 2\sin\theta\cos\theta$$

The last step is true since $\sin^2\theta + \cos^2\theta = 1$. We have verified the identity since we have shown symbolically that the left side of the given equation is equal to the right side.

> **Now Try Exercise 37**

EXAMPLE 5 **Verifying identities**

Verify each identity.

(a) $\dfrac{\sin^2\theta}{1 + \cos\theta} = 1 - \cos\theta$ **(b)** $\dfrac{\csc t}{\cot t} - \dfrac{\cot t}{\csc t} = \tan t \sin t$

SOLUTION

(a) Begin by applying a Pythagorean identity.

$$\frac{\boldsymbol{\sin^2\theta}}{1 + \cos\theta} = \frac{\boldsymbol{1 - \cos^2\theta}}{1 + \cos\theta} \qquad \textcolor{blue}{\sin^2\theta + \cos^2\theta = 1}$$

$$= \frac{(1 - \cos\theta)(\boldsymbol{1 + \cos\theta})}{\boldsymbol{1 + \cos\theta}} \qquad \textcolor{blue}{\text{Factor difference of squares.}}$$

$$= 1 - \cos\theta \qquad \textcolor{blue}{\text{Simplify.}}$$

(b) Begin by finding a common denominator for the expressions on the left side of the equation.

$$\frac{\csc t}{\cot t} - \frac{\cot t}{\csc t} = \frac{\csc t}{\cot t} \cdot \frac{\csc t}{\csc t} - \frac{\cot t}{\csc t} \cdot \frac{\cot t}{\cot t} \qquad \text{Multiply each ratio by 1.}$$

$$= \frac{\csc^2 t}{\cot t \csc t} - \frac{\cot^2 t}{\csc t \cot t} \qquad \text{Simplify.}$$

$$= \frac{\csc^2 t - \cot^2 t}{\cot t \csc t} \qquad \text{Subtract.}$$

$$= \frac{1 + \cot^2 t - \cot^2 t}{\cot t \csc t} \qquad \csc^2 t = 1 + \cot^2 t$$

$$= \frac{1}{\cot t \csc t} \qquad \text{Simplify.}$$

$$= \frac{1}{\cot t} \cdot \frac{1}{\csc t} \qquad \text{Rewrite the expression.}$$

$$= \tan t \sin t \qquad \text{Reciprocal identities}$$

Now Try Exercises 53 and 63

EXAMPLE 6 **Verifying an identity**

Verify that $\frac{\sin t}{1 - \cos t} = \frac{1 + \cos t}{\sin t}$ is an identity.

SOLUTION

Getting Started Earlier, we saw that $(1 - \cos\theta)(1 + \cos\theta) = \sin^2\theta$. This result can be used to verify the given identity by multiplying the numerator and denominator by 1 written as $\frac{1 + \cos t}{1 + \cos t}$. This technique is demonstrated below. ▶

$$\frac{\sin t}{1 - \cos t} = \frac{\sin t}{1 - \cos t} \cdot \frac{1 + \cos t}{1 + \cos t} \qquad \text{Multiply the ratio by 1.}$$

$$= \frac{\sin t (1 + \cos t)}{1 - \cos^2 t} \qquad \text{Simplify.}$$

$$= \frac{\sin t (1 + \cos t)}{\sin^2 t} \qquad \sin^2 t = 1 - \cos^2 t$$

$$= \frac{1 + \cos t}{\sin t} \qquad \text{Simplify.}$$

Now Try Exercise 65

EXAMPLE 7 **Verifying an identity**

Verify that $\frac{\sec x + \tan x}{\sec x - \tan x} = \frac{1 + \sin x}{1 - \sin x}$.

SOLUTION

Getting Started Because $\sec x = \frac{1}{\cos x}$ and $\tan x = \frac{\sin x}{\cos x}$, we start to simplify the expression by multiplying the numerator and denominator by $\cos x$. ▶

$$\frac{\sec x + \tan x}{\sec x - \tan x} = \frac{\sec x + \tan x}{\sec x - \tan x} \cdot \frac{\cos x}{\cos x} \qquad \text{Multiply the ratio by 1.}$$

$$= \frac{\sec x \cos x + \tan x \cos x}{\sec x \cos x - \tan x \cos x} \qquad \text{Distributive property}$$

$$= \frac{\frac{1}{\cos x} \cdot \cos x + \frac{\sin x}{\cos x} \cdot \cos x}{\frac{1}{\cos x} \cdot \cos x - \frac{\sin x}{\cos x} \cdot \cos x} \qquad \text{Reciprocal and quotient identities}$$

$$= \frac{1 + \sin x}{1 - \sin x} \qquad \text{Simplify.}$$

Now Try Exercise 61

In the next example, we show how to give graphical and numerical support.

EXAMPLE 8 Verifying an identity

Verify that $\frac{\tan\theta}{\sec\theta} = \sin\theta$ symbolically. Give graphical and numerical support.

SOLUTION

Symbolic Verification **Getting Started** We will start with the more complicated expression $\frac{\tan\theta}{\sec\theta}$ and simplify it to $\sin\theta$. We begin by writing $\tan\theta$ and $\sec\theta$ in terms of $\sin\theta$ and $\cos\theta$. ▶

$$\frac{\tan\theta}{\sec\theta} = \frac{\sin\theta/\cos\theta}{1/\cos\theta} \qquad \text{Quotient and reciprocal identities}$$

$$= \frac{\sin\theta}{\cos\theta} \cdot \frac{\cos\theta}{1} \qquad \text{Invert and multiply.}$$

$$= \sin\theta \qquad \text{Simplify.}$$

These steps verify symbolically that $\frac{\tan\theta}{\sec\theta} = \sin\theta$ is an identity.

Graphical Support Graph $Y_1 = \tan(X)/(1/\cos(X))$ and $Y_2 = \sin(X)$, as shown below in Figures 7.21 and 7.22. Their graphs appear to be identical. Note that because most calculators do not have a secant button, we use the reciprocal identity $\sec\theta = \frac{1}{\cos\theta}$ to write $y_1 = \frac{\tan\theta}{\sec\theta}$ as $y_1 = \frac{\tan\theta}{1/\cos\theta}$.

Numerical Support See Figure 7.23. Note that when $\theta = \frac{\pi}{2} \approx 1.5708$, the ratio $\frac{\tan\theta}{\sec\theta}$ is undefined, whereas $\sin\theta = 1$. However, the equation $\frac{\tan\theta}{\sec\theta} = \sin\theta$ is nonetheless an identity because $y_1 = y_2$ whenever both expressions are defined. (The step size for this table is $\frac{\pi}{6}$.)

$[-2\pi, 2\pi, \pi/2]$ by $[-2, 2, 1]$ $\qquad$ $[-2\pi, 2\pi, \pi/2]$ by $[-2, 2, 1]$

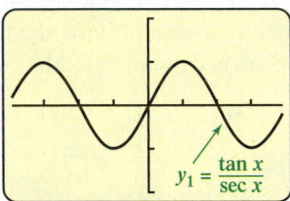

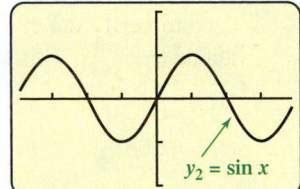

X	Y1	Y2
0	0	0
.5236	.5	.5
1.0472	.86603	.86603
1.5708	ERROR	1
2.0944	.86603	.86603
2.618	.5	.5
3.1416	0	0
X=0		

Figure 7.21 Radian Mode $\qquad$ **Figure 7.22** Radian Mode $\qquad$ **Figure 7.23** Radian Mode

Now Try Exercise 73

7.2 Putting It All Together

Becoming proficient at verifying identities requires practice. Many of the skills learned in algebra can be used to help verify identities. The following table lists some suggestions that may be helpful.

SUGGESTIONS FOR VERIFYING IDENTITIES

1. Become familiar with the fundamental identities found in Section 7.1.
2. Use your knowledge of simplifying algebraic expressions as a guide, particularly when factoring or combining ratios.
3. When verifying an identity, start by simplifying the more complicated side of the equation. Otherwise, choose a side of the equation that you can transform into a different expression.
4. If you are simplifying the left side of the equation, work toward making the left side appear more like the right. For example, if the left side contains an addition sign but the right side does not, add the terms on the left side.
5. If you are uncertain how to proceed, one strategy is to write each trigonometric function in terms of sine and cosine and then simplify. Another strategy is to apply fundamental identities, if possible.
6. If a ratio contains $1 + \sin\theta$, it is sometimes helpful to multiply the numerator and denominator by $1 - \sin\theta$. Then

$$(1 + \sin\theta)(1 - \sin\theta) = 1 - \sin^2\theta = \cos^2\theta.$$

Similar statements can be made for $1 - \sin\theta$, $1 + \cos\theta$, and $1 - \cos\theta$. See Example 6.

7.2 Exercises

Simplifying Expressions

Exercises 1–6: Multiply the algebraic expression. Then multiply the corresponding trigonometric expression. If possible, simplify the resulting trigonometric expression.

1. (a) $(1 + x)(1 - x)$ **(b)** $(1 + \sin\theta)(1 - \sin\theta)$

2. (a) $(x - 1)(x + 1)$ **(b)** $(\csc\theta - 1)(\csc\theta + 1)$

3. (a) $x(x - 1)$ **(b)** $\sec\theta(\sec\theta - 1)$

4. (a) $(x + 1)(2x - 1)$ **(b)** $(\tan\theta + 1)(2\tan\theta - 1)$

5. (a) $\dfrac{x}{1} \cdot \dfrac{y}{x}$ **(b)** $\cos\theta \cdot \tan\theta$

6. (a) $\dfrac{1}{x} \cdot \dfrac{x}{y}$ **(b)** $\csc\theta \cdot \tan\theta$

Exercises 7–12: Factor the algebraic expression. Then factor the corresponding trigonometric expression. If possible, simplify the resulting trigonometric expression.

7. (a) $x^2 + 2x + 1$ **(b)** $\cos^2\theta + 2\cos\theta + 1$

8. (a) $2x^2 - 3x + 1$ **(b)** $2\sin^2 t - 3\sin t + 1$

9. (a) $x^2 - 2x$ **(b)** $\sec^2 t - 2\sec t$

10. (a) $3x - 9x^2$ **(b)** $3\tan\theta - 9\tan^2\theta$

11. (a) $x + x^3$ **(b)** $\tan\theta + \tan^3\theta$

12. (a) $x^2 + x^2y^2$ **(b)** $\sin^2\theta + \sin^2\theta \tan^2\theta$

Exercises 13–20: Simplify the algebraic expression. Then simplify the corresponding trigonometric expression completely.

13. (a) $\dfrac{1}{1 - x} + \dfrac{1}{1 + x}$ **(b)** $\dfrac{1}{1 - \cos\theta} + \dfrac{1}{1 + \cos\theta}$

14. (a) $x + \dfrac{1}{x}$ **(b)** $\tan t + \dfrac{1}{\tan t}$

15. (a) $\dfrac{x}{y} + \dfrac{y}{x}$ **(b)** $\dfrac{\cos t}{\sin t} + \dfrac{\sin t}{\cos t}$

16. (a) $\dfrac{1}{y} - \dfrac{x^2}{y}$ **(b)** $\dfrac{1}{\sin\theta} - \dfrac{\cos^2\theta}{\sin\theta}$

17. (a) $\dfrac{1}{1/y^2} + \dfrac{1}{1/x^2}$ **(b)** $\dfrac{1}{\csc^2 t} + \dfrac{1}{\sec^2 t}$

18. (a) $\left(\dfrac{1}{x} + x\right)^2$ **(b)** $(\cot\theta + \tan\theta)^2$

19. (a) $\dfrac{x/y}{1/y}$ **(b)** $\dfrac{\cot\theta}{\csc\theta}$

20. (a) $\dfrac{1 - x^2}{1 + x}$ **(b)** $\dfrac{1 - \cos^2\theta}{1 + \cos\theta}$

Exercises 21–28: Perform the indicated operations and simplify.

21. $\cos\theta \tan\theta$ **22.** $\sin^2\theta \csc\theta$

23. $\tan\theta(\cos\theta - \csc\theta)$ **24.** $(\sin\theta - \cos\theta)^2$

25. $(1 + \tan t)^2$ **26.** $(\sin t - 1)(\sin t + 1)$

27. $\dfrac{\csc^2\theta - 1}{\csc^2\theta}$ **28.** $\sin^2 t(1 + \cot^2 t)$

Exercises 29–34: Factor the trigonometric expression and simplify, if possible.

29. $1 - \tan^2\theta$ **30.** $\sin^2 t - \cos^2 t$

31. $\sec^2 t - \sec t - 6$ **32.** $\cos\theta \sin^2\theta + \cos^3\theta$

33. $\tan^4\theta + 3\tan^2\theta + 2$ **34.** $\sin^4 t - \cos^4 t$

Verifying Identities

Exercises 35–72: Verify the identity.

35. $\csc^2\theta - \cot^2\theta = 1$ **36.** $\dfrac{\tan^2\theta + 1}{\sec\theta} = \sec\theta$

37. $(1 - \sin t)^2 = 1 - 2\sin t + \sin^2 t$

38. $\dfrac{\sin^2 t}{\cos t} = \sec t - \cos t$ **39.** $\dfrac{\sin t + \cos t}{\sin t} = 1 + \cot t$

40. $\sec^4\theta - \sec^2\theta = \tan^4\theta + \tan^2\theta$

41. $\sec^2\theta - 1 = \tan^2\theta$ **42.** $\dfrac{\csc^2\theta}{\cot\theta} = \csc\theta \sec\theta$

43. $\dfrac{\tan^2 t}{\sec t} = \sec t - \cos t$ **44.** $\dfrac{\sec^2\theta - 1}{\sec^2\theta} = \sin^2\theta$

45. $\cot x + 1 = \csc x(\cos x + \sin x)$

46. $\dfrac{1 + \sin x}{\cos x} = \dfrac{\cos x}{1 - \sin x}$ **47.** $\dfrac{\sec t}{1 + \sec t} = \dfrac{1}{\cos t + 1}$

48. $\sec^2 t + \csc^2 t = \sec^2 t \csc^2 t$

49. $(\sec t - 1)(\sec t + 1) = \tan^2 t$

50. $\csc^4\theta - \cot^4\theta = \csc^2\theta + \cot^2\theta$

51. $\dfrac{1 - \sin^2\theta}{\cos\theta} = \cos\theta$ **52.** $\dfrac{\tan^2 t - 1}{1 + \tan^2 t} = 1 - 2\cos^2 t$

53. $\dfrac{\sec t}{\tan t} - \dfrac{\tan t}{\sec t} = \cos t \cot t$

54. $\dfrac{\sin^4 t - \cos^4 t}{\sin^2 t - \cos^2 t} = 1$ **55.** $\dfrac{\cot^2 t}{\csc t + 1} = \csc t - 1$

56. $\sec\theta - \cos\theta = \tan\theta\sin\theta$

57. $\dfrac{\cot t}{\cot t + 1} = \dfrac{1}{1 + \tan t}$

58. $\cos^4 t - \sin^4 t = 2\cos^2 t - 1$

59. $\dfrac{1}{1 - \sin t} + \dfrac{1}{1 + \sin t} = 2\sec^2 t$

60. $\cot\theta + \tan\theta = \csc\theta\sec\theta$

61. $\dfrac{\csc t + \cot t}{\csc t - \cot t} = (\csc t + \cot t)^2$

62. $\dfrac{\csc t}{1 + \csc t} - \dfrac{\csc t}{1 - \csc t} = 2\sec^2 t$

63. $\dfrac{\cos^2 t}{1 - \sin t} = 1 + \sin t$

64. $\csc t + \dfrac{\sec t}{\tan t} = \dfrac{2}{\sin t}$ **65.** $\dfrac{1}{1 + \sin\theta} = \dfrac{1 - \sin\theta}{\cos^2\theta}$

66. $\dfrac{2\sin^2 t + 3\sin t - 2}{\sin t + 2} = 2\sin t - 1$

67. $\sqrt{1 - \sin^2\theta} = \cos\theta$, where θ is acute

68. $\sqrt{\sec^2\theta - 1} = \tan\theta$, where θ is acute

69. $\dfrac{1 + 2\sin x + \sin^2 x}{\cos^2 x} = \dfrac{1 + \sin x}{1 - \sin x}$

70. $\dfrac{\tan t - \cot t}{\sin t \cos t} = \sec^2 t - \csc^2 t$

71. $(1 - \cos^2 x)(1 + \cos^2 x) = 2\sin^2 x - \sin^4 x$

72. $\sin^4 x - \cos^4 x = 2\sin^2 x - 1$

Exercises 73–80: Verify the identity. If you have a graphing calculator, give graphical or numerical support.

73. $\cot\theta\sin\theta = \cos\theta$ **74.** $\tan\theta\cos\theta = \sin\theta$

75. $(1 - \cos^2\theta)(1 + \tan^2\theta) = \tan^2\theta$

76. $\cos^2\theta(1 + \cot^2\theta) = \cot^2\theta$

77. $\cos t(\tan t - \sec t) = \sin t - 1$

78. $\dfrac{\cos\theta}{1 - \sin\theta} = \sec\theta + \tan\theta$

79. $\dfrac{\tan(-\theta)}{\sin(-\theta)} = \sec\theta$

80. $\tan^2 t - \sin^2 t = \tan^2 t \sin^2 t$

Applications

81. Electromagnets (Refer to Example 3.) Let the wattage consumed by an electromagnet be given by the formula $W(t) = 5\cos^2(120\pi t)$ and the voltage be given by the formula $V(t) = 25\sin(120\pi t)$, where t is in seconds.
    (a) Express $W(t)$ in terms of the sine function. When V is maximum or minimum, what is the value of W?

    (b) Support your answer in part (a) by graphing W and V in $[0, 1/15, 1/60]$ by $[-30, 30, 10]$.

82. An Oscillating Spring The potential energy P of a weight on a spring is given by $P(t) = 5\cos^2(4\pi t)$, and its kinetic energy K is given by $K(t) = 5\sin^2(4\pi t)$, where t is in seconds.
    (a) Express $P(t)$ in terms of the sine function. When K is maximum or minimum, what is the value of P?

    (b) Support your answer in part (a) by graphing P and K in $[0, 0.5, 0.25]$ by $[-1, 5, 1]$. Interpret the graph.

Writing about Mathematics

83. Create a trigonometric identity of your own. Verify the identity symbolically and then give graphical and numerical support.

84. Explain how to show that an equation is not an identity. Give an example.

CHECKING BASIC CONCEPTS FOR SECTIONS 7.1 AND 7.2

1. Determine the quadrant containing θ if $\cot\theta > 0$ and $\sin\theta < 0$.

2. Determine the other trigonometric functions of θ using the given information.
    (a) $\sin\theta = \frac{5}{13}$ and $\cos\theta = -\frac{12}{13}$

    (b) $\sec\theta = \frac{5}{4}$ and $\sin\theta < 0$

    (c) $\tan\theta = -\frac{1}{2}$ and $\cos\theta = \frac{2}{\sqrt{5}}$

3. Simplify each expression.
    (a) $(1 - \sin\theta)(1 + \sin\theta)$

    (b) $\tan^2 t \csc^2 t - 1$

4. Factor the trigonometric expression.
    (a) $\tan^2 t - 1$ **(b)** $3\sin^2 t + \sin t - 2$

5. Verify each identity.
    (a) $(1 - \sin^2\theta)(1 + \cot^2\theta) = \cot^2\theta$

    (b) $\dfrac{\cot^2 t}{\csc t} = \csc t - \sin t$

7.3 Trigonometric Equations

- Find and use reference angles
- Solve trigonometric equations and applications
- Solve inverse trigonometric equations

Introduction

In previous chapters we saw how applications that involve functions result in the need to solve equations. In a similar manner, applications that involve trigonometric functions result in the need to solve trigonometric equations. For example, the number of daylight hours near 30°N latitude can be modeled by

$$f(x) = 1.95 \cos\left(\frac{\pi}{6}(x - 6.6)\right) + 12.15,$$

where $x = 1$ corresponds to January 1, $x = 2$ to February 1, and so on. To estimate when there are 11 hours of daylight, we can solve the *trigonometric equation* $f(x) = 11$.

Trigonometric Equation

$$1.95 \cos\left(\frac{\pi}{6}(x - 6.6)\right) + 12.15 = 11$$

Like other types of equations, trigonometric equations can be solved graphically, numerically, and symbolically. We begin by discussing reference angles, which are used when solving trigonometric equations symbolically.

Reference Angles

A **reference angle** for an angle θ, written θ_R, is the acute angle made by the terminal side of θ and the x-axis. It is assumed that θ is in standard position and its terminal side does not lie on either the x- or the y-axis. Examples of reference angles in the four quadrants are shown in Figures 7.24–7.27.

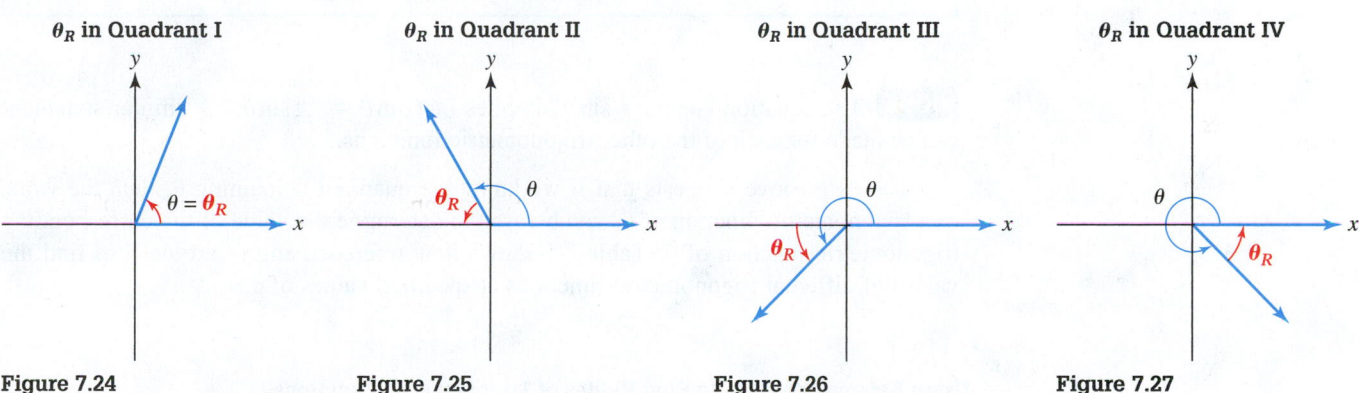

| θ_R in Quadrant I | θ_R in Quadrant II | θ_R in Quadrant III | θ_R in Quadrant IV |

Figure 7.24 Figure 7.25 Figure 7.26 Figure 7.27

EXAMPLE 1 Finding reference angles

Find the reference angle for θ.

(a) $\theta = 43°$ **(b)** $\theta = \dfrac{2\pi}{3}$ **(c)** $\theta = -55°$ **(d)** $\theta = -\dfrac{3\pi}{4}$

SOLUTION

(a) Since θ is in quadrant I, θ and θ_R are equal. Thus $\theta_R = 43°$.

(b) The terminal side of $\theta = \frac{2\pi}{3}$ lies in quadrant II, as in Figure 7.25. In this case, the acute angle θ_R between the terminal side of θ and the x-axis is given by

$$\theta_R = \pi - \theta = \pi - \frac{2\pi}{3} = \frac{\pi}{3}.$$

(c) The terminal side of $\theta = -55°$ lies in quadrant IV, and the acute angle between it and the x-axis is $\theta_R = \mathbf{55°}$. See Figure 7.28.

(d) The terminal side of $\theta = -\frac{3\pi}{4}$ (or $-135°$) lies in quadrant III, and the acute angle between it and the x-axis is $\theta_R = \frac{\pi}{4}$. See Figure 7.29.

CLASS DISCUSSION

Let $0 < \theta < 2\pi$. Find expressions for θ_R given the quadrant containing θ.

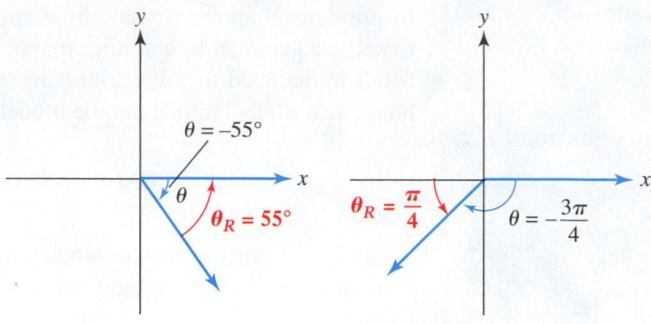

Figure 7.28   **Figure 7.29**

Now Try Exercises 3, 5, 7, and 9

The reference angle is important because it can help determine trigonometric values.

REFERENCE ANGLES AND TRIGONOMETRIC FUNCTIONS

Let θ be an angle in standard position with reference angle θ_R. Then

$$|\sin\theta| = \sin\theta_R \qquad |\cos\theta| = \cos\theta_R \qquad |\tan\theta| = \tan\theta_R$$
$$|\csc\theta| = \csc\theta_R \qquad |\sec\theta| = \sec\theta_R \qquad |\cot\theta| = \cot\theta_R.$$

The signs of the trigonometric functions of θ are determined by the quadrant that contains θ. (See Figure 7.9 in the See the Concept on page 558.)

NOTE The equation $|\sin\theta| = \sin\theta_R$ implies that $\sin\theta = \pm\sin\theta_R$. A similar statement can be made for each of the other trigonometric functions.

The note above suggests that if we know the quadrant containing θ, then the value of a trigonometric function of θ_R can be used to determine the value of the corresponding trigonometric function of θ. Table 7.1 shows how reference angles are used to find the values of different trigonometric functions at specified values of θ.

Using Reference Angles to Find Values of Trigonometric Functions

Desired Function Value	θ	θ_R	Trigonometric Function of θ_R	Quadrant Containing θ	Positive or Negative	Resulting Value
$\cos-\frac{3\pi}{4}$	$-\frac{3\pi}{4}$	$\frac{\pi}{4}$	$\cos\frac{\pi}{4} = \frac{1}{\sqrt{2}}$	III	$\cos\theta < 0$ in quadrant III	$\cos-\frac{3\pi}{4} = -\frac{1}{\sqrt{2}}$
$\csc 150°$	$150°$	$30°$	$\csc 30° = 2$	II	$\csc\theta > 0$ in quadrant II	$\csc 150° = 2$
$\sin\frac{11\pi}{6}$	$\frac{11\pi}{6}$	$\frac{\pi}{6}$	$\sin\frac{\pi}{6} = \frac{1}{2}$	IV	$\sin\theta < 0$ in quadrant IV	$\sin\frac{11\pi}{6} = -\frac{1}{2}$
$\tan-315°$	$-315°$	$45°$	$\tan 45° = 1$	I	$\tan\theta > 0$ in quadrant I	$\tan-315° = 1$

Table 7.1

Basic Trigonometric Equations

In general, trigonometric equations are solved symbolically by using algebraic properties to write the equation as an equivalent *basic trigonometric equation*. For this reason, it is important to be able to solve basic trigonometric equations such as $\cos\theta = -\frac{\sqrt{3}}{2}$, $\sin\theta = -\frac{1}{2}$, and $\tan\theta = 1$. The following See the Concept shows how a reference angle is used as an aid in solving $\cos\theta = -\frac{\sqrt{3}}{2}$ for θ in $[0, 2\pi)$.

See the Concept: Solving a Basic Trigonometric Equation

θ_R in Quadrant II

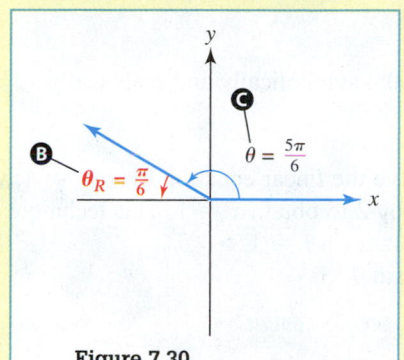

Figure 7.30

θ_R in Quadrant III

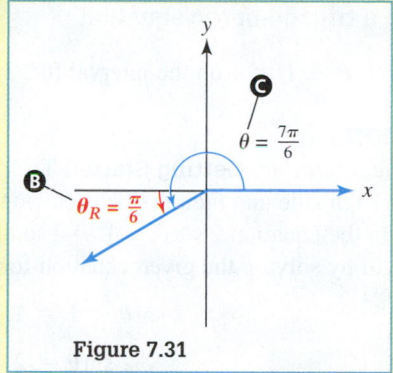

Figure 7.31

Solve $\cos\theta = -\frac{\sqrt{3}}{2}$ for θ in $[0, 2\pi)$.

Ⓐ Begin by finding the reference angle for θ. Since $\cos\theta_R = |\cos\theta|$, we know that $\cos\theta_R = \frac{\sqrt{3}}{2}$. So $\theta_R = \cos^{-1}\frac{\sqrt{3}}{2} = \frac{\pi}{6}$.

Ⓑ Because $\cos\theta < 0$, angle θ is a second or third quadrant angle. Locate θ_R in quadrants II and III as shown in Figures 7.30 and 7.31, respectively.

Ⓒ The angles in $[0, 2\pi)$ with $\cos\theta < 0$ and whose reference angles are $\theta_R = \frac{\pi}{6}$ represent the solutions to the given equation. The solutions are $\frac{5\pi}{6}$ and $\frac{7\pi}{6}$.

> **EXAMPLE 2** Solving trigonometric equations using reference angles

Solve the following equations.
(a) $\sin\theta = -\frac{1}{2}$ for θ in $[0°, 360°)$ **(b)** $\tan\theta = 1$ for θ in $[0, 2\pi)$

SOLUTION
(a) We start by solving the equation $\sin\theta_R = \frac{1}{2}$. The solution to this equation is $\theta_R = \sin^{-1}\frac{1}{2} = 30°$. The sine function is negative in quadrants III and IV. Therefore the solution to $\sin\theta = -\frac{1}{2}$ is an angle θ, located in either quadrant III or quadrant IV, that has a reference angle of **30°**. There are two such angles: 210° and 330°. See Figures 7.32 and 7.33. Thus 210° and 330° are solutions.
(b) The solution to $\tan\theta_R = 1$ is $\theta_R = \tan^{-1} 1 = \frac{\pi}{4}$. The tangent function is positive in quadrants I and III. Thus $\frac{\pi}{4}$ and $\frac{5\pi}{4}$ are solutions. See Figures 7.34 and 7.35.

θ_R in Quadrant III

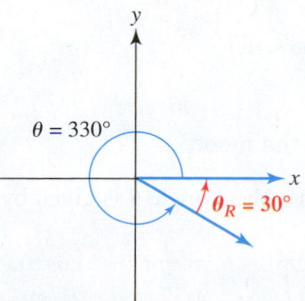

Figure 7.32

θ_R in Quadrant IV

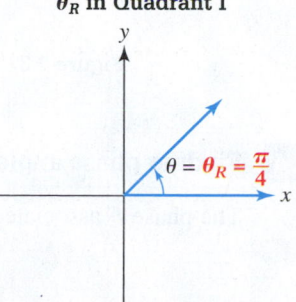

Figure 7.33

θ_R in Quadrant I

Figure 7.34

θ_R in Quadrant III

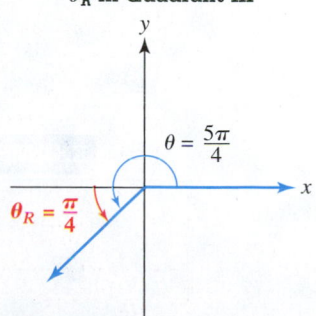

Figure 7.35

> **Now Try Exercises 13 and 15**

Solving Trigonometric Equations

In the previous section we verified trigonometric *identities*. Identities are equations that are true for all meaningful values of the variable. In this section we discuss trigonometric equations that are *conditional*. Conditional equations are satisfied by some but not all values of the variable. For example,

$$\cos \theta = 1$$

is a conditional trigonometric equation since the cosine function equals 1 only for certain values of θ, such as $\theta = 0$ or $\theta = 2\pi$.

EXAMPLE 3 Solving a trigonometric equation

Solve $2 \sin \theta - 1 = 1$ on the interval $[0°, 360°)$ symbolically and graphically.

SOLUTION

Symbolic Solution   **Getting Started** To solve the linear equation $2x - 1 = 1$, we first add 1 to each side and then divide each side by 2 to obtain $x = 1$. This technique can be applied to the equation $2 \sin \theta - 1 = 1$ to obtain $\sin \theta = 1$. ▶

Begin by solving the given equation for $\sin \theta$.

Solving $2 \sin \theta - 1 = 1$
$[0, 360, 90]$ by $[-4, 4, 1]$

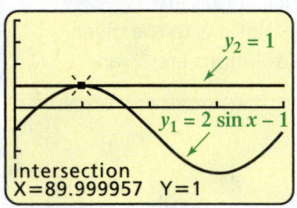

Figure 7.36 Degree Mode

$2 \sin \theta - 1 = 1$	*Given equation*
$2 \sin \theta = 2$	*Add 1 to each side.*
$\sin \theta = 1$	*Divide each side by 2.*

The only solution to $\sin \theta = 1$ on the interval $[0°, 360°)$ is $\theta = \sin^{-1} 1 = 90°$.

Graphical Solution   Graph $y_1 = 2 \sin x - 1$ and $y_2 = 1$, as in Figure 7.36. Their graphs intersect at $x = 90°$.

Now Try Exercise 31

An Application from Astronomy   In the next example, we find all the phase angles associated with a particular phase F of the moon. The fraction of the moon that appears illuminated is called the *phase F*. The *phase angle* θ is shown in Figure 7.37.   (*Source:* M. Zeilik et al., *Introductory Astronomy and Astrophysics.*)

A Phase of the Moon with Phase Angle θ

Figure 7.37 (Not to scale)

EXAMPLE 4 Finding phase angles for the moon

The phase F associated with a phase angle θ is given by

$$F = \frac{1}{2}(1 - \cos \theta).$$

Find all phase angles θ in degrees when $F = 0.75$. (Note that $F = 0.5$ corresponds to a first quarter moon or last quarter moon and $F = 1$ corresponds to a full moon.)

SOLUTION Let $F = 0.75$ and solve the given equation.

$$0.75 = \frac{1}{2}(1 - \cos\theta) \qquad \textcolor{teal}{\text{Let } F = 0.75.}$$

$$1.5 = 1 - \cos\theta \qquad \textcolor{teal}{\text{Multiply each side by 2.}}$$

$$\cos\theta = -0.5 \qquad \textcolor{teal}{\text{Solve for } \cos\theta.}$$

Start by solving the equation $\cos\theta_R = 0.5$. The solution is $\theta_R = \cos^{-1}0.5 = 60°$. The cosine function is negative in quadrants II and III. Angles in these quadrants with a 60° reference angle are 120° and 240°, where $0° \leq \theta \leq 360°$. Verify this fact. Since the cosine function has period 360°, all solutions can be written in the form

$$\theta = 120° + \mathbf{360°} \cdot n \qquad \text{or} \qquad \theta = 240° + \mathbf{360°} \cdot n,$$

> The cosine function has period 360°.

where n is an integer. For example, 120°, 120° ± 360°, and 120° ± 720° are solutions, as well as 240°, 240° ± 360°, and 240° ± 720°.

> **Now Try Exercise 117**

Finding All Solutions Many of the techniques used to solve polynomial equations can be applied to trigonometric equations, as illustrated in the next example.

EXAMPLE 5 **Solving trigonometric equations**

Find all solutions to each equation. Express your results in radians.
(a) $2\cot t + 1 = -1$ **(b)** $2\sin^2 t - 5\sin t + 2 = 0$

SOLUTION
(a) In algebra the equation $2x + 1 = -1$ implies $x = -1$. In a similar manner,

$$\mathbf{2\cot t + 1 = -1} \qquad \text{implies} \qquad \mathbf{\cot t = -1}.$$
$$\textcolor{teal}{\text{Subtract 1 and then divide by 2.}}$$

If $\cot t = -1$, then $\tan t = \frac{1}{\cot t} = -1$ and t has a reference angle of $\tan^{-1}1 = \frac{\pi}{4}$. The cotangent is negative in quadrants II and IV, so the solutions to $\cot t = -1$ in $[0, 2\pi)$ are $t = \frac{3\pi}{4}$ and $t = \frac{7\pi}{4}$. Note that these angles both have a reference angle of $\frac{\pi}{4}$. Since cotangent has a period of π, all solutions can be expressed in the form

$$t = \frac{3\pi}{4} + \pi n \qquad \text{or} \qquad t = \frac{7\pi}{4} + \pi n,$$

> The cotangent function has period π.

where n is an integer. These solutions are equivalent to just $t = \frac{3\pi}{4} + \pi n$ because the difference between $\frac{7\pi}{4}$ and $\frac{3\pi}{4}$ is π.

(b) **Getting Started** In algebra the equation $2x^2 - 5x + 2 = 0$ can be solved by factoring.

$$2x^2 - 5x + 2 = (2x - 1)(x - 2) = 0$$

The solutions are $\frac{1}{2}$ and 2. We can factor a trigonometric expression in the same way. ▶

$$2\sin^2 t - 5\sin t + 2 = (2\sin t - 1)(\sin t - 2) = 0$$
$$\textcolor{teal}{\text{Factor the quadratic expression as the product of two binomials.}}$$

We must solve the equations $\sin t = \frac{1}{2}$ and $\sin t = 2$. If $\sin t = \frac{1}{2}$, the reference angle is $\sin^{-1}\frac{1}{2} = \frac{\pi}{6}$. The sine function is positive in quadrants I and II, so the solutions in $[0, 2\pi)$ are $t = \frac{\pi}{6}$ and $t = \frac{5\pi}{6}$. Since $-1 \leq \sin t \leq 1$ for all t, the equation $\sin t = 2$

has no solutions. The sine function has period 2π, and all solutions to the given equation can be expressed as

$$t = \frac{\pi}{6} + 2\pi n \qquad \text{or} \qquad t = \frac{5\pi}{6} + 2\pi n.$$

The sine function has period 2π.

Now Try Exercises 55 and 59

MAKING CONNECTIONS

Polynomial and Trigonometric Equations A polynomial equation of degree n has at most n solutions. However, a trigonometric equation typically has an infinite number of solutions, as demonstrated in Examples 4 and 5.

Approximating Solutions Sometimes equations have solutions that are not multiples of common angles such as $\frac{\pi}{6}$ or $\frac{\pi}{4}$. In the next example, we find and then approximate solutions to a trigonometric equation.

EXAMPLE 6 **Approximating solutions**

Solve $1.7 \csc t + 2.3 = 0$ to the nearest thousandth for t in $[0, 2\pi)$.

SOLUTION Begin by solving the equation for $\csc t$.

$$1.7 \csc t + 2.3 = 0 \qquad \textit{Given equation}$$

$$1.7 \csc t = -2.3 \qquad \textit{Subtract 2.3 from each side.}$$

$$\csc t = -\frac{2.3}{1.7} \qquad \textit{Divide each side by 1.7.}$$

Because $\csc t = \frac{1}{\sin t}$, it follows that $\sin t = -\frac{1.7}{2.3}$, or $\sin t = -\frac{17}{23}$. The reference number (angle) is $t_R = \sin^{-1}\frac{17}{23} \approx 0.832$ (radian). The sine function is negative in quadrants III and IV. For t in $[0, 2\pi)$, there are two solutions given by

$$\pi + \sin^{-1}\frac{17}{23} \approx 3.973 \qquad \text{and} \qquad 2\pi - \sin^{-1}\frac{17}{23} \approx 5.451.$$

See Figures 7.38 and 7.39. (Note that 0.832 radian is approximately equivalent to 47.7°.)

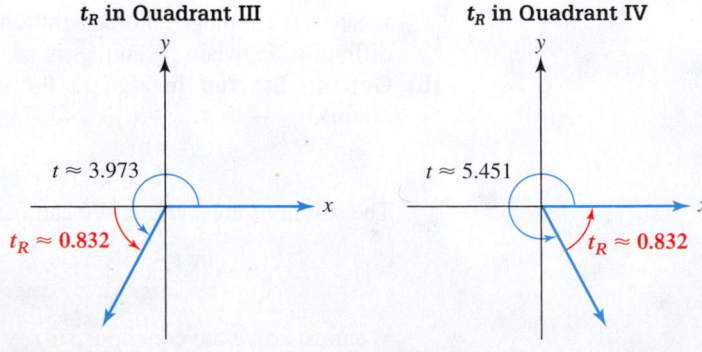

Figure 7.38 **Figure 7.39**

Now Try Exercise 93

An Application   Solar power companies are interested in the number of daylight hours during different times of the year and at different latitudes. In the next example, we solve the trigonometric equation

$$1.95 \cos\left(\frac{\pi}{6}(x - 6.6)\right) + 12.15 = 11$$

presented in the introduction to this section. The solution to this equation tells us when there are 11 hours of daylight at 30°N latitude.

EXAMPLE 7   **Analyzing daylight hours**

The number of daylight hours at 30°N latitude can be modeled by

$$f(x) = 1.95 \cos\left(\frac{\pi}{6}(x - 6.6)\right) + 12.15,$$

where $x = 1$ corresponds to January 1, $x = 2$ to February 1, and so on. Estimate graphically and numerically when there are 11 hours of daylight.

SOLUTION

Graphical Solution   Graph $Y_1 = 1.95 \cos(\pi/6(X - 6.6)) + 12.15$ and $Y_2 = 11$ in radian mode. Their graphs intersect near $x = 2.4$ and $x = 10.8$. See Figures 7.40 and 7.41. These values correspond to about February 11 and October 25. (Note that four-tenths of February is $0.4 \times 28 \approx 11$ days and eight-tenths of October is $0.8 \times 31 \approx 25$ days.)

Numerical Solution   Make a table of Y_1, starting at $x = 2$ and incrementing by 0.1. The table in Figure 7.42 shows that $Y_1 \approx 11$ when $x = 2.4$. Scrolling down the table would also show that $Y_1 \approx 11$ when $x = 10.8$.

[0, 13, 1] by [8, 16, 1]   [0, 13, 1] by [8, 16, 1]

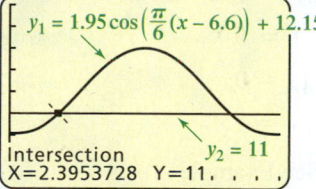   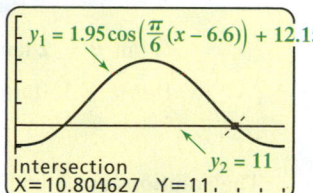

X	Y1	Y2
2	10.701	11
2.1	10.771	11
2.2	10.845	11
2.3	10.923	11
2.4	11.004	11 ←
2.5	11.088	11
2.6	11.175	11
X=2.4		

Figure 7.40 Radian Mode   **Figure 7.41** Radian Mode   **Figure 7.42** Radian Mode

Now Try Exercise 121

More Trigonometric Equations   Some equations contain trigonometric functions such as $\cos 2t$ or $\tan 3\theta$, where the argument is a multiple of t or θ. An additional step is required when solving this type of equation. In the next example, we solve a trigonometric equation for all real numbers t, where the argument of the trigonometric function is $4t$.

EXAMPLE 8   **Solving a trigonometric equation**

Solve $-0.6 \sin 4t = 0.3$, where t is any real number.

SOLUTION   First we let $\theta = 4t$. Then the given equation becomes

$$-0.6 \sin \theta = 0.3.$$

Next we solve this modified equation for all real numbers θ.

$$-0.6 \sin \theta = 0.3 \qquad \text{Let } \theta = 4t.$$

$$\sin \theta = -\frac{1}{2} \qquad \text{Divide each side by } -0.6.$$

From Example 2(a), the solutions to $\sin\theta = -\frac{1}{2}$ on the interval $[0°, 360°)$ are $210°$ and $330°$. In radian measure these solutions are $\theta = \frac{7\pi}{6}$ and $\frac{11\pi}{6}$. Thus all real number solutions to the equation $-0.6\sin\theta = 0.3$ are

$$\theta = \frac{7\pi}{6} + 2\pi n \qquad \text{or} \qquad \theta = \frac{11\pi}{6} + 2\pi n,$$

where n is an integer. Because $\theta = 4t$, we can determine t by substituting $4t$ for θ.

$$4t = \frac{7\pi}{6} + 2\pi n \qquad \text{or} \qquad 4t = \frac{11\pi}{6} + 2\pi n$$

Finally we divide each equation by 4 to obtain

$$t = \frac{7\pi}{24} + \frac{\pi n}{2} \qquad \text{or} \qquad t = \frac{11\pi}{24} + \frac{\pi n}{2}.$$

> **Now Try Exercise 75**

Some equations contain more than one type of trigonometric function. In these situations it is sometimes helpful to use trigonometric identities to rewrite the equation in terms of one trigonometric function. This is illustrated in the next example.

EXAMPLE 9　**Solving a trigonometric equation**

Solve $2\tan\theta = \sec^2\theta$ symbolically on the interval $[0, 2\pi)$.

SOLUTION This equation contains two different trigonometric functions. We begin by applying the identity $1 + \tan^2\theta = \sec^2\theta$ to rewrite the equation only in terms of $\tan\theta$.

$$2\tan\theta = \mathbf{sec^2\,\theta} \qquad \textit{Given equation}$$
$$2\tan\theta = \mathbf{1 + tan^2\,\theta} \qquad \textit{sec}^2\theta = 1 + \textit{tan}^2\theta$$
$$\tan^2\theta - 2\tan\theta + 1 = 0 \qquad \textit{Rewrite the equation.}$$
$$(\tan\theta - 1)(\tan\theta - 1) = 0 \qquad \textit{Factor.}$$
$$\tan\theta = 1 \qquad \textit{Solve for tan }\theta.$$

The solutions are $\frac{\pi}{4}$ and $\frac{5\pi}{4}$. See Example 2(b).

> **Now Try Exercise 45**

In the next example, we solve a trigonometric equation by squaring each side. When squaring each side of an equation, it is important to *check the answers.*

EXAMPLE 10　**Solving a trigonometric equation by squaring**

Solve $\sec t = 1 + \tan t$ for $0 \le t < 2\pi$.

SOLUTION Begin by squaring each side of the equation.

$$\sec t = 1 + \tan t \qquad \textit{Given equation}$$
$$\sec^2 t = (1 + \tan t)^2 \qquad \textit{Square each side}$$
$$\sec^2 t = \mathbf{1} + 2\tan t + \mathbf{tan^2\,t} \qquad \textit{Square the expression.}$$
$$\sec^2 t = \mathbf{sec^2\,t} + 2\tan t \qquad \textit{1 + tan}^2 t = \textit{sec}^2 t$$
$$0 = 2\tan t \qquad \textit{Subtract sec}^2 t \textit{ from each side.}$$
$$\tan t = 0 \qquad \textit{Divide each side by 2 and rewrite.}$$
$$t = 0 \qquad \text{or} \qquad t = \pi \qquad \textit{Solve for t when } 0 \le t < 2\pi.$$

Algebra Review

To review squaring a binomial, see Chapter R (page R-17).

Since we squared each side of the equation, we check $t = 0$ and $t = \pi$ in the given equation.

Solution checks

$$\sec 0 \overset{?}{=} 1 + \tan 0 \qquad \sec \pi \overset{?}{=} 1 + \tan \pi$$

$$1 = 1 + 0 \qquad\qquad -1 \neq 1 + 0$$

Solution does not check

The only solution is 0. The value of π is an *extraneous solution*.

Now Try Exercise 73

More Applications

Highway Curves   Highway curves are sometimes banked so that the outside of the curve is slightly elevated or inclined, as shown in Figure 7.43. This inclination is called the *superelevation*. The relationship among a car's velocity v in feet per second, the safe radius r of the curve in feet, and the superelevation θ in degrees is given by

$$r = \frac{v^2}{4.5 + 32.2 \tan \theta}.$$

(*Source:* F. Mannering and W. Kilareski, *Principles of Highway Engineering and Traffic Analysis.*)

Figure 7.43

EXAMPLE 11   **Determining superelevation for a highway curve**

A highway curve with a radius of 700 feet and a speed limit of 88 feet per second (60 mi/hr) is being designed. Find the appropriate superelevation for the curve.

SOLUTION   Let $r = 700$ and $v = 88$ and then solve the equation for θ.

$$700 = \frac{88^2}{4.5 + 32.2 \tan \theta} \qquad \textit{Let } r = 700 \textit{ and } v = 88.$$

$$4.5 + 32.2 \tan \theta = \frac{88^2}{700} \qquad \textit{Properties of ratios}$$

$$32.2 \tan \theta = \frac{88^2}{700} - 4.5 \qquad \textit{Subtract 4.5 from each side.}$$

$$\tan \theta = \frac{88^2/700 - 4.5}{32.2} \qquad \textit{Divide each side by 32.2.}$$

$$\tan \theta \approx 0.2038 \qquad \textit{Approximate.}$$

$$\theta \approx \tan^{-1} 0.2038 \approx 11.5° \qquad \textit{Apply the inverse tangent.}$$

The superelevation should be about 11.5°.

Now Try Exercise 119

Locating the Position of a Planet   One step in the process used by astronomers to calculate the position of a planet as it orbits the sun involves finding the solution of Kepler's equation. Kepler's equation is a trigonometric equation that *cannot* be solved symbolically. It must be solved either graphically or numerically. In real applications, it is quite common to encounter equations that cannot be solved symbolically.

EXAMPLE 12 Solving Kepler's equation graphically

$[0, 0.2, 0.05]$ by $[-0.2, 0.2, 0.1]$

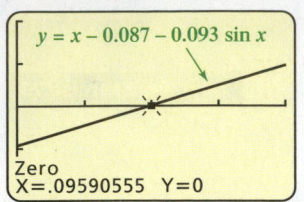

Zero
X=.09590555 Y=0

Figure 7.44 Radian Mode

One example of Kepler's equation is given by $\theta = 0.087 + 0.093 \sin\theta$. Solve this equation graphically. (***Source:*** J. Meeus, *Astronomical Algorithms*.)

SOLUTION The given equation is equivalent to $\theta - 0.087 - 0.093 \sin\theta = 0$. To solve this equation graphically, let $Y_1 = X - 0.087 - 0.093 \sin X$ and graph. (Be sure to use radian mode.) The x-intercept near 0.096 represents the solution. See Figure 7.44.

> **Now Try Exercise 127**

Solving Inverse Trigonometric Equations

Some types of equations contain inverse trigonometric functions. To solve these equations we often make use of the following inverse properties.

Inverse Trigonometric Properties

$$\sin(\sin^{-1}x) = x \qquad \text{for} \qquad -1 \le x \le 1$$
$$\cos(\cos^{-1}x) = x \qquad \text{for} \qquad -1 \le x \le 1$$
$$\tan(\tan^{-1}x) = x \qquad \text{for} \qquad -\infty < x < \infty$$

These types of equations are solved in the next example.

EXAMPLE 13 Solving inverse trigonometric equations

Solve each equation symbolically.

(a) $\cos^{-1}x = \pi$

(b) $\dfrac{\pi}{2} - \tan^{-1}x = \dfrac{\pi}{4}$

(c) $\sin^{-1}2x = \dfrac{\pi}{3}$

SOLUTION

(a) Begin by taking the cosine of each side of the given equation $\cos^{-1}x = \pi$.

$$\cos(\cos^{-1}x) = \cos\pi \qquad \textcolor{blue}{\textit{Take the cosine of each side.}}$$
$$x = -1 \qquad \textcolor{blue}{\textit{Simplify.}}$$

(b) Begin by solving the given equation $\frac{\pi}{2} - \tan^{-1}x = \frac{\pi}{4}$ for $\tan^{-1}x$.

$$\frac{\pi}{4} = \tan^{-1}x \qquad \textcolor{blue}{\textit{Subtract } \frac{\pi}{4}\textit{; add } \tan^{-1}x.}$$

$$\tan\frac{\pi}{4} = \tan(\tan^{-1}x) \qquad \textcolor{blue}{\textit{Take the tangent of each side.}}$$

$$1 = x \qquad \textcolor{blue}{\textit{Simplify.}}$$

(c) Take the sine of each side of the given equation $\sin^{-1}2x = \frac{\pi}{3}$.

$$\sin(\sin^{-1}2x) = \sin\frac{\pi}{3} \qquad \textcolor{blue}{\textit{Take the sine of each side.}}$$

$$2x = \frac{\sqrt{3}}{2} \qquad \textcolor{blue}{\textit{Simplify.}}$$

$$x = \frac{\sqrt{3}}{4} \qquad \textcolor{blue}{\textit{Divide each side by 2.}}$$

> **Now Try Exercises 107, 113, and 115**

7.3 Putting It All Together

$\mathbf{A}$lgebraic skills such as factoring and solving equations can also be used to solve trigonometric equations. Trigonometric equations frequently have infinitely many solutions.

CONCEPT	COMMENTS	EXAMPLES
Reference angles	If an angle θ is in standard position, then its reference angle is the acute angle made by the terminal side of θ and the x-axis.	If $\theta = \frac{11\pi}{6}$, then $\theta_R = \frac{\pi}{6}$.
Trigonometric equations	First, use techniques from algebra to isolate any trigonometric functions. Then solve these simpler equations for the given variable.	$\sqrt{3}\tan\theta = -1$ $\tan\theta = -\frac{1}{\sqrt{3}}$ The reference angle is $\theta_R = \tan^{-1}\frac{1}{\sqrt{3}} = 30°$. Since $\tan\theta$ is negative in quadrants II and IV, the solutions in $[0°, 360°)$ are $150°$ and $330°$.
Inverse trigonometric equations	Equations that involve inverse trigonometric functions can sometimes be solved by using the following properties. $\sin(\sin^{-1}x) = x, \, -1 \le x \le 1$ $\cos(\cos^{-1}x) = x, \, -1 \le x \le 1$ $\tan(\tan^{-1}x) = x, \, -\infty < x < \infty$	$\tan^{-1}x = \frac{\pi}{3}$ $\tan(\tan^{-1}x) = \tan\frac{\pi}{3}$ $x = \sqrt{3}$

7.3 Exercises

Reference Angles

Exercises 1–12: Find the reference angle for θ.

1. $\theta = 120°$

2. $\theta = 230°$

3. $\theta = 85°$

4. $\theta = -130°$

5. $\theta = -65°$

6. $\theta = 340°$

7. $\theta = \frac{5\pi}{6}$

8. $\theta = \frac{7\pi}{4}$

9. $\theta = -\frac{2\pi}{3}$

10. $\theta = -\frac{4\pi}{3}$

11. $\theta = \frac{5\pi}{4}$

12. $\theta = \frac{7\pi}{6}$

Basic Trigonometric Equations

Exercises 13–20: Solve for θ in [0°, 360°) and in [0, 2π).

13. (a) $\sin\theta = 1$ (b) $\sin\theta = -1$

14. (a) $\cos\theta = \frac{1}{2}$ (b) $\cos\theta = -\frac{1}{2}$

15. (a) $\tan\theta = \sqrt{3}$ (b) $\tan\theta = -\sqrt{3}$

16. (a) $\cot\theta = 1$ (b) $\cot\theta = -1$

17. (a) $\sec\theta = 2$ (b) $\sec\theta = -2$

18. (a) $\csc\theta = \sqrt{2}$ (b) $\csc\theta = -\sqrt{2}$

19. (a) $\sin\theta = 3$ (b) $\sin\theta = -3$

20. (a) $\cos\theta = \frac{\sqrt{3}}{2}$ (b) $\cos\theta = -\frac{\sqrt{3}}{2}$

Exercises 21–26: Find all solutions. Express your answer in radians.

21. (a) $\sin t = \frac{1}{2}$ (b) $\sin t = -\frac{1}{2}$

22. (a) $\cos t = 1$ (b) $\cos t = -1$

23. (a) $\tan t = 1$ (b) $\tan t = -1$

24. (a) $\csc t = \frac{1}{4}$ (b) $\csc t = -\frac{1}{4}$

25. (a) $\sec t = 2$ (b) $\sec t = -2$

26. (a) $\cot t = \sqrt{3}$ (b) $\cot t = -\sqrt{3}$

Solving Trigonometric Equations

Exercises 27–30: Use the graph to estimate any solutions to the given equation for $0 \le t < 2\pi$. Then solve the equation symbolically.

27. $\sin t = \cos t$ **28.** $\csc t = \sec t$

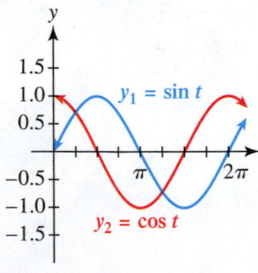 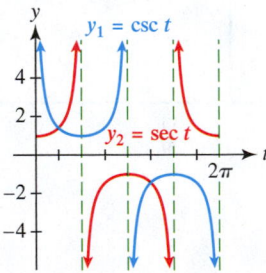

29. $3 \cot t = 2 \sin t$ **30.** $2 \cos^2 t = 1 - \cos t$

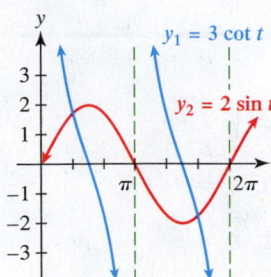 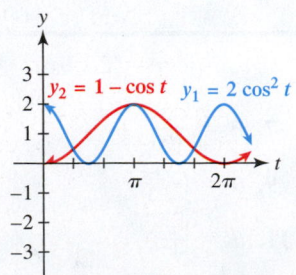

Exercises 31–34: Solve the algebraic equation for x. Then solve the trigonometric equation for $0° \le \theta < 360°$.

31. (a) $2x - 1 = 0$ (b) $2 \sin\theta - 1 = 0$

32. (a) $x - 1 = 0$ (b) $\cot\theta - 1 = 0$

33. (a) $x^2 = x$ (b) $\sin^2\theta = \sin\theta$

34. (a) $x^2 - x = 0$ (b) $\cos^2\theta - \cos\theta = 0$

Exercises 35–38: Solve the algebraic equation for x. Then solve the trigonometric equation for $0 \le t < 2\pi$.

35. (a) $x^2 + 1 = 2$ (b) $\tan^2 t + 1 = 2$

36. (a) $(x - 1)(x + 1) = 0$

     (b) $(\sin t - 1)(\sin t + 1) = 0$

37. (a) $x^2 + x = 2$ (b) $\cos^2 t + \cos t = 2$

38. (a) $2x^2 + 3x = -1$ (b) $2 \sin^2 t + 3 \sin t = -1$

Exercises 39–54: Solve the equation for t in [0, 2π).

39. $\tan^2 t - 3 = 0$ **40.** $2 \sin t = \sqrt{3}$

41. $3 \cos t + 4 = 0$ **42.** $\cos^2 t + \cos t - 6 = 0$

43. $\sin t \cos t = \cos t$ **44.** $\cos^2 t - \sin^2 t = 0$

45. $\csc^2 t = 2 \cot t$ **46.** $2 \sin^2 t - 3 \cos t = 3$

47. $\sin^2 t = \frac{1}{4}$ **48.** $\cos^2 t = -\frac{1}{2}$

49. $\sin t \cos t = 0$ **50.** $\tan t - \cot t = 0$

51. $2 \sec t = \tan^2 t + 1$ **52.** $\cos^2 t = 3 \sin^2 t$

53. $\tan t + \sec t = 1$ **54.** $\cos t - \sin t = 1$

Exercises 55–74: Find all solutions to the equation. Express your results in radians.

55. $2 \tan t - 1 = 1$ **56.** $\sqrt{3} \cot t - 1 = 0$

57. $2 \sin t + 2 = 3$ **58.** $2 \cos t - 1 = 0$

59. $2 \sin^2 t - 3 \sin t = -1$ **60.** $2 \cos^2 t + 3 \cos t + 1 = 0$

61. $\sec^2 t + 3 \sec t + 2 = 0$ **62.** $\csc^2 t - 3 = -1$

63. $\tan^2 t - 1 = 0$ **64.** $2 \cos t = -1$

65. $\sin^2 t + \sin t - 20 = 0$ **66.** $3 \cos t - 5 = 0$

67. $\cos t \sin t = \sin t$ **68.** $2 \cos^2 t - 1 = 0$

69. $\sec^2 t = 2 \tan t$ **70.** $\cos^2 t - 2 \sin t - 1 = 0$

71. $\sin^2 t \cos^2 t = 0$ **72.** $2 \cot^2 t \sin t - \cot^2 t = 0$

73. $\sin t + \cos t = 1$ **74.** $\sin t - \cos t = 1$

Exercises 75–92: (Refer to Example 8.) Solve the equation, where t is any real number. Approximate t to three decimal places when appropriate.

75. $\sin 3t = \frac{1}{2}$ **76.** $\cos 2t = -\frac{1}{2}$

77. $\cos 4t = -\frac{\sqrt{3}}{2}$ **78.** $\sin 4t = \frac{1}{\sqrt{2}}$

79. $\tan 5t = 1$ **80.** $\cot 3t = -\sqrt{3}$

81. $2 \sin 4t = -1$ **82.** $5 \cos 6t = 2.5$

83. $-\sec 4t = \sqrt{2}$ **84.** $\sqrt{3} \csc 3t = -2$

85. $2 \sin 8t - 3 = -1$ **86.** $3 \cos 8t - 4 = -4$

87. $\cot 4t + 5 = 6$ **88.** $\sqrt{3} \tan 2t + 2 = 3$

89. $5 \cos 3t = 1$ **90.** $7 \cos 5t = -2$

91. $\sin 2t = \frac{1}{3}$ **92.** $\frac{1}{7} \sin 7t = \frac{1}{20}$

Approximate Solutions

Exercises 93–98: Solve the equation to the nearest thousandth for t in $[0, 2\pi)$.

93. $2.1 \sec t - 4.5 = 0$ **94.** $2 \csc t + 2 = 8.3$

95. $5.8 \sin t - 3.7 = 0.2$ **96.** $6 \cos t + 2 = 0$

97. $5 \tan^2 t - 3 = 0$ **98.** $7 \cot^2 t + 1.2 = 6$

Exercises 99–104: The following equations cannot be solved symbolically. Approximate to two decimal places any solutions on $[0, 2\pi)$ graphically or numerically.

99. $\tan x = x$ **100.** $x - \cos x = 0$

101. $\sin x = (x - 1)^2$ **102.** $\sin^2 x - \ln x = 0$

103. $2x \cos(x + 1) = \sin(\cos x)$

104. $e^{-0.1x} \cos x = x \sin x$

Numerical Solutions

Exercises 105 and 106: Use the table to find the solutions to the given equation on $[0°, 360°)$. Then write all solutions to the equation.

105. $\tan \theta - \frac{1}{\sqrt{3}} = 0$ **106.** $\tan \theta - \sin \theta = 0$

X	Y1
30	0
90	ERROR
150	−1.155
210	0
270	ERROR
330	−1.155
390	0

Y1 ◼ tan(X)−1/√(3)

X	Y1
0	0
60	.86603
120	−2.598
180	0
240	2.5981
300	−.866
360	0

Y1 ◼ tan(X)−sin(X)

Solving Inverse Trigonometric Equations

Exercises 107–116: Solve the equation.

107. $\sin^{-1} x = \frac{\pi}{2}$ **108.** $\sin^{-1} 2x = -\frac{\pi}{4}$

109. $2 \cos^{-1} x = \frac{5\pi}{3}$ **110.** $\cos^{-1} x = 0$

111. $\pi + \tan^{-1} x = \frac{3\pi}{4}$ **112.** $\tan^{-1} x = -\frac{\pi}{3}$

113. $\tan^{-1}(3x + 1) = \frac{\pi}{4}$ **114.** $\frac{\pi}{4} + \sin^{-1}(x + 1) = \frac{\pi}{2}$

115. $\cos^{-1} x + 3 \cos^{-1} x = \pi$

116. $\frac{\pi}{6} + \sin^{-1} 4x = \frac{\pi}{3}$

Applications

117. First and Third Quarter (Refer to Example 4.) Let $F = 0.25$ and solve the equation $F = \frac{1}{2}(1 - \cos \theta)$ for all θ to determine the phase angles for when 25% of the moon is illuminated.

118. Full Moon (Refer to Example 4.) Let $F = 1$ and solve the equation $F = \frac{1}{2}(1 - \cos \theta)$ for all θ to determine the phase angles for a full moon.

119. Designing Highway Curves (Refer to Example 11.) A highway curve with a radius of $r = 200$ feet and a speed limit of $v = 44$ feet per second (30 mi/hr) is being designed. Use the formula

$$r = \frac{v^2}{4.5 + 32.2 \tan \theta}$$

to find an appropriate superelevation θ for the given curve.

120. Designing Highway Curves (Refer to Example 11.) A highway curve with a radius of $r = 800$ feet and a speed limit of $v = 66$ feet per second (45 mi/hr) is being designed. Find the appropriate superelevation for the curve.

121. Daylight Hours (Refer to Example 7.) The number of daylight hours y at 60°N latitude can be modeled by

$$y = 6.5 \sin\left(\frac{\pi}{6}(x - 3.65)\right) + 12.4,$$

where $x = 1$ corresponds to January 1, $x = 2$ to February 1, and so on. Estimate graphically or numerically when there are 9 hours of daylight. (*Source:* J. Williams.)

122. Average Temperatures The monthly average high temperature y in degrees Fahrenheit at Phoenix, Arizona, can be modeled by

$$y = 20.3 \sin(0.53x - 2.18) + 83.8,$$

where $x = 1$ corresponds to January, $x = 2$ to February, and so on. Estimate graphically or numerically when the monthly average high temperature is 93°F. (*Source:* J. Williams.)

123. Daylight Hours Solve Exercise 121 symbolically.

124. Monthly Average Temperatures Solve Exercise 122 symbolically.

125. Maximum Monthly Sunshine The maximum number of hours of sunshine each month is listed in the table for 50°N latitude.

Month	1	2	3	4	5	6
Hours of Sunshine	261	279	363	407	471	482
Month	7	8	9	10	11	12
Hours of Sunshine	486	442	374	329	267	246

Source: C. Winter, *Solar Power Plants.*

(a) Find a function f that models the data.

(b) Estimate graphically any solutions to the inequality $f(x) \geq 350$ on the interval $[1, 12]$. Interpret the result.

126. Music and Pure Tones A pure tone can be described by a sinusoidal graph. The graph of the function $P = 0.004 \sin(100\pi t)$ shown in the figure represents the pressure of a pure tone on an eardrum in pounds per square foot at time t in seconds. (*Source:* J. Roederer, *Introduction to the Physics and Psychophysics of Music.*)

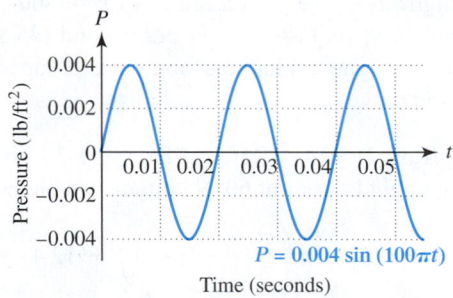

$$P = 0.004 \sin(100\pi t)$$
Time (seconds)

(a) Estimate all solutions to the equation $P = 0.004$ on the interval $[0, 0.05]$.

(b) Interpret these solutions.

Exercises 127 and 128: **Kepler's Equation** *(Refer to Example 12.) Solve Kepler's equation to within two decimal places graphically or numerically.*

127. $\theta = 0.26 + 0.017 \sin\theta$

128. $\theta = 0.18 + 0.249 \sin\theta$

129. Sums of Pure Tones If two loudspeakers located at different positions produce the same pure tone, the human ear may hear one sound that is equal to the sum of the individual tones. Since the sources are at different locations, their sinusoidal waves will have different phase angles. (*Source:* N. Fletcher and T. Rossing, *The Physics of Musical Instruments.*)

(a) Let two musical tones be given by

$$P_1 = 0.003 \sin(880\pi t - 0.7) \quad \text{and}$$
$$P_2 = 0.002 \sin(880\pi t + 0.6).$$

Graph P_1, P_2, and their sum $P = P_1 + P_2$ separately in the viewing rectangle $[0, 0.01, 0.005]$ by $[-0.005, 0.005, 0.001]$.

(b) Determine the maximum pressure for P.

(c) Is the maximum pressure for P equal to the sum of the maximums of P_1 and P_2? Explain.

130. Sound (Refer to Exercise 129.) Suppose that two loudspeakers located at different positions produce pure tones given by

$$P_1 = A_1 \sin(2\pi F t + \alpha) \quad \text{and}$$
$$P_2 = A_2 \sin(2\pi F t + \beta),$$

where F is their common frequency. Then the resulting tone heard by a listener may be written as $P = A \sin(2\pi F t + \theta)$, where

$$A = \sqrt{(A_1 \cos\alpha + A_2 \cos\beta)^2 + (A_1 \sin\alpha + A_2 \sin\beta)^2}$$

and

$$\theta = \arctan\left[\frac{A_1 \sin\alpha + A_2 \sin\beta}{A_1 \cos\alpha + A_2 \cos\beta}\right].$$

(*Source:* N. Fletcher.)

(a) Find A and θ for P if $F = 440$, $A_1 = 0.003$, $\alpha = -0.7$, $A_2 = 0.002$, and $\beta = 0.6$.

(b) Graph $P = A \sin(2\pi F t + \theta)$ and $y = P_1 + P_2$ in $[0, 0.01, 0.005]$ by $[-0.005, 0.005, 0.001]$. Do the graphs appear to be identical?

Writing about Mathematics

131. Explain the difference between a conditional equation and an identity. Give one example of each.

132. Explain why knowledge of algebra is important when solving trigonometric equations. Give one example of how knowledge of algebra can be applied to solving a trigonometric equation.

7.4 Sum and Difference Identities

- Apply the sum and difference identities for cosine
- Apply sum and difference identities for sine and tangent

Introduction

Music is made up of vibrations that create pressure on our eardrums. Musical tones can sometimes be modeled with sinusoidal graphs. When more than one tone is played, the resulting pressure is equal to the sum of the individual pressures. Sum and difference identities are sometimes helpful in the analysis of music. This section introduces several trigonometric identities and some of their applications.

Sum and Difference Identities for Cosine

The graph of $y = \cos\left(t - \frac{\pi}{2}\right)$ is translated to the right $\frac{\pi}{2}$ units compared to the graph of $y = \cos t$, as shown in Figure 7.45. If the graph of $y = \cos t$ is translated right $\frac{\pi}{2}$ units, it coincides with the graph of $y = \sin t$, which is shown in Figure 7.46. This discussion suggests that the equation

$$\cos\left(t - \frac{\pi}{2}\right) = \sin t$$

is true for all real numbers t.

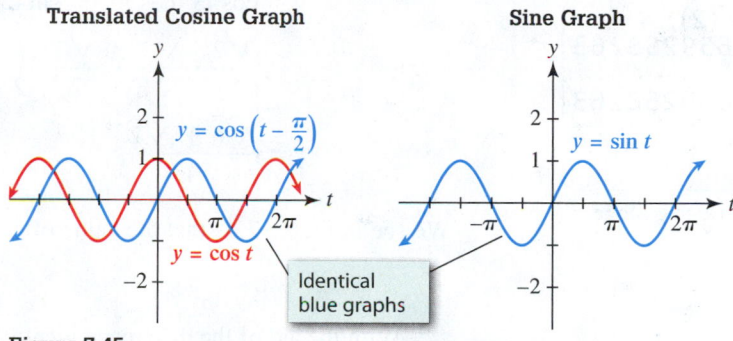

Figure 7.45   **Figure 7.46**

NOTE It is important to understand that

$$\cos\left(t - \frac{\pi}{2}\right) \neq \cos t - \cos\frac{\pi}{2} = \cos t - 0 = \cos t.$$

To verify this result symbolically, a new identity is needed. Suppose that α and β represent any two angles or real numbers. Then the following identity can be used to calculate the cosine of their difference. (Its proof is given at the end of this section.)

$$\cos(\alpha - \beta) = \cos\alpha \cos\beta + \sin\alpha \sin\beta$$

The next example demonstrates how to apply this identity.

EXAMPLE 1   **Using the cosine difference identity**

Verify the identity

$$\cos\left(t - \frac{\pi}{2}\right) = \sin t.$$

SOLUTION

Getting Started Start by letting $\alpha = t$ and $\beta = \frac{\pi}{2}$ in the cosine difference identity

$$\cos(\alpha - \beta) = \cos\alpha \cos\beta + \sin\alpha \sin\beta. \blacktriangleright$$

This substitution gives the following result.

$$\cos\left(t - \frac{\pi}{2}\right) = \cos t \cos \frac{\pi}{2} + \sin t \sin \frac{\pi}{2}$$

$$= \cos t (0) + \sin t (1)$$

$$= \sin t$$

Now Try Exercise 29

In the next example, we use the difference identity

$$\cos(\alpha - \beta) = \cos\alpha \cos\beta + \sin\alpha \sin\beta$$

to find the exact value of cos 15°.

EXAMPLE 2 **Applying the cosine difference identity**

Find the exact value of cos 15°. Use a calculator to support your result.

SOLUTION Since $45° - 30° = 15°$ and the exact trigonometric values for 45° and 30° are known, we proceed as follows.

$$\cos 15° = \cos(45° - 30°) \qquad \text{15° = 45° − 30°}$$

$$= \cos 45° \cos 30° + \sin 45° \sin 30° \qquad \text{Difference identity for cosine}$$

$$= \frac{\sqrt{2}}{2} \cdot \frac{\sqrt{3}}{2} + \frac{\sqrt{2}}{2} \cdot \frac{1}{2} \qquad \text{Evaluate each function.}$$

$$= \frac{\sqrt{6} + \sqrt{2}}{4} \qquad \text{Simplify the exact value.}$$

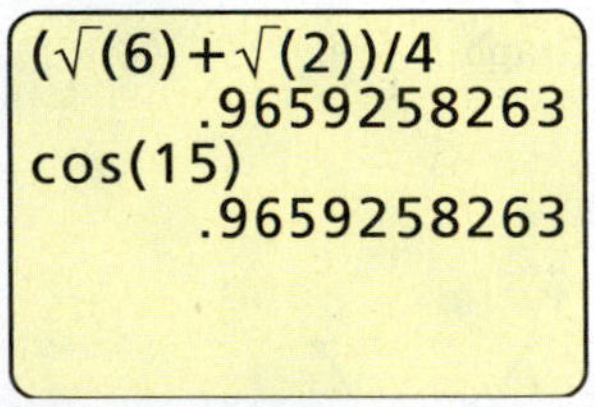

```
(√(6)+√(2))/4
         .9659258263
cos(15)
         .9659258263
```

Figure 7.47 Degree Mode

We see in Figure 7.47 that the value of cos 15° agrees with the symbolic result.

Now Try Exercise 5

With the aid of the difference identity for cosine, we can derive a sum identity.

$$\cos(\alpha + \beta) = \cos(\alpha - (-\beta)) \qquad \alpha + \beta = \alpha - (-\beta)$$

$$= \cos\alpha \cos(-\beta) + \sin\alpha \sin(-\beta) \qquad \text{Difference identity for cosine}$$

$$= \cos\alpha \cos\beta - \sin\alpha \sin\beta \qquad \cos(-\beta) = \cos\beta;$$
$$\sin(-\beta) = -\sin\beta$$

Sum and difference identities for cosine are as follows.

COSINE OF A SUM OR DIFFERENCE

$$\cos(\alpha + \beta) = \cos\alpha \cos\beta - \sin\alpha \sin\beta$$

$$\cos(\alpha - \beta) = \cos\alpha \cos\beta + \sin\alpha \sin\beta$$

NOTE $\cos(\alpha + \beta) \neq \cos\alpha + \cos\beta$ and $\cos(\alpha - \beta) \neq \cos\alpha - \cos\beta$

Cofunction Identities Section 6.2 introduced the cofunction identities for an acute angle θ. These identities are true for any real number t. We verify one of these identities in the next example.

EXAMPLE 3 **Verifying a cofunction identity**

Verify that $\cos\left(\frac{\pi}{2} - t\right) = \sin t$.

SOLUTION Let $\alpha = \frac{\pi}{2}$ and $\beta = t$ in the cosine difference identity.

$$\cos\left(\frac{\pi}{2} - t\right) = \cos\frac{\pi}{2}\cos t + \sin\frac{\pi}{2}\sin t$$

$$= (0)\cos t + (1)\sin t$$

$$= \sin t$$

Now Try Exercise 19

The following cofunction identities are valid for *any* real number t.

COFUNCTION IDENTITIES FOR ANY REAL NUMBER t

$$\cos\left(\frac{\pi}{2} - t\right) = \sin t \qquad \sin\left(\frac{\pi}{2} - t\right) = \cos t$$

$$\cot\left(\frac{\pi}{2} - t\right) = \tan t \qquad \tan\left(\frac{\pi}{2} - t\right) = \cot t$$

$$\csc\left(\frac{\pi}{2} - t\right) = \sec t \qquad \sec\left(\frac{\pi}{2} - t\right) = \csc t$$

Sum and Difference Identities for Sine

There are also sum and difference identities for sine.

$$\sin(\alpha + \beta) = \cos\left(\frac{\pi}{2} - (\alpha + \beta)\right) \qquad \text{Cofunction identity}$$

$$= \cos\left(\left(\frac{\pi}{2} - \alpha\right) - \beta\right) \qquad \text{Associative property}$$

$$= \cos\left(\frac{\pi}{2} - \alpha\right)\cos\beta + \sin\left(\frac{\pi}{2} - \alpha\right)\sin\beta \qquad \text{Difference identity for cosine}$$

$$= \sin\alpha\cos\beta + \cos\alpha\sin\beta \qquad \text{Cofunction identities}$$

The difference identity for sine can be derived in a similar manner.

The following box gives the sum and difference identities for sine.

SINE OF A SUM OR DIFFERENCE

$$\sin(\alpha + \beta) = \sin\alpha\cos\beta + \cos\alpha\sin\beta$$

$$\sin(\alpha - \beta) = \sin\alpha\cos\beta - \cos\alpha\sin\beta$$

NOTE $\sin(\alpha + \beta) \neq \sin\alpha + \sin\beta$ and $\sin(\alpha - \beta) \neq \sin\alpha - \sin\beta$

EXAMPLE 4 **Analyzing an identity graphically and symbolically**

Give graphical support for $\sin(\theta + \pi) = -\sin\theta$. Then verify the identity symbolically.

SOLUTION

Graphical Support Graph $y = \sin(\theta + \pi)$ and $y = -\sin\theta$, as shown in Figures 7.48 and 7.49 on the next page. Their graphs appear to be identical.

Identical Graphs

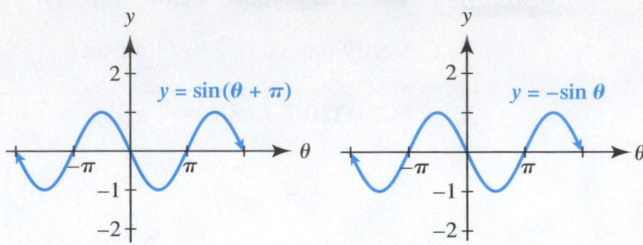

Figure 7.48   **Figure 7.49**

Symbolic Verification   Let $\alpha = \theta$ and $\beta = \pi$ in the sum identity for sine.

$$\sin(\theta + \pi) = \sin\theta \cos\pi + \cos\theta \sin\pi$$
$$= \sin\theta(-1) + \cos\theta(0)$$
$$= -\sin\theta$$

Thus $\sin(\theta + \pi) = -\sin\theta$.

> **Now Try Exercise 11**

EXAMPLE 5   **Applying sum identities for sine and cosine**

Let $\sin\alpha = \frac{4}{5}$ and $\cos\beta = \frac{3}{5}$. If α is in quadrant II and β is in quadrant IV, find each of the following.

(a) $\sin(\alpha + \beta)$ **(b)** $\cos(\alpha + \beta)$ **(c)** $\tan(\alpha + \beta)$
(d) The quadrant containing $\alpha + \beta$

SOLUTION

Getting Started First sketch possible angles for α and for β, as shown in Figures 7.50 and 7.51. We can see that

$$\sin\alpha = \frac{4}{5}, \quad \cos\alpha = -\frac{3}{5}, \quad \sin\beta = -\frac{4}{5}, \quad \text{and} \quad \cos\beta = \frac{3}{5}. \blacktriangleright$$

α in Quadrant II

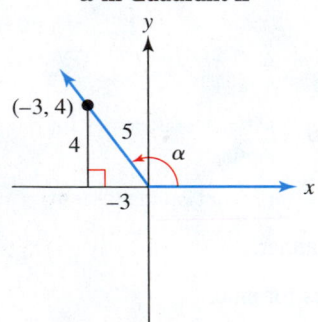

Figure 7.50

(a) To find $\sin(\alpha + \beta)$, apply the sum identity for sine.

$$\sin(\alpha + \beta) = \sin\alpha \cos\beta + \cos\alpha \sin\beta$$
$$= \left(\frac{4}{5}\right)\left(\frac{3}{5}\right) + \left(-\frac{3}{5}\right)\left(-\frac{4}{5}\right)$$
$$= \frac{24}{25}$$

β in Quadrant IV

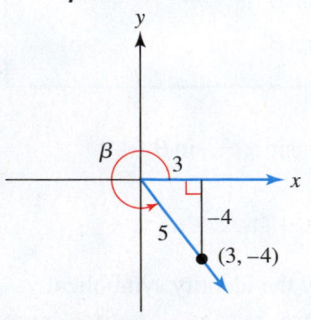

Figure 7.51

(b) To find $\cos(\alpha + \beta)$, apply the sum identity for cosine.

$$\cos(\alpha + \beta) = \cos\alpha \cos\beta - \sin\alpha \sin\beta$$
$$= \left(-\frac{3}{5}\right)\left(\frac{3}{5}\right) - \left(\frac{4}{5}\right)\left(-\frac{4}{5}\right)$$
$$= \frac{7}{25}$$

(c) $\tan(\alpha + \beta) = \dfrac{\sin(\alpha + \beta)}{\cos(\alpha + \beta)} = \dfrac{24/25}{7/25} = \dfrac{24}{7}$

(d) Since both $\sin(\alpha + \beta)$ and $\cos(\alpha + \beta)$ are positive, $\alpha + \beta$ is in quadrant I.

> **Now Try Exercise 21**

EXAMPLE 6   Verifying an identity

Verify the identity $\dfrac{\sin(\alpha - \beta)}{\sin\alpha \sin\beta} = \cot\beta - \cot\alpha$.

SOLUTION Begin by expanding the expression $\sin(\alpha - \beta)$.

$$\dfrac{\sin(\alpha - \beta)}{\sin\alpha \sin\beta} = \dfrac{\sin\alpha \cos\beta - \cos\alpha \sin\beta}{\sin\alpha \sin\beta} \qquad \textit{Difference identity}$$

$$= \dfrac{\sin\alpha \cos\beta}{\sin\alpha \sin\beta} - \dfrac{\cos\alpha \sin\beta}{\sin\alpha \sin\beta} \qquad \tfrac{a-b}{c} = \tfrac{a}{c} - \tfrac{b}{c}$$

$$= \dfrac{\cos\beta}{\sin\beta} - \dfrac{\cos\alpha}{\sin\alpha} \qquad \textit{Simplify each ratio.}$$

$$= \cot\beta - \cot\alpha \qquad \textit{Quotient identity}$$

Now Try Exercise 45

Applications

Back Stress   Because human joints both bend and rotate, trigonometry frequently is applied to human physiology. The next example shows how to calculate the force exerted by a person's back muscles and gives a rather amazing result.

EXAMPLE 7   Analyzing stress on a person's back

If a person with weight W bends at the waist with a straight back, then the force F exerted by the lower back muscles may be approximated using $F = 2.89W \sin\left(\theta + \frac{\pi}{2}\right)$, where θ is the angle between a person's torso and the horizontal. See Figure 7.52.   (*Source:* H. Metcalf, *Topics in Classical Biophysics*.)

(a) Let $W = 155$ pounds. The graph of $F = 2.89W \sin\left(\theta + \frac{\pi}{2}\right)$ with $W = 155$ is shown in Figure 7.53. Interpret the graph.

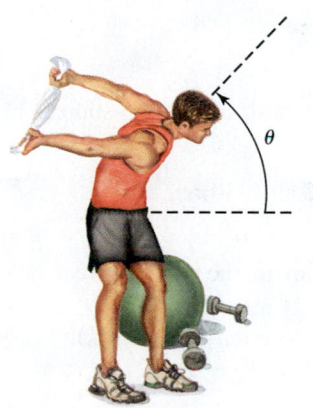

Figure 7.52

Lower Back Force

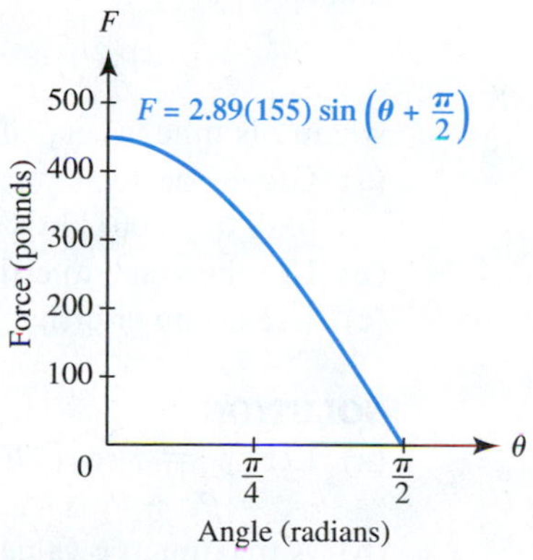

$F = 2.89(155) \sin\left(\theta + \frac{\pi}{2}\right)$

Force (pounds) vs. Angle (radians)

Figure 7.53

(b) Show that $F = 2.89W \cos\theta$.
(c) When $W = 155$ pounds, for what value of θ does F equal 400 pounds?

SOLUTION
(a) When $\theta = 0$, the person's back is parallel to the ground and force F exerted by the back muscles has a maximum of about 450 pounds. This is nearly *three times* the person's weight! As the person straightens up, θ increases, while F decreases. When $\theta = \frac{\pi}{2}$, the person is standing straight up and $F = 0$.

(b) Given $F = 2.89W \sin\left(\theta + \frac{\pi}{2}\right)$, we can show $F = 2.89W \cos\theta$, by applying a sum identity for sine.

$$F = 2.89W \sin\left(\theta + \frac{\pi}{2}\right)$$

$$= 2.89W \left(\sin\theta \cos\frac{\pi}{2} + \cos\theta \sin\frac{\pi}{2}\right)$$

$$= 2.89W \left(\sin\theta (0) + \cos\theta (1)\right)$$

$$= 2.89W \cos\theta$$

(c) To determine θ when the force is 400 pounds for a person weighing 155 pounds, let $F = 400$ and $W = 155$ in $F = 2.89W \cos\theta$, and then solve for θ.

$$400 = 2.89\,(155)\cos\theta \qquad \textcolor{blue}{F = 400;\ W = 155}$$

$$\cos\theta = \frac{400}{2.89\,(155)} \qquad \textcolor{blue}{\text{Solve for } \cos\theta.}$$

$$\theta = \cos^{-1}\left(\frac{400}{2.89\,(155)}\right) \qquad \textcolor{blue}{\text{Solve for } \theta.}$$

$$\theta \approx 0.467, \text{ or } 26.8° \qquad \textcolor{blue}{\text{Approximate } \theta.}$$

> **Now Try Exercise 53**

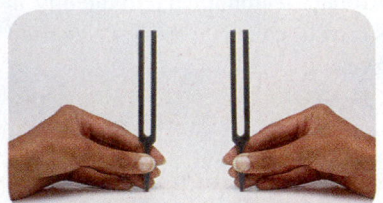

Music and Mathematics Music is composed of tones with various frequencies. Pressure exerted on the eardrum by a pure tone may be modeled by either $P(t) = a \cos bt$ or $P(t) = a \sin bt$, where a and b are constants and t represents time. When two tuning forks produce the same pure tone, the human ear hears only one sound that is equal to the sum of the individual tones. Trigonometry can be used to model this situation. (*Source:* N. Fletcher and T. Rossing, *The Physics of Musical Instruments.*)

EXAMPLE 8 **Modeling musical tones**

Let the pressure P in grams per square meter exerted on the eardrum by two sources be modeled by

$$P_1(t) = 5\cos(440\pi t) \qquad \text{and} \qquad P_2(t) = 3\sin(440\pi t),$$

where t is time in seconds.
(a) Graph the total pressure, $P = P_1 + P_2$, on the eardrum in the viewing rectangle $[0, 0.01, 0.001]$ by $[-8, 8, 1]$.
(b) Use the graph to estimate values for a and k such that $P = a\sin(440\pi t + k)$.
(c) Use a sum or difference identity for sine to verify that $P \approx P_1 + P_2$.

SOLUTION

$[0, 0.01, 0.001]$ by $[-8, 8, 1]$

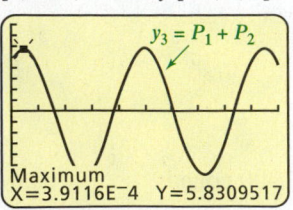

Figure 7.54 Radian Mode

(a) Let $Y_1 = 5\cos(440\pi X)$, $Y_2 = 3\sin(440\pi X)$, and $Y_3 = Y_1 + Y_2$. The graph of $y_3 = P_1 + P_2$ is shown in Figure 7.54.
(b) A maximum y-value occurs near $(0.00039116, 5.831)$. (Note that the x- and y-values shown in Figure 7.54 may vary slightly.) Thus let $a = 5.831$. Since $\sin\theta$ is maximum when $\theta = \frac{\pi}{2}$, we let $t = 0.00039116$ and solve the following equation for k.

$$440\pi\,(0.00039116) + k = \frac{\pi}{2} \qquad \textcolor{blue}{440\pi t + k = \frac{\pi}{2}}$$

$$k = \frac{\pi}{2} - 440\pi\,(0.00039116)$$

$$\approx 1.0301$$

Thus let $P \approx 5.831\sin(440\pi t + 1.0301)$. (Other values for k are possible.)

Calculator Help

To find a maximum point on a graph, see Appendix A (page AP-8).

(c) Apply the sum identity for sine.

$$P \approx 5.831 \sin(440\pi t + 1.0301)$$
$$= 5.831 \left(\sin(440\pi t) \cos(1.0301) + \cos(440\pi t) \sin(1.0301) \right)$$
$$\approx 5.831 \left(\sin(440\pi t)(0.5147) + \cos(440\pi t)(0.8574) \right)$$
$$\approx 3.00 \sin(440\pi t) + 5.00 \cos(440\pi t)$$
$$= P_1 + P_2$$

> **Now Try Exercise 55**

Sum and Difference Identities for Tangent

Sum and difference identities can also be found for the tangent function.

> **TANGENT OF A SUM OR DIFFERENCE**
>
> $$\tan(\alpha + \beta) = \frac{\tan\alpha + \tan\beta}{1 - \tan\alpha \tan\beta}$$
>
> $$\tan(\alpha - \beta) = \frac{\tan\alpha - \tan\beta}{1 + \tan\alpha \tan\beta}$$

NOTE $\tan(\alpha + \beta) \neq \tan\alpha + \tan\beta$ and $\tan(\alpha - \beta) \neq \tan\alpha - \tan\beta$

These identities are a result of the sum and difference identities for sine and cosine. For example, the difference identity for tangent can be verified as follows.

$$\tan(\alpha - \beta) = \frac{\sin(\alpha - \beta)}{\cos(\alpha - \beta)}$$ Use $\tan\theta = \frac{\sin\theta}{\cos\theta}$ with $\theta = \alpha - \beta$.

$$= \frac{\sin\alpha \cos\beta - \cos\alpha \sin\beta}{\cos\alpha \cos\beta + \sin\alpha \sin\beta}$$ Apply difference identities.

$$= \frac{\dfrac{\sin\alpha \cos\beta}{\cos\alpha \cos\beta} - \dfrac{\cos\alpha \sin\beta}{\cos\alpha \cos\beta}}{\dfrac{\cos\alpha \cos\beta}{\cos\alpha \cos\beta} + \dfrac{\sin\alpha \sin\beta}{\cos\alpha \cos\beta}}$$ Divide each term by $\cos\alpha \cos\beta$.

$$= \frac{\tan\alpha - \tan\beta}{1 + \tan\alpha \tan\beta}$$ Simplify.

> **EXAMPLE 9**   **Using the tangent difference identity**

Use Figure 7.55 to find $\tan\gamma$ if $\tan\alpha = \frac{4}{3}$ and $\tan\beta = \frac{3}{4}$.

SOLUTION From Figure 7.55, α and $\beta + \gamma$ are both supplements of angle BAC. Thus $\alpha = \beta + \gamma$ or $\gamma = \alpha - \beta$.

$$\tan\gamma = \tan(\alpha - \beta)$$ $\gamma = \alpha - \beta$

$$= \frac{\tan\alpha - \tan\beta}{1 + \tan\alpha \tan\beta}$$ Tangent difference identity

$$= \frac{\frac{4}{3} - \frac{3}{4}}{1 + \frac{4}{3} \cdot \frac{3}{4}}$$ $\tan\alpha = \frac{4}{3}$; $\tan\beta = \frac{3}{4}$

$$= \frac{7}{24}$$ Simplify.

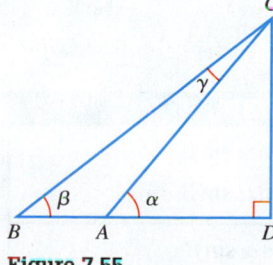

Figure 7.55

> **Now Try Exercise 47**

Derivation of an Identity

We conclude this section by deriving the difference identity for cosine. Begin by considering the angles α and β in standard position and the unit circle, as shown in Figure 7.56. The terminal side of α intersects the unit circle at $(\cos\alpha, \sin\alpha)$, and the terminal side of β intersects the unit circle at $(\cos\beta, \sin\beta)$. The angle formed between the terminal sides of α and β equals $\alpha - \beta$. Now consider angle $\alpha - \beta$ in standard position. Its terminal side intersects the unit circle at the point $(\cos(\alpha - \beta), \sin(\alpha - \beta))$. This is shown in Figure 7.57.

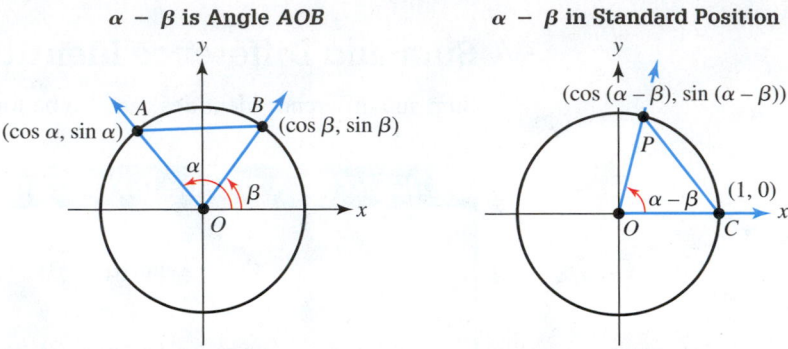

Figure 7.56 Unit Circle   **Figure 7.57** Unit Circle

Since triangles ABO and PCO are congruent, the distance from A to B in Figure 7.56 equals the distance from P to C in Figure 7.57.

$$\sqrt{(\cos\alpha - \cos\beta)^2 + (\sin\alpha - \sin\beta)^2} = \sqrt{(\cos(\alpha - \beta) - 1)^2 + (\sin(\alpha - \beta) - 0)^2}$$

Distance from A to B Distance from P to C

Squaring each side and clearing parentheses produces

$$\cos^2\alpha - 2\cos\alpha\cos\beta + \cos^2\beta + \sin^2\alpha - 2\sin\alpha\sin\beta + \sin^2\beta =$$
$$\cos^2(\alpha - \beta) - 2\cos(\alpha - \beta) + 1 + \sin^2(\alpha - \beta).$$

Since $\sin^2\theta + \cos^2\theta = 1$ for any θ, the above equation simplifies to

$$2 - 2\cos\alpha\cos\beta - 2\sin\alpha\sin\beta = 2 - 2\cos(\alpha - \beta).$$

Solving this equation for $\cos(\alpha - \beta)$ gives the cosine difference identity

$$\cos(\alpha - \beta) = \cos\alpha\cos\beta + \sin\alpha\sin\beta.$$

7.4 Putting It All Together

The following table lists the important identities in this section.

IDENTITIES	GENERAL FORM
Cosine sum and difference	$\cos(\alpha + \beta) = \cos\alpha\cos\beta - \sin\alpha\sin\beta$ $\cos(\alpha - \beta) = \cos\alpha\cos\beta + \sin\alpha\sin\beta$
Sine sum and difference	$\sin(\alpha + \beta) = \sin\alpha\cos\beta + \cos\alpha\sin\beta$ $\sin(\alpha - \beta) = \sin\alpha\cos\beta - \cos\alpha\sin\beta$

IDENTITIES	GENERAL FORM
Tangent sum and difference	$\tan(\alpha + \beta) = \dfrac{\tan\alpha + \tan\beta}{1 - \tan\alpha\tan\beta}$ $\tan(\alpha - \beta) = \dfrac{\tan\alpha - \tan\beta}{1 + \tan\alpha\tan\beta}$
Cofunction	$\cos\left(\frac{\pi}{2} - t\right) = \sin t$ $\qquad$ $\sin\left(\frac{\pi}{2} - t\right) = \cos t$ $\cot\left(\frac{\pi}{2} - t\right) = \tan t$ $\qquad$ $\tan\left(\frac{\pi}{2} - t\right) = \cot t$ $\csc\left(\frac{\pi}{2} - t\right) = \sec t$ $\qquad$ $\sec\left(\frac{\pi}{2} - t\right) = \csc t$

7.4 Exercises

Sum and Difference Identities

Exercises 1–10: Find the exact value for each expression. Use a calculator to support your result numerically.

1. $\sin 15°$ **2.** $\sin 105°$

3. $\tan 15°$ **4.** $\sin 75°$

5. $\cos 75°$ **6.** $\cos 105°$

7. $\sin\frac{\pi}{12}$ $\left(Hint: \frac{\pi}{3} - \frac{\pi}{4} = \frac{\pi}{12}\right)$

8. $\cos\frac{5\pi}{12}$

9. $\sin\frac{5\pi}{12}$ **10.** $\tan\frac{\pi}{12}$

Exercises 11–16: Complete the following for the identity.

(a) Give graphical support for the identity.
(b) Verify the identity symbolically.

11. $\sin\left(t + \frac{\pi}{2}\right) = \cos t$ **12.** $\sin\left(t + \frac{3\pi}{2}\right) = -\cos t$

13. $\cos(t + \pi) = -\cos t$ **14.** $\cos\left(t + \frac{3\pi}{2}\right) = \sin t$

15. $\sec\left(t - \frac{\pi}{2}\right) = \csc t$ **16.** $\tan\left(t + \frac{\pi}{2}\right) = -\cot t$

Exercises 17–20: Use a difference identity to verify the cofunction identity.

17. $\sin\left(\frac{\pi}{2} - t\right) = \cos t$ **18.** $\tan\left(\frac{\pi}{2} - t\right) = \cot t$

19. $\sec\left(\frac{\pi}{2} - t\right) = \csc t$ **20.** $\csc\left(\frac{\pi}{2} - t\right) = \sec t$

Exercises 21–28: (Refer to Example 5.) Find the following.

(a) $\sin(\alpha + \beta)$ *(b)* $\cos(\alpha + \beta)$
(c) $\tan(\alpha + \beta)$ *(d) The quadrant containing $\alpha + \beta$*

21. $\sin\alpha = \frac{3}{5}$ and $\sin\beta = \frac{5}{13}$, α and β in quadrant I

22. $\cos\alpha = -\frac{12}{13}$ and $\cos\beta = -\frac{5}{13}$, α and β in quadrant II

23. $\sin\alpha = -\frac{8}{17}$ and $\cos\beta = \frac{11}{61}$, α in quadrant III and β in quadrant I

24. $\cos\alpha = -\frac{24}{25}$ and $\sin\beta = \frac{4}{5}$, α in quadrant II and β in quadrant I

25. $\cos\alpha = -\frac{3}{5}$ and $\cos\beta = \frac{12}{13}$, α in quadrant III and β in quadrant IV

26. $\tan\alpha = \frac{3}{4}$ and $\cos\beta = -\frac{4}{5}$, α in quadrant I and β in quadrant III

27. $\tan\alpha = -\frac{5}{12}$ and $\sec\beta = -\frac{61}{11}$, α and β in quadrant II

28. $\cot\alpha = \frac{3}{4}$ and $\csc\beta = \frac{25}{24}$, α and β in quadrant I

Exercises 29–46: Verify the identity.

29. $\cos\left(t - \frac{\pi}{4}\right) = \frac{\sqrt{2}}{2}(\cos t + \sin t)$

30. $\sin\left(t + \frac{\pi}{4}\right) = \frac{\sqrt{2}}{2}(\cos t + \sin t)$

31. $\tan\left(t + \frac{\pi}{4}\right) = \dfrac{1 + \tan t}{1 - \tan t}$

32. $\tan(45° - \theta) = \dfrac{1 - \tan\theta}{1 + \tan\theta}$

33. $\dfrac{\cos(x - y)}{\cos(x + y)} = \dfrac{1 + \tan x\tan y}{1 - \tan x\tan y}$

34. $\dfrac{\sin(x - y)}{\sin(x + y)} = \dfrac{\tan x - \tan y}{\tan x + \tan y}$

35. $\dfrac{\cos(\alpha - \beta)}{\cos\alpha \sin\beta} = \tan\alpha + \cot\beta$

36. $\cos(\theta + \theta) = 1 - 2\sin^2\theta$

37. $\sin 2t = 2\sin t \cos t$

38. $\cos 2t = \cos^2 t - \sin^2 t$

39. $\sin(\alpha + \beta) + \sin(\alpha - \beta) = 2\sin\alpha \cos\beta$

40. $\cos(\alpha + \beta) + \cos(\alpha - \beta) = 2\cos\alpha \cos\beta$

41. $\tan(\pi - \theta) = -\tan\theta$ **42.** $\tan(\theta + \pi) = \tan\theta$

43. $\tan(x - y) - \tan(y - x) = \dfrac{2(\tan x - \tan y)}{1 + \tan x \tan y}$

44. $\dfrac{\sin(x - y)}{\sin y} + \dfrac{\cos(x - y)}{\cos y} = \dfrac{\sin x}{\sin y \cos y}$

45. $\dfrac{\sin(x + y)}{\cos x \cos y} = \tan x + \tan y$

46. $\dfrac{\tan(x + y) - \tan y}{1 + \tan(x + y)\tan y} = \tan x$

Exercises 47 and 48: Solve Example 9 by using the given information.

47. $\tan\alpha = \frac{6}{7}$ and $\tan\beta = \frac{5}{7}$

48. $\cot\alpha = \frac{8}{13}$ and $\cot\beta = \frac{11}{13}$

Lines and Slopes

Exercises 49–52: Suppose two lines, l_1 and l_2, intersect the x-axis making angles α and β, as shown in the figure. Then the slopes of l_1 and l_2 satisfy $m_1 = \tan\alpha$ and $m_2 = \tan\beta$, respectively. If l_1 and l_2 intersect with angle θ as shown, then it follows that $\beta = \alpha + \theta$, or equivalently, $\theta = \beta - \alpha$.

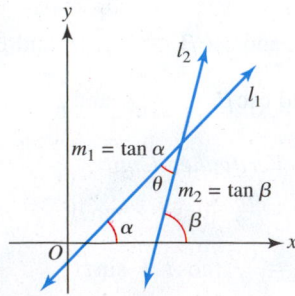

49. Use a difference identity for tangent to show that

$$\tan\theta = \dfrac{m_2 - m_1}{1 + m_1 m_2}.$$

50. Is the formula in Exercise 49 valid if β is an obtuse angle? Explain.

51. Find θ for two intersecting lines given by $y = 2x - 3$ and $y = \frac{3}{5}x + 1$.

52. Find θ for two intersecting lines given by $y = \frac{1}{2}x + 1$ and $y = 3 - x$.

Applications

53. Back Stress (Refer to Example 7.) Answer the following if $F = 2.89W\cos\theta$.
   (a) Suppose a 200-pound person bends at the waist so that $\theta = \frac{\pi}{4}$. Estimate the force F exerted by the person's back muscles.

   (b) For a 200-pound person, approximate the value of θ that results in the back muscles exerting a force F of 400 pounds.

54. Sound Waves Sound is a result of waves applying pressure to a person's eardrum. For a particular sound wave radiating outward, the trigonometric function $P(r) = \frac{a}{r}\cos(\pi r - 1000t)$ can be used to express the pressure at a radius of r feet from the source after t seconds. In this formula, a is the maximum sound pressure at the source, measured in pounds per square foot. (*Source:* L. Beranek, *Noise and Vibration Control.*)
   (a) Let $a = 0.4$, $t = 1$, and graph the sound pressure for $0 \le r \le 20$. What happens to the pressure P as the radius r increases?

   (b) Use a difference identity to simplify the expression for $P(r)$ when r is an even integer.

55. Modeling Musical Tones (Refer to Example 8.) Let the pressure exerted by two sound waves in grams per square meter be given by $P_1(t) = 4\cos(220\pi t)$ and $P_2(t) = 3\sin(220\pi t)$, where t is in seconds.
   (a) Graph the total pressure $P = P_1 + P_2$ in the window $[0, 0.02, 0.001]$ by $[-6, 6, 1]$.

   (b) Use the graph to estimate values for a and k such that $P = a\sin(220\pi t + k)$.

   (c) Use a sum or difference identity for sine to verify that

   $a\sin(220\pi t + k) \approx 4\cos(220\pi t) + 3\sin(220\pi t)$.

56. Electricity When voltages $V_1 = 50\sin(120\pi t)$ and $V_2 = 120\cos(120\pi t)$ are applied to the same circuit, the resulting voltage V is equal to their sum. (*Source:* D. Bell, *Fundamentals of Electric Circuits.*)
   (a) Graph $V = V_1 + V_2$ in the viewing rectangle $[0, 0.05, 0.01]$ by $[-160, 160, 40]$.

   (b) Use the graph to estimate values for a and k so that $V = a\sin(120\pi t + k)$.

   (c) Use a sum or difference identity for sine to verify part (b).

Writing about Mathematics

57. Are $\sin(45° + 30°)$ and $\sin 45° + \sin 30°$ equivalent expressions? Explain your answer.

58. Are the expressions $\cos(\alpha - \beta)$ and $\cos\alpha - \cos\beta$ equal? Give an example to justify your answer.

Extended and Discovery Exercises

Modeling Musical Beats Musicians sometimes tune instruments by playing the same tone on two instruments and listening for a phenomenon known as *beats*. The human ear hears beats because the sound pressure slowly rises and falls when two tones vary slightly. When the two instruments are in tune, the beats will disappear. The pressure P on an eardrum can be modeled by $P = a\sin(2\pi Ft)$, where F is the frequency of the tone, t is time in seconds, and P is in pounds per square foot. (*Source:* J. Pierce, *The Science of Musical Sound.*)

1. Consider two tones with similar frequencies of 440 and 443 cycles per second and pressures

$$P_1 = 0.006\sin(880\pi t) \text{ and } P_2 = 0.004\sin(886\pi t),$$

respectively.

(a) Graph the sum $P = P_1 + P_2$ in $[0.15, 1.15, 0.05]$ by $[-0.01, 0.01, 0.001]$, where P is the total pressure exerted by the tones on an eardrum. How many beats are there in this 1-second interval?

(b) Repeat part (a) with frequencies of 220 and 224.

(c) Determine a way to find the number of beats per second if the frequencies of the tones are F_1 and F_2.

Exercises 2 and 3: **Music and Beats** (*Refer to Exercise 1.*) *Given two musical tones P_1 and P_2, graph their sum in* $[0.2, 1.2, 0.05]$ *by* $[-0.01, 0.01, 0.001.]$ *Count the number of beats in 1 second.*

2. $P_1 = 0.007\sin(450\pi t)$, $P_2 = 0.005\sin(454\pi t)$

3. $P_1 = 0.004\cos(830\pi t)$, $P_2 = 0.005\sin(836\pi t)$

CHECKING BASIC CONCEPTS FOR SECTIONS 7.3 AND 7.4

1. Find the reference angle of each angle.
   (a) $225°$
   (b) $\frac{5\pi}{6}$

2. Solve each equation for θ in $[0°, 360°)$.
   (a) $\cos\theta = \frac{1}{2}$
   (b) $\sin\theta = -\frac{\sqrt{3}}{2}$

3. Find all solutions where t is a real number.
   (a) $\sin t = -\cos t$
   (b) $2\sin^2 t = 1 - \cos t$

4. Use a sum or difference identity to find $\cos\frac{\pi}{12}$.

5. Verify the identity $\sin(t - \pi) = -\sin t$ symbolically. If you have a graphing calculator, give graphical or numerical support.

7.5 Multiple-Angle Identities

- Learn and use the double-angle identities
- Learn and use power-reducing identities
- Learn and use the half-angle formulas
- Solve equations
- Learn and use product-to-sum and sum-to-product identities

Introduction

In 1831, Michael Faraday discovered that when a wire is passed near a magnet, a small electric current is produced in the wire. By rotating thousands of wires near large electromagnets, massive amounts of electricity can be produced. In 1 year, utilities in the United States generate enough electricity to power a 100-watt light bulb for over 3 billion years!

Voltage, amperage, and wattage are quantities that can be modeled by sinusoidal graphs and functions. To model electricity and other phenomena, trigonometric functions and identities are used. This section introduces several important multiple-angle identities. (*Sources:* R. Weidner and R. Sells, *Elementary Classical Physics*, Vol. 2; J. Wright, *The Universal Almanac.*)

Double-Angle Identities

The double-angle identities for sine, cosine, and tangent can be derived using the sum identities with $\alpha = \theta$ and $\beta = \theta$.

Sine Double-Angle Identity

$$\sin 2\theta = \sin(\theta + \theta) \qquad \color{blue}{2\theta = \theta + \theta}$$
$$= \sin\theta\cos\theta + \cos\theta\sin\theta \qquad \color{blue}{\sin(\alpha + \beta) = \sin\alpha\cos\beta + \cos\alpha\sin\beta}$$
$$= 2\sin\theta\cos\theta \qquad \color{blue}{\text{Simplify.}}$$

Cosine Double-Angle Identities

$$\cos 2\theta = \cos(\theta + \theta) \qquad\qquad 2\theta = \theta + \theta$$
$$= \cos\theta\cos\theta - \sin\theta\sin\theta \qquad \cos(\alpha + \beta) = \cos\alpha\cos\beta - \sin\alpha\sin\beta$$
$$= \cos^2\theta - \sin^2\theta \qquad\qquad \text{Simplify.}$$

Applying the Pythagorean identity $\sin^2\theta + \cos^2\theta = 1$, we can write the expression $\cos^2\theta - \sin^2\theta$ as

$$\cos^2\theta - \mathbf{\sin^2\theta} = \cos^2\theta - (\mathbf{1 - \cos^2\theta})$$
$$= 2\cos^2\theta - 1$$

or as

$$\mathbf{\cos^2\theta} - \sin^2\theta = (\mathbf{1 - \sin^2\theta}) - \sin^2\theta$$
$$= 1 - 2\sin^2\theta.$$

Tangent Double-Angle Identity

$$\tan 2\theta = \tan(\theta + \theta) \qquad\qquad 2\theta = \theta + \theta$$
$$= \frac{\tan\theta + \tan\theta}{1 - \tan\theta\tan\theta} \qquad \tan(\alpha + \beta) = \frac{\tan\alpha + \tan\beta}{1 - \tan\alpha\tan\beta}$$
$$= \frac{2\tan\theta}{1 - \tan^2\theta} \qquad\qquad \text{Simplify.}$$

A summary of these double-angle identities is given below.

DOUBLE-ANGLE IDENTITIES

$$\sin 2\theta = 2\sin\theta\cos\theta \qquad \cos 2\theta = \cos^2\theta - \sin^2\theta \qquad \tan 2\theta = \frac{2\tan\theta}{1 - \tan^2\theta}$$
$$\text{or}$$
$$\cos 2\theta = 2\cos^2\theta - 1$$
$$\text{or}$$
$$\cos 2\theta = 1 - 2\sin^2\theta$$

The next example illustrates that $\sin 2\theta \neq 2\sin\theta$.

EXAMPLE 1 **Using double-angle identities**

Verify symbolically and graphically that $\sin 2\theta$ and $2\sin\theta$ are *not* equivalent expressions.

SOLUTION

Symbolic Verification   The expressions $\sin 2\theta$ and $2\sin\theta$ are not equivalent; rather,

$$\sin 2\theta = 2\sin\theta\cos\theta \neq 2\sin\theta.$$

Graphical Verification   Graph $y = \sin 2\theta$ and $y = 2\sin\theta$, as shown in Figures 7.58 and 7.59. Notice that the graphs are different.

Graphs Showing that $\sin 2\theta \neq 2 \sin \theta$

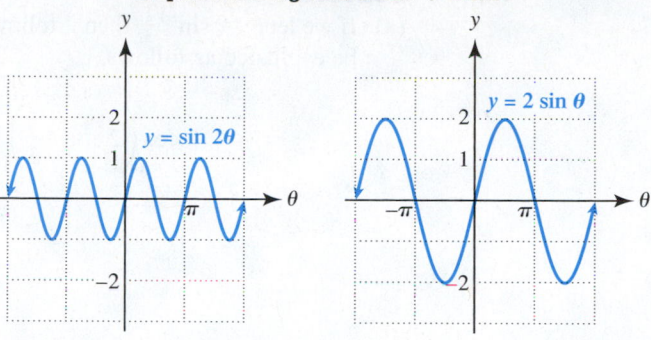

Figure 7.58   **Figure 7.59**

Now Try Exercise 15

EXAMPLE 2   **Using double-angle identities**

Given $\cos \theta = -\frac{12}{13}$ and $\sin \theta > 0$, find $\sin 2\theta$, $\cos 2\theta$, and $\tan 2\theta$. Use a calculator to support your result.

SOLUTION

Symbolic Solution   Since $\cos \theta = -\frac{12}{13} < 0$ and $\sin \theta > 0$, θ is contained in quadrant II. One possibility for θ is shown in Figure 7.60. We see that $\sin \theta = \frac{5}{13}$. Using double-angle identities, we obtain the following results.

$$\sin 2\theta = 2 \sin \theta \cos \theta = 2 \cdot \frac{5}{13} \cdot \left(-\frac{12}{13}\right) = -\frac{120}{169}$$

$$\cos 2\theta = \cos^2 \theta - \sin^2 \theta = \left(-\frac{12}{13}\right)^2 - \left(\frac{5}{13}\right)^2 = \frac{119}{169}$$

$$\tan 2\theta = \frac{\sin 2\theta}{\cos 2\theta} = \frac{-120/169}{119/169} = -\frac{120}{119}$$

Using $\tan \theta = -\frac{5}{12}$ and the double-angle identity for $\tan 2\theta$ gives the same result.

θ in Quadrant II

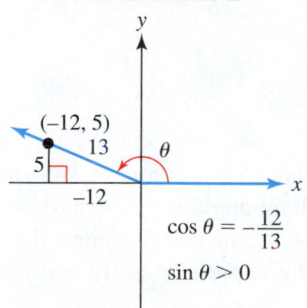

$\cos \theta = -\frac{12}{13}$

$\sin \theta > 0$

Figure 7.60

Calculator Support   Since θ is contained in quadrant II, we can support these results by letting $\theta = \cos^{-1}\left(-\frac{12}{13}\right) \approx 157.38°$ and performing the calculations shown in Figure 7.61 and in Figure 7.62.

Let $\theta = \cos^{-1}(-12/13)$

```
sin(2cos⁻¹(⁻12/13
))▶Frac
              ⁻120/169
cos(2cos⁻¹(⁻12/13
))▶Frac
               119/169
```

```
tan(2cos⁻¹(⁻12/13
))▶Frac
              ⁻120/119
```

Figure 7.61   **Figure 7.62**

Now Try Exercise 19

EXAMPLE 3   **Evaluating expressions with double-angle identities**

Use a double-angle identity to evaluate each expression.

(a) $\cos\left(2 \sin^{-1}\frac{1}{3}\right)$ **(b)** $\sin\left(2 \cos^{-1}\left(-\frac{3}{5}\right)\right)$ **(c)** $\sin(2 \tan^{-1}x), x > 0$

SOLUTION

(a) If we let $\theta = \sin^{-1}\frac{1}{3}$, then it follows that $\sin\theta = \frac{1}{3}$. The expression $\cos\left(2\sin^{-1}\frac{1}{3}\right)$ can be evaluated as follows.

$$\cos\left(2\,\mathbf{sin^{-1}\frac{1}{3}}\right) = \cos(2\theta) \qquad \textcolor{blue}{\theta = \sin^{-1}\frac{1}{3}}$$

$$= 1 - 2\sin^2\theta \qquad \textcolor{blue}{\text{Double-angle identity}}$$

$$= 1 - 2\left(\frac{1}{3}\right)^2 \qquad \textcolor{blue}{\sin\theta = \frac{1}{3}}$$

$$= \frac{7}{9} \qquad \textcolor{blue}{\text{Simplify.}}$$

θ in Quadrant II

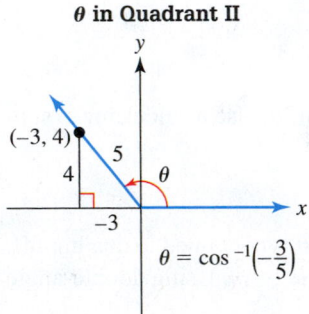

Figure 7.63

(b) Begin by letting $\theta = \cos^{-1}\left(-\frac{3}{5}\right)$ and sketching angle θ in standard position, as shown in Figure 7.63. Notice that $\cos\theta = -\frac{3}{5}$ and that θ is a second quadrant angle. From Figure 7.63 it follows that $\sin\theta = \frac{4}{5}$. The expression $\sin\left(2\cos^{-1}\left(-\frac{3}{5}\right)\right)$ can be evaluated as follows.

$$\sin\left(2\,\mathbf{cos^{-1}\left(-\frac{3}{5}\right)}\right) = \sin(2\theta) \qquad \textcolor{blue}{\theta = \cos^{-1}\left(-\frac{3}{5}\right)}$$

$$= 2\sin\theta\cos\theta \qquad \textcolor{blue}{\text{Double-angle identity}}$$

$$= 2\left(\frac{4}{5}\right)\left(-\frac{3}{5}\right) \qquad \textcolor{blue}{\sin\theta = \frac{4}{5},\ \cos\theta = -\frac{3}{5}}$$

$$= -\frac{24}{25} \qquad \textcolor{blue}{\text{Simplify.}}$$

θ in Quadrant I

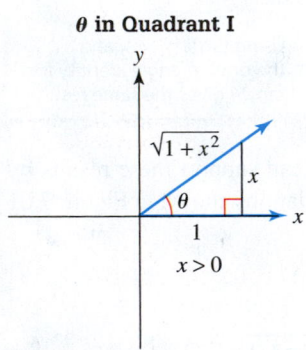

Figure 7.64

(c) Begin by letting $\theta = \tan^{-1}x$ and sketching angle θ in standard position, as shown in Figure 7.64. Since $x > 0$, we have drawn a first quadrant angle θ whose tangent function equals $\frac{x}{1}$. That is, $\tan\theta = x$. From Figure 7.64 it can be concluded that $\sin\theta = \frac{x}{\sqrt{1+x^2}}$ and $\cos\theta = \frac{1}{\sqrt{1+x^2}}$. The trigonometric expression $\sin(2\tan^{-1}x)$ can be evaluated as follows.

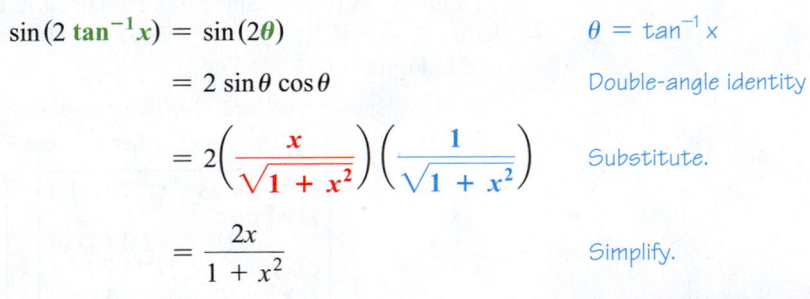

$$\sin(2\,\mathbf{tan^{-1}x}) = \sin(2\theta) \qquad \textcolor{blue}{\theta = \tan^{-1}x}$$

$$= 2\sin\theta\cos\theta \qquad \textcolor{blue}{\text{Double-angle identity}}$$

$$= 2\left(\frac{x}{\sqrt{1+x^2}}\right)\left(\frac{1}{\sqrt{1+x^2}}\right) \qquad \textcolor{blue}{\text{Substitute.}}$$

$$= \frac{2x}{1+x^2} \qquad \textcolor{blue}{\text{Simplify.}}$$

Now Try Exercises 25, 27, and 31

In the next example, we verify an identity by using a double-angle identity.

EXAMPLE 4 Verifying an identity

Verify the identity $\dfrac{\sec^2\theta}{1 - \tan^2\theta} = \sec 2\theta$.

SOLUTION Begin by applying a reciprocal identity.

$$\frac{\sec^2\theta}{1-\tan^2\theta} = \frac{1}{\cos^2\theta\,(1-\tan^2\theta)} \qquad \text{Reciprocal identity}$$

$$= \frac{1}{\cos^2\theta\left(1-\dfrac{\sin^2\theta}{\cos^2\theta}\right)} \qquad \text{Quotient identity}$$

$$= \frac{1}{\cos^2\theta-\sin^2\theta} \qquad \text{Distributive property}$$

$$= \frac{1}{\cos 2\theta} \qquad \text{Double-angle identity}$$

$$= \sec 2\theta \qquad \text{Reciprocal identity}$$

Now Try Exercise 67

EXAMPLE 5 **Deriving a triple-angle identity**

Write $\cos 3\theta$ in terms of $\cos\theta$.

SOLUTION

$$\cos 3\theta = \cos(2\theta + \theta) \qquad 3\theta = 2\theta + \theta$$

$$= \cos 2\theta \cos\theta - \sin 2\theta \sin\theta \qquad \text{Sum identity for cosine}$$

$$= (2\cos^2\theta - 1)\cos\theta - (2\sin\theta\cos\theta)\sin\theta \qquad \text{Double-angle identities}$$

$$= 2\cos^3\theta - \cos\theta - 2\sin^2\theta\cos\theta \qquad \text{Multiply.}$$

$$= 2\cos^3\theta - \cos\theta - 2(1-\cos^2\theta)\cos\theta \qquad \text{Apply } \sin^2\theta + \cos^2\theta = 1.$$

$$= 2\cos^3\theta - \cos\theta - 2\cos\theta + 2\cos^3\theta \qquad \text{Distributive property}$$

$$= 4\cos^3\theta - 3\cos\theta \qquad \text{Combine like terms.}$$

Now Try Exercise 71

Power-Reducing Identities

Power-reducing identities for sine, cosine, and tangent can be derived using the double-angle identities.

Sine Power-Reducing Identity We can solve $\cos 2\theta = 1 - 2\sin^2\theta$ for $\sin^2\theta$ to obtain

$$\sin^2\theta = \frac{1-\cos 2\theta}{2}.$$

Cosine Power-Reducing Identity Solving $\cos 2\theta = 2\cos^2\theta - 1$ for $\cos^2\theta$ gives

$$\cos^2\theta = \frac{1+\cos 2\theta}{2}.$$

Tangent Power-Reducing Identity We can use the power-reducing identities

$$\sin^2\theta = \frac{1-\cos 2\theta}{2} \qquad \text{and} \qquad \cos^2\theta = \frac{1+\cos 2\theta}{2}$$

to derive

$$\tan^2\theta = \frac{\sin^2\theta}{\cos^2\theta} = \frac{(1-\cos 2\theta)/2}{(1+\cos 2\theta)/2} = \frac{1-\cos 2\theta}{1+\cos 2\theta}.$$

A summary of these identities is given in the box below.

POWER-REDUCING IDENTITIES

$$\sin^2\theta = \frac{1 - \cos 2\theta}{2} \qquad \cos^2\theta = \frac{1 + \cos 2\theta}{2} \qquad \tan^2\theta = \frac{1 - \cos 2\theta}{1 + \cos 2\theta}$$

In the next example, we write known identities in terms of inputs other than θ.

EXAMPLE 6 Writing identities

Complete each statement.

(a) $\sin^2\theta = \frac{1 - \cos 2\theta}{2}$, so $\sin^2 4t = $ _____.

(b) $\tan^2\theta = \frac{1 - \cos 2\theta}{1 + \cos 2\theta}$, so $\tan^2\left(\frac{1}{2}x\right) = $ _____.

(c) $\sin 2\theta = 2\sin\theta\cos\theta$, so $\sin 8t = $ _____.

SOLUTION

(a) Substitute $4t$ for θ on each side of the given equation.

$$\sin^2 4t = \frac{1 - \cos(2 \cdot 4t)}{2} \qquad \boxed{\text{Let } \theta = 4t} \qquad sin^2\theta = \frac{1 - \cos 2\theta}{2}$$

$$= \frac{1 - \cos 8t}{2}$$

(b) Substitute $\frac{1}{2}x$ for θ on each side of the given equation.

$$\tan^2\left(\frac{1}{2}x\right) = \frac{1 - \cos\left(2 \cdot \frac{1}{2}x\right)}{1 + \cos\left(2 \cdot \frac{1}{2}x\right)} \qquad \boxed{\text{Let } \theta = \frac{1}{2}x} \qquad tan^2\theta = \frac{1 - \cos 2\theta}{1 + \cos 2\theta}$$

$$= \frac{1 - \cos x}{1 + \cos x}$$

(c) Let $2\theta = 8t$, so $\theta = 4t$. Thus $\sin 2\theta = 2\sin\theta\cos\theta$ implies

$$\boxed{\text{Let } \theta = 4t} \qquad \sin 8t = 2\sin 4t\cos 4t.$$

Now Try Exercises 1, 3, and 5

EXAMPLE 7 Using a power-reducing identity

Find the exact value of $\sin^2 22.5°$. Use a calculator to support your results.

SOLUTION Let $\theta = 22.5°$ and $2\theta = 45°$ and apply the sine power-reducing identity.

$$\sin^2 22.5° = \frac{1 - \cos 45°}{2} \qquad sin^2\theta = \frac{1 - \cos 2\theta}{2}$$

$$= \frac{1 - \sqrt{2}/2}{2} \qquad \cos 45° = \frac{\sqrt{2}}{2}$$

$$= \frac{2 - \sqrt{2}}{4} \qquad \text{Multiply by } \frac{2}{2}.$$

Degree Mode

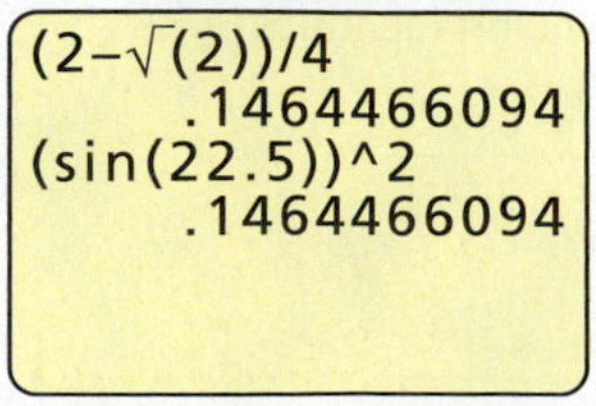

```
(2−√(2))/4
         .1464466094
(sin(22.5))^2
         .1464466094
```

Figure 7.65

Support for this result is shown in Figure 7.65.

Now Try Exercise 45

An Application Next we apply a power-reducing identity to electrical circuits.

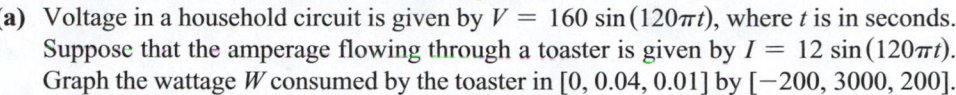

EXAMPLE 8   **Using a power-reducing identity to analyze wattage**

Amperage I is a measure of the amount of electricity passing through a wire, and voltage V is a measure of the force "pushing" the electricity. The wattage W consumed by an electrical device can be calculated using the equation $W = VI$.   (**Source:** G. Wilcox and C. Hesselberth, *Electricity for Engineering Technology*.)

(a)   Voltage in a household circuit is given by $V = 160 \sin(120\pi t)$, where t is in seconds. Suppose that the amperage flowing through a toaster is given by $I = 12 \sin(120\pi t)$. Graph the wattage W consumed by the toaster in [0, 0.04, 0.01] by [−200, 3000, 200].

(b)   Write the wattage as $W = a \cos(k\pi t) + d$, where a, k, and d are constants.

(c)   Compare the periods of the voltage, amperage, and wattage.

(d)   The wattage of this toaster equals half the maximum of W. Find the wattage.

Wattage W

[0, 0.04, 0.01] by [−200, 3000, 200]

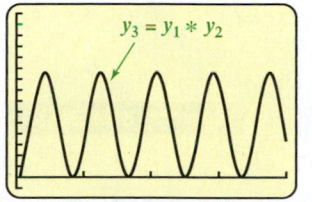

Figure 7.66 Radian Mode

SOLUTION

(a)   Since $W = VI$, graph the equation $Y_3 = Y_1 * Y_2$, where $Y_1 = 160 \sin(120\pi X)$ and $Y_2 = 12 \sin(120\pi X)$, as shown in Figure 7.66, where *radian* mode is used.

(b)
$$W = VI$$

$$= 160 \sin(120\pi t) \cdot 12 \sin(120\pi t) \qquad \text{Substitute for } V \text{ and } I.$$

$$= 1920 \sin^2(120\pi t) \qquad \text{Multiply.}$$

$$= 1920 \cdot \frac{1 - \cos(240\pi t)}{2} \qquad \text{Power-reducing identity}$$

$$= 960 - 960 \cos(240\pi t) \qquad \text{Simplify.}$$

Thus let $a = -960$, $k = 240$, and $d = 960$. Then the wattage can be written as

$$W = -960 \cos(240\pi t) + 960.$$

Voltage V

[0, 0.04, 0.01] by [−200, 200, 50]

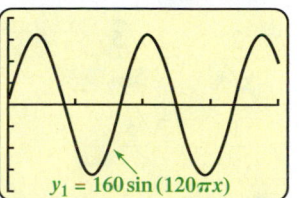

$y_1 = 160 \sin(120\pi x)$

Figure 7.67 Radian Mode

(c)   The period for both V and I is $\frac{2\pi}{120\pi} = \frac{1}{60}$ second, and the period for W is $\frac{2\pi}{240\pi} = \frac{1}{120}$ second. This result is supported in Figure 7.67, where the graph of V requires twice as much time as W to complete one oscillation.

(d)   The maximum wattage is 1920 watts whenever $\cos(240\pi t) = -1$. Half this amount is 960 watts, which is the wattage rating for the toaster.

Now Try Exercise 119

Half-Angle Formulas

We obtain half-angle formulas by using the power-reducing identities. For example,

$$\sin^2 x = \frac{1 - \cos 2x}{2} \qquad \text{Power-reducing identity}$$

$$\sin x = \pm\sqrt{\frac{1 - \cos 2x}{2}} \qquad \text{Square root property}$$

$$\sin\frac{\theta}{2} = \pm\sqrt{\frac{1 - \cos\theta}{2}}. \qquad \text{Let } x = \frac{\theta}{2} \text{ and } 2x = \theta.$$

The following box gives some half-angle formulas. (The second and third half-angle formulas for tangent are verified in Exercises 83 and 84.)

HALF-ANGLE FORMULAS

$$\sin\frac{\theta}{2} = \pm\sqrt{\frac{1 - \cos\theta}{2}} \qquad \cos\frac{\theta}{2} = \pm\sqrt{\frac{1 + \cos\theta}{2}}$$

$$\tan\frac{\theta}{2} = \pm\sqrt{\frac{1 - \cos\theta}{1 + \cos\theta}} \qquad \tan\frac{\theta}{2} = \frac{1 - \cos\theta}{\sin\theta} \qquad \tan\frac{\theta}{2} = \frac{\sin\theta}{1 + \cos\theta}$$

To decide whether a positive or negative sign should be used in a half-angle formula, we must determine the quadrant containing $\frac{\theta}{2}$. This is illustrated in the next example.

EXAMPLE 9 **Using a half-angle formula to find an exact value**

Find the exact value of $\sin(-15°)$.

SOLUTION We use the fact that $\cos(-30°) = \frac{\sqrt{3}}{2}$ to find $\sin(-15°)$. Note if $\theta = -30°$, then $\frac{\theta}{2}$ is in quadrant IV.

$$\sin(-15°) = \sin\left(\frac{-30°}{2}\right) \qquad \text{Let } \frac{\theta}{2} = -15° \text{ and } \theta = -30°.$$

$$= -\sqrt{\frac{1 - \cos(-30°)}{2}} \qquad \sin\frac{\theta}{2} = \pm\sqrt{\frac{1 - \cos\theta}{2}}$$

> Sine is negative in quadrant IV.

$$= -\sqrt{\frac{1 - \sqrt{3}/2}{2}} \qquad \cos(-30°) = \frac{\sqrt{3}}{2}$$

$$= -\sqrt{\frac{2 - \sqrt{3}}{4}} \qquad \begin{array}{l}\text{Multiply numerator}\\ \text{and denominator by 2.}\end{array}$$

$$= -\frac{\sqrt{2 - \sqrt{3}}}{2} \qquad \sqrt{4} = 2$$

> **Now Try Exercise 49**

EXAMPLE 10 **Using half-angle formulas to find exact values**

If $\cos\theta = -\frac{3}{5}$ and $90° \le \theta \le 180°$, find $\sin\frac{\theta}{2}$, $\cos\frac{\theta}{2}$, and $\tan\frac{\theta}{2}$.

SOLUTION Since $90° \le \theta \le 180°$, it follows that $45° \le \frac{\theta}{2} \le 90°$. That is, $\frac{\theta}{2}$ is in quadrant I.

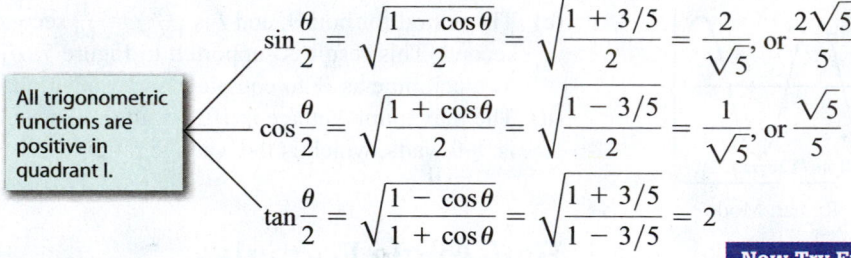

> All trigonometric functions are positive in quadrant I.

$$\sin\frac{\theta}{2} = \sqrt{\frac{1 - \cos\theta}{2}} = \sqrt{\frac{1 + 3/5}{2}} = \frac{2}{\sqrt{5}}, \text{ or } \frac{2\sqrt{5}}{5}$$

$$\cos\frac{\theta}{2} = \sqrt{\frac{1 + \cos\theta}{2}} = \sqrt{\frac{1 - 3/5}{2}} = \frac{1}{\sqrt{5}}, \text{ or } \frac{\sqrt{5}}{5}$$

$$\tan\frac{\theta}{2} = \sqrt{\frac{1 - \cos\theta}{1 + \cos\theta}} = \sqrt{\frac{1 + 3/5}{1 - 3/5}} = 2$$

> **Now Try Exercise 59**

Solving Equations

When trigonometric functions are used in modeling to make predictions, there is often a need to solve trigonometric equations.

EXAMPLE 11 **Solving a trigonometric equation**

Solve the trigonometric equation $\cos\theta - \sin 2\theta = 0$ symbolically, graphically, and numerically for $0° \le \theta < 360°$.

SOLUTION
Symbolic Solution We begin by applying the identity $\sin 2\theta = 2\sin\theta\cos\theta$.

$$\cos\theta - \sin 2\theta = 0 \qquad \text{Given equation}$$

$$\cos\theta - 2\sin\theta\cos\theta = 0 \qquad \text{Double-angle identity}$$

$$\cos\theta(1 - 2\sin\theta) = 0 \qquad \text{Factor out } \cos\theta.$$

$$\cos\theta = 0 \quad \text{or} \quad 1 - 2\sin\theta = 0 \qquad \text{Zero-product property}$$

$$\cos\theta = 0 \quad \text{or} \quad \sin\theta = \frac{1}{2} \qquad \text{Solve for } \sin\theta.$$

$$\theta = 90°, 270° \quad \text{or} \quad \theta = 30°, 150° \qquad \text{Solve for } \theta.$$

On the interval $[0°, 360°)$, the solutions are $30°$, $90°$, $150°$, and $270°$.

Graphical Solution Graphical support is given in Figure 7.68, where the graph of $Y_1 = \cos(X) - \sin(2X)$ is shown. Note that the four *x*-intercepts 30°, 90°, 150°, and 270° correspond to the four symbolic solutions.

Numerical Solution Numerical support is shown in Figure 7.69.

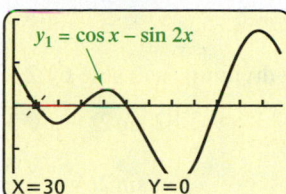

Degree Mode
[0, 352.5, 30] by [−2, 2, 1]

Figure 7.68

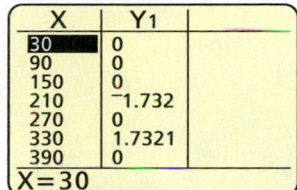

Degree Mode

Figure 7.69

> **Now Try Exercise 95**

EXAMPLE 12   **Solving a trigonometric equation**

Solve the equation $4 \cos\frac{\theta}{2} - 2 = 0$ for $0 \le \theta < 2\pi$.

SOLUTION Begin by solving $4 \cos\frac{\theta}{2} - 2 = 0$ for $\cos\frac{\theta}{2}$.

$$4 \cos\frac{\theta}{2} - 2 = 0 \qquad \textit{Given equation}$$

$$\cos\frac{\theta}{2} = \frac{1}{2} \qquad \textit{Solve for } \cos\frac{\theta}{2}.$$

$$\frac{\theta}{2} = \frac{\pi}{3} \quad \text{or} \quad \frac{\theta}{2} = \frac{5\pi}{3} \qquad \textit{Solve for } \frac{\theta}{2}.$$

$$\theta = \frac{2\pi}{3} \quad \text{or} \quad \theta = \frac{10\pi}{3} \qquad \textit{Multiply by 2.}$$

The only solution on $[0, 2\pi)$ is $\frac{2\pi}{3}$.

> **Now Try Exercise 99**

EXAMPLE 13   **Solving trigonometric equations**

Find all solutions, expressed in radians.
(a) $\cos 2t + 2\cos^2 t = 0$ **(b)** $2 \sin 2t = \sqrt{3}$

SOLUTION
(a) Begin by applying the double-angle identity $\cos 2t = 2\cos^2 t - 1$.

$$\cos 2t + 2\cos^2 t = 0 \qquad \textit{Given equation}$$

$$2\cos^2 t - 1 + 2\cos^2 t = 0 \qquad \textit{Double-angle identity}$$

$$4\cos^2 t - 1 = 0 \qquad \textit{Combine terms.}$$

$$\cos^2 t = \frac{1}{4} \qquad \textit{Solve for } \cos^2 t.$$

$$\cos t = \pm\frac{1}{2} \qquad \textit{Square root property}$$

On the interval $[0, 2\pi)$, there are four angles whose cosines equal either $\frac{1}{2}$ or $-\frac{1}{2}$. They are $\frac{\pi}{3}, \frac{2\pi}{3}, \frac{4\pi}{3}$, and $\frac{5\pi}{3}$. Since $\cos t$ has period 2π, all solutions can be written as

All Solutions Written as Four Expressions

$$\frac{\pi}{3} + 2\pi n, \qquad \frac{2\pi}{3} + 2\pi n, \qquad \frac{4\pi}{3} + 2\pi n, \qquad \text{or} \qquad \frac{5\pi}{3} + 2\pi n,$$

where n is an integer. Note that since $\frac{4\pi}{3} - \frac{\pi}{3} = \pi$, the two solutions $\frac{\pi}{3} + 2\pi n$ and $\frac{4\pi}{3} + 2\pi n$ can be combined and written as $\frac{\pi}{3} + \pi n$. Similarly, the two solutions $\frac{2\pi}{3} + 2\pi n$ and $\frac{5\pi}{3} + 2\pi n$ can be written as $\frac{2\pi}{3} + \pi n$. As a result, all solutions can also be written as

All Solutions Written as Two Expressions

$$\frac{\pi}{3} + \pi n \quad \text{or} \quad \frac{2\pi}{3} + \pi n.$$

(b) Begin by dividing each side by 2.

$$2\sin 2t = \sqrt{3} \qquad \textit{Given equation}$$

$$\sin 2t = \frac{\sqrt{3}}{2} \qquad \textit{Divide each side by 2.}$$

Let $\theta = 2t$ and find all values of θ on the interval $[0, 2\pi)$, where $\sin\theta = \frac{\sqrt{3}}{2}$. Because the reference angle for θ is $\theta_R = \sin^{-1}\frac{\sqrt{3}}{2} = \frac{\pi}{3}$ and the sine function is positive in quadrants I and II, it follows that $\theta = \frac{\pi}{3}$ or $\theta = \frac{2\pi}{3}$. Thus all possible solutions in terms of θ can be written as

$$\theta = \frac{\pi}{3} + 2\pi n \quad \text{or} \quad \theta = \frac{2\pi}{3} + 2\pi n.$$

Since $\theta = 2t$, we can write the solutions to the given equation in terms of t.

$$2t = \frac{\pi}{3} + 2\pi n \quad \text{or} \quad 2t = \frac{2\pi}{3} + 2\pi n \qquad \textit{Let } \theta = 2t.$$

To solve for t, divide each term by 2.

$$t = \frac{\pi}{6} + \pi n \quad \text{or} \quad t = \frac{\pi}{3} + \pi n \qquad \textit{Divide by 2.}$$

Now Try Exercises 101 and 107

Product-to-Sum and Sum-to-Product Identities

The sum and difference identities for sine and cosine can be used to derive several identities that make it possible to rewrite a product as a sum. For example, adding the identities for $\cos(\alpha + \beta)$ and $\cos(\alpha - \beta)$ results in the following.

$$\cos(\alpha + \beta) = \cos\alpha\cos\beta - \sin\alpha\sin\beta$$
$$\cos(\alpha - \beta) = \cos\alpha\cos\beta + \sin\alpha\sin\beta$$
$$\overline{\cos(\alpha + \beta) + \cos(\alpha - \beta) = 2\cos\alpha\cos\beta}$$

Rewriting gives $\cos\alpha\cos\beta = \frac{1}{2}(\cos(\alpha + \beta) + \cos(\alpha - \beta))$. Four product-to-sum identities are given below. The other three identities are derived in a similar manner.

PRODUCT-TO-SUM IDENTITIES

$$\cos\alpha\cos\beta = \frac{1}{2}(\cos(\alpha + \beta) + \cos(\alpha - \beta))$$

$$\sin\alpha\sin\beta = \frac{1}{2}(\cos(\alpha - \beta) - \cos(\alpha + \beta))$$

$$\sin\alpha\cos\beta = \frac{1}{2}(\sin(\alpha + \beta) + \sin(\alpha - \beta))$$

$$\cos\alpha\sin\beta = \frac{1}{2}(\sin(\alpha + \beta) - \sin(\alpha - \beta))$$

EXAMPLE 14 Using a product-to-sum identity

Write the product $\cos 5\theta \cos 3\theta$ as a sum.

SOLUTION We begin by applying the first product-to-sum identity with the substitution $\alpha = 5\theta$ and $\beta = 3\theta$.

$$\cos 5\theta \cos 3\theta = \frac{1}{2}(\cos(5\theta + 3\theta) + \cos(5\theta - 3\theta)) \qquad \alpha = 5\theta, \beta = 3\theta$$

$$= \frac{1}{2}(\cos 8\theta + \cos 2\theta) \qquad \text{Add; subtract.}$$

Now Try Exercise 85

By rewriting the four product-to-sum identities, we can derive four sum-to-product identities. If we let $a = \alpha + \beta$ and $b = \alpha - \beta$, it follows that

$$\frac{a + b}{2} = \frac{\alpha + \beta + \alpha - \beta}{2} = \alpha \quad \text{and} \quad \frac{a - b}{2} = \frac{(\alpha + \beta) - (\alpha - \beta)}{2} = \beta.$$

Now, multiplying both sides of the identity

$$\cos \alpha \cos \beta = \frac{1}{2}(\cos(\alpha + \beta) + \cos(\alpha - \beta))$$

by 2 and substituting yields

$$2 \cos \frac{a + b}{2} \cos \frac{a - b}{2} = \cos a + \cos b.$$

The other three sum-to-product identities can be derived in a similar manner.

SUM-TO-PRODUCT IDENTITIES

$$\cos a + \cos b = 2 \cos \frac{a + b}{2} \cos \frac{a - b}{2}$$

$$\cos a - \cos b = -2 \sin \frac{a + b}{2} \sin \frac{a - b}{2}$$

$$\sin a + \sin b = 2 \sin \frac{a + b}{2} \cos \frac{a - b}{2}$$

$$\sin a - \sin b = 2 \cos \frac{a + b}{2} \sin \frac{a - b}{2}$$

For example, we can write the sum $\cos 70° + \cos 40°$ as follows.

$$\cos 70° + \cos 40° = 2 \cos \frac{70° + 40°}{2} \cos \frac{70° - 40°}{2} = 2 \cos 55° \cos 15°$$

An Application from Telephone Technology Each number on a touch-tone phone produces a unique pair of frequencies. For example, when 1 is pressed, frequencies of 697 hertz and 1209 hertz are simultaneously transmitted. A *hertz* (Hz) is equal to one cycle per second. When 2 is pressed, the pair of frequencies transmitted is 697 hertz and 1336 hertz. As a result, 2 has a different tone from 1. Table 7.2 shows the frequency pairs for the numbers 0 through 9.

Frequencies Used in Touch-Tone Phones

Number	0	1	2	3	4	5	6	7	8	9
Frequency 1 (Hz)	941	697	697	697	770	770	770	852	852	852
Frequency 2 (Hz)	1336	1209	1336	1477	1209	1336	1477	1209	1336	1477

Table 7.2

A tone with frequencies F_1 and F_2 can be modeled by

$$a_1 \cos(2\pi F_1 t) + a_2 \cos(2\pi F_2 t).$$

If both tones have the same intensity, then we can let $a_1 = a_2 = 1$.

EXAMPLE 15 **Analyzing touch-tone phones**

For a touch-tone phone, assume that $a_1 = a_2 = 1$.
(a) Write an expression that models the sound of a 5 on a touch-tone phone.
(b) Rewrite the expression in part (a) as a product of trigonometric expressions.

SOLUTION
(a) From Table 7.2 we can see that for number 5, $F_1 = 770$ and $F_2 = 1336$. Thus

$$y = \cos(1540\pi t) + \cos(2672\pi t).$$

(b) Let $a = 1540\pi t$ and $b = 2672\pi t$ in the appropriate sum-to-product identity.

$$\cos 1540\pi t + \cos 2672\pi t$$

$$= 2\cos\frac{1540\pi t + 2672\pi t}{2}\cos\frac{1540\pi t - 2672\pi t}{2}$$

$$= 2\cos(2106\pi t)\cos(-566\pi t)$$

$$= 2\cos(2106\pi t)\cos(566\pi t) \qquad \boxed{\cos(-\theta) = \cos\theta}$$

Now Try Exercise 125

7.5 Putting It All Together

The multiple-angle identities or formulas presented in this section are summarized in the following table.

IDENTITY OR FORMULA	GENERAL FORM
Double-angle	$\sin 2\theta = 2\sin\theta\cos\theta$ $\cos 2\theta = \cos^2\theta - \sin^2\theta = 2\cos^2\theta - 1 = 1 - 2\sin^2\theta$ $\tan 2\theta = \dfrac{2\tan\theta}{1 - \tan^2\theta}$
Power-reducing	$\sin^2\theta = \dfrac{1 - \cos 2\theta}{2} \qquad \cos^2\theta = \dfrac{1 + \cos 2\theta}{2} \qquad \tan^2\theta = \dfrac{1 - \cos 2\theta}{1 + \cos 2\theta}$

IDENTITY OR FORMULA	GENERAL FORM
Half-angle	$\sin\dfrac{\theta}{2} = \pm\sqrt{\dfrac{1-\cos\theta}{2}}$ $\cos\dfrac{\theta}{2} = \pm\sqrt{\dfrac{1+\cos\theta}{2}}$ $\tan\dfrac{\theta}{2} = \pm\sqrt{\dfrac{1-\cos\theta}{1+\cos\theta}}$ $\tan\dfrac{\theta}{2} = \dfrac{1-\cos\theta}{\sin\theta}$ $\tan\dfrac{\theta}{2} = \dfrac{\sin\theta}{1+\cos\theta}$
Product-to-sum	$\cos\alpha\cos\beta = \dfrac{1}{2}(\cos(\alpha+\beta) + \cos(\alpha-\beta))$ $\sin\alpha\sin\beta = \dfrac{1}{2}(\cos(\alpha-\beta) - \cos(\alpha+\beta))$ $\sin\alpha\cos\beta = \dfrac{1}{2}(\sin(\alpha+\beta) + \sin(\alpha-\beta))$ $\cos\alpha\sin\beta = \dfrac{1}{2}(\sin(\alpha+\beta) - \sin(\alpha-\beta))$
Sum-to-product	$\cos a + \cos b = 2\cos\dfrac{a+b}{2}\cos\dfrac{a-b}{2}$ $\cos a - \cos b = -2\sin\dfrac{a+b}{2}\sin\dfrac{a-b}{2}$ $\sin a + \sin b = 2\sin\dfrac{a+b}{2}\cos\dfrac{a-b}{2}$ $\sin a - \sin b = 2\cos\dfrac{a+b}{2}\sin\dfrac{a-b}{2}$

7.5   Exercises

Writing Identities

Exercises 1–8: Complete each statement.

1. $\sin^2\theta = \frac{1-\cos 2\theta}{2}$, so $\sin^2 10t =$ _____.

2. $\cos^2\theta = \frac{1+\cos 2\theta}{2}$, so $\cos^2 8x =$ _____.

3. $\tan^2\theta = \frac{1-\cos 2\theta}{1+\cos 2\theta}$, so $\tan^2 5t =$ _____.

4. $\tan 2\theta = \frac{2\tan\theta}{1-\tan^2\theta}$, so $\tan t =$ _____.

5. $\sin 2\theta = 2\sin\theta\cos\theta$, so $\sin 20x =$ _____.

6. $\cos 2\theta = \cos^2\theta - \sin^2\theta$, so $\cos 16t =$ _____.

7. $\tan\frac{\theta}{2} = \frac{1-\cos\theta}{\sin\theta}$, so $\tan 5x =$ _____.

8. $\tan\frac{\theta}{2} = \frac{\sin\theta}{1+\cos\theta}$, so $\tan 4t =$ _____.

Double-Angle Identities

Exercises 9–14: If possible, evaluate expressions (a) and (b) and compare their values.

9. (a) $\sin 30° + \sin 30°$ (b) $\sin 60°$

10. (a) $\sin 45° + \sin 45°$ (b) $\sin 90°$

11. (a) $\cos 60° + \cos 60°$ (b) $\cos 120°$

12. (a) $\cos 90° + \cos 90°$ (b) $\cos 180°$

13. (a) $\tan 45° + \tan 45°$ (b) $\tan 90°$

14. (a) $\tan 30° + \tan 30°$ (b) $\tan 60°$

Exercises 15 and 16: (Refer to Example 1.) Verify graphically and symbolically that the two expressions are not equivalent.

15. $\tan 2\theta,\ 2\tan\theta$ **16.** $\cos 3\theta,\ 3\cos\theta$

Exercises 17–24: (Refer to Example 2.) Do the following.

(a) Find $\sin 2\theta$, $\cos 2\theta$, and $\tan 2\theta$.

(b) Use a calculator to support your results.

17. $\cos\theta = \frac{4}{5}$ and $\sin\theta = \frac{3}{5}$

18. $\sin\theta = \frac{12}{13}$ and $\cos\theta = \frac{5}{13}$

19. $\sin\theta = -\frac{24}{25}$ and $\cos\theta > 0$

20. $\cos\theta = -\frac{7}{25}$ and $\tan\theta > 0$

21. $\sin\theta = -\frac{11}{61}$ and $\sec\theta > 0$

22. $\csc\theta = -2$ and $\sec\theta > 0$

23. $\tan\theta = \frac{7}{24}$ and $\cos\theta < 0$

24. $\cot\theta = \frac{5}{12}$ and $\sin\theta > 0$

Exercises 25–34: (Refer to Example 3.) Use an identity to evaluate the expression.

25. $\sin(2\cos^{-1}1)$ **26.** $\cos\left(2\sin^{-1}\frac{1}{2}\right)$

27. $\cos\left(2\sin^{-1}\frac{7}{25}\right)$ **28.** $\sin\left(2\tan^{-1}\frac{3}{4}\right)$

29. $\cos\left(3\sin^{-1}\frac{5}{13}\right)$ **30.** $\tan\left(2\tan^{-1}\frac{1}{2}\right)$

31. $\sin(2\tan^{-1}x),\ x < 0$

32. $\cos(2\sin^{-1}x),\ x > 0$

33. $\cos\left(\sin^{-1}\frac{3}{5} - \sin^{-1}\frac{4}{5}\right)$ (*Hint:* Let $\alpha = \sin^{-1}\frac{3}{5}$, $\beta = \sin^{-1}\frac{4}{5}$, and apply a difference identity.)

34. $\sin\left(\tan^{-1}\frac{12}{5} + \cos^{-1}\frac{12}{13}\right)$

Exercises 35–40: Rewrite using a double-angle identity.

35. $2\cos\theta\sin\theta$ **36.** $2\sin 2\theta\cos 2\theta$

37. $\sin\theta\cos\theta$

38. $(\sin\theta - \cos\theta)(\sin\theta + \cos\theta)$

39. $2\cos^2 2\theta - 1$ **40.** $1 - 2\sin^2 3\theta$

Exercises 41–44: Write the expression as one term.

41. $\sin^2 3\theta + \cos^2 3\theta$ **42.** $1 + \tan^2 2\theta$

43. $\csc^2 5x - 1$ **44.** $\sin^2 8x + \cos^2 8x$

Power-Reducing Identities and Half-Angle Formulas

Exercises 45–48: Find the exact value of the expression. Use a calculator to support your result.

45. $\cos^2 22.5°$ **46.** $\sin^2 15°$

47. $\tan^2 75°$ **48.** $\csc^2 105°$

Exercises 49–52: Use a half-angle formula to find the exact value of the expression. Use a calculator to support your result.

49. (a) $\cos 15°$ (b) $\tan(-15°)$

50. (a) $\sin 67.5°$ (b) $\cos(-67.5°)$

51. (a) $\tan\frac{\pi}{8}$ (b) $\sin\left(-\frac{\pi}{8}\right)$

52. (a) $\cos\frac{5\pi}{12}$ (b) $\cos\left(-\frac{5\pi}{12}\right)$

Exercises 53–58: Use a half-angle formula to simplify the expression. Use a calculator to support your result.

53. $\sqrt{\dfrac{1 - \cos 60°}{2}}$ **54.** $\sqrt{\dfrac{1 + \cos 60°}{2}}$

55. $\sqrt{\dfrac{1 + \cos 50°}{2}}$ **56.** $\sqrt{\dfrac{1 - \cos 50°}{2}}$

57. $\sqrt{\dfrac{1 - \cos 40°}{1 + \cos 40°}}$ **58.** $\sqrt{\dfrac{1 + \cos 26°}{1 - \cos 26°}}$

Exercises 59–64: (Refer to Example 10.) Find $\sin\frac{\theta}{2}$, $\cos\frac{\theta}{2}$, and $\tan\frac{\theta}{2}$.

59. $\cos\theta = \frac{4}{5}$ and $0° < \theta < 90°$

60. $\cos\theta = \frac{1}{3}$ and $0° < \theta < 90°$

61. $\tan\theta = -\frac{5}{12}$ and $-90° < \theta < 0°$

62. $\sec\theta = -2$ and $90° < \theta < 180°$

63. $\csc\theta = \frac{25}{24}$ and $90° < \theta < 180°$

64. $\sin\theta = \frac{4}{5}$ and $0° < \theta < 90°$

Verifying Identities

Exercises 65–74: Verify the identity. Give graphical or numerical support.

65. $4\sin 2x = 8\sin x\cos x$

66. $\cos 4\theta = 1 - 2\sin^2 2\theta$

67. $\dfrac{2 - \sec^2 x}{\sec^2 x} = \cos 2x$

68. $(\sin x + \cos x)^2 = \sin 2x + 1$

69. $\sec 2x = \dfrac{1}{1 - 2\sin^2 x}$

70. $2\csc 2t = \csc t\sec t$

71. $\sin 3\theta = 3\sin\theta - 4\sin^3\theta$

72. $\dfrac{2\tan x}{1 + \tan^2 x} = \sin 2x$

73. $\sin 4\theta = 4\sin\theta\cos\theta\cos 2\theta$

74. $\cos 4t = 8\cos^4 t - 8\cos^2 t + 1$

Exercises 75–84: Verify the identity.

75. $\dfrac{\sin 2\theta}{\sin\theta} = 2\cos\theta$ 　　 **76.** $2\sin^2 4\theta = 1 - \cos 8\theta$

77. $2\cos^2\dfrac{\theta}{2} = 1 + \cos\theta$

78. $\dfrac{\sin^2 2\theta}{1 + \cos 2\theta} = 2\sin^2\theta$ 　 **79.** $\cos^4\theta - \sin^4\theta = \cos 2\theta$

80. $\dfrac{1 - \tan^2 x}{1 + \tan^2 x} = \cos 2x$ 　 **81.** $\csc 2t = \dfrac{\csc t}{2\cos t}$

82. $\tan\theta + \cot\theta = \dfrac{2}{\sin 2\theta}$

83. $\tan\dfrac{x}{2} = \dfrac{\sin x}{1 + \cos x}$ $\left(\textit{Hint: Let } \tan\dfrac{x}{2} = \dfrac{\sin\frac{x}{2}}{\cos\frac{x}{2}}.\right)$

84. $\tan\dfrac{x}{2} = \dfrac{1 - \cos x}{\sin x}$ (*Hint:* Use the identity in Exercise 83.)

Product-to-Sum and Sum-to-Product Identities

Exercises 85–88: Write each expression as a sum or difference of trigonometric functions.

85. (a) $\cos 50°\sin 20°$ 　　 **(b)** $\cos 2x\cos x$

86. (a) $2\sin 74°\sin 24°$ 　 **(b)** $8\sin 18x\cos 13x$

87. (a) $\sin 7\theta\cos 3\theta$ 　　 **(b)** $\sin 8x\sin 4x$

88. (a) $2\cos 5x\cos 7x$ 　　 **(b)** $4\cos 9\theta\sin 2\theta$

Exercises 89–92: Write each expression as a product of trigonometric functions.

89. (a) $\sin 40° + \sin 30°$ 　 **(b)** $\cos 45° + \cos 35°$

90. (a) $\cos 104° - \cos 24°$ 　 **(b)** $\sin 32° - \sin 64°$

91. (a) $\cos 6\theta + \cos 4\theta$ 　 **(b)** $\sin 7x + \sin 4x$

92. (a) $\sin 3x - \sin 5x$ 　　 **(b)** $\cos 3\theta - \cos\theta$

Solving Equations

Exercises 93–96: Find the solutions to the equation in the interval $[0°, 360°)$

(a) symbolically, (b) graphically, and (c) numerically.

93. $\cos 2\theta = 1$ 　　　 **94.** $\sin 2\theta = \dfrac{1}{2}$

95. $\sin 2\theta + \cos\theta = 0$ 　 **96.** $\cos\dfrac{\theta}{2} = -\dfrac{\sqrt{3}}{2}$

Exercises 97–100: Find the solutions to the equation in the interval $[0, 2\pi)$.

97. $\sin\dfrac{\theta}{2} = 1$ 　　　 **98.** $\cos 2\theta + \cos\theta = 0$

99. $\sqrt{2}\sin\dfrac{\theta}{2} - 1 = 0$

100. $2\cos\dfrac{\theta}{2} + 1 = 0$

Exercises 101–116: Find all solutions, expressed in radians.

101. $2\cos 2t = \sqrt{3}$ 　　 **102.** $2\sin 2t = -1$

103. $\sin 2t + \sin t = 0$ 　 **104.** $\sin t - \cos 2t = 0$

105. $2\sin\dfrac{t}{2} - 1 = 0$ 　 **106.** $\sin 2t = 2\cos^2 t$

107. $\cos 2t = \sin t$ 　　 **108.** $\cos 2t - \cos t = 0$

109. $\tan 2t = 1$ 　　　 **110.** $\cot 2t = \sqrt{3}$

111. $2\cos\dfrac{t}{2} = 1$ 　　 **112.** $\tan\dfrac{t}{2} = 1$

113. $\cos 2t = 2\sin t\cos t$

114. $2\cos^2 2t = 1 - \cos 2t$

115. $2\sin^2 2t + \sin 2t - 1 = 0$

116. $2\cos\dfrac{t}{2} + 1 = 0$

Exercises 117 and 118: Approximate to the nearest thousandth all solutions on $[0, 2\pi)$.

117. $\sin t + \sin 2t = \cos t$

118. $\sin 3t + \sin 2t = 2\cos t$

Applications

119. Electricity (Refer to Example 8.) Suppose that the voltage in a 220-volt electrical circuit is modeled by $V(t) = 310\sin(120\pi t)$ and that the amperage flowing through a heater is $I = 7\sin(120\pi t)$. (*Source:* G. Wilcox and C. Hesselberth, *Electricity for Engineering Technology.*)

(a) Graph the wattage $W = VI$ consumed by the heater in $[0, 0.04, 0.01]$ by $[-500, 2500, 500]$.

(b) Find values for the constants a, k, and d so that $W = a\cos(k\pi t) + d$.

120. Electricity If a toaster is plugged into a common household outlet, the wattage W used varies according to the equation $W = \dfrac{V^2}{R}$, where V is the voltage and R is a constant that measures the resistance of the toaster in ohms. (*Source:* D. Bell, *Fundamentals of Electric Circuits.*)

(a) Graph W if $R = 15$ and $V = 163\sin(120\pi t)$ in $[0, 0.05, 0.01]$ by $[-500, 2000, 500]$.

(b) Approximate the maximum wattage consumed by the toaster.

(c) Use a power-reducing identity to express the wattage as $W = a\cos(240\pi t) + d$, where a and d are constants.

Exercises 121 and 122: **Electricity** *Let the voltage in an electrical outlet be given by* $V(t) = 320 \sin(120\pi t)$ *at time t in seconds. Find the times when V equals the following values.*

121. 160 volts

122. $160\sqrt{3}$ volts

123. Highway Curves When an automobile travels along a circular curve, objects like trees and buildings situated on the inside of the curve can obstruct the driver's vision. If the cars in the figure are a safe stopping distance apart, then the distance d that should be cleared on the inside of the curve is $d = r\left(1 - \cos\frac{\beta}{2}\right)$, where r is the radius of the curve and β is the central angle between the cars. (*Source:* F. Mannering and W. Kilareski, *Principles of Highway Engineering and Traffic Analysis.*)

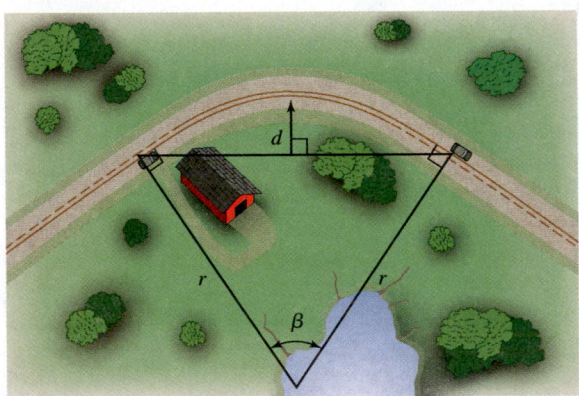

(a) Find d if $\beta = 80°$ and $r = 600$ feet.
(b) Use the figure to justify this formula.

(c) Is the given formula equivalent to the formula $d = r\left(1 - \frac{1}{2}\cos\beta\right)$? Explain.

124. Highway Curves The figure at the top of the next column represents a circular curve with radius r and central angle θ. The tangent length T is an important distance used by surveyors. (*Source:* F. Mannering.)

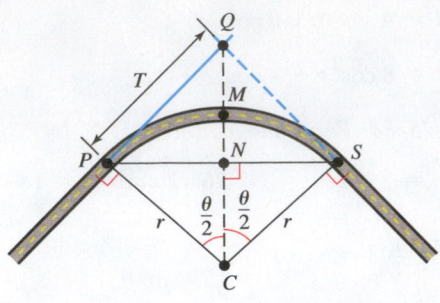

(a) Show that $T = r\tan\frac{\theta}{2}$.

(b) Find T for a curve with a 1500-foot radius and $\theta = 80°$.

125. Touch-Tone Phones (Refer to Example 15.) For the numbers 3 and 4 on a touch-tone phone, complete the following.
(a) Write formulas, f for 3 and g for 4, that are the *sum* of two cosine functions and that model the tone generated by each number.

(b) Write the formulas for each tone from part (a) as a *product* of two trigonometric functions.

126. Musical Tones and Beats If two musical tones with nearly the same frequency are played simultaneously, a phenomenon called *beats* occurs. Let $P_1 = 0.04\cos(110\pi t)$ and $P_2 = 0.04\cos(116\pi t)$ represent these tones.
(a) Graph $P = P_1 + P_2$ in the viewing rectangle $[0.2, 1.2, 0.2]$ by $[-0.08, 0.08, 0.01]$. How many beats are there?

(b) Use a sum-to-product identity to write $P_1 + P_2$ as a product of trigonometric expressions.

Writing about Mathematics

127. Suppose a student believes that an equation is an identity but cannot verify it symbolically. Discuss techniques that the student could use to support this belief.

128. Does the equation $\sin^2 3\theta + \cos^2 2\theta = 1$ represent an identity? Explain your reasoning.

CHECKING BASIC CONCEPTS FOR SECTION 7.5

1. Find values for $\sin 2\theta$ and $\cos 2\theta$ if $\cos\theta = -\frac{7}{25}$ and $\sin\theta = \frac{24}{25}$.

2. Use a half-angle formula to find the exact value of $\sin 22.5°$.

3. Find $\sin\frac{\theta}{2}$ and $\cos\frac{\theta}{2}$ if $\sin\theta = \frac{4}{5}$ and θ is acute.

4. Verify that $\frac{\cos 2\theta}{\cos^2\theta} = 2 - \sec^2\theta$.

5. Solve $\sin 2\theta = 2\cos\theta$ for $0 \le \theta < 2\pi$.

7 Summary

CONCEPT	EXPLANATION AND EXAMPLES

Section 7.1 Fundamental Identities

Reciprocal Identities

$$\sin\theta = \frac{1}{\csc\theta} \qquad \cos\theta = \frac{1}{\sec\theta} \qquad \tan\theta = \frac{1}{\cot\theta}$$

$$\csc\theta = \frac{1}{\sin\theta} \qquad \sec\theta = \frac{1}{\cos\theta} \qquad \cot\theta = \frac{1}{\tan\theta}$$

Example: If $\sin\theta = \frac{3}{5}$, then $\csc\theta = \frac{5}{3}$.

Quotient Identities

$$\tan\theta = \frac{\sin\theta}{\cos\theta} \qquad \cot\theta = \frac{\cos\theta}{\sin\theta}$$

Example: If $\sin\theta = -\frac{3}{5}$ and $\cos\theta = \frac{4}{5}$, then $\tan\theta = \frac{-3/5}{4/5} = -\frac{3}{4}$.

Pythagorean Identities

$$\sin^2\theta + \cos^2\theta = 1 \qquad 1 + \tan^2\theta = \sec^2\theta \qquad 1 + \cot^2\theta = \csc^2\theta$$

Examples: $\sin^2 30° + \cos^2 30° = 1$

$$1 + \tan^2\frac{\pi}{4} = \sec^2\frac{\pi}{4}$$

Negative-Angle Identities

$$\sin(-\theta) = -\sin\theta \qquad \cos(-\theta) = \cos\theta \qquad \tan(-\theta) = -\tan\theta$$
$$\csc(-\theta) = -\csc\theta \qquad \sec(-\theta) = \sec\theta \qquad \cot(-\theta) = -\cot\theta$$

Note: Cosine and secant are even functions, having graphs that are symmetric with respect to the *y*-axis. Sine, cosecant, tangent, and cotangent are odd functions, having graphs that are symmetric with respect to the origin.

Examples: $\cos(-45°) = \cos 45°,\ \sin(-60°) = -\sin 60°$

Section 7.2 Verifying Identities

Verifying Identities

To verify an identity, simplify one side of the equation until it equals the other side of the equation. Make use of the fundamental identities from Section 7.1.

Example: Verify that $\dfrac{1 - \sin^2\theta}{\cos\theta} = \cos\theta$.

$$\frac{1 - \sin^2\theta}{\cos\theta} = \frac{\cos^2\theta}{\cos\theta} \qquad \textcolor{teal}{\sin^2\theta + \cos^2\theta = 1}$$

$$= \cos\theta \qquad \textcolor{teal}{\text{Simplify.}}$$

Section 7.3 Trigonometric Equations

Reference Angle

The reference angle θ_R for an angle θ in standard position is the acute angle between the terminal side of θ and the *x*-axis.

Example: The reference angle for $\theta = 120°$ is $\theta_R = 60°$.

CONCEPT	EXPLANATION AND EXAMPLES

Section 7.3 Trigonometric Equations (CONTINUED)

Trigonometry Equations

Unlike polynomial equations, which have a finite number of solutions, trigonometric equations can have infinitely many solutions.

> **Example:** Find all solutions to $2\sin\theta - 1 = 0$.
>
> $$2\sin\theta - 1 = 0$$
> $$2\sin\theta = 1$$
> $$\sin\theta = \frac{1}{2}$$

Since $\theta_R = \sin^{-1}\frac{1}{2} = 30°$ and the sine function is positive in quadrants I and II, the solutions to $\sin\theta = \frac{1}{2}$ for $0° \le \theta < 360°$ are 30° and 150°.

Since $\sin\theta$ has period 360°, all solutions to the equation can be written as
$\theta = 30° + 360° \cdot n$ or $\theta = 150° + 360° \cdot n$,   where n is an integer.

Section 7.4 Sum and Difference Identities

Cosine Sum and Difference

$$\cos(\alpha + \beta) = \cos\alpha\cos\beta - \sin\alpha\sin\beta$$
$$\cos(\alpha - \beta) = \cos\alpha\cos\beta + \sin\alpha\sin\beta$$

> **Example:** $\cos(60° - 45°) = \cos 60° \cos 45° + \sin 60° \sin 45°$

Sine Sum and Difference

$$\sin(\alpha + \beta) = \sin\alpha\cos\beta + \cos\alpha\sin\beta$$
$$\sin(\alpha - \beta) = \sin\alpha\cos\beta - \cos\alpha\sin\beta$$

> **Example:** $\sin(60° - 45°) = \sin 60° \cos 45° - \cos 60° \sin 45°$

Tangent Sum and Difference

$$\tan(\alpha + \beta) = \frac{\tan\alpha + \tan\beta}{1 - \tan\alpha\tan\beta}, \qquad \tan(\alpha - \beta) = \frac{\tan\alpha - \tan\beta}{1 + \tan\alpha\tan\beta}$$

> **Example:** $\tan(60° - 45°) = \dfrac{\tan 60° - \tan 45°}{1 + \tan 60° \tan 45°}$

Cofunction Identities

$$\cos\left(\frac{\pi}{2} - t\right) = \sin t \qquad \sin\left(\frac{\pi}{2} - t\right) = \cos t$$

$$\cot\left(\frac{\pi}{2} - t\right) = \tan t \qquad \tan\left(\frac{\pi}{2} - t\right) = \cot t$$

$$\csc\left(\frac{\pi}{2} - t\right) = \sec t \qquad \sec\left(\frac{\pi}{2} - t\right) = \csc t$$

> **Example:** $\tan\dfrac{\pi}{3} = \tan\left(\dfrac{\pi}{2} - \dfrac{\pi}{6}\right) = \cot\dfrac{\pi}{6}$

CONCEPT	EXPLANATION AND EXAMPLES

Section 7.5 Multiple-Angle Identities

Double-Angle Identities

$$\sin 2\theta = 2\sin\theta\cos\theta \qquad \cos 2\theta = \cos^2\theta - \sin^2\theta \qquad \tan 2\theta = \frac{2\tan\theta}{1 - \tan^2\theta}$$

$$\text{or}$$
$$\cos 2\theta = 2\cos^2\theta - 1$$
$$\text{or}$$
$$\cos 2\theta = 1 - 2\sin^2\theta$$

Example: $\sin 120° = \sin(2\cdot 60°) = 2\sin 60°\cos 60°$

Power-Reducing Identities

$$\sin^2\theta = \frac{1 - \cos 2\theta}{2} \qquad \cos^2\theta = \frac{1 + \cos 2\theta}{2} \qquad \tan^2\theta = \frac{1 - \cos 2\theta}{1 + \cos 2\theta}$$

Example: $\sin^2 30° = \frac{1 - \cos 60°}{2}$

Half-Angle Formulas

$$\sin\frac{\theta}{2} = \pm\sqrt{\frac{1 - \cos\theta}{2}} \qquad \cos\frac{\theta}{2} = \pm\sqrt{\frac{1 + \cos\theta}{2}}$$

$$\tan\frac{\theta}{2} = \pm\sqrt{\frac{1 - \cos\theta}{1 + \cos\theta}} \qquad \tan\frac{\theta}{2} = \frac{1 - \cos\theta}{\sin\theta} \qquad \tan\frac{\theta}{2} = \frac{\sin\theta}{1 + \cos\theta}$$

Example: $\cos 15° = \cos\frac{30°}{2} = \sqrt{\frac{1 + \cos 30°}{2}}$

Product-to-Sum Identities

$$\cos\alpha\cos\beta = \tfrac{1}{2}(\cos(\alpha + \beta) + \cos(\alpha - \beta))$$
$$\sin\alpha\sin\beta = \tfrac{1}{2}(\cos(\alpha - \beta) - \cos(\alpha + \beta))$$
$$\sin\alpha\cos\beta = \tfrac{1}{2}(\sin(\alpha + \beta) + \sin(\alpha - \beta))$$
$$\cos\alpha\sin\beta = \tfrac{1}{2}(\sin(\alpha + \beta) - \sin(\alpha - \beta))$$

Example: $\cos 45°\cos 60° = \tfrac{1}{2}(\cos(45° + 60°) + \cos(45° - 60°))$

Sum-to-Product Identities

$$\cos a + \cos b = 2\cos\frac{a + b}{2}\cos\frac{a - b}{2}$$

$$\cos a - \cos b = -2\sin\frac{a + b}{2}\sin\frac{a - b}{2}$$

$$\sin a + \sin b = 2\sin\frac{a + b}{2}\cos\frac{a - b}{2}$$

$$\sin a - \sin b = 2\cos\frac{a + b}{2}\sin\frac{a - b}{2}$$

Example: $\cos 60° + \cos 30° = 2\cos\frac{60° + 30°}{2}\cos\frac{60° - 30°}{2}$

7 | Review Exercises

Exercises 1 and 2: Determine the quadrant that contains θ.

1. $\sec\theta < 0$ and $\sin\theta > 0$

2. $\cot\theta > 0$ and $\cos\theta < 0$

Exercises 3–6: Use the given information to find the other trigonometric functions of θ.

3. $\sin\theta = \frac{3}{5}$ and $\cos\theta = -\frac{4}{5}$

4. $\sec\theta = -\frac{13}{12}$ and $\csc\theta = -\frac{13}{5}$

5. $\tan\theta = -\frac{7}{24}$ and $\cos\theta = \frac{24}{25}$

6. $\cot\theta = -\frac{1}{2}$ and $\sin\theta > 0$

Exercises 7–10: Use a negative-angle identity to write an equivalent trigonometric expression involving a positive angle.

7. $\sin(-13°)$ **8.** $\cos(-106°)$

9. $\sec\left(-\frac{3\pi}{7}\right)$ **10.** $\tan\left(-\frac{5\pi}{11}\right)$

Exercises 11–16: Simplify each expression.

11. $\sec\theta\cot\theta\sin\theta$ **12.** $\sin\theta\csc\theta$

13. $(\sec^2 t - 1)(\csc^2 t - 1)$ **14.** $\dfrac{\sec\theta}{\csc\theta} + \dfrac{\sin\theta}{\cos\theta}$

15. $\dfrac{\csc\theta\sin\theta}{\sec\theta}$ **16.** $\dfrac{\cos^2\theta}{1 - \sin\theta}$

Exercises 17–20: Use a calculator to approximate the other trigonometric functions of θ to four decimal places.

17. $\tan\theta = 1.2367$ and θ is acute

18. $\sin\theta = -0.3434$ and θ is in quadrant IV

19. $\cos\theta = -0.4544$ and θ is in quadrant II

20. $\tan\theta = -0.8595$ and θ is in quadrant IV

Exercises 21–24: Factor the trigonometric expression.

21. $\sin^2\theta + 2\sin\theta + 1$

22. $2\cos^2 t - 3\cos t + 1$

23. $\tan^2\theta - 9$

24. $2\sec^2\theta - 3\sec\theta - 5$

Exercises 25–38: Verify the identity.

25. $(\sec\theta - 1)(\sec\theta + 1) = \tan^2\theta$

26. $(\cos\theta + \sin\theta)^2 + (\cos\theta - \sin\theta)^2 = 2$

27. $(1 + \tan t)^2 = \sec^2 t + 2\tan t$

28. $(1 - \cos^2 t)(1 + \tan^2 t) = \tan^2 t$

29. $\sin(x - \pi) = -\sin x$

30. $\cos(\pi + x) = -\cos x$

31. $\sin 8x = 2\sin 4x\cos 4x$

32. $\cos^4 x - \sin^4 x = \cos 2x$

33. $\sec 2x = \dfrac{1}{2\cos^2 x - 1}$

34. $\dfrac{1 + \tan^2 x}{\sin^2 x + \cos^2 x} = \sec^2 x$

35. $\cos^4 x\sin^3 x = (\cos^4 x - \cos^6 x)\sin x$

36. $\sin^4 x = \frac{3}{8} - \frac{1}{2}\cos 2x + \frac{1}{8}\cos 4x$

37. $\sec^4\theta - \tan^4\theta = 1 + 2\tan^2\theta$

38. $\dfrac{1 + \cos\theta}{\sin\theta} + \dfrac{\sin\theta}{1 + \cos\theta} = 2\csc\theta$

Exercises 39–42: Find the reference angle for θ.

39. $\theta = 240°$ **40.** $\theta = 320°$

41. $\theta = \frac{9\pi}{7}$ **42.** $\theta = -\frac{7\pi}{6}$

Exercises 43 and 44: Use the graph to estimate any solutions to the trigonometric equation for $[0, 2\pi)$. Then solve the equation symbolically.

43. $\cos^2\theta - 2\cos\theta = 0$ **44.** $2\sin t\cos t - \cos t = 0$

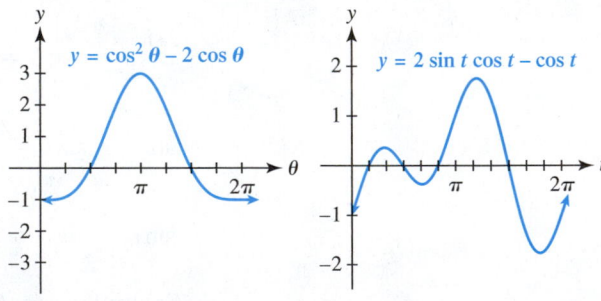

Exercises 45 and 46: Solve the given trigonometric equation for $0° \le \theta < 360°$.

45. **(a)** $\tan\theta = \sqrt{3}$ **(b)** $\cot\theta = -\sqrt{3}$

46. **(a)** $\sin\theta = 1$ **(b)** $\cos\theta = -1$

Exercises 47–52: Solve the equation for $0 \le t < 2\pi$.

47. $2\cos t - 1 = 0$ **48.** $\cot^2 t = 1$

49. $2 \sin^2 t + \sin t - 3 = 0$

50. $\sin^2 t + 2 \cos t = 1$

51. $\tan^2 t - 2 \tan t + 1 = 0$

52. $2 \sin t = \sqrt{3}$

Exercises 53–56: Find all solutions to the equation. Express your results in both degrees and radians.

53. $3 \tan^2 t - 1 = 0$ **54.** $2 \sin^2 t - \sin t - 1 = 0$

55. $\sin 2t + 3 \cos t = 0$ **56.** $\cos 2t = 1$

Exercises 57 and 58: Use a half-angle formula to find the exact value for the expression. Use a calculator to support your result.

57. $\cos 105°$ **58.** $\sin \frac{\pi}{12}$

Exercises 59 and 60: Estimate any solutions in the interval $[-2\pi, 2\pi]$ graphically to two decimal places.

59. $\tan x = x + 1$ **60.** $\sin(\cos x) = \tan x$

Exercises 61 and 62: Find the following.

(a) $\sin(\alpha + \beta)$ *(b)* $\cos(\alpha + \beta)$
(c) $\tan(\alpha + \beta)$ *(d)* *The quadrant containing* $\alpha + \beta$

61. $\cos \alpha = \frac{3}{5}$ and $\cos \beta = \frac{12}{13}$, where α and β are both in quadrant I.

62. $\sin \alpha = -\frac{12}{13}$ and $\tan \beta = -\frac{4}{3}$, where α and β are both in quadrant IV.

Exercises 63 and 64: Complete the following.

(a) *Find* $\sin 2\theta$, $\cos 2\theta$, *and* $\tan 2\theta$.
(b) *Use a calculator to support your results.*

63. $\sin \theta = \frac{4}{5}$ and $\tan \theta = -\frac{4}{3}$

64. $\sin \theta = -\frac{12}{13}$ and $\cos \theta = -\frac{5}{13}$

Exercises 65 and 66: Complete the following.

(a) *Find* $\sin \frac{\theta}{2}$, $\cos \frac{\theta}{2}$, *and* $\tan \frac{\theta}{2}$.
(b) *Use a calculator to support your results.*

65. $\cos \theta = \frac{1}{4}$ and $0° < \theta < 90°$

66. $\tan \theta = -\frac{8}{15}$ and $-90° < \theta < 0°$

Exercises 67 and 68: Evaluate the expression.

67. $\cos\left(2 \tan^{-1} \frac{11}{60}\right)$ **68.** $\sin(2 \sin^{-1} x), x > 0$

Applications

69. Daylight Hours The number of daylight hours y at $20°$ S latitude can be modeled by

$$y = 1.2 \cos\left(\frac{\pi}{6}(x - 0.7)\right) + 12.1,$$

where $x = 1$ corresponds to January 1, $x = 2$ to February 1, and so on. Estimate when the number of daylight hours equals 11.5 hours. (*Source:* J. Williams.)

70. Music and Pure Tones A pure tone is modeled by the graph of $P(t) = 0.006 \cos(50\pi t)$, where P represents the pressure on an eardrum in pounds per square foot at time t in seconds. (*Source:* J. Roederer, *Introduction to the Physics and Psychophysics of Music.*)

(a) Graph P in $[0, 0.1, 0.01]$ by $[-0.008, 0.008, 0.001]$.

(b) Estimate all solutions to the equation $P = 0$ on the interval $0 \leq t \leq 0.1$.

71. Modeling Musical Tones Let the pressure exerted on the eardrum by two sound waves in pounds per square foot be given by

$$P_1(t) = 0.006 \cos(100\pi t) \quad \text{and}$$

$$P_2(t) = 0.008 \sin(100\pi t).$$

(a) Graph the pressure on the eardrum $P = P_1 + P_2$ in $[0, 0.06, 0.01]$ by $[-0.012, 0.012, 0.002]$.

(b) Use the graph to find values for a and k such that $P = a \sin(100\pi t + k)$.

(c) Use a sum or difference identity for sine to verify your result in part (b).

72. Electricity Suppose the voltage in an electrical heater is given by $V(t) = 17 \sin(120\pi t)$ and the amperage flowing through the heater is $I(t) = 2 \sin(120\pi t)$, where t is in seconds. (*Source:* G. Wilcox and C. Hesselberth, *Electricity for Engineering Technology.*)

(a) Graph the wattage $W = VI$ consumed by the heater in $[0, 0.04, 0.01]$ by $[-40, 40, 10]$.

(b) Use a power-reducing identity to express the wattage in the form $W = a \cos(k\pi t) + d$, where a, k, and d are constants.

73. Electromagnets Let the wattage consumed by an electromagnet be given by $W(t) = 7 \cos^2(240\pi t)$ and the voltage be given by $V(t) = 50 \sin(240\pi t)$, where t is in seconds. Express $W(t)$ in terms of the sine function. When V is maximum or minimum, what is the value of W? Explain.

74. Average Temperatures Let the monthly average high temperature y in degrees Fahrenheit at a location be given by $y = 15 \sin\left(\frac{\pi}{6}(x - 4)\right) + 60$, where x is the month and $x = 1$ corresponds to January. Estimate when the monthly average high temperature is $60°$ F.

75. Back Stress The force exerted by the back muscles of a 100-pound person can be estimated by $F = 289 \cos \theta$. Find θ in degrees and radians when $F = 250$ pounds.

76. Electricity Let $V(t) = 80 \sin(120\pi t)$ denote the voltage in an electrical outlet at time t in seconds, where t is a real number. Find all times when the voltage is 40 volts.

Extended and Discovery Exercises

1. Piano Strings If a piano key with a frequency of f_1 is played, then the corresponding string will not only vibrate at f_1 but also vibrate at the higher frequencies of $2f_1$, $3f_1$, $4f_1$, and so on. The *fundamental frequency* of the string is f_1, and the higher frequencies are called the *upper harmonics*. The human ear will hear the sum of these frequencies as one complex tone. (*Source:* J. Roederer, *Introduction to the Physics and Psychophysics of Music.*)

(a) If the A note above middle C is played, its fundamental frequency is $f_1 = 440$ hertz. (One hertz equals one cycle per second.) The piano string also vibrates at frequencies of $f_2 = 2(440) = 880$, $f_3 = 3(440) = 1320$, $f_4 = 4(440) = 1760$, and so on. The pressure for each frequency in pounds per square foot is modeled by

$$P_1 = 0.002 \sin(2\pi(440)t),$$

$$P_2 = \frac{0.002}{2} \sin(2\pi(880)t),$$

$$P_3 = \frac{0.002}{3} \sin(2\pi(1320)t),$$

$$P_4 = \frac{0.002}{4} \sin(2\pi(1760)t), \quad \text{and}$$

$$P_5 = \frac{0.002}{5} \sin(2\pi(2200)t),$$

where t is in seconds. Graph each of the following expressions for P in the viewing rectangle given by $[0, 0.01, 0.002]$ by $[-0.005, 0.005, 0.001]$.

   i. $P = P_1$

   ii. $P = P_1 + P_2$

   iii. $P = P_1 + P_2 + P_3$

   iv. $P = P_1 + P_2 + P_3 + P_4$

   v. $P = P_1 + P_2 + P_3 + P_4 + P_5$

(b) The final graph of P models what the human ear hears. Describe this graph.

(c) Estimate the maximum pressure of

$$P = P_1 + P_2 + P_3 + P_4 + P_5.$$

(d) A pure tone with a frequency of 440 hertz is modeled by $P = P_1$, whereas a piano generates the graph of $P = P_1 + P_2 + P_3 + P_4 + P_5$. Compare and contrast these two graphs.

2. Low Tones and Small Speakers Small speakers often cannot vibrate at frequencies less than 200 hertz. Nonetheless, these tones can still be heard on such speakers. When a piano string creates a tone of 110 hertz, it also creates tones at 220, 330, 440, 550, and 660 hertz. A small speaker cannot reproduce the 110 hertz vibration, but it can reproduce the higher frequencies. The low tones can still be heard because the speaker produces *difference tones* of these *upper harmonics*. The difference between consecutive frequencies is 110 hertz, and this difference tone will be heard on a small speaker even though the speaker cannot vibrate at 110 hertz. This phenomenon can be visualized with a graphing calculator. (*Source:* A. Benade.)

(a) Graph the upper harmonics represented by the pressure wave

$$P = \frac{1}{2}\sin(2\pi(220)t) + \frac{1}{3}\sin(2\pi(330)t) + \frac{1}{4}\sin(2\pi(440)t)$$

in $[0, 0.03, 0.01]$ by $[-1.2, 1.2, 0.5]$.

(b) Estimate the t-values on the interval $0 \le t \le 0.03$ where P is maximum.

(c) Approximate the frequency of these maximum values. What does a person hear in addition to the frequencies of 220, 330, and 440 hertz?

(d) Discuss the advantage of having large speakers instead of smaller ones. (*Hint:* Try graphing the pressure produced by a speaker that can vibrate both at 110 hertz and at the upper harmonics.)

3. Piano Strings When a string is set into vibration by striking it, the amplitude A of the vibrations decreases over time, while the frequency of the vibration remains constant. This phenomenon is called *exponential decay* and can be modeled by $A = A_0 e^{-kt} \sin(2\pi F t)$, where F is the frequency, t is time in seconds, and k and A_0 are positive constants. (*Source:* J. Roederer.)

(a) Graph A when $A_0 = 0.1$, $F = 15$, and $k = 1.2$ in the window $[0, 1, 0.1]$ by $[-0.15, 0.15, 0.05]$.

(b) Now graph the equations $y_1 = -0.1e^{-1.2t}$ and $y_2 = 0.1e^{-1.2t}$ with A. Describe how the graphs of y_1 and y_2 relate to the graph of A.

(c) The *decay half-time* is the time it takes for the maximum amplitude of A_0 to decrease to $\frac{1}{2}A_0$. Estimate this time graphically. (The decay half-time for a typical piano string is about 0.4 second.)

8 Further Topics in Trigonometry

There is no branch of mathematics, however abstract, which may not some day be applied to the real world.

—Nikolai Lobachevsky

Many unusual scientific concepts defy intuition, and mathematics is necessary for their understanding. For example, world technology took a giant step forward when the Large Hadron Collider (LHC) was put into operation in September 2008. The LHC, a 17-mile-long nuclear particle accelerator located 100 meters beneath the border between France and Switzerland, now propels particles near light speed. With the LHC, scientists now believe that they have found a new particle, called the Higgs (or god) particle, which was first predicted by five equations in 1964. Scientists also hope to create tiny black holes that will fortunately disappear rapidly, according to mathematical predictions made by Stephen Hawking. The LHC may even reveal the existence of higher dimensions. Even though there is *no experimental evidence* for many of these phenomena, *mathematical equations predict them*, so scientists are looking. The LHC is one of the most expensive scientific instruments ever built, yet the confidence in its success relies heavily on mathematics.

Angles, triangles, vectors, and complex numbers help make new technologies such as the LHC, the Global Positioning System, aerial photography, and navigation possible. In this chapter, we use these mathematical concepts to solve a variety of applications.

619

8.1 Law of Sines

- Learn about oblique triangles
- Derive the law of sines
- Solve triangles
- Solve the ambiguous case

Introduction

In Chapter 6, we solved right triangles. Solving a triangle involves finding the length of each side and the measure of each angle in the triangle. In many areas of study, triangles without right angles often occur. For example, the height of the Gateway to the West Arch in St. Louis can be found from the ground by using triangles without right angles. (See Exercise 63.) To solve these *oblique triangles*, we will use the law of sines and the law of cosines. In this section, the law of sines is derived and used to solve several problems.

Oblique Triangles

If a triangle is not a right triangle, then it is called an **oblique triangle**. There are four different situations, or cases, that can occur when attempting to solve an oblique triangle.

1. **SSS**: All three sides are given. This situation determines a unique triangle and is referred to as **SSS**. See Figure 8.1. Note that the length of any one side must be less than the sum of the lengths of the other two sides.
2. **SAS**: Two sides and the angle included (between them) are given. This situation determines a unique triangle and is referred to as **SAS**. See Figure 8.2.
3. **AAS or ASA**: One side and two angles are given. This situation determines a unique triangle and is referred to as **AAS or ASA**. See Figure 8.3. Note that whenever two angles of a triangle are known, the third angle can be found by using the fact that the sum of the measures of the angles equals 180°.
4. **SSA**: Two sides and an angle opposite one of the sides are given. This situation does *not always* determine a unique triangle and is referred to as SSA. There may be zero, one, or two triangles that can satisfy SSA. As a result, we call **SSA** the **ambiguous case**. See Figures 8.4–8.6.

Cases 1–3: One Triangle Possible
(Unique)

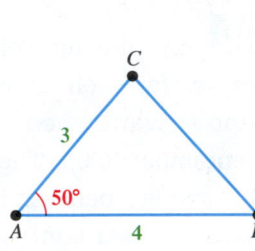

Figure 8.1 Case 1: **SSS**

Figure 8.2 Case 2: **SAS**

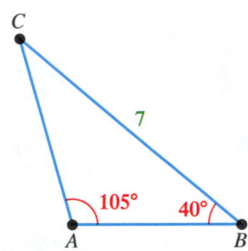

Figure 8.3 Case 3: **AAS or ASA**

Case 4: Zero, One, or Two Triangles Possible
(Ambiguous Case)

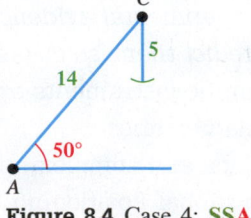

Figure 8.4 Case 4: **SSA**
(No Triangles)

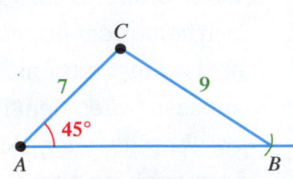

Figure 8.5 Case 4: **SSA**
(One Triangle)

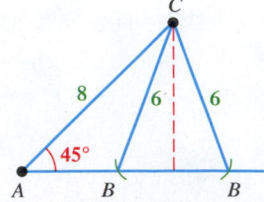

Figure 8.6 Case 4: **SSA**
(Two Triangles)

Cases 1 and 2 are solved in Section 8.2 using the *law of cosines*, and Cases 3 and 4 are solved in this section using the *law of sines*.

Standard Labeling

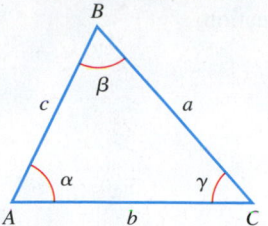

Figure 8.7

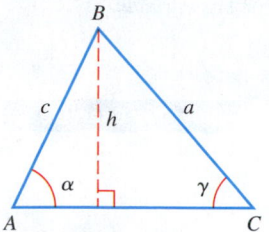

Figure 8.8

Solving Triangles with the Law of Sines

We will label triangles as shown in Figure 8.7 and refer to this labeling as the **standard labeling**. For example, angle α is located at vertex A and side a is opposite angle α. Note that angle γ need *not* be a right angle.

Derivation of Law of Sines   The law of sines can be derived using the oblique triangle shown in Figure 8.8.

$$\sin\alpha = \frac{h}{c}, \quad \text{or} \quad \mathbf{\textit{h} = \textit{c}\sin\alpha}$$

$$\sin\gamma = \frac{h}{a}, \quad \text{or} \quad \mathbf{\textit{h} = \textit{a}\sin\gamma}$$

Since $\mathbf{\textit{h} = \textit{c}\sin\alpha}$ and $\mathbf{\textit{h} = \textit{a}\sin\gamma}$, it follows that

$$\mathbf{\textit{c}\sin\alpha = \textit{a}\sin\gamma,}$$

or, if we divide each side by ac,

$$\frac{\sin\alpha}{a} = \frac{\sin\gamma}{c}.$$

In a similar manner it can be shown that

$$\frac{\sin\alpha}{a} = \frac{\sin\beta}{b}.$$

This discussion supports the following result.

LAW OF SINES

Any triangle with standard labeling satisfies

$$\frac{\sin\alpha}{a} = \frac{\sin\beta}{b} = \frac{\sin\gamma}{c}, \quad \text{or equivalently,} \quad \frac{a}{\sin\alpha} = \frac{b}{\sin\beta} = \frac{c}{\sin\gamma}.$$

We usually use these equations to find angles $\alpha, \beta,$ or γ.

We usually use these equations to find sides $a, b,$ or c.

In the next example we solve a triangle by using the law of sines.

EXAMPLE 1   Solving a triangle (ASA)

Solve the triangle shown in Figure 8.9

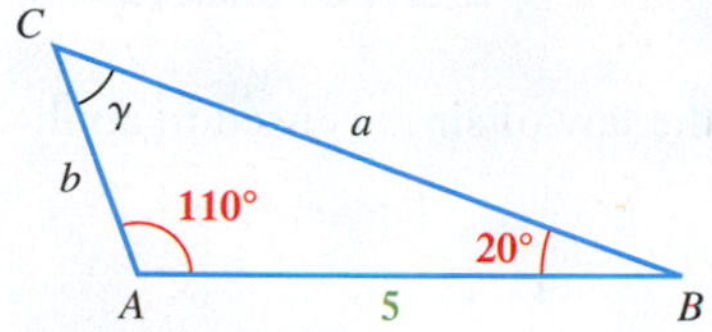

SOLUTION In Figure 8.9, $\alpha = 110°, \beta = 20°,$ and $c = 5$. To solve this triangle we need to find $a, b,$ and angle γ located at vertex C. Because we are given **ASA**, we can apply the law of sines.

Find γ   The sum of the measures of the angles in a triangle equal $180°$, so it follows that $\alpha + \beta + \gamma = 180°$.

$$\gamma = 180° - \alpha - \beta \qquad \text{Solve for } \gamma.$$

$$= 180° - 110° - 20° \qquad \text{Substitute.}$$

$$= 50° \qquad \text{Simplify.}$$

Figure 8.9

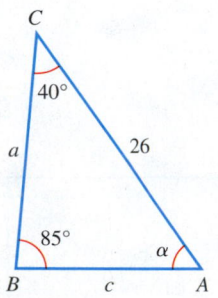

Figure 8.9 (Repeated)

Find a and b We can find a and b by writing the law of sines so that a is the only unknown in one equation and b is the only unknown in the other equation.

Find Side a	Find Side b	
$\dfrac{a}{\sin \alpha} = \dfrac{c}{\sin \gamma}$	$\dfrac{b}{\sin \beta} = \dfrac{c}{\sin \gamma}$	Law of sines
$\dfrac{a}{\sin 110°} = \dfrac{5}{\sin 50°}$	$\dfrac{b}{\sin 20°} = \dfrac{5}{\sin 50°}$	Substitute.
$a = \dfrac{5 \sin 110°}{\sin 50°}$	$b = \dfrac{5 \sin 20°}{\sin 50°}$	Solve for the variable.
$a \approx 6.13$	$b \approx 2.23$	Approximate.

> **Now Try Exercise 3**

EXAMPLE 2 **Solving a triangle (AAS)**

If $\beta = 85°$, $\gamma = 40°$, and $b = 26$, solve triangle ABC.

SOLUTION Sketch triangle ABC as shown in Figure 8.10.

$$\alpha = 180° - \beta - \gamma = 180° - 85° - 40° = 55°$$

Side a can be found by using the law of sines.

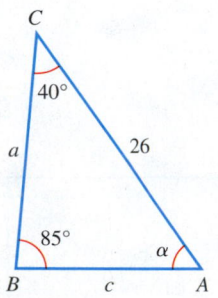

Figure 8.10

> Write the law of sines so that a is the only unknown.

$\dfrac{a}{\sin \alpha} = \dfrac{b}{\sin \beta}$	Law of sines
$\dfrac{a}{\sin 55°} = \dfrac{26}{\sin 85°}$	Substitute.
$a = \dfrac{26 \sin 55°}{\sin 85°}$	Multiply by $\sin 55°$.
$a \approx 21.4$	Approximate.

Side c can be found in a similar manner.

> Write the law of sines so that c is the only unknown.

$\dfrac{c}{\sin \gamma} = \dfrac{b}{\sin \beta}$	Law of sines
$\dfrac{c}{\sin 40°} = \dfrac{26}{\sin 85°}$	Substitute.
$c = \dfrac{26 \sin 40°}{\sin 85°}$	Multiply by $\sin 40°$.
$c \approx 16.8$	Approximate.

> **Now Try Exercise 27**

Applications The next two examples illustrate how the law of sines is used in applications.

EXAMPLE 3 **Using aerial photography to find distances (ASA)**

Figure 8.11 depicts a situation in which a camera lens has an angular coverage of 75°. As a picture is taken over level ground, the airplane's distance is 4800 feet from a house located on the edge of the photograph and the angle of elevation of the airplane from the house is 48°. Find the ground distance a shown in the photograph. (**Source:** F. Moffitt, *Photogrammetry.*)

Aerial Photography

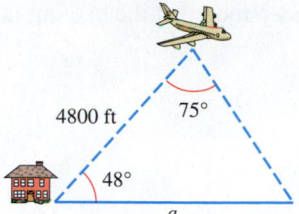

Figure 8.11

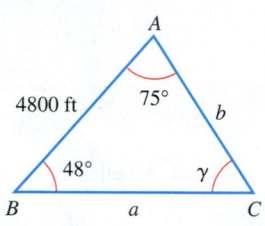

Figure 8.12

SOLUTION In this example we are given two angles and the side included (ASA), so let $\alpha = 75°$, $\beta = 48°$, and $c = 4800$, as shown in Figure 8.12. To find the third angle γ we can use the fact that the angles sum to 180°.

$$\gamma = 180° - \alpha - \beta$$
$$= 180° - 75° - 48°$$
$$= 57°$$

Side a corresponds to the ground distance shown in the photograph and can be found by using the law of sines.

$$\frac{a}{\sin \alpha} = \frac{c}{\sin \gamma} \qquad \text{Law of sines}$$

$$\frac{a}{\sin 75°} = \frac{4800}{\sin 57°} \qquad \text{Substitute.}$$

$$a = \frac{4800 \sin 75°}{\sin 57°} \qquad \text{Multiply by } \sin 75°.$$

$$a \approx 5528 \qquad \text{Approximate.}$$

The photograph will show about 5528 feet of ground distance from one edge of the photograph to the other.

Now Try Exercise 45

EXAMPLE 4   **Estimating the distance to the moon (ASA)**

Since the moon is a relatively close celestial object, its distance can be approximated using trigonometry. To find this distance, two photographs of the moon were taken at precisely the same time from two locations. On April 29, 1976, at 11:35 A.M., the lunar angles of elevation during a partial solar eclipse at Bochum in upper Germany and at Donaueschingen in lower Germany were measured as $\alpha = 52.6997°$ and $\theta = 52.7430°$, respectively. See Figure 8.13. If the two cities are 398.02 kilometers apart, approximate the distance to the moon. Disregard the curvature of Earth in this calculation.   (*Source:* W. Schlosser, *Challenges of Astronomy.*)

Finding the Distance to the Moon

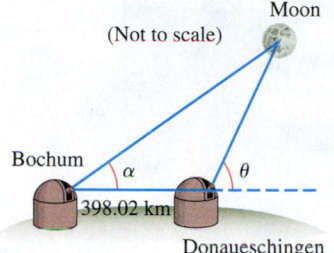

Figure 8.13

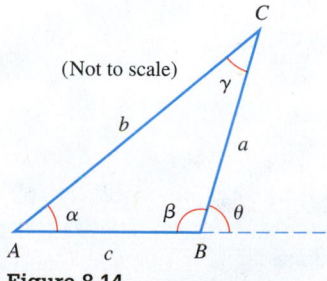

Figure 8.14

SOLUTION

Getting Started Consider triangle ABC shown in Figure 8.14, where $\alpha = 52.6997°$, $\theta = 52.7430°$, and $c = 398.02$. Because $\alpha + \gamma$ and θ are both supplements of angle β, it follows that $\alpha + \gamma = \theta$ and so

$$\gamma = \theta - \alpha = 52.7430° - 52.6997° = 0.0433°.$$

The distance to the moon can be approximated by finding either a or b. Why? ▶
    We find a by applying the law of sines.

$$\frac{a}{\sin \alpha} = \frac{c}{\sin \gamma} \qquad \text{Law of sines}$$

$$\frac{a}{\sin 52.6997°} = \frac{398.02}{\sin 0.0433°} \qquad \text{Substitute.}$$

$$a = \frac{398.02 \sin 52.6997°}{\sin 0.0433°} \qquad \text{Multiply by } \sin 52.6997°.$$

$$a \approx 419,000 \text{ km} \qquad \text{Approximate.}$$

The distance to the moon on that day was about 419,000 kilometers.

Now Try Exercise 47

Bearings Bearings are used in both surveying and aerial navigation to determine directions. If a single angle is used for a **bearing**, then it is understood that the bearing is measured *clockwise* from due north. See Figure 8.15.

Examples of Bearings

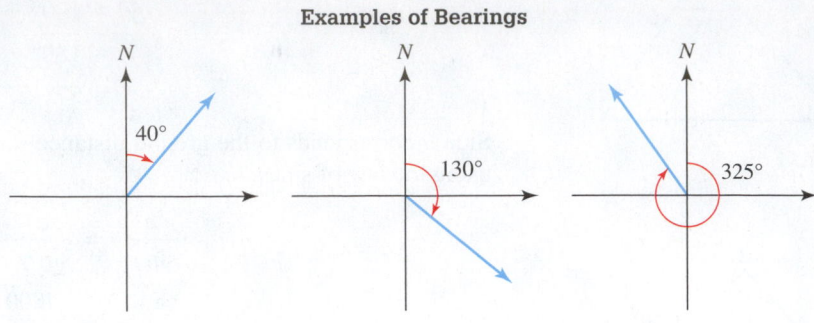

Figure 8.15

In the next example, we locate a fire by using bearings. (*Source:* I. Mueller and K. Ramsayer, *Introduction to Surveying.*)

EXAMPLE 5 **Determining the location of a forest fire**

A fire is spotted from two ranger stations that are 4 miles apart, as illustrated in Figure 8.16. From station *A* the bearing of the fire is 35°, and from station *B* the bearing of the fire is 335°. Find the distance to the nearest hundredth of a mile between the fire and each ranger station if station *A* lies directly west of station *B*.

Finding the Distance to a Fire

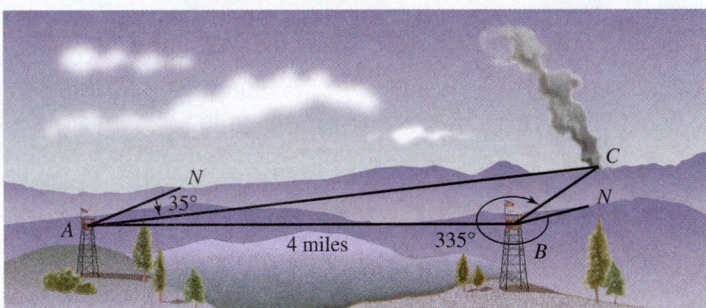

Figure 8.16

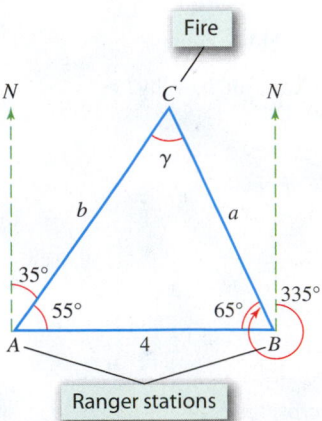

Figure 8.17

SOLUTION See triangle *ABC* in Figure 8.17, where $\alpha = 55°$, $\beta = 335° - 270° = 65°$, and $c = 4$. Thus

$$\gamma = 180° - 55° - 65° = \mathbf{60°}.$$

Using the law of sines, we can find *a*.

$$\frac{a}{\sin \alpha} = \frac{c}{\sin \gamma} \qquad \text{Law of sines}$$

$$\frac{a}{\sin 55°} = \frac{4}{\sin 60°} \qquad \text{Substitute.}$$

$$a = \frac{4 \sin 55°}{\sin 60°} \qquad \text{Multiply by } \sin 55°.$$

$$a \approx 3.78 \text{ mi} \qquad \text{Approximate.}$$

In a similar manner, we can find b.

$$\frac{b}{\sin \beta} = \frac{c}{\sin \gamma} \qquad \textit{Law of sines}$$

$$\frac{b}{\sin 65°} = \frac{4}{\sin 60°} \qquad \textit{Substitute.}$$

$$b = \frac{4 \sin 65°}{\sin 60°} \qquad \textit{Multiply by } \sin 65°.$$

$$b \approx 4.19 \text{ mi} \qquad \textit{Approximate.}$$

The fire is 4.19 miles from station A and 3.78 miles from station B.

Now Try Exercise 49

The Ambiguous Case (SSA)

If we are given two sides and an angle opposite one of the sides (SSA), there may be zero, one, or two triangles that satisfy these conditions. For this reason **SSA** is called the *ambiguous case*.

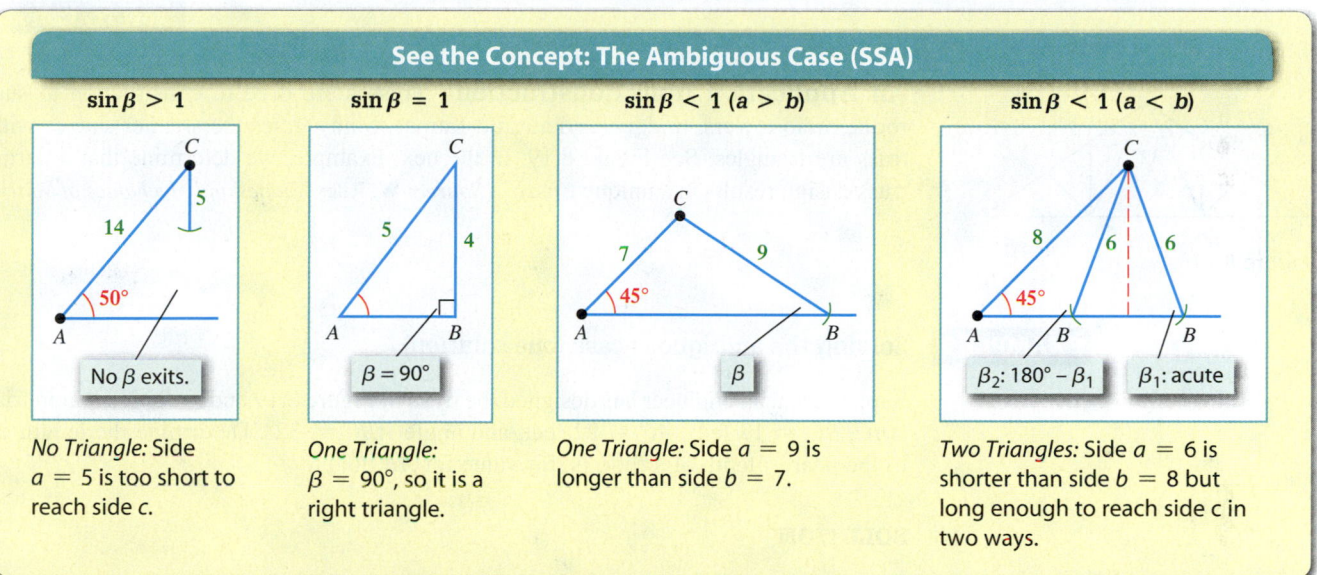

See the Concept: The Ambiguous Case (SSA)

$\sin \beta > 1$

No Triangle: Side $a = 5$ is too short to reach side c.

$\sin \beta = 1$

One Triangle: $\beta = 90°$, so it is a right triangle.

$\sin \beta < 1 \ (a > b)$

One Triangle: Side $a = 9$ is longer than side $b = 7$.

$\sin \beta < 1 \ (a < b)$

Two Triangles: Side $a = 6$ is shorter than side $b = 8$ but long enough to reach side c in two ways.

The following box summarizes the steps needed to recognize and solve each situation of the ambiguous case, where we are given two sides and an angle opposite one side.

SOLVING THE AMBIGUOUS CASE (SSA)

Given SSA, the following steps can be used to determine whether there are zero, one, or two triangles that satisfy the conditions. For simplicity, assume that a, b, and α are given.

STEP 1: Use the law of sines to find $\sin \beta$, where β is the unknown angle opposite b.

    No Solutions: If $\sin \beta > 1$, there are no possible triangles.

    One Solution: If $\sin \beta = 1$, then $\beta = 90°$. Find γ and c.

STEP 2: If $\sin \beta = k$, where $k < 1$, calculate $\beta_1 = \sin^{-1} k$ and $\beta_2 = 180° - \beta_1$. Note that β_1 and β_2 are the two *possible* solutions to $\sin \beta = k$.

    One Solution: If $\alpha + \beta_2 \geq 180°$ (or $a > b$), there is one triangle determined by β_1. Find γ and c.

    Two Solutions: If $\alpha + \beta_2 < 180°$ (or $a < b$), there are two triangles. Let $\gamma_1 = 180° - \alpha - \beta_1$ and $\gamma_2 = 180° - \alpha - \beta_2$. Find c_1 and c_2.

EXAMPLE 6 Solving the ambiguous case (no solutions)

Let $\alpha = 62°$, $a = 6$, and $b = 10$. If possible, solve the triangle.

SOLUTION

STEP 1: We begin by using the law of sines to find $\sin \beta$.

No Triangle Exists

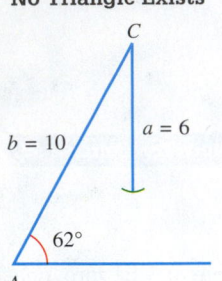

Figure 8.18

> Write the law of sines so that $\sin \beta$ is the only unknown.

$$\frac{\sin \beta}{b} = \frac{\sin \alpha}{a} \qquad \text{Law of sines}$$

$$\frac{\sin \beta}{10} = \frac{\sin 62°}{6} \qquad \text{Substitute.}$$

$$\sin \beta = \frac{10 \sin 62°}{6} \qquad \text{Multiply by 10.}$$

$$\sin \beta \approx 1.47 > 1 \qquad \text{Approximate } \sin \beta.$$

Since the sine function is never greater than 1, there are no solutions for β. No such triangle exists. See Figure 8.18.

Now Try Exercise 37

A Bridge Truss

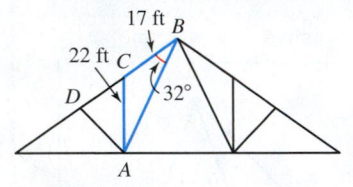

Figure 8.19

An Application from Construction Trusses are used in construction to support roofs, radio towers, bridges, and aircraft frames. Many times the smaller shapes within a truss are triangles. See Figure 8.19. In the next example, we determine that a particular truss design results in a unique truss. (*Source:* W. Riley, *Statics and Mechanics of Materials.*)

EXAMPLE 7 Solving the ambiguous case (one solution)

Suppose that an engineer has designed the truss in Figure 8.19 and specified that in triangle ABC, $BC = 17$ feet, $AC = 22$ feet, and angle $ABC = 32°$. Determine the length of AB to the nearest tenth of a foot. Is this value for AB unique?

SOLUTION

STEP 1: Triangle ABC from the truss in Figure 8.19 is shown in Figure 8.20. Using standard labeling, let $a = 17$, $b = 22$, and $\beta = 32°$. Start by finding $\sin \alpha$.

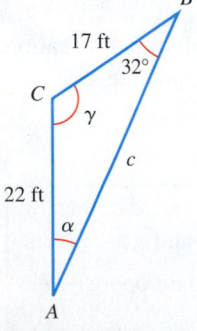

Figure 8.20

$$\frac{\sin \alpha}{a} = \frac{\sin \beta}{b} \qquad \text{Law of sines}$$

$$\frac{\sin \alpha}{17} = \frac{\sin 32°}{22} \qquad \text{Substitute.}$$

$$\sin \alpha = \frac{17 \sin 32°}{22} \qquad \text{Multiply by 17.}$$

$$\sin \alpha \approx 0.4095 \qquad \text{Approximate } \sin \alpha.$$

STEP 2: There are two values possible for angle α if $\sin \alpha \approx 0.4095$. Angle α lies in quadrant I or II with reference angle $\alpha_R \approx \sin^{-1}(0.4095) \approx 24.2°$. Thus

$$\alpha_1 \approx \mathbf{24.2°} \qquad \text{and} \qquad \alpha_2 \approx 180° - 24.2° = \mathbf{155.8°}.$$

However, if $\alpha_2 \approx 155.8°$, then $\alpha_2 + \beta = \mathbf{155.8°} + 32° \geq 180°$, which is not possible in a triangle. Therefore $\alpha_1 \approx \mathbf{24.2°}$ is the only possibility and

$$\gamma \approx 180° - \mathbf{24.2°} - 32° = \mathbf{123.8°}.$$

There is only one triangle possible.

The law of sines allows us to find AB, or side c.

$$\frac{c}{\sin \gamma} = \frac{b}{\sin \beta} \qquad \text{Law of sines}$$

$$\frac{c}{\sin 123.8°} \approx \frac{22}{\sin 32°} \qquad \text{Substitute.}$$

$$c \approx \frac{22 \sin \mathbf{123.8°}}{\sin 32°} \qquad \text{Multiply by } \sin 123.8°.$$

$$c \approx 34.5 \text{ ft} \qquad \text{Approximate } c.$$

Thus AB is about 34.5 feet long, and this value is unique.

Now Try Exercise 33

EXAMPLE 8 | Solving the ambiguous case (two solutions)

Let $\alpha = 55°$, $b = 8.5$, and $a = 7.3$. Solve the triangle. Round to the nearest tenth.

SOLUTION

STEP 1: Begin by finding $\sin \beta$.

$$\frac{\sin \beta}{b} = \frac{\sin \alpha}{a} \qquad \text{Law of sines}$$

$$\frac{\sin \beta}{8.5} = \frac{\sin 55°}{7.3} \qquad \text{Substitute.}$$

$$\sin \beta = \frac{8.5 \sin 55°}{7.3} \qquad \text{Multiply by 8.5.}$$

$$\sin \beta \approx 0.9538 \qquad \text{Approximate } \sin \beta.$$

First Triangle

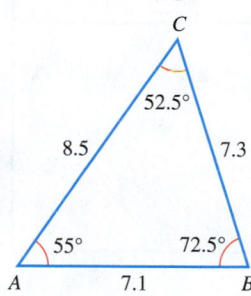

Figure 8.21 Solution 1

STEP 2: Two angles that satisfy $\sin \beta \approx 0.9538$ in quadrants I or II are

$$\beta_1 \approx \sin^{-1}(0.9538) \approx \mathbf{72.5°} \qquad \text{and} \qquad \beta_2 \approx 180° - 72.5° = \mathbf{107.5°}.$$

Both of these values for β are valid, since they do not result in the sum of the angles exceeding 180°. There are *two* solutions. (Note that $a < b$.)

Solution 1 Let $\beta_1 \approx \mathbf{72.5°}$. Then $\gamma_1 \approx 180° - \mathbf{72.5°} - 55° = \mathbf{52.5°}$. Side c_1 can then be found.

$$\frac{c_1}{\sin \gamma_1} = \frac{a}{\sin \alpha} \qquad \text{Law of sines}$$

$$c_1 \approx \frac{7.3 \sin \mathbf{52.5°}}{\sin 55°} \qquad \text{Substitute and solve for } c_1.$$

$$c_1 \approx 7.1 \qquad \text{Approximate } c_1.$$

A sketch of first triangle ABC is shown in Figure 8.21.

Solution 2 Let $\beta_2 \approx \mathbf{107.5°}$. Then $\gamma_2 \approx 180° - \mathbf{107.5°} - 55° = \mathbf{17.5°}$. Then side c_2 can be found.

Second Triangle

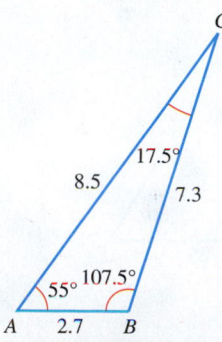

Figure 8.22 Solution 2

$$\frac{c_2}{\sin \gamma_2} = \frac{a}{\sin \alpha} \qquad \text{Law of sines}$$

$$c_2 \approx \frac{7.3 \sin \mathbf{17.5°}}{\sin 55°} \qquad \text{Substitute and solve for } c_2.$$

$$c_2 \approx 2.7 \qquad \text{Approximate } c_2.$$

A sketch of second triangle ABC is shown in Figure 8.22.

Now Try Exercise 29

8.1 Putting It All Together

The law of sines can be expressed as either

$$\frac{\sin \alpha}{a} = \frac{\sin \beta}{b} = \frac{\sin \gamma}{c} \quad \text{or} \quad \frac{a}{\sin \alpha} = \frac{b}{\sin \beta} = \frac{c}{\sin \gamma}.$$

When using the law of sines, a good strategy is to select an equation with the unknown variable in the numerator. We can use the law of sines to solve triangles when we are given ASA, AAS, or SSA. The cases ASA and AAS occur when two angles are given. In these cases a unique triangle is determined. The case where we are given SSA is called the ambiguous case, since there may be zero, one, or two triangles that satisfy the conditions. The ambiguous case is shown in the accompanying table.

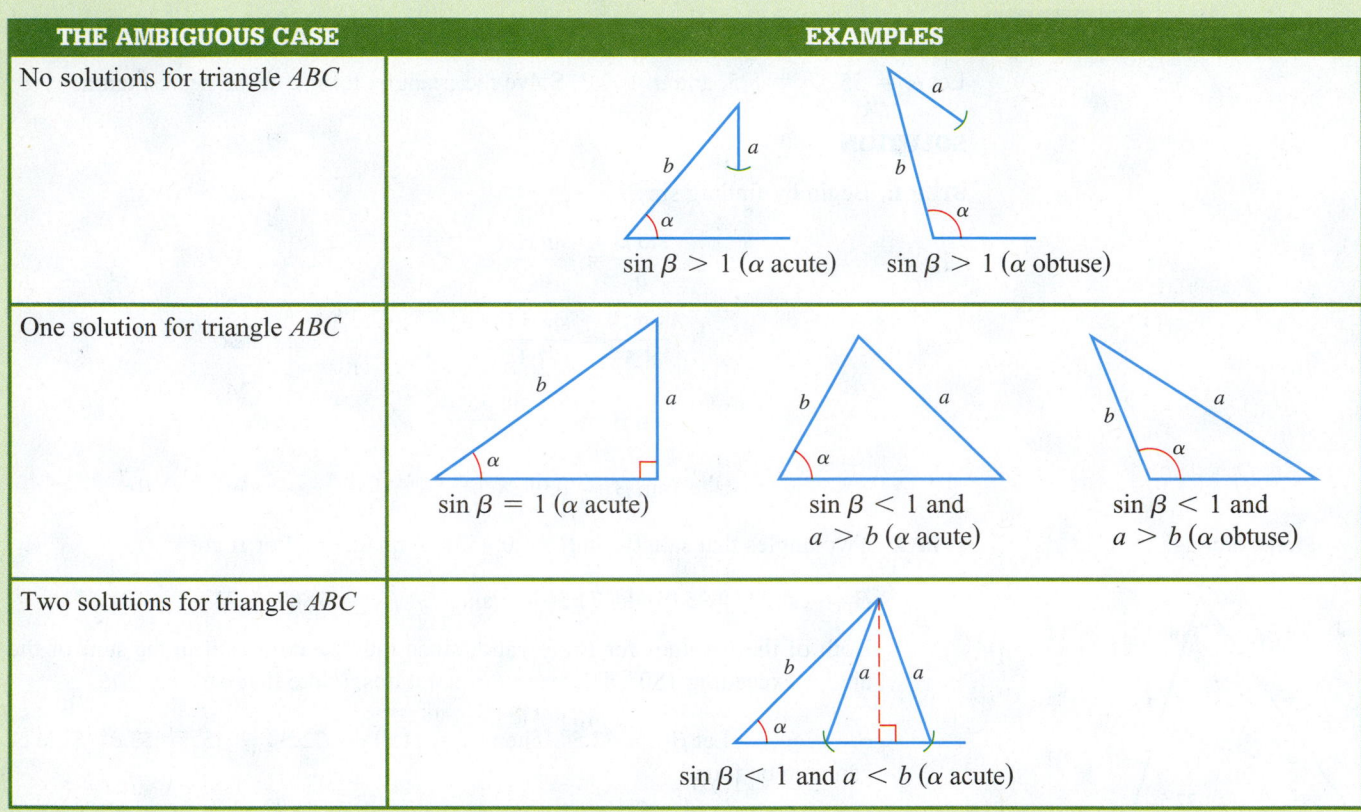

THE AMBIGUOUS CASE	EXAMPLES
No solutions for triangle *ABC*	$\sin \beta > 1$ (α acute) $\sin \beta > 1$ (α obtuse)
One solution for triangle *ABC*	$\sin \beta = 1$ (α acute) $\sin \beta < 1$ and $a > b$ (α acute) $\sin \beta < 1$ and $a > b$ (α obtuse)
Two solutions for triangle *ABC*	$\sin \beta < 1$ and $a < b$ (α acute)

8.1 Exercises

Solving for Unique Triangles

Exercises 1–4: Solve the triangle. Approximate values to the nearest tenth.

1.

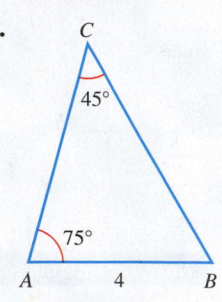

2.

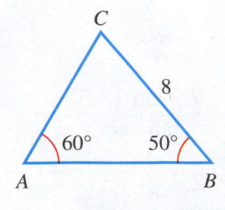

3.

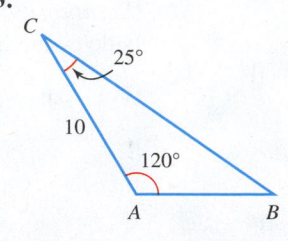

4.

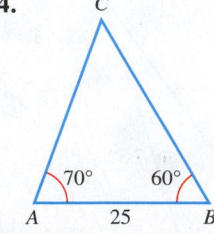

Exercises 5–12: Solve the triangle. Approximate values to the nearest tenth.

5. $\alpha = 40°, \beta = 60°, c = 10$

6. $\alpha = 35°, \beta = 75°, c = 8.1$

7. $\alpha = 25°, \gamma = 40°, a = 9.7$

8. $\beta = 36°, \gamma = 72°, b = 6$

9. $\beta = 40.2°, \gamma = 60.7°, a = 5.5$

10. $\alpha = 15.7°, \gamma = 23°, c = 7.2$

11. $\alpha = 27°, \gamma = 49°, b = 67$

12. $\beta = 39°, \gamma = 67°, a = 79$

Recognizing the Ambiguous Case (SSA)

Exercises 13–20: Let triangle ABC have standard labeling. Given the following angles and sides, decide if solving the triangle results in the ambiguous case.

13. $\alpha, \beta,$ and a

14. $\alpha, \gamma,$ and c

15. $a, b,$ and c

16. $\alpha, a,$ and b

17. $\beta, b,$ and c

18. $\alpha, b,$ and c

19. $\gamma, a,$ and c

20. $\beta, b,$ and α

Solving the Ambiguous Case

Exercises 21–26: Solve the triangle, if possible. Approximate values to the nearest tenth.

21.
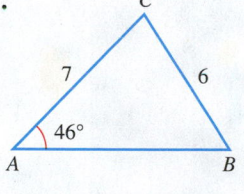

22.
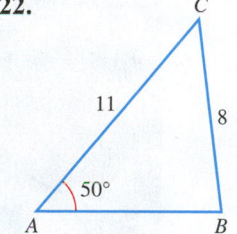

23.
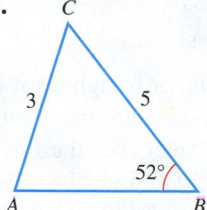

24.
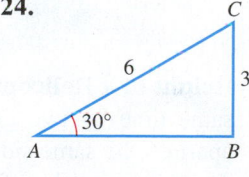

25.
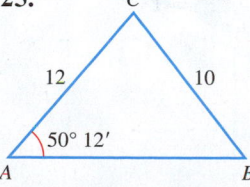

26.
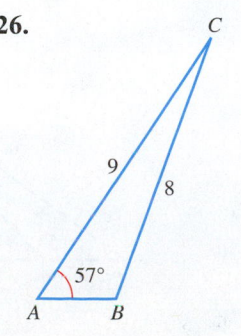

Solving Triangles

Exercises 27–44: Solve the triangle, if possible. Approximate values to the nearest tenth when appropriate.

27. $\alpha = 32°, \beta = 55°, b = 12$

28. $\beta = 20°, \gamma = 67°, c = 9$

29. $\alpha = 20°, b = 9, a = 7$

30. $\alpha = 20°, b = 7, a = 9$

31. $b = 10, \beta = 30°, c = 20$

32. $a = 13.5, \alpha = 46°, c = 27.8$

33. $\gamma = 102°, c = 51.6, a = 42.1$

34. $\beta = 43°, b = 22.1, c = 30.7$

35. $\alpha = 55.2°, \gamma = 114.8°, b = 19.5$

36. $c = 225, \alpha = 103.2°, \beta = 62.5°$

37. $b = 6.2, c = 7.4, \beta = 73°$

38. $\alpha = 45°, a = 5, b = 5\sqrt{2}$

39. $\alpha = 35°15', a = 5, b = 12$

40. $\gamma = 71°35', c = 6, b = 9$

41. $\beta = 46°45', a = 6, b = 5$

42. $\alpha = 54°12', c = 12, a = 10$

43. $\alpha = 56°30', \beta = 23°45', c = 100$

44. $\beta = 56°48', \gamma = 10°12', a = 55$

Applications

45. **Aerial Photography** (Refer to Example 3.) The plane shown in the figure is taking an aerial photograph with a camera lens that has an angular coverage of 70°. The ground below is inclined at 7°. If the angle of elevation of the plane at B is 52° and distance BC is 3500 feet, estimate the ground distance AB (to the nearest foot) that will appear in the picture. (*Source:* F. Moffit, *Photogrammetry.*)

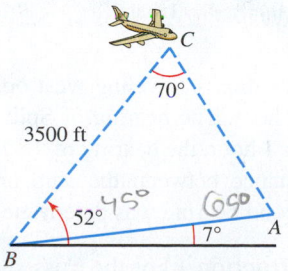

46. Aerial Photography As a picture is taken over level ground, the airplane's distance from a building located at the edge of the photograph is 7500 feet and the angle of depression to the building is 56°. See the figure. Find the ground distance *b* shown in the photograph to the nearest hundred feet.

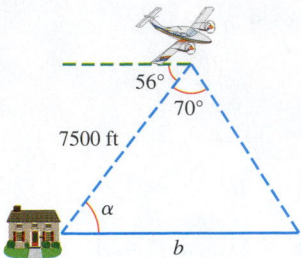

47. Distance to the Moon (Refer to Example 4 and Figure 8.13.) Suppose that the lunar angle at Bochum in upper Germany had been measured as $\alpha = 52.6901°$ instead of $52.6997°$. Determine the effect that this would have on the estimation of the distance to the moon. Interpret the result.

48. Distance to the Moon (Refer to Example 4 and Figure 8.13.) Suppose that the distance between two locations is 452.45 kilometers and that their angles of elevation to the moon are $\alpha = 47.8981°$ and $\theta = 47.9443°$. Estimate the distance to the moon (to the nearest thousand kilometers).

49. Locating a Ship The figure shows the bearings of a ship from two observation points located on a straight shoreline. The bearing from the first observation point is 54.3°, and the bearing from the second observation point is 325.2°. If the distance between these points is 15 miles, how far is it from the ship to shore? Assume that the first observation point is directly west of the second observation point.

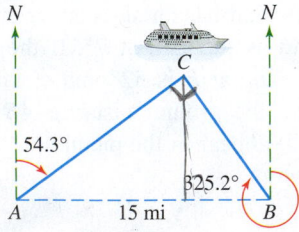

50. Distance A ship is traveling west on Lake Superior at 18 miles per hour. The bearing of Split Rock Lighthouse is 285°. After 1 hour, the bearing of the lighthouse is 340°. Find the distance between the ship and the lighthouse when the second bearing was determined.

51. Truss Construction For the truss shown at the top of the next column, *AB* is 24.2 feet, angle *ABD* is 118°, and angle *BDF* is 28°. Find the length of *BD*.

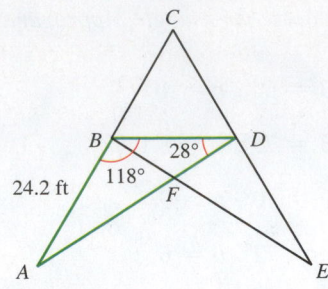

52. Truss Construction Use the results of Example 7 to solve triangle *ACD* in Figure 8.19 if angle *BAD* is 90°.

53. Distance Across a River To find the distance *AB* across a river, a distance *BC* = 354 meters is measured off on one side of the river. See the figure. It is found that angle *ABC* = 112°10′ and that angle *BCA* = 15°20′. Find the distance *AB*.

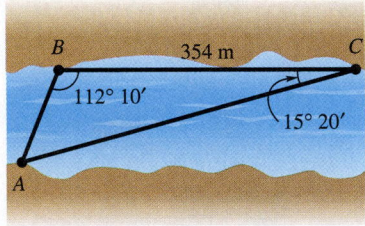

54. Distance Across a Canyon To find the distance *RS* across a canyon, a distance *TR* = 582 yards is measured off on one side of the canyon. See the figure. It is found that angle *TRS* = 102°20′ and that angle *RTS* = 32°50′. Find the distance *RS*.

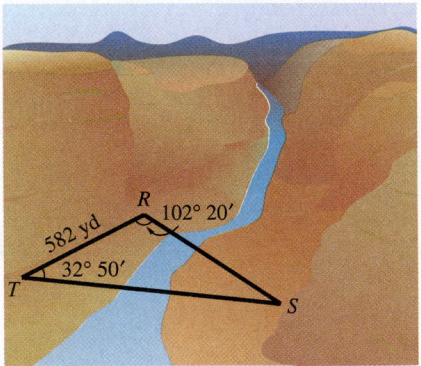

55. Height of a Helicopter A helicopter is sighted at the same time by two ground observers who are 3 miles apart on the same side of the helicopter. See the figure. They report angles of elevation of 20.5° and 27.8°. How high is the helicopter?

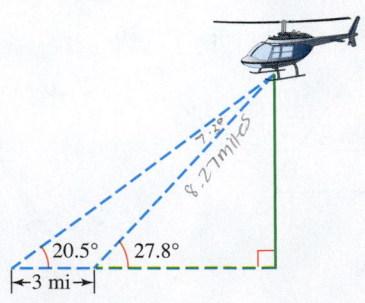

56. Height of a Hot-Air Balloon Two observation points *A* and *B* are 1500 feet apart. From these points the angles of elevation of a hot-air balloon are 43° and 47°, as illustrated in the figure. Find the height of the balloon to the nearest foot.

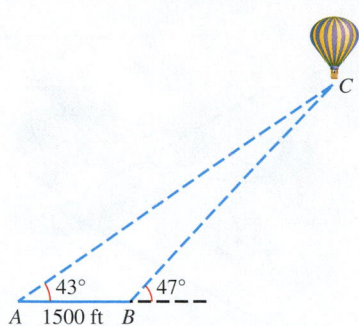

57. Height of a Balloon A balloonist is directly above a straight road 1.5 miles long that joins two towns, as illustrated in the figure. She finds that the town closer to her is at an angle of depression of 35° and the farther town is at an angle of depression of 31°. How high above the ground is the balloon?

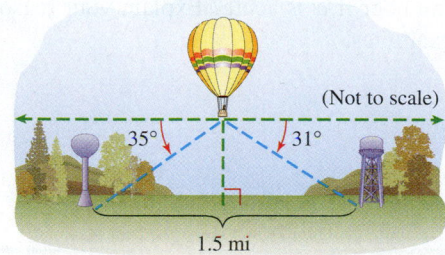

58. Highway Construction In a *reverse curve*, or S-curve, two circular curves are used to connect two straight portions of highway that are offset, as illustrated in the figure. Angles *α* and *β* will not be equal if the two straight portions of highway have different directions. Typically the same radius *r* is used for both portions of the reverse curve. If $r = 480$ feet, $\alpha = 38°$, $\beta = 15°$, and $\theta = 75°$, find the distance between *A* and *B* to the nearest foot. (**Source:** P. Kissam, *Surveying Practice.*)

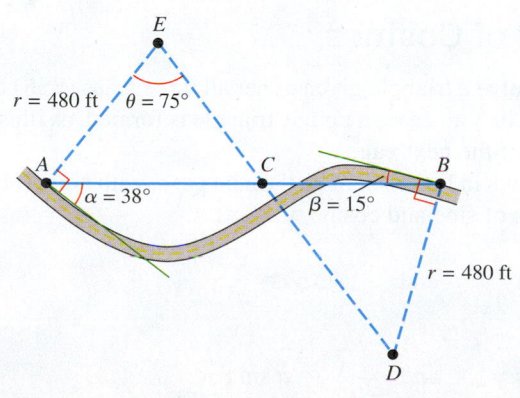

59. Surveying To find the distance between two points *A* and *B* on opposite sides of a small pond, a surveyor determines that *AC* is 97.3 feet, angle *ACB* is 55.1°, and angle *CAB* is 75.7°, as illustrated in the figure. Find the distance between *A* and *B*.

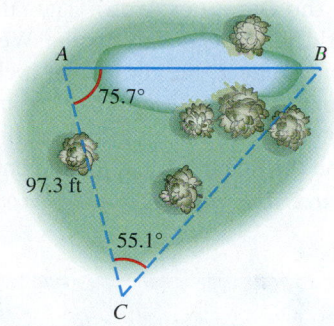

60. Trigonometric Leveling In surveying it is often necessary to determine the height of an inaccessible point *P*, as illustrated in the figure. Points *A*, *B*, and *C* lie on level ground. (**Source:** P. Kissam, *Surveying Practice.*)
(a) If angle *ABP* is 50°, angle *PAB* is 53.3°, and *AB* is 102 feet, find *PB*.

(b) If angle *PBC* is 47° and *C* is directly below *P*, find *PC*.

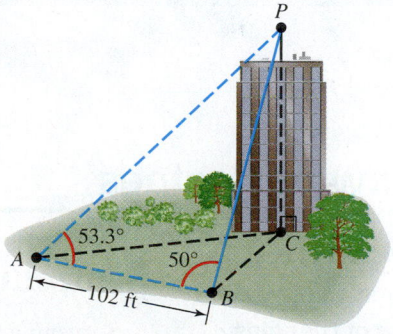

61. Locating a Ship From two observation points *A* and *B*, a sinking ship is spotted at point *C*. Angle *CAB* and angle *ABC* are measured as 28° and 60°, and the distance *AB* is about 4.12 miles, as illustrated in the figure.

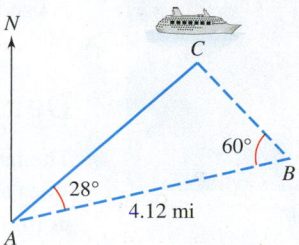

(a) How far is the ship from point *A*, to the nearest hundredth of a mile?

(b) If the coordinates of *A* are (0, 0) and the coordinates of *B* are (4, 1), find the bearing of the ship from point *A* to the nearest degree.

62. Airplane Navigation An airplane takes off with a bearing of 55° and flies 480 miles, after which it changes its course to a bearing of 285°. Finally the airplane flies back to its starting point with a bearing of 180°. Find the total mileage flown by the airplane.

63. Height of the Gateway Arch The tallest monument in the world is the Gateway to the West Arch in St. Louis. From point *A*, the top of the arch has an angle of elevation of 64.91°, and from point *B* the angle of elevation is 60.81°. See the figure. If distance *AB* is 57 feet, find the height of this monument to the nearest foot. (**Source:** *The Guinness Book of Records.*)

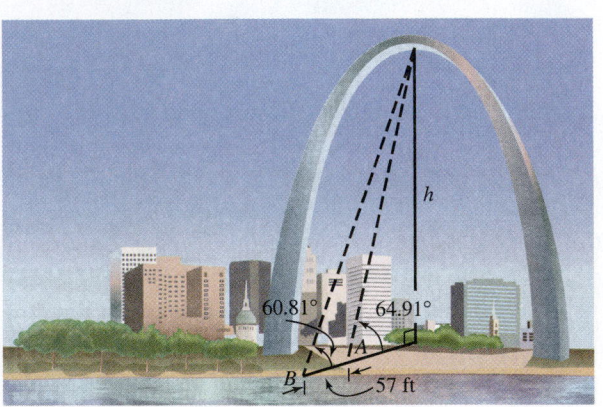

64. Height of a Tower A vertical tower supporting a cable for chairlifts to transport skiers up a mountain is located on a ski slope inclined at 28°, as illustrated in the figure. If the length of the tower's shadow is 21 feet along the mountain side when the angle of the sun is 57° with respect to the ski slope, calculate the height of the tower.

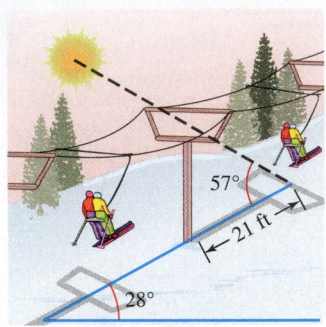

Writing about Mathematics

65. In your own words, describe two situations where the law of sines can be applied. Give an example of each situation.

66. Suppose that you are given α, a, and b for triangle *ABC*. If α is obtuse, what is the maximum number of triangles that could satisfy these conditions? What is the maximum number if α is acute? Explain your reasoning and give examples.

8.2 Law of Cosines

- Derive the law of cosines
- Solve triangles
- Find areas of triangles

Introduction

Surveying has been used for centuries in construction and in the determination of boundaries. Today the Global Positioning System (GPS) is being used to determine distances on Earth. The signal from a GPS satellite contains the information necessary for hand-held receivers to calculate both the position of a GPS satellite and its distance from the receiver. This information can be used to accurately calculate distances and angles between points on the ground. The law of cosines, which is a generalization of the Pythagorean theorem, is used in GPS calculations. In this section, the law of cosines is introduced and used to solve several applications. (**Source:** J. Van Sickle, *GPS for Land Surveyors.*)

Derivation of the Law of Cosines

The law of cosines can be used to solve a triangle given either all three sides (**SSS**) or two sides and the angle included (**SAS**). In both cases a unique triangle is formed, as illustrated in Figures 8.23 and 8.24 at the top of the next page.

Next consider triangle *ABC* shown in Figure 8.25 on the next page, with point *B* having coordinates (x, y). Using definitions of sine and cosine, we find

$$\cos \gamma = \frac{x}{a} \quad \text{and} \quad \sin \gamma = \frac{y}{a},$$

or equivalently,

$$x = a \cos \gamma \quad \text{and} \quad y = a \sin \gamma.$$

Law of Cosines Solves SSS and SAS

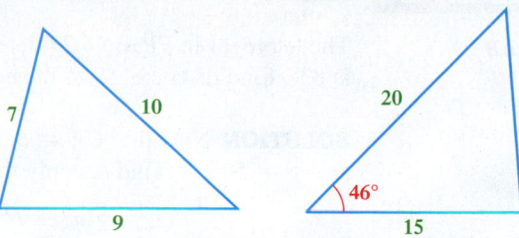

Figure 8.23 Given SSS

Figure 8.24 Given SAS

General Triangle

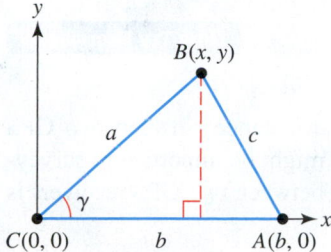

Figure 8.25

As a result, the coordinates of B are $(a \cos \gamma, a \sin \gamma)$. Since the coordinates of point A are $(b, 0)$, the distance c between points A and B can be found.

$$c = \sqrt{(a \cos \gamma - b)^2 + (a \sin \gamma - 0)^2} \quad \text{Distance formula}$$

$$c^2 = (a \cos \gamma - b)^2 + (a \sin \gamma - 0)^2 \quad \text{Square each side.}$$

$$c^2 = a^2 \cos^2 \gamma - 2ab \cos \gamma + b^2 + a^2 \sin^2 \gamma \quad \text{Expand each expression.}$$

$$c^2 = a^2 (\cos^2 \gamma + \sin^2 \gamma) - 2ab \cos \gamma + b^2 \quad \text{Distributive property}$$

$$c^2 = a^2 + b^2 - 2ab \cos \gamma \quad \cos^2 \gamma + \sin^2 \gamma = 1$$

This result is valid for any triangle ABC and is known as the *law of cosines*. Since the vertices in Figure 8.25 could be rearranged, three possible equations are associated with the law of cosines.

CLASS DISCUSSION

Let $\gamma = 90°$ in the formula

$$c^2 = a^2 + b^2 - 2ab \cos \gamma$$

and simplify. Discuss the relationship between the law of cosines and the Pythagorean theorem.

LAW OF COSINES

Any triangle with standard labeling satisfies

$$a^2 = b^2 + c^2 - 2bc \cos \alpha$$

$$b^2 = a^2 + c^2 - 2ac \cos \beta$$

$$c^2 = a^2 + b^2 - 2ab \cos \gamma.$$

Solving Triangles

In the first example we use the law of cosines to find the missing side, given SAS.

EXAMPLE 1 **Applying the law of cosines (SAS)**

Find the missing side in the triangle shown in Figure 8.26.

Figure 8.26

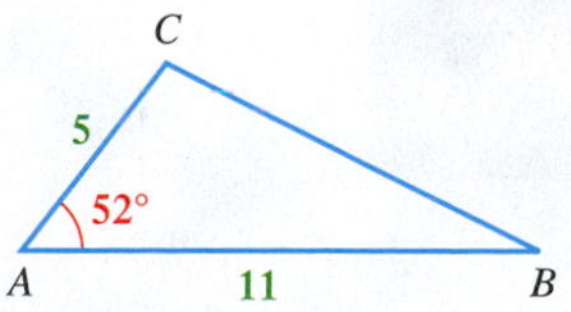

SOLUTION We are given SAS, so let $\alpha = 52°$, $b = 5$, and $c = 11$. The unknown side is side a. We will use the first equation for the law of cosines.

$$a^2 = b^2 + c^2 - 2bc \cos \alpha \quad \text{Law of cosines}$$

$$= 5^2 + 11^2 - 2(5)(11) \cos 52° \quad \text{Substitute.}$$

$$\approx 78.277 \quad \text{Approximate.}$$

Thus $a \approx \sqrt{78.277} \approx 8.85$.

Now Try Exercise 9

Surveying A common problem in surveying is to find the distance between two points A and B situated on opposite sides of a building, as illustrated in Figure 8.27 on the next page. This distance can be found by applying the law of cosines.

EXAMPLE 2 Finding the distance between two points (SAS)

The surveyor in Figure 8.27 determines that CA is 75 feet, CB is 58 feet, and angle ACB is 83°. Find distance AB to the nearest foot.

Finding Distance _AB_

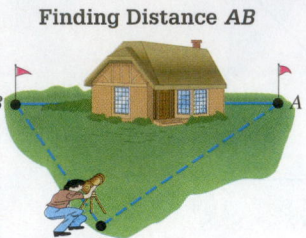

Figure 8.27

SOLUTION Note that $CA = b$, $CB = a$, and angle $ACB = \gamma$. So let $b = 75$, $a = 58$, and $\gamma = 83°$. To find c, apply the law of cosines.

$$
\begin{aligned}
c^2 &= a^2 + b^2 - 2ab \cos \gamma && \text{Law of cosines} \\
&= (58)^2 + (75)^2 - 2(58)(75) \cos 83° && \text{Substitute.} \\
&\approx 7929 && \text{Approximate.} \\
c &\approx 89.04 && \text{Take the square root.}
\end{aligned}
$$

The points A and B are about 89 feet apart.

Now Try Exercise 53

Global Positioning System In the next example, the distance between two GPS receivers is found. Finding the distance between two points might be important to surveyors or to search parties looking for lost hikers. The distance between two GPS receivers is sometimes called the *baseline*.

EXAMPLE 3 Using GPS to find a baseline distance (SAS)

A search party and an injured hiker both have hand-held GPS receivers, as illustrated in Figure 8.28. The distance from the satellite to the search party is $b = 20{,}231.15$ kilometers, and the distance from the satellite to the hiker is $c = 20{,}231.57$ kilometers. If it is determined that $\alpha = 0.01456°$, estimate the baseline a between the search party and the hiker to the nearest hundredth of a kilometer.

Locating an Injured Hiker

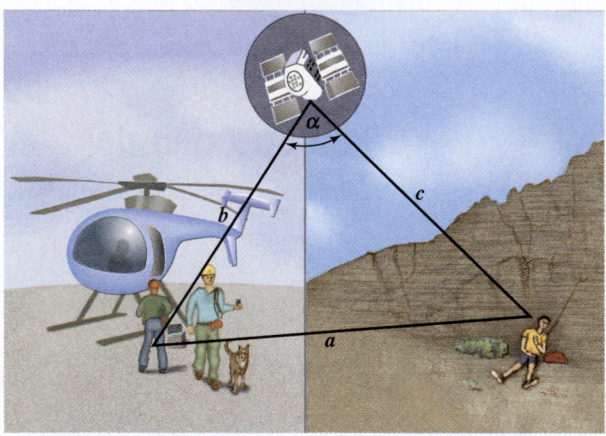

Figure 8.28 (Not to scale)

SOLUTION We can use the law of cosines to find a.

$$
\begin{aligned}
a^2 &= b^2 + c^2 - 2bc \cos \alpha \\
&= (20{,}231.15)^2 + (20{,}231.57)^2 - 2(20{,}231.15)(20{,}231.57) \cos 0.01456° \\
&\approx 26.61 \\
a &\approx 5.16
\end{aligned}
$$

The distance between the search party and the hiker is about 5.16 kilometers.

Now Try Exercise 55

In the next example we use the law of cosines to find a missing angle, given SSS.

EXAMPLE 4 Applying the law of cosines (SSS)

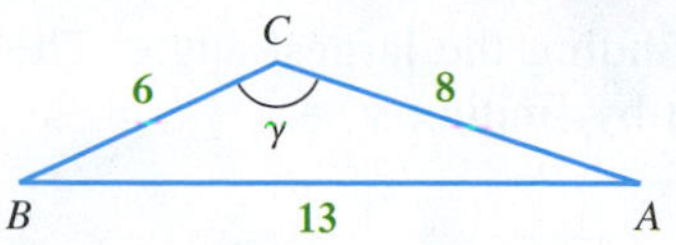

Find angle γ in the triangle shown in Figure 8.29.

Figure 8.29

SOLUTION We are given SSS, so let $a = 6$, $b = 8$, and and $c = 13$. Angle γ is opposite side c. We use the third equation for law of cosines that contains **cos γ** and solve for **cos γ**.

$$c^2 = a^2 + b^2 - 2ab\cos\gamma \qquad \text{Law of cosines}$$

$$2ab\cos\gamma = a^2 + b^2 - c^2 \qquad \text{Rewrite the equation.}$$

$$\cos\gamma = \frac{a^2 + b^2 - c^2}{2ab} \qquad \text{Divide by 2ab.}$$

$$\cos\gamma = \frac{6^2 + 8^2 - 13^2}{2(6)(8)} \qquad \text{Substitute.}$$

$$\cos\gamma = -0.71875 \qquad \text{Simplify.}$$

Thus $\gamma = \cos^{-1}(-0.71875) \approx 136.0°$.

Now Try Exercise 11

An Application from Construction Trusses are frequently used to support roofs on buildings, as illustrated in Figure 8.30. The simplest type of roof truss is a triangle, as shown in Figure 8.31. One basic task when constructing a roof truss is to cut the ends of the rafters so that the roof has the correct slope. (*Source:* W. Riley, *Statics and Mechanics of Materials.*)

Designing a Triangular Roof Truss

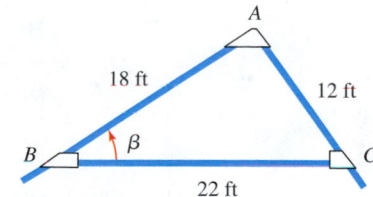

Figure 8.30

Figure 8.31

EXAMPLE 5 Designing a roof truss (SSS)

Find β to the nearest degree for the truss shown in Figure 8.31.

SOLUTION Begin by letting $a = 22$, $b = 12$, and $c = 18$, and then use the law of cosines to find β.

$$b^2 = a^2 + c^2 - 2ac\cos\beta \qquad \text{Law of cosines}$$

$$2ac\cos\beta = a^2 + c^2 - b^2 \qquad \text{Rewrite the equation.}$$

$$\cos\beta = \frac{a^2 + c^2 - b^2}{2ac} \qquad \text{Divide by 2ac.}$$

$$\cos\beta = \frac{22^2 + 18^2 - 12^2}{2(22)(18)} \qquad \text{Substitute.}$$

$$\cos\beta \approx 0.8384 \qquad \text{Approximate.}$$

Thus $\beta \approx \cos^{-1}(0.8384) \approx 33°$.

Now Try Exercise 63

EXAMPLE 6   **Solving a triangle (SSS)**

Find α, β, and γ in triangle ABC to the nearest tenth of a degree if $a = 5$, $b = 6$, and $c = 9$.

SOLUTION

Getting Started When solving SSS, always start by finding the largest angle. The largest angle, γ, is opposite the largest side, c. We start by finding γ. ▶

$$c^2 = a^2 + b^2 - 2ab\cos\gamma \qquad \text{Law of cosines}$$

$$2ab\cos\gamma = a^2 + b^2 - c^2 \qquad \text{Transpose terms.}$$

$$\cos\gamma = \frac{a^2 + b^2 - c^2}{2ab} \qquad \text{Divide by 2ab.}$$

$$\cos\gamma = \frac{5^2 + 6^2 - 9^2}{2(5)(6)} \qquad \text{Let } a = 5, b = 6, \text{ and } c = 9.$$

$$\cos\gamma = -\frac{1}{3} \qquad \text{Simplify.}$$

Thus $\gamma \approx \cos^{-1}\left(-\frac{1}{3}\right) \approx 109.5°$. The law of cosines could be used again to find either α or β. However, we use the law of sines to find α.

$$\frac{\sin\alpha}{a} = \frac{\sin\gamma}{c} \qquad \text{Law of sines}$$

$$\sin\alpha = \frac{a\sin\gamma}{c} \qquad \text{Multiply by a.}$$

$$\sin\alpha \approx \frac{5\sin 109.5°}{9} \qquad \text{Let } a = 5, c = 9, \text{ and } \gamma \approx 109.5°.$$

$$\sin\alpha \approx 0.5237 \qquad \text{Simplify.}$$

Because γ is the largest angle, it follows that α must be an acute angle and that $\alpha \approx \sin^{-1}(0.5237) \approx 31.6°$. To find β we use the fact that the measures of the angles sum to $180°$ in a triangle.

$$\beta \approx 180° - 109.5° - 31.6° = 38.9°$$

Now Try Exercise 21

NOTE   In solving the case SSS, first find the largest angle, which is opposite the longest side. There can be at most one obtuse angle in a triangle, and the law of cosines finds an obtuse angle, if it exists, because the cosine of an obtuse angle is negative, as was the case in Example 6.

See the Concept: Applying the Law of Cosines

SAS

SSS

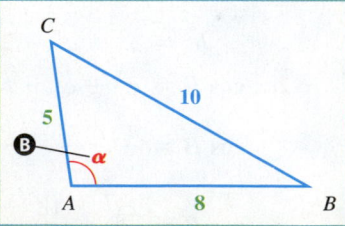

Ⓐ SAS Given two sides and the angle between, find the side opposite the given angle.

$$a^2 = b^2 + c^2 - 2bc\cos\alpha$$

Ⓑ SSS Given three sides, first find the angle opposite the *longest* side.

$$\cos\alpha = \frac{b^2 + c^2 - a^2}{2bc}$$

Ⓐ Find a:

$$a^2 = 5^2 + 8^2 - 2(5)(8)\cos 60°,$$

so $a = \sqrt{49} = 7$.

Ⓑ Find α:

$$\cos\alpha = \frac{5^2 + 8^2 - 10^2}{2(5)(8)},$$

so $\alpha = \cos^{-1}(-0.1375) \approx 97.9°$.

Area Formulas

One task that is frequently performed by surveyors is to find the acreage of a lot using a technique called triangulation. *Triangulation* divides a parcel of land into triangles. The area of the lot equals the sum of the areas of the triangles. We begin our discussion by developing some area formulas for triangles.

The area K of any triangle is given by $K = \frac{1}{2}bh$, where b is its base and h is its height. Using trigonometry, we can find a formula for the area of the triangle shown in Figure 8.32.

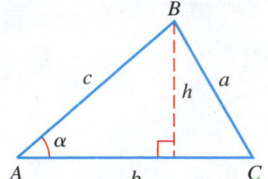

Figure 8.32

$$\sin \alpha = \frac{h}{c}, \quad \text{or} \quad h = c \sin \alpha$$

Thus the area equals

$$K = \frac{1}{2}bh = \frac{1}{2}bc \sin \alpha.$$

Since the labels for the vertices in triangle ABC could be rearranged, three area formulas can be written as follows. Notice that these formulas can be applied when we are given SAS. (We use K for area rather than A so as not to cause confusion with vertex A.)

> **AREA OF A TRIANGLE**
>
> For any triangle with standard labeling, the area K is given by
>
> $$K = \frac{1}{2}ab \sin \gamma, \qquad K = \frac{1}{2}ac \sin \beta, \qquad \text{or} \qquad K = \frac{1}{2}bc \sin \alpha.$$

EXAMPLE 7 Finding the area of a triangle (SAS)

Find the area of triangle ABC in Figure 8.33 to the nearest square foot.

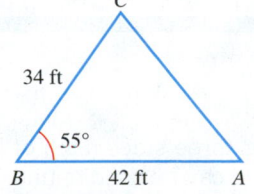

Figure 8.33

SOLUTION We are given $\beta = 55°$, $a = 34$ feet, and $c = 42$ feet. Thus the area K is given by the following.

$$K = \frac{1}{2}ac \sin \beta = \frac{1}{2}(34)(42) \sin 55° \approx 585 \text{ square feet}$$

Now Try Exercise 37

Heron's Formula The next formula can be used to find the area of a triangle when the lengths of three sides are known. It is named after the Greek mathematician Heron.

> **HERON'S FORMULA**
>
> If a triangle has sides with lengths a, b, and c, then its area K is given by
>
> $$K = \sqrt{s(s - a)(s - b)(s - c)},$$
>
> where $s = \frac{1}{2}(a + b + c)$ and s is called the **semiperimeter**.

EXAMPLE 8 Finding the area of a triangle (SSS)

Approximate the area of triangle ABC with sides $a = 4$, $b = 5$, and $c = 7$.

SOLUTION

Getting Started Begin by calculating $s = \frac{1}{2}(4 + 5 + 7) = 8$. ▶

Then the area is

$$K = \sqrt{8(8 - 4)(8 - 5)(8 - 7)} = \sqrt{96} \approx 9.8.$$

Now Try Exercise 39

An Application from Surveying One method for finding the area of a lot is called the *distance method*. This method can be used to find the area of an irregular lot, as illustrated in the next example. The distance method does not measure angles—it measures only distances between points. Triangulation and Heron's formula can be used to find the area of the lot. (***Source:*** I. Mueller and K. Ramsayer, *Introduction to Surveying*.)

EXAMPLE 9 Applying the distance method to find the area of a lot

Find the area of the parcel of land determined by *ABCDE* in Figure 8.34.

SOLUTION For triangle *ABE*, $s = \frac{1}{2}(60.5 + 68.4 + 61.7) = 95.3$, and its area is

$$K_1 = \sqrt{95.3(95.3 - 60.5)(95.3 - 68.4)(95.3 - 61.7)} \approx 1731.$$

For triangle *BCE*, $s = \frac{1}{2}(78.9 + 108.2 + 68.4) = 127.75$, and its area is

$$K_2 = \sqrt{127.75(127.75 - 78.9)(127.75 - 108.2)(127.75 - 68.4)} \approx 2691.$$

For triangle *CDE*, $s = \frac{1}{2}(68.4 + 52.3 + 108.2) = 114.45$, and its area is

$$K_3 = \sqrt{114.45(114.45 - 68.4)(114.45 - 52.3)(114.45 - 108.2)} \approx 1431.$$

The area of the lot is

$$K_1 + K_2 + K_3 \approx 1731 + 2691 + 1431 = 5853 \text{ square feet.}$$

Now Try Exercise 77

Finding Area of a Lot

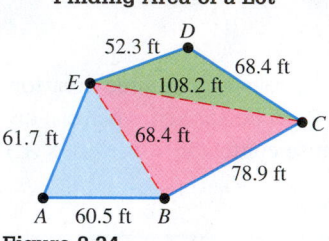

Figure 8.34

8.2 Putting It All Together

The law of cosines can be used to solve triangles when either three sides (SSS) or two sides and the angle included (SAS) are given. Heron's formula can be used to find the area of a triangle given SSS.

CONCEPT	EXPLANATION	EXAMPLES
Law of cosines	$a^2 = b^2 + c^2 - 2bc \cos \alpha$ $b^2 = a^2 + c^2 - 2ac \cos \beta$ $c^2 = a^2 + b^2 - 2ab \cos \gamma$ Can be used to solve a triangle given either SSS or SAS	If $b = 3$, $c = 4$, and $\alpha = 60°$, then $$a^2 = 3^2 + 4^2 - 2(3)(4) \cos 60° = 13$$ and $a = \sqrt{13}$.
Area formulas (SAS)	$K = \frac{1}{2}ab \sin \gamma$ $K = \frac{1}{2}ac \sin \beta$ $K = \frac{1}{2}bc \sin \alpha$ Can be used to find the area of a triangle given SAS	If $a = 2$ feet, $b = 3$ feet, and $\gamma = 30°$, then the area of the triangle is $$K = \frac{1}{2}(2)(3) \sin 30° = 1.5 \text{ ft}^2.$$

CONCEPT	EXPLANATION	EXAMPLES
Heron's formula (SSS)	$K = \sqrt{s(s-a)(s-b)(s-c)}$, where $s = \frac{1}{2}(a+b+c)$ Can be used to find the area of a triangle given SSS	If $a = 3$ feet, $b = 5$ feet, and $c = 4$ feet, then $$s = \frac{1}{2}(3+5+4) = 6,$$ and the area of the triangle is $$K = \sqrt{6(6-3)(6-5)(6-4)} = 6 \text{ ft}^2.$$

8.2 Exercises

Determining a Method to Solve a Triangle

Exercises 1–8: Assume triangle ABC has standard labeling and complete the following.

(a) *Determine if AAS, ASA, SSA, SAS, or SSS is given.*
(b) *Decide if the law of sines or the law of cosines should be used first to solve the triangle.*

1. a, b, and γ

2. α, γ, and c

3. a, b, and α

4. a, b, and c

5. α, β, and c

6. a, c, and α

7. β, a, and γ

8. b, c, and α

Solving Triangles

Exercises 9 and 10: Find the length of the remaining side of each triangle. Approximate to the nearest tenth when appropriate.

9.
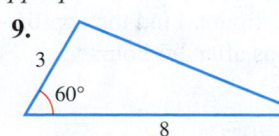

10.
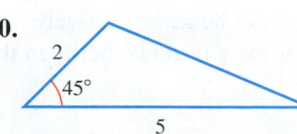

Exercises 11 and 12: Find the value of θ in each triangle. Approximate to the nearest tenth when appropriate.

11.
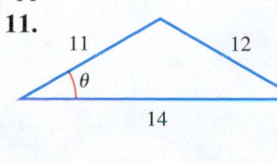

12.
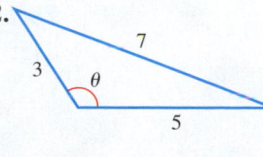

Exercises 13–18: Solve the triangle. Approximate values to the nearest tenth.

13.
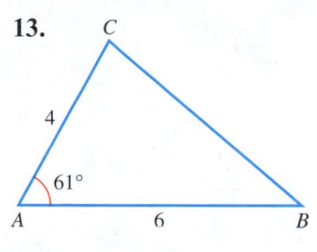

14.
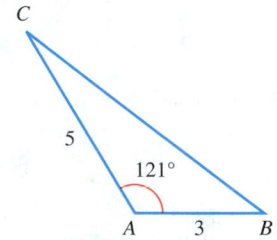

15.
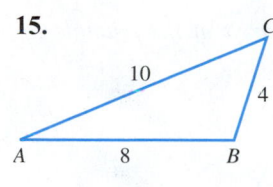

16.
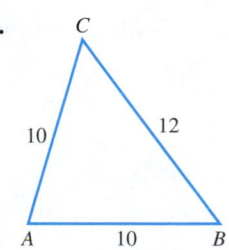

17.
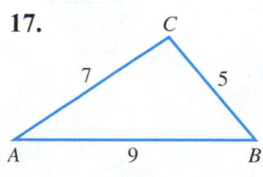

18.
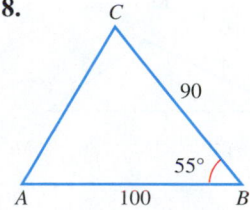

Exercises 19–30: Solve the triangle. Round values to the nearest tenth when appropriate.

19. $a = 45$, $\gamma = 35°$, $b = 24$

20. $c = 7.9$, $\beta = 52°$, $a = 9.6$

21. $a = 2.4, b = 1.7, c = 1.4$

22. $a = 43, b = 41, c = 34$

23. $\alpha = 10°30', b = 24.1, c = 15.8$

24. $a = 12.8, b = 15.8, \gamma = 36°$

25. $a = 10.6, b = 25.8, c = 20.6$

26. $a = 104, b = 121, c = 111$

27. $\beta = 122°10', a = 20, c = 15$

28. $b = 9.1, \alpha = 43°30', c = 12.5$

29. $a = 5.3, b = 6.7, c = 7.1$

30. $a = 4.2, b = 5.1, c = 3.7$

Does This Triangle Exist?

Exercises 31–36: Decide if a triangle exists that satisfies the conditions. Justify your answer.

31. $a = 10, b = 12, c = 25$

32. $a = 10, \beta = 51°, c = 5$

33. $\alpha = 89°, b = 63, \gamma = 112°$

34. $a = 2, b = 10, \alpha = 50°$

35. $\gamma = 54°, b = 63, \alpha = 63°$

36. $a = 5, b = 6, c = 8$

Area of Triangles

Exercises 37–40: Approximate the area of the triangle to the nearest tenth.

37.

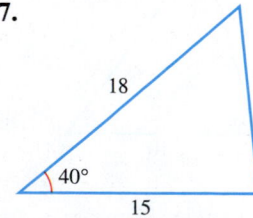

38.

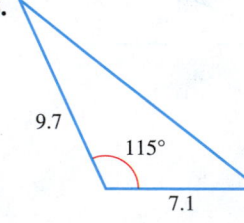

39.

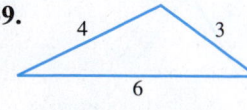

40.

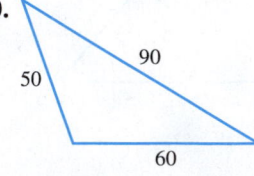

Exercises 41–52: Approximate the area of the triangle to the nearest tenth.

41. $a = 10, b = 12, \gamma = 58°$

42. $\alpha = 40°, b = 5.8, c = 8.8$

43. $\beta = 78°, a = 5.5, c = 6.8$

44. $\alpha = 23°, \gamma = 47°, b = 53$

45. $\beta = 31°, \alpha = 54°, a = 2.6$

46. $a = 7, b = 8, c = 9$

47. $a = 5.5, b = 6.7, c = 9.2$

48. $a = 104, b = 98, c = 112$

49. $a = 11, b = 13, c = 20$

50. $a = 13, b = 14, c = 15$

51. $a = 21, \alpha = 42°, c = 16$

52. $b = 35, c = 38, \gamma = 50°48'$

Applications

53. Obstructed View (Refer to Example 2.) In the figure, a surveyor is attempting to find the distance between points A and B. A grove of trees is obstructing the view, so the surveyor determines that AC is 143 feet, BC is 123 feet, and angle ACB is $78°35'$. Find the distance between A and B to the nearest foot.

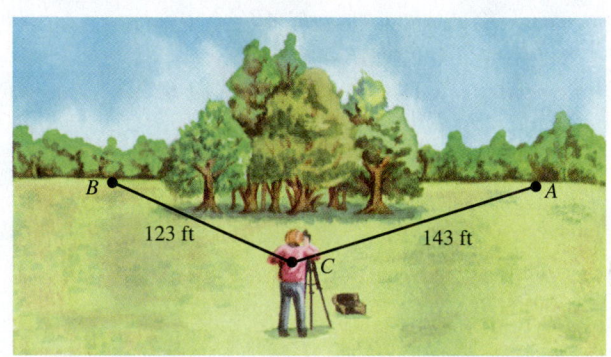

54. Distance Across a Lake Points A and B are on opposite sides of a lake. From a third point, C, the angle between the lines of sight to A and B is $46.3°$. If AC is 350 meters and BC is 286 meters, find AB.

55. Ship Navigation Two ships set sail with bearings of $52°$ and $121°$, traveling at 20 miles per hour and 14 miles per hour, respectively. See the figure. Find the approximate distance between the ships after 1.5 hours.

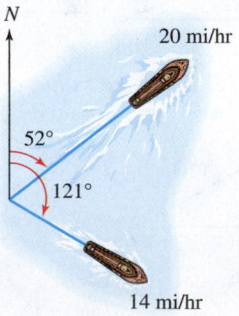

56. Distance Between Two Boats   Two boats leave a dock together. Each travels in a straight line. The angle between their courses measures 54°10′. One boat travels 36.2 kilometers per hour and the other 45.6 kilometers per hour. How far apart will they be after 3 hours?

57. Diagonals of a Parallelogram   One side of a parallelogram is 3.5 feet and another side is 5.2 feet. The angle between these two sides is 56°. Find the lengths of the diagonals to the nearest tenth of a foot.

58. Diagonals of a Parallelogram   The sides of a parallelogram are 4.0 centimeters and 6.0 centimeters. One angle is 58° while another is 122°. Find the lengths of the diagonals.

59. Air Navigation   An airplane flies in the triangular course shown in the figure. To the nearest degree, find the bearings of the plane while traveling from A to B and from B to C.

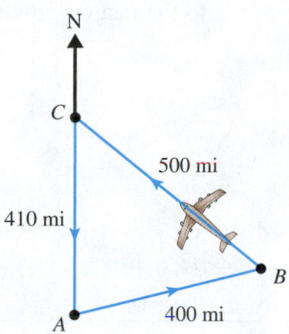

60. Flight Distance   Airports A and B are 450 kilometers apart, on an east-west line. A pilot flies in a northeast direction from A to airport C. From C, the pilot flies 359 kilometers on a bearing of 128°40′ to B. How far is C from A?

61. Area of a Lot   A surveyor measures the sides of a triangular lot to be $a = 145.2$, $b = 136.8$, and $c = 95.3$, where measurements are in feet.
(a) Approximate angles α, β, and γ.

(b) What is the area of the lot to the nearest square foot?

62. Angle in a Parallelogram   One side of a parallelogram is 6.4 yards and another side is 5.3 yards. The shorter diagonal is 3.5 yards. Find the angle opposite the shorter diagonal to the nearest tenth of a degree.

63. Truss Construction   (Refer to Example 5.) Find angle θ for the triangular truss shown in the figure.

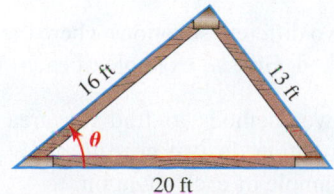

64. Robotics   The figure illustrates the MIT Scheinman robotic arm. Suppose the length of the upper arm is 20 centimeters and the combined length of the forearm and hand is 30 centimeters. If the arm is positioned so that $\theta = 126°$, find the distance between the hand at point A and the shoulder joint at point B.   (***Source:*** G. Beni and S. Hackwood, *Recent Advances in Robotics.*)

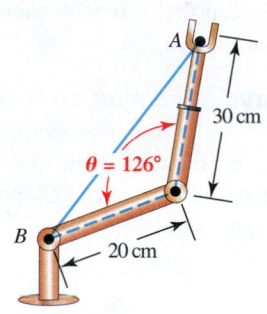

65. Distance Between Airports   Airports A and B are 515 miles apart, and airport A is directly west of airport B. Airport C is located in a northeasterly direction from airport A and is 357 miles from airport B. See the figure. If the bearing from airport C to airport B is 125°, find the distance between airports A and C to the nearest mile.

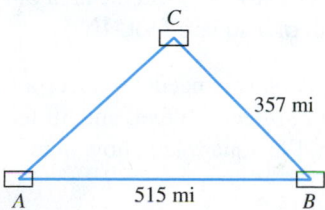

66. Navigation   A ship is sailing east. At one moment, the bearing of a submerged rock is 38°45′. After the ship sails 20.4 miles, the bearing of the rock is 291°15′. Find the distance between the ship and the rock (to the nearest tenth of a mile) when the second bearing is taken.

67. Distance Between a Ship and a Submarine   From an airplane flying over the ocean, the angle of depression to a submarine lying just under the surface is 24°10′. At the same moment, the angle of depression from the airplane to a battleship is 17°30′. See the figure. The distance from the airplane to the battleship is 5120 feet. Find the distance between the battleship and the submarine. (Assume the airplane, submarine, and battleship are in a vertical plane.)

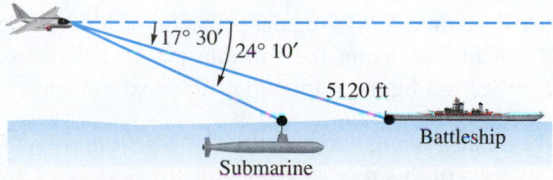

68. Distance Between a Ship and a Rock A ship is sailing east. At one point, the bearing of a submerged rock is 45°20′. After the ship has sailed 15.2 miles, the bearing of the rock has become 308°40′. Find the distance of the ship from the rock at the latter point.

69. Distance Between Two Ships Two ships leave a harbor together, traveling on courses that have an angle of 135°40′ between them. If they each travel 402 miles, how far apart are they?

70. Highway Curve The most common highway curve consists of a circular arc connecting two sections of straight road, as illustrated in the figure. Find the straight-line distance between *PC* (point of curve) and *PT* (point of tangency). (*Source:* P. Kissam, *Surveying Practice.*)

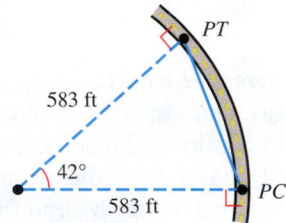

71. Painting A painter needs to cover a triangular region with sides of 25 feet and 15 feet. The angle between these two sides is 128°. Find the area of the region to the nearest tenth of a square foot.

72. Painting A painter needs to cover a triangular region with sides of 30 feet, 40 feet, and 38 feet. If each can of paint covers 125 square feet, how many cans of paint are needed?

73. Area of Regular Polygons If a regular polygon has *n* sides of equal length *L*, then its area *A* is computed by

$$A = \frac{nL^2}{4} \cot \frac{\pi}{n}.$$

(*Source:* M. Mortenson, *Computer Graphics.*)
(a) Using this formula, find the area of an equilateral triangle with sides of 6 inches.

(b) Using Heron's formula, find the area of this triangle. Compare answers.

74. Area of Regular Polygons (Refer to Exercise 73.) The measure of an interior angle in a regular polygon with *n* sides is given by $180°\left(1 - \frac{2}{n}\right)$. For example, a square is a regular polygon with $n = 4$ and each interior angle equals $180°\left(1 - \frac{2}{4}\right) = 90°$.
(a) Find the area of a regular pentagon with sides of length 8 inches, using the formula given in Exercise 73. See the figure at the top of the next column.

(b) Using triangulation and Heron's formula, find the area of this regular pentagon.

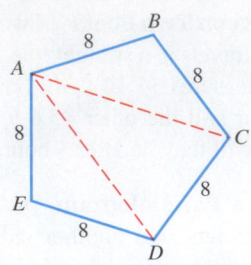

75. Area of a Lot Find the area of the lot in the figure to the nearest square foot.

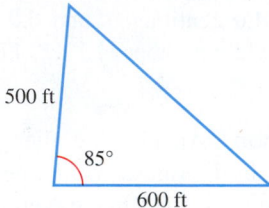

76. Area of a Lot Find the area of the quadrangular lot shown in the figure to the nearest square foot.

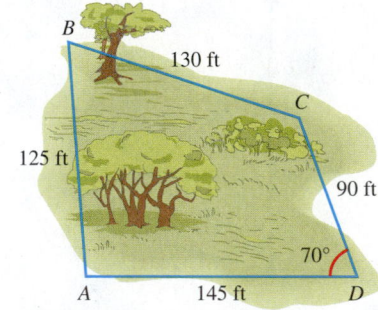

77. Area of a Lot Apply the distance method discussed in Example 9 to find the area of the lot in the figure to the nearest square foot.

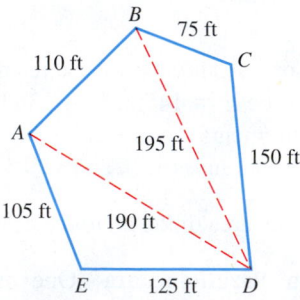

78. Area of the Bermuda Triangle Find the area of the Bermuda Triangle (to the nearest thousand square miles) if the sides of the triangle have approximate lengths of 850 miles, 925 miles, and 1300 miles.

Writing about Mathematics

79. Describe two different situations where the law of cosines can be applied. Give an example of each situation.

80. Describe two methods to find the area of a triangle. What information do you need to apply each method? Give an example of each situation.

Extended and Discovery Exercise

1. **Distance Between a Satellite and a Tracking Station**
A satellite traveling in a circular orbit 1600 kilometers above Earth is due to pass directly over a tracking station at noon. See the figure. Assume that the satellite takes 2 hours to make an orbit and that the radius of Earth is 6400 kilometers. Find the distance between the satellite and the tracking station at 12:03 P.M.   (*Source:* NASA.)

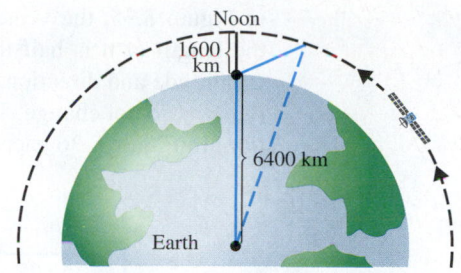

CHECKING BASIC CONCEPTS FOR SECTIONS 8.1 AND 8.2

1. Solve triangle ABC if $\alpha = 44°$, $\gamma = 62°$, and $a = 12$.

2. Solve triangle ABC if $\alpha = 32°$, $a = 6$, and $b = 8$. How many solutions are there?

3. Use the law of cosines to solve the triangles.
   (a)
   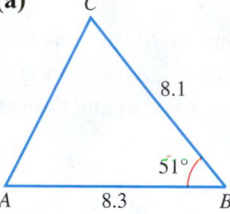

   (b)
   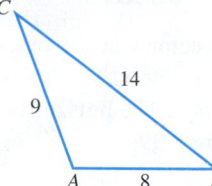

4. Find the area of triangle ABC to the nearest hundredth.
   (a) $a = 4.5$, $b = 5.2$, $\gamma = 55°$

   (b) $a = 6$, $b = 7$, $c = 9$

8.3 Vectors

- **Learn basic concepts about vectors**
- **Learn representations of vectors**
- **Find the magnitude, direction angle, and components of a vector**
- **Perform operations on vectors**
- **Learn to apply the dot product**
- **Use vectors to calculate work**

Introduction

The beginnings of vectors go back centuries to the notion of a directed line segment, but the formal development of vectors occurred during the 19th and 20th centuries, after the invention of complex numbers. It was not until Einstein used vectors in his theory of relativity that their importance became readily accepted.

Vectors are a profound invention. In science and technology they provide a simple model for visualizing difficult concepts such as force, velocity, and electric fields. Vectors are essential to creating today's amazing computer graphics. This section discusses some of the important properties and applications of vectors.   (*Sources: Historical Topics for the Mathematical Classroom, Thirty-first Yearbook, NCTM;* M. Mortenson, *Computer Graphics.*)

Basic Concepts

Many quantities in mathematics can be described using real numbers, or **scalars**. Examples include a person's weight, the cost of a DVD player, and the gas mileage of a car. Other quantities must be represented using vector quantities. A **vector quantity** involves both *magnitude* and *direction*. Magnitude can be interpreted as size or length. For example, if a car is traveling north at 50 miles per hour, then the direction *north* coupled with a *speed* of 50 miles per hour represents a vector quantity called *velocity*. In science a distinction is made between speed and velocity—speed is the magnitude of velocity.

A vector quantity can be represented by a directed line segment called a **vector**. A vector **v** representing the velocity of a car traveling 50 miles per hour north is shown in Figure 8.35; the vector **u** represents a velocity of 25 miles per hour east. Notice that the length of **u** is half the length of **v**. Vectors do *not* have position—rather, they have magnitude and direction. A vector can be translated, provided its direction and magnitude (length) do not change. Two vectors are **equal** if they have the same magnitude and direction. In Figure 8.36 each directed line segment represents the same vector **v**.

Vectors: Magnitude and Direction

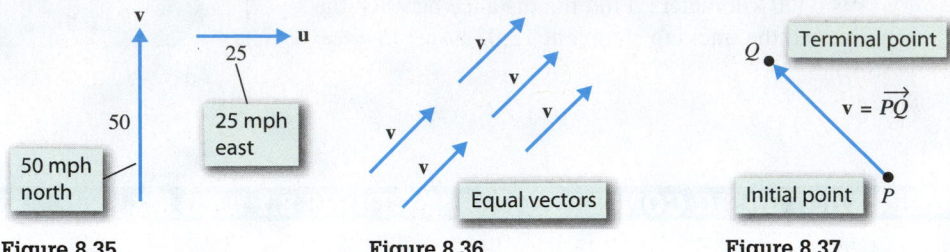

Figure 8.35   **Figure 8.36**   **Figure 8.37**

A vector is usually represented symbolically by a letter printed in boldface type, such as **a**, **b**, **v**, or **F**. A second way to denote a vector is to use two points. If the **initial point** of a vector **v** is P and its **terminal point** is Q, then $\mathbf{v} = \overrightarrow{PQ}$, as illustrated in Figure 8.37.

Representations of Vectors

If we place the initial point of vector **v** at the origin, as in Figure 8.38, then its terminal point (a_1, a_2) can be used to determine **v**. To distinguish the *point* (a_1, a_2) from the *vector* **v**, we use the notation $\mathbf{v} = \langle a_1, a_2 \rangle$. The **horizontal component** of **v** is a_1 and the **vertical component** of **v** is a_2. See Figure 8.39.

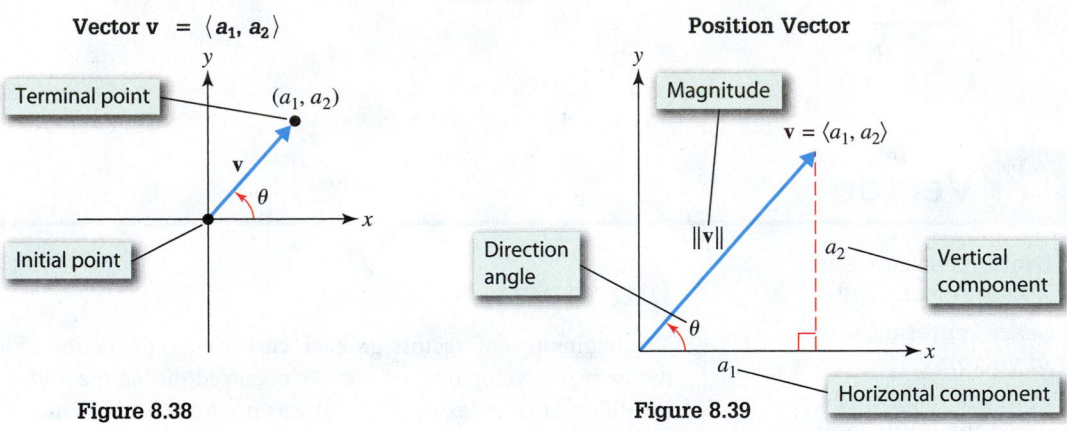

Figure 8.38   **Figure 8.39**

A vector with its initial point at the origin in the rectangular coordinate system is called a **position vector**. Figure 8.39 shows the position vector **v**. The *positive* angle θ ($0° \le \theta < 360°$) between the x-axis and the position vector is called the **direction angle** for the vector. In Figure 8.39, θ is the direction angle for vector **v**. If $\mathbf{v} = \langle a_1, a_2 \rangle$, then its direction angle θ satisfies

$$\tan \theta = \frac{a_2}{a_1}, \quad \text{where} \quad a_1 \ne 0.$$

The length of a vector equals its magnitude. If $\mathbf{v} = \langle a_1, a_2 \rangle$, then the *magnitude* of **v** is denoted $\|\mathbf{v}\|$. Applying the Pythagorean theorem to Figure 8.39 gives $\|\mathbf{v}\| = \sqrt{a_1^2 + a_2^2}$.

MAGNITUDE OF A VECTOR

If $\mathbf{v} = \langle a_1, a_2 \rangle$, then the **magnitude** (or length) of **v** is given by

$$\|\mathbf{v}\| = \sqrt{a_1^2 + a_2^2}.$$

If $\|\mathbf{v}\| = 1$, then **v** is a **unit vector**.

EXAMPLE 1 **Finding magnitude and direction angle**

Finding $\|\mathbf{u}\|$ and θ

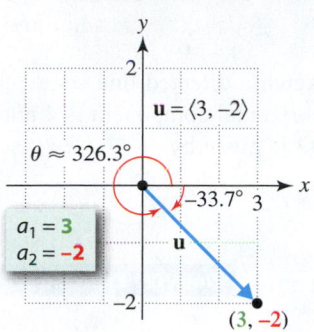

Figure 8.40

Find the magnitude and direction angle of $\mathbf{u} = \langle 3, -2 \rangle$.

SOLUTION Figure 8.40 shows the position vector for $\mathbf{u} = \langle 3, -2 \rangle$. The magnitude of vector $\mathbf{u}$ is

$$\|\mathbf{u}\| = \sqrt{3^2 + (-2)^2} = \sqrt{13}.$$

To find the direction angle θ, start with $\tan \theta = \frac{a_2}{a_1} = \frac{-2}{3} = -\frac{2}{3}$. A calculator reveals that $\tan^{-1}\left(-\frac{2}{3}\right) \approx -33.7°$. Adding 360° yields the positive direction angle $\theta \approx 326.3°$. See Figure 8.40.

Now Try Exercise 27

If the magnitude and the direction angle of a vector $\mathbf{v}$ are known, then horizontal and vertical components of $\mathbf{v}$ can be found by using the following equations. (Explain why.)

HORIZONTAL AND VERTICAL COMPONENTS

The horizontal and vertical components for a vector $\mathbf{v} = \langle a_1, a_2 \rangle$ having direction angle θ are given by

$$a_1 = \|\mathbf{v}\| \cos \theta \qquad \text{and} \qquad a_2 = \|\mathbf{v}\| \sin \theta.$$

That is, $\mathbf{v} = \langle \|\mathbf{v}\| \cos \theta, \|\mathbf{v}\| \sin \theta \rangle$.

EXAMPLE 2 **Finding horizontal and vertical components**

Finding a_1 and a_2

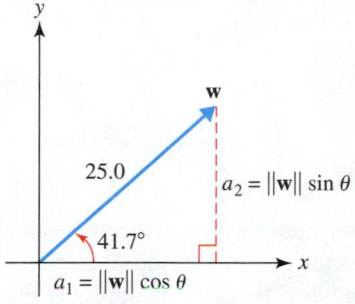

Figure 8.41

Vector $\mathbf{w}$ in Figure 8.41 has magnitude 25.0 and direction angle 41.7°. Find the horizontal and vertical components. Round to the nearest tenth.

SOLUTION Let $\|\mathbf{w}\| = 25.0$ and $\theta = 41.7°$.

$$\begin{array}{c|c}
a_1 = \|\mathbf{w}\| \cos \theta & a_2 = \|\mathbf{w}\| \sin \theta \\
a_1 = 25.0 \cos 41.7° & a_2 = 25.0 \sin 41.7° \\
a_1 \approx 18.7 & a_2 \approx 16.6
\end{array}$$

Therefore $\mathbf{w} \approx \langle 18.7, 16.6 \rangle$. The horizontal component is 18.7, and the vertical component is 16.6 (rounded to the nearest tenth).

Now Try Exercise 35

EXAMPLE 3 **Writing vectors in the form $\langle a_1, a_2 \rangle$**

Finding $\langle a_1, a_2 \rangle$

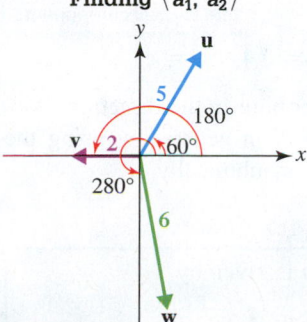

Figure 8.42

Write each vector in Figure 8.42 in the form $\langle a_1, a_2 \rangle$.

SOLUTION

$$\mathbf{u} = \left\langle 5 \cos 60°, 5 \sin 60° \right\rangle = \left\langle 5 \cdot \frac{1}{2}, 5 \cdot \frac{\sqrt{3}}{2} \right\rangle = \left\langle \frac{5}{2}, \frac{5\sqrt{3}}{2} \right\rangle$$

$$\mathbf{v} = \left\langle 2 \cos 180°, 2 \sin 180° \right\rangle = \left\langle 2(-1), 2(0) \right\rangle = \left\langle -2, 0 \right\rangle$$

$$\mathbf{w} = \left\langle 6 \cos 280°, 6 \sin 280° \right\rangle \approx \left\langle 1.04, -5.91 \right\rangle$$

Now Try Exercises 39 and 43

Using Two Points to Determine a Vector If a vector has initial point P with coordinates (a_1, b_1) and terminal point Q with coordinates (a_2, b_2), then vector $\overrightarrow{PQ}$ is given by $\overrightarrow{PQ} = \langle a_2 - a_1, b_2 - b_1 \rangle$. See the next example. Note that $\overrightarrow{QP} = \langle a_1 - a_2, b_1 - b_2 \rangle$.

| EXAMPLE 4 | **Finding a vector graphically and symbolically** |

Let P have coordinates $(-1, 2)$ and Q have coordinates $(3, 4)$. Find vector $\overrightarrow{PQ}$ graphically and symbolically. Calculate the magnitude of $\overrightarrow{PQ}$

Determing $\overrightarrow{PQ}$ from 2 Points

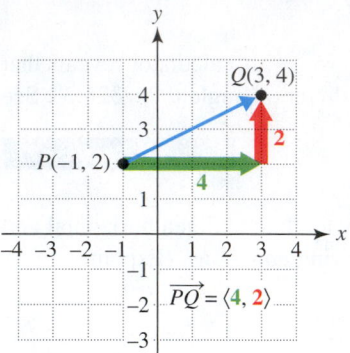

Figure 8.43

SOLUTION To graph $\overrightarrow{PQ}$ plot the points P and Q. Then sketch a directed line segment from P to Q, as shown in Figure 8.43. We can see that the horizontal component is **4** and the vertical component is **2**. A symbolic representation of $\overrightarrow{PQ}$ is given by

$$\overrightarrow{PQ} = \langle 3 - (-1), 4 - 2 \rangle = \langle 4, 2 \rangle.$$

The magnitude, or length, of $\overrightarrow{PQ}$ is

$$\|\overrightarrow{PQ}\| = \sqrt{4^2 + 2^2} = \sqrt{20} \approx 4.47.$$

Now Try Exercise 49

Operations on Vectors

Vector Addition Suppose that a swimmer heads directly across a river at 3 miles per hour. If the current is 4 miles per hour, then the person will be carried a distance downstream before reaching the other side, as illustrated in Figure 8.44. We can use vectors to visually find the direction and speed that the swimmer will travel across the river.

A Swimmer in a Current

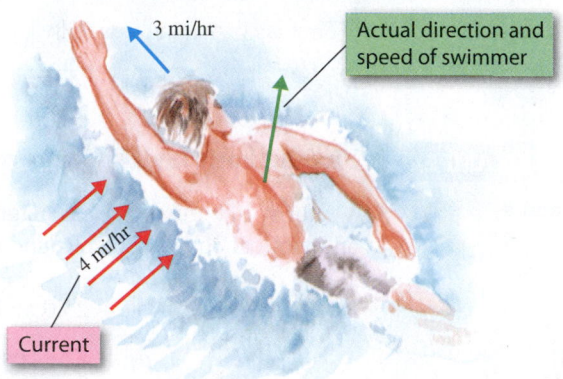

Figure 8.44

Vector Addition: c = a + b

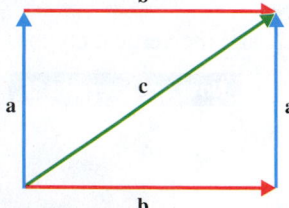

Figure 8.45

Let vector **a** represent the speed and direction of the swimmer with no current, vector **b** represent the direction and speed of the current, and vector **c** represent the final direction and speed of the swimmer. We can find the length and direction of **c** by applying the **parallelogram rule**, as shown in Figure 8.45. The speed and direction of the swimmer are represented by the diagonal **c** of the parallelogram (rectangle), which is determined by **a** and **b**. Vector **c** is called the **sum** or **resultant** of vectors **a** and **b**.

We can represent the velocity of the swimmer with no current by $\mathbf{a} = \langle 0, 3 \rangle$, the velocity of the current by $\mathbf{b} = \langle 4, 0 \rangle$, and the velocity of the swimmer in the current by $\mathbf{c} = \langle 4, 3 \rangle$. Vector **c** is the sum of vectors **a** and **b** and can be found as follows.

The Parallelogram Rule

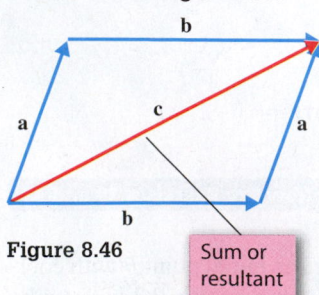

Figure 8.46

Sum or resultant

| Sum horizontal components | | | Sum vertical components |

$$\mathbf{a} + \mathbf{b} = \langle 0, 3 \rangle + \langle 4, 0 \rangle = \langle 0 + 4, 3 + 0 \rangle = \langle 4, 3 \rangle = \mathbf{c}$$

Since $\|\mathbf{c}\| = \sqrt{4^2 + 3^2} = 5$, the swimmer moves 5 miles per hour in the direction of **c**.

Figure 8.46 illustrates graphically how to find $\mathbf{c} = \mathbf{a} + \mathbf{b}$ in general by using the parallelogram rule. The following box defines vector addition symbolically.

VECTOR ADDITION

If $\mathbf{a} = \langle a_1, a_2 \rangle$ and $\mathbf{b} = \langle b_1, b_2 \rangle$, then the **sum** of **a** and **b** is given by

$$\mathbf{a} + \mathbf{b} = \langle a_1, a_2 \rangle + \langle b_1, b_2 \rangle = \langle a_1 + b_1, a_2 + b_2 \rangle.$$

Suppose that vector **a** represents a force of 80 pounds pulling on a water-ski towrope and **b** represents a force of 60 pounds pulling on a second towrope with an angle of 25° between them. See Figure 8.47. The resultant force **c** = **a** + **b** is given by the diagonal of the parallelogram shown in Figure 8.48. Vector **c** represents the net force exerted by the two water skiers.

Force on a Boat

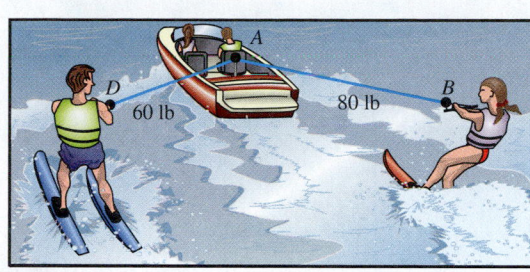

Figure 8.47

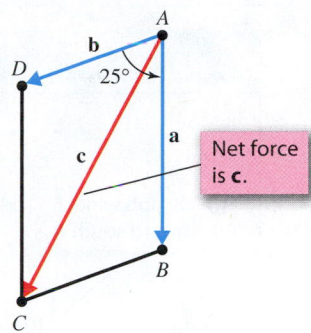

Figure 8.48

EXAMPLE 5 **Applying the parallelogram rule**

Find the magnitude of the resultant force on the ski boat in the preceding discussion.

SOLUTION The magnitude of the force equals the length of the diagonal AC in Figure 8.48. Angle BAD equals 25°. Since angle ABC is $180° - 25° = 155°$, we find AC by applying the law of cosines.

$$AC^2 = 60^2 + 80^2 - 2(60)(80)\cos 155° \approx 18{,}701$$
$$AC \approx 137 \text{ pounds}$$

> **Now Try Exercise 53**

Vector Notation Using i and j A second type of vector notation involves the vectors $\mathbf{i} = \langle 1, 0 \rangle$ and $\mathbf{j} = \langle 0, 1 \rangle$. A vector $\mathbf{a} = \langle a_1, a_2 \rangle$ can be expressed as

$$\mathbf{a} = a_1\mathbf{i} + a_2\mathbf{j}.$$

For example, $\langle 3, -4 \rangle$ and $3\mathbf{i} - 4\mathbf{j}$ represent the same vector.

> **MAKING CONNECTIONS**
>
> **Imaginary unit *i* and unit vector i** Section 3.3 discussed the *imaginary unit i*, where $i = \sqrt{-1}$ and $i^2 = -1$. The *vector* $\mathbf{i} = \langle 1, 0 \rangle$ represents a different concept.

EXAMPLE 6 **Finding resultant forces**

Forces $\mathbf{F}_1 = 5\mathbf{i} + 12\mathbf{j}$ and $\mathbf{F}_2 = 4\mathbf{i} - 3\mathbf{j}$ act at the same point. Find the resultant force $\mathbf{F}$ and its magnitude.

Resultant Force

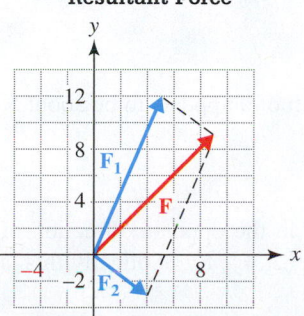

Figure 8.49

SOLUTION The position vectors for $\mathbf{F}_1$ and $\mathbf{F}_2$ and the resultant force $\mathbf{F}$ are shown in Figure 8.49. The resultant force $\mathbf{F}$ equals $\mathbf{F}_1 + \mathbf{F}_2$.

$$\mathbf{F} = \mathbf{F}_1 + \mathbf{F}_2$$
$$= (5\mathbf{i} + 12\mathbf{j}) + (4\mathbf{i} - 3\mathbf{j}) \qquad \text{Add } \mathbf{j} \text{ components}$$
$$= (5 + 4)\mathbf{i} + (12 + (-3))\mathbf{j}$$
$$= 9\mathbf{i} + 9\mathbf{j} \qquad \text{Add } \mathbf{i} \text{ components}$$

It follows that $\|\mathbf{F}\| = \sqrt{9^2 + 9^2} = \sqrt{162} \approx 12.7$.

> **Now Try Exercise 57**

Scalar Multiplication Scalar multiplication occurs when a vector **v** is multiplied by a real number, or *scalar,* k to form $k\mathbf{v}$. Vectors **v** and $k\mathbf{v}$ are parallel if $k \neq 0$. Vector $k\mathbf{v}$ points in the *same* direction as **v** if $k > 0$, and $k\mathbf{v}$ points in the *opposite* direction of **v** if $k < 0$. The magnitude of $k\mathbf{v}$ is $|k|$ times the magnitude of **v**. These ideas are illustrated by Figures 8.50 and 8.51 in the following See the Concept, where vector **v** represents the wind.

See the Concept: Scalar Multiplication

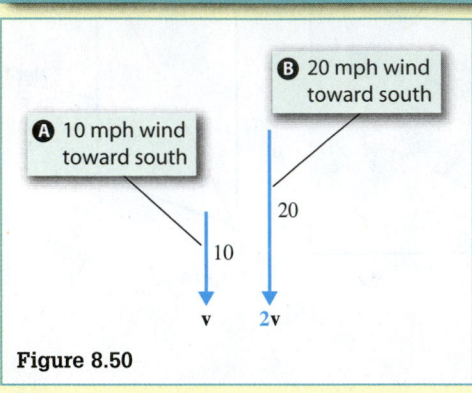

Ⓐ 10 mph wind toward south

Ⓑ 20 mph wind toward south

20

10

v   **2v**

Figure 8.50

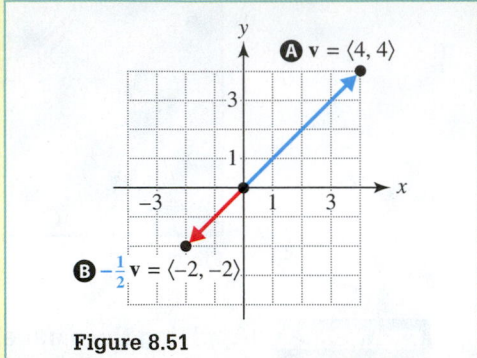

Ⓐ $\mathbf{v} = \langle 4, 4 \rangle$

Ⓑ $-\frac{1}{2}\mathbf{v} = \langle -2, -2 \rangle$

Figure 8.51

Ⓐ Represent the wind by $\mathbf{v} = \langle 0, -10 \rangle$.

Ⓑ If the wind doubles in speed but does not change direction, then the *scalar multiple* 2**v** models this situation.

$$2\mathbf{v} = 2\langle 0, -10 \rangle = \langle 2 \cdot 0, 2 \cdot (-10) \rangle = \langle 0, -20 \rangle$$

Ⓐ A wind toward the northeast is modeled by $\mathbf{v} = \langle 4, 4 \rangle$.

Ⓑ Then a wind in the opposite direction of **v** with half the speed is modeled by

$$-\frac{1}{2}\mathbf{v} = -\frac{1}{2}\langle 4, 4 \rangle = \left\langle -\frac{1}{2} \cdot 4, -\frac{1}{2} \cdot 4 \right\rangle = \langle -2, -2 \rangle.$$

The following is a definition of scalar multiplication.

SCALAR MULTIPLICATION

If $\mathbf{v} = \langle v_1, v_2 \rangle$ and k is a real number, then the **scalar multiple** $k\mathbf{v}$ is given by

$$k\mathbf{v} = k\langle v_1, v_2 \rangle = \langle kv_1, kv_2 \rangle.$$

Sums and scalar multiples can be calculated graphically and symbolically, as illustrated in the next example.

EXAMPLE 7   **Performing operations on vectors**

Find each of the following expressions graphically and symbolically if $\mathbf{a} = \langle -3, 4 \rangle$ and $\mathbf{b} = \langle -1, -2 \rangle$.

(a) $\|\mathbf{a}\|$ (b) $-2\mathbf{b}$ (c) $\mathbf{a} + 2\mathbf{b}$

SOLUTION

(a) Graph $\mathbf{a} = \langle -3, 4 \rangle$, as shown in Figure 8.52. The length of **a** appears to be about 5. This can be verified symbolically.

$$\|\mathbf{a}\| = \sqrt{(-3)^2 + (4)^2} = 5$$

(b) Graph $\mathbf{b} = \langle -1, -2 \rangle$. The scalar multiple $-2\mathbf{b}$ points in the opposite direction of **b** with twice the length. See Figure 8.53, where $-2\mathbf{b} = \langle 2, 4 \rangle$. Symbolically this is given by

$$-2\mathbf{b} = -2\langle -1, -2 \rangle = \langle -2 \cdot (-1), -2 \cdot (-2) \rangle = \langle 2, 4 \rangle.$$

(c) Graph $\mathbf{a} = \langle -3, 4 \rangle$ and $2\mathbf{b} = \langle -2, -4 \rangle$. See Figure 8.54. By the parallelogram rule, the diagonal represents $\mathbf{a} + 2\mathbf{b} = \langle -5, 0 \rangle$. This can be verified symbolically.

$$\mathbf{a} + 2\mathbf{b} = \langle -3, 4 \rangle + 2\langle -1, -2 \rangle = \langle -3, 4 \rangle + \langle -2, -4 \rangle = \langle -5, 0 \rangle$$

Finding $\|\mathbf{a}\|$

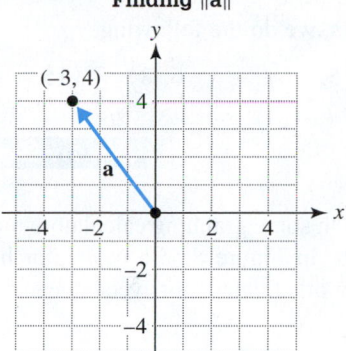

Figure 8.52

Finding $-2\mathbf{b}$

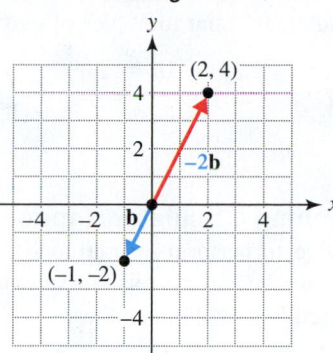

Figure 8.53

Finding $\mathbf{a} + 2\mathbf{b}$

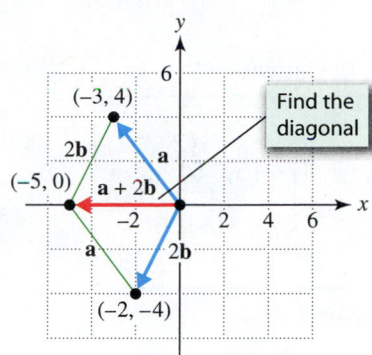

Figure 8.54

Now Try Exercise 79

Vector Subtraction Subtraction can be defined both symbolically and graphically. The difference $\mathbf{a} - \mathbf{b}$ can be thought of as the sum $\mathbf{a} + (-\mathbf{b})$. Then,

$$\mathbf{a} - \mathbf{b} = \mathbf{a} + (-\mathbf{b})$$
$$= \langle a_1, a_2 \rangle + \langle -b_1, -b_2 \rangle$$
$$= \langle a_1 + (-b_1), a_2 + (-b_2) \rangle$$
$$= \langle a_1 - b_1, a_2 - b_2 \rangle.$$

Vector subtraction is defined symbolically as follows.

VECTOR SUBTRACTION

If $\mathbf{a} = \langle a_1, a_2 \rangle$ and $\mathbf{b} = \langle b_1, b_2 \rangle$, then the **difference** of $\mathbf{a}$ and $\mathbf{b}$ is given by

$$\mathbf{a} - \mathbf{b} = \langle a_1, a_2 \rangle - \langle b_1, b_2 \rangle = \langle a_1 - b_1, a_2 - b_2 \rangle.$$

The difference $\mathbf{a} - \mathbf{b}$ is shown graphically in Figure 8.55 in the following See the Concept.

See the Concept: Vector Subtraction (Graphical)

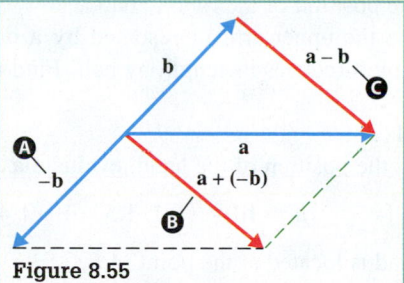

Figure 8.55

To find $\mathbf{a} - \mathbf{b}$:

Ⓐ Find $-\mathbf{b}$.

Ⓑ Add $\mathbf{a} + (-\mathbf{b})$ using the parallelogram rule.

Ⓒ The resultant is $\mathbf{a} - \mathbf{b}$. Notice that by the parallelogram rule $\mathbf{b} + (\mathbf{a} - \mathbf{b}) = \mathbf{a}$.

EXAMPLE 8 Adding and subtracting vectors

Let $\mathbf{a} = \langle -3, 4 \rangle$ and $\mathbf{b} = \langle 5, -6 \rangle$. Find $\mathbf{a} + \mathbf{b}$, $\mathbf{a} - \mathbf{b}$, and $2\mathbf{a} - 3\mathbf{b}$.

SOLUTION To add two vectors, we add corresponding components.

$$\mathbf{a} + \mathbf{b} = \langle -3, 4 \rangle + \langle 5, -6 \rangle = \langle -3 + 5, 4 + (-6) \rangle = \langle 2, -2 \rangle$$

To subtract two vectors, we subtract corresponding components.

$$\mathbf{a} - \mathbf{b} = \langle -3, 4 \rangle - \langle 5, -6 \rangle = \langle -3 - 5, 4 - (-6) \rangle = \langle -8, 10 \rangle$$

Subtract horizontal components   Subtract vertical components

To subtract scalar multiples of two vectors, we do the following.

$$2\mathbf{a} - 3\mathbf{b} = 2\langle -3, 4 \rangle - 3\langle 5, -6 \rangle = \langle -6, 8 \rangle - \langle 15, -18 \rangle$$
$$= \langle -6 - 15, 8 - (-18) \rangle = \langle -21, 26 \rangle$$

Now Try Exercises 63 and 83

Operations on Vectors

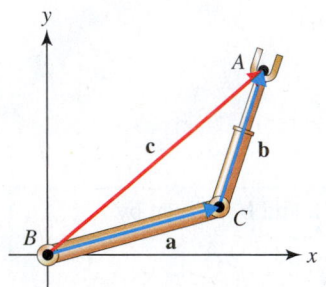

{⁻3,4}+{5,⁻6}
        {2 ⁻2}
{⁻3,4}−{5,⁻6}
        {⁻8 10}
2{⁻3,4}−3{5,⁻6}
        {⁻21 26}

Figure 8.56

Graphing Calculators (optional)   On some graphing calculators the list feature can be used to perform operations on vectors. In Figure 8.56 a calculator has been used to evaluate the expressions in Example 8. On other calculators vectors can be represented by ordered pairs with parentheses.

EXAMPLE 9   **Performing operations on vectors**

Find $-3\mathbf{a} + 5\mathbf{b}$ if $\mathbf{a} = 2\mathbf{i} + 3\mathbf{j}$ and $\mathbf{b} = 6\mathbf{i} - 7\mathbf{j}$.

SOLUTION

$$-3\mathbf{a} + 5\mathbf{b} = -3(2\mathbf{i} + 3\mathbf{j}) + 5(6\mathbf{i} - 7\mathbf{j})$$
$$= (-6\mathbf{i} - 9\mathbf{j}) + (30\mathbf{i} - 35\mathbf{j})$$
$$= (-6 + 30)\mathbf{i} + (-9 - 35)\mathbf{j}$$
$$= 24\mathbf{i} - 44\mathbf{j}$$

Now Try Exercise 81

Robotic Arm

Figure 8.57

An Application from Robotics   Robotic arms are sometimes modeled using vectors. Consider the *planar two-arm manipulator* in Figure 8.57. If $\overrightarrow{BC} = \mathbf{a}$ and $\overrightarrow{CA} = \mathbf{b}$, then the position of the hand is given by $\overrightarrow{BA} = \mathbf{c}$. Since $\mathbf{c} = \mathbf{a} + \mathbf{b}$, we can easily locate the position of the hand if $\mathbf{a}$ and $\mathbf{b}$ are known.   (***Source:*** J. Craig, *Introduction to Robotics.*)

EXAMPLE 10   **Using vectors to locate a robotic hand**

Let $\mathbf{a} = \langle 3.1, 1.5 \rangle$ and $\mathbf{b} = \langle 1.4, 2.4 \rangle$ in Figure 8.57.
(a) Find the position of the robotic hand.
(b) Suppose the upper arm represented by $\mathbf{a}$ doubles its length and the forearm represented by $\mathbf{b}$ reduces its length by half. Find the new position of the hand.

SOLUTION
(a) To find the position of the hand, evaluate $\mathbf{a} + \mathbf{b}$.

$$\mathbf{a} + \mathbf{b} = \langle 3.1, 1.5 \rangle + \langle 1.4, 2.4 \rangle = \langle 4.5, 3.9 \rangle$$

The hand is located at the point (4.5, 3.9).
(b) The new position is represented by

$$2\mathbf{a} + \frac{1}{2}\mathbf{b} = 2\langle 3.1, 1.5 \rangle + \frac{1}{2}\langle 1.4, 2.4 \rangle$$
$$= \langle 6.2, 3.0 \rangle + \langle 0.7, 1.2 \rangle$$
$$= \langle 6.9, 4.2 \rangle.$$

The new coordinates of the robotic hand are (6.9, 4.2).

Now Try Exercise 117

An Application from Navigation Vectors are frequently used to describe air and fluid flow. The next example uses vectors to describe the motion of an airplane.

EXAMPLE 11   Using vectors in navigation

An airplane is flying with an airspeed of 300 miles per hour and a bearing of 40° in a 30-mile-per-hour west wind.
(a) Find vectors **v** and **u** that model the velocity of the airplane and the velocity of the wind, respectively.
(b) Use vectors to determine the groundspeed of the plane.
(c) Find the final bearing of the plane in the wind.

Bearing of an Airplane

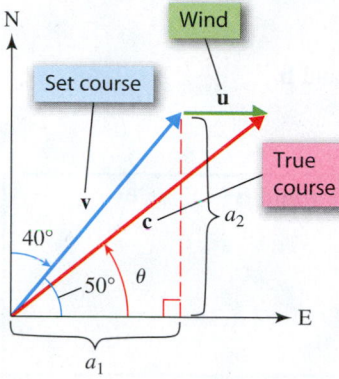

Figure 8.58

SOLUTION
(a) Consider Figure 8.58, which shows vectors **v** and **u** graphically. Since **u** models a west wind, it points to the right with length 30 and can be represented symbolically by $\mathbf{u} = \langle 30, 0 \rangle$. Let a_1 be the horizontal component and a_2 be the vertical component of **v**. Since $\|\mathbf{v}\| = 300$, it follows that $a_1 = 300 \cos 50°$ and $a_2 = 300 \sin 50°$. Thus

$$\mathbf{v} = \langle 300 \cos 50°, 300 \sin 50° \rangle \approx \langle 192.8, 229.8 \rangle.$$

(b) The true course of the plane is given by $\mathbf{c} = \mathbf{v} + \mathbf{u}$.

$$\begin{aligned} \mathbf{c} &= \mathbf{v} + \mathbf{u} \\ &= \langle 300 \cos 50°, 300 \sin 50° \rangle + \langle 30, 0 \rangle \\ &= \langle 300 \cos 50° + 30, 300 \sin 50° \rangle \end{aligned}$$

The groundspeed of the plane equals $\|\mathbf{c}\|$.

$$\|\mathbf{c}\| = \sqrt{(300 \cos 50° + 30)^2 + (300 \sin 50°)^2} \approx 320.1$$

The groundspeed of the airplane is approximately 320 miles per hour.
(c) Since $\mathbf{c} = \langle 300 \cos 50° + 30, 300 \sin 50° \rangle$, the direction angle θ is determined by the vector **c** and the positive x-axis (East). Thus $\tan \theta = \dfrac{300 \sin 50°}{300 \cos 50° + 30} \approx 1.0313$ and $\theta \approx \tan^{-1} 1.0313 \approx 45.9°$. The final bearing of the plane in the wind equals $90° - 45.9° = 44.1°$.

Now Try Exercise 109

The Dot Product

Thus far we have discussed addition, subtraction, and scalar multiplication of vectors. Another operation on vectors, called the *dot product*, is important because it can be used to find angles between vectors. The dot product has applications in computer graphics, solar energy, and physics. We begin by defining the dot product.

> **DOT PRODUCT**
>
> Let $\mathbf{a} = \langle a_1, a_2 \rangle$ and $\mathbf{b} = \langle b_1, b_2 \rangle$. The **dot product** of **a** and **b**, denoted $\mathbf{a} \cdot \mathbf{b}$, is a *real number* given by
>
> $$\mathbf{a} \cdot \mathbf{b} = a_1 b_1 + a_2 b_2.$$

In the next example, we calculate dot products. Notice that the dot product of two vectors is a real number, rather than a vector.

EXAMPLE 12   Calculating dot products

Calculate $\mathbf{a} \cdot \mathbf{b}$.
(a) $\mathbf{a} = \langle 4, -3 \rangle, \mathbf{b} = \langle -1, 2 \rangle$
(b) $\mathbf{a} = 2\mathbf{i} + 5\mathbf{j}, \mathbf{b} = -3\mathbf{i} + 2\mathbf{j}$

SOLUTION

(a) $\mathbf{a} \cdot \mathbf{b} = \langle 4, -3 \rangle \cdot \langle -1, 2 \rangle = (4)(-1) + (-3)(2) = -10$

(b) $\mathbf{a} \cdot \mathbf{b} = (2\mathbf{i} + 5\mathbf{j}) \cdot (-3\mathbf{i} + 2\mathbf{j}) = (2)(-3) + (5)(2) = 4$

Now Try Exercises 87(a) and 91(a)

Angle Between Vectors

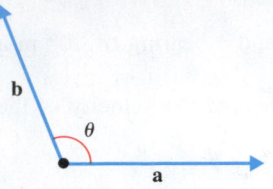

Figure 8.59

In Figure 8.59 the *angle between vectors* **a** *and* **b** is θ, where $0° \le \theta \le 180°$. If $\theta = 90°$ the vectors are **perpendicular**, and if $\theta = 0°$ or $180°$ the vectors are **parallel**. If $\theta = 0°$ the vectors point in the *same* direction, and if $\theta = 180°$ they point in *opposite* directions.

It is shown in Exercise 5 of the Extended Exercises at the end of the chapter that for any two nonzero vectors **a** and **b**,

$$\mathbf{a} \cdot \mathbf{b} = \|\mathbf{a}\| \, \|\mathbf{b}\| \cos \theta.$$

This result can be used to find the angle θ between **a** and **b**.

> **ANGLE BETWEEN TWO VECTORS**
>
> If **a** and **b** are nonzero vectors, then the **angle θ between a and b** is given by
>
> $$\theta = \cos^{-1} \frac{\mathbf{a} \cdot \mathbf{b}}{\|\mathbf{a}\| \, \|\mathbf{b}\|}.$$
>
> Vectors **a** and **b** are perpendicular if and only if $\mathbf{a} \cdot \mathbf{b} = 0$.

EXAMPLE 13 **Finding the angle between two vectors**

Sketch the vectors **a** and **b**. Then find the angle θ between **a** and **b**.

(a) $\mathbf{a} = 2\mathbf{i} - 3\mathbf{j}, \mathbf{b} = 3\mathbf{i} + 2\mathbf{j}$ (b) $\mathbf{a} = \langle -4, 3 \rangle, \mathbf{b} = \langle 1, -2 \rangle$

Perpendicular Vectors

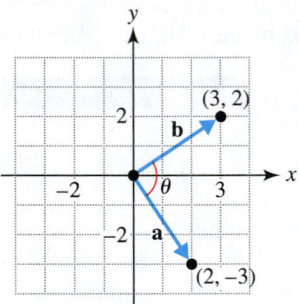

Figure 8.60

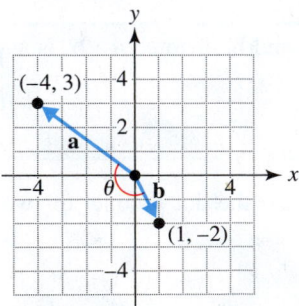

Figure 8.61

SOLUTION

(a) Vectors **a** and **b** appear to be perpendicular in Figure 8.60. Since

$$\mathbf{a} \cdot \mathbf{b} = (2)(3) + (-3)(2) = 0,$$

the vectors are perpendicular and $\theta = 90°$.

(b) A sketch of the vectors is shown in Figure 8.61. They are neither perpendicular nor parallel. Since $\mathbf{a} \cdot \mathbf{b} = (-4)(1) + (3)(-2) = -10$,

$$\|\mathbf{a}\| = \sqrt{(-4)^2 + (3)^2} = 5, \quad \text{and} \quad \|\mathbf{b}\| = \sqrt{(1)^2 + (-2)^2} = \sqrt{5},$$

it follows that

$$\theta = \cos^{-1} \frac{-10}{5\sqrt{5}} \approx 153.4°.$$

Now Try Exercise 87(b) and (c)

Work

In science a force does work only when an object moves. For example, a person pushing against a brick wall does no work, whereas a person lifting a 20-pound weight does work. Work equals force times distance, *provided the force is in the same direction as the movement of the object.* If a 150-pound person climbs up a 20-foot rope, then the work W done is

$$W = 150 \times 20 = 3000 \text{ foot-pounds.}$$

A **foot-pound** equals the work required to lift 1 pound a vertical distance of 1 foot. If the force is not in the same direction as the movement, then we must use trigonometry to determine work.

Consider a person pulling a wagon, as shown in Figure 8.62, where **F** represents the force on the handle and **D** represents the distance and direction that the wagon is pulled. The force **F** can be expressed as the sum of a horizontal vector in the direction of **D** and a vertical vector perpendicular to **D**, as illustrated in Figure 8.63. Using the right triangle in Figure 8.64, we see that the horizontal component of **F** is given by $\|\mathbf{F}\| \cos \theta$ and the vertical component of **F** is given by $\|\mathbf{F}\| \sin \theta$.

Forces Pulling a Wagon

Figure 8.62

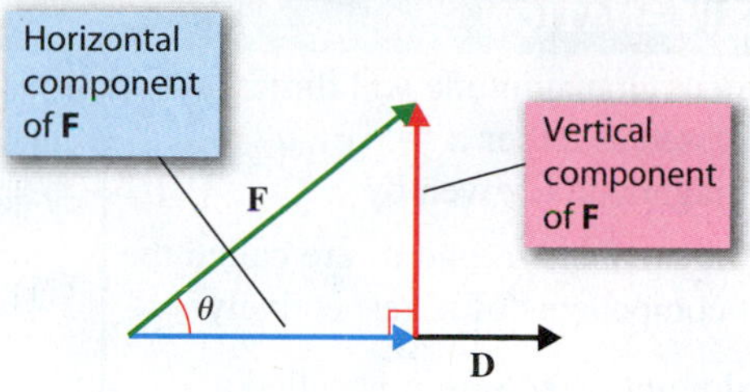

Figure 8.63

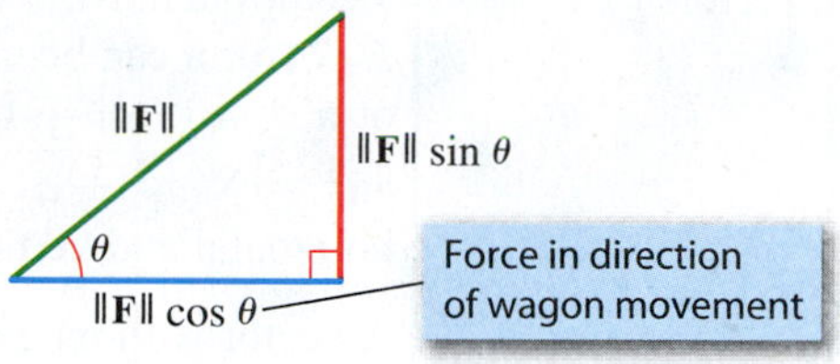

Figure 8.64

CLASS DISCUSSION

Suppose a force vector **F** is perpendicular to **D**. How much work is done? Interpret your answer.

The work W done pulling the wagon is equal to the horizontal component $\|\mathbf{F}\| \cos \theta$ times the distance the wagon moves, which is given by $\|\mathbf{D}\|$. That is,

$$W = \|\mathbf{F}\| \|\mathbf{D}\| \cos \theta = \mathbf{F} \cdot \mathbf{D}.$$

Work equals the dot product of the force vector **F** and the displacement vector **D**.

> **WORK**
>
> If a constant force **F** is applied to an object that moves along a vector **D**, then the work W done is
>
> $$W = \mathbf{F} \cdot \mathbf{D}.$$

EXAMPLE 14 Calculating work

Find the work done when a force $\mathbf{F} = \langle 3, -2 \rangle$ moves an object from point $P = (-2, 1)$ to point $Q = (3, -1)$, where force is measured in pounds and distance in feet.

SOLUTION First we must find the displacement vector $\mathbf{D} = \overrightarrow{PQ}$, where

$$\overrightarrow{PQ} = \langle 3 - (-2), -1 - 1 \rangle = \langle 5, -2 \rangle.$$

The work W done can be calculated as follows.

$$W = \mathbf{F} \cdot \mathbf{D} = \langle 3, -2 \rangle \cdot \langle 5, -2 \rangle = 15 + 4 = 19 \text{ foot-pounds}$$

Now Try Exercise 103

EXAMPLE 15 Calculating work

A 150-pound person walks 500 feet up a hiking trail that is inclined at 20°. Use vectors to compute the work done by the person, as illustrated in Figure 8.65.

SOLUTION Vector **D** is 500 feet long and is given by $\mathbf{D} = \langle 500 \cos 20°, 500 \sin 20° \rangle$. Since gravity pulls downward, the force exerted by the 150-pound person against gravity is given by $\mathbf{F} = \langle 0, 150 \rangle$. The work done is

$$W = \mathbf{F} \cdot \mathbf{D} = (0)(500 \cos 20°) + (150)(500 \sin 20°) \approx 25,650 \text{ foot-pounds}.$$

Now Try Exercise 121

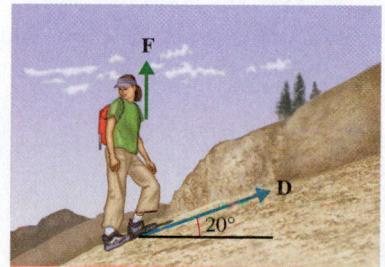

Figure 8.65

Vectors are an important invention of the 19th and 20th centuries that have enabled science and technology to model a wide variety of phenomena. The following table summarizes some basic concepts regarding vectors.

CONCEPT	EXPLANATION	EXAMPLES
Vectors	Vector quantities denote both magnitude and direction. A vector $\mathbf{a}$ can be expressed as either $\mathbf{a} = \langle a_1, a_2 \rangle$ or $\mathbf{a} = a_1\mathbf{i} + a_2\mathbf{j}$. Its magnitude is given by $\|\mathbf{a}\| = \sqrt{a_1^2 + a_2^2}$. The numbers a_1 and a_2 are called the horizontal and vertical components of $\mathbf{a}$, respectively. A vector with its initial point at the origin is called a position vector. The positive angle between the x-axis and the position vector is called the direction angle θ, where $0° \leq \theta < 360°$. If vector $\mathbf{a}$ has direction angle θ, then $\mathbf{a} = \langle \|\mathbf{a}\| \cos \theta, \|\mathbf{a}\| \sin \theta \rangle$.	$\mathbf{a} = \langle 1, 2 \rangle$ and $\mathbf{a} = \mathbf{i} + 2\mathbf{j}$ represent the same vector. The magnitude of $\mathbf{a}$ is given by $\|\mathbf{a}\| = \sqrt{1^2 + 2^2} = \sqrt{5}$. The horizontal component is 1, and the vertical component is 2.
Operations on vectors	Let $\mathbf{a} = \langle a_1, a_2 \rangle$, $\mathbf{b} = \langle b_1, b_2 \rangle$, and k be a real number. $\begin{aligned}\mathbf{a} + \mathbf{b} &= \langle a_1, a_2 \rangle + \langle b_1, b_2 \rangle \\ &= \langle a_1 + b_1, a_2 + b_2 \rangle \quad \text{Sum} \\ \mathbf{a} - \mathbf{b} &= \langle a_1, a_2 \rangle - \langle b_1, b_2 \rangle \\ &= \langle a_1 - b_1, a_2 - b_2 \rangle \quad \text{Difference} \\ k\mathbf{a} &= k\langle a_1, a_2 \rangle = \langle ka_1, ka_2 \rangle \quad \text{Scalar multiple}\end{aligned}$	Let $\mathbf{a} = \langle 4, 1 \rangle$, $\mathbf{b} = \langle 3, 2 \rangle$. $\begin{aligned}\mathbf{a} + \mathbf{b} &= \langle 4, 1 \rangle + \langle 3, 2 \rangle \\ &= \langle 7, 3 \rangle \\ \mathbf{a} - \mathbf{b} &= \langle 4, 1 \rangle - \langle 3, 2 \rangle \\ &= \langle 1, -1 \rangle \\ 3\mathbf{a} &= 3\langle 4, 1 \rangle = \langle 12, 3 \rangle\end{aligned}$
Dot product	If $\mathbf{a} = \langle a_1, a_2 \rangle$ and $\mathbf{b} = \langle b_1, b_2 \rangle$, then $\mathbf{a} \cdot \mathbf{b} = a_1 b_1 + a_2 b_2$.	Let $\mathbf{a} = \langle 2, -2 \rangle$, $\mathbf{b} = \langle 3, 1 \rangle$. $\mathbf{a} \cdot \mathbf{b} = (2)(3) + (-2)(1) = 4$
Angle θ between two vectors	If $\mathbf{a} = \langle a_1, a_2 \rangle$ and $\mathbf{b} = \langle b_1, b_2 \rangle$, then $\theta = \cos^{-1} \dfrac{\mathbf{a} \cdot \mathbf{b}}{\|\mathbf{a}\| \|\mathbf{b}\|}$. Vectors $\mathbf{a}$ and $\mathbf{b}$ are perpendicular ($\theta = 90°$) if and only if $\mathbf{a} \cdot \mathbf{b} = 0$.	If $\mathbf{a} = \langle 1, 0 \rangle$ and $\mathbf{b} = \langle 3, 4 \rangle$, then $\begin{aligned}\theta &= \cos^{-1} \dfrac{\langle 1, 0 \rangle \cdot \langle 3, 4 \rangle}{\|\langle 1, 0 \rangle\| \|\langle 3, 4 \rangle\|} \\ &= \cos^{-1} \dfrac{3}{(1)(5)} \approx 53.1°.\end{aligned}$ If $\mathbf{a} = \langle 4, -3 \rangle$ and $\mathbf{b} = \langle 3, 4 \rangle$, then $\mathbf{a} \cdot \mathbf{b} = (4)(3) + (-3)(4) = 0$. Thus $\mathbf{a}$ and $\mathbf{b}$ are perpendicular.
Work	If a constant force $\mathbf{F}$ is applied to an object that moves along a vector $\mathbf{D}$, then the work done is $W = \mathbf{F} \cdot \mathbf{D}$.	If $\mathbf{F} = 3\mathbf{i} - 4\mathbf{j}$ and $\mathbf{D} = 10\mathbf{i} - 20\mathbf{j}$, then $\begin{aligned}W &= \mathbf{F} \cdot \mathbf{D} \\ &= (3)(10) + (-4)(-20) \\ &= 110 \text{ foot-pounds,}\end{aligned}$ where units are in feet and pounds.

8.3 Exercises

Representing Vectors and Their Magnitudes

*Exercises 1–4: Use the graphical representation of **v** to complete the following.*

(a) *Estimate integer values for components a_1 and a_2 so that $\mathbf{v} = \langle a_1, a_2 \rangle$.*

(b) *Calculate $\|\mathbf{v}\|$.*

1.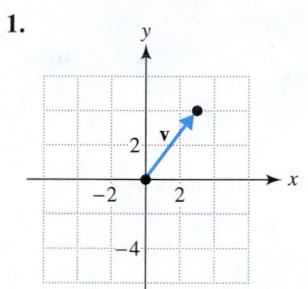

2.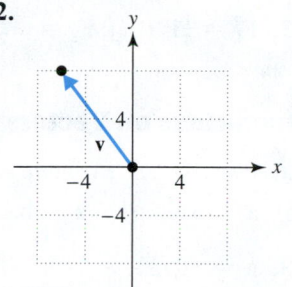

3.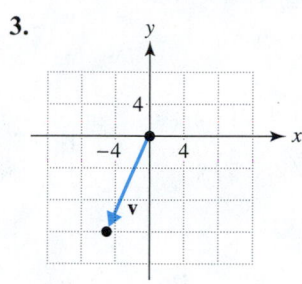

4.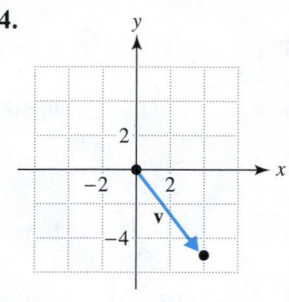

Exercises 5–10: Complete the following.

(a) *Sketch a vector **v** that models the situation.*

(b) *Express **v** as $\langle a_1, a_2 \rangle$.*

(c) *Find $2\mathbf{v}$ and $-\frac{1}{2}\mathbf{v}$. Interpret each result.*

5. A 20-mile-per-hour wind from the north

6. A 10-mile-per-hour wind from the west

7. A 5-mile-per-hour wind from the northeast

8. A 7-mile-per-hour wind from the southeast

9. A 30-pound force upward

10. A 15-pound force pulling at a 45° angle in standard position

*Exercises 11–18: Complete the following for vector **v**.*

(a) *Find the horizontal and vertical components.*

(b) *Calculate $\|\mathbf{v}\|$ and decide if **v** is a unit vector.*

(c) *Graph **v** and interpret $\|\mathbf{v}\|$.*

11. $\mathbf{v} = \langle 1, 1 \rangle$

12. $\mathbf{v} = \langle -1, 0 \rangle$

13. $\mathbf{v} = \langle 3, -4 \rangle$

14. $\mathbf{v} = \langle -2, -2 \rangle$

15. $\mathbf{v} = \mathbf{i}$

16. $\mathbf{v} = -3\mathbf{j}$

17. $\mathbf{v} = 5\mathbf{i} + 12\mathbf{j}$

18. $\mathbf{v} = -\frac{3}{5}\mathbf{i} - \frac{4}{5}\mathbf{j}$

Direction Angles and Components

Exercises 19–30: Find the magnitude and the direction angle θ for the given vector. Let $0° \leq \theta < 360°$ and round θ to the nearest tenth when appropriate.

19. $\langle 3, 0 \rangle$

20. $\langle 0, -2 \rangle$

21. $\langle -1, 1 \rangle$

22. $\langle 2, 2 \rangle$

23. $\langle \sqrt{3}, -1 \rangle$

24. $\langle -1, \sqrt{3} \rangle$

25. $\langle -5, -12 \rangle$

26. $\langle 7, -24 \rangle$

27. $\langle 13, -84 \rangle$

28. $\langle -11, -60 \rangle$

29. $\langle -20, 21 \rangle$

30. $\langle 16, -63 \rangle$

*Exercises 31–38: Find the horizontal and vertical components for **v**, given magnitude $\|\mathbf{v}\|$ and direction angle θ. Round values to the nearest tenth.*

31. $\|\mathbf{v}\| = 4, \theta = 180°$

32. $\|\mathbf{v}\| = 1, \theta = 270°$

33. $\|\mathbf{v}\| = \sqrt{2}, \theta = 135°$

34. $\|\mathbf{v}\| = \sqrt{2}, \theta = 225°$

35. $\|\mathbf{v}\| = 23, \theta = 54°$

36. $\|\mathbf{v}\| = 71, \theta = 163°$

37. $\|\mathbf{v}\| = 34, \theta = 312°$

38. $\|\mathbf{v}\| = 25, \theta = 73°$

Exercises 39–42: Write the vector in the form $\langle a_1, a_2 \rangle$.

39.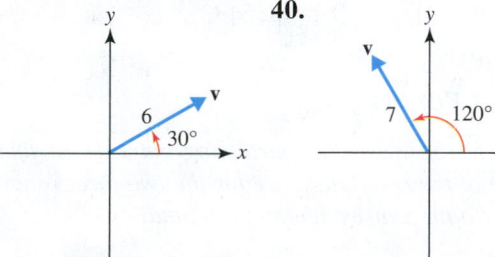

40.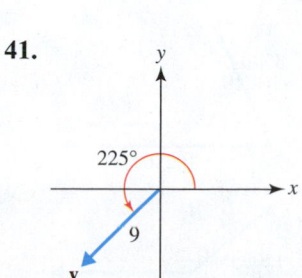

41.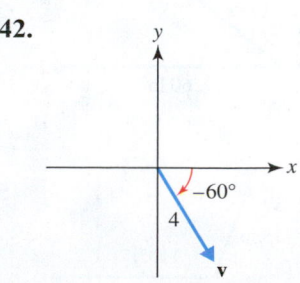

42.

Exercises 43–46: Write the vector in the form $\langle a_1, a_2 \rangle$. Round values to the nearest hundredth.

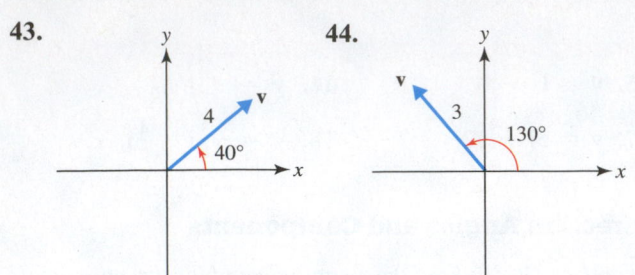

43.

44.

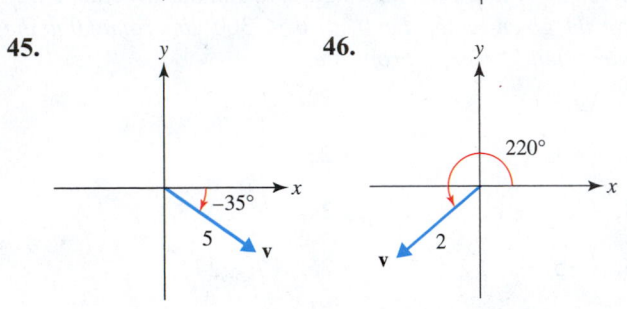

45.

46.

Exercises 47–52: A vector $\mathbf{v}$ has initial point P and terminal point Q.

(a) Graph $\overrightarrow{PQ}$.
(b) Write $\overrightarrow{PQ}$ as $\mathbf{v} = \langle a_1, a_2 \rangle$.
(c) Find the magnitude of $\overrightarrow{PQ}$.

47. $P = (0, 0)$, $Q = (-1, 2)$

48. $P = (0, 0)$, $Q = (4, -6)$

49. $P = (1, 2)$, $Q = (3, 6)$

50. $P = (-1, -2)$, $Q = (4, 4)$

51. $P = (-2, 4)$, $Q = (3, -2)$

52. $P = (0, -4)$, $Q = (1, 3)$

Resultant Forces

Exercises 53–56: Use the parallelogram rule to find the magnitude of the resultant force for the two forces shown in the figure (to the nearest tenth of a pound).

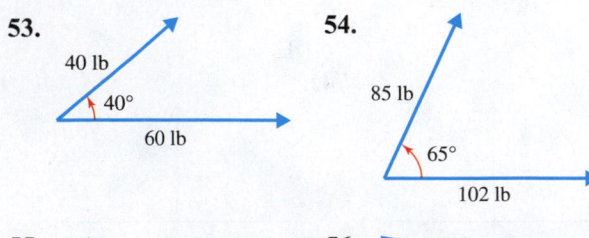

53.

54.

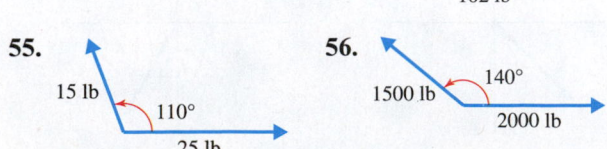

55.

56.

Exercises 57–62: Forces $\mathbf{F}_1$ and $\mathbf{F}_2$ act at the same point. Find the resultant force $\mathbf{F}$ and its magnitude.

57. $\mathbf{F}_1 = 3\mathbf{i} - 4\mathbf{j}$, $\mathbf{F}_2 = -8\mathbf{i} + 16\mathbf{j}$

58. $\mathbf{F}_1 = -7\mathbf{i} + 3\mathbf{j}$, $\mathbf{F}_2 = -\mathbf{i} + 12\mathbf{j}$

59. $\mathbf{F}_1 = \langle 8, 9 \rangle$, $\mathbf{F}_2 = \langle 1, -31 \rangle$

60. $\mathbf{F}_1 = \langle 48, 0 \rangle$, $\mathbf{F}_2 = \langle 0, -55 \rangle$

61. $\mathbf{F}_1 = 0.5\mathbf{i} + 0.7\mathbf{j}$, $\mathbf{F}_2 = -1.5\mathbf{i} - 5.7\mathbf{j}$

62. $\mathbf{F}_1 = \frac{1}{4}\mathbf{i} + \frac{1}{2}\mathbf{j}$, $\mathbf{F}_2 = -\frac{1}{2}\mathbf{i} + \frac{3}{4}\mathbf{j}$

Operations on Vectors

Exercises 63–70: Evaluate each of the following.

(a) $\mathbf{a} + \mathbf{b}$ (b) $\mathbf{a} - \mathbf{b}$

63. $\mathbf{a} = \langle 0, 2 \rangle$, $\mathbf{b} = \langle 3, 0 \rangle$

64. $\mathbf{a} = \langle 1, 1 \rangle$, $\mathbf{b} = \langle -2, 3 \rangle$

65. $\mathbf{a} = 2\mathbf{i} + \mathbf{j}$, $\mathbf{b} = \mathbf{i} - 2\mathbf{j}$

66. $\mathbf{a} = \mathbf{i} + 2\mathbf{j}$, $\mathbf{b} = -2\mathbf{i} + 3\mathbf{j}$

67. $\mathbf{a} = \langle -\sqrt{2}, \frac{1}{2} \rangle$, $\mathbf{b} = \langle \sqrt{2}, -\frac{3}{4} \rangle$

68. $\mathbf{a} = \langle \frac{4}{5}, -\frac{5}{6} \rangle$, $\mathbf{b} = \langle \frac{3}{10}, \frac{2}{3} \rangle$

69. $\mathbf{a} = \left(\cos \frac{\pi}{4} \right)\mathbf{i} + \left(\sin \frac{\pi}{4} \right)\mathbf{j}$,
    $\mathbf{b} = \left(\cos \frac{\pi}{2} \right)\mathbf{i} + \left(\sin \frac{\pi}{2} \right)\mathbf{j}$

70. $\mathbf{a} = \left(\cos \frac{3\pi}{2} \right)\mathbf{i} + \left(\sin \frac{3\pi}{2} \right)\mathbf{j}$,
    $\mathbf{b} = (\cos \pi)\mathbf{i} + (\sin \pi)\mathbf{j}$

Exercises 71–76: Use the figure to evaluate each of the following.

(a) $\mathbf{a} + \mathbf{b}$ (b) $\mathbf{a} - \mathbf{b}$ (c) $-\mathbf{a}$

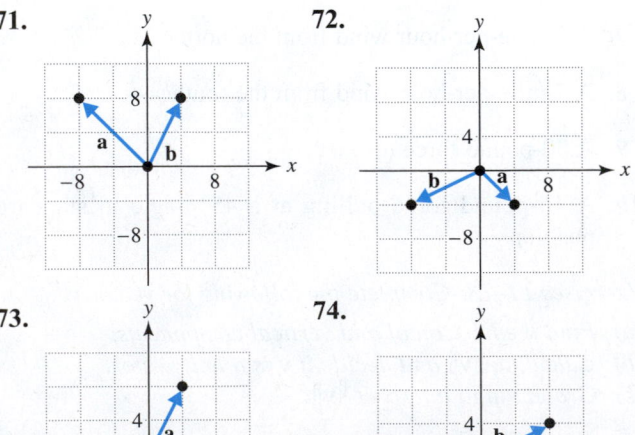

71.

72.

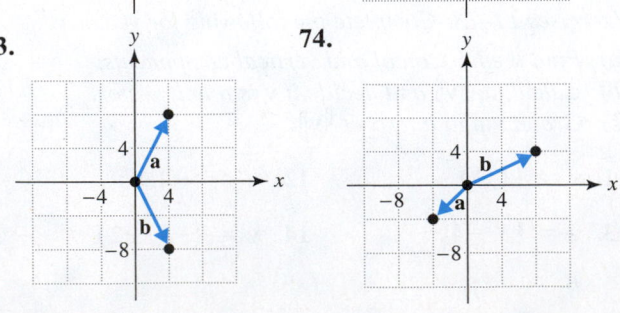

73.

74.

75.

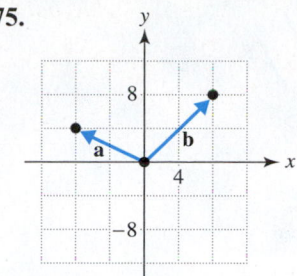

76.

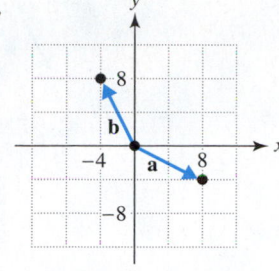

Exercises 77–80: Evaluate each of the following graphically and symbolically.

(a) $\|\mathbf{a}\|$ *(b)* $2\mathbf{a}$ *(c)* $2\mathbf{a} + 3\mathbf{b}$

77. $\mathbf{a} = 2\mathbf{i}$, $\mathbf{b} = \mathbf{i} + \mathbf{j}$

78. $\mathbf{a} = -\mathbf{i} + \mathbf{j}$, $\mathbf{b} = \mathbf{i} - \mathbf{j}$

79. $\mathbf{a} = \langle -1, 2 \rangle$, $\mathbf{b} = \langle 3, 0 \rangle$

80. $\mathbf{a} = \langle -2, -1 \rangle$, $\mathbf{b} = \langle -3, 2 \rangle$

*Exercises 81–86: Given vectors **a** and **b**, evaluate each of the following.*

(a) $-\mathbf{a} + 4\mathbf{b}$ *(b)* $2\mathbf{a} - 3\mathbf{b}$

81. $\mathbf{a} = \mathbf{i} - 2\mathbf{j}$, $\mathbf{b} = -5\mathbf{i} + 2\mathbf{j}$

82. $\mathbf{a} = -6\mathbf{i} + \mathbf{j}$, $\mathbf{b} = 7\mathbf{i} - 3\mathbf{j}$

83. $\mathbf{a} = \langle 1, -4 \rangle$, $\mathbf{b} = \langle -3, 5 \rangle$

84. $\mathbf{a} = \langle -5, 13 \rangle$, $\mathbf{b} = \langle 0, -1 \rangle$

85. $\mathbf{a} = \langle -7, -2 \rangle$, $\mathbf{b} = \langle 9, -1 \rangle$

86. $\mathbf{a} = \langle 5, 1 \rangle$, $\mathbf{b} = \langle -2, -4 \rangle$

Dot Product and Work

*Exercises 87–94: Complete the following for vectors **a** and **b**.*

(a) Find $\mathbf{a} \cdot \mathbf{b}$.

*(b) Approximate the angle θ between **a** and **b** to the nearest tenth of a degree.*

*(c) State if vectors **a** and **b** are perpendicular, parallel, or neither. If **a** and **b** are parallel, state whether they point in the same direction or in opposite directions.*

87. $\mathbf{a} = \langle 1, -2 \rangle$, $\mathbf{b} = \langle 3, 1 \rangle$

88. $\mathbf{a} = \langle 4, -5 \rangle$, $\mathbf{b} = \langle 2, -2 \rangle$

89. $\mathbf{a} = \langle 6, 8 \rangle$, $\mathbf{b} = \langle -4, 3 \rangle$

90. $\mathbf{a} = \langle 1, -2 \rangle$, $\mathbf{b} = \langle -2, 4 \rangle$

91. $\mathbf{a} = 5\mathbf{i} + 6\mathbf{j}$, $\mathbf{b} = 10\mathbf{i} + 12\mathbf{j}$

92. $\mathbf{a} = -2\mathbf{i} + 6\mathbf{j}$, $\mathbf{b} = 3\mathbf{i} + \mathbf{j}$

93. $\mathbf{a} = \mathbf{i} + 3\mathbf{j}$, $\mathbf{b} = 0.5\mathbf{i} - 1.5\mathbf{j}$

94. $\mathbf{a} = -12\mathbf{i} + 16\mathbf{j}$, $\mathbf{b} = -3\mathbf{i} + 4\mathbf{j}$

Exercises 95–98: Find the work done in each situation.

95. Lifting a 30-pound weight 5 feet into the air

96. Lifting a 15-pound bucket 8 feet into the air

97. Pushing a stalled car on level ground with a force of 100 pounds for 1000 feet

98. A 150-pound person running up 5 flights of steps with 10 feet between floors

*Exercises 99–102: Find the work done when a constant force **F** is applied to an object that moves along the vector **D**, where units are in pounds and feet. Find the magnitude of **F**.*

99. $\mathbf{F} = \langle 10, 20 \rangle$, $\mathbf{D} = \langle 15, 22 \rangle$

100. $\mathbf{F} = \langle 64, 36 \rangle$, $\mathbf{D} = \langle 22, -33 \rangle$

101. $\mathbf{F} = 5\mathbf{i} - 3\mathbf{j}$, $\mathbf{D} = 3\mathbf{i} - 4\mathbf{j}$

102. $\mathbf{F} = 7\mathbf{i} - 24\mathbf{j}$, $\mathbf{D} = -2\mathbf{i} - 5\mathbf{j}$

Exercises 103–106: Calculate the work done when the force $\mathbf{F} = 5\mathbf{i} + 3\mathbf{j}$ moves an object from P to Q.

103. $P = (-2, 3)$, $Q = (1, 6)$

104. $P = (-2, -1)$, $Q = (1, 3)$

105. $P = (2, -3)$, $Q = (4, -5)$

106. $P = (1, 1)$, $Q = (-1, 6)$

Applications

107. Swimming in a Current   A swimmer heads directly north across a river at 3 miles per hour, in a current that flows west to east at 2 miles per hour, as illustrated in the figure. Find a vector that models the resulting direction and speed of the swimmer. With what speed is the swimmer moving in the river?

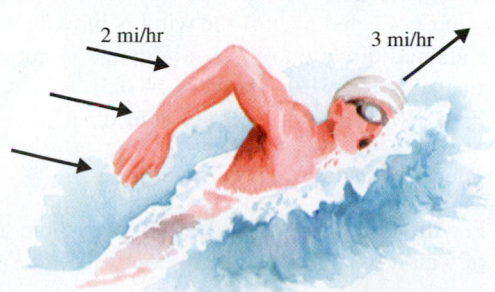

2 mi/hr   3 mi/hr

108. Wind and Vectors A wind can be described by $\mathbf{v} = 6\mathbf{i} + 8\mathbf{j}$, where vector $\mathbf{j}$ points north and represents a south wind of 1 mile per hour.

(a) What is the speed of the wind?

(b) Find $3\mathbf{v}$. Interpret the result.

(c) Interpret the wind if it switches to $\mathbf{u} = -8\mathbf{i} + 8\mathbf{j}$.

109. Air Navigation (Refer to Example 11.) An airplane heads west at 400 miles per hour in a 50-mile-per-hour northwest wind. Find a vector that models the resulting direction and speed of the airplane. Find the groundspeed and bearing of the airplane in the wind.

110. Air Navigation A plane with an airspeed of 240 miles per hour is headed on a bearing of 110°. A wind is blowing from the north at 18 miles per hour. Find the groundspeed and the final bearing of the plane in the wind.

111. Course and Groundspeed A plane flies 450 miles per hour on a bearing of 160°. A 20-mile-per-hour wind is blowing from the south. Find the groundspeed and the final bearing of the plane in the wind.

112. Course and Groundspeed A plane flies on a bearing of 230° at 350 miles per hour. A wind is blowing from the west at 30 miles per hour. Find the groundspeed and the final bearing of the plane in the wind.

113. Airspeed and Groundspeed A pilot wants to fly on a course of 75°. By flying due east, the pilot finds that a 40-mile-per-hour wind, blowing from the south, puts the plane on course. Find the airspeed and the groundspeed.

114. Force and Water-Ski Towropes (Refer to Example 5.) Forces of 65 pounds and 110 pounds are exerted by two water-ski towropes. If the angle between the towropes is 19°, find the magnitude of the resultant force.

115. Measuring Rainfall Suppose that vector $\mathbf{R}$ models the amount of rainfall in inches and the direction it falls, and vector $\mathbf{A}$ models the area in square inches and orientation of the opening of a rain gauge, as illustrated in the figure at the top of next column. The total volume V of water collected in the rain gauge is given by $V = |\mathbf{R} \cdot \mathbf{A}|$. This formula calculates the volume of water collected even if the wind is blowing the rain in a slanted direction or the rain gauge is not exactly vertical. Let $\mathbf{R} = \mathbf{i} - 2\mathbf{j}$ and $\mathbf{A} = 0.5\mathbf{i} + \mathbf{j}$.

(a) Find $\|\mathbf{R}\|$ and $\|\mathbf{A}\|$. Interpret your results.

(b) Calculate V and interpret this result.

(c) For the rain gauge to collect the maximum amount of water, what must be true about vectors $\mathbf{R}$ and $\mathbf{A}$?

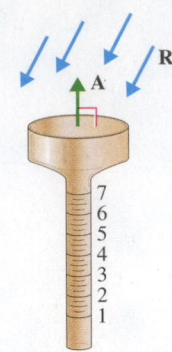

116. Solar Panels Suppose that the sun's intensity (in watts per square centimeter) and direction are given by vector $\mathbf{I}$, and a solar panel's area (in square centimeters) and orientation are given by vector $\mathbf{A}$, as illustrated in the figure. Then the total number of watts W that are collected by the solar panel is given by $W = |\mathbf{I} \cdot \mathbf{A}|$. Let $\mathbf{I} = 0.01\mathbf{i} - 0.02\mathbf{j}$ and $\mathbf{A} = 400\mathbf{i} + 300\mathbf{j}$.

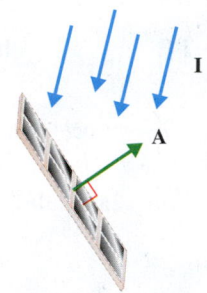

(a) Find $\|\mathbf{I}\|$ and $\|\mathbf{A}\|$. Interpret your results.

(b) Calculate W and interpret this result.

(c) For the solar panel to absorb maximum wattage, what must be true about vectors $\mathbf{I}$ and $\mathbf{A}$?

117. Robotics (Refer to Example 10.) Consider the planar two-arm manipulator shown in the figure. Let the upper arm be modeled by $\mathbf{a} = \langle 3, 2 \rangle$ and the forearm be modeled by $\mathbf{b} = \langle -2, 2 \rangle$, where units are in feet. (*Source:* J. Craig, *Introduction to Robotics.*)

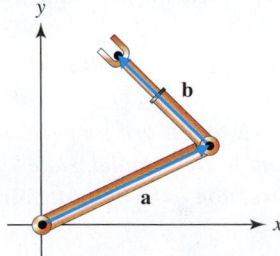

(a) Find a vector $\mathbf{c}$ that represents the position of the hand.

(b) How far is the hand from the origin?

(c) Find the position of the hand if the length of the upper arm triples and the length of the forearm is reduced by half.

118. Robotics A planar three-arm manipulator is shown in the figure, with joint angles measured relative to a positive horizontal axis.

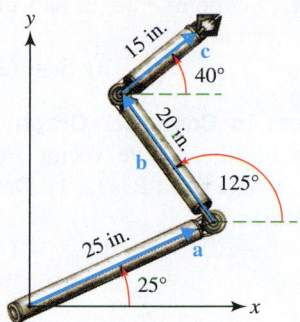

(a) Find vectors **a**, **b**, and **c** that represent each part of the robotic arm.

(b) Find a vector **d** that represents the position of the hand. How far is the hand from the origin?

119. Translations in Computer Graphics Vectors are used in computer graphics to compute translations of points. For example, suppose we would like to translate the point $(-1, 2)$ by $\mathbf{v} = \langle 2, 1 \rangle$, as illustrated in the figure, where the *point* $(-1, 2)$ has been represented by the *vector* $\mathbf{a} = \langle -1, 2 \rangle$. The new location of $(-1, 2)$ is modeled by

$$\mathbf{b} = \mathbf{a} + \mathbf{v} = \langle -1, 2 \rangle + \langle 2, 1 \rangle = \langle 1, 3 \rangle.$$

Thus $(-1, 2)$ has been translated by **v** to $(1, 3)$. (*Source:* J. Foley, *Introduction to Computer Graphics.*)

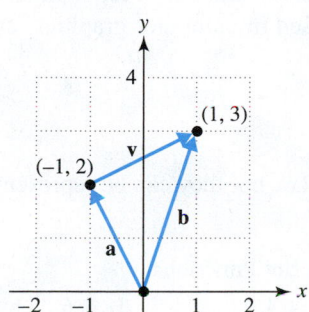

(a) Find the new coordinates of $(-2, 4)$ if it is translated by $\mathbf{v} = \langle 4, -2 \rangle$.

(b) Represent this situation graphically.

120. Translations in Computer Graphics (Refer to Exercise 119.) Let triangle ABC have vertices $(1, 1)$, $(3, 0)$, and $(4, 3)$.

(a) Find the new vertices D, E, and F if the triangle ABC is translated by $\mathbf{v} = \langle -2, 1 \rangle$.

(b) Describe the change in triangle ABC if it were translated by $-2\mathbf{v}$.

121. Work (Refer to Example 15.) A 145-pound person walks 1.5 miles up a hiking trail inclined at 15°. Use a dot product to calculate the work W done (in foot-pounds).

122. Work A wagon is pulled 500 feet using a force of 10 pounds applied to the handle, which makes a 40° angle with the horizontal. See Figure 8.62. Use a dot product to calculate the work W done.

123. Air Navigation A pilot would like to fly to a city that is 200 miles away and has a bearing of 135°. The wind is blowing from the north at 30 miles per hour, and the trip is to take 1 hour. Find the direction and speed that the pilot should adopt to accomplish this.

124. Work Calculate the work required to push an 1800-pound car up a 7° incline for 0.1 mile.

Writing about Mathematics

125. State the basic properties of a vector. Does a vector have position? Explain. Give two examples of how to write a vector.

126. State one application of vectors. Give a specific example of a vector for this application and explain how the vector models that application.

Extended and Discovery Exercises

1. Computer Graphics Vectors frequently are used to determine the color of pixels on computer screens. For example, suppose that we would like to color the right side of the screen blue and the left side yellow, where the boundary is positioned along vector $\overrightarrow{PQ}$, as illustrated in the figure below.

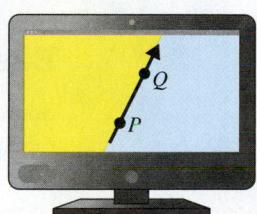

First find $\overrightarrow{PR}$ perpendicular to $\overrightarrow{PQ}$ Then to determine if a pixel at point S should be blue or yellow, consider the angle θ between vectors $\overrightarrow{PS}$ and $\overrightarrow{PR}$ If θ is acute, then S must be on the same side of $\overrightarrow{PQ}$ as R and is colored blue, as illustrated in the left figure below. If angle θ is obtuse, then S is on the opposite side of $\overrightarrow{PQ}$ from R and is colored yellow, as shown in the right figure below.

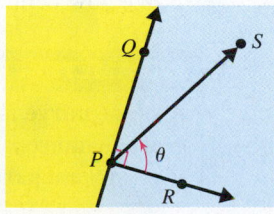
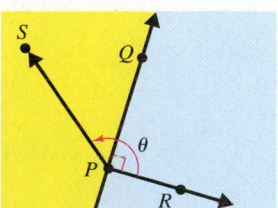

We can determine whether θ is acute or obtuse by calculating the dot product $\overrightarrow{PS} \cdot \overrightarrow{PR}$. If the dot product is positive, then

$$\theta = \cos^{-1}\left(\frac{\overrightarrow{PS} \cdot \overrightarrow{PR}}{\|\overrightarrow{PS}\| \, \|\overrightarrow{PR}\|}\right)$$

is acute and S should be blue. If the dot product is negative, then θ is obtuse and S should be yellow.

Let P be $(-1, 2)$, S be $(-2, -5)$, $\overrightarrow{PQ} = \mathbf{i} + 5\mathbf{j}$, and $\overrightarrow{PR} = 5\mathbf{i} - \mathbf{j}$. Determine the appropriate color at point S. See the figure below.

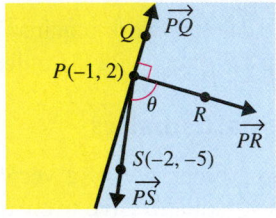

2. **Dot Products in Computer Graphics** A computer screen is gray to the left of vector $\overrightarrow{PQ} = 2\mathbf{i} - 3\mathbf{j}$ and blue to the right, where vector $\overrightarrow{PR} = 3\mathbf{i} + 2\mathbf{j}$ is perpendicular to $\overrightarrow{PQ}$. Let point P be $(2, 1)$. Determine the color of a pixel located at S.
   (a) $S = (3, -2)$ (b) $S = (2, 2)$

3. **Dot Products in Computer Graphics** A computer screen is to be blue above vector $\overrightarrow{PQ} = 5\mathbf{i} - \mathbf{j}$ and white below, where point P is $(2, 1)$. Determine the color of a pixel located at S.
   (a) $S = (100, -10)$ (b) $S = (-500, 50)$

8.4 Parametric Equations

- Learn basic concepts about parametric equations
- Graph parametric equations
- Use parametric equations to solve applications

Introduction

Sometimes a curve cannot be modeled by a function. For example, a circle cannot be described by a single function because a circle fails the vertical line test. *Parametric equations* represent a different approach to describing curves in the *xy*-plane. They are used in industry to draw complicated curves and surfaces, such as the hood of an automobile. Parametric equations are also used in computer graphics, engineering, and physics. (***Source:*** F. Hill, *Computer Graphics.*)

Basic Concepts

Some curves cannot be represented by $y = f(x)$, but they can be represented by parametric equations. See Figures 8.66–8.68.

Curves That Are Not Functions

$[-6, 6, 1]$ by $[-4, 4, 1]$ $[-6, 6, 1]$ by $[-4, 4, 1]$ $[-6, 6, 1]$ by $[-4, 4, 1]$

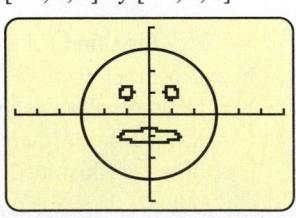

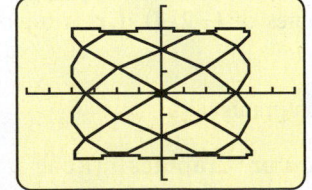

Figure 8.66   **Figure 8.67**   **Figure 8.68**

We now define parametric equations of a plane curve.

PARAMETRIC EQUATIONS OF A PLANE CURVE

A **plane curve** is a set of points (x, y) such that $x = f(t)$ and $y = g(t)$, where f and g are continuous functions on an interval $a \le t \le b$. The equations $x = f(t)$ and $y = g(t)$ are **parametric equations** with **parameter** t.

Parametric equations can be represented symbolically, numerically, graphically, and verbally. This is illustrated in the next example.

EXAMPLE 1   Representing parametric equations

Let $x = t + 3$ and $y = t^2$ for $-3 \le t \le 3$.

(a) Make a table of values for x and y with $t = -3, -2, -1, \ldots, 3$.

(b) Plot the points in the table and graph the curve. Add arrows to show how the curve is traced out.

(c) Describe the curve.

SOLUTION

(a) *Numerical Representation*   A numerical representation of the parametric equations is shown in Table 8.1. For example, if $t = 2$, then $x = 2 + 3 = 5$ and $y = 2^2 = 4$.

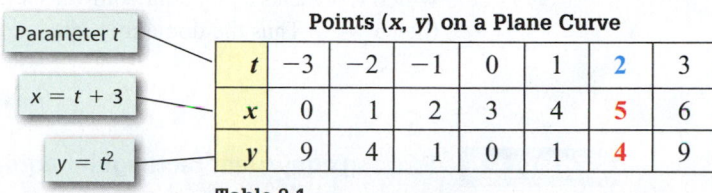

Points (x, y) on a Plane Curve

t	-3	-2	-1	0	1	2	3
x	0	1	2	3	4	5	6
y	9	4	1	0	1	4	9

Table 8.1

(b) *Graphical Representation*   Each ordered pair (x, y) in Table 8.1 is plotted in Figure 8.69, and then the points are connected to obtain the curve.

Curve Defined Parametrically

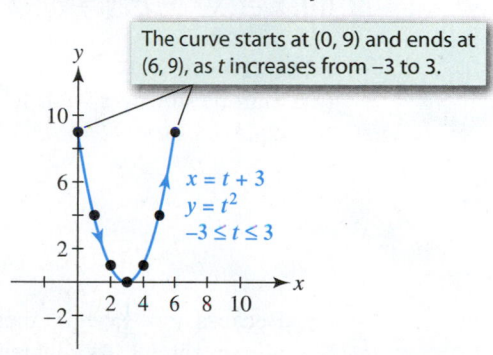

The curve starts at (0, 9) and ends at (6, 9), as t increases from -3 to 3.

$x = t + 3$
$y = t^2$
$-3 \le t \le 3$

Figure 8.69

(c) *Verbal Representation*   The curve in Figure 8.69 appears to be the lower portion of a parabola with vertex $(3, 0)$. See Example 2.

Now Try Exercise 1

Graphing Calculators (Optional)   Graphing calculators are capable of using parametric equations to make tables and graphs. In addition to setting values for the viewing rectangle, we must specify the interval for t. A window setting, table, and graph for the parametric equations in Example 1 are shown in Figures 8.70–8.72. The variable Tstep represents the increment in the parameter t and has a value of 0.1 in this case.

Calculator Help

To create the graph shown in Figure 8.72 with parametric equations, see Appendix A (page AP-18).

Setting a Window

```
WINDOW
 Tmin=-3
 Tmax=3
 Tstep=.1
 Xmin=-2
 Xmax=10
 Xscl=1
↓Ymin=-2
```

Figure 8.70

Table of Points

T	X1T	Y1T
-3	0	9
-2	1	4
-1	2	1
0	3	0
1	4	1
2	5	4
3	6	9

T=-3

Figure 8.71

Plane Curve
$[-2, 10, 1]$ by $[-2, 10, 1]$

$x = t + 3$
$y = t^2$
$-3 \le t \le 3$

Figure 8.72

Converting and Graphing Equations

We can verify symbolically that the curve in Figure 8.69 in Example 1 is indeed a portion of a parabola, as demonstrated in the next example.

EXAMPLE 2 Finding an equivalent rectangular equation

Find an equivalent rectangular equation for $x = t + 3$ and $y = t^2$, where $-3 \leq t \leq 3$. Note that these parametric equations were discussed in Example 1.

SOLUTION Begin by solving $x = t + 3$ for t to obtain $t = x - 3$. Substituting for t in $y = t^2$ results in

$$y = (x - 3)^2,$$

which represents a parabola with vertex $(3, 0)$. When $t = -3$ then $x = 0$, and when $t = 3$ then $x = 6$. Thus the domain is restricted to $0 \leq x \leq 6$. See Figure 8.69.

> **Now Try Exercise 15**

EXAMPLE 3 Finding equivalent rectangular equations

Find an equivalent rectangular equation for each pair of parametric equations. Use the rectangular equation to help graph the parametric equation. Add arrows to show how the curve is traced out.
(a) $x = 4t, y = t - 3; -\infty < t < \infty$
(b) $x = \sqrt{4 - t^2}, y = t; -2 \leq t \leq 2$

A Line

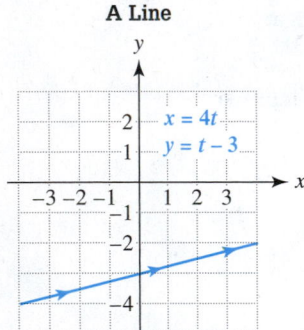

Figure 8.73

SOLUTION
(a) Start by solving $x = 4t$ for t to obtain $t = \frac{1}{4}x$. Substitute for t in the given parametric equation for y.

$$y = t - 3 \qquad \text{Given parametric equation}$$
$$y = \frac{1}{4}x - 3 \qquad \text{Let } t = \tfrac{1}{4}x.$$

Because $y = \frac{1}{4}x - 3$, these parametric equations trace out a line with slope $\frac{1}{4}$ and y-intercept -3. As t increases, x also increases, so this line is traced out from left to right, as illustrated in Figure 8.73. Note that t can be any real number.

(b) Because $y = t$, it follows that a rectangular equation is $x = \sqrt{4 - y^2}$. To determine the graph of this equation, square each side.

$$x = \sqrt{4 - y^2} \qquad \text{Rectangular equation}$$
$$x^2 = 4 - y^2 \qquad \text{Square each side.}$$
$$x^2 + y^2 = 4 \qquad \text{Add } y^2 \text{ to each side.}$$

A Semicircle

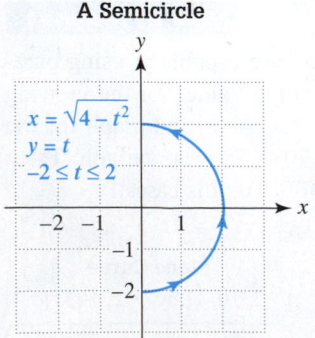

Figure 8.74

The equation $x^2 + y^2 = 4$ is a circle with center $(0, 0)$ and radius 2. Because $\sqrt{4 - y^2}$ is never negative, it follows that $x \geq 0$. Thus the parametric equation traces out only the right half of this circle. See Figure 8.74. Because $y = t$, this semicircle is traced from bottom to top as y increases from -2 to 2.

> **Now Try Exercises 9 and 13**

Parametric equations can model a circle, as shown in the next example.

EXAMPLE 4 Graphing a circle with parametric equations

Graph $x = 2 \cos t$ and $y = 2 \sin t$ for $0 \leq t \leq 2\pi$. Find an equivalent equation by using rectangular coordinates.

SOLUTION Enter and graph these parametric equations, as shown in Figures 8.75 and 8.76. Be sure to have the mode of the calculator set for parametric equations. (Note that Tstep = 0.1.)

Enter Parametric Equations

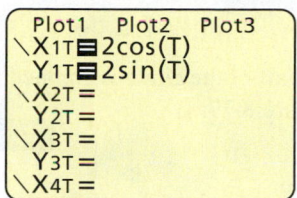

Figure 8.75

Graph Equations to Form a Circle ($r = 2$)
$[-3, 3, 1]$ by $[-2, 2, 1]$

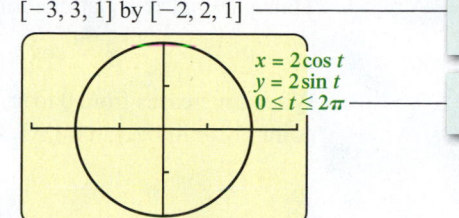

Figure 8.76

The window must be square for a circle to appear circular, rather than elliptical.

t must increase from 0 to 2π for entire circle to appear.

To verify that this is a circle, consider the following.

$$x^2 + y^2 = (2\cos t)^2 + (2\sin t)^2 \qquad x = 2\cos t, y = 2\sin t$$
$$= 4\cos^2 t + 4\sin^2 t \qquad \text{Properties of exponents}$$
$$= 4(\cos^2 t + \sin^2 t) \qquad \text{Distributive property}$$
$$= 4 \qquad \cos^2 t + \sin^2 t = 1$$

The parametric equations are equivalent to $x^2 + y^2 = 4$, which is a circle with its center at $(0, 0)$ and having radius 2.

Now Try Exercise 19

In the next two examples, an equation written in terms of x and y is converted to parametric equations.

EXAMPLE 5 Converting to parametric equations

Convert $x = y^2 - 4y + 4$ to parametric equations.

SOLUTION There is more than one way to convert this equation to parametric equations. One simple way is to let $y = t$ and then write the parametric equations as

$$x = t^2 - 4t + 4, \quad y = t,$$

where t is any real number. To write a different pair of parametric equations, note that

$$x = y^2 - 4y + 4 = (y - 2)^2.$$

Let $t = y - 2$, or $y = t + 2$, and then another pair of parametric equations is

$$x = t^2, y = t + 2.$$

Now Try Exercise 53

EXAMPLE 6 Converting to parametric equations

Given the equation $x^2 + y^2 = 1$, complete the following.
(a) Find parametric equations for this equation.
(b) What portion of the graph appears for $0 \le t \le \pi$?

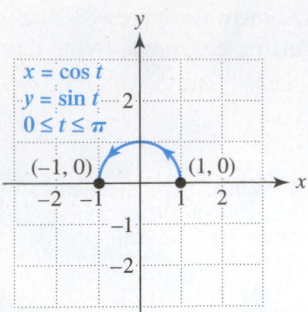

Figure 8.77

SOLUTION

(a) The graph of $x^2 + y^2 = 1$ is the unit circle. From trigonometry we know that on the unit circle $x = \cos t$ and $y = \sin t$. Since $\cos^2 t + \sin^2 t = 1$ for all t, we have the following result.

$$x^2 + y^2 = \cos^2 t + \sin^2 t = 1$$

Thus parametric equations for the unit circle are

$$x = \cos t, \qquad y = \sin t; \qquad 0 \le t \le 2\pi.$$

(b) When t increases from 0 to π, the upper half of the circle is graphed, moving from the point $(1, 0)$ to the point $(-1, 0)$. See Figure 8.77.

> **Now Try Exercise 49**

> **NOTE** The equations $x = a \cos t$, $y = a \sin t$ for $0 \le t \le 2\pi$ trace out a circle with radius $r = a$. If t is limited to an interval that is less than 2π in length, then only a portion of a circle will appear.

Applications of Parametric Equations

Parametric equations are used to simulate motion. If a ball or shot is thrown with a velocity of v feet per second at an angle θ with the horizontal, its flight can be modeled by the parametric equations

$$x = (v \cos \theta)t \qquad \text{and} \qquad y = (v \sin \theta)t - 16t^2 + h,$$

where t is in seconds and h is the initial height above the ground. The term $-16t^2$ occurs because gravity is pulling downward. See Figure 8.78. (These equations ignore air resistance.)

Modeling the Path of a Shot

Figure 8.78

EXAMPLE 7 Simulating motion with parametric equations

Three golf balls are hit simultaneously into the air at 132 feet per second (90 miles per hour), making angles of 30°, 50°, and 70° with the horizontal.
(a) Assuming the ground is level, determine graphically which ball travels the farthest horizontally. Estimate this distance.
(b) Which ball reaches the greatest height? Estimate this height.

Golf Ball Hit at Three Angles
[0, 600, 50] by [0, 400, 50]

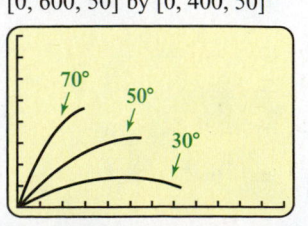

Figure 8.79 Degree Mode

SOLUTION

(a) The three sets of parametric equations determined by the three golf balls are as follows. Since $h = 0$, the only difference between the three balls is the angle of elevation.

$$X_1 = 132 \cos (30)T, \qquad Y_1 = 132 \sin (30)T - 16T^2 \quad \text{— Hit at 30°}$$
$$X_2 = 132 \cos (50)T, \qquad Y_2 = 132 \sin (50)T - 16T^2 \quad \text{— Hit at 50°}$$
$$X_3 = 132 \cos (70)T, \qquad Y_3 = 132 \sin (70)T - 16T^2 \quad \text{— Hit at 70°}$$

Angles Affect the Distance
[0, 600, 50] by [0, 400, 50]

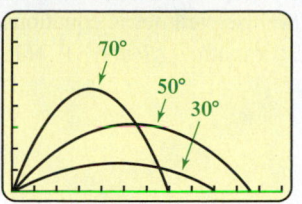

Figure 8.80 Degree Mode

The graphs of the three sets of parametric equations are shown in Figures 8.79 and 8.80, where $0 \le t \le 9$. A graphing calculator in *simultaneous mode* has been used so that we can view all three balls in flight at the same time. From the second graph we can see that the ball hit at 50° travels the farthest distance. Using the trace feature, we estimate this horizontal distance to be about 540 feet.

(b) Using the trace feature, the ball hit at 70° reaches the greatest height of about 240 feet.

Now Try Exercise 71

CLASS DISCUSSION

If a golf ball is hit at 88 feet per second (60 mi/hr), use trial and error to find the angle θ that results in a maximum distance for the ball.

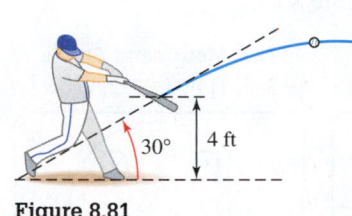

Figure 8.81

EXAMPLE 8 | **Modeling the flight of a baseball**

A baseball is hit from a height of 4 feet at a 30° angle above the horizontal. Its initial velocity is 128 feet per second. See Figure 8.81.

(a) Write parametric equations that model the flight of the baseball.
(b) Determine the horizontal distance that the ball travels in the air, assuming that the ground is level.
(c) What is the maximum height of the baseball?
(d) Would the ball clear a 4-foot-high fence that is 400 feet from the batter?

SOLUTION

(a) Let $v = 128$, $\theta = 30°$, and $h = 4$. Then the parametric equations become

$$x = (128 \cos 30°)t \quad \text{and} \quad y = (128 \sin 30°)t - 16t^2 + 4.$$

Since $\cos 30° = \frac{\sqrt{3}}{2}$ and $\sin 30° = \frac{1}{2}$, these equations can be rewritten as

$$x = (64\sqrt{3})t \quad \text{and} \quad y = 64t - 16t^2 + 4.$$

(b) To find how far the ball travels, we first determine the length of time that the ball is in flight. The ball hits the ground when $y = 0$.

$$64t - 16t^2 + 4 = 0 \qquad \text{Substitute for } y.$$

$$16t^2 - 64t - 4 = 0 \qquad \text{Rewrite quadratic equation.}$$

$$t = \frac{64 \pm \sqrt{(-64)^2 - 4(16)(-4)}}{2(16)} \qquad \text{Quadratic formula}$$

$$t \approx 4.0616 \qquad \text{or} \qquad t \approx -0.0616 \qquad \text{Approximate.}$$

After **4.0616** seconds, the ball traveled *horizontally* $x = 64\sqrt{3}(4.0616) \approx 450.2$ feet.

(c) The graph of $y = 64t - 16t^2 + 4$ is a parabola that opens downward. Using the *vertex formula*, we find that the maximum height of the ball occurs after

$$t = -\frac{b}{2a} = -\frac{64}{2(-16)} = 2 \text{ seconds.}$$

The maximum height is $y = 64(2) - 16(2)^2 + 4 = 68$ feet.

(d) Because $x = (64\sqrt{3})t$, we can determine how long it takes the ball to reach the fence by solving the equation $(64\sqrt{3})t = 400$.

$$(64\sqrt{3})t = 400, \qquad \text{or} \qquad t = \frac{400}{64\sqrt{3}} \approx 3.61 \text{ seconds}$$

After 3.61 seconds, the ball has traveled horizontally 400 feet and is

$$y = 64(3.61) - 16(3.61)^2 + 4 \approx 27 \text{ feet}$$

high. The baseball easily clears the 4-foot fence.

Now Try Exercise 75

An Application from Computer Graphics Parametric equations are used frequently in computer graphics to design a variety of figures and letters. Computer fonts are sometimes designed using parametric equations. In the next example, we use parametric equations to design a "smiley" face consisting of a head, two eyes, and a mouth. (*Source:* F. Hill, *Computer Graphics.*)

EXAMPLE 9 **Creating drawings with parametric equations**

Graph a "smiley" face using parametric equations. Answers may vary.

SOLUTION

Head We can use a circle centered at the origin for the head. If the radius is 2, then let $x = 2 \cos t$ and $y = 2 \sin t$ for $0 \le t \le 2\pi$. These equations are graphed in Figure 8.82.

Eyes For the eyes we can use two small circles. The eye in the first quadrant can be modeled by $x = 1 + 0.3 \cos t$ and $y = 1 + 0.3 \sin t$ for $0 \le t \le 2\pi$. This represents a circle centered at $(1, 1)$ with radius 0.3. The eye in the second quadrant can be modeled by $x = -1 + 0.3 \cos t$ and $y = 1 + 0.3 \sin t$ for $0 \le t \le 2\pi$, which is a circle centered at $(-1, 1)$ with radius 0.3. These equations are graphed in Figure 8.83.

Start with the Head	Add Eyes	Add Mouth and Pupils
$[-3, 3, 1]$ by $[-2, 2, 1]$	$[-3, 3, 1]$ by $[-2, 2, 1]$	$[-3, 3, 1]$ by $[-2, 2, 1]$
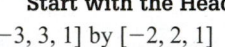	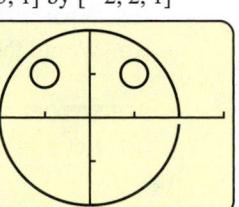	
Figure 8.82	**Figure 8.83**	**Figure 8.84**

CLASS DISCUSSION

Modify the face in Example 9 so that it is frowning. Try to find a way to make the right eye shut rather than open.

Mouth For the smile we can use the lower half of a circle. Using trial and error, we might arrive at $x = 0.5 \cos \frac{1}{2}t$ and $y = -0.5 - 0.5 \sin \frac{1}{2}t$. This is a semicircle centered at $(0, -0.5)$ with radius 0.5. Since we are letting $0 \le t \le 2\pi$, the term $\frac{1}{2}t$ ensures that only half a circle (a semicircle) is drawn. The minus sign before $0.5 \sin \frac{1}{2}t$ in the y-equation causes the lower half of the semicircle to be drawn rather than the upper half. The final result is shown in Figure 8.84. The pupils have been added by plotting the points $(1, 1)$ and $(-1, 1)$, and the coordinate axes have been turned off.

Now Try Exercise 69

8.4 Putting It All Together

Parametric equations can be used to model a wide variety of curves in the xy-plane that cannot be represented by a single function.

CONCEPT	EXPLANATION	EXAMPLES
Plane curve and parametric equations	A plane curve is a set of points (x, y) such that $x = f(t)$ and $y = g(t)$, where f and g are continuous on an interval $a \le t \le b$. The equations $x = f(t)$ and $y = g(t)$ are parametric equations with parameter t.	If $x = 2t$ and $y = t^2$ for $-1 \le t \le 2$, then the resulting graph is a portion of a parabola.

CONCEPT	EXPLANATION	EXAMPLES
Writing parametric equations in terms of x and y	Solve one of the parametric equations for t and substitute into the second equation.	If $x = t^3$ and $y = t^2 - 2$, solve $x = t^3$ for t to obtain $t = x^{1/3}$. Substituting gives $y = x^{2/3} - 2$.
Converting to parametric equations	If possible, solve the equation for one of the variables. Let the other variable equal t. Now write the first variable in terms of t. Answers may vary.	If $4x = y^2$, solve the equation for x to obtain $x = \frac{1}{4}y^2$. Let $y = t$. Then $$x = \tfrac{1}{4}t^2 \quad \text{and} \quad y = t.$$ Another possibility is $$x = t^2 \quad \text{and} \quad y = 2t.$$

8.4 Exercises

Graphs of Parametric Equations

Exercises 1–8: Use the parametric equations to complete the following.

(a) *Make a table of values for* $t = 0, 1, 2, 3$.

(b) *Plot the points from the table and graph the curve for* $0 \le t \le 3$. *Add arrows to show how the curve is traced out.*

(c) *Describe the curve.*

1. $x = t - 1,$ $\qquad$ $y = 2t$

2. $x = t + 1,$ $\qquad$ $y = t - 2$

3. $x = t + 2,$ $\qquad$ $y = (t - 2)^2$

4. $x = \frac{1}{3}t^2,$ $\qquad$ $y = t - 1$

5. $x = \sqrt{9 - t^2},$ $\qquad$ $y = t$

6. $x = t^2,$ $\qquad$ $y = 2t + 1$

7. $x = t,$ $\qquad$ $y = \sqrt{9 - t^2}$

8. $x = 3t,$ $\qquad$ $y = t^2 + 2$

Exercises 9–24: Find a rectangular equation for each curve and describe the curve. Support your result by graphing the parametric equations.

9. $x = 3t,$ $\qquad$ $y = t - 1;$ $\qquad$ $-\infty < t < \infty$

10. $x = t + 3,$ $\qquad$ $y = 2t;$ $\qquad$ $-\infty < t < \infty$

11. $x = 3t^2,$ $\qquad$ $y = t + 1;$ $\qquad$ $-\infty < t < \infty$

12. $x = t^2 - 2t + 1,$ $\qquad$ $y = t - 1;$ $\qquad$ $-\infty < t < \infty$

13. $x = \sqrt{1 - t^2},$ $\qquad$ $y = t;$ $\qquad$ $-1 \le t \le 1$

14. $x = t,$ $\qquad$ $y = \sqrt{9 - t^2};$ $\qquad$ $-3 \le t \le 3$

15. $x = t,$ $\qquad$ $y = \frac{1}{2}t^2;$ $\qquad$ $-2 \le t \le 2$

16. $x = \sqrt[3]{t},$ $\qquad$ $y = t;$ $\qquad$ $-2 \le t \le 2$

17. $x = t - 2,$ $\qquad$ $y = t^2 + 1;$ $\qquad$ $-1 \le t \le 2$

18. $x = 2t,$ $\qquad$ $y = t^2 + 1;$ $\qquad$ $-1 \le t \le 2$

19. $x = 3 \sin t,$ $\qquad$ $y = 3 \cos t;$ $\qquad$ $-\pi \le t \le \pi$

20. $x = 4 \cos t,$ $\qquad$ $y = 4 \sin t;$ $\qquad$ $0 \le t \le 2\pi$

21. $x = 2 \sin t,$ $\qquad$ $y = -2 \cos t;$ $\qquad$ $0 \le t \le 2\pi$

22. $x = \cos 2t,$ $\qquad$ $y = \sin 2t;$ $\qquad$ $0 \le t \le \pi$

23. $x = 3 \cos 2t,$ $\qquad$ $y = 3 \sin 2t;$ $\qquad$ $0 \le t \le \pi$

24. $x = 2 \cos^2 t,$ $\qquad$ $y = 2 \sin^2 t;$ $\qquad$ $0 \le t \le \frac{\pi}{2}$

Exercises 25–42: Graph the parametric equations.

25. $x = \frac{1}{3}t,$ $\qquad$ $y = \frac{2}{3}t + 1;$ $\qquad$ $-\infty < t < \infty$

26. $x = t + 3,$ $\qquad$ $y = 2t - 1;$ $\qquad$ $-\infty < t < \infty$

27. $x = t^2,$ $\qquad$ $y = 2t;$ $\qquad$ $-\infty < t < \infty$

28. $x = \frac{1}{2}(t + 2)^2,$ $\qquad$ $y = t + 2;$ $\qquad$ $-\infty < t < \infty$

29. $x = \cos t,$ $\qquad$ $y = \sin t;$ $\qquad$ $0 \le t \le \pi$

30. $x = 2 \sin t,$ $\qquad$ $y = 2 \cos t;$ $\qquad$ $-\pi \le t \le 0$

31. $x = t^3$, $\qquad\qquad y = t^2$; $\qquad\qquad -2 \le t \le 2$

32. $x = e^t$, $\qquad\qquad y = t - 1$; $\qquad\qquad -2 \le t \le 2$

33. $x = t^2$, $\qquad\qquad y = \ln t$; $\qquad\qquad 0 < t \le 2$

34. $x = t^3 - t$, $\qquad y = e^t$; $\qquad\qquad -1.5 \le t \le 1.5$

35. $x = t - \sin t$, $\qquad y = 1 - \cos t$; $\qquad 0 \le t \le 6\pi$

36. $x = t^3 + 3t$, $\qquad y = 2 \cos t$; $\qquad -1 \le t \le 1$

37. $x = 2 + \cos t$, $\qquad y = \sin t - 1$; $\qquad 0 \le t \le 2\pi$

38. $x = -2 + \cos t$, $\qquad y = \sin t + 1$; $\qquad 0 \le t \le 2\pi$

39. $x = \cos^3 t$, $\qquad y = \sin^3 t$; $\qquad 0 \le t \le 2\pi$

40. $x = \cos^5 t$, $\qquad y = \sin^5 t$; $\qquad 0 \le t \le 2\pi$

41. $x = |3 \sin t|$, $\qquad y = |3 \cos t|$; $\qquad 0 \le t \le \pi$

42. $x = 3 \sin 2t$, $\qquad y = 3 \cos t$; $\qquad 0 \le t \le 2\pi$

Exercises 43–54: Convert the given equation to parametric equations. Answers may vary.

43. $2x + y = 4$ $\qquad\qquad$ **44.** $5x - 4y = 20$

45. $y = 4 - x^2$ $\qquad\qquad$ **46.** $x = y^2 - 2$

47. $x = y^2 + y - 3$ $\qquad$ **48.** $5x = y^3 + 1$

49. $x^2 + y^2 = 4$ $\qquad\qquad$ **50.** $x^2 + y^2 = 9$

51. $\ln y = 0.1x^2$ $\qquad\qquad$ **52.** $e^x = |1 - y|$

53. $x = y^2 - 2y + 1$ $\qquad$ **54.** $x = 4y^2 + 4y + 1$

Exercises 55–60: Graph each pair of parametric equations for $0 \le t \le 2\pi$. Describe any differences in the two graphs.

55. (a) $x = 3 \cos t$, $\qquad\qquad y = 3 \sin t$
$\qquad$ (b) $x = 3 \cos 2t$, $\qquad\qquad y = 3 \sin 2t$

56. (a) $x = 2 \cos t$, $\qquad\qquad y = 2 \sin t$
$\qquad$ (b) $x = 2 \cos t$, $\qquad\qquad y = -2 \sin t$

57. (a) $x = 3 \cos t$, $\qquad\qquad y = 3 \sin t$
$\qquad$ (b) $x = 3 \sin t$, $\qquad\qquad y = 3 \cos t$

58. (a) $x = t$, $\qquad\qquad y = t^2$
$\qquad$ (b) $x = t^2$, $\qquad\qquad y = t$

59. (a) $x = -1 + \cos t$, $\qquad y = 2 + \sin t$
$\qquad$ (b) $x = 1 + \cos t$, $\qquad y = 2 + \sin t$

60. (a) $x = 2 \cos\frac{1}{2}t$, $\qquad y = 2 \sin\frac{1}{2}t$
$\qquad$ (b) $x = 2 \cos t$, $\qquad\qquad y = 2 \sin t$

Designing Shapes and Figures

Exercises 61–64: Graph the following set of parametric equations for $0 \le t \le 2\pi$ in the viewing rectangle $[0, 6, 1]$ by $[0, 4, 1]$. Identify the letter of the alphabet that is graphed.

61. $x_1 = 1$, $\qquad\qquad y_1 = 1 + t/\pi$
$\qquad x_2 = 1 + t/(3\pi)$, $\qquad y_2 = 2$
$\qquad x_3 = 1 + t/(2\pi)$, $\qquad y_3 = 3$

62. $x_1 = 1$, $\qquad\qquad y_1 = 1 + t/\pi$
$\qquad x_2 = 1 + t/(3\pi)$, $\qquad y_2 = 2$
$\qquad x_3 = 1 + t/(2\pi)$, $\qquad y_3 = 3$
$\qquad x_4 = 1 + t/(2\pi)$, $\qquad y_4 = 1$

63. $x_1 = 1$, $\qquad\qquad y_1 = 1 + t/\pi$
$\qquad x_2 = 1 + 1.3 \sin(0.5t)$, $\qquad y_2 = 2 + \cos(0.5t)$

64. $x_1 = 2 + 0.8 \cos(0.85t)$, $\qquad y_1 = 2 + \sin(0.85t)$
$\qquad x_2 = 1.2 + t/(1.3\pi)$, $\qquad y_2 = 2$

Exercises 65–68: **Designing Letters** *Find a set of parametric equations that results in a letter similar to the one shown in the figure. Use the viewing rectangle given by $[-4.7, 4.7, 1]$ by $[-3.1, 3.1, 1]$ and turn off the coordinate axes. Answers may vary.*

65. $\qquad\qquad\qquad$ **66.**

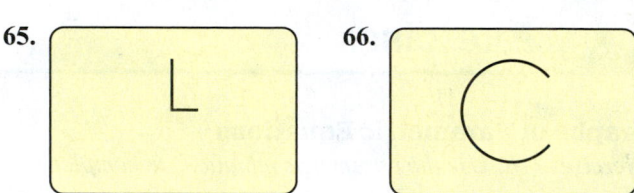

67. $\qquad\qquad\qquad$ **68.**

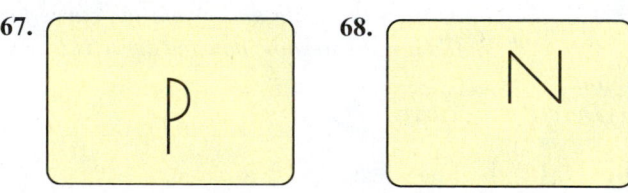

69. Designing a Face (Refer to Example 9.) Use parametric equations to create your own "smiley" face. This face should have a head, a mouth, and eyes.

70. Designing a Face Add a nose to the face that you designed in Exercise 69.

Applications

71. Flight of a Golf Ball (Refer to Example 7.) Two golf balls are hit into the air at 66 feet per second (45 mi/hr), making angles of 35° and 50° with the horizontal. If the ground is level, estimate the horizontal distance traveled by each golf ball.

72. Flight of a Golf Ball Solve Exercise 71 if, instead of the ground being level, the ground is inclined with a slope of $m = 0.1$.

73. Flight of a Golf Ball If a golf ball is hit at 88 feet per second (60 mi/hr), making an angle of 45° with the horizontal, will it go over a fence 10 feet high that is 200 feet away on level ground?

74. Simulating Gravity on the Moon   (Refer to Example 7.) If an object is thrown on the moon, the parametric equations of flight are

$$x = (v \cos\theta)t \quad \text{and} \quad y = (v \sin\theta)t - 2.66t^2 + h.$$

Estimate the horizontal distance that a golf ball hit 88 feet per second (60 mi/hr) at an angle of 45° with the horizontal travels on the moon if the moon's surface is level.

75. Flight of a Baseball   (Refer to Example 8.) A baseball is hit with an angle of elevation of 45°, from the top of a ridge that is 50 feet above an area of level ground. The initial velocity of the ball is 88 feet per second, or 60 miles per hour. Find the horizontal distance traveled by the ball in the air.

76. Flight of a Baseball   A baseball is hit from a height of 3 feet at a 60° angle above the horizontal. Its initial velocity is 64 feet per second.
(a) Write parametric equations that model the flight of the baseball.

(b) Determine the horizontal distance traveled by the ball in the air. Assume that the ground is level.

(c) What is the maximum height of the baseball? At that time, how far has the ball traveled horizontally?

(d) Would the ball clear a 5-foot high fence that is 100 feet from the batter?

Exercises 77–80: **Lissajous Figures**   *Lissajous figures occur in electronics and may be used to find the frequency of an unknown voltage. (See Extended and Discovery Exercises 1–4.) Graph the Lissajous figure for $0 \le t \le 6.5$ in the viewing rectangle $[-6, 6, 1]$ by $[-4, 4, 1]$.*

77. $x = 2 \cos t, \qquad y = 3 \sin 2t$

78. $x = 3 \cos 2t, \qquad y = 3 \sin 3t$

79. $x = 3 \sin 4t, \qquad y = 3 \cos 3t$

80. $x = 4 \sin 4t, \qquad y = 3 \sin 5t$

Writing about Mathematics

81. Describe the basic form of parametric equations. Give an example. Explain how graphs of parametric equations can differ from graphs of functions.

82. Suppose that a function is defined by $y = f(x)$, where the domain of f is $a \le x \le b$. Explain how we could represent f with parametric equations. Apply your method to $f(x) = x^2 + 1$, where $-2 \le x \le 2$.

Extended and Discovery Exercises

Exercises 1–4: **Electronic Technology**   *Parametric equations have applications in electricity. If two sinusoidal voltages, denoted by $x = V_1(t)$ and $y = V_2(t)$, are applied to an oscilloscope, a stationary pattern called a Lissajous figure may appear, as shown in the figure below. If the frequency F_1 of V_1 is known and the frequency F_2 of V_2 is unknown, then a Lissajous figure may be used to find F_2. The ratio $\frac{F_1}{F_2}$ is equal to the ratio of the corresponding number of tangents to the enclosing rectangle. The number of tangents along a vertical side of the rectangle corresponds to F_1, and the number of tangents along a horizontal side corresponds to F_2. In this figure $\frac{F_1}{F_2} = \frac{3}{2}$. Therefore $F_2 = \frac{2}{3}F_1$. Determine F_2 given the Lissajous figure and the frequency F_1 in cycles per second.* (**Source:** R. Smith and R. Dorf, Circuits, Devices, and Systems.)

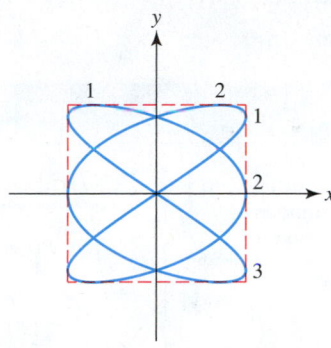

1. $F_1 = 150$

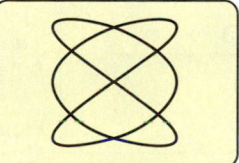

2. $F_1 = 60$

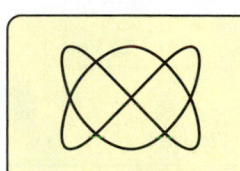

3. $F_1 = 400$

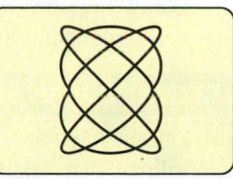

4. $F_1 = 1200$

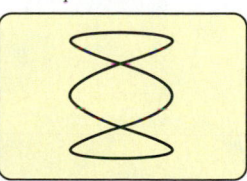

CHECKING BASIC CONCEPTS FOR SECTIONS 8.3 AND 8.4

1. Let the point P be $(-1, 3)$ and the point Q be $(3, 7)$. Find the following.
(a) $\mathbf{v} = \vec{PQ}$ (b) $\|\mathbf{v}\|$ (c) $\vec{PQ} + \vec{QP}$

2. Let $\mathbf{v} = 2\mathbf{i} - \mathbf{j}$ and $\mathbf{u} = -3\mathbf{i} + 2\mathbf{j}$. Find the following graphically and symbolically.
(a) $2\mathbf{v} + \mathbf{u}$ (b) $2\mathbf{v}$ (c) $\mathbf{v} - 3\mathbf{u}$

3. Let $\mathbf{a} = \langle 3, -2 \rangle$ and $\mathbf{b} = \langle -1, 3 \rangle$. Find the following.
(a) $\mathbf{a} \cdot \mathbf{b}$

(b) The angle θ between $\mathbf{a}$ and $\mathbf{b}$ rounded to the nearest tenth of a degree

4. Graph the parametric equations given by $x = t + 1$ and $y = (t - 1)^2$ for $-1 \le t \le 5$. Write these parametric equations in terms of x and y.

8.5 Polar Equations

- Learn the polar coordinate system
- Graph polar equations
- Graph polar equations with graphing calculators (optional)
- Solve polar equations

Introduction

Many times a change in a frame of reference can have a profound effect on the solution of a problem. Thus far we have graphed functions only in the xy-plane. Many interesting curves, such as a spiral, cannot be represented by a function since these curves fail the vertical line test. Creating a new coordinate system makes some types of equations simpler. For example, the equation $x^2 + y^2 = 1$ describes the unit circle in the rectangular coordinate system. Every point lying on the unit circle is 1 unit from the origin. If we specify a new variable r that represents the radius of the circle, then $r = 1$ also describes the unit circle in the *polar coordinate system*. A change of variable has resulted in a simpler equation.

The Polar Coordinate System

In the xy-plane we are accustomed to identifying points using (x, y), where x and y are real numbers. However, using the xy-plane is not the only way to locate a point in a plane. The **polar coordinate system** uses r and θ instead of x and y to locate a point P, as shown in Figures 8.85 and 8.86 in the following See the Concept.

See the Concept: Polar Coordinates

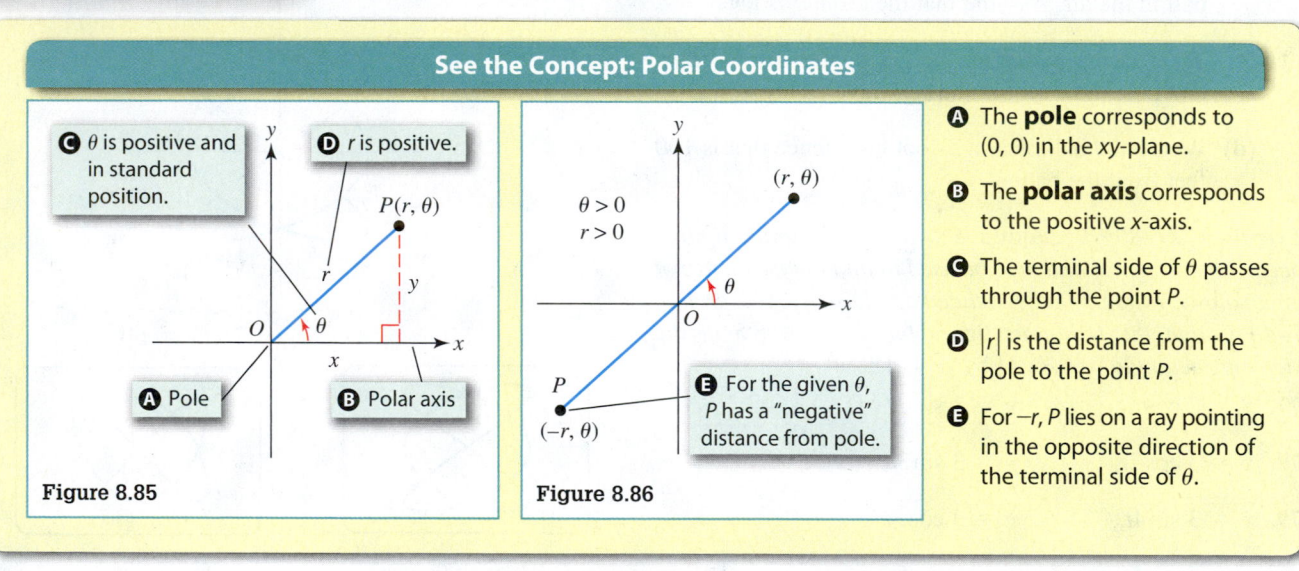

C θ is positive and in standard position.

D r is positive.

$P(r, \theta)$

r

θ

A Pole

B Polar axis

Figure 8.85

$\theta > 0$
$r > 0$

(r, θ)

θ

P

$(-r, \theta)$

E For the given θ, P has a "negative" distance from pole.

Figure 8.86

A The **pole** corresponds to $(0, 0)$ in the xy-plane.

B The **polar axis** corresponds to the positive x-axis.

C The terminal side of θ passes through the point P.

D $|r|$ is the distance from the pole to the point P.

E For $-r$, P lies on a ray pointing in the opposite direction of the terminal side of θ.

Using Figure 8.85 and trigonometry, we can establish the following relationships between rectangular and polar coordinates.

RECTANGULAR AND POLAR COORDINATES

If a point has rectangular coordinates (x, y) and polar coordinates (r, θ), then these coordinates are related as follows. (Both r and θ can be negative.)

$$x = r \cos\theta, \qquad y = r \sin\theta$$

$$r^2 = x^2 + y^2, \qquad \tan\theta = \frac{y}{x} \ (x \neq 0)$$

EXAMPLE 1 Plotting points in polar coordinates

Plot the points (r, θ) on a polar grid.

(a) $(2, 45°)$ **(b)** $(-3, 150°)$ **(c)** $\left(3.5, -\dfrac{\pi}{3}\right)$

SOLUTION

(a) Let $r = 2$ and $\theta = 45°$. Plot a point 2 units from the pole on the terminal side of $\theta = 45°$, as shown in Figure 8.87. Note that θ is in standard position.

(b) Since $r = -3 < 0$ and $\theta = 150°$, begin by locating the terminal side of θ in the second quadrant. Next plot a point 3 units from the pole in the *opposite* direction of the terminal side of θ, as shown in Figure 8.88.

(c) Since $r = 3.5$ and $\theta = -\frac{\pi}{3}$ (radians), the point is in the fourth quadrant with a distance of 3.5 from the pole. See Figure 8.89.

$r > 0$ and $\theta > 0$	$r < 0$ and $\theta > 0$	$r > 0$ and $\theta < 0$
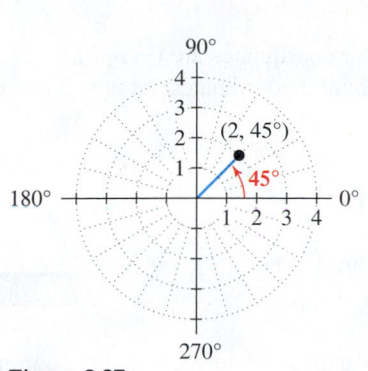	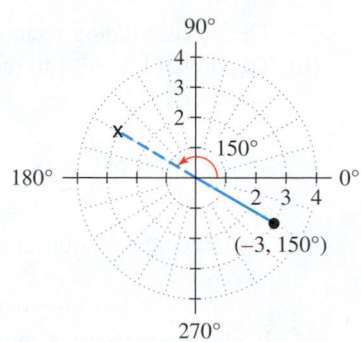	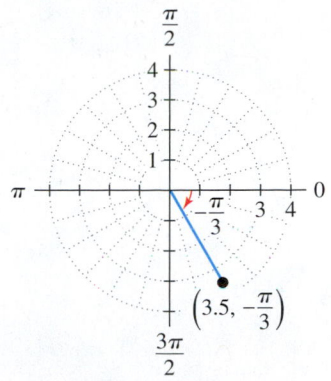
Figure 8.87	**Figure 8.88**	**Figure 8.89**

Now Try Exercises 1 and 3

MAKING CONNECTIONS

Vectors and Polar Coordinates If a vector $\mathbf{v} = \langle a, b \rangle$ has magnitude $\|\mathbf{v}\|$ and direction angle θ, then

$$a = \|\mathbf{v}\| \cos \theta \qquad \text{and} \qquad b = \|\mathbf{v}\| \sin \theta,$$

where $\|\mathbf{v}\| = \sqrt{a^2 + b^2}$ and $\tan \theta = \frac{b}{a}$, $a \neq 0$. See Figure 8.90. Similarly, if a point has rectangular coordinates (a, b) and polar coordinates (r, θ) with $r > 0$, then

$$a = r \cos \theta \qquad \text{and} \qquad b = r \sin \theta,$$

where $r = \sqrt{a^2 + b^2}$ and $\tan \theta = \frac{b}{a}$, $a \neq 0$. See Figure 8.91.

Vector **v**	Polar Coordinates (r, θ)
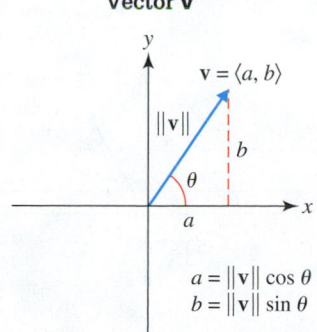	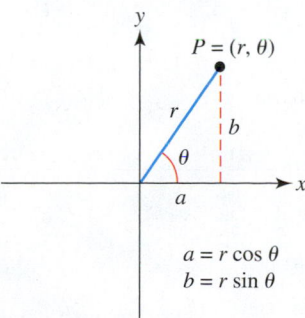
Figure 8.90	**Figure 8.91**

CLASS DISCUSSION

What can be said about a point (r, θ) if $r = 0$?

NOTE Unlike xy-coordinates, polar coordinates are *not* unique. For example, the $r\theta$-coordinates of $(2, 0°)$, $(2, 360°)$, $(2, -360°)$, and $(-2, 180°)$ all represent the same point.

In the next example, we convert polar coordinates to rectangular coordinates.

EXAMPLE 2 **Converting to rectangular coordinates**

Convert each point from polar coordinates to rectangular coordinates.

(a) $(5, 180°)$ (b) $\left(3, -\dfrac{\pi}{3}\right)$

SOLUTION

(a) To convert $(5, 180°)$ to rectangular coordinates, use the equations $x = r\cos\theta$ and $y = r\sin\theta$ with $r = 5$ and $\theta = 180°$.

$$x = 5\cos(180°) = -5$$
$$y = 5\sin(180°) = 0$$

The corresponding rectangular coordinates are $(-5, 0)$.

(b) To convert $\left(3, -\dfrac{\pi}{3}\right)$ to rectangular coordinates, let $r = 3$ and $\theta = -\dfrac{\pi}{3}$, where θ is in radians.

$$x = 3\cos\left(-\frac{\pi}{3}\right) = \frac{3}{2}, \qquad y = 3\sin\left(-\frac{\pi}{3}\right) = -\frac{3\sqrt{3}}{2} \approx -2.6$$

The rectangular coordinates are $\left(\frac{3}{2}, -\frac{3\sqrt{3}}{2}\right)$.

> **Now Try Exercises 11 and 15**

In the next example, points expressed in rectangular coordinates are converted to polar coordinates. However, this conversion is *not* unique because, unlike rectangular coordinates, polar coordinates are not unique.

EXAMPLE 3 **Expressing a point in polar coordinates**

Given the point $\left(1, \sqrt{3}\right)$ in rectangular coordinates, find polar coordinates (r, θ) that satisfy each condition.

(a) $r > 0,$ $0° \le \theta < 360°$
(b) $r > 0,$ $-360° \le \theta < 0°$
(c) $r < 0,$ $0° \le \theta < 360°$

SOLUTION

(a) $r > 0,\ 0° \le \theta < 360°$ Because $r^2 = x^2 + y^2$, let $r = \sqrt{1 + 3} = 2$. The point $\left(1, \sqrt{3}\right)$ is located in the first quadrant of the xy-plane. Therefore $\tan\theta = \frac{y}{x} = \frac{\sqrt{3}}{1}$ and $\theta = \tan^{-1}\sqrt{3} = 60°$. Let $(r, \theta) = (2, 60°)$. See Figure 8.92.

Polar Coordinates Are Not Unique

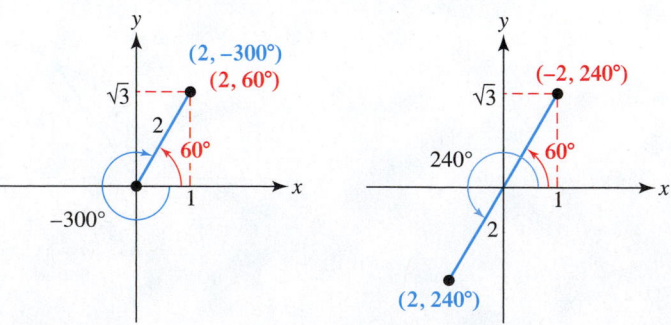

Figure 8.92   Figure 8.93

(b) $r > 0,\ -360° \le \theta < 0°$ Rather than let $\theta = 60°$, we can use $\theta = -300°$, so $(r, \theta) = (2, -300°)$. See Figure 8.92. Notice that $60°$ and $-300°$ are coterminal angles.

(c) $r < 0,\ 0° \le \theta < 360°$ We can let $r = -2$, but then we need to let angle $\theta = 60° + 180° = 240°$. Thus $(r, \theta) = (-2, 240°)$, as illustrated in Figure 8.93.

> **Now Try Exercise 23**

Graphs of Polar Equations

When we graph $y = f(x)$ in the *xy*-coordinate system, we are graphing a function. A vertical line can intersect the graph of a function at most once. As a result, many shapes such as circles, hearts, and leaves cannot be represented by a function. Like parametric equations, polar equations can be valuable for representing a variety of curves.

EXAMPLE 4   Representing polar equations

Make a table of values and graph each curve. Then describe the curve.

(a) $\theta = \dfrac{\pi}{4}$ (b) $r = 3$ (c) $r = 2 + 2\cos\theta$

SOLUTION

(a) *Numerical Representation* Since $\theta = \frac{\pi}{4}$, every point lying on this graph is of the form $\left(r, \frac{\pi}{4}\right)$, where r can be any real number. Five of these points are listed in Table 8.2.

$$\theta = \frac{\pi}{4}$$

θ	$\frac{\pi}{4}$	$\frac{\pi}{4}$	$\frac{\pi}{4}$	$\frac{\pi}{4}$	$\frac{\pi}{4}$
r	-2	-1	0	1	2

Table 8.2

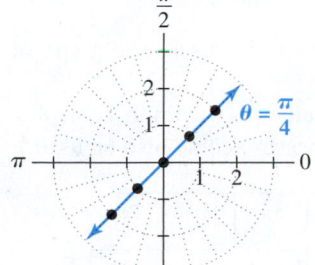

Line

Figure 8.94

Graphical Representation A graph of $\theta = \frac{\pi}{4}$ and the points in Table 8.2 are shown in Figure 8.94.

Verbal Representation The graph is a line with slope 1 passing through the pole.

(b) *Numerical Representation* A table of values is shown in Table 8.3.

$$r = 3$$

θ	0°	60°	120°	180°	240°	300°
r	3	3	3	3	3	3

Table 8.3

Circle

Figure 8.95

Graphical Representation A graph, with these points, is shown in Figure 8.95.

Verbal Representation The polar equation represents a circle with radius 3.

(c) *Numerical Representation* A table of values for $r = 2 + 2\cos\theta$ is shown in Table 8.4. For example, when $\theta = \mathbf{60°}$, then

$$r = 2 + 2\cos\mathbf{60°} = 2 + 2(0.5) = \mathbf{3}.$$

$$r = 2 + 2\cos\theta$$

θ	0°	**60°**	120°	180°	240°	300°	360°
r	4	**3**	1	0	1	3	4

Table 8.4

Cardioid

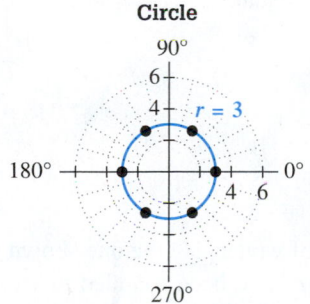

Figure 8.96

Graphical Representation To help graph the equation we can plot the points in Table 8.4. See Figure 8.96. Notice that since the cosine function has period 360°, the graph repeats after 360°.

Verbal Representation The polar equation represents a heart-shaped graph called a **cardioid**.

Now Try Exercises 31, 33, and 37

EXAMPLE 5 **Graphing in polar coordinates**

Graph each polar equation.
(a) $r = 2\sin\theta$ **(b)** $r = 2 + \cos\theta$

SOLUTION

(a) The polar equation $r = 2\sin\theta$ can be converted to rectangular coordinates by first multiplying each side of the equation by r.

$$r = 2\sin\theta \qquad \text{Given polar equation}$$
$$r^2 = 2r\sin\theta \qquad \text{Multiply by } r.$$
$$x^2 + y^2 = 2y \qquad \text{Convert to xy-coordinates.}$$
$$x^2 + y^2 - 2y = 0 \qquad \text{Subtract 2y.}$$
$$x^2 + y^2 - 2y + 1 = 1 \qquad \text{Complete the square by adding 1.}$$
$$x^2 + (y - 1)^2 = 1 \qquad \text{Perfect square trinomial}$$

This final equation represents a circle with center $(0, 1)$ and radius 1. The graph of $r = 2\sin\theta$ is shown in Figure 8.97.

Circle

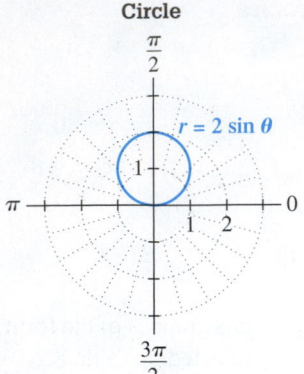

Figure 8.97

(b) To graph this polar equation, start by making a table of values as shown in Table 8.5, where values are rounded to the nearest tenth. (Degree measure could also be used.)

θ	0	$\frac{\pi}{4}$	$\frac{\pi}{2}$	$\frac{3\pi}{4}$	π	$\frac{5\pi}{4}$	$\frac{3\pi}{2}$	$\frac{7\pi}{4}$
$r = 2 + \cos\theta$	3	2.7	2	1.3	1	1.3	2	2.7

Table 8.5

The points in Table 8.5 and the equation $r = 2 + \cos\theta$ are graphed in Figure 8.98. The graph is called a **limaçon** *without an inner loop*.

Now Try Exercises 47 and 51

Limaçon

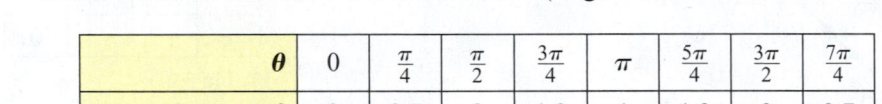

Figure 8.98

Rose Curves Polar equations in the form $r = a\sin n\theta$ or $r = a\cos n\theta$ result in graphs of rose curves. It can be shown that when n is odd there are n leaves and when n is even there are $2n$ leaves. In the next example, we graph this type of equation.

EXAMPLE 6 **Graphing a four-leaved rose**

Graph $r = 3\cos 2\theta$.

SOLUTION

To graph this polar equation by hand, start by making a table of values like the one shown in Table 8.6. Degree measure has been used, and values for r have been rounded to the nearest tenth. Notice that the values repeat themselves starting at $180°$.

θ	0°	15°	30°	45°	60°	75°	90°
$r = 3\cos 2\theta$	3	2.6	1.5	0	-1.5	-2.6	-3

θ	105°	120°	135°	150°	165°	180°
$r = 3\cos 2\theta$	-2.6	-1.5	0	1.5	2.6	3

Table 8.6

Four-Leaved Rose

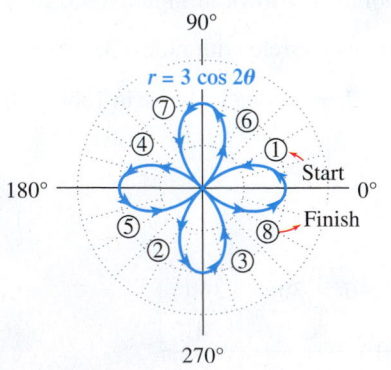

Figure 8.99

Plotting these points in order gives the graph, called a **four-leaved rose**. Note in Figure 8.99 that the graph is developed with a continuous curve, beginning with the upper half of the right horizontal leaf and ending with the lower half of that leaf. As the graph is traced, the curve passes through the pole four times. Each leaf has length 3.

Now Try Exercise 55

EXAMPLE 7   **Writing polar equations in rectangular form**

Write the polar equation in terms of x and y. Describe its graph.

(a) $r = 3 \csc \theta$ **(b)** $r = \dfrac{2}{4 \cos \theta - 3 \sin \theta}$

SOLUTION

(a) Begin by applying a reciprocal identity.

$$r = 3 \csc \theta \qquad \textit{Given equation}$$

$$r = \frac{3}{\sin \theta} \qquad \textit{Reciprocal identity: } \csc \theta = \frac{1}{\sin \theta}$$

$$\boldsymbol{r \sin \theta = 3} \qquad \textit{Multiply by } \sin \theta.$$

$$\boldsymbol{y = 3} \qquad \textit{y} = r \sin \theta$$

Its graph is a horizontal line.

(b) Start by cross multiplying.

$$r = \frac{2}{4 \cos \theta - 3 \sin \theta} \qquad \textit{Given equation}$$

$$\boldsymbol{4r \cos \theta - 3r \sin \theta = 2} \qquad \textit{Cross multiply.}$$

$$\boldsymbol{4x - 3y = 2} \qquad \textit{Substitute.}$$

$$y = \frac{4}{3}x - \frac{2}{3} \qquad \textit{Solve for y.}$$

Its graph is a line with slope $\frac{4}{3}$ and y-intercept $-\frac{2}{3}$.

Now Try Exercises 69 and 71

Logarithmic Spiral

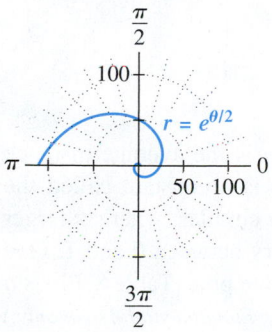

Figure 8.100

Logarithmic Spiral   In 1638, René Descartes described a *logarithmic spiral* using the complicated equation $y = x \tan (\ln (x^2 + y^2))$. This curve, shown in Figure 8.100, cannot be represented by a function. With polar coordinates, this rectangular equation reduces to the much simpler equation $r = e^{\theta/2}$. Johann Bernoulli was so entranced by this remarkable curve that he ordered it carved on his tombstone.   (***Source:*** H. Resnikoff and R. Wells, *Mathematics in Civilization.*)

Graphing Calculators and Polar Equations (Optional)

Technology can be used to make tables and graphs in polar coordinates. The table and graph in Example 4(c) are shown in Figures 8.101 and 8.102.

Cardioid

$[-6, 6, 1]$ by $[-4, 4, 1]$

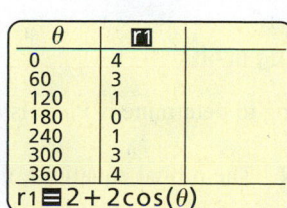

Figure 8.101 Degree Mode

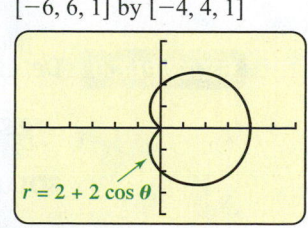

Figure 8.102

Calculator Help

To create the graph shown in Figure 8.102 with a polar equation, see Appendix A (page AP-18).

As is the case with rectangular and parametric equations, a viewing rectangle must be selected before graphing a polar equation. First choose whether the graph should be plotted in radian or degree mode. Next determine an interval for θ. Many times the interval $0° \le \theta \le 360°$ is sufficient to have the entire graph generated. Then select a square viewing rectangle, if possible.

EXAMPLE 8 Representing polar equations

Graph each curve. Then describe the curve.
(a) $r = 3\cos 2\theta$ **(b)** $r = 1 - 2\sin\theta$

Four-Leaved Rose

$[-6, 6, 1]$ by $[-4, 4, 1]$

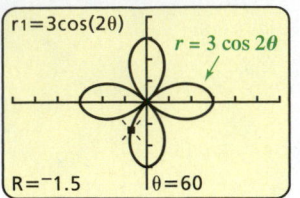

Figure 8.103

SOLUTION

(a) This equation was graphed by hand in Example 6. In Figure 8.103 a graphing calculator has been used to create a graph of $r = 3\cos 2\theta$. Turn on the polar grid and try tracing the graph to see how each leaf of the rose is generated by the polar equation. Notice, for example, the location of the point $(-1.5, 60°)$ in Figure 8.103.

Limaçon with Inner Loop

$[-6, 6, 1]$ by $[-4, 4, 1]$

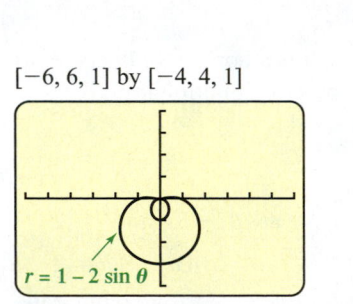

Figure 8.104

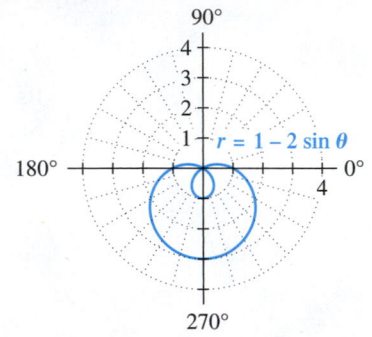

Figure 8.105

(b) A graph of $r = 1 - 2\sin\theta$ is shown in Figure 8.104. The graph is called a limaçon *with an inner loop*. A graph with better resolution is shown in Figure 8.105. Notice that the inner loop occurs when $30° \le \theta \le 150°$ and $r \le 0$.

> **Now Try Exercises 79 and 83**

Polar Equations of Conics   The polar equation

$$r = \frac{a(1 - e^2)}{1 + e\cos\theta}$$

can be used to model the orbits of planets and comets, where a is the average distance of the celestial body from the sun in astronomical units and e is a constant called the *eccentricity*. (Smaller values of e indicate that an orbit is more circular, whereas larger values indicate a more elliptical orbit. Note that values for e vary between 0 and 1. One astronomical unit equals 93 million miles.) The sun is located at the pole. Table 8.7 lists a and e for the outer planets and Pluto. (***Source:*** H. Karttunen et al., *Fundamental Astronomy*.)

Distances and Eccentricities

Planet	a	e
Jupiter	5.20	0.048
Saturn	9.54	0.056
Uranus	19.2	0.047
Neptune	30.1	0.009
Pluto	39.4	0.249

Table 8.7

EXAMPLE 9 Determining orbits

Use graphing to determine if Pluto is always farther from the sun than Neptune is.

SOLUTION   The orbital equations, given below, are graphed in Figure 8.106.

$$\text{Neptune: } r_1 = \frac{30.1(1 - 0.009^2)}{1 + 0.009\cos\theta} \qquad a = 30.1, e = 0.009$$

$$\text{Pluto: } r_2 = \frac{39.4(1 - 0.249^2)}{1 + 0.249\cos\theta} \qquad a = 39.4, e = 0.249$$

The graph shows that their orbits pass near each other. By zooming in we can determine that the orbit of Pluto actually passes inside the orbit of Neptune. See Figure 8.107.

Therefore there are times when Neptune—not Pluto—is farther from the sun. However, Pluto's average distance from the sun is considerably greater than Neptune's average distance. Neptune was farther from the sun than Pluto was for a 20-year period that ended in 1999.

Orbits of Neptune and Pluto

$[-60, 60, 10]$ by $[-40, 40, 10]$

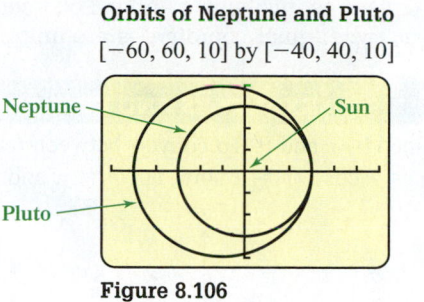

Figure 8.106

A Magnified View

$[27, 33, 1]$ by $[-2, 2, 1]$

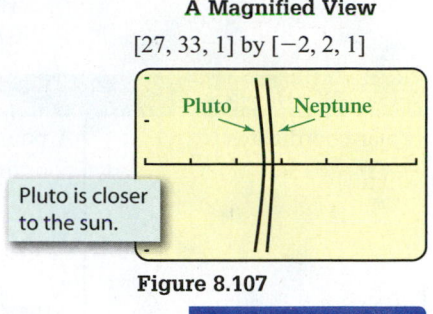

Figure 8.107

Now Try Exercise 95

Solving Polar Equations

We can solve polar equations symbolically, graphically, and numerically.

EXAMPLE 10   **Solving a polar equation**

Find values for θ where the circle $r = 3$ intersects the cardioid $r = 2 + 2\cos\theta$. Assume that $0° \le \theta \le 360°$.

SOLUTION

Symbolic Solution Begin by setting the two equations equal.

$$3 = 2 + 2\cos\theta \qquad \textit{Set equations equal.}$$

$$\cos\theta = \frac{1}{2} \qquad \textit{Solve for cos } \theta.$$

$\theta = \cos^{-1}\dfrac{1}{2} = 60°$

$$\theta = 60° \text{ or } 300° \qquad \textit{The angle in quadrant IV with a reference angle of 60° is 300°.}$$

(Be sure to check symbolic solutions.)

Graphical Solution Using the intersection-of-graphs method, let $r_1 = 3$ and let $r_2 = 2 + 2\cos(\theta)$. Their graphs intersect when $\theta = 60°$ and $300°$, as shown in Figures 8.108 and 8.109.

Solving 2 + 2 cos θ = 3 Graphically and Numerically

$[-6, 6, 1]$ by $[-4, 4, 1]$ $\qquad$ $[-6, 6, 1]$ by $[-4, 4, 1]$

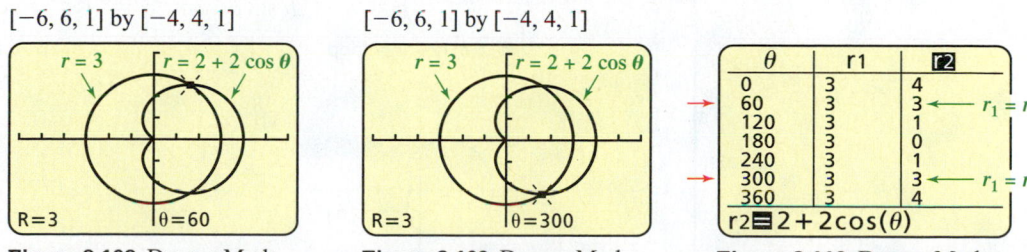

Figure 8.108 Degree Mode $\qquad$ **Figure 8.109** Degree Mode $\qquad$ **Figure 8.110** Degree Mode

Numerical Solution Numerical support is shown in Figure 8.110, where $r_1 = r_2 = 3$ when $\theta = 60°$ and $300°$.

Now Try Exercise 91

8.5 Putting It All Together

Polar equations and graphs can be used to describe curves that are not easily represented by equations involving x and y. Polar coordinates are not unique for a given point, whereas rectangular coordinates are unique for a given point.

CONCEPT	EXAMPLES AND EXPLANATIONS
Polar coordinates	A point is determined by r and θ. To convert between rectangular and polar coordinates, use $x = r\cos\theta$, $y = r\sin\theta$, $\tan\theta = \frac{y}{x}$, and $r^2 = x^2 + y^2$.

Graphs in polar coordinates	

Circle

$r = a$

Cardioid

$r = a \pm a\sin\theta$ or $r = a \pm a\cos\theta$

Rose Curve

$r = a\sin n\theta$ or $r = a\cos n\theta$

The number of leaves equals n when n is odd and is twice n when n is even.

$n = 3$

$n = 4$

Limaçon ($b \neq a$)

$r = a \pm b\sin\theta$ or $r = a \pm b\cos\theta$

$a > b$; no inner loop

$a < b$; one inner loop

8.5 Exercises

Polar Coordinates

Exercises 1–4: Plot the points (r, θ).

1. (a) $(2, 0°)$ (b) $(3, 120°)$ (c) $(-1, 135°)$

2. (a) $(-2, 60°)$ (b) $(1, 120°)$ (c) $(2, 270°)$

3. (a) $\left(2, \frac{\pi}{3}\right)$ (b) $\left(-3, -\frac{\pi}{6}\right)$ (c) $\left(0, \frac{3\pi}{4}\right)$

4. (a) $(4, \pi)$ (b) $\left(1, \frac{\pi}{4}\right)$ (c) $\left(-3, -\frac{3\pi}{2}\right)$

Exercises 5–10: Determine if the pair of polar coordinates represents the same point.

5. $(2, 180°), (2, -180°)$ 6. $(1, 90°), (-1, -90°)$

7. $(3, 45°), (3, -45°)$ 8. $(-2, 135°), (2, -135°)$

9. $(0, 40°), (0, 50°)$ 10. $(-3, 30°), (3, 210°)$

Exercises 11–20: Change the polar coordinates (r, θ) to rectangular coordinates (x, y).

11. $(3, 45°)$ 12. $(-4, 225°)$

13. $(10, 90°)$ 14. $\left(-1, \frac{\pi}{3}\right)$

15. $(5, 2\pi)$ 16. $\left(-3, -\frac{\pi}{2}\right)$

17. $(-3, 60°)$ 18. $(4, \pi)$

19. $\left(-2, -\frac{3\pi}{2}\right)$ 20. $\left(10, \frac{2\pi}{3}\right)$

Exercises 21–26: For the point given in rectangular coordinates, find equivalent polar coordinates (r, θ) that satisfy the conditions.

(a) $r > 0$, $0° \leq \theta < 360°$
(b) $r < 0$, $-180° < \theta \leq 180°$

21. $(0, 3)$ 22. $(-3, 0)$

23. $(-1, -\sqrt{3})$ 24. $(\sqrt{3}, -1)$

25. $(3, -3)$ 26. $(2, 2)$

Exercises 27–30: For the point given in rectangular coordinates, find equivalent polar coordinates (r, θ) that satisfy $r > 0$ and $0 \leq \theta < 2\pi$. Approximate θ to the nearest hundredth of a radian.

27. $(7, 24)$ 28. $(3, -4)$

29. $(-5, 12)$ 30. $(11, -60)$

Graphs of Polar Equations

Exercises 31 and 32: Graph the equation.

31. $\theta = 60°$ 32. $\theta = -135°$

Exercises 33–40: Complete the following.

(a) Make a table of values with $\theta = 0°, 90°, 180°, 270°$.

(b) Plot the points from the table and graph the curve for $0° \leq \theta \leq 360°$.

33. $r = 2$ 34. $r = 1$

35. $r = 3 \sin \theta$ 36. $r = 2 - 2 \sin \theta$

37. $r = 2 + 2 \sin \theta$ 38. $r = 3 - 2 \cos \theta$

39. $r = 2 - \cos \theta$ 40. $r = 2 + \sin \theta$

Exercises 41–58: Graph the polar equation by hand.

41. $\theta = 60°$ 42. $\theta = -\frac{\pi}{4}$

43. $r = 3$ 44. $r = \cos \theta$

45. $r \sin \theta = 3$ 46. $r \cos \theta = -2$

47. $r = 2 \cos \theta$ 48. $r = -2 \sin \theta$

49. $r = \sin 2\theta$ 50. $r = \cos 2\theta$

51. $r = 3 + \cos \theta$ 52. $r = 3 - \sin \theta$

53. $r = 1 - 2 \sin \theta$ 54. $r = 1 + 2 \cos \theta$

55. $r = 3 \sin 2\theta$ 56. $r = 2 \cos 2\theta$

57. $r = 2 \cos 3\theta$ 58. $r = 4 \sin 3\theta$

Exercises 59–66: Write the equation in polar form.

59. $y = 3$ 60. $x = -5$

61. $y = x$ 62. $y = -\sqrt{3}$

63. $x^2 + y^2 = 9$ 64. $x^2 + y^2 = 36$

65. $x^2 + y^2 = 2x$ 66. $x^2 + y^2 = -4y$

Exercises 67–74: (Refer to Example 7.) Write the polar equation in terms of x and y.

67. $r = 3$ 68. $r = 5$

69. $r = 2 \sec \theta$ 70. $r = 2 \csc \theta$

71. $r = \dfrac{3}{2 \cos \theta + 4 \sin \theta}$ 72. $r = \dfrac{2}{5 \cos \theta - \sin \theta}$

73. $r = \cos \theta$ 74. $r = 2 \sin \theta$

Exercises 75–88: Graph the curve.

75. $r = 3 + 3 \cos \theta$ (cardioid)

76. $r = 2 - 2 \sin \theta$ (cardioid)

77. $r = 3 - 2 \sin \theta$ (limaçon)

78. $r = 4 + \cos \theta$ (limaçon)

79. $r = 2 - 4\cos\theta$ (limaçon with a loop)

80. $r = 1 + 2\sin\theta$ (limaçon with a loop)

81. $r = 4\sin\theta$ (circle)

82. $r = 2\cos 3\theta$ (three-leaved rose)

83. $r = 2\cos 5\theta$ (five-leaved rose)

84. $r = 3\sin 4\theta$ (eight-leaved rose)

85. $r = \frac{\theta}{2}$ (spiral)

86. $r = e^{\theta/4}$ (logarithmic spiral)

87. $r^2 = 2\sin 2\theta$ (lemniscate)

88. $r^2 = 4\cos 2\theta$ (lemniscate)

Solving Equations in Polar Coordinates

Exercises 89 and 90: Find values for θ that satisfy the equation $r_1 = r_2$, where $0° \leq \theta \leq 360°$. Check any solutions.

89. $r_1 = 3, r_2 = 2 + 2\sin\theta$

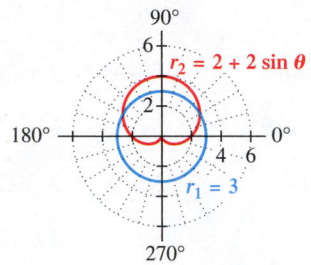

90. $r_1 = 1, r_2 = 2\cos\theta$

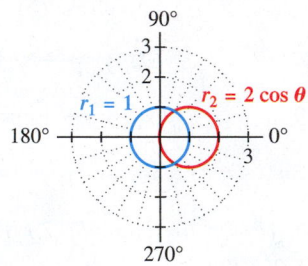

Exercises 91–94: Solve the polar equation $r_1 = r_2$, where $0° \leq \theta \leq 360°$,

(a) symbolically, (b) graphically, and (c) numerically.

91. $r_1 = 3, r_2 = 2 - 2\sin\theta$

92. $r_1 = 3, r_2 = 2 - \sin\theta$ **93.** $r_1 = 1, r_2 = 2\sin\theta$

94. $r_1 = 2 - \sin\theta, r_2 = 2 + \cos\theta$

Applications

*Exercises 95 and 96: **Planetary Orbits** (Refer to Example 9 and Table 8.7.) Graph the planetary orbits in a square viewing rectangle using polar coordinates.*

95. Saturn and Uranus   **96.** Jupiter and Neptune

*Exercises 97 and 98: **Planetary Orbits** (Refer to Example 9.)*

Planet	a	e
Mercury	0.39	0.206
Venus	0.78	0.007
Earth	1.00	0.017
Mars	1.52	0.093

Source: H. Karttunen et al., *Fundamental Astronomy.*

97. Graph the orbits of the four inner planets in the same square viewing rectangle.

98. NASA is planning future missions to Mars. Estimate graphically the closest distance possible between Earth and Mars.

*Exercises 99 and 100: **Broadcasting Patterns** Many times radio stations do not broadcast in all directions with the same intensity. To avoid interference with an existing station to the north, a new station may be licensed to broadcast only east and west. To create an east–west signal, two radio towers are sometimes used, as illustrated in the figure. Locations where the radio signal is received correspond to the interior of the curve defined by $r^2 = 40{,}000\cos 2\theta$, where the polar axis (or positive x-axis) points east. (**Source:** R. Weidner and R. Sells, Elementary Classical Physics, Vol. 2.)*

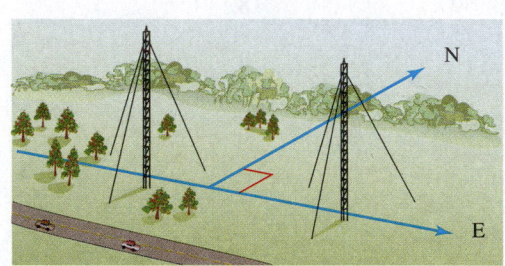

99. Graph $r^2 = 40{,}000\cos 2\theta$ for $0° \leq \theta \leq 360°$, where units are in miles. Assuming the radio towers are located near the pole, use the graph to describe the regions where the signal can be received and where the signal cannot be received.

100. (Refer to Exercise 99.) Suppose a radio signal pattern is given by $r^2 = 22{,}500\sin 2\theta$. Graph this pattern and interpret the results.

Writing about Mathematics

101. Explain why (r, θ) and $(-r, \theta + 180°)$ represent the same points in polar coordinates. Give two examples.

102. Give an example of a curve other than a circle that is more convenient to express in polar coordinates than in rectangular coordinates. Give an example of a curve that is more convenient to express using rectangular coordinates. Explain your reasoning.

Extended and Discovery Exercises

1. **Logarithmic Spiral**   Figure 8.100 shows a logarithmic spiral that can be described in rectangular coordinates by $y = x \tan (\ln (x^2 + y^2))$ and can be described in polar coordinates by the simpler equation $r = e^{\theta/2}$. Show that the first equation reduces to the second equation by assuming that $-\frac{\pi}{2} < \theta < \frac{\pi}{2}$. (*Hint:* Let $\tan \theta = \frac{y}{x}$ and $r^2 = x^2 + y^2$.)

2. **Polar Graphs**   Consider the graphs of $r = a \cos n\theta$ and $r = a \sin n\theta$, where n is a positive integer and θ is given by $0° \le \theta < 360°$.
   (a) How many times are these graphs traced over when n is even?

   (b) How many times are these graphs traced over when n is odd?

8.6 Trigonometric Form and Roots of Complex Numbers

- Learn trigonometric form
- Find products and quotients of complex numbers
- Apply De Moivre's theorem
- Find roots of complex numbers

Introduction

One of the earliest encounters with the square root of a negative number was in A.D. 50, when Heron of Alexandria derived the expression $\sqrt{81 - 144}$. Square roots of negative numbers resulted in the invention (or discovery) of complex numbers. As late as the 16th and 17th centuries, mathematicians felt uneasy about negative numbers and square roots of negative numbers. The famous mathematician René Descartes rejected complex numbers and coined the term "imaginary" numbers.

In the historical development of our present-day number system, the introduction of new numbers was often met with resistance. Today, complex numbers are readily accepted and play an important role in the design of airplanes, ships, fractals, and electrical circuits. See Exercises 67 and 68.   (*Sources:* M. Kline, *Mathematics: The Loss of Certainty; Historical Topics for the Mathematics Classroom, Thirty-first Yearbook,* NCTM.)

Trigonometric Form

Complex numbers were introduced in Section 3.3. Any complex number can be expressed in *standard form* as $a + bi$, where a and b are real numbers. The *real part* is a and the *imaginary part* is b.

Real numbers can be plotted on a number line. Since complex numbers are determined by both the real part a and the imaginary part b, we use the **complex plane** to plot complex numbers. The horizontal axis is the **real axis** and the vertical axis is the **imaginary axis**. For example, $2 + 3i$ can be plotted in the complex plane as the point $(2, 3)$, and $3 - 4i$ can be plotted as the point $(3, -4)$. See Figure 8.111.

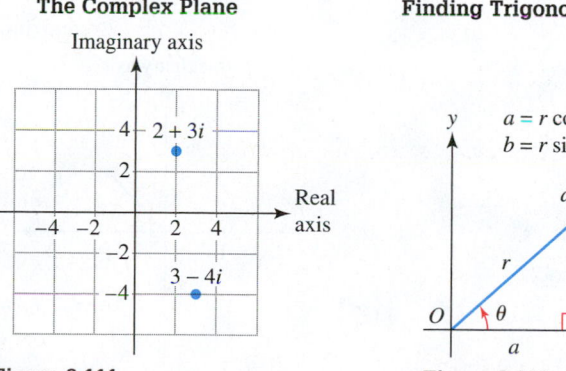

Figure 8.111

Figure 8.112

A second form for complex numbers is called *trigonometric form*, which uses the variables r and θ to locate a complex number. In Figure 8.112 we see that

$$\cos \theta = \frac{a}{r} \quad \text{and} \quad \sin \theta = \frac{b}{r}.$$

Finding Trigonometric Form

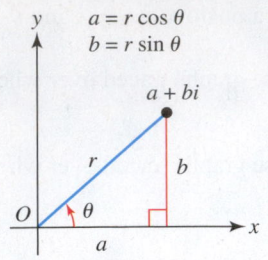

Figure 8.112 (Repeated)

Solving $\cos\theta = \frac{a}{r}$ and $\sin\theta = \frac{b}{r}$ for a and b gives

$$a = r\cos\theta \quad \text{and} \quad b = r\sin\theta. \qquad \textit{Solve for a and b.}$$

As a result, we can write the complex number $a + bi$ as follows.

$$a + bi = r\cos\theta + (r\sin\theta)i \qquad \textit{Substitute.}$$

Trigonometric form
$$= r(\cos\theta + i\sin\theta) \qquad \textit{Factor out r.}$$

By the Pythagorean theorem, $r = \sqrt{a^2 + b^2}$. It also follows that

$$\tan\theta = \frac{b}{a} \quad (a \neq 0).$$

TRIGONOMETRIC FORM OF A COMPLEX NUMBER

The expression

$$r(\cos\theta + i\sin\theta)$$

is called a **trigonometric form** of the complex number $a + bi$, where $a = r\cos\theta$ and $b = r\sin\theta$. The number $r = \sqrt{a^2 + b^2}$ is the **modulus** of $a + bi$, and θ is the **argument** of $a + bi$.

NOTE The expression $\cos\theta + i\sin\theta$ can be written as cis θ. The expression $|a + bi|$ is sometimes used to denote the *modulus* of the complex number $a + bi$. The modulus equals the distance in the complex plane between the point $a + bi$ and the origin.

EXAMPLE 1 **Converting standard form to trigonometric form**

Find the trigonometric form for each complex number, where $0° \leq \theta < 360°$.
(a) $1 + i$
(b) $-1 - i\sqrt{3}$

SOLUTION
(a) Plot $1 + i$ in the complex plane, as shown in Figure 8.113. The modulus r is

$$r = \sqrt{1^2 + 1^2} = \sqrt{2}.$$

We can see that $\tan\theta = \frac{b}{a} = \frac{1}{1}$. Therefore $\theta = \tan^{-1} 1 = 45°$. The trigonometric form is

$$\sqrt{2}\,(\cos 45° + i\sin 45°).$$

Converting to Trigonometric Form

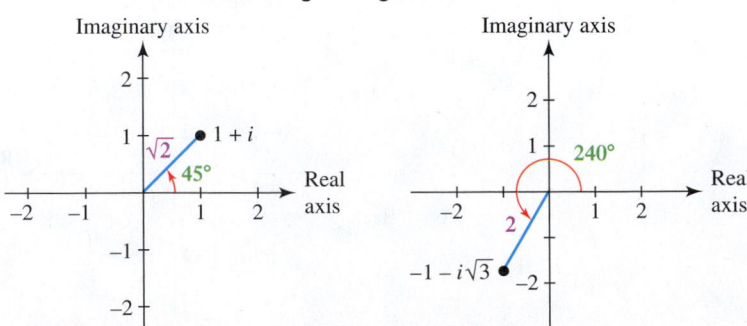

Figure 8.113   **Figure 8.114**

(b) Plot $-1 - i\sqrt{3}$, as shown in Figure 8.114. The modulus r is

$$r = \sqrt{(-1)^2 + (-\sqrt{3})^2} = 2.$$

The argument θ is in quadrant III and satisfies $\tan \theta = \dfrac{-\sqrt{3}}{-1} = \sqrt{3}$. The reference angle for θ is $\theta_R = \tan^{-1}(\sqrt{3}) = 60°$. Thus $\theta = 60° + 180° = \mathbf{240°}$ and the trigonometric form is $\mathbf{2}(\cos \mathbf{240°} + i \sin \mathbf{240°})$.

Now Try Exercises 13 and 19

Graphing Calculators (Optional)   Some calculators have the capability to convert the complex number $a + bi$ to trigonometric or **polar form**. This is illustrated in Figures 8.115 and 8.116, where the first computation gives the modulus and the second gives the argument. See Example 1. Notice that angles of 240° and −120° are coterminal angles. The value of θ is *not unique* in a trigonometric form. If θ_1 and θ_2 are *coterminal angles*, then

$$r(\cos \theta_1 + i \sin \theta_1) \qquad \text{and} \qquad r(\cos \theta_2 + i \sin \theta_2)$$

represent equivalent trigonometric forms.

Converting to Trigonometric Form

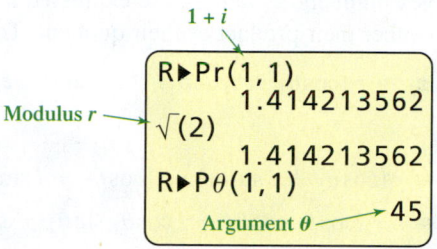

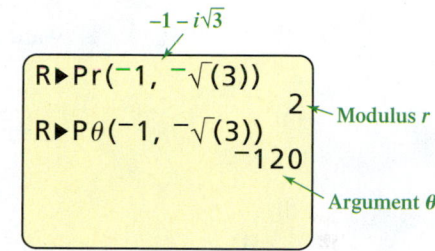

Figure 8.115 Degree Mode   **Figure 8.116** Degree Mode

MAKING CONNECTIONS

Vectors, Polar Coordinates, and Trigonometric Forms   The following are equivalent concepts.

1. Finding the magnitude $\|\mathbf{v}\|$ and direction angle θ for $\mathbf{v} = \langle a, b \rangle$
2. Finding polar coordinates (r, θ), where $r > 0$, for the point (a, b)
3. Finding the trigonometric form $r(\cos \theta + i \sin \theta)$ for $a + bi$

For example:

1. Given $\mathbf{v} = \langle 3, 4 \rangle$, then $\|\mathbf{v}\| = \sqrt{3^2 + 4^2} = 5$ and $\theta = \tan^{-1} \frac{4}{3} \approx 53.1°$. It follows that $\mathbf{v} \approx \langle 5 \cos 53.1°, 5 \sin 53.1° \rangle$. See Figure 8.117.
2. Given the point $(3, 4)$, then $r = \sqrt{3^2 + 4^2} = 5$ and $\theta = \tan^{-1} \frac{4}{3} \approx 53.1°$. It follows that $P \approx (5, 53.1°)$ in polar coordinates. See Figure 8.118.
3. Given $3 + 4i$, then $r = \sqrt{3^2 + 4^2} = 5$ and $\theta = \tan^{-1} \frac{4}{3} \approx 53.1°$. It follows that $3 + 4i \approx 5(\cos 53.1° + i \sin 53.1°)$ in trigonometric form. See Figure 8.119.

See also Extended and Discovery Exercises 1–4 at the end of this section.

Vector

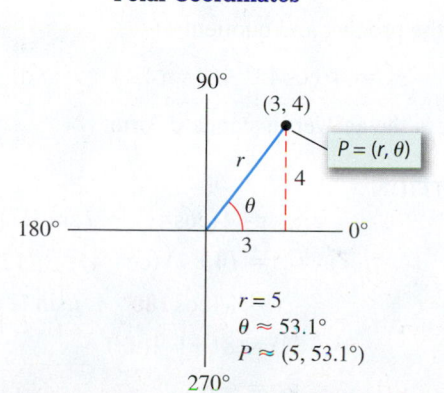

$\|\mathbf{v}\| = 5$
$\theta \approx 53.1°$
$\mathbf{v} \approx \langle 5 \cos 53.1°, 5 \sin 53.1° \rangle$

Figure 8.117

Polar Coordinates

$r = 5$
$\theta \approx 53.1°$
$P \approx (5, 53.1°)$

Figure 8.118

Trigonometric Form

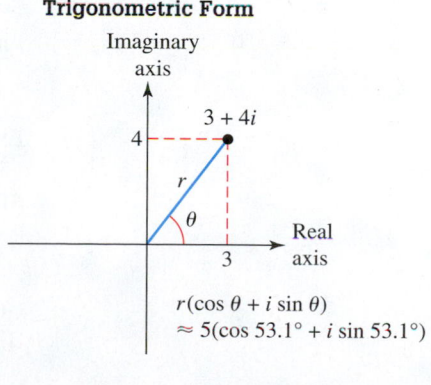

$r(\cos \theta + i \sin \theta)$
$\approx 5(\cos 53.1° + i \sin 53.1°)$

Figure 8.119

EXAMPLE 2 Converting trigonometric form to standard form

Write the complex number as $a + bi$, where a and b are real numbers.

(a) $4\left(\cos\frac{\pi}{2} + i\sin\frac{\pi}{2}\right)$

(b) $\sqrt{3}(\cos 150° + i\sin 150°)$

SOLUTION

(a) $4\left(\cos\frac{\pi}{2} + i\sin\frac{\pi}{2}\right) = 4(0 + i(1)) = 4i$

(b) $\sqrt{3}(\cos 150° + i\sin 150°) = \sqrt{3}\left(-\frac{\sqrt{3}}{2} + \frac{1}{2}i\right) = -\frac{3}{2} + \frac{\sqrt{3}}{2}i$

> **Now Try Exercises 27 and 31**

Products and Quotients of Complex Numbers

If two complex numbers, z_1 and z_2, are expressed in trigonometric form, it is straightforward to find either their product or their quotient. To see this, let

$$z_1 = r_1(\cos\theta_1 + i\sin\theta_1) \quad \text{and} \quad z_2 = r_2(\cos\theta_2 + i\sin\theta_2).$$

Then

$$z_1 z_2 = r_1(\cos\theta_1 + i\sin\theta_1) \cdot r_2(\cos\theta_2 + i\sin\theta_2)$$
$$= r_1 r_2(\cos\theta_1\cos\theta_2 + i\cos\theta_1\sin\theta_2 + i\sin\theta_1\cos\theta_2 + i^2\sin\theta_1\sin\theta_2)$$
$$= r_1 r_2((\cos\theta_1\cos\theta_2 - \sin\theta_1\sin\theta_2) + i(\cos\theta_1\sin\theta_2 + \sin\theta_1\cos\theta_2)).$$

Using the sum identities for cosine and sine, we can simplify the last expression to

$$z_1 z_2 = r_1 r_2(\cos(\theta_1 + \theta_2) + i\sin(\theta_1 + \theta_2)).$$

Using similar reasoning, it can be shown that

$$\frac{z_1}{z_2} = \frac{r_1}{r_2}(\cos(\theta_1 - \theta_2) + i\sin(\theta_1 - \theta_2)).$$

These results are summarized in the following box.

PRODUCTS AND QUOTIENTS OF COMPLEX NUMBERS

Let $z_1 = r_1(\cos\theta_1 + i\sin\theta_1)$ and $z_2 = r_2(\cos\theta_2 + i\sin\theta_2)$. Then

$$z_1 z_2 = r_1 r_2(\cos(\theta_1 + \theta_2) + i\sin(\theta_1 + \theta_2))$$

$$\frac{z_1}{z_2} = \frac{r_1}{r_2}(\cos(\theta_1 - \theta_2) + i\sin(\theta_1 - \theta_2)), \quad r_2 \neq 0.$$

EXAMPLE 3 Finding products and quotients

Find the product and quotient of

$$z_1 = 4(\cos 45° + i\sin 45°) \quad \text{and} \quad z_2 = 2(\cos 135° + i\sin 135°).$$

Express the answer in standard form.

SOLUTION

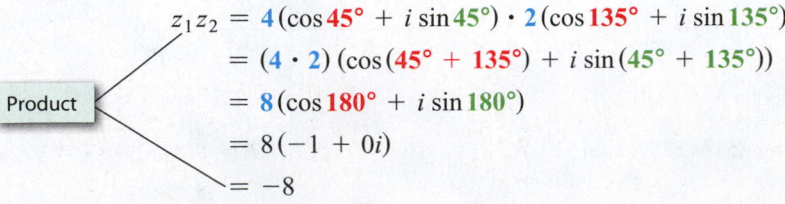

$$z_1 z_2 = 4(\cos 45° + i\sin 45°) \cdot 2(\cos 135° + i\sin 135°)$$
$$= (4 \cdot 2)(\cos(45° + 135°) + i\sin(45° + 135°))$$
$$= 8(\cos 180° + i\sin 180°)$$
$$= 8(-1 + 0i)$$
$$= -8$$

Product

$$\frac{z_1}{z_2} = \frac{4(\cos 45° + i \sin 45°)}{2(\cos 135° + i \sin 135°)}$$

$$= \frac{4}{2}(\cos(45° - 135°) + i \sin(45° - 135°))$$

Quotient

$$= 2(\cos(-90°) + i \sin(-90°))$$

$$= 2(0 + -1i)$$

$$= -2i$$

Now Try Exercise 35

Fractals and Complex Numbers   During the past 30 years, computer graphics and complex numbers have made it possible to produce many beautiful fractals. In 1977, Benoit B. Mandelbrot first used the term *fractal*. Largely because of his efforts, fractal geometry has become a new field of study. A fractal is an enchanting geometric figure with an endless self-similarity property, repeating itself infinitely with ever decreasing dimensions. If you look at smaller and smaller portions of the figure, you will continue to see the whole—much like when you look into two parallel mirrors that are facing each other. Not only do fractals have aesthetic appeal, they also have applications in science. An example of a fractal is the *Mandelbrot set* shown in Figure 8.120.   (*Source:* B. Mandelbrot, *The Fractal Geometry of Nature*.)

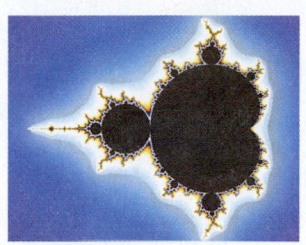

Figure 8.120

EXAMPLE 4   Analyzing the Mandelbrot set

The fractal called the Mandelbrot set is shown in Figure 8.120. To determine if a complex number $z = a + bi$ is in the Mandelbrot set, we can perform the following sequence of calculations. Let

$$z_0 = z$$
$$z_1 = z_0^2 + z_0$$
$$z_2 = z_1^2 + z_0$$
$$z_3 = z_2^2 + z_0$$

and so on. If the modulus of any z_k ever exceeds 2, then z is not in the Mandelbrot set; otherwise, z is in the Mandelbrot set. Determine if the complex number belongs to the Mandelbrot set.   (*Source:* F. Hill, *Computer Graphics*.)
(a) $z = 1 + i$ **(b)** $z = 0.5i$

SOLUTION
(a) Let $z_0 = 1 + i$. Then

$$z_1 = (1 + i)^2 + (1 + i) = 1 + 3i. \qquad z_1 = z_0^2 + z_0$$

Since the modulus of z_1 is

$$|1 + 3i| = \sqrt{1^2 + 3^2} = \sqrt{10} > 2,$$

the complex number $1 + i$ is not in the Mandelbrot set.
(b) Let $z_0 = 0.5i$. Then

$$z_1 = (0.5i)^2 + 0.5i = -0.25 + 0.5i$$
$$z_2 = (-0.25 + 0.5i)^2 + 0.5i = -0.1875 + 0.25i$$
$$z_3 = (-0.1875 + 0.25i)^2 + 0.5i \approx -0.0273 + 0.406i.$$

The modulus of each consecutive z_k never exceeds 2. Thus $0.5i$ is in the Mandelbrot set. You may find it helpful to use a calculator to perform these calculations. See Figure 8.121.

Now Try Exercises 63 and 65

$z_0^2 + z_0$

```
(.5i)^2+(.5i)
            ⁻.25+.5i
Ans^2+(.5i)
            ⁻.1875+.25i
⁻.02734375+.406...
```

Figure 8.121

De Moivre's Theorem

If a complex number z is expressed in trigonometric form, then z^n for any positive integer n can be computed easily. Let $z = r(\cos\theta + i\sin\theta)$ and consider the following.

$$z^2 = r(\cos\theta + i\sin\theta) \cdot r(\cos\theta + i\sin\theta)$$
$$= r^2(\cos(\theta + \theta) + i\sin(\theta + \theta))$$
$$= r^2(\cos 2\theta + i\sin 2\theta)$$
$$z^3 = zz^2$$
$$= r(\cos\theta + i\sin\theta) \cdot r^2(\cos 2\theta + i\sin 2\theta)$$
$$= r^3(\cos 3\theta + i\sin 3\theta)$$

In general it can be shown that

$$z^n = r^n(\cos n\theta + i\sin n\theta).$$

This result is summarized in the following theorem, which is due to Abraham De Moivre (1667–1754), a French Huguenot who was a close friend of Isaac Newton. This theorem has become a keystone of analytic trigonometry. (***Source:*** H. Eves, *An Introduction to the History of Mathematics*.)

DE MOIVRE'S THEOREM

Let $z = r(\cos\theta + i\sin\theta)$ and n be a positive integer. Then

$$z^n = r^n(\cos n\theta + i\sin n\theta).$$

EXAMPLE 5   **Finding a power of a complex number**

Use De Moivre's theorem to evaluate $(1 + i)^8$ and express the result in standard form.

SOLUTION   From Example 1(a), the trigonometric form of $z = 1 + i$ is

$$z = \sqrt{2}(\cos 45° + i\sin 45°).$$

By De Moivre's theorem,

$$z^8 = (\sqrt{2})^8(\cos(8 \cdot 45°) + i\sin(8 \cdot 45°))$$
$$= 16(\cos 360° + i\sin 360°)$$
$$= 16(1 + 0i)$$
$$= 16.$$

Now Try Exercise 47

Roots of Complex Numbers

A number w is the ***n*th root** of a number z if $w^n = z$. De Moivre's theorem can be used to find roots of complex numbers. To see this, let the trigonometric forms of w and z be

$$w = s(\cos\alpha + i\sin\alpha) \qquad \text{and} \qquad z = r(\cos\theta + i\sin\theta).$$

Then, by De Moivre's theorem, $w^n = z$ implies that

$$s^n(\cos n\alpha + i\sin n\alpha) = r(\cos\theta + i\sin\theta).$$

Thus $s^n = r$, or $s = \sqrt[n]{r}$. Furthermore, the following two equations must be satisfied.

$$\cos n\alpha = \cos\theta \qquad \text{and} \qquad \sin n\alpha = \sin\theta$$

Since the cosine and sine functions have period 360°, $n\alpha = \theta + 360° \cdot k$ for some integer k, or

$$\alpha = \frac{\theta + 360° \cdot k}{n}.$$

Substituting these results in the trigonometric form for w gives

$$w_k = \sqrt[n]{r}\left(\cos\frac{\theta + 360° \cdot k}{n} + i\sin\frac{\theta + 360° \cdot k}{n}\right).$$

We obtain a unique value of w_k for $k = 0, 1, 2, \ldots, n - 1$. This discussion is summarized in the following box.

ROOTS OF A COMPLEX NUMBER

Let $z = r(\cos\theta + i\sin\theta)$ be a nonzero complex number and n be any positive integer. Then z has exactly n distinct nth roots given by

$$w_k = \sqrt[n]{r}\left(\cos\frac{\theta + 360° \cdot k}{n} + i\sin\frac{\theta + 360° \cdot k}{n}\right),$$

where $k = 0, 1, 2, \ldots, n - 1$. If radian measure is used, then let

$$w_k = \sqrt[n]{r}\left(\cos\frac{\theta + 2\pi k}{n} + i\sin\frac{\theta + 2\pi k}{n}\right).$$

EXAMPLE 6 Finding cube roots of a complex number

Find the three cube roots of $8i$. Check your results with a calculator.

SOLUTION

Getting Started First write the complex number $8i$ in trigonometric form.

$$8i = 8(\cos 90° + i\sin 90°)$$

The three cube roots of $8i$ can be found by letting $n = 3, r = 8, \theta = 90°$, and $k = 0, 1, 2.$ ▶

$$w_0 = \sqrt[3]{8}\left(\cos\frac{90° + 360° \cdot 0}{3} + i\sin\frac{90° + 360° \cdot 0}{3}\right)$$

$k=0$
$$= 2(\cos 30° + i\sin 30°)$$
$$= 2\left(\frac{\sqrt{3}}{2} + \frac{1}{2}i\right)$$
$$= \sqrt{3} + i$$

$$w_1 = \sqrt[3]{8}\left(\cos\frac{90° + 360° \cdot 1}{3} + i\sin\frac{90° + 360° \cdot 1}{3}\right)$$

$k=1$
$$= 2(\cos 150° + i\sin 150°)$$
$$= 2\left(-\frac{\sqrt{3}}{2} + \frac{1}{2}i\right)$$
$$= -\sqrt{3} + i$$

$$w_2 = \sqrt[3]{8}\left(\cos\frac{90° + 360° \cdot 2}{3} + i\sin\frac{90° + 360° \cdot 2}{3}\right)$$

$k=2$
$$= 2(\cos 270° + i\sin 270°)$$
$$= 2(0 - i)$$
$$= -2i$$

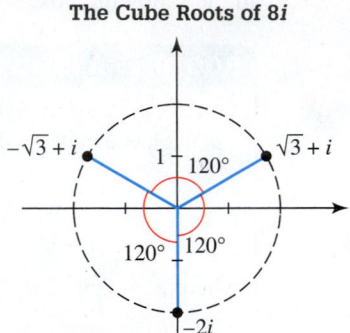

$(\sqrt{(3)}+i)^\wedge 3$

$\qquad 8i$

$(-\sqrt{(3)}+i)^\wedge 3$

$\qquad 8i$

$(-2i)^\wedge 3$

$\qquad 8i$

Figure 8.122

One cube root of the complex number $z = -1$ is $w = -1$. Find the other two cube roots of z graphically.

The three cube roots of $8i$ are $\sqrt{3} + i$, $-\sqrt{3} + i$, and $-2i$. These can be checked using a calculator, as shown in Figure 8.122.

Now Try Exercise 59

Graphical Interpretation If the three cube roots of $8i$ are plotted in the complex plane, they lie on a circle of radius 2, equally spaced 120° apart, as shown in Figure 8.123. In general, the nth roots of a complex number $z = r(\cos\theta + i\sin\theta)$ will lie equally spaced on a circle of radius $\sqrt[n]{r}$.

The Cube Roots of 8i

Figure 8.123

EXAMPLE 7 **Finding square roots of a complex number**

Find the two square roots of $1 + i\sqrt{3}$.

SOLUTION First write the complex number $1 + i\sqrt{3}$ in trigonometric form.

$$1 + i\sqrt{3} = 2\left(\cos\frac{\pi}{3} + i\sin\frac{\pi}{3}\right)$$

The two square roots of $1 + i\sqrt{3}$ can be found by letting $n = 2$, $r = 2$, $\theta = \frac{\pi}{3}$, and $k = 0, 1$.

$k = 0$

$$w_0 = \sqrt{2}\left(\cos\frac{\frac{\pi}{3} + 2\pi \cdot 0}{2} + i\sin\frac{\frac{\pi}{3} + 2\pi \cdot 0}{2}\right)$$

$$= \sqrt{2}\left(\cos\frac{\pi}{6} + i\sin\frac{\pi}{6}\right)$$

$$= \sqrt{2}\left(\frac{\sqrt{3}}{2} + \frac{1}{2}i\right)$$

$$= \frac{\sqrt{6}}{2} + \frac{\sqrt{2}}{2}i$$

$k = 1$

$$w_1 = \sqrt{2}\left(\cos\frac{\frac{\pi}{3} + 2\pi \cdot 1}{2} + i\sin\frac{\frac{\pi}{3} + 2\pi \cdot 1}{2}\right)$$

$$= \sqrt{2}\left(\cos\frac{7\pi}{6} + i\sin\frac{7\pi}{6}\right)$$

$$= \sqrt{2}\left(-\frac{\sqrt{3}}{2} - \frac{1}{2}i\right)$$

$$= -\frac{\sqrt{6}}{2} - \frac{\sqrt{2}}{2}i$$

Thus the two square roots of $1 + i\sqrt{3}$ are $\frac{\sqrt{6}}{2} + \frac{\sqrt{2}}{2}i$ and $-\frac{\sqrt{6}}{2} - \frac{\sqrt{2}}{2}i$.

Now Try Exercise 55

8.6 Putting It All Together

$\mathbf{T}$he following table summarizes some of the important topics in this section.

CONCEPT	EXPLANATION	EXAMPLES
Trigonometric form	If $z = a + bi$, then its trigonometric form is $z = r(\cos\theta + i\sin\theta)$, where $r = \sqrt{a^2 + b^2}$ and $\tan\theta = \dfrac{b}{a} (a \neq 0)$. The modulus is r and the argument is θ.	If $z = 1 + i\sqrt{3}$, then $$r = \sqrt{1^2 + (\sqrt{3})^2} = 2$$ and $\theta = \tan^{-1}\dfrac{\sqrt{3}}{1} = 60°$. Thus $z = 2(\cos 60° + i\sin 60°)$.
Products and quotients	Let $z_1 = r_1(\cos\theta_1 + i\sin\theta_1)$ and $z_2 = r_2(\cos\theta_2 + i\sin\theta_2)$. Then $$z_1 z_2 = r_1 r_2(\cos(\theta_1 + \theta_2) + i\sin(\theta_1 + \theta_2))$$ and for $r_2 \neq 0$, $$\frac{z_1}{z_2} = \frac{r_1}{r_2}(\cos(\theta_1 - \theta_2) + i\sin(\theta_1 - \theta_2)).$$	If $z_1 = 3(\cos 66° + i\sin 66°)$ and $z_2 = 2(\cos 22° + i\sin 22°)$, then $$z_1 z_2 = 6(\cos 88° + i\sin 88°)$$ and $$\frac{z_1}{z_2} = \frac{3}{2}(\cos 44° + i\sin 44°).$$
De Moivre's theorem	Let $z = r(\cos\theta + i\sin\theta)$. Then $$z^n = r^n(\cos n\theta + i\sin n\theta).$$	If $z = 2(\cos 7° + i\sin 7°)$, then $$z^3 = 8(\cos 21° + i\sin 21°).$$
Roots of complex numbers	Let $z = r(\cos\theta + i\sin\theta)$ and n be any positive integer. Then the nth roots of z are given by $$w_k = \sqrt[n]{r}\left(\cos\frac{\theta + 360° \cdot k}{n} + i\sin\frac{\theta + 360° \cdot k}{n}\right),$$ where $k = 0, 1, 2, \ldots, n - 1$.	The three cube roots of $8 = 8(\cos 0° + i\sin 0°)$ are as follows. $$w_0 = \sqrt[3]{8}\left(\cos\frac{0° + 360° \cdot 0}{3} + i\sin\frac{0° + 360° \cdot 0}{3}\right)$$ $$= 2$$ $$w_1 = \sqrt[3]{8}\left(\cos\frac{0° + 360° \cdot 1}{3} + i\sin\frac{0° + 360° \cdot 1}{3}\right)$$ $$= 2(\cos 120° + i\sin 120°) = -1 + i\sqrt{3}$$ $$w_2 = \sqrt[3]{8}\left(\cos\frac{0° + 360° \cdot 2}{3} + i\sin\frac{0° + 360° \cdot 2}{3}\right)$$ $$= 2(\cos 240° + i\sin 240°) = -1 - i\sqrt{3}$$

8.6 Exercises

The Complex Plane

Exercises 1–4: Plot the numbers in the complex plane.

1. (a) $3 + 2i$ (b) $-1 + i$ (c) $3i$

2. (a) $-2i$ (b) $2 + 2i$ (c) $2 - 2i$

3. (a) -3 (b) $4 - 2i$ (c) $-1 - 3i$

4. (a) $-1 - i$ (b) $4 + 3i$ (c) 4

Trigonometric Form

Exercises 5–12: Find the modulus of the number.

5. $1 + i$

6. $3 - 4i$

7. $12 - 5i$

8. $-24 + 7i$

9. -6

10. $15i$

11. $2 - 3i$

12. $11 - 60i$

Exercises 13–22: Write the number in trigonometric form. Let $0° \le \theta < 360°$.

13. $-1 + i$

14. $1 - i$

15. 5

16. -3

17. $4i$

18. $-i$

19. $-1 + i\sqrt{3}$

20. $-\sqrt{2} - i\sqrt{2}$

21. $\sqrt{3} + i$

22. $-\frac{\sqrt{3}}{2} + \frac{1}{2}i$

Exercises 23–26: Write the number in trigonometric form. Let $0 \le \theta < 2\pi$.

23. -2

24. $4i$

25. $-2 + 2i$

26. $1 + i\sqrt{3}$

Exercises 27–34: Write the number in standard form.

27. $5(\cos 180° + i \sin 180°)$

28. $3(\cos 90° + i \sin 90°)$

29. $2(\cos 45° + i \sin 45°)$

30. $\cos 150° + i \sin 150°$

31. $2\left(\cos \frac{\pi}{6} + i \sin \frac{\pi}{6}\right)$

32. $4\left(\cos \frac{3\pi}{2} + i \sin \frac{3\pi}{2}\right)$

33. $3(\cos 2\pi + i \sin 2\pi)$

34. $5\left(\cos \frac{3\pi}{4} + i \sin \frac{3\pi}{4}\right)$

Exercises 35–40: Find $z_1 z_2$ and $\frac{z_1}{z_2}$. Express your answer in standard form.

35. $z_1 = 9(\cos 45° + i \sin 45°)$,
$z_2 = 3(\cos 15° + i \sin 15°)$

36. $z_1 = 5(\cos 90° + i \sin 90°)$,
$z_2 = 2(\cos 30° + i \sin 30°)$

37. $z_1 = 6\left(\cos \frac{3\pi}{4} + i \sin \frac{3\pi}{4}\right)$,
$z_2 = \cos \frac{\pi}{4} + i \sin \frac{\pi}{4}$

38. $z_1 = 4(\cos 300° + i \sin 300°)$,
$z_2 = 2(\cos 60° + i \sin 60°)$

39. $z_1 = \cos 15° + i \sin 15°$,
$z_2 = \cos \left(-\frac{\pi}{4}\right) + i \sin \left(-\frac{\pi}{4}\right)$

40. $z_1 = 11\left(\cos \frac{2\pi}{3} + i \sin \frac{2\pi}{3}\right)$,
$z_2 = 22(\cos 30° + i \sin 30°)$

Powers of Complex Numbers

Exercises 41–46: Use De Moivre's theorem to evaluate the expression. Write the result in standard form.

41. $(2(\cos 30° + i \sin 30°))^3$

42. $(3(\cos 45° + i \sin 45°))^4$

43. $(\cos 10° + i \sin 10°)^{36}$

44. $(\cos 1° + i \sin 1°)^{90}$

45. $(5(\cos 60° + i \sin 60°))^2$

46. $(2(\cos 90° + i \sin 90°))^5$

Exercises 47–50: Use De Moivre's theorem to evaluate the expression. Write the result in standard form and check it using a calculator.

47. $(1 + i)^3$

48. $(3i)^4$

49. $(\sqrt{3} + i)^5$

50. $(2 - 2i)^6$

Roots of Complex Numbers

Exercises 51–62: Find the following roots and express them in standard form. Check your results with a calculator.

51. The square roots of $4(\cos 120° + i \sin 120°)$

52. The cube roots of $27(\cos 180° + i \sin 180°)$

53. The cube roots of $\cos 180° + i \sin 180°$

54. The fourth roots of $16(\cos 240° + i \sin 240°)$

55. The square roots of i

56. The cube roots of 1

57. The cube roots of -8

58. The square roots of $-4i$

59. The cube roots of $64i$

60. The fourth roots of -1

61. The fourth roots of 81

62. The square roots of $-1 + i\sqrt{3}$

Fractals

Exercises 63–66: **Mandelbrot Set** *(Refer to Example 4.) Determine if the complex number belongs to the Mandelbrot set.*

63. $-0.4i$

64. $0.5 + i$

65. $1 + i$

66. $-0.2 + 0.2i$

Applications

67. Electrical Circuits *Impedance* is a measure of the opposition to the flow of current in an electrical circuit. It consists of two parts called the *resistance* and the *reactance*. Light bulbs add resistance to an electrical circuit, and reactance occurs when electricity passes through coils of wire like those found in electric motors. Impedance Z in ohms (Ω) may be expressed as a complex number, where the real part represents the resistance and the imaginary part represents the reactance. For example, if the resistive part is 3 ohms and the reactive part is 4 ohms, then the impedance could be described by the complex number $Z = 3 + 4i$. The modulus of Z gives the total impedance in ohms. In a series circuit like the one shown in the figure, the total impedance is the sum of the individual impedances. (**Source:** R. Smith and R. Dorf, *Circuits, Devices and Systems.*)

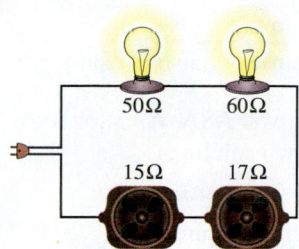

(a) The circuit contains two light bulbs and two electric motors. If it is assumed that the light bulbs represent resistance and the motors represent reactance, express impedance as $Z = a + bi$.

(b) Find total impedance in ohms by calculating the modulus of Z.

68. Electrical Circuits (Continuation of Exercise 67.) In the parallel electrical circuit shown in the figure, impedance Z is given by

$$Z = \frac{1}{\frac{1}{Z_1} + \frac{1}{Z_2}},$$

where Z_1 and Z_2 represent the impedances for the two branches of the circuit. (**Source:** G. Wilcox and C. Hesselberth, *Electricity for Engineering Technology.*)

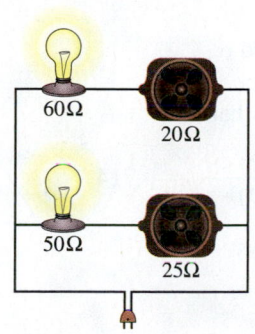

(a) Find Z.

(b) Find total impedance to the nearest tenth of an ohm by calculating the modulus of Z.

Writing about Mathematics

69. Explain how to find a trigonometric form of a complex number $a + bi$. Give an example. Is trigonometric form unique for a given complex number z? Explain.

70. Suppose that one fourth root w of a complex number z is known. Explain how to find the other fourth roots of z graphically.

Extended and Discovery Exercises

Exercises 1–4: **Making Connections** *Complete the following. Choose angles in $[0°, 360°)$ and round to the nearest tenth.*

(a) Write the vector in terms of its magnitude and direction angle.
(b) Write the point (a, b) in polar coordinates.
(c) Write the complex number $a + bi$ in trigonometric form.

1. $\langle \sqrt{3}, 1 \rangle, (\sqrt{3}, 1), \sqrt{3} + i$

2. $\langle 5, 12 \rangle, (5, 12), 5 + 12i$

3. $\langle 4, -3 \rangle, (4, -3), 4 - 3i$

4. $\langle -7, 24 \rangle, (-7, 24), -7 + 24i$

CHECKING BASIC CONCEPTS FOR SECTIONS 8.5 AND 8.6

1. Plot the following points (r, θ) on a polar grid.
  (a) $(2, 30°)$ **(b)** $(3, -60°)$ **(c)** $(-4, 120°)$

2. Graph the equation.
  (a) $r = 2$ **(b)** $\theta = -\frac{\pi}{4}$
  (c) $r = 3 + 3\cos\theta$ **(d)** $r = 3\cos 2\theta$

3. Plot the numbers in the complex plane.
  (a) $-3 + 2i$ **(b)** $-4 - 3i$

4. Find the trigonometric form of $1 + i\sqrt{3}$.

5. Find the three cube roots of i.

8 Summary

CONCEPT	EXPLANATION AND EXAMPLES

Section 8.1 Law of Sines

Law of Sines

$$\frac{\sin \alpha}{a} = \frac{\sin \beta}{b} = \frac{\sin \gamma}{c}, \quad \text{or} \quad \frac{a}{\sin \alpha} = \frac{b}{\sin \beta} = \frac{c}{\sin \gamma}$$

The law of sines can be used to solve triangles given ASA, AAS, or SSA. SSA is called the ambiguous case and can have zero, one, or two solutions.

Example: Given $\beta = 32°$, $\gamma = 46°$, and $c = 10$, find b.
We are given AAS, as illustrated in the figure.

$$\frac{b}{\sin \beta} = \frac{c}{\sin \gamma} \quad \text{implies that} \quad b = \frac{10 \sin 32°}{\sin 46°} \approx 7.37.$$

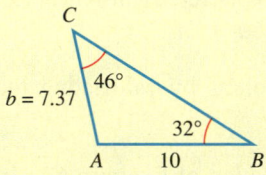

Section 8.2 Law of Cosines

Law of Cosines

$$a^2 = b^2 + c^2 - 2bc \cos \alpha$$
$$b^2 = a^2 + c^2 - 2ac \cos \beta$$
$$c^2 = a^2 + b^2 - 2ab \cos \gamma$$

The law of cosines can be used to solve triangles given SAS or SSS. Each situation results in a unique solution.

Example: Given $a = 5$, $b = 6$, and $c = 7$, find α.
We are given SSS, as illustrated in the figure.
$a^2 = b^2 + c^2 - 2bc \cos \alpha$ implies that

$$\cos \alpha = \frac{b^2 + c^2 - a^2}{2bc} = \frac{6^2 + 7^2 - 5^2}{2(6)(7)} \approx 0.714.$$

Thus $\alpha \approx \cos^{-1}(0.714) \approx 44.4°$.

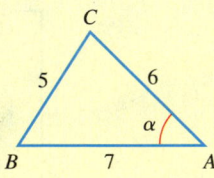

Section 8.3 Vectors

Vectors

A vector is a directed line segment that has both magnitude (length) and direction. Three different representations for a vector are

$$\mathbf{v} = \langle a_1, a_2 \rangle, \quad \mathbf{v} = a_1 \mathbf{i} + a_2 \mathbf{j}, \quad \text{and} \quad \mathbf{v} = \overrightarrow{PQ}.$$

Horizontal component $= a_1$; *vertical component* $= a_2$

Magnitude: $\|\mathbf{v}\| = \sqrt{a_1{}^2 + a_2{}^2}$

Direction angle: The positive angle θ $(0° \le \theta < 360°)$ between the x-axis and the position vector $\mathbf{v}$, where

$$\mathbf{v} = \langle \|\mathbf{v}\| \cos \theta, \|\mathbf{v}\| \sin \theta \rangle.$$

The horizontal component is $\|\mathbf{v}\| \cos \theta$ and the vertical component is $\|\mathbf{v}\| \sin \theta$.

CONCEPT	EXPLANATION AND EXAMPLES

Section 8.3 Vectors (CONTINUED)

Calculations Involving Vectors

Let $\mathbf{a} = \langle a_1, a_2 \rangle$ and $\mathbf{b} = \langle b_1, b_2 \rangle$.

Sum: $\mathbf{a} + \mathbf{b} = \langle a_1 + b_1, a_2 + b_2 \rangle$
Difference: $\mathbf{a} - \mathbf{b} = \langle a_1 - b_1, a_2 - b_2 \rangle$
Scalar Multiple: $k\mathbf{a} = \langle ka_1, ka_2 \rangle$
Dot Product: $\mathbf{a} \cdot \mathbf{b} = a_1 b_1 + a_2 b_2$

Angle θ between $\mathbf{a}$ and $\mathbf{b}$: $\theta = \cos^{-1}\left(\dfrac{\mathbf{a} \cdot \mathbf{b}}{\|\mathbf{a}\| \, \|\mathbf{b}\|} \right)$

Work

If a constant force $\mathbf{F}$ is applied to an object that moves along a vector $\mathbf{D}$, then the work done is $W = \mathbf{F} \cdot \mathbf{D}$.

Section 8.4 Parametric Equations

Plane Curve and Parametric Equations

A plane curve can be defined by the parametric equations $x = f(t)$ and $y = g(t)$, where f and g are continuous and t is the parameter.

Example: $x = \cos t, y = \sin t; 0 \le t \le 2\pi$
Since $x^2 + y^2 = \cos^2 t + \sin^2 t = 1$,
this curve describes the unit circle.
See the figure.

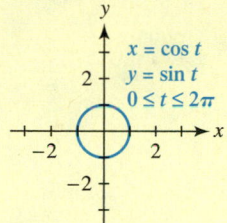

Section 8.5 Polar Equations

Polar Coordinates and Equations

Points are identified using r and θ instead of x and y. Polar equations are plotted in the polar plane, where the pole corresponds to the origin and the polar axis corresponds to the positive x-axis. Polar coordinates are not unique.

Example: $r = \cos 5\theta$ (rose curve with 5 leaves)

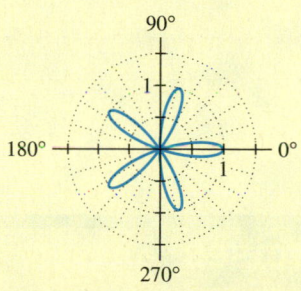

Section 8.6 Trigonometric Form and Roots of Complex Numbers

Trigonometric Form and Complex Numbers

The expression $r(\cos\theta + i\sin\theta)$ is the trigonometric form of $a + bi$, where $a = r\cos\theta$ and $b = r\sin\theta$.

Modulus: $|z| = r = \sqrt{a^2 + b^2}$; *argument*: θ

Example: $z = 2(\cos 30° + i\sin 30°) = \sqrt{3} + i$
Modulus: $r = 2$; argument: $\theta = 30°$

CONCEPT	EXPLANATION AND EXAMPLES

Section 8.6 Trigonometric Form and Roots of Complex Numbers (CONTINUED)

Operations on Complex Numbers and Trigonometric Form

Let $z_1 = r_1(\cos\theta_1 + i\sin\theta_1)$ and $z_2 = r_2(\cos\theta_2 + i\sin\theta_2)$.

$$z_1 z_2 = r_1 r_2(\cos(\theta_1 + \theta_2) + i\sin(\theta_1 + \theta_2))$$

$$\frac{z_1}{z_2} = \frac{r_1}{r_2}(\cos(\theta_1 - \theta_2) + i\sin(\theta_1 - \theta_2))$$

$$z_1{}^n = r_1{}^n(\cos(n\theta_1) + i\sin(n\theta_1)) \qquad \text{De Moivre's theorem}$$

Example: Let $z_1 = 6(\cos 120 + i\sin 120°)$ and $z_2 = 2(\cos 80° + i\sin 80°)$.

$$z_1 z_2 = 12(\cos 200° + i\sin 200°)$$

$$\frac{z_1}{z_2} = 3(\cos 40° + i\sin 40°)$$

$$z_1{}^4 = 6^4(\cos(4 \cdot 120°) + i\sin(4 \cdot 120°))$$

Roots of Complex Numbers

If $z = r(\cos\theta + i\sin\theta)$, then the nth roots of z are given by

$$w_k = \sqrt[n]{r}\left(\cos\frac{\theta + 360° \cdot k}{n} + i\sin\frac{\theta + 360° \cdot k}{n}\right),$$

where $k = 0, 1, 2, \ldots, n-1$.

Example: Let $z = 4i = 4(\cos 90° + i\sin 90°)$.
The square roots of z are as follows.

$$w_k = \sqrt{4}\left(\cos\frac{90° + 360° \cdot k}{2} + i\sin\frac{90° + 360° \cdot k}{2}\right) \quad \text{for } k = 0 \text{ and } 1$$

Simplifying gives

$$w_0 = 2(\cos 45° + i\sin 45°) = \sqrt{2} + i\sqrt{2} \quad \text{and}$$
$$w_1 = 2(\cos 225° + i\sin 225°) = -\sqrt{2} - i\sqrt{2}.$$

8 Review Exercises

Exercises 1–4: Solve the triangle. Approximate values to the nearest tenth.

1.
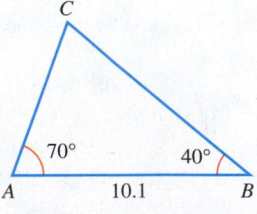

2.
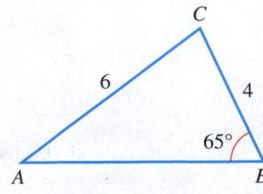

3.
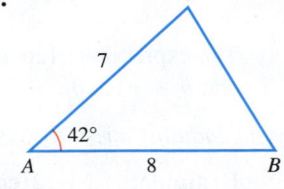

4.
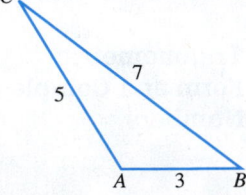

Exercises 5–10: Solve the triangle. Approximate values to the nearest tenth.

5. $\alpha = 19°, \beta = 46°, b = 13$

6. $\alpha = 30°, b = 10, a = 8$

7. $\gamma = 20°, b = 8, c = 11$

8. $\alpha = 70°, b = 17, a = 5$

9. $b = 23, \gamma = 35°, a = 18$

10. $a = 65, b = 45, c = 32$

Exercises 11–14: Approximate the area of the triangle to the nearest tenth.

11. $a = 12.3, b = 13.7, \gamma = 39°$

12. $\alpha = 40°, \beta = 55°, c = 67$

13. $a = 34, b = 67, c = 53$

14. $a = 2.1, b = 1.7, c = 2.2$

*Exercises 15 and 16: Complete the following for vector **v**.*

(a) Give the horizontal and vertical components.
(b) Find $\|\mathbf{v}\|$.
*(c) Graph **v** and interpret $\|\mathbf{v}\|$.*

15. $\mathbf{v} = \langle 3, 4 \rangle$ **16.** $\mathbf{v} = -5\mathbf{i} + 12\mathbf{j}$

*Exercises 17 and 18: A vector **v** has initial point P and terminal point Q.*

(a) Graph $\overrightarrow{PQ}$.
(b) Write $\overrightarrow{PQ}$ as $\mathbf{v} = a_1\mathbf{i} + a_2\mathbf{j}$.
(c) Find $\|\overrightarrow{PQ}\|$.

17. $P = (0, 0), Q = (-2, -4)$

18. $P = (3, 2), Q = (-3, -1)$

Exercises 19–22: Find each of the following.

(a) $2\mathbf{a}$ (b) $\mathbf{a} - 3\mathbf{b}$ (c) $\mathbf{a} \cdot \mathbf{b}$
*(d) The angle θ between **a** and **b** rounded to a tenth of a degree*

19. $\mathbf{a} = \langle 3, -2 \rangle, \mathbf{b} = \langle 1, 1 \rangle$

20. $\mathbf{a} = \langle 3, 2 \rangle, \mathbf{b} = \langle -2, -3 \rangle$

21. $\mathbf{a} = 2\mathbf{i} + 2\mathbf{j}, \mathbf{b} = \mathbf{i} + \mathbf{j}$

22. $\mathbf{a} = \mathbf{i} - 2\mathbf{j}, \mathbf{b} = 2\mathbf{i} + \mathbf{j}$

23. Resultant Force   Use the parallelogram rule to find the magnitude of the resultant force of the two forces shown in the figure.

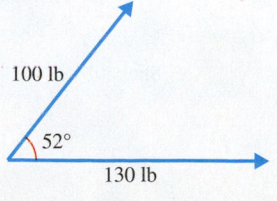

24. Work   Find the work done when $\mathbf{F} = 300\mathbf{i} + 400\mathbf{j}$ is applied to an object that moves along the vector $\mathbf{D} = 10\mathbf{i} - 2\mathbf{j}$, where units are in pounds and feet. Find the magnitude of $\mathbf{F}$ and interpret the result.

Exercises 25–27: Graph the parametric equations.

25. $x = t + 2, \quad y = t^2 - 3; \quad -2 \le t \le 2$

26. $x = t^3 - 4, \quad y = t - 1; \quad 0 \le t \le 2$

27. $x = 2\cos t, \quad y = -2\sin t; \quad 0 \le t \le 2\pi$

28. Change the polar coordinates (r, θ) to rectangular coordinates (x, y).

  (a) $(2, 135°)$ **(b)** $(-1, 60°)$

Exercises 29–34: Graph the polar equation.

29. $r = 1 + \cos\theta$ **30.** $r = \sin\theta$

31. $r = 3\sin 3\theta$ **32.** $r = 2 - \cos\theta$

33. $r = 3 + 3\sin\theta$ **34.** $r = 1 - 2\sin\theta$

35. Plot each number in the complex plane.

  (a) $4 - i$ **(b)** $-2 + 2i$

  (c) $-2i$ **(d)** -4

36. Write each complex number in trigonometric form. Let θ satisfy $0° \le \theta < 360°$.

  (a) $-2 + 2i$ **(b)** $\sqrt{3} + i$

  (c) $5i$ **(d)** -6

37. Find $z_1 z_2$ and $\dfrac{z_1}{z_2}$ in standard form, if

$$z_1 = 4(\cos 150° + i \sin 150°) \text{ and}$$
$$z_2 = 2(\cos 30° + i \sin 30°).$$

38. Use De Moivre's theorem to evaluate z^4 if the trigonometric form of z is $z = 2(\cos 45° + i \sin 45°)$. Write the result in standard form.

Exercises 39 and 40: Find the following roots.

39. The square roots of $4(\cos 60° + i \sin 60°)$

40. The cube roots of $27i$

Applications

41. Airplane Navigation   An airplane takes off with a bearing of 130° and flies 350 miles. Then it changes its course to a bearing of 60° and flies for 500 miles. Determine how far the plane is from its takeoff point.

42. Obstructed View   To find the distance between two points A and B on opposite sides of a small building, a surveyor measures AC as 63.15 feet, angle ACB as 43.56°, and CB as 103.53 feet. Find the distance between A and B to the nearest tenth of a foot.

43. Height of an Airplane Two observation points A and B are 950 feet apart. From these points the angles of elevation of an airplane are 52° and 57°, as illustrated in the figure. Find the height of the airplane to the nearest foot.

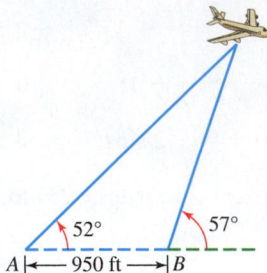

44. Area of a Lot A surveyor measures two sides and the included angle of a triangular lot as $a = 93.6$ feet, $b = 110.6$ feet, and $\gamma = 51.8°$. Find the area of the lot.

45. Area of a Lot Find the area of the quadrangular lot shown in the figure to the nearest square foot.

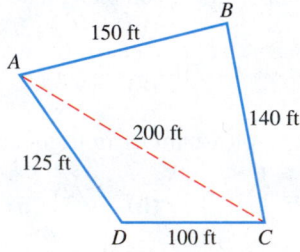

46. Interpreting a Vector A boat is heading west at 20 miles per hour in a current that is flowing south at 6 miles per hour. Find a vector **v** that models the direction and speed of the boat. What does $\|\mathbf{v}\|$ represent?

47. Robotics Consider the planar two-arm manipulator shown in the figure, where units are in centimeters. Let the upper arm be modeled by the vector $\mathbf{a} = 40\mathbf{i} - 20\mathbf{j}$ and the forearm be modeled by the vector $\mathbf{b} = 20\mathbf{i} + 30\mathbf{j}$. (**Source:** J. Craig, *Introduction to Robotics.*)

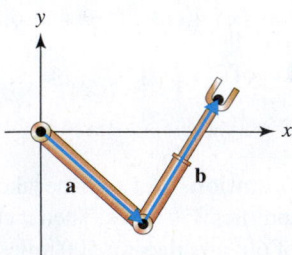

(a) Find a vector **c** that gives the position of the hand.

(b) How far is the hand from the origin?

(c) Find the position of the hand if the length of the forearm doubles.

48. Work A 200-pound person walks 0.75 mile up a hiking trail inclined at 15°. Use a dot product to compute the work done in foot-pounds.

49. Flight of a Golf Ball A golf ball is hit at 50 feet per second, making an angle of 45° with the horizontal as it leaves the club. If the ground is level, estimate the horizontal distance traveled by the golf ball in the air.

50. Aerial Photography A camera lens has an angular coverage of 86°. Suppose an aerial photograph is taken vertically with no tilt at an altitude of 3500 feet over ground with an increasing slope of 5°, as shown in the figure. Calculate the ground distance CB that would appear in the resulting photograph (to the nearest foot). (**Source:** F. Moffitt, *Photogrammetry.*)

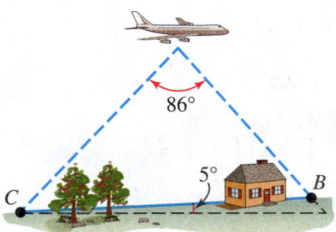

Extended and Discovery Exercises

1. Velocity of a Star The velocity vector **v** of a star relative to the sun can be expressed as the resultant vector of two perpendicular vectors—the radial velocity $\mathbf{v}_r$ and the tangential velocity $\mathbf{v}_t$, where $\mathbf{v} = \mathbf{v}_r + \mathbf{v}_t$, as illustrated in the figure. If a star is located near the sun and its velocity is large, then its motion across the sky will also be large. Barnard's Star is relatively close to the sun with a distance of 35 trillion miles. Relative to the sun, it moves across the sky through an angle of 10.34″ per year, which is the largest of any known star. Its radial velocity is $\mathbf{v}_r = 67$ miles per second toward the sun. (**Sources:** A. Acker and C. Jaschek, *Astronomical Methods and Calculations;* M. Zeilik, *Introductory Astronomy and Astrophysics.*)

(a) Approximate $\|\mathbf{v}_t\|$ for Barnard's Star in miles per second. (*Hint:* Use $s = r\theta$.)

(b) Compute $\|\mathbf{v}\|$.

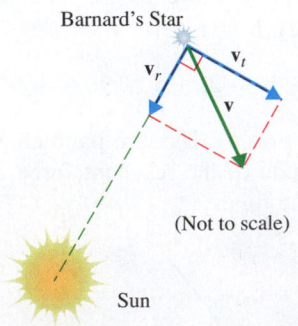

(Not to scale)

2. Fractals The fractal called the *Julia set* is shown in the figure. To determine if a complex number $z = a + bi$ belongs to this set, repeatedly compute the sequence of values

$$z_1 = z^2 - 1, \quad z_2 = z_1^2 - 1, \quad z_3 = z_2^2 - 1,$$

and so on. If the modulus of any of the resulting complex numbers exceeds 2, then the complex number z is not in the Julia set. Otherwise z is in this set. Determine if the complex numbers belong to the Julia set.

(a) $z = 0 + 0i$

(b) $z = 1 + i$

(c) $z = -0.2i$

3. Aerial Photography Aerial photography from satellites and planes has become important to many applications such as map-making, national security, and surveying. If a photograph is taken from a plane with the camera tilted at an angle θ, then trigonometry can be used to find the ground coordinates of the object, as illustrated in the figure. If an object's photographic coordinates in inches are (x, y), then its ground coordinates (X, Y) in feet can be computed using the formulas

$$X = \frac{ax}{f \sec\theta - y \sin\theta},$$

$$Y = \frac{ay \cos\theta}{f \sec\theta - y \sin\theta},$$

where f is the focal length of the camera in inches and a is the altitude of the airplane in feet. Suppose the photographic coordinates of a house and nearby forest fire are $(x_H, y_H) = (0.9, 3.5)$ and $(x_F, y_F) = (2.1, -2.4)$, respectively. (**Source:** F. Moffitt, *Photogrammetry*.)

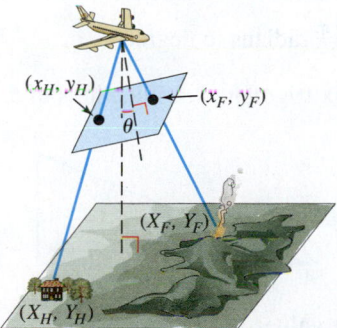

(a) Find the distance between the house and the fire on the photograph to the nearest hundredth of an inch.

(b) If the photograph was taken at 7400 feet by a camera with a focal length of 6 inches and a tilt of $\theta = 4.1°$, find the ground distance in feet between the house and the fire.

4. Shadows in Computer Graphics Vectors are used frequently in computer graphics to simulate realistic shadows. For example, suppose an airplane is taking off from a runway, as illustrated in the figure. Let the length and direction of the airplane at takeoff be given by vector **L**. If the sunlight is assumed to be perpendicular to the runway, then the length of the airplane's shadow cast on the runway equals $\|\mathbf{L}\| \cos\theta$. From previous work, we know that if vector **R** points in the direction of the runway, then

$$\mathbf{L} \cdot \mathbf{R} = \|\mathbf{L}\| \, \|\mathbf{R}\| \cos\theta.$$

Solving for $\|\mathbf{L}\| \cos\theta$ results in

$$\|\mathbf{L}\| \cos\theta = \frac{\mathbf{L} \cdot \mathbf{R}}{\|\mathbf{R}\|}.$$

The expression $\dfrac{\mathbf{L} \cdot \mathbf{R}}{\|\mathbf{R}\|}$ represents the **component of L in the direction of R**. Find the length of the shadow on the runway for each **L** and **R**. Assume units are in feet. (**Source:** C. Pokorny and C. Gerald, *Computer Graphics*.)

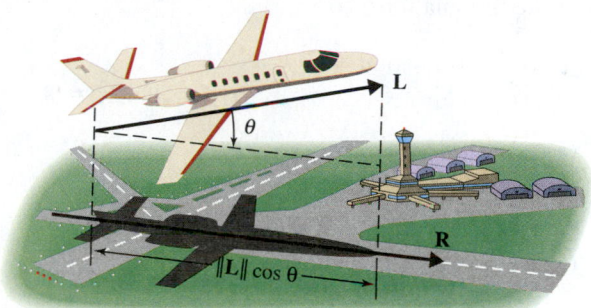

(a) $\mathbf{L} = 40\mathbf{i} + 10\mathbf{j}, \qquad \mathbf{R} = \mathbf{i}$

(b) $\mathbf{L} = 35\mathbf{i} + 5\mathbf{j}, \qquad \mathbf{R} = 10\mathbf{i} + \mathbf{j}$

(c) $\mathbf{L} = 100\mathbf{i} + 8\mathbf{j}, \qquad \mathbf{R} = 30\mathbf{i} + 2\mathbf{j}$

5. The Dot Product In the figure,

$$\mathbf{a} = \langle a_1, a_2 \rangle, \qquad \mathbf{b} = \langle b_1, b_2 \rangle, \qquad \text{and}$$

$$\mathbf{a} - \mathbf{b} = \langle a_1 - b_1, a_2 - b_2 \rangle.$$

Apply the law of cosines to the triangle and derive the equation $\mathbf{a} \cdot \mathbf{b} = \|\mathbf{a}\| \, \|\mathbf{b}\| \cos\theta$.

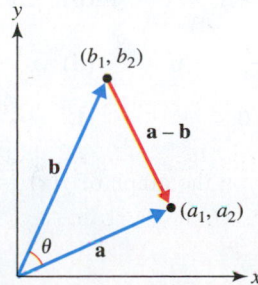

Cumulative Review Exercises

1. Find the exact distance between $(3, -2)$ and $(7, -9)$.

2. Graph $y = g(x)$ by hand.
   (a) $g(x) = \frac{1}{2}x - 1$ (b) $g(x) = |x + 2|$
   (c) $g(x) = \sqrt{x + 1}$ (d) $g(x) = \frac{1}{x + 1}$
   (e) $g(x) = -3\cos\left(2\left(x - \frac{\pi}{2}\right)\right)$

3. Find the domain of $f(x) = \sqrt{4 - x}$ and evaluate $f(-5)$.

4. Find the average rate of change of $f(x) = 3x^2 - 2x$ from $x = 1$ to $x = 3$.

5. Find the difference quotient for $f(x) = 4x^2$.

6. Write the slope-intercept form for a line that passes through $(-1, 4)$ and $(1, -3)$.

7. Determine the x- and y-intercepts on the graph of $4x + 3y = -12$. Graph the equation.

8. If $G(t) = 300 - 10t$ calculates the gallons of water in a tank after t seconds, interpret the numbers 300 and -10 in the formula for $G(t)$.

9. Solve each equation.
   (a) $|2x - 5| = 6$ (b) $6x^2 + 22x = 8$
   (c) $x^3 = x$ (d) $x^4 - 2x^2 - 3 = 0$
   (e) $2e^{3x} - 1 = 50$
   (f) $3x^{2/3} = 12$
   (g) $\sin t = \frac{1}{2}, 0 \le t < 2\pi$
   (h) $\tan 2t = -\sqrt{3}$
   (i) $2\cos^2 t + \cos t = 1$

10. Evaluate $f(-1)$ and graph $y = f(x)$. Is f continuous on its domain?
$$f(x) = \begin{cases} x + 2 & \text{if } -3 \le x \le -1 \\ -2x - 1 & \text{if } -1 < x < 1 \\ x^2 - 4 & \text{if } 1 \le x \le 3 \end{cases}$$

11. Solve each inequality. Use interval notation.
    (a) $4(x - 3) > 1 - x$ (b) $|2x - 1| \le 3$
    (c) $x^2 - 2x - 3 > 0$ (d) $x^3 - 4x > 0$
    (e) $\frac{x}{x - 1} \le 0$ (f) $-4 \le 4 - 3x \le 12$

12. Find the vertex on the graph of $f(x) = 2x^2 - 4x + 1$.

13. Divide each expression.
    (a) $\frac{3x^3 - x + 2}{x + 2}$ (b) $\frac{2x^3 - 3x^2 + x - 1}{2x - 1}$

14. A cubic function f has zeros $-2, 3$, and 5 and leading coefficient 4. Write the complete factored form of $f(x)$.

15. State the domain of $f(x) = \frac{3x + 4}{2 - 3x}$. Find any vertical or horizontal asymptotes on the graph of $y = f(x)$.

16. Let $f(x) = \frac{1}{x^2 - 3}$ and $g(x) = 2x + 1$. Find each of the following.
    (a) $(f + g)(2)$ (b) $(g \circ f)(2)$
    (c) $(g/f)(x)$ (d) $(f \circ g)(x)$

17. Find $f^{-1}(x)$ if $f(x) = 3x - 2$.

18. Find an exponential function given by $f(x) = Ca^x$ that models the data in the table.

x	0	1	2	3
$f(x)$	2	6	18	54

19. There are initially 5000 bacteria, and this number doubles in size every 1.5 hours. Find C and a so that $f(t) = Ca^t$ models the number of bacteria after t hours.

20. One thousand dollars is deposited in an account that pays 7% annual interest compounded quarterly. Find the amount in the account after 8 years.

21. Simplify each logarithm by hand.
    (a) $\log_3 \frac{1}{27}$ (b) $\ln \frac{1}{e^3}$
    (c) $\log \sqrt[3]{10}$ (d) $\log_4 32 - \log_4 \frac{1}{2}$

22. Expand the expression $\ln \sqrt[3]{\frac{x^3 y}{z^2}}$.

23. Convert $225°$ to radians.

24. Convert $\frac{11\pi}{6}$ radians to degrees.

25. Find the six trigonometric functions of θ.

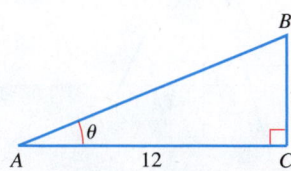

26. Find exact values of the six trigonometric functions of $\theta = \frac{2\pi}{3}$.

27. Find the values of the six trigonometric functions of θ if $\sin \theta = -\frac{7}{25}$ and $\sec \theta < 0$.

28. Evaluate $\tan^{-1} \sqrt{3}$.

29. Solve the right triangle shown in Exercise 25.

30. Simplify $(1 - \cos t)(1 + \cos t)$.

31. Factor $\cot^2 \theta - 2 \cot \theta + 1$.

32. Verify each identity.
   (a) $(1 - \cos^2 \theta)(1 + \tan^2 \theta) = \tan^2 \theta$
   (b) $\dfrac{\sin(\alpha + \beta)}{\cos \alpha \cos \beta} = \tan \alpha + \tan \beta$
   (c) $\dfrac{\csc^2 \theta}{1 - \cot^2 \theta} = -\sec 2\theta$

33. Solve triangle ABC. Approximate to the nearest tenth.
   (a) $\alpha = 31°, \gamma = 53°, b = 15$
   (b) $\alpha = 31°, a = 6, b = 5$
   (c) $\beta = 56°, a = 6, c = 8$
   (d) $a = 6, b = 7, c = 8$

34. Find the area of a triangle with sides of length 7, 10, and 15 feet.

35. Let $\mathbf{a} = \langle -5, 12 \rangle$ and $\mathbf{b} = \langle 7, -24 \rangle$. Find the following.
   (a) $\|\mathbf{b}\|$ (b) $2\mathbf{a} - 3\mathbf{b}$ (c) $\mathbf{a} \cdot \mathbf{b}$
   (d) The angle between $\mathbf{a}$ and $\mathbf{b}$

36. Graph the parametric equations $x = \frac{1}{2}t$, $y = (t - 1)^2$ for any real number t.

37. Graph the polar equation $r = 3 - 2 \sin \theta$ for θ satisfying $0° \le \theta < 360°$.

38. Find the three cube roots of $27i$.

Applications

39. **Construction**   A box is being constructed by cutting 3-inch squares from the corners of a rectangular sheet of metal that is 4 inches longer than it is wide. If the box is to have a volume of 351 cubic inches, find the dimensions of the metal sheet.

40. **Modeling Data**   Find a quadratic function in the form $f(x) = a(x - h)^2 + k$ that models the data in the table at the top of the next column *exactly*.

x	-1	0	1	2	3	4
y	-26	-11	-2	1	-2	-11

41. **Designing a Box**   A box with rectangular sides and a top is being designed to hold 288 cubic inches and to have a surface area of 288 square inches. If the width is half the length, find possible dimensions for the box.

42. **Inverse Variation**   The force of gravity F varies inversely with the square of the distance r from the *center* of the moon. If a rock weighs 50 pounds on the surface of the moon ($r = 1750$ kilometers), how much would this rock weigh at a distance of 7000 kilometers from the center of the moon?

43. **Length of a Shadow**   The angle of elevation of the sun is 63°. Find the length of the shadow cast by a person who is 5 feet tall. Round to the nearest tenth of a foot.

44. **Modeling Temperature**   The monthly average high temperatures for a location are shown in the table. Model these data using $f(x) = a \sin(b(x - c)) + d$.

Month	1	2	3	4	5	6
Temperature (°F)	25	28	37	50	63	72

Month	7	8	9	10	11	12
Temperature (°F)	75	72	62	50	38	28

45. **Angle of Elevation**   An 85-foot tree casts a 57-foot shadow. Estimate the angle of elevation of the sun to the nearest tenth of a degree.

46. **Distance**   An ore ship is traveling east at 20 miles per hour. The bearing of a submerged rock is 75°. After 2 hours, the bearing of the rock is 305°. Find the distance between the ship and the rock when the second bearing is determined.

47. **Surveyor**   A surveyor measures two sides of a triangular lot to be $a = 242$ feet and $b = 165$ feet. The angle between these sides is $\gamma = 72°$.
   (a) Find the length of the third side c.
   (b) Estimate the area of the lot.

48. **Flight of a Golf Ball**   A golf ball is hit into the air at 96 feet per second, making an angle of 60° with the horizontal. Use parametric equations to estimate the horizontal distance traveled by the golf ball before it strikes the ground.

9 Systems of Equations and Inequalities

In 2000, less than 6% of the world's population had Internet access. Today, the majority of people on Earth are able to get online. It would be impossible for billions of people to download, post, tweet, and stream data, without the mathematics used to create and manage Internet networks. Systems of equations and matrices are vitally important to the success of social networks such as Facebook, Twitter, Spotify, and Pinterest.

Special types of graphs can be used to represent simple networks, and matrices can be used to summarize these graphs. With matrices it is possible to identify the connections between friends in a social network or to analyze web page links. (See Examples 2 and 9 in Section 9.5.)

In this chapter, we will see that mathematics can be used to solve systems of equations, compute movement in computer graphics, analyze web page links, and even represent social networks. Throughout history, many important discoveries have been based on the insights of a few people. The mathematicians who first worked with matrices and systems of equations could not have imagined the profound impact their work would have in the 21st century.

> Go deep enough into anything and you will find mathematics.
>
> —Dean Schlicter

Source: *Internet World Stats;* R. Hanneman and M. Riddle, *Introduction to Social Network Methods.*

9.1 Functions and Systems of Equations in Two Variables

- Evaluate functions of two variables
- Understand basic concepts about systems of equations
- Recognize types of linear systems
- Apply the method of substitution
- Apply the elimination method
- Apply graphical and numerical methods
- Solve problems involving joint variation

Introduction

Many quantities in everyday life depend on more than one variable.

- Finding the area of a rectangular room requires both its *length* and *width*.
- The heat index is a function of *temperature* and *humidity*.
- Grade point average is computed using *grades* and *credit hours*.

Quantities determined by more than one variable often are computed by a *function* of more than one variable. The mathematical concepts that we have already studied concerning functions of one input also apply to functions of more than one input. One unifying concept about every function is that it produces *at most one output* each time it is evaluated.

Functions of Two Variables

In order to perform addition, two numbers must be provided. The addition of x and y results in one output, z. The addition function f can be represented symbolically by

$$f(x, y) = x + y, \text{ where } z = f(x, y).$$

For example, the addition of 3 and 4 can be written as

$$z = f(3, 4) = 3 + 4 = 7.$$

In this case, $f(x, y)$ is a **function of two inputs** or a **function of two variables**. The **independent variables** are x and y, and z is the **dependent variable**. The output z depends on the inputs x and y. Other arithmetic operations can be defined similarly. For example, a division function can be defined by $g(x, y) = \frac{x}{y}$, where $z = g(x, y)$.

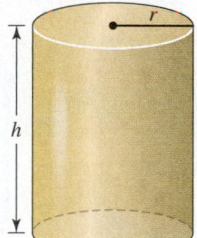

Figure 9.1

EXAMPLE 1 Evaluating functions of more than one input

For each function, evaluate the expression and interpret the result.
(a) $f(3, -4)$, where $f(x, y) = xy$ represents the multiplication function
(b) $M(120, 5)$, where $M(m, g) = \frac{m}{g}$ computes the gas mileage when traveling m miles on g gallons of gasoline
(c) $V(0.5, 2)$, where $V(r, h) = \pi r^2 h$ calculates the volume of a cylindrical barrel with radius r feet and height h feet. (See Figure 9.1.)

SOLUTION
(a) $f(3, -4) = (3)(-4) = -12$. The product of 3 and -4 is -12.
(b) $M(120, 5) = \frac{120}{5} = 24$. If a car travels 120 miles on 5 gallons of gasoline, its gas mileage is 24 miles per gallon.
(c) $V(0.5, 2) = \pi(0.5)^2(2) = 0.5\pi \approx 1.57$. If a barrel has a radius of 0.5 foot and a height of 2 feet, it holds about 1.57 cubic feet of liquid.

Now Try Exercises 1, 3, and 5

Systems of Equations in Two Variables

A **linear equation in two variables** can be written in the form

$$ax + by = k,$$

where a, b, and k are constants and a and b are not equal to 0. Examples of linear equations in two variables include

$$2x - 3y = 4, \qquad -x - 5y = 0, \qquad \text{and} \qquad 5x - y = 10.$$

Many situations involving two variables result in the need to determine values for x and y that satisfy *two* equations. For example, suppose that we would like to find a pair of numbers whose average is 10 and whose difference is 2. The function $f(x, y) = \frac{x + y}{2}$ calculates the average of two numbers, and $g(x, y) = x - y$ computes their difference. The solution can be found by solving two linear equations $f(x, y) = 10$ and $g(x, y) = 2$.

$$\frac{x + y}{2} = 10$$
$$x - y = 2$$

> System of linear equations

This pair of equations is called a **system of linear equations** because we are solving more than one linear equation at once. A **solution** to a system of equations in two variables consists of an x-value *and* a y-value that satisfy *both* equations simultaneously. The set of all solutions is called the **solution set**. Using trial and error, we see that $x = $ **11** and $y = $ **9** satisfy both equations. This is the only solution and it can be expressed as the *ordered pair* (**11**, **9**).

Systems of equations that have at least one nonlinear equation are called **nonlinear systems of equations**.

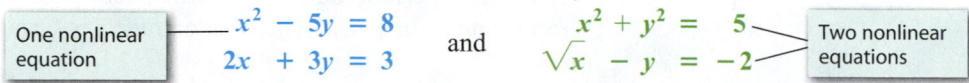

Nonlinear Systems of Equations

One nonlinear equation
$$x^2 - 5y = 8$$
$$2x + 3y = 3$$

and

$$x^2 + y^2 = 5$$
$$\sqrt{x} - y = -2$$
Two nonlinear equations

Types of Linear Systems in Two Variables

Any system of linear equations in two variables can be written in the form

$$a_1 x + b_1 y = c_1$$
$$a_2 x + b_2 y = c_2,$$

where $a_1, b_1, c_1, a_2, b_2,$ and c_2 are constants. The graph of this system consists of *two* lines in the xy-plane. The following See the Concept summarizes the three possible types of linear systems. Note that **coincident lines** are identical lines and indicate that the two equations are equivalent and have the same graph. (See **B** below.)

CLASS DISCUSSION

Explain why a system of linear equations in two variables cannot have two or three solutions.

See the Concept: Three Types of Linear Systems

One Solution	Infinitely Many Solutions	No Solutions

B Coincident lines

C Parallel lines

A Intersecting lines

Consistent System
Independent Equations

Consistent System
Dependent Equations

Inconsistent System

A The solution is given by the coordinates of the point of intersection.

B Every point on the coincident lines represents a solution.

C The distinct parallel lines have no points in common.

A **consistent system** of linear equations has either one solution, meaning the equations are **independent**, or infinitely many solutions, meaning the equations are **dependent.** An **inconsistent system** has no solutions.

EXAMPLE 2 **Recognizing types of linear systems**

Graph each system of equations and find any solutions. Identify the system as consistent or inconsistent. If the system is consistent, state whether the equations are dependent or independent.

(a) $\begin{aligned} x - y &= 2 \\ -x + y &= 1 \end{aligned}$ (b) $\begin{aligned} 4x - y &= 2 \\ x - 2y &= -3 \end{aligned}$ (c) $\begin{aligned} 2x - y &= 1 \\ -4x + 2y &= -2 \end{aligned}$

SOLUTION

Getting Started Start by solving each equation for y. Use the resulting slope-intercept form to graph each line. Determine any points of intersection. ▶

(a) Graph $y = x - 2$ and $y = x + 1$, as shown in Figure 9.2. Their graphs (parallel lines) do not intersect, so there are no solutions. The system is inconsistent.

(b) Graph $y = 4x - 2$ and $y = \frac{1}{2}x + \frac{3}{2}$, as shown in Figure 9.3. Their graphs intersect at $(1, 2)$. There is one solution, so the system is consistent and the equations are independent. Because graphical solutions can be approximate, we check this solution by substituting 1 for x and 2 for y in the given system.

$$4(1) - 2 = 2 \checkmark \quad \text{True}$$
$$(1) - 2(2) = -3 \checkmark \quad \text{True}$$

Because $(1, 2)$ satisfies *both* equations, it is the solution.

(c) Solving the equations $2x - y = 1$ and $-4x + 2y = -2$ for y results in the same equation: $y = 2x - 1$. Therefore their graphs coincide, as shown in Figure 9.4. (Note that the second equation results when the first equation is multiplied by -2, so the equations are equivalent.) Any point on this line is a solution to both equations. Thus the system has infinitely many solutions of the form $\{(x, y) \mid 2x - y = 1\}$ and is consistent. The equations are dependent.

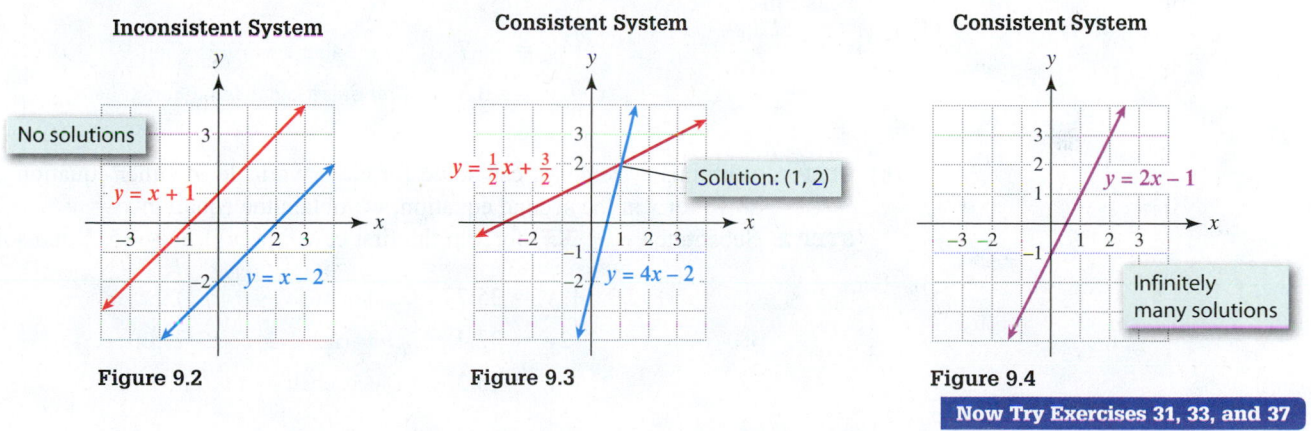

| Figure 9.2 | Figure 9.3 | Figure 9.4 |

Now Try Exercises 31, 33, and 37

The Method of Substitution

The **method of substitution** is often used to solve systems of equations symbolically. It is summarized by the following steps.

> **THE METHOD OF SUBSTITUTION**
>
> To use the method of substitution to solve a system of two equations in two variables, perform the following steps.
>
> **STEP 1:** Choose a variable in one of the two equations. Solve the equation for that variable.
>
> **STEP 2:** Substitute the result from Step 1 into the other equation and solve for the remaining variable.
>
> **STEP 3:** Use the value of the variable from Step 2 to determine the value of the other variable. To do this, you may want to use the equation you found in Step 1.
>
> To check your answer, substitute the value of each variable into the *given* equations. These values should satisfy *both* equations.

An Application In the next example, we use the method of substitution to solve an example involving real data.

EXAMPLE 3 Applying the method of substitution

In the first quarter of 2011, Apple Corporation sold a combined total of 35.7 million iPods and iPhones. There were 3.3 million more iPods sold than iPhones. (*Source:* Apple Corporation.)

(a) Write a system of equations whose solution gives the individual sales of iPods and of iPhones.

(b) Solve the system of equations. Interpret the results.

(c) Is your system consistent or inconsistent? If it is consistent, state whether the equations are dependent or independent.

SOLUTION

Getting Started When setting up a system of equations, it is important to identify what each variable represents. Then express the situation with equations. Finally, apply the method of substitution. ▶

(a) Let x be the number of iPods sold in millions and y be the number of iPhones sold in millions. The combined total is 35.7 million, so let $x + y = 35.7$. Because iPod sales exceeded iPhone sales by 3.3 million, let $x - y = 3.3$. Thus the system of equations is as follows.

$$x + y = 35.7 \qquad \text{Total sales are 35.7 million.}$$
$$x - y = 3.3 \qquad \text{iPod sales exceeded iPhone sales by 3.3 million.}$$

(b) **STEP 1:** With this system, we can solve for either variable in either equation. If we solve for x in the second equation, we obtain the equation $x = y + 3.3$.

 STEP 2: Substitute $(y + 3.3)$ for x in the first equation, $x + y = 35.7$, and solve.

$$(y + 3.3) + y = 35.7 \qquad \text{Substitute } (y + 3.3) \text{ for } x.$$
$$2y = 32.4 \qquad \text{Subtract 3.3; combine terms.}$$
$$y = 16.2 \qquad \text{Divide each side by 2.}$$

 STEP 3: To find x, substitute 16.2 for y in the equation $x = y + 3.3$ from Step 1 to obtain $x = 16.2 + 3.3 = 19.5$. The solution is $(19.5, 16.2)$.

Thus, there were 19.5 million iPods and 16.2 million iPhones sold.

(c) There is one solution, so the system is consistent and the equations are independent.

Now Try Exercise 113

In the next example, we solve a system and check our result.

EXAMPLE 4   **Using the method of substitution**

Solve the system symbolically. Check your answer.

$$5x - 2y = -16$$
$$x + 4y = -1$$

SOLUTION

STEP 1: Begin by solving one of the equations for one of the variables. One possibility is to solve the second equation for x.

$$x + 4y = -1 \qquad \text{Second equation}$$
$$x = -4y - 1 \qquad \text{Subtract 4y from each side.}$$

STEP 2: Next, substitute $(-4y - 1)$ for x in the first equation and solve the resulting equation for y.

$$5x - 2y = -16 \qquad \text{First equation}$$
$$5(-4y - 1) - 2y = -16 \qquad \text{Let x = -4y - 1.}$$
$$-20y - 5 - 2y = -16 \qquad \text{Distributive property}$$
$$-5 - 22y = -16 \qquad \text{Combine like terms.}$$
$$-22y = -11 \qquad \text{Add 5 to each side.}$$
$$y = \frac{1}{2} \qquad \text{Divide each side by -22; Simplify.}$$

STEP 3: Now find the value of x by using the equation $x = -4y - 1$ from Step 1. Since $y = \frac{1}{2}$, it follows that $x = -4\left(\frac{1}{2}\right) - 1 = -3$. The solution can be written as an ordered pair: $\left(-3, \frac{1}{2}\right)$.

Check: Substitute $x = -3$ and $y = \frac{1}{2}$ in both given equations.

$$5(-3) - 2\left(\frac{1}{2}\right) \stackrel{?}{=} -16 \checkmark \quad \text{True}$$

$$-3 + 4\left(\frac{1}{2}\right) \stackrel{?}{=} -1 \checkmark \quad \text{True}$$

Both equations are satisfied, so the solution is $\left(-3, \frac{1}{2}\right)$.

Now Try Exercise 39

Nonlinear Systems of Equations   The method of substitution can also be used to solve nonlinear systems of equations. In the next example, we solve a nonlinear system of equations having two solutions. In general, a nonlinear system of equations can have *any number of solutions*.

EXAMPLE 5   **Solving a nonlinear system of equations**

Solve the system symbolically.

$$6x + 2y = 10$$
$$2x^2 - 3y = 11$$

SOLUTION

STEP 1: Begin by solving one of the equations for one of the variables. One possibility is to solve the first equation for y.

$$6x + 2y = 10 \qquad \text{First equation}$$
$$2y = 10 - 6x \qquad \text{Subtract 6x from each side.}$$
$$y = 5 - 3x \qquad \text{Divide each side by 2.}$$

STEP 2: Next, substitute $(5 - 3x)$ for y in the second equation and solve the resulting quadratic equation for x.

Algebra Review
To review factoring, see Chapter R (pages R-22–R-23).

$$2x^2 - 3y = 11 \qquad \text{Second equation}$$
$$2x^2 - 3(5 - 3x) = 11 \qquad \text{Let } y = 5 - 3x.$$
$$2x^2 - 15 + 9x = 11 \qquad \text{Distributive property}$$
$$2x^2 + 9x - 26 = 0 \qquad \text{Subtract 11 from each side; rewrite.}$$
$$(2x + 13)(x - 2) = 0 \qquad \text{Factor.}$$
$$x = -\frac{13}{2} \quad \text{or} \quad x = 2 \qquad \text{Zero-product property}$$

STEP 3: Now find the corresponding y-values for each x-value. From Step 1 we know that $y = 5 - 3x$, so it follows that $y = 5 - 3\left(-\frac{13}{2}\right) = \frac{49}{2}$ or $y = 5 - 3(2) = -1$. Thus the solutions are $\left(-\frac{13}{2}, \frac{49}{2}\right)$ and $(2, -1)$.

Now Try Exercise 51

EXAMPLE 6 Solving a nonlinear system of equations

Use the method of substitution to determine the points where the line $y = 2x$ intersects the circle $x^2 + y^2 = 5$. Sketch a graph that illustrates the solutions.

SOLUTION Substitute $(2x)$ for y in the equation $x^2 + y^2 = 5$.

Two Solutions

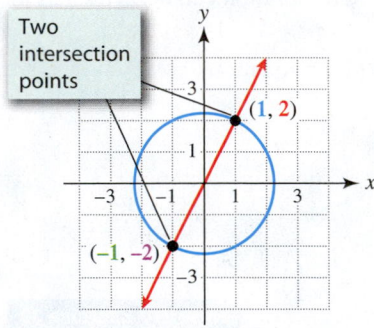

Two intersection points

$(1, 2)$

$(-1, -2)$

Figure 9.5

$$x^2 + y^2 = 5 \qquad \text{Second equation}$$
$$x^2 + (2x)^2 = 5 \qquad y = 2x$$
$$x^2 + 4x^2 = 5 \qquad \text{Square the expression.}$$
$$5x^2 = 5 \qquad \text{Add like terms.}$$
$$x^2 = 1 \qquad \text{Divide each side by 5.}$$
$$x = \pm 1 \qquad \text{Square root property}$$

Since $y = 2x$ we see that when $x = 1$, $y = 2$ and when $x = -1$, $y = -2$. The graphs of $x^2 + y^2 = 5$ and $y = 2x$ intersect at the points $(1, 2)$ and $(-1, -2)$. This nonlinear system has two solutions, which are shown in Figure 9.5.

Now Try Exercise 99

EXAMPLE 7 Identifying a system with zero or infinitely many solutions

If possible, solve each system of equations.
(a) $x^2 + y = 1$ **(b)** $2x - 4y = 5$
      $x^2 - y = -2$ $-x + 2y = -\frac{5}{2}$

SOLUTION
(a) STEP 1: Solve the second equation for y, which gives $y = x^2 + 2$.

STEP 2: Substitute $(x^2 + 2)$ for y in the first equation and then solve for x, if possible.

$$x^2 + y = 1 \qquad \text{First equation}$$
$$x^2 + (x^2 + 2) = 1 \qquad \text{Let } y = x^2 + 2.$$
$$2x^2 + 2 = 1 \qquad \text{Combine like terms.}$$
$$2x^2 = -1 \qquad \text{Subtract 2 from each side.}$$

Because $2x^2 \geq 0$, it follows that there are no real solutions and the system is inconsistent.

No Solutions

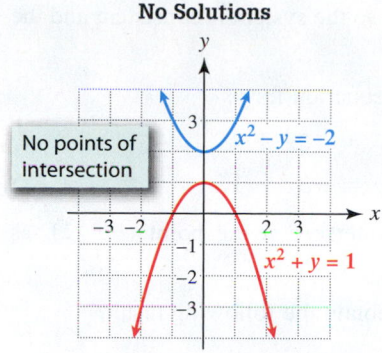

Figure 9.6

Infinitely Many Solutions

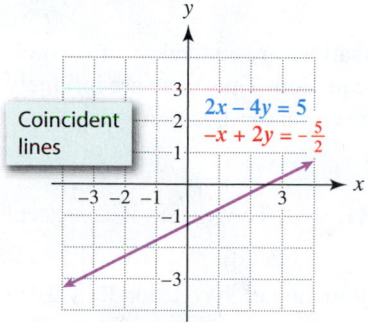

Figure 9.7

STEP 3: This step is not necessary because there are no real solutions for x. Figure 9.6 shows that the graphs, which are parabolas, do not intersect.

(b) **STEP 1:** First solve the second equation for x to obtain $x = 2y + \frac{5}{2}$.

STEP 2: Substitute $\left(2y + \frac{5}{2}\right)$ for x in the first equation and then solve for y.

$$2x - 4y = 5 \qquad \text{First equation}$$

$$2\left(2y + \frac{5}{2}\right) - 4y = 5 \qquad \text{Let } x = 2y + \tfrac{5}{2}.$$

$$4y + 5 - 4y = 5 \qquad \text{Distributive property}$$

$$5 = 5 \qquad \text{Combine like terms.}$$

The equation $5 = 5$ is an identity and indicates that there are infinitely many solutions. The system is consistent and the equations are dependent. Note that we can multiply each side of the second equation by -2 to obtain the first equation.

$$-2(-x + 2y) = -2\left(-\frac{5}{2}\right) \qquad \text{Multiply second equation by } -2.$$

$$2x - 4y = 5 \qquad \text{Distributive property}$$

STEP 3: In Figure 9.7 the graphs of the equations are identical lines. The solution set is $\{(x, y) \mid 2x - 4y = 5\}$ and includes all points on this line; $\left(0, -\frac{5}{4}\right)$ and $\left(2, -\frac{1}{4}\right)$ are examples of solutions to this system.

> **Now Try Exercises 45 and 57**

The Elimination Method

The **elimination method** is another way to solve systems of equations symbolically. This method is based on the property that *equals added to equals are equal*. That is, if

$$a = b \quad \text{and} \quad c = d, \qquad \text{then} \qquad a + c = b + d.$$

The goal of this method is to obtain an equation where one of the two variables has been eliminated. This task is sometimes accomplished by adding two equations. The elimination method is demonstrated in the next example for three types of linear systems.

EXAMPLE 8 Using elimination to solve a system

Use elimination to solve each system of equations, if possible. Identify the system as consistent or inconsistent. If the system is consistent, state whether the equations are dependent or independent. Support your results graphically.

(a) $2x - y = -4$ **(b)** $4x - y = 10$ **(c)** $x - y = 6$
      $3x + y = -1$ $-4x + y = -10$ $x - y = 3$

SOLUTION

(a) *Symbolic Solution*   We can eliminate the y-variable by adding the equations.

$$2x - y = -4 \qquad \text{First equation}$$

$$\underline{3x + y = -1} \qquad \text{Second equation}$$

$$5x \quad\;\; = -5 \quad \text{or} \quad x = -1 \qquad \text{Add equations.}$$

Now the y-variable can be determined by substituting $x = -1$ in either equation.

$$2x - y = -4 \qquad \text{First equation}$$

$$2(-1) - y = -4 \qquad \text{Let } x = -1.$$

$$-y = -2 \qquad \text{Add 2.}$$

$$y = 2 \qquad \text{Multiply by } -1.$$

One Solution

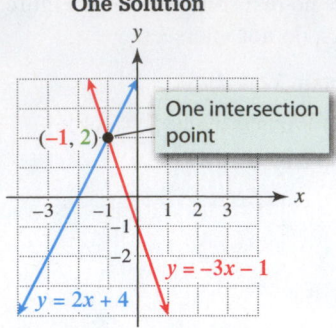

Figure 9.8

Infinitely Many Solutions

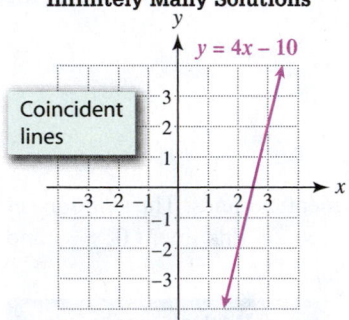

Coincident lines

Figure 9.9

No Solutions

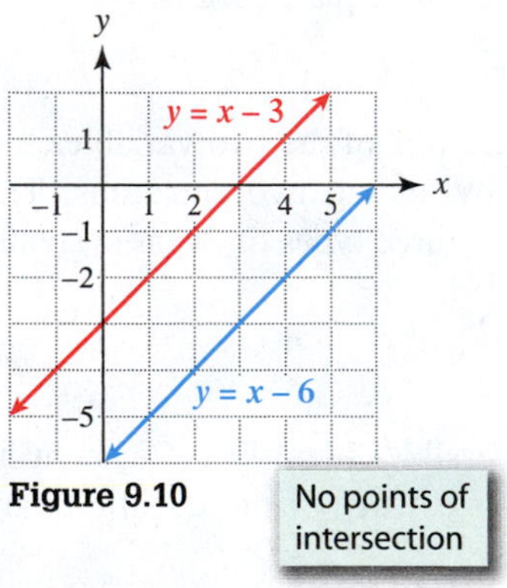

No points of intersection

Figure 9.10

The solution is $(-1, 2)$. There is a unique solution so the system is consistent and the equations are independent.

Graphical Solution Start by solving each given equation for y.

$$2x - y = -4 \quad \text{is equivalent to} \quad y = 2x + 4.$$
$$3x + y = -1 \quad \text{is equivalent to} \quad y = -3x - 1.$$

The graphs of $y = 2x + 4$ and $y = -3x - 1$ intersect at the point $(-1, 2)$, as shown in Figure 9.8.

(b) ***Symbolic Solution*** If we add the equations, we obtain the following result.

$$\begin{array}{rll} 4x - y = & 10 & \textcolor{blue}{\text{First equation}} \\ -4x + y = & -10 & \textcolor{blue}{\text{Second equation}} \\ \hline 0 = & 0 & \textcolor{blue}{\text{Add equations.}} \end{array}$$

The equation $0 = 0$ is an identity. The two given equations are equivalent: if we multiply the first equation by -1, we obtain the second equation. Thus there are infinitely many solutions, and we can write the solution set in set-builder notation.

$$\{(x, y) \mid 4x - y = 10\}$$

Some examples of solutions are $(3, 2)$, $(4, 6)$, and $(1, -6)$. The system is consistent and the equations are dependent.

Graphical Solution For a graphical solution, start by solving each equation for y. Both equations are equivalent to $y = 4x - 10$. Their graphs are identical and coincide. The graph of $y = 4x - 10$ is shown in Figure 9.9. Any point on the line represents a solution to the system. For example, $(3, 2)$ is a solution.

(c) ***Symbolic Solution*** If we subtract the second equation from the first, we obtain the following result. (Note that subtracting the second equation from the first is equivalent to multiplying the second equation by -1 and then adding it to the first.)

$$\begin{array}{rll} x - y = & 6 & \\ x - y = & 3 & \\ \hline 0 = & 3 & \textcolor{blue}{\text{Subtract.}} \end{array}$$

The equation $0 = 3$ is a contradiction. Therefore there are no solutions, and the system is inconsistent.

Graphical Solution For a graphical solution, start by solving each equation for y to obtain $y = x - 6$ and $y = x - 3$. The graphs of $y = x - 6$ and $y = x - 3$, shown in Figure 9.10, are parallel lines that never intersect, so there are no solutions.

Now Try Exercises 71, 75, and 77

Sometimes multiplication is performed before elimination is used, as illustrated in the next example.

EXAMPLE 9 Multiplying before using elimination

Solve each system of equations by using elimination.

(a) $2x - 3y = 18$ **(b)** $5x + 10y = 10$
      $5x + 2y = 7$ $x + 2y = 2$

SOLUTION

(a) If we multiply the first equation by 2 and the second equation by 3, then the y-coefficients become -6 and 6. Addition eliminates the y-variable.

$$\begin{array}{rll} 4x - 6y = 36 & \textcolor{blue}{\text{Multiply first equation by 2.}} \\ 15x + 6y = 21 & \textcolor{blue}{\text{Multiply second equation by 3.}} \\ \hline 19x \quad\quad = 57, \text{ or } x = 3 & \textcolor{blue}{\text{Add equations.}} \end{array}$$

Substituting $x = 3$ in $2x - 3y = 18$ (first equation) results in

$$2(3) - 3y = 18, \text{ or } y = -4.$$

The solution is $(3, -4)$.

(b) If the second equation is multiplied by -5, addition eliminates both variables.

$$
\begin{array}{ll}
5x + 10y = 10 & \text{First equation} \\
-5x - 10y = -10 & \text{Multiply second equation by } -5. \\
\hline
0 = 0 & \text{Add equations.}
\end{array}
$$

The statement $0 = 0$ is an identity. The equations are dependent and there are infinitely many solutions. The solution set is $\{(x, y) \mid x + 2y = 2\}$.

> **Now Try Exercises 81 and 83**

Elimination and Nonlinear Systems   Elimination can also be used to solve some nonlinear systems of equations, as illustrated in the next example.

EXAMPLE 10   **Using elimination to solve a nonlinear system**

Solve the system of equations.

$$
\begin{aligned}
x^2 + y^2 &= 4 \\
2x^2 - y &= 7
\end{aligned}
$$

SOLUTION   If we multiply each side of the first equation by 2, multiply each side of the second equation by -1, and then add the equations, the x variable is eliminated.

$$
\begin{array}{ll}
2x^2 + 2y^2 = 8 & \text{Multiply first equation by 2.} \\
-2x^2 + y = -7 & \text{Multiply second equation by } -1. \\
\hline
2y^2 + y = 1 & \text{Add equations.}
\end{array}
$$

Next we solve $2y^2 + y - 1 = 0$ for y.

$$
\begin{array}{ll}
(2y - 1)(y + 1) = 0 & \text{Factor.} \\[4pt]
y = \dfrac{1}{2} \text{ or } y = -1 & \text{Solve.}
\end{array}
$$

Solving $x^2 + y^2 = 4$ for x results in $x = \pm\sqrt{4 - y^2}$. If $y = \frac{1}{2}$, then $x = \pm\sqrt{\frac{15}{4}}$, which can be written as $\pm\frac{\sqrt{15}}{2}$. If $y = -1$, then $x = \pm\sqrt{3}$. Thus there are four solutions: $\left(\pm\frac{\sqrt{15}}{2}, \frac{1}{2}\right)$ and $(\pm\sqrt{3}, -1)$.

A graph of the system of equations is shown in Figure 9.11. The four points of intersection correspond to the four solutions. In Figure 9.12 the four points of intersection are labeled.

Four Solutions to a Nonliner System

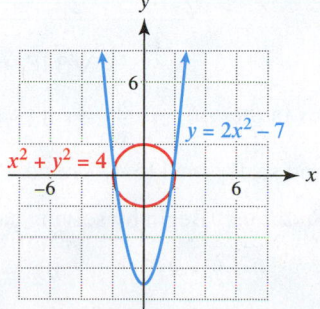

Figure 9.11

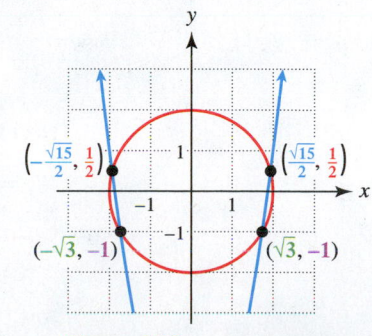

Figure 9.12

> **Now Try Exercise 93**

Graphical and Numerical Methods

An Application   The next example illustrates how a system with two variables can be solved symbolically, graphically, and numerically.

EXAMPLE 11   **Modeling roof trusses**

Linear systems occur in the design of roof trusses for homes and buildings. See Figure 9.13. One of the simplest types of roof trusses is an equilateral triangle. If a 200-pound force is applied to the peak of a truss, as shown in Figure 9.14, then the weights W_1 and W_2 exerted on each rafter of the truss are determined by the following system of linear equations. (**Source:** R. Hibbeler, *Structural Analysis.*)

$$W_1 - W_2 = 0$$
$$\frac{\sqrt{3}}{2}(W_1 + W_2) = 200$$

Estimate the solution symbolically, graphically, and numerically.

Equilateral Triangle Roof Trusses

Figure 9.13

Figure 9.14

SOLUTION

Symbolic Solution   The system of equations can be written as follows.

$$W_1 - W_2 = 0 \qquad \textcolor{blue}{\text{First equation}}$$
$$\frac{\sqrt{3}}{2}W_1 + \frac{\sqrt{3}}{2}W_2 = 200 \qquad \textcolor{blue}{\text{Distributive property}}$$

We can apply elimination by multiplying the first equation by $\frac{\sqrt{3}}{2}$ and then adding.

$$\frac{\sqrt{3}}{2}W_1 - \frac{\sqrt{3}}{2}W_2 = 0 \qquad \textcolor{blue}{\text{Multiply by } \frac{\sqrt{3}}{2}.}$$
$$\frac{\sqrt{3}}{2}W_1 + \frac{\sqrt{3}}{2}W_2 = 200$$
$$\overline{}$$
$$\sqrt{3}\,W_1 = 200 \qquad \textcolor{blue}{\text{Add equations.}}$$

Dividing by $\sqrt{3}$ gives $W_1 = \frac{200}{\sqrt{3}} \approx 115.47$ pounds. From the first equation, it follows that $W_1 = W_2$, and so $W_2 \approx 115.47$ pounds.

Graphical Solution   Begin by solving each equation for the variable W_2.

$$W_2 = W_1$$
$$W_2 = \frac{400}{\sqrt{3}} - W_1$$

> Solve each equation for W_2.

Graph the equations $y_1 = x$ and $y_2 = \dfrac{400}{\sqrt{3}} - x$. Their graphs intersect near the point

(115.47, 115.47), as shown in Figure 9.15. This means that each rafter supports a weight of approximately 115 pounds.

Graphical Solution   **Numerical Solution**

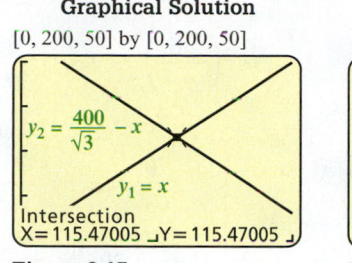

Figure 9.15   **Figure 9.16**

Numerical Solution   In Figure 9.16, $y_1 \approx y_2$ for $x = 115$.

Now Try Exercise 115

EXAMPLE 12   **Determining the dimensions of a cylinder**

The volume V of a cylindrical container with a radius r and height h is computed by $V(r, h) = \pi r^2 h$. See Figure 9.17. The lateral surface area S of the container, *excluding* the circular top and bottom, is computed by $S(r, h) = 2\pi rh$.

Volume and Lateral Surface Area of a Cylinder

Geometry Review
To review formulas related to cylinders, see Chapter R (page R-4).

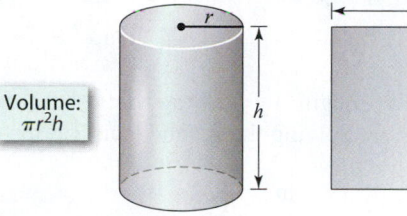 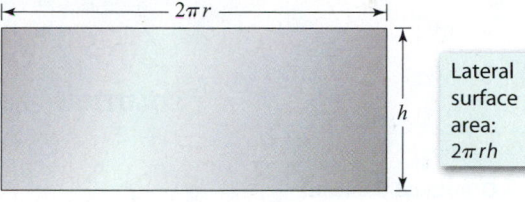

Figure 9.17

(a) Write a system of equations whose solution is the dimensions for a cylinder with a volume of 38 cubic inches and a lateral surface area of 63 square inches.
(b) Solve the system of equations graphically and symbolically.

SOLUTION
(a) The equations $V(r, h) = 38$ and $S(r, h) = 63$ must be satisfied. This results in the following system of nonlinear equations.

$$\pi r^2 h = 38 \qquad \text{Volume}$$

$$2\pi rh = 63 \qquad \text{Lateral surface area}$$

(b) ***Graphical Solution***   To find the solution graphically, we can solve each equation for h and then apply the intersection-of-graphs method.

Graphical Solution
[0, 4, 1] by [0, 20, 5]

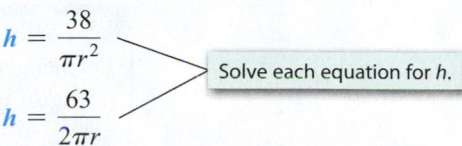

Figure 9.18

$$h = \frac{38}{\pi r^2}$$

$$h = \frac{63}{2\pi r}$$

Solve each equation for h.

Let r correspond to x and h to y. Graph $y_1 = \dfrac{38}{\pi x^2}$ and $y_2 = \dfrac{63}{2\pi x}$. Their graphs intersect near the point (1.206, 8.312), as shown in Figure 9.18. Therefore a cylinder with a radius of $r \approx 1.206$ inches and height of $h \approx 8.312$ inches has a volume of 38 cubic inches and lateral surface area of 63 square inches.

Symbolic Solution   Because $h = \frac{38}{\pi r^2}$ and $h = \frac{63}{2\pi r}$, we can determine r by solving the following equation.

$$\frac{38}{\pi r^2} = \frac{63}{2\pi r}$$ Equation to be solved

$$2\pi r^2\left(\frac{38}{\pi r^2}\right) = 2\pi r^2\left(\frac{63}{2\pi r}\right)$$ Multiply each side by the LCD, $2\pi r^2$.

$$76 = 63r$$ Simplify.

$$\frac{76}{63} = r$$ Divide each side by 63.

Because $r = \frac{76}{63} \approx 1.206$, $h = \frac{63}{2\pi r} = \frac{63}{2\pi(76/63)} \approx 8.312$; the symbolic result verifies our graphical result.

Now Try Exercise 117

An Example That Requires a Graphical Solution   Sometimes it is either difficult or impossible to solve a nonlinear system of equations symbolically. However, it might be possible to solve such a system graphically.

EXAMPLE 13   Solving a nonlinear system of equations graphically

Solve the system graphically to the nearest thousandth.

$$2x^3 - y = 2$$
$$\ln x^2 - 3y = -1$$

SOLUTION   Begin by solving both equations for y. The first equation becomes $y = 2x^3 - 2$. Solving the second equation for y gives the following results.

$$\ln x^2 - 3y = -1$$ Second equation

$$\ln x^2 + 1 = 3y$$ Add $3y$ and 1 to each side.

$$\frac{\ln x^2 + 1}{3} = y$$ Divide each side by 3.

Graphical Solution

$[-6, 6, 1]$ by $[-4, 4, 1]$

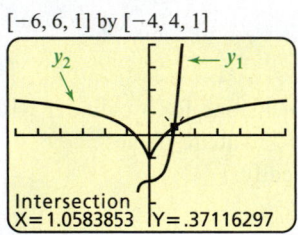

Intersection
X=1.0583853 Y=.37116297

Figure 9.19

The graphs of $y_1 = 2x^3 - 2$ and $y_2 = \frac{\ln x^2 + 1}{3}$ in Figure 9.19 intersect at one point. To the nearest thousandth, the solution is $(1.058, 0.371)$.

Now Try Exercise 107

Joint Variation

A quantity may depend on more than one variable. For example, the volume V of a cylinder is given by $V = \pi r^2 h$. We say that V *varies jointly* with h and the square of r. The *constant of variation* is π.

JOINT VARIATION

Let m and n be real numbers. Then z **varies jointly** with the mth power of x and the nth power of y if a nonzero real number k exists such that

$$z = kx^m y^n.$$

In the following example we use joint variation to determine the amount of timber in a tree with a specified diameter and height.

EXAMPLE 14   Modeling the amount of wood in a tree

To estimate the volume of timber in a given area of forest, formulas have been developed to find the amount of wood contained in a tree with height h in feet and diameter d in inches. See Figure 9.20. One study concluded that the volume V of wood in a tree varies jointly with the 1.12 power of h and the 1.98 power of d. (The diameter is measured 4.5 feet above the ground.)   (*Source:* B. Ryan, B. Joiner, and T. Ryan, *Minitab Handbook.*)

(a) Write an equation that relates V, h, and d.

(b) A tree with a 13.8-inch diameter and a 64-foot height has a volume of 25.14 cubic feet. Estimate the constant of variation k.

(c) Estimate the volume of wood in a tree with $d = 11$ inches and $h = 47$ feet.

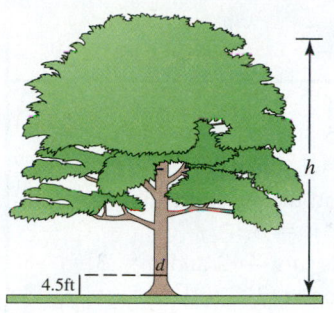

Figure 9.20

SOLUTION

(a) $V = kh^{1.12}d^{1.98}$, where k is the constant of variation.

(b) Substitute $d = 13.8$, $h = 64$, and $V = 25.14$ into the equation and solve for k.

$$25.14 = k(64)^{1.12}(13.8)^{1.98}$$

$$k = \frac{25.14}{(64)^{1.12}(13.8)^{1.98}} \approx 0.00132$$

Thus let $V = 0.00132\,h^{1.12}d^{1.98}$.

(c) $V = 0.00132(47)^{1.12}(11)^{1.98} \approx 11.4$ cubic feet

Now Try Exercise 143

9.1 Putting It All Together

The following table summarizes some mathematical concepts involved with functions and equations in two variables.

CONCEPT	COMMENTS	EXAMPLE
Function of two inputs or variables	$z = f(x, y)$ where x and y are inputs and z is the output.	$f(x, y) = x^2 + 5y$ $f(2, 3) = 2^2 + 5(3) = 19$
System of two linear equations	The equations can be written as $ax + by = k$. A solution is an ordered pair (x, y) that satisfies both equations.	$2x - 3y = 6$ $5x + 4y = -8$ Solution: $(0, -2)$
Nonlinear system of two equations	A system of equations that has at least one nonlinear equation is a nonlinear system. A solution is an ordered pair (x, y) that satisfies both equations. A nonlinear system of equations can have any number of solutions.	$5x^2 - 4xy = -3$ $\dfrac{5}{x} - 2y = 1$ Solutions: $(1, 2), \left(-\frac{7}{5}, -\frac{16}{7}\right)$
Consistent system of linear equations in two variables	A consistent linear system has either one or infinitely many solutions. Its graph is either distinct, intersecting lines or identical lines.	$\begin{array}{r} x + y = 10 \\ x - y = 4 \\ \hline 2x \quad\quad = 14 \end{array}$ Solution is given by $x = 7$ and $y = 3$. The equations are independent.
System of dependent linear equations in two variables	A system of dependent linear equations has infinitely many solutions. The graph consists of two identical lines.	$2x + 2y = 2$ and $x + y = 1$ are equivalent (dependent) equations. The solution set is $\{(x, y) \mid x + y = 1\}$.

continued on next page

CONCEPT	COMMENTS	EXAMPLE
Inconsistent system of linear equations in two variables	An inconsistent linear system has no solutions. The graph is two parallel lines.	$\begin{aligned} x + y &= 1 \\ x + y &= 2 \quad \text{Subtract.} \\ \hline 0 &= -1 \quad \text{Always false} \end{aligned}$
Method of substitution	Solve one equation for a variable. Then substitute the result in the second equation and solve.	$x - y = 1$ $x + y = 5$ If $x - y = 1$, then $x = 1 + y$. Substitute in the second equation: $(1 + y) + y = 5$. This results in $y = 2$ and $x = 3$. The solution is $(3, 2)$.
Elimination method	By performing arithmetic operations on a system, a variable is eliminated.	$\begin{aligned} 2x + y &= 5 \\ x - y &= 1 \quad \text{Add.} \\ \hline 3x \quad\;\; &= 6 \end{aligned}$ so $x = 2$ and $y = 1$.
Graphical method for two equations	Solve both equations for the same variable. Then apply the intersection-of-graphs method.	If $x + y = 3$, then $y = 3 - x$. If $4x - y = 2$, then $y = 4x - 2$. Graph and locate the point of intersection at $(1, 2)$.

9.1 Exercises

Functions of More Than One Input

Exercises 1 and 2: Evaluate the function for the indicated inputs and interpret the result.

1. $A(5, 8)$, where $A(b, h) = \frac{1}{2}bh$ (A computes the area of a triangle with base b and height h.)

2. $A(20, 35)$, where $A(w, l) = wl$ (A computes the area of a rectangle with width w and length l.)

Exercises 3–8: Evaluate the expression for the given $f(x, y)$.

3. $f(2, -3)$ if $f(x, y) = x^2 + y^2$

4. $f(-1, 3)$ if $f(x, y) = 2x^2 - y^2$

5. $f(-2, 3)$ if $f(x, y) = 3x - 4y$

6. $f(5, -2)$ if $f(x, y) = 6y - \frac{1}{2}x$

7. $f\left(\frac{1}{2}, -\frac{7}{4}\right)$ if $f(x, y) = \frac{2x}{y + 3}$

8. $f(0.2, 0.5)$ if $f(x, y) = \frac{5x}{2y + 1}$

Exercises 9–12: Write a symbolic representation for $f(x, y)$ if the function f computes the following quantity.

9. The sum of y and twice x

10. The product of x^2 and y^2

11. The product of x and y divided by $1 + x$

12. The square root of the sum of x and y

Exercises 13–18: Solve the equation for x and then solve it for y.

13. $3x - 4y = 7$

14. $-x - 5y = 4$

15. $x - y^2 = 5$

16. $2x^2 + y = 4$

17. $\dfrac{2x - y}{3y} = 1$

18. $\dfrac{x + y}{x - y} = 2$

Solutions to Systems of Equations

Exercises 19–22: Determine which ordered pairs are solutions to the given system of equations. State whether the system is linear or nonlinear.

19. $(2, 1), (-2, 1), (1, 0)$
$2x + y = 5$
$x + y = 3$

20. $(3, 2), (3, -4), (5, 0)$
$x - y = 5$
$2x + y = 10$

21. $(4, -3), (0, 5), (4, 3)$
$x^2 + y^2 = 25$
$2x + 3y = -1$

22. $(4, 8), (8, 4), (-4, -8)$
$xy = 32$
$x + y = 12$

Exercises 23–26: The figure shows the graph of a system of two linear equations. Use the graph to estimate the solution to this system of equations. Then solve the system symbolically.

23.

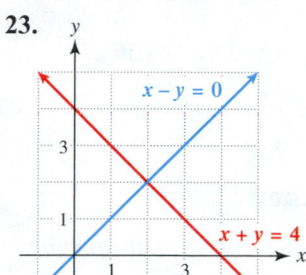

24.

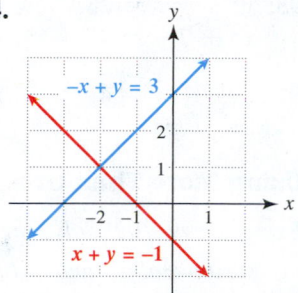

25.

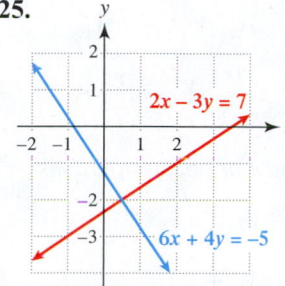

26.

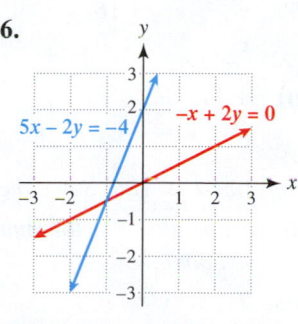

Consistent and Inconsistent Linear Systems

Exercises 27–30: The figure represents a system of linear equations. Classify the system as consistent or inconsistent. Solve the system graphically and symbolically, if possible.

27.

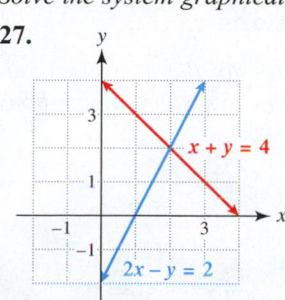

28.

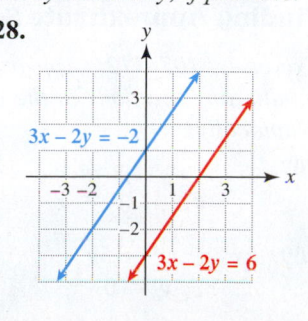

29.

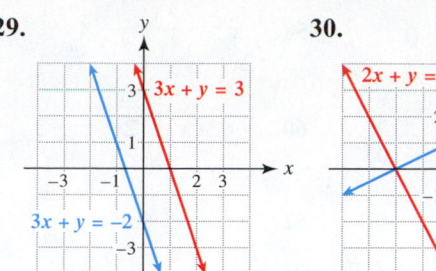

30.

Exercises 31–38: Graph each system of equations and find any solutions. Check your answers. Identify the system as consistent or inconsistent. If the system is consistent, state whether the equations are dependent or independent.

31. $2x + y = 3$
$-2x - y = 4$

32. $x - 4y = 4$
$2x - 8y = 4$

33. $3x - y = 7$
$-2x + y = -5$

34. $-x + 2y = 3$
$3x - y = 1$

35. $x - 2y = -6$
$-2x + y = 6$

36. $2x - 3y = 1$
$x + y = -2$

37. $2x - y = -4$
$-4x + 2y = 8$

38. $3x - y = -2$
$-3x + y = 2$

The Method of Substitution

Exercises 39–50: If possible, solve the system of linear equations and check your answer.

39. $x + 2y = 0$
$3x + 7y = 1$

40. $-2x - y = -2$
$3x + 4y = -7$

41. $2x - 9y = -17$
$8x + 5y = 14$

42. $3x + 6y = 0$
$4x - 2y = -5$

43. $\frac{1}{2}x - y = -5$
$x + \frac{1}{2}y = 10$

44. $-x - \frac{1}{3}y = -4$
$\frac{1}{3}x + 2y = 7$

45. $3x - 2y = 5$
$-6x + 4y = -10$

46. $\frac{1}{2}x - \frac{3}{4}y = \frac{1}{2}$
$\frac{1}{5}x - \frac{3}{10}y = \frac{1}{5}$

47. $2x - 7y = 8$
$-3x + \frac{21}{2}y = 5$

48. $0.6x - 0.2y = 2$
$-1.2x + 0.4y = 3$

49. $0.2x - 0.1y = 0.5$
$0.4x + 0.3y = 2.5$

50. $100x + 200y = 300$
$200x + 100y = 0$

Exercises 51–64: If possible, solve the nonlinear system of equations.

51. $x^2 - y = 0$
$2x + y = 0$

52. $x^2 - y = 3$
$x + y = 3$

53. $xy = 8$
$x + y = 6$

54. $2x - y = 0$
$2xy = 4$

55. $x^2 + y^2 = 20$
$y = 2x$

56. $x^2 + y^2 = 9$
$x + y = 3$

57. $\sqrt{x} - 2y = 0$
$\quad x - y = -2$

58. $x^2 + y^2 = 4$
$\quad 2x^2 + y = -3$

59. $2x^2 - y = 5$
$\quad -4x^2 + 2y = -10$

60. $-6\sqrt{x} + 2y = -3$
$\quad 2\sqrt{x} - \frac{2}{3}y = 1$

61. $x^2 - y = 4$
$\quad x^2 + y = 4$

62. $x^2 + x = y$
$\quad 2x^2 - y = 2$

63. $x^3 - x = 3y$
$\quad x - y = 0$

64. $x^4 + y = 4$
$\quad 3x^2 - y = 0$

Exercises 65–68: Write a system of linear equations with two variables whose solution satisfies the problem. State what each variable represents. Then solve the system.

65. Screen Dimensions The screen of a rectangular television set is 2 inches wider than it is high. If the perimeter of the screen is 38 inches, find its dimensions.

66. Numbers The sum of two numbers is 300 and their difference is 8. Find the two numbers.

67. Tickets Admission prices to a movie are $4 for children and $7 for adults. If 75 tickets were sold for $456, how many of each type of ticket were sold?

68. Coins A sample of 16 dimes and quarters has a value of $2.65. How many of each type of coin are there?

Exercises 69 and 70: **Area and Perimeter** *The area of a rectangle with length l and width w is computed by $A(l, w) = lw$, and its perimeter is calculated by $P(l, w) = 2l + 2w$. Assume that $l > w$ and use the method of substitution to solve the system of equations for l and w.*

69. $A(l, w) = 35$
$\quad P(l, w) = 24$

70. $A(l, w) = 300$
$\quad P(l, w) = 70$

The Elimination Method

Exercises 71–80: Use elimination to solve the system of equations, if possible. Identify the system as consistent or inconsistent. If the system is consistent, state whether the equations are dependent or independent. Support your results graphically or numerically.

71. $x + y = 20$
$\quad x - y = 8$

72. $2x + y = 15$
$\quad x - y = 0$

73. $x + 3y = 10$
$\quad x - 2y = -5$

74. $4x + 2y = 10$
$\quad -2x - y = 10$

75. $x + y = 500$
$\quad -x - y = -500$

76. $2x + 3y = 5$
$\quad 5x - 2y = 3$

77. $2x + 4y = 7$
$\quad -x - 2y = 5$

78. $4x - 3y = 5$
$\quad 3x + 4y = 2$

79. $2x + 3y = 2$
$\quad x - 2y = -5$

80. $x - 3y = 1$
$\quad 2x - 6y = 2$

Exercises 81–92: Solve the system, if possible.

81. $\frac{1}{2}x - y = 5$
$\quad x - \frac{1}{2}y = 4$

82. $\frac{1}{2}x - \frac{1}{3}y = 1$
$\quad \frac{1}{3}x - \frac{1}{2}y = 1$

83. $7x - 3y = -17$
$\quad -21x + 9y = 51$

84. $-\frac{1}{3}x + \frac{1}{6}y = -1$
$\quad 2x - y = 6$

85. $\frac{2}{3}x + \frac{4}{3}y = \frac{1}{3}$
$\quad -2x - 4y = 5$

86. $5x - 2y = 7$
$\quad 10x - 4y = 6$

87. $0.2x + 0.3y = 8$
$\quad -0.4x + 0.2y = 0$

88. $2x - 3y = 1$
$\quad 3x - 2y = 2$

89. $2x + 3y = 7$
$\quad -3x + 2y = -4$

90. $5x + 4y = -3$
$\quad 3x - 6y = -6$

91. $7x - 5y = -15$
$\quad -2x + 3y = -2$

92. $-5x + 3y = -36$
$\quad 4x - 5y = 34$

Exercises 93–98: Use elimination to solve the nonlinear system of equations.

93. $x^2 + y = 12$
$\quad x^2 - y = 6$

94. $x^2 + 2y = 15$
$\quad 2x^2 - y = 10$

95. $x^2 + y^2 = 25$
$\quad x^2 + 7y = 37$

96. $x^2 + y^2 = 36$
$\quad x^2 - 6y = 36$

97. $x^2 + y^2 = 4$
$\quad 2x^2 + y^2 = 8$

98. $x^2 + y^2 = 4$
$\quad x^2 - y^2 = 4$

Using More Than One Method

Exercises 99–102: Solve the nonlinear system of equations
(a) symbolically and (b) graphically.

99. $x^2 + y^2 = 16$
$\quad x - y = 0$

100. $x^2 - y = 1$
$\quad 3x + y = -1$

101. $xy = 12$
$\quad x - y = 4$

102. $x^2 + y^2 = 2$
$\quad x^2 - y = 0$

Exercises 103–106: Solve the system of linear equations
(a) graphically, (b) numerically, and
(c) symbolically.

103. $2x + y = 1$
$\quad x - 2y = 3$

104. $3x + 2y = -2$
$\quad 2x - y = -6$

105. $-2x + y = 0$
$\quad 7x - 2y = 3$

106. $x - 4y = 15$
$\quad 3x - 2y = 15$

Finding Approximate Solutions

Exercises 107–112: Approximate, to the nearest thousandth, any solutions to the nonlinear system of equations graphically.

107. $x^3 - 3x + y = 1$
$\quad x^2 + 2y = 3$

108. $x^2 + y = 5$
$\quad x + y^2 = 6$

109. $2x^3 - x^2 = 5y$
$\quad 2^{-x} - y = 0$

110. $x^4 - 3x^3 = y$
$\quad \log x^2 - y = 0$

111. $e^{2x} + y = 4$
$\ln x - 2y = 0$

112. $3x^2 + y = 3$
$(0.3)^x + 4y = 1$

Applications

113. Population In 2010, the combined population of Minneapolis/St. Paul, Minnesota, was 670,000. The population of Minneapolis was 98,000 greater than the population of St. Paul. (*Source:* Census Bureau.)
  (a) Write a system of equations whose solution gives the population of each city in thousands.

  (b) Solve the system of equations.

  (c) Is your system consistent or inconsistent? If it is consistent, state whether the equations are dependent or independent.

114. U.S. Energy Consumption In 2010, the United States consumed 94.58 quadrillion (10^{15}) Btu of energy from renewable and nonrenewable sources. It used 79.44 quadrillion Btu more from nonrenewable sources than from renewable sources. (*Source:* Department of Energy.)
  (a) Write a system of equations whose solution gives the consumption of energy from renewable and nonrenewable sources (in quadrillion Btu).

  (b) Solve the system of equations.

  (c) Is your system consistent or inconsistent? If it is consistent, state whether the equations are dependent or independent.

115. Roof Truss (Refer to Example 11.) The weights W_1 and W_2 exerted on each rafter for the roof truss shown in the figure are determined by the system of linear equations. Solve the system.

$$W_1 + \sqrt{2}W_2 = 300$$
$$\sqrt{3}W_1 - \sqrt{2}W_2 = 0$$

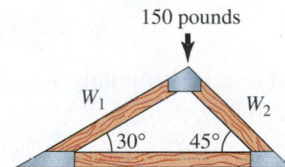

150 pounds

W_1 W_2

30°   45°

116. Time on the Internet From 2001 to 2010 the average number of hours that a user spent on the Internet each week increased by 180%. This percent increase amounted to 8 hours. Find the average number of hours that a user spent on the Internet each week in 2001 and 2010. (*Source:* eMarketer.)

117. Geometry (Refer to Example 12.) Find the radius and height of a cylindrical container with a volume of 50 cubic inches and a lateral surface area of 65 square inches.

118. Geometry (Refer to Example 12.) Determine if it is possible to construct a cylindrical container, *including* the top and bottom, with a volume of 38 cubic inches and a surface area of 38 square inches.

119. Dimensions of a Box A box has an *open* top, rectangular sides, and a square base. Its volume is 576 cubic inches, and its outside surface area is 336 square inches. Find the dimensions of the box.

120. Dimensions of a Box A box has rectangular sides, and its rectangular top and base are twice as long as they are wide. Its volume is 588 cubic inches, and its outside surface area is 448 square inches. Find its dimensions.

121. Bank Theft The total incidences of bank theft in 2009 and 2010 was 11,693. There were 437 fewer incidences in 2009 than in 2010. (*Source:* FBI.)
  (a) Write a system of equations whose solution represents the incidences of bank theft in each of these years.

  (b) Solve the system symbolically.

  (c) Solve the system graphically.

122. e-Waste The United States and China together produce 5.9 million tons of e-waste each year. About 0.7 million more tons are produced in the United States than in China.
  (a) Write a system of equations whose solution represents the amount of e-waste produced in each country.

  (b) Solve the system symbolically.

  (c) Solve the system graphically.

123. Student Loans A student takes out two loans totaling $3000 to help pay for college expenses. One loan is at 8% interest, and the other is at 10%. Interest for both loans is compounded annually.
  (a) If the first-year interest is $264, write a system of equations whose solution is the amount of each loan.

  (b) Find the amount of each loan.

124. Student Loans (Refer to Exercise 123.) Suppose that both loans have an interest rate of 10% and the total first-year interest is $300. If possible, determine the amount of each loan. Interpret your results.

125. **Student Loans** (Refer to Exercises 123 and 124.) Suppose that both loans are at 10% and the total annual interest is $264. If possible, determine the amount of each loan. Interpret your results.

126. **Investments** A student invests $5000 at two annual interest rates, 5% and 7%. After 1 year the student receives a total of $325 in interest. How much did the student invest at each interest rate?

127. **Air Speed** A jet airliner travels 1680 miles in 3 hours with a tail wind. The return trip, into the wind, takes 3.5 hours. Find both the speed of the jet with no wind and the wind speed. (*Hint:* First find the ground speed of the airplane in each direction.)

128. **River Current** A tugboat can pull a barge 60 miles upstream in 15 hours. The same tugboat and barge can make the return trip downstream in 6 hours. Determine the speed of the current in the river.

129. **Maximizing Area** Suppose a rectangular pen for a pet is to be made using 40 feet of fence. Let l represent its length and w its width, with $l \geq w$.
(a) Find l and w if the area is 91 square feet.

(b) Write a formula for the area A in terms of w.

(c) What is the maximum area possible for the pen? Interpret this result.

130. **The Toll of War** American battlefield deaths in World Wars I and II totaled about 345,000. There were about 5.5 times as many deaths in World War II as World War I. Find the number of American battlefield deaths in each war. Round your answers to the nearest whole number. (*Source:* Defense Department.)

131. **Height and Weight** The relationship between a professional basketball player's height h in inches and weight w in pounds was modeled using two samples of players. The resulting modeling equations for the two samples were $w = 7.46h - 374$ and $w = 7.93h - 405$. Assume that $65 \leq h \leq 85$.
(a) Use each equation to predict the weight of a professional basketball player who is 6'11".

(b) Determine graphically the height where the two models give the same weight.

(c) For each model, what change in weight is associated with a 1-inch increase in height?

132. **Heart Rate** In one study a group of athletes were exercised to exhaustion. Let x and y represent an athlete's heart rate 5 seconds and 10 seconds after stopping exercise, respectively. It was found that the maximum heart rate H for these athletes satisfied the following two equations.

$$H = \quad 0.491x + 0.468y + 11.2$$
$$H = -0.981x + 1.872y + 26.4$$

If an athlete had a maximum heart rate of $H = 180$, determine x and y graphically. Interpret your answer. (*Source:* V. Thomas, *Science and Sport.*)

133. **Surface Area and the Human Body** The surface area of the skin covering the human body is a function of more than one variable. A taller person tends to have a larger surface area, as does a heavier person. Both height and weight influence the surface area of a person's body. A formula used to determine the surface area of a person's body in square meters is given by

$$S(w, h) = 0.007184w^{0.425}h^{0.725},$$

where w is weight in kilograms and h is height in centimeters. Use S to estimate the surface area of a person who is 65 inches (165.1 centimeters) tall and weighs 154 pounds (70 kilograms). (*Source:* H. Lancaster, *Quantitative Methods in Biological and Medical Sciences.*)

Exercises 134–136: **Skin and the Human Body** *(Refer to Exercise 133.) Estimate, to the nearest tenth, the surface area of a person with weight w and height h.*

134. $w = 86$ kilograms, $h = 185$ centimeters

135. $w = 132$ pounds, $h = 62$ inches

136. $w = 220$ pounds, $h = 75$ inches

Joint Variation

Exercises 137 and 138: Approximate the constant of variation to the nearest hundredth.

137. The variable z varies jointly with the second power of x and the third power of y. When $x = 2$ and $y = 2.5$, $z = 31.9$.

138. The variable z varies jointly with the 1.5 power of x and the 2.1 power of y. When $x = 4$ and $y = 3.5$, $z = 397$.

139. The variable z varies jointly with the square root of x and the cube root of y. If $z = 10.8$ when $x = 4$ and $y = 8$, find z when $x = 16$ and $y = 27$.

140. The variable z varies jointly with the third powers of x and y. If $z = 2160$ when $x = 3$ and $y = 4$, find z when $x = 2$ and $y = 5$.

141. **Wind Power** The electrical power generated by a windmill varies jointly with the square of the diameter of the area swept out by the blades and the cube of the wind velocity. If a windmill with an 8-foot diameter and a 10-mile-per-hour wind generates 2405 watts, how much power would be generated if the blades swept out an area 6 feet in diameter and the wind was 20 miles per hour?

142. **Strength of a Beam** The strength of a rectangular beam varies jointly with its width and the square of its thickness. If a beam 5.5 inches wide and 2.5 inches thick supports 600 pounds, how much can a similar beam that is 4 inches wide and 1.5 inches thick support?

143. **Volume of Wood** (Refer to Example 14.) One cord of wood contains 128 cubic feet. Estimate the number of cords in a tree that is 105 feet tall and has a diameter of 38 inches.

144. **Carpeting** The cost of carpet for a rectangular room varies jointly with its width and length. If a room 10 feet wide and 12 feet long costs $1560 to carpet, find the cost to carpet a room that is 11 feet by 23 feet. Interpret the constant of variation.

145. **Surface Area** Use the results of Exercise 133 to find a formula for $S(w, h)$ that calculates the surface area of a person if w is given in *pounds* and h is given in *inches*.

146. **Surface Area** Use the results of Exercise 145 to solve Exercises 135 and 136.

Writing about Mathematics

147. Give an example of a quantity occurring in everyday life that can be computed by a function of more than one input. Identify the inputs and the output.

148. Give an example of a system of linear equations with two variables. Explain how to solve the system graphically and symbolically

---

9.2 Systems of Inequalities in Two Variables

- Solve inequalities in two variables graphically
- Solve systems of inequalities in two variables
- Learn basic properties of linear programming in two variables

Introduction

For people who regularly consume caffeinated beverages, too much caffeine may cause "caffeine jitters," while too little caffeine may bring on a caffeine withdrawal headache. These caffeine amounts vary depending on an individual's weight. Figure 9.21 shows one possible relationship between a person's weight and the effects of caffeine, while Table 9.1 gives the caffeine content of selected beverages. To describe the middle shaded region in the figure, we need a *system of linear inequalities*. See Exercises 37–40. (*Source:* Mayo Clinic)

Effects of Caffeine

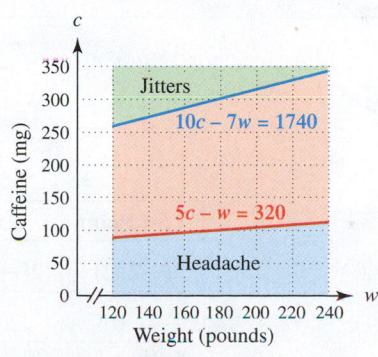

Weight (pounds)

Figure 9.21

Beverage Caffeine Content

Beverage	Caffeine
7-Up: 12 oz	0 mg
Mt Dew: 12 oz	54 mg
Red Bull: 8.4 oz	80 mg
Brewed Coffee: 8 oz	108 mg
Monster: 16 oz	160 mg
Starbucks Tall Coffee: 12 oz	260 mg
All City NRG: 16 oz	300 mg

Source: Energy Fiend
Table 9.1

Systems of Linear and Nonlinear Inequalities

A linear inequality in two variables can be written as

$$ax + by \le c,$$

where a, b, and c are constants with a and b not equal to zero. (The symbol $\le$ can be replaced by $\ge$, $<$, or $>$.) If an ordered pair (x, y) makes the inequality a true statement, then (x, y) is a solution. The set of all solutions is called the *solution set*. The graph of an inequality includes all points (x, y) in the solution set.

The graph of a linear inequality is a (shaded) **half-plane,** which may include the boundary. To determine which half-plane to shade, select a **test point** that is not on the boundary. If the test point satisfies the given inequality, then shade the half-plane containing the test point. Otherwise, shade the other half-plane. For example, the following See the Concept demonstrates how to graph the solution to the inequality $3x - 2y \le 6$.

See the Concept: Graphing a Linear Inequality

To graph $3x - 2y \le 6$:

A Solve the inequality for y.

$$3x - 2y \le 6$$
$$-2y \le -3x + 6$$
$$y \ge \tfrac{3}{2}x - 3$$

B Graph the *equation* $y = \tfrac{3}{2}x - 3$.

C Choose a test point that is not on the line.

D Substitute the test point (**0, 0**) in the given inequality. Since $3(\mathbf{0}) - 2(\mathbf{0}) \le 6$ is true, shade the half-plane containing the test point.

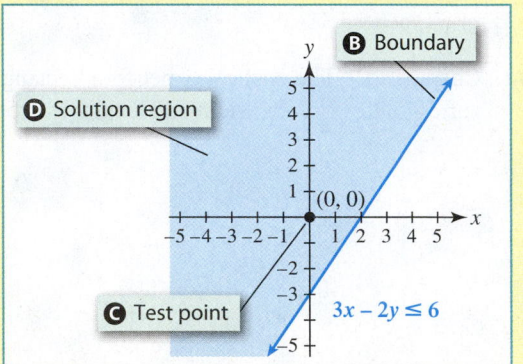

B Boundary
D Solution region
C Test point
$3x - 2y \le 6$

EXAMPLE 1   Graphing inequalities

Graph the solution set to each inequality.
(a) $2x - 3y \le -6$ **(b)** $x^2 + y^2 < 9$

SOLUTION
(a) For $2x - 3y \le -6$ start by graphing the line $2x - 3y = -6$, as in Figure 9.22. Note that this line is solid because equality is included. We can determine which side of the line to shade by using test points. For example, the test point $(-2, 2)$ lies above the line and the test point $(0, 0)$ lies below the line.

The Boundary

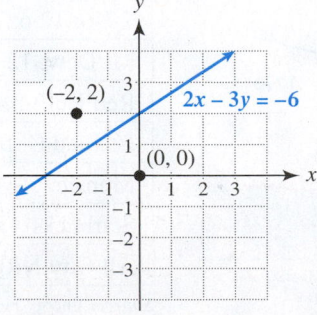

Figure 9.22

Checking Test Points

Test Point	$2x - 3y \le -6$	True or False?
$(-2, 2)$	$2(-2) - 3(2) \overset{?}{\le} -6$	True
$(0, 0)$	$2(0) - 3(0) \overset{?}{\le} -6$	False

Table 9.2

In Table 9.2, the test point $(-2, 2)$ satisfies the given inequality, so shade the region above the line that contains the point $(-2, 2)$. See Figure 9.23.

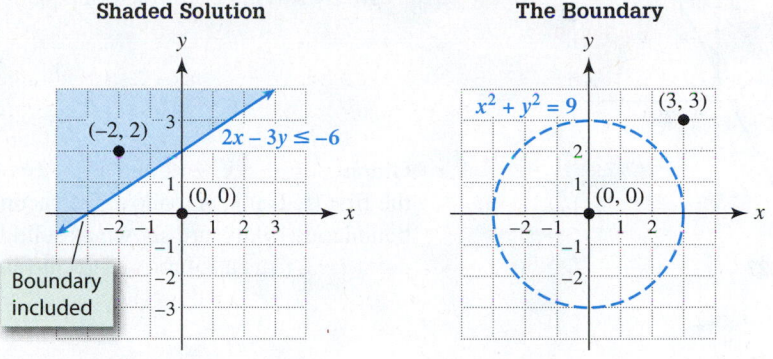

Figure 9.23   **Figure 9.24**

Shaded Solution

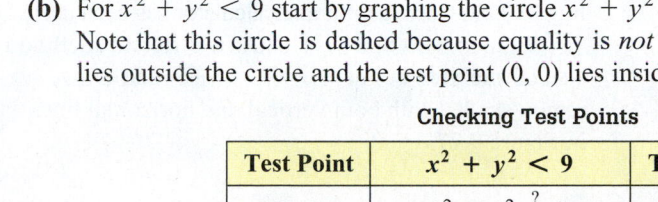

Figure 9.25

(b) For $x^2 + y^2 < 9$ start by graphing the circle $x^2 + y^2 = 9$, as shown in Figure 9.24. Note that this circle is dashed because equality is *not* included. The test point $(3, 3)$ lies outside the circle and the test point $(0, 0)$ lies inside the circle.

Checking Test Points

Test Point	$x^2 + y^2 < 9$	True or False?
$(3, 3)$	$3^2 + 3^2 \overset{?}{<} 9$	**False**
$(0, 0)$	$0^2 + 0^2 \overset{?}{<} 9$	**True**

Table 9.3

In Table 9.3, the test point $(0, 0)$ satisfies the given inequality, so shade the region inside the circle. The actual circle is not part of the solution set. See Figure 9.25.

> **Now Try Exercises 7 and 9**

In Section 9.1, we saw that systems of equations could be linear or nonlinear. Similary, systems of inequalities can be linear or nonlinear. The next example illustrates a system of each type. Both are solved graphically.

EXAMPLE 2 **Solving systems of inequalities graphically**

Solve each system of inequalities by shading the solution set. Identify one solution.
(a) $y > x^2$ **(b)** $x + 3y \le 9$
        $x + y < 4$ $2x - y \le -1$

SOLUTION
(a) This is a nonlinear system. Graph the parabola $y = x^2$ and the line $y = 4 - x$. Since $y > x^2$ and $y < 4 - x$, the region satisfying the system lies **above** the parabola and **below** the line. It does not include the boundaries, which are shown using a dashed line and curve. See Figure 9.26.

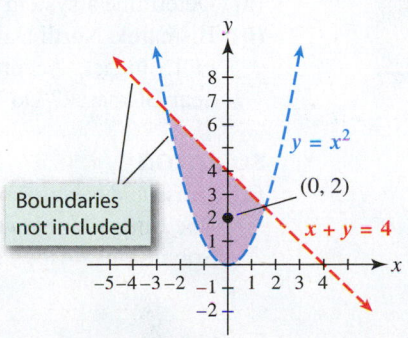

Figure 9.26

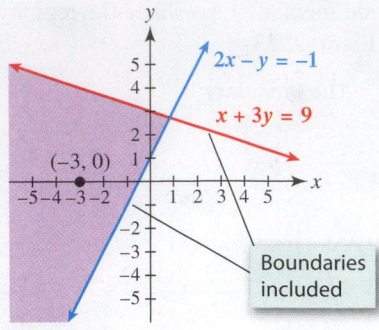

Figure 9.27

Any point in the shaded region represents a solution. For example, $(0, 2)$ lies in the shaded region and is a solution, since $x = 0$ and $y = 2$ satisfy *both* inequalities.

(b) Begin by solving each linear inequality for y.

$$y \leq -\frac{1}{3}x + 3$$

$$y \geq 2x + 1$$

Graph $y = -\frac{1}{3}x + 3$ and $y = 2x + 1$. The region satisfying the system is below the first (red) line and above the second (blue) line. Because equality is included, the boundaries, which are shown as solid lines in Figure 9.27, are part of the region. The point $(-3, 0)$ is a solution, since it satisfies both inequalities.

Now Try Exercises 17 and 19

Calculator Help

To shade a graph, see Appendix A (pages AP-10 and AP-12).

Graphing Calculators (Optional)   Graphing calculators can be used to shade regions in the xy-plane. See Figure 9.28. The solution set shown in Figure 9.26 is also shown in Figure 9.29, where a graphing calculator has been used. However, the boundary is not dashed.

Figures 9.30 and 9.31 show a different method of shading a solution set; we have shaded the area below $y_1 = -\frac{1}{3}x + 3$ and above $y_2 = 2x + 1$. The solution set is the region shaded with both vertical *and* horizontal lines and corresponds to the shaded region in Figure 9.27.

Shading using the "Shade" Function

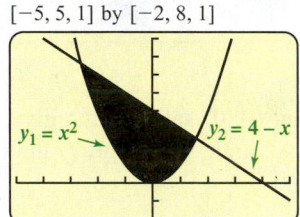

Figure 9.28   **Figure 9.29**

Shading using the Y= Menu

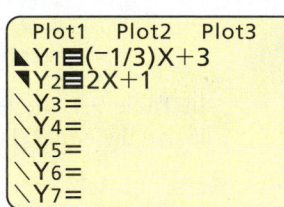

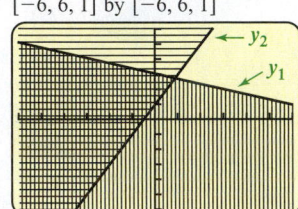

Figure 9.30   **Figure 9.31**

An Application of Inequalities   The next example discusses how a system of inequalities can be used to determine where forests, grasslands, and deserts will occur.

EXAMPLE 3   **Modeling plant growth**

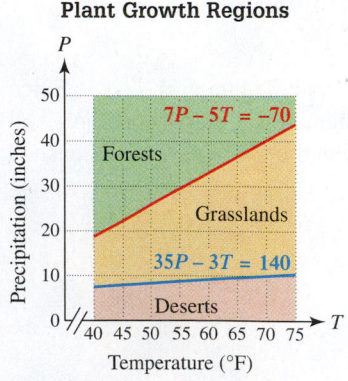

Figure 9.32

If a region has too little precipitation, it will be a desert. Forests tend to exist in regions where temperatures are relatively low and there is sufficient rainfall. At other levels of precipitation and temperature, grasslands may prevail. Figure 9.32 illustrates the relationship among forests, grasslands, and deserts, as suggested by annual average temperature T in degrees Fahrenheit and precipitation P in inches.   (**Source:** A. Miller and J. Thompson, *Elements of Meteorology.*)

(a) Determine a system of linear inequalities that describes where grasslands occur.

(b) Bismarck, North Dakota, has an annual average temperature of 40°F and precipitation of 15 inches. According to the graph, what type of plant growth would you expect near Bismarck? Do these values satisfy the system of inequalities from part (a)?

SOLUTION

(a) Grasslands occur for ordered pairs (T, P) lying between the two lines in Figure 9.32. The boundary between deserts and grasslands is determined by $35P - 3T = 140$. Solving for P (the variable on the vertical axis) results in

$$P = \frac{3}{35}T + \frac{140}{35}.$$

Grasslands grow where values of P are above the line. This region is described by $P > \frac{3}{35}T + \frac{140}{35}$, or equivalently, $35P - 3T > 140$. In a similar manner, the region below the boundary between grasslands and forests is represented by the inequality $7P - 5T < -70$. Thus grasslands satisfy the following system of inequalities.

$$35P - 3T > 140$$
$$7P - 5T < -70$$

Grasslands satisfy both inequalities.

(b) For Bismarck, $T = 40$ and $P = 15$. Figure 9.32 shows that the point $(40, 15)$ lies between the two lines, so the graph predicts that grasslands will exist around Bismarck. Substituting these values for T and P into the system of inequalities results in the following true statements.

$$35(15) - 3(40) = 405 > 140 \; \checkmark \quad \text{True}$$
$$7(15) - 5(40) = -95 < -70 \; \checkmark \quad \text{True}$$

The temperature and precipitation values for Bismarck satisfy the system of inequalities for grasslands.

Now Try Exercises 43 and 45

Linear Programming

Linear programming is a procedure used to optimize quantities such as cost and profit. It was developed during World War II as a method of efficiently allocating supplies. Linear programming applications frequently contain thousands of variables and are solved by computers. However, here we focus on problems involving two variables.

A linear programming problem consists of a linear **objective function** and a system of linear inequalities called **constraints**. The solution set for the system of linear inequalities is called the set of **feasible solutions**. The objective function describes a quantity that is to be optimized. For example, linear programming is often used to maximize profit or minimize cost. The following example illustrates these concepts.

EXAMPLE 4 **Finding maximum profit**

Suppose a small company manufactures two products—car radios and stereos. Each radio results in a profit of $15, and each stereo provides a profit of $35. Due to demand, the company must produce at least 5 and not more than 25 radios per day. The number of radios cannot exceed the number of stereos, and the number of stereos cannot exceed 30. How many of each should the company manufacture to obtain maximum profit?

SOLUTION Let x be the number of car radios produced daily and y be the number of stereos produced daily. Since the profit from x radios is $15x$ dollars and the profit from y stereos is $35y$ dollars, the total daily profit P is given by

$$P = 15x + 35y.$$

The company produces from 5 to 25 radios per day, so the inequalities

$$x \geq 5 \quad \text{and} \quad x \leq 25$$

must be satisfied. The requirements that the number of radios cannot exceed the number of stereos and the number of stereos cannot exceed 30 indicate that

$$x \leq y \quad \text{and} \quad y \leq 30.$$

Since the numbers of radios and stereos cannot be negative, we have

$$x \geq 0 \quad \text{and} \quad y \geq 0.$$

Listing all the constraints on production gives

$$x \geq 5, \quad x \leq 25, \quad y \leq 30, \quad x \leq y, \quad x \geq 0, \quad \text{and} \quad y \geq 0.$$

Graphing these constraints results in the shaded region shown in Figure 9.33. This shaded region is the set of feasible solutions. The vertices (or corners) of this region are $(5, 5)$, $(25, 25)$, $(25, 30)$, and $(5, 30)$.

It can be shown that maximum profit occurs at a vertex of the region of feasible solutions. Thus we evaluate P at each vertex, as shown in Table 9.4.

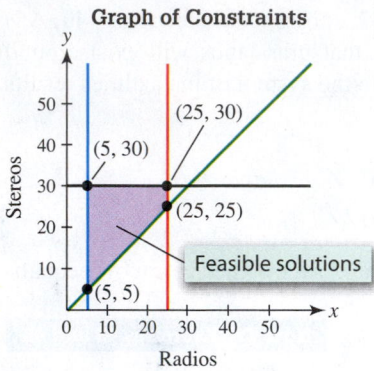

Graph of Constraints

Figure 9.33

Checking Vertices in the Profit Equation

Vertex	$P = 15x + 35y$
$(5, 5)$	$15(5) + 35(5) = 250$
$(25, 25)$	$15(25) + 35(25) = 1250$
$(25, 30)$	$15(25) + 35(30) = \mathbf{1425}$ — Maximum profit
$(5, 30)$	$15(5) + 35(30) = 1125$

Table 9.4

The maximum value of P is **1425** at vertex $(25, 30)$. Thus the maximum profit is \$1425, and it occurs when 25 car radios and 30 stereos are manufactured.

Now Try Exercise 63

The following theorem holds for linear programming problems.

FUNDAMENTAL THEOREM OF LINEAR PROGRAMMING

If the optimal value for a linear programming problem exists, then it occurs at a vertex of the region of feasible solutions.

Justification of the Fundamental Theorem   To better understand the fundamental theorem of linear programming, consider the following example. Suppose that we want to maximize $P = 30x + 70y$ subject to the following four constraints:

$$x \geq 10, \quad x \leq 50, \quad y \geq x, \quad \text{and} \quad y \leq 60.$$

The corresponding region of feasible solutions is shown in Figure 9.34.

Each value of P determines a unique line. For example, if $P = 7000$, then the equation for P becomes $30x + 70y = 7000$. The resulting line, shown in Figure 9.35, does not intersect the region of feasible solutions. Thus there are no values for x and y that lie in this region and result in a profit of 7000. Figure 9.35 also shows the lines that result from letting $P = 0, 1000,$ and 3000. If $P = 1000$, then the line intersects the region of feasible solutions only at the vertex $(10, 10)$. This means that if $x = 10$ and $y = 10$, then $P = 30(10) + 70(10) = 1000$. If $P = 3000$, then the line $30x + 70y = 3000$ intersects the region of feasible solutions infinitely many times. However, it appears that values greater than 3000 are possible for P.

In Figure 9.36 lines are drawn for $P = 5700, 6300,$ and 7000. Notice that there are no points of intersection for $P = 6300$ or $P = 7000$, but there is one vertex in the region of feasible solutions at $(50, 60)$ that gives $P = 5700$. Thus the maximum value of P is 5700 and this maximum occurs at a vertex of the region of feasible solutions. The fundamental theorem of linear programming generalizes this result.

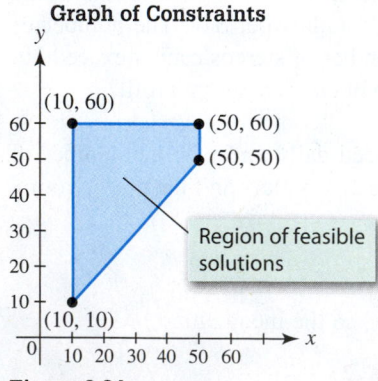

Graph of Constraints

Figure 9.34

CLASS DISCUSSION

What is the minimum value for P subject to the given constraints?

How the Objective Function Intersects the Region of Feasible Solutions

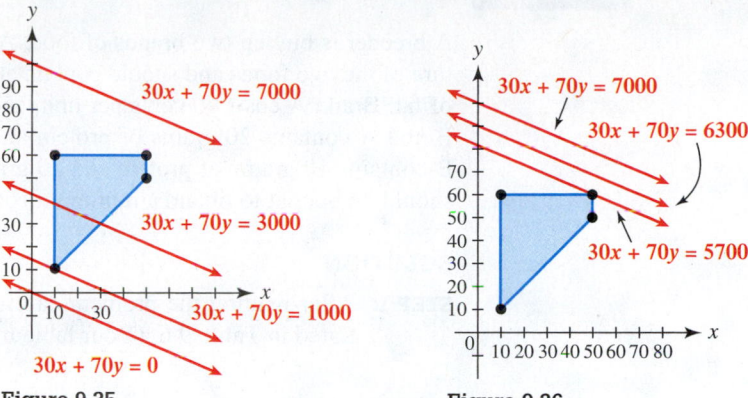

Figure 9.35

Figure 9.36

EXAMPLE 5 Finding the minimum of an objective function

Find the minimum value of $C = 2x + 3y$ subject to the following constraints.

$$x + y \geq 4$$
$$2x + y \leq 8$$
$$x \geq 0, \quad y \geq 0$$

SOLUTION Sketch the region determined by the constraints and find all vertices, as shown in Figure 9.37.

Evaluate the objective function C at each vertex, as shown in Table 9.5.

Graph of Constraints

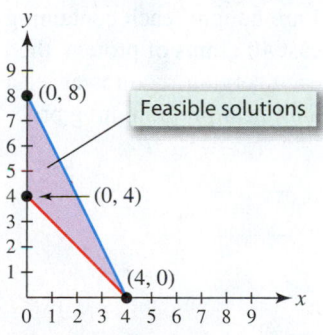

Figure 9.37

Checking Vertices in the Objective Function

Vertex	$C = 2x + 3y$	
$(4, 0)$	$2(4) + 3(0) = \mathbf{8}$	Minimum value
$(0, 8)$	$2(0) + 3(8) = 24$	
$(0, 4)$	$2(0) + 3(4) = 12$	

Table 9.5

The minimum value for C is **8** and it occurs at vertex $(4, 0)$, or when $x = 4$ and $y = 0$.

Now Try Exercise 59

The following procedure describes how to solve a linear programming problem.

SOLVING A LINEAR PROGRAMMING PROBLEM

STEP 1: Read the problem carefully. Consider making a table to display the given information.

STEP 2: Use the table to write the objective function and all the constraints.

STEP 3: Sketch a graph of the region of feasible solutions. Identify all vertices, or corner points.

STEP 4: Evaluate the objective function at each vertex. A maximum (or a minimum) occurs at a vertex. If the region is unbounded, a maximum (or minimum) may not exist.

EXAMPLE 6   **Minimizing cost**

A breeder is buying two brands of food, A and B, for her animals. Each serving is a mixture of the two foods and should contain at least 40 grams of protein and at least 30 grams of fat. Brand A costs 90 cents per unit, and Brand B costs 60 cents per unit. Each unit of Brand A contains 20 grams of protein and 10 grams of fat, whereas each unit of Brand B contains 10 grams of protein and 10 grams of fat. Determine how much of each brand should be bought to obtain a minimum cost per serving.

SOLUTION

STEP 1: After reading the problem carefully, begin by listing the information, as illustrated in Table 9.6. (Your table may be different.)

Protein and Fat Content by Brand, with Cost

Brand	Units	Protein/Unit	Total Protein	Fat/Unit	Total Fat	Cost
A	x	20	$20x$	10	$10x$	$90x$
B	y	10	$10y$	10	$10y$	$60y$

Minimum Total Protein —— **40**   **30** —— Minimum Total Fat

Table 9.6

Graph of Constraints

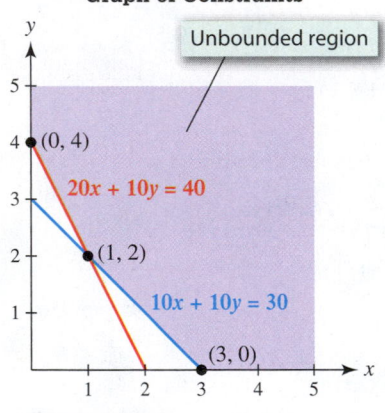

Figure 9.38

STEP 2: If x units of Brand A are purchased at 90¢ per unit and y units of Brand B are purchased at 60¢ per unit, then the cost C is given by $C = 90x + 60y$. Each serving requires at least 40 grams of protein. If x units of Brand A are bought (each containing 20 grams of protein), y units of Brand B are bought (each containing 10 grams of protein), and each serving requires at least 40 grams of protein, then we can write $20x + 10y \geq 40$. Similarly, since each serving requires at least 30 grams of fat, we can write $10x + 10y \geq 30$. The linear programming problem can be written as follows.

$$\begin{aligned} \text{Minimize:} \quad & C = 90x + 60y & \text{Cost} \\ \text{Subject to:} \quad & 20x + 10y \geq 40 & \text{Protein} \\ & 10x + 10y \geq 30 & \text{Fat} \\ & x \geq 0, \quad y \geq 0 \end{aligned}$$

STEP 3: The region of feasible solutions is unbounded and shown in Figure 9.38. The vertices for this region are $(0, 4)$, $(1, 2)$, and $(3, 0)$.

STEP 4: Evaluate the objective function C at each vertex, as shown in Table 9.7.

Checking Vertices in the Cost Equation

Vertex	$C = 90x + 60y$
$(0, 4)$	240
$(1, 2)$	**210** —— Minimum cost
$(3, 0)$	270

Table 9.7

The minimum cost occurs when 1 unit of Brand A and 2 units of Brand B are mixed, at a cost of **$2.10** per serving.

Now Try Exercise 65

9.2 Putting It All Together

The following table summarizes some important mathematical concepts from this section.

CONCEPT	COMMENTS	EXAMPLE
Linear inequality in two variables	$ax + by \leq c$ ($\leq$ may be replaced by $<$, $>$, or $\geq$) The solution set is typically a shaded region in the xy-plane.	$2x - 3y \leq 12$ $-2x + y \leq 4$
Linear programming	In a linear programming problem, the maximum or minimum of an objective function is found, subject to constraints. If a solution exists, it occurs at a vertex of the region of feasible solutions.	Maximize the objective function $P = 2x + 3y$ subject to the following constraints. $2x + y \leq 6$ $x + 2y \leq 6$ $x \geq 0, \, y \geq 0$ The maximum of $P = 10$ occurs at vertex $(2, 2)$.

9.2 Exercises

Inequalities

Exercises 1–12: Graph the solution set to the inequality.

1. $x \geq y$

2. $y > -3$

3. $x < 1$

4. $y > 2x$

5. $x + y \leq 2$

6. $x + y > -3$

7. $2x + y > 4$

8. $2x + 3y \leq 6$

9. $x^2 + y^2 > 4$

10. $x^2 + y^2 \leq 1$

11. $x^2 + y \leq 2$

12. $2x^2 - y < 1$

Exercises 13–16: Match the system of inequalities with the appropriate graph (a–d). Use the graph to identify one solution.

13. $x + y \geq 2$
$x - y \leq 1$

14. $2x - y > 0$
$x - 2y \leq 1$

15. $\frac{1}{2}x^3 - y > 0$
$2x - y \leq 1$

16. $x^2 + y \leq 4$
$x^2 - y \leq 2$

a.

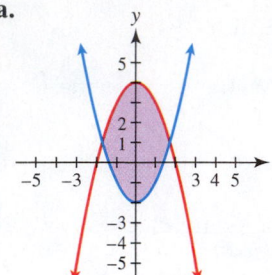

b.

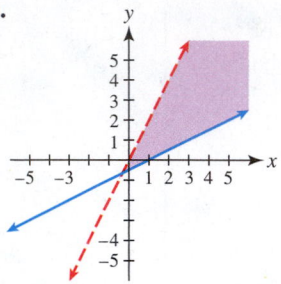

c.

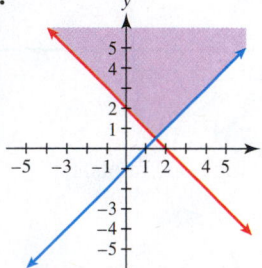

d.

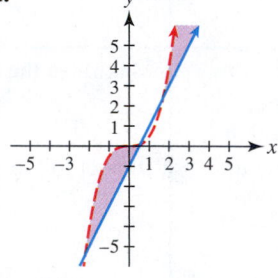

Exercises 17–24: Graph the solution set to the system of inequalities. Use the graph to identify one solution.

17. $y \geq x^2$
$x + y \leq 6$

18. $y \leq \sqrt{x}$
$y \geq 1$

19. $x + 2y > -2$
$x + 2y < 5$

20. $x - y \leq 3$
$x + y \leq 3$

21. $x^2 + y^2 \leq 16$
$x + y < 2$

22. $x^2 + y \leq 4$
$x^2 - y \leq 3$

23. $x^2 + y > 2$
$x^2 + y^2 \leq 9$

24. $x^2 + y^2 > 4$
$x^2 + y^2 < 16$

Exercises 25–36: Graph the solution set to the system of inequalities.

25. $x + 2y \leq 4$
$2x - y \geq 6$

26. $3x - y \leq 3$
$x + 2y \leq 2$

27. $3x + 2y < 6$
$x + 3y \leq 6$

28. $4x + 3y \geq 12$
$2x + 6y \geq 4$

29. $x - 2y \geq 0$
$x - 3y \leq 3$

30. $2x - 4y \geq 4$
$x + y \leq 0$

31. $x^2 + y^2 \leq 4$
$y \geq 1$

32. $x^2 - y \leq 0$
$x^2 + y^2 \leq 6$

33. $2x^2 + y \leq 0$
$x^2 - y \leq 3$

34. $x^2 + 2y \leq 4$
$x^2 - y \leq 0$

35. $x^2 + y^2 \leq 4$
$x^2 + 2y \leq 2$

36. $2x + 3y \leq 6$
$\frac{1}{2}x^2 - y \leq 2$

Applications

Exercises 37–40: **Caffeine Consumption** *(Refer to the introduction to this section.) The following graph shows one possible relationship between a person's weight and the effects of caffeine.* (**Source:** Mayo Clinic)

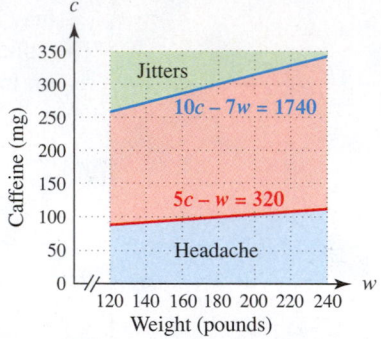

37. What does the graph indicate about the effects of caffeine on 140-pound person who has consumed 335 mg of caffeine?

38. Suppose a 180-pound person wishes to avoid both a headache and the jitters. What range of caffeine consumption is suggested?

39. (Refer to Table 9.1.) For what weights could a person drink a 16-ounce can of All City NRG without experiencing the jitters?

40. (Refer to Table 9.1.) According to this graph, does a single 12-ounce can of Mountain Dew contain enough caffeine for most people to avoid a headache?

41. Traffic Control The figure shows two intersections, labeled A and B, that involve one-way streets. The numbers and variables represent the average traffic flow rates measured in vehicles per hour. For example, an average of 500 vehicles per hour enter intersection A from the west, whereas 150 vehicles per hour enter this intersection from the north. A stoplight will control the unknown traffic flow denoted by the variables x and y. Use the fact that the number of vehicles entering an intersection must equal the number leaving to determine x and y.

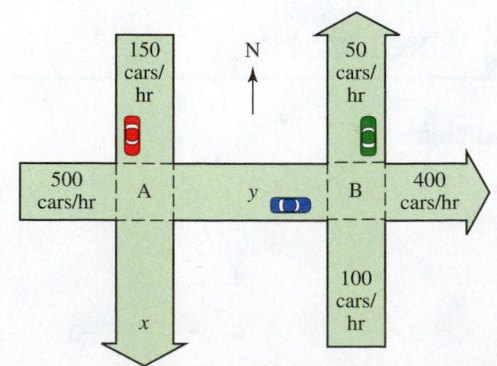

42. Traffic Control   (Refer to Exercise 41.) Suppose that the number of vehicles entering intersection A from the west varies between 400 and 600. If all other traffic flows remain the same as in the figure, what effect does this have on the ranges of the values for x and y?

Exercises 43–46: **Weight and Height**   *The following graph shows a weight and height chart. The weight w is listed in pounds and the height h in inches. The shaded area is a recommended region. (**Source:** Department of Agriculture.)*

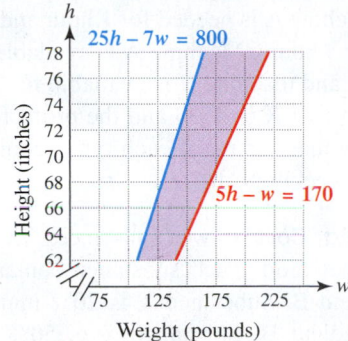

43. What does this chart indicate about an individual who weighs 125 pounds and is 70 inches tall?

44. Use the graph to estimate the recommended weight range for a person 74 inches tall.

45. Use the graph to find a system of linear inequalities that describes the recommended region.

46. Explain why inequalities are more appropriate than equalities for describing recommended weight and height combinations.

Linear Programming

Exercises 47–50: Shade the region of feasible solutions for the following constraints.

47. $x + y \leq 4$
$x + y \geq 1$
$x \geq 0, y \geq 0$

48. $x + 2y \leq 8$
$2x + y \geq 2$
$x \geq 0, y \geq 0$

49. $3x + 2y \leq 12$
$2x + 3y \leq 12$
$x \geq 0, y \geq 0$

50. $x + y \leq 4$
$x + 4y \geq 4$
$x \geq 0, y \geq 0$

Exercises 51 and 52: The graph shows a region of feasible solutions for P. Find the maximum and minimum values of P.

51. $P = 3x + 5y$

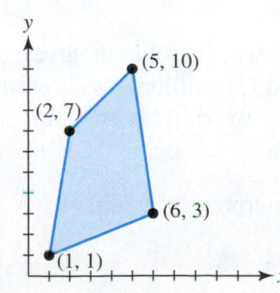

52. $P = 6x + y$

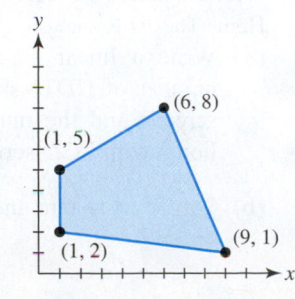

Exercises 53–56: The graph shows a region of feasible solutions for C. Find the maximum and minimum values of C.

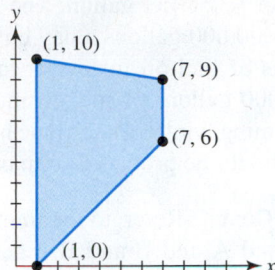

53. $C = 3x + 5y$

54. $C = 5x + 5y$

55. $C = 10y$

56. $C = 3x - y$

Exercises 57 and 58: Write a system of linear inequalities that describes the shaded region.

57.

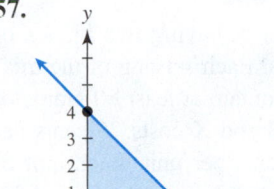

58.

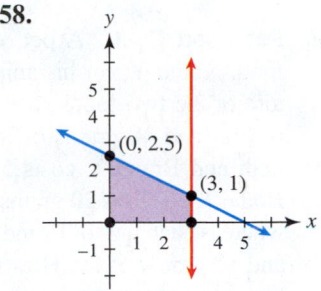

59. Find the minimum value of $C = 4x + 2y$ subject to the following constraints.

$$x + y \geq 3$$
$$2x + 3y \leq 12$$
$$x \geq 0, y \geq 0$$

60. Find the maximum value of $P = 3x + 5y$ subject to the following constraints.

$$3x + y \leq 8$$
$$x + 3y \leq 8$$
$$x \geq 0, y \geq 0$$

Exercises 61 and 62: If possible, maximize and minimize z subject to the given constraints.

61. $z = 7x + 6y$

$$x + y \leq 8$$
$$x + y \geq 4$$
$$x \geq 0, y \geq 0$$

62. $z = 8x + 3y$

$$4x + y \geq 12$$
$$x + 2y \geq 6$$
$$x \geq 0, y \geq 0$$

63. Maximizing Profit   Rework Example 4 if the profit from each radio is $20 and the profit from each CD player is $15.

64. Maximizing Revenue A refinery produces both gasoline and fuel oil, and sells gasoline for $4.00 per gallon and fuel oil for $3.60 per gallon. The refinery can produce at most 600,000 gallons a day but must produce at least 2 gallons of fuel oil for every gallon of gasoline. At least 150,000 gallons of fuel oil must be made each day for the coming winter. Determine how much of each type of fuel should be produced to maximize revenue.

65. Minimizing Cost (Refer to Example 6.) A breeder is mixing Brand A and Brand B. Each serving should contain at least 60 grams of protein and 30 grams of fat. Brand A costs 80 cents per unit, and Brand B costs 50 cents per unit. Each unit of Brand A contains 15 grams of protein and 10 grams of fat, whereas each unit of Brand B contains 20 grams of protein and 5 grams of fat. Determine how much of each food should be bought to achieve a minimum cost per serving.

66. Pet Food Cost A pet owner is buying two brands of food, X and Y, for his animals. Each serving of the mixture of the two foods should contain at least 60 grams of protein and 40 grams of fat. Brand X costs 75 cents per unit, and Brand Y costs 50 cents per unit. Each unit of Brand X contains 20 grams of protein and 10 grams of fat, whereas each unit of Brand Y contains 10 grams of protein and 10 grams of fat. How much of each brand should be bought to obtain a minimum cost per serving?

67. Raising Animals A breeder can raise no more than 50 hamsters and mice and no more than 20 hamsters. If she sells the hamsters for $15 each and the mice for $10 each, find the maximum revenue produced.

68. Maximizing Storage A manager wants to buy filing cabinets. Cabinet X costs $100, requires 6 square feet of floor space, and holds 8 cubic feet. Cabinet Y costs $200, requires 8 square feet of floor space, and holds 12 cubic feet. No more than $1400 can be spent, and the office has room for no more than 72 square feet of cabinets. The manager wants the maximum storage capacity within the limits imposed by funds and space. How many of each type of cabinet should be bought?

69. Maximizing Profit A business manufactures two parts, X and Y. Machines A and B are needed to make each part. To make part X, machine A is needed for 4 hours and machine B is needed for 2 hours. To make part Y, machine A is needed for 1 hour and machine B is needed for 3 hours. Machine A is available for 40 hours each week and machine B is available for 30 hours. The profit from part X is $500 and the profit from part Y is $600. How many parts of each type should be made to maximize weekly profit?

70. Minimizing Cost Two substances, X and Y, are found in pet food. Each substance contains the ingredients A and B. Substance X is 20% ingredient A and 50% ingredient B. Substance Y is 50% ingredient A and 30% ingredient B. The cost of substance X is $2 per pound, and the cost of substance Y is $3 per pound. The pet store needs at least 251 pounds of ingredient A and at least 200 pounds of ingredient B. If cost is to be minimal, how many pounds of each substance should be ordered? Find the minimum cost.

Writing about Mathematics

71. Give the general form of a system of linear inequalities in two variables. Discuss what distinguishes a system of linear inequalities from a nonlinear system of inequalities.

72. Discuss how to use test points to solve a linear inequality. Give an example.

CHECKING BASIC CONCEPTS FOR SECTIONS 9.1 AND 9.2

1. Evaluate $d(13, 18)$ if
$$d(x, y) = \sqrt{(x - 1)^2 + (y - 2)^2}.$$

2. Solve the nonlinear system of equations using the method of substitution.
$$2x^2 - y = 0$$
$$3x + 2y = 7$$

3. Solve $z = x^2 + y^2$ for y.

4. Solve the system of linear equations by using elimination.
$$3x - 2y = 4$$
$$-x + 6y = 8$$

5. Graph the solution set to $3x - 2y \leq 6$.

6. Graph the solution set to the system of inequalities. Use the graph to identify one solution.
$$x^2 - y < 3$$
$$x - y \geq 1$$

7. HDTV Sales In 2010 there were 220 million HDTVs sold worldwide. For every 2 HDTVs sold with plasma screens, 9 HDTVs were sold with LCD screens. (*Source:* Home Theater Review.)
   (a) Write a linear system whose solution gives the number of HDTVs sold (in millions) with plasma screens and the number of HDTVs sold (in millions) with LCD screens.

   (b) Solve the system and interpret the result.

9.3 Systems of Linear Equations in Three Variables

- Learn basic concepts about systems in three variables
- Solve systems using elimination and substitution
- Identify systems with no solutions
- Solve systems with infinitely many solutions

Introduction

In Section 9.1 we discussed how to solve systems of linear equations in two variables. Systems of linear equations can have any number of variables. For example, Internet search sites such as Google, Yahoo!, and Bing use algorithms that involve linear systems with millions of variables. Computers are necessary to solve these systems efficiently. However, in this section we discuss solving systems of linear equations containing three variables by hand.

Basic Concepts

When writing systems of linear equations in three variables it is common, but not necessary, to use the variables x, y, and z. For example,

$$\left.\begin{array}{rcl} 2x - 3y + 4z &=& 4 \\ -y + 2z &=& 0 \\ x + 5y - 6z &=& 7 \end{array}\right\}$$

Linear System with three variables

represents a system of linear equations in three variables. The solution to this system is given by $x = 3$, $y = 2$, and $z = 1$ because each equation is satisfied when these values are substituted for the variables in the system of linear equations.

$$2(3) - 3(2) + 4(1) \stackrel{?}{=} 4 \; \checkmark \quad \text{True}$$
$$-(2) + 2(1) \stackrel{?}{=} 0 \; \checkmark \quad \text{True}$$
$$(3) + 5(2) - 6(1) \stackrel{?}{=} 7 \; \checkmark \quad \text{True}$$

The solution to this system can be written as the **ordered triple** $(3, 2, 1)$. This system of linear equations has exactly one solution. In general, systems of linear equations can have zero, one, or infinitely many solutions.

EXAMPLE 1   Checking for solutions

Determine whether $(-1, -3, 2)$ or $(1, -10, -13)$ is a solution to the system of equations.

$$\begin{array}{rcl} x - 4y + 2z &=& 15 \\ 4x - y + z &=& 1 \\ 6x - 2y - 3z &=& -6 \end{array}$$

SOLUTION   First substitute $x = -1$, $y = -3$, and $z = 2$ in the system of linear equations, and then substitute $x = 1$, $y = -10$, and $z = -13$.

$$(-1) - 4(-3) + 2(2) \stackrel{?}{=} 15 \quad \text{True} \qquad (1) - 4(-10) + 2(-13) \stackrel{?}{=} 15 \quad \text{True}$$
$$4(-1) - (-3) + (2) \stackrel{?}{=} 1 \quad \text{True} \qquad 4(1) - (-10) + (-13) \stackrel{?}{=} 1 \quad \text{True}$$
$$6(-1) - 2(-3) - 3(2) \stackrel{?}{=} -6 \quad \text{True} \qquad 6(1) - 2(-10) - 3(-13) \stackrel{?}{=} -6 \quad \text{False}$$

The ordered triple $(-1, -3, 2)$ satisfies all three equations, so it is a solution to the system of equations. The ordered triple $(1, -10, -13)$ is not a solution to the system of equations because it satisfies only two of the three equations.

Now Try Exercise 7

Solving with Elimination and Substitution

We can solve systems of linear equations in three variables by hand. The following procedure uses substitution and elimination and assumes that the variables are x, y, and z.

> **SOLVING A SYSTEM OF LINEAR EQUATIONS IN THREE VARIABLES**
>
> **STEP 1:** Eliminate one variable, such as x, from two of the equations.
>
> **STEP 2:** Apply the techniques discussed in Section 9.1 to solve the two resulting equations in two variables from Step 1. If x is eliminated, then solve these equations to find y and z.
>
> If there are no solutions for y and z, then the given system also has no solutions. If there are infinitely many solutions for y and z, then write y in terms of z and proceed to Step 3.
>
> **STEP 3:** Substitute the values for y and z in one of the given equations to find x. The solution is (x, y, z). If possible, check your answer as in Example 1.

EXAMPLE 2 Solving a linear system in three variables

Solve the following system.

$$\begin{aligned} x - y + 2z &= 6 \\ 2x + y - 2z &= -3 \\ -x - 2y + 3z &= 7 \end{aligned}$$

SOLUTION

STEP 1: We begin by eliminating the variable x from the second and third equations. To eliminate x from the second equation, we multiply the first equation by -2 and then add it to the second equation. To eliminate x from the third equation, we add the first and third equations.

$$\begin{array}{ll} -2x + 2y - 4z = -12 & \text{First equation times } -2 \\ \underline{2x + y - 2z = -3} & \text{Second equation} \\ 3y - 6z = -15 & \text{Add.} \end{array}$$

$$\begin{array}{ll} x - y + 2z = 6 & \text{First equation} \\ \underline{-x - 2y + 3z = 7} & \text{Third equation} \\ -3y + 5z = 13 & \text{Add.} \end{array}$$

STEP 2: Take the two resulting equations from Step 1 and eliminate either variable. Here we add the two equations to eliminate y.

$$\begin{array}{ll} 3y - 6z = -15 & \\ \underline{-3y + 5z = 13} & \\ -z = -2 & \text{Add the equations.} \\ z = 2 & \text{Multiply each side by } -1. \end{array}$$

Now we can use substitution to find the value of y. We let $z = 2$ in *either* equation used in Step 2 to find y.

$$\begin{array}{ll} 3y - 6z = -15 & \text{Equation from Step 2} \\ 3y - 6(2) = -15 & \text{Substitute } z = 2. \\ 3y - 12 = -15 & \text{Multiply.} \\ 3y = -3 & \text{Add 12 to each side.} \\ y = -1 & \text{Divide each side by 3.} \end{array}$$

STEP 3: Finally, we substitute $y = -1$ and $z = 2$ in any of the *given* equations to find x.

$$\begin{array}{ll} x - y + 2z = 6 & \text{First given equation} \\ x - (-1) + 2(2) = 6 & \text{Let } y = -1 \text{ and } z = 2. \\ x + 1 + 4 = 6 & \text{Simplify.} \\ x = 1 & \text{Subtract 5 from each side.} \end{array}$$

The solution is $(1, -1, 2)$. Check this solution.

Now Try Exercise 11

In the next example, we determine numbers of tickets sold for a play.

EXAMPLE 3 **Finding numbers of tickets sold**

One thousand tickets were sold for a play, generating \$3800 in revenue. The prices of the tickets were \$3 for children, \$4 for students, and \$5 for adults. There were 100 fewer student tickets sold than adult tickets. Find the number of each type of ticket sold.

SOLUTION Let x be the number of tickets sold to children, y be the number of tickets sold to students, and z be the number of tickets sold to adults. The total number of tickets sold was 1000, so

$$x + y + z = 1000.$$

Each child's ticket costs \$3, so the revenue generated from selling x tickets is $3x$. Similarly, the revenue generated from students is $4y$, and the revenue from adults is $5z$. Total ticket sales were \$3800, so

$$3x + 4y + 5z = 3800.$$

The equation $z - y = 100$, or $y - z = -100$, must also be satisfied, because 100 fewer tickets were sold to students than adults.

To find the price of a ticket, we need to solve the following system of linear equations.

$$
\begin{array}{ll}
x + y + z = 1000 & \text{Total number of tickets is 1000.} \\
3x + 4y + 5z = 3800 & \text{Total revenue is \$3800.} \\
y - z = -100 & \text{100 fewer student tickets than adult tickets}
\end{array}
$$

STEP 1: We begin by eliminating the variable x from the first equation. To do this, we multiply the first equation by 3 and subtract the second equation.

$$
\begin{array}{ll}
3x + 3y + 3z = 3000 & \text{First given equation times 3} \\
\underline{3x + 4y + 5z = 3800} & \text{Second equation} \\
-y - 2z = -800 & \text{Subtract.}
\end{array}
$$

STEP 2: We then use the equation that resulted from Step 1 and the third given equation to eliminate y.

$$
\begin{array}{ll}
-y - 2z = -800 & \text{Equation from Step 1} \\
\underline{y - z = -100} & \text{Third given equation} \\
-3z = -900 & \text{Add the equations.}
\end{array}
$$

Thus $z = 300$. To find y, we can substitute $z = 300$ in the third equation.

$$
\begin{array}{ll}
y - z = -100 & \text{Third given equation} \\
y - 300 = -100 & \text{Let } z = 300. \\
y = 200 & \text{Add 300.}
\end{array}
$$

STEP 3: Finally, we substitute $y = 200$ and $z = 300$ in the first equation.

$$
\begin{array}{ll}
x + y + z = 1000 & \text{First given equation} \\
x + 200 + 300 = 1000 & \text{Let } y = 200 \text{ and } z = 300. \\
x = 500 & \text{Subtract 500.}
\end{array}
$$

Thus **500** tickets were sold to children, **200** to students, and **300** to adults. Check this answer.

Now Try Exercise 33

Systems with No Solutions

Regardless of the number of variables, a system of linear equations can have zero, one, or infinitely many solutions. In the next example, a system of linear equations has no solutions.

EXAMPLE 4 Identifying a system with no solutions

Three students buy lunch in the cafeteria. One student buys 2 hamburgers, 1 order of fries, and 1 soda for $9. Another student buys 1 hamburger, 2 orders of fries, and 1 soda for $8. The third student buys 3 hamburgers, 3 orders of fries, and 2 sodas for $18. If possible, find the cost of each item. Interpret the results.

SOLUTION Let x be the cost of a hamburger, y be the cost of an order of fries, and z be the cost of a soda. Then the purchases of the three students can be expressed as a system of linear equations.

$$
\begin{aligned}
2x + y + z &= 9 && \text{2 burgers, 1 order of fries, and 1 soda for \$9} \\
x + 2y + z &= 8 && \text{1 burger, 2 orders of fries, and 1 soda for \$8} \\
3x + 3y + 2z &= 18 && \text{3 burgers, 3 orders of fries, and 2 sodas for \$18}
\end{aligned}
$$

STEP 1: We can eliminate z in the first equation by subtracting the second equation from the first equation. We can eliminate z in the third equation by subtracting twice the second equation from the third equation.

$$
\begin{array}{ll}
2x + y + z = 9 & \text{First equation} \\
\underline{x + 2y + z = 8} & \text{Second equation} \\
x - y \quad\quad = 1 & \text{Subtract.}
\end{array}
\qquad
\begin{array}{ll}
3x + 3y + 2z = 18 & \text{Third equation} \\
\underline{2x + 4y + 2z = 16} & \text{Twice second equation} \\
x - y \quad\quad = 2 & \text{Subtract.}
\end{array}
$$

STEP 2: The equations $x - y = 1$ and $x - y = 2$ are *inconsistent* because the difference between two numbers cannot be both 1 and 2. Step 3 is not necessary—the system of equations has no solutions.

NOTE In this problem the third student bought the same amount of food as the first and second students bought together. Therefore the third student should have paid $9 + $8 = $17 rather than $18. *Inconsistent pricing* led to an *inconsistent system* of linear equations.

Now Try Exercise 35

Systems with Infinitely Many Solutions

Some systems of linear equations have infinitely many solutions. In this case, we say that the system of linear equations is consistent, but the equations are dependent. A system of dependent equations is solved in the next example.

EXAMPLE 5 Solving a system with infinitely many solutions

Solve the following system of linear equations.

$$
\begin{aligned}
x + y - z &= -2 \\
x + 2y - 2z &= -3 \\
y - z &= -1
\end{aligned}
$$

SOLUTION

STEP 1: Because x does not appear in the third equation, begin by eliminating x from the first equation. To do this, subtract the second equation from the first equation.

$$
\begin{array}{ll}
x + y - z = -2 & \text{First equation} \\
\underline{x + 2y - 2z = -3} & \text{Second equation} \\
-y + z = 1 & \text{Subtract.}
\end{array}
$$

STEP 2: Adding the resulting equation from Step 1 and the third given equation gives the equation $0 = 0$, which indicates that there are infinitely many solutions.

$$-y + z = 1$$
$$\underline{y - z = -1}$$
$$0 = 0 \quad \text{Add.}$$

> Dependent equations:
> Solve either equation for y.

CLASS DISCUSSION

Three students buy lunch in the cafeteria. One student buys 1 hamburger, 1 order of fries, and 1 soda for $5. Another student buys 2 hamburgers, 2 orders of fries, and 2 sodas for $10. The third student buys 3 hamburgers, 3 orders of fries, and 3 sodas for $15. Can you find the cost of each item? Interpret your answer.

The variable y can be written in terms of z as $y = z - 1$.

STEP 3: To find x, substitute the results from Step 2 in the first given equation.

$$x + y - z = -2 \quad \text{First given equation}$$
$$x + (z - 1) - z = -2 \quad \text{Let } y = z - 1.$$
$$x = -1 \quad \text{Solve for } x.$$

Solutions to the given system are of the form $(-1, z - 1, z)$, where z is any real number. For example, if $z = 2$, then $(-1, 1, 2)$ is one possible solution.

> **Now Try Exercise 17**

9.3 Putting It All Together

In this section we discussed how to solve a system of three linear equations in three variables by hand. Systems of linear equations can have no solutions, one solution, or infinitely many solutions. The following table summarizes some of the important concepts presented in this section.

CONCEPT	EXPLANATION
System of linear equations in three variables	The following is a system of three linear equations in three variables. $$x - 2y + z = 0$$ $$-x + y + z = 4$$ $$-y + 4z = 10$$
Solution to a linear system in three variables	The solution to a linear system in three variables is an ordered triple, expressed as (x, y, z). The solution to the preceding system is $(1, 2, 3)$ because substituting $x = 1$, $y = 2$, and $z = 3$ in each equation results in a true statement. $$(1) - 2(2) + (3) = 0 \checkmark \quad \text{True}$$ $$-(1) + (2) + (3) = 4 \checkmark \quad \text{True}$$ $$-(2) + 4(3) = 10 \checkmark \quad \text{True}$$
Solving a linear system with substitution and elimination	Refer to Example 2. **STEP 1:** Eliminate one variable, such as x, from two of the equations. **STEP 2:** Apply the techniques discussed in Section 9.1 to solve the two resulting equations in two variables from Step 1. If x is eliminated, then solve these equations to find y and z. If there are no solutions for y and z, then the given system also has no solutions. If there are infinitely many solutions for y and z, then write y in terms of z and proceed to Step 3. **STEP 3:** Substitute the values for y and z in one of the given equations to find x. The solution is (x, y, z). If possible, check your answer.

9.3 Exercises

1. Can a system of linear equations have exactly three solutions?

2. Does the ordered triple $(1, 2, 3)$ satisfy the equation $3x + 2y + z = 10$?

3. To solve a system of linear equations in two variables, how many equations do you usually need?

4. To solve a system of linear equations in three variables, how many equations do you usually need?

5. If a system of linear equations has infinitely many solutions, are the equations dependent or independent?

6. If a system of linear equations is inconsistent, how many solutions does it have?

Exercises 7–10: Determine whether each ordered triple is a solution to the system of linear equations.

7. $(0, 2, -2), (-1, 3, -2)$

$$x + y - z = 4$$
$$-x + y + z = 2$$
$$x + y + z = 0$$

8. $(5, 2, 2), (2, -1, 1)$

$$2x - 3y + 3z = 10$$
$$x - 2y - 3z = 1$$
$$4x - y + z = 10$$

9. $\left(-\frac{5}{11}, \frac{20}{11}, -2\right), (1, 2, -1)$

$$x + 3y - 2z = 9$$
$$-3x + 2y + 4z = -3$$
$$-2x + 5y + 2z = 6$$

10. $(1, 2, 3), (11, 16, -3)$

$$4x - 2y + 2z = 6$$
$$2x - 4y - 6z = -24$$
$$-3x + 3y + 2z = 9$$

Exercises 11–32: If possible, solve the system.

11.
$$x + y + z = 6$$
$$-x + 2y + z = 6$$
$$y + z = 5$$

12.
$$x - y + z = -2$$
$$x - 2y + z = 0$$
$$y - z = 1$$

13.
$$x + 2y + 3z = 4$$
$$2x + y + 3z = 5$$
$$x - y + z = 2$$

14.
$$x - y + z = 2$$
$$3x - 2y + z = -1$$
$$x + y = -3$$

15.
$$3x + y + z = 0$$
$$4x + 2y + z = 1$$
$$2x - 2y - z = 2$$

16.
$$-x - 5y + 2z = 2$$
$$x + y + 2z = 2$$
$$3x + y - 4z = -10$$

17.
$$x + 3y + z = 6$$
$$3x + y - z = 6$$
$$x - y - z = 0$$

18.
$$2x - y + 2z = 6$$
$$-x + y + z = 0$$
$$-x - 3z = -6$$

19.
$$x - 4y + 2z = -2$$
$$x + 2y - 2z = -3$$
$$x - y = 4$$

20.
$$2x + y + 3z = 4$$
$$-3x - y - 4z = 5$$
$$x + y + 2z = 0$$

21.
$$4a - b + 2c = 0$$
$$2a + b - c = -11$$
$$2a - 2b + c = 3$$

22.
$$a - 4b + 3c = 2$$
$$-a - 2b + 5c = 9$$
$$a + 2b + c = 6$$

23.
$$a + b + c = 0$$
$$a - b - c = 3$$
$$a + 3b + 3c = 5$$

24.
$$a - 2b + c = -1$$
$$a + 5b = -3$$
$$2a + 3b + c = -2$$

25.
$$3x + 2y + z = -1$$
$$3x + 4y - z = 1$$
$$x + 2y + z = 0$$

26.
$$x - 2y + z = 1$$
$$x + y + 2z = 2$$
$$2x + 3y + z = 6$$

27.
$$-x + 3y + z = 3$$
$$2x + 7y + 4z = 13$$
$$4x + y + 2z = 7$$

28.
$$x + 2y + z = 0$$
$$3x + 2y - z = 4$$
$$-x + 2y + 3z = -4$$

29.
$$-x + 2z = -9$$
$$y + 4z = -13$$
$$3x + y = 13$$

30.
$$x + y + z = -1$$
$$2x + z = -6$$
$$2y + 3z = 0$$

31.
$$\tfrac{1}{2}x - y + \tfrac{1}{2}z = -4$$
$$x + 2y - 3z = 20$$
$$-\tfrac{1}{2}x + 3y + 2z = 0$$

32.
$$\tfrac{3}{4}x + y + \tfrac{1}{2}z = -3$$
$$x + y - z = -8$$
$$\tfrac{1}{4}x - 2y + z = -4$$

Applications

33. **Tickets Sold** Five hundred tickets were sold for a play, generating \$3560. The prices of the tickets were \$5 for children, \$7 for students, and \$10 for adults. There were 180 more student tickets sold than adult tickets. Find the number of each type of ticket sold.

34. **Tickets Sold** One thousand tickets were sold for a baseball game. There were one hundred more adult tickets sold than student tickets, and there were four times as many tickets sold to students as to children. How many of each type of ticket were sold?

35. **Buying Lunch** Three students buy lunch in the cafeteria. One student buys 2 hamburgers, 2 orders of fries, and 1 soda for \$9. Another student buys 1 hamburger, 1 order of fries, and 1 soda for \$5. The third student buys 1 hamburger and 1 order of fries for \$5. If possible, find the cost of each item. Interpret the results.

36. Cost of DVDs The table shows the total cost of purchasing various combinations of differently priced DVDs. The types of DVDs are labeled A, B, and C.

A	B	C	Total Cost
2	1	1	$48
3	2	1	$71
1	1	2	$53

(a) Let a be the cost of a DVD of type A, b be the cost of a DVD of type B, and c be the cost of a DVD of type C. Write a system of three linear equations whose solution gives the cost of each type of DVD.

(b) Solve the system of equations and check your answer.

37. Geometry The largest angle in a triangle is 25° more than the smallest angle. The sum of the measures of the two smaller angles is 30° more than the measure of the largest angle.

(a) Let x, y, and z be the measures of the three angles from largest to smallest. Write a system of three linear equations whose solution gives the measure of each angle.

(b) Solve the system of equations and check your answer.

38. Geometry The perimeter of a triangle is 105 inches. The longest side is 22 inches longer than the shortest side. The sum of the lengths of the two shorter sides is 15 inches more than the length of the longest side. Find the lengths of the sides of the triangle.

39. Investment Mixture A sum of $20,000 is invested in three mutual funds. In one year the first fund grew by 5%, the second by 7%, and the third by 10%. Total earnings for the year were $1650. The amount invested in the third fund was four times the amount invested in the first fund. Find the amount invested in each fund.

40. Home Prices Prices of homes can depend on several factors such as size and age. The table shows the selling prices for three homes. In this table, price P is given in thousands of dollars, age A in years, and home size S in thousands of square feet. These data may be modeled by $P = a + bA + cS$.

Price (P)	Age (A)	Size (S)
190	20	2
320	5	3
50	40	1

(a) Write a system of linear equations whose solution gives a, b, and c.

(b) Solve this system of linear equations.

(c) Predict the price of a home that is 10 years old and has 2500 square feet.

41. Mixture Problem One type of lawn fertilizer consists of a mixture of nitrogen, N; phosphorus, P; and potassium, K. An 80-pound sample contains 8 more pounds of nitrogen and phosphorus than of potassium. There is nine times as much potassium as phosphorus.

(a) Write a system of three equations whose solution gives the amount of nitrogen, phosphorus, and potassium in this sample.

(b) Solve the system of equations.

42. Business Production A business has three machines that manufacture containers. Together they can make 100 containers per day, whereas the two fastest machines can make 80 containers per day. The fastest machine makes 34 more containers per day than the slowest machine.

(a) Let x, y, and z be the numbers of containers that the machines make from fastest to slowest. Write a system of three equations whose solution gives the number of containers each machine can make.

(b) Solve the system of equations.

Writing about Mathematics

43. When using elimination and substitution, explain how to recognize a system of linear equations that has no solutions.

44. When using elimination and substitution, explain how to recognize a system of linear equations that has infinitely many solutions.

9.4 Solutions to Linear Systems Using Matrices

- **Represent systems of linear equations with matrices**
- **Learn row-echelon form**
- **Perform Gaussian elimination**
- **Learn reduced row-echelon form**
- **Perform Gauss-Jordan elimination**
- **Solve systems of linear equations with technology (optional)**

Introduction

After its release in 2010, the iPad recorded remarkable quarterly sales growth. The three points plotted in Figure 9.39 give the *cumulative* sales y in millions of units, sold x quarters after the iPad's release. For example, the point (5, 29) indicates that Apple sold 29 million iPads by the end of the 5th quarter after its release. Because three distinct points (that are not collinear) determine the graph of a quadratic function, we can model these data by finding a unique parabola that passes though the given points, as illustrated in Figure 9.39. (*Source:* Apple Corporation.)

iPad Sales

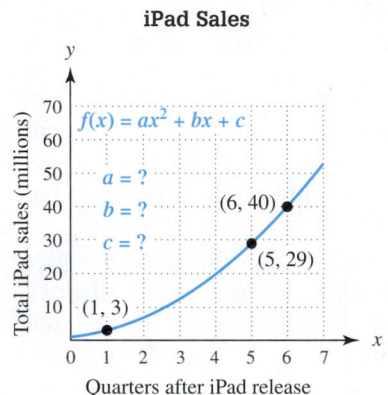

Figure 9.39

One way to accomplish this task is to set up a linear system of equations in three variables and solve it by using a matrix. This section discusses matrices and how they can be used to solve systems of linear equations. (See Example 11 and Exercise 87.)

Representing Systems of Linear Equations with Matrices

Arrays of numbers occur frequently in many different situations. Spreadsheets often make use of arrays, where data are displayed in a tabular format. A **matrix** is a rectangular array of elements. The following are examples of matrices whose elements are real numbers.

Examples of Matrices

$$\begin{bmatrix} 4 & -7 \\ -2 & 9 \end{bmatrix} \quad \begin{bmatrix} -1 & -5 & 3 \\ 1.2 & 0 & -1.3 \\ 4.1 & 5 & 7 \end{bmatrix} \quad \begin{bmatrix} -3 & -6 & 9 & 5 \\ \sqrt{2} & -8 & -8 & 0 \\ 3 & 0 & 19 & -7 \\ -11 & -3 & 7 & 8 \end{bmatrix} \quad \begin{bmatrix} 5 & -2 \\ -2 & \pi \\ 1 & -1 \end{bmatrix} \quad \begin{bmatrix} 1 & -0.5 & 9 \\ 5 & 0.4 & -3 \end{bmatrix}$$

$$2 \times 2 \qquad\qquad 3 \times 3 \qquad\qquad\qquad 4 \times 4 \qquad\qquad\qquad 3 \times 2 \qquad\qquad 2 \times 3$$

The dimension of a matrix is given much like the dimensions of a rectangular room. We might say a room is m feet long and n feet wide. The **dimension** of a matrix is $m \times n$ (m by n) if it has m rows and n columns. For example, the last matrix has a dimension of 2×3 because it has 2 rows and 3 columns. If the numbers of rows and columns are equal, the matrix is a **square matrix.** The first three matrices are square matrices.

CLASS DISCUSSION

Give a general form of a system of linear equations with four equations and four variables. Write its augmented matrix.

Matrices are frequently used to represent systems of linear equations.

See the Concept: Representing a Linear System with a Matrix

Ⓐ

System of Three Equations

$a_1 x + b_1 y + c_1 z = d_1$
$a_2 x + b_2 y + c_2 z = d_2$
$a_3 x + b_3 y + c_3 z = d_3$

Ⓑ

Coefficient Matrix

$$\begin{bmatrix} a_1 & b_1 & c_1 \\ a_2 & b_2 & c_2 \\ a_3 & b_3 & c_3 \end{bmatrix}$$

Ⓒ

Augmented Matrix

$$\begin{bmatrix} a_1 & b_1 & c_1 & | & d_1 \\ a_2 & b_2 & c_2 & | & d_2 \\ a_3 & b_3 & c_3 & | & d_3 \end{bmatrix}$$

Ⓐ The a_k, b_k, c_k, and d_k are constants and x, y, and z are variables.

Ⓑ The coefficients of the variables are represented in a square matrix called the **coefficient matrix** of the linear system.

Ⓒ The matrix is enlarged to include the constants d_k. The vertical line in this matrix corresponds to where the equals sign occurs in each equation. This matrix is commonly called an **augmented matrix.**

EXAMPLE 1 **Representing a linear system with an augmented matrix**

Express each linear system with an augmented matrix. State the dimension of the matrix.

(a) $3x - 4y = 6$
$-5x + y = -5$

(b) $2x - 5y + 6z = -3$
$3x + 7y - 3z = 8$
$x + 7y = 5$

SOLUTION

(a) This system has two equations with two variables. It can be represented by an augmented matrix having dimension 2×3.

$$\begin{bmatrix} 3 & -4 & | & 6 \\ -5 & 1 & | & -5 \end{bmatrix} \qquad \begin{aligned} 3x - 4y &= 6 \\ -5x + y &= -5 \end{aligned}$$

(b) This system has three equations with three variables. Note that variable z does not appear in the third equation. A value of 0 is inserted for its coefficient.

$$\begin{bmatrix} 2 & -5 & 6 & | & -3 \\ 3 & 7 & -3 & | & 8 \\ 1 & 7 & 0 & | & 5 \end{bmatrix} \qquad \begin{aligned} 2x - 5y + 6z &= -3 \\ 3x + 7y - 3z &= 8 \\ x + 7y &= 5 \end{aligned}$$

This matrix has dimension 3×4.

Now Try Exercises 7 and 9

EXAMPLE 2 **Converting an augmented matrix into a linear system**

Write the linear system represented by the augmented matrix. Let the variables be x, y, and z.

(a) $$\begin{bmatrix} 1 & 0 & 2 & | & -3 \\ 2 & 2 & 10 & | & 3 \\ -1 & 2 & 3 & | & 5 \end{bmatrix}$$
(b) $$\begin{bmatrix} 1 & 2 & 3 & | & -4 \\ 0 & 1 & -6 & | & 7 \\ 0 & 0 & 1 & | & 8 \end{bmatrix}$$

SOLUTION

Getting Started The first column corresponds to x, the second to y, and the third to z. When a 0 appears, the variable for that column does not appear in the equation. The vertical line gives the location of the equals sign. The last column represents the constant terms. ▶

$$\begin{bmatrix} 1 & 0 & 2 & | & -3 \\ 2 & 2 & 10 & | & 3 \\ -1 & 2 & 3 & | & 5 \end{bmatrix}$$

(a) The augmented matrix (repeated in the margin) represents the following linear system.

$$\begin{array}{rcl} x \quad\quad + 2z &=& -3 \quad\quad \text{First row in the matrix} \\ 2x + 2y + 10z &=& 3 \quad\quad \text{Second row in the matrix} \\ -x + 2y + 3z &=& 5 \quad\quad \text{Third row in the matrix} \end{array}$$

$$\begin{bmatrix} 1 & 2 & 3 & | & -4 \\ 0 & 1 & -6 & | & 7 \\ 0 & 0 & 1 & | & 8 \end{bmatrix}$$

(b) The augmented matrix (repeated in the margin) represents the following linear system.

$$\begin{array}{rcl} x + 2y + 3z &=& -4 \quad\quad \text{First row in the matrix} \\ y - 6z &=& 7 \quad\quad \text{Second row in the matrix} \\ z &=& 8 \quad\quad \text{Third row in the matrix} \end{array}$$

Now Try Exercises 11 and 13

Row-Echelon Form

To solve a linear system with an augmented matrix, it is convenient to get the matrix in **row-echelon form.** The following matrices are in row-echelon form.

$$\begin{bmatrix} 1 & 3 & 0 & -1 \\ 0 & 1 & -6 & 1 \\ 0 & 0 & 1 & -2 \end{bmatrix} \quad \begin{bmatrix} 1 & 2 & 0 \\ 0 & 1 & 4 \end{bmatrix} \quad \begin{bmatrix} 1 & 3 & -1 & 5 \\ 0 & 1 & -1 & 3 \\ 0 & 0 & 1 & 0 \end{bmatrix} \quad \begin{bmatrix} 1 & 3 & -1 & 5 \\ 0 & 0 & 1 & 3 \\ 0 & 0 & 0 & 0 \end{bmatrix} \quad \begin{bmatrix} 1 & 3 & 5 \\ 0 & 0 & 1 \end{bmatrix}$$

The elements of the **main diagonal** are blue in each matrix. Scanning down the main diagonal of a matrix in row-echelon form, we see that this diagonal first contains only 1's, and then possibly 0's. The first nonzero element in any row is 1. Rows containing only 0's occur at the bottom of the matrix. All elements below the main diagonal are 0.

The next example demonstrates a technique called **backward substitution.** It can be used to solve linear systems represented by an augmented matrix in row-echelon form.

EXAMPLE 3 **Solving a linear system with backward substitution**

Solve the system of linear equations represented by the augmented matrix.

(a) $$\begin{bmatrix} 1 & 1 & 3 & | & 12 \\ 0 & 1 & -2 & | & -4 \\ 0 & 0 & 1 & | & 3 \end{bmatrix}$$
(b) $$\begin{bmatrix} 1 & -1 & 5 & | & 5 \\ 0 & 1 & 3 & | & 3 \\ 0 & 0 & 0 & | & 0 \end{bmatrix}$$

SOLUTION
(a) The matrix represents the following linear system.

$$\begin{array}{rcl} x + y + 3z &=& 12 \quad\quad \text{First row in the matrix} \\ y - 2z &=& -4 \quad\quad \text{Second row in the matrix} \\ z &=& 3 \quad\quad \text{Third row in the matrix} \end{array}$$

Since $z = 3$, substitute this value in the second equation to find y.

Second equation with $z = 3$ ——— $y - 2(3) = -4$, or $y = 2$

Then $y = 2$ and $z = 3$ can be substituted in the first equation to determine x.

First equation with $y = 2$ and $z = 3$ ——— $x + 2 + 3(3) = 12$, or $x = 1$

The solution is given by $x = 1, y = 2$, and $z = 3$ and can be expressed as the *ordered triple* $(1, 2, 3)$.

(b) The matrix represents the following linear system.

$$\begin{array}{rcl} x - y + 5z &=& 5 \quad\quad \text{First row in the matrix} \\ y + 3z &=& 3 \quad\quad \text{Second row in the matrix} \\ 0 &=& 0 \quad\quad \text{Third row in the matrix} \end{array}$$

The last equation, $0 = 0$, is an identity. Its presence usually indicates infinitely many solutions. Use the second equation to write y in terms of z.

> Solve second equation for y. ——— $y = 3 - 3z$

Next, substitute $(3 - 3z)$ for y in the first equation and write x in terms of z.

$x - (3 - 3z) + 5z = 5$	Substitute $(3 - 3z)$ for y
$x - 3 + 3z + 5z = 5$	Distributive property
$x = 8 - 8z$	Solve for x.

All solutions can be written as the ordered triple $(8 - 8z, 3 - 3z, z)$, where z is any real number. There are infinitely many solutions. Sometimes we say that all solutions can be written in terms of the **parameter** z, where z is any real number. For example, if we let $z = 1$, then $y = 3 - 3(1) = 0$ and $x = 8 - 8(1) = 0$. Thus one solution to the system is $(0, 0, 1)$.

> **Now Try Exercises 21 and 23**

Gaussian Elimination

The methods of elimination and substitution from Section 9.1 can be combined to create a state-of-the-art numerical method capable of solving systems of linear equations that contain thousands of variables. Even though this method, called *Gaussian elimination with backward substitution*, dates back to Carl Friedrich Gauss (1777–1855), it continues to be one of the most efficient methods for solving systems of linear equations.

If an augmented matrix is not in row-echelon form, it can be transformed into row-echelon form using *Gaussian elimination*. This method uses the following three basic matrix row transformations.

MATRIX ROW TRANSFORMATIONS

For any augmented matrix representing a system of linear equations, the following row transformations result in an equivalent system of linear equations.
1. Any two rows may be interchanged.
2. The elements of any row may be multiplied by a nonzero constant.
3. Any row may be changed by adding to (or subtracting from) its elements a multiple of the corresponding elements of another row.

When we transform a matrix into row-echelon form, we also are transforming a system of linear equations. The next two examples illustrate how Gaussian elimination with backward substitution is performed.

EXAMPLE 4   Transforming a matrix into row-echelon form

Use Gaussian elimination with backward substitution to solve the linear system of equations.

$$x + y + z = 1$$
$$-x + y + z = 5$$
$$y + 2z = 5$$

SOLUTION

Getting Started The goal is to apply matrix row transformations that transform the given matrix into row-echelon form. Then we can perform backward substitution to determine the solution. ▶

The linear system is written to the right to illustrate how each row transformation affects the corresponding system of linear equations. Note that it is *not* necessary to write the system of equations to the right of the augmented matrix.

Augmented Matrix	*Linear System*

$$\begin{bmatrix} 1 & 1 & 1 & | & 1 \\ -1 & 1 & 1 & | & 5 \\ 0 & 1 & 2 & | & 5 \end{bmatrix} \qquad \begin{matrix} x + y + z = 1 \\ -x + y + z = 5 \\ y + 2z = 5 \end{matrix}$$

We can add the first equation to the second equation to obtain a 0 where the coefficient of x in the second row is highlighted. This row operation is denoted $R_2 + R_1$, and the result becomes the new row 2.

The row that is changing is written first. $\quad R_2 + R_1 \rightarrow \begin{bmatrix} 1 & 1 & 1 & | & 1 \\ 0 & 2 & 2 & | & 6 \\ 0 & 1 & 2 & | & 5 \end{bmatrix} \qquad \begin{matrix} x + y + z = 1 \\ 2y + 2z = 6 \\ y + 2z = 5 \end{matrix}$

To have the matrix in row-echelon form, we need the highlighted 2 in the second row to be a 1. Multiply each element in row 2 by $\frac{1}{2}$ and denote the operation $\frac{1}{2}R_2$.

Row 2 is changing. $\quad \frac{1}{2}R_2 \rightarrow \begin{bmatrix} 1 & 1 & 1 & | & 1 \\ 0 & 1 & 1 & | & 3 \\ 0 & 1 & 2 & | & 5 \end{bmatrix} \qquad \begin{matrix} x + y + z = 1 \\ y + z = 3 \\ y + 2z = 5 \end{matrix}$

Next, we need a 0 where the 1 is highlighted in row 3. Subtract row 2 from row 3 and denote the operation $R_3 - R_2$.

Row 3 is changing. $\quad R_3 - R_2 \rightarrow \begin{bmatrix} 1 & 1 & 1 & | & 1 \\ 0 & 1 & 1 & | & 3 \\ 0 & 0 & 1 & | & 2 \end{bmatrix} \qquad \begin{matrix} x + y + z = 1 \\ y + z = 3 \\ z = 2 \end{matrix}$

Because we have a 1 in the highlighted box, the matrix is now in row-echelon form, and we see that $z = 2$. Backward substitution may be applied now to find the solution. Substituting $z = 2$ in the second equation gives

$$y + 2 = 3, \quad \text{or} \quad y = 1.$$

Finally, let $y = 1$ and $z = 2$ in the first equation to determine x.

$$x + 1 + 2 = 1, \quad \text{or} \quad x = -2$$

The solution to the system is given by $x = -2, y = 1$, and $z = 2$, or $(-2, 1, 2)$.

Now Try Exercise 33

EXAMPLE 5 **Transforming a matrix into row-echelon form**

Use Gaussian elimination with backward substitution to solve the linear system of equations.

$$2x + 4y + 4z = 4$$
$$x + 3y + z = 4$$
$$-x + 3y + 2z = -1$$

SOLUTION The initial linear system and augmented matrix are written first.

Augmented Matrix	*Linear System*

$$\begin{bmatrix} 2 & 4 & 4 & | & 4 \\ 1 & 3 & 1 & | & 4 \\ -1 & 3 & 2 & | & -1 \end{bmatrix} \qquad \begin{matrix} 2x + 4y + 4z = 4 \\ x + 3y + z = 4 \\ -x + 3y + 2z = -1 \end{matrix}$$

First we obtain a 1 where the x-coefficient of 2 in the first row is highlighted. This can be accomplished by either multiplying the first equation by $\frac{1}{2}$ or interchanging rows 1 and 2. We multiply row 1 by $\frac{1}{2}$. This operation is denoted $\frac{1}{2}R_1$.

$$\frac{1}{2}R_1 \rightarrow \begin{bmatrix} 1 & 2 & 2 & | & 2 \\ 1 & 3 & 1 & | & 4 \\ -1 & 3 & 2 & | & -1 \end{bmatrix} \qquad \begin{matrix} x + 2y + 2z = 2 \\ x + 3y + z = 4 \\ -x + 3y + 2z = -1 \end{matrix}$$

The next step is to eliminate the x-variable in rows 2 and 3 by obtaining zeros in the highlighted positions. To do this, subtract row 1 from row 2, and add row 1 to row 3.

$$\begin{matrix} \\ R_2 - R_1 \rightarrow \\ R_3 + R_1 \rightarrow \end{matrix} \begin{bmatrix} 1 & 2 & 2 & | & 2 \\ 0 & 1 & -1 & | & 2 \\ 0 & 5 & 4 & | & 1 \end{bmatrix} \qquad \begin{matrix} x + 2y + 2z = 2 \\ y - z = 2 \\ 5y + 4z = 1 \end{matrix}$$

Since we have a 1 for the y-coefficient in the second row, the next step is to eliminate the y-variable in row 3 and obtain a zero where the y-coefficient of 5 is highlighted. Multiply row 2 by 5, and subtract the result from row 3.

$$\begin{matrix} \\ \\ R_3 - 5R_2 \rightarrow \end{matrix} \begin{bmatrix} 1 & 2 & 2 & | & 2 \\ 0 & 1 & -1 & | & 2 \\ 0 & 0 & 9 & | & -9 \end{bmatrix} \qquad \begin{matrix} x + 2y + 2z = 2 \\ y - z = 2 \\ 9z = -9 \end{matrix}$$

Finally, make the z-coefficient of 9 in the third row equal 1 by multiplying row 3 by $\frac{1}{9}$.

$$\begin{matrix} \\ \\ \frac{1}{9}R_3 \rightarrow \end{matrix} \begin{bmatrix} 1 & 2 & 2 & | & 2 \\ 0 & 1 & -1 & | & 2 \\ 0 & 0 & 1 & | & -1 \end{bmatrix} \qquad \begin{matrix} x + 2y + 2z = 2 \\ y - z = 2 \\ z = -1 \end{matrix}$$

The final matrix is in row-echelon form. Backward substitution may be applied to find the solution. Substituting $z = -1$ in the second equation gives

$$y - (-1) = 2, \qquad \text{or} \qquad y = 1.$$

Next, substitute $y = 1$ and $z = -1$ in the first equation to determine x.

$$x + 2(1) + 2(-1) = 2, \qquad \text{or} \qquad x = 2$$

The solution to the system is $(2, 1, -1)$.

> **Now Try Exercise 37**

EXAMPLE 6   **Transforming a system that has no solutions**

Solve the system of linear equations, if possible.

$$x - 2y + 3z = 2$$
$$2x + 3y + 2z = 7$$
$$4x - y + 8z = 8$$

SOLUTION   Because it is not necessary to write the linear system next to the matrix, we write the matrices in a horizontal format in this example.

$$\begin{bmatrix} 1 & -2 & 3 & | & 2 \\ 2 & 3 & 2 & | & 7 \\ 4 & -1 & 8 & | & 8 \end{bmatrix} \begin{matrix} \\ R_2 - 2R_1 \rightarrow \\ R_3 - 4R_1 \rightarrow \end{matrix} \begin{bmatrix} 1 & -2 & 3 & | & 2 \\ 0 & 7 & -4 & | & 3 \\ 0 & 7 & -4 & | & 0 \end{bmatrix} \begin{matrix} \\ \\ R_3 - R_2 \rightarrow \end{matrix} \begin{bmatrix} 1 & -2 & 3 & | & 2 \\ 0 & 7 & -4 & | & 3 \\ 0 & 0 & 0 & | & -3 \end{bmatrix}$$

The last row of the last matrix represents $0x + 0y + 0z = -3$, which has no solutions because $0 \neq -3$. There are no solutions.

> **Now Try Exercise 69**

A Geometric Interpretation   The graph of a single linear equation in three variables is a plane in three-dimensional space. For a system of three equations in three variables, the possible intersections of the planes are illustrated in Figure 9.40. The solution set of such a system may be either a single ordered triple (x, y, z), an infinite set of ordered triples (dependent equations), or the empty set (an inconsistent system).

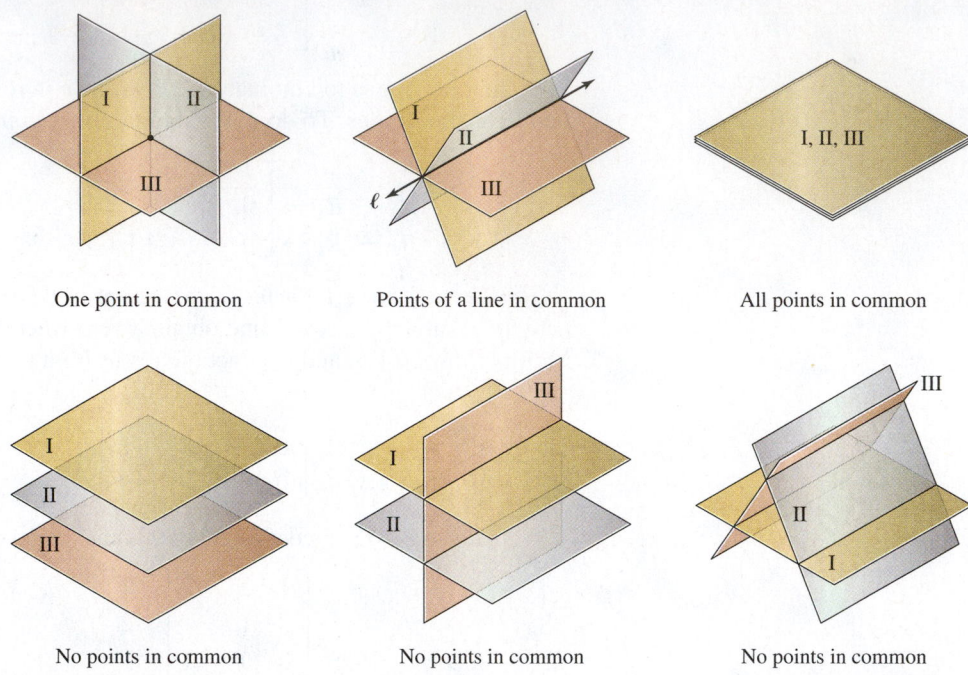

| One point in common | Points of a line in common | All points in common |

| No points in common | No points in common | No points in common |

Figure 9.40

Reduced Row-Echelon Form   Sometimes it is convenient to express a matrix in *reduced* row-echelon form. A matrix in row-echelon form is in **reduced row-echelon form** if every element above and below a 1 on the main diagonal is 0. The following matrices are examples of reduced row-echelon form.

$$\begin{bmatrix} 1 & 0 \\ 0 & 1 \end{bmatrix} \quad \begin{bmatrix} 1 & 0 \\ 0 & 0 \end{bmatrix} \quad \begin{bmatrix} 1 & 0 & 0 \\ 0 & 1 & 0 \\ 0 & 0 & 1 \end{bmatrix} \quad \begin{bmatrix} 1 & 0 & 3 \\ 0 & 1 & -2 \end{bmatrix} \quad \begin{bmatrix} 1 & 0 & 0 & 3 \\ 0 & 1 & 0 & 1 \\ 0 & 0 & 1 & -1 \end{bmatrix} \quad \begin{bmatrix} 1 & 0 & 4 & 8 \\ 0 & 1 & -1 & 2 \\ 0 & 0 & 0 & 0 \end{bmatrix}$$

If an augmented matrix is in reduced row-echelon form, solving the system of linear equations is often straightforward.

EXAMPLE 7   **Determining a solution from a matrix in reduced row-echelon form**

Each matrix represents a system of linear equations. Find the solution.

(a) $\begin{bmatrix} 1 & 0 & | & 6 \\ 0 & 1 & | & -5 \end{bmatrix}$

(b) $\begin{bmatrix} 1 & 0 & 0 & | & 3 \\ 0 & 1 & 0 & | & -1 \\ 0 & 0 & 1 & | & 2 \end{bmatrix}$

(c) $\begin{bmatrix} 1 & 0 & 0 & | & 4 \\ 0 & 1 & 0 & | & 3 \\ 0 & 0 & 0 & | & 2 \end{bmatrix}$

(d) $\begin{bmatrix} 1 & 0 & -2 & | & -3 \\ 0 & 1 & 2 & | & 1 \\ 0 & 0 & 0 & | & 0 \end{bmatrix}$

SOLUTION

(a) To see how the matrix in reduced row-echelon form provides immediate access to the solution to the related system of linear equations, we write the corresponding system of equations next to the given matrix.

Given Matrix   *Linear System*

$$\begin{bmatrix} 1 & 0 & | & 6 \\ 0 & 1 & | & -5 \end{bmatrix} \qquad \begin{aligned} 1x + 0y &= 6 \quad \text{or} \quad x = 6 \\ 0x + 1y &= -5 \quad \text{or} \quad y = -5 \end{aligned}$$

The solution is $(6, -5)$.

(b) The top row represents $1x + 0y + 0z = 3$, or $x = 3$. Using similar reasoning for the second and third rows yields $y = -1$ and $z = 2$. The solution is $(3, -1, 2)$.

(c) The last row represents $0x + 0y + 0z = 2$, which has no solutions because $0 \neq 2$. Therefore there are no solutions to the system of equations.

(d) The last row simplifies to $0 = 0$, which is an identity and is always true. The second row gives $y + 2z = 1$, or $y = -2z + 1$. The first row represents $x - 2z = -3$, or $x = 2z - 3$. Thus this system of linear equations has infinitely many solutions. Every solution can be written as an ordered triple in the form $(2z - 3, -2z + 1, z)$, where z can be any real number.

> **Now Try Exercises 55, 57, 59, and 61**

Gauss-Jordan Elimination   Matrix row transformations can be used to transform an augmented matrix into reduced row-echelon form. This approach requires more effort than transforming a matrix into row-echelon form, but often eliminates the need for backward substitution. The technique is sometimes called **Gauss-Jordan elimination.**

EXAMPLE 8   **Transforming a matrix into reduced row-echelon form**

Use Gauss-Jordan elimination to solve the linear system.

$$2x + y + 2z = 10$$
$$x \qquad + 2z = 5$$
$$x - 2y + 2z = 1$$

SOLUTION   The linear system has been written to the right for illustrative purposes.

Augmented Matrix $\qquad\qquad$ *Linear System*

$$\begin{bmatrix} 2 & 1 & 2 & | & 10 \\ 1 & 0 & 2 & | & 5 \\ 1 & -2 & 2 & | & 1 \end{bmatrix} \qquad \begin{aligned} 2x + y + 2z &= 10 \\ x \quad\;\; + 2z &= 5 \\ x - 2y + 2z &= 1 \end{aligned}$$

Obtain a 1 in the highlighted position in row 1 by interchanging rows 1 and 2.

$$\begin{matrix} R_2 \to \\ R_1 \to \\ {} \end{matrix} \begin{bmatrix} 1 & 0 & 2 & | & 5 \\ 2 & 1 & 2 & | & 10 \\ 1 & -2 & 2 & | & 1 \end{bmatrix} \qquad \begin{aligned} x \quad\;\; + 2z &= 5 \\ 2x + y + 2z &= 10 \\ x - 2y + 2z &= 1 \end{aligned}$$

Next subtract 2 times row 1 from row 2. Then subtract row 1 from row 3. This eliminates the x-variable from the second and third equations.

$$\begin{matrix} {} \\ R_2 - 2R_1 \to \\ R_3 - R_1 \to \end{matrix} \begin{bmatrix} 1 & 0 & 2 & | & 5 \\ 0 & 1 & -2 & | & 0 \\ 0 & -2 & 0 & | & -4 \end{bmatrix} \qquad \begin{aligned} x \qquad\;\; + 2z &= 5 \\ y - 2z &= 0 \\ -2y \qquad &= -4 \end{aligned}$$

To eliminate the y-variable in row 3, add 2 times row 2 to row 3.

$$\begin{matrix} {} \\ {} \\ R_3 + 2R_2 \to \end{matrix} \begin{bmatrix} 1 & 0 & 2 & | & 5 \\ 0 & 1 & -2 & | & 0 \\ 0 & 0 & -4 & | & -4 \end{bmatrix} \qquad \begin{aligned} x \qquad\;\; + 2z &= 5 \\ y - 2z &= 0 \\ -4z &= -4 \end{aligned}$$

To obtain a 1 in the highlighted position in row 3, multiply row 3 by $-\frac{1}{4}$.

$$\begin{matrix} {} \\ {} \\ -\dfrac{1}{4}R_3 \to \end{matrix} \begin{bmatrix} 1 & 0 & 2 & | & 5 \\ 0 & 1 & -2 & | & 0 \\ 0 & 0 & 1 & | & 1 \end{bmatrix} \qquad \begin{aligned} x \qquad\;\; + 2z &= 5 \\ y - 2z &= 0 \\ z &= 1 \end{aligned}$$

Finally, the matrix can be transformed into reduced row-echelon form by subtracting 2 times row 3 from row 1, and adding 2 times row 3 to row 2.

$$\begin{matrix} R_1 - 2R_3 \rightarrow \\ R_2 + 2R_3 \rightarrow \\ \\ \end{matrix} \begin{bmatrix} 1 & 0 & 0 & | & 3 \\ 0 & 1 & 0 & | & 2 \\ 0 & 0 & 1 & | & 1 \end{bmatrix} \qquad \begin{matrix} x = 3 \\ y = 2 \\ z = 1 \end{matrix}$$

This final matrix is in reduced row-echelon form. The solution is (3, 2, 1).

Now Try Exercise 65

Solving Systems of Linear Equations with Technology (Optional)

If the arithmetic at each step of Gaussian elimination is done exactly, then it may be thought of as an exact symbolic procedure. However, when calculators and computers are used to solve systems of equations, their solutions often are approximate. The next three examples use a graphing calculator to solve systems of linear equations.

EXAMPLE 9 Solving a system of equations using technology

Use a graphing calculator to solve the system of linear equations in Example 8.

SOLUTION To solve this system, enter the augmented matrix

$$A = \begin{bmatrix} 2 & 1 & 2 & | & 10 \\ 1 & 0 & 2 & | & 5 \\ 1 & -2 & 2 & | & 1 \end{bmatrix},$$

as shown in Figures 9.41–9.43.

Calculator Help

To enter the elements of a matrix, see Appendix A (page AP-12).

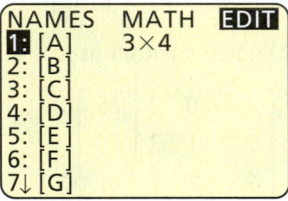

Figure 9.41

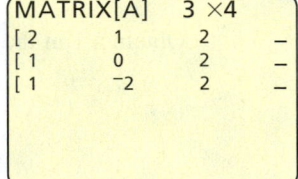

Figure 9.42

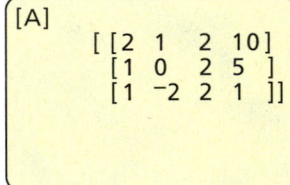

Figure 9.43

A graphing calculator can transform matrix A into reduced row-echelon form, as illustrated in Figures 9.44 and 9.45. Notice that the reduced row-echelon form obtained from the graphing calculator agrees with our results from Example 8. The solution is (3, 2, 1).

Calculator Help

To transform a matrix into reduced row-echelon form, see Appendix A (page AP-13).

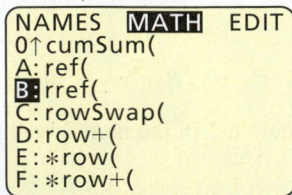

Figure 9.44

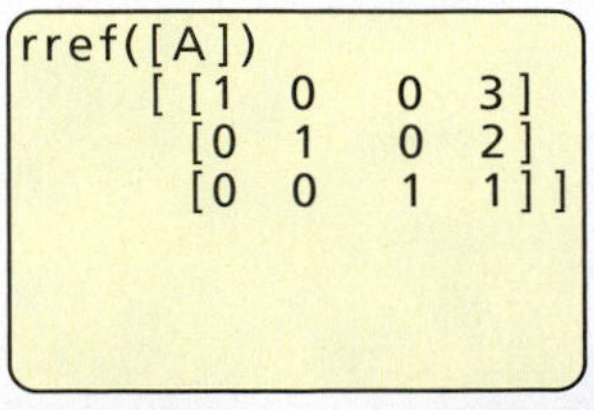

Figure 9.45

Now Try Exercise 71

EXAMPLE 10   Transforming a matrix into reduced row-echelon form

For three food shelters operated by a charitable organization, three different quantities are computed: monthly food costs F in dollars, number of people served per month N, and monthly charitable receipts R in dollars. The data are shown in Table 9.8.

Food Shelter Operations

Food Costs (F)	Number Served (N)	Charitable Receipts (R)
3000	2400	8000
4000	2600	10,000
8000	5900	14,000

Table 9.8

(a) Model these data by using $F = aN + bR + c$, where a, b, and c are constants.
(b) Predict the food costs for a shelter that serves 4000 people and receives charitable receipts of $12,000. Round your answer to the nearest hundred dollars.

SOLUTION
(a) Getting Started Table 9.8 provides several values for F, N, and R in the equation $F = aN + bR + c$. The goal is to write a system of linear equations whose solution gives the values of a, b, and c. ▶

Since $F = aN + bR + c$, the constants a, b, and c satisfy the following equations.

$$3000 = a(2400) + b(8000) \quad + c$$
$$4000 = a(2600) + b(10,000) + c$$
$$8000 = a(5900) + b(14,000) + c$$

This system can be rewritten as

$$2400a + \quad 8000b + c = 3000$$
$$2600a + 10,000b + c = 4000$$
$$5900a + 14,000b + c = 8000.$$

The associated augmented matrix is

$$A = \begin{bmatrix} 2400 & 8000 & 1 & 3000 \\ 2600 & 10,000 & 1 & 4000 \\ 5900 & 14,000 & 1 & 8000 \end{bmatrix}.$$

Figure 9.46 shows the matrix A. The fourth column of A may be viewed by using the arrow keys. In Figure 9.47, A has been transformed into reduced row-echelon form where $a \approx 0.6897$, $b \approx 0.4310$, and $c \approx -2103$. Thus let $F = 0.6897N + 0.431R - 2103$.

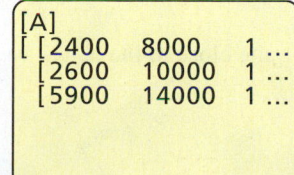

Figure 9.46

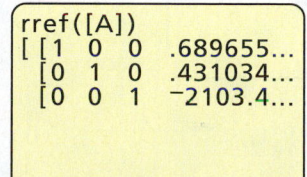

Figure 9.47

(b) To predict the food costs for a shelter that serves 4000 people and receives charitable receipts of $12,000, let $N = \textbf{4000}$ and $R = \textbf{12,000}$ and evaluate F.

$$F = 0.6897(\textbf{4000}) + 0.431(\textbf{12,000}) - 2103 = 5827.8.$$

This model predicts monthly food costs of about $5800.

Now Try Exercise 79

Determining a Quadratic Function The introduction to this section discussed how three points can be used to determine a quadratic function whose graph passes through these points. The next example illustrates this method.

EXAMPLE 11 Determining a quadratic function

More than half of private-sector employees cannot carry vacation days into a new year. The average number y of paid days off for full-time workers at medium to large companies after x years of employment is listed in Table 9.9.

Paid Days Off

x (years)	1	15	30
y (days)	9.4	18.8	21.9

Source: Bureau of Labor Statistics.

Table 9.9

(a) Determine the coefficients for $f(x) = ax^2 + bx + c$ so that f models these data.
(b) Graph f with the data in $[-4, 32, 5]$ by $[8, 23, 2]$.
(c) Estimate the number of paid days off after 3 years of employment. Compare it to the actual value of 11.2 days.

Calculator Help

To plot data and to graph an equation, see Appendix A (page AP-6).

```
rref([A])
...1  0  0  ⁻.016026...
...0  1  0  .9278489...
...0  0  1  8.488177...
```

Figure 9.48

$[-4, 32, 5]$ by $[8, 23, 2]$

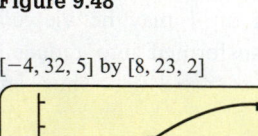

Figure 9.49

SOLUTION

(a) For f to model the data, the equations $f(\textbf{1}) = 9.4$, $f(\textbf{15}) = 18.8$, and $f(\textbf{30}) = 21.9$ must be satisfied. See Table 9.9.

$$f(\textbf{1})\ = a(\textbf{1})^2\ + b(\textbf{1})\ + c =\ \textbf{9.4}$$
$$f(\textbf{15}) = a(\textbf{15})^2 + b(\textbf{15}) + c = \textbf{18.8}$$
$$f(\textbf{30}) = a(\textbf{30})^2 + b(\textbf{30}) + c = \textbf{21.9}$$

The associated augmented matrix is

$$\begin{bmatrix} \textbf{1}^2 & \textbf{1} & 1 & \textbf{9.4} \\ \textbf{15}^2 & \textbf{15} & 1 & \textbf{18.8} \\ \textbf{30}^2 & \textbf{30} & 1 & \textbf{21.9} \end{bmatrix}.$$

Figure 9.48 shows a portion of the matrix represented in reduced row-echelon form. The solution is $a \approx -0.016026$, $b \approx 0.92785$, and $c \approx 8.4882$.

(b) Graph $y_1 = -0.016026x^2 + 0.92785x + 8.4882$ together with the points $(1, 9.4)$, $(15, 18.8)$, and $(30, 21.9)$. The graph of f passes through the points. See Figure 9.49.

(c) To estimate the number of paid days off after 3 years of employment, evaluate $f(3)$.

$$f(3) = -0.016026(3)^2 + 0.92785(3) + 8.4882 \approx 11.1$$

This is quite close to the actual value of 11.2 days.

Now Try Exercise 87

9.4 Putting It All Together

Through a sequence of matrix row operations known as Gaussian elimination, an augmented matrix can be transformed into row-echelon form or reduced row-echelon form. Backward substitution is frequently used to find the solution when a matrix is in row-echelon form.

AUGMENTED MATRIX

A linear system can be represented by an augmented matrix.

$$\left[\begin{array}{ccc|c} 2 & 0 & -3 & 2 \\ -1 & 2 & -2 & -5 \\ 1 & -2 & -1 & 7 \end{array}\right] \qquad \begin{array}{rcr} 2x \quad\;\; - 3z &=& 2 \\ -x + 2y - 2z &=& -5 \\ x - 2y - \;\,z &=& 7 \end{array}$$

ROW-ECHELON FORM

The following matrices are in row-echelon form. They represent three possible situations: no solutions, one solution, and infinitely many solutions.

$$\left[\begin{array}{ccc|c} 1 & -2 & 1 & 0 \\ 0 & 1 & 2 & 3 \\ 0 & 0 & 0 & 1 \end{array}\right] \qquad \left[\begin{array}{ccc|c} 1 & -2 & 1 & 0 \\ 0 & 1 & 2 & 3 \\ 0 & 0 & 1 & 1 \end{array}\right] \qquad \left[\begin{array}{ccc|c} 1 & -2 & 1 & 0 \\ 0 & 1 & 2 & 3 \\ 0 & 0 & 0 & 0 \end{array}\right]$$

No solutions   *One solution*   *Infinitely many solutions*

                     (1, 1, 1)   *(6 − 5z, 3 − 2z, z)*

BACKWARD SUBSTITUTION

Backward substitution can be used to solve a system of linear equations represented by an augmented matrix in row-echelon form.

Augmented Matrix

$$\left[\begin{array}{ccc|c} 1 & -2 & 1 & 3 \\ 0 & 1 & -2 & -3 \\ 0 & 0 & 1 & 2 \end{array}\right]$$

From the last row, $z = 2$.

Substitute $z = 2$ in the second row: $y - 2(2) = -3$, or $y = 1$.

Let $z = 2$ and $y = 1$ in the first row: $x - 2(1) + 2 = 3$, or $x = 3$.

The solution is $(3, 1, 2)$.

REDUCED ROW-ECHELON FORM

The Gauss-Jordan elimination method can be used to transform an augmented matrix into reduced row-echelon form, which often eliminates the need for backward substitution. The following matrices are in reduced row-echelon form. The solution is given below the matrix.

$$\left[\begin{array}{ccc|c} 1 & 0 & 0 & 4 \\ 0 & 1 & 0 & 5 \\ 0 & 0 & 0 & 2 \end{array}\right] \qquad \left[\begin{array}{ccc|c} 1 & 0 & 0 & 1 \\ 0 & 1 & 0 & 2 \\ 0 & 0 & 1 & 3 \end{array}\right] \qquad \left[\begin{array}{ccc|c} 1 & 0 & 3 & 2 \\ 0 & 1 & 2 & 3 \\ 0 & 0 & 0 & 0 \end{array}\right]$$

No solutions   *One solution*   *Infinitely many solutions*

                     (1, 2, 3)   *(2 − 3z, 3 − 2z, z)*

9.4 Exercises

Dimensions of Matrices and Augmented Matrices

Exercises 1–6: State the dimension of each matrix.

1. $\begin{bmatrix} 1 \\ 2 \\ 3 \end{bmatrix}$
2. $\begin{bmatrix} a & b & c \\ d & e & b \end{bmatrix}$

3. $\begin{bmatrix} 3 & 0 \\ 1 & -4 \end{bmatrix}$
4. $\begin{bmatrix} -1 & 1 \end{bmatrix}$

5. $\begin{bmatrix} 1 & -1 \\ 7 & 5 \\ -4 & 0 \end{bmatrix}$
6. $\begin{bmatrix} 1 & 3 & 8 & -3 \\ 1 & -1 & 1 & -2 \\ 4 & 5 & 0 & -1 \end{bmatrix}$

Exercises 7–10: Represent the linear system by an augmented matrix, and state the dimension of the matrix.

7. $5x - 2y = 3$
$-x + 3y = -1$

8. $3x + y = 4$
$-x + 4y = 5$

9. $-3x + 2y + z = -4$
$5x \quad - z = 9$
$x - 3y - 6z = -9$

10. $x + 2y - z = 2$
$-2x + y - 2z = -3$
$7x + y - z = 7$

Exercises 11–14: Write the system of linear equations that the augmented matrix represents.

11. $\begin{bmatrix} 3 & 2 & | & 4 \\ 0 & 1 & | & 5 \end{bmatrix}$
12. $\begin{bmatrix} -2 & 1 & | & 5 \\ 7 & 9 & | & 2 \end{bmatrix}$

13. $\begin{bmatrix} 3 & 1 & 4 & | & 0 \\ 0 & 5 & 8 & | & -1 \\ 0 & 0 & -7 & | & 1 \end{bmatrix}$
14. $\begin{bmatrix} 1 & -1 & 3 & | & 2 \\ -2 & 1 & 1 & | & -2 \\ -1 & 0 & -2 & | & 1 \end{bmatrix}$

Row-Echelon Form

Exercises 15 and 16: Is the matrix in row-echelon form?

15. (a) $\begin{bmatrix} 1 & 3 & | & 2 \\ 0 & 1 & | & -1 \end{bmatrix}$
(b) $\begin{bmatrix} 1 & 4 & -1 & | & 0 \\ 0 & -1 & 1 & | & 3 \\ 0 & 2 & 1 & | & 7 \end{bmatrix}$

(c) $\begin{bmatrix} 1 & 6 & -8 & | & 5 \\ 0 & 1 & 7 & | & 9 \\ 0 & 0 & 1 & | & 11 \end{bmatrix}$

16. (a) $\begin{bmatrix} 1 & 3 & | & 2 \\ 0 & -1 & | & -1 \end{bmatrix}$
(b) $\begin{bmatrix} 1 & 3 & -1 & | & 8 \\ 0 & 1 & 5 & | & 3 \\ 0 & 0 & 0 & | & 0 \end{bmatrix}$

(c) $\begin{bmatrix} 0 & 0 & 1 & | & 1 \\ 0 & 1 & 7 & | & 9 \\ 1 & 2 & -1 & | & 11 \end{bmatrix}$

Exercises 17–26: The augmented matrix is in row-echelon form and represents a linear system. Solve the system by using backward substitution, if possible. Write the solution as either an ordered pair or an ordered triple.

17. $\begin{bmatrix} 1 & 2 & | & 3 \\ 0 & 1 & | & -1 \end{bmatrix}$
18. $\begin{bmatrix} 1 & -5 & | & 6 \\ 0 & 0 & | & 1 \end{bmatrix}$

19. $\begin{bmatrix} 1 & -1 & | & 2 \\ 0 & 1 & | & 0 \end{bmatrix}$
20. $\begin{bmatrix} 1 & 4 & | & -2 \\ 0 & 1 & | & 3 \end{bmatrix}$

21. $\begin{bmatrix} 1 & 1 & -1 & | & 4 \\ 0 & 1 & -1 & | & 2 \\ 0 & 0 & 1 & | & 1 \end{bmatrix}$
22. $\begin{bmatrix} 1 & -2 & -1 & | & 0 \\ 0 & 1 & -3 & | & 1 \\ 0 & 0 & 1 & | & 2 \end{bmatrix}$

23. $\begin{bmatrix} 1 & 2 & -1 & | & 5 \\ 0 & 1 & -2 & | & 1 \\ 0 & 0 & 0 & | & 0 \end{bmatrix}$
24. $\begin{bmatrix} 1 & -1 & 2 & | & 8 \\ 0 & 1 & -4 & | & 2 \\ 0 & 0 & 0 & | & 0 \end{bmatrix}$

25. $\begin{bmatrix} 1 & 2 & 1 & | & -3 \\ 0 & 1 & -3 & | & \frac{1}{2} \\ 0 & 0 & 0 & | & 4 \end{bmatrix}$
26. $\begin{bmatrix} 1 & 0 & -4 & | & \frac{3}{4} \\ 0 & 1 & 2 & | & 1 \\ 0 & 0 & 0 & | & -3 \end{bmatrix}$

Solving Systems with Gaussian Elimination

Exercises 27–30: Perform each row operation on the given matrix by completing the matrix at the right.

27. $\begin{bmatrix} 2 & -4 & 6 & | & 10 \\ -3 & 5 & 3 & | & 2 \\ 4 & 8 & 4 & | & -8 \end{bmatrix} \begin{matrix} \frac{1}{2}R_1 \to \\ \\ \frac{1}{4}R_3 \to \end{matrix} \begin{bmatrix} 1 & & & | & \\ -3 & 5 & 3 & | & 2 \\ & & 1 & | & \end{bmatrix}$

28. $\begin{bmatrix} 1 & -2 & 1 & | & 3 \\ 1 & 4 & 0 & | & -1 \\ 2 & 0 & 1 & | & 5 \end{bmatrix} \begin{matrix} \\ R_2 - R_1 \to \\ R_3 - 2R_1 \to \end{matrix} \begin{bmatrix} 1 & -2 & 1 & | & 3 \\ & 6 & & | & \\ & & & | & -1 \end{bmatrix}$

29. $\begin{bmatrix} 1 & -1 & 1 & | & 2 \\ -1 & 2 & -2 & | & 0 \\ 1 & 7 & 0 & | & 5 \end{bmatrix} \begin{matrix} \\ R_2 + R_1 \to \\ R_3 - R_1 \to \end{matrix} \begin{bmatrix} 1 & -1 & 1 & | & 2 \\ & & & | & \\ & & & | & \end{bmatrix}$

30. $\begin{bmatrix} 1 & -2 & 3 & | & 6 \\ 2 & 1 & 4 & | & 5 \\ -3 & 5 & 3 & | & 2 \end{bmatrix} \begin{matrix} \\ R_2 - 2R_1 \to \\ R_3 + 3R_1 \to \end{matrix} \begin{bmatrix} 1 & -2 & 3 & | & 6 \\ & & & | & \\ & & & | & \end{bmatrix}$

Exercises 31–42: Use Gaussian elimination with backward substitution to solve the system of linear equations. Write the solution as an ordered pair or an ordered triple whenever possible.

31. $x + 2y = 3$
$-x - y = 7$

32. $2x + 4y = 10$
$x - 2y = -3$

33. $\begin{aligned} x + 2y + z &= 3 \\ x + y - z &= 3 \\ -x - 2y + z &= -5 \end{aligned}$ **34.** $\begin{aligned} x + y + z &= 6 \\ 2x + 3y - z &= 3 \\ x + y + 2z &= 10 \end{aligned}$

35. $\begin{aligned} x + 2y - z &= -1 \\ 2x - y + z &= 0 \\ -x - y + 2z &= 7 \end{aligned}$ **36.** $\begin{aligned} x + 3y - 2z &= -4 \\ 2x + 6y + z &= -3 \\ x + y - 4z &= -2 \end{aligned}$

37. $\begin{aligned} 3x + y + 3z &= 14 \\ x + y + z &= 6 \\ -2x - 2y + 3z &= -7 \end{aligned}$ **38.** $\begin{aligned} x + 3y - 2z &= 3 \\ -x - 2y + z &= -2 \\ 2x - 7y + z &= 1 \end{aligned}$

39. $\begin{aligned} 2x + 5y + z &= 8 \\ x + 2y - z &= 2 \\ 3x + 7y &= 5 \end{aligned}$ **40.** $\begin{aligned} x + y + z &= 3 \\ x + y + 2z &= 4 \\ 2x + 2y + 3z &= 7 \end{aligned}$

41. $\begin{aligned} -x + 2y + 4z &= 10 \\ 3x - 2y - 2z &= -12 \\ x + 2y + 6z &= 8 \end{aligned}$ **42.** $\begin{aligned} 4x - 2y + 4z &= 8 \\ 3x - 7y + 6z &= 4 \\ -x - 5y + 2z &= 7 \end{aligned}$

Exercises 43–54: Solve the system, if possible.

43. $\begin{aligned} x - y + z &= 1 \\ x + 2y - z &= 2 \\ y - z &= 0 \end{aligned}$ **44.** $\begin{aligned} x - y - 2z &= -11 \\ x - 2y - z &= -11 \\ -x + y + 3z &= 14 \end{aligned}$

45. $\begin{aligned} 2x - 4y + 2z &= 11 \\ x + 3y - 2z &= -9 \\ 4x - 2y + z &= 7 \end{aligned}$ **46.** $\begin{aligned} x - 4y + z &= 9 \\ 3y - 2z &= -7 \\ -x + z &= 0 \end{aligned}$

47. $\begin{aligned} 3x - 2y + 2z &= -18 \\ -x + 2y - 4z &= 16 \\ 4x - 3y - 2z &= -21 \end{aligned}$ **48.** $\begin{aligned} 2x - y - z &= 0 \\ x - y - z &= -2 \\ 3x - 2y - 2z &= -2 \end{aligned}$

49. $\begin{aligned} x - 4y + 3z &= 26 \\ -x + 3y - 2z &= -19 \\ -y + z &= 10 \end{aligned}$ **50.** $\begin{aligned} 4x - y - z &= 0 \\ 4x - 2y &= 0 \\ 2x + z &= 1 \end{aligned}$

51. $\begin{aligned} 5x + 4z &= 7 \\ 2x - 4y &= 6 \\ 3y + 3z &= 3 \end{aligned}$ **52.** $\begin{aligned} y + 2z &= -5 \\ 3x - 2z &= -6 \\ -x - 4y &= 11 \end{aligned}$

53. $\begin{aligned} 5x - 2y + z &= 5 \\ x + y - 2z &= -2 \\ 4x - 3y + 3z &= 7 \end{aligned}$ **54.** $\begin{aligned} 2x - 4y - z &= 2 \\ x + y - 3z &= 10 \\ -x - 7y + 8z &= 2 \end{aligned}$

Exercises 55–62: (Refer to Example 7.) The augmented matrix is in reduced row-echelon form and represents a system of linear equations. If possible, solve the system.

55. $\left[\begin{array}{cc|c} 1 & 0 & 12 \\ 0 & 1 & 3 \end{array}\right]$

56. $\left[\begin{array}{cc|c} 1 & -1 & 1 \\ 0 & 0 & 0 \end{array}\right]$

57. $\left[\begin{array}{ccc|c} 1 & 0 & 0 & -2 \\ 0 & 1 & 0 & 4 \\ 0 & 0 & 1 & \frac{1}{2} \end{array}\right]$

58. $\left[\begin{array}{ccc|c} 1 & 0 & 0 & 7 \\ 0 & 1 & 0 & -9 \\ 0 & 0 & 1 & 3 \end{array}\right]$

59. $\left[\begin{array}{ccc|c} 1 & 0 & 2 & 4 \\ 0 & 1 & -1 & -3 \\ 0 & 0 & 0 & 0 \end{array}\right]$

60. $\left[\begin{array}{ccc|c} 1 & 0 & 1 & -2 \\ 0 & 1 & 3 & 5 \\ 0 & 0 & 0 & 0 \end{array}\right]$

61. $\left[\begin{array}{ccc|c} 1 & 0 & 0 & \frac{3}{4} \\ 0 & 1 & 0 & -1 \\ 0 & 0 & 0 & \frac{2}{3} \end{array}\right]$

62. $\left[\begin{array}{ccc|c} 1 & 0 & 0 & 10 \\ 0 & 1 & 0 & 21 \\ 0 & 0 & 0 & -2 \end{array}\right]$

Exercises 63–70: **Reduced Row-Echelon Form** *Use Gauss-Jordan elimination to solve the system of equations.*

63. $\begin{aligned} x - y &= 1 \\ x + y &= 5 \end{aligned}$ **64.** $\begin{aligned} 2x + 3y &= 1 \\ x - 2y &= -3 \end{aligned}$

65. $\begin{aligned} x + 2y + z &= 3 \\ y - z &= -2 \\ -x - 2y + 2z &= 6 \end{aligned}$ **66.** $\begin{aligned} x + z &= 2 \\ x - y - z &= 0 \\ -2x + y &= -2 \end{aligned}$

67. $\begin{aligned} x - y + 2z &= 7 \\ 2x + y - 4z &= -27 \\ -x + y - z &= 0 \end{aligned}$ **68.** $\begin{aligned} 2x - 4y - 6z &= 2 \\ x - 3y + z &= 12 \\ 2x + y + 3z &= 5 \end{aligned}$

69. $\begin{aligned} 2x + y - z &= 2 \\ x - 2y + z &= 0 \\ x + 3y - 2z &= 4 \end{aligned}$ **70.** $\begin{aligned} -2x - y + z &= 3 \\ x + y - 3z &= 1 \\ x - 2y - 4z &= 2 \end{aligned}$

Exercises 71–76: **Technology** *Use technology to find the solution. Approximate values to the nearest thousandth.*

71. $\begin{aligned} 5x - 7y + 9z &= 40 \\ -7x + 3y - 7z &= 20 \\ 5x - 8y - 5z &= 15 \end{aligned}$

72. $\begin{aligned} 12x - 4y - 7z &= 8 \\ -8x - 6y + 9z &= 7 \\ 34x + 6y - 2z &= 5 \end{aligned}$

73. $\begin{aligned} 2.1x + 0.5y + 1.7z &= 4.9 \\ -2x + 1.5y - 1.7z &= 3.1 \\ 5.8x - 4.6y + 0.8z &= 9.3 \end{aligned}$

74. $\begin{aligned} 53x + 95y + 12z &= 108 \\ 81x - 57y - 24z &= -92 \\ -9x + 11y - 78z &= 21 \end{aligned}$

75. $\begin{aligned} 0.1x + 0.3y + 1.7z &= 0.6 \\ 0.6x + 0.1y - 3.1z &= 6.2 \\ 2.4y + 0.9z &= 3.5 \end{aligned}$

76. $\begin{aligned} 103x - 886y + 431z &= 1200 \\ -55x + 981y &= 1108 \\ -327x + 421y + 337z &= 99 \end{aligned}$

Applications

77. Pumping Water Three pumps are being used to empty a small swimming pool. The first pump is twice as fast as the second pump. The first two pumps can empty the pool in 8 hours, while all three pumps can empty it in 6 hours. How long would it take each pump to empty the pool individually? (*Hint:* Let x represent the fraction of the pool that the first pump can empty in 1 hour. Let y and z represent this fraction for the second and third pumps, respectively.)

78. Pumping Water Suppose in Exercise 77 that the first pump is three times as fast as the third pump, the first and second pumps can empty the pool in 6 hours, and all three pumps can empty the pool in 8 hours.
(a) Are these data realistic? Explain your reasoning.

(b) Make a conjecture about a solution to these data.

(c) Test your conjecture by solving the problem.

79. Food Shelters (Refer to Example 10.) For three food shelters, monthly food costs F in dollars, number of people served per month N, and monthly charitable receipts R in dollars are as shown in the table.

Food Costs (F)	Number Served (N)	Charitable Receipts (R)
1300	1800	5000
5300	3200	12,000
6500	4500	13,000

(a) Model these data using $F = aN + bR + c$, where a, b, and c are constants.

(b) Predict the food costs for a shelter that serves 3500 people and receives charitable receipts of $12,500. Round your answer to the nearest hundred dollars.

80. Estimating the Weight of a Bear The following table shows the weight W, neck size N, and chest size C for a representative sample of black bears.

W (pounds)	N (inches)	C (inches)
100	17	27
272	25	36
381	30	43

Source: M. Triola, *Elementary Statistics.*

(a) Find values for a, b, and c so that the equation $W = a + bN + cC$ models these data.

(b) Estimate the weight of a bear with a 20-inch neck and a 31-inch chest size.

(c) Explain why it is reasonable for the coefficients b and c to be positive.

81. Electricity In the study of electrical circuits, the application of Kirchoff's rules frequently results in systems of linear equations. To determine the current I (in amperes) in each branch of the circuit shown in the figure, solve the system of linear equations. Round values to the nearest hundredth.

$$I_1 = I_2 + I_3$$
$$15 + 4I_3 = 14I_2$$
$$10 + 4I_3 = 5I_1$$

82. Electricity (Refer to Exercise 81.) Find the current (in amperes) in each branch of the circuit shown in the figure by solving the system of linear equations. Round values to the nearest hundredth.

$$I_1 = I_2 + I_3$$
$$20 = 4I_1 + 7I_3$$
$$10 + 7I_3 = 6I_2$$

83. Investment A sum of $5000 is invested in three mutual funds that pay 8%, 11%, and 14% annual interest rates. The amount of money invested in the fund paying 14% equals the total amount of money invested in the other two funds, and the total annual interest from all three funds is $595.
(a) Write a system of equations whose solution gives the amount invested in each mutual fund. Be sure to state what each variable represents.

(b) Solve the system of equations.

84. Investment A sum of $10,000 is invested in three accounts that pay 3%, 4%, and 5% interest. Twice as much money is invested in the account paying 5% as in the account paying 3%, and the total annual interest from all three accounts is $421.
(a) Write a system of equations whose solution gives the amount invested in each account. Be sure to state what each variable represents.

(b) Solve the system of equations.

Exercises 85 and 86: **Traffic Flow** *The figure shows three one-way streets with intersections A, B, and C. Numbers indicate the average traffic flow in vehicles per minute. The variables x, y, and z denote unknown traffic flows that need to be determined for timing of stoplights.*
(a) If the number of vehicles per minute entering an intersection must equal the number exiting an intersection, verify that the accompanying system of linear equations describes the traffic flow.
(b) Rewrite the system and solve.
(c) Interpret your solution.

85. A: $x + 5 = y + 7$
    B: $z + 6 = x + 3$
    C: $y + 3 = z + 4$

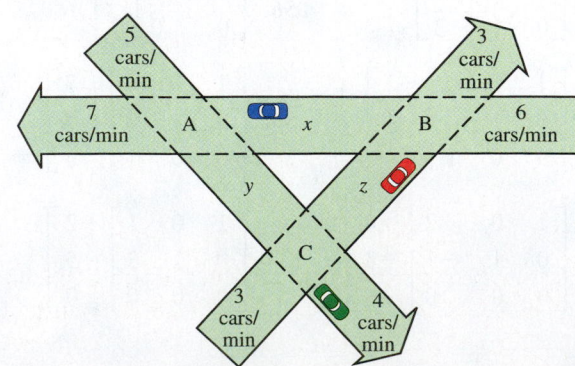

86. A: $x + 7 = y + 4$
B: $4 + 5 = x + z$
C: $y + 8 = 9 + 4$

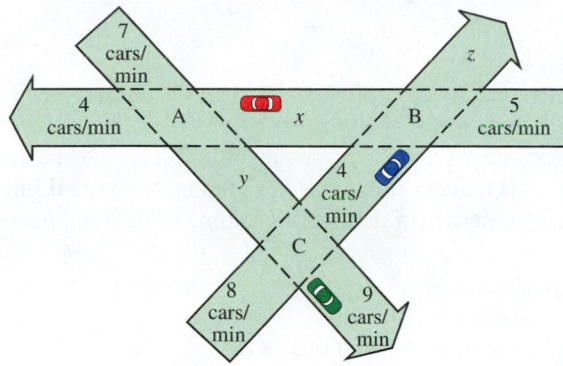

Exercises 87–90: Each set of data can be modeled by the quadratic function $f(x) = ax^2 + bx + c$.

(a) *Write a linear system whose solution represents values of a, b, and c.*
(b) *Use technology to find the solution.*
(c) *Graph f and the data in the same viewing rectangle.*
(d) *Make your own prediction using f.*

87. Estimating iPad Sales (Refer to the introduction to this section.) The table lists total iPad sales y in millions x quarters after its release.

x	1	5	6
y	3	29	40

Source: Apple Corporation.

88. Head Start Enrollment The table lists annual enrollment in thousands for the Head Start program x years after 1980.

x	0	10	26
y	376	541	909

Source: Dept. of Health and Human Services.

89. Chronic Health Care A large percentage of the U.S. population will require chronic health care in the coming decades. The average age of caregivers is 50–64, while the typical person needing chronic care is 85 or older. The ratio y of potential caregivers to those needing chronic health care will shrink in the coming years x, as shown in the table.

x	1990	2010	2030
y	11	10	6

Source: Robert Wood Johnson Foundation, *Chronic Care in America: A 21st Century Challenge.*

90. Carbon Dioxide Levels Carbon dioxide (CO_2) is a greenhouse gas. The table lists its concentration y in parts per million (ppm) measured at Mauna Loa, Hawaii, for three selected years x.

x	1958	1973	2003
y	315	325	376

Source: A. Nilsson, *Greenhouse Earth.*

Writing about Mathematics

91. A linear equation in three variables can be represented by a flat plane. Describe geometrically situations that can occur when a system of three linear equations has either no solution or an infinite number of solutions.

92. Give an example of an augmented matrix in row-echelon form that represents a system of linear equations that has no solution. Explain your reasoning.

Extended and Discovery Exercises

Exercises 1 and 2: Solve the system of four equations with four variables.

1. $\begin{aligned} w + x + 2y - z &= 4 \\ 2w + x + 2y + z &= 5 \\ -w + 3x + y - 2z &= -2 \\ 3w + 2x + y + 3z &= 3 \end{aligned}$

2. $\begin{aligned} 2w - 5x + 3y - 2z &= -13 \\ 3w + 2x + 4y - 9z &= -28 \\ 4w + 3x - 2y - 4z &= -13 \\ 5w - 4x - 3y + 3z &= 0 \end{aligned}$

CHECKING BASIC CONCEPTS FOR SECTIONS 9.3 AND 9.4

1. If possible, solve the system of linear equations.

(a) $\begin{aligned} x - 2y + z &= -2 \\ x + y + 2z &= 3 \\ 2x - y - z &= 5 \end{aligned}$

(b) $\begin{aligned} x - 2y + z &= -2 \\ x + y + 2z &= 3 \\ 2x - y + 3z &= 1 \end{aligned}$

(c) $\begin{aligned} x - 2y + z &= -2 \\ x + y + 2z &= 3 \\ 2x - y + 3z &= 5 \end{aligned}$

2. Tickets Sold Two thousand tickets were sold for a play, generating $19,700. The prices of the tickets were $5 for children, $10 for students, and $12 for adults. There were 100 more adult tickets sold than student tickets. Find the number of each type of ticket sold.

3. Solve the system of linear equations using Gaussian elimination and backward substitution.

$$\begin{aligned} x \qquad + z &= 2 \\ x + y - z &= 1 \\ -x - 2y - z &= 0 \end{aligned}$$

4. Use technology to solve the system of linear equations in Exercise 3.

9.5 Properties and Applications of Matrices

- **Learn matrix notation**
- **Learn how matrices are used in social networks**
- **Find sums, differences, and scalar multiples of matrices**
- **Find matrix products**
- **Use technology (optional)**

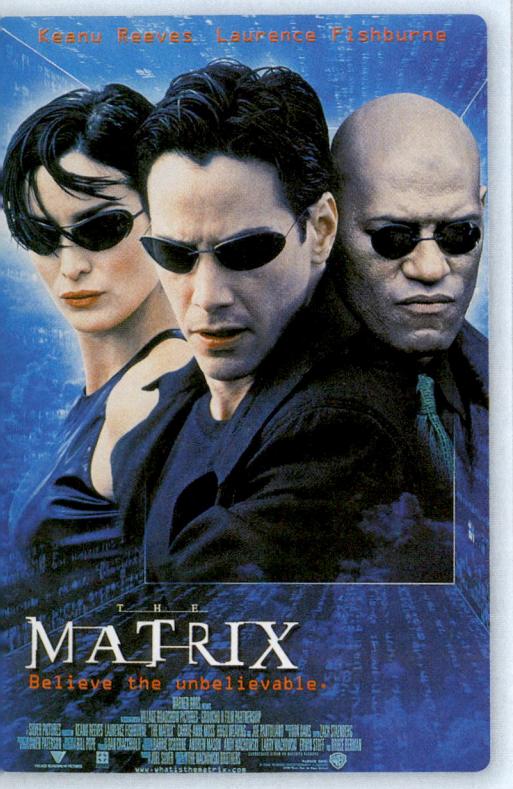

Introduction

In the movie trilogy *The Matrix,* the reality that most humans perceive is nothing more than a simulated reality constructed by machines. Although these films are works of fiction, matrices make it possible for programmers to create popular, multiplayer virtual reality games for the Internet. Matrices are also vitally important in social networks and Internet browsing. In this section we discuss properties of matrices and some of their applications.

Matrix Notation

The following notation is used to denote elements in a matrix A.

$$\begin{bmatrix} a_{11} & a_{12} \\ a_{21} & a_{22} \end{bmatrix} \quad \begin{bmatrix} a_{11} & a_{12} & a_{13} \\ a_{21} & a_{22} & a_{23} \\ a_{31} & a_{32} & a_{33} \end{bmatrix} \quad \begin{bmatrix} a_{11} & a_{12} & a_{13} & a_{14} \\ a_{21} & a_{22} & a_{23} & a_{24} \\ a_{31} & a_{32} & a_{33} & a_{34} \\ a_{41} & a_{42} & a_{43} & a_{44} \end{bmatrix} \quad \begin{bmatrix} a_{11} & a_{12} \\ a_{21} & a_{22} \\ a_{31} & a_{32} \end{bmatrix} \quad \begin{bmatrix} a_{11} & a_{12} & a_{13} \\ a_{21} & a_{22} & a_{23} \end{bmatrix}$$

A general element is denoted by a_{ij}. This refers to the element in the ith row, jth column. For example, a_{23} would be the element of A located in the second row, third column. Two m by n matrices A and B are **equal** if corresponding elements are equal. If A and B have different dimensions, they cannot be equal. For example,

$$\begin{bmatrix} 3 & -3 & 7 \\ 2 & 6 & -2 \\ 4 & 2 & 5 \end{bmatrix} = \begin{bmatrix} 3 & -3 & 7 \\ 2 & 6 & -2 \\ 4 & 2 & 5 \end{bmatrix} \qquad \text{Equal matrices}$$

because *all* corresponding elements are equal. However,

$$\begin{bmatrix} 1 & 4 \\ -3 & 2 \\ 4 & -7 \end{bmatrix} \neq \begin{bmatrix} 1 & 4 \\ -3 & 2 \\ 5 & -7 \end{bmatrix} \qquad \text{Unequal matrices:}\ a_{31} \neq b_{31}$$

because $4 \neq 5$ in row 3 and column 1, and

$$\underset{2 \times 3}{\begin{bmatrix} 1 & 2 & 3 \\ 4 & 5 & 6 \end{bmatrix}} \neq \underset{2 \times 2}{\begin{bmatrix} 1 & 2 \\ 4 & 5 \end{bmatrix}} \qquad \text{Unequal matrices: different dimensions}$$

because the matrices have different dimensions.

EXAMPLE 1   Determining matrix elements

Let a_{ij} denote a general element in A and b_{ij} a general element in B, where

$$A = \begin{bmatrix} 3 & -3 & 7 \\ 1 & 6 & -2 \\ 4 & 2 & 5 \end{bmatrix} \quad \text{and} \quad B = \begin{bmatrix} 3 & x & 7 \\ 1 & 6 & -2 \\ 4 & 5 & 2 \end{bmatrix}.$$

(a) Identify a_{12}, b_{32}, and a_{13}.
(b) Compute $a_{31} b_{13} + a_{32}b_{23} + a_{33}b_{33}$.
(c) Is there a value for x that will make the statement $A = B$ true?

SOLUTION
(a) The element a_{12} is located in the first row, second column of A. Thus, $a_{12} = -3$. In a similar manner, we find that $b_{32} = 5$ and $a_{13} = 7$.

(b) $a_{31}b_{13} + a_{32}b_{23} + a_{33}b_{33} = (4)(7) + (2)(-2) + (5)(2) = 34$

(c) No, since $a_{32} = 2 \neq 5 = b_{32}$ and $a_{33} = 5 \neq 2 = b_{33}$. Even if we let $x = -3$, other corresponding elements in A and B are not equal.

Now Try Exercise 1

Matrices and Social Networks

People with Internet access often choose to participate in at least one social network such as Facebook, Pinterest, or Twitter. Mathematics is essential to the success of these social networks, and matrices play an important role in processing social network data. Consider the diagram in Figure 9.50, which represents a simple social network of four people.

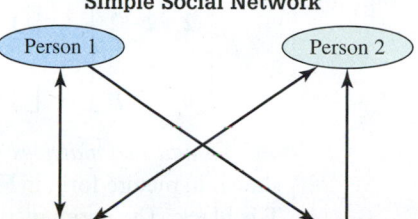

Simple Social Network

Figure 9.50

The arrows in Figure 9.50 show the social relationships among these people. For example, an arrow from Person 1 to Person 4 indicates that Person 1 likes Person 4. But Person 4 does not like Person 1 because there is no arrow pointing in the opposite direction. On the other hand Person 2 and Person 3 like each other, which is indicated by a double arrow between them. In the next example we see how a matrix can represent this social network.

EXAMPLE 2   **Representing a social network with a matrix**

Use a matrix to represent the social network shown in Figure 9.50.

SOLUTION A social network with four people can be represented by a 4×4 square matrix. Because **Person 1** likes **Person 4**, we put a **1** in **row 1 column 4**. Similarly, **Person 4** likes **Person 2**, so we put a **1** in **row 4 column 2**. When no arrow exists to indicate that one person likes another, we place a 0 in the appropriate row and column of the matrix. Using this process results in the following matrix.

$$\begin{bmatrix} 0 & 0 & 1 & 1 \\ 0 & 0 & 1 & 0 \\ 1 & 1 & 0 & 0 \\ 0 & 1 & 1 & 0 \end{bmatrix}$$

Now Try Exercise 53

Sums, Differences, and Scalar Multiples of Matrices

An Application   As a result of an FCC mandate, all television stations are now required to broadcast digital signals. HDTVs can display digital images with a resolution up to 1920×1080 pixels. Matrices play an important role in processing digital images.

Matrix Addition and Subtraction   To simplify the concept of a digital image, we reduce the resolution to 3×3 pixels and have just four gray levels, rather than colors.

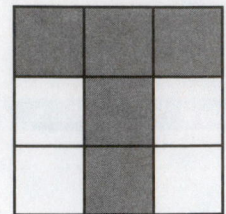

Figure 9.51

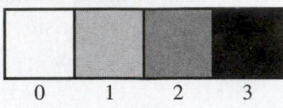

0   1   2   3
Figure 9.52 Gray Levels

Figure 9.53

We will let 0 represent white, 1 light gray, 2 dark gray, and 3 black. Suppose that we would like to digitize the letter T shown in Figure 9.51, using the gray levels shown in Figure 9.52. Since the T is dark gray and the background is white, Figure 9.51 can be represented by

$$A = \begin{bmatrix} 2 & 2 & 2 \\ 0 & 2 & 0 \\ 0 & 2 & 0 \end{bmatrix}.$$

Suppose that we want to make the entire picture darker. If we changed every element in A to 3, the entire picture would be black. A more acceptable solution would be to darken each pixel by one gray level. This corresponds to adding 1 to each element in the matrix A and can be accomplished efficiently using matrix notation.

Matrix Addition: Add Corresponding Elements

$$\begin{bmatrix} 2 & 2 & 2 \\ 0 & 2 & 0 \\ 0 & 2 & 0 \end{bmatrix} + \begin{bmatrix} 1 & 1 & 1 \\ 1 & 1 & 1 \\ 1 & 1 & 1 \end{bmatrix} = \begin{bmatrix} 2+1 & 2+1 & 2+1 \\ 0+1 & 2+1 & 0+1 \\ 0+1 & 2+1 & 0+1 \end{bmatrix} = \begin{bmatrix} 3 & 3 & 3 \\ 1 & 3 & 1 \\ 1 & 3 & 1 \end{bmatrix}$$

To add two matrices of equal dimension, add corresponding elements. The result is shown in picture form in Figure 9.53. Notice that the background is now light gray and the T is black. The entire picture is darker.

To lighten the picture in Figure 9.53, subtract 1 from each element. *To subtract two matrices of equal dimension, subtract corresponding elements.*

Matrix Subtraction: Subtract Corresponding Elements

$$\begin{bmatrix} 3 & 3 & 3 \\ 1 & 3 & 1 \\ 1 & 3 & 1 \end{bmatrix} - \begin{bmatrix} 1 & 1 & 1 \\ 1 & 1 & 1 \\ 1 & 1 & 1 \end{bmatrix} = \begin{bmatrix} 3-1 & 3-1 & 3-1 \\ 1-1 & 3-1 & 1-1 \\ 1-1 & 3-1 & 1-1 \end{bmatrix} = \begin{bmatrix} 2 & 2 & 2 \\ 0 & 2 & 0 \\ 0 & 2 & 0 \end{bmatrix}$$

EXAMPLE 3 **Adding and subtracting matrices**

If $A = \begin{bmatrix} 7 & 8 & -1 \\ 0 & -1 & 6 \end{bmatrix}$ and $B = \begin{bmatrix} 5 & -2 & 10 \\ -3 & 2 & 4 \end{bmatrix}$, find the following.

(a) $A + B$ **(b)** $B + A$ **(c)** $A - B$

SOLUTION

(a) $A + B = \begin{bmatrix} 7 & 8 & -1 \\ 0 & -1 & 6 \end{bmatrix} + \begin{bmatrix} 5 & -2 & 10 \\ -3 & 2 & 4 \end{bmatrix}$

$$= \begin{bmatrix} 7+5 & 8+(-2) & -1+10 \\ 0+(-3) & -1+2 & 6+4 \end{bmatrix}$$

$$= \begin{bmatrix} 12 & 6 & 9 \\ -3 & 1 & 10 \end{bmatrix}$$

(b) $B + A = \begin{bmatrix} 5 & -2 & 10 \\ -3 & 2 & 4 \end{bmatrix} + \begin{bmatrix} 7 & 8 & -1 \\ 0 & -1 & 6 \end{bmatrix}$

$$= \begin{bmatrix} 5+7 & -2+8 & 10+(-1) \\ -3+0 & 2+(-1) & 4+6 \end{bmatrix}$$

$$= \begin{bmatrix} 12 & 6 & 9 \\ -3 & 1 & 10 \end{bmatrix}$$

Notice that $A + B = B + A$. The commutative property for matrix addition holds in general, provided that A and B have the same dimension.

CLASS DISCUSSION

If matrices A and B have the same dimension, does $A - B = B - A$?

(c) $A - B = \begin{bmatrix} 7 & 8 & -1 \\ 0 & -1 & 6 \end{bmatrix} - \begin{bmatrix} 5 & -2 & 10 \\ -3 & 2 & 4 \end{bmatrix}$

$ = \begin{bmatrix} 7 - 5 & 8 - (-2) & -1 - 10 \\ 0 - (-3) & -1 - 2 & 6 - 4 \end{bmatrix}$

$ = \begin{bmatrix} 2 & 10 & -11 \\ 3 & -3 & 2 \end{bmatrix}$

Now Try Exercise 9

An Application   Increasing the contrast in a digital image causes light areas to become lighter and dark areas to become darker. As a result, there are fewer pixels with intermediate gray levels. Changing contrast is different from making the entire picture lighter or darker.

EXAMPLE 4	**Applying matrix addition to a digital image**

Less Contrast

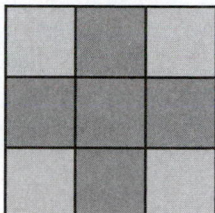

Figure 9.54

Increase the contrast of the + sign in Figure 9.54 by changing light gray to white and dark gray to black. Use matrices to represent this computation.

SOLUTION   Figure 9.54 can be represented by the matrix A.

$$A = \begin{bmatrix} 1 & 2 & 1 \\ 2 & 2 & 2 \\ 1 & 2 & 1 \end{bmatrix}$$

To change the contrast, we reduce each 1 in matrix A to 0 and increase each 2 to 3. The addition of matrix B can accomplish this task.

Matrix Addition for Changing Contrast

$$A + B = \begin{bmatrix} 1 & 2 & 1 \\ 2 & 2 & 2 \\ 1 & 2 & 1 \end{bmatrix} + \begin{bmatrix} -1 & 1 & -1 \\ 1 & 1 & 1 \\ -1 & 1 & -1 \end{bmatrix} = \begin{bmatrix} 0 & 3 & 0 \\ 3 & 3 & 3 \\ 0 & 3 & 0 \end{bmatrix}$$

The picture corresponding to $A + B$ is shown in Figure 9.55.

More Contrast

Figure 9.55

Now Try Exercises 23 and 25

Multiplication of a Matrix by a Scalar   The matrix

$$B = \begin{bmatrix} 1 & 1 & 1 \\ 1 & 1 & 1 \\ 1 & 1 & 1 \end{bmatrix}$$

can be used to darken a digital picture. Suppose that a photograph is represented by a matrix A with gray levels 0 through 11. Every time matrix B is added to A, the picture becomes slightly darker. For example, if

$$A = \begin{bmatrix} 0 & 5 & 0 \\ 5 & 5 & 5 \\ 0 & 5 & 0 \end{bmatrix}$$

then the addition of $A + B + B$ would darken the picture by two gray levels and could be computed by

$$A + B + B = \begin{bmatrix} 0 & 5 & 0 \\ 5 & 5 & 5 \\ 0 & 5 & 0 \end{bmatrix} + \begin{bmatrix} 1 & 1 & 1 \\ 1 & 1 & 1 \\ 1 & 1 & 1 \end{bmatrix} + \begin{bmatrix} 1 & 1 & 1 \\ 1 & 1 & 1 \\ 1 & 1 & 1 \end{bmatrix} = \begin{bmatrix} 2 & 7 & 2 \\ 7 & 7 & 7 \\ 2 & 7 & 2 \end{bmatrix}.$$

A simpler way to write the expression $A + B + B$ is $A + 2B$. Multiplying B by 2 to obtain $2B$ is called **scalar multiplication.**

Scalar Multiplication: Multiply Each Element by the Scalar

Scalar

$$2B = 2 \begin{bmatrix} 1 & 1 & 1 \\ 1 & 1 & 1 \\ 1 & 1 & 1 \end{bmatrix} = \begin{bmatrix} 2(1) & 2(1) & 2(1) \\ 2(1) & 2(1) & 2(1) \\ 2(1) & 2(1) & 2(1) \end{bmatrix} = \begin{bmatrix} 2 & 2 & 2 \\ 2 & 2 & 2 \\ 2 & 2 & 2 \end{bmatrix}$$

Every element of B is multiplied by the real number (scalar) 2.

Sometimes a matrix B is denoted $B = [b_{ij}]$, where b_{ij} represents the element in the ith row, jth column. In this way, we could write $2B$ as $2[b_{ij}] = [2b_{ij}]$. This indicates that to calculate $2B$, multiply each b_{ij} by 2. In a similar manner, a matrix A is sometimes denoted by $[a_{ij}]$.

Some operations on matrices are now summarized.

OPERATIONS ON MATRICES

Matrix Addition

The sum of two $m \times n$ matrices A and B is the $m \times n$ matrix $A + B$, in which each element is the sum of the corresponding elements of A and B. This is written as $A + B = [a_{ij}] + [b_{ij}] = [a_{ij} + b_{ij}]$. If A and B have different dimensions, then $A + B$ is undefined.

Matrix Subtraction

The difference of two $m \times n$ matrices A and B is the $m \times n$ matrix $A - B$, in which each element is the difference of the corresponding elements of A and B. This is written as $A - B = [a_{ij}] - [b_{ij}] = [a_{ij} - b_{ij}]$. If A and B have different dimensions, then $A - B$ is undefined.

Multiplication of a Matrix by a Scalar

The product of a scalar (real number) k and an $m \times n$ matrix A is the $m \times n$ matrix kA, in which each element is k times the corresponding element of A. This is written as $kA = k[a_{ij}] = [ka_{ij}]$.

EXAMPLE 5 Performing scalar multiplication

If $A = \begin{bmatrix} 2 & 7 & 11 \\ -1 & 3 & -5 \\ 0 & 9 & -12 \end{bmatrix}$, find $-4A$.

SOLUTION

$$-4A = -4 \begin{bmatrix} 2 & 7 & 11 \\ -1 & 3 & -5 \\ 0 & 9 & -12 \end{bmatrix} = \begin{bmatrix} -4(2) & -4(7) & -4(11) \\ -4(-1) & -4(3) & -4(-5) \\ -4(0) & -4(9) & -4(-12) \end{bmatrix} = \begin{bmatrix} -8 & -28 & -44 \\ 4 & -12 & 20 \\ 0 & -36 & 48 \end{bmatrix}$$

Now Try Exercise 11(b)

EXAMPLE 6   **Performing operations on matrices**

If possible, perform the indicated operations using

$$A = \begin{bmatrix} 4 & -2 \\ 3 & 5 \end{bmatrix}, B = \begin{bmatrix} 0 & 1 \\ -2 & 3 \end{bmatrix}, C = \begin{bmatrix} 1 & -1 \\ 0 & 7 \\ -4 & 2 \end{bmatrix}, \text{ and } D = \begin{bmatrix} -1 & -3 \\ 9 & -7 \\ 1 & 8 \end{bmatrix}.$$

(a) $A + 3B$ **(b)** $A - C$ **(c)** $-2C - 3D$

SOLUTION

(a) $A + 3B = \begin{bmatrix} 4 & -2 \\ 3 & 5 \end{bmatrix} + 3 \begin{bmatrix} 0 & 1 \\ -2 & 3 \end{bmatrix}$

$$= \begin{bmatrix} 4 & -2 \\ 3 & 5 \end{bmatrix} + \begin{bmatrix} 0 & 3 \\ -6 & 9 \end{bmatrix} = \begin{bmatrix} 4 & 1 \\ -3 & 14 \end{bmatrix}$$

(b) $A - C$ is undefined because the dimension of A is 2×2 and unequal to the dimension of C, which is 3×2.

(c) $-2C - 3D = -2 \begin{bmatrix} 1 & -1 \\ 0 & 7 \\ -4 & 2 \end{bmatrix} - 3 \begin{bmatrix} -1 & -3 \\ 9 & -7 \\ 1 & 8 \end{bmatrix}$

$$= \begin{bmatrix} -2 & 2 \\ 0 & -14 \\ 8 & -4 \end{bmatrix} - \begin{bmatrix} -3 & -9 \\ 27 & -21 \\ 3 & 24 \end{bmatrix} = \begin{bmatrix} 1 & 11 \\ -27 & 7 \\ 5 & -28 \end{bmatrix}$$

Now Try Exercises 13 and 15

Matrix Products

Addition, subtraction, and multiplication can be performed on numbers, variables, and functions. The same operations apply to matrices. Matrix multiplication is different from scalar multiplication.

An Application   Suppose two students are taking day classes at one college and night classes at another, in order to graduate on time. Tables 9.10 and 9.11 list the number of credits taken by the students and the cost per credit at each college.

The cost of tuition is computed by multiplying the number of credits and the cost of each credit. Student 1 is taking 10 credits at $60 each and 7 credits at $80 each. The total tuition for Student 1 is $10(\$60) + 7(\$80) = \$1160$. In a similar manner, the tuition for Student 2 is given by $11(\$60) + 4(\$80) = \$980$. The information in these tables can be represented by matrices. Let A represent Table 9.10 and B represent Table 9.11. B is called a **column matrix** because it has exactly one column.

Credits Taken

	College A	College B
Student 1	10	7
Student 2	11	4

Table 9.10

Credit Cost

	Cost per Credit
College A	$60
College B	$80

Table 9.11

$$A = \begin{bmatrix} 10 & 7 \\ 11 & 4 \end{bmatrix} \quad \text{and} \quad B = \begin{bmatrix} 60 \\ 80 \end{bmatrix} \text{ — Column matrix}$$

The matrix product AB calculates total tuition for each student.

Matrix Multiplication

$$AB = \begin{bmatrix} 10 & 7 \\ 11 & 4 \end{bmatrix} \begin{bmatrix} 60 \\ 80 \end{bmatrix} = \begin{bmatrix} 10(60) + 7(80) \\ 11(60) + 4(80) \end{bmatrix} = \begin{bmatrix} 1160 \\ 980 \end{bmatrix}$$

Generalizing from this example provides the following definition of matrix multiplication.

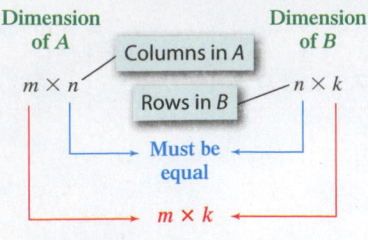

Dimension of A $\quad$ Dimension of B

$m \times n$ $\quad$ $n \times k$

Columns in A / Rows in B — Must be equal

$m \times k$ — Dimension of AB

Figure 9.56

MATRIX MULTIPLICATION

The **product** of an $m \times n$ matrix A and an $n \times k$ matrix B is the $m \times k$ matrix AB, which is computed as follows. To find the element of AB in the ith row and jth column, multiply each element in the ith row of A by the corresponding element in the jth column of B. The sum of these products will give the element in row i, column j of AB.

NOTE In order to compute the product of two matrices, the number of columns in the first matrix must equal the number of rows in the second matrix, as illustrated in Figure 9.56.

EXAMPLE 7 Multiplying matrices

If possible, compute each product using

$$A = \begin{bmatrix} 1 & -1 \\ 0 & 3 \\ 4 & -2 \end{bmatrix}, B = \begin{bmatrix} -1 \\ -2 \end{bmatrix}, C = \begin{bmatrix} 1 & 2 & 3 \\ 4 & 5 & 6 \end{bmatrix}, \text{ and } D = \begin{bmatrix} 1 & -1 & 2 \\ 0 & 3 & -2 \\ -3 & 4 & 5 \end{bmatrix}.$$

(a) AB **(b)** CA **(c)** DC **(d)** CD

SOLUTION

(a) The dimension of A is 3×2 and the dimension of B is 2×1. The dimension of AB is 3×1, as shown in Figure 9.57. The product AB is found as follows.

$$AB = \begin{bmatrix} 1 & -1 \\ 0 & 3 \\ 4 & -2 \end{bmatrix} \begin{bmatrix} -1 \\ -2 \end{bmatrix} = \begin{bmatrix} (1)(-1) + (-1)(-2) \\ (0)(-1) + (3)(-2) \\ (4)(-1) + (-2)(-2) \end{bmatrix} = \begin{bmatrix} 1 \\ -6 \\ 0 \end{bmatrix}$$

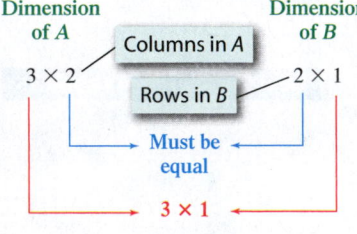

Dimension of A $\quad$ Dimension of B

3×2 $\quad$ 2×1

Columns in A / Rows in B — Must be equal

3×1 — Dimension of AB

Figure 9.57

(b) The dimension of C is 2×3 and the dimension of A is 3×2. Thus CA is 2×2.

$$CA = \begin{bmatrix} 1 & 2 & 3 \\ 4 & 5 & 6 \end{bmatrix} \begin{bmatrix} 1 & -1 \\ 0 & 3 \\ 4 & -2 \end{bmatrix}$$

$$= \begin{bmatrix} 1(1) + 2(0) + 3(4) & 1(-1) + 2(3) + 3(-2) \\ 4(1) + 5(0) + 6(4) & 4(-1) + 5(3) + 6(-2) \end{bmatrix}$$

$$= \begin{bmatrix} 13 & -1 \\ 28 & -1 \end{bmatrix}$$

(c) The dimension of D is 3×3 and the dimension of C is 2×3. Therefore DC is undefined. Note that D has 3 columns and C has only 2 rows.

(d) The dimension of C is 2×3 and the dimension of D is 3×3. Thus CD is 2×3.

$$CD = \begin{bmatrix} 1 & 2 & 3 \\ 4 & 5 & 6 \end{bmatrix} \begin{bmatrix} 1 & -1 & 2 \\ 0 & 3 & -2 \\ -3 & 4 & 5 \end{bmatrix}$$

$$= \begin{bmatrix} 1(1) + 2(0) + 3(-3) & 1(-1) + 2(3) + 3(4) & 1(2) + 2(-2) + 3(5) \\ 4(1) + 5(0) + 6(-3) & 4(-1) + 5(3) + 6(4) & 4(2) + 5(-2) + 6(5) \end{bmatrix}$$

$$= \begin{bmatrix} -8 & 17 & 13 \\ -14 & 35 & 28 \end{bmatrix}$$

Now Try Exercises 27, 31, 33, and 43

> **MAKING CONNECTIONS**
>
> **The Commutative Property and Matrix Multiplication** Example 7 shows that $CD \neq DC$. Unlike multiplication of numbers, variables, and functions, matrix multiplication is *not* commutative. Instead, matrix multiplication is similar to function composition, where for a general pair of functions $f \circ g \neq g \circ f$.

Square matrices have the same number of rows as columns and have dimension $n \times n$ for some natural number n. When we multiply two square matrices, both having dimension $n \times n$, the resulting matrix also has dimension $n \times n$.

EXAMPLE 8   Multiplying square matrices

If $A = \begin{bmatrix} 1 & 0 & 7 \\ 3 & 2 & -1 \\ -5 & -2 & 5 \end{bmatrix}$ and $B = \begin{bmatrix} 4 & -6 & 7 \\ 8 & 9 & 10 \\ 0 & 1 & -3 \end{bmatrix}$, find AB.

SOLUTION

$$AB = \begin{bmatrix} 1 & 0 & 7 \\ 3 & 2 & -1 \\ -5 & -2 & 5 \end{bmatrix} \begin{bmatrix} 4 & -6 & 7 \\ 8 & 9 & 10 \\ 0 & 1 & -3 \end{bmatrix}$$

$$= \begin{bmatrix} 1(4) + 0(8) + 7(0) & 1(-6) + 0(9) + 7(1) & 1(7) + 0(10) + 7(-3) \\ 3(4) + 2(8) - 1(0) & 3(-6) + 2(9) - 1(1) & 3(7) + 2(10) - 1(-3) \\ -5(4) - 2(8) + 5(0) & -5(-6) - 2(9) + 5(1) & -5(7) - 2(10) + 5(-3) \end{bmatrix}$$

$$= \begin{bmatrix} 4 & 1 & -14 \\ 28 & -1 & 44 \\ -36 & 17 & -70 \end{bmatrix}$$

> **Now Try Exercise 37**

An Application   People can navigate from one web page to another by clicking a link. Figure 9.58 shows the links connecting four web pages. An arrow from one web page to another indicates a link. For example, it is possible to navigate from Page 1 to Page 3 in a single click, but it is not possible to navigate from Page 2 to Page 4 in a single click.

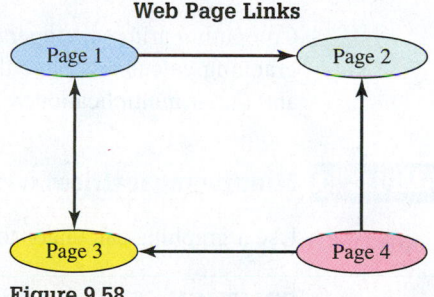

Web Page Links

Figure 9.58

These web page links can be represented by a 4×4 square matrix. Because there is a link from **Page 1** to **Page 2**, we put a **1** in **row 1 column 2**. Similarly, a link exists from **Page 4** to **Page 3**, so we put a **1** in **row 4 column 3**. When no link exists from one web page to another, we place a 0 in the appropriate row and column of the matrix. Using this process results in the following matrix.

$$A = \begin{bmatrix} 0 & 1 & 1 & 0 \\ 0 & 0 & 0 & 0 \\ 1 & 0 & 0 & 0 \\ 0 & 1 & 1 & 0 \end{bmatrix}$$

In the next example we use matrix multiplication to find all of the 2-click paths between web pages.

EXAMPLE 9 Finding 2-click paths between web pages

Use matrix multiplication to find all 2-click paths among the web pages in Figure 9.58 on the previous page.

SOLUTION

The computation A^2 can be used to determine if it is possible to get from web page i to web page j in 2 clicks (links).

$$A^2 = \begin{bmatrix} 0 & 1 & 1 & 0 \\ 0 & 0 & 0 & 0 \\ 1 & 0 & 0 & 0 \\ 0 & 1 & 1 & 0 \end{bmatrix} \cdot \begin{bmatrix} 0 & 1 & 1 & 0 \\ 0 & 0 & 0 & 0 \\ 1 & 0 & 0 & 0 \\ 0 & 1 & 1 & 0 \end{bmatrix} = \begin{bmatrix} 1 & 0 & 0 & 0 \\ 0 & 0 & 0 & 0 \\ 0 & 1 & 1 & 0 \\ 1 & 0 & 0 & 0 \end{bmatrix}$$

The **1** in row 3 column 2 of A^2 indicates that there is a 2-click path from Page 3 to Page 2 (Page 3 to Page 1 to Page 2.) The other 1's in A^2 can be interpreted similarly.

Now Try Exercises 77, 79

NOTE In Example 9, computing A^3 would give all **3**-click paths between the web pages. Similar statements can be made for A^n, where n is a positive integer.

Real numbers satisfy the commutative, associative, and distributive properties for various arithmetic operations. Matrices also satisfy some of these properties, provided that their dimensions are valid so the resulting expressions are defined.

PROPERTIES OF MATRICES

Let A, B, and C be matrices. Assume that each matrix operation is defined.

1. $A + B = B + A$ Commutative property for matrix addition (No commutative property for matrix multiplication)
2. $(A + B) + C = A + (B + C)$ Associative property for matrix addition
3. $(AB)C = A(BC)$ Associative property for matrix multiplication
4. $A(B + C) = AB + AC$ Distributive property

Technology and Matrices (Optional)

Computing arithmetic operations on large matrices by hand can be a difficult task. Many graphing calculators have the capability to perform addition, subtraction, multiplication, and scalar multiplication with matrices, as the next two examples demonstrate.

EXAMPLE 10 Multiplying matrices with technology

Use a graphing calculator to find the product AB from Example 8.

SOLUTION First enter the matrices A and B into your calculator, as illustrated in Figures 9.59 and 9.60. Then find their product on the home screen, as shown in Figure 9.61. Notice that the answer agrees with our results from Example 8.

Calculator Help

To enter the elements of a matrix, see Appendix A (page AP-12). To multiply two matrices, see Appendix A (page AP-13).

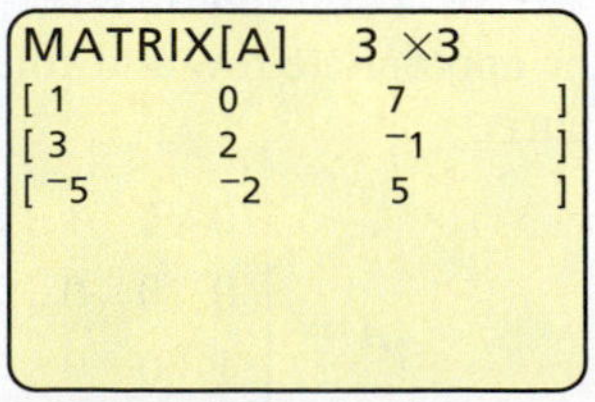

```
MATRIX[A]   3×3
[1     0     7    ]
[3     2    -1    ]
[-5   -2     5    ]
```

Figure 9.59

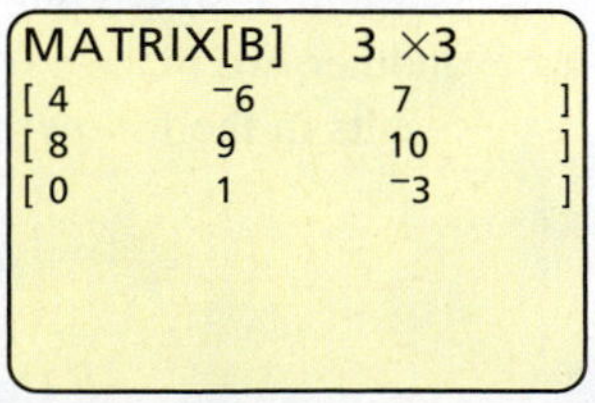

```
MATRIX[B]   3×3
[4    -6     7    ]
[8     9    10    ]
[0     1    -3    ]
```

Figure 9.60

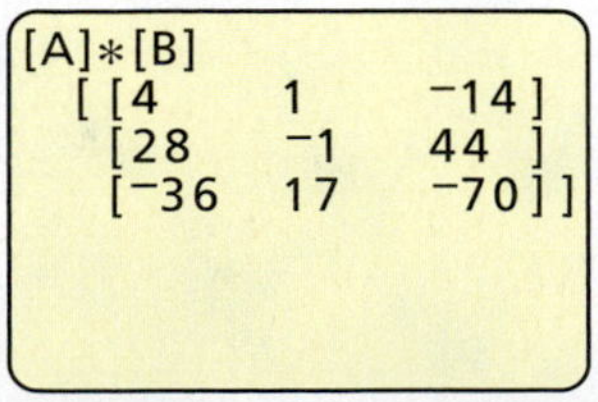

```
[A]*[B]
[[4     1    -14]
 [28   -1    44 ]
 [-36  17   -70]]
```

Figure 9.61

Now Try Exercise 45

EXAMPLE 11   **Using technology to evaluate a matrix expression**

Evaluate the expression $2A + 3B^3$, where

$$A = \begin{bmatrix} 3 & -1 & 2 \\ -1 & 6 & -1 \\ 2 & -1 & 9 \end{bmatrix} \quad \text{and} \quad B = \begin{bmatrix} 1 & -2 & 5 \\ 3 & 1 & -1 \\ 5 & 2 & 1 \end{bmatrix}.$$

```
2[A]+3[B]^3
[[300   -32   322]
 [133    63    73 ]
 [358    28   348]]
```

Figure 9.62

SOLUTION   In the expression $2A + 3B^3$, B^3 is equal to BBB. Enter each matrix into a calculator and evaluate the expression. Figure 9.62 shows the result of this computation.

Now Try Exercise 47

9.5 Putting It All Together

Addition, subtraction, and multiplication can be performed on numbers, variables, and functions. In this section we looked at how these operations also apply to matrices. The following table provides examples of these operations.

MATRIX ADDITION

$$\begin{bmatrix} 1 & 2 & 3 \\ 5 & 6 & 7 \end{bmatrix} + \begin{bmatrix} -1 & 0 & 8 \\ 9 & -2 & 10 \end{bmatrix} = \begin{bmatrix} 1 + (-1) & 2 + 0 & 3 + 8 \\ 5 + 9 & 6 + (-2) & 7 + 10 \end{bmatrix} = \begin{bmatrix} 0 & 2 & 11 \\ 14 & 4 & 17 \end{bmatrix}$$

The matrices must have the same dimension for their sum to be defined.

MATRIX SUBTRACTION

$$\begin{bmatrix} 1 & -4 \\ -3 & 4 \\ 2 & 7 \end{bmatrix} - \begin{bmatrix} 5 & 1 \\ 3 & 6 \\ 8 & -9 \end{bmatrix} = \begin{bmatrix} 1 - 5 & -4 - 1 \\ -3 - 3 & 4 - 6 \\ 2 - 8 & 7 - (-9) \end{bmatrix} = \begin{bmatrix} -4 & -5 \\ -6 & -2 \\ -6 & 16 \end{bmatrix}$$

The matrices must have the same dimension for their difference to be defined.

SCALAR MULTIPLICATION

$$3 \begin{bmatrix} 3 & -2 \\ 0 & 1 \end{bmatrix} = \begin{bmatrix} 3(3) & 3(-2) \\ 3(0) & 3(1) \end{bmatrix} = \begin{bmatrix} 9 & -6 \\ 0 & 3 \end{bmatrix}$$

MATRIX MULTIPLICATION

$$\begin{bmatrix} 0 & 1 \\ 2 & -3 \end{bmatrix} \begin{bmatrix} 3 & -5 \\ 4 & 6 \end{bmatrix} = \begin{bmatrix} 0(3) + 1(4) & 0(-5) + 1(6) \\ 2(3) + (-3)(4) & 2(-5) + (-3)(6) \end{bmatrix} = \begin{bmatrix} 4 & 6 \\ -6 & -28 \end{bmatrix}$$

For a matrix product to be defined, the number of columns in the first matrix must equal the number of rows in the second matrix. Matrix multiplication is not commutative. That is, $AB \neq BA$ in general.

9.5 Exercises

Elements of Matrices

Exercises 1 and 2: Let a_{ij} and b_{ij} be general elements for the given matrices A and B.

(a) *Identify a_{12}, b_{32}, and b_{22}.*
(b) *Compute $a_{11}b_{11} + a_{12}b_{21} + a_{13}b_{31}$.*
(c) *If possible, find a value for x that makes $A = B$.*

1. $A = \begin{bmatrix} 1 & 3 & -4 \\ 3 & 0 & 7 \\ x & 1 & -1 \end{bmatrix}$, $B = \begin{bmatrix} 1 & x & -4 \\ 3 & 0 & 7 \\ 3 & 1 & -1 \end{bmatrix}$

2. $A = \begin{bmatrix} 0 & -1 & 6 \\ 2 & x & -1 \\ 9 & -2 & 1 \end{bmatrix}$, $B = \begin{bmatrix} 0 & -1 & x \\ 2 & 6 & -1 \\ 7 & -2 & 1 \end{bmatrix}$

Exercises 3–6: If possible, find values for x and y so that the matrices A and B are equal.

3. $A = \begin{bmatrix} x & 2 \\ -2 & 1 \end{bmatrix}$, $B = \begin{bmatrix} 1 & 2 \\ -2 & y \end{bmatrix}$

4. $A = \begin{bmatrix} 1 & x+y & 3 \\ 4 & -1 & 6 \\ 3 & 7 & -2 \end{bmatrix}$, $B = \begin{bmatrix} 1 & 2 & 3 \\ 4 & -1 & 6 \\ 3 & y & -2 \end{bmatrix}$

5. $A = \begin{bmatrix} x & 3 \\ 6 & -2 \end{bmatrix}$, $B = \begin{bmatrix} 1 & y & 0 \\ 6 & -2 & 0 \\ 0 & 0 & 0 \end{bmatrix}$

6. $A = \begin{bmatrix} 4 & -2 \\ 3 & -4 \\ x & y \end{bmatrix}$, $B = \begin{bmatrix} 4 & -2 & -2 \\ 3 & -4 & -4 \\ 7 & 8 & 8 \end{bmatrix}$

Addition, Subtraction, and Scalar Multiplication

Exercises 7–10: For the given matrices A and B find each of the following.

(a) $A + B$ (b) $B + A$ (c) $A - B$

7. $A = \begin{bmatrix} 4 & -1 \\ -1 & 4 \end{bmatrix}$, $B = \begin{bmatrix} -1 & 4 \\ 4 & -1 \end{bmatrix}$

8. $A = \begin{bmatrix} 2 & -4 \\ -1 & \frac{1}{2} \\ 3 & -2 \end{bmatrix}$, $B = \begin{bmatrix} 5 & 0 \\ 3 & \frac{1}{2} \\ -1 & 1 \end{bmatrix}$

9. $A = \begin{bmatrix} 3 & 4 & -1 \\ 0 & -3 & 2 \\ -2 & 5 & 10 \end{bmatrix}$, $B = \begin{bmatrix} 11 & 5 & -2 \\ 4 & -7 & 12 \\ 6 & 6 & 6 \end{bmatrix}$

10. $A = \begin{bmatrix} 1 & 6 & 1 & -2 \\ 0 & 1 & 3 & 5 \\ 0 & 0 & 1 & -2 \end{bmatrix}$, $B = \begin{bmatrix} 1 & 0 & 0 & 9 \\ 3 & 1 & 0 & 3 \\ -1 & 4 & 1 & -2 \end{bmatrix}$

Exercises 11–16: If possible, find each of the following.

(a) $A + B$ (b) $3A$ (c) $2A - 3B$

11. $A = \begin{bmatrix} 2 & -6 \\ 3 & 1 \end{bmatrix}$, $B = \begin{bmatrix} -1 & 0 \\ -2 & 3 \end{bmatrix}$

12. $A = \begin{bmatrix} 1 & -2 & 5 \\ 3 & -4 & -1 \end{bmatrix}$, $B = \begin{bmatrix} 0 & -1 & -5 \\ -3 & 1 & 2 \end{bmatrix}$

13. $A = \begin{bmatrix} 1 & -1 & 0 \\ 1 & 5 & 9 \\ -4 & 8 & -5 \end{bmatrix}$, $B = \begin{bmatrix} 2 & 8 & -1 \\ 6 & -1 & 3 \end{bmatrix}$

14. $A = \begin{bmatrix} 6 & 2 & 9 \\ 3 & -2 & 0 \\ -1 & 4 & 8 \end{bmatrix}$, $B = \begin{bmatrix} 1 & 0 & -1 \\ 3 & 0 & 7 \\ 0 & -2 & -5 \end{bmatrix}$

15. $A = \begin{bmatrix} -2 & -1 \\ -5 & 1 \\ 2 & -3 \end{bmatrix}$, $B = \begin{bmatrix} 2 & -1 \\ 3 & 1 \\ 7 & -5 \end{bmatrix}$

16. $A = \begin{bmatrix} 0 & 1 \\ 3 & 2 \\ 4 & -9 \end{bmatrix}$, $B = \begin{bmatrix} 5 & 2 & -7 \\ 8 & -2 & 0 \end{bmatrix}$

Exercises 17–22: Evaluate the matrix expression.

17. $2\begin{bmatrix} 2 & -1 \\ 5 & 1 \\ 0 & 3 \end{bmatrix} + \begin{bmatrix} 5 & 0 \\ 7 & -3 \\ 1 & 1 \end{bmatrix} - \begin{bmatrix} 9 & -4 \\ 4 & 4 \\ 1 & 6 \end{bmatrix}$

18. $-3\begin{bmatrix} 3 & 8 \\ -1 & -9 \end{bmatrix} + 5\begin{bmatrix} 4 & -8 \\ 1 & 6 \end{bmatrix}$

19. $\begin{bmatrix} 4 & 6 \\ 3 & -7 \end{bmatrix} - 2\begin{bmatrix} 1 & 0 \\ -4 & 1 \end{bmatrix}$

20. $\begin{bmatrix} 5 & -1 & 6 \\ -2 & 10 & 12 \\ 5 & 2 & 9 \end{bmatrix} - \begin{bmatrix} -1 & 2 & 2 \\ 2 & -1 & 2 \\ 2 & 2 & -1 \end{bmatrix}$

21. $2\begin{bmatrix} 2 & -1 & -1 \\ -1 & 2 & -1 \\ -1 & -1 & 2 \end{bmatrix} + 3\begin{bmatrix} 1 & 2 & 3 \\ 2 & 1 & 3 \\ 2 & 3 & 1 \end{bmatrix}$

22. $3\begin{bmatrix} 1 & 0 & 3 & -1 \\ 0 & 1 & 2 & -1 \\ 1 & 0 & -3 & 1 \end{bmatrix} - 4\begin{bmatrix} -1 & 0 & 0 & 4 \\ 0 & -1 & 3 & 2 \\ 2 & 0 & 1 & -1 \end{bmatrix}$

Matrices and Digital Photography

Exercises 23–26: **Digital Photography** *(Refer to the discussion of digital images in this section.) Consider the following simplified digital image, which has a 3 × 3 grid with four gray levels numbered from 0 to 3. It shows the number* 1 *in dark gray on a light gray background. Let A be the* 3 × 3 *matrix that represents this image digitally.*

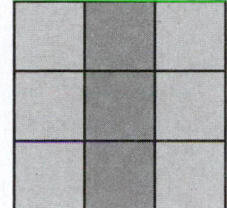

23. Find the matrix A.

24. Find a matrix B such that adding B to A will cause the entire image to become one gray level darker. Evaluate the expression $A + B$.

25. (Refer to Example 4.) Find a matrix B such that adding B to A will enhance the contrast of A by one gray level. Evaluate $A + B$.

26. Find a matrix B such that subtracting B from A will cause the entire image to become lighter by one gray level. Evaluate the expression $A - B$.

Matrix Multiplication

Exercises 27–44: If possible, find AB and BA.

27. $A = \begin{bmatrix} 1 & -1 \\ 2 & 0 \end{bmatrix}$, $\quad B = \begin{bmatrix} -2 & 3 \\ 1 & 2 \end{bmatrix}$

28. $A = \begin{bmatrix} -3 & 5 \\ 2 & 7 \end{bmatrix}$, $\quad B = \begin{bmatrix} -1 & 2 \\ 0 & 7 \end{bmatrix}$

29. $A = \begin{bmatrix} 5 & -7 & 2 \\ 0 & 1 & 5 \end{bmatrix}$, $\quad B = \begin{bmatrix} 9 & 8 & 7 \\ 1 & -1 & -2 \end{bmatrix}$

30. $A = \begin{bmatrix} 2 & 1 & -1 \\ 0 & 2 & 1 \\ 3 & 2 & -1 \end{bmatrix}$, $\quad B = \begin{bmatrix} 1 & 0 \\ 2 & -1 \\ 3 & 1 \end{bmatrix}$

31. $A = \begin{bmatrix} 3 & -1 \\ 1 & 0 \\ -2 & -4 \end{bmatrix}$, $\quad B = \begin{bmatrix} -2 & 5 & -3 \\ 9 & -7 & 0 \end{bmatrix}$

32. $A = \begin{bmatrix} -1 & 0 & -2 \\ 4 & -2 & 1 \end{bmatrix}$, $\quad B = \begin{bmatrix} 2 & -2 \\ 5 & -1 \\ 0 & 1 \end{bmatrix}$

33. $A = \begin{bmatrix} 1 & -1 & 0 \\ 2 & -1 & 5 \\ 6 & 1 & -4 \end{bmatrix}$, $\quad B = \begin{bmatrix} -1 & 3 & -1 \\ 7 & -7 & 1 \end{bmatrix}$

34. $A = \begin{bmatrix} 2 & -1 & -5 \\ 4 & -1 & 6 \\ -2 & 0 & 9 \end{bmatrix}$, $\quad B = \begin{bmatrix} 1 & 2 \\ -1 & -1 \\ 2 & 0 \end{bmatrix}$

35. $A = \begin{bmatrix} 2 & -3 \\ 5 & 3 \end{bmatrix}$, $\quad B = \begin{bmatrix} -3 \\ 4 \\ 1 \end{bmatrix}$

36. $A = \begin{bmatrix} 3 & -1 \\ 2 & -2 \\ 0 & 4 \end{bmatrix}$, $\quad B = \begin{bmatrix} 1 & -4 & 0 \\ -1 & 3 & 2 \end{bmatrix}$

37. $A = \begin{bmatrix} 2 & -1 & 3 \\ 0 & 1 & 0 \\ 2 & -2 & 3 \end{bmatrix}$, $\quad B = \begin{bmatrix} 1 & 5 & -1 \\ 0 & 1 & 3 \\ -1 & 2 & 1 \end{bmatrix}$

38. $A = \begin{bmatrix} 1 & -2 & 5 \\ 1 & 0 & -2 \\ 1 & 3 & 2 \end{bmatrix}$, $\quad B = \begin{bmatrix} -1 & 4 & 2 \\ -3 & 0 & 1 \\ 5 & 1 & 0 \end{bmatrix}$

39. $A = \begin{bmatrix} 2 & -1 \\ 3 & 1 \end{bmatrix}$, $\quad B = \begin{bmatrix} 1 \\ 3 \end{bmatrix}$

40. $A = \begin{bmatrix} 5 & -3 \end{bmatrix}$, $\quad B = \begin{bmatrix} 1 \\ 3 \end{bmatrix}$

41. $A = \begin{bmatrix} -3 & 1 \\ 2 & -4 \end{bmatrix}$, $\quad B = \begin{bmatrix} 1 & 0 & -2 \\ -4 & 8 & 1 \end{bmatrix}$

42. $A = \begin{bmatrix} 6 & 1 & 0 \\ -2 & 5 & 1 \\ 4 & -7 & 10 \end{bmatrix}$, $\quad B = \begin{bmatrix} 10 \\ 20 \\ 30 \end{bmatrix}$

43. $A = \begin{bmatrix} 1 & 0 & -2 \\ 3 & -4 & 1 \\ 2 & 0 & 5 \end{bmatrix}$, $\quad B = \begin{bmatrix} 1 \\ -1 \\ 3 \end{bmatrix}$

44. $A = \begin{bmatrix} 1 & -1 & 3 & -2 \\ 1 & 0 & 3 & 4 \\ 2 & -2 & 0 & 8 \end{bmatrix}$, $\quad B = \begin{bmatrix} 1 & -1 \\ 0 & 5 \\ 2 & 3 \\ -5 & 4 \end{bmatrix}$

Technology and Matrices

Exercises 45–48: Use the given A and B to evaluate each expression.

$$A = \begin{bmatrix} 3 & -2 & 4 \\ 5 & 2 & -3 \\ 7 & 5 & 4 \end{bmatrix}, \quad B = \begin{bmatrix} 1 & 1 & -5 \\ -1 & 0 & -7 \\ -6 & 4 & 3 \end{bmatrix}$$

45. AB

46. BA

47. $3A^2 + 2B$

48. $B^2 - 3A$

Exercises 49–52: **Properties of Matrices** *Use a graphing calculator to evaluate the expression with the given matrices A, B, and C. Compare your answers for parts (a) and (b). Then interpret the results.*

$$A = \begin{bmatrix} 2 & -1 & 3 \\ 1 & 3 & -5 \\ 0 & -2 & 1 \end{bmatrix}, B = \begin{bmatrix} 6 & 2 & 7 \\ 3 & -4 & -5 \\ 7 & 1 & 0 \end{bmatrix},$$

$$C = \begin{bmatrix} 1 & 4 & -3 \\ 8 & 1 & -1 \\ 4 & 6 & -2 \end{bmatrix}$$

49. (a) $A(B + C)$ (b) $AB + AC$

50. (a) $(A - B)C$ (b) $AC - BC$

51. (a) $(A - B)^2$ (b) $A^2 - AB - BA + B^2$

52. (a) $(AB)C$ (b) $A(BC)$

Applications

Exercises 53–56: **Social Networks** *(Refer to Example 2.) The following graph shows a simple social network.*

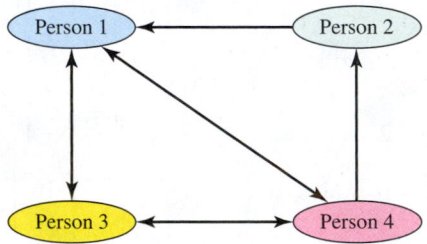

53. Use a matrix to represent this social network.

54. Which person is the most liked person in the network?

55. Which person is the least liked person in the network?

56. Which person likes the most people in the network?

Exercises 57–60: **Social Networks** *(Refer to the previous four exercises.) The following matrix represents a simple social network.*

$$\begin{bmatrix} 0 & 0 & 1 & 0 \\ 0 & 0 & 0 & 0 \\ 1 & 1 & 0 & 0 \\ 1 & 0 & 1 & 0 \end{bmatrix}$$

57. Draw a graph of this network.

58. Row 2 in the matrix contains only 0's. What does this tell us about Person 2?

59. Column 4 in the matrix contains only 0's. What does this tell us about Person 4?

60. If a column of a social network matrix contains only 1's (except on the main diagonal), what can be said about the person represented by that column?

61. **Negative Image** The negative image of a picture interchanges black and white. The number 1 is represented by the matrix A. Determine a matrix B such that $B - A$ represents the negative image of the picture represented by A. Evaluate $B - A$.

$$A = \begin{bmatrix} 0 & 3 & 0 \\ 0 & 3 & 0 \\ 0 & 3 & 0 \end{bmatrix}$$

62. **Negative Image** (Refer to the previous exercise.) Matrix A represents a digital photograph. Find a matrix B that represents the negative image of this picture.

$$A = \begin{bmatrix} 0 & 3 & 0 \\ 1 & 3 & 1 \\ 2 & 3 & 2 \end{bmatrix}$$

Exercises 63 and 64: **Digital Photography** *The digital image represents the letter F using 20 pixels in a 5 × 4 grid. Assume that there are four gray levels from 0 to 3.*

63. Find a matrix A that represents this digital image of the letter F.

64. (Continuation of Exercise 63)
(a) Find a matrix B such that $B - A$ represents the negative image of the picture represented by A.

(b) Find a matrix C such that $A + C$ represents a decrease in the contrast of A by one gray level.

Exercises 65–68: **Digitizing Letters** *(Refer to Exercise 61.) Complete the following.*

(a) *Design a matrix A with dimension 4 × 4 that represents a digital image of the given letter. Assume that there are four gray levels from 0 to 3.*

(b) *Find a matrix B such that B − A represents the negative image of the picture represented by matrix A from part (a).*

65. Z   **66.** N

67. L   **68.** O

Exercises 69–72: **Tuition Costs** *(Refer to the discussion after Example 6.)*

(a) *Find a matrix A and a column matrix B that describe the following tables.*

(b) *Find the matrix product AB, and interpret the result.*

69.

	College A	College B
Student 1	12	4
Student 2	8	7

	Cost per Credit
College A	$55
College B	$70

70.

	College A	College B
Student 1	15	2
Student 2	12	4

	Cost per Credit
College A	$90
College B	$75

71.

	College A	College B
Student 1	10	5
Student 2	9	8
Student 3	11	3

	Cost per Credit
College A	$60
College B	$70

72.

	College A	College B	College C
Student 1	6	0	3
Student 2	11	3	0
Student 3	0	12	3

	Cost per Credit
College A	$50
College B	$65
College C	$60

73. Auto Parts A store owner makes two separate orders for three types of auto parts: I, II, and III. The numbers of parts ordered are represented by the matrix A.

$$A = \begin{matrix} & \text{I} \quad \text{II} \quad \text{III} \\ \begin{bmatrix} 3 & 4 & 8 \\ 5 & 6 & 2 \end{bmatrix} & \begin{matrix} \text{Order 1} \\ \text{Order 2} \end{matrix} \end{matrix}$$

For example, Order 1 called for 4 parts of type II. The cost in dollars of each part can be represented by the matrix B.

$$B = \begin{matrix} \text{Cost} \\ \begin{bmatrix} 10 \\ 20 \\ 30 \end{bmatrix} & \begin{matrix} \text{Part I} \\ \text{Part II} \\ \text{Part III} \end{matrix} \end{matrix}$$

Find AB and interpret the result.

74. Car Sales Two car dealers buy four different makes of cars: I, II, III, and IV. The number of each make of automobile bought by each dealer is represented by the matrix A.

$$A = \begin{matrix} & \text{I} \quad \text{II} \quad \text{III} \quad \text{IV} \\ \begin{bmatrix} 1 & 3 & 8 & 4 \\ 3 & 5 & 7 & 0 \end{bmatrix} & \begin{matrix} \text{Dealer 1} \\ \text{Dealer 2} \end{matrix} \end{matrix}$$

For example, Dealer 2 bought 7 cars of type III. The cost in thousands of dollars of each type of car can be represented by the matrix B.

$$B = \begin{matrix} \text{Cost} \\ \begin{bmatrix} 15 \\ 21 \\ 28 \\ 38 \end{bmatrix} & \begin{matrix} \text{Make I} \\ \text{Make II} \\ \text{Make III} \\ \text{Make IV} \end{matrix} \end{matrix}$$

Find AB and interpret the result.

Exercises 75–80: **Web Page Links** *(Refer to Example 9 and the application preceding it.) The following graph shows web page links.*

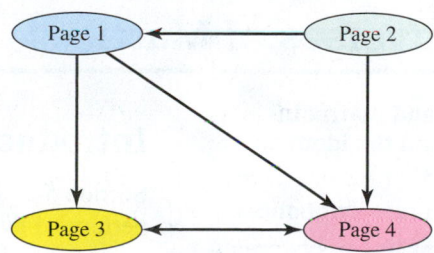

75. Create a matrix A that represents this situation.

76. Which page can be reached in a single click from every other page in the network?

77. Compute A^2.

78. Which two pages cannot be reached using a 2-click path from any other page in the network?

79. There is a 2 in row 2 column 3 of A^2. What does this tell us?

80. There is a 1 in row 4 column 4 of A^2. What does this tell us?

Writing about Mathematics

81. Discuss whether matrix multiplication is more like multiplication of functions or composition of functions. Explain your reasoning.

82. Describe one application of matrices.

Extended and Discovery Exercises

Exercises 1–4: **Representing Colors** *Colors for computer monitors are often described using ordered triples. One model, called the RGB system, uses red, green, and blue to generate all colors. The figure describes the relationships of these colors in this system. Red is* $(1, 0, 0)$*, green is* $(0, 1, 0)$*, and blue is* $(0, 0, 1)$*. Since equal amounts of red and green combine to form yellow, yellow is represented by* $(1, 1, 0)$*. Similarly, magenta (a deep reddish purple) is a mixture of blue and red and is represented by* $(1, 0, 1)$*. Cyan is* $(0, 1, 1)$*, since it is a mixture of blue and green.*

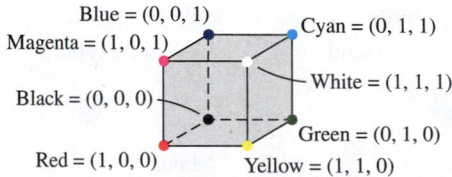

Blue = (0, 0, 1)
Magenta = (1, 0, 1)
Cyan = (0, 1, 1)
White = (1, 1, 1)
Black = (0, 0, 0)
Green = (0, 1, 0)
Red = (1, 0, 0)
Yellow = (1, 1, 0)

Another color model uses cyan, magenta, and yellow. Referred to as the CMY model, it is used in the four-color printing process for textbooks like this one. In this system, cyan is $(1, 0, 0)$*, magenta is* $(0, 1, 0)$*, and yellow is* $(0, 0, 1)$*.*

In the CMY model, red is created by mixing magenta and yellow. Thus, red is $(0, 1, 1)$ *in this system. To convert ordered triples in the RGB model to ordered triples in the CMY model, we can use the following matrix equation. In both of these systems, color intensities vary between 0 and 1.* (**Sources:** *I. Kerlow,* The Art of 3-D Computer Animation and Imaging*; R. Wolff.*)

$$\begin{bmatrix} C \\ M \\ Y \end{bmatrix} = \begin{bmatrix} 1 \\ 1 \\ 1 \end{bmatrix} - \begin{bmatrix} R \\ G \\ B \end{bmatrix}$$

1. In the RGB model, aquamarine is $(0.631, 1, 0.933)$. Use the matrix equation to determine the mixture of cyan, magenta, and yellow that makes aquamarine in the CMY model.

2. In the RGB model, rust is $(0.552, 0.168, 0.066)$. Use the matrix equation to determine the mixture of cyan, magenta, and yellow that makes rust in the CMY model.

3. Use the given matrix equation to find a matrix equation that changes colors represented by ordered triples in the CMY model into ordered triples in the RGB model.

4. In the CMY model, $(0.012, 0, 0.597)$ is a cream color. Use the matrix equation from Exercise 3 to determine the mixture of red, green, and blue that makes a cream color in the RGB model.

9.6 Inverses of Matrices

- **Understand matrix inverses and the identity matrices**
- **Find inverses symbolically**
- **Represent linear systems with matrix equations**
- **Solve linear systems with matrix inverses**

Introduction

Section 5.2 discussed how the inverse function f^{-1} will undo or cancel the computation performed by the function f. Like functions, some matrices have inverses. The inverse of a matrix A will undo or cancel the computation performed by A. For example, in computer graphics, if a matrix A rotates a figure on the screen 90° clockwise, then the inverse matrix will rotate the figure 90° counterclockwise. Similarly, if a matrix B translates a figure 3 units right, then B^{-1} will restore the figure to its original position by translating it 3 units left. This section discusses matrix inverses and some of their applications.

Understanding Matrix Inverses

An Application In computer graphics, the matrix

$$A = \begin{bmatrix} 1 & 0 & h \\ 0 & 1 & k \\ 0 & 0 & 1 \end{bmatrix} \qquad \text{Matrix for translating a point}$$

is used to translate a point (x, y) horizontally h units and vertically k units. The translation is to the right if $h > 0$ and to the left if $h < 0$. Similarly, the translation is upward if

$k > 0$ and downward if $k < 0$. A point (x, y) is represented by the 3×1 *column matrix*

$$X = \begin{bmatrix} x \\ y \\ 1 \end{bmatrix}. \quad \text{Column matrix}$$

The third element in X is always equal to 1. For example, the point $(-1, 2)$ could be translated **3 units right** and **4 units downward** by computing the following matrix product.

Translating $(-1, 2)$ Right 3 Units and Downward 4 Units

$$AX = \begin{bmatrix} 1 & 0 & 3 \\ 0 & 1 & -4 \\ 0 & 0 & 1 \end{bmatrix} \begin{bmatrix} -1 \\ 2 \\ 1 \end{bmatrix} = \begin{bmatrix} 2 \\ -2 \\ 1 \end{bmatrix} = Y$$

Its new location is $(2, -2)$. In the matrix A, $h = 3$ and $k = -4$. See Figure 9.63. (*Source:* C. Pokorny and C. Gerald, *Computer Graphics.*)

If A translates a point 3 units right and 4 units downward, then the inverse matrix translates a point 3 units left and 4 units upward. This would return a point to its original position after being translated by A. Therefore the *inverse matrix of A*, denoted A^{-1}, is given by

$$A^{-1} = \begin{bmatrix} 1 & 0 & -3 \\ 0 & 1 & 4 \\ 0 & 0 & 1 \end{bmatrix}. \quad \text{Inverse matrix}$$

In A^{-1}, $h = -3$ and $k = 4$. The matrix product $A^{-1}Y$ results in

Translating (2, –2) Left 3 Units and Upward 4 Units

$$A^{-1}Y = \begin{bmatrix} 1 & 0 & -3 \\ 0 & 1 & 4 \\ 0 & 0 & 1 \end{bmatrix} \begin{bmatrix} 2 \\ -2 \\ 1 \end{bmatrix} = \begin{bmatrix} -1 \\ 2 \\ 1 \end{bmatrix} = X.$$

The matrix A^{-1} translates $(2, -2)$ to its original coordinates of $(-1, 2)$. The two translations acting on the point $(-1, 2)$ can be represented by the following computation.

$$A^{-1}AX = \begin{bmatrix} 1 & 0 & -3 \\ 0 & 1 & 4 \\ 0 & 0 & 1 \end{bmatrix} \begin{bmatrix} 1 & 0 & 3 \\ 0 & 1 & -4 \\ 0 & 0 & 1 \end{bmatrix} \begin{bmatrix} -1 \\ 2 \\ 1 \end{bmatrix} = \begin{bmatrix} 1 & 0 & 0 \\ 0 & 1 & 0 \\ 0 & 0 & 1 \end{bmatrix} \begin{bmatrix} -1 \\ 2 \\ 1 \end{bmatrix} = \begin{bmatrix} -1 \\ 2 \\ 1 \end{bmatrix} = X$$

That is, the action of A followed by A^{-1} on the point $(-1, 2)$ results in $(-1, 2)$. In a similar manner, if we reverse the order of A^{-1} and A to compute $AA^{-1}X$, the result is again X.

$$AA^{-1}X = \begin{bmatrix} 1 & 0 & 3 \\ 0 & 1 & -4 \\ 0 & 0 & 1 \end{bmatrix} \begin{bmatrix} 1 & 0 & -3 \\ 0 & 1 & 4 \\ 0 & 0 & 1 \end{bmatrix} \begin{bmatrix} -1 \\ 2 \\ 1 \end{bmatrix} = \begin{bmatrix} 1 & 0 & 0 \\ 0 & 1 & 0 \\ 0 & 0 & 1 \end{bmatrix} \begin{bmatrix} -1 \\ 2 \\ 1 \end{bmatrix} = \begin{bmatrix} -1 \\ 2 \\ 1 \end{bmatrix} = X$$

Notice that both matrix products $A^{-1}A$ and AA^{-1} resulted in a matrix with 1's on its main diagonal and 0's elsewhere.

The Identity Matrix

An $n \times n$ matrix with 1's on its main diagonal and 0's elsewhere is called the $n \times n$ *identity matrix*. This matrix is important because its product with any $n \times n$ matrix A always equals A.

Translating a Point

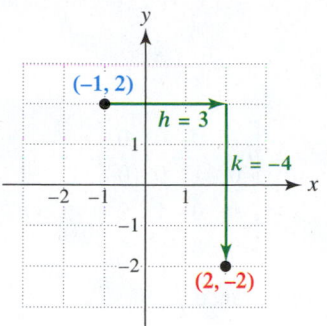

Figure 9.63

> **THE $n \times n$ IDENTITY MATRIX**
>
> The $n \times n$ **identity matrix,** denoted I_n, has only 1's on its main diagonal and 0's elsewhere.

Some examples of identity matrices are shown here.

Identity Matrices

$$I_2 = \begin{bmatrix} 1 & 0 \\ 0 & 1 \end{bmatrix}, \qquad I_3 = \begin{bmatrix} 1 & 0 & 0 \\ 0 & 1 & 0 \\ 0 & 0 & 1 \end{bmatrix}, \qquad \text{and} \qquad I_4 = \begin{bmatrix} 1 & 0 & 0 & 0 \\ 0 & 1 & 0 & 0 \\ 0 & 0 & 1 & 0 \\ 0 & 0 & 0 & 1 \end{bmatrix}$$

If A is any $n \times n$ matrix, then $I_n A = A$ and $AI_n = A$. For instance, if

$$A = \begin{bmatrix} 2 & 3 \\ 4 & 5 \end{bmatrix},$$

then

$$I_2 A = \begin{bmatrix} 1 & 0 \\ 0 & 1 \end{bmatrix}\begin{bmatrix} 2 & 3 \\ 4 & 5 \end{bmatrix} = \begin{bmatrix} 2 & 3 \\ 4 & 5 \end{bmatrix} = A \qquad \text{and}$$

$$AI_2 = \begin{bmatrix} 2 & 3 \\ 4 & 5 \end{bmatrix}\begin{bmatrix} 1 & 0 \\ 0 & 1 \end{bmatrix} = \begin{bmatrix} 2 & 3 \\ 4 & 5 \end{bmatrix} = A.$$

Matrix Inverses

Next we formally define the inverse of an $n \times n$ matrix A, whenever it exists.

> **INVERSE OF A SQUARE MATRIX**
>
> Let A be an $n \times n$ matrix. If there exists an $n \times n$ matrix, denoted A^{-1}, that satisfies
>
> $$A^{-1}A = I_n \qquad \text{and} \qquad AA^{-1} = I_n,$$
>
> then A^{-1} is the **inverse** of A.

If A^{-1} exists, then A is **invertible** or **nonsingular.** On the other hand, if a matrix A is not invertible, then it is **singular.** Not every matrix has an inverse. For example, the **zero matrix** with dimension 3×3 is given by

$$O_3 = \begin{bmatrix} 0 & 0 & 0 \\ 0 & 0 & 0 \\ 0 & 0 & 0 \end{bmatrix}. \qquad \textit{Zero matrix}$$

The matrix O_3 does not have an inverse. The product of O_3 with any 3×3 matrix B would be O_3, rather than the identity matrix I_3.

EXAMPLE 1 **Verifying an inverse**

Determine if B is the inverse of A, where

$$A = \begin{bmatrix} 5 & 3 \\ -3 & -2 \end{bmatrix} \qquad \text{and} \qquad B = \begin{bmatrix} 2 & 3 \\ -3 & -5 \end{bmatrix}.$$

SOLUTION For B to be the inverse of A, it must satisfy $AB = I_2$ and $BA = I_2$.

$$AB = \begin{bmatrix} 5 & 3 \\ -3 & -2 \end{bmatrix}\begin{bmatrix} 2 & 3 \\ -3 & -5 \end{bmatrix} = \begin{bmatrix} 1 & 0 \\ 0 & 1 \end{bmatrix} = I_2$$

$$BA = \begin{bmatrix} 2 & 3 \\ -3 & -5 \end{bmatrix}\begin{bmatrix} 5 & 3 \\ -3 & -2 \end{bmatrix} = \begin{bmatrix} 1 & 0 \\ 0 & 1 \end{bmatrix} = I_2$$

Thus B is the inverse of A. That is, $B = A^{-1}$.

> **Now Try Exercise 1**

An Application   The next example discusses the significance of an inverse matrix in computer graphics.

EXAMPLE 2   Interpreting an inverse matrix

The matrix A can be used to rotate a point 90° clockwise about the origin, where

$$A = \begin{bmatrix} 0 & 1 & 0 \\ -1 & 0 & 0 \\ 0 & 0 & 1 \end{bmatrix} \quad \text{and} \quad A^{-1} = \begin{bmatrix} 0 & -1 & 0 \\ 1 & 0 & 0 \\ 0 & 0 & 1 \end{bmatrix}.$$

(a) Use A to rotate the point $(-2, 0)$ clockwise 90° about the origin.
(b) Make a conjecture about the effect of A^{-1} on the resulting point.
(c) Test this conjecture.

SOLUTION

(a) First, let the point $(-2, 0)$ be represented by the column matrix

$$X = \begin{bmatrix} -2 \\ 0 \\ 1 \end{bmatrix}.$$

Then compute

$$AX = \begin{bmatrix} 0 & 1 & 0 \\ -1 & 0 & 0 \\ 0 & 0 & 1 \end{bmatrix}\begin{bmatrix} -2 \\ 0 \\ 1 \end{bmatrix} = \begin{bmatrix} 0 \\ 2 \\ 1 \end{bmatrix} = Y.$$

If the point $(-2, 0)$ is rotated 90° clockwise about the origin, its new location is $(0, 2)$. See Figure 9.64.

(b) Since A^{-1} represents the inverse operation of A, A^{-1} will rotate the point located at $(0, 2)$ counterclockwise 90°, back to $(-2, 0)$.

(c) This conjecture is correct, since

$$A^{-1}Y = \begin{bmatrix} 0 & -1 & 0 \\ 1 & 0 & 0 \\ 0 & 0 & 1 \end{bmatrix}\begin{bmatrix} 0 \\ 2 \\ 1 \end{bmatrix} = \begin{bmatrix} -2 \\ 0 \\ 1 \end{bmatrix} = X.$$

> **Now Try Exercise 67**

Rotating a Point About the Origin

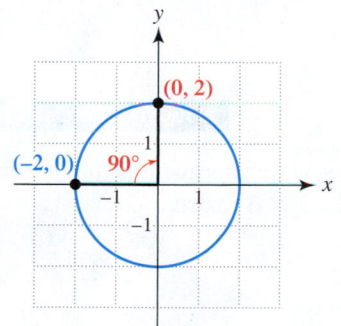

Figure 9.64

CLASS DISCUSSION

What will the results be of the computations AAX and $A^{-1}A^{-1}X$?

Finding Inverses Symbolically

The inverse matrix of an $n \times n$ matrix A can be found symbolically by first forming the augmented matrix $[A \,|\, I_n]$ and then performing matrix row operations, until the left side of the augmented matrix becomes the identity matrix. The resulting augmented matrix can be written as $[I_n \,|\, A^{-1}]$, where the right side of the matrix is A^{-1}.

In the next example, we find A^{-1} from Example 2 by hand.

EXAMPLE 3 Finding an inverse symbolically

Find A^{-1} if

$$A = \begin{bmatrix} 0 & 1 & 0 \\ -1 & 0 & 0 \\ 0 & 0 & 1 \end{bmatrix}.$$

SOLUTION

Getting Started Begin by forming the following 3×6 augmented matrix with the 3×3 identity matrix on the right half.

Augmented Matrix $[A \mid I_3]$

$$\begin{bmatrix} 0 & 1 & 0 & 1 & 0 & 0 \\ -1 & 0 & 0 & 0 & 1 & 0 \\ 0 & 0 & 1 & 0 & 0 & 1 \end{bmatrix}$$

Next use row transformations to obtain the 3×3 identity on the left side. ▶

To obtain a 1 in the first row and first column, we negate the elements in row 2 and then interchange row 1 and row 2. The same row transformations are also applied to the right side of the augmented matrix.

$$\begin{bmatrix} 0 & 1 & 0 & 1 & 0 & 0 \\ -1 & 0 & 0 & 0 & 1 & 0 \\ 0 & 0 & 1 & 0 & 0 & 1 \end{bmatrix} \begin{array}{c} -R_2 \to \\ R_1 \to \\ R_3 \to \end{array} \begin{bmatrix} 1 & 0 & 0 & 0 & -1 & 0 \\ 0 & 1 & 0 & 1 & 0 & 0 \\ 0 & 0 & 1 & 0 & 0 & 1 \end{bmatrix}$$

Because the left side of the augmented matrix is now the 3×3 identity, we stop. The right side of the augmented matrix is A^{-1}. Thus

$$A^{-1} = \begin{bmatrix} 0 & -1 & 0 \\ 1 & 0 & 0 \\ 0 & 0 & 1 \end{bmatrix}, \qquad \text{Inverse matrix}$$

and our result agrees with the information in Example 2.

Now Try Exercise 21

Many times finding inverses requires several steps of row transformations. In the next two examples, we find the inverse of a 2×2 matrix and a 3×3 matrix.

EXAMPLE 4 Finding the inverse of a 2×2 matrix symbolically

Find A^{-1} if

$$A = \begin{bmatrix} 1 & 4 \\ 2 & 9 \end{bmatrix}.$$

SOLUTION Begin by forming a 2×4 augmented matrix. Perform matrix row operations to obtain the identity matrix on the left side, and perform the same operation on the right side of this matrix.

$$\begin{bmatrix} 1 & 4 & 1 & 0 \\ 2 & 9 & 0 & 1 \end{bmatrix} \begin{array}{c} \\ R_2 - 2R_1 \to \end{array} \begin{bmatrix} 1 & 4 & 1 & 0 \\ 0 & 1 & -2 & 1 \end{bmatrix} \begin{array}{c} R_1 - 4R_2 \to \\ \end{array} \begin{bmatrix} 1 & 0 & 9 & -4 \\ 0 & 1 & -2 & 1 \end{bmatrix}$$

Since the 2×2 identity matrix appears on the left side, it follows that the right side equals A^{-1}. That is,

$$A^{-1} = \begin{bmatrix} 9 & -4 \\ -2 & 1 \end{bmatrix}.$$

Furthermore, it can be verified that $A^{-1}A = I_2 = AA^{-1}$.

Now Try Exercise 15

EXAMPLE 5 Finding the inverse of a 3 × 3 matrix symbolically

Find A^{-1} if

$$A = \begin{bmatrix} 1 & 0 & 1 \\ 2 & 1 & 3 \\ -1 & 1 & 1 \end{bmatrix}.$$

SOLUTION Begin by forming the following 3 × 6 augmented matrix. Perform matrix row operations to obtain the identity matrix on the left side, and perform the same operation on the right side of this matrix.

$$\begin{bmatrix} 1 & 0 & 1 & 1 & 0 & 0 \\ 2 & 1 & 3 & 0 & 1 & 0 \\ -1 & 1 & 1 & 0 & 0 & 1 \end{bmatrix} \begin{matrix} \\ R_2 - 2R_1 \to \\ R_3 + R_1 \to \end{matrix} \begin{bmatrix} 1 & 0 & 1 & 1 & 0 & 0 \\ 0 & 1 & 1 & -2 & 1 & 0 \\ 0 & 1 & 2 & 1 & 0 & 1 \end{bmatrix}$$

$$\begin{matrix} \\ \\ R_3 - R_2 \to \end{matrix} \begin{bmatrix} 1 & 0 & 1 & 1 & 0 & 0 \\ 0 & 1 & 1 & -2 & 1 & 0 \\ 0 & 0 & 1 & 3 & -1 & 1 \end{bmatrix} \begin{matrix} R_1 - R_3 \to \\ R_2 - R_3 \to \\ \end{matrix} \begin{bmatrix} 1 & 0 & 0 & -2 & 1 & -1 \\ 0 & 1 & 0 & -5 & 2 & -1 \\ 0 & 0 & 1 & 3 & -1 & 1 \end{bmatrix}$$

The right side is equal to A^{-1}. That is,

$$A^{-1} = \begin{bmatrix} -2 & 1 & -1 \\ -5 & 2 & -1 \\ 3 & -1 & 1 \end{bmatrix}.$$

It can be verified that $A^{-1}A = I_3 = AA^{-1}$.

Now Try Exercise 25

NOTE If it is not possible to obtain the identity matrix on the left side of the augmented matrix by using matrix row operations, then A^{-1} does *not* exist.

Representing Linear Systems with Matrix Equations

In Section 9.4 linear systems were solved using Gaussian elimination with backward substitution. This method used an augmented matrix to represent a system of linear equations. A system of linear equations can also be represented by a matrix equation.

$$3x - 2y + 4z = 5$$
$$2x + y + 3z = 9$$
$$-x + 5y - 2z = 5$$

Let A, X, and B be matrices defined as

|Coefficient Matrix|Variable Matrix|Constant Matrix|

$$A = \begin{bmatrix} 3 & -2 & 4 \\ 2 & 1 & 3 \\ -1 & 5 & -2 \end{bmatrix}, \quad X = \begin{bmatrix} x \\ y \\ z \end{bmatrix}, \quad \text{and} \quad B = \begin{bmatrix} 5 \\ 9 \\ 5 \end{bmatrix}.$$

The matrix product AX is given by

$$AX = \begin{bmatrix} 3 & -2 & 4 \\ 2 & 1 & 3 \\ -1 & 5 & -2 \end{bmatrix} \begin{bmatrix} x \\ y \\ z \end{bmatrix} = \begin{bmatrix} 3x + (-2)y + 4z \\ 2x + 1y + 3z \\ (-1)x + 5y + (-2)z \end{bmatrix} = \begin{bmatrix} 3x - 2y + 4z \\ 2x + y + 3z \\ -x + 5y - 2z \end{bmatrix}.$$

Thus the matrix equation $AX = B$ simplifies to

$$\begin{bmatrix} 3x - 2y + 4z \\ 2x + y + 3z \\ -x + 5y - 2z \end{bmatrix} = \begin{bmatrix} 5 \\ 9 \\ 5 \end{bmatrix}.$$

This matrix equation $AX = B$ is equivalent to the original system of linear equations. Any system of *linear* equations can be represented by a matrix equation.

EXAMPLE 6 **Representing linear systems with matrix equations**

Represent each system of linear equations in the form $AX = B$.

(a) $\begin{aligned} 3x - 4y &= 7 \\ -x + 6y &= -3 \end{aligned}$

(b) $\begin{aligned} x - 5y &= 2 \\ -3x + 2y + z &= -7 \\ 4x + 5y + 6z &= 10 \end{aligned}$

SOLUTION

(a) This linear system comprises two equations and two variables. The equivalent matrix equation is

$$AX = \begin{bmatrix} 3 & -4 \\ -1 & 6 \end{bmatrix} \begin{bmatrix} x \\ y \end{bmatrix} = \begin{bmatrix} 7 \\ -3 \end{bmatrix} = B.$$

(b) The equivalent matrix equation is

$$AX = \begin{bmatrix} 1 & -5 & 0 \\ -3 & 2 & 1 \\ 4 & 5 & 6 \end{bmatrix} \begin{bmatrix} x \\ y \\ z \end{bmatrix} = \begin{bmatrix} 2 \\ -7 \\ 10 \end{bmatrix} = B.$$

Now Try Exercises 39 and 43

Solving Linear Systems with Inverses

The matrix equation $AX = B$ can be solved by using A^{-1}, if it exists.

$$\begin{aligned} AX &= B && \text{Linear system} \\ A^{-1}AX &= A^{-1}B && \text{Multiply each side by } A^{-1}. \\ I_nX &= A^{-1}B && A^{-1}A = I_n \\ X &= A^{-1}B && I_nX = X \text{ for any } n \times 1 \text{ matrix } X \end{aligned}$$

To solve a linear system, multiply each side of the matrix equation $AX = B$ by A^{-1}, if it exists. The solution to the system is unique and can be written as $X = A^{-1}B$.

NOTE Since matrix multiplication is not commutative, it is essential to multiply each side of the equation on the *left* by A^{-1}. That is, $X = A^{-1}B \neq BA^{-1}$ in general.

EXAMPLE 7 **Solving a linear system using the inverse of a 2 × 2 matrix**

Write the linear system as the matrix equation $AX = B$. Find A^{-1} and solve for X.

$$\begin{aligned} x + 4y &= 3 \\ 2x + 9y &= 5 \end{aligned}$$

SOLUTION The linear system can be written as

$$AX = \begin{bmatrix} 1 & 4 \\ 2 & 9 \end{bmatrix} \begin{bmatrix} x \\ y \end{bmatrix} = \begin{bmatrix} 3 \\ 5 \end{bmatrix} = B.$$

The matrix A^{-1} was found in Example 4. Thus we can solve for X as follows.

$$X = A^{-1}B = \begin{bmatrix} 9 & -4 \\ -2 & 1 \end{bmatrix} \begin{bmatrix} 3 \\ 5 \end{bmatrix} = \begin{bmatrix} 7 \\ -1 \end{bmatrix}$$

The solution to the system is $(7, -1)$. Check this.

Now Try Exercise 47

Technology and Inverse Matrices (Optional)   In the next two examples, we use technology to solve the system of linear equations. Technology is especially helpful when finding A^{-1}.

EXAMPLE 8   **Solving a linear system using the inverse of a 3 × 3 matrix**

Write the linear system as the matrix equation $AX = B$. Find A^{-1} and solve for X.

$$\begin{aligned} x + 3y - z &= 6 \\ -2y + z &= -2 \\ -x + y - 3z &= 4 \end{aligned}$$

SOLUTION   The linear system can be written as

$$AX = \begin{bmatrix} 1 & 3 & -1 \\ 0 & -2 & 1 \\ -1 & 1 & -3 \end{bmatrix} \begin{bmatrix} x \\ y \\ z \end{bmatrix} = \begin{bmatrix} 6 \\ -2 \\ 4 \end{bmatrix} = B.$$

The matrix A^{-1} can be found by hand or with a graphing calculator, as shown in Figure 9.65. The solution to the system is given by $x = 4.5$, $y = -0.5$, and $z = -3$. See Figure 9.66.

Calculator Help

To find the inverse of a matrix, see Appendix A (page AP-14). To solve a linear system with a matrix inverse, see Appendix A (page AP-14).

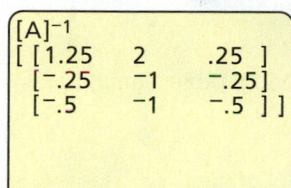

Figure 9.65

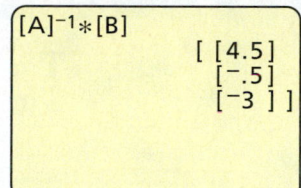

Figure 9.66

Now Try Exercise 59

EXAMPLE 9   **Modeling blood pressure**

In one study of adult males, the effect of both age A in years and weight W in pounds on systolic blood pressure P was found to be modeled by $P(A, W) = a + bA + cW$, where a, b, and c are constants. Table 9.12 lists three individuals with representative blood pressures.

(a) Use Table 9.12 to approximate values for the constants a, b, and c.
(b) Estimate a typical systolic blood pressure for an individual who is 55 years old and weighs 175 pounds.

P	A	W
113	39	142
138	53	181
152	65	191

Table 9.12

SOLUTION

(a) Determine the constants a, b, and c in $P(A, W) = a + bA + cW$ by solving the following three equations.

$$\begin{aligned} P(39, 142) &= a + b(39) + c(142) = 113 \\ P(53, 181) &= a + b(53) + c(181) = 138 \\ P(65, 191) &= a + b(65) + c(191) = 152 \end{aligned}$$

These three equations can be rewritten as follows.

$$\begin{aligned} a + 39b + 142c &= 113 \\ a + 53b + 181c &= 138 \\ a + 65b + 191c &= 152 \end{aligned}$$

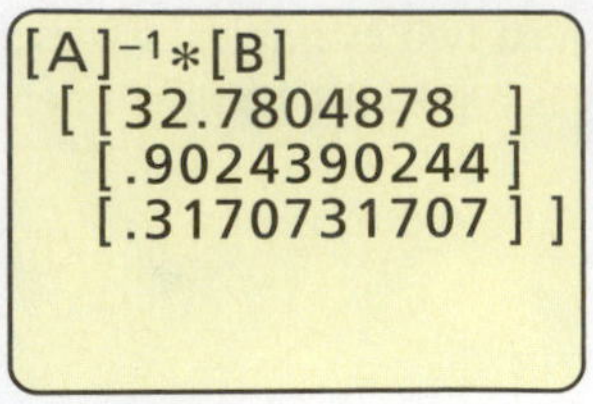

```
[A]⁻¹*[B]
 [[32.7804878  ]
  [.9024390244 ]
  [.3170731707 ]]
```

Figure 9.67

This system can be represented by the matrix equation $AX = B$.

$$AX = \begin{bmatrix} 1 & 39 & 142 \\ 1 & 53 & 181 \\ 1 & 65 & 191 \end{bmatrix} \begin{bmatrix} a \\ b \\ c \end{bmatrix} = \begin{bmatrix} 113 \\ 138 \\ 152 \end{bmatrix} = B$$

The solution, $X = A^{-1}B$, is shown in Figure 9.67. The values for the constants are $a \approx 32.78$, $b \approx 0.9024$, and $c \approx 0.3171$. Thus it follows that P is given by the equation $P(A, W) = 32.78 + 0.9024A + 0.3171W$.

(b) Evaluate $P(55, 175) = 32.78 + 0.9024(55) + 0.3171(175) \approx 137.9$. This model predicts that a typical (male) individual 55 years old, weighing 175 pounds, has a systolic blood pressure of approximately 138. Clearly, this could vary greatly among individuals.

Now Try Exercise 73

9.6 Putting It All Together

The following table summarizes some of the mathematical concepts presented in this section.

CONCEPT	COMMENTS	EXAMPLES
Identity matrix	The $n \times n$ identity matrix I_n has only 1's on the main diagonal and 0's elsewhere. When it is multiplied by any $n \times n$ matrix A, the result is A.	$\begin{bmatrix} 1 & 0 \\ 0 & 1 \end{bmatrix}\begin{bmatrix} 2 & 3 \\ 4 & 5 \end{bmatrix} = \begin{bmatrix} 2 & 3 \\ 4 & 5 \end{bmatrix}$ and $\begin{bmatrix} 2 & 3 \\ 4 & 5 \end{bmatrix}\begin{bmatrix} 1 & 0 \\ 0 & 1 \end{bmatrix} = \begin{bmatrix} 2 & 3 \\ 4 & 5 \end{bmatrix}$, where $I_2 = \begin{bmatrix} 1 & 0 \\ 0 & 1 \end{bmatrix}$.
Matrix inverse	If an $n \times n$ matrix A has an inverse, it is unique, is denoted A^{-1}, and satisfies the equations $AA^{-1} = I_n$ and $A^{-1}A = I_n$. Matrix inverses can be found by using technology. They can also be found with pencil and paper by performing matrix row operations on the augmented matrix $[A\,|\,I_n]$ until it is transformed to $[I_n\,|\,A^{-1}]$.	If $A = \begin{bmatrix} 2 & 3 \\ 3 & 5 \end{bmatrix}$, then $A^{-1} = \begin{bmatrix} 5 & -3 \\ -3 & 2 \end{bmatrix}$ because $AA^{-1} = \begin{bmatrix} 2 & 3 \\ 3 & 5 \end{bmatrix}\begin{bmatrix} 5 & -3 \\ -3 & 2 \end{bmatrix} = \begin{bmatrix} 1 & 0 \\ 0 & 1 \end{bmatrix} = I_2$ $A^{-1}A = \begin{bmatrix} 5 & -3 \\ -3 & 2 \end{bmatrix}\begin{bmatrix} 2 & 3 \\ 3 & 5 \end{bmatrix} = \begin{bmatrix} 1 & 0 \\ 0 & 1 \end{bmatrix} = I_2$.
Matrix equations	Systems of linear equations can be written by using the matrix equation $AX = B$. If A is invertible, then there will be a unique solution given by $X = A^{-1}B$. If A is not invertible, then there could be either no solution or infinitely many solutions. In the latter case, Gaussian elimination should be applied.	The linear system $2x - y = 3$ $x + 2y = 4$ can be written as $AX = B$, where $A = \begin{bmatrix} 2 & -1 \\ 1 & 2 \end{bmatrix}, X = \begin{bmatrix} x \\ y \end{bmatrix}$, and $B = \begin{bmatrix} 3 \\ 4 \end{bmatrix}$. The solution to the system is given by $X = A^{-1}B = \begin{bmatrix} 0.4 & 0.2 \\ -0.2 & 0.4 \end{bmatrix}\begin{bmatrix} 3 \\ 4 \end{bmatrix} = \begin{bmatrix} 2 \\ 1 \end{bmatrix}$. The solution is $(2, 1)$.

9.6 Exercises

Inverse and Identity Matrices

Exercises 1–6: Determine if B is the inverse matrix of A by calculating AB and BA.

1. $A = \begin{bmatrix} 4 & 3 \\ 5 & 4 \end{bmatrix}$, $\qquad B = \begin{bmatrix} 4 & -3 \\ -5 & 4 \end{bmatrix}$

2. $A = \begin{bmatrix} -1 & 2 \\ -3 & 8 \end{bmatrix}$, $\qquad B = \begin{bmatrix} -4 & 1 \\ -2 & 0.5 \end{bmatrix}$

3. $A = \begin{bmatrix} 1 & -1 & 2 \\ 0 & 1 & -1 \\ 1 & 0 & 2 \end{bmatrix}$, $B = \begin{bmatrix} 2 & 2 & -1 \\ -1 & 0 & 1 \\ -1 & -1 & 1 \end{bmatrix}$

4. $A = \begin{bmatrix} 2 & 1 & 1 \\ -1 & 0 & -1 \\ 0 & 2 & -1 \end{bmatrix}$, $B = \begin{bmatrix} 2 & 3 & -1 \\ -1 & -2 & 1 \\ -2 & -4 & 1 \end{bmatrix}$

5. $A = \begin{bmatrix} 2 & 1 & -1 \\ 3 & 0 & 2 \\ -1 & 0 & 1 \end{bmatrix}$, $B = \begin{bmatrix} 0 & 1 & -2 \\ 1 & -3 & 7 \\ 0 & -1 & 3 \end{bmatrix}$

6. $A = \begin{bmatrix} 1 & -1 & 1 \\ 0 & 1 & 0 \\ 1 & 1 & 2 \end{bmatrix}$, $B = \begin{bmatrix} 2 & 3 & -1 \\ 0 & 1 & 0 \\ -1 & -2 & 1 \end{bmatrix}$

Exercises 7–10: Find the value of the constant k in A^{-1}.

7. $A = \begin{bmatrix} 1 & 1 \\ 1 & 2 \end{bmatrix}$, $\qquad A^{-1} = \begin{bmatrix} 2 & -1 \\ -1 & k \end{bmatrix}$

8. $A = \begin{bmatrix} -2 & 2 \\ 1 & -2 \end{bmatrix}$, $\qquad A^{-1} = \begin{bmatrix} -1 & k \\ -0.5 & -1 \end{bmatrix}$

9. $A = \begin{bmatrix} 1 & 3 \\ -1 & -5 \end{bmatrix}$, $\qquad A^{-1} = \begin{bmatrix} k & 1.5 \\ -0.5 & -0.5 \end{bmatrix}$

10. $A = \begin{bmatrix} -2 & 5 \\ -3 & 4 \end{bmatrix}$, $\qquad A^{-1} = \begin{bmatrix} \frac{4}{7} & -\frac{5}{7} \\ k & -\frac{2}{7} \end{bmatrix}$

Exercises 11–14: Predict the results of $I_n A$ and $A I_n$. Then verify your prediction.

11. $I_2 = \begin{bmatrix} 1 & 0 \\ 0 & 1 \end{bmatrix}$, $\qquad A = \begin{bmatrix} 1 & -2 \\ 4 & 3 \end{bmatrix}$

12. $I_3 = \begin{bmatrix} 1 & 0 & 0 \\ 0 & 1 & 0 \\ 0 & 0 & 1 \end{bmatrix}$, $\qquad A = \begin{bmatrix} 1 & -4 & 3 \\ 1 & 9 & 5 \\ 3 & -5 & 0 \end{bmatrix}$

13. $I_3 = \begin{bmatrix} 1 & 0 & 0 \\ 0 & 1 & 0 \\ 0 & 0 & 1 \end{bmatrix}$, $\qquad A = \begin{bmatrix} 0 & 0 & 0 \\ 0 & 0 & 0 \\ 0 & 0 & 0 \end{bmatrix}$

14. $I_4 = \begin{bmatrix} 1 & 0 & 0 & 0 \\ 0 & 1 & 0 & 0 \\ 0 & 0 & 1 & 0 \\ 0 & 0 & 0 & 1 \end{bmatrix}$, $\qquad A = \begin{bmatrix} 5 & -2 & 6 & -3 \\ 0 & 1 & 4 & -1 \\ -5 & 7 & 9 & 8 \\ 0 & 0 & 3 & 1 \end{bmatrix}$

Calculating Inverses

Exercises 15–28: (Refer to Examples 3–5.) Let A be the given matrix. Find A^{-1} without a calculator.

15. $\begin{bmatrix} 1 & 2 \\ 1 & 3 \end{bmatrix}$

16. $\begin{bmatrix} 1 & 0 \\ 1 & -1 \end{bmatrix}$

17. $\begin{bmatrix} -1 & 2 \\ 3 & -5 \end{bmatrix}$

18. $\begin{bmatrix} 1 & 3 \\ 2 & 5 \end{bmatrix}$

19. $\begin{bmatrix} 8 & 5 \\ 2 & 1 \end{bmatrix}$

20. $\begin{bmatrix} -2 & 4 \\ -5 & 9 \end{bmatrix}$

21. $\begin{bmatrix} 0 & 0 & 1 \\ 1 & 0 & 0 \\ 0 & 1 & 0 \end{bmatrix}$

22. $\begin{bmatrix} 1 & 0 & 0 \\ 1 & 1 & 0 \\ 0 & 1 & 1 \end{bmatrix}$

23. $\begin{bmatrix} 1 & 0 & 1 \\ 2 & 1 & 3 \\ -1 & 1 & 1 \end{bmatrix}$

24. $\begin{bmatrix} -2 & 1 & 0 \\ 1 & 0 & 1 \\ -1 & 1 & 0 \end{bmatrix}$

25. $\begin{bmatrix} 1 & 2 & -1 \\ 2 & 5 & 0 \\ -1 & -1 & 2 \end{bmatrix}$

26. $\begin{bmatrix} 2 & -2 & 1 \\ 1 & 3 & 2 \\ 4 & -2 & 4 \end{bmatrix}$

27. $\begin{bmatrix} -2 & 1 & -3 \\ 0 & 1 & 2 \\ 1 & -2 & 1 \end{bmatrix}$

28. $\begin{bmatrix} 1 & -1 & 1 \\ -1 & 2 & 1 \\ 0 & 2 & 1 \end{bmatrix}$

Exercises 29–38: Let A be the given matrix. Find A^{-1}.

29. $\begin{bmatrix} 0.5 & -1.5 \\ 0.2 & -0.5 \end{bmatrix}$

30. $\begin{bmatrix} -0.5 & 0.5 \\ 3 & 2 \end{bmatrix}$

31. $\begin{bmatrix} 1 & 2 & 0 \\ -1 & 4 & -1 \\ 2 & -1 & 0 \end{bmatrix}$

32. $\begin{bmatrix} -2 & 0 & 1 \\ 5 & -4 & 1 \\ 1 & -2 & 0 \end{bmatrix}$

33. $\begin{bmatrix} 2 & -2 & 1 \\ 0 & 5 & 8 \\ 0 & 0 & -1 \end{bmatrix}$

34. $\begin{bmatrix} 2 & 0 & 2 \\ 1 & 5 & 0 \\ -1 & 0 & 2 \end{bmatrix}$

35. $\begin{bmatrix} 3 & -1 & -1 \\ -1 & 3 & -1 \\ -1 & -1 & 3 \end{bmatrix}$

36. $\begin{bmatrix} 2 & -3 & 1 \\ 5 & -6 & 3 \\ 3 & 2 & 0 \end{bmatrix}$

37. $\begin{bmatrix} 1 & -1 & 0 & 0 \\ -1 & 5 & -1 & 0 \\ 0 & -1 & 5 & -1 \\ 0 & 0 & -1 & 1 \end{bmatrix}$ **38.** $\begin{bmatrix} 3 & 1 & 0 & 0 \\ 1 & 3 & 1 & 0 \\ 0 & 1 & 3 & 1 \\ 0 & 0 & 1 & 3 \end{bmatrix}$

Matrices and Linear Systems

Exercises 39–46: Represent the system of linear equations in the form AX = B.

39. $\begin{aligned} 2x - 3y &= 7 \\ -3x - 4y &= 9 \end{aligned}$ **40.** $\begin{aligned} -x + 3y &= 10 \\ 2x - 6y &= -1 \end{aligned}$

41. $\begin{aligned} \tfrac{1}{2}x - \tfrac{3}{2}y &= \tfrac{1}{4} \\ -x + 2y &= 5 \end{aligned}$ **42.** $\begin{aligned} -1.1x + 3.2y &= -2.7 \\ 5.6x - 3.8y &= -3.0 \end{aligned}$

43. $\begin{aligned} x - 2y + z &= 5 \\ 3y - z &= 6 \\ 5x - 4y - 7z &= 0 \end{aligned}$ **44.** $\begin{aligned} 4x - 3y + 2z &= 8 \\ -x + 4y + 3z &= 2 \\ -2x \quad\quad - 5z &= 2 \end{aligned}$

45. $\begin{aligned} 4x - y + 3z &= -2 \\ x + 2y + 5z &= 11 \\ 2x - 3y \quad\quad &= -1 \end{aligned}$ **46.** $\begin{aligned} x - 2y + z &= 12 \\ 4y + 3z &= 13 \\ -2x + 7y \quad\quad &= -2 \end{aligned}$

Solving Linear Systems

Exercises 47–54: Complete the following.

(a) *Write the system in the form AX = B.*

(b) *Solve the system by finding A^{-1} and then using the equation $X = A^{-1}B$. (Hint: Some of your answers from Exercises 15–28 may be helpful.)*

47. $\begin{aligned} x + 2y &= 3 \\ x + 3y &= 6 \end{aligned}$ **48.** $\begin{aligned} 2x + y &= 4 \\ -x + 2y &= -1 \end{aligned}$

49. $\begin{aligned} -x + 2y &= 5 \\ 3x - 5y &= -2 \end{aligned}$ **50.** $\begin{aligned} x + 3y &= -3 \\ 2x + 5y &= -2 \end{aligned}$

51. $\begin{aligned} x + z &= -7 \\ 2x + y + 3z &= -13 \\ -x + y + z &= -4 \end{aligned}$ **52.** $\begin{aligned} -2x + y &= -5 \\ x + z &= -5 \\ -x + y &= -4 \end{aligned}$

53. $\begin{aligned} x + 2y - z &= 2 \\ 2x + 5y &= -1 \\ -x - y + 2z &= 0 \end{aligned}$ **54.** $\begin{aligned} 2x - 2y + z &= 1 \\ x + 3y + 2z &= 3 \\ 4x - 2y + 4z &= 4 \end{aligned}$

Exercises 55–62: Complete the following for the given system of linear equations.

(a) *Write the system in the form AX = B.*

(b) *Solve the linear system by computing $X = A^{-1}B$ with a calculator. Approximate the solution to the nearest hundredth when appropriate.*

55. $\begin{aligned} 1.5x + 3.7y &= 0.32 \\ -0.4x - 2.1y &= 0.36 \end{aligned}$ **56.** $\begin{aligned} 31x + 18y &= 64.1 \\ 5x - 23y &= -59.6 \end{aligned}$

57. $\begin{aligned} 0.08x - 0.7y &= -0.504 \\ 1.1x - 0.05y &= 0.73 \end{aligned}$

58. $\begin{aligned} -231x + 178y &= -439 \\ 525x - 329y &= 2282 \end{aligned}$

59. $\begin{aligned} 3.1x + 1.9y - z &= 1.99 \\ 6.3x \quad\quad - 9.9z &= -3.78 \\ -x + 1.5y + 7z &= 5.3 \end{aligned}$

60. $\begin{aligned} 17x - 22y - 19z &= -25.2 \\ 3x + 13y - 9z &= 105.9 \\ x - 2y + 6.1z &= -23.55 \end{aligned}$

61. $\begin{aligned} 3x - y + z &= 4.9 \\ 5.8x - 2.1y \quad\quad &= -3.8 \\ -x \quad\quad + 2.9z &= 3.8 \end{aligned}$

62. $\begin{aligned} 1.2x - 0.3y - 0.7z &= -0.5 \\ -0.4x + 1.3y + 0.4z &= 0.9 \\ 1.7x + 0.6y + 1.1z &= 1.3 \end{aligned}$

Interpreting Inverses

Exercises 63 and 64: **Translations** *(Refer to the discussion in this section about translating a point.) The matrix product AX performs a translation on the point (x, y), where*

$$A = \begin{bmatrix} 1 & 0 & h \\ 0 & 1 & k \\ 0 & 0 & 1 \end{bmatrix} \quad \text{and} \quad X = \begin{bmatrix} x \\ y \\ 1 \end{bmatrix}.$$

(a) *Predict the new location of the point (x, y) when it is translated by A. Compute $Y = AX$ to verify your prediction.*

(b) *Make a conjecture as to what $A^{-1}Y$ represents. Find A^{-1} and calculate $A^{-1}Y$ to test your conjecture.*

(c) *What will AA^{-1} and $A^{-1}A$ equal?*

63. $A = \begin{bmatrix} 1 & 0 & 2 \\ 0 & 1 & 3 \\ 0 & 0 & 1 \end{bmatrix}$, $(x, y) = (0, 1)$, and $X = \begin{bmatrix} 0 \\ 1 \\ 1 \end{bmatrix}$

64. $A = \begin{bmatrix} 1 & 0 & -4 \\ 0 & 1 & 5 \\ 0 & 0 & 1 \end{bmatrix}$, $(x, y) = (4, 2)$, and $X = \begin{bmatrix} 4 \\ 2 \\ 1 \end{bmatrix}$

Exercises 65 and 66: **Translations** *(Refer to the discussion in this section about translating a point.) Find a 3×3 matrix A that performs the following translation of a point (x, y) represented by X. Find A^{-1} and describe what it computes.*

65. 3 units to the left and 5 units downward

66. 6 units to the right and 1 unit upward

67. **Rotation** (Refer to Example 2.) The matrix *B* rotates the point (x, y) clockwise about the origin 45°, where

$$B = \begin{bmatrix} \frac{1}{\sqrt{2}} & \frac{1}{\sqrt{2}} & 0 \\ -\frac{1}{\sqrt{2}} & \frac{1}{\sqrt{2}} & 0 \\ 0 & 0 & 1 \end{bmatrix} \quad \text{and} \quad B^{-1} = \begin{bmatrix} \frac{1}{\sqrt{2}} & -\frac{1}{\sqrt{2}} & 0 \\ \frac{1}{\sqrt{2}} & \frac{1}{\sqrt{2}} & 0 \\ 0 & 0 & 1 \end{bmatrix}.$$

(a) Let *X* represent the point $(-\sqrt{2}, -\sqrt{2})$. Compute $Y = BX$.

(b) Find $B^{-1}Y$. Interpret what B^{-1} computes.

68. Rotation (Refer to Exercise 67.) Predict the result of the computations $BB^{-1}X$ and $B^{-1}BX$ for any point (x, y) represented by X. Explain this result geometrically.

69. Translations The matrix A translates a point to the right 4 units and downward 2 units, and the matrix B translates a point to the left 3 units and upward 3 units, where

$$A = \begin{bmatrix} 1 & 0 & 4 \\ 0 & 1 & -2 \\ 0 & 0 & 1 \end{bmatrix} \quad \text{and} \quad B = \begin{bmatrix} 1 & 0 & -3 \\ 0 & 1 & 3 \\ 0 & 0 & 1 \end{bmatrix}.$$

(a) Let X represent the point $(1, 1)$. Predict the result of $Y = ABX$. Check your prediction.

(b) Find AB mentally, and then compute AB.

(c) Would you expect $AB = BA$? Verify your answer.

(d) Find $(AB)^{-1}$ mentally. Explain your reasoning.

70. Rotation (Refer to Exercises 63 and 67 for A and B.)
(a) Let X represent the point $(0, \sqrt{2})$. If this point is rotated about the origin $45°$ clockwise and then translated 2 units to the right and 3 units upward, determine its new coordinates geometrically.

(b) Compute $Y = ABX$, and explain the result.

(c) Is ABX equal to BAX? Interpret your answer.

(d) Find a matrix that translates Y back to X.

Applications

71. Cost of CDs A music store marks its compact discs A, B, or C to indicate one of three selling prices. The last column in the table shows the total cost of a purchase. Use this information to determine the cost of one CD of each type by setting up a matrix equation and solving it with an inverse.

A	B	C	Total
2	3	4	$120.91
1	4	0	$62.95
2	1	3	$79.94

72. Traffic Flow (Refer to Exercises 85 and 86 in Section 9.4.) The figure at the top of the next column shows four one-way streets with intersections A, B, C, and D. Numbers indicate the average traffic flow in vehicles per minute. The variables $x_1, x_2, x_3,$ and x_4 denote unknown traffic flows.

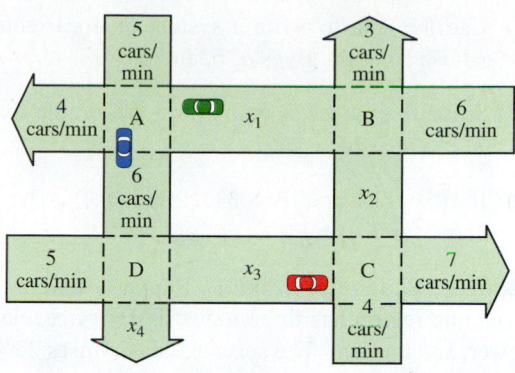

(a) The number of vehicles per minute entering an intersection equals the number exiting an intersection. Verify that the given system of linear equations describes the traffic flow.

A: $x_1 + 5 = 4 + 6$ B: $x_2 + 6 = x_1 + 3$
C: $x_3 + 4 = x_2 + 7$ D: $6 + 5 = x_3 + x_4$

(b) Write the system as $AX = B$ and solve using A^{-1}.

(c) Interpret your results.

73. Home Prices The table contains data on sales of three homes. Price P is measured in thousands of dollars, home size S is in square feet, and condition C is rated on a scale from 1 to 10, where 10 represents excellent condition. The variables were found to be related by the equation $P = a + bS + cC$.

P	S	C
122	1500	8
130	2000	5
158	2200	10

(a) Use the table to write a system of linear equations whose solution gives a, b, and c. Solve this system of linear equations.

(b) Estimate the selling price of a home with 1800 square feet and a condition of 7.

74. Tire Sales A study investigated the relationship among annual tire sales T in thousands, automobile registrations A in millions, and personal disposable income I in millions of dollars. Representative data for three different years are shown in the table. The data were modeled by $T = aA + bI + c$, where a, b, and c are constants. (*Source:* J. Jarrett, *Business Forecasting Methods.*)

T	A	I
10,170	113	308
15,305	133	622
21,289	155	1937

continued on next page

(a) Use the data to write a system of linear equations whose solution gives a, b, and c.

(b) Solve this linear system. Write a formula for T, as $T = aA + bI + c$.

(c) If $A = 118$ and $I = 311$, predict T. (The actual value for T was 11,314.)

75. Leontief Economic Model Suppose that a closed economic region has three industries: service, electrical power, and tourism. The service industry uses 20% of its own production, 40% of the electrical power, and 80% of the tourism. The power company uses 40% of the service industry, 20% of the electrical power, and 10% of the tourism. The tourism industry uses 40% of the service industry, 40% of the electrical power, and 10% of the tourism.

(a) Let S, E, and T be the numbers of units produced by the service, electrical, and tourism industries, respectively. The following system of linear equations can be used to determine the relative number of units each industry needs to produce. (This model assumes that all production is consumed by the region.)

$$0.2S + 0.4E + 0.8T = S$$

$$0.4S + 0.2E + 0.1T = E$$

$$0.4S + 0.4E + 0.1T = T$$

Solve the system and write the solution in terms of T.

(b) If tourism produces 60 units, how many units should the service and electrical industries produce?

76. Plate Glass Sales Plate glass sales G can be affected by the number of new building contracts B issued and the number of automobiles A produced, since plate glass is used in buildings and cars. To forecast sales, a plate glass company in California collected data for three consecutive years, shown in the table. All units are in millions. The data were modeled by $G = aA + bB + c$, where a, b, and c are constants. (*Source:* S. Makridakis and S. Wheelwright, *Forecasting Methods for Management.*)

G	A	B
603	5.54	37.1
657	6.93	41.3
779	7.64	45.6

(a) Write a system of linear equations whose solution gives a, b, and c.

(b) Solve this linear system. Write a formula for G.

(c) For the following year, it was estimated that $A = 7.75$ and $B = 47.4$. Predict G. (The actual value for G was 878.)

Writing about Mathematics

77. Discuss how to solve the matrix equation $AX = B$ if A^{-1} exists.

78. Give an example of a 2×2 matrix A with only nonzero elements that does not have an inverse. Explain what happens if one attempts to find A^{-1} symbolically.

CHECKING BASIC CONCEPTS FOR SECTIONS 9.5 AND 9.6

1. Perform the operations on the given matrices A and B.

$$A = \begin{bmatrix} 1 & 0 & 1 \\ -1 & 1 & 2 \\ 1 & 3 & 0 \end{bmatrix}, \quad B = \begin{bmatrix} -1 & 1 & 2 \\ 0 & 4 & 1 \\ 1 & -2 & 0 \end{bmatrix}$$

(a) $A + B$ **(b)** $2A - B$ **(c)** AB

2. Find the inverse of the matrix A by hand.

$$A = \begin{bmatrix} 0 & 0 & 1 \\ 1 & 1 & 0 \\ 1 & 0 & 1 \end{bmatrix}$$

3. Write each system of linear equations as a matrix equation $AX = B$. Solve the system utilizing A^{-1}.

(a) $\quad x - 2y = 13$
$\qquad 2x + 3y = 5$

(b) $\quad x - y + z = 2$
$\qquad -x + y + z = 4$
$\qquad\qquad\quad y - z = -1$

(c) $\quad 3.1x - 5.3y = -2.682$
$\qquad -0.1x + 1.8y = 0.787$

4. Find A^{-1} if $A = \begin{bmatrix} 2 & -3 & 5 \\ 4 & -3 & 2 \\ 1 & 5 & -4 \end{bmatrix}$.

9.7 Determinants

- **Define and calculate determinants**
- **Apply Cramer's rule**
- **Use determinants to find areas of regions**

Introduction

Determinants are used in mathematics for theoretical purposes. However, they also are used to test if a matrix is invertible and to find the area of certain geometric figures, such as triangles. A *determinant* is a real number associated with a square matrix. We begin our discussion by defining a determinant for a 2 × 2 matrix.

Definition and Calculation of Determinants

Finding the determinant of a matrix with dimension 2 × 2 is a straightforward arithmetic calculation.

DETERMINANT OF A 2 × 2 MATRIX

The **determinant** of

$$A = \begin{bmatrix} a & b \\ c & d \end{bmatrix}$$

is a real number defined by

$$\det A = ad - cb.$$

Later we define determinants for any $n \times n$ matrix. The following theorem can be used to determine if a matrix has an inverse.

INVERTIBLE MATRIX

A square matrix A is invertible if and only if $\det A \neq 0$.

EXAMPLE 1   Determining if a 2 × 2 matrix is invertible

Determine if A^{-1} exists by computing the determinant of the matrix A.

(a) $A = \begin{bmatrix} 3 & -4 \\ -5 & 9 \end{bmatrix}$ **(b)** $A = \begin{bmatrix} 52 & -32 \\ 65 & -40 \end{bmatrix}$

SOLUTION
(a) The determinant of the 2 × 2 matrix A is calculated as follows.

$$\det A = \det \begin{bmatrix} 3 & -4 \\ -5 & 9 \end{bmatrix} = (3)(9) - (-5)(-4) = 7$$

Since $\det A = 7 \neq 0$, the matrix A is invertible and A^{-1} exists.
(b) Similarly,

$$\det A = \det \begin{bmatrix} 52 & -32 \\ 65 & -40 \end{bmatrix} = (52)(-40) - (65)(-32) = 0.$$

Since $\det A = 0$, A^{-1} does not exist. Try finding A^{-1}. What happens?

Now Try Exercises 1 and 3

We can use determinants of 2×2 matrices to find determinants of larger square matrices. In order to do this, we first define the concepts of a *minor* and a *cofactor*.

MINORS AND COFACTORS

The **minor**, denoted by M_{ij}, for element a_{ij} in the square matrix A is the real number computed by performing the following steps.

STEP 1: Delete the ith row and jth column from the matrix A.

STEP 2: Compute the determinant of the resulting matrix, which is equal to M_{ij}.

The **cofactor**, denoted A_{ij}, for a_{ij} is defined by $A_{ij} = (-1)^{i+j} M_{ij}$.

EXAMPLE 2 **Calculating minors and cofactors**

Find the following minors and cofactors for the matrix A.

$$A = \begin{bmatrix} 2 & -3 & 1 \\ -2 & 1 & 0 \\ 0 & -1 & 4 \end{bmatrix}$$

(a) M_{11} and M_{21} **(b)** A_{11} and A_{21}

SOLUTION

(a) To obtain the minor M_{11}, begin by crossing out the first row and first column of A.

$$A = \begin{bmatrix} 2 & -3 & 1 \\ -2 & 1 & 0 \\ 0 & -1 & 4 \end{bmatrix}$$

For M_{11}, cross out row **1** and column **1**.

The remaining elements form the 2×2 matrix

$$B = \begin{bmatrix} 1 & 0 \\ -1 & 4 \end{bmatrix}.$$

The minor M_{11} is equal to $\det B = (1)(4) - (-1)(0) = 4$.

M_{21} is found by crossing out the second row and first column of A.

$$A = \begin{bmatrix} 2 & -3 & 1 \\ -2 & 1 & 0 \\ 0 & -1 & 4 \end{bmatrix}$$

For M_{21}, cross out row **2** and column **1**.

The resulting matrix is

$$B = \begin{bmatrix} -3 & 1 \\ -1 & 4 \end{bmatrix}.$$

Thus $M_{21} = \det B = (-3)(4) - (-1)(1) = -11$.

(b) Since $A_{ij} = (-1)^{i+j} M_{ij}$, A_{11} and A_{21} can be computed as follows.

$$A_{11} = (-1)^{1+1} M_{11} = (-1)^2 (4) = 4$$
$$A_{21} = (-1)^{2+1} M_{21} = (-1)^3 (-11) = 11$$

Now Try Exercise 5

Using the concept of a cofactor, we can calculate the determinant of *any* square matrix.

> **DETERMINANT OF A SQUARE MATRIX USING COFACTORS**
>
> For a square matrix A, multiply each element in any row or column of the matrix by its cofactor. The sum of the products is equal to the determinant of A.

To compute the determinant of a 3×3 matrix A, begin by selecting either a row or a column.

$$A = \begin{bmatrix} a_{11} & a_{12} & a_{13} \\ a_{21} & a_{22} & a_{23} \\ a_{31} & a_{32} & a_{33} \end{bmatrix}$$

For example, if the *second row* of A is selected, the elements are a_{21}, a_{22}, and a_{23}. Then

$$\det A = a_{21}A_{21} + a_{22}A_{22} + a_{23}A_{23}.$$

On the other hand, utilizing the elements of a_{11}, a_{21}, and a_{31} in the *first column* gives

$$\det A = a_{11}A_{11} + a_{21}A_{21} + a_{31}A_{31}.$$

Regardless of the row or column selected, the value of $\det A$ is the same. The calculation is easier if some elements in the selected row or column equal 0.

EXAMPLE 3 Evaluating the determinant of a 3×3 matrix

Find $\det A$ if

$$A = \begin{bmatrix} 2 & -3 & 1 \\ -2 & 1 & 0 \\ 0 & -1 & 4 \end{bmatrix}.$$

SOLUTION To find the determinant of A, we can select any row or column. If we begin *expanding* about the first column of A, then

$$\det A = a_{11}A_{11} + a_{21}A_{21} + a_{31}A_{31}.$$

In the first column, $a_{11} = 2$, $a_{21} = -2$, and $a_{31} = 0$. In Example 2, the cofactors A_{11} and A_{21} were computed as 4 and 11, respectively. Since A_{31} is multiplied by $a_{31} = 0$, we do not need to calculate its value. Thus

$$\begin{aligned}
\det &= a_{11}A_{11} + a_{21}A_{21} + a_{31}A_{31} \\
&= 2(4) + (-2)(11) + (0)A_{31} \qquad \text{First column of A} \\
&= -14.
\end{aligned}$$

We could also have expanded about the second row.

$$\begin{aligned}
\det A &= a_{21}A_{21} + a_{22}A_{22} + a_{23}A_{23} \\
&= (-2)A_{21} + (1)A_{22} + (0)A_{23} \qquad \text{Second row of A}
\end{aligned}$$

CLASS DISCUSSION

If a row or column in matrix A contains only zeros, what is $\det A$?

To complete this computation we need to determine only A_{22}, since A_{21} is known to be 11 and A_{23} is multiplied by 0. To compute A_{22}, delete the second row and column of A to obtain M_{22}.

$$M_{22} = \det \begin{bmatrix} 2 & 1 \\ 0 & 4 \end{bmatrix} = 8 \qquad \text{and} \qquad A_{22} = (-1)^{2+2}(8) = 8$$

Thus $\det A = (-2)(11) + (1)(8) + (0)A_{23} = -14$. The same value for $\det A$ is obtained in both calculations.

Now Try Exercise 17

Instead of calculating $(-1)^{i+j}$ for each cofactor, we can use the following **sign matrix** to find determinants of 3×3 matrices. The checkerboard pattern can be expanded to include larger square matrices.

Sign Matrix

$$\begin{bmatrix} + & - & + \\ - & + & - \\ + & - & + \end{bmatrix}$$

For example, if

$$A = \begin{bmatrix} 2 & 3 & 7 \\ -3 & -2 & -1 \\ 4 & 0 & 2 \end{bmatrix},$$

we can compute det A by expanding about the second column to take advantage of the 0. The second column contains $-$, $+$, and $-$ signs. Therefore

$$\det A = -(3) \det \begin{bmatrix} -3 & -1 \\ 4 & 2 \end{bmatrix} + (-2) \det \begin{bmatrix} 2 & 7 \\ 4 & 2 \end{bmatrix} - (0) \det \begin{bmatrix} 2 & 7 \\ -3 & -1 \end{bmatrix}$$

$$= -3(-2) + (-2)(-24) - (0)(19)$$

$$= 54.$$

NOTE We could have computed det A by expanding about *any* row or column. However, computation can be simplified by taking advantage of any 0's in the matrix. For the matrix above, the 0 can be used by expanding about either the second column or the third row.

Graphing calculators can evaluate determinants, as shown in the next example.

EXAMPLE 4 Using technology to find a determinant

Find the determinant of A.

(a) $A = \begin{bmatrix} 2 & -3 & 1 \\ -2 & 1 & 0 \\ 0 & -1 & 4 \end{bmatrix}$ (b) $A = \begin{bmatrix} 2 & -3 & 1 & 5 \\ 7 & 1 & -8 & 0 \\ 5 & 4 & 9 & 7 \\ -2 & 3 & 3 & 0 \end{bmatrix}$

SOLUTION

(a) The determinant of this matrix was calculated in Example 3 by hand. To use technology, enter the matrix and evaluate its determinant, as shown in Figure 9.68. The result is det $A = -14$, which agrees with our earlier calculation.

Calculator Help

To calculate a determinant, see Appendix A (page AP-15).

Finding Determinants with a Calculator

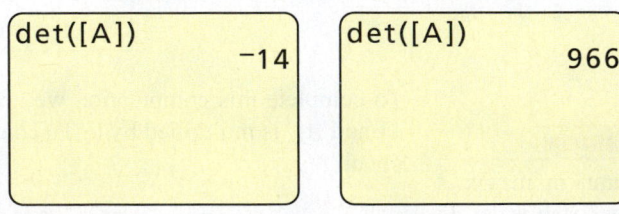

Figure 9.68   Figure 9.69

(b) The determinant of a 4×4 matrix can be computed using cofactors. However, it is considerably easier to use technology. From Figure 9.69 we see that det $A = 966$.

Now Try Exercises 23 and 24

Cramer's Rule

We can solve *linear* systems in two variables using determinants and a method called **Cramer's rule.** Cramer's rule for linear systems in *three* variables is discussed in the Extended and Discovery Exercises at the end of this section. Although Cramer's rule can be used to solve linear systems with more than three variables, it is not practical to do so.

> ### CRAMER'S RULE FOR LINEAR SYSTEMS IN TWO VARIABLES
>
> The solution to the linear system
>
> $$a_1 x + b_1 y = c_1$$
> $$a_2 x + b_2 y = c_2$$
>
> is given by $x = \frac{E}{D}$ and $y = \frac{F}{D}$, where
>
> $$E = \det \begin{bmatrix} c_1 & b_1 \\ c_2 & b_2 \end{bmatrix}, \quad F = \det \begin{bmatrix} a_1 & c_1 \\ a_2 & c_2 \end{bmatrix}, \quad \text{and} \quad D = \det \begin{bmatrix} a_1 & b_1 \\ a_2 & b_2 \end{bmatrix} \neq 0.$$

NOTE If $D = 0$, then the system does not have a unique solution. There are either no solutions or infinitely many solutions.

EXAMPLE 5 Using Cramer's rule to solve a linear system in two variables

Use Cramer's rule to solve the linear system

$$4x + y = 146$$
$$9x + y = 66.$$

SOLUTION In this system $a_1 = 4$, $b_1 = 1$, $c_1 = 146$, $a_2 = 9$, $b_2 = 1$, and $c_2 = 66$. By Cramer's rule, the solution can be found as follows.

$$E = \det \begin{bmatrix} c_1 & b_1 \\ c_2 & b_2 \end{bmatrix} = \det \begin{bmatrix} 146 & 1 \\ 66 & 1 \end{bmatrix} = (146)(1) - (66)(1) = \mathbf{80}$$

$$F = \det \begin{bmatrix} a_1 & c_1 \\ a_2 & c_2 \end{bmatrix} = \det \begin{bmatrix} 4 & 146 \\ 9 & 66 \end{bmatrix} = (4)(66) - (9)(146) = \mathbf{-1050}$$

$$D = \det \begin{bmatrix} a_1 & b_1 \\ a_2 & b_2 \end{bmatrix} = \det \begin{bmatrix} 4 & 1 \\ 9 & 1 \end{bmatrix} = (4)(1) - (9)(1) = \mathbf{-5}$$

The solution is

$$x = \frac{E}{D} = \frac{80}{-5} = -16 \quad \text{and} \quad y = \frac{F}{D} = \frac{-1050}{-5} = 210.$$

Now Try Exercise 25

Area of Regions

Determinants may be used to find the area of a triangle. If a triangle has vertices (a_1, b_1), (a_2, b_2), and (a_3, b_3), as shown in Figure 9.70, then its area is equal to the *absolute value* of D, where

$$D = \frac{1}{2} \det \begin{bmatrix} a_1 & a_2 & a_3 \\ b_1 & b_2 & b_3 \\ 1 & 1 & 1 \end{bmatrix}.$$

If the vertices are entered into the columns of D in a *counterclockwise* direction, then D will be positive. (***Source:*** W. Taylor, *The Geometry of Computer Graphics.*)

Triangular Region

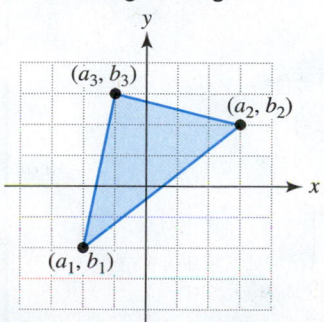

Figure 9.70

EXAMPLE 6 Computing the area of a parallelogram

Use determinants to calculate the area of the parallelogram in Figure 9.71.

SOLUTION To find the area of the parallelogram, we view the parallelogram as comprising two triangles. One triangle has vertices at $(\mathbf{0}, \mathbf{0})$, $(\mathbf{4}, \mathbf{2})$, and $(\mathbf{1}, \mathbf{2})$, and the other triangle has vertices at $(4, 2)$, $(5, 4)$, and $(1, 2)$. The area of the parallelogram is equal to the sum of the areas of the two triangles. Since these triangles are congruent, we can calculate the area of one triangle and double it. The area of one triangle is equal to D.

$$D = \frac{1}{2}\det\begin{bmatrix} \mathbf{0} & \mathbf{4} & \mathbf{1} \\ \mathbf{0} & \mathbf{2} & \mathbf{2} \\ 1 & 1 & 1 \end{bmatrix} = \frac{1}{2}(6) = 3$$

Since the vertices were entered in a counterclockwise direction, D is positive. The area of one triangle is equal to 3 square units. Therefore the area of the parallelogram is twice this value, or 6 square units.

Now Try Exercise 35

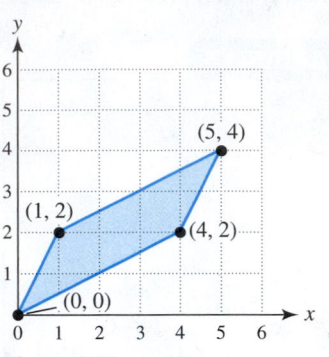

Figure 9.71

CLASS DISCUSSION

Suppose we are given three distinct vertices and $D = 0$. What must be true about the three points?

9.7 Putting It All Together

The determinant of a square matrix A is a real number, denoted det A. If det $A \neq 0$, then the matrix A is invertible. The following table summarizes the calculation of 2×2 and 3×3 determinants by hand.

DETERMINANTS OF 2 × 2 MATRICES

The determinant of a 2×2 matrix A is given by

$$\det A = \det\begin{bmatrix} a & b \\ c & d \end{bmatrix} = ad - cb.$$

Example: $\det\begin{bmatrix} 6 & -2 \\ 3 & 7 \end{bmatrix} = (6)(7) - (3)(-2) = 48$

DETERMINANTS OF 3 × 3 MATRICES

Finding the determinant of a 3×3 matrix A can be reduced to calculating the determinants of three 2×2 matrices. This calculation can be performed using cofactors.

$$\det A = \det\begin{bmatrix} a_1 & b_1 & c_1 \\ a_2 & b_2 & c_2 \\ a_3 & b_3 & c_3 \end{bmatrix}$$

$$= a_1\det\begin{bmatrix} b_2 & c_2 \\ b_3 & c_3 \end{bmatrix} - a_2\det\begin{bmatrix} b_1 & c_1 \\ b_3 & c_3 \end{bmatrix} + a_3\det\begin{bmatrix} b_1 & c_1 \\ b_2 & c_2 \end{bmatrix}$$

Example: $\det\begin{bmatrix} 1 & -2 & 3 \\ 4 & 5 & -1 \\ -3 & 7 & 8 \end{bmatrix} = (1)\det\begin{bmatrix} 5 & -1 \\ 7 & 8 \end{bmatrix} - (4)\det\begin{bmatrix} -2 & 3 \\ 7 & 8 \end{bmatrix} + (-3)\det\begin{bmatrix} -2 & 3 \\ 5 & -1 \end{bmatrix}$

$$= (1)(47) - 4(-37) - 3(-13) = 234$$

9.7 Exercises

Calculating Determinants

Exercises 1–4: Determine if the matrix A is invertible by calculating det A.

1. $A = \begin{bmatrix} 4 & 3 \\ 5 & 4 \end{bmatrix}$

2. $A = \begin{bmatrix} 1 & -3 \\ 2 & 6 \end{bmatrix}$

3. $A = \begin{bmatrix} -4 & 6 \\ -8 & 12 \end{bmatrix}$

4. $A = \begin{bmatrix} 10 & -20 \\ -5 & 10 \end{bmatrix}$

Exercises 5–8: Find the specified minor and cofactor for A.

5. M_{12} and A_{12} if $A = \begin{bmatrix} 1 & -1 & 3 \\ 2 & 3 & -2 \\ 0 & 1 & 5 \end{bmatrix}$

6. M_{23} and A_{23} if $A = \begin{bmatrix} 1 & 2 & -1 \\ 4 & 6 & -3 \\ 2 & 3 & 9 \end{bmatrix}$

7. M_{22} and A_{22} if $A = \begin{bmatrix} 7 & -8 & 1 \\ 3 & -5 & 2 \\ 1 & 0 & -2 \end{bmatrix}$

8. M_{31} and A_{31} if $A = \begin{bmatrix} 0 & 0 & -1 \\ 6 & -7 & 1 \\ 8 & -9 & -1 \end{bmatrix}$

Exercises 9–12: Let A be the given matrix. Find det A by expanding about the first column. State whether A^{-1} exists.

9. $\begin{bmatrix} 1 & 4 & -7 \\ 0 & 2 & -3 \\ 0 & -1 & 3 \end{bmatrix}$

10. $\begin{bmatrix} 0 & 2 & 8 \\ -1 & 3 & 5 \\ 0 & 4 & 1 \end{bmatrix}$

11. $\begin{bmatrix} 5 & 1 & 6 \\ 0 & -2 & 0 \\ 0 & 4 & 0 \end{bmatrix}$

12. $\begin{bmatrix} 3 & 2 & 3 \\ 2 & 2 & 2 \\ 1 & 3 & 1 \end{bmatrix}$

Exercises 13–20: Let A be the given matrix. Find det A by using the method of cofactors.

13. $\begin{bmatrix} 2 & 0 & 0 \\ 0 & 3 & 0 \\ 0 & 0 & 5 \end{bmatrix}$

14. $\begin{bmatrix} 0 & 0 & 2 \\ 0 & 3 & 0 \\ 5 & 0 & 0 \end{bmatrix}$

15. $\begin{bmatrix} 0 & 0 & 0 \\ -8 & 3 & -9 \\ 15 & 5 & 9 \end{bmatrix}$

16. $\begin{bmatrix} 1 & 1 & 5 \\ -3 & -3 & 0 \\ 7 & 0 & 0 \end{bmatrix}$

17. $\begin{bmatrix} 3 & -1 & 2 \\ 0 & 5 & 7 \\ 1 & 0 & -1 \end{bmatrix}$

18. $\begin{bmatrix} 3 & 0 & -1 \\ 2 & 3 & -4 \\ 6 & -5 & 1 \end{bmatrix}$

19. $\begin{bmatrix} 1 & -5 & 2 \\ -7 & 1 & 3 \\ 0 & 4 & -2 \end{bmatrix}$

20. $\begin{bmatrix} 1 & -1 & 2 \\ -2 & 0 & 1 \\ 1 & 1 & -1 \end{bmatrix}$

Exercises 21–24: Let A be the given matrix. Use technology to calculate det A.

21. $\begin{bmatrix} 11 & -32 \\ 1.2 & 55 \end{bmatrix}$

22. $\begin{bmatrix} 17 & -4 & 3 \\ 11 & 5 & -15 \\ 7 & -9 & 23 \end{bmatrix}$

23. $\begin{bmatrix} 2.3 & 5.1 & 2.8 \\ 1.2 & 4.5 & 8.8 \\ -0.4 & -0.8 & -1.2 \end{bmatrix}$

24. $\begin{bmatrix} 1 & -1 & 3 & 7 \\ 9 & 2 & -7 & -4 \\ 5 & -7 & 1 & -9 \\ 7 & 1 & 3 & 6 \end{bmatrix}$

Cramer's Rule

Exercises 25–32: Use Cramer's rule to solve the system of linear equations.

25. $\begin{aligned} -x + 2y &= 5 \\ 3x + 3y &= 1 \end{aligned}$

26. $\begin{aligned} 2x + y &= -3 \\ -4x - 6y &= -7 \end{aligned}$

27. $\begin{aligned} -2x + 3y &= 8 \\ 4x - 5y &= 3 \end{aligned}$

28. $\begin{aligned} 5x - 3y &= 4 \\ -3x - 7y &= 5 \end{aligned}$

29. $\begin{aligned} 7x + 4y &= 23 \\ 11x - 5y &= 70 \end{aligned}$

30. $\begin{aligned} -7x + 5y &= 8.2 \\ 6x + 4y &= -0.4 \end{aligned}$

31. $\begin{aligned} 1.7x - 2.5y &= -0.91 \\ -0.4x + 0.9y &= 0.423 \end{aligned}$

32. $\begin{aligned} -2.7x + 1.5y &= -1.53 \\ 1.8x - 5.5y &= -1.68 \end{aligned}$

Calculating Area

Exercises 33–36: Use a determinant to find the area of the shaded region.

33.

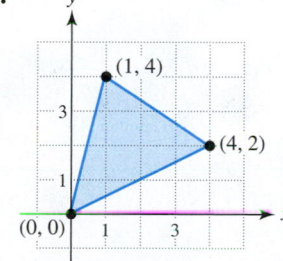

34.

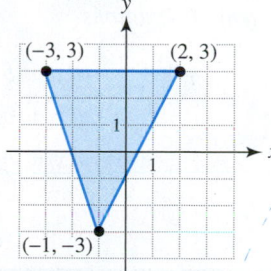

35.

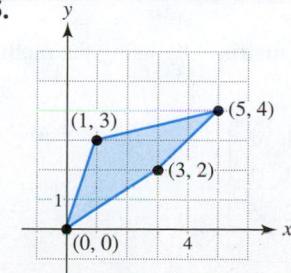

36.

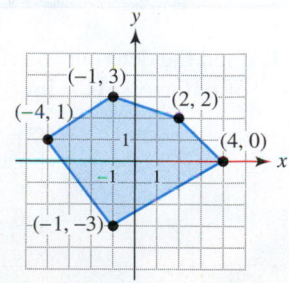

Applying a Concept

Exercises 37–40: Use the concept of the area of a triangle to determine if the three points are collinear.

37. $(1, 3), (-3, 11), (2, 1)$

38. $(3, 6), (-1, -6), (5, 11)$

39. $(-2, -5), (4, 4), (2, 3)$

40. $(4, -5), (-2, 10), (6, -10)$

Equations of Lines

Exercises 41–44: If a line passes through the points (x_1, y_1) and (x_2, y_2), then an equation of this line can be found by calculating the determinant.

$$\det \begin{bmatrix} x & y & 1 \\ x_1 & y_1 & 1 \\ x_2 & y_2 & 1 \end{bmatrix} = 0$$

Find the standard form $ax + by = c$ of the line passing through the given points.

41. $(2, 1)$ and $(-1, 4)$ **42.** $(-1, 3)$ and $(4, 2)$

43. $(6, -7)$ and $(4, -3)$ **44.** $(5, 1)$ and $(2, -2)$

Writing about Mathematics

45. Choose two matrices A and B with dimension 2×2. Calculate det A, det B, and det (AB). Repeat this process until you are able to discover how these three determinants are related. Summarize your results.

46. Calculate both det A and det A^{-1} for several different matrices. Compare the determinants. Try to generalize your results.

Extended and Discovery Exercises

Exercises 1–6: **Cramer's Rule**   *Cramer's rule can be applied to systems of three linear equations in three variables. For the system of equations*

$$a_1 x + b_1 y + c_1 z = d_1$$
$$a_2 x + b_2 y + c_2 z = d_2$$
$$a_3 x + b_3 y + c_3 z = d_3,$$

the solution can be written as follows.

$$D = \det \begin{bmatrix} a_1 & b_1 & c_1 \\ a_2 & b_2 & c_2 \\ a_3 & b_3 & c_3 \end{bmatrix}, \quad E = \det \begin{bmatrix} d_1 & b_1 & c_1 \\ d_2 & b_2 & c_2 \\ d_3 & b_3 & c_3 \end{bmatrix}$$

$$F = \det \begin{bmatrix} a_1 & d_1 & c_1 \\ a_2 & d_2 & c_2 \\ a_3 & d_3 & c_3 \end{bmatrix}, \quad G = \det \begin{bmatrix} a_1 & b_1 & d_1 \\ a_2 & b_2 & d_2 \\ a_3 & b_3 & d_3 \end{bmatrix}$$

If $D \neq 0$, a unique solution exists and is given by

$$x = \frac{E}{D}, \quad y = \frac{F}{D}, \quad z = \frac{G}{D}.$$

Use Cramer's rule to solve the system of equations.

1. $x + y + z = 6$
$\quad 2x + y + 2z = 9$
$\quad\quad\quad y + 3z = 9$

2. $\quad\quad y + z = 1$
$\quad 2x - y - z = -1$
$\quad x + y - z = 3$

3. $x \quad\quad + z = 2$
$\quad x + y \quad\quad = 0$
$\quad\quad\quad y + 2z = 1$

4. $x + y + 2z = 1$
$\quad -x - 2y - 3z = -2$
$\quad\quad\quad y - 3z = 5$

5. $x \quad\quad + 2z = 7$
$\quad -x + y + z = 5$
$\quad 2x - y + 2z = 6$

6. $x + 2y + 3z = -1$
$\quad 2x - 3y - z = 12$
$\quad x + 4y - 2z = -12$

Exercises 7–10: **Equations of Circles**   *Given three distinct points on a circle (x_1, y_1), (x_2, y_2), and (x_3, y_3), we can find the equation of the circle by using the following determinant.*

$$\det \begin{bmatrix} x^2 + y^2 & x & y & 1 \\ x_1^2 + y_1^2 & x_1 & y_1 & 1 \\ x_2^2 + y_2^2 & x_2 & y_2 & 1 \\ x_3^2 + y_3^2 & x_3 & y_3 & 1 \end{bmatrix} = 0$$

Find the equation of the circle through the given points.

7. $(0, 2)$ $(2, 0)$, and $(-2, 0)$

8. $(0, 0), (4, 0)$, and $(2, -2)$

9. $(0, 1), (1, -1)$, and $(2, 2)$

10. $(1, 0), (-1, 2)$, and $(3, 2)$

CHECKING BASIC CONCEPTS FOR SECTION 9.7

1. Find the determinant of the matrix A by using the method of cofactors. Is A invertible?

$$A = \begin{bmatrix} 1 & -1 & 2 \\ 2 & 3 & 1 \\ 0 & -2 & 5 \end{bmatrix}$$

2. Use Cramer's rule to solve the system of equations.
$$3x - 4y = 7$$
$$-4x + 3y = 5$$

9 Summary

CONCEPT	EXPLANATION AND EXAMPLES

Section 9.1 Functions and Systems of Equations in Two Variables

Functions of Two Variables

$z = f(x, y)$, where x and y are inputs to f

Example: $f(x, y) = 2x - 3y$
$f(4, -1) = 2(4) - 3(-1) = 11$

System of Linear Equations in Two Variables

General form: $a_1x + b_1y = c_1$
                  $a_2x + b_2y = c_2$

A linear system can have zero, one, or infinitely many solutions. A solution can be written as an ordered pair. A linear system may be solved symbolically, graphically, or numerically.

Example: $x - y = 2$
               $2x + y = 7$ Solution: $(3, 1)$

Types of Linear Systems with Two Variables

Consistent system: Has either one solution (independent equations) or infinitely many solutions (dependent equations)

Inconsistent system: Has no solutions

One Solution

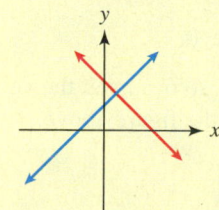

Consistent System
Independent Equations

Infinitely Many Solutions

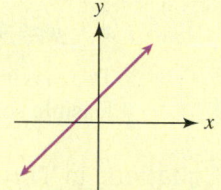

Consistent System
Dependent Equations

No Solution

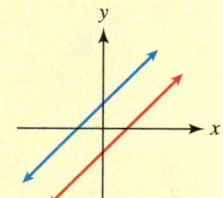

Inconsistent System

Method of Substitution for Two Equations

Can be used to solve systems of linear or nonlinear equations

Example: $x - y = -3$
               $x + 4y = 17$

Solve the first equation for x to obtain $x = y - 3$. Substitute this result in the second equation and solve for y.

$$(y - 3) + 4y = 17 \quad \text{implies that} \quad y = 4.$$

Then $x = 4 - 3 = 1$ and the solution is $(1, 4)$.

Method of Elimination

Can be used to solve systems of linear or nonlinear equations

Example: $2x - 3y = 4$
               $x + 3y = 11$
               $\overline{3x \qquad = 15}$,   or $x = 5$ Add.

Substituting $x = 5$ in the first equation gives $y = 2$.
The solution is $(5, 2)$.

CONCEPT	EXPLANATION AND EXAMPLES

Section 9.1 Functions and Systems of Equations in Two Variables (CONTINUED)

Joint Variation

Let m and n be real numbers. Then z *varies jointly* with the mth power of x and the nth power of y if a nonzero real number k exists such that $z = kx^m y^n$.

Example: The area of a triangle varies jointly with the base b and the height h because $A = \frac{1}{2}bh$. Note that $k = \frac{1}{2}$, $m = 1$, and $n = 1$ in this example.

Section 9.2 Systems of Inequalities in Two Variables

System of Inequalities in Two Variables

The solution set is often a shaded region in the xy-plane.

Example: $x + y \leq 4$
$y \geq 0$
$x \geq 0$

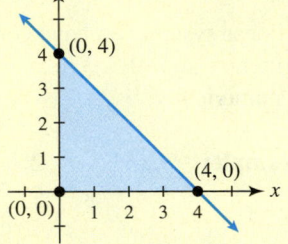

Linear Programming

Method for maximizing (or minimizing) an objective function subject to a set of constraints

Example: Maximize $P = 2x + 4y$, subject to

$x + y \leq 4, x \geq 0, y \geq 0$.

The maximum of $P = 16$ occurs at the vertex $(0, 4)$, in the region of feasible solutions. See the figure above.

Section 9.3 Systems of Linear Equations in Three Variables

Solution to a System of Linear Equations in Three Variables

An ordered triple (x, y, z) that satisfies *every* equation

Example: $x - 2y + 3z = 6$
$-x + 3y + 4z = 17$
$3x + 4y - 5z = -4$

The solution is $(1, 2, 3)$ because the values $x = 1$, $y = 2$, and $z = 3$ satisfy all three equations. (Check this fact.)

Elimination and Substitution

Systems of linear equations in three variables can be solved by using elimination and substitution. The following three steps outline this process.

STEP 1: Eliminate one variable, such as x, from two of the equations.

STEP 2: Apply the techniques discussed in Section 9.1 to solve the two equations in two variables resulting from Step 1. If x is eliminated, then solve these equations to find y and z.

If there are no solutions for y and z, then the given system has no solutions. If there are infinitely many solutions for y and z, then write y in terms of z and go to Step 3.

STEP 3: Substitute the values for y and z in one of the given equations to find x. The solution is (x, y, z). If possible, check your solution.

CONCEPT	EXPLANATION AND EXAMPLES

Section 9.4 Solutions to Linear Systems Using Matrices

Matrices and Systems of Linear Equations

An augmented matrix can be used to represent a system of linear equations.

Example:

$$\begin{array}{rcl} x - 2y + z &=& 0 \\ -x + 4y - z &=& 4 \\ 2x + y - 3z &=& -5 \end{array}$$

Augmented Matrix

$$\left[\begin{array}{ccc|c} 1 & -2 & 1 & 0 \\ -1 & 4 & -1 & 4 \\ 2 & 1 & -3 & -5 \end{array}\right]$$

Row-Echelon Form

Examples:

$$\begin{bmatrix} 1 & 2 & -1 \\ 0 & 1 & 2 \end{bmatrix} \qquad \left[\begin{array}{ccc|c} 1 & 3 & -2 & 7 \\ 0 & 1 & 4 & 5 \\ 0 & 0 & 1 & -3 \end{array}\right]$$

Gaussian Elimination with Backward Substitution

Gaussian elimination can be used to transform a matrix representing a system of linear equations into row-echelon form. Then backward substitution can be used to solve the resulting system of linear equations. (Graphing calculators can also be used to solve systems of equations.)

Section 9.5 Properties and Applications of Matrices

Operations on Matrices

Matrices can be added, subtracted, and multiplied, but there is *no* division of matrices.

Addition

$$\begin{bmatrix} 2 & 4 \\ 5 & 6 \end{bmatrix} + \begin{bmatrix} -2 & 1 \\ 7 & 3 \end{bmatrix} = \begin{bmatrix} 0 & 5 \\ 12 & 9 \end{bmatrix}$$

Subtraction

$$\begin{bmatrix} -3 & 0 \\ 4 & -4 \end{bmatrix} - \begin{bmatrix} 1 & 2 \\ 6 & -7 \end{bmatrix} = \begin{bmatrix} -4 & -2 \\ -2 & 3 \end{bmatrix}$$

Scalar Multiplication

$$3\begin{bmatrix} 5 & 1 & 6 & -1 \\ 0 & -2 & 3 & 2 \end{bmatrix} = \begin{bmatrix} 15 & 3 & 18 & -3 \\ 0 & -6 & 9 & 6 \end{bmatrix}$$

Multiplication

$$\begin{bmatrix} 2 & -1 \\ 0 & 3 \\ -7 & 1 \end{bmatrix}\begin{bmatrix} 1 & -1 & 0 \\ 3 & -5 & -4 \end{bmatrix} = \begin{bmatrix} -1 & 3 & 4 \\ 9 & -15 & -12 \\ -4 & 2 & -4 \end{bmatrix}$$

Section 9.6 Inverses of Matrices

Matrix Inverses

The inverse of an $n \times n$ matrix A, denoted A^{-1}, satisfies $A^{-1}A = I_n$ and $AA^{-1} = I_n$, where I_n is the $n \times n$ identity matrix. The inverse of a matrix can be found by hand or with technology.

Example: $A = \begin{bmatrix} 5 & 2 \\ 2 & 1 \end{bmatrix}$ and $A^{-1} = \begin{bmatrix} 1 & -2 \\ -2 & 5 \end{bmatrix}$

$$\begin{bmatrix} 5 & 2 \\ 2 & 1 \end{bmatrix}\begin{bmatrix} 1 & -2 \\ -2 & 5 \end{bmatrix} = \begin{bmatrix} 1 & 0 \\ 0 & 1 \end{bmatrix} = I_2$$

$$\begin{bmatrix} 1 & -2 \\ -2 & 5 \end{bmatrix}\begin{bmatrix} 5 & 2 \\ 2 & 1 \end{bmatrix} = \begin{bmatrix} 1 & 0 \\ 0 & 1 \end{bmatrix} = I_2$$

CONCEPT	EXPLANATION AND EXAMPLES
	Section 9.6 Inverses of Matrices (CONTINUED)
Matrix Equations	A system of linear equations can be written as the matrix equation $AX = B$. **Example:** $2x - 2y = 3$ $-3x + 4y = 2$ $$AX = \begin{bmatrix} 2 & -2 \\ -3 & 4 \end{bmatrix} \begin{bmatrix} x \\ y \end{bmatrix} = \begin{bmatrix} 3 \\ 2 \end{bmatrix} = B$$ The solution can be found as follows. $$X = A^{-1}B = \begin{bmatrix} 2 & 1 \\ 1.5 & 1 \end{bmatrix} \begin{bmatrix} 3 \\ 2 \end{bmatrix} = \begin{bmatrix} 8 \\ 6.5 \end{bmatrix}$$ The solution is $(8, 6.5)$.
	Section 9.7 Determinants
Determinant of a 2 × 2 Matrix	$$\det A = \det \begin{bmatrix} a & b \\ c & d \end{bmatrix} = ad - cb$$ **Example:** $\det \begin{bmatrix} 1 & 4 \\ 3 & 5 \end{bmatrix} = (1)(5) - (3)(4) = -7$
Determinant of a 3 × 3 Matrix	Finding a 3 × 3 determinant can be reduced to calculating the determinants of three 2 × 2 matrices by using cofactors. If $\det A \neq 0$, then A^{-1} exists. **Example:** $\det \begin{bmatrix} 3 & 1 & -1 \\ 2 & 2 & 0 \\ 0 & 1 & -3 \end{bmatrix}$ $$= 3\begin{bmatrix} 2 & 0 \\ 1 & -3 \end{bmatrix} - 2\begin{bmatrix} 1 & -1 \\ 1 & -3 \end{bmatrix} + 0\begin{bmatrix} 1 & -1 \\ 2 & 0 \end{bmatrix}$$ $$= 3(-6) - 2(-2) + 0(2)$$ $$= -14$$
Cramer's Rule	Cramer's rule makes use of determinants to solve systems of linear equations. However, Gaussian elimination with backward substitution is usually more efficient.

9 Review Exercises

Exercises 1 and 2: Evaluate the function for the inputs.

1. $A(3, 6)$, where $A(b, h) = \frac{1}{2}bh$

2. $V(2, 5)$, where $V(r, h) = \pi r^2 h$

Exercises 3 and 4: The figure in the next column shows the graph of a system of two linear equations. Use the graph to estimate the solution to the system of equations. Then solve the system symbolically.

3.

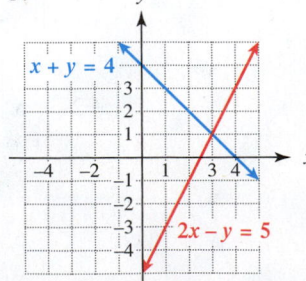

4.

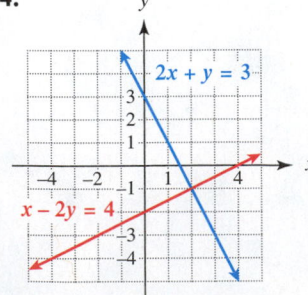

Exercises 5 and 6: Solve the system of equations

(a) *graphically and* (b) *symbolically.*

5. $3x + y = 1$
$2x - 3y = 8$

6. $x^2 - y = 1$
$x + y = 1$

Exercises 7–10: Use the elimination method to solve each system of linear equations, if possible. Identify the system as consistent or inconsistent.

7. $2x + y = 7$
$x - 2y = -4$

8. $3x + 3y = 15$
$-x - y = -4$

9. $6x - 15y = 12$
$-4x + 10y = -8$

10. $3x - 4y = -10$
$4x + 3y = -30$

Exercises 11 and 12: Use elimination to solve the nonlinear system of equations.

11. $x^2 - 3y = 3$
$x^2 + 2y^2 = 5$

12. $2x - 3y = 1$
$2x^2 + y = 1$

Exercises 13 and 14: Graph the solution set to the inequality.

13. $y \geq -1$

14. $2x - y < 4$

Exercises 15 and 16: Graph the solution set to the system of inequalities. Use the graph to find one solution.

15. $x^2 + y^2 < 9$
$x + y > 3$

16. $x + 3y \geq 3$
$x + y \leq 4$

Exercises 17–20: If possible, solve the system.

17. $x - y + z = -2$
$x + 2y - z = 2$
$2y + 3z = 7$

18. $x - 3y + 2z = -10$
$2x - y + 3z = -9$
$-x - y + z = -1$

19. $-x + 2y + 2z = 9$
$x + y - 3z = 6$
$3y - z = 8$

20. $2x - y - 3z = -9$
$x - 8z = -23$
$-3x + 2y - 2z = -5$

Exercises 21–24: The augmented matrix represents a system of linear equations. Solve the system.

21. $\begin{bmatrix} 1 & 5 & | & 6 \\ 0 & 1 & | & 3 \end{bmatrix}$

22. $\begin{bmatrix} 1 & 2 & -2 & | & 8 \\ 0 & 1 & 1 & | & 5 \\ 0 & 0 & 0 & | & 0 \end{bmatrix}$

23. $\begin{bmatrix} 1 & 0 & 0 & | & -2 \\ 0 & 1 & 0 & | & 3 \\ 0 & 0 & 1 & | & 0 \end{bmatrix}$

24. $\begin{bmatrix} 1 & 0 & 0 & | & -5 \\ 0 & 1 & -4 & | & 1 \\ 0 & 0 & 0 & | & 5 \end{bmatrix}$

Exercises 25 and 26: Use Gaussian elimination with backward substitution to solve the system of linear equations.

25. $2x - y + 2z = 10$
$x - 2y + z = 8$
$3x - y + 2z = 11$

26. $x - 2y + z = 1$
$2x - 5y + 3z = 4$
$2x - 3y + z = 0$

Exercises 27 and 28: Let a_{ij} denote the general term for the matrix A. Find each of the following.

(a) $a_{12} + a_{22}$ (b) $a_{11} - 2a_{23}$

27. $A = \begin{bmatrix} -2 & 3 & -1 \\ 5 & 2 & 4 \end{bmatrix}$

28. $\begin{bmatrix} -1 & 2 & 5 \\ 1 & -3 & 7 \\ 0 & 7 & -2 \end{bmatrix}$

Exercises 29 and 30: Evaluate the following.

(a) $A + 2B$ (b) $A - B$ (c) $-4A$

29. $A = \begin{bmatrix} 1 & -3 \\ 2 & -1 \end{bmatrix}$, $B = \begin{bmatrix} 3 & 2 \\ -5 & 1 \end{bmatrix}$

30. $A = \begin{bmatrix} 4 & 0 & 1 \\ -2 & 8 & 9 \end{bmatrix}$, $B = \begin{bmatrix} -5 & 3 & 2 \\ -4 & 0 & 7 \end{bmatrix}$

Exercises 31–34: If possible, find AB and BA.

31. $A = \begin{bmatrix} 2 & 0 \\ -5 & 3 \end{bmatrix}$, $B = \begin{bmatrix} -1 & -2 \\ 4 & 7 \end{bmatrix}$

32. $A = \begin{bmatrix} 1 & -2 \\ 2 & 3 \end{bmatrix}$, $B = \begin{bmatrix} 1 & 0 & 2 \\ -1 & 3 & 4 \end{bmatrix}$

33. $A = \begin{bmatrix} 2 & -1 & 3 \\ 2 & 4 & 0 \end{bmatrix}$, $B = \begin{bmatrix} 1 & 0 \\ -1 & 2 \\ 0 & 3 \end{bmatrix}$

34. $A = \begin{bmatrix} 1 & -1 & 2 \\ 0 & 3 & 4 \\ 1 & 0 & 2 \end{bmatrix}$, $B = \begin{bmatrix} -1 & 0 & 0 \\ 2 & 0 & -1 \\ 1 & 4 & 2 \end{bmatrix}$

Exercises 35 and 36: Determine if B is the inverse matrix of A by evaluating AB and BA.

35. $A = \begin{bmatrix} 8 & 5 \\ 6 & 4 \end{bmatrix}$, $B = \begin{bmatrix} 2 & -2.5 \\ -3 & 4 \end{bmatrix}$

36. $A = \begin{bmatrix} -1 & 1 & 2 \\ 1 & 0 & -1 \\ 0 & 1 & 2 \end{bmatrix}$, $B = \begin{bmatrix} -1 & 0 & 1 \\ 2 & 2 & -1 \\ -1 & -1 & -1 \end{bmatrix}$

Exercises 37 and 38: Let A be the given matrix. Find A^{-1}.

37. $\begin{bmatrix} 1 & -2 \\ -1 & 1 \end{bmatrix}$

38. $\begin{bmatrix} 1 & 0 & 1 \\ 1 & 1 & 1 \\ 0 & 1 & -1 \end{bmatrix}$

Exercises 39 and 40: Complete the following.

(a) *Write the system in the form $AX = B$.*

(b) *Solve the linear system by computing $X = A^{-1}B$.*

39. $x - 3y = 4$
$2x - y = 3$

40. $x - 2y + z = 0$
$2x + y + 2z = 10$
$y + z = 3$

41. Solve the system using technology.
$$12x + 7y - 3z = 14.6$$
$$8x - 11y + 13z = -60.4$$
$$-23x + 9z = -14.6$$

42. If possible, graphically approximate the solution of each system of equations to the nearest thousandth. Identify each system as consistent or inconsistent. If the system is consistent, determine if the equations are dependent or independent.

(a) $3.1x + 4.2y = 6.4$
  $1.7x - 9.1y = 1.6$

(b) $6.3x - 5.1y = 9.3$
  $4.2x - 3.4y = 6.2$

(c) $0.32x - 0.64y = 0.96$
  $-0.08x + 0.16y = -0.72$

Exercises 43 and 44: Let A be the given matrix. Find det A by using the method of cofactors.

43. $\begin{bmatrix} 2 & 1 & 3 \\ 0 & 3 & 4 \\ 1 & 0 & 5 \end{bmatrix}$ **44.** $\begin{bmatrix} 3 & 0 & 2 \\ 1 & 3 & 5 \\ -5 & 2 & 0 \end{bmatrix}$

Exercises 45 and 46: Let A be the given matrix. Use technology to find det A. State whether A is invertible.

45. $\begin{bmatrix} 13 & 22 \\ 55 & -57 \end{bmatrix}$ **46.** $\begin{bmatrix} 6 & -7 & -1 \\ -7 & 3 & -4 \\ 23 & 54 & 77 \end{bmatrix}$

Applications

47. Area and Perimeter Let l represent the length of a rectangle and w its width, where $l \geq w$. Then its area can be computed by $A(l, w) = lw$ and its perimeter by $P(l, w) = 2l + 2w$. Solve the system of equations determined by $A(l, w) = 77$ and $P(l, w) = 36$.

48. Cylinder Approximate the radius r and height h of a cylindrical container with a volume V of 30 cubic inches and a lateral (side) surface area S of 45 square inches.

49. Student Loans A student takes out two loans totaling $2000 to help pay for college expenses. One loan is at 7% interest, and the other is at 9%. Interest for both loans is compounded annually.
(a) If the combined total interest for the first year is $156, find the amount of each loan symbolically.

(b) Determine the amount of each loan graphically or numerically.

50. Dimensions of a Screen The screen of a rectangular television set is 3 inches wider than it is high. If the perimeter of the screen is 42 inches, find its dimensions by writing a system of linear equations and solving.

51. CD Prices A music store marks its compact discs A or B to indicate one of two selling prices. Each row in the table represents a purchase. Determine the cost of each type of CD by using a matrix inverse.

A	B	Total
1	2	$37.47
2	3	$61.95

52. Digital Photography Design a 3×3 matrix A that represents a digital photograph of the letter T in black on a white background. Find a matrix B such that adding B to A darkens only the white background by one gray level.

53. Area Use a determinant to find the area of the triangle whose vertices are $(0, 0)$, $(5, 2)$, and $(2, 5)$.

54. Voter Turnout The table shows the percent y of voter turnout in the United States for the presidential election in year x, where $x = 0$ corresponds to 1900. Find a quadratic function defined by $f(x) = ax^2 + bx + c$ that models these data. Graph f together with the data.

x	24	60	96
y	48.9	62.8	48.8

Source: Committee for the Study of the American Electorate.

55. Joint Variation Suppose P varies jointly with the square of x and the cube of y. If $P = 432$ when $x = 2$ and $y = 3$, find P when $x = 3$ and $y = 5$.

56. Linear Programming Find the maximum value of $P = 3x + 4y$ subject to the following constraints.
$$x + 3y \leq 12$$
$$3x + y \leq 12$$
$$x \geq 0, y \geq 0$$

Extended and Discovery Exercises

1. To form the **transpose** of a matrix A, denoted A^T, let the first row of A be the first column of A^T, the second row of A be the second column of A^T, and so on, for each row of A. The following are examples of A and A^T. If A has dimension $m \times n$, then A^T has dimension $n \times m$.

$$A = \begin{bmatrix} 3 & -3 & 7 \\ 1 & 6 & -2 \\ 4 & 2 & 5 \end{bmatrix}, \quad A^T = \begin{bmatrix} 3 & 1 & 4 \\ -3 & 6 & 2 \\ 7 & -2 & 5 \end{bmatrix}$$

$$A = \begin{bmatrix} 1 & 2 \\ 3 & 4 \\ 5 & 6 \end{bmatrix}, \quad A^T = \begin{bmatrix} 1 & 3 & 5 \\ 2 & 4 & 6 \end{bmatrix}$$

Find the transpose of each matrix A.

(a) $A = \begin{bmatrix} 3 & -3 \\ 2 & 6 \\ 4 & 2 \end{bmatrix}$ **(b)** $A = \begin{bmatrix} 0 & 1 & -2 \\ 2 & 5 & 4 \\ -4 & 3 & 9 \end{bmatrix}$

(c) $A = \begin{bmatrix} 5 & 7 \\ 1 & -7 \\ 6 & 3 \\ -9 & 2 \end{bmatrix}$

Exercises 2 and 3: **Least-Square Models**   *The table shows the average cost of tuition and fees y in dollars at 4-year public colleges. In this table x = 0 represents 1980 and x = 20 corresponds to 2000.*

x	0	5	10	15	20
y	804	1318	1908	2860	3487

Source: The College Board.

These data can be modeled by using linear regression. Ideally, we would like f(x) = ax + b to satisfy the following five equations.

$$f(0) = a(0) + b = 804$$
$$f(5) = a(5) + b = 1318$$
$$f(10) = a(10) + b = 1908$$
$$f(15) = a(15) + b = 2860$$
$$f(20) = a(20) + b = 3487$$

Since the data points are not collinear, it is impossible for the graph of a line to pass through all five points. These five equations can be written as

$$AX = \begin{bmatrix} 0 & 1 \\ 5 & 1 \\ 10 & 1 \\ 15 & 1 \\ 20 & 1 \end{bmatrix} \begin{bmatrix} a \\ b \end{bmatrix} = \begin{bmatrix} 804 \\ 1318 \\ 1908 \\ 2860 \\ 3487 \end{bmatrix} = B.$$

The least-squares solution is found by solving the **normal equations**

$$A^{T}AX = A^{T}B$$

for X. The solution is $X = (A^T A)^{-1} A^T B$. Using technology, we find a = 138.16 and b = 693.8. Thus f is given by the formula f(x) = 138.16x + 693.8. The function f and the data can be graphed. See the figure.

$[-5, 25, 5]$ by $[0, 4000, 1000]$

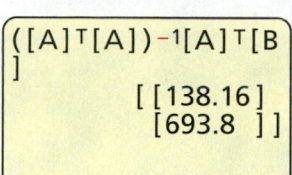

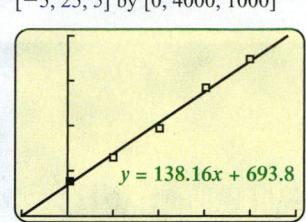

Solve the normal equations to model the data with the line determined by f(x) = ax + b. Plot the data and f in the same viewing rectangle.

2. **Tuition and Fees**   The table shows average cost of tuition and fees y in dollars at private 4-year colleges. In this table $x = 0$ corresponds to 1980 and $x = 20$ to 2000.

x	0	5	10	15	20
y	3617	6121	9340	12,216	16,233

Source: The College Board.

3. **Early Satellite TV**   The table lists the number of satellite television subscribers y in millions. In this table $x = 0$ corresponds to 1995 and $x = 5$ to the year 2000.

x	0	1	2	3	4	5
y	2.2	4.5	7.9	10.5	13	15

Source: USA Today.

10 Conic Sections

Throughout history, people have been fascinated by the universe around them and compelled to try to understand it. Conic sections have played an important role in gaining this understanding. In the sixteenth century Tycho Brahe, the greatest observational astronomer of the age, recorded precise data on planetary movement in the sky. In 1619, using Brahe's data, Johannes Kepler determined that planets move in elliptical orbits around the sun. Later, Newton used Kepler's work to show that elliptical orbits are the result of his famous theory of gravitation. We now know that all celestial objects—including planets, comets, asteroids, and satellites—travel in paths described by conic sections.

Parabolas, ellipses, and hyperbolas have had a profound influence on our understanding of ourselves and the cosmos around us. In this chapter we consider these age-old curves.

Source: Historical Topics for the Mathematics Classroom, Thirty-first Yearbook, NCTM.

> I want to put a ding in the universe.
> —Steve Jobs

10.1 Parabolas

- **Find equations of parabolas**
- **Graph parabolas**
- **Learn the reflective property of parabolas**
- **Translate parabolas**

Introduction

Conic sections are named after the different ways in which a plane can intersect a cone. See Figure 10.1. Three basic conic sections are parabolas, ellipses, and hyperbolas. A circle is also an example of a conic section.

Examples of Conic Sections

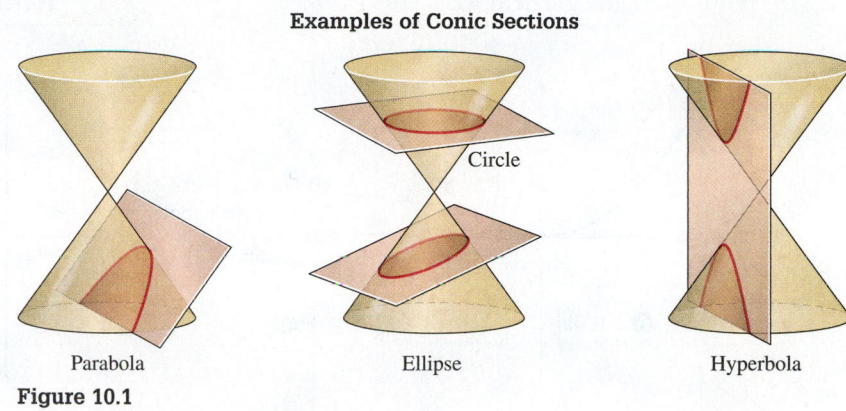

Parabola   Ellipse   Hyperbola

Figure 10.1

Conic sections can be graphed in the xy-plane as shown in Figures 10.2–10.4.

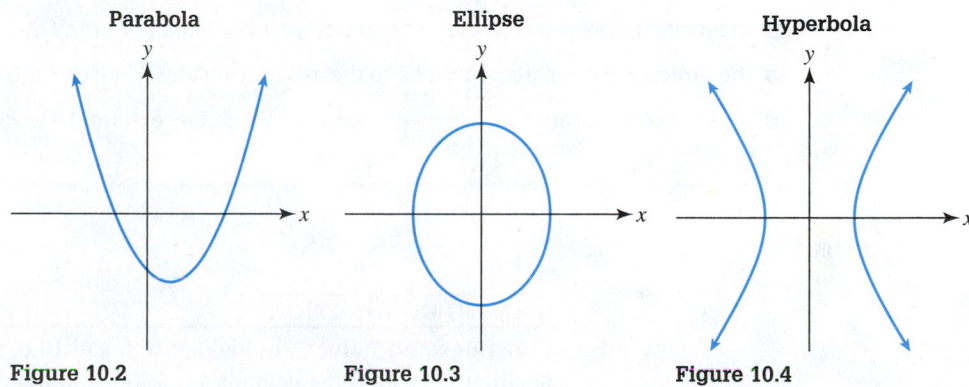

Parabola   Ellipse   Hyperbola

Figure 10.2   **Figure 10.3**   **Figure 10.4**

Ellipses will be discussed in Section 10.2 and hyperbolas will be discussed in Section 10.3. In this section we focus on parabolas.

Equations and Graphs of Parabolas

In Chapter 3 we saw that a parabola with vertex $(0, 0)$ can be represented symbolically by the equation $y = ax^2$. With this representation, a parabola can open either upward when $a > 0$ or downward when $a < 0$. The following definition of a parabola allows it to open in *any* direction.

PARABOLA

A **parabola** is the set of points in a plane that are equidistant from a fixed point and a fixed line. The fixed point is called the **focus** and the fixed line is called the **directrix** of the parabola.

The following See the Concept shows some of the important features of parabolas. Note that the blue parabola in Figure 10.5 passes the vertical line test and can be represented by a function. However, the blue parabola in Figure 10.6 fails the vertical line test and cannot be represented by a function.

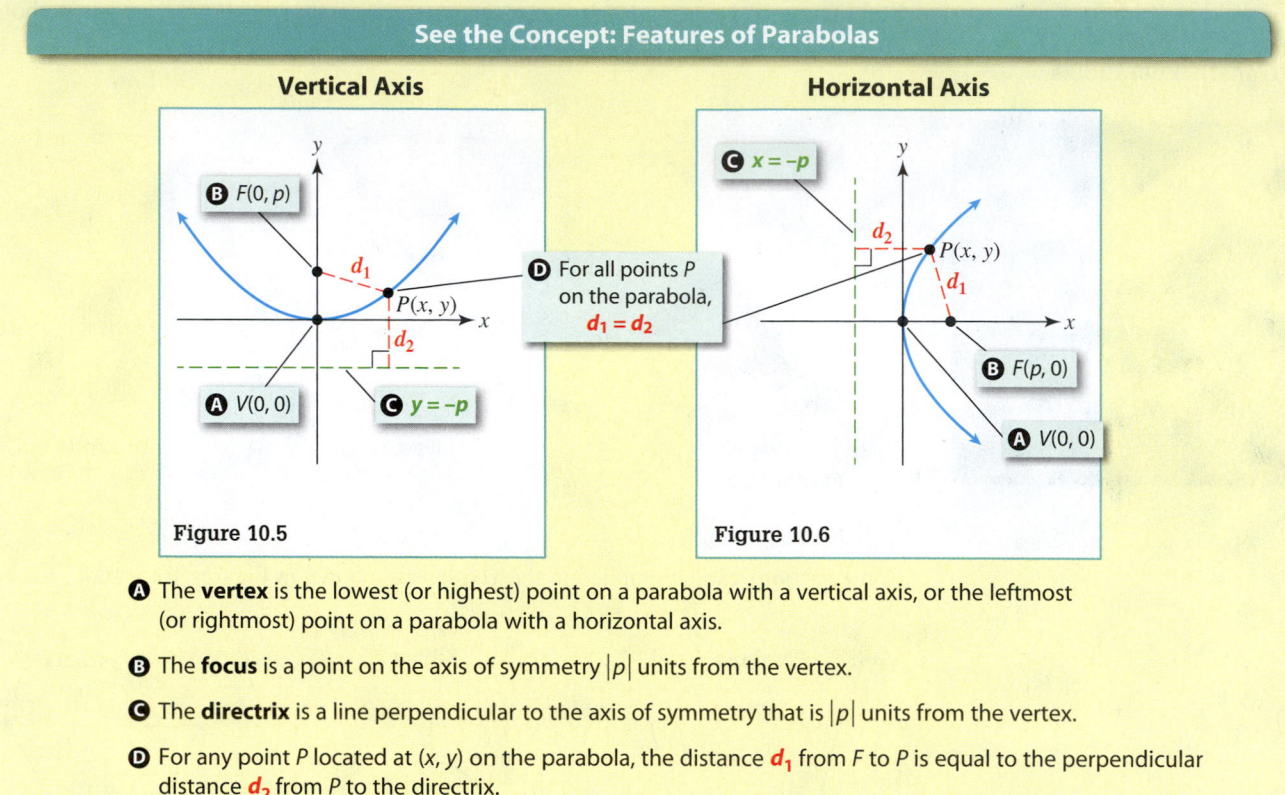

See the Concept: Features of Parabolas

Vertical Axis   **Horizontal Axis**

Figure 10.5   Figure 10.6

A The **vertex** is the lowest (or highest) point on a parabola with a vertical axis, or the leftmost (or rightmost) point on a parabola with a horizontal axis.

B The **focus** is a point on the axis of symmetry $|p|$ units from the vertex.

C The **directrix** is a line perpendicular to the axis of symmetry that is $|p|$ units from the vertex.

D For any point P located at (x, y) on the parabola, the distance d_1 from F to P is equal to the perpendicular distance d_2 from P to the directrix.

MAKING CONNECTIONS

Functions and Points   In Figures 10.5 and 10.6, the point P is labeled $P(x, y)$. This notation resembles the notation for a function involving two inputs, since the point P is determined by x and y.

We can derive an equation of the parabola shown in Figure 10.5. Since $d_1 = d_2$, the distance formula can be used to express the variables x, y, and p in an equation.

$$d_1 = d_2$$

$$\sqrt{(x - 0)^2 + (y - p)^2} = \sqrt{(x - x)^2 + (y - (-p))^2} \qquad \text{Distance formula}$$

$$x^2 + (y - p)^2 = 0^2 + (y + p)^2 \qquad \text{Square each side.}$$

$$x^2 + y^2 - 2py + p^2 = y^2 + 2py + p^2 \qquad \text{Expand binomials.}$$

$$x^2 - 2py = 2py \qquad \text{Subtract } y^2 \text{ and } p^2.$$

$$x^2 = 4py \qquad \text{Add 2py.}$$

If the value of p is known, then the equation of a parabola with vertex $(0, 0)$ can be found using one of the following equations.

EQUATION OF A PARABOLA WITH VERTEX (0, 0)

Vertical Axis

The parabola with a focus at $(0, p)$ and directrix $y = -p$ has equation

$$x^2 = 4py.$$

The parabola opens upward if $p > 0$ and downward if $p < 0$.

Horizontal Axis

The parabola with a focus at $(p, 0)$ and directrix $x = -p$ has equation

$$y^2 = 4px.$$

The parabola opens to the right if $p > 0$ and to the left if $p < 0$.

EXAMPLE 1 **Sketching graphs of parabolas**

Sketch a graph of each parabola. Label the vertex, focus, and directrix.
(a) $x^2 = 8y$ **(b)** $y^2 = -2x$

SOLUTION

(a) The equation $x^2 = 8y$ is in the form $x^2 = 4py$, where $8 = 4p$. Therefore the parabola has a vertical axis with $p = 2$. Since $p > 0$, the parabola opens **upward**. The focus is $(0, 2)$ and the directrix is $y = -2$. See Figure 10.7.

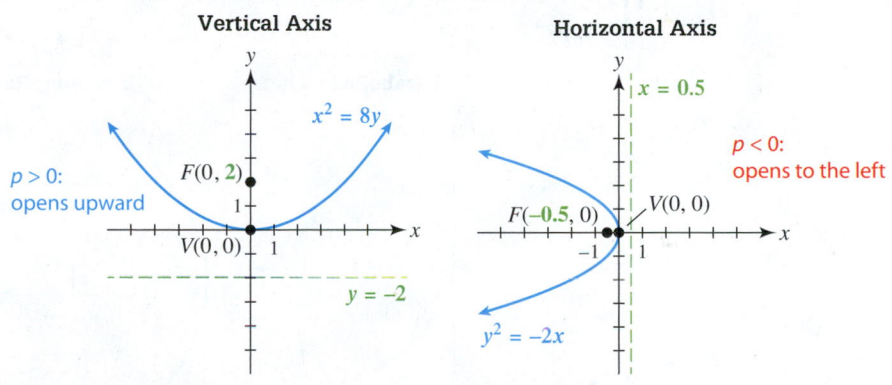

Figure 10.7   **Figure 10.8**

(b) The equation $y^2 = -2x$ has the form $y^2 = 4px$, where $-2 = 4p$. Therefore the parabola has a horizontal axis with $p = -0.5$. Since $p < 0$, the parabola opens to the **left**. The focus is $(-0.5, 0)$ and the directrix is $x = 0.5$. See Figure 10.8.

Now Try Exercises 23 and 27

EXAMPLE 2 **Finding the equation of a parabola**

Find the equation of the parabola with focus $(-1.5, 0)$ and directrix $x = 1.5$, as shown in Figure 10.9. Sketch a graph of the parabola.

SOLUTION A parabola always *opens toward the focus* and *away from the directrix*. From Figure 10.9 we see that the parabola should open to the left. It follows that $p < 0$ in the equation $y^2 = 4px$. The distance between the focus at $(-1.5, 0)$ and the vertex at $(0, 0)$ is 1.5, and so $p = -1.5 < 0$. (Note that the vertex of the parabola is $(0, 0)$ because the vertex always lies *midway* between the focus and the directrix.) The equation of the parabola is $y^2 = 4(-1.5)x$, or $y^2 = -6x$, and a graph of the parabola is shown in Figure 10.10 at the top of the next page.

Figure 10.9

Horizontal Axis

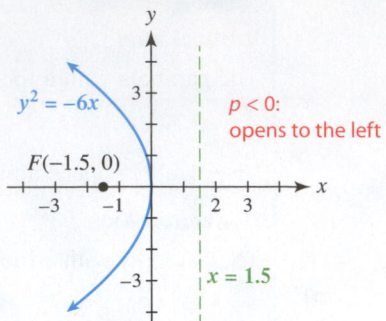

Figure 10.10

Now Try Exercise 35

Reflective Property of Parabolas

When a parabola is rotated about its axis, it sweeps out a shape called a **paraboloid**, as shown in Figure 10.11. Paraboloids have a special reflective property. When incoming, parallel rays of light from the sun or distant stars strike the surface of a paraboloid, each ray is reflected toward the focus. See Figure 10.12. This property of a paraboloid can also be used in reverse. If a light source is placed at the focus, then the light is reflected straight ahead, as shown in Figure 10.13. Searchlights, flashlights, and car headlights make use of this property.

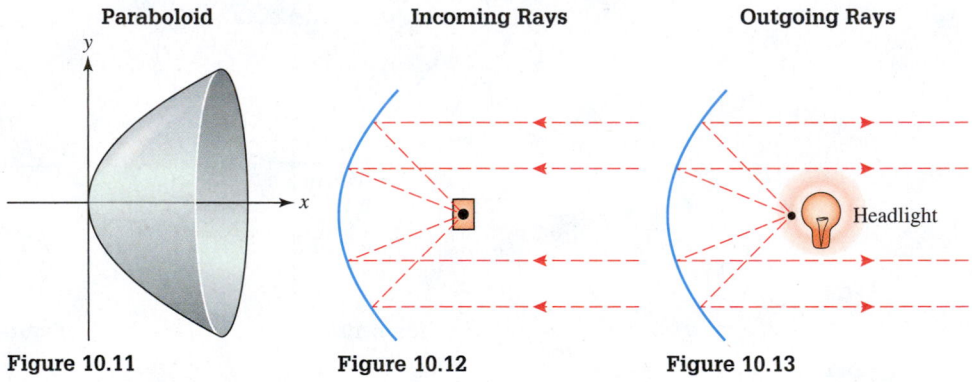

| Paraboloid | Incoming Rays | Outgoing Rays |

Figure 10.11   Figure 10.12   Figure 10.13

An Application   The next example illustrates how this reflective property is used in the construction of a telescope in the shape of a large parabolic dish.

EXAMPLE 3   Locating the receiver for a radio telescope

The giant Arecibo telescope in Puerto Rico has a parabolic dish with a diameter of 300 meters and a depth of 50 meters. See Figure 10.14. (*Source:* National Astronomy and Ionosphere Center, Arecibo Observatory.)
(a) Find an equation in the form $y = ax^2$ that describes a cross section of this dish.
(b) If the receiver is located at the focus, how far should it be from the vertex?

SOLUTION
(a) A parabola that passes through $(-150, 50)$ and $(150, 50)$, as shown in Figure 10.15, has a diameter of 300 meters and a depth of 50 meters. Substitute either point into $y = ax^2$.

Figure 10.14

Telescope Cross Section

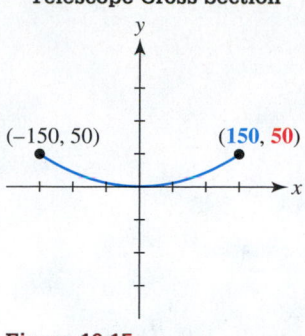

Figure 10.15

$$y = ax^2 \qquad \text{Equation of parabola}$$
$$50 = a(150)^2 \qquad \text{Substitute.}$$
$$a = \frac{50}{150^2} = \frac{1}{450} \qquad \text{Solve for } a.$$

The equation of the parabola is $y = \frac{1}{450}x^2$, where $-150 \le x \le 150$.

(b) The value of p represents the distance from the vertex to the focus. To determine p, write the equation in the form $x^2 = 4py$. Then

$$y = \frac{1}{450}x^2 \qquad \text{is equivalent to} \qquad x^2 = 450y.$$

It follows that $4p = 450$, or $p = 112.5$. Therefore the receiver should be located about 112.5 meters from the vertex.

> **Now Try Exercise 95**

Translations of Parabolas

If the equation of a parabola is either $x^2 = 4py$ or $y^2 = 4px$, then its vertex is $(0, 0)$. We can use translations of graphs to find the equation of a parabola with vertex (h, k). This translation can be obtained by replacing x with $(x - h)$ and y with $(y - k)$.

$$(x - h)^2 = 4p(y - k) \qquad \text{Vertex } (h, k); \text{ vertical axis}$$
$$(y - k)^2 = 4p(x - h) \qquad \text{Vertex } (h, k); \text{ horizontal axis}$$

These two parabolas with $p > 0$ are shown in Figures 10.16 and 10.17, respectively.

Vertex (h, k); Vertical Axis

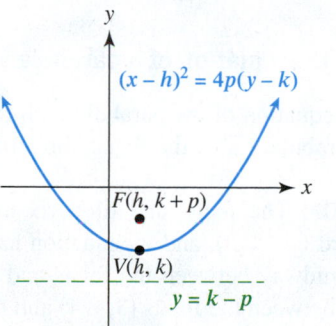

Figure 10.16

Vertex (h, k); Horizontal Axis

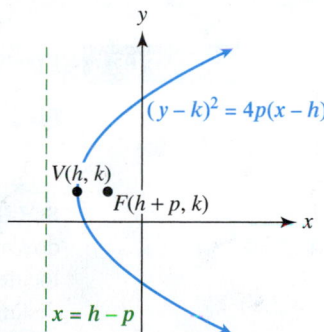

Figure 10.17

EQUATION OF A PARABOLA WITH VERTEX (h, k)

$(x - h)^2 = 4p(y - k)$ Vertical axis; vertex: (h, k)
         $p > 0$: opens upward; $p < 0$: opens downward
         Focus: $(h, k + p)$; directrix: $y = k - p$

$(y - k)^2 = 4p(x - h)$ Horizontal axis; vertex: (h, k)
         $p > 0$: opens to the right; $p < 0$: opens to the left
         Focus: $(h + p, k)$; directrix: $x = h - p$

EXAMPLE 4   Graphing a parabola with vertex (h, k)

Graph the parabola $x = -\frac{1}{8}(y + 3)^2 + 2$. Label the vertex, focus, and directrix.

Algebra Review
To review translations, or shifts, see Section 3.5.

SOLUTION Rewrite the equation in the form $(y - k)^2 = 4p(x - h)$.

$$x = -\frac{1}{8}(y + 3)^2 + 2 \qquad \textit{Given equation}$$

$$x - 2 = -\frac{1}{8}(y + 3)^2 \qquad \textit{Subtract 2.}$$

$$-8(x - 2) = (y + 3)^2 \qquad \textit{Multiply by } -8.$$

$$(y + 3)^2 = -8(x - 2) \qquad \textit{Rewrite equation.}$$

$$(y - (-3))^2 = 4(-2)(x - 2) \qquad (y - k)^2 = 4p(x - h)$$

The vertex is $(2, -3)$, $p = -2$, and the parabola opens to the left. The focus is 2 units left of the vertex, and the directrix is 2 units right of the vertex. Thus the focus is $(0, -3)$, and the directrix is $x = 4$. See Figure 10.18.

Translated Parabola

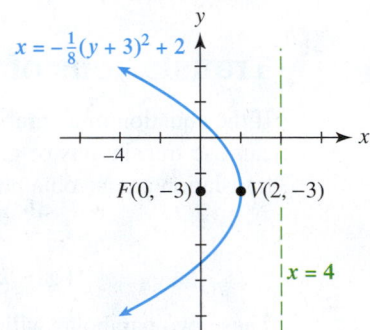

Figure 10.18

Now Try Exercise 61

EXAMPLE 5 Finding the equation of a parabola with vertex (h, k)

Find the equation of the parabola with focus $(3, -4)$ and directrix $y = 2$. Sketch a graph of the parabola. Label the focus, directrix, and vertex.

SOLUTION The focus and directrix are shown in Figure 10.19. The parabola opens downward ($p < 0$), and its equation has the form $(x - h)^2 = 4p(y - k)$. The vertex is located midway between the focus and the directrix, so its coordinates are $(3, -1)$. The distance between the focus $(3, -4)$ and the vertex $(3, -1)$ is 3, so $p = -3$. The equation of the parabola is

$$(x - 3)^2 = -12(y + 1),$$

and its graph is shown in Figure 10.20.

Focus and Directrix

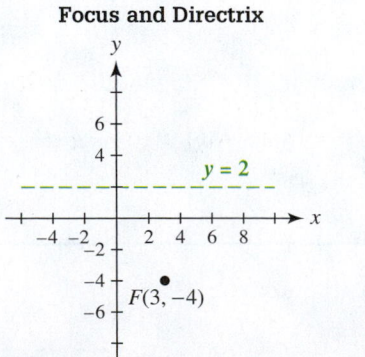

Figure 10.19

Translated Parabola

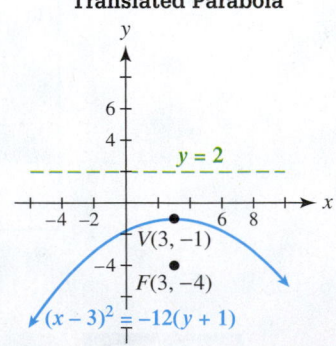

Figure 10.20

Now Try Exercise 67

EXAMPLE 6   **Finding the equation of a parabola**

Write $2x = y^2 + 4y + 12$ in the form $(y - k)^2 = a(x - h)$.

SOLUTION Write the given equation in the required form by completing the square.

$$2x = y^2 + 4y + 12 \qquad \textit{Given equation}$$
$$2x - 12 = y^2 + 4y \qquad \textit{Subtract 12 from each side.}$$

To complete the square on the right side of the equation, add $\left(\frac{4}{2}\right)^2 = 4$ to each side.

$$2x - 12 + 4 = y^2 + 4y + 4 \qquad \textit{Add 4 to each side.}$$
$$2x - 8 = (y + 2)^2 \qquad \textit{Perfect square trinomial}$$
$$2(x - 4) = (y + 2)^2 \qquad \textit{Factor out 2.}$$

The given equation is equivalent to $(y + 2)^2 = 2(x - 4)$.

Now Try Exercise 73

CLASS DISCUSSION

Sketch a graph of the parabola in Example 6. Identify the vertex, focus, and directrix.

Using Technology Graphing calculators can graph parabolas with horizontal axes, as illustrated in the next example.

EXAMPLE 7   **Graphing a parabola with technology**

Graph the equation $(y - 1)^2 = -0.5(x - 2)$ with a graphing calculator.

SOLUTION Begin by solving the equation for y.

$$(y - 1)^2 = -0.5(x - 2) \qquad \textit{Given equation}$$
$$y - 1 = \pm\sqrt{-0.5(x - 2)} \qquad \textit{Square root property}$$
$$y = 1 \pm \sqrt{-0.5(x - 2)} \qquad \textit{Add 1.}$$

Let $y_1 = 1 + \sqrt{-0.5(x - 2)}$ and $y_2 = 1 - \sqrt{-0.5(x - 2)}$. The graph of y_1 creates the upper portion of the parabola, and the graph of y_2 creates the lower portion of the parabola, as shown in Figure 10.21.

Now Try Exercise 83

$[-4.7, 4.7, 1]$ by $[-3.1, 3.1, 1]$

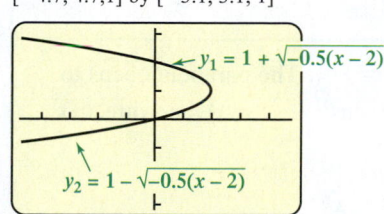

Figure 10.21

10.1 Putting It All Together

The following table summarizes some important concepts about parabolas.

CONCEPT	EQUATION	EXAMPLE
Parabola with vertex $(0, 0)$ and vertical axis	$x^2 = 4py$ $p > 0$: opens upward $p < 0$: opens downward Focus: $(0, p)$ Directrix: $y = -p$	$x^2 = -2y$ has $4p = -2$, or $p = -\frac{1}{2}$. The parabola opens downward with vertex $(0, 0)$, focus $\left(0, -\frac{1}{2}\right)$, and directrix $y = \frac{1}{2}$.

continued on next page

CONCEPT	EQUATION	EXAMPLE
Parabola with vertex $(0, 0)$ and horizontal axis	$y^2 = 4px$ $p > 0$: opens to the right $p < 0$: opens to the left Focus: $(p, 0)$ Directrix: $x = -p$	$y^2 = 4x$ has $4p = 4$, or $p = 1$. The parabola opens to the right with vertex $(0, 0)$, focus $(1, 0)$, and directrix $x = -1$.
Parabola with vertex (h, k) and vertical axis	$(x - h)^2 = 4p(y - k)$ $p > 0$: opens upward $p < 0$: opens downward Focus: $(h, k + p)$ Directrix: $y = k - p$	$(x - 1)^2 = 8(y - 3)$ has $p = 2$. The parabola opens upward with vertex $(1, 3)$, focus $(1, 5)$, and directrix $y = 1$.
Parabola with vertex (h, k) and horizontal axis	$(y - k)^2 = 4p(x - h)$ $p > 0$: opens to the right $p < 0$: opens to the left Focus: $(h + p, k)$ Directrix: $x = h - p$	$(y + 1)^2 = -2(x + 2)$ has $p = -\frac{1}{2}$. The parabola opens to the left with vertex $(-2, -1)$, focus $\left(-\frac{5}{2}, -1\right)$, and directrix $x = -\frac{3}{2}$.

10.1 Exercises

Basic Concepts

1. A parabola always opens toward its _____.

2. A parabola always opens away from its _____.

3. The parabola $x^2 = 4py$ opens _____ if $p > 0$ and it opens _____ if $p < 0$.

4. The parabola $y^2 = 4px$ opens _____ if $p > 0$ and it opens _____ if $p < 0$.

5. The parabola $x^2 = 4py$ has a vertical/horizontal axis of symmetry.

6. The parabola $y^2 = 4px$ has a vertical/horizontal axis of symmetry.

Parabolas with Vertex (0, 0)

Exercises 7–16: Sketch a graph of the parabola.

7. $x^2 = y$ **8.** $x^2 = -y$

9. $y^2 = -x$ **10.** $y^2 = x$

11. $4x^2 = -2y$ **12.** $y^2 = -3x$

13. $y^2 = -4x$ **14.** $x^2 = 4y$

15. $y^2 = -\frac{1}{2}x$ **16.** $8x = y^2$

Exercises 17–22: Match the equation with its graph (a–f).

17. $x^2 = 2y$ **18.** $x^2 = -2y$

19. $y^2 = -8x$ **20.** $y^2 = 4x$

21. $x = \frac{1}{2}y^2$ **22.** $y = -2x^2$

a.
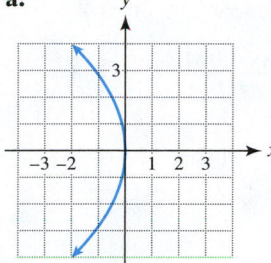

b.
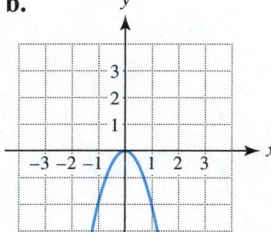

c.
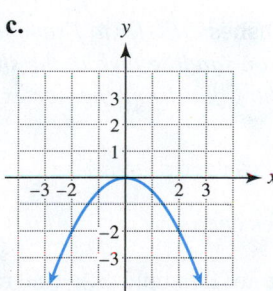

d.
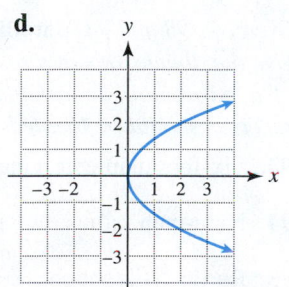

e.
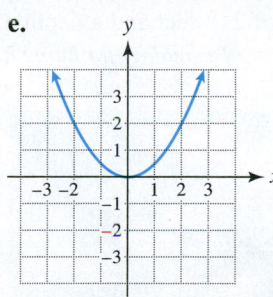

f.
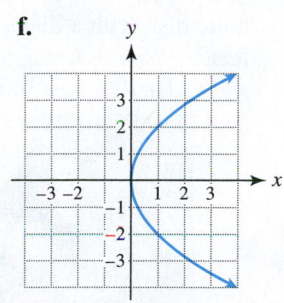

Exercises 23–32: Graph the parabola. Label the vertex, focus, and directrix.

23. $16y = x^2$ **24.** $y = -2x^2$

25. $x = \frac{1}{8}y^2$ **26.** $-y^2 = 6x$

27. $-4x = y^2$ **28.** $\frac{1}{2}y^2 = 3x$

29. $x^2 = -8y$ **30.** $x^2 = -4y$

31. $2y^2 = -8x$ **32.** $-3x = \frac{1}{4}y^2$

Exercises 33–36: Sketch a parabola with focus and directrix as shown in the figure. Find an equation of the parabola.

33.
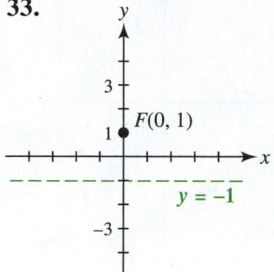

34.
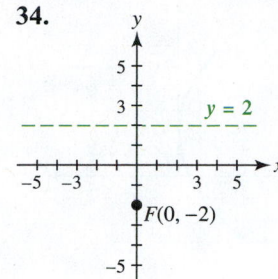

35.
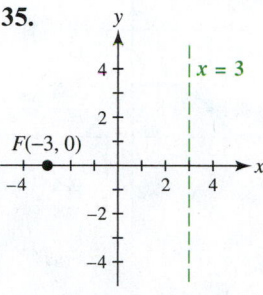

36.
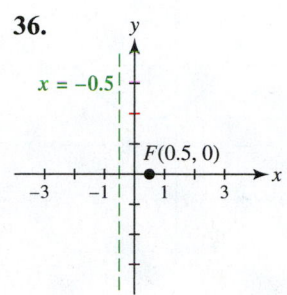

Exercises 37–46: Find an equation of the parabola with vertex $(0, 0)$ that satisfies the given conditions.

37. Focus $\left(0, \frac{3}{4}\right)$ **38.** Directrix $y = 2$

39. Directrix $x = 2$ **40.** Focus $(-1, 0)$

41. Focus $(1, 0)$ **42.** Focus $\left(0, -\frac{1}{2}\right)$

43. Directrix $x = \frac{1}{4}$ **44.** Directrix $y = -1$

45. Horizontal axis, passing through $(1, -2)$

46. Vertical axis, passing through $(-2, 3)$

Exercises 47–50: Find an equation of a parabola that satisfies the given conditions.

47. Focus $(0, -3)$ and directrix $y = 3$

48. Focus $(0, 2)$ and directrix $y = -2$

49. Focus $(-1, 0)$ and directrix $x = 1$

50. Focus $(3, 0)$ and directrix $x = -3$

Parabolas with Vertex (h, k)

Exercises 51–54: Sketch a graph of the parabola.

51. $(x - 1)^2 = (y - 2)$ **52.** $(x - 2)^2 = -(y + 1)$

53. $(y - 1)^2 = -(x + 1)$ **54.** $(y + 2)^2 = 2x$

Exercises 55–58: Match the equation with its graph (a–d).

55. $(x - 1)^2 = 4(y - 1)$ **56.** $(x + 1)^2 = -4(y - 2)$

57. $(y - 2)^2 = -8x$ **58.** $(y + 1)^2 = 8(x + 3)$

a.
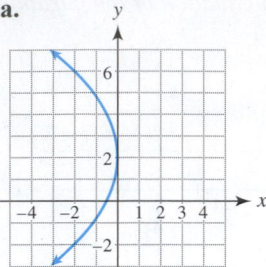

b.
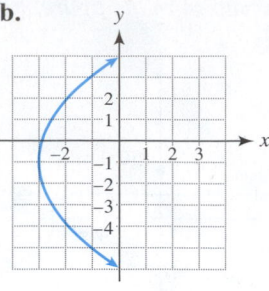

c.
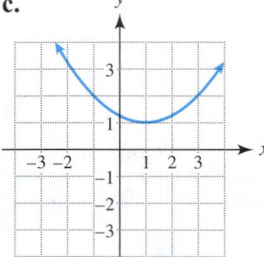

d.
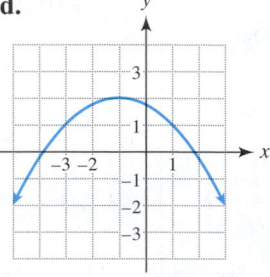

Exercises 59–64: Graph the parabola. Label the vertex, focus, and directrix.

59. $(x - 2)^2 = 8(y + 2)$ **60.** $\frac{1}{16}(x + 4)^2 = -(y - 4)$

61. $x = -\frac{1}{4}(y + 3)^2 + 2$ **62.** $x = 2(y - 2)^2 - 1$

63. $y = -\frac{1}{4}(x + 2)^2$ **64.** $-2(y + 1) = (x + 3)^2$

Exercises 65–68: Find an equation of a parabola that satisfies the given conditions. Sketch a graph of the parabola. Label the focus, directrix, and vertex.

65. Focus $(0, 2)$ and vertex $(0, 1)$

66. Focus $(-1, 2)$ and vertex $(3, 2)$

67. Focus $(0, 0)$ and directrix $x = -2$

68. Focus $(2, 1)$ and directrix $x = -1$

Exercises 69–72: Find an equation of a parabola that satisfies the given conditions.

69. Focus $(-1, 3)$ and directrix $y = 7$

70. Focus $(1, 2)$ and directrix $y = 4$

71. Horizontal axis, vertex $(-2, 3)$, passing through $(-4, 0)$

72. Horizontal axis, vertex $(-1, 2)$, passing through $(2, 3)$

Exercises 73–80: Write the given equation either in the form $(y - k)^2 = a(x - h)$ or in the form $(x - h)^2 = a(y - k)$.

73. $-2x = y^2 + 6x + 10$ **74.** $y^2 + 8x - 8 = 4x$

75. $x = 2y^2 + 4y - 1$ **76.** $x = 3y^2 - 6y - 2$

77. $x^2 - 3x + 4 = 2y$ **78.** $-3y = -x^2 + 4x - 6$

79. $4y^2 + 4y - 5 = 5x$ **80.** $-2y^2 + 5y + 1 = -x$

Graphing Parabolas with Technology

 Exercises 81–86: Graph the parabola.

81. $(y + 0.75)^2 = -3x$ **82.** $(y - 3)^2 = \frac{1}{7}x$

83. $(y - 0.5)^2 = 3.1(x + 1.3)$

84. $1.4(y - 1.5)^2 = 0.5(x + 2.1)$

85. $x = 2.3(y + 1)^2$

86. $(y - 2.5)^2 = 4.1(x + 1)$

Solving Nonlinear Systems

Exercises 87–92: Solve each system.

87. $x^2 = 2y$
$\ x^2 = y + 1$

88. $x^2 = -3y$
$\ -x^2 = 2y - 2$

89. $\frac{1}{3}y^2 = -3x$
$\ y^2 = x + 1$

90. $-2y^2 = x - 5$
$\ y^2 = 2x$

91. $(y - 1)^2 = x + 1$
$\ (y + 2)^2 = -x + 4$

92. $(y + 1)^2 = -x$
$\ -(y - 1)^2 = x + 4$

Applications

Exercises 93 and 94: **Satellite Dishes** *(Refer to Example 3.) Use the dimensions of a television satellite dish in the shape of a paraboloid to calculate how far from the vertex the receiver should be located.*

93. Six-foot diameter, nine inches deep

94. Nine-inch radius, two inches deep

95. **Radio Telescope** (Refer to Example 3.) The radio telescope shown in the figure has the shape of a parabolic dish with a diameter of 210 feet and a depth of 32 feet. (*Source:* J. Mar, *Structure Technology for Large Radio and Radar Telescope Systems.*)

(a) Determine an equation of the form $y = ax^2$ with $a > 0$ describing a cross section of the dish.

(b) The receiver is placed at the focus. How far from the vertex is the receiver located?

96. Radio Telescope (Refer to Example 3.) A radio telescope is being designed in the shape of a parabolic dish with a diameter of 180 feet and a depth of 25 feet.
(a) Determine an equation of the form $x = ay^2$ with $a > 0$ describing a cross section of the dish.

(b) The receiver is placed at the focus. How far from the vertex should the receiver be located?

97. Comets A comet sometimes travels along a parabolic path as it passes the sun. In this case the sun is located at the focus of the parabola and the comet passes the sun once, rather than orbiting the sun. Suppose the path of a comet is given by $y^2 = 100x$, where units are in millions of miles.
(a) Find the coordinates of the sun.

(b) Find the minimum distance between the sun and the comet.

98. Headlight A headlight is being constructed in the shape of a paraboloid with a depth of 4 inches and a diameter of 5 inches, as illustrated in the figure. Find the distance d that the bulb should be from the vertex in order to have the beam of light shine straight ahead.

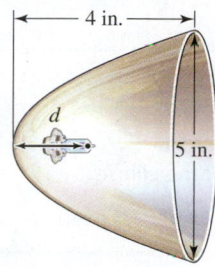

99. Solar Heater A solar heater is being designed to heat a pipe that will contain water, as illustrated in the figure. A cross section of the heater is described by the equation $x^2 = ky$, where k is a constant and all units are in feet. If the pipe is to be placed 18 inches from the vertex of this cross section, find the value of k.

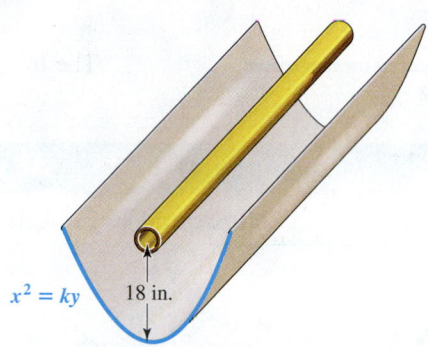

100. Solar Heater If the figure in Exercise 99 is rotated 90° clockwise, then the front edge of the reflector can be described by the equation $y^2 = kx$. If the pipe is placed 2 feet from the vertex of this cross section, find the value of k.

Writing about Mathematics

101. Explain how the distance between the focus and the vertex of a parabola affects the shape of the parabola.

102. Explain how to determine the direction that a parabola opens, given the focus and the directrix.

10.2 Ellipses

- **Find equations of ellipses**
- **Graph ellipses**
- **Learn the reflective property of ellipses**
- **Translate ellipses**
- **Solve nonlinear systems of equations and inequalities**

Introduction

When planets travel around the sun, they travel in elliptical orbits. This discovery by Johannes Kepler made it possible for astronomers to determine the precise positions of all types of celestial objects, such as asteroids, comets, and moons, and made it easier to predict both solar and lunar eclipses. Ellipses are also used in construction and medicine. In this section we examine the basic properties of ellipses.

Equations and Graphs of Ellipses

One method for sketching an ellipse is to tie a string to two nails driven into a flat board. If a pencil is placed inside the loop formed by the string, the curve it traces (shown in Figure 10.22 on the next page) is an ellipse. The sum of the distances d_1 and d_2 between the pencil and each of the nails is always fixed by the string. The locations of the nails correspond to the *foci* of the ellipse. If the two nails coincide, the ellipse becomes a circle. As the nails spread farther apart, the ellipse becomes more elongated, or *eccentric*.

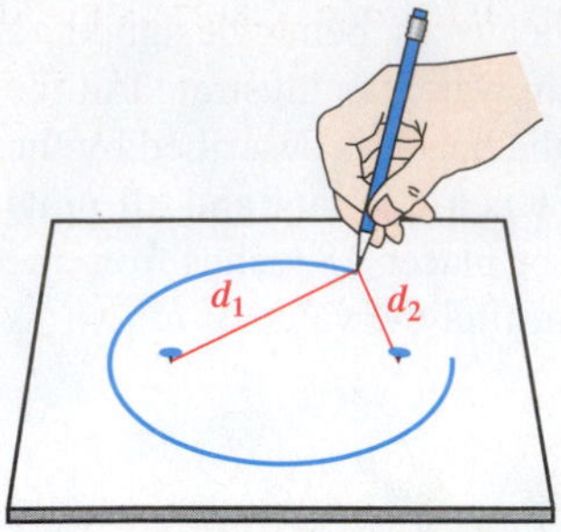

This method of sketching an ellipse suggests the following definition.

Figure 10.22

> **ELLIPSE**
>
> An **ellipse** is the set of points in a plane, the sum of whose distances from two fixed points is constant. Each fixed point is called a **focus** (plural, **foci**) of the ellipse.

The following See the Concept shows some of the important features of ellipses.

> ### See the Concept: Features of Ellipses
>
> **Horizontal Major Axis**   **Vertical Major Axis** 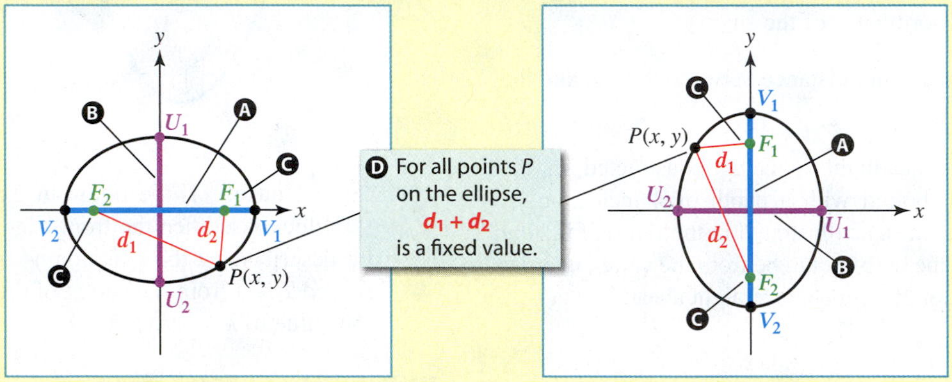
>
>
> **D** For all points P on the ellipse, $d_1 + d_2$ is a fixed value.
>
> **A** The **major axis** is the longer line segment connecting the **vertices** V_1 and V_2.
>
> **B** The **minor axis** is the shorter line segment connecting U_1 and U_2.
>
> **C** The **foci** F_1 and F_2 (singular, focus) are points located on the major axis of the ellipse.
>
> **D** For any point P located at (x, y) on the ellipse, the sum of the distances d_1 and d_2 is fixed.

NOTE Since a vertical line can intersect the graph of an ellipse more than once, an ellipse cannot be described by a function.

Some ellipses can be represented by the following equations.

> **STANDARD EQUATIONS FOR ELLIPSES CENTERED AT (0, 0)**
>
> The ellipse with center at the origin, *horizontal* major axis, and equation
>
> $$\frac{x^2}{a^2} + \frac{y^2}{b^2} = 1 \qquad (a > b > 0)$$
>
> has vertices $(\pm a, 0)$, endpoints of the minor axis $(0, \pm b)$, and foci $(\pm c, 0)$, where $c^2 = a^2 - b^2$ and $c \geq 0$.
>
> The ellipse with center at the origin, *vertical* major axis, and equation
>
> $$\frac{x^2}{b^2} + \frac{y^2}{a^2} = 1 \qquad (a > b > 0)$$
>
> has vertices $(0, \pm a)$, endpoints of the minor axis $(\pm b, 0)$, and foci $(0, \pm c)$, where $c^2 = a^2 - b^2$ and $c \geq 0$.

NOTE If $a = b$, then the ellipse becomes a circle with radius $r = a$ and center $(0, 0)$.

Figures 10.23 and 10.24 show two ellipses. The first has a horizontal major axis and the second has a vertical major axis. The coordinates of the vertices V_1 and V_2, foci F_1 and F_2, and endpoints of the minor axis U_1 and U_2 are labeled. In each figure $a > b > 0$.

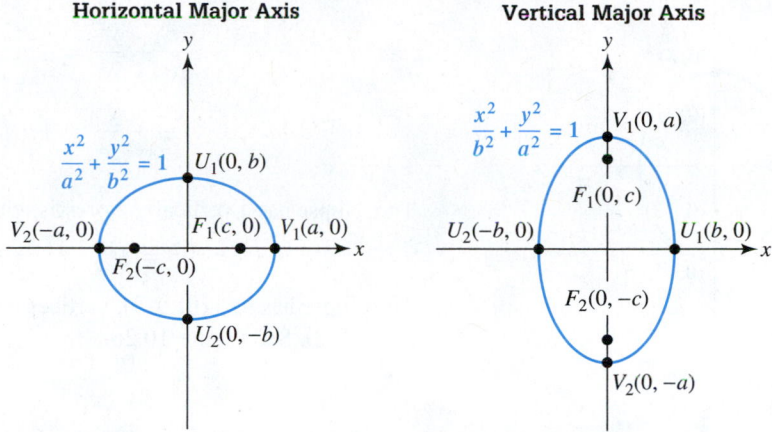

Figure 10.23   **Figure 10.24**

NOTE In every ellipse the major and minor axes are perpendicular bisectors of each other.

MAKING CONNECTIONS

Intercept Form of a Line and Standard Form for an Ellipse   If the equation of a line can be written as

$$\frac{x}{a} + \frac{y}{b} = 1, \qquad \text{Intercept form}$$

then the x-intercept is a and the y-intercept is b. If the equation of an ellipse centered at (0, 0) can be written as

$$\frac{x^2}{a^2} + \frac{y^2}{b^2} = 1, \qquad \text{Standard form}$$

then the x-intercepts are $\pm a$ and the y-intercepts are $\pm b$. See Extended and Discovery Exercises 1–6 at the end of this section.

EXAMPLE 1 **Sketching graphs of ellipses**

Sketch a graph of each ellipse. Label the vertices, foci, and endpoints of the minor axis.

(a) $\dfrac{x^2}{9} + \dfrac{y^2}{4} = 1$ **(b)** $25x^2 + 16y^2 = 400$

SOLUTION

(a) The equation $\frac{x^2}{9} + \frac{y^2}{4} = 1$ can be written as $\frac{x^2}{3^2} + \frac{y^2}{2^2} = 1$ and describes an ellipse with $a = 3$ and $b = 2$. The ellipse has a horizontal major axis with vertices $(\pm 3, 0)$. The endpoints of the minor axis are $(0, \pm 2)$. To locate the foci, find c.

$$c^2 = a^2 - b^2 = 3^2 - 2^2 = 5, \quad \text{or} \quad c = \sqrt{5} \approx 2.24.$$

The foci are located on the major axis, with coordinates $\left(\pm \sqrt{5}, 0\right)$. See Figure 10.25.

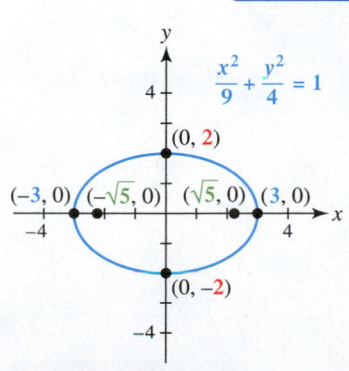

Figure 10.25

(b) The given equation can be written in standard form by dividing each side by **400**.

$$25x^2 + 16y^2 = 400 \qquad \textit{Given equation}$$

$$\frac{25x^2}{400} + \frac{16y^2}{400} = \frac{400}{400} \qquad \textit{Divide each side by 400.}$$

$$\frac{x^2}{16} + \frac{y^2}{25} = 1 \qquad \textit{Simplify each fraction.}$$

$$\frac{x^2}{4^2} + \frac{y^2}{5^2} = 1 \qquad \textit{Rewrite the equation.}$$

This ellipse has a vertical major axis with $a = 5$ and $b = 4$. The value of c is given by

$$c^2 = 5^2 - 4^2 = 9, \qquad \text{or} \qquad c = 3.$$

The ellipse has foci $(0, \pm 3)$, vertices $(0, \pm 5)$, and endpoints of the minor axis located at $(\pm 4, 0)$. See Figure 10.26.

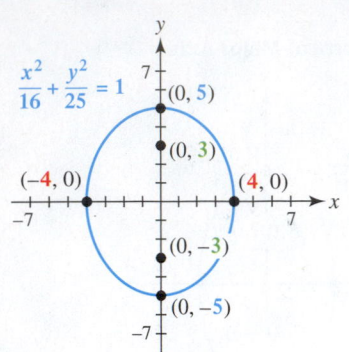

Figure 10.26

> **Now Try Exercises 5 and 9**

In the next two examples, we find standard equations for ellipses.

EXAMPLE 2 Finding the equation of an ellipse

Find the standard equation of the ellipse shown in Figure 10.27. Identify the coordinates of the vertices and the foci.

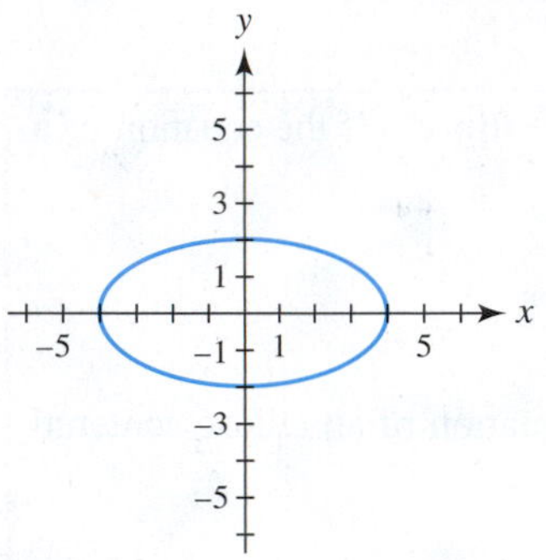

SOLUTION The ellipse is centered at $(0, 0)$ and has a horizontal major axis. Its standard equation has the form

$$\frac{x^2}{a^2} + \frac{y^2}{b^2} = 1.$$

Figure 10.27

The endpoints of the major axis are $(\pm 4, 0)$, and the endpoints of the minor axis are $(0, \pm 2)$. It follows that $a = 4$ and $b = 2$, and the standard equation is

$$\frac{x^2}{4^2} + \frac{y^2}{2^2} = 1 \qquad \text{or} \qquad \frac{x^2}{16} + \frac{y^2}{4} = 1.$$

The foci lie on the horizontal *major* axis and can be determined as follows.

$$c^2 = a^2 - b^2 = 4^2 - 2^2 = 12$$

Thus $c = \sqrt{12} \approx 3.46$, and the coordinates of the foci are $\left(\pm \sqrt{12}, 0 \right)$. A graph of the ellipse with the vertices and foci plotted is shown in Figure 10.28.

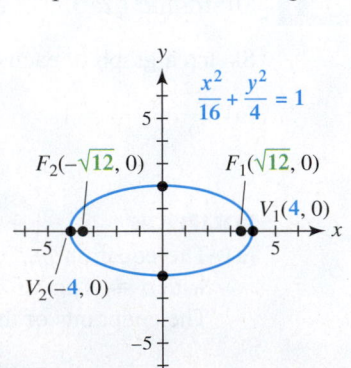

Figure 10.28

> **Now Try Exercise 17**

EXAMPLE 3 Finding the equation of an ellipse

Find the standard equation of the ellipse with foci $(0, \pm 1)$ and vertices $(0, \pm 3)$.

SOLUTION Since the foci and vertices lie on the y-axis, the ellipse has a vertical major axis. Its standard equation has the form

$$\frac{x^2}{b^2} + \frac{y^2}{a^2} = 1.$$

Because the foci are $(0, \pm 1)$ and the vertices are $(0, \pm 3)$, it follows that $c = 1$ and $a = 3$. The value of b^2 can be found by rewriting the equation $c^2 = a^2 - b^2$.

$$b^2 = a^2 - c^2 = 3^2 - 1^2 = 8 \qquad \text{or} \qquad b = \sqrt{8}$$

Thus the equation of the ellipse is $\frac{x^2}{(\sqrt{8})^2} + \frac{y^2}{3^2} = 1$ or $\frac{x^2}{8} + \frac{y^2}{9} = 1$. Its graph is shown in Figure 10.29.

Now Try Exercise 25

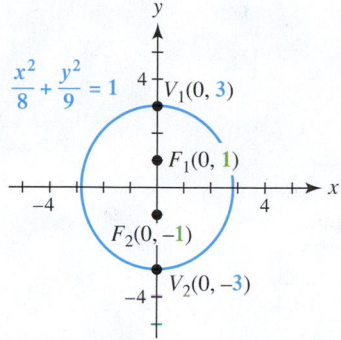

Figure 10.29

Eccentricity and Applications The planets travel around the sun in elliptical orbits. Although their orbits are nearly circular, many have a slight eccentricity to them. The **eccentricity** e of an ellipse is defined by

$$e = \frac{\sqrt{a^2 - b^2}}{a} = \frac{c}{a}. \qquad a > b > 0$$

Since the foci of an ellipse lie *inside* the ellipse, $0 < c < a$ and $0 < \frac{c}{a} < 1$. Therefore the eccentricity e of an ellipse satisfies $0 < e < 1$. If $a = b$, then $e = 0$ and the ellipse becomes a circle. See Figure 10.30. As e increases, the foci spread apart and the ellipse becomes more elongated. See Figures 10.31 and 10.32.

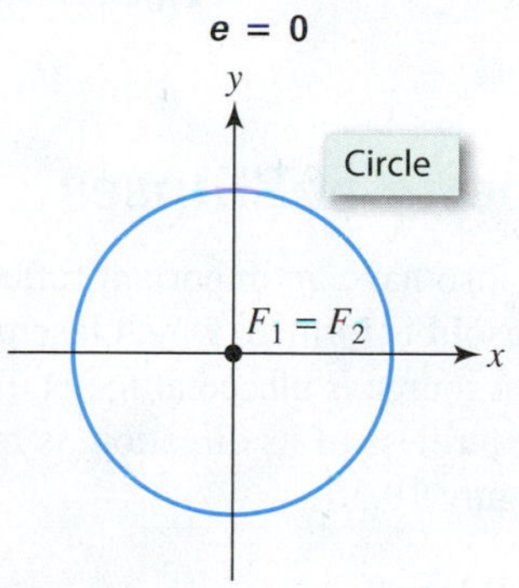

$e = 0$

Circle

$F_1 = F_2$

Figure 10.30

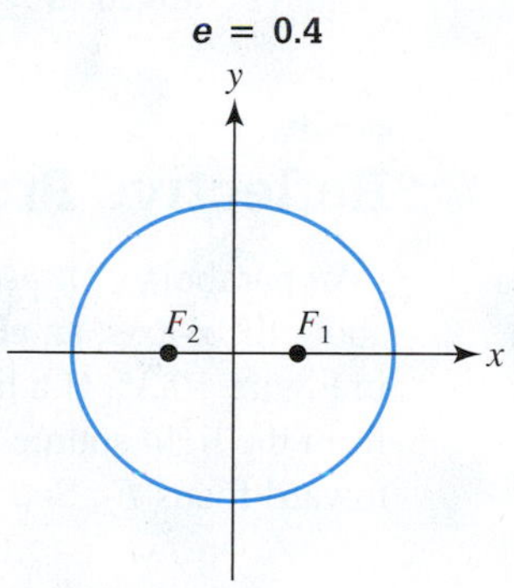

$e = 0.4$

F_2 F_1

Figure 10.31

$e = 0.8$

F_2 F_1

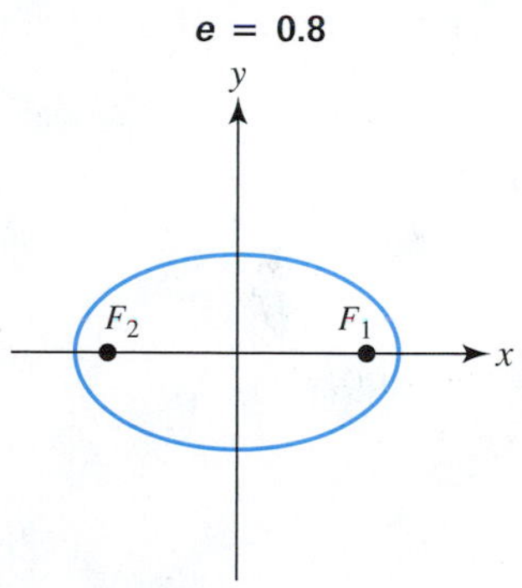

Figure 10.32

MAKING CONNECTIONS

The Number e and the Variable e Do not confuse the *variable e*, which is used to denote the eccentricity of an ellipse, with the irrational *number $e \approx 2.72$*, which is the base of the natural exponential function $f(x) = e^x$ and of the natural logarithmic function $g(x) = \ln x$.

Astronomers have measured values of a and e for the eight major planets and Pluto. With this information and the fact that the sun is located at one focus of the ellipse, the equations of their orbits can be found.

EXAMPLE 4 Finding the orbital equation for Pluto

Pluto has $a = 39.44$ and $e = 0.249$. (For Earth, $a = 1$.) Graph the orbit of Pluto and the position of the sun in $[-60, 60, 10]$ by $[-40, 40, 10]$. (*Source:* M. Zeilik, *Introductory Astronomy and Astrophysics.*)

SOLUTION Let the orbit of Pluto be given by $\frac{x^2}{a^2} + \frac{y^2}{b^2} = 1$. Then

$$e = \frac{c}{a} = 0.249 \qquad \text{implies} \qquad c = 0.249a = 0.249(39.44) \approx \mathbf{9.821}.$$

To find b, solve the equation $c^2 = a^2 - b^2$ for b.

$$b = \sqrt{a^2 - c^2}$$
$$= \sqrt{39.44^2 - 9.821^2} \approx 38.20$$

Pluto's orbit is modeled by $\dfrac{x^2}{39.44^2} + \dfrac{y^2}{38.20^2} = 1$. Since $c \approx 9.821$, the foci are $(\pm 9.821, 0)$. The sun could be located at either focus. We locate the sun at $(9.821, 0)$.

To graph this ellipse on a graphing calculator, we must solve the equation for y.

$$\frac{x^2}{39.44^2} + \frac{y^2}{38.20^2} = 1$$
$$\frac{y^2}{38.20^2} = 1 - \frac{x^2}{39.44^2}$$
$$\frac{y}{38.20} = \pm\sqrt{1 - \frac{x^2}{39.44^2}}$$
$$y = \pm 38.20\sqrt{1 - \frac{x^2}{39.44^2}}$$

See Figures 10.33 and 10.34.

Graphing Pluto's Orbit

$[-60, 60, 10]$ by $[-40, 40, 10]$

Calculator Help

To learn how to access the variable Y_1, see Appendix A (page AP-10).

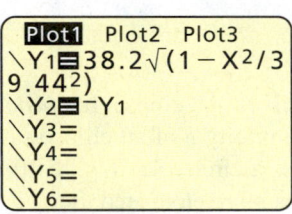

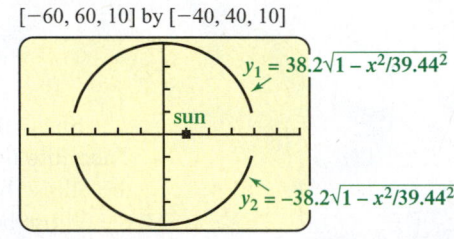

$y_1 = 38.2\sqrt{1 - x^2/39.44^2}$

sun

$y_2 = -38.2\sqrt{1 - x^2/39.44^2}$

Figure 10.33   **Figure 10.34**

Now Try Exercise 93

Reflective Property of Ellipses

Like parabolas, ellipses also have an important reflective property. If an ellipse is rotated about the x-axis, an **ellipsoid** is formed, which resembles the shell of an egg, as illustrated in Figure 10.35. If a light source is placed at focus F_1, then every beam of light emanating from the light source, regardless of its direction, is reflected at the surface of the ellipsoid toward focus F_2. See Figure 10.36

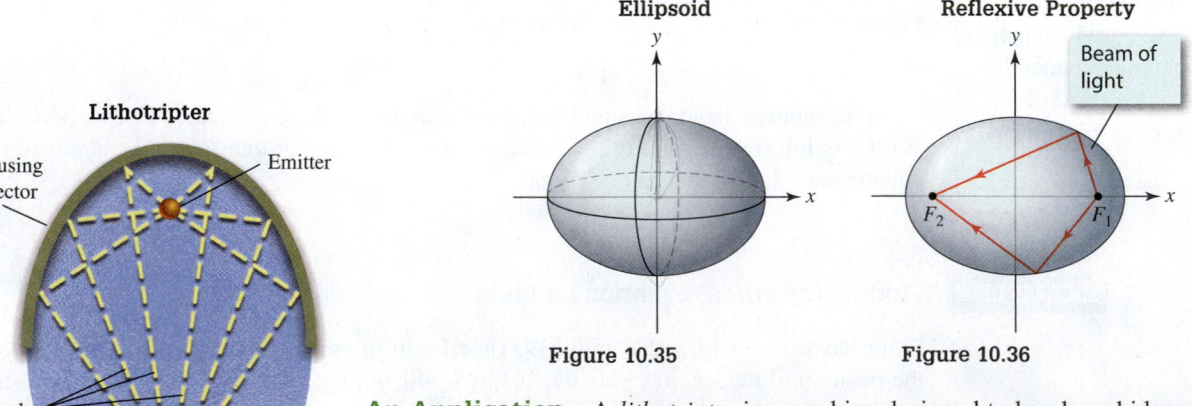

Ellipsoid

Reflexive Property

Beam of light

F_2 F_1

Figure 10.35   **Figure 10.36**

Lithotripter

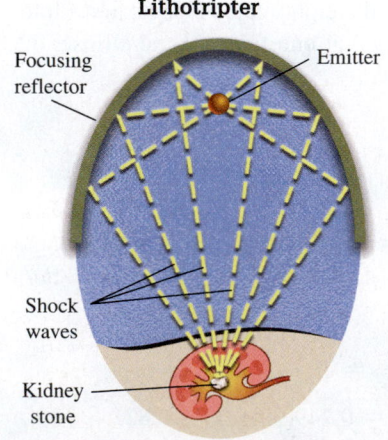

Focusing reflector

Emitter

Shock waves

Kidney stone

Figure 10.37

An Application   A *lithotripter* is a machine designed to break up kidney stones without surgery, by using the reflexive property of ellipses to focus powerful shock waves. See Figure 10.37. A patient is carefully positioned so that the kidney stone is at one focus, while the source of the shock waves is located at the other focus. The kidney stone absorbs all of the energy from the shock wave and is broken up without harming the patient. In Exercises 95 – 97 this reflective property of ellipses is applied.

Translations of Ellipses

If the equation of an ellipse is given by either $\frac{x^2}{a^2} + \frac{y^2}{b^2} = 1$ or $\frac{x^2}{b^2} + \frac{y^2}{a^2} = 1$, then the center of the ellipse is $(0, 0)$. We can use translations of graphs to find the equation of an ellipse centered at (h, k) by replacing x with $(x - h)$ and y with $(y - k)$.

> **STANDARD EQUATIONS FOR ELLIPSES CENTERED AT (h, k)**
>
> An ellipse with center (h, k) and either a horizontal or a vertical major axis satisfies one of the following equations, where $a > b > 0$ and $c^2 = a^2 - b^2$ with $c \geq 0$.
>
> $$\frac{(x - h)^2}{a^2} + \frac{(y - k)^2}{b^2} = 1 \qquad \text{Horizontal major axis; foci: } (h \pm c, k); \\ \text{Vertices: } (h \pm a, k)$$
>
> $$\frac{(x - h)^2}{b^2} + \frac{(y - k)^2}{a^2} = 1 \qquad \text{Vertical major axis; foci: } (h, k \pm c); \\ \text{Vertices: } (h, k \pm a)$$

EXAMPLE 5 Translating an ellipse

Translate the ellipse with equation $\frac{x^2}{9} + \frac{y^2}{4} = 1$ so that it is centered at $(-1, 2)$. Find the new equation and sketch its graph.

SOLUTION To translate the center from $(0, 0)$ to $(-1, 2)$, replace x with $(x - (-1))$ or $(x + 1)$ and replace y with $(y - 2)$. The new equation is

$$\frac{(x + 1)^2}{9} + \frac{(y - 2)^2}{4} = 1.$$

The given ellipse is shown in Figure 10.38, and the translated ellipse is shown in Figure 10.39.

Given Ellipse

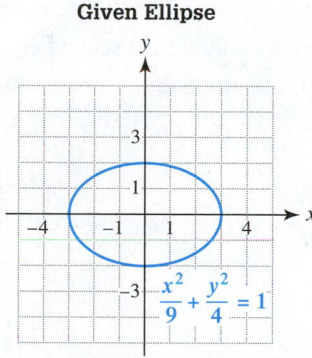

Figure 10.38

Translated Ellipse

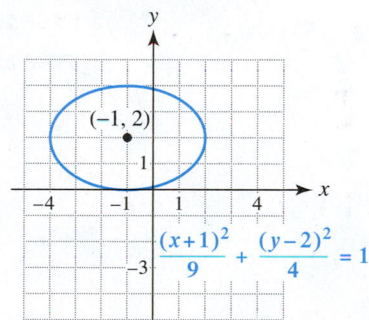

Figure 10.39

> **Now Try Exercise 33**

EXAMPLE 6 Graphing an ellipse with center (h, k)

Graph the ellipse given by $\frac{(x + 2)^2}{16} + \frac{(y - 2)^2}{25} = 1$. Label the vertices and the foci.

SOLUTION The ellipse has a vertical major axis, and its center is $(-2, 2)$. Since $a^2 = 25$ and $b^2 = 16$, it follows that $c^2 = a^2 - b^2 = 25 - 16 = 9$. Thus $a = 5$, $b = 4$, and $c = 3$. The vertices are located **5** units above and below the center of the ellipse, and the foci are located **3** units above and below the center of the ellipse. That is, the vertices are $(-2, 2 \pm 5)$, or $(-2, 7)$ and $(-2, -3)$, and the foci are $(-2, 2 \pm 3)$, or $(-2, 5)$ and $(-2, -1)$. A graph of the ellipse is shown in Figure 10.40

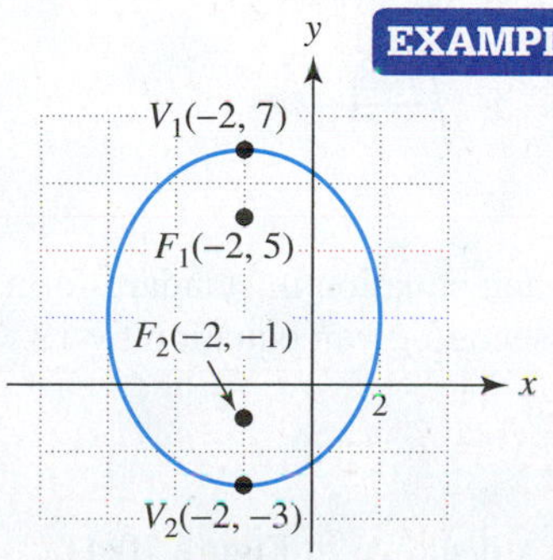

> **Now Try Exercise 47**

$V_1(-2, 7)$

$F_1(-2, 5)$

$F_2(-2, -1)$

$V_2(-2, -3)$

Figure 10.40

EXAMPLE 7 **Finding the standard equation of an ellipse**

Write $4x^2 - 16x + 9y^2 + 54y + 61 = 0$ in the standard form for an ellipse centered at (h, k). Identify the center and the vertices.

SOLUTION We can write the given equation in standard form by completing the square.

$$4x^2 - 16x + 9y^2 + 54y + 61 = 0 \qquad \text{Given equation}$$
$$4(x^2 - 4x + \underline{}) + 9(y^2 + 6y + \underline{}) = -61 \qquad \text{Distributive property}$$
$$4(x^2 - 4x + \underline{\mathbf{4}}) + 9(y^2 + 6y + \underline{\mathbf{9}}) = -61 + \mathbf{16} + \mathbf{81} \qquad \text{Complete the square.}$$
$$4(x - 2)^2 + 9(y + 3)^2 = 36 \qquad \text{Perfect square trinomials}$$
$$\frac{(x - 2)^2}{9} + \frac{(y + 3)^2}{4} = 1 \qquad \text{Divide each side by 36.}$$

The center is $(\mathbf{2}, -\mathbf{3})$. Because the major axis is horizontal and $a = \mathbf{3}$, the vertices of the ellipse are $(\mathbf{2} \pm \mathbf{3}, -\mathbf{3})$, or $(5, -3)$ and $(-1, -3)$.

Now Try Exercise 57

More Nonlinear Systems of Equations

In Sections 9.1 and 9.2 we discussed systems of equations and inequalities. In this and the next subsections we revisit these topics.

EXAMPLE 8 **Solving a nonlinear system of equations**

Use substitution to solve the following system of equations. Give graphical support.

$$9x^2 + 4y^2 = 36$$
$$12x^2 + y^2 = 12$$

SOLUTION

STEP 1: Begin by solving the second equation for y^2 to obtain $y^2 = 12 - 12x^2$.

STEP 2: Next, substitute $(\mathbf{12 - 12x^2})$ for $\mathbf{y^2}$ in the first equation and solve for x.

$$9x^2 + 4\mathbf{y^2} = 36 \qquad \text{First equation}$$
$$9x^2 + 4(\mathbf{12 - 12x^2}) = 36 \qquad \text{Let } y^2 = 12 - 12x^2.$$
$$9x^2 + 48 - 48x^2 = 36 \qquad \text{Distributive property}$$
$$-39x^2 = -12 \qquad \text{Subtract 48; simplify.}$$
$$x^2 = \frac{4}{13} \qquad \text{Divide by } -39; \text{ simplify.}$$
$$x = \pm\sqrt{\frac{4}{13}} \qquad \text{Square root property}$$

Four Solutions

$$\left(-\sqrt{\tfrac{4}{13}}, \sqrt{\tfrac{108}{13}}\right) \qquad \left(\sqrt{\tfrac{4}{13}}, \sqrt{\tfrac{108}{13}}\right)$$

$$\left(-\sqrt{\tfrac{4}{13}}, -\sqrt{\tfrac{108}{13}}\right) \qquad \left(\sqrt{\tfrac{4}{13}}, -\sqrt{\tfrac{108}{13}}\right)$$

Figure 10.41

STEP 3: To determine the y-values, substitute $x^2 = \frac{4}{13}$ in $y^2 = 12 - 12\mathbf{x^2}$.

$$y^2 = 12 - 12\left(\frac{4}{13}\right) = \frac{108}{13}, \quad \text{or} \quad y = \pm\sqrt{\frac{108}{13}}$$

There are four solutions: $\left(\pm\sqrt{\frac{4}{13}}, \pm\sqrt{\frac{108}{13}}\right)$.

To graph the system of equations by hand, put each equation in standard form by dividing the first given equation by 36 and the second given equation by 12.

$$\frac{x^2}{4} + \frac{y^2}{9} = 1 \qquad \text{and} \qquad x^2 + \frac{y^2}{12} = 1$$

The graphs of these ellipses and the four solutions are shown in Figure 10.41.

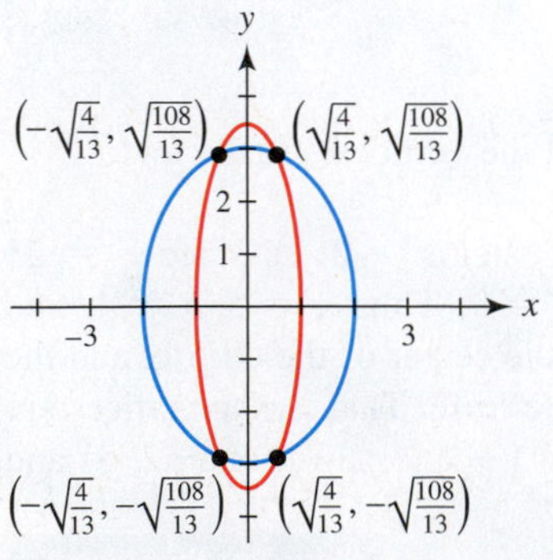

Now Try Exercise 71

NOTE The system of equations in Example 8 could also be solved by elimination. To do this, multiply the second equation by -4 and then add the equations.

More Nonlinear Systems of Inequalities

In the next example, we solve a nonlinear system of inequalities whose graph involves a parabola and an ellipse.

EXAMPLE 9   Solving a nonlinear inequality

Shade the region in the xy-plane that satisfies the system of inequalities.

$$36x^2 + 25y^2 \le 900$$
$$x + (y + 2)^2 \le 4$$

SOLUTION   Before sketching a graph, rewrite these two inequalities as follows.

First Inequality: $\dfrac{36}{900}x^2 + \dfrac{25}{900}y^2 \le \dfrac{900}{900}$ *Divide each term by 900.*

$\dfrac{x^2}{25} + \dfrac{y^2}{36} \le 1$ *Simplify to standard form.*

Second Inequality: $(y + 2)^2 \le -x + 4$ *Subtract x.*

$(y + 2)^2 \le -(x - 4)$ *Distributive property*

The first inequality represents the region inside an ellipse, as shown in Figure 10.42. The second inequality represents the region left of a parabola that opens to the left with vertex $(4, -2)$, as shown in Figure 10.43. The solution set for the system, which satisfies both inequalities, is shaded in Figure 10.44. (To verify this, try the test point $(-2, -2)$.)

First Inequality

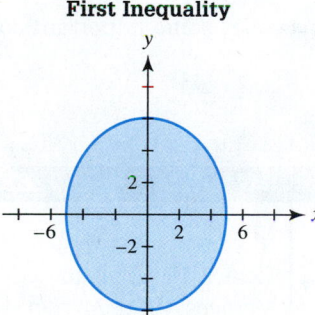

Figure 10.42

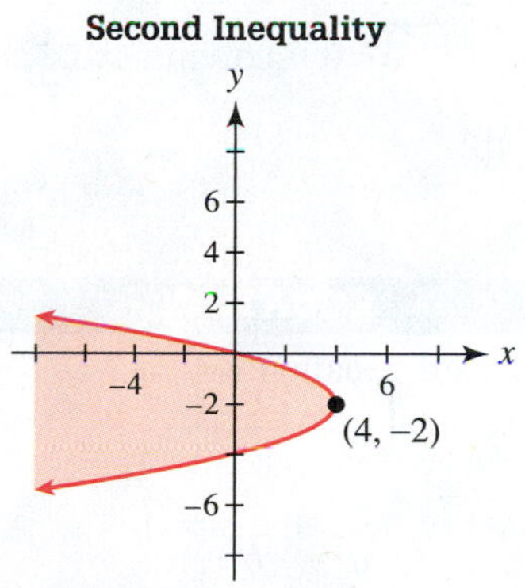

Second Inequality

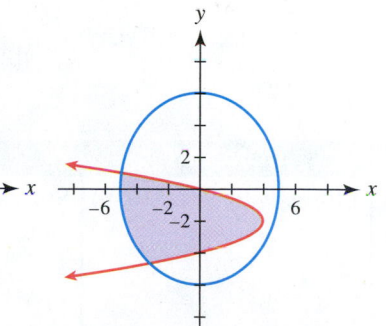

Figure 10.43

System of Inequalities

Figure 10.44

Now Try Exercise 87

Area Inside an Ellipse   The following formula can be used to calculate the area inside an ellipse.

CLASS DISCUSSION
Explain why the area formula for an ellipse is a generalization of the area formula for a circle.

AREA INSIDE AN ELLIPSE

The area A of the region contained inside an ellipse is given by $A = \pi ab$, where a and b are from the standard equation of the ellipse.

This formula is applied in the next example.

EXAMPLE 10 **Finding the area inside an ellipse**

Shade the region in the xy-plane that satisfies the inequality $x^2 + 4y^2 \le 4$. Find the area of this region if units are in inches.

SOLUTION Begin by dividing each term in the given inequality by **4**.

$$x^2 + 4y^2 \le 4 \qquad \textcolor{blue}{\text{Given inequality}}$$

$$\frac{x^2}{4} + \frac{4y^2}{4} \le \frac{4}{4} \qquad \textcolor{blue}{\text{Divide by 4.}}$$

$$\textcolor{red}{\begin{array}{l} a = 2 \\ b = 1 \end{array}} \qquad \frac{x^2}{4} + \frac{y^2}{1} \le 1 \qquad \textcolor{blue}{\text{Simplify the fractions.}}$$

The boundary of the region is the ellipse $\frac{x^2}{4} + \frac{y^2}{1} = 1$. The region *inside* the ellipse satisfies the inequality. To verify this fact, note that the test point $(0, 0)$, which is located inside the ellipse, satisfies the inequality. The solution set is shaded in Figure 10.45. The area of this elliptical region with $a = 2$ and $b = 1$ is

$$A = \pi ab = \pi(2)(1) = 2\pi \approx 6.28 \text{ square inches.}$$

Now Try Exercise 89

Elliptical Region

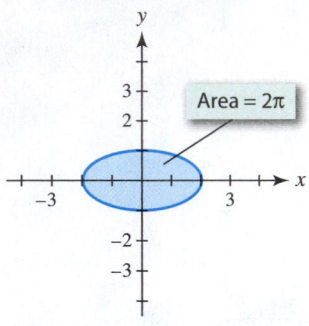

Figure 10.45

10.2 Putting It All Together

The following table summarizes some important concepts about ellipses.

CONCEPT	EQUATION	EXAMPLE
Ellipse with center $(0, 0)$	Standard equation with $a > b > 0$ Horizontal major axis: $$\frac{x^2}{a^2} + \frac{y^2}{b^2} = 1$$ Vertical major axis: $$\frac{x^2}{b^2} + \frac{y^2}{a^2} = 1$$ See the box on page 808.	$\frac{x^2}{4} + \frac{y^2}{9} = 1$; $a = 3, b = 2$ Center: $(0, 0)$; major axis: vertical Vertices; $(0, \pm 3)$; foci: $\left(0, \pm\sqrt{5}\right)$ $\left(c^2 = a^2 - b^2 = 9 - 4 = 5, \text{ so } c = \sqrt{5}.\right)$

CONCEPT	EQUATION	EXAMPLE
Ellipse with center (h, k)	Standard equation with $a > b > 0$ Horizontal major axis: $$\frac{(x - h)^2}{a^2} + \frac{(y - k)^2}{b^2} = 1$$ Vertical major axis: $$\frac{(x - h)^2}{b^2} + \frac{(y - k)^2}{a^2} = 1$$ See the box on page 813.	$\dfrac{(x - 1)^2}{4} + \dfrac{(y + 1)^2}{9} = 1;\ a = 3, b = 2$ Center: $(1, -1)$; major axis: vertical Vertices: $(1, -1 \pm 3)$; foci: $\left(1, -1 \pm \sqrt{5}\right)$ $\left(c^2 = a^2 - b^2 = 9 - 4 = 5, \text{ so } c = \sqrt{5}.\right)$
Area inside an ellipse	$A = \pi ab$	The area inside the ellipse given by $\frac{x^2}{49} + \frac{y^2}{9} = 1$ is $$A = \pi(7)(3) = 21\pi \text{ square units.}$$

10.2 Exercises

Basic Concepts

1. The endpoints of the major axis of an ellipse are called the _____ of the ellipse.

2. The foci of an ellipse are located on the _____ axis.

3. An ellipse with equation $\dfrac{x^2}{a^2} + \dfrac{y^2}{b^2} = 1$ $(a > b > 0)$ has a(n) __vertical/horizontal__ major axis.

4. An ellipse with equation $\dfrac{x^2}{b^2} + \dfrac{y^2}{a^2} = 1$ $(a > b > 0)$ has a(n) __vertical/horizontal__ major axis.

Ellipses with Center (0, 0)

Exercises 5–12: Graph the ellipse. Label the foci and the endpoints of each axis.

5. $\dfrac{x^2}{4} + \dfrac{y^2}{9} = 1$

6. $\dfrac{x^2}{9} + \dfrac{y^2}{4} = 1$

7. $\dfrac{x^2}{36} + \dfrac{y^2}{16} = 1$

8. $x^2 + \dfrac{y^2}{4} = 1$

9. $x^2 + 4y^2 = 400$

10. $9x^2 + 5y^2 = 45$

11. $25x^2 + 9y^2 = 225$

12. $5x^2 + 4y^2 = 20$

Exercises 13–16: Match the equation with its graph (a–d).

13. $\dfrac{x^2}{16} + \dfrac{y^2}{36} = 1$

14. $\dfrac{x^2}{4} + y^2 = 1$

15. $\dfrac{x^2}{16} + \dfrac{y^2}{4} = 1$

16. $\dfrac{x^2}{9} + \dfrac{y^2}{9} = 1$

a.

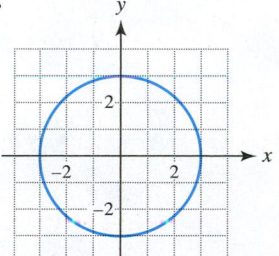

b.

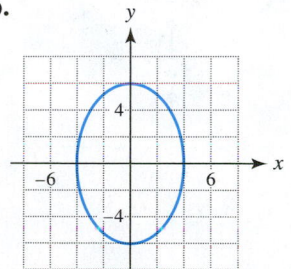

c.

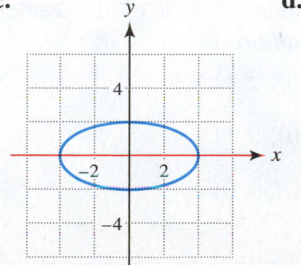

d.

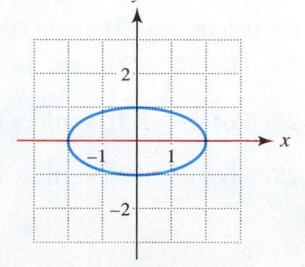

Exercises 17–20: Find the standard equation of the ellipse shown in the figure. Identify the coordinates of the vertices, endpoints of the minor axis, and the foci.

17.

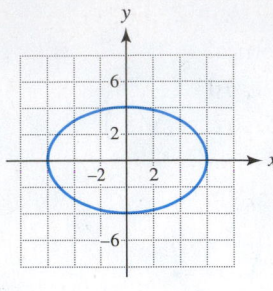

18.

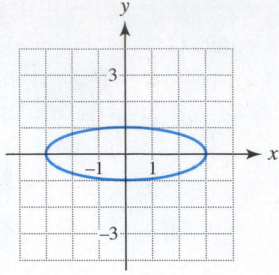

19.

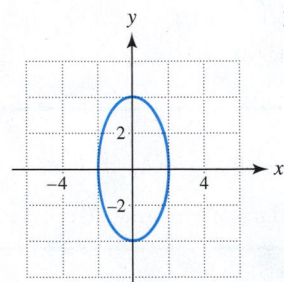

20.

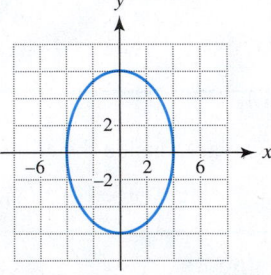

Exercises 21–24: The foci F_1 and F_2, vertices V_1 and V_2, and endpoints U_1 and U_2 of the minor axis of an ellipse are labeled in the figure. Graph the ellipse and find its standard equation. (The coordinates of V_1, V_2, F_1, and F_2 are integers.)

21.

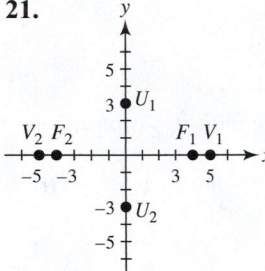

22.

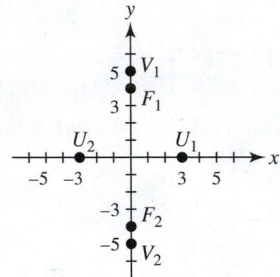

23.

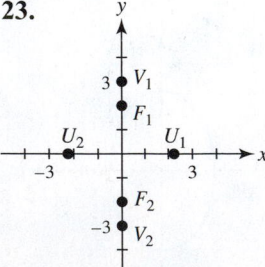

24.

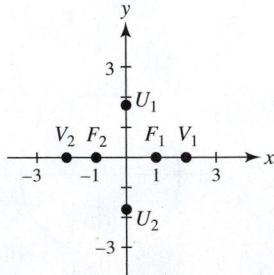

Exercises 25–32: Find an equation of the ellipse, centered at the origin, satisfying the conditions.

25. Foci $(0, \pm 2)$, vertices $(0, \pm 4)$

26. Foci $(0, \pm 3)$, vertices $(0, \pm 5)$

27. Foci $(\pm 5, 0)$, vertices $(\pm 6, 0)$

28. Foci $(\pm 4, 0)$, vertices $(\pm 6, 0)$

29. Horizontal major axis of length 8, minor axis of length 6

30. Vertical major axis of length 12, minor axis of length 8

31. Eccentricity $\frac{2}{3}$, horizontal major axis of length 6

32. Eccentricity $\frac{3}{4}$, vertices $(0, \pm 8)$

Ellipses with Center (*h*, *k*)

Exercises 33–36: Translate the ellipse with the given equation so that it is centered at the given point. Find the new equation and sketch its graph.

33. $\dfrac{x^2}{4} + \dfrac{y^2}{3} = 1; (2, -1)$ **34.** $\dfrac{x^2}{9} + \dfrac{y^2}{2} = 1; (-3, 7)$

35. $\dfrac{x^2}{2} + \dfrac{y^2}{9} = 1; (-3, -4)$ **36.** $\dfrac{x^2}{15} + \dfrac{y^2}{16} = 1; (5, -6)$

Exercises 37–42: Sketch a graph of the ellipse.

37. $\dfrac{(x - 2)^2}{4} + \dfrac{(y - 1)^2}{9} = 1$

38. $\dfrac{(x + 1)^2}{16} + \dfrac{(y + 3)^2}{9} = 1$

39. $\dfrac{(x + 1)^2}{16} + \dfrac{(y + 2)^2}{25} = 1$

40. $\dfrac{(x - 4)^2}{9} + \dfrac{y^2}{4} = 1$ **41.** $\dfrac{(x + 2)^2}{4} + y^2 = 1$

42. $x^2 + \dfrac{(y - 3)^2}{4} = 1$

Exercises 43–46: Match the equation with its graph (a–d).

43. $\dfrac{(x - 2)^2}{16} + \dfrac{(y + 4)^2}{36} = 1$

44. $\dfrac{(x + 1)^2}{4} + \dfrac{y^2}{9} = 1$

45. $\dfrac{(x + 1)^2}{9} + \dfrac{(y - 1)^2}{4} = 1$

46. $\dfrac{x^2}{25} + \dfrac{(y + 1)^2}{10} = 1$

a.

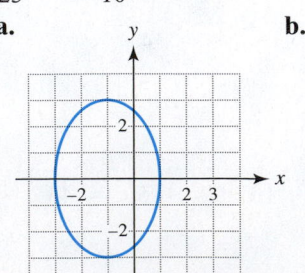

b.

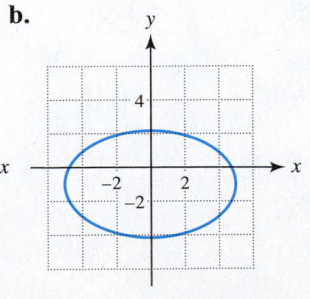

c. 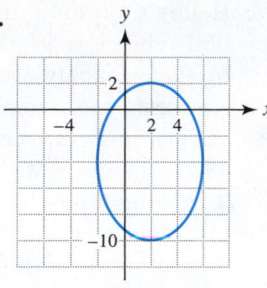 **d.**

Exercises 47–50: *Sketch a graph of the ellipse. Identify the foci and vertices.*

47. $\dfrac{(x-1)^2}{9} + \dfrac{(y-1)^2}{25} = 1$

48. $\dfrac{(x+2)^2}{25} + \dfrac{(y+1)^2}{16} = 1$

49. $\dfrac{(x+4)^2}{16} + \dfrac{(y-2)^2}{9} = 1$

50. $\dfrac{x^2}{4} + \dfrac{(y-1)^2}{9} = 1$

Exercises 51–54: *Find an equation of an ellipse that satisfies the given conditions.*

51. Center (2, 1), focus (2, 3), and vertex (2, 4)

52. Center $(-3, -2)$, focus $(-1, -2)$, and vertex $(1, -2)$

53. Vertices $(\pm 3, 2)$ and foci $(\pm 2, 2)$

54. Vertices $(-1, \pm 3)$ and foci $(-1, \pm 1)$

Exercises 55 and 56: *Find an (approximate) equation of the ellipse shown in the figure.*

55. 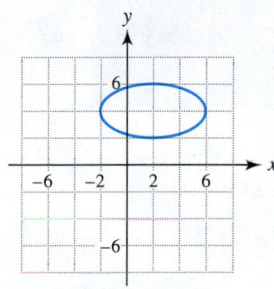   **56.**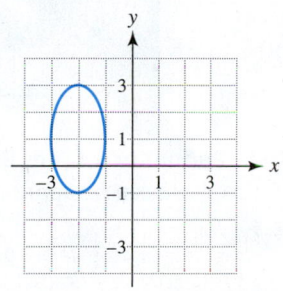

Exercises 57–64: *(Refer to Example 7.) Write the equation in standard form for an ellipse centered at (h, k). Identify the center and the vertices.*

57. $9x^2 + 18x + 4y^2 - 8y - 23 = 0$

58. $9x^2 - 36x + 16y^2 - 64y - 44 = 0$

59. $4x^2 + 8x + y^2 + 2y + 1 = 0$

60. $x^2 - 6x + 9y^2 = 0$

61. $4x^2 + 16x + 5y^2 - 10y + 1 = 0$

62. $2x^2 + 4x + 3y^2 - 18y + 23 = 0$

63. $16x^2 - 16x + 4y^2 + 12y = 51$

64. $16x^2 + 48x + 4y^2 - 20y + 57 = 0$

Graphing Ellipses with Technology

Exercises 65–68: *Graph the ellipse.*

65. $\dfrac{x^2}{15} + \dfrac{y^2}{10} = 1$ **66.** $\dfrac{(x-1.2)^2}{7.1} + \dfrac{y^2}{3.5} = 1$

67. $4.1x^2 + 6.3y^2 = 25$ **68.** $\frac{1}{2}x^2 + \frac{1}{3}y^2 = \frac{1}{6}$

Solving Equations and Inequalities

Exercises 69–74: *Solve the system of equations. Give graphical support by making a sketch.*

69. $\dfrac{x^2}{4} + \dfrac{y^2}{9} = 1$ **70.** $\dfrac{x^2}{16} + \dfrac{y^2}{25} = 1$
$\quad\ \ x + y = 3$ $\quad -2x + y = 5$

71. $4x^2 + 16y^2 = 64$ **72.** $4x^2 + y^2 = 4$
$\quad\ x^2 + \ \ y^2 = 9$ $\quad\ x^2 + y^2 = 2$

73. $\ x^2 + \ y^2 = 9$ **74.** $\qquad x^2 + y^2 = 4$
$\ 2x^2 + 3y^2 = 18$ $\quad (x-1)^2 + y^2 = 4$

Exercises 75–80: *Solve the system of equations.*

75. $\dfrac{x^2}{2} + \dfrac{y^2}{4} = 1$ **76.** $x^2 + \frac{1}{9}y^2 = 1$
$\ -x^2 + 2y = 4$ $\quad x\ +\ y\ = 3$

77. $\dfrac{x^2}{2} + \dfrac{y^2}{4} = 1$ **78.** $\dfrac{x^2}{5} + \dfrac{y^2}{10} = 1$
$\ \dfrac{x^2}{4} + \dfrac{y^2}{2} = 1$ $\ \dfrac{x^2}{10} + \dfrac{y^2}{5} = 1$

79. $(x-2)^2 + y^2 = 9$ **80.** $(x-2) - y^2 = 0$
$\quad x^2 + y^2 = 9$ $\quad \dfrac{x^2}{4} + \dfrac{y^2}{9} = 1$

Exercises 81–88: *Shade the solution set to the system.*

81. $(x-1)^2 + (y+1)^2 < 4$
$\ \ (x+1)^2 + y^2 \qquad\ > 1$

82. $\dfrac{x^2}{16} + \dfrac{y^2}{25} < 1$
$\ \dfrac{x^2}{4} + \dfrac{y^2}{9} > 1$

83. $\dfrac{x^2}{4} + \dfrac{y^2}{9} \le 1$ **84.** $\dfrac{x^2}{16} + \dfrac{y^2}{25} \le 1$
$\ x\ +\ y\ \ge 2$ $\ -x + y\ \ \le 4$

85. $\quad x^2 + y^2 \leq 4$
$\quad x^2 + (y - 2)^2 \leq 4$

86. $\quad x^2 + (y + 1)^2 \leq 9$
$\quad (x + 1)^2 + y^2 \leq 9$

87. $\quad x^2 + y^2 \leq 4$
$\quad (x + 1)^2 - y \leq 0$

88. $\quad 4x^2 + 9y^2 \leq 36$
$\quad x - (y - 2)^2 \geq 0$

Exercises 89–92: (Refer to Example 10.) Shade the region in the xy-plane that satisfies the given inequality. Find the area of this region if units are in feet.

89. $4x^2 + 9y^2 \leq 36$

90. $9x^2 + y^2 \leq 9$

91. $\dfrac{(x - 1)^2}{25} + \dfrac{(y + 2)^2}{16} \leq 1$

92. $\dfrac{(x + 3)^2}{4} + \dfrac{(y - 2)^2}{8} \leq 1$

Applications

Exercises 93 and 94: **Orbits of Planets** *(Refer to Example 4.) Find an equation for the orbit of the planet. Graph its orbit and the location of the sun at a focus on the positive x-axis.*

93. Mercury: $e = 0.206$, $a = 0.387$

94. Mars: $e = 0.093$, $a = 1.524$

95. Lithotripter (Refer to the discussion in this section.) The source of a shock wave is placed at one focus of an ellipsoid with a major axis of 8 inches and a minor axis of 5 inches. Estimate, to the nearest thousandth of an inch, how far a kidney stone should be positioned from the source.

96. Shape of a Lithotripter A patient's kidney stone is placed 12 units away from the source of the shock waves of a lithotripter. The lithotripter is based on an ellipse with a minor axis that measures 16 units. Find the equation of an ellipse that would satisfy this situation.

97. Whispering Gallery A large room constructed in the shape of the upper half of an ellipsoid has a unique property. Any sound emanating from one focus is reflected directly toward the other focus. See the figure. If the foci are 100 feet apart and the maximum height of the ceiling is 40 feet, estimate the area of the floor of the room.

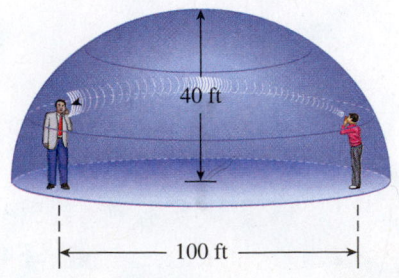

98. Halley's Comet Halley's comet travels in an elliptical orbit with $a = 17.95$ and $b = 4.44$ and passes by Earth roughly every 76 years. Note that each unit represents one astronomical unit, or 93 million miles. The comet most recently passed by Earth in February 1986. (*Source: M. Zeilik, Introductory Astronomy and Astrophysics.*)

(a) Write an equation for this orbit, centered at (0, 0) with major axis on the x-axis.

(b) If the sun lies (at the focus) on the positive x-axis, approximate its coordinates.

(c) Determine the maximum and minimum distances between Halley's comet and the sun.

99. The Roman Colosseum The perimeter of the Roman Colosseum is an ellipse with major axis 620 feet and minor axis 513 feet. Find the distance between the foci of this ellipse.

100. Orbit of Earth (Refer to Example 4.) Earth has a nearly circular orbit with $e \approx 0.0167$ and $a = 93$ million miles. Approximate the minimum and maximum distances between Earth and the sun. (*Source: M. Zeilik, Introductory Astronomy and Astrophysics.*)

101. Arch Bridge An elliptical arch under a bridge is constructed so that it is 60 feet wide and has a maximum height of 25 feet, as illustrated in the figure. Find the height of the arch 15 feet from the center of the arch.

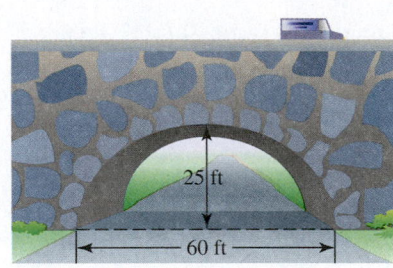

102. Perimeter of an Ellipse The perimeter P of an ellipse can be approximated by

$$P \approx 2\pi \sqrt{\frac{a^2 + b^2}{2}}.$$

(a) Approximate the distance in miles that Mercury travels in one orbit of the sun if $a = 36.0$, $b = 35.2$, and the units are in millions of miles.

(b) If a planet has a circular orbit, does this formula give the *exact* perimeter? Explain.

103. Satellite Orbit The orbit of *Explorer VII* and the outline of Earth's surface are shown in the figure at the top of the next column. This orbit can be described by the equation $\dfrac{x^2}{a^2} + \dfrac{y^2}{b^2} = 1$, where $a = 4464$ and $b = 4462$. The surface of Earth can be described by $(x - 164)^2 + y^2 = 3960^2$. Find the maximum and minimum heights of the satellite above Earth's surface if all units are in miles. (*Sources:* W. Loh; W. Thomson.)

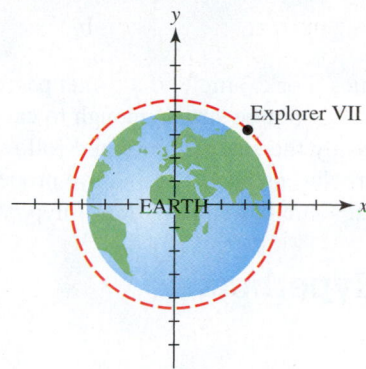

Explorer VII

EARTH

104. Orbital Velocity The maximum and minimum velocities in kilometers per second of a celestial body moving in an elliptical orbit can be calculated by

$$v_{max} = \frac{2\pi a}{P}\sqrt{\frac{1+e}{1-e}} \quad \text{and} \quad v_{min} = \frac{2\pi a}{P}\sqrt{\frac{1-e}{1+e}}.$$

In these equations, a is half the length of the major axis of the orbit in kilometers, P is the orbital period in seconds, and e is the eccentricity of the orbit. (*Source:* M. Zeilik.)

(a) Find v_{max} and v_{min} for Pluto if $a = 5.913 \times 10^9$ kilometers, the period is $P = 2.86 \times 10^{12}$ seconds, and the eccentricity is $e = 0.249$.

(b) If a planet has a circular orbit, what can be said about its orbital velocity?

Writing about Mathematics

105. Explain how the distance between the foci of an ellipse affects the shape of the ellipse.

106. Given the standard equation of an ellipse, explain how to determine the length of the major axis. How can you determine whether the major axis is vertical or horizontal?

Extended and Discovery Exercises

Exercises 1–4: (Refer to the first Making Connections in this Section.) Write in intercept form the equation of the line satisfying the given conditions. Then find the x- and y-intercepts of the line.

1. Passing through $(-2, 6)$ and $(4, -3)$

2. Passing through $(-6, -4)$ and $(3, 8)$

3. Slope -2, passing through $(3, -1)$

4. Slope 4, passing through $(-2, 1)$

Exercises 5 and 6: (Refer to the first Making Connections in this Section.) Find the x- and y-intercepts of the ellipse.

5. $\dfrac{x^2}{25} + \dfrac{y^2}{9} = 1$

6. Vertices $(0, \pm 13)$ and foci $(0, \pm 12)$

CHECKING BASIC CONCEPTS FOR SECTIONS 10.1 AND 10.2

1. Graph the parabola defined by $x = \frac{1}{2}y^2$. Include the focus and directrix.

2. Find an equation of the parabola with focus $(-1, 0)$ and directrix $y = 3$.

3. Graph the ellipse defined by $\frac{x^2}{36} + \frac{y^2}{100} = 1$. Include the foci and label the major and minor axes.

4. Find an equation of the ellipse centered at $(3, -2)$ with a vertical major axis of length 6 and minor axis of length 4. What are the coordinates of the foci?

5. A parabolic reflector for a searchlight has a diameter of 4 feet and a depth of 1 foot. How far from the vertex should the filament of the light bulb be located?

6. Solve the nonlinear system of equations.

$$x^2 + y^2 = 10$$
$$2x^2 + 3y^2 = 29$$

7. Write $x^2 - 4x + 4y^2 + 8y - 8 = 0$ in the standard form for an ellipse centered at (h, k). Identify the center and the vertices.

10.3 Hyperbolas

- **Find equations of hyperbolas**
- **Graph hyperbolas**
- **Learn the reflective property of hyperbolas**
- **Translate hyperbolas**

Introduction

Hyperbolas have several interesting properties. For example, if a comet passes by the sun with a high velocity, then the sun's gravity may not be strong enough to cause the comet to go into orbit; instead, the comet will pass by the sun just once and follow a trajectory that can be described by a hyperbola. Hyperbolas also have a reflective property, which is used in telescopes. In this section we look at some basic properties of hyperbolas.

Equations and Graphs of Hyperbolas

A third type of conic section is a hyperbola.

> **HYPERBOLA**
>
> A **hyperbola** is the set of points in a plane, the difference of whose distances from two fixed points is constant. Each fixed point is called a **focus** of the hyperbola.

The following See the Concept shows some of the important features of hyperbolas.

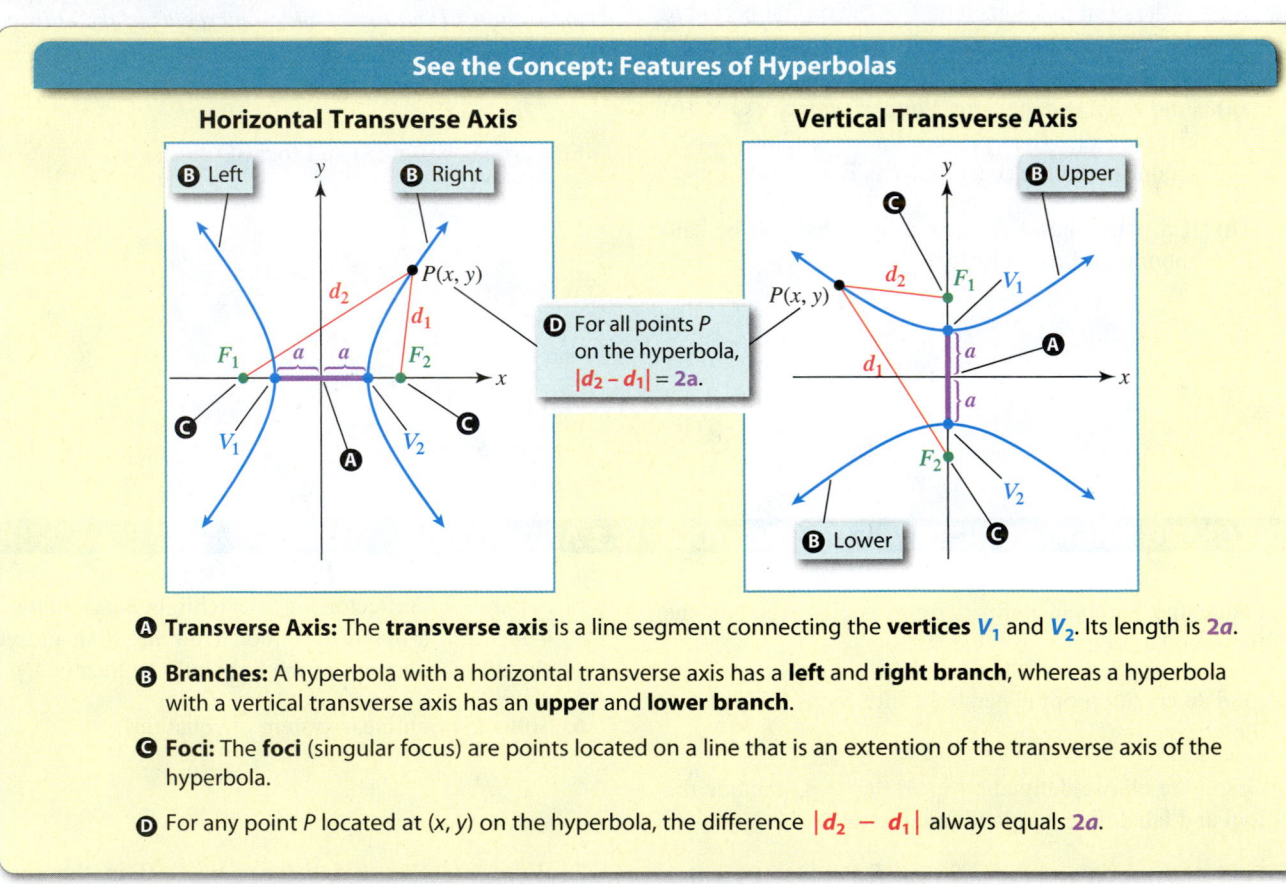

See the Concept: Features of Hyperbolas

Horizontal Transverse Axis

Vertical Transverse Axis

D For all points P on the hyperbola, $|d_2 - d_1| = 2a$.

A Transverse Axis: The **transverse axis** is a line segment connecting the **vertices** V_1 and V_2. Its length is **2a**.

B Branches: A hyperbola with a horizontal transverse axis has a **left** and **right branch**, whereas a hyperbola with a vertical transverse axis has an **upper** and **lower branch**.

C Foci: The **foci** (singular focus) are points located on a line that is an extention of the transverse axis of the hyperbola.

D For any point P located at (x, y) on the hyperbola, the difference $|d_2 - d_1|$ always equals **2a**.

Two hyperbolas are shown in Figures 10.46 and 10.47. The vertices, $(\pm a, 0)$ or $(0, \pm a)$, and foci, $(\pm c, 0)$ or $(0, \pm c)$, are labeled. A line segment connecting the points $(0, \pm b)$ in Figure 10.46 and $(\pm b, 0)$ in Figure 10.47 is the **conjugate axis**. The dashed lines $y = \pm \frac{b}{a}x$ and $y = \pm \frac{a}{b}x$ are **asymptotes** for the respective hyperbolas. They can be used as an aid in graphing. The dashed rectangle is called the **fundamental rectangle**.

Horizontal Transverse Axis   **Vertical Transverse Axis**

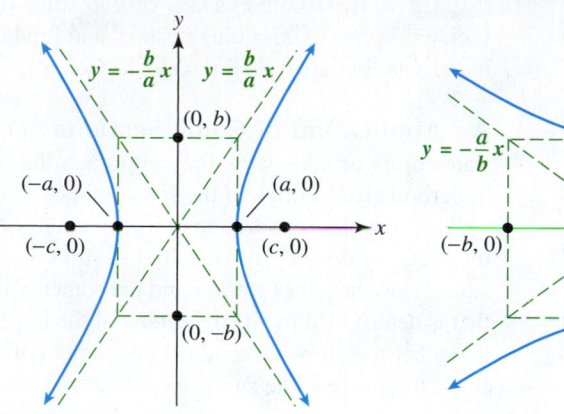

Figure 10.46   Figure 10.47

By the vertical line test, a hyperbola cannot be represented by a function, but many can be described by the following equations. The constants a, b, and c are *positive*.

STANDARD EQUATIONS FOR HYPERBOLAS CENTERED AT (0, 0)

The hyperbola with center at the origin, *horizontal* transverse axis, and equation

$$\frac{x^2}{a^2} - \frac{y^2}{b^2} = 1$$

has asymptotes $y = \pm \frac{b}{a}x$, vertices $(\pm a, 0)$, and foci $(\pm c, 0)$, where $c^2 = a^2 + b^2$. The hyperbola with center at the origin, *vertical* transverse axis, and equation

$$\frac{y^2}{a^2} - \frac{x^2}{b^2} = 1$$

has asymptotes $y = \pm \frac{a}{b}x$, vertices $(0, \pm a)$, and foci $(0, \pm c)$, where $c^2 = a^2 + b^2$.

EXAMPLE 1   **Sketching the graph of a hyperbola**

Sketch a graph of $\frac{x^2}{4} - \frac{y^2}{9} = 1$. Label the vertices, foci, and asymptotes.

SOLUTION   The equation is in standard form with $a = 2$ and $b = 3$. It has a horizontal transverse axis with vertices $(\pm 2, 0)$. The endpoints of the conjugate axis are $(0, \pm 3)$. To locate the foci, find c.

$$c^2 = a^2 + b^2 = 2^2 + 3^2 = 13, \qquad \text{or} \qquad c = \sqrt{13} \approx 3.61.$$

The foci are $\left(\pm \sqrt{13}, 0 \right)$. The asymptotes are $y = \pm \frac{b}{a}x$, or $y = \pm \frac{3}{2}x$. See Figure 10.48.

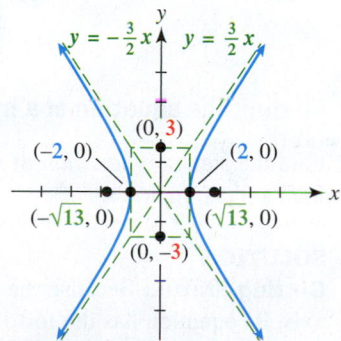

Figure 10.48

Now Try Exercise 5

NOTE A hyperbola consists of two solid (blue) curves, or branches. The asymptotes, foci, transverse axis, conjugate axis, and fundamental (dashed) rectangle are not part of the hyperbola, but are aids for sketching its graph.

An Application of Conic Sections One interpretation of an asymptote relates to trajectories of comets as they approach the sun. Comets travel in parabolic, elliptic, or hyperbolic trajectories. If the speed of a comet is too slow, the gravitational pull of the sun will capture the comet in an elliptical orbit. See Figure 10.49. If the speed of the comet is too fast, the comet will pass by the sun once in a hyperbolic trajectory; farther from the sun, gravity becomes weaker and the comet will eventually return to a straight-line trajectory that is determined by the *asymptote* of the hyperbola. See Figure 10.50. Finally, if the speed is neither too slow nor too fast, the comet will travel in a parabolic path. See Figure 10.51. In all three cases, the sun is located at a focus of the conic section.

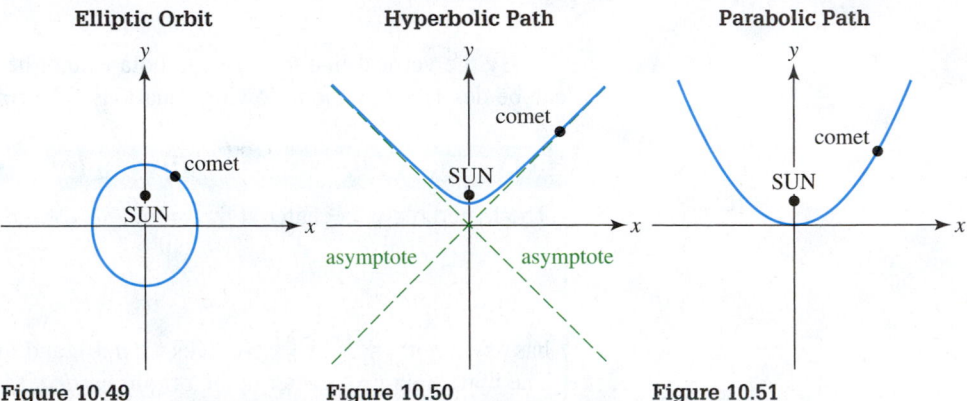

Figure 10.49   Figure 10.50   Figure 10.51

EXAMPLE 2 **Finding the equation of a hyperbola**

Find the equation of the hyperbola centered at the origin with a vertical transverse axis of length 6 and focus (0, 5). Also find the equations of its asymptotes.

SOLUTION Since the hyperbola is centered at the origin with a vertical transverse axis, its equation is $\frac{y^2}{a^2} - \frac{x^2}{b^2} = 1$. The transverse axis has length $6 = 2a$, so $a = 3$. Since one focus is located at (0, 5), it follows that $c = 5$. We can find b by using the following equation.

$$b^2 = c^2 - a^2$$
$$b = \sqrt{c^2 - a^2}$$
$$b = \sqrt{5^2 - 3^2} = 4$$

The standard equation of this hyperbola is $\frac{y^2}{9} - \frac{x^2}{16} = 1$. Its asymptotes are $y = \pm\frac{a}{b}x$, or $y = \pm\frac{3}{4}x$.

Now Try Exercise 23

EXAMPLE 3 **Finding the equation of a hyperbola**

Find the standard equation of the hyperbola shown in Figure 10.52. Identify the vertices, foci, and asymptotes.

SOLUTION
Getting Started Because the hyperbola is centered at (0, 0) with a horizontal transverse axis, its equation has the form

$$\frac{x^2}{a^2} - \frac{y^2}{b^2} = 1.$$

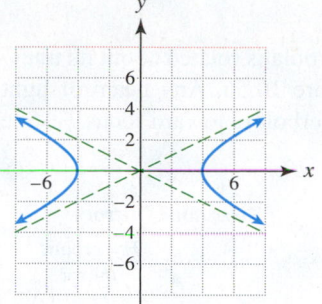

Figure 10.52

Sketch the fundamental rectangle first. Half of its length equals a, and half of its width equals b. ▶

In Figure 10.53 the fundamental rectangle is determined by the four points $(\pm 4, 0)$ and $(0, \pm 2)$, and its diagonals correspond to the asymptotes. It follows that $a = 4$ and $b = 2$. Thus the standard equation of the hyperbola is

$$\frac{x^2}{4^2} - \frac{y^2}{2^2} = 1 \quad \text{or} \quad \frac{x^2}{16} - \frac{y^2}{4} = 1.$$

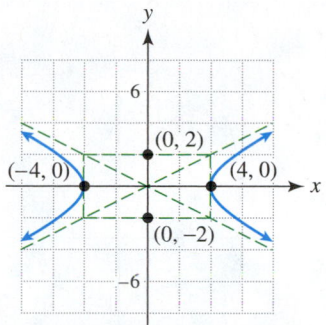

Figure 10.53

The vertices are $(\pm 4, 0)$, and the asymptotes are $y = \pm\frac{1}{2}x$. To find the foci, find c.

$$c^2 = a^2 + b^2 = 4^2 + 2^2 = 20, \quad \text{or} \quad c = \sqrt{20}.$$

The foci are $\left(\pm\sqrt{20}, 0\right)$.

Now Try Exercise 41

Using Technology   In the next example, a graphing calculator is used to graph a hyperbola.

EXAMPLE 4   **Graphing a hyperbola with technology**

Use a graphing calculator to graph $\frac{y^2}{4.2} - \frac{x^2}{8.4} = 1$.

SOLUTION   Begin by solving the given equation for y.

$$\frac{y^2}{4.2} = 1 + \frac{x^2}{8.4} \qquad \text{Add } \tfrac{x^2}{8.4}.$$

$$y^2 = 4.2\left(1 + \frac{x^2}{8.4}\right) \qquad \text{Multiply by 4.2.}$$

$$y = \pm\sqrt{4.2\left(1 + \frac{x^2}{8.4}\right)} \qquad \text{Square root property}$$

Graph $Y_1 = \sqrt{(4.2(1 + X^2/8.4))}$ and $Y_2 = -\sqrt{(4.2(1 + X^2/8.4))}$. See Figures 10.54 and 10.55.

$[-10, 10, 1]$ by $[-10, 10, 1]$

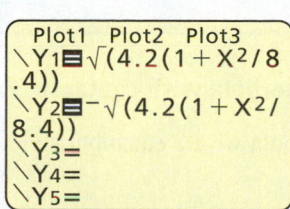

```
Plot1  Plot2  Plot3
\Y1■√(4.2(1+X²/8
.4))
\Y2■-√(4.2(1+X²/
8.4))
\Y3=
\Y4=
\Y5=
```

$y_1 = \sqrt{4.2(1 + x^2/8.4)}$

$y_2 = -\sqrt{4.2(1 + x^2/8.4)}$

Figure 10.54   **Figure 10.55**

Now Try Exercise 59

Reflective Property of Hyperbolas

Hyperbolas have an important reflective property. If a hyperbola is rotated about its transverse axis, a **hyperboloid** is formed, as illustrated in Figure 10.56. Any beam of light that is directed toward focus F_1 will be reflected by the hyperboloid toward focus F_2. See Figure 10.57.

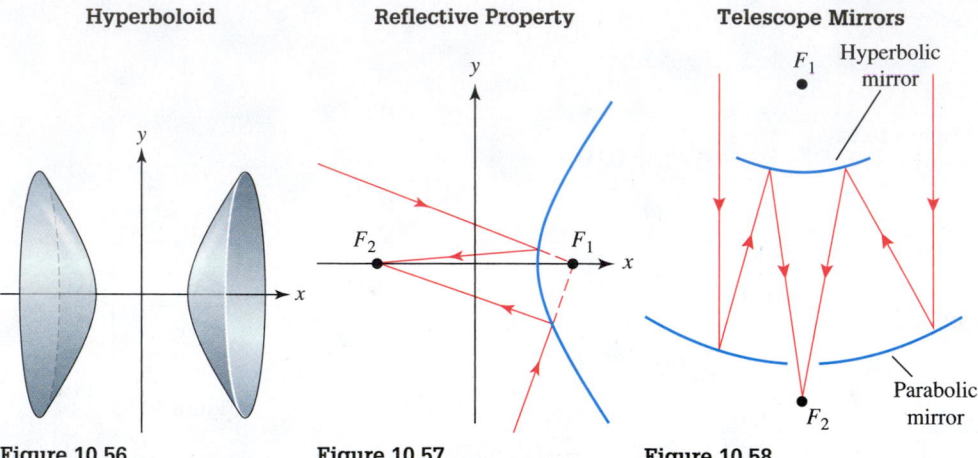

| Hyperboloid | Reflective Property | Telescope Mirrors |

Figure 10.56   Figure 10.57   Figure 10.58

An Application Telescopes sometimes make use of both parabolic and hyperbolic mirrors, as shown in Figure 10.58. When parallel rays of light from distant stars strike the large parabolic (primary) mirror, they are reflected toward its focus, F_1. A smaller hyperbolic (secondary) mirror is placed so that its focus is also located at F_1. Light rays striking the hyperbolic mirror are reflected toward its other focus, F_2, through a small hole in the parabolic mirror, and into an eye piece. See Exercise 72.

Translations of Hyperbolas

If the equation of a hyperbola is either $\frac{x^2}{a^2} - \frac{y^2}{b^2} = 1$ or $\frac{y^2}{a^2} - \frac{x^2}{b^2} = 1$ then the center of the hyperbola is $(0, 0)$. We can use translations of graphs to find the equation of a hyperbola centered at (h, k) by replacing x with $(x - h)$ and y with $(y - k)$. The constants a, b, and c are positive.

STANDARD EQUATIONS FOR HYPERBOLAS CENTERED AT (h, k)

A hyperbola with center (h, k) and either a horizontal or a vertical transverse axis satisfies one of the following equations, where $c^2 = a^2 + b^2$.

$$\frac{(x - h)^2}{a^2} - \frac{(y - k)^2}{b^2} = 1$$

Transverse axis: horizontal
Vertices: $(h \pm a, k)$; foci: $(h \pm c, k)$
Asymptotes: $y = \pm \frac{b}{a}(x - h) + k$

$$\frac{(y - k)^2}{a^2} - \frac{(x - h)^2}{b^2} = 1$$

Transverse axis: vertical
Vertices: $(h, k \pm a)$; foci: $(h, k \pm c)$
Asymptotes: $y = \pm \frac{a}{b}(x - h) + k$

EXAMPLE 5   Graphing a hyperbola with center (h, k)

Graph the hyperbola whose equation is $\frac{(y + 2)^2}{9} - \frac{(x - 2)^2}{16} = 1$. Label the vertices, foci, and asymptotes.

SOLUTION   The hyperbola has a vertical transverse axis, and its center is $(2, -2)$. Since $a^2 = 9$ and $b^2 = 16$, it follows that $c^2 = a^2 + b^2 = 9 + 16 = 25$. Thus $a = \mathbf{3}$, $b = \mathbf{4}$,

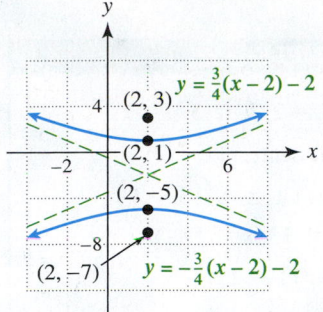

Figure 10.59

and $c = 5$. The vertices are located **3** units above and below the center of the hyperbola, and the foci are located **5** units above and below the center of the hyperbola. That is, the vertices are $(2, -2 \pm 3)$, or $(2, 1)$ and $(2, -5)$, and the foci are $(2, -2 \pm 5)$, or $(2, 3)$ and $(2, -7)$. The asymptotes are given by

$$y = \pm \frac{a}{b}(x - h) + k, \quad \text{or} \quad y = \pm \frac{3}{4}(x - 2) - 2.$$

A graph of the hyperbola is shown in Figure 10.59.

> **Now Try Exercise 45**

EXAMPLE 6   **Finding the standard equation of a hyperbola**

Write $9x^2 - 18x - 4y^2 - 16y = 43$ in the standard form for a hyperbola centered at (h, k). Identify the center and the vertices.

SOLUTION   We can write the given equation in standard form by completing the square.

$9x^2 - 18x - 4y^2 - 16y = 43$	Given equation
$9(x^2 - 2x + \underline{}) - 4(y^2 + 4y + \underline{}) = 43$	Distributive property
$9(x^2 - 2x + \underline{\ 1\ }) - 4(y^2 + 4y + \underline{\ 4\ }) = 43 + 9 - 16$	Complete the square.
$9(x - 1)^2 - 4(y + 2)^2 = 36$	Perfect square trinomials
$\dfrac{(x - 1)^2}{4} - \dfrac{(y + 2)^2}{9} = 1$	Divide each side by 36.

The center is $(1, -2)$. Because $a = 2$ and the transverse axis is horizontal, the vertices of the hyperbola are $(h \pm a, k) = (1 \pm 2, -2)$.

> **Now Try Exercise 51**

10.3 Putting It All Together

The following table summarizes some important concepts about hyperbolas.

CONCEPT	EQUATION	EXAMPLE
Hyperbola with center $(0, 0)$	Standard equation Transverse axis: horizontal $\dfrac{x^2}{a^2} - \dfrac{y^2}{b^2} = 1$ Transverse axis: vertical $\dfrac{y^2}{a^2} - \dfrac{x^2}{b^2} = 1$ See the box on page 823.	$\dfrac{y^2}{4} - \dfrac{x^2}{9} = 1$; $a = 2, b = 3$ Transverse axis: vertical Vertices: $(0, \pm 2)$; foci: $\left(0, \pm\sqrt{13}\right)$ $\left(c^2 = a^2 + b^2 = 4 + 9 = 13, \text{ so } c = \sqrt{13}.\right)$ Asymptotes: $y = \pm\frac{2}{3}x$

continued on next page

CONCEPT	EQUATION	EXAMPLE
Hyperbola with center (h, k)	Standard equation Transverse axis: horizontal $$\frac{(x-h)^2}{a^2} - \frac{(y-k)^2}{b^2} = 1$$ Transverse axis: vertical $$\frac{(y-k)^2}{a^2} - \frac{(x-h)^2}{b^2} = 1$$ See the box on page 826.	$\frac{(x-1)^2}{4} - \frac{(y+1)^2}{9} = 1; a = 2, b = 3$ Transverse axis: horizontal; center: $(1, -1)$ Vertices: $(1 \pm 2, -1)$; foci: $\left(1 \pm \sqrt{13}, -1\right)$ $\left(c^2 = a^2 + b^2 = 4 + 9 = 13, \text{ so } c = \sqrt{13}.\right)$ Asymptotes: $y = \pm\frac{3}{2}(x-1) - 1$

10.3 Exercises

Basic Concepts

1. The _____ are the endpoints of the transverse axis of a hyperbola.

2. The (dashed) rectangle used as an aid in graphing a hyperbola is called the _____ rectangle.

3. A hyperbola with an equation of the form $\frac{x^2}{a^2} - \frac{y^2}{b^2} = 1$

has a(n) _vertical/horizontal_ transverse axis.

4. A hyperbola with an equation of the form $\frac{y^2}{a^2} - \frac{x^2}{b^2} = 1$

has a(n) _vertical/horizontal_ transverse axis.

Hyperbolas with Center (0, 0)

Exercises 5–12: Sketch a graph of the hyperbola, including the asymptotes. Give the coordinates of the foci and vertices.

5. $\frac{x^2}{9} - \frac{y^2}{49} = 1$ **6.** $\frac{x^2}{16} - \frac{y^2}{4} = 1$

7. $\frac{y^2}{36} - \frac{x^2}{16} = 1$ **8.** $\frac{y^2}{4} - \frac{x^2}{4} = 1$

9. $x^2 - y^2 = 9$ **10.** $49y^2 - 25x^2 = 1225$

11. $9y^2 - 16x^2 = 144$ **12.** $4x^2 - 4y^2 = 100$

Exercises 13–16: Match the equation with its graph (a–d).

13. $\frac{x^2}{4} - \frac{y^2}{6} = 1$ **14.** $\frac{x^2}{9} - y^2 = 1$

15. $\frac{y^2}{9} - \frac{x^2}{16} = 1$ **16.** $\frac{y^2}{4} - \frac{x^2}{4} = 1$

a.
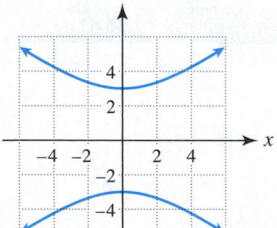

b.
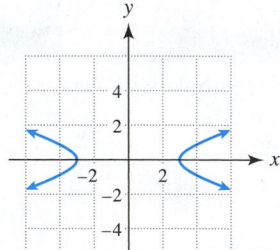

c.
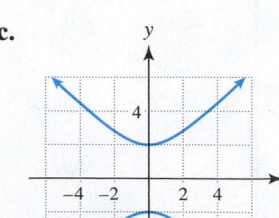

d.
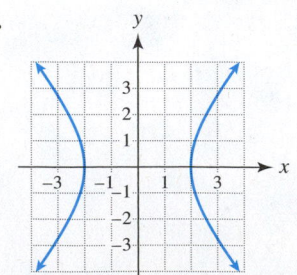

Exercises 17–20: Sketch a graph of a hyperbola, centered at the origin, with the foci, vertices, and asymptotes shown in the figure. Find an equation of the hyperbola. (The coordinates of the foci and vertices are integers.)

17.
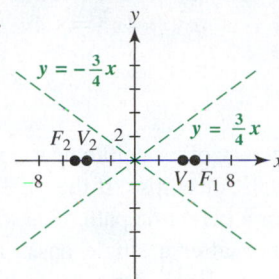

18.
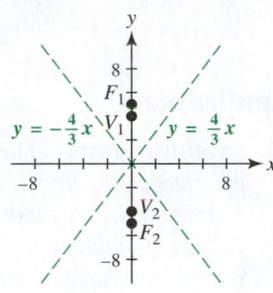

19.
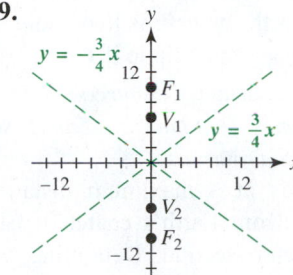

20.
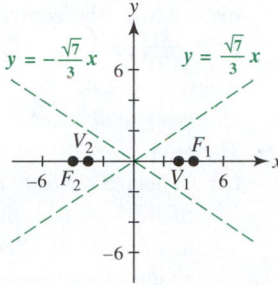

Exercises 21–30: Determine an equation of the hyperbola, centered at the origin, satisfying the conditions. Give the equations of its asymptotes.

21. Foci $(0, \pm 13)$, vertices $(0, \pm 12)$

22. Foci $(\pm 13, 0)$, vertices $(\pm 5, 0)$

23. Vertical transverse axis of length 4, foci $(0, \pm 5)$

24. Horizontal transverse axis of length 12, foci $(\pm 10, 0)$

25. Vertices $(\pm 3, 0)$, asymptotes $y = \pm\frac{2}{3}x$

26. Vertices $(0, \pm 4)$, asymptotes $y = \pm\frac{1}{2}x$

27. Endpoints of conjugate axis $(0, \pm 3)$, vertices $(\pm 4, 0)$

28. Endpoints of conjugate axis $(\pm 4, 0)$, vertices $(0, \pm 2)$

29. Vertices $\left(\pm\sqrt{10}, 0\right)$, passing through $(10, 9)$

30. Vertices $\left(0, \pm\sqrt{5}\right)$, passing through $(4, 5)$

Hyperbolas with Center (*h*, *k*)

Exercises 31–36: Sketch a graph of the hyperbola. Identify the vertices, foci, and asymptotes.

31. $\dfrac{(x-1)^2}{16} - \dfrac{(y-2)^2}{4} = 1$

32. $\dfrac{(y+1)^2}{16} - \dfrac{(x+3)^2}{9} = 1$

33. $\dfrac{(y-2)^2}{36} - \dfrac{(x+2)^2}{4} = 1$

34. $\dfrac{(x+1)^2}{4} - \dfrac{(y-1)^2}{4} = 1$

35. $\dfrac{x^2}{4} - (y-1)^2 = 1$

36. $(y+1)^2 - \dfrac{(x-3)^2}{4} = 1$

Exercises 37–40: Match the equation with its graph (a–d).

37. $\dfrac{(x-2)^2}{4} - \dfrac{(y+4)^2}{4} = 1$ **38.** $\dfrac{(x+1)^2}{4} - \dfrac{y^2}{9} = 1$

39. $\dfrac{(y+1)^2}{16} - \dfrac{(x-2)^2}{16} = 1$ **40.** $\dfrac{y^2}{25} - \dfrac{(x+1)^2}{9} = 1$

a.
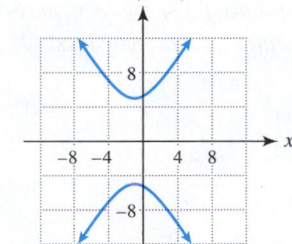

b.
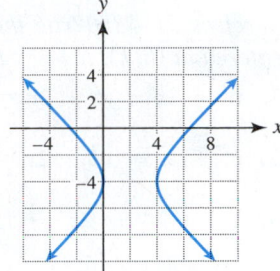

c.
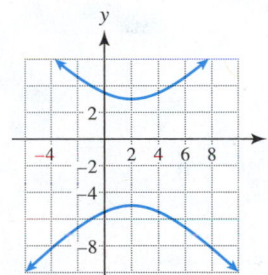

d.
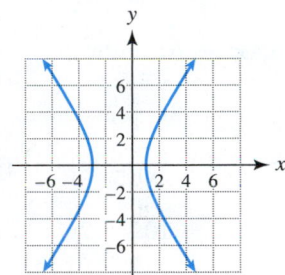

Exercises 41 and 42: Find an (approximate) equation of the hyperbola shown in the graph. Identify the vertices, foci, and asymptotes.

41.
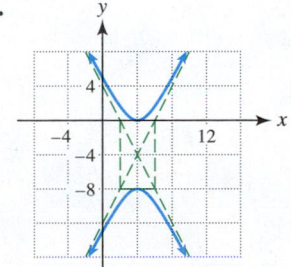

42.
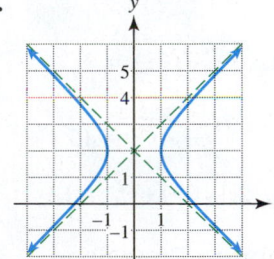

Exercises 43–46: Sketch a graph of the hyperbola, including the asymptotes. Give the coordinates of the vertices and foci.

43. $\dfrac{(x-1)^2}{4} - \dfrac{(y-1)^2}{4} = 1$

44. $\dfrac{(x+2)^2}{4} - \dfrac{(y+1)^2}{16} = 1$

45. $\dfrac{(y+1)^2}{16} - \dfrac{(x-1)^2}{9} = 1$

46. $y^2 - \dfrac{(x-2)^2}{4} = 1$

Exercises 47–50: Find the standard equation of a hyperbola with center (h, k) that satisfies the given conditions.

47. Center $(2, -2)$, focus $(4, -2)$, and vertex $(3, -2)$

48. Center $(-1, 1)$, focus $(-1, 4)$, and vertex $(-1, 3)$

49. Vertices $(-1, \pm 1)$ and foci $(-1, \pm 3)$

50. Vertices $(2 \pm 1, 1)$ and foci $(2 \pm 3, 1)$

Exercises 51–58: Write the standard form for a hyperbola centered at (h, k). Identify the center and the vertices.

51. $x^2 - 2x - y^2 + 2y = 4$

52. $y^2 + 4y - x^2 + 2x = 6$

53. $3y^2 + 24y - 2x^2 + 12x + 24 = 0$

54. $4x^2 + 16x - 9y^2 + 18y = 29$

55. $x^2 - 6x - 2y^2 + 7 = 0$

56. $y^2 + 8y - 3x^2 + 13 = 0$

57. $4y^2 + 32y - 5x^2 - 10x + 39 = 0$

58. $5x^2 + 10x - 7y^2 + 28y = 58$

Graphing Hyperbolas with Technology

Exercises 59–62: Graph the hyperbola.

59. $\dfrac{(y-1)^2}{11} - \dfrac{x^2}{5.9} = 1$

60. $\dfrac{x^2}{5.3} - \dfrac{y^2}{6.7} = 1$

61. $3y^2 - 4x^2 = 15$

62. $2.1x^2 - 6y^2 = 12$

Solving Equations

Exercises 63–70: Solve the system of equations.

63. $x^2 - y^2 = 4$
$x^2 + y^2 = 9$

64. $x^2 - 4y^2 = 16$
$x^2 + 4y^2 = 16$

65. $\dfrac{x^2}{4} - \dfrac{y^2}{9} = 1$
$x + y = 2$

66. $x^2 - y^2 = 4$
$x + y = 2$

67. $8x^2 - 6y^2 = 24$
$5x^2 + 3y^2 = 24$

68. $3y^2 - 4x^2 = 12$
$y^2 + 2x^2 = 34$

69. $\dfrac{y^2}{3} - \dfrac{x^2}{4} = 1$
$3x - y = 0$

70. $x^2 - 4y^2 = 16$
$y^2 - 4x^2 = 4$

Applications

71. Satellite Orbits The trajectory of a satellite near Earth can trace a hyperbola, parabola, or ellipse. If the satellite follows either a hyperbolic or a parabolic path, it escapes Earth's gravitational influence after a single pass. The path that a satellite travels near Earth depends on both its velocity V in meters per second and its distance D in meters from the center of Earth. Its path is hyperbolic if $V > \dfrac{k}{\sqrt{D}}$, parabolic if $V = \dfrac{k}{\sqrt{D}}$, and elliptic if $V < \dfrac{k}{\sqrt{D}}$, where $k = 2.82 \times 10^7$ is a constant. (*Sources*: W. Loh, *Dynamics and Thermodynamics of Planetary Entry*; W. Thomson, *Introduction to Space Dynamics*.)

(a) When *Explorer IV* was at a maximum distance of 42.5×10^6 meters from Earth's center, it had a velocity of 2090 meters/second. Determine the shape of its trajectory.

(b) If an orbiting satellite is to escape Earth's gravity so that it can travel to another planet, its velocity must be increased so that its trajectory changes from elliptic to hyperbolic. What range of velocities would allow *Explorer IV* to leave Earth's influence when it is at a maximum distance?

(c) Explain why it is easier to change a satellite's trajectory from an ellipse to a hyperbola when D is maximum rather than minimum.

72. Telescopes (Refer to Figure 10.58.) Suppose that the coordinates of F_1 are $(0, 5.2)$ and the coordinates of F_2 are $(0, -5.2)$. If the coordinates of the vertex of the hyperbolic mirror are $(0, 4.1)$, find the standard equation of a hyperbola whose upper branch coincides with the hyperbolic mirror.

Writing about Mathematics

73. Explain how the center, vertices, and asymptotes of a hyperbola are related to the fundamental rectangle.

74. Given the standard equation of a hyperbola, explain how to determine the length of the transverse axis. How can you determine whether the transverse axis is vertical or horizontal?

Extended and Discovery Exercises

1. **Structure of an Atom** In 1911, Ernest Rutherford discovered the basic structure of the atom by "shooting" positively charged alpha particles with a speed of 10^7 meter per second at a piece of gold foil 6×10^{-7} meter thick. Only a small percentage of the alpha particles struck a gold nucleus head on and were deflected directly back toward their source. The rest of the particles often followed a *hyperbolic* trajectory because they were repelled by positively charged gold nuclei. The figure shows the (blue) path of an alpha particle A initially approaching a gold nucleus N and being deflected at an angle $\theta = 90°$. N is located at a focus of the hyperbola, and the trajectory of A passes through a vertex of the hyperbola. (*Source:* H. Semat and J. Albright, *Introduction to Atomic and Nuclear Physics.*)

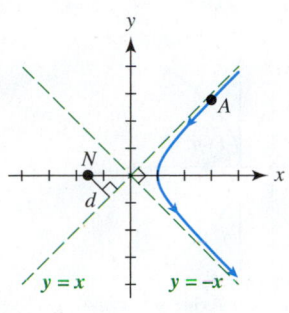

(a) Determine the equation of the trajectory of the alpha particle if $d = 5 \times 10^{-14}$ meter.

(b) What was the minimum distance between the centers of the alpha particle and the gold nucleus?

2. **Sound Detection** Microphones are placed at points $(-c, 0)$ and $(c, 0)$. An explosion occurs at point $P(x, y)$, which has a positive x-coordinate. See the figure.

   The sound is detected at the closer microphone t seconds before being detected at the farther microphone. Assume that sound travels at a speed of 330 meters per second, and show that P must be on the hyperbola

$$\frac{x^2}{330^2 t^2} - \frac{y^2}{4c^2 - 330^2 t^2} = \frac{1}{4}.$$

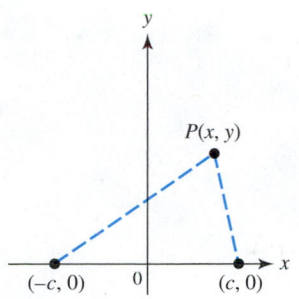

CHECKING BASIC CONCEPTS FOR SECTION 10.3

1. Use the graph to find the standard form of the hyperbola.

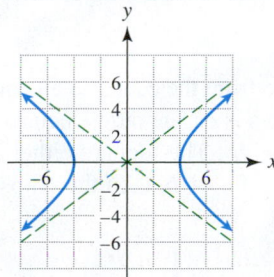

2. Graph the hyperbola defined by $\frac{x^2}{9} - \frac{y^2}{16} = 1$. Include the foci and asymptotes.

3. Find an equation of the hyperbola centered at $(1, 3)$ with a horizontal transverse axis of length 6 and a conjugate axis of length 4. Identify the foci.

4. Write $9y^2 - 54y - 16x^2 - 32x = 79$ in the standard form for a hyperbola centered at (h, k). Identify the center and the vertices.

10 Summary

CONCEPT	EXPLANATION AND EXAMPLES

Section 10.1 Parabolas

Conic Sections

Basic types: parabola, ellipse, circle, and hyperbola.

Parabolas with Vertex (0, 0)

Standard Forms

$x^2 = 4py$ (vertical axis)　　or　　$y^2 = 4px$ (horizontal axis)

Meaning of p

Both the vertex–focus distance and the vertex–directrix distance are p. The sign of p determines if the parabola opens upward or downward—or left or right. The focus is either $(0, p)$ or $(p, 0)$, and the directrix is either $y = -p$ or $x = -p$. In the figures below $p > 0$, but $p < 0$ is also possible.

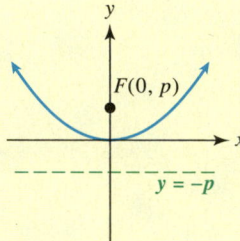 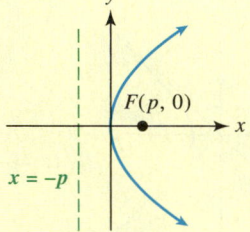

Vertical Axis: $x^2 = 4py$　　Horizontal Axis: $y^2 = 4px$

Parabolas with Vertex (h, k)

$(x - h)^2 = 4p(y - k)$　　Vertical axis; Focus: $(h, k + p)$
　　　　　　　　　　　　　　Directrix: $y = k - p$

$(y - k)^2 = 4p(x - h)$　　Horizontal axis; Focus: $(h + p, k)$
　　　　　　　　　　　　　　Directrix: $x = h - p$

Section 10.2 Ellipses

Ellipses with Center (0, 0)

Standard Forms with a > b > 0

$$\frac{x^2}{a^2} + \frac{y^2}{b^2} = 1 \text{ (horizontal major axis)} \quad \text{or} \quad \frac{x^2}{b^2} + \frac{y^2}{a^2} = 1 \text{ (vertical major axis)}$$

Meaning of a, b, and c

The distance from the center to a vertex is a, the distance from the center to an endpoint of the minor axis is b, and the distance from the center to a focus is c. The ratio $\frac{c}{a}$ equals the eccentricity e. The values of a, b, and c are related by $c^2 = a^2 - b^2$. The foci are either $(\pm c, 0)$ or $(0, \pm c)$, and the vertices are either $(\pm a, 0)$ or $(0, \pm a)$. If $a = b$, then the ellipse becomes a circle.

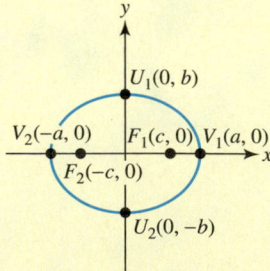 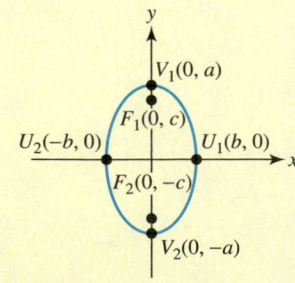

Horizontal Major Axis　　　　　Vertical Major Axis

CONCEPT	EXPLANATION AND EXAMPLES

Section 10.2 Ellipses (CONTINUED)

Ellipses with Center (h, k)

$a > b > 0; c^2 = a^2 - b^2$

$$\frac{(x - h)^2}{a^2} + \frac{(y - k)^2}{b^2} = 1$$

Major axis: horizontal; Foci: $(h \pm c, k)$
Vertices: $(h \pm a, k)$

$$\frac{(x - h)^2}{b^2} + \frac{(y - k)^2}{a^2} = 1$$

Major axis: vertical; Foci: $(h, k \pm c)$
Vertices: $(h, k \pm a)$

Area Inside an Ellipse

The area A of the region contained inside an ellipse is given by $A = \pi ab$, where a and b are from the standard equation of the ellipse.

Section 10.3 Hyperbolas

Hyperbolas with Center (0, 0)

Standard Forms with both a and b positive

$$\frac{x^2}{a^2} - \frac{y^2}{b^2} = 1 \text{ (horizontal transverse axis)} \quad \text{or} \quad \frac{y^2}{a^2} - \frac{x^2}{b^2} = 1 \text{ (vertical transverse axis)}$$

Meaning of a, b, and c

The distance from the center to a vertex is a, and the distance from the center to a focus is c. The asymptotes are $y = \pm \frac{b}{a}x$ if the transverse axis is horizontal and $y = \pm \frac{a}{b}x$ if the transverse axis is vertical. The values of a, b, and c are related by $c^2 = a^2 + b^2$. The foci are either $(\pm c, 0)$ or $(0, \pm c)$, and the vertices are either $(\pm a, 0)$ or $(0, \pm a)$.

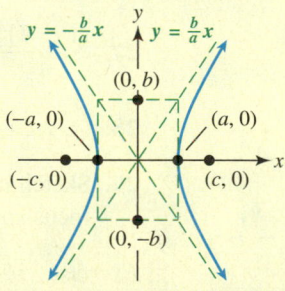

Horizontal Transverse Axis

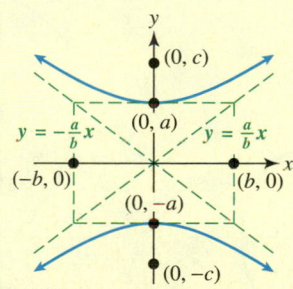

Vertical Transverse Axis

Hyperbolas with Center (h, k)

$a > 0, b > 0; c^2 = a^2 + b^2$

$$\frac{(x - h)^2}{a^2} - \frac{(y - k)^2}{b^2} = 1$$

Transverse axis: horizontal
Foci: $(h \pm c, k)$; Vertices: $(h \pm a, k)$
Asymptotes: $y = \pm \frac{b}{a}(x - h) + k$

$$\frac{(y - k)^2}{a^2} - \frac{(x - h)^2}{b^2} = 1$$

Transverse axis: vertical
Foci: $(h, k \pm c)$; Vertices: $(h, k \pm a)$
Asymptotes: $y = \pm \frac{a}{b}(x - h) + k$

10 Review Exercises

Exercises 1–6: Sketch a graph of the equation.

1. $-x^2 = y$ **2.** $y^2 = 2x$

3. $\dfrac{x^2}{25} + \dfrac{y^2}{49} = 1$ **4.** $\dfrac{y^2}{4} + \dfrac{x^2}{2} = 1$

5. $\dfrac{y^2}{4} - \dfrac{x^2}{9} = 1$ **6.** $x^2 - y^2 = 4$

Exercises 7–12: Match the equation with its graph (a–f).

7. $x^2 = 2y$ **8.** $y^2 = -3x$

9. $x^2 + y^2 = 4$ **10.** $\dfrac{x^2}{36} + \dfrac{y^2}{49} = 1$

11. $\dfrac{x^2}{4} - \dfrac{y^2}{9} = 1$ **12.** $\dfrac{y^2}{36} - \dfrac{x^2}{25} = 1$

a.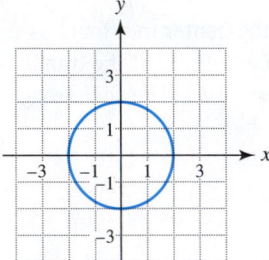

b.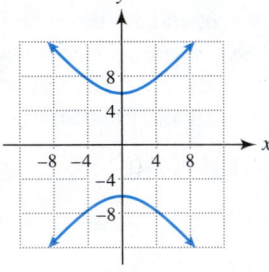

c.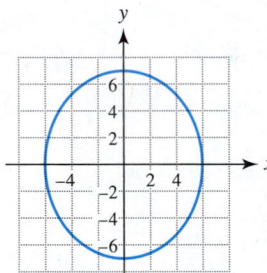

d.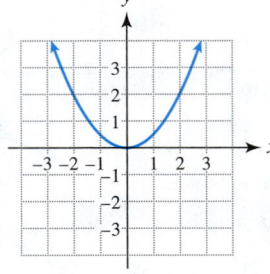

e.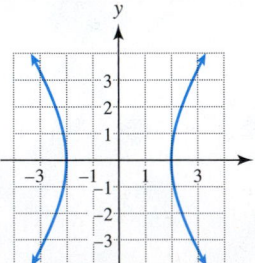

f.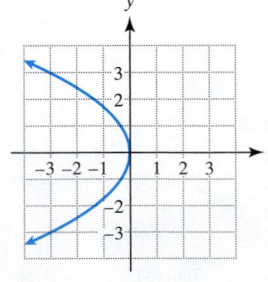

Exercises 13–18: Determine an equation of the conic section that satisfies the given conditions.

13. Parabola with focus $(2, 0)$ and vertex $(0, 0)$

14. Parabola with vertex $(5, 2)$ and focus $(5, 0)$

15. Ellipse with foci $(\pm 4, 0)$ and vertices $(\pm 5, 0)$

16. Ellipse centered at the origin with vertical major axis of length 14 and minor axis of length 8

17. Hyperbola with foci $(0, \pm 10)$ and endpoints of the conjugate axis $(\pm 6, 0)$

18. Hyperbola with vertices $(-2 \pm 3, 3)$ and foci given by $(-2 \pm 4, 3)$

Exercises 19–25: Sketch a graph of the conic section. Give the coordinates of any foci.

19. $-4y = x^2$ **20.** $y^2 = 8x$

21. $\dfrac{x^2}{25} + \dfrac{y^2}{4} = 1$ **22.** $49x^2 + 36y^2 = 1764$

23. $\dfrac{x^2}{16} - \dfrac{y^2}{9} = 1$ **24.** $\dfrac{y^2}{4} - x^2 = 1$

25. $(x - 3)^2 + (y + 1)^2 = 9$

Exercises 26–28: Sketch a graph of the conic section. Identify the coordinates of its center when appropriate.

26. $\dfrac{(y - 2)^2}{4} + \dfrac{(x + 1)^2}{16} = 1$

27. $\dfrac{(x - 1)^2}{4} - \dfrac{(y + 1)^2}{4} = 1$

28. $(x + 2) = 4(y - 1)^2$

29. Sketch a graph of $(y - 4)^2 = -8(x - 8)$. Include the focus and the directrix.

Exercises 30–32: Graph the equation.

30. $y^2 = \dfrac{3}{4}x$ **31.** $7.1x^2 + 8.2y^2 = 60$

32. $\dfrac{(y - 1.4)^2}{7} - \dfrac{(x + 2.3)^2}{11} = 1$

Exercises 33 and 34: Write the equation in the form given by $(y - k)^2 = a(x - h)$.

33. $-2x = y^2 + 8x + 14$ **34.** $2y^2 - 12y + 16 = x$

Exercises 35–38: Write the equation in standard form for an ellipse or a hyperbola centered at (h, k). Identify the center and the vertices.

35. $4x^2 + 8x + 25y^2 - 250y = -529$

36. $5x^2 + 20x + 2y^2 - 8y = -18$

37. $x^2 + 4x - 4y^2 + 24y = 36$

38. $4y^2 + 8y - 3x^2 + 6x = 11$

Exercises 39 and 40: Solve the system of equations.

39. $\dfrac{x^2}{4} + \dfrac{y^2}{4} = 1$

$\dfrac{x^2}{8} + \dfrac{y^2}{2} = 1$

40. $x^2 - y^2 = 1$

$x + y = 2$

Exercises 41 and 42: Shade the solution set to the system.

41. $\dfrac{x^2}{9} + \dfrac{y^2}{4} \le 1$

$x + y \le 3$

42. $y^2 - x^2 \le 9$

$y - x \le 0$

Applications

43. Comets A comet travels along an elliptical orbit around the sun. Its path can be described by the equation

$$\frac{x^2}{70^2} + \frac{y^2}{500^2} = 1,$$

where units are in millions of miles.

(a) What are the comet's minimum and maximum distances from the sun?

(b) Estimate the distance that the comet travels in one orbit around the sun. (Refer to Exercise 102 from Section 10.2.)

44. Searchlight A searchlight is constructed in the shape of a paraboloid with a depth of 7 inches and a diameter of 20 inches, as illustrated in the figure. Determine the distance d that the bulb should be from the vertex in order to have the beam of light shine straight ahead.

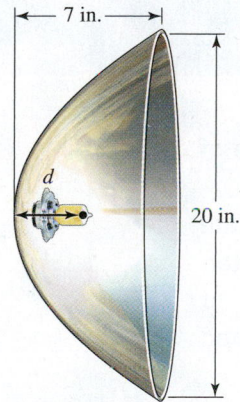

45. Arch Bridge An elliptical arch under a bridge is constructed so that it is 80 feet wide and has a maximum height of 30 feet, as illustrated in the figure. Find the height of the arch 10 feet from the center of the arch.

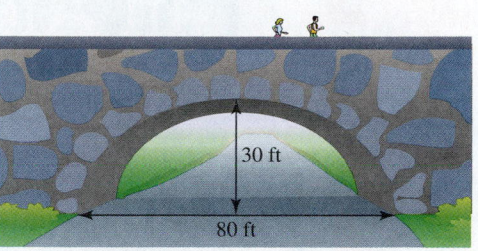

Extended and Discovery Exercises

Exercises 1–5: **Neptune and Pluto** *Both Neptune and Pluto travel around the sun in elliptical orbits. For Neptune's orbit, $a = 30.10$, and for Pluto's orbit, $a = 39.44$, where the variable a represents their average distances from the sun in astronomical units. (One astronomical unit equals 93 million miles.) The value of the variable a also corresponds to half the length of the major axis. Pluto has a highly eccentric orbit with $e = 0.249$, and Neptune has a nearly circular orbit with $e = 0.009$.*

1. Calculate the value of c for Neptune and Pluto.

2. Position the sun at the *origin* of the xy-plane. Find the coordinates of the center of Neptune's orbit and the coordinates of the center of Pluto's orbit. Assume that the centers lie on the positive x-axis.

3. Find equations for Neptune's orbit and for Pluto's orbit.

4. Graph both orbits in the same xy-plane.

5. Is Pluto always farther from the sun than Neptune? Explain.

11 Further Topics in Algebra

Mathematics permeates the fabric of modern society. It is the language of technology and allows society to quantify its experiences.

While technology and visualization provide new ways for us to investigate mathematical problems, they are *not* replacements for mathematical understanding. Computers and calculators are incapable of mathematical insight, but are excellent at performing arithmetic and other routine computation. Together, technology and the human mind can create amazing inventions.

In previous chapters we saw hundreds of examples of the use of mathematics to describe physical and social phenomena and events. Mathematics is diverse in its ability to adapt to new situations and solve complex problems. If any subject area is studied in enough detail, mathematics usually appears.

This chapter introduces further topics in mathematics. It represents only a small fraction of the topics found in mathematics. Although it may be difficult to predict exactly what the future will bring, one thing is certain—mathematics will continue to play a very important role.

> The art of asking the right questions in mathematics is more important than the art of solving them.
> —Georg Cantor

11.1 Sequences

- Understand basic concepts about sequences
- Learn how to represent sequences
- Identify and use arithmetic sequences
- Identify and use geometric sequences

Introduction

A sequence is a *function* that computes an ordered list. For example, the average person in the United States uses about 100 gallons of water each day. The function $f(n) = 100n$ generates the terms of the *sequence*

$$100, 200, 300, 400, 500, 600, 700, \ldots$$

when $n = 1, 2, 3, 4, 5, 6, 7, \ldots$. This *ordered list* represents the gallons of water used by the average person after n days. Sequences are a fundamental concept in mathematics and have many applications.

Basic Concepts

A second example of a sequence involves investing money. If \$100 is deposited into a savings account paying 5% interest compounded annually, then the function defined by $g(n) = 100(1.05)^n$ calculates the account balance after n years, which is given by

$$g(1), g(2), g(3), g(4), g(5), g(6), g(7), \ldots.$$

Terms of the sequence

These terms can be approximated as

$$105, 110.25, 115.76, 121.55, 127.63, 134.01, 140.71, \ldots.$$

We now define a sequence formally.

> **SEQUENCE**
>
> An **infinite sequence** is a function that has the set of natural numbers as its domain. A **finite sequence** is a function with domain $D = \{1, 2, 3, \ldots, n\}$ for some fixed natural number n.

Since sequences are functions, many of the concepts discussed in previous chapters apply to sequences. Instead of letting y represent the output, it is common to write $a_n = f(n)$, where n is a natural number in the domain of the sequence. The **terms** of a sequence are

$$a_1, a_2, a_3, \ldots, a_n, \ldots.$$

The first term is $a_1 = f(1)$, the second term is $a_2 = f(2)$, and so on. The **nth term**, or **general term**, of a sequence is $a_n = f(n)$.

NOTE The nth term, a general term, and a symbolic representation (formula) of a sequence are equivalent concepts.

EXAMPLE 1 Finding terms of sequences

Write the first four terms $a_1, a_2, a_3,$ and a_4 of each sequence, where $a_n = f(n)$.

(a) $f(n) = 2n - 5$ **(b)** $f(n) = 4(2)^{n-1}$ **(c)** $f(n) = (-1)^n \left(\dfrac{n}{n+1} \right)$

SOLUTION
(a) Evaluate $f(n) = 2n - 5$ as follows.

$$a_1 = f(1) = 2(1) - 5 = -3$$

$$a_2 = f(2) = 2(2) - 5 = -1$$

In a similar manner, $a_3 = f(3) = 1$ and $a_4 = f(4) = 3$.

(b) Since $f(n) = 4(2)^{n-1}$,

$$a_1 = f(1) = 4(2)^{1-1} = 4.$$

Similarly, $a_2 = 8$, $a_3 = 16$, and $a_4 = 32$.

(c) Let $f(n) = (-1)^n \left(\frac{n}{n+1} \right)$, and substitute $n = 1, 2, 3,$ and 4.

$$a_1 = f(1) = (-1)^1 \left(\frac{1}{1+1} \right) = -\frac{1}{2}$$

$$a_2 = f(2) = (-1)^2 \left(\frac{2}{2+1} \right) = \frac{2}{3}$$

$$a_3 = f(3) = (-1)^3 \left(\frac{3}{3+1} \right) = -\frac{3}{4}$$

$$a_4 = f(4) = (-1)^4 \left(\frac{4}{4+1} \right) = \frac{4}{5}$$

> Note the alternating signs.

Note that the factor $(-1)^n$ causes the terms of the sequence to alternate signs.

> **Now Try Exercises 1, 3, and 9**

Calculator Help

To learn how to generate a sequence, see Appendix A (page AP-15).

Graphing Calculators and Sequences Graphing calculators may be used to calculate the terms of a sequence. Figures 11.1–11.3 show the first four terms of each sequence in Example 1. The graphing calculator is in sequence mode.

```
seq(2n-5,n,1,4)
      {-3 -1 1 3}
```

Figure 11.1

```
seq(4(2)^(n-1),n
,1,4)
     {4 8 16 32}
```

Figure 11.2

```
seq((-1)^n*(n/(n
+1)),n,1,4)▶Frac
{-1/2 2/3 -3/4 ...
```

Figure 11.3

Recursive Sequences Some sequences are not defined using a general term. Instead they are defined *recursively*. With a **recursive sequence**, we must find terms a_1 through a_{n-1} before we can find a_n. For example, we need to find a_1 before we can find a_2, a_2 before we can find a_3, and so on.

EXAMPLE 2 **Finding the terms of a recursive sequence**

Find the first four terms of the recursive sequence that is defined by

$$a_n = 2a_{n-1} + 1; \qquad a_1 = 3.$$

SOLUTION The sequence is defined recursively, so we must find the terms in order.

> We substitute the value of a_1 to find the value of a_2.

$$a_1 = 3$$
$$a_2 = 2a_1 + 1 = 2(3) + 1 = 7$$
$$a_3 = 2a_2 + 1 = 2(7) + 1 = 15$$
$$a_4 = 2a_3 + 1 = 2(15) + 1 = 31$$

The first four terms are 3, 7, 15, and 31.

> **Now Try Exercise 15(a)**

An Application A population model for an insect with a life span of 1 year can be described using a recursive sequence. Suppose each adult female insect produces r female offspring that survive to reproduce the following year. Let $f(n)$ calculate the female insect population during year n. Then the number of female insects is given recursively by

$$f(n) = rf(n - 1) \qquad \text{for } n > 1.$$

Females in year n ⟋   ⟍ Females in previous year

The number of female insects in the year n is equal to r times the number of female insects in the previous year $n - 1$. Note that this function f is defined in terms of *itself*. To evaluate $f(n)$, we evaluate $f(n - 1)$. To evaluate $f(n - 1)$, we evaluate $f(n - 2)$, and so on. If we know the number of adult female insects during the first year, then we can determine the sequence. That is, if $f(1)$ is given, we can determine $f(n)$ by first computing

$$f(1), f(2), f(3), \ldots, f(n - 1).$$

The next example illustrates a similar recursively defined sequence where $f(n)$ gives the female population density during year n.   (**Source:** D. Brown and P. Rothery, *Models in Biology: Mathematics, Statistics and Computing.*)

EXAMPLE 3   **Modeling an insect population**

Suppose that the initial density of adult female insects is 1000 per acre and $r = 1.1$. Then the density of female insects during year n is described by

$$f(1) = 1000$$

$$f(n) = 1.1f(n - 1), \qquad n > 1.$$

(a) Rewrite this symbolic representation in terms of a_n.
(b) Find a_4 and interpret the result. Is the density of female insects increasing or decreasing?
(c) A general term for this sequence is given by $f(n) = 1000(1.1)^{n-1}$. Use this representation to find a_4.

SOLUTION
(a) Since $a_n = f(n)$ for all n, $a_{n-1} = f(n - 1)$. The sequence can be expressed as

$$a_1 = 1000$$

$$a_n = 1.1a_{n-1}, \qquad n > 1.$$

CLASS DISCUSSION

How does the value of r in Example 3 affect the population density in future years?

(b) In order to calculate the fourth term, a_4, we must first determine a_1, a_2, and a_3.

$$a_1 = \mathbf{1000}$$

$$a_2 = 1.1a_1 = 1.1(\mathbf{1000}) = \mathbf{1100}$$

$$a_3 = 1.1a_2 = 1.1(\mathbf{1100}) = \mathbf{1210}$$

$$a_4 = 1.1a_3 = 1.1(\mathbf{1210}) = \mathbf{1331}$$

The fourth term is $a_4 = 1331$. The female population density is increasing and reaches 1331 per acre during the fourth year.
(c) Since $a_4 = f(4)$,

$$a_4 = \mathbf{1000}(1.1)^{4-1} = \mathbf{1331}.$$

It is less work to find a_n using a formula for a general term rather than a recursive formula—particularly if n is large.

Now Try Exercise 89

Representations of Sequences

Sequences are functions. Therefore they have graphical, numerical, and symbolic representations. In the next example we make a table of values and a graph for a sequence.

EXAMPLE 4 Representing a sequence numerically and graphically

Let a recursive sequence be defined as follows.

$$a_1 = 3$$
$$a_n = 2a_{n-1} - 2, \quad n > 1$$

(a) Give a numerical representation (list each term in a table) for $n = 1, 2, 3, 4, 5$.
(b) Graph the first five terms of this sequence.

SOLUTION
(a) *Numerical Representation* Start by calculating the first five terms of the sequence.

$$a_1 = 3$$
$$a_2 = 2a_1 - 2 = 2(3) - 2 = 4$$
$$a_3 = 2a_2 - 2 = 2(4) - 2 = 6$$
$$a_4 = 2a_3 - 2 = 2(6) - 2 = 10$$
$$a_5 = 2a_4 - 2 = 2(10) - 2 = 18$$

n	a_n
1	3
2	4
3	6
4	10
5	18

Table 11.1

The first five terms are 3, 4, 6, 10, and 18. A numerical representation of the sequence is shown in Table 11.1.

(b) *Graphical Representation* To represent these terms graphically, plot the points $(1, 3)$, $(2, 4)$, $(3, 6)$, $(4, 10)$, and $(5, 18)$, as shown in Figure 11.4. Because the domain of a sequence contains only natural numbers, the graph of a sequence is a scatterplot.

Figure 11.4

Now Try Exercise 19

NOTE A graphing calculator set in sequence mode may be used to calculate the terms of the recursive sequence in Example 4. See Figures 11.5 and 11.6.

Calculator Help

To make a table or graph of a sequence, see Appendix A (page AP-16).

Table of a Sequence

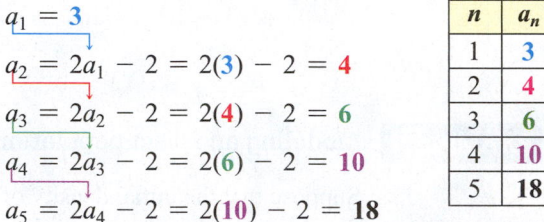

$$u(n) = 2u(n-1) - 2$$

Figure 11.5

Graph of a Sequence
[0, 6, 1] by [0, 20, 4]

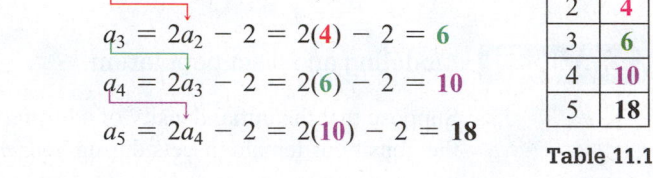

Figure 11.6 Dot Mode

An Application The next example illustrates numerical and graphical representations for a sequence involving population growth.

EXAMPLE 5 Representing a sequence numerically and graphically

Frequently the population of a particular insect does not continue to grow indefinitely, as it does in Example 3. Instead, the population grows rapidly at first and then levels off because of competition for limited resources. In one study, the population of the winter moth was modeled with a sequence similar to the following, where a_n represents the

population density in thousands per acre at the beginning of year n.   (*Source*: G. Varley and G. Gradwell, "Population models for the winter moth.")

$$a_1 = 1$$

$$a_n = 2.85a_{n-1} - 0.19a_{n-1}^2, \qquad n \geq 2$$

(a) Make a table of values for $n = 1, 2, 3, \ldots, 10$. Describe what happens to the population density of the winter moth.

(b) Use the table to graph the sequence.

SOLUTION

(a) *Numerical Representation* Evaluate $a_1, a_2, a_3, \ldots, a_{10}$ recursively. Since $a_1 = 1$,

$$a_2 = 2.85a_1 - 0.19a_1^2 = 2.85(1) - 0.19(1)^2 = 2.66 \qquad \text{and}$$

$$a_3 = 2.85a_2 - 0.19a_2^2 = 2.85(2.66) - 0.19(2.66)^2 \approx 6.24.$$

Approximate values for other terms are shown in Table 11.2. Figure 11.7 shows the sequence computed using a calculator, where the sequence is denoted $u(n)$ rather than a_n.

Calculating Table 11.2

n	$u(n)$	
1	1	
2	2.66	
3	6.2366	
4	10.384	
5	9.1069	
6	10.197	
7	9.3056	
$n = 1$		

Figure 11.7

Winter Moth Population Density

n	1	2	3	4	5	6	7	8	9	10
a_n	1	2.66	6.24	10.4	9.11	10.2	9.31	10.1	9.43	9.98

Table 11.2

(b) *Graphical Representation* The graph of a sequence is a scatterplot. Plot the points

$$(1, 1), (2, 2.66), (3, 6.24), \ldots, (10, 9.98),$$

as shown in Figure 11.8. The insect population increases rapidly at first and then oscillates about the line $y = 9.7$. (See the Class Discussion in the margin.) The oscillations become smaller as n increases, indicating that the population density may stabilize near 9.7 thousand per acre. Some calculators can plot sequences, as shown in Figure 11.9. In this figure, the first 20 terms have been plotted in dot mode.

CLASS DISCUSSION

In Example 5, the insect population stabilizes near the value $k = 9.74$ thousand. This value of k can be found by solving the quadratic equation

$$k = 2.85k - 0.19k^2.$$

Try to explain why this is true.

Graphical Representations of an Insect Population

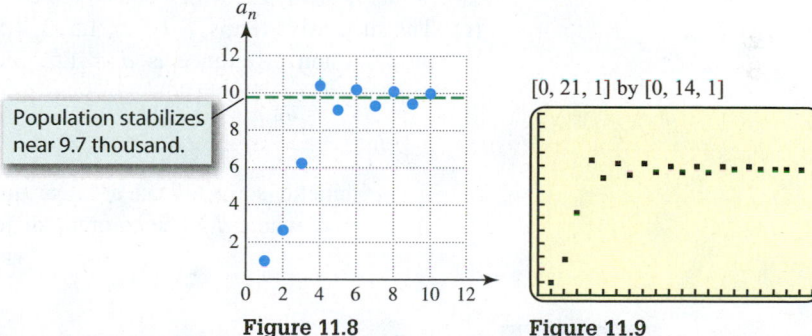

Population stabilizes near 9.7 thousand.

[0, 21, 1] by [0, 14, 1]

Figure 11.8   **Figure 11.9**

Now Try Exercise 91

Arithmetic Sequences

Suppose that a person receives a starting salary of $30,000 per year and a $1000 raise each year. The salary *after n* years of experience is represented by

$$f(n) = 1000n + 30,000, \quad \boxed{\text{Linear Function}}$$

where f is a linear function. After **10** years of experience, the annual salary would be

$$f(10) = 1000(10) + 30,000 = \$40,000.$$

If a sequence can be defined by a linear function, it is an *arithmetic sequence*. Its formula is given by $f(n) = dn + c$.

> **INFINITE ARITHMETIC SEQUENCE**
>
> An **infinite arithmetic sequence** is a linear function f whose domain is the set of natural numbers. The general term can be written as $a_n = dn + c$ where d and c are constants and $a_n = f(n)$.

An arithmetic sequence can be defined recursively by $a_n = a_{n-1} + d$, where d is a constant. Since $d = a_n - a_{n-1}$ for each valid n, d is called the **common difference**. If $d = 0$, then the sequence is a **constant sequence**. A **finite arithmetic sequence** is similar to an infinite arithmetic sequence except that its domain is $D = \{1, 2, 3, \ldots, n\}$, where n is a fixed natural number.

EXAMPLE 6 Determining arithmetic sequences

Determine if f is an arithmetic sequence for each situation.
(a) $f(n) = n^2 + 3n$
(b) f as graphed in Figure 11.10
(c) f as given in Table 11.3

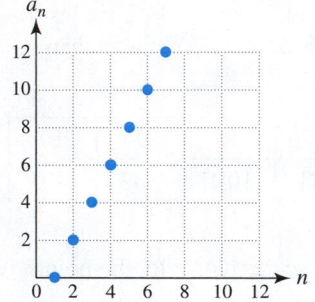

Figure 11.10

n	1	2	3	4	5	6	7
$f(n)$	-1.5	0	1.5	3	4.5	6	7.5

Table 11.3

SOLUTION

Getting Started The formula for an arithmetic sequence is given by $f(n) = dn + c$, its graph lies on a line with slope d, and consecutive terms in its numerical representation always change by a common difference d. ▶

(a) This sequence is not arithmetic because $f(n) = n^2 + 3n$ is nonlinear.
(b) The sequence in Figure 11.10 is an arithmetic sequence because the points lie on a line. A linear function could generate these points. Notice that the slope between points is always 2, which equals the common difference d.
(c) The successive terms, $-1.5, 0, 1.5, 3, 4.5, 6, 7.5$, increase by precisely 1.5. Therefore the common difference is $d = 1.5$. Since $a_n = a_{n-1} + 1.5$ for each valid n, the sequence is arithmetic.

Now Try Exercises 61, 65, and 67

An arithmetic sequence is a linear function and can always be represented by $f(n) = dn + c$, where d is the common difference and c is a constant.

EXAMPLE 7 Finding general terms for arithmetic sequences

Find a general term $a_n = f(n)$ for each arithmetic sequence.
(a) $a_1 = 3$ and $d = -2$
(b) $a_3 = 4$ and $a_9 = 17$

SOLUTION

(a) Let $f(n) = dn + c$. Since $d = -2$, $f(n) = -2n + c$. Since $a_1 = 3$,

$$a_1 = f(1) = -2(1) + c = 3, \quad \text{or} \quad c = 5.$$

Thus $a_n = -2n + 5$.

(b) Since $a_3 = 4$ and $a_9 = 17$, we find a linear function $f(n) = dn + c$ that satisfies the equations $f(3) = 4$ and $f(9) = 17$. The common difference is equal to the slope between the points $(3, 4)$ and $(9, 17)$.

MAKING CONNECTIONS

Linear Functions and Arithmetic Sequences

In Chapter 2 we discussed several techniques for finding a formula for linear functions. These methods can be applied to finding a general term for arithmetic sequences. It is important to realize that the mathematical concept of linear functions is simply being applied to the new topic of sequences by restricting their domains to the natural numbers.

$$d = \frac{17 - 4}{9 - 3} = \frac{13}{6}$$ Common difference = slope

It follows that $f(n) = \frac{13}{6}n + c$.

$$a_3 = f(3) = \frac{13}{6}(3) + c = 4, \quad \text{or} \quad c = -\frac{5}{2}$$ Use one point and d to find c.

Thus $a_n = \frac{13}{6}n - \frac{5}{2}$.

Now Try Exercises 41 and 45

Finding the nth Term If a_1 is the first term of an arithmetic sequence and d is the common difference, then consecutive terms of the sequence are given by

$$a_2 \qquad\qquad = a_1 + d$$
$$a_3 = a_2 + d = a_1 + 1d + d = a_1 + 2d$$
$$a_4 = a_3 + d = a_1 + 2d + d = a_1 + 3d \qquad (n-1)d$$
$$a_5 = a_4 + d = a_1 + 3d + d = a_1 + 4d$$

and, in general, $a_n = a_1 + (n - 1)d$. This result is summarized in the following box.

nTH TERM OF AN ARITHMETIC SEQUENCE

In an arithmetic sequence with first term a_1 and common difference d, the nth term, a_n, is given by

$$a_n = a_1 + (n - 1)d.$$

EXAMPLE 8 Finding the nth term of an arithmetic sequence

(a) Find a symbolic representation for the nth term of the arithmetic sequence

$$9, 8.5, 8, 7.5, 7, 6.5, 6, \ldots.$$

(b) Find the 12th term in the sequence using your formula for a_n.

SOLUTION

(a) The first term is 9. Successive terms can be found by subtracting 0.5 from (or adding -0.5 to) the previous term. Therefore $a_1 = 9$ and $d = -0.5$, and it follows that

$$a_n = a_1 + (n - 1)d \qquad \text{General formula}$$
$$= 9 + (n - 1)(-0.5) \qquad \text{Substitute.}$$
$$= -0.5n + 9.5. \qquad \text{Simplify.}$$

(b) $a_{12} = -0.5(12) + 9.5 = 3.5$

Now Try Exercise 31(c)

Geometric Sequences

Suppose that a person with a starting salary of \$30,000 per year receives a 5% raise each year. If $a_n = f(n)$ computes this salary at the *beginning* of the nth year, then

$$f(1) = 30,000$$
$$f(2) = 30,000(1.05) = 31,500$$
$$f(3) = 31,500(1.05) = 33,075$$
$$f(4) = 33,075(1.05) = 34,728.75.$$

Previous year's salary   Salary in year n

Terms of Geometric Sequences

c	r	a_1, a_2, a_3, a_4
1	2	$1, 2, 4, 8$
1	$\frac{1}{2}$	$1, \frac{1}{2}, \frac{1}{4}, \frac{1}{8}$
2	-4	$2, -8, 32, -128$
3	$\frac{1}{10}$	$3, 0.3, 0.03, 0.003$

Table 11.4

Each salary results from multiplying the previous salary by **1.05**. A general term in the sequence can be written as

$$f(n) = \mathbf{30{,}000}(\mathbf{1.05})^{n-1}.$$

During the 10th year, the annual salary is

$$f(\mathbf{10}) = \mathbf{30{,}000}(\mathbf{1.05})^{\mathbf{10}-1} \approx \$46{,}540.$$

This type of sequence is a *geometric sequence* given by $f(n) = \boldsymbol{c}r^{n-1}$, where $\boldsymbol{c}$ and $\boldsymbol{r}$ are constants, as shown in our example above. Geometric sequences are capable of either rapid growth or rapid decay. The first four terms from some geometric sequences are shown in Table 11.4. The corresponding values of c and r have been included.

The terms of a geometric sequence can be found by multiplying the previous term by r. In our example we multiplied the previous term by 1.05, indicating a 5% raise each year. We now define a geometric sequence formally.

INFINITE GEOMETRIC SEQUENCE

An **infinite geometric sequence** is a function defined by $f(n) = cr^{n-1}$, where c and r are nonzero constants. The domain of f is the set of natural numbers.

A geometric sequence can be defined recursively by $a_n = ra_{n-1}$, where $a_n = f(n)$ and the first term is $a_1 = c$. Since $r = \frac{a_n}{a_{n-1}}$ for each valid n, r is called the **common ratio**.

The next example illustrates how to recognize symbolic, graphical, and numerical representations of geometric sequences.

EXAMPLE 9 Determining geometric sequences

Decide which of the following represents a geometric sequence.
(a) The sequence defined by $a_n = 4(0.5)^n$
(b) The sequence a_n in Figure 11.11
(c) The sequence a_n in Table 11.5

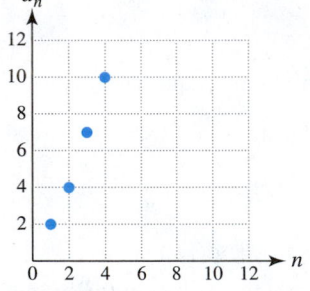

Figure 11.11

n	1	2	3	4	5	6
a_n	1	-3	9	-27	81	-243

Table 11.5

SOLUTION

Getting Started The formula for a geometric sequence is given by $a_n = cr^{n-1}$, and consecutive terms on its graph change by a common ratio r, as do consecutive terms in its numerical representation. ▶

(a) The formula for a_n can be written as

$$a_n = 4(0.5)^n = 4(0.5)(0.5)^{n-1} = \mathbf{2}(\mathbf{0.5})^{n-1}.$$

Thus a_n represents a geometric sequence with $c = \mathbf{2}$ and $r = \mathbf{0.5}$.

(b) The points on the graph are $(1, 2)$, $(2, 4)$, $(3, 7)$, and $(4, 10)$. Thus $a_1 = 2$, $a_2 = 4$, $a_3 = 7$, and $a_4 = 10$. Taking ratios of successive terms results in

$$\frac{a_2}{a_1} = \frac{4}{2} = 2, \qquad \frac{a_3}{a_2} = \frac{7}{4}, \qquad \text{and} \qquad \frac{a_4}{a_3} = \frac{10}{7}.$$

Since these ratios are not equal, there is no *common* ratio. The sequence is *not* geometric.

(c) The terms in Table 11.5 are $a_1 = 1$, $a_2 = -3$, $a_3 = 9$, $a_4 = -27$, $a_5 = 81$, and $a_6 = -243$. Note that these terms result from multiplying the previous term by -3. This sequence can be written either as

---

MAKING CONNECTIONS

Exponential Functions and Geometric Sequences

If the domain of an exponential function $f(x) = Ca^x$ is restricted to the natural numbers, then f represents a geometric sequence. For example, $f(x) = 3(2)^x$ generates the geometric sequence

$$6, 12, 24, 48, 96, \ldots$$

when

$$x = 1, 2, 3, 4, 5, \ldots.$$

$$a_n = -3a_{n-1} \quad \text{with} \quad a_1 = 1$$

or as

$$a_n = (-3)^{n-1}.$$

Therefore the sequence is geometric.

Now Try Exercises 69, 73, and 75

EXAMPLE 10 **Finding general terms for geometric sequences**

Find a general term a_n for each geometric sequence.
(a) $a_1 = 5$ and $r = 1.12$ **(b)** $a_2 = 8$ and $a_5 = 512$

SOLUTION
(a) Since $a_1 = c = 5$ and the common ratio is $r = 1.12$, $a_n = 5(1.12)^{n-1}$.
(b) We need to find $a_n = cr^{n-1}$ so that $a_2 = 8$ and $a_5 = 512$. Start by determining the common ratio r. Since

$$\frac{a_5}{a_2} = \frac{cr^{5-1}}{cr^{2-1}} = \frac{r^4}{r^1} = r^3 \quad \text{and} \quad \frac{a_5}{a_2} = \frac{512}{8} = 64,$$

it follows that $r^3 = 64$, or $r = 4$. So $a_n = c(4)^{n-1}$. Now

$$a_2 = c(4)^{2-1} = 8, \quad \text{or} \quad c = 2.$$

Thus $a_n = 2(4)^{n-1}$.

Now Try Exercises 51 and 55

11.1 Putting It All Together

The following table summarizes some fundamental concepts about sequences.

CONCEPT	EXPLANATION	EXAMPLE
Infinite sequence	A function f whose domain is the set of natural numbers; denoted $a_n = f(n)$; the terms are $a_1, a_2, a_3,\ldots$.	$f(n) = n^2 - 2n$, where $a_n = f(n)$ The first three terms are $a_1 = 1^2 - 2(1) = -1$ $a_2 = 2^2 - 2(2) = 0$ $a_3 = 3^2 - 2(3) = 3$. Graphs of sequences are scatterplots.
Recursive sequence	Defined in terms of previous terms; a_1 through a_{n-1} must be calculated before a_n can be found.	$a_n = 2a_{n-1}, a_1 = 1$ $a_1 = 1, a_2 = 2, a_3 = 4$, and $a_4 = 8$. A new term is found by multiplying the previous term by 2.
Arithmetic sequence	A *linear* function whose domain is the natural numbers; $a_n = dn + c$ or $a_n = a_{n-1} + d$, with common difference d. General term is $a_n = a_1 + (n-1)d$.	$f(n) = 2n - 1$, where $a_n = f(n)$ $a_1 = 1, a_2 = 3, a_3 = 5$, and $a_4 = 7$. Consecutive terms increase by the common difference $d = 2$. The points on the graph of this sequence lie on a line with slope 2.
Geometric sequence	$f(n) = cr^{n-1}$, where c is a nonzero constant and r is the nonzero common ratio; may also be written as $a_n = ra_{n-1}$	$f(n) = 2(3)^{n-1}$, where $a_n = f(n)$ $a_1 = 2, a_2 = 6, a_3 = 18$, and $a_4 = 54$. Consecutive terms are found by multiplying the previous term by the common ratio $r = 3$.

11.1 Exercises

Finding Terms of Sequences

Exercises 1–12: Find the first four terms of the sequence.

1. $a_n = 2n + 1$

2. $a_n = 3(n - 1) + 5$

3. $a_n = 4(-2)^{n-1}$

4. $a_n = 2(3)^n$

5. $a_n = \dfrac{n}{n^2 + 1}$

6. $a_n = 5 - \dfrac{1}{n^2}$

7. $a_n = (-1)^n \left(\dfrac{1}{2}\right)^n$

8. $a_n = (-1)^n \left(\dfrac{1}{n}\right)$

9. $a_n = (-1)^{n-1}\left(\dfrac{2^n}{1 + 2^n}\right)$

10. $a_n = (-1)^{n-1}\left(\dfrac{1}{3^n}\right)$

11. $a_n = 2^n + n^2$

12. $a_n = \dfrac{1}{n} + \dfrac{1}{3n}$

Exercises 13 and 14: Use the graphical representation to list the terms of the sequence.

13.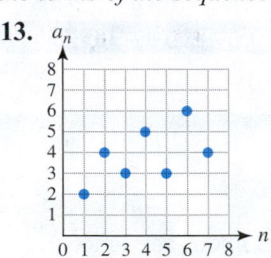

14.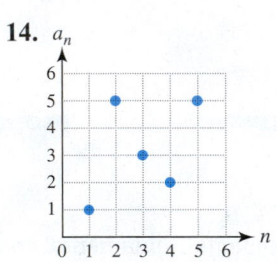

Exercises 15–28: Complete the following for the recursively defined sequence.

(a) Find the first four terms.
(b) Graph these terms.

15. $a_n = 2a_{n-1}; a_1 = 1$

16. $a_n = a_{n-1} + 5; a_1 = -4$

17. $a_n = a_{n-1} + 3; a_1 = -3$

18. $a_n = 2a_{n-1} + 1; a_1 = 1$

19. $a_n = 3a_{n-1} - 1; a_1 = 2$

20. $a_n = \frac{1}{2}a_{n-1}; a_1 = 16$

21. $a_n = a_{n-1} - a_{n-2}; a_1 = 2, a_2 = 5$

22. $a_n = 2a_{n-1} + a_{n-2}; a_1 = 0, a_2 = 1$

23. $a_n = a_{n-1}^2; a_1 = 2$

24. $a_n = \frac{1}{2}a_{n-1}^3 + 1; a_1 = 0$

25. $a_n = a_{n-1} + n; a_1 = 1$

26. $a_n = 3a_{n-1}^2; a_1 = 2$

27. $a_n = a_{n-1}a_{n-2}; a_1 = 2, a_2 = 3$

28. $a_n = 2a_{n-1}^2 + a_{n-2}; a_1 = 2, a_2 = 1$

Representations of Sequences

Exercises 29–34: The first five terms of an arithmetic sequence are given. Find

(a) numerical, (b) graphical, and (c) symbolic

representations of the sequence. Include at least eight terms for the graphical and numerical representations.

29. $1, 3, 5, 7, 9$

30. $4, 1, -2, -5, -8$

31. $7.5, 6, 4.5, 3, 1.5$

32. $5.1, 5.5, 5.9, 6.3, 6.7$

33. $\frac{1}{2}, 2, \frac{7}{2}, 5, \frac{13}{2}$

34. $2, 4, 6, 8, 10$

Exercises 35–40: The first five terms of a geometric sequence are given. Find

(a) numerical, (b) graphical, and (c) symbolic

representations of the sequence. Include at least eight terms for the graphical and numerical representations.

35. $8, 4, 2, 1, \frac{1}{2}$

36. $32, -8, 2, -\frac{1}{2}, \frac{1}{8}$

37. $\frac{3}{4}, \frac{3}{2}, 3, 6, 12$

38. $\frac{1}{27}, \frac{1}{9}, \frac{1}{3}, 1, 3$

39. $-\frac{1}{4}, -\frac{1}{2}, -1, -2, -4$

40. $9, 6, 4, \frac{8}{3}, \frac{16}{9}$

Exercises 41–50: Find a general term a_n for the arithmetic sequence.

41. $a_1 = 5, d = -2$

42. $a_1 = -3, d = 5$

43. $a_3 = 1, d = 3$

44. $a_4 = 12, d = -10$

45. $a_2 = 5, a_6 = 13$ **46.** $a_3 = 22, a_{17} = -20$

47. $a_1 = 8, a_4 = 17$ **48.** $a_1 = -2, a_5 = 8$

49. $a_5 = -4, a_8 = -2.5$ **50.** $a_3 = 10, a_7 = -4$

Exercises 51–60: Find a general term a_n for the geometric sequence.

51. $a_1 = 2, r = \frac{1}{2}$ **52.** $a_1 = 0.8, r = -3$

53. $a_3 = \frac{1}{32}, r = -\frac{1}{4}$ **54.** $a_4 = 3, r = 3$

55. $a_3 = 2, a_6 = \frac{1}{4}$ **56.** $a_2 = 6, a_4 = 24, r > 0$

57. $a_1 = -5, a_3 = -125, r < 0$

58. $a_1 = 10, a_2 = 2$

59. $a_2 = -1, a_7 = -32$ **60.** $a_2 = \frac{9}{4}, a_4 = \frac{81}{4}, r < 0$

Identifying Types of Sequences

Exercises 61–68: Determine if f is an arithmetic sequence.

61. $f(n) = 4 - 3n^3$ **62.** $f(n) = 2(n - 1)$

63. $f(n) = 4n - (3 - n)$ **64.** $f(n) = n^2 - n + 2$

65.

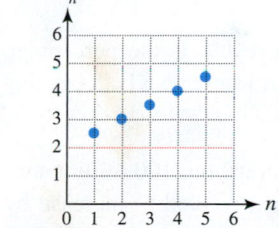

66.

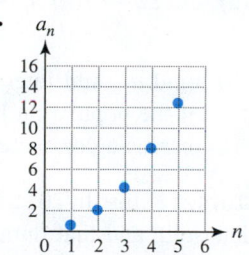

67.

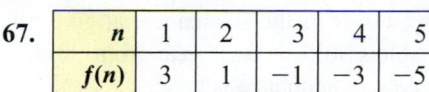

n	1	2	3	4	5
$f(n)$	3	1	−1	−3	−5

68.

n	1	2	3	4	5
$f(n)$	1	4	9	16	25

Exercises 69–76: Determine if f is a geometric sequence.

69. $f(n) = 4(2)^{n-1}$ **70.** $f(n) = -3(0.25)^n$

71. $f(n) = -3(n)^2$ **72.** $f(n) = 2(n - 1)^n$

73.

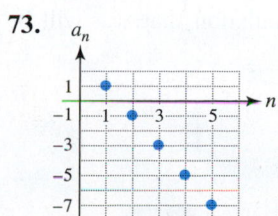

74.

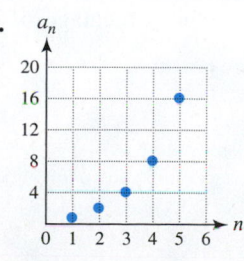

75.

n	1	2	3	4	5
$f(n)$	$\frac{1}{2}$	$\frac{3}{4}$	1	$\frac{5}{4}$	$\frac{5}{2}$

76.

n	1	2	3	4	5
$f(n)$	9	3	1	$\frac{1}{3}$	$\frac{1}{9}$

Exercises 77–82: Given the terms of a finite sequence, classify it as arithmetic, geometric, or neither.

77. $-5, 2, 9, 16, 23, 30$ **78.** $5, 2, -2, -6, -11$

79. $2, 8, 32, 128, 512$ **80.** $5.75, 5.5, 5.25, 5, 4.75, 4.5$

81. $100, 110, 130, 160, 200$

82. $0.7, 0.21, 0.063, 0.0189, 0.00567$

Exercises 83–86: Use the graph to determine if the sequence is arithmetic or geometric. If the sequence is arithmetic, state the sign of the common difference d and estimate its value. If the sequence is geometric, give the sign of the common ratio r and state if $|r| < 1$ or $|r| > 1$.

83.

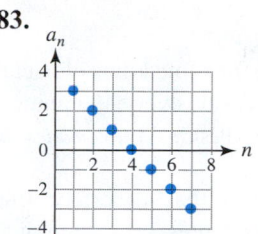

84.

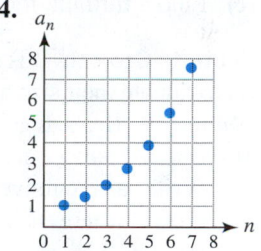

85.

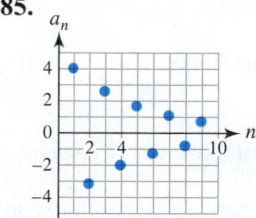

86.

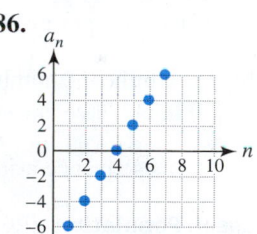

Modeling Insect and Bacteria Populations

Exercises 87 and 88: **Insect Population** *The annual population density of a species of insect after n years is modeled by a sequence. Use the graph to discuss trends in the insect population.*

87.

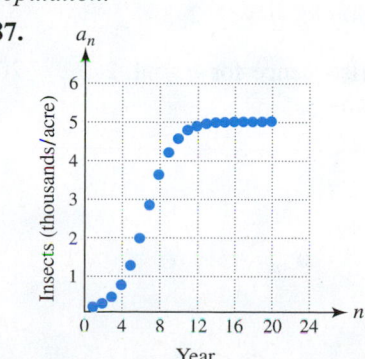

88.

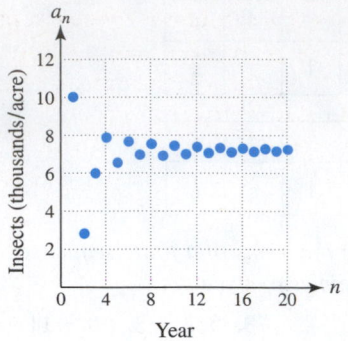

Year

89. Insect Population (Refer to Example 3.) Suppose that the density of female insects during the first year is 500 per acre with $r = 0.8$.
   (a) Write a recursive sequence that describes these data, where a_n denotes the female insect density during year n.

   (b) Find the six terms $a_1, a_2, a_3, \ldots, a_6$. Interpret the results.

   (c) Find a formula for a_n.

90. Bacteria Growth It is possible for some kinds of bacteria to double their size and then divide every 40 minutes. (*Source:* F. Hoppensteadt and C. Peskin, *Mathematics in Medicine and the Life Sciences.*)
   (a) Write a recursive sequence that describes this growth where each value of n represents a 40-minute interval. Let $a_1 = 300$ represent the initial number of bacteria per milliliter. Find the first five terms.

   (b) Determine the number of bacteria per milliliter after 10 hours have elapsed.

   (c) Is this sequence arithmetic or geometric? Explain.

91. Insect Population (Refer to Example 5.) Suppose an insect population density in thousands per acre at the beginning of year n can be modeled by the following recursive sequence.

$$a_1 = 8$$
$$a_n = 2.9a_{n-1} - 0.2a_{n-1}^2, \quad n > 1$$

   (a) Find the population for $n = 1, 2, 3$.

   (b) Graph the given sequence for $n = 1, 2, 3, \ldots, 20$. Interpret the graph.

92. Bacteria Growth (Refer to Exercise 90.) If bacteria are cultured in a medium with limited nutrients, competition ensues and growth slows. According to *Verhulst's model,* the number of bacteria at 40-minute intervals is given by

$$a_n = \left(\frac{2}{1 + a_{n-1}/K} \right) a_{n-1},$$

where K is a constant.

   (a) Let $a_1 = 200$ and $K = 10,000$. Graph the sequence for $n = 1, 2, 3, \ldots, 20$.
   (b) Describe the growth of these bacteria.
   (c) Trace the graph of the sequence. Make a conjecture as to why K is called the **saturation constant**. Test your conjecture by changing the value of K.

Applications

93. Digital Waste On average, the United States throws out 100 million cell phones per year. Write a general term a_n for a sequence that gives the number of cell phones thrown out after n years. Find a_5 and interpret the result. (*Source:* EPA.)

94. Global Poverty Between 1981 and 2008, the percent of the world population living on less than \$1.25 per day can be modeled by the function $f(n) = -1.1n + 53.1$, where $n = 1$ corresponds to 1981, $n = 2$ to 1982, and so on. (*Source: World Bank.*)
   (a) Let $a_n = f(n)$. Find a_1 and interpret the result.

   (b) Find a_{28} and interpret the result. Is the percentage of the population living on less than \$1.25 per day increasing or decreasing?

95. Tablet Sales In 2011 there were about 100 million tablets sold, and that number was expected to increase by 94 million per year until 2015. (*Source: Business Insider.*)
   (a) Write the five terms of the sequence that gives the number of tablets sold in each year from 2011 to 2015. What type of sequence is this?

   (b) Give a graphical representation of these terms.
   (c) Find a general term a_n.

96. Overweight in the U.S. In 2010, 68% of the U.S. population was overweight. This number is expected to increase by 0.7% per year until 2020. (*Source: New York Times.*)
   (a) Write the first six terms of the sequence that gives the percentage of the population that was/will be

overweight in each year from 2010 to 2015. What type of sequence is this?

(b) Find the general term a_n

97. Fibonacci Sequence The *Fibonacci sequence* dates back to 1202. It is one of the most famous sequences in mathematics and can be defined recursively.

$$a_1 = 1, a_2 = 1$$

$$a_n = a_{n-1} + a_{n-2} \quad \text{for } n > 2$$

(a) Find the first 12 terms of this sequence.

(b) Compute $\frac{a_n}{a_{n-1}}$ when $n = 2, 3, 4, \ldots, 12$. What happens to this ratio?

(c) Show that for $n = 2$, 3, and 4 the terms of the Fibonacci sequence satisfy the equation

$$a_{n-1} \cdot a_{n+1} - a_n^2 = (-1)^n.$$

98. Bouncing Ball If a tennis ball is dropped, it bounces, or rebounds, to 80% of its initial height.
(a) Write the first five terms of a sequence that gives the maximum height attained by the tennis ball on each rebound when it is dropped from an initial height of 5 feet. Let $a_1 = 5$. What type of sequence is this?

(b) Graph these terms.

(c) Find a general term a_n.

99. Salary Increases Suppose an employee's initial salary is $30,000.
(a) If this person receives a $2000 raise for each year of experience, determine a sequence that gives the salary at the beginning of the nth year. What type of sequence is this?

(b) Suppose another employee has the same starting salary and receives a 5% raise after each year. Find a sequence that computes the salary at the beginning of the nth year. What type of sequence is this?

(c) Which salary is higher at the beginning of the 10th year and the 20th year?

(d) Graph both sequences in the same viewing rectangle. Compare the two salaries.

100. Area A sequence of smaller squares is formed by connecting the midpoints of the sides of a larger square, as shown in the figure. If the area of the largest square is one square unit, give the first five terms of a sequence that describes the area of each successive square. What type of sequence is this? Write an expression for the area of the nth square.

Exercises 101–104: **Computing Square Roots** *The following recursively defined sequence can be used to compute $\sqrt{k}$ for any positive number k.*

$$a_1 = k; \ a_n = \frac{1}{2}\left(a_{n-1} + \frac{k}{a_{n-1}}\right)$$

This sequence was known to Sumerian mathematicians 4000 years ago, but it is still used today. Use this sequence to approximate the given square root by finding a_6. Compare your result with the actual value. (**Source:** P. Heinz-Otto, *Chaos and Fractals.*)

101. $\sqrt{2}$ **102.** $\sqrt{11}$

103. $\sqrt{21}$ **104.** $\sqrt{41}$

105. Suppose that a_n and b_n represent arithmetic sequences. Show that their sum, $c_n = a_n + b_n$, is also an arithmetic sequence.

106. Explain why the sequence

$$\log 2, \log 4, \log 8, \log 16, \ldots$$

is an arithmetic sequence.

Writing about Mathematics

107. Explain how we can distinguish between an arithmetic and a geometric sequence. Give examples.

108. Compare a sequence whose nth term is given by $a_n = f(n)$ to a sequence that is defined recursively. Give examples. Which symbolic representation for defining a sequence is usually more convenient to use? Explain why.

11.2 Series

Introduction

Although the terms *sequence* and *series* are sometimes used interchangeably in everyday English, they represent different mathematical concepts. In mathematics, a sequence is an *ordered list* (a function whose domain is the set of natural numbers), whereas a series is a summation of the terms in a sequence. Series have played a central role in the development of modern mathematics. Today series are often used to approximate functions that are too complicated to have simple formulas. Series are also instrumental in calculating approximations of numbers like π and e.

Basic Concepts

Suppose a person has a starting salary of \$30,000 per year and receives a \$2000 raise each year. Then

A Finite Sequence

$$30{,}000, \ 32{,}000, \ 34{,}000, \ 36{,}000, \ 38{,}000$$

5 terms of a sequence

are terms of the finite sequence that describe this person's salaries over a 5-year period. The total amount earned is given by the finite *series*

A Finite Series

$$30{,}000 \ + \ 32{,}000 \ + \ 34{,}000 \ + \ 36{,}000 \ + \ 38{,}000,$$

Sum of 5 terms of a sequence

whose sum is \$170,000. Any sequence can be used to define a series. For example, the *infinite sequence*

An Infinite Sequence

$$1, \frac{1}{3}, \frac{1}{9}, \frac{1}{27}, \frac{1}{81}, \frac{1}{243}, \ldots$$

defines the terms of the *infinite series*

An Infinite Series

Add the terms of a sequence to define a series.

$$1 + \frac{1}{3} + \frac{1}{9} + \frac{1}{27} + \frac{1}{81} + \frac{1}{243} + \cdots.$$

Finite Series We now define the concept of a finite series, where $a_1, a_2, a_3, \ldots, a_n$ represent terms of a sequence.

> **FINITE SERIES**
>
> A **finite series** is the sum of the first n terms of a sequence and can be written as
>
> $$a_1 + a_2 + a_3 + \cdots + a_n.$$

EXAMPLE 1 Writing sequences and series

Complete the following.
(a) Write a sequence that is the first 5 even natural numbers.
(b) Write a series that sums the terms of the sequence in part (a).
(c) Find the sum of the series in part (b).

SOLUTION
(a) A sequence is a list of values, so the terms are 2, 4, 6, 8, 10.
(b) The corresponding series is $2 + 4 + 6 + 8 + 10$.
(c) Because $2 + 4 + 6 + 8 + 10 = 30$, the sum of the series is 30.

Now Try Exercise 1

See the Concept: Sequences and Series

(A) Sequence —— 1, 3, 5, 7, 9, 11, 13   **(A)** A sequence is an ordered list.

(B) Series —— $1 + 3 + 5 + 7 + 9 + 11 + 13$ **(B)** A series is the sum of the terms of a sequence.

(C) Sum of the series —— 49   **(C)** When we evaluate the sum in **(B)**, we say that we are finding the "sum of the series."

Infinite Series and Partial Sums In the following box we define an infinite series.

INFINITE SERIES

An **infinite series** is the sum of the terms of an infinite sequence and can be written as

$$a_1 + a_2 + a_3 + \cdots + a_n + \cdots .$$

An infinite series contains infinitely many terms. Since a series represents a sum, we must define what is meant by finding the sum of an *infinite* series. Let the following be a **sequence of partial sums**.

$$S_1 = a_1 \qquad \text{Sum of the first term}$$

$$S_2 = a_1 + a_2 \qquad \text{Sum of the first two terms}$$

$$S_3 = a_1 + a_2 + a_3 \qquad \text{Sum of the first three terms}$$

$$\vdots$$

$$S_n = a_1 + a_2 + a_3 + \cdots + a_n \qquad \text{Sum of the first } n \text{ terms}$$

If the sequence of partial sums $S_1, S_2, S_3, \ldots, S_n$ approaches a real number S as $n \to \infty$, then the sum of the infinite series is S.

For example, let $S_1 = 0.3$, $S_2 = 0.3 + 0.03$, $S_3 = 0.3 + 0.03 + 0.003$, and so on. Then, as $n \to \infty$, $S_n \to \frac{1}{3}$. We say that the infinite series

> As the number of terms n approaches infinity, the sum approaches $\frac{1}{3}$.

$$0.3 + 0.03 + 0.003 + 0.0003 + \cdots = 0.3333\ldots = 0.\overline{3}$$

has sum $\frac{1}{3}$. Some infinite series do not have a sum S. For example, the series given by $1 + 2 + 3 + 4 + 5 + \cdots$ would have an **unbounded**, or "infinite," sum.

EXAMPLE 2   Finding partial sums

For each a_n, calculate S_4.
(a) $a_n = 2n + 1$ **(b)** $a_n = n^2$

SOLUTION
(a) Because $S_4 = a_1 + a_2 + a_3 + a_4$, start by calculating the first four terms of the sequence $a_n = 2n + 1$.

$$a_1 = 2(1) + 1 = 3; \qquad a_2 = 2(2) + 1 = 5;$$

$$a_3 = 2(3) + 1 = 7; \qquad a_4 = 2(4) + 1 = 9$$

Thus $S_4 = \underbrace{3 + 5 + 7 + 9}_{4 \text{ terms}} = 24$.

(b) $a_1 = 1^2 = 1; \qquad a_2 = 2^2 = 4; \qquad a_3 = 3^2 = 9; \qquad a_4 = 4^2 = 16$

Thus $S_4 = 1 + 4 + 9 + 16 = 30$.

Now Try Exercises 7 and 11

Calculating π Since π is an irrational number, it cannot be represented exactly by a fraction. Its decimal expansion neither repeats nor has a discernible pattern. The ability to compute π was essential to the development of every modern society, because π appears in formulas used in construction, surveying, and geometry. It was not until the discovery of series that exceedingly accurate decimal approximations of π were possible. In 2002, after 400 hours of supercomputer time, π was computed to 1.24 trillion digits. Why would anyone want to compute π to so many decimal places? One practical reason is to test electrical circuits in new computers. If a computer has a small defect in its hardware, there is a good chance that an error will appear after trillions of arithmetic calculations are performed during the computation of π. (*Sources:* P. Beckmann, *A History of PI;* P. Heinz-Otto, *Chaos and Fractals.*)

EXAMPLE 3 **Computing π with a series**

The infinite series given by

$$\frac{\pi^4}{90} = \frac{1}{1^4} + \frac{1}{2^4} + \frac{1}{3^4} + \frac{1}{4^4} + \frac{1}{5^4} + \cdots + \frac{1}{n^4} + \cdots$$

can be used to estimate π.
(a) Approximate π by finding the sum of the first four terms.
(b) Use technology to approximate π by summing the first 50 terms. Compare the result to the actual value of π.

SOLUTION
(a) Summing the first four terms results in the following approximation.

$$\frac{\pi^4}{90} \approx \frac{1}{1^4} + \frac{1}{2^4} + \frac{1}{3^4} + \frac{1}{4^4} \approx 1.078751929$$

This approximation can be solved for π by multiplying by 90 and then taking the fourth root. Thus

$$\pi \approx \sqrt[4]{90(1.078751929)} \approx 3.139.$$

```
sum(seq(1/n^4,n,
1,50))
        1.082320646
(90*Ans)^(1/4)
        3.141590776
π
        3.141592654
```

Figure 11.12

(b) Some calculators are capable of summing the terms of a sequence, as shown in Figure 11.12. (Summing the terms of a sequence is equivalent to finding the sum of a series.) The first 50 terms of the series provide an approximation of $\pi \approx 3.141590776$. This computation matches the actual value of π for the first five decimal places.

Now Try Exercise 107

Calculator Help
To find the sum of a series, see Appendix A (page AP-17).

Arithmetic Series

Summing the terms of an arithmetic sequence results in an **arithmetic series**. For example, the sequence defined by $a_n = 2n - 1$ for $n = 1, 2, 3, \ldots, 7$ is the arithmetic sequence

$$1, 3, 5, 7, 9, 11, 13. \qquad \boxed{\text{Arithmetic sequence}}$$

The corresponding arithmetic series is

$$1 + 3 + 5 + 7 + 9 + 11 + 13. \qquad \boxed{\text{Arithmetic series}}$$

The following formula can be used to sum a finite arithmetic series. (For a proof, see Exercise 3 in the Extended and Discovery Exercises at the end of this chapter.)

> **SUM OF A FINITE ARITHMETIC SERIES**
>
> The sum of the first n terms of an arithmetic series, denoted S_n, is found by averaging the first and nth terms and then multiplying by n. That is,
>
> $$S_n = a_1 + a_2 + a_3 + \cdots + a_n = n\left(\frac{a_1 + a_n}{2}\right).$$

From Section 11.1, the general term a_n for an arithmetic sequence can be written as $a_n = a_1 + (n - 1)d$, so S_n can also be written as follows.

$$S_n = n\left(\frac{a_1 + a_n}{2}\right)$$

$$= \frac{n}{2}(a_1 + a_1 + (n - 1)d)$$

$$= \frac{n}{2}(2a_1 + (n - 1)d) \quad \longleftarrow \begin{array}{l}\text{Sum of a finite arithmetic} \\ \text{series with } n \text{ terms}\end{array}$$

EXAMPLE 4 **Finding the sum of a finite arithmetic series**

Use a formula to find the sum of the arithmetic series

$$2 + 4 + 6 + 8 + \cdots + 100.$$

SOLUTION The series $2 + 4 + 6 + 8 + \cdots + 100$ has $n = 50$ terms with $a_1 = 2$ and $a_{50} = 100$. We can use the formula

$$S_n = n\left(\frac{a_1 + a_n}{2}\right)$$

to find its sum.

$$S_{50} = 50\left(\frac{2 + 100}{2}\right) = 2550$$

We can also use the formula

$$S_n = \frac{n}{2}(2a_1 + (n - 1)d)$$

with common difference $d = 2$ to find this sum.

$$S_{50} = \frac{50}{2}(2(2) + (50 - 1)2) = 2550$$

The two answers agree, as expected.

> **Now Try Exercise 15**

EXAMPLE 5 **Finding the sum of a finite arithmetic series**

A person has a starting annual salary of \$30,000 and receives a \$1500 raise each year.
(a) Calculate the total amount earned over 10 years.
(b) Verify this value using a calculator.

SOLUTION
(a) The arithmetic sequence describing the salary during year n is computed by

$$a_n = 30,000 + 1500(n - 1).$$

Because $a_n = 30,000 + 1500(n - 1)$, the first and tenth year's salaries are

$$a_1 = 30,000 + 1500(1 - 1) = \mathbf{30,000}$$

$$a_{10} = 30,000 + 1500(10 - 1) = \mathbf{43,500}.$$

Thus the total amount earned during this 10-year period is

$$S_{10} = 10\left(\frac{\mathbf{30,000} + \mathbf{43,500}}{2}\right) = \$367,500.$$

This sum can also be found using $\boldsymbol{S_n = \frac{n}{2}(2a_1 + (n - 1)d)}$.

$$S_{10} = \frac{10}{2}(2 \cdot 30,000 + (10 - 1)1500) = \$367,500$$

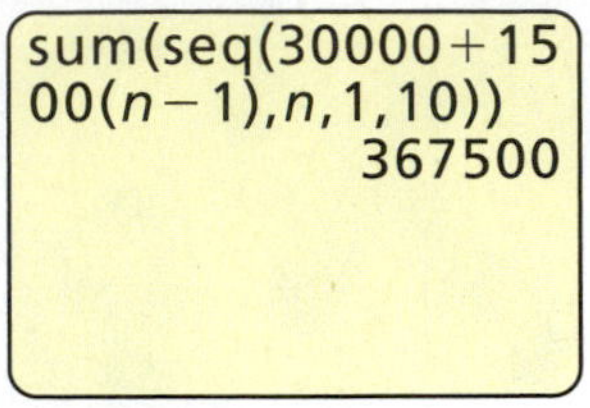

(b) To verify this result with a calculator, compute the sum

$$a_1 + a_2 + a_3 + \cdots + a_{10},$$

where $a_n = 30,000 + 1500(n - 1)$. This calculation is shown in Figure 11.13. The result of 367,500 agrees with part (a).

> **Now Try Exercise 95**

```
sum(seq(30000+15
00(n-1),n,1,10))
                  367500
```

Figure 11.13

EXAMPLE 6 **Finding a term of an arithmetic series**

The sum of an arithmetic series with 15 terms is 285. If $a_{15} = 40$, find a_1.

SOLUTION To find a_1, we apply the sum formula

$$S_n = n\left(\frac{a_1 + a_n}{2}\right) = a_1 + a_2 + a_3 + \cdots + a_n$$

with $n = 15$ and $a_{15} = 40$.

> Apply the formula and solve for a_1.

$$15\left(\frac{a_1 + 40}{2}\right) = 285$$

$$15(a_1 + 40) = 570 \qquad \text{Multiply by 2.}$$

$$a_1 + 40 = 38 \qquad \text{Divide by 15.}$$

$$a_1 = -2 \qquad \text{Subtract 40.}$$

> **Now Try Exercise 23**

CLASS DISCUSSION

Explain why a formula for the sum of an infinite arithmetic series is not given.

Geometric Series

What will happen if we attempt to find the sum of an infinite geometric series? For example, suppose that a person walked 1 mile on the first day, $\frac{1}{2}$ mile the second day, $\frac{1}{4}$ mile the third day, and so on. How far down the road would this person travel? This distance is described by the infinite series

$$1 + \frac{1}{2} + \frac{1}{4} + \frac{1}{8} + \frac{1}{16} + \frac{1}{32} + \frac{1}{64} + \cdots . \qquad \boxed{\text{Total distance traveled}}$$

Does the sum of an infinite number of positive values always become infinitely large? We answer this question later in Example 9.

Finite Geometric Series In a manner similar to the way an arithmetic series was defined, a **geometric series** is defined as the sum of the terms of a geometric sequence.

In order to calculate sums of infinite geometric series, we begin by finding sums of finite geometric series. Any finite geometric sequence can be written as

Geometric Sequence

$$a_1, a_1r, a_1r^2, a_1r^3, \ldots, a_1r^{n-1}.$$

The summation of these n terms is a finite geometric series. Its sum S_n is expressed by

Geometric Series

$$S_n = a_1 + a_1r + a_1r^2 + a_1r^3 + \cdots + a_1r^{n-1}. \qquad \text{Equation 1}$$

To find the value of S_n, multiply this equation by r.

$$rS_n = a_1r + a_1r^2 + a_1r^3 + \cdots + a_1r^{n-1} + a_1r^n \qquad \text{Equation 2}$$

Subtracting equation 2 from equation 1 results in

$$S_n - rS_n = a_1 - a_1r^n$$

$$S_n(1 - r) = a_1(1 - r^n)$$

Sum of a finite geometric series with n terms

$$S_n = a_1\left(\frac{1 - r^n}{1 - r}\right), \qquad \text{provided } r \neq 1.$$

SUM OF A FINITE GEOMETRIC SERIES

If a geometric sequence has first term a_1 and common ratio r, then the sum of the first n terms is given by

$$S_n = a_1\left(\frac{1 - r^n}{1 - r}\right), \qquad r \neq 1.$$

EXAMPLE 7   **Finding the sums of finite geometric series**

Approximate the sum S_n for the given values of n.
(a) $1 + \frac{1}{2} + \frac{1}{4} + \cdots + \left(\frac{1}{2}\right)^{n-1}$; $n = 5, 10$, and 20
(b) $3 - 6 + 12 - 24 + 48 - \cdots + 3(-2)^{n-1}$; $n = 3, 8$, and 13

SOLUTION
(a) This geometric series has $a_1 = 1$ and $r = \frac{1}{2} = 0.5$.

$$S_5 = 1\left(\frac{1 - 0.5^5}{1 - 0.5}\right) = 1.9375$$

$$S_{10} = 1\left(\frac{1 - 0.5^{10}}{1 - 0.5}\right) \approx 1.998047$$

$$S_{20} = 1\left(\frac{1 - 0.5^{20}}{1 - 0.5}\right) \approx 1.999998$$

(b) This geometric series has $a_1 = 3$ and $r = -2$.

$$S_3 = 3\left(\frac{1 - (-2)^3}{1 - (-2)}\right) = 9$$

$$S_8 = 3\left(\frac{1 - (-2)^8}{1 - (-2)}\right) = -255$$

$$S_{13} = 3\left(\frac{1 - (-2)^{13}}{1 - (-2)}\right) = 8193$$

Now Try Exercises 41 and 43

Annuities With an **annuity**, an individual often makes a sequence of deposits at equal time intervals. Suppose A_0 dollars is deposited at the end of each year into an account that pays an annual interest rate i compounded annually. At the end of the first year, the account contains A_0 dollars. At the end of the second year, A_0 dollars would be deposited again. In addition, the first deposit of A_0 dollars would have received interest during the second year. Therefore the value of the annuity after 2 years is

$$A_0 + A_0(1 + i).$$

2nd year deposit | 1st year deposit plus interest

After 3 years the balance is

$$A_0 + A_0(1 + i) + A_0(1 + i)^2,$$

By the end of the 3rd year the 1st year deposit has earned **2** years of interest.

and after n years this amount is given by

$$A_0 + A_0(1 + i) + A_0(1 + i)^2 + \cdots + A_0(1 + i)^{n-1}.$$

This is a geometric series with first term $a_1 = A_0$ and common ratio $r = (1 + i)$. The sum of the first n terms is given by

$$S_n = A_0\left(\frac{1 - (1 + i)^n}{1 - (1 + i)}\right) = A_0\left(\frac{(1 + i)^n - 1}{i}\right). \qquad S_n = a_1\left(\frac{1 - r^n}{1 - r}\right)$$

EXAMPLE 8 **Finding the future value of an annuity**

Suppose that a 20-year-old worker deposits $1000 into an account at the end of each year until age 65. If the interest rate is 4%, find the future value of the annuity.

SOLUTION Let $A_0 = 1000$, $i = 0.04$, and $n = 45$. The future value of the annuity is

$$S_n = A_0\left(\frac{(1 + i)^n - 1}{i}\right)$$

$$= 1000\left(\frac{(1 + 0.04)^{45} - 1}{0.04}\right)$$

$$\approx \$121,029.39.$$

Now Try Exercise 97

Infinite Geometric Series The absolute value of r affects the sum of a finite geometric series.

- If $|r| > 1$, then $|r^n|$ becomes large as n increases and the sum of the series S_n also becomes large in *absolute value*. See Example 7, part (b) where $r = -2$.
- If $|r| < 1$, then $|r^n|$ becomes closer to 0. So as n increases, S_n approaches a number

$$S_n = a_1\left(\frac{1 - r^n}{1 - r}\right) \approx a_1\left(\frac{1 - 0}{1 - r}\right) = \frac{a_1}{1 - r}.$$

See Example 7, part (a), where $r = \frac{1}{2}$ and the values appear to approach 2. This result is summarized in the following box.

SUM OF AN INFINITE GEOMETRIC SERIES

The sum of the infinite geometric series with first term a_1 and common ratio r is given by

$$S = \frac{a_1}{1 - r},$$

provided $|r| < 1$. If $|r| \geq 1$, then this sum does not exist.

Infinite series can be used to describe repeating decimals. For example, the fraction $\frac{1}{3}$ can be written as the repeating decimal $0.333333\ldots$. This decimal can be expressed as an infinite series.

Writing a Fraction as a Series

$$\frac{1}{3} = 0.3 + 0.03 + 0.003 + 0.0003 + \cdots + 0.3(0.1)^{n-1} + \cdots$$

In this series $a_1 = \mathbf{0.3}$ and $r = \mathbf{0.1}$. Since $|r| < 1$, the sum exists and is given by

Summing the Series

$$S = \frac{\mathbf{0.3}}{1 - \mathbf{0.1}} = \frac{3}{9} = \frac{1}{3}$$

as expected.

We are now able to answer the question concerning how far a person will walk if he or she travels 1 mile on the first day, $\frac{1}{2}$ mile the second day, $\frac{1}{4}$ mile the third day, and so on.

EXAMPLE 9   Finding the sum of an infinite geometric series

Find the sum of the infinite geometric series

$$1 + \frac{1}{2} + \frac{1}{4} + \frac{1}{8} + \cdots.$$

SOLUTION In this series, the first term is $a_1 = 1$ and the common ratio is $\frac{1}{2} = 0.5$. Its sum is

$$S = \frac{a_1}{1 - r} = \frac{1}{1 - \mathbf{0.5}} = 2.$$

If it were possible to walk in the prescribed manner, the total distance traveled after many days would always be slightly less than 2 miles.

Now Try Exercise 45

Summation Notation

Summation notation is used to write series efficiently. The symbol Σ, the uppercase Greek letter *sigma*, indicates a sum.

SUMMATION NOTATION

$$\sum_{k=1}^{n} a_k = a_1 + a_2 + a_3 + \cdots + a_n$$

The letter k is called the **index of summation**. The numbers 1 and n represent the subscripts of the first and last terms in the series. They are called the **lower limit** and **upper limit** of the summation, respectively.

EXAMPLE 10   Using summation notation

Evaluate each series.

(a) $\displaystyle\sum_{k=1}^{5} k^2$ **(b)** $\displaystyle\sum_{k=1}^{4} 5$ **(c)** $\displaystyle\sum_{k=3}^{6} (2k - 5)$

SOLUTION

(a) $\displaystyle\sum_{k=1}^{5} k^2 = 1^2 + 2^2 + 3^2 + 4^2 + 5^2 = 55$ $k = 1, 2, 3, 4, 5$

(b) $\displaystyle\sum_{k=1}^{4} 5 = 5 + 5 + 5 + 5 = 20$ $\quad k = 1, 2, 3, 4$

(c) $\displaystyle\sum_{k=3}^{6} (2k - 5) = (2(3) - 5) + (2(4) - 5) + (2(5) - 5) + (2(6) - 5)$

$\qquad\qquad\qquad\quad k = 3 \qquad\quad k = 4 \qquad\quad k = 5 \qquad\quad k = 6$

$\qquad\qquad\qquad = 1 + 3 + 5 + 7 = 16$

> **Now Try Exercise 61, 63, and 67**

EXAMPLE 11 Writing a series in summation notation

Write the series using summation notation. Let the lower limit equal 1.

(a) $\dfrac{1}{2^3} + \dfrac{1}{3^3} + \dfrac{1}{4^3} + \dfrac{1}{5^3} + \dfrac{1}{6^3} + \dfrac{1}{7^3} + \dfrac{1}{8^3}$

(b) $\dfrac{1}{2} + \dfrac{2}{3} + \dfrac{3}{4} + \dfrac{4}{5} + \dfrac{5}{6} + \dfrac{6}{7} + \dfrac{7}{8}$

SOLUTION

(a) The terms of the series can be written as $\dfrac{1}{(k+1)^3}$ for $k = 1, 2, 3, \ldots, 7$. Thus

$$\frac{1}{2^3} + \frac{1}{3^3} + \frac{1}{4^3} + \frac{1}{5^3} + \frac{1}{6^3} + \frac{1}{7^3} + \frac{1}{8^3} = \sum_{k=1}^{7} \frac{1}{(k+1)^3}.$$

MAKING CONNECTIONS

Notation The expressions

$$\sum_{k=1}^{n} a_k \text{ and } \Sigma_{k=1}^{n}\, a_k$$

are equivalent.

(b) The terms of the series can be written as $\dfrac{k}{k+1}$ for $k = 1, 2, 3, \ldots, 7$. Thus

$$\frac{1}{2} + \frac{2}{3} + \frac{3}{4} + \frac{4}{5} + \frac{5}{6} + \frac{6}{7} + \frac{7}{8} = \sum_{k=1}^{7} \frac{k}{k+1}.$$

> **Now Try Exercises 69 and 71**

EXAMPLE 12 Shifting the index of a series

Rewrite each summation so that the index starts with $n = 1$.

(a) $\displaystyle\sum_{k=4}^{7} k^2$ **(b)** $\displaystyle\sum_{k=8}^{30} (2k - 3)$

SOLUTION

(a) **Getting Started** Because $\sum_{k=4}^{7} k^2 = 4^2 + 5^2 + 6^2 + 7^2$, the summation has four terms. Instead of letting $k = 4, 5, 6, 7$, we must rewrite the sum so that $n = 1, 2, 3, 4$. It follows that $n + 3 = k$, so substitute $n + 3$ for k in the expression. ▶

$$\sum_{k=4}^{7} k^2 = \sum_{n=1}^{4} (n + 3)^2$$

$$= (1 + 3)^2 + (2 + 3)^2 + (3 + 3)^2 + (4 + 3)^2$$

$$= 4^2 + 5^2 + 6^2 + 7^2$$

(b) Instead of letting $k = 8, 9, 10, \ldots, 30$, we must rewrite the given sum so that $n = 1, 2, 3, \ldots, 23$. Thus $n + 7 = k$.

$$\sum_{k=8}^{30} (2k - 3) = \sum_{n=1}^{23} (2(n + 7) - 3)$$

$$= \sum_{n=1}^{23} (2n + 11)$$

> **Now Try Exercises 75 and 77**

The following box lists properties for summation notation.

PROPERTIES FOR SUMMATION NOTATION

Let $a_1, a_2, a_3, \ldots, a_n$ and $b_1, b_2, b_3, \ldots, b_n$ be sequences, and c be a constant.

1. $\displaystyle\sum_{k=1}^{n} ca_k = c\sum_{k=1}^{n} a_k$

2. $\displaystyle\sum_{k=1}^{n} (a_k + b_k) = \sum_{k=1}^{n} a_k + \sum_{k=1}^{n} b_k$

3. $\displaystyle\sum_{k=1}^{n} (a_k - b_k) = \sum_{k=1}^{n} a_k - \sum_{k=1}^{n} b_k$

4. $\displaystyle\sum_{k=1}^{n} c = nc$

5. $\displaystyle\sum_{k=1}^{n} k = \frac{n(n+1)}{2}$

6. $\displaystyle\sum_{k=1}^{n} k^2 = \frac{n(n+1)(2n+1)}{6}$

These properties can be used to find sums, as illustrated in the next example.

EXAMPLE 13   Applying summation notation

Use properties for summation notation to find each sum.

(a) $\displaystyle\sum_{k=1}^{40} 5$

(b) $\displaystyle\sum_{k=1}^{22} 2k$

(c) $\displaystyle\sum_{k=1}^{14} (2k^2 - 3)$

SOLUTION

(a) $\displaystyle\sum_{k=1}^{40} 5 = 40(5) = 200$ *Property 4 with $n = 40$ and $c = 5$*

(b) $\displaystyle\sum_{k=1}^{22} 2k = 2\sum_{k=1}^{22} k$ *Property 1 with $c = 2$ and $a_k = k$*

$\displaystyle\qquad\quad = 2 \cdot \frac{22(22+1)}{2}$ *Property 5 with $n = 22$*

$\displaystyle\qquad\quad = 506$ *Simplify.*

(c) $\displaystyle\sum_{k=1}^{14} (2k^2 - 3) = \sum_{k=1}^{14} 2k^2 - \sum_{k=1}^{14} 3$ *Property 3 with $a_k = 2k^2$ and $b_k = 3$*

$\displaystyle\qquad\qquad\quad = 2\sum_{k=1}^{14} k^2 - \sum_{k=1}^{14} 3$ *Property 1 with $c = 2$ and $a_k = k^2$*

$\displaystyle\qquad\qquad\quad = 2 \cdot \frac{14(14+1)(2 \cdot 14 + 1)}{6} - 14(3)$ *Properties 6 and 4*

$\displaystyle\qquad\qquad\quad = 1988$ *Simplify.*

Now Try Exercises 81, 83, and 89

NOTE   The infinite series $a_1 + a_2 + a_3 + \ldots + a_k + \ldots$ is sometimes written in summation notation as

$$\sum_{k=1}^{\infty} a_k, \text{ where the upper limit is infinity.}$$

11.2 Putting It All Together

The following table summarizes concepts related to series.

CONCEPT	EXPLANATION	EXAMPLE
Series	A series is the summation of the terms of a sequence. A finite series always has a sum, but an infinite series may not have a sum.	$2 + 4 + 6 + 8 + \cdots + 20$ $S_4 = a_1 + a_2 + a_3 + a_4$ **(partial sum)** $\quad = 2 + 4 + 6 + 8$ $\quad = 20$
Arithmetic series	An arithmetic series is the summation of the terms of an arithmetic sequence. The sum of the first n terms is given by $$S_n = n\left(\frac{a_1 + a_n}{2}\right)$$ or $$S_n = \frac{n}{2}(2a_1 + (n-1)d).$$	$1 + 4 + 7 + 10 + 13 + 16 + 19$ **(7 terms)** $S_7 = 7\left(\dfrac{a_1 + a_7}{2}\right) = 7\left(\dfrac{1 + 19}{2}\right) = 70$ or $S_7 = \dfrac{7}{2}(2a_1 + 6d) = \dfrac{7}{2}(2(1) + 6(3)) = 70$
Geometric series	A geometric series is the summation of the terms of a geometric sequence. The sum of the first n terms is given by $$S_n = a_1\left(\frac{1 - r^n}{1 - r}\right).$$ If $|r| < 1$, then an infinite geometric series has the sum $$S = \frac{a_1}{1 - r}.$$	$3 + 6 + 12 + 24 + 48 + 96$ **(6 terms)** $$S_6 = 3\left(\frac{1 - 2^6}{1 - 2}\right) = 189$$ The infinite geometric series $$1 + \frac{1}{4} + \frac{1}{16} + \frac{1}{64} + \cdots$$ has a sum $$S = \frac{1}{1 - \frac{1}{4}} = \frac{4}{3}.$$
Summation notation	The series $a_1 + a_2 + \cdots + a_n$ can be written with summation notation as $$\sum_{k=1}^{n} a_k.$$	$1^2 + 2^2 + 3^2 + 4^2 + \cdots + 10^2$ can be written as $$\sum_{k=1}^{10} k^2.$$

11.2 Exercises

Concepts

Exercises 1–4: Complete the following.

(a) *Write the described sequence.*

(b) *Write a series that sums the terms of the sequence in part (a).*

(c) *Find the sum of the series in part (b).*

1. The counting numbers from 1 to 5

2. The first seven positive odd integers

3. The integers counting down from 1 to -3

4. The integers counting down from 10 to 6

Exercises 5 and 6: Let A_n represent the number of U.S. AIDS deaths reported n years after 2000.

5. Write a series whose sum gives the cumulative number of AIDS deaths from 2005 to 2009.

6. Explain what S_6 represents.

Finding Sums of Series

Exercises 7–14: For the given a_n, calculate S_5.

7. $a_n = 3n$

8. $a_n = n + 4$

9. $a_n = 2n - 1$

10. $a_n = 4n + 1$

11. $a_n = n^2 + 1$

12. $a_n = 2n^2$

13. $a_n = \dfrac{n}{n + 1}$

14. $a_n = \dfrac{1}{2n}$

Exercises 15–22: Use a formula to find the sum of the arithmetic series.

15. $3 + 5 + 7 + 9 + 11 + 13 + 15 + 17$

16. $7.5 + 6 + 4.5 + 3 + 1.5 + 0 + (-1.5)$

17. $1 + 2 + 3 + 4 + \cdots + 50$

18. $1 + 3 + 5 + 7 + \cdots + 97$

19. $-7 + (-4) + (-1) + 2 + 5 + \cdots + 98 + 101$

20. $89 + 84 + 79 + 74 + \cdots + 9 + 4$

21. The first 40 terms of the series defined by $a_n = 5n$

22. The first 50 terms of the series defined by $a_n = 1 - 3n$

23. The sum of an arithmetic series with 15 terms is 255. If $a_1 = 3$, find a_{15}.

24. The sum of an arithmetic series with 20 terms is 610. If $a_{20} = 59$, find a_1.

Exercises 25–34: Use a formula to find the sum of the first 20 terms for the arithmetic sequence.

25. $a_1 = 4, d = 2$

26. $a_1 = -3, d = \frac{2}{3}$

27. $a_1 = 10, d = -\frac{1}{2}$

28. $a_1 = 0, d = -4$

29. $a_1 = 4, a_{20} = 190.2$

30. $a_1 = -4, a_{20} = 15$

31. $a_1 = -2, a_{11} = 50$

32. $a_1 = 6, a_5 = -30$

33. $a_2 = 6, a_{12} = 31$

34. $a_8 = 4, a_{10} = 14$

Exercises 35–40: Use a formula to find the sum of the finite geometric series.

35. $1 + 2 + 4 + 8 + 16 + 32 + 64 + 128$

36. $2 + \frac{1}{2} + \frac{1}{8} + \frac{1}{32} + \frac{1}{128} + \frac{1}{512}$

37. $0.5 + 1.5 + 4.5 + 13.5 + 40.5 + 121.5 + 364.5$

38. $0.6 + 0.3 + 0.15 + 0.075 + 0.0375$

39. The first 20 terms of the series defined by $a_n = 3(2)^{n-1}$

40. The first 15 terms of the series defined by $a_n = 2\left(\frac{1}{3}\right)^n$

Exercises 41–44: Use a formula to approximate the sum for $n = 4, 7,$ and 10.

41. $1 - \frac{1}{2} + \frac{1}{4} - \frac{1}{8} + \cdots + \left(-\frac{1}{2}\right)^{n-1}$

42. $3 - 1 + \frac{1}{3} - \frac{1}{9} + \cdots + 3\left(-\frac{1}{3}\right)^{n-1}$

43. $\frac{1}{3} + \frac{2}{3} + \frac{4}{3} + \frac{8}{3} + \cdots + \frac{1}{3}(2)^{n-1}$

44. $4 + \frac{8}{3} + \frac{16}{9} + \frac{32}{27} + \cdots + 4\left(\frac{2}{3}\right)^{n-1}$

Exercises 45–50: Find the sum of the infinite geometric series.

45. $1 + \frac{1}{3} + \frac{1}{9} + \frac{1}{27} + \frac{1}{81} + \cdots$

46. $5 + \frac{5}{2} + \frac{5}{4} + \frac{5}{8} + \frac{5}{16} + \cdots$

47. $6 - 4 + \frac{8}{3} - \frac{16}{9} + \frac{32}{27} - \frac{64}{81} + \cdots$

48. $-2 + \frac{1}{2} - \frac{1}{8} + \frac{1}{32} - \frac{1}{128} + \cdots$

49. $1 - \frac{1}{10} + \frac{1}{100} - \frac{1}{1000} + \cdots + \left(-\frac{1}{10}\right)^{n-1} + \cdots$

50. $25 - 5 + 1 - \frac{1}{5} + \cdots + 25\left(-\frac{1}{5}\right)^{n-1} + \cdots$

Decimal Numbers and Geometric Series

Exercises 51–56: Write each rational number in the form of an infinite geometric series.

51. $\frac{2}{3}$ **52.** $\frac{1}{9}$

53. $\frac{9}{11}$ **54.** $\frac{14}{33}$

55. $\frac{1}{7}$ **56.** $\frac{23}{99}$

Exercises 57–60: Write the sum of each geometric series as a rational number.

57. $0.8 + 0.08 + 0.008 + 0.0008 + \cdots$

58. $0.9 + 0.09 + 0.009 + 0.0009 + \cdots$

59. $0.45 + 0.0045 + 0.000045 + \cdots$

60. $0.36 + 0.0036 + 0.000036 + \cdots$

Summation Notation

Exercises 61–68: Write out the terms of the series and then evaluate it.

61. $\sum_{k=1}^{4} (k + 1)$ **62.** $\sum_{k=1}^{6} (3k - 1)$

63. $\sum_{k=1}^{8} 4$ **64.** $\sum_{k=2}^{6} (5 - 2k)$

65. $\sum_{k=1}^{7} k^3$ **66.** $\sum_{k=1}^{4} 5(2)^{k-1}$

67. $\sum_{k=4}^{5} (k^2 - k)$ **68.** $\sum_{k=1}^{5} \log k$

Exercises 69–74: Write the series with summation notation. Let the lower limit equal 1.

69. $1^4 + 2^4 + 3^4 + 4^4 + 5^4 + 6^4$

70. $1 + \frac{1}{5} + \frac{1}{25} + \frac{1}{125} + \frac{1}{625}$

71. $1 + \frac{4}{3} + \frac{6}{4} + \frac{8}{5} + \frac{10}{6} + \frac{12}{7} + \frac{14}{8}$

72. $2 + \frac{5}{8} + \frac{10}{27} + \frac{17}{64} + \frac{26}{125} + \frac{37}{216}$

73. $1 + \frac{1}{2^2} + \frac{1}{3^2} + \frac{1}{4^2} + \frac{1}{5^2} + \cdots$

74. $1 + \frac{1}{10} + \frac{1}{100} + \frac{1}{1000} + \frac{1}{10,000} + \cdots$

Exercises 75–80: (Refer to Example 12.) Rewrite each summation so that the index starts with $n = 1$.

75. $\sum_{k=6}^{9} k^3$ **76.** $\sum_{k=5}^{10} (k^2 - 2)$

77. $\sum_{k=9}^{32} (3k - 2)$ **78.** $\sum_{k=8}^{21} (4k + 1)$

79. $\sum_{k=16}^{52} (k^2 - 3k)$ **80.** $\sum_{k=25}^{59} (k^2 + 4k)$

Exercises 81–92: (Refer to Example 13.) Use properties for summation notation to find the sum.

81. $\sum_{k=1}^{60} 9$ **82.** $\sum_{k=1}^{43} -4$

83. $\sum_{k=1}^{15} 5k$ **84.** $\sum_{k=1}^{22} -2k$

85. $\sum_{k=1}^{31} (3k - 3)$ **86.** $\sum_{k=1}^{17} (1 - 4k)$

87. $\sum_{k=1}^{25} k^2$ **88.** $\sum_{k=1}^{12} 3k^2$

89. $\sum_{k=1}^{16} (k^2 - k)$ **90.** $\sum_{k=1}^{18} (k^2 - 4k + 3)$

91. $\sum_{k=5}^{24} k$ **92.** $\sum_{k=7}^{19} (k^2 + 1)$

93. Verify the formula $\sum_{k=1}^{n} k = \frac{n(n+1)}{2}$ by using the formula for the sum of the first n terms of a finite arithmetic sequence.

94. Use Exercise 93 to find the sum of the series $\sum_{k=1}^{200} k$.

Applications

Exercises 95 and 96: **Salaries** *A person has the given starting salary S and receives a raise R each year thereafter.*

(a) *Use a formula to calculate the total amount earned over 15 years.*

(b) *Use a calculator to verify this value.*

95. $S = \$42,000$, $R = \$1800$

96. $S = \$35,000$, $R = \$2500$

Exercises 97–100: **Annuities** *(Refer to Example 8.) Determine the future value of each annuity.*

97. $A_0 = \$2000,\quad i = 0.08, n = 20$

98. $A_0 = \$500,\quad i = 0.15, n = 10$

99. $A_0 = \$10{,}000,\quad i = 0.11, n = 5$

100. $A_0 = \$3000,\quad i = 0.19, n = 45$

101. **Stacking Logs** Logs are stacked in layers, with one fewer log in each layer. See the figure. If the top layer has 7 logs and the bottom layer has 15 logs, what is the total number of logs in the pile? Use a formula to find the sum.

102. **Stacking Logs** (Refer to Exercise 101.) Suppose a stack of logs has 13 logs in the top layer and a total of 7 layers. How many logs are in the stack?

103. **Filter** Suppose that one filter removes half of the impurities in a water supply.
   (a) Find a series that represents the amount of impurities removed by a sequence of n filters. Express your answer in summation notation.

   (b) How many filters would be necessary to remove all of the impurities?

104. **Walking** Suppose that a person walks 1 mile on the first day, $\frac{1}{3}$ mile on the second day, $\frac{1}{9}$ mile on the third day, and so on. Assuming that a person could walk each distance precisely, estimate how far the person would have traveled after a very long time.

105. **Area** (Refer to Exercise 100, Section 11.1.) Use a geometric series to find the sum of the areas of the squares if they continue indefinitely.

106. **Perimeter** (Refer to Exercise 100, Section 11.1.) Use a geometric series to find the sum of the perimeters of the squares if they continue indefinitely.

Exercises 107 and 108: **The Natural Exponential Function** *The following series can be used to estimate the value of e^a for any real number a:*

$$e^a \approx 1 + a + \frac{a^2}{2!} + \frac{a^3}{3!} + \cdots + \frac{a^n}{n!},$$

where $n! = 1 \cdot 2 \cdot 3 \cdot 4 \cdots \cdots n$. Use the first eight terms of this series to approximate the given expression. Compare this estimate with the actual value.

107. e

108. e^{-1}

Computing Partial Sums

Exercises 109–112: Use a_k and n to find $S_n = \sum_{k=1}^{n} a_k$. (Refer to Example 7.) Then evaluate the infinite geometric series $S = \sum_{k=1}^{\infty} a_k$. Compare S to the values for S_n.

109. $a_k = \left(\frac{1}{3}\right)^{k-1}; n = 2, 4, 8, 16$

110. $a_k = 3\left(\frac{1}{2}\right)^{k-1}; n = 5, 10, 15, 20$

111. $a_k = 4\left(-\frac{1}{10}\right)^{k-1}; n = 1, 2, 3, 4, 5, 6$

112. $a_k = 2(-0.02)^{k-1}; n = 1, 2, 3, 4, 5, 6$

Writing about Mathematics

113. Discuss the difference between a sequence and a series. Give examples.

114. Under what circumstances can we find the sum of a geometric series? Give examples.

115. Explain how to write the series

$$\log 1 + \log 2 + \log 3 + \cdots + \log n$$

as one term.

116. Explain how to write the series

$$\log 2 - \log 4 + \log 6 - \log 8 + \cdots + (-1)^{n-1}\log 2n$$

as one term. Assume n is even.

CHECKING BASIC CONCEPTS FOR SECTIONS 11.1 AND 11.2

1. Give graphical and numerical representations of the sequence defined by $a_n = -2n + 3$, where $a_n = f(n)$. Include the first six terms.

2. Determine if the sequence is arithmetic or geometric. If it is arithmetic, state the common difference; if it is geometric, give the common ratio.
   (a) $2, -4, 8, -16, 32, -64, 128, \ldots$

   (b) $-3, 0, 3, 6, 9, 12, \ldots$

   (c) $4, 2, 1, \frac{1}{2}, \frac{1}{4}, \frac{1}{8}, \frac{1}{16}, \ldots$

3. Determine if the series is arithmetic or geometric. Use a formula to find its sum.
   (a) $1 + 5 + 9 + 13 + \cdots + 37$

   (b) $3 + 1 + \frac{1}{3} + \frac{1}{9} + \frac{1}{27} + \frac{1}{81}$

   (c) $2 + \frac{1}{2} + \frac{1}{8} + \frac{1}{32} + \cdots$

   (d) $0.9 + 0.09 + 0.009 + 0.0009 + \cdots$

4. Write each series in Exercise 3 in summation notation.

5. Use properties of summation notation to find the sum.
   (a) $\sum_{k=1}^{15}(k + 2)$ (b) $\sum_{k=1}^{21}2k^2$

6. **Bouncing Ball** A ball is dropped from a height of 6 feet. On each bounce the ball returns to $\frac{2}{3}$ of its previous height. How far does the ball travel (up and down) before it comes to rest?

11.3 Counting

- Apply the fundamental counting principle
- Calculate and apply permutations
- Calculate and apply combinations

Introduction

The notion of *counting* in mathematics includes much more than simply counting from 1 to 100. It also includes determining the number of ways that an event can occur. For example, how many ways are there to answer a true-false quiz with ten questions? The answer involves counting the different ways that a student could answer such a quiz. Counting is an important concept that is used to calculate probabilities. Probability is discussed in Section 11.6.

Fundamental Counting Principle

Suppose that a quiz has only two questions. The first is a multiple-choice question with four choices, A, B, C, or D, and the second is a true-false (T-F) question. The **tree diagram** in Figure 11.14 can be used to count the ways that this quiz can be answered.

Different Ways to Answer a Quiz

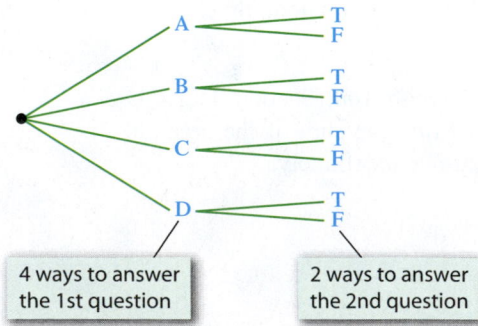

4 ways to answer the 1st question

2 ways to answer the 2nd question

Figure 11.14

A tree diagram is a systematic way of listing every possibility. From Figure 11.14 we can see that there are eight ways to answer the quiz. They are

AT, AF, BT, BF, CT, CF, DT, and DF.   8 ways to answer the quiz

For instance, CF indicates a quiz with answers of C on the first question and F on the second question.

A tree diagram is not always practical, because it can quickly become very large. For this reason mathematicians have developed more efficient ways of counting. Since the multiple-choice question has four possible answers, after which the true-false question has two possible answers, there are $4 \cdot 2 = 8$ possible ways of answering the test. This is an application of the *fundamental counting principle*, which applies to independent events. Two events are **independent** if neither event influences the outcome of the other.

> **FUNDAMENTAL COUNTING PRINCIPLE**
>
> Let $E_1, E_2, E_3, \ldots, E_n$ be a sequence of n independent events. If event E_k can occur m_k ways for $k = 1, 2, 3, \ldots, n$, then there are
>
> $$m_1 \cdot m_2 \cdot m_3 \cdot \cdots \cdot m_n$$
>
> ways for all n events to occur.

EXAMPLE 1 Counting ways to answer an exam

An exam contains four true-false questions and six multiple-choice questions. Each multiple-choice question has five possible answers. Count the number of ways that the exam can be answered.

SOLUTION

Getting Started Answering these ten questions can be thought of as a sequence of ten independent events. There are two ways to answer each of the first **four** questions and five ways of answering each of the next **six** questions. ▶

The number of ways to answer the exam is

Each factor represents the number of ways to answer the given question.

$$\underbrace{2 \cdot 2 \cdot 2 \cdot 2}_{\textbf{4 factors}} \cdot \underbrace{5 \cdot 5 \cdot 5 \cdot 5 \cdot 5 \cdot 5}_{\textbf{6 factors}} = 2^4 5^6 = 250,000.$$

Now Try Exercise 3

In the next example, we count the number of different license plates possible with a given format.

EXAMPLE 2 Counting license plates

Sometimes a license plate is limited to three uppercase letters (A through Z) followed by three digits (0 through 9). For example, ABB 112 would be a valid license plate. Would this format provide enough license plates for a state with 8 million vehicles?

SOLUTION Since there are 26 letters of the alphabet, it follows that there are 26 ways to choose each of the first three letters of the license plate. Similarly, there are 10 digits, so there are 10 ways to choose each of the three digits in the license plate. By the fundamental counting principle, there are

$$26 \cdot 26 \cdot 26 \cdot 10 \cdot 10 \cdot 10 = 17,576,000$$

unique license plates that could be issued. This format for license plates could accommodate more than 8 million vehicles.

Now Try Exercise 5

An Application At one time, toll-free numbers always began with 800. Because there were not enough toll-free 800 numbers to meet demand, phone companies started to use toll-free numbers that began with 888 and 877. In the next example, we count the number of valid 800 numbers. (*Source:* Database Services Management.)

EXAMPLE 3 Counting toll-free telephone numbers

Count the total number of 800 numbers if the local portion of a telephone number (the last seven digits) does not start with a 0 or 1.

SOLUTION A toll-free 800 number assumes the following form.

We can think of choosing the remaining digits for the local number as seven independent events. Since the local number cannot begin with a 0 or 1, there are **eight possibilities** (2 to 9) for the first digit. The remaining six digits can be any number from 0 to 9, so there are **ten possibilities** for each of these digits. The total is given by

$$8 \cdot 10 \cdot 10 \cdot 10 \cdot 10 \cdot 10 \cdot 10 = 8 \cdot 10^6 = 8{,}000{,}000.$$

First digit   Last 6 digits

> **Now Try Exercise 47**

Permutations

A **permutation** is an *ordering* or *arrangement*. For example, suppose that three students are scheduled to give speeches in a class. The different arrangements in which these speeches can be ordered are called permutations. Initially, any one of the three students could give the first speech. After the first speech, there are two students remaining for the second speech. For the third speech there is only one possibility. By the fundamental counting principle, the total number of permutations is equal to

$$3 \cdot 2 \cdot 1 = 6. \quad \boxed{\text{Six possible orderings}}$$

If the students are denoted as A, B, and C, then these six permutations are ABC, ACB, BAC, BCA, CAB, and CBA. In a similar manner, if there were ten students scheduled to give speeches, the number of permutations would increase to

$$\boxed{\begin{array}{c}\text{Ten possible}\\\text{choices to go first}\end{array}}\!\!-10 \cdot 9 \cdot 8 \cdot 7 \cdot 6 \cdot 5 \cdot 4 \cdot 3 \cdot 2 \cdot 1 = 3{,}628{,}800.$$

A more efficient way of writing the previous two products is to use *factorial notation*. The number $n!$ (read "n-factorial") is defined as follows.

Calculator Help

To calculate $n!$, see Appendix A (page AP-17).

n-FACTORIAL

For any natural number n,

$$n! = n(n - 1)(n - 2) \cdots (3)(2)(1)$$

and

$$0! = 1.$$

EXAMPLE 4 Calculating factorials

Compute $n!$ for $n = 0, 1, 2, 3, 4,$ and 5 by hand. Use a calculator to find 8!, 13!, and 25!.

SOLUTION The values for $n!$ can be calculated as

$$0! = 1, \quad 1! = 1, \quad 2! = 2 \cdot 1 = 2, \quad 3! = 3 \cdot 2 \cdot 1 = 6,$$

$$4! = 4 \cdot 3 \cdot 2 \cdot 1 = 24, \quad \text{and} \quad 5! = 5 \cdot 4 \cdot 3 \cdot 2 \cdot 1 = 120.$$

```
8!
              40320
13!
         6227020800
25!
    1.551121004E25
```

Figure 11.15

Figure 11.15 shows the values of 8!, 13!, and 25!. The value for 25! is an approximation. Notice how rapidly $n!$ increases.

> **Now Try Exercise 27**

A Famous Unsolved Problem One of the most famous unanswered questions in computing today is called the *traveling salesperson problem*. It is a relatively simple problem to state, but if someone could design a procedure to solve this problem *efficiently*, he or she would not only become famous but also provide a valuable way for businesses to save millions of dollars on scheduling problems, such as assigning bus routes and truck deliveries.

One example of the traveling salesperson problem can be stated as follows. A salesperson must begin and end at home and travel to three cities. Assuming that the salesperson can travel between any pair of cities, what route would minimize the salesperson's mileage? In Figure 11.16, the four cities are labeled A, B, C, and D. Let the salesperson live in city A. There are six routes that could be tried. They are listed in Table 11.6 with the appropriate mileage for each.

Traveling Salesperson Problem

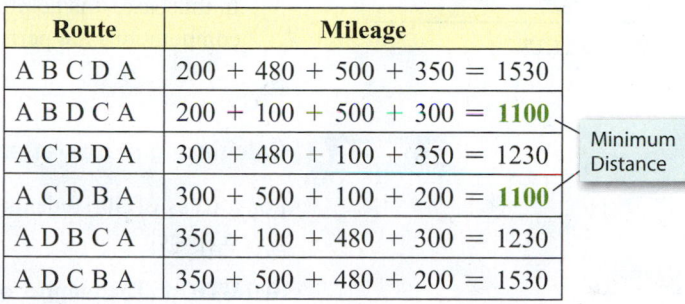

Route	Mileage
A B C D A	200 + 480 + 500 + 350 = 1530
A B D C A	200 + 100 + 500 + 300 = **1100**
A C B D A	300 + 480 + 100 + 350 = 1230
A C D B A	300 + 500 + 100 + 200 = **1100**
A D B C A	350 + 100 + 480 + 300 = 1230
A D C B A	350 + 500 + 480 + 200 = 1530

Minimum Distance

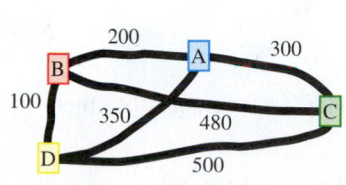

Figure 11.16   **Table 11.6**

The shortest route of **1100** miles occurs when the salesperson either (1) starts at A and travels through B, D, and C and back to A or (2) reverses this route. Currently, this method of listing all possible routes to find the minimum distance is the only known way to consistently find the optimal solution for any general map containing n cities. In fact, people have not been able to determine whether a *significantly* faster method even exists.

Counting the number of routes involves the fundamental counting principle. At the first step, the salesperson can travel to any one of three cities. Once this city has been selected, there are two possible cities to choose, and then one. Finally the salesperson returns home. The total number of routes is given by

$$3! = 3 \cdot 2 \cdot 1 = 6.$$

If the salesperson must travel to 30 cities, then there are

$$30! \approx 2.7 \times 10^{32}$$

routes to check, far too many to check even with the largest supercomputers.

Next suppose that a salesperson must visit three of eight possible cities. At first there are eight cities to choose. After the first city has been visited, there are seven cities to select. Since the salesperson travels to only three of the eight cities, there are

$$8 \cdot 7 \cdot 6 = 336$$

possible routes. This number of permutations is denoted $P(8, 3)$. It represents the number of arrangements that can be made using three elements taken from a sample of eight.

PERMUTATIONS OF n ELEMENTS TAKEN r AT A TIME

If $P(n, r)$ denotes the number of permutations of n elements taken r at a time, with $r \leq n$, then

$$P(n, r) = \frac{n!}{(n - r)!} = \underbrace{n(n - 1)(n - 2) \cdots (n - r + 1)}_{r \text{ factors}}.$$

Calculator Help

To calculate $P(n, r)$, see Appendix A (page AP-17).

EXAMPLE 5 Calculating $P(n, r)$

Calculate each of the following by hand. Then support your answers by using a calculator.
(a) $P(7, 3)$ **(b)** $P(100, 2)$

SOLUTION
(a) $P(7, 3) = \underbrace{7 \cdot 6 \cdot 5}_{3 \text{ factors}} = 210$

Permutations: $P(n, r)$

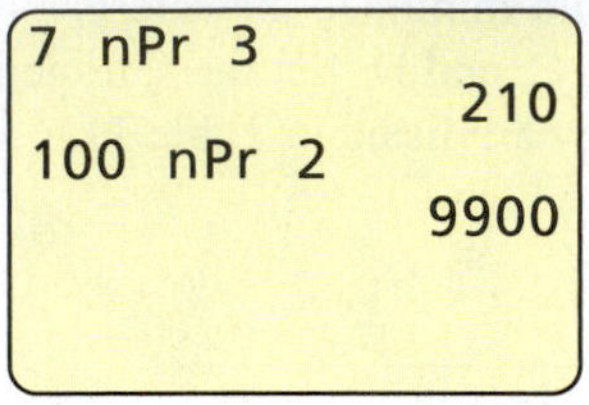

```
7 nPr 3
              210
100 nPr 2
             9900
```

Figure 11.17

(b) $P(100, 2) = 100 \cdot 99 = 9900$. We also can compute this number as follows.

$$P(100, 2) = \frac{100!}{(100 - 2)!} = \frac{100 \cdot 99 \cdot 98!}{98!} = 100 \cdot 99 = 9900$$

In this case, it is helpful to cancel 98! before performing the arithmetic. Both of these computations are performed by a calculator in Figure 11.17.

> **Now Try Exercise 31 and 37**

EXAMPLE 6 Calculating permutations

For a class of 30 students, how many arrangements are there in which 4 students each give a speech?

SOLUTION The number of permutations of 30 elements taken 4 at a time is given by

$$P(30, 4) = \underbrace{30 \cdot 29 \cdot 28 \cdot 27}_{4 \text{ factors}} = 657{,}720.$$

Thus there are 657,720 ways to arrange the four speeches.

> **Now Try Exercise 43**

In the next example, we determine the number of ways that four people can have *different* birthdays. For example, if the birthdays of four people are February 29, May 2, June 30, and July 11, then these dates would be one way that four people could have different birthdays.

EXAMPLE 7 Counting birthdays

Count the possible ways that four people can have different birthdays.

SOLUTION Counting February 29, there are 366 possible birthdays. The first person could have any of the 366 possible birthdays. The second person could have any of 365 birthdays, because the first person's birthday cannot be duplicated. Similarly, the third person could have any of 364 birthdays, and the fourth person could have any of 363 birthdays. The total number of ways that four people could have different birthdays equals

$$P(366, 4) = \underbrace{366 \cdot 365 \cdot 364 \cdot 363}_{4 \text{ factors}} \approx 1.77 \times 10^{10}.$$

> **Now Try Exercise 53**

CLASS DISCUSSION

Count the number of arrangements of 52 cards in a standard deck. Is it likely that there are arrangements that no one has ever shuffled at any time in the history of the world? Explain.

Combinations

Unlike a permutation, a **combination** is not an ordering or arrangement, but rather a subset of a set of elements. Order is unimportant in finding combinations. For example, suppose we want to select a tennis team of two players from four people. The order in which the selection is made does not affect the final team of two players. From a set of four people, we select a subset of two players. The number of possible subsets, or combinations, is denoted either $C(4, 2)$ or $\binom{4}{2}$.

To calculate $C(4, 2)$, we first consider $P(4, 2)$. If we denote the four players by the letters A, B, C, and D, there are $P(4, 2) = 4 \cdot 3 = 12$ permutations given by

Equivalent teams   AB, BA, AC, CA, AD, DA, BC, CB, BD, DB, CD, DC.

However, the team that comprises person A and person B is equivalent to the team with person B and person A. The sets {AB} and {BA} are equal. The valid combinations are the following **two**-element subsets of {A, B, C, D}.

$$\{AB\} \quad \{AC\} \quad \{AD\} \quad \{BC\} \quad \{BD\} \quad \{CD\}$$

That is, $C(4, \textbf{2}) = \dfrac{P(4, 2)}{2!} = 6$. The relationship between $P(n, r)$ and $C(n, r)$ is given below.

COMBINATIONS OF *n* ELEMENTS TAKEN *r* AT A TIME

If $C(n, r)$ denotes the number of combinations of n elements taken r at a time, with $r \leq n$, then

$$C(n, r) = \frac{P(n, r)}{r!} = \frac{n!}{(n - r)! \, r!}.$$

Calculator Help

To calculate $C(n, r)$, see Appendix A (page AP-17).

EXAMPLE 8   Calculating $C(n, r)$

Calculate each of the following. Support your answers by using a calculator.

(a) $C(7, 3)$ (b) $\binom{50}{47}$

SOLUTION

(a) $C(7, \textbf{3}) = \dfrac{7!}{(7 - \textbf{3})!\textbf{3}!} = \dfrac{7!}{4!3!} = \dfrac{7 \cdot 6 \cdot 5 \cdot 4!}{4!3!} = \dfrac{7 \cdot 6 \cdot 5}{3!} = \dfrac{210}{6} = 35$

$\quad\quad\;\; n\;\; r \quad\quad n - r$

Combinations: $C(n, r)$

7 nCr 3	
	35
50 nCr 47	
	19600

Figure 11.18

(b) The notation $\binom{50}{47}$ is equivalent to $C(50, 47)$.

$$\binom{50}{47} = \frac{50!}{(50 - 47)! \, 47!} = \frac{50!}{3! \, 47!} = \frac{50 \cdot 49 \cdot 48 \cdot 47!}{3! \, 47!}$$

$$= \frac{50 \cdot 49 \cdot 48}{3!} = \frac{117,600}{6} = 19,600$$

These computations are performed by a calculator in Figure 11.18.

Now Try Exercise 59 and 65

EXAMPLE 9   Counting combinations

A college student has five courses left in her major and plans to take two of them this semester. Assuming that this student has the prerequisites for all five courses, determine how many ways these two courses can be selected.

SOLUTION   The order in which the courses are selected is unimportant. From a set of **five** courses, the student selects a subset of **two** courses. The number of subsets is

$$C(5, \textbf{2}) = \frac{5!}{(5 - \textbf{2})! \, \textbf{2}!} = \frac{5!}{3! \, 2!} = 10.$$

There are 10 ways to select two courses from a set of five.

Now Try Exercise 67

EXAMPLE 10 Calculating the number of ways to play the lottery

To win the jackpot in a lottery, a person must select five different numbers from 1 to 59 and then pick the powerball, which has a number from 1 to 35. Count the ways to play the game. (*Source:* Minnesota State Lottery.)

SOLUTION From 59 numbers a player picks five numbers. There are $C(59, 5)$ ways of doing this. There are 35 ways to choose the powerball. By the fundamental counting principle, the number of ways to play the game equals $C(59, 5) \cdot 35 = 175{,}223{,}510$.

> **Now Try Exercise 71**

MAKING CONNECTIONS

Permutations and Combinations

Permutation: An arrangement (or list) in which ordering of the objects *is* important. For example, $P(20, 9)$ would give the number of possible batting orders for 9 players from a team of 20.

Combination: A subset of a set of objects where the ordering of the objects *is not* important. For example, $C(20, 9)$ would give the number of committees possible when 9 people are selected from a group of 20. The order in which members are selected does not affect the resulting committee.

EXAMPLE 11 Counting committees

How many committees of six people can be selected from six women and three men if a committee must have at least two men?

SOLUTION

Getting Started Because there are three men and each committee must have at least two men, a committee can include either two or three men. The order of selection is not important. Therefore we need to consider combinations of committee members rather than permutations. ▶

Two Men: This committee would consist of four women and two men. We can select two men from a group of three in $C(3, 2) = 3$ ways. Four women can be selected from a group of six in $C(6, 4) = 15$ ways. By the fundamental counting principle, a total of

$$C(3, 2) \cdot C(6, 4) = 3 \cdot 15 = \mathbf{45}$$

committees have two men.

Three Men: This committee would have three women and three men. We can select three men from a group of three in $C(3, 3) = 1$ way. We can select three women from a group of six in $C(6, 3) = 20$ ways. By the fundamental counting principle, a total of

$$C(3, 3) \cdot C(6, 3) = 1 \cdot 20 = \mathbf{20}$$

committees have three men.

The total number of possible committees would be $\mathbf{45} + \mathbf{20} = 65$.

> **Now Try Exercise 69**

11.3 Putting It All Together

$\mathbf{T}$he fundamental counting principle can be used to determine the number of ways a sequence of independent events can occur. Permutations are arrangements or listings, whereas combinations are subsets of a set of events.

The following table summarizes some concepts related to counting in mathematics.

NOTATION	EXPLANATION	EXAMPLES
n-factorial: $n!$	$n!$ represents the product $$n(n-1)\cdots(3)(2)(1).$$	$6! = 6 \cdot 5 \cdot 4 \cdot 3 \cdot 2 \cdot 1 = 720$ $0! = 1$
$P(n, r) = \dfrac{n!}{(n-r)!}$	$P(n, r)$ represents the number of permutations of n elements taken r at a time.	The number of two-letter strings that can be formed using the four letters A, B, C, and D with no letter repeated is given by $$P(4, 2) = \frac{4!}{(4-2)!} = \frac{4!}{2!} = 12,$$ or $P(4, 2) = 4 \cdot 3 = 12.$
$C(n, r) = \dfrac{n!}{(n-r)!\,r!}$	$C(n, r)$ represents the number of combinations of n elements taken r at a time.	The number of committees of three people that can be formed from five people is $$C(5, 3) = \frac{5!}{(5-3)!\,3!} = \frac{5!}{2!3!} = 10.$$

11.3 Exercises

Counting

Exercises 1–4: **Exam Questions** *Count the number of ways that the questions on an exam could be answered.*

1. Ten true-false questions

2. Ten multiple-choice questions with five choices each

3. Five true-false questions and ten multiple-choice questions with four choices each

4. One question involving matching ten items in one column with ten items in another column, using a one-to-one correspondence

Exercises 5–8: **License Plates** *Count the number of possible license plates with the given constraints.*

5. Three digits followed by three letters

6. Two letters followed by four digits

7. Three letters followed by three digits or letters

8. Two letters followed by either three or four digits

Exercises 9–12: **Counting Strings** *Count the number of five-letter strings that can be formed with the given letters, assuming a letter can be used more than once.*

9. A, B, C

10. W, X, Y, Z

11. D, E, F, G, H

12. A, C

Exercises 13–16: **Counting Strings** *Count the number of strings that can be formed with the given letters, assuming each letter is used exactly once.*

13. A, B

14. A, B, C

15. W, X, Y, Z

16. V, W, X, Y, Z

17. Combination Lock A briefcase has two locks. The combination to each lock consists of a three-digit number, in which digits may be repeated. See the figure. How many combinations are possible? (*Hint:* The word *combination* is a misnomer. Lock combinations are permutations in which the arrangement of the numbers is important.)

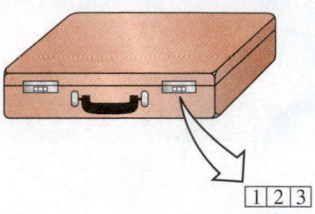

18. Combination Lock A typical combination for a padlock consists of three numbers from 0 to 39. Count the combinations possible with this type of lock if a number may be repeated.

19. Garage Door Openers The code for some garage door openers consists of 12 electrical switches that can be set to either 0 or 1 by the owner. With this type of opener, how many codes are possible? (*Source:* Promax.)

20. Lottery To win the jackpot in a lottery game, a person must pick three numbers from 0 to 9 in the correct order. If a number can be repeated, how many ways are there to play the game?

21. Radio Stations Call letters for a radio station usually begin with either a K or a W, followed by three letters. In 2012, there were 14,952 radio stations on the air. Is there any shortage of call letters for new radio stations? (*Source:* M. Street Corporation.)

22. Access Codes An ATM access code often consists of a four-digit number. How many codes are possible without giving two accounts the same access code?

23. Computer Package A computer store offers a package in which buyers choose one of two monitors, one of three printers, and one of four types of software. How many different packages can be purchased?

24. Dice A red die and a blue die are thrown. How many ways are there for both dice to show an even number?

25. Telephone Numbers How many different 7-digit telephone numbers are possible if the first digit cannot be a 0 or a 1?

26. Dinner Choices A menu offers 5 different salads, 10 different entrées, and 4 different desserts. How many ways are there to order a salad, an entrée, and a dessert?

Exercises 27–30: Evaluate the expression.

27. 6!   **28.** 0!

29. 10!   **30.** 7!

Permutations

Exercises 31–40: Evaluate the expression.

31. $P(5, 3)$ **32.** $P(10, 2)$

33. $P(8, 1)$ **34.** $P(6, 6)$

35. $P(7, 3)$ **36.** $P(12, 3)$

37. $P(25, 2)$ **38.** $P(20, 1)$

39. $P(10, 4)$ **40.** $P(34, 2)$

41. Standing in Line How many ways can four people stand in a line?

42. Books on a Shelf How many arrangements are there of six different books on a shelf?

43. Giving a Speech In how many arrangements can 3 students from a class of 15 each give a speech?

44. Introductions How many ways could five basketball players be introduced at a game?

45. Traveling Salesperson (Refer to the discussion after Example 4.) A salesperson must travel to three of seven cities. Direct travel is possible between every pair of cities. How many arrangements are there in which the salesperson could visit these three cities? Assume that traveling a route in reverse order constitutes a different arrangement.

46. Traveling Salesperson A salesperson must start at city A, stop in each city once, and then return to city A. Use the figure to find the route with the least mileage.

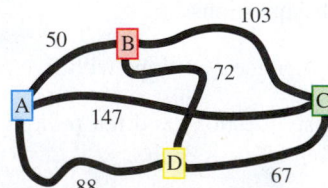

47. Phone Numbers How many seven-digit phone numbers are there if the first three numbers must be 387, 388, or 389?

48. Keys How many distinguishable ways can four keys be put on a key ring?

49. **Sitting at a Round Table** How many ways can seven people sit at a round table? (For a way to be different, at least one person must be sitting next to someone different.)

50. **Batting Orders** A softball team has 10 players. How many batting orders are possible?

51. **Baseball Positions** In how many ways can nine players be assigned to the nine positions on a baseball team, assuming that any player can play any position?

52. **Musical Chairs Seating** In a game of musical chairs, seven children will sit in six chairs arranged in a circle. One child will be left out. How many (different) ways can the children sit in the chairs? (For a way to be different, at least one child must be sitting next to someone different.)

53. **Birthdays** In how many ways can five people have different birthdays?

54. **Course Schedule** A scheduling committee has one room in which to offer five mathematics courses. In how many ways can the committee arrange the five courses over the day?

55. **Telephone Numbers** How many 10-digit telephone numbers are there if the first digit and the fourth digit cannot be a 0 or a 1?

56. **Car Designs** There are 10 basic colors available for a new car, along with 5 basic styles of trim. In how many ways can a person pick the color and trim?

Combinations

Exercises 57–66: Evaluate the expression.

57. $C(3, 1)$

58. $C(4, 3)$

59. $C(6, 3)$

60. $C(7, 5)$

61. $C(5, 0)$

62. $C(10, 2)$

63. $\begin{pmatrix} 8 \\ 2 \end{pmatrix}$

64. $\begin{pmatrix} 9 \\ 4 \end{pmatrix}$

65. $\begin{pmatrix} 20 \\ 18 \end{pmatrix}$

66. $\begin{pmatrix} 100 \\ 2 \end{pmatrix}$

67. **Lottery** To win the jackpot in a lottery, one must select five different numbers from 1 to 39. How many ways are there to play this game?

68. **Selecting a Committee** How many ways can a committee of five be selected from eight people?

69. **Selecting a Coed Team** How many teams of four people can be selected from five women and three men if a team must have two people of each sex on it?

70. **Essay Questions** On a test with six essay questions, students are asked to answer four questions. How many ways can the essay questions be selected?

71. **Test Questions** A test consists of two parts. In the first part a student must choose three of five essay questions, and in the second part a student must choose four of five essay questions. How many ways can the essay questions be selected?

72. **Cards** How many ways are there to draw a 5-card hand from a 52-card deck?

73. **Selecting Marbles** How many ways are there to draw 3 red marbles and 2 blue marbles from a jar that contains 10 red marbles and 12 blue marbles?

74. **Book Arrangements** A professor has three copies of an algebra book and four copies of a calculus text. How many distinguishable ways can the books be placed on a shelf?

75. **Peach Samples** How many samples of 3 peaches can be drawn from a crate of 24 peaches? (Assume that the peaches are distinguishable.)

76. **Flower Samples** A bouquet of flowers contains three red roses, four yellow roses, and five white roses. In how many ways can a person choose one flower of each type? (Assume that the flowers are distinguishable.)

77. **Permutations** Show that $P(n, n - 1) = P(n, n)$. Give an example that supports your result.

78. **Combinations** Show that $\begin{pmatrix} n \\ r \end{pmatrix} = \begin{pmatrix} n \\ n - r \end{pmatrix}$. Give an example that supports your result.

Writing about Mathematics

79. Explain the difference between a permutation and a combination. Give examples.

80. Explain what counting is, as presented in this section.

11.4 The Binomial Theorem

- Derive the binomial theorem
- Use the binomial theorem
- Apply Pascal's triangle

Introduction

In this section we discuss how to expand expressions in the form $(a + b)^n$, where n is a natural number. Some examples include the following.

$$(a + b)^2, (a + b)^5, (2x + 1)^3, \text{ and } (x - y)^4$$

These expressions occur in statistics, finite mathematics, computer science, and calculus.

Derivation of the Binomial Theorem

Combinations play a central role in the development of the binomial theorem. The binomial theorem can be used to expand expressions of the form $(a + b)^n$. Before stating the binomial theorem, we begin by counting the number of strings of a given length that can be formed with only the variables a and b.

EXAMPLE 1 Calculating distinguishable strings

Count the number of distinguishable strings that can be formed with the given number of a's and b's. List these strings.

(a) Two a's, one b **(b)** Two a's, three b's   **(c)** Four a's, no b's

SOLUTION

(a) Using **two a's** and **one b**, we can form strings of length three. Once the b has been positioned, the string is determined. For example, if the b is placed in the middle position,

$$\boxed{}\;\boxed{b}\;\boxed{},$$

then the string must be ***aba***. From a set of three slots, we choose one slot in which to place the b. This is computed by $C(3, \mathbf{1}) = 3$. The strings are *aab*, *aba*, and *baa*.

(b) With **two a's** and **three b's** we can form strings of length five. Once the locations of the three b's have been selected, the string is determined. For instance, if the b's are placed in the first, third, and fifth positions,

$$\boxed{b}\;\boxed{}\;\boxed{b}\;\boxed{}\;\boxed{b}$$

then the string becomes *babab*. From a set of five slots, we select three in which to place the b's. This is computed by $C(5, \mathbf{3}) = 10$. The 10 strings are

bbbaa, *bbaba*, *bbaab*, *babba*, *babab*, *baabb*, *abbba*, *abbab*, *ababb*, and *aabbb*.

(c) There is only one string of length four that contains **no b's**. This is *aaaa* and is computed by

$$C(4, \mathbf{0}) = \frac{4!}{(4 - 0)!0!} = \frac{24}{24(1)} = 1.$$

Now Try Exercise 9, 11, and 13

Next we expand $(a + b)^n$ for a few values of n, without simplifying.

$$(a + b)^1 = a + b$$

$$(a + b)^2 = (a + b)(a + b)$$

$$= aa + \mathbf{ab} + \mathbf{ba} + bb$$

$$(a + b)^3 = (a + b)(a + b)^2$$

$$= (a + b)(aa + ab + ba + bb)$$

$$= aaa + \mathbf{aab} + \mathbf{aba} + \mathbf{abb} + \mathbf{baa} + \mathbf{bab} + \mathbf{bba} + bbb$$

Notice that $(a + b)^1$ is the sum of all possible strings of length one that can be formed using a and b. The only possibilities are a and b. The expression $(a + b)^2$ is the sum of all possible strings of length two using a and b. The strings are aa, ab, ba, and bb. Similarly, $(a + b)^3$ is the sum of all possible strings of length three using a and b. This pattern continues for higher powers of $(a + b)$.

Strings with equal numbers of a's and equal numbers of b's can be combined into one term. For example, in $(a + b)^2$ the terms ab and ba can be combined as $2ab$. Notice that there are $C(2, 1) = 2$ distinguishable strings of length two containing one b. Similarly, in the expansion of $(a + b)^3$, the terms containing one a and two b's can be combined as

$$abb + bab + bba = 3ab^2.$$

There are $C(3, 2) = 3$ strings of length three that contain two b's.

We can use these concepts to expand $(a + b)^4$. The expression $(a + b)^4$ consists of the sum of all strings of length four using only the letters a and b: $C(4, 0) = 1$ string containing no b's, $C(4, 1) = 4$ strings containing one b, and so on, up to $C(4, 4) = 1$ string containing four b's. Thus

$$(a + b)^4 = \binom{4}{0}a^4b^0 + \binom{4}{1}a^3b^1 + \binom{4}{2}a^2b^2 + \binom{4}{3}a^1b^3 + \binom{4}{4}a^0b^4$$

$$= a^4 + 4a^3b + 6a^2b^2 + 4ab^3 + b^4.$$

These results are summarized by the **binomial theorem**.

BINOMIAL THEOREM

For any positive integer n and numbers a and b,

$$(a + b)^n = \binom{n}{0}a^n + \binom{n}{1}a^{n-1}b^1 + \cdots + \binom{n}{n-1}a^1b^{n-1} + \binom{n}{n}b^n.$$

We can use the binomial theorem to expand $(a + b)^n$. For example,

$$(a + b)^3 = \binom{3}{0}a^3 + \binom{3}{1}a^2b^1 + \binom{3}{2}a^1b^2 + \binom{3}{3}b^3$$

$$= 1a^3 + 3a^2b + 3ab^2 + 1b^3$$

$$= a^3 + 3a^2b + 3ab^2 + b^3.$$

Since $\binom{n}{r} = C(n, r)$, we can use the combination formula $C(n, r) = \frac{n!}{(n - r)!\, r!}$ to evaluate the binomial coefficients.

EXAMPLE 2 Applying the binomial theorem

Use the binomial theorem to expand the expression $(2a + 1)^5$.

SOLUTION Using the binomial theorem, we arrive at the following result.

$$(2a + 1)^5 = \binom{5}{0}(2a)^5 + \binom{5}{1}(2a)^4 1^1 + \binom{5}{2}(2a)^3 1^2 + \binom{5}{3}(2a)^2 1^3 + \binom{5}{4}(2a)^1 1^4 + \binom{5}{5}1^5$$

$$= \frac{5!}{5!0!}(32a^5) + \frac{5!}{4!1!}(16a^4) + \frac{5!}{3!2!}(8a^3) + \frac{5!}{2!3!}(4a^2) + \frac{5!}{1!4!}(2a) + \frac{5!}{0!5!}$$

$$= 32a^5 + 80a^4 + 80a^3 + 40a^2 + 10a + 1$$

Now Try Exercise 19

Pascal's Triangle

Expanding $(a + b)^n$ for increasing values of n gives the following results.

Pascal's Triangle

```
          1
        1   1
      1   2   1
    1   3   3   1
  1   4   6   4   1
1   5   10   10   5   1
```

Figure 11.19

$$(a + b)^0 = \qquad\qquad 1$$
$$(a + b)^1 = \qquad\qquad 1a + 1b$$
$$(a + b)^2 = \qquad\qquad 1a^2 + 2ab + 1b^2$$
$$(a + b)^3 = \qquad\qquad 1a^3 + 3a^2b + 3ab^2 + 1b^3$$
$$(a + b)^4 = \qquad\qquad 1a^4 + 4a^3b + 6a^2b^2 + 4ab^3 + 1b^4$$
$$(a + b)^5 = \qquad 1a^5 + 5a^4b + 10a^3b^2 + 10a^2b^3 + 5ab^4 + 1b^5$$

| Exponent on a decreases by 1 each term. | | Exponent on b increases by 1 each term. |

Notice that $(a + b)^1$ has two terms starting with a and ending with b, $(a + b)^2$ has three terms starting with a^2 and ending with b^2, and in general $(a + b)^n$ has $n + 1$ terms starting with a^n and ending with b^n. The exponent on a decreases by 1 each successive term, and the exponent on b increases by 1 each successive term.

The triangle formed by the highlighted numbers is called **Pascal's triangle**. It can be used to efficiently compute the binomial coefficients, $C(n, r)$. The triangle consists of 1's along the sides. Each element inside the triangle is the sum of the two numbers above it. Pascal's triangle is usually written without variables, as in Figure 11.19. It can be extended to include as many rows as needed.

We can use this triangle to expand powers of binomials in the form $(a + b)^n$, where n is a natural number. For example, the expression $(m + n)^4$ consists of five terms written as follows.

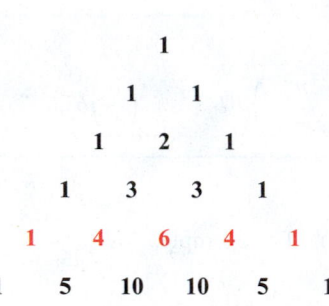

$$(m + n)^4 = \underline{}m^4 + \underline{}m^3n^1 + \underline{}m^2n^2 + \underline{}m^1n^3 + \underline{}n^4$$

Because there are five terms, the coefficients can be found in the fifth row of Pascal's triangle, which from the figure in the margin is

$$1 \quad 4 \quad 6 \quad 4 \quad 1.$$

Thus

$$(m + n)^4 = \mathbf{1}\,m^4 + \mathbf{4}\,m^3n^1 + \mathbf{6}\,m^2n^2 + \mathbf{4}\,m^1n^3 + \mathbf{1}\,n^4$$
$$= m^4 + 4m^3n + 6m^2n^2 + 4mn^3 + n^4.$$

EXAMPLE 3 **Expanding expressions with Pascal's triangle**

Expand each of the following.
(a) $(2x + 1)^5$ **(b)** $(3x - y)^3$

SOLUTION
(a) To expand $(2x + 1)^5$, let $a = 2x$ and $b = 1$ in the binomial theorem. We can use the sixth row of Pascal's triangle to obtain the coefficients **1, 5, 10, 10, 5,** and **1.** Compare this solution with the solution for Example 2.

$$(2x + 1)^5 = \mathbf{1}(2x)^5 + \mathbf{5}(2x)^4(1)^1 + \mathbf{10}(2x)^3(1)^2 + \mathbf{10}(2x)^2(1)^3 + \mathbf{5}(2x)^1(1)^4 + \mathbf{1}(1)^5$$
$$= 32x^5 + 80x^4 + 80x^3 + 40x^2 + 10x + 1$$

(b) Let $a = 3x$ and $b = -y$ in the binomial theorem. Use the coefficients **1, 3, 3,** and **1** from the fourth row of Pascal's triangle.

$$(3x - y)^3 = \mathbf{1}(3x)^3 + \mathbf{3}(3x)^2(-y)^1 + \mathbf{3}(3x)^1(-y)^2 + \mathbf{1}(-y)^3$$
$$= 27x^3 - 27x^2y + 9xy^2 - y^3$$

Now Try Exercise 33 and 35

Finding the *k*th Term The binomial theorem gives *all* of the terms of $(a + b)^n$. However, we can find any individual term by noting that the $(r + 1)$st term in the binomial expansion for $(a + b)^n$ is given by the formula $\binom{n}{r}a^{n-r}b^r$ for $0 \le r \le n$. The next example shows how to use this formula to find the $(r + 1)$st term of $(a + b)^n$.

EXAMPLE 4 Finding the *k*th term in a binomial expansion

Find the third term of $(x - y)^5$.

SOLUTION In this example the $(r + 1)$st term is the *third* term in the expansion of $(x - y)^5$. That is, $r + 1 = 3$, or $r = 2$. Also, the exponent in the expression is $n = 5$. To get this binomial into the form $(a + b)^n$, we note that the first term in the binomial is $a = x$ and that the second term in the binomial is $b = -y$. Substituting the values for r, n, a, and b in the formula for the $(r + 1)$st term yields

$$\binom{5}{2}(x)^{5-2}(-y)^2 = 10x^3y^2.$$

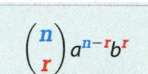
$\binom{n}{r}a^{n-r}b^r$

The third term in the binomial expansion of $(x - y)^5$ is $10x^3y^2$.

Now Try Exercise 45

11.4 Putting It All Together

The following table summarizes topics related to the binomial theorem.

NOTATION	EXPLANATION	EXAMPLES
Binomial coefficient	$\binom{n}{r} = C(n, r) = \dfrac{n!}{(n - r)!\, r!}$	$\binom{5}{3} = \dfrac{5!}{(5 - 3)!\, 3!} = \dfrac{120}{2 \cdot 6} = 10$ $\binom{4}{0} = \dfrac{4!}{(4 - 0)!\, 0!} = \dfrac{24}{24 \cdot 1} = 1$
Binomial theorem	$(a + b)^n =$ $\binom{n}{0}a^n + \binom{n}{1}a^{n-1}b + \cdots + \binom{n}{n}b^n$ for any positive integer n and real numbers a and b.	$(a + b)^3 = \binom{3}{0}a^3 + \binom{3}{1}a^2b + \binom{3}{2}ab^2 + \binom{3}{3}b^3$ $= a^3 + 3a^2b + 3ab^2 + b^3$ The binomial coefficients can also be found using the fourth row of Pascal's triangle, which is shown below.
Pascal's triangle	A triangle of numbers that can be used to find the binomial coefficients needed to expand an expression of the form $(a + b)^n$. To expand $(a + b)^n$, use row $n + 1$ of Pascal's triangle.	$\begin{array}{ccccccccccc} & & & & & 1 & & & & & \\ & & & & 1 & & 1 & & & & \\ & & & 1 & & 2 & & 1 & & & \\ & & \mathbf{1} & & \mathbf{3} & & \mathbf{3} & & \mathbf{1} & & \\ & 1 & & 4 & & 6 & & 4 & & 1 & \\ 1 & & 5 & & 10 & & 10 & & 5 & & 1 \end{array}$
Finding the $(r + 1)$st term of $(a + b)^n$	The $(r + 1)$st term of $(a + b)^n$ is given by $\binom{n}{r}a^{n-r}b^r$ for $0 \le r \le n$.	To find the fifth term of $(x + y)^6$, let $r + 1 = 5$, or $r = 4$, and $n = 6$. $\binom{n}{r}a^{n-r}b^r = \binom{6}{4}x^{6-4}y^4$ $= 15x^2y^4$

11.4 Exercises

Binomial Coefficients

Exercises 1–8: Evaluate the expression.

1. $\binom{5}{4}$

2. $\binom{6}{2}$

3. $\binom{4}{0}$

4. $\binom{4}{2}$

5. $\binom{6}{5}$

6. $\binom{6}{3}$

7. $\binom{3}{3}$

8. $\binom{5}{2}$

Binomial Theorem

Exercises 9–16: (Refer to Example 1.) Calculate the number of distinguishable strings that can be formed with the given number of a's and b's.

9. Three a's, two b's

10. Five a's, three b's

11. Four a's, four b's

12. One a, five b's

13. Five a's, no b's

14. No a's, three b's

15. Four a's, one b

16. Four a's, two b's

Exercises 17–30: Use the binomial theorem to expand each expression.

17. $(x + y)^2$

18. $(x + y)^4$

19. $(m + 2)^3$

20. $(m + 2n)^5$

21. $(2x - 3)^3$

22. $(x + y^2)^3$

23. $(p - q)^6$

24. $(p^2 - 3)^4$

25. $(2m + 3n)^3$

26. $(3a - 2b)^5$

27. $(1 - x^2)^4$

28. $(2 + 3x^2)^3$

29. $(2p^3 - 3)^3$

30. $(2r + 3t)^4$

Pascal's Triangle

Exercises 31–44: Use Pascal's triangle to help expand the expression.

31. $(x + y)^2$

32. $(m + n)^3$

33. $(3x + 1)^4$

34. $(2x - 1)^4$

35. $(2 - x)^5$

36. $(2a + 3b)^3$

37. $(x^2 + 2)^4$

38. $(5 - x^2)^3$

39. $(4x - 3y)^4$

40. $(3 - 2x)^5$

41. $(m + n)^6$

42. $(2m - n)^4$

43. $(2x^3 - y^2)^3$

44. $(3x^2 + y^3)^4$

Finding the kth Term

Exercises 45–52: Find the specified term.

45. The fourth term of $(a + b)^9$

46. The second term of $(m - n)^9$

47. The fifth term of $(x + y)^8$

48. The third term of $(a + b)^7$

49. The fourth term of $(2x + y)^5$

50. The eighth term of $(2a - b)^9$

51. The sixth term of $(3x - 2y)^6$

52. The seventh term of $(2a + b)^9$

Writing about Mathematics

53. Explain how to find the numbers in Pascal's triangle.

54. Compare the expansion of $(a + b)^n$ with the expansion of $(a - b)^n$. Give an example.

CHECKING BASIC CONCEPTS FOR SECTIONS 11.3 AND 11.4

1. Count the ways to answer a quiz that consists of eight true-false questions.

2. Count the number of 5-card poker hands that can be dealt using a standard deck of 52 cards.

3. How many distinct license plates could be made using a letter followed by five digits or letters?

4. Expand each expression.
   (a) $(2x + 1)^4$
   (b) $(4 - 3x)^3$

11.5 Mathematical Induction

- Learn basic concepts about mathematical induction
- Use mathematical induction to prove statements
- Apply the generalized principle of mathematical induction

Introduction

The brilliant mathematician Carl Friedrich Gauss (1777–1855) proved the fundamental theorem of algebra at age 20. When he was a young child, he amazed his teacher by showing that

$$1 + 2 + 3 + 4 + \cdots + 100 = \frac{100(101)}{2}.$$

With *mathematical induction* we will be able to show, more generally, that

$$1 + 2 + 3 + 4 + \cdots + n = \frac{n(n + 1)}{2}.$$

Mathematical induction is a powerful method of proof. It is used not only in mathematics; it is also used in computer science to prove that programs and basic concepts are correct.

Mathematical Induction

Many results in mathematics are claimed true for every positive integer. Any of these results could be checked for $n = 1$, $n = 2$, $n = 3$, and so on, but since the set of positive integers is infinite, it would be impossible to check every number. For example, let S_n represent the statement that the sum of the first n positive integers is $\frac{n(n + 1)}{2}$; that is,

$$S_n: 1 + 2 + 3 + \cdots + n = \frac{n(n + 1)}{2}.$$

The truth of this statement can be checked quickly for the first few values of n.

If $n = \mathbf{1}$, S_1 is $1 = \frac{1(1 + 1)}{2}$, a true statement, since $1 = 1$.

If $n = \mathbf{2}$, S_2 is $1 + 2 = \frac{2(2 + 1)}{2}$, a true statement, since $3 = 3$.

If $n = \mathbf{3}$, S_3 is $1 + 2 + 3 = \frac{3(3 + 1)}{2}$, a true statement, since $6 = 6$.

If $n = \mathbf{4}$, S_4 is $1 + 2 + 3 + 4 = \frac{4(4 + 1)}{2}$, a true statement, since $10 = 10$.

Since the statement is true for $n = 1, 2, 3, 4$, can we conclude that the statement is true for all positive integers? The answer is *no*. To prove that such a statement is true for every positive integer, we use the following principle.

> **PRINCIPLE OF MATHEMATICAL INDUCTION**
>
> Let S_n be a statement concerning the positive integer n. Suppose that
>
> 1. S_1 is true;
> 2. for any positive integer k, if S_k is true, then S_{k+1} is also true.
>
> Then S_n is true for every positive integer n.

A proof by mathematical induction can be explained as follows. By assumption (1), the statement is true when $n = 1$. By assumption (2), the fact that the statement is true for $n = 1$ implies that it is true for $n = 1 + 1 = 2$. Using (2) again, the statement is thus true for $2 + 1 = 3$, for $3 + 1 = 4$, for $4 + 1 = 5$, and so on. Continuing in this way shows that the statement must be true for *every* positive integer.

The situation is similar to that of an infinite number of dominoes lined up. If the first domino is pushed over, it pushes the next, which pushes the next, and so on, indefinitely.

Another example of the principle of mathematical induction might be an infinite ladder. Suppose the rungs are spaced so that, whenever you are on a rung, you know you can move to the next rung. Then *if* you can get to the first rung, you can go as high up the ladder as you wish.

Two separate steps are required for a proof by mathematical induction.

PROOF BY MATHEMATICAL INDUCTION

STEP 1: Prove that the statement is true for $n = 1$.

STEP 2: Show that for any positive integer k, if S_k is true, then S_{k+1} is also true.

Proving Statements

Mathematical induction is used in the next example to prove the statement S_n discussed earlier.

EXAMPLE 1 **Proving an equality statement**

Let S_n represent the statement

$$1 + 2 + 3 + \cdots + n = \frac{n(n + 1)}{2}.$$

Prove that S_n is true for every positive integer n.

SOLUTION The proof by mathematical induction is as follows.

STEP 1: Show that the statement is true when $n = 1$. If $n = 1$, S_1 becomes

$$1 = \frac{1(1 + 1)}{2},$$

which is true.

STEP 2: Show that if S_k is true, then S_{k+1} is also true, where S_k is the statement

Statement S_k $\quad 1 + 2 + 3 + \cdots + k = \dfrac{k(k + 1)}{2}$

and S_{k+1} is the statement

Statement S_{k+1} $\quad 1 + 2 + 3 + \cdots + k + (k + 1) = \dfrac{(k + 1)[(k + 1) + 1]}{2}.$

Start with S_k and assume it is a true statement.

$$1 + 2 + 3 + \cdots + k = \frac{k(k + 1)}{2} \qquad S_k \text{ is true.}$$

Add $k + 1$ to each side of equation S_k and simplify to obtain S_{k+1}.

$$1 + 2 + 3 + \cdots + k + (k + 1) = \frac{k(k + 1)}{2} + (k + 1)$$

$$= (k + 1)\left(\frac{k}{2} + 1\right) \qquad \text{Factor out } k + 1.$$

Statement S_{k+1} is true.

$$= (k + 1)\left(\frac{k + 2}{2}\right)$$

$$= \frac{(k + 1)[(k + 1) + 1]}{2}$$

This final result is the statement S_{k+1}. Therefore, if S_k is true, then S_{k+1} is also true. The two steps required for a proof by mathematical induction have been completed, so the statement S_n is true for every positive integer n.

Now Try Exercise 1

EXAMPLE 2 **Proving an equality statement**

Let S_n represent the statement

$$2^1 + 2^2 + 2^3 + 2^4 + \cdots + 2^n = 2^{n+1} - 2.$$

Prove that S_n is true for every positive integer n.

SOLUTION

STEP 1: Show that the statement S_1 is true, where S_1 is

$$2^1 = 2^{1+1} - 2.$$

Since $2 = 4 - 2$, S_1 is a true statement.

STEP 2: Show that if S_k is true, then S_{k+1} is also true, where S_k is

Statement S_k $2^1 + 2^2 + 2^3 + \cdots + 2^k = 2^{k+1} - 2$

and S_{k+1} is

Statement S_{k+1} $2^1 + 2^2 + 2^3 + \cdots + 2^k + 2^{k+1} = 2^{(k+1)+1} - 2.$

Start with the given equation S_k and add 2^{k+1} to each side of this equation. Then algebraically change the right side to look like the right side of S_{k+1}.

$$2^1 + 2^2 + 2^3 + \cdots + 2^k + 2^{k+1} = 2^{k+1} - 2 + 2^{k+1}$$

Statement S_{k+1} is true.

$$= 2 \cdot 2^{k+1} - 2$$
$$= 2^{(k+1)+1} - 2$$

The final result is the statement S_{k+1}. Therefore, if S_k is true, then S_{k+1} is also true. The two steps required for a proof by mathematical induction have been completed, so the statement S_n is true for every positive integer n.

Now Try Exercise 5

EXAMPLE 3 **Proving an inequality statement**

Prove that if x satisfies $0 < x < 1$, then for every positive integer n,

$$0 < x^n < 1.$$

SOLUTION

STEP 1: Here S_1 is the statement

$$\text{if } 0 < x < 1, \text{ then } 0 < x^1 < 1,$$

which is true.

STEP 2: S_k is the statement

$$\text{if } 0 < x < 1, \text{ then } 0 < x^k < 1. \quad \text{Statement } S_k$$

To show that S_k implies that S_{k+1} is true, multiply all three parts of $0 < x^k < 1$ by x to get

$$x \cdot 0 < x \cdot x^k < x \cdot 1.$$

(Here the fact that $0 < x$ is used. Why?) Simplify to obtain

$$0 < x^{k+1} < x.$$

Because $x < 1$,

$$0 < x^{k+1} < 1,$$

Statement S_{k+1} is true.

which implies that S_{k+1} is true. Therefore, if S_k is true, then S_{k+1} is true. Since both steps for a proof by mathematical induction have been completed, the given statement is true for every positive integer n.

Now Try Exercise 23

Generalized Principle of Mathematical Induction

Some statements S_n are not true for the first few values of n, but are true for all values of n that are greater than or equal to some fixed integer j. The following slightly generalized form of the principle of mathematical induction addresses these cases.

> **GENERALIZED PRINCIPLE OF MATHEMATICAL INDUCTION**
>
> Let S_n be a statement concerning the positive integer n. Let j be a fixed positive integer. Suppose that
>
> **1.** S_j is true;
> **2.** for any positive integer k, $k \geq j$, S_k implies S_{k+1}.
>
> Then S_n is true for all positive integers n, where $n \geq j$.

EXAMPLE 4 **Using the generalized principle**

Let S_n represent the statement $2^n > 2n + 1$. Show that S_n is true for all values of n such that $n \geq 3$.

SOLUTION (Check that S_n is false for $n = 1$ and $n = 2$.)

STEP 1: Show that S_n is true for $n = 3$. If $n = 3$, then S_3 is

$$2^3 > 2 \cdot 3 + 1, \quad \text{or}$$

$$8 > 7.$$

Thus S_3 is true.

STEP 2: Now show that S_k implies S_{k+1} for $k \geq 3$, where

$$S_k \text{ is } 2^k > 2k + 1 \qquad \text{and} \qquad \text{Assume } S_k \text{ is true.}$$

$$S_{k+1} \text{ is } 2^{k+1} > 2(k + 1) + 1. \qquad \text{Show } S_{k+1} \text{ is true.}$$

Multiply each side of $2^k > 2k + 1$ by 2, obtaining

$$2 \cdot 2^k > 2(2k + 1), \quad \text{or}$$

$$2^{k+1} > 4k + 2.$$

Rewrite $4k + 2$ as $2(k + 1) + 2k$, to get

$$2^{k+1} > 2(k + 1) + 2k.$$

Since k is a positive integer greater than or equal to 3,

$$2k > 1.$$

It follows that

$$2^{k+1} > 2(k + 1) + 2k > 2(k + 1) + 1, \quad \text{or}$$

Statement S_{k+1} is true.

$$2^{k+1} > 2(k + 1) + 1,$$

as required. Thus S_k implies S_{k+1}, and this, together with the fact that S_3 is true, shows that S_n is true for every positive integer n greater than or equal to 3.

Now Try Exercise 27

11.5 Putting It All Together

Some important concepts about mathematical induction are summarized in the following table.

CONCEPT	EXPLANATION
Principle of mathematical induction	Let S_n be a statement concerning the positive integer n. Suppose that 1. S_1 is true; 2. for any positive integer k, if S_k is true, then S_{k+1} is also true. Then S_n is true for every positive integer n.
Proof by mathematical induction	**STEP 1:** Prove that the statement is true for $n = 1$. **STEP 2:** Show that for any positive integer k, if S_k is true, then S_{k+1} is also true.
Generalized principle of mathematical induction	Let S_n be a statement concerning the positive integer n. Let j be a fixed positive integer. Suppose that 1. S_j is true; 2. for any positive integer k, $k \geq j$, S_k implies S_{k+1}. Then S_n is true for all positive integers n, where $n \geq j$.

11.5 Exercises

Mathematical Induction

Exercises 1–14: Use mathematical induction to prove the statement. Assume that n is a positive integer.

1. $3 + 6 + 9 + \cdots + 3n = \dfrac{3n(n + 1)}{2}$

2. $1 + 3 + 5 + \cdots + (2n - 1) = n^2$

3. $5 + 10 + 15 + \cdots + 5n = \dfrac{5n(n + 1)}{2}$

4. $4 + 7 + 10 + \cdots + (3n + 1) = \dfrac{n(3n + 5)}{2}$

5. $3 + 3^2 + 3^3 + \cdots + 3^n = \dfrac{3(3^n - 1)}{2}$

6. $1^2 + 2^2 + 3^2 + \cdots + n^2 = \dfrac{n(n + 1)(2n + 1)}{6}$

7. $1^3 + 2^3 + 3^3 + \cdots + n^3 = \dfrac{n^2(n + 1)^2}{4}$

8. $5 \cdot 6 + 5 \cdot 6^2 + 5 \cdot 6^3 + \cdots + 5 \cdot 6^n = 6(6^n - 1)$

9. $\dfrac{1}{1 \cdot 2} + \dfrac{1}{2 \cdot 3} + \dfrac{1}{3 \cdot 4} + \cdots + \dfrac{1}{n(n + 1)} = \dfrac{n}{n + 1}$

10. $7 \cdot 8 + 7 \cdot 8^2 + 7 \cdot 8^3 + \cdots + 7 \cdot 8^n = 8(8^n - 1)$

11. $\dfrac{4}{5} + \dfrac{4}{5^2} + \dfrac{4}{5^3} + \cdots + \dfrac{4}{5^n} = 1 - \dfrac{1}{5^n}$

12. $\dfrac{1}{2} + \dfrac{1}{2^2} + \dfrac{1}{2^3} + \cdots + \dfrac{1}{2^n} = 1 - \dfrac{1}{2^n}$

13. $\dfrac{1}{1 \cdot 4} + \dfrac{1}{4 \cdot 7} + \cdots + \dfrac{1}{(3n - 2)(3n + 1)} = \dfrac{n}{3n + 1}$

14. $x^{2n} + x^{2n-1}y + \cdots + xy^{2n-1} + y^{2n} = \dfrac{x^{2n+1} - y^{2n+1}}{x - y}$

Exercises 15–18: Find all positive integers n for which the given statement is not true.

15. $3^n > 6n$

16. $3^n > 2n + 1$

17. $2^n > n^2$

18. $n! > 2n$

Exercises 19–28: Prove the statement by mathematical induction.

19. $(a^m)^n = a^{mn}$ (Assume a and m are constants.)

20. $(ab)^n = a^n b^n$ (Assume a and b are constants.)

21. $2^n > 2n$ if $n \geq 3$

22. $3^n > 2n + 1$, if $n \geq 2$

23. If $a > 1$, then $a^n > 1$.

24. If $a > 1$, then $a^n > a^{n-1}$.

25. If $0 < a < 1$, then $a^n < a^{n-1}$.

26. $2^n > n^2$ for $n > 4$

27. If $n \geq 4$, then $n! > 2^n$, where

$$n! = n(n-1)(n-2) \cdots (3)(2)(1).$$

28. $4^n > n^4$ for $n \geq 5$

Applications

29. Number of Handshakes Suppose that each of the n ($n \geq 2$) people in a room shakes hands with everyone else, but not with himself. Show that the number of handshakes is $\frac{n^2 - n}{2}$.

30. Sides of a Polygon The series of sketches starts with an equilateral triangle having sides of length 1. In the following steps, equilateral triangles are constructed on each side of the preceding figure. The length of the sides of each new triangle is $\frac{1}{3}$ the length of the sides of the preceding triangles. Develop a formula for the number of sides of the nth figure. Use mathematical induction to prove your answer.

31. Perimeter Find the perimeter of the nth figure in Exercise 30.

32. Area Show that the area of the nth figure in Exercise 30 is given by

$$\sqrt{3}\left[\frac{2}{5} - \frac{3}{20}\left(\frac{4}{9}\right)^{n-1}\right].$$

33. Tower of Hanoi A pile of n rings, each ring smaller than the one below it, is on a peg. Two other pegs are attached to the same board as this peg. In a game called the *Tower of Hanoi* puzzle, all the rings must be moved to a different peg, with only one ring moved at a time and with no ring ever placed on top of a smaller ring. Find the least number of moves required. Prove your result with mathematical induction.

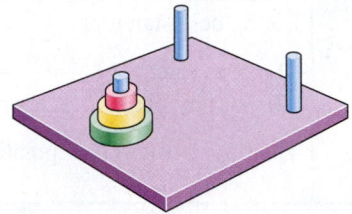

Writing about Mathematics

34. Explain the principle of mathematical induction.

35. Explain how the generalized principle of mathematical induction differs from the principle of mathematical induction.

36. When using mathematical induction, why is it important to prove that the statement holds for $n = 1$?

11.6 Probability

- Learn the basic concepts about probability
- Calculate the probability of compound events
- Calculate the probability of independent and dependent events

Introduction

Questions of chance have no doubt engaged the minds of people since antiquity. However, the mathematical treatment of probability did not begin until the 15th century. The birth of probability theory as a mathematical discipline occurred in the 17th century with the work of Blaise Pascal and Pierre Fermat. Today probability pervades society. It is used not only to determine outcomes in gambling, but also to predict weather, genetic outcomes, and the risk involved with various types of substances and behaviors.

Probability provides us with a measure of the likelihood that an event will occur. Risk is the chance, or probability, that a harmful event will occur. The following activities increase the annual risk of death by one chance in a million: flying 1000 miles in a jet, traveling 300 miles in a car, riding 10 miles on a bicycle, smoking 1.4 cigarettes, living 2 days in New York City, having one chest X-ray, or living 2 months with a cigarette smoker. Knowledge about probability allows individuals to make informed decisions about their lives. (**Sources:** NCTM, *Historical Topics for the Mathematics Classroom, Thirty-first Yearbook;* J. Rodricks, *Calculated Risk.*)

Definition of Probability

In the study of probability, experiments often are performed. The following explains terminology about experiments and gives an example.

Experiment Terminology	Example: Rolling a Die
Outcome: A result from an experiment	*Outcome*: The number showing on the die
Sample Space: The set of all possible outcomes S	*Sample Space*: $S = \{1, 2, 3, 4, 5, 6\}$
	Two outcomes in E Six outcomes in S
Event: Any subset E of a sample space	*Event*: $E = \{1, 6\}$ contains the outcomes of either 1 or 6 showing on the die

If $n(E)$ and $n(S)$ denote the number of outcomes in E and S, then $n(E) = 2$ and $n(S) = 6$ in the above experiment of rolling a die. The probability of rolling a 1 or a 6 is given by $P(E) = \frac{2}{6}$. That is, the likelihood of event E occurring is **two** chances in **six**. These concepts are summarized in the following box.

> **PROBABILITY OF AN EVENT**
>
> If the outcomes of a finite sample space S are equally likely and if E is an event in S, then the **probability of E** is given by
> $$P(E) = \frac{n(E)}{n(S)},$$
> where $n(E)$ and $n(S)$ represent the number of outcomes in E and S, respectively.

Since $n(E) \leq n(S)$, the probability of an event E satisfies $0 \leq P(E) \leq 1$. If $P(E) = 1$, then event E is *certain* to occur. If $P(E) = 0$, then event E is *impossible*.

EXAMPLE 1   Drawing a card

One card is drawn at random from a standard deck of 52 cards. Find the probability that the card is an ace.

SOLUTION The sample space S consists of 52 outcomes that correspond to drawing any one of 52 cards. Each outcome is equally likely. Let E represent the event of drawing

an ace. There are four aces in the deck, so event E contains four outcomes. Therefore $n(S) = 52$ and $n(E) = 4$. The probability of drawing an ace is given by

$$P(E) = \frac{n(E)}{n(S)} = \frac{4}{52} = \frac{1}{13}.$$

Now Try Exercise 15

EXAMPLE 2   Estimating probability of organ transplants

Waiting List

Heart	3186
Kidney	92,346
Liver	16,082
Lung	1645

Table 11.7
Source: UNOS.

In 2012, there were 114,448 patients waiting for an organ transplant. Table 11.7 lists the numbers of patients waiting for the *most common types* of transplants. None of these people need two or more transplants. Approximate the probability that a transplant patient chosen at random will need

(a) a kidney or a heart,   **(b)** neither a kidney nor a heart.

SOLUTION

(a) Let each patient represent an outcome in a sample space S. The event E of a transplant patient needing either a kidney or a heart contains $92{,}346 + 3186 = 95{,}532$ outcomes. The desired probability is

$$P(E) = \frac{n(E)}{n(S)} = \frac{95{,}532}{114{,}448} \approx 0.83.$$

In 2012, about 83% of transplant patients needed either a kidney or a heart.

(b) Let F be the event of a patient waiting for an organ other than a kidney or a heart. Then

$$n(F) = n(S) - n(E) = 114{,}448 - 95{,}532 = 18{,}916.$$

The probability of F is

$$P(F) = \frac{n(F)}{n(S)} = \frac{18{,}916}{114{,}448} \approx 0.17, \quad \text{or} \quad 17\%.$$

Now Try Exercise 43

Union and Intersection   The **union** of set A and set B is the set of elements that belong to either A or B, and is denoted $A \cup B$ and sometimes read "A or B". For example, if

$$A = \{1, 2, 3\} \text{ and } B = \{3, 4, 5\}, \text{ then}$$

$$A \cup B = \{1, 2, 3, 4, 5\}.$$

> Appears in both A and B, but is only listed once

The **intersection** of A and B is the set of elements that belong to *both A and B*, and is denoted $A \cap B$ and sometimes read "A and B." Using the sets A and B above,

$$A \cap B = \{3\}, \text{ as it is the only element that is in both } A \text{ and } B.$$

Notice that $P(E) + P(F) = 1$ in Example 2. The events E and F are **complements** because $E \cap F = \varnothing$ and $E \cup F = S$, where $\varnothing$ denotes the *empty set*. That is, a transplant patient is either waiting for a kidney or a heart (event E) or not waiting for a kidney or a heart (event F). The complement of E may be denoted by E'.

Venn Diagrams   Probability concepts can be illustrated using **Venn diagrams**. In Figure 11.20 the sample space S of an experiment is the union of the disjoint sets E and its complement E'. That is, $E \cup E' = S$ and $E \cap E' = \varnothing$.

If $P(E)$ is known, then $P(E')$ can be calculated as follows.

Complements E and E'

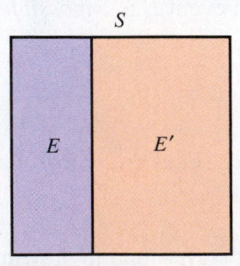

Figure 11.20 $E \cup E' = S$

$$P(E') = \frac{n(E')}{n(S)} = \frac{n(S) - n(E)}{n(S)} = 1 - \frac{n(E)}{n(S)} = 1 - P(E)$$

In Example 2(b), the probability of $F = E'$ could also have been calculated by using

$$P(F) = 1 - P(E) \approx 1 - 0.83 = 0.17.$$

> **PROBABILITY OF A COMPLEMENT**
>
> Let E be an event and E' be its complement. If the probability of E is $P(E)$, then the probability of its complement is given by
>
> $$P(E') = 1 - P(E).$$

EXAMPLE 3 **Finding probabilities of human eye color**

In 1865, Gregor Mendel performed important research in genetics. His work led to a better understanding of dominant and recessive genetic traits. According to Mendel's research, brown eye color B is dominant over blue eye color b. A person receives one gene (B or b) from each parent. If a person has the genotype BB, Bb, or bB, he or she will have brown eyes. The genotype bb will result in blue eyes. Table 11.8 shows how these two genes can be paired. (**Source:** H. Lancaster, *Quantitative Methods in Biology and Medical Sciences.*)
(a) Assuming that each genotype is equally likely, find the probability of blue eyes.
(b) What is the probability that a person has brown eyes?

Eye Color Outcomes

	B	b
B	BB	Bb
b	bB	bb

Table 11.8

SOLUTION
(a) The sample space S consists of **four** equally likely outcomes denoted BB, Bb, bB, and bb. The event E of blue eye color (bb) occurs **once**. Therefore

$$P(E) = \frac{n(E)}{n(S)} = \frac{1}{4} = 0.25.$$

(b) In Table 11.8 brown eyes are the complement of blue eyes. The probability of brown eyes is

$$1 - P(\text{blue eyes}) = P(\text{brown eyes})$$

$$P(E') = 1 - P(E) = 1 - 0.25 = 0.75.$$

This probability also could be computed as

$$P(E') = \frac{n(E')}{n(S)} = \frac{3}{4} = 0.75,$$

since there are three genotypes that result in brown eye color.

Now Try Exercise 47

Compound Events

Student Enrollment

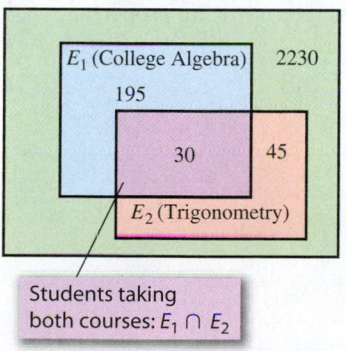

Figure 11.21

Frequently the probability of more than one event is needed. For example, suppose a college with a total of 2500 students has 225 students enrolled in college algebra, 75 in trigonometry, and 30 in both. Let E_1 denote the event that a student is enrolled in college algebra and E_2 the event that a student is enrolled in trigonometry. Then the Venn diagram in Figure 11.21 visually describes the situation.

In this Venn diagram, it is important that the **30** students taking both courses not be counted twice. Set E_1 has a total of $195 + 30 = 225$ students, and set E_2 contains $45 + 30 = 75$ students.

EXAMPLE 4 **Calculating the probability of a union**

In the preceding scenario, suppose a student is selected at random. What is the probability that this student is enrolled in college algebra, trigonometry, or both?

Student Enrollment

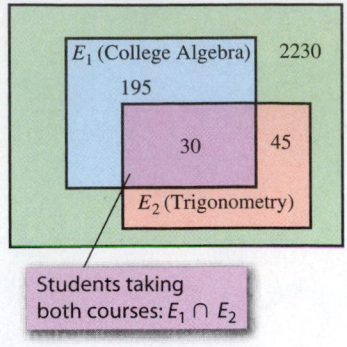

Students taking both courses: $E_1 \cap E_2$

Figure 11.21 (Repeated)

SOLUTION We would like to find the probability $P(E_1 \text{ or } E_2)$, which is usually denoted $P(E_1 \cup E_2)$. Since $n(E_1 \cup E_2) = 195 + 30 + 45 = 270$ and $n(S) = 2500$,

$$P(E_1 \cup E_2) = \frac{n(E_1 \cup E_2)}{n(S)} = \frac{270}{2500} = \mathbf{0.108}.$$

There is a **10.8%** chance that a student selected at random will be taking college algebra, trigonometry, or both.

Now Try Exercise 41

In the previous example, it would have been incorrect to simply add the probability of a student taking algebra and the probability of a student taking trigonometry. Their sum would be

$$P(E_1) + P(E_2) = \frac{n(E_1)}{n(S)} + \frac{n(E_2)}{n(S)}$$

$$= \frac{225}{2500} + \frac{75}{2500}$$

$$= \frac{300}{2500}$$

$$= 0.12, \quad \text{or} \quad 12\%. \ \text{✗}$$

This sum is greater than $P(E_1 \cup E_2)$ because the 30 students taking both courses are counted *twice* in this calculation. In order to find the correct probability for $P(E_1 \cup E_2)$, **we must subtract the probability of the intersection $E_1 \cap E_2$.**

$$P(E_1 \cup E_2) = P(E_1) + P(E_2) - \mathbf{P(E_1 \cap E_2)}$$

$$= \frac{n(E_1)}{n(S)} + \frac{n(E_2)}{n(S)} - \frac{n(E_1 \cap E_2)}{n(S)}$$

$$= \frac{225}{2500} + \frac{75}{2500} - \frac{30}{2500}$$

$$= \frac{270}{2500}$$

$$= \mathbf{0.108}, \quad \text{or} \quad \mathbf{10.8\%} \ \text{✓}$$

This is the same result obtained in Example 4 and suggests the following property.

PROBABILITY OF THE UNION OF TWO EVENTS

For any two events E_1 and E_2,

$$P(E_1 \cup E_2) = P(E_1) + P(E_2) - P(E_1 \cap E_2).$$

EXAMPLE 5 **Rolling dice**

Suppose two dice are rolled. Find the probability that the dice show either a sum of 8 or a pair.

SOLUTION In Table 11.9 the roll of the dice is represented by an ordered pair. For example, the ordered pair (3, 6) repre4sents the first die showing 3 and the second die 6.

Possibilities when Rolling Two Dice

(1, 1)	(1, 2)	(1, 3)	(1, 4)	(1, 5)	(1, 6)
(2, 1)	**(2, 2)**	(2, 3)	(2, 4)	(2, 5)	**(2, 6)**
(3, 1)	(3, 2)	**(3, 3)**	(3, 4)	**(3, 5)**	(3, 6)
(4, 1)	(4, 2)	(4, 3)	**(4, 4)**	(4, 5)	(4, 6)
(5, 1)	(5, 2)	**(5, 3)**	(5, 4)	**(5, 5)**	(5, 6)
(6, 1)	**(6, 2)**	(6, 3)	(6, 4)	(6, 5)	**(6, 6)**

Table 11.9

Because each die can show six different outcomes, there are a total of $6 \cdot 6 = 36$ outcomes in the sample space S. Let E_1 denote the event of rolling a sum of 8 and E_2 the event of rolling a pair. Then

> 5 possible ways to roll a sum of 8

$$E_1 = \{(6, 2), (5, 3), \mathbf{(4, 4)}, (3, 5), (2, 6)\} \quad \text{and}$$

$$E_2 = \{(1, 1), (2, 2), (3, 3), \mathbf{(4, 4)}, (5, 5), (6, 6)\}.$$

> 6 possible ways to roll a pair

The intersection of E_1 and E_2 is

$$E_1 \cap E_2 = \{\mathbf{(4, 4)}\}.$$

Since $n(S) = 36$, $n(E_1) = \mathbf{5}$, $n(E_2) = \mathbf{6}$, and $n(E_1 \cap E_2) = 1$, the following can be computed.

$$P(E_1 \cup E_2) = P(E_1) + P(E_2) - P(E_1 \cap E_2)$$

> Subtract the probability of the intersection.

$$= \frac{n(E_1)}{n(S)} + \frac{n(E_2)}{n(S)} - \frac{n(E_1 \cap E_2)}{n(S)}$$

$$= \frac{5}{36} + \frac{6}{36} - \frac{1}{36}$$

$$= \frac{10}{36}, \quad \text{or} \quad \frac{5}{18}$$

This result can be verified by counting the number of boldfaced outcomes in Table 11.9. Of the 36 possible outcomes, 10 satisfy the conditions, so the probability is $\frac{10}{36}$, or $\frac{5}{18}$.

> **Now Try Exercise 49**

Mutually Exclusive Events If $E_1 \cap E_2 = \varnothing$, then the events E_1 and E_2 are **mutually exclusive**. Mutually exclusive events have no outcomes in common, so $P(E_1 \cap E_2) = 0$. In this case, $P(E_1 \cup E_2) = P(E_1) + P(E_2)$.

EXAMPLE 6 **Drawing cards**

Find the probability of drawing either an ace or a king from a standard deck of 52 cards.

SOLUTION The event E_1 of drawing an ace and the event E_2 of drawing a king are mutually exclusive. No card can be both an ace and a king. Therefore $P(E_1 \cap E_2) = 0$. The probability of drawing an ace is $P(E_1) = \frac{4}{52}$, since there are 4 aces in 52 cards. Similarly, the probability of drawing a king is $P(E_2) = \frac{4}{52}$.

$$P(E_1 \cup E_2) = P(E_1) + P(E_2)$$

$$= \frac{4}{52} + \frac{4}{52}$$

$$= \frac{8}{52}, \quad \text{or} \quad \frac{2}{13}$$

Thus the probability of drawing either an ace or a king is $\frac{2}{13}$.

> **Now Try Exercise 51**

The next example uses concepts from both counting and probability.

EXAMPLE 7 Drawing a poker hand

A standard deck of cards contains 52 cards, consisting of 13 different cards from each of four suits: hearts, diamonds, spades, and clubs. A poker hand consists of 5 cards drawn from a standard deck of cards, and a flush occurs when the 5 cards are all from the same suit. Find an expression for the probability of drawing 5 cards of the same suit in one try. Assume that the cards are not replaced.

SOLUTION Let E be the event of drawing 5 cards of the same suit. To determine $n(E)$, start by calculating the number of ways to draw a flush in a particular suit, such as hearts. From a set of 13 hearts, 5 hearts need to be drawn. There are $\binom{13}{5}$ ways to draw this hand. Because there are 4 suits, there are $4\binom{13}{5}$ ways to draw 5 cards of the same suit. Thus $n(E) = 4\binom{13}{5}$. The sample space S consists of all 5-card poker hands that can be drawn from a deck of 52 cards. There are $\binom{52}{5}$ different poker hands. Thus $n(S) = \binom{52}{5}$. The probability of a flush can now be calculated as follows.

> 4(13 nCr 5)/(52
> nCr 5)
>   .0019807923

Figure 11.22

$$P(E) = \frac{n(E)}{n(S)} = \frac{4\dbinom{13}{5}}{\dbinom{52}{5}}$$

— Number of ways to draw a flush

— Number of ways to draw any 5-card poker hand

In Figure 11.22, we see that $P(E) \approx 0.00198$. Thus there is about a 0.2% chance of drawing 5 cards of the same suit (in one try) from a standard deck of 52 cards.

Now Try Exercise 37

Independent Events

Two events are **independent** if they do not influence each other. Otherwise they are **dependent**. An example of independent events would be one coin being tossed twice. The result of the first toss does not affect the second toss. The following shows how to calculate the probability of two independent events both occurring.

> **INDEPENDENT EVENTS**
>
> If E_1 and E_2 are independent events, then
>
> $$P(E_1 \cap E_2) = P(E_1) \cdot P(E_2).$$

EXAMPLE 8 Tossing a coin

Suppose a coin is tossed twice. Determine the probability that the result is two heads.

SOLUTION Let E_1 be the event of a head on the first toss, and let E_2 be the event of a head on the second toss. Then $P(E_1) = P(E_2) = \frac{1}{2}$. The two events are independent. The probability of two heads occurring is

$$P(E_1 \cap E_2) = P(E_1) \cdot P(E_2) = \frac{1}{2} \cdot \frac{1}{2} = \frac{1}{4}.$$

This probability of $\frac{1}{4}$ also can be found using a tree diagram, as shown in Figure 11.23. There are four equally likely outcomes. Tosses resulting in two heads occur once.

Now Try Exercise 27

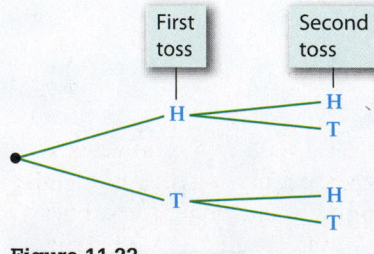

Figure 11.23

EXAMPLE 9   **Rolling dice**

What is the probability of rolling a sum of 12 with two dice?

SOLUTION The roll of one die does not influence the roll of the other. They are independent events. To obtain a sum of 12, both dice must show a 6. Let E_1 be the event of rolling a 6 with the first die and E_2 the event of rolling a 6 with the second die. Then $P(E_1) = P(E_2) = \frac{1}{6}$. The probability of rolling a 12 is

$$P(E_1 \cap E_2) = P(E_1) \cdot P(E_2) = \frac{1}{6} \cdot \frac{1}{6} = \frac{1}{36}.$$

CLASS DISCUSSION

When a pair of dice are rolled, what sum is most likely to appear?

This result can be verified by using Table 11.9. There is only 1 outcome out of 36 that results in a sum of 12.

Now Try Exercise 31

Conditional Probability and Dependent Events

The **conditional probability** of an event E_2 is the probability that the event will occur, given the knowledge that event E_1 already occurred. For example, we might want to find the probability of attending college, given that graduating from high school already occurred. $P(E_2$, given E_1 occurred$)$ is called the conditional probability of E_2, given E_1.

When events E_1 and E_2 are independent, $P(E_2$, given that E_1 occurred$) = P(E_2)$, as event E_1 does not influence event E_2. For example, P(tossing heads given that the last toss was tails) is simply P(tossing heads) $= \frac{1}{2}$, since two separate coin flips are independent events. However, we often use conditional probability to find the probability of dependent events.

If events E_1 and E_2 influence each other, they are *dependent*. The probability of dependent events both occurring is given as follows.

PROBABILITY OF DEPENDENT EVENTS

If E_1 and E_2 are dependent events, then

$$P(E_1 \cap E_2) = P(E_1) \cdot P(E_2, \text{ given that } E_1 \text{ occurred}).$$

EXAMPLE 10   **Drawing cards**

Find the probability of drawing two hearts from a standard deck of 52 cards when the first card is
(a) replaced before drawing the second card,   **(b)** not replaced.

SOLUTION
(a) Let E_1 denote the event of the first card being a heart and E_2 the event of the second card being a heart. If the first card is replaced before the second card is drawn, the two events are *independent*. Since there are 13 hearts in a standard deck of 52 cards, the probability of two hearts being drawn is

$$P(E_1 \cap E_2) = P(E_1) \cdot P(E_2) = \frac{13}{52} \cdot \frac{13}{52} = \frac{1}{16}.$$

(b) If the first card is not replaced, then the outcome for the second card is influenced by the first card. Therefore the events E_1 and E_2 are dependent and we need to find the conditional probability of E_2, given E_1. The probability of drawing a second heart, given that the first card is a heart, is $P(E_2$, given E_1 has occurred$) = \frac{12}{51}$. That is, if the first card drawn were a heart and removed from the deck, then there would be 12 hearts in a sample space of 51 cards.

$$P(E_1 \cap E_2) = P(E_1) \cdot P(E_2, \text{ given that } E_1 \text{ occurred}) = \frac{13}{52} \cdot \frac{12}{51} = \frac{1}{17}$$

Thus the probability of drawing two hearts is slightly less if the first card is not replaced.

Now Try Exercise 61

EXAMPLE 11 **Calculating the probability of dependent events**

Table 11.10 shows the number of students (by gender) registered for either a Spanish class or a French class. No student is taking both languages.

	Spanish	French	Totals
Females	20	25	45
Males	40	25	65
Totals	60	50	110

Table 11.10

If one student is selected at random, calculate each of the following.
(a) The probability that the student is female
(b) The probability that the student is taking Spanish, given that the student is female
(c) The probability that the student is female and taking Spanish

SOLUTION
(a) Let F represent the event that the student is female. Since 45 of the 110 students are female, $P(F) = \frac{45}{110} = \frac{9}{22}$.
(b) Let S be the event that the student is taking Spanish. The probability that the student is taking Spanish, given that the student is female, is $P(S, \text{ given } F) = \frac{20}{45} = \frac{4}{9}$, because 20 of the 45 female students are taking Spanish.
(c) The probability that the student is female *and* taking Spanish is calculated by

$$P(F \cap S) = P(F) \cdot P(S, \text{ given } F) = \frac{45}{110} \cdot \frac{20}{45} = \frac{2}{11}.$$

Table 11.10 shows that 20 of the 110 students are female and taking Spanish. Thus the required probability is $\frac{20}{110} = \frac{2}{11}$, which is in agreement with our previous calculation.

Now Try Exercise 71

EXAMPLE 12 **Analyzing a polygraph test**

Suppose there is a 6% chance that a polygraph test will incorrectly say a person is lying when he or she is actually telling the truth. If a person tells the truth 95% of the time, what percentage of the time will the polygraph test incorrectly indicate a lie for this person?

SOLUTION Let E_1 be the event that the person is telling the truth and E_2 be the event that the polygraph test is incorrect. Then

$$P(E_1 \cap E_2) = P(E_1) \cdot P(E_2, \text{ given the person is telling the truth)}$$

$$= (0.95)(0.06)$$

$$= 0.057, \quad \text{or} \quad 5.7\%.$$

Now Try Exercise 67

11.6 Putting It All Together

The following table summarizes some concepts about probability. In this table, S denotes a finite sample space of equally likely outcomes, and E, E_1, and E_2 denote events in the sample space S.

CONCEPT	EXPLANATION	EXAMPLE
Probability P	A number P satisfying $0 \leq P \leq 1$	$P(E) = 1$ indicates that an event E is certain to occur, $P(E) = 0$ indicates that an event E is impossible, and $P(E) = 0.3$ indicates that an event E has a 30% chance of occurring.
Probability of an event	The probability of event E is $$P(E) = \frac{n(E)}{n(S)},$$ where $n(E)$ and $n(S)$ denote the number of equally likely outcomes in E and S, respectively.	The probability of rolling two dice and their sum being 3 is $\frac{2}{36}$. This is because there are $6 \cdot 6 = 36$ ways to roll two dice and only 2 ways to roll a sum of 3. (They are 1 and 2 or 2 and 1.)
Probability of a complement	If E and E' are complementary events, then $E \cap E' = \varnothing$, $E \cup E' = S$, and $P(E') = 1 - P(E)$.	The probability of rolling a 6 with one die is $\frac{1}{6}$. Therefore the probability of *not* rolling a 6 is $1 - \frac{1}{6} = \frac{5}{6}$.
Probability of the union of two events	The probability of E_1 or E_2 (or both) occurring is $P(E_1 \cup E_2) = P(E_1) + P(E_2) - P(E_1 \cap E_2)$.	If $P(E_1) = 0.5$, $P(E_2) = 0.2$, and $P(E_1 \cap E_2) = 0.1$, then $P(E_1 \cup E_2) = 0.5 + 0.2 - 0.1 = 0.6$.
Probability of independent events	Two events E_1 and E_2 are independent if they do not influence one another. $$P(E_1 \cap E_2) = P(E_1) \cdot P(E_2)$$	If E_1 represents a head on the first toss of a coin and E_2 represents a tail on the second toss, then E_1 and E_2 are independent events and $$P(E_1 \cap E_2) = \frac{1}{2} \cdot \frac{1}{2} = \frac{1}{4}.$$
Conditional probability	The conditional probability of E_2 is the probability the event will occur, given that E_1 already occurred. $P(E_2$, given that E_1 occurred)	An example would be the probability of rain, given that it's cloudy.
Probability of dependent events	Two events E_1 and E_2 are dependent if they are not independent. $P(E_1 \cap E_2) = P(E_1) \cdot P(E_2$, given that E_1 occurred)	Drawing two hearts without replacement from a standard deck of cards is an example of dependent events. See Example 10.

11.6   Exercises

Probability of an Event

Exercises 1–8: Does the number represent a probability?

1. $\frac{11}{13}$ 2. 0.995

3. 2.5   4. 1

5. 0   6. 110%

7. −0.375   8. $\frac{9}{8}$

Exercises 9–18: Find the probability of each event.

9. Tossing a head with a fair coin

10. Tossing a tail with a fair coin

11. Rolling a 2 with a fair die

12. Rolling a 5 or 6 with a fair die

13. Guessing the correct answer to a true-false question

14. Guessing the correct answer to a multiple-choice question with five choices

15. Randomly drawing a king from a standard deck of 52 cards

16. Randomly drawing a club from a standard deck of 52 cards

17. Randomly guessing a four-digit ATM access code

18. Randomly picking the winning team at a basketball game

19. The following table shows people's favorite pizza toppings.

Pepperoni	43%
Sausage	19%
Mushrooms	14%
Vegetables	13%

Source: USA Today

(a) If a person is selected at random, what is the probability that pepperoni is not his or her favorite topping?

(b) Find the probability that a person's favorite topping is either mushrooms or sausage.

20. (Refer to Example 2.) Find the probability that a transplant patient in 2012 was waiting for the following.
    (a) A lung

    (b) A heart or a liver

Union and Intersection

Exercises 21–26: Find the union and the intersection of events A and B.

21. $A = \{10, 25, 26\}$; $B = \{25, 26, 35\}$

22. $A = \{100, 200, 300\}$; $B = \{100, 500, 1000\}$

23. $A = \{1, 3, 5, 7\}$; $B = \{9, 11\}$

24. $A = \{2, 4, 6\}$; $B = \{2, 4, 6, 8\}$

25. $A = \{\text{Heads}\}$; $B = \{\text{Tails}\}$

26. $A = \{\text{Rain}\}$; $B = \{\text{No rain}\}$

Probability of Compound Events

Exercises 27–36: Find the probability of the compound event.

27. Tossing a coin twice with the outcomes of two tails

28. Tossing a coin three times with the outcomes of three heads

29. Rolling a die three times and obtaining a 5 or 6 on each roll

30. Rolling a sum of 7 with two dice

31. Rolling a sum of 2 with two dice

32. Rolling a sum other than 7 with two dice

33. Rolling a die four times without obtaining a 6

34. Rolling a die four times and obtaining at least one 6

35. Drawing four consecutive aces from a standard deck of 52 cards without replacement

36. Drawing a pair (two cards with the same value) from a standard deck of 52 cards without replacement

37. **Poker Hands** (Refer to Example 7.) Calculate the probability of drawing 3 hearts and 2 diamonds in a 5-card poker hand. Assume that drawn cards are not replaced and that the 5 cards are drawn only once.

38. **Poker Hands** (Refer to Example 7.) Calculate the probability of drawing 3 kings and 2 queens in a 5-card poker hand. Assume that drawn cards are not replaced and that the 5 cards are drawn only once.

39. **Quality Control** A quality-control experiment involves selecting 1 string of decorative lights from a box of 20.

If the string is defective, the entire box of 20 is rejected. Suppose the box contains 4 defective strings of lights. What is the probability of rejecting the box?

40. Quality Control (Refer to Exercise 39.) Suppose 3 strings of lights are tested. If any of the strings are defective, the entire box of 20 is rejected. What is the probability of rejecting a box if there are 4 defective strings of lights in the box? (*Hint:* Start by finding the probability that the box is not rejected.)

41. Entrance Exams A group of students is preparing for college entrance exams. It is estimated that 50% need help with mathematics, 45% with English, and 25% with both.

(a) Draw a Venn diagram representing these data.

(b) Use this diagram to find the probability that a student needs help with mathematics, English, or both.

(c) Solve part (b) symbolically by applying a probability formula.

42. College Classes In a college of 5500 students, 950 students are enrolled in English classes, 1220 in business classes, and 350 in both. If a student is chosen at random, find the probability that he or she is enrolled in an English class, a business class, or both.

43. Pinterest Categories The table shows the most popular Pinterest categories in 2012 and the percentage of total pins associated with them. Find the probability that a randomly selected pin will be as specified.

Home	18%
Arts/Crafts	11%
Food	11%
Style/Fashion	11%
Inspiration/Education	9%
Holiday/Seasonal	4%

Source: Analytics by RJ Metrics

(a) In the Home category

(b) In none of the categories in the table

(c) In neither the Home nor the Style/Fashion category

44. Death Rates In 2012, the U.S. death rate per 100,000 people was 838. What is the probability that a person selected at random died during 2012? (*Source:* Department of Health and Human Services.)

45. Death Rates In 2010, the death rate per 100,000 females was 634. What is the probability that a person selected at random from this gender group died during 2010? (*Source:* Department of Health and Human Services.)

46. Tossing a Coin Find the probability of tossing a coin n times and obtaining n heads. What happens to this probability as n increases? Does this agree with your intuition? Explain.

47. U.S. HIV By 2008, a total of 679,590 people were living with an HIV diagnoses. The table lists HIV cases diagnosed in certain cities. Estimate the probability that a person diagnosed with HIV satisfied the following conditions.

New York	223,508
Los Angeles	65,947
San Francisco	44,422
Miami	64,573

Source: Department of Health and Human Services.

(a) Resided in New York

(b) Did not reside in New York

(c) Resided in Los Angeles or Miami

48. Rolling Dice Find the probability of rolling a die five times and obtaining a 6 on the first two rolls, a 5 on the third roll, and a 1, 2, 3, or 4 on the last two rolls.

49. Rolling Dice (Refer to Example 5.) Two dice are rolled. Find the probability that the dice show either a pair or a sum of 6.

50. Rolling Dice Two dice are rolled. Find the probability that the dice show a sum other than 7 or 11.

51. Drawing Cards (Refer to Example 6.) Find the probability of drawing a 2, 3, or 4 from a standard deck of 52 playing cards.

52. Drawing Cards Find the probability of drawing two cards, neither of which is an ace or a queen.

53. Unfair Die Suppose a die is not fair, but instead the probability P of each number n is as listed in the table. Find the probability of each event.

n	1	2	3	4	5	6
P	0.1	0.1	0.1	0.2	0.2	0.3

(a) Rolling a number that is 4 or higher

(b) Rolling a 6 twice on consecutive rolls

54. Unfair Coin Suppose a coin is not fair, but instead the probability of obtaining a head (H) is $\frac{3}{4}$ and a tail (T) is $\frac{1}{4}$. What is the probability of each event?

(a) HT   (b) HH

(c) HHT   (d) THT

55. Dice Suppose there are two dice, one red and one blue, having the probabilities shown in the table in Exercise 53. If both dice are rolled, find the probability of the given sum.

(a) 12   (b) 11

56. Garage Door Code The code for some garage door openers consists of 12 electrical switches that can be set to either 0 or 1 by the owner. Each setting represents a different code. What is the probability of guessing someone's code at random? (*Source:* Promax.)

57. Lottery To win a lottery, a person must pick three numbers from 0 to 9 in the correct order. If a number may be repeated, what is the probability of winning this game with one play?

58. Lottery To win the jackpot in a lottery, a person must pick five numbers from 1 to 59 and then pick the powerball, which has a number from 1 to 35. If the numbers are picked at random, what is the probability of winning this game with one play?

59. Marbles A jar contains 22 red marbles, 18 blue marbles, and 10 green marbles. If a marble is drawn from the jar at random, find the probability that the color is the following.

(a) Red   (b) Not red

(c) Blue or green

60. Marbles A jar contains 55 red marbles and 45 blue marbles. If 2 marbles are drawn from the jar at random without replacement, find the probability that the marbles satisfy the following.

(a) Both are blue

(b) Neither is blue

(c) The first marble is red and the second marble is blue

Conditional Probability and Dependent Events

61. Drawing a Card Find the probability of drawing a queen from a standard deck of cards given that one card, a queen, has already been drawn and not replaced.

62. Drawing a Card Find the probability of drawing a king from a standard deck of cards given that two cards, both kings, have already been drawn and not replaced.

63. Drawing a Card A card is drawn from a standard deck of 52 cards. Given that the card is a face card, what is the probability that the card is a king? (*Hint:* A face card is a jack, queen, or king.)

64. Drawing a Card Three cards are drawn from a deck without replacement. Find the probability that the three cards are an ace, king, and queen in that order.

65. Drawing Marbles A jar initially contains 10 red marbles and 23 blue marbles. What is the probability of drawing a blue marble, given that 2 red marbles and 4 blue marbles have already been drawn?

66. Tennis Serve The probability that the first serve of a tennis ball is out of bounds is 0.3, and the probability that the second serve of a tennis ball is in bounds, given that the first serve was out of bounds, is 0.8. Find the probability that the first serve is out of bounds and the second serve is in bounds.

67. Cloudy and Windy The probability of a day being cloudy is 30%, and the probability of it being cloudy and windy is 12%. Given that the day is cloudy, what is the probability that it will be windy?

68. Rainy and Windy The probability of a day being rainy is 80%, and the probability of it being windy and rainy is 72%. Given that the day is rainy, what is the probability that it will be windy?

69. Rolling Dice Two dice are rolled. If the first die shows a 2, find the probability that the sum of the dice is 7 or more.

70. Rolling Dice Three dice are rolled. If the first die shows a 4, find the probability that the sum of the three dice is less than 12.

71. Defective Parts The table shows numbers of automobile parts that are either defective or not defective.

	Type A	Type B	Totals
Defective	7	11	18
Not defective	123	94	217
Totals	130	105	235

If one part is selected at random, calculate each of the following.

(a) The probability that the part is defective

(b) The probability that the part is type A, given that it is defective

(c) The probability that the part is type A and defective

72. Health The table shows numbers of patients with two different diseases by gender. (Assume that a person does not have both diseases.)

	Disease A	Disease B	Totals
Females	145	851	996
Males	256	355	611
Totals	401	1206	1607

If one patient is selected at random, find the probability that the patient is female and has disease B.

73. Prime Numbers Suppose a number from 1 to 15 is selected at random. Find the probability of each event.
(a) The number is odd

(b) The number is even

(c) The number is prime (*Hint:* A natural number greater than 1 that has only itself and 1 as factors is called a **prime number**.)

(d) The number is prime and odd

(e) The number is prime and even

74. Students and Classes (Refer to Example 11.) If one student is selected at random, use Table 11.10 to calculate each of the following.
(a) The probability that the student is male

(b) The probability that the student is taking French, given that the student is male

(c) The probability that the student is male and taking French

Writing about Mathematics

75. What values are possible for a probability? Interpret different probabilities and give examples.

76. Discuss the difference between dependent and independent events. How are their probabilities calculated?

CHECKING BASIC CONCEPTS FOR SECTIONS 11.5 AND 11.6

1. Use mathematical induction to prove that

$$4 + 8 + 12 + 16 + \cdots + 4n = 2n(n + 1).$$

2. Use mathematical induction to prove that $n^2 \leq 2^n$ for $n \geq 4$.

3. Find the probability of tossing a coin four times and obtaining a head every time.

4. Find the probability of rolling a sum of 11 with two dice.

5. Find the probability of drawing four aces and a queen from a standard deck of 52 cards.

6. Electronic Waste Worldwide electronic waste is increasing at about 40 million tons per year, with China producing 2.6 million tons per year. Estimate the probability that a given ton of electronic waste is *not* produced by China. (*Source:* EPA)

11 Summary

CONCEPT	EXPLANATION AND EXAMPLES
Section 11.1 Sequences	
Sequences	An infinite sequence is a function whose domain is the natural numbers. Its graph is a scatterplot.

Example: $a_n = \frac{1}{2}n^2 - 2$; the first 4 terms a_1, a_2, a_3, a_4 are as follows.

$$a_1 = \tfrac{1}{2}(1)^2 - 2 = -\tfrac{3}{2}, \qquad a_2 = \tfrac{1}{2}(2)^2 - 2 = 0$$

$$a_3 = \tfrac{1}{2}(3)^2 - 2 = \tfrac{5}{2}, \qquad a_4 = \tfrac{1}{2}(4)^2 - 2 = 6$$

CONCEPT	EXPLANATION AND EXAMPLES

Section 11.1 Sequences (CONTINUED)

Arithmetic Sequence

Recursive Definition:
$a_n = a_{n-1} + d$, where d is the common difference

Function Definition:
$f(n) = dn + c$, or equivalently, $f(n) = a_1 + d(n - 1)$, where $a_n = f(n)$ and d is the common difference

Example: $a_n = a_{n-1} + 3$, $a_1 = 4$ and $f(n) = 3n + 1$ describe the same sequence. The common difference is $d = 3$. The terms of the sequence are

$$4, 7, 10, 13, 16, 19, 22, \ldots.$$

Geometric Sequence

Recursive Definition:
$a_n = ra_{n-1}$, where r is the common ratio

Function Definition:
$f(n) = cr^{n-1}$, where $c = a_1$ and r is the common ratio

Example: $a_n = -2a_{n-1}$, $a_1 = 3$ and $f(n) = 3(-2)^{n-1}$ describe the same sequence. The common ratio is $r = -2$. The terms of the sequence are

$$3, -6, 12, -24, 48, -96, 192, \ldots.$$

Section 11.2 Series

Series

A series is the summation of the terms of a sequence.

Examples: $1 + \dfrac{1}{2} + \dfrac{1}{4} + \dfrac{1}{8} + \cdots + \dfrac{1}{2^{n-1}} + \cdots$ *Infinite series*

$$\sum_{k=1}^{5} 2k = 2 + 4 + 6 + 8 + 10 \qquad \textit{Finite series}$$

Arithmetic Series

Finite Arithmetic Series:

$$\sum_{k=1}^{n} a_k = a_1 + a_2 + a_3 + \cdots + a_n,$$

where $a_k = dk + c$ for some constants c and d and d is the common difference. Summing the terms of an arithmetic *sequence* results in an arithmetic *series*.

Sum of the First n Terms:

$$S_n = n\left(\frac{a_1 + a_n}{2}\right) \qquad \text{or} \qquad S_n = \frac{n}{2}(2a_1 + (n - 1)d)$$

Example: The series $4 + 7 + 10 + 13 + 16 + 19 + 22$ is defined by $a_k = 3k + 1$. Its sum is $S_7 = 7\left(\frac{4 + 22}{2}\right) = 91$.

CONCEPT	EXPLANATION AND EXAMPLES

Section 11.2 Series (CONTINUED)

Geometric Series

Infinite Geometric Series:

$$\sum_{k=1}^{\infty} a_k = a_1 + a_2 + a_3 + \cdots + a_n + \cdots,$$

where $a_k = a_1 r^{k-1}$ for some nonzero constants a_1 and r.
Summing the terms of a geometric *sequence* results in a geometric *series*.

Sum of First n Terms:

$S_n = a_1\left(\frac{1 - r^n}{1 - r}\right)$, where a_1 is the first term and r is the common ratio

Example: The series $3 + 6 + 12 + 24 + 48 + 96$ has $a_1 = 3$ and $r = 2$.
Its sum is $S_6 = 3\left(\frac{1 - 2^6}{1 - 2}\right) = 189$.

Sum of Infinite Geometric Series:

$S = \frac{a_1}{1 - r}$, if $|r| < 1$. The sum S does not exist if $|r| \geq 1$.

Example: $4 + 1 + \frac{1}{4} + \frac{1}{16} + \frac{1}{64} + \cdots$ has sum $S = \frac{4}{1 - \frac{1}{4}} = \frac{16}{3}$.

Section 11.3 Counting

Fundamental Counting Principle

Let E_1 and E_2 be independent events. If event E_1 can occur m_1 ways and if event E_2 can occur m_2 ways, then there are $m_1 \cdot m_2$ ways for both events to occur.

Example: If two multiple-choice questions have 5 choices each, then there are $5 \cdot 5 = 25$ ways to answer the two questions.

Factorial Notation

$n! = n(n - 1)(n - 2) \cdots (3)(2)(1)$

Examples: $0! = 1$; $1! = 1$; $5! = 5 \cdot 4 \cdot 3 \cdot 2 \cdot 1 = 120$

Permutations

$P(n, r) = \frac{n!}{(n - r)!}$ represents the number of permutations, or arrangements, of n elements taken r at a time. Order is important in calculating a permutation.

Example: $P(5, 3) = \frac{5!}{(5 - 3)!} = \frac{120}{2} = 60$

Combinations

$C(n, r) = \binom{n}{r} = \frac{n!}{(n - r)!\, r!}$ represents the number of combinations of n elements taken r at a time. Order is unimportant in calculating a combination.

Example: $C(5, 3) = \binom{5}{3} = \frac{5!}{(5 - 3)!3!} = \frac{120}{2 \cdot 6} = 10$

CONCEPT	EXPLANATION AND EXAMPLES

Section 11.4 The Binomial Theorem

Binomial Theorem

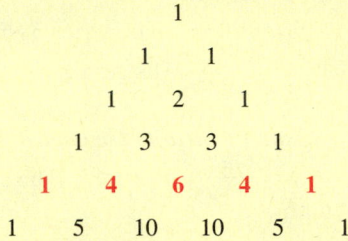

$$(a + b)^n = \binom{n}{0}a^n + \binom{n}{1}a^{n-1}b + \cdots + \binom{n}{n-1}ab^{n-1} + \binom{n}{n}b^n$$

Example: $(x + y)^4 = \mathbf{1}x^4 + \mathbf{4}x^3y + \mathbf{6}x^2y^2 + \mathbf{4}xy^3 + \mathbf{1}y^4$
The coefficients can also be found in the fifth row of Pascal's triangle.

Pascal's Triangle

Pascal's triangle can be used to calculate the binomial coefficients when expanding the expression $(a + b)^n$.

$$
\begin{array}{ccccccccccc}
 & & & & & 1 & & & & & \\
 & & & & 1 & & 1 & & & & \\
 & & & 1 & & 2 & & 1 & & & \\
 & & 1 & & 3 & & 3 & & 1 & & \\
 & \mathbf{1} & & \mathbf{4} & & \mathbf{6} & & \mathbf{4} & & \mathbf{1} & \\
1 & & 5 & & 10 & & 10 & & 5 & & 1
\end{array}
$$

Section 11.5 Mathematical Induction

Principle of Mathematical Induction

Let S_n be a statement concerning the positive integer n. Suppose that

1. S_1 is true;

2. for any positive integer k, if S_k is true, then S_{k+1} is also true.

Then S_n is true for every positive integer n.

Section 11.6 Probability

Probability

$P(E) = \frac{n(E)}{n(S)}$, where $n(E)$ is the number of outcomes in event E and $n(S)$ is the number of equally likely outcomes in the finite sample space S. Note that $0 \le P(E) \le 1$.

Compound Events

Probability of either E_1 or E_2 (or both) occurring:

$$P(E_1 \cup E_2) = P(E_1) + P(E_2) - P(E_1 \cap E_2)$$

If E_1 and E_2 are *mutually exclusive*, then $E_1 \cap E_2 = \varnothing$ and

$$P(E_1 \cup E_2) = P(E_1) + P(E_2).$$

Probability of *both* E_1 and E_2 occurring:
If E_1 and E_2 are *independent*, then

$$P(E_1 \cap E_2) = P(E_1) \cdot P(E_2).$$

If E_1 and E_2 are *dependent*, then

$$P(E_1 \cap E_2) = P(E_1) \cdot P(E_2, \text{ given that } E_1 \text{ occurred}),$$

where $P(E_2$, given that E_1 occurred) is called the ***conditional probability*** of event E_2, given E_1.

11 Review Exercises

Exercises 1–4: Find the first four terms of the sequence.

1. $a_n = -3n + 2$

2. $a_n = n^2 + n$

3. $a_n = 2a_{n-1} + 1; a_1 = 0$

4. $a_n = a_{n-1} + 2a_{n-2}; a_1 = 1, a_2 = 4$

Exercises 5 and 6: Use the graphical representation to identify the terms of the finite sequence.

5. 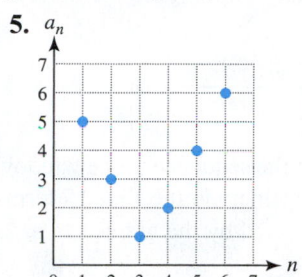 **6.**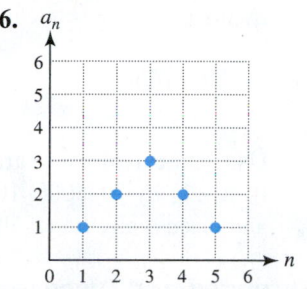

Exercises 7 and 8: The first five terms of an infinite arithmetic or geometric sequence are given. Find

(a) numerical, (b) graphical, and (c) symbolic

representations of the sequence. Include at least the first eight terms of the sequence in your table and graph.

7. $3, 1, -1, -3, -5$

8. $1.5, -3, 6, -12, 24$

9. Find a general term a_n for the arithmetic sequence with $a_3 = -3$ and $d = 4$.

10. Find a general term a_n for the geometric sequence with $a_1 = 2.5$ and $a_6 = -80$.

Exercises 11–14: Determine if the sequence is arithmetic, geometric, or neither.

11. $f(n) = 5 - 2n$ **12.** $f(n) = 3n^2$

13. $f(n) = 3(2)^n$ **14.** $f(n) = n + \left(\frac{1}{2}\right)^{n-1}$

Exercises 15 and 16: For the given a_n, calculate S_5.

15. $a_n = 4n + 1$ **16.** $a_n = 3(4)^{n-1}$

Exercises 17–20: Use a formula to find the sum of the series.

17. $-2 + 1 + 4 + 7 + 10 + 13 + 16 + 19 + 22$

18. $2 + 4 + 6 + 8 + \cdots + 98 + 100$

19. $1 + 3 + 9 + 27 + 81 + 243 + 729 + 2187$

20. $64 + 16 + 4 + 1 + \frac{1}{4} + \frac{1}{16}$

Exercises 21 and 22: Find the sum of the infinite geometric series.

21. $4 - \frac{4}{3} + \frac{4}{9} - \frac{4}{27} + \frac{4}{81} - \frac{4}{243} + \cdots$

22. $0.2 + 0.02 + 0.002 + 0.0002 + 0.00002 + \cdots$

Exercises 23 and 24: Write out the terms of the series.

23. $\sum_{k=1}^{5}(5k + 1)$ **24.** $\sum_{k=1}^{4}(2 - k^2)$

Exercises 25 and 26: Write the series using summation notation.

25. $1^3 + 2^3 + 3^3 + 4^3 + 5^3 + 6^3$

26. $1 + \frac{1}{10} + \frac{1}{100} + \frac{1}{1000} + \frac{1}{10,000}$

Exercises 27 and 28: Use a formula to find the sum of the first 30 terms of the arithmetic sequence.

27. $a_1 = 5, d = -3$ **28.** $a_1 = -2, a_{10} = 16$

29. Write $\frac{2}{11}$ as an infinite geometric series.

30. Write the infinite series

$$0.23 + 0.0023 + 0.000023 + \cdots$$

as a rational number.

Exercises 31 and 32: Evaluate the expression.

31. $P(6, 3)$ **32.** $C(7, 4)$

Exercises 33 and 34: Use mathematical induction to prove that the statement is true for every positive integer.

33. $1 + 3 + 5 + 7 + \cdots + (2n - 1) = n^2$

34. $2 + 2^2 + 2^3 + \cdots + 2^n = 2(2^n - 1)$

Exercises 35 and 36: Find the probability of each event.

35. A 1, 2, or 3 appears when a die is rolled

36. Tossing three heads in a row using a fair coin

Applications

37. Standing in Line   In how many arrangements can five people stand in a line?

38. Giving a Speech   How many arrangements are possible in which 4 students out of a class of 15 each give a speech?

39. Exam Questions   Count the ways that an exam, consisting of 20 multiple-choice questions with four choices each, could be answered.

40. License Plates   Count the different license plates having four numeric digits followed by two letters.

41. Combination Lock A combination lock consists of four numbers from 0 to 49. If a number may be repeated, find the number of possible combinations.

42. Dice A red die and a blue die are rolled. How many ways are there for the sum to equal 4?

43. Height of a Ball When a Ping-Pong ball is dropped, it rebounds to 90% of its initial height.
  (a) Write the first five terms of a sequence that gives the maximum height attained by the ball on each rebound when it is dropped from an initial height of 4 feet. Let $a_1 = 4$. What type of sequence describes these maximum heights?

  (b) Give a graphical representation of this sequence for the first five terms.

  (c) Find a formula for a_n.

44. Falling Object If air resistance is ignored, an object falls 16, 48, 80, and 112 feet during each successive 1-second interval.
  (a) What type of sequence describes these distances?

  (b) Determine how far an object falls during the sixth second.

  (c) Find a formula for the nth term of this sequence.

45. Committees In how many ways can a committee of three be selected from six people?

46. Committees Find the number of committees with three women and three men that can be selected from a group of seven women and five men.

47. Test Questions On a test with 10 essay questions, students are asked to answer 6 questions. How many ways can the essay questions be selected?

48. Binomial Theorem Use the binomial theorem to expand the expression $(2x - y)^4$.

49. Quality Control A quality-control experiment involves selecting 2 batteries from a pack of 16. If either battery is defective, the entire pack of 16 is rejected. Suppose a pack contains 2 defective batteries. What is the probability that this pack will not be rejected?

50. Venn Diagram Of a group of 82 students, 19 are enrolled in music, 22 in art, and 10 in both.
  (a) Draw a Venn diagram representing these data.

  (b) Use this diagram to determine the probability that a student selected at random is enrolled in music, art, or both.

  (c) Solve part (b) symbolically by applying a probability formula.

51. Marbles A jar contains 13 red, 27 blue, and 20 green marbles. If a marble is drawn from the jar at random, find the probability that the marble is the following.
  (a) Blue

  (b) Not blue

  (c) Red

52. Cards Find the probability of drawing 2 diamonds from a standard deck of 52 cards without replacing the first card.

53. Insect Populations The monthly density of an insect population, measured in thousands per acre, is described by the recursive sequence

$$a_1 = 100; \quad a_n = \frac{2a_{n-1}}{1 + (a_{n-1}/4000)}, \quad n > 1.$$

Use your calculator to graph the sequence in the window $[0, 16, 1]$ by $[0, 5000, 1000]$. Include the first 15 terms and discuss any trends illustrated by the graph.

Extended and Discovery Exercises

Exercises 1 and 2: **Antibiotic Resistance** *Because of the frequent use of antibiotics in society, many strains of bacteria are becoming resistant. Some types of haploid bacteria contain genetic material called plasmids. Plasmids are capable of making a strain of bacterium resistant to antibiotic drugs. Genetic engineers want to predict the resistance of various bacteria after many generations.*

1. Suppose a strain of bacterium contains two plasmids R_1 and R_2. Plasmid R_1 is resistant to the antibiotic ampicillin, whereas plasmid R_2 is resistant to the antibiotic tetracycline. When bacteria reproduce through cell division, the type of plasmids passed on to each new cell is random. For example, a daughter cell could have two plasmids of type R_1 and no plasmid of type R_2, one of each type, or no plasmid of type R_1 and two plasmids of type R_2. The probability $P_{k,j}$ that a mother cell with k plasmids of type R_1 produces a daughter cell with j plasmids of type R_1 can be calculated by the formula

$$P_{k,j} = \frac{\binom{2k}{j}\binom{4-2k}{2-j}}{\binom{4}{2}}.$$

(*Source:* F. Hoppensteadt and C. Peskin, *Mathematics in Medicine and the Life Sciences.*)
  (a) Compute $P_{k,j}$ for $0 \le k, \; j \le 2$. Assume that $\binom{0}{0} = 1$ and $\binom{k}{j} = 0$ whenever $k < j$. Record your results in the matrix

$$P = \begin{bmatrix} P_{00} & P_{01} & P_{02} \\ P_{10} & P_{11} & P_{12} \\ P_{20} & P_{21} & P_{22} \end{bmatrix}.$$

(b) Which elements in P are the greatest? Interpret the result.

2. (Continuation of Exercise 1.) The genetic makeup of future generations of the haploid bacterium can be modeled using matrices. Let $A = [a_1, a_2, a_3]$ be a 1×3 matrix containing three probabilities. The value of a_1 is the probability that it has two R_1 plasmids and no R_2 plasmid; a_2 is the probability that it has one R_1 plasmid and one R_2 plasmid; a_3 is the probability that it has no R_1 plasmid and two R_2 plasmids. If an entire generation of the bacterium has one plasmid of each type, then $A_1 = [0, 1, 0]$. In this case the bacterium is resistant to both antibiotics. The probabilities A_n for plasmids R_1 and R_2 in the nth generation of the bacterium can be calculated with the matrix recurrence equation $A_n = A_{n-1}P$, where $n > 1$ and P is the 3×3 matrix determined in Exercise 1. The resulting phenomenon

was not well understood until recently. It is now used in the genetic engineering of plasmids.

(a) If an entire strain of the bacterium is resistant to both the antibiotics ampicillin and tetracycline, make a conjecture as to the drug resistance of future generations of this bacterium.

(b) Test your conjecture by repeatedly computing the matrix product $A_n = A_{n-1}P$. Let $A_1 = [0, 1, 0]$ and $n = 2, 3, \ldots, 12$. Interpret the result. (It may surprise you.)

3. The sum of the first n terms of an arithmetic series is given by $S_n = n\left(\frac{a_1 + a_n}{2}\right)$. Justify this formula using a geometric discussion. (*Hint:* Start by graphing an arithmetic sequence where n is an odd number.)*

1-11 Cumulative Review Exercises

1. Write 34,500 in scientific notation and 1.52×10^{-4} in standard form.

2. Evaluate $\dfrac{5 - \sqrt[3]{4}}{\pi^2 - (\sqrt{3} + 1)}$. Round your answer to the nearest hundredth.

3. Find the exact distance between $(-4, 2)$ and $(1, -2)$.

4. Graph f.
   (a) $f(x) = 4x - 2$ **(b)** $f(x) = |2x - 1|$
   (c) $f(x) = x^2 + 2x$ **(d)** $f(x) = \sqrt{x - 1}$
   (e) $f(x) = \dfrac{1}{x + 2}$ **(f)** $f(x) = x^3 - 2$
   (g) $f(x) = \log_2 x$ **(h)** $f(x) = 3\left(\frac{1}{2}\right)^x$

Exercises 5 and 6: Complete the following.
(a) Evaluate $f(-3)$ and $f(a + 1)$.
(b) Determine the domain of f.

5. $f(x) = \sqrt{1 - x}$ 6. $f(x) = \dfrac{1}{x^2 - 4}$

7. Find the average rate of change of $f(x) = x^3 - 4$ from $x = -2$ to $x = -1$.

8. Find the difference quotient for $f(x) = x^2 - 3x$.

Exercises 9 and 10: Write an equation of a line satisfying the conditions. Use slope-intercept form whenever possible.

9. Passing through $(2, -4)$ and $(-3, 2)$

10. Passing through the point $(-1, 3)$ and perpendicular to the line $y = -\frac{3}{4}x + 1$

11. The graph of a linear function f is shown.
    (a) Identify the slope, y-intercept, and x-intercept.
    (b) Write a formula for f.
    (c) Find any zeros of f.

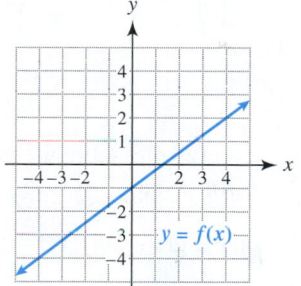

12. Determine the x- and y-intercepts on the graph of $-3x + 4y = 12$. Graph the equation.

13. Solve each equation.
    (a) $4(x - 2) + 1 = 3 - \frac{1}{2}(2x + 3)$
    (b) $6x^2 = 13x + 5$ **(c)** $x^2 - x - 3 = 0$
    (d) $x^3 + x^2 = 4x + 4$ **(e)** $x^4 - 4x^2 + 3 = 0$
    (f) $\dfrac{1}{x - 3} = \dfrac{4}{x + 5}$ **(g)** $3e^{2x} - 5 = 23$
    (h) $2\log(x + 1) - 1 = 2$
    (i) $\sqrt{x + 3} + 4 = x + 1$
    (j) $|3x - 1| = 5$

14. Solve each inequality. Write your answer in set-builder or interval notation.

(a) $3x - 5 < x + 1$ (b) $x^2 - 4x - 5 \leq 0$

(c) $(x + 1)(x - 2)(x - 3) > 0$

(d) $\dfrac{2}{x - 1} < 0$ (e) $|3x - 5| \leq 4$

(f) $|4 - x| > 0$ (g) $\dfrac{3}{4} \leq \dfrac{1 - 2x}{3} < \dfrac{5}{2}$

15. Graph f. Is f continuous on its domain?

$$f(x) = \begin{cases} 2x + 3 & \text{if } -3 \leq x < -1 \\ x^2 & \text{if } -1 \leq x < 1 \\ 2 - x & \text{if } 1 \leq x \leq 3 \end{cases}$$

16. Solve $-2.3x + 3.4 = \sqrt{2x^2 - 1}$ graphically. Round your answers to the nearest tenth.

17. The graph of a nonlinear function f is shown. Solve each equation or inequality. Write the solution set to each inequality in set-builder or interval notation.

(a) $f(x) = 0$ (b) $f(x) > 0$ (c) $f(x) \leq 0$

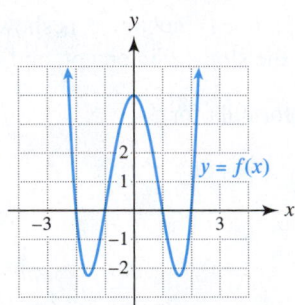

18. Write $f(x) = 3x^2 + 24x + 43$ in the vertex form given by $f(x) = a(x - h)^2 + k$.

19. Find the vertex on the graph of $f(x) = -3x^2 + 9x + 1$.

20. Use the given graph of $y = f(x)$ to sketch a graph of each equation.

(a) $y = f(x - 1) + 2$ (b) $y = \frac{1}{2}f(x)$

(c) $y = -f(x)$ (d) $y = f(2x)$

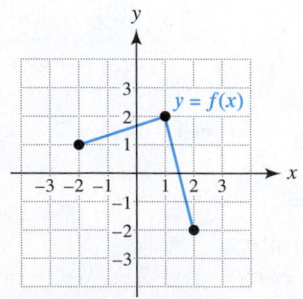

21. Use the graph of f to estimate each of the following.

(a) Intervals where f is increasing or decreasing

(b) The zeros of f

(c) The coordinates of any turning points

(d) Any local extrema

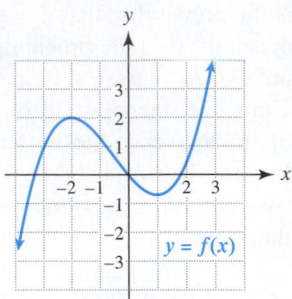

22. Determine if the function defined by $f(x) = x^3 - 5x$ is even, odd, or neither.

23. Divide each expression.

(a) $\dfrac{6x^4 - 2x^2 + 1}{2x^2}$ (b) $\dfrac{2x^4 - 3x^3 - x + 2}{x + 1}$

24. A degree 4 function f has zeros $-2, -1, 1,$ and 2 and leading coefficient 6. Write the complete factored form of $f(x)$.

25. A degree 3 function f with real coefficients has leading coefficient 3 and zeros -1 and $3i$. Write $f(x)$ in complete factored form and expanded form.

26. Write $(2 - i)(2 + 3i)$ in standard form.

27. Find all solutions, real or imaginary, to the quadratic equation $x^2 + 2x + 5 = 0$.

28. State the domain of $f(x) = \dfrac{2x - 5}{x + 5}$. Find any vertical or horizontal asymptotes for the graph of f.

29. Write $\sqrt[5]{(x + 1)^3}$ using rational exponents. Evaluate the expression for $x = 31$.

30. Use the tables to evaluate each expression, if possible.

x	0	1	2	3
$f(x)$	1	2	4	5

x	0	1	2	3
$g(x)$	4	3	2	1

(a) $(f - g)(1)$ (b) $(f/g)(2)$

(c) $(g \circ f)(3)$ (d) $(g \circ f^{-1})(5)$

31. Use the graphs of f and g to complete the following.

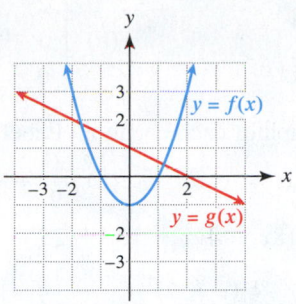

(a) $(f + g)(2)$ **(b)** $(fg)(0)$

(c) $(g \circ f)(1)$ **(d)** $(g^{-1} \circ f)(-2)$

32. Let $f(x) = \dfrac{1}{x + 2}$ and $g(x) = x^2 + x - 4$. Find each of the following.

(a) $(f - g)(0)$ **(b)** $(g \circ f)(-1)$

(c) $(fg)(x)$ **(d)** $(g \circ f)(x)$

33. Find $f^{-1}(x)$ if $f(x) = \dfrac{x}{x + 1}$.

34. Graph $f(x) = x^3 - 1$, $y = x$, and $y = f^{-1}(x)$ on the same axes.

35. Find either a linear or an exponential function f that models the data in the table.

x	0	1	2	3
$f(x)$	2	6	18	54

36. There are initially 1000 bacteria in a sample, and the number doubles every 2 hours.
(a) Find C and a so that $f(x) = Ca^x$ models the number of bacteria after x hours.

(b) Estimate the number of bacteria after 5.2 hours.

(c) When are there 9000 bacteria in the sample?

37. Five hundred dollars is deposited in an account that pays 6% annual interest compounded quarterly. Find the amount in the account after 15 years.

38. Simplify each logarithm by hand.
(a) $\log_2 \frac{1}{16}$ **(b)** $\log \sqrt{10}$

(c) $\ln e^4$ **(d)** $\log_4 2 + \log_4 32$

39. Find the domain and range of each function f.
(a) $f(x) = x^2 - 2x + 1$

(b) $f(x) = 10^x$ **(c)** $f(x) = \ln x$

(d) $f(x) = \dfrac{1}{x}$

40. Expand the expression $\log \sqrt{\frac{x + 1}{yz}}$.

41. Write $2 \log x + 3 \log y - \frac{1}{3} \log z$ as a log single expression.

42. Approximate $\log_4 52$ to three decimal places.

43. Graph $y = f(x)$.
(a) $f(x) = \csc x$

(b) $f(x) = -2 \sin \frac{1}{2}x$

(c) $f(x) = 3 \cos \left(2 \left(x - \frac{\pi}{2} \right) \right)$

(d) $f(x) = \tan (\pi x)$

44. Find the domain of $f(x) = \sec 2x$.

45. Solve each equation for all real numbers t.
(a) $\cos t = -\dfrac{\sqrt{3}}{2}$

(b) $\tan^2 t - 1 = 0$

(c) $\sin^2 t + \frac{1}{2} \sin t = 0$

(d) $\cos 2t = -1$

46. Find the complementary angle α and the supplementary angle β to $54°35'12''$.

47. Convert $5.54°$ to degrees, minutes, and seconds.

48. Convert $135°$ to radians.

49. Convert $\frac{5\pi}{4}$ radians to degrees.

50. Approximate $\cot \frac{2\pi}{9}$ to the nearest hundredth.

51. Find the values of the six trigonometric functions of an angle θ in standard position having its terminal side pass through the point $(-7, 24)$,

52. Find the exact values of the six trigonometric functions of $\theta = \frac{5\pi}{6}$.

53. Find the values of the six trigonometric functions of θ if $\cos \theta = -\frac{11}{61}$ and $\cos \theta < 0$.

54. Evaluate $\cos^{-1} \left(\frac{1}{2} \right)$.

55. Solve the right triangle shown in the figure.

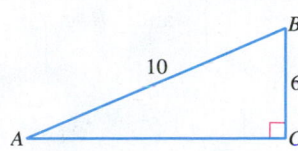

56. Simplify $(\sec t - 1)(\sec t + 1)$.

57. Verify the identity.
$$1 - \sin^2 \theta + \cot^2 \theta - \sin^2 \theta \cot^2 \theta = \cot^2 \theta$$

58. Solve $2 \sin 2x = \ln x$ graphically for $0 \le x < 2\pi$.

59. Solve triangle ABC. Approximate values to the nearest tenth.

(a) $\beta = 42°, \gamma = 31°, a = 22$

(b) $\gamma = 50°, c = 7, b = 8$

(c) $\alpha = 44°, b = 7, c = 8$

(d) $a = 10, b = 11, c = 12$

60. Find the area of a triangle with $\alpha = 30°, b = 15$ feet, and $c = 20$ feet.

61. Let $\mathbf{a} = \langle 3, -4 \rangle$ and $\mathbf{b} = \langle -5, 12 \rangle$. Find the following.

(a) $\|\mathbf{a}\|$

(b) $4\mathbf{b} - 2\mathbf{a}$

(c) $\mathbf{a} \cdot \mathbf{b}$

(d) The angle between $\mathbf{a}$ and $\mathbf{b}$

62. Graph the parametric equations $x = 2\cos t, y = 2\sin t$ for $0 \le t \le 2\pi$.

63. Graph the polar equation given by $r = 1 + \sin\theta$ for $0 \le \theta < 2\pi$.

64. Find the two square roots of $-16i$.

65. Solve the system of equations.

(a) $2x + 3y = 4$
$2x - 5y = -12$

(b) $-2x + \frac{1}{2}y = 1$
$4x - y = -2$

(c) $x^2 + y^2 = 16$
$2x^2 - y^2 = 11$

(d) $x + y - 2z = -6$
$2x - y - 3z = -18$
$3y - z = 6$

66. The variable z varies inversely with the square of x. If $z = 8$ when $x = 50$, find z when $x = 36$.

67. Graph the solution set to each inequality or system of inequalities.*

(a) $-2x + 3y < 6$

(b) $x + y \le 4$
$x - 2y > 6$

68. Find $2A + B$ and AB if

$$A = \begin{bmatrix} -1 & 0 & 2 \\ 1 & -3 & 1 \\ 0 & -3 & 4 \end{bmatrix} \quad \text{and} \quad B = \begin{bmatrix} 1 & 5 & 1 \\ -2 & 2 & 1 \\ 0 & 1 & -2 \end{bmatrix}.$$

69. Find A^{-1} if $A = \begin{bmatrix} 1 & -2 \\ -3 & 4 \end{bmatrix}$.

70. Solve by using A^{-1}.

$$x - y - 2z = 5$$
$$-x + 2y + 3z = -7$$
$$2y + z = -2$$

71. Calculate the determinant of each matrix.

(a) $\begin{bmatrix} -1 & 4 \\ 2 & 3 \end{bmatrix}$

(b) $\begin{bmatrix} 2 & 3 & -1 \\ 3 & -1 & 5 \\ 0 & 0 & -2 \end{bmatrix}$

72. Sketch a graph of each equation. Label any foci.

(a) $y^2 = 2x$

(b) $\dfrac{x^2}{9} + \dfrac{y^2}{25} = 1$

(c) $9x^2 - 18x + 4y^2 + 16y + 25 = 36$

(d) $\dfrac{(x + 1)^2}{16} - \dfrac{(y - 2)^2}{9} = 1$

73. Find an equation of a parabola with vertex $(0, 0)$ and focus $\left(\frac{3}{4}, 0\right)$.

74. Find an equation of an ellipse with vertices $(\pm 3, 1)$ and foci $(\pm 2, 1)$.

75. Find an equation of a hyperbola with foci $(0, \pm 13)$ and vertices $(0, \pm 5)$.

76. Find the first four terms of each sequence.

(a) $a_n = (-1)^{n-1}(3)^n$

(b) $a_n = a_{n-1}a_{n-2}; a_1 = 2, a_2 = 3$

77. Find a general term for the arithmetic sequence given that $a_1 = 4$ and $a_3 = 12$.

78. Find a general term for the geometric sequence given that $a_2 = 6$ and $r = \frac{1}{2}$.

79. Use a formula to find the sum of each series.

(a) $2 + 5 + 8 + 11 + \cdots + 74$

(b) $0.2 + 0.02 + 0.002 + 0.0002 + \cdots$

80. Find the sum $\sum_{k=1}^{7}(k^2 + k)$.

81. Count the number of license plates that can be formed by three letters followed by four digits.

82. Evaluate $P(4, 2)$ and $\binom{6}{3}$.

83. Expand $(2x - 1)^4$.

84. Use mathematical induction to show that

$$5 + 7 + 9 + 11 + \cdots + (2n + 3) = n(n + 4).$$

85. Find the probability of drawing a heart or an ace from a standard deck of 52 cards.

86. Find the probability of rolling a sum of 7 with two dice.

87. A number from 1 to 20 is drawn at random. Find the probability that the number is prime.

88. The probability of a day being cloudy is 40%, and the probability of it being cloudy *and* windy is 15%. Given that the day is cloudy, what is the probability that it will be windy?

Applications

89. Distance At noon, car A is traveling north at 50 miles per hour and is located 30 miles north of car B. Car B is traveling east at 50 miles per hour. Approximate the distance between the cars at 1:45 P.M. to the nearest tenth of a mile.

90. Distance from Home A driver is initially 240 miles from home, traveling toward home on a straight interstate at 60 miles per hour.
  (a) Write a formula for a linear function D that models the distance between the driver and home after x hours.

  (b) What is an appropriate domain for D?

  (c) Graph D.

  (d) Identify the x- and y-intercepts. Interpret each.

91. Running An athlete traveled 10.5 miles in 1.3 hours, jogging at 7 miles per hour and 9 miles per hour. How long did the athlete run at each speed?

92. Average Rate of Change The total distance D in feet traveled by a racehorse after t seconds is given by $D(t) = 3t^2$ for $0 \leq t \leq 5$.
  (a) Find the average rate of change of D from 0 to 1 and 3 to 4.

  (b) Interpret these average rates of change.*

93. Working Together Suppose one person can mow a lawn in 5 hours and another person can mow the same lawn in 4 hours. How long will it take to mow the lawn if they work together?

94. Maximum Height A stone is shot upward with a slingshot. Its height s in feet after t seconds is given by $s(t) = -16t^2 + 96t + 4$. Find its maximum height.

95. Airline Tickets Tickets for a charter flight are regularly $400, but for each additional ticket bought the cost of each ticket is reduced by $5.
  (a) Write a quadratic function C that gives the total cost of buying x tickets.

  (b) Solve $C(x) = 7000$ and interpret the result.

  (c) Find the absolute maximum for C and interpret your result. Assume that x is an integer.

96. Dimensions of a Rectangle A rectangle has a perimeter of 48 inches and an area of 143 square inches. Find its dimensions.

97. Satellite Dish A parabolic satellite dish has a 3-foot diameter and is 6 inches deep. How far from the vertex should the receiver be so that it is located at the focus?

98. Bouncing Ball A Ping-Pong ball is dropped from 4 feet. Each time the ball bounces, it rebounds to 75% of its previous height. Use an infinite geometric series to estimate the *total* distance that the ball travels before coming to rest on the floor.

99. Marbles A jar contains 15 red, 28 blue, and 34 green marbles. If a marble is drawn at random, find the probability that the marble is not blue.

R Reference: Basic Concepts from Algebra and Geometry

Throughout the text there are algebra and geometry review notes that direct students to "see Chapter R." This reference chapter contains seven sections, which provide a review of important topics from algebra and geometry. Students can refer to these sections for more explanation or extra practice. Instructors can use these sections to emphasize a variety of mathematical skills.

R.1 Formulas from Geometry

- Use formulas for shapes in a plane
- Find sides of right triangles by applying the Pythagorean theorem
- Apply formulas to three-dimensional objects
- Use similar triangles to solve problems

Geometric Shapes in a Plane

This subsection discusses formulas related to rectangles, triangles, and circles.

Rectangles   The distance around the boundary of a geometric shape in a plane is called its **perimeter**. The perimeter of a rectangle equals the sum of the lengths of its four sides. For example, the perimeter of the rectangle shown in Figure R.1 is $5 + 4 + 5 + 4 = 18$ feet. The perimeter P of a rectangle with length L and width W is $P = 2L + 2W$.

Rectangle

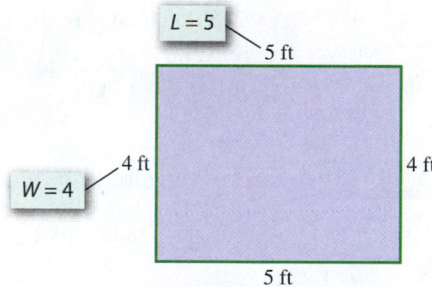

Figure R.1

The area A of a rectangle equals the product of its length and width: $A = LW$. So the rectangle in Figure R.1 has an area of $5 \cdot 4 = 20$ square feet.

Many times the perimeter or area of a rectangle is written in terms of variables, as demonstrated in the next example.

EXAMPLE 1   Finding the perimeter and area of a rectangle

The length of a rectangle is three times greater than its width. If the width is x inches, write expressions that give the perimeter and area.

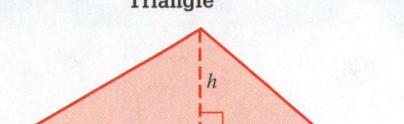

Figure R.2

SOLUTION The **width** of the rectangle is x inches, so its **length** is $3x$ inches. A sketch is shown in Figure R.2. The perimeter is

$$P = 2L + 2W$$
$$= 2(3x) + 2(x)$$
$$= 8x \text{ inches.}$$

> Write the length in terms of the width.

The area is $A = LW = 3x \cdot x = 3x^2$ square inches.

Now Try Exercise 7

Triangle

Figure R.3

Triangles If the base of a triangle is b and its height is h, as illustrated in Figure R.3, then the area A of the triangle is given by

$$A = \frac{1}{2}bh.$$

EXAMPLE 2 Finding the area of a triangle

Calculate the area of the triangle.

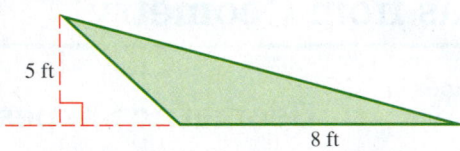

SOLUTION The triangle has a base of 8 feet and a height of 5 feet. Therefore its area is

$$A = \frac{1}{2}bh = \frac{1}{2} \cdot 8 \cdot 5 = 20 \text{ square feet.}$$

Now Try Exercise 11

Circle

$C = 2\pi r$
$A = \pi r^2$

Figure R.4

Circles The perimeter of a circle is called its **circumference** C and is given by $C = 2\pi r$, where r is the radius of the circle. The area A of a circle is $A = \pi r^2$. See Figure R.4.

EXAMPLE 3 Finding the circumference and area of a circle

A circle has a radius of 12.5 inches. Approximate its circumference and area.

SOLUTION
Circumference: $C = 2\pi r = 2\pi(12.5) = 25\pi \approx 78.5$ inches
Area: $A = \pi r^2 = \pi(12.5)^2 = 156.25\pi \approx 490.9$ square inches

Now Try Exercise 21

Pythagorean Theorem

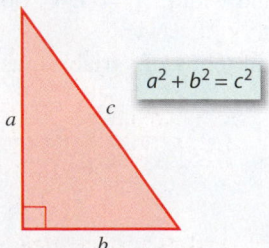

Figure R.5

The Pythagorean Theorem

One of the most famous theorems in mathematics is the Pythagorean theorem. It states that a triangle with legs a and b and hypotenuse c is a right triangle if and only if

$$a^2 + b^2 = c^2,$$

as illustrated in Figure R.5.

EXAMPLE 4 Finding the perimeter of a right triangle

One Side Unknown

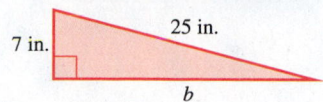

Figure R.6

Find the perimeter of the triangle shown in Figure R.6.

SOLUTION Given one leg and the hypotenuse of a right triangle, we can use the Pythagorean theorem to find the other leg. Let $a = 7$, $c = 25$, and find b.

$$a^2 + b^2 = c^2 \qquad \text{Pythagorean theorem}$$
$$b^2 = c^2 - a^2 \qquad \text{Subtract } a^2.$$
$$b^2 = 25^2 - 7^2 \qquad \text{Let } a = 7 \text{ and } c = 25.$$
$$b^2 = 576 \qquad \text{Simplify.}$$
$$b = 24 \qquad \text{Solve for } b > 0.$$

The perimeter of the triangle is
$$a + b + c = 7 + 24 + 25 = 56 \text{ inches.}$$

Now Try Exercise 29

Three-Dimensional Objects

Rectangular Box

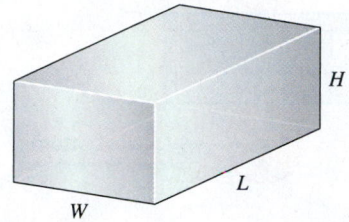

Figure R.7

Objects that occupy space have both volume and surface area. This subsection discusses rectangular boxes, spheres, cylinders, and cones.

Rectangular Boxes The volume V of a rectangular box with length L, width W, and height H equals $V = LWH$. See Figure R.7. The surface area S of the box equals the sum of the areas of the six sides: $S = 2LW + 2WH + 2LH$.

EXAMPLE 5 Finding the volume and surface area of a box

The box in Figure R.8 has dimensions x by $2x$ by y. Find its volume and surface area.

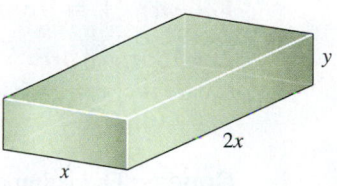

Figure R.8

SOLUTION

Volume: $LWH = 2x \cdot x \cdot y = 2x^2y$ cubic units

Surface Area: Base and top: $2x \cdot x + 2x \cdot x = 4x^2$
Front and back: $xy + xy = 2xy$
Left and right sides: $2xy + 2xy = 4xy$
Total surface area: $4x^2 + 2xy + 4xy = 4x^2 + 6xy$ square units

> Find the area of each side and add.

Now Try Exercise 41

Sphere

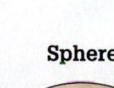

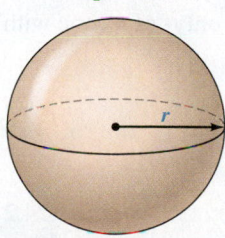

Figure R.9

Spheres The volume V of a sphere with radius r is $V = \frac{4}{3}\pi r^3$, and its surface area S is $S = 4\pi r^2$. See Figure R.9.

EXAMPLE 6 Finding the volume and surface area of a sphere

Estimate, to the nearest tenth, the volume and surface area of a sphere with a radius of 5.1 feet.

SOLUTION

Volume: $V = \frac{4}{3}\pi r^3 = \frac{4}{3}\pi(5.1)^3 \approx 555.6$ cubic feet

Surface Area: $S = 4\pi r^2 = 4\pi(5.1)^2 \approx 326.9$ square feet

> **Now Try Exercise 47**

Cylinders The volume of a cylinder with radius r and height h is $V = \pi r^2 h$. See Figure R.10. To find the total surface area of a cylinder, we add the area of the top and bottom to the area of the side. Figure R.11 illustrates a cylinder cut open to determine its surface area. The top and bottom are circular with areas of πr^2 each, and the side has a surface area of $2\pi r h$. The total surface area is $S = 2\pi r^2 + 2\pi r h$. The side surface area is called the **lateral surface area**.

Cylinder   **A Cylinder Cut Open**

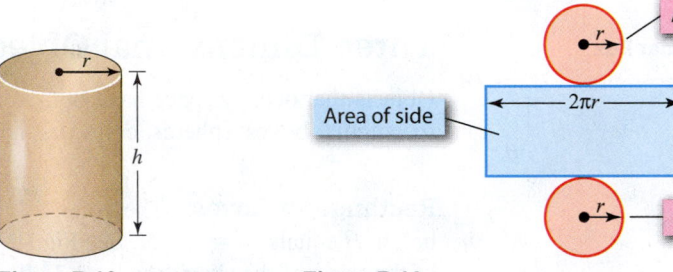

Figure R.10   **Figure R.11**

EXAMPLE 7 Finding the volume and surface area of a cylinder

A cylinder has radius $r = 3$ inches and height $h = 2.5$ feet. Find its volume and total surface area to the nearest tenth.

SOLUTION

Begin by changing 2.5 feet to 30 inches so that all units are in inches.

Volume: $V = \pi r^2 h = \pi(3)^2(30) = 270\pi \approx 848.2$ cubic inches

Total Surface Area: $S = 2\pi r^2 + 2\pi r h = 2\pi(3)^2 + 2\pi(3)(30) = 198\pi \approx 622.0$ square inches

> **Now Try Exercise 51**

Cone

Figure R.12

Cones The volume of a cone with radius r and height h is $V = \frac{1}{3}\pi r^2 h$, as shown in Figure R.12. (Compare this formula with the formula for the volume of a cylinder.) Excluding the bottom of the cone, the side (or lateral) surface area is $S = \pi r \sqrt{r^2 + h^2}$. The bottom of the cone is circular and has a surface area of πr^2.

EXAMPLE 8 Finding the volume and surface area of a cone

Approximate, to the nearest tenth, the volume and surface area (side only) of a cone with a radius of 1.45 inches and a height of 5.12 inches.

SOLUTION

Volume: $V = \frac{1}{3}\pi r^2 h = \frac{1}{3}\pi(1.45)^2(5.12) \approx 11.3$ cubic inches

Surface Area (side only): $S = \pi r \sqrt{r^2 + h^2} = \pi(1.45)\sqrt{(1.45)^2 + (5.12)^2} \approx 24.2$ square inches

> **Now Try Exercise 55**

Similar Triangles

The corresponding angles of **similar triangles** have equal measure, but similar triangles are not necessarily the same size. Two similar triangles are shown in Figure R.13. Notice that both triangles have angles of 30°, 60°, and 90°. Corresponding sides are not equal in length; however, corresponding ratios are equal. For example, in triangle ABC the ratio of the shortest leg to the hypotenuse equals $\frac{2}{4} = \frac{1}{2}$, and in triangle DEF this ratio is $\frac{3}{6} = \frac{1}{2}$.

Similar Triangles

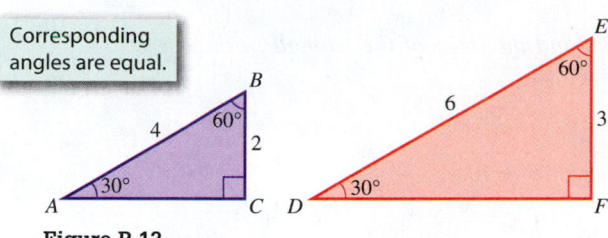

Figure R.13

EXAMPLE 9 **Using similar triangles**

Find the length of BC in Figure R.14.

SOLUTION Notice that triangle ABC and triangle ADE are both right triangles. These triangles share an angle at vertex A. Therefore triangles ABC and ADE have two corresponding angles that are congruent. Because the sum of the angles in a triangle equals 180°, all three corresponding angles in these two triangles are congruent. Thus triangles ABC and ADE are similar.

Since corresponding ratios are equal, we can find BC as follows.

$$\frac{BC}{AC} = \frac{DE}{AE}$$

$$\frac{BC}{5} = \frac{6}{7}$$

Corresponding ratios are equal.

Figure R.14

Solving this equation for BC gives $BC = \frac{30}{7} \approx 4.3$.

Now Try Exercise 61

NOTE For a summary of these formulas, see the back endpapers of this text.

R.1 Exercises

Rectangles

Exercises 1–6: Find the area and perimeter of the rectangle with length L and width W.

1. $L = 15$ feet, $W = 7$ feet

2. $L = 16$ inches, $W = 10$ inches

3. $L = 100$ meters, $W = 35$ meters

4. $L = 80$ yards, $W = 13$ yards

5. $L = 3x$, $W = y$ **6.** $L = a + 5$, $W = a$

Exercises 7–10: Find the area and perimeter of the rectangle in terms of the width W.

7. The width W is half the length.

8. Triple the width W minus 3 equals the length.

9. The length equals the width W plus 5.

10. The length is 2 less than twice the width W.

Triangles

Exercises 11 and 12: Find the area of the triangle shown in the figure.

11.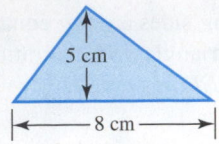

12.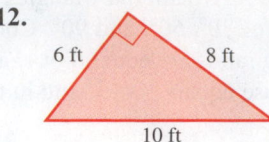

Exercises 13–20: Find the area of the triangle with base b and height h.

13. $b = 5$ inches, $h = 8$ inches

14. $b = 24$ inches, $h = 9$ feet

15. $b = 10.1$ meters, $h = 730$ meters

16. $b = 52$ yards, $h = 102$ feet

17. $b = 2x, h = 6x$ **18.** $b = x, h = x + 4$

19. $b = z, h = 5z$ **20.** $b = y + 1, h = 2y$

Circles

Exercises 21–26: Find the circumference and area of the circle. Approximate each value to the nearest tenth when appropriate.

21. $r = 4$ meters   **22.** $r = 1.5$ feet

23. $r = 19$ inches   **24.** $r = 22$ miles

25. $r = 2x$ **26.** $r = 5z$

Pythagorean Theorem

Exercises 27–32: Use the Pythagorean theorem to find the missing side of the right triangle with legs a and b and hypotenuse c. Then calculate the perimeter. Approximate values to the nearest tenth when appropriate.

27. $a = 60$ feet, $b = 11$ feet

28. $a = 21$ feet, $b = 11$ yards

29. $a = 5$ centimeters, $c = 13$ centimeters

30. $a = 6$ meters, $c = 15$ meters

31. $b = 7$ millimeters, $c = 10$ millimeters

32. $b = 1.2$ miles, $c = 2$ miles

Exercises 33–36: Find the area of the right triangle that satisfies the conditions. Approximate values to the nearest tenth when appropriate.

33. Legs with lengths 3 feet and 6 feet

34. Hypotenuse 10 inches and leg 6 inches

35. Hypotenuse 15 inches and leg 11 inches

36. Shorter leg 40 centimeters and hypotenuse twice the shorter leg

Rectangular Boxes

Exercises 37–44: Find the volume and surface area of a rectangular box with length L, width W, and height H.

37. $L = 4$ feet, $W = 3$ feet, $H = 2$ feet

38. $L = 6$ meters, $W = 4$ meters, $H = 1.5$ meters

39. $L = 4.5$ inches, $W = 4$ inches, $H = 1$ foot

40. $L = 9.1$ yards, $W = 8$ yards, $H = 6$ feet

41. $L = 3x, W = 2x, H = x$

42. $L = 6z, W = 5z, H = 7z$

43. $L = x, W = 2y, H = 3z$

44. $L = 8x, W = y, H = z$

Exercises 45 and 46: Find the volume of the rectangular box in terms of the width W.

45. The length is twice the width W, and the height is half the width.

46. The width W is three times the height and one-third of the length.

Spheres

Exercises 47–50: Find the volume and surface area of the sphere satisfying the given condition, where r is the radius and d is the diameter. Approximate values to the nearest tenth.

47. $r = 3$ feet   **48.** $r = 4.1$ inches

49. $d = 6.4$ meters   **50.** $d = 16$ feet

Cylinders

Exercises 51–54: Find the volume, the surface area of the side, and the total surface area of the cylinder that satisfies the given conditions, where r is the radius and h is the height. Approximate values to the nearest tenth.

51. $r = 0.5$ foot, $h = 2$ feet

52. $r = \frac{1}{3}$ yard, $h = 2$ feet

53. $r = 12$ millimeters, and h is twice r

54. r is one-fourth of h, and $h = 2.1$ feet

Cones

Exercises 55–60: Approximate, to the nearest tenth, the volume and surface area (side only) of the cone satisfying the given conditions, where r is the radius and h is the height.

55. $r = 5$ centimeters, $h = 6$ centimeters

56. $r = 8$ inches, $h = 30$ inches

57. $r = 24$ inches, $h = 3$ feet

58. $r = 100$ centimeters, $h = 1.3$ meters

59. Three times r equals h, and $r = 2.4$ feet

60. Twice h equals r, and $h = 3$ centimeters

Similar Triangles

Exercises 61–64: Use the fact that triangles ABC and DEF are similar to find the value of x.

61.

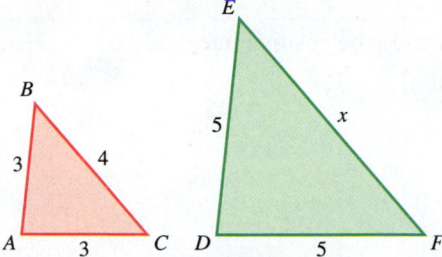

62.

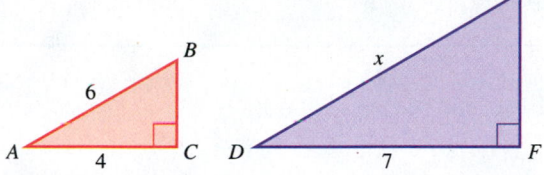

63.

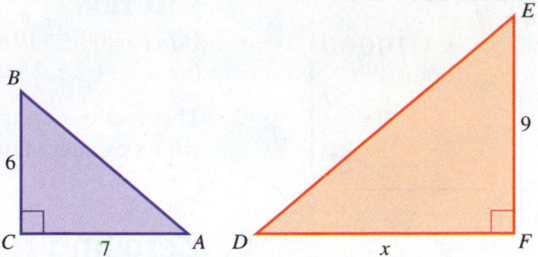

64.

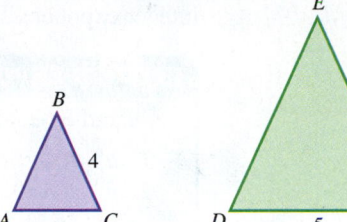

- **Use bases and exponents**
- **Use zero and negative exponents**
- **Apply the product, quotient, and power rules**

Bases and Positive Exponents

The expression 8^2 is an exponential expression with base **8** and exponent **2**. Table R.1 contains examples of other exponential expressions.

Exponential Expressions

Equal Expressions	Base	Exponent
$2 \cdot 2 \cdot 2 = 2^3$	2	3
$6 \cdot 6 \cdot 6 \cdot 6 = 6^4$	6	4
$7 = 7^1$	7	1
$0.5 \cdot 0.5 = 0.5^2$	0.5	2
$x \cdot x \cdot x = x^3$	x	3

Table R.1

We read 0.5^2 as "0.5 squared," 2^3 as "2 cubed," and 6^4 as "6 to the fourth power."

EXAMPLE 1 **Writing numbers in exponential notation**

Use the given base to write each number as an exponential expression. Check your results with a calculator.

(a) 10,000 (base 10)   **(b)** 27 (base 3)   **(c)** 32 (base 2)

```
10^4
            10000
3^3
               27
2^5
               32
```

Figure R.15

SOLUTION

(a) $10{,}000 = 10 \cdot 10 \cdot 10 \cdot 10 = 10^4$ **(b)** $27 = 3 \cdot 3 \cdot 3 = 3^3$
(c) $32 = 2 \cdot 2 \cdot 2 \cdot 2 \cdot 2 = 2^5$

These values are supported in Figure R.15, where exponential expressions are evaluated with a calculator, using the ⌃ key.

> **Now Try Exercises 11 and 13**

Zero and Negative Exponents

Exponents can be defined for any integer. The following box lists some properties for integer exponents.

INTEGER EXPONENTS

Let a and b be nonzero real numbers and m and n be positive integers. Then

1. $a^n = a \cdot a \cdot a \cdot \cdots \cdot a$ (n factors of a)
2. $a^0 = 1$ (*Note:* 0^0 is undefined.)
3. $a^{-n} = \dfrac{1}{a^n}$ and $\dfrac{1}{a^{-n}} = a^n$
4. $\dfrac{a^{-n}}{b^{-m}} = \dfrac{b^m}{a^n}$
5. $\left(\dfrac{a}{b}\right)^{-n} = \left(\dfrac{b}{a}\right)^n$

EXAMPLE 2   **Evaluating expressions**

Evaluate each expression.

(a) 3^{-4} **(b)** $\dfrac{1}{2^{-3}}$ **(c)** $\left(\dfrac{5}{7}\right)^{-2}$ **(d)** $\dfrac{1}{(xy)^{-1}}$ **(e)** $\dfrac{2^{-2}}{3t^{-3}}$

SOLUTION

(a) $3^{-4} = \dfrac{1}{3^4} = \dfrac{1}{3 \cdot 3 \cdot 3 \cdot 3} = \dfrac{1}{81}$ **(b)** $\dfrac{1}{2^{-3}} = 2^3 = 2 \cdot 2 \cdot 2 = 8$

(c) $\left(\dfrac{5}{7}\right)^{-2} = \left(\dfrac{7}{5}\right)^2 = \dfrac{7}{5} \cdot \dfrac{7}{5} = \dfrac{49}{25}$ **(d)** $\dfrac{1}{(xy)^{-1}} = (xy)^1 = xy$

> Base is xy.

(e) Note that only t, not $3t$, is raised to the power of -3.

$$\dfrac{2^{-2}}{3t^{-3}} = \dfrac{t^3}{3(2^2)} = \dfrac{t^3}{3 \cdot 4} = \dfrac{t^3}{12}$$

> Base is t.

> **Now Try Exercises 27, 29, 31, 65, and 67**

Product, Quotient, and Power Rules

We can calculate products and quotients of exponential expressions *provided their bases are the same.* For example,

> 5 factors of 3

$$3^2 \cdot 3^3 = \overbrace{(3 \cdot 3) \cdot (3 \cdot 3 \cdot 3)} = 3^5.$$

This expression has a total of $2 + 3 = 5$ factors of 3, so the result is 3^5. To multiply exponential expressions with *like bases*, add exponents.

THE PRODUCT RULE

For any nonzero number a and integers m and n,

$$a^m \cdot a^n = a^{m+n}.$$

The product rule holds for negative exponents. For example,

$$10^5 \cdot 10^{-2} = 10^{5+(-2)} = 10^3.$$

> Product rule: add exponents.

EXAMPLE 3   Using the product rule

Multiply and simplify.

(a) $7^3 \cdot 7^{-4}$ (b) $x^3 x^{-2} x^4$ (c) $(3y^2)(2y^{-4})$

SOLUTION

(a) $7^3 \cdot 7^{-4} = 7^{3+(-4)} = 7^{-1} = \dfrac{1}{7}$ (b) $x^3 x^{-2} x^4 = x^{3+(-2)+4} = x^5$

(c) $(3y^2)(2y^{-4}) = 3 \cdot 2 \cdot y^2 \cdot y^{-4} = 6y^{2+(-4)} = 6y^{-2} = \dfrac{6}{y^2}$

Note that 6 is not raised to the power of -2 in the expression $6y^{-2}$.

> **Now Try Exercises 35, 37, and 39**

Consider division of exponential expressions using the following example.

$$\frac{6^5}{6^3} = \frac{6 \cdot 6 \cdot \cancel{6} \cdot \cancel{6} \cdot \cancel{6}}{\cancel{6} \cdot \cancel{6} \cdot \cancel{6}} = 6 \cdot 6 = 6^2$$

After simplifying, there are two 6s left in the numerator. The result is $6^{5-3} = 6^2 = 36$. To divide exponential expressions with *like bases*, subtract exponents.

THE QUOTIENT RULE

For any nonzero number a and integers m and n,

$$\frac{a^m}{a^n} = a^{m-n}.$$

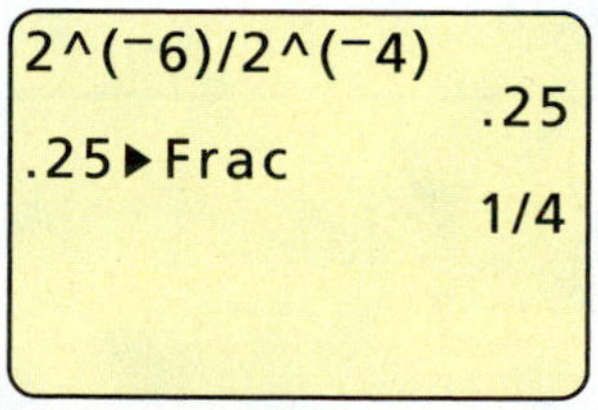

The quotient rule holds true for negative exponents. For example,

$$\frac{2^{-6}}{2^{-4}} = 2^{-6-(-4)} = 2^{-2} = \frac{1}{2^2} = \frac{1}{4}.$$

> Quotient rule: subtract exponents.

This result is supported by Figure R.16.

```
2^(-6)/2^(-4)
                    .25
.25▶Frac
                    1/4
```

Figure R.16

EXAMPLE 4   Using the quotient rule

Simplify the expression. Use positive exponents.

(a) $\dfrac{10^4}{10^6}$ (b) $\dfrac{x^5}{x^2}$ (c) $\dfrac{15x^2y^3}{5x^4y}$

SOLUTION

(a) $\dfrac{10^4}{10^6} = 10^{4-6} = 10^{-2} = \dfrac{1}{10^2} = \dfrac{1}{100}$ (b) $\dfrac{x^5}{x^2} = x^{5-2} = x^3$

(c) $\dfrac{15x^2y^3}{5x^4y} = \dfrac{15}{5} \cdot \dfrac{x^2}{x^4} \cdot \dfrac{y^3}{y^1} = 3 \cdot x^{(2-4)}y^{(3-1)} = 3x^{-2}y^2 = \dfrac{3y^2}{x^2}$

> **Now Try Exercises 45, 49, and 51**

How should we evaluate $(4^3)^2$? To answer this question, consider

$$(4^3)^2 = \underbrace{4^3 \cdot 4^3}_{\text{2 factors of } 4^3} = 4^{3+3} = 4^6.$$

Similarly,

$$(x^4)^3 = \underbrace{x^4 \cdot x^4 \cdot x^4}_{\text{3 factors of } x^4} = x^{4+4+4} = x^{12}.$$

These results suggest that to raise a power to a power, we must multiply the exponents.

> **RAISING POWERS TO POWERS**
>
> For any nonzero real number a and integers m and n,
>
> $$(a^m)^n = a^{mn}.$$

EXAMPLE 5 Raising powers to powers

Simplify each expression. Use positive exponents.
(a) $(5^2)^3$ **(b)** $(2^4)^{-2}$ **(c)** $(b^{-7})^5$

SOLUTION

(a) $(5^2)^3 = 5^{2 \cdot 3} = 5^6 = 15{,}625$ **(b)** $(2^4)^{-2} = 2^{4(-2)} = 2^{-8} = \dfrac{1}{2^8} = \dfrac{1}{256}$

(c) $(b^{-7})^5 = b^{-7 \cdot 5} = b^{-35} = \dfrac{1}{b^{35}}$

Multiply exponents.

Now Try Exercises 55 and 57

How can we simplify the expression $(2x)^3$? Consider the following.

$$(2x)^3 = 2x \cdot 2x \cdot 2x = \underbrace{(2 \cdot 2 \cdot 2)}_{\text{3 factors}} \cdot \underbrace{(x \cdot x \cdot x)}_{\text{3 factors}} = 2^3 x^3$$

This result suggests that to cube a product, we can cube each factor.

> **RAISING PRODUCTS TO POWERS**
>
> For any nonzero real numbers a and b and integer n,
>
> $$(ab)^n = a^n b^n.$$

EXAMPLE 6 Raising products to powers

Simplify each expression. Use positive exponents.
(a) $(6y)^2$ **(b)** $(x^2 y)^{-2}$ **(c)** $(2xy^3)^4$

SOLUTION

(a) $(6y)^2 = 6^2 y^2 = 36y^2$ **(b)** $(x^2 y)^{-2} = (x^2)^{-2} y^{-2} = x^{-4} y^{-2} = \dfrac{1}{x^4 y^2}$

(c) $(2xy^3)^4 = 2^4 x^4 (y^3)^4 = 16x^4 y^{12}$

Now Try Exercises 59 and 73

To simplify a power of a quotient, use the following rule.

> **RAISING QUOTIENTS TO POWERS**
>
> For nonzero numbers a and b and any integer n,
>
> $$\left(\frac{a}{b}\right)^n = \frac{a^n}{b^n}.$$

EXAMPLE 7 Raising quotients to powers

Simplify each expression. Use positive exponents.

(a) $\left(\dfrac{3}{x}\right)^3$ **(b)** $\left(\dfrac{1}{2^3}\right)^{-2}$ **(c)** $\left(\dfrac{3x^{-3}}{y^2}\right)^4$

SOLUTION

(a) $\left(\dfrac{3}{x}\right)^3 = \dfrac{3^3}{x^3} = \dfrac{27}{x^3}$ (b) $\left(\dfrac{1}{2^3}\right)^{-2} = \dfrac{1^{-2}}{(2^3)^{-2}} = \dfrac{1}{2^{-6}} = 2^6 = 64$

(c) $\left(\dfrac{3x^{-3}}{y^2}\right)^4 = \dfrac{3^4(x^{-3})^4}{(y^2)^4} = \dfrac{81x^{-12}}{y^8} = \dfrac{81}{x^{12}y^8}$

> **Now Try Exercises 61, 63, and 81**

In the next example, we use several properties of exponents to simplify expressions.

EXAMPLE 8 **Simplifying expressions**

Write each expression using positive exponents. Simplify the result completely.

(a) $\left(\dfrac{x^2 y^{-3}}{3z^{-4}}\right)^{-2}$ (b) $\dfrac{(rt^3)^{-3}}{(r^2 t^3)^{-2}}$

SOLUTION

(a) $\left(\dfrac{x^2 y^{-3}}{3z^{-4}}\right)^{-2} = \left(\dfrac{3z^{-4}}{x^2 y^{-3}}\right)^2$ (b) $\dfrac{(rt^3)^{-3}}{(r^2 t^3)^{-2}} = \dfrac{(r^2 t^3)^2}{(rt^3)^3}$

$= \left(\dfrac{3y^3}{x^2 z^4}\right)^2$ $= \dfrac{r^4 t^6}{r^3 t^9}$

$= \dfrac{9y^6}{x^4 z^8}$ $= \dfrac{r}{t^3}$

> **Now Try Exercises 77 and 79**

R.2 Exercises

Concepts

1. Are the expressions 2^3 and 3^2 equal? Explain your answer.

2. Are the expressions -4^2 and $(-4)^2$ equal? Explain your answer.

3. $7^{-n} = $ _____

4. $6^m \cdot 6^n = $ _____

5. $\dfrac{5^m}{5^n} = $ _____

6. $(3x)^k = $ _____

7. $(2^m)^k = $ _____

8. $\left(\dfrac{x}{y}\right)^m = $ _____

9. $5 \times 10^3 = $ _____

10. $5 \times 10^{-3} = $ _____

Properties of Exponents

Exercises 11–16: (Refer to Example 1.) Write the number as an exponential expression, using the base shown. Check your result with a calculator.

11. 8 (base 2)

12. 1000 (base 10)

13. 256 (base 4)

14. $\frac{1}{64}$ (base 4)

15. 1 (base 3)

16. $\frac{1}{49}$ (base 7)

Exercises 17–34: Evaluate the expression by hand. Check your result with a calculator.

17. 5^3

18. 5^{-3}

19. -2^4

20. $(-2)^4$

21. 5^0

22. $\left(-\frac{2}{3}\right)^{-3}$

23. $\left(\frac{2}{3}\right)^3$

24. $\frac{1}{4^{-2}}$

25. $\left(-\frac{1}{2}\right)^4$

26. $\left(-\frac{3}{4}\right)^3$

27. 4^{-3}

28. 10^{-4}

29. $\frac{1}{2^{-4}}$

30. $\frac{1}{3^{-2}}$

31. $\left(\frac{3}{4}\right)^{-3}$

32. $\left(\frac{1}{2}\right)^0$

33. $\frac{3^{-2}}{2^{-3}}$

34. $\frac{10^{-4}}{4^{-3}}$

Exercises 35–44: Use the product rule to simplify.

35. $6^3 \cdot 6^{-4}$

36. $10^2 \cdot 10^5 \cdot 10^{-3}$

37. $2x^2 \cdot 3x^{-3} \cdot x^4$

38. $3y^4 \cdot 6y^{-4} \cdot y$

39. $10^0 \cdot 10^6 \cdot 10^2$

40. $y^3 \cdot y^{-5} \cdot y^4$

41. $5^{-2} \cdot 5^3 \cdot 2^{-4} \cdot 2^3$ **42.** $2^{-3} \cdot 3^4 \cdot 3^{-2} \cdot 2^5$

43. $(2a^3)(b^2)(a^{-4})(4b^{-5})$ **44.** $(3x^{-4})(2x^2)(5y^4)(y^{-3})$

Exercises 45–54: Use the quotient rule to simplify the expression. Use positive exponents to write your answer.

45. $\dfrac{5^4}{5^2}$ **46.** $\dfrac{6^2}{6^{-7}}$

47. $\dfrac{a^{-3}}{a^2 \cdot a}$ **48.** $\dfrac{y^0 \cdot y \cdot y^5}{y^{-2} \cdot y^{-3}}$

49. $\dfrac{24x^3}{6x}$ **50.** $\dfrac{10x^5}{5x^{-3}}$

51. $\dfrac{12a^2b^3}{18a^4b^2}$ **52.** $\dfrac{-6x^7y^3}{3x^2y^{-5}}$

53. $\dfrac{21x^{-3}y^4}{7x^4y^{-2}}$ **54.** $\dfrac{32x^3y}{-24x^5y^{-3}}$

Exercises 55–64: Use the power rules to simplify the expression. Use positive exponents to write your answer.

55. $(5^{-1})^3$ **56.** $(-4^2)^3$

57. $(y^4)^{-2}$ **58.** $(x^2)^4$

59. $(4y^2)^3$ **60.** $(-2xy^3)^{-4}$

61. $\left(\dfrac{4}{x}\right)^3$ **62.** $\left(\dfrac{-3}{x^3}\right)^2$

63. $\left(\dfrac{2x}{z^4}\right)^{-5}$ **64.** $\left(\dfrac{2xy}{3z^5}\right)^{-1}$

Exercises 65–90: Use rules of exponents to simplify the expression. Use positive exponents to write your answer.

65. $\dfrac{2}{(ab)^{-1}}$ **66.** $\dfrac{5a^2}{(xy)^{-1}}$

67. $\dfrac{2^{-3}}{2t^{-2}}$ **68.** $\dfrac{t^{-3}}{2t^{-1}}$

69. $\dfrac{6a^2b^{-3}}{4ab^{-2}}$ **70.** $\dfrac{20a^{-2}b}{4a^{-2}b^{-1}}$

71. $\dfrac{5r^2st^{-3}}{25rs^{-2}t^2}$ **72.** $\dfrac{36r^{-1}(st)^2}{9(rs)^2t^{-1}}$

73. $(3x^2y^{-3})^{-2}$ **74.** $(-2x^{-3}y^{-2})^3$

75. $\dfrac{(d^3)^{-2}}{(d^{-2})^3}$ **76.** $\dfrac{(b^2)^{-1}}{(b^{-4})^3}$

77. $\left(\dfrac{3t^2}{2t^{-1}}\right)^3$ **78.** $\left(\dfrac{-2t}{4t^{-2}}\right)^{-1}$

79. $\dfrac{(-m^2n^{-1})^{-2}}{(mn)^{-1}}$ **80.** $\dfrac{(-mn^4)^{-1}}{(m^2n)^{-3}}$

81. $\left(\dfrac{2a^3}{6b}\right)^4$ **82.** $\left(\dfrac{-3a^2}{9b^3}\right)^4$

83. $\dfrac{8x^{-3}y^{-2}}{4x^{-2}y^{-4}}$ **84.** $\dfrac{6x^{-1}y^{-1}}{9x^{-2}y^3}$

85. $\dfrac{(r^2t^2)^{-2}}{(r^3t)^{-1}}$ **86.** $\dfrac{(2rt)^2}{(rt^4)^{-2}}$

87. $\dfrac{4x^{-2}y^3}{(2x^{-1}y)^2}$ **88.** $\dfrac{(ab)^3}{a^4b^{-4}}$

89. $\left(\dfrac{15r^2t}{3r^{-3}t^4}\right)^3$ **90.** $\left(\dfrac{4(xy)^2}{(2xy^{-2})^3}\right)^{-2}$

R.3 Polynomial Expressions

- Perform addition and subtraction on monomials
- Perform addition and subtraction on polynomials
- Apply the distributive property
- Perform multiplication on polynomials
- Find the product of a sum and difference
- Square a binomial

Addition and Subtraction of Monomials

A **term** is a number, a variable, or a *product* of numbers and variables raised to powers. Examples of terms include

$$-15, \quad y, \quad x^4, \quad 3x^3z, \quad x^{-1/2}y^{-2}, \quad \text{and} \quad 6x^{-1}y^3.$$
$$\text{Terms}$$

If the variables in a term have only *nonnegative integer* exponents, the term is called a *monomial*. Examples of monomials include

$$-4, \quad 5y, \quad x^2, \quad 5x^2z^6, \quad -xy^7, \quad \text{and} \quad 6xy^3.$$
$$\text{Monomials}$$

If two terms contain the same variables raised to the same powers, we call them **like terms**. We can add or subtract *like* terms, but not *unlike* terms. For example, the terms

$$x^3 + y^3 \quad \boxed{\text{Unlike terms cannot be combined.}}$$

cannot be combined because they are *unlike terms.* However,

$$x^3 + 3x^3 = (1 + 3)x^3 = 4x^3 \quad \boxed{\text{Like terms can be combined.}}$$

because x^3 and $3x^3$ are *like terms.* To add or subtract monomials, we simply combine like terms, as illustrated in the next example.

EXAMPLE 1   **Adding and subtracting monomials**

Simplify each of the following expressions by combining like terms.
(a) $8x^2 - 4x^2 + x^3$ **(b)** $9x - 6xy^2 + 2xy^2 + 4x$

SOLUTION
(a) The terms $8x^2$ and $-4x^2$ are like terms, so they may be combined.

$$8x^2 - 4x^2 + x^3 = (8 - 4)x^2 + x^3 \quad \text{Combine like terms.}$$
$$= 4x^2 + x^3 \quad \text{Subtract.}$$

$\boxed{\text{Unlike terms cannot be combined.}}$

(b) The terms $9x$ and $4x$ may be combined, as may $-6xy^2$ and $2xy^2$.

$$9x - 6xy^2 + 2xy^2 + 4x = 9x + 4x - 6xy^2 + 2xy^2 \quad \text{Commutative property}$$
$$= (9 + 4)x + (-6 + 2)xy^2 \quad \text{Combine like terms.}$$
$$= 13x - 4xy^2 \quad \text{Add.}$$

$\boxed{\text{Now Try Exercises 5 and 11}}$

Addition and Subtraction of Polynomials

A polynomial is either a monomial or a sum (or difference) of monomials. Examples of polynomials include

$$5x^4z^2, \qquad 9x^4 - 5, \qquad 4x^2 + 5xy - y^2, \qquad \text{and} \qquad 4 - y^2 + 5y^4 + y^5.$$

$\quad$ 1 term $\qquad\quad$ 2 terms $\qquad\quad$ 3 terms $\qquad\qquad\qquad\qquad$ 4 terms

Polynomials containing one variable are called polynomials of one variable. The second and fourth polynomials shown above are examples of polynomials of one variable. The leading coefficient of a polynomial of one variable is the coefficient of the monomial with highest degree. The degree of a polynomial with one variable equals the exponent of the monomial with the highest power. Table R.2 shows several polynomials of one variable along with their degrees and leading coefficients. A polynomial of degree 1 is a *linear* polynomial, a polynomial of degree 2 is a *quadratic* polynomial, and a polynomial of degree 3 is a *cubic* polynomial.

Characterizing Polynomials

Polynomial	Degree	Leading Coefficient	Type
-98	0	-98	Constant
$2x - 7$	1	2	Linear
$-5z + 9z^2 + 7$	2	9	Quadratic
$-2x^3 + 4x^2 + x - 1$	3	-2	Cubic
$7 - x + 4x^2 + x^5$	5	1	Fifth Degree

Table R.2

To add two polynomials, we combine like terms.

EXAMPLE 2 Adding polynomials

Simplify.
(a) $(2x^2 - 3x + 7) + (3x^2 + 4x - 2)$ (b) $(z^3 + 4z + 8) + (4z^2 - z + 6)$

SOLUTION
(a) $(2x^2 - 3x + 7) + (3x^2 + 4x - 2) = 2x^2 + 3x^2 - 3x + 4x + 7 - 2$
$$= (2 + 3)x^2 + (-3 + 4)x + (7 - 2)$$
$$= 5x^2 + x + 5$$

(b) $(z^3 + 4z + 8) + (4z^2 - z + 6) = z^3 + 4z^2 + 4z - z + 8 + 6$
$$= z^3 + 4z^2 + (4 - 1)z + (8 + 6)$$
$$= z^3 + 4z^2 + 3z + 14$$

Now Try Exercises 21 and 27

To subtract integers, we add the first integer and the **opposite, or additive inverse,** of the second integer. For example, to evaluate $3 - 5$, we perform the following operations.

Subtracting integers ——— $3 - 5 = 3 + (-5)$ Add the opposite.
$$= -2$$ Simplify.

Similarly, to subtract two polynomials, we add the first polynomial and the opposite, or additive inverse, of the second polynomial. To find the opposite of a polynomial, we simply negate each term. Table R.3 shows three polynomials and their opposites.

Finding the Additive Inverse

Polynomial	Opposite
$9 - x$	$-9 + x$
$5x^2 + 4x - 1$	$-5x^2 - 4x + 1$
$-x^4 + 5x^3 - x^2 + 5x - 1$	$x^4 - 5x^3 + x^2 - 5x + 1$

Table R.3

EXAMPLE 3 Subtracting polynomials

Simplify.
(a) $(y^5 + 3y^3) - (-y^4 + 2y^3)$ (b) $(5x^3 + 9x^2 - 6) - (5x^3 - 4x^2 - 7)$

SOLUTION
(a) The opposite of $(-y^4 + 2y^3)$ is $(y^4 - 2y^3)$.

Add the opposite.

$$(y^5 + 3y^3) - (-y^4 + 2y^3) = (y^5 + 3y^3) + (y^4 - 2y^3)$$
$$= y^5 + y^4 + (3 - 2)y^3$$
$$= y^5 + y^4 + y^3$$

(b) The opposite of $(5x^3 - 4x^2 - 7)$ is $(-5x^3 + 4x^2 + 7)$.

$$(5x^3 + 9x^2 - 6) - (5x^3 - 4x^2 - 7) = (5x^3 + 9x^2 - 6) + (-5x^3 + 4x^2 + 7)$$
$$= (5 - 5)x^3 + (9 + 4)x^2 + (-6 + 7)$$
$$= 0x^3 + 13x^2 + 1$$
$$= 13x^2 + 1$$

Now Try Exercises 39 and 41

Distributive Properties

Distributive properties are used frequently in the multiplication of polynomials. For all real numbers a, b, and c,

$$a(b + c) = ab + ac \qquad \text{and}$$

$$a(b - c) = ab - ac.$$

In the next example, we use these distributive properties to multiply expressions.

| EXAMPLE 4 | Using distributive properties |

$4(5 + x) = 20 + 4x$

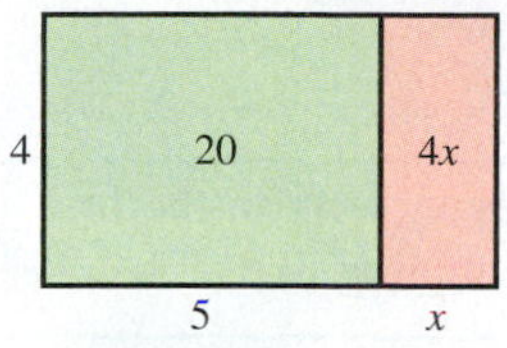

4	20	4x
	5	x

Visualize the solution by using areas.

Figure R.17

Multiply.
(a) $4(5 + x)$ **(b)** $-3(x - 4y)$ **(c)** $(2x - 5)(6)$

SOLUTION

(a) $4(5 + x) = 4 \cdot 5 + 4 \cdot x = 20 + 4x$

(b) $-3(x - 4y) = -3 \cdot x - (-3) \cdot (4y) = -3x + 12y$

(c) $(2x - 5)(6) = 2x \cdot 6 - 5 \cdot 6 = 12x - 30$ **Now Try Exercises 47, 51, and 53**

You can visualize the product in part (a) of Example 4 by using areas of rectangles. If a rectangle has width 4 and length $5 + x$, its area is $20 + 4x$, as shown in Figure R.17.

Multiplying Polynomials

A polynomial with two terms is a **binomial**, and a polynomial with three terms is a **trinomial**. Examples are shown in Table R.4.

Types of Polynomials

One term ⎯	**Monomials**	$2x^2$	$-3x^4$	9
Two terms ⎯	**Binomials**	$3x - 1$	$2x^3 - x$	$x^2 + 5$
Three terms ⎯	**Trinomials**	$x^2 - 3x + 5$	$5x^2 - 2x + 10$	$2x^3 - x^2 - 2$

Table R.4

In the next example, we multiply two binomials.

| EXAMPLE 5 | Multiplying binomials |

Multiply $(x + 1)(x + 3)$.

SOLUTION   To multiply $(x + 1)(x + 3)$, we apply the distributive property.

$$(x + 1)(x + 3) = (x + 1)(x) + (x + 1)(3)$$
$$= x \cdot x + 1 \cdot x + x \cdot 3 + 1 \cdot 3$$
$$= x^2 + x + 3x + 3$$
$$= x^2 + 4x + 3$$

Now Try Exercise 55

To multiply the bionomials $(x + 1)$ and $(x + 3)$, we multiplied every term in $x + 1$ by every term in $x + 3$. That is,

$$(x + 1)(x + 3) = x^2 + 3x + x + 3$$
$$= x^2 + 4x + 3.$$

NOTE This process of multiplying binomials is called *FOIL*. You may use the name to remind yourself to multiply the first terms (F), outside terms (O), inside terms (I), and last terms (L).

Multiply the *First terms* to obtain x^2. $(x + 1)(x + 3)$ F

Multiply the *Outside terms* to obtain $3x$. $(x + 1)(x + 3)$ O

Multiply the *Inside terms* to obtain x. $(x + 1)(x + 3)$ I

Multiply the *Last terms* to obtain 3. $(x + 1)(x + 3)$ L

The following box summarizes how to multiply two polynomials in general.

> **MULTIPLYING POLYNOMIALS**
>
> The product of two polynomials may be found by multiplying every term in the first polynomial by every term in the second polynomial and then combining like terms.

EXAMPLE 6 **Multiplying binomials**

Multiply each binomial.
(a) $(2x - 1)(x + 2)$ **(b)** $(1 - 3x)(2 - 4x)$ **(c)** $(x^2 + 1)(5x - 3)$

SOLUTION

(a) $\begin{aligned}(2x - 1)(x + 2) &= 2x \cdot x + 2x \cdot 2 - 1 \cdot x - 1 \cdot 2 \\ &= 2x^2 + 4x - x - 2 \\ &= 2x^2 + 3x - 2\end{aligned}$

(b) $\begin{aligned}(1 - 3x)(2 - 4x) &= 1 \cdot 2 - 1 \cdot 4x - 3x \cdot 2 + 3x \cdot 4x \\ &= 2 - 4x - 6x + 12x^2 \\ &= 2 - 10x + 12x^2\end{aligned}$

(c) $\begin{aligned}(x^2 + 1)(5x - 3) &= x^2 \cdot 5x - x^2 \cdot 3 + 1 \cdot 5x - 1 \cdot 3 \\ &= 5x^3 - 3x^2 + 5x - 3\end{aligned}$

Now Try Exercises 57, 59, and 77

EXAMPLE 7 **Multiplying polynomials**

Multiply each expression.
(a) $3x(x^2 + 5x - 4)$ **(b)** $-x^2(x^4 - 2x + 5)$ **(c)** $(x + 2)(x^2 + 4x - 3)$

SOLUTION

(a) $\begin{aligned}3x(x^2 + 5x - 4) &= 3x \cdot x^2 + 3x \cdot 5x - 3x \cdot 4 \\ &= 3x^3 + 15x^2 - 12x\end{aligned}$

(b) $\begin{aligned}-x^2(x^4 - 2x + 5) &= -x^2 \cdot x^4 + x^2 \cdot 2x - x^2 \cdot 5 \\ &= -x^6 + 2x^3 - 5x^2\end{aligned}$

(c) $\begin{aligned}(x + 2)(x^2 + 4x - 3) &= x \cdot x^2 + x \cdot 4x - x \cdot 3 + 2 \cdot x^2 + 2 \cdot 4x - 2 \cdot 3 \\ &= x^3 + 4x^2 - 3x + 2x^2 + 8x - 6 \\ &= x^3 + 6x^2 + 5x - 6\end{aligned}$

Now Try Exercises 67, 69, and 73

Some Special Products

Sum and Difference   The following special product often occurs in mathematics.

Product of a sum of two numbers and their difference

$$(a + b)(a - b) = a \cdot a - a \cdot b + b \cdot a - b \cdot b$$
$$= a^2 - ab + ba - b^2$$
$$= a^2 - b^2$$

That is, the product of a sum of two numbers and their difference equals the difference of their squares: $(a + b)(a - b) = a^2 - b^2$.

EXAMPLE 8   **Finding the product of a sum and difference**

Multiply.
(a) $(x - 3)(x + 3)$ **(b)** $(5 + 4x^2)(5 - 4x^2)$

SOLUTION
(a) If we let $a = x$ and $b = 3$, we can apply $(a - b)(a + b) = a^2 - b^2$. Thus
$$(x - 3)(x + 3) = (x)^2 - (3)^2$$
$$= x^2 - 9.$$

Product is the difference of the squares of x and 3.

(b) Similarly,
$$(5 + 4x^2)(5 - 4x^2) = (5)^2 - (4x^2)^2$$
$$= 25 - 16x^4.$$

Now Try Exercises 79 and 95

Squaring a Binomial   Two other special products involve *squaring a binomial:*
$$(a + b)^2 = (a + b)(a + b)$$
$$= a^2 + ab + ba + b^2$$
$$= a^2 + 2ab + b^2$$

and

$$(a - b)^2 = (a - b)(a - b)$$
$$= a^2 - ab - ba + b^2$$
$$= a^2 - 2ab + b^2.$$

Note that to obtain the middle term, we multiply the two terms in the binomial and double the result.

EXAMPLE 9   **Squaring a binomial**

Multiply.
(a) $(x + 5)^2$ **(b)** $(3 - 2x)^2$

SOLUTION
(a) If we let $a = x$ and $b = 5$, we can apply $(a + b)^2 = a^2 + 2ab + b^2$. Thus
$$(x + 5)^2 = (x)^2 + 2(x)(5) + (5)^2$$
$$= x^2 + 10x + 25.$$

(b) Applying the formula $(a - b)^2 = a^2 - 2ab + b^2$, we find
$$(3 - 2x)^2 = (3)^2 - 2(3)(2x) + (2x)^2$$
$$= 9 - 12x + 4x^2.$$

Now Try Exercises 85 and 91

R.3 Exercises

Monomials and Polynomials

Exercises 1–12: Combine like terms whenever possible.

1. $3x^3 + 5x^3$

2. $-9z + 6z$

3. $5y^7 - 8y^7$

4. $9x - 7x$

5. $5x^2 + 8x + x^2$

6. $5x + 2x + 10x$

7. $9x^2 - x + 4x - 6x^2$

8. $-y^2 - \frac{1}{2}y^2$

9. $x^2 + 9x - 2 + 4x^2 + 4x$

10. $6y + 4y^2 - 6y + y^2$

11. $7y + 9x^2y - 5y + x^2y$

12. $5ab - b^2 + 7ab + 6b^2$

Exercises 13–18: Identify the degree and leading coefficient of the polynomial.

13. $5x^2 - 4x + \frac{3}{4}$

14. $-9y^4 + y^2 + 5$

15. $5 - x + 3x^2 - \frac{2}{5}x^3$

16. $7x + 4x^4 - \frac{4}{3}x^3$

17. $8x^4 + 3x^3 - 4x + x^5$

18. $5x^2 - x^3 + 7x^4 + 10$

Exercises 19–28: Add the polynomials.

19. $(5x + 6) + (-2x + 6)$

20. $(5y^2 + y^3) + (12y^2 - 5y^3)$

21. $(2x^2 - x + 7) + (-2x^2 + 4x - 9)$

22. $(x^3 - 5x^2 + 6) + (5x^2 + 3x + 1)$

23. $(4x) + (1 - 4.5x)$

24. $(y^5 + y) + \left(5 - y + \frac{1}{3}y^2\right)$

25. $(x^4 - 3x^2 - 4) + \left(-8x^4 + x^2 - \frac{1}{2}\right)$

26. $(3z + z^4 + 2) + (-3z^4 - 5 + z^2)$

27. $(2z^3 + 5z - 6) + (z^2 - 3z + 2)$

28. $(z^4 - 6z^2 + 3) + (5z^3 + 3z^2 - 3)$

Exercises 29–34: Find the opposite of the polynomial.

29. $7x^3$

30. $-3z^8$

31. $19z^5 - 5z^2 + 3z$

32. $-x^2 - x + 6$

33. $z^4 - z^2 - 9$

34. $1 - 8x + 6x^2 - \frac{1}{6}x^3$

Exercises 35–42: Subtract the polynomials.

35. $(5x - 3) - (2x + 4)$

36. $(10x + 5) - (-6x - 4)$

37. $(x^2 - 3x + 1) - (-5x^2 + 2x - 4)$

38. $(-x^2 + x - 5) - (x^2 - x + 5)$

39. $(4x^4 + 2x^2 - 9) - (x^4 - 2x^2 - 5)$

40. $(8x^3 + 5x^2 - 3x + 1) - (-5x^3 + 6x - 11)$

41. $(x^4 - 1) - (4x^4 + 3x + 7)$

42. $(5x^4 - 6x^3 + x^2 + 5) - (x^3 + 11x^2 + 9x - 3)$

Exercises 43–54: Apply the distributive property.

43. $5x(x - 5)$

44. $3x^2(-2x + 2)$

45. $-5(3x + 1)$

46. $-(-3x + 1)$

47. $5(y + 2)$

48. $4(x - 7)$

49. $-2(5x + 9)$

50. $-3x(5 + x)$

51. $(y - 3)6y$

52. $(2x - 5)8x^3$

53. $-4(5x - y)$

54. $-6(3y - 2x)$

Exercises 55–66: Multiply the binomials.

55. $(y + 5)(y - 7)$

56. $(3x + 1)(2x + 1)$

57. $(3 - 2x)(3 + x)$

58. $(7x - 3)(4 - 7x)$

59. $(-2x + 3)(x - 2)$

60. $(z - 2)(4z + 3)$

61. $\left(x - \frac{1}{2}\right)\left(x + \frac{1}{4}\right)$

62. $\left(z - \frac{1}{3}\right)\left(z - \frac{1}{6}\right)$

63. $(x^2 + 1)(2x^2 - 1)$

64. $(x^2 - 2)(x^2 + 4)$

65. $(x + y)(x - 2y)$

66. $(x^2 + y^2)(x - y)$

Exercises 67–78: Multiply the polynomials.

67. $3x(2x^2 - x - 1)$

68. $-2x(3 - 2x + 5x^2)$

69. $-x(2x^4 - x^2 + 10)$

70. $-2x^2(5x^3 + x^2 - 2)$

71. $(2x^2 - 4x + 1)(3x^2)$

72. $(x - y + 5)(xy)$

73. $(x + 1)(x^2 + 2x - 3)$

74. $(2x - 1)(3x^2 - x + 6)$

75. $(2 - 3x)(5 - 2x)(x^2 - 1)$

76. $(3 + z)(6 - 4z)(4 + 2z^2)$

77. $(x^2 + 2)(3x - 2)$

78. $(4 + x)(2x^2 - 3)$

Exercises 79–96: Multiply the expressions.

79. $(x - 7)(x + 7)$

80. $(x + 9)(x - 9)$

81. $(3x + 4)(3x - 4)$

82. $(9x - 4)(9x + 4)$

83. $(2x - 3y)(2x + 3y)$

84. $(x + 2y)(x - 2y)$

85. $(x + 4)^2$

86. $(z + 9)^2$

87. $(2x + 1)^2$

88. $(3x + 5)^2$

89. $(x - 1)^2$

90. $(x - 7)^2$

91. $(2 - 3x)^2$

92. $(5 - 6x)^2$

93. $3x(x + 1)(x - 1)$

94. $-4x(3x - 5)^2$

95. $(2 - 5x^2)(2 + 5x^2)$

96. $(6y - x^2)(6y + x^2)$

R.4 Factoring Polynomials

- **Use common factors**
- **Factor by grouping**
- **Factor $x^2 + bx + c$**
- **Factor trinomials by grouping**
- **Factor trinomials with FOIL**
- **Factor the difference of two squares**
- **Factor perfect square trinomials**
- **Factor the sum and difference of two cubes**

Common Factors

When factoring a polynomial, we first look for factors that are common to each term in an expression. By applying a distributive property, we can write a polynomial as two factors. For example, each term in $2x^2 + 4x$ contains a factor of $2x$.

$$2x^2 = \mathbf{2x} \cdot \mathbf{x}$$
$$4x = \mathbf{2x} \cdot \mathbf{2}$$

> Factor 2x out of each term.

Thus the polynomial $2x^2 + 4x$ can be factored as follows.

$$2x^2 + 4x = \mathbf{2x}(x + 2)$$

> Apply a distributive property to factor.

EXAMPLE 1 Finding common factors

Factor.
(a) $6z^3 - 2z^2 + 4z$ **(b)** $4x^3y^2 + x^2y^3$

SOLUTION
(a) Each of the terms $6z^3$, $2z^2$, and $4z$ contains a common factor of $2z$. That is,

$$6z^3 = \mathbf{2z} \cdot \mathbf{3z^2}, \quad 2z^2 = \mathbf{2z} \cdot \mathbf{z}, \quad \text{and} \quad 4z = \mathbf{2z} \cdot \mathbf{2}.$$

Thus $6z^3 - 2z^2 + 4z = \mathbf{2z}(3z^2 - z + 2)$.
(b) Both $4x^3y^2$ and x^2y^3 contain a common factor of x^2y^2. That is,

$$4x^3y^2 = \mathbf{x^2y^2} \cdot \mathbf{4x} \quad \text{and} \quad x^2y^3 = \mathbf{x^2y^2} \cdot \mathbf{y}.$$

Thus $4x^3y^2 + x^2y^3 = \mathbf{x^2y^2}(4x + y)$.

Now Try Exercises 5 and 15

Many times we factor out the *greatest common factor*. For example, the polynomial $15x^4 - 5x^2$ has a common factor of $5x$. We could write this polynomial as

$$15x^4 - 5x^2 = 5x(3x^3 - x).$$

However, we can also factor out $5x^2$ to obtain

$$15x^4 - 5x^2 = \mathbf{5x^2}(3x^2 - 1).$$

Because $\mathbf{5x^2}$ is the common factor with the highest degree and largest coefficient, we say that $5x^2$ is the **greatest common factor** (GCF) of $15x^4 - 5x^2$.

EXAMPLE 2 Factoring greatest common factors

Factor.
(a) $6m^3n^2 - 3mn^2 + 9m$ **(b)** $-9x^4 + 6x^3 - 3x^2$

SOLUTION
(a) The GCF of $6m^3n^2$, $3mn^2$, and $9m$ is $3m$.

$$6m^3n^2 = \mathbf{3m} \cdot \mathbf{2m^2n^2}, \quad 3mn^2 = \mathbf{3m} \cdot \mathbf{n^2}, \quad \text{and} \quad 9m = \mathbf{3m} \cdot \mathbf{3}$$

Thus $6m^3n^2 - 3mn^2 + 9m = \mathbf{3m}(\mathbf{2m^2n^2} - \mathbf{n^2} + \mathbf{3})$.

(b) Rather than factoring out $3x^2$, we can factor out $-3x^2$ and make the leading coefficient of the remaining expression positive.

$$-9x^4 = \mathbf{-3x^2} \cdot \mathbf{3x^2}, \quad 6x^3 = \mathbf{-3x^2} \cdot \mathbf{-2x}, \quad \text{and} \quad -3x^2 = \mathbf{-3x^2} \cdot \mathbf{1}$$

Thus $-9x^4 + 6x^3 - 3x^2 = \mathbf{-3x^2}(\mathbf{3x^2} - \mathbf{2x} + \mathbf{1})$.

> **Now Try Exercises 11 and 17**

Factoring by Grouping

Factoring by grouping is a technique that makes use of the associative and distributive properties. The next example illustrates the first step in this factoring technique.

Consider the cubic polynomial

$$3t^3 + 6t^2 + 2t + 4.$$

We can factor this polynomial by first grouping it into two binomials.

$$(3t^3 + 6t^2) + (2t + 4) \qquad \text{Associative property}$$
$$\mathbf{3t^2(t + 2)} + \mathbf{2(t + 2)} \qquad \text{Factor out common factors.}$$
$$\mathbf{(3t^2 + 2)(t + 2)}. \qquad \text{Factor out } (t + 2).$$

The following steps summarize factoring four terms by grouping.

FACTORING BY GROUPING

STEP 1: Use parentheses to group the terms into binomials with common factors. Begin by writing the expression with a plus sign between the binomials.

STEP 2: Factor out the common factor in each binomial.

STEP 3: Factor out the common binomial. If there is no common binomial, try a different grouping or a different method of factoring.

EXAMPLE 3 Factoring by grouping

Factor each polynomial.
(a) $12x^3 - 9x^2 - 8x + 6$ **(b)** $2x - 2y + ax - ay$

SOLUTION

(a) $12x^3 - 9x^2 - 8x + 6 = (12x^3 - 9x^2) + (-8x + 6)$ *Write with a plus sign between binomials.*

$\qquad\qquad\qquad\qquad = \mathbf{3x^2(4x - 3)} - \mathbf{2(4x - 3)}$ *Factor out $3x^2$ and -2.*

$\qquad\qquad\qquad\qquad = \mathbf{(3x^2 - 2)(4x - 3)}$ *Factor out $4x - 3$.*

(b) $2x - 2y + ax - ay = (2x - 2y) + (ax - ay)$ *Group terms.*

$\qquad\qquad\qquad\qquad = \mathbf{2(x - y)} + \mathbf{a(x - y)}$ *Factor out 2 and a.*

$\qquad\qquad\qquad\qquad = \mathbf{(2 + a)(x - y)}$ *Factor out $x - y$.*

> **Now Try Exercises 21 and 31**

Factoring $x^2 + bx + c$

The product $(x + 3)(x + 4)$ can be found as follows.

Multiply using FOIL.

$$(x + 3)(x + 4) = x^2 + 4x + 3x + 12$$
$$= x^2 + 7x + 12$$

The middle term $7x$ is found by calculating the sum $4x + 3x$, and the last term is found by calculating the product $3 \cdot 4 = 12$.

When we factor polynomials, we are *reversing* the process of multiplication. To factor $x^2 + 7x + 12$, we must find m and n that satisfy

$$x^2 + 7x + 12 = (x + m)(x + n).$$

Because

$$(x + m)(x + n) = x^2 + (m + n)x + mn,$$

it follows that $mn = 12$ and $m + n = 7$. To determine m and n, we list factors of 12 and their sum, as shown in Table R.5.

Because $3 \cdot 4 = 12$ and $3 + 4 = 7$, we can write the factored form as

$$x^2 + 7x + 12 = (x + 3)(x + 4).$$

This result can always be checked by multiplying the two binomials.

$$(x + 3)\ (x + 4) = x^2 + 7x + 12$$

$$3x$$
$$+4x$$
$$\overline{7x} \longleftarrow \text{The middle term checks.}$$

Factor Pairs for 12

Factors	1, 12	2, 6	3, 4
Sum	13	8	7

Table R.5

FACTORING $x^2 + bx + c$

To factor the trinomial $x^2 + bx + c$, find integers m and n that satisfy

$$m \cdot n = c \qquad \text{and} \qquad m + n = b.$$

Then $x^2 + bx + c = (x + m)(x + n)$.

EXAMPLE 4   Factoring the form $x^2 + bx + c$

Factor each trinomial.
(a) $x^2 + 10x + 16$ (b) $x^2 + 7x - 30$

SOLUTION
(a) We need to find a factor pair for 16 whose sum is **10**. From Table R.6 the required factor pair is $m = 2$ and $n = 8$. Thus

$$x^2 + 10x + 16 = (x + 2)(x + 8).$$

(b) Factors of -30 whose sum equals **7** are -3 and **10**. Thus

$$x^2 + 7x - 30 = (x - 3)(x + 10).$$

Now Try Exercises 33 and 37

Factor Pairs for 16

Factors	1, 16	2, 8	4, 4
Sum	17	10	8

Table R.6

EXAMPLE 5   Removing common factors first

Factor completely.
(a) $3x^2 + 15x + 18$ (b) $5x^3 + 5x^2 - 60x$

SOLUTION
(a) If we first factor out the common factor of 3, the resulting trinomial is easier to factor.

$$3x^2 + 15x + 18 = 3(x^2 + 5x + 6)$$

Now we find m and n such that $mn = 6$ and $m + n = 5$. These numbers are 2 and 3.

$$3x^2 + 15x + 18 = 3(x^2 + 5x + 6)$$
$$= 3(x + 2)(x + 3)$$

(b) First, we factor out the common factor of $5x$. Then we factor the resulting trinomial.

$$5x^3 + 5x^2 - 60x = 5x(x^2 + x - 12)$$
$$= 5x(x - 3)(x + 4)$$

> **Now Try Exercises 51 and 53**

Factoring Trinomials by Grouping

In this subsection we use grouping to factor trinomials in the form $ax^2 + bx + c$ with $a \neq 1$. For example, one way to factor $3x^2 + 14x + 8$ is to find two numbers m and n such that $mn = 3 \cdot 8 = 24$ and $m + n = 14$. Because $2 \cdot 12 = 24$ and $2 + 12 = 14$, we let $m = 2$ and $n = 12$. Using grouping, we can now factor this trinomial.

$$\begin{aligned}
3x^2 + 14x + 8 &= 3x^2 + 2x + 12x + 8 && \text{Write } 14x \text{ as } 2x + 12x. \\
&= (3x^2 + 2x) + (12x + 8) && \text{Associative property} \\
&= x(3x + 2) + 4(3x + 2) && \text{Factor out } x \text{ and } 4. \\
&= (x + 4)(3x + 2) && \text{Distributive property}
\end{aligned}$$

Writing the polynomial as $3x^2 + 12x + 2x + 8$ would also work.

FACTORING $ax^2 + bx + c$ BY GROUPING

To factor $ax^2 + bx + c$, perform the following steps. (Assume that a, b, and c have no factor in common.)

1. Find numbers m and n such that $mn = ac$ and $m + n = b$. This step may require trial and error.
2. Write the trinomial as $ax^2 + mx + nx + c$.
3. Use grouping to factor this expression as two binomials.

EXAMPLE 6 **Factoring $ax^2 + bx + c$ by grouping**

Factor each trinomial.
(a) $12y^2 + 5y - 3$ **(b)** $6r^2 - 19r + 10$

SOLUTION
(a) In this trinomial $a = 12$, $b = 5$, and $c = -3$. Because $mn = ac$ and $m + n = b$, the numbers m and n satisfy $mn = -36$ and $m + n = 5$. Thus $m = 9$ and $n = -4$.

$$\begin{aligned}
12y^2 + 5y - 3 &= 12y^2 + 9y - 4y - 3 && \text{Write } 5y \text{ as } 9y - 4y. \\
&= (12y^2 + 9y) + (-4y - 3) && \text{Associative property} \\
&= 3y(4y + 3) - 1(4y + 3) && \text{Factor out } 3y \text{ and } -1. \\
&= (3y - 1)(4y + 3) && \text{Distributive property}
\end{aligned}$$

(b) In this trinomial $a = 6$, $b = -19$, and $c = 10$. Because $mn = ac$ and $m + n = b$, the numbers m and n satisfy $mn = 60$ and $m + n = -19$. Thus $m = -4$ and $n = -15$.

$$\begin{aligned}
6r^2 - 19r + 10 &= 6r^2 - 4r - 15r + 10 && \text{Write } -19r \text{ as } -4r - 15r. \\
&= (6r^2 - 4r) + (-15r + 10) && \text{Associative property} \\
&= 2r(3r - 2) - 5(3r - 2) && \text{Factor out } 2r \text{ and } -5. \\
&= (2r - 5)(3r - 2) && \text{Distributive property}
\end{aligned}$$

> **Now Try Exercises 41 and 43**

Factoring Trinomials with FOIL

An alternative to factoring trinomials by grouping is to use FOIL in *reverse*. For example, the factors of $3x^2 + 7x + 2$ are two binomials.

$$3x^2 + 7x + 2 \stackrel{?}{=} (\underline{} + \underline{})(\underline{} + \underline{})$$

The expressions to be placed in the four blanks are yet to be found. By the FOIL method, we know that the product of the first terms is $3x^2$. Because $3x^2 = 3x \cdot x$, we can write

$$3x^2 + 7x + 2 \stackrel{?}{=} (\underline{3x} + \underline{})(\underline{x} + \underline{}).$$

The product of the last terms in each binomial must equal 2. Because $2 = 1 \cdot 2$, we can put the 1 and 2 in the blanks, but we must be sure to place them correctly so that the product of the *outside terms* plus the product of the *inside terms* equals $7x$.

Determine where to place the 1 and 2 and then check the middle term.

$$(3x + 1)\,(x + 2) = 3x^2 + 7x + 2$$

1x
+6x
7x ⟵ Middle term checks. ⟶ Correct

If we had interchanged the 1 and 2, we would have obtained an incorrect result.

$$(3x + 2)\,(x + 1) = 3x^2 + \mathbf{5x} + 2$$

2x
+3x
5x ⟵ Middle term is *not* 7x. ⟶ Incorrect

In the next example, we factor expressions of the form $ax^2 + bx + c$, where $a \neq 1$. In this situation, we may need to *guess and check* or use *trial and error* a few times to find the correct factors.

EXAMPLE 7 **Factoring the form $ax^2 + bx + c$**

Factor each trinomial.
(a) $6x^2 - x - 2$ (b) $4x^3 - 14x^2 + 6x$

SOLUTION
(a) The factors of $6x^2$ are either $2x$ and $3x$ or $6x$ and x. The factors of -2 are either -1 and 2 or 1 and -2. To obtain a middle term of $-x$, we use the following factors.

$$(3x - 2)\,(2x + 1) = 6x^2 - x - 2$$

−4x
+ 3x
−x ⟵ It checks.

To find the correct factorization, we may need to guess and check a few times.
(b) Each term contains a common factor of $2x$, so we do the following step first.

$$4x^3 - 14x^2 + 6x = 2x(2x^2 - 7x + 3)$$

Next we factor $2x^2 - 7x + 3$. The factors of $2x^2$ are $2x$ and x. Because the middle term is negative, we use -1 and -3 for factors of 3.

$$4x^3 - 14x^2 + 6x = 2x(2x^2 - 7x + 3)$$
$$= 2x(2x - 1)(x - 3)$$

Now Try Exercises 55 and 57

Difference of Two Squares

When we factor polynomials, we are *reversing* the process of multiplying polynomials. In Section R.3 we discussed the equation

$$(a - b)(a + b) = a^2 - b^2.$$

We can use this equation to factor a difference of two squares.

DIFFERENCE OF TWO SQUARES

For any real numbers a and b,

$$a^2 - b^2 = (a - b)(a + b).$$

NOTE The sum of two squares *cannot* be factored (using real numbers). For example, $x^2 + y^2$ cannot be factored, whereas $x^2 - y^2$ can be factored. It is important to remember that $x^2 + y^2 \neq (x + y)^2$.

EXAMPLE 8 Factoring the difference of two squares

Factor each polynomial, if possible.
(a) $9x^2 - 64$ **(b)** $4x^2 + 9y^2$ **(c)** $4a^3 - 4a$

SOLUTION
(a) Note that $9x^2 = (3x)^2$ and $64 = 8^2$.

$$9x^2 - 64 = (3x)^2 - (8)^2 \qquad \text{Factor difference of two squares.}$$
$$= (3x - 8)(3x + 8)$$

(b) Because $4x^2 + 9y^2$ is the *sum* of two squares, it *cannot* be factored.
(c) Start by factoring out the common factor of $4a$.

$$4a^3 - 4a = 4a(a^2 - 1)$$
$$= 4a(a - 1)(a + 1)$$

Now Try Exercises 61, 65, and 69

EXAMPLE 9 Applying the difference of two squares

Factor each expression.
(a) $x^4 - y^4$ **(b)** $6r^2 - 24t^4$

SOLUTION
(a) Use $a^2 - b^2 = (a - b)(a + b)$, with $a = x^2$ and $b = y^2$.

$$x^4 - y^4 = (x^2)^2 - (y^2)^2 \qquad \text{Write as difference of squares.}$$
$$= (x^2 - y^2)(x^2 + y^2) \qquad \text{Difference of squares}$$
$$= (x - y)(x + y)(x^2 + y^2) \qquad \text{Difference of squares}$$

(b) Start by factoring out the common factor of 6.

$$6r^2 - 24t^4 = 6(r^2 - 4t^4) \qquad \text{Factor out 6.}$$
$$= 6\left(r^2 - (2t^2)^2\right) \qquad \text{Write as difference of squares.}$$
$$= 6(r - 2t^2)(r + 2t^2) \qquad \text{Difference of squares}$$

Now Try Exercises 66 and 67

Perfect Square Trinomials

In Section R.3 we expanded $(a + b)^2$ and $(a - b)^2$ as follows.

$$(a + b)^2 = a^2 + 2ab + b^2 \quad \text{and} \quad (a - b)^2 = a^2 - 2ab + b^2$$

The expressions $a^2 + 2ab + b^2$ and $a^2 - 2ab + b^2$ are called **perfect square trinomials**. We can use the following formulas to factor them.

> **PERFECT SQUARE TRINOMIALS**
>
> For any real numbers a and b,
> $$a^2 + 2ab + b^2 = (a + b)^2 \quad \text{and}$$
> $$a^2 - 2ab + b^2 = (a - b)^2.$$

EXAMPLE 10 Factoring perfect square trinomials

Factor each expression.
(a) $x^2 + 6x + 9$ **(b)** $81x^2 - 72x + 16$

SOLUTION
(a) Let $a^2 = x^2$ and $b^2 = 3^2$. In a perfect square trinomial, the middle term is $2ab$.
$$2ab = 2(x)(3) = 6x,$$
which equals the given middle term. Thus $a^2 + 2ab + b^2 = (a + b)^2$ implies
$$x^2 + 6x + 9 = (x + 3)^2.$$

Perfect square trinomial

(b) Let $a^2 = (9x)^2$ and $b^2 = 4^2$. In a perfect square trinomial, the middle term is $2ab$.
$$2ab = 2(9x)(4) = 72x,$$
which equals the given middle term. Thus $a^2 - 2ab + b^2 = (a - b)^2$ implies
$$81x^2 - 72x + 16 = (9x - 4)^2.$$

Now Try Exercises 77 and 81

EXAMPLE 11 Factoring perfect square trinomials

Factor $25a^3 + 10a^2b + ab^2$.

SOLUTION Start by factoring out the common factor of a. Then factor the resulting perfect square trinomial.
$$25a^3 + 10a^2b + ab^2 = a(25a^2 + 10ab + b^2)$$

Perfect square trinomial

$$= a(5a + b)^2$$

Now Try Exercise 89

Sum and Difference of Two Cubes

The sum or difference of two cubes may be factored. This fact is justified by the following two equations.

$$(a + b)(a^2 - ab + b^2) = a^3 + b^3 \quad \text{and}$$
$$(a - b)(a^2 + ab + b^2) = a^3 - b^3$$

These equations can be verified by multiplying the left side to obtain the right side. For example,

$$(a + b)(a^2 - ab + b^2) = a \cdot a^2 - a \cdot ab + a \cdot b^2 + b \cdot a^2 - b \cdot ab + b \cdot b^2$$
$$= a^3 - a^2b + ab^2 + a^2b - ab^2 + b^3$$
$$= a^3 + b^3.$$

> **SUM AND DIFFERENCE OF TWO CUBES**
>
> For any real numbers a and b,
> $$a^3 + b^3 = (a + b)(a^2 - ab + b^2) \quad \text{and}$$
> $$a^3 - b^3 = (a - b)(a^2 + ab + b^2).$$

EXAMPLE 12 Factoring the sum and difference of two cubes

Factor each polynomial.
(a) $x^3 + 8$ **(b)** $27x^3 - 64y^3$ **(c)** $27p^9 - 8q^6$

SOLUTION
(a) Because $x^3 = (x)^3$ and $8 = 2^3$, we let $a = x, b = 2$ and factor. Substituting in
$$a^3 + b^3 = (a + b)(a^2 - ab + b^2)$$
gives $\quad x^3 + 2^3 = (x + 2)(x^2 - x \cdot 2 + 2^2)$
$$= (x + 2)(x^2 - 2x + 4).$$

Note that the quadratic expression does not factor further.
(b) Here, $27x^3 = (3x)^3$ and $64y^3 = (4y)^3$, so
$$27x^3 - 64y^3 = (3x)^3 - (4y)^3.$$

Substituting $a = 3x$ and $b = 4y$ in
$$a^3 - b^3 = (a - b)(a^2 + ab + b^2)$$
gives $\quad (3x)^3 - (4y)^3 = (3x - 4y)((3x)^2 + 3x \cdot 4y + (4y)^2)$
$$= (3x - 4y)(9x^2 + 12xy + 16y^2).$$

(c) Let $a^3 = (3p^3)^3$ and $b^3 = (2q^2)^3$. Then $a^3 - b^3 = (a - b)(a^2 + ab + b^2)$ implies
$$27p^9 - 8q^6 = (3p^3 - 2q^2)(9p^6 + 6p^3q^2 + 4q^4).$$

> **Now Try Exercises 93, 95, and 99**

R.4 Exercises

Greatest Common Factor
Exercises 1–18: Factor out the greatest common factor.

1. $10x - 15$
2. $32 - 16x$

3. $2x^3 - 5x$
4. $3y - 9y^2$

5. $8x^3 - 4x^2 + 16x$
6. $-5x^3 + x^2 - 4x$

7. $5x^4 - 15x^3 + 15x^2$
8. $28y + 14y^3 - 7y^5$

9. $15x^3 + 10x^2 - 30x$
10. $14a^4 - 21a^2 + 35a$

11. $6r^5 - 8r^4 + 12r^3$
12. $15r^6 + 20r^4 - 10r^3$

13. $8x^2y^2 - 24x^2y^3$
14. $36xy - 24x^3y^3$

15. $18mn^2 - 12m^2n^3$
16. $24m^2n^3 + 12m^3n^2$

17. $-4a^2 - 2ab + 6ab^2$
18. $-5a^2 + 10a^2b^2 - 15ab$

Factoring by Grouping
Exercises 19–32: Use grouping to factor the polynomial.

19. $x^3 + 3x^2 + 2x + 6$ **20.** $4x^3 + 3x^2 + 8x + 6$

21. $6x^3 - 4x^2 + 9x - 6$ **22.** $x^3 - 3x^2 - 5x + 15$

23. $z^3 - 5z^2 + z - 5$ **24.** $y^3 - 7y^2 + 8y - 56$

25. $y^4 + 2y^3 - 5y^2 - 10y$ **26.** $4z^4 + 4z^3 + z^2 + z$

27. $2x^3 - 3x^2 + 2x - 3$ **28.** $8x^3 - 2x^2 + 12x - 3$

29. $2x^4 - x^3 + 4x - 2$ **30.** $2x^4 - 5x^3 + 10x - 25$

31. $ab - 3a + 2b - 6$ **32.** $2ax - 6bx - ay + 3by$

Factoring Trinomials
Exercises 33–58: Factor the expression completely.

33. $x^2 + 7x + 10$ **34.** $x^2 + 3x - 10$

35. $x^2 + 8x + 12$ **36.** $x^2 - 8x + 12$

37. $z^2 + z - 42$ **38.** $z^2 - 9z + 20$

39. $z^2 + 11z + 24$ **40.** $z^2 + 15z + 54$

41. $24x^2 + 14x - 3$ **42.** $25x^2 - 5x - 6$

43. $6x^2 - x - 2$ **44.** $10x^2 + 3x - 1$

45. $1 + x - 2x^2$ **46.** $3 - 5x - 2x^2$

47. $20 + 7x - 6x^2$ **48.** $4 + 13x - 12x^2$

49. $5x^3 + x^2 - 6x$ **50.** $2x^3 + 8x^2 - 24x$

51. $3x^3 + 12x^2 + 9x$ **52.** $12x^3 - 8x^2 - 20x$

53. $2x^2 - 14x + 20$ **54.** $7x^2 + 35x + 42$

55. $60t^4 + 230t^3 - 40t^2$ **56.** $24r^4 + 8r^3 - 80r^2$

57. $4m^3 + 10m^2 - 6m$ **58.** $30m^4 + 3m^3 - 9m^2$

Difference of Two Squares
Exercises 59–76: Factor the expression completely, if possible.

59. $x^2 - 25$ **60.** $z^2 - 169$

61. $4x^2 - 25$ **62.** $36 - y^2$

63. $36x^2 - 100$ **64.** $9x^2 - 4y^2$

65. $64z^2 - 25z^4$ **66.** $100x^3 - x$

67. $16x^4 - y^4$ **68.** $x^4 - 9y^2$

69. $a^2 + 4b^2$ **70.** $9r^4 + 25t^4$

71. $4 - r^2t^2$ **72.** $25 - x^4y^2$

73. $(x - 1)^2 - 16$ **74.** $(y + 2)^2 - 1$

75. $4 - (z + 3)^2$ **76.** $64 - (t - 3)^2$

Perfect Square Trinomials
Exercises 77–90: Factor the expression.

77. $x^2 + 2x + 1$ **78.** $x^2 - 6x + 9$

79. $4x^2 + 20x + 25$ **80.** $x^2 + 10x + 25$

81. $x^2 - 12x + 36$ **82.** $16z^4 - 24z^3 + 9z^2$

83. $9z^3 - 6z^2 + z$ **84.** $49y^2 + 42y + 9$

85. $9y^3 + 30y^2 + 25y$ **86.** $25y^3 - 20y^2 + 4y$

87. $4x^2 - 12xy + 9y^2$ **88.** $25a^2 + 60ab + 36b^2$

89. $9a^3b - 12a^2b + 4ab$ **90.** $16a^3 + 8a^2b + ab^2$

Sum and Difference of Two Cubes
Exercises 91–102: Factor the expression.

91. $x^3 - 1$ **92.** $x^3 + 1$

93. $y^3 + z^3$ **94.** $y^3 - z^3$

95. $8x^3 - 27$ **96.** $8 - z^3$

97. $x^4 + 125x$ **98.** $3x^4 - 81x$

99. $8r^6 - t^3$ **100.** $125r^6 + 64t^3$

101. $10m^9 - 270n^6$ **102.** $5t^6 + 40r^3$

General Factoring
Exercises 103–158: Factor the expression completely.

103. $16x^2 - 25$ **104.** $25x^2 - 30x + 9$

105. $x^3 - 64$ **106.** $1 + 8y^3$

107. $x^2 + 16x + 64$ **108.** $12x^2 + x - 6$

109. $5x^2 - 38x - 16$ **110.** $125x^3 - 1$

111. $x^4 + 8x$ **112.** $2x^3 - 12x^2 + 18x$

113. $64x^3 + 8y^3$ **114.** $54 - 16x^3$

115. $3x^2 - 5x - 8$ **116.** $15x^2 - 11x + 2$

117. $7a^3 + 20a^2 - 3a$ **118.** $b^3 - b^2 - 2b$

119. $2x^3 - x^2 + 6x - 3$ **120.** $3x^3 - 5x^2 + 3x - 5$

121. $2x^4 - 5x^3 - 25x^2$ **122.** $10x^3 + 28x^2 - 6x$

123. $2x^4 + 5x^2 + 3$ **124.** $2x^4 + 2x^2 - 4$

125. $x^3 + 3x^2 + x + 3$ **126.** $x^3 + 5x^2 + 4x + 20$

127. $5x^3 - 5x^2 + 10x - 10$

128. $5x^4 - 20x^3 + 10x - 40$

129. $ax + bx - ay - by$ **130.** $ax - bx - ay + by$

131. $18x^2 + 12x + 2$ **132.** $-3x^2 + 30x - 75$

133. $-4x^3 + 24x^2 - 36x$ **134.** $18x^3 - 60x^2 + 50x$

135. $27x^3 - 8$ **136.** $27x^3 + 8$

137. $-x^4 - 8x$ **138.** $x^5 - 27x^2$

139. $x^4 - 2x^3 - x + 2$ **140.** $x^4 + 3x^3 + x + 3$

141. $r^4 - 16$ **142.** $r^4 - 81$

143. $25x^2 - 4a^2$ **144.** $9y^2 - 16z^2$

145. $2x^4 - 2y^4$ **146.** $a^4 - b^4$

147. $9x^3 + 6x^2 - 3x$ **148.** $8x^3 + 28x^2 - 16x$

149. $(z - 2)^2 - 9$ **150.** $(y + 2)^2 - 4$

151. $3x^5 - 27x^3 + 3x^2 - 27$

152. $2x^5 - 8x^3 - 16x^2 + 64$

153. $(x + 2)^2(x + 4)^4 + (x + 2)^3(x + 4)^3$

154. $(x - 3)(2x + 1)^3 + (x - 3)^2(2x + 1)^2$

155. $(6x + 1)(8x - 3)^4 - (6x + 1)^2(8x - 3)^3$

156. $(2x + 3)^4(x + 1)^4 - (2x + 3)^3(x + 1)^5$

157. $4x^2(5x - 1)^5 + 2x(5x - 1)^6$

158. $x^4(7x + 3)^3 + x^5(7x + 3)^2$

R.5 Rational Expressions

- Simplify rational expressions
- Multiply and divide fractions
- Perform multiplication and division on rational expressions
- Find least common multiples and denominators
- Add and subtract fractions
- Perform addition and subtraction on rational expressions
- Clear fractions from equations
- Simplify complex fractions

Simplifying Rational Expressions

When simplifying fractions, we sometimes use the **basic principle of fractions**, which states that

$$\frac{a \cdot c}{b \cdot c} = \frac{a}{b}.$$

This principle holds because $\frac{c}{c} = 1$ and $\frac{a}{b} \cdot 1 = \frac{a}{b}$. It can be used to simplify a fraction.

$$\frac{6}{44} = \frac{3 \cdot 2}{22 \cdot 2} = \frac{3}{22}$$

> Factor out 2 in the numerator and denominator and simplify.

This same principle can also be used to simplify rational expressions.

> **SIMPLIFYING RATIONAL EXPRESSIONS**
>
> The following principle can be used to simplify rational expressions, where A, B, and C are polynomials.
>
> $$\frac{A \cdot C}{B \cdot C} = \frac{A}{B}, \quad B \text{ and } C \text{ are nonzero.}$$

EXAMPLE 1   Simplifying rational expressions

Simplify each expression.

(a) $\dfrac{9x}{3x^2}$ **(b)** $\dfrac{2z^2 - 3z - 9}{z^2 + 2z - 15}$ **(c)** $\dfrac{a^2 - b^2}{a + b}$

SOLUTION

(a) First factor out the greatest common factor, $3x$, in the numerator and denominator.

$$\frac{9x}{3x^2} = \frac{3 \cdot 3x}{x \cdot 3x} = \frac{3}{x} \cdot 1 = \frac{3}{x}$$

(b) Start by factoring the numerator and denominator.

$$\frac{2z^2 - 3z - 9}{z^2 + 2z - 15} = \frac{(2z + 3)(z - 3)}{(z + 5)(z - 3)} = \frac{2z + 3}{z + 5}$$

(c) Start by factoring the numerator as the difference of squares.

$$\frac{a^2 - b^2}{a + b} = \frac{(a - b)(a + b)}{a + b} = a - b$$

> Now Try Exercises 1, 5, and 11

Review of Multiplication and Division of Fractions

Recall that to multiply two fractions we use the property

$$\frac{a}{b} \cdot \frac{c}{d} = \frac{ac}{bd}.$$

For example, $\frac{2}{5} \cdot \frac{3}{7} = \frac{2 \cdot 3}{5 \cdot 7} = \frac{6}{35}$.

EXAMPLE 2 Multiplying fractions

Multiply and simplify the product.

(a) $\dfrac{4}{9} \cdot \dfrac{3}{8}$ **(b)** $\dfrac{2}{3} \cdot \dfrac{3}{4} \cdot \dfrac{5}{6}$

SOLUTION

(a) $\dfrac{4}{9} \cdot \dfrac{3}{8} = \dfrac{4 \cdot 3}{9 \cdot 8} = \dfrac{12}{72} = \dfrac{1 \cdot 12}{6 \cdot 12} = \dfrac{1}{6}$ **(b)** $\dfrac{2}{3} \cdot \dfrac{3}{4} \cdot \dfrac{5}{6} = \dfrac{6 \cdot 5}{12 \cdot 6} = \dfrac{5}{12}$

> Now Try Exercises 15 and 17

Recall that to divide two fractions we "invert and multiply." That is, we change a division problem to a multiplication problem. For example,

$$\frac{3}{4} \div \frac{5}{4} = \frac{3}{4} \cdot \frac{4}{5} = \frac{3 \cdot 4}{5 \cdot 4} = \frac{3}{5}.$$

Multiplication and Division of Rational Expressions

Multiplying and dividing rational expressions is similar to multiplying and dividing fractions.

PRODUCTS AND QUOTIENTS OF RATIONAL EXPRESSIONS

To multiply two rational expressions, multiply numerators and multiply denominators.

$$\frac{A}{B} \cdot \frac{C}{D} = \frac{AC}{BD}, \qquad B \text{ and } D \text{ are nonzero.}$$

To divide two rational expressions, multiply by the reciprocal of the divisor.

$$\frac{A}{B} \div \frac{C}{D} = \frac{A}{B} \cdot \frac{D}{C}, \qquad B, C, \text{ and } D \text{ are nonzero.}$$

EXAMPLE 3 Multiplying rational expressions

Multiply.

(a) $\dfrac{1}{x} \cdot \dfrac{2x}{x + 1}$ **(b)** $\dfrac{x - 1}{x} \cdot \dfrac{x - 1}{x + 2}$

SOLUTION

(a) $\dfrac{1}{x} \cdot \dfrac{2x}{x + 1} = \dfrac{1 \cdot 2x}{x(x + 1)} = \dfrac{2}{x + 1}$ **(b)** $\dfrac{x - 1}{x} \cdot \dfrac{x - 1}{x + 2} = \dfrac{(x - 1)(x - 1)}{x(x + 2)}$

> Now Try Exercises 33 and 37

EXAMPLE 4 Dividing two rational expressions

Divide and simplify.

(a) $\dfrac{2}{x} \div \dfrac{2x - 1}{4x}$ (b) $\dfrac{x^2 - 1}{x^2 + x - 6} \div \dfrac{x - 1}{x + 3}$

SOLUTION

(a)
$$\frac{2}{x} \div \frac{2x - 1}{4x} = \frac{2}{x} \cdot \frac{4x}{2x - 1}$$ "Invert and multiply."

$$= \frac{8x}{x(2x - 1)}$$ Multiply.

$$= \frac{8}{2x - 1}$$ Simplify.

(b)
$$\frac{x^2 - 1}{x^2 + x - 6} \div \frac{x - 1}{x + 3} = \frac{x^2 - 1}{x^2 + x - 6} \cdot \frac{x + 3}{x - 1}$$ "Invert and multiply."

$$= \frac{(x + 1)(x - 1)}{(x - 2)(x + 3)} \cdot \frac{x + 3}{x - 1}$$ Factor.

$$= \frac{(x + 1)(x - 1)(x + 3)}{(x - 2)(x - 1)(x + 3)}$$ Commutative property

$$= \frac{x + 1}{x - 2}$$ Simplify.

Now Try Exercises 45 and 53

Least Common Multiples and Denominators

To add or subtract fractions and rational expressions, we need to find a common denominator. The **least common denominator** (LCD) is equivalent to the **least common multiple** (LCM) of the denominators. The following procedure can be used to find the least common multiple.

FINDING THE LEAST COMMON MULTIPLE

The least common multiple (LCM) of two polynomials can be found as follows.

STEP 1: Factor each polynomial completely.

STEP 2: List each factor the greatest number of times that it occurs in either factorization.

STEP 3: Find the product of this list of factors. The result is the LCM.

The next example illustrates how to use this procedure.

EXAMPLE 5 Finding least common multiples

Find the least common multiple for each pair of expressions.

(a) $4x, 5x^3$ (b) $x^2 + 4x + 4, x^2 + 3x + 2$

SOLUTION

(a) **STEP 1:** Factor each polynomial completely.

$$4x = 2 \cdot 2 \cdot x \quad \text{and} \quad 5x^3 = 5 \cdot x \cdot x \cdot x$$

STEP 2: The factor 2 occurs twice, the factor 5 occurs once, and the factor x occurs at most three times. The list then is $2, 2, 5, x, x,$ and x.

STEP 3: The LCM is the product $2 \cdot 2 \cdot 5 \cdot x \cdot x \cdot x$, or $20x^3$.

(b) **STEP 1:** Factor each polynomial as follows.

$$x^2 + 4x + 4 = (x + 2)(x + 2) \quad \text{and} \quad x^2 + 3x + 2 = (x + 1)(x + 2)$$

STEP 2: The factor $(x + 1)$ occurs once, and $(x + 2)$ occurs at most twice.

STEP 3: The LCM is the product $(x + 1)(x + 2)^2$, which is left in factored form.

> **Now Try Exercises 61 and 65**

EXAMPLE 6 **Finding a least common denominator**

Find the LCD for the expressions $\dfrac{1}{x^2 + 4x + 4}$ and $\dfrac{5}{x^2 + 3x + 2}$.

SOLUTION From Example 5(b), the LCM for $x^2 + 4x + 4$ and $x^2 + 3x + 2$ is

$$(x + 1)(x + 2)^2.$$

> The LCD is the same as the LCM of the denominators.

Therefore the LCD is *also* $(x + 1)(x + 2)^2$.

> **Now Try Exercise 69**

Review of Addition and Subtraction of Fractions

Recall that to add two fractions we use the property $\dfrac{a}{c} + \dfrac{b}{c} = \dfrac{a + b}{c}$. This property requires *like* denominators. For example, $\dfrac{1}{5} + \dfrac{3}{5} = \dfrac{1 + 3}{5} = \dfrac{4}{5}$. When the denominators are not alike, we must find a common denominator. Before adding two fractions, such as $\dfrac{2}{3}$ and $\dfrac{1}{4}$, we write them with 12 as their common denominator. That is, we multiply each fraction by 1 written in an appropriate form. For example, to write $\dfrac{2}{3}$ with a denominator of 12, we multiply $\dfrac{2}{3}$ by 1, written as $\dfrac{4}{4}$.

$$\frac{2}{3} = \frac{2}{3} \cdot \frac{4}{4} = \frac{8}{12} \quad \text{and} \quad \frac{1}{4} = \frac{1}{4} \cdot \frac{3}{3} = \frac{3}{12}$$

> 1. Write each fraction with a common denominator.

Once the fractions have a common denominator, we can add them, as in

$$\frac{2}{3} + \frac{1}{4} = \frac{8}{12} + \frac{3}{12} = \frac{11}{12}.$$

> 2. Add numerators. Keep the common denominator.

The *least common denominator* (LCD) for $\dfrac{2}{3}$ and $\dfrac{1}{4}$ is equal to the *least common multiple* (LCM) of 3 and 4. Thus the least common denominator is 12.

EXAMPLE 7 **Adding fractions**

Simplify $\dfrac{3}{5} + \dfrac{2}{7}$.

SOLUTION The LCD is 35.

$$\frac{3}{5} + \frac{2}{7} = \frac{3}{5} \cdot \frac{7}{7} + \frac{2}{7} \cdot \frac{5}{5} = \frac{21}{35} + \frac{10}{35} = \frac{31}{35}$$

> **Now Try Exercise 27**

Recall that subtraction is similar to addition. To subtract two fractions with *like* denominators, we use the property $\dfrac{a}{c} - \dfrac{b}{c} = \dfrac{a - b}{c}$. For example, $\dfrac{3}{11} - \dfrac{7}{11} = \dfrac{3 - 7}{11} = -\dfrac{4}{11}$.

EXAMPLE 8 **Subtracting fractions**

Simplify $\dfrac{3}{8} - \dfrac{5}{6}$.

> Multiply each fraction by **1** in the appropriate form.

SOLUTION The LCD is 24.

$$\frac{3}{8} - \frac{5}{6} = \frac{3}{8} \cdot \frac{3}{3} - \frac{5}{6} \cdot \frac{4}{4} = \frac{9}{24} - \frac{20}{24} = -\frac{11}{24}$$

> **Now Try Exercise 29**

Addition and Subtraction of Rational Expressions

Addition and subtraction of rational expressions with like denominators are performed in the following manner.

> **SUMS AND DIFFERENCES OF RATIONAL EXPRESSIONS**
>
> To add (or subtract) two rational expressions with like denominators, add (or subtract) their numerators. The denominator does not change.
>
> $$\frac{A}{C} + \frac{B}{C} = \frac{A + B}{C}$$
>
> $$\frac{A}{C} - \frac{B}{C} = \frac{A - B}{C}, \quad C \neq 0$$

NOTE If the denominators are not alike, begin by writing each rational expression, using a common denominator. The LCD equals the LCM of the denominators.

EXAMPLE 9 Adding rational expressions

Add the expressions.

(a) $\dfrac{x}{x + 2} + \dfrac{3x + 1}{x + 2}$ **(b)** $\dfrac{1}{x - 1} + \dfrac{2x}{x + 1}$

SOLUTION

(a) The denominators are alike, so we add the numerators and keep the same denominator.

$$\frac{x}{x + 2} + \frac{3x + 1}{x + 2} = \frac{x + 3x + 1}{x + 2} \qquad \text{Add numerators.}$$

$$= \frac{4x + 1}{x + 2} \qquad \text{Combine like terms.}$$

(b) The LCM for $x - 1$ and $x + 1$ is their product, $(x - 1)(x + 1)$.

$$\frac{1}{x - 1} + \frac{2x}{x + 1} = \frac{1}{x - 1} \cdot \frac{x + 1}{x + 1} + \frac{2x}{x + 1} \cdot \frac{x - 1}{x - 1} \qquad \begin{array}{l}\text{Change to a common} \\ \text{denominator.}\end{array}$$

$$= \frac{x + 1}{(x - 1)(x + 1)} + \frac{2x(x - 1)}{(x + 1)(x - 1)} \qquad \text{Multiply.}$$

$$= \frac{x + 1 + 2x^2 - 2x}{(x - 1)(x + 1)} \qquad \begin{array}{l}\text{Add numerators;} \\ \text{distributive property}\end{array}$$

$$= \frac{2x^2 - x + 1}{(x - 1)(x + 1)} \qquad \text{Combine like terms.}$$

Now Try Exercises 73 and 81

Subtraction of rational expressions is similar.

EXAMPLE 10 Subtracting rational expressions

Subtract the expressions: $\dfrac{x - 1}{x} - \dfrac{5}{x + 5}$.

SOLUTION The LCD is $x(x + 5)$.

$$\frac{x - 1}{x} - \frac{5}{x + 5} = \frac{x - 1}{x} \cdot \frac{x + 5}{x + 5} - \frac{5}{x + 5} \cdot \frac{x}{x} \qquad \begin{array}{l}\text{Change to a common} \\ \text{denominator.}\end{array}$$

$$= \frac{(x - 1)(x + 5)}{x(x + 5)} - \frac{5x}{x(x + 5)} \qquad \text{Multiply.}$$

$$= \frac{(x - 1)(x + 5) - 5x}{x(x + 5)} \qquad \text{Subtract numerators.}$$

$$= \frac{x^2 + 4x - 5 - 5x}{x(x + 5)} \qquad \text{Multiply binomials.}$$

$$= \frac{x^2 - x - 5}{x(x + 5)} \qquad \text{Combine like terms.}$$

Now Try Exercise 79

Clearing Fractions

To solve rational equations, it is sometimes advantageous to multiply each side by the LCD to clear fractions. For example, the LCD for the equation $\frac{1}{x+2} + \frac{1}{x-2} = 0$ is $(x+2)(x-2)$. Multiplying each side by the LCD results in the following.

$$(x+2)(x-2)\left(\frac{1}{x+2} + \frac{1}{x-2}\right) = 0 \qquad \text{Multiply each side by LCD.}$$

$$\frac{(x+2)(x-2)}{x+2} + \frac{(x+2)(x-2)}{x-2} = 0 \qquad \text{Distributive property}$$

$$(x-2) + (x+2) = 0 \qquad \text{Simplify.}$$

$$x = 0 \qquad \text{Combine like terms and solve.}$$

This technique is applied in the next example.

EXAMPLE 11 Clearing fractions

Clear fractions from the equation and solve.

$$\frac{3}{x} + \frac{x}{x^2-1} - \frac{4}{x+1} = 0$$

SOLUTION The LCD is $x(x^2-1) = x(x-1)(x+1)$.

$$x(x^2-1)\left(\frac{3}{x} + \frac{x}{x^2-1} - \frac{4}{x+1}\right) = x(x^2-1)\cdot 0$$

$$\frac{3x(x^2-1)}{x} + \frac{x(x)(x^2-1)}{x^2-1} - \frac{4x(x^2-1)}{x+1} = 0$$

$$\frac{3x(x^2-1)}{x} + \frac{x^2(x^2-1)}{x^2-1} - \frac{4x(x-1)(x+1)}{x+1} = 0$$

$$3(x^2-1) + x^2 - 4x(x-1) = 0$$

$$3x^2 - 3 + x^2 - 4x^2 + 4x = 0$$

$$4x - 3 = 0$$

$$x = \frac{3}{4}$$

The solution is $\frac{3}{4}$. Check this answer.

Now Try Exercise 111

Complex Fractions

A complex fraction is a rational expression that contains fractions in its numerator, denominator, or both. One strategy for simplifying a complex fraction is to multiply the numerator and denominator by the LCD of the fractions in the numerator and denominator. For example, the LCD for the complex fraction

$$\frac{1 - \dfrac{1}{x}}{1 + \dfrac{1}{2x}}$$

is $2x$. To simplify, multiply the complex fraction by $\mathbf{1}$, expressed in the form $\frac{2x}{2x}$.

$$\frac{\left(1 - \dfrac{1}{x}\right)\cdot 2x}{\left(1 + \dfrac{1}{2x}\right)\cdot 2x} = \frac{2x - \dfrac{2x}{x}}{2x + \dfrac{2x}{2x}} \qquad \text{Distributive property}$$

$$= \frac{2x - 2}{2x + 1} \qquad \text{Simplify.}$$

In the next example, we simplify a complex fraction.

EXAMPLE 12 Simplifying a complex fraction

Simplify the complex fraction.

$$\frac{\dfrac{3}{x-1} - \dfrac{2}{x}}{\dfrac{1}{x-1} + \dfrac{3}{x}}$$

SOLUTION The LCD is the product, $x(x-1)$. Multiply the expression by $\dfrac{x(x-1)}{x(x-1)}$.

$$\frac{\left(\dfrac{3}{x-1} - \dfrac{2}{x}\right)}{\left(\dfrac{1}{x-1} + \dfrac{3}{x}\right)} \cdot \frac{x(x-1)}{x(x-1)} = \frac{\dfrac{3x(x-1)}{x-1} - \dfrac{2x(x-1)}{x}}{\dfrac{x(x-1)}{x-1} + \dfrac{3x(x-1)}{x}}$$ Distributive property

$$= \frac{3x - 2(x-1)}{x + 3(x-1)}$$ Simplify.

$$= \frac{3x - 2x + 2}{x + 3x - 3}$$ Distributive property

$$= \frac{x + 2}{4x - 3}$$ Combine like terms.

Now Try Exercise 115

R.5 Exercises

Simplifying Rational Expressions

Exercises 1–14: Simplify the expression.

1. $\dfrac{10x^3}{5x^2}$

2. $\dfrac{24t^3}{6t^2}$

3. $\dfrac{(x-5)(x+5)}{x-5}$

4. $-\dfrac{5-a}{a-5}$

5. $\dfrac{x^2-16}{x-4}$

6. $\dfrac{(x+5)(x-4)}{(x+7)(x+5)}$

7. $\dfrac{x+3}{2x^2+5x-3}$

8. $\dfrac{2x^2-9x+4}{6x^2+7x-5}$

9. $-\dfrac{z+2}{4z+8}$

10. $\dfrac{x^2-25}{x^2+10x+25}$

11. $\dfrac{x^2+2x}{x^2+3x+2}$

12. $\dfrac{x^2-3x-10}{x^2-6x+5}$

13. $\dfrac{a^3+b^3}{a+b}$

14. $\dfrac{a^3-b^3}{a-b}$

Review of Fractions

Exercises 15–32: Simplify.

15. $\dfrac{5}{8} \cdot \dfrac{4}{15}$

16. $\dfrac{7}{2} \cdot \dfrac{4}{21}$

17. $\dfrac{5}{6} \cdot \dfrac{3}{10} \cdot \dfrac{8}{3}$

18. $\dfrac{9}{5} \cdot \dfrac{10}{3} \cdot \dfrac{1}{27}$

19. $\dfrac{4}{7} \div \dfrac{8}{7}$

20. $\dfrac{5}{12} \div \dfrac{10}{9}$

21. $\dfrac{1}{2} \div \dfrac{3}{4} \div \dfrac{5}{6}$

22. $\dfrac{3}{4} \div \dfrac{7}{8} \div \dfrac{5}{14}$

23. $\dfrac{3}{8} + \dfrac{5}{8}$

24. $\dfrac{5}{9} + \dfrac{2}{9}$

25. $\dfrac{3}{7} - \dfrac{4}{7}$

26. $\dfrac{8}{11} - \dfrac{9}{11}$

27. $\dfrac{2}{3} + \dfrac{5}{11}$

28. $\dfrac{9}{13} + \dfrac{3}{2}$

29. $\dfrac{4}{5} - \dfrac{1}{10}$

30. $\dfrac{3}{4} - \dfrac{7}{12}$

31. $\dfrac{1}{3} + \dfrac{3}{4} - \dfrac{3}{7}$

32. $\dfrac{6}{11} - \dfrac{1}{2} + \dfrac{3}{8}$

Multiplication and Division of Rational Expressions

Exercises 33–58: Simplify. Leave numerators and denominators in factored form when appropriate.

33. $\dfrac{1}{x^2} \cdot \dfrac{3x}{2}$

34. $\dfrac{6a}{5} \cdot \dfrac{5}{12a^2}$

35. $\dfrac{5x}{3} \div \dfrac{10x}{6}$

36. $\dfrac{2x^2 + x}{3x + 9} \div \dfrac{x}{x + 3}$

37. $\dfrac{x + 1}{2x - 5} \cdot \dfrac{x}{x + 1}$

38. $\dfrac{4x + 8}{2x} \cdot \dfrac{x^2}{x + 2}$

39. $\dfrac{(x - 5)(x + 3)}{3x - 1} \cdot \dfrac{x(3x - 1)}{(x - 5)}$

40. $\dfrac{b^2 + 1}{b^2 - 1} \cdot \dfrac{b - 1}{b + 1}$

41. $\dfrac{x^2 - 2x - 35}{2x^3 - 3x^2} \cdot \dfrac{x^3 - x^2}{2x - 14}$

42. $\dfrac{2x + 4}{x + 1} \cdot \dfrac{x^2 + 3x + 2}{4x + 2}$

43. $\dfrac{6b}{b + 2} \div \dfrac{3b^4}{2b + 4}$

44. $\dfrac{5x^5}{x - 2} \div \dfrac{10x^3}{5x - 10}$

45. $\dfrac{3a + 1}{a^7} \div \dfrac{a + 1}{3a^8}$

46. $\dfrac{x^2 - 16}{x + 3} \div \dfrac{x + 4}{x^2 - 9}$

47. $\dfrac{x + 5}{x^3 - x} \div \dfrac{x^2 - 25}{x^3}$

48. $\dfrac{x^2 + x - 12}{2x^2 - 9x - 5} \div \dfrac{x^2 + 7x + 12}{2x^2 - 7x - 4}$

49. $\dfrac{x - 2}{x^3 - x} \div \dfrac{x^2 - 2x}{x^2 - 1}$

50. $\dfrac{x^2 + 3x + 2}{2x^2 + 7x + 3} \div \dfrac{x^2 - 4}{2x^2 - x - 1}$

51. $\dfrac{x^2 - 3x + 2}{x^2 + 5x + 6} \div \dfrac{x^2 + x - 2}{x^2 + 2x - 3}$

52. $\dfrac{2x^2 + x - 1}{6x^2 + x - 2} \div \dfrac{2x^2 + 5x + 3}{6x^2 + 13x + 6}$

53. $\dfrac{x^2 - 4}{x^2 + x - 2} \div \dfrac{x - 2}{x - 1}$

54. $\dfrac{x^2 + 2x + 1}{x - 2} \div \dfrac{x + 1}{2x - 4}$

55. $\dfrac{3y}{x^2} \div \dfrac{y^2}{x} \div \dfrac{y}{5x}$

56. $\dfrac{x + 1}{y - 2} \div \dfrac{2x + 2}{y - 2} \div \dfrac{x}{y}$

57. $\dfrac{x - 3}{x - 1} \div \dfrac{x^2}{x - 1} \div \dfrac{x - 3}{x}$

58. $\dfrac{2x}{x - 2} \div \dfrac{x + 2}{x} \div \dfrac{7x}{x^2 - 4}$

Least Common Multiples

Exercises 59–66: Find the least common multiple.

59. $12, 18$

60. $9, 15$

61. $5a^3, 10a$

62. $6a^2, 9a^5$

63. $z^2 - 4z, (z - 4)^2$

64. $z^2 - 1, z^2 + 2z + 1$

65. $x^2 - 6x + 9, x^2 - 5x + 6$

66. $x^2 - 4, x^2 - 4x + 4$

Common Denominators

Exercises 67–72: Find the LCD for the rational expressions.

67. $\dfrac{1}{x + 1}, \dfrac{1}{7}$

68. $\dfrac{1}{2x - 1}, \dfrac{1}{x + 1}$

69. $\dfrac{1}{x + 4}, \dfrac{1}{x^2 - 16}$

70. $\dfrac{4}{2x^2}, \dfrac{1}{2x + 2}$

71. $\dfrac{3}{2}, \dfrac{x}{2x + 1}, \dfrac{x}{2x - 4}$

72. $\dfrac{1}{x}, \dfrac{1}{x^2 - 4x}, \dfrac{1}{2x}$

Addition and Subtraction of Rational Expressions

Exercises 73–102: Simplify. Leave numerators and denominators in factored form when appropriate.

73. $\dfrac{4}{x + 1} + \dfrac{3}{x + 1}$

74. $\dfrac{2}{x^2} + \dfrac{5}{x^2}$

75. $\dfrac{2}{x^2 - 1} - \dfrac{x + 1}{x^2 - 1}$

76. $\dfrac{2x}{x^2 + x} - \dfrac{2x}{x + 1}$

77. $\dfrac{x}{x + 4} - \dfrac{x + 1}{x(x + 4)}$

78. $\dfrac{4x}{x + 2} + \dfrac{x - 5}{x - 2}$

79. $\dfrac{2}{x^2} - \dfrac{4x - 1}{x}$

80. $\dfrac{2x}{x - 5} - \dfrac{x}{x + 5}$

81. $\dfrac{x + 3}{x - 5} + \dfrac{5}{x - 3}$

82. $\dfrac{x}{2x - 1} + \dfrac{1 - x}{3x}$

83. $\dfrac{3}{x-5} - \dfrac{1}{x-3} - \dfrac{2x}{x-5}$

84. $\dfrac{2x+1}{x-1} - \dfrac{3}{x+1} + \dfrac{x}{x-1}$

85. $\dfrac{x}{x^2-9} + \dfrac{5x}{x-3}$ **86.** $\dfrac{a^2+1}{a^2-1} + \dfrac{a}{1-a^2}$

87. $\dfrac{b}{2b-4} - \dfrac{b-1}{b-2}$ **88.** $\dfrac{y^2}{2-y} - \dfrac{y}{y^2-4}$

89. $\dfrac{2x}{x-5} + \dfrac{2x-1}{3x^2-16x+5}$

90. $\dfrac{x+3}{2x-1} + \dfrac{3}{10x^2-5x}$

91. $\dfrac{x}{(x-1)^2} - \dfrac{1}{(x-1)(x+3)}$

92. $\dfrac{3}{x^2-x-6} - \dfrac{2}{x^2+5x+6}$

93. $\dfrac{x}{x^2-5x+4} + \dfrac{2}{x^2-2x-8}$

94. $\dfrac{3}{x^2-2x+1} + \dfrac{1}{x^2-3x+2}$

95. $\dfrac{x}{x^2-4} - \dfrac{1}{x^2+4x+4}$

96. $\dfrac{3x}{x^2+2x-3} + \dfrac{1}{x^2-2x+1}$

97. $\dfrac{3x}{x-y} - \dfrac{3y}{x^2-2xy+y^2}$

98. $\dfrac{4c}{ab} + \dfrac{3b}{ac} - \dfrac{2a}{bc}$ **99.** $x + \dfrac{1}{x-1} - \dfrac{1}{x+1}$

100. $5 - \dfrac{6}{n^2-36} + \dfrac{3}{n-6}$

101. $\dfrac{6}{t-1} + \dfrac{2}{t-2} + \dfrac{1}{t}$

102. $\dfrac{3}{x-5} - \dfrac{1}{x-3} - \dfrac{2x}{x-5}$

Clearing Fractions

Exercises 103–112: (Refer to Example 11.) Clear fractions and solve. Check your answers.

103. $\dfrac{1}{x} + \dfrac{3}{x^2} = 0$ **104.** $\dfrac{1}{x-2} + \dfrac{3}{x+1} = 0$

105. $\dfrac{1}{x} + \dfrac{3x}{2x-1} = 0$ **106.** $\dfrac{x}{2x-5} + \dfrac{4}{x} = 0$

107. $\dfrac{2x}{9-x^2} + \dfrac{1}{3-x} = 0$

108. $\dfrac{1}{1-x^2} + \dfrac{1}{1+x} = 0$

109. $\dfrac{1}{2x} + \dfrac{1}{2x^2} - \dfrac{1}{x^3} = 0$

110. $\dfrac{1}{x^2-16} + \dfrac{4}{x+4} - \dfrac{5}{x-4} = 0$

111. $\dfrac{1}{x} - \dfrac{2}{x+5} + \dfrac{1}{x-5} = 0$

112. $\dfrac{1}{x-2} + \dfrac{1}{x-3} - \dfrac{2}{x} = 0$

Complex Fractions

Exercises 113–124: Simplify. Leave numerators and denominators in factored form when appropriate.

113. $\dfrac{1 + \dfrac{1}{x}}{1 - \dfrac{1}{x}}$ **114.** $\dfrac{\dfrac{1}{2} - x}{\dfrac{1}{x} - 2}$

115. $\dfrac{\dfrac{1}{x-5}}{\dfrac{4}{x} - \dfrac{1}{x-5}}$ **116.** $\dfrac{1 + \dfrac{1}{x-3}}{\dfrac{1}{x-3} - 1}$

117. $\dfrac{\dfrac{1}{x} + \dfrac{2-x}{x^2}}{\dfrac{3}{x^2} - \dfrac{1}{x}}$ **118.** $\dfrac{\dfrac{1}{x-1} + \dfrac{2}{x}}{2 - \dfrac{1}{x}}$

119. $\dfrac{\dfrac{1}{x+3} + \dfrac{2}{x-3}}{2 - \dfrac{1}{x-3}}$ **120.** $\dfrac{\dfrac{1}{x} + \dfrac{2}{x}}{\dfrac{1}{x-1} + \dfrac{x}{2}}$

121. $\dfrac{\dfrac{4}{x-5}}{\dfrac{1}{x+5} + \dfrac{1}{x}}$ **122.** $\dfrac{\dfrac{2}{x-4}}{1 - \dfrac{1}{x+4}}$

123. $\dfrac{\dfrac{1}{2a} - \dfrac{1}{2b}}{\dfrac{1}{a^2} - \dfrac{1}{b^2}}$ **124.** $\dfrac{\dfrac{1}{2x^2} - \dfrac{1}{2y^2}}{\dfrac{1}{3y^2} + \dfrac{1}{3x^2}}$

R.6 Radical Notation and Rational Exponents

- Use radical notation
- Apply rational exponents
- Use properties of rational exponents

Radical Notation

Square Root Recall the definition of the **square root** of a number a.

> **SQUARE ROOT**
>
> The number b is a *square root* of a if $b^2 = a$.

Every positive number a has two square roots, one positive and one negative. For example, the square roots of 100 are 10 and -10. Recall that the *positive square root* is called the **principal square root** and is denoted $\sqrt{a}$. The *negative square root* is denoted $-\sqrt{a}$. To identify both square roots, we write $\pm\sqrt{a}$. The symbol $\pm$ is read "plus or minus." The symbol $\sqrt{}$ is called the **radical sign**. The expression under the radical sign is called the **radicand**, and an expression containing a radical sign is called a **radical expression**. Examples of radical expressions include

Radical sign ⟶ $\sqrt{6},$ $5 + \sqrt{x + 1},$ and $\sqrt{\dfrac{3x}{2x - 1}}.$

Radicand (over the 6)

Radical Expressions

EXAMPLE 1 **Finding principal square roots**

Find the principal square root of each expression.

(a) 25   **(b)** 17   **(c)** $\dfrac{4}{9}$ **(d)** $c^2, c > 0$

SOLUTION

(a) Because $5 \cdot 5 = 25$, the principal, or positive, square root of 25 is $\sqrt{25} = 5$.

(b) The principal square root of 17 is $\sqrt{17}$. This value is not an integer, but we can approximate it. Figure R.18 shows that $\sqrt{17} \approx 4.12$, rounded to the nearest hundredth. Note that calculators do not give *exact* answers when approximating many radical expressions; they give decimal *approximations*.

(c) Because $\frac{2}{3} \cdot \frac{2}{3} = \frac{4}{9}$, the principal square root of $\frac{4}{9}$ is $\sqrt{\frac{4}{9}} = \frac{2}{3}$.

(d) The principal square root of c^2 is $\sqrt{c^2} = c$, as it is given that c is positive.

> **Now Try Exercises 7, 9, 11 and 13**

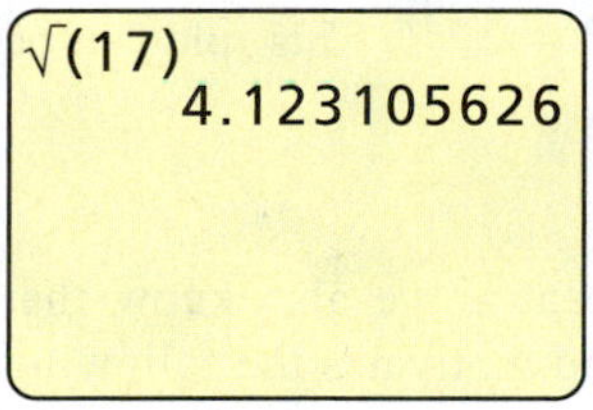

```
√(17)
        4.123105626
```

Figure R.18

Cube Root Another common radical expression is the **cube root** of a number a, denoted $\sqrt[3]{a}$.

> **CUBE ROOT**
>
> The number b is a *cube root* of a if $b^3 = a$.

Although the square root of a negative number is not a real number, the cube root of a negative number is a negative real number. *Every real number has one real cube root.*

EXAMPLE 2 **Finding cube roots**

Find the cube root of each expression.

(a) 8   **(b)** -27 **(c)** 16   **(d)** $\dfrac{1}{64}$ **(e)** d^6

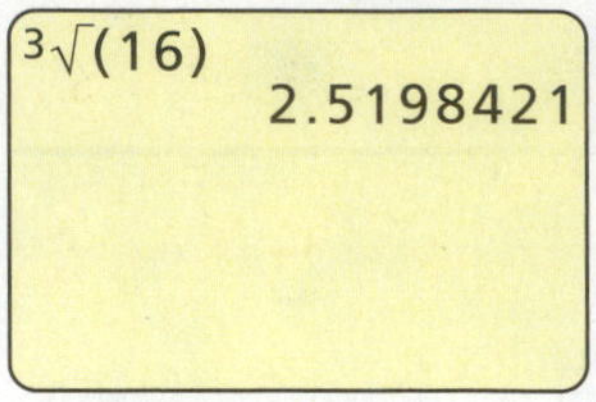

$^3\sqrt{(16)}$
2.5198421

Figure R.19

SOLUTION

(a) $\sqrt[3]{8} = 2$ because $2^3 = 2 \cdot 2 \cdot 2 = 8$.

(b) $\sqrt[3]{-27} = -3$ because $(-3)^3 = (-3)(-3)(-3) = -27$.

(c) $\sqrt[3]{16}$ is not an integer. Figure R.19 shows that $\sqrt[3]{16} \approx 2.52$.

(d) $\sqrt[3]{\frac{1}{64}} = \frac{1}{4}$ because $\left(\frac{1}{4}\right)^3 = \frac{1}{4} \cdot \frac{1}{4} \cdot \frac{1}{4} = \frac{1}{64}$.

(e) $\sqrt[3]{d^6} = d^2$ because $(d^2)^3 = d^2 \cdot d^2 \cdot d^2 = d^{2+2+2} = d^6$.

Now Try Exercises 15, 17, 19, 21, and 31

nth Root We can generalize square roots and cube roots to include the *n*th root of a number *a*. The number *b* is an **nth root** of *a* if $b^n = a$, where *n* is a positive integer, and the **principal nth root** is denoted $\sqrt[n]{a}$. The number *n* is called the **index**. For the square root, the index is 2, although we usually write $\sqrt{a}$ rather than $\sqrt[2]{a}$. When *n* is odd, we are finding an **odd root**, and when *n* is even, we are finding an **even root**. The square root $\sqrt{a}$ is an example of an even root, and the cube root $\sqrt[3]{a}$ is an example of an odd root.

NOTE An odd root of a negative number is a negative number, but the even root of a negative number is *not* a real number. For example, $\sqrt[3]{-8} = -2$, whereas $\sqrt[4]{-81}$ is *not* a real number.

EXAMPLE 3 Finding *n*th roots

Find each root.

(a) $\sqrt[4]{16}$ (b) $\sqrt[5]{-32}$

SOLUTION

(a) $\sqrt[4]{16} = 2$ because $2^4 = 2 \cdot 2 \cdot 2 \cdot 2 = 16$. Note that when *n* is even the principal *n*th root is positive.

(b) $\sqrt[5]{-32} = -2$ because $(-2)^5 = (-2)(-2)(-2)(-2)(-2) = -32$.

Now Try Exercises 37 and 38

Rational Exponents

When *m* and *n* are integers, the product rule states that $a^m a^n = a^{m+n}$. This rule can be extended to include exponents that are fractions. For example,

$$4^{1/2} \cdot 4^{1/2} = 4^{1/2+1/2} = 4^1 = 4.$$

That is, if we multiply $4^{1/2}$ by itself, the result is 4. Because we also know that $\sqrt{4} \cdot \sqrt{4} = 4$, this discussion suggests that $4^{1/2} = \sqrt{4}$ and motivates the following definition.

THE EXPRESSION $a^{1/n}$

If *n* is an integer greater than 1, then

$$a^{1/n} = \sqrt[n]{a}.$$

If $a < 0$ and *n* is an even positive integer, then $a^{1/n}$ is not a real number.

The next two examples show how to interpret rational exponents.

EXAMPLE 4 Interpreting rational exponents

Write each expression in radical notation. Then evaluate the expression (to the nearest hundredth when appropriate).

(a) $36^{1/2}$ (b) $23^{1/5}$ (c) $(5x)^{1/2}$

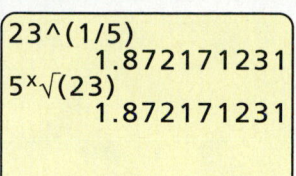

Figure R.20

SOLUTION

(a) The exponent $\frac{1}{2}$ indicates a square root. Thus $36^{1/2} = \sqrt{36}$, which evaluates to 6.

(b) The exponent $\frac{1}{5}$ indicates a fifth root. Thus $23^{1/5} = \sqrt[5]{23}$, which is not an integer. Figure R.20 shows this expression approximated in both exponential and radical notation. In either case $23^{1/5} \approx 1.87$.

(c) The exponent $\frac{1}{2}$ indicates a square root, so $(5x)^{1/2} = \sqrt{5x}$.

> **Now Try Exercises 39, 43, and 55**

Suppose that we want to define the expression $8^{2/3}$. On the one hand, using properties of exponents we have

$$8^{1/3} \cdot 8^{1/3} = 8^{1/3+1/3} = 8^{2/3}.$$

On the other hand, we have

$$8^{1/3} \cdot 8^{1/3} = \sqrt[3]{8} \cdot \sqrt[3]{8} = 2 \cdot 2 = 4.$$

Thus $8^{2/3} = 4$, and that value is obtained whether we interpret $8^{2/3}$ as either

$$8^{2/3} = (8^2)^{1/3} = \sqrt[3]{8^2} = \sqrt[3]{64} = 4$$

or

$$8^{2/3} = (8^{1/3})^2 = (\sqrt[3]{8})^2 = 2^2 = 4.$$

This result illustrates that $8^{2/3} = \sqrt[3]{8^2} = (\sqrt[3]{8})^2 = 4$ and suggests the following definition.

> **THE EXPRESSION $a^{m/n}$**
>
> If m and n are positive integers with $\frac{m}{n}$ in lowest terms, then
> $$a^{m/n} = \sqrt[n]{a^m} = (\sqrt[n]{a})^m.$$
> If $a < 0$ and n is an even integer, then $a^{m/n}$ is not a real number.

EXAMPLE 5 **Interpreting rational exponents**

Write each expression in radical notation. Then evaluate the expression when the result is an integer.

(a) $(-27)^{2/3}$ **(b)** $12^{3/5}$

SOLUTION

(a) The exponent $\frac{2}{3}$ indicates either that we take the **cube root** of -27 and then **square** it or that we **square** -27 and then take the cube root. In either case the result will be the same. Thus

$$(-27)^{2/3} = (\sqrt[3]{-27})^2 = (-3)^2 = 9$$

or

$$(-27)^{2/3} = \sqrt[3]{(-27)^2} = \sqrt[3]{729} = 9.$$

(b) The exponent $\frac{3}{5}$ indicates either that we take the fifth root of 12 and then cube it or that we cube 12 and then take the fifth root. Thus

$$12^{3/5} = (\sqrt[5]{12})^3 \quad \text{or} \quad 12^{3/5} = \sqrt[5]{12^3}.$$

This result is not an integer.

> **Now Try Exercises 47 and 61**

From properties of exponents we know that $a^{-n} = \frac{1}{a^n}$, where n is a positive integer. We now define this property for negative rational exponents.

> **THE EXPRESSION $a^{-m/n}$**
>
> If m and n are positive integers with $\frac{m}{n}$ in lowest terms, then
> $$a^{-m/n} = \frac{1}{a^{m/n}}, \qquad a \neq 0.$$

EXAMPLE 6 **Interpreting negative rational exponents**

Write each expression in radical notation and then evaluate.
(a) $(64)^{-1/3}$ **(b)** $(81)^{-3/4}$

SOLUTION

(a) $(64)^{-1/3} = \dfrac{1}{64^{1/3}} = \dfrac{1}{\sqrt[3]{64}} = \dfrac{1}{4}$

(b) $(81)^{-3/4} = \dfrac{1}{81^{3/4}} = \dfrac{1}{\left(\sqrt[4]{81}\right)^3} = \dfrac{1}{3^3} = \dfrac{1}{27}$

Now Try Exercises 51 and 53

Properties of Rational Exponents

Any rational number can be written as a ratio of two integers. That is, if p is a rational number, then $p = \frac{m}{n}$, where m and n are integers. Properties for integer exponents also apply to rational exponents—with one exception. If n is even in the expression $a^{m/n}$ and $\frac{m}{n}$ is written in lowest terms, then a must be nonnegative (not negative) for the result to be a real number.

PROPERTIES OF EXPONENTS

Let p and q be rational numbers written in lowest terms. For all real numbers a and b for which the expressions are real numbers, the following properties hold.

1. $a^p \cdot a^q = a^{p+q}$ Product rule for exponents

2. $a^{-p} = \dfrac{1}{a^p}, \quad \dfrac{1}{a^{-p}} = a^p$ Negative exponents

3. $\left(\dfrac{a}{b}\right)^{-p} = \left(\dfrac{b}{a}\right)^p$ Negative exponents for quotients

4. $\dfrac{a^p}{a^q} = a^{p-q}$ Quotient rule for exponents

5. $(a^p)^q = a^{pq}$ Power rule for exponents

6. $(ab)^p = a^p b^p$ Power rule for products

7. $\left(\dfrac{a}{b}\right)^p = \dfrac{a^p}{b^p}$ Power rule for quotients

---

In the next example, we apply these properties.

EXAMPLE 7 **Applying properties of exponents**

Write each expression using rational exponents and simplify. Write the answer with a positive exponent. Assume that all variables are positive numbers.

(a) $\sqrt{x} \cdot \sqrt[3]{x}$ **(b)** $\sqrt[3]{27x^2}$ **(c)** $\left(\dfrac{x^2}{81}\right)^{-1/2}$

SOLUTION

(a) $\sqrt{x} \cdot \sqrt[3]{x} = x^{1/2} \cdot x^{1/3}$ Use rational exponents.

$\qquad\qquad = x^{1/2+1/3}$ Product rule for exponents

$\qquad\qquad = x^{5/6}$ Simplify.

(b) $\sqrt[3]{27x^2} = (27x^2)^{1/3}$ Use rational exponents.

$\qquad\qquad = 27^{1/3}(x^2)^{1/3}$ Power rule for products

$\qquad\qquad = 3x^{2/3}$ Power rule for exponents

(c) $\left(\dfrac{x^2}{81}\right)^{-1/2} = \left(\dfrac{81}{x^2}\right)^{1/2}$ Negative exponents for quotients

$\qquad\qquad = \dfrac{(81)^{1/2}}{(x^2)^{1/2}}$ Power rule for quotients

$\qquad\qquad = \dfrac{9}{x}$ Power rule for exponents; simplify.

Now Try Exercises 81, 89, and 99

R.6 Exercises

Square Roots and Cube Roots

Exercises 1–6: Find the square roots of the number. Approximate your answers to the nearest hundredth whenever appropriate.

1. 25

2. 49

3. $\dfrac{16}{25}$

4. $\dfrac{64}{81}$

5. 11

6. 17

Exercises 7–14: Find the principal square root of the number. Approximate your answer to the nearest hundredth whenever appropriate.

7. 144

8. 100

9. 23

10. 45

11. $\dfrac{4}{49}$

12. $\dfrac{16}{121}$

13. $b^2, b < 0$

14. $(xy)^2, xy > 0$

Exercises 15–22: Find the cube root of the number.

15. 27

16. 64

17. -8

18. -125

19. $\dfrac{1}{27}$

20. $-\dfrac{1}{64}$

21. b^9

22. $8x^6$

Radical Notation

Exercises 23–40: If possible, simplify the expression by hand. If you cannot, approximate the answer to the nearest hundredth. Variables represent any real number.

23. $\sqrt{9}$

24. $\sqrt{121}$

25. $-\sqrt{5}$

26. $\sqrt{11}$

27. $\sqrt[3]{27}$

28. $\sqrt[3]{64}$

29. $\sqrt[3]{-64}$

30. $-\sqrt[3]{-1}$

31. $\sqrt[3]{5}$

32. $\sqrt[3]{-13}$

33. $-\sqrt[3]{x^9}$

34. $\sqrt[3]{(x+1)^6}$

35. $\sqrt[3]{(2x)^6}$

36. $\sqrt[3]{9x^3}$

37. $\sqrt[4]{81}$

38. $\sqrt[5]{-1}$

39. $\sqrt[5]{-7}$

40. $\sqrt[4]{6}$

Rational Exponents

Exercises 41–46: Write the expression in radical notation.

41. $6^{1/2}$

42. $7^{1/3}$

43. $(xy)^{1/2}$

44. $x^{2/3}y^{1/5}$

45. $y^{-1/5}$

46. $\left(\dfrac{x}{y}\right)^{-2/7}$

Exercises 47–54: Write the expression in radical notation. Then evaluate the expression when the result is an integer.

47. $27^{2/3}$

48. $8^{4/3}$

49. $(-1)^{4/3}$

50. $81^{3/4}$

51. $8^{-1/3}$

52. $16^{-3/4}$

53. $13^{-3/5}$

54. $23^{-1/2}$

Exercises 55–76: Evaluate the expression by hand. Approximate the answer to the nearest hundredth when appropriate.

55. $16^{1/2}$

56. $8^{1/3}$

57. $256^{1/4}$

58. $4^{3/2}$

59. $32^{1/5}$

60. $(-32)^{1/5}$

61. $(-8)^{4/3}$

62. $(-1)^{3/5}$

63. $2^{1/2} \cdot 2^{2/3}$

64. $5^{3/5} \cdot 5^{1/10}$

65. $\left(\dfrac{4}{9}\right)^{1/2}$

66. $\left(\dfrac{27}{64}\right)^{1/3}$

67. $\dfrac{4^{2/3}}{4^{1/2}}$

68. $\dfrac{6^{1/5} \cdot 6^{3/5}}{6^{2/5}}$

69. $4^{-1/2}$

70. $9^{-3/2}$

71. $(-8)^{-1/3}$

72. $49^{-1/2}$

73. $\left(\frac{1}{16}\right)^{-1/4}$

74. $\left(\frac{16}{25}\right)^{-3/2}$

75. $(2^{1/2})^3$

76. $(5^{6/5})^{-1/2}$

Exercises 77–106: Simplify the expression and write it with rational exponents. Assume that all variables are positive.

77. $(x^2)^{3/2}$

78. $(y^4)^{1/2}$

79. $(x^2 y^8)^{1/2}$

80. $(y^{10} z^4)^{1/4}$

81. $\sqrt[3]{x^3 y^6}$

82. $\sqrt{16x^4}$

83. $\sqrt{\dfrac{y^4}{x^2}}$

84. $\sqrt[3]{\dfrac{x^{12}}{z^6}}$

85. $\sqrt{y^3} \cdot \sqrt[3]{y^2}$

86. $\left(\dfrac{x^6}{81}\right)^{1/4}$

87. $\left(\dfrac{x^6}{27}\right)^{2/3}$

88. $\left(\dfrac{1}{x^8}\right)^{-1/4}$

89. $\left(\dfrac{x^2}{y^6}\right)^{-1/2}$

90. $\dfrac{\sqrt{x}}{\sqrt[3]{27x^6}}$

91. $\sqrt{\sqrt{y}}$

92. $\sqrt{\sqrt[3]{(3x)^2}}$

93. $(a^{-1/2})^{4/3}$

94. $(x^{-3/2})^{2/3}$

95. $(a^3 b^6)^{1/3}$

96. $(64x^3 y^{18})^{1/6}$

97. $\dfrac{(k^{1/2})^{-3}}{(k^2)^{1/4}}$

98. $\dfrac{(b^{3/4})^4}{(b^{4/5})^{-5}}$

99. $\sqrt{b} \cdot \sqrt[4]{b}$

100. $\sqrt[3]{t} \cdot \sqrt[5]{t}$

101. $\sqrt{z} \cdot \sqrt[3]{z^2} \cdot \sqrt[4]{z^3}$

102. $\sqrt{b} \cdot \sqrt[3]{b} \cdot \sqrt[5]{b}$

103. $p^{1/2}(p^{3/2} + p^{1/2})$

104. $d^{3/4}(d^{1/4} - d^{-1/4})$

105. $\sqrt[3]{x}(\sqrt{x} - \sqrt[3]{x^2})$

106. $\frac{1}{2}\sqrt{x}(\sqrt{x} + \sqrt[4]{x^2})$

R.7 Radical Expressions

- Apply the product rule
- Simplify radical expressions
- Apply the quotient rule
- Perform addition, subtraction, and multiplication on radical expressions
- Rationalize the denominator

Product Rule for Radical Expressions

The product of two (like) roots is equal to the root of their product.

PRODUCT RULE FOR RADICAL EXPRESSIONS

Let a and b be real numbers, where $\sqrt[n]{a}$ and $\sqrt[n]{b}$ are both defined. Then

$$\sqrt[n]{a} \cdot \sqrt[n]{b} = \sqrt[n]{a \cdot b}.$$

NOTE The product rule works only when the radicals have the *same* index.

We apply the product rule in the next two examples.

EXAMPLE 1 Multiplying radical expressions

Multiply each pair of radical expressions.

(a) $\sqrt{5} \cdot \sqrt{20}$ **(b)** $\sqrt[3]{-3} \cdot \sqrt[3]{9}$

SOLUTION

(a) $\sqrt{5} \cdot \sqrt{20} = \sqrt{5 \cdot 20} = \sqrt{100} = 10$

(b) $\sqrt[3]{-3} \cdot \sqrt[3]{9} = \sqrt[3]{-3 \cdot 9} = \sqrt[3]{-27} = -3$

Now Try Exercises 3 and 5

EXAMPLE 2 Multiplying radical expressions containing variables

Multiply each pair of radical expressions. Assume that all variables are positive.

(a) $\sqrt{x} \cdot \sqrt{x^3}$ **(b)** $\sqrt[3]{2a} \cdot \sqrt[3]{5a}$ **(c)** $\sqrt[5]{\dfrac{2x}{y}} \cdot \sqrt[5]{\dfrac{16y}{x}}$

SOLUTION

(a) $\sqrt{x} \cdot \sqrt{x^3} = \sqrt{x \cdot x^3} = \sqrt{x^4} = x^2$

(b) $\sqrt[3]{2a} \cdot \sqrt[3]{5a} = \sqrt[3]{2a \cdot 5a} = \sqrt[3]{10a^2}$

(c) $\sqrt[5]{\dfrac{2x}{y}} \cdot \sqrt[5]{\dfrac{16y}{x}} = \sqrt[5]{\dfrac{2x}{y} \cdot \dfrac{16y}{x}}$ Product rule

$\qquad\qquad\qquad = \sqrt[5]{\dfrac{32xy}{xy}}$ Multiply fractions.

$\qquad\qquad\qquad = \sqrt[5]{32}$ Simplify.

$\qquad\qquad\qquad = 2$ $2^5 = 32$

> **Now Try Exercises 27, 33, and 35**

Simplifying Radicals An integer a is a **perfect nth power** if there exists an integer b such that $b^n = a$. Thus 36 is a **perfect square** because $6^2 = 36$, 8 is a **perfect cube** because $2^3 = 8$, and 81 is a *perfect fourth power* because $3^4 = 81$.

The product rule for radicals can be used to simplify radical expressions. For example, because the largest perfect square factor of 50 is 25, the expression $\sqrt{50}$ can be simplified as

> 25 is a perfect square.

$$\sqrt{50} = \sqrt{25 \cdot 2} = \sqrt{25} \cdot \sqrt{2} = 5\sqrt{2}.$$

This procedure is generalized as follows.

SIMPLIFYING RADICALS (nth ROOTS)

STEP 1: Determine the largest perfect nth power factor of the radicand.

STEP 2: Use the product rule to factor out and simplify this perfect nth power.

EXAMPLE 3 **Simplifying radical expressions**

Simplify each expression.

(a) $\sqrt{300}$ (b) $\sqrt[3]{16}$ (c) $\sqrt[4]{512}$

SOLUTION

(a) First note that $300 = 100 \cdot 3$ and that 100 is the largest perfect square factor of 300.

$$\sqrt{300} = \sqrt{100} \cdot \sqrt{3} = 10\sqrt{3}$$

(b) The largest perfect cube factor of 16 is 8. Thus $\sqrt[3]{16} = \sqrt[3]{8} \cdot \sqrt[3]{2} = 2\sqrt[3]{2}$.

(c) $\sqrt[4]{512} = \sqrt[4]{256} \cdot \sqrt[4]{2} = 4\sqrt[4]{2}$ because $4^4 = 256$.

> **Now Try Exercises 37, 39, and 41**

> **NOTE** To simplify a cube root of a negative number, we usually factor out the negative of the largest perfect cube factor. For example, because $-16 = -8 \cdot 2$, it follows that $\sqrt[3]{-16} = \sqrt[3]{-8} \cdot \sqrt[3]{2} = -2\sqrt[3]{2}$. This procedure can be used with any odd root of a negative number.

EXAMPLE 4 **Simplifying radical expressions**

Simplify each expression. Assume that all variables are positive.

(a) $\sqrt{25x^4}$ (b) $\sqrt{32n^3}$ (c) $\sqrt[3]{-16x^3y^5}$ (d) $\sqrt[3]{2a} \cdot \sqrt[3]{4a^2b}$

SOLUTION

(a) $\sqrt{25x^4} = 5x^2$ $\qquad$ $(5x^2)^2 = 25x^4$

(b) $\sqrt{32n^3} = \sqrt{(16n^2)2n}$ $\qquad$ $16n^2$ is the largest perfect square factor.

$\qquad\qquad = \sqrt{16n^2} \cdot \sqrt{2n}$ $\qquad$ Product rule

$\qquad\qquad = 4n\sqrt{2n}$ $\qquad$ $(4n)^2 = 16n^2$

(c) $\sqrt[3]{-16x^3y^5} = \sqrt[3]{(-8x^3y^3)2y^2}$ $\qquad$ $8x^3y^3$ is the largest perfect cube factor.

$\qquad\qquad = \sqrt[3]{-8x^3y^3} \cdot \sqrt[3]{2y^2}$ $\qquad$ Product rule

$\qquad\qquad = -2xy\sqrt[3]{2y^2}$ $\qquad$ $(-2xy)^3 = -8x^3y^3$

(d) $\sqrt[3]{2a} \cdot \sqrt[3]{4a^2b} = \sqrt[3]{(2a)(4a^2b)}$ $\qquad$ Product rule

$\qquad\qquad = \sqrt[3]{(8a^3)b}$ $\qquad$ $8a^3$ is the largest perfect cube factor.

$\qquad\qquad = \sqrt[3]{8a^3} \cdot \sqrt[3]{b}$ $\qquad$ Product rule

$\qquad\qquad = 2a\sqrt[3]{b}$ $\qquad$ $(2a)^3 = 8a^3$

Now Try Exercises 45, 47, 49, and 51

The product rule for radical expressions cannot be used if the radicals do not have the same indexes. In this case we use rational exponents, as illustrated in the next example.

EXAMPLE 5 **Multiplying radicals with different indexes**

Simplify each expression. Write your answer in radical notation.

(a) $\sqrt{2} \cdot \sqrt[3]{4}$ $\qquad$ (b) $\sqrt[3]{x} \cdot \sqrt[4]{x}$

SOLUTION

(a) First note that $\sqrt[3]{4} = \sqrt[3]{2^2} = 2^{2/3}$. Thus

$$\sqrt{2} \cdot \sqrt[3]{4} = 2^{1/2} \cdot 2^{2/3} = 2^{1/2+2/3} = 2^{7/6}.$$

In radical notation, $2^{7/6} = \sqrt[6]{2^7} = \sqrt[6]{2^6 \cdot 2^1} = \sqrt[6]{2^6} \cdot \sqrt[6]{2} = 2\sqrt[6]{2}$.

(b) $\sqrt[3]{x} \cdot \sqrt[4]{x} = x^{1/3} \cdot x^{1/4} = x^{7/12} = \sqrt[12]{x^7}$ $\qquad$ **Now Try Exercises 57 and 59**

Quotient Rule for Radical Expressions

The root of a quotient is equal to the quotient of the roots.

> **QUOTIENT RULE FOR RADICAL EXPRESSIONS**
>
> Let a and b be real numbers, where $\sqrt[n]{a}$ and $\sqrt[n]{b}$ are both defined and $b \neq 0$. Then
>
> $$\sqrt[n]{\frac{a}{b}} = \frac{\sqrt[n]{a}}{\sqrt[n]{b}}.$$

EXAMPLE 6 **Simplifying quotients**

Simplify each radical expression. Assume that all variables are positive.

(a) $\sqrt[3]{\dfrac{5}{8}}$ $\qquad$ (b) $\sqrt{\dfrac{16}{y^2}}$

SOLUTION

(a) $\sqrt[3]{\dfrac{5}{8}} = \dfrac{\sqrt[3]{5}}{\sqrt[3]{8}} = \dfrac{\sqrt[3]{5}}{2}$ $\qquad$ Quotient rule

(b) $\sqrt{\dfrac{16}{y^2}} = \dfrac{\sqrt{16}}{\sqrt{y^2}} = \dfrac{4}{y}$ because $y > 0$. $\qquad$ **Now Try Exercises 7 and 21**

EXAMPLE 7 Simplifying radical expressions

Simplify each radical expression. Assume that all variables are positive.

(a) $\dfrac{\sqrt{40}}{\sqrt{10}}$ **(b)** $\sqrt[4]{\dfrac{16x^3}{y^4}}$ **(c)** $\sqrt{\dfrac{5a^2}{8}} \cdot \sqrt{\dfrac{5a^3}{2}}$

SOLUTION

(a) $\dfrac{\sqrt{40}}{\sqrt{10}} = \sqrt{\dfrac{40}{10}} = \sqrt{4} = 2$

(b) $\sqrt[4]{\dfrac{16x^3}{y^4}} = \dfrac{\sqrt[4]{16x^3}}{\sqrt[4]{y^4}} = \dfrac{\sqrt[4]{16} \cdot \sqrt[4]{x^3}}{\sqrt[4]{y^4}} = \dfrac{2\sqrt[4]{x^3}}{y}$

(c) To simplify this expression, we use both the product and quotient rules.

$$\sqrt{\dfrac{5a^2}{8}} \cdot \sqrt{\dfrac{5a^3}{2}} = \sqrt{\dfrac{25a^5}{16}} \qquad \textcolor{blue}{\text{Product rule}}$$

$$= \dfrac{\sqrt{25a^5}}{\sqrt{16}} \qquad \textcolor{blue}{\text{Quotient rule}}$$

$$= \dfrac{\sqrt{25a^4} \cdot \sqrt{a}}{\sqrt{16}} \qquad \textcolor{blue}{\text{Factor out largest perfect square.}}$$

$$= \dfrac{5a^2\sqrt{a}}{4} \qquad \textcolor{blue}{(5a^2)^2 = 25a^4}$$

> **Now Try Exercises 13, 19, and 53**

Addition and Subtraction

We can add $2x^2$ and $5x^2$ to obtain $7x^2$ because they are *like* terms. That is,

$$2x^2 + 5x^2 = (2 + 5)x^2 = 7x^2.$$

We can add and subtract **like radicals**, which have the same index and the same radicand. For example, we can add $3\sqrt{2}$ and $5\sqrt{2}$ because they are like radicals.

$$3\sqrt{2} + 5\sqrt{2} = (3 + 5)\sqrt{2} = 8\sqrt{2}$$

Sometimes two radical expressions that are not alike can be added by changing them to like radicals. For example, $\sqrt{20}$ and $\sqrt{5}$ are unlike radicals. However,

$$\sqrt{20} = \sqrt{4 \cdot 5} = \sqrt{4} \cdot \sqrt{5} = 2\sqrt{5},$$

so it follows that

> Write $\sqrt{20}$ as a multiple of $\sqrt{5}$.

$$\sqrt{20} + \sqrt{5} = 2\sqrt{5} + \sqrt{5} = 3\sqrt{5}.$$

We cannot combine $x + x^2$ because they are unlike terms. Similarly, we cannot combine $\sqrt{2} + \sqrt{5}$ because they are unlike radicals.

EXAMPLE 8 Adding radical expressions

Add the expressions and simplify.

(a) $10\sqrt{11} + 4\sqrt{11}$ **(b)** $5\sqrt[3]{6} + \sqrt[3]{6}$ **(c)** $\sqrt{12} + 7\sqrt{3}$

SOLUTION

(a) $10\sqrt{11} + 4\sqrt{11} = (10 + 4)\sqrt{11} = 14\sqrt{11}$
(b) $5\sqrt[3]{6} + \sqrt[3]{6} = (5 + 1)\sqrt[3]{6} = 6\sqrt[3]{6}$
(c) $\sqrt{12} + 7\sqrt{3} = \sqrt{4 \cdot 3} + 7\sqrt{3}$

$$= \sqrt{4} \cdot \sqrt{3} + 7\sqrt{3}$$

$$= 2\sqrt{3} + 7\sqrt{3}$$

$$= 9\sqrt{3}$$

> **Now Try Exercises 63, 67, and 69**

EXAMPLE 9 Adding radical expressions

Add the expressions and simplify. Assume that all variables are positive.

(a) $-2\sqrt{4x} + \sqrt{x}$ **(b)** $3\sqrt{3k} + 5\sqrt{12k} + 9\sqrt{48k}$

SOLUTION

(a) Note that $\sqrt{4x} = \sqrt{4} \cdot \sqrt{x} = 2\sqrt{x}$.

$$-2\sqrt{4x} + \sqrt{x} = -2(2\sqrt{x}) + \sqrt{x} = -4\sqrt{x} + \sqrt{x} = -3\sqrt{x}$$

(b) Note that $\sqrt{12k} = \sqrt{4} \cdot \sqrt{3k} = 2\sqrt{3k}$ and that $\sqrt{48k} = \sqrt{16} \cdot \sqrt{3k} = 4\sqrt{3k}$.

$$3\sqrt{3k} + 5\sqrt{12k} + 9\sqrt{48k} = 3\sqrt{3k} + 5(2\sqrt{3k}) + 9(4\sqrt{3k})$$
$$= (3 + 10 + 36)\sqrt{3k}$$
$$= 49\sqrt{3k}$$

Now Try Exercises 77 and 83

Subtraction of radical expressions is similar to addition of radical expressions, as illustrated in the next example.

EXAMPLE 10 Subtracting radical expressions

Subtract and simplify. Assume that all variables are positive.

(a) $3\sqrt[3]{xy^2} - 2\sqrt[3]{xy^2}$ **(b)** $\sqrt{16x^3} - \sqrt{x^3}$

SOLUTION

(a) $3\sqrt[3]{xy^2} - 2\sqrt[3]{xy^2} = (3 - 2)\sqrt[3]{xy^2} = \sqrt[3]{xy^2}$

(b) $\sqrt{16x^3} - \sqrt{x^3} = \sqrt{16} \cdot \sqrt{x^3} - \sqrt{x^3}$
$$= 4\sqrt{x^3} - \sqrt{x^3}$$
$$= 3\sqrt{x^3}$$
$$= 3x\sqrt{x}$$

Now Try Exercises 75 and 81

Multiplication

Some types of radical expressions can be multiplied like binomials. The next example demonstrates this technique.

EXAMPLE 11 Multiplying radical expressions

Multiply and simplify.

(a) $(\sqrt{b} - 4)(\sqrt{b} + 5)$ **(b)** $(4 + \sqrt{3})(4 - \sqrt{3})$

SOLUTION

(a) This expression can be multiplied and then simplified.

$$(\sqrt{b} - 4)(\sqrt{b} + 5) = \sqrt{b} \cdot \sqrt{b} + 5\sqrt{b} - 4\sqrt{b} - 4 \cdot 5$$
$$= b + \sqrt{b} - 20$$

(Compare this product to $(b - 4)(b + 5) = b^2 + b - 20$.)

(b) This expression is in the form $(a + b)(a - b)$, which equals $a^2 - b^2$.

$$(4 + \sqrt{3})(4 - \sqrt{3}) = (4)^2 - (\sqrt{3})^2$$
$$= 16 - 3$$
$$= 13$$

Now Try Exercises 89 and 95

Rationalizing the Denominator

Quotients containing radical expressions can appear to be different but actually be equal. For example, $\frac{1}{\sqrt{3}}$ and $\frac{\sqrt{3}}{3}$ represent the same real number even though they do not look equal. To show this fact, we multiply the first quotient by **1** in the form $\frac{\sqrt{3}}{\sqrt{3}}$.

$$\frac{1}{\sqrt{3}} \cdot \frac{\sqrt{3}}{\sqrt{3}} = \frac{1 \cdot \sqrt{3}}{\sqrt{3} \cdot \sqrt{3}} = \frac{\sqrt{3}}{3}$$

NOTE $\sqrt{b} \cdot \sqrt{b} = \sqrt{b^2} = b$ for any *positive* number b.

One way to standardize radical expressions is to remove any radical expressions from the denominator. This process is called **rationalizing the denominator**. The next example demonstrates how to rationalize the denominator of two quotients.

EXAMPLE 12 **Rationalizing the denominator**

Rationalize each denominator. Assume that all variables are positive.

(a) $\dfrac{3}{5\sqrt{3}}$ (b) $\sqrt{\dfrac{x}{24}}$

SOLUTION

(a) We multiply this expression by **1** in the form $\frac{\sqrt{3}}{\sqrt{3}}$.

$$\frac{3}{5\sqrt{3}} \cdot \frac{\sqrt{3}}{\sqrt{3}} = \frac{3\sqrt{3}}{5\sqrt{9}} = \frac{3\sqrt{3}}{5 \cdot 3} = \frac{\sqrt{3}}{5}$$

(b) Because $\sqrt{24} = \sqrt{4} \cdot \sqrt{6} = 2\sqrt{6}$, we start by simplifying the expression.

$$\sqrt{\frac{x}{24}} = \frac{\sqrt{x}}{\sqrt{24}} = \frac{\sqrt{x}}{2\sqrt{6}}$$

To rationalize the denominator, we multiply this expression by **1** in the form $\frac{\sqrt{6}}{\sqrt{6}}$.

$$\frac{\sqrt{x}}{2\sqrt{6}} = \frac{\sqrt{x}}{2\sqrt{6}} \cdot \frac{\sqrt{6}}{\sqrt{6}} = \frac{\sqrt{6x}}{12}$$

Now Try Exercises 99 and 101

If the denominator consists of two terms, at least one of which contains a radical expression, then the **conjugate** of the denominator is found by changing a $+$ sign to a $-$ sign or vice versa. For example, the conjugate of $\sqrt{2} + \sqrt{3}$ is $\sqrt{2} - \sqrt{3}$, and the conjugate of $\sqrt{3} - 1$ is $\sqrt{3} + 1$. In the next example, we multiply the numerator and denominator by the conjugate of the *denominator* to rationalize the denominator of fractions that contain radicals.

EXAMPLE 13 **Rationalizing the denominator**

Rationalize the denominator. Assume that all variables are positive.

(a) $\dfrac{3 + \sqrt{5}}{2 - \sqrt{5}}$ (b) $\dfrac{\sqrt{x}}{\sqrt{x} - 2}$

SOLUTION

(a) The conjugate of the denominator $2 - \sqrt{5}$ is $2 + \sqrt{5}$.

$$\frac{3 + \sqrt{5}}{2 - \sqrt{5}} = \frac{(3 + \sqrt{5})}{(2 - \sqrt{5})} \cdot \frac{(2 + \sqrt{5})}{(2 + \sqrt{5})} \qquad \text{Multiply by } \frac{conjugate}{conjugate}.$$

$$= \frac{6 + 3\sqrt{5} + 2\sqrt{5} + (\sqrt{5})^2}{(2)^2 - (\sqrt{5})^2} \qquad \text{Multiply.}$$

$$= \frac{11 + 5\sqrt{5}}{4 - 5} \qquad \text{Combine terms.}$$

$$= -11 - 5\sqrt{5} \qquad \text{Simplify.}$$

(b) The conjugate of the denominator $\sqrt{x} - 2$ is $\sqrt{x} + 2$.

$$\frac{\sqrt{x}}{\sqrt{x} - 2} = \frac{\sqrt{x}}{(\sqrt{x} - 2)} \cdot \frac{(\sqrt{x} + 2)}{(\sqrt{x} + 2)} \qquad \text{Multiply by } \frac{\text{conjugate}}{\text{conjugate}}.$$

$$= \frac{x + 2\sqrt{x}}{x - 4} \qquad \text{Multiply.}$$

Now Try Exercises 103 and 109

R.7 Exercises

Multiplying and Dividing

Exercises 1–36: Simplify the expression. Assume that all variables are positive.

1. $\sqrt{3} \cdot \sqrt{3}$

2. $\sqrt{2} \cdot \sqrt{18}$

3. $\sqrt{2} \cdot \sqrt{50}$

4. $\sqrt[3]{-2} \cdot \sqrt[3]{-4}$

5. $\sqrt[3]{4} \cdot \sqrt[3]{16}$

6. $\sqrt[3]{x} \cdot \sqrt[3]{x^2}$

7. $\sqrt{\dfrac{9}{25}}$

8. $\sqrt[3]{\dfrac{x}{8}}$

9. $\sqrt{\dfrac{1}{2}} \cdot \sqrt{\dfrac{1}{8}}$

10. $\sqrt{\dfrac{5}{3}} \cdot \sqrt{\dfrac{1}{3}}$

11. $\sqrt{\dfrac{x}{2}} \cdot \sqrt{\dfrac{x}{8}}$

12. $\sqrt{\dfrac{4}{y}} \cdot \sqrt{\dfrac{y}{5}}$

13. $\dfrac{\sqrt{45}}{\sqrt{5}}$

14. $\dfrac{\sqrt{7}}{\sqrt{28}}$

15. $\sqrt[4]{9} \cdot \sqrt[4]{9}$

16. $\sqrt[5]{16} \cdot \sqrt[5]{-2}$

17. $\dfrac{\sqrt[5]{64}}{\sqrt[5]{-2}}$

18. $\dfrac{\sqrt[4]{324}}{\sqrt[4]{4}}$

19. $\sqrt{\dfrac{36}{z^4}}$

20. $\dfrac{\sqrt{4xy^2}}{\sqrt{x}}$

21. $\sqrt[3]{\dfrac{x^3}{8}}$

22. $\dfrac{\sqrt{a^2 b}}{\sqrt{b}}$

23. $\sqrt{4x^4}$

24. $\sqrt[3]{-8y^3}$

25. $\sqrt[4]{16x^4 y}$

26. $\sqrt[3]{8xy^3}$

27. $\sqrt{3x} \cdot \sqrt{12x}$

28. $\sqrt{6x^5} \cdot \sqrt{6x}$

29. $\sqrt[3]{8x^6 y^3 z^9}$

30. $\sqrt{16x^4 y^6}$

31. $\sqrt[4]{\dfrac{3}{4}} \cdot \sqrt[4]{\dfrac{27}{4}}$

32. $\sqrt[5]{\dfrac{4}{-9}} \cdot \sqrt[5]{\dfrac{8}{-27}}$

33. $\sqrt[4]{25z} \cdot \sqrt[4]{25z}$

34. $\sqrt[5]{3z^2} \cdot \sqrt[5]{7z}$

35. $\sqrt[5]{\dfrac{7a}{b^2}} \cdot \sqrt[5]{\dfrac{b^2}{7a^6}}$

36. $\sqrt[3]{\dfrac{8m}{n}} \cdot \sqrt[3]{\dfrac{n^4}{m^2}}$

Exercises 37–54: Simplify the radical expression by factoring out the largest perfect nth power. Assume that all variables are positive.

37. $\sqrt{200}$

38. $\sqrt{72}$

39. $\sqrt[3]{81}$

40. $\sqrt[3]{256}$

41. $\sqrt[4]{64}$

42. $\sqrt[5]{27 \cdot 81}$

43. $\sqrt[5]{-64}$

44. $\sqrt[3]{-81}$

45. $\sqrt{8n^3}$

46. $\sqrt{32a^2}$

47. $\sqrt{12a^2 b^5}$

48. $\sqrt{20a^3 b^2}$

49. $\sqrt[3]{-125x^4 y^5}$

50. $\sqrt[3]{-81a^5 b^2}$

51. $\sqrt[3]{5t} \cdot \sqrt[3]{125t}$

52. $\sqrt[4]{4bc^3} \cdot \sqrt[4]{64ab^3 c^2}$

53. $\sqrt[4]{\dfrac{9t^5}{r^8}} \cdot \sqrt[4]{\dfrac{9r}{5t}}$

54. $\sqrt[5]{\dfrac{4t^6}{r}} \cdot \sqrt[5]{\dfrac{8t}{r^6}}$

Exercises 55–62: Simplify the expression. Assume that all variables are positive and write your answer in radical notation.

55. $\sqrt{3} \cdot \sqrt[3]{3}$

56. $\sqrt{5} \cdot \sqrt[3]{5}$

57. $\sqrt[4]{8} \cdot \sqrt[3]{4}$

58. $\sqrt[5]{16} \cdot \sqrt{2}$

59. $\sqrt[4]{x^3} \cdot \sqrt[3]{x}$

60. $\sqrt[4]{x^3} \cdot \sqrt{x}$

61. $\sqrt[4]{rt} \cdot \sqrt[3]{r^2 t}$

62. $\sqrt[3]{a^3 b^2} \cdot \sqrt{a^2 b}$

Exercises 63–88: Simplify the expression. Assume that all variables are positive.

63. $2\sqrt{3} + 7\sqrt{3}$

64. $8\sqrt{7} + 2\sqrt{7}$

65. $\sqrt{x} + \sqrt{x} - \sqrt{y}$

66. $\sqrt{xy^2} - \sqrt{x}$

67. $2\sqrt[3]{6} + 7\sqrt[3]{6}$

68. $18\sqrt[3]{3} + 3\sqrt[3]{3}$

69. $3\sqrt{28} + 3\sqrt{7}$ **70.** $9\sqrt{18} - 2\sqrt{8}$

71. $\sqrt{44} - 4\sqrt{11}$ **72.** $\sqrt[4]{5} + 2\sqrt[4]{5}$

73. $2\sqrt[3]{16} + \sqrt[3]{2} - \sqrt{2}$

74. $5\sqrt[3]{x} - 3\sqrt[3]{x}$ **75.** $\sqrt[3]{xy} - 2\sqrt[3]{xy}$

76. $3\sqrt{x^3} - \sqrt{x}$ **77.** $\sqrt{4x + 8} + \sqrt{x + 2}$

78. $\sqrt{2a + 1} + \sqrt{8a + 4}$

79. $\dfrac{15\sqrt{8}}{4} - \dfrac{2\sqrt{2}}{5}$ **80.** $\dfrac{23\sqrt{11}}{2} - \dfrac{\sqrt{44}}{8}$

81. $20\sqrt[3]{b^4} - 4\sqrt[3]{b}$

82. $2\sqrt[4]{64} - \sqrt[4]{324} + \sqrt[4]{4}$

83. $2\sqrt{3z} + 3\sqrt{12z} + 3\sqrt{48z}$

84. $\sqrt{64x^3} - \sqrt{x} + 3\sqrt{x}$

85. $\sqrt[4]{81a^5b^5} - \sqrt[4]{ab}$ **86.** $\sqrt[4]{xy^5} + \sqrt[4]{x^5y}$

87. $5\sqrt[3]{\dfrac{n^4}{125}} - 2\sqrt[3]{n}$ **88.** $\sqrt[3]{\dfrac{8x}{27}} - \dfrac{2\sqrt[3]{x}}{3}$

Exercises 89–96: Multiply and simplify.

89. $(3 + \sqrt{7})(3 - \sqrt{7})$

90. $(5 - \sqrt{5})(5 + \sqrt{5})$

91. $(\sqrt{x} + 8)(\sqrt{x} - 8)$

92. $(\sqrt{ab} - 3)(\sqrt{ab} + 3)$

93. $(\sqrt{ab} - \sqrt{c})(\sqrt{ab} + \sqrt{c})$

94. $(\sqrt{2x} + \sqrt{3y})(\sqrt{2x} - \sqrt{3y})$

95. $(\sqrt{x} - 7)(\sqrt{x} + 8)$

96. $(\sqrt{ab} - 1)(\sqrt{ab} - 2)$

Exercises 97–112: Rationalize the denominator.

97. $\dfrac{4}{\sqrt{3}}$ **98.** $\dfrac{8}{\sqrt{2}}$

99. $\dfrac{5}{3\sqrt{5}}$ **100.** $\dfrac{6}{11\sqrt{3}}$

101. $\sqrt{\dfrac{b}{12}}$ **102.** $\sqrt{\dfrac{5b}{72}}$

103. $\dfrac{1}{3 - \sqrt{2}}$ **104.** $\dfrac{1}{\sqrt{3} - 2}$

105. $\dfrac{\sqrt{2}}{\sqrt{5} + 2}$ **106.** $\dfrac{\sqrt{3} - 1}{\sqrt{3} + 1}$

107. $\dfrac{1}{\sqrt{7} - \sqrt{6}}$ **108.** $\dfrac{1}{\sqrt{8} - \sqrt{7}}$

109. $\dfrac{\sqrt{z}}{\sqrt{z} - 3}$ **110.** $\dfrac{2\sqrt{z}}{2 - \sqrt{z}}$

111. $\dfrac{\sqrt{a} + \sqrt{b}}{\sqrt{a} - \sqrt{b}}$ **112.** $\dfrac{1}{\sqrt{a + 1} + \sqrt{a}}$

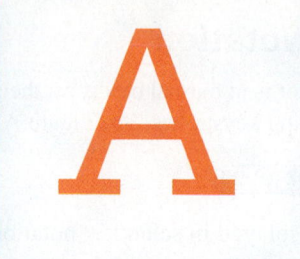

Appendix A: Using the Graphing Calculator

Overview of the Appendix

The intent of this appendix is to provide instruction in the TI-83, TI-83 Plus, and TI-84 Plus graphing calculators that may be used in conjunction with this textbook. It includes specific keystrokes needed to work several examples from the text. Students are also advised to consult the *Graphing Calculator Guidebook* provided by the manufacturer.

The following is a listing of the topics covered in this appendix.

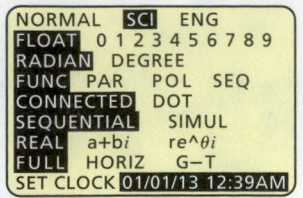

Figure A.1

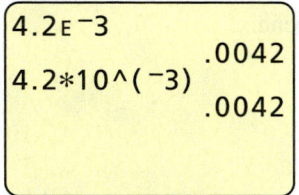

Figure A.2

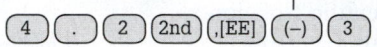

Figure A.3 Normal Mode

Displaying Numbers in Scientific Notation

To display numbers in scientific notation when the calculator is in normal mode, set the graphing calculator in scientific mode (SCI) by using the following keystrokes. See Figure A.1.

$$\boxed{\text{MODE}} \boxed{\triangleright} \boxed{\text{ENTER}} \boxed{\text{2nd}} \boxed{\text{MODE [QUIT]}}$$

Figure A.2 shows the numbers 5432 and 0.00001234 displayed in scientific notation.

> **SUMMARY: SETTING SCIENTIFIC MODE**
>
> If your calculator is in normal mode, it can be set in scientific mode by pressing
>
> $$\boxed{\text{MODE}} \boxed{\triangleright} \boxed{\text{ENTER}} \boxed{\text{2nd}} \boxed{\text{MODE [QUIT]}}.$$
>
> These keystrokes return the graphing calculator to the home screen.

Entering Numbers in Scientific Notation

Numbers can be entered in scientific notation. For example, to enter 4.2×10^{-3} in scientific notation, use the following keystrokes. (Be sure to use the negation key ($-$) rather than the subtraction key.)

> Use the negation key not the subtraction key.

$$\boxed{4} \boxed{.} \boxed{2} \boxed{\text{2nd}} \boxed{\text{,[EE]}} \boxed{(-)} \boxed{3}$$

This number can also be entered using the following keystrokes. See Figure A.3.

$$\boxed{4} \boxed{.} \boxed{2} \boxed{\times} \boxed{1} \boxed{0} \boxed{\wedge} \boxed{(} \boxed{(-)} \boxed{3} \boxed{)}$$

> **SUMMARY: ENTERING NUMBERS IN SCIENTIFIC NOTATION**
>
> One way to enter a number in scientific notation is to use the keystrokes
>
> $$\boxed{\text{2nd}} \boxed{\text{,[EE]}}$$
>
> to access an exponent (EE) of 10.

Entering Mathematical Expressions

Several expressions are evaluated in Example 7, Section 1.1. To evaluate $\sqrt[3]{131}$, use the following keystrokes from the home screen.

$$\boxed{\text{MATH}} \boxed{4} \boxed{1} \boxed{3} \boxed{1} \boxed{)} \boxed{\text{ENTER}}$$

To calculate $\pi^3 + 1.2^2$, use the following keystrokes. (Do *not* use 3.14 for π.)

$$\boxed{\text{2nd}} \boxed{\wedge[\pi]} \boxed{\wedge} \boxed{3} \boxed{+} \boxed{1} \boxed{.} \boxed{2} \boxed{x^2} \boxed{\text{ENTER}}$$

To calculate $|\sqrt{3} - 6|$, use the following keystrokes.

$$\boxed{\text{MATH}} \boxed{\triangleright} \boxed{1} \boxed{\text{2nd}} \boxed{x^2[\sqrt{\ }]} \boxed{3} \boxed{)} \boxed{-} \boxed{6} \boxed{)} \boxed{\text{ENTER}}$$

> **SUMMARY: ENTERING COMMON MATHEMATICAL EXPRESSIONS**
>
> To calculate a cube root, use the keystrokes $\boxed{\text{MATH}} \boxed{4}$.
>
> To access the number π, use the keystrokes $\boxed{\text{2nd}} \boxed{\wedge[\pi]}$.
>
> To access the absolute value, use the keystrokes $\boxed{\text{MATH}} \boxed{\triangleright} \boxed{1}$.
>
> To access the square root, use the keystrokes $\boxed{\text{2nd}} \boxed{x^2[\sqrt{\ }]}$.

```
ZOOM MEMORY
1:ZBox
2:Zoom In
3:Zoom Out
4:ZDecimal
5:ZSquare
6:ZStandard
7↓ZTrig
```
Figure A.4

```
WINDOW
 Xmin=⁻10
 Xmax=10
 Xscl=1
 Ymin=⁻10
 Ymax=10
 Yscl=1
 Xres=1
```
Figure A.5

```
WINDOW
 Xmin=⁻30
 Xmax=40
 Xscl=10
 Ymin=⁻400
 Ymax=800
 Yscl=100
 Xres=1
```
Figure A.6

```
L1    L2    L3    1
1     4     ------
2     5
3     6
------
L1={1,2,3}
```
Figure A.7

```
L1    L2    L3    1
⁻5    ⁻5    ------
⁻2    3
1     ⁻7
4     8
      ------
L1(5)=
```
Figure A.8

```
STAT PLOTS
1:Plot1...Off
    L1   L2   □
2:Plot2...Off
    L1   L2   □
3:Plot3...Off
    L1   L2   □
4↓PlotsOff
```
Figure A.9

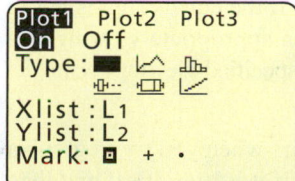
Figure A.10

Setting the Viewing Rectangle

In Example 12, Section 1.2, there are at least two ways to set the standard viewing rectangle to $[-10, 10, 1]$ by $[-10, 10, 1]$. The first method involves pressing ZOOM followed by 6. (See Figure A.4.) The second method is to press WINDOW and enter the following keystrokes. (See Figure A.5.)

Use the negation key.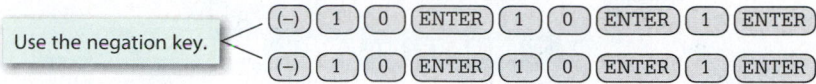

(Be sure to use the negation key $(-)$ rather than the subtraction key.) The viewing rectangle $[-30, 40, 10]$ by $[-400, 800, 100]$ can be set in a similar manner, as shown in Figure A.6. To see the viewing rectangle, press GRAPH.

> **SUMMARY: SETTING THE VIEWING RECTANGLE**
>
> To set the standard viewing rectangle, press ZOOM 6. To set any viewing rectangle, press WINDOW and enter the necessary values. To see the viewing rectangle, press GRAPH.
>
> *Note:* You do not need to change "Xres".

Making a Scatterplot or a Line Graph

In Example 13, Section 1.2, we are asked to make a scatterplot with $(-5, -5)$, $(-2, 3)$, $(1, -7)$, and $(4, 8)$. Begin this task by following these steps.

1. Press STAT followed by 1.
2. If list L1 is not empty, use the arrow keys to place the cursor on L1, as shown in Figure A.7. Then press CLEAR followed by ENTER. This deletes all elements in the list. Similarly, if L2 is not empty, clear the list.
3. Input each x-value into list L1, followed by ENTER. Input each y-value into list L2, followed by ENTER. See Figure A.8.

It is essential that both lists have the same number of values—otherwise an error message will appear when a scatterplot is attempted. Before these four points can be plotted, STATPLOT must be turned on. It is accessed by pressing

$$\boxed{2\text{nd}}\ \boxed{\text{Y}=[\text{STAT PLOT}]}$$

as shown in Figure A.9.

There are three possible STATPLOTS, numbered 1, 2, and 3. Any one of the three can be selected. The first plot is selected by pressing 1. Next, place the cursor over "On" and press ENTER to turn Plot1 on. There are six types of plots that can be selected. The first type is a *scatterplot* and the second type is a *line graph*, so place the cursor over the first type of plot and press ENTER to select a scatterplot. (To make the line graph in Example 14, Section 1.2, be sure to select the line graph.) The x-values are stored in list L1, so select L1 for "Xlist" by pressing 2nd 1. Similarly, press 2nd 2 for "Ylist," since the y-values are stored in list L2. Finally, there are three styles of marks that can be used to show data points in the graph. We usually use the first, because it is largest and shows up the best. Make the screen appear as in Figure A.10. Before plotting the four data points, be sure to set an appropriate viewing rectangle. Then press GRAPH. The data points will appear as in Figure A.11 on the next page.

NOTE 1 A fast way to set the viewing rectangle for any scatterplot is to select the ZOOMSTAT feature by pressing ZOOM 9. This feature automatically scales the viewing rectangle so that all data points are shown.

[−10, 10, 1] by [−10, 10, 1]

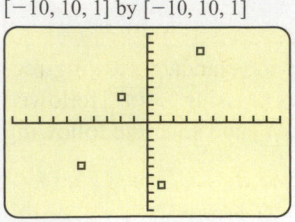

Figure A.11

NOTE 2 If an equation has been entered into the (Y =) menu and selected, it will be graphed with the data. Throughout this textbook, this feature is used frequently in modeling data.

NOTE 3 When the error message "ERR: DIM MISMATCH" appears, it usually means that there are not equal numbers of x-values and y-values in lists L1 and L2.

SUMMARY: MAKING A SCATTERPLOT OR A LINE GRAPH

The following are basic steps necessary to make either a scatterplot or a line graph.

STEP 1: Use (STAT)(1) to access lists L1 and L2.

STEP 2: If list L1 is not empty, place the cursor on L1 and press (CLEAR)(ENTER). Repeat for list L2 if it is not empty.

STEP 3: Enter the x-values into list L1 and the y-values into list L2.

STEP 4: Use (2nd)(Y = [STAT PLOT]) to set the appropriate parameters for the scatterplot or line graph.

STEP 5: Either set an appropriate viewing rectangle or press (ZOOM)(9). This feature automatically sets the viewing rectangle and plots the data.

Note: (ZOOM)(9) *cannot* be used to set a viewing rectangle for the graph of a function.

Deleting and Inserting a List A list, such as L2, can be deleted. Press (STAT)(1) and then place the cursor on L2 and press (DEL). If you want to insert a deleted list, press (STAT)(1) and then place the cursor where you want to insert the list. For example, to insert L2, place the cursor on L3. Press (2nd)(DEL [INS])(2nd)(2[L2])(ENTER).

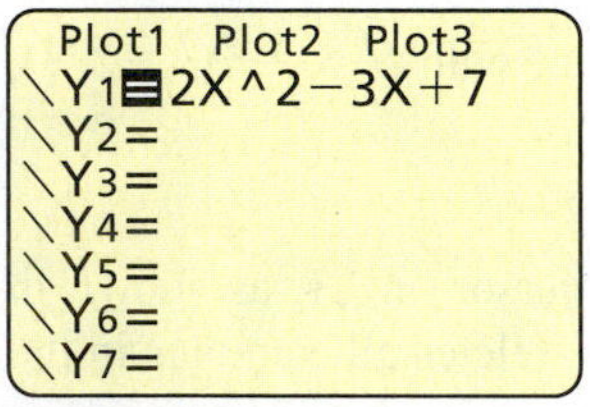

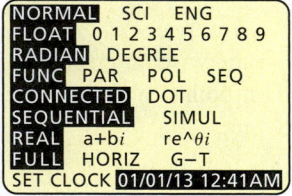

Figure A.12

Entering a Formula for a Function

To enter the formula for a function f, press (Y =). For example, use the following keystrokes after "$Y_1 = $" to enter $f(x) = 2x^2 - 3x + 7$. See Figure A.12.

$$(Y =)(CLEAR)(2)(X,T,\theta,n)(\wedge)(2)(-)(3)(X,T,\theta,n)(+)(7)$$

Note that there is a built-in key for entering the variable X. If "$Y_1 = $" does not appear after you press (Y =), press (MODE) and make sure the calculator is set in function mode, denoted "Func". See Figure A.13.

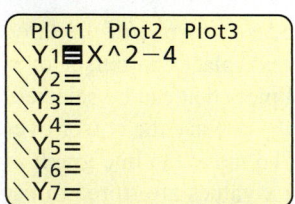

Figure A.13

SUMMARY: ENTERING A FORMULA FOR A FUNCTION

To enter the formula for a function, press (Y =). To delete an existing formula, press (CLEAR). Then enter the symbolic representation for the function.

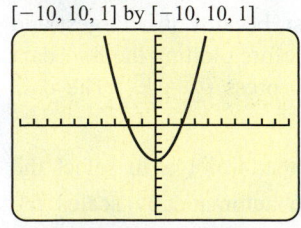

Figure A.14

[−10, 10, 1] by [−10, 10, 1]

Graphing a Function

To graph a function such as $f(x) = x^2 - 4$, start by pressing (Y =) and then enter $Y_1 = X^2 - 4$. If there is an equation already entered, remove it by pressing (CLEAR). The equals sign in "$Y_1 = $" should be in reverse video (a dark rectangle surrounding a white equals sign), which indicates that the equation will be graphed. If the equals sign is not in reverse video, place the cursor over it and press (ENTER). Set an appropriate viewing rectangle and then press (GRAPH). The graph of f will appear in the specified viewing rectangle. See Figures A.14 and A.15.

NOTE If the error message "ERR: DIM MISMATCH" appears when you try to graph a function, check to see if one of the STATPLOTS is turned on. If it is, turn it off and then try graphing the function.

Figure A.15

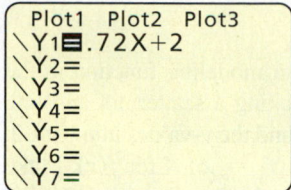

Figure A.16

Figure A.17

SUMMARY: GRAPHING A FUNCTION

Use the $\boxed{Y =}$ menu to enter the formula for the function and the $\boxed{\text{WINDOW}}$ menu to set an appropriate viewing rectangle. Then press $\boxed{\text{GRAPH}}$.

ZoomFit The ZoomFit feature can be used to find an appropriate window when graphing a function. ZoomFit leaves the current Xmin and Xmax settings unchanged and adjusts the current Ymin and Ymax values so that they are equal to the smallest and largest y-values on the graph of the function between Xmin and Xmax. To use ZoomFit, press $\boxed{\text{ZOOM}}$ $\boxed{0}$.

Making a Table

To make a table of values for a function, such as $f(x) = 0.72x + 2$, start by pressing $\boxed{Y =}$ and then entering the formula $Y_1 = .72X + 2$, as shown in Figure A.16. To set the table parameters, use the following keystrokes. (See Figure A.17.)

$$\boxed{\text{2nd}}\ \boxed{\text{WINDOW [TBLSET]}}\ \boxed{6}\ \boxed{0}\ \boxed{\text{ENTER}}\ \boxed{1}$$

These keystrokes specify a table that starts at $x = 60$ and increments the x-values by 1. Therefore the values of Y_1 at $x = 60, 61, 62, \ldots$ appear in the table. To create this table, press the following keys.

$$\boxed{\text{2nd}}\ \boxed{\text{GRAPH [TABLE]}}$$

X	Y1
60	45.2
61	45.92
62	46.64
63	47.36
64	48.08
65	48.8
66	49.52

Y1■.72X+2

Figure A.18

One can scroll through x- and y-values by using the arrow keys. See Figure A.18. Note that there is no first or last x-value in the table.

SUMMARY: MAKING A TABLE OF A FUNCTION

Enter the formula for the function using $\boxed{Y =}$. Then press

$$\boxed{\text{2nd}}\ \boxed{\text{WINDOW [TBLSET]}}$$

to set the starting x-value and the increment between x-values appearing in the table. Create the table by pressing

$$\boxed{\text{2nd}}\ \boxed{\text{GRAPH [TABLE]}}.$$

Squaring a Viewing Rectangle

In a square viewing rectangle, the graph of $y = x$ is a line that makes a 45° angle with the positive x-axis, a circle appears circular, and all sides of a square have the same length. An approximately square viewing rectangle can be set if the distance along the x-axis is 1.5 times the distance along the y-axis. Examples of viewing rectangles that are (approximately) square include

$$[-6, 6, 1]\ \text{by}\ [-4, 4, 1] \quad \text{and} \quad [-9, 9, 1]\ \text{by}\ [-6, 6, 1].$$

Square viewing rectangles can be set automatically by pressing either

$$\boxed{\text{ZOOM}}\ \boxed{4} \quad \text{or} \quad \boxed{\text{ZOOM}}\ \boxed{5}.$$

ZOOM 4 provides a decimal window, which is discussed later. See Figure A.19.

```
ZOOM MEMORY
1:ZBox
2:Zoom In
3:Zoom Out
4:ZDecimal
5:ZSquare
6:ZStandard
7↓ZTrig
```

Figure A.19

SUMMARY: SQUARING A VIEWING RECTANGLE

Either $\boxed{\text{ZOOM}}$ $\boxed{4}$ or $\boxed{\text{ZOOM}}$ $\boxed{5}$ may be used to produce a square viewing rectangle. An (approximately) square viewing rectangle has the form

$$[-1.5k, 1.5k, 1]\ \text{by}\ [-k, k, 1],$$

where k is a positive number.

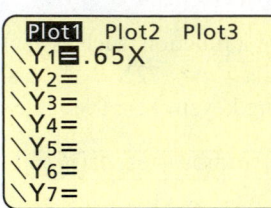

Figure A.20

Plotting Data and an Equation

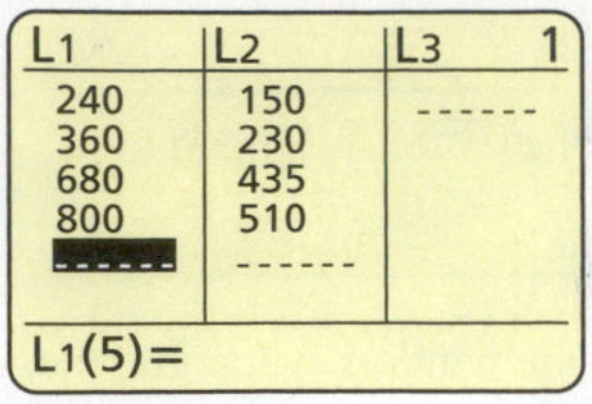

In Example 3, Section 2.4, we are asked to plot data and graph a modeling function in the same xy-plane. (You may want to refer to the subsection on making a scatterplot and line graph in this appendix.) Start by entering the x-values into list L1 and the y-values into list L2, as shown in Figure A.20. Then press $\boxed{Y=}$ and enter the formula $Y_1 = .65X$ for $f(x)$. Make sure that STATPLOT is on, and set an appropriate viewing rectangle. See Figures A.21 and A.22, and note that Figure A.21 shows "Plot1" in reverse video, which indicates that the scatterplot is on. Now press $\boxed{\text{GRAPH}}$ to have both the scatterplot and the graph of Y_1 appear in the same viewing rectangle, as shown in Figure A.23.

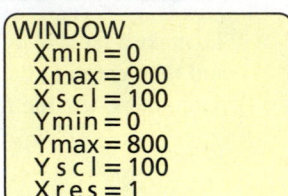

Figure A.21

Figure A.22

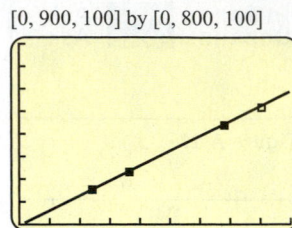

Figure A.23

SUMMARY: PLOTTING DATA AND AN EQUATION

STEP 1: Enter the x-values into list L1 and the y-values into list L2 using the STAT EDIT menu. Turn on Plot1 so that the scatterplot appears.

STEP 2: Use the $\boxed{Y=}$ menu to enter the equation to be graphed.

STEP 3: Use $\boxed{\text{WINDOW}}$ or $\boxed{\text{ZOOM}}$ to set an appropriate viewing rectangle.

STEP 4: Press $\boxed{\text{GRAPH}}$ to graph both the scatterplot and the equation in the same viewing rectangle.

Figure A.24

Accessing the Greatest Integer Function

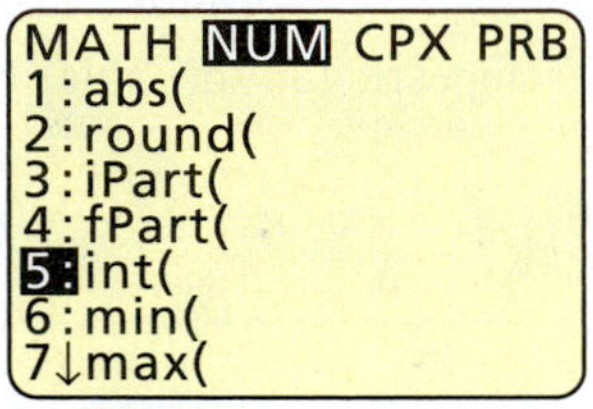

To access the greatest integer function, enter the following keystrokes from the home screen.

$$\boxed{\text{MATH}} \quad \boxed{\triangleright} \quad \boxed{5}$$

See Figure A.24.

SUMMARY: ACCESSING THE GREATEST INTEGER FUNCTION

STEP 1: Press $\boxed{\text{MATH}}$.

STEP 2: Position the cursor over "NUM".

STEP 3: Press $\boxed{5}$ to select the greatest integer function, which is denoted "int(".

Finding the Line of Least-Squares Fit

In Example 11, Section 2.1, the line of least-squares fit for the points (1, 1), (2, 3), and (3, 4) is found. Begin by entering the points in the same way as for a scatterplot. See Figure 2.14, where the x-values are in list L1 and the y-values are in list L2.

After the data have been entered, perform the following keystrokes from the home screen.

$$\boxed{\text{CLEAR}} \quad \boxed{\text{STAT}} \quad \boxed{\triangleright} \quad \boxed{4}$$

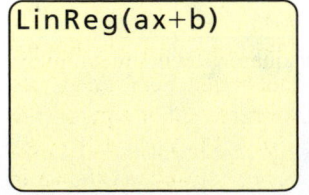

Figure A.25

(See Figure 2.15.) This causes "LinReg(ax+b)" to appear on the home screen, as shown in Figure A.25. The graphing calculator assumes that the x-values are in list L1 and the y-values are in list L2. Now press $\boxed{\text{ENTER}}$. The result is shown in Figure 2.16.

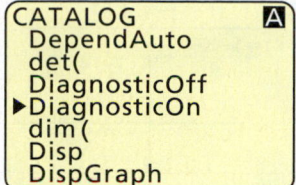

Figure A.26

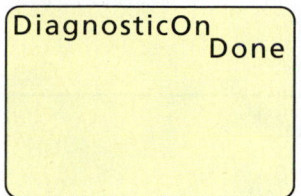

Figure A.27

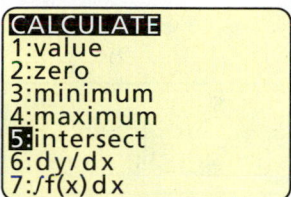

Figure A.28

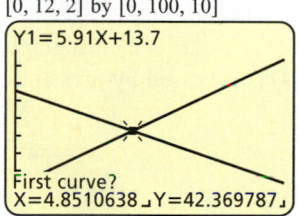

Figure A.29

[0, 12, 2] by [0, 100, 10]

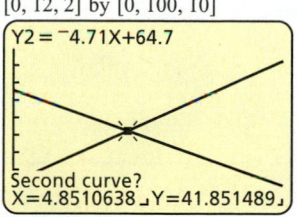

Figure A.30

[0, 12, 2] by [0, 100, 10]

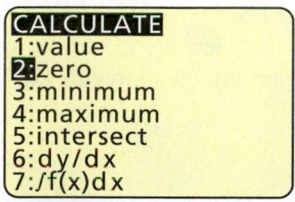

Figure A.31

Figure A.32

If the correlation coefficient r does not appear, enter the keystrokes

$$\boxed{\text{2nd}}\ \boxed{\text{0 [CATALOG]}}$$

and scroll down until you find "DiagnosticsOn". Press $\boxed{\text{ENTER}}$ twice. See Figures A.26 and A.27. The graphs of the data and the least-squares regression line are shown in Figure 2.17.

SUMMARY: LINEAR LEAST-SQUARES FIT

STEP 1: Enter the data using $\boxed{\text{STAT}}\ \boxed{1}$, as is done for a scatterplot. Input the x-values into list L1 and the y-values into list L2.

STEP 2: Press $\boxed{\text{STAT}}\ \boxed{\triangleright}\ \boxed{4}$ from the home screen to access the least-squares regression line. Press $\boxed{\text{ENTER}}$ to start the computation. See page AP-11 to learn how to copy a regression equation into Y_1.

Locating a Point of Intersection

In Example 8, Section 2.2, we are asked to find the point of intersection for two lines. To find the point of intersection for the graphs of

$$f(x) = 5.91x + 13.7 \quad \text{and} \quad g(x) = -4.71x + 64.7,$$

start by entering Y_1 and Y_2, as shown in Figure A.28. Set the viewing rectangle to [0, 12, 2] by [0, 100, 10], and graph both equations in the same viewing rectangle, as shown in Figure 2.24. Then press the following keys to find the intersection point.

$$\boxed{\text{2nd}}\ \boxed{\text{TRACE [CALC]}}\ \boxed{5}$$

See Figure A.29, where the intersect utility is being selected. The calculator prompts for the first curve, as shown in Figure A.30. Use the arrow keys to locate the cursor near the point of intersection and press $\boxed{\text{ENTER}}$. Repeat these steps for the second curve, as shown in Figure A.31. Finally the calculator prompts for a guess. For each of the three prompts, place the free-moving cursor near the point of intersection and press $\boxed{\text{ENTER}}$. The approximate coordinates of the point of intersection are shown in Figure 2.25.

SUMMARY: FINDING A POINT OF INTERSECTION

STEP 1: Graph the two functions in an appropriate viewing rectangle.

STEP 2: Press $\boxed{\text{2nd}}\ \boxed{\text{TRACE [CALC]}}\ \boxed{5}$.

STEP 3: Use the arrow keys to select an approximate location for the point of intersection. Press $\boxed{\text{ENTER}}$ to make the three selections for "First curve?", "Second curve?", and "Guess?". (If the cursor is near the point of intersection, you usually do not need to move the cursor for each selection. Just press $\boxed{\text{ENTER}}$ three times.)

Locating a Zero of a Function

In Example 4, Section 2.3, we are asked to locate an x-intercept, or *zero*, of the function f given by $f(x) = 1 - x - \frac{1}{2}x + 2$. Start by entering $Y_1 = 1 - X - .5X + 2$ into the $\boxed{\text{Y =}}$ menu. Set the viewing rectangle to [−6, 6, 1] by [−4, 4, 1] and graph Y_1. Afterwards, press the following keys to invoke the zero finder. (See Figure A.32.)

$$\boxed{\text{2nd}}\ \boxed{\text{TRACE [CALC]}}\ \boxed{2}$$

The calculator prompts for a left bound. Use the arrow keys to set the cursor to the left of the x-intercept and press $\boxed{\text{ENTER}}$. The calculator then prompts for a right bound. Set the cursor to the right of the x-intercept and press $\boxed{\text{ENTER}}$. Finally the calculator prompts for a guess. Set the cursor roughly at the x-intercept and press $\boxed{\text{ENTER}}$. See Figures A.33–A.35 on the next page. The calculator then approximates the x-intercept, or zero, automatically, as shown in Figure 2.35(b).

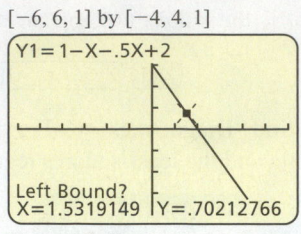

Figure A.33

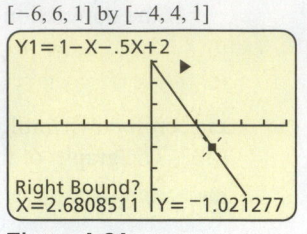

Figure A.34

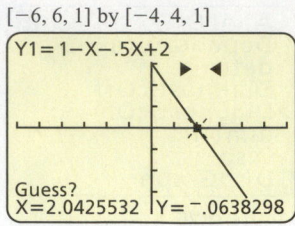

Figure A.35

SUMMARY: LOCATING A ZERO OF A FUNCTION

STEP 1: Graph the function in an appropriate viewing rectangle.

STEP 2: Press (2nd) (TRACE [CALC]) (2).

STEP 3: Select the left and right bounds, followed by a guess. Press (ENTER) after each selection. The calculator then approximates the zero.

Setting Connected and Dot Mode

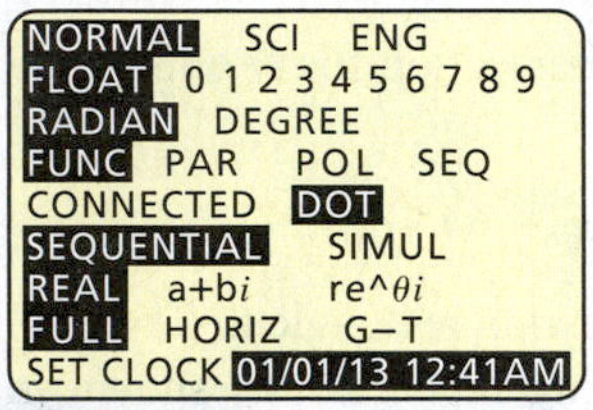

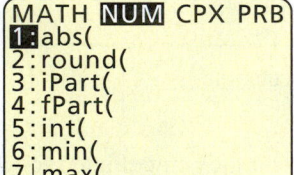

Figure A.36

In Figure 2.48 of Section 2.4 a form of the greatest integer function is graphed in dot mode, and in Figure 2.49 it is graphed in connected mode. To set your graphing calculator in dot mode, press (MODE), position the cursor over "Dot", and press (ENTER). See Figure A.36. Graphs will now appear in dot mode rather than connected mode.

SUMMARY: SETTING CONNECTED OR DOT MODE

STEP 1: Press (MODE).

STEP 2: Position the cursor over "Connected" or "Dot". Press (ENTER).

Accessing the Absolute Value

Figure A.37

In Example 1, Section 2.5, the absolute value is used to graph $f(x) = |x + 2|$. To graph f, begin by entering $Y_1 = abs(X + 2)$. The absolute value (abs) is accessed by pressing

(MATH) (▷) (1).

See Figure A.37.

SUMMARY: ACCESSING THE ABSOLUTE VALUE

STEP 1: Press (MATH).

STEP 2: Position the cursor over "NUM".

STEP 3: Press (1) to select the absolute value.

Finding Extrema (Minima and Maxima)

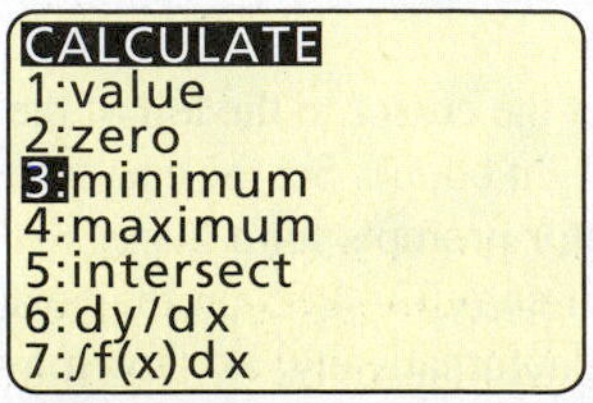

Figure A.38

To find a minimum point (or vertex) on a graph, such as $f(x) = 1.5x^2 - 6x + 4$, start by entering $Y_1 = 1.5X^2 - 6X + 4$ into the (Y=) menu. Set the viewing rectangle to $[-4.7, 4.7, 1]$ by $[-3.1, 3.1, 1]$ by entering (ZOOM) (4). Then perform the following keystrokes to find the minimum y-value.

(2nd) (TRACE [CALC]) (3)

See Figure A.38.

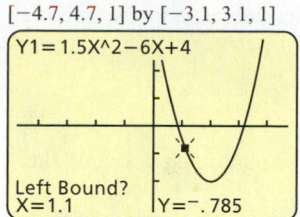

[−4.7, 4.7, 1] by [−3.1, 3.1, 1]

Y1=1.5X^2−6X+4

Left Bound?
X=1.1   Y=−.785

Figure A.39

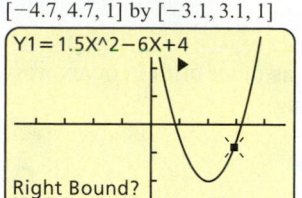

[−4.7, 4.7, 1] by [−3.1, 3.1, 1]

Y1=1.5X^2−6X+4

Right Bound?
X=2.9   Y=−.785

Figure A.40

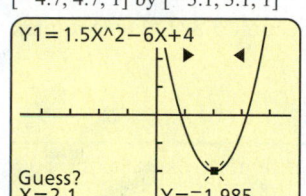

[−4.7, 4.7, 1] by [−3.1, 3.1, 1]

Y1=1.5X^2−6X+4

Guess?
X=2.1   Y=−1.985

Figure A.41

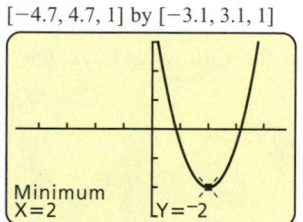

[−4.7, 4.7, 1] by [−3.1, 3.1, 1]

Minimum
X=2   Y=−2

Figure A.42

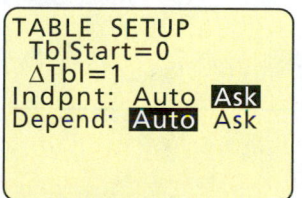

TABLE SETUP
 TblStart=0
 ΔTbl=1
Indpnt: Auto **Ask**
Depend: **Auto** Ask

Figure A.43

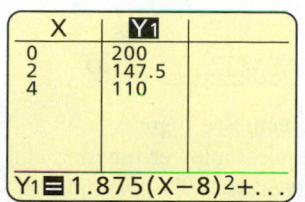

X	Y1
0	200
2	147.5
4	110

Y1 ▪ 1.875(X−8)2+...

Figure A.44

The calculator prompts for a left bound. Use the arrow keys to position the cursor to the left of the vertex and press (ENTER). Similarly, position the cursor to the right of the vertex for the right bound and press (ENTER). Finally the calculator asks for a guess between the left and right bounds. Place the cursor near the minimum point and press (ENTER). See Figures A.39–A.41. The minimum point (or vertex) is shown in Figure A.42.

To find a maximum of the function f on an interval, use a similar approach, except enter

$$\boxed{2nd}\ \boxed{TRACE\ [CALC]}\ \boxed{4}.$$

The calculator prompts for left and right bounds, followed by a guess. Press (ENTER) after the cursor has been located appropriately for each prompt. An example of a maximum point is displayed in Figure 3.17 in Section 3.1.

SUMMARY: FINDING EXTREMA (MAXIMA AND MINIMA)

STEP 1: Graph the function in an appropriate viewing rectangle.

STEP 2: Press (2nd) (TRACE [CALC]) (3) to find a minimum point or (2nd) (TRACE [CALC]) (4) to find a maximum point.

STEP 3: Use the arrow keys to locate the left and right x-bounds, followed by a guess. Press (ENTER) to select each position of the cursor.

Using the Ask Table Feature

In Example 10, Section 3.1, a table with x-values of 0, 2, 4, 6, and 8 is created. Start by entering $Y_1 = 1.875(X - 8)^2 + 80$. To obtain the table shown in Figure 3.20, use the Ask feature rather than the Auto feature for the independent variable (Indpnt:). Press (2nd) (GRAPH [TABLE]). Whenever an x-value is entered, the corresponding y-value is calculated automatically. See Figures A.43 and A.44.

SUMMARY: USING THE ASK FEATURE FOR A TABLE

STEP 1: Enter the formula for $f(x)$ into Y_1 by using the (Y =) menu.

STEP 2: Press (2nd) (WINDOW [TBLSET]) to access "TABLE SETUP" and then select "Ask" for the independent variable (Indpnt:). "TblStart" and "ΔTbl" do not need to be set.

STEP 3: Enter x-values of your choice. The corresponding y-values will be calculated automatically.

Finding a Nonlinear Function of Least-Squares Fit

In Example 11, Section 3.1, a quadratic function of least-squares fit is found in a manner similar to the way a linear function of least-squares fit is found. To solve Example 11, start by pressing (STAT) (1) and then enter the data points from Table 3.4, as shown in Figure 3.22. Input the x-values into list L1 and the y-values into list L2. To find the equation for a quadratic polynomial of least-squares fit, perform the following keystrokes from the home screen.

$$\boxed{CLEAR}\ \boxed{STAT}\ \boxed{\triangleright}\ \boxed{5}$$

This causes "Quadreg" to appear on the home screen. The calculator assumes that the x-values are in list L1 and the y-values are in list L2, unless otherwise designated. Press (ENTER) to obtain the quadratic regression equation, as shown in Figure 3.24. Graphs of the data and the regression equation are shown in Figure 3.25.

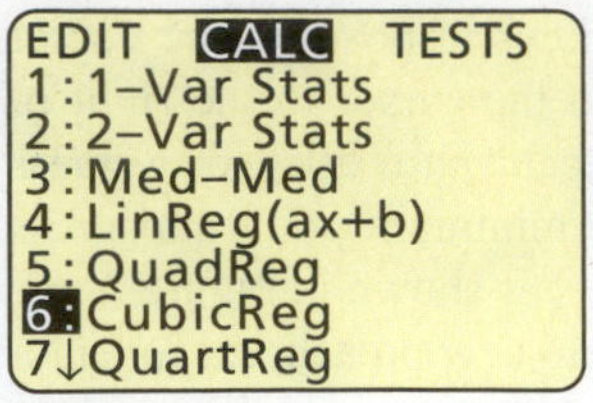

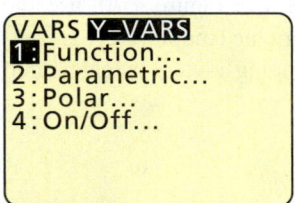

Figure A.45

Other types of regression equations, such as cubic, quartic, power, and exponential, can be selected from the STAT CALC menu. See Figure A.45.

> **SUMMARY: NONLINEAR LEAST-SQUARES FIT**
>
> **STEP 1:** Enter the data using (STAT)(1). Input the *x*-values into list L1 and the *y*-values into list L2, as is done for a scatterplot.
>
> **STEP 2:** From the home screen, press (STAT)(▷) and select a type of least-squares modeling function from the menu. Press (ENTER) to initiate the computation.

Evaluating Complex Arithmetic

Complex arithmetic can be performed in much the same way as other arithmetic expressions are evaluated. The imaginary unit *i* is obtained by entering

(2nd)(. [i])

from the home screen. For example, to add the numbers $(-2 + 3i) + (4 - 6i)$, perform the following keystrokes on the home screen.

The result is shown in the first two lines of Figure 3.37 in Section 3.3. Other complex arithmetic operations are done similarly.

> **SUMMARY: EVALUATING COMPLEX ARITHMETIC**
>
> Enter a complex expression in the same way as any other arithmetic expression. To obtain the complex number *i*, use (2nd)(. [i]).

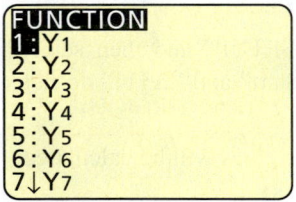

Figure A.46

Accessing the Variable Y_1

In Figure 3.78, Section 3.5, the expressions $-Y_1$ and $Y_1(-X)$ in the (Y =) menu are used to graph reflections. The Y_1 variable can be found by pressing the following keys. (See Figures A.46 and A.47.)

> **SUMMARY: ACCESSING THE VARIABLE Y_1**
>
> **STEP 1:** Press (VARS).
>
> **STEP 2:** Position the cursor over "Y-VARS".
>
> **STEP 3:** Press (1) twice.
>
> These keystrokes will make Y_1 appear on the screen.

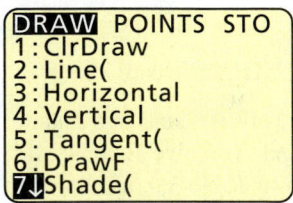

Figure A.47

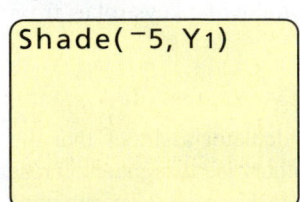

Figure A.48

Shade(⁻5, Y1)

Figure A.49

Shading between Two Graphs

In Example 9, Section 3.5, the region below the graph of $f(x) = -0.4x^2 + 4$ is shaded to make it look like a mountain, as illustrated in Figure 3.97. One way to shade below the graph of *f* is to begin by entering $Y_1 = -.4X{^\wedge}2 + 4$ after pressing (Y =). Then use the following keystrokes from the home screen.

The expression Shade(-5, Y_1) should appear on your home screen. See Figures A.48 and A.49. The shading utility, accessed from the DRAW menu, requires a lower function and then an upper function, separated by a comma. When (ENTER) is pressed, the graphing calculator shades between the graph of the lower function and the graph of the upper function.

For the lower function we have arbitrarily selected $y = -5$ because its graph lies below the graph of f and does not appear in the viewing rectangle in Figure 3.97. Instead of entering the variable Y_1, we could enter the formula $-.4X^2 + 4$ for the upper function.

SUMMARY: SHADING A GRAPH

STEP 1: Press (2nd) (PRGM [DRAW]) (7) from the home screen.

STEP 2: Enter a formula or a variable such as Y_1 for the lower function, followed by a comma.

STEP 3: Enter a formula or a variable such as Y_2 for the upper function, followed by a right parenthesis.

STEP 4: Set an appropriate viewing rectangle.

STEP 5: Press (ENTER). The region between the two graphs will be shaded.

Copying a Regression Equation into $Y_1 =$

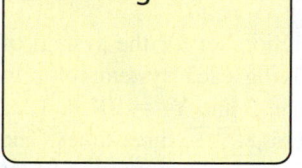

Figure A.50

In Example 8, Section 4.2, we are asked to use cubic regression to model real data. The resulting formula for the cubic function, shown in Figure 4.54, is quite complicated and tedious to enter into $Y_1 =$ by hand. A graphing calculator has the capability to copy this equation into Y_1 automatically. To do this, clear the equation for $Y_1 =$. Then enter Y_1 after "CubicReg", as shown in Figure A.50. When (ENTER) is pressed, the regression equation will be calculated and then copied into $Y_1 =$, as shown in Figure A.51. The following keystrokes may be used from the home screen. (Be sure to enter the data into lists L1 and L2.)

$$\boxed{\text{STAT}} \ \boxed{\triangleright} \ \boxed{6} \ \boxed{\text{VARS}} \ \boxed{\triangleright} \ \boxed{1} \ \boxed{1} \ \boxed{\text{ENTER}}$$

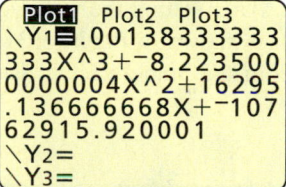

Figure A.51

SUMMARY: COPYING A REGRESSION EQUATION INTO $Y_1 =$

STEP 1: Clear Y_1 in the (Y =) menu if an equation is present. Return to the home screen.

STEP 2: Select a type of regression from the STAT CALC menu.

STEP 3: Press (VARS) (▷) (1) (1) (ENTER).

Setting a Decimal Window

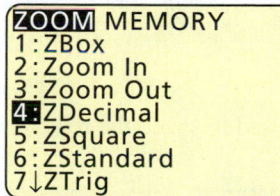

Figure A.52

In Example 1, Section 4.7, a decimal (or friendly) window is used to trace the graph of f. With a decimal window, the cursor stops on convenient x-values. In the decimal window $[-9.4, 9.4, 1]$ by $[-6.2, 6.2, 1]$, the cursor stops on x-values that are multiples of 0.2. If we reduce the viewing rectangle to $[-4.7, 4.7, 1]$ by $[-3.1, 3.1, 1]$, the cursor stops on x-values that are multiples of 0.1. To set this smaller window automatically, press (ZOOM) (4). See Figure A.52. Decimal windows are useful when graphing rational functions with asymptotes in connected mode.

SUMMARY: SETTING A DECIMAL WINDOW

Press (ZOOM) (4) to set the viewing rectangle $[-4.7, 4.7, 1]$ by $[-3.1, 3.1, 1]$. A convenient larger decimal window is $[-9.4, 9.4, 1]$ by $[-6.2, 6.2, 1]$.

Graphing an Inverse Function

In Example 7, Section 5.2, the inverse function of $f(x) = x^3 + 2$ is graphed. A graphing calculator can graph the inverse of a function without a formula for $f^{-1}(x)$. Begin by entering $Y_1 = X^3 + 2$ into the (Y =) menu. Then return to the home screen by pressing

$$\boxed{\text{2nd}} \ \boxed{\text{MODE [QUIT]}} \ .$$

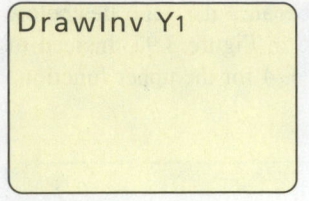

Figure A.53

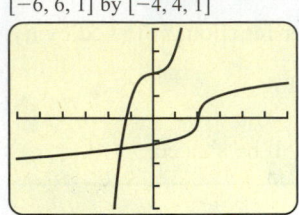

$[-6, 6, 1]$ by $[-4, 4, 1]$

Figure A.54

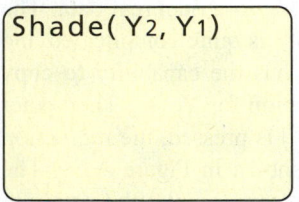

Figure A.55

$[-6, 6, 1]$ by $[-6, 6, 1]$

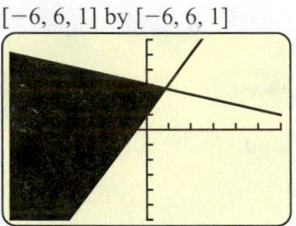

Figure A.56

The DrawInv utility may be accessed by pressing

$$\boxed{2nd} \ \boxed{PRGM\ [DRAW]} \ \boxed{8} \ ,$$

followed by

$$\boxed{VARS} \ \boxed{\triangleright} \ \boxed{1} \ \boxed{1}$$

to obtain the variable Y_1. See A.53. Pressing $\boxed{ENTER}$ causes both Y_1 and its inverse to be graphed, as shown in A.54.

SUMMARY: GRAPHING AN INVERSE FUNCTION

STEP 1: Enter the formula for $f(x)$ into Y_1 using the $\boxed{Y=}$ menu.

STEP 2: Set an appropriate viewing rectangle by pressing $\boxed{WINDOW}$.

STEP 3: Return to the home screen by pressing $\boxed{2nd}$ $\boxed{MODE\ [QUIT]}$.

STEP 4: Press $\boxed{2nd}$ $\boxed{PRGM\ [DRAW]}$ $\boxed{8}$ $\boxed{VARS}$ $\boxed{\triangleright}$ $\boxed{1}$ $\boxed{1}$ $\boxed{ENTER}$ to create the graphs of f and f^{-1}.

Shading a System of Inequalities

In Example 2(b), Section 9.2, we are asked to shade the solution set for the system of linear inequalities $x + 3y \leq 9, 2x - y \leq -1$. Begin by solving each system for y to obtain $y \leq -\frac{1}{3}x + 3$ and $y \geq 2x + 1$. Then let $Y_1 = -X/3 + 3$ and $Y_2 = 2X + 1$, as shown in Figure 9.30. Position the cursor to left of Y_1 and press $\boxed{ENTER}$ three times. The triangle that appears indicates that the calculator will shade the region below the graph of Y_1. Next locate the cursor to the left of Y_2 and press $\boxed{ENTER}$ twice. This triangle indicates that the calculator will shade the region above the graph of Y_2. After setting the viewing rectangle to $[-6, 6, 1]$ by $[-6, 6, 1]$, press $\boxed{GRAPH}$. The result is shown in Figure 9.31. The solution set could also be shaded using Shade(Y_2, Y_1) from the home screen. See Figures A.55 and A.56.

SUMMARY: SHADING A SYSTEM OF INEQUALITIES

STEP 1: Solve each inequality for y.

STEP 2: Enter the formulas as Y_1 and Y_2 in the $\boxed{Y=}$ menu.

STEP 3: Locate the cursor to the left of Y_1 and press $\boxed{ENTER}$ two or three times, to shade either above or below the graph of Y_1. Repeat for Y_2.

STEP 4: Set an appropriate viewing rectangle.

STEP 5: Press $\boxed{GRAPH}$.

Note: The Shade utility under the DRAW menu can also be used to shade the region *between* two graphs.

Entering the Elements of a Matrix

In Example 9, Section 9.4, the augmented matrix A is given by

$$A = \begin{bmatrix} 2 & 1 & 2 & 10 \\ 1 & 0 & 2 & 5 \\ 1 & -2 & 2 & 1 \end{bmatrix}.$$

On the TI-83 Plus and TI-84 Plus, use the following keystrokes to define a matrix A with dimension 3×4. (On the TI-83 graphing calculator, the matrix menu is found by pressing $\boxed{MATRIX}$.)

See Figures 9.41 and 9.42. Then input the 12 elements of the matrix A, row by row, as shown in Figure 9.42. Finish each entry by pressing (ENTER). After these elements have been entered, press

$$\boxed{\text{2nd}}\ \boxed{\text{MODE [QUIT]}}$$

to return to the home screen. To display the matrix A, press

$$\boxed{\text{2nd}}\ \boxed{x^{-1}\ [\text{MATRIX}]}\ \boxed{1}\ \boxed{\text{ENTER}}.$$

See Figure 9.43.

SUMMARY: ENTERING THE ELEMENTS OF A MATRIX A

STEP 1: Begin by accessing the matrix A by pressing $\boxed{\text{2nd}}\ \boxed{x^{-1}\ [\text{MATRIX}]}\ \boxed{\triangleright}\ \boxed{\triangleright}\ \boxed{1}$.

STEP 2: Enter the dimension of A by pressing $\boxed{m}\ \boxed{\text{ENTER}}\ \boxed{n}\ \boxed{\text{ENTER}}$, where the dimension of the matrix is $m \times n$.

STEP 3: Input each element of the matrix, row by row. Finish each entry by pressing $\boxed{\text{ENTER}}$. Use $\boxed{\text{2nd}}\ \boxed{\text{MODE [QUIT]}}$ to return to the home screen.

Note: On the TI-83, replace the keystrokes $\boxed{\text{2nd}}\ \boxed{x^{-1}\ [\text{MATRIX}]}$ with $\boxed{\text{MATRIX}}$.

Reduced Row-Echelon Form

In Example 9, Section 9.4, the reduced row-echelon form of a matrix is found. To find this reduced row-echelon form on the TI-83 Plus and TI-84 Plus, use the following keystrokes from the home screen. (See Figure 9.44.)

$$\boxed{\text{2nd}}\ \boxed{x^{-1}\ [\text{MATRIX}]}\ \boxed{\triangleright}\ \boxed{\text{ALPHA}}\ \boxed{\text{APPS [B]}}\ \boxed{\text{2nd}}\ \boxed{x^{-1}\ [\text{MATRIX}]}\ \boxed{1}\ \boxed{)}\ \boxed{\text{ENTER}}$$

The resulting matrix is shown in Figure 9.45. On the TI-83 graphing calculator, use the following keystrokes to find the reduced row-echelon form.

$$\boxed{\text{MATRIX}}\ \boxed{\triangleright}\ \boxed{\text{ALPHA}}\ \boxed{\text{MATRIX [B]}}\ \boxed{\text{MATRIX}}\ \boxed{1}\ \boxed{)}\ \boxed{\text{ENTER}}$$

SUMMARY: FINDING THE REDUCED ROW-ECHELON FORM OF A MATRIX

STEP 1: To make rref([A]) appear on the home screen, use the following keystrokes.

$$\boxed{\text{2nd}}\ \boxed{x^{-1}\ [\text{MATRIX}]}\ \boxed{\triangleright}\ \boxed{\text{ALPHA}}\ \boxed{\text{APPS [B]}}\ \boxed{\text{2nd}}\ \boxed{x^{-1}\ [\text{MATRIX}]}\ \boxed{1}\ \boxed{)}\ \boxed{\text{ENTER}}$$

STEP 2: Press $\boxed{\text{ENTER}}$ to calculate the reduced row-echelon form. Use arrow keys to access elements that do not appear on the screen.

Note: On the TI-83, replace the keystrokes $\boxed{\text{2nd}}\ \boxed{x^{-1}\ [\text{MATRIX}]}$ with $\boxed{\text{MATRIX}}$ and $\boxed{\text{APPS [B]}}$ with $\boxed{\text{MATRIX [B]}}$.

Performing Arithmetic Operations on Matrices

In Example 10, Section 9.5, the matrices A and B are multiplied. Begin by entering the elements for the matrices A and B. The following keystrokes can be used to define a matrix A with dimension 3×3.

$$\boxed{\text{2nd}}\ \boxed{x^{-1}\ [\text{MATRIX}]}\ \boxed{\triangleright}\ \boxed{\triangleright}\ \boxed{1}\ \boxed{3}\ \boxed{\text{ENTER}}\ \boxed{3}\ \boxed{\text{ENTER}}$$

Next input the 9 elements in the matrix A, row by row. Finish each entry by pressing $\boxed{\text{ENTER}}$. See Figure 9.59. Repeat this process to define a matrix B with dimension 3×3.

$$\boxed{\text{2nd}}\ \boxed{x^{-1}\ [\text{MATRIX}]}\ \boxed{\triangleright}\ \boxed{\triangleright}\ \boxed{2}\ \boxed{3}\ \boxed{\text{ENTER}}\ \boxed{3}\ \boxed{\text{ENTER}}$$

Enter the 9 elements in B. See Figure 9.60. After the elements of A and B have been entered, press

$$\boxed{\text{2nd}}\ \boxed{\text{MODE [QUIT]}}$$

to return to the home screen. To multiply the expression AB, use the following keystrokes from the home screen.

$$\boxed{\text{2nd}}\ \boxed{x^{-1}\ \text{[MATRIX]}}\ \boxed{1}\ \boxed{\times}\ \boxed{\text{2nd}}\ \boxed{x^{-1}\ \text{[MATRIX]}}\ \boxed{2}\ \boxed{\text{ENTER}}$$

The result is shown in Figure 9.61.

SUMMARY: PERFORMING ARITHMETIC OPERATIONS ON MATRICES

STEP 1: Enter the elements of each matrix, beginning with the keystrokes

$$\boxed{\text{2nd}}\ \boxed{x^{-1}\ \text{[MATRIX]}}\ \boxed{\triangleright}\ \boxed{\triangleright}\ \boxed{k}\ \boxed{m}\ \boxed{\text{ENTER}}\ \boxed{n}\ \boxed{\text{ENTER}},$$

where k is the menu number of the matrix and the dimension of the matrix is $m \times n$.

STEP 2: Return to the home screen by pressing $\boxed{\text{2nd}}\ \boxed{\text{MODE [QUIT]}}$.

STEP 3: Enter the matrix expression, followed by $\boxed{\text{ENTER}}$. Use the keystrokes

$$\boxed{\text{2nd}}\ \boxed{x^{-1}\ \text{[MATRIX]}}\ \boxed{k}$$

to access the matrix with menu number k.

Note: On the TI-83, replace the keystrokes $\boxed{\text{2nd}}\ \boxed{x^{-1}\ \text{[MATRIX]}}$ with $\boxed{\text{MATRIX}}$.

Finding the Inverse of a Matrix

In Example 8, Section 9.6, the inverse of A, denoted A^{-1}, is displayed in Figure 9.65. To calculate A^{-1}, start by entering the elements of the matrix A, as shown in Figure A.57. To compute A^{-1}, perform the following keystrokes from the home screen.

$$\boxed{\text{2nd}}\ \boxed{x^{-1}\ \text{[MATRIX]}}\ \boxed{1}\ \boxed{x^{-1}}\ \boxed{\text{ENTER}}$$

The results are shown in Figure 9.65.

SUMMARY: FINDING THE INVERSE OF A SQUARE MATRIX

STEP 1: Enter the elements of the square matrix A.

STEP 2: Return to the home screen by pressing

$$\boxed{\text{2nd}}\ \boxed{\text{MODE [QUIT]}}.$$

STEP 3: Perform the following keystrokes from the home screen to display A^{-1}.

$$\boxed{\text{2nd}}\ \boxed{x^{-1}\ \text{[MATRIX]}}\ \boxed{1}\ \boxed{x^{-1}}\ \boxed{\text{ENTER}}$$

Note: On the TI-83, replace the keystrokes $\boxed{\text{2nd}}\ \boxed{x^{-1}\ \text{[MATRIX]}}$ with $\boxed{\text{MATRIX}}$.

Solving a Linear System with a Matrix Inverse

In Example 8, Section 9.6, the solution to a system of equations is found. The matrix equation $AX = B$ has the solution $X = A^{-1}B$, provided A^{-1} exists, and is given by

$$AX = \begin{bmatrix} 1 & 3 & -1 \\ 0 & -2 & 1 \\ -1 & 1 & -3 \end{bmatrix} \begin{bmatrix} x \\ y \\ z \end{bmatrix} = \begin{bmatrix} 6 \\ -2 \\ 4 \end{bmatrix} = B.$$

```
MATRIX[A]   3 ×3
[ 1      3     ⁻1    ]
[ 0     ⁻2     1     ]
[ ⁻1     1    ⁻3     ]
```

Figure A.57

To solve this equation, start by entering the elements of the matrices A and B. To compute the solution $A^{-1}B$, perform the following keystrokes from the home screen.

The results are shown in Figure 9.66.

SUMMARY: SOLVING A LINEAR SYSTEM WITH A MATRIX INVERSE

STEP 1: Write the system of equations as $AX = B$.

STEP 2: Enter the elements of the matrices A and B.

STEP 3: Return to the home screen by pressing

$\boxed{\text{2nd}}\ \boxed{\text{MODE [QUIT]}}$.

STEP 4: Perform the following keystrokes.

$\boxed{\text{2nd}}\ \boxed{x^{-1}\ [\text{MATRIX}]}\ \boxed{1}\ \boxed{x^{-1}}\ \boxed{\times}\ \boxed{\text{2nd}}\ \boxed{x^{-1}\ [\text{MATRIX}]}\ \boxed{2}\ \boxed{\text{ENTER}}$

Note: On the TI-83, replace the keystrokes with $\boxed{\text{2nd}}\boxed{x^{-1}\ [\text{MATRIX}]}$ with $\boxed{\text{MATRIX}}$.

Evaluating a Determinant

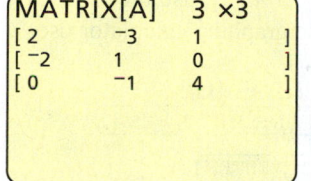

Figure A.58

In Example 4(a), Section 9.7, a graphing calculator is used to evaluate a determinant of a matrix. Start by entering the 9 elements of the 3×3 matrix A, as shown in Figure A.58. To compute det A, perform the following keystrokes from the home screen.

$\boxed{\text{2nd}}\ \boxed{x^{-1}\ [\text{MATRIX}]}\ \boxed{\triangleright}\ \boxed{1}\ \boxed{\text{2nd}}\ \boxed{x^{-1}\ [\text{MATRIX}]}\ \boxed{1}\ \boxed{)}\ \boxed{\text{ENTER}}$

The results are shown in Figure 9.68.

SUMMARY: EVALUATING A DETERMINANT OF A SQUARE MATRIX

STEP 1: Enter the elements of the matrix A.

STEP 2: Return to the home screen by pressing

$\boxed{\text{2nd}}\ \boxed{\text{MODE [QUIT]}}$.

STEP 3: Perform the following keystrokes.

$\boxed{\text{2nd}}\ \boxed{x^{-1}\ [\text{MATRIX}]}\ \boxed{\triangleright}\ \boxed{1}\ \boxed{\text{2nd}}\ \boxed{x^{-1}\ [\text{MATRIX}]}\ \boxed{1}\ \boxed{)}\ \boxed{\text{ENTER}}$

Note: On the TI-83, replace the keystrokes $\boxed{\text{2nd}}\boxed{x^{-1}\ [\text{MATRIX}]}$ with $\boxed{\text{MATRIX}}$.

Creating a Sequence

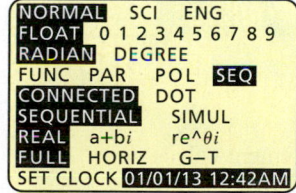

Figure A.59

A graphing calculator can be used to calculate the terms of the sequence given by $f(n) = 2n - 5$ for $n = 1, 2, 3, 4$. See Example 1(a), Section 11.1. Start by setting the mode of the calculator to sequence ("Seq") using the following keystrokes. (See Figure A.59.)

$\boxed{\text{MODE}}\ \boxed{\triangledown}\ \boxed{\triangledown}\ \boxed{\triangledown}\ \boxed{\triangleright}\ \boxed{\triangleright}\ \boxed{\triangleright}\ \boxed{\text{ENTER}}\ \boxed{\text{2nd}}\ \boxed{\text{MODE [QUIT]}}$

Then enter the following from the home screen.

$\boxed{\text{2nd}}\ \boxed{\text{STAT [LIST]}}\ \boxed{\triangleright}\ \boxed{5}$

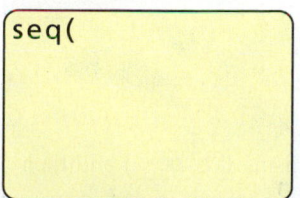

Figure A.60

On the home screen, "seq(" will appear, as shown in Figure A.60. This sequence utility requires that four things be entered—all separated by commas. They are the formula, the variable, the subscript of the first term, and the subscript of the last term. Use the following keystrokes to obtain the first four terms (a_1, a_2, a_3, a_4) of the sequence $a_n = 2n - 5$, as shown in Figure 11.1.

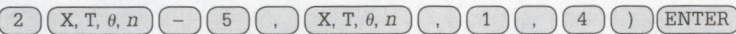

SUMMARY: CREATING A SEQUENCE

STEP 1: To create a sequence, use the keystrokes

$$\boxed{\text{2nd}}\ \boxed{\text{STAT [LIST]}}\ \boxed{\triangleright}\ \boxed{5}\ .$$

STEP 2: Enter the formula, the variable, the subscript of the first term, and the subscript of the last term—all separated by commas. For example, if you want the first 10 terms $(a_1, a_2, a_3, \dots, a_{10})$ of $a_n = n^2$, enter seq$(n^2, n, 1, 10)$. Be sure to set your calculator in sequence mode.

STEP 3: Press $\boxed{\text{ENTER}}$ to get the terms of the sequence to appear.

Entering, Tabling, and Graphing a Sequence

In Example 5, Section 11.1, a table and a graph of a sequence are created with a graphing calculator. The calculator should be set to sequence mode by entering the following keystrokes.

$$\boxed{\text{MODE}}\ \boxed{\triangledown}\ \boxed{\triangledown}\ \boxed{\triangledown}\ \boxed{\triangleright}\ \boxed{\triangleright}\ \boxed{\triangleright}\ \boxed{\text{ENTER}}$$

To enter the formula for a sequence, press $\boxed{\text{Y =}}$. See Figure A.61. Let $n\text{Min} = 1$, since the initial value of n is equal to 1. To enter $a_n = 2.85a_{n-1} - .19a_{n-1}^2$, use the following keystrokes, after clearing out any old formula. (Notice that the graphing calculator uses u instead of a to denote a term of the sequence.)

$$\boxed{2}\ \boxed{.}\ \boxed{8}\ \boxed{5}\ \boxed{\text{2nd}}\ \boxed{7[u]}\ \boxed{(}\ \boxed{\text{X, T, }\theta\text{, }n}\ \boxed{-}\ \boxed{1}\ \boxed{)}\ \boxed{-}\ \boxed{.}\ \boxed{1}\ \boxed{9}$$

$$\boxed{\text{2nd}}\ \boxed{7[u]}\ \boxed{(}\ \boxed{\text{X, T, }\theta\text{, }n}\ \boxed{-}\ \boxed{1}\ \boxed{)}\ \boxed{\wedge}\ \boxed{2}\ \boxed{\text{ENTER}}$$

Since $a_1 = 1$, let $u(n\text{Min}) = \{1\}$. This can be done as follows. See Figure A.61.

$$\boxed{\text{CLEAR}}\ \boxed{\text{2nd}}\ \boxed{(}\ \boxed{1}\ \boxed{\text{2nd}}\ \boxed{)}$$

To create a table for this sequence, starting with a_1 and incrementing n by 1, perform the following keystrokes. See Figure A.62 and Figure 11.7.

$$\boxed{\text{2nd}}\ \boxed{\text{WINDOW [TBLSET]}}\ \boxed{1}\ \boxed{\text{ENTER}}\ \boxed{1}\ \boxed{\text{2nd}}\ \boxed{\text{GRAPH [TABLE]}}$$

To graph the first 20 terms of this sequence, start by selecting $\boxed{\text{WINDOW}}$. Since we want the first 20 terms plotted, let $n\text{Min} = 1$, $n\text{Max} = 20$, PlotStart $= 1$, and PlotStep $= 1$. The window can be set as $[0, 21, 1]$ by $[0, 14, 1]$. See Figure A.63. To graph the sequence, press $\boxed{\text{GRAPH}}$. The resulting graph uses dot mode and is shown in Figure 11.9.

SUMMARY: ENTERING, TABLING, AND GRAPHING A SEQUENCE

STEP 1: Set the mode to "Seq" by using the $\boxed{\text{MODE}}$ menu.

STEP 2: Enter the formula for the sequence by pressing $\boxed{\text{Y =}}$.

STEP 3: To create a table of a sequence, set the start and increment values with

$$\boxed{\text{2nd}}\ \boxed{\text{WINDOW [TBLSET]}}$$

and then press

$$\boxed{\text{2nd}}\ \boxed{\text{GRAPH [TABLE]}}\ .$$

STEP 4: To graph a sequence, set the viewing rectangle by using $\boxed{\text{WINDOW}}$ and then press $\boxed{\text{GRAPH}}$. Be sure to use dot mode.

Figure A.61

```
Plot1  Plot2  Plot3
nMin=1
∴u(n)⬚2.85u(n−1)
−.19u(n−1)^2
 u(nMin)⬚{1}
\v(n)=
 v(nMin)=
\w(n)=
```

Figure A.62

```
TABLE SETUP
 TblStart=1
 ΔTbl=1
Indpnt: Auto  Ask
Depend: Auto  Ask
```

Figure A.63

```
WINDOW
 nMin=1
 nMax=20
 PlotStart=1
 PlotStep=1
 Xmin=0
 Xmax=21
↓Xscl=1
```

Summing a Series

In Example 3, Section 11.2, the sum of the series $\sum_{n=1}^{50} \left(\frac{1}{n^4}\right)$ is found by using a graphing calculator. Use the following keystrokes from the home screen.

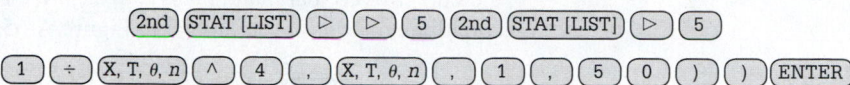

The results are shown in the first three lines of Figure 11.12.

SUMMARY: SUMMING A SERIES

STEP 1: Use (2nd) (STAT [LIST]) (▷) (▷) (5) to access the sum utility.

STEP 2: Use (2nd) (STAT [LIST]) (▷) (5) to access the sequence utility. (To use the sequence utility, see "Creating a Sequence" in this appendix.)

Calculating Factorial Notation

In Example 4, Section 11.3, factorial notation is evaluated with a graphing calculator. The factorial utility is found under the MATH PRB menus. To calculate 8!, use the following keystrokes from the home screen.

The results are shown in the first two lines of Figure 11.15.

SUMMARY: CALCULATING FACTORIAL NOTATION

To calculate n factorial, use the following keystrokes.

(n) (MATH) (▷) (▷) (▷) (4) (ENTER)

The value of n should be entered as a number, not a variable.

Calculating Permutations and Combinations

In Example 5(a), Section 11.3, the permutation $P(7, 3)$ is evaluated. To perform this calculation, use the following keystrokes from the home screen.

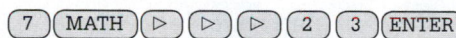

The results are shown in the first two lines of Figure 11.17.

In Example 8(a), Section 11.3, the combination $C(7, 3)$ can be calculated by using the following keystrokes.

The results are shown in the first two lines of Figure 11.18.

SUMMARY: CALCULATING PERMUTATIONS AND COMBINATIONS

STEP 1: To calculate $P(n, r)$, use (MATH) and select "PRB" followed by (2).

STEP 2: To calculate $C(n, r)$, use (MATH) and select "PRB" followed by (3).

Graphing Parametric Equations

In Figure 8.72, Section 8.4, the parametric equations $x = t + 3$, $y = t^2$ for $-3 \le t \le 3$ are graphed. To set your graphing calculator in parametric mode, press (MODE), position the cursor over "Par", and press (ENTER). See Figure A.64. Next Press (Y =) and enter the equations for x and y, as shown in Figure A.65.

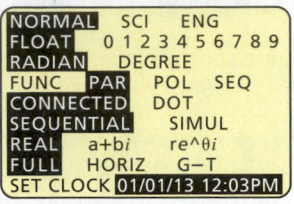

Figure A.64

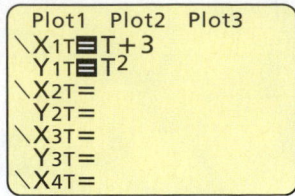

Figure A.65

To set a viewing rectangle, press (WINDOW). In addition to setting Xmin, Xmax, Xscl, Ymin, Ymax, and Yscl, you must set values for Tmin, Tmax, and Tstep. Tmin refers to the minimum value of t in the graph, and Tmax refers to the maximum value of t. It is given that $-3 \le t \le 3$, so it follows that Tmin $= -3$ and Tmax $= 3$. However, when an interval for t is not given, it may take a little experimentation to determine an appropriate interval for t. Tstep represents the increment between consecutive t-values on the graph. If Tstep is too large, the graph appears more like a line graph than a smooth curve. If Tstep is too small, the graphing calculator will take a long time to create the graph. Many times a reasonable value is Tstep $= 0.1$. See Figure 8.70. A parametric graph can be created by pressing (GRAPH).

Tables for parametric equations can be created. Press

(2nd) (WINDOW [TBLSET])

and proceed in the usual manner. Note that the variables TblStart and ΔTbl refer to t and not x. See Figure 8.71.

SUMMARY: GRAPHING PARAMETRIC EQUATIONS

1. Press (MODE), move the cursor to "Par", and press (ENTER).
2. Press (Y =) and enter the equations for x and y.
3. Press (WINDOW) and set the viewing rectangle. Be sure to set Tmin, Tmax, and Tstep. When in doubt, let Tstep $= 0.1$.
4. To make the graph appear, press (GRAPH).

Graphing in Polar Coordinates

In Figure 8.102, Section 8.5, the polar equation $r = 2 + 2 \cos \theta$ for $0° \le \theta \le 360°$ is graphed. To set your graphing calculator in polar coordinate mode, press (MODE), position the cursor over "Pol", and press (ENTER). See Figure A.66. Polar equations can be graphed in either degree or radian mode. To set your calculator in degree mode, position the cursor over "Degree" and press (ENTER). See Figure A.67. Next press (Y =) and enter the equation for "$r_1 = $", as shown in Figure A.68. Note that the polar equation must be solved for the variable r.

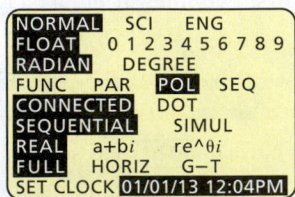

Figure A.66

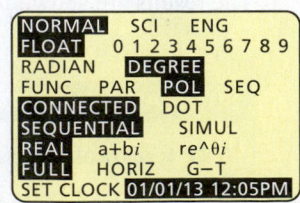

Figure A.67

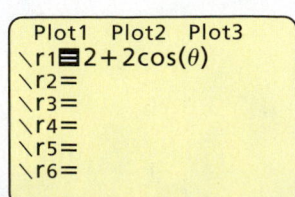

Figure A.68

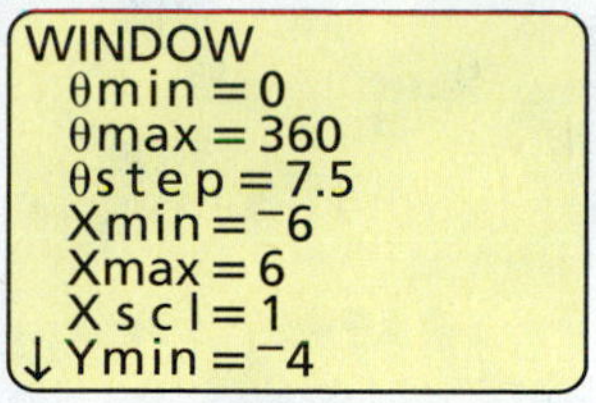

WINDOW
 θmin = 0
 θmax = 360
 θstep = 7.5
 Xmin = ⁻6
 Xmax = 6
 Xscl = 1
↓Ymin = ⁻4

Figure A.69 Degree Mode

To set a viewing rectangle, press (WINDOW). In addition to setting Xmin, Xmax, Xscl, Ymin, Ymax, and Yscl, you must set values for θmin, θmax, and θstep. The variable θmin refers to the minimum value of θ, and θmax refers to the maximum value of θ. Since cos θ is periodic with 360°, the entire graph will appear if we let $0° \leq \theta \leq 360°$. Let θmin = 0 and θmax = 360. The variable θstep represents the increment between consecutive θ-values on the polar graph. If θstep is too large, the graph appears more like a line graph than a smooth curve. If θstep is too small, the graphing calculator will take a long time to create the graph. In degree mode a reasonable value for θstep is 7.5°, and in radian mode a reasonable value for θstep is 0.1 radian. See Figure A.69. A polar graph can be created by pressing (GRAPH).

Tables for polar coordinates can be created. Press

(2nd) (WINDOW [TBLSET])

and proceed in the usual manner. Note that the variables TblStart and ΔTbl refer to θ and not x. See Figure 8.101.

> **SUMMARY: GRAPHING IN POLAR COORDINATES**
>
> 1. Press (MODE), move the cursor to "Pol", and press (ENTER). Set the calculator to either degree or radian mode.
> 2. Press (Y =) and enter the polar equation.
> 3. Press (WINDOW) and set the viewing rectangle. Be sure to set θmin, θmax, and θstep. When in doubt, let θstep = 7.5 in degree mode and θstep = 0.1 in radian mode.
> 4. To make the graph appear, press (GRAPH).

Appendix B:
A Library of Functions

Basic Functions

The following are symbolic, numerical, and graphical representations of several functions used in algebra and trigonometry. Their domains D and ranges R are given.

Identity Function: $f(x) = x$

x	-2	-1	0	1	2
$y = x$	-2	-1	0	1	2

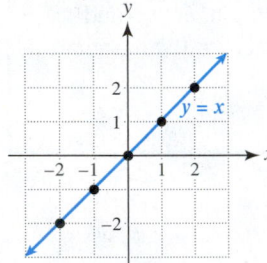

$D = (-\infty, \infty)$
$R = (-\infty, \infty)$

Absolute Value Function: $f(x) = |x|$

x	-2	-1	0	1	2
$y = |x|$	2	1	0	1	2

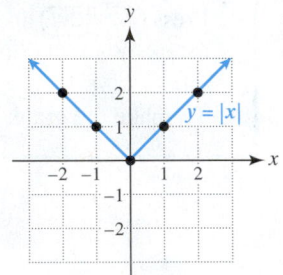

$D = (-\infty, \infty)$
$R = [0, \infty)$

Square Function: $f(x) = x^2$

x	-2	-1	0	1	2
$y = x^2$	4	1	0	1	4

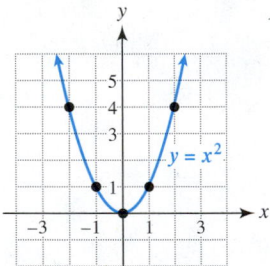

$D = (-\infty, \infty)$
$R = [0, \infty)$

Cube Function: $f(x) = x^3$

x	-2	-1	0	1	2
$y = x^3$	-8	-1	0	1	8

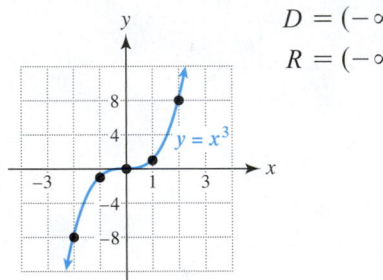

$D = (-\infty, \infty)$
$R = (-\infty, \infty)$

Square Root Function: $f(x) = \sqrt{x}$

x	0	1	4	9
$y = \sqrt{x}$	0	1	2	3

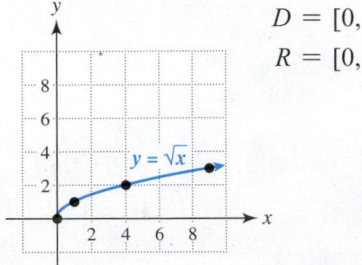

$D = [0, \infty)$
$R = [0, \infty)$

Cube Root Function: $f(x) = \sqrt[3]{x}$

x	-8	-1	0	1	8
$y = \sqrt[3]{x}$	-2	-1	0	1	2

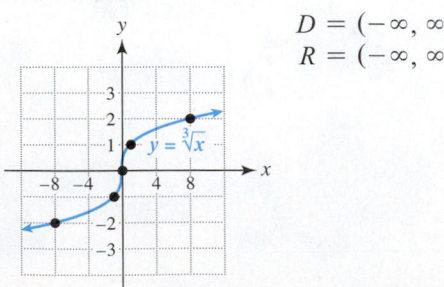

$D = (-\infty, \infty)$
$R = (-\infty, \infty)$

Greatest Integer Function: $f(x) = [x]$

x	-2.5	-1.5	0	1.5	2.5
$y = [x]$	-3	-2	0	1	2

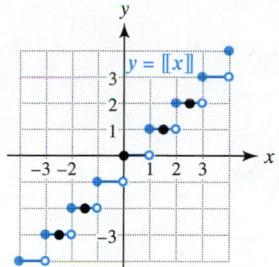

$$D = (-\infty, \infty)$$
$$R = \text{Integers}$$

Reciprocal Function: $f(x) = \frac{1}{x}$

x	-2	-1	0	1	2
$y = \frac{1}{x}$	$-\frac{1}{2}$	-1	—	1	$\frac{1}{2}$

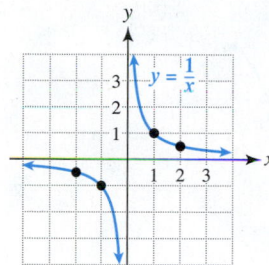

$$D = (-\infty, 0) \cup (0, \infty)$$
$$R = (-\infty, 0) \cup (0, \infty)$$

Base-2 Exponential Function: $f(x) = 2^x$

x	-2	-1	0	1	2
$y = 2^x$	$\frac{1}{4}$	$\frac{1}{2}$	1	2	4

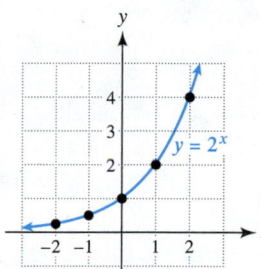

$$D = (-\infty, \infty)$$
$$R = (0, \infty)$$

Natural Exponential Function: $f(x) = e^x$

x	-2	-1	0	1	2
$y = e^x$	e^{-2}	e^{-1}	1	e^1	e^2

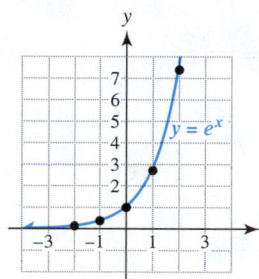

$$D = (-\infty, \infty)$$
$$R = (0, \infty)$$

Common Logarithmic Function: $f(x) = \log x$

x	0.1	1	4	7	10
$y = \log x$	-1	0	$\log 4$	$\log 7$	1

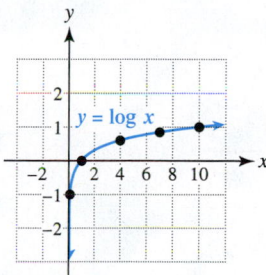

$$D = (0, \infty)$$
$$R = (-\infty, \infty)$$

Natural Logarithmic Function: $f(x) = \ln x$

x	$\frac{1}{2}$	1	2	e	e^2
$y = \ln x$	$\ln \frac{1}{2}$	0	$\ln 2$	1	2

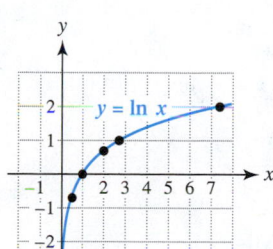

$$D = (0, \infty)$$
$$R = (-\infty, \infty)$$

Sine Function: $f(x) = \sin x$

x	0	$\frac{\pi}{2}$	π	$\frac{3\pi}{2}$	2π
$y = \sin x$	0	1	0	-1	0

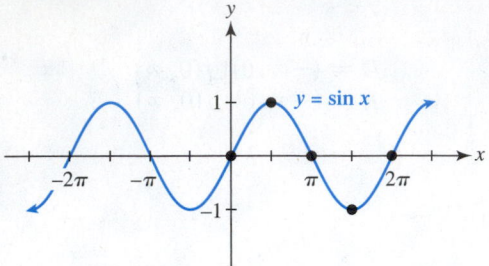

$D = (-\infty, \infty), R = [-1, 1]$

Cosine Function: $f(x) = \cos x$

x	0	$\frac{\pi}{2}$	π	$\frac{3\pi}{2}$	2π
$y = \cos x$	1	0	-1	0	1

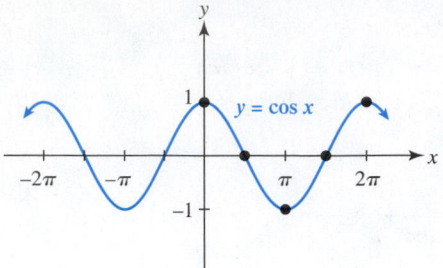

$D = (-\infty, \infty), R = [-1, 1]$

Tangent Function: $f(x) = \tan x$

x	$-\frac{\pi}{3}$	$-\frac{\pi}{4}$	0	$\frac{\pi}{4}$	$\frac{\pi}{3}$
$y = \tan x$	$-\sqrt{3}$	-1	0	1	$\sqrt{3}$

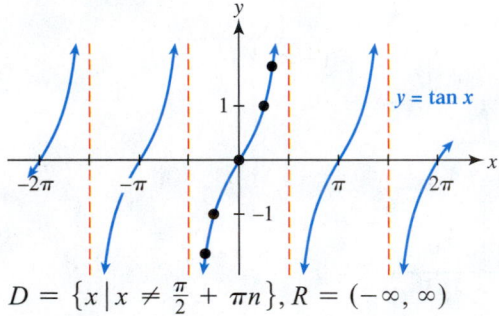

$D = \left\{ x \mid x \neq \frac{\pi}{2} + \pi n \right\}, R = (-\infty, \infty)$

Cotangent Function: $f(x) = \cot x$

x	$\frac{\pi}{6}$	$\frac{\pi}{4}$	$\frac{\pi}{2}$	$\frac{3\pi}{4}$	$\frac{5\pi}{6}$
$y = \cot x$	$\sqrt{3}$	1	0	-1	$-\sqrt{3}$

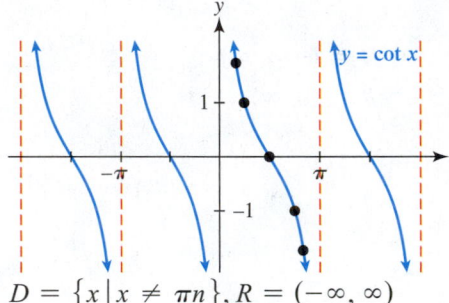

$D = \{ x \mid x \neq \pi n \}, R = (-\infty, \infty)$

Cosecant Function: $f(x) = \csc x$

x	$\frac{\pi}{6}$	$\frac{\pi}{4}$	$\frac{\pi}{2}$	$\frac{3\pi}{4}$	$\frac{5\pi}{6}$
$y = \csc x$	2	$\sqrt{2}$	1	$\sqrt{2}$	2

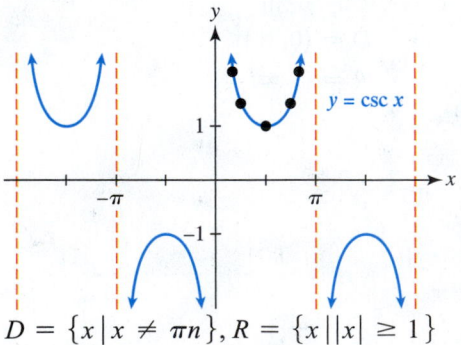

$D = \{ x \mid x \neq \pi n \}, R = \{ x \mid |x| \geq 1 \}$

Secant Function: $f(x) = \sec x$

x	$-\frac{\pi}{3}$	$-\frac{\pi}{4}$	0	$\frac{\pi}{4}$	$\frac{\pi}{3}$
$y = \sec x$	2	$\sqrt{2}$	1	$\sqrt{2}$	2

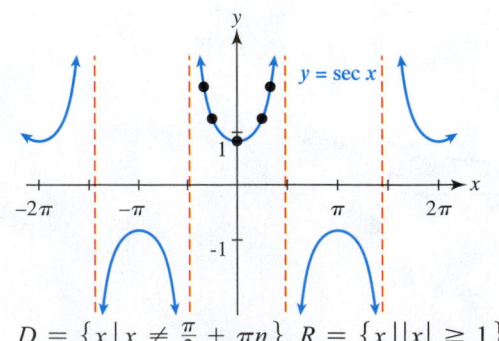

$D = \left\{ x \mid x \neq \frac{\pi}{2} + \pi n \right\}, R = \{ x \mid |x| \geq 1 \}$

Families of Functions

This subsection shows the formulas and graphs of some families of functions, such as linear, quadratic, and exponential. Notice that the appearance of the graphs of these functions depends on the value of k, m, or a.

Constant Functions: $f(x) = k$

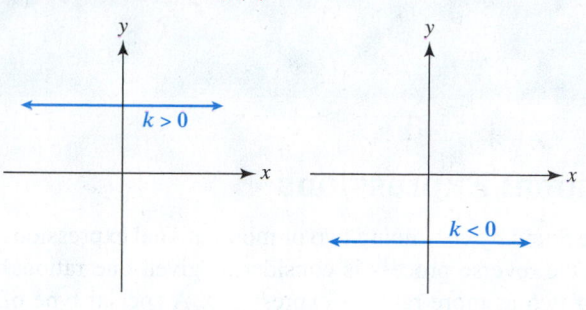

Linear Functions: $f(x) = mx + b$

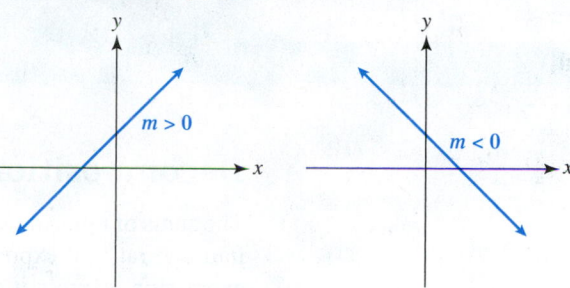

Quadratic Functions: $f(x) = ax^2 + bx + c$

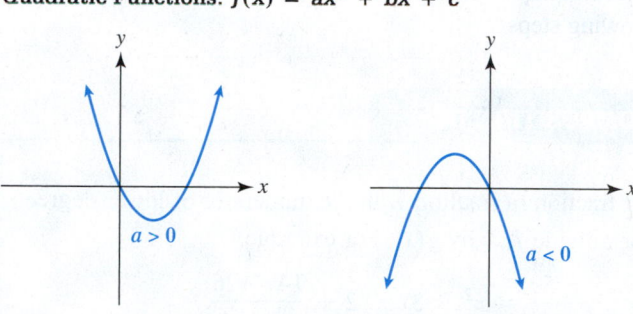

Cubic Functions: $f(x) = ax^3 + bx^2 + cx + d$

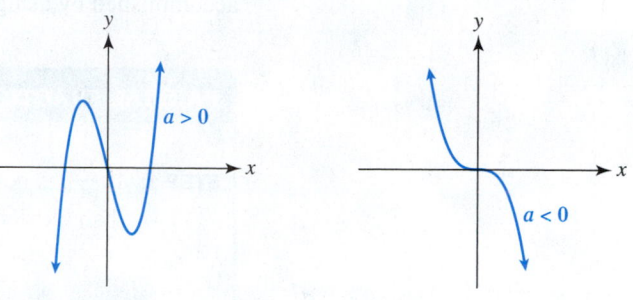

Power Functions: $f(x) = x^a, x > 0$

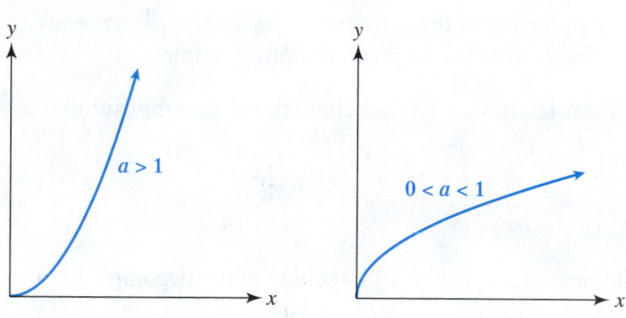

Sinusoidal Functions: $f(x) = a \sin(b(x - c)) + d$ or
$$f(x) = a \cos(b(x - c)) + d$$

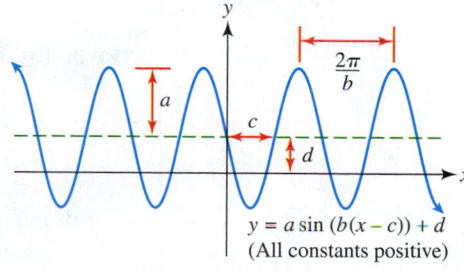

$y = a \sin(b(x - c)) + d$
(All constants positive)

Exponential Functions: $f(x) = Ca^x, C > 0$

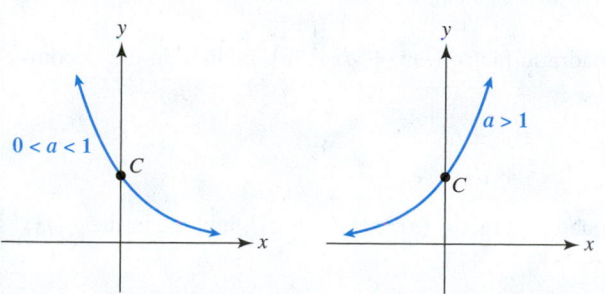

Logarithmic Functions: $f(x) = \log_a x$

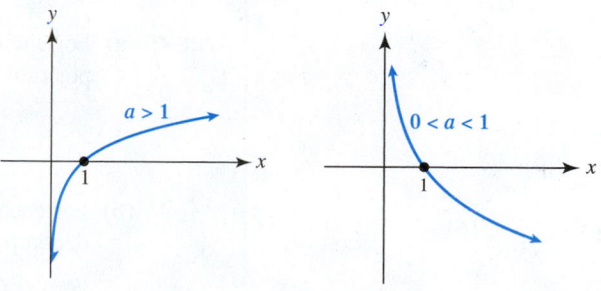

C Appendix C: Partial Fractions

Decomposition of Rational Expressions

The sums of rational expressions are found by combining two or more rational expressions into one rational expression. Here, the reverse process is considered: given one rational expression, express it as the sum of two or more rational expressions. A special type of sum of rational expressions is called the **partial fraction decomposition**; each term in the sum is a **partial fraction**. The technique of finding partial fraction decompositions can be accomplished by using the following steps.

PARTIAL FRACTION DECOMPOSITION OF $\dfrac{f(x)}{g(x)}$

STEP 1: If $\dfrac{f(x)}{g(x)}$ is not a proper fraction (a fraction with the numerator of lower degree than the denominator), divide $f(x)$ by $g(x)$. For example,

$$\frac{x^4 - 3x^3 + x^2 + 5x}{x^2 + 3} = x^2 - 3x - 2 + \frac{14x + 6}{x^2 + 3}.$$

Then apply the following steps to the remainder, which is a proper fraction.

STEP 2: Factor $g(x)$ completely into factors of the form $(ax + b)^m$ or $(cx^2 + dx + e)^n$, where $cx^2 + dx + e$ is *irreducible* and m and n are positive integers.

STEP 3: **(a)** For each distinct linear factor $(ax + b)$, include in the decomposition the term

$$\frac{A}{ax + b}.$$

(b) For each repeated linear factor $(ax + b)^m$, include in the decomposition the terms

$$\frac{A_1}{ax + b} + \frac{A_2}{(ax + b)^2} + \cdots + \frac{A_m}{(ax + b)^m}.$$

STEP 4: **(a)** For each distinct quadratic factor $(cx^2 + dx + e)$, include in the decomposition the term

$$\frac{Bx + C}{cx^2 + dx + e}.$$

(b) For each repeated quadratic factor $(cx^2 + dx + e)^n$, include in the decomposition the terms

$$\frac{B_1 x + C_1}{cx^2 + dx + e} + \frac{B_2 x + C_2}{(cx^2 + dx + e)^2} + \cdots + \frac{B_n x + C_n}{(cx^2 + dx + e)^n}.$$

STEP 5: Use algebraic techniques to solve for the constants in the numerators.

To find the constants in Step 5, the goal is to get a system of equations with as many equations as there are unknowns in the numerators. One method for finding these equations is to substitute values for x on each side of the rational equation formed in Steps 3 and 4.

Distinct Linear Factors

Finding a partial fraction decomposition

Find the partial fraction decomposition of

$$\frac{2x^4 - 8x^2 + 5x - 2}{x^3 - 4x}.$$

SOLUTION The given fraction is not a proper fraction; the numerator has higher degree than the denominator. Perform the following division.

$$\begin{array}{r} 2x \\ x^3 - 4x \overline{)\,2x^4 - 8x^2 + 5x - 2} \\ \underline{2x^4 - 8x^2 } \\ 5x - 2 \end{array}$$

Algebra Review

To review clearing fractions from rational equations, see chapter R (page R-33).

The result is $2x + \frac{5x - 2}{x^3 - 4x}$. Now work with the remainder fraction. Factor the denominator as $x^3 - 4x = x(x + 2)(x - 2)$. Since the factors are distinct linear factors, use Step 3(a) to write the decomposition as

$$\frac{5x - 2}{x^3 - 4x} = \frac{A}{x} + \frac{B}{x + 2} + \frac{C}{x - 2}, \qquad \text{Equation 1}$$

where A, B, and C are constants that need to be found. Multiply each side of equation 1 by $x(x + 2)(x - 2)$ to clear fractions and get

$$5x - 2 = A(x + 2)(x - 2) + Bx(x - 2) + Cx(x + 2). \qquad \text{Equation 2}$$

Equation 1 is an identity, since both sides represent the same rational expression. Thus equation 2 is also an identity. Equation 1 holds for all values of x except 0, -2, and 2. However, equation 2 holds for all values of x. In particular, substituting 0 for x in equation 2 gives $-2 = -4A$, so $A = \frac{1}{2}$. Similarly, choosing $x = -2$ gives $-12 = 8B$, so $B = -\frac{3}{2}$. Finally, choosing $x = 2$ gives $8 = 8C$, so $C = 1$. The remainder rational expression can be written as the sum of partial fractions

$$\frac{5x - 2}{x^3 - 4x} = \frac{1}{2x} + \frac{-3}{2(x + 2)} + \frac{1}{x - 2},$$

and the given rational expression can be written as

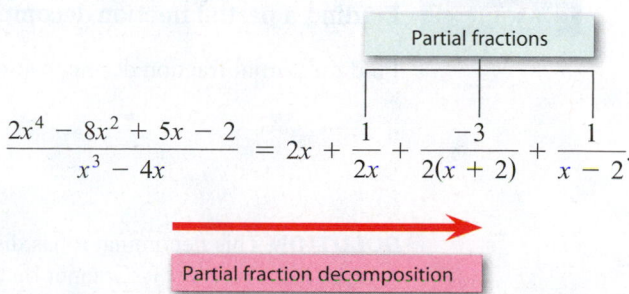

Partial fractions

$$\frac{2x^4 - 8x^2 + 5x - 2}{x^3 - 4x} = 2x + \frac{1}{2x} + \frac{-3}{2(x + 2)} + \frac{1}{x - 2}.$$

Partial fraction decomposition

Check the work by combining the terms on the right.

Repeated Linear Factors

EXAMPLE 2 Finding a partial fraction decomposition

Find the partial fraction decomposition of

$$\frac{2x}{(x-1)^3}.$$

SOLUTION This is a proper fraction. The denominator is already factored with repeated linear factors. We write the decomposition as shown, using Step 3(b).

$$\frac{2x}{(x-1)^3} = \frac{A}{x-1} + \frac{B}{(x-1)^2} + \frac{C}{(x-1)^3}$$

We clear denominators by multiplying each side of this equation by $(x-1)^3$.

$$2x = A(x-1)^2 + B(x-1) + C$$

Substituting 1 for x leads to $C = \mathbf{2}$, so

$$2x = A(x-1)^2 + B(x-1) + 2. \qquad \text{Equation 1}$$

We found C and we still need to find values for A and B. *Any* number can be substituted for x. For example, when we choose $x = -1$ (because it is easy to substitute), equation 1 becomes

$$-2 = 4A - 2B + 2$$
$$-4 = 4A - 2B$$
$$-2 = 2A - B. \qquad \text{Equation 2}$$

Substituting 0 for x in equation 1 gives

$$0 = A - B + 2$$
$$2 = -A + B. \qquad \text{Equation 3}$$

Now we solve the system of equations 2 and 3 to get $A = \mathbf{0}$ and $B = \mathbf{2}$. Since $A = 0$, the term $\frac{A}{x-1}$ is not used. The partial fraction decomposition is

$$\frac{2x}{(x-1)^3} = \frac{\mathbf{2}}{(x-1)^2} + \frac{\mathbf{2}}{(x-1)^3}.$$

We needed three substitutions because there were three constants to find: A, B, and C. To check this result, we could combine the terms on the right.

Now Try Exercise 13

Distinct Linear and Quadratic Factors

EXAMPLE 3 Finding a partial fraction decomposition

Find the partial fraction decomposition of

$$\frac{x^2 + 3x - 1}{(x+1)(x^2+2)}.$$

SOLUTION This denominator has distinct linear and quadratic factors, neither of which is repeated. Since $x^2 + 2$ cannot be factored, it is *irreducible*. The partial fraction decomposition is

$$\frac{x^2 + 3x - 1}{(x+1)(x^2+2)} = \frac{A}{x+1} + \frac{Bx + C}{x^2+2}.$$

Multiply each side by $(x + 1)(x^2 + 2)$ to get

$$x^2 + 3x - 1 = A(x^2 + 2) + (Bx + C)(x + 1). \qquad \text{Equation 1}$$

First substitute -1 for x to get

$$(-1)^2 + 3(-1) - 1 = A((-1)^2 + 2) + 0$$

$$-3 = 3A$$

$$A = -1.$$

Replace A with -1 in equation 1 and substitute any value for x. If $x = 0$, then

$$0^2 + 3(0) - 1 = -1(0^2 + 2) + (B \cdot 0 + C)(0 + 1)$$

$$-1 = -2 + C$$

$$C = 1.$$

Now, letting $A = -1$ and $C = 1$, substitute again in equation 1, using another number for x. For $x = 1$,

$$3 = -3 + (B + 1)(2)$$

$$6 = 2B + 2$$

$$B = 2.$$

With $A = -1$, $B = 2$, and $C = 1$, the partial fraction decomposition is

$$\frac{x^2 + 3x - 1}{(x + 1)(x^2 + 2)} = \frac{-1}{x + 1} + \frac{2x + 1}{x^2 + 2}.$$

This work can be checked by combining terms on the right. **Now Try Exercise 21**

For fractions with denominators that have quadratic factors, another method is often more convenient. The system of equations is formed by equating coefficients of like terms on each side of the partial fraction decomposition. For instance, in Example 3, after each side was multiplied by the common denominator, equation 1 was

$$x^2 + 3x - 1 = A(x^2 + 2) + (Bx + C)(x + 1). \qquad \text{Equation 1}$$

Multiplying on the right and collecting like terms, we have

$$x^2 + 3x - 1 = Ax^2 + 2A + Bx^2 + Bx + Cx + C$$

$$x^2 + 3x - 1 = (A + B)x^2 + (B + C)x + (C + 2A).$$

Now equating the coefficients of like powers of x gives three equations:

$$1 = A + B$$

$$3 = B + C$$

$$-1 = C + 2A.$$

Solving this system of equations for A, B, and C would give the partial fraction decomposition. The next example uses a combination of the two methods.

Repeated Quadratic Factors

EXAMPLE 4 **Finding a partial fraction decomposition**

Find the partial fraction decomposition of

$$\frac{2x}{(x^2 + 1)^2 (x - 1)}.$$

SOLUTION This expression has both a linear factor and a repeated quadratic factor. By Steps 3(a) and 4(b),

$$\frac{2x}{(x^2 + 1)^2(x - 1)} = \frac{Ax + B}{x^2 + 1} + \frac{Cx + D}{(x^2 + 1)^2} + \frac{E}{x - 1}.$$

Multiplying each side by $(x^2 + 1)^2(x - 1)$ leads to

$$2x = (Ax + B)(x^2 + 1)(x - 1) + (Cx + D)(x - 1) + E(x^2 + 1)^2. \qquad \text{Equation 1}$$

If $x = 1$, equation 1 reduces to $2 = 4E$, or $E = \frac{1}{2}$. Substituting $\frac{1}{2}$ for E in equation 1 and combining terms on the right gives

$$2x = \left(A + \frac{1}{2}\right)x^4 + (-A + B)x^3 + (A - B + C + 1)x^2$$

$$+ (-A + B + D - C)x + \left(-B - D + \frac{1}{2}\right). \qquad \text{Equation 2}$$

To get additional equations involving the unknowns, equate the coefficients of like powers of x on each side of equation 2. Setting corresponding coefficients of x^4 equal gives $0 = A + \frac{1}{2}$, or $A = -\frac{1}{2}$. From the corresponding coefficients of x^3, $0 = -A + B$, which means that since $A = -\frac{1}{2}$, $B = -\frac{1}{2}$. From the coefficients of x^2, $0 = A - B + C + 1$. Since $A = -\frac{1}{2}$ and $B = -\frac{1}{2}$, it follows that $C = -1$. Finally, from the coefficients of x, $2 = -A + B + D - C$. Substituting for A, B, and C gives $D = 1$. With

$$A = -\frac{1}{2}, \quad B = -\frac{1}{2}, \quad C = -1, \quad D = 1, \quad \text{and} \quad E = \frac{1}{2},$$

the given fraction has the partial fraction decomposition

$$\frac{2x}{(x^2 + 1)^2(x - 1)} = \frac{-\frac{1}{2}x - \frac{1}{2}}{x^2 + 1} + \frac{-x + 1}{(x^2 + 1)^2} + \frac{\frac{1}{2}}{x - 1},$$

or

$$\frac{2x}{(x^2 + 1)^2(x - 1)} = \frac{-(x + 1)}{2(x^2 + 1)} + \frac{-x + 1}{(x^2 + 1)^2} + \frac{1}{2(x - 1)}.$$

Now Try Exercise 25

In summary, to solve for the constants in the numerators of a partial fraction decomposition, use either of the following methods or a combination of the two.

TECHNIQUES FOR DECOMPOSITION INTO PARTIAL FRACTIONS

Method 1 for Linear Factors

STEP 1: Multiply each side of the rational expression by the common denominator.

STEP 2: Substitute the zero of each factor in the resulting equation. For repeated linear factors, substitute as many other numbers as necessary to find all the constants in the numerators. The number of substitutions required will equal the number of constants.

Method 2 for Quadratic Factors

STEP 1: Multiply each side of the rational expression by the common denominator.

STEP 2: Collect terms on the right side of the resulting equation.

STEP 3: Equate the coefficients of like terms to get a system of equations.

STEP 4: Solve the system to find the constants in the numerators.

C Exercises

Exercises 1–30: Find the partial fraction decomposition for the rational expression.

1. $\dfrac{5}{3x(2x + 1)}$

2. $\dfrac{3x - 1}{x(x + 1)}$

3. $\dfrac{4x + 2}{(x + 2)(2x - 1)}$

4. $\dfrac{x + 2}{(x + 1)(x - 1)}$

5. $\dfrac{x}{x^2 + 4x - 5}$

6. $\dfrac{5x - 3}{(x + 1)(x - 3)}$

7. $\dfrac{2x}{(x + 1)(x + 2)^2}$

8. $\dfrac{2}{x^2(x + 3)}$

9. $\dfrac{4}{x(1 - x)}$

10. $\dfrac{4x^2 - 4x^3}{x^2(1 - x)}$

11. $\dfrac{4x^2 - x - 15}{x(x + 1)(x - 1)}$

12. $\dfrac{2x + 1}{(x + 2)^3}$

13. $\dfrac{x^2}{x^2 + 2x + 1}$

14. $\dfrac{3}{x^2 + 4x + 3}$

15. $\dfrac{2x^5 + 3x^4 - 3x^3 - 2x^2 + x}{2x^2 + 5x + 2}$

16. $\dfrac{6x^5 + 7x^4 - x^2 + 2x}{3x^2 + 2x - 1}$

17. $\dfrac{x^3 + 4}{9x^3 - 4x}$

18. $\dfrac{x^3 + 2}{x^3 - 3x^2 + 2x}$

19. $\dfrac{-3}{x^2(x^2 + 5)}$

20. $\dfrac{2x + 1}{(x + 1)(x^2 + 2)}$

21. $\dfrac{3x - 2}{(x + 4)(3x^2 + 1)}$

22. $\dfrac{3}{x(x + 1)(x^2 + 1)}$

23. $\dfrac{1}{x(2x + 1)(3x^2 + 4)}$

24. $\dfrac{x^4 + 1}{x(x^2 + 1)^2}$

25. $\dfrac{3x - 1}{x(2x^2 + 1)^2}$

26. $\dfrac{3x^4 + x^3 + 5x^2 - x + 4}{(x - 1)(x^2 + 1)^2}$

27. $\dfrac{-x^4 - 8x^2 + 3x - 10}{(x + 2)(x^2 + 4)^2}$

28. $\dfrac{x^2}{x^4 - 1}$

29. $\dfrac{5x^5 + 10x^4 - 15x^3 + 4x^2 + 13x - 9}{x^3 + 2x^2 - 3x}$

30. $\dfrac{3x^6 + 3x^4 + 3x}{x^4 + x^2}$

Percent Change and Exponential Functions

Percentages and Percent Change

Percentages A percentage can be written either in **percent form** or in **decimal form**. For example, the percent form of 15% can also be written in decimal form as 0.15. To change a percent form $R\%$ to a decimal form r we divide R by 100. That is, $r = \frac{R}{100}$.

EXAMPLE 1 **Writing percentages as decimals**

Write each percentage as a decimal.
(a) 45% **(b)** 0.03% **(c)** 420% **(d)** −1.45% **(e)** $\frac{2}{5}\%$

SOLUTION
(a) Let $R = 45$. Then $r = \frac{45}{100} = 0.45$.

(b) Let $R = 0.03$. Then $r = \frac{0.03}{100} = 0.0003$.

(c) Let $R = 420$. Then $r = \frac{420}{100} = 4.2$.

(d) Let $R = -1.45$. Then $r = \frac{-1.45}{100} = -0.0145$. A negative percentage generally corresponds to a quantity decreasing rather than increasing.

(e) Let $R = \frac{2}{5}$. Then $r = \frac{\frac{2}{5}}{100} = \frac{2}{500} = 0.004$.

Now Try Exercise 1

NOTE Dividing a number by 100 is equivalent to moving the decimal point two places to the *left*.

In a similar manner a decimal form r can be changed to a percent form R by using the formula $R = 100r$. For example, the decimal form 0.047 has the percent form

$$R = 100r = 100(0.047) = 4.7\%.$$

In this calculation the decimal point is moved 2 places to the *right*.

Percent Change When an amount A_1 changes to a new amount A_2, then the **percent change** is

$$\frac{A_2 - A_1}{A_1} \times 100. \qquad \textit{Percent change}$$

We multiply by 100 to change decimal form to percent form.

EXAMPLE 2 **Finding percent change**

Complete the following.
(a) Find the percent change if an account increases from $1200 to $1500.
(b) Find the percent change if an account decreases from $1500 to $1200.
(c) Comment on your results from parts (a) and (b).

SOLUTION

(a) Let $A_1 = 1200$ and $A_2 = 1500$.

$$\frac{1500 - 1200}{1200} \times 100 = \frac{300}{1200} \times 100$$

$$= \frac{1}{4} \times 100$$

$$= 25\%$$

The percent change (increase) is 25%.

(b) Let $A_1 = 1500$ and $A_2 = 1200$.

$$\frac{1200 - 1500}{1500} \times 100 = -\frac{300}{1500} \times 100$$

$$= -\frac{1}{5} \times 100$$

$$= -20\%$$

The percent change (decrease) is −20%.

(c) Notice that the account increased by 25% and then decreased by 20% to return to its initial value. Because the initial amount of $A_1 = \$1500$ in part (b) is larger than the initial amount $A_1 = \$1200$ in part (a), the amount of $1500 only needs to decrease by 20% or $300, to return to the original $1200.

> **Now Try Exercise 9**

Suppose a child's weight increases from 20 pounds to 60 pounds over a period of years. The percent increase is

$$\frac{60 - 20}{20} \times 100 = 2 \times 100 = 200\%.$$

Notice that the child's weight *tripled* and the percent change is 200%, *not* 300%. The actual *increase* in weight is 40 pounds and can be found by taking 200% of 20 pounds.

$$200\% \text{ of } 20 = 2.00 \times 20 = 40 \text{ pounds} \qquad \textit{Change 200\% to decimal form.}$$

If we want to find the percent change, expressed in *decimal form*, of an amount A_1 changing to an amount A_2, then we do not need to multiply by 100. Thus

$$r = \frac{A_2 - A_1}{A_1}.$$

We can solve this equation for A_2.

$$r = \frac{A_2 - A_1}{A_1} \qquad \textit{Percent change in decimal form}$$

$$rA_1 = A_2 - A_1 \qquad \textit{Multiply each side by } A_1.$$

$$A_1 + rA_1 = A_2 \qquad \textit{Add } A_1 \textit{ to each side.}$$

$$A_2 = A_1 + rA_1 \qquad \textit{Rewrite the equation.}$$

Thus, if the percent increase in an amount A_1 is given by r in *decimal form*, then the *increase* (or *decrease*) in A_1 is given by rA_1 and the *final amount* is given by $A_1 + rA_1$, or $A_1(1 + r)$. The initial amount A_1 changes by the *factor* $1 + r$.

For example, if a $100,000 budget decreases by 12%, then

$$rA_1 = -0.12(100{,}000) = -\$12{,}000$$

and the budget decrease is $12,000. Also,

$$A_1 + rA_1 = 100{,}000 + (-0.12)(100{,}000) = \$88{,}000$$

and the new budget decreased to $88,000. The budget changed by a factor of

$$1 + r = 1 + (-0.12) = 0.88,$$

or the budget is now 88% of the original budget.

EXAMPLE 3 **Analyzing the increase in an account**

An account that contains $5000 increases in value by 150%.
(a) Find the increase in value of the account.
(b) Find the final value of the account.
(c) By what factor did the account increase?

SOLUTION
(a) Let $A_1 = 5000$ and $r = 1.50$ (150% in decimal form). The increase is

$$rA_1 = 1.50(5000) = 7500.$$

The account increased in value by $7500.
(b) The final value of the account is $A_1 + rA_1 = 5000 + 7500 = \$12,500$.
(c) The account increased in value by a factor of $1 + r = 1 + 1.50 = 2.5$. Note that $5000(2.5) = 12,500$.

> **Now Try Exercise 15**

More Exponential Functions and Models

Section 5.3 discussed how an exponential function results when the *initial value C* is multiplied by a *constant factor a* for each unit increase in x. For example, if an initial value of $C = 3$ is multiplied by a *constant growth factor* of $a = 2$ for each unit increase in x, then the exponential function

$$f(x) = 3(2)^x \qquad \text{Initial value = 3, growth factor = 2}$$

models this growth. This concept can be used to describe exponential functions and models in terms of *constant percent change*.

Suppose that an initial population of a country is $P_0 = 10$ million and the population increases by 1.2% in 1 year. Then the increase is

$$rP_0 = 0.012(10) = 0.12 \text{ million}, \qquad \text{1.2\% equals 0.012 in decimal form.}$$

and after 1 year the new population is

$$P_0(1 + r)^1 = 10(1.012)^1 = 10.12 \text{ million.}$$

After 1 year the population has increased by a *growth factor* of $1 + r$, or 1.012. If the rate of growth were to remain constant in future years, then after x years the population would be

$$P_0(1 + r)^x = 10(1.012)^x$$

with initial value $C = 10$ and *growth* factor $a = 1.012$.

EXAMPLE 4 **Finding exponential models**

A sample of 10,000 insects is decreasing in number by 8% per week. Find an exponential model $f(x)$ that describes this population after x weeks.

SOLUTION The initial value is $C = 10,000$, the rate of decrease is $r = -0.08$, and the *decay* factor is

$$a = 1 + r = 1 + (-0.08) = 0.92.$$

Thus the sample of insects contains

$$f(x) = 10,000(0.92)^x$$

insects after x weeks.

Now Try Exercise 21

These concepts are summarized in the following box.

PERCENT CHANGE AND EXPONENTIAL FUNCTIONS

Suppose that an amount A changes by R percent (or r expressed in decimal form) for each unit increase in x. Then the following hold.

1. $r = \dfrac{R}{100}$ and $R = 100r$.
2. If $r > 0$, the **constant growth factor** is $a = 1 + r$ and $a > 1$.
3. If $r < 0$, the **constant decay factor** is $a = 1 + r$ and $0 < a < 1$.
4. If the initial amount is C, then the amount A after an x-unit increase in time is given by the exponential model

$$A(x) = C(1 + r)^x, \quad \text{or} \quad A(x) = Ca^x.$$

EXAMPLE 5 | **Analyzing constant percent change**

For each $f(x)$, give the initial value, the growth or decay factor, and percent change for each unit increase in x.

(a) $f(x) = 5(1.034)^x$ **(b)** $f(x) = 10(0.45)^x$ **(c)** $f(x) = 3^x$

SOLUTION

(a) For $f(x) = 5(1.034)^x$ the initial value is $C = 5$ and the growth factor is $a = 1.034$. Because $a = 1 + r$, it follows that

$$r = a - 1 = 1.034 - 1 = 0.034.$$

The percent change for each unit increase in x is 3.4%.

(b) For $f(x) = 10(0.45)^x$ the initial value is $C = 10$ and the decay factor is $a = 0.45$. The percent change for each unit increase in x is

$$r = a - 1 = 0.45 - 1 = -0.55, \text{ or } -55\%.$$

(c) For $f(x) = 3^x$ the initial value is $C = 1$ and the growth factor is $a = 3$. The percent change for each unit increase in x is $r = a - 1 = 3 - 1 = 2$, or 200%.

Now Try Exercises 27, 29, and 31

Growth and Decay Models If an initial quantity A_0 either grows or decays by a factor of b each k units of time, then the amount A after t units of time is given by the exponential model

$$A(t) = Ab^{t/k}.$$

For example, if **700** bacteria **triple** every **5** days, the formula

$$A(t) = \mathbf{700}(\mathbf{3})^{t/5}$$

gives the number of bacteria after t days.

EXAMPLE 6 Applying an exponential model

The population of a city is currently 239,000 and is increasing at a constant rate of 8.5% every 4 years. Find the population of this city after 7 years.

SOLUTION The population of the city is 239,000 and increasing by a factor of 1.085 every 4 years. Thus $A_0 = \textbf{239,000}$, $b = \textbf{1.085}$, $k = \textbf{4}$, and

$$A(t) = \textbf{239,000}(\textbf{1.085})^{t/4}.$$

After **7** years the population is

$$A(7) = 239,000(1.085)^{7/4} \approx 275,677.$$

Now Try Exercise 37

Rule of 70 The **rule of 70** can be used to quickly estimate the number of years it takes for an investment to double. If R is the interest rate (in percent form) and T is the number of years for a quantity to double, then

$$RT = 70. \qquad \textit{Rule of 70}$$

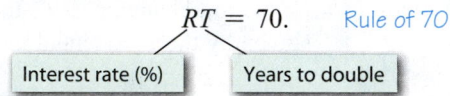

This formula is most accurate for continuous compounding, but it can also be applied to other types of compound interest. For example, if we deposit an amount of money at 5% interest compounded continuously, then it will require about

$$T = \frac{70}{R} = \frac{70}{5} = 14 \text{ years}$$

to double. Similarly, if a city's population doubles in 35 years, then its annual growth rate is about

$$R = \frac{70}{T} = \frac{70}{35} = 2\%.$$

See Exercises 45–50.

D Exercises

Percentages

Exercises 1–4: Write each percentage in decimal form.

1. (a) 35%   (b) −0.07%   (c) 721%   (d) $\frac{3}{10}$%

2. (a) 95%   (b) 0.321%   (c) −175%   (d) $\frac{4}{5}$%

3. (a) −5.5%   (b) −1.54%   (c) 120%   (d) $\frac{3}{20}$%

4. (a) −4.7%   (b) −0.01%   (c) 500%   (d) $\frac{1}{40}$%

Exercises 5–8: Write each decimal form in percent form.

5. (a) 0.37   (b) −0.095   (c) 1.9   (d) $\frac{7}{20}$

6. (a) 0.97   (b) −0.04   (c) 10   (d) $\frac{9}{10}$

7. (a) −0.121   (b) 1.4   (c) 3.2   (d) $-\frac{1}{4}$

8. (a) 0.001   (b) 12   (c) 1.01   (d) $-\frac{1}{8}$

Percent Change

Exercises 9–14: For the given amounts A and B, find each of the following. Round values to the nearest hundredth when appropriate.

(a) *The percent change if A changes to B*
(b) *The percent change if B changes to A*

 9. $A = \$500, B = \$1000$ **10.** $A = \$500, B = \$200$

11. $A = \$1.27, B = \$1.30$ **12.** $A = 15, B = 5$

13. $A = 45, B = 65$ **14.** $A = 75, B = 50$

Exercises 15–20: An account that initially contains A dollars increases/decreases by R percent. For each A and R, complete the following.

(a) *Find the increase/decrease in value of the account.*
(b) *Find the final value of the account.*
(c) *By what factor did the account value increase/decrease?*

15. $A = \$1500, R = 120\%$

16. $A = \$3500, R = 210\%$

17. $A = \$4000, R = -55\%$

18. $A = \$6000, R = -75\%$

19. $A = \$7500, R = -60\%$

20. $A = \$9000, R = 85\%$

Exponential Models

Exercises 21–26: Find an exponential model f(x) that describes each situation.

21. A sample of 9500 insects decreases in number by 35% per week

22. A sample of 5000 insects increases in number by 120% per day

23. A sample of 2500 fish increases in number by 5% per month

24. A sample of 152 birds decreases in number by 3.4% per week

25. A mutual fund account contains $1000 and decreases by 6.5% per year

26. A mutual fund account contains $2500 and increases by 2.1% per year

Exercises 27–36: For the given f(x), state the initial value, the growth or decay factor, and percent change for each unit increase in x.

27. $f(x) = 8(1.12)^x$ **28.** $f(x) = 9(1.005)^x$

29. $f(x) = 1.5(0.35)^x$

30. $f(x) = 100(1.23)^x$

31. $f(x) = 0.55^x$

32. $f(x) = 0.4^x$

33. $f(x) = 7e^x$

34. $f(x) = 91e^{x/2}$

35. $f(x) = 6(3^{-x})$

36. $f(x) = 9(4^{-x})$

Exercises 37–44: (Refer to Example 6.) Write a formula for f(t) that models the situation and then answer the question.

37. The population of a city is currently 35,000 and is increasing at a constant rate of 9.8% every 2 years. What is the population after 5 years?

38. A savings account contains $2500 and increases by 10% in 3 years. How much is in the account after 8 years?

39. A sample of 1000 bacteria triples in number every 7 hours. How many bacteria are there after 11 hours?

40. A sample of 5 million insects decreases in number by $\frac{2}{3}$ every 10 days. In millions, how many insects are there after 65 days?

41. The intensity I_0 of a light passing through colored glass decreases by $\frac{1}{3}$ for each 2 millimeters in thickness of the glass. What is the intensity of the light in terms of I_0 after passing through 4.3 millimeters of colored glass?

42. The intensity I_0 of a sound passing through the atmosphere decreases 20% for each 100 feet of distance. What is the intensity of the sound in terms of I_0 after traveling a distance of 450 feet?

43. An investment of $5000 will quadruple every 35 years. How much is the investment worth after 8 years?

44. An investment of $2500 increases by a factor of 1.2 every 4 years. How much is the investment worth after 9 years?

Exercises 45–50: **Rule of 70** *Use the rule of 70 to estimate the time required for the given principal P to double at the annual percent interest rate R. Check your answer by using the continuously compounded interest formula.*

45. $P = \$2000, R = 7\%$ **46.** $P = \$1200, R = 14\%$

47. $P = \$500, R = 20\%$ **48.** $P = \$9000, R = 10\%$

49. $P = \$1500, R = 25\%$ **50.** $P = \$5000, R = 8\%$

Exercises 51–56: **Rule of 70** *Use the rule of 70 to estimate the annual percent rate of growth for a city whose population P doubles in time T.*

51. $P = 150,000, T = 40$ years

52. $P = 400,000, T = 25$ years

53. $P = 1,500,000, T = 35$ years

54. $P = 20,000, T = 10$ years

55. $P = 750,000, T = 70$ years

56. $P = 80,000, T = 50$ years

Applications

57. Bacteria Growth A population of bacteria increases by 6% every 8 hours. By what percentage does the sample increase in 3 hours?

58. Bacteria Growth A population of bacteria decreases by 40% every 4 hours. By what percentage does the sample decrease in 7 hours?

59. Percent Change The number of cell phone subscribers to a company increases by 25% during the first year and then decreases by 20% the second year. Compare the number of subscribers at the beginning of the first year with the number of subscribers at the end of the second year.

60. Wage Increase If your wages are $8 per hour and you receive a 300% raise for excellent work, determine your new wages.

61. Wages If a wage of $9.81 decreases by 9% each year, what is the new wage after 3 years?

62. Pollution A pollutant in a river has an initial concentration of 3 parts per million and degrades at a rate of 3% every 2 years. Approximate its concentration after 20 years.

63. Radioactive Half-Life A radioactive element decays to 40% of its original amount every 2 years. Approximate the percentage that remains after 8 years.

64. Radioactive Half-Life A radioactive element decays to 80% of its original amount every 3 years. Approximate the percentage that remains after 8 years.

Appendix E: Rotation of Axes

Derivation of Rotation Equations

If we begin with an xy-coordinate system having origin O and rotate the axes about O through an angle θ, the new coordinate system is called a **rotation** of the xy-system. Trigonometric identities can be used to obtain equations for converting the coordinates of a point from the xy-system to the rotated $x'y'$-system. Let P be any point other than the origin, with coordinates (x, y) in the xy-system and (x', y') in the $x'y'$-system. See Figure E.1. Let $OP = r$, and let α represent the angle made by OP and the x'-axis. As shown in Figure E.1,

$$\cos(\theta + \alpha) = \frac{OA}{r} = \frac{x}{r}, \quad \sin(\theta + \alpha) = \frac{AP}{r} = \frac{y}{r},$$

$$\cos\alpha = \frac{OB}{r} = \frac{x'}{r}, \quad \sin\alpha = \frac{BP}{r} = \frac{y'}{r}.$$

These four statements can be rewritten as

$$x = r\cos(\theta + \alpha), \quad y = r\sin(\theta + \alpha), \quad x' = r\cos\alpha, \quad y' = r\sin\alpha.$$

Using the trigonometric identity for the cosine of the sum of two angles gives

$$x = r\cos(\theta + \alpha)$$

$$= r(\cos\theta\cos\alpha - \sin\theta\sin\alpha)$$

$$= (\boldsymbol{r\cos\alpha})\cos\theta - (\boldsymbol{r\sin\alpha})\sin\theta$$

$$= \boldsymbol{x'}\cos\theta - \boldsymbol{y'}\sin\theta.$$

Using the identity for the sine of the sum of two angles in the same way gives $y = x' \sin\theta + y'\cos\theta$. This proves the following result.

ROTATION EQUATIONS

If the rectangular coordinate axes are rotated about the origin through an angle θ and if the coordinates of a point P are (x, y) and (x', y') with respect to the xy-system and the $x'y'$-system, respectively, then the rotation equations are

$$x = x'\cos\theta - y'\sin\theta \quad \text{and} \quad y = x'\sin\theta + y'\cos\theta.$$

Applying a Rotation Equation

EXAMPLE 1 Finding an equation after a rotation

The equation of a curve is $x^2 + y^2 + 2xy + 2\sqrt{2}x - 2\sqrt{2}y = 0$. Find the resulting equation if the axes are rotated $45°$. Graph the equation.

Rotation of xy-Plane

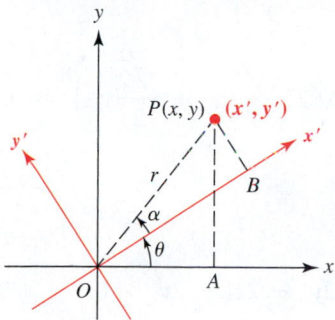

Figure E.1

SOLUTION   If $\theta = 45°$, then $\sin \theta = \frac{\sqrt{2}}{2}$ and $\cos \theta = \frac{\sqrt{2}}{2}$, and the rotation equations become

$$x = \frac{\sqrt{2}}{2}x' - \frac{\sqrt{2}}{2}y' \quad \text{and} \quad y = \frac{\sqrt{2}}{2}x' + \frac{\sqrt{2}}{2}y'.$$

Substituting these values into the given equation yields

$$x^2 + y^2 + 2xy + 2\sqrt{2}x - 2\sqrt{2}y = 0$$

$$\left(\frac{\sqrt{2}}{2}x' - \frac{\sqrt{2}}{2}y'\right)^2 + \left(\frac{\sqrt{2}}{2}x' + \frac{\sqrt{2}}{2}y'\right)^2$$

$$+ 2\left(\frac{\sqrt{2}}{2}x' - \frac{\sqrt{2}}{2}y'\right)\left(\frac{\sqrt{2}}{2}x' + \frac{\sqrt{2}}{2}y'\right)$$

$$+ 2\sqrt{2}\left(\frac{\sqrt{2}}{2}x' - \frac{\sqrt{2}}{2}y'\right) - 2\sqrt{2}\left(\frac{\sqrt{2}}{2}x' + \frac{\sqrt{2}}{2}y'\right) = 0.$$

Expanding these terms yields

$$\frac{1}{2}x'^2 - x'y' + \frac{1}{2}y'^2 + \frac{1}{2}x'^2 + x'y' + \frac{1}{2}y'^2 + x'^2 - y'^2$$

$$+ 2x' - 2y' - 2x' - 2y' = 0.$$

Collecting terms gives

$$2x'^2 - 4y' = 0$$

$$x'^2 - 2y' = 0 \qquad \textcolor{blue}{\text{Divide by 2.}}$$

or, finally,

$$x'^2 = 2y',$$

the equation of a parabola. The graph is shown in Figure E.2.

Now Try Exercise 13

Rotation of a Parabola

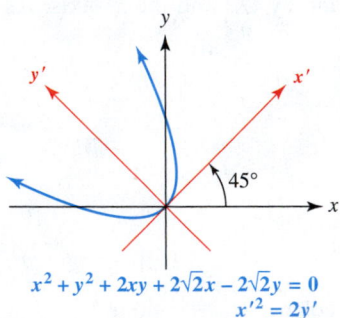

$x^2 + y^2 + 2xy + 2\sqrt{2}x - 2\sqrt{2}y = 0$
$x'^2 = 2y'$

Figure E.2

We have graphed equations written in the form $Ax^2 + Cy^2 + Dx + Ey + F = 0$. As we saw in the preceding example, the rotation of axes eliminated the xy-term. Thus, to graph by hand an equation that has an xy-term, it is necessary to find an appropriate **angle of rotation** to eliminate the xy-term. The necessary angle of rotation can be determined by using the following result. The proof is quite lengthy and is not presented here.

ANGLE OF ROTATION

The xy-term is removed from the general equation

$$Ax^2 + Bxy + Cy^2 + Dx + Ey + F = 0$$

by a rotation of the axes through an angle θ, $0° < \theta < 90°$, where

$$\cot 2\theta = \frac{A - C}{B}.$$

This result can be used to find the appropriate angle of rotation, θ. To find the rotation equations, first find $\sin \theta$ and $\cos \theta$. The following example illustrates a way to obtain $\sin \theta$ and $\cos \theta$ from $\cot 2\theta$ without first identifying the angle θ.

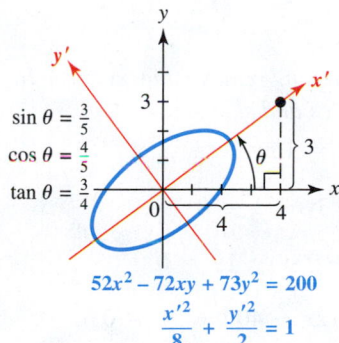

Figure E.3

EXAMPLE **Rotating and graphing**

Rotate the axes and graph $52x^2 - 72xy + 73y^2 = 200$.

SOLUTION Here $A = 52$, $B = -72$, and $C = 73$. By substitution,

$$\cot 2\theta = \frac{52 - 73}{-72} = \frac{-21}{-72} = \frac{7}{24}.$$

To find $\sin \theta$ and $\cos \theta$, use the trigonometric identities

$$\sin \theta = \sqrt{\frac{1 - \cos 2\theta}{2}} \quad \text{and} \quad \cos \theta = \sqrt{\frac{1 + \cos 2\theta}{2}}.$$

Sketch a right triangle and label it as in Figure E.3, to see that $\cos 2\theta = \frac{7}{25}$. (Recall that in the two quadrants with which we are concerned, $0° \le 2\theta \le 180°$, cosine and cotangent have the same sign.) Then

$$\sin \theta = \sqrt{\frac{1 - \frac{7}{25}}{2}} = \sqrt{\frac{9}{25}} = \frac{3}{5} \quad \text{and} \quad \cos \theta = \sqrt{\frac{1 + \frac{7}{25}}{2}} = \sqrt{\frac{16}{25}} = \frac{4}{5}.$$

Use these values for $\sin \theta$ and $\cos \theta$ to obtain

$$x = \frac{4}{5}x' - \frac{3}{5}y' \quad \text{and} \quad y = \frac{3}{5}x' + \frac{4}{5}y'.$$

Substituting these expressions for x and y into the original equation yields

$$52\left(\frac{4}{5}x' - \frac{3}{5}y'\right)^2 - 72\left(\frac{4}{5}x' - \frac{3}{5}y'\right)\left(\frac{3}{5}x' + \frac{4}{5}y'\right) + 73\left(\frac{3}{5}x' + \frac{4}{5}y'\right)^2 = 200.$$

This becomes

$$52\left(\frac{16}{25}x'^2 - \frac{24}{25}x'y' + \frac{9}{25}y'^2\right) - 72\left(\frac{12}{25}x'^2 + \frac{7}{25}x'y' - \frac{12}{25}y'^2\right)$$
$$+ 73\left(\frac{9}{25}x'^2 + \frac{24}{25}x'y' + \frac{16}{25}y'^2\right) = 200.$$

Combining terms gives

$$25x'^2 + 100y'^2 = 200.$$

Divide each side by 200 to get

$$\frac{x'^2}{8} + \frac{y'^2}{2} = 1,$$

> New equation after rotation

an equation of an ellipse having x'-intercepts $\pm\sqrt{8}$ and y'-intercepts $\pm\sqrt{2}$. The graph is shown in Figure E.4. To find θ, use the fact that

$$\frac{\sin \theta}{\cos \theta} = \frac{\frac{3}{5}}{\frac{4}{5}} = \frac{3}{4} = \tan \theta,$$

from which $\theta = \tan^{-1}\frac{3}{4} \approx 37°$.

Now Try Exercise 17

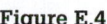

$\sin \theta = \frac{3}{5}$
$\cos \theta = \frac{4}{5}$
$\tan \theta = \frac{3}{4}$

$52x^2 - 72xy + 73y^2 = 200$
$\frac{x'^2}{8} + \frac{y'^2}{2} = 1$

Figure E.4

Summary of Conics with an *xy*-Term

The following summary enables us to use the general equation to decide on the type of graph to expect.

EQUATIONS OF CONICS WITH AN *XY*-TERM

If the general second-degree equation

$$Ax^2 + Bxy + Cy^2 + Dx + Ey + F = 0$$

has a graph, it will be one of the following:

(a) a circle or an ellipse (or a point) if $B^2 - 4AC < 0$;
(b) a parabola (or one line or two parallel lines) if $B^2 - 4AC = 0$;
(c) a hyperbola (or two intersecting lines) if $B^2 - 4AC > 0$;
(d) a straight line if $A = B = C = 0$ and $D \neq 0$ or $E \neq 0$.

E   Exercises

Exercises 1–6: Use the summary in this section to predict the graph of the second-degree equation.

1. $4x^2 + 3y^2 + 2xy - 5x = 8$

2. $x^2 + 2xy - 3y^2 + 2y = 12$

3. $2x^2 + 3xy - 4y^2 = 0$

4. $x^2 - 2xy + y^2 + 4x - 8y = 0$

5. $4x^2 + 4xy + y^2 + 15 = 0$

6. $-x^2 + 2xy - y^2 + 16 = 0$

Exercises 7–12: Find the angle of rotation θ that will remove the xy-term in the equation.

7. $2x^2 + \sqrt{3}xy + y^2 + x = 5$

8. $4\sqrt{3}x^2 + xy + 3\sqrt{3}y^2 = 10$

9. $3x^2 + \sqrt{3}xy + 4y^2 + 2x - 3y = 12$

10. $4x^2 + 2xy + 2y^2 + x - 7 = 0$

11. $x^2 - 4xy + 5y^2 = 18$

12. $3\sqrt{3}x^2 - 2xy + \sqrt{3}y^2 = 25$

Exercises 13–16: Use the given angle of rotation to remove the xy-term and graph the equation.

13. $x^2 - xy + y^2 = 6; \theta = 45°$

14. $2x^2 - xy + 2y^2 = 25; \theta = 45°$

15. $8x^2 - 4xy + 5y^2 = 36; \sin\theta = \dfrac{2}{\sqrt{5}}$

16. $5y^2 + 12xy = 10; \sin\theta = \dfrac{3}{\sqrt{13}}$

Exercises 17–24: Remove the xy-term from the equation by performing a suitable rotation. Graph the equation.

17. $3x^2 - 2xy + 3y^2 = 8$ **18.** $x^2 + xy + y^2 = 3$

19. $x^2 - 4xy + y^2 = -5$

20. $x^2 + 2xy + y^2 + 4\sqrt{2}x - 4\sqrt{2}y = 0$

21. $7x^2 + 6\sqrt{3}xy + 13y^2 = 64$

22. $7x^2 + 2\sqrt{3}xy + 5y^2 = 24$

23. $3x^2 - 2\sqrt{3}xy + y^2 - 2x - 2\sqrt{3}y = 0$

24. $2x^2 + 2\sqrt{3}xy + 4y^2 = 5$

Exercises 25–30: In the equation, remove the xy-term by rotation. Then translate the axes and sketch the graph.

25. $x^2 + 3xy + y^2 - 5\sqrt{2}y = 15$

26. $x^2 - \sqrt{3}xy + 2\sqrt{3}x - 3y - 3 = 0$

27. $4x^2 + 4xy + y^2 - 24x + 38y - 19 = 0$

28. $12x^2 + 24xy + 19y^2 - 12x - 40y + 31 = 0$

29. $16x^2 + 24xy + 9y^2 - 130x + 90y = 0$

30. $9x^2 - 6xy + y^2 - 12\sqrt{10}x - 36\sqrt{10}y = 0$

Bibliography

Acker, A., and C. Jaschek. *Astronomical Methods and Calculations*. New York: John Wiley and Sons, 1986.

Battan, L. *Weather in Your Life*. San Francisco: W. H. Freeman, 1983.

Beckmann, P. *A History of Pi*. New York: Barnes and Noble, 1993.

Bell, D. *Fundamentals of Electric Circuits,* Reston. Va.: Reston Publishing Company. 1981.

Benade, A. *Fundamentals of Musical Acoustics*. New York: Oxford University Press, 1976.

Beni, G., and S. Hackwood. *Recent Advances in Robotics*. New York: John Wiley and Sons, 1985.

Beranek, L. *Noise and Vibration Control*. Washington, D.C.: Institute of Noise Control Engineering, 1988.

Brearley, J., and A. Nicholas. *This Is the Bichon Frise*. Hong Kong: TFH Publication, 1973.

Brown, D., and P. Rothery. *Models in Biology: Mathematics, Statistics and Computing*. West Sussex, England: John Wiley and Sons, 1993.

Brown, F., J. Hastings, and J. Palmer. *The Biological Clock*. New York: Academic Press, 1970.

Carlson, T. "Über Geschwindigkeit und Grösse der Hefevermehrung in Würze." *Biochem. A.* 57:313–334.

Cheney, W., and D. Kincaid. *Numerical Mathematics and Computing*. 3rd ed. Pacific Grove, Calif.: Brooks/Cole Publishing Company, 1994.

Clime, W. *The Economics of Global Warming*. Washington, D.C.: Institute for International Economics, 1992.

Cole, F. *Introduction to Meteorology*. New York: Wiley, 1980.

Cooper, J., and R. Glassow. *Kinesiology*. 2nd ed. St. Louis: The C. V. Mosby Company, 1968.

Cotton, W., and R. Pielke. *Human Impacts on Weather and Climate*. Geophysical Science Series, vol. 2. Fort Collins, Colo.: *ASTeR Press, 1992.

Craig, J. *Introduction to Robotics: Mechanics and Control*. Reading, Mass.: Addison-Wesley Publishing Company, 1989.

Crownover, R. *Introduction to Fractals and Chaos*. Boston: Jones and Bartlett, 1995.

Duffet-Smith, P. *Practical Astronomy with Your Calculator*. New York: Cambridge University Press, 1988.

Eves, H. *An Introduction to the History of Mathematics*. 5th ed. Philadelphia: Saunders College Publishing, 1983.

Fletcher, N., and T. Rossing. *The Physics of Musical Instruments*. New York: Springer-Verlag, 1991.

Foley, J., A. van Dam, S. Feiner, J. Hughes, and R. Phillips. *Introduction to Computer Graphics*. Reading, Mass.: Addison-Wesley Publishing Company, 1994.

Foster, R., and J. Bates. "Use of mussels to monitor point source industrial discharges." *Environ. Sci. Technol.* 12:958–962.

Freebury, H. *A History of Mathematics*. New York: Macmillan, 1961.

Freedman, B. *Environmental Ecology: The Ecological Effects of Pollution, Disturbance, and Other Stresses*. 2nd ed. San Diego: Academic Press, 1995.

Garber, N., and L. Hoel. *Traffic and Highway Engineering*. Boston, Mass.: PWS Publishing Co., 1997.

Good, I. J. "What is the most amazing approximate integer in the universe?" *Pi Mu Epsilon Journal* 5 (1972): 314–315.

Grigg, D. *The World Food Problem*. Oxford: Blackwell Publishers, 1993.

Haber-Schaim, U., J. Cross, G. Abegg, J. Dodge, and J. Walter. *Introductory Physical Science*. Englewood Cliffs, N.J.: Prentice Hall, 1972.

Haefner, L. *Introduction to Transportation Systems*. New York: Holt, Rinehart and Winston, 1986.

Harker, J. *The Physiology of Diurnal Rhythms*. New York: Cambridge University Press, 1964.

Harrison, F., F. Hills, J. Paterson, and R. Saunders. "The measurement of liver blood flow in conscious calves." *Quarterly Journal of Experimental Physiology* 71:235–247.

Hartman, D. *Global Physical Climatology*. San Diego: Academic Press, 1994.

Heinz-Otto, P., H. Jürgens, and D. Saupe. *Chaos and Fractals: New Frontiers in Science*. New York: Springer-Verlag, 1993.

Hibbeler, R. *Structural Analysis*. Englewood Cliffs, N.J.: Prentice-Hall, 1995.

Hill, F. *Computer Graphics*. New York: Macmillan Publishing Company, 1990.

Hines, A., T. Ghosh, S. Loyalka, and R. Warder, Jr. *Indoor Air Quality and Control*. Englewood Cliffs, N.J.: Prentice-Hall, 1993.

Hoggar, S. *Mathematics for Computer Graphics*. New York: Cambridge University Press, 1993.

Hoppensteadt, F., and C. Peskin. *Mathematics in Medicine and the Life Sciences*. New York: Springer-Verlag, 1992.

Hosmer, D., and S. Lemeshow. *Applied Logistic Regression*. New York: John Wiley and Sons, 1989.

Howatson, A. *Electrical Circuits and Systems*. New York: Oxford University Press, 1996.

Howells, G. *Acid Rain and Acid Waters*. 2nd ed. New York: Ellis Horwood, 1995.

Huffman, R. *Atmospheric Ultraviolet Remote Sensing*. San Diego: Academic Press, 1992.

Huxley, J. *Problems of Relative Growth*. London: Methuen and Co., 1932.

Jarrett, J. *Business Forecasting Methods*. Oxford: Basil Blackwell, 1991.

Karttunen, H., P. Kroger, H. Oja, M. Poutanen, and K. Donner, eds. *Fundamental Astronomy*. 2nd ed. New York: Springer-Verlag, 1994.

Kerlow, I. *The Art of 3-D Computer Animation and Imaging*. New York: Van Nostrand Riehold, 1996.

Kissam, P. *Surveying Practice*. 3rd ed. New York: McGraw-Hill, 1978.

Kline, M. *The Loss of Certainty*. New York: Oxford University Press, 1980.

Kraljic, M. *The Greenhouse Effect*. New York: The H. W. Wilson Company, 1992.

Kress, S. *Bird Life—A Guide to the Behavior and Biology of Birds*. Racine, Wisc.: Western Publishing Company, 1991.

Lack, D. *The Life of a Robin*. London: Collins, 1965.

Lancaster, H. *Quantitative Methods in Biological and Medical Sciences: A Historical Essay*. New York: Springer-Verlag, 1994.

Loh, W. *Dynamics and Thermodynamics of Planetary Entry*. Englewood Cliffs, N.J.: Prentice-Hall, 1963.

Makridakis, S., and S. Wheelwright. *Forecasting Methods for Management*. New York: John Wiley and Sons, 1989.

Mandelbrot, B. *The Fractal Geometry of Nature*. New York: W. H. Freeman Company, 1982.

Mannering, F., and W. Kilareski. *Principles of Highway Engineering and Traffic Analysis*. New York: John Wiley and Sons, 1990.

Mar, J., and H. Liebowitz. *Structure Technology for Large Radio and Radar Telescope Systems*. Cambridge, Mass.: The MIT Press, 1969.

Mason, C. *Biology of Freshwater Pollution*. New York: Longman and Scientific and Technical, John Wiley and Sons, 1991.

Meeus, J. *Astronomical Algorithms*. Richmond, Va.: Willman-Bell, 1991.

Mehrotra, A. *Cellular Radio: Analog and Digital Systems*. Boston: Artech House, 1994.

Metcalf, H. *Topics in Classical Biophysics*. Englewood Cliffs, N.J.: Prentice-Hall, 1980.

Miller, A., and J. Thompson. *Elements of Meteorology.* 2nd ed. Columbus, Ohio: Charles E. Merrill Publishing Company, 1975.

Miller, A., and R. Anthes. *Meteorology.* 5th ed. Columbus, Ohio: Charles E. Merrill Publishing Company, 1985.

Moffitt, F. *Photogrammetry.* Scranton, Pa.: International Textbook Company, 1967.

Mortenson, M. *Computer Graphics: An Introduction to Mathematics and Geometry.* New York: Industrial Press Inc., 1989.

Motz, L., and J. Weaver. *The Story of Mathematics.* New York: Plenum Press, 1993.

Mueller, I., and K. Ramsayer. *Introduction to Surveying.* New York: Frederick Ungar Publishing Company, 1979.

National Council of Teachers of Mathematics. *Historical Topics for the Mathematics Classroom, Thirty-first Yearbook,* 1969.

Navarra, J. *Atmosphere, Weather and Climate.* Philadelphia: W. B. Saunders, 1979.

Nilsson, A. *Greenhouse Earth.* New York: John Wiley and Sons, 1992.

Pennycuick, C. *Newton Rules Biology.* New York: Oxford University Press, 1992.

Pielou, E. *Population and Community Ecology: Principles and Methods.* New York: Gordon and Breach Science Publishers, 1974.

Pierce, J. *The Science of Musical Sound.* New York: W. H. Freeman, 1992.

Pokorny, C., and C. Gerald. *Computer Graphics: The Principles behind the Art and Science.* Irvine, Calif.: Franklin, Beedle, and Associates, 1989.

Raggett, G. "Modeling the Eyam plague." *The Institute of Mathematics and Its Applications* 18: 221–226.

Resnikoff, H., and R. Wells, Jr. *Mathematics in Civilization.* New York: Dover Publications, Inc., 1984.

Riley, W., L. Sturges, and D. Morris. *Statics and Mechanics of Materials: An Integrated Approach.* New York: John Wiley and Sons, Inc., 1995.

Rist, Curtis. "The Physics of Foul Shots." *Discover,* October 2000.

Robert Wood Johnson Foundation. *Chronic Care in America: A 21st Century Challenge,* 1996.

Rodricks, J. *Calculated Risk.* New York: Cambridge University Press, 1992.

Roederer, J. *Introduction to the Physics and Psychophysics of Music.* New York: Springer-Verlag, 1973.

Rogers, E., and T. Kostigen. *The Green Book.* New York: Random House, 2007.

Ronan, C. *The Natural History of the Universe.* New York: MacMillan Publishing Company, 1991.

Ryan, B., B. Joiner, and T. Ryan. *Minitab Handbook.* Boston: Duxbury Press, 1985.

Sanders, D. *Statistics: A First Course.* 5th ed. New York: McGraw-Hill, 1995.

Schlosser, W. *Challenges of Astronomy.* New York: Springer-Verlag, 1991.

Semat, H., and J. Albright. *Introduction to Atomic and Nuclear Physics.* Austin, Tex.: Holt, Rinehart and Winston, 1972.

Sharov, A., and I. Novikov. *Edwin Hubble: The Discoverer of the Big Bang Universe.* New York: Cambridge University Press, 1993.

Sinkov, A. *Elementary Cryptanalysis: A Mathematical Approach.* New York: Random House, 1968.

Smith, R., and R. Dorf. *Circuits, Devices and Systems.* 5th ed. New York: John Wiley and Sons, Inc., 1992.

Socolow, R., and S. Pacala. "A Plan to Keep Carbon in Check." *Scientific American,* September 2006.

Stadler, W. *Analytical Robotics and Mechatronics.* New York: McGraw-Hill, Inc., 1995.

Stent, G. S. *Molecular Biology of Bacterial Viruses.* San Francisco: W. H. Freeman, 1963.

Thomas, D. *Swimming Pool Operators Handbook.* Washington, D.C.: National Swimming Pool Foundation, 1972.

Thomas, R. *The Old Farmer's 2012 Almanac.* Dublin, N.H.: The Old Farmer's Almanac, 2011.

Thomas, V. *Science and Sport.* London: Faber and Faber, 1970.

Thomson, W. *Introduction to Space Dynamics.* New York: John Wiley and Sons, 1961.

Triola, M. *Elementary Statistics.* Pearson Education, 2012.

Tucker, A., A. Bernat, W. Bradley, R. Cupper, and G. Scragg. *Fundamentals of Computing 1. Logic: Problem Solving, Programs, and Computers.* New York: McGraw-Hill, 1995.

Turner, R. K., D. Pierce, and I. Bateman. *Environmental Economics: An Elementary Approach.* Baltimore: The Johns Hopkins University Press, 1993.

Van Sickle, J. *GPS for Land Surveyors.* Chelsea, Mich.: Ann Arbor Press, 1996.

Varley, G., and G. Gradwell. "Population models for the winter moth." *Symposium of the Royal Entomological Society of London* 4:132–142.

Walker, A. *Observation and Inference: An Introduction to the Methods of Epidemiology.* Newton Lower Falls, Mass.: Epidemiology Resources, 1991.

Wang, Z. "Self-Powered Nanotech." *Scientific American*, January 2008.

Watt, A. *3D Computer Graphics.* Reading, Mass.: Addison-Wesley Publishing Company, 1993.

Webb, T. *Celestial Objects for Common Telescopes.* New York: Dover Publications Inc., 1962.

Weidner, R., and R. Sells. *Elementary Classical Physics,* vol. 2. Boston: Allyn and Bacon, 1965.

Wilcox, G., and C. Hesselberth. *Electricity for Engineering Technology.* Boston: Allyn and Bacon, 1970.

Williams, J. *The Weather Almanac 1995.* New York: Vintage Books, 1994.

Winter, C. *Solar Power Plants.* New York: Springer-Verlag, 1991.

Wolff, R., and L. Yaeger. *Visualization of Natural Phenomena.* New York: Springer-Verlag, 1993.

Wuebbles, D., and J. Edmonds. *Primer on Greenhouse Gases.* Chelsea, Mich.: Lewis Publishers, 1991.

Zeilik, M., S. Gregory, and D. Smith. *Introductory Astronomy and Astrophysics.* 3rd ed. Philadelphia: Saunders College Publishers, 1992.

Zhao, Y. *Vehicle Location and Navigation Systems.* Boston, Mass.: Artech House, 1997.

Answers to Selected Exercises

CHAPTER 1: Introduction to Functions and Graphs

SECTION 1.1 (pp. 8–10)

1. Rational number, real number **3.** Rational number, real number **5.** Real number **7.** Natural number: $\sqrt{9}$; integers: $-3, \sqrt{9}$; rational numbers: $-3, \frac{2}{9}, \sqrt{9}, 1.\overline{3}$; irrational numbers: $\pi, -\sqrt{2}$ **9.** Natural numbers: none; integer: $-\sqrt{4}$; rational numbers: $\frac{1}{3}, 5.1 \times 10^{-6}, -2.33, 0.\overline{7}, -\sqrt{4}$; irrational number: $\sqrt{13}$ **11.** Rational numbers
13. Natural numbers **15.** Integers **17.** 51 **19.** -84
21. 0 **23.** 5 **25.** 8 **27.** -32 **29.** 4×10^1
31. 3.65×10^{-3} **33.** 2.45×10^3 **35.** 5.6×10^{-1}
37. -8.7×10^{-3} **39.** 2.068×10^2 **41.** 0.000001
43. 200,000,000 **45.** 156.7 **47.** 500,000 **49.** 4500
51. 67,000 **53.** 8×10^8; 800,000,000 **55.** 3.5×10^{-1}; 0.35
57. 2.1×10^{-3}; 0.0021 **59.** 5×10^{-3}; 0.005
61. 4.24×10^{19} **63.** 8.72×10^4 **65.** 7.67×10^{11}
67. 5.769 **69.** 0.058 **71.** 0.419 **73.** -1.235 **75.** 15.819
77. 1.4×10^{-1} watt **79.** About 53,794 miles per hour
81. (a) 1.4% **(b)** 3.9% **83. (a)** 45,000,000 ft **(b)** Yes
85. (a) $7.436\pi \approx 23.4$ in^3 **(b)** Yes

1.1 EXTENDED AND DISCOVERY EXERCISES (p. 11)

1. 2.9×10^{-4} cm **3.** 0.25 feet, or 3 inches

SECTION 1.2 (pp. 22–25)

1. (a)
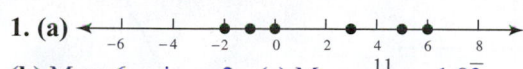
(b) Max: 6; min: -2 **(c)** Mean: $\frac{11}{6} = 1.8\overline{3}$

3. (a)

(b) Max: 30; min: -20 **(c)** Mean: 5

5.

-30	-30	-10	5	15	25	45	55	61

(a) Max: 61; min: -30
(b) Mean: $\frac{136}{9} \approx 15.11$; median: 15
7. $\sqrt{15} \approx 3.87$, $2^{2.3} \approx 4.92$, $\sqrt[3]{69} \approx 4.102$, $\pi^2 \approx 9.87$, $2^{\pi} \approx 8.82$, 4.1

$\sqrt{15}$	4.1	$\sqrt[3]{69}$	$2^{2.3}$	2^{π}	π^2

(a) Max: π^2; min: $\sqrt{15}$ **(b)** Mean: 5.95; median: 4.51
9. (a)

(b) Mean: 23.5; median: 23.95
The average area of the six largest lakes is 23,500 square miles. Half of these lakes have areas below 23,950 square miles and half have areas above. **(c)** Lake Superior
11. 16, 18, 26; no

13. (a) $S = \{(-1, 5), (2, 2), (3, -1), (5, -4), (9, -5)\}$
(b) $D = \{-1, 2, 3, 5, 9\}, R = \{-5, -4, -1, 2, 5\}$
15. (a) $S = \{(1, 5), (4, 5), (5, 6), (4, 6), (1, 5)\}$
(b) $D = \{1, 4, 5\}, R = \{5, 6\}$
17. (a) $D = \{-3, -2, 0, 7\}, R = \{-5, -3, 0, 4, 5\}$
(b) x-min: -3; x-max: 7; y-min: -5; y-max: 5
(c) & (d)

19. (a) $D = \{-4, -3, -1, 0, 2\}, R = \{-2, -1, 1, 2, 3\}$
(b) x-min: -4; x-max: 2; y-min: -2; y-max: 3
(c) & (d)

21. (a) $D = \{-35, -25, 0, 10, 75\}$,
$R = \{-55, -25, 25, 45, 50\}$
(b) x-min: -35; x-max: 75; y-min: -55; y-max: 50
(c) & (d)

23. Scatterplot   Line graph

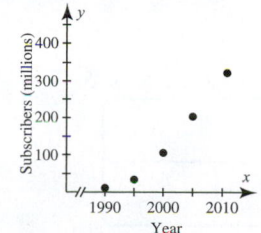
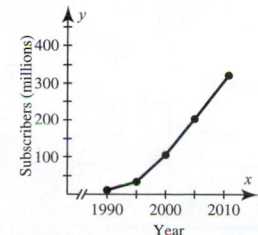

25. 5 **27.** $\sqrt{29} \approx 5.39$ **29.** $\sqrt{41.49} \approx 6.44$ **31.** 8
33. $\frac{\sqrt{17}}{4} \approx 1.03$ **35.** $\frac{\sqrt{2}}{2} \approx 0.71$ **37.** 130 **39.** $\sqrt{a^2 + b^2}$
41. Yes

43. (a)

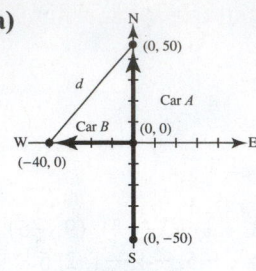

(b) $d = \sqrt{4100} \approx 64.0$ miles
45. 27 million units **47.** 9.5 seconds **49.** $(3, -0.5)$
51. $(10, 10)$ **53.** $(-2.1, -0.35)$ **55.** $(\sqrt{2}, 0)$ **57.** $(0, 2b)$
59. Center: $(0, 0)$; radius: 5 **61.** Center: $(0, 0)$; radius: $\sqrt{7}$
63. Center: $(2, -3)$; radius: 3 **65.** Center: $(0, -1)$; radius: 10
67. $(x - 1)^2 + (y + 2)^2 = 1$
69. $(x + 2)^2 + (y - 1)^2 = 4$
71. $(x - 3)^2 + (y + 5)^2 = 64$ **73.** $(x - 3)^2 + y^2 = 49$
75. $(x - 3)^2 + (y + 5)^2 = 50$
77. $(x + 2)^2 + (y + 3)^2 = 25$
79. $\left(x - \frac{7}{2}\right)^2 + (y - 3)^2 = \frac{25}{4}$

81. 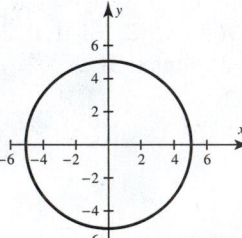 **83.**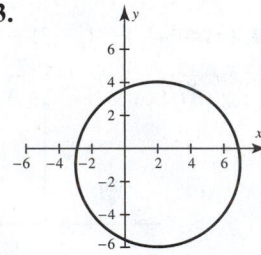

85. 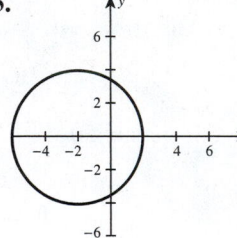 **87.**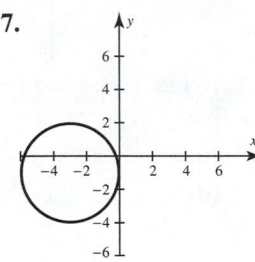

89.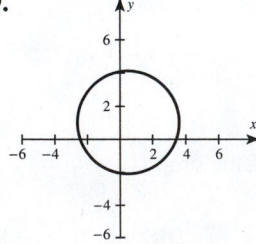

91. x-axis: 10; y-axis: 10 **93.** x-axis: 10; y-axis: 5

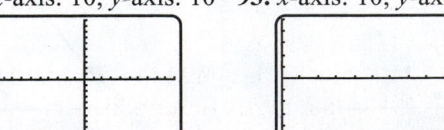

95. x-axis: 16; y-axis: 5

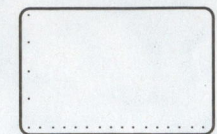

97. b **99.** a
101. $[-5, 5, 1]$ by $[-5, 5, 1]$ **103.** $[-100, 100, 10]$ by $[-100, 100, 10]$

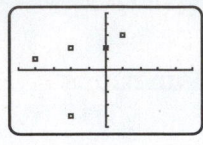 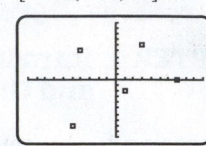

105. (a) x-min: 2006; x-max: 2010; y-min: 6.1; y-max: 19.4
(b) $[2005, 2011, 1]$ by $[5, 20, 5]$ (answers may vary)
(c) 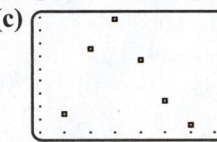 **(d)**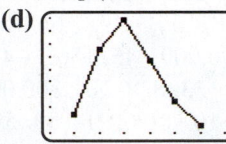

107. (a) x-min: 2006; x-max: 2011; y-min: 180; y-max: 590
(b) $[2005, 2012, 1]$ by $[150, 600, 50]$ (answers may vary)
(c) 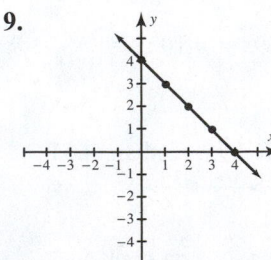 **(d)**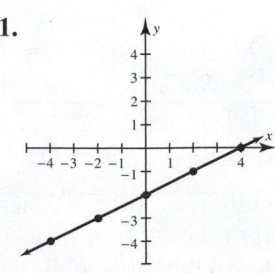

CHECKING BASIC CONCEPTS FOR SECTIONS 1.1 AND 1.2 (p. 25)

1. (a) 9.88 **(b)** 1.28
3. (a) 3.485×10^8 **(b)** -1.2374×10^3 **(c)** 1.98×10^{-3}
5. $\left(1, \frac{5}{2}\right)$ **7.** Mean $= 10{,}762.75$; median $= 12{,}941.5$

SECTION 1.3 (pp. 38–41)

1. $(-2, 3)$ **3.** $f(7) = 8$
5. **7.**

9. **11.**

13.
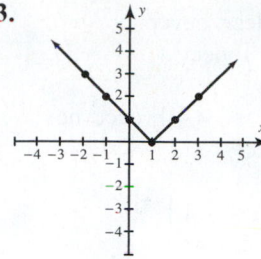

15.
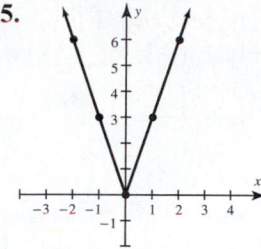

17.
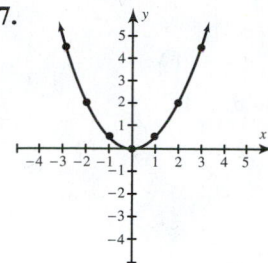

19.
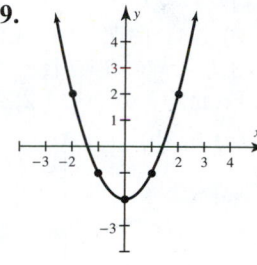

21. (a) $g = \{(-1, 0), (2, -2), (5, 7)\}$
(b) $D = \{-1, 2, 5\}, R = \{-2, 0, 7\}$
23. (a) $g = \{(1, 8), (2, 8), (3, 8)\}$
(b) $D = \{1, 2, 3\}, R = \{8\}$
25. (a) $g = \{(-1, 2), (0, 4), (1, -3), (2, 2)\}$
(b) $D = \{-1, 0, 1, 2\}, R = \{-3, 2, 4\}$
27. (a) $f(-2) = -8, f(5) = 125$ **(b)** All real numbers
29. (a) $f(-1)$ is undefined, $f(a + 1) = \sqrt{a + 1}$
(b) Nonnegative real numbers
31. (a) $f(-1) = 9, f(a + 1) = 3 - 3a$
(b) All real numbers
33. (a) $f(-1) = -2, f(a) = \frac{3a - 5}{a + 5}$ **(b)** $x \neq -5$
35. (a) $f(4) = \frac{1}{16}, f(-7) = \frac{1}{49}$ **(b)** $x \neq 0$
37. (a) $D = $ All real numbers; $R = $ All real numbers
(b) $g(-1) = -3; g(2) = 3$ **(c)** $g(-1) = -3; g(2) = 3$
39. (a) $D = $ All real numbers; $R = \{y \mid y \leq 2\}$
(b) $g(-1) = 1; g(2) = -2$ **(c)** $g(-1) = 1; g(2) = -2$
41. (a) $D = \{x \mid -2 \leq x \leq 2\}; R = \{y \mid -3 \leq y \leq 1\}$
(b) $g(-1) = -2; g(2) = 1$ **(c)** $g(-1) = -2; g(2) = 1$
43. $D = \{x \mid -3 \leq x \leq 3\}, R = \{y \mid 0 \leq y \leq 3\};$
$f(0) = 3$
45. $D = $ all real numbers, $R = \{y \mid y \leq 2\}; f(0) = 2$
47. $D = \{x \mid x \geq -1\}, R = \{y \mid y \leq 2\}; f(0) = 0$
49. (a) $f(2) = 7$ **(b)** $f = \{(1, 7), (2, 7), (3, 8)\}$
(c) $D = \{1, 2, 3\}, R = \{7, 8\}$
51. (a) $[-4.7, 4.7, 1]$ by $[-3.1, 3.1, 1]$ **(b)** $f(2) = 1$
(c)
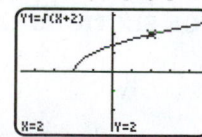

53. (a) $[-4.7, 4.7, 1]$ by $[-3.1, 3.1, 1]$ **(b)** $f(2) = 2$
(c)
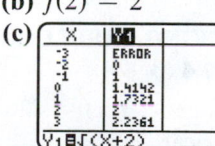

55. Verbal: Square the input x.
Graphical:
$[-10, 10, 1]$ by $[-10, 10, 1]$
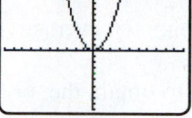
Numerical:

x	-2	-1	0	1	2
y	4	1	0	1	4

$f(2) = 4$

57. Verbal: Multiply the input by 2, add 1, and then take the absolute value.
Graphical:
$[-6, 6, 1]$ by $[-4, 4, 1]$
Numerical:

x	-2	-1	0	1	2
y	3	1	1	3	5

$f(2) = 5$

59. Verbal: Subtract x from 5.
Graphical:
$[-10, 10, 1]$ by $[-10, 10, 1]$
Numerical:

x	-2	-1	0	1	2
y	7	6	5	4	3

$f(2) = 3$

61. Verbal: Add 1 to the input and then take the square root of the result.
Graphical:
$[-6, 6, 1]$ by $[-4, 4, 1]$
Numerical:

x	-2	-1	0	1	2
y	—	0	1	$\sqrt{2}$	$\sqrt{3}$

$f(2) = \sqrt{3}$

63. Symbolic: $f(x) = 0.50x$
Graphical:
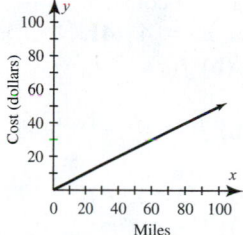
Numerical:

Miles	1	2	3	4	5	6
Cost	0.50	1.00	1.50	2.00	2.50	3.00

65. Yes. Domain and range include all real numbers.
67. No
69. Yes. $D:\{x \mid -4 \leq x \leq 4\}; R:\{y \mid 0 \leq y \leq 4\}$
71. (a) Yes
(b) Each real number has exactly one real cube root.
73. (a) No
(b) More than one student could have score x.
75. Yes, because IDs are unique.   **77.** No   **79.** Yes   **81.** No
83. No   **85.** Yes   **87.** No   **89.** Yes
91. $g(x) = 12x; g(10) = 120$; there are 120 inches in 10 feet.
93. $g(x) = \frac{x}{4}; g(10) = 2.5$; there are 2.5 dollars in 10 quarters.
95. $g(x) = 86{,}400x; g(10) = 864{,}000$; there are 864,000 seconds in 10 days.

97. (a) $V = \{(R, 37), (N, 30), (S, 17)\}$
(b) $D = \{N, R, S\}$, $R = \{17, 30, 37\}$
99. 200; 200 million tons of electronic waste will accumulate after 5 years.
101. $N(x) = 2200x$; $N(3) = 6600$; in 3 years the average person uses 6600 napkins.
103. Verbal: Multiply the input x by -5.8 to obtain the change in temperature.
Symbolic: $f(x) = -5.8x$.

Graphical: Numerical:

$[0, 3, 1]$ by $[-20, 20, 5]$

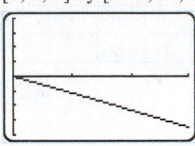

X	Y1
0	0
.5	-2.9
1	-5.8
1.5	-8.7
2	-11.6
2.5	-14.5
3	-17.4

$Y_1 = -5.8X$

SECTION 1.4 (pp. 55–59)

1. $m = -2$, $b = 5$ **3.** $m = -8$, $b = 0$ **5.** $\frac{1}{2}$ **7.** -1
9. 0   **11.** -1 **13.** -8 **15.** Undefined
17. $-\frac{39}{35} \approx -1.1143$
19. Slope $= 2$; the graph rises 2 units for every unit increase in x.
21. Slope $= -\frac{3}{4}$; the graph falls $\frac{3}{4}$ unit for every unit increase in x, or equivalently, the line falls 3 units for every 4-unit increase in x.
23. Slope $= -1$; the graph falls 1 unit for every unit increase in x.
25. (a) Zero square feet of carpet would cost \$0.
(b) Slope $= 20$ **(c)** The carpet costs \$20 per square foot.
27. (a) $D(5) = 50$; after 5 hours the train is 50 miles from the station   **(b)** -20; the train is traveling *toward* the station at 20 miles per hour.
29. (a) 150 miles   **(b)** Slope $= 75$; the car is traveling *away* from the rest stop at 75 miles per hour.
31. Linear, but not constant   **33.** Linear and constant
35. Nonlinear   **37.** Nonlinear   **39.** Yes, $m = 4$ **41.** No
43. (a) Slope: 2; y-int: -1; x-int: 0.5   **(b)** $f(x) = 2x - 1$
(c) 0.5
45. (a) Slope: $-\frac{1}{3}$; y-int: 2; x-int: 6   **(b)** $f(x) = -\frac{1}{3}x + 2$
(c) 6
47. $f(x) = -\frac{3}{4}x + \frac{1}{3}$ **49.** $f(x) = 15x$ **51.** $[5, \infty)$
53. $[4, 19)$ **55.** $[-1, \infty)$ **57.** $(-\infty, 1) \cup [3, \infty)$
59. $(-3, 5]$ **61.** $(-\infty, -2)$ **63.** $(-\infty, -2) \cup [1, \infty)$
65. Incr: never; decr: $(-\infty, \infty)$, $(\{x \mid -\infty < x < \infty\})$
67. Incr: $(2, \infty)$, $(\{x \mid x > 2\})$; decr: $(-\infty, 2)$, $(\{x \mid x < 2\})$
69. Incr: $(-\infty, -2)$, $(1, \infty)$, $(\{x \mid x < -2\}, \{x \mid x > 1\})$; decr: $(-2, 1)$, $(\{x \mid -2 < x < 1\})$
71. Incr: $(-8, 0)$, $(8, \infty)$, $(\{x \mid -8 < x < 0\}, \{x \mid x > 8\})$; decr: $(-\infty, -8)$, $(0, 8)$, $(\{x \mid x < -8\}, \{x \mid 0 < x < 8\})$
73. Incr: $(-\infty, \infty)$, $(\{x \mid -\infty < x < \infty\})$; decr: never
75. Incr: $(0, \infty)$, $(\{x \mid x > 0\})$; decr: $(-\infty, 0)$, $(\{x \mid x < 0\})$
77. Incr: $(-\infty, 1)$, $(\{x \mid x < 1\})$; decr: $(1, \infty)$, $(\{x \mid x > 1\})$

79. Incr: $(1, \infty)$, $(\{x \mid x > 1\})$; decr: never
81. Incr: $(-3, \infty)$, $(\{x \mid x > -3\})$; decr: $(-\infty, -3)$, $(\{x \mid x < -3\})$
83. Incr: $(-\infty, \infty)$, $(\{x \mid -\infty < x < \infty\})$; decr: never
85. Incr: $(-\infty, -2)$, $(2, \infty)$, $(\{x \mid x < -2\}, \{x \mid x > 2\})$; decr: $(-2, 2)$, $(\{x \mid -2 < x < 2\})$
87. Incr: $(-\infty, -1)$, $(0, 2)$, $(\{x \mid x < -1\}, \{x \mid 0 < x < 2\})$; decr: $(-1, 0), (2, \infty)$, $(\{x \mid -1 < x < 0\}, \{x \mid x > 2\})$
89. $(0, 2.4)$, $(8.7, 14.7)$, $(21, 27)$, $(\{x \mid 0 < x < 2.4\}, \{x \mid 8.7 < x < 14.7\}, \{x \mid 21 < x < 27\})$
91. From -3 to -1: 1.2; from 1 to 3: -1.2
93. (a) 3   **(b)**

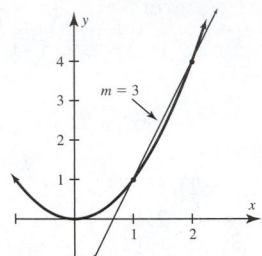

95. 7   **97.** 0.62   **99. (a)** From 1900 to 1940: 4.475; from 1940 to 1980: 11.25; from 1980 to 2010: about -10.57.
(b) From 1900 to 1940 cigarette consumption increased by 4.475 billion cigarettes per year, on average. The other rates may be interpreted similarly.
101.   **103.** Answers may vary.

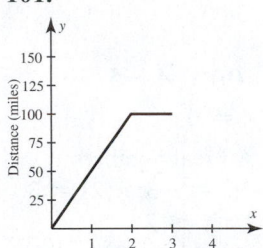

105. (a) 3   **(b)** 0   **107. (a)** $2x + 2h + 1$ **(b)** 2
109. (a) $3x^2 + 6xh + 3h^2 + 1$ **(b)** $6x + 3h$
111. (a) $-x^2 - 2xh - h^2 + 2x + 2h$ **(b)** $-2x - h + 2$
113. (a) $2x^2 + 4xh + 2h^2 - x - h + 1$ **(b)** $4x + 2h - 1$
115. (a) $x^3 + 3x^2h + 3h^2x + h^3$ **(b)** $3x^2 + 3hx + h^2$
117. (a) $8t^2 + 16th + 8h^2$ **(b)** $16t + 8h$
(c) 64.4; the average speed of the car from 4 to 4.05 seconds is 64.4 feet per second.

1.4 EXTENDED AND DISCOVERY EXERCISES (p. 59)

1. (a) Yes; 2π inches per second   **(b)** No; because the area function depends on the radius squared, the area function is not a linear function and does not increase at a constant rate.

CHECKING BASIC CONCEPTS FOR SECTIONS 1.3 AND 1.4 (p. 59)

1. Symbolic: $f(x) = 5280x$
Numerical:

x	1	2	3	4	5
$f(x)$	5280	10,560	15,840	21,120	26,400

Graphical:

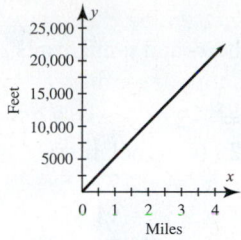

3. $-\frac{3}{2}$ **5. (a)** $(-\infty, 5]$ **(b)** $[1, 6)$ **7.** -7

CHAPTER 1 REVIEW EXERCISES (pp. 63–65)

1. Natural number: $\sqrt{16}$; integers: $-2, 0, \sqrt{16}$;
rational numbers: $-2, \frac{1}{2}, 0, 1.23, \sqrt{16}$; real numbers:
$-2, \frac{1}{2}, 0, 1.23, \sqrt{7}, \sqrt{16}$
3. 1.891×10^6 **5.** 15,200 **7. (a)** 32.07 **(b)** 2.62
(c) 5.21 **(d)** 49.12 **9.** -41
11.

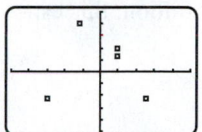

| -23 | -5 | 8 | 19 | 24 |

(a) Max: 24; min: -23
(b) Mean: 4.6; median: 8
13. (a) $S = \{(-15, -3), (-10, -1), (0, 1), (5, 3), (20, 5)\}$
(b) $D = \{-15, -10, 0, 5, 20\}$, $R = \{-3, -1, 1, 3, 5\}$
15. Not a function
$[-50, 50, 10]$ by $[-50, 50, 10]$

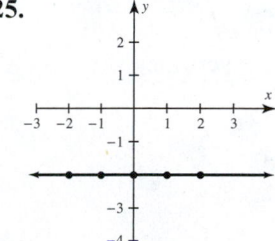

17. 10 **19.** $\left(2, -\frac{3}{2}\right)$ **21.** Yes
23. $(x - 2)^2 + (y - 5)^2 = 17$
25.

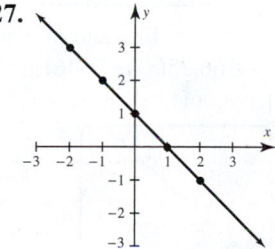

27.

29.

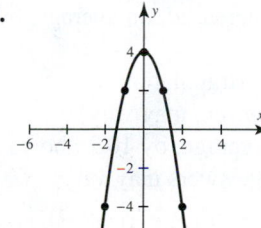

31.

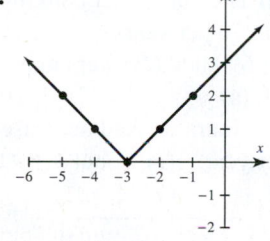

33. $f(x) = 16x$
$[0, 100, 10]$ by $[0, 1800, 300]$

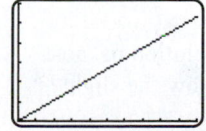

| x | 0 | 25 | 50 | 75 | 100 |
| $f(x)$ | 0 | 400 | 800 | 1200 | 1600 |

35. (a) $f(-3) = 5$, $f(1.5) = 5$ **(b)** All real numbers
37. (a) $f(-10) = 97$, $f(a + 2) = a^2 + 4a + 1$
(b) All real numbers

39. (a) $f(-3) = -\frac{1}{7}$, $f(a + 1) = \frac{1}{a - 3}$
(b) $D = \{x \mid x \neq 4\}$
41. No **43. (a)** Slope: -2; y-int: 6; x-int: 3
(b) $f(x) = -2x + 6$ **(c)** 3
45. Yes **47.** Yes **49.** 0 **51.** $-\frac{3}{4}$ **53.** 0 **55.** Linear
57. Nonlinear
59.

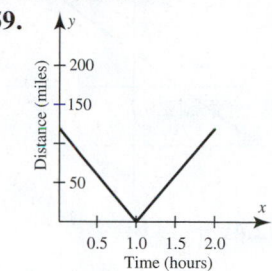

61. Yes; -4 **63.** 5 **65.** 760 seconds
67. (a) 198 feet **(b)** 1044 sq ft
69. (a) The data decrease rapidly, indicating a very high
mortality rate during the first year.
$[-1, 5, 1]$ by $[0, 110, 10]$

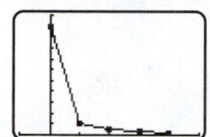

(b) Yes **(c)** From 0 to 1: -90; from 1 to 2: -4; from 2
to 3: -3; from 3 to 4: -1. During the first year the population decreased, on average, by 90 birds. The other average
rates of change can be interpreted similarly.
71. (a) $[1, 5, 1]$ by $[40, 70, 5]$

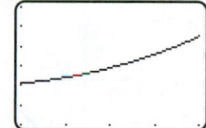

Nonlinear
(b) 2.5 **(c)** The average rate of change in outside temperature from 1 P.M. to 4 P.M. was 2.5°F per hour.

**CHAPTER 1 EXTENDED AND DISCOVERY EXERCISES
(pp. 65–66)**

1. About 7.17 km
3. About 3.862 **5.** About 4.039
$[-3, 3, 1]$ by $[-2, 2, 1]$ $[-3, 3, 1]$ by $[-1, 3, 1]$

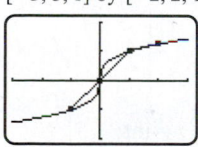 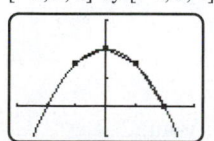

7. About 3.16; estimates will be less than the true value.
9. (a) Determine the number of square miles of Earth's surface that are covered by the oceans. Then divide the total
volume of the water from the ice cap by the surface area of
the oceans to get the rise in sea level. **(b)** About 25.7 feet
(c) Since the average elevations of Boston, New Orleans,
and San Diego are all less than 25 feet, these cities would
be under water without some type of dike system.
(d) About 238 feet

CHAPTER 2: Linear Functions and Equations

SECTION 2.1 (pp. 80–85)

1. $y = -2(x - 1) + 2$

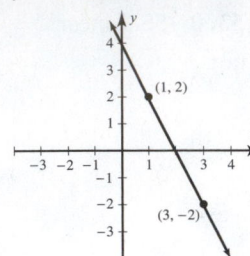

3. $y = \frac{3}{4}(x + 3) - 1$

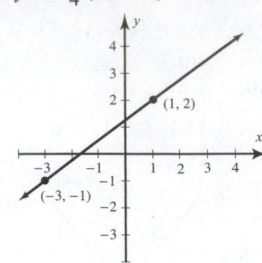

5. $y = -2.4(x - 4) + 5$; $y = -2.4x + 14.6$; $f(x) = -2.4x + 14.6$

7. $y = -\frac{1}{2}(x - 1) - 2$; $y = -\frac{1}{2}x - \frac{3}{2}$; $f(x) = -\frac{1}{2}x - \frac{3}{2}$

9. $y = \frac{3}{4}(x - 4) + 0$; $y = \frac{3}{4}x - 3$; $f(x) = \frac{3}{4}x - 3$

11. $y = \frac{2}{3}x - 1$ **13.** $y = -\frac{3}{5}x + \frac{3}{5}$ **15.** c

17. b **19.** e **21.** $y = 3x - 1$ **23.** $y = \frac{8}{3}x - \frac{17}{3}$

25. $y = -7.8x + 5$ **27.** $y = -\frac{1}{2}x + 45$

29. $y = -3x + 5$ **31.** $y = \frac{3}{2}x - 6$ **33.** $y = \frac{5}{18}x + \frac{11}{18}$

35. $y = 4x + 9$ **37.** $y = \frac{3}{2}x - 2960$ **39.** $y = \frac{2}{3}x - 2.1$

41. $y = \frac{1}{2}x + 6$ **43.** $y = x - 20$

45. $y = \frac{1}{2}x + \frac{9}{2}$ **47.** $y = -12x - 20$

49. $x = -5$; not possible **51.** $y = 6$, $f(x) = 6$

53. $x = 4$; not possible **55.** $x = 19$; not possible

57. x-int: 5; y-int: -4

59. x-int: 7; y-int: -7

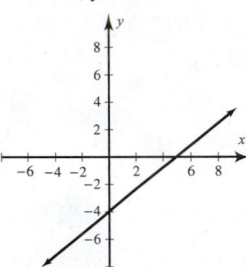

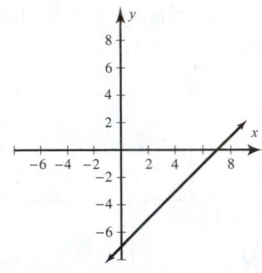
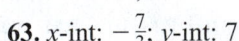

61. x-int: -7; y-int: 6

63. x-int: $-\frac{7}{3}$; y-int: 7

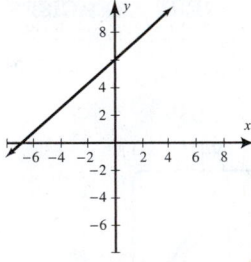

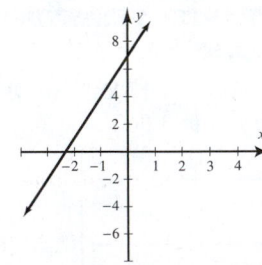

65. x-int: 4; y-int: 2

67. x-int: $\frac{5}{8}$; y-int: -5

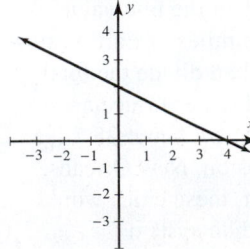

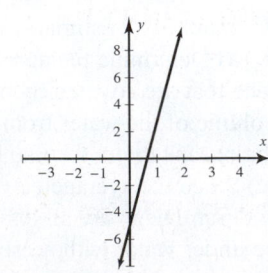

69. x-int: 5; y-int: 7; a and b represent the x- and y-intercepts, respectively.

71. x-int: $\frac{3}{2}$; y-int: $\frac{5}{4}$; a and b represent the x- and y-intercepts, respectively. **73.** $\frac{x}{5} + \frac{y}{9} = 1$

75. (a) $y = 1.5x - 3.2$ **(b)** When $x = -2.7$, $y = -7.25$ (interpolation); when $x = 6.3$, $y = 6.25$ (extrapolation)

77. (a) $y = -2.1x + 105.2$ **(b)** When $x = -2.7$, $y = 110.87$ (extrapolation); when $x = 6.3$, $y = 91.97$ (interpolation).

79. (a) $f(x) = 7(x - 2008) + 3$, or $f(x) = 7x - 14,053$ (answers may vary); approximate **(b)** -4%
(c) Extapolation; the result is negative, which is not possible.

81. (a) $y = \frac{12{,}000}{7}(x - 2003) + 25{,}000$, or $y = \frac{12{,}000}{7}(x - 2010) + 37{,}000$; the cost of attending a private college or university is increasing by $\frac{12{,}000}{7} \approx \$1714$ per year, on average. **(b)** About $31,857

83. (a) Leaving; 70 gallons **(b)** x-int: 10, the tank is empty after 10 minutes; y-int: 100, the tank held 100 gallons initially. **(c)** $y = -10x + 100$; the slope is -10, so the water is being drained at a rate of 10 gal/min. **(d)** 5

85. (a) $y = -\frac{3}{7}(x - 1999) + 15$ or $y = -\frac{3}{7}(x - 2013) + 9$ **(b)** Sales decreased, on average, by $\frac{3}{7} \approx \$0.43$ billion per year. **(c)** $11.1 billion; this estimate is $0.7 billion high; interpolation

87. (a) $[2005, 2011, 1]$ by $[150, 450, 50]$

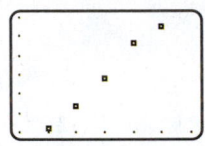

(b) Using the first and last points gives $f(x) = 66.25(x - 2006) + 160$ (answers may vary).
(c) $[2005, 2011, 1]$ by $[150, 450, 50]$

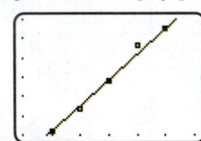

(d) The number of bankruptcies increased, on average, by 66,250 per year.
(e) 690,000 (answers may vary); extrapolation

89. (a) $y = -10.25x + 2000$ (answers may vary)
(b) Hours worked decreased, on average, by 10.25 hours per year. **(c)** About 1549 hours (answers may vary)

91. $[0, 3, 1]$ by $[-2, 2, 1]$

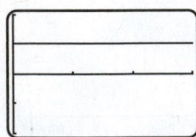

(a) No, the slope is not zero. **(b)** The resolution of most calculator screens is not high enough to show the slight increase in the y-values.

93. (a) They do not appear to be perpendicular in the standard viewing rectangle (answers may vary for different calculators).

(b) In $[-15, 15, 1]$ by $[-10, 10, 1]$ and $[-3, 3, 1]$ by $[-2, 2, 1]$ they appear to be perpendicular.

$[-10, 10, 1]$ by $[-10, 10, 1]$

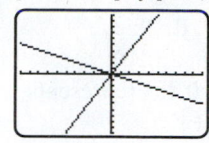

$[-15, 15, 1]$ by $[-10, 10, 1]$

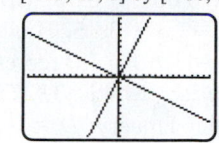

$[-10, 10, 1]$ by $[-3, 3, 1]$

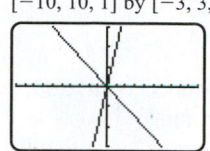

$[-3, 3, 1]$ by $[-2, 2, 1]$

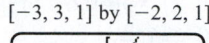
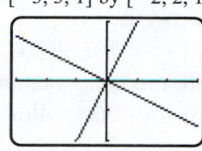

(c) The lines appear perpendicular when the distance shown along the x-axis is approximately 1.5 times the distance along the y-axis.
95. $y_1 = x, y_2 = -x, y_3 = x + 2, y_4 = -x + 4$
97. $y_1 = x + 4, y_2 = x - 4, y_3 = -x + 4, y_4 = -x - 4$
99. $y \approx -0.789x + 0.526; r \approx -0.993$

$[-3, 4, 1]$ by $[-3, 3, 1]$

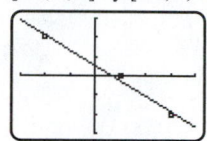

101. (a) Positive **(b)** $y = ax + b$, where $a = 3.25$ and $b = -2.45; r \approx 0.9994$ **(c)** $y = 5.35$
103. (a) Negative **(b)** $y = ax + b$, where $a \approx -3.8857$ and $b \approx 9.3254; r \approx -0.9996$ **(c)** $y \approx -0.00028$ (because of rounding, answers may vary slightly)
105. (a) $[-100, 1800, 100]$ by $[-1000, 28000, 1000]$

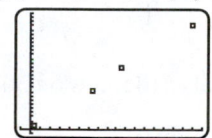

(b) $y = ax + b$, where $a \approx 14.680$ and $b \approx 277.82$
(c) 2500 light-years
107. (a) Positive

$[-5, 45, 5]$ by $[0, 6, 1]$

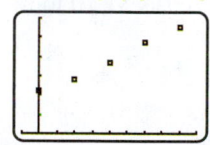

(b) $f(x) \approx 0.085x + 2.08$
(c) The number of passenger miles increased by about 0.085 trillion per year, on average.

$[-5, 45, 5]$ by $[0, 6, 1]$

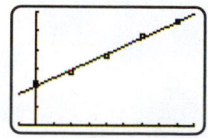

(d) About 5.9 trillion

SECTION 2.2 (pp. 99–103)

1. One **3.** $4x - 1$ **5.** They are equal. **7.** Linear
9. Nonlinear **11.** Linear **13.** 4 **15.** -4

17. $\frac{32}{3}$ **19.** 4 **21.** 3 **23.** $\frac{1}{3}$ **25.** $\frac{4}{7}$ **27.** $-\frac{2}{19}$
29. -5 **31.** $\frac{17}{10}$ **33.** $\frac{7}{32}$ **35.** $\frac{5}{17}$ **37.** $\frac{400}{7}$
39. (a) No solutions **(b)** Contradiction
41. (a) $\frac{8}{3}$ **(b)** Conditional
43. (a) All real numbers **(b)** Identity
45. (a) No solutions **(b)** Contradiction
47. (a) All real numbers **(b)** Identity
49. 3 **51. (a)** 4 **(b)** 2 **(c)** -2
53. -1 **55.** 1 **57.** 4 **59.** 1.3 **61.** 0.675
63. 3.621 **65.** 2.294 **67.** 3 **69.** 8.6
71. 0.2 **73.** -1.2 **75.** 1 **77.** -4 **79.** 0.8
81. 2 **83.** $W = \frac{4}{L}$ **85.** $L = \frac{1}{2}P - W$
87. $y = 4 - \frac{3}{2}x$ **89.** $x = \frac{1}{4}y + \frac{3}{2}$
91. (a) $y = -2x + 8$ **(b)** $f(x) = -2x + 8$
93. (a) $y = \frac{1}{2}x + \frac{1}{4}$ **(b)** $f(x) = \frac{1}{2}x + \frac{1}{4}$
95. (a) $y = \frac{9}{8}x + \frac{9}{8}$ **(b)** $f(x) = \frac{9}{8}x + \frac{9}{8}$
97. 1989 **99.** About 1987
101. (a) For the graph of A, the percentage of people *against* legalization was decreasing at a rate of 1.6% per year. For the graph of F, the percentage of people *for* legalization was increasing at a rate of 1.5% per year.
(b) About 2011
103. $f(x) = 0.75x$; \$42.18
105. (a) $0.048x$ **(b)** About 1,583,000
107. (a) About 2 hours (answers may vary)
(b) $\frac{15}{8} = 1.875$ hours
109. 3.2 hours at 55 mi/hr and 2.8 hours at 70 mi/hr
111. $\frac{1}{9}$ hour, or $6\frac{2}{3}$ minutes **113.** 41.25 feet
115. About 36.4 cubic feet **117.** About 8.33 liters
119. 36 inches by 54 inches
121. (a) $S(x) = 19x - 38,017$ **(b)** Sales increased, on average, by \$19 billion per year. **(c)** 2013
123. $-40°F$ is equivalent to $-40°C$.
125. (a) f is linear because the amount of oil is mixed at a constant rate. **(b)** 0.48 pint; 0.48 pint of oil should be added to 3 gallons of gasoline to get the correct mixture.
(c) 12.5 gallons **127.** $\frac{80}{9} \approx 8.89$

2.2 EXTENDED AND DISCOVERY EXERCISES (p. 103)

1. (a) Yes; since multiplication distributes over addition, doubling the lengths gives double the sum of the lengths.
(b) No; for example, in the case of a square (a type of rectangle), the square of twice a side is four times the square of the side.
3. (a) $f(x) = 14,000x$ **(b)** About 1.9 hours

CHECKING BASIC CONCEPTS FOR SECTIONS 2.1 AND 2.2 (p. 104)

1. $y = -\frac{3}{4}x + \frac{7}{4}$; parallel: $y = -\frac{3}{4}x$;
perpendicular: $y = \frac{4}{3}x$ (answers may vary)
3. $y = -\frac{3}{2}x + \frac{1}{2}$ **5.** $-\frac{5}{3}$
7. (a) 2.5 **(b)** 2.5 **(c)** 2.5; The results are the same.

SECTION 2.3 (pp. 114–117)

1. $(-\infty, 2)$ **3.** $[-1, \infty)$ **5.** $[1, 8)$ **7.** $(-\infty, 1]$
9. $\{x | x \geq 2\}$, or $[2, \infty)$

11. $\{x\,|\,x < 10.5\}$, or $(-\infty, 10.5)$
13. $\{x\,|\,x \geq 13\}$, or $[13, \infty)$ 15. $\{x\,|\,x < 0\}$, or $(-\infty, 0)$
17. $\{x\,|\,x \geq -10\}$, or $[-10, \infty)$
19. $\{x\,|\,x > 1\}$, or $(1, \infty)$
21. $\{x\,|\,x > \frac{7}{3}\}$, or $\left(\frac{7}{3}, \infty\right)$
23. $\{x\,|\,\frac{3}{2} < x \leq 3\}$, or $\left(\frac{3}{2}, 3\right]$
25. $\{x\,|\,-16 \leq x \leq 1\}$, or $[-16, 1]$
27. $\{x\,|\,-20.75 < x \leq 12.5\}$, or $(-20.75, 12.5]$
29. $\{x\,|\,-4 < x < 1\}$, or $(-4, 1)$
31. $\{x\,|\,\frac{9}{2} \leq x \leq \frac{21}{2}\}$, or $\left[\frac{9}{2}, \frac{21}{2}\right]$
33. $\{x\,|\,x \geq \frac{5}{3}\}$, or $\left[\frac{5}{3}, \infty\right)$
35. $\{x\,|\,-\frac{1}{2} < x \leq -\frac{1}{4}\}$, or $\left(-\frac{1}{2}, -\frac{1}{4}\right]$
37. $\{z\,|\,z \leq \frac{21}{19}\}$, or $\left(-\infty, \frac{21}{19}\right]$
39. $\{x\,|\,x \leq 2\}$ 41. $\{x\,|\,x > 3\}$
43. $\{x\,|\,0 \leq x \leq 2\}$ 45. $\{x\,|\,-1 < x \leq 4\}$
47. (a) 2   (b) $\{x\,|\,x < 2\}$, or $(-\infty, 2)$
(c) $\{x\,|\,x \geq 2\}$, or $[2, \infty)$
49. (a) -2 (b) $\{x\,|\,x > -2\}$, or $(-2, \infty)$
(c) $\{x\,|\,x \leq -2\}$, or $(-\infty, -2]$
51. $\{x\,|\,x \leq 2\}$, or $(-\infty, 2]$
53. $\{x\,|\,x > 1\}$, or $(1, \infty)$
55. $\{x\,|\,x > 2.8\}$ 57. $\{x\,|\,x \leq 1987.5\}$
59. $\{x\,|\,x > -1.82\}$
61. $\{x\,|\,4 \leq x < 6.4\}$, or $[4, 6.4)$
63. $\{x\,|\,4.6 \leq x \leq 15.2\}$, or $[4.6, 15.2]$
65. $\{x\,|\,1 < x < 5.5\}$, or $(1, 5.5)$
67. (a) 8   (b) $\{x\,|\,x < 8\}$
69. $\{x\,|\,x < 4\}$; $\{x\,|\,x \geq 4\}$
71. $\{x\,|\,x < -\frac{3}{2}\}$, or $\left(-\infty, -\frac{3}{2}\right)$
73. $\{x\,|\,1 \leq x \leq 4\}$, or $[1, 4]$
75. $\{x\,|\,-\frac{1}{20} \leq x < \frac{17}{20}\}$, or $\left[-\frac{1}{20}, \frac{17}{20}\right)$
77. $\{x\,|\,x \leq 31.4\}$, or $(-\infty, 31.4]$
79. $\{x\,|\,x > \frac{13}{2}\}$, or $\left(\frac{13}{2}, \infty\right)$
81. $\{x\,|\,x \leq 1.534\}$, or $(-\infty, 1.534]$
83. (a) Car A is traveling faster since its graph has the greater slope.   (b) 2.5 hours; 225 miles
(c) $0 \leq x < 2.5$
85. (a) $T(x) = 65 - 19x$ (b) $D(x) = 50 - 5.8x$
(c) Below 1.14 miles (approximately)
(d) Below 1.14 miles (approximately)
87. (a) Revenue increased, on average, by \$0.86 billion per year.   (b) From 2012 to 2015
89. (a) $U(x) = 225x + 100$ (b) 2010 and after
91. (a) $V(x) = 7.6x - 15,212.6$
(b) About 2007 to 2009
93. About day 110 (April 19) to day 119 (April 28)
95. $3.98\pi \leq C \leq 4.02\pi$
97. (a) $f(x) = 3x - 1.5$ (b) $x > 1.25$
99. (a) $P(x) = 0.658x - 1290.76$ (b) Between 1989 and 2005   (c) Interpolation

2.3 EXTENDED AND DISCOVERY EXERCISES (p. 117)

1. $a < b \Rightarrow 2a < a + b < 2b \Rightarrow a < \dfrac{a + b}{2} < b$

SECTION 2.4 (pp. 127–133)

1. (a) $f(x) = \frac{x}{16}$ (b) $f(x) = 10x$
(c) $f(x) = 0.06x + 6.50$ (d) $f(x) = 500$
3. $f(x) = -\frac{1}{2}x + 3$ 5. $f(x) = 2x + 5$ 7. d   9. c
11. $B(t) = 1.2t + 27$; t represents years after 2010;
$D = \{t\,|\,0 \leq t \leq 4\}$ 13. $T(t) = 0.5t + 0.4$; t represents months after January; $D = \{t\,|\,0 \leq t \leq 11\}$
15. $P(t) = 21.5 + 0.6t$; t represents years after 1900;
$D = \{t\,|\,0 \leq t \leq 111\}$
17. (a) $W(t) = -10t + 300$ (b) 230 gal
(c) x-int: 30, the tank is empty after 30 minutes; y-int: 300, the tank contains 300 gallons of water initially.

(d) $D = \{t\,|\,0 \leq t \leq 30\}$

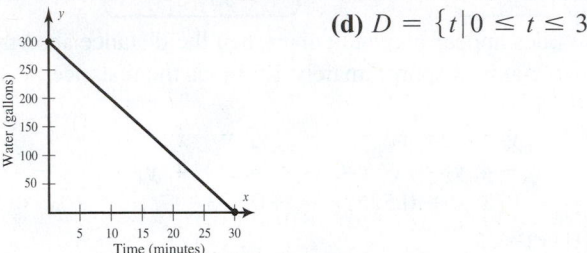

19. (a) $f(x) = 0.04x + 1.2$ (b) 1,480,000
21. (a) $f(x) = 0.25x + 0.5$ (b) 1.125 inches
23. (a) 17; 17.2; 16.6   (b) $f(x) = 17x$
(c) The vehicle is getting 17 miles per gallon.

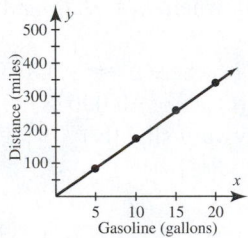

(d) $f(30) = 510$; the vehicle travels 510 miles on 30 gallons of gas.
25. (a) Max: 55 mi/hr; min: 30 mi/hr   (b) 12 miles
(c) $f(4) = 40$, $f(12) = 30$, $f(18) = 55$
(d) $x = 4, 6, 8, 12,$ and 16. The speed limit changes at each discontinuity.
27. (a) $P(1.5) = 1.10$, $P(3) = 1.30$; it costs \$1.10 to mail 1.5 ounces and \$1.30 to mail 3 ounces.
(b) $D = \{x\,|\,0 < x \leq 5\}$

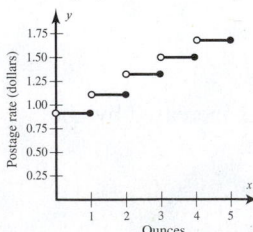

(c) $x = 1, 2, 3, 4$
29. (a) $f(1.5) = 30$; $f(4) = 10$
(b) $m_1 = 20$ indicates that the car is moving away from home at 20 mi/hr; $m_2 = -30$ indicates that the car is moving toward home at 30 mi/hr; $m_3 = 0$ indicates that the car is not moving; $m_4 = -10$ indicates that the car is moving toward home at 10 mi/hr.   (c) The driver starts at home and

drives away from home at 20 mi/hr for 2 hours until the car is 40 miles from home. The driver then travels toward home at 30 mi/hr for 1 hour until the car is 10 miles from home. Then the car does not move for 1 hour. Finally, the driver returns home in 1 hour at 10 mi/hr.
(d) Incr: $0 < x < 2$; decr: $2 < x < 3$ or $4 < x < 5$; const: $3 < x < 4$

31. (a) $-5 \le x \le 5$
(b) $f(-2) = 2, f(0) = 3, f(3) = 6$
(c)   **(d)** f is continuous.

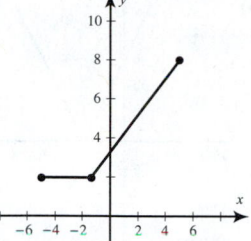

33. (a) $-1 \le x \le 2$
(b) $f(-2) =$ undefined, $f(0) = 0, f(3) =$ undefined
(c)   **(d)** f is not continuous.

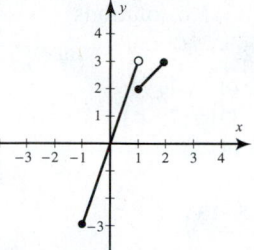

35. (a) $-3 \le x \le 3$
(b) $f(-2) = -2, f(0) = 1, f(3) = -1$
(c)   **(d)** f is not continuous.

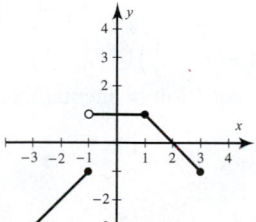

37.

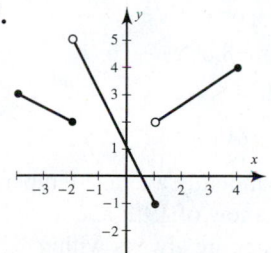

39. (a) $f(-3) = -10, f(1) = 4, f(2) = 4, f(5) = 1$
(b) $[1, 3]$ **(c)** f is not continuous.

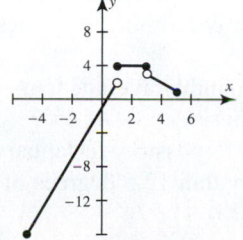

41. $xf(x) = \begin{cases} \frac{3}{4}x + 3, & \text{if } -4 \le x < 0 \\ -\frac{2}{3}x + 2, & \text{if } 0 \le x \le 3 \end{cases}$

43. (a)

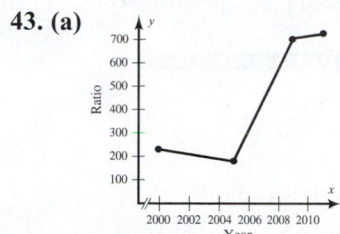

(b) 700; there were 700 people for each housing start in 2009; small values   **(c)** Yes
(d) Increasing: $(2005, 2011)$; decreasing: $(2000, 2005)$; constant: never
(e) $R(x) = \begin{cases} -9x + 18{,}225, & \text{if } 2000 \le x \le 2005 \\ 130x - 260{,}470, & \text{if } 2005 < x \le 2009 \\ 13.5x - 26{,}421.5, & \text{if } 2009 < x \le 2011 \end{cases}$

45. (a) $[-10, 10, 1]$ by $[-10, 10, 1]$

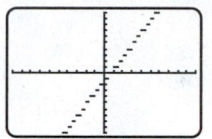

(b) $f(-3.1) = -8, f(1.7) = 2$
47. (a) $[-10, 10, 1]$ by $[-10, 10, 1]$

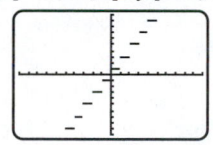

(b) $f(-3.1) = -7, f(1.7) = 3$
49. (a) $f(x) = 0.8\lfloor x/2 \rfloor$ for $6 \le x \le 18$
(b) $[6, 18, 1]$ by $[0, 8, 1]$

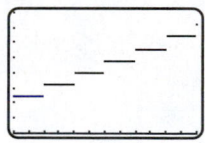

(c) $f(8.5) = \$3.20, f(15.2) = \$5.60$
51. 2.5   **53.** $\frac{9}{8}$ **55.** $k = 2.5, y = 20$ when $x = 8$
57. $k = 0.06, x = \$85$ when $y = \$5.10$
59. $\$1048, k = 65.5$ **61. (a)** $k = 0.01$ **(b)** 1.1 mm
63. (a) $k = \frac{15}{8}$ **(b)** $13\frac{1}{3}$ inches
65. (a) For $(150, 26), \frac{F}{x} \approx 0.173$; for $(180, 31)$, $\frac{F}{x} \approx 0.172$; for $(210, 36), \frac{F}{x} \approx 0.171$; for $(320, 54)$, $\frac{F}{x} \approx 0.169$; the ratios give the force needed to push a 1-pound box.   **(b)** $k \approx 0.17$ (answers may vary)
(c) $[125, 350, 25]$ by $[0, 75, 5]$

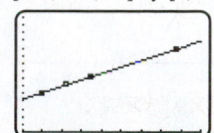

(d) 46.75 pounds

67. (a) $S(x) \approx 3.974x - 14.479$ (answers may vary)
(b) About 5.15 cm
69. (a) $f(x) \approx 0.12331x - 244.75$ (answers may vary)
(b) \$3.0 million; interpolation   **(c)** About 2017

2.4 EXTENDED AND DISCOVERY EXERCISES
(p. 133)

1. About 615 fish
3. Answers will vary.
5. (a) [1.580, 1.584, 0.001] by [−6.252, −6.248, 0.001]

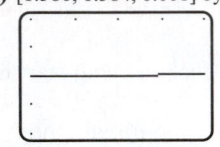

(b) A linear function would be a good approximation over a small interval.

CHECKING BASIC CONCEPTS FOR SECTIONS 2.3 AND 2.4 (p. 133)

1. $\{x \mid x > 3\}$
3. (a) 3   **(b)** $\{x \mid x > 3\}$ or $(3, \infty)$
(c) $\{x \mid x \le 3\}$ or $(-\infty, 3]$
5. $f(t) = 60t + 50$, where t is in hours

SECTION 2.5 (pp. 142–144)

1. $-3, 3$ **3.** $x < -3$ or $x > 3$, or $(-\infty, -3) \cup (3, \infty)$
5. It is V-shaped with the vertex on the x-axis.   **7.** $|6a|$
9.

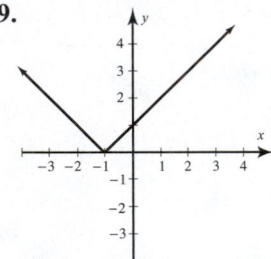

(a) -1 **(b)** Incr: $x > -1$, or $(-1, \infty)$; decr: $x < -1$, or $(-\infty, -1)$

11.

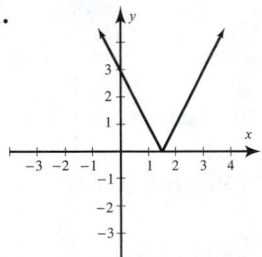

(a) $\frac{3}{2}$ **(b)** Incr: $x > \frac{3}{2}$, or $\left(\frac{3}{2}, \infty\right)$; decr: $x < \frac{3}{2}$, or $\left(-\infty, \frac{3}{2}\right)$

13. (a)   **(b)**

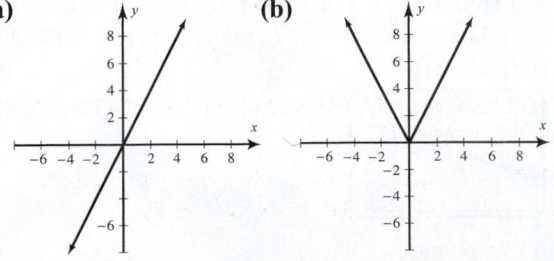

(c) 0

15. (a)   **(b)**

(c) 1
17. (a)   **(b)**

(c) 3
19. $-2, 2$ **21.** $1, \frac{9}{5}$ **23.** $-\frac{1}{2}, 2$ **25.** $-\frac{1}{3}$ **27.** $\frac{7}{16}$
29. No solutions   **31.** $-\frac{23}{12}, \frac{19}{4}$ **33.** No solutions
35. $-1, \frac{17}{5}$ **37.** $-1, 1$ **39.** $-3, 2$
41. (a) $-1, 7$ **(b)** $-1 < x < 7$, or $(-1, 7)$
(c) $x < -1$ or $x > 7$, or $(-\infty, -1) \cup (7, \infty)$
43. (a) $1, 2$ **(b)** $1 < x < 2$, or $(1, 2)$
(c) $x < 1$ or $x > 2$, or $(-\infty, 1) \cup (2, \infty)$
45. $-\frac{5}{2}, \frac{15}{2}; -\frac{5}{2} < x < \frac{15}{2}$, or $\left(-\frac{5}{2}, \frac{15}{2}\right)$
47. $\frac{7}{3}, 1; x < 1$ or $x > \frac{7}{3}$, or $(-\infty, 1) \cup \left(\frac{7}{3}, \infty\right)$
49. $-\frac{17}{21}, \frac{31}{21}; x \le -\frac{17}{21}$ or $x \ge \frac{31}{21}$, or
$\left(-\infty, -\frac{17}{21}\right] \cup \left[\frac{31}{21}, \infty\right)$
51. $-\frac{1}{3}, \frac{1}{3}; x < -\frac{1}{3}$ or $x > \frac{1}{3}$, or $\left(-\infty, -\frac{1}{3}\right) \cup \left(\frac{1}{3}, \infty\right)$
53. There are no solutions for the equation or inequality.
55. $\left(-\frac{7}{3}, 3\right)$, or $-\frac{7}{3} < x < 3$
57. $\left[-1, \frac{9}{2}\right]$, or $-1 \le x \le \frac{9}{2}$
59. $\left(-\frac{5}{2}, \frac{11}{2}\right)$, or $-\frac{5}{2} < x < \frac{11}{2}$
61. $(-\infty, 1) \cup (2, \infty)$, or $x < 1$ or $x > 2$
63. $\left(-\infty, \frac{5}{3}\right] \cup \left[\frac{11}{3}, \infty\right)$, or $x \le \frac{5}{3}$ or $x \ge \frac{11}{3}$
65. $(-\infty, -8) \cup (16, \infty)$, or $x < -8$ or $x > 16$
67. 6   **69.** $[-2, 4]$ **71.** $[0, \infty)$
73. Max: 75 mi/hr; min: 40 mi/hr
75. (a) $\frac{48}{19} \le x \le \frac{80}{19}$ **(b)** $\left| x - \frac{64}{19} \right| \le \frac{16}{19}$
77. (a) $19 \le T \le 67$ **(b)** The monthly average temperatures in Marquette vary between a low of 19°F and a high of 67°F. The monthly averages are always within 24 degrees of 43°F.
79. (a) $28 \le T \le 72$ **(b)** The monthly average temperatures in Boston vary between a low of 28°F and a high of 72°F. The monthly averages are always within 22 degrees of 50°F.
81. (a) $49 \le T \le 74$ **(b)** The monthly average temperatures in Buenos Aires vary between a low of 49°F (possibly in July) and a high of 74°F (possibly in January). The monthly averages are always within 12.5 degrees of 61.5°F.

83. (a) $|T - 10.5| < 0.5$ **(b)** $10.45 < T < 10.55$; The actual thickness must be greater than 10.45 mm and less than 10.55 mm.
85. (a) $|D - 2.118| \le 0.007$ **(b)** $2.111 \le D \le 2.125$; D must be greater than or equal to 2.111 inches and less than or equal to 2.125 inches.
87. $34.3 \le Q \le 35.7$

2.5 EXTENDED AND DISCOVERY EXERCISES (p. 144)

1. $|x - c| < \delta$

CHECKING BASIC CONCEPTS FOR SECTION 2.5 (p. 145)

1. $|2x|$ **3. (a)** $-2, 3$ **(b)** $[-2, 3]$, or $-2 \le x \le 3$; $(-\infty, -2) \cup (3, \infty)$, or $x < -2$ or $x > 3$ **5.** $-\frac{1}{3}, 1$

CHAPTER 2 REVIEW EXERCISES (pp. 148–152)

1. $y = \frac{1}{5}(x + 3) + 4$
3. $y = -\frac{7}{5}(x + 5) + 6$; $y = -\frac{7}{5}x - 1$; $f(x) = -\frac{7}{5}x - 1$
5. $y = \frac{2}{5}(x - 5) + 0$; $y = \frac{2}{5}x - 2$; $f(x) = \frac{2}{5}x - 2$
7. $f(x) = -2x - 1$
9. $y = 7x + 30$ **11.** $y = -3x + 2$
13. $y = -\frac{31}{57}x - \frac{368}{57}$
15. $x = 6$ **17.** $y = 3$ **19.** $x = 2.7$
21. x-int: 4; y-int: -5

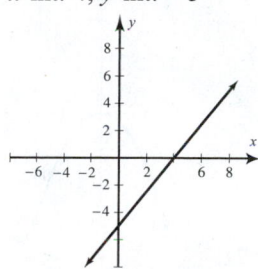

23. 6.4 **25.** $\frac{15}{7} \approx 2.143$ **27.** $\frac{5}{\pi} \approx 1.592$ **29.** -2.9
31. (a) All real numbers **(b)** Identity
33. (a) -3 **(b)** Conditional
35. $(-3, \infty)$ **37.** $\left[-2, \frac{3}{4}\right)$
39. $\{x \mid x \le 3\}$, or $(-\infty, 3]$
41. $\{x \mid -\infty < x < \infty\}$, or $(-\infty, \infty)$
43. $\{x \mid -1 < x \le \frac{7}{2}\}$, or $\left(-1, \frac{7}{2}\right]$
45. $\{x \mid x > -1\}$, or $(-1, \infty)$
47. (a) 2 **(b)** $x > 2$ **(c)** $x < 2$
49. (a) $f(-2) = 4$, $f(-1) = 6$, $f(2) = 3$, $f(3) = 4$
(b) f is continuous. **(c)** $x = -\frac{5}{2}$ or 2

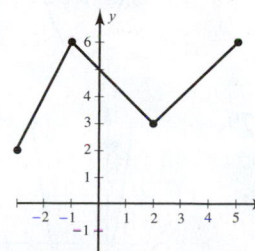

51. $-2, 7$ **53.** No solutions
55. ± 3; $x < -3$ or $x > 3$, or $(-\infty, -3) \cup (3, \infty)$
57. $\frac{17}{3}, -1$; $x < -1$ or $x > \frac{17}{3}$, or $(-\infty, -1) \cup \left(\frac{17}{3}, \infty\right)$

59. $-3 < x < 6$, or $(-3, 6)$
61. $x \le -\frac{5}{2}$ or $x \ge \frac{7}{2}$, or $\left(-\infty, -\frac{5}{2}\right] \cup \left[\frac{7}{2}, \infty\right)$
63. (a) $f(x) = 1060x + 17{,}700$ **(b)** Slope 1060 means that median income increased, on average, by about $1060 per year; y-intercept 17,700 means that in 1980 median income was $17,700 **(c)** $30,420; they are approximately equal. **(d)** About 2020; extrapolation
65. (a) $f(x) = 42.5x + 524$ **(b)** Slope 42.5 means that spending increased, on average, by about $42.5 billion per year; y-intercept 524 means that in 2010 spending was $524 billion. **(c)** $779 billion; interpolation **(d)** From 2014 to 2018
67. Initially the car is at home. After traveling 30 mi/hr for 1 hour, the car is 30 miles away from home. During the second hour the car travels 20 mi/hr until it is 50 miles away. During the third hour the car travels toward home at 30 mi/hr until it is 20 miles away. During the fourth hour the car travels away from home at 40 mi/hr until it is 60 miles away from home. During the last hour the car travels 60 miles at 60 mi/hr until it arrives home.
69. 155,590 **71.** 18.75 minutes
73. 0.9 hour at 7 mi/hr and 0.9 hour at 8 mi/hr
75. (a) $y = -1.2x + 3$
(b) $y = 4.8$ when $x = -1.5$, interpolation; $y = -1.2$ when $x = 3.5$, extrapolation **(c)** $\frac{17}{12}$
77. When $0 \le x \le 3$, the slope is 5, which means the inlet pipe is open and the outlet pipe is closed; when $3 < x \le 5$, the slope is 2, which means both pipes are open; when $5 < x \le 8$, the slope is 0, which means both pipes are closed; when $8 < x \le 10$, the slope is -3, which means the inlet pipe is closed and the outlet pipe is open.
79. The distance above the ground is between $1\frac{2}{3}$ kilometers and $3\frac{1}{3}$ kilometers.
81. Between 52.1431 feet and 52.4569 feet

CHAPTER 2 EXTENDED AND DISCOVERY EXERCISES (pp. 152)

1. (a) 62.8 inches
(b)

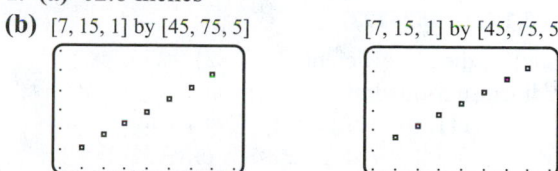

[7, 15, 1] by [45, 75, 5]   [7, 15, 1] by [45, 75, 5]

Both sets of data are linear.
(c) Female: 3.1 inches; male: 3.0 inches
(d) $f(x) = 3.1(x - 8) + 50.4$; $g(x) = 3.0(x - 8) + 53$
(e) $55.67 \le$ female height ≤ 56.91; $58.1 \le$ male height ≤ 59.3
3. 3 miles

CHAPTERS 1–2 CUMULATIVE REVIEW EXERCISES (pp. 152–154)

1. 1.23×10^5; 5.1×10^{-3} **3.** 2.09
5. $(x + 2)^2 + (y - 3)^2 = 49$ **7.** $\sqrt{89}$

9. (a) $D = \{x|-\infty < x < \infty\}, R = \{y|y \geq -2\}$; $f(-1) = -1$
(b) $D = \{x|-3 \leq x \leq 3\}, R = \{y|-3 \leq y \leq 2\}$; $f(-1) = -\frac{1}{2}$
11. (a) $f(2) = 7$; $f(a - 1) = 5a - 8$
(b) $D = \{x|-\infty < x < \infty\}$
13. No. The graph does not pass the vertical line test.
15. 1 **17. (a)** $\frac{2}{3}$; -2; 3 **(b)** $f(x) = \frac{2}{3}x - 2$ **(c)** 3
19. $y = -\frac{11}{8}x - \frac{29}{8}$
21. $x = -1$ **23.** $y = 2x + 11$
25. x-int: -3; y-int: 2

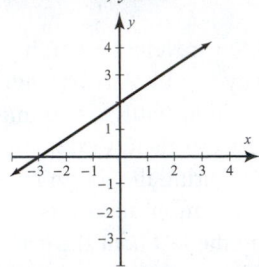

27. 1 **29.** $-\frac{24}{17}$ **31.** 3 **33.** $(-\infty, 5)$
35. $(-\infty, -2) \cup (2, \infty)$
37. $\{x|x \leq \frac{5}{8}\}$, or $(-\infty, \frac{5}{8}]$
39. (a) 2 **(b)** $x < 2$ **(c)** $x \geq 2$
41. $-6, 4$ **43.** $-7, 7$
45. $\{x|0 \leq x \leq 5\}$, or $[0, 5]$
47. (a) 770,000; it costs \$770,000 to manufacture 1500 computers. **(b)** 500; each additional computer costs \$500 to manufacture; fixed costs are \$20,000.
49. (a) 9°F per hour **(b)** On average, the temperature increased by 9°F per hour over this 2-hour period.
51. $\frac{60}{17} \approx 3.53$ hours
53. (a) $f(x) = \frac{28}{9}(x - 2001) + 56$ or $f(x) = \frac{28}{9}(x - 2010) + 84$ **(b)** About 100 pounds

CHAPTER 3: Quadratic Functions and Equations

SECTION 3.1 (pp. 167–172)

1. Quadratic; leading coefficient: 3; $f(-2) = 17$
3. Neither linear nor quadratic **5.** Linear
7. (a) $a > 0$ **(b)** $(1, 0)$ **(c)** $x = 1$
(d) Incr: $x > 1$, or $(1, \infty)$; decr: $x < 1$, or $(-\infty, 1)$
(e) $D = (-\infty, \infty)$; $R = [0, \infty)$
9. (a) $a < 0$ **(b)** $(-3, -2)$ **(c)** $x = -3$
(d) Incr: $x < -3$, or $(-\infty, -3)$; decr: $x > -3$, or $(-3, \infty)$
(e) $D = (-\infty, \infty)$; $R = (-\infty, -2]$
11. The graph of g is narrower than the graph of f.
13. The graph of g is wider than the graph of f and opens downward rather than upward.
15. Vertex: $(1, 2)$; leading coefficient: -3; $f(x) = -3x^2 + 6x - 1$
17. Vertex: $(4, 5)$; leading coefficient: -2; $f(x) = -2x^2 + 16x - 27$
19. Vertex: $\left(-5, -\frac{7}{4}\right)$; leading coefficient: $\frac{3}{4}$; $f(x) = \frac{3}{4}x^2 + \frac{15}{2}x + 17$

21. $f(x) = (x - 2)^2 - 2$ **23.** $f(x) = \frac{1}{2}(x - 2)^2 - 3$
25. $f(x) = -2(x + 1)^2 + 3$ **27.** $f(x) = -3(x - 2)^2 + 6$
29. $f(x) = (x + 2)^2 - 9$; vertex: $(-2, -9)$
31. $f(x) = \left(x - \frac{3}{2}\right)^2 - \frac{9}{4}$; vertex: $\left(\frac{3}{2}, -\frac{9}{4}\right)$
33. $f(x) = 2\left(x - \frac{5}{4}\right)^2 - \frac{1}{8}$; vertex: $\left(\frac{5}{4}, -\frac{1}{8}\right)$
35. $f(x) = \frac{1}{3}\left(x + \frac{3}{2}\right)^2 + \frac{1}{4}$; vertex: $\left(-\frac{3}{2}, \frac{1}{4}\right)$
37. $f(x) = 2(x - 2)^2 - 9$; vertex: $(2, -9)$
39. $f(x) = -3(x + 1.5)^2 + 8.75$; vertex: $(-1.5, 8.75)$
41. (a) $(0, 6)$ **(b)** Incr: $x < 0$, or $(-\infty, 0)$; decr: $x > 0$, or $(0, \infty)$
43. (a) $(3, -9)$ **(b)** Incr: $x > 3$, or $(3, \infty)$; decr: $x < 3$, or $(-\infty, 3)$
45. (a) $(1, -1)$ **(b)** Incr: $x > 1$, or $(1, \infty)$; decr: $x < 1$, or $(-\infty, 1)$
47. (a) $(0, 10)$ **(b)** Incr: $x > 0$, or $(0, \infty)$; decr: $x < 0$, or $(-\infty, 0)$
49. (a) $\left(\frac{1}{3}, -\frac{35}{12}\right)$ **(b)** Incr: $x < \frac{1}{3}$, or $\left(-\infty, \frac{1}{3}\right)$; decr: $x > \frac{1}{3}$, or $\left(\frac{1}{3}, \infty\right)$
51. (a) $\left(-\frac{1}{4}, \frac{15}{8}\right)$ **(b)** Incr: $x < -\frac{1}{4}$, or $\left(-\infty, -\frac{1}{4}\right)$; decr: $x > -\frac{1}{4}$, or $\left(-\frac{1}{4}, \infty\right)$
53. -6 **55.** $\frac{2}{3}$ **57.** $\frac{11}{4}$ **59.** $\frac{1}{4}$ **61.** $\frac{25}{4}$ **63.** $\frac{1}{3}$

65.

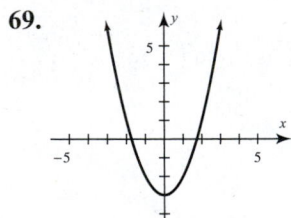

67.

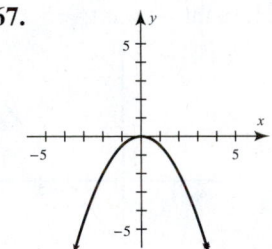

69.

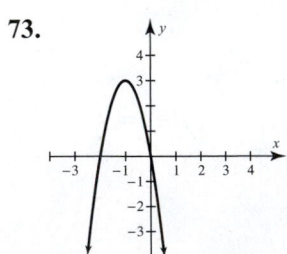

71.

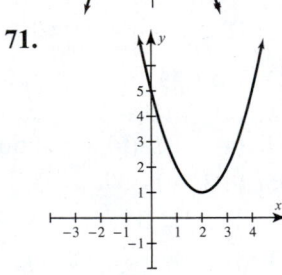

73.

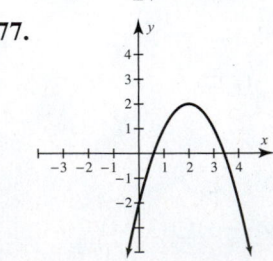

75.

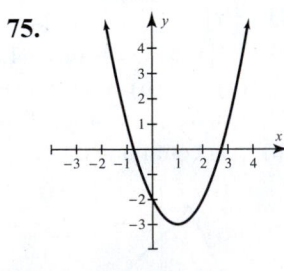

77.

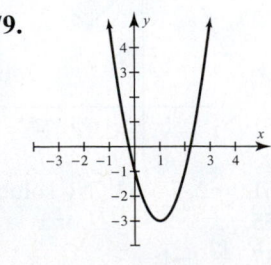

79.

81.

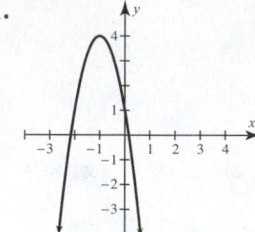

83.

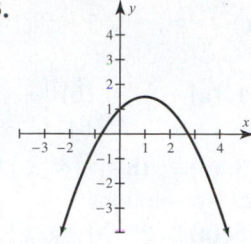

85. -7 **87.** $6x + 3h - 2$ **89.** $f(x) = 2(x - 3)^2 + 1$
91. d **93.** a **95.** 250 ft by 250 ft
97. (a) $R(2) = 72$; the company receives $72,000 for producing 2000 DVD players. **(b)** 10,000 **(c)** $200,000
99. (a) It is a parabola that opens upward. **(b)** About 3.5¢ per copy when 7796 copies are made.
101. (a) 32 ft **(b)** 34.25 ft
103. (a) $s(t) = -16t^2 - 66t + 120$
(b) Yes, because $s(2) < 0$.
105. 40 ft by 80 ft **107.** 146 ft after 2.75 sec
109. 323 ft after 6.77 sec
111. $f(x) = \frac{1}{225}x^2 + 20$, or $f(x) \approx 0.0044x^2 + 20$
113. $f(x) = 2(x - 1)^2 - 3$
115. (a) $H(t) = 2(t - 4)^2 + 90; D = \{t \mid 0 \le t \le 4\}$
(b) $H(1.5) = 102.5$ beats per minute
117. (a) $V(x) = 3.66x^2 + 1.5$ **(b)** 93 million
119. (a) $f(x) = 3.06(x - 1982)^2 + 1.6$ (answers may vary)
(b) [1980, 1996, 2] by [-50, 500, 100]

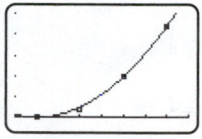

(c) About 250; by 1991 a total of about 250 thousand AIDS cases had been reported.
121. $f(x) = 3.125x^2 + 2.05x - 0.9; f(3.5) \approx 44.56$
123. (a) $f(x) \approx 0.055562x^2 - 220.05x + 217{,}886$
(b) About 49 thouand, which is 2 thousand too low
125. (a) $f(x) \approx 0.59462x^2 - 2350.82x + 2{,}323{,}895$
(b) About 454 thousand

3.1 EXTENDED AND DISCOVERY EXERCISES (p. 172)

1. (a)

x	1	2	3	4	5
$f(x)$	-2	1	6	13	22

(b) 3; 5; 7; 9 **(c)** $2x + h$; $2x + 1$
(d) 3; 5; 7; 9; the results are the same.
3. (a)

x	1	2	3	4	5
$f(x)$	0	-3	-10	-21	-36

(b) -3; -7; -11; -15
(c) $-4x + 3 - 2h$; $-4x + 1$
(d) -3; -7; -11; -15; the results are the same.

SECTION 3.2 (pp. 183–187)

1. $-4, 3$ **3.** $0, 2$ **5.** $0, \frac{7}{3}$ **7.** $-1, \frac{15}{2}$ **9.** -3
11. $\frac{1}{2}$ **13.** $-5, \frac{1}{3}$ **15.** $\frac{1}{2}, \frac{5}{6}$ **17.** $-3 \pm \sqrt{5}$
19. $\pm \frac{\sqrt{13}}{2}$ **21.** No real solutions **23.** $3 \pm 2\sqrt{2}$

25. $\frac{-1 \pm \sqrt{13}}{3}$ **27.** $\frac{1}{5}$ **29.** $\frac{5 \pm \sqrt{85}}{30}$ **31.** $-2, -1$
33. x-intercepts: $-\frac{5}{2}, \frac{1}{3}$; y-intercept: -5
35. x-intercept: $\frac{3}{2}$; y-intercept: -9
37. x-intercepts: $\frac{2}{3}, 3$; y-intercept: -6
39. $-2, 0$ **41.** $-2, 3$ **43.** $\pm\sqrt{3} \approx \pm 1.7$ **45.** 1.5
47. $-0.75, 0.2$ **49.** $0.7, 1.2$ **51.** $-2 \pm \sqrt{10}$
53. $-\frac{5}{2} \pm \frac{1}{2}\sqrt{41}$ **55.** $1 \pm \frac{\sqrt{15}}{3}$ **57.** $4 \pm \sqrt{26}$
59. $\frac{3 \pm \sqrt{17}}{2}$ **61.** $\frac{3 \pm \sqrt{17}}{4}$ **63.** $\frac{-1 \pm \sqrt{97}}{12}$
65. $\{x \mid x \ne \sqrt{5}, x \ne -\sqrt{5}\}$
67. $\{t \mid t \ne -1, t \ne 2\}$ **69.** $y = \frac{-12x^2 + 1}{8}$; yes
71. $y = \frac{x}{5}$; yes **73.** $y = 3 \pm \sqrt{9 - x^2}$; no
75. $y = \pm\frac{\sqrt{12 - 3x^2}}{2}$; no **77.** $r = \pm\sqrt{\frac{3V}{\pi h}}$
79. $v = \pm\sqrt{\frac{2K}{m}}$ **81.** $b = \pm\sqrt{c^2 - a^2}$
83. $t = \frac{25 \pm \sqrt{625 - 4s}}{8}$
85. (a) $3x^2 - 12 = 0$ **(b)** $b^2 - 4ac = 144 > 0$.
There are two real solutions. **(c)** ± 2
87. (a) $x^2 - 2x + 1 = 0$ **(b)** $b^2 - 4ac = 0$.
There is one real solution. **(c)** 1
89. (a) $x^2 - 4x = 0$ **(b)** $b^2 - 4ac = 16 > 0$.
There are two real solutions. **(c)** 0, 4
91. (a) $x^2 - x + 1 = 0$ **(b)** $b^2 - 4ac = -3 < 0$.
There are no real solutions.
93. (a) $2x^2 + 5x - 12 = 0$ **(b)** $b^2 - 4ac = 121 > 0$.
There are two real solutions. **(c)** $-4, \frac{3}{2}$
95. (a) $9x^2 - 36x + 36 = 0$ **(b)** $b^2 - 4ac = 0$.
There is one real solution. **(c)** 2
97. (a) $\frac{1}{2}x^2 + x + \frac{13}{2} = 0$ **(b)** $b^2 - 4ac = -12 < 0$.
There are no real solutions.
99. (a) $3x^2 + x - 1 = 0$
(b) $b^2 - 4ac = 13 > 0$. There are two real solutions.
(c) $\frac{-1 \pm \sqrt{13}}{6}$
101. (a) $a > 0$ **(b)** $-6, 2$ **(c)** Positive
103. (a) $a > 0$ **(b)** -4 **(c)** Zero
105. 2.2 seconds **107.** 2009 **109.** 8.5 in. by 11 in.; yes
111. 15 in. by 25 in. **113.** About 1.49 in.
115. About 18.23 in.
117. 86 shirts **119. (a)** $s(t) = -16t^2 + 160t + 32$
(b) About 10.2 sec
121. (a) $B(x) = 5.2(x - 1995)^2 + 180$ **(b)** 2005
123. (a) $I(x) = -2.277x^2 + 11.71x + 40.4$
(b) About 2005 and 2012
125. (a) $E(15) = 1.4$; in 2002 there were 1.4 million Walmart employees.
(b) $f(x) \approx 0.00474x^2 + 0.00554x + 0.205$
(answers may vary)
(c) [0, 25, 5] by [0, 2.6, 0.2]

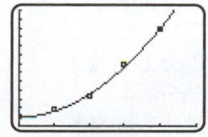

(d) About 2011 (answers may vary)

3.2 EXTENDED AND DISCOVERY EXERCISES (pp. 187–188)

1. $676 = 26^2$; yes; $-\frac{5}{2}, \frac{3}{4}$ **3.** 69; no; $\frac{3 \pm \sqrt{69}}{10}$
7. $x^2 = k \Rightarrow \sqrt{x^2} = \sqrt{k} \Rightarrow |x| \Rightarrow \sqrt{k} \Rightarrow x = \pm\sqrt{k}$

CHECKING BASIC CONCEPTS FOR SECTIONS 3.1 AND 3.2 (p. 188)

1. Vertex; $(1, -4)$; axis of symmetry: $x = 1$; x-intercepts: $-1, 3$

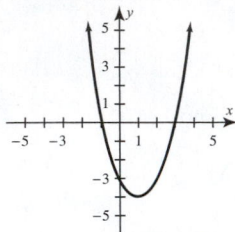

3. $f(x) = 2(x + 1)^2 + 3$
5. $f(x) = (x + 2)^2 - 7$; $(-2, -7)$; -7
7. 11 in. by 15 in.

SECTION 3.3 (pp. 195–196)

1. $2i$ **3.** $10i$ **5.** $i\sqrt{23}$ **7.** $2i\sqrt{3}$ **9.** $3i\sqrt{6}$
11. $2 \pm 2i$ **13.** $-2 \pm 2i\sqrt{2}$ **15.** -5 **17.** -6
19. $-3\sqrt{2}$ **21.** $8i$ **23.** $-2 - i$ **25.** $5 - 21i$
27. $-1 + 6i$ **29.** $4 + 8i$ **31.** $5 - i$ **33.** $4 - 7i$
35. $-5 - 12i$ **37.** 4 **39.** $\frac{1}{2} - \frac{1}{2}i$ **41.** $\frac{19}{26} + \frac{9}{26}i$
43. $-\frac{2}{25} + \frac{4}{25}i$ **45.** $3i$ **47.** $\frac{1}{2} + i$
49. $-18.5 + 87.4i$ **51.** $8.7 - 6.7i$
53. $-117.27 + 88.11i$ **55.** $-0.921 - 0.236i$
57. -1 **59.** $-i$ **61.** 1 **63.** i
65. $\pm i\sqrt{5}$ **67.** $\pm i\sqrt{\frac{1}{2}}$ **69.** $\frac{3}{10} \pm \frac{i\sqrt{11}}{10}$ **71.** $2 \pm i$
73. $\frac{3}{2} \pm \frac{i\sqrt{11}}{2}$ **75.** $-1 \pm i\sqrt{3}$ **77.** $1 \pm \frac{i\sqrt{2}}{2}$
79. $\frac{5}{4} \pm \frac{i\sqrt{7}}{4}$ **81.** $\frac{11}{8} \pm \frac{i\sqrt{7}}{8}$
83. (a) Two real zeros **(b)** $-1, \frac{3}{2}$
85. (a) Two imaginary zeros **(b)** $-\frac{1}{2} \pm \frac{i\sqrt{7}}{2}$
87. (a) Two imaginary zeros **(b)** $\pm i\sqrt{2}$
89. $Z = 10 + 6i$ **91.** $V = 11 + 2i$ **93.** $I = 1 + i$

3.3 EXTENDED AND DISCOVERY EXERCISE (p. 196)

1. (a) $i^1 = i, i^2 = -1, i^3 = -i, i^4 = 1, i^5 = i, i^6 = -1,$ $i^7 = -i, i^8 = 1$, and so on. **(b)** Divide n by 4. If the remainder is r, then $i^n = i^r$, where $i^0 = 1, i^1 = i,$ $i^2 = -1,$ and $i^3 = -i.$

SECTION 3.4 (pp. 203–206)

1. (a) $(-\infty, -1) \cup (1, \infty)$, or $\{x \mid x < -1 \text{ or } x > 1\}$
(b) $(-1, 1)$, or $\{x \mid -1 < x < 1\}$
3. (a) $[-4, 4]$, or $\{x \mid -4 \le x \le 4\}$
(b) $(-\infty, -4] \cup [4, \infty)$, or $\{x \mid x \le -4 \text{ or } x \ge 4\}$
5. (a) $(-2, 2)$, or $\{x \mid -2 < x < 2\}$
(b) $(-\infty, -2) \cup (2, \infty)$, or $\{x \mid x < -2 \text{ or } x > 2\}$
7. (a) $-3, 4$ **(b)** $(-3, 4)$, or $\{x \mid -3 < x < 4\}$
(c) $(-\infty, -3) \cup (4, \infty)$, or $\{x \mid x < -3 \text{ or } x > 4\}$
9. (a) $\pm\sqrt{5}$ **(b)** $[-\sqrt{5}, \sqrt{5}]$, or $\{x \mid -\sqrt{5} \le x \le \sqrt{5}\}$

(c) $(-\infty, -\sqrt{5}] \cup [\sqrt{5}, \infty)$, or $\{x \mid x \le -\sqrt{5} \text{ or } x \ge \sqrt{5}\}$
11. (a) $-\frac{8}{3}, 0$ **(b)** $[-\frac{8}{3}, 0]$, or $\{x \mid -\frac{8}{3} \le x \le 0\}$
(c) $(-\infty, -\frac{8}{3}] \cup [0, \infty)$, or $\{x \mid x \le -\frac{8}{3} \text{ or } x \ge 0\}$
13. (a) $\frac{3}{2}$ **(b)** $(-\infty, \frac{3}{2}) \cup (\frac{3}{2}, \infty)$, or $\{x \mid x < \frac{3}{2} \text{ or } x > \frac{3}{2}\}$
(c) No solutions
15. (a) $\frac{2}{3}, \frac{5}{4}$ **(b)** $[\frac{2}{3}, \frac{5}{4}]$, or $\{x \mid \frac{2}{3} \le x \le \frac{5}{4}\}$
(c) $(-\infty, \frac{2}{3}] \cup [\frac{5}{4}, \infty)$, or $\{x \mid x \le \frac{2}{3} \text{ or } x \ge \frac{5}{4}\}$
17. (a) $-1 \pm \sqrt{2}$ **(b)** $(-1 - \sqrt{2}, -1 + \sqrt{2})$, or $\{x \mid -1 - \sqrt{2} < x < -1 + \sqrt{2}\}$
(c) $(-\infty, -1 - \sqrt{2}) \cup (-1 + \sqrt{2}, \infty)$, or $\{x \mid x < -1 - \sqrt{2} \text{ or } x > -1 + \sqrt{2}\}$
19. (a) $-3 < x < 2$ **(b)** $x \le -3 \text{ or } x \ge 2$
21. (a) $x = -2$ **(b)** $x \ne -2$
23. (a) No solutions **(b)** All real numbers
25. (a) $-\frac{5}{2}, -\frac{1}{2}$ **(b)** $(-\frac{5}{2}, -\frac{1}{2})$, or $\{x \mid -\frac{5}{2} < x < -\frac{1}{2}\}$
(c) $(-\infty, -\frac{5}{2}) \cup (-\frac{1}{2}, \infty)$, or $\{x \mid x < -\frac{5}{2} \text{ or } x > -\frac{1}{2}\}$
27. (a) $-1, \frac{7}{5}$ **(b)** $(-\infty, -1) \cup (\frac{7}{5}, \infty)$, or $\{x \mid x < -1 \text{ or } x > \frac{7}{5}\}$ **(c)** $(-1, \frac{7}{5})$, or $\{x \mid -1 < x < \frac{7}{5}\}$
29. (a) $x < -1 \text{ or } x > 1$ **(b)** $-1 \le x \le 1$
31. (a) $-6 < x < -2$ **(b)** $x \le -6 \text{ or } x \ge -2$
33. $-2 \le x \le -0.5$ **35.** $x < -3 \text{ or } x > 2$
37. $-2 \le x \le 2$ **39.** All real numbers
41. $x \le -2 \text{ or } x \ge 3$ **43.** $-\frac{1}{3} < x < \frac{1}{2}$
45. $-4 \le x \le 10$ **47.** No solutions
49. All real numbers except $\frac{2}{3}$ **51.** $x \le 0 \text{ or } x \ge 1$
53. $x \le -2 \text{ or } x \ge 3$ **55.** $-\sqrt{5} \le x \le \sqrt{5}$
57. $x \le \frac{23}{7} \text{ or } x \ge 22.4$ **59.** $2 \le x \le 7$
61. $x \le -2 \text{ or } x \ge 5$ **63.** All real numbers
65. $x < -2 - \sqrt{7} \text{ or } x > -2 + \sqrt{7}$
67. About 35 mi/hr, but not more than 38 mi/hr
69. $2 \le r \le 3$ (inches)
71. (a) $f(0) = 160, f(2) = 131.2$; initially the heart rate is 160 bpm, and after 2 minutes it is about 131 bpm.
(b) About $2.9 \le x \le 5$ (minutes)
73. From 1989 to 1992
75. From 2007 to 2008 and from 2009 to 2010

CHECKING BASIC CONCEPTS FOR SECTIONS 3.3 AND 3.4 (p. 206)

1. (a) $5i$ **(b)** $-3\sqrt{6}$ **(c)** $\frac{1}{2} \pm \frac{i\sqrt{2}}{2}$
3. (a) $[-3, 0]$, or $\{x \mid -3 \le x \le 0\}$; $(-\infty, -3) \cup (0, \infty)$, or $\{x \mid x < -3 \text{ or } x > 0\}$
(b) $(-\infty, \infty)$, or $\{x \mid -\infty < x < \infty\}$; no solutions
5. (a) $(-\infty, -6] \cup [6, \infty)$, or $\{x \mid x \le -6 \text{ or } x \ge 6\}$
(b) $(-\infty, \infty)$, or $\{x \mid -\infty < x < \infty\}$
(c) $\left[\frac{1 - \sqrt{5}}{2}, \frac{1 + \sqrt{5}}{2}\right]$, or $\left\{x \mid \frac{1 - \sqrt{5}}{2} \le x \le \frac{1 + \sqrt{5}}{2}\right\}$

SECTION 3.5 (pp. 220–223)

1. $y = (x + 2)^2$ **3.** $y = \sqrt{x + 3}$
5. $y = |x + 2| - 1$
7. $y = \sqrt{x + 2} - 3$

9. $y = (x - 2)^2 - 3$
$[-10, 10, 1]$ by $[-10, 10, 1]$

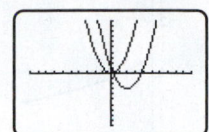

11. $y = (x + 6)^2 - 4(x + 6) + 5$
$[-10, 10, 1]$ by $[-10, 10, 1]$

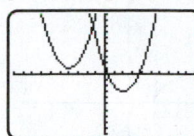

13. $y = \frac{1}{2}(x + 3)^2 + 2(x + 3) - 3$
$[-10, 10, 1]$ by $[-10, 10, 1]$

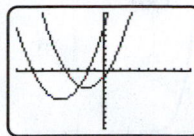

15. (a) $g(x) = 3(x + 3)^2 + 2(x + 3) - 5$
(b) $g(x) = 3x^2 + 2x - 9$
17. (a) $g(x) = 2(x - 2)^2 + 4$
(b) $g(x) = 2(x + 8)^2 - 5$
19. (a) $g(x) = 3(x - 2000)^2 - 3(x - 2000) + 72$
(b) $g(x) = 3(x + 300)^2 - 3(x + 300) - 28$
21. (a) $g(x) = -\sqrt{x - 4}$ **(b)** $g(x) = \sqrt{-x + 2}$
23. $(x - 3)^2 + (y + 4)^2 = 4$; center: $(3, -4)$; $r = 2$
25. $(x + 5)^2 + (y - 3)^2 = 5$;
center: $(-5, 3)$; $r = \sqrt{5}$

27. (a)

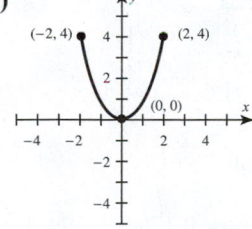

(b)

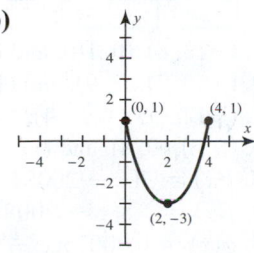

(c)

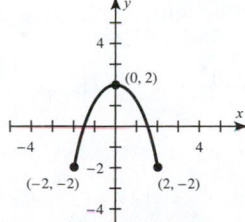

29. (a)

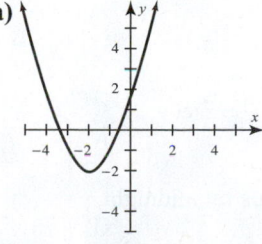

(b)

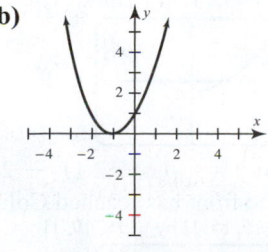

(c)

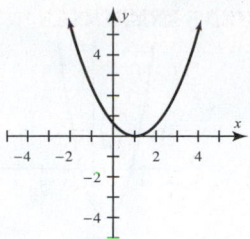

31. (a)

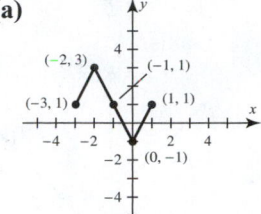

(b)

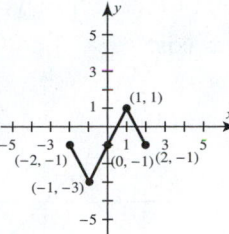

(c)

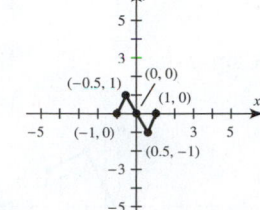

33. (a)

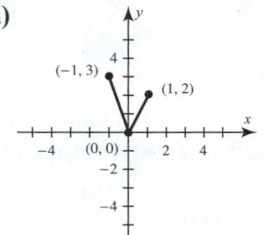

(b)

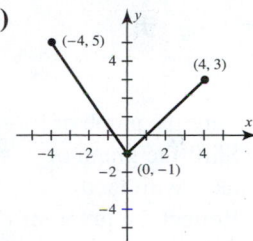

(c)

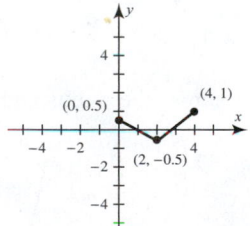

35. $y = -\sqrt{x} - 1$

Given function   x-axis reflection

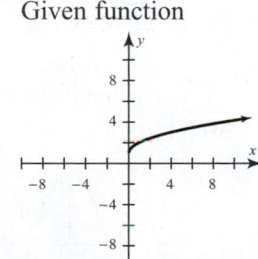

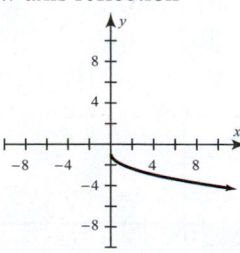

37. $y = x^2 + x$

Given function   y-axis reflection

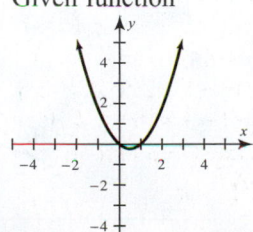

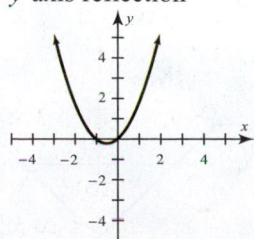

39. *x*-axis: *y*-axis:

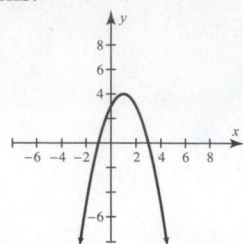   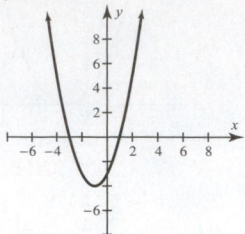

41. *x*-axis: *y*-axis:

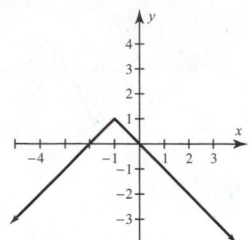   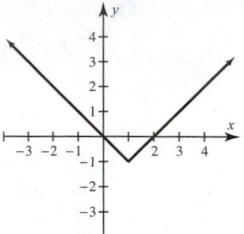

43. *x*-axis: *y*-axis:

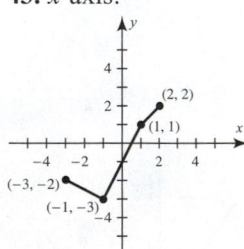   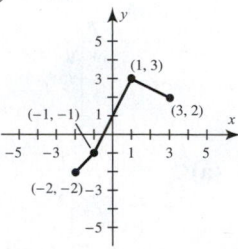

45. Shift the graph of $y = x^2$ right 3 units and upward 1 unit.

47. Shift the graph of $y = x^2$ left 1 unit and vertically shrink it with factor $\frac{1}{4}$.

49. Reflect the graph of $y = \sqrt{x}$ across the *x*-axis and shift it left 5 units.

51. Reflect the graph of $y = \sqrt{x}$ across the *y*-axis and vertically stretch it with factor 2.

53. Reflect the graph of $y = |x|$ across the *y*-axis and shift it left 1 unit.

55.   **57.**

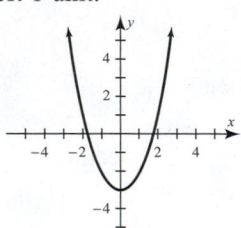   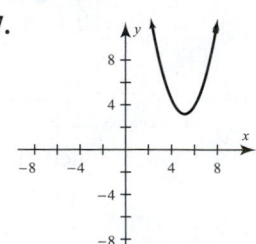

59.   **61.**

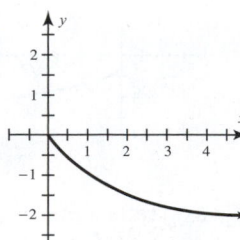   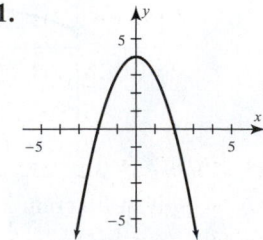

63.   **65.**

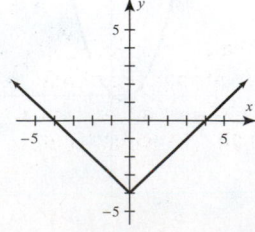   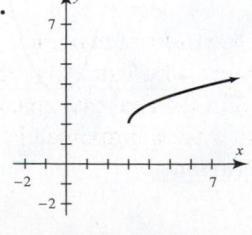

67.   **69.**

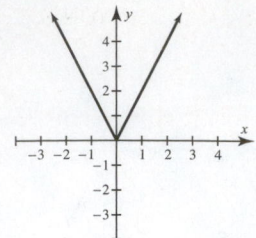   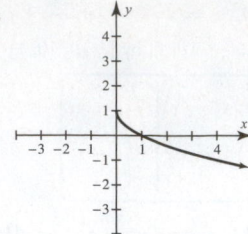

71.   **73.**

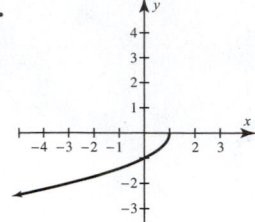   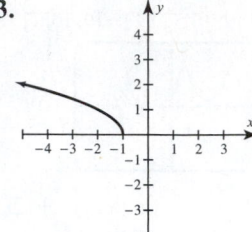

75.   **77.**

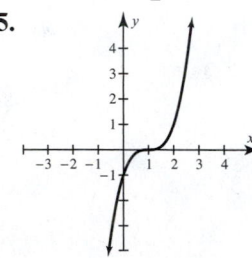   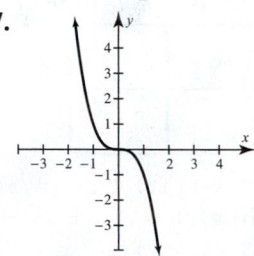

79.

x	1	2	3	4	5	6
$g(x)$	12	8	13	9	14	16

81.

x	−2	0	2	4	6
$g(x)$	5	2	−3	−5	−9

83.

x	0	1	2	3	4	5
$g(x)$	0	2	1	5	6	8

85.

x	−2	−1	0	1	2
$g(x)$	0	3	6	9	12

87. $(-12, 8)$, $(0, 10)$, and $(8, -2)$

89. $(-10, 7)$, $(2, 9)$, and $(10, -3)$

91. $(-12, -3)$, $(0, -4)$, and $(8, 2)$

93. $(6, 6)$, $(0, 8)$, and $(-4, -4)$

95. $f(x) = 7(x - 2008)^2 + 11.6$

97. $f(x) = 1.8(x - 2008)^2 + 22$

99. $g(x) = 0.00075(x - 1990)^2 + 0.17(x - 1990) + 44$

101. $y = -0.4(x - 3)^2 + 4$ (mountain)
$[-4, 4, 1]$ by $[0, 6, 1]$

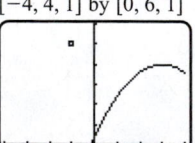

103. (a) $y = \frac{1}{20}x^2 - 1.6$
$[-15, 15, 1]$ by $[-10, 10, 1]$

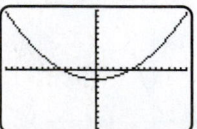

(b) $y = \frac{1}{20}(x - 2.1)^2 - 2.5$.
The front has reached Columbus by midnight.
$[-15, 15, 1]$ by $[-10, 10, 1]$

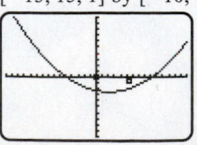

CHECKING BASIC CONCEPTS FOR SECTION 3.5 (p. 224)

1. (a) Shifted 4 units to the left
(b) Shifted 3 units down
(c) Shifted 5 units to the right and 3 units up
3. (a) $y = (x - 3)^2 - 4$ **(b)** $y = -x^2$
(c) $y = (-x + 6)^2$ **(d)** $y = (-(x + 6))^2$

5. (a)

x	-2	0	2	4	6
$g(x)$	4	6	9	11	12

(b)

x	-5	-3	-1	1	3
$h(x)$	-2	-6	-12	-16	-18

CHAPTER 3 REVIEW EXERCISES (pp. 228–229)

1. (a) $a < 0$ **(b)** $(2, 4)$ **(c)** $x = 2$
(d) Incr: $x < 2$, or $(-\infty, 2)$; Decr: $x > 2$, or $(2, \infty)$
3. $f(x) = -2x^2 + 20x - 49$; leading coefficient: -2
5. $f(x) = -(x + 1)^2 + 2$
7. $f(x) = (x + 3)^2 - 10$; vertex: $(-3, -10)$ **9.** $\left(\frac{1}{3}, -\frac{11}{3}\right)$

11.

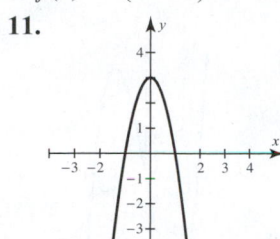

13.

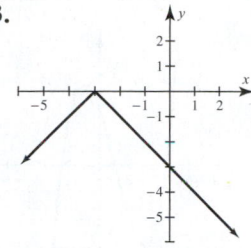

15. -29 **17.** $-4, 5$ **19.** $\pm\frac{\sqrt{7}}{2}$ **21.** $-\frac{7}{2}, 2$ **23.** $-2, 5$
25. $-1 \pm \sqrt{6}$ **27.** $\frac{3 \pm \sqrt{11}}{2}$ **29.** $y = \pm\sqrt{\frac{2x^2 - 6}{3}}$; no
31. (a) $4i$ **(b)** $4i\sqrt{3}$ **(c)** $-5\sqrt{3}$
33. (a) $-\frac{5}{2}, \frac{1}{2}$ **(b)** $-\frac{5}{2}, \frac{1}{2}$ **35.** $\pm\frac{3}{2}i$
37. (a) $-3 < x < 2$, or $(-3, 2)$
(b) $x \leq -3$ or $x \geq 2$, or $(-\infty, -3] \cup [2, \infty)$
39. $\{x \mid 1 \leq x \leq 2\}$, or $[1, 2]$
41. $\{x \mid x \leq -3$ or $x \geq 5\}$, or $(-\infty, -3] \cup [5, \infty)$
43. $y = -f(x)$ $\qquad$ $y = f(-x)$

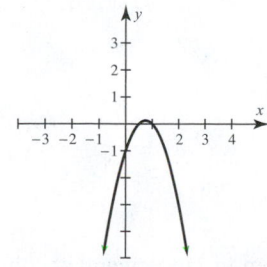 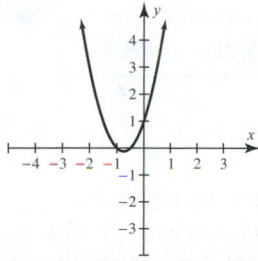

45. $\qquad$ **47.**

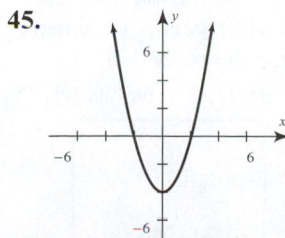 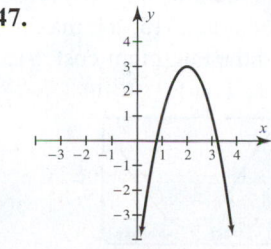

49. 11 ft by 22 ft

51. (a) $h(0) = 5$; the stone was 5 ft above the ground when it was released. **(b)** 117 ft **(c)** 126 ft
(d) After 2 seconds and 3.5 seconds
53. 11 in. by 15 in.
55. $M(x) = 30(x - 2007)^2 + 60$

CHAPTER 3 EXTENDED AND DISCOVERY EXERCISES (p. 230)

1. (a) 23.32 ft/sec
(b) $[-1, 16, 1]$ by $[-1, 16, 1]$ Yes **(c)** 12.88 ft

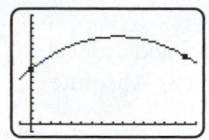

3. (a) $[-1, 8, 1]$ by $[-4, 4, 1]$

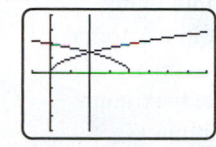

(b) The graph of $y = f(2k - x) = f(4 - x)$ is a reflection of $y = f(x)$ across the line $x = 2$.
5. (a) $[-15, 3, 1]$ by $[-3, 9, 1]$

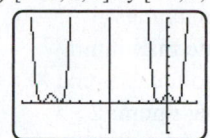

(b) The graph of $f(2k - x) = f(-12 - x)$ is a reflection of $y = f(x)$ across the line $x = -6$.
7. (a) $[0, 1200, 100]$ by $[-800, 0, 100]$

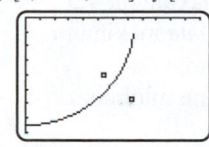

The front reached St. Louis, but not Nashville.
(b) $g(x) = -\sqrt{750^2 - (x - 160)^2} - 110$
(c) $[0, 1200, 100]$ by $[-800, 0, 100]$

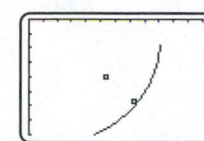

The cold front reached both cities in less than 12 hours.

CHAPTER 4: More Nonlinear Functions and Equations

SECTION 4.1 (pp. 238–242)

1. Yes; degree: 3; a: 2 **3.** No
5. Yes; degree: 4; a: -5 **7.** No
9. Yes; degree: 0; a: 22
11. (a) Local maximum: approximately 5.5; local minimum: approximately -5.5 **(b)** No absolute extrema
13. (a) Local maxima: approximately 17 and 27; local minima: approximately -10 and 24 **(b)** No absolute extrema

15. (a) Local maxima: approximately 0.5 and 2.8; local minimum: approximately 0 **(b)** Absolute maximum: 2.8; no absolute minimum
17. (a) Local maximum: 0; local minimum: approximately −1000 **(b)** No absolute maximum; absolute minimum: −1000
19. (a) Local maximum: 1; local minimum: −1 **(b)** Absolute maximum: 1; absolute minimum: −1
21. (a) No local maxima; local minimum: approximately −3.2 **(b)** Absolute maximum: 3; absolute minimum: approximately −3.2
23. (a) Local maxima: approximately 0.5 and 2; local minima: approximately −2 and −0.5 **(b)** Absolute maximum: 2; absolute minimum: −2
25. (a) No local maxima; local minimum: −2 **(b)** No absolute maximum; absolute minimum: −2
27. (a) No local extrema **(b)** No absolute extrema
29. (a) Local minimum: 1; no local maxima **(b)** Absolute minimum: 1; no absolute maximum
31. (a) Local maximum: 4; no local minima **(b)** Absolute maximum: 4; no absolute minimum
33. (a) Local minimum: $-\frac{1}{8}$; no local maxima **(b)** Absolute minimum: $-\frac{1}{8}$; no absolute maximum
35. (a) Local minimum: 0; no local maxima **(b)** Absolute minimum: 0; no absolute maximum
37. (a) No local extrema **(b)** No absolute extrema
39. (a) Local minimum: −2; local maximum: 2 **(b)** No absolute extrema
41. (a) Local maxima: 19, −8; local minimum: −13 **(b)** Absolute maximum: 19; no absolute minimum
43. (a) Local minimum: −8; local maximum: 4.5 **(b)** Absolute minimum: −8; no absolute maximum
45. (a) Local maximum: 8; no local minima **(b)** Absolute maximum: 8; no absolute minimum
47. Neither **49.** Even
51. Odd **53.** Neither **55.** Even **57.** Even
59. Odd **61.** Neither **63.** Even
65. Even **67.** Even **69.** Neither **71.** Odd
73. Note that $f(0)$ can be any number.

x	−3	−2	−1	0	1	2	3
$f(x)$	21	−12	−25	1	−25	−12	21

75. $f(5) = 6$; $f(3) = -4$
77. Answers may vary.

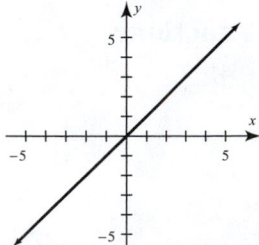

79. No. If (2, 5) is on the graph of an odd function f, then so is (−2, −5). Since f would pass through (−3, −4) and then (−2, −5), it could not always be increasing.

81. Answers may vary.

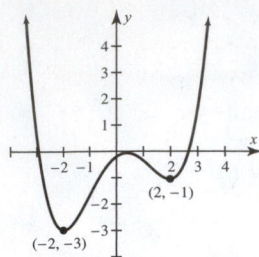

83. Answers may vary; yes, but it does not have to be quadratic.

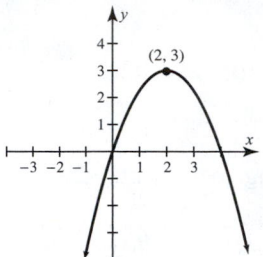

85.

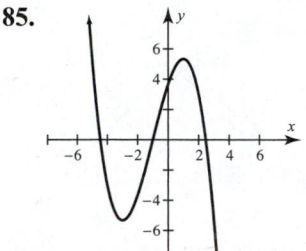

87.

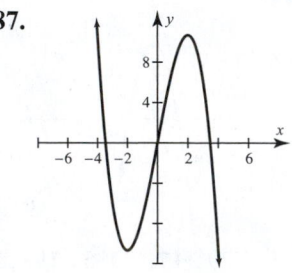

89. On (0, 3); on (3, 6)
91. (a) Absolute maximum: 84°F; absolute minimum: 63°F; the high temperature was 84°F and the low was 63°F. **(b)** Local maxima: approximately 78°F and 84°F; local minima: approximately 63°F and 72°F
(c) $1.6 < x < 2.9$; $3.8 < x < 5$ (approximate)
93. (a) $F(1) \approx 69$; in January 2009, Facebook had about 69 million unique users. $F(12) \approx 132$; in December 2009, Facebook had about 132 million unique users. **(b)** No, F has no real extrema.
$[0, 13, 2]$ by $[0, 150, 25]$

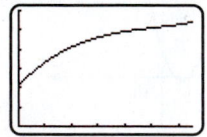

(c) (1, 12) or $\{x \mid 1 < x < 12\}$
95. (a) Possible absolute maximum in January and absolute minimum in July **(b)** Absolute maximum: $140; absolute minimum: $15; the maximum cost, $140, occurs in January and the minimum cost, $15, occurs in July.
$[1, 12, 1]$ by $[0, 150, 10]$ $[1, 12, 1]$ by $[0, 150, 10]$

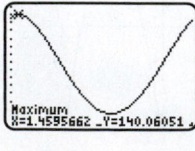
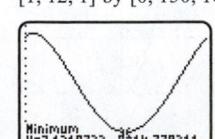

97. (a) Even **(b)** 83°F **(c)** They are equal. **(d)** Monthly average temperatures are symmetric about July. July has the highest average and January the lowest. The pairs June-August, May-September, April-October, March-November, and February-December have approximately the same average temperatures.

4.1 EXTENDED AND DISCOVERY EXERCISES (pp. 242–243)

1. The maximum area occurs when the figure is a rectangle with length about 4.24 and height about 2.12.
3. (a) About 1 hour 54 minutes **(b)** About 2 hours 8 minutes **(c)** About 1 hour 46 minutes

SECTION 4.2 (pp. 253–258)

1. (a) The turning points are approximately $(1.6, 3.6)$, $(3, 1.2)$, $(4.4, 3.6)$. **(b)** After 1.6 minutes, the runner is 360 feet from the starting line. The runner turns and jogs toward the starting line. After 3 minutes, the runner is 120 feet from the starting line. The runner turns and jogs away from the starting line. After 4.4 minutes, the runner is again 360 feet from the starting line. The runner turns and jogs back to the starting line.
3. (a) 0; 0.5 **(b)** $a > 0$ **(c)** 1
5. (a) 3; $-6, -1$, and 6 **(b)** $a < 0$ **(c)** 4
7. (a) 4; $-3, -1, 0, 1$, and 2 **(b)** $a > 0$ **(c)** 5
9. (a) 2; -3 **(b)** $a > 0$ **(c)** 3
11. (a) 1; -1 and 2 **(b)** $a > 0$ **(c)** 2
13. (a) d **(b)** $(1, 0)$ **(c)** 1 **(d)** No local maxima; local minimum: 0 **(e)** No absolute maxima; absolute minimum: 0
15. (a) b **(b)** $(-3, 27), (1, -5)$ **(c)** $x \approx -4.9$, $x = 0, x \approx 1.9$ **(d)** Local maximum: 27; local minimum: -5 **(e)** No absolute maximum; no absolute minimum
17. (a) a **(b)** $(-2, 16), (0, 0), (2, 16)$ **(c)** $x \approx -2.8$, $x = 0, x \approx 2.8$ **(d)** Local maximum: 16; local minimum: 0 **(e)** Absolute maximum: 16; no absolute minimum
19. (a) $[-10, 10, 1]$ by $[-10, 10, 1]$

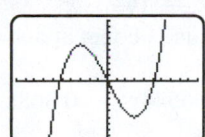

(b) $(-3, 6), (3, -6)$ **(c)** Local minimum: -6; local maximum: 6
21. (a) $[-10, 10, 1]$ by $[-10, 10, 1]$

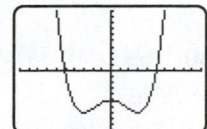

(b) There are three turning points located at $(-3, -7.025), (0, -5)$, and $(3, -7.025)$.
(c) Local minimum: -7.025; local maximum: -5
23. (a) $[-10, 10, 1]$ by $[-10, 10, 1]$

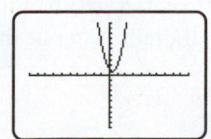

(b) $\left(\frac{1}{3}, \frac{2}{3}\right) \approx (0.333, 0.667)$
(c) Local minimum: $\frac{2}{3} \approx 0.667$; no local maximum
25. (a) $[-10, 10, 1]$ by $[-10, 10, 1]$

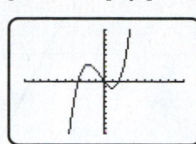

(b) $\left(-2, \frac{10}{3}\right) \approx (-2, 3.333)$, $\left(1, -\frac{7}{6}\right) \approx (1, -1.167)$
(c) Local minimum: $-\frac{7}{6} \approx -1.167$; local maximum: $\frac{10}{3} \approx 3.333$
27. (a) Degree: 1; leading coefficient: -2
(b) Up on left end, down on right end; $f(x) \to \infty$ as $x \to -\infty$, $f(x) \to -\infty$ as $x \to \infty$
29. (a) Degree: 2; leading coefficient: 1 **(b)** Up on both ends; $f(x) \to \infty$ as $x \to -\infty$, $f(x) \to \infty$ as $x \to \infty$
31. (a) Degree: 3; leading coefficient: -2
(b) Up on left end, down on right end; $f(x) \to \infty$ as $x \to -\infty$, $f(x) \to -\infty$ as $x \to \infty$
33. (a) Degree: 3; leading coefficient: -1
(b) Up on left end, down on right end; $f(x) \to \infty$ as $x \to -\infty$, $f(x) \to -\infty$ as $x \to \infty$
35. (a) Degree: 5; leading coefficient: 0.1
(b) Down on left end, up on right end; $f(x) \to -\infty$ as $x \to -\infty$, $f(x) \to \infty$ as $x \to \infty$
37. (a) Degree: 2; leading coefficient: $-\frac{1}{2}$ **(b)** Down on both ends; $f(x) \to -\infty$ as $x \to -\infty$, $f(x) \to -\infty$ as $x \to \infty$
39. (a)

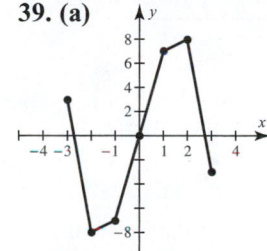

(b) Degree 3 **(c)** $a < 0$ **(d)** $f(x) = -x^3 + 8x$
41. (a)

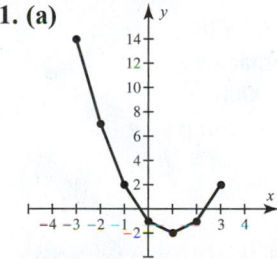

(b) Degree 2 **(c)** $a > 0$ **(d)** $f(x) = x^2 - 2x - 1$
43. (a)

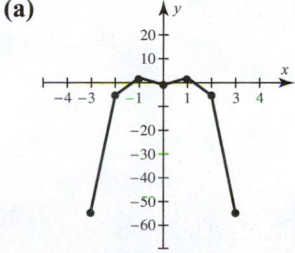

(b) Degree 4 **(c)** $a < 0$ **(d)** $f(x) = -x^4 + 3x^2 - 1$

45. Answers may vary.

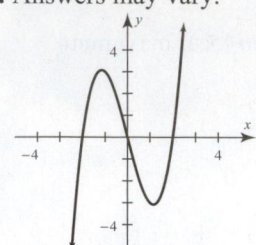

47. Answers may vary.

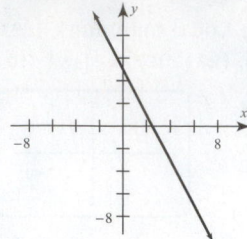

79. (a)

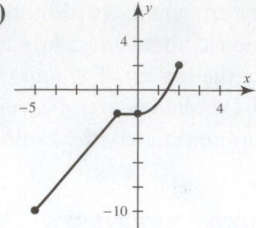

(b) f is continuous.   **(c)** $\sqrt{2}$

81. (a)

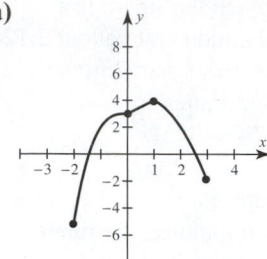

(b) f is continuous.   **(c)** $-\sqrt[3]{3}, \dfrac{\sqrt{17}+1}{2}$

49. Not possible

51. Not possible

53. Answers may vary.

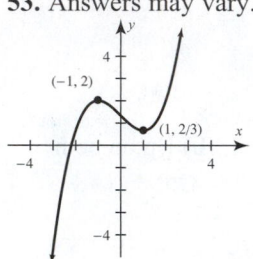

55.

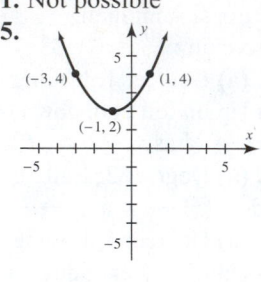

83. (a) $F(3) = 6$; 300 boats are able to harvest 6 thousand tons of fish.   **(b)** Absolute max: 12; 12 thousand tons of fish is the maximum that can be caught.

85. (a) $H(-2) = 0$; $H(0) = 1$; $H(3.5) = 1$

(b)

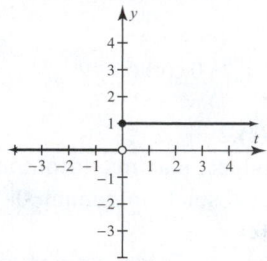

57. As the viewing rectangle increases in size, the graphs begin to look alike. Each formula contains the term $2x^4$, which determines the end behavior of the graph for large values of $|x|$.

(a) $[-4, 4, 1]$ by $[-4, 4, 1]$

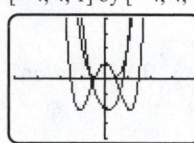

(b) $[-10, 10, 1]$ by $[-100, 100, 10]$

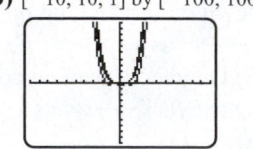

(c) $[-100, 100, 10]$ by $[-10^6, 10^6, 10^5]$

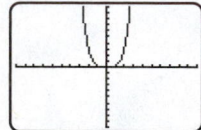

59. For f: 1; for g: 0.5; for h: 0.25. On the interval $[0, 0.5]$, the higher the degree of the function, the smaller the average rate of change.

61. (a) 12.01   **(b)** 12.0001   **(c)** 12.000001
The average rate of change is approaching 12.

63. (a) $4.01\overline{6}$ **(b)** $4.0001\overline{6}$ **(c)** $4.000001\overline{6}$
The average rate of change is approaching 4.

65. $9x^2 + 9xh + 3h^2$ **67.** $-3x^2 - 3xh - h^2 + 1$

69. $f(-2) \approx 5$, $f(1) \approx 0$

71. $f(-1) \approx -1$, $f(1) \approx 1$, $f(2) \approx -2$

73. $f(-3) = -63$, $f(1) = 3$, $f(4) = 10$

75. $f(-2) = 6$, $f(1) = 7$, $f(2) = 9$

77. (a)

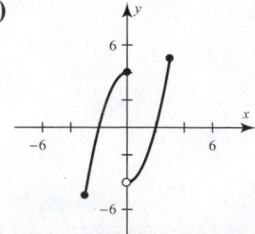

(b) f is not continuous.   **(c)** ± 2

87. (a) Approximately $(1, 13)$ and $(7, 72)$ **(b)** The low monthly average temperature of 13°F occurs in January. The high monthly average temperature of 72°F occurs in July.

89. (a) 4 seconds   **(b)** From 0 to 4 seconds; from 4 to 7 seconds   **(c)** $m = 36$, $a = -16$, $b = 144$
(d) $x \approx 2.8$ or $x \approx 5.7$; the height is 100 feet at about 2.8 seconds and at about 5.7 seconds.

91. (a) $f(x) \approx -0.0006296x^3 + 0.06544x^2 - 0.368x + 2.8$

(b) $[-5, 60, 10]$ by $[0, 80, 20]$

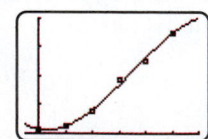

(c) $f(34) \approx 41.2$; $f(60) \approx 80.3$ **(d)** 1994 ($x = 34$), interpolation; 2020 ($x = 60$), extrapolation

4.2 EXTENDED AND DISCOVERY EXERCISES (pp. 258–259)

1. (a) $D = [0, 10]$, or $\{x \mid 0 \le x \le 10\}$
(b) $A(1) = 405$; after 1 min, the tank contains 405 gal of water.   **(c)** Degree: 2; leading coefficient: 5   **(d)** No, more than half; yes, because the water will drain faster at first.

3. Conc. down: $(-\infty, \infty)$
5. Conc. up: $(1, \infty)$; conc. down: $(-\infty, 1)$
7. Conc. up: $(-2, 2)$; conc. down: $(-\infty, -2)$, $(2, \infty)$

CHECKING BASIC CONCEPTS FOR SECTIONS 4.1 AND 4.2 (p. 259)

1. (a) Incr: $(-2, 1)$, $(3, \infty)$, or $\{x \mid -2 < x < 1\}$, $\{x \mid x > 3\}$; decr: $(-\infty, -2)$, $(1, 3)$, or $\{x \mid x < -2\}$, $\{x \mid 1 < x < 3\}$
(b) Local maximum: approximately 3; local minima: approximately -13 and -2
(c) No absolute maximum; absolute minimum: approximately -13 **(d)** Approximately -3.1, 0, 2.2, and 3.6; they are the same values.
3. (a) Not possible
(b) Answers may vary. **(c)** Answers may vary.

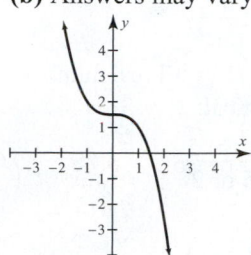

(d) Not possible
5. $f(x) \approx -1.01725x^4 + 10.319x^2 - 10$

SECTION 4.3 (pp. 266–267)

1. $\frac{x^3}{2} - \frac{3}{2x}$ **3.** $x - \frac{2}{3x} - \frac{1}{3x^3}$ **5.** $-\frac{1}{x^3} + \frac{1}{4}$ **7.** $5x - 10 + \frac{5}{3x}$
9. Quotient: $x^2 + x - 2$; remainder: 0
11. Quotient: $2x^3 - 9x^2 + 4x - 23$; remainder: 40
13. Quotient: $3x^2 + 3x - 4$; remainder: 6
15. $x^3 - 1$ **17.** $4x^2 + 3x - 2 + \frac{4}{x - 1}$ **19.** $x^2 - x + 1$
21. $3x^2 + 4x - 2 + \frac{2}{2x - 1}$ **23.** $x^3 + 2 + \frac{-2}{3x - 7}$
25. $5x^2 - 12 + \frac{30}{x^2 + 2}$ **27.** $4x + 5$
29. $x^2 - 2x + 4 + \frac{-1}{2x^2 + 3x + 2}$ **31.** $x^3 - 8x^2 + 15x - 6$
33. $(x - 2)(x - 1) - 1$ **35.** $(2x + 1)(x^2 - 1) + 1$
37. $(x^2 + 1)(x - 1) + 2$ **39.** $x^2 - 3x - 2$
41. $3x^2 + 4x + \frac{3}{x - 5}$ **43.** $x^3 - x^2 - 6x$
45. $2x^4 - 2x^3 + 4 + \frac{1}{x + 0.5}$ **47.** 3 **49.** -42
51. $L = 4x + 3$; 43 ft

SECTION 4.4 (pp. 278–282)

1. $(x + 2)$, $(x + 1)$, $(x - 1)$
3. $(x + 2)$, $(x + 1)$, $(x - 1)$, $(x - 2)$
5. $f(x) = 2\left(x - \frac{11}{2}\right)(x - 7)$
7. $f(x) = (x + 2)(x - 1)(x - 3)$
9. $f(x) = -2(x + 5)\left(x - \frac{1}{2}\right)(x - 6)$
11. $f(x) = 7(x + 3)(x - 2)$
13. $f(x) = -2x(x + 1)(x - 1)$
15. $f(x) = (x + 4)(x - 2)(x - 8)$
17. $f(x) = -1(x + 8)(x + 4)(x + 2)(x - 4)$
19. $f(x) = \frac{1}{2}(x + 1)(x - 2)(x - 3)$
21. $f(x) = \frac{1}{2}(x + 1)(x - 1)(x - 2)$
23. $f(x) = -2(x + 2)(x + 1)(x - 1)(x - 2)$

25. $f(x) = 10(x + 2)\left(x - \frac{3}{10}\right)$
27. $f(x) = -3x(x - 2)(x + 3)$
29. $f(x) = x(x + 1)(x + 3)\left(x - \frac{3}{2}\right)$
31. $f(x) = (x - 1)(x - 3)(x - 5)$
33. $f(x) = -4(x + 4)\left(x - \frac{3}{4}\right)(x - 3)$
35. $f(x) = 2x(x + 2)\left(x + \frac{1}{2}\right)(x - 3)$
37. Yes **39.** No
41. -2 (odd), 4 (even); minimum degree: 5
43. $f(x) = (x + 1)^2(x - 6)$
45. $f(x) = (x - 2)^3(x - 6)$
47. $f(x) = (x + 2)^2(x - 4)$
49. $f(x) = -1(x + 3)^2(x - 3)^2$
51. $f(x) = 2(x + 1)^2(x - 1)^3$
53. (a) x-int: -2, -1; y-int: 4
(b) -2 has multiplicity 1; -1 has multiplicity 2
(c)

55. (a) x-int: -2, 0, 2; y-int: 0
(b) 0 has multiplicity 2; -2 and 2 each have multiplicity 1
(c)

57. (a) -3, $\frac{1}{2}$, 1 **(b)** $f(x) = 2(x + 3)\left(x - \frac{1}{2}\right)(x - 1)$
59. (a) -2, -1, 1, $\frac{3}{2}$
(b) $f(x) = 2(x + 2)(x + 1)(x - 1)\left(x - \frac{3}{2}\right)$
61. (a) $\frac{1}{3}$, 1, 4 **(b)** $f(x) = 3\left(x - \frac{1}{3}\right)(x - 1)(x - 4)$
63. (a) 1 **(b)** $f(x) = \left(x + \sqrt{7}\right)(x - 1)\left(x - \sqrt{7}\right)$
65. Possible: 0 or 2 positive, 1 negative; actual: 0 positive, 1 negative
67. Possible: 1 positive, 1 negative; actual: 1 positive, 1 negative
69. Possible: 0 or 2 positive, 1 or 3 negative; actual: 0 positive, 1 negative
71. -3, 0, 2 **73.** -1, 1 **75.** -2, 0, 2 **77.** -5, 0, 5
79. ± 2 **81.** -3, 0, 6 **83.** 0, 1 **85.** $-\frac{1}{4}$, 0, $\frac{5}{3}$
87. $\pm \frac{2}{3}$, ± 1 **89.** -1, $\pm \frac{\sqrt{3}}{2}$ **91.** ± 2, $\frac{1}{2}$ **93.** $\pm \frac{3}{2}$, $\pm \frac{\sqrt{6}}{2}$
95. -2, 3 **97.** -2.01, 0.12, 2.99 **99.** -4.05, -0.52, 1.71
101. -2.69, -1.10, 0.55, 3.98
103. Because $f(2) = -1 < 0$ and $f(3) = 4 > 0$, the intermediate value property states that there exists an x-value between 2 and 3 where $f(x) = 0$.
105. Because $f(0) = -1 < 0$ and $f(1) = 1 > 0$, the intermediate value property states that there exists an x-value between 0 and 1 where $f(x) = 0$.

107. 4, 32; yes, by the intermediate value property
109. 12 A.M., 2 A.M., 4 A.M.
111. Approximately 11.34 cm
113. June 2, June 22, and July 12
115. (a) $f(x) \approx -0.184(x + 6.01)(x - 2.15)(x - 11.7)$
(b) The zero of -6.01 has no significance. The zeros of $2.15 \approx 2$ and $11.7 \approx 12$ indicate that during February and December the average temperature is 0°F.
117. (a) As x increases, C decreases.
(b) $C(x) \approx -0.000068x^3 + 0.0099x^2 - 0.653x + 23$
(c) [0, 70, 10] by [0, 22, 5]

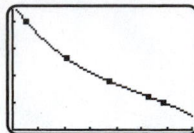

(d) $0 \le x < 32.1$ (approximately)

4.4 EXTENDED AND DISCOVERY EXERCISES (p. 282)

1. Dividing $P(x)$ by $x - 2$ synthetically results in the following bottom row: 1 1 5 2 12.
Since $2 > 0$ and the bottom row values are all nonnegative, there is no real zero greater than 2.
3. Dividing $P(x)$ by $x + 2$ synthetically results in the following bottom row: 1 -1 1 -2 7.
Since $-2 < 0$ and the bottom row values alternate in sign, there is no real zero less than -2.
5. Dividing $P(x)$ by $x - 1$ synthetically results in the following bottom row: 3 5 1 2 1.
Since $1 > 0$ and the bottom row values are all nonnegative, there is no real zero greater than 1.

CHECKING BASIC CONCEPTS FOR SECTIONS 4.3 AND 4.4 (p. 283)

1. $x^2 - 2x + 1$ **3.** $f(x) = -\frac{1}{2}(x + 2)^2(x - 1)$;
-2 has multiplicity 2; 1 has multiplicity 1.
5. The zeros are -4, -1, 2, and 4;
$f(x) = (x + 4)(x + 1)(x - 2)(x - 4)$.

SECTION 4.5 (pp. 288–289)

1. Two imaginary zeros
3. One real zero; two imaginary zeros
5. Two real zeros; two imaginary zeros
7. Three real zeros; two imaginary zeros
9. (a) $f(x) = (x - 6i)(x + 6i)$ **(b)** $f(x) = x^2 + 36$
11. (a) $f(x) = -1(x + 1)(x - 2i)(x + 2i)$
(b) $f(x) = -x^3 - x^2 - 4x - 4$
13. (a) $f(x) = 10(x - 1)(x + 1)(x - 3i)(x + 3i)$
(b) $f(x) = 10x^4 + 80x^2 - 90$
15. (a) $f(x) = \frac{1}{2}(x + i)(x - i)(x + 2i)(x - 2i)$
(b) $f(x) = \frac{1}{2}x^4 + \frac{5}{2}x^2 + 2$
17. (a) $f(x) = -2(x - (1 - i))(x - (1 + i))(x - 3)$
(b) $f(x) = -2x^3 + 10x^2 - 16x + 12$
19. $\frac{5}{3}, \pm 5i$ **21.** $\pm 3i, \frac{1}{4} \pm \frac{i\sqrt{7}}{4}$
23. (a) $\pm 5i$ **(b)** $f(x) = (x - 5i)(x + 5i)$
25. (a) $0, \pm i$ **(b)** $f(x) = 3(x - 0)(x - i)(x + i)$, or $f(x) = 3x(x - i)(x + i)$

27. (a) $\pm i, \pm 2i$
(b) $f(x) = (x - i)(x + i)(x - 2i)(x + 2i)$
29. (a) $-2, \pm 4i$ **(b)** $f(x) = (x + 2)(x + 4i)(x - 4i)$
31. $0, \pm i$ **33.** $2, \pm i\sqrt{7}$ **35.** $0, \pm i\sqrt{5}$
37. $0, \frac{1}{2} \pm \frac{i\sqrt{15}}{2}$ **39.** $-2, 1, \pm i\sqrt{8}$ **41.** $-2, \frac{1}{3} \pm \frac{i\sqrt{8}}{3}$

SECTION 4.6 (pp. 302–305)

1. Yes; $D = \left\{ x \mid x \ne \frac{5}{4} \right\}$
3. Yes; $D =$ all real numbers
5. No; $D = \{ x \mid x \ne -1 \}$
7. Yes; $D =$ all real numbers
9. No; $D = \{ x \mid x \ne -1, x \ne 0 \}$
11. Yes; $D = \{ x \mid x \ne -1 \}$
13. Horizontal: $y = 4$; vertical: $x = 2$; $D = \{ x \mid x \ne 2 \}$
15. Horizontal: $y = -4$; vertical: $x = \pm 2$;
$D = \{ x \mid x \ne 2, x \ne -2 \}$
17. Horizontal: $y = 0$; vertical: none; $D =$ all real numbers
19. $y = 3$ **21.** Horizontal: $y = 2$; vertical: $x = 3$
23. Horizontal: $y = 0$; vertical: $x = \pm\sqrt{5}$
25. Horizontal: none; vertical: $x = -5$ or 2
27. Horizontal: $y = \frac{1}{2}$; vertical: $x = \frac{5}{2}$
29. Horizontal: $y = 3$; vertical: $x = 1$
31. Horizontal: none; vertical: none, since $f(x) = x - 3$ for $x \ne -3$ **33.** b **35.** d
37. $f(x) = \frac{x + 1}{x + 3}$ (answers may vary)
39. $f(x) = \frac{1}{x^2 - 9}$ (answers may vary)
41. Horizontal: $y = 0$;   **43.** Horizontal: $y = 0$
vertical: $x = 0$ vertical: $x = 0$

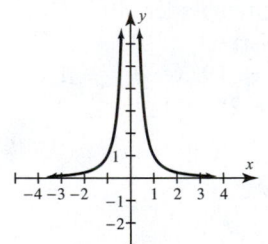
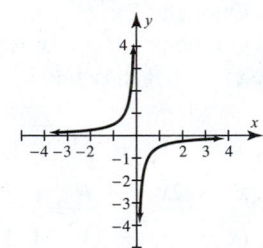

45. $g(x) = f(x - 3)$ **47.** $g(x) = f(x) + 2$

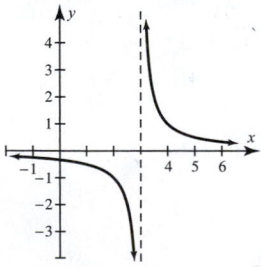
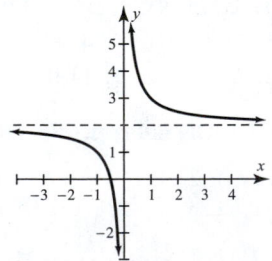

49. $g(x) = f(x + 1) - 2$ **51.** $g(x) = -2h(x - 1)$

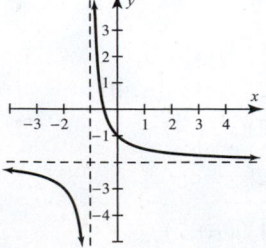
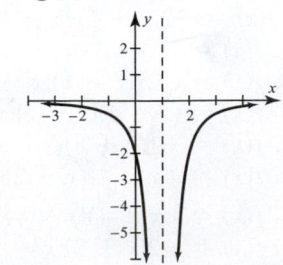

53. $g(x) = h(x + 1) - 2$

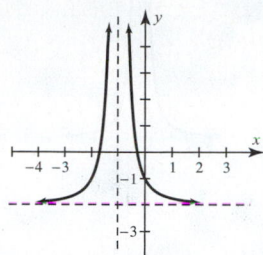

55. (a) $D = \{x \mid x \neq 2\}$
(b) $[-9.4, 9.4, 1]$ by
$[-6.2, 6.2, 1]$

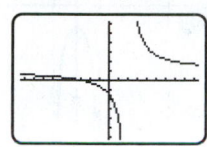

(c) Horizontal: $y = 1$;
vertical: $x = 2$
(d)

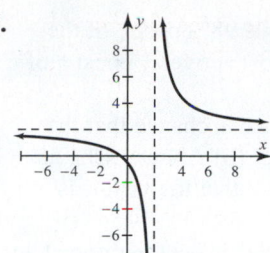

67.

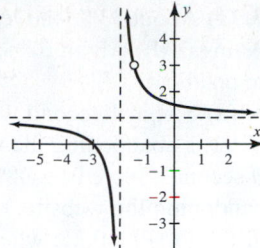

69.

71.

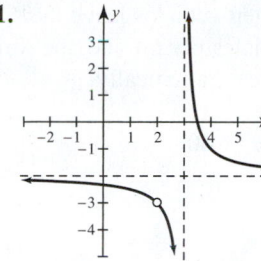

73. (a) Slant: $y = x - 1$;
vertical: $x = -1$
(b)

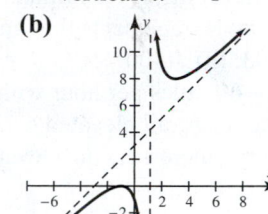

57. (a) $D = \{x \mid x \neq 2,$
$x \neq -2\}$
(b) $[-9.4, 9.4, 1]$ by
$[-6.2, 6.2, 1]$

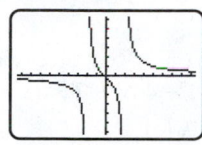

(c) Horizontal: $y = 0$;
vertical: $x = \pm 2$
(d)

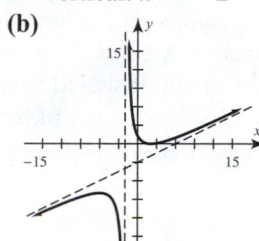

75. (a) Slant: $y = \frac{1}{2}x - 3$;
vertical: $x = -2$
(b)

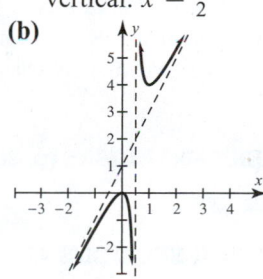

77. (a) Slant: $y = x + 3$;
vertical: $x = 1$
(b)

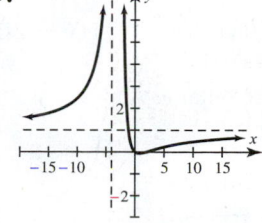

59. (a) $D = \{x \mid x \neq 2,$
$x \neq -2\}$
(b) $[-9.4, 9.4, 1]$ by
$[-9.3, 9.3, 1]$

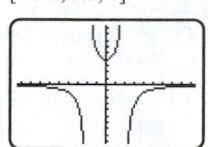

(c) Horizontal: $y = 0$;
vertical: $x = \pm 2$
(d)

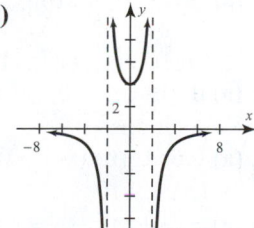

79. (a) Slant: $y = 2x + 1$;
vertical: $x = \frac{1}{2}$
(b)

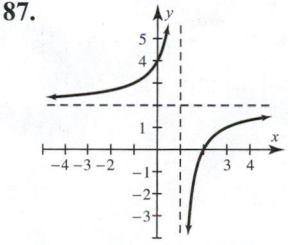

81. -3 **83.** 1 **85.** $-\frac{1}{2}$

61. (a) $D = \{x \mid x \neq 2\}$
(b) $[-4.7, 4.7, 1]$ by
$[-6.2, 6.2, 1]$

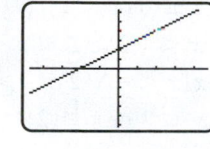

(c) Horizontal: none;
vertical: none, since
$f(x) = x + 2$ for $x \neq 2$
(d)

87.

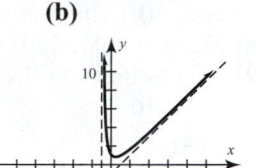

89.

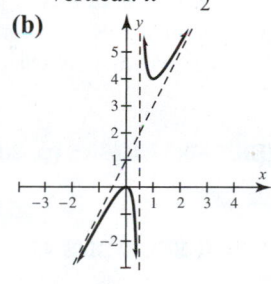

63.

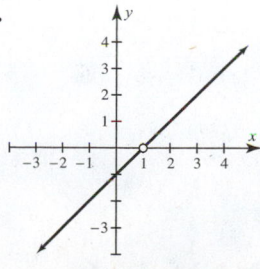

65.

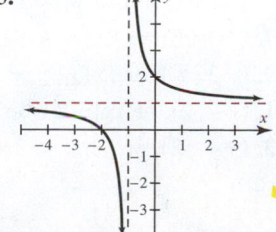

91.

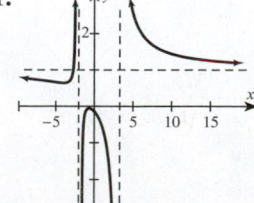

93. (a) About 33%; the least active 98% post $\frac{1}{3}$ of the postings. **(b)** About 67%; the most active 2% post $\frac{2}{3}$ of the postings.
95. (a) 5%; at 1 second, there is a 5% chance that the visitor is abandoning the website. **(b)** About 1.8%; at 60 seconds, there is a 1.8% chance that the visitor is abandoning the website.
97. (a) $T(4) = 0.25$; when vehicles leave the ramp at an average rate of 4 vehicles per minute, the wait is 0.25 minute or 15 seconds. $T(7.5) = 2$; when vehicles leave the ramp at an average rate of 7.5 vehicles per minute, the wait is 2 minutes. **(b)** The wait increases dramatically.
99. (a) $N(20) = 0.5$, $N(39) \approx 38$
(b) It increases dramatically. **(c)** $x = 40$
101. (a) $y = 10$

[0, 14, 1] by [0, 14, 1]

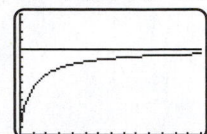

(b) When $x = 0$, there are 1 million insects. **(c)** It starts to level off at 10 million. **(d)** The horizontal asymptote $y = 10$ represents the limiting population after a long time.
103. (a) $f(400) = \frac{2540}{400} = 6.35$ inches. A curve designed for 60 miles per hour with a radius of 400 ft should have the outer rail elevated 6.35 in. **(b)** As the radius x of the curve increases, the elevation of the outer rail decreases.

[0, 600, 100] by [0, 50, 5]

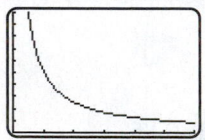

(c) The horizontal asymptote is $y = 0$. As the radius of the curve increases without bound ($x \to \infty$), the tracks become straight and no elevation or banking ($y \to 0$) is necessary.
(d) 200 ft

4.6 EXTENDED AND DISCOVERY EXERCISES (p. 305)

1. $-\frac{1}{3}$; $-\frac{1}{x(x+h)}$ **3.** $-\frac{1}{2}$; $-\frac{3}{2x(x+h)}$

CHECKING BASIC CONCEPTS FOR SECTIONS 4.5 AND 4.6 (pp. 305–306)

1. $f(x) = 3(x - 4i)(x + 4i) = 3x^2 + 48$
3. $f(x) = (x - 1)(x - 2i)(x + 2i)$
5. (a) $D = \{x \mid x \neq 1\}$ **(b)** Vertical asymptote: $x = 1$; horizontal asymptote: $y = 2$
(c)

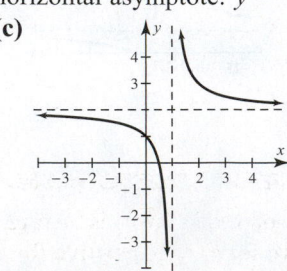

7. (a)

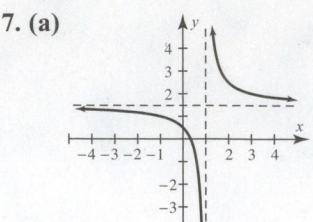

(b)

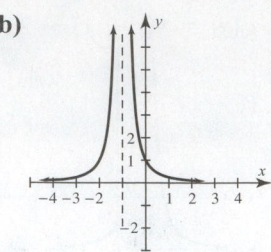

(c)

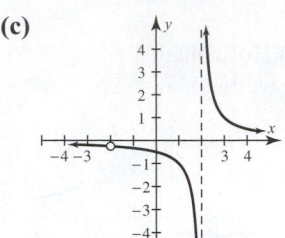

(d)

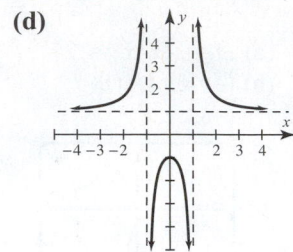

SECTION 4.7 (pp. 317–322)

1. -3 **3.** $\frac{1}{2}, 2$ **5.** $\frac{1}{2}$ **7.** -1 **9.** $\frac{13}{8}$ **11.** $\pm\sqrt{2}$ **13.** No real solutions **15.** 0, ± 2 **17.** $-\frac{5}{3}, \frac{7}{5}$ **19.** -14
21. No real solutions (extraneous: 2)
23. -3 (extraneous: 1) **25.** 1 (extraneous: -2)
27. No real solutions (extraneous: 1)
29. (a) -4, -2, or 2
(b) $(-4, -2) \cup (2, \infty)$, or $\{x \mid -4 < x < -2 \text{ or } x > 2\}$
(c) $(-\infty, -4) \cup (-2, 2)$, or $\{x \mid x < -4 \text{ or } -2 < x < 2\}$
31. (a) -4, -2, 0, or 2
(b) $(-4, -2) \cup (0, 2)$, or $\{x \mid -4 < x < -2 \text{ or } 0 < x < 2\}$
(c) $(-\infty, -4) \cup (-2, 0) \cup (2, \infty)$, or $\{x \mid x < -4 \text{ or } -2 < x < 0 \text{ or } x > 2\}$
33. (a) -2, 1, or 2 **(b)** $(-\infty, -2) \cup (-2, 1)$, or $\{x \mid x < -2 \text{ or } -2 < x < 1\}$
(c) $(1, 2) \cup (2, \infty)$, or $\{x \mid 1 < x < 2 \text{ or } x > 2\}$
35. (a) 0 **(b)** $(-\infty, 0) \cup (0, \infty)$, or $\{x \mid x < 0 \text{ or } x > 0\}$
(c) No solutions
37. (a) 0 or 1 **(b)** $(-\infty, 0) \cup (1, \infty)$, or $\{x \mid x < 0 \text{ or } x > 1\}$
(c) $(0, 1)$, or $\{x \mid 0 < x < 1\}$
39. (a) -2, 0, or 2
(b) $(-\infty, -2) \cup (2, \infty)$, or $\{x \mid x < -2 \text{ or } x > 2\}$
(c) $(-2, 0) \cup (0, 2)$, or $\{x \mid -2 < x < 0 \text{ or } 0 < x < 2\}$
41. $(-1, 0) \cup (1, \infty)$, or $\{x \mid -1 < x < 0 \text{ or } x > 1\}$
43. $[-2, 0] \cup [1, \infty)$, or $\{x \mid -2 \leq x \leq 0 \text{ or } x \geq 1\}$
45. $(-3, -2) \cup (2, 3)$, or $\{x \mid -3 < x < -2 \text{ or } 2 < x < 3\}$
47. $(-\infty, -\sqrt{2}) \cup (\sqrt{2}, \infty)$, or $\{x \mid x < -\sqrt{2} \text{ or } x > \sqrt{2}\}$
49. $[-2, 1] \cup [2, \infty)$, or $\{x \mid -2 \leq x \leq 1 \text{ or } x \geq 2\}$
51. $[-3, 2]$, or $\{x \mid -3 \leq x \leq 2\}$
53. $(-\infty, 1] \cup [2, 4]$, or $\{x \mid x \leq 1 \text{ or } 2 \leq x \leq 4\}$
55. $(-\infty, -1) \cup \left(0, \frac{4}{3}\right) \cup (2, \infty)$, or $\left\{x \mid x < -1 \text{ or } 0 < x < \frac{4}{3} \text{ or } x > 2\right\}$
57. $(-\infty, 0)$, or $\{x \mid x < 0\}$
59. $(-3, \infty)$, or $\{x \mid x > -3\}$
61. $(-2, 2)$, or $\{x \mid -2 < x < 2\}$
63. $(-\infty, 2)$, or $\{x \mid x < 2\}$

65. $(-\infty, -1) \cup \left(\frac{3}{2}, \infty\right)$, or $\left\{x \mid x < -1 \text{ or } x > \frac{3}{2}\right\}$

67. $(-\infty, -3) \cup (-1, 2)$, or $\{x \mid x < -3 \text{ or } -1 < x < 2\}$

69. $(-1, 1) \cup \left[\frac{5}{2}, \infty\right)$, or $\left\{x \mid -1 < x < 1 \text{ or } x \geq \frac{5}{2}\right\}$

71. $(3, \infty)$, or $\{x \mid x > 3\}$

73. $(-\infty, 0) \cup \left(0, \frac{1}{2}\right] \cup [2, \infty)$, or
$\left\{x \mid x < 0 \text{ or } 0 < x \leq \frac{1}{2} \text{ or } x \geq 2\right\}$

75. $(-2, 0) \cup [2, \infty)$, or $\{x \mid -2 < x < 0 \text{ or } x \geq 2\}$

77. (a) About 12.4 cars per minute (b) 3

79. Two possible solutions: width $= 7$ in., length $= 14$ in., height $= 2$ in.; width ≈ 2.266 in., length ≈ 4.532 in., height ≈ 19.086 in.

81. (a) $A(x) = \frac{x^2}{144} + \frac{3}{x}$ (b) $C(x) = 0.1\left(\frac{x^2}{144} + \frac{3}{x}\right)$
(c) 6 in. $\times$ 6 in. $\times$ 3 in.

83. (a) $D(0.05) \approx 238$; the braking distance for a car traveling at 50 miles per hour on a 5% uphill grade is about 238 ft.
(b) As the uphill grade x increases, the braking distance decreases, which agrees with driving experience.
(c) $x = \frac{13}{165} \approx 0.079$, or 7.9%

85. (a) $x \leq 36$ (approximately) (b) The average line length is less than or equal to 8 cars when the average arrival rate is about 36 cars per hour or less.

87. (a) The braking distance increases. (b) $0 < x \leq 0.3$

89. $\sqrt[3]{212.8} \leq x \leq \sqrt[3]{213.2}$, or (approximately)
$5.97022 \leq x \leq 5.97396$ inches

91. $k = 6$ 93. $k = 8$ 95. $T = 160$ 97. $y = 2$

99. Becomes half as much 101. Becomes 27 times as much

103. $k = 0.5, n = 2$ 105. $k = 3, n = 1$

107. 1.18 grams 109. $\sqrt{50} \approx 7$ times as far

111. $\frac{2}{9}$ ohm 113. F decreases by a factor of $\frac{\sqrt{2}}{2}$

SECTION 4.8 (pp. 333–336)

1. 4 3. $\frac{1}{8}$ 5. -9 7. 27 9. 2 11. 9 13. $-\frac{1}{243}$

15. 16 17. $\frac{9}{4}$ 19. $(2x)^{1/2}$ 21. $z^{5/3}$ 23. $\frac{1}{y^{3/4}}$

25. $x^{5/6}$ 27. $y^{3/4}$ 29. $\frac{\sqrt{b}}{\sqrt[4]{a^3}}$ 31. $\sqrt{\sqrt{a} + \sqrt{b}}$

33. 4 35. 64 37. 5 39. -2

41.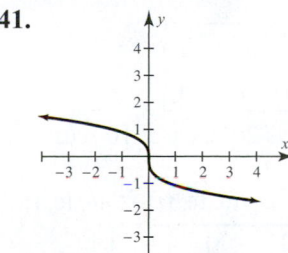

43.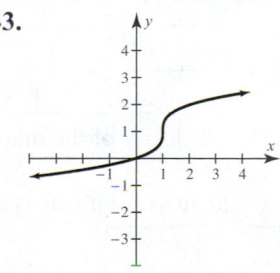

45.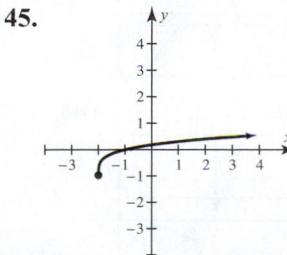

47.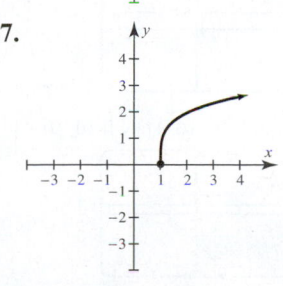

49.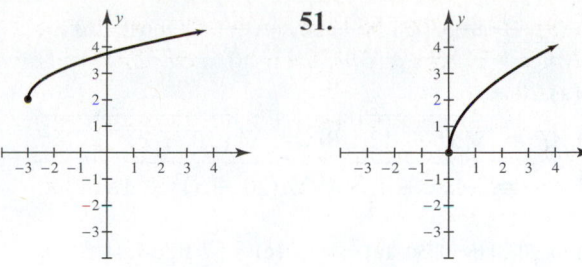

51.

53. 7 55. -1 57. 2, 3 59. 4 61. 15 63. 8

65. -28 67. 2 69. 65,538

71. $50^{3/2} - 50^{1/2} \approx 346.48$ 73. b

75.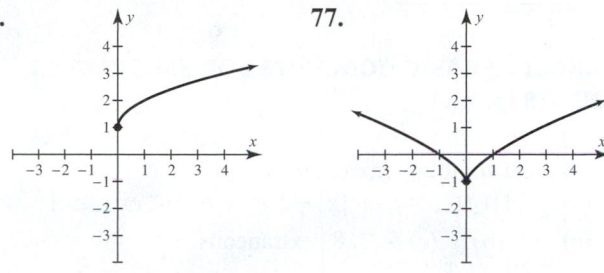

77.

79.

81. 2 83. 81 85. ± 32 87. ± 8 89. $\sqrt[3]{16}$

91. $-1, -\frac{1}{2}$ 93. $-\frac{1}{4}, \frac{5}{7}$ 95. $-8, 27$ 97. $\frac{1}{8}, \frac{64}{27}$

99. 1 101. $-1, \frac{1}{27}$

103. About 3.5; average speed is about 3.5 mi/hr.

105. $w \approx 3.63$ lb 107. About 58.1 yr

109. (a) $a = 1960$ (b) $b \approx -1.2$
(c) $f(4) = 1960(4)^{-1.2} \approx 371$. If the zinc ion concentration reaches 371 milligrams per liter, a rainbow trout will survive, on average, 4 minutes.

111. (a) $f(2) \approx 1.06$ grams
(b) & (c) Approximately 1.1 grams

113. $a \approx 3.20, b \approx 0.20$

[1, 9, 1] by [0, 6, 1]

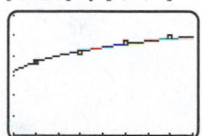

115. (a) $f(x) = 0.005192x^{1.7902}$ (Answers may vary.)
(b) About 2.6 million; too high; extrapolation
(c) About 1999

117. $a \approx 874.54, b \approx -0.49789$

4.8 EXTENDED AND DISCOVERY EXERCISES (p. 336)

1. The graph of an odd root function (not shown) is always increasing; the function is negative for $x < 0$, positive for $x > 0$, and zero at $x = 0$. It is an odd function.

3. $\dfrac{1}{\sqrt{x + h} + \sqrt{x}}$ **5.** $\dfrac{x + 1}{x^{5/3}}$ **7.** $\dfrac{2 - x}{3x^{1/3}(x + 1)^2}$

9. (a) $20 - x$

(b) $AP = \sqrt{x^2 + 12^2}$; $BP = \sqrt{(20 - x)^2 + 16^2}$

(c) $f(x) = \sqrt{x^2 + 12^2} + \sqrt{(20 - x)^2 + 16^2}$, $0 < x < 20$

(d) [0, 20, 5] by [0, 50, 10]   **(e)** 8.57 ft; 34.41 ft

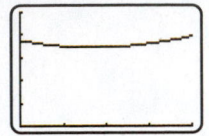

CHECKING BASIC CONCEPTS FOR SECTIONS 4.7 AND 4.8 (p. 336)

1. (a) $\frac{1}{2}$ **(b)** $-5, \frac{7}{3}$
(c) No real solutions (extraneous: −2)
3. $(-2, -1] \cup [1, \infty)$, or $\{x \mid -2 < x \le -1 \text{ or } x \ge 1\}$
5. (a) -8 **(b)** $\frac{1}{4}$ **(c)** 9 **7.** 8 (extraneous: 1)
9. $a = 2, b = \frac{1}{2}$

CHAPTER 4 REVIEW EXERCISES (pp. 341–344)

1. Degree: 3; leading coefficient: -7
3. (a) Local minimum: -2; local maximum: 4.1
(b) No absolute minimum; no absolute maximum
5. Even **7.** Odd **9.** Odd
11. Answers may vary.

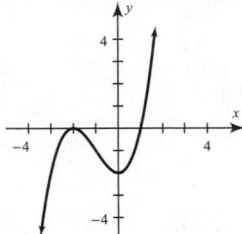

13. (a) 2; $-2, 0, 1$ **(b)** $a < 0$ **(c)** 3
15. Up on left end, down on right end;
$f(x) \to \infty$ as $x \to -\infty$; $f(x) \to -\infty$ as $x \to \infty$
17. 7
19. (a)

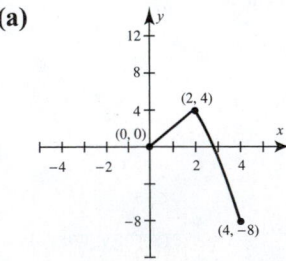

f is continuous.

(b) $f(1) = 2$; $f(3) = -1$ **(c)** $1, \sqrt{6}$
21. $2x^2 - 3x - 1$ **23.** $2x^2 - 3x + 1 + \dfrac{1}{2x + 3}$
25. $f(x) = \frac{1}{2}(x - 1)(x - 2)(x - 3)$
27. $f(x) = \frac{1}{2}(x + 2)(x - 1)(x - 3)$
29. $-3, \frac{1}{2}, 2$ **31.** $0, \pm\sqrt{3}$ **33.** $-1.88, 0.35, 1.53$
35. $0, \pm i$ **37.** $f(x) = 4(x - 1)(x - 3i)(x + 3i)$;
$f(x) = 4x^3 - 4x^2 + 36x - 36$;

39. $f(x) = (x + i)(x - i)\left(x - \left(-\frac{1}{2} + \frac{i\sqrt{3}}{2}\right)\right) \times$
$\left(x - \left(-\frac{1}{2} - \frac{i\sqrt{3}}{2}\right)\right)$

41. Horizontal: $y = \frac{2}{3}$; vertical: $x = \frac{1}{3}$

43. 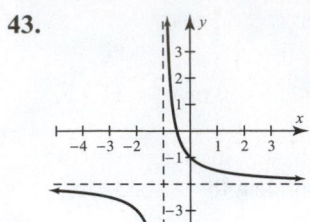   **45.**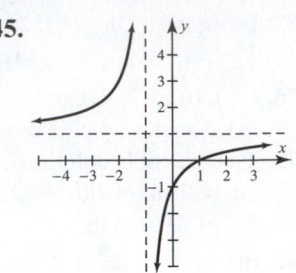

47. Answers may vary.

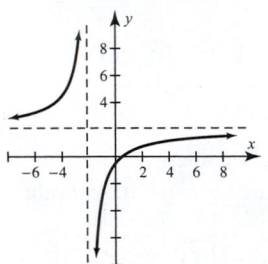

49. 4 **51.** No real solutions (extraneous: 2)
53. (a) $(-\infty, -4) \cup (-2, 3)$, or
$\{x \mid x < -4 \text{ or } -2 < x < 3\}$
(b) $(-4, -2) \cup (3, \infty)$, or $\{x \mid -4 < x < -2 \text{ or } x > 3\}$
55. $(-3, 0) \cup (2, \infty)$, or $\{x \mid -3 < x < 0 \text{ or } x > 2\}$
57. $(-\infty, -2) \cup \left(\frac{1}{2}, \infty\right)$, or $\left\{x \mid x < -2 \text{ or } x > \frac{1}{2}\right\}$
59. 216 **61.** 8 **63.** $x^{4/3}$ **65.** $y^{1/2}$
67. $D = \{x \mid x \ge 0\}$; $f(3) \approx 15.59$
69. 4 **71.** 6 **73.** $\frac{81}{16}$ **75.** 15 **77.** $-2, \frac{1}{3}$ **79.** $-1, 125$
81. (a) Dog: 148; person: 69 **(b)** 6.4 in.
83. (a) 81.5°F **(b)** 87.3; the ocean reaches a maximum temperature of about 87.3°F in late July. **85.** 4 sec

CHAPTER 4 EXTENDED AND DISCOVERY EXERCISES (pp. 344–345)

1. (a)

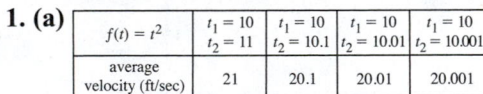

$f(t) = t^2$	$t_1 = 10$ $t_2 = 11$	$t_1 = 10$ $t_2 = 10.1$	$t_1 = 10$ $t_2 = 10.01$	$t_1 = 10$ $t_2 = 10.001$
average velocity (ft/sec)	21	20.1	20.01	20.001

(b) The velocity of the bike rider is 20 ft/sec at 10 sec.
3. (a)

f_1 in [−10, 10, 1] by [−10, 10, 1] f_2 in [−10, 10, 1] by [−10, 10, 1]

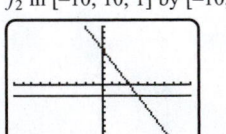

f_3 in [−10, 10, 1] by [−10, 10, 1] f_4 in [−10, 10, 1] by [−10, 10, 1]

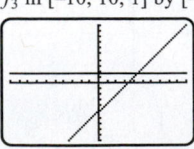   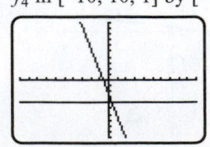

(b) Neither the graph of each linear function nor the graph of its average rate of change has any turning points.

(c) For any linear function, the graph of its (average) rate of change is a constant function whose value is equal to the slope of the graph of the linear function.

5. (a)

f_1 in $[-10, 10, 1]$ by $[-10, 10, 1]$ f_2 in $[-10, 10, 1]$ by $[-10, 10, 1]$

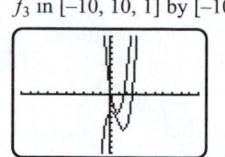   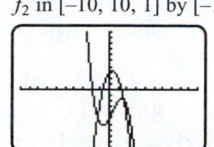

f_3 in $[-10, 10, 1]$ by $[-10, 10, 1]$ f_4 in $[-10, 10, 1]$ by $[-10, 10, 1]$

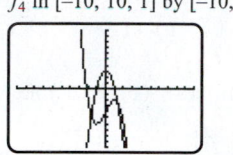

(b) The graph of a cubic function has two turning points or none; the graph of its average rate of change has one turning point. The x-coordinate of a turning point of the function corresponds to an x-intercept on the graph of the average rate of change.

(c) For any cubic function, the graph of its average rate of change is a quadratic function. The leading coefficients of a cubic function and its average rate of change have the same sign.

CHAPTERS 1–4 CUMULATIVE REVIEW EXERCISES (pp. 345–347)

1. (a) $D = \{-3, -1, 0, 1\}$; $R = \{-2, 4, 5\}$ **(b)** No
3. $D = \{x \mid -2 \le x \le 2\}$; $R = \{x \mid 0 \le x \le 3\}$; $f(0) = 3$
5. $C(x) = 0.25x + 200$; $C(2000) = 700$; the cost of driving 2000 miles in one month is $700. **7.** 18
9. $y = -\frac{9}{5}x + \frac{7}{5}$ **11.** $y = -5$ **13.** Each radio costs $15 to manufacture. The fixed cost is $2000. **15.** $\frac{3}{5}$ **17.** $0, \frac{8}{3}$
19. $-2, \frac{5}{7}$ **21.** $\frac{1}{27}$; -8 **23.** $\frac{4}{5}$ **25.** No solutions; contradiction
27. $\left(-\infty, -\frac{1}{2}\right)$, or $\left\{x \mid x < -\frac{1}{2}\right\}$
29. $\left(-\infty, \frac{4}{5}\right] \cup [2, \infty)$, or $\left\{x \mid x \le \frac{4}{5} \text{ or } x \ge 2\right\}$
31. $(-\infty, -3] \cup [0, 3]$, or $\{x \mid x \le -3 \text{ or } 0 \le x \le 3\}$
33. (a) $-3, -1, 1, 2$ **(b)** $(-\infty, -3) \cup (-1, 1) \cup (2, \infty)$, or $\{x \mid x < -3 \text{ or } -1 < x < 1 \text{ or } x > 2\}$
(c) $[-3, -1] \cup [1, 2]$, or $\{x \mid -3 \le x \le -1 \text{ or } 1 \le x \le 2\}$
35. (a)   **(b)**

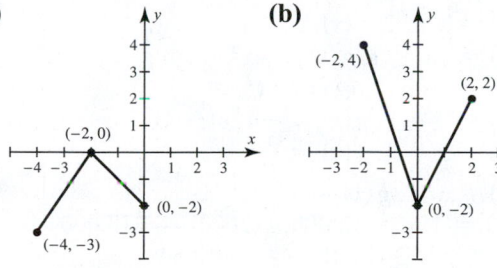

(c)   **(d)**

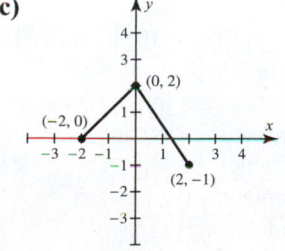   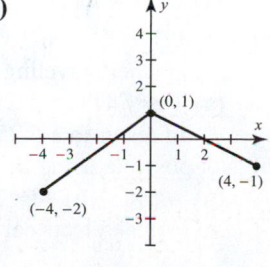

37. (a) Incr: $(-\infty, -2)$, $(1, \infty)$, or $\{x \mid x < -2\}$, $\{x \mid x > 1\}$; decr: $(-2, 1)$, or $\{x \mid -2 < x < 1\}$
(b) Approximately -3.3, 0, and 1.8
(c) $(-2, 3)$ and $(1, -1)$ **(d)** Local maximum: 3; local minimum: -1
39. Answers may vary.

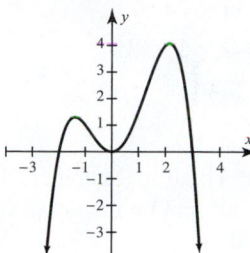

41. (a) $a - 2 + \frac{3}{a^2}$ **(b)** $2x^2 + 2x - 2 + \frac{-1}{x - 1}$
43. $f(x) = 4(x + 3)(x - 1)^2(x - 4)^3$ **45.** $-\frac{1}{2} + \frac{7}{2}i$
47. $D = \{x \mid x \ne -1, x \ne 4\}$; vertical: $x = -1$, $x = 4$; horizontal: $y = 0$
49. $m_1 = 20$ thousand: the pool is being filled at a rate of 20 thousand gallons per hour. $m_2 = 10$ thousand: the pool is being filled at a rate of 10 thousand gallons per hour. $m_3 = 0$: the pool is neither being filled nor being emptied. $m_4 = -15$ thousand: the pool is being emptied at a rate of 15 thousand gallons per hour.
51. 3.75 liters **53.** 400 thousand toy figures
55. (a) $C(t) = t(805 - 5t) = 805t - 5t^2$ **(b)** 25, 136; the cost is $17,000 when either 25 or 136 tickets are purchased. **(c)** $32,400 when 80 or 81 tickets are sold
57. $r \approx 1.7$ in., $h \approx 3.5$ in.

CHAPTER 5: Exponential and Logarithmic Functions

SECTION 5.1 (pp. 359–365)

1. 7 **3.** $4x^3$ **5.** The cost of x square yards of carpet
7. (a) -5 **(b)** -5 **(c)** -3 **(d)** $-\frac{1}{3}$
9. (a) $\frac{11}{2}$ **(b)** 0 **(c)** $\frac{9}{4}$ **(d)** Undefined
11. (a) $(f + g)(x) = 2x + x^2$; all real numbers
(b) $(f - g)(x) = 2x - x^2$; all real numbers
(c) $(fg)(x) = 2x^3$; all real numbers
(d) $(f/g)(x) = \frac{2}{x}$; $D = \{x \mid x \ne 0\}$
13. (a) $(f + g)(x) = 2x^2$; all real numbers
(b) $(f - g)(x) = -2$; all real numbers
(c) $(fg)(x) = x^4 - 1$; all real numbers
(d) $(f/g)(x) = \frac{x^2 - 1}{x^2 + 1}$; all real numbers
15. (a) $(f + g)(x) = 2x$; $D = \{x \mid x \ge 1\}$
(b) $(f - g)(x) = -2\sqrt{x - 1}$; $D = \{x \mid x \ge 1\}$
(c) $(fg)(x) = x^2 - x + 1$; $D = \{x \mid x \ge 1\}$
(d) $(f/g)(x) = \frac{x - \sqrt{x - 1}}{x + \sqrt{x - 1}}$; $D = \{x \mid x \ge 1\}$
17. (a) $(f + g)(x) = 2\sqrt{x}$; $D = \{x \mid x \ge 0\}$
(b) $(f - g)(x) = -2$; $D = \{x \mid x \ge 0\}$
(c) $(fg)(x) = x - 1$; $D = \{x \mid x \ge 0\}$
(d) $(f/g)(x) = \frac{\sqrt{x} - 1}{\sqrt{x} + 1}$; $D = \{x \mid x \ge 0\}$

19. (a) $(f + g)(x) = \frac{4}{x + 1}$; $D = \{x \mid x \neq -1\}$

(b) $(f - g)(x) = -\frac{2}{x + 1}$; $D = \{x \mid x \neq -1\}$

(c) $(fg)(x) = \frac{3}{(x + 1)^2}$; $D = \{x \mid x \neq -1\}$

(d) $(f/g)(x) = \frac{1}{3}$; $D = \{x \mid x \neq -1\}$

21. (a) $(f + g)(x) = \frac{x + 1}{2x - 4}$; $D = \{x \mid x \neq 2\}$

(b) $(f - g)(x) = \frac{1 - x}{2x - 4}$; $D = \{x \mid x \neq 2\}$

(c) $(fg)(x) = \frac{x}{(2x - 4)^2}$; $D = \{x \mid x \neq 2\}$

(d) $(f/g)(x) = \frac{1}{x}$; $D = \{x \mid x \neq 0, x \neq 2\}$

23. (a) $(f + g)(x) = x^2 - 1 + |x + 1|$; all real numbers

(b) $(f - g)(x) = x^2 - 1 - |x + 1|$; all real numbers

(c) $(fg)(x) = (x^2 - 1)|x + 1|$; all real numbers

(d) $(f/g)(x) = \frac{x^2 - 1}{|x + 1|}$; $D = \{x \mid x \neq -1\}$

25. (a) $(f + g)(x) = \frac{(x - 1)(2x^2 - 2x + 5)}{(x + 1)(x - 2)}$;

$D = \{x \mid x \neq -1, x \neq 2\}$

(b) $(f - g)(x) = \frac{-3(x - 1)(2x - 1)}{(x + 1)(x - 2)}$;

$D = \{x \mid x \neq -1, x \neq 2\}$

(c) $(fg)(x) = (x - 1)^2$; $D = \{x \mid x \neq -1, x \neq 2\}$

(d) $(f/g)(x) = \frac{(x - 2)^2}{(x + 1)^2}$;

$D = \{x \mid x \neq 1, x \neq -1, \text{ and } x \neq 2\}$

27. (a) $(f + g)(x) = \frac{x^2 + 4x - 1}{(x - 1)^2(x + 1)}$;

$D = \{x \mid x \neq 1, x \neq -1\}$

(b) $(f - g)(x) = \frac{-x^2 - 3}{(x - 1)^2(x + 1)}$;

$D = \{x \mid x \neq 1, x \neq -1\}$

(c) $(fg)(x) = \frac{2}{(x - 1)^3}$; $D = \{x \mid x \neq 1, x \neq -1\}$

(d) $(f/g)(x) = \frac{2(x - 1)}{(x + 1)^2}$; $D = \{x \mid x \neq 1, x \neq -1\}$

29. (a) $(f + g)(x) = x^{1/2}(x^2 - x + 1)$; $D = \{x \mid x \geq 0\}$

(b) $(f - g)(x) = x^{1/2}(x^2 - x - 1)$; $D = \{x \mid x \geq 0\}$

(c) $(fg)(x) = x^2(x - 1)$; $D = \{x \mid x \geq 0\}$

(d) $(f/g)(x) = x(x - 1)$; $D = \{x \mid x > 0\}$

31. (a) 2   **(b)** 4   **(c)** 0   **(d)** $-\frac{1}{3}$

33. (a) 2   **(b)** -3 **(c)** 2   **(d)** -2

35. (a) -5 **(b)** -2 **(c)** 0   **(d)** Undefined

37. (a) 5   **(b)** 5   **(c)** 0   **(d)** Undefined

39.

x	-2	0	2	4
$(f + g)(x)$	6	5	5	15
$(f - g)(x)$	-6	5	9	5
$(fg)(x)$	0	0	-14	50
$(f/g)(x)$	0	—	-3.5	2

41. (a) $g(-3) = -5$ **(b)** $g(b) = 2b + 1$

(c) $g(x^3) = 2x^3 + 1$ **(d)** $g(2x - 3) = 4x - 5$

43. (a) $g(-3) = -4$ **(b)** $g(b) = 2(b + 3)^2 - 4$

(c) $g(x^3) = 2(x^3 + 3)^2 - 4$

(d) $g(2x - 3) = 8x^2 - 4$

45. (a) $g(-3) = -\frac{11}{2}$ **(b)** $g(b) = \frac{1}{2}b^2 + 3b - 1$

(c) $g(x^3) = \frac{1}{2}x^6 + 3x^3 - 1$ **(d)** $g(2x - 3) = 2x^2 - \frac{11}{2}$

47. (a) $g(-3) = 1$ **(b)** $g(b) = \sqrt{b + 4}$

(c) $g(x^3) = \sqrt{x^3 + 4}$ **(d)** $g(2x - 3) = \sqrt{2x + 1}$

49. (a) $g(-3) = 14$ **(b)** $g(b) = |3b - 1| + 4$

(c) $g(x^3) = |3x^3 - 1| + 4$

(d) $g(2x - 3) = |6x - 10| + 4$

51. (a) $g(-3)$ is undefined.   **(b)** $g(b) = \frac{4b}{b + 3}$

(c) $g(x^3) = \frac{4x^3}{x^3 + 3}$ **(d)** $g(2x - 3) = \frac{2(2x - 3)}{x}$

53. (a) 3   **(b)** 4   **55. (a)** 18   **(b)** 23

57. (a) $(f \circ g)(x) = (x^2 + 3x - 1)^3$; all real numbers

(b) $(g \circ f)(x) = x^6 + 3x^3 - 1$; all real numbers

(c) $(f \circ f)(x) = x^9$; all real numbers

59. (a) $(f \circ g)(x) = x^4 + x^2 - 3x - 2$; all real numbers

(b) $(g \circ f)(x) = (x + 2)^4 + (x + 2)^2 - 3(x + 2) - 4$; all

real numbers   **(c)** $(f \circ f)(x) = x + 4$; all real numbers

61. (a) $(f \circ g)(x) = 2 - 3x^3$; all real numbers

(b) $(g \circ f)(x) = (2 - 3x)^3$; all real numbers

(c) $(f \circ f)(x) = 9x - 4$; all real numbers

63. (a) $(f \circ g)(x) = \frac{1}{5x + 1}$; $D = \{x \mid x \neq -\frac{1}{5}\}$

(b) $(g \circ f)(x) = \frac{5}{x + 1}$; $D = \{x \mid x \neq -1\}$

(c) $(f \circ f)(x) = \frac{x + 1}{x + 2}$; $D = \{x \mid x \neq -1, x \neq -2\}$

65. (a) $(f \circ g)(x) = \sqrt{4 - x^2} + 4$;

$D = \{x \mid -2 \leq x \leq 2\}$

(b) $(g \circ f)(x) = \sqrt{4 - (x + 4)^2}$;

$D = \{x \mid -6 \leq x \leq -2\}$

(c) $(f \circ f)(x) = x + 8$; all real numbers

67. (a) $(f \circ g)(x) = \sqrt{3x - 1}$; $D = \{x \mid x \geq \frac{1}{3}\}$

(b) $(g \circ f)(x) = 3\sqrt{x - 1}$; $D = \{x \mid x \geq 1\}$

(c) $(f \circ f)(x) = \sqrt{\sqrt{x - 1} - 1}$; $D = \{x \mid x \geq 2\}$

69. (a) $(f \circ g)(x) = x$; all real numbers

(b) $(g \circ f)(x) = x$; all real numbers

(c) $(f \circ f)(x) = 25x - 4$; all real numbers

71. (a) $(f \circ g)(x) = x$; $D = \{x \mid x \neq 0\}$

(b) $(g \circ f)(x) = x$; $D = \{x \mid x \neq 0\}$

(c) $(f \circ f)(x) = x$; $D = \{x \mid x \neq 0\}$

73. (a) -4 **(b)** 2   **(c)** -4

75. (a) -3 **(b)** -2 **(c)** 0

77. (a) 5   **(b)** Undefined   **(c)** 4   **79.** 4; 2

Answers may vary for Exercises 81–93.

81. $f(x) = x - 2, g(x) = \sqrt{x}$

83. $f(x) = x + 2, g(x) = \frac{1}{x}$

85. $f(x) = 2x + 1, g(x) = 4x^3$

87. $f(x) = x^3 - 1, g(x) = x^2$

89. $f(x) = x + 2, g(x) = -4|x| - 3$

91. $f(x) = x - 1, g(x) = \frac{1}{x^2}$

93. $f(x) = x^{1/4}, g(x) = x^3 - x$

95. $P(x) = 10x - 12,000$; $P(3000) = \$18,000$

97. (a) $I(x) = 36x$ **(b)** $C(x) = 2.54x$

(c) $F(x) = (C \circ I)(x)$ **(d)** $F(x) = 91.44x$

99. (a) $s(x) = \frac{11}{6}x + \frac{1}{9}x^2$ **(b)** $s(60) = 510$; it takes 510

feet to stop when traveling 60 mi/hr.

101. (a) $(g \circ f)(1) = 5.25$; a 1% decrease in the ozone

layer could result in a 5.25% increase in skin cancer.

(b) Not possible using the given tables

103. (a) 4.5%   **(b)** $(g \circ f)(x)$ computes the percent

increase in peak electrical demand during year x.

105. (a) $(g \circ f)(1960) = 1.98$
(b) $(g \circ f)(x) = 0.165(x - 1948)$
(c) f, g, and $g \circ f$ are all linear functions.
107. (a) $(g \circ f)(2) \approx 25°C$ **(b)** $(g \circ f)(x)$ computes the Celsius temperature after x hours.
109. (a) $A(4s) = 16\frac{\sqrt{3}}{4}s^2 = 16A(s)$; if the length of a side is quadrupled, the area increases by a factor of 16.
(b) $A(s + 2) = \frac{\sqrt{3}}{4}(s^2 + 4s + 4) = A(s) + \sqrt{3}(s + 1)$; if the length of a side increases by 2, the area increases by $\sqrt{3}(s + 1)$.
111. $A = 36\pi t^2$
113. (a)

x	1990	2000	2010	2020	2030
$h(x)$	32	35.5	39	42.5	46

(b) $h(x) = f(x) + g(x)$
115. $h(x) = 0.35x - 664.5$
117. (a) $M(20) = 50$; $N(20) = 140$; the median income was $50 thousand in 2010, while the 90th percentile was $140 thousand.
(b) $D(x) = -0.15x^2 + 3.9x + 72$; $D(x)$ gives the household income difference between the 90th percentile and the median.
(c) $D(20) = 90$; the household income difference between the 90th percentile and the median in 2010 was $90 thousand.
119. Let $f(x) = ax + b$ and $g(x) = cx + d$. Then $f(x) + g(x) = (ax + b) + (cx + d) = (a + c)x + (b + d)$, which is linear.
121. (a) $(f \circ g)(x) = k$; a constant function
(b) $(g \circ f)(x) = ak + b$; a constant function

SECTION 5.2 (pp. 375–379)

1. Closing a window
3. Closing a book, standing up, and walking out of the classroom
5. Subtract 2 from x; $x + 2$ and $x - 2$
7. Divide x by 3 and then add 2; $3(x - 2)$ and $\frac{x}{3} + 2$
9. Subtract 1 from x and cube the result; $\sqrt[3]{x} + 1$ and $(x - 1)^3$ **11.** Take the reciprocal of x; $\frac{1}{x}$ and $\frac{1}{x}$
13. One-to-one   **15.** Not one-to-one
17. Not one-to-one
19. Not one-to-one; does not have an inverse
21. One-to-one; does have an inverse
23. One-to-one   **25.** Not one-to-one
27. Not one-to-one   **29.** Not one-to-one
31. Not one-to-one   **33.** Not one-to-one
35. One-to-one   **37.** No   **39.** Yes
41. $f^{-1}(x) = x^3$ **43.** $f^{-1}(x) = -\frac{1}{2}x + 5$
45. $f^{-1}(x) = \frac{x + 1}{3}$ **47.** $f^{-1}(x) = \sqrt[3]{\frac{x + 5}{2}}$
49. $f^{-1}(x) = \sqrt{x + 1}$ **51.** $f^{-1}(x) = \frac{1}{2x}$
53. $f^{-1}(x) = -\frac{2(x - 3)}{5}$ **55.** $f^{-1}(x) = -\frac{2x}{x - 1}$
57. $f^{-1}(x) = \frac{x + 1}{x - 2}$ **59.** $f^{-1}(x) = \frac{1}{x + 3}$
61. $f^{-1}(x) = \sqrt[3]{\frac{1 + x}{x}}$
63. If the domain of f is restricted to $x \geq 0$, then $f^{-1}(x) = \sqrt{4 - x}$.

65. If the domain of f is restricted to $x \geq 2$, then $f^{-1}(x) = 2 + \sqrt{x - 4}$.
67. If the domain of f is restricted to $x \geq 0$, then $f^{-1}(x) = (x - 1)^{3/2}$.
69. If the domain of f is restricted to $0 \leq x \leq \frac{3}{\sqrt{2}}$, then $f^{-1}(x) = \sqrt{\frac{9 - x^2}{2}}$.
71. $f^{-1}(x) = \frac{x + 15}{5}$; D and R are all real numbers.
73. $f^{-1}(x) = x^3 + 5$; D and R are all real numbers.
75. $f^{-1}(x) = 4x + 5$; D and R are all real numbers.
77. $f^{-1}(x) = x^2 + 5$; $D = \{x \mid x \geq 0\}$ and $R = \{y \mid y \geq 5\}$
79. $f^{-1}(x) = \frac{1}{x} - 3$; $D = \{x \mid x \neq 0\}$ and $R = \{y \mid y \neq -3\}$
81. $f^{-1}(x) = \sqrt[3]{\frac{x}{2}}$; D and R are all real numbers.
83. $f^{-1}(x) = \sqrt{x}$; D and R include all nonnegative real numbers.
85.

x	5	7	9
$f^{-1}(x)$	1	2	3

For f: $D = \{1, 2, 3\}$, $R = \{5, 7, 9\}$;
for f^{-1}: $D = \{5, 7, 9\}$, $R = \{1, 2, 3\}$
87.

x	0	4	16
$f^{-1}(x)$	0	2	4

For f: $D = \{0, 2, 4\}$, $R = \{0, 4, 16\}$;
for f^{-1}: $D = \{0, 4, 16\}$, $R = \{0, 2, 4\}$
89.

x	0	2	4	6
$f^{-1}(x)$	0	$\frac{1}{2}$	1	$\frac{3}{2}$

91. 1   **93.** 3   **95.** 5   **97.** 1
99. (a) $f(1) \approx 110$ dollars   **(b)** $f^{-1}(110) \approx 1$ year
(c) $f^{-1}(160) \approx 5$ years; $f^{-1}(x)$ computes the years necessary for the account to accumulate x dollars.
101. (a) 2   **(b)** 3   **(c)** 1   **(d)** 3
103. (a) 4   **(b)** 0   **(c)** 9   **(d)** 4
105. 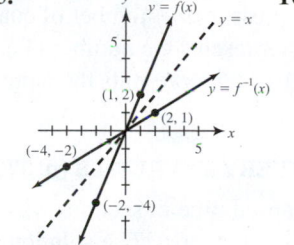   **107.**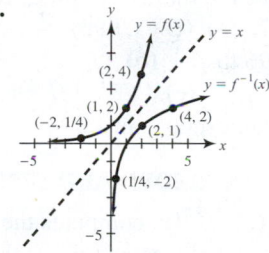

109.   **111.**
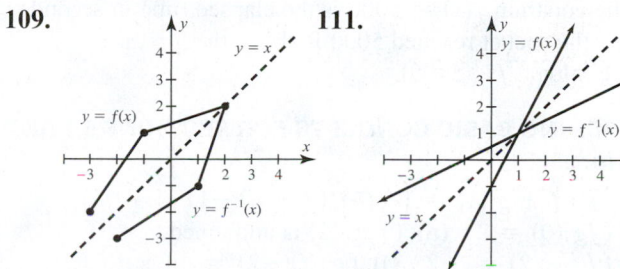

113. 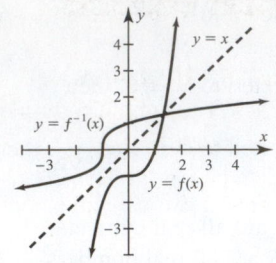 **115.**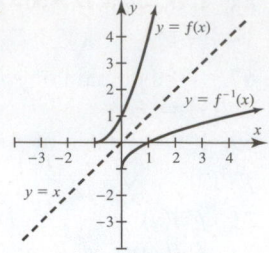

117. $Y_1 = 3X - 1$, $Y_2 = (X + 1)/3$, $Y_3 = X$
$[-4.7, 4.7, 1]$ by $[-3.1, 3.1, 1]$

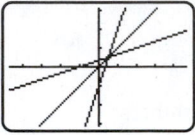

119. $Y_1 = X^3/3 - 1$, $Y_2 = \sqrt[3]{(3X + 3)}$, $Y_3 = X$
$[-4.7, 4.7, 1]$ by $[-3.1, 3.1, 1]$

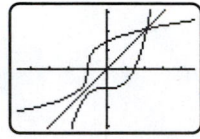

121. (a) Yes **(b)** The radius r of a sphere with volume V
(c) $r = \sqrt[3]{\frac{3V}{4\pi}}$ **(d)** No. If V and r were interchanged, then r would represent the volume and V would represent the radius. **123. (a)** 135.7 pounds **(b)** Yes
(c) $h = \frac{7}{25}\left(W + \frac{800}{7}\right) = \frac{7}{25}W + 32$ **(d)** 74; the maximum recommended height for a person weighing 150 lb is 74 in. **(e)** The inverse formula computes the maximum recommended height for a person of a given weight.
125. (a) $(F \circ Y)(2) = 10,560$ represents the number of feet in 2 miles. **(b)** $F^{-1}(26,400) = 8800$ represents the number of yards in 26,400 ft. **(c)** $(Y^{-1} \circ F^{-1})(21,120) = 4$ represents the number of miles in 21,120 ft.
127. (a) $(Q \circ C)(96) = 1.5$ represents the number of quarts in 96 tbsp. **(b)** $Q^{-1}(2) = 8$ represents the number of cups in 2 qt. **(c)** $(C^{-1} \circ Q^{-1})(1.5) = 96$ represents the number of tablespoons in 1.5 qt.

5.2 EXTENDED AND DISCOVERY EXERCISES (p. 379)

1. (a) $f^{-1}(x)$ computes the elapsed time in seconds when the rocket was x ft above the ground. **(b)** The solution to the equation $f(x) = 5000$ is the elapsed time in seconds when the rocket reached 5000 ft above the ground.
(c) Evaluate $f^{-1}(5000)$.

CHECKING BASIC CONCEPTS FOR SECTIONS 5.1 AND 5.2 (p. 380)

1. (a) $(f + g)(1) = 1$ **(b)** $(f - g)(-1) = 3$
(c) $(fg)(0) = 2$ **(d)** $(f/g)(2)$ is undefined.
(e) $(f \circ g)(2) = -2$ **(f)** $(g \circ f)(-2) = -1$
3. (a) $(f + g)(x) = x^2 + 6x - 3$
(b) $(f/g)(x) = \frac{x^2 + 3x - 2}{3x - 1}, x \neq \frac{1}{3}$
(c) $(f \circ g)(x) = 9x^2 + 3x - 4$

5. (a) Yes; yes; $f^{-1}(x) = x - 1$ **(b)** No; no
7. (a) 0 **(b)** 2

SECTION 5.3 (p. 393–398)
1. $\frac{1}{8}$ **3.** 6 **5.** -18 **7.** 2 **9.** 1 **11.** 5
13. Linear; $f(x) = -1.2x + 2$
15. Exponential; $f(x) = 8(\frac{1}{2})^x$
17. Exponential; $f(x) = 5(2^x)$
19. Linear; $f(x) = 3.5x - 2$
21. Exponential; $f(x) = 81(\frac{1}{3})^x$
23. $f(x) = 2^x$ **25.** $f(x) = 2x + 1$
27. $C = 5, a = 1.5$ **29.** $C = 10, a = 2$
31. $C = 3, a = 3$ **33.** $C = \frac{1}{2}, a = \frac{1}{3}$
35. $f(x) = 4x + 4; g(x) = 4(2)^x$
37. $f(x) = -\frac{7}{2}x + 5; g(x) = 3(\frac{1}{2})^x$
39. $C = 5000, a = 2$; x represents time in hours.
41. $C = 200,000, a = 0.95$; x represents the number of years after 2000.
43. About 11 pounds per square inch
45. 71.2571 **47.** -0.7586

49.  **51.**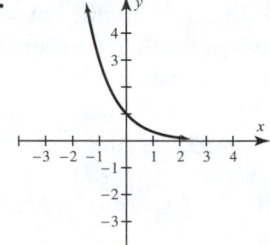

53. 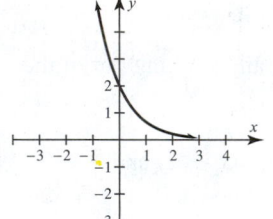 **55.**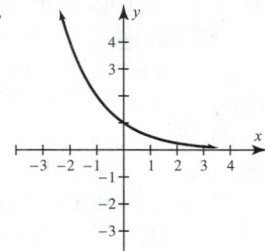

57. $C = 1, a = \frac{1}{2}$ **59.** $C = \frac{1}{2}, a = 4$
61. (a) $D: (-\infty, \infty)$ or $\{x \mid -\infty < x < \infty\}$; $R: (0, \infty)$ or $\{x \mid x > 0\}$ **(b)** Decreasing **(c)** $y = 0$
(d) y-intercept: 7; no x-intercept **(e)** Yes; yes
63. (i) b **(ii)** d **(iii)** a **(iv)** c

65. (a) 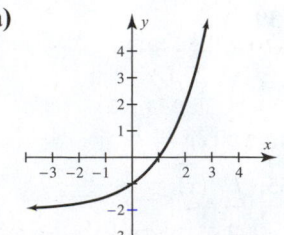 **(b)**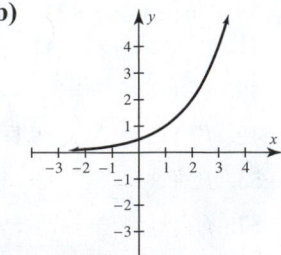

(c)

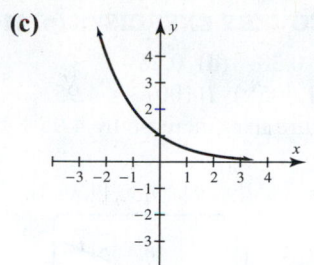

(d)

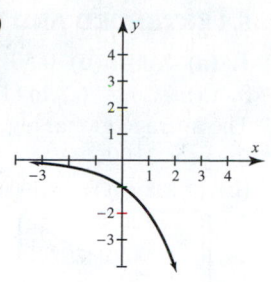

67.

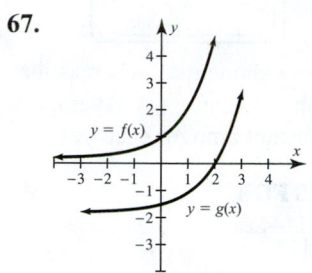

69.

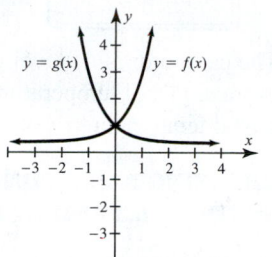

71.

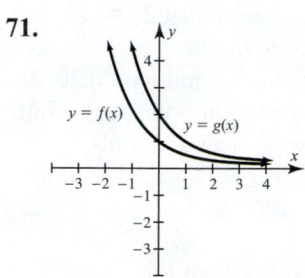

73. $841.53   **75.** $1730.97   **77.** $4451.08
79. $1652.83   **81.** $14,326.78   **83.** Annually
85. (a) $B(x) = 2.5(3)^x$
(b) About 13 million per milliliter
87. (a) About 1.5 watts per square meter
(b) The y-intercept indicates that the intensity is 300 watts per square meter at the surface.
$[-5, 50, 5]$ by $[0, 350, 50]$

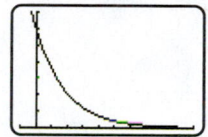

89. About 7.2 million
91. (a) $C \approx 0.72, a \approx 1.041$ (answers may vary slightly)
(b) 1.21 ppb (answers may vary slightly)
93. (a) About 6,214,000 bacteria per milliliter
(b) There will be 10 million *E. coli* bacteria per milliliter after about 214 min, or about 3.6 hr.
95. (a) About 92%   **(b)** About 0.83 min
97. $T(t) = 100\left(\frac{1}{2}\right)^{t/2.8}; T(5.5) \approx 25.6\%$
99. (a) $P(x) = 95(0.9)^x$ **(b)** About 11.5%
(c) About 6.6 feet   **101.** About 18,935 yr
103. About 27.3%
105. (a) $C = 2.5, a = 0.7$ **(b)** 1.225 ppm
107. (a) $H(30) \approx 0.42$; about 0.42 horsepower is required for each ton pulled at 30 mi/hr.
(b) About 2100 horsepower   **(c)** 2

5.3 EXTENDED AND DISCOVERY EXERCISES (p. 398)

1. The formulas are equivalent.   **3.** $12,914.64
5. (a) About 1.0005   **(b)** 1   **(c)** They are very similar.
7. (a) About 0.6068   **(b)** About 0.6065
(c) They are very similar.
9. The average rate of change near x and the value of the function at x are approximately equal.

SECTION 5.4 (pp. 410–414)

1.

x	10^0	10^4	10^{-8}	$10^{1.26}$
$\log x$	0	4	-8	1.26

3. (a) Undefined   **(b)** -2 **(c)** $-\frac{1}{2}$ **(d)** 0
5. (a) 1   **(b)** 4   **(c)** -20 **(d)** -2
7. (a) 2   **(b)** $\frac{1}{2}$ **(c)** 3   **(d)** Undefined
9. (a) $n = 1, \log 79 \approx 1.898$
(b) $n = 2, \log 500 \approx 2.699$ **(c)** $n = 0, \log 5 \approx 0.699$
(d) $n = -1, \log 0.5 \approx -0.301$
11. (a) $\frac{3}{2}$ **(b)** $\frac{1}{3}$ **(c)** $-\frac{1}{5}$ **(d)** -1
13. $\{x \mid x > -3\}$, or $(-3, \infty)$
15. $\{x \mid x < -1 \text{ or } x > 1\}$, or $(-\infty, -1) \cup (1, \infty)$
17. $\{x \mid -\infty < x < \infty\}$, or $(-\infty, \infty)$
19. $\{x \mid x < 2\}$, or $(-\infty, 2)$
21. -5.7 **23.** $2x, x > 0$ **25.** 64
27. -4 **29.** π **31.** $x - 1$ for $x > 1$
33. 6   **35.** $\frac{1}{2}$ **37.** -3 **39.** 2   **41.** 2   **43.** -2
45. -1 **47.** 0   **49.** -4
51.

x	6	7	21
$f(x)$	0	2	8

53. (a) $4x = \log_7 4$ **(b)** $x = \ln 7$ **(c)** $x = \log_c b$
55. (a) $x = 8^3$ **(b)** $2 + x = 9^5$ **(c)** $b = k^c$
57. (a) -2 **(b)** $\log 7 \approx 0.85$ **(c)** No solutions
59. (a) -2 **(b)** $\ln 2 \approx 0.69$ **(c)** 3
61. (a) 0   **(b)** $\frac{1}{2}$ **(c)** $\frac{1}{3}$
63. $-\ln 3 \approx -1.10$ **65.** 2   **67.** $\frac{2}{3}$
69. $\frac{\log 13}{4} \approx 0.28$ **71.** $\log_3 4 \approx 1.26$ **73.** $\ln 23 \approx 3.14$
75. $\log_2 14 \approx 3.81$ **77.** $\ln\left(\frac{18}{5}\right) \approx 1.28$ **79.** 8
81. (a) $10^2 = 100$ **(b)** $10^{-3} = 0.001$
(c) $10^{1.2} \approx 15.8489$
83. (a) $2^6 = 64$ **(b)** $3^{-2} = \frac{1}{9}$ **(c)** $e^2 \approx 7.3891$
85. $2^{1.2} \approx 2.2974$ **87.** $\frac{49}{2}$ **89.** 1000   **91.** 20
93. $e^{3/4} \approx 2.1170$ **95.** $e^{7/5} \approx 4.0552$ **97.** 16
99. $\frac{1}{2}e^{6/5} \approx 1.6601$ **101.** 8   **103.** $a = 5, b = 2$
105. $f^{-1}(x) = \ln x$ **107.** $f^{-1}(x) = 2^x$

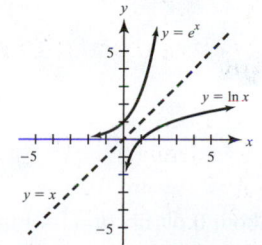

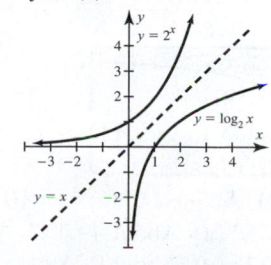

109. Decreasing

111. (a)

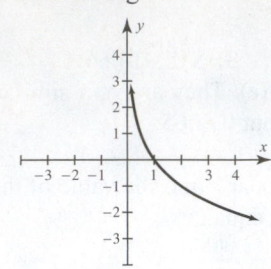
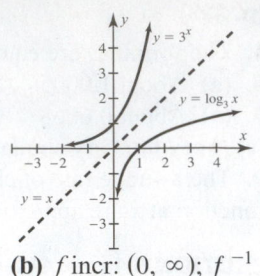

(b) f incr: $(0, \infty)$; f^{-1}
incr: $(-\infty, \infty)$

113. $x = 2$

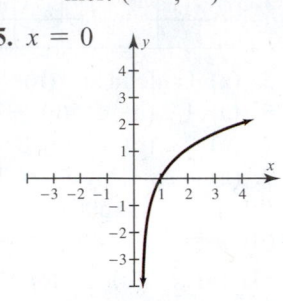

115. $x = 0$

117. $x = 0$

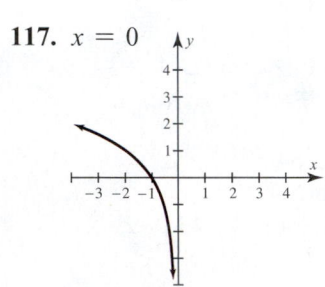

119. $x = 1$

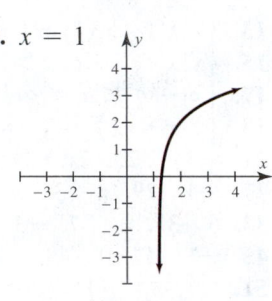

121. $D = \{x \mid x > -1\}$
$[-6, 6, 1]$ by $[-4, 4, 1]$

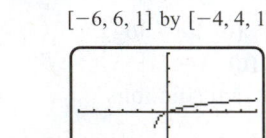

123. $D = \{x \mid x < 0\}$
$[-6, 6, 1]$ by $[-4, 4, 1]$

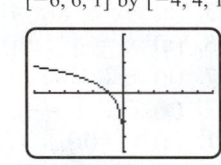

125. 120 dB   **127. (a)** $L(100) = 6$; a 100-thousand-pound plane needs 6000 feet of runway.   **(b)** No. It increases by 3000 ft.   **(c)** If the weight increases tenfold, the runway length increases by 3000 ft.
129. (a) 4.7   **(b)** $10^{3.5} \approx 3162$
131. (a) Yugoslavia: $x = 1,000,000$;
Indonesia: $x = 100,000,000$ **(b)** 100
133. (a) $f(0) = 27$, $f(100) \approx 29.2$. At the eye, the barometric air pressure is 27 in.; 100 mi away, it is 29.2 in.
(b) The air pressure increases rapidly at first and then starts to level off.

$[0, 250, 50]$ by $[25, 30, 1]$

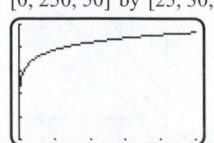

(c) About 7 miles
135. (a) About $49.4^\circ C$ **(b)** About 0.69 hr, or 41.4 min
137. (a) About 0.189, or an 18.9% chance
(b) $x = -3 \ln (0.3) \approx 3.6$ min
139. $a = 7$, $b = 4$; about 178 km^2
141. (a) $C = 3$, $a = 2$ **(b)** After about 2.4 days

5.4 EXTENDED AND DISCOVERY EXERCISES (p. 414)

1. (a) 1.00   **(b)** 0.50   **(c)** 0.33   **(d)** 0.25
3. (a) $T(x) = 6.5 \ln (1.3 \cdot 1.005^x)$; $T(100) \approx 4.95$.
The average global temperature may increase by 4.95°F by the year 2100.
(b) $[0, 200, 50]$ by $[0, 1000, 100]$ $[0, 200, 50]$ by $[0, 10, 1]$

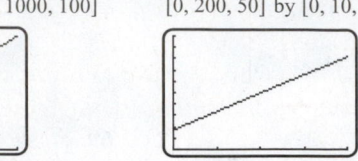

The carbon dioxide level graph is exponential, whereas the average global temperature graph is linear.   **(c)** Average global temperature rises by a constant amount each year.

CHECKING BASIC CONCEPTS FOR SECTIONS 5.3 AND 5.4 (pp. 414–415)

1. \$1752.12; \$1754.74
3. $\log_2 15$ represents the exponent k such that $2^k = 15$. No.
5. (a) $\ln 5 \approx 1.609$ **(b)** $\log 25 \approx 1.398$
(c) $10^{1.5} \approx 31.623$ **7. (a)** About 38.1 million; in 2012 California's population was about 38.1 million.   **(b)** 37.3; in 2010 California's population was about 37.3 million.   **(c)** About 1%   **(d)** 2017

SECTION 5.5 (pp. 421–423)

1. 1.447; Product rule   **3.** 2.197; Quotient rule
5. 3.010; Power rule   **7.** $\log_2 a + \log_2 b$
9. $\ln 7 + 4 \ln a$ **11.** $\log 6 - \log z$
13. $2 \log x - \log 3$ **15.** $\ln 2 + 7 \ln x - \ln 3 - \ln k$
17. $2 + 2 \log_2 k + 3 \log_2 x$
19. $2 + 3 \log_5 x - 4 \log_5 y$
21. $4 \ln x - 2 \ln y - \frac{3}{2} \ln z$
23. $-1 + 3 \log_4 (x + 2)$ **25.** $3 \log_5 x - 4 \log_5 (x - 4)$
27. $\frac{1}{2} \log_2 x - 2 \log_2 z$ **29.** $\frac{1}{3} \ln (2x + 6) - \frac{5}{3} \ln (x + 1)$
31. $\frac{1}{3} \log_2 (x^2 - 1) - \frac{1}{2} \log_2 (1 + x^2)$
33. $\log 6$ **35.** $\ln 5^{-3/2}$ **37.** $\log 2$
39. $\log 6$ **41.** $\log_7 5k^2$ **43.** $\ln x^3$ **45.** $\log x^{3/2}$
47. $\ln \dfrac{x^3 z^4}{\sqrt{y^3}}$ **49.** $\ln \frac{2}{e}$ **51.** $\ln \dfrac{x^2 \sqrt{z}}{y^4}$
53. $\log (4 \sqrt{x^5})$ **55.** $\log \dfrac{(x^2 - 1)^2 (x - 2)^4}{\sqrt{y}}$
57. (a) Yes   **(b)** By the product rule:
$\log 3x + \log 2x = \log (3x \cdot 2x) = \log 6x^2$
59. (a) Yes   **(b)** By the quotient rule:
$\ln 2x^2 - \ln x = \ln \left(\frac{2x^2}{x} \right) = \ln 2x$
61. (a) Yes   **(b)** By the power rule:
$\ln x^4 - \ln x^2 = 4 \ln x - 2 \ln x = 2 \ln x$
63.   **65.**

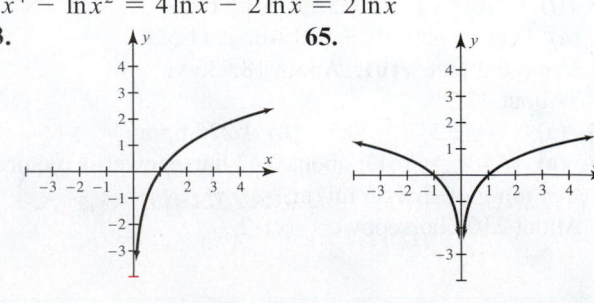

67. $\frac{\log 25}{\log 2} \approx 4.644$ **69.** $\frac{\log 130}{\log 5} \approx 3.024$

71. $\frac{\log 5}{\log 2} + \frac{\log 7}{\log 2} \approx 5.129$ **73.** $\sqrt{\frac{\log 46}{\log 4}} \approx 1.662$

75. $\frac{\log 12/\log 2}{\log 3/\log 2} = \frac{\log 12}{\log 3} \approx 2.262$ **77.** 4.714

79. ± 2.035 **81.** $L(x) = \frac{3\ln x}{\ln 10}$; $L(50) \approx 5.097$; yes

83. 10 decibels **85.** $I = I_0\, e^{-kx}$

87. (a) $x = 100\ln\frac{P}{37.3}$ **(b)** About 7; the population is expected to reach 40 million during 2017.

89. $t = \dfrac{\ln\frac{A}{P}}{r}$

91. $\log(1 \cdot 2^2 \cdot 3^3 \cdot 4^4 \cdot 5^5) = \log 86{,}400{,}000$

SECTION 5.6 (pp. 431–435)

1. (a) About 2 **(b)** $\ln 7.5 \approx 2.015$
3. (a) About 2 **(b)** $5\log 2.5 \approx 1.990$
5. $\ln 1.25 \approx 0.2231$ **7.** $\log 20 \approx 1.301$
9. $-\frac{5}{6}\ln 0.4 \approx 0.7636$ **11.** $\frac{\ln 0.5}{\ln 0.9} \approx 6.579$
13. $1 + \frac{\log 4}{\log 1.1} \approx 15.55$ **15.** 1 **17.** $\frac{1}{5}$ **19.** 1, 2
21. No solutions **23.** $2 + \frac{\log(1/3)}{\log(2/5)} \approx 3.199$
25. $\frac{\log 4}{\log 4 - 2\log 3} \approx -1.710$ **27.** $\frac{3}{1 - 3\ln 2} \approx -2.779$
29. $\frac{\log(64/3)}{\log 1.4} \approx 9.095$ **31.** $1980 + \frac{\log(8/5)}{\log 1.015} \approx 2012$
33. -3 **35.** $0, -1$ **37.** 1 **39.** $\frac{6}{7}$ **41.** $-\frac{12}{5}$
43. $10^{2/3} \approx 4.642$ **45.** $\frac{1}{2}e^5 \approx 74.207$
47. $\pm\sqrt{50} \approx \pm 7.071$ **49.** 6 **51.** -39
53. 10^{-11} **55.** $e \approx 2.718$ **57.** $2^{2.1} \approx 4.287$
59. $\sqrt{50} \approx 7.071$ **61.** $-\frac{998}{3}$ **63.** $2\left(\text{extraneous } -\frac{5}{3}\right)$
65. $1 + \sqrt{2} \approx 2.414$ (extraneous $1 - \sqrt{2}$)
67. 2 (extraneous -2) **69.** 2 (extraneous -4)
71. 4 (extraneous -4) **73.** 0.31 **75.** 1.71 **77.** 2.10
79. 2018 **81.** $\frac{\ln 0.25}{-0.12} \approx 11.55$ ft
83. (a) $T(x) = 2300(2)^{x/2}$ **(b)** $T(40) \approx 2{,}411{,}724{,}800$; in 2011 there were about 2.4 billion transistors on an integrated circuit. **(c)** About 1995 **85. (a)** About 0.88; the Human Development Index is 0.88. **(b)** 3.9
87. About 1981 **89. (a)** $C = 1, a \approx 1.01355$
(b) $P(2010) \approx 1.14$ billion **(c)** About 2030
91. (a) $T_0 = 32, D = 180, a \approx 0.045$ **(b)** About 139°F
(c) About 1 hr **93. (a)** About 16.7°C **(b)** About 1.09 hr
95. About 2.8 acres **97.** $100 - 10^{1.16} \approx 85.5\%$
99. (a) $\frac{\ln 1.15}{6} \approx 0.0233$ **(b)** About 1.45 billion per liter
(c) About 2.36% **101.** 8.25 yr **103. (a)** $\frac{\ln 1.5}{0.03} \approx 13.5$
(b) $500 invested at 3% compounded continuously results in $750 after 13.5 years. **105.** About 8633 yr
107. $-2\ln 0.5 \approx 1.39$ min **109.** About 3.6 hr
111. About 13.5 yr **113. (a)** 11 milligrams/liter
(b) About 2.11 hr
115. $t = \dfrac{\log(A/P)}{n\log(1 + r/n)}$

5.6 EXTENDED AND DISCOVERY EXERCISE (p. 435)

1. $f(x) = Ce^{x\ln a}$ and $g(x) = e^{x\ln 2}$; that is, $k = \ln a$.

CHECKING BASIC CONCEPTS FOR SECTIONS 5.5 AND 5.6 (p. 435)

1. $2\log x + 3\log y - \frac{1}{3}\log z$ **3. (a)** $\frac{\log(29/5)}{\log(1.4)} \approx 5.224$
(b) $\frac{1}{3}$ **5. (a)** It levels off at 80°F. **(b)** 17 min

SECTION 5.7 (pp. 441–443)

1. Logarithmic **3.** Exponential
5. Exponential: $f(x) = 1.2(1.7)^x$
7. Logarithmic: $f(x) = 1.088 + 2.937\ln x$
9. Logistic: $f(x) = \frac{9.96}{1 + 30.6e^{-1.51x}}$
11. (a) Quadratic: $V(x) = 3.55x^2 + 0.21x + 1.9$
(b) 91.7 million **13.** $f(x) = 9.02 + 1.03\ln x$
15. (a) $f(x) = 1.4734(0.99986)^x$
(b) Approximately 0.55 kg/m³
17. (a) $f(x) = \frac{4.9955}{1 + 49.7081e^{-0.6998x}}$
(b) About 5 thousand per acre
19. (a) $C(x) = \frac{0.94}{1 + 17.97e^{-0.862x}}$ **(b)** About 0.93 billion
21. (a) $A(w) = 101w^{0.662}$
(b) [0, 20, 2] by [100, 700, 50]

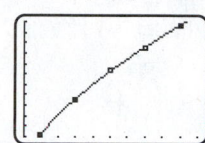

(c) About 11.2 lb
23. (a) 3 ft; after 5 years, the tree is 3 feet tall.
(b) $H(x) = \frac{50.2}{1 + 47.4e^{-0.221x}}$
(c) [0, 45, 5] by [0, 55, 5]

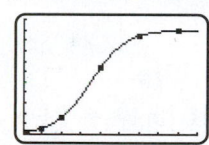

(d) About 17.4 yr **(e)** Interpolation
25. (a) The data are not linear.
$[-2, 32, 5]$ by $[0, 80, 10]$

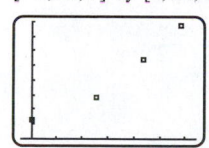

(b) $f(x) = 12.42(1.066)^x$ **(c)** $f(39) \approx 150$; fertilizer use increased but at a slower rate than predicted by f.

5.7 EXTENDED AND DISCOVERY EXERCISE (p. 443)

1. (a) $f(x) = 0.0904(0.844)^x$ (regression) or $f(x) \approx 0.128(0.777)^x$ (trial and error)
(b) About 4.6 min **(c)** Answers may vary.

CHECKING BASIC CONCEPTS FOR SECTION 5.7 (p. 444)

1. Exponential: $f(x) = 0.5(1.2)^x$
3. Logistic: $f(x) = \frac{4.5}{1 + 277e^{-1.4x}}$

CHAPTER 5 REVIEW EXERCISES (pp. 447–450)

1. (a) 8 **(b)** 0 **(c)** -6 **(d)** Undefined

3. (a) 7 **(b)** 1 **(c)** 0 **(d)** $-\frac{9}{2}$

5. (a) 2 **(b)** 1 **(c)** 2 **7. (a)** $\sqrt{6}$ **(b)** 12

9. $(f \circ g)(x) = \left(\frac{1}{x}\right)^3 - \left(\frac{1}{x}\right)^2 + 3\left(\frac{1}{x}\right) - 2; D = \{x \mid x \neq 0\}$

11. $(f \circ g)(x) = x$; all real numbers

13. $f(x) = x^2 + 3, g(x) = \sqrt{x}$ (answers may vary)

15. Subtract 6 from x and then multiply the result by 10; $\frac{x}{10} + 6$ and $10(x - 6)$ **17.** f is one-to-one. **19.** f is not one-to-one.

21.

x	6	4	3	1
$f^{-1}(x)$	-1	0	4	6

For f: $D = \{-1, 0, 4, 6\}$ and $R = \{1, 3, 4, 6\}$; for f^{-1}: $D = \{1, 3, 4, 6\}$ and $R = \{-1, 0, 4, 6\}$

23. $f^{-1}(x) = \frac{x + 5}{3}$

25. $(f \circ f^{-1})(x) = f\left(\frac{x + 1}{2}\right) = 2\left(\frac{x + 1}{2}\right) - 1 = x;$
$(f^{-1} \circ f)(x) = f^{-1}(2x - 1) = \frac{(2x - 1) + 1}{2} = x$

27. 1 **29.** $f^{-1}(x) = x^2 - 1, x \geq 0$; for f: $D = \{x \mid x \geq -1\}$ and $R = \{y \mid y \geq 0\}$; for f^{-1}: $D = \{x \mid x \geq 0\}$ and $R = \{y \mid y \geq -1\}$

31. $C = 3, a = 2$

33. 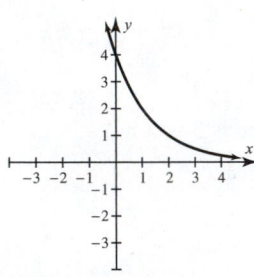 **35.**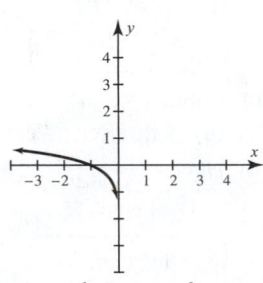

$D = $ all real numbers $\qquad D = \{x \mid x < 0\}$

37. $C = 2, a = 2$ **39.** \$1562.71 **41.** $\ln 19 \approx 2.9444$

43. 3 **45.** -21 **47.** 2 **49.** 1 **51.** 2.631

53. $\log 125 \approx 2.097$ **55.** $10 \ln 5.2 \approx 16.49$

57. $-\frac{1}{\log 5} \approx -1.431$ **59.** $10 + \frac{\log(29/3)}{\log(0.78)} \approx 0.869$

61. $f(x) = 1.5(2)^x$ **63.** $10^{1.5} \approx 31.62$

65. $e^{3.4} \approx 29.96$ **67.** $\log 30x$ **69.** $\ln y - 2 \ln x$

71. $10^{1/4} \approx 1.778$ **73.** $\frac{100,000}{3} \approx 33,333$

75. 1 (extraneous -4)

77. The x-intercept is b. If $(0, b)$ is on the graph of f, then $(b, 0)$ is on the graph of f^{-1}.

79. (a) $11,022/\text{mL}$ **(b)** About 1.86 hr **81.** $h(x) = 15x$

83. $x \approx 2.74$; the fish weighs 50 milligrams at about 3 weeks. **85.** About 3.8 min

87. Logistic: $f(x) = \frac{171.4}{1 + 18.4e^{-0.0744x}}$

89. $a \approx 3.50, b \approx 0.74$

$[0, 5, 1]$ by $[0, 3, 1]$

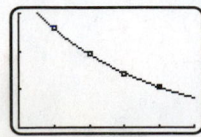

CHAPTER 5 EXTENDED AND DISCOVERY EXERCISES (pp. 450–451)

1. The data points $(\ln x, \ln y)$ seem to be almost linear. Using linear regression gives the equation $y = 1.5x - 8.5$. Since $b = 1.5$ and $a = e^{-8.5} \approx 0.0002$, we get the power function $f(x) = 0.0002x^{1.5}$.

3. (a) $163.0000000000000000000000000000000232$; it is not an integer.

CHAPTER 6: Trigonometric Functions

SECTION 6.1 (pp. 462–465)

1. (a) 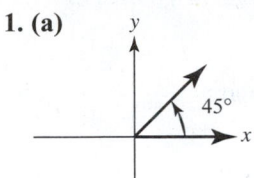 **(b)**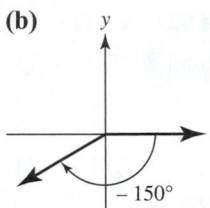

(c) 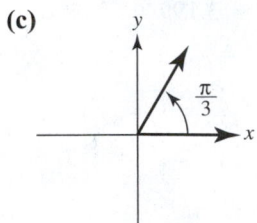 **(d)**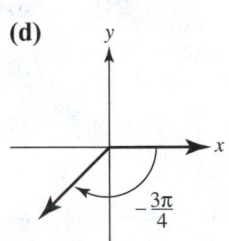

3. Answers may vary. **5.**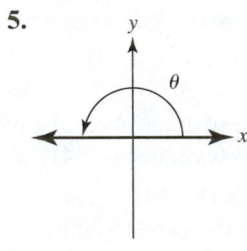

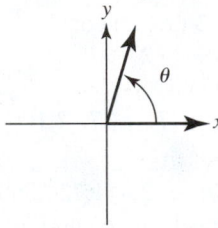

7. Answers may vary. **9.** Answers may vary.

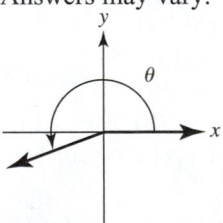 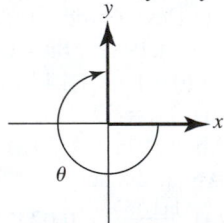

11. (a) $\frac{1}{4}$ **(b)** $\frac{1}{12}$ **(c)** $\frac{1}{6}$ **(d)** $\frac{1}{8}$

13. $510°, -210°$ (answers may vary)

15. $288°, -432°$ (answers may vary)

17. $\frac{5\pi}{2}, -\frac{3\pi}{2}$ (answers may vary)

19. $\frac{9\pi}{5}, -\frac{11\pi}{5}$ (answers may vary)

21. $125.25°$ **23.** $108.76°$ **25.** $125°18'$

27. $51°21'36''$ **29.** $\alpha = 34.1°, \beta = 124.1°$

31. $\alpha = 4°36'15'', \beta = 94°36'15''$

33. $\alpha = 66°19'25'', \beta = 156°19'25''$

35. $\theta = 2$ radians; $\theta \approx 114.6°$

37. $\theta = 1.3$ radians; $\theta \approx 74.5°$

39. (a) $\frac{\pi}{4}$ **(b)** $\frac{3\pi}{4}$ **(c)** $-\frac{2\pi}{3}$ **(d)** $-\frac{7\pi}{6}$

41. (a) $\frac{37\pi}{180}$ **(b)** 2.15 **(c)** -1.61 **(d)** 4.02

43. (a) 30° **(b)** 12° **(c)** $-300°$ **(d)** $-210°$

45. (a) 45° **(b)** 25.71° **(c)** 177.62° **(d)** $-143.24°$

47. $s = \frac{2\pi}{3}$ in. **49.** $\theta = \frac{12}{5}$ radians **51.** $r = \frac{5}{\pi}$ ft

53. $\frac{\pi}{4}$ m **55.** π ft **57.** $\frac{7\pi}{360}$ mi **59.** 2π in.; $\frac{2\pi}{15}$ in./min

61. 10π in.; $\frac{2\pi}{15}$ in./min **63.** 1.5π in^2 **65.** 4.5π in^2

67. $\frac{17,161\pi}{3000} \approx 5.72\pi$ cm^2 **69.** $\frac{3\pi}{16} \approx 0.59$ ft^2

71. 240π in^2 **73.** 4292π cm^2 **75.** 16.25 ft/sec

77. (a) $\frac{\pi}{210} \approx 0.015$ radian/sec **(b)** $\frac{\pi}{3} \approx 1.05$ ft/sec

79. (a) $1000\pi \approx 3141.6$ radians/min

(b) $15,000\pi \approx 47,123.89$ in./min, or about 65.4 ft/sec

81. 810 mi **83.** About 69 ft/sec **85.** 14.5 in.

87. (a) 2.5 revolutions **(b)** $\frac{65\pi}{6} \approx 34$ ft/sec

89. (a) 78,370 mi/hr **(b)** 66,630 mi/hr

(c) 29,250 mi/hr **(d)** 12,160 mi/hr

Planets farther from the sun have slower orbital velocities.

91. 137.2 m **93. (a)** 388.8 m **(b)** 881.8 m

6.1 EXTENDED AND DISCOVERY EXERCISES (p. 465)

1. $s = r\theta\left(\frac{\pi}{180°}\right)$, where θ is in degrees. The formula for radian measure is simpler.

SECTION 6.2 (pp. 475–479)

1.

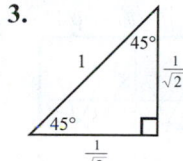

3.

5.

7. $\frac{\sqrt{3}}{2}$ **9.** $\frac{\sqrt{3}}{2}$ **11.** 2 **13.** 1 **15.** 1 **17.** $\frac{1}{\sqrt{2}}$

19. $\sin\theta = \frac{4}{5}$, $\cos\theta = \frac{3}{5}$, $\tan\theta = \frac{4}{3}$,

  $\csc\theta = \frac{5}{4}$, $\sec\theta = \frac{5}{3}$, $\cot\theta = \frac{3}{4}$

21. $\sin\theta = \frac{12}{13}$, $\cos\theta = \frac{5}{13}$, $\tan\theta = \frac{12}{5}$,

  $\csc\theta = \frac{13}{12}$, $\sec\theta = \frac{13}{5}$, $\cot\theta = \frac{5}{12}$

23. $\sin 60° \approx 0.866$, $\cos 60° \approx 0.5$, $\tan 60° \approx 1.732$,

  $\csc 60° \approx 1.155$, $\sec 60° \approx 2$, $\cot 60° \approx 0.577$

25. $\sin 25° \approx 0.423$, $\cos 25° \approx 0.906$, $\tan 25° \approx 0.466$,

  $\csc 25° \approx 2.366$, $\sec 25° \approx 1.103$, $\cot 25° \approx 2.145$

27. $\sin 5°35' \approx 0.097$, $\cos 5°35' \approx 0.995$,

  $\tan 5°35' \approx 0.098$, $\csc 5°35' \approx 10.278$,

  $\sec 5°35' \approx 1.005$, $\cot 5°35' \approx 10.229$

29. $\sin 13°45'30'' \approx 0.238$, $\cos 13°45'30'' \approx 0.971$,

  $\tan 13°45'30'' \approx 0.245$, $\csc 13°45'30'' \approx 4.205$,

  $\sec 13°45'30'' \approx 1.030$, $\cot 13°45'30'' \approx 4.084$

31. $\sin 1.05° \approx 0.018$, $\cos 1.05° \approx 1.000$,

  $\tan 1.05° \approx 0.018$, $\csc 1.05° \approx 54.570$,

  $\sec 1.05° \approx 1.000$, $\cot 1.05° \approx 54.561$

33. 3 **35.** $\frac{13}{12}$ **37.** $\frac{24}{7}$ **39.** $a \approx 13.86$, $b = 8$

41. $b \approx 5.03$, $c \approx 7.83$ **43.** $a \approx 5.25$, $b \approx 6.04$

45. $a \approx 16.82$, $c \approx 23.28$

47. $a = 12$, $b = 12\sqrt{3}$, $c = 12\sqrt{6}$, $d = 12\sqrt{3}$

49. $a = \frac{14\sqrt{3}}{3}$, $b = \frac{7\sqrt{3}}{3}$, $c = \frac{14\sqrt{3}}{3}$, $d = \frac{14\sqrt{6}}{3}$

51. $a \approx 20.78$ **53.** $c \approx 168.98$

55. (a) $\cos 20° \approx 0.9397$ **(b)** $\sin 50° \approx 0.7660$

57. (a) $\sec 41° \approx 1.3250$ **(b)** $\csc 27° \approx 2.2027$

59. $1500 \tan 37°30' \approx 1151$ ft **61.** 7.8 ft

63. 39.2 ft **65.** 52,000 ft **67.** About 128.2 ft

69. About 194.5 ft **71.** 19,600 ft **73.** 114 ft

75. Barnard's Star: 3.5×10^{13} mi, 5.9 light-years;

Sirius: 5.1×10^{13} mi, 8.6 light-years;

61 Cygni: 6.6×10^{13} mi, 11.1 light-years;

Procyon: 6.7×10^{13} mi, 11.3 light-years

77. Min: 2.9×10^7 mi; max: 4.4×10^7 mi

79. 12,534 mi

81. (a) About 704 ft **(b)** About 595 ft

  (c) Increasing θ decreases r.

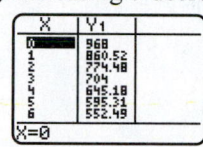

83. $d = 625\left(\frac{1}{\cos 54°} - 1\right) \approx 438$ ft **85.** $A = \frac{\sqrt{3}}{4}s^2$

CHECKING BASIC CONCEPTS FOR SECTIONS 6.1 AND 6.2 (p. 480)

1. (a) $\frac{\pi}{4}$ **(b)** $\frac{5\pi}{12}$ **3.** $s = 2\pi$ in.; $A = 12\pi$ in^2

5. $\sin\theta = \frac{5}{13}$, $\cos\theta = \frac{12}{13}$, $\tan\theta = \frac{5}{12}$, $\csc\theta = \frac{13}{5}$,

  $\sec\theta = \frac{13}{12}$, $\cot\theta = \frac{12}{5}$

SECTION 6.3 (pp. 494–497)

1. (a) 13 **(b)** $\sin\theta = \frac{5}{13}$, $\cos\theta = \frac{12}{13}$

3. (a) 17 **(b)** $\sin\theta = \frac{8}{17}$, $\cos\theta = -\frac{15}{17}$

5. $\sin\theta = \frac{3}{5}$, $\cos\theta = \frac{4}{5}$

7. $\sin\theta = -\frac{2}{\sqrt{5}}$, $\cos\theta = \frac{1}{\sqrt{5}}$

9. $\sin\theta = \frac{2}{\sqrt{5}}$, $\cos\theta = \frac{1}{\sqrt{5}}$

11. $\sin\theta = -\frac{3}{\sqrt{10}}$, $\cos\theta = \frac{1}{\sqrt{10}}$

13. $\sin 45° = \frac{1}{\sqrt{2}}$, $\cos 45° = \frac{1}{\sqrt{2}}$

15. $\sin(-30°) = -\frac{1}{2}$, $\cos(-30°) = \frac{\sqrt{3}}{2}$

17. $\sin 225° = -\frac{1}{\sqrt{2}}$, $\cos 225° = -\frac{1}{\sqrt{2}}$

19. $\sin(-420°) = -\frac{\sqrt{3}}{2}$, $\cos(-420°) = \frac{1}{2}$

21. $\sin\frac{\pi}{3} = \frac{\sqrt{3}}{2}$, $\cos\frac{\pi}{3} = \frac{1}{2}$

23. $\sin\left(-\frac{\pi}{2}\right) = -1$, $\cos\left(-\frac{\pi}{2}\right) = 0$

25. $\sin\frac{7\pi}{6} = -\frac{1}{2}$, $\cos\frac{7\pi}{6} = -\frac{\sqrt{3}}{2}$

27. $\sin\left(-\frac{9\pi}{4}\right) = -\frac{1}{\sqrt{2}}$, $\cos\left(-\frac{9\pi}{4}\right) = \frac{1}{\sqrt{2}}$

29. $\sin 93.2° \approx 0.9984$, $\cos 93.2° \approx -0.0558$

31. $\sin 123°50' \approx 0.8307$, $\cos 123°50' \approx -0.5568$

33. $\sin(-4) \approx 0.7568$, $\cos(-4) \approx -0.6536$

35. $\sin\frac{11\pi}{7} \approx -0.9749$, $\cos\frac{11\pi}{7} \approx 0.2225$

37. $\sin\theta = \frac{3}{5}$, $\cos\theta = \frac{4}{5}$

39. $\sin\theta = -\frac{5}{13}$, $\cos\theta = \frac{12}{13}$

41. $\sin \frac{\pi}{2} = 1$, $\cos \frac{\pi}{2} = 0$

43. $\sin \frac{7\pi}{6} = -\frac{1}{2}$, $\cos \frac{7\pi}{6} = -\frac{\sqrt{3}}{2}$

45. $\sin \left(-\frac{3\pi}{4}\right) = -\frac{1}{\sqrt{2}}$, $\cos \left(-\frac{3\pi}{4}\right) = -\frac{1}{\sqrt{2}}$

47. $\sin \frac{5\pi}{2} = 1$, $\cos \frac{5\pi}{2} = 0$

49. $\sin \left(-\frac{\pi}{3}\right) = -\frac{\sqrt{3}}{2}$, $\cos \left(-\frac{\pi}{3}\right) = \frac{1}{2}$

51. $(-1, 0)$ **53.** $(1, 0)$ **55.** $(0, -1)$ **57.** $(0, -1)$

59. $\left(-\frac{1}{\sqrt{2}}, -\frac{1}{\sqrt{2}}\right)$ **61.** $\left(-\frac{1}{\sqrt{2}}, \frac{1}{\sqrt{2}}\right)$ **63.** $\left(\frac{\sqrt{3}}{2}, -\frac{1}{2}\right)$

65. $\left(\frac{1}{2}, -\frac{\sqrt{3}}{2}\right)$ **67.** $\sin 3\pi = 0$, $\cos 3\pi = -1$

69. $\sin(-2\pi) = 0$, $\cos(-2\pi) = 1$

71. $\sin \frac{3\pi}{2} = -1$, $\cos \frac{3\pi}{2} = 0$

73. $\sin \left(-\frac{5\pi}{2}\right) = -1$, $\cos \left(-\frac{5\pi}{2}\right) = 0$

75. $\sin \frac{5\pi}{4} = -\frac{1}{\sqrt{2}}$, $\cos \frac{5\pi}{4} = -\frac{1}{\sqrt{2}}$

77. $\sin \left(-\frac{5\pi}{4}\right) = \frac{1}{\sqrt{2}}$, $\cos \left(-\frac{5\pi}{4}\right) = -\frac{1}{\sqrt{2}}$

79. $\sin \frac{11\pi}{6} = -\frac{1}{2}$, $\cos \frac{11\pi}{6} = \frac{\sqrt{3}}{2}$

81. $\sin \left(-\frac{7\pi}{3}\right) = -\frac{\sqrt{3}}{2}$, $\cos \left(-\frac{7\pi}{3}\right) = \frac{1}{2}$

83.

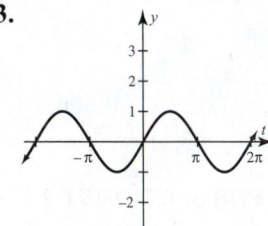

85. (a) $[-2\pi, 2\pi, \pi/2]$ by $[-4, 4, 1]$

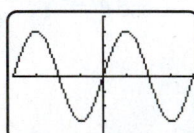

$R = \{y \,|\, -3 \le y \le 3\}$; $f\left(\frac{3\pi}{2}\right) = -3$
(b) $[-2\pi, 2\pi, \pi/2]$ by $[-4, 4, 1]$

$R = \{y \,|\, -1 \le y \le 1\}$; $f\left(\frac{3\pi}{2}\right) = 1$

87. (a) $[-2\pi, 2\pi, \pi/2]$ by $[-4, 4, 1]$

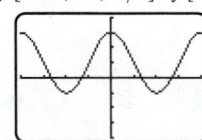

$R = \{y \,|\, -1 \le y \le 3\}$; $f\left(\frac{3\pi}{2}\right) = 1$
(b) $[-2\pi, 2\pi, \pi/2]$ by $[-4, 4, 1]$

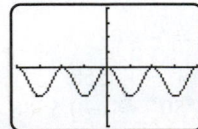

$R = \{y \,|\, -2 \le y \le 0\}$; $f\left(\frac{3\pi}{2}\right) = -2$

89. (a) 3% **(b)** $25{,}000 \left(\dfrac{3}{\sqrt{10{,}009}}\right) \approx 750$ lb

91. First quarter: $\frac{\pi}{2} + 2\pi n$

93. (a) $[0, 24, 2]$ by $[0, 100, 10]$

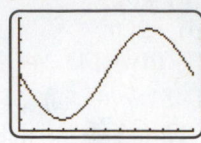

(b) 6:00 P.M. **(c)** 6:00 A.M.
95. (a) V is sinusoidal and varies between -310 volts and 310 volts. **(b)** $V(1/120) = 0$; after $1/120$ second, the voltage is 0. **(c)** $310/\sqrt{2} \approx 219$ volts
97. (a) 201 ft **(b)** 258 ft **(c)** It is easier to stop going uphill $(\theta > 0)$ than downhill $(\theta < 0)$.
99. (a) 31.4 ft **(b)** 78.3 ft **(c)** When the speed limit is higher, more land needs to be cleared on the inside of the curve.
101. (a) $[0, 13, 1]$ by $[3, 8, 1]$ **(b)** $[0, 13, 1]$ by $[3, 8, 1]$

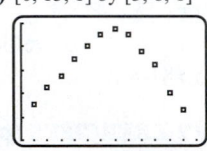 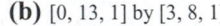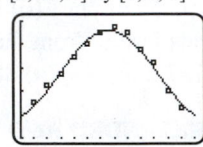

From the close data fit shown, we may conclude that flying squirrels become active near sunset.

6.3 EXTENDED AND DISCOVERY EXERCISES (p. 497)

1. $[-2\pi, 2\pi, \pi/2]$ by $[-2, 2, 1]$

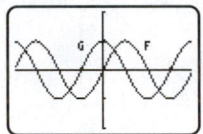

The translated graph and the graph of g are identical.

SECTION 6.4 (pp. 510–513)

1. $\sin \theta = \frac{12}{13}$, $\cos \theta = \frac{5}{13}$, $\tan \theta = \frac{12}{5}$, $\csc \theta = \frac{13}{12}$, $\sec \theta = \frac{13}{5}$, $\cot \theta = \frac{5}{12}$ **3.** $\sin \theta = -\frac{24}{25}$, $\cos \theta = \frac{7}{25}$, $\tan \theta = -\frac{24}{7}$, $\csc \theta = -\frac{25}{24}$, $\sec \theta = \frac{25}{7}$, $\cot \theta = -\frac{7}{24}$

5. $\sin 90° = 1$, $\cos 90° = 0$, $\tan 90°$ is undefined, $\csc 90° = 1$, $\sec 90°$ is undefined, $\cot 90° = 0$

7. $\sin(-45°) = -\frac{1}{\sqrt{2}}$, $\cos(-45°) = \frac{1}{\sqrt{2}}$, $\tan(-45°) = -1$, $\csc(-45°) = -\sqrt{2}$, $\sec(-45°) = \sqrt{2}$, $\cot(-45°) = -1$

9. $\sin \pi = 0$, $\cos \pi = -1$, $\tan \pi = 0$, $\csc \pi$ is undefined, $\sec \pi = -1$, $\cot \pi$ is undefined

11. $\sin \left(-\frac{\pi}{3}\right) = -\frac{\sqrt{3}}{2}$, $\cos \left(-\frac{\pi}{3}\right) = \frac{1}{2}$, $\tan \left(-\frac{\pi}{3}\right) = -\sqrt{3}$, $\csc \left(-\frac{\pi}{3}\right) = -\frac{2}{\sqrt{3}}$, $\sec \left(-\frac{\pi}{3}\right) = 2$, $\cot \left(-\frac{\pi}{3}\right) = -\frac{1}{\sqrt{3}}$

13. $\sin \left(-\frac{\pi}{2}\right) = -1$, $\cos \left(-\frac{\pi}{2}\right) = 0$, $\tan \left(-\frac{\pi}{2}\right)$ is undefined, $\csc \left(-\frac{\pi}{2}\right) = -1$, $\sec \left(-\frac{\pi}{2}\right)$ is undefined, $\cot \left(-\frac{\pi}{2}\right) = 0$

15. $\sin 360° = 0$, $\cos 360° = 1$, $\tan 360° = 0$, $\csc 360°$ is undefined, $\sec 360° = 1$, $\cot 360°$ is undefined

17. $\sin \frac{\pi}{6} = \frac{1}{2}$, $\cos \frac{\pi}{6} = \frac{\sqrt{3}}{2}$, $\tan \frac{\pi}{6} = \frac{1}{\sqrt{3}}$, $\csc \frac{\pi}{6} = 2$, $\sec \frac{\pi}{6} = \frac{2}{\sqrt{3}}$, $\cot \frac{\pi}{6} = \sqrt{3}$

19. $\sin \frac{4\pi}{3} = -\frac{\sqrt{3}}{2}$, $\cos \frac{4\pi}{3} = -\frac{1}{2}$, $\tan \frac{4\pi}{3} = \sqrt{3}$, $\csc \frac{4\pi}{3} = -\frac{2}{\sqrt{3}}$, $\sec \frac{4\pi}{3} = -2$, $\cot \frac{4\pi}{3} = \frac{1}{\sqrt{3}}$

21. $\sin(-225°) = \frac{1}{\sqrt{2}}$, $\cos(-225°) = -\frac{1}{\sqrt{2}}$, $\tan(-225°) = -1$, $\csc(-225°) = \sqrt{2}$, $\sec(-225°) = -\sqrt{2}$, $\cot(-225°) = -1$

23. $\sin\left(-\frac{13\pi}{6}\right) = -\frac{1}{2}$, $\cos\left(-\frac{13\pi}{6}\right) = \frac{\sqrt{3}}{2}$, $\tan\left(-\frac{13\pi}{6}\right) = -\frac{1}{\sqrt{3}}$, $\csc\left(-\frac{13\pi}{6}\right) = -2$, $\sec\left(-\frac{13\pi}{6}\right) = \frac{2}{\sqrt{3}}$, $\cot\left(-\frac{13\pi}{6}\right) = -\sqrt{3}$

25. $\sin\theta = \frac{4}{\sqrt{17}}$, $\cos\theta = -\frac{1}{\sqrt{17}}$, $\tan\theta = -4$, $\csc\theta = \frac{\sqrt{17}}{4}$, $\sec\theta = -\sqrt{17}$, $\cot\theta = -\frac{1}{4}$
The slope of the line equals $\tan\theta$.

27. $\sin\theta = -\frac{6}{\sqrt{37}}$, $\cos\theta = -\frac{1}{\sqrt{37}}$, $\tan\theta = 6$, $\csc\theta = -\frac{\sqrt{37}}{6}$, $\sec\theta = -\sqrt{37}$, $\cot\theta = \frac{1}{6}$
The slope of the line equals $\tan\theta$.

29. $\tan\theta = \frac{3}{4}$, $\cot\theta = \frac{4}{3}$, $\csc\theta = \frac{5}{3}$, $\sec\theta = \frac{5}{4}$

31. $\sin\theta = -\frac{15}{17}$, $\cos\theta = -\frac{8}{17}$, $\tan\theta = \frac{15}{8}$, $\cot\theta = \frac{8}{15}$

33. $\sin\theta = \frac{5}{13}$, $\csc\theta = \frac{13}{5}$, $\cot\theta = \frac{12}{5}$, $\sec\theta = \frac{13}{12}$

35. $\cos\theta = \frac{4}{5}$, $\tan\theta = -\frac{3}{4}$, $\csc\theta = -\frac{5}{3}$, $\sec\theta = \frac{5}{4}$, $\cot\theta = -\frac{4}{3}$

37. $\sin\theta = -\frac{3}{5}$, $\tan\theta = \frac{3}{4}$, $\csc\theta = -\frac{5}{3}$, $\sec\theta = -\frac{5}{4}$, $\cot\theta = \frac{4}{3}$

39. $\sin(-7\pi) = 0$, $\cos(-7\pi) = -1$, $\tan(-7\pi) = 0$, $\csc(-7\pi)$ is undefined, $\sec(-7\pi) = -1$, $\cot(-7\pi)$ is undefined

41. $\sin\frac{7\pi}{2} = -1$, $\cos\frac{7\pi}{2} = 0$, $\tan\frac{7\pi}{2}$ is undefined, $\csc\frac{7\pi}{2} = -1$, $\sec\frac{7\pi}{2}$ is undefined, $\cot\frac{7\pi}{2} = 0$

43. $\sin\left(-\frac{3\pi}{4}\right) = -\frac{1}{\sqrt{2}}$, $\cos\left(-\frac{3\pi}{4}\right) = -\frac{1}{\sqrt{2}}$, $\tan\left(-\frac{3\pi}{4}\right) = 1$, $\csc\left(-\frac{3\pi}{4}\right) = -\sqrt{2}$, $\sec\left(-\frac{3\pi}{4}\right) = -\sqrt{2}$, $\cot\left(-\frac{3\pi}{4}\right) = 1$

45. $\sin\frac{7\pi}{6} = -\frac{1}{2}$, $\cos\frac{7\pi}{6} = -\frac{\sqrt{3}}{2}$, $\tan\frac{7\pi}{6} = \frac{1}{\sqrt{3}}$, $\csc\frac{7\pi}{6} = -2$, $\sec\frac{7\pi}{6} = -\frac{2}{\sqrt{3}}$, $\cot\frac{7\pi}{6} = \sqrt{3}$

47. $\sin\theta = \frac{1}{\sqrt{2}}$, $\cos\theta = \frac{1}{\sqrt{2}}$, $\tan\theta = 1$, $\csc\theta = \sqrt{2}$, $\sec\theta = \sqrt{2}$, $\cot\theta = 1$

49. $\sin\theta = -\frac{12}{13}$, $\cos\theta = \frac{5}{13}$, $\tan\theta = -\frac{12}{5}$, $\csc\theta = -\frac{13}{12}$, $\sec\theta = \frac{13}{5}$, $\cot\theta = -\frac{5}{12}$

51. -1 **53.** 2 **55.** $\frac{2}{\sqrt{3}}$

57. $\sin\frac{9\pi}{2} = 1$, $\cos\frac{9\pi}{2} = 0$, $\tan\frac{9\pi}{2}$ is undefined, $\csc\frac{9\pi}{2} = 1$, $\sec\frac{9\pi}{2}$ is undefined, $\cot\frac{9\pi}{2} = 0$

59. $\sin(-\pi) = 0$, $\cos(-\pi) = -1$, $\tan(-\pi) = 0$, $\csc(-\pi)$ is undefined, $\sec(-\pi) = -1$, $\cot(-\pi)$ is undefined

61. $\sin\frac{4\pi}{3} = -\frac{\sqrt{3}}{2}$, $\cos\frac{4\pi}{3} = -\frac{1}{2}$, $\tan\frac{4\pi}{3} = \sqrt{3}$, $\csc\frac{4\pi}{3} = -\frac{2}{\sqrt{3}}$, $\sec\frac{4\pi}{3} = -2$, $\cot\frac{4\pi}{3} = \frac{1}{\sqrt{3}}$

63. (a) 0.9984 (b) 1.0016 **65.** (a) 1.4045 (b) 0.7120
67. (a) -0.8637 (b) -1.1578
69. (a) 0.2225 (b) 4.4940
71. 0.900 **73.** 0.527

75. Origin symmetry

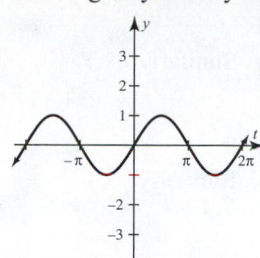

77. Origin symmetry

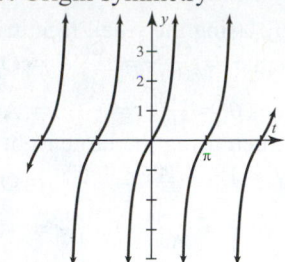

79. y-axis symmetry

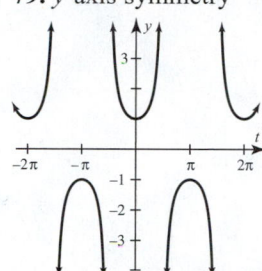

81. D = all real numbers, $R = \{y \mid -1 \le y \le 1\}$, period $= 2\pi$

83. $D = \left\{t \mid t \ne \pm\frac{\pi}{2}, \pm\frac{3\pi}{2}, \pm\frac{5\pi}{2}, \ldots\right\}$, R = all real numbers, period $= \pi$

85. $D = \left\{t \mid t \ne \pm\frac{\pi}{2}, \pm\frac{3\pi}{2}, \pm\frac{5\pi}{2}, \ldots\right\}$, $R = \{y \mid |y| \ge 1\}$, period $= 2\pi$

87. (a)

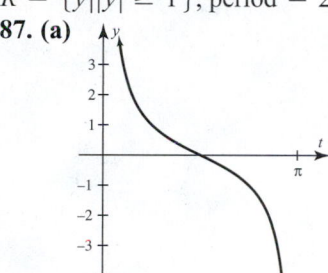

(b) Just after sunrise ($t = 0$), the shadow is very long. As the elevation of the sun increases, the shadow decreases in length until it is 0 when $t = \pi/2$. In the afternoon, the shadow increases in length in the opposite direction until sunset ($t = \pi$).

89. About 1300 ft **91.** About $17.95''$; $\theta = \frac{57.3\sin\alpha}{\cos\alpha}$

93. About 41%

95. As θ increases, the values of y_1 and y_2 become closer together. When $\theta = 20°$, the difference is only about 0.02, or 2%.

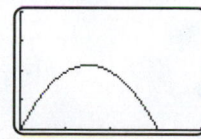

97. [0, 20000, 5000] by [0, 4000, 1000]

(a) About (7612, 2197) (b) About 15,223 ft

6.4 EXTENDED AND DISCOVERY EXERCISE (p. 513)

1. Using the small right triangle yields
$\sin \theta = \frac{\text{Opp.}}{\text{Hyp.}} = \frac{\text{Opp.}}{1} =$ Opposite side. Similarly,

$\cos \theta = \frac{\text{Adj.}}{\text{Hyp.}} = \frac{\text{Adj.}}{1} =$ Adjacent side.
Then using the large right triangle yields
$\tan \theta = \frac{\text{Opp.}}{\text{Adj.}} = \frac{\text{Opp.}}{1} =$ Opposite side. Similarly,

$\sec \theta = \frac{\text{Hyp.}}{\text{Adj.}} = \frac{\text{Hyp.}}{1} =$ Hypotenuse.

CHECKING BASIC CONCEPTS FOR SECTIONS 6.3 AND 6.4 (p. 513)

1. $\sin \theta = \frac{6}{\sqrt{85}}$, $\cos \theta = -\frac{7}{\sqrt{85}}$, $\tan \theta = -\frac{6}{7}$,

$\csc \theta = \frac{\sqrt{85}}{6}$, $\sec \theta = -\frac{\sqrt{85}}{7}$, $\cot \theta = -\frac{7}{6}$

3. Sine   Cosine

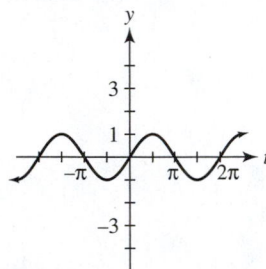 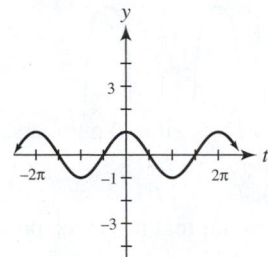

Tangent

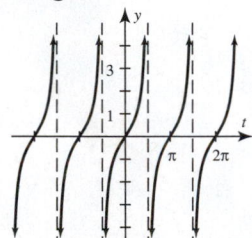

SECTION 6.5 (pp. 524–529)

1. Period: 4π;
amplitude: 3

3. Period: $\frac{2\pi}{3}$;
amplitude: 2

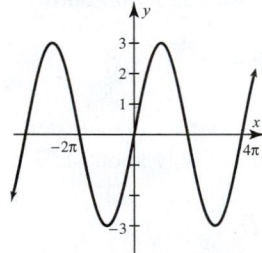 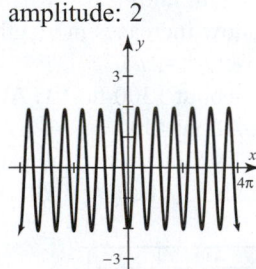

5. Period: 2; amplitude: $\frac{1}{2}$

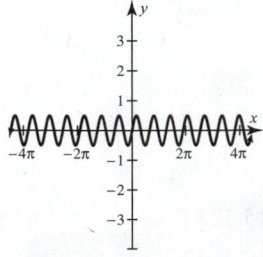

7. Shorten the period of the sine graph to π, increase the amplitude to 3, and shift the graph downward 1 unit.

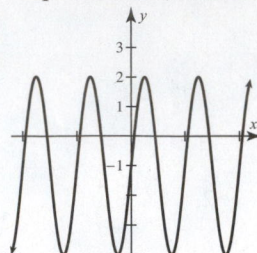

9. Shift the cosine graph $\frac{\pi}{2}$ units to the left, increase the amplitude to 2, and reflect the graph across the x-axis.

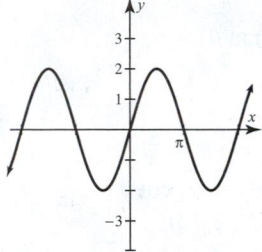

11. Shift the cosine graph $\frac{1}{\pi}$ unit to the right, shorten the period to 2, and decrease the amplitude to $\frac{1}{2}$.

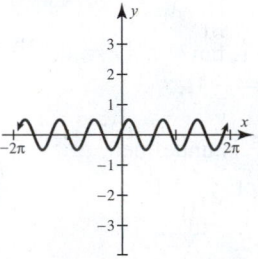

13. Amplitude: 3; period: $\frac{\pi}{2}$; phase shift: $\frac{\pi}{4}$; vertical shift: -4
15. Amplitude: 4; period: 4; phase shift: 1; vertical shift: 6
17. Amplitude: 20; period: 3π; phase shift: $-\frac{3\pi}{2}$; vertical shift: 2
19. $a = 3$, $b = 2$
21. Amplitude: 3; period: 4π; phase shift: 0
23. c   **25.** d   **27.** a   **29.** $y = 3 \sin \left(\frac{1}{2} x \right)$
31. Amplitude: 2;   **33.** Amplitude: 1;
period: 2π;   period: 4π;
phase shift: 0   phase shift: 0

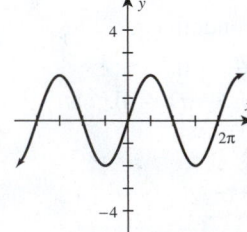 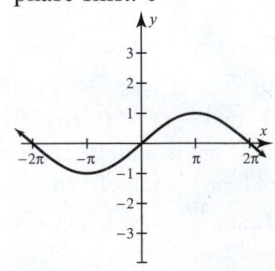

35. Amplitude: 1;
period: 2π;
phase shift: 0

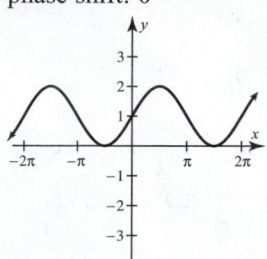

37. Amplitude: 1;
period: 2;
phase shift: 0

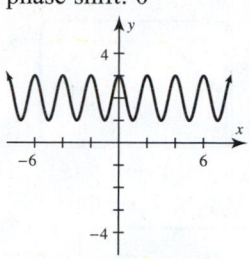

53. Amplitude = 2.5;
period = π;
phase shift = $-\frac{\pi}{4}$
$[-2\pi, 2\pi, \pi/2]$ by $[-4, 4, 1]$

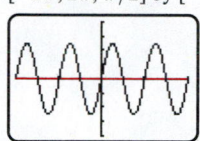

55. Amplitude = 2;
period = 1;
phase shift = $-\frac{1}{8}$
$[-2\pi, 2\pi, \pi/2]$ by $[-4, 4, 1]$

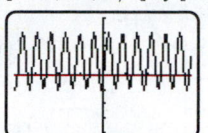

57. Period = $\frac{\pi}{2}$; phase shift = 0
Asymptotes: $x = \pm\frac{\pi}{4}, \pm\frac{3\pi}{4}, \pm\frac{5\pi}{4}, \pm\frac{7\pi}{4}$

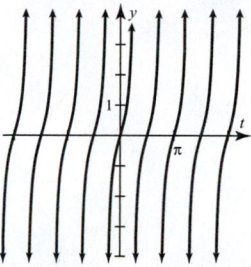

39. Amplitude: 1;
period: π;
phase shift: $-\pi$

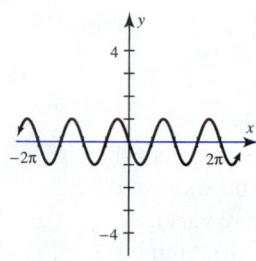

41. Amplitude: 1;
period: 2π;
phase shift: $\frac{\pi}{2}$

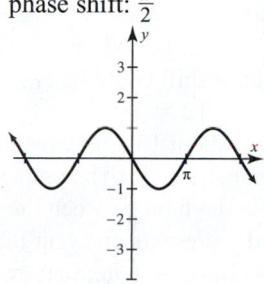

59. Period = π; phase shift = $\frac{\pi}{2}$
Asymptotes: $x = 0, \pm\pi, \pm2\pi$

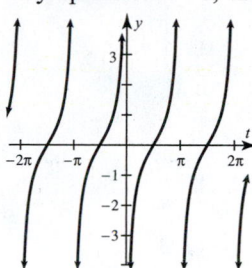

43. Amplitude: $\frac{1}{2}$;
period: π;
phase shift: 0

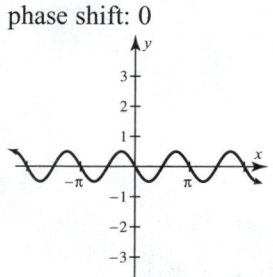

45. Amplitude: 2;
period: π;
phase shift: $-\frac{\pi}{4}$

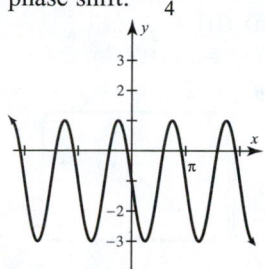

61. Period = $\frac{\pi}{2}$; phase shift = 0
Asymptotes: $x = 0, \pm\frac{\pi}{2}, \pm\pi, \pm\frac{3\pi}{2}, \pm2\pi$

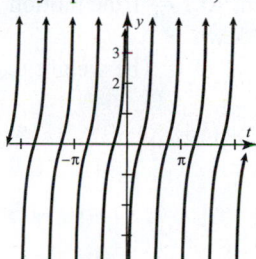

47. Amplitude: 1; period: π; phase shift: $\frac{\pi}{2}$

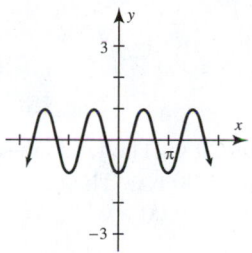

63. Period = $\frac{\pi}{2}$; phase shift = $\frac{\pi}{4}$
Asymptotes: $x = \pm\frac{\pi}{4}, \pm\frac{3\pi}{4}, \pm\frac{5\pi}{4}, \pm\frac{7\pi}{4}$

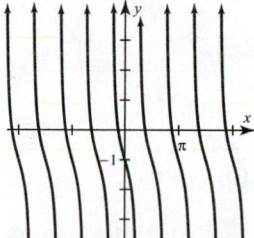

49. Amplitude = 2;
period = π;
phase shift = 0
$[-2\pi, 2\pi, \pi/2]$ by $[-4, 4, 1]$

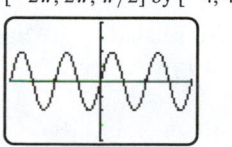

51. Amplitude = $\frac{1}{2}$;
period = $\frac{2\pi}{3}$;
phase shift = $-\frac{\pi}{3}$
$[-2\pi, 2\pi, \pi/2]$ by $[-4, 4, 1]$

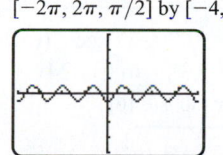

65. Period $= 4\pi$; phase shift $= 0$
Asymptotes: $x = \pm\pi$

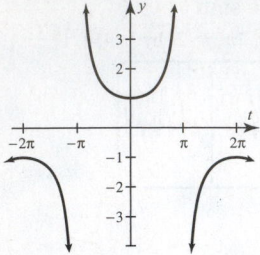

67. Period $= 2\pi$; phase shift $= \pi$
Asymptotes: $x = 0, \pm\pi, \pm2\pi$

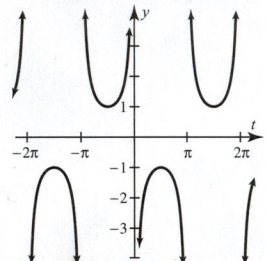

69. Period $= 6\pi$; phase shift $= \frac{\pi}{6}$
Asymptotes: $x = -\frac{4\pi}{3}, \frac{5\pi}{3}$

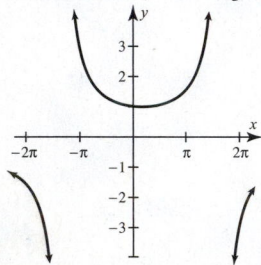

71. (a) $s(t) = 2\cos(4\pi t)$ **(b)** $s(1) = 2$. The weight is moving neither upward nor downward. At $t = 1$ the motion of the weight is changing from up to down.
73. (a) $s(t) = -3\cos(2.5\pi t)$ **(b)** $s(1) = 0$. The weight is moving upward.
75. $a = 0.21$, $b = 55\pi$; $Y_1 = 0.21\cos(55\pi X)$
$[0, 0.05, 0.01]$ by $[-0.3, 0.3, 0.1]$

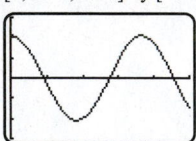

77. $a = 0.14$, $b = 110\pi$; $Y_1 = 0.14\cos(110\pi X)$
$[0, 0.05, 0.01]$ by $[-0.3, 0.3, 0.1]$

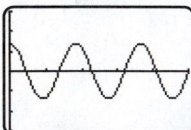

79. (a) $P(x) = 40\sin\left(\frac{\pi}{12}x\right) + 50$ **(b)** Noon
(c) Midnight
81. (a) $40°F$, $-40°F$ **(b)** Amplitude $= 40$, half the difference between the maximum and minimum monthly average temperatures; period $= 12$ means that the temperature pattern repeats every 12 months. **(c)** The months when the average temperature is $0°F$

83. (a) Amplitude $= 34$; period $= 12$; phase shift $= 4.3$
(b) $f(5) \approx 12.2°F$; $f(12) \approx -26.4°F$ **(c)** About $0°F$
85. (a) $[0, 25, 2]$ by $[0, 80, 10]$

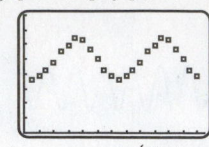

(b) $f(x) = 14\sin\left(\frac{\pi}{6}(x - 4)\right) + 50$
(c) $[0, 25, 2]$ by $[0, 80, 10]$

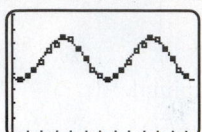

87. (a) $f(x) = 17\cos\left(\frac{\pi}{6}(x - 7)\right) + 75$
(b) Yes. The period of the graph is 12, so, for example, the phase shift could be $c = 7 + 12 = 19$ or $c = 7 - 12 = -5$.
89. (a) About 18.5 hr; June 21 **(b)** About 6 hr; December 21 **(c)** The amplitude represents half the difference in daylight between the longest and shortest day; the period represents one year (answers may vary).
91. (a) Max ≈ 8 in.; min ≈ 0.5 in. **(b)** Amp ≈ 3.75. The amplitude represents half the difference between the maximum and minimum monthly average precipitations.
(c) $f(x) = 3.75\cos\left(\frac{\pi}{6}x\right) + 4.25$
93. (a) $f(t) = 2\cos\left(\frac{\pi}{6}(t - 1)\right) + 4$
(b) $f(1) = 6$, $f(7) = 2$
95. (a) $[0, 25, 2]$ by $[60, 90, 5]$

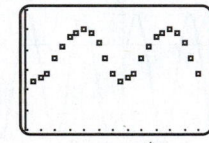

(b) $f(x) = 6.5\sin\left(\frac{\pi}{6}(x - 4)\right) + 78.5$
97. $[0, 12, 3]$ by $[0, 50, 10]$

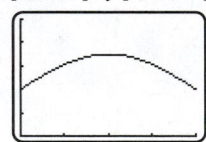

At 9:00 A.M. the outdoor temperature is $20°C$. The temperature increases to a maximum of $35°C$ at 3:00 P.M. Then it begins to fall until it reaches $20°C$ again at 9:00 P.M.
99. (a) $[0, 1/100, 1/880]$ by $[-1.5, 1.5, 0.5]$

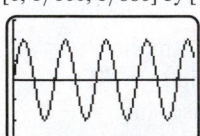

(b) $\frac{1}{440} \approx 0.00227$ sec **(c)** 440 cycles per second
101. $y \approx 13.2\sin(0.524x - 2.18) + 49.7$
$[0, 25, 2]$ by $[30, 80, 10]$

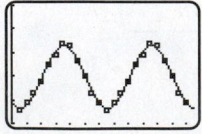

103. $y \approx 16.9 \sin(0.522x - 2.09) + 75.4$
$[0, 25, 2]$ by $[50, 100, 10]$

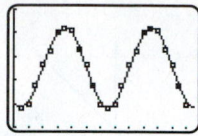

SECTION 6.6 (pp. 540–543)

1. one-to-one **3.** π **5.** 0 **7.** $\frac{\pi}{2}$ **9.** $-\frac{\pi}{4}$ **11.** $\frac{2\pi}{3}$
13. (a) $\frac{\pi}{2}$, or $90°$ (b) Undefined (c) $-\frac{\pi}{3}$, or $-60°$
15. (a) $\frac{\pi}{2}$, or $90°$ (b) π, or $180°$ (c) $\frac{\pi}{3}$, or $60°$
17. (a) $\frac{\pi}{4}$, or $45°$ (b) $-\frac{\pi}{4}$, or $-45°$ (c) $\frac{\pi}{3}$, or $60°$
19. (a) Undefined (b) 1.47, or $84.3°$ (c) 1.82, or $104.5°$
21. (a) β (b) α (c) β **23.** 1 **25.** $\frac{3\pi}{4}$ **27.** -3
29. (a) $90°$ (b) $90°$ (c) $90°$; $\sin^{-1} x + \cos^{-1} x = 90°$
whenever $-1 \le x \le 1$.
31. *Verbal:* Determine the angle (or real number) θ such that $\sin \theta = 2x$ and $-\frac{\pi}{2} \le \theta \le \frac{\pi}{2}$.
Numerical:

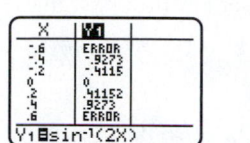

Graphical:
$[-1, 1, 1]$ by $[-2, 2, 1]$

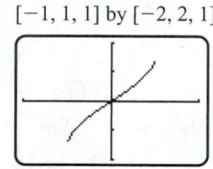

33. *Verbal:* Determine the angle (or real number) θ such that $\cos \theta = \frac{1}{2}x$ and $0 \le \theta \le \pi$.
Numerical:

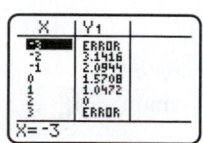

Graphical:
$[-3, 3, 1]$ by $[-1, 4, 1]$

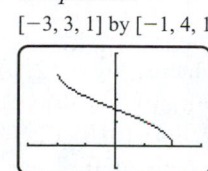

35. $\tan \theta = \frac{x}{\sqrt{1 - x^2}}$ **37.** $\cos \theta = \frac{1}{\sqrt{1 + x^2}}$
39. $\sqrt{1 - u^2}$ **41.** $\frac{\sqrt{1 - u^2}}{u}$ **43.** u
45. $\alpha = \tan^{-1} \frac{7}{24} \approx 16.3°$, $\beta = \tan^{-1} \frac{24}{7} \approx 73.7°$, $c = 25$
47. $\alpha = \sin^{-1} \frac{6}{10} \approx 36.9°$, $\beta = \cos^{-1} \frac{6}{10} \approx 53.1°$, $b = 8$
49. $\beta = 35°$, $b = \frac{5}{\tan 55°} \approx 3.5$, $c = \frac{5}{\sin 55°} \approx 6.1$
51. $90°$ **53.** $45°$ **55.** $90°$ **57.** $82.8°$ **59.** $80.5°$
61. $68.9°$ **63.** -0.197 **65.** 1.102 **67.** -0.282
69. 1.369, 1.772 **71.** π **73.** $-\frac{\pi}{4}$ **75.** $\frac{3\pi}{4}$ **77.** 1.231
79. 0.197 **81.** 0.588 **83.** $\tan^{-1} \frac{50}{85} \approx 30.5°$
85. (a) $4.6°$ (b) $2.1°$ (c) $-2.9°$ **87.** $\tan^{-1} \frac{4}{10} \approx 21.8°$
89. (a) $\tan^{-1} \frac{3}{12} \approx 14.0°$ (b) $\tan^{-1} \frac{4}{12} \approx 18.4°$
(c) $\tan^{-1} \frac{6}{12} \approx 26.6°$ (d) $45°$
91. $\theta \approx 41.9°$ **93.** (a) 14.9 hr (b) 13.8 hr (c) 14.6 hr
95. $\theta_1 \approx 35.8°$, $\theta_2 \approx 54.2°$
97. Let α and β represent the angles of elevation from the shrub to the shorter and taller buildings, respectively. The distance from the shrub to the shorter building is $100 - x$; thus $\alpha = \arctan \frac{75}{100 - x}$. Similarly, $\beta = \arctan \frac{150}{x}$. Because the angles α, θ, and β form a straight angle,

$\theta = \pi - \alpha - \beta$. That is,
$\theta = \pi - \arctan \frac{75}{100 - x} - \arctan \frac{150}{x}$.

6.6 EXTENDED AND DISCOVERY EXERCISE (p. 543)

1. The maximum value of θ is $38.7°$ when $x \approx 6.24$.
$[0, 50, 10]$ by $[0, 50, 10]$

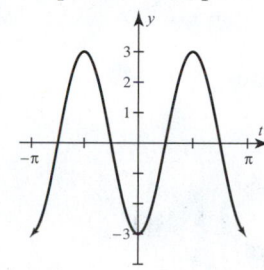

CHECKING BASIC CONCEPTS FOR SECTIONS 6.5 AND 6.6 (p. 543)

1. Amplitude $= 3$; period $= \frac{2\pi}{2} = \pi$; phase shift $= \frac{\pi}{4}$

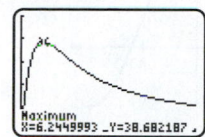

3. (a) $\sin^{-1} 0 = 0°$ (b) $\cos^{-1}(-1) = 180°$
(c) $\tan^{-1}(-1) = -45°$ (d) $\sin^{-1} \frac{1}{2} = 30°$
(e) $\tan^{-1} \sqrt{3} = 60°$ (f) $\cos^{-1} \frac{1}{2} = 60°$
5. (a) $\sin^{-1} 0.55 \approx 0.582$ (b) $\cos^{-1} (-0.35) \approx 1.93$
(c) $\tan^{-1} (-2.9) \approx -1.24$

CHAPTER 6 REVIEW EXERCISES (pp. 547–549)

1. (a) (b)

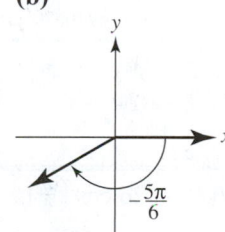

3. (a) $60°$ (b) $5°$ (c) $-150°$ (d) $-315°$
5. 2π ft **7.** $\frac{1}{2}$ **9.** $\frac{1}{\sqrt{3}}$ **11.** $\sqrt{2}$
13. $\sin \theta = \frac{8}{\sqrt{145}}$, $\cos \theta = \frac{9}{\sqrt{145}}$, $\tan \theta = \frac{8}{9}$,
$\csc \theta = \frac{\sqrt{145}}{8}$, $\sec \theta = \frac{\sqrt{145}}{9}$, $\cot \theta = \frac{9}{8}$
15. $\csc \theta = 3$
17. $\sin 25° \approx 0.423$, $\cos 25° \approx 0.906$, $\tan 25° \approx 0.466$, $\csc 25° \approx 2.366$, $\sec 25° \approx 1.103$, $\cot 25° \approx 2.145$
19. $\sin \theta = -\frac{2}{\sqrt{5}}$, $\cos \theta = \frac{1}{\sqrt{5}}$, $\tan \theta = -2$,
$\csc \theta = -\frac{\sqrt{5}}{2}$, $\sec \theta = \sqrt{5}$, $\cot \theta = -\frac{1}{2}$
21. $\sin \theta = \frac{\sqrt{3}}{2}$, $\cos \theta = -\frac{1}{2}$, $\tan \theta = -\sqrt{3}$,
$\csc \theta = \frac{2}{\sqrt{3}}$, $\sec \theta = -2$, $\cot \theta = -\frac{1}{\sqrt{3}}$
23. -1 **25.** 0
27. $\tan \theta = -\frac{4}{3}$, $\csc \theta = -\frac{5}{4}$, $\sec \theta = \frac{5}{3}$, $\cot \theta = -\frac{3}{4}$

29. Amplitude = 3;
period = π;
phase shift = 0

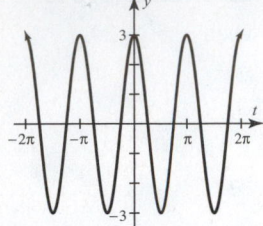

31. Amplitude = 3;
period = 4π;
phase shift = π

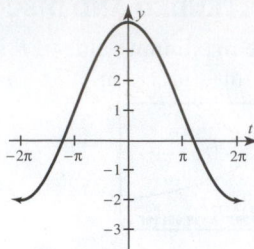

33. $a = 2, b = 3$
35. Period = $\frac{\pi}{2}$; phase shift = 0

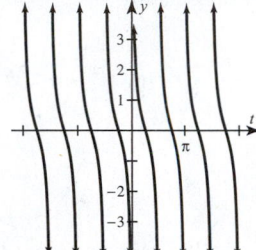

37. (a) $-\frac{\pi}{2}$, or $-90°$ **(b)** $\frac{\pi}{3}$, or $60°$ **(c)** $\frac{\pi}{4}$, or $45°$
39. (a) -0.64, or $-36.9°$ **(b)** 1.37, or $78.7°$
(c) 1.45, or $83.1°$
41. $\alpha = \tan^{-1}\frac{5}{3} \approx 59.0°$, $\beta = \tan^{-1}\frac{3}{5} \approx 31.0°$, $c = \sqrt{34}$
43. $30°$ **45.** $78.5°$ **47.** -0.6435
49. (a) $\frac{\pi}{25}$ radian/sec **(b)** $\pi \approx 3.14$ ft/sec
51. 89 ft **53. (a)** $\sin^{-1}\frac{350}{6000} \approx 3.3°$ **(b)** $\sin^{-1}\frac{160}{4500} \approx 2.0°$
55. 161 mi
57. (a) [0, 25, 2] by [0, 70, 10] **(c)** [0, 25, 2] by [0, 70, 10]

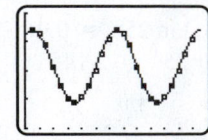

(b) $f(x) = 26\cos\left(\frac{\pi}{6}(x-7)\right) + 32$

CHAPTER 6 EXTENDED AND DISCOVERY EXERCISES (p. 549)

1. (a) $x_Q = d\sin\theta + x_p$ and $y_Q = d\cos\theta + y_p$
(b) Approximately $(233.9, 377.2)$
3. (a) $f(x) = 12.5\cos\left(\frac{\pi}{6}(x-1)\right) + 61.5$
(b) [0, 25, 2] by [40, 80, 10]

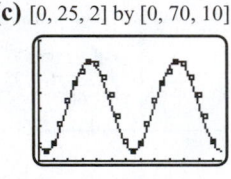

Answers may vary.
(c) The high temperatures occur in January in the Southern Hemisphere as opposed to July in the Northern Hemisphere. This affects the phase shift of f.

CHAPTERS 1–6 CUMULATIVE REVIEW EXERCISES (pp. 550–552)

1. 1.25×10^5; 0.00467
3. (a) **(b)**

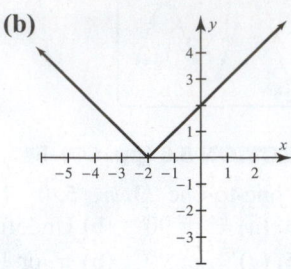

(c) **(d)**

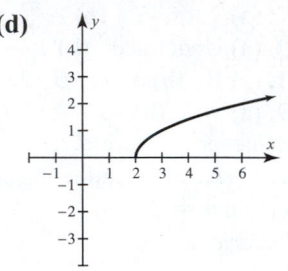

5. (a) $D = \{x \mid x \neq 2, x \neq -2\}$
(b) $f(-1) = \frac{1}{4}$; $f(2a) = \frac{1}{8(a+1)}$
7. 90 **9.** $y = -\frac{2}{3}x - \frac{5}{3}$
11. (a) $-\frac{4}{5}, \frac{12}{5}$ **(b)** $\ln 14 \approx 2.64$ **(c)** $0, 1, 2$
(d) 5 **(e)** $-1, 2$ **(f)** 65,535
13. (a) $\left(-\infty, \frac{9}{5}\right)$ **(b)** $\left(-\frac{2}{3}, \frac{7}{3}\right]$ **(c)** $\left(-\infty, -\frac{3}{2}\right] \cup [3, \infty)$
(d) $[1, 4]$ **(e)** $(-1, 0) \cup (1, \infty)$
15. $\frac{-2 \pm \sqrt{6}}{2}$
17. (a) Increasing: $(-2, 0), (2, \infty)$; decreasing: $(-\infty, -2), (0, 2)$ **(b)** $-2.8, 0, 2.8$ (approximate)
(c) $(-2, -4), (0, 0), (2, -4)$ **(d)** Local minimum: -4; local maximum: 0
19. (a) $\frac{5a^2}{2} - 1 + \frac{2}{a^2}$ **(b)** $x - 3 + \frac{4}{x^2+1}$
21. $D = \{x \mid x \neq \frac{7}{3}\}$; vertical: $x = \frac{7}{3}$; horizontal: $y = \frac{2}{3}$
23. (a) 5 **(b)** Undefined **(c)** 2 **(d)** 0
25. (a) 8 **(b)** 0 **(c)** $(f-g)(x) = x^2 + 2x$
(d) $(f \circ g)(x) = x^2 - x - 4$ **27.** $C = \frac{1}{2}, a = 2$
29. (a) 2 **(b)** 4 **(c)** -2 **(d)** 2 **31.** 4.395
33. $\frac{5\pi}{6}$ **35.** $\frac{\pi}{4} \approx 0.79$ ft **37.** $\frac{9}{\tan 30°} \approx 15.6$
39. $\sin\theta = -\frac{1}{\sqrt{2}}$, $\cos\theta = -\frac{1}{\sqrt{2}}$, $\tan\theta = 1$,
$\csc\theta = -\sqrt{2}$, $\sec\theta = -\sqrt{2}$, $\cot\theta = 1$
41. $\sin\theta = -\frac{12}{13}$, $\cos\theta = \frac{5}{13}$, $\tan\theta = -\frac{12}{5}$,
$\csc\theta = -\frac{13}{12}$, $\sec\theta = \frac{13}{5}$, $\cot\theta = -\frac{5}{12}$
43. (a) **(b)**

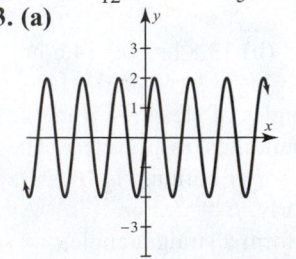
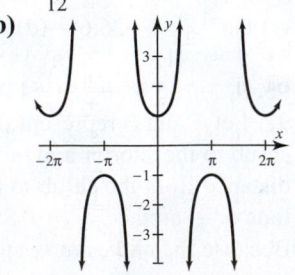

(c)

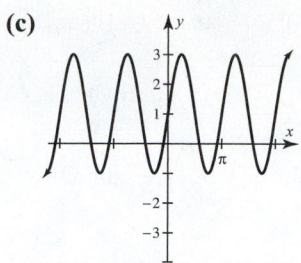

45. $\alpha \approx 33.4°; \beta \approx 56.6°; b \approx 16.7$ **47.** 24 lb

49. (a) $f^{-1}(x) = \frac{9}{5}x + 32$

(b) f^{-1} converts degrees Celsius to degrees Fahrenheit.

51. (a) $N(t) = 200,000e^{0.135t}$

(b) $N(5) \approx 392,807$; after 5 hours, there are about 393,000 bacteria per mL.   **(c)** About 6.8 hours

53. $f(x) \approx 14.5 \sin\left(\frac{\pi}{6}(x - 4)\right) + 75.5$

CHAPTER 7: Trigonometric Identities and Equations

SECTION 7.1 (pp. 563–565)

1. $\cot\theta = 2$ **3.** $\sec\theta = \frac{7}{2}$ **5.** $\cos\theta = -\frac{1}{4}$

7. Undefined   **9.** $\frac{\sin^2\theta}{\cos^2\theta}$ **11.** $\cos^2\theta$ **13.** $\sin\theta + \cos\theta$

15. $\tan\theta = -\frac{3}{4}, \csc\theta = \frac{5}{3}, \sec\theta = -\frac{5}{4}, \cot\theta = -\frac{4}{3}$

17. $\cos\theta = -\frac{7}{25}, \tan\theta = \frac{24}{7}, \csc\theta = -\frac{25}{24}, \sec\theta = -\frac{25}{7}$

19. $\tan\theta = \frac{60}{11}, \cot\theta = \frac{11}{60}, \csc\theta = -\frac{61}{60}, \sec\theta = -\frac{61}{11}$

21. $\sin\theta = \frac{1}{\sqrt{2}}, \cos\theta = -\frac{1}{\sqrt{2}}, \tan\theta = -1, \cot\theta = -1$

23. 1   **25.** 1   **27.** 1   **29.** $\cos^2\theta$ **31.** $\tan^2\theta$

33. $-\tan\theta$ **35.** $\sec\theta$ **37.** $\tan^2\theta$ **39.** $\cos x$ **41.** 1

43. $\sec x - \csc x$ **45.** Yes   **47.** No   **49.** Quadrant IV

51. Quadrant III   **53.** Quadrant II

55. $\sin\theta = -\frac{\sqrt{3}}{2}, \tan\theta = -\sqrt{3}, \csc\theta = -\frac{2}{\sqrt{3}}, \sec\theta = 2, \cot\theta = -\frac{1}{\sqrt{3}}$

57. $\sin\theta = -\frac{11}{61}, \cos\theta = \frac{60}{61}, \cot\theta = -\frac{60}{11}, \csc\theta = -\frac{61}{11}, \sec\theta = \frac{61}{60}$

59. $\cos\theta = \frac{24}{25}, \tan\theta = \frac{7}{24}, \cot\theta = \frac{24}{7}, \csc\theta = \frac{25}{7}, \sec\theta = \frac{25}{24}$

61. $\cos\theta = -\frac{\sqrt{8}}{3}, \tan\theta = \frac{1}{\sqrt{8}}, \cot\theta = \sqrt{8}, \csc\theta = -3, \sec\theta = -\frac{3}{\sqrt{8}}$

63. $\sin\theta = -\frac{35}{37}, \cos\theta = \frac{12}{37}, \tan\theta = -\frac{35}{12}, \csc\theta = -\frac{37}{35}, \cot\theta = -\frac{12}{35}$

65. $\sin\theta = \frac{3}{7}, \cos\theta = -\frac{\sqrt{40}}{7}, \tan\theta = -\frac{3}{\sqrt{40}}, \sec\theta = -\frac{7}{\sqrt{40}}, \cot\theta = -\frac{\sqrt{40}}{3}$

67. $\sin x = \sqrt{1 - \cos^2 x}; \tan x = \frac{\sqrt{1 - \cos^2 x}}{\cos x}$

69. $\sin x = -\frac{\tan x}{\sqrt{1 + \tan^2 x}}; \sec x = -\sqrt{1 + \tan^2 x}$

71. $\cot x = -\sqrt{\csc^2 x - 1}; \cos x = -\frac{\sqrt{\csc^2 x - 1}}{\csc x}$

73. $\cos\theta = \sqrt{1 - x^2}, 0.8586$

75. $\sin\theta = -\frac{1}{\sqrt{1 + x^2}}, -0.8899$ **77.** $\tan\theta = \frac{x}{\sqrt{1 - x^2}}$

79. $-\sin 13°$ **81.** $-\tan\frac{\pi}{11}$ **83.** $\sec\frac{2\pi}{5}$

85. (a) I is a maximum when $\theta = 0$.

$[-90, 90, 45]$ by $[-1, 2, 1]$

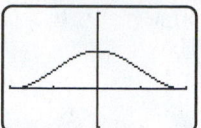

(b) $I = k(1 - \sin^2\theta)$

87. (a) The sum of L and C equals 3.

$[0, 10^{-6}, 10^{-7}]$ by $[-1, 4, 1]$

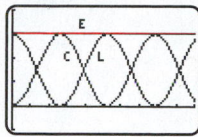

(b) Let $Y_1 = L(t), Y_2 = C(t),$ and $Y_3 = E(t)$.
$E(t) = 3$ for all inputs

(c) $E(t) = L(t) + C(t)$
$= 3\cos^2(6,000,000t) + 3\sin^2(6,000,000t)$
$= 3(\cos^2(6,000,000t) + \sin^2(6,000,000t))$
$= 3(1) = 3$

89. (a) The graph has y-axis symmetry. The monthly high temperatures x months before and x months after July are equal.

$[-6, 6, 1]$ by $[0, 100, 10]$ **(b)** f is even.

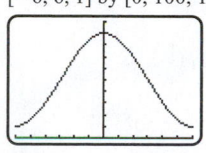   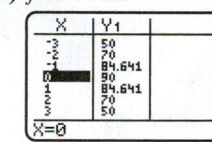

(c) $f(-x) = f(x)$

91. (a) $P = 16k\cos^2(2\pi t)$

(b) Let $Y_1 = 32(\cos(2\pi X))^2$. Y_1 has a maximum value of 32 when $t = 0, 0.5, 1, 1.5, 2.0,$ and Y_1 has a minimum value of 0 when $t = 0.25, 0.75, 1.25, 1.75$. The spring is either stretched or compressed the most when Y_1 is maximum.

$[0, 2, 0.5]$ by $[-1, 40, 8]$

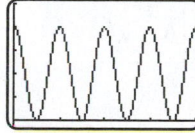

(c) $P = 16k(1 - \sin^2(2\pi t))$

SECTION 7.2 (pp. 571–572)

1. (a) $1 - x^2$ **(b)** $\cos^2\theta$

3. (a) $x^2 - x$ **(b)** $\sec^2\theta - \sec\theta$ **5. (a)** y **(b)** $\sin\theta$

7. (a) $(x + 1)(x + 1)$ **(b)** $(\cos\theta + 1)(\cos\theta + 1)$

9. (a) $x(x - 2)$ **(b)** $\sec t(\sec t - 2)$

11. (a) $x(1 + x^2)$ **(b)** $\tan\theta\sec^2\theta$

13. (a) $\frac{2}{1 - x^2}$ **(b)** $2\csc^2\theta$

15. (a) $\frac{x^2 + y^2}{xy}$ **(b)** $\sec t\csc t$ **17. (a)** $y^2 + x^2$ **(b)** 1

19. (a) x **(b)** $\cos\theta$ **21.** $\sin\theta$ **23.** $\sin\theta - \sec\theta$

25. $2\tan t + \sec^2 t$ **27.** $\cos^2\theta$

29. $(1 - \tan\theta)(1 + \tan\theta)$ **31.** $(\sec t - 3)(\sec t + 2)$

33. $\sec^2\theta\,(\tan^2\theta + 2)$

35. $\csc^2\theta - \cot^2\theta = 1 + \cot^2\theta - \cot^2\theta = 1$

37. $(1 - \sin t)^2 = 1 - 2\sin t + \sin^2 t$ (FOIL)

39. $\dfrac{\sin t + \cos t}{\sin t} = \dfrac{\sin t}{\sin t} + \dfrac{\cos t}{\sin t} = 1 + \cot t$

41. $\sec^2\theta - 1 = (1 + \tan^2\theta) - 1$
$= \tan^2\theta$

43. $\dfrac{\tan^2 t}{\sec t} = \dfrac{\sec^2 t - 1}{\sec t}$
$= \sec t - \dfrac{1}{\sec t}$
$= \sec t - \cos t$

45. $\cot x + 1 = \dfrac{\cos x}{\sin x} + \dfrac{\sin x}{\sin x}$
$= \dfrac{1}{\sin x}(\cos x + \sin x)$
$= \csc x(\cos x + \sin x)$

47. $\dfrac{\sec t}{1 + \sec t} = \dfrac{\sec t}{1 + \sec t} \cdot \dfrac{\cos t}{\cos t}$
$= \dfrac{1}{\cos t + 1}$

49. $(\sec t - 1)(\sec t + 1) = \sec^2 t - 1$
$= \tan^2 t$

51. $\dfrac{1 - \sin^2\theta}{\cos\theta} = \dfrac{\cos^2\theta}{\cos\theta} = \cos\theta$

53. $\dfrac{\sec t}{\tan t} - \dfrac{\tan t}{\sec t} = \dfrac{\sec^2 t - \tan^2 t}{\sec t\,\tan t}$
$= \dfrac{(1 + \tan^2 t) - \tan^2 t}{\sec t\,\tan t}$
$= \dfrac{1}{\sec t\,\tan t}$
$= \cos t\,\cot t$

55. $\dfrac{\cot^2 t}{\csc t + 1} = \dfrac{\csc^2 t - 1}{\csc t + 1}$
$= \dfrac{(\csc t + 1)(\csc t - 1)}{\csc t + 1}$
$= \csc t - 1$

57. $\dfrac{\cot t}{\cot t + 1} = \dfrac{\cot t}{\cot t + 1} \cdot \dfrac{\tan t}{\tan t}$
$= \dfrac{\cot t\,\tan t}{\cot t\,\tan t + \tan t}$
$= \dfrac{1}{1 + \tan t}$

59. $\dfrac{1}{1 - \sin t} + \dfrac{1}{1 + \sin t} = \dfrac{(1 + \sin t) + (1 - \sin t)}{(1 - \sin t)(1 + \sin t)}$
$= \dfrac{2}{1 - \sin^2 t}$
$= \dfrac{2}{\cos^2 t}$
$= 2\sec^2 t$

61. $\dfrac{\csc t + \cot t}{\csc t - \cot t} = \dfrac{\csc t + \cot t}{\csc t - \cot t} \cdot \dfrac{\csc t + \cot t}{\csc t + \cot t}$
$= \dfrac{(\csc t + \cot t)^2}{\csc^2 t - \cot^2 t}$
$= \dfrac{(\csc t + \cot t)^2}{\csc^2 t - (\csc^2 t - 1)}$
$= (\csc t + \cot t)^2$

63. $\dfrac{\cos^2 t}{1 - \sin t} = \dfrac{\cos^2 t}{1 - \sin t} \cdot \dfrac{1 + \sin t}{1 + \sin t}$
$= \dfrac{\cos^2 t(1 + \sin t)}{1 - \sin^2 t}$
$= \dfrac{\cos^2 t(1 + \sin t)}{\cos^2 t}$
$= 1 + \sin t$

65. $\dfrac{1}{1 + \sin\theta} = \dfrac{1}{1 + \sin\theta} \cdot \dfrac{1 - \sin\theta}{1 - \sin\theta}$
$= \dfrac{1 - \sin\theta}{1 - \sin^2\theta}$
$= \dfrac{1 - \sin\theta}{\cos^2\theta}$

67. $\sqrt{1 - \sin^2\theta} = \sqrt{\cos^2\theta}$
$= |\cos\theta|$
$= \cos\theta$, where θ is acute

69. $\dfrac{1 + 2\sin x + \sin^2 x}{\cos^2 x} = \dfrac{(1 + \sin x)^2}{1 - \sin^2 x}$
$= \dfrac{(1 + \sin x)(1 + \sin x)}{(1 - \sin x)(1 + \sin x)}$
$= \dfrac{1 + \sin x}{1 - \sin x}$

71. $(1 - \cos^2 x)(1 + \cos^2 x) = \sin^2 x\,(1 + (1 - \sin^2 x))$
$= \sin^2 x\,(2 - \sin^2 x)$
$= 2\sin^2 x - \sin^4 x$

73. $\cot\theta\,\sin\theta = \dfrac{\cos\theta}{\sin\theta} \cdot \sin\theta$
$= \cos\theta$

75. $(1 - \cos^2\theta)(1 + \tan^2\theta) = \sin^2\theta\,\sec^2\theta$
$= \dfrac{\sin^2\theta}{\cos^2\theta}$
$= \tan^2\theta$

77. $\cos t(\tan t - \sec t) = \cos t\left(\dfrac{\sin t}{\cos t} - \dfrac{1}{\cos t}\right)$
$= \sin t - 1$

79. $\dfrac{\tan(-\theta)}{\sin(-\theta)} = \dfrac{-\tan\theta}{-\sin\theta}$
$= \dfrac{1}{\cos\theta}$
$= \sec\theta$

81. (a) $W(t) = 5(1 - \sin^2(120\pi t))$; $W = 0$
(b) Whenever there is a peak or valley on the graph of V, the graph of W intersects the x-axis, which corresponds to a zero of W.
$[0, 1/15, 1/60]$ by $[-30, 30, 10]$

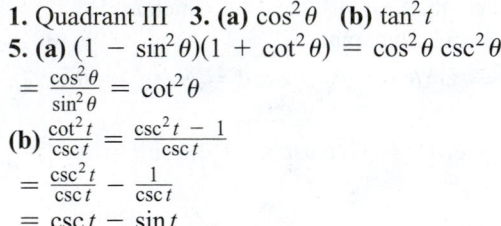

CHECKING BASIC CONCEPTS FOR SECTIONS 7.1 AND 7.2 (p. 572)

1. Quadrant III   **3. (a)** $\cos^2\theta$ **(b)** $\tan^2 t$
5. (a) $(1 - \sin^2\theta)(1 + \cot^2\theta) = \cos^2\theta\,\csc^2\theta$
$= \dfrac{\cos^2\theta}{\sin^2\theta} = \cot^2\theta$
(b) $\dfrac{\cot^2 t}{\csc t} = \dfrac{\csc^2 t - 1}{\csc t}$
$= \dfrac{\csc^2 t}{\csc t} - \dfrac{1}{\csc t}$
$= \csc t - \sin t$

SECTION 7.3 (pp. 583–586)

1. $60°$ **3.** $85°$ **5.** $65°$ **7.** $\dfrac{\pi}{6}$ **9.** $\dfrac{\pi}{3}$ **11.** $\dfrac{\pi}{4}$

13. (a) $90°$; $\dfrac{\pi}{2}$ **(b)** $270°$; $\dfrac{3\pi}{2}$

15. (a) $60°, 240°$; $\dfrac{\pi}{3}, \dfrac{4\pi}{3}$ **(b)** $120°, 300°$; $\dfrac{2\pi}{3}, \dfrac{5\pi}{3}$

17. (a) $60°, 300°$; $\dfrac{\pi}{3}, \dfrac{5\pi}{3}$ **(b)** $120°, 240°$; $\dfrac{2\pi}{3}, \dfrac{4\pi}{3}$

19. (a) No solutions   **(b)** No solutions

21. (a) $\dfrac{\pi}{6} + 2\pi n, \dfrac{5\pi}{6} + 2\pi n$ **(b)** $\dfrac{7\pi}{6} + 2\pi n, \dfrac{11\pi}{6} + 2\pi n$

23. (a) $\dfrac{\pi}{4} + \pi n$ **(b)** $\dfrac{3\pi}{4} + \pi n$

25. (a) $\dfrac{\pi}{3} + 2\pi n, \dfrac{5\pi}{3} + 2\pi n$ **(b)** $\dfrac{2\pi}{3} + 2\pi n, \dfrac{4\pi}{3} + 2\pi n$

27. $\dfrac{\pi}{4}, \dfrac{5\pi}{4}$ **29.** $\dfrac{\pi}{3}, \dfrac{5\pi}{3}$ **31. (a)** $\dfrac{1}{2}$ **(b)** $30°, 150°$

33. (a) $0, 1$ **(b)** $0°, 90°, 180°$

35. (a) $-1, 1$ **(b)** $\dfrac{\pi}{4}, \dfrac{3\pi}{4}, \dfrac{5\pi}{4}, \dfrac{7\pi}{4}$

37. (a) $-2, 1$ **(b)** 0 **39.** $\dfrac{\pi}{3}, \dfrac{2\pi}{3}, \dfrac{4\pi}{3}, \dfrac{5\pi}{3}$ **41.** No solutions

43. $\frac{\pi}{2}, \frac{3\pi}{2}$ **45.** $\frac{\pi}{4}, \frac{5\pi}{4}$ **47.** $\frac{\pi}{6}, \frac{5\pi}{6}, \frac{7\pi}{6}, \frac{11\pi}{6}$ **49.** $0, \frac{\pi}{2}, \pi, \frac{3\pi}{2}$

51. $\frac{\pi}{3}, \frac{5\pi}{3}$ **53.** 0 **55.** $\frac{\pi}{4} + \pi n$ **57.** $\frac{\pi}{6} + 2\pi n, \frac{5\pi}{6} + 2\pi n$

59. $\frac{\pi}{6} + 2\pi n, \frac{5\pi}{6} + 2\pi n, \frac{\pi}{2} + 2\pi n$

61. $\frac{2\pi}{3} + 2\pi n, \frac{4\pi}{3} + 2\pi n, \pi + 2\pi n$ **63.** $\frac{\pi}{4} + \frac{\pi n}{2}$

65. No solutions **67.** πn **69.** $\frac{\pi}{4} + \pi n$ **71.** $\frac{\pi n}{2}$

73. $2\pi n, \frac{\pi}{2} + 2\pi n$ **75.** $\frac{\pi}{18} + \frac{2\pi}{3} n, \frac{5\pi}{18} + \frac{2\pi}{3} n$

77. $\frac{5\pi}{24} + \frac{\pi}{2} n, \frac{7\pi}{24} + \frac{\pi}{2} n$ **79.** $\frac{\pi}{20} + \frac{\pi}{5} n$

81. $\frac{7\pi}{24} + \frac{\pi}{2} n, \frac{11\pi}{24} + \frac{\pi}{2} n$ **83.** $\frac{3\pi}{16} + \frac{\pi}{2} n, \frac{5\pi}{16} + \frac{\pi}{2} n$

85. $\frac{\pi}{16} + \frac{\pi}{4} n$ **87.** $\frac{\pi}{16} + \frac{\pi}{4} n$

89. $0.456 + \frac{2\pi}{3} n, 1.638 + \frac{2\pi}{3} n$

91. $0.170 + \pi n, 1.401 + \pi n$ **93.** 1.085, 5.198

95. 0.737, 2.404 **97.** 0.659, 2.483, 3.801, 5.624

99. 0, 4.49 **101.** 0.39, 1.96 **103.** 3.60

105. $30°, 210°; 30° + 180° \cdot n$ **107.** 1

109. $-\frac{\sqrt{3}}{2} \approx -0.866$ **111.** -1 **113.** 0

115. $\frac{1}{\sqrt{2}} \approx 0.707$ **117.** $60° + 360° \cdot n, 300° + 360° \cdot n$

119. About 9.1°

121. 2.6, 10.7; near February 17 and October 22

123. 2.6, 10.7; near February 17 and October 22

125. (a) $f(x) = 122.3 \sin(0.524x - 1.7) + 367$ (answers may vary)

(b) $2.98 \le x \le 9.51$ (approximately). At 50°N latitude, the maximum number of monthly hours of sunshine is greater than or equal to 350 hours roughly from March through September.

127. 0.26

129. (a) [0, 0.01, 0.005] by [−0.005, 0.005, 0.001]

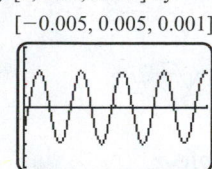

[0, 0.01, 0.005] by [−0.005, 0.005, 0.001]

[0, 0.01, 0.005] by [−0.005, 0.005, 0.001]

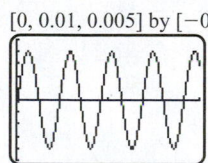

(b) Maximum: $P \approx 0.004$

(c) No; the maximum of P_1 is 0.003, and the maximum of P_2 is 0.002.

SECTION 7.4 (pp. 595–597)

1. $\frac{\sqrt{6} - \sqrt{2}}{4}$ **3.** $2 - \sqrt{3}$ **5.** $\frac{\sqrt{6} - \sqrt{2}}{4}$ **7.** $\frac{\sqrt{6} - \sqrt{2}}{4}$

9. $\frac{\sqrt{6} + \sqrt{2}}{4}$ **11. (a)** The graphs of $y = \sin\left(t + \frac{\pi}{2}\right)$ and $y = \cos t$ are the same.

(b) $\sin\left(t + \frac{\pi}{2}\right) = \sin t \cos\frac{\pi}{2} + \cos t \sin\frac{\pi}{2}$
$= \sin t\,(0) + \cos t\,(1)$
$= \cos t$

13. (a) The graphs of $y = \cos(t + \pi)$ and $y = -\cos t$ are the same.

(b) $\cos(t + \pi) = \cos t \cos\pi - \sin t \sin\pi$
$= \cos t\,(-1) - \sin t\,(0)$
$= -\cos t$

15. (a) The graphs of $y = \sec\left(t - \frac{\pi}{2}\right)$ and $y = \csc t$ are the same.

(b) $\sec\left(t - \frac{\pi}{2}\right) = \dfrac{1}{\cos\left(t - \frac{\pi}{2}\right)}$

$= \dfrac{1}{\cos t \cos\frac{\pi}{2} + \sin t \sin\frac{\pi}{2}}$

$= \dfrac{1}{\cos t\,(0) + \sin t\,(1)}$

$= \dfrac{1}{\sin t}$

$= \csc t$

17. $\sin\left(\frac{\pi}{2} - t\right) = \sin\frac{\pi}{2}\cos t - \cos\frac{\pi}{2}\sin t$
$= (1)\cos t - (0)\sin t$
$= \cos t$

19. $\sec\left(\frac{\pi}{2} - t\right) = \dfrac{1}{\cos\left(\frac{\pi}{2} - t\right)}$

$= \dfrac{1}{\cos\frac{\pi}{2}\cos t + \sin\frac{\pi}{2}\sin t}$

$= \dfrac{1}{(0)\cos t + (1)\sin t}$

$= \dfrac{1}{\sin t}$

$= \csc t$

21. (a) $\frac{56}{65}$ **(b)** $\frac{33}{65}$ **(c)** $\frac{56}{33}$ **(d)** I

23. (a) $-\frac{988}{1037}$ **(b)** $\frac{315}{1037}$ **(c)** $-\frac{988}{315}$ **(d)** IV

25. (a) $-\frac{33}{65}$ **(b)** $-\frac{56}{65}$ **(c)** $\frac{33}{56}$ **(d)** III

27. (a) $-\frac{775}{793}$ **(b)** $-\frac{168}{793}$ **(c)** $\frac{775}{168}$ **(d)** III

29. $\cos\left(t - \frac{\pi}{4}\right) = \cos t \cos\frac{\pi}{4} + \sin t \sin\frac{\pi}{4}$
$= \cos t\left(\frac{\sqrt{2}}{2}\right) + \sin t\left(\frac{\sqrt{2}}{2}\right)$
$= \frac{\sqrt{2}}{2}(\cos t + \sin t)$

31. $\tan\left(t + \frac{\pi}{4}\right) = \dfrac{\tan t + \tan\frac{\pi}{4}}{1 - \tan t \tan\frac{\pi}{4}}$

$= \dfrac{\tan t + 1}{1 - \tan t\,(1)}$

$= \dfrac{1 + \tan t}{1 - \tan t}$

33. $\dfrac{\cos(x - y)}{\cos(x + y)} = \dfrac{\cos x \cos y + \sin x \sin y}{\cos x \cos y - \sin x \sin y}$

$= \dfrac{\dfrac{\cos x \cos y}{\cos x \cos y} + \dfrac{\sin x \sin y}{\cos x \cos y}}{\dfrac{\cos x \cos y}{\cos x \cos y} - \dfrac{\sin x \sin y}{\cos x \cos y}}$

$= \dfrac{1 + \tan x \tan y}{1 - \tan x \tan y}$

35. $\dfrac{\cos(\alpha - \beta)}{\cos\alpha \sin\beta} = \dfrac{\cos\alpha \cos\beta + \sin\alpha \sin\beta}{\cos\alpha \sin\beta}$

$= \dfrac{\cos\alpha \cos\beta}{\cos\alpha \sin\beta} + \dfrac{\sin\alpha \sin\beta}{\cos\alpha \sin\beta}$

$= \dfrac{\cos\beta}{\sin\beta} + \dfrac{\sin\alpha}{\cos\alpha}$

$= \cot\beta + \tan\alpha$

$= \tan\alpha + \cot\beta$

37. $\sin 2t = \sin(t + t)$
$= \sin t \cos t + \cos t \sin t$
$= 2 \sin t \cos t$

39. $\sin(\alpha + \beta) + \sin(\alpha - \beta)$
$= \sin\alpha \cos\beta + \cos\alpha \sin\beta + \sin\alpha \cos\beta$
$\quad - \cos\alpha \sin\beta$
$= \sin\alpha \cos\beta + \sin\alpha \cos\beta$
$= 2 \sin\alpha \cos\beta$

41. $\tan(\pi - \theta) = \dfrac{\tan \pi - \tan \theta}{1 + \tan \pi \tan \theta}$

$= \dfrac{0 - \tan \theta}{1 + (0)\tan \theta}$

$= -\tan \theta$

43. $\tan(x - y) - \tan(y - x)$

$= \dfrac{\tan x - \tan y}{1 + \tan x \tan y} - \dfrac{\tan y - \tan x}{1 + \tan y \tan x}$

$= \dfrac{\tan x - \tan y - \tan y + \tan x}{1 + \tan x \tan y}$

$= \dfrac{2(\tan x - \tan y)}{1 + \tan x \tan y}$

45. $\dfrac{\sin(x + y)}{\cos x \cos y} = \dfrac{\sin x \cos y + \cos x \sin y}{\cos x \cos y}$

$= \dfrac{\sin x \cos y}{\cos x \cos y} + \dfrac{\cos x \sin y}{\cos x \cos y}$

$= \dfrac{\sin x}{\cos x} + \dfrac{\sin y}{\cos y}$

$= \tan x + \tan y$

47. $\dfrac{7}{79}$

49. $\tan \theta = \tan(\beta - \alpha)$

$= \dfrac{\tan \beta - \tan \alpha}{1 + \tan \alpha \tan \beta}$

$= \dfrac{m_2 - m_1}{1 + m_1 m_2}$

51. $\theta = \tan^{-1}\dfrac{7}{11} \approx 32.5°$

53. (a) $F \approx 409$ lb   **(b)** About 0.81, or 46.2°

55. (a) $[0, 0.02, 0.001]$ by $[-6, 6, 1]$

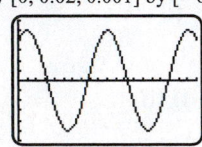

(b) $a = 5, k \approx 0.9272$

(c) $5 \sin(220\pi t + 0.9272)$

$= 5(\sin(220\pi t)\cos(0.9272)$

$\quad + \cos(220\pi t)\sin(0.9272))$

$\approx 5(0.6 \sin(220\pi t) + 0.8 \cos(220\pi t))$

$= 3 \sin(220\pi t) + 4 \cos(220\pi t)$

7.4 EXTENDED AND DISCOVERY EXERCISES (p. 597)

1. (a) $[0.15, 1.15, 0.05]$ by $[-0.01, 0.01, 0.001]$

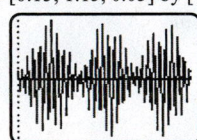

There are 3 beats in 1 second.

(b) $[0.15, 1.15, 0.05]$ by $[-0.01, 0.01, 0.001]$

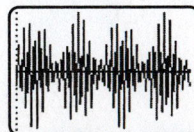

There are 4 beats in 1 second.

(c) When the frequencies are F_1 and F_2, the rate of beats per second is given by $|F_2 - F_1|$.

3. $[0.2, 1.2, 0.05]$ by $[-0.01, 0.01, 0.001]$

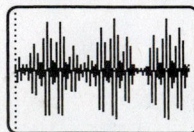

There are 3 beats in 1 second.

CHECKING BASIC CONCEPTS FOR SECTIONS 7.3 AND 7.4 (p. 597)

1. (a) 45°   **(b)** $\dfrac{\pi}{6}$

3. (a) $\dfrac{3\pi}{4} + \pi n$ **(b)** $\dfrac{2\pi}{3} + 2\pi n; \dfrac{4\pi}{3} + 2\pi n; 2\pi n$

5. $\sin(t - \pi) = \sin t \cos \pi - \cos t \sin \pi$

$= \sin t(-1) - \cos t(0)$

$= -\sin t$

SECTION 7.5 (pp. 609–612)

1. $\dfrac{1 - \cos 20t}{2}$ **3.** $\dfrac{1 - \cos 10t}{1 + \cos 10t}$

5. $2 \sin 10x \cos 10x$ **7.** $\dfrac{1 - \cos 10x}{\sin 10x}$

9. (a) 1   **(b)** $\dfrac{\sqrt{3}}{2}$; not equal

11. (a) 1   **(b)** $-\dfrac{1}{2}$; not equal

13. (a) 2   **(b)** Undefined; not equal

15. *Graphical:* Graphs of $y = \tan 2\theta$ and $y = 2 \tan \theta$ are not the same.

Symbolic: $\tan 2\theta = \dfrac{2 \tan \theta}{1 - \tan^2 \theta} \neq 2 \tan \theta$, unless $\tan \theta = 0$.

17. $\sin 2\theta = \dfrac{24}{25}, \cos 2\theta = \dfrac{7}{25}, \tan 2\theta = \dfrac{24}{7}$

19. $\sin 2\theta = -\dfrac{336}{625}, \cos 2\theta = -\dfrac{527}{625}, \tan 2\theta = \dfrac{336}{527}$

21. $\sin 2\theta = -\dfrac{1320}{3721}, \cos 2\theta = \dfrac{3479}{3721}, \tan 2\theta = -\dfrac{1320}{3479}$

23. $\sin 2\theta = \dfrac{336}{625}, \cos 2\theta = \dfrac{527}{625}, \tan 2\theta = \dfrac{336}{527}$

25. 0   **27.** $\dfrac{527}{625}$ **29.** $\dfrac{828}{2197}$ **31.** $\dfrac{2x}{x^2 + 1}$ **33.** $\dfrac{24}{25}$

35. $\sin 2\theta$ **37.** $\dfrac{1}{2}\sin 2\theta$ **39.** $\cos 4\theta$ **41.** 1

43. $\cot^2 5x$ **45.** $\dfrac{2 + \sqrt{2}}{4}$ **47.** $\dfrac{2 + \sqrt{3}}{2 - \sqrt{3}}$

49. (a) $\dfrac{\sqrt{2 + \sqrt{3}}}{2}$

(b) $-\sqrt{\dfrac{2 - \sqrt{3}}{2 + \sqrt{3}}}$, or $-\dfrac{1}{2 + \sqrt{3}}$, or $\sqrt{3} - 2$

51. (a) $\sqrt{\dfrac{2 - \sqrt{2}}{2 + \sqrt{2}}}$, or $\dfrac{\sqrt{2}}{2 + \sqrt{2}}$, or $\sqrt{2} - 1$

(b) $-\dfrac{\sqrt{2 - \sqrt{2}}}{2}$

53. $\sin 30°$ **55.** $\cos 25°$ **57.** $\tan 20°$

59. $\sin\dfrac{\theta}{2} = \dfrac{1}{\sqrt{10}}, \cos\dfrac{\theta}{2} = \dfrac{3}{\sqrt{10}}, \tan\dfrac{\theta}{2} = \dfrac{1}{3}$

61. $\sin\dfrac{\theta}{2} = -\dfrac{1}{\sqrt{26}}, \cos\dfrac{\theta}{2} = \dfrac{5}{\sqrt{26}}, \tan\dfrac{\theta}{2} = -\dfrac{1}{5}$

63. $\sin\dfrac{\theta}{2} = \dfrac{4}{5}, \cos\dfrac{\theta}{2} = \dfrac{3}{5}, \tan\dfrac{\theta}{2} = \dfrac{4}{3}$

65. $4 \sin 2x = 4(2 \sin x \cos x)$

$= 8 \sin x \cos x$

67. $\dfrac{2 - \sec^2 x}{\sec^2 x} = \dfrac{2}{\sec^2 x} - 1$

$= 2 \cos^2 x - 1$

$= \cos 2x$

69. $\sec 2x = \dfrac{1}{\cos 2x}$

$= \dfrac{1}{1 - 2 \sin^2 x}$

71. $\sin 3\theta = \sin(2\theta + \theta)$

$= \sin 2\theta \cos \theta + \cos 2\theta \sin \theta$

$= (2 \sin \theta \cos \theta)\cos \theta + (1 - 2 \sin^2 \theta)\sin \theta$

$= 2 \sin \theta \cos^2 \theta + \sin \theta - 2 \sin^3 \theta$

$= 2 \sin \theta(1 - \sin^2 \theta) + \sin \theta - 2 \sin^3 \theta$

$= 2 \sin \theta - 2 \sin^3 \theta + \sin \theta - 2 \sin^3 \theta$

$= 3 \sin \theta - 4 \sin^3 \theta$

73. $\sin 4\theta = 2 \sin 2\theta \cos 2\theta$

$= 2(2 \sin \theta \cos \theta)\cos 2\theta$

$= 4 \sin \theta \cos \theta \cos 2\theta$

75. $\dfrac{\sin 2\theta}{\sin \theta} = \dfrac{2 \sin \theta \cos \theta}{\sin \theta}$
$= 2 \cos \theta$

77. $2 \cos^2 \dfrac{\theta}{2} = 2\left(\dfrac{1 + \cos \theta}{2}\right)$
$= 1 + \cos \theta$

79. $\cos^4 \theta - \sin^4 \theta = (\cos^2 \theta - \sin^2 \theta)(\cos^2 \theta + \sin^2 \theta)$
$= (\cos 2\theta)(1)$
$= \cos 2\theta$

81. $\csc 2t = \dfrac{1}{\sin 2t}$
$= \dfrac{1}{2 \sin t \cos t}$
$= \dfrac{\csc t}{2 \cos t}$

83. $\tan \dfrac{x}{2} = \dfrac{\sin \frac{x}{2}}{\cos \frac{x}{2}}$
$= \dfrac{2 \sin \frac{x}{2} \cos \frac{x}{2}}{2 \cos \frac{x}{2} \cos \frac{x}{2}}$
$= \dfrac{\sin x}{2 \cos^2 \frac{x}{2}}$
$= \dfrac{\sin x}{1 + \cos x}$

85. (a) $\frac{1}{2}(\sin 70° - \sin 30°)$ (b) $\frac{1}{2}(\cos 3x + \cos x)$
87. (a) $\frac{1}{2}(\sin 10\theta + \sin 4\theta)$ (b) $\frac{1}{2}(\cos 4x - \cos 12x)$
89. (a) $2 \sin 35° \cos 5°$ (b) $2 \cos 40° \cos 5°$
91. (a) $2 \cos 5\theta \cos \theta$ (b) $2 \sin \frac{11x}{2} \cos \frac{3x}{2}$ 93. 0°, 180°
95. 90°, 210°, 270°, 330°   97. π 99. $\frac{\pi}{2}, \frac{3\pi}{2}$
101. $\frac{\pi}{12} + \pi n, \frac{11\pi}{12} + \pi n$ 103. $\pi n, \frac{2\pi}{3} + 2\pi n, \frac{4\pi}{3} + 2\pi n$
105. $\frac{\pi}{3} + 4\pi n, \frac{5\pi}{3} + 4\pi n$
107. $\frac{\pi}{6} + 2\pi n, \frac{5\pi}{6} + 2\pi n, \frac{3\pi}{2} + 2\pi n$
109. $\frac{\pi}{8} + \pi n, \frac{5\pi}{8} + \pi n$, or equivalently, $\frac{\pi}{8} + \frac{\pi n}{2}$
111. $\frac{2\pi}{3} + 4\pi n, \frac{10\pi}{3} + 4\pi n$
113. $\frac{\pi}{8} + \pi n, \frac{5\pi}{8} + \pi n$, or equivalently, $\frac{\pi}{8} + \frac{\pi n}{2}$
115. $\frac{\pi}{12} + \pi n, \frac{5\pi}{12} + \pi n, \frac{\pi}{4} + \pi n$ 117. 0.333, 4.379
119. (a) $[0, 0.04, 0.01]$ by $[-500, 2500, 500]$

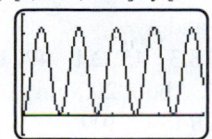

(b) $a = -1085, k = 240, d = 1085$
121. $\frac{1}{720} + \frac{n}{60}, \frac{5}{720} + \frac{n}{60}$ sec
123. (a) $d = 600\left(1 - \cos \frac{80°}{2}\right) \approx 140.4$ ft
(b) *Hint:* First show that $r = d + r \cos \frac{\beta}{2}$. See Student's Solutions Manual.
(c) No, since $\cos \frac{\beta}{2} \neq \frac{1}{2} \cos \beta$ in general
125. Let $f(t)$ model the tone for the number 3 and $g(t)$ model the tone for number 4.
(a) $f(t) = \cos(1394\pi t) + \cos(2954\pi t)$
$g(t) = \cos(1540\pi t) + \cos(2418\pi t)$
(b) $f(t) = 2 \cos(2174\pi t) \cos(780\pi t)$
$g(t) = 2 \cos(1979\pi t) \cos(439\pi t)$

CHECKING BASIC CONCEPTS FOR SECTION 7.5 (p. 612)

1. $\sin 2\theta = -\frac{336}{625}, \cos 2\theta = -\frac{527}{625}$
3. $\sin \frac{\theta}{2} = \frac{1}{\sqrt{5}}, \cos \frac{\theta}{2} = \frac{2}{\sqrt{5}}$ 5. $\frac{\pi}{2}, \frac{3\pi}{2}$

CHAPTER 7 REVIEW EXERCISES (pp. 616–618)

1. Quadrant II
3. $\tan \theta = -\frac{3}{4}, \csc \theta = \frac{5}{3}, \sec \theta = -\frac{5}{4}, \cot \theta = -\frac{4}{3}$
5. $\sin \theta = -\frac{7}{25}, \csc \theta = -\frac{25}{7}, \sec \theta = \frac{25}{24}, \cot \theta = -\frac{24}{7}$
7. $-\sin 13°$ 9. $\sec \frac{3\pi}{7}$ 11. 1   13. 1   15. $\cos \theta$
17. $\sin \theta \approx 0.7776, \cos \theta \approx 0.6288, \csc \theta \approx 1.2860,$
$\sec \theta \approx 1.5904, \cot \theta \approx 0.8086$
19. $\sin \theta \approx 0.8908, \tan \theta \approx -1.9604, \csc \theta \approx 1.1226,$
$\sec \theta \approx -2.2007, \cot \theta \approx -0.5101$
21. $(\sin \theta + 1)(\sin \theta + 1)$ 23. $(\tan \theta + 3)(\tan \theta - 3)$
25. $(\sec \theta - 1)(\sec \theta + 1) = \sec^2 \theta - 1$
$= (1 + \tan^2 \theta) - 1$
$= \tan^2 \theta$
27. $(1 + \tan t)^2 = 1 + 2 \tan t + \tan^2 t$
$= \sec^2 t + 2 \tan t$
29. $\sin(x - \pi) = \sin x \cos \pi - \cos x \sin \pi$
$= \sin x(-1) - \cos x(0)$
$= -\sin x$
31. $\sin 8x = \sin(2 \cdot 4x)$
$= 2 \sin 4x \cos 4x$
33. $\sec 2x = \dfrac{1}{\cos 2x}$
$= \dfrac{1}{2 \cos^2 x - 1}$
35. $\cos^4 x \sin^3 x = \cos^4 x \sin^2 x \sin x$
$= \cos^4 x(1 - \cos^2 x) \sin x$
$= (\cos^4 x - \cos^6 x) \sin x$
37. $\sec^4 \theta - \tan^4 \theta = (\sec^2 \theta - \tan^2 \theta)(\sec^2 \theta + \tan^2 \theta)$
$= (1)(1 + \tan^2 \theta + \tan^2 \theta)$
$= 1 + 2 \tan^2 \theta$
39. 60°   41. $\frac{2\pi}{7}$ 43. $\frac{\pi}{2}, \frac{3\pi}{2}$
45. (a) 60°, 240°   (b) 150°, 330°
47. $\frac{\pi}{3}, \frac{5\pi}{3}$ 49. $\frac{\pi}{2}$ 51. $\frac{\pi}{4}, \frac{5\pi}{4}$
53. $\frac{\pi}{6} + \pi n, -\frac{\pi}{6} + \pi n$; $30° + 180° \cdot n, -30° + 180° \cdot n$
55. $\frac{\pi}{2} + \pi n$; $90° + 180° \cdot n$
57. $-\frac{\sqrt{2 - \sqrt{3}}}{2}$ 59. $-4.43, 1.13, 4.53$
61. (a) $\sin(\alpha + \beta) = \frac{63}{65}$ (b) $\cos(\alpha + \beta) = \frac{16}{65}$,
(c) $\tan(\alpha + \beta) = \frac{63}{16}$ (d) Quadrant I
63. $\sin 2\theta = -\frac{24}{25}, \cos 2\theta = -\frac{7}{25}, \tan 2\theta = \frac{24}{7}$
65. $\sin \frac{\theta}{2} = \sqrt{\frac{3}{8}}, \cos \frac{\theta}{2} = \sqrt{\frac{5}{8}}, \tan \frac{\theta}{2} = \sqrt{\frac{3}{5}}$
67. $\frac{3479}{3721}$ 69. $x = 4.7, 8.7$, about April 21 and August 22
71. (a) $[0, 0.06, 0.01]$ by $[-0.012, 0.012, 0.002]$

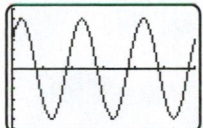

(b) $a = 0.01, k \approx 0.6435$
(c) $0.01 \sin(100\pi t + 0.6435)$
$= 0.01(\sin(100\pi t) \cos(0.6435)$
$+ \cos(100\pi t) \sin(0.6435))$
$\approx 0.01(0.8 \sin(100\pi t) + 0.6 \cos(100\pi t))$
$\approx 0.008 \sin(100\pi t) + 0.006 \cos(100\pi t)$
73. $W(t) = 7 - 7 \sin^2(240\pi t)$; when V is maximum or minimum, $W = 0$.
75. $\theta \approx 30.11°$, or 0.53 radian

CHAPTER 7 EXTENDED AND DISCOVERY EXERCISES (p. 618)

1. (a) i. [0, 0.01, 0.002] by
[−0.005, 0.005, 0.001]

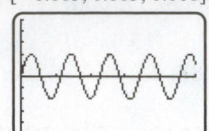

ii. [0, 0.01, 0.002] by
[−0.005, 0.005, 0.001]

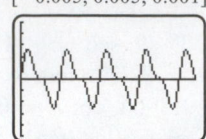

iii. [0, 0.01, 0.002] by
[−0.005, 0.005, 0.001]

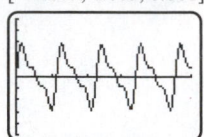

iv. [0, 0.01, 0.002] by
[−0.005, 0.005, 0.001]

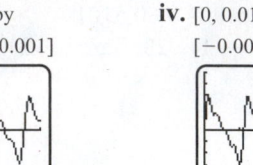

v. [0, 0.01, 0.002] by [−0.005, 0.005, 0.001]

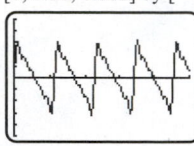

(b) The graph approximates a saw-tooth shape.
(c) The maximum pressure of P is approximately 0.00317.
(d) The pure tone is modeled by a smooth graph, whereas the piano tone is modeled by a saw-tooth shape.
3. (a) [0, 1, 0.1] by [−0.15, 0.15, 0.05]

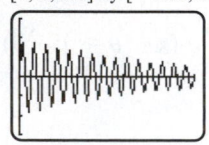

(b) These graphs bound the changing amplitude of A.
[0, 1, 0.1] by [−0.15, 0.15, 0.05]

(c) $t \approx 0.55$ second (answers may vary slightly)

CHAPTER 8: Further Topics in Trigonometry

SECTION 8.1 (PP. 628–632)

1. $\beta = 60°, a \approx 5.5, b \approx 4.9$
3. $\beta = 35°, a \approx 15.1, c \approx 7.4$
5. $\gamma = 80°, a \approx 6.5, b \approx 8.8$
7. $\beta = 115°, b \approx 20.8, c \approx 14.8$
9. $\alpha = 79.1°, b \approx 3.6, c \approx 4.9$
11. $\beta = 104°, a \approx 31.3, c \approx 52.1$
13. No **15.** No **17.** Yes **19.** Yes
21. $\beta_1 \approx 57.1°, \gamma_1 \approx 76.9°, c_1 \approx 8.1$
     $\beta_2 \approx 122.9°, \gamma_2 \approx 11.1°, c_2 \approx 1.6$
23. There are no solutions.
25. $\beta_1 \approx 67°13', \gamma_1 \approx 62°35', c_1 \approx 11.6$
     $\beta_2 \approx 112°48', \gamma_2 \approx 17°0', c_2 \approx 3.8$
27. $\gamma = 93°, a \approx 7.8, c \approx 14.6$

29. $\beta_1 \approx 26.1°, \gamma_1 \approx 133.9°, c_1 \approx 14.7$
     $\beta_2 \approx 153.9°, \gamma_2 \approx 6.1°, c_2 \approx 2.2$
31. $\gamma = 90°, \alpha = 60°, a = 10\sqrt{3} \approx 17.3$
33. $\alpha \approx 52.9°, \beta \approx 25.1°, b \approx 22.4$
35. $\beta = 10°, a \approx 92.2, c \approx 101.9$
37. There are no solutions. **39.** There are no solutions.
41. $\alpha_1 \approx 60°56', \gamma_1 \approx 72°19', c_1 \approx 6.5$
     $\alpha_2 \approx 119°4', \gamma_2 \approx 14°11', c_2 \approx 1.7$
43. $\gamma = 99°45', a \approx 84.6, b \approx 40.9$ **45.** 3629 ft
47. The calculated distance to the moon changes to about 343,000 km, a difference of about 76,000 km. A small error in measuring the lunar angle could result in large errors in calculating the distance to the moon.
49. $d \approx 7.2$ mi **51.** About 28.8 ft
53. About 118.0 m **55.** About 3.86 mi
57. About 0.49 mi **59.** $AB \approx 105.4$ ft
61. (a) 3.57 mi **(b)** 48° **63.** 630 ft

SECTION 8.2 (pp. 639–642)

1. (a) SAS **(b)** Law of cosines
3. (a) SSA **(b)** Law of sines
5. (a) ASA **(b)** Law of sines
7. (a) ASA **(b)** Law of sines
9. 7 **11.** 55.8° **13.** $a \approx 5.4, \beta \approx 40.7°, \gamma \approx 78.3°$
15. $\alpha \approx 22.3°, \beta \approx 108.2°, \gamma \approx 49.5°$
17. $\alpha \approx 33.6°, \beta \approx 50.7°, \gamma \approx 95.7°$
19. $c \approx 28.8, \alpha \approx 116.5°, \beta \approx 28.5°$
21. $\alpha \approx 101.0°, \beta \approx 44.0°, \gamma \approx 34.9°$
Angles do not sum to 180° because of rounding.
23. $a \approx 9.0, \beta \approx 150.9°, \gamma \approx 18.6°$
25. $\alpha \approx 23.1°, \beta \approx 107.2°, \gamma \approx 49.7°$
27. $b \approx 30.7, \alpha \approx 33°26', \gamma \approx 24°24'$
29. $\alpha \approx 45.1°, \beta \approx 63.5°, \gamma \approx 71.5°$
Angles do not sum to 180° because of rounding.
31. No, since $a + b < c$
33. No, since $89° + 112° > 180°$
35. Yes, since we are given ASA and $\alpha + \gamma < 180°$
37. 86.8 **39.** 5.3 **41.** 50.9 **43.** 18.3 **45.** 2.1 **47.** 18.3
49. 66 **51.** 160.4 **53.** 169 ft **55.** About 29.8 mi
57. 4.4 ft; 7.7 ft **59.** A to B: 76°; B to C: 309°
61. (a) $\alpha \approx 75.1°, \beta \approx 65.6°, \gamma \approx 39.4°$
Angles do not sum to 180° because of rounding.
(b) 6299 ft^2
63. $\theta \approx 40.5°$ **65.** 302 mi **67.** About 1452 ft
69. About 745 mi **71.** 147.8 ft^2
73. (a) $9\sqrt{3} \approx 15.6$ in^2 **(b)** The results are equal.
75. 149,429 ft^2 **77.** 21,309 ft^2

8.2 EXTENDED AND DISCOVERY EXERCISES (p. 643)

1. About 2000 km

CHECKING BASIC CONCEPTS FOR SECTIONS 8.1 AND 8.2 (p. 643)

1. $\beta = 74°, b \approx 16.6, c \approx 15.3$
3. (a) $b \approx 7.1, \alpha \approx 63.0°, \gamma \approx 66.0°$
(b) $\alpha \approx 110.7°, \beta \approx 37.0°, \gamma \approx 32.3°$

SECTION 8.3 (pp. 655–659)

1. (a) $a_1 \approx 3, a_2 \approx 4$ **(b)** $\|\mathbf{v}\| = 5$
3. (a) $a_1 \approx -5, a_2 \approx -12$ **(b)** $\|\mathbf{v}\| = 13$
5. (a) 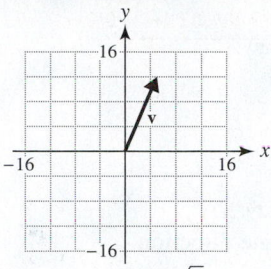 v (20 mi/hr) **(b)** $\mathbf{v} = \langle 0, -20 \rangle$

(c) $2\mathbf{v} = \langle 0, -40 \rangle$; this represents a 40-mi/hr north wind.
$-\frac{1}{2}\mathbf{v} = \langle 0, 10 \rangle$; this represents a 10-mi/hr south wind.
7. (a) v (5 mi/hr)

(b) $\mathbf{v} = \langle \frac{5}{\sqrt{2}}, -\frac{5}{\sqrt{2}} \rangle$, or $\langle \frac{5}{2}\sqrt{2}, -\frac{5}{2}\sqrt{2} \rangle$
(c) $2\mathbf{v} = \langle \frac{10}{\sqrt{2}}, -\frac{10}{\sqrt{2}} \rangle$, or $\langle 5\sqrt{2}, -5\sqrt{2} \rangle$; this represents a
10-mi/hr northwest wind. $-\frac{1}{2}\mathbf{v} = \langle -\frac{5}{4}\sqrt{2}, \frac{5}{4}\sqrt{2} \rangle$; this
represents a 2.5-mi/hr southeast wind.
9. (a) ↑ v (30 lb) **(b)** $\mathbf{v} = \langle 0, 30 \rangle$

(c) $2\mathbf{v} = \langle 0, 60 \rangle$; this represents a 60-lb force upward.
$-\frac{1}{2}\mathbf{v} = \langle 0, -15 \rangle$; this represents a 15-lb force downward.
11. (a) Horizontal $= 1$, vertical $= 1$
(b) $\|\mathbf{v}\| = \sqrt{2}$; $\mathbf{v}$ is not a unit vector.
(c) $\|\mathbf{v}\|$ represents the length of $\mathbf{v}$.

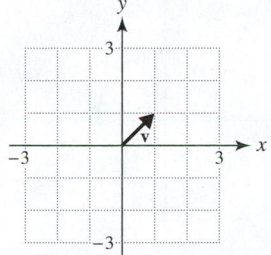

13. (a) Horizontal $= 3$, vertical $= -4$
(b) $\|\mathbf{v}\| = 5$; $\mathbf{v}$ is not a unit vector.
(c) $\|\mathbf{v}\|$ represents the length of $\mathbf{v}$.

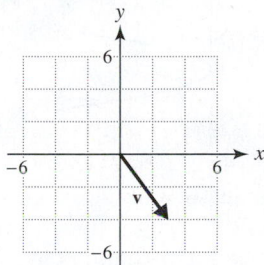

15. (a) Horizontal $= 1$, vertical $= 0$
(b) $\|\mathbf{v}\| = 1$; $\mathbf{v}$ is a unit vector.
(c) $\|\mathbf{v}\|$ represents the length of $\mathbf{v}$.

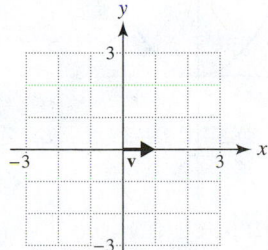

17. (a) Horizontal $= 5$, vertical $= 12$
(b) $\|\mathbf{v}\| = 13$; $\mathbf{v}$ is not a unit vector.
(c) $\|\mathbf{v}\|$ represents the length of $\mathbf{v}$.

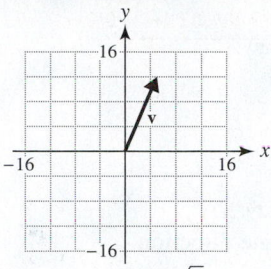

19. $3; 0°$ **21.** $\sqrt{2}; 135°$ **23.** $2; 330°$ **25.** $13; 247.4°$
27. $85; 278.8°$ **29.** $29; 133.6°$ **31.** $-4, 0$ **33.** $-1, 1$
35. $13.5, 18.6$ **37.** $22.8, -25.3$ **39.** $\langle 3\sqrt{3}, 3 \rangle$
41. $\langle -\frac{9\sqrt{2}}{2}, -\frac{9\sqrt{2}}{2} \rangle$ **43.** $\langle 3.06, 2.57 \rangle$ **45.** $\langle 4.10, -2.87 \rangle$
47. (a)

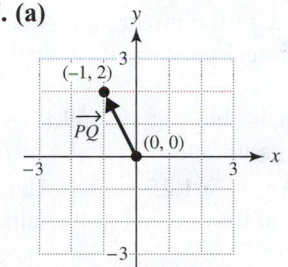

(b) $\overrightarrow{PQ} = \langle -1, 2 \rangle$
(c) $\|\overrightarrow{PQ}\| = \sqrt{5}$
49. (a)

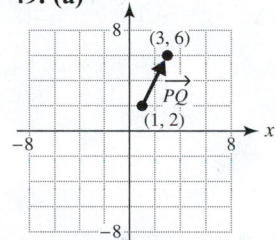

(b) $\overrightarrow{PQ} = \langle 2, 4 \rangle$
(c) $\|\overrightarrow{PQ}\| = \sqrt{20}$
51. (a)

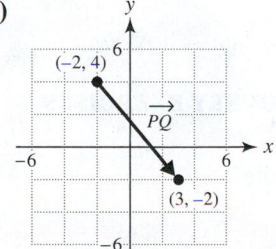

(b) $\overrightarrow{PQ} = \langle 5, -6 \rangle$ **(c)** $\|\overrightarrow{PQ}\| = \sqrt{61}$
53. 94.2 lb **55.** 24.4 lb **57.** $\mathbf{F} = -5\mathbf{i} + 12\mathbf{j}, \|\mathbf{F}\| = 13$
59. $\mathbf{F} = \langle 9, -22 \rangle, \|\mathbf{F}\| = \sqrt{565}$
61. $\mathbf{F} = -\mathbf{i} - 5\mathbf{j}, \|\mathbf{F}\| = \sqrt{26}$
63. (a) $\langle 3, 2 \rangle$ **(b)** $\langle -3, 2 \rangle$
65. (a) $3\mathbf{i} - \mathbf{j}$ **(b)** $\mathbf{i} + 3\mathbf{j}$
67. (a) $\langle 0, -\frac{1}{4} \rangle$ **(b)** $\langle -2\sqrt{2}, \frac{5}{4} \rangle$
69. (a) $\frac{\sqrt{2}}{2}\mathbf{i} + \frac{\sqrt{2}+2}{2}\mathbf{j}$ **(b)** $\frac{\sqrt{2}}{2}\mathbf{i} + \frac{\sqrt{2}-2}{2}\mathbf{j}$

71. (a) $\langle -4, 16 \rangle$ **(b)** $\langle -12, 0 \rangle$ **(c)** $\langle 8, -8 \rangle$
73. (a) $\langle 8, 0 \rangle$ **(b)** $\langle 0, 16 \rangle$ **(c)** $\langle -4, -8 \rangle$
75. (a) $\langle 0, 12 \rangle$ **(b)** $\langle -16, -4 \rangle$ **(c)** $\langle 8, -4 \rangle$
77. (a) 2 **(b)** $4\mathbf{i}$ **(c)** $7\mathbf{i} + 3\mathbf{j}$
79. (a) $\sqrt{5}$ **(b)** $\langle -2, 4 \rangle$ **(c)** $\langle 7, 4 \rangle$
81. (a) $-21\mathbf{i} + 10\mathbf{j}$ **(b)** $17\mathbf{i} - 10\mathbf{j}$
83. (a) $\langle -13, 24 \rangle$ **(b)** $\langle 11, -23 \rangle$
85. (a) $\langle 43, -2 \rangle$ **(b)** $\langle -41, -1 \rangle$
87. (a) 1 **(b)** $81.9°$ **(c)** Neither
89. (a) 0 **(b)** $90°$ **(c)** Perpendicular
91. (a) 122 **(b)** $0°$ **(c)** Parallel, same direction
93. (a) -4 **(b)** $143.1°$ **(c)** Neither
95. 150 ft-lb **97.** 100,000 ft-lb
99. Work $= 590$ ft-lb, $\|\mathbf{F}\| = \sqrt{500} \approx 22.4$ lb
101. Work $= 27$ ft-lb, $\|\mathbf{F}\| = \sqrt{34} \approx 5.8$ lb
103. 24 **105.** 4
107. $\mathbf{v} = \langle 2, 3 \rangle$, speed $= \sqrt{13} \approx 3.6$ mi/hr
109. $\mathbf{v} \approx \langle -364.6, -35.4 \rangle$, groundspeed ≈ 366.3 mi/hr, bearing $\approx 264.5°$
111. Ground speed ≈ 431.3 mi/hr, bearing $\approx 159.1°$
113. Airspeed ≈ 149.3 mi/hr, groundspeed ≈ 154.6 mi/hr
115. (a) $\|\mathbf{R}\| = \sqrt{5} \approx 2.2$, $\|\mathbf{A}\| = \sqrt{1.25} \approx 1.1$. About 2.2 inches of rain fell. The area of the opening of the rain gauge is about 1.1 square inches.
(b) $V = 1.5$; the volume of rain collected in the gauge was 1.5 cubic inches. **(c)** $\mathbf{R}$ and $\mathbf{A}$ must be parallel and point in opposite directions.
117. (a) $\mathbf{c} = \mathbf{a} + \mathbf{b} = \langle 1, 4 \rangle$ **(b)** $\sqrt{17} \approx 4.1$ ft
(c) $3\mathbf{a} + \frac{1}{2}\mathbf{b} = \langle 8, 7 \rangle$
119. (a) $(2, 2)$ **(b)**

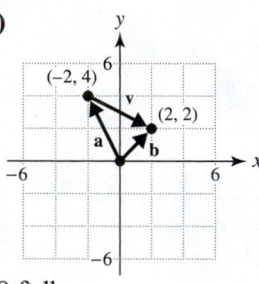

121. $W \approx 297{,}228$ ft-lb
123. Speed ≈ 180 mi/hr, bearing $\approx 128.2°$

8.3 EXTENDED AND DISCOVERY EXERCISES (pp. 659–660)

1. Blue **3. (a)** Blue **(b)** White

SECTION 8.4 (pp. 667–669)

1. (a)

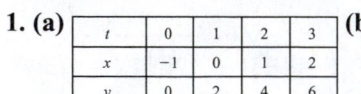

t	0	1	2	3
x	-1	0	1	2
y	0	2	4	6

(b)

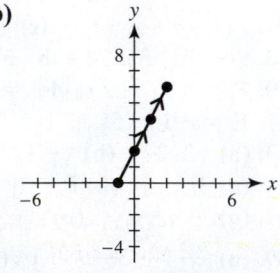

(c) Line segment

3. (a)

t	0	1	2	3
x	2	3	4	5
y	4	1	0	1

(b)

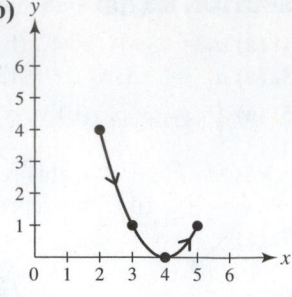

(c) Lower portion of a parabola

5. (a)

t	0	1	2	3
x	3	$\sqrt{8}$	$\sqrt{5}$	0
y	0	1	2	3

(b)

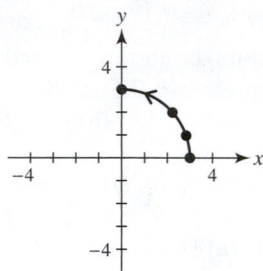

(c) Portion of a circle with radius 3

7. (a)

t	0	1	2	3
x	0	1	2	3
y	3	$\sqrt{8}$	$\sqrt{5}$	0

(b)

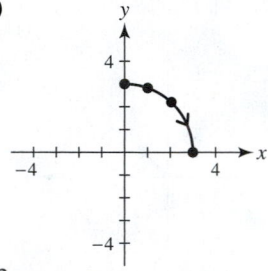

(c) Portion of a circle with radius 3

9. $y = \frac{1}{3}x - 1$; line

11. $x = 3(y - 1)^2$; parabola

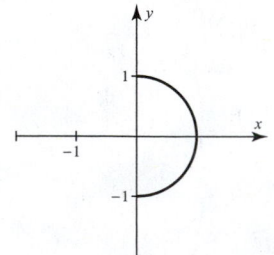

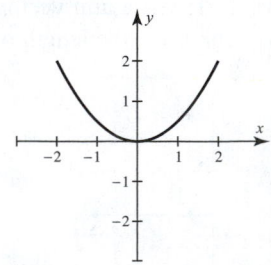

13. $x = \sqrt{1 - y^2}$; portion of a circle with radius 1

15. $y = \frac{1}{2}x^2$; portion of a parabola

17. $y = x^2 + 4x + 5$;
portion of a parabola

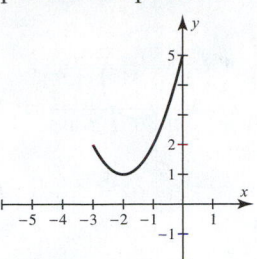

19. $x^2 + y^2 = 9$;
circle with radius 3

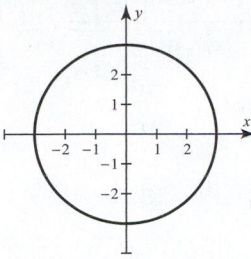

21. $x^2 + y^2 = 4$;
circle with radius 2

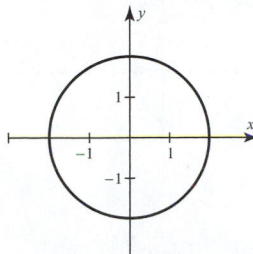

23. $x^2 + y^2 = 9$;
circle with radius 3

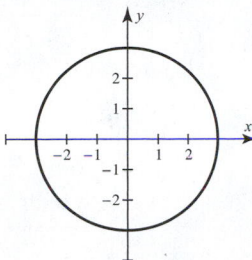

25.

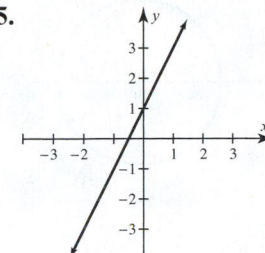

27.

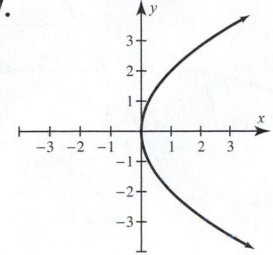

29.

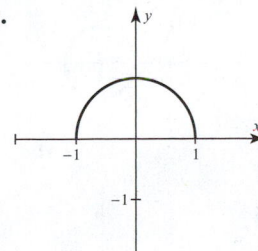

31.

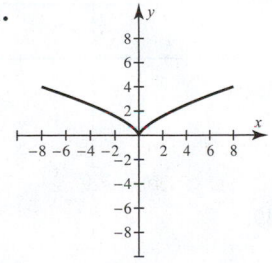

33. $[0, 6, 1]$ by $[-2, 2, 1]$

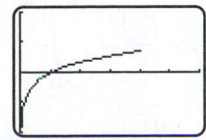

35. $[-2, 20, 2]$ by $[-1, 3, 1]$

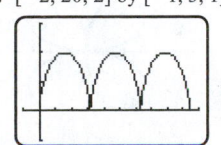

37. $[-4.7, 4.7, 1]$ by
$[-3.1, 3.1, 1]$

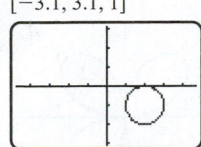

39. $[-1.5, 1.5, 0.5]$ by
$[-1, 1, 0.5]$

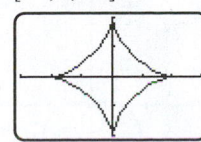

41. $[-4.7, 4.7, 1]$ by $[-3.1, 3.1, 1]$

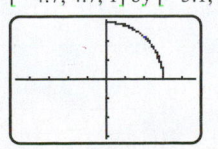

43. $x = t, y = 4 - 2t$ **45.** $x = t, y = 4 - t^2$
47. $x = t^2 + t - 3, y = t$ **49.** $x = 2\cos t, y = 2\sin t$
51. $x = t, y = e^{0.1t^2}$ **53.** $x = t^2 - 2t + 1, y = t$
55. (a) The curve traces a
circle of radius 3 once.
$[-4.7, 4.7, 1]$ by $[-3.1, 3.1, 1]$

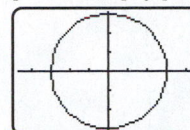

(b) The curve traces a circle
of radius 3 twice.
$[-4.7, 4.7, 1]$ by $[-3.1, 3.1, 1]$

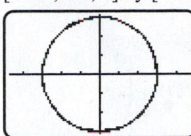

57. (a) The curve traces a circle of radius 3 once counterclockwise, starting at $(3, 0)$.
$[-4.7, 4.7, 1]$ by $[-3.1, 3.1, 1]$

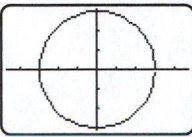

(b) The curve traces a circle of radius 3 once clockwise,
starting at $(0, 3)$.
$[-4.7, 4.7, 1]$ by $[-3.1, 3.1, 1]$

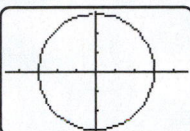

59. (a) Circle of radius 1
centered at $(-1, 2)$
$[-4.7, 4.7, 1]$ by $[-3.1, 3.1, 1]$

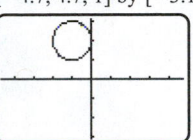

(b) Circle of radius 1
centered at $(1, 2)$
$[-4.7, 4.7, 1]$ by $[-3.1, 3.1, 1]$

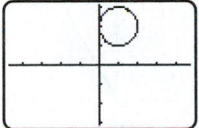

61. F
$[0, 6, 1]$ by $[0, 4, 1]$

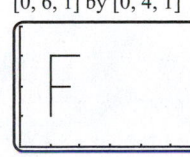

63. D
$[0, 6, 1]$ by $[0, 4, 1]$

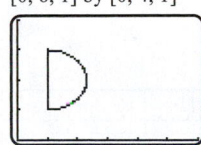

65. $x_1 = 0, y_1 = 2t; x_2 = t, y_2 = 0; 0 \le t \le 1$
67. $x_1 = \sin t, y_1 = \cos t; x_2 = 0, y_2 = t - 2; 0 \le t \le \pi$
69. Answers may vary.
71. The ball hit at $35°$ travels about 128 feet. The ball hit at
$50°$ travels about 134 feet.
73. Yes **75.** About 285 feet
77. $[-6, 6, 1]$ by $[-4, 4, 1]$

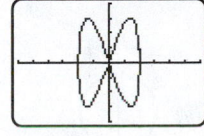

79. $[-6, 6, 1]$ by $[-4, 4, 1]$

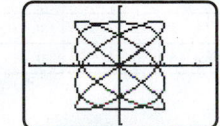

**8.4 EXTENDED AND DISCOVERY
EXERCISES (p. 669)**

1. $F_2 = 100$ **3.** $F_2 = 300$

CHECKING BASIC CONCEPTS FOR SECTIONS 8.3 AND 8.4 (p. 669)

1. (a) $v = \langle 4, 4 \rangle$ **(b)** $\|v\| = 4\sqrt{2}$ **(c)** $\langle 0, 0 \rangle$
3. (a) -9 **(b)** $142.1°$

SECTION 8.5 (pp. 679–680)

1.

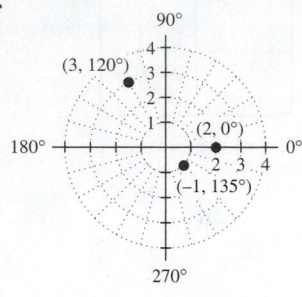

3.

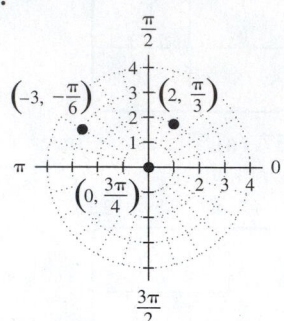

5. Yes **7.** No **9.** Yes

11. $\left(\dfrac{3}{\sqrt{2}}, \dfrac{3}{\sqrt{2}} \right)$, or $\left(\dfrac{3\sqrt{2}}{2}, \dfrac{3\sqrt{2}}{2} \right)$

13. $(0, 10)$ **15.** $(5, 0)$

17. $\left(-\dfrac{3}{2}, -\dfrac{3\sqrt{3}}{2} \right)$ **19.** $(0, -2)$

21. (a) $(3, 90°)$ **(b)** $(-3, -90°)$
23. (a) $(2, 240°)$ **(b)** $(-2, 60°)$
25. (a) $(\sqrt{18}, 315°)$ **(b)** $(-\sqrt{18}, 135°)$
27. $(25, 1.29)$ **29.** $(13, 1.97)$

31.

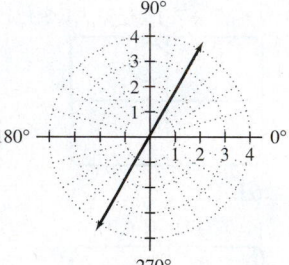

33. (a)

θ	0°	90°	180°	270°
r	2	2	2	2

(b)

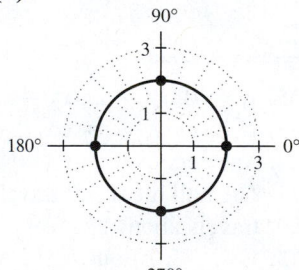

35. (a)

θ	0°	90°	180°	270°
r	0	3	0	-3

(b)

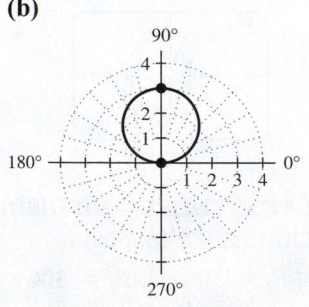

37. (a)

θ	0°	90°	180°	270°
r	2	4	2	0

(b)

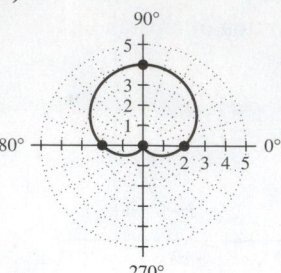

39. (a)

θ	0°	90°	180°	270°
r	1	2	3	2

(b)

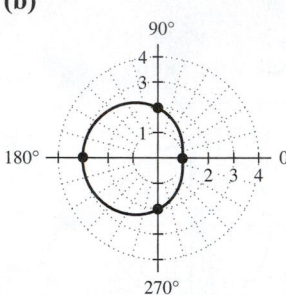

41.

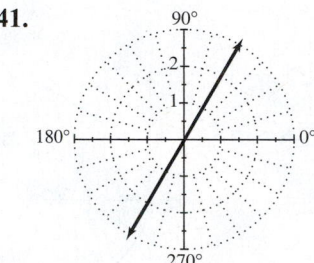

43.

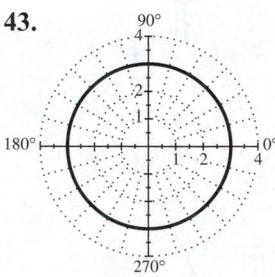

45.

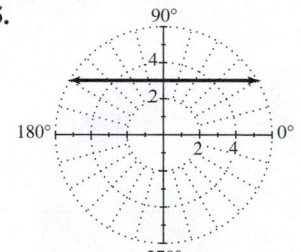

47.

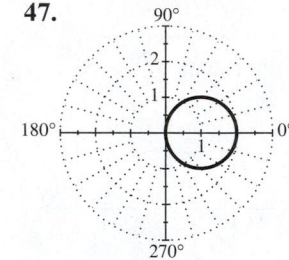

49.

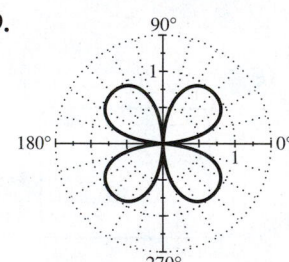

51.

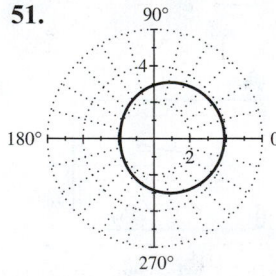

53.

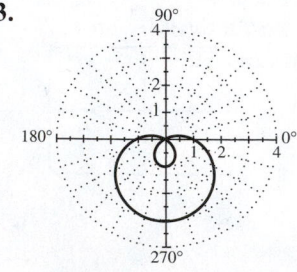

55.

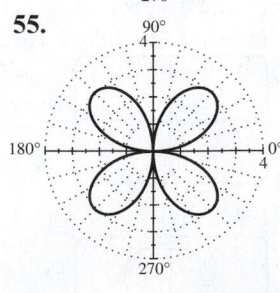

57.

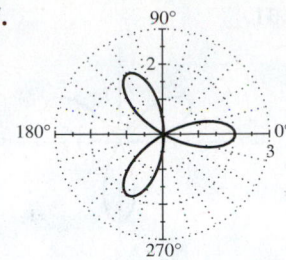

59. $r = 3 \csc \theta$ **61.** $\theta = \frac{\pi}{4}$ **63.** $r = 3$
65. $r = 2 \cos \theta$ **67.** $x^2 + y^2 = 9$ **69.** $x = 2$
71. $2x + 4y = 3$ **73.** $x^2 + y^2 = x$

75.
$[-9.4, 9.4, 1]$ by $[-6.2, 6.2, 1]$

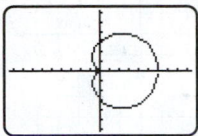

77.
$[-9.4, 9.4, 1]$ by $[-6.2, 6.2, 1]$

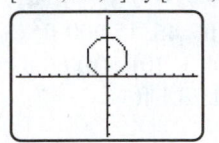

79.
$[-9.4, 9.4, 1]$ by $[-6.2, 6.2, 1]$

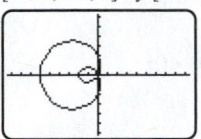

81.
$[-9.4, 9.4, 1]$ by $[-6.2, 6.2, 1]$

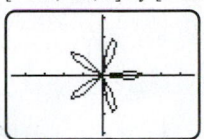

83. $[-4.7, 4.7, 1]$ by $[-3.1, 3.1, 1]$

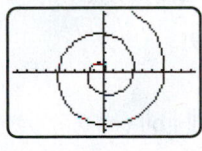

85. Use radian mode
with $0 \leq \theta \leq \frac{9\pi}{2}$.
$[-9.4, 9.4, 1]$ by $[-6.2, 6.2, 1]$

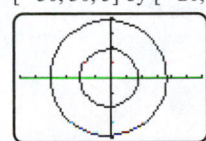

87. Let $r_1 = \sqrt{2 \sin (2\theta)}$
and $r_2 = -\sqrt{2 \sin (2\theta)}$.
$[-3, 3, 1]$ by $[-2, 2, 1]$

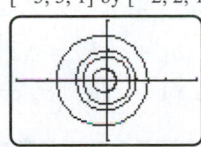

89. $30°, 150°$ **91.** $210°, 330°$ **93.** $30°, 150°$

95. $[-30, 30, 5]$ by $[-20, 20, 5]$

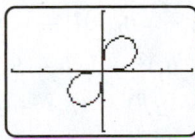

97. $[-3, 3, 1]$ by $[-2, 2, 1]$

99. The radio signal can be received inside the "figure eight."
This region is generally in an east–west direction from the
two towers, with a maximum distance of 200 miles.
$[-300, 300, 100]$ by $[-200, 200, 100]$

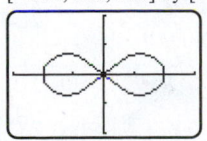

8.5 EXTENDED AND DISCOVERY EXERCISES (p. 681)

1.
$$y = x \tan (\ln (x^2 + y^2))$$
$$\frac{y}{x} = \tan (\ln r^2)$$
$$\tan \theta = \tan (\ln r^2)$$
$$\theta = \ln r^2$$
$$\theta = 2 \ln r$$
$$\frac{\theta}{2} = \ln r$$
$$e^{\theta/2} = r$$

SECTION 8.6 (pp. 689–691)

1.

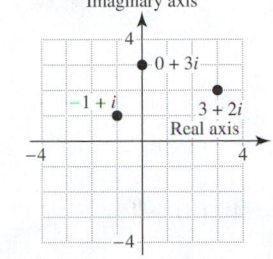

3.

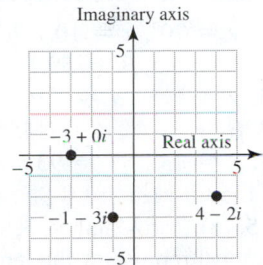

5. $\sqrt{2}$ **7.** 13 **9.** 6 **11.** $\sqrt{13}$
13. $\sqrt{2} (\cos 135° + i \sin 135°)$
15. $5(\cos 0° + i \sin 0°)$ **17.** $4(\cos 90° + i \sin 90°)$
19. $2(\cos 120° + i \sin 120°)$ **21.** $2(\cos 30° + i \sin 30°)$
23. $2(\cos \pi + i \sin \pi)$ **25.** $\sqrt{8} \left(\cos \frac{3\pi}{4} + i \sin \frac{3\pi}{4} \right)$
27. -5 **29.** $\sqrt{2} + i\sqrt{2}$ **31.** $\sqrt{3} + i$ **33.** 3
35. $z_1 z_2 = \frac{27}{2} + \frac{27\sqrt{3}}{2} i, \frac{z_1}{z_2} = \frac{3\sqrt{3}}{2} + \frac{3}{2}i$
37. $z_1 z_2 = -6, \frac{z_1}{z_2} = 6i$
39. $z_1 z_2 = \frac{\sqrt{3}}{2} - \frac{1}{2}i, \frac{z_1}{z_2} = \frac{1}{2} + \frac{\sqrt{3}}{2}i$ **41.** $8i$ **43.** 1
45. $-\frac{25}{2} + \frac{25\sqrt{3}}{2}i$ **47.** $-2 + 2i$ **49.** $-16\sqrt{3} + 16i$
51. $1 + i\sqrt{3}, -1 - i\sqrt{3}$ **53.** $-1, \frac{1}{2} + \frac{\sqrt{3}}{2} i, \frac{1}{2} - \frac{\sqrt{3}}{2} i$
55. $\frac{\sqrt{2}}{2} + \frac{\sqrt{2}}{2} i, -\frac{\sqrt{2}}{2} - \frac{\sqrt{2}}{2} i$ **57.** $-2, 1 + i\sqrt{3}, 1 - i\sqrt{3}$
59. $-4i, 2\sqrt{3} + 2i, -2\sqrt{3} + 2i$
61. $\pm 3, \pm 3i$ **63.** Yes   **65.** No
67. (a) $Z = 110 + 32i$ **(b)** $\sqrt{13,124} \approx 114.6$ ohms

8.6 EXTENDED AND DISCOVERY EXERCISES (p. 691)

1. (a) $\langle 2 \cos 30°, 2 \sin 30° \rangle$ **(b)** $(2, 30°)$
(c) $2(\cos 30° + i \sin 30°)$
3. (a) $\langle 5 \cos 323.1°, 5 \sin 323.1° \rangle$ **(b)** $(5, 323.1°)$
(c) $5(\cos 323.1° + i \sin 323.1°)$

CHECKING BASIC CONCEPTS FOR SECTIONS 8.5 AND 8.6 (p. 691)

1.
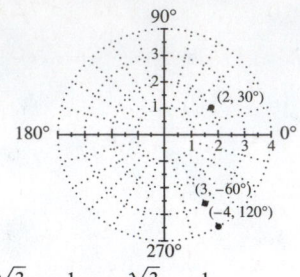

3.
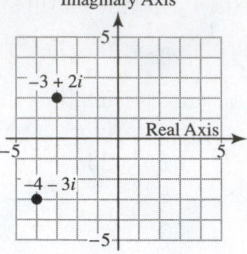

5. $\frac{\sqrt{3}}{2} + \frac{1}{2}i, -\frac{\sqrt{3}}{2} + \frac{1}{2}i, -i$

CHAPTER 8 REVIEW EXERCISES (pp. 694–696)

1. $\gamma = 70°, a = 10.1, b \approx 6.9$
3. $a \approx 5.5, \beta \approx 59.1°, \gamma \approx 78.9°$
5. $\gamma = 115°, a \approx 5.9, c \approx 16.4$
7. $\beta \approx 14.4°, \alpha \approx 145.6°, a \approx 18.2$
9. $c \approx 13.2, \beta \approx 93.6°, \alpha \approx 51.4°$
11. 53.0 **13.** 891.4
15. (a) Horizontal = 3, vertical = 4 (b) $\|\mathbf{v}\| = 5$
(c) $\|\mathbf{v}\|$ represents the length of $\mathbf{v}$.

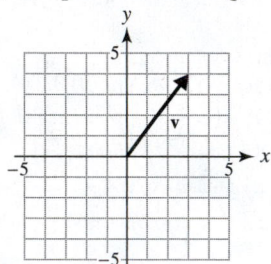

17. (a)
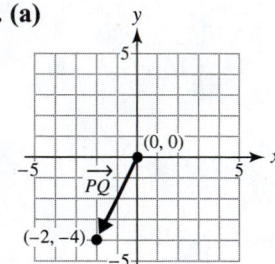

(b) $\overrightarrow{PQ} = -2\mathbf{i} - 4\mathbf{j}$ (c) $\|\overrightarrow{PQ}\| = \sqrt{20}$
19. (a) $2\mathbf{a} = \langle 6, -4 \rangle$ (b) $\mathbf{a} - 3\mathbf{b} = \langle 0, -5 \rangle$
(c) $\mathbf{a} \cdot \mathbf{b} = 1$ (d) $\theta \approx 78.7°$
21. (a) $2\mathbf{a} = 4\mathbf{i} + 4\mathbf{j}$ (b) $\mathbf{a} - 3\mathbf{b} = -\mathbf{i} - \mathbf{j}$
(c) $\mathbf{a} \cdot \mathbf{b} = 4$ (d) $\theta = 0°$
23. About 207.1 lb
25. $[-4.7, 4.7, 1]$ by $[-3.1, 3.1, 1]$ **27.** $[-4.7, 4.7, 1]$ by $[-3.1, 3.1, 1]$

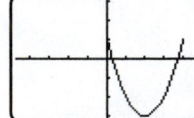 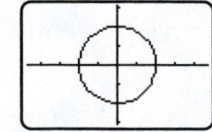

29.
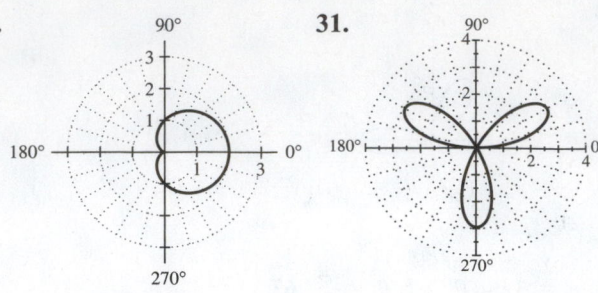

31.

33.
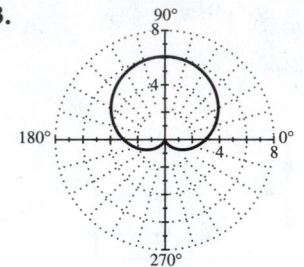

35.
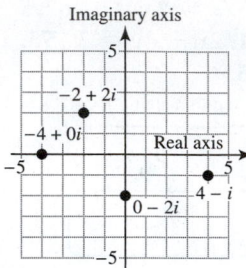

37. $z_1 z_2 = -8, \frac{z_1}{z_2} = -1 + i\sqrt{3}$
39. $\sqrt{3} + i, -\sqrt{3} - i$ **41.** About 701.6 mi
43. 7204 ft **45.** 15,600 ft^2
47. (a) $60\mathbf{i} + 10\mathbf{j}$ (b) 60.8 cm (c) $80\mathbf{i} + 40\mathbf{j}$
49. About 78.1 ft

CHAPTER 8 EXTENDED AND DISCOVERY EXERCISES (pp. 696–697)

1. (a) About 56 miles per second
(b) About 87 miles per second
3. (a) 6.02 in. (b) About 7470 ft
5. $\|\mathbf{a} - \mathbf{b}\|^2 = \|\mathbf{a}\|^2 + \|\mathbf{b}\|^2 - 2\|\mathbf{a}\| \|\mathbf{b}\| \cos\theta$
$\Rightarrow \|\mathbf{a} - \mathbf{b}\|^2 - \|\mathbf{a}\|^2 - \|\mathbf{b}\|^2 = -2\|\mathbf{a}\| \|\mathbf{b}\| \cos\theta$
$\Rightarrow \left(\sqrt{(a_1 - b_1)^2 + (a_2 - b_2)^2}\right)^2 - \left(\sqrt{a_1^2 + a_2^2}\right)^2$
$- \left(\sqrt{b_1^2 + b_2^2}\right)^2 = -2\|\mathbf{a}\| \|\mathbf{b}\| \cos\theta$
$\Rightarrow (a_1 - b_1)^2 + (a_2 - b_2)^2$
$- (a_1^2 + a_2^2) - (b_1^2 + b_2^2) = -2\|\mathbf{a}\| \|\mathbf{b}\| \cos\theta$
$\Rightarrow a_1^2 - 2a_1 b_1 + b_1^2 + a_2^2 - 2a_2 b_2$
$+ b_2^2 - a_1^2 - a_2^2 - b_1^2 - b_2^2 = -2\|\mathbf{a}\| \|\mathbf{b}\| \cos\theta$
$\Rightarrow -2a_1 b_1 - 2a_2 b_2 = -2\|\mathbf{a}\| \|\mathbf{b}\| \cos\theta$
$\Rightarrow a_1 b_1 + a_2 b_2 = \|\mathbf{a}\| \|\mathbf{b}\| \cos\theta$
$\Rightarrow \mathbf{a} \cdot \mathbf{b} = \|\mathbf{a}\| \|\mathbf{b}\| \cos\theta$

CHAPTERS 1–8 CUMULATIVE REVIEW EXERCISES (pp. 698–699)

1. $\sqrt{65}$ **3.** $\{x \mid x \le 4\}$; 3 **5.** $8x + 4h$
7. x-intercept: -3; y-intercept: -4

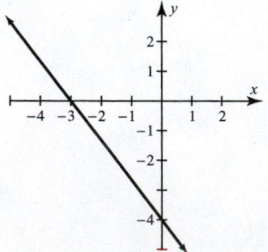

9. (a) $-\frac{1}{2}, \frac{11}{2}$ **(b)** $-4, \frac{1}{3}$ **(c)** $-1, 0, 1$
(d) $\pm i, \pm \sqrt{3}$ **(e)** $\frac{\ln(51/2)}{3} \approx 1.08$ **(f)** 8 **(g)** $\frac{\pi}{6}, \frac{5\pi}{6}$
(h) $\frac{\pi}{3} + \frac{\pi}{2}n$ **(i)** $\frac{\pi}{3} + 2\pi n, \pi + 2\pi n, \frac{5\pi}{3} + 2\pi n$
11. (a) $\left(\frac{13}{5}, \infty\right)$ **(b)** $[-1, 2]$
(c) $(-\infty, -1) \cup (3, \infty)$ **(d)** $(-2, 0) \cup (2, \infty)$
(e) $[0, 1)$ **(f)** $\left[-\frac{8}{3}, \frac{8}{3}\right]$
13. (a) $3x^2 - 6x + 11 + \frac{-20}{x+2}$ **(b)** $x^2 - x + \frac{-1}{2x-1}$
15. $D = \left\{x \mid x \neq \frac{2}{3}\right\}$; vertical asymptote: $x = \frac{2}{3}$; horizontal asymptote: $y = -1$
17. $f^{-1}(x) = \frac{x+2}{3}$ **19.** $C = 5000; a = 2^{2/3}$
21. (a) -3 **(b)** -3 **(c)** $\frac{1}{3}$ **(d)** 3 **23.** $\frac{5\pi}{4}$
25. $\sin\theta = \frac{5}{13}, \cos\theta = \frac{12}{13}, \tan\theta = \frac{5}{12}, \csc\theta = \frac{13}{5},$
$\sec\theta = \frac{13}{12}, \cot\theta = \frac{12}{5}$
27. $\sin\theta = -\frac{7}{25}, \cos\theta = -\frac{24}{25}, \tan\theta = \frac{7}{24}, \csc\theta = -\frac{25}{7},$
$\sec\theta = -\frac{25}{24}, \cot\theta = \frac{24}{7}$
29. $c = 13, \theta \approx 22.6°, \beta \approx 67.4°$ **31.** $(\cot\theta - 1)^2$
33. (a) $\beta = 96°, a \approx 7.8, c \approx 12.0$
(b) $\beta \approx 25.4°, \gamma \approx 123.6°, c \approx 9.7$
(c) $\alpha \approx 47.0°, \gamma \approx 77.0°, b \approx 6.8$
(d) $\alpha \approx 46.6°, \beta \approx 57.9°, \gamma \approx 75.5°$
35. (a) 25 **(b)** $\langle -31, 96 \rangle$ **(c)** -323
(d) About 173.6°
37.

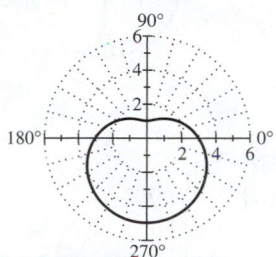

39. 15 by 19 inches
41. 12 by 6 by 4 inches or 7.4 by 3.7 by 10.5 inches
43. 2.5 feet **45.** 56.2°
47. (a) About 247 feet **(b)** About 18,988 ft^2

CHAPTER 9: Systems of Equations and Inequalities

SECTION 9.1 (pp. 714–719)

1. 20; the area of a triangle with base 5 and height 8 is 20.
3. 13 **5.** -18 **7.** $\frac{4}{5}$ **9.** $f(x, y) = y + 2x$
11. $f(x, y) = \frac{xy}{1+x}$ **13.** $x = \frac{4y+7}{3}; y = \frac{3x-7}{4}$
15. $x = y^2 + 5; y = \pm\sqrt{x-5}$ **17.** $x = 2y; y = \frac{x}{2}$
19. $(2, 1)$; linear **21.** $(4, -3)$; nonlinear **23.** $(2, 2)$
25. $\left(\frac{1}{2}, -2\right)$ **27.** Consistent with solution $(2, 2)$
29. Inconsistent; no solutions

31.

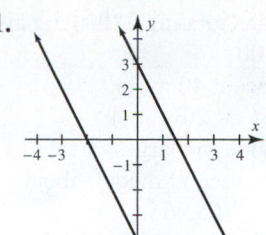

No solutions; inconsistent

33.

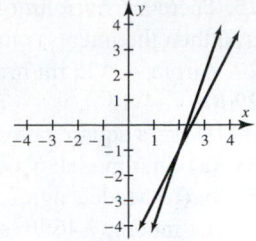

$(2, -1)$; consistent, independent

35.

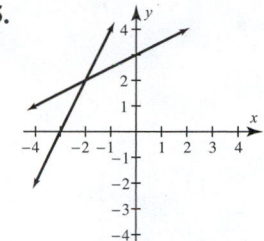

$(-2, 2)$; consistent, independent

37.

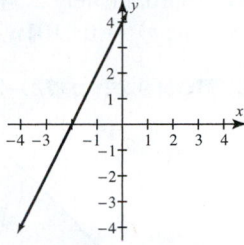

$\{(x, y) \mid 2x - y = -4\}$; consistent, dependent

39. $(-2, 1)$ **41.** $\left(\frac{1}{2}, 2\right)$ **43.** $(6, 8)$
45. $\{(x, y) \mid 3x - 2y = 5\}$ **47.** No solutions **49.** $(4, 3)$
51. $(-2, 4), (0, 0)$ **53.** $(2, 4), (4, 2)$ **55.** $(2, 4), (-2, -4)$
57. No real solutions **59.** $\{(x, y) \mid 2x^2 - y = 5\}$
61. $(-2, 0), (2, 0)$ **63.** $(-2, -2), (0, 0), (2, 2)$
65. $x - y = 2, 2x + 2y = 38$, where x is width and y is height; 10.5 in. wide, 8.5 in. high
67. $x + y = 75, 4x + 7y = 456$, where x is the number of child tickets and y is the number of adult tickets; 23 child tickets, 52 adult tickets **69.** $l = 7, w = 5$ **71.** $(14, 6)$; consistent, independent **73.** $(1, 3)$; consistent, independent
75. $\{(x, y) \mid x + y = 500\}$; consistent, dependent
77. No solutions; inconsistent
79. $\left(-\frac{11}{7}, \frac{12}{7}\right)$; consistent, independent
81. $(2, -4)$ **83.** $\{(x, y) \mid 7x - 3y = -17\}$
85. No solutions **87.** $(10, 20)$ **89.** $(2, 1)$
91. $(-5, -4)$ **93.** $(3, 3), (-3, 3)$
95. $(4, 3), (-4, 3), (3, 4), (-3, 4)$ **97.** $(2, 0), (-2, 0)$
99. $(-\sqrt{8}, -\sqrt{8}), (\sqrt{8}, \sqrt{8})$ **101.** $(6, 2), (-2, -6)$
103. $(1, -1)$ **105.** $(1, 2)$
107. $(-1.588, 0.239), (0.164, 1.487), (1.924, -0.351)$
109. $(1.220, 0.429)$ **111.** $(0.714, -0.169)$
113. (a) $x + y = 670, x - y = 98$
(b) $(384, 286)$ **(c)** Consistent; independent
115. $W_1 = \frac{300}{1 + \sqrt{3}} \approx 109.8$ lb,

$W_2 = \frac{300\sqrt{3}}{\sqrt{6} + \sqrt{2}} \approx 134.5$ lb

117. $r \approx 1.538$ in., $h \approx 6.724$ in.
119. 12 by 12 by 4 in. or 9.10 by 9.10 by 6.96 in.
121. (a) $x + y = 11,693, x - y = 437$
(b) & (c) $(6065, 5628)$
123. (a) $x + y = 3000, 0.08x + 0.10y = 264$
(b) \$1800 at 8%; \$1200 at 10%

125. There are no solutions. If loans totaling $3000 are at 10%, then the interest must be $300.
127. Airplane: 520 mi/hr; wind speed: 40 mi/hr
129. (a) $l = 13$ ft, $w = 7$ ft **(b)** $A = 20w - w^2$
(c) 100 ft^2; a square pen will provide the largest area.
131. (a) First model: about 245 lb; second model: about 253 lb **(b)** Models agree when $h \approx 65.96$ in.
(c) First model: 7.46 lb; second model: 7.93 lb
133. About 1.77 m^2 **135.** $S(60, 157.48) \approx 1.6$ m^2
137. 0.51 **139.** 32.4 **141.** Approximately 10,823 watts
143. Approximately 2.54 cords
145. $S(w, h) = 0.0101w^{0.425}h^{0.725}$

SECTION 9.2 (pp. 727–730)

1.
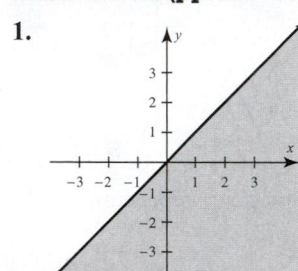

3.
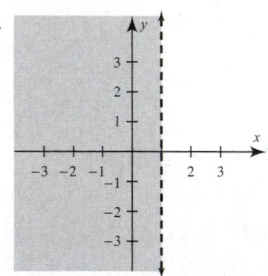

5.
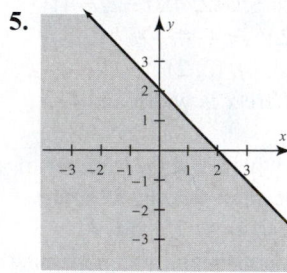

7.
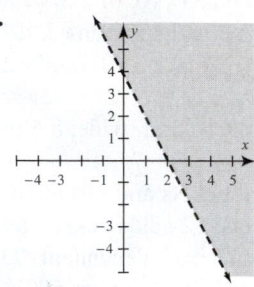

9.
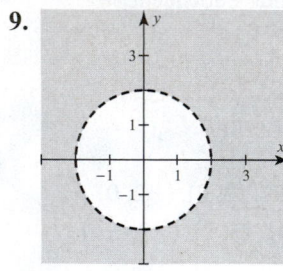

11.
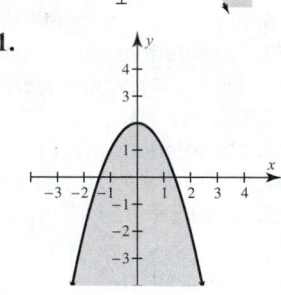

13. c; (2, 3) (answers may vary)
15. d; (−1, −1) (answers may vary)
17. (0, 2) (answers may vary)
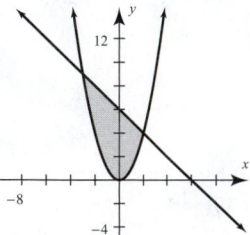

19. (0, 0) (answers may vary)
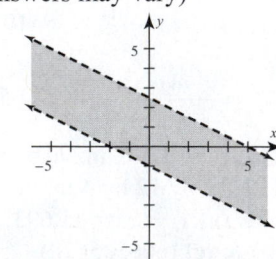

21. (−1, 1) (answers may vary)
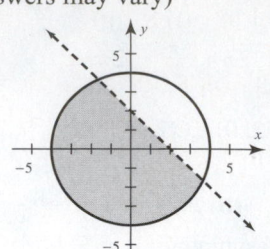

23. (2, 1) (answers may vary)
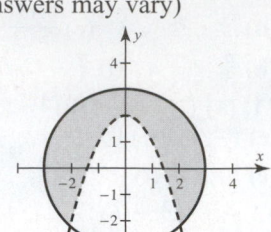

25.
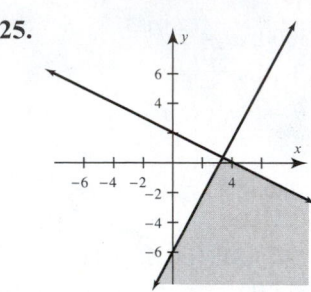

27.
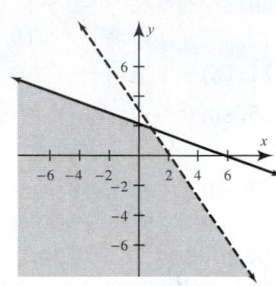

29.
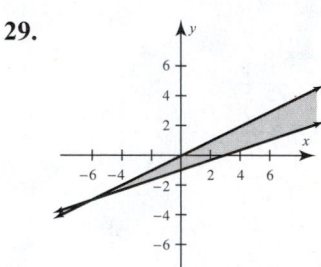

31.
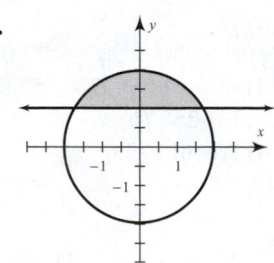

33.
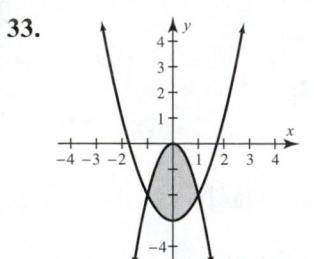

35.
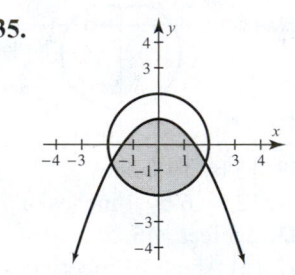

37. Jitters **39.** 180 pounds or more
41. $x = 300, y = 350$
43. This individual weighs less than recommended for his or her height.
45. $25h - 7w \le 800, 5h - w \ge 170$
47.
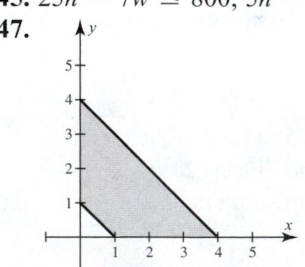

49.
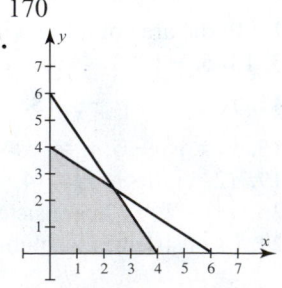

51. Maximum: 65; minimum: 8
53. Maximum: 66; minimum: 3
55. Maximum: 100; minimum: 0
57. $x + y \le 4, x \ge 0, y \ge 0$
59. Minimum: 6
61. Maximum: $z = 56$; minimum: $z = 24$

63. 25 radios, 30 CD players
65. 2.4 units of Brand A, 1.2 units of Brand B
67. $600 **69.** Part X: 9, part Y: 4

CHECKING BASIC CONCEPTS FOR SECTIONS 9.1 AND 9.2 (p. 730)

1. $d(13, 18) = 20$ **3.** $y = \pm\sqrt{z - x^2}$
5.

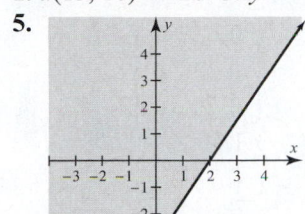

7. (a) $x + y = 220, 4.5x - y = 0$
(b) $(40, 180)$; in 2010 40 million HDTVs were sold with plasma screens and 180 million were sold with LCD screens.

SECTION 9.3 (pp. 736–737)

1. No **3.** 2 **5.** Dependent
7. $(0, 2, -2)$ is not, but $(-1, 3, -2)$ is a solution.
9. Both are solutions. **11.** $(1, 2, 3)$ **13.** $(1, 0, 1)$
15. $\left(\frac{1}{2}, \frac{1}{2}, -2\right)$ **17.** $\left(\frac{z+3}{2}, \frac{-z+3}{2}, z\right)$
19. No solutions **21.** $\left(-\frac{5}{2}, -2, 4\right)$ **23.** No solutions
25. $\left(-\frac{1}{2}, \frac{1}{2}, -\frac{1}{2}\right)$ **27.** $\left(\frac{-5z + 18}{13}, \frac{-6z + 19}{13}, z\right)$
29. $\left(8, -11, -\frac{1}{2}\right)$ **31.** $(2, 3, -4)$
33. 120 child, 280 student, and 100 adult tickets
35. No solutions; at least one student was charged incorrectly.
37. (a)
$$\begin{aligned} x + y + z &= 180 \\ x \quad\;\; - z &= 25 \\ -x + y + z &= 30 \end{aligned}$$
(b) 75°, 55°, 50°
39. $2500 at 5%, $7500 at 7%, $10,000 at 10%
41. (a)
$$\begin{aligned} N + P + K &= 80 \\ N + P - K &= 8 \\ 9P - K &= 0 \end{aligned}$$
(b) $(40, 4, 36)$; 40 lb of nitrogen, 4 lb of phosphorus, 36 lb of potassium

SECTION 9.4 (pp. 750–753)

1. 3×1 **3.** 2×2 **5.** 3×2
7. Dimension: 2×3
$$\left[\begin{array}{cc|c} 5 & -2 & 3 \\ -1 & 3 & -1 \end{array}\right]$$

9. Dimension: 3×4
$$\left[\begin{array}{ccc|c} -3 & 2 & 1 & -4 \\ 5 & 0 & -1 & 9 \\ 1 & -3 & -6 & -9 \end{array}\right]$$

11. $\begin{aligned} 3x + 2y &= 4 \\ y &= 5 \end{aligned}$

13. $\begin{aligned} 3x + y + 4z &= 0 \\ 5y + 8z &= -1 \\ - 7z &= 1 \end{aligned}$

15. (a) Yes **(b)** No **(c)** Yes **17.** $(5, -1)$ **19.** $(2, 0)$
21. $(2, 3, 1)$ **23.** $(3 - 3z, 1 + 2z, z)$ **25.** No solutions

27. $\left[\begin{array}{ccc|c} 1 & -2 & 3 & 5 \\ -3 & 5 & 3 & 2 \\ 1 & 2 & 1 & -2 \end{array}\right]$ **29.** $\left[\begin{array}{ccc|c} 1 & -1 & 1 & 2 \\ 0 & 1 & -1 & 2 \\ 0 & 8 & -1 & 3 \end{array}\right]$

31. $(-17, 10)$ **33.** $(0, 2, -1)$ **35.** $(-1, 2, 4)$
37. $(3, 2, 1)$ **39.** No solutions
41. $\left(-1 - z, \frac{9 - 5z}{2}, z\right)$ **43.** $(1, 1, 1)$
45. $\left(\frac{1}{2}, -\frac{1}{2}, 4\right)$ **47.** $(-2, 5, -1)$
49. No solutions **51.** $(-1, -2, 3)$
53. $\left(\frac{3z + 1}{7}, \frac{11z - 15}{7}, z\right)$ **55.** $(12, 3)$
57. $\left(-2, 4, \frac{1}{2}\right)$ **59.** $(4 - 2z, z - 3, z)$
61. No solutions **63.** $(3, 2)$ **65.** $(-2, 1, 3)$
67. $(-2, 5, 7)$ **69.** No solutions
71. $(-9.226, -9.167, 2.440)$
73. $(5.211, 3.739, -4.655)$
75. $(7.993, 1.609, -0.401)$
77. Pump 1: 12 hours; pumps 2 and 3: 24 hours
79. (a) $F = 0.5714N + 0.4571R - 2014$ **(b)** $5700
81. $(3.53, 1.62, 1.91)$
83. (a)
$$\begin{aligned} x + y + z &= 5000 \\ x + y - z &= 0 \\ 0.08x + 0.11y + 0.14z &= 595, \end{aligned}$$
where x is amount invested at 8%, y is amount invested at 11%, and z is amount invested at 14%
(b) $1000 at 8%; $1500 at 11%; $2500 at 14%
85. (a) At intersection A, incoming traffic is equal to $x + 5$. The outgoing traffic is given by $y + 7$. Therefore, $x + 5 = y + 7$. The other equations can be justified in a similar way.
(b) The three equations can be written as
$$\begin{aligned} x - y &= 2 \\ x - z &= 3 \\ y - z &= 1 \end{aligned}$$
The solution can be written as $\{(z + 3, z + 1, z) \mid z \geq 0\}$.
(c) There are infinitely many solutions, since some cars could be driving around the block continually.
87. (a)
$$\begin{aligned} a + b + c &= 3 \\ 25a + 5b + c &= 29 \\ 36a + 6b + c &= 40 \end{aligned}$$
(b) $f(x) = 0.9x^2 + 1.1x + 1$
(c) $[-0.5, 10, 1]$ by $[-5, 90, 10]$

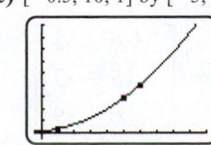

(d) After 9 quarters predicted sales are $f(9) = 83.8$ million (answers may vary).
89. (a)
$$\begin{aligned} 1990^2a + 1990b + c &= 11 \\ 2010^2a + 2010b + c &= 10 \\ 2030^2a + 2030b + c &= 6 \end{aligned}$$

(b) $f(x) = -0.00375x^2 + 14.95x - 14,889.125$

(c) [1985, 2035, 5] by [5, 12, 1]

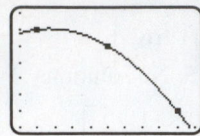

(d) In 2015 the predicted ratio is $f(2015) \approx 9.3$ (answers may vary).

9.4 EXTENDED AND DISCOVERY EXERCISES (p. 753)

1. $(1, -1, 2, 0)$

CHECKING BASIC CONCEPTS FOR SECTIONS 9.3 AND 9.4 (p. 753)

1. (a) $(3, 2, -1)$ **(b)** $\left(\frac{4 - 5z}{3}, \frac{5 - z}{3}, z\right)$ **(c)** No solutions

3. $(2, -1, 0)$

SECTION 9.5 (pp. 764–768)

1. (a) $a_{12} = 3, b_{32} = 1, b_{22} = 0$ **(b)** -2 **(c)** $x = 3$

3. $x = 1, y = 1$ **5.** Not possible

7. (a) $A + B = \begin{bmatrix} 3 & 3 \\ 3 & 3 \end{bmatrix}$ **(b)** $B + A = \begin{bmatrix} 3 & 3 \\ 3 & 3 \end{bmatrix}$

(c) $A - B = \begin{bmatrix} 5 & -5 \\ -5 & 5 \end{bmatrix}$

9. (a) $A + B = \begin{bmatrix} 14 & 9 & -3 \\ 4 & -10 & 14 \\ 4 & 11 & 16 \end{bmatrix}$

(b) $B + A = \begin{bmatrix} 14 & 9 & -3 \\ 4 & -10 & 14 \\ 4 & 11 & 16 \end{bmatrix}$

(c) $A - B = \begin{bmatrix} -8 & -1 & 1 \\ -4 & 4 & -10 \\ -8 & -1 & 4 \end{bmatrix}$

11. (a) $A + B = \begin{bmatrix} 1 & -6 \\ 1 & 4 \end{bmatrix}$ **(b)** $3A = \begin{bmatrix} 6 & -18 \\ 9 & 3 \end{bmatrix}$

(c) $2A - 3B = \begin{bmatrix} 7 & -12 \\ 12 & -7 \end{bmatrix}$

13. (a) $A + B$ is undefined.

(b) $3A = \begin{bmatrix} 3 & -3 & 0 \\ 3 & 15 & 27 \\ -12 & 24 & -15 \end{bmatrix}$

(c) $2A - 3B$ is undefined.

15. (a) $A + B = \begin{bmatrix} 0 & -2 \\ -2 & 2 \\ 9 & -8 \end{bmatrix}$ **(b)** $3A = \begin{bmatrix} -6 & -3 \\ -15 & 3 \\ 6 & -9 \end{bmatrix}$

(c) $2A - 3B = \begin{bmatrix} -10 & 1 \\ -19 & -1 \\ -17 & 9 \end{bmatrix}$

17. $\begin{bmatrix} 0 & 2 \\ 13 & -5 \\ 0 & 1 \end{bmatrix}$ **19.** $\begin{bmatrix} 2 & 6 \\ 11 & -9 \end{bmatrix}$

21. $\begin{bmatrix} 7 & 4 & 7 \\ 4 & 7 & 7 \\ 4 & 7 & 7 \end{bmatrix}$ **23.** $A = \begin{bmatrix} 1 & 2 & 1 \\ 1 & 2 & 1 \\ 1 & 2 & 1 \end{bmatrix}$

25. $B = \begin{bmatrix} -1 & 1 & -1 \\ -1 & 1 & -1 \\ -1 & 1 & -1 \end{bmatrix}, A + B = \begin{bmatrix} 0 & 3 & 0 \\ 0 & 3 & 0 \\ 0 & 3 & 0 \end{bmatrix}$

27. $AB = \begin{bmatrix} -3 & 1 \\ -4 & 6 \end{bmatrix}, BA = \begin{bmatrix} 4 & 2 \\ 5 & -1 \end{bmatrix}$

29. AB and BA are undefined.

31. $AB = \begin{bmatrix} -15 & 22 & -9 \\ -2 & 5 & -3 \\ -32 & 18 & 6 \end{bmatrix}, BA = \begin{bmatrix} 5 & 14 \\ 20 & -9 \end{bmatrix}$

33. AB is undefined. $BA = \begin{bmatrix} -1 & -3 & 19 \\ -1 & 1 & -39 \end{bmatrix}$

35. AB and BA are undefined.

37. $AB = \begin{bmatrix} -1 & 15 & -2 \\ 0 & 1 & 3 \\ -1 & 14 & -5 \end{bmatrix}, BA = \begin{bmatrix} 0 & 6 & 0 \\ 6 & -5 & 9 \\ 0 & 1 & 0 \end{bmatrix}$

39. $AB = \begin{bmatrix} -1 \\ 6 \end{bmatrix}, BA$ is undefined.

41. $AB = \begin{bmatrix} -7 & 8 & 7 \\ 18 & -32 & -8 \end{bmatrix}, BA$ is undefined.

43. $AB = \begin{bmatrix} -5 \\ 10 \\ 17 \end{bmatrix}, BA$ is undefined.

45. $\begin{bmatrix} -19 & 19 & 11 \\ 21 & -7 & -48 \\ -22 & 23 & -58 \end{bmatrix}$ **47.** $\begin{bmatrix} 83 & 32 & 92 \\ 10 & -63 & -8 \\ 210 & 56 & 93 \end{bmatrix}$

49. They both equal $\begin{bmatrix} 36 & 36 & 8 \\ -15 & -38 & -4 \\ -11 & 13 & 10 \end{bmatrix}$. The distributive property appears to hold for matrices.

51. They both equal $\begin{bmatrix} 50 & 3 & 12 \\ -6 & 55 & 8 \\ 27 & -3 & 29 \end{bmatrix}$. Matrices appear to conform to rules of algebra except that $AB \neq BA$.

53. $\begin{bmatrix} 0 & 0 & 1 & 1 \\ 1 & 0 & 0 & 0 \\ 1 & 0 & 0 & 1 \\ 1 & 1 & 1 & 0 \end{bmatrix}$

55. Person 2

57. Person 1 → ... Person 2, Person 3 ↔ ... Person 4

59. No one likes Person 4.

61. $B = \begin{bmatrix} 3 & 3 & 3 \\ 3 & 3 & 3 \\ 3 & 3 & 3 \end{bmatrix}$, $B - A = \begin{bmatrix} 3 & 0 & 3 \\ 3 & 0 & 3 \\ 3 & 0 & 3 \end{bmatrix}$

63. $A = \begin{bmatrix} 3 & 3 & 3 & 3 \\ 3 & 0 & 0 & 0 \\ 3 & 3 & 3 & 0 \\ 3 & 0 & 0 & 0 \\ 3 & 0 & 0 & 0 \end{bmatrix}$

65. (a) One possibility is $A = \begin{bmatrix} 3 & 3 & 3 & 3 \\ 0 & 0 & 3 & 0 \\ 0 & 3 & 0 & 0 \\ 3 & 3 & 3 & 3 \end{bmatrix}$.

(b) $B = \begin{bmatrix} 3 & 3 & 3 & 3 \\ 3 & 3 & 3 & 3 \\ 3 & 3 & 3 & 3 \\ 3 & 3 & 3 & 3 \end{bmatrix}$

67. (a) One possibility is $A = \begin{bmatrix} 3 & 0 & 0 & 0 \\ 3 & 0 & 0 & 0 \\ 3 & 0 & 0 & 0 \\ 3 & 3 & 3 & 3 \end{bmatrix}$.

(b) $B = \begin{bmatrix} 3 & 3 & 3 & 3 \\ 3 & 3 & 3 & 3 \\ 3 & 3 & 3 & 3 \\ 3 & 3 & 3 & 3 \end{bmatrix}$

69. (a) $A = \begin{bmatrix} 12 & 4 \\ 8 & 7 \end{bmatrix}$, $B = \begin{bmatrix} 55 \\ 70 \end{bmatrix}$

(b) $AB = \begin{bmatrix} 940 \\ 930 \end{bmatrix}$. Tuition for Student 1 is \$940, and tuition for Student 2 is \$930.

71. (a) $A = \begin{bmatrix} 10 & 5 \\ 9 & 8 \\ 11 & 3 \end{bmatrix}$, $B = \begin{bmatrix} 60 \\ 70 \end{bmatrix}$

(b) $AB = \begin{bmatrix} 950 \\ 1100 \\ 870 \end{bmatrix}$. Tuition for Student 1 is \$950, for Student 2 it is \$1100, and for Student 3 it is \$870.

73. $AB = \begin{bmatrix} 350 \\ 230 \end{bmatrix}$. The total cost of order 1 is \$350, and the total cost of order 2 is \$230.

75. $\begin{bmatrix} 0 & 0 & 1 & 1 \\ 1 & 0 & 0 & 1 \\ 0 & 0 & 0 & 1 \\ 0 & 0 & 1 & 0 \end{bmatrix}$ **77.** $\begin{bmatrix} 0 & 0 & 1 & 1 \\ 0 & 0 & 2 & 1 \\ 0 & 0 & 1 & 0 \\ 0 & 0 & 0 & 1 \end{bmatrix}$

79. There are two different 2-click paths from Page 2 to Page 3

9.5 EXTENDED AND DISCOVERY
EXERCISES (p. 768)

1. Aquamarine is represented by (0.369, 0, 0.067) in CMY.

3. $\begin{bmatrix} R \\ G \\ B \end{bmatrix} = \begin{bmatrix} 1 \\ 1 \\ 1 \end{bmatrix} - \begin{bmatrix} C \\ M \\ Y \end{bmatrix}$

SECTION 9.6 (pp. 777–780)

1. Yes **3.** Yes **5.** No **7.** $k = 1$ **9.** $k = 2.5$ **11.** A **13.** A

15. $\begin{bmatrix} 3 & -2 \\ -1 & 1 \end{bmatrix}$ **17.** $\begin{bmatrix} 5 & 2 \\ 3 & 1 \end{bmatrix}$ **19.** $\begin{bmatrix} -\frac{1}{2} & \frac{5}{2} \\ 1 & -4 \end{bmatrix}$

21. $\begin{bmatrix} 0 & 1 & 0 \\ 0 & 0 & 1 \\ 1 & 0 & 0 \end{bmatrix}$ **23.** $\begin{bmatrix} -2 & 1 & -1 \\ -5 & 2 & -1 \\ 3 & -1 & 1 \end{bmatrix}$

25. $\begin{bmatrix} -10 & 3 & -5 \\ 4 & -1 & 2 \\ -3 & 1 & -1 \end{bmatrix}$ **27.** $\begin{bmatrix} -1 & -1 & -1 \\ -\frac{2}{5} & -\frac{1}{5} & -\frac{4}{5} \\ \frac{1}{5} & \frac{3}{5} & \frac{2}{5} \end{bmatrix}$

29. $\begin{bmatrix} -10 & 30 \\ -4 & 10 \end{bmatrix}$ **31.** $\begin{bmatrix} 0.2 & 0 & 0.4 \\ 0.4 & 0 & -0.2 \\ 1.4 & -1 & -1.2 \end{bmatrix}$

33. $\begin{bmatrix} 0.5 & 0.2 & 2.1 \\ 0 & 0.2 & 1.6 \\ 0 & 0 & -1 \end{bmatrix}$ **35.** $\begin{bmatrix} 0.5 & 0.25 & 0.25 \\ 0.25 & 0.5 & 0.25 \\ 0.25 & 0.25 & 0.5 \end{bmatrix}$

37. $\begin{bmatrix} 1.2\overline{6} & 0.2\overline{6} & 0.0\overline{6} & 0.0\overline{6} \\ 0.2\overline{6} & 0.2\overline{6} & 0.0\overline{6} & 0.0\overline{6} \\ 0.0\overline{6} & 0.0\overline{6} & 0.2\overline{6} & 0.2\overline{6} \\ 0.0\overline{6} & 0.0\overline{6} & 0.2\overline{6} & 1.2\overline{6} \end{bmatrix}$

39. $AX = \begin{bmatrix} 2 & -3 \\ -3 & -4 \end{bmatrix} \begin{bmatrix} x \\ y \end{bmatrix} = \begin{bmatrix} 7 \\ 9 \end{bmatrix} = B$

41. $AX = \begin{bmatrix} \frac{1}{2} & -\frac{3}{2} \\ -1 & 2 \end{bmatrix} \begin{bmatrix} x \\ y \end{bmatrix} = \begin{bmatrix} \frac{1}{4} \\ 5 \end{bmatrix} = B$

43. $AX = \begin{bmatrix} 1 & -2 & 1 \\ 0 & 3 & -1 \\ 5 & -4 & -7 \end{bmatrix} \begin{bmatrix} x \\ y \\ z \end{bmatrix} = \begin{bmatrix} 5 \\ 6 \\ 0 \end{bmatrix} = B$

45. $AX = \begin{bmatrix} 4 & -1 & 3 \\ 1 & 2 & 5 \\ 2 & -3 & 0 \end{bmatrix} \begin{bmatrix} x \\ y \\ z \end{bmatrix} = \begin{bmatrix} -2 \\ 11 \\ -1 \end{bmatrix} = B$

47. (a) $AX = \begin{bmatrix} 1 & 2 \\ 1 & 3 \end{bmatrix} \begin{bmatrix} x \\ y \end{bmatrix} = \begin{bmatrix} 3 \\ 6 \end{bmatrix} = B$ **(b)** $X = \begin{bmatrix} -3 \\ 3 \end{bmatrix}$

49. (a) $AX = \begin{bmatrix} -1 & 2 \\ 3 & -5 \end{bmatrix} \begin{bmatrix} x \\ y \end{bmatrix} = \begin{bmatrix} 5 \\ -2 \end{bmatrix} = B$

(b) $X = \begin{bmatrix} 21 \\ 13 \end{bmatrix}$

51. (a) $AX = \begin{bmatrix} 1 & 0 & 1 \\ 2 & 1 & 3 \\ -1 & 1 & 1 \end{bmatrix} \begin{bmatrix} x \\ y \\ z \end{bmatrix} = \begin{bmatrix} -7 \\ -13 \\ -4 \end{bmatrix} = B$

(b) $X = \begin{bmatrix} 5 \\ 13 \\ -12 \end{bmatrix}$

53. (a) $AX = \begin{bmatrix} 1 & 2 & -1 \\ 2 & 5 & 0 \\ -1 & -1 & 2 \end{bmatrix} \begin{bmatrix} x \\ y \\ z \end{bmatrix} = \begin{bmatrix} 2 \\ -1 \\ 0 \end{bmatrix} = B$

(b) $X = \begin{bmatrix} -23 \\ 9 \\ -7 \end{bmatrix}$

55. (a) $AX = \begin{bmatrix} 1.5 & 3.7 \\ -0.4 & -2.1 \end{bmatrix} \begin{bmatrix} x \\ y \end{bmatrix} = \begin{bmatrix} 0.32 \\ 0.36 \end{bmatrix} = B$

(b) $X = \begin{bmatrix} 1.2 \\ -0.4 \end{bmatrix}$

57. (a) $AX = \begin{bmatrix} 0.08 & -0.7 \\ 1.1 & -0.05 \end{bmatrix} \begin{bmatrix} x \\ y \end{bmatrix} = \begin{bmatrix} -0.504 \\ 0.73 \end{bmatrix} = B$

(b) $X = \begin{bmatrix} 0.7 \\ 0.8 \end{bmatrix}$

59. (a) $AX = \begin{bmatrix} 3.1 & 1.9 & -1 \\ 6.3 & 0 & -9.9 \\ -1 & 1.5 & 7 \end{bmatrix} \begin{bmatrix} x \\ y \\ z \end{bmatrix} = \begin{bmatrix} 1.99 \\ -3.78 \\ 5.3 \end{bmatrix} = B$

(b) $X = \begin{bmatrix} 0.5 \\ 0.6 \\ 0.7 \end{bmatrix}$

61. (a) $AX = \begin{bmatrix} 3 & -1 & 1 \\ 5.8 & -2.1 & 0 \\ -1 & 0 & 2.9 \end{bmatrix} \begin{bmatrix} x \\ y \\ z \end{bmatrix} = \begin{bmatrix} 4.9 \\ -3.8 \\ 3.8 \end{bmatrix} = B$

(b) $X \approx \begin{bmatrix} 9.26 \\ 27.39 \\ 4.50 \end{bmatrix}$

63. (a) $(2, 4)$ **(b)** It will translate $(2, 4)$ to the left 2 units and downward 3 units, back to $(0, 1)$;

$A^{-1} = \begin{bmatrix} 1 & 0 & -2 \\ 0 & 1 & -3 \\ 0 & 0 & 1 \end{bmatrix}$ **(c)** I_3

65. $A = \begin{bmatrix} 1 & 0 & -3 \\ 0 & 1 & -5 \\ 0 & 0 & 1 \end{bmatrix}$ and $A^{-1} = \begin{bmatrix} 1 & 0 & 3 \\ 0 & 1 & 5 \\ 0 & 0 & 1 \end{bmatrix}$.

A^{-1} will translate a point 3 units to the right and 5 units upward.

67. (a) $BX = \begin{bmatrix} -2 \\ 0 \\ 1 \end{bmatrix} = Y$ **(b)** $B^{-1}Y = \begin{bmatrix} -\sqrt{2} \\ -\sqrt{2} \\ 1 \end{bmatrix} = X.$

B^{-1} rotates the point represented by Y counterclockwise $45°$ about the origin.

69. (a) $ABX = \begin{bmatrix} 2 \\ 2 \\ 1 \end{bmatrix} = Y$ **(b)** The net result of A and B

is to translate a point 1 unit to the right and 1 unit upward.

$AB = \begin{bmatrix} 1 & 0 & 1 \\ 0 & 1 & 1 \\ 0 & 0 & 1 \end{bmatrix}$. **(c)** Yes **(d)** Since AB translates

a point 1 unit right and 1 unit upward, the inverse of AB would translate a point 1 unit left and 1 unit downward. Therefore

$(AB)^{-1} = \begin{bmatrix} 1 & 0 & -1 \\ 0 & 1 & -1 \\ 0 & 0 & 1 \end{bmatrix}.$

71. A: $10.99; B: $12.99; C: $14.99
73. (a) $\begin{aligned} a + 1500b + 8c &= 122 \\ a + 2000b + 5c &= 130 \\ a + 2200b + 10c &= 158 \\ a = 30, b = 0.04, c &= 4 \end{aligned}$

(b) $130,000 **75. (a)** $\left(\frac{17}{12}T, \frac{5}{6}T, T\right)$
(b) Service: 85 units; electrical: 50 units

CHECKING BASIC CONCEPTS FOR SECTIONS 9.5 AND 9.6 (p. 780)

1. (a) $A + B = \begin{bmatrix} 0 & 1 & 3 \\ -1 & 5 & 3 \\ 2 & 1 & 0 \end{bmatrix}$

(b) $2A - B = \begin{bmatrix} 3 & -1 & 0 \\ -2 & -2 & 3 \\ 1 & 8 & 0 \end{bmatrix}$

(c) $AB = \begin{bmatrix} 0 & -1 & 2 \\ 3 & -1 & -1 \\ -1 & 13 & 5 \end{bmatrix}$

3. (a) $AX = \begin{bmatrix} 1 & -2 \\ 2 & 3 \end{bmatrix} \begin{bmatrix} x \\ y \end{bmatrix} = \begin{bmatrix} 13 \\ 5 \end{bmatrix} = B; X = \begin{bmatrix} 7 \\ -3 \end{bmatrix}$

(b) $AX = \begin{bmatrix} 1 & -1 & 1 \\ -1 & 1 & 1 \\ 0 & 1 & -1 \end{bmatrix} \begin{bmatrix} x \\ y \\ z \end{bmatrix} = \begin{bmatrix} 2 \\ 4 \\ -1 \end{bmatrix} = B; X = \begin{bmatrix} 1 \\ 2 \\ 3 \end{bmatrix}$

(c) $AX = \begin{bmatrix} 3.1 & -5.3 \\ -0.1 & 1.8 \end{bmatrix} \begin{bmatrix} x \\ y \end{bmatrix} = \begin{bmatrix} -2.682 \\ 0.787 \end{bmatrix} = B;$

$X = \begin{bmatrix} -0.13 \\ 0.43 \end{bmatrix}$

SECTION 9.7 (pp. 787–788)

1. det $A = 1 \neq 0$. A is invertible.
3. det $A = 0$. A is not invertible.
5. $M_{12} = 10, A_{12} = -10$
7. $M_{22} = -15, A_{22} = -15$
9. det $A = 3 \neq 0$. A^{-1} exists.
11. det $A = 0$. A^{-1} does not exist.
13. 30 **15.** 0 **17.** -32 **19.** 0
21. 643.4 **23.** -4.484 **25.** $\left(-\frac{13}{9}, \frac{16}{9}\right)$ **27.** $\left(\frac{49}{2}, 19\right)$
29. $(5, -3)$ **31.** $(0.45, 0.67)$ **33.** 7 square units
35. 6.5 square units **37.** The points are collinear.
39. The points are not collinear.
41. $x + y = 3$ **43.** $2x + y = 5$

9.7 EXTENDED AND DISCOVERY EXERCISES (p. 788)

1. $(1, 3, 2)$ **3.** $(1, -1, 1)$ **5.** $(-1, 0, 4)$
7. $x^2 + y^2 - 4 = 0$ **9.** $5x^2 + 5y^2 - 15x - 5y = 0$

CHECKING BASIC CONCEPTS FOR SECTION 9.7 (p. 788)

1. det $A = 19$; A is invertible.

CHAPTER 9 REVIEW EXERCISES (pp. 792–794)

1. $A(3, 6) = 9$ **3.** $(3, 1)$ **5.** $(1, -2)$
7. $(2, 3)$; consistent
9. $\{(x, y) \mid 2x - 5y = 4\}$; consistent
11. $\left(\frac{3\sqrt{2}}{2}, \frac{1}{2}\right), \left(-\frac{3\sqrt{2}}{2}, \frac{1}{2}\right)$

13.

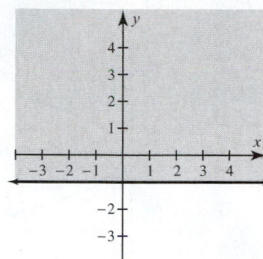

15. $(2, 2)$ (answers may vary)

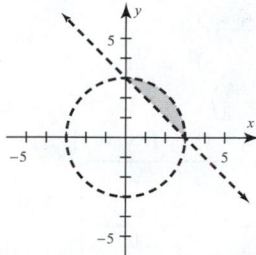

17. $(-1, 2, 1)$ **19.** No solutions **21.** $(-9, 3)$ **23.** $(-2, 3, 0)$
25. $(1, -2, 3)$ **27. (a)** 5 **(b)** -10

29. (a) $A + 2B = \begin{bmatrix} 7 & 1 \\ -8 & 1 \end{bmatrix}$

(b) $A - B = \begin{bmatrix} -2 & -5 \\ 7 & -2 \end{bmatrix}$ **(c)** $-4A = \begin{bmatrix} -4 & 12 \\ -8 & 4 \end{bmatrix}$

31. $AB = \begin{bmatrix} -2 & -4 \\ 17 & 31 \end{bmatrix}, BA = \begin{bmatrix} 8 & -6 \\ -27 & 21 \end{bmatrix}$

33. $AB = \begin{bmatrix} 3 & 7 \\ -2 & 8 \end{bmatrix}, BA = \begin{bmatrix} 2 & -1 & 3 \\ 2 & 9 & -3 \\ 6 & 12 & 0 \end{bmatrix}$ **35.** Yes

37. $\begin{bmatrix} -1 & -2 \\ -1 & -1 \end{bmatrix}$

39. (a) $AX = \begin{bmatrix} 1 & -3 \\ 2 & -1 \end{bmatrix} \begin{bmatrix} x \\ y \end{bmatrix} = \begin{bmatrix} 4 \\ 3 \end{bmatrix} = B$

(b) $X = \begin{bmatrix} 1 \\ -1 \end{bmatrix}$

41. $(-0.5, 1.7, -2.9)$
43. det $A = 25$ **45.** det $A = -1951 \neq 0$. A is invertible.
47. $l = 11, w = 7$
49. Both methods yield $1200 at 7%, $800 at 9%.
51. A: $11.49; B: $12.99
53. 10.5 square units **55.** 4500

CHAPTER 9 EXTENDED AND DISCOVERY EXERCISES (pp. 794–795)

1. (a) $A^T = \begin{bmatrix} 3 & 2 & 4 \\ -3 & 6 & 2 \end{bmatrix}$ **(b)** $A^T = \begin{bmatrix} 0 & 2 & -4 \\ 1 & 5 & 3 \\ -2 & 4 & 9 \end{bmatrix}$

(c) $A^T = \begin{bmatrix} 5 & 1 & 6 & -9 \\ 7 & -7 & 3 & 2 \end{bmatrix}$

3. $f(x) = 2.6314x + 2.2714$
$[-1, 6, 1]$ by $[0, 18, 2]$

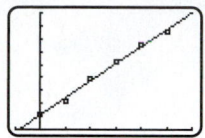

CHAPTER 10: Conic Sections

SECTION 10.1 (pp. 804–807)

1. focus **3.** upward; downward **5.** vertical

7.

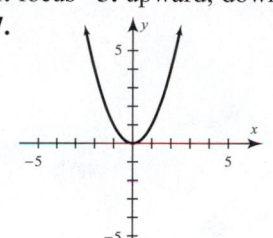

9.

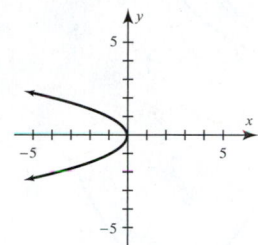

11.

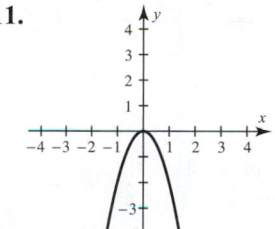

13.

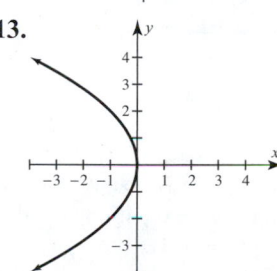

15.

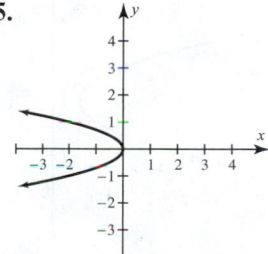

17. e **19.** a **21.** d
23. Vertex: $V(0, 0)$;
focus: $F(0, 4)$;
directrix: $y = -4$

25. Vertex: $V(0, 0)$;
focus: $F(2, 0)$;
directrix: $x = -2$

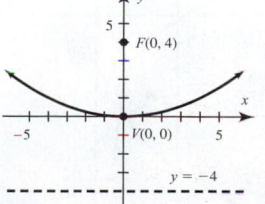

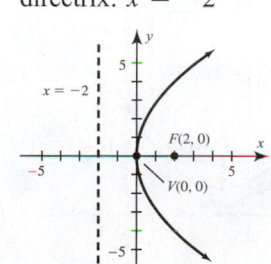

27. Vertex: $V(0, 0)$; focus: $F(-1, 0)$; directrix: $x = 1$

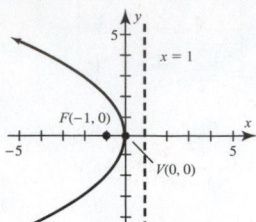

29. Vertex: $V(0, 0)$; focus: $F(0, -2)$; directrix: $y = 2$

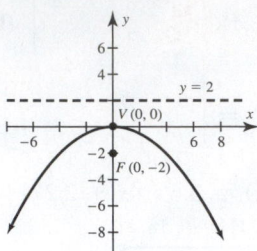

31. Vertex: $V(0, 0)$; focus: $F(-1, 0)$; directrix: $x = 1$

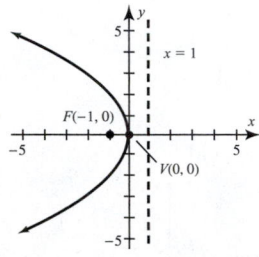

33. $x^2 = 4y$

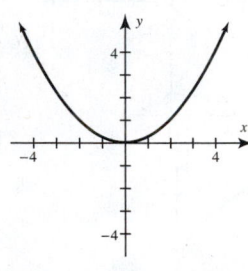

35. $y^2 = -12x$

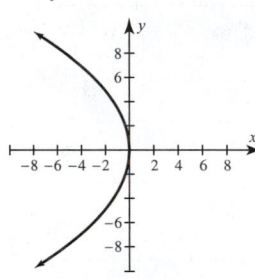

37. $x^2 = 3y$ **39.** $y^2 = -8x$ **41.** $y^2 = 4x$
43. $y^2 = -x$ **45.** $y^2 = 4x$ **47.** $x^2 = -12y$
49. $y^2 = -4x$

51.

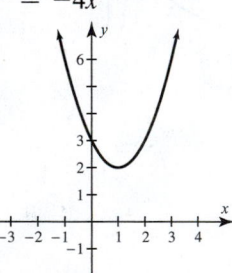

53.

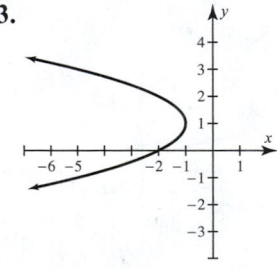

55. c   **57.** a
59. Vertex: $V(2, -2)$; focus: $F(2, 0)$; directrix; $y = -4$

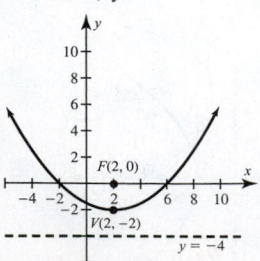

61. Vertex: $V(2, -3)$; focus: $F(1, -3)$; directrix: $x = 3$

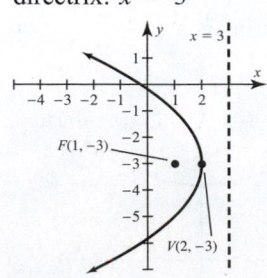

63. Vertex: $V(-2, 0)$; focus: $F(-2, -1)$; directrix: $y = 1$

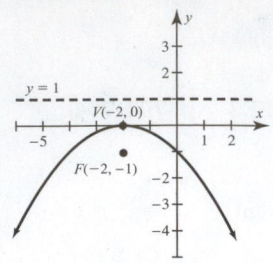

65. $x^2 = 4(y - 1)$

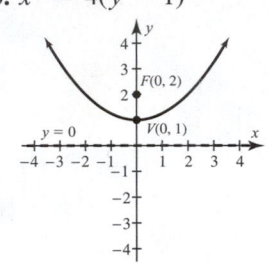

67. $y^2 = 4(x + 1)$

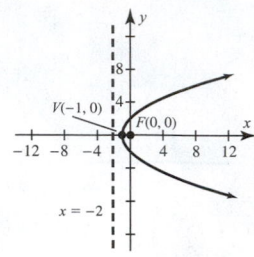

69. $(x + 1)^2 = -8(y - 5)$ **71.** $(y - 3)^2 = -\frac{9}{2}(x + 2)$
73. $(y - 0)^2 = -8\left(x + \frac{5}{4}\right)$ **75.** $(y + 1)^2 = \frac{1}{2}(x + 3)$
77. $\left(x - \frac{3}{2}\right)^2 = 2\left(y - \frac{7}{8}\right)$ **79.** $\left(y + \frac{1}{2}\right)^2 = \frac{5}{4}\left(x + \frac{6}{5}\right)$
81. $[-6, 6, 1]$ by $[-4, 4, 1]$ **83.** $[-9, 9, 1]$ by $[-6, 6, 1]$

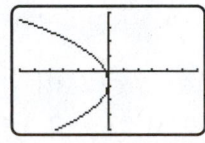   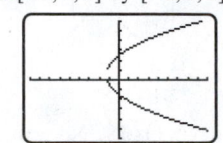

Note: If a break in the graph appears near the vertex, it should not be there. It is a result of the low resolution of the graphing calculator screen.
85. $[-6, 6, 1]$ by $[-4, 4, 1]$

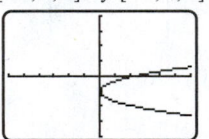

87. $\left(\pm\sqrt{2}, 1\right)$ **89.** $\left(-0.1, \pm\sqrt{0.9}\right)$
91. $(0, 0), (3, -1)$ **93.** $p = 3$ ft
95. (a) $y = \frac{32}{11{,}025}x^2$ **(b)** About 86.1 ft
97. (a) $(25, 0)$ **(b)** 25 million mi   **99.** $k = 6$

SECTION 10.2 (pp. 817–821)

1. vertices   **3.** horizontal
5. Foci: $F\left(0, \pm\sqrt{5}\right)$; vertices: $V(0, \pm3)$; endpoints of the minor axis: $U(\pm2, 0)$

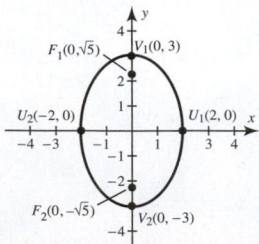

7. Foci: $F\left(\pm\sqrt{20}, 0\right)$; vertices: $V(\pm6, 0)$; endpoints of the minor axis: $U(0, \pm4)$

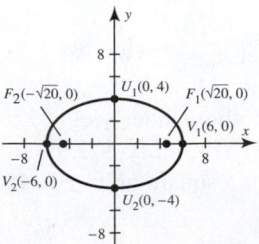

9. Foci: $F\left(\pm\sqrt{300}, 0\right)$;
vertices: $V(\pm 20, 0)$;
endpoints of the
minor axis: $U(0, \pm 10)$

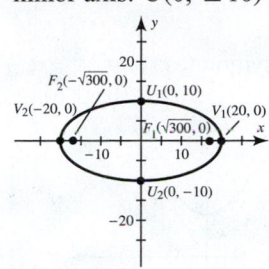

11. Foci: $F(0, \pm 4)$;
vertices: $V(0, \pm 5)$;
endpoints of the
minor axis: $U(\pm 3, 0)$

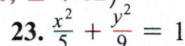

13. b **15.** c

17. $\dfrac{x^2}{36} + \dfrac{y^2}{16} = 1$; vertices: $V(\pm 6, 0)$; endpoints of the minor
axis: $U(0, \pm 4)$; foci: $F\left(\pm\sqrt{20}, 0\right)$

19. $\dfrac{x^2}{4} + \dfrac{y^2}{16} = 1$; vertices: $V(0, \pm 4)$; endpoints of the
minor axis: $U(\pm 2, 0)$; foci: $F\left(0, \pm\sqrt{12}\right)$

21. $\dfrac{x^2}{25} + \dfrac{y^2}{9} = 1$ **23.** $\dfrac{x^2}{5} + \dfrac{y^2}{9} = 1$

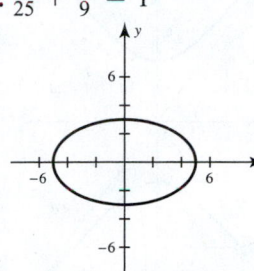 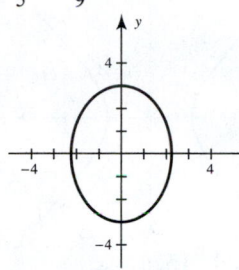

25. $\dfrac{x^2}{12} + \dfrac{y^2}{16} = 1$ **27.** $\dfrac{x^2}{36} + \dfrac{y^2}{11} = 1$

29. $\dfrac{x^2}{16} + \dfrac{y^2}{9} = 1$ **31.** $\dfrac{x^2}{9} + \dfrac{y^2}{5} = 1$

33. $\dfrac{(x-2)^2}{4} + \dfrac{(y+1)^2}{3} = 1$ **35.** $\dfrac{(x+3)^2}{2} + \dfrac{(y+4)^2}{9} = 1$

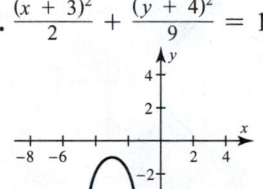

37.

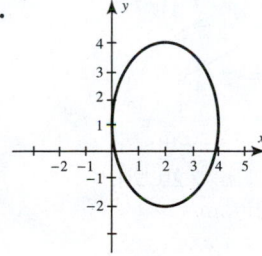

39.

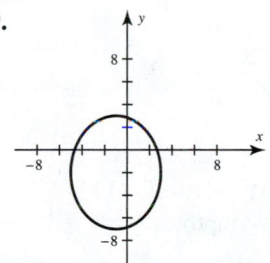

41.

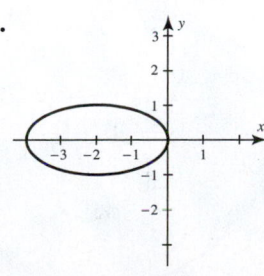

43. d   **45.** c

47. Foci: $F(1, 1 \pm 4)$;
vertices: $V(1, 1 \pm 5)$

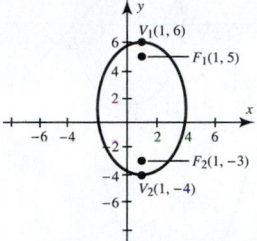

49. Foci: $F(-4 \pm \sqrt{7}, 2)$;
vertices: $V(-4 \pm 4, 2)$

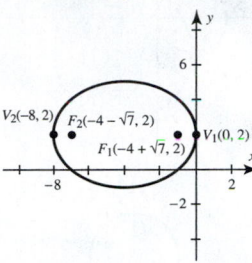

51. $\dfrac{(x-2)^2}{5} + \dfrac{(y-1)^2}{9} = 1$ **53.** $\dfrac{x^2}{9} + \dfrac{(y-2)^2}{5} = 1$

55. $\dfrac{(x-2)^2}{16} + \dfrac{(y-4)^2}{4} = 1$

57. $\dfrac{(x+1)^2}{4} + \dfrac{(y-1)^2}{9} = 1$; center: $C(-1, 1)$;
vertices: $V(-1, -2), V(-1, 4)$

59. $\dfrac{(x+1)^2}{1} + \dfrac{(y+1)^2}{4} = 1$; center: $C(-1, -1)$;
vertices: $V(-1, -3), V(-1, 1)$

61. $\dfrac{(x+2)^2}{5} + \dfrac{(y-1)^2}{4} = 1$; center: $C(-2, 1)$;
vertices: $V\left(-2 - \sqrt{5}, 1\right), V\left(-2 + \sqrt{5}, 1\right)$

63. $\dfrac{\left(x - \frac{1}{2}\right)^2}{4} + \dfrac{\left(y + \frac{3}{2}\right)^2}{16} = 1$; center: $C\left(\frac{1}{2}, -\frac{3}{2}\right)$;
vertices: $V\left(\frac{1}{2}, \frac{5}{2}\right), V\left(\frac{1}{2}, -\frac{11}{2}\right)$

65. $[-6, 6, 1]$ by $[-4, 4, 1]$ **67.** $[-4.7, 4.7, 1]$ by $[-3.1, 3.1, 1]$

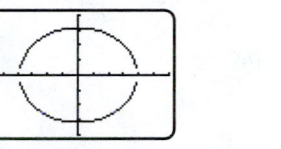 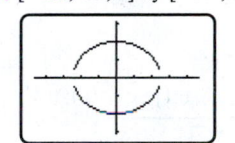

69. $(0, 3), \left(\frac{24}{13}, \frac{15}{13}\right)$

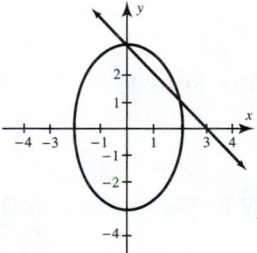

71. Four solutions:
$\left(\pm\sqrt{\dfrac{20}{3}}, \pm\sqrt{\dfrac{7}{3}}\right)$

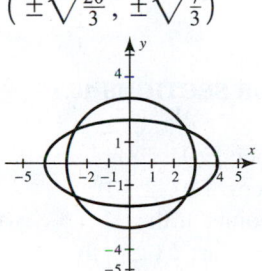

73. $(3, 0), (-3, 0)$

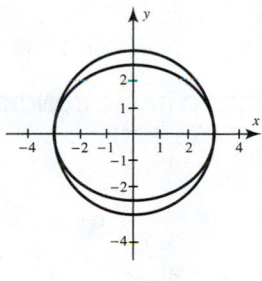

75. $(0, 2)$ **77.** Four solutions: $\left(\pm\sqrt{\dfrac{4}{3}}, \pm\sqrt{\dfrac{4}{3}}\right)$

79. $\left(1, \sqrt{8}\right), \left(1, -\sqrt{8}\right)$

81.
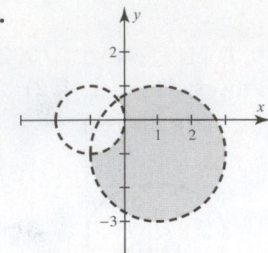

83.
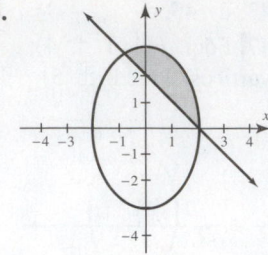

85.
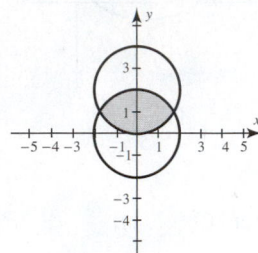

87.
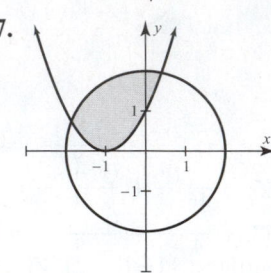

89. $A = 6\pi \approx 18.85$ ft^2 **91.** $A = 20\pi \approx 62.83$ ft^2

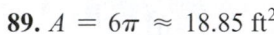
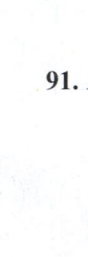
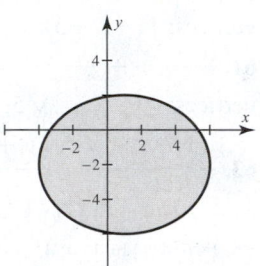

93. $\dfrac{x^2}{0.387^2} + \dfrac{y^2}{0.379^2} = 1$; sun: $(0.0797, 0)$
$[-0.6, 0.6, 0.1]$ by $[-0.4, 0.4, 0.1]$

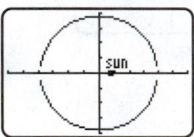

95. 6.245 in.   **97.** $A = \pi(64.03)(40) \approx 8046.25$ ft^2
99. 348.2 ft   **101.** About 21.65 ft
103. Maximum: 668 mi; minimum: 340 mi

10.2 EXTENDED AND DISCOVERY EXERCISES (p. 821)

1. $\dfrac{x}{2} + \dfrac{y}{3} = 1$; x-int: 2, y-int: 3
3. $\dfrac{x}{2.5} + \dfrac{y}{5} = 1$; x-int: 2.5, y-int: 5
5. x-int: ± 5, y-int: ± 3

CHECKING BASIC CONCEPTS FOR SECTIONS 10.1 AND 10.2 (p. 821)

1. Focus: $F\left(\dfrac{1}{2}, 0\right)$;
directrix: $x = -\dfrac{1}{2}$

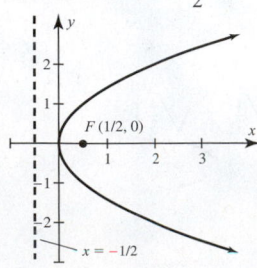

3. Foci: $F(0, \pm 8)$;
vertices: $V(0, \pm 10)$;
endpoints of the
minor axis: $U(\pm 6, 0)$

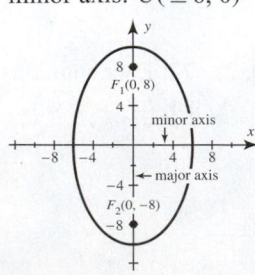

5. 1 ft   **7.** $\dfrac{(x - 2)^2}{16} + \dfrac{(y + 1)^2}{4} = 1$; center: $C(2, -1)$;
vertices: $V(-2, -1)$, $V(6, -1)$

SECTION 10.3 (pp. 828–830)

1. vertices   **3.** horizontal
5. Asymptotes: $y = \pm\dfrac{7}{3}x$;
$F\left(\pm\sqrt{58}, 0\right)$; $V(\pm 3, 0)$

7. Asymptotes: $y = \pm\dfrac{3}{2}x$;
$F\left(0, \pm\sqrt{52}\right)$; $V(0, \pm 6)$

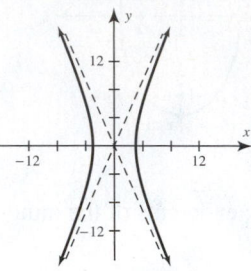
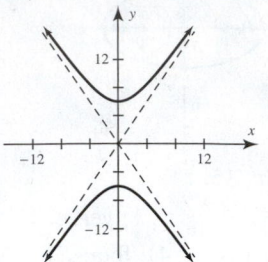

9. Asymptotes: $y = \pm x$;
$F\left(\pm\sqrt{18}, 0\right)$; $V(\pm 3, 0)$

11. Asymptotes: $y = \pm\dfrac{4}{3}x$;
$F(0, \pm 5)$; $V(0, \pm 4)$

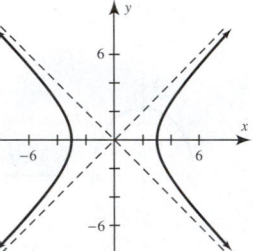
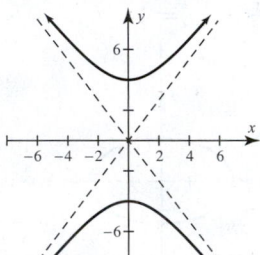

13. d   **15.** a
17. $\dfrac{x^2}{16} - \dfrac{y^2}{9} = 1$

19. $\dfrac{y^2}{36} - \dfrac{x^2}{64} = 1$

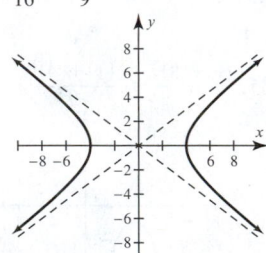
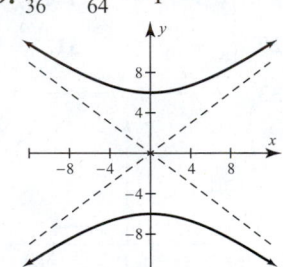

21. $\dfrac{y^2}{144} - \dfrac{x^2}{25} = 1$; asymptotes: $y = \pm\dfrac{12}{5}x$

23. $\dfrac{y^2}{4} - \dfrac{x^2}{21} = 1$; asymptotes: $y = \pm\dfrac{2}{\sqrt{21}}x$

25. $\dfrac{x^2}{9} - \dfrac{y^2}{4} = 1$; asymptotes: $y = \pm\dfrac{2}{3}x$

27. $\dfrac{x^2}{16} - \dfrac{y^2}{9} = 1$; asymptotes: $y = \pm\dfrac{3}{4}x$

29. $\dfrac{x^2}{10} - \dfrac{y^2}{9} = 1$; asymptotes: $y = \pm\dfrac{3}{\sqrt{10}}x$

31. Vertices: $V(1 \pm 4, 2)$; foci: $F\left(1 \pm \sqrt{20}, 2\right)$;
asymptotes: $y = \pm\dfrac{1}{2}(x - 1) + 2$

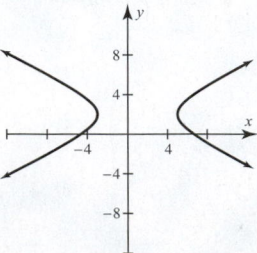

33. Vertices: $V(-2, 2 \pm 6)$; foci: $F(-2, 2 \pm \sqrt{40})$; asymptotes: $y = \pm 3(x + 2) + 2$

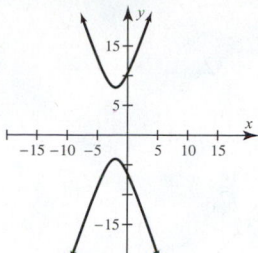

35. Vertices: $V(\pm 2, 1)$; foci: $F(\pm \sqrt{5}, 1)$; asymptotes: $y = \pm\frac{1}{2}x + 1$

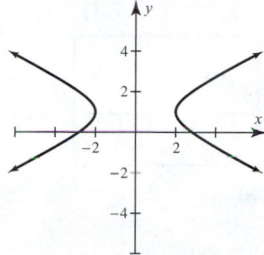

37. b   **39.** c

41. $\frac{(y + 4)^2}{16} - \frac{(x - 4)^2}{4} = 1$; vertices: $V(4, -4 \pm 4)$; foci: $F(4, -4 \pm \sqrt{20})$; asymptotes: $y = \pm 2(x - 4) - 4$

43. Vertices: $V(1 \pm 2, 1)$; foci: $F(1 \pm \sqrt{8}, 1)$; asymptotes: $y = \pm(x - 1) + 1$

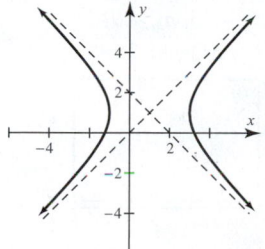

45. Vertices: $V(1, -1 \pm 4)$; foci: $F(1, -1 \pm 5)$; asymptotes: $y = \pm\frac{4}{3}(x - 1) - 1$

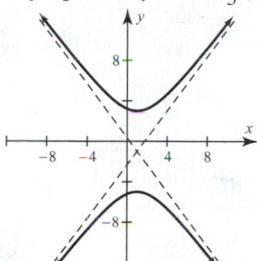

47. $(x - 2)^2 - \frac{(y + 2)^2}{3} = 1$ **49.** $y^2 - \frac{(x + 1)^2}{8} = 1$

51. $\frac{(x - 1)^2}{4} - \frac{(y - 1)^2}{4} = 1$; center: $C(1, 1)$; vertices: $V(-1, 1), V(3, 1)$

53. $\frac{(y + 4)^2}{2} - \frac{(x - 3)^2}{3} = 1$; center: $C(3, -4)$; vertices: $V(3, -4 - \sqrt{2}), V(3, -4 + \sqrt{2})$

55. $\frac{(x - 3)^2}{2} - \frac{(y - 0)^2}{1} = 1$; center: $C(3, 0)$; vertices: $V(3 - \sqrt{2}, 0), V(3 + \sqrt{2}, 0)$

57. $\frac{(y + 4)^2}{5} - \frac{(x + 1)^2}{4} = 1$; center: $C(-1, -4)$; vertices: $V(-1, -4 - \sqrt{5}), V(-1, -4 + \sqrt{5})$

59. $[-15, 15, 5]$ by $[-10, 10, 5]$ **61.** $[-9, 9, 1]$ by $[-6, 6, 1]$

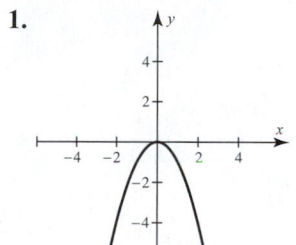 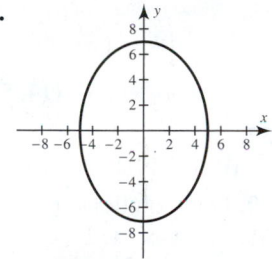

63. Four solutions: $\left(\pm\sqrt{\frac{13}{2}}, \pm\sqrt{\frac{5}{2}}\right)$

65. $(2, 0), (-5.2, 7.2)$ **67.** Four solutions: $\left(\pm 2, \pm\frac{2}{\sqrt{3}}\right)$

69. $\left(\frac{2}{\sqrt{11}}, \frac{6}{\sqrt{11}}\right), \left(-\frac{2}{\sqrt{11}}, -\frac{6}{\sqrt{11}}\right)$

71. (a) Elliptic   **(b)** Its speed should be 4326 m/sec or greater.   **(c)** If D is larger, then $\frac{k}{\sqrt{D}}$ is smaller, so smaller values for V satisfy $V > \frac{k}{\sqrt{D}}$.

10.3 EXTENDED AND DISCOVERY EXERCISES (p. 831)

1. (a) $x = \sqrt{y^2 + 2.5 \times 10^{-27}}$; this equation represents the right half of the hyperbola.   **(b)** About 1.2×10^{-13} m

CHECKING BASIC CONCEPTS FOR SECTION 10.3 (p. 831)

1. $\frac{x^2}{16} - \frac{y^2}{9} = 1$

3. $\frac{(x - 1)^2}{9} - \frac{(y - 3)^2}{4} = 1$; $F(1 \pm \sqrt{13}, 3)$

CHAPTER 10 REVIEW EXERCISES (pp. 834–835)

1.

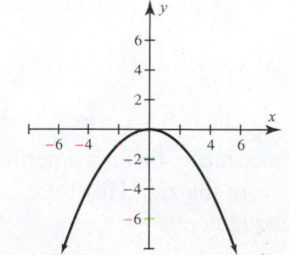

3.

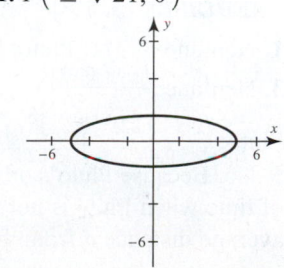

5.

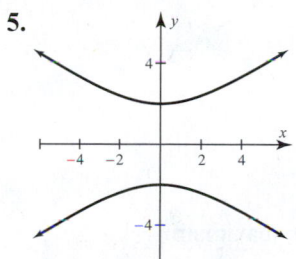

7. d   **9.** a   **11.** e

13. $y^2 = 8x$ **15.** $\frac{x^2}{25} + \frac{y^2}{9} = 1$ **17.** $\frac{y^2}{64} - \frac{x^2}{36} = 1$

19. $F(0, -1)$ **21.** $F(\pm\sqrt{21}, 0)$

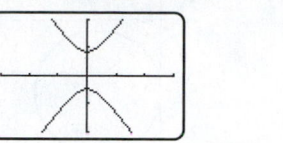

23. $F(\pm 5, 0)$

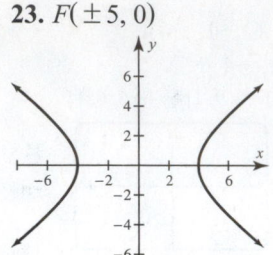

25. Both foci are located at $(3, -1)$.

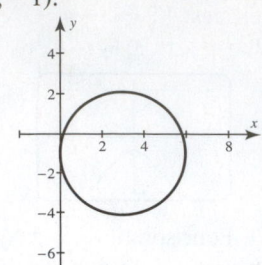

27. Center: $C(1, -1)$

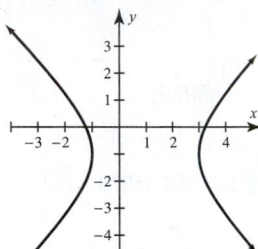

29.

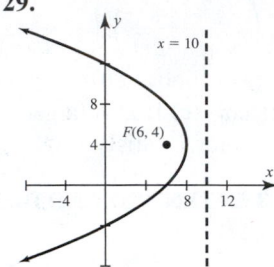

31. $[-5, 5, 1]$ by $[-5, 5, 1]$

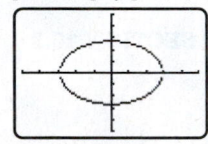

Note: If breaks in the graph appear near the vertices, they should not be there. It is a result of the low resolution of the graphing calculator screen.

33. $(y - 0)^2 = -10\left(x + \frac{7}{5}\right)$

35. $\dfrac{(x + 1)^2}{25} + \dfrac{(y - 5)^2}{4} = 1$; center: $C(-1, 5)$; vertices: $V(-6, 5)$, $V(4, 5)$

37. $\dfrac{(x + 2)^2}{4} - \dfrac{(y - 3)^2}{1} = 1$; center: $C(-2, 3)$; vertices: $V(-4, 3)$, $V(0, 3)$

39. Four solutions: $\left(\pm\sqrt{\tfrac{8}{3}}, \pm\sqrt{\tfrac{4}{3}}\right)$

41.

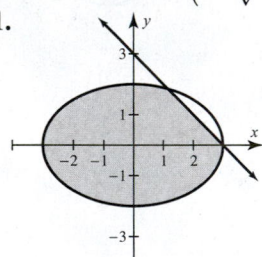

43. (a) Minimum: 4.92 million mi; maximum: 995.08 million mi **(b)** $2\pi\sqrt{\dfrac{500^2 + 70^2}{2}} \approx 2243$ million miles, or 2.243 billion miles **45.** About 29.05 ft

CHAPTER 10 EXTENDED AND DISCOVERY EXERCISES (p. 835)

1. Neptune: 0.271; Pluto: 9.82

3. Neptune: $\dfrac{(x - 0.271)^2}{30.10^2} + \dfrac{y^2}{30.10^2} = 1$;

Pluto: $\dfrac{(x - 9.82)^2}{39.44^2} + \dfrac{y^2}{38.20^2} = 1$

5. No. Because Pluto's orbit is so eccentric, there is a period of time when Pluto is not farther from the sun. However, its average distance a from the sun is greater.

CHAPTER 11: Further Topics in Algebra

SECTION 11.1 (pp. 846–849)

1. $a_1 = 3, a_2 = 5, a_3 = 7, a_4 = 9$

3. $a_1 = 4, a_2 = -8, a_3 = 16, a_4 = -32$

5. $a_1 = \frac{1}{2}, a_2 = \frac{2}{5}, a_3 = \frac{3}{10}, a_4 = \frac{4}{17}$

7. $a_1 = -\frac{1}{2}, a_2 = \frac{1}{4}, a_3 = -\frac{1}{8}, a_4 = \frac{1}{16}$

9. $a_1 = \frac{2}{3}, a_2 = -\frac{4}{5}, a_3 = \frac{8}{9}, a_4 = -\frac{16}{17}$

11. $a_1 = 3, a_2 = 8, a_3 = 17, a_4 = 32$

13. 2, 4, 3, 5, 3, 6, 4

15. (a) $a_1 = 1, a_2 = 2, a_3 = 4, a_4 = 8$
(b) $[0, 5, 1]$ by $[0, 9, 1]$

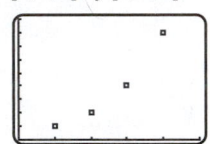

17. (a) $a_1 = -3, a_2 = 0, a_3 = 3, a_4 = 6$
(b) $[0, 5, 1]$ by $[-4, 7, 1]$

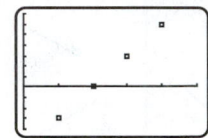

19. (a) $a_1 = 2, a_2 = 5, a_3 = 14, a_4 = 41$
(b) $[0, 5, 1]$ by $[0, 45, 5]$

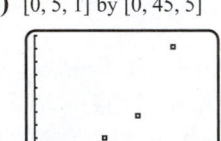

21. (a) $a_1 = 2, a_2 = 5, a_3 = 3, a_4 = -2$
(b) $[0, 5, 1]$ by $[-3, 6, 1]$

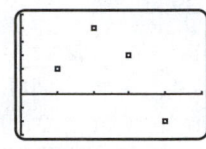

23. (a) $a_1 = 2, a_2 = 4, a_3 = 16, a_4 = 256$
(b) $[0, 5, 1]$ by $[0, 300, 50]$

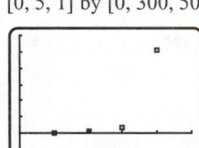

25. (a) $a_1 = 1, a_2 = 3, a_3 = 6, a_4 = 10$
(b) $[0, 5, 1]$ by $[0, 12, 1]$

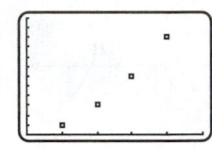

27. (a) $a_1 = 2, a_2 = 3, a_3 = 6, a_4 = 18$
(b) $[0, 5, 1]$ by $[0, 20, 2]$

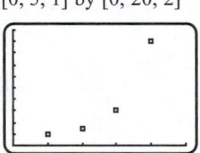

29. (a)

n	1	2	3	4	5	6	7	8
a_n	1	3	5	7	9	11	13	15

(b) $[0, 10, 1]$ by $[0, 16, 1]$ **(c)** $a_n = 2n - 1$

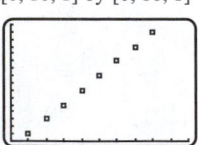

31. (a)

n	1	2	3	4	5	6	7	8
a_n	7.5	6	4.5	3	1.5	0	-1.5	-3

(b) $[0, 12, 1]$ by $[-4, 8, 1]$ **(c)** $a_n = -1.5n + 9$

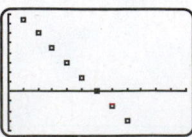

33. (a)

n	1	2	3	4	5	6	7	8
a_n	$\frac{1}{2}$	2	$\frac{7}{2}$	5	$\frac{13}{2}$	8	$\frac{19}{2}$	11

(b) $[0, 9, 1]$ by $[0, 12, 1]$ **(c)** $a_n = \frac{3}{2}n - 1$

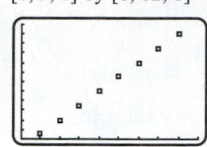

35. (a)

n	1	2	3	4	5	6	7	8
a_n	8	4	2	1	$\frac{1}{2}$	$\frac{1}{4}$	$\frac{1}{8}$	$\frac{1}{16}$

(b) $[0, 10, 1]$ by $[-1, 9, 1]$ **(c)** $a_n = 8\left(\frac{1}{2}\right)^{n-1}$

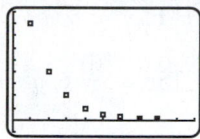

37. (a)

n	1	2	3	4	5	6	7	8
a_n	$\frac{3}{4}$	$\frac{3}{2}$	3	6	12	24	48	96

(b) $[0, 10, 1]$ by $[-10, 110, 10]$ **(c)** $a_n = \frac{3}{4}(2)^{n-1}$

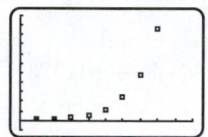

39. (a)

n	1	2	3	4	5	6	7	8
a_n	$-\frac{1}{4}$	$-\frac{1}{2}$	-1	-2	-4	-8	-16	-32

(b) $[0, 9, 1]$ by $[-36, 4, 4]$ **(c)** $a_n = -\frac{1}{4}(2)^{n-1}$

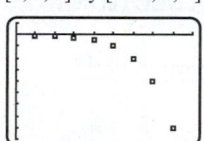

41. $a_n = -2n + 7$ **43.** $a_n = 3n - 8$
45. $a_n = 2n + 1$ **47.** $a_n = 3n + 5$
49. $a_n = 0.5n - 6.5$ **51.** $a_n = 2\left(\frac{1}{2}\right)^{n-1}$
53. $a_n = \frac{1}{2}\left(-\frac{1}{4}\right)^{n-1}$ **55.** $a_n = 8\left(\frac{1}{2}\right)^{n-1}$
57. $a_n = -5(-5)^{n-1}$ **59.** $a_n = -\frac{1}{2}(2)^{n-1}$
61. No   **63.** Yes   **65.** Yes   **67.** Yes
69. Yes   **71.** No   **73.** No   **75.** No
77. Arithmetic   **79.** Geometric   **81.** Neither
83. Arithmetic, $d < 0$, $d = -1$
85. Geometric, $r < 0$, $|r| < 1$
87. The insect population density increases rapidly and then levels off near 5000 per acre.
89. (a) $a_n = 0.8a_{n-1}$, $a_1 = 500$ **(b)** $a_1 = 500$, $a_2 = 400$, $a_3 = 320$, $a_4 = 256$, $a_5 = 204.8$, and $a_6 = 163.84$. The population density decreases by 20% each year.   **(c)** $a_n = 500(0.8)^{n-1}$
91. (a) $a_1 = 8$, $a_2 = 10.4$, $a_3 = 8.528$
(b) $[0, 21, 1]$ by $[0, 14, 1]$

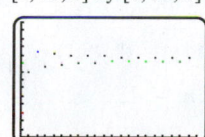

The population density oscillates above and below approximately 9.5.

93. $a_n = 100n$; $a_5 = 500$; after 5 years, 500 million cell phones have been thrown out.
95. (a) 100, 194, 288, 382, 476; arithmetic
(b) $[0, 6, 1]$ by $[0, 500, 100]$

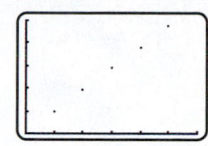

(c) $a_n = 94(n - 1) + 100$
97. (a) 1, 1, 2, 3, 5, 8, 13, 21, 34, 55, 89, 144
(b) $\frac{a_2}{a_1} = 1$, $\frac{a_3}{a_2} = 2$, $\frac{a_4}{a_3} = 1.5$, $\frac{a_5}{a_4} = \frac{5}{3} \approx 1.6667$, $\frac{a_6}{a_5} = \frac{8}{5} = 1.6$, $\frac{a_7}{a_6} = \frac{13}{8} = 1.625$, $\frac{a_8}{a_7} = \frac{21}{13} \approx 1.6154$, $\frac{a_9}{a_8} = \frac{34}{21} \approx 1.6190$, $\frac{a_{10}}{a_9} = \frac{55}{34} \approx 1.6176$, $\frac{a_{11}}{a_{10}} = \frac{89}{55} \approx 1.6182$, and $\frac{a_{12}}{a_{11}} = \frac{144}{89} \approx 1.6180$.
The ratio appears to approach a number near 1.618.
(c) $n = 2$: $a_1 \cdot a_3 - a_2^2 = (1)(2) - (1)^2 = 1 = (-1)^2$
$n = 3$: $a_2 \cdot a_4 - a_3^2 = (1)(3) - (2)^2 = -1 = (-1)^3$
$n = 4$: $a_3 \cdot a_5 - a_4^2 = (2)(5) - (3)^2 = 1 = (-1)^4$
99. (a) $a_n = 2000n + 28{,}000$, or $a_n = 30{,}000 + 2000(n - 1)$; arithmetic
(b) $b_n = 30{,}000(1.05)^{n-1}$; geometric
(c) Since $a_{10} = \$48{,}000 > b_{10} \approx \$46{,}540$, the first salary is higher after 10 years. Since $a_{20} = \$68{,}000 < b_{20} \approx \$75{,}809$, the second salary is higher after 20 years.
(d) With time, the geometric sequence with $r > 1$ overtakes the arithmetic sequence.
$[0, 30, 10]$ by $[0, 150000, 50000]$

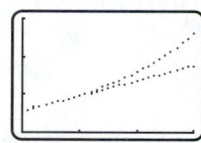

101. $a_6 \approx 1.414213562$, $\sqrt{2} \approx 1.414213562$
103. $a_6 = 4.582581971$, $\sqrt{21} \approx 4.582575695$
105. By definition,
$a_n = a_1 + (n - 1)d_1$ and $b_n = b_1 + (n - 1)d_2$.
Then $c_n = a_n + b_n$
$= [a_1 + (n - 1)d_1] + [b_1 + (n - 1)d_2]$
$= (a_1 + b_1) + [(n - 1)d_1 + (n - 1)d_2]$
$= (a_1 + b_1) + (n - 1)(d_1 + d_2)$
$= c_1 + (n - 1)d$,
where $c_1 = a_1 + b_1$ and $d = d_1 + d_2$.

SECTION 11.2 (pp. 861–863)

1. (a) 1, 2, 3, 4, 5   **(b)** $1 + 2 + 3 + 4 + 5$ **(c)** 15
3. (a) 1, 0, -1, -2, -3
(b) $1 + 0 + (-1) + (-2) + (-3)$ **(c)** -5
5. $A_5 + A_6 + A_7 + A_8 + A_9$
7. 45   **9.** 25   **11.** 60   **13.** $\frac{71}{20}$ **15.** 80   **17.** 1275
19. 1739   **21.** 4100   **23.** 31   **25.** 460   **27.** 105
29. 1942   **31.** 948   **33.** 545   **35.** 255   **37.** 546.5
39. 3,145,725   **41.** 0.625; 0.671875; 0.666015625
43. 5; 42.333333333; 341
45. $\frac{3}{2}$ **47.** $\frac{18}{5}$ **49.** $\frac{10}{11}$
51. $\frac{2}{3} = 0.6 + 0.06 + 0.006 + 0.0006 + \cdots$

53. $\frac{9}{11} = 0.81 + 0.0081 + 0.000081 + \cdots$

55. $\frac{1}{7} = 0.142857 + 0.000000142857 + \cdots$

57. $\frac{8}{9}$ **59.** $\frac{5}{11}$ **61.** $2 + 3 + 4 + 5 = 14$

63. $4 + 4 + 4 + 4 + 4 + 4 + 4 + 4 = 32$

65. $1 + 8 + 27 + 64 + 125 + 216 + 343 = 784$

67. $12 + 20 = 32$

69. $\sum_{k=1}^{6} k^4$ **71.** $\sum_{k=1}^{7} \left(\frac{2k}{k + 1} \right)$

73. $\sum_{k=1}^{\infty} \left(\frac{1}{k^2} \right)$ **75.** $\sum_{n=1}^{4} (n + 5)^3$ **77.** $\sum_{n=1}^{24} (3n + 22)$

79. $\sum_{n=1}^{37} (n^2 + 27n + 180)$

81. 540 **83.** 600 **85.** 1395 **87.** 5525

89. 1360 **91.** 290

93. $\sum_{k=1}^{n} k = 1 + 2 + 3 + \cdots + n$ is an arithmetic series with $a_1 = 1$ and $a_n = n$. Its sum equals $S_n = n\left(\frac{a_1 + a_n}{2} \right) = n\left(\frac{1 + n}{2} \right) = \frac{n(n + 1)}{2}$.

95. (a) & (b) $819,000 **97.** $91,523.93 **99.** $62,278.01

101. $S_9 = 9\left(\frac{7 + 15}{2} \right) = 99$ logs

103. (a) $\sum_{k=1}^{n} 0.5(0.5)^{k-1}$ **(b)** Infinitely many filters

105. 2

107. $1 + 1 + \frac{1}{2} + \frac{1}{6} + \frac{1}{24} + \frac{1}{120} + \frac{1}{720} + \frac{1}{5040}$
≈ 2.718254; $e \approx 2.718282$

109. $S_2 = \frac{4}{3} \approx 1.3333$, $S_4 = \frac{40}{27} \approx 1.4815$, $S_8 \approx 1.49977$, $S_{16} \approx 1.49999997$; $S = 1.5$. As n increases, the partial sums approach S.

111. $S_1 = 4$, $S_2 = 3.6$, $S_3 = 3.64$, $S_4 = 3.636$, $S_5 = 3.6364$, $S_6 = 3.63636$; $S = \frac{40}{11} = 3.\overline{63}$. As n increases, the partial sums approach S.

CHECKING BASIC CONCEPTS FOR SECTIONS 11.1 AND 11.2 (p. 864)

1. $[0, 7, 1]$ by $[-10, 2, 1]$

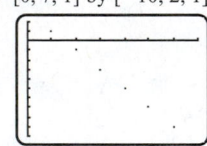 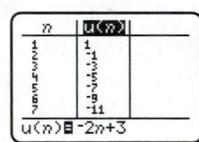

3. (a) Arithmetic; $S_{10} = 190$

(b) Geometric; $S_6 = \frac{364}{81} \approx 4.494$

(c) Geometric; $S = \frac{8}{3} \approx 2.667$ **(d)** Geometric; $S = 1$

5. (a) 150 **(b)** 6622

SECTION 11.3 (pp. 871–873)

1. $2^{10} = 1024$

3. $2^5 4^{10} = 33,554,432$

5. $10^3 \cdot 26^3 = 17,576,000$

7. $26^3 \cdot 36^3 = 820,025,856$

9. $3^5 = 243$ **11.** $5^5 = 3125$ **13.** 2 **15.** 24

17. 1,000,000 **19.** $2^{12} = 4096$

21. No; there are 35,152 call letters possible.

23. 24 **25.** 8,000,000 **27.** 720 **29.** 3,628,800

31. 60 **33.** 8 **35.** 210 **37.** 600 **39.** 5040

41. 24 **43.** 2730 **45.** 210 **47.** 30,000

49. 360 **51.** 362,880 **53.** About 6.39×10^{12}

55. 6,400,000,000 **57.** 3 **59.** 20 **61.** 1

63. 28 **65.** 190 **67.** 575,757 **69.** 30

71. 50 **73.** 7920 **75.** 2024

77. $P(n, n - 1) = \frac{n!}{(n - (n - 1))!} = \frac{n!}{1} = n!$ and $P(n, n) = \frac{n!}{(n - n)!} = \frac{n!}{0!} = \frac{n!}{1} = n!$. For example, $P(7, 6) = 5040 = P(7, 7)$.

SECTION 11.4 (p. 878)

1. 5 **3.** 1 **5.** 6 **7.** 1 **9.** 10 **11.** 70 **13.** 1 **15.** 5

17. $x^2 + 2xy + y^2$ **19.** $m^3 + 6m^2 + 12m + 8$

21. $8x^3 - 36x^2 + 54x - 27$

23. $p^6 - 6p^5q + 15p^4q^2 - 20p^3q^3 + 15p^2q^4 - 6pq^5 + q^6$

25. $8m^3 + 36m^2n + 54mn^2 + 27n^3$

27. $1 - 4x^2 + 6x^4 - 4x^6 + x^8$

29. $8p^9 - 36p^6 + 54p^3 - 27$

31. $x^2 + 2xy + y^2$

33. $81x^4 + 108x^3 + 54x^2 + 12x + 1$

35. $32 - 80x + 80x^2 - 40x^3 + 10x^4 - x^5$

37. $x^8 + 8x^6 + 24x^4 + 32x^2 + 16$

39. $256x^4 - 768x^3y + 864x^2y^2 - 432xy^3 + 81y^4$

41. $m^6 + 6m^5n + 15m^4n^2 + 20m^3n^3 + 15m^2n^4 + 6mn^5 + n^6$

43. $8x^9 - 12x^6y^2 + 6x^3y^4 - y^6$

45. $84a^6b^3$ **47.** $70x^4y^4$ **49.** $40x^2y^3$ **51.** $-576xy^5$

CHECKING BASIC CONCEPTS FOR SECTIONS 11.3 AND 11.4 (p. 878)

1. $2^8 = 256$ **3.** $26 \cdot 36^5 = 1,572,120,576$

SECTION 11.5 (pp. 883–884)

1. $3 + 6 + 9 + \cdots + 3n = \frac{3n(n + 1)}{2}$

(i) Show that the statement is true for $n = 1$:
$3(1) = \frac{3(1)(2)}{2}$
$3 = 3$

(ii) Assume that S_k is true:
$3 + 6 + 9 + \cdots + 3k = \frac{3k(k + 1)}{2}$
Show that S_{k+1} is true:
$3 + 6 + \cdots + 3(k + 1) = \frac{3(k + 1)(k + 2)}{2}$
Add $3(k + 1)$ to each side of S_k:
$3 + 6 + 9 + \cdots + 3k + 3(k + 1)$
$= \frac{3k(k + 1)}{2} + 3(k + 1)$
$= \frac{3k(k + 1) + 6(k + 1)}{2}$
$= \frac{(k + 1)(3k + 6)}{2}$
$= \frac{3(k + 1)(k + 2)}{2}$
Since S_k implies S_{k+1}, the statement is true for every positive integer n.

3–13. See the Student's Solutions Manual.

15. 1, 2 **17.** 2, 3, 4

19. $(a^m)^n = a^{mn}$

(i) Show that the statement is true for $n = 1$:
$(a^m)^1 = a^{m \cdot 1}$
$a^m = a^m$

(ii) Assume that S_k is true:
$(a^m)^k = a^{mk}$
Show that S_{k+1} is true:
$(a^m)^{k+1} = a^{m(k+1)}$
Multiply each side of S_k by a^m:
$(a^m)^k \cdot (a^m)^1 = a^{mk} \cdot a^m$
$(a^m)^{k+1} = a^{mk+m}$
$(a^m)^{k+1} = a^{m(k+1)}$
Since S_k implies S_{k+1}, the statement is true for every positive integer n.

21–29. See the Student's Solutions Manual.

31. $P = 3\left(\frac{4}{3}\right)^{n-1}$ **33.** $2^n - 1$

SECTION 11.6 (pp. 894–897)

1. Yes   **3.** No   **5.** Yes   **7.** No   **9.** $\frac{1}{2}$ **11.** $\frac{1}{6}$
13. $\frac{1}{2}$ **15.** $\frac{4}{52} = \frac{1}{13}$ **17.** $\frac{1}{10,000}$
19. **(a)** 0.57, or 57%   **(b)** 0.33, or 33%
21. $A \cup B = \{10, 25, 26, 35\}$; $A \cap B = \{25, 26\}$
23. $A \cup B = \{1, 3, 5, 7, 9, 11\}$; $A \cap B = \varnothing$
25. $A \cup B = \{\text{Heads, Tails}\}$; $A \cap B = \varnothing$
27. $\frac{1}{4}$ **29.** $\frac{1}{27}$ **31.** $\frac{1}{36}$ **33.** $\frac{625}{1296} \approx 0.482$ **35.** $\frac{1}{270,725}$

37. $\dfrac{\binom{13}{3} \cdot \binom{13}{2}}{\binom{52}{5}} \approx 0.0086$, or a 0.86% chance

39. $\frac{4}{20} = 0.2$

41. **(a)**

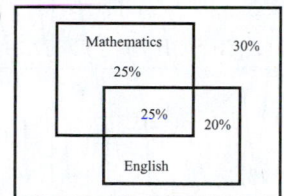

Mathematics 25%   30%
25%   20%
English

(b) 0.7, or 70%
(c) Let M denote the event of needing help with math and E the event of needing help with English. Then $P(M \cup E) = P(M) + P(E) - P(M \cap E)$
$= 0.5 + 0.45 - 0.25 = 0.7$.

43. **(a)** $\frac{9}{50}$ **(b)** $\frac{9}{25}$ **(c)** $\frac{71}{100}$

45. $\frac{634}{100,000} = 0.00634$

47. **(a)** $\frac{223,508}{679,590} \approx 0.329$ **(b)** $\frac{456,082}{679,590} \approx 0.671$
(c) $\frac{130,520}{679,590} \approx 0.192$

49. $\frac{10}{36} = \frac{5}{18}$ **51.** $\frac{12}{52} = \frac{3}{13}$ **53.** **(a)** 0.7   **(b)** 0.09
55. **(a)** 0.09   **(b)** 0.12   **57.** $\frac{1}{1000}$
59. **(a)** $\frac{22}{50} = 0.44$ **(b)** $\frac{28}{50} = 0.56$ **(c)** $\frac{28}{50} = 0.56$
61. $\frac{3}{51} = \frac{1}{17}$ **63.** $\frac{1}{3}$ **65.** $\frac{19}{27}$ **67.** 40%   **69.** $\frac{1}{3}$
71. **(a)** $\frac{18}{235}$ **(b)** $\frac{7}{18}$ **(c)** $\frac{7}{235}$
73. **(a)** $\frac{8}{15}$ **(b)** $\frac{7}{15}$ **(c)** $\frac{2}{5}$ **(d)** $\frac{1}{3}$ **(e)** $\frac{1}{15}$

CHECKING BASIC CONCEPTS FOR SECTIONS 11.5 AND 11.6 (p. 897)

1. $4 + 8 + 12 + \cdots + 4n = 2n(n + 1)$
(i) Show that the statement is true for $n = 1$:
$4(1) = 2(1)(1 + 1)$
$4 = 4$

(ii) Assume that S_k is true:
$4 + 8 + 12 + \cdots + 4k = 2k(k + 1)$
Show that S_{k+1} is true:
$4 + 8 + \cdots + 4(k + 1) = 2(k + 1)(k + 2)$
Add $4(k + 1)$ to each side of S_k:
$4 + 8 + 12 + \cdots + 4k + 4(k + 1)$
$= 2k(k + 1) + 4(k + 1)$
$= 2k^2 + 6k + 4$
$= 2(k + 1)(k + 2)$
Since S_k implies S_{k+1}, the statement is true for every positive integer n.

3. $\frac{1}{16}$ **5.** $\dfrac{\binom{4}{4} \cdot \binom{4}{1}}{\binom{52}{5}} = \dfrac{4}{2,598,960} \approx 0.0000015$

CHAPTER 11: REVIEW EXERCISES (pp. 901–902)

1. $-1, -4, -7, -10$ **3.** $0, 1, 3, 7$
5. 5, 3, 1, 2, 4, 6
7. **(a)**

n	1	2	3	4	5	6	7	8
a_n	3	1	−1	−3	−5	−7	−9	−11

(b) [0, 10, 1] by [−12, 4, 1]

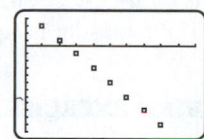

(c) $a_n = -2n + 5$
9. $a_n = 4n - 15$ **11.** Arithmetic   **13.** Geometric
15. 65   **17.** 90   **19.** 3280   **21.** 3
23. $6 + 11 + 16 + 21 + 26$
25. $\displaystyle\sum_{k=1}^{6} k^3$ **27.** -1155
29. $0.18 + 0.0018 + 0.000018 + \cdots$
31. 120
33. $1 + 3 + 5 + \cdots + (2n - 1) = n^2$
(i) Show that the statement is true for $n = 1$:
$2(1) - 1 = 1^2$
$1 = 1$
(ii) Assume that S_k is true:
$1 + 3 + 5 + \cdots + (2k - 1) = k^2$
Show that S_{k+1} is true:
$1 + 3 + \cdots + (2(k + 1) - 1) = (k + 1)^2$
Add $2k + 1$ to each side of S_k:
$1 + 3 + 5 + \cdots + (2k - 1) + (2k + 1)$
$= k^2 + 2k + 1$
$= (k + 1)^2$
Since S_k implies S_{k+1}, the statement is true for every positive integer n.

35. $\frac{1}{2}$ **37.** 120   **39.** $4^{20} \approx 1.1 \times 10^{12}$ **41.** 6,250,000
43. **(a)** 4, 3.6, 3.24, 2.916, 2.6244; geometric
(b) [0, 6, 1] by [0, 6, 1]   **(c)** $a_n = 4(0.9)^{n-1}$

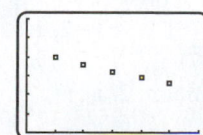

45. 20   **47.** 210   **49.** $\frac{91}{120} \approx 0.758$

51. (a) $\frac{27}{60} = 0.45$ **(b)** $\frac{33}{60} = 0.55$ **(c)** $\frac{13}{60} \approx 0.217$

53. The population density grows slowly initially, then increases rapidly, and finally levels off near 4,000,000 per acre.

[0, 16, 1] by [0, 5000, 1000]

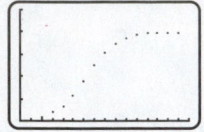

CHAPTER 11 EXTENDED AND DISCOVERY EXERCISES (pp. 902–903)

1. (a) $P = \begin{bmatrix} 1 & 0 & 0 \\ \frac{1}{6} & \frac{2}{3} & \frac{1}{6} \\ 0 & 0 & 1 \end{bmatrix}$

(b) The greatest probabilities lie on the main diagonal: $1, \frac{2}{3}, 1$; this means that a mother cell is most likely to produce a daughter cell like itself (answers will vary).

3. The quantity $\frac{a_1 + a_n}{2}$ represents not only the average of a_1 and a_n but also the average of the terms $a_1, a_2, a_3, \ldots, a_n$. This is true whether n is odd or even. The total sum is equal to n times the average of the terms.

CHAPTERS 1–11 CUMULATIVE REVIEW EXERCISES (pp. 903–907)

1. $3.45 \times 10^4; 0.000152$ **3.** $\sqrt{41}$

5. (a) $2; \sqrt{-a}$ **(b)** $D = \{x \mid x \leq 1\}$, or $(-\infty, 1]$

7. 7 **9.** $y = -\frac{6}{5}x - \frac{8}{5}$

11. (a) $\frac{3}{4}; -1; \frac{4}{3}$ **(b)** $f(x) = \frac{3}{4}x - 1$ **(c)** $\frac{4}{3}$

13. (a) $\frac{17}{10}$ **(b)** $-\frac{1}{3}, \frac{5}{2}$ **(c)** $\frac{1 \pm \sqrt{13}}{2}$ **(d)** $-2, -1, 2$

(e) $\pm\sqrt{3}, \pm 1$ **(f)** $\frac{17}{3}$ **(g)** $\frac{\ln(28/3)}{2} \approx 1.117$

(h) $10^{3/2} - 1 \approx 30.623$ **(i)** 6 **(j)** $-\frac{4}{3}, 2$

15. f is continuous.

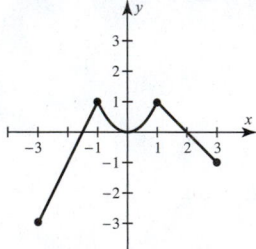

17. (a) $-2, -1, 1, 2$

(b) $\{x \mid x < -2 \text{ or } -1 < x < 1 \text{ or } x > 2\}$, or $(-\infty, -2) \cup (-1, 1) \cup (2, \infty)$

(c) $\{x \mid -2 \leq x \leq -1 \text{ or } 1 \leq x \leq 2\}$, or $[-2, -1] \cup [1, 2]$

19. $\left(\frac{3}{2}, \frac{31}{4}\right)$

21. (a) Incr: $\{x \mid x < -2\}, \{x \mid x > 1\}$, or $(-\infty, -2), (1, \infty)$; decr: $\{x \mid -2 < x < 1\}$, or $(-2, 1)$

(b) $-3.3, 0, 1.8$ **(c)** $(-2, 2), (1, -0.7)$

(d) Local minimum: -0.7, local maximum: 2

23. (a) $3x^2 - 1 + \frac{1}{2x^2}$

(b) $2x^3 - 5x^2 + 5x - 6 + \frac{8}{x+1}$

25. $f(x) = 3(x + 1)(x - 3i)(x + 3i)$; $f(x) = 3x^3 + 3x^2 + 27x + 27$

27. $-1 \pm 2i$ **29.** $(x + 1)^{3/5}; 8$

31. (a) 3 **(b)** -1 **(c)** 1 **(d)** -4

33. $f^{-1}(x) = -\frac{x}{x - 1}$ **35.** $f(x) = 2(3)^x$

37. \$1221.61

39. (a) $D = \{x \mid -\infty < x < \infty\}$, or $(-\infty, \infty)$; $R = \{x \mid x \geq 0\}$, or $[0, \infty)$

(b) $D = \{x \mid -\infty < x < \infty\}$, or $(-\infty, \infty)$; $R = \{x \mid x > 0\}$, or $(0, \infty)$

(c) $D = \{x \mid x > 0\}$, or $(0, \infty)$; $R = \{x \mid -\infty < x < \infty\}$, or $(-\infty, \infty)$

(d) $D = \{x \mid x \neq 0\}$, or $(-\infty, 0) \cup (0, \infty)$; $R = \{x \mid x \neq 0\}$, or $(-\infty, 0) \cup (0, \infty)$

41. $\log \dfrac{x^2 y^3}{\sqrt[3]{z}}$

43. (a) (b) (c) (d)

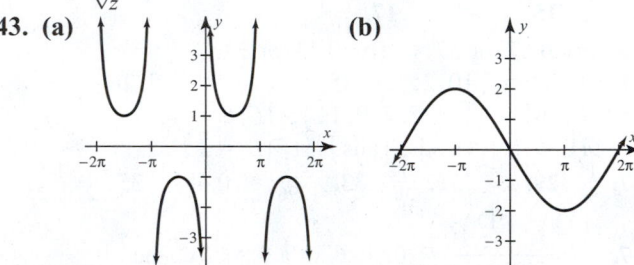

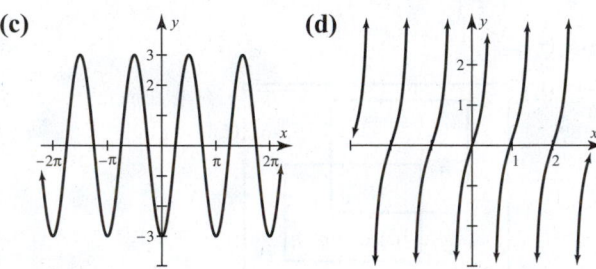

45. (a) $\dfrac{5\pi}{6} + 2\pi n, \dfrac{7\pi}{6} + 2\pi n$ **(b)** $\dfrac{\pi}{4} + \dfrac{\pi}{2}n$

(c) $\pi n, \dfrac{7\pi}{6} + 2\pi n, \dfrac{11\pi}{6} + 2\pi n$ **(d)** $\dfrac{\pi}{2} + \pi n$

47. $5°32'24''$ **49.** $225°$

51. $\sin\theta = \frac{24}{25}, \cos\theta = -\frac{7}{25}, \tan\theta = -\frac{24}{7}, \csc\theta = \frac{25}{4},$ $\sec\theta = -\frac{25}{7}, \cot\theta = -\frac{7}{24}$

53. $\sin\theta = -\frac{60}{61}, \cos\theta = -\frac{11}{61}, \tan\theta = \frac{60}{11}, \csc\theta = -\frac{61}{60},$ $\sec\theta = -\frac{61}{11}, \cot\theta = \frac{11}{60}$

55. $b = 8; \alpha \approx 36.9°; \beta \approx 53.1°$

57. $1 - \sin^2\theta + \cot^2\theta - \sin^2\theta \cot^2\theta$
$= 1 - \sin^2\theta - \sin^2\theta \cot^2\theta + \cot^2\theta$
$= 1 - \sin^2\theta(1 + \cot^2\theta) + \cot^2\theta$
$= 1 - \sin^2\theta(\csc^2\theta) + \cot^2\theta$
$= 1 - 1 + \cot^2\theta$
$= \cot^2\theta$

59. (a) $b \approx 15.4; c \approx 11.8; \alpha = 107°$

(b) $a_1 \approx 8.5; \alpha_1 \approx 68.9°; \beta_1 \approx 61.1°$ or $a_2 \approx 1.8; \alpha_2 \approx 11.1°; \beta_2 \approx 118.9°$

(c) $a \approx 5.7; \beta \approx 58.6°; \gamma \approx 77.4°$

(d) $\alpha \approx 51.3°; \beta \approx 59.2°; \gamma \approx 69.5°$

61. (a) 5 **(b)** $\langle -26, 56 \rangle$ **(c)** -63
(d) About $165.75°$

63.

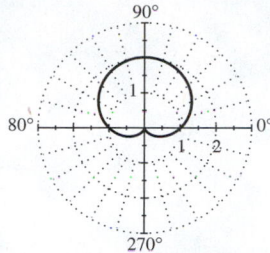

65. (a) $(-1, 2)$ **(b)** $\{(x, y) \mid 4x - y = -2\}$
(c) Four solutions: $\left(\pm 3, \pm \sqrt{7} \right)$ **(d)** $\left(\frac{5z - 24}{3}, \frac{z + 6}{3}, z \right)$

67. (a)   **(b)**

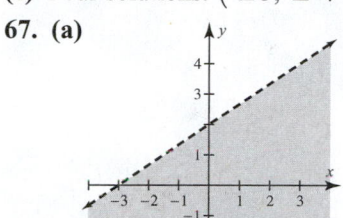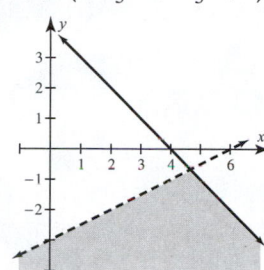

69. $\begin{bmatrix} -2 & -1 \\ -1.5 & -0.5 \end{bmatrix}$

71. (a) -11 **(b)** 22 **73.** $y^2 = 3x$

75. $\frac{y^2}{25} - \frac{x^2}{144} = 1$ **77.** $a_n = 4n$

79. (a) 950 **(b)** $\frac{2}{9}$ **81.** 175,760,000

83. $16x^4 - 32x^3 + 24x^2 - 8x + 1$

85. $\frac{16}{52} = \frac{4}{13}$ **87.** $\frac{8}{20} = \frac{2}{5}$ **89.** 146.5 mi

91. 0.6 hr at 7 mi/hr; 0.7 hr at 9 mi/hr

93. $\frac{20}{9} \approx 2.22$ hr

95. (a) $C(x) = x(405 - 5x)$ **(b)** 25 or 56; the cost is $7000 when 25 or 56 tickets are purchased.
(c) $8200; the cost is $8200 when 40 or 41 tickets are purchased.

97. 1.125 ft **99.** $\frac{49}{77} = \frac{7}{11}$

APPENDIX C: Partial Fractions (p. AP-29)

1. $\frac{5}{3x} + \frac{-10}{3(2x + 1)}$ **3.** $\frac{6}{5(x + 2)} + \frac{8}{5(2x - 1)}$

5. $\frac{5}{6(x + 5)} + \frac{1}{6(x - 1)}$

7. $\frac{-2}{x + 1} + \frac{2}{x + 2} + \frac{4}{(x + 2)^2}$ **9.** $\frac{4}{x} + \frac{4}{1 - x}$

11. $\frac{15}{x} + \frac{-5}{x + 1} + \frac{-6}{x - 1}$ **13.** $1 + \frac{-2}{x + 1} + \frac{1}{(x + 1)^2}$

15. $x^3 - x^2 + \frac{-1}{3(2x + 1)} + \frac{2}{3(x + 2)}$

17. $\frac{1}{9} + \frac{-1}{x} + \frac{25}{18(3x + 2)} + \frac{29}{18(3x - 2)}$

19. $\frac{-3}{5x^2} + \frac{3}{5(x^2 + 5)}$ **21.** $\frac{-2}{7(x + 4)} + \frac{6x - 3}{7(3x^2 + 1)}$

23. $\frac{1}{4x} + \frac{-8}{19(2x + 1)} + \frac{-9x - 24}{76(3x^2 + 4)}$

25. $\frac{-1}{x} + \frac{2x}{2x^2 + 1} + \frac{2x + 3}{(2x^2 + 1)^2}$

27. $\frac{-1}{x + 2} + \frac{3}{(x^2 + 4)^2}$

29. $5x^2 + \frac{3}{x} + \frac{-1}{x + 3} + \frac{2}{x - 1}$

APPENDIX D: Percent Change and Exponential Functions (pp. AP-34–AP-36)

1. (a) 0.35 **(b)** -0.0007 **(c)** 7.21 **(d)** 0.003
3. (a) -0.055 **(b)** -0.0154 **(c)** 1.2 **(d)** 0.0015
5. (a) 37% **(b)** -9.5% **(c)** 190% **(d)** 35%
7. (a) -12.1% **(b)** 140% **(c)** 320% **(d)** -25%
9. (a) 100% **(b)** -50%
11. (a) 2.36% **(b)** -2.31%
13. (a) 44.44% **(b)** -30.77%
15. (a) $1800 **(b)** $3300 **(c)** 2.2
17. (a) $-$2200 **(b)** $1800 **(c)** 0.45
19. (a) $-$4500 **(b)** $3000 **(c)** 0.4
21. $f(x) = 9500(0.65)^x$ **23.** $f(x) = 2500(1.05)^x$
25. $f(x) = 1000(0.935)^x$
27. $C = 8, a = 1.12, r = 0.12$ or 12%
29. $C = 1.5, a = 0.35, r = -0.65$ or -65%
31. $C = 1, a = 0.55, r = -0.45$ or -45%
33. $C = 7, a = e, r = e - 1$ or about 171.8%
35. $C = 6, a = \frac{1}{3}, r = -\frac{2}{3}$ or about -66.7%
37. $f(t) = 35,000(1.098)^{t/2}$; about 44,215
39. $f(t) = 1000(3)^{t/7}$; about 5620
41. $f(t) = I_0 \left(\frac{2}{3} \right)^{t/2}$; about $0.418 I_0$
43. $f(t) = 5000(4)^{t/35}$; $6864.10
45. 10 years; $4027.51 **47.** 3.5 years; $1006.88
49. 2.8 years; $3020.63 **51.** 1.75% **53.** 2% **55.** 1%
57. About 2.21% **59.** They are equal. **61.** $7.39
63. About 2.56%

APPENDIX E: Rotation of Axes (p. AP-40)

1. Circle or ellipse or a point
3. Hyperbola or two intersecting lines
5. Parabola or one line or two parallel lines
7. 30° **9.** 60° **11.** 22.5°

13.   **15.**

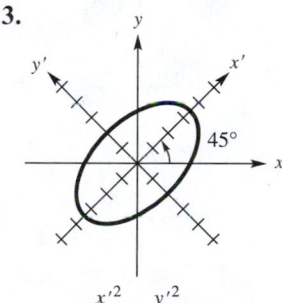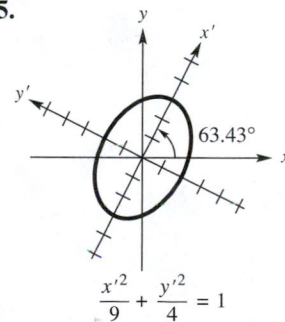

$\frac{x'^2}{12} + \frac{y'^2}{4} = 1$ $\frac{x'^2}{9} + \frac{y'^2}{4} = 1$

17.   **19.**

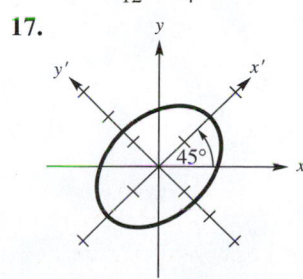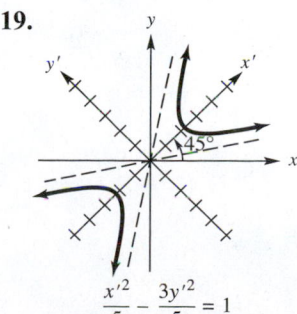

$\frac{x'^2}{4} + \frac{y'^2}{2} = 1$ $\frac{x'^2}{5} - \frac{3y'^2}{5} = 1$

21.

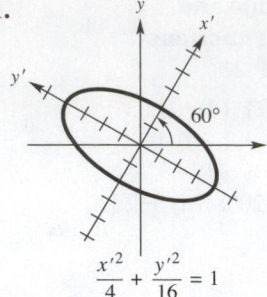

$$\frac{x'^2}{4} + \frac{y'^2}{16} = 1$$

23.

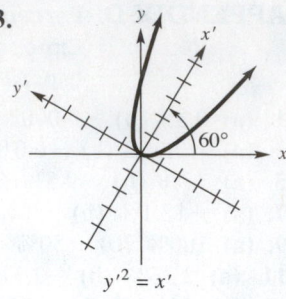

$$y'^2 = x'$$

25.

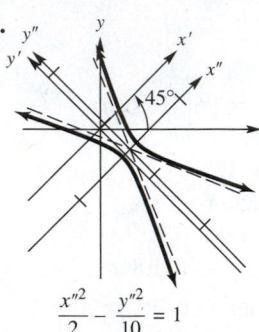

$$\frac{x''^2}{2} - \frac{y''^2}{10} = 1$$

27.

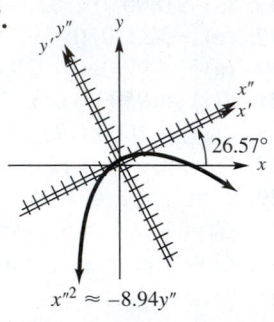

$$x''^2 \approx -8.94y''$$

29.

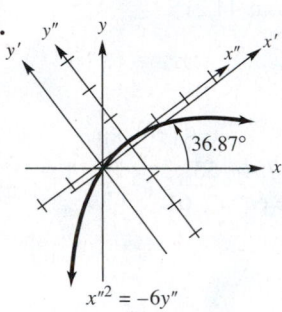

$$x''^2 = -6y''$$

Photo Credits

Index of Applications

Index

Trigonometric Functions and Identities

Right Triangle Trigonometry

$$\sin\theta = \frac{\text{side opposite}}{\text{hypotenuse}} \qquad \cos\theta = \frac{\text{side adjacent}}{\text{hypotenuse}}$$

$$\csc\theta = \frac{\text{hypotenuse}}{\text{side opposite}} \qquad \sec\theta = \frac{\text{hypotenuse}}{\text{side adjacent}}$$

$$\tan\theta = \frac{\text{side opposite}}{\text{side adjacent}}$$

$$\cot\theta = \frac{\text{side adjacent}}{\text{side opposite}}$$

Hypotenuse   Side opposite

A   Side adjacent   C

The Unit Circle

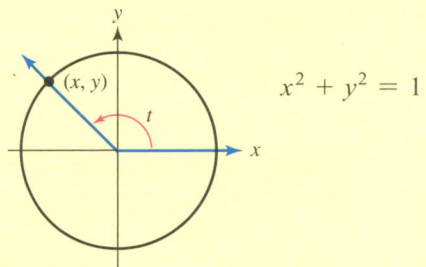

$$x^2 + y^2 = 1$$

$$\sin t = y \qquad \cos t = x \qquad \tan t = \frac{y}{x}$$

$$\csc t = \frac{1}{y} \qquad \sec t = \frac{1}{x} \qquad \cot t = \frac{x}{y}$$

The Unit Circle and Special Angles

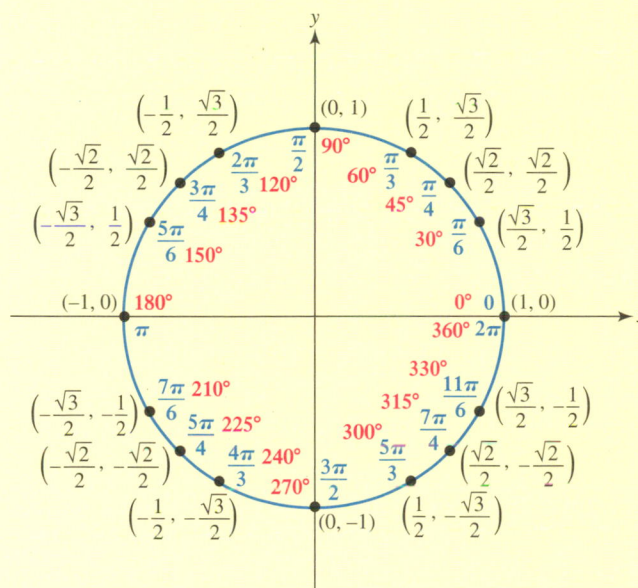

Pythagorean

$$\sin^2\theta + \cos^2\theta = 1$$

$$1 + \tan^2\theta = \sec^2\theta$$

$$1 + \cot^2\theta = \csc^2\theta$$

Sum and Difference

$$\cos(\alpha + \beta) = \cos\alpha\cos\beta - \sin\alpha\sin\beta$$

$$\cos(\alpha - \beta) = \cos\alpha\cos\beta + \sin\alpha\sin\beta$$

$$\sin(\alpha + \beta) = \sin\alpha\cos\beta + \cos\alpha\sin\beta$$

$$\sin(\alpha - \beta) = \sin\alpha\cos\beta - \cos\alpha\sin\beta$$

$$\tan(\alpha + \beta) = \frac{\tan\alpha + \tan\beta}{1 - \tan\alpha\tan\beta}$$

$$\tan(\alpha - \beta) = \frac{\tan\alpha - \tan\beta}{1 + \tan\alpha\tan\beta}$$

Reciprocal

$$\sin\theta = \frac{1}{\csc\theta} \qquad \cos\theta = \frac{1}{\sec\theta} \qquad \tan\theta = \frac{1}{\cot\theta}$$

$$\csc\theta = \frac{1}{\sin\theta} \qquad \sec\theta = \frac{1}{\cos\theta} \qquad \cot\theta = \frac{1}{\tan\theta}$$

Quotient

$$\tan\theta = \frac{\sin\theta}{\cos\theta} \qquad \cot\theta = \frac{\cos\theta}{\sin\theta}$$

Negative-Angle

$$\sin(-\theta) = -\sin\theta \qquad \csc(-\theta) = -\csc\theta$$

$$\cos(-\theta) = \cos\theta \qquad \sec(-\theta) = \sec\theta$$

$$\tan(-\theta) = -\tan\theta \qquad \cot(-\theta) = -\cot\theta$$

Double-Angle

$$\sin 2\theta = 2\sin\theta\cos\theta$$

$$\cos 2\theta = \cos^2\theta - \sin^2\theta$$

$$= 2\cos^2\theta - 1$$

$$= 1 - 2\sin^2\theta$$

$$\tan 2\theta = \frac{2\tan\theta}{1 - \tan^2\theta}$$

Power-Reducing

$$\sin^2\theta = \frac{1 - \cos 2\theta}{2}$$

$$\cos^2\theta = \frac{1 + \cos 2\theta}{2}$$

$$\tan^2\theta = \frac{1 - \cos 2\theta}{1 + \cos 2\theta}$$

Half-Angle

$$\sin\frac{\theta}{2} = \pm\sqrt{\frac{1 - \cos\theta}{2}} \qquad \cos\frac{\theta}{2} = \pm\sqrt{\frac{1 + \cos\theta}{2}}$$

$$\tan\frac{\theta}{2} = \frac{1 - \cos\theta}{\sin\theta} \qquad \tan\frac{\theta}{2} = \frac{\sin\theta}{1 + \cos\theta}$$

Formulas from Geometry

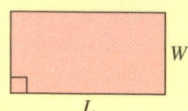

Rectangle

$A = LW$

$P = 2L + 2W$

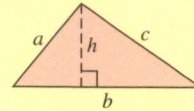

Triangle

$A = \frac{1}{2}bh$

$P = a + b + c$

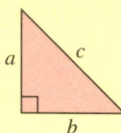

Pythagorean Theorem

$c^2 = a^2 + b^2$

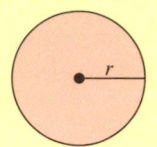

Circle

$C = 2\pi r$

$A = \pi r^2$

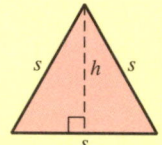

Equilateral Triangle

$A = \dfrac{\sqrt{3}}{4}s^2$

$P = 3s$

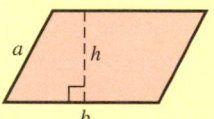

Parallelogram

$A = bh$

$P = 2a + 2b$

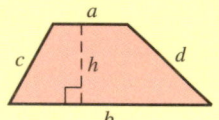

Trapezoid

$A = \frac{1}{2}(a + b)h$

$P = a + b + c + d$

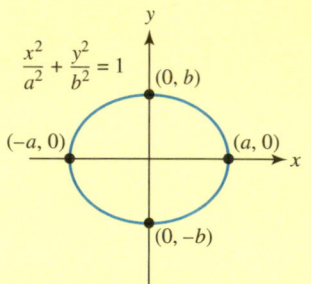

Ellipse

$A = \pi ab$

$P \approx 2\pi\sqrt{\dfrac{a^2 + b^2}{2}}$

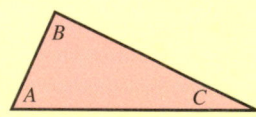

Sum of the Angles in a Triangle

$A + B + C = 180°$

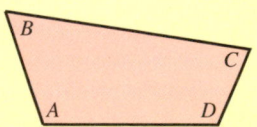

Sum of the Angles in a Quadrilateral

$A + B + C + D = 360°$

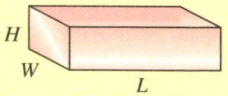

Rectangular (Parallelepiped) Box

$V = LWH$

$S = 2LW + 2LH + 2WH$

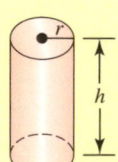

Cylinder

$V = \pi r^2 h$

$S = 2\pi rh + 2\pi r^2$

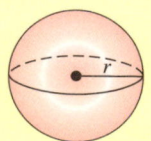

Sphere

$V = \frac{4}{3}\pi r^3$

$S = 4\pi r^2$

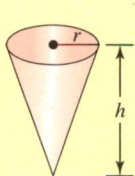

Cone

$V = \frac{1}{3}\pi r^2 h$

$S = \pi r^2 + \pi r\sqrt{r^2 + h^2}$